Forms Containing $u^2 \pm a^2$

38. $\displaystyle\int \frac{du}{a^2 - u^2} = \frac{1}{2a} \ln\left|\frac{a + u}{a - u}\right| + C$

39. $\displaystyle\int \frac{du}{u^2 - a^2} = \frac{1}{2a} \ln\left|\frac{u - a}{u + a}\right| + C$

40. $\displaystyle\int \frac{du}{\sqrt{u^2 \pm a^2}} = \ln\left|u + \sqrt{u^2 \pm a^2}\right| + C$

41. $\displaystyle\int \frac{du}{u\sqrt{a^2 \pm u^2}} = -\frac{1}{a} \ln\left|\frac{a + \sqrt{a^2 \pm u^2}}{u}\right| + C$

42. $\displaystyle\int \frac{du}{au^2 + c} = \frac{1}{\sqrt{ac}} \tan^{-1}\left(u\sqrt{\frac{a}{c}}\right) + C, \quad ac > 0$

43. $\displaystyle\int \frac{du}{c^2 - u^2} = \frac{1}{2c} \ln\left|\frac{c + u}{c - u}\right| + C, \quad c^2 > u^2$

44. $\displaystyle\int \sqrt{u^2 \pm a^2}\, du = \frac{1}{2}\left[u\sqrt{u^2 \pm a^2} \pm a^2 \ln(u + \sqrt{u^2 \pm a^2})\right] + C$

45. $\displaystyle\int u\sqrt{u^2 \pm a^2}\, du = \frac{1}{3}(u^2 \pm a^2)^{3/2} + C$

46. $\displaystyle\int \frac{du}{\sqrt{u^2 \pm a^2}} = \ln(u + \sqrt{u^2 \pm a^2}) + C$

47. $\displaystyle\int \frac{du}{u\sqrt{u^2 - a^2}} = \frac{1}{a} \sec^{-1}\left(\frac{u}{a}\right) + C$

48. $\displaystyle\int \frac{du}{u\sqrt{u^2 + a^2}} = -\frac{1}{a} \ln\left(\frac{a + \sqrt{u^2 + a^2}}{u}\right) + C$

49. $\displaystyle\int \frac{\sqrt{u^2 + a^2}}{u}\, du = \sqrt{u^2 + a^2} - a \ln\left(\frac{a + \sqrt{u^2 + a^2}}{u}\right) \quad C$

50. $\displaystyle\int \frac{\sqrt{u^2 - a^2}}{u}\, du = \sqrt{u^2 - a^2} - a \sec^{-1}\left(\frac{u}{a}\right) + C, \quad a > 0$

51. $\displaystyle\int \frac{du}{\sqrt{(u^2 \pm a^2)^3}} = \frac{\pm u}{a^2\sqrt{u^2 \pm a^2}} + C$

52. $\displaystyle\int \frac{\sqrt{u^2 \pm a^2}}{u^2}\, du = -\frac{\sqrt{u^2 \pm a^2}}{u} + \ln(u + \sqrt{u^2 \pm a^2}) + C$

53. $\displaystyle\int \frac{\sqrt{u^2 - a^2}}{u^3}\, du = -\frac{\sqrt{u^2 - a^2}}{2u^2} + \frac{1}{2a} \sec^{-1}\left(\frac{u}{a}\right) + C, a > 0$

54. $\displaystyle\int \sqrt{a^2 - u^2}\, du = \frac{1}{2}\left[u\sqrt{a^2 - u^2} + a^2 \sin^{-1}\left(\frac{u}{a}\right)\right] + C, a > 0$

55. $\displaystyle\int \frac{du}{\sqrt{a^2 - u^2}} = \sin^{-1}\left(\frac{u}{a}\right) + C$

56. $\displaystyle\int \frac{du}{u\sqrt{a^2 - u^2}} = -\frac{1}{a} \ln\left(\frac{a + \sqrt{a^2 - u^2}}{u}\right) + C$

57. $\displaystyle\int \frac{du}{\sqrt{(a^2 - u^2)^3}} = \frac{u}{a^2\sqrt{a^2 - u^2}} + C$

58. $\displaystyle\int u^2\sqrt{a^2 - u^2}\, du = -\frac{u}{4}\sqrt{(a^2 - u^2)^3}$

$\displaystyle+ \frac{a^2}{8}\left(u\sqrt{a^2 - u^2} + a^2 \sin^{-1}\frac{u}{a}\right) + C, \quad a > 0$

59. $\displaystyle\int \frac{u^2\, du}{\sqrt{a^2 - u^2}} = -\frac{u}{2}\sqrt{a^2 - u^2} + \frac{a^2}{2} \sin^{-1}\left(\frac{u}{a}\right) + C, \quad a > 0$

60. $\displaystyle\int \frac{du}{u^2\sqrt{a^2 - u^2}} = -\frac{\sqrt{a^2 - u^2}}{a^2 u} + C$

61. $\displaystyle\int \frac{\sqrt{a^2 - u^2}}{u^2}\, du = -\frac{\sqrt{a^2 - u^2}}{u} - \sin^{-1}\left(\frac{u}{a}\right) + C, \quad a > 0$

Forms Containing $a + bu^2$

62. $\displaystyle\int \frac{du}{a + bu^2} = \frac{1}{\sqrt{ab}} \tan^{-1} \frac{u\sqrt{ab}}{a} + C$

63. $\displaystyle\int \frac{du}{(u^2 + a^2)^2} = \frac{1}{2a^3} \tan^{-1} \frac{u}{a} + \frac{u}{2a^2(u^2 + a^2)} + C$

64. $\displaystyle\int \frac{du}{u(a + bu^2)} = \frac{1}{2a} \ln\left|\frac{u^2}{a + bu^2}\right| + C$

Forms Involving $2au - u^2$

65. $\displaystyle\int \sqrt{2au - u^2}\, du = \frac{1}{2}\left[(u - a)\sqrt{2au - u^2} + a^2 \sin^{-1}\left(\frac{u - a}{a}\right)\right] + C$

66. $\displaystyle\int \frac{du}{\sqrt{2au - u^2}} = 2 \sin^{-1}\sqrt{\frac{u}{2a}} + C, \quad a > 0$

67. $\displaystyle\int \frac{du}{(2au - u^2)^{3/2}} = \frac{u - a}{a^2\sqrt{2au - u^2}} + C$

68. $\displaystyle\int \frac{du}{\sqrt{2au + u^2}} = \ln\left|u + a + \sqrt{2au + u^2}\right| + C$

(Continued on back endsheet)

Calculus

Calculus

DENNIS D. BERKEY
Boston University

SAUNDERS COLLEGE PUBLISHING
Philadelphia New York Chicago
San Francisco Montreal Toronto
London Sydney Tokyo Mexico City
Rio de Janeiro Madrid

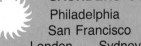

Address orders to:
383 Madison Avenue
New York, NY 10017

Address editorial correspondence to:
West Washington Square
Philadelphia, PA 19105

Text Typeface: Times Roman
Compositor: York Graphic Services, Inc.
Acquisitions Editor: Leslie Hawke
Developmental Editor: Jay Freedman
Project Editor: Sally Kusch
Copyeditor: Charlotte Nelson
Index: Eleanor Kuljian
Managing Editor & Art Director: Richard L. Moore
Art/Design Assistant: Virginia A. Bollard
Text Design: Emily Harste
Cover Design: Lawrence R. Didona
Text Artwork: J & R Technical Services, Inc.
Production Manager: Tim Frelick
Assistant Production Manager: Maureen Iannuzzi

Cover credit: Photograph of French horn. © Pete Turner/THE IMAGE BANK.

Library of Congress Cataloging in Publication Data

Berkey, Dennis D.
Calculus.

Includes index.

1. Calculus. I. Title
QA303.B48 1983 515 83-20046
ISBN 0-03-059522-3

CALCULUS

ISBN 0-03-059522-3

Library of Congress catalog card number 83-20046

56 032 9876543

CBS COLLEGE PUBLISHING
Saunders College Publishing
Holt, Rinehart and Winston
The Dryden Press

PREFACE

This text is intended for use in a traditional three-semester or four-quarter sequence of courses on the calculus, populated principally by mathematics, science and engineering students. It reflects my own philosophy that freshman calculus should be taught so as to produce skilled practitioners who have some real feeling for the mathematical issues underlying the techniques they have acquired. It was written in order to provide students of widely varying interests and abilities with a highly readable exposition of the principal results, including ample motivation, numerous well-articulated examples, a rich discussion of applications, and a nontrivial description and use of numerical techniques. In particular, I have tried to indicate, in an unobtrusive manner, how computers can be used both to illustrate the theory and to provide approximate solutions to problems for which more elegant techniques break down.

LEVEL AND RIGOR: Nearly all topics in the traditional calculus curriculum are treated. However, the discussion is informal, and geometric arguments are used wherever possible. I do not prove that a continuous function assumes its extreme values on a closed bounded interval. Limits are first presented quite informally in the context of the tangent line problem. The more formal δ-ϵ definition appears briefly at the end of Chapter 2, along with formal proofs of several limit theorems. When the topic of limits appears later in the chapters on infinite series, it is treated more formally, reflecting the increased maturity of the reader. The notion of differentiability for a function of several variables is discussed following the development of partial differentiation and the gradient, and the theorems of Green and Stokes, as well as the Divergence Theorem, are fully discussed.

STRATEGY/SOLUTION FORMAT IN EXAMPLES: An unusually large number of examples (more than 700) has been included. In many of these, especially in the early parts of the text, the solutions have been written in a two-column format, with one of the columns labelled ''strategy.'' In this column the student will find, in very abbreviated form, a description of the principal steps involved in the fuller solution. Here I have attempted to help students identify the more general aspects of the particular solution, and to develop problem-solving strategies of their own.

ORGANIZATION: The order of topics is consistent with those of most popular texts. The trigonometric functions are introduced as part of the review material in Chapter 1, and are used frequently throughout the text. Applications of the derivative are organized into two successive chapters, with the simpler notions of related rates and extrema on closed bounded intervals (Chapter 4) followed by the more general discussions on increasing and decreasing functions, relative extrema, asymptotes, and graphing techniques (Chapter 5). The applications of the definite integral are split into two chapters as well (Chapters 7, 8), although the rationale for this is simply to break up what otherwise would be a long and less than coherent chapter. The unit on infinite series (Chapters 12–14) may be postponed without consequence.

ANTIDERIVATIVES, INTEGRALS, AND DIFFERENTIAL EQUATIONS: The organization of Chapters 5 and 6 reflects a concern I have held for some time, that the definite integral is often presented to students almost simultaneously with the notion of an antiderivative and the statement of the Fundamental Theorem of Calculus. As a result, students can leave their first calculus course thinking of a definite integral only as a difference between two values of an antiderivative that, coincidentally, might also be identified with the area of a certain planar region. I have addressed this concern in two ways. First, antiderivatives are introduced as one of the applications of the derivative, followed by a section on separable differential equations. This serves to introduce and apply the notion of antidifferentiation as a topic that is independent of the definite integral. Second, the definite integral is introduced as the limit of approximating sums, which result from a discussion of approximating planar regions by rectangles. This approach should help the student to grasp more quickly the notion of approximate integration (including the role computers can play), as well as making the development of the various applications of the integral more accessible. Further topics on differential equations appear in optional sections at the ends of chapters throughout the text, rather than in a single final chapter, in the hope that they will be included at points in the development where they occur naturally.

EXERCISES: More than 6,000 exercises are included, ranging from drill to challenging in type, and including many applied exercises from a broad range of disciplines. The extensive review exercises at the end of each chapter reflect the range of topics included in that chapter.

While many theorems, particularly those with instructive proofs, can and should be presented and proved, the time available to most of us for this task is not sufficient to allow a careful treatment of the least upper bound axiom, uniform continuity, differentiability of power series, or several other topics where statements of fact must simply be made. Honesty about these omissions, together with the right picture here and a good heuristic discussion there, can result in a presentation that is both factual and effective, and one that allows us to succeed in sharing with students the excitement of the triumphs of this classic subject.

ACKNOWLEDGMENTS: Many individuals played instrumental roles in the development of this text. It is my pleasure to acknowledge some of these here.

Twenty-seven teachers of the calculus scrutinized one or more drafts of the manuscript, both on matters of content and to identify errors. These were

Paul Baum, Brown University
David Bellamy, University of Delaware
George Blakley, Texas A&M University
Jan List Boal, Georgia State University

Alan Candiotti, Drew University
George Feissner, SUNY at Cortland
Herbert Gindler, San Diego State University
Arthur Goodman, Queen's College, CUNY
Douglas Hall, Michigan State University
Peter Herron, Suffolk Community College
Laurence Hoffman, Claremont Men's College
Adam Hulin, University of New Orleans
Frank Kocher, Pennsylvania State University
Carlon Krantz, Kean College
Paul Kumpel, SUNY at Stony Brook
Ed Landesman, University of California, Santa Cruz
Robert Lohman, Kent State University
Eldon Miller, University of Mississippi
Daniel Moran, Michigan State University
Thomas Rishel, Cornell University
Rainer Sachs, University of California, Berkeley
Philip Schaefer, University of Tennessee
Mark Schilling, University of Southern California
Donald Sherbert, University of Illinois
John Thorpe, SUNY at Stony Brook
William Wheeler, Indiana University
Robert Zink, Purdue University

Professor Paul Baum checked both the final manuscript and the exercise solutions for accuracy, and prepared the solutions manual. Professors Barry Granoff, of Boston University, and Richard Porter, of Northeastern University, read the entire text in galley form, rechecking all examples, exercises, and mathematical content. Professor Granoff also reviewed the artwork as it came from the artist, and scrutinized all page proofs. Each of these individuals worked meticulously to ensure the accuracy of the text. Whatever errors might remain are, of course, the sole responsibility of the author. Any comments on correcting or improving the text will be gratefully acknowledged.

At Boston University I am indebted to Tom Orowan for typing near-perfect drafts of the manuscript, and to Lisa Doherty for managing the flow of materials between Boston and Philadelphia. The BASIC programs included in the appendix were used by students on the University's IBM 3081 based time-sharing system, as well as on the author's personal computer. The computer-generated graphs of quadric surfaces were produced by the Graphics Laboratory of the University's Academic Computing Center.

Forewarned about hazards in dealings with publishers, I was delighted to experience a warm, professional, and highly supportive relationship with each of several key individuals at Saunders College Publishing: Mathematics Editor Leslie Hawke, Developmental Editor Jay Freedman, Project Editor Sally Kusch, and Publisher Don Jackson. Their commitment to excellence was a strong guiding force throughout the development of this text.

The historical notes, which provide an important human contrast, were written by Professor Duane Deal of Ball State University.

On a more general and personal level, the writing of this text was supported by three very special groups of people. First, my students at Boston University, who have encouraged this project and helped sharpen my thinking about teaching and the calculus for the past decade. Second, my colleagues in the faculty and administration of Boston University, who really do believe in the importance of effective teaching. And, most importantly, my family, who understood my need to write this book and shared fully and willingly in the sacrifices that were required. To all I am truly grateful.

Dennis D. Berkey

Boston, Massachusetts

CONTENTS OVERVIEW

CONTENTS

Isaac Newton

Gottfried Leibniz

UNIT 1

PRELIMINARY NOTIONS

Newton and Leibniz—The Unification of Calculus

More than once in history, a significant mathematical development has been independently discovered by two mathematicians widely separated by geography and having no contact with one another. Perhaps the greatest of these discoveries was that of the calculus. Isaac Newton (1642–1727) in England, and Gottfried Wilhelm Leibniz (1646–1716) in Germany, were completely unaware of each other's work. Newton actually developed his version of calculus some ten years before Leibniz, but Newton was generally reluctant to publish his results, and Leibniz presented his own version to the world about twenty years before Newton's first publication. These time differences ultimately resulted in an extended controversy between English and Continental mathematicians concerning priority and the possibility of plagiarism. To their credit, however, Newton and Leibniz did not attack one another.

Isaac Newton was still an undergraduate at Cambridge University when he essentially created differential calculus, or *fluxions* as he named derivatives. When bubonic plague swept over England in 1665–66, the university was closed for nearly two years, and Newton spent the time at his home. In this fruitful period, he developed differential and integral calculus, made the first observations that culminated in his theory of gravitation, made his first experiments in optics and the theory of color, and stated the binomial theorem for general exponents.

Although the binomial theorem for positive integral exponents had long been known, no one had thought to apply it to negative and fractional exponents before Newton. His discoveries concerning the infinite series generated by such exponents greatly intrigued him, and led to his using infinite series as a basis for expressing functions for his calculus. From this time on, the use of infinite processes in mathematics came to be considered legitimate. Newton himself thought infinite series and rates of change to be inextricably linked, and referred to them together as "my method."

Newton's ideas came from considering a curve as the result of a continuously moving point. The changing quantity he called a *fluent,* and its rate of change he named a *fluxion* (from the Latin *fluere,* to flow). The fluxion was thus what we shall refer to as a derivative, denoted $\dot{y}$ if y were the original fluent. Similarly, Newton saw that the original fluent y could be thought of as the fluxion of another function, designated $\overset{\square}{y}$ or $\overset{|}{y}$. The theory of the integral, which we shall develop in Unit III, explains Newton's claim that the fluent is the integral of the fluxion.

For some twenty years Newton made significant discoveries in many fields, but then the light of creative genius flickered. He suffered a long illness that affected his ability and, while he continued to work, the results were not of the quality of his earlier period. He served in Parliament and was appointed Warden (and later Master) of the Mint, in which administrative capacity he spent his last quarter century checking the quality of the metal in British coins.

Gottfried Leibniz was the son of a university professor who died when the boy was only six years old. Young Leibniz had access to his father's extensive library and read broadly. A brilliant scholar, he received his bachelor's degree at age 17, and his doctorate (in law) at 20. He then became a diplomat by profession, but soon became interested in mathematics as a consuming avocation. By the time Leibniz was 30 he had invented his calculus, which is in many ways *our* calculus. About 1673 he came to the realization that areas can be calculated by *summing* "infinitely thin" rectangles, and that tangents to curves involve *differences* in the x and y coordinates. This led him to suspect that the two processes are inverses of one another. In doing these calculations, he soon found himself immersed in infinite series representing functions, as had Newton some years before. Realizing the power of his discoveries, Leibniz set to work developing terminology and notation. He selected the integral sign $\int$ as representing the Latin word *summa,* or sum, and the derivative notation $\dfrac{dy}{dx}$.

Leibniz's, and the world's, first paper on differential calculus was published in 1684 with the title (translated) *A New Method for Maxima and Minima, and also for Tangents, which is not Obstructed by Irrational Quantities*. He gave formulas for differentials of powers, products, and quotients of functions—the same formulas we learn today. Two years later he published his integral calculus, in which quadratures, or the calculation of areas, are shown to be the inverse operation of finding tangents.

Newton and Leibniz both came to realize that the calculus is a much more general tool than merely a method for finding tangent lines or areas under curves. It was this generalization that made them such important people in the development of mathematics. Others had differentiated before them, had found areas, and had even realized to some extent the relation between the two operations. But these two, independently, unified the calculus and made it applicable to many fields, and to a much broader range of functions.

Newton was respected and revered in his old age. He was reelected President of the Royal Society of London for 25 years, and was knighted by Queen Anne in 1705. He was buried in Westminster Abbey with great honor. Leibniz, on the other hand, was neglected and became embittered before his death. Only one person, his secretary, was in attendance at his funeral.

Newton and Leibniz were both creative mathematicians of the first rank. The genius of neither is diminished by that of the other.

Pronunciation Guide

Isaac Newton (Eye′zak New′ton)

Gottfried Wilhelm Leibniz (Got′freed Vil′helm Libe′nitz)

(Photographs from the David Eugene Smith Papers, Rare Book and Manuscript Library, Columbia University.)

CHAPTER 1

REVIEW OF PRECALCULUS CONCEPTS

1.1 INTRODUCTION

This chapter summarizes many of the concepts and techniques of precalculus mathematics used throughout the text. You will find little here which you have not seen before. However, there are several reasons why you should examine this chapter carefully.

First, you need to become familiar with the notation and style in which this text is written. Studying mathematics is similar to studying a foreign language in that a vocabulary must be mastered, as well as a set of concepts. In mathematics, the vocabulary is quite formal, consisting of symbols and terms that have very precise meanings. However, different authors may not use exactly the same notation. Let's agree on notation and basic concepts before discussing new ideas. This will make the material which follows seem less intimidating.

Second, as mathematics has come to be more widely used as a tool in the natural and social sciences, more students than ever before are studying the calculus. This means that the typical calculus class contains students with quite diverse backgrounds in mathematics. This chapter defines the level and scope of precalculus mathematics assumed throughout the text. It therefore provides a common base from which to proceed.

Finally, this chapter provides a gauge by which you can determine whether you are sufficiently well prepared to begin a course on the calculus. If, after carefully reviewing the chapter, you are unable to work most of the exercises successfully, you probably need to consider enrolling in a course on basic mathematical skills. On the other hand, many students will already have mastered the concepts presented here. If this is your situation, you should quickly survey the chapter to observe the notation, and then proceed into Chapter 2.

1.2 THE REAL NUMBER SYSTEM

Real Numbers

The **real numbers** may be thought of as the set of all points lying on an infinite line. To do so we must first select a point 0 on the line that we refer to as the **origin** (Figure 2.1). We then select a **unit** of measurement and locate the point on the line lying one unit to the right of 0. This point is labeled 1 (one). This procedure establishes the **scale** for the **number line** and also uniquely determines the location

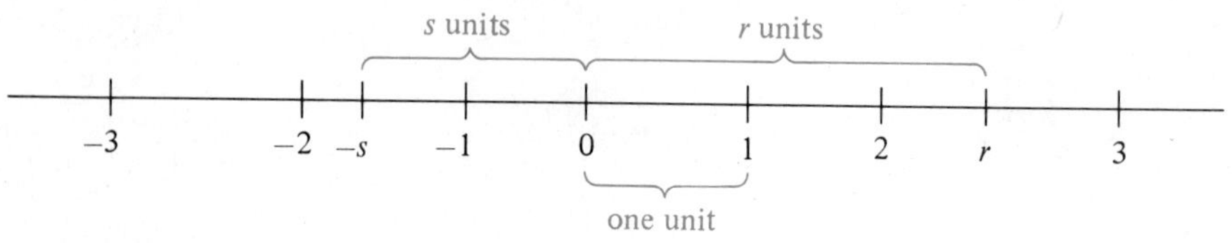

Figure 2.1 The real number line.

of each real number. Each positive real number r lies r units to the right of 0 and each negative real number, $-s$, lies s units to the left of 0.

You have previously learned that every real number has a **decimal expansion,** such as

$$\frac{4}{3} = 1.33333\ldots,$$

$$-\frac{3}{2} = -1.50000\ldots, \text{ and}$$

$$\pi = 3.14159\ldots.$$

There are three important special types of real numbers that can be identified by their decimal forms. The **integers** are the real numbers . . . −2, −1, 0, 1, 2, ... Integers may be expressed in decimal form with only zeros to the right of the decimal point. The **rational** numbers are those real numbers that can be expressed as quotients of integers. In other words, if r is a rational number then $r = p/q$ for some integers p and q ($q \neq 0$). The decimal forms of rational numbers repeat. For example, using a bar to identify the digits that repeat, we may write

$$\frac{2}{9} = 0.222\overline{2}\ldots,$$

$$-\frac{25}{11} = -2.2727\overline{27}\ldots, \text{ and}$$

$$\frac{20}{13} = 1.538461\overline{538461}\ldots.$$

See Exercise 83 at the end of this section.)

Those real numbers whose decimal forms do not repeat are referred to as **irrational numbers.** The discovery nearly 2500 years ago that not every real number is also a rational number was both profound and traumatic, since the geometric mathematics of the early Greeks assumed that every length could be expressed as the ratio of two integral lengths. Most high school students today study the proof that the real number $\sqrt{2}$ is irrational, and the irrational number π* plays a central role in Euclidean geometry. Later in this text we will encounter the irrational number $e \approx 2.71828\ldots$ that will appear as the base for the natural logarithmic and exponential functions. Besides these two prominent examples, there are infinitely many irrational real numbers.

*The number $\pi = 3.14159\ldots$ is the ratio of the circumference of a circle to its diameter. Since this irrational number is often *approximated* by the fraction $\frac{22}{7}$, many students mistakenly believe that $\pi = \frac{22}{7}$ and, therefore, that π is rational. The proof that π is irrational is not simple.

Sets of Real Numbers

There are several types of notation that we will use to specify sets of real numbers and operations on these sets. The most familiar notation to you is probably **set builder notation** in which one specifies a choice of variable and also a rule by which values of that variable are determined. For example, $\{x \mid 2 \leq x \leq 4\}$ is read, "the set of all real numbers equal to or greater than 2 and equal to or less than 4." We will frequently refer to **intervals** of the real number line with the following notation:

$$\begin{aligned} [a, b] &= \{x \mid a \leq x \leq b\}, \\ [a, b) &= \{x \mid a \leq x < b\}, \\ (a, b] &= \{x \mid a < x \leq b\}, \\ (a, b) &= \{x \mid a < x < b\}. \end{aligned}$$

An interval of the form $[a, b]$ is called a **closed** interval because it includes both its endpoints. An interval of the form (a, b) is called an **open** interval because it includes neither of its endpoints. Intervals of the form $[a, b)$ or $(a, b]$ are referred to as either **half-open** or **half-closed.**

We can denote intervals that extend indefinitely in one direction by use of the infinity symbol, ∞, as follows:

$$\begin{aligned} [a, \infty) &= \{x \mid a \leq x\}, \\ (-\infty, b) &= \{x \mid x < b\}. \end{aligned}$$

There are two other cases of intervals that extend indefinitely in one direction, namely, (a, ∞) and $(-\infty, b]$; the interval $(-\infty, \infty)$ is the entire number line. It must be clearly understood that ∞ is not a number: it is shorthand for the phrase "continuing infinitely through positive values" (or negative values for $-\infty$). In particular, ∞ and $-\infty$ cannot be endpoints for closed intervals. In sketching intervals, we use a square bracket to indicate an endpoint that is included and a round bracket to indicate an endpoint that is excluded, as illustrated in Figures 2.2 and 2.3.

Figure 2.2 The interval $[a, b)$, or $\{x \mid a \leq x < b\}$.

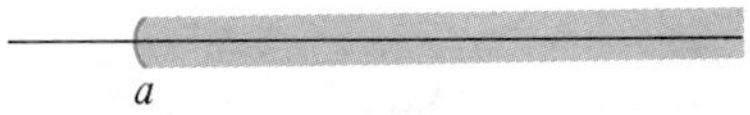

Figure 2.3 The interval (a, ∞), or $\{x \mid a < x < \infty\}$.

If x is a number and A is a set of numbers (not necessarily an interval), the statement "$x \in A$," read "x is an element of set A," means that x is one of the numbers in set A. The negation of this statement is "$x \notin A$," read "x is not an element of set A." For example,

$$3 \in \{1, 3, 5\}, \qquad 2 \in [0, 7), \qquad \text{but} \qquad 6 \notin \{x \mid -2 \leq x \leq 2\}.$$

If A and B are sets of numbers, we say that A is a **subset** of B if every element of A is also an element of B. The notation for this is $A \subseteq B$. When both A and B are intervals, we say that A is a subinterval of B if $A \subseteq B$. For example

$$\{1, 3\} \subseteq \{1, 2, 3, 4\}, \qquad \text{and} \qquad [-1, 1] \subseteq (-\infty, 3)$$

The symbol for the empty set is $\varnothing$. It is the set with no elements.

There are two **operations** on sets that we shall use frequently. Given two sets, A and B, the **union** of the two sets is the set containing precisely those numbers which are elements of either set A or set B, or both. The symbol for the union of A and B is $A \cup B$. In other words,

$$A \cup B = \{x \mid x \in A \text{ or } x \in B\}^*$$

---

*In mathematics, the usage of the word "or" is *inclusive*. This means that the statement "U or V" is interpreted as "either U or V, or *both*." This is in contrast to the *exclusive* usage, in which "U or V" would be interpreted as "either U or V, but *not* both."

For example,

$$\{1, 3\} \cup \{2, 3, 5\} = \{1, 2, 3, 5\}, \qquad \text{and} \qquad [-3, 4] \cup (2, 6) = [-3, 6).$$

The **intersection** of two sets A and B is the set containing precisely those numbers which are elements of *both* set A and set B. The symbol for the intersection of A and B is $A \cap B$. Thus,

$$A \cap B = \{x \mid x \in A \text{ and } x \in B\}.$$

For example,

$$\{1, 3\} \cap \{2, 3, 5\} = \{3\}, \qquad \text{and} \qquad [-3, 4] \cap (2, 6) = (2, 4].$$

Figure 2.4 illustrates these two set operations for the interval examples above.

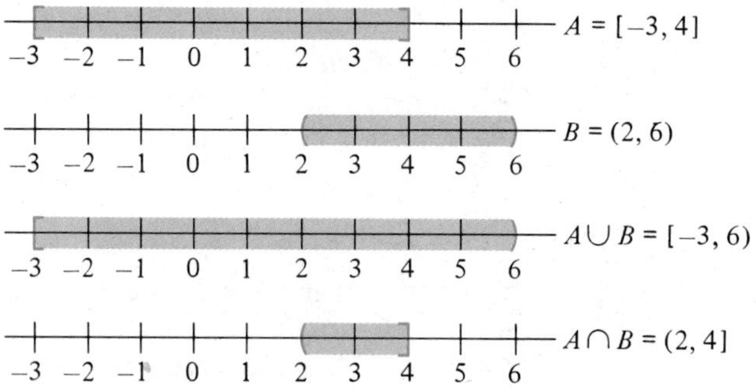

Figure 2.4 Set union and intersection.

The notation, terminology and operations for sets enables us to ensure precision in solving mathematical problems, as we shall see in Example 3.

Inequalities

There is a natural ordering for the real numbers. If b lies to the right of a on the number line, we say that b is larger than a, and we write $b > a$. Of course, this is the same as saying that a is smaller than b and writing $a < b$ (Figures 2.5 and 2.6).

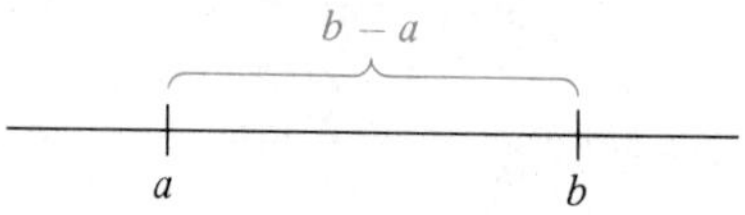

Figure 2.5 $a < b$, or $(b - a) > 0$.

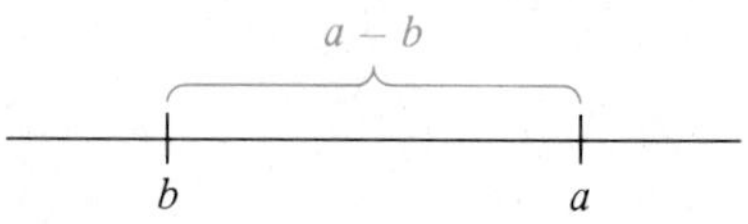

Figure 2.6 $a > b$, or $(a - b) > 0$.

A more formal way to say that b lies to the right of a is to say that the difference $(b - a)$ is positive. The formal definition of inequality is therefore the following.

DEFINITION 1

The **inequality** $a < b$ means $(b - a)$ is positive. The inequality $a \leq b$ means that either $a = b$ or $a < b$.

We will often use the properties given by the following theorem. Their proofs are quite simple, and you are asked to provide them in Exercises 70, 71, and 74.

THEOREM 1
Properties of Inequality

Let x, y, z and c be real numbers. Then

(i) Either $x = y$, $x < y$, or $x > y$. (trichotomy)
(ii) If $x < y$ and $y < z$, then $x < z$. (transitivity)
(iii) If $x < y$, then $x + c < y + c$ for any c.
(iv) If $x < y$ and $c > 0$, then $cx < cy$.
(v) If $x < y$ and $c < 0$, then $cx > cy$.

Properties (iii)–(v) are frequently used in ''solving'' inequalities, that is, finding the numbers for which the inequalities are true statements. Property (iii) says that one can add a number to both sides of an inequality without changing the sense of the inequality. However, properties (iv) and (v) show that in multiplying both sides of an inequality by c, the sense of the inequality will be retained if $c > 0$ and reversed if $c < 0$. The incorrect use of property (v) is a frequent source of error.

Example 1
(a) To solve the inequality

$$x + 2 < 5$$

we add the constant -2 to both sides, using property (iii) of Theorem 1, to obtain the inequality

$$(x + 2) + (-2) < 5 + (-2),$$

or

$$x < 3.$$

The *solution set* is, therefore, $\{x \mid x < 3\} = (-\infty, 3)$.
(b) To solve the inequality

$$-\frac{3}{2}x \geq 9,$$

we multiply both sides by $-\frac{2}{3}$, using property (v). This gives the inequality

$$\left(-\frac{2}{3}\right)\left(-\frac{3}{2}x\right) \leq \left(-\frac{2}{3}\right) \cdot 9,$$

or

$$x \leq -6.$$

The solution set is $\{x \mid x \leq -6\} = (-\infty, -6]$. ■

The gist of ''solving inequalities'' is that we attempt to isolate the variable on one side of the inequality, just as we do in solving equations containing a variable. However, you must remember one critical difference—multiplying (or dividing) by a negative number reverses the sense of the inequality.

Example 2 Solve the inequality $7x - 4 \leq 3x + 8$.

Solution: Adding 4 to both sides of the inequality gives

$$7x \leq 3x + 12.$$

We next add $-3x$ to both sides (or, subtract $3x$ from both sides) to obtain

$$4x \leq 12.$$

Finally, we divide both sides by 4 $\left(\text{or, multiply both sides by } \frac{1}{4}\right)$ to conclude that

$$x \leq 3.$$

The solution set is therefore $\{x \mid x \leq 3\} = (-\infty, 3]$. ■

Quadratic inequalities are more difficult to solve than the simple linear inequalities of Examples 1 and 2, since they involve x^2 terms. The following example demonstrates how quadratic inequalities can be solved when we are fortunate enough to be able to "factor" the quadratic expression. Note the careful use of the notions of set unions and intersections. (See Exercises 84–88 for a discussion of how to handle inequalities involving quadratic expressions which do not factor.)

Example 3 Solve the inequality $x^2 + x + 1 > 7$.

Strategy

Move all terms to the left side, leaving zero on the right.

Factor the left-hand side (if possible) and find its zeros.

These zeros divide the number line into intervals, on which $x^2 + x - 6$ will have constant sign.

Determine the sign of each factor on each interval.

From the signs of the factors, determine the sign of the quadratic expression on each interval.

Solution

Adding -7 to both sides of the inequality gives

$$x^2 + x - 6 > 0.$$

Since $x^2 + x - 6 = (x + 3)(x - 2)$, the inequality may be written as

$$(x + 3)(x - 2) > 0.$$

From this factored form we can see that $(x + 3)(x - 2) = 0$ if $x = -3$ or if $x = 2$. We may therefore determine the sign of the left hand side on the three resulting intervals $(-\infty, -3)$, $(-3, 2)$, and $(2, \infty)$ by examining the signs of the individual factors:

(i) For $x \in (-\infty, -3)$, $x < -3$, so both

$$x + 3 < 0 \quad \text{and} \quad x - 2 < 0.$$

Thus, $(x + 3)(x - 2) > 0$ if $x \in (-\infty, -3)$.

(ii) For $x \in (-3, 2)$,

(a) $x > -3$, so $x + 3 > 0$.
(b) $x < 2$, so $x - 2 < 0$.

Thus, $(x + 3)(x - 2) < 0$ if $x \in (-3, 2)$.

(iii) For $x \in (2, \infty)$, $x > 2$, so both

$$x + 3 > 0 \quad \text{and} \quad x - 2 > 0.$$

Thus, $(x + 3)(x - 2) > 0$ if $x \in (2, \infty)$.

From cases (i)–(iii), we conclude that $(x + 3)(x - 2) > 0$ if $x \in (-\infty, -3) \cup (2, \infty)$. ■

REMARK: Figure 2.7 suggests a simple graphical device for visualizing the procedure in Example 3. Using one number line for each factor and one for the quadratic expression, mark the zeros, and then mark the signs of each factor on each interval. The signs for the quadratic on each interval are then easily recognized.

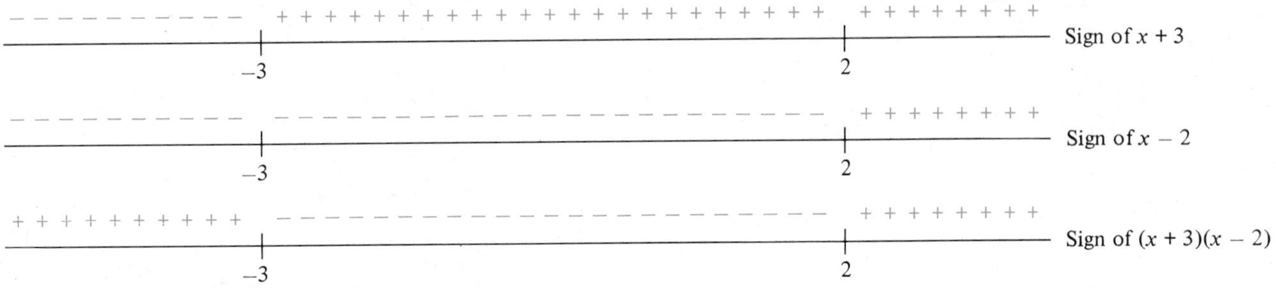

Figure 2.7 Graphical device for determining sign of a quadratic expression.

Absolute Value and Distance

In order to speak about the distance between two points on the number line, we need a way to refer to the magnitude of a number, regardless of its sign. This is the concept of the **absolute value** of a number x, denoted by $|x|$.

DEFINITION 2

$$|x| = \begin{cases} x & \text{if} \quad x \geq 0 \\ -x & \text{if} \quad x < 0. \end{cases}$$

For example, $|3| = 3$, $|-7| = 7$, and $|\pi - 4| = 4 - \pi$. (You should be cautioned that the absolute value of a number is *always* nonnegative. Even though the expression $-x$ appears in the second line of the definition of $|x|$, x itself is negative in this case, so $-x$ is positive.)

Example 4 Solve the equation $|2x - 3| = 6$.

Strategy
Write down the two cases from Definition 2.

Solution
From Definition 2 either

$$2x - 3 = 6, \quad \text{or} \quad -(2x - 3) = 6,$$

so either

$$2x = 9, \quad \text{or} \quad -2x = 3.$$

Solve the two equations that result.

These two equations produce the solutions

$$x_1 = \frac{9}{2}, \quad \text{and} \quad x_2 = -\frac{3}{2}.$$

■

Using the concept of absolute value, we can now give a precise definition for the distance between two numbers (points) on the number line.

DEFINITION 3

The **distance** between the numbers a and b is the number $|b - a|$.

Figure 2.8 illustrates the meaning of Definition 3. The distance between the numbers a and b is just the absolute value of their difference.

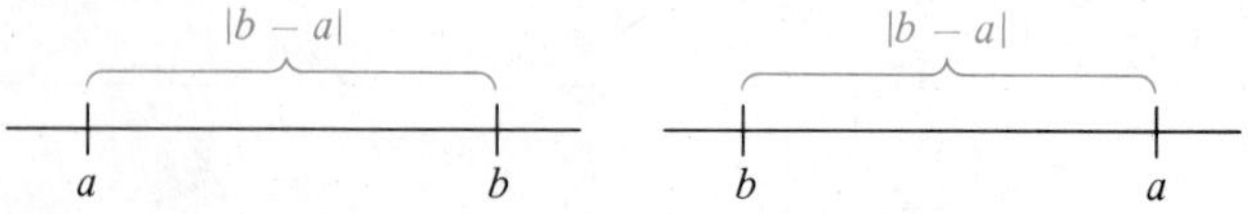

Figure 2.8 Distance on the number line.

The following example shows how we may use the notion of distance to interpret certain inequalities.

Example 5 Find all numbers x which satisfy the inequality

$$|x - 2| < 5.$$

Strategy
Interpret the statement geometrically, using the definition of distance.

Solution
Since $|x - 2|$ is the distance between the numbers x and 2, the inequality is describing precisely those numbers x which lie at a distance less than 5 from the number 2. These numbers are described by the inequality

$$-3 < x < 7.$$

Provide a sketch.

The solution set is $\{x \mid -3 < x < 7\} = (-3, 7)$.
See Figure 2.9. ■

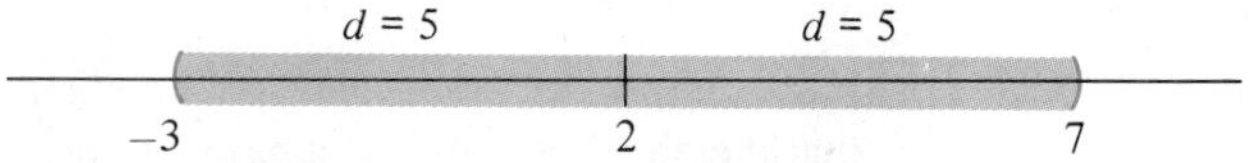

Figure 2.9 Solution of the inequality $|x - 2| < 5$.

You will need to be able to solve inequalities such as that in Example 5 throughout this text. From the solution of Example 5 we can see that the inequality

$$|x| < a$$

can always be rewritten as the inequality

$$-a < x < a$$

according to the interpretation of $|x|$ as $|x - 0|$, the distance between x and the origin.

Example 6 Solve the inequality $|7 - 2x| \leq 5$

Strategy
Use the above remark to rewrite the inequality.

Solution
By the preceding remark we can rewrite the inequality as

$$-5 \leq 7 - 2x \leq 5,$$

Subtract 7 from all terms.

so

$$-12 \leq -2x \leq -2,$$

Divide all terms by −2, remembering to reverse sense of all inequalities.

and

$$6 \geq x \geq 1.$$

The solution set is therefore $\{x \mid 1 \leq x \leq 6\}$, or $[1, 6]$. ■

We shall make frequent use of the following properties of absolute value.

THEOREM 2
Properties of Absolute Value

For any real numbers x and y,

(i) $|x| \geq 0$,
(ii) $|xy| = |x| \cdot |y|$,
(iii) $|x + y| \leq |x| + |y|$.

Statement (i) simply reminds us that absolute values are never negative. Statement (ii) says that the absolute value of a product is the same as the product of the absolute values. Statement (iii) is referred to as the **triangle inequality** (for reasons we shall see later) and states that the absolute value of the sum of two numbers can never exceed the sum of their absolute values. Notice that statement (iii) is an inequality rather than an equation. An example where equality fails is $|3 + (-5)| = |-2| = 2 < |3| + |-5|$.

Laws of Exponents

Finally, we recall the laws of exponents, which are based on the notion of repeated multiplications.

DEFINITION 4

Let x and y be real numbers, and let n and m be positive integers.

(a) x^n means $x \cdot x \cdot x \cdot \cdots \cdot x$. ($n$ factors).

(b) x^{-n} means $\dfrac{1}{x^n}$. ($x \neq 0$).

(c) $x^{1/n} = y$ means $y^n = x$.
(d) $x^{m/n}$ means $(x^{1/n})^m$.
(e) x^0 means 1 if $x \neq 0$.

Example 7

(a) $2^5 = 2 \cdot 2 \cdot 2 \cdot 2 \cdot 2 = 32$

(b) $6^{-2} = \dfrac{1}{6^2} = \dfrac{1}{36}$

(c) $16^{3/4} = (16^{1/4})^3 = 2^3 = 8$
(d) $(-27)^{2/3} = ((-27)^{1/3})^2 = (-3)^2 = 9$
(e) But, $(-27)^{3/2}$ is not defined, since $(-27)^{1/2} = y$ implies $-27 = y^2$, which has no solution. ■

The following properties of exponents follow from Definition 4.

THEOREM 3
Laws of Exponents

Let x and y be real numbers, and let m and n be integers. Then

(a) $x^n \cdot x^m = x^{n+m}$,

(b) $\dfrac{x^n}{x^m} = x^{n-m}$, $x \neq 0$,

(c) $(x^n)^m = x^{nm}$,
(d) $x^{m/n} = (x^{1/n})^m = (x^m)^{1/n}$.

Example 8 The expression $\dfrac{(4x^3y^{2/3})^2}{x^{1/3}y^2}$ may be simplified as follows:

$$\begin{aligned}\frac{(4x^3y^{2/3})^2}{x^{1/3}y^2} &= \frac{4^2\cdot(x^3)^2\cdot(y^{2/3})^2}{x^{1/3}y^2}\\ &= \frac{16x^6y^{4/3}}{x^{1/3}y^2}\\ &= 16x^{\left(6-\frac{1}{3}\right)}y^{\left(\frac{4}{3}-2\right)}\\ &= 16x^{17/3}y^{-2/3}.\end{aligned}$$

■

Exercise Set 1.2

In each of Exercises 1–9, label the real number as either a rational number, an irrational number, or an integer.

1. $22/7$ **2.** $3.141\overline{414}\ldots$ **3.** $2.101001001\ldots$

4. $\pi + 3$ **5.** $\dfrac{51}{17}$ **6.** $-6.\overline{163163}\ldots$

7. $\sqrt{256}$ **8.** $\sqrt{2}$ **9.** $1 + \sqrt{2}$

In Exercises 10–15, use the intervals $A = [-2, 5]$, $B = (-1, 6)$, and $C = (-\infty, 0)$ to find the indicated intervals.

10. $A \cup B$ **11.** $A \cap B$ **12.** $A \cup C$

13. $A \cap C$ **14.** $B \cup C$ **15.** $B \cap C$

16. Let $A = [2, 4)$. Find the interval B so that $A \cup B = [2, 6]$ and $A \cap B = (3, 4)$.

17. True or false? For a given circle, the radius and the circumference cannot both be rational numbers. What about area and radius? Area and circumference?

In each of Exercises 18–31, solve the given inequality.

18. $2x + 3 < 9$ **19.** $x + 4 \le 3x + \pi$

20. $2x + 7 > 4x - 5$ **21.** $6x - 6 \le 8x + 8$

22. $-4 \le 2(x + 2) < 10$ **23.** $(x - 3)(x + 1) < 0$

24. $(x + 7)(2x - 4) > 0$ **25.** $x(x + 6) > -8$

26. $\dfrac{x}{x - 1} > 0,\ x \ne 1$ **27.** $\dfrac{2x - 3}{3} < 3$

28. $x^2 + x < 0$ **29.** $x^3 + x^2 - 2x > 0$

30. $x^2 + x + 7 > 19$ **31.** $x^4 - 9x^2 < 0$

In Exercises 32–43, simplify the given expression as much as possible.

32. $8^{4/3}$ **33.** $81^{3/4}$ **34.** $121^{-3/2}$

35. $\left(\dfrac{1}{4}\right)^{-3/2}$ **36.** $(-32)^{-1/5}$ **37.** $(36^{-1/2})\cdot(9^{3/2})$

38. $((x^2y)y^2)^2$ **39.** $\dfrac{\sqrt{x^3y^4}}{xy^3}$ **40.** $(x^2y^6z^{-2})^{-3/2}$

41. $\dfrac{4^{3/2}\cdot 16^{-3/4}\cdot 32^{2/5}}{8\cdot 2^{-3}}$ **42.** $\sqrt{2\sqrt{4\sqrt{16}}}$ **43.** $\dfrac{\sqrt{xy^{2/3}z}}{(x^2\sqrt{y})^2z^{-3/2}}$

In Exercises 44–53, solve the given equations.

44. $|x - 7| = 2$ **45.** $|5x - 2| = 0$

46. $x + |x| = 0$ **47.** $2|3x - 1| = 22$

48. $x + |-3| = 7$ **49.** $(|x| + 6)^2 = 49$

50. $x - 6 = 2x - |\pi - 6|$ **51.** $x + 3|x| = 8$

52. $|(x + 2)^2 + 3| = 12$ **53.** $|x - 4| = |x - 7|$

54. True or false? $|a - b| = |b - a|$ for all real numbers a and b.

In Exercises 55–62, solve for x and sketch the solution set.

55. $|x - 3| \le 2$ **56.** $|x + 5| < 4$

57. $|x + 2| > 1$ **58.** $3|x - 6| \ge 12$

59. $|2x - 7| \le 3$ **60.** $|8 - 3x| \ge 5$

61. $|x - 1| = x - 1$ **62.** $x > |x|$

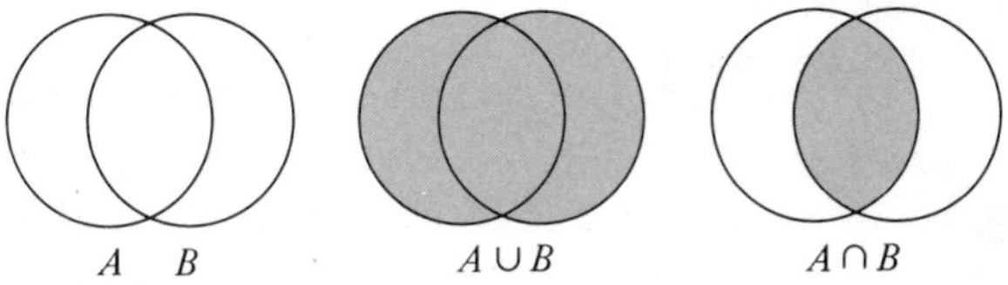

In Exercises 63–66, solve the inequality geometrically by describing in words the meaning of the inequality in terms of distances.

63. $|x - 5| > 2$ **64.** $|x - 1| = |x + 3|$

65. $|x - 1| = 2|x - 3|$ **66.** $2 < |x - 4| < 5$

67. Write down an inequality whose solution is the set of all numbers lying at a distance greater than 4 from the number -5.

68. Write down an equation whose solution is the number whose distance from -2 is twice its distance from 12.

69. The terminology and operations described for sets of real numbers may be applied to more general kinds of sets. **Venn diagrams** may be used to illustrate these concepts, as shown below. The idea is simply that the region inside the figure represents the elements of the particular set.

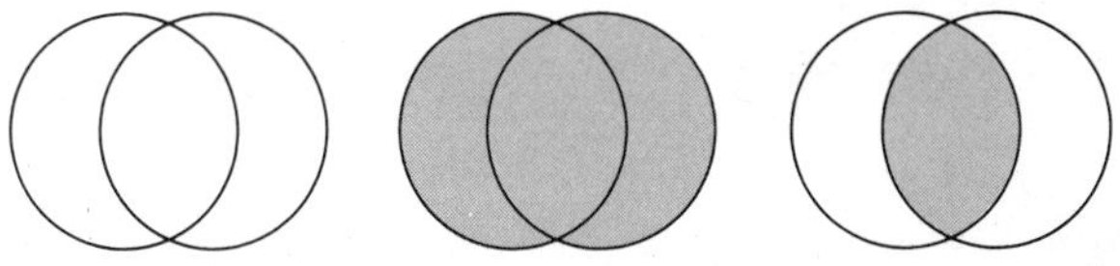

$A \quad B$ $\qquad A \cup B$ $\qquad A \cap B$

Draw Venn diagrams illustrating three sets A, B, and C, so that each pair shares a common area, for the following sets.
a. $A \cup B \cup C$ **b.** $A \cap B \cap C$ **c.** $A \cap (B \cup C)$

70. Prove Theorem 1, part (i), by examining the sign of $(y - x)$.

71. Prove Theorem 1, part (ii) by writing $(z - x) = (z - y) + (y - x)$ and applying Definition 1.

72. True or false? If $a^2 < b^2$ then $a < b$.

73. Prove that if $a^2 < b^2$ and $b > 0$ then $-b < a < b$. (*Hint:* Factor $b^2 - a^2$ as $(b - a) \cdot (b + a)$.)

74. Prove properties (iii)–(v) of Theorem 1.

75. Use Exercise 73 to solve the inequality $(x + 6)^2 < 49$.

76. Prove that $\sqrt{2}$ is not a rational number by filling in the details of the following argument:

a. Argue by contradiction. This means, assume that $\sqrt{2}$ *is* rational and seek a contradiction.

b. If $\sqrt{2}$ is rational, $\sqrt{2} = p/q$ where p and q have no common factors.

c. Thus $\sqrt{2}q = p$, so $2q^2 = p^2$.

d. This means p is divisible by 2 (why?), so $p = 2k$. Thus,

$$2q^2 = (2k)^2, \text{ so } q^2 = 2k^2.$$

e. This means q is divisible by 2.

f. Statements d. and e. contradict the last phrase of statement b. The assumption that $\sqrt{2}$ is rational is therefore false.

77. Prove that $|x| = \sqrt{x^2}$ for all real numbers x.

78. Prove the triangle inequality by noticing that

$$-|x| \le x \le |x| \quad \text{and} \quad -|y| \le y \le |y|$$

and adding to obtain $-(|x| + |y|) \le x + y \le |x| + |y|$.

79. True or false? Taking reciprocals reverses the sense of an inequality, i.e., if $a < b$ then $\dfrac{1}{a} > \dfrac{1}{b}$.

80. State and prove the correct version of the statement in Exercise 79.

81. Prove the inequality $|x| - |y| \le |x - y|$ by writing $y = x + (y - x)$ and applying the triangle inequality.

82. Does the same basic argument as Exercise 76 show that $\sqrt{3}$ is not rational?

83. The following demonstration shows how to recover the fractional form of a repeating decimal (rational number) $2.6363\overline{63}\ldots$.

$$x = 2.6363\overline{63}\ldots, \quad \text{so} \quad 100x = 263.6363\overline{63}\ldots,$$

so

$$99x = (100x - x) = 261.00\overline{00}\ldots.$$

Thus

$$x = \frac{261}{99}.$$

a. Use a similar procedure to find a fractional form for the rational number $1.341341\overline{341}\ldots$.

b. Generalize your findings to a statement about how to recover the fractional form for the repeating decimal $x = 0.\overline{a_1a_2\ldots a_n}$.

84. The method of Example 3 for solving quadratic inequalities assumed that the quadratic expression could be easily factored. Explain how the quadratic formula

$$x = \frac{-b \pm \sqrt{b^2 - 4ac}}{2a}$$

may be used to solve the quadratic inequality

$$ax^2 + bx + c > 0 \qquad (\text{or}, < 0)$$

when the expression $ax^2 + bx + c$ "does not factor."

In Exercises 85–88, use the method of Exercise 84 to solve the quadratic inequality.

85. $x^2 + x - 3 > 0$ **86.** $x^2 + 5x - 5 < 0$

87. $4 - x - x^2 > 0$ **88.** $2x^2 - 8x + 3 > 0$

1.3 THE COORDINATE PLANE, DISTANCE, AND CIRCLES

The theory and techniques that we shall develop in the first two thirds of this text are designed for analyzing situations involving two variables. For example, we will consider the relationship between time and distance for particles moving along a line, the relationship between selling price and profit for an item being marketed in a monopoly, and the relationship between pressure and temperature for a quantity of an ideal gas.

The Coordinate Plane

In order to illustrate these relationships graphically, we make use of a two-dimensional coordinate system constructed as follows. If x and y represent the two variables in question, we construct axes for x and for y (known as the **x-axis** and **y-axis,** respectively) so that the axes are perpendicular and so that the two origins lie at the same point. (Also, we follow the convention that the positive half of the horizontal axis extends to the right and the positive half of the vertical axis extends upward.) The plane determined by these two axes is called the **xy-coordinate plane.**

Each point P in the xy-plane is associated with a pair of real numbers, (x_0, y_0), called its coordinates. These coordinates are found by constructing lines through P parallel to both axes. The number x_0 on the x-axis where the line parallel to the y-axis crosses is called the **x-coordinate** for P. Similarly, the number y_0 where the line parallel to the x-axis crosses the y-axis is called the **y-coordinate** for P. The coordinate axes divide the plane into four quadrants, numbered as in Figure 3.1. Figure 3.2 shows the coordinates of several points in the plane.

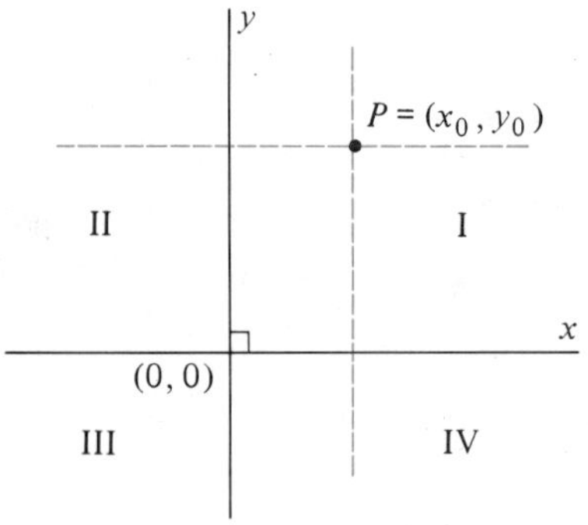

Figure 3.1 The xy-coordinate plane.

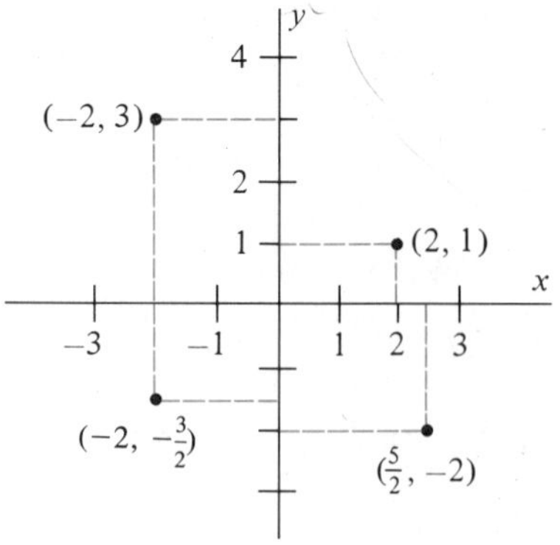

Figure 3.2 Points in the xy-plane.

To find the point Q in the xy-plane corresponding to the given coordinates (x_1, y_1), we reverse the procedure described above, constructing a vertical line through the point $x = x_1$ on the x-axis and a horizontal line through $y = y_1$ on the y-axis. The point where these lines cross is the desired point.

Distance

To calculate the distance between the points $P = (x_1, y_1)$ and $Q = (x_2, y_2)$ in the plane, we make use of the point $R = (x_2, y_1)$. By the way R is chosen, the points P, Q, and R are vertices of a right triangle.

It then follows from the way we have defined distance on the number line, and from the Pythagorean theorem, that

$$\begin{aligned} d^2 &= |x_2 - x_1|^2 + |y_2 - y_1|^2 \\ &= (x_2 - x_1)^2 + (y_2 - y_1)^2. \end{aligned}$$

(See Figure 3.3.) This establishes our definition of distance in the plane.

DEFINITION 5

The distance $d(P, Q)$ between the points $P = (x_1, y_1)$ and $Q = (x_2, y_2)$ in the xy-plane is the number

$$d(P, Q) = \sqrt{(x_2 - x_1)^2 + (y_2 - y_1)^2}.$$

(Note that $d(P, Q)$ cannot be negative.)

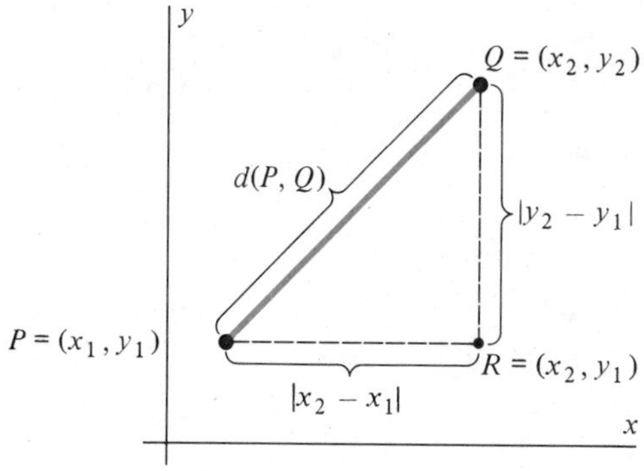

Figure 3.3 Distance between P and Q is $d(P, Q) = \sqrt{(x_2 - x_1)^2 + (y_2 - y_1)^2}$.

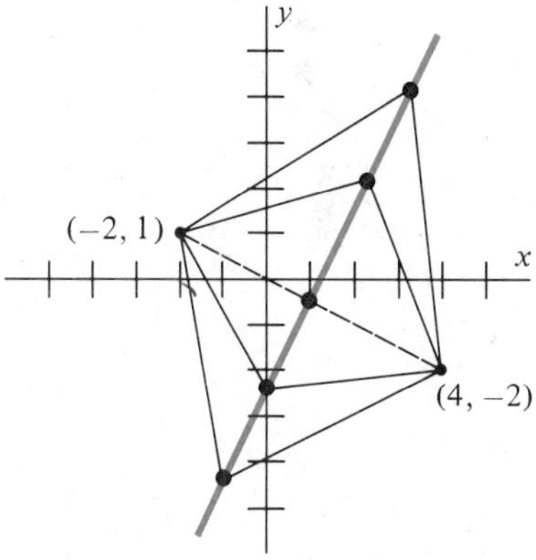

Figure 3.4 Points equidistant from two points lie along a line.

Example 1 The distance between the points $P = (-4, 1)$ and $Q = (1, 3)$ is

$$\begin{aligned} d(P, Q) &= \sqrt{(1 - (-4))^2 + (3 - 1)^2} \\ &= \sqrt{5^2 + 2^2} \\ &= \sqrt{29}. \end{aligned}$$

Example 2 Find the set of all points lying equidistant from the points $P = (-2, 1)$ and $Q = (4, -2)$.

Strategy

Select an arbitrary point satisfying the condition.

Solution

We let $T = (x, y)$ be any point lying equidistant from the points $P = (-2, 1)$ and $Q = (4, -2)$. This means that

$$d(P, T) = d(T, Q)$$

Use Definition 5 to convert the condition of equal distances to an equation in x and y.

or

$$\sqrt{(x - (-2))^2 + (y - 1)^2} = \sqrt{(x - 4)^2 + (y - (-2))^2}.$$

Squaring both sides of this equation gives

$$(x + 2)^2 + (y - 1)^2 = (x - 4)^2 + (y + 2)^2,$$

Simplify the resulting equation as much as possible.

so

$$(x^2 + 4x + 4) + (y^2 - 2y + 1) = (x^2 - 8x + 16) + (y^2 + 4y + 4).$$

This equation simplifies to the equation

$$12x = 6y + 15,$$

or

$$y = 2x - \frac{5}{2}.$$

The solution is therefore the set

$$\left\{(x, y) \,\middle|\, y = 2x - \frac{5}{2}\right\}.$$

This set of points is the line sketched in Figure 3.4. ■

Circles

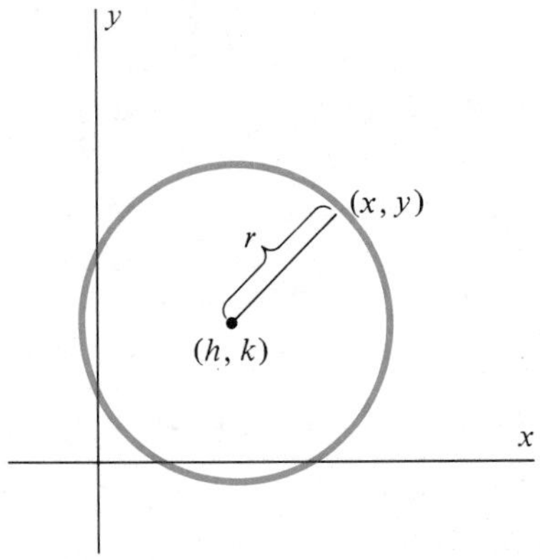

Figure 3.5 The circle $(x - h)^2 + (y - k)^2 = r^2$.

Geometrically, a **circle** may be defined as the set of all points lying at a fixed distance (the **radius**) from a fixed point (the **center**). According to Definition 5, this means that the point (x, y) lies on the circle with radius r and center (h, k) if and only if

$$\sqrt{(x - h)^2 + (y - k)^2} = r \tag{1}$$

Squaring both sides of equation (1), we obtain the **standard** form for the equation of the circle with radius r and center (h, k):

$$\boxed{(x - h)^2 + (y - k)^2 = r^2} \tag{2}$$

(See Figure 3.5.)

Example 3 The equation for the circle with center $(1, -2)$ and radius $r = 4$ is, according to equation (2),

$$(x - 1)^2 + (y - (-2))^2 = 4^2,$$

which simplifies to

$$(x - 1)^2 + (y + 2)^2 = 16,$$

or

$$x^2 - 2x + y^2 + 4y - 11 = 0.$$ ■

Example 4 Find an equation for the circle(s) of radius $r = 5$ that contains the points $(1, 2)$ and $(-1, -2)$.

Strategy

Begin with the standard equation for a circle with r known.

Solution

According to equation (2), a circle of radius 5 must have an equation of the form

$$(x - h)^2 + (y - k)^2 = 25$$

Substitute given values for x and y to obtain equations for h and k.

where (h, k) is the center. Substituting $x = 1$ and $y = 2$ into this equation gives

$$(1 - h)^2 + (2 - k)^2 = 25$$

which simplifies to the equation

$$h^2 - 2h + k^2 - 4k = 20. \tag{3}$$

Similarly, the fact that the point $(-1, -2)$ lies on the circle gives the equation

$$(-1 - h)^2 + (-2 - k)^2 = 25$$

which simplifies to the equation

$$h^2 + 2h + k^2 + 4k = 20. \tag{4}$$

Use algebra to solve the resulting two equations:

(a) Subtract, to eliminate the h^2 and k^2 terms.

(b) Solve for h in terms of k.

(c) Substitute back into (3) or (4) to obtain an equation in just one variable.

(d) Solve the resulting equation for k. Then obtain h from the substitution.

Note that *two* solutions result.

Next, we subtract the corresponding sides of equation (3) from those of equation (4) to find that h and k satisfy the equation

$$4h + 8k = 0,$$

or

$$h = -2k. \tag{5}$$

Finally, we use this substitution in equation (4) to obtain an equation in k alone:

$$(-2k)^2 + 2(-2k) + k^2 + 4k = 20,$$

or

$$5k^2 = 20.$$

Thus, $k = -2$ or 2. From this information and equation (5), we conclude that the two possible centers for the circle are

$$(h, k) = (4, -2) \text{ and } (h, k) = (-4, 2).$$

The equations for the two circles are

$$(x - 4)^2 + (y + 2)^2 = 25$$

and

$$(x + 4)^2 + (y - 2)^2 = 25.$$

(See Figure 3.6.) ■

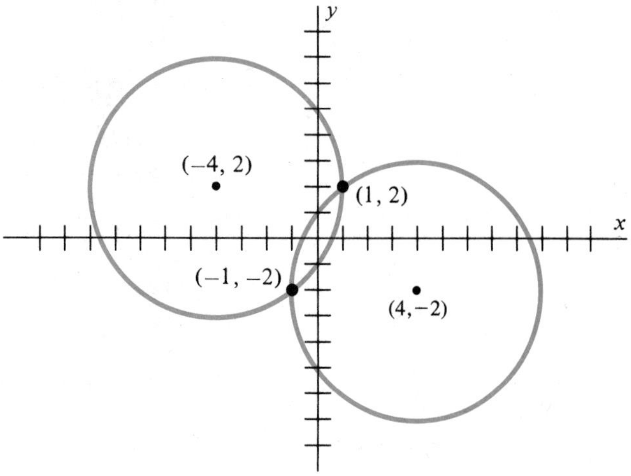

Figure 3.6 Two circles of radius 5 contain the points (1, 2) and (−1, −2).

Examples 3 and 4 had to do with finding an equation for a circle, given certain data. Example 6 involves the reverse problem—identifying a circle from its equation. The procedure for doing so involves **completing the square,** a technique from elementary algebra that goes as follows.

Recall, the result of squaring the binomial $(x + a)$ is the quadratic expression

$$(x + a)^2 = x^2 + 2ax + a^2. \tag{6}$$

The technique of completing the square answers the opposite question: What constant, when added to the expression $x^2 + bx$, will produce a perfect square, as in equation (6)? By examining the right side of equation (6) you can see that *the constant term* (a^2) *in a perfect square is the square of half the coefficient of* x. We therefore "complete the square on $x^2 + bx$" as follows:

$$\begin{aligned} x^2 + bx &= \left[x^2 + bx + \left(\frac{b}{2}\right)^2\right] - \left(\frac{b}{2}\right)^2 \\ &= \left(x + \frac{b}{2}\right)^2 - \frac{b^2}{4}. \end{aligned}$$

It is important to note that whatever constant is added to the expression $x^2 + bx$ must also be subtracted from the result, so that the values of the expression are not changed.

Example 5 Completing the square.

(a) $$\begin{aligned} x^2 + bx &= \left[x^2 + bx + \left(\frac{b}{2}\right)^2\right] - \left(\frac{b}{2}\right)^2 \\ &= \left(x + \frac{b}{2}\right)^2 - \frac{b^2}{4} \end{aligned}$$

(b) $$\begin{aligned} x^2 + 6x &= [x^2 + 6x + 3^2] - 3^2 \\ &= (x + 3)^2 - 9 \end{aligned}$$

(c) $$\begin{aligned} x^2 - 8x &= [x^2 - 8x + (-4)^2] - (-4)^2 \\ &= (x - 4)^2 - 16 \end{aligned}$$

(d) $$\begin{aligned} x^2 + \frac{3}{4}x &= \left[x^2 + \frac{3}{4}x + \left(\frac{3}{8}\right)^2\right] - \left(\frac{3}{8}\right)^2 \\ &= \left(x + \frac{3}{8}\right)^2 - \frac{9}{64} \end{aligned}$$

(e) $$\begin{aligned} 3x^2 - 12x &= 3[x^2 - 4x] \\ &= 3[x^2 - 4x + (-2)^2] - 3\cdot(-2)^2 \\ &= 3(x - 2)^2 - 12. \end{aligned}$$ (*Note:* factor of 3.) ■

Example 6 Describe and sketch the graph of the equation

$$x^2 + y^2 + 6x - 2y + 6 = 0.$$

Strategy

Complete the square on x and y terms to try to bring equation into form of equation (1).

Solution

We write the given equation as

$$(x^2 + 6x) + (y^2 - 2y) + 6 = 0. \tag{7}$$

Completing the square in x.

To complete the square on the first term, we add the square of half the coefficient of x, that is, we add $\left(\frac{6}{2}\right)^2 = 9$. This gives

$$(x^2 + 6x) = (x^2 + 6x + 9) - 9 = (x + 3)^2 - 9. \tag{8}$$

Completing the square in y.

Similarly, we complete the square on the second term as

$$(y^2 - 2y) = (y^2 - 2y + 1) - 1 = (y - 1)^2 - 1. \tag{9}$$

Combining the results.

Substituting the expressions in equations (8) and (9) into equation (7) gives the equation

$$[(x + 3)^2 - 9] + [(y - 1)^2 - 1] + 6 = 0$$

Compare the equation obtained with equation (2).

or

$$(x + 3)^2 + (y - 1)^2 = 4.$$

The graph is therefore a circle with center $(-3, 1)$ and radius $r = \sqrt{4} = 2$ (see Figure 3.7). ■

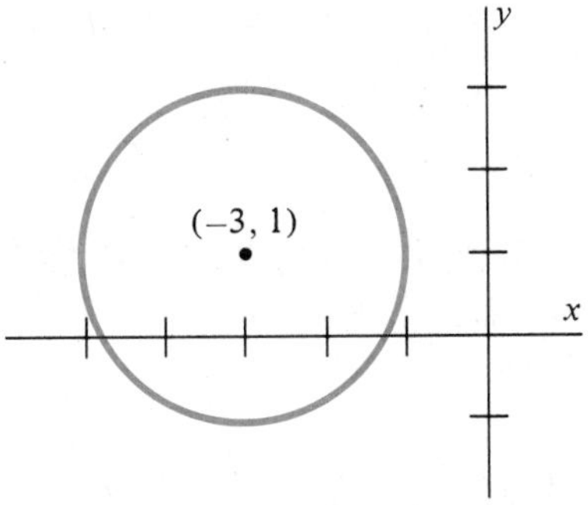

Figure 3.7 Circle $x^2 + y^2 + 6x - 2y = -6$.

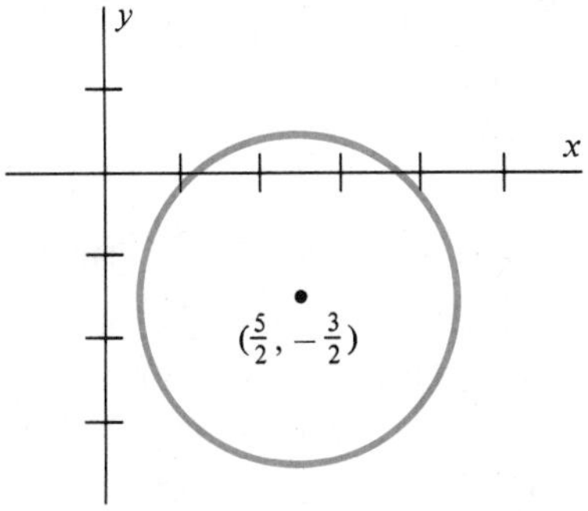

Figure 3.8 Circle $2x^2 - 10x + 2y^2 + 6y = -9$.

Example 7 Describe the circle with equation

$$2x^2 - 10x + 2y^2 + 6y + 9 = 0.$$

Solution: Grouping x and y terms and completing squares we obtain

$$2x^2 - 10x + 2y^2 + 6y + 9 = 0$$
$$2(x^2 - 5x) + 2(y^2 + 3y) = -9$$
$$2\left(x^2 - 5x + \frac{25}{4}\right) - \frac{25}{2} + 2\left(y^2 + 3y + \frac{9}{4}\right) - \frac{9}{2} = -9$$
$$2\left(x - \frac{5}{2}\right)^2 + 2\left(y + \frac{3}{2}\right)^2 = -9 + \frac{25}{2} + \frac{9}{2}$$
$$\left(x - \frac{5}{2}\right)^2 + \left(y + \frac{3}{2}\right)^2 = 4.$$

The graph is therefore a circle with center $\left(\frac{5}{2}, -\frac{3}{2}\right)$ with radius 2. (Figure 3.8.) ■

Exercise Set 1.3

1. Plot the given points on the xy-coordinate plane.
   a. (4, 2) **c.** (−2, 0) **e.** (−3, −3)
   b. (−6, 3) **d.** (1, −3) **f.** (6, −2)

2. Find the distance between the following pairs of points.
   a. (2, −1) and (0, 2) **d.** (1, −3) and (6, 6)
   b. (3, 1) and (1, 3) **e.** (1, 1) and (−1, −1)
   c. (0, 2) and (1, −9) **f.** (−2, −2) and (1, 2)

3. Find two points (h, k) so that $h = 5$ and so that the distance between (h, k) and (1, 3) equals 5.

4. Find two points in the xy-coordinate plane, each of which lies a distance of 4 units from both (−2, 3) and (2, 3).

5. True or false? *Every* point in the plane lying at a distance r from point P must lie on the circle with center P and radius r.

6. Find a so that the triangle with vertices (−1, 0), (2, 3), and $(a, 0)$ is isosceles.

7. Find b so that the triangle with vertices (0, 0), (2, 0), and $(1, b)$ is equilateral.

8. Find an equation for the set of all points lying equidistant from the points (−1, −1) and (3, 1).

9. Find an equation for the set of all points lying equidistant from the points (1, 2) and (5, −1).

10. Verify that the coordinates of the midpoint of the line segment joining the points (x_1, y_1) and (x_2, y_2) are $\left(\frac{x_1 + x_2}{2}, \frac{y_1 + y_2}{2}\right)$. (*Hint:* Use the distance formula.)

11. Find the midpoints of the line segments joining the following pairs of points. (See Exercise 10.)
    a. (0, 0) and (0, 6) **c.** (−1, 2) and (6, 4)
    b. (1, 3) and (5, 5) **d.** (1, 1) and (−7, −3)

In Exercises 12–15, use set builder notation to identify the points in the xy-plane satisfying the given property, and sketch the set.

12. The set of all points lying more than $\sqrt{2}$ units from the point (−3, 1).

13. The set of all points lying at least 3 units to the right of the y-axis and at least 3 units from the x-axis.

14. The set of all points lying no more than 2 units from either axis.

15. The set of all points lying more than 3 but less than 5 units from the point (1, −2).

In Exercises 16–20, write the equation for the circle with the stated properties.

16. The circle with center (0, 0) and radius 3.

17. The circle with center (0, 0) and radius 5.

18. The circle with center (1, 3) and radius 2.

19. The circle with center (−2, 4) and radius 3.

20. The circle with center (−6, −4) and radius 5.

21. Find an equation for the circle with center (2, 3) that contains the point (2, −1).

22. Find equations for the circles with radius 5 and containing the points (−1, −4) and (4, 1).

In Exercises 23–28, complete the square.

23. $x^2 - 3x$ **24.** $x^2 + 4x$ **25.** $3x^2 - 9x$

26. $t - 2t^2$ **27.** $ax^2 + bx$ **28.** $ax - bx^2$

In each of Exercises 29–36, find the center and radius of the circle whose equation is given, and sketch the circle.

29. $x^2 - 2x + y^2 - 8 = 0$

30. $x^2 - 2x + y^2 + 6y - 12 = 0$

31. $x^2 + y^2 + 4x + 2y - 11 = 0$

32. $x^2 + 14x + y^2 - 10y + 70 = 0$

33. $x^2 - 6x + y^2 - 4y + 8 = 0$

34. $x^2 + y^2 - 2x - 6y + 3 = 0$

35. $x^2 - 2ax + y^2 + 4ay + 5a^2 - 1 = 0$

36. $x^2 + y^2 - 2by + b^2 - a^2 = 0$

37. Find an equation for the circle with center (3, 2) which is tangent to the y-axis. (That is, which intersects the y-axis in a single point.)

38. Find the points of intersection of the line $x - y - 5 = 0$ with the circle $x^2 - 8x + y^2 - 4y + 11 = 0$.

39. True or false? Three points determine a circle. Why?

40. Show that the set of all points whose distance from the point (−2, 1) is twice the distance from the point (4, −2) is a circle. Find its radius and center.

41. Find an equation for the circle containing the points (−2, 1), (1, 4), and (4, 1).

42. Find the family of *all* circles containing the two points (1, 2) and (−1, 2). (See Example 4.)

43. Show that the equation $(x - h)^2 + (y - k)^2 = d$ will have (1) infinitely many solutions (i.e., the points on a circle), (2) exactly one solution, or (3) no solutions, depending on whether d is positive, zero, or negative. (The last two cases are referred to as **degenerate** circles.)

1.4 LINEAR EQUATIONS

Equations whose graphs are straight lines are very important in the calculus. Indeed, one of the most fundamental problems in the entire subject is that of finding the equation of the line tangent to a given curve at a given point. More generally, lines are important because they represent the simplest relationship between two variables. For example, we shall see that for a freely falling body the relationship between velocity and time is a line. Even when the relationship between two variables is more complicated than simply a straight line, scientists often "linearize" their models. That is, they find the straight line which "best approximates" the true relationship, usually out of a desire to keep their models as simple as possible. While all these issues are well down the road, it is important to develop sharp skills in working with lines before more complex situations arise.

Slope

If $P_1 = (x_1, y_1)$ and $P_2 = (x_2, y_2)$ are points on a nonvertical line ℓ, the **slope** of ℓ is defined by the ratio

$$m = \frac{y_2 - y_1}{x_2 - x_1} \tag{1}$$

as illustrated in Figure 4.1.

For a given line, the value of m is independent of the particular choices for (x_1, y_1) and (x_2, y_2). (See Exercise 46.) Notice that expression (1) for slope is not defined if $x_1 = x_2$. For this reason, we say that a vertical line has no slope. For all other lines slope is a real number, and every real number can occur as the slope of some line. Some illustrations of lines with various slopes appear in Figure 4.2.

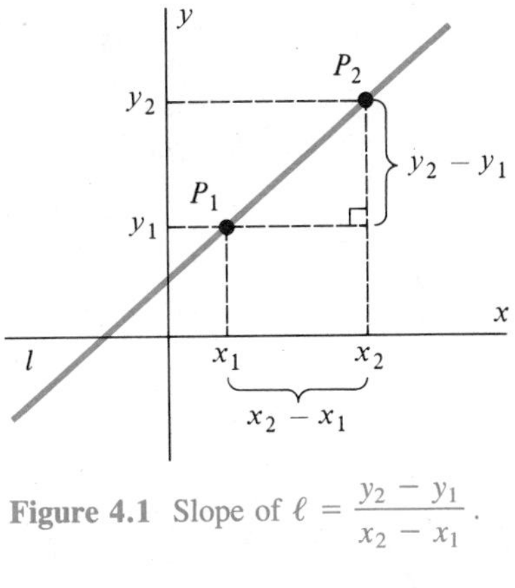

Figure 4.1 Slope of $\ell = \dfrac{y_2 - y_1}{x_2 - x_1}$.

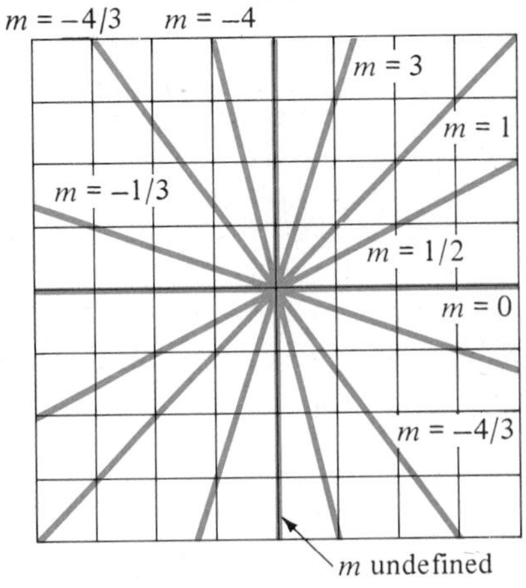

Figure 4.2 Some slopes.

Example 1 Find the slope of the line through the points

(a) $(-7, 2)$ and $(3, -3)$, and
(b) $(-1, -4)$ and $(7, 16)$.

Solution: From equation (1) we have

(a) $m = \dfrac{-3 - 2}{3 - (-7)} = \dfrac{-5}{10} = -\dfrac{1}{2}$, and

(b) $m = \dfrac{16 - (-4)}{7 - (-1)} = \dfrac{20}{8} = \dfrac{5}{2}$. ■

Equations for Lines

Suppose ℓ is a line with slope m which contains the point (x_1, y_1). To find the equation for ℓ we let $P = (x, y)$ be an arbitrary point on ℓ. Then, by equation (1), we obtain

$$m = \frac{y - y_1}{x - x_1}$$

so

$$\boxed{y - y_1 = m(x - x_1),} \tag{2a}$$

or

$$\boxed{y = mx + b} \tag{2b}$$

where $b = y_1 - mx_1$.

Equation (2a) is called the **point-slope** equation for the line ℓ. Equation (2b) is called the **slope-intercept** form of the equation for ℓ. This is because the constants m and b appearing in equation (2b) are the slope and y-intercept for ℓ, respectively. To see this, notice that if we set $x = 0$ in equation (2b), we obtain the statement $y = b$. This means that the point $(0, b)$ is on the graph of ℓ. (See Figure 4.3.)

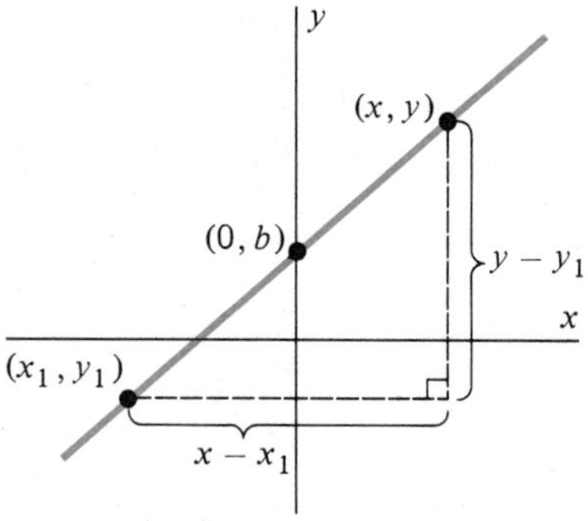

Figure 4.3 $y = mx + b$ has y intercept b.

Thus, every nonvertical line can be described by an equation of the form (2b). Conversely, if (x_1, y_1) and (x_2, y_2) are any two distinct points whose coordinates satisfy equation (2b), then the slope of the line determined by these two points is

$$\text{Slope} = \frac{y_2 - y_1}{x_2 - x_1} = \frac{(mx_2 + b) - (mx_1 + b)}{x_2 - x_1} = \frac{m(x_2 - x_1)}{x_2 - x_1} = m.$$

It follows that any point (x, y) that satisfies equation (2b) lies on the line with slope m and y-intercept b. The following theorem summarizes our discussion.

THEOREM 4 The graph of the equation $y = mx + b$ is a line with slope m and y-intercept b. Conversely, the line with slope m and y-intercept b has an equation of this form.

Example 2 Find an equation for the line through (2, 5) with slope $m = 3$.

Solution: Substituting directly into equation (2a) gives

$$y - 5 = 3(x - 2),$$

so

$$y = 3x - 1$$

is an equation for the desired line. ■

Example 3 Find an equation for the line through $(-3, 4)$ and $(1, -2)$.

Strategy
First, find the slope using (1).

Solution
By equation (1), the slope of the line is

$$m = \frac{-2 - 4}{1 - (-3)} = -\frac{6}{4} = -\frac{3}{2}.$$

Then, use slope and one point to write an equation for the line.

Using the point $(-3, 4)$ and the slope $m = -\frac{3}{2}$ in equation (2a) then gives

$$y - 4 = \left(-\frac{3}{2}\right)(x + 3),$$

or

$$y = -\frac{3}{2}x - \frac{1}{2}.$$

(See Figure 4.4.) ■

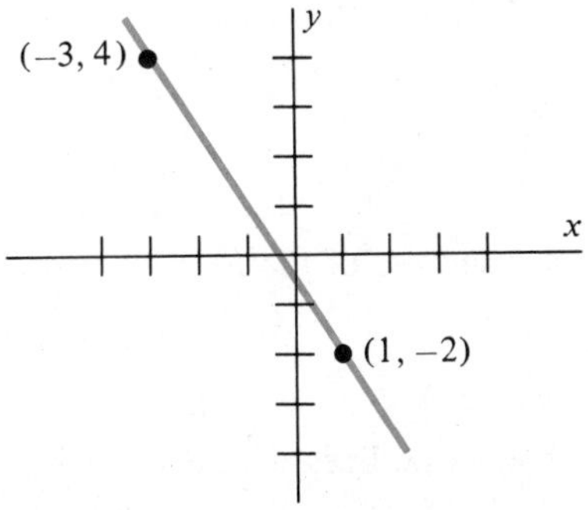

Figure 4.4 Line $y = -\frac{3}{2}x - \frac{1}{2}$.

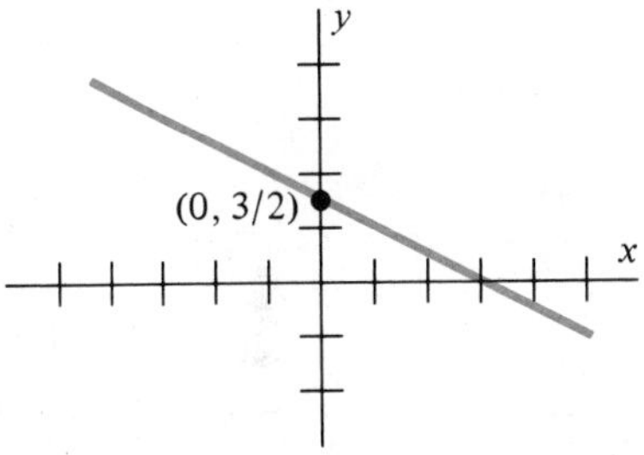

Figure 4.5 Line $y = -\frac{1}{2}x + \frac{3}{2}$.

Example 4 Find the slope and y-intercept for the line with equation

$$2x + 4y - 6 = 0.$$

Strategy
Bring equation to slope-intercept form of equation (2b).

Solution
We put the given equation in slope-intercept form by solving for y. We obtain

$$4y = -2x + 6,$$

so

$$y = -\frac{1}{2}x + \frac{3}{2}.$$

Read off m and b.

The line therefore has slope $-\frac{1}{2}$ and y-intercept $\frac{3}{2}$.

(See Figure 4.5.) ■

Vertical and Horizontal Lines

A vertical line has the property that all x-coordinates of points of the line are the same, while the values of the y-coordinate are unrestricted. The equation of a vertical line therefore has the form

$$x = a \qquad \text{(vertical line).}$$

A horizontal line has the property that its slope is zero. Setting $m = 0$ in equation (2b) shows that a horizontal line has an equation of the form

$$y = b \qquad \text{(horizontal line).}$$

Figure 4.6 shows the graphs of the vertical line $x = 2$ and the horizontal line $y = 1$.

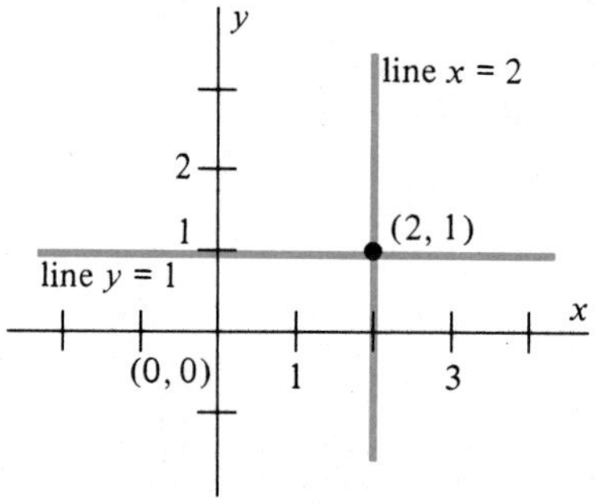

Figure 4.6 Vertical and horizontal lines.

Parallel and Perpendicular Lines

We will make use of the following theorem.

THEOREM 5

Let ℓ_1 and ℓ_2 be lines with equations

$$\ell_1\colon y = m_1x + b_1$$
$$\ell_2\colon y = m_2x + b_2$$

where neither m_1 nor m_2 are zero. Then

(a) ℓ_1 and ℓ_2 are **parallel** if and only if $m_1 = m_2$.
(b) ℓ_1 and ℓ_2 are **perpendicular** if $m_1m_2 = -1$.

Theorem 5 states that parallel lines have the same slope, while the slopes of perpendicular lines are negative reciprocals.

Example 5 For the line ℓ_1: $3y - 9x = 12$, find

(a) an equation of the line ℓ_2 which is parallel to ℓ_1 and contains the point (2, 0),
(b) an equation of the line ℓ_3 which is perpendicular to ℓ_1 and contains the point (3, −2).

Strategy
(a) Find the slope of ℓ_1 by putting equation in slope-intercept form.

Solution
(a) We first bring the equation for ℓ_1 into slope-intercept form by solving for y:

$$y = 3x + 4.$$

Take $m_2 = m_1$.

The slope of ℓ_1 is therefore $m_1 = 3$, so the slope of ℓ_2 must also be $m_2 = 3$. Since ℓ_2 contains the point (2, 0), we obtain from the definition of slope that

Use definition of slope to find equation of point (x, y) on ℓ_2.

$$\frac{y - 0}{x - 2} = 3,$$

or

$$\ell_2\text{: } y = 3(x - 2).$$

(b) Slope of ℓ_3 is negative reciprocal of m_1.

(b) Line ℓ_3 is perpendicular to ℓ_1, so $m_3 = -\dfrac{1}{m_1} = -\dfrac{1}{3}$. Since ℓ_3 contains the point $(3, -2)$ we obtain

Use definition of slope to find equation for ℓ_3.

$$\frac{y - (-2)}{x - 3} = -\frac{1}{3},$$

so

$$y + 2 = -\frac{1}{3}(x - 3),$$

or

$$\ell_3\text{: } y = -\frac{1}{3}x - 1.$$

(See Figure 4.7.) ■

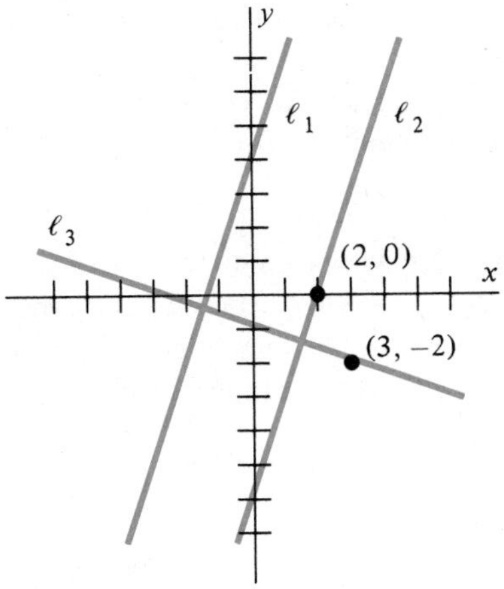

Figure 4.7

Points of Intersection

Lines in the plane which are not parallel must meet. If the lines are distinct (that is, if they are not the same line), they meet at a single point. Since the coordinates of this point of intersection must satisfy the equations for both lines, we can find the point of intersection by solving both equations for y and then equating these two expressions. The result is a first degree equation in x that is easily solved.

Example 6 Find the point of intersection for the lines

$$\ell_1: 4x - 2y - 4 = 0$$
$$\ell_2: x + y - 4 = 0$$

Strategy
Solve both equations for y.

Solution
Solving both equations for y gives

$$\ell_1: y = 2x - 2$$
$$\ell_2: y = -x + 4$$

Equate expressions for y. Solve for x.

Equating expressions for y gives

$$2x - 2 = -x + 4$$

or

$$3x = 6.$$

Substitute value for x into ℓ_1 or ℓ_2 to find y.

Thus, $x = 2$. From the equation for ℓ_1 we find that $y = 2(2) - 2 = 2$. The point of intersection is therefore (2, 2). ■

Linear equations arise in scientific applications when it is desired that the ratio of respective changes in two variables remain the same, regardless of the particular values. A typical example is the relationship between various temperature scales, the subject of the next example.

Example 7 The most frequently used temperature scale in science is the Celsius scale, named in honor of the Swedish astronomer Anders Celsius. (The Celsius scale is sometimes referred to as the **centigrade** scale.) On this scale the freezing point of water is 0°C and the boiling point is 100°C. In everyday use, however, the Fahrenheit scale, named for the German scientist G. D. Fahrenheit, is still common. On this scale the freezing point of water is 32°F and the boiling point is 212°F. Both scales are divided into degrees evenly. Find an equation by which Fahrenheit temperatures may be converted to Celsius temperatures.

Strategy
Verify that the desired relationship is linear.

Identify and name the variables.

Solution
Since both scales are divided evenly into degrees, the ratio between temperature changes on the two scales will be constant. The relationship is, therefore, linear.

Let F represent temperature in degrees Fahrenheit, and let C represent temperature in degrees centigrade. We seek an equation of the form

State the form of the desired equation.

$$C = aF + b. \tag{3}$$

Substitute given data into the equation to obtain equations for the unknown constants.

To find the constants a and b, we note that $C = 0°$ when $F = 32°$ and, also, $C = 100°$ when $F = 212°$. Substituting these two pairs of values into equation (3) gives the equations

$$0 = 32a + b$$
$$100 = 212a + b$$

Solve the resulting equations to determine the constants.

Subtracting corresponding sides of the top equation from the bottom equation gives

$$180a = 100, \qquad \text{or} \qquad a = \frac{5}{9}.$$

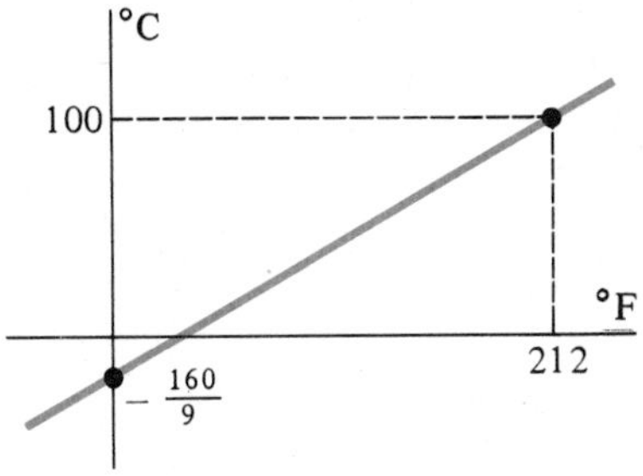

Figure 4.8 Linear relationship between Fahrenheit and Celsius temperatures.

Substituting $a = \frac{5}{9}$ into the top equation then gives

$$\begin{aligned} 0 &= 32\left(\frac{5}{9}\right) + b \\ &= \frac{160}{9} + b, \end{aligned}$$

so

$$b = -\frac{160}{9}.$$

The desired equation is therefore

$$C = \frac{5}{9}\text{F} - \frac{160}{9}.$$

(See Figure 4.8.) ■

Proportionality

Biologists often write expressions of the form "$y \propto x$", meaning "y is proportional to x." The mathematical meaning of this phrase is that all corresponding values of x and y satisfy the equation $y = kx$ for some (fixed) constant k, called the **constant of proportionality.** For example, the equation $y = 7x$ is interpreted by saying "y is proportional to x with proportionality constant $k = 7$."

Linear relationships of proportionality occur in many areas of science, such as Charles' Law in chemistry, $V = kT$, which states that the volume of a gas is proportional to its absolute (Kelvin) temperature. However, there are other types of relationships of proportionality which are not linear, such as $K = \left(\frac{1}{2}m\right)v^2$ in physics, the statement that the kinetic energy of a particle is proportional to the *square* of its velocity. Only when both variables appear to the first power will a relationship of proportionality be a linear relationship.

Exercise Set 1.4

1. Find the slope of the line through the following pairs of points
   a. (1, 2) and (0, 2)
   b. (3, 1) and (−1, 2)
   c. (6, −2) and (−1, −1)
   d. (1, 1) and (−1, −5)
   e. (a, b) and (b, a)
   f. (1, 1) and (a, a)
2. True or false? Every line has a slope.
3. True or false? Every slope determines one and only one line.
4. True or false? If ℓ_1 and ℓ_2 are perpendicular lines then ℓ_1 and ℓ_2 have precisely one point of intersection.
5. True or false? If ℓ_1 and ℓ_2 are parallel and distinct, they do not intersect.
6. Find a if the line through (2, 4) and $(-2, a)$ has slope 3.
7. Find b if the line through $(b, 1)$ and $(1, -5)$ has slope 2.

In Exercises 8–22, find an equation for the line determined by the given information.

8. Slope 4 and y-intercept −2.
9. Slope −2 and y-intercept 5.

10. Slope zero and y-intercept -5.

11. Through $(-1, 6)$ and $(4, 12)$.

12. Through $(4, 1)$ and with slope 7.

13. Through $(1, 3)$ and with slope -3.

14. x-intercept -3 and y-intercept 6.

15. Through $(-3, 5)$ and vertical.

16. Through $(-3, 5)$ and horizontal.

17. Through $(-2, 4)$ and $(-6, 8)$.

18. Through $(0, 2)$ and $(-1, -4)$.

19. Through $(1, 4)$ and parallel to the line with equation $2x - 6y + 5 = 0$.

20. Through $(5, -2)$ and parallel to the line with equation $x - y = 2$.

21. Through $(1, 3)$ and perpendicular to the line with equation $3x + y = 7$.

22. Through $(4, -1)$ and perpendicular to the line through the points $(-2, 5)$ and $(-1, 9)$.

23. True or false? If x and y satisfy a linear equation, then y is proportional to x.

24. True or false? If x and y satisfy a linear equation, then y is proportional to $x + a$ for some constant a.

In Exercises 25–34, find the slope, x-intercept, and y-intercept of the line determined by the given equation. Graph the line.

25. $x = 7 - y$

26. $3x - 2y = 6$

27. $x + y + 3 = 0$

28. $2x = 10 - 3y$

29. $y = 5$

30. $y - 2x = 9$

31. $x = 4$

32. $y = x$

33. $3y - \pi x = 4$

34. $9x - 9y = 27$

In each of Exercises 35–40, find the point(s) of intersection of the two lines, if any.

35. $3x - y - 1 = 0$
$x + y - 3 = 0$

36. $x - 2y + 3 = 0$
$x - 2y - 7 = 0$

37. $x - 3y + 3 = 0$
$2x - 3y + 6 = 0$

38. $x - 2y + 4 = 0$
$3x + 6y - 12 = 0$

39. $2x - y - 2 = 0$
$8x - 4y + 2 = 0$

40. $3x - 2y + 2 = 0$
$3x - y = 0$

41. Explain why any equation of the form $ax + by + c = 0$, where a and b are not both zero, must have a graph which is a line. (*Hint:* Show that if $b \neq 0$, the equation can be put in slope-intercept form. What about the remaining case, $b = 0$?)

42. Because of the result in Exercise 41, the equation $ax + by + c = 0$ is often referred to as the **general linear equation.** If only two points are required to determine a line, why does this equation have three constants?

43. By checking slopes, determine whether the following sets of points lie on a common line.
a. $(1, 3)$, $(-2, 0)$, and $(4, 6)$
b. $(2, -7)$, $(-2, -3)$, and $(-1, -4)$
c. $(5, 15)$, $(0, 3)$, and $(-2, -7)$

44. Determine whether or not the points $(1, 3)$, $(3, 5)$, and $(4, 0)$ form the vertices of a right triangle.

45. The points $(-2, 2)$, $(4, 4)$, and $(0, a)$ are the vertices of a right triangle.
a. For how many distinct values of a is this condition satisfied?
b. Find these values of a.

46. Use information about similar triangles to explain why, for a given line, the calculation of slope is independent of the particular points chosen.

47. A market research firm determines that the demand (d) for a certain product, in terms of purchasers per thousand population, is 12 when the product is priced at $P = 20$ dollars, but that the demand is only 6 when the item is priced at \$60. Write down a linear equation which models the relationship between demand (d) and price (P) according to these data. At what price level will demand reach zero?

48. A state has an income tax of 5% on all income over \$5,000.
a. Write a linear equation which determines a person's tax, T, in terms of income, i, for a person earning more than \$5,000.
b. At what income level will a person owe \$800 in state tax?

49. Find a linear equation which determines temperatures in degrees Fahrenheit from temperatures in degrees Celsius.

50. Find the distance from the point $(4, -2)$ to the line ℓ with equation $2x - y + 1 = 0$ as follows:
a. Find the slope m of the line ℓ.
b. The slope of the perpendicular from $(4, -2)$ to ℓ is then $-\dfrac{1}{m}$.
c. Write the equation for this line.
d. Find its point of intersection with ℓ.
e. Compute the desired distance.

51. As a beaker of water is heated the following temperatures are noted:

time (minutes)	0	10	20
temperature (°C)	22	42	62

a. Find a linear equation which relates time t to temperature T.

b. According to this equation what will the temperature of the water be after 25 minutes? 35 minutes? (Using an equation obtained from data to predict values of one variable from stated values of the other variable in this manner is called **extrapolation.**)

52. Temperatures T on the absolute, or Kelvin, scale are related to temperatures t on the Celsius scale by the equation $T = t + 273°C$.
At the Kelvin temperature 273° (0°C) the volume of a certain quantity of gas is 40 liters.

a. Find a linear relationship for volume in terms of temperature, according to Charles' Law: $V = kT = k(t + 273)$.

b. What is the volume of the gas at temperature 323°K?

c. What is the volume of the gas at temperature 100°C?

53. The length of a rod l is given by the linear equation $l = l_0(1 + at)$, where a is a constant called the coefficient of thermal expansion, l_0 is the length of the rod at 0°C, and t is the temperature of the rod. Find a if the rod is 100 cm long at temperature 0°C and 100.2 cm long at temperature 50°C.

54. Find an equation for the line through the points of intersection of the circles $x^2 + y^2 - 2y = 0$ and $x^2 - 2x + y^2 = 0$.

55. Find an equation for the line through the centers of the circles with equations $x^2 - 2x + y^2 - 4y + 1 = 0$ and $x^2 + y^2 + 2y = 0$.

1.5 FUNCTIONS

A linear equation in the form $y = mx + b$ may be thought of as a rule which assigns a *unique* number y to each number x. We want to generalize this concept by saying that a *function* is any rule which uniquely assigns elements of one set to elements of another set. For example,

(a) The equation $y = x^3$ assigns to each number x its (unique) cube. Thus, $y = x^3$ is a function.

(b) The rule, "measure his or her height," assigns a unique number (height) to each member of your calculus class. It is also a function.

(c) But, the equation $x^2 + y^2 = 1$ does *not* determine a function, since it does not determine a unique value of y for a given value of x. This is most readily seen by writing the equation in the form $y = \pm\sqrt{1 - x^2}$. It is then clear, for example, that both $y = 1$ and $y = -1$ are assigned to $x = 0$, violating the property of uniqueness.

Here is a more precise definition of what we are talking about.

DEFINITION 6

A **function** from set A to set B is a rule that assigns to each element $x \in A$ a **unique** element $y \in B$. We write $y = f(x)$ to indicate that the element y is the value assigned by the function f to the element x. The set A is called the **domain** of the function f. The set of all values $\{f(x) \mid x \in A\}$ is called the **range** of the function.

Thus, in order to specify a function, we must indicate both the domain of the function and the rule by which values of the function are determined. We will almost always do this by writing an equation of the form $y = f(x)$, where $f(x)$ is an expression involving x's and constants. If we wish to specify a special subset of the real numbers as the domain of the function, we do so by writing a statement to this effect following the equation $y = f(x)$. Otherwise, we shall mean that the domain of the function consists of all numbers x for which the expression $f(x)$ is defined.

Example 1 For the function $f(x) = x^2$,

(a) the domain is the whole real line, since nothing has been indicated to restrict the domain;
(b) the range is the interval $[0, \infty)$, since all nonnegative numbers are squares (that is, have square roots), and all squares must be nonnegative.

(See Figure 5.1.) ■

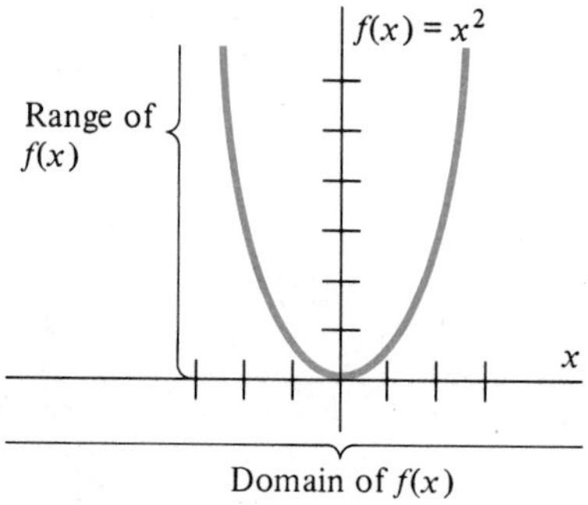

Figure 5.1 For the function $f(x) = x^2$, domain of $f = (-\infty, \infty)$ and range of $f = [0, \infty)$.

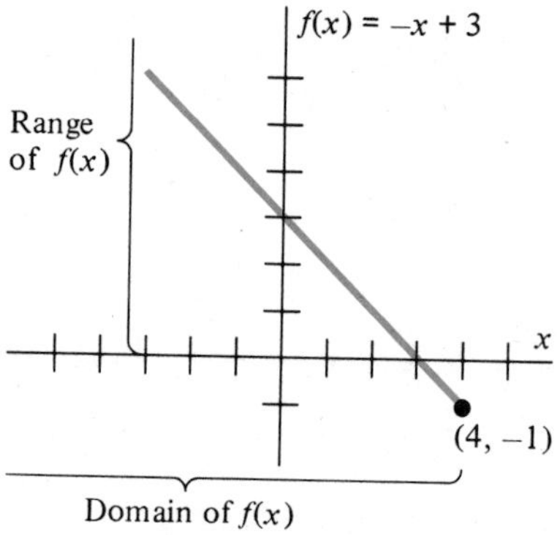

Figure 5.2 For $f(x) = -x + 3$, $x \leq 4$, domain of $f = (-\infty, 4]$ and range of $f = [-1, \infty)$.

Example 2 For the function $f(x) = -x + 3$, $x \leq 4$,

(a) the domain is specified to be the interval $(-\infty, 4]$ by the inequality $x \leq 4$;
(b) the range is the interval $[-1, \infty)$, since $f(x) = -x + 3 \geq -1$ whenever $x \leq 4$, and $y = -x + 3$ has a solution in $(-\infty, 4]$ whenever $y \geq -1$.

(See Figure 5.2.) ■

Example 3 For the function $f(x) = \dfrac{1}{x - 2}$,

(a) the domain is the set $(-\infty, 2) \cup (2, \infty)$. The number $x = 2$ is excluded from the domain, since the expression $\dfrac{1}{x - 2}$ is undefined for $x = 2$.
(b) The range is the set $(-\infty, 0) \cup (0, \infty)$, since every value of $f(x)$ is a real number other than zero, and the equation $y = \dfrac{1}{x - 2}$ has a solution for every value of y except $y = 0$ (see Figure 5.3). ■

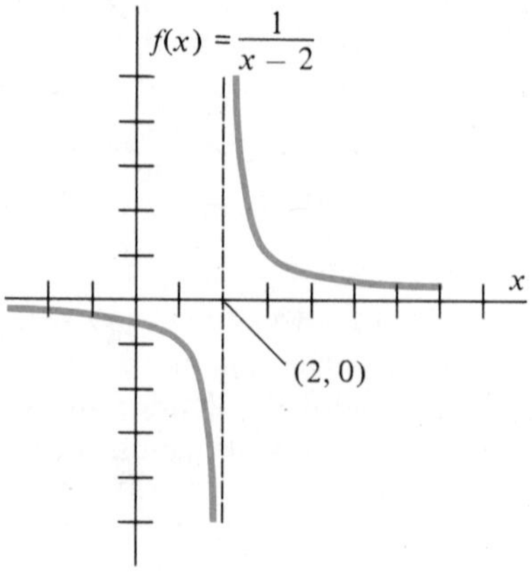

Figure 5.3 For $f(x) = \dfrac{1}{x - 2}$, domain $= (-\infty, 2) \cup (2, \infty)$ and range $= (-\infty, 0) \cup (0, \infty)$.

Example 3 points out one reason for which certain values of x must be excluded from the domain of a function, even though a domain is not explicitly stated: a denominator becomes zero. Example 4 presents another situation of this type, this time involving an even root, which exists only for nonnegative numbers.

Example 4 Find the domain and range of the function

$$f(x) = (x^2 - x - 2)^{3/2}.$$

(See Figure 5.4.)

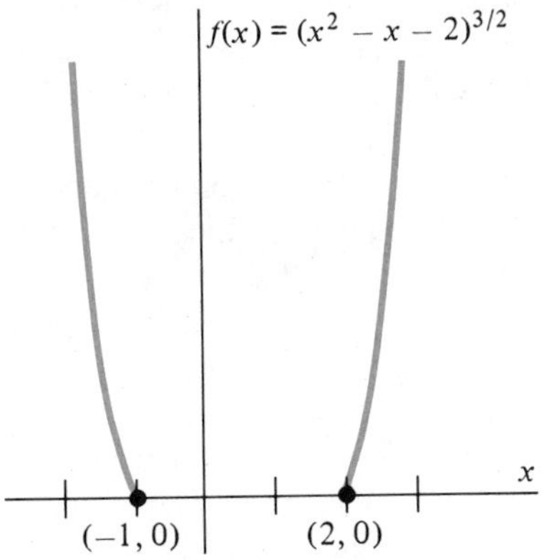

Figure 5.4 For $f(x) = (x^2 - x - 2)^{3/2}$, domain $= (-\infty, -1] \cup [2, \infty)$ and range $= [0, \infty)$.

Strategy

Note that an even root requires a nonnegative base. Write this as an inequality.

Factor the left-hand side and find its zeros.

Examine the signs of individual factors between zeros to determine sign of $x^2 - x - 2$ on each of these intervals.

The domain of $f(x)$ is the solution set of inequality (1).

Determine the range by asking, "What values of $f(x)$ occur when x is in the domain of $f(x)$?"

Solution

Since the indicated root is even, the function is defined only when

$$x^2 - x - 2 \geq 0. \tag{1}$$

To determine the solution set of inequality (1), we write it in factored form as

$$(x - 2)(x + 1) \geq 0.$$

Since $(x - 2)(x + 1) = 0$ for $x = -1$ and $x = 2$, we must examine the signs of the factors on the intervals $(-\infty, -1)$, $(-1, 2)$, and $(2, \infty)$.

(i) If $x < -1$, then $x - 2 < 0$ and $x + 1 < 0$, so $(x - 2)(x + 1) > 0$.
(ii) If $-1 < x < 2$, then $x - 2 < 0$ but $x + 1 > 0$, so $(x - 2)(x + 1) < 0$.
(iii) If $x > 2$, both $x - 2 > 0$ and $x + 1 > 0$, so $(x - 2)(x + 1) > 0$.

From statements (i)–(iii), it follows that the solution set of inequality (1) is $(-\infty, -1] \cup [2, \infty)$. This is the domain. To determine the range we can write $f(x)$ as

$$f(x) = [(x - 2)(x + 1)]^{3/2}$$

Now think of what happens to this expression as x starts at $x = 2$ and increases. The product inside the braces starts at zero and increases through all positive numbers. Since cubes and square roots of positive numbers are positive numbers, $f(x)$ can equal any nonnegative number. Since even roots cannot be negative, the range of $f(x)$ is simply $[0, \infty)$. ■

Graphs of Functions: The Vertical Line Property

For a function of the form $y = f(x)$, the **graph** of the function is the set of all points (x, y) in the xy-coordinate plane satisfying the equation $y = f(x)$. That is, the graph is the set

$$\{(x, y) \mid x \text{ is in the domain of } f, \text{ and } y = f(x)\}.$$

The ability to sketch an accurate graph for a given function is an important and powerful tool in analyzing the nature of that function. Indeed, curve sketching is a major topic pursued throughout much of this text, one which requires the theory of the derivative (Chapters 3–5) to master. However, the technique of simply selecting a few values of x, calculating the corresponding values of $y = f(x)$, plotting the resulting points $(x, f(x))$, and making a "best guess" at sketching a curve through these points is always available to you. Program 1 in Appendix I is a BASIC program that will help you do this, if you have access to a computer. (See Exercises 37–42.)

There is a simple but important property that distinguishes graphs of functions from graphs of equations which are not functions. It is called the **vertical line property,** and it means that *any vertical line* (that is, a line parallel to the y-axis) *can intersect the graph of a function at most once*. The reason for this is simple: a vertical line has equation $x = a$; but the graph of the *function* $y = f(x)$ has at most one point, $(a, f(a))$, whose x-coordinate is a. Figure 5.5 illustrates our previous conclusion, that $x^2 + y^2 = 1$ cannot be the graph of a function—the vertical line $x = a$ can intersect the graph twice. Figure 5.6 illustrates the vertical line property for an arbitrary function.

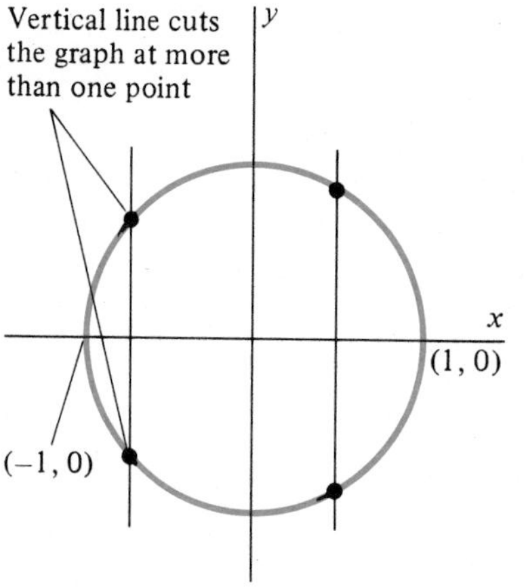

Figure 5.5 Graph of $x^2 + y^2 = 1$ does not have the vertical line property.

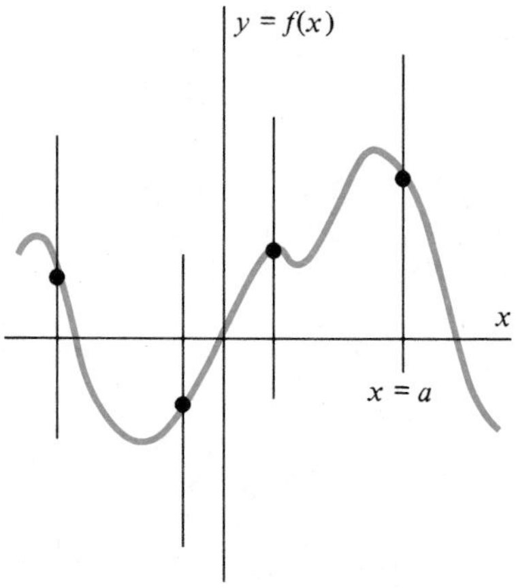

Figure 5.6 Vertical lines intersect the graph of a *function at most* once (Vertical Line Property).

Power and Quadratic Functions

Among the simplest types of functions are the power functions. These have the form

$$f(x) = x^n$$

where n is a positive integer. Graphs of several power functions appear in Figure 5.7.

Figures 5.8 through 5.10 show the effect of introducing various constants into the equation $y = x^2$.

(a) The graph of $y = ax^2$, $a > 0$, opens upward, but does so more or less quickly depending on whether the constant a is large or small (Figure 5.8).
(b) The graph of $y = ax^2$, $a < 0$, opens downward (Figure 5.8).
(c) The graph of $y = ax^2 + b$ is like the graph of $y = ax^2$, except that it is "shifted" b units upward ($b > 0$) or $|b|$ units downward ($b < 0$, Figure 5.9).

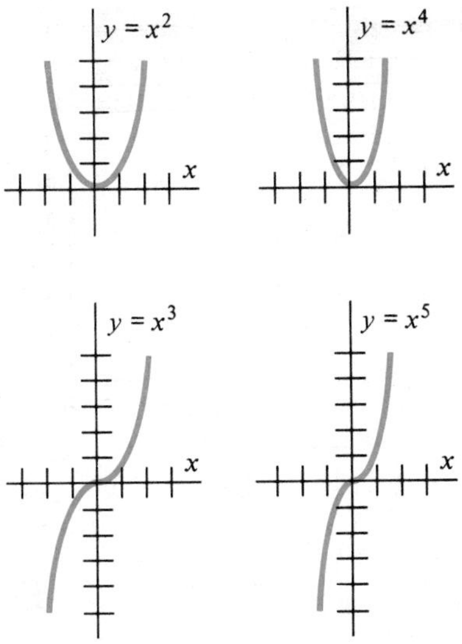

Figure 5.7 Some power functions.

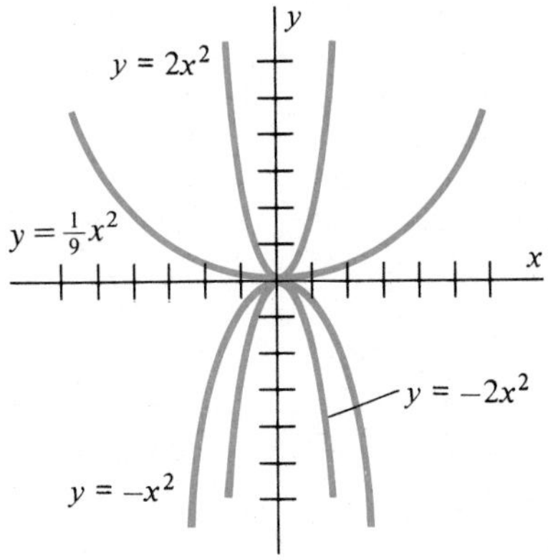

Figure 5.8 Functions of the form $f(x) = ax^2$.

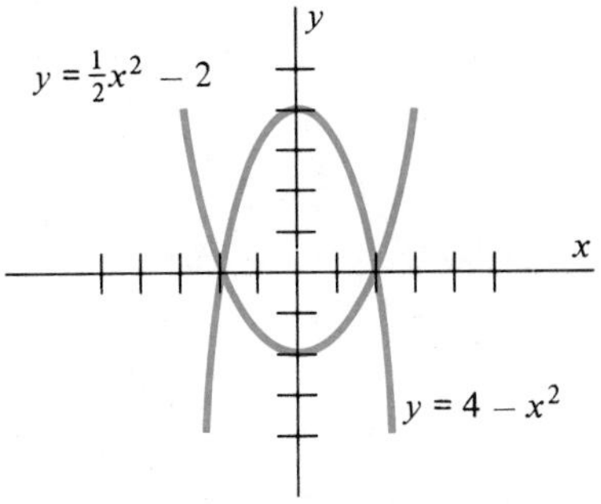

Figure 5.9 Functions of the form $h(x) = ax^2 + b$.

(d) The graph of $y = a(x - c)^2 + b$ is like the graph of $y = ax^2 + b$, except that it is symmetric with respect to the vertical line $x = c$ rather than with respect to the y-axis (Figure 5.10).

A **quadratic function** is any function that can be written in the form

$$f(x) = Ax^2 + Bx + C. \tag{2}$$

By completing the square on the right hand side of equation (2), we can bring the equation into the form $f(x) = a(x - c)^2 + b$. (See Section 1.3.) Thus, the graph of any quadratic function will be of the form illustrated in Figure 5.10. (Such curves are called **parabolas,** and they are discussed in detail in Chapter 15.) The following example is typical.

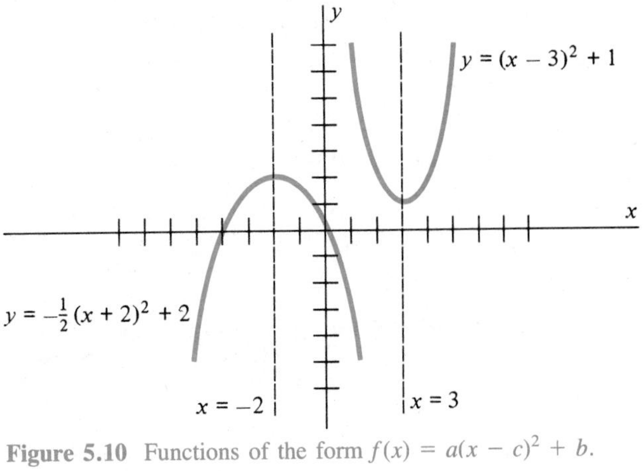

Figure 5.10 Functions of the form $f(x) = a(x - c)^2 + b$.

Example 5 Sketch the graph of the quadratic function $f(x) = \frac{1}{4}x^2 - x$.

Strategy

Complete the square in x by

(1) factoring by $\frac{1}{4}$ and

(2) adding the square of half the coefficient of x.

Solution

We have

$$f(x) = \frac{1}{4}x^2 - x = \frac{1}{4}[x^2 - 4x]$$
$$= \frac{1}{4}[x^2 - 4x + 4] - \frac{1}{4}(4)$$
$$= \frac{1}{4}(x - 2)^2 - 1,$$

so

$$f(x) = \frac{1}{4}(x - 2)^2 - 1.$$

The graph appears in Figure 5.11. ■

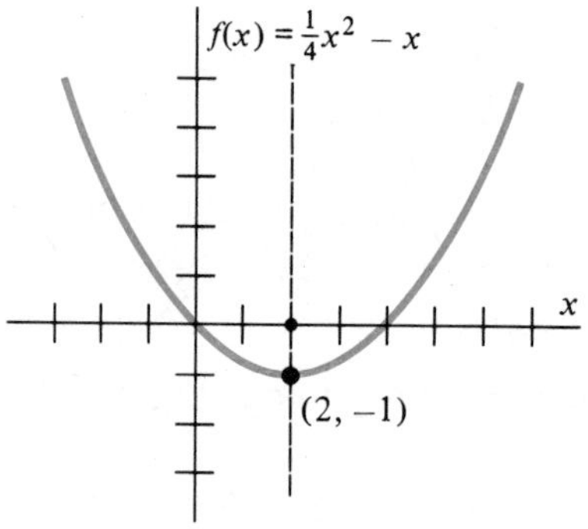

Figure 5.11 Graph of quadratic function $f(x) = \frac{1}{4}x^2 - x$.

Polynomial Functions

A **polynomial function** is a function that can be written in the form

$$f(x) = a_nx^n + a_{n-1}x^{n-1} + \cdots + a_1x + a_0$$

where $a_0, a_1, \ldots, a_n$ are constants and n is a positive integer. Note that every function that is linear or quadratic is a polynomial and that all polynomial functions are defined for all values of x. The integer n is called the **degree** of the polynomial, provided that $a_n \neq 0$. A polynomial function of degree n can have at most n zeros, although it actually may have fewer; for instance, the polynomial function $f(x) = x^2 + 1$ has no real zeros. The following are examples of polynomial functions:

$$f(x) = x^3 - 7x + 2 \qquad \text{(degree 3)}$$
$$f(x) = 1 - x^{10} \qquad \text{(degree 10)}$$
$$y = (x - 3)^4 - 2(x + 5)^3 \qquad \text{(degree 4)}$$

Rational Functions

A **rational function** is a quotient of two polynomial functions. Thus, the general form of a rational function is

$$f(x) = \frac{a_nx^n + a_{n-1}x^{n-1} + \cdots + a_1x + a_0}{b_mx^m + b_{m-1}x^{m-1} + \cdots + b_1x + b_0}.$$

A rational function is not defined for values of x for which its denominator equals zero. For example, the rational function

$$f(x) = \frac{x^2 + 4}{x^3 + x^2 - 12x}$$

is not defined for $x = 0$, 3, or -4 since the denominator $x^3 + x^2 - 12x = x(x - 3)(x + 4)$ equals zero for these values of x.

Absolute Value Function

The definition of the absolute value* of x, $|x|$, gives rise to the absolute value function $f(x) = |x|$. The graph of $f(x) = |x|$ appears in Figure 5.12.

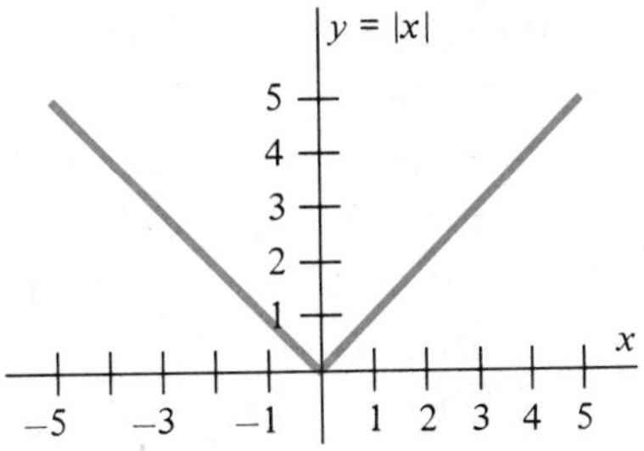

Figure 5.12 Graph of absolute value function.

Example 6 Sketch the graph of the function $f(x) = |4 - x^2|$ and state its domain and range.

Strategy

Because of the absolute value signs, we must determine the sign of $4 - x^2$. Begin by factoring $4 - x^2$ to find its zeros.

Solution

The zeros of $4 - x^2$ are found by setting

$$4 - x^2 = 0$$

and factoring. We obtain

$$(2 - x)(2 + x) = 0,$$

which has solutions $x = -2$ and $x = 2$.

Examine the sign of each factor on each interval to determine the sign of $4 - x^2$.

(i) For $x \in (-\infty, -2)$, $x < -2$, so

$$2 - x > 0 \text{ and } 2 + x < 0.$$

Thus, $(2 - x)(2 + x) < 0$ if $x \in (-\infty, -2)$.

(ii) For $x \in (-2, 2)$

(a) $x > -2$, so $2 + x > 0$;
(b) $x < 2$, so $2 - x > 0$.

Thus, $(2 - x)(2 + x) > 0$ if $x \in (-2, 2)$.

(iii) For $x \in (2, \infty)$, $x > 2$, so

$$2 - x < 0 \text{ and } 2 + x > 0.$$

---

*See Section 1.2 for the definition of absolute value.

Use the definition of absolute value to state the explicit form of $|4 - x^2|$ on each interval.

(Specifically,

$$|4 - x^2| = (4 - x^2) = x^2 - 4$$

whenever $4 - x^2 < 0$.)

Thus, $(2 - x)(2 + x) < 0$ if $x \in (2, \infty)$.

From (i)–(iii) we may now conclude that

$$|4 - x^2| = \begin{cases} x^2 - 4 \text{ if } -\infty < x < -2 \\ 4 - x^2 \text{ if } -2 \le x \le 2 \\ x^2 - 4 \text{ if } 2 < x < \infty. \end{cases}$$

Since both the polynomial $4 - x^2$ and the absolute value function are defined for all x, the domain is $(-\infty, \infty)$. The range is $[0, \infty)$, as you can see from Figure 5.13. ■

REMARK: You can see the effect of the absolute value signs in the function $f(x) = |4 - x^2|$ by comparing Figure 5.13 with the graph of $g(x) = 4 - x^2$ in Figure 5.14. The "legs" of the graph of $4 - x^2$, which otherwise extend below the x-axis, have been turned upward.

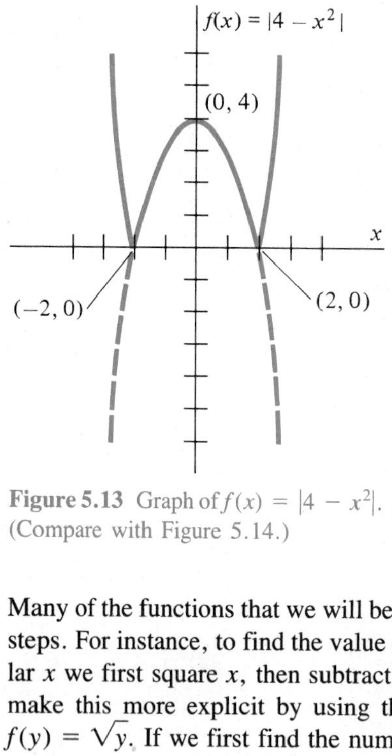

Figure 5.13 Graph of $f(x) = |4 - x^2|$. (Compare with Figure 5.14.)

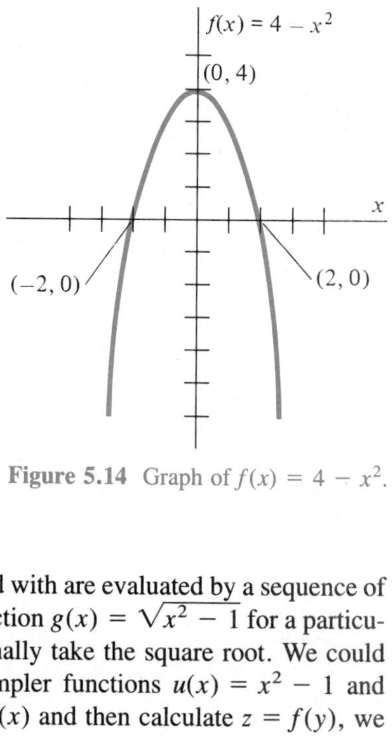

Figure 5.14 Graph of $f(x) = 4 - x^2$.

Composite Functions

Many of the functions that we will be concerned with are evaluated by a sequence of steps. For instance, to find the value of the function $g(x) = \sqrt{x^2 - 1}$ for a particular x we first square x, then subtract 1, and finally take the square root. We could make this more explicit by using the two simpler functions $u(x) = x^2 - 1$ and $f(y) = \sqrt{y}$. If we first find the number $y = u(x)$ and then calculate $z = f(y)$, we must obtain the same number $z = g(x)$ as before, since the two sequences of steps are the same. We could write this observation more compactly as simply $g(x) = f(u(x))$.

A function such as this is called a **composite function.** We say that $g(x)$ is the *composition of f with u,* written

$$g(x) = f(u(x)) \qquad \text{or} \qquad g(x) = (f \circ u)(x).$$

The domain of $g(x)$ consists of all elements of the domain of $u(x)$ for which $y = u(x)$ is in the domain of $f(y)$. In the case of $g(x) = \sqrt{x^2 - 1}$, we see that the domain of $u(x) = x^2 - 1$ is all real numbers. However, the domain of $f(y)$ is $[0, \infty)$, so f can accept the output of u only when $y = u(x)$ is nonnegative. This is true only for the subset $(-\infty, -1] \cup [1, \infty)$ of the domain of u; therefore, the domain of $g(x)$ is $(-\infty, -1] \cup [1, \infty)$.

The following example demonstrates a fact about composition of functions: The order in which the functions are composed generally cannot be changed without altering the result. That is, $(f \circ u)(x) \neq (u \circ f)(x)$.

Example 7 For $f(x) = \dfrac{1}{x}$, $g(x) = x^3 - 7$ and $u(x) = \sqrt{x}$,

(a) $(f \circ u)(x) = f(u(x)) = f(\sqrt{x}) = \dfrac{1}{\sqrt{x}}$; Domain of $f \circ u$ is $(0, \infty)$.

(b) $(f \circ g)(x) = f(g(x)) = f(x^3 - 7) = \dfrac{1}{x^3 - 7}$; Domain of $f \circ g$ is $(-\infty, \sqrt[3]{7}) \cup (\sqrt[3]{7}, \infty)$.

(c) $g(u(x)) = [u(x)]^3 - 7 = (\sqrt{x})^3 - 7 = x^{3/2} - 7$; Domain of $g \circ u$ is $[0, \infty)$.

(d) $g(f(x)) = \left(\dfrac{1}{x}\right)^3 - 7 = \dfrac{1}{x^3} - 7$; Domain of $g \circ f$ is $(-\infty, 0) \cup (0, \infty)$.

(e) $u(f(g(x))) = u(f(x^3 - 7)) = u\left(\dfrac{1}{x^3 - 7}\right) = \dfrac{1}{\sqrt{x^3 - 7}}$; Domain of $u \circ f \circ g$ is $(\sqrt[3]{7}, \infty)$. ■

The last example shows that composite functions often can be decomposed in several ways.

Example 8 The function $g(x) = \sin \sqrt{x + 5}$ is an example of a composite function that may be decomposed in various ways.

(a) It can be viewed as $f(u(x))$, with

$$f(x) = \sin x, \qquad u(x) = \sqrt{x + 5}.$$

(b) It can also be viewed as $f(u(x))$, with

$$f(x) = \sin \sqrt{x}, \qquad u(x) = x + 5.$$

(c) Yet another way to view $g(x)$ is as $g(x) = f(u(h(x)))$, with

$$f(x) = \sin x, \qquad u(x) = \sqrt{x}, \qquad \text{and} \qquad h(x) = x + 5.$$ ■

Exercise Set 1.5

1. Complete the following chart.

$f(x)$	$f(-2)$	$f(0)$	$f(4)$	$f(5)$	$\lvert f(3)\rvert$
$1 - 3x^2$	-11	1	-47	-74	26
$\frac{1}{x+2}$	undefined	$\frac{1}{2}$			
$\frac{(x-3)^2}{x^2+1}$	5				
$\sqrt{x+4}$					
$\frac{1}{\sqrt{16-x^2}}$					
$\begin{cases} 1-x, & x<-3 \\ x-1, & x>1 \end{cases}$					

In Exercises 2–7, determine whether the given equation determines y as a function of x.

2. $2x - y = 7$

3. $xy = 5$

4. $xy^2 + 2x = 6$

5. $x^2 + 2x + y^2 = 8$

6. $x - y = x + y$

7. $y = 5$

8. True or false? The functions $f(x) = \dfrac{x^2-4}{x+2}$ and $g(x) = x - 2$ are equal. (*Hint:* Do they have the same equations? Domains?)

In Exercises 9–20, state the domain of the given function.

9. $f(x) = x^5 - 5x^3$

10. $f(x) = \dfrac{x-1}{x+1}, x > 3$

11. $g(x) = \sqrt{x+4}$

12. $h(x) = \sqrt{x(x-1)}$

13. $f(t) = 1 + t^2, t \geq 0$

14. $f(x) = \dfrac{1}{1 - |x|}$

15. $h(s) = \sqrt{16 - s^2}$

16. $g(t) = \dfrac{1}{t^2 + 2t - 35}$

17. $f(x) = \begin{cases} \dfrac{1}{x}, & x > 0 \\ -\dfrac{1}{x}, & x < 0 \end{cases}$

18. $f(s) = s(s+3)\sqrt{1 - s^2}$

19. $h(x) = \dfrac{\sqrt{x^2 + 2x}}{x - 2}$

20. $f(x) = \sqrt{6 - |x + 2|}$

21. True or false? For the rational function $f(x) = \dfrac{g(x)}{h(x)}$, if $h(x)$ is a polynomial of degree n then $f(x)$ will be undefined for n distinct values of x.

22. True or false? If a horizontal line intersects the graph of the equation $y = f(x)$ in more than one point, the equation $y = f(x)$ cannot determine y as a function of x. Explain.

For each of the equations in Exercises 23–28, y is determined as a quadratic function of x. Graph each by first completing the square.

23. $3x^2 - 2y - 8 = 0$

24. $3x^2 - y - 7 = 0$

25. $y - 3x^2 + x - \dfrac{1}{2} = 0$

26. $y - 2x^2 + 4x - 5 = 0$

27. $4x^2 + y - 24x + 34 = 0$

28. $y - \dfrac{1}{2}x^2 - \dfrac{1}{2}x - \dfrac{11}{24} = 0$

An equation of the form $x = ky^2$ determines x as a quadratic function of y. The graph will therefore be a parabola opening in either the positive x direction ($k > 0$) or the negative x direction ($k < 0$). In each of the Exercises 29–34, graph the given parabola.

29. $y^2 + x - 2 = 0$

30. $3y^2 - x - 4 = 0$

31. $3x + 2y^2 - 3 = 0$

32. $x - 3y^2 + 6y - 4 = 0$

33. $x - 2y^2 + 4y - 2 = 0$

34. $4y^2 - x - 24y + 34 = 0$

35. Find the point(s) of intersection of the line $x - y - 2 = 0$ and the parabola $y = 4 - x^2$.

36. Find the points of intersection of the parabolas $y = 6 - x^2$ and $y = x^2 - 2$.

In each of Exercises 37–42, graph the given function by plotting points. (Program 1 in Appendix I is a BASIC computer program that will help you calculate values of these functions.)

37. $f(x) = \dfrac{1}{x - 1}$

38. $f(x) = \sqrt{2x + 1}$

39. $f(x) = \begin{cases} 7 - x, & x \le 2 \\ 2x + 1, & x > 2 \end{cases}$

40. $f(x) = x(4 - 2x)$

41. $f(x) = \dfrac{x^2 + 1}{\sqrt{3 + x^2}}$

42. $f(x) = \begin{cases} \sqrt{x + 2}, & x \le 2 \\ 4 - x, & x > 2 \end{cases}$

In Exercises 43–48, sketch the graph by first noting where the expression inside the absolute value signs changes sign.

43. $f(x) = x + |x|$

44. $y = |1 - x^2|$

45. $y = \dfrac{x + 1}{|x - 1|}$

46. $f(x) = |x^2 - x + 2|$

47. $f(t) = \dfrac{1}{|t|}$

48. $g(s) = |s^2 - 1|$

49. A function $f(x)$ is called **even** if whenever x is in the domain of f, so is $-x$, and $f(-x) = f(x)$. It is called **odd** if whenever x is in the domain of f, so is $-x$, and $f(-x) = -f(x)$. Determine whether $f(x)$ is even, odd, or neither, and sketch the graph.

a. $f(x) = x^2$

b. $f(x) = 2 - x^2$

c. $f(x) = x^3$

d. $f(x) = 1 - x^3$

e. $f(x) = x^3 + x$

f. $f(x) = 2x^4 + x^2$

g. $f(x) = |x| + 2$

h. $f(x) = x^2 + x$

50. What symmetry properties does the graph of an even function possess? An odd function?

In Exercises 51–58, use the functions

$$f(x) = 3x + 1 \qquad h(x) = \frac{1}{x + 1}$$

$$g(x) = x^3 \qquad u(x) = \sqrt{x}$$

to form the indicated composite function.

51. $f(u(x))$

52. $u(h(x))$

53. $g(f(x))$

54. $u(g(x))$

55. $h(u(x))$

56. $h(f(u(x)))$

57. $f(g(u(x)))$

58. $g(h(u(x)))$

59. True or false? Every polynomial is a rational function.

60. A rectangle has area 36 cm^2. Find a function that expresses the width in terms of the length, ℓ, of the rectangle. What is the domain of this function?

61. A water tank has the shape of a right circular cylinder. If the radius of the cylinder is fixed, express its volume as a function of its height.

62. Find two nonconstant functions $f(x)$ and $u(x)$ so that the composite function $f(u(x))$ is constant.

63. Find a function $f(e)$ giving the volume of a cube in terms of the length of one of its edges.

64. Find a function $e(v)$ giving the length of an edge of a cube in terms of its volume.

65. The frequency of x-rays emitted from a chemical element varies with the atomic number Z according to the equation $\gamma = a(z - b)^2$, where a and b are constants. Discuss the nature of this graph.

66. For a moving automobile, air resistance is proportional to the square of the vehicle's speed. Write this statement in function form and discuss the nature of its graph.

67. Show that the quadratic function $f(x) = Ax^2 + Bx + C$ may be written as $f(x) = A(x - D)^2 + E$ with $D = \dfrac{-B}{2A}$ and $E = C - \dfrac{B^2}{4A}$. (*Hint:* Complete the square.)

68. What is the range of the quadratic function $f(x) = ax^2 + bx + c$ if $a > 0$? What if $a < 0$?

69. Find the range of the function;

a. In Exercise 11

b. In Exercise 13

c. In Exercise 17

d. In Exercise 20

e. In Exercise 38

70. Use what you know about the graph of the quadratic function $f(x) = ax^2 + bx + c$, and the method of completing the square, to prove the quadratic formula for finding the zeros of the quadratic polynomial $ax^2 + bx + c$.

$$x = \frac{-b \pm \sqrt{b^2 - 4ac}}{2a}$$

1.6 TRIGONOMETRIC FUNCTIONS

Many problems in mathematics, science, and engineering have to do with the measurement of angles. For example, in the study of optics a fundamental fact is that a ray of light will bend when passing from one medium into another (such as passing from water into air). This phenomenon is called *refraction*. Figure 6.1 is meant to illustrate the fact that the sizes of the angles associated with the refraction phenomenon are intimately related to the lengths of the sides of certain triangles. The subject of *trigonometry* addresses these kinds of questions, and the relationships that are obtained give rise to the six *trigonometric functions*.

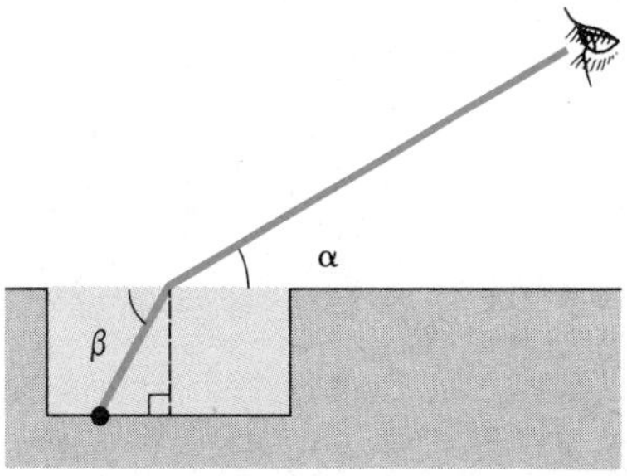

Figure 6.1 Ray of light from coin in wishing well to observer's eye "bends" at water surface.

Radian Measure

You are already familiar with the measurement of angles in degrees, and you know that one complete revolution equals 360°. However, the degree system (which originated in ancient Babylonia) is poorly suited to the needs of calculus. We therefore define a more natural unit of angular measurement, the *radian*.

Consider a circle located in the *xy*-coordinate plane, with its center at the origin. Imagine a particle that travels along the circumference of the circle, starting on the positive x-axis and carrying with it one end of a line segment whose other end is pivoted at the origin (see Figure 6.2). The angle θ is defined by the positive x-axis and the rotating line segment. When the distance traveled by the particle is exactly equal to the length of the circle's radius, the angle θ is defined as one **radian.**

This construction method works for a circle of any size, but it is easiest to visualize on a circle of radius $r = 1$, called a **unit circle** (see Figure 6.3). In plane geometry we define the number π to be the ratio of a circle's circumference to its diameter, $\pi = C/d = C/2r$, so the circumference of the unit circle is 2π. When our hypothetical point has traveled one complete revolution around the circle, it has gone a distance of 2π times the radius (or just 2π on the *unit* circle), and the associated angle is 2π radians. This leads to the general relation between radians and degrees:

$$2\pi \text{ radians} = 360°.$$

The relation between any angle θ_r measured in radians and the same angle θ_d measured in degrees is therefore

$$\frac{\theta_r}{2\pi} = \frac{\theta_d}{360} \quad \text{or} \quad \theta_r = \frac{2\pi}{360}\theta_d.$$

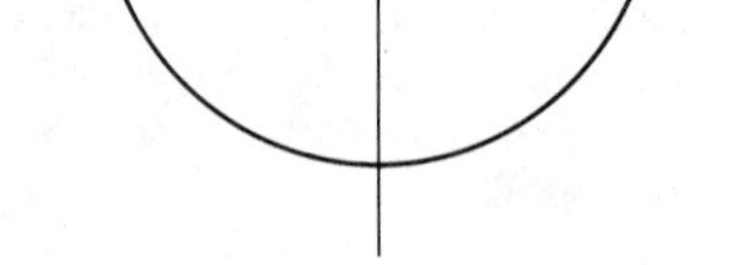

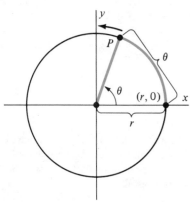

Figure 6.2 Circle of arbitrary radius—angle measure in radians equals arc length in radii.

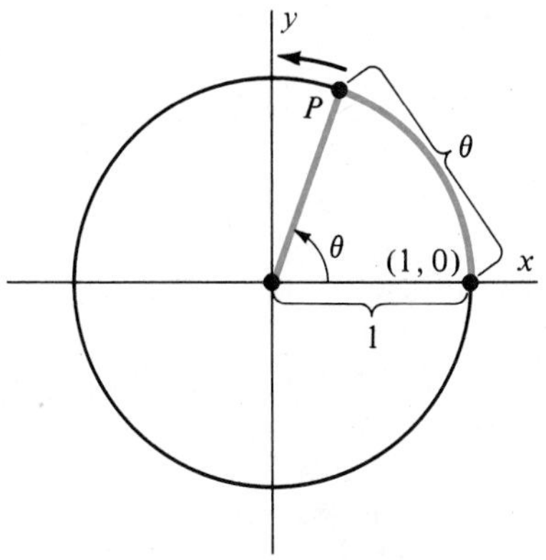

Figure 6.3 Unit circle—angle measure, in radians, equals arc length.

Example 1 We have the following relationships.

θ_d	0°	30°	45°	60°	90°	120°	135°	180°	210°	270°	315°	360°
θ_r	0	$\frac{\pi}{6}$	$\frac{\pi}{4}$	$\frac{\pi}{3}$	$\frac{\pi}{2}$	$\frac{2\pi}{3}$	$\frac{3\pi}{4}$	π	$\frac{7\pi}{6}$	$\frac{3\pi}{2}$	$\frac{7\pi}{4}$	2π

■

We can imagine the point and line segment continuing to rotate for more than one complete revolution, defining angles greater than 360° or 2π. Notice, however, that the final position of the line segment after such a rotation is the same as its position after some rotation of *less than* 2π. In order to avoid confusion, we define the **principal angle** of any rotation θ to be the angle θ_p such that

$$\theta_p + 2n\pi = \theta \quad \text{for} \quad n = 0, \pm1, \pm2, \pm3, \ldots$$
$$\text{and} \quad 0 \le \theta_p < 2\pi. \tag{1}$$

From now on we will discard the subscript p and consider all angles to be expressed as their principal values unless stated otherwise.

By use of equation (1) we can identify each real number t with a corresponding point on the unit circle as follows. If $t > 0$, we allow our traveling point to go a distance t counterclockwise along the unit circle, starting from $(1, 0)$. Let P denote the point on the circle where this motion ends. Then the number t corresponds to the angle formed between the positive x-axis and the rotating line segment, *and* to the length of the arc along the circle between $(1, 0)$ and P. In other words, we associate t with the number θ that satisfies the equation

$$t = \theta + 2n\pi \quad \text{for} \quad 0 \le \theta < 2\pi \tag{2}$$

and some integer $n \ge 0$. If $t < 0$, we use the same procedure, except that the point travels in the clockwise direction and n is a negative integer. This procedure of "wrapping the real number line around the unit circle" is useful in expressing arbitrary angles in terms of their principal angle equivalents, and it permits the definition of the trigonometric *functions* for all real numbers.

Example 2 The following real numbers t correspond to principal angles θ as indicated. The value of n for which equation (2) holds is also indicated.

t	0	π	2π	$\frac{5\pi}{2}$	3π	$\frac{7\pi}{2}$	4π	$\frac{13\pi}{3}$	$-\frac{\pi}{4}$	$-\pi$	-2π	$-\frac{7\pi}{2}$
θ	0	π	0	$\frac{\pi}{2}$	π	$\frac{3\pi}{2}$	0	$\frac{\pi}{3}$	$\frac{7\pi}{4}$	π	0	$\frac{\pi}{2}$
n	0	0	1	1	1	1	2	2	-1	-1	-1	-2

■

Sine and Cosine Functions

Let θ be an angle, measured in radians, satisfying the inequality

$$0 \le \theta < 2\pi.$$

Let P be the point on the unit circle so that the radius OP forms an angle θ with the positive half of the x-axis. We define the **sine** of the angle θ, written $\sin\theta$, to be the

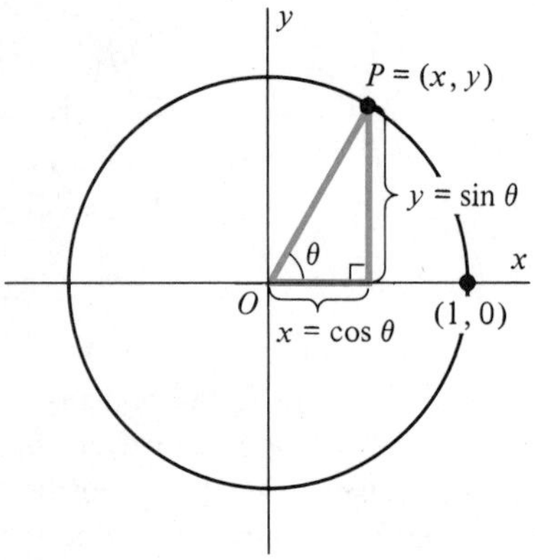

Figure 6.4 Definition of $\sin\theta$ and $\cos\theta$.

y-coordinate of P. Similarly, the cosine function, $\cos\theta$, is defined to be the x-coordinate of P. In other words,

$$\cos\theta = x$$
$$\sin\theta = y$$

(See Figure 6.4.)

In general, values of the sine and cosine are difficult to compute. (We will develop the ability to do this as part of our work on infinite series later in this text.) However, values of $\sin\theta$ and $\cos\theta$ have been tabulated for many angles (tables of these values appear at the end of this text). Moreover, most hand calculators are preprogrammed to give values of $\sin\theta$ and $\cos\theta$ with high degrees of precision. There are certain angles, however, for which values of $\sin\theta$ and $\cos\theta$ are easy to compute. These include multiples of $\pi/2$, and the angles involved in 30°-60°-90° and isosceles right triangles. This is because the ratios of sides in such triangles are easily found from elementary trigonometry (see Figure 6.5).

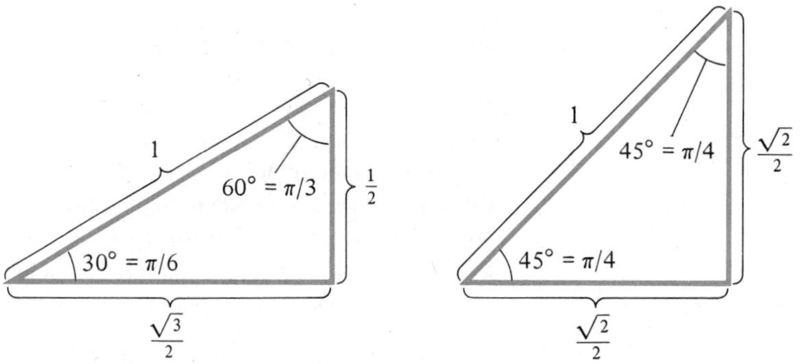

Figure 6.5 Ratios of sides in 30°-60°-90° and 45°-45°-90° triangles.

Example 3 Values of $\sin\theta$ and $\cos\theta$ for various angles θ are shown. Several such angles are sketched in Figure 6.6.

θ	0	$\frac{\pi}{6}$	$\frac{\pi}{4}$	$\frac{\pi}{3}$	$\frac{\pi}{2}$	$\frac{2\pi}{3}$	$\frac{3\pi}{4}$	$\frac{5\pi}{6}$	π	$\frac{7\pi}{6}$	$\frac{5\pi}{4}$	$\frac{4\pi}{3}$	$\frac{3\pi}{2}$	$\frac{5\pi}{3}$	$\frac{7\pi}{4}$	$\frac{11\pi}{6}$
$\sin\theta$	0	$\frac{1}{2}$	$\frac{\sqrt{2}}{2}$	$\frac{\sqrt{3}}{2}$	1	$\frac{\sqrt{3}}{2}$	$\frac{\sqrt{2}}{2}$	$\frac{1}{2}$	0	$-\frac{1}{2}$	$-\frac{\sqrt{2}}{2}$	$-\frac{\sqrt{3}}{2}$	-1	$-\frac{\sqrt{3}}{2}$	$-\frac{\sqrt{2}}{2}$	$-\frac{1}{2}$
$\cos\theta$	1	$\frac{\sqrt{3}}{2}$	$\frac{\sqrt{2}}{2}$	$\frac{1}{2}$	0	$-\frac{1}{2}$	$-\frac{\sqrt{2}}{2}$	$-\frac{\sqrt{3}}{2}$	-1	$-\frac{\sqrt{3}}{2}$	$-\frac{\sqrt{2}}{2}$	$-\frac{1}{2}$	0	$\frac{1}{2}$	$\frac{\sqrt{2}}{2}$	$\frac{\sqrt{3}}{2}$

■

We may extend the definitions of sine and cosine to apply to all real numbers (in radian units) by use of equation (2). Given any number t, we find the principal angle θ corresponding to t by equation (2). We then define $\sin t = \sin\theta$, and $\cos t = \cos\theta$. For example, $\sin\left(\frac{9\pi}{2}\right) = \sin\left(\frac{\pi}{2}\right) = 1$, and $\cos\left(\frac{11\pi}{4}\right) = \cos\left(\frac{3\pi}{4}\right) = -\frac{\sqrt{2}}{2}$. In other words, we have the identities

$$\sin t = \sin(t + 2n\pi), \qquad n = \pm 1, \pm 2, \ldots. \tag{3a}$$
$$\cos t = \cos(t + 2n\pi), \qquad n = \pm 1, \pm 2, \ldots. \tag{3b}$$

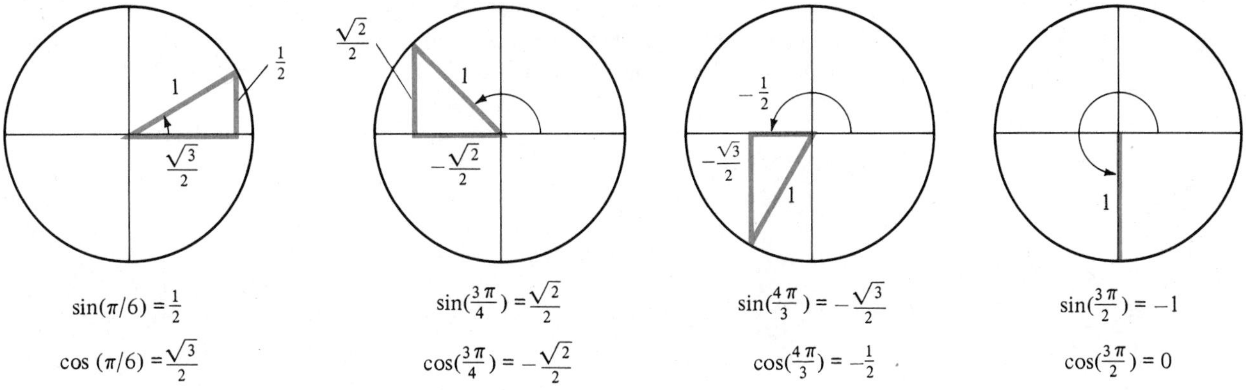

Figure 6.6 Sin θ and cos θ for various angles.

Identities (3a) and (3b) say that the values of the sine and cosine are the same at any number located a multiple of 2π radians away from t as they are at the number t. For this reason, we say that these functions are **periodic,** with period $T = 2\pi$.

Having extended the definitions of $\sin t$ and $\cos t$ to all real numbers t, we may regard $\sin t$ and $\cos t$ as functions defined on $(-\infty, \infty)$. Graphs of these two functions appear in Figure 6.7. Notice that, for both functions, the maximum value is 1 and the minimum value is -1. That is, $|\sin t| \le 1$ and $|\cos t| \le 1$.

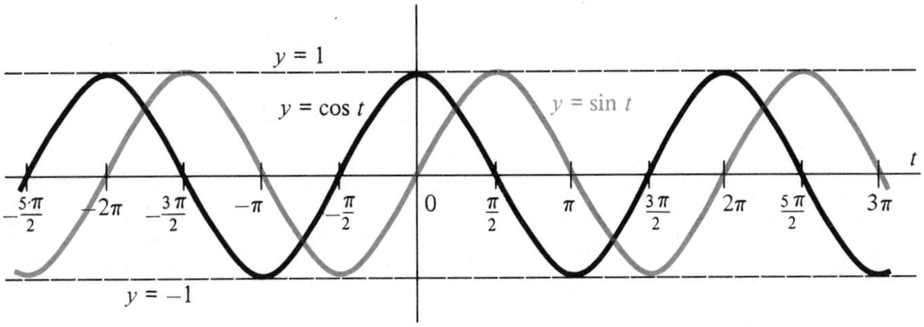

Figure 6.7 Graphs of functions $y = \sin t$ and $y = \cos t$.

The Other Trigonometric Functions

Four additional trigonometric functions are defined as the quotients and reciprocals of the sine and cosine functions. They are as follows:

(i) tangent: $\tan t = \dfrac{\sin t}{\cos t}, \quad t \neq \dfrac{\pi}{2} + n\pi$

(ii) cotangent: $\cot t = \dfrac{\cos t}{\sin t}, \quad t \neq n\pi$

(iii) secant: $\sec t = \dfrac{1}{\cos t}, \quad t \neq \dfrac{\pi}{2} + n\pi$

(iv) cosecant: $\csc t = \dfrac{1}{\sin t}, \quad t \neq n\pi$

Graphs of these functions appear in Figure 6.8. (Note that you can sketch these graphs geometrically, using Figure 6.7. For example, the y-coordinates of the points on the graph of $y = \tan t$ are just the quotients of the y-coordinates of the points on the sine and cosine curves.) You can see from the graphs that the tangent and cotangent functions are periodic with period $T = \pi$, whereas the other four trigonometric functions have a period of $T = 2\pi$. (See Exercise 37.)

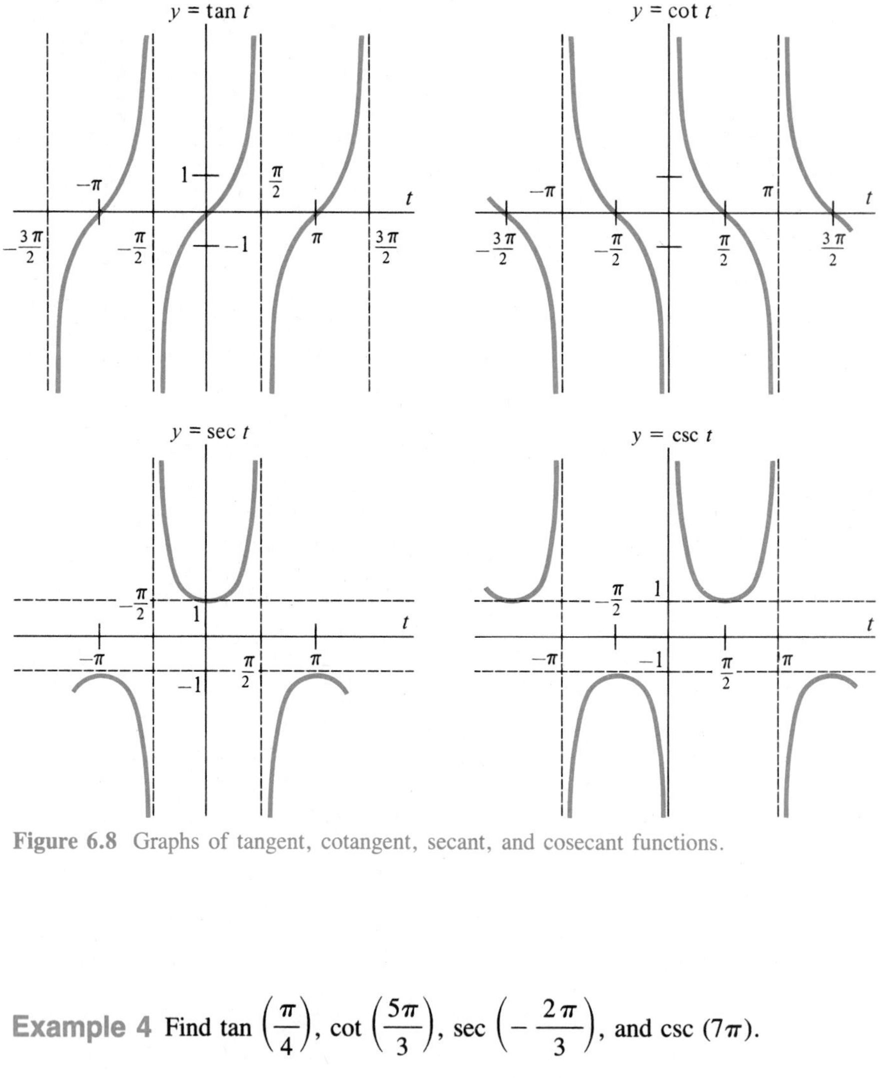

Figure 6.8 Graphs of tangent, cotangent, secant, and cosecant functions.

Example 4 Find $\tan\left(\frac{\pi}{4}\right)$, $\cot\left(\frac{5\pi}{3}\right)$, $\sec\left(-\frac{2\pi}{3}\right)$, and $\csc(7\pi)$.

Strategy

Convert angles to principal angles.

Solution

$$\tan\left(\frac{\pi}{4}\right) = \frac{\sin(\pi/4)}{\cos(\pi/4)} = \frac{\sqrt{2}/2}{\sqrt{2}/2} = 1$$

Then use definitions (i)–(iv).

$$\cot\left(\frac{5\pi}{3}\right) = \frac{\cos(5\pi/3)}{\sin(5\pi/3)} = \frac{1/2}{-\sqrt{3}/2} = -\frac{1}{\sqrt{3}}$$

$\frac{-2\pi}{3} = \frac{4\pi}{3} - 2\pi$

$$\sec\left(-\frac{2\pi}{3}\right) = \sec\left(\frac{4\pi}{3}\right) = \frac{1}{\cos(4\pi/3)} = \frac{1}{-1/2} = -2$$

Denominator undefined!

$\csc(7\pi) = \csc(\pi) = \dfrac{1}{\sin\pi}$, which is *undefined* since $\sin\pi = 0$. ■

Trigonometric Identities

Because $\sin\theta$ and $\cos\theta$ correspond to the legs of a right triangle with hypotenuse equal to one, the Pythagorean theorem gives the identity

$$\sin^2\theta + \cos^2\theta = 1. \tag{4}$$

If $\cos\theta \neq 0$, we may divide each term of equation (4) by $\cos^2\theta$ to obtain the identity

$$\tan^2\theta + 1 = \sec^2\theta.$$

Similarly, dividing (4) through by $\sin^2\theta \neq 0$ gives

$$1 + \cot^2\theta = \csc^2\theta.$$

Here are some other useful identities:

$$\sin(\theta \pm \phi) = \sin\theta\cos\phi \pm \cos\theta\sin\phi$$
$$\cos(\theta \pm \phi) = \cos\theta\cos\phi \mp \sin\theta\sin\phi$$
$$\tan(\theta \pm \phi) = \frac{\tan\theta \pm \tan\phi}{1 \mp \tan\theta\tan\phi}, \quad 1 \mp \tan\theta\tan\phi \neq 0.$$
$$\sin^2\theta = \frac{1}{2}(1 - \cos 2\theta)$$
$$\cos^2\theta = \frac{1}{2}(1 + \cos 2\theta)$$
$$\sin 2\theta = 2\sin\theta\cos\theta$$
$$\cos 2\theta = \cos^2\theta - \sin^2\theta$$
$$\sin\theta = \cos(\pi/2 - \theta)$$
$$\cos\theta = \sin(\pi/2 - \theta)$$
$$\sin(-\theta) = -\sin\theta$$
$$\cos(-\theta) = \cos\theta$$

Finally, for any triangle with interior angles α, β, and γ and corresponding opposite sides of length a, b, and c, we have the Law of Sines:

$$\frac{\sin\alpha}{a} = \frac{\sin\beta}{b} = \frac{\sin\gamma}{c}$$

and the Law of Cosines:

$$c^2 = a^2 + b^2 - 2ab\cos\gamma.$$

(See Figure 6.9.)

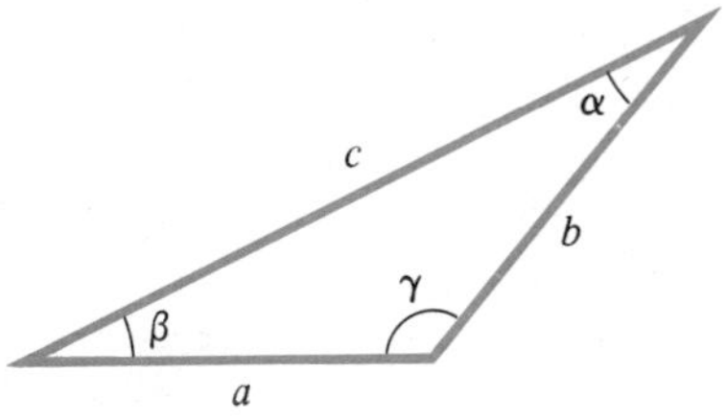

Figure 6.9

Exercise Set 1.6

1. Convert the following angles from degrees to radians.
   a. 30° b. 75° c. −15° d. 315° e. $(x + 30)°$ f. 2910°

2. Convert the following angles in radians to angles in degrees.
   a. π b. 1 c. $9\pi/2$ d. $-21\pi/6$ e. $a + \pi$ f. $7\pi/16$

3. Write a linear equation which converts angles θ_d in degrees to angles θ_r in radians. What is the slope of this line?

4. Find $\tan x$, $\cot x$, $\sec x$, and $\csc x$ for each of the following values of x.
   a. $x = \pi/6$ b. $x = 5\pi/2$ c. $x = -13\pi/3$ d. $x = 9\pi$ e. $x = -9\pi/4$ f. $x = 7\pi/6$

5. Find all principal values of x, in radians, so that

a. $\sin x = 0$
b. $\sin x = \cos x$
c. $\cos x = \sqrt{2}/2$
d. $\tan x = -1$
e. $\csc x = \sqrt{2}$
f. $\sin 3x = 0$
g. $\cos(\pi/2 + 2x) = 1$
h. $\sin 2x = \cos x$

In Exercises 6–11, find all principal angles, in radians, satisfying all stated properties.

6. $\begin{cases} \sin t = \cos t \\ \tan t > 0 \end{cases}$ **7.** $\begin{cases} \sec t = 2 \\ \cot(t) < 0 \end{cases}$

8. $\begin{cases} \sin t > 0 \\ \tan t > 0 \end{cases}$ **9.** $\begin{cases} \sin t \geq \tan t \\ \sin t \geq 0 \end{cases}$

10. $\begin{cases} \cos t \geq 0 \\ \sec t \leq 0 \end{cases}$ **11.** $\begin{cases} \cos t < 0 \\ \tan t < 0 \end{cases}$

In each of Exercises 12–17, graph the given function by comparing it with the function $y = \sin x$.

12. $y = 2 \sin x$ **13.** $f(x) = -2 \sin 3x$

14. $y = \pi \sin(\pi x)$ **15.** $f(x) = \sin(2x + \pi/2)$

16. $y = A \sin(\pi + x)$ **17.** $y = 4 \sin(2x + \pi)$

In each of Exercises 18–25, sketch the graph of the given function by plotting "convenient" points and by comparing with Figures 6.7 and 6.8.

18. $y = \sin x + \cos x$ **19.** $y = \sin x \cos x$

20. $y = 2 + 3 \sin x$ **21.** $y = x \tan x$

22. $y = |\sec x|$ **23.** $y = |\tan x| + 1$

24. $y = \tan(2x + \pi)$ **25.** $y = \dfrac{\sec x}{\sin x}$

26. State the domains and ranges for each of the six trigonometric functions.

In Exercises 27–32, state whether the given function is even, odd, or neither. (See Exercise 49, Section 1.5.)

27. $y = \sin x$ **28.** $y = \cos x$ **29.** $y = \tan x$

30. $y = \dfrac{\sin x}{x}$ **31.** $y = 3 \sec x$ **32.** $y = \sin^2 x$

33. True or false? If the size of an angle is the same in both radians and degrees then that angle must be zero.

34. For acute angles θ (i.e., for $0 \leq \theta \leq 90°$), the six trigonometric functions may be defined as follows. First, construct a right triangle with θ as one of the three angles (Figure 6.10). Label the side opposite θ as y, the side adjacent to θ as x, and the hypotenuse as h. Then,

$$\sin \theta = \frac{y}{h} \qquad \cos \theta = \frac{x}{h}$$

$$\tan \theta = \frac{y}{x} \qquad \cot \theta = \frac{x}{y}$$

$$\sec \theta = \frac{h}{x} \qquad \csc \theta = \frac{h}{y}$$

a. Show that these definitions are entirely equivalent to the definitions given in this section. (*Hint:* Divide all dimensions by h, construct coordinate axes.)

b. Explain how the above definitions can be extended to apply to angles of arbitrary size.

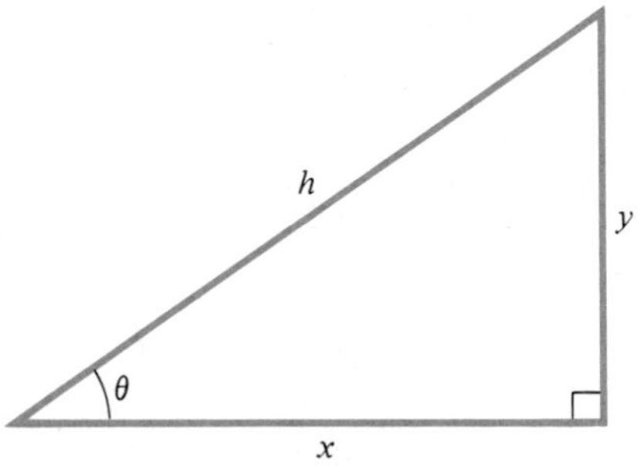

Figure 6.10 x = side adjacent θ; y = side opposite θ; h = hypotenuse.

35. Use the results of Exercise 34 to explain why the following formula for the area of the triangle in Figure 6.10 is valid.

$$A = \frac{1}{2} xh \cdot \sin\theta$$

36. Use the identity $\tan(\theta - \phi) = \dfrac{\tan \theta - \tan \phi}{1 + \tan \theta \tan \phi}$ to prove that if l_1 and l_2 are two nonvertical and nonperpendicular lines with slopes m_1 and m_2, respectively, then the tangent of the angle θ between l_1 and l_2 is

$$\tan \theta = \frac{m_2 - m_1}{1 + m_1 m_2}.$$

37. Use the definitions of the tangent and cotangent functions, together with the periodicity relations of the sine and cosine (equations (3)), to show that the period of the tangent and cotangent is $T = \pi$.

38. True or false? $\sin(x + y) = \sin x + \sin y$.

39. A pendulum consists of a bob hanging at the end of a cord 2 meters long. If the maximum angle of displacement for the pendulum is $\theta = \pi/6$, how far does the bob travel in one complete cycle? (Figure 6.11.)

40. Points A and B are on opposite banks of a straight riverbed, as in Figure 6.12. Point A is directly opposite point B. What is the distance between points B and C if $\theta = \pi/6$?

41. What is the answer to the question in Exercise 40 if $\theta = \pi/3$?

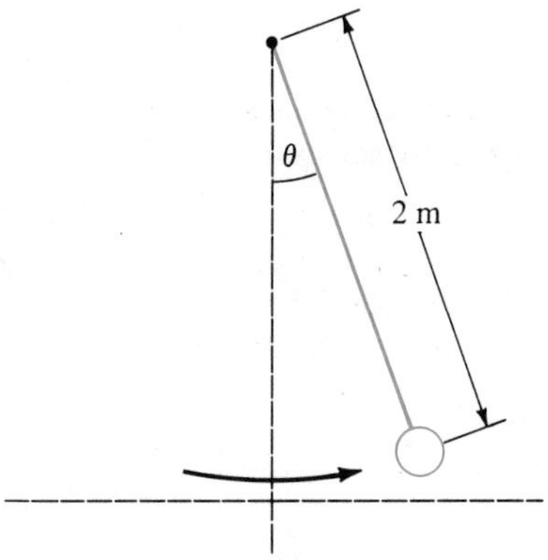

Figure 6.11 Pendulum in Exercise 39.

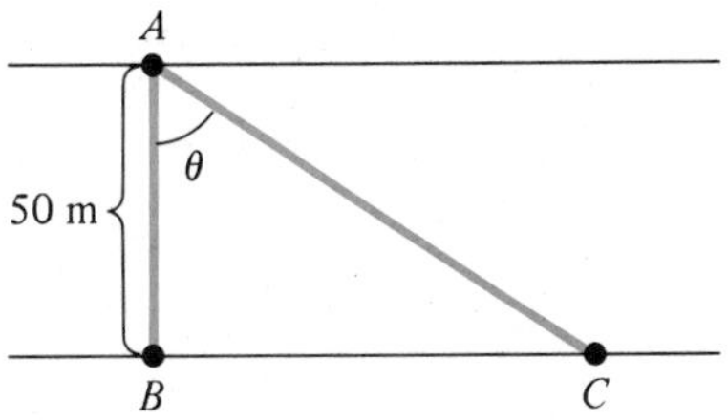

Figure 6.12 Points on opposite sides of a straight riverbed.

42. A child stands 30 meters from a point directly below the location of an airplane. If the angle between ground level and the child's line of sight to the airplane is $\pi/6 = 30°$, what is the altitude of the airplane?

43. What is the altitude of the airplane in Exercise 42 if the angle is $\theta = \pi/4$?

44. What is the period of the function $f(t) = 3\sin\left(\frac{3t}{2} + \frac{\pi}{6}\right)$?

45. Find a constant a so that $\sin\left(at + \frac{\pi}{2}\right) = 2\sin\left(\frac{t}{4} + \frac{\pi}{4}\right)\sin\left(\frac{\pi}{4} - \frac{t}{4}\right)$.

SUMMARY OUTLINE OF CHAPTER 1

- The **real numbers** correspond to points on a line. The **integers** are the numbers $\ldots -2, -1, 0, 1, 2, \ldots$. **Rational numbers** are quotients of integers. **Decimal expansions** for rational numbers terminate or repeat. Decimal expansions for **irrational** numbers neither repeat nor terminate.

- The **inequality** $x < y$ means $y - x$ is positive.

- ***Theorem 1*** (Properties of Inequality)

  (i) Either $x = y$, $x < y$, or $x > y$.
  (ii) If $x < y$ and $y < z$, then $x < z$.
  (iii) If $x < y$, then $x + c < y + c$ for any c.
  (iv) If $x < y$ and $c > 0$, then $cx < cy$.
  (v) If $x < y$ and $c < 0$, then $cx > cy$.

- ***Interval Notation***

  $[a, b) = \{x \mid a \le x < b\}$; $(a, b) = \{x \mid a < x < b\}$

  $[a, \infty) = \{x \mid a \le x\}$; $[a, b] = \{x \mid a \le x \le b\}$

- ***Set Operations***

  $A \cup B = \{x \mid x \in A \text{ or } x \in B\}$
  $A \cap B = \{x \mid x \in A \text{ and } x \in B\}$

- ***Absolute Value***

  $$|x| = \begin{cases} x & \text{if } x \ge 0 \\ -x & \text{if } x < 0. \end{cases}$$

- ***Theorem 2*** (Properties of absolute value): For all x

  (i) $|x| \ge 0$,
  (ii) $|xy| = |x| \cdot |y|$,
  (iii) $|x + y| \le |x| + |y|$.

- The **distance** D between (x_1, y_1) and (x_2, y_2) in the xy plane is the nonnegative number

  $$D = \sqrt{(x_2 - x_1)^2 + (y_2 - y_1)^2}.$$

- The **circle** with center (x_0, y_0) and radius r has equation

  $$(x - x_0)^2 + (y - y_0)^2 = r^2.$$

- The **slope** of the **line** containing (x_1, y_1) and (x_2, y_2) is

  $$m = \frac{y_2 - y_1}{x_2 - x_1}, \quad x_2 \ne x_1.$$

- An **equation** for the line through (x_1, y_1) with slope m is

  $$(y - y_1) = m(x - x_1).$$

- The **slope-intercept** form for this line is

$$y = mx + b, \qquad b = y\text{-intercept}.$$

- A **vertical** line has no slope, and equation $x = a$. A **horizontal** line has slope zero, and equation $y = b$.

- Two lines with slope m_1 and m_2 are (a) **parallel** if $m_1 = m_2$, (b) **perpendicular** if $m_1 m_2 = -1$.

- The variables x and y are **proportional** if $y = kx$ for some constant k and all values of x and y.

- The equation $y = f(x)$ is a **function** if there is a **unique** value of y corresponding to each value of x in the **domain** of f.

- The **domain** of $f(x)$, if not explicitly stated, is the largest set for which $f(x)$ is defined. The **graph** of the function f is the graph of the equation $y = f(x)$.

- The graph of an equation of the form $y = a(x - b)^2 + c$ is a **parabola.** The graph of every **quadratic function** $f(x) = Ax^2 + Bx + C$ is a parabola. The quadratic function can be graphed by **completing the square** in x.

- **Radian** measure is defined by the equation 2π radians $= 360°$. If $P = (x, y)$ is a point on the unit circle and θ is the angle formed between the radius through P and the positive x-axis, then $\sin\theta = y$ and $\cos\theta = x$. Also,

$$\tan\theta = \frac{\sin\theta}{\cos\theta} = \frac{y}{x} \qquad \sec\theta = \frac{1}{\cos\theta} = \frac{1}{x}$$

$$\cot\theta = \frac{\cos\theta}{\sin\theta} = \frac{x}{y} \qquad \csc\theta = \frac{1}{\sin\theta} = \frac{1}{y}$$

when these expressions are defined. The functions $\sin\theta$, $\cos\theta$, $\sec\theta$, and $\csc\theta$ are **periodic** with period $T = 2\pi$. The functions $\tan\theta$ and $\cot\theta$ are periodic with period π.

REVIEW EXERCISES—CHAPTER 1

1. Give an example of a set with no smallest element.

2. For $A = \{1, 2, 3, 5, 7\}$ and $B = \{3, 5, 7, 9, 10\}$, find
a. $A \cup B$ **b.** $A \cap B$

3. For $A = [2, 7)$ and $B = (3, 9)$, find
a. $A \cup B$ **b.** $A \cap B$

4. Which of the following sets have largest elements?
a. $\{1, 2, -3, 6\}$
b. $[0, 5)$
c. $\{x \mid 3 \le x \le 5\}$
d. $\{x \mid x \text{ is a rational number}\}$
e. $\{x \mid 0 \le x^2 \le 2 \text{ and } x \text{ is rational}\}$

5. List all subsets of the set $\{1, 7, 19, \pi\}$.

6. Find two sets of numbers A and B so that neither A nor B has a largest element but so that $A \cap B$ has a largest element.

7. Solve the inequality $2x - 7 \ge 9$.

8. Find $\{x \mid 6 \le x^2 - 3 \le 22\}$.

9. Let $A = \{x \mid x^2 \ge 4\}$, $B = \{x \mid -3 \le x \le 3\}$, and $C = \{-4, -3, -2, -1, 0, 1, 2, 3, 4\}$: find
a. $A \cap B$ **d.** $B \cap C$
b. $A \cup B$ **e.** $A \cup (B \cap C)$
c. $A \cap C$ **f.** $A \cap (B \cup C)$

10. Solve the inequality $4 \le (x + 2)^2 \le 36$.

11. Solve the inequality $|x - 7| \le 12$.

12. Find $\{x \mid 2 \le |x^2 - 2| \le 7\}$.

13. Let $A = \{x \mid |x| \le 9\}$ and $B = \{x \mid -2 < x < 2\}$: find $A \cup B$ and $A \cap B$.

Solve the following inequalities.

14. $|x - 3| \le 5$ **15.** $-3 \le |2x + 1| \le 11$

16. $|x + 2| \ge 6$ **17.** $|2x^2 + 1| \le 9$

18. $|9 - x^2| \ge 0$ **19.** $(x + 1)(x - 1) \ge 0$

20. $(2x - 1)(x + 3) \le 0$ **21.** $x(x + 1)(x + 2) \ge 0$

22. Graph the inequality $y \ge 3$ in the xy-plane.

23. Graph the inequality $x + y \ge 0$.

24. Find the solution set for the equation $|x - 2| = |x + 2|$.

25. Determine which sets of points lie on a common line.
a. $\{(0, 0), (-3, 6), (2, 4)\}$
b. $\{(-4, 2), (-1, 5), (6, 10)\}$
c. $\{(-3, 2), (-1, 1), (2, -3)\}$
d. $\{(1, 3), (2, 4), (-2, 0)\}$

26. Find the area of the triangle with vertices $(0, 0)$, $(5, 0)$, and $(2, 3)$.

27. Find a if a right triangle has vertices $(0, 0)$, $(2, 4)$, and $(a, 0)$, and
a. the right angle is at vertex $(a, 0)$
b. the right angle is at vertex $(2, 4)$.

28. Find the distance between the following pairs of points.
a. $(0, 4)$ and $(2, 7)$
b. $(-3, 9)$ and $(3, 3)$
c. $(-\sqrt{2}, 0)$ and $(0, \sqrt{2})$

29. Find an equation for the circle with center (0, 0) and radius $r = 3$.

30. Find an equation for the circle with center at (2, −4) and diameter $d = 10$.

Find the center and radius of each of the following circles.

31. $x^2 + y^2 = 49$

32. $x^2 - 4x + y^2 = 0$

33. $x^2 - 2x + y^2 + 6y = -9$

34. $x^2 - 2x + y^2 + 2y = 14$

35. Graph the equation $2x - y = 9$.

For each of the following lines find the slope, y-intercept, and x-intercept, if possible.

36. $x + y = 1$

37. $x - 3y + 4 = 0$

38. $y = 4$

39. $7x - 7y + 21 = 0$

40. $x - y = 5$

41. $3x = 6$

42. Find an equation for the line through the points (−2, 1) and (3, 3).

43. Find an equation for the line with slope 6 and y-intercept 7.

44. Find the line through the point (2, −4) and parallel to the line with equation $2x - 4y = 14$.

45. The lines $y - ax = 1$ and $3y = 6x + 12$ are parallel. Find a.

46. Find an equation for the line perpendicular to the line with equation $3x - 6y = 8$ and passing through the origin.

47. Find an equation for the line through (−1, 3) parallel to the line through the points (1, 3) and (6, −4).

48. Where does the line $ax + by + c = 0$ cross the x-axis?

49. The line $y = ax + b$ passes through the points (2, 2) and (4, −5). Find a and b.

50. Graph the inequality $y \le 4x + 4$.

51. Graph the inequality $3y \ge -6x - 9$.

52. Graph the region bounded by the graph of $y = |4 - x^2|$ and the x-axis.

53. Graph the region R where

$$R = \{(x, y) \mid y \le x + 4\} \cap \{(x, y) \mid y \le -2x + 8\}.$$

54. Sketch the region in the plane whose points satisfy all of the following inequalities.
a. $x + y \ge 0$
b. $2y \le 4x + 4$
c. $x \le 4$

55. Sketch the region R consisting of all points satisfying both of the following inequalities
a. $x^2 + y^2 \le 4$, and
b. $x + y \ge 2$.

Convert the following decimals to quotients of integers.

56. $32.6161\overline{61}\ldots$

57. $-6.214\overline{214}\ldots$

58. Draw Venn diagrams to illustrate the following statements.
(a) $A \cap (B \cup C) = (A \cap B) \cup (A \cap C)$
(b) $A \cup (B \cap C) = (A \cup B) \cap (A \cup C)$

59. Find the coordinates of the point of intersection of the lines with equations $x + 3y - 6 = 0$ and $x - y = 2$.

60. A beaker of water is being heated. The water is initially at 10°C. After 3 minutes its temperature is 25°C. Using this data, find a linear equation which will predict the temperature of the water at a later time.
a. What is the predicted temperature after 5 minutes?
b. What is wrong with extrapolating to predict the temperature after 30 minutes?

61. Find a function which converts temperatures in *absolute* (Kelvin) degrees to temperature in Celsius degrees, given that 100°C = 373°K and 0°C = 273°K.

62. Find an equation which converts temperatures in absolute degrees to temperatures in degrees Fahrenheit, given that 212°F = 373°K and 32°F = 273°K.

63. Find the point on the line $x + y = 2$ nearest the point (4, 1).

64. When two chemicals combine to form a solution, the chemical used in greater quantity is called the *solvent. Raoult's law* states that in a dilute solution, the vapor pressure of the solvent, P_s, is proportional to the vapor pressure of the solvent when pure, P_p. Moreover, the constant of proportionality is the ratio of the number of moles of the solvent present, n_s, to the total number of moles in the solution, n_T.
a. Write an equation expressing Raoult's law and sketch its graph.
b. In a certain solution, the number of moles of solvent is 85% of the total present. Write the equation for Raoult's law in this case, and find the vapor pressure of the solvent in solution if its vapor pressure when pure is 0.25 atmospheres.

In Exercises 65–70, state the domain of the given function.

65. $y = \sqrt{x^2 - 1}$

66. $y = \dfrac{1}{x + 2}$

67. $f(x) = \sin(1 - x^2)$

68. $f(x) = \dfrac{1}{1 + \cos x}$

69. $y = \dfrac{1}{\sqrt{1 - \sin^2 x}}$

70. $f(x) = \dfrac{1}{\sqrt{x(x + 2)}}$

In Exercises 71–74, state whether the given function is odd, even, or neither.

71. $f(x) = x^3 \sin x$

72. $f(x) = \dfrac{1 - \cos x}{x^2}$

73. $f(x) = (x - 1) \cdot (x + 1)$ **74.** $f(x) = \dfrac{x \tan x}{1 + x^3}$

In Exercises 75–78, use the technique of completing the square to graph the function.

75. $y - 2x^2 + 2x - \dfrac{7}{2} = 0$ **76.** $y^2 - 2x + 4y + 6 = 0$

77. $y^2 + 2x - y + \dfrac{3}{4} = 0$

78. $y - \dfrac{1}{6}x^2 - x - \dfrac{5}{2} = 0$

In Exercises 79–82, find the values of the six trigonometric functions for the indicated angles.

79. $3\pi/4 = 135°$ **80.** $17\pi/6 = 510°$

81. $-7\pi/3 = -420°$ **82.** $7\pi/3 = 420°$

83. Find θ if
 a. $\sin\theta = \sqrt{3}/2$ and $\cos\theta < 0$
 b. $\sec\theta = \sqrt{2}$ and $\tan\theta = -1$
 c. $\sin 2\theta = \sin\theta$ and $\sin\theta \neq 0$

84. Sketch a right triangle containing an angle θ for which
 a. $\sec\theta = 1$
 b. $\tan\theta = \sqrt{3}$
 c. $\csc\theta = -\sqrt{2}$

In Exercises 85–90, graph the given function.

85. $f(x) = 3 \tan 2x$ **86.** $y = \sin(x - \pi/4)$

87. $y = 2\sin(2x + \pi)$ **88.** $y = \sec 2x$

89. $f(x) = 4\cos(2x - \pi/4)$ **90.** $f(x) = x + \sin x$

In Exercises 91–96, let

$$f(x) = x^3 - x, \quad g(x) = \frac{1}{1 - x}, \quad \text{and} \quad u(x) = \sin x.$$

Form the indicated composite function.

91. $g(u(x))$ **92.** $f(g(x))$ **93.** $u(g(x))$

94. $f(g(u(x)))$ **95.** $g(f(u(x)))$ **96.** $g(u(g(x)))$

97. Prove that the product of two odd functions is an even function.

98. Prove that the product of an odd function and an even function is an odd function.

CHAPTER 2

LIMITS OF FUNCTIONS

2.1 INTRODUCTION

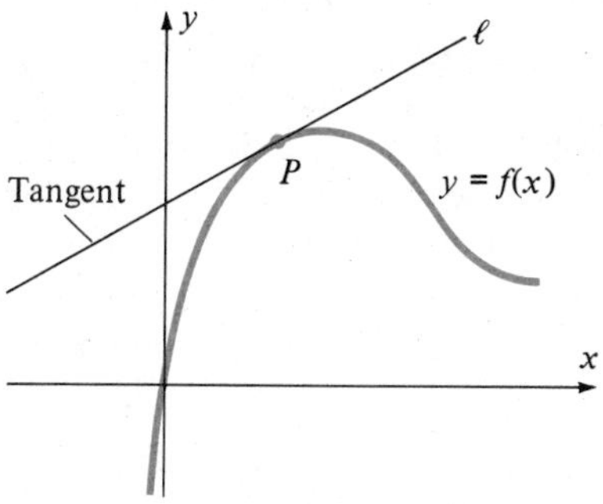

Figure 1.1 The Tangent Line Problem: Find the slope of ℓ.

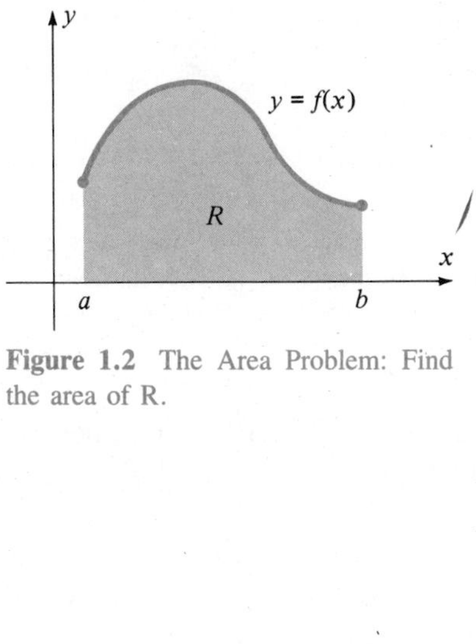

Figure 1.2 The Area Problem: Find the area of R.

Two general problems characterize much of what we shall do in this text:

THE TANGENT LINE PROBLEM: Find the slope of the line tangent to the graph of the function $y = f(x)$ at point P (Figure 1.1).

THE AREA PROBLEM: Determine the area of the region R bounded by the graph of the function $y = f(x)$ and the x-axis for $a \leq x \leq b$ (Figure 1.2).

Both of these problems may be viewed as generalizations of simpler problems that you already know how to solve. For example, if the function $f(x)$ is the linear function $f(x) = mx + b$, then we know that the graph of $f(x)$ is itself a line, so the slope of its tangent at each point is just the number m. If the graph of $f(x)$ is an arc of a circle, we can find the slope of the radius through P and then use the fact that the radius and the tangent are perpendicular to find the slope of the tangent. But if the graph of $f(x)$ is a more general curve, there is no easy technique from algebra or geometry that will enable us to address the Tangent Line Problem.

Similarly, the Area Problem is easy to solve if the region R is a rigid figure from plane geometry, such as a square, a rectangle, a triangle, a trapezoid (each of which can occur when $f(x)$ is linear), or a circle. But plane geometry does not address the Area Problem for more general regions, as illustrated in Figure 1.2.

The **differential calculus** resolves the Tangent Line Problem, together with a host of related issues, all having to do with the rate of change of a function with respect to changes in its independent variable. The **integral calculus** addresses the Area Problem and other related issues. This text follows the conventional format of treating the differential calculus first.

In developing both the differential and the integral calculus, we will make heavy use of the notation, theory, and techniques of elementary algebra and plane geometry, together with the powerful techniques of analytic geometry that synthesize these two subjects. (The development of analytic geometry is due primarily to René Descartes, 1596–1650, a French mathematician and philosopher.)

However, we shall see immediately that something else is required. Whether one begins a study of the calculus with the Tangent Line Problem or the Area Problem, the notion of the **limit** of a function unavoidably surfaces as a fundamental issue.

Because this concept underlies every aspect of the calculus, we need to develop it carefully at the outset. Our plan, therefore, is to begin the discussion of the Tangent Line Problem (Section 2.2), and then to devote the remaining portion of this chapter to developing a rather thorough understanding of the limit concept that arises in Section 2.2. We will then be armed for a vigorous study of the differential calculus, beginning in Chapter 3.

2.2 THE TANGENT LINE PROBLEM: A NEED FOR LIMITS

The first step in attacking the Tangent Line Problem is to clearly define what we mean by "the line tangent to the graph of $y = f(x)$ at point P." From geometry, we know that if the graph of $y = f(x)$ is an arc of a circle, then the tangent at point P may be defined as the (unique) line that intersects the circle only at point P. This is a fine definition for circles, but it fails for more general curves. For example, Figure 2.1 shows that several lines (none of which we would wish to call a tangent) can intersect the graph of $y = f(x)$ only at point P.

However, there is another way of defining the tangent to a circle at point P that does have a satisfactory generalization to more general curves. As Figure 2.2 illustrates, a second point Q on the circle determines a **secant** line through P and Q. As the point Q moves toward P along the circle, these secant lines rotate toward a fixed line through point P. This fixed line is the tangent to the circle at P. This is how we shall define tangents to more general curves.

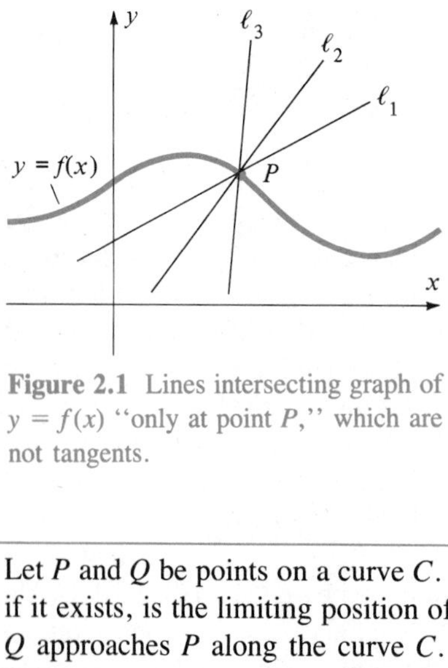

Figure 2.1 Lines intersecting graph of $y = f(x)$ "only at point P," which are not tangents.

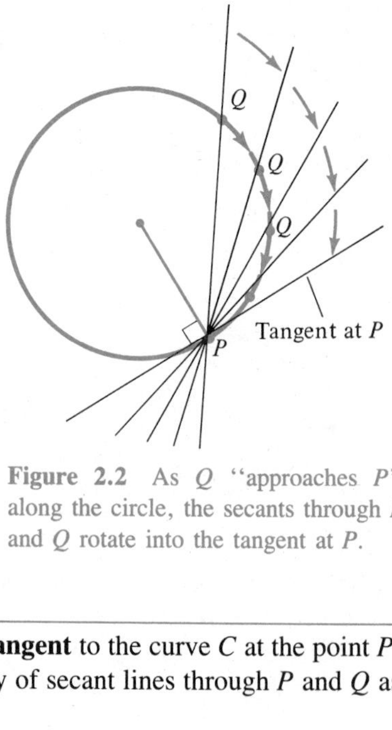

Figure 2.2 As Q "approaches P" along the circle, the secants through P and Q rotate into the tangent at P.

DEFINITION 1

Let P and Q be points on a curve C. The line **tangent** to the curve C at the point P, if it exists, is the limiting position of the family of secant lines through P and Q as Q approaches P along the curve C.

To interpret this definition, we imagine the point Q sliding along the curve C toward the point P (see Figure 2.3). As it does so, the secant through P and Q rotates into the position of line ℓ (Figure 2.4). This is the tangent to C at point P in the definition.

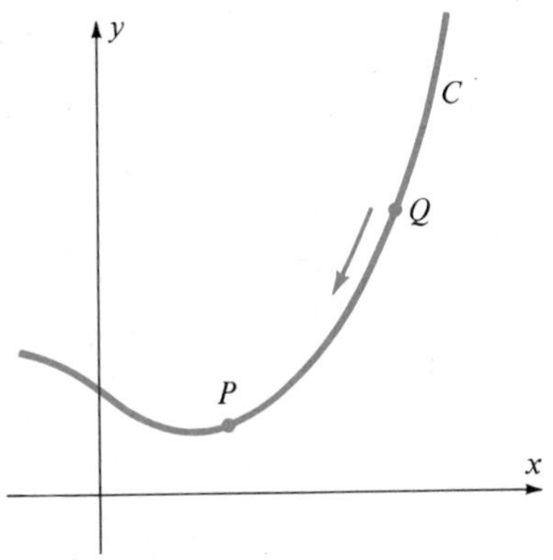

Figure 2.3 Curve of more general type.

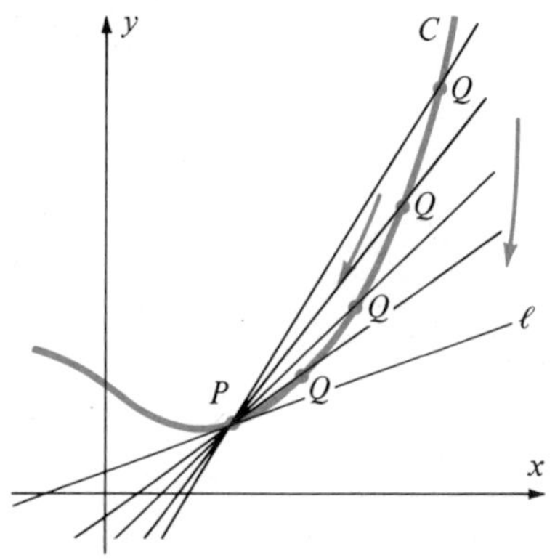

Figure 2.4 Tangent is limiting position of secants as $Q \to P$.

Solving the Tangent Line Problem

Armed with this definition, we can now say how to find the slope of the line tangent to the graph of $y = f(x)$ at the point P. First, we let x_0 be the x-coordinate of P. Then $P = (x_0, f(x_0))$, since P is on the graph of $f(x)$. To find a second point Q on the graph of $f(x)$, we choose any number $\Delta x \neq 0$ and let $Q = (x_0 + \Delta x, f(x_0 + \Delta x))$ (see Figure 2.5). Then

$$\left\{\begin{matrix}\text{Slope of secant}\\ \text{through } P \text{ and } Q\end{matrix}\right\} = m_{\text{sec}} = \frac{f(x_0 + \Delta x) - f(x_0)}{\Delta x}. \tag{1}$$

Assuming for the moment that $f(x)$ is defined for all x, we note that the point Q lies on the graph of $f(x)$ for all values of Δx. Thus, as the number Δx "shrinks" toward zero, the point Q will approach P along the graph of $f(x)$. Since the tangent ℓ to the graph of $f(x)$ at the point P is the "limiting position" of the secants ℓ_{sec} through P and Q as Q approaches P (i.e., as Δx approaches zero), the slopes of these secants, m_{sec}, should approach the slope of the tangent, $m_{\tan}$. That is,

$$m_{\tan} = \left\{\begin{matrix}\text{limiting value of } m_{\text{sec}} \text{ as } Q\\ \text{approaches } P \text{ (i.e., as } \Delta x \text{ approaches zero)}\end{matrix}\right\}. \tag{2}$$

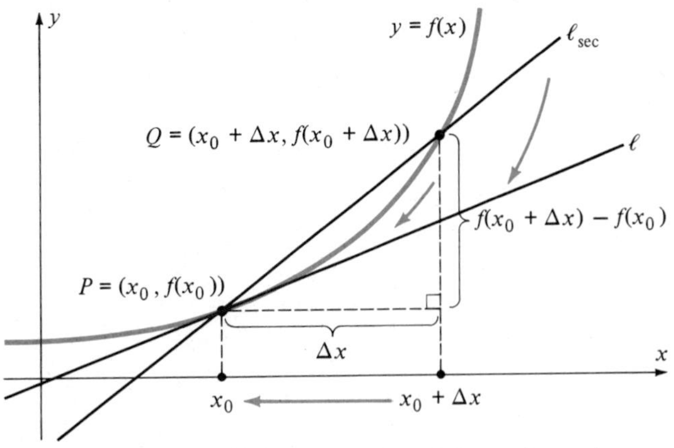

Figure 2.5 Slope of ℓ is limit of slope of ℓ_{sec} as $\Delta x \to 0$.

Using equation (1) we write the statement (2) more formally as follows:

$$m_{\tan} = \lim_{\Delta x \to 0} \frac{f(x_0 + \Delta x) - f(x_0)}{\Delta x} \tag{3}$$

The remaining sections of this chapter are devoted to the issue of making the limit concept, represented by the symbol "$\lim\limits_{\Delta x \to 0}$" in equation (3), precise. For now, let's simply agree that it represents the informal notion embodied by the discussion leading up to equation (3), and take a look at several examples in which slopes are actually calculated in this way.

Example 1 For a linear function $f(x) = mx + b$, the secant through two points P and Q on the graph will be just the graph (line) itself. Thus, the line $y = mx + b$ is its own tangent, so the slope given by equation (3) should be the number m. To verify this, we let $(x_0, f(x_0))$ be any point on the graph of $f(x) = mx + b$. Then

$$f(x_0 + \Delta x) = m(x_0 + \Delta x) + b, \qquad \text{and} \qquad f(x_0) = mx_0 + b,$$

so, by equation (3),

$$\begin{aligned} m_{\tan} &= \lim_{\Delta x \to 0} \frac{[m(x_0 + \Delta x) + b] - [mx_0 + b]}{\Delta x} \\ &= \lim_{\Delta x \to 0} \frac{mx_0 + m\,\Delta x + b - mx_0 - b}{\Delta x} \\ &= \lim_{\Delta x \to 0} \frac{m\,\Delta x}{\Delta x}. \end{aligned}$$

Now, as long as $\Delta x \neq 0$, we may divide both the numerator and denominator in the last line by Δx. (That is, we can "cancel" the common factor Δx.) Since this leaves only m, which is unaffected by changes in Δx, we conclude that

$$m_{\tan} = m,$$

as expected. ■

The next example shows how we can put equation (3) to work on a graph that is not itself linear. Note that we follow the same strategy as in Example 1 for evaluating the limit that results: factor Δx from both the numerator and denominator and cancel.

Example 2 Find the slope of the line tangent to the graph of $f(x) = x^2$ at the point $(1, 1)$.

Strategy

Identify x_0.

Find $f(x_0 + \Delta x)$ by setting $x = x_0 + \Delta x$ in the equation for $f(x)$.

Solution

Here $x_0 = 1$ and $f(x) = x^2$, so

$$f(x_0 + \Delta x) = f(1 + \Delta x) = (1 + \Delta x)^2$$

and

$$f(x_0) = f(1) = 1.$$

Thus,

Set up the expression in equation (3) for finding the slope.

Substitute explicit forms for $f(x_0 + \Delta x)$ and $f(x_0)$.

By algebra, simplify the resulting quotient.

Factor Δx in the numerator.

Divide by the common factor of Δx.

$$\begin{aligned} m_{\tan} &= \lim_{\Delta x \to 0} \frac{f(x_0 + \Delta x) - f(x_0)}{\Delta x} \\ &= \lim_{\Delta x \to 0} \frac{(1 + \Delta x)^2 - 1}{\Delta x} \\ &= \lim_{\Delta x \to 0} \frac{(1 + 2\,\Delta x + \Delta x^2) - 1}{\Delta x} \\ &= \lim_{\Delta x \to 0} \frac{2\,\Delta x + \Delta x^2}{\Delta x} \\ &= \lim_{\Delta x \to 0} \frac{\Delta x(2 + \Delta x)}{\Delta x} \\ &= \lim_{\Delta x \to 0} (2 + \Delta x) \\ &= 2. \end{aligned}$$

See Figure 2.6. ■

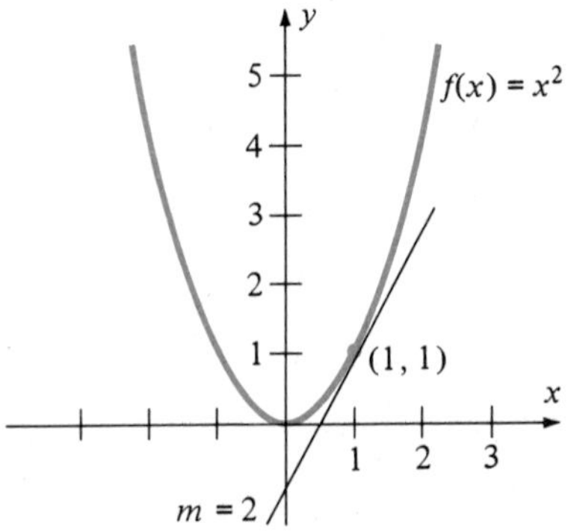

Figure 2.6 Slope of tangent to graph of $f(x) = x^2$ at (1, 1) is $m = 2$.

Difference Quotients

The quotient

$$\frac{f(x_0 + \Delta x) - f(x_0)}{\Delta x} \tag{4}$$

in equation (3) that we use to find the slope of a tangent to the graph of the function $f(x)$ is called the **difference quotient.** We may summarize the process of finding the slope of the tangent to the graph of $f(x)$ by saying that we "find the limit of the difference quotient for $f(x)$ at $x = x_0$ as Δx approaches zero." It is important to note that *we cannot evaluate this limit by simply setting* $\Delta x = 0$ *in the difference quotient* (4), since this would produce a zero in the denominator (as well as in the numerator!). For the functions in the examples of this section, we are able to overcome this difficulty by using algebra to factor a single Δx from the numerator, thereby enabling us to divide the factor of Δx from both the numerator and denominator. The resulting limit is then evaluated by setting $\Delta x = 0$, which causes no further difficulties. However, this strategy will not work for all functions, and the theory of limits developed later in this chapter will be required to handle more general problems of this type.

In the following example, the strategy is the same as in Example 2, although the algebra is a bit trickier.

Example 3 Find the slope of the line tangent to the graph of $f(x) = \sqrt{x}$ at the point (4, 2).

Strategy

Identify x_0, $f(x_0)$, and $f(x_0 + \Delta x)$.

Solution

Here $x_0 = 4$ and $f(x) = \sqrt{x}$, so

$$f(x_0 + \Delta x) = \sqrt{4 + \Delta x}; \qquad f(x_0) = \sqrt{4} = 2.$$

Thus,

Set up the difference quotient in equation (3).

To obtain a factor of Δx in the numerator, eliminate the radical using the fact that

$$\sqrt{a} - b = (\sqrt{a} - b)\frac{\sqrt{a} + b}{\sqrt{a} + b} = \frac{a - b^2}{\sqrt{a} + b}.$$

Divide common factor of Δx from numerator and denominator.

$$\begin{aligned}
m_{\tan} &= \lim_{\Delta x \to 0} \frac{f(x_0 + \Delta x) - f(x_0)}{\Delta x} \\
&= \lim_{\Delta x \to 0} \frac{\sqrt{4 + \Delta x} - 2}{\Delta x} \\
&= \lim_{\Delta x \to 0} \left(\frac{\sqrt{4 + \Delta x} - 2}{\Delta x}\right) \cdot \left(\frac{\sqrt{4 + \Delta x} + 2}{\sqrt{4 + \Delta x} + 2}\right) \\
&= \lim_{\Delta x \to 0} \frac{(4 + \Delta x) - 4}{\Delta x(\sqrt{4 + \Delta x} + 2)} \\
&= \lim_{\Delta x \to 0} \frac{\Delta x}{\Delta x(\sqrt{4 + \Delta x} + 2)} \\
&= \lim_{\Delta x \to 0} \frac{1}{\sqrt{4 + \Delta x} + 2} \\
&= \frac{1}{\sqrt{4} + 2} \\
&= \frac{1}{4}.
\end{aligned}$$

See Figure 2.7. ■

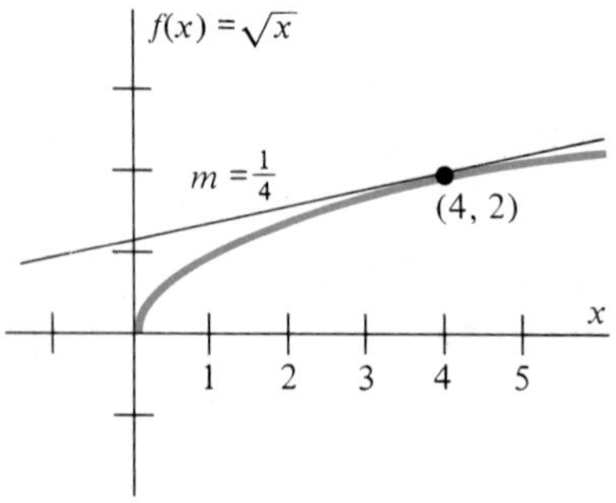

Figure 2.7 Slope of tangent to graph of $f(x) = \sqrt{x}$ at (4, 2) is $m = 1/4$.

The Derivative of a Function

There is some notation and terminology relevant to the preceding discussion. The limit of the difference quotient in equation (3) is called the **derivative** of the function $f(x)$ at $x = x_0$. It is denoted by the expression $f'(x_0)$. That is,

$$f'(x_0) = \lim_{\Delta x \to 0} \frac{f(x_0 + \Delta x) - f(x_0)}{\Delta x} \tag{5}$$

whenever this limit exists. Although we shall find a broader interpretation for the derivative than that of the slope of a tangent, we may combine the statements of equations (3) and (5) to write

$$f'(x_0) = \text{slope of tangent to graph of } y = f(x) \text{ at } (x_0, f(x_0)) \tag{6}$$

Thus, we may summarize the results of Examples 1–3 by saying that

(a) for $f(x) = mx + b$, $\quad f'(x) = m$;
(b) for $f(x) = x^2$, $\quad f'(1) = 2$;
(c) for $f(x) = \sqrt{x}$, $\quad f'(4) = \dfrac{1}{4}$.

Figures 2.8 and 2.9 suggest a broader interpretation of the derivative. Since $f'(x)$ is the slope of the tangent to the graph of $y = f(x)$ at $(x, f(x))$, the derivative is itself a function of x. It is often referred to as the "slope function" for the function $f(x)$.

Given a function $f(x)$, we may find an explicit formulation for the derivative, $f'(x)$, by the method of Examples 1–3, except that we do not specify a particular value $x = x_0$.

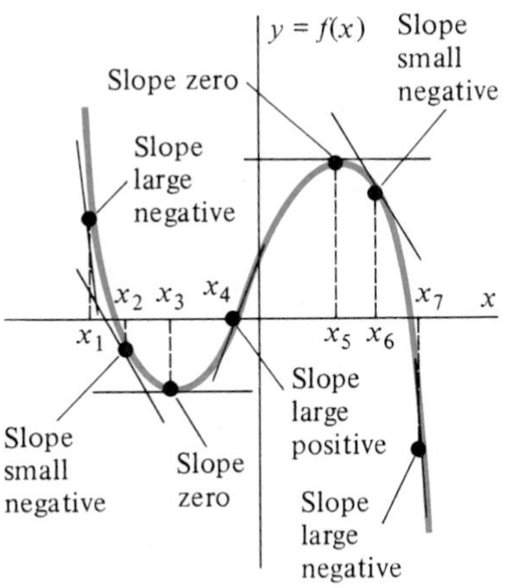

Figure 2.8 Graph of a function $f(x)$, showing several tangents.

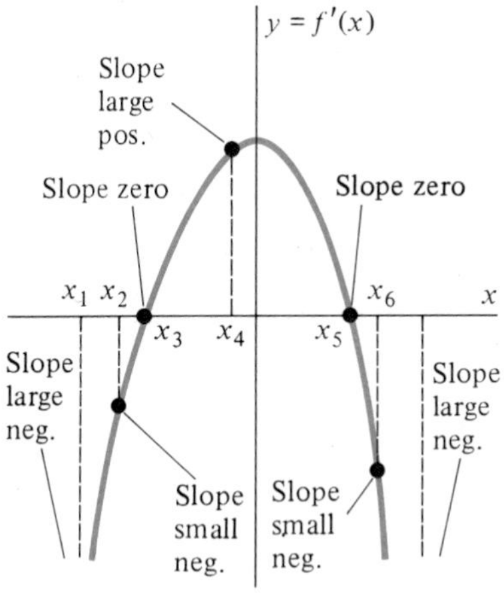

Figure 2.9 Graph of the "slope function" (derivative), $f'(x)$.

Example 4 For the linear function $f(x) = mx + b$, the derivative is

$$f'(x) = \lim_{\Delta x \to 0} \frac{f(x + \Delta x) - f(x)}{\Delta x}$$

$$= \lim_{\Delta x \to 0} \frac{[m(x + \Delta x) + b] - [mx + b]}{\Delta x}$$
$$= \lim_{\Delta x \to 0} \frac{(mx + m\,\Delta x + b) - (mx + b)}{\Delta x}$$
$$= \lim_{\Delta x \to 0} \frac{m\,\Delta x}{\Delta x}$$
$$= m.$$

See Figure 2.10. ■

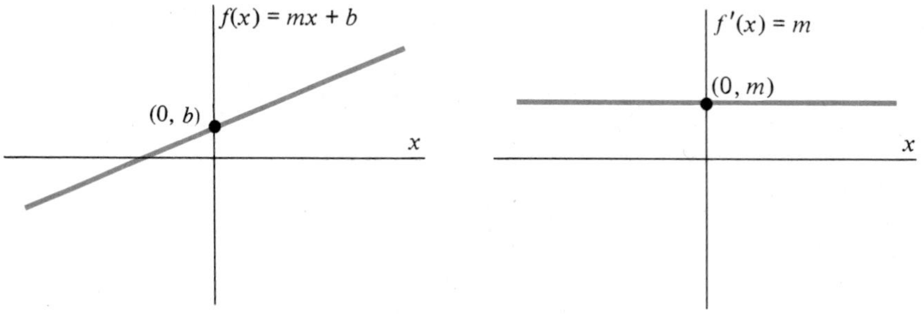

Figure 2.10 For the linear function $f(x) = mx + b$, the slope function (derivative) is $f'(x) = m$.

Example 5 The derivative of the function $f(x) = x^2$ is found as in Example 2.

$$f'(x) = \lim_{\Delta x \to 0} \frac{f(x + \Delta x) - f(x)}{\Delta x}$$
$$= \lim_{\Delta x \to 0} \frac{(x + \Delta x)^2 - x^2}{\Delta x}$$
$$= \lim_{\Delta x \to 0} \frac{(x^2 + 2x\,\Delta x + \Delta x^2) - x^2}{\Delta x}$$
$$= \lim_{\Delta x \to 0} \frac{2x\,\Delta x + \Delta x^2}{\Delta x}$$
$$= \lim_{\Delta x \to 0} (2x + \Delta x)$$
$$= 2x.$$

Thus, for $f(x) = x^2$, $f'(0) = 0$, $f'(-3) = -6$, $f'(2) = 4$, and $f'(10) = 20$ (see Figure 2.11). ■

Example 6 For $f(x) = \sqrt{x}$, find $f'(x)$.

Solution: The algebra is the same as in Example 3, except that we are not working with an explicit value for x_0:

$$f'(x) = \lim_{\Delta x \to 0} \frac{f(x + \Delta x) - f(x)}{\Delta x}$$
$$= \lim_{\Delta x \to 0} \frac{\sqrt{x + \Delta x} - \sqrt{x}}{\Delta x}$$

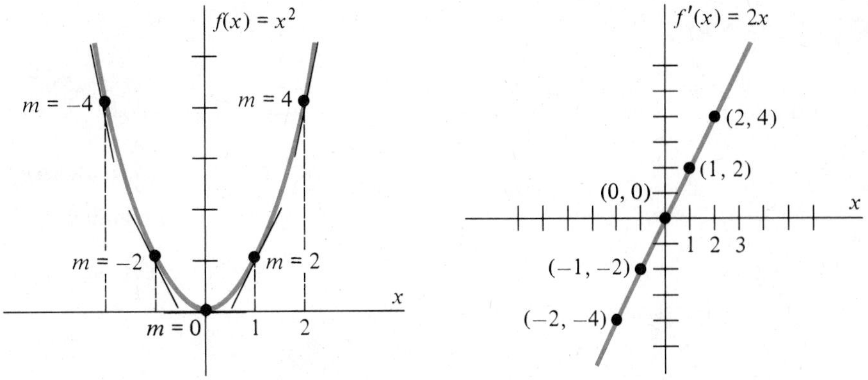

Figure 2.11 For the quadratic function $f(x) = x^2$, the slope function (derivative) is $f'(x) = 2x$.

$$= \lim_{\Delta x \to 0} \left(\frac{\sqrt{x + \Delta x} - \sqrt{x}}{\Delta x} \right) \cdot \left(\frac{\sqrt{x + \Delta x} + \sqrt{x}}{\sqrt{x + \Delta x} + \sqrt{x}} \right)$$

$$= \lim_{\Delta x \to 0} \frac{(x + \Delta x) - x}{\Delta x(\sqrt{x + \Delta x} + \sqrt{x})}$$

$$= \lim_{\Delta x \to 0} \frac{\Delta x}{\Delta x(\sqrt{x + \Delta x} + \sqrt{x})}$$

$$= \lim_{\Delta x \to 0} \frac{1}{(\sqrt{x + \Delta x} + \sqrt{x})}$$

$$= \frac{1}{2\sqrt{x}}.$$

That is, for $f(x) = \sqrt{x}$, $f'(x) = \dfrac{1}{2\sqrt{x}}$. Thus, $f'(4) = \dfrac{1}{4}$, $f'(9) = \dfrac{1}{6}$, $f'(27) = \dfrac{1}{2\sqrt{27}}$, but neither $f'(0)$ nor $f'(-4)$ are defined (see Figure 2.12). ■

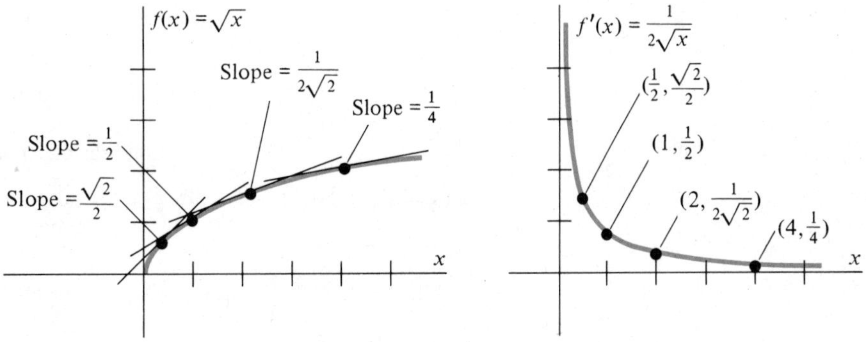

Figure 2.12 For the function $f(x) = \sqrt{x}$, the slope function (derivative) is $f'(x) = 1/2\sqrt{x}$.

Exercise Set 2.2

1. Each of the Figures (i)–(v) is the graph of the slope function for one of the functions whose graph is one of Figures (a)–(e). Match the corresponding functions and derivatives.

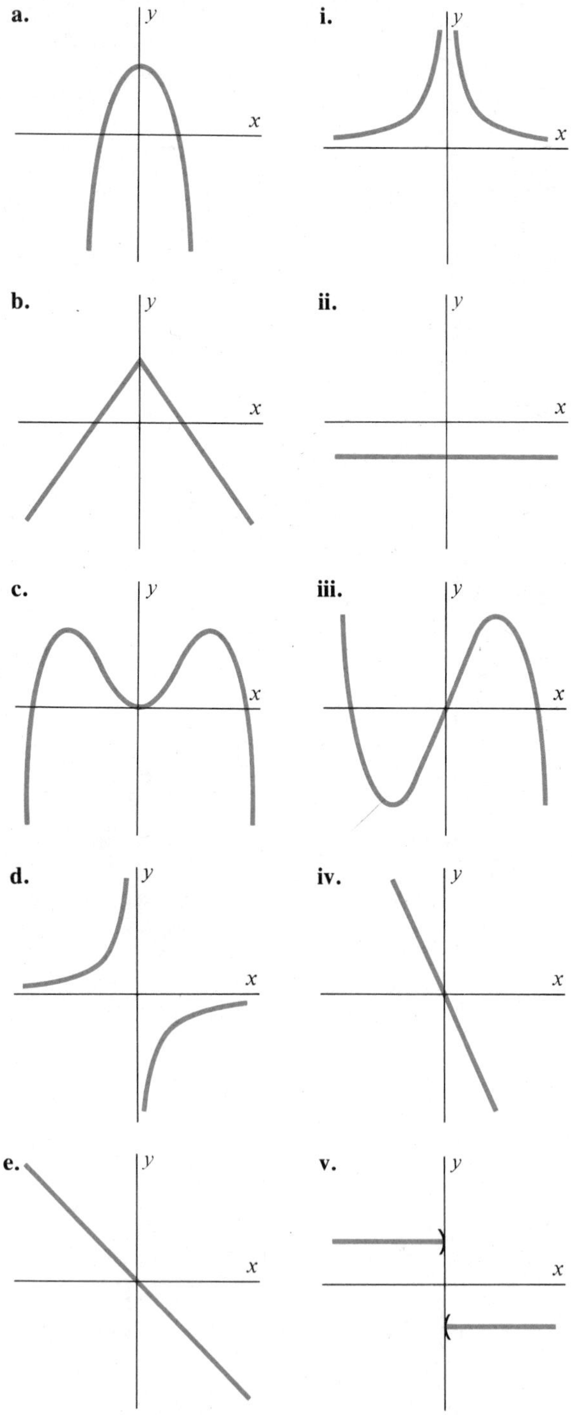

2. Sketch examples to show why each of the following "definitions" of tangent is inadequate in describing the line tangent to a curve at a given point. Figure 2.1 gives one such example for Definition A.

Definition A: The line tangent to the curve C at point P is the line that meets C only at point P.

Definition B: The line tangent to the curve C at point P is the line that meets C only at point P and that lies entirely on one side of C.

In Exercises 3–24, use equation (5) to find the derivative $f'(x)$.

3. $f(x) = x$
4. $f(x) = 3x - 2$
5. $f(x) = 7 - 3x$
6. $f(x) = 2x^2$
7. $f(x) = 9x^2$
8. $f(x) = ax^2$
9. $f(x) = x^2 + 3$
10. $f(x) = 5 - x^2$
11. $f(x) = 3x^2 + 2$
12. $f(x) = x^2 + x$
13. $f(x) = x^2 + 6x + 1$
14. $f(x) = x^3$
15. $f(x) = 2x^3 + x$
16. $f(x) = x^3 - 2x + 6$
17. $f(x) = x^4$
18. $f(x) = ax^3 + bx^2 + c$
19. $f(x) = \sqrt{x + 4}$
20. $f(x) = \sqrt{4 - x}$
21. $f(x) = \dfrac{1}{x}$
22. $f(x) = \dfrac{1}{x + 3}$
23. $f(x) = \dfrac{1}{\sqrt{x}}$
24. $f(x) = \dfrac{1}{\sqrt{x + 2}}$

25. Find the slope of the line tangent to the graph of the function $f(x) = 2x^2$ at the point $(2, 8)$.

26. Find the slope of the line tangent to the graph of $f(x) = ax^2$ at the point $(1, a)$.

27. Find the slope of the line tangent to the graph of $f(x) = 5 - x^2$ at the point $(-2, 1)$.

28. Find the slope of the line tangent to the graph of $f(x) = 2x^3 + x$ at the point where $x = 1$.

29. Find the slope of the line tangent to the graph of $f(x) = \sqrt{x + 4}$ at the point where this line is parallel to the graph of the equation $4y - x - 8 = 0$.

30. True or false? A tangent to the graph of $f(x) = \sqrt{4 - x}$ cannot be horizontal. Why or why not?

In Exercises 31–35, find an equation for the tangent line determined in the given exercise.

31. Exercise 25.

32. Exercise 26.

33. Exercise 27.

34. Exercise 28.

35. Exercise 29.

36. *(Calculator)* By use of a calculator we can approximate the slope of the line tangent to the graph of $y = f(x)$ at $(x_0, f(x_0))$ by computing values of the difference quotient $\dfrac{f(x_0 + \Delta x) - f(x_0)}{\Delta x}$ for small values of Δx. Approximate the slope of the line tangent to the graph of $f(x) = x^3 - 3x$ at the point (2, 2) by completing Table 2.1.

Table 2.1

x	Δx	$\dfrac{[(2 + \Delta x)^3 - 3(2 + \Delta x)] - 2}{\Delta x}$
2	2	
2	1	
2	0.5	
2	0.2	
2	0.1	
2	0.05	
2	0.01	
2	0.005	

37. Use the method of Examples 2 and 3 to find the slope of the line tangent to the graph of $y = x^3 - 3x$ at the point (2, 2). Does this agree with the results of Exercise 36?

38. Consider the graph of $f(x) = x^3 - 3x$ in Figure 2.13.

a. Sketch tangents at the points with x-coordinates -2, $-\frac{3}{2}$, -1, $-\frac{1}{2}$, 0, $\frac{1}{2}$, 1, $\frac{3}{2}$, and 2.

b. From your sketches estimate the slopes of each of these tangents.

c. Compare the values of $f'(x)$ at each of the x values in part (a) with your slope estimates. How closely do they agree? Should they?

d. What does $f'(x)$ tell you about the high and low points for values of x between -2 and 2?

39. *(Calculator)* Use the method of Exercise 36 to approximate the slope of the line tangent to the graph of $f(x) = \sin x$ at the point $(\pi/4, \sqrt{2}/2)$.

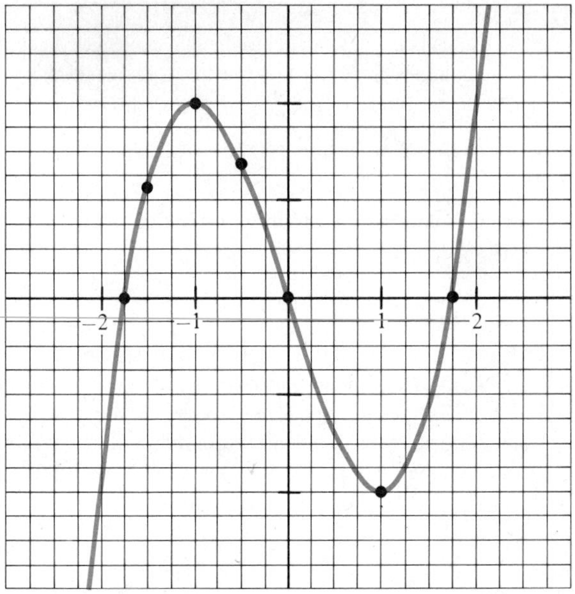

Figure 2.13

40. Find two points where the tangents to the graph of $f(x) = x^2$ are perpendicular (see Figure 2.14).

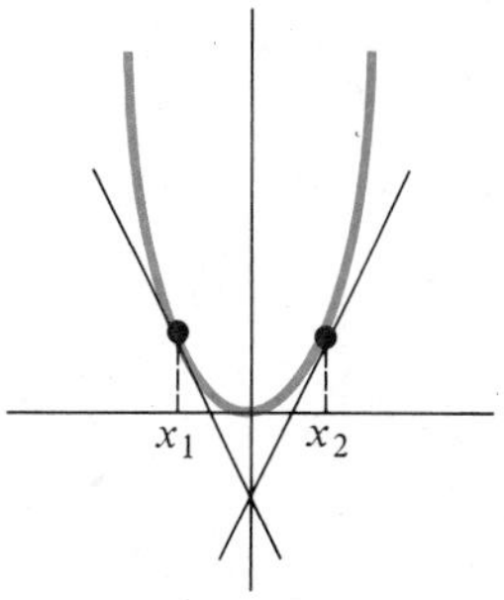

Figure 2.14

41. Find numbers x_1 and x_2 so that the triangle formed by the x-axis and the lines tangent to the graph of $f(x) = 1 - x^2$ at the points $(x_1, f(x_1))$ and $(x_2, f(x_2))$ is equilateral (see Figure 2.15).

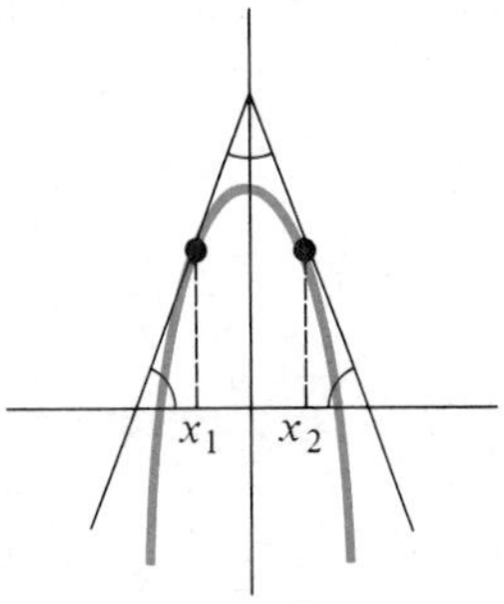

Figure 2.15

42. Prove that the line tangent to the graph of $f(x) = x^3$ at $(x_0, f(x_0))$ is always parallel to the line tangent to this graph at the point $(-x_0, f(-x_0))$. Is this true for any odd function? Why or why not?

43. True or false? If $f'(x) = g'(x)$, then $f(x) = g(x)$.

44. Give a geometric argument to show that if $f(x) = g(x) + C$ for some constant C, and if the derivatives $f'(x)$ and $g'(x)$ exist, then $f'(x) = g'(x)$.

45. Knowledge of the derivative of a function often aids in understanding its behavior, even when the exact expression for the function is unknown. Use the indicated values of $f'(x)$ in Table 2.2 to sketch tangents at each of the points in Figure 2.16. Then sketch the graph of $f(x)$ as best you can. (Notice that the graph you sketch differs from that which you might have sketched without the data given in Table 2.2.)

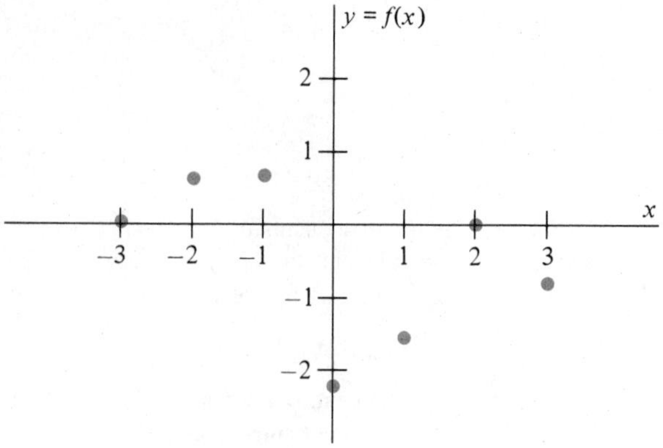

Figure 2.16

Table 2.2

x	-3	-2	-1	0	1	2	3
$f'(x)$	-2	4	0	-1	6	-2	2

2.3 AN INFORMAL DEFINITION OF LIMIT

In Section 2.2 we used a rather ad hoc method in obtaining the derivative

$$f'(x) = \lim_{\Delta x \to 0} \frac{f(x + \Delta x) - f(x)}{\Delta x} \tag{1}$$

as the limit of a difference quotient as $\Delta x \to 0$. Since this method will not handle all types of functions, we need to develop a more precise notion of limit and a set of rules by which limits may be evaluated. That is the goal of the remaining sections of this chapter.

The derivative in line (1) is a special case of the more general statement

$$L = \lim_{x \to a} f(x), \tag{2}$$

read, "The number L is the limit of the function $f(x)$ as x approaches the number a." An informal working definition of what we mean by this statement follows. (A more rigorous definition of limit will be presented in Section 2.7, by which point we will have developed a better understanding of the relevant issues.)

DEFINITION 2

The statement $L = \lim_{x \to a} f(x)$ means that the values $f(x)$ will approach the number L as the values of x approach the number a from either direction.

Figure 3.1 illustrates the geometric interpretation of our working definition of limit. Notice that the point $(a, f(a))$ is not assumed to belong to the graph of

$y = f(x)$. This emphasizes the fact that $L = \lim_{x \to a} f(x)$ depends only on the behavior of $f(x)$ *near* $x = a$, *but not on the number* $f(a)$ *itself.*

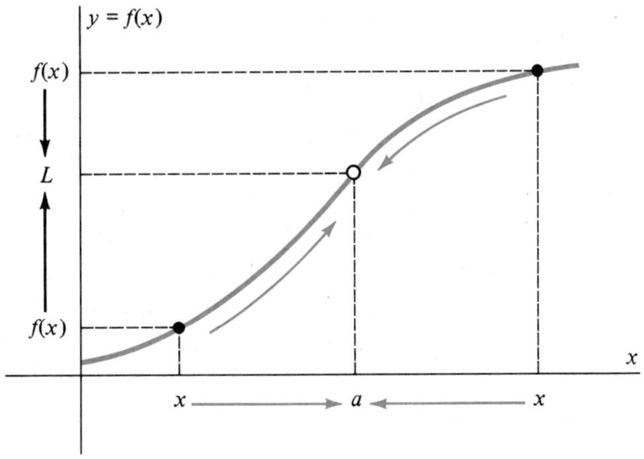

Figure 3.1 Working definition of $L = \lim_{x \to a} f(x)$ is that $f(x) \to L$ as $x \to a$. The concept is independent of $(a, f(a))$.

Continuous Functions

For many functions, the limit $\lim_{x \to a} f(x)$ is just the number $L = f(a)$, even though we have emphasized that this is not necessarily the case. Such functions are called **continuous** functions, and they will be discussed in some detail in Section 2.6. The following are examples of continuous functions.

Example 1 From our knowledge of the graphs of the following functions, together with Definition 2, we conclude that

(a) for $f(x) = 3x - 4$, $\quad f(2) = 3 \cdot 2 - 4 = 2$. Thus,

$$\lim_{x \to 2} (3x - 4) = 2;$$

(b) for $f(x) = \sqrt{x + 6}$, $\quad f(10) = \sqrt{10 + 6} = 4$. Thus,

$$\lim_{x \to 10} \sqrt{x + 6} = 4;$$

(c) for $f(x) = \sin x$, $\quad f(\pi/4) = \sin(\pi/4) = \dfrac{\sqrt{2}}{2}$. Thus,

$$\lim_{x \to \pi/4} \sin x = \frac{\sqrt{2}}{2}. \qquad \blacksquare$$

Limits Where $f(a)$ is Undefined

If every function were continuous, there would be no need to worry about limits at this point. However, we have already encountered an important limit (the derivative) where the function is undefined at precisely the value $x = a$ for which we wish to obtain the limit. This situation usually occurs when $f(x)$ is a quotient of the form $f(x) = \dfrac{g(x)}{h(x)}$, with both $g(a) = 0$ and $h(a) = 0$. The following example is typical.

Example 2 Find $\lim_{x \to 0} \dfrac{(x+1)^3 - 1}{x}$.

Solution: We cannot simply set $x = 0$ to obtain this limit, since this produces the quotient $\dfrac{(0+1)^3 - 1}{0} = \dfrac{0}{0}$, which is undefined. However, using a calculator or a computer, we can obtain values of the function $f(x) = \dfrac{(x+1)^3 - 1}{x}$ for x's near $a = 0$. Table 3.1 contains several such values.

Table 3.1
$f(x) = \dfrac{(x+1)^3 - 1}{x}$

x	$y = f(x)$
−2.0	1.0000
−1.5	0.7500
−1.0	1.0000
−0.5	1.7500
−0.2	2.4400
−0.1	2.7100
−0.01	2.9701
−0.001	2.9970
0.001	3.0030
0.01	3.0301
0.1	3.3100
0.2	3.6400
0.5	4.7500
1.0	7.000
1.5	9.7500
2.0	13.000

From the entries of Table 3.1 we might conclude that $f(x)$ approaches $L = 3$ as x approaches 0, that is, that

$$\lim_{x \to 0} \frac{(x+1)^3 - 1}{x} = 3.$$

We can verify this using simple algebra. Since

$$\begin{aligned} \frac{(x+1)^3 - 1}{x} &= \frac{(x^3 + 3x^2 + 3x + 1) - 1}{x} \\ &= \frac{x^3 + 3x^2 + 3x}{x} \\ &= x^2 + 3x + 3, \ x \neq 0, \end{aligned}$$

we may conclude that the functions

$$f(x) = \frac{(x+1)^3 - 1}{x} \qquad \text{and} \qquad g(x) = x^2 + 3x + 3$$

have the same values *except* at $x = 0$ where $f(x)$ is undefined (see Figures 3.2 and 3.3). Thus, the limits as x approaches zero for these two functions must be the same. (Remember, we do not care about $f(0)$, only the limit of $f(x)$ as x *approaches* 0.)

Our limit is therefore calculated as follows:

$$\lim_{x \to 0} \frac{(x+1)^3 - 1}{x} = \lim_{x \to 0} (x^2 + 3x + 3) = 0 + 0 + 3 = 3.$$

Of course, the technique used here is that of Section 2.2: Factor the term causing the zero from both the numerator and denominator, and divide. The next example uses this same idea. ■

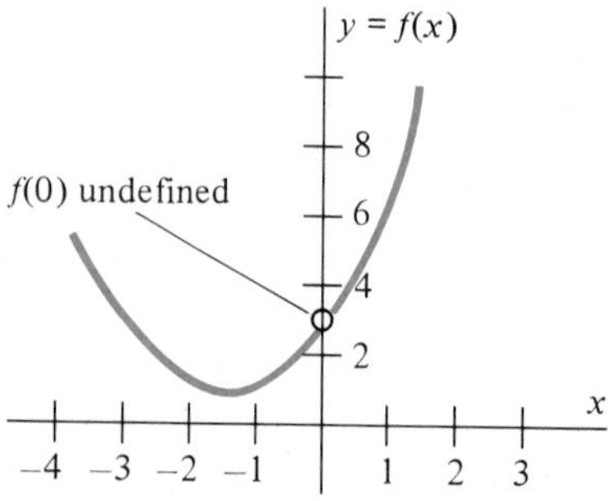

Figure 3.2 $f(x) = \dfrac{(x+1)^3 - 1}{x}$; $f(0)$ is undefined.

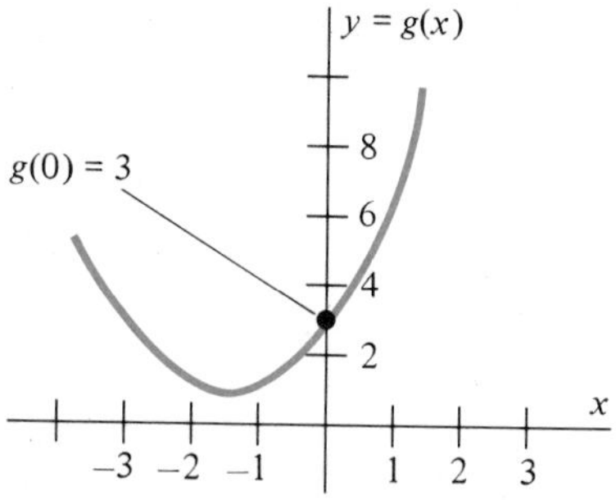

Figure 3.3 $g(x) = x^2 + 3x + 3$; $g(0) = 3$.

Example 3 Find $\lim_{x\to 2} \frac{x^2 - 3x + 2}{x^2 + x - 6}$.

Strategy
Determine whether limit can be obtained by setting $x = 2$. (This fails.)

Solution
We first note that on setting $x = 2$ we obtain the quotient

$$\frac{2^2 - 3 \cdot 2 + 2}{2^2 + 2 - 6} = \frac{4 - 6 + 2}{4 + 2 - 6} = \frac{0}{0}$$

which is undefined.

We therefore factor the numerator and denominator, obtaining

Factor both numerator and denominator.

Divide both by common factor of $x - 2$.

The limit equals the limit of the resulting expression.

$$\begin{aligned}\lim_{x\to 1} \frac{x^2 - 3x + 2}{x^2 + x - 6} &= \lim_{x\to 2} \frac{(x-2)\cdot(x-1)}{(x-2)\cdot(x+3)}\\ &= \lim_{x\to 2} \frac{x-1}{x+3}\\ &= \frac{2-1}{2+3}\\ &= \frac{1}{5}.\end{aligned}$$

■

Difference Quotients Revisited

Derivatives are sometimes written as limits of a slightly different form than was presented in Section 2.2. If, in the expression

$$f'(x_0) = \lim_{\Delta x\to 0} \frac{f(x_0 + \Delta x) - f(x_0)}{\Delta x} \tag{3}$$

we use the notation $x = x_0 + \Delta x$, then $\Delta x = x - x_0$, and $f'(x_0)$ in (3) becomes

$$f'(x_0) = \lim_{x\to x_0} \frac{f(x) - f(x_0)}{x - x_0}, \tag{4}$$

since $x \to x_0$ as $\Delta x \to 0$. Note that the difference quotient in (4) is undefined for $x = x_0$, just as the difference quotient in (3) is undefined for $\Delta x = 0$.

Example 4 Find $\lim_{x\to 3} \frac{x^2 - 9}{x - 3}$.

Solution: First, note that this limit arises when we compute $f'(3)$ for the function $f(x) = x^2$, according to equation (4). To evaluate this limit we factor the numerator:

$$\begin{aligned}\lim_{x\to 3} \frac{x^2 - 9}{x - 3} &= \lim_{x\to 3} \frac{(x-3)(x+3)}{x-3}\\ &= \lim_{x\to 3} (x + 3)\\ &= 6.\end{aligned}$$

Figures 3.4 and 3.5 illustrate that in making this calculation we used the fact that the functions $f(x) = \frac{(x-3)(x+3)}{x-3}$ and $g(x) = x + 3$ have the same limit at $a = 3$, even though $f(3)$ is undefined while $g(3) = 6$.

■

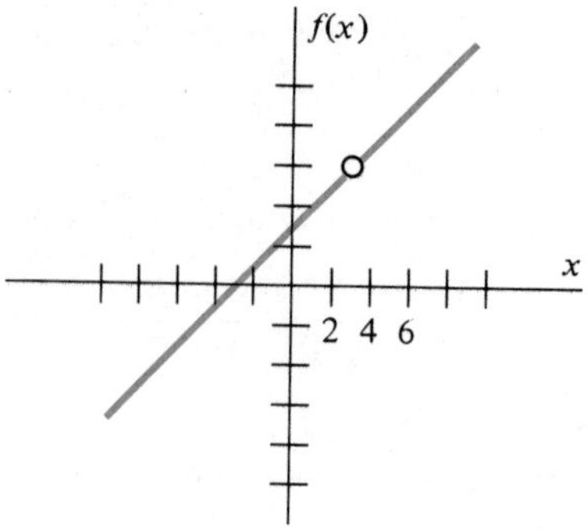

Figure 3.4 $f(x) = \dfrac{(x-3)(x+3)}{x-3}$; $f(3)$ undefined, but $\lim_{x\to 3} f(x) = 6$.

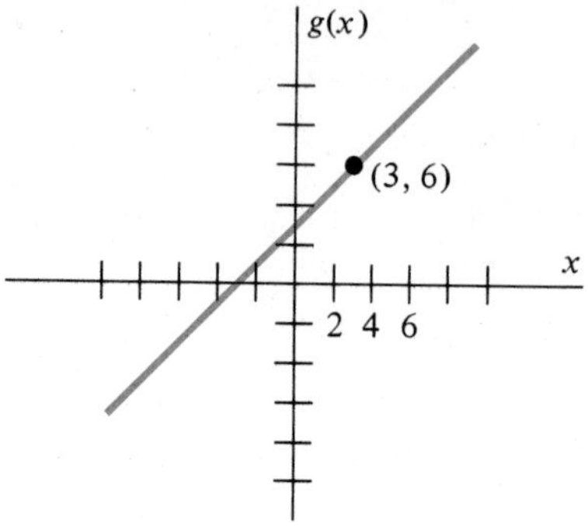

Figure 3.5 $g(x) = x + 3$; $g(3) = 6 = \lim_{x\to 3} g(x)$.

Example 5 Find $\displaystyle\lim_{x\to 1} \frac{x^3 + x - 2}{x - 1}$.

Solution: We leave it for you to verify that this limit results from calculating $f'(1)$ for the function $f(x) = x^3 + x + 2$, according to equation (4). The only strategy available to us in evaluating this limit is to attempt to factor $x - 1$ from the numerator. To determine whether $x - 1$ is indeed a factor of $x^3 + x - 2$, we can perform a "polynomial long division," which yields the following result:

$$
\begin{array}{r|l}
 & x^2 + x + 2 \\
\hline
x - 1 & x^3 \qquad\quad + x - 2 \\
 & \underline{x^3 - x^2} \\
 & \qquad x^2 + x \\
 & \qquad \underline{x^2 - x} \\
 & \qquad\qquad 2x - 2 \\
 & \qquad\qquad \underline{2x - 2} \\
 & \qquad\qquad\qquad 0
\end{array}
$$

This shows that $x^3 + x - 2 = (x - 1)(x^2 + x + 2)$.

We may now proceed as in Example 4:

$$
\begin{aligned}
\lim_{x\to 1} \frac{x^3 + x - 2}{x - 1} &= \lim_{x\to 1} \frac{(x-1)(x^2 + x + 2)}{x - 1} \\
&= \lim_{x\to 1} (x^2 + x + 2) \\
&= 1^2 + 1 + 2 \\
&= 4.
\end{aligned}
$$

■

The following example shows that the strategy of factoring the denominator from the numerator will not always succeed.

Example 6 For the function $f(x) = \sin x$, find

$$f'(0) = \lim_{\Delta x\to 0} \frac{f(0 + \Delta x) - f(0)}{\Delta x}.$$

Strategy	*Solution*
Substitute explicit form of $f(x)$.	$f'(0) = \lim_{\Delta x \to 0} \frac{f(0 + \Delta x) - f(0)}{\Delta x}$
Use the fact that sin (0) = 0.	$= \lim_{\Delta x \to 0} \frac{\sin(\Delta x) - \sin(0)}{\Delta x}$
Trouble!	$= \lim_{\Delta x \to 0} \frac{\sin \Delta x}{\Delta x}.$

Here we face a real difficulty. We cannot factor Δx from the numerator as in the preceding examples. At the moment the best we can do is calculate values of $\frac{\sin \Delta x}{\Delta x}$ for Δx near zero and guess the limit from this "circumstantial" evidence.

In Section 2.4, we will obtain a theorem which will enable us to conclude that, in fact,

$$\lim_{\Delta x \to 0} \frac{\sin \Delta x}{\Delta x} = 1.$$

For the moment we can only take this result as that suggested by the entries of Table 3.2 below. ■

Table 3.2

Δx	$\frac{\sin \Delta x}{\Delta x}$
±1.0	.841471
±0.8	.896695
±0.5	.958851
±0.2	.993347
±0.08	.998933
±0.05	.999583
±0.02	.999933
±0.005	.999996
±0.002	.999999

Limits that Fail to Exist

The next three examples describe distinct ways in which $\lim_{x \to a} f(x)$ can fail to exist. In each case note carefully the reason for which Definition 2 fails.

Example 7 Find $\lim_{x \to 0} \frac{|x|}{x}$.

Strategy

Use definition of $|x|$ to simplify statement of the function.

Solution

Since $|x| = \begin{cases} x & \text{if} \quad x \geq 0 \\ -x & \text{if} \quad x < 0 \end{cases}$

we can write the given function as

$$\frac{|x|}{x} = \begin{cases} \dfrac{x}{x} = 1 & \text{if} \quad x > 0 \\ \dfrac{-x}{x} = -1 & \text{if} \quad x < 0. \end{cases}$$

Graph the function.

Note that the values $f(x)$ do *not* approach a unique number as x approaches zero.

From the graph of $f(x) = \dfrac{|x|}{x}$ in Figure 3.6 we can see that as x approaches zero from the right all values of the function are $f(x) = 1$, while as x approaches zero from the left, all values of the function are $f(x) = -1$. For the limit at $a = 0$ to exist, $f(x)$ must approach the *same* number as x approaches zero from *either* direction. Since this does not happen, the limit does not exist. ■

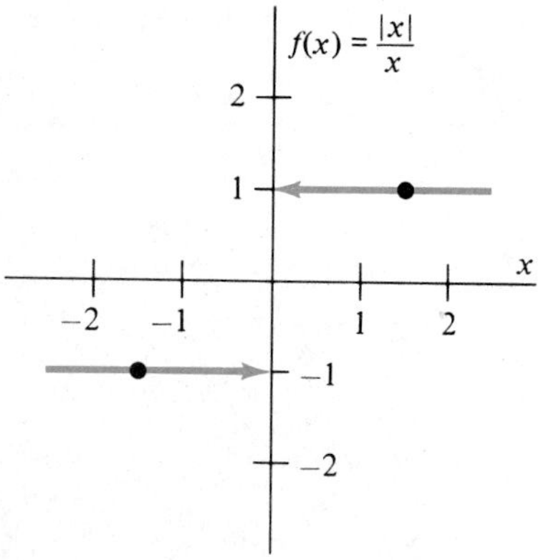

Figure 3.6 $\lim_{x\to 0} \dfrac{|x|}{x}$ does not exist.

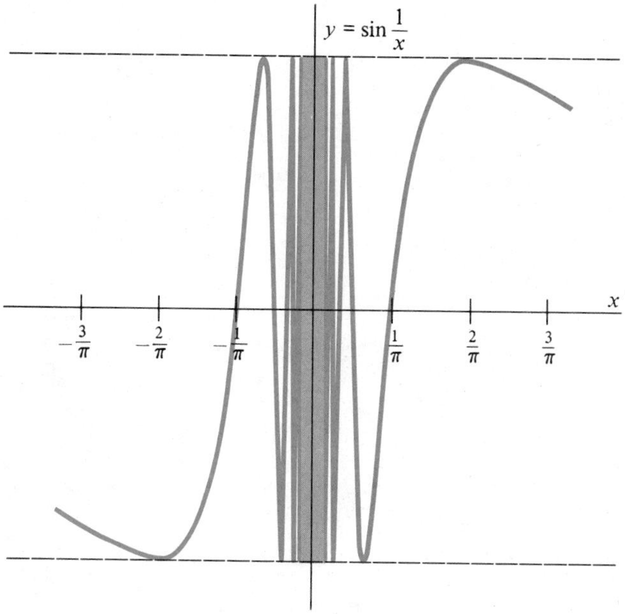

Figure 3.7 $\lim_{x\to 0} \sin\left(\dfrac{1}{x}\right)$ does not exist.

Example 8 A second example of a limit that fails to exist is $\lim_{x\to 0} \sin(1/x)$. Figure 3.7 shows that as x approaches zero from either direction the values of $f(x) = \sin(1/x)$ oscillate between 1 and -1 at an increasingly rapid rate. This is because $1/x$ is increasing rapidly as x approaches zero. Since the numbers $f(x) = \sin(1/x)$ are not approaching any *particular* number L, the limit fails to exist. ■

Example 9 Another way that $\lim_{x \to a} f(x)$ can fail to exist occurs when $f(x)$ becomes infinite as x approaches a. This happens for the function $f(x) = 1/x$ at $a = 0$ (Figure 3.8). The reason why $f(x) = 1/x$ "blows up" as $x \to 0$ is that the denominator becomes zero while the numerator does not. You should always keep this possibility in mind when calculating a limit of a rational function. For example, the limit

$$\lim_{x \to 3} \frac{x^4 - 7x^2 + x - 12}{x^2 + x - 12}$$

does not exist, since as $x \to 3$ the numerator approaches $3^4 - 7 \cdot 3^2 + 3 - 12 = 9$, while the denominator approaches $3^2 + 3 - 12 = 0$. ■

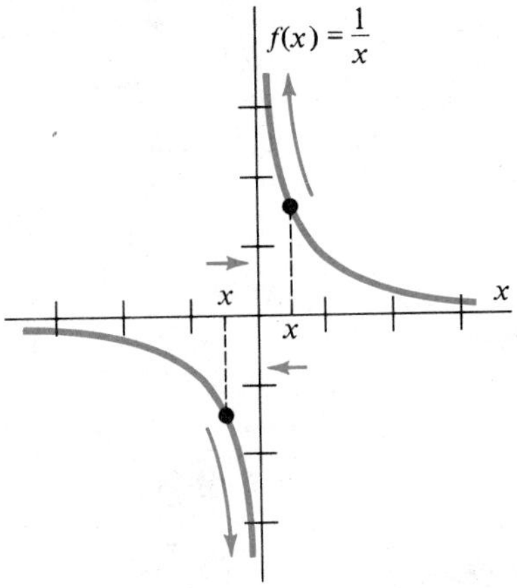

Figure 3.8 $\lim_{x \to 0} \frac{1}{x}$ does not exist.

Exercise Set 2.3

For each of the functions whose graphs are sketched in Exercises 1–5, indicate whether

(i) $\lim_{x \to a} f(x)$ exists and equals $f(a)$,

(ii) $\lim_{x \to a} f(x)$ exists but does not equal $f(a)$, or

(iii) $\lim_{x \to a} f(x)$ does not exist.

1.

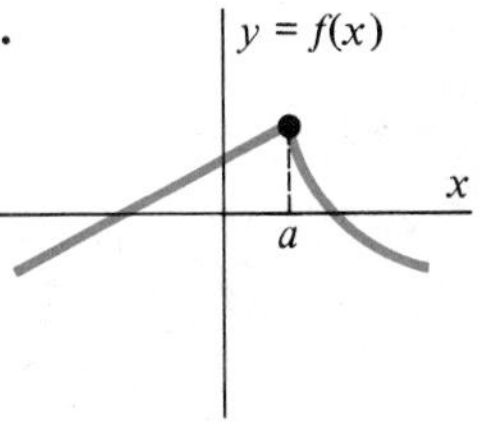

2.

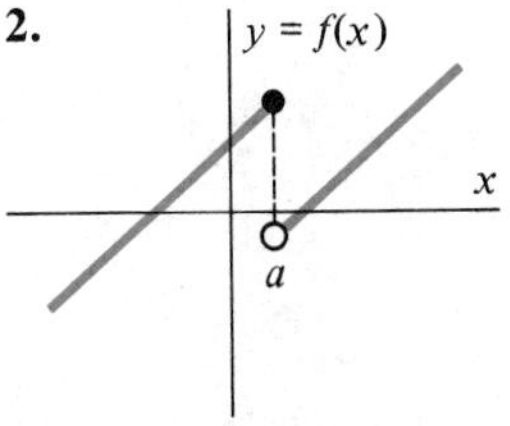

3.

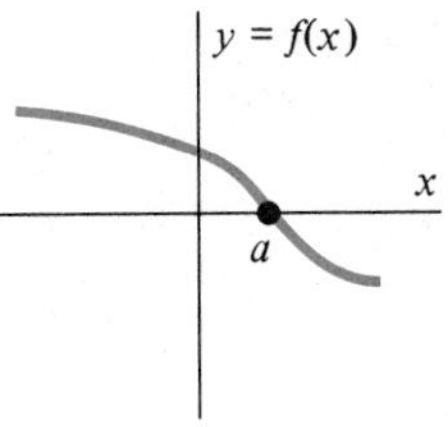

4.

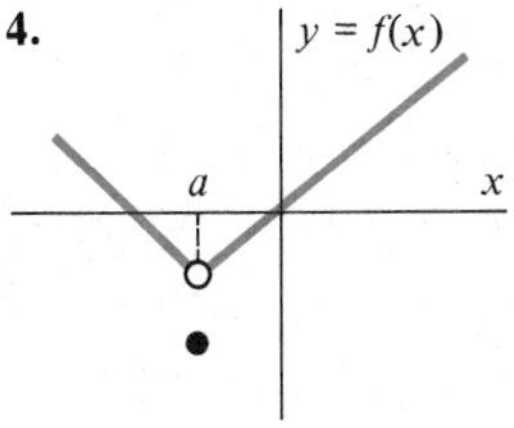

5.

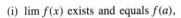

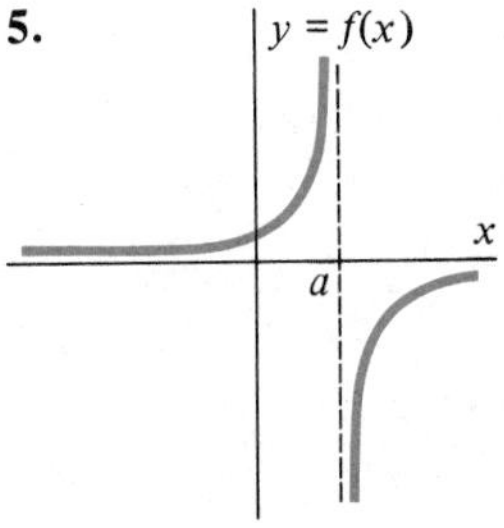

In each of Exercises 6–23, find the indicated limit, if it exists.

6. $\lim_{x\to 0} (3 + 7x)$

7. $\lim_{x\to 6} \dfrac{4 + x^2}{2 + x}$

8. $\lim_{x\to 0} \sqrt{4 + x^2}$

9. $\lim_{\Delta x\to 0} \sin(\pi/4 - \Delta x)$

10. $\lim_{\Delta x\to 0} \dfrac{3\,\Delta x + \Delta x^2}{\Delta x}$

11. $\lim_{x\to 3} \dfrac{x^2 - 9}{x - 3}$

12. $\lim_{x\to 1} \dfrac{x^2 - 1}{x - 1}$

13. $\lim_{x\to -2} \dfrac{x^2 - x - 6}{x + 2}$

14. $\lim_{\Delta x\to 0} \dfrac{(4 + \Delta x)^2 - 16}{\Delta x}$

15. $\lim_{h\to 0} \dfrac{(7 + h)^2 - 49}{h}$

16. $\lim_{h\to 0} \dfrac{(3 + h)^3 - 27}{h}$

17. $\lim_{\theta\to 2} \dfrac{\theta^4 - 2^4}{\theta - 2}$

18. $\lim_{x\to 0} \dfrac{\sin 2x}{\sin x}$ [*Hint:* $\sin 2x = 2 \sin x \cos x$]

19. $\lim_{\Delta x\to 0} \dfrac{(1 + \Delta x)^2 + 2(1 + \Delta x) - 3}{\Delta x}$

20. $\lim_{x\to -1} \dfrac{x^2 - 2x - 3}{x + 1}$

21. $\lim_{x\to 4} \dfrac{x^2 - 2x - 8}{x - 4}$

22. $\lim_{x\to 1} \dfrac{\sqrt{1 + x} - \sqrt{2}}{x - 1}$

23. $\lim_{x\to 0} \dfrac{\tan x}{\sin x}$

In Exercises 24–27, use a hand calculator or a computer to construct a table of values similar to Table 3.1 to develop a guess as to the value of the limit.

24. $\lim_{\Delta x\to 0} \dfrac{\cos \Delta x - 1}{\Delta x}$

25. $\lim_{\Delta x\to 0} \dfrac{\sqrt{\Delta x^2 + 1} - 1}{\Delta x}$

26. $\lim_{\Delta x\to 0} \dfrac{\sin\left(\dfrac{\pi}{4} + \Delta x\right) - \dfrac{\sqrt{2}}{2}}{\Delta x}$

27. $\lim_{\Delta x\to 0} \dfrac{\cos 2\left(\dfrac{\pi}{4} + \Delta x\right)}{\Delta x}$

Several of the limits in Exercises 6–23 arise from the definition of the derivative of the function $y = f(x)$ at $x = a$. For example, in Exercise 14 if we let $f(x) = x^2$ and $a = 4$ we obtain

$$\lim_{\Delta x\to 0} \frac{(4 + \Delta x)^2 - 16}{\Delta x} = \lim_{\Delta x\to 0} \frac{f(4 + \Delta x) - f(4)}{\Delta x} = f'(4).$$

In each of Exercises 28–34, identify the function $f(x)$ and the number $x = a$ for which the given limit is the derivative.

28. The limit in Exercise 11.

29. The limit in Exercise 12.

30. The limit in Exercise 15.

31. The limit in Exercise 16.

32. The limit in Exercise 17.

33. The limit $\lim_{x\to 2} \dfrac{x^3 + x - 10}{x - 2}$.

34. The limit $\lim_{x\to 0} \dfrac{\sqrt{x^2 + 4} - 2}{x}$.

35. True or false? If $f(a)$ does not exist, then neither does $\lim_{x\to a} f(x)$.

36. According to our working definition of limit, $\lim_{x\to 1} \sqrt{x - 1}$ does not exist. Why?

37. Show that $\dfrac{x^3 - 7x + 6}{x^2 + 2x - 3} = x - 2$ if $x \neq 1, -3$. Then find

a. $\lim_{x\to 1} \dfrac{x^3 - 7x + 6}{x^2 + 2x - 3}$

b. $\lim_{x\to -3} \dfrac{x^3 - 7x + 6}{x^2 + 2x - 3}$

38. For the function $f(x) = \dfrac{x^{5/2} - x^{1/2}}{x^{3/2} - x^{1/2}}$, find

a. $\lim_{x\to 0} f(x)$

b. $\lim_{x\to 1} f(x)$

In Exercises 39–44, find the slope of the tangent to the graph of the given function at the given point, and find an equation for this tangent.

39. The graph of $y = 3x^2 + 4$ at $(1, 7)$.

40. The graph of $y = \sqrt{x + 1}$ at $(3, 2)$.

41. The graph of $y = x - \sqrt{x}$ at $(1, 0)$.

42. The graph of $y = \dfrac{1}{x}$ at $(2, 1/2)$.

43. The graph of $y = (x + 1)^4$ at $(1, 16)$.

44. The graph of $y = \dfrac{1}{\sqrt{x + 2}}$ at $(2, 1/2)$.

45. Let $f(x) = |x|$. Note that as x approaches 0, the values $f(x)$ "approach" the number $y = -1$ in a certain sense (just as these values "approach," or "move in the direction of,"

any negative number or zero). Use this observation to comment on the deficiencies of Definition 2.

46. Figure 3.9 is a graph of the function $f(x) = x \sin\left(\frac{1}{x}\right)$. Does $\lim_{x\to 0} x \sin\left(\frac{1}{x}\right)$ exist? What can you say about $\lim_{x\to 0} x^2 \sin\left(\frac{1}{x}\right)$? (Compare these results with those for $\lim_{x\to 0} \sin\left(\frac{1}{x}\right)$ in Example 8.)

47. Assuming that $\lim_{\theta\to 0} \frac{\sin\theta}{\theta} = 1$, find

a. $\lim_{\theta\to 0} \frac{\sin\theta}{3\theta}$ **b.** $\lim_{\theta\to 0} \frac{\cos\theta - 1}{\theta}$

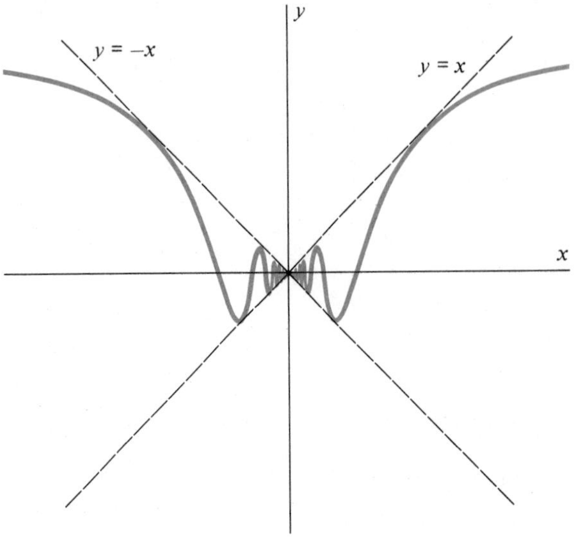

Figure 3.9 $f(x) = x \sin\left(\frac{1}{x}\right)$.

2.4 PROPERTIES OF LIMITS

From our working definition of limit, it is clear that the limit of a constant function $f(x) \equiv c$ as x approaches any particular number a is just the constant c. That is, $\lim_{x\to a} c = c$. This is because the value of $f(x)$ is c for all x. It should also be obvious to you that $\lim_{x\to a} x = a$ for any value of a, since the linear function $f(x) = x$ always has the same value as its independent variable x.

The following theorem establishes several properties of limits, by which we can evaluate more complicated limits in terms of the simple observations above. It will also be used to establish certain basic properties of the derivative in Chapter 3.

THEOREM 1

Suppose that $\lim_{x\to a} f(x) = L$ and $\lim_{x\to a} g(x) = M$. Let c be any constant. Then

(i) $\lim_{x\to a} [f(x) + g(x)] = [\lim_{x\to a} f(x)] + [\lim_{x\to a} g(x)] = L + M;$

(ii) $\lim_{x\to a} c[f(x)] = c[\lim_{x\to a} f(x)] = cL;$

(iii) $\lim_{x\to a} [f(x)g(x)] = [\lim_{x\to a} f(x)] \cdot [\lim_{x\to a} g(x)] = LM;$

(iv) $\lim_{x\to a} \frac{f(x)}{g(x)} = \frac{[\lim_{x\to a} f(x)]}{[\lim_{x\to a} g(x)]} = \frac{L}{M}, \qquad M \neq 0.$

We can paraphrase Theorem 1 by saying that

(i) *the limit of a sum is the sum of the limits;*
(ii) *the limit of a multiple of a function is that multiple of the limit;*

(iii) *the limit of a product is the product of the limits;*
(iv) *the limit of a quotient is the quotient of the limits.*

The proof of Theorem 1 requires a formal definition for the limit concept and will be given in Section 2.8. The following examples demonstrate the use of Theorem 1. In each example, the strategy is to use Theorem 1 to "decompose" the function into simpler parts whose limits are obvious.

Example 1 Find $\lim_{x\to 2} 3(x^2 + 6)$.

Strategy

Theorem 1, part (ii).

Theorem 1, part (i).

Theorem 1, part (iii).

Solution

$$\begin{aligned}\lim_{x\to 2} 3(x^2+6) &= 3[\lim_{x\to 2}(x^2+6)]\\ &= 3[(\lim_{x\to 2} x^2) + \lim_{x\to 2} 6]\\ &= 3[\lim_{x\to 2} x)(\lim_{x\to 2} x) + \lim_{x\to 2} 6]\\ &= 3[2\cdot 2 + 6]\\ &= 30.\end{aligned}$$

■

Example 2 Find $\lim_{x\to -1}\left(\dfrac{2x+3}{1+x^2}\right)$.

Strategy

Theorem 1, part (iv) (after verifying that $\lim_{x\to -1}(1+x^2) \neq 0$).

Theorem 1, part (i).

Theorem 1, parts (ii) and (iii).

Solution

$$\begin{aligned}\lim_{x\to -1}\left(\frac{2x+3}{1+x^2}\right) &= \frac{\lim_{x\to -1}(2x+3)}{\lim_{x\to -1}(1+x^2)}\\ &= \frac{\lim_{x\to -1}(2x) + \lim_{x\to -1} 3}{\lim_{x\to -1} 1 + \lim_{x\to -1} x^2}\\ &= \frac{2\lim_{x\to -1} x + \lim_{x\to -1} 3}{\lim_{x\to -1} 1 + [\lim_{x\to -1} x][\lim_{x\to -1} x]}\\ &= \frac{2(-1)+3}{1+(-1)(-1)}\\ &= \frac{1}{2}.\end{aligned}$$

■

Example 1 suggests the following result.

THEOREM 2

For the polynomial function

$$P_n(x) = a_nx^n + a_{n-1}x^{n-1} + \cdots + a_1x + a_0,$$
$$\lim_{x\to c} P_n(x) = a_nc^n + a_{n-1}c^{n-1} + \cdots + a_1c + a_0 = P_n(c).$$

In other words, the limit of the polynomial $P_n(x)$ as $x \to c$ is just $P_n(c)$, the polynomial evaluated at $x = c$. To see this, we note first that since, for any $k = 1, 2, \ldots, n$,

$$x^k = x\cdot x\cdot \cdots \cdot x \qquad (k \text{ factors}), \qquad 1 \le k \le n$$

we can apply Theorem 1, part (iii) $(k-1)$ times to conclude that

$$\begin{aligned}\lim_{x\to c} x^k &= [\lim_{x\to c} x]\cdot[\lim_{x\to c} x]\cdot\cdots\cdot[\lim_{x\to c} x] \qquad (k \text{ factors})\\ &= [\lim_{x\to c} x]^k\\ &= c^k, \qquad k = 1, 2, \ldots, n.\end{aligned}$$

Using this observation together with parts (i) and (ii) of Theorem 1, we obtain

$$\begin{aligned}&\lim_{x\to c} [a_n x^n + a_{n-1}x^{n-1} + \cdots + a_1 x + a_0]\\ &= a_n \lim_{x\to c} x^n + a_{n-1}\lim_{x\to c} x^{n-1} + \cdots + a_1 \lim_{x\to c} x + a_0\\ &= a_1 c^n + a_{n-1}c^{n-1} + \cdots + a_1 c + a_0 = P_n(c),\end{aligned}$$

which proves Theorem 2.

Example 3

$$\lim_{x\to 3} (x^5 - 3x^3 + 4x^2 - x + 5) = 3^5 - 3(3^3) + 4(3^2) - 3 + 5 = 200. \blacksquare$$

A **rational function** is a quotient of two polynomials. In other words

$$r(x) = \frac{p(x)}{q(x)} \tag{1}$$

is a rational function if $p(x)$ and $q(x)$ are polynomials. If we apply Theorem 1, part (iv), to equation (1) we find that

$$\lim_{x\to c} r(x) = \lim_{x\to c}\frac{p(x)}{q(x)} = \frac{\lim_{x\to c} p(x)}{\lim_{x\to c} q(x)}, \qquad \text{as long as } \lim_{x\to c} q(x) \neq 0. \tag{2}$$

Applying Theorem 2 on the right side of equation (2) then gives

$$\lim_{x\to c} r(x) = \frac{\lim_{x\to c} p(x)}{\lim_{x\to c} q(x)} = \frac{p(c)}{q(c)}, \qquad q(c) \neq 0.$$

Since $\dfrac{p(c)}{q(c)} = r(c)$, this proves the following theorem.

THEOREM 3

For the rational function $r(x) = \dfrac{p(x)}{q(x)}$,

$$\lim_{x\to c} r(x) = r(c)$$

if $q(c) \neq 0$.

In Sections 2.2 and 2.3 we dealt extensively with rational functions for which $\lim_{x\to c} q(x) = 0$. We saw that evaluating limits of functions of this type required special care in overcoming the difficulty caused by a zero in the denominator. Theorem 3 addresses the other (easier) case. It says that the limit of a rational function $r(x)$ as $x \to c$ is just $r(c)$, the function evaluated at c, provided the denominator is not zero.

Example 4 $\lim_{x\to 3} \dfrac{7x - x^2}{3x + 6} = \dfrac{7 \cdot 3 - 3^2}{3 \cdot 3 + 6} = \dfrac{4}{5}$. ■

Example 5 Find $\lim_{x\to 2} \dfrac{x^3 - 3x^2 + 5x - 6}{x^2 - 2x}$.

Solution: Here Theorem 1 does not apply, since $\lim_{x\to 2} (x^2 - 2x) = 2^2 - 2 \cdot 2 = 0$ in the denominator. We must resort to the technique of attempting to divide both numerator and denominator by the factor causing the zero. Since the denominator factors as $x^2 - 2x = x(x - 2)$, we check (by long division, if necessary) to see if $x - 2$ is a factor of the numerator. The result is the following:

$$\begin{aligned} \lim_{x\to 2} \frac{x^3 - 3x^2 + 5x - 6}{x^2 - 2x} &= \lim_{x\to 2} \frac{(x^2 - x + 3)(x - 2)}{x(x - 2)} \\ &= \lim_{x\to 2} \frac{x^2 - x + 3}{x} \\ &= \frac{2^2 - 2 + 3}{2} \\ &= \frac{5}{2}. \end{aligned}$$ ■

The next theorem asserts that the limit of the nth root of x as x approaches a is simply the nth root of a. It has an immediate extension to fractional powers. The proof of Theorem 4 is found in Appendix II.

THEOREM 4

Let n be a positive integer.

(i) If n is even, $\lim_{x\to a} \sqrt[n]{x} = \sqrt[n]{a}$, $\quad 0 \le a < \infty$

(ii) If n is odd, $\lim_{x\to a} \sqrt[n]{x} = \sqrt[n]{a}$, $\quad -\infty < a < \infty$.

If $\dfrac{n}{m} > 0$, we may combine part (iii) of Theorem 1 and Theorem 4 to conclude that

$$\begin{aligned} \lim_{x\to a} x^{n/m} &= \lim_{x\to a} (\sqrt[m]{x})^n \\ &= [\lim_{x\to a} \sqrt[m]{x}]^n \\ &= (\sqrt[m]{a})^n \\ &= a^{n/m}. \end{aligned}$$

(The result is still subject to the restriction that $0 \le x < \infty$ if m is even.) This proves the following corollary in the case $\dfrac{n}{m} > 0$. The case $\dfrac{n}{m} < 0$ is handled in a similar way (see Exercise 52).

COROLLARY 1

Let n and m be positive integers with $m \neq 0$. Then

(i) If m is even, $\lim_{x\to a} x^{n/m} = a^{n/m}, \quad 0 < a < \infty$

(ii) If m is odd, $\lim_{x\to a} x^{n/m} = a^{n/m}, \quad -\infty < a < \infty.$

Example 6 Find $\lim_{x\to 4} \dfrac{x^{5/2} + 5\sqrt{x}}{x^2 - \dfrac{8}{x}}$.

Strategy

Solution

Theorem 1, part (iv).

$$\lim_{x\to 4} \frac{x^{5/2} + 5\sqrt{x}}{x^2 - \dfrac{8}{x}} = \frac{\lim_{x\to 4}(x^{5/2} + 5\sqrt{x})}{\lim_{x\to 4}\left(x^2 - \dfrac{8}{x}\right)}$$

Theorem 1, parts (i) and (ii).

$$= \frac{\lim_{x\to 4} x^{5/2} + 5 \cdot \lim_{x\to 4} \sqrt{x}}{\lim_{x\to 4} x^2 - 8 \cdot \lim_{x\to 4}\left(\dfrac{1}{x}\right)}$$

Theorems 2 and 4.

$(4^{5/2} = (\sqrt{4})^5 = 2^5 = 32)$

$$= \frac{4^{5/2} + 5\sqrt{4}}{4^2 - 8\left(\dfrac{1}{4}\right)}$$

$$= \frac{32 + 5 \cdot 2}{14}$$

$$= \frac{21}{7}$$

$$= 3. \qquad \blacksquare$$

Our final theorem on limits is sometimes called the "Pinching Theorem" or "Sandwich Theorem." These names arise from its geometric interpretation, as illustrated in Figure 4.1. The proof of Theorem 5 is given in Section 2.8.

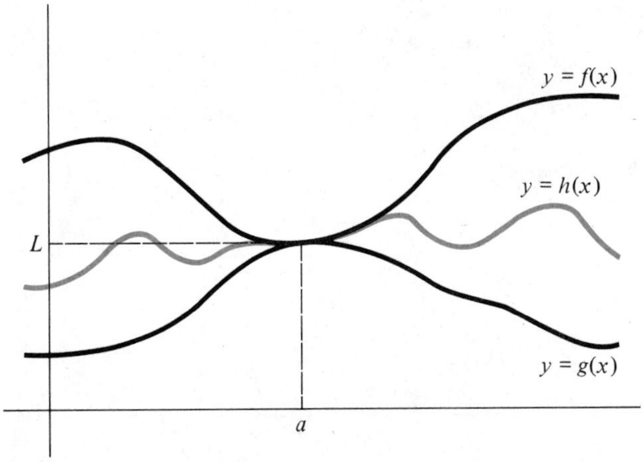

Figure 4.1 The Pinching Theorem.

THEOREM 5 Let $\lim_{x\to a} f(x) = L = \lim_{x\to a} g(x)$. If the function $y = h(x)$ satisfies the inequality

$$f(x) \le h(x) \le g(x)$$

for all x in an interval containing a, then $\lim_{x\to a} h(x) = L$, also.

The following example is a straightforward application of the Pinching Theorem.

Example 7 Suppose that the function $f(x)$ is defined on the interval $[-\pi/4, \pi/4]$, and that

$$\frac{1}{1+x^2} \le f(x) \le \sec x$$

for all $x \in [-\pi/4, \pi/4]$. Since

$$\lim_{x\to 0} \frac{1}{1+x^2} = 1 = \lim_{x\to 0} \sec x$$

we may conclude from the Pinching Theorem that

$$\lim_{x\to 0} f(x) = 1.$$

See Figure 4.2. ■

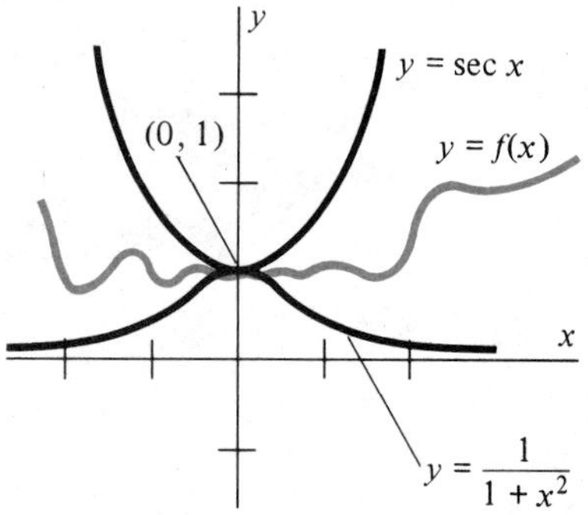

Figure 4.2 $\lim_{x\to 1} f(x) = 1$.

In our next example, we return to the problem of calculating $\lim_{x\to 0} \frac{\sin x}{x}$ (see Example 6, Section 2.3). The details of this example require an understanding of Figure 4.3, which concerns an arc of the unit circle. The following observations concerning Figure 4.3 will help you with these details:

(a) The length of OC is 1, since it's a radius for the unit circle.

(b) $\sin x = \dfrac{\text{length of } AC}{\text{length of } OC} = \text{length of } AC.$

(c) $\cos x = \dfrac{\text{length of } OA}{\text{length of } OC} = \text{length of } OA.$

(d) $\tan x = \dfrac{\text{length of } DB}{\text{length of } OB} = \text{length of } DB$, since length of $OB = 1$.

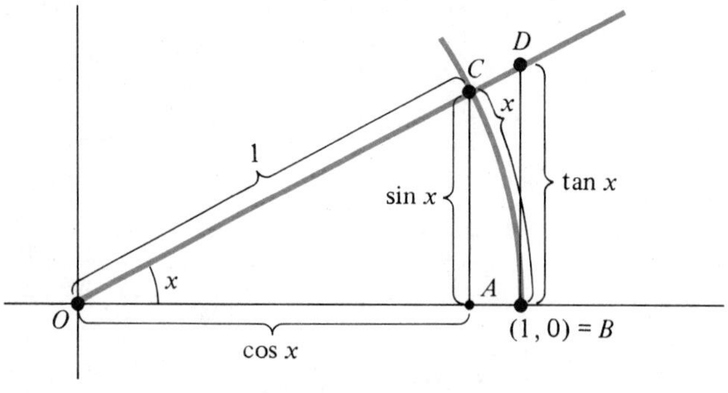

Figure 4.3

Example 8 Find $\lim_{x\to 0} \frac{\sin x}{x}$.

Solution: We assume first that $x > 0$. From Figure 4.3 we observe that

$$\text{Area triangle } BOD \geq \text{Area sector } BOC \geq \text{Area triangle } AOC. \tag{3}$$

Since

$$\text{Area triangle } AOC = \frac{1}{2}\cos x \cdot \sin x$$

$$\text{Area sector } BOC = \frac{1}{2}x$$

$$\text{Area triangle } BOD = \frac{1}{2}\tan x$$

inequality (3) becomes

$$\tan x \geq x \geq \sin x \cdot \cos x.$$

Dividing by $\sin x$ and inverting all terms gives

$$\cos x \leq \frac{\sin x}{x} \leq \frac{1}{\cos x}.$$

Now, from knowledge of the graph of $y = \cos x$ we know that

$$\lim_{x\to 0} \cos x = 1 = \lim_{x\to 0} \frac{1}{\cos x}.$$

Thus, by the Pinching Theorem, $\lim_{x\to 0} \frac{\sin x}{x} = 1$.

(The case $x < 0$ is the subject of Exercise 49.) ■

Example 9 Find $\lim_{x\to 0} \frac{3\sin 2x}{x}$.

Strategy

Bring quotient to the form $\frac{\sin u}{u}$ by letting $u = 2x$.

Use fact that $\lim_{u\to 0} \frac{\sin u}{u} = 1$.

Solution

$$\lim_{x\to 0} \frac{3\sin 2x}{x} = \lim_{x\to 0} \frac{2\cdot 3\sin 2x}{2x}$$

$$= 6\lim_{x\to 0} \frac{\sin 2x}{2x}$$

Now let $u = 2x$. Then $u \to 0$ as $x \to 0$, so we have that

$$\lim_{x\to 0} \frac{3\sin 2x}{x} = 6\cdot \lim_{u\to 0} \frac{\sin u}{u} = 6\cdot 1 = 6.$$ ■

Example 10 Find $\lim_{x\to 0} x\cot x$.

Strategy

Try to obtain form $\frac{\sin x}{x}$.

Solution

$$\lim_{x\to 0} x\cot x = \lim_{x\to 0} \frac{x\cdot \cos x}{\sin x}$$

$$= \lim_{x \to 0} \left(\frac{x}{\sin x}\right) \cdot \lim_{x \to 0} \cos x$$

Use Theorem 1, (iv) and the fact that $\lim_{x \to 0} \cos x = 1$.

$$= \lim_{x \to 0} \frac{1}{\left(\frac{\sin x}{x}\right)} \cdot \lim_{x \to 0} \cos x$$

$$= 1 \cdot 1 = 1.$$

■

Exercise Set 2.4

In Exercises 1–26, use Theorems 1–3 to calculate the limit.

1. $\lim_{x \to 3} (3x - 7)$

2. $\lim_{x \to 4} (1 - x^3)$

3. $\lim_{x \to 2} \frac{x^2 + 3x}{x - 3}$

4. $\lim_{x \to 5} \frac{x^2 - 5x}{x - 5}$

5. $\lim_{x \to 3} \sqrt{x}(1 - x^2)$

6. $\lim_{x \to 2} \sqrt{x} \sin(\pi/x)$

7. $\lim_{x \to -1} \sec(\pi x) \tan(\pi x)$

8. $\lim_{x \to 2} (4\sqrt{x} - 5x^2)$

9. $\lim_{x \to 7} (\sqrt{x - 3} + \sqrt{x - 7})$

10. $\lim_{x \to 1} (3x^9 - \sqrt[3]{x})^6$

11. $\lim_{x \to 2} \frac{x^3 - 6x + 5}{x^2 + 2x + 2}$

12. $\lim_{x \to \pi} \frac{(x^2 - \pi)(x + \pi)}{(x - 2\pi)}$

13. $\lim_{x \to 4} \frac{x^{3/2} + 2\sqrt{x}}{x^{5/2} + \sqrt{x}}$

14. $\lim_{x \to 3} \frac{x^2 - 10x + 21}{x - 3}$

15. $\lim_{x \to 1} \frac{x^3 + 2x^2 + 2x - 5}{x^2 - 1}$

16. $\lim_{x \to 0} \frac{\sin 2x}{\sin x}$

17. $\lim_{x \to -8} \frac{x^{2/3} - x}{x^{5/3}}$

18. $\lim_{x \to 4} \frac{(x^{3/2} - x^{5/2})}{(4 + \sqrt{x})^2}$

19. $\lim_{x \to 4} \frac{x - 4}{\sqrt{x} - 2}$

20. $\lim_{x \to 3} \frac{x^4 - 3x^3 + x - 3}{x - 3}$

21. $\lim_{x \to -8} \frac{x^{2/3} - 4}{x^{1/3} + 2}$

22. $\lim_{x \to -3} \frac{x^2 - 6x - 27}{x^2 + 3x}$

23. $\lim_{x \to -1} (x^{7/3} - 2x^{2/3})^2$

24. $\lim_{x \to -2} \frac{\sqrt{4 - x}}{\sqrt{x + 2}}$

25. $\lim_{x \to 1} \frac{x^2 - 1}{x - 1}$

26. $\lim_{x \to -1} \frac{x^3 + x^2 - x - 1}{x^3 + 2x^2 + 2x + 1}$

In Exercises 27–31, use the facts that $\lim_{x \to a} f(x) = 2$, $\lim_{x \to a} g(x) = -3$, and $\lim_{x \to a} h(x) = 5$ to find the limit.

27. $\lim_{x \to a} 3f(x) \cdot g(x)$

28. $\lim_{x \to a} \frac{f(x) - g(x)}{[h(x)]^2}$

29. $\lim_{x \to a} [f(x) + g(x)]^{-2/3}$

30. $\lim_{x \to a} f(x)[g(x)]^2 \cdot [h(x)]^3$

31. $\lim_{x \to a} \frac{6h(x) - 7\sqrt{8f(x)}}{g(x) - 2f(x)}$

In Exercises 32–41, use the result of Example 8 to find the limit.

32. $\lim_{x \to 0} \frac{\sin x}{2x}$

33. $\lim_{x \to 0} \frac{3 \sin x}{x}$

34. $\lim_{x \to 0} \frac{\tan x}{x}$

35. $\lim_{x \to 0} \frac{\sin 3x}{5x}$

36. $\lim_{x \to 0} \frac{\sin 4x}{2x}$

37. $\lim_{x \to 0} \frac{3x}{\sin x}$

38. $\lim_{x \to 0} x \csc x$

39. $\lim_{x \to 0} \frac{\sin^2 x}{x}$

40. $\lim_{x \to 0} \frac{\sin ax}{\sin bx}$

41. $\lim_{x \to 0} \frac{\sin^4 2x}{4x}$

42. Suppose that $-|x| \le f(x) \le |x|$ for all x. Find $\lim_{x \to 0} f(x)$.

43. Suppose that $1 - x^2 \le f(x) \le 1 + x^2$ for all x. Find $\lim_{x \to 0} f(x)$.

44. Suppose that $4x - x^2 \le f(x) \le |x - 2| + 4$ for all x. Find $\lim_{x \to 2} f(x)$.

45. Use Theorem 1, part (i), to show that

$$\lim_{x \to a} [f(x) + g(x) + h(x)] = \lim_{x \to a} f(x) + \lim_{x \to a} g(x) + \lim_{x \to a} h(x)$$

provided each of these limits exists. (*Hint:* Apply Theorem 1, part (i), twice.)

46. Use Theorem 1, part (iii) to show that

$$\lim_{x\to a} [f(x)g(x)h(x)] = [\lim_{x\to a} f(x)] \cdot [\lim_{x\to a} g(x)] \cdot [\lim_{x\to a} h(x)]$$

provided each of these limits exists.

47. Apply Theorem 1, part (iii) to show that

$$\lim_{x\to a} [f(x)]^2 = [\lim_{x\to a} f(x)]^2$$

if $\lim_{x\to a} f(x)$ exists.

48. Generalize the result of Exercise 47 to show that

$$\lim_{x\to a} [f(x)]^k = [\lim_{x\to a} f(x)]^k,$$

k a positive integer, provided $\lim_{x\to a} f(x)$ exists. Conclude that $\lim_{x\to c} x^k = c^k$, as claimed in the proof of Theorem 2.

49. Comment on the proof in Example 8 for the case $x < 0$.

50. Show that $\lim_{x\to 0} \dfrac{1 - \cos x}{x} = 0$ as follows:

a. Write

$$\frac{1-\cos x}{x} = \frac{1-\cos x}{x} \cdot \left(\frac{1+\cos x}{1+\cos x}\right) = \frac{1-\cos^2 x}{x(1+\cos x)}.$$

b. Show that $\dfrac{1-\cos x}{x} = \dfrac{\sin^2 x}{x(1+\cos x)}$, $x \neq 0$.

c. Use the facts that $\lim_{x\to 0} \sin x = 0$ and $\lim_{x\to 0} \cos x = 1$ to obtain the desired limit.

51. Show where the result of Exercise 48 was assumed in the proof of Corollary 1.

52. Prove Corollary 1 in the case $\dfrac{n}{m} < 0$.

2.5 ONE-SIDED LIMITS

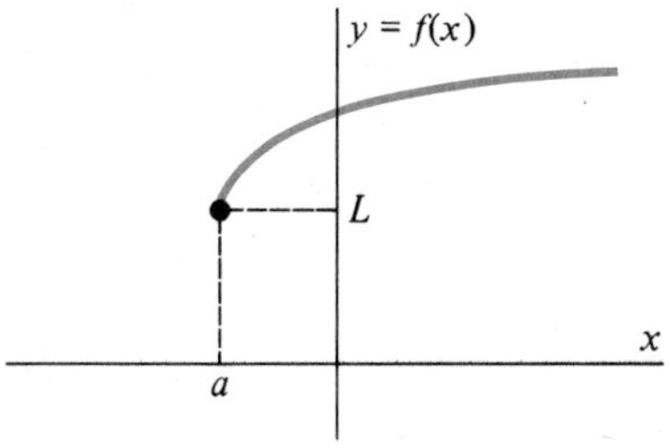

Figure 5.1 $f(x)$ undefined for $x < a$.

Our working definition of limit (Section 2.3) requires that for the statement

$$L = \lim_{x\to a} f(x) \tag{1}$$

to hold, $f(x)$ must approach L as x approaches a *from either direction* along the x-axis. Thus, statement (1) fails in the following cases, among others:

(i) $f(x)$ approaches L as x approaches a *from the right*, but $f(x)$ is not defined for x to the left of a (see Figure 5.1). A similar situation occurs when $f(x)$ approaches L as x approaches a from the left, but $f(x)$ is not defined for x to the right of a.

(ii) $f(x)$ approaches a number L as x approaches a from the right, $f(x)$ approaches a number M as x approaches a from the left, but $L \neq M$ (Figure 5.2).

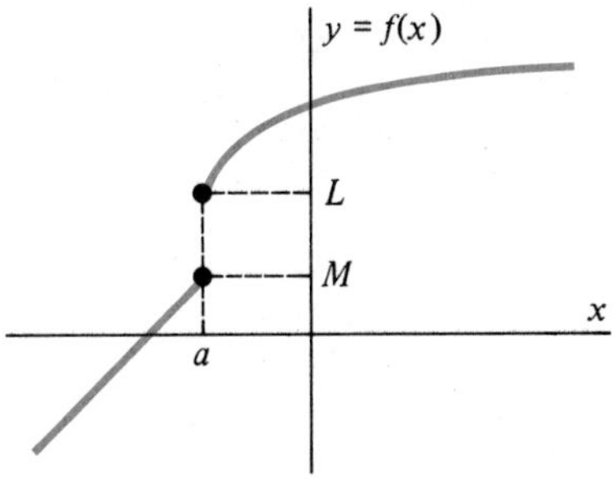

Figure 5.2 $f(x) \to L$ for $x > a$; $f(x) \to M$ for $x < a$.

We say that our working definition describes a **two-sided limit** since x must be allowed to approach a from either direction with the same result. However, in situations (i) and (ii) above there is certain information to be conveyed about the graph of $y = f(x)$ near $(a, f(a))$ even though the two-sided limit fails to exist. For this reason we define the left-hand and right-hand limits of $f(x)$ at $x = a$ as follows.

DEFINITION 3

The statement $L = \lim_{x\to a^+} f(x)$ means that the values $f(x)$ approach the number L as x approaches the number a *from the right,* that is, for $a < x$. The statement $M = \lim_{x\to a^-} f(x)$ means that the values $f(x)$ approach the number M as x approaches the number a *from the left,* that is, for $x < a$ (Figure 5.2).

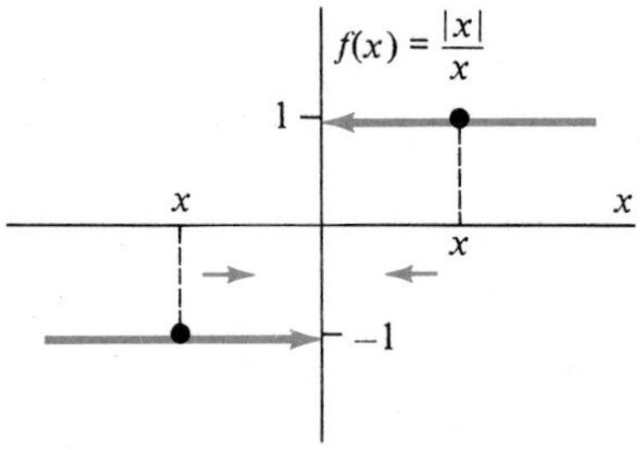

Figure 5.3 $\lim_{x\to 0^+} \frac{|x|}{x} = 1;$ $\lim_{x\to 0^-} \frac{|x|}{x} = -1.$

In the above definition the number L is referred to as the **right-hand limit** of $f(x)$ as x approaches a, and the number M is called the **left-hand limit.**

Example 1 For the function $f(x) = \frac{|x|}{x}$ in Figure 5.3 we have

$$\lim_{x\to 0^+} \frac{|x|}{x} = 1, \qquad \text{and} \qquad \lim_{x\to 0^-} \frac{|x|}{x} = -1.$$

See Figure 5.3. ■

Example 2 The **greatest integer** function is defined by

$$[x] = \text{the largest integer } n \text{ with } n \le x.$$

(See Figure 5.4.) We have

$$\lim_{x\to 2^+} [x] = 2; \qquad \lim_{x\to 2^-} [x] = 1$$

$$\lim_{x\to -2^-} [x] = -3; \qquad \lim_{x\to -2^+} [x] = -2.$$ ■

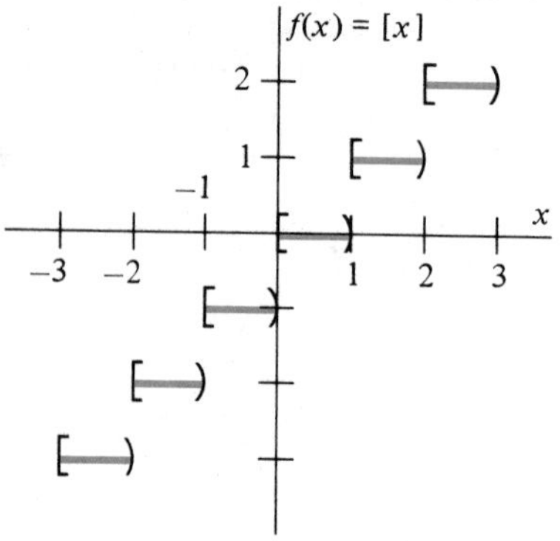

Figure 5.4 Graph of greatest integer function.

Example 3 For the function $f(x) = \sqrt{1 - x^2}$, we have $\lim_{x \to 1^-} \sqrt{1 - x^2} = 0$, but $\lim_{x \to 1^+} \sqrt{1 - x^2}$ is not defined (see Figure 5.5). ■

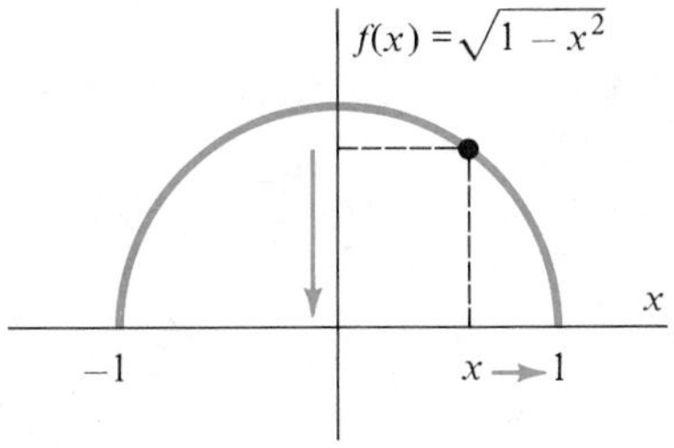

Figure 5.5 $\lim_{x \to 1^-} \sqrt{1 - x^2} = 0.$

Example 4 For the function $f(x) = \begin{cases} 1/x & \text{if} \quad x > 0 \\ -x^2 & \text{if} \quad x \le 0 \end{cases}$

the left-hand limit at $a = 0$ is

$$\lim_{x \to 0^-} f(x) = \lim_{x \to 0^-} -x^2 = 0.$$

However, the right-hand limit fails to exist since the values $f(x)$ become increasingly large as x approaches zero from the right (see Figure 5.6). ■

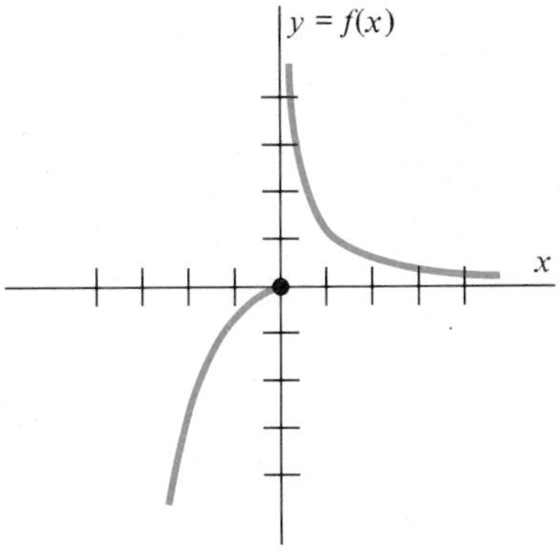

Figure 5.6 $f(x) = \begin{cases} 1/x, & x > 0 \\ -x^2, & x \le 0 \end{cases}$

The relationship between two-sided and one-sided limits may be summarized as follows

(i) $\lim_{x \to a} f(x) = L$ if and only if *both* $\lim_{x \to a^+} f(x) = L$ and $\lim_{x \to a^-} f(x) = L$.

(ii) However, if $\lim_{x \to a} f(x)$ fails to exist, either or both of $\lim_{x \to a^+} f(x)$ and $\lim_{x \to a^-} f(x)$ can nonetheless exist (Example 1).

Examples of each of these situations are to be found in the exercise set.

Exercise Set 2.5

1. For the function $y = f(x)$ in Figure 5.7 find
   a. $f(1)$
   b. $\lim_{x \to 1^-} f(x)$
   c. $\lim_{x \to 1} f(x)$
   d. $\lim_{x \to 0} f(x)$

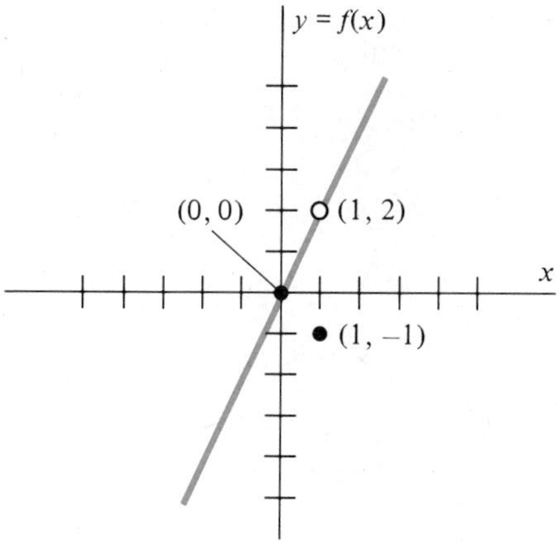

Figure 5.7

2. For the function $y = f(x)$ in Figure 5.8 find
a. $f(1)$
b. $\lim_{x\to 1^-} f(x)$
c. $\lim_{x\to 1^+} f(x)$
d. $\lim_{x\to 1} f(x)$

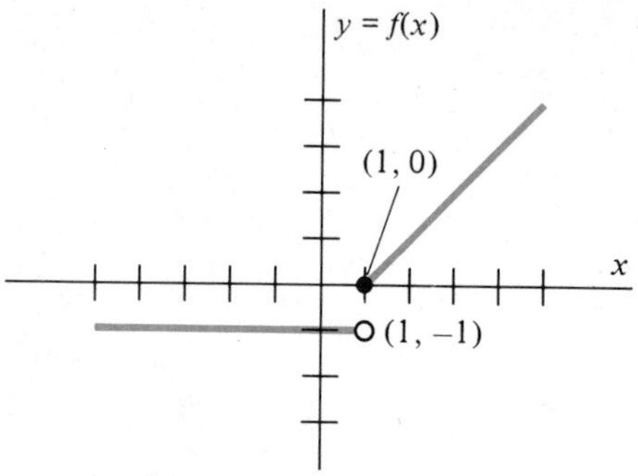

Figure 5.8

3. For the function $y = f(x)$ in Figure 5.9 find
a. $\lim_{x\to 0^-} f(x)$
b. $\lim_{x\to 0^+} f(x)$
c. $\lim_{x\to -2} f(x)$

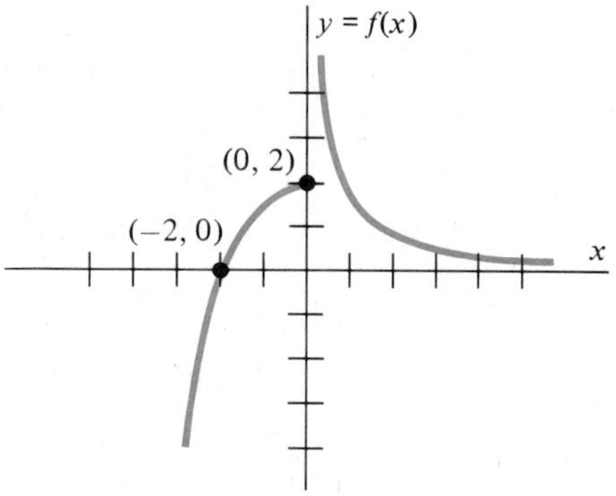

Figure 5.9

4. For the function $y = f(x)$ in Figure 5.10 find
a. $\lim_{x\to -2} f(x)$
b. $\lim_{x\to 2^-} f(x)$
c. $\lim_{x\to 2^+} f(x)$
d. $f(2)$

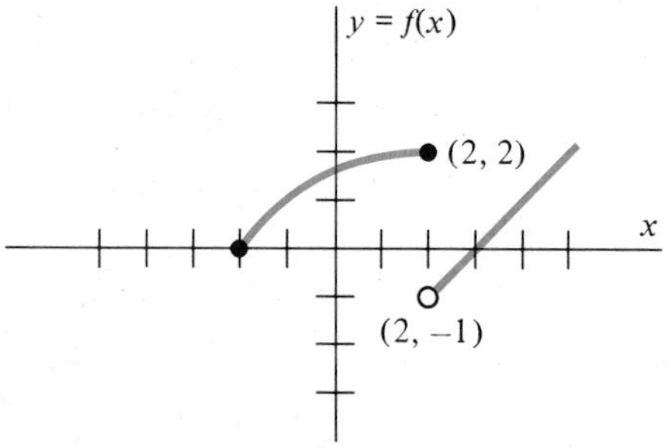

Figure 5.10

In Exercises 5–26, find the limit, if it exists.

5. $\lim_{x\to 0^+} \sqrt{x}$

6. $\lim_{x\to 2^-} \dfrac{|x-2|}{x-2}$

7. $\lim_{x\to 0^+} (x^{5/2} - 5x^{3/2})$

8. $\lim_{x\to 1^-} \sqrt{1-x^2}$

9. $\lim_{x\to 1^+} \sqrt{1-x^2}$

10. $\lim_{x\to 2^+} \sqrt{x^2-x-2}$

11. $\lim_{x\to -1^-} \sqrt{x^2-x-2}$

12. $\lim_{x\to 2^-} \dfrac{x^2-4x+4}{x-2}$

13. $\lim_{x\to 3^+} [1+x]$

14. $\lim_{x\to 2^-} [1-x^2]$

15. $\lim_{x\to 2^+} \sqrt{x^2+x-6}$

16. $\lim_{x\to 3^-} x[x]$

17. $\lim_{x\to 0} \dfrac{|x^2|}{x^2}$

18. $\lim_{x\to -2^-} \dfrac{x^2}{1+x^2}$

19. $\lim_{x\to 3^+} \sqrt{x^2+2x-15}$

20. $\lim_{x\to -5^+} \sqrt{x^2+2x-15}$

21. $\lim_{x\to -2^-} \dfrac{x^2-4}{x-2}$

22. $\lim_{x\to 2^-} \{7-(2-x)^{3/2}\}$

23. $\lim_{x\to -1^-} \dfrac{\sqrt{x^2-5x-6}}{x+3}$

24. $\lim_{x\to 6^-} \dfrac{\sqrt{x^2-5x-6}}{x-3}$

25. $\lim_{x\to 2^-} \frac{[x+2]}{x}$

26. $\lim_{x\to -3^+} \frac{[x-7]}{[x+4]}$

27. Let $f(x) = \begin{cases} x+2, & x \le -1 \\ -x, & x > -1 \end{cases}$

a. Find $\lim_{x\to -1^-} f(x)$.

b. Find $\lim_{x\to -1^+} f(x)$.

c. Does $\lim_{x\to -1} f(x)$ exist?

28. Let $f(x) = \begin{cases} \cos x, & x \le 0 \\ 1-x, & x > 0 \end{cases}$

a. Find $\lim_{x\to 0^-} f(x)$.

b. Find $\lim_{x\to 0^+} f(x)$.

c. Does $\lim_{x\to 0} f(x)$ exist?

29. Let $f(x) = \begin{cases} x^2, & x \le 1 \\ x^5, & x > 1 \end{cases}$

Show that $\lim_{x\to 1} f(x)$ exists.

30. Let $f(x) = \begin{cases} 2, & x \le -1 \\ -x, & -1 < x \le 1 \\ -x^2, & 1 < x \end{cases}$

a. Sketch the graph of $y = f(x)$.

b. Find $\lim_{x\to -1^+} f(x)$.

c. Find $\lim_{x\to 1^-} f(x)$.

d. Find $\lim_{x\to 1} f(x)$.

31. Let $f(x) = \begin{cases} x+2, & x \le -1 \\ cx^2, & x > -1 \end{cases}$

Find c so that $\lim_{x\to -1} f(x)$ exists.

32. Let $f(x) = \begin{cases} 3-x^2, & x \le -2 \\ ax+b, & -2 < x < 2 \\ \frac{1}{2}x^2, & 2 \le x \end{cases}$

Find a and b so that $\lim_{x\to -2} f(x)$ and $\lim_{x\to 2} f(x)$ exist.

2.6 CONTINUITY

In making our informal definition of $L = \lim_{x\to a} f(x)$, we have repeatedly emphasized that $\lim_{x\to a} f(x)$ and $f(a)$ need not be equal. (Indeed, one or both might even fail to exist.) We wish now to turn our attention to the case in which $f(a) = \lim_{x\to a} f(x)$. When this happens, we say that the function $f(x)$ is *continuous* at $x = a$. Here is a more precise definition of this concept.

DEFINITION 4

The function $y = f(x)$ is said to be **continuous at $x = a$** if each of the following is true:

(a) $f(a)$ exists,

(b) $\lim_{x\to a} f(x)$ exists, and

(c) $f(a) = \lim_{x\to a} f(x)$.

Since our definition of limit is an informal one, the above definition of continuity must also be regarded as informal. (A rigorous definition will be given in Section 2.7.) If $f(x)$ is not continuous at $x = a$, we say that it is **discontinuous** at $x = a$. This is the case for each of the functions whose graphs appear in Figures 6.1–6.3. In Figure 6.1, part (a) of Definition 3 fails; in Figure 6.2, (b) fails; and in Figure

6.3, (c) fails. Geometrically, continuity is a property that assures the graph of $y = f(x)$ will not have a hole or otherwise be broken at $(a, f(a))$, by ruling out each of the possibilities represented by Figures 6.1–6.3.

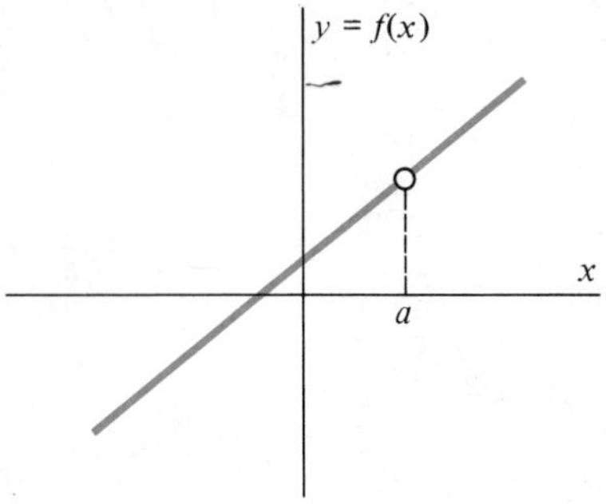

Figure 6.1 $f(a)$ fails to exist.

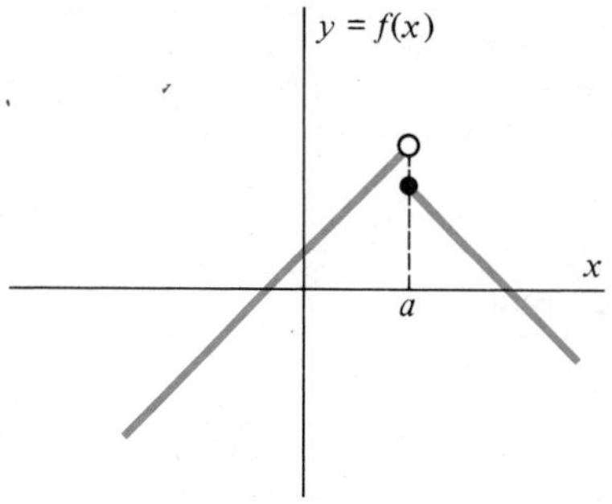

Figure 6.2 $\lim_{x\to a^-} f(x) \neq \lim_{x\to a^+} f(x)$; $\lim_{x\to a} f(x)$ does not exist.

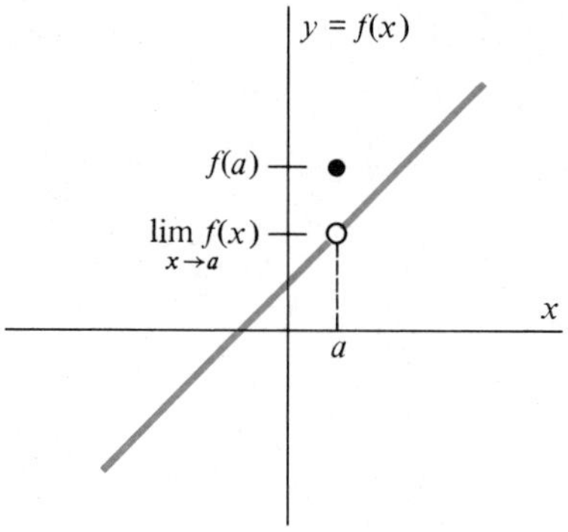

Figure 6.3 $f(a) \neq \lim_{x\to a} f(x)$.

Example 1 Find the numbers where $f(x) = \dfrac{x^2 - 4}{x - 2}$ is continuous.

Strategy

Find numbers x for which the denominator equals zero, since $f(x)$ will be undefined for such numbers.

Solution

Since $x - 2 = 0$ when $x = 2$, the function $f(x) = \dfrac{x^2 - 4}{x - 2}$ is not defined for $x = 2$, that is, $f(2)$ does not exist. The function is therefore *discontinuous* at $x = 2$.

$f(x)$ is continuous wherever it is not discontinuous.

Since $f(x)$ is a rational function, at all other values of a we have $\lim_{x\to a} f(x) = f(a)$, by Theorem 3. Thus $f(x)$ is *continuous* for all $x \neq 2$ (Figure 6.4). ■

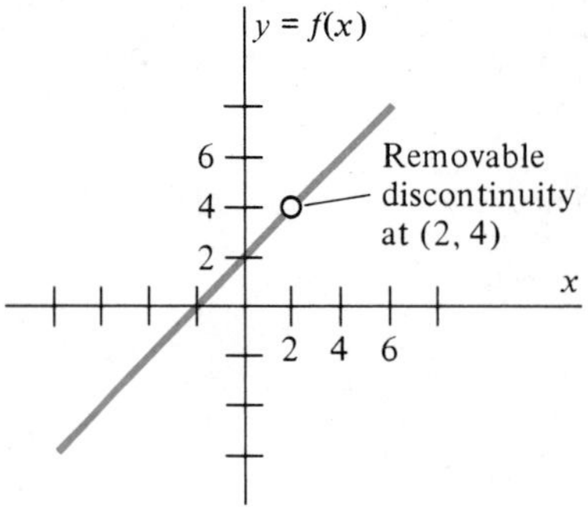

Figure 6.4 $f(x) = \dfrac{x^2 - 4}{x - 2}$ is discontinuous for $x = 2$.

The discontinuity in Example 1 is called a **removable** discontinuity since we can eliminate (i.e., remove) the discontinuity at $x = 2$ by defining $f(2) = \lim_{x\to 2} f(x) = 4$

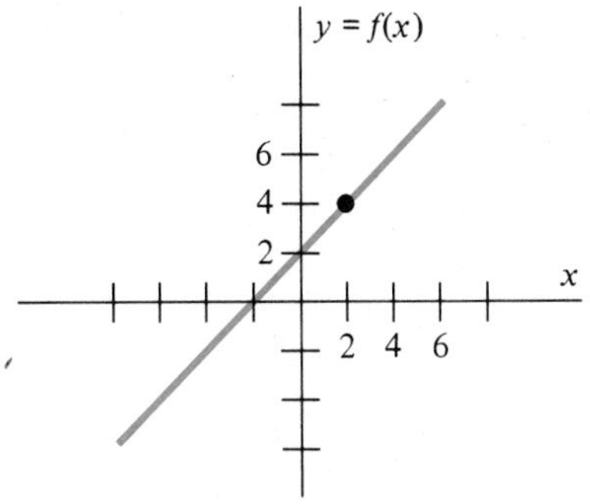

Figure 6.5 Discontinuity in Figure 6.4 is *removed* by defining $f(2) = \lim_{x\to 2} f(x) = 4$.

(see Figure 6.5). In other words, we add $x = 2$ to the domain of $f(x)$ by defining the new function

$$\hat{f}(x) = \begin{cases} \dfrac{x^2 - 4}{x - 2} & \text{if} \quad x \neq 2 \\ 4 & \text{if} \quad x = 2. \end{cases}$$

The function $\hat{f}(x)$ is then continuous for all x, and agrees with $f(x)$ for all $x \neq 2$.

Example 2 Find the values of x for which $f(x) = \dfrac{|x - 1|}{x^2 - 1}$ is discontinuous.

Solution: As in Example 1, we note that the denominator has factorization $x^2 - 1 = (x - 1)(x + 1)$, and equals zero for $x = -1$ and $x = 1$. The function $f(x)$ is therefore undefined, hence discontinuous, for $x = -1$ and $x = 1$. Since the function $f(x)$ can be written

$$\frac{|x - 1|}{x^2 - 1} = \begin{cases} \dfrac{x - 1}{x^2 - 1}, & x > 1 \\ \dfrac{1 - x}{x^2 - 1}, & x < 1, \end{cases}$$

it is a rational function on each of the intervals $(-\infty, -1)$, $(-1, 1)$, and $(1, \infty)$. It is therefore continuous for all x in these intervals by Theorem 3 (see Figure 6.6). ■

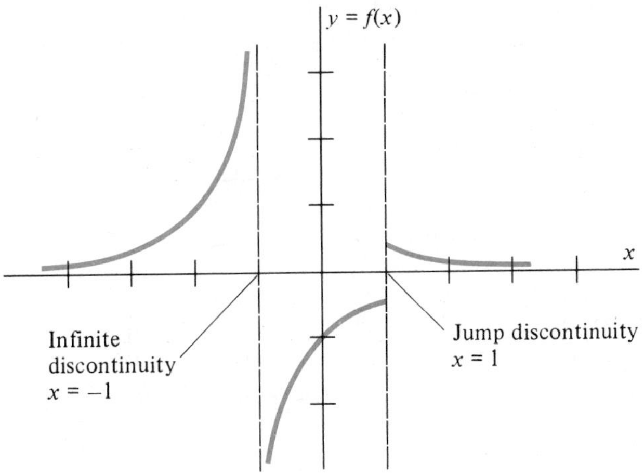

Figure 6.6 Graph of $f(x) = \dfrac{|x - 1|}{x^2 - 1}$.

The function $f(x)$ in Example 2 exhibits two additional types of discontinuities. The discontinuity at $x = 1$ is called a **jump** discontinuity, for obvious reasons, since both one-sided limits exist but are unequal. The discontinuity at $x = -1$ is called an **infinite** discontinuity since one (in this case, both) of the one-sided limits is infinite.

Example 3 The trigonometric functions $f(\theta) = \sin\theta$ and $g(\theta) = \cos\theta$ are continuous for all values of θ. Although we will not formally prove this statement, a geometric argument is compelling: The functions $\cos\theta$ and $\sin\theta$ are defined to be the x- and y-coordinates of a point on the unit circle. This point is determined by the radius making an angle of θ radians with the positive x-axis. Thus, these functions change "uniformly" rather than "abruptly" as θ changes. There can be no "breaks" or "jumps" in their graphs, since small changes in the location of a point on the unit circle can result only in small changes in its x- and y-coordinates. ■

Example 4 The trigonometric function $\tan\theta = \dfrac{\sin\theta}{\cos\theta}$ is discontinuous where $\cos\theta = 0$ (i.e., for $\theta = \dfrac{\pi}{2} + n\pi$, $n = 0, \pm 1, \pm 2, \ldots$). Similarly, the function $\csc\theta = \dfrac{1}{\sin\theta}$ is discontinuous where $\sin\theta = 0$, (i.e., for $\theta = n\pi$, $n = 0, \pm 1, \pm 2, \ldots$). More generally, we can say that the six trigonometric functions are continuous whenever they are defined. ■

Enough of this dwelling on discontinuities! Let's now define what we mean by saying that $y = f(x)$ is continuous on an **interval.** (Definition 6 is slightly more subtle than you might suspect.)

DEFINITION 5 The function $y = f(x)$ is continuous on the **open** interval (a, b) if $f(x)$ is continuous at each $x \in (a, b)$.

DEFINITION 6 The function $y = f(x)$ is continuous on the **closed** interval $[a, b]$ if
(a) $f(x)$ is continuous on (a, b),
(b) $f(a)$ and $f(b)$ both exist, and
(c) both $f(a) = \lim_{x\to a^+} f(x)$ and $f(b) = \lim_{x\to b^-} f(x)$.

Figure 6.7 illustrates why we must add the condition that the "inside limits" equal the function values at the endpoints when defining continuity on closed intervals.

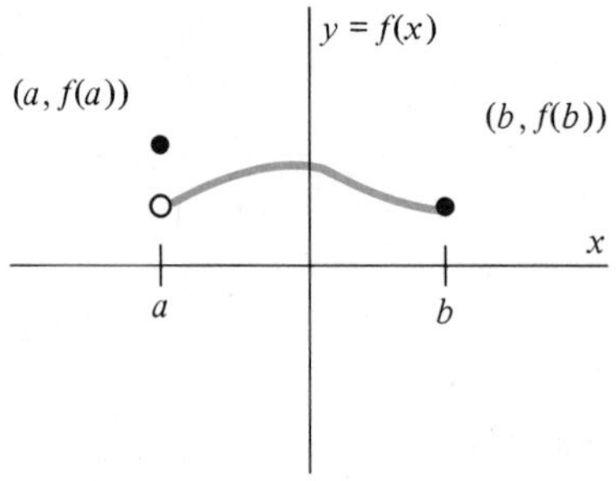

Figure 6.7 $y = f(x)$ is continuous on $(a, b]$ but not on $[a, b]$.

Example 5 The function $f(x) = \sqrt{x+2}$ is continuous on the interval $[-2, \infty)$ since $f(x)$ satisfies our informal notion of continuity on $(-2, \infty)$ and $\lim_{x\to -2^+} f(x) = \lim_{x\to -2^+} \sqrt{x+2} = 0 = f(-2)$ (Figure 6.8). ■

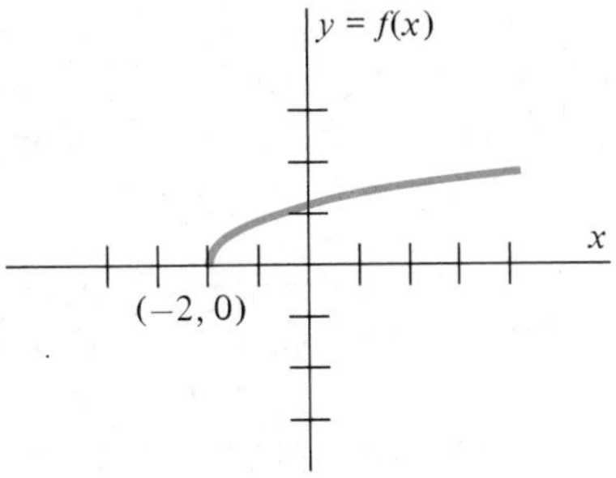

Figure 6.8 $f(x) = \sqrt{x+2}$ is continuous on $[-2, \infty)$.

Example 6 The function $f(x) = \dfrac{|x-1|}{x^2-1}$ is continuous on $(-1, 1)$ but not on $[-1, 1]$. This is because neither $f(-1)$ nor $f(1)$ is defined (Figure 6.6). It is also continuous on the intervals $(-\infty, -1)$ and $(1, \infty)$. ■

The property of continuity is one of the most important concepts in the calculus. It will lurk somewhere near almost every major idea we pursue in this text. In fact, in solving the area problem (Chapter 6) we shall do so only for continuous functions, although a solution for a larger class of functions is possible. In other words, the property of continuity is broad enough to encompass most of the functions with which we must deal, yet continuity is a sufficiently powerful notion to allow "good things" to happen in its presence.

One of these good things is the delightfully simple property that continuous functions cannot "skip" any numbers in passing from one function value to another. A more formal statement of this property is the following:

THEOREM 6
Intermediate Value Theorem

Let $f(x)$ be continuous for $x \in [a, b]$ and let d be any number between $f(a)$ and $f(b)$. Then there is at least one number $c \in (a, b)$ for which $f(c) = d$.

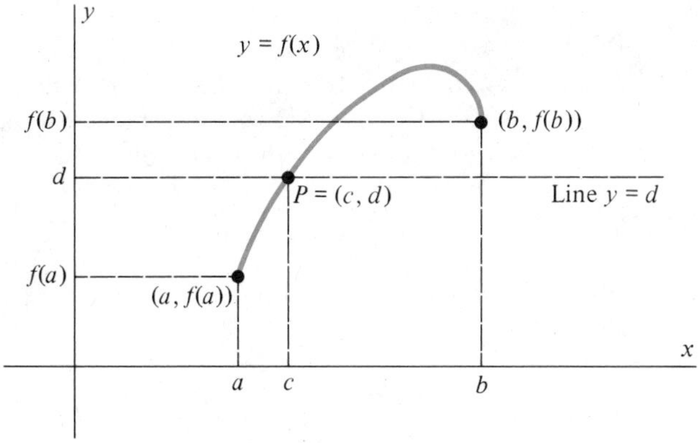

Figure 6.9 The Intermediate Value Theorem.

Figure 6.9 shows the graph of a typical continuous function and the interpretation of the Intermediate Value Theorem. Although a formal proof of this theorem is beyond the scope of this text, a geometric argument for its validity is suggested by

Figure 6.9 in the case $f(a) < f(b)$. Since $f(x)$ is continuous, there cannot be any jump discontinuities between a and b. Since $f(a) < d < f(b)$, the graph of $f(x)$ must intersect the horizontal line $y = d$ at some point P. Thus, P has coordinates (c, d), with $f(c) = d$.

Example 7 For some (but not all) functions, we can easily find the number c satisfying the Intermediate Value Theorem. For example, for the function $f(x) = \sqrt{x^3 + 1}$ and the interval $[a, b] = [0, 2]$, we have $f(a) = f(0) = 1$ and $f(b) = f(2) = 3$. Thus, any number d with $1 \le d \le 3$ is an intermediate value for $f(x)$ on $[0, 2]$. To find $x = c$ for which $f(c) = d$, we simply solve the equation $d = \sqrt{x^3 + 1}$ for x:

$$\begin{aligned} d = \sqrt{x^3 + 1} &\Rightarrow d^2 = x^3 + 1 \\ &\Rightarrow x^3 = d^2 - 1 \\ &\Rightarrow x = \sqrt[3]{d^2 - 1}. \end{aligned}$$

In particular, if $d = 2$, we obtain $c = x = \sqrt[3]{2^2 - 1} = \sqrt[3]{3}$. ■

Although we were able to find the number c satisfying the Intermediate Value Theorem in Example 7, the theorem itself is only an **existence theorem.** It tells us nothing about how to find c, only that such a number exists. However, existence theorems are very important in mathematics because often we must first determine whether or not a problem *has* a solution before we set out to solve it. For example, the student faced with the problem of calculating the real zeros of the polynomial $x^6 + 4x^4 + x^2 + 4 = 0$ should not immediately grab a hand calculator and start blissfully proceeding by trial and error to find such zeros. Indeed, by writing

$$x^6 + 4x^4 + x^2 + 4 = 0 \tag{1}$$

as

$$x^6 + 4x^4 + x^2 = -4 \tag{2}$$

we can see that the left side of equation (2) cannot be negative. Since $-4 < 0$, equation (2) has no solutions. Thus, neither does (1), so the polynomial has no real zeros. Often, determining whether or not a problem *has* a solution is the major part of the task.

Example 8 Use the Intermediate Value Theorem to show that the polynomial $f(x) = 4x^3 - x + 2$ has at least one real zero in the interval $[-1, 1]$. Assume $f(x)$ to be continuous on this interval.

Strategy

Check function values at endpoints. If they are of opposite sign, we must have $f(c) = 0$ somewhere in $[-1, 1]$, since either

$f(a) < 0 < f(b)$

or

$f(b) < 0 < f(a)$.

Solution

We are seeking a number $c \in [-1, 1]$ with $f(c) = 0$.

Notice that

$$f(-1) = 4(-1)^3 - (-1) + 2 = -1 < 0$$

while

$$f(1) = 4(1)^3 - (1) + 2 = 5 > 0.$$

Since $f(-1) < 0 < f(1)$, the Intermediate Value Theorem guarantees the existence of a number $c \in (-1, 1)$ with $f(c) = 0$. ■

Example 8 highlights the importance of knowing which functions are continuous and where this is true. The following two theorems assure that polynomials and rational functions are continuous throughout their domains.

THEOREM 7

Let $p(x)$ be the polynomial

$$p(x) = a_n x^n + a_{n-1}x^{n-1} + \cdots + a_1 x + a_0.$$

Then $p(x)$ is continuous for all values of x.

Strategy
Show that the definition of continuity is fulfilled. Use Theorem 2 (about limits of polynomials) to do this.

Proof of Theorem 7
We must show that the three requirements of Definition 3 are met at $x = a$:

(a) $p(a)$ exists, since polynomials are defined for all values of x.
(b) $\lim_{x \to a} p(x)$ exists, by Theorem 2.
(c) $p(a) = \lim_{x \to a} p(x)$, also by Theorem 2.

Thus, $p(x)$ is continuous at $x = a$. Since a is arbitrary, the statement is proved for all x. ■

THEOREM 8

Let $p(x)$ and $q(x)$ be polynomials with $q(a) \neq 0$. Then the rational function

$$r(x) = \frac{p(x)}{q(x)}$$

is continuous at $x = a$.

Strategy
Show that the definition of continuity is fulfilled. Use Theorem 3 (about limits of rational functions) to do this.

Proof of Theorem 8
We must show that the requirements of Definition 3 are met:

(a) $r(a)$ exists since $p(a)$ and $q(a)$ exist and $q(a) \neq 0$.
(b) $\lim_{x \to a} r(x)$ exists, since $q(a) \neq 0$, by Theorem 3.
(c) $r(a) = \lim_{x \to a} r(x)$, also by Theorem 3.

Thus $r(x)$ is continuous at any number $a \neq 0$. ■

The following theorem asserts the continuity of fractional power functions. Its proof follows from Theorem 4 and Corollary 1 and is left as an exercise.

THEOREM 9

Let n and m be positive integers with $m \neq 0$. Then the function $f(x) = x^{n/m}$ is continuous where defined.

Example 9 By Theorems 7, 8, and 9, each of the following functions is continuous on the indicated interval:

(a) $f(x) = x^3 + \pi x$, $(-\infty\ \infty)$
(b) $f(x) = x^{-n}$, n a positive integer, $(-\infty, 0)$ and $(0, \infty)$

(c) $f(x) = x^{7/2}$, $[0, \infty)$

(d) $f(x) = \dfrac{x^3 + x + 6}{x(x-2)(x+3)}$, $(-\infty, -3)$, $(-3, 0)$, $(0, 2)$, and $(2, \infty)$. ■

Theorem 10 asserts that the algebra of continuous functions works just like the algebra of limits. That is, sums, multiples, products, and quotients of continuous functions are again continuous. It should not be surprising that this theorem is a direct consequence of Theorem 1.

THEOREM 10

Let the functions $f(x)$ and $g(x)$ be continuous at $x = a$ and let c be a real number. Then the following functions are continuous at $x = a$:

(i) $(f + g)(x) = [f(x) + g(x)]$,

(ii) $(cf)(x) = c[f(x)]$,

(iii) $(fg)(x) = [f(x)][g(x)]$, and

(iv) $\left(\dfrac{f}{g}\right)(x) = \left[\dfrac{f(x)}{g(x)}\right]$, $g(x) \neq 0$.

Strategy

Show that the definition of continuity is fulfilled using Theorem 1 (about the algebra of limits) to do this.

Proof of Theorem 10

To prove part (i) we will prove that $[f + g](x)$ is continuous at $x = a$. Since both $f(x)$ and $g(x)$ are assumed continuous at $x = a$, we know that $f(a)$ and $g(a)$ exist, and that

$$f(a) = \lim_{x\to a} f(x), \qquad g(a) = \lim_{x\to a} g(x).$$

We may therefore apply Theorem (1), part (i), to conclude that

$$\begin{aligned}\lim_{x\to a} [f(x) + g(x)] &= \lim_{x\to a} f(x) + \lim_{x\to a} g(x)\\ &= f(a) + g(a).\end{aligned}$$

Thus, $\lim_{x\to a} (f + g)(x) = (f + g)(a)$, so $(f + g)(x)$ is continuous at $x = a$. The remaining cases also follow from Theorem 1 and are left as exercises. ■

Composite Functions

Recall, from Section 1.5, that a composite function has the form $y = f(u(x))$. For example, the function

$$y = \sqrt{x^3 + 4}$$

is composite. It has the form $f(u(x))$ where the "outside" function is the square root function $f(u) = \sqrt{u}$, and the "inside" function is the polynomial $u(x) = x^3 + 4$.

The following theorem will be used throughout the text in evaluating limits of composite functions. Its proof will be given in Section 2.8.

THEOREM 11

Let $f(x)$ be a continuous function on an open interval containing L, and let $\lim_{x\to a} g(x) = L$. Then

$$\lim_{x\to a} f(g(x)) = f(L).$$

We may paraphrase Theorem 11 by saying that

$$\lim_{x\to a} f(g(x)) = f(\lim_{x\to a} g(x))$$

if f is continuous "near" $\lim_{x\to a} g(x)$. That is, we may "pass the limit inside" a continuous function f. This theorem justifies the calculation of limits such as those of Example 10.

Example 10 According to Theorem 11,

(a) $\lim_{x\to 5} \sqrt{34 - x^2} = \sqrt{\lim_{x\to 5}(34 - x^2)} = \sqrt{34 - 25} = 3,$
since $f(u) = \sqrt{u}$ is continuous near $u = 9$.

(b) $\lim_{x\to -2} (6x + 4)^{-2/3} = [\lim_{x\to -2} (6x + 4)]^{-2/3} = (-8)^{-2/3} = \dfrac{1}{4},$
since $f(u) = u^{-2/3}$ is continuous near $u = -8$.

(c) $\lim_{x\to \pi/4} \sin\left(x + \dfrac{\pi}{4}\right) = \sin\left[\lim_{x\to \pi/4}\left(x + \dfrac{\pi}{4}\right)\right] = \sin\left(\dfrac{\pi}{2}\right) = 1,$
since $f(u) = \sin u$ is continuous for all u. ■

Here is a special case of Theorem 11 that we shall use frequently in establishing the theory of the derivative in Chapter 4:

> If $f(x)$ is continuous on an open interval containing x_0, then (3)
>
> $$\lim_{\Delta x\to 0} f(x_0 + \Delta x) = f(x_0)$$

To see why this is true, note simply that

$$\lim_{\Delta x\to 0} f(x_0 + \Delta x) = f[\lim_{\Delta x\to 0} (x_0 + \Delta x)] = f(x_0 + 0) = f(x_0).$$

Finally, we can use Theorem 11 to prove that the composition of two continuous functions is again continuous.

THEOREM 12 Let $f(u)$ be continuous on an open interval containing $u = u(a)$, and let $u(x)$ be continuous on an open interval containing $x = a$. Then the composite function $f(u(x))$ is continuous at $x = a$.

Proof: We must show that $\lim_{x\to a} f(u(x)) = f(u(a))$. Since $u(x)$ is continuous at $x = a$, we know that $\lim_{x\to a} u(x) = u(a)$. Using this fact, the continuity of f, and Theorem 11, we conclude that

$$\lim_{x\to a} f(u(x)) = f(\lim_{x\to a} u(x)) = f(u(a)),$$

as required. ■

Figure 6.10 suggests a way of thinking about composite functions that might help you get a better feel for Theorems 11 and 12. The idea is simply that if f and u are continuous, $u(x_0)$ is close to $u(a)$ when x_0 is close to a, so $f(u(x_0))$ will be close to $f(u(a))$ when x_0 is close to a.

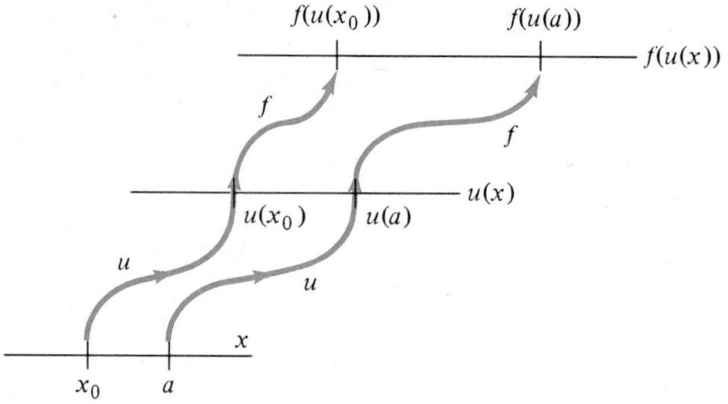

Figure 6.10 The composite function $f(u(x))$.

Example 11 The functions $f(x) = \sqrt{x}$, $g(x) = \sin x$, and $h(x) = x^3 + 1$ satisfy our informal notion of continuity on their respective domains. Thus, by Theorem 11,

(a) $y = f(g(x)) = \sqrt{\sin x}$ is continuous on $[0, \pi]$
(b) $y = g(f(x)) = \sin \sqrt{x}$ is continuous on $[0, \infty)$
(c) $y = f(h(x)) = \sqrt{x^3 + 1}$ is continuous on $[-1, \infty)$
(d) $y = h(f(x)) = (\sqrt{x})^3 + 1 = x^{3/2} + 1$ is continuous on $[0, \infty)$
(e) $y = h(g(x)) = \sin^3 x + 1$ is continuous on $(-\infty, \infty)$
(f) $y = g(h(x)) = \sin(x^3 + 1)$ is continuous on $(-\infty, \infty)$. ■

You should verify that the interval of continuity quoted for each composition is correct, and be sure to understand *why* it is correct.

Exercise Set 2.6

In Exercises 1–12, find the number(s) x at which the given function is discontinuous.

1. $f(x) = \dfrac{\sin x}{x}$

2. $f(x) = \dfrac{1}{1 - x^2}$

3. $f(x) = \cot x$

4. $f(x) = \sec x$

5. $f(x) = \dfrac{x^2 - 1}{x - 1}$

6. $f(x) = \dfrac{x}{\sin x}$

7. $y = \dfrac{x + 2}{x^2 - x - 2}$

8. $y = \dfrac{x^2 + x + 1}{x^3 + 2x^2 - 3x}$

9. $f(x) = \begin{cases} 1 - x, & x \le 2 \\ x - 1, & x > 2 \end{cases}$

10. $y = \begin{cases} x^2, & x \ge 0 \\ 3x, & x < 0 \end{cases}$

11. $f(x) = \begin{cases} -x, & x \le -1 \\ 4 - x^2, & -1 < x \le 2 \\ \frac{1}{2}x - 1, & 2 < x \end{cases}$

12. $y = \begin{cases} ax^2, & x \le 0 \\ bx^3, & 0 < x \le 1 \\ cx^4, & 1 < x \end{cases}$

In Exercises 13–26, state the intervals on which the function is continuous.

13. $y = \dfrac{x^2 - 36}{x - 6}$

14. $f(x) = \dfrac{x^3 - 9x}{x^3 - 3x^2}$

15. $y = \csc x$

16. $y = \dfrac{1}{\sqrt{x^2 + 3x - 10}}$

17. $f(x) = \dfrac{1}{\sqrt{x^3 - 4x}}$

18. $y = \dfrac{1}{\sqrt{\sin x}}$

19. $y = x^{2/3} - x^{5/3}$

20. $y = \dfrac{1}{\sec x \tan x}$

21. $g(x) = \sqrt{\tan x}$

22. $f(x) = \sqrt{8 - x^{3/2}}$

23. $f(x) = \dfrac{\tan x}{x^2 - x - 2}$

24. $f(x) = \dfrac{1}{x^{5/2} - x^{3/2}}$

25. $f(x) = \begin{cases} \sqrt{x + x^2}, & x \geq 0 \\ \sin x, & x < 0 \end{cases}$

26. $f(x) = \begin{cases} x^{4/3} - 1, & x < -1 \\ \sqrt{1 + x^2}, & -1 \leq x \leq \sqrt{3} \\ x^2 - 1, & x > \sqrt{3} \end{cases}$

In Exercises 27–30, the function given has a removable discontinuity at $x = a$. Determine how to define $f(a)$ so that the function is continuous at a.

27. $f(x) = \dfrac{x^2 - 1}{x - 1}, \quad a = 1$

28. $f(x) = \begin{cases} x^2 + 1, & x < 1 \\ \sqrt{3 + x}, & x > 1 \end{cases} \quad a = 1$

29. $f(x) = \dfrac{1 - \cos^2 x}{\sin x}, \quad a = 0$

30. $f(x) = \dfrac{x^2 + x - 2}{x^3 - x^2 - 6x}, \quad a = -2$

In Exercises 31–33, find the constant k which makes the function continuous at $x = a$.

31. $y = \begin{cases} x^k, & x \leq 2 \\ 10 - x, & x > 2 \end{cases} \quad a = 2$

32. $y = \begin{cases} k, & x \geq 1 \\ \dfrac{1}{\sqrt{kx^2 + k}}, & x < 1 \end{cases} \quad a = 1$

33. $y = \begin{cases} (x - k)(x + k), & x \leq 2 \\ kx + 5, & x > 2 \end{cases} \quad a = 2$

In Exercises 34–36, use the Intermediate Value Theorem to determine whether the given function has a zero in the given interval.

34. $f(x) = x^3 + 3x^2 - x - 3$
a. $[-4, -2]$
b. $[-2, 0]$
c. $[2, 4]$

35. $f(x) = x^2 - 4\sqrt{x} + 1$
a. $[0, 1]$
b. $[1, 3]$
c. $[3, 5]$

36. *(Calculator)* $f(x) = \sqrt{x} - x \sin x$.
a. $[0, 1]$
b. $[1, 2]$
c. $[2, 3]$
d. $[3, 4]$
e. $[5, 9]$

In Exercises 37–42, find a number c in the interval $[a, b]$ for which $f(c) = d$, according to the Intermediate Value Theorem.

37. $f(x) = 3x - 4, \quad a = -2, \quad b = 2, \quad d = -1$

38. $f(x) = x^2, \quad a = 0, \quad b = 3, \quad d = 5$

39. $f(x) = \sqrt{9 - x^2}, \quad a = -3, \quad b = 0, \quad d = \sqrt{5}$

40. $f(x) = \sin x, \quad a = 0, \quad b = \dfrac{\pi}{2}, \quad d = \dfrac{\sqrt{2}}{2}$

41. $f(x) = \dfrac{1}{x^2 + 1}, \quad a = -4, \quad b = -1, \quad d = \dfrac{1}{4}$

42. $f(x) = \dfrac{x}{x + 3}, \quad a = -6, \quad b = -4, \quad d = 3$

43. True or false? If $f(x)$ is continuous on $[a, b]$, $f(a) < 0$, and $f(b) > 0$, then $f(x)$ cannot have exactly two zeros in $[a, b]$. Why?

44. Give an example to show that the composite function $f(u(x))$ might be continuous at $x = a$ although $\lim_{x \to a} u(x) \neq u(a)$. Does this contradict Theorem 12?

45. Use Theorem 4 and Corollary 1 to prove Theorem 9.

46. Prove Theorem 10, part (ii).

47. Prove Theorem 10, part (iii).

48. Prove Theorem 10, part (iv).

2.7 A FORMAL DEFINITION OF LIMIT (Optional)

From the viewpoint of precise mathematics, our working definition of limit is problematic. The difficulty lies in the use of the "dynamic" concept of "*f*(*x*) *approaching L as x approaches a*." A precise mathematical statement can involve constants, variables, equal signs, inequalities, arithmetic operations, absolute value signs, and so forth, but not vague references to motion.

In order to capture our intuitive notion of $L = \lim_{x\to a} f(x)$, we use the following elements of precise language.

1. In place of the phrase "$f(x)$ approaches L," we use the inequality

$$|f(x) - L| < \epsilon \tag{1}$$

where ϵ is an arbitrarily small positive number. This means that $f(x)$ will be found as close to the number L as you wish, although we will not necessarily have $f(x) = L$.

2. In place of the phrase "as x approaches a," we use a similar inequality:

$$0 < |x - a| < \delta \tag{2}$$

where δ is a (small) positive constant. (Remember, we do not want to allow $x = a$. That is the reason for the left-hand part of this inequality.)

3. In order to connect these two phrases to reflect the notion that "$f(x)$ approaches L as x approaches a," we say that, no matter what constant ϵ is chosen for inequality (1), we can find an appropriate constant δ so that the validity of inequality (2) will insure the validity of inequality (1). In other words, we want to say that

$|x - a|$ small guarantees $|f(x) - L|$ small.

These conventions bring us to our formal definition of limit.

DEFINITION 7

We say that the number L is the **limit** of the function $f(x)$ as x approaches a, written

$$L = \lim_{x\to a} f(x), \qquad \text{if and only if:}$$

for every $\epsilon > 0$ there exists a corresponding $\delta > 0$ so that

$$|f(x) - L| < \epsilon \qquad \text{whenever} \qquad 0 < |x - a| < \delta.$$

In other words, $L = \lim_{x\to a} f(x)$ means that we can cause $f(x)$ to be as close to L as desired (ϵ) simply by choosing x sufficiently close (δ) to a (see Figure 7.1). This formal definition agrees with our intuitive definition of limit since as $x \to a$, $(x - a) \to 0$. Thus, $|x - a|$ becomes less than δ, so $|f(x) - L|$ becomes less than ϵ. Since ϵ can be arbitrarily small, this means that $|f(x) - L| \to 0$, or $f(x) \to L$. Thus, when Definition 7 holds, $f(x) \to L$ as $x \to a$.

The advantage of Definition 7 over our intuitive definition is that Definition 7 provides us with a formal criterion that we can attempt to verify. In other words, we can actually prove statements of the form "$L = \lim_{x\to a} f(x)$" using Definition 7. To do so we use the following procedure.

Procedure P for Proving $L = \lim_{x\to a} f(x)$

(i) Assume $\epsilon > 0$ is an arbitrary real number.
(ii) Find a relationship between $|f(x) - L|$ and $|x - a|$.
(iii) Using the relationship found in step (ii), show how we may select a number $\delta > 0$ so that

$$|f(x) - L| < \epsilon \qquad \text{whenever} \qquad |x - a| < \delta.$$

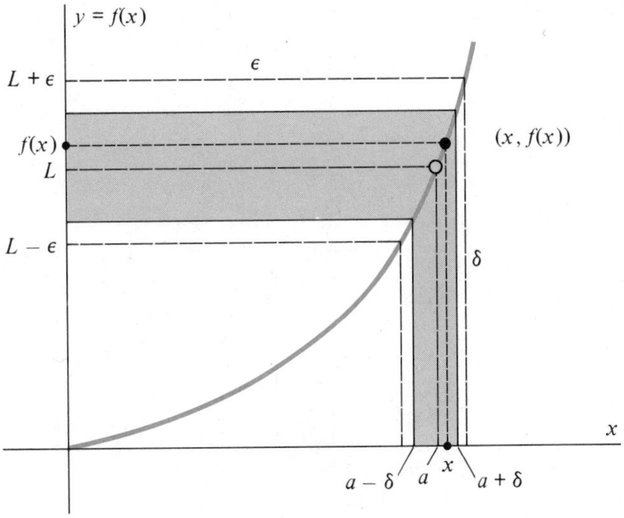

Figure 7.1 $|f(x) - L| < \varepsilon$ if $0 < |x - a| < \delta$.

The following examples show how procedure P is used for several particular functions. Before studying these examples you should note the limitation of Definition 7 and procedure P: *given* the supposed limit L, a procedure is available for proving that $L = \lim\limits_{x \to a} f(x)$. However, neither Definition 7 nor procedure P provide any information as to how the number L is found. Finding the correct value of L is necessary before the statement $L = \lim\limits_{x \to a} f(x)$ can be proved. The techniques for finding L are largely those discussed in Sections 2.2–2.5.

Example 1 Prove that $\lim\limits_{x \to 3} (2x + 1) = 7$.

Strategy

Write down $|f(x) - 7|$ to determine how it is related to $|x - 3|$.

Combine terms and factor out 2.

Use the relationship between $|f(x) - 7|$ and $|x - 3|$ to state the relationship between ϵ and δ, that is, obtain δ as a function of ϵ.

The formal proof:

(i) take $\epsilon > 0$ arbitrary
(ii) choose $\delta = \epsilon/2$
(iii) show that

$$0 < |x - 3| < \delta$$

guarantees that

$$|f(x) - 7| < \epsilon.$$

Solution

Here $f(x) = 2x + 1$ and $L = 7$, so

$$\begin{aligned}
|f(x) - L| &= |(2x + 1) - 7| \\
&= |2x - 6| \\
&= 2|x - 3|.
\end{aligned}$$

This shows that the distance between $(2x + 1)$ and 7 is always *twice* the distance between x and 3. Thus, we can guarantee that $|(2x + 3) - 7| < \epsilon$ by requiring that $0 < |x - 3| < \epsilon/2$. In other words, we use $\delta = \epsilon/2$ in Definition 7.

Formally, the proof goes as follows:

Let $\epsilon > 0$ be arbitrary, and let $\delta = \epsilon/2$. Then if $0 < |x - 3| < \delta$ we have that

$$\begin{aligned}
|(2x + 1) - 7| &= |2x - 6| \\
&= 2|x - 3| \\
&< 2 \cdot \delta \\
&= 2\left(\frac{\epsilon}{2}\right) \\
&= \epsilon.
\end{aligned}$$

Thus $|(2x + 1) - 7| < \epsilon$ whenever $0 < |x - 3| < \delta$, as required by Definition 7 (see Figure 7.2). ∎

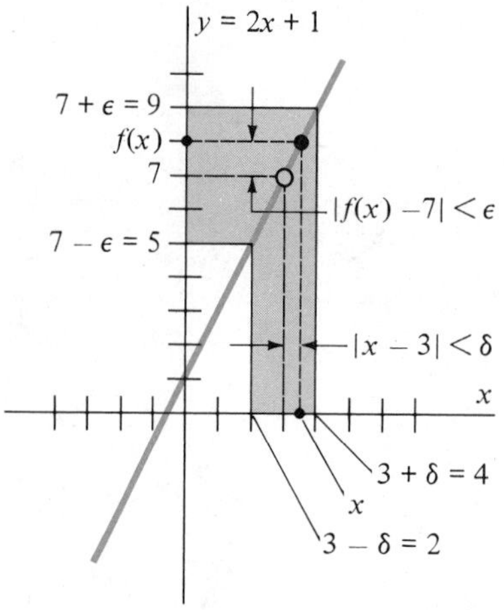

Figure 7.2 $\lim_{x\to3} (2x + 1) = 7$. Case $\varepsilon = 2$, $\delta = \varepsilon/2 = 1$.

The relationship $\delta = \epsilon/2$ in Example 1 should not have been surprising since $\Delta y = 2\,\Delta x$ for any two points on the graph of $y = 2x + 1$. In other words, $\epsilon = 2\delta$ since the slope of the line $y = 2x + 1$ equals 2. A similar relationship results for any linear function. For functions which are not linear, we must make use of particular behavior of the function near the value of x in question. Note how this happens in the following example.

Example 2 Prove that $\lim_{x\to2} x^2 = 4$.

Strategy
Examine $|x^2 - 4|$ to try to find a relationship with $|x - 2|$.

Solution
Here $f(x) = x^2$ and $L = 4$, so we examine

$$\begin{aligned}|x^2 - 4| &= |(x + 2)(x - 2)| \\ &= |x + 2| \cdot |x - 2|.\end{aligned}$$

Try to bound the factor $|x + 2|$ by some constant.

Now $|x - 2|$ is the desired factor $|x - a|$ in Definition 7. But what shall we do about the factor $|x + 2|$? Simply this: If x is close to 2, $x + 2$ is close to 4, so we should be able to replace $|x + 2|$ by a constant close to 4.

$|x + 2| < 5$ if $|x - 2| < 1$

To make this more precise, we observe that if $|x - 2| < 1$, then $|x + 2| < 5$. In this case we will have

$$\begin{aligned}|x^2 - 4| &= |x + 2| \cdot |x - 2| \\ &< 5|x - 2|,\end{aligned}$$

Take $\delta = \min\left\{\frac{\epsilon}{5}, 1\right\}$

that is, $|x^2 - 4|$ will be less than 5 times $|x - 2|$. We therefore will use $\delta = \epsilon/5$ and recall also that we need to have $|x - 2| < 1$. The formal proof is the following:

Apply Definition 7.

Given $\epsilon > 0$, let δ be the smaller of the numbers $\{1, \epsilon/5\}$. Then if $0 < |x - 2| < \delta$, we have both

$$|x - 2| < \epsilon/5 \qquad \text{and} \qquad |x + 2| < 5.$$

Thus,

$$\begin{aligned}|x^2 - 4| &= |x + 2| \cdot |x - 2| \\ &< 5 \cdot |x - 2| \\ &< 5 \cdot (\epsilon/5) \\ &= \epsilon.\end{aligned}$$

This shows that

$$|x^2 - 4| < \epsilon \qquad \text{whenever} \qquad 0 < |x - 2| < \delta, \text{ so}$$

$$\lim_{x \to 2} x^2 = 4.$$

See Figure 7.3. ■

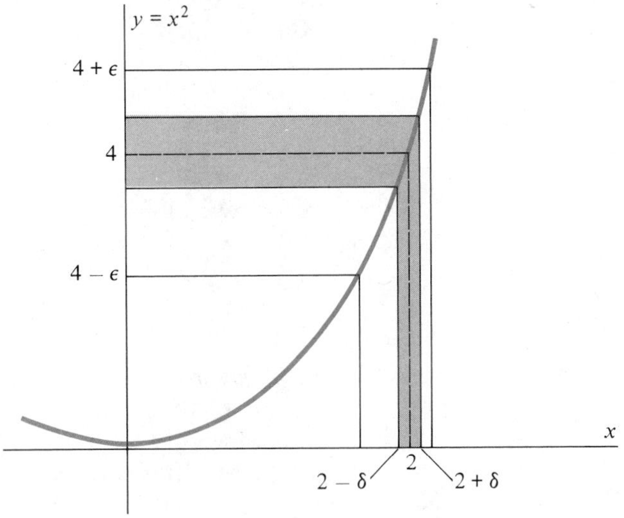

Figure 7.3 $|x^2 - 4| < \varepsilon$ if $0 < |x - 2| < \delta$ for $\delta = \min\{1, \varepsilon/5\}$.

Example 3 Prove that $\lim_{x \to 2} \dfrac{1}{x + 1} = \dfrac{1}{3}$.

Strategy

Examine $\left|\dfrac{1}{x+1} - \dfrac{1}{3}\right|$ to find a relationship with $|x - 2|$.

Solution

Here $f(x) = \dfrac{1}{x + 1}$ and $L = \dfrac{1}{3}$, so

$$\begin{aligned}|f(x) - L| &= \left|\frac{1}{x + 1} - \frac{1}{3}\right| \\ &= \left|\frac{3 - (x + 1)}{3(x + 1)}\right| \\ &= \frac{1}{3} \cdot \frac{|x - 2|}{|x + 1|}.\end{aligned}$$

Find a bound on factors other than $|x - 2|$. Since x approaches 2, we can arbitrarily assume $|x - 2| < 1$.

Now $|x - 2|$ is the desired factor $|x - a|$, so we must find a bound on the factor $|x + 1|^{-1}$. We do this by noting that if $|x - 2| < 1$, then

$$-1 < x - 2 < 1,$$

so

$$2 < x + 1 < 4.$$

We must bound $|x + 1|$ from below since it is a factor of the denominator.

Thus

$$|x + 1| > 2 \quad \text{if} \quad |x - 2| < 1,$$

so

$$\frac{1}{|x + 1|} < \frac{1}{2} \quad \text{when} \quad |x - 2| < 1.$$

In this case we have

$$\begin{aligned} |f(x) - L| &= \frac{1}{3}\frac{|x - 2|}{|x + 1|} \\ &< \frac{1}{3}\cdot\frac{|x - 2|}{2} \\ &= \frac{1}{6}|x - 2|. \end{aligned}$$

We need to take δ less than *both* 6ϵ and 1, since we assumed $|x - 2| < 1$.

Thus, given $\epsilon > 0$, we take $\delta = \min\{6\cdot\epsilon, 1\}$.
Then if $0 < |x - 2| < \delta$ we have

$$\left|\frac{1}{x + 1} - \frac{1}{3}\right| = \frac{1}{3}\frac{|x - 2|}{|x + 1|} < \frac{1}{6}|x - 2| \le \frac{1}{6}\cdot(6\cdot\epsilon) = \epsilon,$$

as required by Definition 7. ■

Example 3 illustrates the fact that proving limits is, in general, a difficult exercise. Determining the relationship between δ and ϵ in Definition 7 often depends on both the function $f(x)$ and the number $x = a$ in complicated ways. We will not dwell at length on this issue in this text, primarily because a well developed understanding of the technical formalities of limits is not essential to the mastery of what follows. However, we do need to verify the algebra of limits developed in Sections 2.4 and 2.6. Before doing this in Section 2.8, we note the formal definitions associated with one-sided limits as introduced in Section 3.3.

DEFINITION 8
One-Sided Limits

We say that $L = \lim_{x\to a^+} f(x)$ if and only if: for every $\epsilon > 0$ there exists a number $\delta > 0$ so that

$$|f(x) - L| < \epsilon \quad \text{whenever} \quad a < x < a + \delta.$$

Similarly, we say that $L = \lim_{x\to a^-} f(x)$ if and only if: for every $\epsilon > 0$ there exists a number $\delta > 0$ so that

$$|f(x) - L| < \epsilon \quad \text{whenever} \quad a - \delta < x < a.$$

Note the differences between Definition 7 and Definition 8. In $\lim_{x\to a^+} f(x)$, the inequality $a < x < a + \delta$ requires that $x > a$, that is, that x approach a from the right only (Figure 7.4). In $\lim_{x\to a^-} f(x)$, the inequality $a - \delta < x < a$ requires that $x < a$, that is, that x approach a from the left only (Figure 7.5).

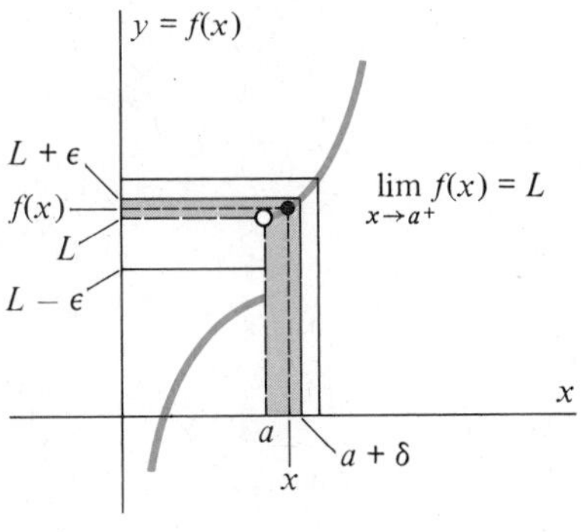

Figure 7.4 If $a < x < a + \delta$, then $|f(x) - L| < \varepsilon$.

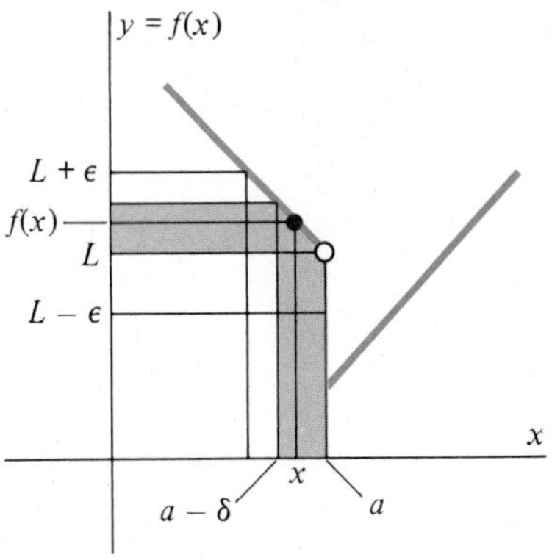

Figure 7.5 If $a - \delta < x < a$, then $|f(x) - L| < \varepsilon$.

Example 4 Prove that $\lim_{x\to 0^+} \sqrt{x} = 0$.

Strategy
Examine $|\sqrt{x} - 0|$ to find relationship with $|x - 0|$.

Solution
Here $f(x) = \sqrt{x}$ and $L = 0$. Thus

$$\begin{aligned}|f(x) - L| &= |\sqrt{x} - 0|\\ &= \sqrt{x},\end{aligned}$$

so

$$|\sqrt{x} - 0| < \epsilon \qquad \text{if} \qquad 0 < x < \epsilon^2.$$

Take $\delta = \epsilon^2$ since
$|\sqrt{x} - 0| < \epsilon$
if $x < \epsilon^2 = \delta$.

We therefore take $\delta = \epsilon^2$ in Definition 8. Then, given $\epsilon > 0$ we have

$$|\sqrt{x} - 0| = \sqrt{x} < \sqrt{\delta} = \sqrt{\epsilon^2} = \epsilon$$

whenever $0 < x < \delta$, which proves that $\lim_{x\to 0^+} \sqrt{x} = 0$

(Figure 7.6). ■

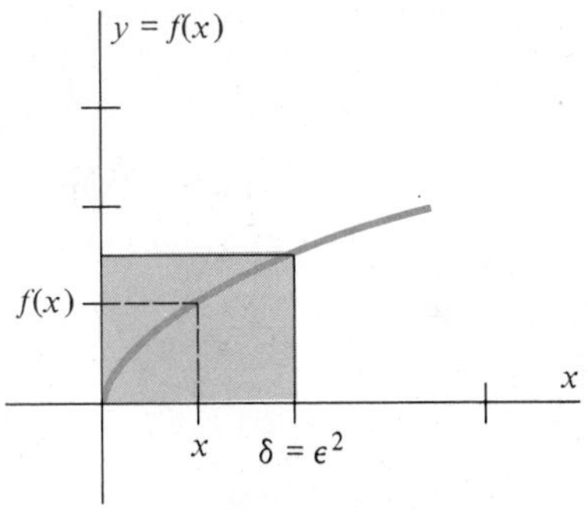

Figure 7.6 $\sqrt{x} < \varepsilon$ if $0 < x < \delta$.

Example 5 For $f(x) = \begin{cases} 4 - x^2, & x \le -1 \\ x + 1, & x > -1 \end{cases}$, find $\lim_{x \to -1^-} f(x)$.

Strategy

Find a candidate for $L = \lim_{x \to -1^-} f(x)$.

Examine $|f(x) - L|$.

Solution

Our informal notion of limit suggests that $f(x) \to 3$ as $x \to -1^-$. To prove this we observe that for $x < -1$

$$\begin{aligned} |f(x) - 3| &= |(4 - x^2) - 3| \\ &= |1 - x^2| \\ &= |x - 1| \cdot |x - (-1)|. \end{aligned}$$

Find a constant to bound the term $|x - 1|$.

We proceed as in Example 2: If x is within one unit of -1, that is, if $|x - (-1)| < 1$, then $|x - 1| < 3$. Using this inequality the above equation becomes

$$\begin{aligned} |f(x) - 3| &= |x - 1| \cdot |x - (-1)| \\ &< 3|x - (-1)|. \end{aligned}$$

Take $\delta = \epsilon/3$.
Apply Definition 8.

Now for $\epsilon > 0$, we take $\delta = \min\{1, \epsilon/3\}$.

Then if $-1 - \delta < x < -1$, we have $|x - (-1)| < \delta$, so the above inequality shows that

$$\begin{aligned} |f(x) - 3| &< 3|x - (-1)| \\ &< 3 \cdot \delta \\ &= 3 \cdot \epsilon/3 \\ &= \epsilon, \end{aligned}$$

as required by Definition 8

(Figure 7.7). ■

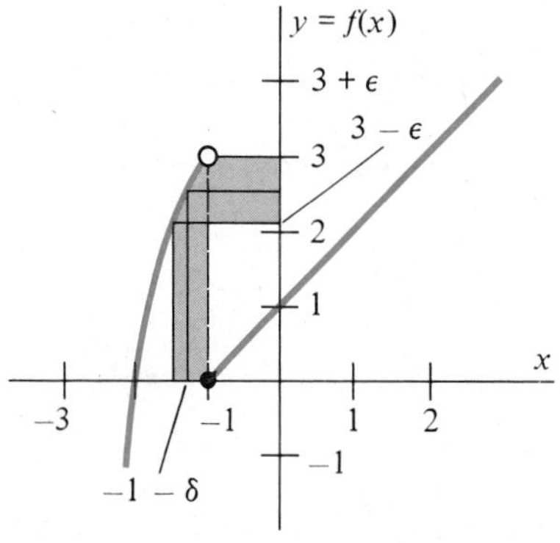

Figure 7.7 $|f(x) - 3| < \varepsilon$ if $-1 - \delta < x < -1$.

Finally, we introduce a formal definition of continuity. Recall (Definition 4) that the function $y = f(x)$ was said to be continuous at $x = a$ if

(i) $\lim_{x \to a} f(x)$ exists,
(ii) $f(a)$ exists, and
(iii) $\lim_{x \to a} f(x) = f(a)$.

A single modification of Definition 7 (limits) incorporates all three of these requirements.

DEFINITION 9 The function $f(x)$ is continuous at $x = a$ if, given $\epsilon > 0$, there can be found a number $\delta > 0$ so,

$$|f(x) - f(a)| < \epsilon \quad \text{whenever} \quad 0 \leq |x - a| < \delta.$$

The differences between Definition 7 and Definition 9 are:

(a) the number $L = \lim_{x \to a} f(x)$ in Definition 7 is replaced by the number $f(a)$ in Definition 9. This assures that (iii) holds.

(b) We can allow $|x - a| = 0$ in Definition 9 since, in this case, we simply have $|f(a) - f(a)| = 0 < \epsilon$.

When used in conjunction with Definition 7 (limits), our original definition of continuity, Definition 4, becomes mathematically precise. However, we will find Definition 9 more convenient to use in proving Theorem 11 in Section 2.8.

Exercise Set 2.7

Each of Exercises 1–7, presents a limit of the form $L = \lim_{x \to a} f(x)$.

For each given ϵ, find a number δ so that $|f(x) - L| < \epsilon$ whenever $0 < |x - a| < \delta$.

1. $\lim_{x \to 2} (2x + 5) = 9$

a. $\epsilon = 2$
b. $\epsilon = 0.4$
c. $\epsilon = 0.05$

2. $\lim_{x \to -3} (1 - 4x) = 13$

a. $\epsilon = 2$
b. $\epsilon = 0.4$
c. $\epsilon = 0.1$

3. $\lim_{x \to 2} (x^2 + 3) = 7$

a. $\epsilon = 2$
b. $\epsilon = 1$
c. $\epsilon = 0.3$

4. $\lim_{x \to -2} \dfrac{1}{x - 2} = -\dfrac{1}{4}$

a. $\epsilon = 2$
b. $\epsilon = 1$
c. $\epsilon = 0.5$

5. *(Calculator)* $\lim_{x \to \pi/2} \sin x = 1$

a. $\epsilon = 0.5$
b. $\epsilon = 0.2$
c. $\epsilon = 0.1$

6. $\lim_{x \to 1} \dfrac{x^2 - 1}{x - 1} = 2$

a. $\epsilon = 2$
b. $\epsilon = 0.8$
c. $\epsilon = 0.05$

7. *(Calculator)* $\lim_{x \to 4} \sqrt{2x + 1} = 3$

a. $\epsilon = 0.5$
b. $\epsilon = 0.2$
c. $\epsilon = 0.05$

In Exercises 8–19, use Definition 7 to prove the stated limit.

8. $\lim_{x \to 3} (x + 3) = 6$

9. $\lim_{x \to 1} (2x - 3) = -1$

10. $\lim_{x \to 4} (7 - 3x) = -5$

11. $\lim_{x \to 2} (3x + 5) = 11$

12. $\lim_{x \to 3} x^2 = 9$

13. $\lim_{x \to 2} 3x^2 = 12$

14. $\lim_{x \to 3} \dfrac{x^2 - 9}{x - 3} = 6$

15. $\lim_{x \to 2} \dfrac{x^2 - 4}{x - 2} = 4$

16. $\lim_{x \to 4} \dfrac{1}{x} = \dfrac{1}{4}$

17. $\lim_{x \to 2} \dfrac{1}{x + 3} = \dfrac{1}{5}$

18. $\lim_{x \to 4} \sqrt{x} = 2$

19. $\lim_{x \to 3} |x - 3| = 0$

In Exercises 20–23, use Definition 8 to prove the stated one-sided limit.

20. $\lim_{x \to 0^+} |x| = 0$

21. $\lim_{x \to 2^+} \sqrt{x - 2} = 0$

22. $\lim_{x \to 1^+} f(x) = 3$, where $f(x) = \begin{cases} 2x + 1, & x \geq 1 \\ -x, & x < 1 \end{cases}$

23. $\lim_{x \to -3^-} \dfrac{2(x - 3)}{|x - 3|} = -2$

2.8 PROOFS OF SEVERAL LIMIT THEOREMS (Optional)

The formal definition of limit (Definition 7) allows us to give rigorous proofs of all theorems of Sections 2.4–2.6 yet unproved. Several of these proofs are given here. Those remaining appear in Appendix II.

Theorem 1

(We prove parts (i) and (ii) here. Proof of parts (iii) and (iv) appear in Appendix II.)

Strategy

Proof of Theorem 1, part (i)

To show that

$$\lim_{x \to a} [f(x) + g(x)] = L + M$$

we must show that we can make the difference $|(f(x) + g(x)) - (L + M)|$ small, (i.e., less than ϵ). The idea of the proof is to use the fact that we can make both $|f(x) - L|$ and $|g(x) - M|$ small, since

$$L = \lim_{x \to a} f(x) \qquad \text{and} \qquad M = \lim_{x \to a} g(x).$$

We do this in the following way:

Break the difference $|(f(x) + g(x)) - (L + M)|$ into two pieces, each of which becomes small as $x \to a$.

$$\begin{aligned} |(f(x) + g(x)) - (L + M)| &= |(f(x) - L) + (g(x) - M)|. \\ &\le |f(x) - L| + |g(x) - M|. \end{aligned}$$

This inequality shows that if each of the differences on the right is less than $\epsilon/2$, the difference on the left will be less than ϵ. We therefore proceed formally as follows:

Use triangle inequality to bound the magnitude of the difference.

Given $\epsilon > 0$, choose δ_1 so that

$$|f(x) - L| < \epsilon/2 \qquad \text{whenever} \qquad 0 < |x - a| < \delta_1 \tag{1}$$

and choose δ_2 so that

$$|g(x) - M| < \epsilon/2 \qquad \text{whenever} \qquad 0 < |x - a| < \delta_2. \tag{2}$$

Then, if we take $\delta = \min\{\delta_1, \delta_2\}$ both (1) and (2) will hold when $0 < |x - a| < \delta$. Thus, we will have that

Apply Definition 7.

$$\begin{aligned} |(f(x) + g(x)) - (L + M)| &= |(f(x) - L) + (g(x) - M)| \\ &\le |f(x) - L| + |g(x) - M| \\ &< \epsilon/2 + \epsilon/2 = \epsilon \end{aligned}$$

whenever $0 < |x - a| < \epsilon$. This proves statement (i) according to Definition 7.

Strategy

Proof of Theorem 1, part (ii)

To show that

$$\lim_{x \to a} cf(x) = cL$$

we must show that we can make the difference $|cf(x) - cL|$ small. The idea here is to note that

Write the difference $|cf(x) - cL|$ as $|c| \cdot |f(x) - L|$.

$$\begin{aligned} |cf(x) - cL| &= |c(f(x) - L)| \\ &= |c| \cdot |f(x) - L| \end{aligned}$$

Use fact that

$|f(x) - L| \to 0$ as $x \to a$.

Since

$$L = \lim_{x \to a} f(x),$$

we can make the factor $|f(x) - L|$ small by requiring $|x - a|$ to be small. To handle the constant $|c|$ we need only require that $|f(x) - L| < \dfrac{\epsilon}{|c|}$. We proceed as follows:

Given $\epsilon > 0$, since

$$L = \lim_{x \to a} f(x)$$

Pick δ small enough to obtain

$|f(x) - L| < \dfrac{\epsilon}{|c|}$.

Apply Definition 7.

we may choose $\delta > 0$ sufficiently small so that $|f(x) - L| < \dfrac{\epsilon}{|c|}$ whenever $0 < |x - a| < \delta$. In this case we will have

$$\begin{aligned} |cf(x) - cL| &= |c(f(x) - L)| \\ &= |c| \cdot |f(x) - L| \\ &< |c| \cdot \left(\frac{\epsilon}{|c|}\right) \\ &= \epsilon. \end{aligned}$$

This shows that $cL = \lim_{x \to a} cf(x)$. ■

To prove the Pinching Theorem (Theorem 5), we use the fact that the inequality $|f(x) - L| < \epsilon$ is equivalent to the double inequality

$$-\epsilon < f(x) - L < \epsilon$$

for any $\epsilon > 0$.

Strategy

Since $f(x) \le h(x) \le g(x)$, we try to find a lower bound on $f(x)$ and an upper bound on $g(x)$.

Proof of Theorem 5

Since

$$\lim_{x \to a} f(x) = L,$$

given $\epsilon > 0$ there is a number δ_1 so that when $0 < |x - a| < \delta_1$,

$$-\epsilon < f(x) - L < \epsilon$$

or, adding L to each term, so that

$$L - \epsilon < f(x) < L + \epsilon. \tag{3}$$

Similarly, since

$$\lim_{x \to a} g(x) = L$$

there is a (possibly different) number δ_2 so that when $0 < |x - a| < \delta_2$,

$$L - \epsilon < g(x) < L + \epsilon. \tag{4}$$

Now since $f(x) \le h(x) \le g(x)$, we may use the left half of inequality (3) and the right half of inequality (4) to conclude that

This provides a two-sided bound on $h(x)$.

$$L - \epsilon < f(x) \le h(x) \le g(x) < L + \epsilon$$

when both $0 < |x - a| < \delta_1$ and $0 < |x - a| < \delta_2$.

This shows that if we take $\delta = \min\{\delta_1, \delta_2\}$ we will have

$-\epsilon < (h(x) - L) < \epsilon$ means $|h(x) - L| < \epsilon$.

$$L - \epsilon < h(x) < L + \epsilon$$

or,

$$-\epsilon < h(x) - L < \epsilon$$

whenever $0 < |x - a| < \delta$, as required. ■

The basic idea behind the proof of Theorem 11 is this: Since $f(u)$ is continuous at $u = L$, $f(u(x))$ is close to $f(L)$ whenever $u(x)$ is close to L, that is, when x is close to a. The situation is illustrated in Figure 8.1.

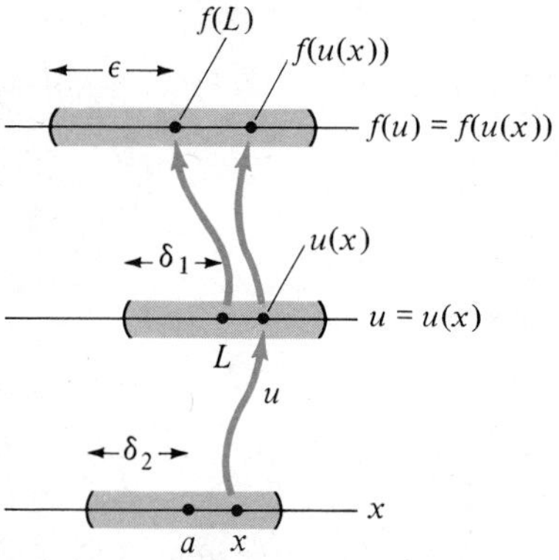

Figure 8.1 Composite function $f(u(x))$.

Strategy
The continuity of f guarantees $|f(u(x)) - f(L)|$ is small when $|u(x) - L|$ is small.

Proof of Theorem 11
Since f is continuous at L, given $\epsilon > 0$ there exists $\delta_1 > 0$ so that

$$|f(u) - f(L)| < \epsilon \qquad \text{whenever} \qquad |u - L| < \delta_1. \tag{5}$$

Also, since

$$\lim_{x\to a} u(x) = L,$$

The fact that $L = \lim_{x\to a} u(x)$ guarantees that $|u(x) - L|$ is small when $|x - a|$ is small.

there exists $\delta_2 > 0$ so that

$$|u(x) - L| < \delta_1 \qquad \text{whenever} \qquad 0 < |x - a| < \delta_2. \tag{6}$$

Combine these two observations.

Combining inequalities (5) and (6), with $u = u(x)$, shows that

$$|f(u(x)) - f(L)| < \epsilon \qquad \text{whenever} \qquad 0 < |x - a| < \delta_2.$$

This proves that $\lim_{x\to a} f(u(x)) = f(L)$. ■

SUMMARY OUTLINE OF CHAPTER 2

- Intuitively,

  $L = \lim_{x\to a} f(x)$ means that $f(x) \to L$ as $x \to a$;

  $L = \lim_{x\to a^+} f(x)$ means that $f(x) \to L$ as $x \to a$ from the right;

  $L = \lim_{x\to a^-} f(x)$ means that $f(x) \to L$ as $x \to a$ from the left.

- ***Theorem 1:***

  (i) $\lim_{x\to a} [f(x) + g(x)] = \lim_{x\to a} f(x) + \lim_{x\to a} g(x)$

  (ii) $\lim_{x\to a} [cf(x)] = c \cdot \lim_{x\to a} f(x)$

  (iii) $\lim_{x\to a} [f(x)g(x)] = [\lim_{x\to a} f(x)][\lim_{x\to a} g(x)]$

  (iv) $\lim_{x\to a} \dfrac{f(x)}{g(x)} = \dfrac{\lim_{x\to a} f(x)}{\lim_{x\to a} g(x)}, \quad \lim_{x\to a} g(x) \neq 0$

- ***Theorem 2:*** $\lim_{x\to a} p(x) = p(a)$ if $p(x)$ is a polynomial

- ***Theorem 3:*** $\lim_{x\to a} r(x) = r(a)$ if $r(x)$ is a rational function

- ***Theorem 4:*** $\lim_{x\to a} \sqrt[n]{x} = \sqrt[n]{a}$

- ***Theorem 5:*** If $f(x) \leq h(x) \leq g(x)$ and $\lim_{x\to a} f(x) = \lim_{x\to a} g(x) = L$, then $\lim_{x\to a} h(x) = L$.

- The function $y = f(x)$ is continuous at $x = a$ if

  (i) $f(a)$ exists,
  (ii) $\lim_{x\to a} f(x)$ exists, and
  (iii) $f(a) = \lim_{x\to a} f(x)$.

- ***Theorem 6:*** (Intermediate Value Theorem): If $f(x)$ is continuous and if $f(a) < d < f(b)$, then there is a number c between a and b with $f(c) = d$.

- ***Theorem 7:*** Every polynomial is continuous for all values of x.

- ***Theorem 8:*** Each rational function $r(x) = \dfrac{p(x)}{q(x)}$ is continuous for all values of x with $q(x) \neq 0$.

- ***Theorem 9:*** The function $f(x) = x^{n/m}$ is continuous, where defined.

- ***Theorem 10:*** Sums, multiples, products, and quotients of continuous functions are continuous (where denominators are not zero).

- ***Theorem 11:*** $\lim_{x\to a} f(u(x)) = f(\lim_{x\to a} u(x))$ if f is continuous.

- ***Theorem 12:*** Compositions of continuous functions are continuous.

- ***The formal definition of limit:*** $L = \lim_{x\to a} f(x)$ if and only if given $\epsilon > 0$ there exists a number $\delta > 0$ so that

  $|f(x) - L| < \epsilon \quad$ whenever $\quad 0 < |x - a| < \delta.$

- ***The formal definition of continuity:*** The function $f(x)$ is continuous at $x = a$ if given $\epsilon > 0$ there exists a number $\delta > 0$ so that

  $|f(x) - f(a)| < \epsilon \quad$ whenever $\quad |x - a| < \delta.$

REVIEW EXERCISES—CHAPTER 2

In Exercises 1–32, find the indicated limit.

1. $\lim_{x\to 2} (3x - 1)$

2. $\lim_{x\to 2} \dfrac{2x - 1}{x + 6}$

3. $\lim_{x\to 3} (x^4 - 4)$

4. $\lim_{x\to -5} \dfrac{x^2 - 25}{x + 5}$

5. $\lim_{x\to 3/2} \dfrac{4x^2 - 9}{2x - 3}$

6. $\lim_{x\to 3} \dfrac{x - 3}{x^2 - 9}$

7. $\lim_{x\to -2} \dfrac{x^2 + x - 2}{x + 2}$

8. $\lim_{x\to 1} \dfrac{x^2 + 6x - 7}{x - 1}$

9. $\lim_{x\to 1} \dfrac{x^2 + 2x - 3}{x^2 + x - 2}$

10. $\lim_{x\to -2} \dfrac{x^2 + 2x - 3}{x^2 + x - 2}$

11. $\lim_{t\to -1} \sqrt{\dfrac{1 - t^2}{1 + t}}$

12. $\lim_{x\to 8} \sqrt{\dfrac{x - 7}{x + 2}}$

13. $\lim_{x\to 0} \dfrac{\tan x}{\sin x}$

14. $\lim_{x\to \pi/2} \dfrac{1 - \sin^2 x}{\cos x}$

15. $\lim_{x\to 0} \dfrac{(2 + x)^2 - 4}{x}$

16. $\lim_{x\to 0} \dfrac{\sqrt{4 + x} - 2}{x}$

17. $\lim_{x\to 0} \dfrac{\sqrt[3]{x + 1} - 1}{x}$

18. $\lim_{x\to 4} \dfrac{x^2 - 16}{x - 4}$

19. $\lim_{x\to 2} \dfrac{3x^3 + x + 1}{1 - x^3}$

20. $\lim_{x\to 0} \dfrac{\sqrt{4x^4 + x}}{x^2 + 2}$

21. $\lim_{x\to 0} \dfrac{3x + 5x^2}{x}$

22. $\lim_{x\to 1} \dfrac{1 - x^3}{1 - x}$

23. $\lim_{x\to -1} \dfrac{1 - x}{1 + x}$

24. $\lim_{x\to 0} \sqrt{4x^4 + x}$

25. $\lim_{x\to 3^+} \frac{x^2}{x-3}$

26. $\lim_{x\to 3^-} \frac{x^2}{x-3}$

27. $\lim_{x\to 1^+} \frac{|x-1|}{x-1}$

28. $\lim_{x\to 1^-} \frac{x-1}{|x-1|}$

29. $\lim_{x\to 0^+} \frac{2+\sqrt{x}}{2-\sqrt{x}}$

30. $\lim_{x\to 1^-} \sqrt{1-[x]}$

31. $\lim_{x\to 1^+} \sqrt{1-[x]}$

32. $\lim_{x\to 0^+} \frac{x-[x]}{x}$

In Exercises 33–42, find the intervals on which the given function is continuous.

33. $y = \frac{x^2-7}{x-2}$

34. $f(x) = \frac{x+2}{x^2-x-6}$

35. $f(x) = \frac{x}{|x|}$

36. $y = \sec x$

37. $y = \frac{\sin x}{x}$

38. $f(x) = \begin{cases} x - x^2, & x \le 2 \\ -x, & x > 2 \end{cases}$

39. $y = \begin{cases} \sin x, & x \le \pi/4 \\ \cos x, & x > \pi/4 \end{cases}$

40. $f(x) = \begin{cases} \sqrt{x}, & x \le 4 \\ x - 1, & x > 4 \end{cases}$

41. $f(x) = \begin{cases} 1 - t, & t < -2 \\ -t, & -2 \le t \le 1 \\ 1 - 2t^2, & 1 < t \end{cases}$

42. $f(x) = \begin{cases} \frac{1}{1-x}, & x \ne 1 \\ 0, & x = 1 \end{cases}$

43. Prove that there exists a number x_0 so that $x_0^2 - 7x_0 + 1 = 3$.

44. Prove that there exist two distinct real numbers x_0 and x_1 so that $3 - x_0^6 = 0.5 = 3 - x_1^5$.

45. The function $f(x) = \begin{cases} 4 - x + x^3, & x \le 1 \\ 9 - ax^2, & x > 1 \end{cases}$ is continuous at $x = 1$. Find a.

In Exercises 46–51, let $f(x) = \begin{cases} 2, & x \le -1 \\ \sqrt{x+2}, & -1 < x < 2 \\ \cos \pi x, & 2 < x \end{cases}$

$$g(x) = \begin{cases} \tan x, & x \le 0 \\ x \sin x, & 0 \le x \le \pi/2 \\ \frac{2x}{\pi}, & \pi/2 < x \end{cases}$$

Find:

46. $\lim_{x\to -1^-} f(x)$

47. $\lim_{x\to 2^-} f(x)$

48. $\lim_{x\to 2^+} f(x)$

49. $\lim_{x\to 0^-} g(x)$

50. $\lim_{x\to 0^+} g(x)$

51. $\lim_{x\to \pi/2^-} g(x)$

52. Suppose $1 + 4x - x^2 \le f(x) \le x^2 - 4x + 9$ for $x \ne 2$. Find $\lim_{x\to 2} f(x)$.

53. Suppose $1 - |x| \le f(x) \le \sec x$ for $-1/2 \le x \le 1/2$. Find $\lim_{x\to 0} f(x)$.

In Exercises 54–59, prove that the stated limit is correct.

54. $\lim_{x\to 3} 2x + 3 = 9$

55. $\lim_{x\to 1} 4x - 3 = 1$

56. $\lim_{x\to 2} x^2 + 3 = 7$

57. $\lim_{x\to 4} \frac{1}{x} = \frac{1}{4}$

58. $\lim_{x\to 2} x^2 + 4x = 12$

59. $\lim_{x\to 3} \sqrt{x+1} = 2$

60. True or false? If $f(x) \le h(x) \le g(x)$ and both $f(x)$ and $g(x)$ are continuous at $x = a$, then $h(x)$ is continuous at $x = a$.

61. Prove that if $\lim_{x\to a} f(x)$ exists, it is unique. In other words, prove that if $\lim_{x\to a} f(x) = L$ and $\lim_{x\to a} f(x) = M$, then $L = M$.

62. Prove that if $f(x) \ge g(x)$ for all x and if $\lim_{x\to a} g(x) = L$, then $\lim_{x\to a} f(x) \ge L$ if this limit exists.

63. Prove that if $\lim_{x\to a} f(x) = L$, then $\lim_{x\to a^+} f(x) = L$.

64. Prove that if $\lim_{x\to a^+} f(x) = L$ and $\lim_{x\to a^-} f(x) = L$, then $\lim_{x\to a} f(x) = L$.

65. If $f(x)$ is continuous at $x = a$ but $(f + g)(x)$ is discontinuous at $x = a$, what can you say about the continuity of $g(x)$ at $x = a$?

Pierre de Fermat

Isaac Barrow

Marquis de l'Hôpital

Augustin-Louis Cauchy

UNIT 2

DIFFERENTIATION

The Origins of Differentiation

Isaac Newton wrote, "If I seem to have seen further than others, it is because I stood on the shoulders of giants." He was referring to the several mathematicians who actually were performing differentiations and integrations in the years before Newton and Leibniz unified calculus.

One of the principal contributors was Pierre de Fermat (1601–1665). A lawyer by profession, he was an amateur mathematician in the true sense of the word: a lover of mathematics. He was a co-inventor with René Descartes (1596–1650) of analytic geometry, and made many important discoveries in the theory of numbers. He was the first person to develop a procedure for solving maximum and minimum problems that amounted to finding a derivative and setting this equal to zero, as we shall be doing in Chapters 4 and 5. His work was a direct inspiration to Newton.

Laplace called Fermat the discoverer of differential calculus. As early as 1629 he was thinking of the process. Although he was hampered by a lack of the limit concept and of logical procedures, and although the notation was cumbersome, he essentially differentiated. This process of determining extreme values is sometimes called Fermat's Method. Fermat also found the tangent lines to curves of the form $y = f(x)$, noting that the process was similar to his maximum-minimum method.

Fermat determined volumes, centers of gravity, and the lengths of curves, but he did not realize something that we shall uncover in Chapter 6—the inverse nature of the differentiation and integration processes. He was also very reluctant to publish his results, often scribbling them in the margins of books, or sending important material to friends without keeping a copy. He thus lost out as an acknowledged creator of the calculus.

Isaac Barrow (1630–1677) was a teacher of Newton who came closer than anyone else before Newton and Leibniz to the complete concept of calculus. In his *Lectiones opticae et geometricae* (1669) he performed differentiations much as we do today, but with differential triangles. His results were correct, but he did not have a theory of limits to put his work on a sound basis. He was the first Lucasian Professor of Mathematics at Cambridge, but he resigned after six years in favor of his own student, Isaac Newton, whom Barrow recognized to be an even greater mathematician. Newton was then 27 years old.

Rolle's Theorem, met in this unit, is named after Michel Rolle (1652–1719). This French mathematician was at first a strong antagonist to the "new" calculus of Newton and Leibniz. He asserted that it was not based on logical grounds, and that it gave erroneous results. He called the calculus "a collection of ingenious fallacies." In later life, however, he became convinced of its validity. The theorem that bears his name, and which is used as a basis for proving the Mean Value Theorem, appears only coincidentally in his writings about finding approximate solutions to equations; Rolle used it only in connection with polynomial functions, although we now know that it is much more widely applicable.

A very useful theorem encountered in this unit, l'Hôpital's Rule, is named after a French nobleman, Guillaume François Antoine de l'Hôpital (1661–1704). The Marquis de l'Hôpital wrote the first textbook on differential calculus, *Analyse des infiniment petits pour l'intelligence des lignes courbés,* in 1696. The so-called rule, which enables one to find the limit of a quotient whose numerator and denominator both tend to zero, appears in his book. However, the rule and much of the other material in the book was actually the work of Johann Bernoulli, l'Hôpital's teacher. They had an interesting agreement according to which, in return for a regular salary, Bernoulli agreed to turn over to l'Hôpital all of his mathematical discoveries, which l'Hôpital was entitled to claim as his own. L'Hôpital had plans for writing what would also have been the first text in integral calculus, but abandoned this project when Leibniz told him that he was planning such a work.

A nineteenth century French mathematician, Augustin-Louis Cauchy (1789–1857), is credited with bringing present-day standards of rigor to calculus. In particular, he defined "limit" in a more precise way than had previously been done, and formally defined "continuous function." Cauchy took the lead in defining the integral as the limit of a sum, rather than as the inverse of the differentiation process. He also first formally defined the derivative as the limit of the difference quotient, as it is done in today's textbooks. Cauchy was a brilliant and prolific creator of mathematics. He published nearly eight hundred papers and several books, including a work on integrals that he wrote at age 25, which served as the basis for modern complex variable theory.

Pronunciation Guide

Pierre de Fermat (Pee-air' day Fair-mah)

René Descartes (Reh-nay' Day-cart')

Michel Rolle (Mi-shell' Roll)

Guillaume de l'Hôpital (Zhe-yome' day Lo' pi-tahl')

Johann Bernoulli (Yo'hahn Burr-noo'lee)

Augustin-Louis Cauchy (Aw-gus-tahn'-Loo-ee' Ko-she)

(Photographs from the David Eugene Smith Papers, Rare Book and Manuscript Library, Columbia University.)

CHAPTER 3

THE DERIVATIVE

3.1 INTRODUCTION

The statement of the Tangent Line Problem in Chapter 2 led to the definition of the **derivative** of a function, which is defined as follows.

DEFINITION 1

The **derivative** of the function $f(x)$ at $x = x_0$ is the number

$$f'(x_0) = \lim_{\Delta x \to 0} \frac{f(x_0 + \Delta x) - f(x_0)}{\Delta x}$$

provided this limit exists.

We say that the function $f(x)$ is **differentiable** at x_0 if the limit $f'(x_0)$ in Definition 1 exists. We may summarize the discussion of Section 2.2 by saying that if $f(x)$ is differentiable at $x = x_0$, the slope of the line tangent to the graph of $y = f(x)$ at $(x_0, f(x_0))$ is the derivative, $f'(x_0)$. More generally, the derivative $f'(x)$ may be interpreted as a **function** whose value at each number x in its domain is the slope of the line tangent to the graph of $y = f(x)$ at $(x, f(x))$.

Throughout Chapter 2, the only method available to you for calculating derivatives was to evaluate the limit of the difference quotient in Definition 1. This method is both tedious and time-consuming, and the chance for error in the lengthy algebraic manipulations is considerable. That is why our next objective is to establish a list of rules by which derivatives can be more easily calculated. We do this in Chapter 3, as well as developing a fuller theory of the derivative, with which we will be able to solve a wide variety of problems in Chapters 4 and 5.

Throughout this chapter we shall make frequent use of the relationship between differentiability and continuity, as described in the following theorem.

THEOREM 1

Let the function $f(x)$ be defined on an open interval containing x_0. If $f'(x_0)$ exists, then $f(x)$ is continuous at $x = x_0$.

Proof: To prove that $f(x)$ is continuous at x_0, we must show that $\lim_{x \to x_0} f(x) = f(x_0)$ (see Definition 3, Chapter 2). If we let $\Delta x = x - x_0$, then $x = x_0 + \Delta x$, so this is the same as showing that

$$\lim_{\Delta x \to 0} f(x_0 + \Delta x) = f(x_0). \tag{1}$$

Using the hypothesis that $f'(x_0)$ exists, and the "Product Rule" for limits (Theorem 1, part [iii], Chapter 2), we can write that

$$\begin{aligned}\lim_{\Delta x \to 0} f(x_0 + \Delta x) - f(x_0) &= \lim_{\Delta x \to 0} [f(x_0 + \Delta x) - f(x_0)] \\ &= \lim_{\Delta x \to 0} \left[\frac{f(x_0 + \Delta x) - f(x_0)}{\Delta x}\right](\Delta x) \\ &= \left(\lim_{\Delta x \to 0} \left[\frac{f(x_0 + \Delta x) - f(x_0)}{\Delta x}\right]\right) \cdot \left(\lim_{\Delta x \to 0} \Delta x\right) \\ &= f'(x_0) \cdot 0 \\ &= 0,\end{aligned}$$

which proves that equation (1) is true. ■

Exercise Set 3.1

In Exercises 1–20, use Definition 1 to find $f'(x)$.

1. $f(x) = 3x + 2$
2. $f(x) = x^2 - 7$
3. $f(x) = 2x^3 + 3$
4. $f(x) = 1 - x^3$
5. $f(x) = ax^3 + bx^2 + cx + d$
6. $f(x) = \dfrac{1}{x - 1}$
7. $f(x) = \dfrac{1}{2x + 3}$
8. $f(x) = \dfrac{1}{x^2 - 9}$
9. $f(x) = \sqrt{x + 1}$
10. $f(x) = \sqrt{2x + 3}$
11. $f(x) = \dfrac{1}{\sqrt{x + 1}}$
12. $f(x) = \dfrac{1}{ax + b}$
13. $f(x) = \dfrac{1}{x^2}$
14. $f(x) = x^4$
15. $f(x) = (x + 3)^3$
16. $f(x) = (x - 1)^2$
17. $f(x) = \dfrac{1}{\sqrt{x + 5}}$
18. $f(x) = \dfrac{1}{(x + 2)^2}$
19. $f(x) = \dfrac{3}{(x - 1)^2}$
20. $f(x) = \dfrac{-2}{\sqrt{x + 1}}$
21. Find an equation for the line tangent to the graph of $y = \dfrac{1}{2x + 3}$ at the point $(0, 1/3)$.
22. Find an equation for the line tangent to the graph of $f(x) = \sqrt{x + 1}$ at the point $(3, 2)$.
23. A **normal** to a curve at point P is a line through P perpendicular to the tangent. Find an equation for the normal to the graph of $f(x) = x^2 - 7$ at the point $P = (3, 2)$.
24. The graph of $y = x^2$ has two tangents that pass through $(0, -4)$. Find equations for these lines.
25. Find the point where the tangent to the graph of $y = x^2 + 4x + 4$ has slope 4.
26. Find a so that the graph of $y = 2 - ax^2$ has a tangent with slope 6 at $x = -1$.
27. Let $f(x) = ax^2 + bx + 3$. Find a and b so that the tangent to the graph of $y = f(x)$ at $(1, 5)$ has slope 1.
28. Find the constants a and b so that the graph of $y = ax^2 + b$ has tangent $y = 4x$ at the point $P = (1, 4)$.
29. *(Calculator)* Sketch the graph of $f(x) = \sin x$ for $0 \le x \le 2\pi$. At each of the values $x = 0, \pi/4, \pi/2, 3\pi/4, \ldots, 2\pi$ estimate the slope of the tangent by evaluating the difference quotient with $\Delta x = \pm 0.5, \pm 0.1, \pm 0.05$, and ± 0.01. Using these results sketch the graph of the "slope function" $f'(x)$ (see Section 2.2).
30. *(Calculator)* For $f(x) = \cos x$, estimate the value of $f'(x)$ at $x = 0, \pi/4, \pi/2, 3\pi/4, \ldots, 2\pi$ by evaluating the difference quotient with $\Delta x = \pm 0.5, \pm 0.1, \pm 0.05$, and ± 0.01. Compare these estimates with the values of $g(x) = \sin x$ at the same angles. What relationship do you observe?

31. *(Computer)* Program 2 in Appendix I is a BASIC Program which prints out values of the difference quotient

$$D = \frac{\sin(y) - \sin(x)}{y - x}$$

for the sine function. Use this program to approximate $f'(x)$ for $f(x) = \sin x$ at various values of x.

32. *(Computer)* Note that Program 2 computes difference quotients as $y \to x$ from the right only. How could Program 2 be modified so as to compute difference quotients as $y \to x$ from the left?

33. *(Computer)* Modify Program 2 to compute difference quotients for the function $f(x) = \sqrt{1 - x^2}$.

34. Show that the function $f(x) = |x|$ is continuous at $x = 0$, but that $f'(0)$ fails to exist. Explain why this shows that the converse of Theorem 1 is false.

3.2 RULES FOR CALCULATING DERIVATIVES

In this section we begin the task of developing a list of rules by which derivatives may be calculated. Before doing so, however, we introduce an additional symbol for the derivative:

$$\frac{d}{dx} f(x) \quad \text{or} \quad \frac{df}{dx} \quad \text{means} \quad f'(x)$$

In other words, the symbol $\frac{d}{dx}$ means "the derivative with respect to x of." We shall see later why the notation suggests a ratio, but for now we shall be concerned only with the above interpretation. The notation $\frac{d}{dx}$ is usually credited to the mathematician Gottfried Leibniz (1646–1716).

Example 1 Using Leibniz notation, we may rewrite two previous results as

(a) $\frac{d}{dx}(x^2 - 7) = 2x$ (Exercise 2, Section 3.1.)

(b) $\frac{d}{dx}\left(\frac{1}{x-1}\right) = \frac{-1}{(x-1)^2}$ (Exercise 6, Section 3.1.) ■

Our first two results formalize what we have already discovered in Section 2.2: the derivative of a constant function is always zero, and the derivative of the linear function $f(x) = x$ is just the slope of its graph, 1.

THEOREM 2

The constant function $f(x) \equiv c$ is differentiable for all x, and $f'(x) = 0$. That is,

$$\frac{d}{dx}(c) = 0.$$

Proof: By hypothesis, $f(x + \Delta x) = c = f(x)$, so

$$f'(x) = \lim_{\Delta x \to 0} \frac{f(x + \Delta x) - f(x)}{\Delta x} = \lim_{\Delta x \to 0} \frac{c - c}{\Delta x} = \lim_{\Delta x \to 0} (0) = 0.$$ ■

THEOREM 3

The linear function $f(x) = x$ is differentiable for all x, and $f'(x) = 1$. That is,

$$\frac{d}{dx}(x) = 1.$$

Proof: For $f(x) = x$, $f(x + \Delta x) = x + \Delta x$, so

$$\begin{aligned} f'(x) = \lim_{\Delta x \to 0} \frac{f(x + \Delta x) - f(x)}{\Delta x} &= \lim_{\Delta x \to 0} \frac{(x + \Delta x) - x}{\Delta x} \\ &= \lim_{\Delta x \to 0} \left(\frac{\Delta x}{\Delta x}\right) \\ &= \lim_{\Delta x \to 0} (1) \\ &= 1. \end{aligned}$$

■

Proceeding as in the proof of Theorem 3 we can demonstrate that

for $f(x) = x^2$, $f'(x) = 2x$ (Example 5, Section 2.2),
for $f(x) = x^3$, $f'(x) = 3x^2$ (Exercise 14, Section 2.2),
for $f(x) = x^4$, $f'(x) = 4x^3$ (Exercise 17, Section 2.2),
for $f(x) = x^5$, $f'(x) = 5x^4$ (Left to you), and so forth.

We handle all these cases at once with the following theorem.

THEOREM 4
Power Rule

Let n be a positive integer. The function $f(x) = x^n$ is differentiable for all x, and $f'(x) = nx^{n-1}$. In Leibniz notation,

$$\frac{d}{dx}x^n = nx^{n-1}, \qquad n = 1, 2, \ldots .$$

Proof: We recall that by the Binomial Theorem we can write

$$(x + \Delta x)^n = x^n + nx^{n-1}\,\Delta x + \frac{n(n-1)}{2}x^{n-2}\,\Delta x^2 + \cdots + \frac{n(n-1)}{2}x^2\,\Delta x^{n-2} + nx\,\Delta x^{n-1} + \Delta x^n.$$

We can therefore write the required difference quotient as follows:

$$\frac{f(x + \Delta x) - f(x)}{\Delta x} = \frac{(x + \Delta x)^n - x^n}{\Delta x} = nx^{n-1} + \{\text{terms with factors of } \Delta x^P,\ P \geq 1\}.$$

We now observe that in the limit as $\Delta x \to 0$, all terms involving factors of Δx^P will approach zero. Thus, we conclude that

$$f'(x) = \lim_{\Delta x \to 0} \frac{(x + \Delta x)^n - x^n}{\Delta x} = nx^{n-1}.$$

(This proof must be regarded as intuitive rather than formal, since we have been vague about the nature of the "terms with factors of Δx" and their limits as $\Delta x \to 0$. A formal proof using the Principle of Mathematical Induction is outlined in Exercise 76.) ■

Example 2

(a) For $f(x) = x^{27}$, $\quad f'(x) = 27x^{26}$.

(b) $\dfrac{d}{dx}(x^{3n}) = 3nx^{3n-1}$, $\quad n$ a positive integer. ■

The next several results will enable us to calculate derivatives for functions that are "built up" from simpler pieces, each of whose derivative is known. Since the derivative is a limit, it should not be surprising that the following theorem guarantees that derivatives have two properties in common with limits (see Theorem 1, Chapter 2).

THEOREM 5

If $f(x)$ and $g(x)$ are differentiable at x and c is any constant, then $(f + g)(x)$ and $cf(x)$ are differentiable at x, and

(i) $(f + g)'(x) = f'(x) + g'(x)$, $\quad$ and
(ii) $(cf)'(x) = cf'(x)$.

In Leibniz notation the statements become

(i) $\dfrac{d}{dx}(f + g) = \dfrac{df}{dx} + \dfrac{dg}{dx}$.

(ii) $\dfrac{d}{dx}(cf) = c\dfrac{df}{dx}$.

Strategy

Proof of Theorem 5

Write down definition of $(f + g)'$, using Definition 1.

$$(f + g)'(x) = \lim_{\Delta x \to 0} \frac{(f + g)(x + \Delta x) - (f + g)(x)}{\Delta x}$$

Apply definition that $(f + g)(a) = f(a) + g(a)$.

$$= \lim_{\Delta x \to 0} \frac{[f(x + \Delta x) + g(x + \Delta x)] - [f(x) + g(x)]}{\Delta x}$$

Collect terms in f and terms in g.

$$= \lim_{\Delta x \to 0} \frac{[f(x + \Delta x) - f(x)] + [g(x + \Delta x) - g(x)]}{\Delta x}$$

Apply Theorem 1, Chapter 2.

$$= \lim_{\Delta x \to 0} \frac{f(x + \Delta x) - f(x)}{\Delta x} + \lim_{\Delta x \to 0} \frac{g(x + \Delta x) - g(x)}{\Delta x}$$

Apply Definition 1.

$$= f'(x) + g'(x).$$

This proves part (i). The proof of part (ii) is similar and is left as an exercise. ■

Example 3 Using the Power Rule and Theorem 5 we find that

(a)
$$\begin{aligned}
\frac{d}{dx}(3x + 4) &= \frac{d}{dx}(3x) + \frac{d}{dx}(4) && \text{(Theorem 5, part (i))}\\
&= 3\frac{d}{dx}(x) + \frac{d}{dx}(4) && \text{(Theorem 5, part (ii))}\\
&= 3 \cdot 1 + 0 && \text{(Power Rule)}\\
&= 3.
\end{aligned}$$

That is, for $f(x) = 3x + 4$, $\quad f'(x) = 3$.

(b) $\frac{d}{dx}(7x^4 - 5x^3) = 7 \cdot \frac{d}{dx}(x^4) - 5\frac{d}{dx}(x^3)$ (Theorem 5, parts (i) and (ii))

$= 7(4x^3) - 5(3x^2)$ (Power Rule)

$= 28x^3 - 15x^2.$

That is, for $f(x) = 7x^4 - 5x^3$, $f'(x) = 28x^3 - 15x^2$. ■

As Example 3 suggests, the Power Rule may be combined with Theorem 5 to produce a rule for differentiating any polynomial.

THEOREM 6
Polynomial Rule

The function $f(x) = a_nx^n + a_{n-1}x^{n-1} + \cdots + a_1x + a_0$ is differentiable for all values of x, and

$$f'(x) = na_nx^{n-1} + (n-1)a_{n-1}x^{n-2} + \cdots + 2a_2x + a_1.$$

Proof: This is left for you as Exercise 75. ■

Example 4 For $f(x) = 6x^5 - 3x^4 - 2x^3 + 4x^2 - 6x + 5$,

$$\begin{aligned} f'(x) &= (5 \cdot 6)x^4 - (4 \cdot 3)x^3 - (3 \cdot 2)x^2 + (2 \cdot 4)x - (1)(6) + 0 \\ &= 30x^4 - 12x^3 - 6x^2 + 8x - 6. \end{aligned}$$ ■

We can paraphrase Theorem 5, part (i), by saying that the derivative of the sum is the sum of the derivatives. This observation might lead you to suspect also that the derivative of a product of two functions would be the product of the individual derivatives. However, this is not true. A simple example is found by letting

$$f(x) = x \quad \text{and} \quad g(x) = x^3.$$

Then,

$$f'(x) = 1 \quad \text{and} \quad g'(x) = 3x^2.$$

Now the product is $h(x) = f(x)g(x) = x \cdot x^3 = x^4$, which has derivative $h'(x) = (f(x)g(x))' = 4x^3$. However, the product of the individual derivatives is $f'(x)g'(x) = (1)(3x^2) = 3x^2$, not $4x^3$.

The correct procedure for differentiating a product is given by the following theorem.

THEOREM 7
Product Rule

Let $f(x)$ and $g(x)$ be differentiable at x. Then the product function $f(x)g(x)$ is differentiable at x, and

$$[f(x)g(x)]' = f'(x)g(x) + f(x)g'(x)$$

or, in Leibniz notation,

$$\frac{d}{dx}(fg) = \frac{df}{dx} \cdot g + f \cdot \frac{dg}{dx}.$$

Proof: By the definition of the derivative

$$[f(x)g(x)]' = \lim_{\Delta x \to 0} \frac{f(x + \Delta x)g(x + \Delta x) - f(x)g(x)}{\Delta x}.$$

In order to factor the numerator above, we subtract and add the quantity $f(x)g(x + \Delta x)$ and use Theorem 1, Chapter 2 (on the algebra of limits) to obtain

$$\begin{aligned}
&[f(x)g(x)]' \\
&= \lim_{\Delta x \to 0} \frac{[f(x + \Delta x)g(x + \Delta x) - f(x)g(x + \Delta x)] + [f(x)g(x + \Delta x) - f(x)g(x)]}{\Delta x} \\
&= \lim_{\Delta x \to 0} \left[\left(\frac{f(x + \Delta x) - f(x)}{\Delta x}\right) g(x + \Delta x) + f(x)\left(\frac{g(x + \Delta x) - g(x)}{\Delta x}\right)\right] \\
&= \left[\lim_{\Delta x \to 0} \frac{f(x + \Delta x) - f(x)}{\Delta x}\right][\lim_{\Delta x \to 0} g(x + \Delta x)] + \\
&\qquad [\lim_{\Delta x \to 0} f(x)]\left[\lim_{\Delta x \to 0} \frac{g(x + \Delta x) - g(x)}{\Delta x}\right].
\end{aligned}$$

Now since the function $g(x)$ is differentiable at x it must be continuous at x, by Theorem 1. Thus $\lim_{\Delta x \to 0} g(x + \Delta x) = g(x)$. Since x is fixed in this argument, $\lim_{\Delta x \to 0} f(x) = f(x)$. We therefore conclude that

$$[f(x)g(x)]' = f'(x)g(x) + f(x)g'(x)$$

and the proof is complete. ■

Example 5 For $f(x) = (2x + 7)(x - 9)$, find $f'(x)$.

Solution: One way to work this problem is to first multiply the two binomial terms,

$$f(x) = (2x + 7)(x - 9) = 2x^2 - 11x - 63,$$

and then apply the Polynomial Rule:

$$f'(x) = 4x - 11.$$

Another approach is to begin by applying the Product Rule:

$$\begin{aligned}
f'(x) &= \left[\frac{d}{dx}(2x + 7)\right](x - 9) + (2x + 7)\left[\frac{d}{dx}(x - 9)\right] \\
&= 2(x - 9) + (2x + 7)(1) \\
&= 4x - 11.
\end{aligned}$$

Of course, the result of either approach must be the same. ■

Example 6 For the function $f(x) = (3x^3 - 6x)(9x^4 + 3x^3 + 3)$, it is easier to use the Product Rule to calculate the derivative:

$$\begin{aligned}
f'(x) &= \left[\frac{d}{dx}(3x^3 - 6x)\right](9x^4 + 3x^3 + 3) + (3x^3 - 6x)\left[\frac{d}{dx}(9x^4 + 3x^3 + 3)\right] \\
&= (9x^2 - 6)(9x^4 + 3x^3 + 3) + (3x^3 - 6x)(36x^3 + 9x^2) \\
&= 189x^6 + 54x^5 - 270x^4 - 72x^3 + 27x^2 - 18.
\end{aligned}$$

■

Our final theorem for Section 3.2 is the rule for differentiating quotients. It is even more surprising than the rule for products.

THEOREM 8
Quotient Rule

Let $f(x)$ and $g(x)$ be differentiable at x with $g(x) \neq 0$. Then the quotient $f(x)/g(x)$ is differentiable at x and

$$\left[\frac{f(x)}{g(x)}\right]' = \frac{f'(x)g(x) - f(x)g'(x)}{[g(x)]^2}$$

or, in Leibniz notation,

$$\frac{d}{dx}\left(\frac{f}{g}\right) = \frac{g\dfrac{df}{dx} - f\dfrac{dg}{dx}}{g^2}.$$

Proof: By the definition of the derivative we have

$$\begin{aligned}\left[\frac{f(x)}{g(x)}\right]' &= \lim_{\Delta x \to 0} \frac{\dfrac{f(x+\Delta x)}{g(x+\Delta x)} - \dfrac{f(x)}{g(x)}}{\Delta x} \\ &= \lim_{\Delta x \to 0} \frac{f(x+\Delta x)g(x) - f(x)g(x+\Delta x)}{\Delta x g(x)g(x+\Delta x)}\end{aligned}$$

We proceed to subtract and add the expression $f(x)g(x)$ in the numerator so as to be able to factor and again use the algebra of limits, to obtain

$$\left[\frac{f(x)}{g(x)}\right]'$$

$$= \lim_{\Delta x \to 0} \frac{[f(x+\Delta x)g(x) - f(x)g(x)] + [f(x)g(x) - f(x)g(x+\Delta x)]}{\Delta x g(x)g(x+\Delta x)}$$

$$= \lim_{\Delta x \to 0} \frac{\left[\dfrac{f(x+\Delta x) - f(x)}{\Delta x}\right]g(x) - f(x)\left[\dfrac{g(x+\Delta x) - g(x)}{\Delta x}\right]}{g(x)g(x+\Delta x)}$$

$$= \frac{\left[\displaystyle\lim_{\Delta x \to 0} \frac{f(x+\Delta x) - f(x)}{\Delta x}\right]g(x) - f(x)\left[\displaystyle\lim_{\Delta x \to 0} \frac{g(x+\Delta x) - g(x)}{\Delta x}\right]}{g(x)[\displaystyle\lim_{\Delta x \to 0} g(x+\Delta x)]}$$

$$= \frac{f'(x)g(x) - f(x)g'(x)}{[g(x)]^2}$$

(Notice that we have used the continuity of $g(x)$ in the denominator in the step above, just as we did in the proof of the Product Rule.) ■

Example 7 Find $f'(x)$ for $f(x) = \dfrac{3x^2 + 7x + 1}{9 - x^3}$.

Solution: By the Quotient Rule we have

$$f'(x) = \frac{\left[\dfrac{d}{dx}(3x^2 + 7x + 1)\right](9 - x^3) - (3x^2 + 7x + 1)\left[\dfrac{d}{dx}(9 - x^3)\right]}{(9 - x^3)^2}$$

$$= \frac{(6x + 7)(9 - x^3) - (3x^2 + 7x + 1)(-3x^2)}{(9 - x^3)^2}$$

$$= \frac{3x^4 + 14x^3 + 3x^2 + 54x + 63}{x^6 - 18x^3 + 81}.$$ ■

Example 8 Find a rule for differentiating $f(x) = x^{-n}$ where n is a positive integer.

Solution: We write $f(x) = \dfrac{1}{x^n}$ and apply the Quotient Rule and the Power Rule to obtain

$$f'(x) = \frac{0 \cdot x^n - 1 \cdot nx^{n-1}}{x^{2n}}$$
$$= -nx^{n-1-2n}$$
$$= -nx^{-n-1}.$$ ■

Thus, Example 8 shows that the Power Rule holds for any integer n regardless of sign, that is,

$$\text{If } f(x) = x^n \text{ then } f'(x) = nx^{n-1} \text{ for any integer } n. \tag{1}$$

Example 9 Find the equation of the line tangent to the graph of $y = x^2 + x^{-2}$ at the point $(1, 2)$.

Strategy

Find $f'(x)$ for $f(x) = x^2 + x^{-2}$ using the Power Rule in (1).

Slope is $f'(1)$.

Solution

For

$$f(x) = x^2 + x^{-2},$$
$$f'(x) = 2x - 2x^{-3}.$$

Then

$$f'(1) = 2 - 2 = 0$$

so the desired line is horizontal. Since this line has slope zero and contains $(1, 2)$ its equation is $y = 2$. ■

Exercise Set 3.2

In Exercises 1–34, find the derivative of the given function.

1. $f(x) = 8x^3 - x^2$

2. $f(x) = x - x^5$

3. $f(x) = ax^3 + bx$

4. $f(x) = a^5 + 3a^2x^2 + x^3$

5. $f(x) = \dfrac{x^2 + 5}{2}$

6. $f(x) = \dfrac{2}{3}x^3 + \dfrac{1}{2}x^2 + x$

7. $f(x) = (x - 1)(x + 2)$

8. $f(x) = (x^2 - 1)(2 - x)$

9. $f(x) = (3x^2 - 8x)(x^2 + 2)$

10. $f(x) = (x^2 + x + 1)(x + 1)$

11. $f(x) = (x^3 - x)^2$

12. $f(x) = \left(x^2 - \frac{3}{x^2}\right)^2$

13. $f(x) = x(x^2 - 1)(3 - x^2)$

14. $f(x) = \frac{x + 2}{x - 2}$

15. $f(x) = \frac{x^2 - 6}{3 - x}$

16. $f(x) = \frac{(8x + 2)(x + 1)}{x - 3}$

17. $f(x) = \frac{x^4 + 4x + 4}{1 - x^3}$

18. $f(x) = \frac{(2x + 1)(3x + 2)}{(x + 1)(x - 1)}$

19. $f(x) = \frac{(x^2 + 7)(x + 2)}{x(x - 3)}$

20. $f(x) = \frac{x^2 - 4}{x + 2}$

21. $f(x) = (x - 2)\left(x + \frac{1}{x}\right)$

22. $f(x) = \frac{1}{(x - 3)^2}$

23. $f(x) = \left(1 + \frac{3}{x}\right)^2$

24. $f(x) = \left(\frac{x - 1}{x}\right)^3$

25. $g(t) = \frac{t}{t^2 + t + 1}$

26. $f(s) = \frac{1 - s}{(1 + s)^2}$

27. $f(x) = \left[\frac{x + 1}{x - 1}\right]^2$

28. $f(t) = \frac{t - 4 + t^2}{t^3 + 3t^2 + 3}$

29. $f(\theta) = \frac{\theta(\theta + 1)(\theta + 2)}{\theta + 3}$

30. $f(x) = \frac{1}{(x - 6)^3}$

31. $f(x) = \frac{ax + b}{cx^2 + d}$

32. $f(u) = (u^2 + 4)^3$

33. $f(x) = \frac{(3x^2 + 4)(6 - x^2)}{(x - 7)(3 - x)}$

34. $f(x) = ((x^2 + 1)^2 + 1)^2$

In each of Exercises 35–44, find $f'(x)$ in two ways—first by the Product Rule, then by first multiplying to eliminate the parentheses.

35. $f(x) = x(x + 1)$

36. $f(x) = (x + 2)(x - 1)^2$

37. $f(x) = (x^2 + 2)(x^2 - 2)$

38. $f(x) = \left(\frac{1}{x} + 1\right)\left(3 - \frac{2}{x^2}\right)$

39. $f(x) = (x^3 + 7)(3x^4 + x + 9)$

40. $f(x) = (1 - x^2)(1 + x^2)$

41. $f(x) = \left(\frac{1}{x + 1}\right)\left(\frac{2}{x + 2}\right)$

42. $f(s) = \left(s - \frac{3}{s^2}\right)\left(s + \frac{5}{s^2}\right)$

43. $f(u) = (u^2 + u + 1)(u^2 - u - 1)$

44. $f(x) = \left(\frac{x^2 + 7}{3 - x}\right)\left(\frac{x^4 + 6x - 2}{9 - x^3}\right)$

45. Use the Product Rule to establish the following formula for the derivative of the product of three functions:

$$(f(x)g(x)h(x))' = f'(x)g(x)h(x) + f(x)g'(x)h(x) + f(x)g(x)h'(x).$$

46. Write the differentiation rule in Exercise 45 in Leibniz notation.

In Exercises 47–52, use the result of Exercise 45 to find the derivative.

47. $f(x) = x(x + 1)(x + 2)$

48. $f(s) = (2s - 1)(s - 3)(s^2 + 4)$

49. $f(t) = (t^2 - 7)(3t^5 + t)(t^3 - 9)$

50. $f(x) = \left(\frac{1}{x}\right)\left(\frac{1}{x + 1}\right)\left(\frac{1}{x + 2}\right)$

51. $f(u) = (u^2 - 4)^3$

52. $f(x) = (2x^3 - 6x + 9)^3$

53. Use the Product Rule and the result of Exercise 45 to establish the following differentiation rules.
    a. $[f^2(x)]' = 2f(x)f'(x)$
    b. $[f^3(x)]' = 3f^2(x)f'(x)$

    Can you find a rule for differentiating $f^4(x)$? What about $f^n(x)$?

54. Find $f'(2)$ for $f(x)$ in Exercise 11.

55. Find $f'(-1)$ for $f(x)$ in Exercise 20.

56. Find $f'(3)$ for $f(x)$ in Exercise 24.

In Exercises 57–64, assume that $f(x)$ and $g(x)$ are differentiable at $x = 2$, and that $f(2) = 3$, $f'(2) = 1$, $g(2) = 7$, and $g'(2) = -3$. Find the value of the indicated derivative at $x = 2$.

57. $\frac{d}{dx}[f(x)g(x)]$

58. $\frac{d}{dx}[f^2(x)g(x)]$

59. $\frac{d}{dx}[f(x) + g(x)]$

60. $\frac{d}{dx}[f^3(x)]$

61. $\frac{d}{dx}\left[\frac{f(x)}{3g(x)}\right]$

62. $\frac{d}{dx}\left[\frac{2g(x) - f(x)}{f^2(x)}\right]$

63. $\frac{d}{dx}[3f(x) - g^3(x)]$

64. $\frac{d}{dx}\left[\frac{g^2(x)}{f(x) - g(x)}\right]$

In each of Exercises 65–70, find an equation for the line tangent to the graph of the given function at the given point.

65. $f(x) = 3x^3 - 7$ at the point $(1, -4)$

66. $f(x) = \frac{x-1}{x+1}$ at the point $(1, 0)$

67. $f(x) = (x^2 + x)(1 - x^3)$ at the point $(2, -42)$

68. $f(x) = \left(1 - \frac{1}{x}\right)^2$ at the point $(1, 0)$

69. $f(x) = \frac{1}{x^3 + x}$ at the point $\left(2, \frac{1}{10}\right)$

70. $f(x) = \left(x + \frac{1}{x}\right)^3$ at the point $\left(2, \frac{125}{8}\right)$

71. Find the constant a if the graph of $y = \frac{1}{ax + 2}$ has tangent $4y + 3x - 2 = 0$ at $(0, 1/2)$.

72. Find the constant b if the graph of $y = \frac{b}{x^2}$ has tangent $4y - bx - 21 = 0$ when $x = -2$.

73. Use the Product Rule to find a formula for the derivative of $f(x) = \sqrt[3]{x}$. (*Hint:* Use the fact that $f(x)f(x)f(x) = x$.)

74. Use the method of Exercise 73 to find a formula for the derivative of $f(x) = x^{1/n}$, $x \neq 0$, where n is a positive integer.

75. Prove the Polynomial Rule, Theorem 6.

76. Complete the steps in the following formal proof of Theorem 4 (The Power Rule):
   a. For $n = 1$, show that the function $f(x) = x$ is differentiable for all x and satisfies the equation $f'(x) = 1 \cdot x^{1-1} = x^0 = 1$. This shows that Theorem 4 holds for $n = 1$.
   b. Assume that Theorem 4 holds for $n = k$ where k is a positive integer greater than 1.
   c. Write $f(x) = x^{k+1} = x(x^k)$ and use assumption (b) and the Product Rule to show that $f(x)$ is differentiable for all x.
   d. Use assumption (b) and the Product Rule to show that $f'(x) = (k + 1)x^k$ for all x.
   e. Apply the Principle of Mathematical Induction to conclude that Theorem 4 holds for all integers $n \geq 1$.

3.3 THE DERIVATIVE AS VELOCITY

Before proceeding to develop additional rules for calculating derivatives, we pause here to note another interpretation of the derivative. In addition to giving the slope of the line tangent to the graph of a function, the derivative may be used to define the *velocity* of a moving object.

Imagine an object moving along a line, such as a jogger on a footpath or an automobile on a highway. In physics, we define the **velocity** of such an object by the equation

$$\text{velocity} = \frac{\text{change in distance}}{\text{change in time}}. \qquad (1)$$

Of course, what is meant by equation (1) is really an *average* velocity for the time period in question. For example, if a jogger wearing both a pedometer and a watch finds that she has traveled 12 kilometers in 45 minutes, the velocity associated with this time interval is

$$\text{velocity} = \frac{12 \text{ kilometers}}{3/4 \text{ hour}} = 16 \text{ kilometers per hour.}$$

However, at various times during the run the jogger will quite likely have been moving both faster and slower than 16 kilometers per hour.

There is a special setting in which we can use the theory of the derivative to define the velocity of an object *at each instant,* rather than having to settle for the average velocity over a finite time interval. First, the motion of the object must be along a line (we call this **rectilinear** motion), rather than along a general curve. Second, we must have available a **position function,** $s(t)$, giving the location of the object along the line at each time t (see Figure 3.1).

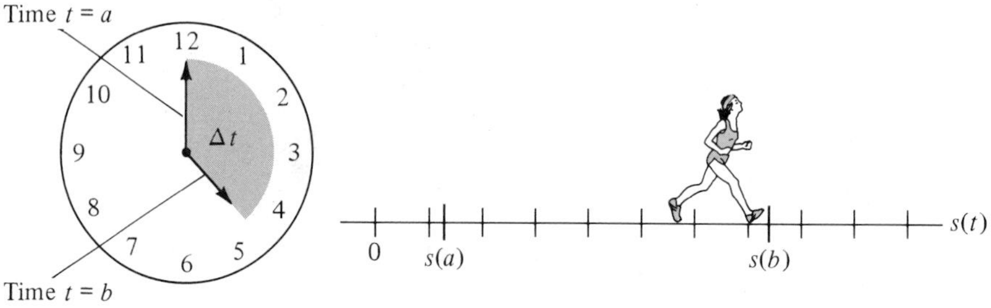

Figure 3.1 A *position function* $s(t)$ gives the location of an object along a (number) line at time t.

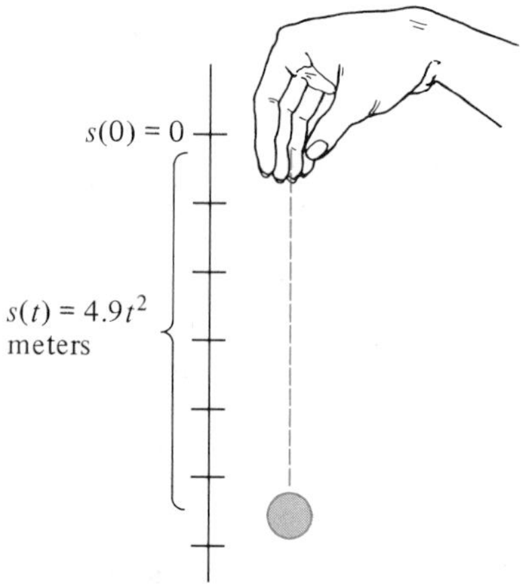

Figure 3.2 A freely falling body falls $4.9t^2$ meters in t seconds.

For example, in physics we are told that, near the surface of the earth, a freely falling body will have fallen $4.9t^2$ meters t seconds after its release. The motion of such an object is therefore along a (vertical) line, and its position function is $s(t) = 4.9t^2$ (Figure 3.2).

To define the velocity of an object at time t_0 (sometimes referred to as the **instantaneous** velocity at time t_0), we begin by observing that if $\Delta t \neq 0$, then $s(t_0 + \Delta t) - s(t_0)$ is the change in the position of the object corresponding to the time interval with endpoints t_0 and $t_0 + \Delta t$. Thus, the expression

$$\left\{\begin{matrix}\text{average}\\ \text{velocity}\end{matrix}\right\} = \frac{s(t_0 + \Delta t) - s(t_0)}{\Delta t} \tag{2}$$

is precisely the concept of (average) velocity as defined by equation (1).

We now argue that as $\Delta t \to 0$, the average velocity corresponding to the (shrinking) time interval with endpoints t_0 and $t_0 + \Delta t$ should provide an increasingly accurate measure of the velocity *at the instant* $t = t_0$. For this reason, we define the velocity at time $t = t_0$ to be the limiting value of these average velocities. That is, we *define* the velocity, $v(t_0)$, as

$$\boxed{v(t_0) = \lim_{\Delta t \to 0} \frac{s(t_0 + \Delta t) - s(t_0)}{\Delta t}} \tag{3}$$

whenever this limit exists. Of course, equation (3) simply states that we have defined velocity as the derivative of the position function.

DEFINITION 2

If the differentiable function $s(t)$ gives the position at time t of an object moving along a line, then the velocity $v(t)$ at time t is the derivative

$$v(t) = s'(t).$$

That is,

$$v(t) = \frac{d}{dt} s(t).$$

Example 1 Starting at time $t = 0$, a particle moves along a line so that its position after t seconds is $s(t) = t^2 - 6t + 8$ meters.

(a) Find its velocity at time t.
(b) When is its velocity zero?

Solution: According to Definition 2, the velocity at time t is

$$v(t) = s'(t) = 2t - 6.$$

Setting $v(t) = 0$ gives $2t - 6 = 0$, so $t = 3$. The particle has zero velocity after 3 seconds (see Figures 3.3 and 3.4). ■

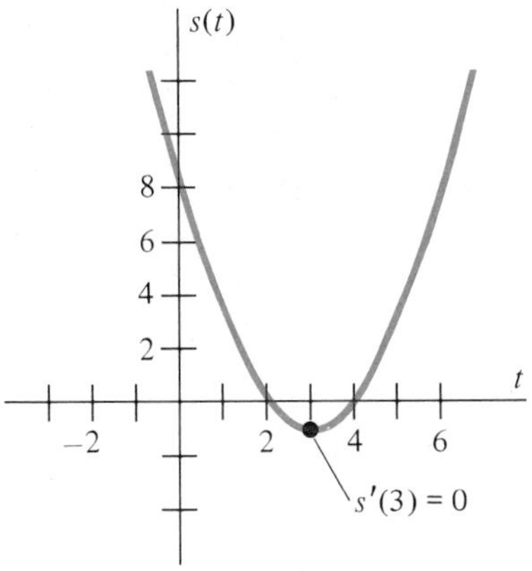

Figure 3.3 $s(t) = t^2 - 6t + 8$

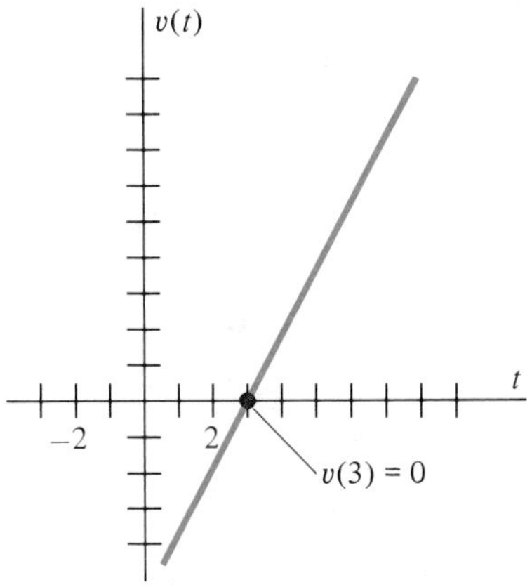

Figure 3.4 $v(t) = s'(t) = 2t - 6$

Notice that the particle in Example 1 lies $s(0) = 8$ meters from the origin at time $t = 0$, but that as t increases from 0 to 3, $s(t)$ decreases from 8 meters to $s(3) = -1$ meter. If we think of the particle as moving along a number line, this simply says that the particle moves to the left for $0 \le t < 3$ (Figure 3.5). This corresponds to the observation that $v(t)$ is negative for $t < 3$. For $t > 3$, the distance $s(t)$ increases as t increases, so $v(t) > 0$ for $t > 3$.

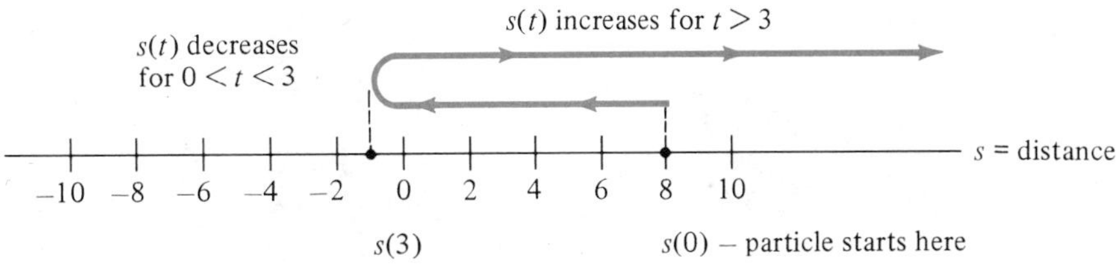

Figure 3.5 For $s(t) = t^2 - 6t + 8$ m/sec, velocity is negative for $0 < t < 3$ and positive for $t > 3$.

Velocity indicates both speed and direction along a line. It is often left up to you to decide which direction along the line will be called positive and which negative. But once this assignment is made, careful attention must be paid to the sign of $v(t)$. Positive velocities indicate motion in the positive direction along the line.

Example 2 A pebble is dropped from an open window 100 meters above the ground. With what velocity does it strike the ground?

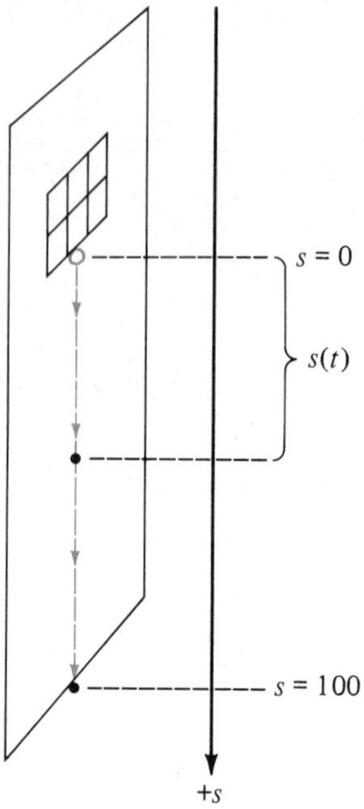

Figure 3.6

Solution: The motion of the pebble is along a vertical line. Since the motion is downward, we take the downward direction as positive. Also, we take the origin $s = 0$ to be at the height of the window (Figure 3.6). As previously stated, after t seconds the freely falling pebble will have fallen $s(t) = 4.9t^2$ meters. It will strike the ground when

$$s(t_0) = 4.9t_0^2 = 100 \text{ meters},$$

or when

$$t_0 = \sqrt{\frac{100}{4.9}} \approx 4.5 \text{ seconds}.$$

Since the corresponding velocity function is

$$v(t) = s'(t) = 2(4.9)t = 9.8t \text{ m/sec},$$

the velocity at the time of impact will be approximately

$$v(t_0) = v(4.5) = (9.8)(4.5) = 44.1$$

meters per second (downward). ■

Example 3 In Chapter 5 we shall show that an object launched vertically upward from ground level with an initial velocity of 72 m/sec will be located

$$s(t) = 72t - (4.9)t^2$$

meters above ground level after t seconds. According to this position function,

(a) when does the object stop rising?
(b) what is its maximum height?

Strategy

Establish an orientation for the line of motion and locate the origin.

Solution

From the statement of the problem it is evident that the motion is vertical, the origin is at ground level, and the positive direction is upward. The velocity function is

Calculate $v(t) = s'(t)$.

$$v(t) = \frac{d}{dt}(72t - 4.9t^2)$$
$$= 72 - 9.8t.$$

The object stops rising when $v(t_0) = 0$. Solve this equation for t_0.

Setting $v(t_0) = 0$ gives

$$72 - 9.8t_0 = 0$$

or

$$t_0 = \frac{72}{9.8} \approx 7.35 \text{ seconds}.$$

Maximum height is $s(t_0)$.

This is the time at which the object is at maximum height. The maximum height is approximately

$$s(t_0) = 72(7.35) - (4.9)(7.35)^2$$
$$= 264.5 \text{ meters}.$$

■

Exercise Set 3.3

In Exercises 1–10, the function $s(t)$ gives the position of a particle moving along a line. Find **a.** the velocity function $v(t)$, **b.** the time(s) $t_0 \geq 0$ at which velocity is zero, and **c.** the intervals in $[0, \infty)$ on which velocity is positive.

1. $s(t) = 3t - 2$

2. $s(t) = t^2 - 6t + 4$

3. $s(t) = \dfrac{1}{1 + t}$

4. $s(t) = t^2 - 3t + 5$

5. $s(t) = t^3 - 9t^2 + 24t + 10$

6. $s(t) = t(t - 1)(t + 2)$

7. $s(t) = \dfrac{1 - t}{1 + t}$

8. $s(t) = \dfrac{t^2}{a^2 + t^2}$

9. $s(t) = t^3 - 6t^2 + 9t + 7$

10. $s(t) = t^4 - 4t + 4$

11. A particle moves along a line so that its position at time t is $s(t) = \dfrac{t^2 + 2}{t + 1}$ units. Find its velocity at time $t = 3$.

12. If a particle is projected vertically upward with an initial velocity v_0, its height after t seconds is $s(t) = v_0 t - 4.9t^2$ meters. Suppose $v_0 = 98$ meters per second.

a. What is the velocity of the particle at time t?

b. At what time does the particle reach its maximum height?

c. What is the maximum height? (*Hint:* What is its velocity at maximum height?)

d. How fast is it moving when it strikes the ground?

13. A particle moves along a line so that after t seconds its position is $s(t) = 6 + 5t - t^2$. Find its maximum distance from the origin during the time interval $[0, 6]$.

14. A particle moves along a line so that at time t its position is $s(t) = 6t - t^2$.

a. What is its initial velocity, that is, v_0?

b. When does it change direction?

c. How fast is it moving when it reaches the origin the second time?

15. Show that Definition 2 agrees with the definition of velocity in equation (1) for motion at a constant speed.

3.4 THE CHAIN RULE

If $u(x)$ is a differentiable function, the Product Rule may be applied to determine the derivative of its square:

$$[u^2(x)]' = [u(x)u(x)]' = u'(x)u(x) + u(x)u'(x) = 2u(x)u'(x).$$

Similarly, for $n \geq 3$ we can show that

$$[u^n(x)]' = n \cdot u^{n-1}(x) \cdot u'(x). \tag{1}$$

(See Exercise 53, Section 3.2.) Equation (1) is called the **Power Rule for Functions.** In Leibniz notation it may be written

$$\frac{d}{dx} u^n = nu^{n-1} \cdot \frac{du}{dx}. \tag{2}$$

The Power Rule for Functions says this: "To find $[u^n(x)]'$, first differentiate $u^n = u^n(x)$ as if u were the independent variable, according to the ordinary Power Rule. Then multiply the result by the derivative of u."

Example 1 Find $f'(x)$ for $f(x) = (x^4 - 6x^2)^3$.

Solution: If we define $u = x^4 - 6x^2$, the function $f(x)$ can be written $f(x) = (x^4 - 6x^2)^3 = u^3$. Since

$$\frac{du}{dx} = \frac{d}{dx}(x^4 - 6x^2) = 4x^3 - 12x,$$

the Power Rule for Functions, equation (2), gives

$$\begin{aligned} f'(x) &= 3u^2(4x^3 - 12x) \\ &= 3(x^4 - 6x^2)^2(4x^3 - 12x). \end{aligned}$$

Of course, we could also obtain $f'(x)$ by first writing $f(x)$ as the polynomial

$$f(x) = x^{12} - 18x^{10} + 108x^8 - 216x^6$$

and then applying the Polynomial Rule

$$f'(x) = 12x^{11} - 180x^9 + 864x^7 - 1296x^5.$$

We leave it for you to verify that these two results are the same. ■

In Example 1 the advantage of the Power Rule solution is that the factors of $f'(x)$ are more easily recognized than in the Polynomial Rule solution, although both methods apply. In Example 2 the Power Rule is the only reasonable method available for finding $f'(x)$.

Example 2 For $f(x) = \left(\dfrac{x-3}{x^3+7}\right)^9$, find $f'(x)$.

Strategy

Identify the "inside" function u.

Solution

Here we define u to be the function

$$u = \frac{x-3}{x^3+7}.$$

Find $\dfrac{du}{dx}$.

Then, by the Quotient Rule,

$$\begin{aligned} \frac{du}{dx} &= \frac{(x^3+7)(1) - (x-3)(3x^2)}{(x^3+7)^2} \\ &= \frac{-2x^3 + 9x^2 + 7}{(x^3+7)^2}. \end{aligned}$$

Apply the Power Rule (2) with $n = 9$.

Then, by the Power Rule for Functions,

$$\begin{aligned} f'(x) &= 9u^8 \cdot \frac{du}{dx} \\ &= 9u^8\left[\frac{-2x^3 + 9x^2 + 7}{(x^3+7)^2}\right] \end{aligned}$$

Substitute back for u in final answer.

$$\begin{aligned} &= 9\left(\frac{x-3}{x^3+7}\right)^8\left[\frac{-2x^3 + 9x^2 + 7}{(x^3+7)^2}\right] \\ &= \frac{9(-2x^3 + 9x^2 + 7)(x-3)^8}{(x^3+7)^{10}}. \end{aligned}$$

■

Composite Functions

The power function $y = u^n(x)$ is a special case of the more general notion of a *composite function* (see Sections 1.5 and 2.6 for a review of composite functions). Composite functions have the form

$$(f \circ u)(x) = f(u(x)) \tag{3}$$

with an "outside" function $f(u)$ acting on values of an "inside" function $u(x)$. In the power function $y = u^n(x)$, the outside function is $f(u) = u^n$.

By way of review, here is a list of composite functions together with an indication of how they may be viewed as having the form (3).

$f(u(x))$	$f(u)$	$u(x)$
$\sqrt{x^2+3}$	$f(u) = \sqrt{u}$	$u(x) = x^2 + 3$
$\dfrac{1}{7-x}$	$f(u) = \dfrac{1}{u}$	$u(x) = 7 - x$
$\dfrac{3}{(\sqrt{x}+2)^4}$	$f(u) = \dfrac{3}{u^4}$	$u(x) = \sqrt{x} + 2$
$\left(\dfrac{2x-9}{2x+9}\right)^3$	$f(u) = u^3$	$u(x) = \dfrac{2x-9}{2x+9}$

It is important to keep in mind that the composite function $f(u(x))$ is defined only when x is in the domain of u *and* $u(x)$ is in the domain of f.

Since the power function $u^n(x)$ is a special case of a composite function, we would expect the derivative of a composite function $f(u(x))$ to be calculated in a manner consistent with the Power Rule for Functions. This is indeed the case, and the result is called the **Chain Rule.**

THEOREM 9
Chain Rule

If $u(x)$ is differentiable at x and $f(u)$ is differentiable at $u(x)$, then the composition $f(u(x))$ is differentiable at x and

$$[f(u(x))]' = f'(u(x))u'(x).$$

In Leibniz notation, the rule is stated

$$\frac{d}{dx}f(u) = \frac{df}{du}\cdot\frac{du}{dx}.$$

The Chain Rule says this: "To differentiate the composite function $f(u(x))$, first differentiate f as a function of u, as if u were the independent variable. Then multiply the result by the derivative of u."

Before concerning ourselves with a proof for Theorem 9 we shall apply it in several examples.

Example 3 For $f(x) = \dfrac{1}{(6x^3 - x)^4}$, find $f'(x)$.

Solution: Here we let u be the "inside" function $u(x) = 6x^3 - x$. Then

$$f(u) = \frac{1}{(6x^3 - x)^4} = \frac{1}{u^4} = u^{-4},$$

so

$$\frac{df}{du} = -4u^{-5}, \quad \text{and} \quad \frac{du}{dx} = 18x^2 - 1.$$

Then, by the Chain Rule

$$\begin{aligned} f'(x) &= \frac{df}{du} \cdot \frac{du}{dx} \\ &= -4u^{-5}(18x^2 - 1) \\ &= -4(6x^3 - x)^{-5}(18x^2 - 1). \end{aligned}$$

■

Example 4 The result of Example 6, Section 2.2, is that $\frac{d}{dx}(\sqrt{x}) = \frac{1}{2\sqrt{x}}$. Use this information to find $f'(t)$ for

$$f(t) = \sqrt{\frac{t^2 - 1}{t^2 + 1}}.$$

Strategy

Identify the inside function u.

Write f as a function of u.

Calculate $\frac{df}{du}$ and $\frac{du}{dt}$.

Apply Chain Rule:

$f'(t) = \frac{df}{du} \cdot \frac{du}{dt}$.

Solution

If we let $u(t) = \frac{t^2 - 1}{t^2 + 1}$, then

$$f(u) = \sqrt{\frac{t^2 - 1}{t^2 + 1}} = \sqrt{u},$$

so

$$\frac{df}{du} = \frac{1}{2\sqrt{u}}$$

and

$$\begin{aligned} \frac{du}{dt} = \frac{d}{dt}\left(\frac{t^2 - 1}{t^2 + 1}\right) &= \frac{(t^2 + 1)(2t) - (t^2 - 1)(2t)}{(t^2 + 1)^2} \\ &= \frac{4t}{(t^2 + 1)^2}. \end{aligned}$$

By the Chain Rule,

$$\begin{aligned} f'(t) &= \frac{1}{2\sqrt{u}} \cdot \left[\frac{4t}{(t^2 + 1)^2}\right] \\ &= \frac{4t}{2(t^2 + 1)^2\sqrt{\frac{t^2 - 1}{t^2 + 1}}} \\ &= \frac{2t}{(t^2 + 1)^{3/2}(t^2 - 1)^{1/2}} \end{aligned}$$

■

By repeated application of the Chain Rule we can differentiate compositions involving more than two functions. For compositions of three functions we have

$$[g(f(u(x)))]' = g'(f(u(x))) \cdot f'(u(x)) \cdot u'(x)$$

or, in Leibniz notation,

$$\frac{d}{dx} g(f(u)) = \frac{dg}{df} \cdot \frac{df}{du} \cdot \frac{du}{dx}. \tag{4}$$

Example 5 Find $\frac{d}{dx}\left(\frac{1}{\sqrt{x^3 - 6x + 2}}\right)$.

Solution: If we let $u(x) = x^3 - 6x + 2$,

$$f(u) = \sqrt{u}, \quad \text{and}$$

$$g(f) = \frac{1}{f}$$

we can write $\frac{1}{\sqrt{x^3 - 6x + 2}} = g(f(u))$. Since

$$\frac{du}{dx} = 3x^2 - 6, \qquad \frac{df}{du} = \frac{1}{2\sqrt{u}}, \qquad \text{and} \qquad \frac{dg}{df} = \frac{-1}{f^2},$$

an application of equation (4) gives

$$\begin{aligned} \frac{d}{dx}\left(\frac{1}{\sqrt{x^3 - 6x + 2}}\right) &= \left(\frac{-1}{f^2}\right)\left(\frac{1}{2\sqrt{u}}\right)(3x^2 - 6) \\ &= \left[\frac{-1}{(\sqrt{u})^2}\right]\left(\frac{1}{2\sqrt{u}}\right)(3x^2 - 6) \\ &= \frac{-(3x^2 - 6)}{2u^{3/2}} \\ &= \frac{-3x^2 + 6}{2(x^3 - 6x + 2)^{3/2}}. \end{aligned}$$

■

Justifying the Chain Rule

Here is an informal argument for the validity of the Chain Rule. Recall that the derivative measures the rate of change of a function. Since f acts on u and u acts on x, the rate of change of f with respect to change in x should be the product of the rate of change of f with respect to change in u multiplied by the rate of change of u with respect to x. In fact, if we begin with the definition:

$$[f(u(x))]' = \lim_{\Delta x \to 0} \frac{f(u(x + \Delta x)) - f(u(x))}{\Delta x}$$

and use some elementary algebra, we obtain the statement

$$[f(u(x))]' = \lim_{\Delta x \to 0} \left[\frac{f(u(x + \Delta x)) - f(u(x))}{u(x + \Delta x) - u(x)}\right] \cdot \left[\frac{u(x + \Delta x) - u(x)}{\Delta x}\right]. \tag{5}$$

If we next let $\Delta u = u(x + \Delta x) - u(x)$, we can rewrite equation (5) as

$$[f(u(x))]' = \lim_{\Delta x \to 0} \left[\frac{f(u + \Delta u) - f(u)}{\Delta u}\right] \cdot \left[\frac{u(x + \Delta x) - u(x)}{\Delta x}\right]. \tag{6}$$

Since $u(x)$ is differentiable at x, it is continuous at x (Theorem 1). Thus, according to equation (3), Section 2.6,

$$\lim_{\Delta x \to 0} \Delta u = \lim_{\Delta x \to 0} [u(x + \Delta x) - u(x)] = 0,$$

that is, $\Delta u \to 0$ as $\Delta x \to 0$. Using this observation, equation (6), and part (iii) of Theorem 1 (Chapter 2) we have

$$[f(u(x))]' = \left[\lim_{\Delta u \to 0} \frac{f(u + \Delta u) - f(u)}{\Delta u}\right] \cdot \left[\lim_{\Delta x \to 0} \frac{u(x + \Delta x) - u(x)}{\Delta x}\right]$$
$$= f'(u) \cdot u'(x)$$

as desired. However, this argument is not entirely rigorous in that closer attention needs to be paid to the question of whether or not the denominator Δu in equation (6) might equal zero for $\Delta x \neq 0$. A formal proof may be found in Appendix II.

Exercise Set 3.4

In each of Exercises 1–30, find $f'(x)$.

1. $f(x) = (x + 4)^3$

2. $f(x) = (3x - 2)^4$

3. $f(x) = (x^2 - 7x)^6$

4. $f(x) = (x^2 + 8x + 8)^9$

5. $f(x) = x(x^4 - 5)^3$

6. $f(x) = (x^2 - 7)^3(5 - x^3)^2$

7. $f(x) = \dfrac{1}{(x^2 - 9)^3}$

8. $f(x) = \dfrac{x + 3}{(x^2 - 6x + 2)^2}$

9. $f(x) = (3\sqrt{x} - 2)^4$

10. $f(x) = (x^6 - x^2 + 2)^{-4}$

11. $f(x) = \left(\dfrac{x - 3}{x + 3}\right)^4$

12. $f(x) = \left(\dfrac{1 + \sqrt{x}}{1 - \sqrt{x}}\right)^{-6}$

13. $f(x) = (4x^2 + 7)^5$

14. $f(x) = (3 - 2x - x^4)^3$

15. $f(x) = \dfrac{(x^3 + 1)^3 + 1}{1 - x}$

16. $f(x) = \dfrac{x + 2}{3 + (x^2 + 1)^3}$

17. $f(x) = \dfrac{1}{(x^2 + x + 1)^6}$

18. $f(x) = \left(\dfrac{1 + x^2}{1 - x^2}\right)^5$

19. $f(x) = \dfrac{(2x^3 + x)^{-3}}{(x^4 - 2x^2)^{-5}}$

20. $f(x) = \left(x + \dfrac{1}{x}\right)^3$

21. $f(x) = \left(\dfrac{ax + b}{cx + d}\right)^3$

22. $f(x) = \sqrt{ax^2 + bx + c}$

23. $f(x) = [(x^4 - 4x^2)(2x + 5)]^3$

24. $f(x) = \left[\dfrac{x^2 + x + 7}{(x + 2)^2 - 4}\right]^3$

25. $f(x) = \left(\sqrt{x} - \dfrac{1}{\sqrt{x}}\right)^3$

26. $f(x) = \dfrac{1}{(1 + \sqrt{x})^4}$

27. $f(x) = \dfrac{\sqrt{x^2 + 1}}{(x + 5)^3}$

28. $f(x) = \dfrac{(x - 4)^5}{\sqrt{x^3 + 6}}$

29. $f(x) = \dfrac{(\sqrt{x} + 7)^3}{\sqrt{9 - x^2}}$

30. $f(x) = \dfrac{1}{\left(1 + \dfrac{1}{x}\right)^3}$

In each of Exercises 31–40, form the composite function $f(u(x))$. Then find $f(u(x))'$ by the Chain Rule.

31. $f(u) = u^3 + 1$
$u(x) = 1 - x^2$

32. $f(u) = \dfrac{1}{u^2 + u}$
$u(x) = 3x^2 - 7$

33. $f(u) = u(1 - u^2)$
$u(x) = \dfrac{1}{x}$

34. $f(u) = \dfrac{u + 1}{u - 1}$
$u(x) = \dfrac{1}{x + 1}$

35. $f(u) = (u + 1)(u - 1)$
$u(x) = x^5 + 5x + 5$

36. $f(u) = (3u^2 - 1)^5$
$u(x) = \dfrac{1}{x + 5}$

37. $f(u) = \sqrt{3 + u}$
$u(x) = 7 - \sqrt{x}$

38. $f(u) = (u - 1) \cdot (u + 1)$
$u(x) = (1 - x) \cdot (1 + x)$

39. $f(u) = u^{-3}$
$u(x) = \dfrac{x}{6 - x^3}$

40. $f(u) = u^3 - u + 2$
$u(x) = \sqrt{x}$

Suppose that $f(x)$ is a differentiable function and that $f'(x) = \dfrac{x}{1 + x}$. Find the derivative of each of the following functions:

41. $f(x + 2)$

42. $f(x^2)$

43. $f(\sqrt{x})$

44. $f\left(\dfrac{3 - x}{2 + x}\right)$

In each of Exercises 45–53, use equation (4) to find the derivative.

45. $y = ((x^4 + 1)^3 + 1)^5$

46. $y = \sqrt{1 + (x^2 + 1)^3}$

47. $y = \sqrt{a + (bx + c)^4}$

48. $y = \dfrac{1}{(\sqrt{x^2 + 1} + 1)^2}$

49. $y = ((x^2 + 1)^2 + 1)^2$

50. $y = (1 - \sqrt{x^2 + 1})^6$

51. $y = \sqrt{1 + \sqrt{x}}$

52. $y = \dfrac{1}{\sqrt{x^2 + 6x + 6}}$

53. $y = \sqrt{\dfrac{1}{\sqrt{x + 3}}}$

In each of Exercises 54–59, find an equation for the line tangent to the graph of $y = f(x)$ at the indicated point P.

54. $y = (1 - x^3)^3$; $P = (1, 0)$

55. $y = \left(\dfrac{x}{x + 1}\right)$; $P = (0, 0)$

56. $y = \sqrt{1 + x^3}$; $P = (2, 3)$

57. $y = \left(1 - \dfrac{3}{x + 2}\right)^3$; $P = (1, 0)$

58. $y = \left(\dfrac{3x^2 + 1}{x + 3}\right)^2$; $P = (-1, 4)$

59. $y = \sqrt{\dfrac{1 - x}{1 + x}}$; $P = (0, 1)$

60. Let $h(x) = f(u(x))$ where $u(2) = 3$, $u'(2) = 7$, and $f'(3) = -2$. Find $h'(2)$.

61. Let $y = [u(t)^2 + 1]^3$. Find $\dfrac{dy}{dt}$ for $t = 3$ if $u(3) = -2$ and $u'(3) = -1$.

62. *(Calculator)* Let $f(u) = 3u^2 + 1$ and $u(x) = 2x + 1$.

a. Estimate the difference quotient for $f(u)$ at $u = 3$ using $\Delta u = 0.2$, 0.08, and 0.02.

b. Estimate the difference quotient for $u(x)$ at $x = 1$ using $\Delta x = 0.1$, 0.04, and 0.01.

c. Estimate the difference quotient for $f(u(x)) = 3(2x + 1)^2 + 1$ at $x = 1$ using $\Delta x = 0.1$, 0.04, and 0.01.

d. Compare the estimates obtained in part (c) with the respective products of the estimates obtained in parts (a) and (b). What do you observe? What *should* you observe? Comment on how these observations relate to the remarks concerning the proof of the Chain Rule.

63. Find $a > 0$ so that the graph of $y = \sqrt{ax^2 + 4}$ has tangent with equation $2y - 3x - b = 0$ at the point where $x = 2$. What is b?

64. Verify equation (1) in the following way.

a. For $n = 1$, $[u(x)^n]' = n[u(x)^{n-1}]u'(x)$ since $u'(x) = 1[u(x)^0] \cdot u'(x)$.

b. Assume that equation (1) holds for the integer n.

c. Write $[u(x)^{n+1}]' = [u(x)u(x)^n]'$ and use the Product Rule to verify that $[u(x)^{n+1}]' = (n + 1)[u(x)^n]u'(x)$.

d. Use parts (a)–(c) and the Principal of Mathematical Induction to conclude that equation (1) holds for all $n = 1, 2, 3, \ldots$.

65. *(Computer)* Program 3 in Appendix I is a BASIC program which prints out difference quotients for the composite function $f(u(x)) = \sqrt{1 + x^2}$ as the products

$$\frac{FY - FX}{UY - UX} \cdot \frac{UY - UX}{Y - X}$$

where X is the input and

$$\begin{aligned} \Delta X &= (0.5)^n, \qquad n = 1, 2, \ldots, 10 \\ Y &= X + \Delta X \\ UX &= 1 + X^2 \\ UY &= 1 + Y^2 \\ FX &= \sqrt{UX} \\ FY &= \sqrt{UY} \end{aligned}$$

Each approximation to the difference quotient is listed alongside the number $\dfrac{x}{\sqrt{1 + x^2}}$, the correct value of the derivative. For example, the following results are obtained for $x = 2$ (Table 4.1):

Table 4.1

N	Difference quotient	Derivative
1	.91302884	.89442720
2	.90458587	.89442720
3	.89975039	.89442720
4	.89715390	.89442720
5	.89580724	.89442720
6	.89512185	.89442720
7	.89477490	.89442720
8	.89459941	.89442720
9	.89451237	.89442720
10	.89446629	.89442720

Use Program 3 to obtain difference quotient approximations to the derivative of $f(x) = \sqrt{1 + x^2}$ for various values of x. For which values of x are these approximations

a. more accurate?

b. less accurate?

3.5 DERIVATIVES OF THE TRIGONOMETRIC FUNCTIONS

The six trigonometric functions are reviewed briefly in Section 1.6. Our purpose here is to determine whether these functions are differentiable and to calculate their derivatives. We will need to make use of two limits, both of which were established in Chapter 2:

$$\lim_{x\to 0} \frac{\sin x}{x} = 1 \qquad \text{(Example 8, Section 2.4)} \tag{1}$$

$$\lim_{x\to 0} \frac{1-\cos x}{x} = 0 \qquad \text{(Exercise 50, Section 2.4)} \tag{2}$$

Since the derivative of a function measures its slope, we begin by sketching several tangents to the graph of $y = \sin x$ and estimating their slopes (by guessing). This gives us a rough idea of several points lying on the graph of $g(x) = (\sin x)'$, the slope function for $f(x) = \sin x$ (Figure 5.1).

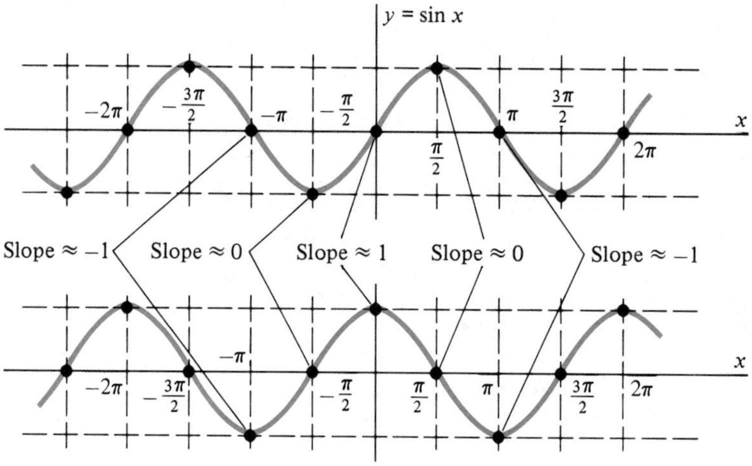

Figure 5.1 Graph of $f(x) = \sin x$ (above) and its slope function $g(x) = f'(x)$ (below).

Notice that since $y = \sin x$ is periodic with period 2π (that is, $f(x + 2\pi)$ is the same as $f(x)$ for all x), the slope function for $\sin x$ must also be 2π-periodic. Furthermore, the slope function will oscillate between a maximum and minimum value in much the same way as does $\sin x$. The following theorem, therefore, should not surprise you.

THEOREM 10 The function $f(x) = \sin x$ is differentiable for all x, and $f'(x) = \cos x$. That is,

$$\frac{d}{dx} \sin x = \cos x.$$

Strategy

Write definition

$$f'(x) = \lim_{\Delta x\to 0} \frac{f(x+\Delta x) - f(x)}{\Delta x}.$$

Proof

$$f'(x) = \lim_{\Delta x\to 0} \frac{\sin(x+\Delta x) - \sin x}{\Delta x}$$

Use identity $\sin(\alpha + \beta) = \sin\alpha\cos\beta + \sin\beta\cos\alpha$.

$$= \lim_{\Delta x\to 0} \frac{[\sin x\cos\Delta x + \cos x\sin\Delta x] - \sin x}{\Delta x}$$

Factor $\sin x$, $\cos x$.

$$= \lim_{\Delta x\to 0}\left\{\sin x\left[\frac{\cos\Delta x - 1}{\Delta x}\right] + \cos x\left[\frac{\sin\Delta x}{\Delta x}\right]\right\}$$

Apply Theorem 1, Chapter 2 and limits (1) and (2).

$$= (\sin x)\left[\lim_{\Delta x\to 0}\left(\frac{\cos\Delta x - 1}{\Delta x}\right)\right] + (\cos x)\left[\lim_{\Delta x\to 0}\left(\frac{\sin\Delta x}{\Delta x}\right)\right]$$

$$= (\sin x)(0) + (\cos x)(1)$$

$$= \cos x.$$

Since this limit exists for all x, the function $f(x) = \sin x$ is differentiable for all x. ■

If $y = u(x)$ is a differentiable function of x, Theorem 10 and the Chain Rule together imply that

For $f(x) = \sin u(x)$, $\quad f'(x) = \cos u(x)\cdot u'(x)$, **(3)**

or

$$\frac{d}{dx}\sin u = \cos u\cdot\frac{du}{dx}. \quad (4)$$

Example 1 Find $f'(x)$ for $f(x) = \sin(1 + x^2)$.

Solution: Here $u(x) = 1 + x^2$, so $u'(x) = 2x$. By (3) we have

$$f'(x) = \cos(1 + x^2)\cdot(2x) = 2x\cdot\cos(1 + x^2)$$ ■

Example 2 Find $\frac{dy}{dx}$ for $y = \sqrt{x}\sin(x^3 + x)$.

Strategy

Solution

Apply Product Rule first.

$$\frac{dy}{dx} = \left[\frac{d}{dx}\sqrt{x}\right]\sin(x^3 + x) + \sqrt{x}\left[\frac{d}{dx}\sin(x^3 + x)\right]$$

Then apply equation (4) in the second term.

$$= \frac{1}{2\sqrt{x}}\sin(x^3 + x) + \sqrt{x}[\cos(x^3 + x)]\frac{d}{dx}(x^3 + x)$$

$$= \frac{1}{2\sqrt{x}}\sin(x^3 + x) + \sqrt{x}(3x^2 + 1)\cos(x^3 + x).$$ ■

Example 3 Use the identity $\cos x = \sin(\pi/2 - x)$ to find the derivative of the function $y = \cos x$.

Strategy

Solution

$\cos x = \sin(\pi/2 - x)$

$$\frac{d}{dx}\cos x = \frac{d}{dx}\sin(\pi/2 - x)$$

Apply equation (4).

$$= \cos(\pi/2 - x)\cdot\frac{d}{dx}(\pi/2 - x)$$

$$= -\cos(\pi/2 - x)$$

$\cos(\pi/2 - x) = \sin x$.

$$= -\sin x.$$ ■

Since the function $y = \sin x$ is differentiable for all x and since the identities $\cos x = \sin(\pi/2 - x)$ and $\sin x = \cos(\pi/2 - x)$ are valid for all x, the result of Example 3 is that the function $y = \cos x$ is differentiable for all values of x, and $\frac{d}{dx}\cos x = -\sin x$.

For a differentiable function $y = u(x)$, we may combine this result with the Chain Rule to obtain the following differentiation formula for $\cos u(x)$.

For $f(x) = \cos u(x), \qquad f'(x) = -\sin u(x) \cdot u'(x),$ (5)

or

$$\frac{d}{dx}\cos u = -\sin u \cdot \frac{du}{dx}. \tag{6}$$

Example 4 Find $\frac{dy}{dx}$ for $y = \frac{2 - \cos x}{2 + \cos x}$.

Strategy

Solution

Apply Quotient Rule first.

$$\frac{dy}{dx} = \frac{(2 + \cos x)\left[\frac{d}{dx}(2 - \cos x)\right] - \left[\frac{d}{dx}(2 + \cos x)\right](2 - \cos x)}{(2 + \cos x)^2}$$

Then apply equation (6).

$$= \frac{(2 + \cos x)[-(-\sin x)] - (-\sin x)(2 - \cos x)}{(2 + \cos x)^2}$$

$$= \frac{4 \sin x}{(2 + \cos x)^2}.$$

■

Example 5 Find the derivative of $y = \tan x$, where it is defined, and determine for which values of x the function $y = \tan x$ is differentiable.

Strategy

Solution

Apply definition of $\tan x$.

$$\frac{d}{dx}\tan x = \frac{d}{dx}\left(\frac{\sin x}{\cos x}\right)$$

Apply Quotient Rule.

$$= \frac{\cos x \cdot \frac{d}{dx}(\sin x) - \sin x \cdot \frac{d}{dx}(\cos x)}{\cos^2 x}$$

$$= \frac{\cos^2 x + \sin^2 x}{\cos^2 x}$$

$\cos^2 x + \sin^2 x = 1$

$$= \left(\frac{1}{\cos x}\right)^2$$

$\frac{1}{\cos x} = \sec x$

$$= \sec^2 x,$$

that is,

$$\frac{d}{dx}\tan x = \sec^2 x.$$

Both $\tan x$ and $\sec x$ are defined for all x for which $\cos x \neq 0$. Since $\cos x$ is the only factor of the denominator in the above calculations, $\tan x$ is differentiable at all values of x with $\cos x \neq 0$ (i.e., except at odd multiples of $\pi/2$). ■

Since each of the remaining trigonometric functions is defined as a ratio involving $\sin x$, $\cos x$, or both, we may proceed as in Example 5 to determine their derivatives (see Exercise 48). The results of doing so are the following, where we assume $y = u(x)$ to be a differentiable function of x:

$$\frac{d}{dx}\tan x = \sec^2 x; \qquad \frac{d}{dx}\tan u = \sec^2 u \cdot \frac{du}{dx} \tag{7}$$

$$\frac{d}{dx}\cot x = -\csc^2 x; \qquad \frac{d}{dx}\cot u = -\csc^2 u \cdot \frac{du}{dx} \tag{8}$$

$$\frac{d}{dx}\sec x = \sec x \tan x; \qquad \frac{d}{dx}\sec u = \sec u \cdot \tan u \cdot \frac{du}{dx} \tag{9}$$

$$\frac{d}{dx}\csc x = -\csc x \cot x; \qquad \frac{d}{dx}\csc u = -\csc u \cdot \cot u \cdot \frac{du}{dx} \tag{10}$$

Example 6 Find the equation of the line tangent to the graph of $y = \sec x \tan x$ at the point $(\pi/4, \sqrt{2})$.

Strategy

Find $\frac{dy}{dx}$ using Product Rule and Equations (7) and (9).

Solution

$$\begin{aligned}\frac{dy}{dx} &= \frac{d}{dx}(\sec x \tan x)\\ &= \sec x\left(\frac{d}{dx}\tan x\right) + \left(\frac{d}{dx}\sec x\right)\tan x\\ &= \sec x(\sec^2 x) + (\sec x \tan x)\tan x\\ &= \sec^3 x + \sec x \tan^2 x.\end{aligned}$$

Slope of line is $\frac{dy}{dx}$.

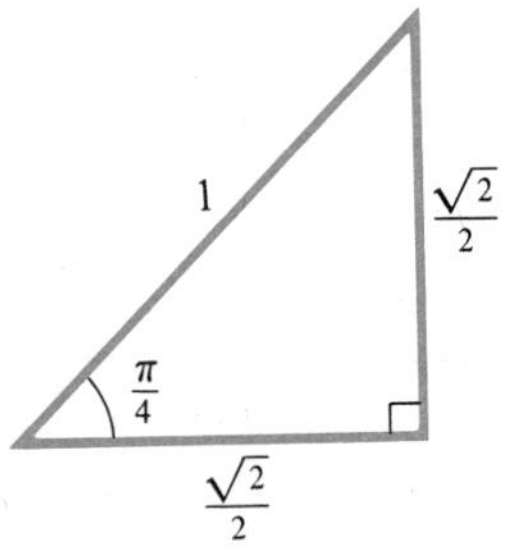

$\sec \pi/4 = \sqrt{2}, \qquad \tan \pi/4 = 1$

At the point $(\pi/4, \sqrt{2})$ the slope of the tangent is

$$\begin{aligned}m &= \sec^3(\pi/4) + \sec(\pi/4)\tan^2(\pi/4)\\ &= \left(\frac{2}{\sqrt{2}}\right)^3 + \frac{2}{\sqrt{2}}(1)^2\\ &= \frac{8}{2\sqrt{2}} + \frac{2}{\sqrt{2}}\\ &= \frac{6}{\sqrt{2}}\\ &= 3\sqrt{2}.\end{aligned}$$

Line containing (a, b) with slope m is $y - b = m(x - a)$.

The equation of the tangent is therefore

$$(y - \sqrt{2}) = 3\sqrt{2}(x - \pi/4),$$

or

$$y = 3\sqrt{2}x + \sqrt{2}\left(1 - \frac{3\pi}{4}\right).$$ ■

Exercise Set 3.5

In each of Exercises 1–38, find $f'(x)$.

1. $f(x) = \sin(3x)$
2. $f(x) = x \sin x$
3. $f(x) = \cos\left(\frac{\pi}{2} - x\right)$
4. $f(x) = \sin x \cdot \cos x$
5. $f(x) = \sin \sqrt{1 + x^2}$
6. $f(x) = \tan\left(x + \frac{\pi}{2}\right)$
7. $f(x) = \cos(2x) - \sin(3x)$
8. $f(x) = \dfrac{\sin x}{x}$
9. $f(x) = x \cos^2 x$
10. $f(x) = \sin^3 x \cdot \tan x$
11. $f(x) = \sec x \cdot \tan x$
12. $f(x) = \dfrac{\sin x}{1 + \cos^2 x}$
13. $f(x) = \sin(2x) - 2 \sin x \cdot \cos x$
14. $f(x) = \dfrac{\csc x}{\cot x}$
15. $f(x) = \dfrac{1}{\sec x}$
16. $f(x) = \sqrt{\sin x + 1}$
17. $f(x) = \dfrac{1}{\cos(2x)}$
18. $f(x) = x^3 \sin^2 x$
19. $f(x) = \cos^2 x^{\ 2}$
20. $f(x) = \dfrac{\cos^2 x + 1}{1 - \sin x}$
21. $f(x) = \cos(\sin\sqrt{1 - x^2})$
22. $f(x) = \dfrac{1 - \sin x}{\tan^2 x + 1}$
23. $f(x) = \cot^3\left(\frac{x}{2}\right)$
24. $f(x) = \sin^3 x \cdot \cos^5 x$
25. $f(x) = x \sin\left(\frac{1}{x}\right)$
26. $f(x) = \dfrac{\tan \sqrt{2x + 1}}{x}$
27. $f(x) = \sec(x^3)$
28. $f(x) = x^2 \sin^2 \left(\frac{1}{x}\right)$
29. $f(x) = \sec^3 \sqrt{x + 1}$
30. $f(x) = \dfrac{\cos \sqrt{x} \cdot \cot \sqrt{x}}{1 + x}$
31. $f(x) = \sqrt{\dfrac{4 + \sin 2x}{2 + \cos 2x}}$
32. $f(x) = \sin^3\left(x - \frac{\pi}{2}\right)\sec \sqrt{x}$
33. $f(x) = \dfrac{\sqrt{x} \cot\left(3x - \frac{\pi}{4}\right)}{\sec x \tan^2 x}$
34. $f(x) = \dfrac{\sqrt{\sec^2 2x + \tan^2 2x}}{(\cos x + \sin^2 x)^3}$
35. $f(x) = \dfrac{1}{1 + \sec^4(x^2 - \pi)}$
36. $f(x) = (\sqrt{x} + x^3 \sec \sqrt{x})^4$
37. $f(x) = \dfrac{1}{\sqrt{x \tan x \cos^3 x}}$
38. $f(x) = \left(\dfrac{\sqrt{\cos^3 x - \sin^3 x}}{x \cos x^2 - x^2 \sec x}\right)^2$
39. Find an equation for the line tangent to the graph of $y = \sin 2x$ at the point $(\pi/6, \sqrt{3}/2)$.
40. Find the tangent of the angle at which the graphs of $y_1 = \sin x$ and $y_2 = \sin 2x$ intersect for $0 \le x \le \pi/2$.
41. A particle moves along a line so that after t seconds its position is $s(t) = \dfrac{\sin 2t}{3 + \cos^2 t}$. Find its velocity at time $t = \pi/4$.
42. For which values of x are the tangents to the graph of $y = \sec^2 x$ horizontal?
43. A particle moves along a line so that at time t its position is $s(t) = 6 \sin 3t$.
    a. At which values of time does it change direction?
    b. What is its maximum velocity?
    c. What is its maximum distance from the origin?
44. Show that $\dfrac{d}{dx} \tan^2 x = \dfrac{d}{dx} \sec^2 x$. Can you give a geometric argument for this result?
45. Find an equation for the line tangent to the graph of $y = \cot \sqrt{x}$ at the point $(\pi^2/16, 1)$.
46. Do there exist numbers x for which the tangents to the graph of $f(x) = \sin x$ and $g(x) = \cos x$ are
    a. parallel? If so, which?
    b. perpendicular? If so, which?
47. Why do the functions $f(x) = \sin^2 x$ and $g(x) = 1 - \cos^2 x$ have the same derivative? What about $f(x) = 1 + \sin^2 x$ and $g(x) = \cos^2 x$?
48. Derive equations (8), (9), and (10).
49. Find the slope of the line normal to the graph of $y = \tan x$ at the point $(\pi/4, 1)$.
50. Find all values of x where $f(x)$ has a horizontal tangent (slope zero) for
    a. $f(x) = \cos x$ **b.** $f(x) = \sec x$ **c.** $f(x) = \csc x$.
51. Determine where the trigonometric functions $\cot x$, $\sec x$, and $\csc x$ are differentiable.

52. Figure 5.2 and the rule for differentiating $f(x) = \sec x$ suggest the identity

$$\sec\left(\frac{\pi}{2} - x\right)\tan\left(\frac{\pi}{2} - x\right) = \sec\left(\frac{\pi}{2} + x\right)\tan\left(\frac{\pi}{2} + x\right)$$

for $0 < x < \frac{\pi}{2}$. Is this true? Why or why not?

53. Explain why the graphs of $f(x) = \tan x$ and $g(x) = \cot x$ suggest the trigonometric identity $\sec^2 x = \csc^2\left(\frac{\pi}{2} - x\right)$ for $-\frac{\pi}{2} < x < \frac{\pi}{2}$. (Sketch a graph as part of your answer.)

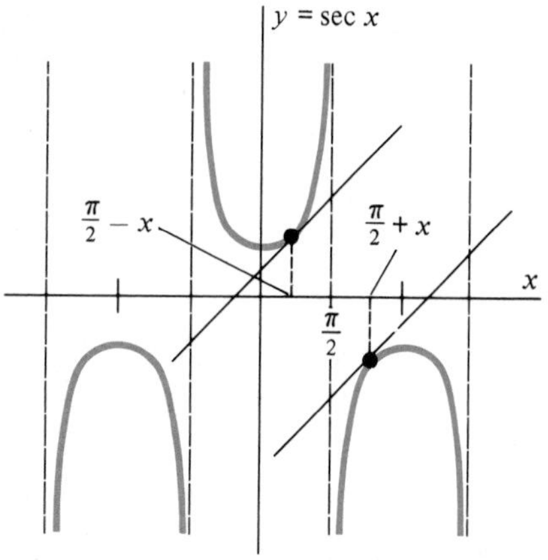

Figure 5.2

3.6 HIGHER ORDER DERIVATIVES

Since the process of differentiation produces a new function $f'(x)$ from a given function $f(x)$, it is entirely reasonable to ask what happens if we differentiate the function $f'(x)$. The answer is simply that if $f'(x)$ itself is a differentiable function we will obtain its derivative, $(f'(x))'$. We call this the **second derivative** of $f(x)$, and we denote it by $f''(x)$. That is,

$$f''(x) = [f'(x)]'.$$

For $y = f(x)$ the Leibniz notation for the second derivative of $y = f(x)$ is

$$\frac{d^2}{dx^2}f(x) = \frac{d}{dx}\left(\frac{df}{dx}\right), \qquad \text{or} \qquad \frac{d^2y}{dx^2} = \frac{d}{dx}\left(\frac{dy}{dx}\right).$$

Similarly, differentiating $f(x)$ three times produces the third derivative, provided that the corresponding limit exists, which is denoted by

$$f'''(x), \qquad \text{or} \qquad \frac{d^3y}{dx^3}, \qquad \text{or} \qquad \frac{d^3}{dx^3}f(x),$$

and so on for higher order derivatives.

Example 1 For the polynomial function

$$f(x) = x^4 - 4x^3 + 3x^2 - 5x + 3$$

we have

$$\begin{aligned} f'(x) &= 4x^3 - 12x^2 + 6x - 5, \\ f''(x) &= 12x^2 - 24x + 6, \\ f'''(x) &= 24x - 24, \\ f^{iv}(x) &= 24, \end{aligned}$$

and

$$f^{v}(x) = 0 = f^{vi}(x) = \dots .$$

■

Example 2 Find the first three derivatives for the function

$$f(x) = x^3 + x \sin x.$$

Solution:

$$\begin{aligned} f'(x) &= 3x^2 + \sin x + x \cos x \\ f''(x) &= 6x + \cos x + \cos x - x \sin x \\ &= 6x + 2 \cos x - x \sin x \\ f'''(x) &= 6 - 2 \sin x - \sin x - x \cos x \\ &= 6 - 3 \sin x - x \cos x \end{aligned}$$

■

It is an important property of the functions $f(x) = \sin x$ and $g(x) = \cos x$ that they reappear in the list of their own derivatives:

$$\begin{array}{ll} f(x) = \sin x & g(x) = \cos x \\ f'(x) = \cos x & g'(x) = -\sin x \\ f''(x) = -\sin x & g''(x) = -\cos x \\ f'''(x) = -\cos x & g'''(x) = \sin x \\ f^{\text{iv}}(x) = \sin x & g^{\text{iv}}(x) = \cos x \\ f^{\text{v}}(x) = \cos x & g^{\text{v}}(x) = \sin x \\ \text{etc.} & \text{etc.} \end{array}$$

From these lists you can see that both $\sin x$ and $\cos x$ satisfy the *differential* (or, *derivative*) equations

(i) $f''(x) = -f(x)$
(ii) $f^{\text{iv}}(x) = f(x)$.

Equation (i) is important in the modelling of oscillatory phenomena (such as problems in engineering involving springs), while equation (ii) occurs in problems involving the deflection of beams under heavy loading. We will have much more to say about differential equations later in the text.

Example 3 Find a number k so that the function $f(x) = \sin kx$ satisfies the equation

$$f''(x) = -9f(x). \tag{1}$$

Strategy
Calculate $f''(x)$.

Solution
Here $f(x) = \sin kx$, so

$$f'(x) = k \cos kx$$

and

$$f''(x) = -k^2 \sin kx.$$

Set $f''(x) = -9f(x)$ and solve for k.

Equation (1) therefore becomes

$$-k^2 \sin kx = -9 \sin kx,$$

so

$$k^2 = 9$$

and

$$k = \pm 3.$$

■

ACCELERATION: For a particle moving along a line, acceleration is defined to be the (instantaneous) rate of change of velocity. That is, if $v(t)$ denotes velocity at time t, the acceleration $a(t)$ is defined as

$$\begin{aligned} a(t) &= \lim_{\Delta t \to 0} \frac{v(t + \Delta t) - v(t)}{\Delta t} \\ &= v'(t). \end{aligned}$$

Recall that if $s(t)$ is the position function for the particle, then $v(t) = s'(t)$, so acceleration is the second derivative of position:

$$a(t) = v'(t) = s''(t),$$

or

$$a = \frac{dv}{dt} = \frac{d^2s}{dt^2}.$$

Example 4 A particle moves along a line so that at time t seconds its position is $s(t) = t^3 - 6t^2 + 7t - 2$.

(a) Find the initial acceleration, $a(0)$.
(b) Find the time t_0 at which acceleration equals zero.

Solution: We have $v(t) = s'(t) = 3t^2 - 12t + 7$, so

$$a(t) = s''(t) = 6t - 12.$$

Thus the initial acceleration is $a(0) = -12 \text{ m/sec}^2$, and

$$a(t) = 0 \quad \text{when} \quad t = 2 \text{ seconds.}$$ ■

Example 5 A projectile is fired vertically upward from ground level with an initial velocity of 100 meters per second. In such a case the distance of the particle above ground level is given by the function $s(t) = 100t - 4.9t^2$ meters. Find the acceleration of the particle.

Solution: Here $v(t) = s'(t) = 100 - 9.8t$ m/sec, so

$$a(t) = v'(t) = -9.8 \text{ m/sec}^2.$$

(Note that the acceleration is both constant and downward. It is referred to as the **acceleration due to gravity.**) ■

Exercise Set 3.6

In each of Exercises 1–14, find $f''(x)$.

1. $f(x) = 2x^3 - 6x$
2. $f(x) = a - bx - cx^2$
3. $f(x) = \dfrac{1}{1 + x}$
4. $f(x) = \dfrac{ax + b}{cx + d}$
5. $f(x) = \sqrt{x^2 + 7}$
6. $f(x) = \sqrt{x + 2}(x^3 - 6)$
7. $f(x) = \sin^2 x$
8. $f(x) = \sin x \cdot \tan x$
9. $f(x) = \dfrac{1}{1 + \cos x}$
10. $f(x) = \cos \sqrt{x}$
11. $f(x) = \sec(1 - x^2)$
12. $f(x) = x \cos(1 + \sqrt{x})$
13. $f(x) = \sqrt{x^3 + 1}$

14. $f(x) = (x - a)^{3/2} + (x + a)^{5/3}$

In each of Exercises 15–22, find $\dfrac{d^2y}{dx^2}$.

15. $y = kx\sqrt{1 - x^2}$

16. $y = x^2 + \dfrac{1}{x^2}$

17. $\dfrac{dy}{dx} = \sec x \cdot \tan x$

18. $\dfrac{dy}{dx} = (1 - x^3)\cos(x^2 + 1)$

19. $y = \sec\sqrt{1 - x^2}$

20. $\dfrac{dy}{dx} = x^4(1 - x^4)$

21. $y = x^{2/3}(1 + x^2)$

22. $y = x^m + x^n$

In each of Exercises 23–26, find the indicated derivative.

23. $f''(x)$ for $f(x) = \dfrac{1}{x + 2}$

24. $f''(x)$ for $f(x) = \sqrt{x}\sin x$

25. $\dfrac{d^4y}{dx^4}$ for $y = (x^3 - 1)^4$

26. $\dfrac{d^3y}{dx^3}$ for $y = x^{4/3} + x^{1/4}$

In Exercises 27–30, find a function of the form $f(x) = \sin kx$ satisfying the given equation.

27. $f''(x) + 4f(x) = 0$

28. $f''(x) = -16f(x)$

29. $y'' + y = 0$

30. $\dfrac{d^2y}{dx^2} + 5y = 0$

31. Explain why a polynomial $p(x) = a_nx^n + \cdots + a_1x + a_0$ has derivatives of all orders (i.e., why $p(x)$ is **infinitely differentiable**).

32. A particle moves along a line so that at time t seconds its position is $s(t) = t^4 - 8t^2 + 2$.
a. What is the velocity function?
b. For what time intervals is velocity positive?
c. What is the acceleration function?
d. When is the acceleration positive?

33. A particle moves along a line with acceleration $a(t) = 2 - t$ for $0 \le t \le 4$.
a. Sketch the graph of $a(t)$.
b. If $v(t)$ = velocity, what can you say about $v(1)$ if $v(0) > 0$?
c. What can you say about $v(4)$ if $v(3) < 0$?
d. Must the particle necessarily change direction at some time t_0 where $0 \le t_0 \le 4$?
e. Find an explicit position function $s(t)$ to support your answer to part (d).

34. Find a formula for $\dfrac{d^ny}{dx^n}$ if $y = (1 + x)^{-1}$.

35. Find a formula for $\dfrac{d^ny}{dx^n}$ if $y = \sqrt{x}$.

36. Find a second degree polynomial $f(x)$ so that $f(2) = 2$, $f'(2) = 4$, and $f''(2) = 6$.

37. A particle moves with constant acceleration along a line. If $v(2) = 6$ meters per second and $v(5) = 15$ meters per second, find the acceleration.

38. Examine the first four derivatives of the trigonometric functions $\tan x$, $\cot x$, $\sec x$, and $\csc x$. Do you note any recurring pattern or otherwise see any pattern among these derivatives as happens for $\sin x$ and $\cos x$?

3.7 IMPLICIT DIFFERENTIATION

When a function is specified by an equation of the form $y = f(x)$ we say that we have y determined as an **explicit** function of x. This is because for each value of x precisely one value of y is determined by substituting x into the right-hand side of the equation $y = f(x)$. However, some equations involving the two variables x and y determine interesting relationships between these variables even though the equations can not be brought into the explicit form $y = f(x)$. For example, the equation

$$\sin(xy) = 1 \tag{1}$$

determines a seemingly complex relationship between the variables x and y that, at the moment, we can study only by plotting pairs (x, y) where $\sin(xy) = 1$ (see Figure 7.1).

Similarly, the equation

$$\frac{x^2}{16} + \frac{y^2}{9} = 1 \tag{2}$$

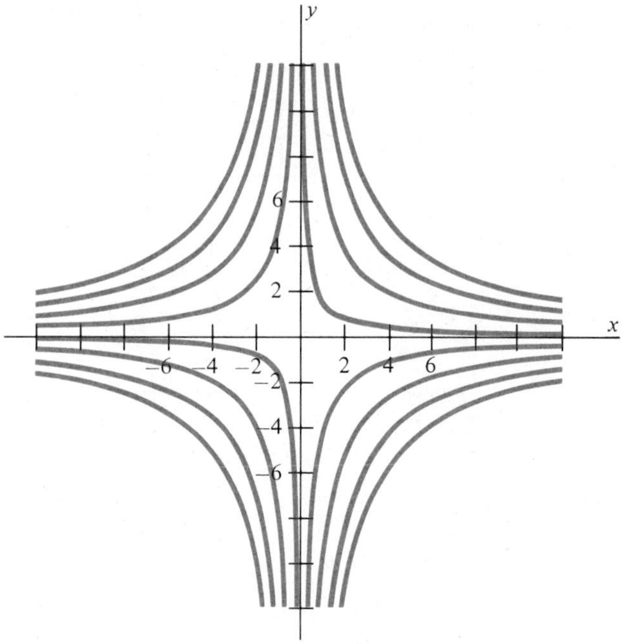

Figure 7.1 Some arcs of the graph of the relation $\sin(xy) = 1$.

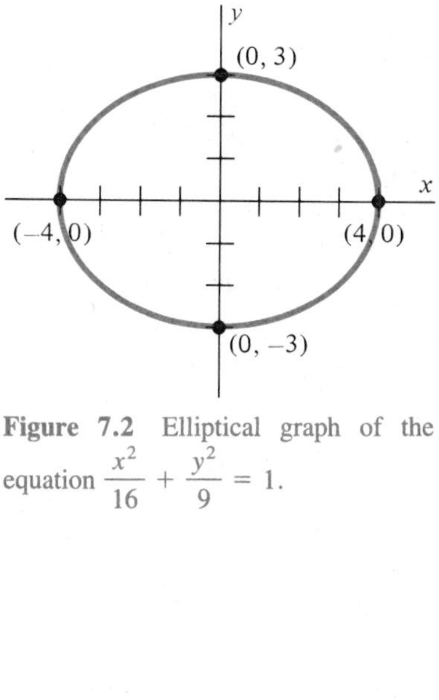

Figure 7.2 Elliptical graph of the equation $\frac{x^2}{16} + \frac{y^2}{9} = 1$.

determines the familiar ellipse in the plane (see Figure 7.2) although it cannot be brought into the form $y = f(x)$. The best we can do is solve for y as

$$y = \pm 3\sqrt{1 - x^2/16}.$$

(Because of the $\pm$ sign, the right-hand side does not determine a function of x since more than one value of y can correspond to a given value of x.)

Both equation (1) and equation (2) are examples of **relations** between the variables x and y that can be written in the form

$$f(x, y) = 0. \tag{3}$$

In particular, we can bring equations (1) and (2) into this form by writing

$$\sin(xy) - 1 = 0$$

and

$$\frac{x^2}{16} + \frac{y^2}{9} - 1 = 0.$$

Regardless of whether a relation between x and y of the form of equation (3) defines a function in the proper sense, near any particular point (x, y) on the graph of that equation we can ask about the slope of the line tangent to the graph at (x, y) (see Figure 7.3). If the tangent at this point exists and is not vertical, the slope $\frac{dy}{dx}$ can often be obtained from equation (3) by application of the Chain Rule. The reason behind this observation is illustrated in Figure 7.4: Near any particular point on the graph of the relation (3), the given equation may indeed define a differentiable function whose derivative is the desired slope.

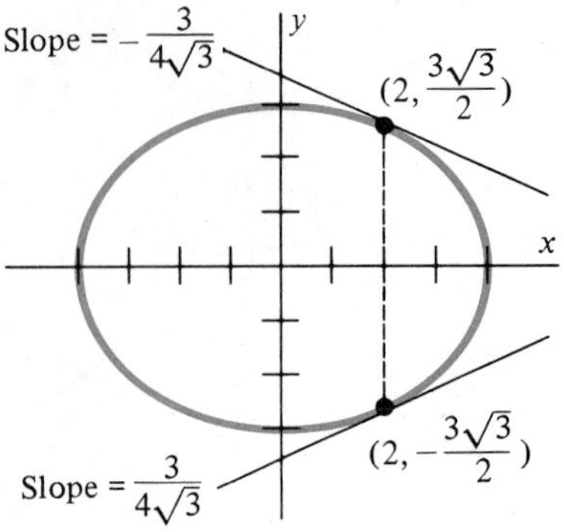

Figure 7.3 Although equation (2) does not define a function, tangents exist at all points on the graph.

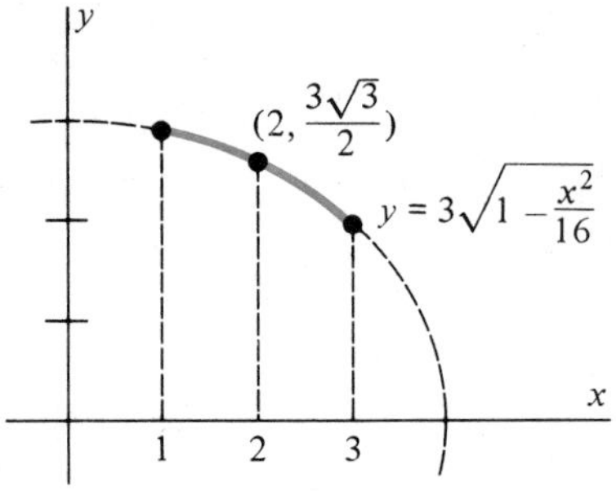

Figure 7.4 "Near" $\left(2, \frac{3\sqrt{3}}{2}\right)$ equation (2) does define a differentiable function of x.

Example 1 Find the slope of the line tangent to the graph of the ellipse

$$\frac{x^2}{16} + \frac{y^2}{9} = 1$$

at the point $\left(2, \frac{3\sqrt{3}}{2}\right)$ (Figure 7.3).

Strategy

Force y to be a differentiable function of x by imposing the restriction that $y > 0$.

Solution

We see from Figure 7.3 that y can be considered a function of x if we impose the condition $y \geq 0$. This function is also **differentiable** if $y > 0$ (i.e., we exclude the points $(-4, 0)$ and $(4, 0)$ where the tangent is vertical). Since the left and right sides of equation (2) are then merely different names for the same functions of x, the derivatives of these two sides must be equal. In other words, we may simply differentiate both sides of the equation. Doing so, we obtain

Differentiate both sides of equation (2), remembering that

$$\frac{d}{dx}\left(\frac{y^2}{9}\right) = \frac{2y}{9} \cdot \frac{dy}{dx},$$

according to the Chain Rule.

$$\frac{2x}{16} + \frac{2y}{9} \cdot \frac{dy}{dx} = 0,$$

so

$$\frac{dy}{dx} = -\frac{9x}{16y}, \qquad y \neq 0. \tag{4}$$

At the point $\left(2, \frac{3\sqrt{3}}{2}\right)$, we have $x = 2$ and $y = \frac{3\sqrt{3}}{2}$

so

Solve for $\frac{dy}{dx}$ and substitute values for x and y.

$$\frac{dy}{dx} = -\frac{9(2)}{16\left(\frac{3\sqrt{3}}{2}\right)} = -\frac{\sqrt{3}}{4},$$

which is the desired slope. (Alternatively, we could have been asked for the slope of the tangent at the point $(2, -3\sqrt{3}/2)$. Then we would have used the restriction $y < 0$ to make y a function of x. The result would be $dy/dx = \sqrt{3}/4$ [see Figure 7.3]). ■

The technique used in the solution of Example 1 is called **implicit differentiation.** It can be summarized as follows:

(i) Impose a condition on y so that the given equation implicitly defines y as a differentiable function of x.
(ii) Differentiate both sides of the given equation, remembering that the factor dy/dx will occur in any differentiation of a term involving the function y, according to the Chain Rule.
(iii) Solve the resulting equation for dy/dx.

Although we can always carry out this procedure symbolically to obtain an expression for dy/dx (assuming that the necessary differentiation formulas are known), there is a hazard with this procedure. The problem lies in the fact that the given equation need not define a differentiable function of x. Indeed, in Example 1 we could not obtain a value of dy/dx at the point $(4, 0)$ since the expression $\frac{dy}{dx} = -\frac{9x}{16y}$ is undefined when $y = 0$. Geometrically, this corresponds to the observation that the tangent to the graph of equation (2) at the point $(4, 0)$ is vertical.

A careful resolution of this difficulty requires a theorem (called the Implicit Function Theorem) that gives precise conditions under which an equation of the form $f(x, y) = 0$ determines y as a differentiable function of x near a particular point (x_0, y_0). Developing this theorem here would take us far astray from our main objectives, so we simply refer you to more advanced texts for a discussion of this theorem. (See, for example, *Mathematical Analysis,* 2nd ed., by Tom Apostol, Addison-Wesley, 1974.) We will resolve this concern here by cautioning that you should always check to see that both the given equation and the resulting expression for dy/dx are defined at the point (x_0, y_0) of interest.

Example 2 The equation

$$y^2 + x^2y = 3x^2$$

determines y as a differentiable function of x near $(2, 2)$. (You may verify this by plotting a portion of the graph of the equation. We obtain y as a differentiable function of x in the region where $1 < x < 3$ and $1 < y < 3$.) Find an expression for dy/dx near $(2, 2)$, and the slope of the line tangent to the graph of the equation at $(2, 2)$.

Strategy

Solution

Differentiate both sides with respect to x.

$$\underbrace{2y \cdot \frac{dy}{dx}}_{\frac{d}{dx}(y^2)} + \underbrace{2xy + x^2 \cdot \frac{dy}{dx}}_{\frac{d}{dx}(x^2y)} = 6x$$

Collect terms involving $\frac{dy}{dx}$ and solve for $\frac{dy}{dx}$.

Thus,

$$(x^2 + 2y)\frac{dy}{dx} = 6x - 2xy,$$

so

$$\frac{dy}{dx} = \frac{6x - 2xy}{x^2 + 2y}.$$

Substitute given values for x and y to obtain the slope.

For $x = 2$, $y = 2$ we have

$$\frac{dy}{dx} = \frac{6 \cdot 2 - 2 \cdot 2 \cdot 2}{2^2 + 2 \cdot 2} = \frac{1}{2},$$

which is the desired slope. ■

Example 3 Find $\frac{dy}{dx}$ if $\sqrt[3]{x} + \sqrt[3]{y} = 1$, using the fact that $\frac{d}{dx}\sqrt[3]{x} = \frac{1}{3}x^{-2/3}$.

Solution: Differentiating implicitly we obtain

$$\frac{1}{3}x^{-2/3} + \frac{1}{3}y^{-2/3} \cdot \frac{dy}{dx} = 0$$

so

$$\frac{dy}{dx} = -\frac{x^{-2/3}}{y^{-2/3}} = -\left(\frac{y}{x}\right)^{2/3}.$$ ■

Example 4 Find the equation of the line tangent to the graph of

$$y^3 - x^2 = 7 \tag{5}$$

at the point (1, 2).

Solution: We could begin by solving equation (5) for y. We would obtain

$$y = (x^2 + 7)^{1/3}$$

so we could obtain dy/dx by the Power Rule. However, we prefer to proceed by differentiating equation (5) implicitly. We obtain

$$3y^2 \cdot \frac{dy}{dx} - 2x = 0$$

so

$$\frac{dy}{dx} = \frac{2x}{3y^2}.$$

At the point (1, 2) the slope of the desired line is therefore

$$m = \frac{2 \cdot 1}{3 \cdot 2^2} = \frac{1}{6}.$$

The equation for the line is

$$y - 2 = \frac{1}{6}(x - 1)$$

or

$$6y - x - 11 = 0.$$

(See Figure 7.5.) ■

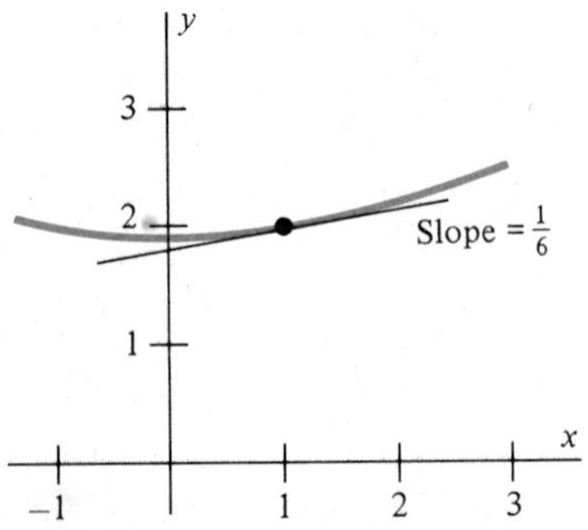

Figure 7.5 Tangent to graph of $y^3 - x^2 = 7$ at (1, 2).

Given an equation in x and y we will sometimes need to compute d^2y/dx^2 as well as dy/dx. If the equation defines y implicitly as a twice differentiable function of x, we can often compute d^2y/dx^2 by applying the method of implicit differentiation twice.

Example 5 The equation $9x^2 + 4y^2 = 36$ defines y as a twice differentiable function of x near the point $(0, 3)$. Find $\dfrac{d^2y}{dx^2}$ for this function.

Strategy

Solution

Differentiate both sides of given equation and solve for $\dfrac{dy}{dx}$.

$$18x + 8y \cdot \frac{dy}{dx} = 0$$

$$\frac{dy}{dx} = -\frac{9x}{4y}.$$

Differentiate resulting equation (using the Quotient Rule).

Differentiating both sides of this equation gives

$$\frac{d^2y}{dx^2} = -\frac{(4y)9 - (9x) \cdot 4 \cdot \dfrac{dy}{dx}}{16y^2}$$

$$= -\frac{36y - 36x \cdot \dfrac{dy}{dx}}{16y^2}$$

Substitute for $\dfrac{dy}{dx}$ as found above.

$$= -\frac{36y - 36x\left(-\dfrac{9x}{4y}\right)}{16y^2}$$

$$= -\frac{9}{4y} - \frac{81x^2}{16y^3}.$$

■

Exercise Set 3.7

In Exercises 1–20, find $\dfrac{dy}{dx}$ by implicit differentiation.

1. $x^2 + y^2 = 25$
2. $x = \sin y$
3. $x^{1/2} + y^{1/2} = 4$
4. $x^2 + 2xy + y^2 = 8$
5. $x = \tan y$
6. $x^2y + xy^2 = 0$
7. $x \sin y = y \cos x$
8. $x = y(y - 1)$
9. $(xy)^{1/2} = xy - x$
10. $y^2 = \sin^2 x - \cos^2 2x$
11. $x^3 + x^2y + xy^2 + y^3 = 0$
12. $\cos(x + y) = y \sin x$
13. $\sqrt{x + y} = xy - x$
14. $y^2 = \dfrac{x + 1}{x^2 + 1}$
15. $\cot y = 3x^2 + \cot(x + y)$
16. $x^2y^2 = 4$
17. $\sin(xy) = 1/2$
18. $\dfrac{1}{x} + \dfrac{1}{y} + \dfrac{1}{4} = 0$
19. $y^4 = x^5$
20. $\sin(x + y) + \cos(x - y) = 1$

In Exercises 21–24, find $\dfrac{dy}{dx}$ and $\dfrac{d^2y}{dx^2}$:

21. $y^2 = 4x$
22. $x^2 + y^2 = 1$
23. $y^2 - xy = 4$
24. $\sqrt{x} + \sqrt{y} = 1$

In each of Exercises 25–30, find an equation for the line tangent to the graph determined by the given equation at the given point.

25. $xy = 9 \quad (3, 3)$
26. $x^2 + y^2 = 4 \quad (\sqrt{2}, \sqrt{2})$

27. $x^3 + y^3 = 16 \quad (2, 2)$

28. $x^2y^2 = 16 \quad (-1, 4)$

29. $\dfrac{x+y}{x-y} = 4 \quad (5, 3)$ **30.** $(y - x)^2 = x \quad (9, 12)$

31. Assume that the equation $f(x, y) = 0$ determines y as a differentiable function of x. Explain why $\dfrac{dy}{dx} = 0$ at points where the graph has a horizontal tangent.

32. Assume that the equation $f(x, y) = 0$ determines x as a differentiable function of y. Explain why $dx/dy = 0$ at points where the graph has a vertical tangent. (We assume the y-axis to be vertical in both Exercises 31 and 32.)

In Exercises 33–38, use the results of Exercises 31 and 32 to find all horizontal and vertical tangents.

33. $x^2 + y^2 = 2$ **34.** $x^2 + 4y^2 = 4$

35. $x^2 - y^2 = 1$ **36.** $(x - 1)^2 + (y + 2)^2 = 9$

37. $x^2 + y^2 + 2x + 4y = -4$

38. $(x + 1)(y - 1) = 1$

39. Use implicit differentiation to show that $\dfrac{d}{dx}(x^{p/q}) = \dfrac{p}{q}x^{p/q-1}$ for $x > 0$ and p, q integers. (*Hint:* If $y = x^{p/q}$, $y^q = x^p$. Assume y to be differentiable.)

40. Find the value of a so that the circles with equations $(x - a)^2 + y^2 = 2$ and $(x + a)^2 + y^2 = 2$ intersect at points where their tangents are perpendicular.

41. Find an equation for the line normal to the curve $9x^2 + 16y^2 = 144$ at the point $(2, 3\sqrt{3}/2)$.

42. Prove that all normals to the curve $x^2 + y^2 = a^2$ pass through the origin.

43. Explain why the equation $4 + x^2 + y^2 = 2xy$ does not define a function $y = f(x)$ for *any* values of x and y. (*Hint:* Rewrite as $x^2 - 2xy + y^2 = -4$. What is the sign of the left-hand side?)

44. Some of the following equations do not determine y as a function of x. Others do, although we may not be able to solve them in practice. Which determine y as a function of x?

a. $x = y^2 - 7$ **b.** $x = y^3$
c. $x = y + y^3$ **d.** $x = y^3 + y^5 - 7$
e. $x = y^3 + y^4 - y^5$ **f.** $x = 3 - y^4 + y$

45. Note that the graph of the equation $xy^2 + x^2y = 4x^2$ consists of the vertical line $x = 0$ (y-axis) in addition to the graph of the equation $y^2 + xy = 4x$. Why is this so? What does this say about the existence of a derivative dy/dx "near" $x = 0$?

3.8 RATIONAL EXPONENTS AND INVERSE FUNCTIONS

Up to this point we have said very little about differentiating the power function $f(x) = x^n$ when n is not an integer. Using the basic definition of the derivative we were able to show that

$$\frac{d}{dx}\sqrt{x} = \frac{d}{dx}(x^{1/2}) = \frac{1}{2}x^{-1/2} = \frac{1}{2\sqrt{x}} \tag{1}$$

and

$$\frac{d}{dx}\sqrt[3]{x} = \frac{d}{dx}(x^{1/3}) = \frac{1}{3}x^{-2/3} = \frac{1}{3x^{2/3}}. \tag{2}$$

However, none of the rules that we have developed so far in this chapter speak to the issue of differentiating nonintegral powers of x.

No doubt you have noticed that both differentiation formulas (1) and (2) correspond to the Power Rule, $\dfrac{d}{dx}(x^n) = nx^{n-1}$, with $n = \dfrac{1}{2}$ and $n = \dfrac{1}{3}$, respectively. The following theorem tells us that the Power Rule extends to *all* power functions, $f(x) = x^r$, with r rational.

THEOREM 11

Let $f(x) = x^{p/q}$, where p and q are integers and $q \neq 0$. Then $f(x)$ is differentiable and

$$f'(x) = \frac{p}{q} x^{p/q-1} \tag{3}$$

whenever $f(x)$ and the right side of equation (3) are defined. In Leibniz notation,

$$\frac{d}{dx}(x^{p/q}) = \left(\frac{p}{q}\right) x^{p/q-1}. \tag{4}$$

If we write $r = \dfrac{p}{q}$, then statements (3) and (4) are just statements of the Power Rule formula, $\dfrac{d}{dx}x^r = rx^{r-1}$. We will therefore simply refer to the Power Rule (which now applies to *all* rational powers of x) when using Theorem 11.

The tricky part of proving Theorem 11 is showing that the power functions $f(x) = x^{p/q}$ are indeed differentiable on the stated intervals. In order to show this we need to develop the notion of the *inverse* of a function, which we do later in this section. Setting the differentiability issue aside for the moment, we can establish equation (3) by implicit differentiation as follows. Let $x \neq 0$ be in the domain of the function

$$y = x^{p/q}$$

where p and $q \neq 0$ are integers. Then $y \neq 0$, and

$$y^q = (x^{p/q})^q = x^p. \tag{5}$$

If we assume y to be differentiable, then since q is an integer, y^q is differentiable. Applying the Power Rule (for integral powers) to both sides of equation (5) gives

$$qy^{q-1} \cdot \frac{dy}{dx} = px^{p-1}.$$

Recalling that $y = x^{p/q} \neq 0$, we can now solve for $\dfrac{dy}{dx}$ as

$$\begin{aligned}
\frac{dy}{dx} &= \left(\frac{p}{q}\right) x^{p-1}y^{1-q} \\
&= \left(\frac{p}{q}\right) x^{p-1}(x^{p/q})^{1-q} \\
&= \left(\frac{p}{q}\right) x^{p-1} \cdot x^{p/q-p} \\
&= \left(\frac{p}{q}\right) x^{(p-1+p/q-p)} \\
&= \left(\frac{p}{q}\right) x^{p/q-1},
\end{aligned}$$

which is the formula in Theorem 11. Before turning to the issue of differentiability, we consider several examples making use of Theorem 11.

Example 1 For $f(x) = \sqrt[4]{x^3 - x^2 + 3}$, find $f'(x)$.

Strategy

Replace the radical sign with a fractional exponent.

Solution

Since $f(x) = (x^3 - x^2 + 3)^{1/4}$,

$$f'(x) = \frac{1}{4}(x^3 - x^2 + 3)^{-3/4} \cdot \frac{d}{dx}(x^3 - x^2 + 3)$$

Apply Theorem 11 together with the Chain Rule.

$$= \frac{1}{4}(x^3 - x^2 + 3)^{-3/4}(3x^2 - 2x).$$

■

Example 2 Find an equation for the line tangent to the graph of the function

$$f(x) = \frac{\sin(x^{4/3} - x^{3/2})}{(x^{7/2} + 4)^{3/2}}$$

at the point (1, 0).

Solution: Applying the Quotient Rule, Chain Rule, and Power Rule (Theorem 11), we find that

$$f'(x) = \frac{(x^{7/2} + 4)^{3/2} \cdot \frac{d}{dx} \sin(x^{4/3} - x^{3/2}) - \sin(x^{4/3} - x^{3/2}) \cdot \frac{d}{dx}(x^{7/2} + 4)^{3/2}}{[(x^{7/2} + 4)^{3/2}]^2}$$

$$= \frac{(x^{7/2} + 4)^{3/2} \cos(x^{4/3} - x^{3/2}) \cdot \left[\frac{4}{3}x^{1/3} - \frac{3}{2}x^{1/2}\right]}{(x^{7/2} + 4)^3}$$

$$- \frac{\sin(x^{4/3} - x^{3/2})\left[\frac{3}{2}(x^{7/2} + 4)^{1/2} \cdot \left(\frac{7}{2}x^{5/2}\right)\right]}{(x^{7/2} + 4)^3}$$

Substituting $x = 1$ gives

$$f'(1) = \frac{(1 + 4)^{3/2}\cos(1 - 1)\left(\frac{4}{3} - \frac{3}{2}\right) - \sin(1 - 1)\left[\frac{3}{2}(1 + 4)^{1/2} \cdot \left(\frac{7}{2}\right)\right]}{(1 + 4)^3}$$

$$= \frac{5^{3/2} \cdot 1 \cdot \left(\frac{4}{3} - \frac{3}{2}\right) - 0}{5^3} = -\frac{1}{6 \cdot 5^{3/2}}.$$

This is the slope. Since the line contains the point (1, 0), an equation for the line is

$$y = -\frac{1}{6 \cdot 5^{3/2}}(x - 1).$$

■

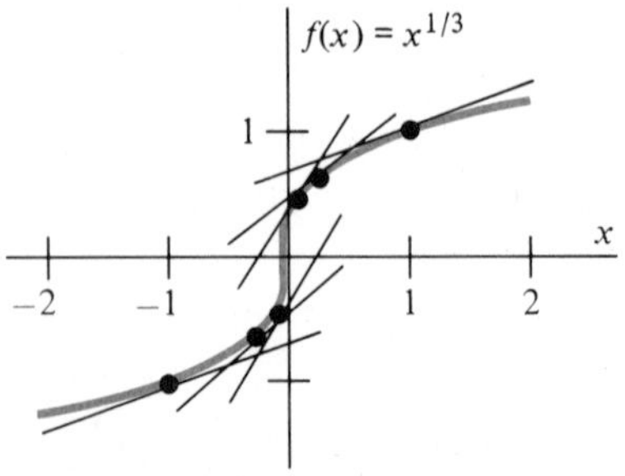

Figure 8.1 Graph of $f(x) = x^{1/3}$ and several tangents.

Example 3 Figure 8.1 shows the graph of $f(x) = x^{1/3}$, with several tangents sketched in. According to the Power Rule, the derivative (slope function) for $f(x) = x^{1/3}$ is the function $f'(x) = \frac{1}{3}x^{-2/3}$. The graph of $f'(x)$ appears in Figure

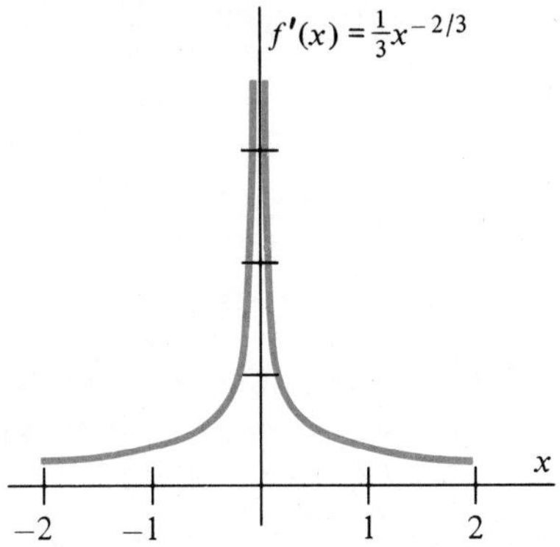

Figure 8.2 Graph of "slope function" $f'(x) = \dfrac{1}{3}x^{-2/3}$.

8.2. Note that $f'(0)$ is undefined and that the slopes of the tangents approach $+\infty$ as x approaches zero from either direction. This illustrates the fact that $f(x) = x^{1/3}$ is not differentiable at $x = 0$.

However, the graph of $f(x) = x^{1/3}$ *does* have a tangent at $x = 0$ in the sense in which we defined tangents in Section 2.2. As Figure 8.3 illustrates, the slope of the secant line through (0, 0) and $(\Delta x, f(\Delta x)) = (\Delta x, \Delta x^{1/3})$ is

$$\begin{aligned} m_{sec} &= \frac{f(0 + \Delta x) - f(0)}{\Delta x} \\ &= \frac{\Delta x^{1/3} - 0}{\Delta x} \\ &= \frac{1}{\Delta x^{2/3}}. \end{aligned}$$

Thus, m_{sec} approaches $+\infty$ as $\Delta x \to 0$ from either direction, and the secant through (0, 0) and $(\Delta x, \Delta x^{1/3})$ rotates into the position of the vertical line through (0, 0). That is, the tangent to the graph of $f(x) = x^{1/3}$ at (0, 0) is just the y-axis. This shows that the graph of $y = f(x)$ may have a *vertical* tangent when $f'(a)$ fails to exist. ■

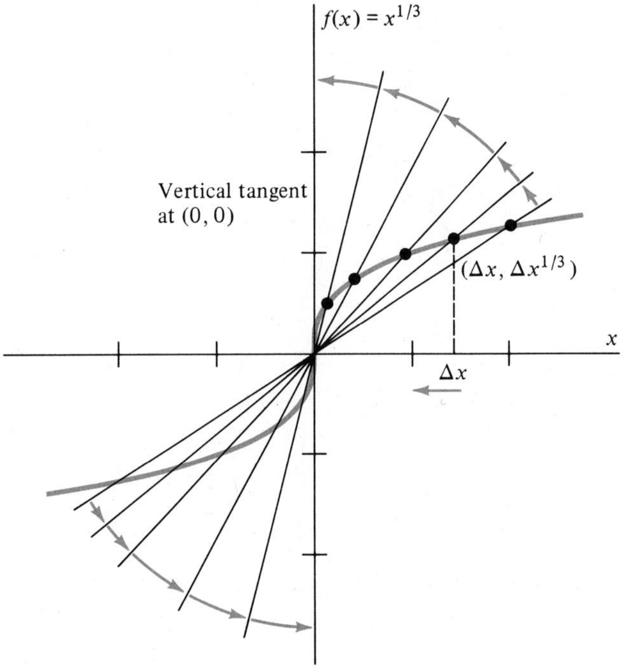

Figure 8.3 Secants through (0, 0) approach the y-axis as $\Delta x \to 0$. Tangent to $f(x) = x^{1/3}$ at (0, 0) is vertical.

Example 4 The graph of $f(x) = x^{2/3}$ appears in Figure 8.4. Since $f'(x) = \dfrac{2}{3}x^{-1/3}$, $f'(0)$ is undefined. However, the situation on the graph of

$f(x) = x^{2/3}$ at (0, 0) is similar to that of Example 3. The secants through (0, 0) have slope

$$m_{\text{sec}} = \frac{(0 + \Delta x)^{2/3} - 0}{\Delta x} = \frac{1}{\Delta x^{1/3}},$$

so $m_{\text{sec}} \to +\infty$ as Δx approaches zero from the right, and $m_{\text{sec}} \to -\infty$ as Δx approaches zero from the left. Again, the secants rotate into the y-axis as $\Delta x \to 0$. Thus, the graph of $f(x) = x^{2/3}$ has a vertical tangent at $x = 0$, even though $f'(0)$ is undefined.

Note, however, that the graph of $f(x) = x^{2/3}$ looks quite different from the graph of $g(x) = x^{1/3}$ near (0, 0). This is because $f(x) = x^{2/3} = (x^{1/3})^2$ is an *even* function, while $g(x) = x^{1/3}$ is an *odd* function. Because the tangent to the graph of $f(x) = x^{2/3}$ at (0, 0) is vertical, and because the branch of the graph to the left of $x = 0$ must be the mirror image of the branch to the right of $x = 0$, a "cusp" (sharp point) exists at (0, 0). Can you explain why the fact that $g(x) = x^{1/3}$ is an odd function rules out the existence of a cusp on its graph at (0, 0)? ■

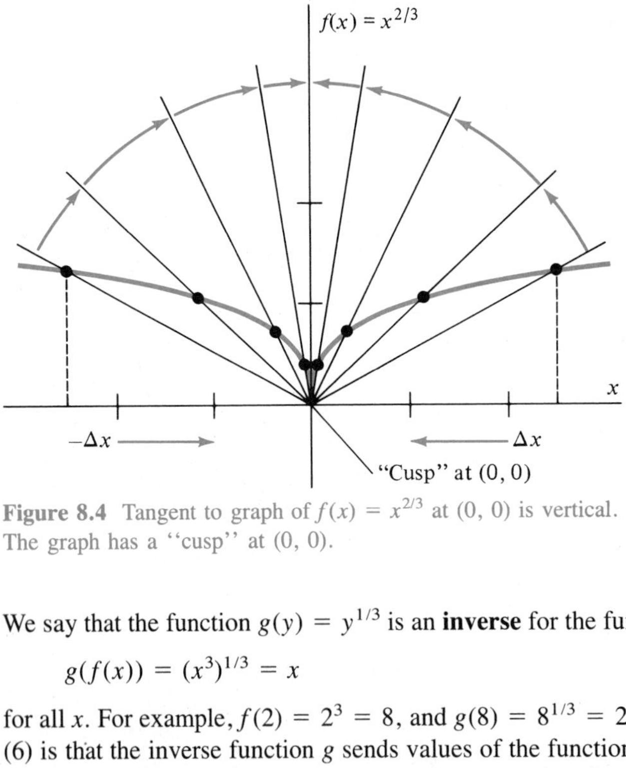

Figure 8.4 Tangent to graph of $f(x) = x^{2/3}$ at (0, 0) is vertical. The graph has a "cusp" at (0, 0).

Inverse Functions

We say that the function $g(y) = y^{1/3}$ is an **inverse** for the function $f(x) = x^3$ since

$$g(f(x)) = (x^3)^{1/3} = x \tag{6}$$

for all x. For example, $f(2) = 2^3 = 8$, and $g(8) = 8^{1/3} = 2$. The point of equation (6) is that the inverse function g sends values of the function $f(x)$ right back where they came from—to the number x. Another way of saying this is that the inverse function g **returns** the value $f(x)$ to the number x (see Figure 8.5).

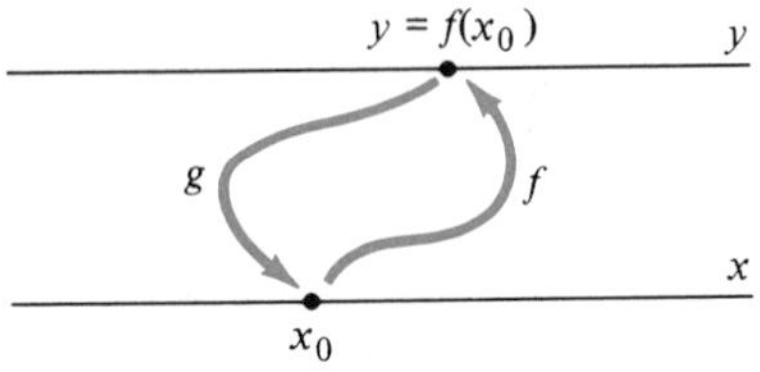

Figure 8.5 g is an inverse for f if $g(y_0) = x_0$ when $y_0 = f(x_0)$.

Example 5 Here are some other examples of inverse functions.

$y = f(x)$	$g(y)$ = inverse of $f(x)$	$g(f(x)) = x$
$f(x) = x$	$g(y) = y$	$g(f(x)) = g(x) = x$
$f(x) = 4x$	$g(y) = \frac{1}{4}y$	$g(f(x)) = \frac{1}{4}(4x) = x$
$f(x) = x + 1$	$g(y) = y - 1$	$g(f(x)) = (x + 1) - 1 = x$
$f(x) = x^5$	$g(y) = y^{1/5}$	$g(f(x)) = (x^5)^{1/5} = x$
$f(x) = 2x - 1$	$g(y) = \frac{y + 1}{2}$	$g(f(x)) = \frac{(2x - 1) + 1}{2} = x$

■

If g is an inverse for a given function $f(x)$, then $g(y) = x$ whenever $f(x) = y$. Applying f to both sides of the first equation and using the second shows that

$$f(g(y)) = f(x) = y.$$

That is, $f(g(y)) = y$, which says that g is an inverse for f if f is an inverse for g. We can often obtain an explicit formula for the inverse function g by solving this equation for $g(y)$. The next example shows how this can be done.

Example 6 Find an inverse for the function $f(x) = \frac{1}{2}x - 2$.

Strategy

Set up the equation

$f(g(y)) = y$.

Solution

The equation $f(g(y)) = y$ becomes

$$\frac{1}{2}g(y) - 2 = y,$$

so

Solve for $g(y)$.

$$\frac{1}{2}g(y) = y + 2.$$

Thus,

$$g(y) = 2y + 4$$

is the inverse for $f(x) = \frac{1}{2}x - 2$. To verify this, we note that

Verify the answer by showing that $g(f(x)) = x$.

$$\begin{aligned} g(f(x)) &= g\left(\frac{1}{2}x - 2\right) \\ &= 2\left(\frac{1}{2}x - 2\right) + 4 \\ &= x - 4 + 4 \\ &= x. \end{aligned}$$

■

Domains and Ranges of Inverses

We have been making use of two basic equations in the case where g is an inverse for f:

$$g(f(x)) = x \qquad (7)$$

$$f(g(y)) = y \qquad (8)$$

Equation (7) is what we mean by saying that g is an inverse for f. Obviously, the function g must be defined for the number $y = f(x)$ if Equation (7) is to make sense. That is, we must have the *range* of f contained within the *domain* of g. Similarly, we must caution you that Equation (8) is defined only for those numbers y in the domain of g for which $g(y)$ lies in the domain of f.

In order to graph both a function and its inverse on the same pair of axes, we will often write both functions in terms of the same independent variable x. Using this convention, we would write the inverse of the function $f(x) = \frac{1}{2}x - 2$ in Example 6 as simply $g(x) = 2x + 4$.

NOTATION: The notation $g(x) = f^{-1}(x)$ is used to denote the inverse of the function $f(x)$. Thus we can write

(i) for $f(x) = x^5$, $\quad f^{-1}(x) = x^{1/5}$ $\quad$ (Example 5)

(ii) for $f(x) = \frac{1}{2}x - 2$, $\quad f^{-1}(x) = 2x + 4$ $\quad$ (Example 6)

and so forth. *It is very important that you do not confuse the notation for inverses with the notation for reciprocals.* That is,

$f^{-1}(x)$ means the inverse for the function,

$[f(x)]^{-1}$ means $\dfrac{1}{f(x)}$, $\quad f(x) \neq 0$,

and

$$f^{-1}(x) \neq [f(x)]^{-1}$$

in general.

Existence of Inverses

Unfortunately, not every function has an inverse, and Figures 8.6 and 8.7 show why. If the function $g = f^{-1}$ is to return the value $y = f(x)$ to the number x, there must be only one value $f(x)$ associated with each number x. Otherwise, the "return function" $g = f^{-1}$ will not be a *function,* since it will be assigning more than one value $g(f(x)) = f^{-1}(f(x)) = f^{-1}(y)$ to the number $y = f(x)$.

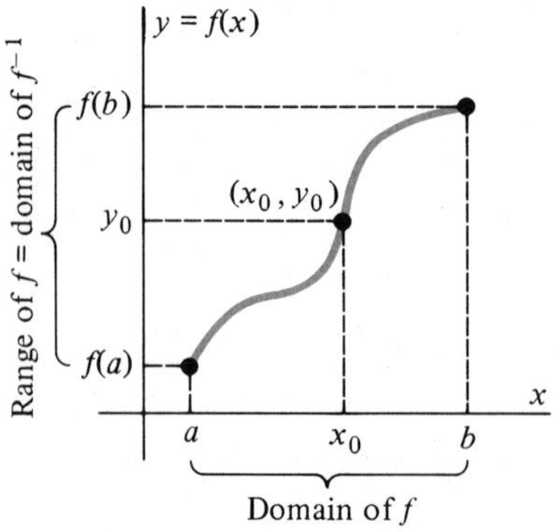

Figure 8.6 This function has an inverse since "returns are unique." There is only one x_0 with $f(x_0) = y_0$, so $f^{-1}(y_0) = x_0$ is unique.

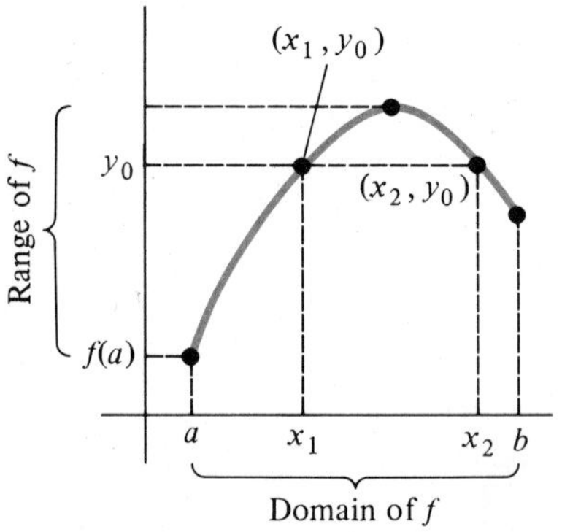

Figure 8.7 This function has no inverse since "returns are not unique." Since both $y_0 = f(x_1)$ and $y_0 = f(x_2)$, $f^{-1}(y_0)$ is not a *single* number.

Functions that have the property of Figure 8.6 are called *one-to-one* functions. Here is a more precise definition.

DEFINITION 3

The function $y = f(x)$ is called **one-to-one** if, whenever y is in the range of $f(x)$, $y = f(x)$ for only one value of x. In other words, $y = f(x)$ is one-to-one if $f(x_1) = f(x_2)$ implies that $x_1 = x_2$.

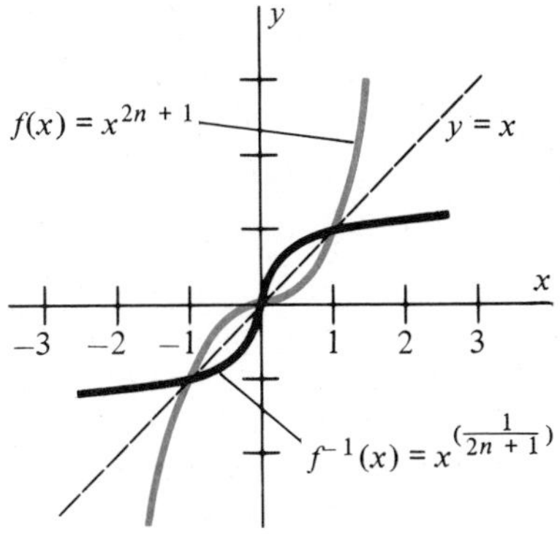

Figure 8.8 Odd power function $f(x) = x^{2n+1}$ has inverse $f^{-1}(x) = x^{1/(2n+1)}$. (Graph is for case $2n + 1 > 0$.)

Example 7 Since odd roots are unique, the odd power functions

$$f(x) = x^{2n+1}, \qquad n \text{ an integer} \tag{9}$$

are one-to-one functions. For example, if $f(x) = x^3$, then the condition that $f(x_1) = f(x_2)$ becomes $x_1^3 = x_2^3$, so $x_1 = x_2$. The inverse for the odd power function in (9) is the odd root function

$$f^{-1}(x) = x^{1/(2n+1)}$$

since $f^{-1}(f(x)) = (x^{2n+1})^{1/(2n+1)} = x$. Both $f(x)$ and $f^{-1}(x)$ are defined for all values of x (except $x = 0$, if $2n + 1 < 0$) (see Figure 8.8). ■

Example 8 The even power functions

$$f(x) = x^{2n}, \qquad n \text{ an integer}$$

are *not* one-to-one, since $f(-x) = (-x)^{2n} = x^{2n} = f(x)$ for all x. For example, if $f(x) = x^2$, the equation $f(x) = 4$ has two solutions: $x = 2$ and $x = -2$. However, Figure 8.9 shows that if $n \geq 0$ and we restrict the domain of the even power functions to $[0, \infty)$, the even power functions $f(x) = x^{2n}$, $x \geq 0$, become one-to-one. (If $n < 0$ we must further restrict the domain to $(0, \infty)$, since $f(0)$ is undefined.) When so restricted, the even power functions have inverses $f^{-1}(x) = x^{1/(2n)}$. ■

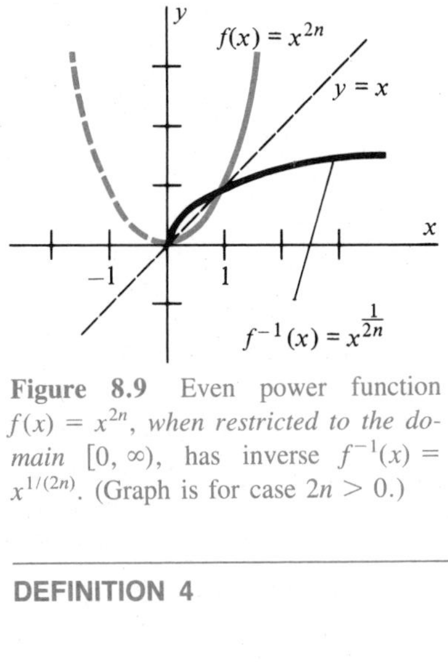

Figure 8.9 Even power function $f(x) = x^{2n}$, *when restricted to the domain* $[0, \infty)$, has inverse $f^{-1}(x) = x^{1/(2n)}$. (Graph is for case $2n > 0$.)

Geometrically, the test for determining whether a function is one-to-one is simple: no horizontal line can intersect its graph in more than one point (see Exercise 37).

As Examples 7 and 8 illustrate, the relationship between $f(x)$ being one-to-one and the existence of $f^{-1}(x)$ is this: If $f(x)$ is one-to-one on its domain D, then f^{-1}, the "return function," exists. Otherwise, returns are not unique, so f^{-1} cannot exist. This is why we formally define f^{-1} as follows.

DEFINITION 4

Let $f(x)$ be a one-to-one function with domain D. The **inverse** of $f(x)$, denoted by f^{-1}, is the function defined by the statement

$$f^{-1}(y) = x \qquad \text{if} \qquad f(x) = y, \qquad x \in D.$$

The domain of f^{-1} is the range of f.

THE GRAPH OF f^{-1}: Note in both Figure 8.8 and Figure 8.9 that the graph of $y = f^{-1}(x)$ appears to be the reflection in the line $y = x$ of the graph of $y = f(x)$. Indeed, this will always be the case since if $b = f(a)$, then $a = f^{-1}(b)$. This means that the point (b, a) will lie on the graph of $y = f^{-1}(x)$ whenever (a, b) lies on the graph of $y = f(x)$, and conversely (see Figure 8.10).

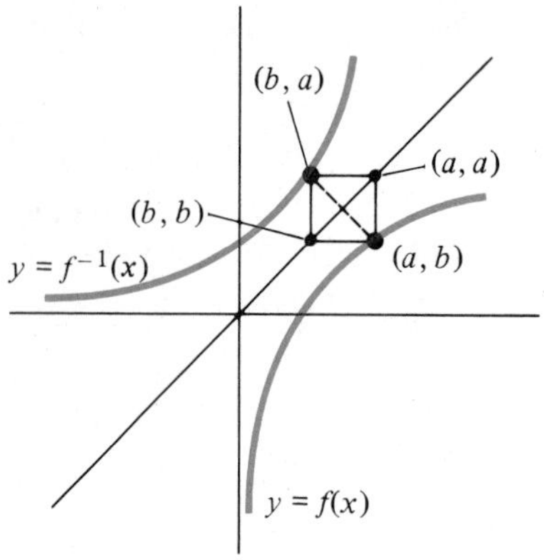

Figure 8.10 Graphs of $y = f(x)$ and $y = f^{-1}(x)$ are reflections in line $y = x$.

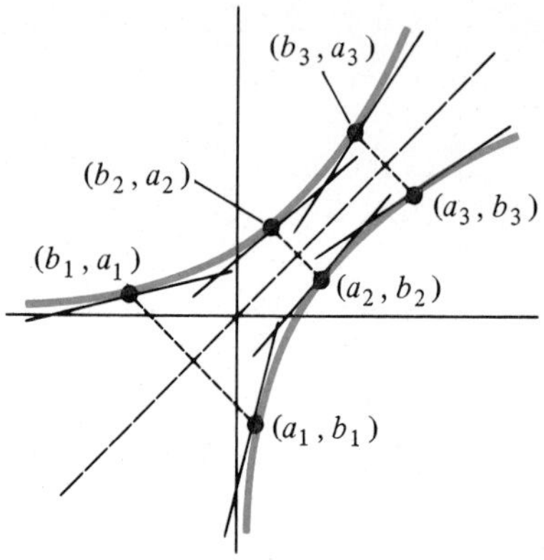

Figure 8.11 Slope of $y = f(x)$ at (a, b) is reciprocal of slope of $y = f^{-1}(x)$ at (b, a).

Continuity of Inverses

Since the graph of $f^{-1}(x)$ can be obtained from the graph of $f(x)$ by simply reflecting the latter in the line $y = x$, the continuity properties of the two graphs should correspond. That is, if the graph of $f(x)$ is "unbroken" at (a, b), the graph of $f^{-1}(x)$ should be "unbroken" at (b, a). The following theorem verifies our intuition, although we shall not give a formal proof.

THEOREM 12

If the function $y = f(x)$ has an inverse function $y = f^{-1}(x)$ on an interval containing the number a, and if $f(x)$ is continuous at $x = a$, then the inverse function $f^{-1}(x)$ is continuous at $b = f(a)$.

In other words, the inverse of a continuous function is a continuous function, where defined.

Differentiability of Inverses

Figure 8.11 suggests that a simple relationship should exist between the derivative $f'(a)$, if it exists, and the derivative $f^{-1}(x)$ at $x = b = f(a)$. Since $f'(a)$ is the slope of the line tangent to the graph of $y = f(x)$ at (a, b), and since the graph of $y = f^{-1}(x)$ is obtained from the graph of $y = f(x)$ by interchanging x- and y-coordinates of points on the graph, the slope of the line tangent to the graph of $y = f^{-1}(x)$ at (b, a) should be the reciprocal of the slope of the tangent to the graph of $y = f(x)$ at (a, b).

The following theorem supports this observation. Its proof is outlined in Exercise 40.

THEOREM 13

Suppose that $f(x)$ is continuous on an open interval containing $x = a$, that $f(x)$ has an inverse $f^{-1}(x)$, that $f'(a)$ exists, and that $f'(a) \neq 0$. Let $b = f(a)$, and assume that $f^{-1}(x)$ is defined on an open interval containing b. Then $(f^{-1})'(b)$ exists, and

$$(f^{-1})'(b) = \frac{1}{f'(a)}$$

That is,

$$\frac{d}{dx} f^{-1}(x) = \frac{1}{f'(f^{-1}(x))}.$$

Actually, we shall make little use of the differentiation formula in Theorem 13, relying instead on the method of implicit differentiation to obtain formulas for derivatives of various inverse functions, as we did at the beginning of this section. However, at several points in the text we shall need to call on Theorem 13 to guarantee that inverse functions indeed are differentiable before we go hunting for their derivatives.

The Differentiability of Power Functions

Theorem 13 enables us to verify the assertion of Theorem 11 that the power functions $f(x) = x^{p/q}$ are indeed differentiable. The argument goes like this:

(i) If q is an odd integer, the power function $g(x) = x^q$ is one-to-one where defined, and its inverse is $f(x) = g^{-1}(x) = x^{1/q}$. Since $g'(x) = qx^{q-1}$ exists, by the Power Rule, the inverse function $f(x) = x^{1/q}$ is differentiable, for $x \neq 0$, by Theorem 13.

(ii) If q is an even integer, the power function $g(x) = x^q$ is one-to-one when restricted to the domain $(0, \infty)$. The same argument as in part (i) shows that its inverse, $f(x) = x^{1/q}$, is differentiable for all $x > 0$.

(iii) The power function $f(x) = x^{p/q}$ may be written as a composite function

$$f(x) = (x^{1/q})^p = u^p, \qquad u = x^{1/q}.$$

Since the function $g(u) = u^p$ is differentiable for all u (by the Power Rule), the Chain Rule and the conclusions in (ii) and (iii) above guarantee that the power function $f(x) = x^{p/q}$ is differentiable whenever $u = x^{1/q}$ is differentiable, that is, for $x \neq 0$ if q is odd, and for $x > 0$ if q is even. In other words, the function $f(x) = x^{p/q}$ is differentiable wherever both sides of the formula

$$\frac{d}{dx}(x^{p/q}) = \frac{p}{q} x^{p/q-1}$$

are defined.

Exercise Set 3.8

In Exercises 1–22, find $f'(x)$.

1. $f(x) = (x + 2)^{4/3}$

2. $f(x) = \sqrt[3]{x^2 + 1}$

3. $f(x) = \sqrt{x} + \dfrac{1}{\sqrt{x}}$

4. $f(x) = \dfrac{\sqrt[3]{x + 2}}{1 + \sqrt{x}}$

5. $f(x) = x^{2/3} + x^{-2/3}$

6. $f(x) = (3x^4 + 4x^3)^{-2}$

7. $f(x) = (x^2 - 1)^{-2}(x^2 + 1)^2$

8. $f(x) = (7 - x^2)^{2/3}$

9. $f(x) = \sqrt{1 + \sqrt[3]{x}}$

10. $f(x) = x^{5/2} + 3x^{2/3} + x^{-4/3}$

11. $f(x) = \dfrac{\sqrt[3]{x}}{\sqrt[3]{x} + x}$

12. $f(x) = \sqrt{x^3} \cdot (x^{-2} + 2x^{-1} + 1)^4$

13. $f(x) = \sqrt{x} + \sqrt[4]{x} + \sqrt[8]{x}$

14. $f(x) = \dfrac{x^{2/3}}{\sqrt[3]{x^2 + 1}}$

15. $f(x) = \dfrac{(x^2 + 1)^{2/3}}{(1 - x^2)^{4/3}}$

16. $f(x) = \dfrac{(x-1)^{1/2} + (x+1)^{1/3}}{(x+2)^{1/4}}$

17. $f(x) = \sqrt{\dfrac{ax^2+b}{cx+d}}$

18. $f(x) = \sqrt[3]{\sin x^2}$

19. $f(x) = \dfrac{x^{1/4}(x^{4/3} - x)}{(\sin^3 x - \sqrt{x})^{1/3}}$

20. $f(x) = \left[\dfrac{\tan(x^{3/2} - x^{1/2})}{1 + \sec(x^3 - \sqrt{x})}\right]^{4/3}$

21. $f(x) = \sin^2(x^{2/3} + x^{-4/3})^{5/2}$

22. $f(x) = (x^{-2/3} - x^{3/2})^{3/5} \cdot (6 - x^{5/3})^{-2/5}$

In Exercises 23–26, find an expression for $(f^{-1})'(b)$ given that $f(x)$ is one-to-one and $f^{-1}(b) = a$.

23. $f(x) = x^3 + x$ **24.** $f(x) = x + \sqrt[3]{x}$

25. $f(x) = \sqrt[3]{1+x}$ **26.** $f(x) = \sqrt[3]{x} + \sqrt[3]{x}$

In Exercises 27–32, find an expression for $f^{-1}(x)$.

27. $f(x) = x + 1$ **28.** $f(x) = 3x - 5$

29. $f(x) = \dfrac{x}{x+3}$ **30.** $f(x) = \dfrac{3x-1}{x+2}, \quad x > -2$

31. $f(x) = \sqrt[3]{x+2}$ **32.** $f(x) = \sqrt[5]{x+1}$

33. Find an equation for the line tangent to the graph of the inverse of the function $f(x) = \dfrac{x}{x-2}$ at the point (2, 4).

34. True or false? A function cannot be its own inverse.

35. Find the value of x for which the graph of $f(x) = \dfrac{x}{1+x}$ and its inverse have parallel tangents.

36. Find the value of x for which the tangents to the graphs of $f(x) = \sqrt{x}$ and its inverse are perpendicular.

37. Explain why no horizontal line can intersect the graph of a one-to-one function more than once (Horizontal Line Property).

38. Since every function has the property that no vertical line can intersect its graph more than once, a one-to-one function must have both the Horizontal and Vertical Line Properties. Use this fact to give a geometric argument for the fact that the inverse of a one-to-one function must also be one-to-one.

39. What is the hazard in writing Theorem 13 in the form

$$\frac{dx}{dy} = \frac{1}{\dfrac{dy}{dx}}?$$

40. Prove Theorem 13 as follows:

a. Set up $(f^{-1})'(b) = \lim\limits_{y \to b} \dfrac{f^{-1}(y) - f^{-1}(b)}{y - b}$.

b. Use the continuity of f^{-1} to conclude that $x \to a$ as $y \to b$.

c. Use (b) to rewrite $(f^{-1})'(b)$ in (a) as

$$(f^{-1})'(b) = \lim_{x \to a} \frac{f^{-1}(f(x)) - f^{-1}(f(a))}{f(x) - f(a)}.$$

d. Explain why we may conclude that $f(x) \neq f(a)$ for x near, but not equal to, a.

e. Use the algebra of limits (Theorem 1, Chapter 2) to show that

$$\lim_{x \to a} \frac{f^{-1}(f(x)) - f^{-1}(f(a))}{f(x) - f(a)} = \frac{1}{\lim\limits_{x \to a}\left(\dfrac{f(x) - f(a)}{x - a}\right)}$$

f. Conclude that $(f^{-1})'(b) = \dfrac{1}{f'(a)}$.

41. Which of the following functions have graphs with vertical tangents at (0, 0)?

a. $f(x) = x^{1/5}$ **b.** $f(x) = x^{7/3}$
c. $f(x) = x^{1/3} + x^2$ **d.** $f(x) = x^{1/3} + x^{2/3}$

42. Find a condition under which the graph of $f(x) = x^{p/q}$ will have a vertical tangent at (0, 0).

3.9 DIFFERENTIALS

In the design and use of photographic equipment a very important formula is the following, which gives the amount of light L entering through a circular aperture of radius r:

$$L = \pi \rho r^2 \tag{1}$$

Here ρ is a constant determined by the intensity of the light in the surrounding environment. Of course, equation (1) is just the formula for the area of the circular opening multiplied by the constant ρ. Let us consider the problem of determining the effect on L of a small change in r. In other words, we want to know what sort of change in light to expect from a given change in the radius of the opening.

One way to answer this question is simply to recompute L from equation (1) for each value of r of interest. For example, let us denote by r_0 the original value of the radius and by L_0 the original amount of light. Then if we desire to increase r by Δr cm to $r_1 = r_0 + \Delta r$ cm, the corresponding amount of light will be

$$L_1 = \pi\rho(r_0 + \Delta r)^2$$

and the increase in light corresponding to Δr is

$$\Delta L = L_1 - L_0 = \pi\rho(r_0 + \Delta r)^2 - \pi\rho r_0^2. \tag{2}$$

Although equation (2) gives us a precise answer for each value of Δr, it does not reveal much about the relationship between the variables Δr and ΔL. To explore this relationship more carefully, we consider the graph of the function L (see Figure 9.1).

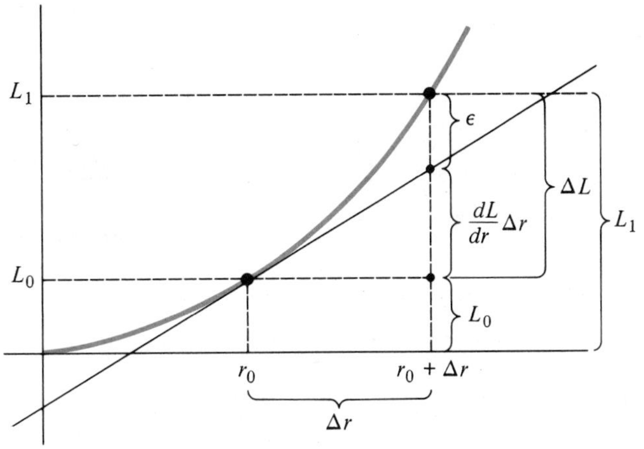

Figure 9.1 Approximating values of the function $L = \pi\rho r^2$.

Since the line tangent to the graph of $L = \pi\rho r^2$ at the point (r_0, L_0) has slope $\dfrac{dL}{dr}$, we can see that a good approximation to the value L_1 is given by

$$L_1 \approx L_0 + \frac{dL}{dr}\Delta r \tag{3}$$

where $\dfrac{dL}{dr}$ is the derivative of the function $L = \pi\rho r^2$ evaluated at $r = r_0$. (The symbol $\approx$ is read "is approximately equal to.") Since $\dfrac{dL}{dr} = 2\pi\rho r$, approximation (3) can be written as

$$L_1 \approx L_0 + 2\pi\rho r_0 \Delta r,$$

or

$$\Delta L \approx (2\pi\rho r_0)\Delta r, \tag{4}$$

where $\Delta L = L_1 - L_0$.

In terms of our original question, approximation (4) is a more satisfying answer than is equation (2) since we have the direct (i.e., *linear*) relationship, "ΔL is approximately the product of the constant $(2\pi p r_0)$ multiplied by the change in r, Δr." Of course, the trade-off is that equation (2) is exact while the simpler approximation (4) is not. The error ϵ in approximation (4) is illustrated in Figure 9.1.

The preceding example suggests a powerful method for approximating the value of a function $f(x)$ at a number $x = x_0 + \Delta x$ when the function value $f(x_0)$ and the derivative $f'(x_0)$ are both known. The method is simply to follow the tangent at $(x_0, f(x_0))$ to the point with x-coordinate $(x_0 + \Delta x)$ rather than following the actual graph of the function $y = f(x)$. As Figure 9.2 illustrates, in so doing we obtain the following approximation formula.

$$f(x_0 + \Delta x) \approx f(x_0) + f'(x_0)\Delta x \tag{5}$$

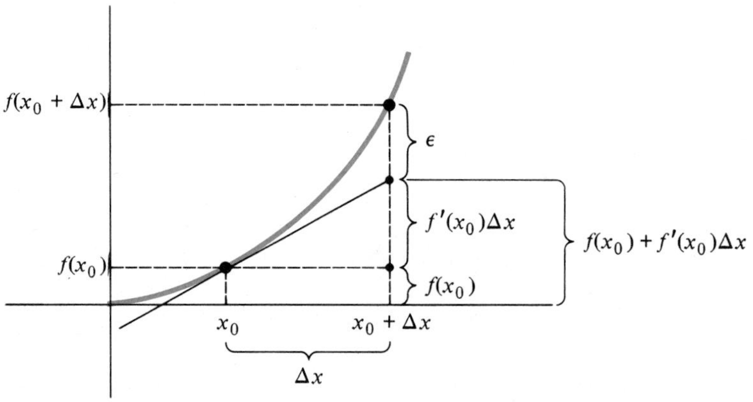

Figure 9.2 $f(x_0) + f'(x_0)\Delta x$ approximates $f(x_0 + \Delta x)$.

Example 1 Use approximation (5) to obtain an approximation for the quantity $\sqrt{36 + \Delta x}$.

Strategy

Identify x_0, and $f(x)$.

Compute $f'(x)$.

Insert results in approximation (5).

Solution

Here we take $x_0 = 36$ and $f(x) = \sqrt{x}$.

Then $f'(x) = \dfrac{1}{2\sqrt{x}}$, so approximation (5) gives

$$\begin{aligned}\sqrt{36 + \Delta x} = f(36 + \Delta x) &\approx f(36) + f'(36)\Delta x \\ &= \sqrt{36} + \frac{1}{2\sqrt{36}}\Delta x \\ &= 6 + \frac{\Delta x}{12},\end{aligned}$$

so

$$\sqrt{36 + \Delta x} \approx 6 + \frac{\Delta x}{12}.$$

Table 9.1 displays values of this approximation for various values of Δx and compares them with the actual values of $\sqrt{36 + \Delta x}$. Notice that the approximation improves as $\Delta x \to 0$. ■

Table 9.1 Approximations to $\sqrt{36 + \Delta x}$

(a) Δx	(b) Actual value $\sqrt{36 + \Delta x}$	(c) Approximation $6 + \frac{\Delta x}{12}$	(d) Absolute error $\lvert(b) - (c)\rvert$	(e) Relative error $\left\lvert\frac{(b) - (c)}{(b)}\right\rvert$
4.0	6.324555	6.333333	.008778	.001388
2.0	6.164414	6.166667	.002253	.000365
1.5	6.123724	6.125000	.001276	.000208
1.0	6.082762	6.083333	.000571	.000094
0.5	6.041523	6.041667	.000144	.000024
0.1	6.008328	6.008333	.000006	.000001

Example 2 Find an approximation to sin 42°.

Strategy

Choose a convenient x_0 near 42°.

Solution

We take $x_0 = 45° = \pi/4$ since we are familiar with the fact that

$$\sin(\pi/4) = \frac{\sqrt{2}}{2}.$$

This means that $42° = x_0 + \Delta x$, so

Identify Δx.

$$\Delta x = 42° - 45° = -3°.$$

Convert Δx to radians.

We must convert Δx to radians, however:

$$\Delta x = \frac{-3°}{360°}(2\pi) = -.05236 \text{ radians.}$$

Calculate $f'(x)$.

Since $f(x) = \sin x$, $f'(x) = \cos x$. Applying approximation (5) we obtain

Apply (5).

$$\begin{aligned}\sin(42°) &\approx \sin(\pi/4) + \cos(\pi/4)(-.05236)\\ &= \frac{\sqrt{2}}{2} + \frac{\sqrt{2}}{2}(-.05236)\\ &= .6701\end{aligned}$$

The actual value is 0.6691.

Table 9.2 presents other approximations to $\sin(45° + \Delta x)$ for various positive values of Δx (in radians). ■

You might wonder why we are concerned with approximating expressions such as $\sqrt{x_0 + \Delta x}$ when inexpensive calculators are available with which this sort of expression can be determined precisely. One answer is simply that you cannot determine the *relationship* between changes in x and the corresponding changes in $f(x)$ without calculating a considerable number of examples. Approximation (5) displays this relationship in a simple manner and is easy to obtain.

Another important aspect of approximation (5) has to do with the problem of **error propagation.** When a scientist or engineer attempts to measure a physical

Table 9.2 Approximations to $\sin\left(\frac{\pi}{4} + \Delta x\right)$

(a) Δx (in radians)	(b) Actual value $\sin\left(\frac{\pi}{4} + \Delta x\right)$	(c) Approximation $\frac{\sqrt{2}}{2}(1 + \Delta x)$	(d) Absolute error $\lvert(b) - (c)\rvert$	(e) Relative error $\left\lvert\frac{(b) - (c)}{(b)}\right\rvert$
1.0	0.977062	1.41421	0.437151	0.447414
0.5	0.959549	1.06066	0.101111	0.105374
0.1	0.774165	0.777817	.003652	0.004718
0.05	0.741562	0.742462	.000901	0.000121
0.01	0.714140	0.714178	.000038	0.000053
0.005	0.710631	0.710642	.000011	0.000016

quantity, the measurement obtained, say x, will vary from the true quantity, say x_0, by an amount determined by the precision of the measuring instrument. In other words, the scientist will observe $x = x_0 + \Delta x$ when the true value is x_0, and Δx is an amount called the **measurement error.** If the researcher then calculates another quantity from the measured quantity x by an equation of the form $y = f(x)$, the result obtained will be $y = f(x_0 + \Delta x)$, which will differ from the true value $y_0 = f(x_0)$. We can use approximation (5) to determine the error propagating through the calculations.

Example 3 A ball bearing is to be machined in the shape of a perfect sphere with radius 3 cm. The bearing is made out of a metal alloy weighing 8 grams per cubic centimeter.

(a) What is the mass of the bearing, assuming its radius is precisely 3 cm?

(b) If the actual radius is in error by not more than $\Delta r = 0.05$ cm, what is the resulting error in the calculation for mass?

Solution:

(a) The volume of the perfect bearing will be

$$v_0 = \frac{4}{3}\pi \cdot 3^3 = 36\pi \text{ cm}^3.$$

Therefore, its mass will be

$$m_0 = (36\pi \text{ cm}^3)(8 \text{ grams/cm}^3) = 288\pi \text{ grams.}$$

(b) Since $M(r) = V(r) \cdot 8 \text{ grams/cm}^3 = \frac{32}{3}\pi r^3$ grams, we have

$$M'(r) = 32\pi r^2 \text{ grams/cm.}$$

Using $r_0 = 3$ cm and $\Delta r = 0.05$ cm we obtain from approximation (5) that

$$\begin{aligned}\text{Error} = M(r_0 + \Delta r) - M(r_0) &\approx M'(r_0)\Delta r\\ &= 32\pi \cdot 3^2(0.05) \text{ grams}\\ &= 14.4\pi \text{ grams.}\end{aligned}$$

The actual mass of the bearings manufactured will therefore fall within the range $(288 \pm 14.4)\pi$ grams. In this case we say that the *relative error* in the calculation of the mass is $\frac{14.4}{288} = 0.05$, and the *percentage error* is 5.0% (see Exercises 11 and 12 in this section). In comparison, the relative error in the radius is only $\frac{0.05}{3} = 0.0167$, so the mass calculation has roughly tripled the error—it has propagated with a vengeance! ■

Geometrically, the error in using approximation (5) to calculate $f(x_0 + \Delta x)$ is represented by the vertical distance labelled ϵ in Figure 9.2. It is

$$\epsilon(\Delta x) = f(x_0 + \Delta x) - [f(x_0) + f'(x_0)\Delta x]. \tag{6}$$

We can rewrite equation (6) in the form

$$\frac{f(x_0 + \Delta x) - f(x_0)}{\Delta x} = f'(x_0) + \frac{\epsilon(\Delta x)}{\Delta x} \tag{7}$$

Now assuming that $f(x)$ is differentiable at x_0, as $\Delta x \to 0$ the limit of the left-hand side of equation (7) is $f'(x_0)$. Thus we may conclude that not only does $\epsilon \to 0$ as $\Delta x \to 0$ but in fact

$$\lim_{\Delta x \to 0} \frac{\epsilon(\Delta x)}{\Delta x} = 0. \tag{8}$$

We may interpret equation (8) by saying that as $\Delta x \to 0$ the error approaches zero faster than Δx.

The preceding discussion is so important in understanding the nature of a differentiable function that we summarize it as follows. (Theorem 14 will provide an important link between the theory developed in this chapter and the question of differentiability for functions of several variables in Chapter 19.)

THEOREM 14

Let $f(x)$ be a differentiable function for all x in an open interval containing x_0. Then for all $\Delta x \neq 0$

$$f(x_0 + \Delta x) = f(x_0) + f'(x_0)\Delta x + \epsilon(\Delta x) \tag{9}$$

where $\lim_{\Delta x \to 0} \frac{\epsilon(\Delta x)}{\Delta x} = 0$.

DIFFERENTIALS: We can characterize the discussion of this section as having to do with small changes in an independent variable and the corresponding change in the value of a dependent variable or function value. We can formalize this discussion now by introducing the terminology of *differentials*.

DEFINITION 5

Let $y = f(x)$ be a function which is differentiable at x. Let Δx be a given number, either positive or negative. Then the **differentials** dx and dy are defined by

$$dx = \Delta x, \qquad \text{and}$$
$$dy = f'(x)\,dx.$$

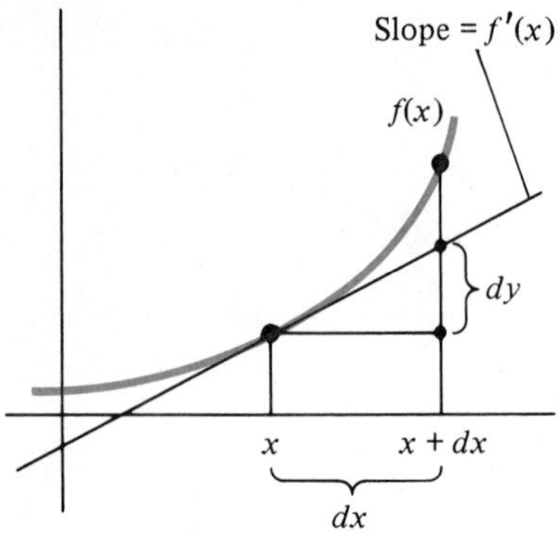

Figure 9.3

Thus, dx is just any change in x, and dy is the corresponding change in y obtained by allowing the point (x, y) to move along the *tangent* to the graph of $f(x)$ until it reaches the point with x-coordinate $(x + dx)$ (see Figure 9.3).

This terminology gives added meaning to the Leibniz notation for the derivative since we obtain

$$\frac{dy}{dx} = \frac{f'(x)\,dx}{dx} = f'(x),$$

provided $dx \neq 0$. This is entirely consistent with our earlier use of this notation and our interpretation of $f'(x)$ as the slope of the line tangent to the graph of $y = f(x)$ at $(x, f(x))$. Thus, $\frac{dy}{dx}$ can actually be viewed as a quotient, and we may legitimately write

$$dy = f'(x)\,dx \qquad \text{for} \qquad \frac{dy}{dx} = f'(x).$$

Example 4 Find differentials dy for (a) $y = ax^2 + bx + c$, and (b) $y = \sin(kx)$.

Solution: We apply Definition 5:

(a) $dy = d(ax^2 + bx + c) = (2ax + b)\,dx$
(b) $dy = d(\sin(kx)) = k\cos(kx) \cdot dx$ ■

Using Definition 5 we can restate all our differentiation formulas in the notation of differentials. For example, the Product Rule

$$[f(x)g(x)]' = f(x)g'(x) + f'(x)g(x)$$

or

$$\frac{d}{dx}(fg) = f \cdot \frac{dg}{dx} + g\frac{df}{dx}$$

becomes

$$d(fg) = fdg + gdf.$$

A complete list of these formulas appears in the Chapter Summary.

Exercise Set 3.9

In each of Exercises 1–10, use approximation (5) of Section 3.9 to approximate the given quantity.

1. $\sqrt{37}$

2. $\sqrt{48}$

3. $\frac{1}{\sqrt{17}}$

4. $\sqrt[3]{124}$

5. $\sin(2°)$

6. $\cos(88°)$

7. $\sqrt[3]{1004}$

8. $(31)^{3/5}$

9. $\frac{1}{\sqrt{26}}$

10. $\sin^2(46°)$

11. *(Calculator)* Use a calculator to find the exact value of the quantities in Exercises 1–10. For each problem compute the relative error by dividing the error by the actual value. Then find the **percentage error** by multiplying the relative error by 100%.

12. *(Calculator)* Find the relative and percentage errors for the approximation in Example 2.

In Exercises 13–18, find dy.

13. $y = \cos(1 - x^2)$

14. $y = \dfrac{x - 1}{x + 1}$

15. $y = x \sec(\pi - x)$

16. $y = \sqrt[3]{x^2 + x}$

17. $y = \dfrac{1 - x^3}{(2 + x)^2}$

18. $y = \dfrac{\sqrt{t}}{t^{2/3} - t}$

19. Find an approximation to $f(9)$ if $f(10) = 6$ and $f'(10) = -2$.

20. Find an approximation to $f(36)$ if $f(39) = 7$ and $f'(39) = 6.5$.

21. Water flows through a pipe at the rate of 3 liters per minute per square centimeter of cross-sectional area.
a. Find the volume per minute if the diameter of the pipe is 10 cm.
b. Find the possible error in the calculation in part (a) if the true diameter of the pipe varies from 9.6 to 10.4 cm.

22. In calculating the relative error in $y = f(x)$ due to a measurement error in x, the actual value of y_0 is usually unknown. Thus in calculating the **relative error** in y (see Exercise 11) we must use the calculated value of y rather than the actual value. Calculate the relative error and the percentage error for the calculation in Exercise 21.

23. A roller bearing has the shape of a cylinder. If the radius of the base is to be 0.5 cm and the length is to be 4.5 cm, find the relative error in the calculated volume if the length is precise but the radius is only accurate to within 0.05 cm.

24. Find the allowable error in the size of the radius of the ball bearing in Example 3 so that the resulting percentage error in the weight calculation will be less than 2%.

25. For the roller bearing in Exercise 23, find the accuracy needed in the radius of the bearing if the percentage error in the volume is to be less than 1%.

26. A square plot of ground is measured to be 200 meters by 200 meters. Find the percentage error in the calculation of the area of the plot if both measurements are known to be accurate to within 1%.

27. How accurate must the surveyor be in laying out the plot of ground in Exercise 26 so that the actual area will differ from 40,000 square meters by less than 1%?

28. Assume that the acceleration of a pendulum bob due to gravity is given by $a(\theta) = -\sin \theta$ where θ is the angle between the pendulum cord and the vertical (see Figure 9.4). Find the relative error in the calculation of the acceleration if θ is measured to be $\pi/4$ and this measurement is known to be correct only to within 20%.

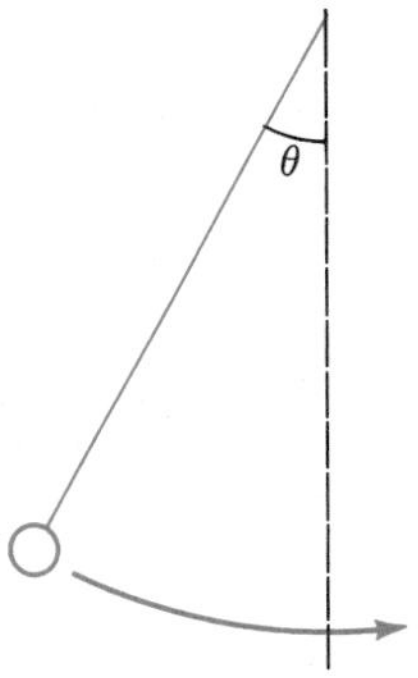

Figure 9.4

29. Students in a physics course conducted a rather naive experiment by dropping a lead weight from the top of a tall building and measuring the time of its fall to the ground with a stopwatch. They observed a value of t equal to 5 seconds. Using the formula

$$s = 4.9t^2 \text{ meters}$$

they calculated the height of the building to be 122.5 meters. If the time measurement is known to be accurate only to within 0.5 second, what are the relative and percentage errors in the calculation of the height?

SUMMARY OUTLINE OF CHAPTER 3

- The **derivative** of $f(x)$ is defined by

$$f'(x) = \lim_{\Delta x \to 0} \frac{f(x + \Delta x) - f(x)}{\Delta x}.$$

The derivative of $f(x)$ at $x = a$ is the slope of the line tangent to the graph of $y = f(x)$ at $(a, f(a))$. The equation of this line is

$$y - f(a) = f'(a)(x - a).$$

- If $s(t)$ denotes the location of a particle moving along a line, then the **velocity** of the particle is $v(t) = s'(t)$.
- ***Theorem:*** If $f(x)$ is differentiable at $x = a$, then $f(x)$ is continuous at $x = a$.
- The function g is the inverse of the function f if $g(f(x)) = x$ for all x in the domain of f.
- ***Notation:*** $\dfrac{df}{dx}$ means $f'(x)$; $\dfrac{d^2f}{dx^2}$ means $f''(x)$; and so forth.

- When y is not an explicit function of x we can often calculate dy/dx by the technique of **implicit differentiation.**
- By using the tangent to $y = f(x)$ at $(x, f(x))$, we **approximate changes** in $f(x)$ with the formula $f(x + \Delta x) \approx f(x) + f'(x)\Delta x$.
- We have the following differentiation formulas:

$f(x)$ notation	Leibniz notation	Differential notation
If $h(x) = f(x) + g(x)$, $\quad h'(x) = f'(x) + g'(x)$	$\dfrac{d}{dx}[f + g] = \dfrac{df}{dx} + \dfrac{dg}{dx}$	$d(f + g) = df + dg$
If $h(x) = cf(x)$, $\quad h'(x) = cf'(x)$	$\dfrac{d}{dx}[cf] = c\dfrac{df}{dx}$	$d(cf) = cdf$
If $h(x) = f(x)g(x)$, $\quad h'(x) = f'(x)g(x) + f(x)g'(x)$	$\dfrac{d}{dx}[fg] = \dfrac{df}{dx}g + f\dfrac{dg}{dx}$	$d(fg) = fdg + gdf$
If $h(x) = \dfrac{f(x)}{g(x)}$, $\quad h'(x) = \dfrac{f'(x)g(x) - f(x)g'(x)}{g^2(x)}$	$\dfrac{d}{dx}[f/g] = \dfrac{g\dfrac{df}{dx} - f\dfrac{dg}{dx}}{g^2}$	$d(f/g) = \dfrac{gdf - fdg}{g^2}$
If $h(x) = f(x)^n$, $\quad h'(x) = nf(x)^{n-1}f'(x)$	$\dfrac{d}{dx}[f(x)^n] = n[f(x)^{n-1}]\dfrac{df}{dx}$	$d(f^n) = nf^{n-1}df$
If $h(x) = f(u(x))$, $\quad h'(x) = f'(u(x))u'(x)$	$\dfrac{dy}{dx} = \dfrac{dy}{du} \cdot \dfrac{du}{dx}$	$dy = \dfrac{dy}{du}du$
If $h(x) = \sin x$, $\quad h'(x) = \cos x$	$\dfrac{d}{dx}\sin x = \cos x$;	$d(\sin x) = \cos x\, dx$
If $h(x) = \cos x$, $\quad h'(x) = -\sin x$	$\dfrac{d}{dx}\cos x = -\sin x$	$d(\cos x) = -\sin x\, dx$
If $h(x) = f^{-1}(x)$, $\quad h'(x) = \dfrac{1}{f'(f^{-1}(x))}$.	$\dfrac{dy}{dx} = \dfrac{1}{\dfrac{dx}{dy}}$	

REVIEW EXERCISES—CHAPTER 3

In Exercises 1–28, find the derivative of the given function.

1. $f(x) = \dfrac{x}{3x - 7}$

2. $y = \dfrac{\sqrt{x}}{1 + \sqrt{x}}$

3. $y = \dfrac{6x^3 - x^2 + x}{3x^5 + x^3}$

4. $f(x) = \dfrac{x^3 - 1}{x^3 + 1}$

5. $f(t) = \sqrt{t}\sin t$

6. $y = (6x - 4)^{-3}$

7. $y = \dfrac{1}{x + x^2 + x^3}$

8. $f(s) = \left(s + \dfrac{1}{s}\right)^5$

9. $g(t) = (t^{-2} - t^{-3})^{-1}$

10. $y = \left(\dfrac{4x^3 + 3}{x + 2}\right)^4$

11. $y = x^2 \sin x^2$

12. $f(x) = \tan^2 x^2$

13. $y = \dfrac{1}{(x^4 + 4x^2)^3}$

14. $f(t) = \tan t \cdot \csc t$

15. $f(x) = \dfrac{2x - 7}{\sqrt{x^2 + 1}}$

16. $g(t) = 4t^2\sqrt{t} + t^3\sqrt{t}$

17. $y = ((x^2 + 1)^3 - 7)^5$

18. $y = [\sin(1 + x^2) + x]^3$

19. $f(x) = \sqrt{\dfrac{3x - 9}{x^2 + 3}}$

20. $y = \csc(\cot(3x))$

21. $f(s) = \tan^2 s \cdot \cot^5 s$

22. $y = \sqrt[3]{\sin x}$

23. $y = \dfrac{\sqrt{1 + \sin x}}{\cos x}$

24. $f(x) = \dfrac{x + \sqrt{x}}{\sqrt{1 + x^3}}$

25. $f(t) = t^{9/2} - 6t^{5/2}$

26. $y = \dfrac{\sqrt[3]{x}}{x + \sqrt[3]{x}}$

27. $y = \sqrt{x^3}(x^{-3} - 2x^{-1} + 2)^2$

28. $f(x) = x|x|$

Find $f''(x)$ and $f'''(x)$ for

29. $f(x) = x \sin x$

30. $f(x) = \dfrac{x}{x + 1}$

31. $f(x) = \sec x \cdot \tan x$

32. $f(x) = \dfrac{1}{\sqrt{1 + x^2}}$

Find the velocity and acceleration of a particle moving along a line if its position along a line after t seconds is

33. $s(t) = \dfrac{\sqrt{t+2}}{1+\sqrt{t}}$

34. $s(t) = \sqrt{t}\sin t$

35. $s(t) = \sqrt{t}(t^2+1)^{3/2}$

36. $s(t) = \dfrac{\sin t}{1+\sqrt{t}}$

Use only the definition of the derivative to find $\dfrac{dy}{dx}$ for

37. $y = 6x - 2$

38. $y = \dfrac{1}{x+1}$

39. $y = \sqrt{x+1}$

40. $y = \dfrac{1}{\sqrt{2x+1}}$

Find $\dfrac{dy}{dx}$ if

41. $6x^2 - xy - 4y^2 = 0$

42. $6x^3 - 2y^2 = x$

43. $x = y + y^2 + y^3$

44. $\sin\sqrt{x} + \sin\sqrt{y} = 1$

45. $xy = \cot xy$

46. $(2x^2y^3)^{1/3} = 1$

47. $x\sin y = y$

48. $\sqrt{x} + \sqrt{y} = K$

Find $\dfrac{d^2y}{dx^2}$ if

49. $x^2 + y^2 = 4$

50. $\sin xy = 1/2$

Find an equation for the line tangent to the graph at the indicated point.

51. $\dfrac{1}{x} + \dfrac{1}{y} = 4,\quad \left(\dfrac{1}{2}, \dfrac{1}{2}\right)$

52. $y = \tan x - x,\quad \left(\dfrac{\pi}{4}, 1 - \dfrac{\pi}{4}\right)$

53. $f(x) = \dfrac{\cos x}{\sin x},\quad \left(\dfrac{\pi}{4}, 1\right)$

54. $\sqrt{x} + \sqrt{y} = 1,\quad \left(\dfrac{1}{4}, \dfrac{1}{4}\right)$

55. $f(x) = \sec x \tan x,\quad \left(\dfrac{\pi}{4}, \sqrt{2}\right)$

Find the differential dy:

56. $y = ax + b$

57. $y = x\sin x$

58. $y = \dfrac{x+1}{x+2}$

59. $y = \sqrt{1+x^2}$

60. $y = \dfrac{\sqrt{x+1}}{\sqrt{x+2}}$

61. $y = \cos\sqrt{x}$

Use differentials to approximate

62. $\sqrt{65}$

63. $(8.2)^{2/3}$

64. $\cos 44°$

65. $(0.96)^3$

Find the inverses of the following functions:

66. $f(x) = 2x + 3$

67. $f(x) = \dfrac{2}{x} + 3,\quad x > 0$

68. $f(x) = \dfrac{x+1}{x+2},\quad x > -2$

69. $f(x) = ax^3$

Determine which of the following functions is differentiable at $x = 2$.

70. $f(x) = \begin{cases} x - 2, & x \le 2 \\ 2x - 4, & x > 2 \end{cases}$

71. $f(x) = \begin{cases} x^2, & x \le 2 \\ 4x - 4, & x > 2 \end{cases}$

72. $f(x) = \begin{cases} x^3 - 8, & x \le 2 \\ 6x - 12, & x > 2 \end{cases}$

73. $f(x) = \begin{cases} 4 - x^2, & x \le 2 \\ 2x - 4, & x > 2 \end{cases}$

74. A particle moves along a line so that its location at time t is given by a polynomial of degree 2. Prove that it can change direction at most once. Must it change direction?

75. The radius of a sphere is increased from 20 cm to 20.5 cm. Use differentials to approximate the change in volume. What is the relative error in this calculation? What is the percentage error?

76. Find $h'(3)$ if $h(x) = f(g(x))$, $g(3) = 4$, $g'(3) = 2$, and $f'(4) = 5$.

77. An isosceles triangle has base 10 cm and base angles $\pi/4$. Approximate the change in the area of the triangle, using differentials, if the base angles are increased 0.05 radian.

78. Prove that all normals to a circle meet at the center. What about ellipses?

79. The adiabatic law for the expansion of air is $PV^{1.4} = C$ where P = pressure, V = volume, and C is a constant. Approximate the percentage change in P caused by a 1% change in V.

80. At what points does the line tangent to the graph of $y = 1 - x^3$ have slope $m = -6$?

81. Find the slope of the line tangent to the circle $(x+2)^2 + (y-3)^2 = 4$ at the point $(-1, 3 + \sqrt{3})$?

82. At what point(s) is the line tangent to the graph of $y = 4x^3 - 4x + 4$ parallel to the line with equation $y - 8x + 6 = 0$?

83. Prove that the derivative of an even function is an odd function.

84. Prove that the derivative of an odd function is an even function.

85. A particle moves along a line so that after t seconds it is at $s(t) = t^3 + 3t + 3$ on the number line. Find the time at which its velocity and acceleration are of equal magnitude.

86. If the composite function $y = f(g(x))$ is differentiable at $x = a$ must both $f(x)$ and $g(x)$ be differentiable at $g(a)$ and a respectively?

CHAPTER 4

APPLICATIONS OF THE DERIVATIVE: RATES AND OPTIMIZATION

4.1 INTRODUCTION

Up to this point we have used the derivative to answer questions about slopes of tangents to curves and about velocity and acceleration in rectilinear motion. Chapters 4 and 5 present a wide variety of additional applications of the derivative.

We have organized these topics into two separate chapters for several reasons. One reason is that you have just completed three very mathematical chapters, with little discussion of how the ideas being developed are applied outside the subject of the calculus itself. Expecting you to be ready for a "change of scenery," we have concentrated the nonmathematical applications as much as possible in Chapter 4. In Chapter 5 we return to the business of using the derivative to develop further mathematical results.

A second reason for organizing these applications into two chapters is that we have put the simpler, more straightforward topics in Chapter 4 and the more complex generalizations of these ideas in Chapter 5. For example, the problem of finding the maximum or minimum value of a differentiable function on a finite closed interval is relatively straightforward, and appears in Chapter 4 (Sections 4.3 and 4.4). But the problem of finding *all relative* maxima and minima for a function throughout its domain, and otherwise analyzing the nature of its graph, is much more complex, calling for generalizations of the simpler techniques developed in Chapter 4. These are among the topics treated in Chapter 5.

Finally, there is the reason of length. Much can be said about applications of the derivative; we want to say most of it, but we do not want to bog you down with a single lengthy chapter. It is likely that your instructor will cover most, but not all, of Chapters 4 and 5. This is both reasonable and expected. The important point is that you experience a variety of applications so that you both increase your computational skills and gain deeper insights into the nature of the derivative.

4.2 RATES OF CHANGE PER UNIT TIME

Often one encounters problems in which two or more variables are functions of time. For example, as an ice cube melts, its volume, its weight, and each of its dimensions change continuously as time passes. The objective of this section is to determine how to calculate and compare the rates at which the variables in such a problem are changing.

If $Q(t)$ denotes a quantity that is a function of t = time, then the **average rate of change** in Q from time t to time $t + \Delta t$ is given by the quotient

$$\frac{\Delta Q}{\Delta t} = \frac{Q(t + \Delta t) - Q(t)}{\Delta t} = \frac{\text{Change in } Q}{\text{Change in time}}. \tag{1}$$

If the quotient in equation (1) is the same for all values of t and Δt, then $Q(t)$ is said to be changing at a constant rate. However, if the rate given by equation (1) depends upon t or Δt (or both), a more precise meaning for the rate of change $Q(t)$ is needed. We therefore define the **instantaneous rate of change of $Q(t)$** to be the limiting value of the average rate of change from time t to time $t + \Delta t$ as $\Delta t \to 0$, that is,

$$\left\{\begin{matrix}\text{Rate of change of} \\ Q(t) \text{ at time } t\end{matrix}\right\} = \lim_{\Delta t \to 0} \frac{Q(t + \Delta t) - Q(t)}{\Delta t}. \tag{2}$$

From the theory developed in Chapter 3, we know that the limit in equation (2) may or may not exist. If this limit exists, the right side of equation (2) is the derivative, that is, we have

$$\left\{\begin{matrix}\text{Rate of change of} \\ Q(t) \text{ at time } t\end{matrix}\right\} = Q'(t) = \frac{dQ}{dt}. \tag{3}$$

(The units associated with the above rates (1), (2), and (3) will be units for Q per unit time.)

Example 1 For the function $f(t) = \sin \pi t$, the average rate of change over the interval [1/4, 1/2] is

$$\frac{\Delta f}{\Delta t} = \frac{\sin(\pi/2) - \sin(\pi/4)}{1/2 - 1/4} = \frac{1 - \sqrt{2}/2}{1/4} = 4 - 2\sqrt{2} \approx 1.17,$$

while the instantaneous rate of change at $t = 1/4$ is

$$f'(\pi/4) = \pi \cos(\pi/4) = \frac{\pi\sqrt{2}}{2} \approx 2.22.$$

(See Figure 2.1.) ■

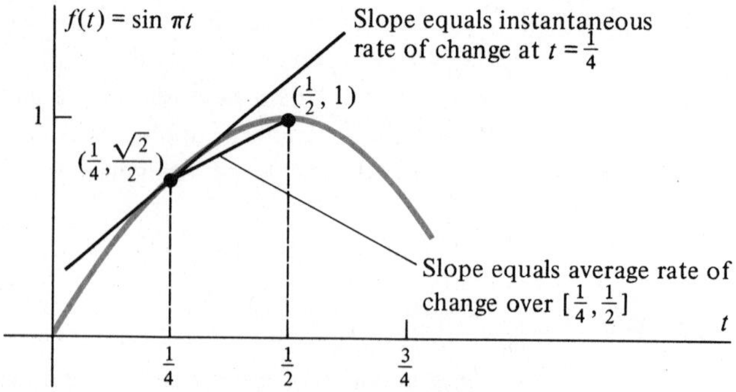

Figure 2.1 Average rate of change versus instantaneous rate of change.

We have already encountered rates such as that in equation (3) in the position/velocity problem. There $v(t)$ was defined as the rate of change of position, $s(t)$, per unit time. It was shown that $v(t) = s'(t)$ when $s(t)$ was differentiable, as equation (3) above states. The following example illustrates another situation in which equations (1) and (3) are used to calculate rates of change per unit time.

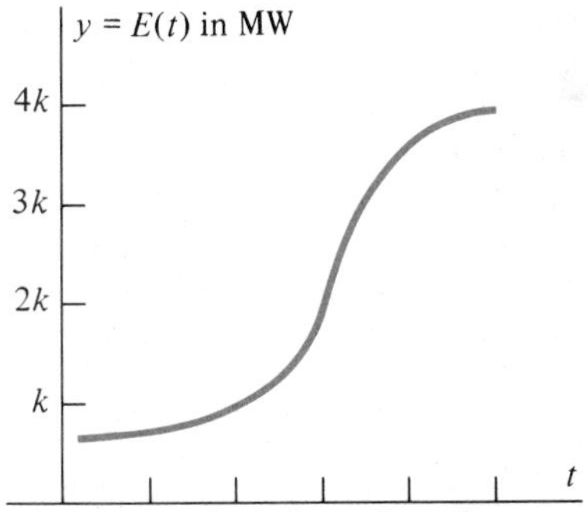

Figure 2.2 Demand for electricity.

Example 2 An engineer at an urban power plant is concerned with developing a mathematical model for the city's electricity demand during the period from 4:00 a.m. to 9:00 a.m. on Mondays. After studying data covering many Mondays, the engineer proposes the following equation as an appropriate mathematical model:

$$E(t) = k(2 + \sqrt[3]{t - 7}) \text{ mW}. \tag{4}$$

In equation (4), $E(t)$ represents the demand for electricity in megawatts, k is a constant, and t denotes time in hours after midnight. The graph associated with this equation appears in Figure 2.2.

In this model, the **average rate of increase** in $E(t)$ from 5:00 a.m. to 7:00 a.m. can be found by using equation (1). We obtain

$$\left\{\begin{array}{l}\text{Average rate of increase in}\\ \text{demand from 5:00 a.m. to}\\ \text{7:00 a.m.}\end{array}\right\} = \frac{[E(7) - E(5)]}{(7 - 5)\text{ hr}} \text{ mW}$$
$$= \frac{2k - .74k}{2 \text{ hr}} \text{ mW}$$
$$= .63k \text{ mW/hr}.$$

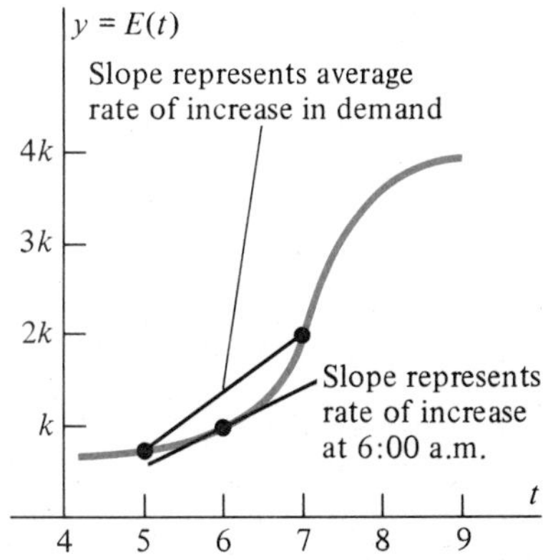

Figure 2.3 Average rate of increase versus (instantaneous) rate of increase.

Geometrically, this average rate of increase is represented by the secant from the point $(5, .74k)$ to the point $(7, 2k)$ in Figure 2.3.

To obtain the **rate of increase** of $E(t)$, we differentiate the function $E(t)$ to obtain

$$\text{Rate of increase of } E(t) = E'(t) = \frac{d}{dt}[k(2 + \sqrt[3]{t - 7})] \text{ mW}$$
$$= \frac{k}{3}(t - 7)^{-2/3} \text{ mW/hr}.$$

The rate at which demand for electricity is increasing at 6:00 a.m., for example, is $E'(6) = (k/3)(6 - 7)^{-2/3} = k/3$ mW/hr. Geometrically, this rate is represented by the slope of the tangent at the point $(6, k)$ in Figure 2.3. ■

REMARK ON MATHEMATICAL MODELS: You should realize that equation (4) in Example 2 was proposed as a **model** for the actual demand function. This does not mean that the *actual* demand function is that of equation (4) but rather that equation (4) produces values of demand for given values of time that correspond very closely with observed data. While the actual demand function will fluctuate from day to day, equation (4) represents a sort of average or "typical" situation.

The value of mathematical models lies in their use as predictors. Their limitations are that they may possess mathematical properties not shared by the actual phenomenon they are modelling. For example, the model in equation (4) is differentiable everywhere except at $t = 7$. It is unlikely, however, that the actual demand function is differentiable at all such values or that the actual rate of increase in demand becomes infinite at 7:00 a.m.—perhaps only very large. A goal of this text is to develop techniques for analyzing models such as that of equation (4). The question of how such models are developed from actual observations is a topic of courses on mathematical statistics and operations research.

Related Rate Problems

When two or more differentiable functions of time are related by a single equation, we can often obtain the relationship between their respective rates of change by differentiating both sides of that equation with respect to t. In so doing we are making use of the fact that the equation $f(t) = g(t)$ means that $f(t)$ and $g(t)$ *are the same function,* since the equation is assumed true for all values of t.* We may therefore conclude that $f'(t) = g'(t)$, since we have merely differentiated two different forms for the same function. Problems of this general type are referred to as **related rate problems.**

Example 3 A balloon in the shape of a sphere is being inflated so that the volume is increasing by 100 cubic centimeters per second. At what rate is the radius increasing when the radius is 9 cm?

Strategy

Name all variables.

Solution

We let V = volume, r = radius, and t = time. We know that

Find an equation relating these variables.

$$V = \frac{4}{3}\pi r^3.$$

Differentiate both sides of the equation, using the Chain Rule where necessary.

We assume that both V and r are differentiable functions of t. Then

$$\frac{dV}{dt} = 4\pi r^2 \cdot \frac{dr}{dt},$$

so

Solve for the desired rate, dr/dt.

$$\frac{dr}{dt} = \frac{1}{4\pi r^2} \cdot \frac{dV}{dt}.$$

Substitute in given values of other variables.

We are given that $dV/dt = 100 \text{ cm}^3/\text{sec}$. When $r = 9$ cm,

$$\frac{dr}{dt} = \frac{1}{4\pi (9 \text{ cm})^2} \cdot 100 \frac{\text{cm}^3}{\text{sec}}$$

$$= \frac{25}{81\pi} \approx .098 \frac{\text{cm}}{\text{sec}}.$$

■

In Example 3, there was no difficulty in determining the basic equation relating the variables. Often, however, related rate problems do not clearly indicate what the basic relationship between the variables might be. We therefore suggest the following *procedure for solving related rate problems:*

(i) If the problem is geometric in nature (or can be interpreted geometrically), draw a sketch.

*For example, if $ax^2 + b = 3x^2 + 4$ we may conclude, by differentiating both sides, that $2ax = 6x$. But we cannot differentiate both sides of the equation $x^2 = 4$ to conclude that $2x = 0$, since the equation $x^2 = 4$ is true only for the two numbers $x = 2$ and $x = -2$. It is not an equation involving *functions*.

(ii) Label the important lengths and quantities as variables or constants, according to the statement of the problem.
(iii) From the sketch, together with known relationships among the variables (either given in the problem or known from geometry or trigonometry), write down an equation relating the relevant variables.
(iv) Differentiate both sides of the equation, using the Chain Rule if necessary, to obtain an equation relating the rates.
(v) Solve for the desired rate. After doing so, substitute given data into this expression to obtain the solution.

Whether or not you actually write out each of these steps, you should learn to reason through problems of this type by an "algorithm" similar to the above procedure. We do not say this because we are trying to tell you "how to think." Rather, we are trying to emphasize the importance of learning how to attack a problem by first planning a strategy that will lead to a solution. Once this strategy is clearly in mind, the details of the particular steps may be pursued with less risk of confusion than is the case when one plunges too quickly into calculations.

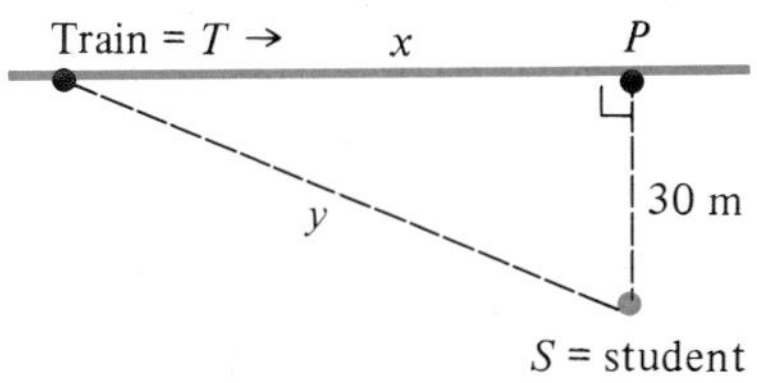

Figure 2.4 Sketch for Doppler effect experiment.

Example 4 A physics student is standing 30 meters from a straight section of railroad track in order to perform an experiment on the Doppler effect. A train is approaching, moving along the track at 90 kilometers per hour. How fast is the distance between the train and the student decreasing when the train is 50 meters from the student?

Strategy

Draw a sketch.

Solution

The situation is sketched in Figure 2.4. We let T denote the location of the train, S the location of the student, and P the point on the tracks nearest the student. Also, we let

x = distance from T to P, and
y = distance from T to S.

The problem is, therefore, to find dy/dt when $y = 50$ m.

Use Pythagorean Theorem to find an equation in x and y.

Since T, P, and S are vertices of a right triangle, we have the equation

$$y^2 = x^2 + 30^2. \tag{5}$$

Differentiate both sides of (5).

Assuming both x and y to be differentiable functions of t, we differentiate both sides of (5) to obtain

$$2y \cdot \frac{dy}{dt} = 2x \cdot \frac{dx}{dt},$$

so

Solve for $\frac{dy}{dt}$.

$$\frac{dy}{dt} = \frac{x}{y} \cdot \frac{dx}{dt}. \tag{6}$$

Find x when $y = 50$.

Now when $y = 50$ m, we obtain from equation (5)

$$x = \sqrt{50^2 - 30^2} = \sqrt{40^2} = 40 \text{ m}.$$

Insert given values for all variables and rates.

We are given that $dx/dt = -90$ km/hr, so when $y = 50$ m we have from equation (6)

$$\frac{dy}{dt} = \left(\frac{40 \text{ m}}{50 \text{ m}}\right)(-90 \text{ km/hr})$$
$$= -72 \text{ km/hr}.$$

■

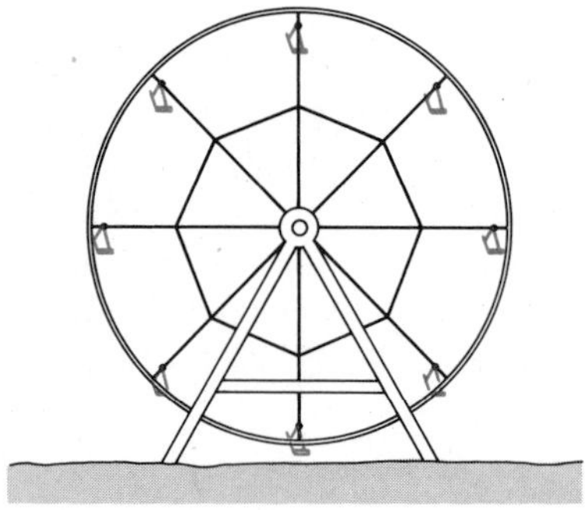

Figure 2.5 Ferris wheel.

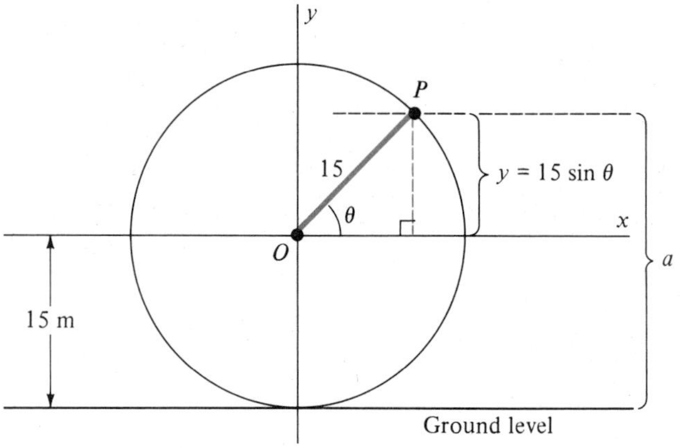

Figure 2.6 Idealized Ferris wheel.

Example 5 A ferris wheel 30 meters high turns at a uniform rate of 3 revolutions per minute (see Figure 2.5). Find the rate at which the altitude of a passenger changes.

Strategy

Solution

Draw a sketch.

The idealized ferris wheel is sketched in Figure 2.6. In order to label the relevant variables, we superimpose a rectangular coordinate system with its origin lying at the center of the ferris wheel.

Label variables.

Since the wheel is 30 m high, its radius is $r = 15$ m. We let P denote a point on the circumference of the wheel, we let θ be the angle between the ray OP and the positive x-axis, and we let y denote the vertical distance from the x-axis to the point P. Then, from the definition of the sine function we have

Use definition of $\sin\theta$ to obtain equation in θ, y, and r.

$$\sin\theta = \frac{y}{r} = \frac{y}{15},$$

Solve for y.

so

$$y = 15 \sin\theta \quad \text{meters}.$$

Add y to 15 m to find altitude.

Finally, since the x-axis lies 15 meters above ground level, we see that the altitude h of the point P is given by the equation

$$h = 15 + y$$
$$= 15 + 15\sin\theta \quad \text{meters}.$$

Assuming both h and θ to be differentiable functions of t, we differentiate both sides of this equation to obtain

Differentiate both sides.

$$\frac{dh}{dt} = 15\cos\theta \cdot \frac{d\theta}{dt} \text{ m/min.} \tag{7}$$

Convert $d\theta/dt$ from rev/min to rad/min and substitute in equation (7).

We are given that $\dfrac{d\theta}{dt} = 3$ revolutions per minute. Since θ is measured in radians, we must convert revolutions to radians (1 revolution $= 2\pi$ radians). We obtain

$$\frac{d\theta}{dt} = 3 \cdot 2\pi = 6\pi \quad \text{rad/min.}$$

Substituting this value into equation (7) then gives

$$\begin{aligned}\frac{dh}{dt} &= 90\pi \cos\theta \quad \text{m/min} \qquad (8)\\ &= 1.5\pi \cos\theta \quad \text{m/sec.}\end{aligned}$$

■

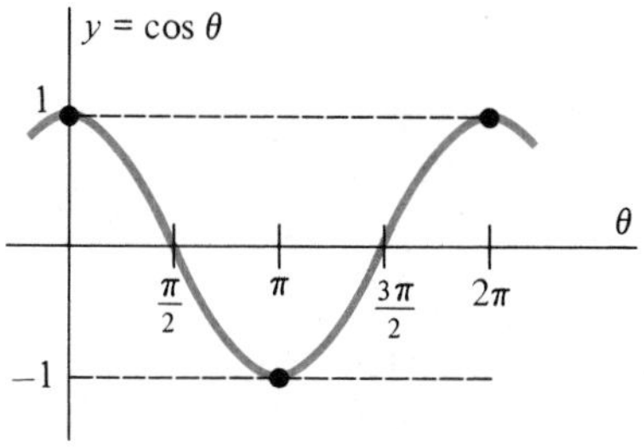

Figure 2.7 cos θ achieves its maximum value at $\theta = 0$ and its minimum value at $\theta = \pi$.

As you have probably experienced by riding a ferris wheel, the rate at which you are rising or falling, dh/dt, depends upon your position. Since the only variable on the right-hand side of equation (8) in Example 5 is θ, we can easily see that dh/dt will achieve its largest and smallest values at angles where cos θ achieves its largest and smallest values. Since $\cos(0) = 1$, $\cos(\pi) = -1$, and $|\cos\theta| < 1$ for all other values of θ in $[0, 2\pi]$ (see Figure 2.7), we conclude that

(i) $\left.\dfrac{dh}{dt}\right|_{\theta=0} = \dfrac{3\pi}{2}$ m/sec is the maximum rate at which you will rise, and this occurs when $\theta = 0$ (i.e., halfway up).

(ii) $\left.\dfrac{dh}{dt}\right|_{\theta=\pi} = -\dfrac{3\pi}{2}$ m/sec is the maximum rate at which you will fall, and this occurs when $\theta = \pi$ (i.e., halfway down). Notice that the sign of dh/dt indicates whether you are rising or falling (see Figure 2.8).

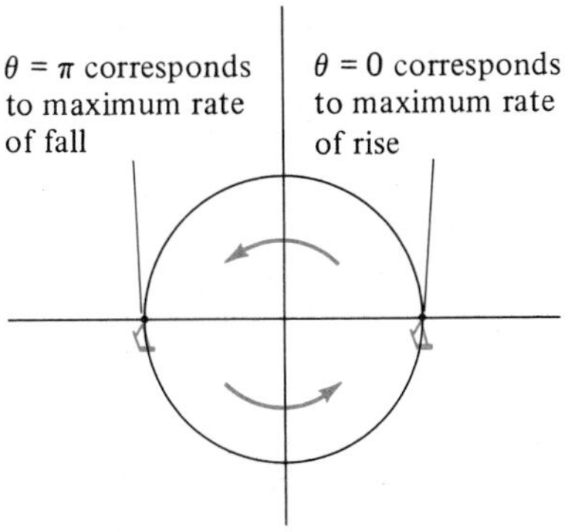

Figure 2.8 Maximum rates of rise and fall occur at halfway points.

The above remarks illustrate the importance of knowing how to find the maximum and minimum values of a given function. This will be the topic of the next two sections.

Example 6 A coffee maker has the shape of a double cone 20 cm high. The radii at both the top and the base are 4 cm. Coffee is flowing from the top section into the bottom section at a rate of 4 cm^3/sec. At what rate is the level of coffee in the top section falling when the coffee in the top section is 4 cm deep?

Strategy

Draw a sketch.

Solution

The coffee maker is sketched in Figure 2.9. An idealized sketch of the top portion appears in Figure 2.10.

Label variables.

We let h denote the depth of the coffee in the top section and we let r denote the radius of its surface.

Use equation for volume of a cone.

The formula for the volume of coffee in the upper cone is therefore

$$V = \frac{1}{3}\pi r^2 h. \qquad (9)$$

No data given on dr/dt, so r must be eliminated from (9) before differentiation.

Now if we proceed directly to differentiate both sides of equation (9) with respect to t, we will obtain an equation involving dV/dt, dr/dt, and dh/dt, since each of the variables, V, h, and r is a function of time. However, we have no given

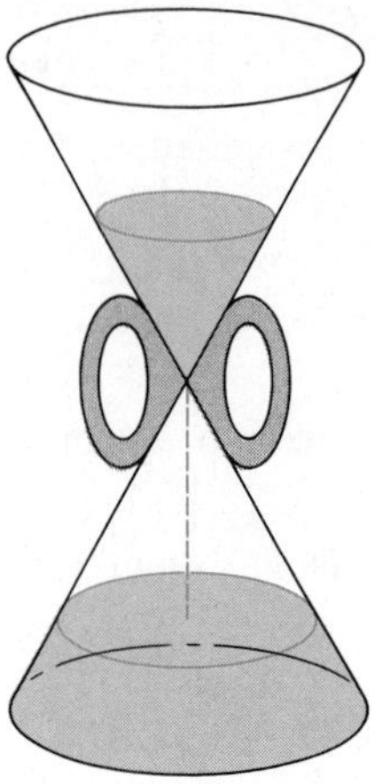
Figure 2.9 Coffee maker in shape of double cone.

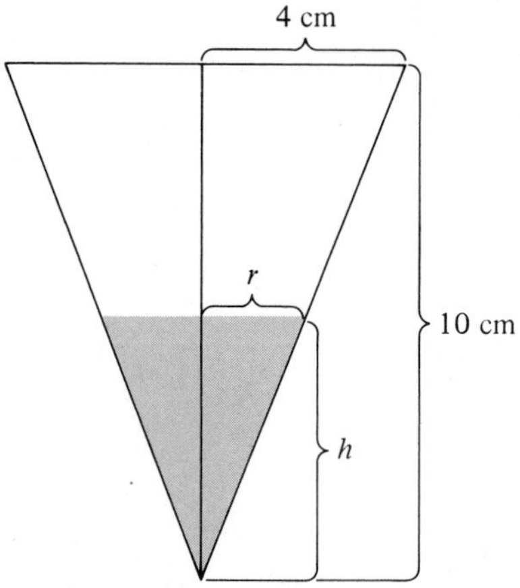

Figure 2.10 Top section of coffee maker.

information on dr/dt. We therefore attempt to find an equation expressing r in terms of V and/or h so that r can be eliminated from equation (9). To do so we observe from Figure 2.10 that, by similar triangles,

Use similar triangles to find equation relating r and h.

$$\frac{r}{h} = \frac{4}{10},$$

so

$$r = \frac{2}{5}h \tag{10}$$

Insert expression for r in (9).

is the desired equation. Substituting this expression for r in (9) gives

$$V = \frac{1}{3}\pi\left(\frac{2}{5}h\right)^2 h$$

$$= \frac{4\pi}{75}h^3. \tag{11}$$

Differentiating both sides of this equation with respect to t, we obtain

Differentiate both sides.

$$\frac{dV}{dt} = \frac{4\pi}{25}h^2 \cdot \frac{dh}{dt},$$

so

Solve for dh/dt.

$$\frac{dh}{dt} = \frac{25}{4\pi h^2} \cdot \frac{dV}{dt}.$$

Insert given information.

(Note that the sign of dV/dt is negative.)

Substituting the given information $dV/dt = -4\ \text{cm}^3/\text{sec}$ and $h = 4$ cm then gives

$$\frac{dh}{dt} = \frac{25}{4\pi(4\ \text{cm})^2}\left(-4\frac{\text{cm}^2}{\text{sec}}\right)$$

$$= -\frac{25}{16\pi} \approx -0.497\frac{\text{cm}}{\text{sec}}.$$

■

REMARK 1: Equation (10) above is an example of an **auxiliary** equation. Its purpose is to allow one of the variables in the principal equation (9) to be eliminated. The strategy in the above example was to use the auxiliary equation to eliminate the variable r since no information about r was given. In general, you should look for auxiliary equations in problems of this type when you encounter more variables than can be evaluated with the given data.

REMARK 2: In the solution to Example 6, it was important that the two sides of equation (11) were differentiated *before* the particular values for the variables were substituted. Otherwise the desired rates would not have appeared. A common error of students encountering problems of this kind for the first time is to substitute the given values of the variables into the principal equation before differentiating. Be very careful not to make this mistake.

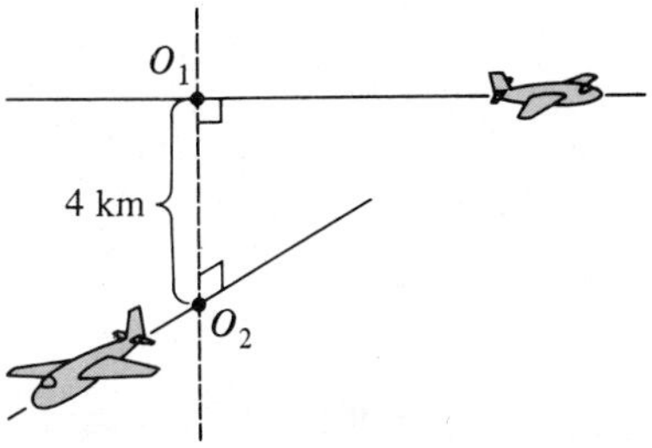

Figure 2.11 Two airplanes.

Example 7 An airplane is flying a level course due east at a speed of 3 kilometers per minute. A second airplane is flying a level course due south at a speed of 2 kilometers per minute at an altitude 4 kilometers below that of the first plane. At one instant the second airplane is directly beneath the first. At what rate is the distance between the two airplanes increasing one minute later?

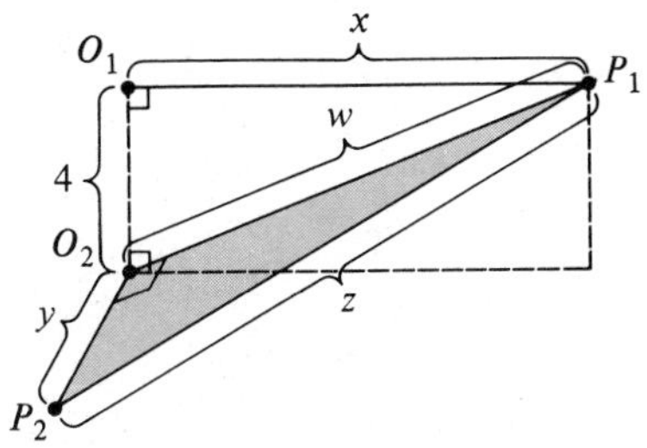

Figure 2.12 P_1 and P_2 denote the two airplanes.

Solution: The situation is sketched in Figure 2.11. In Figure 2.12, points P_1 and P_2 represent the first and second airplanes, respectively, and O_1 and O_2 are the locations of the planes when the first is directly above the second. We name variables as follows:

x = distance between O_1 and P_1
y = distance between O_2 and P_2
w = distance between O_2 and P_1
z = distance between P_1 and P_2

The rate to be calculated is dz/dt. Since z is the length of the hypotenuse of the right triangle $P_1O_2P_2$, we have the equation

$$z = \sqrt{w^2 + y^2}. \tag{12}$$

We are given information about dy/dt, the speed of the second airplane, but we do not have direct information about the variable w. We must therefore find an auxiliary equation relating w to one or more of the other variables. From Figure 2.12 we can see that w is the length of the hypotenuse of triangle $P_1O_1O_2$, so we obtain the auxiliary equation

$$w^2 = x^2 + 4^2.$$

Substituting this expression for w^2 into the principal equation (12), we obtain the equation

$$z = \sqrt{x^2 + 4^2 + y^2}.$$

We now differentiate with respect to t. This gives

$$\frac{dz}{dt} = \frac{x \cdot \dfrac{dx}{dt} + y \cdot \dfrac{dy}{dt}}{\sqrt{x^2 + 4^2 + y^2}}. \tag{13}$$

We are given

$$\frac{dx}{dt} = 3\frac{\text{km}}{\text{min}} \quad \text{and} \quad \frac{dy}{dt} = 2\frac{\text{km}}{\text{min}}.$$

Therefore, one minute after crossing, we will have $x = 3$ km and $y = 2$ km. Substituting these values into equation (3) gives

$$\frac{dz}{dt} = \frac{(3 \text{ km})(3 \text{ km/min}) + (2 \text{ km})(2 \text{ km/min})}{\sqrt{(3 \text{ km})^2 + (4 \text{ km})^2 + (2 \text{ km})^2}}$$

$$= \frac{13}{\sqrt{29}} \approx 2.414 \text{ km/min}.$$

■

Exercise Set 4.2

In each of Exercises 1–10, find the average rate of change of the given function for the given time interval. Then find the instantaneous rate of change per unit time. Note any values of t where the instantaneous rate is undefined.

1. $s(t) = (1 + t)^2, \quad t \in [2, 3]$
2. $s(t) = \dfrac{1}{1 - t}, \quad t \in [2, 2.5]$
3. $f(t) = \sqrt{t}(1 - t^3), \quad t \in [0, 1]$
4. $g(t) = \sin(\pi - t^2), \quad t \in [-\pi/2, \pi/2]$
5. $r(t) = t^{5/2} - 2t^{3/2}, \quad t \in [1, 4]$
6. $w(t) = \dfrac{\sqrt{t}}{1 + t^2}, \quad t \in [0, 5]$
7. $k(t) = \tan(\pi t), \quad t \in \left[\dfrac{1}{6}, \dfrac{1}{3}\right]$
8. $v(t) = (t^{-2} + 2)^3, \quad t \in [1, 2]$
9. $g(t) = \dfrac{1 - \sec^2 t}{1 + \sec^2 t}, \quad t \in [0, \pi/4]$
10. $p(t) = t^{2/3} \sin\sqrt{1 + t}, \quad t \in [0, 1]$
11. A spherical balloon is being inflated so that the radius is increasing at a rate of 3 cm/sec. Find the rate at which the volume is increasing when $r = 10$ cm.
12. At what rate is the diagonal of a cube increasing if the sides are increasing at a rate of 2 cm/sec?
13. When a pebble is tossed into a still pond, ripples move out from the point where the stone hits in the form of concentric circles. Find the rate at which the area of the disturbed water is increasing when the radius of the outermost circle equals 10 meters if this radius is increasing at a rate of 2 meters per second.
14. A snowball, in the shape of a sphere, is melting so that the radius is decreasing at a uniform rate of 1 centimeter per second. How fast is the volume decreasing when the radius equals 6 centimeters?
15. A point moves along the graph of $y = x^{5/2}$ so that its x-coordinate increases at the constant rate of 2 units per second. Find the rate at which its y-coordinate is increasing as it passes the point (4, 32).
16. A conical water tank has a radius of 20 meters and a depth of 20 meters. Water is being pumped into the tank at the rate of 40 cubic meters per minute. How fast is the level of the water rising when the water is 8 meters deep?
17. A radio transmitter is located 3 kilometers from a fairly straight section of interstate highway. A truck is travelling away from the transmitter along the highway at a speed of 80 kilometers per hour. How fast is the distance between the truck and the transmitter increasing when they are 5 kilometers apart?
18. The area of a rectangle, whose length is twice its width, is increasing at the rate of 8 cm^2/sec. Find the rate at which the length is increasing when the width is 5 cm.
19. A ladder 5 meters long is leaning against a wall. The base of the ladder is sliding away from the wall at a rate of 1 meter per second. How fast is the top of the ladder sliding down the wall at the instant when the base is 3 meters from the wall?
20. A boat is being pulled to shore by a rope attached to a windlass atop a pier. The height of the windlass above the water is 6 meters, and the rope is being wound in at the rate

of 5 meters per minute. How fast is the boat approaching the shore when it is 8 meters away?

21. The law of cosines, relating the lengths of the three sides of a triangle, is

$$C^2 = A^2 + B^2 - 2AB \cos \theta$$

where θ is the angle opposite side C. Sides $A = 1$ and $B = 2$ are of fixed length, and the angle θ is increasing at the rate of 0.2 radians/minute. Find the rate at which C is increasing at the instant when $C = \sqrt{3}$.

22. A woman 1.6 meters tall is walking away from a lamp post 10 meters tall. If the woman is walking at a speed of 1.2 meters per second, how fast is her shadow increasing when she is 15 meters from the lamp post?

23. The lengths of the two equal sides in an isosceles triangle are increasing at the rate of 2 cm/sec. The base of the triangle has fixed length 18 cm. Find the rate at which the area of the triangle is increasing when its height equals 12 cm.

24. Following the outbreak of an epidemic, a population of $N(t)$ people can be regarded as being made up of immunes, $I(t)$, and susceptibles, $S(t)$, that is,

$$N(t) = I(t) + S(t).$$

$I(t)$ includes both those who have contracted the disease and those who cannot contract the disease. If the rate of decrease of susceptibles is 20 persons per day, and the rate of increase of immunes is 24 persons per day, how fast is the population growing?

25. Gravel is being poured onto a pile at the rate of 180 cubic meters per minute. The pile is in the shape of an inverted cone whose diameter is always three times its height. Find the rate at which the diameter of the base is increasing when the pile is 6 meters high.

26. At noon a truck leaves Columbus, Ohio, driving east at a speed of 40 kilometers per hour. An hour later a second truck leaves Columbus driving north at a speed of 60 kilometers per hour. At what rate is the distance between the two trucks increasing at 2:00 p.m. that day?

27. A water trough has end pieces in the shape of inverted isosceles triangles with bases 60 cm and heights 40 cm. The trough is 4 meters long. Water is being pumped into the trough at a rate of 9000 cm^3 per second. How fast is the level of the water in the trough rising when the water is 10 cm deep?

28. Work Exercise 27 if the ends of the trough have the shape of equilateral triangles whose sides have length 60 cm.

29. An observer stands 150 meters from a fireworks display rocket which is fired directly upward. When the rocket reaches a height of 200 meters, it is travelling at a speed of 12 meters per second. At what rate is the angle of elevation formed with the observer increasing at that instant?

30. A point moves along the unit circle at a constant speed. Find the points at which the x- and y-coordinates are
a. changing at the same rates,
b. changing at opposite rates.

31. For a body moving through air, **drag** is defined as the force opposing the motion of the body. For such a body, the drag D is jointly proportional to the squares of its velocity V and its surface area S, that is,

$$D = kV^2S^2.$$

Find the rate at which drag is increasing if the body is undergoing an acceleration of 8 meters/sec^2 at the instant when the velocity equals 30 meters/sec.

32. On a particular autumn day, the sun moves across the sky at the rate of 15° per hour. How fast is the shadow cast by a building 30 meters high increasing at the moment when the evening sun is 30° above the horizon?

33. For a gas at temperature T confined within a container of volume V at a pressure P, the gas law states that the ratio $\frac{PV}{T}$ is a constant, that is, $PV = kT$. A 1000 cm^3 tank of oxygen is being heated so that the temperature of the oxygen is increasing at a rate of 2°C per minute. Find the rate at which the pressure inside the tank is increasing.

34. How fast is the level of the coffee in the bottom of the coffee pot in Example 6 rising at the instant when the coffee in the bottom of the pot is 4 cm deep?

35. The gravitational force of attraction between two bodies of mass m_1 and m_2 is given by Newton's Law of Gravitation as

$$F = G\frac{m_1m_2}{r^2}$$

where G is the gravitational constant and r is the distance between the two bodies. Find the rate at which the distance between the bodies is changing if the force is changing at the rate α.

36. A lighthouse lies 200 meters from a straight shore. The light rotates at a rate of 2 revolutions per minute. A piling marks the spot on the beach nearest the lighthouse. Find the speed at which the light beam is moving along the shore at a point 200 meters from the piling.

37. An *idealized heat engine,* designed to provide a model for studying the question of maximum possible efficiency, was proposed by the French engineer Sadi Carnot in 1824. For the Carnot engine the efficiency is defined by the equation

$$E = 1 - \frac{T_c}{T_h}$$

where T_h is the intake temperature and T_c is the exhaust temperature in degrees Kelvin.

a. If T_c is increasing by 3°K per minute and the efficiency is not changing, what is the rate of change of T_h?

b. If the efficiency is decreasing by 2% per hour and T_c is fixed, what is the rate of change of T_h?

38. If a ray of monochromatic light travelling in a vacuum makes an angle of incidence α with the normal to a surface

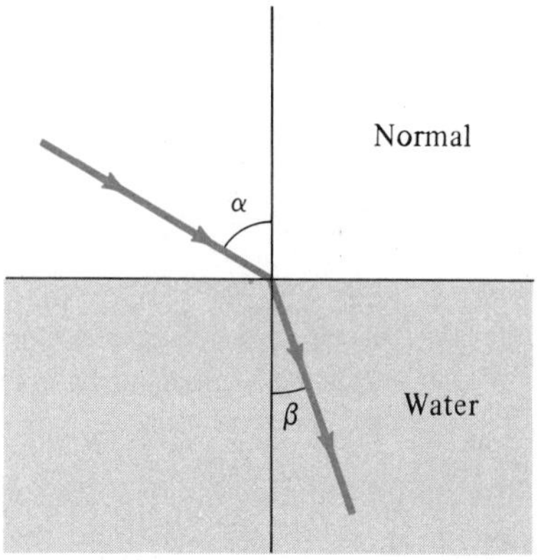

of a substance a and an angle of refraction β in the substance, then **Snell's Law** states that

$$\frac{\sin \alpha}{\sin \beta} = C_a$$

where C_a is a constant, called the index of refraction for the substance a. For water, the index of refraction is $C_w = 1.33$. If the angle of incidence of a light ray striking water is decreasing at the rate of 0.2 radians per second, find an expression for the rate at which the angle of refraction is decreasing.

39. A submarine running due north at a depth of 1 kilometer and a speed of 60 kilometers per hour passes directly beneath a ship sailing due west at 30 kilometers per hour. At what rate is the distance between the two vessels increasing 20 minutes later?

40. The velocity of viscous fluid flowing through a circular tube is not the same at all points of a cross section. Provided the velocity is not too great, the flow has maximum velocity at the center and decreases to zero at the walls. For a point at a radial distance r from the center of the tube, the velocity of the flow is given by

$$v = \frac{\alpha}{L}(R^2 - r^2)$$

where R is the radius of the tube, L is its length, and α is a constant. Find the acceleration of the fluid moving at the center of a tube if

a. $L = 25$ cm is fixed and R is increasing at a rate of 0.2 cm/min at the instant when $R = 10$ cm,

b. $R = 10$ cm is fixed and L is increasing at a rate of 0.5 cm/sec at the instant when $L = 25$ cm.

4.3 AN OPTIMIZATION PROBLEM

In the ferris wheel problem (Example 5) of Section 4.2, we raised the issue of finding the maximum and minimum values of the function $y = \cos x$ on the interval $[0, 2\pi]$. Because $y = \cos x$ is a familiar function, its maximum and minimum values are not hard to find. However, we shall soon encounter problems involving a wide variety of less familiar functions for which we will need to find maximum and minimum values on certain finite closed intervals. (Doing so is often referred to as **optimizing** the given function on the given interval.) The objective of this section is to develop a theory by which problems of the following general type can be solved.

OPTIMIZATION PROBLEM: Given the continuous function $y = f(x)$ and the closed finite interval $[a, b]$, find the maximum and minimum values of $f(x)$ for $x \in [a, b]$ and the corresponding numbers x.

A first step in attacking a general problem of this kind is to ensure that all terms are clearly defined. A testable condition that defines what we mean by maximum and minimum values follows (see Figure 3.1).

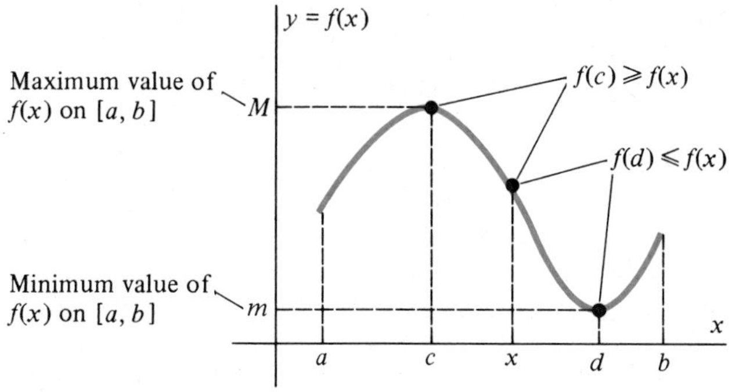

Figure 3.1 Maximum value of $f(x)$ on $[a, b]$ is $f(c) = M$. Minimum value is $f(d) = m$.

DEFINITION 1

By **the maximum value of $f(x)$ on $[a, b]$** we mean the number M so that

(i) $f(c) = M$ for at least one number $c \in [a, b]$, and
(ii) $f(x) \leq M$ for all $x \in [a, b]$.

(Statement (i) says that M is indeed a function value for at least one c in $[a, b]$. Statement (ii) says that no function value is larger than M.) Similarly, the *minimum* value of $f(x)$ on $[a, b]$ is the number m so that

(i) $f(d) = m$ for at least one number $d \in [a, b]$, and
(ii) $f(x) \geq m$ for all $x \in [a, b]$.

Although we have achieved a precise definition of terms for the Optimization Problem, we are not yet ready to take up the question of finding the maximum and minimum values for $f(x)$. The intermediate step that remains is to determine whether or not such values actually exist. This question is resolved by the following **existence theorem.**

THEOREM 1

If the function $f(x)$ is continuous on the closed interval $[a, b]$ then $f(x)$ has both a maximum value and a minimum value on $[a, b]$.

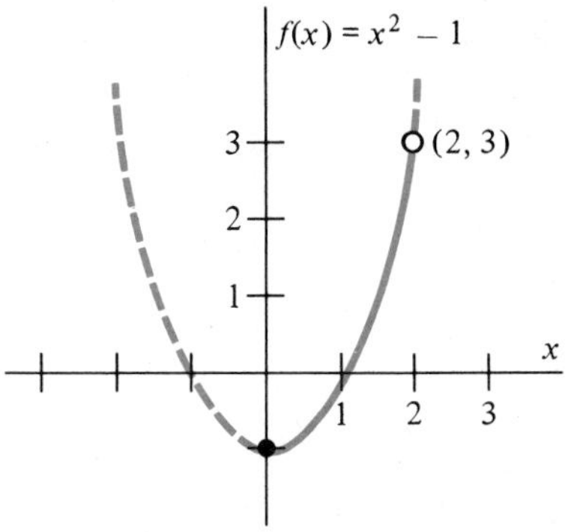

Figure 3.2 $f(x) = x^2 - 1$ has no maximum on $[0, 2)$ because $[0, 2)$ is not *closed*.

A rigorous proof of this theorem calls upon rather subtle properties of the real number system and is beyond the scope of this text. (This theorem is proved in courses entitled *Advanced Calculus* or *Introductory Analysis*.) However, it is instructive to observe several examples in which functions fail to possess maximum or minimum values on particular intervals. In each case, one of the two hypotheses of Theorem 1 is violated.

Example 1 The function $f(x) = x^2 - 1$ does not have a maximum value on the interval $[0, 2)$. This is because the interval $[0, 2)$ is not closed—its right endpoint is excluded. Thus $f(2) = 3$ is *not* a value of the function on $[0, 2)$, yet all numbers $y \in [-1, 3)$ *are* values of the function $x^2 - 1$ for $x \in [0, 2)$. Since $[-1, 3)$ does not have a largest element, $f(x)$ does not have a maximum value on $[0, 2)$ (see Figure 3.2). ■

Example 2 The function $f(x) = x - [x]$, where $[x]$ denotes the greatest integer function (see Section 2.5), has no maximum value on the closed finite interval $[0, 4]$. This is because $f(x)$ is discontinuous at each integer (Figure 3.3). Its minimum value, zero, is achieved four times in the interval. ■

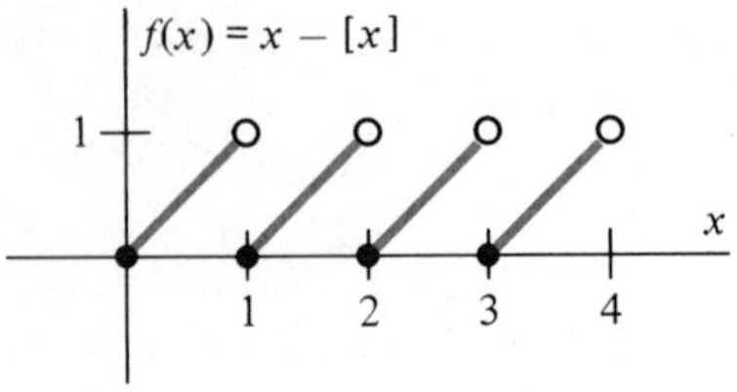

Figure 3.3 $f(x) = x - [x]$ has no maximum on $[0, 4)$. It is not *continuous* on $[0, 4]$.

Given the assurance of Theorem 1 that the Optimization Problem does have solutions, we are now prepared to pursue the question of how to find the maximum and minimum values for $f(x)$ on $[a, b]$. A fruitful way to begin this investigation is to try to identify geometric properties of the graph of $f(x)$ that might be associated with these maximum and minimum values. Figures 3.4–3.6 illustrate three geometric properties associated with particular maxima and minima: a **horizontal tangent** (derivative zero), a derivative that fails to exist (in this case, resulting in a **cusp**), and the existence of **endpoint** maxima and minima.

Further curve sketching fails to identify additional phenomena by which maximum and minimum values for $f(x)$ might occur. However, before formulating a procedure for finding maxima and minima by these three criteria, we should attempt to find a rigorous argument assuring that these are the only ways that maxima and minima occur. The role of the following theorem is to formalize and establish these observations.

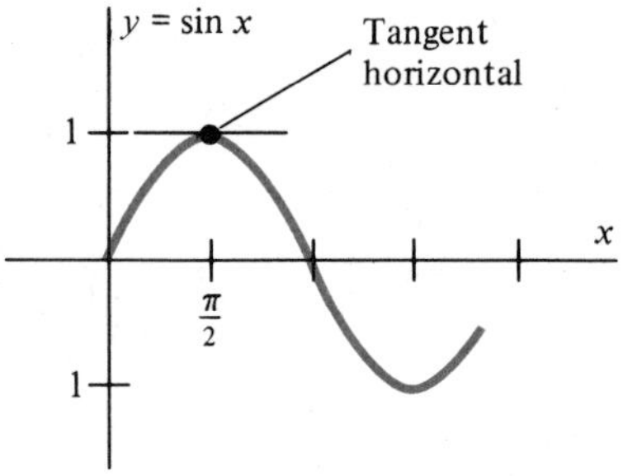

Figure 3.4 Maximum value for $f(x) = \sin x$ on $[0, \pi]$ occurs at $\pi/2$, where tangent is horizontal, that is, $f'(\pi/2) = \cos(\pi/2) = 0$.

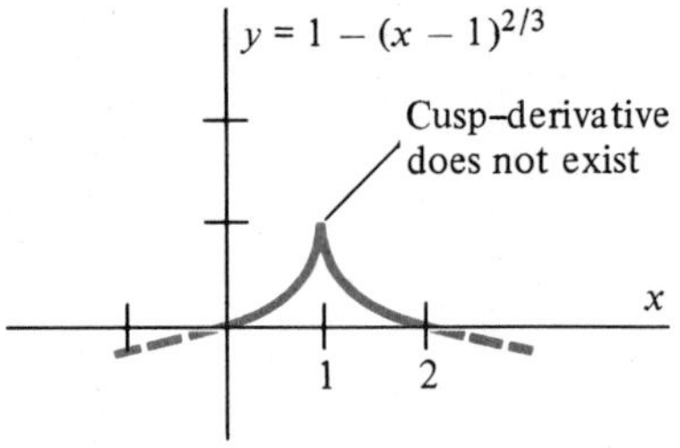

Figure 3.5 Maximum value for $f(x) = 1 - (x - 1)^{2/3}$ on $[0, 2]$ occurs at $x = 1$, where

$$f'(1) = \left.\frac{2}{3\sqrt[3]{x-1}}\right|_{x=1}$$

is undefined.

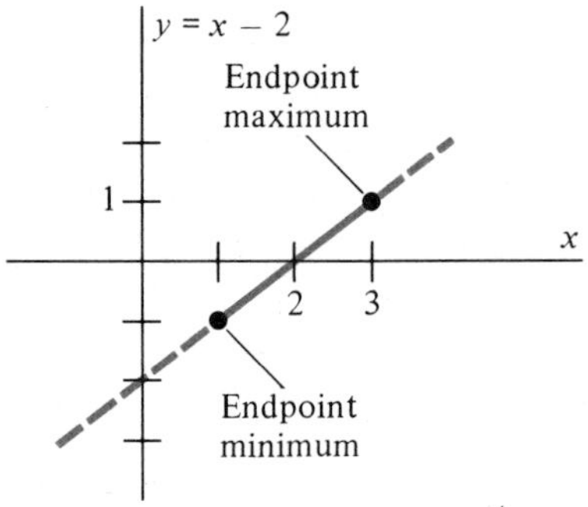

Figure 3.6 Maximum value for $f(x) = x - 2$ on $[1, 3]$ occurs at endpoint $x = 3$. Minimum occurs at endpoint $x = 1$.

THEOREM 2

Let $f(x)$ be a continuous function on the closed interval $[a, b]$. Let $c \in [a, b]$. If $f(c)$ is a maximum or minimum value for $f(x)$ on $[a, b]$, then one of the following statements must hold:

(i) c is an endpoint (that is, $c = a$, or $c = b$),
(ii) $f'(c)$ fails to exist, or
(iii) $f'(c) = 0$ (horizontal tangent).

Proof: We will prove the theorem for the case where $f(c)$ is a maximum. The proof for the case where $f(c)$ is a minimum is analogous. Let c be an arbitrary number in $[a, b]$. Our technique of proof will be to show that if neither (i) nor (ii) holds then

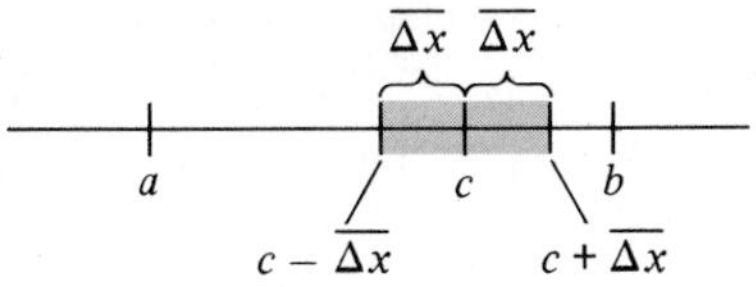

Figure 3.7 The interval $[c - \overline{\Delta x}, c + \overline{\Delta x}]$ lies within the interval $[a, b]$.

(iii) must hold,* i.e., we assume that c is *not* an endpoint and that $f'(c)$ exists, and we attempt to prove that $f'(c) = 0$.

Since c is not an endpoint of $[a, b]$, we can find a number $\overline{\Delta x} > 0$ sufficiently small that the interval $[c - \overline{\Delta x}, c + \overline{\Delta x}]$ is contained within the interval $[a, b]$ (see Figure 3.7). Throughout the remainder of the proof, we shall assume that all values of Δx satisfy the inequality $|\Delta x| \leq \overline{\Delta x}$. This assures that the number $(x + \Delta x)$ belongs to the interval $[a, b]$.

From the definition of the derivative and the assumption that $f'(c)$ exists, we know that

$$f'(c) = \lim_{\Delta x \to 0} \frac{f(c + \Delta x) - f(c)}{\Delta x}$$

exists. The above limit is a two-sided limit. We will complete the proof by considering each of the one-sided limits individually.

Case i: Let $\Delta x > 0$. Since $f(c)$ is the maximum value of $f(x)$ on $[a, b]$ we must have

$$f(c) \geq f(c + \Delta x),$$

so

$$f(c + \Delta x) - f(c) \leq 0.$$

Since $\Delta x > 0$, it follows that

$$\frac{f(c + \Delta x) - f(c)}{\Delta x} \leq 0. \tag{1}$$

Since inequality (1) holds for all $\Delta x > 0$, regardless of how small Δx might be, we must have

$$\lim_{\Delta x \to 0^+} \frac{f(c + \Delta x) - f(c)}{\Delta x} \leq 0. \tag{2}$$

Case ii: Let $\Delta x < 0$. Then, as above,

$$f(c) \geq f(c + \Delta x)$$

so

$$f(c + \Delta x) - f(c) \leq 0.$$

But $\Delta x < 0$, so in this case we obtain the inequality

$$\frac{f(c + \Delta x) - f(c)}{\Delta x} \geq 0.$$

Thus,

$$\lim_{\Delta x \to 0^-} \frac{f(c + \Delta x) - f(c)}{\Delta x} \geq 0. \tag{3}$$

---

*Since the statement of the theorem is that *one* (but not *all*) of the statements (i)–(iii) hold, we do not prove that all hold, or even that each holds when the other two fail. (Indeed, one or more of these statements might be entirely superfluous.) What we must show is that among these three statements there is *one* which holds whenever the other two fail. This guarantees that, among the three statements, at least one is always true.

Combining inequalities (2) and (3) with the fact that the two-sided limit must equal each of the one-sided limits, we obtain the statement

$$0 \le \lim_{\Delta x \to 0^-} \frac{f(c + \Delta x) - f(c)}{\Delta x} = \lim_{\Delta x \to 0} \frac{f(c + \Delta x) - f(c)}{\Delta x}$$
$$= \lim_{\Delta x \to 0^+} \frac{f(c + \Delta x) - f(c)}{\Delta x} \le 0.$$

By the Pinching Theorem we conclude that

$$f'(c) = \lim_{\Delta x \to 0} \frac{f(c + \Delta x) - f(c)}{\Delta x} = 0,$$

which shows that statement (iii) holds. The proof is complete. ■

Theorem 2 establishes a procedure by which we will determine the maximum and minimum values of $f(x)$ on $[a, b]$. We will simply inspect the values $f(c)$ for all numbers c in $[a, b]$ that satisfy one of three conditions stated in Theorem 2. Among these, the largest will be the maximum value of $f(x)$ on $[a, b]$ and the smallest will be the minimum.

Thus, we have solved the Optimization Problem in principle. Before turning to particular examples, we will state the procedure implied by Theorem 2 for finding maxima and minima. Such procedures are referred to in mathematics as **algorithms.** The value in formulating algorithms as solutions to generally posed problems, such as the Optimization Problem, is that we obtain a general strategy by which the solutions to particular problems may be executed.

Algorithm for Finding the Maximum and Minimum Values of a Continuous Function $f(x)$ on a Closed Interval $[a, b]$:

1. Find all numbers $c \in [a, b]$ for which either $f'(c) = 0$ or $f'(c)$ does not exist. (These numbers are called **critical numbers.**)
2. Compute $f(a), f(b)$, and all values $f(c)$ where c is a critical number. (That is, check $f(x)$ at the *endpoints* and at all *critical numbers.*)
3. Select the largest and smallest values computed in step 2. These are the maximum and minimum values, respectively.

Example 3 Find the maximum and minimum values of $f(x) = x^3 - 3x$ on the interval $[-2, 2]$ and the corresponding numbers x.

Strategy

Verify that $f(x)$ is continuous on $[-2, 2]$.

Set $f'(x) = 0$ to find critical numbers.

Find remaining critical numbers where $f'(x)$ is undefined. (In this case, there are none.)

Find $f(x)$ for all critical numbers and endpoints.

Solution

Since $f(x) = x^3 - 3x$ is a polynomial, we know that $f(x)$ is continuous on $[-2, 2]$ by Theorem 7, Chapter 2. We have

$$f'(x) = 3x^2 - 3,$$

so setting $f'(x) = 0$ gives the equation

$$3x^2 - 3 = 0,$$

which has solutions $x = \pm 1$. There are no numbers x for which $f'(x)$ is undefined. The critical numbers are, therefore, $x = -1$ and $x = 1$. The values of $f(x)$ at the critical numbers and endpoints are as follows:

$$f(-2) = (-2)^3 - 3(-2) = -2$$
$$f(-1) = (-1)^3 - 3(-1) = 2$$

$$f(1) = (1)^3 - 3(1) = -2$$
$$f(2) = 2^3 - 3(2) = 2$$

Find max and min by inspecting these values. Note the corresponding numbers x.

Thus, the maximum value of $f(x)$ on $[-2, 2]$ is 2, which occurs both at $x = -1$ and at $x = 2$. The minimum is -2, which occurs both at $x = -2$ and at $x = 1$. The graph of $y = f(x)$ appears in Figure 3.8. ■

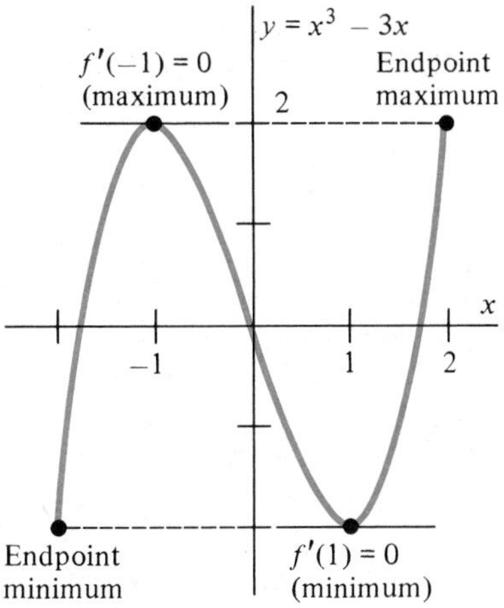

Figure 3.8 For $f(x) = x^3 - 3x$ on $[-2, 2]$, maximum value = 2 and minimum value = -2.

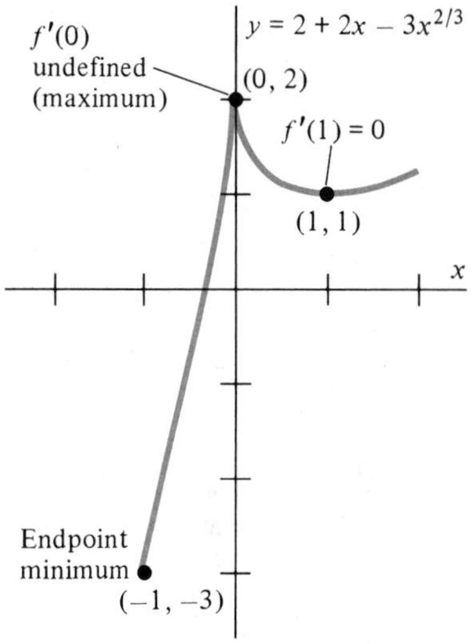

Figure 3.9 For $f(x) = 2 + 2x - 3x^{2/3}$ on $[-1, 2]$, maximum value = 2 and minimum value = -3.

Example 4 Find the maximum and minimum values of the function $f(x) = 2 + 2x - 3x^{2/3}$ on the interval $[-1, 2]$.

Strategy

Verify $f(x)$ continuous.

Solution

The function $f(x)$ is continuous at $[-1, 2]$ since it is the sum of the polynomial $2 + 2x$ and a multiple of the rational power function $x^{2/3}$. Since $f'(x) = 2 - 2x^{-1/3}$, setting $f'(x) = 0$ gives the equation

Set $f'(x) = 0$ and solve to find critical number(s).

$$2 - 2x^{-1/3} = 0,$$

so

$$x^{1/3} = 1.$$

Determine where

$$f'(x) = 2 - \frac{2}{\sqrt[3]{x}}$$

is undefined to find remaining critical numbers.

The only solution of the equation $f'(x) = 0$ is therefore $x = 1$. Also, $f'(x)$ is undefined at $x = 0$. The critical numbers are therefore $x = 0$ and $x = 1$. Examining $f(x)$ at the critical numbers and endpoints we see that

Inspect $f(x)$ at critical numbers and endpoints to identify max and min.

$$f(-1) = 2 + 2(-1) - 3(-1)^{2/3} = -3$$
$$f(0) = 2 + 2(0) - 3(0)^{2/3} = 2$$
$$f(1) = 2 + 2(1) - 3(1)^{2/3} = 1$$
$$f(2) = 2 + 2(2) - 3(2)^{2/3} = 6 - 3\sqrt[3]{4} \approx 1.2378.$$

We conclude that the maximum value of $f(x)$ on $[-1, 2]$ is $f(0) = 2$ and that the minimum is $f(-1) = -3$. The graph of $y = f(x)$ appears in Figure 3.9. ■

Example 5 A large suburban community receives its electrical power supply from an urban power plant. A local engineer, after studying data on energy usage,

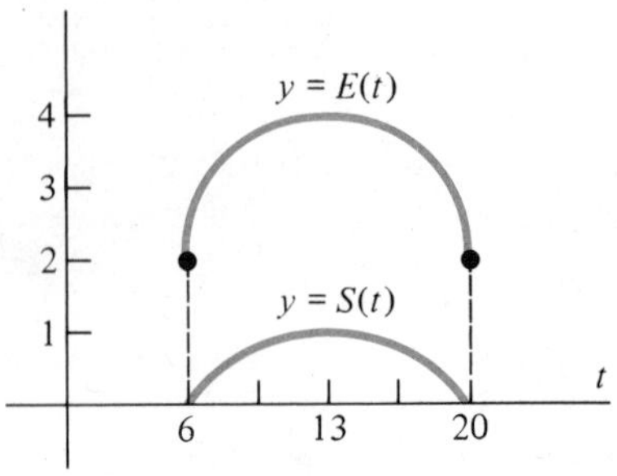

Figure 3.10 Models for energy demand $E(t)$ and solar energy supply $S(t)$.

determines that a good model for daily energy usage between 6:00 a.m. and 8:00 p.m. is given by the function

$$E(t) = 4 - 2\left(\frac{t-13}{7}\right)^4, \qquad t \in [6, 20].$$

Here t represents time in hours after midnight. The engineer argues that through the use of solar generators the community could reduce its energy demands from the urban power plant by the amount $S(t)$ where

$$S(t) = 1 - \left(\frac{t-13}{7}\right)^2, \qquad t \in [6, 20].$$

The graphs of both functions appear in Figure 3.10.

(a) Find the peak demand for energy according to the model for present usage (i.e., the maximum value of $E(t)$) and the time at which it occurs.
(b) Find the resulting peak demand according to the model for the contribution due to solar power and the time(s) at which it occurs.
(c) Calculate the amount by which peak demand from the power plant can be reduced according to the model.

Solution: (a) The derivative of $E(t)$ is

$$E'(t) = -8\left(\frac{t-13}{7}\right)^3 \cdot \left(\frac{1}{7}\right).$$

Setting $E'(t) = 0$ gives the equation

$$\frac{-8}{7} \cdot \left(\frac{t-13}{7}\right)^3 = 0, \qquad \text{or} \qquad t = 13 \qquad \text{(1:00 p.m.)}.$$

Since $E'(t)$ is defined for all $t \in [6, 20]$, the only critical number is $t = 13$. The values of $E(t)$ corresponding to this critical number and the endpoints are

$$E(6) = 4 - 2\left(\frac{6-13}{7}\right)^4 = 2$$

$$E(13) = 4 - 2\left(\frac{13-13}{7}\right)^4 = 4$$

$$E(20) = 4 - 2\left(\frac{20-13}{7}\right)^4 = 2.$$

The peak demand is therefore $E(13) = 4$ units, which occurs at $t = 13 = $ 1:00 p.m.

(b) According to the engineer's model, if solar generators are installed, the resulting demand for energy from the urban power plant will be given by the function

$$\begin{aligned} R(t) &= E(t) - S(t) \\ &= \left[4 - 2\left(\frac{t-13}{7}\right)^4\right] - \left[1 - \left(\frac{t-13}{7}\right)^2\right] \\ &= 3 + \left(\frac{t-13}{7}\right)^2 - 2\left(\frac{t-13}{7}\right)^4. \end{aligned}$$

Here

$$R'(t) = \frac{2}{7}\cdot\left(\frac{t-13}{7}\right) - \frac{8}{7}\left(\frac{t-13}{7}\right)^3$$
$$= \frac{2}{7}\cdot\left(\frac{t-13}{7}\right)\left[1 - 4\left(\frac{t-13}{7}\right)^2\right].$$

The solutions of the equation $R'(t) = 0$ occur when

$$\frac{t-13}{7} = 0 \qquad \text{or} \qquad 1 - 4\left(\frac{t-13}{7}\right)^2 = 0.$$

The first equation has the single solution $t = 13$. The second equation can be written as

$$4\cdot\left(\frac{t-13}{7}\right)^2 = 1$$

so

$$\left(\frac{t-13}{7}\right)^2 = \frac{1}{4}.$$

Taking square roots of both sides gives

$$\frac{t-13}{7} = \pm\frac{1}{2}$$

so

$$t = 13 \pm \frac{7}{2}.$$

Since $R'(t)$ is defined for all $t \in [6, 20]$, we have found the critical numbers for $R(t)$ to be $t = 13$ (1:00 p.m.), $t = 13 - \frac{7}{2} = \frac{19}{2}$ (9:30 a.m.), and $t = 13 + \frac{7}{2} = \frac{33}{2}$ (4:30 p.m.). The values of $R(t)$ corresponding to these critical numbers and to the endpoints are given in Table 3.1.

Table 3.1

t	6	$\frac{19}{2}$	13	$\frac{33}{2}$	20
$R(t)$	2	$\frac{25}{8}$	3	$\frac{25}{8}$	2

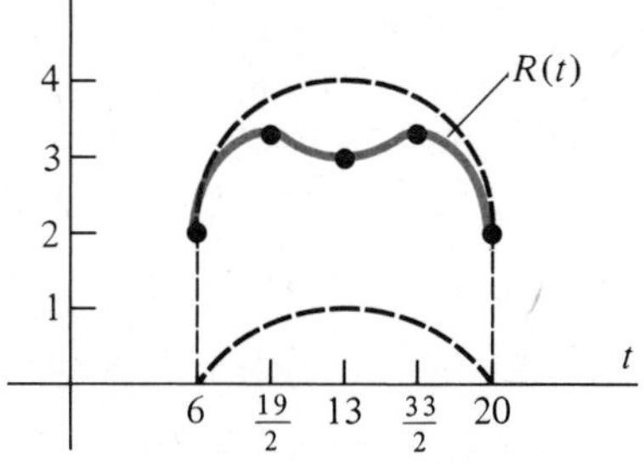

Figure 3.11 Model for resulting demand $R(t) = E(t) - S(t)$.

The peak demand is therefore $\frac{25}{8}$ units, and this peak demand occurs at $t = \frac{19}{2}$ = 9:30 a.m. and again at $t = \frac{33}{2}$ = 4:30 p.m. The graph of $R(t)$ appears in Figure 3.11.

(c) The reduction in peak demand, according to the proposed model, is

$$E(13) - R\left(\frac{19}{2}\right) = \left(4 - \frac{25}{8}\right) \text{ units}$$
$$= \frac{7}{8} \text{ units.}$$

Relative to the original peak demand of 4 units this is a percentage decrease of

$$\frac{\frac{7}{8}\text{ units}}{4\text{ units}} \times 100\% \approx 21.9\%$$

■

REMARK: Again, you should be cautioned that the calculated reduction in peak demand and the corresponding flattening of the power demand function (Figure 3.11) are both consequences of the engineer's proposed models. The question of whether or not these conclusions will actually result if the solar generators are employed depends on the accuracy of these models.

Exercise Set 4.3

In Exercises 1–27, find all critical numbers in the given interval. Then find the maximum and minimum values for $f(x)$ on the given interval and the corresponding numbers x.

1. $f(x) = x^2(x - 1), \quad x \in [0, 3]$
2. $f(x) = |x - 2|, \quad x \in [0, 5]$
3. $f(x) = \dfrac{1}{x(x-4)}, \quad x \in [1, 3]$
4. $f(x) = x(x^2 - 2), \quad x \in [-1, 2]$
5. $f(x) = x^2(x - 3), \quad x \in [-1, 2]$
6. $f(x) = \sin\left(\dfrac{x}{2}\right), \quad x \in [0, \pi]$
7. $f(x) = 1 - \tan^2 x, \quad x \in [-\pi/4, \pi/4]$
8. $f(x) = \sin x \cos x, \quad x \in [-\pi/2, \pi/2]$
9. $f(x) = x + \dfrac{1}{x}, \quad x \in [1/2, 2]$
10. $f(x) = (\sqrt{x} - x)^2, \quad x \in [0, 4]$
11. $f(x) = x^{2/3} + 2, \quad x \in [-2, 1]$
12. $f(x) = \sin x + \cos x, \quad x \in [0, \pi]$
13. $f(x) = \left(\dfrac{x+1}{x-1}\right)^2, \quad x \in [-3, 0]$
14. $f(x) = \sec x, \quad x \in [-\pi/4, \pi/4]$
15. $f(x) = \dfrac{x^2}{1+x}, \quad x \in [-1/2, 2]$
16. $f(x) = \dfrac{x^3}{1+x}, \quad x \in [0, 2]$
17. $f(x) = x \sin x, \quad x \in [-\pi/2, \pi/2]$
18. $f(x) = 8x^{1/3} - 2x^{4/3}, \quad x \in [-1, 8]$
19. $f(x) = 3x^5 - 5x^3, \quad x \in [-2, 2]$
20. $f(x) = \sqrt{x}(1 - x^2), \quad x \in [0, 4]$
21. $f(x) = \dfrac{x^{2/3}}{2 + \sqrt[3]{x}}, \quad x \in [-1, 8]$
22. $f(x) = x - \sin x, \quad x \in [\pi/2, 3\pi/2]$
23. $f(x) = \dfrac{\sqrt{x}}{1+x}, \quad x \in [0, 4]$
24. $f(x) = \sec x \tan x, \quad x \in [-\pi/4, \pi/4]$
25. $f(x) = \dfrac{\sqrt[3]{1-x}}{4+x^2}, \quad x \in [-1, 0]$
26. $f(x) = \sec x - \tan x, \quad x \in [0, \pi/4]$
27. $f(x) = \dfrac{\sqrt{x}}{1 + \sqrt[3]{x}}, \quad x \in [0, 8]$
28. In Example 5, calculate the peak demand for the function $R(t) = E(t) - S(t)$ if $E(t)$ is as given in Example 5 and $S(t) = \dfrac{1}{2} - \dfrac{1}{2}\left(\dfrac{t-13}{7}\right)^2$.
29. If $f(x)$ has maximum value M on $[a, b]$ and $g(x)$ has maximum value L on $[a, b]$, is $(L + M)$ necessarily the maximum value of $h(x) = f(x) + g(x)$ on $[a, b]$? Under what conditions will this be true? What relationship will always hold between the maximum value of $h(x)$ on $[a, b]$ and the number $L + M$ (see Example 5)?
30. Find the numbers b and c if the function $f(x) = x^2 + bx + c$ has minimum value $f(3) = -7$ on the interval $[0, 5]$.
31. Let $f(x) = ax^3 - bx$. Find a and b if $f(2) = 4$ is the maximum value of $f(x)$ on $[0, 4]$.

32. Assume that the velocity at which automobiles will travel along a certain section of highway is modeled by the function $v(\rho) = \frac{1}{4}(\rho - 2)^2$ where ρ is the density of automobiles (in units of automobiles per kilometer) and $\rho \in [0, 2]$.
a. Find the density for which velocity will be a maximum.
b. If the flow rate is defined to be $q = \rho v(\rho)$, find the density at which flow will be a maximum for the velocity function $v(\rho)$ given above.

33. Let c be a real number. What is the relationship between the critical numbers for the function $f(x)$ on a given interval $[a, b]$ and those of the functions
a. $g(x) = cf(x)$?
b. $h(x) = f(x + c)$?
c. $k(x) = f(cx)$?

34. To find the maximum and minimum values of the *periodic* function $f(x) = \sin x$ on the entire number line $(-\infty, \infty)$, we need only find these numbers on $[0, 2\pi]$ (or on *any* interval of length 2π). This is because the graph of $f(x) = \sin x$ on $(-\infty, \infty)$ consists of copies of the graph on the closed interval $[0, 2\pi]$. Generalize this observation to find the maximum and minimum values of the following periodic functions on $(-\infty, \infty)$.
a. $f(x) = \sin x - \cos x$
b. $f(x) = 2\sin(\pi/2 - x)$
c. $f(x) = 2\sin x \cos x$.

35. Figure 3.12 shows the graph of $f(x) = |x^2 - 5x + 4|$, which has maximum value $f(0) = 4 = f(5)$ and minimum value $f(1) = f(4) = 0$ for $x \in [0, 5]$. By comparing the graph of $f(x)$ with the graph of $g(x) = x^2 - 5x + 4$ in Figure 3.13, you can see that the effect of the absolute value sign is to "turn the portion of the graph extending below the x-axis upward."
a. Show that $f(x)$ may be rewritten as

$$f(x) = \begin{cases} x^2 - 5x + 4, & x \le 1 \\ -x^2 + 5x - 4, & 1 < x < 4 \\ x^2 - 5x + 4, & 4 \le x. \end{cases}$$

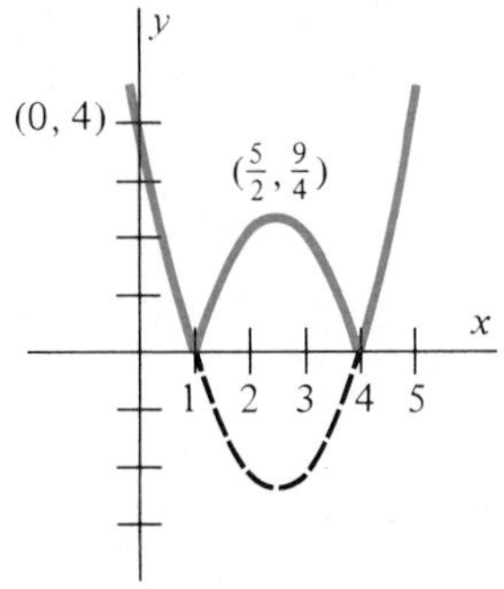

Figure 3.12 Graph of $f(x) = |x^2 - 5x + 4|$.

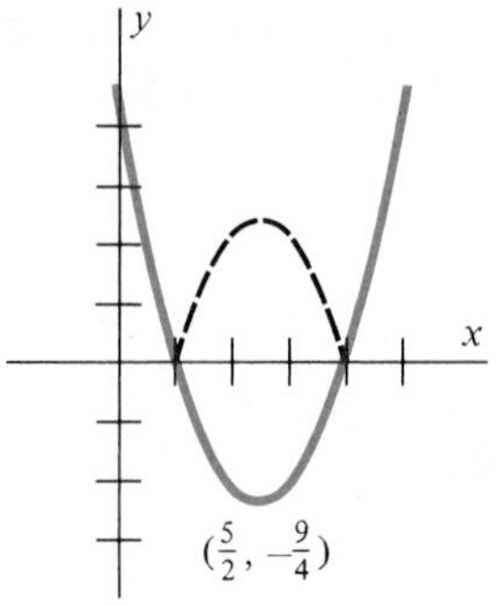

Figure 3.13 Graph of $g(x) = x^2 - 5x + 4$.

b. Show that $f(x)$ is differentiable for $x \neq 1$ and $x \neq 4$, and that

$$f'(x) = \begin{cases} 2x - 5, & x < 1 \\ 5 - 2x, & 1 < x < 4 \\ 2x - 5, & 4 < x. \end{cases}$$

c. By examining the difference quotient for $f(x)$, show that $f'(1)$ and $f'(4)$ do not exist.
d. Generalize your findings to a rule for finding the maximum and minimum values for the function $|h(x)|$, $x \in [a, b]$.

In Exercises 36–40, use the results of Exercise 35 to find the maximum and minimum values for the given function on the given interval and the corresponding numbers x.

36. $f(x) = |x - 3|, \quad x \in [-2, 6]$

37. $f(x) = |3x - 5|, \quad x \in [0, 4]$

38. $f(x) = |x^2 - x - 6|, \quad x \in [-3, 5]$

39. $f(x) = |\cos x|, \quad x \in [\pi/4, \pi]$

40. $f(x) = \dfrac{|x|}{4 + x}, \quad x \in [-2, 4]$.

4.4 APPLIED MAXIMUM-MINIMUM PROBLEMS

One of the most important (and difficult) skills needed in becoming a successful practitioner of the calculus is the ability to formulate a given problem in precise mathematical terms. In this section, we encounter optimization problems described in paragraph form rather than in the precise mathematical formulation of the problems in Section 4.3. For example, suppose that a person interested in building a single-story home containing 180 square meters of floor space wishes to find the dimensions that minimize the perimeter of the building, because of energy considerations. Suppose further that esthetic considerations limit the dimensions for length

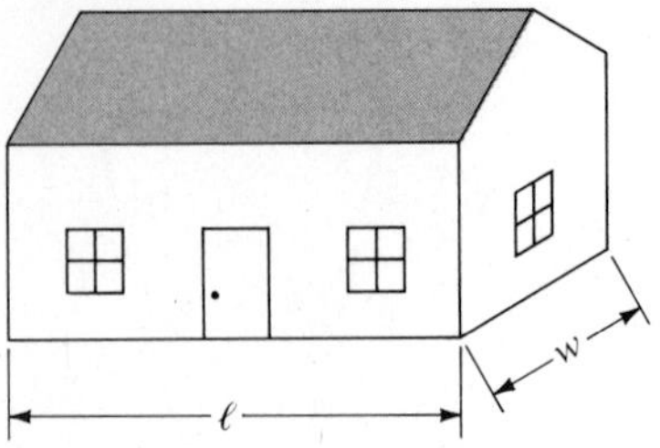

Figure 4.1

and width, say, to a minimum of 10 meters. The problem of finding the dimensions, subject to the given constraints, that minimize the perimeter is an example of an optimization problem. However, before the theory and techniques of Section 4.3 can be applied, the problem must be given precise mathematical formulation.

A first step in developing a mathematical formulation of the above problem is to label the relevant variables (Figure 4.1). We do so by letting

ℓ = length of the house
w = width of the house
P = perimeter of the house
A = area of the floor space

We next identify all known relationships among the variables and write down any given information concerning these variables. We obtain

$$P = 2\ell + 2w \qquad \text{(principal equation)} \tag{1}$$

$$A = \ell w = 180 \qquad \text{(auxiliary equation)} \tag{2}$$

$$10 \le \ell \qquad \text{(constraint)} \tag{3}$$

$$10 \le w \qquad \text{(constraint)} \tag{4}$$

We can now give a precise formulation of our problem: "Find the values ℓ and w that minimize the expression $P = 2\ell + 2w$ given the fact that $\ell w = 180$ and the constraints $\ell \ge 10$ and $w \ge 10$."

Although we have now achieved a precise mathematical formulation, our problem is not yet in the form to which the techniques of Section 4.3 can be applied. The difficulty is that the variable P to be minimized is not a function of a single independent variable. Indeed, the *principal equation* (1) gives P as a function of *two* independent variables, ℓ and w. We therefore use the *auxiliary equation* (2) to solve for one of these variables as a function of the other. Solving equation (2) for w gives

$$w = \frac{180}{\ell}. \tag{5}$$

Substituting this expression for w in equation (1) then gives the variable P as a function of ℓ alone:

$$P(\ell) = 2\ell + 2\left(\frac{180}{\ell}\right)$$
$$= 2\ell + \frac{360}{\ell}.$$

Finally, we observe that since neither variable ℓ or w can be less than 10 (inequalities (3), (4)), neither can be greater than 18, by equation (2). The independent variable ℓ must therefore lie within the closed interval [10, 18]. We can now state our original problem in a form to which the techniques of Section 4.3 directly apply:

"Find the minimum value of the function

$$P(\ell) = 2\ell + \frac{360}{\ell} \qquad \text{for} \qquad \ell \in [10, 18]."$$

We solve this problem, as in Section 4.3, by finding the critical numbers for P. Setting $P'(\ell) = 0$ gives

$$2 - \frac{360}{\ell^2} = 0,$$

so

$$\ell^2 = 180, \qquad \text{or} \qquad \ell = \pm 6\sqrt{5}.$$

Now $P'(\ell)$ is undefined for $\ell = 0$, but 0 lies outside the interval $[10, 18]$, as does the critical number $-6\sqrt{5}$. Thus the only critical number in $[10, 18]$ is $\ell = 6\sqrt{5}$. The minimum value for P must therefore lie among the three values

$$P(10) = 2(10) + \frac{360}{10} = 56 \text{ meters} \qquad \text{(endpoint)}$$

$$P(6\sqrt{5}) = 2(6\sqrt{5}) + \frac{360}{6\sqrt{5}} \approx 53.67 \text{ meters} \qquad \text{(critical number)}$$

$$P(18) = 2(18) + \frac{360}{18} = 56 \text{ meters} \qquad \text{(endpoint)}$$

The dimensions that minimize the perimeter are therefore $\ell = 6\sqrt{5}$ and $w = \dfrac{180}{6\sqrt{5}} = \dfrac{30}{\sqrt{5}} = 6\sqrt{5}$. In other words, the house should be designed in the shape of a square.

The following diagram illustrates the major conceptual steps involved in applying mathematics to solve practical problems:

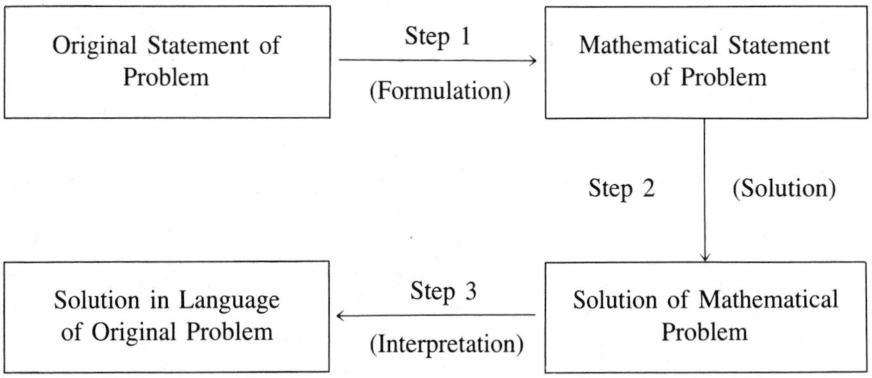

Diagram 4.1 Steps in solving applied problems.

The objective of this section is to develop skills in all three of the general steps represented in Diagram 4.1. Each of the problems encountered will reduce to an optimization problem of the type discussed in Section 4.3 in its mathematical formulation. However, the task here is more complex, in that the steps involved in actually calculating the maximum and minimum value of the appropriate function will be preceded by steps on formulation and followed by steps on interpretation of the results. The following algorithm breaks the three conceptual steps of the diagram down into finer detail, corresponding more closely to the actual steps involved in solving such problems.

Algorithm for Solving Applied Max-Min Problems:

Step 1: Formulation

a. Draw a sketch, if possible, illustrating all variables and relevant constant quantities.
b. Label all variables.
c. Identify the variable to be optimized.
d. Find an equation expressing the variable to be optimized in terms of other variables and constants (principal equation).
e. Find all other equations relating variables in the problem (auxiliary equations).
f. Write down any constraints (i.e., intervals within which variables must lie).

Step 2: Mathematical Solution

a. Use auxiliary equations to substitute for variables in principal equation until the principal equation contains only one independent variable.
b. Determine the interval within which the independent variable must lie.
c. Solve the resulting optimization problem by the techniques of Section 4.3.

Step 3: Interpretation

a. From the auxiliary equations find the values of all remaining variables corresponding to the solution of the optimization problem.
b. Describe the solution in the language of the original statement of the problem.

The remaining pages of this section consist of examples of problems solved by the procedure we have just described. The strategy for solving each one is outlined by this algorithm. While they are obviously contrived, these examples and the exercises at the end of the section provide a ''laboratory'' setting in which you can gain experience with the several steps in problem solving outlined above.

Example 1 Find two nonnegative numbers whose sum is 10 and the sum of whose squares is a maximum.

Strategy

Label variables.

Solution

Let x and y be the two numbers and let S be the desired sum. We want to maximize

Find equation for S by summing squares.

$$S = x^2 + y^2 \qquad \text{(principal equation)} \tag{6}$$

subject to the conditions that

Auxiliary equation states that sum is 10.

Write down constraints.

$$x + y = 10, \qquad \text{(auxiliary equation)} \tag{7}$$

$$x \geq 0, \qquad \text{(constraint)}$$

and

$$y \geq 0. \qquad \text{(constraint)}$$

Eliminate y in (6), using auxiliary equation.

To eliminate the variable y in equation (6), we solve equation (7) for y to obtain

$$y = 10 - x$$

and substitute into equation (6), which gives

$$S(x) = x^2 + (10 - x)^2. \tag{8}$$

Find domain of $S(x)$ in the form $[a, b]$ from auxiliary equation and constraints.

Equation (7) together with the constraints imply that

$$0 \leq x \leq 10.$$

We therefore seek to find the maximum value of the function $S(x)$ in equation (8) on the interval [0, 10].

Set $S'(x) = 0$ and solve to find critical numbers.

Setting $S'(x) = 0$ gives the equation

$$S'(x) = 2x - 2(10 - x) = 0,$$

so

$$x = 5.$$

Identify those critical numbers lying in the domain [0, 10].

Since $S'(x)$ is defined for all $x \in [0, 10]$, the only critical number in [0, 10] is $x = 5$. The maximum value of $S(x)$ will therefore occur among the following values:

Inspect $S(x)$ at critical number and endpoints to find maximum.

$$\begin{aligned} S(0) &= 0^2 + 10^2 = 100 \quad \text{(endpoint)} \\ S(5) &= 5^2 + 5^2 = 50 \quad \text{(critical number)} \\ S(10) &= 10^2 + 0^2 = 100 \quad \text{(endpoint)} \end{aligned}$$

The maximum occurs at the endpoints 0 and 10, both of which give the solution as the pair of numbers 0 and 10. ■

Example 2 A water trough is to be constructed from three metal sheets 1 meter wide and 6 meters long plus end panels in the shape of trapezoids. Find the angle at which the long panels should be joined so as to provide a trough of maximum volume (see Figures 4.2 and 4.3).

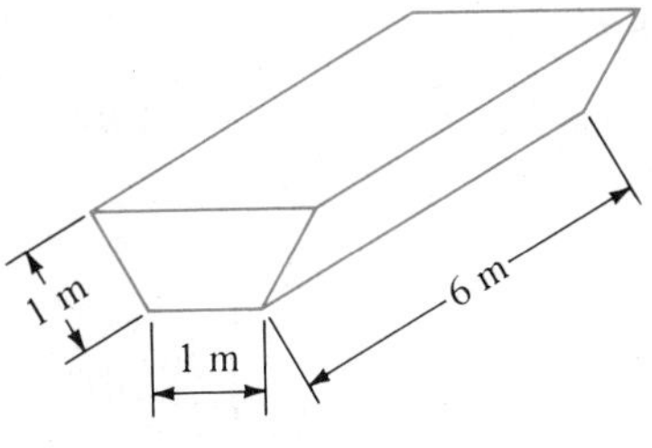

Figure 4.2 Water trough.

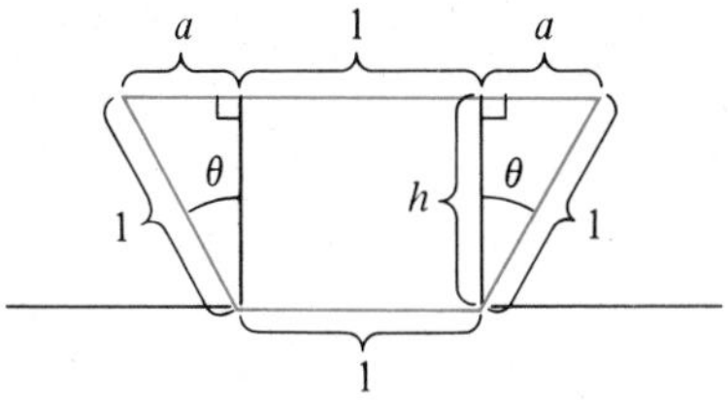

Figure 4.3 End view of the trough.

Strategy

Label variables.

Solution

Consider the end view as shown in Figure 4.2. Let θ be the angle between the side panel and the vertical, h the height, and a the length of the side of the triangle opposite the angle θ.

Use

$$h = \frac{h}{1} = \frac{\text{adjacent}}{\text{hypotenuse}} = \cos\theta.$$

Find equation for V using formula for area of a trapezoid.

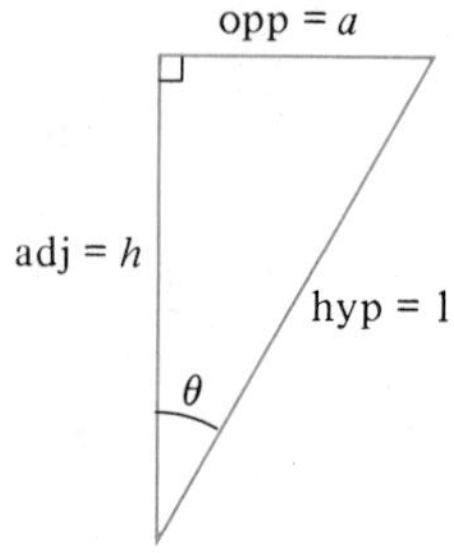

Then

$$h = \cos\theta, \qquad \text{and} \qquad a = \sin\theta.$$

The end panel is a trapezoid with small base $b = 1$ and large base $B = 1 + 2a$. The volume of the trough is therefore

$$\begin{aligned} V &= \frac{1}{2}(B + b)\cdot h \cdot 6 \\ &= \frac{1}{2}[(1 + 2a) + 1]\cdot h \cdot 6 \\ &= 6(1 + \sin\theta)\cos\theta. \end{aligned}$$

Domain of V is $[0, \pi/2]$, from sketch.

Clearly θ must lie within the interval $[0, \pi/2]$. Since V is expressed as a function of θ alone, we may proceed to find the critical numbers for θ:

$$\begin{aligned} \frac{dV}{d\theta} &= -6\sin\theta + 6\cos^2\theta - 6\sin^2\theta \\ &= -6\sin\theta + 6(1 - \sin^2\theta) - 6\sin^2\theta \\ &= -12\sin^2\theta - 6\sin\theta + 6 \\ &= -6(2\sin\theta - 1)(\sin\theta + 1). \end{aligned}$$

Set $V'(\theta) = 0$ to find critical numbers.

The equation $dV/d\theta = 0$ therefore yields two critical numbers since

$$2\sin\theta - 1 = 0 \text{ implies } \sin\theta = \frac{1}{2}, \qquad \text{or} \qquad \theta = \frac{\pi}{6}.$$

and

$$\sin\theta + 1 = 0 \text{ implies } \sin\theta = -1, \qquad \text{or} \qquad \theta = -\frac{\pi}{2}.$$

Remaining critical numbers occur when $V'(\theta) = 0$.

Determine which critical numbers lie in domain $[0, \pi/2]$.

Inspect $V(\theta)$ at critical numbers and endpoints to determine maximum.

Since $dV/d\theta$ is defined for all θ, and the critical number $\theta = -\pi/2$ does not lie within $[0, \pi/2]$, the only critical number for θ in $[0, \pi/2]$ is $\theta = \pi/6$. The maximum value for V must lie among the numbers

$$\begin{aligned} V(0) &= 6(1 + 0)(1) = 6 \\ V(\pi/6) &= 6\left(1 + \frac{1}{2}\right)\left(\frac{\sqrt{3}}{2}\right) = \frac{9\sqrt{3}}{2} \approx 7.79 \\ V(\pi/2) &= 6(1 + 1)(0) = 0 \end{aligned}$$

The angle corresponding to the maximum volume is therefore $\theta = \pi/6 = 30°$. ■

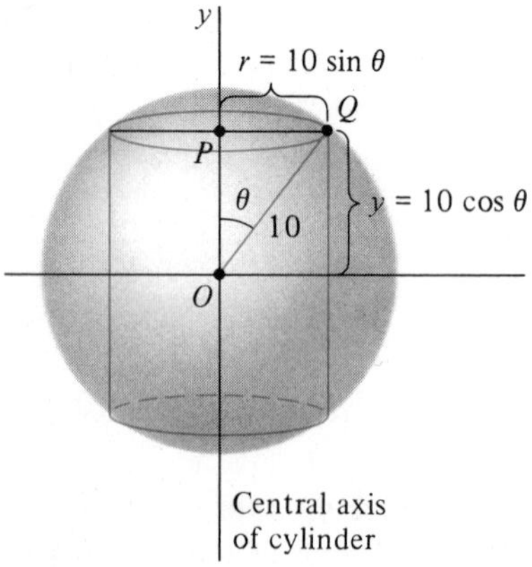

Figure 4.4 Cylinder within a sphere.

Example 3 Find the right circular cylinder of maximum volume that can be inscribed in a sphere of radius 10 cm.

Solution: As illustrated in Figure 4.4, let O be the center of the sphere, P the point lying on both the top of the cylinder and the central axis for the cylinder, and Q a point at which the top of the cylinder meets the sphere. Then since the top of the cylinder is perpendicular to its central axis, triangle OPQ is a right triangle. Let θ be the angle formed between OP and OQ. Since OQ is a radius of the sphere,

$OQ = 10$ cm. Let $y = OP$ and $r = PQ$. Then

$$\frac{y}{10} = \cos\theta, \qquad \text{so} \qquad y = 10\cos\theta$$

and

$$\frac{r}{10} = \sin\theta, \qquad \text{so} \qquad r = 10\sin\theta.$$

Since y is half the height of the cylinder and r is the radius of its base, we can express the volume of the cylinder as

$$\begin{aligned} V &= \pi r^2(2y) \\ &= 2\pi(10\sin\theta)^2(10\cos\theta) \\ &= 2000\pi\sin^2\theta\cos\theta. \end{aligned}$$

The function V to be maximized is now a function of $\theta \in [0, \pi/2]$, so we proceed to compute

$$\begin{aligned} V'(\theta) &= 2000\pi(2\sin\theta\cos^2\theta - \sin^3\theta] \\ &= 2000\pi \cdot \sin\theta \cdot [2\cos^2\theta - \sin^2\theta] \\ &= 2000\pi \cdot \sin\theta \cdot [2\cos^2\theta - (1 - \cos^2\theta)] \\ &= 2000\pi\sin\theta[3\cos^2\theta - 1]. \end{aligned}$$

We will have $V'(\theta) = 0$ when $\sin\theta = 0$ and when $\cos\theta = 1/\sqrt{3}$. We need not solve for θ in either of these equations since the dimensions of the cylinder can be computed directly from this information.

If $\sin\theta = 0$, then $\cos\theta = \sqrt{1 - \sin^2\theta} = 1$, so

$$r = 10\sin\theta = 0, \qquad \text{and}$$
$$h = 2y = 2 \cdot 10\cos\theta = 20.$$

In this case, $V = \pi r^2 h = \pi \cdot 0^2 \cdot 20 = 0$.

If $\cos\theta = \dfrac{1}{\sqrt{3}}$, then

$$\sin\theta = \sqrt{1 - \left(\frac{1}{\sqrt{3}}\right)^2} = \sqrt{\frac{2}{3}}, \qquad \text{so}$$

$$r = 10\sin\theta = \frac{10\sqrt{2}}{\sqrt{3}}, \qquad \text{and} \tag{9}$$

$$h = 2y = 2 \cdot 10 \cdot \frac{1}{\sqrt{3}} = \frac{20}{\sqrt{3}}. \tag{10}$$

Hence

$$V = \pi r^2 h = \pi \cdot \left(\frac{10^2 \cdot 2}{3}\right)\left(\frac{20}{\sqrt{3}}\right) = \frac{4000\pi\sqrt{3}}{9}\ \text{cm}^3. \tag{11}$$

Since $V'(\theta)$ is defined for all $\theta \in [0, \pi/2]$ and the endpoint $\theta = 0$ corresponds to the case $\sin\theta = 0$ above, we need only to check the value of V for $\theta = \pi/2$. But here $h = 2y = 20\cos(\pi/2) = 0$, so $V = 0$. We can therefore now conclude that the cylinder of maximum volume occurs when $\cos\theta = 1/\sqrt{3}$. The dimensions and volume of this cylinder are as given in equations (9), (10), and (11). ■

Example 4 Researchers interested in modelling the rate at which animals (including humans) grow know that growth is not uniform. Periods of rapid growth often occur between periods of very slow growth.

A scientist interested in modelling the growth pattern for a particular species wishes to use an equation of the form

$$y = \alpha t + \beta \sin(\gamma \pi t) + p$$

as a growth model, since this function possesses the growth characteristics described above (see Figure 4.5). Here t represents time in months after birth and y represents height in centimeters. The parameters (constants) α, β, γ and p are positive numbers calculated by the scientist on the basis of observed data. (This process, called "fitting" the model to the data, is a subject in courses on mathematical statistics.) For a model of this type the growth rate is defined to be the derivative dy/dt.

If $\gamma = 1/4$ and $\beta = \alpha = 1$, find the maximum and minimum rates of growth and the times at which they occur during the first year of growth according to the proposed model.

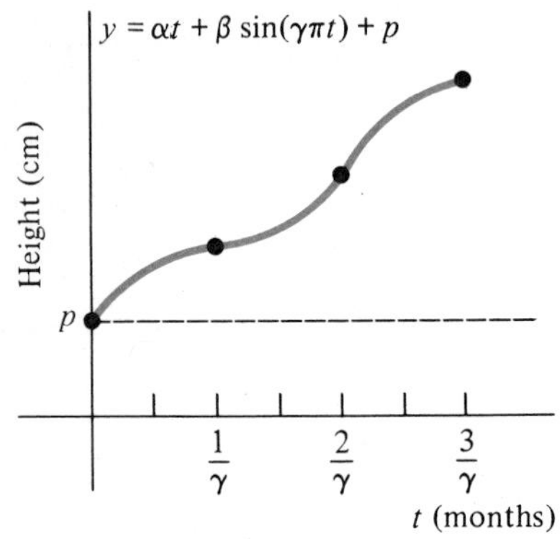

Figure 4.5 A model for animal growth.

Solution: The growth rate is

$$\frac{dy}{dt} = 1 + \frac{\pi}{4}\cos\left(\frac{\pi t}{4}\right).$$

The mathematical problem is to find the extreme values for this function on the interval [0, 12].* To obtain the critical numbers for dy/dt we differentiate, obtaining

$$\frac{d^2y}{dt^2} = -\frac{\pi^2}{16}\sin\left(\frac{\pi t}{4}\right).$$

This derivative equals zero when $\sin\left(\frac{\pi t}{4}\right) = 0$, that is, when

$$\frac{\pi t}{4} = 0, \pm\pi, \pm 2\pi, \ldots$$

or

$$t = 0, \pm 4, \pm 8, \pm 12, \ldots .$$

Now only the critical numbers $t = 0, 4, 8$, and 12 lie within the interval [0, 12]. Since d^2y/dt^2 is never undefined, there are no other critical numbers. Table 4.2 shows the growth rate at each critical number and endpoint.

Table 4.2

t	0	4	8	12
$\frac{dy}{dt}$	$1 + \frac{\pi}{4}$	$1 - \frac{\pi}{4}$	$1 + \frac{\pi}{4}$	$1 - \frac{\pi}{4}$

*Note here that we are seeking the extreme values of the *derivative* dy/dt rather than the extreme values of y. We therefore examine $\frac{d}{dt}\left(\frac{dy}{dt}\right) = \frac{d^2y}{dt^2}$ for critical numbers.

The maximum growth rate is observed to be $1 + \frac{\pi}{4} \approx 1.785$ cm/month, and this rate occurs at birth and at age 8 months. The minimum growth rate $1 - \frac{\pi}{4} \approx .215$ cm/month occurs at ages 4 months and 12 months (see Figure 4.6). ■

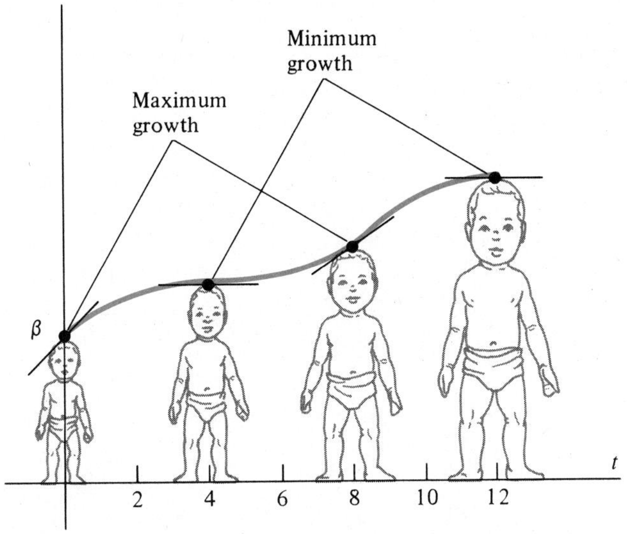

Figure 4.6 Times of minimum and maximum growth.

Example 5 Suppose that a traffic engineer has sampled speed and density of automobiles along a particular section of highway and determined that a good model for the relationship between velocity and density is given by the equation

$$v(\rho) = \frac{100}{1 + \rho^2} \text{ km/hr}, \qquad 0 \le \rho \le 3.$$

Here ρ represents density in units of 100 automobiles per kilometer, and v represents velocity. Find the density at which traffic *flow* $q(\rho) = \rho v(\rho)$ will be a maximum according to this model (see Figures 4.7 and 4.8).

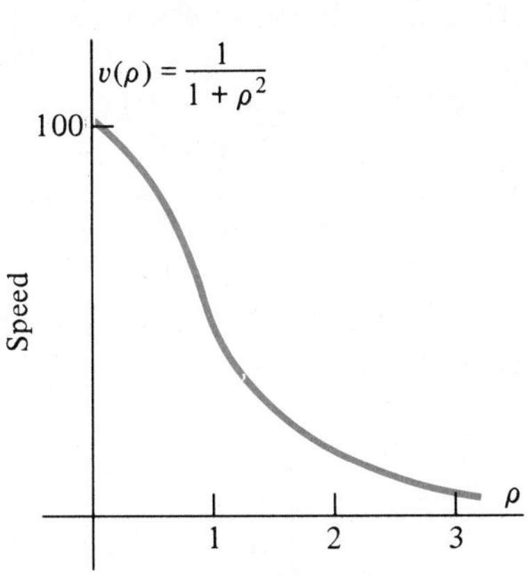

Figure 4.7 Velocity as a function of density.

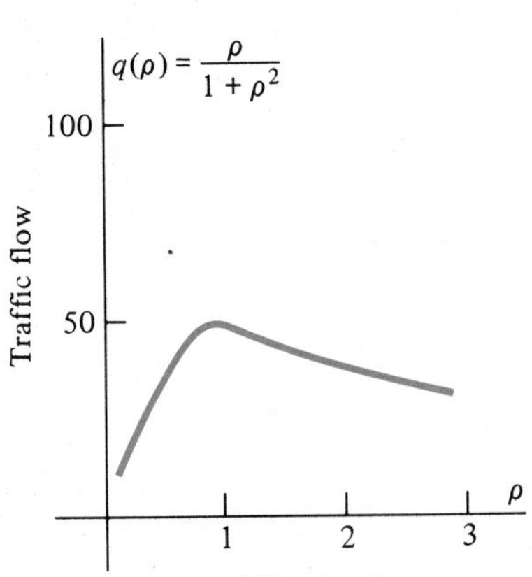

Figure 4.8 Traffic flow as a function of density.

Solution: For the above model we have

$$q(\rho) = \rho v(\rho) = \frac{100\rho}{1 + \rho^2} \text{ hundred automobiles/hr.}$$

The mathematical problem here is to find the maximum value of the function $q(\rho)$ for $\rho \in [0, 3]$. We find

$$\frac{dq}{d\rho} = \frac{100(1 + \rho^2) - (100\rho)(2\rho)}{[1 + \rho^2]^2}$$
$$= \frac{100 - 100\rho^2}{[1 + \rho^2]^2}.$$

Since $dq/d\rho = 0$ implies $100 - 100\rho^2 = 0$, or $\rho^2 = 1$, it follows that the only critical numbers for q are $\rho = -1$ and $\rho = 1$. Since only $\rho = 1$ lies within the interval $[0, 3]$, the maximum flow lies among the values

$$q(0) = \frac{100(0)}{1 + 0^2} = 0 \qquad \text{(endpoint)}$$

$$q(1) = \frac{100}{1 + 1^2} = 50 \qquad \text{(critical number)}$$

$$q(3) = \frac{100(3)}{1 + 3^2} = 30 \qquad \text{(endpoint).}$$

The maximum flow is therefore $50 \times 100 = 5000$ automobiles per hour, which corresponds to a density of $\rho = 100$ automobiles per kilometer and a speed of $v(1) = 50$ kilometers per hour. ■

Example 6 Two laws in optics can be established using the calculus and a principle discovered by Pierre de Fermat, the brilliant seventeenth century French mathematician. Fermat's principle states that a ray of light traveling in a uniform medium will follow the path of minimum time.

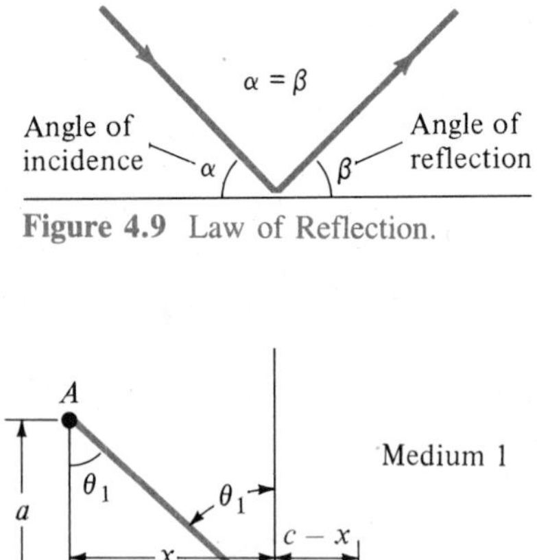

Figure 4.9 Law of Reflection.

The first law is the Law of Reflection, which states that a ray of light which strikes a flat reflecting surface at an angle α will be reflected away at that same angle, i.e., that the angle of incidence equals the angle of reflection (see Figure 4.9). You are asked to show this in Exercise 34.

The second law, first discovered by the seventeenth century Dutch mathematician Willebrord Snell, concerns a ray of light that passes from one medium, in which light travels at a velocity v_1, into a second medium, in which light travels at a velocity v_2. Within each medium the path of minimum time is a straight line. However, if $v_1 \neq v_2$, the path followed between a point A in medium 1 and a point B in medium 2 will consist of two line segments meeting at a point P on the surface separating the media (see Figure 4.10). This "bending" of the light ray is called refraction, and Snell's Law of Refraction states that

$$\frac{\sin\theta_1}{v_1} = \frac{\sin\theta_2}{v_2}$$

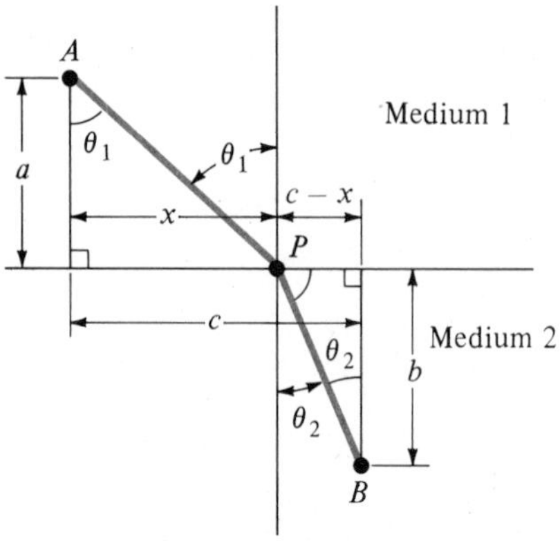

Figure 4.10 Snell's Law of Refraction.

where θ_1 and θ_2 are the angles formed by the ray and the lines normal to the surface at P in each of the media.

To establish this law, we let a and b denote the lengths of the normals from points A and B to the surface, we let c be the distance between these normals, and we let x be the distance along the surface from the normal through A to point P.

Since the time required for the light to travel from point A to point P is

$$T_1 = \frac{\sqrt{a^2 + x^2}}{v_1} \qquad \left(\text{time} = \frac{\text{distance}}{\text{velocity}}\right)$$

and the time required for light to travel from point P to point B is

$$T_2 = \frac{\sqrt{(c - x)^2 + b^2}}{v_2},$$

we seek to minimize the total time

$$T = T_1 + T_2 = \frac{\sqrt{a^2 + x^2}}{v_1} + \frac{\sqrt{(c - x)^2 + b^2}}{v_2}. \tag{12}$$

It is reasonable to assume that x lies within the closed interval $[0, c]$. Since equation (12) for T contains only one independent variable x, we may proceed to calculate $T'(x)$. We obtain

$$T'(x) = \frac{x}{v_1\sqrt{a^2 + x^2}} - \frac{c - x}{v_2\sqrt{(c - x)^2 + b^2}}.$$

This derivative is defined for all x, so the only way in which a critical number can arise is from the equation $T'(x) = 0$. This equation gives the condition that

$$\frac{x}{v_1\sqrt{a^2 + x^2}} = \frac{c - x}{v_2\sqrt{(c - x)^2 + b^2}}. \tag{13}$$

From Figure 4.10 we can see that

$$\frac{x}{\sqrt{a^2 + x^2}} = \sin\theta_1; \qquad \frac{c - x}{\sqrt{(c - x)^2 + b^2}} = \sin\theta_2.$$

Using these equations, we obtain from equation (13) that

$$\frac{\sin\theta_1}{v_1} = \frac{\sin\theta_2}{v_2},$$

which is Snell's Law. (The verification that this condition actually corresponds to a minimum rather than a maximum is less straightforward here than in the other examples. One reasons on physical grounds that a minimum must exist for $x \in (0, c)$. Since the condition of Snell's Law arises from the only critical number in $(0, c)$, it must correspond to the minimum travel time.) ■

Exercise Set 4.4

1. The sum of two nonnegative numbers is 10. Find these numbers if
   a. their product is as small as possible,
   b. the sum of their squares is as small as possible,
   c. the sum of their squares is as large as possible.

2. The sum of two nonnegative numbers is 36. Find these numbers if the first plus the square of the second is
   a. a maximum,
   b. a minimum.

3. A rectangular play yard is to be constructed along the side of a house by erecting a fence on three sides, using the house wall as the fourth wall of the play yard. Find the dimensions that produce the play yard of maximum area if 20 meters of fence is available for the project.

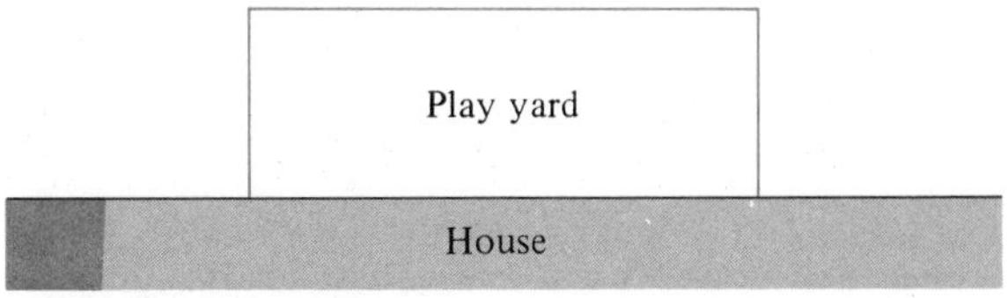

4. A model for the spread of disease assumes that the rate at which a disease spreads is proportional to the product of the number of people infected and the number not infected.

Assume the size of the population to be a constant N. When is the disease spreading most rapidly?

5. A farmer has 120 meters of fencing with which he plans to make a rectangular pig pen. The pen is to have one internal fence running parallel to the end fences that divides the pen into two sections. Find the dimensions that produce the pen of maximum area if the length of the larger section is to be twice the length of the smaller section.

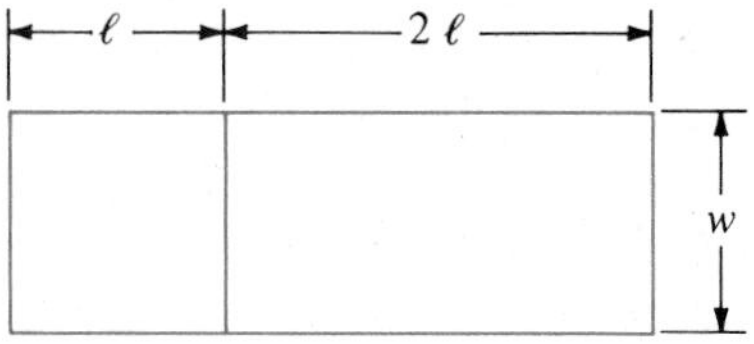

6. Find the minimum and maximum values of the slopes of the lines tangent to the graph

$$y = x^3 - 9x^2 + 7x - 6, \qquad 1 \leq x \leq 4,$$

and the points where these slopes occur.

7. An open box is to be made from a square sheet of cardboard by cutting out squares of equal size from each of the four corners and bending up the flaps. The sheet of cardboard measures 20 cm on each side. Find the dimensions of the box of largest volume which can be made in this way.

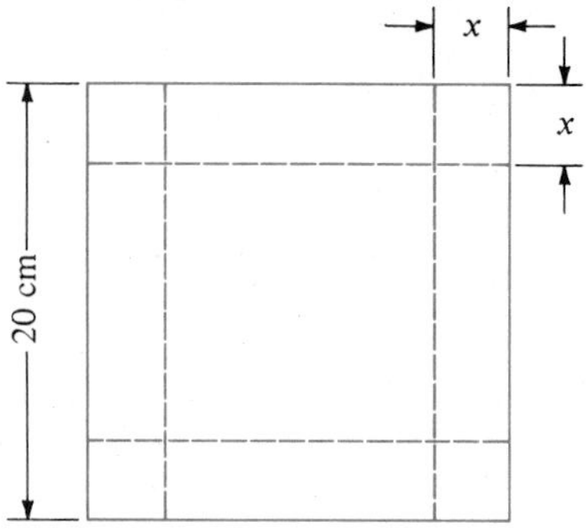

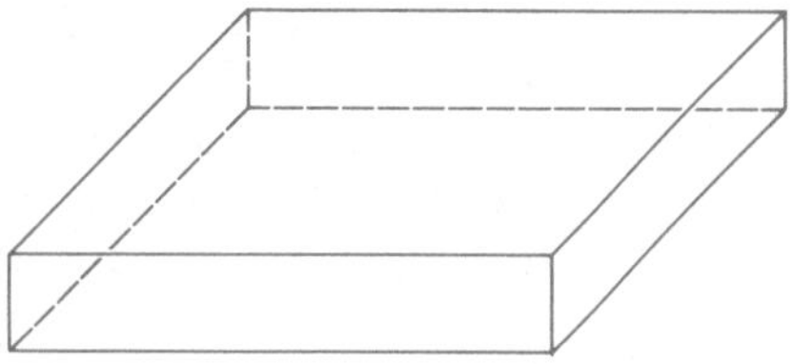

8. A rectangle is inscribed in a triangle with sides of length 6 cm, 8 cm, and 10 cm, respectively. Find the dimensions of the rectangle of maximum area if two sides of the rectangle lie along two sides of the triangle.

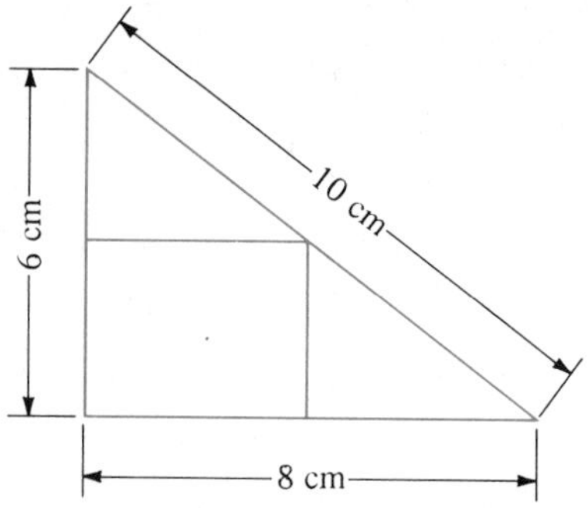

9. A pattern for a rectangular box with a top is to be cut from a sheet of cardboard measuring 10 cm by 16 cm. Find the dimensions of the box for which volume is a maximum.

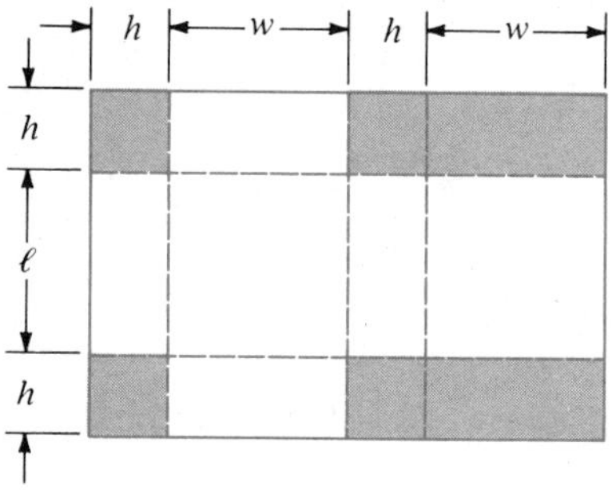

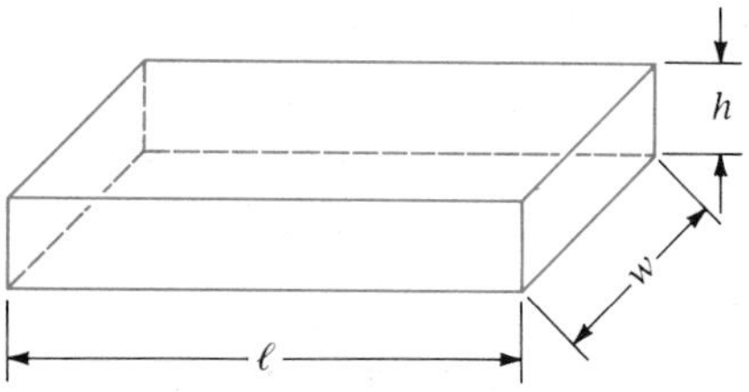

10. A window has the shape of a rectangle surmounted by a semicircle. Find the dimensions that provide maximum area if the perimeter of the window is 10 meters.

11. A rectangle is inscribed in an isosceles triangle with base 6 cm and height 4 cm. Find the dimensions of the rectangle of maximum area if one side of the rectangle lies along the base of the triangle.

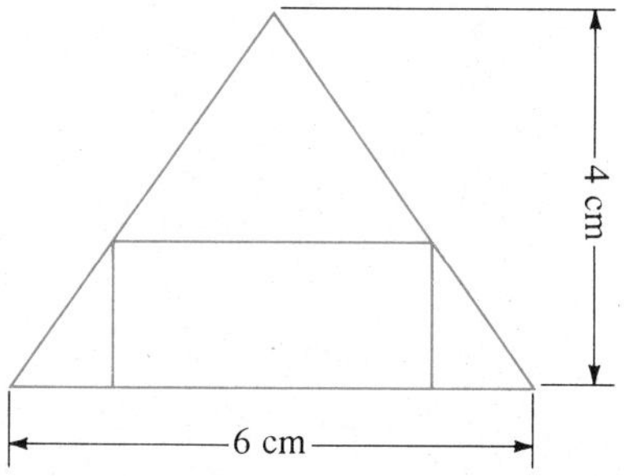

12. A right triangle with hypotenuse 10 cm is rotated about one of its legs to sweep out a right circular cone. Of all such triangles, which generates the cone of maximum volume?

13. A rectangle has two vertices on the x-axis and the other two above the x-axis and on the graph of the equation $y = 4 - x^2$. Find the dimensions for which the area of such a rectangle is a maximum.

14. A triangle has two legs of length a and b. The angle at the vertex where these legs meet is θ. Find the value of θ for which the area of the triangle is a maximum.

15. A sector of a circle of radius r and angle θ is to have fixed perimeter P. Find the dimensions r and θ that maximize the area.

16. Find the dimensions of the right triangle with hypotenuse $h = 2$ and maximum area.

17. An orchard presently has 25 trees per acre. The average yield has been calculated to be 495 apples per tree. It is predicted that for each additional tree planted per acre the yield will be reduced by 15 apples per tree. According to this information, how many additional trees per acre should be planted in order to maximize yield?

18. Find the dimensions of the rectangle of largest area that can be inscribed in a semicircle of radius 8 cm.

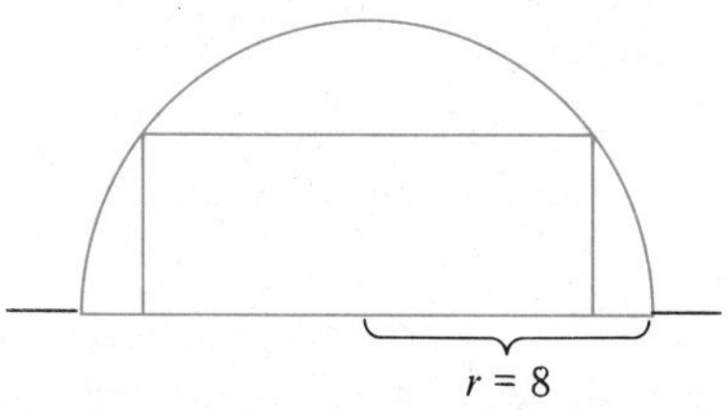

19. Find the dimensions and the area for the rectangle of maximum area that can be inscribed in a circle of radius 4.

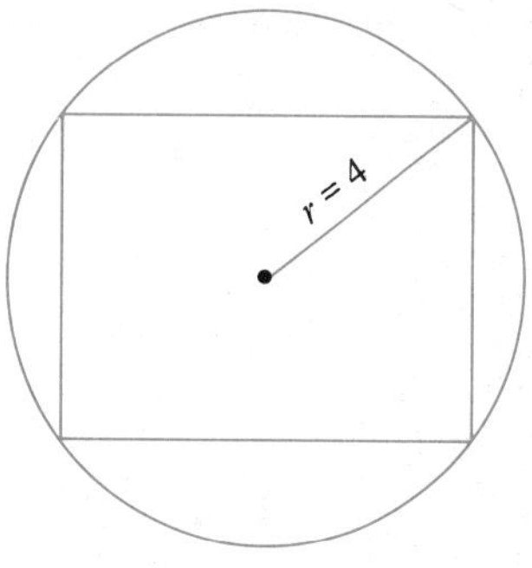

20. A rectangular beam is to be cut from a round log 20 cm in diameter. If the strength of the beam is proportional to the product of its width and the square of its depth, find the dimensions of the cross section for the beam of maximum strength.

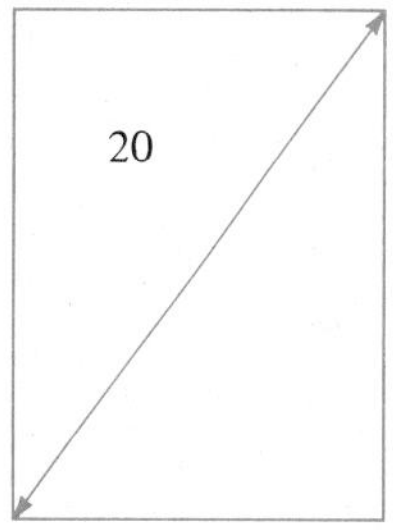

21. A format for textbook page layouts is to be chosen so that each printed page has a 4-cm margin at top and bottom and a 2-cm margin on the left and right sides. The rectangular region of printed matter is to have area 800 cm^2. Find the dimensions for the textbook pages that minimize their areas.

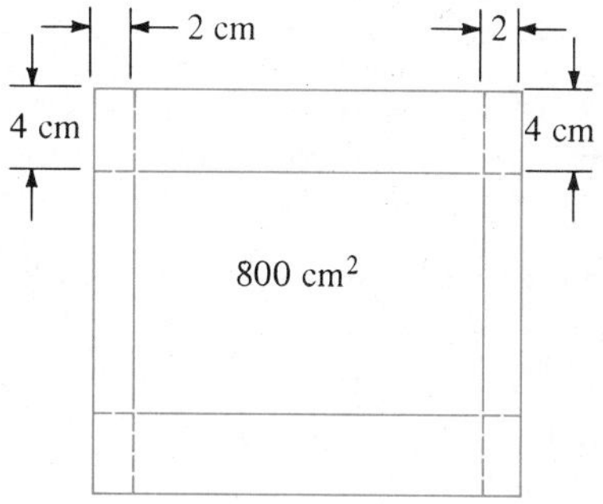

22. Find the rectangle of maximum area that can be inscribed in the ellipse $x^2 + 4y^2 = 4$.

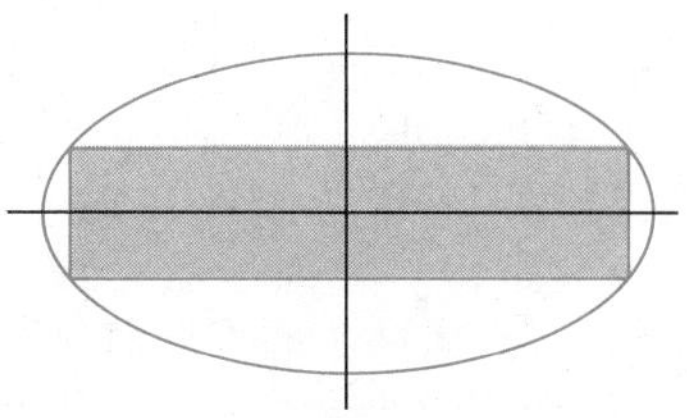

23. A right circular cylinder is inscribed in a right circular cone of radius 3 cm and height 5 cm. Find the dimensions of the cylinder of maximum volume.

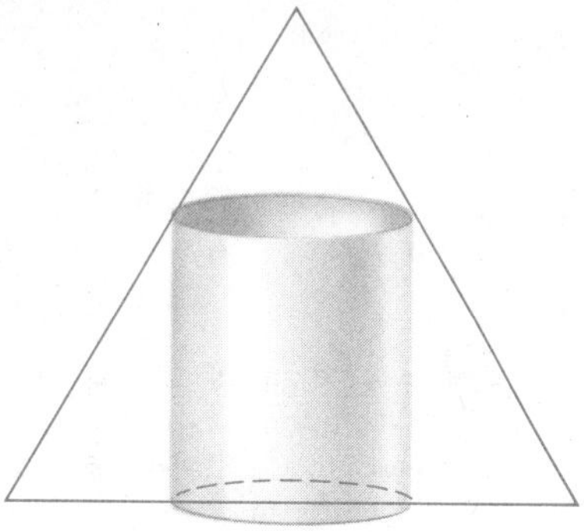

24. A right circular cone is inscribed in a sphere of radius 10 cm. Find the dimensions for which the volume of the cone is a maximum.

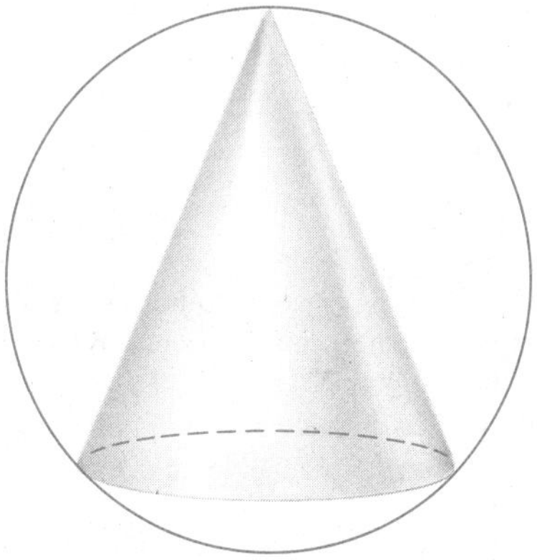

25. For the growth model of Example 4, find the time at which minimum and maximum growth occurs if $\alpha = 3$, $\beta = 2$, $\gamma = 1/3$, and $0 \le t \le 9$.

26. Suppose that the density of automobiles along a section of highway between the hours of $t = 3$ p.m. and $t = 7$ p.m. is given by the function

$$\rho(t) = -10(t-3)(t-7) \qquad \text{automobiles/kilometer.}$$

If the velocity $v(\rho)$ as a function of density is as given in Example 5, find the maximum velocity and the time at which it occurs.

27. Prove that $\sin x \le x$ for all $x \ge 0$. (*Hint:* On the interval $[0, 1]$ find the minimum value of the function $f(x) = x - \sin x$. Then argue the case on $[1, \infty)$ from known properties of $\sin x$.)

28. Find the point on the graph of $y = \sqrt{x}$ nearest the point $(2, 0)$. (*Hint:* Assume x lies in an interval of the form $[0, a]$ where a is large. Then minimize the *square* of the distance.)

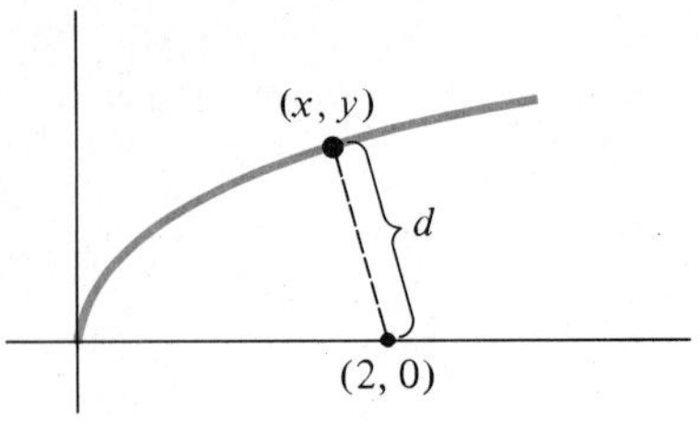

29. Find the point on the ellipse $x^2 + 4y^2 = 4$ nearest the point $(1, 0)$.

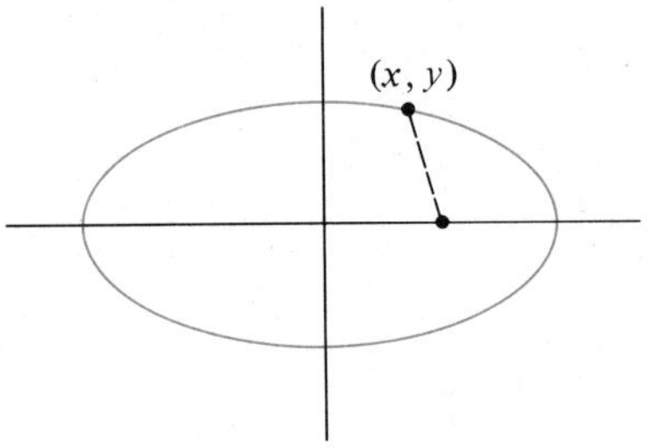

30. A wire 50 cm long is to be cut into two pieces, one of which is to be bent into the shape of a circle and the other of which is to be bent into the shape of a square. Find where the wire should be cut so that the area of the resulting figures is

a. a maximum,

b. a minimum.

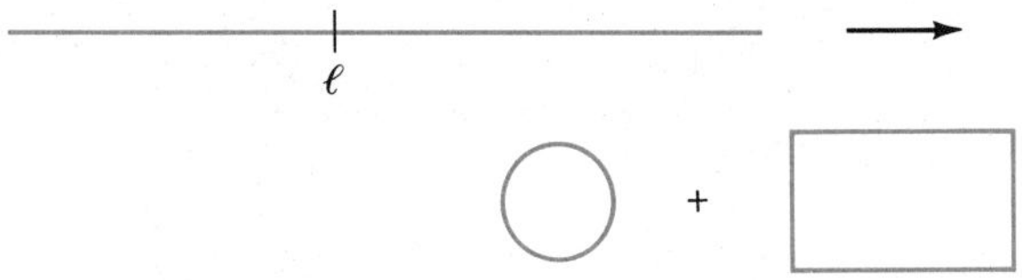

31. United States postal regulations limit the size of a parcel to be mailed parcel post at second class post offices, according to the rule "length plus girth not to exceed 100 inches." (Girth is the largest circumference perpendicular to the length.) Find the dimensions of the rectangular box of maximum volume that can meet this restriction. (Neglect the thickness of the material. Then attack the problem in two steps: (1) For fixed girth, find the relationship between width and height that maximizes the area of the cross section. (2) Then find the relationship between girth and length that maximizes volume.)

32. Work Problem 30 if one of the two pieces of wire is bent into the shape of an equilateral triangle rather than a circle.

33. The illumination at point Q from a source of light at point P is proportional to the intensity of the light source at P and inversely proportional to the square of the distance from P to Q. Two light sources are 20 meters apart. Find the point on the line joining the two points of minimum illumination if one source is twice as strong as the other.

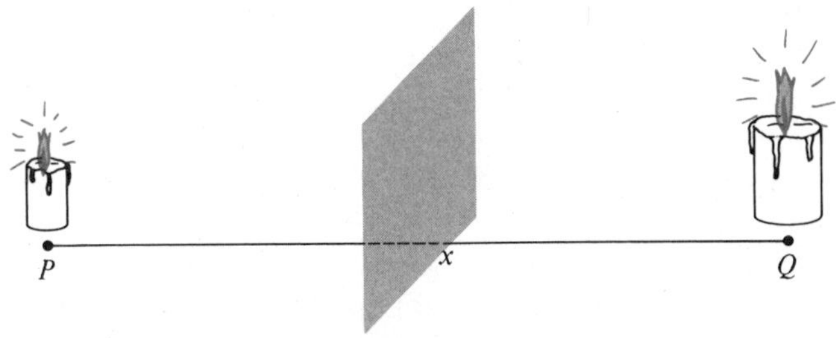

34. Prove the Law of Reflection: A ray of light striking a flat reflecting surface at an angle α will be reflected away at the same angle, i.e., that the angle of incidence equals the angle of reflection (see Example 6).

35. A swimmer is in the ocean 100 meters from a straight shoreline. A person in distress is located on the shoreline 300 meters from the point on the shoreline closest to the swimmer. If the swimmer can swim 3 meters per second and run 5 meters per second, what path should the swimmer follow in order to reach the distressed person as quickly as possible?

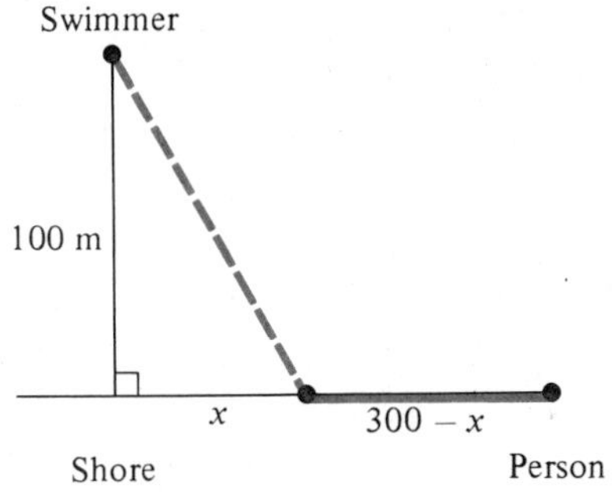

36. An underground telephone cable is to be laid between two boat docks on opposite banks of a straight river. One boathouse is 600 meters downstream from the other. The river is 200 meters wide. If the cost of laying the cable is \$50 per meter under water and \$30 per meter on land, how should the cable be laid to minimize cost?

37. Observations by plant biologists support the thesis that the survival rate for seedlings in the vicinity of a parent tree is proportional to the product of the density of seeds on the ground and their probability of survival against herbivores. (The density of herbivores tends to decrease as distance from the tree increases since the density of the food supply also decreases.) Let x denote the distance in meters from the trunk of the tree. The results of sampling indicate that the density of seeds on the ground for $0 \leq x \leq 10$ is given by

$$d(x) = \frac{1}{1 + (0.2x)^2}$$

and the probability of survival against herbivores is

$$p(x) = (0.1)x.$$

Find the distance, according to the model proposed above, at which the survival rate is a maximum.

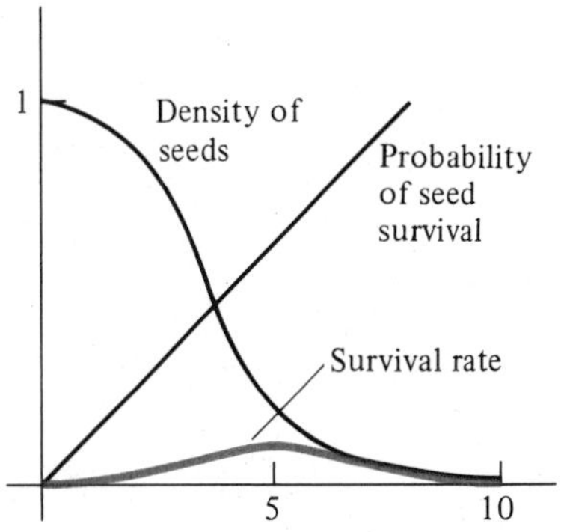

38. In Exercise 37, suppose that the density of seeds for $1 \leq x \leq 9$ is given by the function

$$d(x) = \frac{1}{x}$$

and that the probability of a seed surviving against herbivores is

$$p(x) = \frac{1}{(x - 10)^2}.$$

Find the distance from the parent tree at which seed survival will be a minimum.

39. Use the fact that the lateral surface area of a right circular cone is $S = \pi r\sqrt{r^2 + h^2}$ to find the dimensions of a drinking cup in this shape if the volume of the cup is 64π cm^3 and the surface area is a minimum.

40. A rectangle is inscribed inside the ellipse $\dfrac{x^2}{4} + \dfrac{y^2}{9} = 1$ with its sides parallel to the coordinate axes. Find the dimensions for which the area of the rectangle is a maximum.

41. Find the length of the longest rod which can be carried horizontally around a (square) corner from a corridor 4 ft wide into another corridor 4 ft wide.

42. Work Problem 41 if the corridors are each 8 ft high and the rod need not be carried horizontally.

43. Work Problem 41 for a 3-ft wide corridor meeting a 6-ft wide corridor, and a horizontal rod.

4.5 APPLICATIONS IN ECONOMICS (Optional)

This text has been developed primarily for use by students in the natural sciences, mathematics, and engineering. However, as quantitative methods find increasing use in the social sciences, students of science and mathematics are finding familiar mathematical concepts arising in a broad range of subject areas. It is therefore appropriate to devote several pages here to some of the ways in which applications of the differential calculus arise in economics, the most quantitative of the social sciences.

In a broad sense the discipline of economics concerns itself with the production, distribution, and consumption of goods and services, and how policy decisions are related to these processes. The quantitative economist attempts to develop a mathematical model whose behavior reflects, and therefore predicts, a phenomenon of interest. We will restrict our considerations to rather simple situations involving relationships between the cost of producing a particular good or service, its selling price, the demand for that good or service, the revenue and profit generated by its sale, and the effect of taxation on these relationships. In reading these examples you should be cautioned that precise mathematical models in business and economics are very difficult to achieve, primarily because the basic laws governing the relationships between variables are usually vague and frequently change with time. The examples that appear here are overly specific, by design, so that you may clearly see the role of the derivative in analyzing these models.

Maximizing Revenue

Let p represent the selling price of a good or service and let x represent the number of such items sold. Then the product $R = px$ is called **total revenue.** When the selling price is a function of x, the revenue equation takes the form

$$R(x) = xp(x). \tag{1}$$

In reality, equation (1) can be defined only for nonnegative *integer* values of x. However, in developing models of the sort appearing in this section, the economist takes the view that equation (1) represents a differentiable function defined for all x whose values at nonnegative integers correspond to revenue. With this interpretation, we may apply the theory of extrema to problems involving revenue functions in the form suggested by equation (1).

Example 1 A catering service will serve a particular dinner on its menu to groups of between 20 and 50 people. For groups of size 20 the price charged for the meal is \$12 per person. For each additional person beyond 20 the price is reduced by 20¢ per person. What size group provides the service with maximum revenue?

Strategy
Name variables.

Solution
We let x represent the number of persons in the group being served. Then $(x - 20)$ is the number of persons by which the group exceeds 20. The price per person is therefore

Find equation for $p(x)$ from given information.

$$p(x) = 12 - (x - 20)(0.20) \text{ dollars.}$$

Find equation for $R(x)$ using equation (1).

Using the above and equation (1), we see that total revenue is given by the equation

$$\begin{aligned} R(x) &= xp(x) \\ &= x[12 - (x - 20)(0.20)] \\ &= 16x - (0.2)x^2 \text{ dollars.} \end{aligned}$$

Set $R'(x) = 0$ to find critical numbers.

We seek the maximum value of this function for x in the interval [20, 50]. As in Sections 4.3 and 4.4, we do this by first calculating the derivative $R'(x)$ and finding the critical numbers. We obtain

$$R'(x) = 16 - (0.4)x,$$

so the equation $R'(x) = 0$ gives

$$(0.4)x = 16, \qquad \text{or} \qquad x = \frac{16}{0.4} = 40.$$

Inspect $R(x)$ at critical numbers and endpoints to find maximum.

There are no values of x for which $R'(x)$ fails to exist, so the only critical number for R is $x = 40$. The maximum revenue therefore lies among the values:

$$\begin{aligned} R(20) &= 12 \text{ dollars} \\ R(40) &= 16 \cdot 40 - (0.2)(40)^2 = 320 \text{ dollars} \\ R(50) &= 16 \cdot 50 - (0.2)(50)^2 = 300 \text{ dollars} \end{aligned}$$

Interpret results.

The maximum revenue is 320 dollars, achieved for groups of size 40. ■

Minimizing Costs

Merchants encounter both fixed and variable costs associated with purchasing, storing, and selling their wares. Careful accounting procedures can often give fairly accurate estimates of the costs attributable to the various aspects of such an enterprise. If a differentiable function $C(x)$ can be found that models cost as a function of a single independent variable x then the theory of the derivative can be applied to find the value(s) of x for which cost is minimized.

Example 2 A merchant sells brooms rather uniformly at a rate of about 1600 brooms per year. Accounting procedures determine that the cost of carrying a broom in stock is \$2 per year. When ordering brooms, the merchant experiences a fixed order cost of \$25 plus a variable order cost of 10¢ per broom. Assuming that orders can be placed so that delivery occurs precisely when stock is depleted, how often should the merchant order brooms so as to minimize yearly costs?

Strategy

Write equation for cost in words. Then translate words into mathematical expressions.

Solution

Our merchant faces a tradeoff between ordering brooms frequently, thereby reducing storage costs but increasing ordering costs, or ordering infrequently with the opposite consequences. To analyze this cost problem, we first identify the contributing factors of cost, as described above. We find that

$$\begin{aligned} \text{Yearly cost} &= (\text{storage costs}) + (\text{fixed ordering costs}) \\ &\quad + (\text{variable ordering costs}). \end{aligned} \tag{2}$$

Label independent variable.

The only independent variable present is the number of times the merchant orders brooms per year, which we denote by x. Then the fraction of a year each order lasts in stock is $1/x$. Since brooms are sold at a uniform rate, individual brooms remain in stock half this long, $1/2x$, on the average. The merchant's annual storage costs are therefore as follows:

Write equation for storage costs in words. Then convert to mathematical expressions.

$$\begin{aligned}\text{storage costs} &= (\text{number of brooms}) \cdot \\ &\quad (\text{average storage time per broom}) \cdot \\ &\quad (\text{annual storage cost per broom}) \\ &= (1600)\left(\frac{1}{2x}\right)(2) \\ &= \frac{1600}{x} \text{ dollars per year}\end{aligned}$$

Write equation for fixed ordering costs in words. Then convert to mathematical expressions.

The fixed ordering costs will be:

$$\begin{aligned}\text{fixed ordering costs} &= (\text{orders per year}) \cdot \\ &\quad (\$25 \text{ per order}) \\ &= 25x \text{ dollars per year.}\end{aligned}$$

Find equation for variable ordering costs.

Obtain equation (2) in mathematical terms.

Finally, the variable ordering costs will be $(1600) \cdot (.10) = 160$ dollars per year regardless of the frequency with which orders are placed. With these observations, we can return to equation (2) and write down the annual cost $C(x)$ as

$$C(x) = \frac{1600}{x} + 25x + 160 \text{ dollars per year.}$$

Now the range of feasible values for x will be limited to an interval, say $[1, 52]$. Then the problem of minimizing cost is simply the problem of finding the minimum value for the function $C(x)$ on the interval $[1, 52]$. We find that

Set $C'(x) = 0$ to find critical numbers.

$$C'(x) = -\frac{1600}{x^2} + 25$$

Remaining critical numbers occur where $C'(x) = 0$.

so setting $C'(x) = 0$ gives the equation

$$x^2 = \frac{1600}{25} = \left(\frac{40}{5}\right)^2 = 8^2,$$

which yields the critical numbers $x = \pm 8$. In addition, $x = 0$ is also a critical number since $C'(0)$ is undefined. However, among these critical numbers only $x = 8$ lies within the interval of feasible solutions $[1, 52]$. Since

Examine $C(x)$ at critical points and endpoints to find minimum.

$$\begin{aligned}C(1) &= 1600 + 25 + 160 = 1785 \text{ dollars,} \\ C(8) &= 200 + 400 + 160 = 760 \text{ dollars,} \qquad \text{and} \\ C(52) &\approx 31 + 1300 + 160 = 1491 \text{ dollars}\end{aligned}$$

the minimum cost occurs when brooms are ordered eight times per year. ■

Maximizing Profits

The **profit** made on the sale of an item is the difference between the selling price (revenue) and the cost of producing the item for sale. If $P(x)$, $R(x)$, and $C(x)$ represent the profit, revenue, and cost associated with the production and sale of x items, the basic equation among these three quantities is

$$P(x) = R(x) - C(x), \qquad x = \text{output} \tag{3}$$

If $R(x)$ and $C(x)$ are modelled by differentiable functions of x, we may apply our optimization techniques as before to find the output x which maximizes profit.

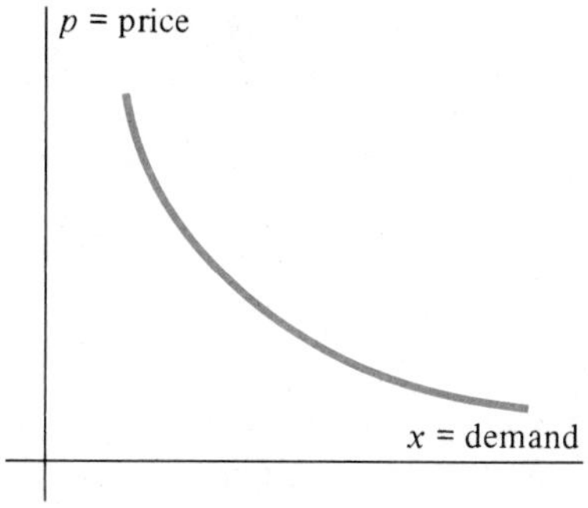

Figure 5.1 Typical demand curve.

Example 3 A monopoly exists when a single firm is the sole producer of a particular good or service. In such a situation there is a direct relationship between the price the monopolist charges for the item (price) and the number of items the public will purchase (demand). Decreasing the price increases sales, and conversely. A typical demand curve appears in Figure 5.1.

Suppose that a monopolist can manufacture at most 175 items per week and that market research shows that the price at which the monopolist can sell x items per week is roughly $p(x) = (500 - x)$ dollars. Furthermore, suppose that the monopolist estimates the cost of producing x items per week to be approximately $C(x) = 150 + 4x + x^2$ dollars. Then,

(a) the revenue obtained from selling x items per week is

$$R(x) = xp(x) = 500x - x^2,$$

(b) the profit obtained from selling x items per week is

$$\begin{aligned} P(x) &= R(x) - C(x) \\ &= [500x - x^2] - [150 + 4x + x^2] \\ &= -2x^2 + 496x - 150 \text{ dollars}, \qquad \text{and} \end{aligned}$$

(c) the maximum possible weekly profit corresponds to the maximum value of the function

$$P(x) = -2x^2 + 496x - 150, \qquad \text{for} \qquad x \in [0, 175].$$

To find this maximum, we set

$$P'(x) = -4x + 496 = 0$$

and obtain the single critical number $x = 496/4 = 124$. The maximum profit thus occurs among the values

$$\begin{aligned} P(0) &= -150 \\ P(124) &= 30{,}602 \\ P(175) &= 25{,}400. \end{aligned}$$

The maximum weekly profit of \$30,602 is therefore achieved at a production level of 124 items per week and a selling price of $(500 - 124) = 376$ dollars per item. ■

Marginal Revenue and Cost

In each of the preceding examples the variables of interest (revenue, cost, and profit) were all known as explicit functions of a single independent variable. In practice, economists often encounter situations in which explicit relationships between variables are not known. Rather, information is limited to observations concerning the effect on a particular dependent variable (say, cost) of a change in the independent variable (say, output level). Economists refer to these kinds of information as **marginal rates.** More specifically, if the cost function $C(x)$ is a differentiable function of x, the **marginal cost** function $MC(x)$ is defined to be the derivative of $C(x)$, that is,

$$MC(x) = C'(x) = \lim_{\Delta x \to 0} \frac{C(x + \Delta x) - C(x)}{\Delta x}. \qquad (4)$$

Similarly, marginal revenue, MR, is defined as

$$MR(x) = R'(x) = \lim_{\Delta x \to 0} \frac{R(x + \Delta x) - R(x)}{\Delta x}. \qquad (5)$$

Since the derivatives appearing in equations (4) and (5) are the limits of the difference quotients $\Delta C/\Delta x$ and $\Delta R/\Delta x$, respectively, we see that the marginal rates MC and MR measure the rates at which C and R change with respect to changes in the independent variable x. In particular, when $\Delta x = 1$ we obtain the approximation

$$MC = C'(x) \approx \left.\frac{\Delta C}{\Delta x}\right|_{\Delta x=1} = \Delta C.$$

For this reason the economist interprets the marginal cost $MC(x)$ at output level x as the cost of producing one additional item. A similar interpretation is placed on $MR(x)$.

Example 4 For the cost and revenue functions of Example 3, we obtain

$$MC(x) = C'(x) = \frac{d}{dx}[150 + 4x + x^2] = 4 + 2x \text{ dollars,}$$

and

$$MR(x) = R'(x) = \frac{d}{dx}[500x - x^2] = 500 - 2x \text{ dollars.}$$

Notice that at the optimum profit level, $x = 124$, we have

$$MR(124) = 500 - 2(124) = 252 \text{ dollars,}$$

and

$$MC(124) = 4 + 2(124) = 252 \text{ dollars.}$$

This is an example of the general principle governing situations in which equation (3) holds: *marginal cost will equal marginal revenue at maximum profit.*

The proof of this statement is straightforward—differentiate both sides of equation (3) to obtain the equation

$$\begin{aligned} P'(x) &= R'(x) - C'(x) \\ &= MR(x) - MC(x). \end{aligned}$$

When $P(x)$ is a maximum, we will have $P'(x) = 0$, and the principle follows. (Of course, all this assumes that $P(x)$, $R(x)$, and $C(x)$ are differentiable at the output level x corresponding to maximum profit.) ■

Taxation and Other Policy Questions

In addition to serving the fundamental capitalist objective of maximizing profits, models such as those of the preceding examples can be helpful in analyzing policy questions such as the effect on output of taxation. For instance, if a government decides to impose a tax of t dollars on each item produced, the basic profit equation (3) becomes

$$P(x) = R(x) - C(x) - tx, \qquad x = \text{output.} \tag{6}$$

If, on the other hand, the tax is imposed as a percentage α of profits, the profit equation (3) becomes

$$P(x) = (1 - \alpha)[R(x) - C(x)]. \tag{7}$$

(In practice, a tax on profits rarely occurs, mainly because of the difficulty in establishing accurate figures on profit levels.)

Example 5 Suppose that the city in which the monopolist in Example 3 is located is contemplating a tax of \$20 to be imposed on each item manufactured by the monopolist. How will the monopolist respond, given the objective of profit maximization?

Solution: Applying the taxed profit equation (6) to the calculations of Example 3 gives the monopolist's revised profit equation as

$$\begin{aligned} P(x) &= R(x) - C(x) - 20x \\ &= (-2x^2 + 496x - 150) - 20x \\ &= -2x^2 + 476x - 150 \text{ dollars.} \end{aligned}$$

The new maximum is obtained by setting

$$P'(x) = -4x + 476 = 0,$$

which gives the new critical number $x = 119$. Inspection of endpoint values $P(0)$ and $P(175)$ shows that, indeed, the new maximum value of profit is $P(119) = 28{,}172$ dollars, a reduction in profit of \$2,430 per week. Notice that the sale price corresponding to the new output level is $p = (500 - 119) = 381$ dollars per item, as opposed to $p = 376$ dollars per item at the original output level of 124 items per week. The results of the proposed tax will therefore be that

(a) the monopolist will reduce output from 124 to 119 items per week, achieving a maximum profit of \$28,172 weekly;
(b) the monopolist will sell the items at a price of \$381, thus passing \$5 of the \$20 tax along to the consumer and absorbing the remaining \$15;
(c) the tax will generate a weekly revenue of $119 \times 20 = 2{,}380$ dollars to the city government. ■

Elasticity of Demand

In situations involving monopolies or near monopolies, economists often express the demand Q for a good or service as a function of price, p. The **elasticity, $E(p)$, of the associated demand curve** at price level p is defined to be the negative of the percentage change in demand divided by the associated percentage change in price as $\Delta p \to 0$, that is,

$$E(p) = \lim_{\Delta p \to 0} -\frac{\left[\dfrac{Q(p + \Delta p) - Q(p)}{Q(p)}\right]}{\dfrac{\Delta p}{p}}.$$

If $Q(p)$ is a differentiable function of p, this definition becomes

$$\begin{aligned} E(p) &= \lim_{\Delta p \to 0} -\left[\frac{Q(p + \Delta p) - Q(p)}{\Delta p}\right]\frac{p}{Q(p)} \\ &= -\frac{pQ'(p)}{Q(p)}. \end{aligned}$$

The definition of $E(p)$ is written with a negative sign by convention so that $E(p)$ is always a positive quantity. (Both p and $Q(p)$ are positive, and $Q'(p)$ is negative

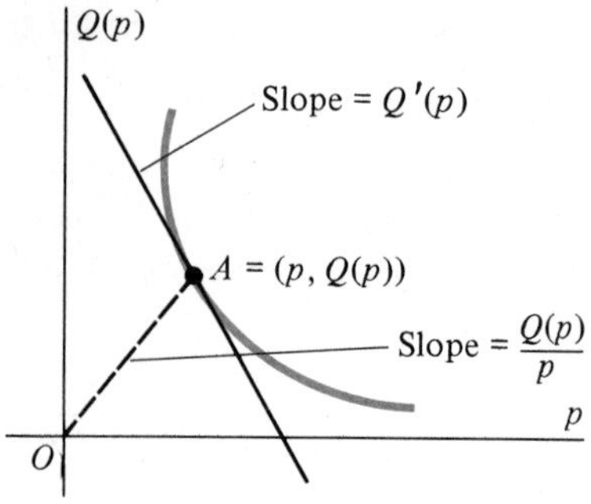

Figure 5.2 Demand vs. supply curve is *elastic* at A if the tangent at A is steeper than OA $\left(\text{i.e., if } |Q'(p)| > \dfrac{Q(p)}{p}\right)$.

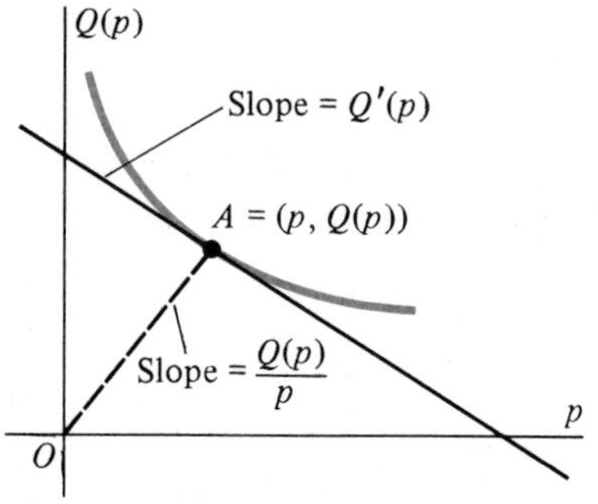

Figure 5.3 Demand vs. supply curve is *inelastic* at A if OA is steeper than the tangent at A.

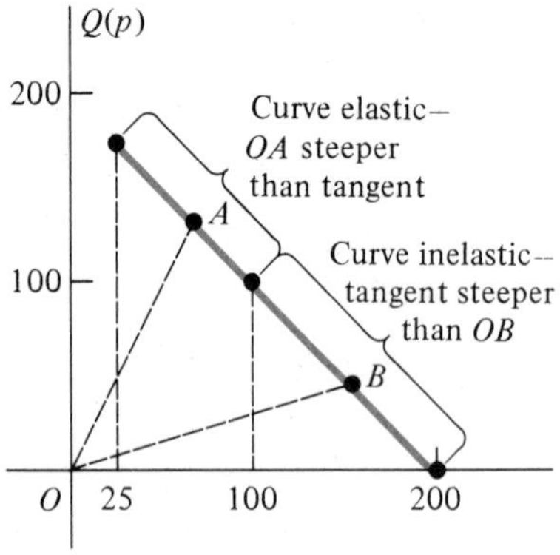

Figure 5.4 Elasticity analysis for $Q(p) = 200 - p$.

since demand decreases as price increases.) When $E(p) > 1$, the demand curve is called **elastic** at p. When $E(p) < 1$, the curve is called **inelastic** at p.

We can gain some geometric insight into elasticity by writing $E(p)$ as

$$E(p) = \frac{-Q'(p)}{\left(\dfrac{Q(p)}{p}\right)} \tag{8}$$

and considering arcs of typical demand-versus-supply curves in Figures 5.2 and 5.3. If $A = (p, Q(p))$ is a point on this curve, then the slope of the line segment OA, from the origin to A, is $\dfrac{\Delta Q}{\Delta p} = \dfrac{Q(p)}{p}$. Note that this is the denominator of $E(p)$ in equation (8). Moreover, the slope of the tangent to this curve at A is just $Q'(p)$. Since $Q'(p)$ is negative, $-Q'(p) = |Q'(p)|$. Thus, the curve will be *elastic* at A if the absolute value of the slope of the tangent is greater than the slope of OA. We paraphrase this by saying that the demand curve is elastic at A if "the tangent is steeper than the line segment OA" (see Figure 5.2). Similarly, the demand curve is *inelastic* at A if the line segment OA is steeper than the tangent at A (see Figure 5.3).

Example 6 The demand for an item is found to be roughly

$$Q(p) = 200 - p$$

for prices between $p = 25$ and $p = 200$ dollars. For which prices is this demand curve elastic? Inelastic?

Solution: Since $Q'(p) = -1$, we have from (8) that

$$E(p) = \frac{-(-1)}{\left(\dfrac{200 - p}{p}\right)} = \frac{p}{200 - p}.$$

Since $E(p) = 1$ has solution $P = 100$, we can see by inspection that

$$E(p) \text{ is } \begin{cases} < 1 \text{ for } 25 \le p < 100 \\ > 1 \text{ for } 100 < p < 200. \end{cases}$$

Thus, the curve is inelastic for $25 \le p < 100$ and elastic for $100 < p < 200$. Figure 5.4 illustrates this conclusion. Since the slope of $Q(p) = 200 - p$ is $m = -1$, the tangent (i.e., the curve itself) will be steeper than the line from the origin to $(p, Q(p))$ when $p > Q(p) = 200 - p$ (i.e., when $p > 100$), and the reverse comparison will hold when $p < 100$. ■

Interest in the concept of elasticity has to do in part with its relation to marginal revenue. If $Q(p)$ denotes the demand for an item at price p, then the revenue function is $R(p) = pQ(p)$. The marginal revenue is therefore

$$\begin{aligned} MR(p) = R'(p) &= \frac{d}{dp}[pQ(p)] \\ &= Q(p) + pQ'(p) \\ &= Q(p)\left[1 + p\frac{Q'(p)}{Q(p)}\right] \\ &= Q(p)[1 - E(p)]. \end{aligned}$$

From this calculation we can conclude that

(1) If the demand curve is *elastic* at p (i.e., if $E(p) > 1$), revenue will decrease if price is increased since in this case $MR(p) = R'(p) < 0$.
(2) If the demand curve is *inelastic* at p, meaning $E(p) < 1$, then we will have $MR(p) = R'(p) > 0$ so revenue will *increase* if price is increased.

These observations illustrate the type of conclusion that can be drawn from general models such as the revenue equation (3) and the definition of elasticity above, even though explicit equations for the demand function and its derivative are not known. The determining factor in the above observations is the magnitude of elasticity, $E(p)$, which in turn depends on the relationship between the variables p and Q and the derivative $Q'(p)$. The study of relationships between marginal rates constitutes a significant part of modern economic theory, and the concept of the derivative clearly plays a central role in this subject.

Exercise Set 4.5

In Exercises 1–4, the cost $C(x)$ and revenue $R(x)$ from the manufacture and sale of x items per month for a company is given, along with the range of possible production levels. In each case find the production level which most nearly maximizes profits, $P(x) = R(x) - C(x)$.

1. $C(x) = 500 + 4x$

$R(x) = 100x - \frac{1}{4}x^2$

$0 \le x \le 200$

2. $C(x) = 1000 + 30x + \frac{1}{2}x^2$

$R(x) = 80x$

$0 \le x \le 500$

3. $C(x) = 500 + 20x + x^2$
$R(x) = 30x$
$0 \le x \le 100$

4. $C(x) = 200 + 50x + \frac{1}{10}x^3$

$R(x) = 100x - \frac{1}{20}x^2$

$0 \le x \le 40$

5. Not all of the companies in Exercises 1–4 are profitable, even at the output level that maximizes profit. Which are profitable ($P(x_0) > 0$)?

6. The cost of operating a small aircraft is calculated to be \$200 per hour plus the cost of fuel. Fuel costs are $s/100$ dollars per kilometer, where s represents speed in kilometers per hour. Find the speed that minimizes the cost per kilometer.

7. An automobile parts store sells 3000 headlight bulbs per year at a uniform rate. Orders can be placed from the distributor so that shipments arrive just as supplies run out. In ordering bulbs the parts store encounters costs of \$20 per order plus 5¢ per bulb. How frequently should the parts store order bulbs if the cost of maintaining one bulb in stock for one year is estimated to be 48¢?

8. A chartered cruise requires a minimum of 100 persons. If 100 people sign up, the cost is \$400 per person. For each additional person the cost per person decreases by \$1.50. Find the number of passengers that maximizes revenue.

9. A merchant sells shirts uniformly at a rate of about 2000 per year. The cost of carrying a shirt in stock is \$3 per year. The cost of ordering shirts is \$50 per order plus 20¢ per shirt. Approximately how many times per year should the merchant order shirts to minimize his total yearly costs? (Assume that orders can be placed so that shipment arrives precisely when stock is depleted.)

10. Find the effect on the broom merchant's reorder scheme in Example 2 if sales fall to 1200 brooms per year and reorder costs increase by 20%.

11. A manufacturing company determines that the revenue obtained from the sale of x items will be $R(x) = 200x - 4x^2$ dollars for $0 \le x \le 25$ and that the cost of producing these x items will be $C(x) = 900 + 40x$. Find the output x at which profits will be a maximum.

12. A publisher can produce a certain book at a net cost of \$4 per book. Market research indicates that 20,000 copies of the book can be sold at a price of \$16 per book and that sales can be increased by 2000 copies for each \$1 reduction in price.

a. What price maximizes revenue?
b. What is the marginal cost per book?
c. What price maximizes profit?

13. A manufacturer of dishwashers can produce up to 100 dishwashers per week. Sales experience indicates that the manufacturer can sell x dishwashers per week at price p where $p + 3x = 600$ dollars. Production records show that the cost of producing x dishwashers per week is

$$C(x) = 4000 + 150x + 0.5x^2.$$

a. Find the weekly revenue function $R(x)$.
b. Find the weekly profit function.
c. Find the marginal cost and marginal revenue functions.
d. Find the weekly production level for which profit is a maximum.
e. What is the relationship between marginal cost and marginal revenue for the production level found in part d?

14. Suppose that the supply of uranium available on world markets is related to the price per pound by the function

$$S(p) = \frac{100}{200 - p}$$

where p is the price per pound, $0 \le p < 180$.

a. Find the price at which the marginal supply equals 0.25 ton.
b. Explain why $S(p)$ is an increasing function.
c. What would be unrealistic about this model if p were to have the range $0 \le p \le 200$?

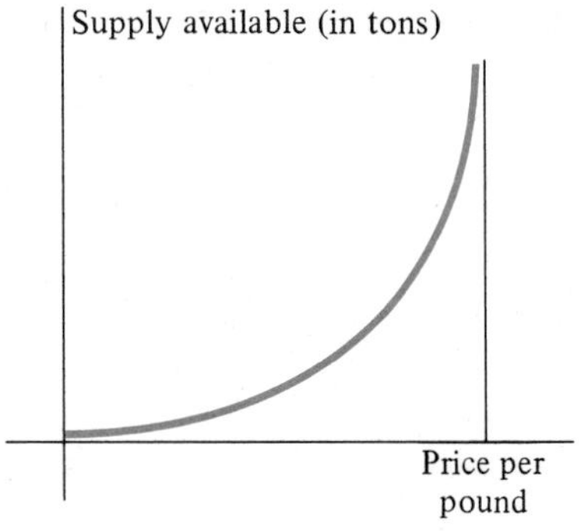

15. Let $C(x)$ denote the cost of producing x items. Then the **average cost** of production for these items is $A(x) = C(x)/x$. Assuming $C(x)$ and $A(x)$ to be differentiable functions of x, find the relationship between average cost and marginal cost when average cost is a minimum.

16. From equation (6), find the relationship between marginal cost and marginal revenue, in the presence of a tax of t dollars per item, when profit is a maximum.

17. A firm is called a **perfect competitor** if the price it can receive for its items is independent of production level (in other words, price is a fixed constant p beyond the control of the firm). In this case the revenue obtained from the sale of x items is simply $R(x) = px$. Assuming that marginal cost is an increasing function of output (i.e., $C'(x) > 0$), show that for a perfect competitor subject to a per item tax, an increase in the tax rate will cause a decrease in output level.

18. Suppose a tax of \$30 per dishwasher is imposed on the manufacturer in Exercise 13. Find

a. the output level that maximizes profit,
b. the resulting change in price per dishwasher,
c. the amount of tax absorbed by the manufacturer.

19. In Exercise 13, find the resulting production level, assuming the manufacturer will always schedule production so as to maximize profit, of the granting of a tax credit of \$35 per dishwasher to the manufacturer.

20. For the demand curve $Q = 40 + 6p - p^2$, $0 < p < 10$, find

a. the elasticity of demand $E(p)$,
b. $E(4)$.

21. A producer in a monopoly faces a demand curve $Q(p) = 200 - p$ where p is the price per item of the manufactured item. Suppose that the cost of manufacturing Q items is given by the function $C(Q) = 2Q + 100$.

a. Find the output Q that maximizes profits and the corresponding price.
b. Find the elasticity of demand at the output that maximizes profit.

For each of Exercises 22–25, determine for which numbers p the given demand curve is elastic or inelastic.

22. $Q(p) = \dfrac{1}{\sqrt{p}}$

23. $Q(p) = \dfrac{1}{p^2}$

24. $Q(p) = 500 - \dfrac{1}{2}p$

25. $Q(p) = \dfrac{\sqrt{p}}{1 + p}, \quad p > 1$

4.6 NEWTON'S METHOD

In solving the optimization problems of Sections 4.3 through 4.5, it is essential to be able to find the critical numbers that arise as solutions of the equation $f'(x) = 0$, that is, to find the *zeros* of the derivative $f'(x)$. This is only one of many circumstances in mathematics in which one needs to find the zeros of a particular function.

In some cases, finding the zeros of a given function is not difficult—simple algebra suffices for linear functions, the quadratic formula handles the case of second degree polynomials, and elegant algorithms exist for computing the zeros of third and fourth degree polynomials (although, sadly, the latter have almost completely disappeared from modern mathematics curricula). However, no such procedures can be found for polynomials of degree five or higher (a fact that you can learn more about in a course on modern algebra), and the situation for equations involving functions other than polynomials is, in general, even more difficult.

In light of this situation, the brilliant mathematician and physicist Isaac Newton (1642–1727) was interested in finding a simple procedure by which zeros of functions could be approximated to within any desired accuracy. He succeeded handsomely, and his procedure made very simple use of tangents and, therefore, of derivatives. We discuss his method briefly, as the last of the rather pragmatic applications of the derivative presented in this chapter. Because the observations on which the method is based issue so directly from properties of the graph of the given function, this topic also serves as a rather natural bridge to Chapter 5, where the more geometric applications of the derivative are discussed.

Let us consider a differentiable function $y = f(x)$ and the problem of approximating a zero c of $f(x)$ which is known to lie between the numbers $x = a$ and $x = b$ (Figure 6.1). Newton's method for approximating c calls for an initial guess, x_1, to be made. The line ℓ_1 tangent to the graph of $y = f(x)$ at $(x_1, f(x_1))$ is then constructed. By finding the x-intercept, x_2, of ℓ_1, we obtain a second approximation to c. Newton's famous observation is simply that for many functions the second approximation, x_2, is better than the first, x_1. If the procedure is then repeated by finding the line ℓ_2 tangent to the graph of $f(x)$ at $(x_2, f(x_2))$, the x-intercept of ℓ_2 (say, x_3) provides an even closer approximation to c than x_2. The procedure is repeated again and again, until an approximation of sufficiently high accuracy is obtained.

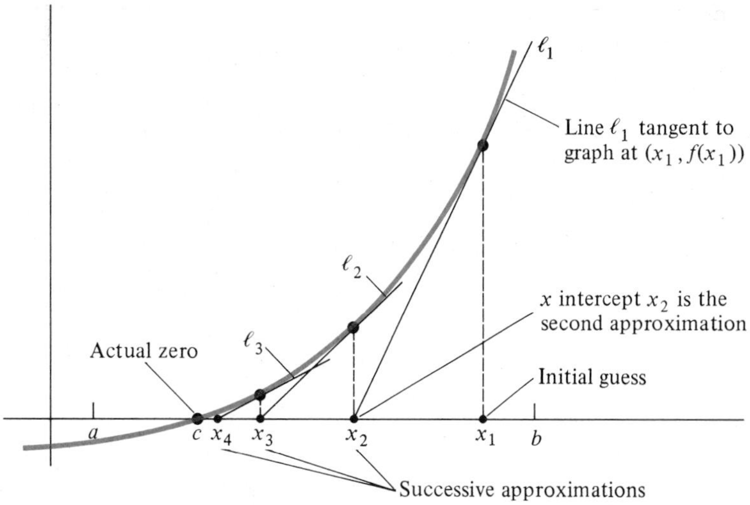

Figure 6.1 Newton's Method for approximating zero of $y = f(x)$.

We can obtain a simple equation for obtaining the $(n + 1)$st approximation, x_{n+1}, from the nth approximation, x_n, by considering Figure 6.2. Since the slope of line ℓ_n tangent to the graph of $y = f(x)$ at $(x_n, f(x_n))$ is given by the derivative

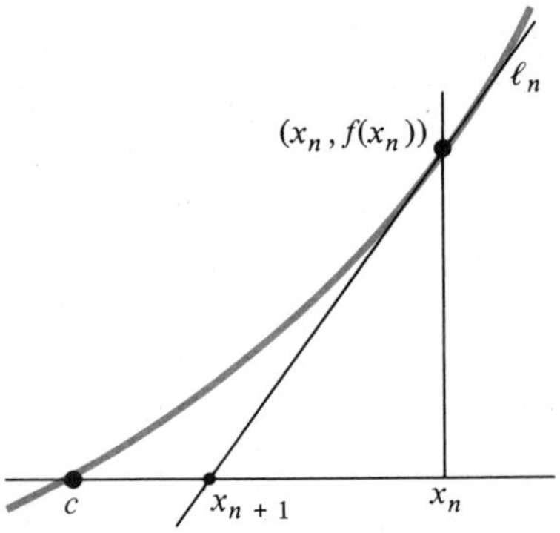

Figure 6.2 Obtaining the approximation x_{n+1} from x_n.

$f'(x_n)$, the equation for ℓ_n can be written as

$$y - f(x_n) = f'(x_n)(x - x_n). \tag{1}$$

To find x_{n+1}, the x-intercept of ℓ_n, we set $y = 0$ and solve for x in equation (1). We obtain

$$x = x_n - \frac{f(x_n)}{f'(x_n)}.$$

Since this is the desired approximation $x = x_{n+1}$, we have **the approximation scheme for Newton's Method:**

$$x_{n+1} = x_n - \frac{f(x_n)}{f'(x_n)} \tag{2}$$

Example 1 Use Newton's Method to approximate the zero of the function $f(x) = x^3 - 10$ lying between $x = 0$ and $x = 4$.

Strategy
First, verify that a zero exists in [0, 4] by applying the Intermediate Value Theorem.

Solution
Before applying the method we should verify that the given function indeed has a zero between the given numbers. Since $f(0) = -10$ and $f(4) = 54$ are of opposite sign and $f(x)$ is continuous, the Intermediate Value Theorem guarantees that $f(x)$ must equal zero for some $x \in (0, 4)$.

Make an initial guess for the first approximation.

We begin by making an initial guess for c, say $x_1 = 3$. Then, from equation (2), with $n = 1$, we obtain the second approximation

Apply (2), using $f'(x) = 3x^2$.

$$\begin{aligned} x_2 &= 3 - \frac{f(3)}{f'(3)} \\ &= 3 - \frac{(3^3 - 10)}{3(3)^2} \\ &= 3 - \frac{17}{27} \approx 2.37. \end{aligned}$$

Apply (2) again and so on.

The next approximation, x_3, is obtained from equation (2) by using $x_2 = 2.37$ and $n = 2$:

$$\begin{aligned} x_3 &= 2.37 - \frac{f(2.37)}{f'(2.37)} \\ &= 2.37 - \frac{(2.37)^3 - 10}{3(2.37)^2} \\ &\approx 2.17, \end{aligned}$$

and so on. ■

Although the formula for Newton's Method is simple to state, the hand calculations quickly become tedious and the chance for error grows rapidly. However, this is precisely the sort of problem that is easy to implement on a hand calculator or on a computer. (A BASIC program for implementing Newton's Method is listed in Appendix I.) Table 6.1 contains the results obtained by using a computer to continue the calculations of this example through $n = 4$ iterations.

Table 6.1

n	x_n	$f(x_n)$	$f'(x_n)$	x_{n+1}
1	3.	17.	27.	2.370370
2	2.370370	3.318295	16.855967	2.173509
3	2.173509	0.267958	14.172419	2.154602
4	2.154602	0.002324	13.926924	2.154435

The advantage of listing partial calculations in a table such as this is that each approximation to the zero, x_n (column 2), can be compared directly to the function value (column 3) at each step. The results obtained in the second and fifth column of Table 6.1 suggest that $x_4 = 2.154602$ is a good approximation to the desired zero.

Example 2 Discuss the accuracy of the approximation obtained in Example 1.

Solution: A precise and thorough analysis of the error associated with Newton's Method is beyond the objectives of this text. The student interested in how such error questions are analyzed as well as other issues in approximation theory is encouraged to pursue these topics in courses on numerical analysis. However, we can give a reasonably good estimate for the error in our approximation by use of the data appearing in columns 3 and 4 of Table 6.1. Recall from the theory of the differential (Chapter 3, Section 9) that when Δx is small we have the approximation

$$\frac{f(x + \Delta x) - f(x)}{\Delta x} \approx f'(x). \tag{3}$$

If c denotes the zero in question and we let $x = c$ and $x + \Delta x = x_n$, then $\Delta x = x_n - x = x_n - c$. Approximation (3) then becomes

$$\frac{f(x_n) - f(c)}{x_n - c} \approx f'(c).$$

Since $f(c) = 0$, we may solve for the quantity $x_n - c$ obtaining

$$x_n - c \approx \frac{f(x_n)}{f'(c)}. \tag{4}$$

Now Table 6.1 contains values for x_n and $f(x_n)$, but not for $f'(c)$. However, if the derivative, $f'(x)$ is a continuous function (as is certainly the case), we know that the value $f'(c)$ will be close to $f'(x_n)$ when x_n is close to c. We therefore approximate $f'(c)$ by $f'(x_n)$, which gives the approximation

$$x_n - c \approx \frac{f(x_n)}{f'(x_n)}.$$

From the values in row 4 of Table 6.1 we obtain the estimate

$$x_4 - c \approx \frac{.002324}{13.926924} \approx 0.00017,$$

suggesting that the approximation $x_4 = 2.154602$ is accurate to three decimal places. ■

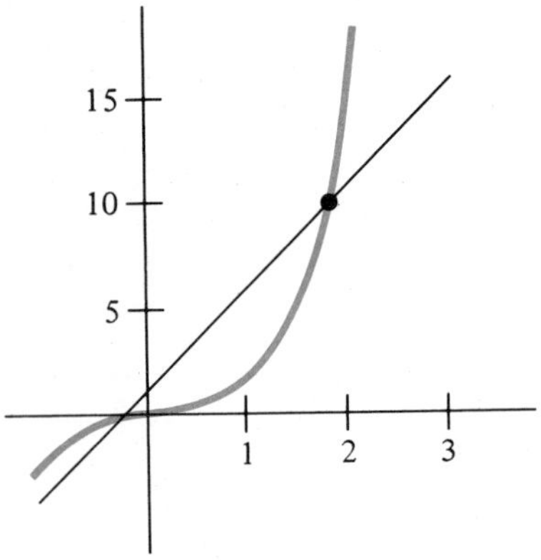

Figure 6.3

In Section 3.9 we saw that the theory of differentials could be used to approximate square roots of numbers lying close to perfect squares. Newton's Method can be used to approximate roots regardless of where the particular number lies.

Example 3 Use Newton's Method to approximate $\sqrt[3]{a}$.

Solution: Finding the number x for which $\sqrt[3]{a} = x$ is equivalent to solving the equation $a = x^3$, which in turn is equivalent to finding a zero for the function $f(x) = x^3 - a$. This is the problem treated in Example 1 with $a = 10$. ■

Example 4 Use Newton's Method to find the point in the right half plane where the graphs of $f(x) = 2x^3$ and $g(x) = 5x + 1$ intersect (see Figure 6.3).

Solution: Finding a value of x for which

$$2x^3 = 5x + 1$$

is equivalent to finding a zero of the function $h(x) = 2x^3 - (5x + 1)$.

For this function the approximation scheme (2) becomes

$$x_{n+1} = x_n - \frac{2x_n^3 - (5x_n + 1)}{6x_n^2 - 5}.$$

Since $h(0) = -1$ and $h(2) = 5$, we know by the Intermediate Value Theorem that a zero must lie within the interval [0, 2]. Using a first approximation $x_1 = 2$, we obtain the information contained in Table 6.2 for four iterations of Newton's Method.

Table 6.2

n	x_n	$f(x_n)$	$f'(x_n)$	x_{n+1}
1	2.0	5.0	19.0	1.736842
2	1.736842	0.794576	13.099723	1.676186
3	1.676186	0.037894	11.857599	1.672990
4	1.672990	0.000103	11.793380	1.672982

The desired point of intersection is approximately (1.672982, 9.364910). ■

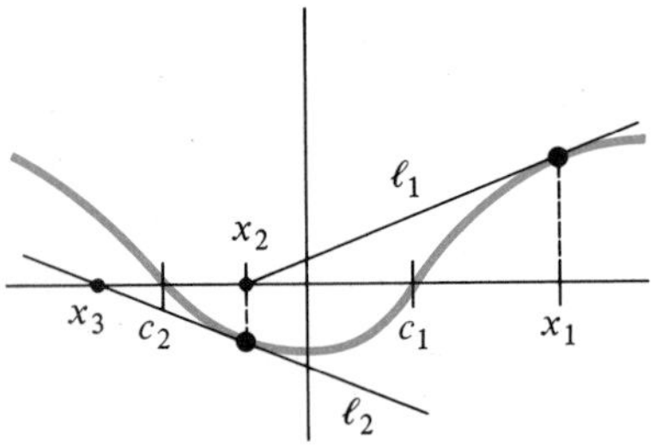

Figure 6.4 Initial approximation x_1 leads to zero c_2 where zero c_1 was desired.

When the approximations produced by Newton's Method approach the desired zero, we say that the method **converges** to that zero. Unfortunately, depending on the function and the initial approximation, Newton's Method may not converge to the desired zero. We comment on some of the reasons for this here, and pursue some of the details in the Exercise Set.

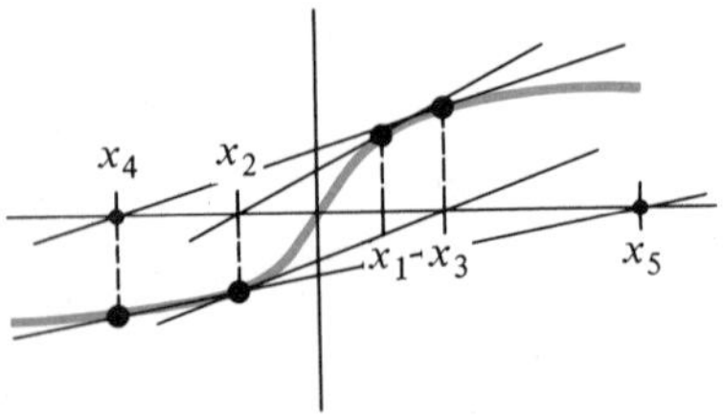

Figure 6.5 Iterates move away from zero rather than converging.

(1) The function may have more than one zero. Depending on the initial approximation x_1, the method may converge to a zero other than the one desired (see Figure 6.4).
(2) An approximation x_k may be obtained for which $f'(x_k) = 0$. In this case the denominator in the expression for x_{k+1} is zero, so the method fails.
(3) The slope of the graph of $f(x)$ may be such that the approximations simply do not converge, but instead move away from the desired zero (see Figure 6.5 and Exercise 15) or simply oscillate between two or more distinct approximations.

Newton's Method is not a theorem but rather a procedure that produces useful approximations in a large number of cases and that, unfortunately, fails in other cases. The fact that the method does not converge for all functions is not a criticism of Newton's idea, but rather an invitation to you to pursue a deeper study of the geometric and analytic properties of functions that allow techniques such as Newton's Method to succeed. The relevant courses in which to pursue these issues are those on advanced calculus, numerical analysis, and modern analysis.

Exercise Set 4.6

In Exercises 1–8, use Newton's Method to approximate the zero of the given function lying between the given values $x = a$ and $x = b$. In each case, first verify that such a zero indeed exists. If a calculator or computer is not available to you, only two or three iterations should be performed. If a calculator or computer is available, a table such as Table 6.1 should be constructed.

1. $f(x) = x^2 - \frac{7}{2}x + \frac{3}{4}$, $a = 0$, $b = 2$

2. $f(x) = 1 - x - x^2$, $a = -2$, $b = -1$

3. $f(x) = 1 - x - x^2$, $a = 0$, $b = 1$

4. $f(x) = x^3 + x - 3$, $a = 1$, $b = 2$

5. $f(x) = \sqrt{x + 3} - x$, $a = 1$, $b = 3$

6. $f(x) = x^3 + x^2 + 3$, $a = -3$, $b = -1$

7. $f(x) = x^4 - 5$, $a = -2$, $b = -1$

8. $f(x) = 5 - x^4$, $a = 1$, $b = 2$

9. Use the results of Exercise 5 to find a point of intersection between the graphs of $f(x) = \sqrt{x + 3}$ and $g(x) = x$.

10. Use Newton's Method to approximate
a. $\sqrt{40}$ **b.** $\sqrt[3]{49}$
c. $\sqrt[5]{18}$ **d.** $\sqrt{37}$

11. Show how the formula for the approximation scheme in Newton's Method can be obtained from the formula for approximation by differentials in Section 3.9. (Newton's Method can therefore be viewed as repeated application of the idea of approximation by differentials.)

12. What can you say about the accuracy of the approximation obtained in Example 4? (*Hint:* See Example 2.)

13. What results from applying Newton's Method if the (lucky) first guess is precisely the desired zero?

14. Use a calculator or computer to find an approximation to the point where the graphs of $f(x) = 2x$ and $g(x) = \tan x$ intersect for $0 < x \leq \pi/2$.

15. Attempt to use Newton's Method to find the zero of $f(x) = (x - 2)^{1/3}$ with initial guess $x_1 = 3$. What happens? Why?

16. The graphs of the functions $f(x)$ and $g(x)$ in Example 4 cross at two other points. Find them.

SUMMARY OUTLINE OF CHAPTER 4

- If $f(t)$ is a differentiable function of time, the (instantaneous) rate of change of $f(t)$ per unit time is the derivative $f'(t)$.
- If two differentiable functions $f(t)$ and $g(t)$ are related by the equation $f(t) = g(t)$, we may differentiate both sides of the equation to obtain the equation $f'(t) = g'(t)$ relating their rates of change.
- To find the maximum and minimum values of a differentiable function $f(x)$ on the interval $[a, b]$ we

  (i) find the critical numbers by finding all values of x for which $f'(x) = 0$ or $f'(x)$ is undefined, and
  (ii) inspect $f(x)$ for all critical numbers and both endpoints.
- The largest of the values observed in (ii) is the **maximum value of $f(x)$** on $[a, b]$, the smallest is the **minimum.**
- In solving applied max-min problems, one often needs to make use of **auxiliary equations** to obtain the function to be optimized as a function of a single independent variable. Also, one needs to determine the interval $[a, b]$ within which the independent variable must lie by examining the **constraints.**
- If $p(x)$ is the price at which x items of a particular commodity will sell, the revenue generated by the sale of x such items is $R(x) = xp(x)$. If $C(x)$ denotes the cost of producing

these x items, the profit resulting from the sale is

$$P(x) = R(x) - C(x) = xp(x) - C(x).$$

- For differentiable revenue functions $R(x)$ and cost functions $C(x)$, **marginal revenue, $MR(x)$,** and **marginal cost, $MC(x)$,** are defined by the equations

$$MR(x) = R'(x) = \frac{dR}{dx}; \qquad MC(x) = C'(x) = \frac{dC}{dx}.$$

- If $Q(p)$ measures the demand for a good or service at price level p, the **elasticity of demand, $E(p)$** is

$$E(p) = -\frac{pQ'(p)}{Q(p)}.$$

- **Newton's Method** is an iterative scheme for generating successive approximations to a zero of $f(x)$ via the formula

$$x_{n+1} = x_n - \frac{f(x_n)}{f'(x_n)}.$$

REVIEW EXERCISES—CHAPTER 4

In Exercises 1–16, find the maximum and minimum values of the function on the given interval and the corresponding values of x.

1. $f(x) = \dfrac{1}{x^2 - 4}, \quad x \in [-1, 1]$

2. $f(x) = \sin x + \cos x, \quad x \in [-\pi, \pi]$

3. $y = x^3 - 3x^2 + 1, \quad x \in [-1, 1]$

4. $f(x) = x\sqrt{1 - x}, \quad x \in [-3, 0]$

5. $f(x) = x + x^{2/3}, \quad x \in [-1, 1]$

6. $y = x - 2|x|, \quad x \in [-3, 2]$

7. $y = x - \sqrt{1 - x^2}, \quad x \in [-1, 1]$

8. $f(x) = \sqrt[3]{x} - x, \quad x \in [-1, 1]$

9. $f(x) = x - 4x^{-2}, \quad x \in [-3, -1]$

10. $y = \dfrac{x}{1 + x^2}, \quad x \in [-2, 2]$

11. $y = x^4 - 2x^2, \quad x \in [-2, 2]$

12. $f(x) = |x + \sin x|, \quad x \in [-\pi/2, \pi/2]$

13. $f(x) = \dfrac{2 + \cos x}{2 - \cos x}, \quad x \in [-\pi, \pi]$

14. $y = \dfrac{x - 1}{x^2 + 3}, \quad x \in [-2, 4]$

15. $y = \dfrac{x + 2}{\sqrt{x^2 + 1}}, \quad x \in [0, 2]$

16. $f(x) = |x^2 - 1|, \quad x \in [-3, 3]$

17. The function $f(x) = \dfrac{x}{1 + ax^2}$ has minimum values at $x = \pm 3$ for $x \in [-5, 5]$. Find a.

18. Find the number most exceeded by its square root.

19. The sides of a regular hexagon are increasing at a rate of 2 cm/sec. At what rate is the area increasing when the length of a side is 10 cm?

20. Approximate the maximum value of the function $f(x) = x^2 \sin x$ for $x \in [1, 3]$. (*Hint:* Do not overlook the zero of the derivative lying between 2 and 3.)

21. Prove that the function $f(x) = x^n + ax + b$ cannot have more than 2 zeros when n is even. What if n is odd?

22. A peach orchard has 25 trees per acre, and the average yield is 300 peaches per tree. For each additional tree planted per acre, the average yield per tree will be reduced by approximately 10 peaches. How many trees per acre will give the largest peach crop?

23. Use Newton's Method to approximate the solution of $x^2 - x - 1 = 0$ lying in the interval $[0, 2]$.

24. Use Newton's Method to approximate the solution of $\cot x = x$ lying in the interval $(0, \pi/2]$.

25. Use Newton's Method to approximate $\sqrt[3]{175}$.

26. Use Newton's Method to approximate the maximum value of the function $f(x) = x \sin x$ for $x \in [0, 3]$ (see hint to Exercise 20).

27. The volume of a sphere is increasing at a rate of 500 cm^3 per second at the instant when the radius is 25 cm. At what rate is the radius increasing?

28. A swimming pool is to be constructed in the shape of a sector of a circle with radius r and central angle θ. Find r and θ if the area of the water surface is the constant S and the perimeter is to be a minimum.

29. A volume V of oil is spilled at sea. The spill takes the shape of a disc whose radius is increasing at a rate of $a/\sqrt{t}$ m/sec where a is constant. Find the rate at which the thickness of the layer of oil is decreasing after 9 seconds, if the radius at that time is $r(9) = 6a$.

30. A ladder 10 meters high leans against a wall. The bottom of the ladder is sliding away from the wall at a rate of 1 meter per second. At what rate is the area of the triangle formed by the ladder, wall, and ground changing when the base of the ladder is 6 meters from the wall?

31. A pebble dropped in still water sends out ripples in the shape of concentric circles. If the radius of the outer ripple increases at a rate of 2 meters per second, how fast is the area of the disturbance increasing after 5 seconds?

32. A fertilizer company can sell x pounds of fertilizer per week at a price of $p(x)$ cents per pound, where

$$p(x) = 90 - \frac{x}{500}.$$

a. Find the total weekly revenue.
b. Find the marginal revenue.
c. If weekly production is limited to 50 tons, what sales level maximizes revenue?

33. Find the rectangle of maximum area that can be inscribed within the equilateral triangle of side 10.

34. The cost of producing x wastebaskets is $C(x) = 5000 + 2x$ dollars. Market research shows that the relationship between the number x of wastebaskets which can be sold and the selling price p is $p(x) = 5 + \dfrac{200}{x}$. If production is limited to 1000 wastebaskets per week, find the production level that maximizes profit.

35. At what rate is the ladder in Exercise 30 rotating at the instant when $x = 6$?

36. The cost per hour of driving a particular automobile at speed v is calculated to be $c(v) = 10 + .004v^2$ dollars per hour. What is the most economical speed at which to operate this automobile? (Assume v to be in units of miles per hr.)

37. Find the points on the graph of the hyperbola $x^2 - x + \dfrac{5}{4} - y^2 = 0$ nearest the origin.

38. Find the dimensions of the rectangle of maximum area which can be inscribed in the ellipse $x^2 + \dfrac{y^2}{4} = 1$.

39. A truck traveling 80 kilometers per hour is heading north on highway X which crosses east-west highway Y at point P. A car traveling east on highway Y at 60 kilometers per hour passed point P when the truck was 20 kilometers south of point P. At what rate is the distance between the truck and the car increasing when the truck passes point P?

40. A spherical snowball melts at a rate proportional to its surface area. Show that its radius decreases at a constant rate.

41. The strength of a wooden beam cut from a log is proportional to the product of its width and the square of its depth. Find the ratio of depth to width of the strongest beam that can be cut from a circular log.

42. Find the maximum and minimum values of the functions $f(x) = x^3 - x^2 - x + 1$ and $g(x) = 4 + 2x - x^2$ for $x \in [-1, 5]$. Then find the maximum and minimum values for the function $h(x) = f(x) + g(x)$.

43. A student 1.6 meters tall walks directly away from a street light 8 meters above the ground at a rate of 1.2 meters per second. Find the rate at which the student's shadow is increasing when the student is 20 meters from the point directly beneath the light.

44. A propane storage tank is to have the shape of a cylinder capped at both ends by hemispheres. If the hemispherical caps are three times as expensive to construct, per square unit of surface area, as the cylindrical walls, find the dimensions that minimize construction costs for a given volume.

45. An appliance dealer sells 150 dishwashers per year at a uniform rate. The storage costs are \$15 per year per dishwasher. The ordering costs are \$2 per dishwasher plus an additional \$25 per order. How many times per year should the store order dishwashers so as to minimize costs?

46. What is the maximum possible area of an isosceles triangle that can be inscribed in a circle of radius r?

47. Mathematical statisticians often "estimate" the true value x of a parameter in a large population by extracting a sample from the population and calculating the value $\bar{x}$ of the parameter in that sample. For example, we could estimate the average age of all citizens of Chicago by calculating the average age for a sample of 100 citizens, assuming the sample to be representative of the population. Given sample values $x_1, x_2, \ldots, x_n$ of the parameter of interest, the estimate $\bar{x}$ which statisticians prefer to use is the one which minimizes the sum of the squares of the errors $(\bar{x} - x_1)^2 + (\bar{x} - x_2)^2 + \cdots + (\bar{x} - x_n)^2$. Show that the number $\bar{x}$ which satisfies this criterion is the simple average

$$\bar{x} = \frac{x_1 + x_2 + \cdots + x_n}{n}.$$

48. When a set of data $(x_1, y_1), (x_2, y_2), \ldots, (x_n, y_n)$ appears to be related by an equation of the form $y = mx$, the statistician defines the "best" straight line modelling this relationship as the line $y = mx$ that minimizes the sum of the squares of the errors

$$(mx_1 - y_1)^2 + (mx_2 - y_2)^2 + \cdots + (mx_n - y_n)^2.$$

Show that the value of m producing this "best" straight line is

$$m = \frac{x_1y_1 + x_2y_2 + \cdots + x_ny_n}{x_1^2 + x_2^2 + \cdots + x_n^2}.$$

49. Use Newton's Method to approximate the number $x \in (0, \pi/2)$ where $\tan x = 3x$.

50. A monopolist can manufacture at most 200 items per week. The price at which the monopolist can sell x items per week is $p(x) = 800 - 2x$ dollars, while the cost of producing x items per week is $C(x) = 100 + 6x + x^2$ dollars. Find the output level x_0 that maximizes profits. What is $P(x_0)$, the weekly profit, at this level?

51. How would the imposition of a \$30 per item manufacturing tax change the optimal output level in Exercise 50?

52. Determine the elasticity of the demand curve

$$Q(p) = \frac{1}{1 + p} \quad \text{for } 1 \leq p \leq 10.$$

53. A rectangular box is to be constructed so as to hold 1024 cm^3. Its base is to have length twice its width. Material for the top costs 12¢ per square centimeter, and material for the sides and bottom cost 6¢ per square centimeter. Find the dimensions that will minimize cost.

CHAPTER 5

GEOMETRIC APPLICATIONS OF THE DERIVATIVE

5.1 INTRODUCTION

In Chapter 4, we saw glimpses of ways in which the derivative reveals certain information about the graph of the function $y = f(x)$. For example, the maximum value of the differentiable function $f(x)$ on $[a, b]$ can occur only at critical numbers or at endpoints. However, a function can have critical numbers that do not correspond to maxima or minima. What can we learn from such numbers? We will learn that the derivative yields enough information to specify completely the *shape* of the graph of $f(x)$, leaving only its *position* in the coordinate plane to be determined.

The objective of this chapter is to take the more general approach of asking what the derivative can tell us about the nature of the function $f(x)$ throughout its natural domain, rather than just the nature of $f(x)$ when it is restricted to a closed interval $[a, b]$. The answers to this question will provide us with a set of techniques for analyzing the graph of $y = f(x)$. Moreover, the process of drawing conclusions about the nature of $f(x)$ from observed properties of $f'(x)$ will lead us naturally into the topics of antidifferentiation and differential equations. These last issues form the bridge to the integral calculus, the second major topic of the text.

5.2 THE MEAN VALUE THEOREM

The Mean Value Theorem plays a central role in the development of the calculus. Although it appears on first reading to be little more than a geometric observation about slopes, you will see in the chapters which follow that this theorem finds wide application in the proofs of important theorems concerning antiderivatives, area calculations, error estimates for approximation procedures, limits of indeterminate forms, convergence of infinite series, and other topics. At the moment, our interest lies in applying this theorem to determine the intervals on which a function increases or decreases (Section 5.3).

Before beginning a careful development of this theorem, we provide a quick overview: the Mean Value Theorem states that for a differentiable function $f(x)$, the slope of the secant through $(a, f(a))$ and $(b, f(b))$ equals the slope of the tangent at some point c between a and b (Figure 2.1). Another way to say this is that the

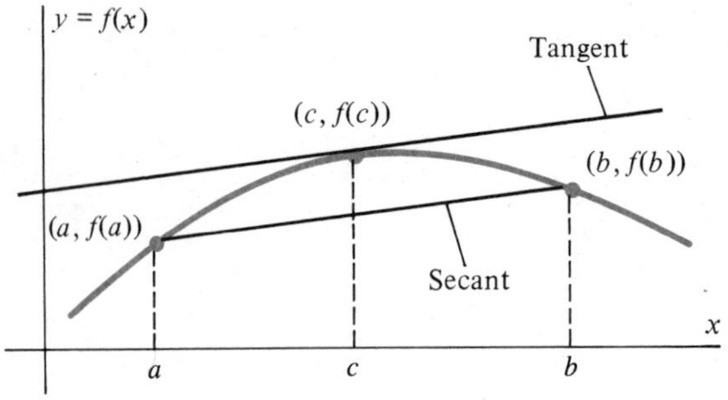

Figure 2.1 Secant through $(a, f(a))$ and $(b, f(b))$ parallel to tangent through $(c, f(c))$.

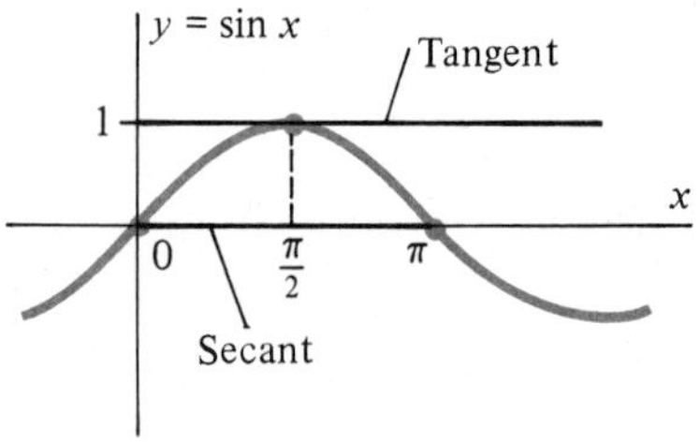

Figure 2.2 $f(x) = \sin x$.

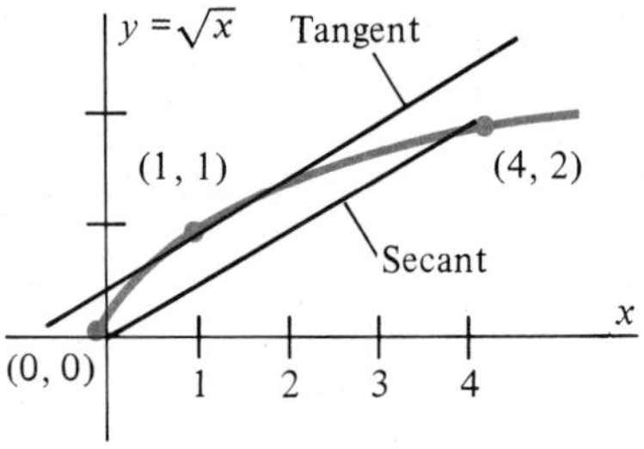

Figure 2.3 $f(x) = \sqrt{x}$.

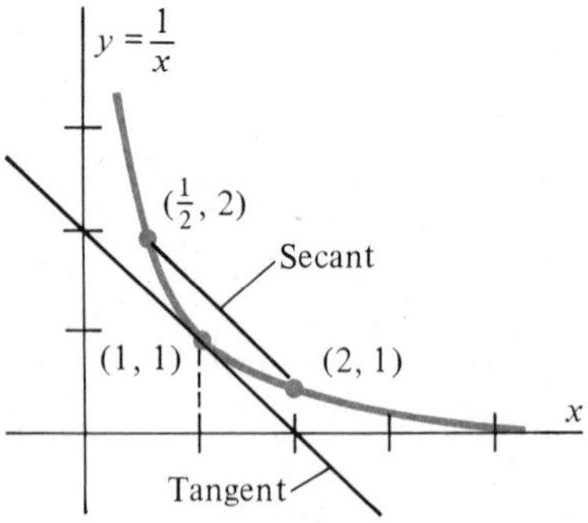

Figure 2.4 $f(x) = \dfrac{1}{x}$.

average change of $f(x)$ over $[a, b]$ equals the instantaneous rate of change at some point $c \in (a, b)$. Illustrations of this theorem are the following:

(i) For $f(x) = \sin x$ on the interval $[0, \pi]$, the average change in f is

$$\frac{\sin(\pi) - \sin(0)}{\pi - 0} = \frac{0}{\pi} = 0,$$

which equals

$$f'(\pi/2) = \cos(\pi/2)$$

(Figure 2.2).

(ii) For $f(x) = \sqrt{x}$ on the interval $[0, 4]$, the average change in f is

$$\frac{\sqrt{4} - \sqrt{0}}{4 - 0} = \frac{2}{4} = \frac{1}{2},$$

which equals

$$f'(1) = \frac{1}{2\sqrt{1}}$$

(Figure 2.3).

(iii) For $f(x) = \dfrac{1}{x}$ on the interval $\left[\dfrac{1}{2}, 2\right]$, the average change in f is

$$\frac{1/2 - 2}{2 - 1/2} = \frac{-3/2}{3/2} = -1$$

which equals

$$f'(1) = \frac{-1}{(1)^2}$$

(Figure 2.4).

The geometry of illustrations (i)–(iii) explains the terminology "mean value." The *average* (mean) *change in $f(x)$* is given by the *derivative*, $f'(c)$. Note that although c may be the midpoint of the interval $[a, b]$ (as it is in illustration (i)), this need *not* be true in general (see Figures 2.3 and 2.4).

Before discussing a precise formulation of the Mean Value Theorem, we state and prove a special case, known as Rolle's Theorem, that we shall use in proving the general case.

THEOREM 1
Rolle's Theorem

Let $f(x)$ be a continuous function for x in the closed interval $[a, b]$ and assume that $f'(x)$ exists for each x in the open interval (a, b). If $f(a) = f(b) = 0$, there exists a number $c \in (a, b)$ for which $f'(c) = 0$.

In other words, Rolle's Theorem asserts that if function is continuous on a closed interval and differentiable everywhere except possibly at endpoints and if the function equals zero at both endpoints, then the graph must have at least one horizontal tangent.

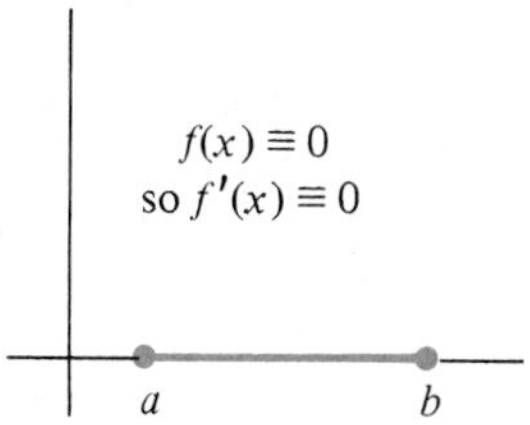

Figure 2.5 Rolle's Theorem for constant $f(x)$.

Proof of Rolle's Theorem: Since $f(x)$ is continuous on the closed interval $[a, b]$, $f(x)$ must have both a maximum and a minimum value on the interval $[a, b]$ (Theorem 1, Chapter 4). If both these values occur at endpoints, then $f(a) = 0$ and $f(b) = 0$ are both the maximum and minimum values for $f(x)$ on $[a, b]$. Obviously, this means that $f(x) = 0$ for *all* $x \in [a, b]$, so $f(x)$ is the constant function $f(x) \equiv 0$.* Then $f'(c) = 0$ for all $c \in [a, b]$, so any such c satisfies the theorem (see Figure 2.5).

On the other hand, if $f(x)$ has a maximum or minimum value at a number x between a and b then, since $f'(x)$ exists, we must have $f'(x) = 0$ (by Theorem 2, Chapter 4). Taking $c = x$ then satisfies the statement of the theorem (see Figure 2.6). Since these two cases exhaust all possibilities, the proof of Rolle's Theorem is complete. ■

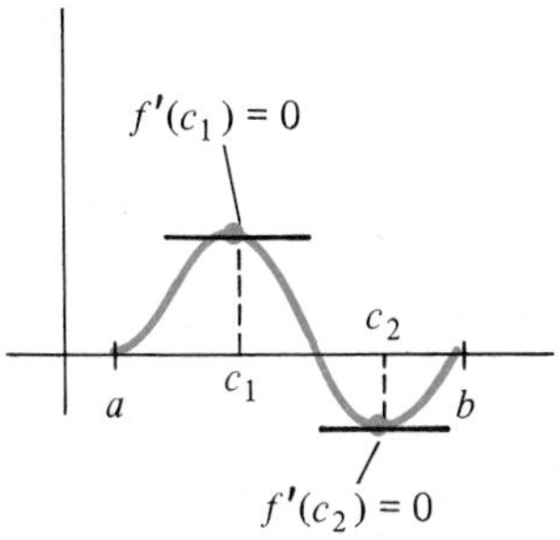

Figure 2.6 Rolle's Theorem—general case.

We are now prepared to discuss the Mean Value Theorem. Recall that if the function $f(x)$ is defined on $[a, b]$, the slope of the secant joining $(a, f(a))$ and $(b, f(b))$ is given by the quotient

$$m = \frac{f(b) - f(a)}{b - a}.$$

The statement that the slope of this secant equals the slope of the tangent at $x = c$ is simply $f'(c) = m$. Of course, we must be careful to include hypotheses which guarantee that both of these numbers will exist.

THEOREM 2
Mean Value Theorem

Let $f(x)$ be continuous on the closed interval $[a, b]$ and let $f'(x)$ exist for each x in the open interval (a, b). Then there exists at least one number $c \in (a, b)$ so that

$$f'(c) = \frac{f(b) - f(a)}{b - a}.$$

*The statement $f(x) \equiv 0$ is read "$f(x)$ is identically equal to the zero function" and means that $f(x) = 0$ for all x in the interval $[a, b]$.

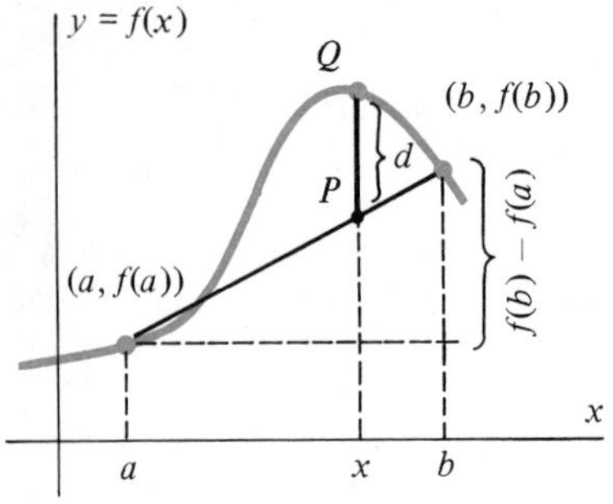

Figure 2.7 Function $d = d(x)$ in proof of Mean Value Theorem.

Proof: The proof is every bit as geometric as is the statement of the theorem. We begin by describing the vertical distance between a point on the secant, say P, and the corresponding point Q on the graph of $f(x)$ (see Figure 2.7). Since the secant has slope $\dfrac{f(b) - f(a)}{b - a}$ and passes through the point $(a, f(a))$, the equation for the secant can be obtained using the point-slope form for straight lines (see Chapter 1). If (x, y) denotes an arbitrary point on the secant, we must have

$$\frac{y - f(a)}{x - a} = \underbrace{\left(\frac{f(b) - f(a)}{b - a}\right)}_{\text{slope}},$$

or

$$y = f(a) + (x - a) \cdot \left[\frac{f(b) - f(a)}{b - a}\right]. \tag{1}$$

Thus, if point P on the secant has coordinates (x, y), x and y must satisfy equation (1). Since the y-coordinate of the point Q is simply $f(x)$, the (vertical) distance between P and Q is

$$d(x) = f(x) - \left\{f(a) + (x - a)\left[\frac{f(b) - f(a)}{b - a}\right]\right\}. \tag{2}$$

(We have written the distance $d(x)$ as a function of x since the preceding discussion holds for each x in $[a, b]$).

We now make several observations concerning the function $d(x)$:

(a) Since the expression in braces in equation (2) is a polynomial (and, therefore, continuous and differentiable), the function $d(x)$ satisfies the hypotheses of Rolle's Theorem.

(b) $d(a) = f(a) - \left\{f(a) + (a - a)\left[\dfrac{f(b) - f(a)}{b - a}\right]\right\} = f(a) - f(a) = 0.$

(c) $$\begin{aligned} d(b) &= f(b) - \left\{f(a) + (b - a)\left[\frac{f(b) - f(a)}{b - a}\right]\right\} \\ &= f(b) - \{f(a) + [f(b) - f(a)]\} \\ &= 0. \end{aligned}$$

Thus, the function $d(x)$ satisfies the hypotheses of Rolle's Theorem, including the endpoint condition $d(a) = d(b) = 0$. Therefore, by Rolle's Theorem, there is a number $c \in (a, b)$ for which $d'(c) = 0$. But differentiating both sides of equation (2) shows that

$$\begin{aligned} d'(x) &= f'(x) - \left\{0 + (1 - 0)\left[\frac{f(b) - f(a)}{b - a}\right]\right\} \\ &= f'(x) - \frac{f(b) - f(a)}{b - a}. \end{aligned}$$

Setting $x = c$ thus gives the statement

$$0 = d'(c) = f'(c) - \frac{f(b) - f(a)}{b - a},$$

or

$$f'(c) = \frac{f(b) - f(a)}{b - a}.$$

■

Notice that Rolle's Theorem can be viewed as a special case of the Mean Value Theorem and that both theorems are statements about averages. The great applicability of the Mean Value Theorem derives from its role of obtaining information about a function over an entire interval from the value of its derivative at a single point.

Example 1 Verify that the Mean Value Theorem holds for the function $f(x) = -x^2 + 6x - 6$ on the interval $[1, 3]$.

Strategy

Verify hypotheses.

Solution

Since $f(x)$ is a polynomial, $f(x)$ is differentiable and, therefore, continuous for all values of x. For $a = 1$, $b = 3$ we have

Find $m = \dfrac{f(b) - f(a)}{b - a}$.

$$\frac{f(b) - f(a)}{b - a} = \frac{f(3) - f(1)}{3 - 1} = \frac{3 - (-1)}{2} = 2.$$

Set $f'(c) = m$ and solve for c.

We must therefore find c so that

$$f'(c) = -2(c) + 6 = 2.$$

This gives $-2c = -4$, or $c = 2$ (Figure 2.8). ■

Example 2 The function $f(x) = x^{2/3}$ fails to satisfy the hypothesis of the Mean Value Theorem on the interval $[-1, 1]$, since $f'(0) = \frac{2}{3}x^{-1/3}\Big|_{x=0}$ is not defined (see Example 4, Section 3.8). As a result we have

$$\frac{f(1) - f(-1)}{1 - (-1)} = \frac{1 - 1}{2} = 0,$$

but $f'(x) \neq 0$ for all $x \neq 0$ in the interval $[-1, 1]$ (see Figure 2.9). ■

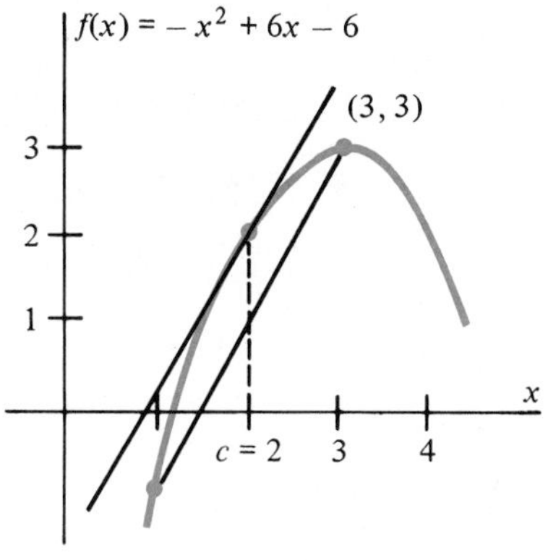

Figure 2.8 $f'(2) = \dfrac{f(3) - f(1)}{3 - 1}$.

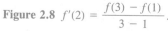

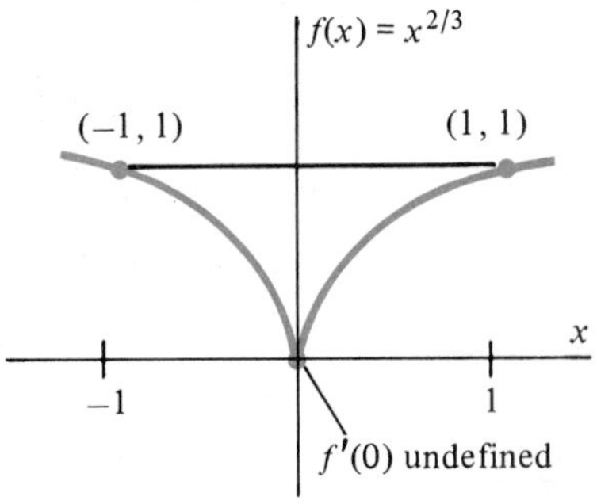

Figure 2.9 Mean Value Theorem fails for $f(x) = x^{2/3}$ on $[-1, 1]$ since $f'(0)$ does not exist.

Exercise Set 5.2

1. Sketch the graph of $f(x) = |x|$ for $-1 \le x \le 1$. Does the Mean Value Theorem hold for $f(x)$ on $[-1, 1]$. Why or why not? (See Exercise 35, Section 4.3.)

2. The function $f(x) = \dfrac{1}{x-3}$ is differentiable on the open interval $(0, 3)$, yet the Mean Value Theorem does not hold on $[0, 3]$. Why?

In each of Exercises 3–8, verify that the given function satisfies the Mean Value Theorem on the stated interval.

3. $f(x) = 2x - 7, \quad x \in [-1, 5]$

4. $f(x) = 4 - x^2, \quad x \in [0, 2]$

5. $f(x) = x^2 + 2x, \quad x \in [0, 4]$

6. $f(x) = 3 \sin 2x, \quad x \in [0, \pi/2]$

7. $f(x) = x^2 + 2x - 3, \quad x \in [-3, 0]$

8. $f(x) = x^3 - 2x + 4, \quad x \in [0, 2]$

In Exercises 9–14, find the number that satisfies the Mean Value Theorem on the given interval or state why the theorem does not apply.

9. $y = x^{2/5}, \quad x \in [0, 8]$

10. $y = \dfrac{1}{(x-2)^2}, \quad x \in [2, 5]$

11. $y = |a - x^2|, \quad x \in [-2, 2], \quad a > 4$

12. $y = x^{1/2} + 2(x-2)^{1/3}, \quad x \in [1, 9]$

13. $f(x) = x + \dfrac{1}{x}, \quad x \in [1, 3]$

14. $f(x) = \dfrac{x-1}{x+1}, \quad x \in [0, 1]$

15. Show that if $f(x)$ satisfies the hypotheses of the Mean Value Theorem on $[a, b]$ and $|f'(x)| \le M$ for all $x \in (a, b)$ then

$$|f(b) - f(a)| \le M(b - a).$$

16. Show that if $f(x)$ satisfies the hypotheses of Exercise 15 then

$$f(x) \le f(a) + M(b - a)$$

for all $x \in [a, b]$.

17. Use the results of Exercise 15 to show that

$$|\sin x - \sin y| \le |x - y|$$

for any numbers x and y.

18. Use the results of Exercise 16 to show that

$$6 < \sqrt{36.2} < 6.02$$

19. Rolle's Theorem shows that between any two zeros of a differentiable function there must always be a zero of the derivative. Is the reverse true, that is, if $f(x)$ is differentiable for all x and if $f'(x_1) = f'(x_2) = 0$, must $f(x)$ have a zero in (x_1, x_2)?

20. Use the Mean Value Theorem to show that if $f(x)$ and $g(x)$ are differentiable on (a, b) and continuous on $[a, b]$, if $f(a) = g(a)$, and if $f'(x) < g'(x)$ for all $x \in (a, b)$, then $f(b) < g(b)$. (*Hint:* If the two graphs meet on (a, b), a contradiction is generated.)

21. Extend the results of Exercise 15 to the case $m \le f'(x) \le M$, $x \in [a, b]$ and obtain the inequality

$$m(b - a) \le |f(b) - f(a)| \le M(b - a)$$

for differentiable functions $f(x)$.

22. Suppose an automobile travels at speeds between 75 and 90 kilometers per hour for 6 hours. From Exercise 21 what can you conclude about the distance travelled during this time period?

(Calculator/Computer) In each of Exercises 23–25, use Newton's Method to approximate the number c satisfying the Mean Value Theorem for the given function on the given interval.

23. $f(x) = x^4 - 3x^2 + 2, \quad x \in [1, 4]$

24. $f(x) = \tan x, \quad x \in \left[0, \dfrac{\pi}{4}\right]$

25. $f(x) = \sin \sqrt{x}, \quad x \in \left[\dfrac{\pi^2}{16}, \dfrac{\pi^2}{4}\right].$

26. Prove that if $f(x)$ is a quadratic function $f(x) = ax^2 + bx + c$ then the number c in the interval $[x_1, x_2]$ satisfying the Mean Value Theorem can be assumed to be the midpoint $c = \dfrac{x_1 + x_2}{2}$.

5.3 INCREASING AND DECREASING FUNCTIONS

The objective of this section is to establish a relationship between the sign of the derivative and the question of whether the graph of the function $y = f(x)$ is rising or falling as the independent variable x increases. As Figure 3.1 suggests, a rising graph corresponds to tangents with positive slope. Similarly, a falling graph suggests negative slopes. Although these relationships may seem clear from the geometry of Figure 3.1, we need to make the observations both precise and legitimate with the following definition and theorem.

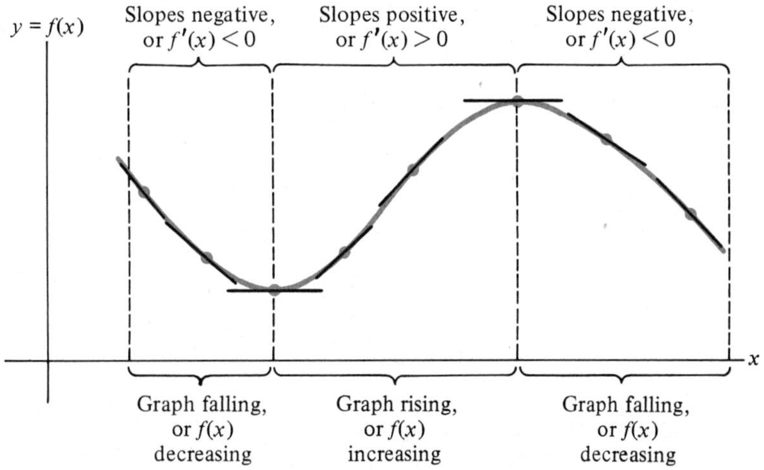

Figure 3.1 Relationship between rising graph and sign of $f'(x)$.

DEFINITION 1

The function $f(x)$ is said to be **increasing** on an interval I if $f(x)$ increases as x increases, that is, if, whenever $x_1, x_2 \in I$,

$$x_2 > x_1 \text{ implies } f(x_2) > f(x_1).$$

Similarly, $f(x)$ is said to be **decreasing** on I if, whenever $x_1, x_2 \in I$,

$$x_2 > x_1 \text{ implies } f(x_2) < f(x_1).$$

(See Figures 3.2 and 3.3.)

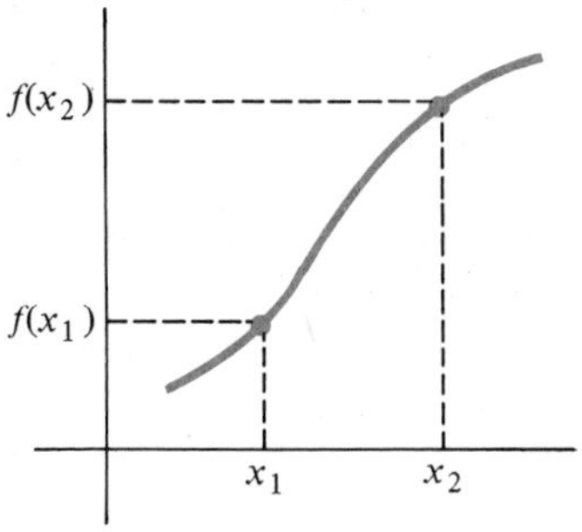

Figure 3.2 $f(x)$ increasing: $x_2 > x_1$ implies $f(x_2) > f(x_1)$.

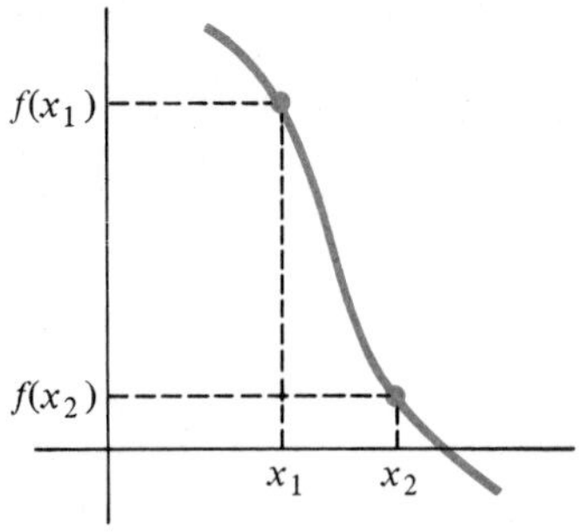

Figure 3.3 $f(x)$ decreasing: $x_2 > x_1$ implies $f(x_2) < f(x_1)$.

The following theorem establishes the relationship suggested by Figures 3.1–3.3.

THEOREM 3

Let I be an interval and let I^o be the open interval obtained from I by excluding its endpoints. Let $f(x)$ be continuous on I and differentiable on I^o.

(i) If $f'(x) > 0$ for all $x \in I^o$, then $f(x)$ is increasing on I.
(ii) If $f'(x) < 0$ for all $x \in I^o$, then $f(x)$ is decreasing on I.

Don't let the distinction between I and I^o throw you—we are just trying to be careful about the fact that I may or may not have endpoints. Let's look at two simple examples before taking up the proof of Theorem 3.

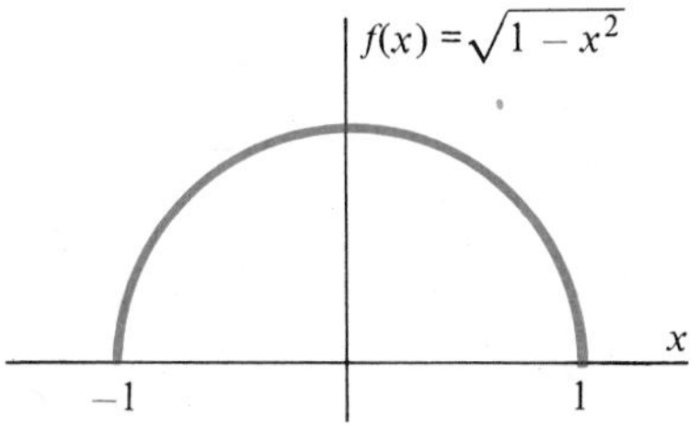

Figure 3.4 $f(x) = \sqrt{1 - x^2}$ is increasing on $[-1, 0]$ and decreasing on $[0, 1]$.

Example 1 The function $f(x) = \sqrt{1 - x^2}$ is increasing on $I_1 = [-1, 0]$ and decreasing on $I_2 = [0, 1]$. This is easy to see from its graph (Figure 3.4), and since $f'(x) = \dfrac{-x}{\sqrt{1 - x^2}}$ it is just as easy to conclude by applying Theorem 3. Here $f'(x) > 0$ for $x \in I_1^o = (-1, 0)$ and $f'(x) < 0$ for $x \in I_2^o = (0, 1)$. (Note, however, that $f'(-1)$ and $f'(1)$ are not defined. That is why we are being careful about endpoints.) ■

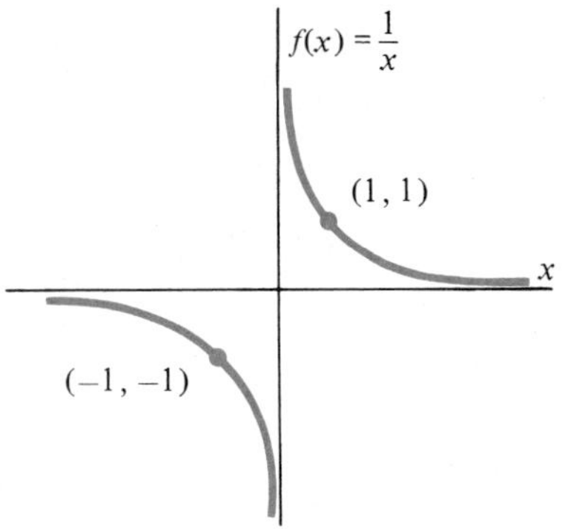

Figure 3.5 $f(x) = \dfrac{1}{x}$ is decreasing, both on $(-\infty, 0)$ and on $(0, \infty)$.

Example 2 The function $f(x) = \dfrac{1}{x}$ is decreasing both on $I_1 = (-\infty, 0)$ and on $I_2 = (0, \infty)$. Again, this is obvious from its graph (Figure 3.5). It also follows easily from Theorem 3, since $f'(x) = \dfrac{-1}{x^2}$. Thus, $f'(x) < 0$ for $x \in (-\infty, 0) = I_1 = I_1^o$, and $f'(x) < 0$ for $x \in (0, \infty) = I_2 = I_2^o$. Note, however, that we do *not* say that $f(x)$ is decreasing on $(-\infty, \infty)$, since $f(0)$ is not defined. ■

Proof of Theorem 3: The argument is a straightforward application of the Mean Value Theorem. To prove statement (i), we assume that $f'(x) > 0$ for all $x \in I^o$ and show that $f(x)$ is increasing on I. To do this we must let x_1 and x_2 denote any two numbers in I with $x_1 < x_2$ and show that $f(x_2) > f(x_1)$. The key observation in doing this is that since the interval $[x_1, x_2]$ lies in I, $f(x)$ is continuous on $[x_1, x_2]$ and differentiable on (x_1, x_2). Thus, the Mean Value Theorem guarantees that there is a number $c \in (x_1, x_2)$ with

$$\frac{f(x_2) - f(x_1)}{x_2 - x_1} = f'(c). \tag{1}$$

Now since we are assuming $f'(x) > 0$ for all $x \in I^o$, we must have $f'(c) > 0$. Thus, the quotient on the left-hand side of equation (1) is positive. Since $x_1 < x_2$, the denominator of this quotient is positive and, therefore, so is the numerator. That is, $f(x_2) - f(x_1)$ is positive. This shows that $f(x_2) > f(x_1)$, which is what we wanted to prove. The proof of statement (ii) is similar and is left as Exercise 26. ■

Theorem 3 provides a simple *procedure for determining the intervals on which a function is increasing and decreasing:*

(i) Find those numbers x in the domain of f for which $f'(x) = 0$ or for which $f'(x)$ is undefined (i.e., the critical numbers for $f(x)$).

(ii) Determine the sign of $f'(x)$ on the intervals determined by the critical numbers and by any numbers x for which $f(x)$ is undefined.
(iii) Apply Theorem 3 in each such interval.

Example 3 Find the intervals on which the function

$$f(x) = \frac{1}{3}x^3 - x^2 - 3x + 4$$

is increasing and decreasing.

Strategy
Find $f'(x)$.

Solution
Here $f'(x) = x^2 - 2x - 3 = (x - 3)(x + 1)$.

Find values of x for which $f'(x) = 0$ or $f'(x)$ is undefined (critical numbers).

Setting $f'(x) = (x - 3)(x + 1) = 0$ gives $x = -1$ and $x = 3$. There are no values of x for which $f'(x)$ is undefined or for which $f(x)$ is undefined.

Determine the sign of $f'(x)$ on each of the resulting intervals.

From the factored form of $f'(x)$, we can see (Figure 3.6) that

(i) If $x < -1$, then both $(x - 3) < 0$ and $(x + 1) < 0$. Thus,

$$f'(x) > 0 \text{ for } x \in (-\infty, -1).$$

(ii) If $-1 < x < 3$, then $(x - 3) < 0$, but $(x + 1) > 0$. Thus,

$$f'(x) < 0 \text{ for } x \in (-1, 3).$$

(iii) If $x > 3$, then both $(x - 3) > 0$ and $(x + 1) > 0$. Thus,

$$f'(x) > 0 \text{ for } x \in (3, \infty).$$

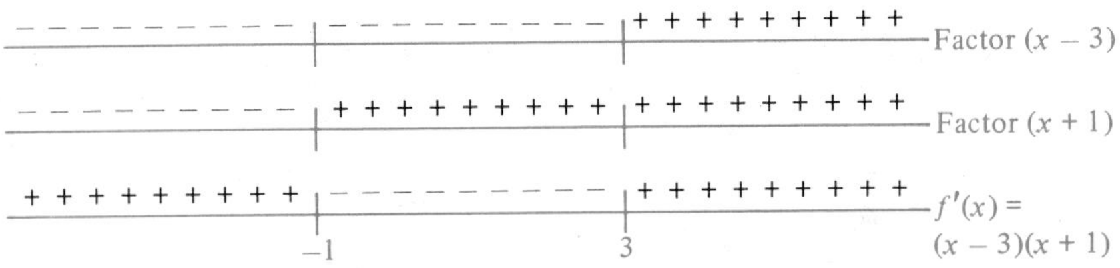

Figure 3.6 Sign analysis for $f'(x) = (x - 3)(x + 1)$ in Example 3.

Apply Theorem 3.

By Theorem 3 we conclude that

(i) $f(x)$ is increasing on $(-\infty, -1]$,
(ii) $f(x)$ is decreasing on $[-1, 3]$,
(iii) $f(x)$ is increasing on $[3, \infty)$.

(See Figure 3.7.) ■

Often it is difficult to determine the sign of $f'(x)$ on various intervals by simply inspecting the expression for $f'(x)$, especially when $f(x)$ is not easily factored. When $f'(x)$ is a continuous function, a simple procedure based on the Intermediate Value Theorem may be used. The idea is simply that if a and b are successive critical numbers for $f(x)$, then $f'(x)$ cannot change sign on the interval (a, b). (Otherwise the Intermediate Value Theorem would imply the existence of another zero for $f'(x)$ between a and b.) We can therefore determine the sign of $f'(x)$ on all of (a, b) by checking the sign of $f'(t)$ at any particular $t \in (a, b)$. We use this idea in the following example.

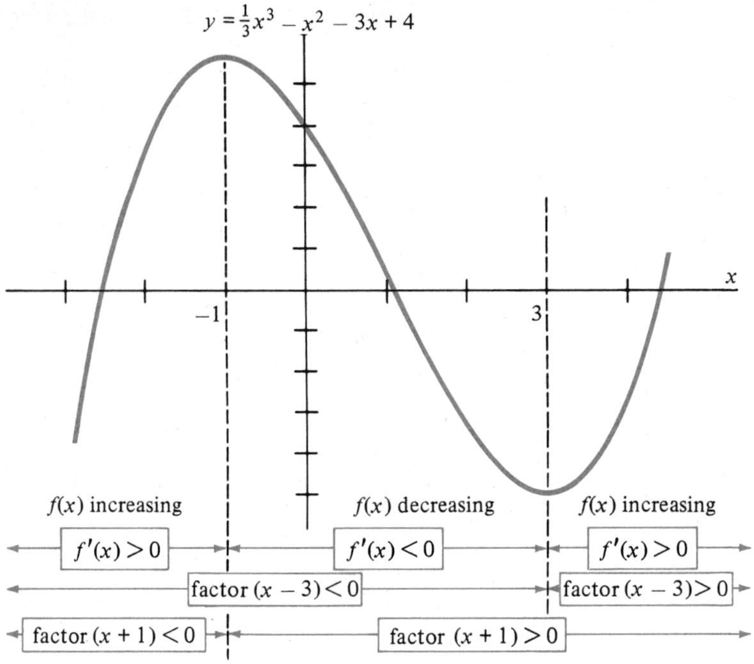

Figure 3.7 Graph of $f(x) = \dfrac{1}{3}x^3 - x^2 - 3x + 4$ and analysis of signs of factors of $f'(x) = x^2 - 2x - 3$.

Example 4 Determine the intervals on which $f(x) = \dfrac{x^2 + 1}{x}$ is increasing and decreasing.

Solution: Here

$$f'(x) = \frac{x(2x) - (x^2 + 1)}{x^2} = \frac{x^2 - 1}{x^2},$$

so $f'(x) = 0$ for $x = -1$ and $x = 1$. The critical numbers are $x = -1$ and 1. Also, $f'(x)$ is undefined for $x = 0$, as is $f(x)$. We must therefore inspect $f'(x)$ on the intervals $(-\infty, -1)$, $(-1, 0)$, $(0, 1)$, and $(1, \infty)$. Since $f'(x)$ is continuous on each of these open intervals, we may determine the sign of $f'(x)$ on each interval by checking a single point. We arbitrarily choose $-2 \in (-\infty, -1)$, $-\dfrac{1}{2} \in (-1, 0)$, $\dfrac{1}{2} \in (0, 1)$, and $2 \in (1, \infty)$. We find that

$f'(-2) = 3/4 > 0$, so $f(x)$ increases on $(-\infty, -1]$;
$f'(-1/2) = -3 < 0$, so $f(x)$ decreases on $[-1, 0)$;
$f(1/2) = -3 < 0$, so $f(x)$ decreases on $(0, 1]$;
$f(2) = 3/4 > 0$, so $f(x)$ increases on $[1, \infty)$.

(See Figure 3.8.)

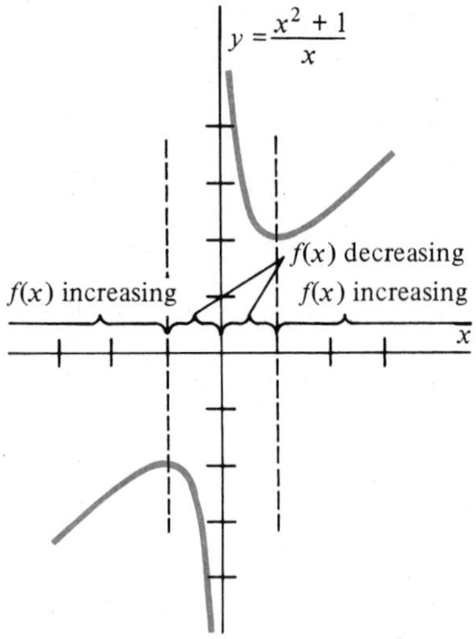

Figure 3.8

You may find a table such as Table 3.1 helpful in organizing the information leading to these conclusions. ■

Table 3.1 Analysis of $f(x) = \dfrac{x^2+1}{x}$ for increase and decrease

Interval I	Arbitrary test number $t \in I$	Value of $f'(t)$	Sign of $f'(t)$	Conclusion about $f(x)$
$(-\infty, -1)$	$t = -2$	$f'(2) = \dfrac{3}{4}$	+	$f(x)$ increasing on $(-\infty, -1]$
$(-1, 0)$	$t = -\dfrac{1}{2}$	$f'\left(-\dfrac{1}{2}\right) = -3$	−	$f(x)$ decreasing on $[-1, 0)$
$(0, 1)$	$t = \dfrac{1}{2}$	$f'\left(\dfrac{1}{2}\right) = -3$	−	$f(x)$ decreasing on $(0, 1]$
$(1, \infty)$	$t = 2$	$f'(2) = \dfrac{3}{4}$	+	$f(x)$ increasing $[1, \infty)$

Example 5 Determine the intervals on which the function $f(x) = \tan^2 x$ is increasing and decreasing for $-\pi/2 < x < 3\pi/2$.

Strategy

Find $f'(x)$.

Find critical numbers for $f(x)$ and the numbers where $f(x)$ is undefined.

Select one number c in each of the resulting intervals and check the sign of $f'(c)$.

Tabulate the results and apply Theorem 3.

Solution

Here

$$f'(x) = 2\tan x \cdot \sec^2 x = 2\,\frac{\sin x}{\cos^3 x}.$$

Now $f'(x) = 0$ if $\sin x = 0$, which occurs for $x = 0$ and $x = \pi$. Also, $f'(x)$ and $f(x)$ are undefined if $\cos x = 0$, that is, at $x = \pi/2$. We must therefore determine the sign of $f'(x)$ on the intervals $(-\pi/2, 0)$, $(0, \pi/2)$, $(\pi/2, \pi)$, and $(\pi, 3\pi/2)$. To do so we arbitrarily choose the midpoint of each interval and examine the sign. The results are given in Table 3.2 and Figure 3.9. ■

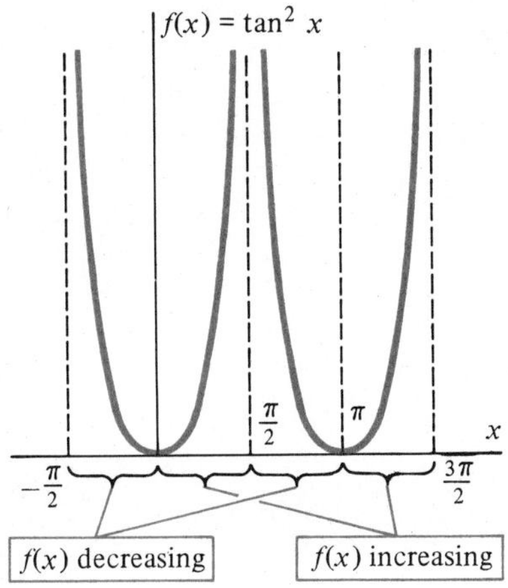

Figure 3.9

Table 3.2 Analysis of $f(x) = \tan^2 x$ for increase and decrease

Interval I	Arbitrary test number $t \in I$	Value of $f'(t) = \dfrac{2 \sin t}{\cos^3 t}$	Sign of $f'(t)$	Conclusion
$\left(-\frac{\pi}{2}, 0\right)$	$t = -\frac{\pi}{4}$	$f'\left(-\frac{\pi}{4}\right) = \dfrac{2\left(-\frac{\sqrt{2}}{2}\right)}{\left(\frac{\sqrt{2}}{2}\right)^3} = -4$	$-$	$f(x)$ decreasing on $\left(-\frac{\pi}{2}, 0\right]$
$\left(0, \frac{\pi}{2}\right)$	$t = \frac{\pi}{4}$	$f'\left(\frac{\pi}{4}\right) = \dfrac{2\left(\frac{\sqrt{2}}{2}\right)}{\left(\frac{\sqrt{2}}{2}\right)^3} = 4$	$+$	$f(x)$ increasing on $\left[0, \frac{\pi}{2}\right)$
$\left(\frac{\pi}{2}, \pi\right)$	$t = \frac{3\pi}{4}$	$f'\left(\frac{3\pi}{4}\right) = \dfrac{2\left(\frac{\sqrt{2}}{2}\right)}{\left(-\frac{\sqrt{2}}{2}\right)^3} = -4$	$-$	$f(x)$ decreasing on $\left(\frac{\pi}{2}, \pi\right]$
$\left(\pi, \frac{3\pi}{2}\right)$	$t = \frac{5\pi}{4}$	$f'\left(\frac{5\pi}{4}\right) = \dfrac{2\left(-\frac{\sqrt{2}}{2}\right)}{\left(-\frac{\sqrt{2}}{2}\right)^3} = 4$	$+$	$f(x)$ increasing on $\left[\pi, \frac{3\pi}{2}\right)$

Exercise Set 5.3

In each of Exercises 1–20, find the intervals on which $f(x)$ increases and decreases by finding the critical numbers.

1. $f(x) = x^2 - 2x + 2$

2. $f(x) = 1 - x^3$

3. $f(x) = \dfrac{x}{1 + x}$

4. $f(x) = \sqrt{x + 2}$

5. $f(x) = \sin x$

6. $f(x) = \tan x$

7. $f(x) = \dfrac{1}{x^2 + 1}$

8. $f(x) = \dfrac{x}{x^2 + 1}$

9. $f(x) = x + \sin x$

10. $f(x) = (x + 3)^{2/3}$

11. $f(x) = \dfrac{1}{3}x^3 - 3x^2 - 7x + 5$

12. $f(x) = x^4 - 1$

13. $f(x) = \dfrac{1}{1 - x}$

14. $f(x) = \sec^2 x$

15. $f(x) = \dfrac{x^{1/3}}{x^{2/3} - 4}$

16. $f(x) = x^{2/3}(x + 8)^2$

17. $f(x) = \sqrt[3]{8 - x^3}$

18. $f(x) = \sqrt[3]{\dfrac{2x + 1}{3x - 4}}$

19. $f(x) = |3x - 7|$

20. $f(x) = |9 - x^2|$

21. The function $f(x) = 2x^3 - 3ax^2 + 6$ decreases only on the interval $(0, 3)$. Find a.

22. Profits for the XYZ Corporation for the 10 years of its existence were given by the function $P(t) = \dfrac{50}{1 + (t - 6)^2}$. During which periods were profits **(a)** increasing, and **(b)** decreasing?

23. The sum of two numbers is 50. If x denotes one of the numbers, for what values of x is the product of the two numbers increasing?

24. Is the converse of Theorem 3 true? In other words, if $f(x)$ is a differentiable function that is increasing on all intervals must $f'(x)$ be positive for all x? (*Hint:* Consider $f(x) = x^3$ and $x = 0$.)

25. Find a function $f(x)$ that is neither increasing nor decreasing on any interval. Can $f(x)$ be differentiable? Does this violate Theorem 3?

26. Prove part (ii) of Theorem 3.

27. Use the result of Example 4 to conclude that

$$x + \frac{1}{x} \geq 2$$

for all $x > 0$. Can you establish this inequality without using calculus? (*Hint:* Begin with the inequality $(x-1)^2 \geq 0$.) What can you say about the case $x < 0$?

5.4 RELATIVE EXTREMA

Recall from Section 4.3 that the function $f(x)$ has a *maximum value* at $c \in [a, b]$ if $f(c) \geq f(x)$ for all $x \in [a, b]$. This definition was appropriate for addressing the Optimization Problem, since we were seeking the *single* largest value of the given function on a *closed, bounded* interval. However, our interest here is in using the theory of the derivative to study the graph of the function $y = f(x)$ for all x in the domain of $f(x)$. Although the (absolute) maximum and minimum values for $f(x)$ remain of interest, we also want to know how to locate the *relative* maxima and minima as illustrated in Figure 4.1. We define such *relative extrema* as follows.

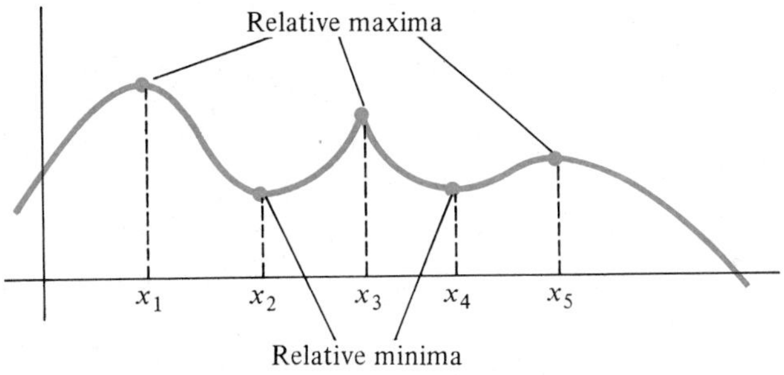

Figure 4.1

DEFINITION 2

The number $f(c)$ is a **relative maximum** for the function $f(x)$ if there exists a number $h > 0$ so that the interval $[c-h, c+h]$ is in the domain of $f(x)$, and $f(c)$ is the maximum value for $f(x)$ on $[c-h, c+h]$, that is, so that

$$f(c) \geq f(x) \qquad \text{for all} \qquad x \in [c-h, c+h].$$

Similarly, the number $f(c)$ is a **relative minimum** if there exists a number $k > 0$ so that the interval $[c-k, c+k]$ is in the domain of $f(x)$, and $f(c)$ is the minimum value for $f(x)$ on $[c-k, c+k]$, that is, so that

$$f(c) \leq f(x) \qquad \text{for all} \qquad x \in [c-k, c+k].$$

The term **relative extrema** refers to both relative maxima and relative minima.

Intuitively, this definition states that a relative maximum occurs when the number $f(c)$ is larger than $f(x)$ for all x "near" c. The number h is a measure of how close we must take x to c in order to observe this relationship. Figure 4.2 illustrates this concept for the relative maximum which occurs at $c = x_3$ in Figure 4.1. In Figure 4.3, the relative minimum $f(x_2)$ is shown.

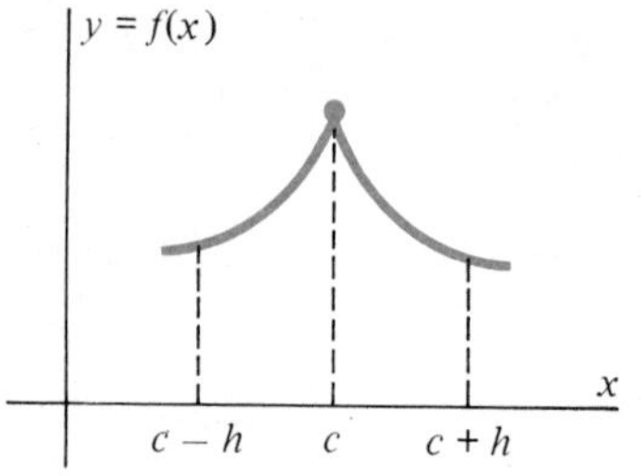

Figure 4.2 Relative maximum is absolute maximum on some interval about c.

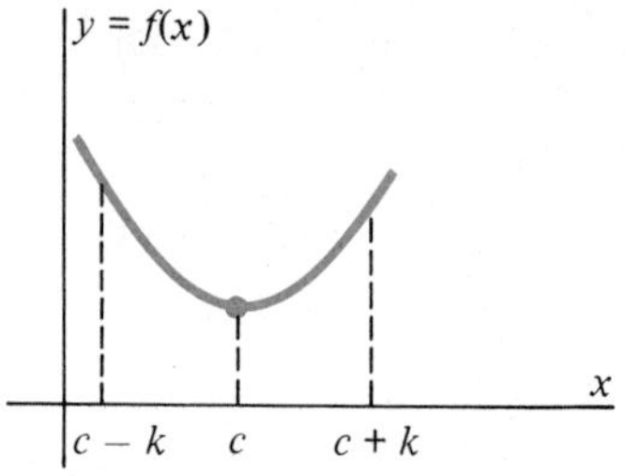

Figure 4.3 Relative minimum is absolute minimum on some interval about c.

Definition 2 is stated in a manner that emphasizes the relationship between *relative extrema* and *extrema on closed intervals*. A relative extremum $f(c)$ becomes an extremum if we restrict the interval on which the function is defined to a ''small'' interval containing $x = c$. Conversely, a value $f(c)$ may be an extremum for $f(x)$ on a closed, finite interval $[a, b]$, but only one of several relative extrema if $f(x)$ is actually defined on a larger set. (The case of endpoint extrema is particularly tenuous—these may fail to correspond to relative extrema when the domain of $f(x)$ is enlarged—see Figure 4.4.)

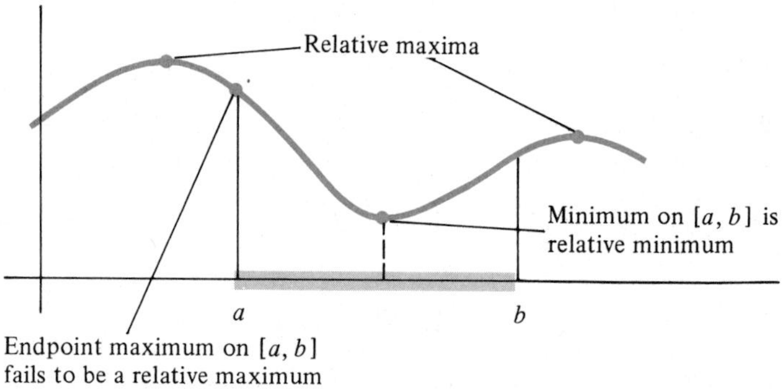

Figure 4.4 Relationships between extrema on $[a, b]$ and relative extrema.

If a function $f(x)$ has a largest value on its domain, say $f(x) = M$, we will continue to refer to M as the maximum or **absolute maximum** for $f(x)$, and similarly for **absolute minima.** However, functions whose domains are not restricted to finite closed intervals need not have **absolute extrema,** as Example 1 shows.

Example 1 For $f(x) = \frac{3}{2}x^{2/3} - x$, find

(a) the maximum and minimum values for $x \in [-1, 3]$,
(b) the relative extrema for $x \in (-\infty, \infty)$,
(c) the (absolute) maximum and minimum values for $x \in (-\infty, \infty)$.

Solution: Part (a) is an optimization problem on a closed, bounded interval. We therefore proceed as in Section 4.3 to find the critical numbers for $f(x)$. Since

$$f'(x) = \frac{2}{3} \cdot \frac{3}{2} x^{-1/3} - 1 = \frac{1}{\sqrt[3]{x}} - 1,$$

we know that $f'(x) = 0$ if $x = 1$ and $f'(x)$ is undefined for $x = 0$. The critical numbers are therefore $x = 0$ and $x = 1$. Computing the function values at these critical numbers and endpoints we find that

$$f(-1) = \frac{3}{2}(-1)^{2/3} - (-1) = \frac{3}{2} + 1 = \frac{5}{2} \quad \text{(endpoint maximum)}$$

$$f(0) = \frac{3}{2}(0)^{2/3} - (0) = 0 \quad \text{(minimum-critical number)}$$

$$f(1) = \frac{3}{2}(1)^{2/3} - 1 = \frac{3}{2} - 1 = \frac{1}{2} \quad \text{(not an extremum)}$$

$$f(3) = \frac{3}{2}(3)^{2/3} - 3 \approx 3.12 - 3 = 0.12 \quad \text{(not an extremum).}$$

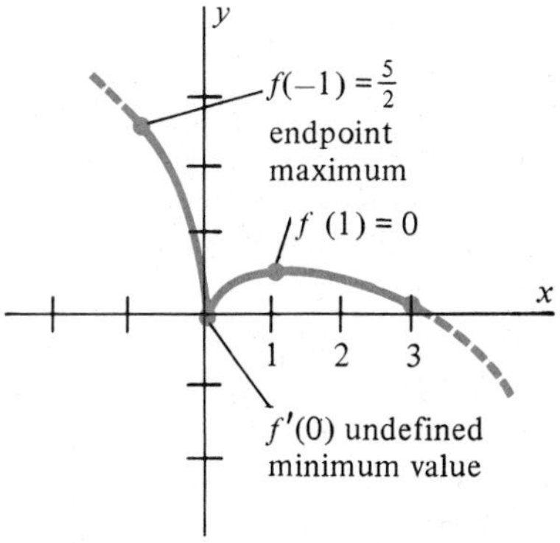

Figure 4.5 Maximum and minimum values of $f(x) = \frac{3}{2}x^{2/3} - x$ on $[-1, 3]$.

The maximum is $f(-1) = \frac{5}{2}$, which occurs at the endpoint $x = -1$, and the minimum is $f(0) = 0$, occurring at the critical number $x = 0$ (see Figure 4.5).

(b) The situation for relative extrema on the larger domain $(-\infty, \infty)$ is quite different, however, as illustrated in Figure 4.6. Note the following:

(i) Since $x = -1$ is no longer an endpoint and $f'(-1)$ exists and is not zero, $f(-1)$ is *not* a relative extremum.
(ii) $f(0) = 0$ is a relative minimum.
(iii) $f(1)$ is a relative maximum.
(iv) Like $f(-1)$, $f(3)$ is not a relative extremum.

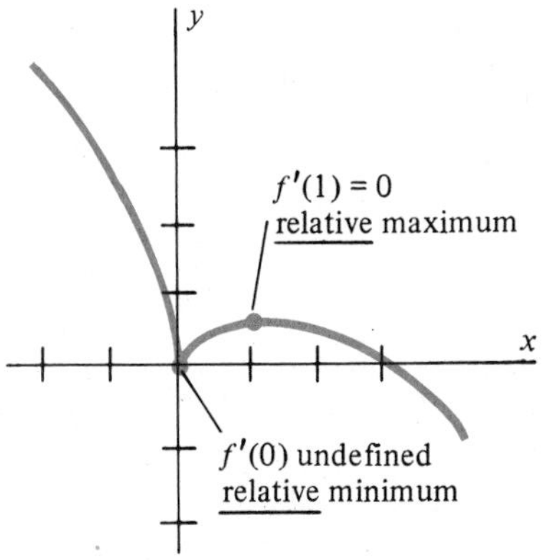

Figure 4.6 *Relative* extrema for $f(x) = \frac{3}{2}x^{2/3} - x$ on $(-\infty, \infty)$.

(c) The function $f(x)$ has no (absolute) maximum or minimum value on $(-\infty, \infty)$. To see this we write $f(x)$ in the form

$$f(x) = x \cdot \left(\frac{3}{2\sqrt[3]{x}} - 1\right).$$

As $|x|$ becomes large, the term $\frac{3}{2\sqrt[3]{x}}$ approaches zero, so the second factor approaches -1. Thus, $f(x)$ becomes increasingly large and positive as $x \to -\infty$, and $f(x)$ becomes increasingly large and negative as $x \to \infty$. (We will speak with more precision about infinite limits in Section 5.6.) This shows that a continuous function need not have a maximum or minimum value if its domain is not a closed, bounded interval, as we have seen before. ■

Finding Relative Extrema

Example 1 and the discussion preceding it suggest that there is a strong relationship between relative extrema and the conditions of Chapter 4 for identifying extrema on closed intervals. The next theorem verifies this suggestion. (Note the similarity between Theorem 4 below and Theorem 2 of Section 4.3.)

THEOREM 4

Let $f(x)$ be a continuous function. If $f(c)$ is a relative extremum for $f(x)$, then either $f'(c) = 0$ or else $f'(c)$ fails to exist.

Proof: If $f(c)$ is a relative extremum for $f(x)$ then, according to Definition 2, there is a number $h > 0$ so that $f(x)$ is defined on the closed bounded interval

$[c - h, c + h]$ and $f(c)$ is either the maximum or minimum value of $f(x)$ on this interval. Thus, by Theorem 2, Section 4.3, either

(i) $f'(c) = 0$,
(ii) $f'(c)$ fails to exist, or
(iii) c is an endpoint of $[c - h, c + h]$.

Since $h > 0$, c cannot be an endpoint. Thus, either (i) or (ii) must hold. ■

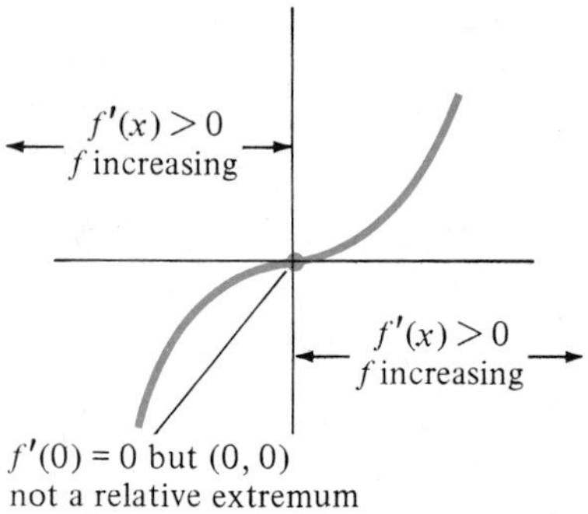

Figure 4.7 Critical number for $f(x) = x^3$ does not yield a relative extremum.

Recall (Section 4.3) that the number c is called a **critical number** for $f(x)$ if either $f'(c) = 0$ or $f'(c)$ fails to exist. Thus, Theorem 4 states that **relative extrema can occur only at critical numbers.**

However, not all critical numbers correspond to relative extrema. A classic example of this is found in the function $f(x) = x^3$. Here $f'(x) = 3x^2$, so $f'(0) = 0$, but $f(0)$ is neither a relative maximum nor a relative minimum (see Figure 4.7). The difficulty in this example is that $f(x) = x^3$ is an increasing function for x on either side of the critical number $x = 0$. For a critical number to yield a relative extremum, the function must *change* from increasing to decreasing, or vice versa, at the critical number.

Since the sign of the derivative $f'(x)$ determines whether $f(x)$ is increasing or decreasing, we can use $f'(x)$ to identify relative maxima and minima, as well as critical numbers that yield neither. The procedure for doing this is called the *First Derivative Test*.

THEOREM 5
First Derivative Test

Let c be a critical number for $f(x)$ and let $f'(x)$ exist for all x in an interval $[c - h, c + h]$, $h > 0$, except possibly at $x = c$. Then $f(c)$ is a relative extremum for $f(x)$ if and only if $f'(x)$ changes sign at $x = c$. Specifically, for $x \in [c - h, c + h]$,

(i) if $f'(x) > 0$ for $x < c$ and $f'(x) < 0$ for $x > c$, then $f(c)$ is a relative maximum;
(ii) if $f'(x) < 0$ for $x < c$ and $f'(x) > 0$ for $x > c$, then $f(c)$ is a relative minimum;
(iii) if $f'(x)$ does not change sign at $x = c$, then $f(c)$ is neither a relative maximum nor a relative minimum.

(See Figures 4.8 and 4.9.)

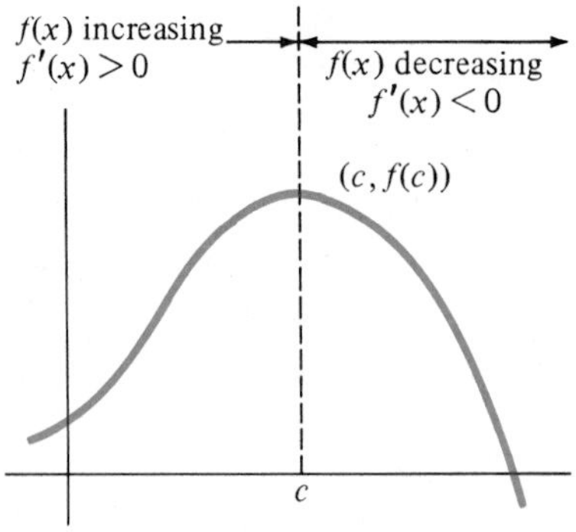

Figure 4.8 $f(x)$ changes from increasing to decreasing at a relative maximum.

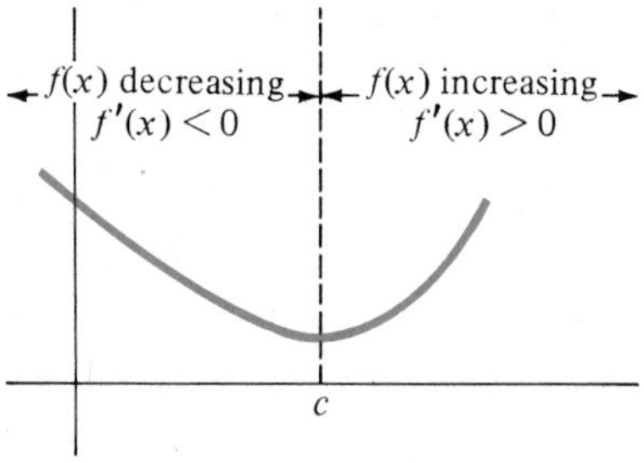

Figure 4.9 $f(x)$ changes from decreasing to increasing at a relative minimum.

Proof of Theorem 5: To prove statement (i) we note that if $f'(x) > 0$ for $x < c$ and $f'(x) < 0$ for $x > c$, then $f(x)$ is *increasing* for $x < c$ and *decreasing* for $x > c$. This guarantees that $f(c) \geq f(x)$ for all $x \in [c - h, c + h]$. That is, $f(c)$ is a relative maximum (see Figure 4.8). Similar observations establish statements (ii) and (iii). We leave these to you as Exercise 40. ■

Taken together, Theorems 4 and 5 give us a complete procedure for finding and classifying relative extrema. The following examples illustrate this procedure.

Example 2 For $f(x) = 4x^2(1 - x^2)$, find all relative and absolute extrema.

Solution: We first find the relative extrema using Theorems 4 and 5. The derivative for $f(x)$ is

$$\begin{aligned} f'(x) &= 8x(1 - x^2) + 4x^2(-2x) \\ &= 8x - 16x^3 \\ &= 8x(1 - 2x^2). \end{aligned}$$

The equation $f'(x) = 0$ therefore yields the critical numbers $x = 0$, $x = -\sqrt{2}/2$, and $x = \sqrt{2}/2$. There are no other critical numbers since $f'(x)$ is defined for all x. We must therefore examine the sign of $f'(x)$ on each of the intervals $(-\infty, -\sqrt{2}/2)$, $(-\sqrt{2}/2, 0)$, $(0, \sqrt{2}/2)$, and $(\sqrt{2}/2, \infty)$. To do so we simply select a convenient number a in each interval and examine the sign of $f'(a)$. The results of doing so appear in Table 4.1.

Table 4.1

Interval I	Test number $t \in I$	Value of $f'(t)$	Sign of $f'(t)$
$\left(-\infty, -\frac{\sqrt{2}}{2}\right)$	$t = -1$	$f'(-1) = 8$	+
$\left(-\frac{\sqrt{2}}{2}, 0\right)$	$t = -\frac{1}{2}$	$f'\left(-\frac{1}{2}\right) = -2$	−
$\left(0, \frac{\sqrt{2}}{2}\right)$	$t = \frac{1}{2}$	$f'\left(\frac{1}{2}\right) = 2$	+
$\left(\frac{\sqrt{2}}{2}, \infty\right)$	$t = 1$	$f'(1) = -8$	−

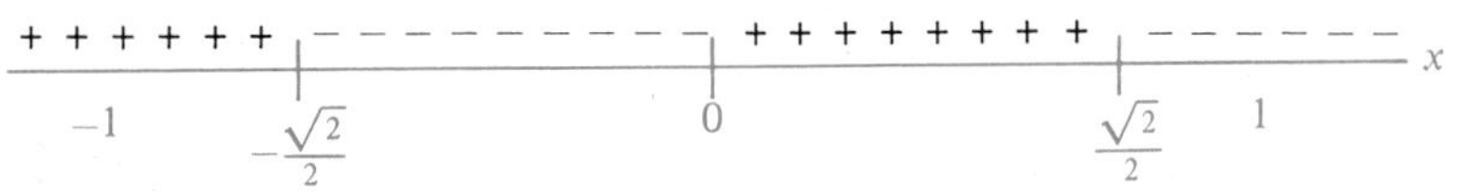

Figure 4.10 Sign of $f'(x)$ in Example 2.

From the results of Table 4.1, as illustrated in Figure 4.10, and the First Derivative Test, we conclude that $f(-\sqrt{2}/2) = 1$ and $f(\sqrt{2}/2) = 1$ are relative maxima, and that $f(0) = 0$ is a relative minima. The graph of $f(x)$ appears in Figure 4.11. Note that $f(-\sqrt{2}/2)$ and $f(\sqrt{2}/2)$ are indeed absolute maxima, but that no absolute minima exists. ■

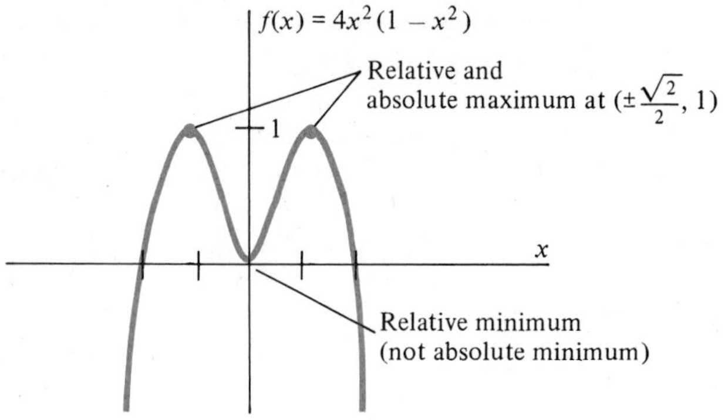

Figure 4.11 Relative extrema for $f(x) = 4x^2(1 - x^2)$.

Example 3 Find all relative extrema for the function

$$f(x) = \sin x + \cos x.$$

Solution: Although the domain of $f(x)$ is $(-\infty, \infty)$, $f(x)$ is periodic, of period 2π. This means that $f(x + 2n\pi) = f(x)$ for all integers $n = \pm 1, \pm 2, \ldots$. Thus, if we can identify the relative extrema in the interval $[0, 2\pi]$, we can find all others by simply adding $2n\pi$ to the ones found in $[0, 2\pi]$. Setting $f'(x) = 0$ gives

$$f'(x) = \cos x - \sin x = 0,$$

or

$$\sin x = \cos x.$$

The only solutions of this equation in $[0, 2\pi]$ are $x = \pi/4$ and $x = 5\pi/4$. Since $f'(x)$ is defined for all x this means that the only critical numbers in $[0, 2\pi]$ are $x = \pi/4$ and $x = 5\pi/4$. Table 4.2 shows the result of selecting one "test number" in each of the resulting intervals and checking the sign of $f'(x)$. ■

Table 4.2

Interval I	Arbitrary test number $t \in I$	$f'(t) = \cos t - \sin t$	Sign of $f'(t)$
$\left[0, \frac{\pi}{4}\right)$	$t = \frac{\pi}{6}$	$f'\left(\frac{\pi}{6}\right) = \frac{\sqrt{3}}{2} - \frac{1}{2}$	+
$\left(\frac{\pi}{4}, \frac{5\pi}{4}\right)$	$t = \pi$	$f'(\pi) = -1 - 0$	−
$\left(\frac{5\pi}{4}, 2\pi\right)$	$t = \frac{3\pi}{2}$	$f'\left(\frac{3\pi}{2}\right) = 0 - (-1)$	+

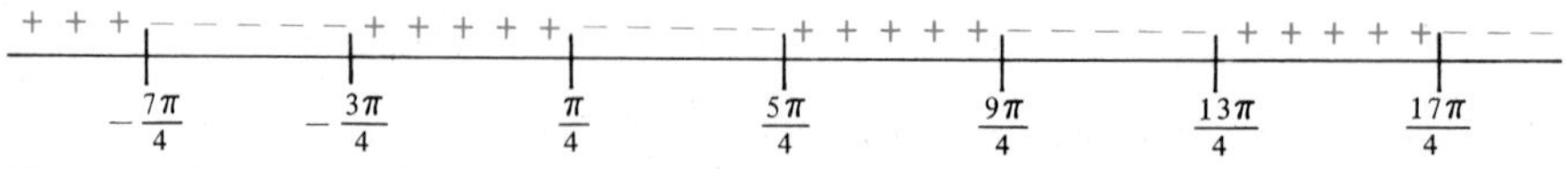

Figure 4.12 Sign of $f'(x)$ for $f(x) = \sin x + \cos x$.

From the results of Table 4.2, together with the First Derivative Test, we may conclude that the sign of $f'(x)$ is as illustrated in Figure 4.12 and that

(i) $f\left(\dfrac{\pi}{4} + 2n\pi\right) = \sqrt{2}$ is a relative maximum for $n = 0, \pm 1, \pm 2, \ldots$,

(ii) $f\left(\dfrac{5\pi}{4} + 2n\pi\right) = -\sqrt{2}$ is a relative minimum for $n = 0, \pm 1, \pm 2, \ldots$.

(See Figure 4.13.)

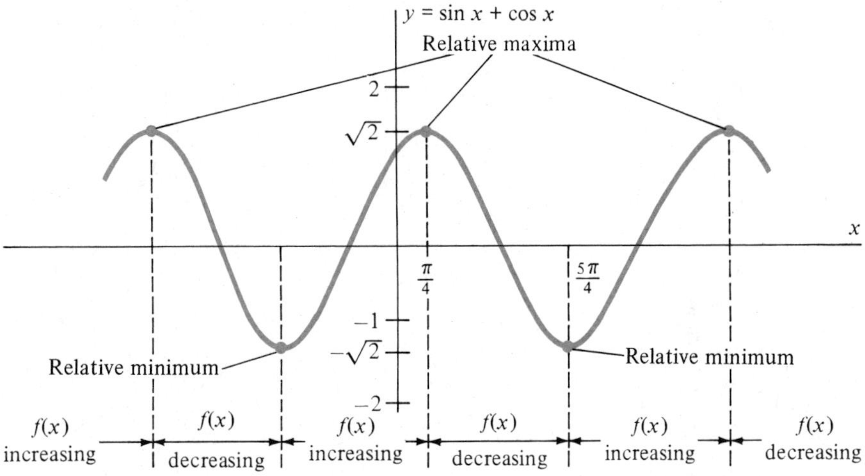

Figure 4.13 Relative extrema for $y = \sin x + \cos x$.

Example 4 Find all relative extrema for the function

$$f(x) = \frac{\sqrt[3]{1 - x}}{1 + x^2}.$$

Strategy

Note domain of $f(x)$.

Calculate $f'(x)$ using the Quotient Rule.

Multiply by

$$\frac{(1 - x)^{2/3}}{(1 - x)^{2/3}}$$

to simplify $f'(x)$.

Solution

Since $\sqrt[3]{x}$ is defined for all x, the domain of $f(x)$ is $(-\infty, \infty)$. Using the Quotient Rule, we find the derivative to be

$$f'(x) = \frac{(1 + x^2)\left[\dfrac{1}{3}(1 - x)^{-2/3}(-1)\right] - 2x(1 - x)^{1/3}}{[1 + x^2]^2}$$

$$= \frac{\left[-\dfrac{1}{3}(1 + x^2)(1 - x)^{-2/3} - 2x(1 - x)^{1/3}\right]}{[1 + x^2]^2} \cdot \left[\frac{(1 - x)^{2/3}}{(1 - x)^{2/3}}\right]$$

$$= \frac{-(1 + x^2) - 6x(1 - x)}{3[1 + x^2]^2(1 - x)^{2/3}}$$

$$= \frac{5x^2 - 6x - 1}{3[1 + x^2]^2(1 - x)^{2/3}}.$$

Find the zeros of the derivative by finding the zeros of its numerator.

Using the quadratic formula, we find that the zeros of the numerator are

$$x = \frac{6 \pm \sqrt{6^2 - 4(5)(-1)}}{2 \cdot 5} = \frac{6 \pm \sqrt{56}}{10} = \frac{3}{5} \pm \frac{\sqrt{14}}{5}.$$

Note that denominator equals zero for $x = 1$.

Construct a table showing sign of $f'(x)$ on each of the intervals determined by the critical numbers.

Also, $f'(x)$ is undefined for $x = 1$. The critical numbers are, therefore

$$c_1 = \frac{3}{5} - \frac{\sqrt{14}}{5} \approx -0.148,$$

$$c_2 = 1, \quad \text{and}$$

$$c_3 = \frac{3}{5} + \frac{\sqrt{14}}{5} \approx 1.348.$$

Table 4.3 shows the result of checking the sign of $f'(x)$ on each of the resulting intervals.

Table 4.3

Interval I	Test Number $t \in I$	Value of $f'(t)$	Sign of $f'(x)$ on I
$(-\infty, -0.148)$	-1	0.524	+
$(-0.148, 1)$	0	-0.333	−
$(1, 1.348)$	1.2	-0.084	−
$(1.348, \infty)$	2	0.093	+

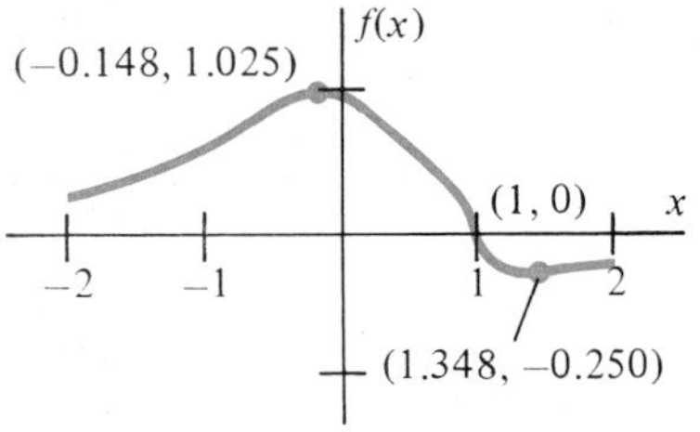

Figure 4.14 Graph of $f(x) = \dfrac{\sqrt[3]{1-x}}{1+x^2}$.

From the results of Table 4.3, and the First Derivative Test, it follows that

(i) $f(c_1) = f(-0.148) = 1.025$ is a relative maximum.
(ii) $f(c_2) = f(1) = 0$ is neither a relative maximum nor a relative minimum.
(iii) $f(c_3) = f(1.348) = -0.250$ is a relative minimum.

The graph of $f(x)$ appears in Figure 4.14. ■

A type of optimization problem similar to those of Sections 4.4 and 4.5 involves finding the maximum or minimum value of a quantity when the variable on which the quantity depends is not constrained to lie within a closed interval. In this case the solution is obtained by finding the absolute extrema of the function in question. One often needs then to argue on physical or geometric grounds why a particular relative extremum is indeed the desired absolute extremum.

Example 5 A rectangular container is to be designed so as to have a square base and a volume of 4000 cm^3. The container will not have a top. Find the dimensions that require the least amount of material.

Strategy

Clearly state the problem.

Draw a sketch.

Label variables.

Solution

The problem is to find the rectangular box of minimum surface area. If we denote one dimension of the base by x, then since the base is square the other dimension of the base is also x. Let y denote the height of the box (see Figure 4.15). Then

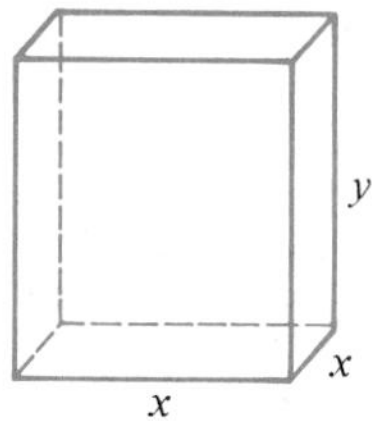

Figure 4.15 Box surface area to be minimized.

Find an equation for the variable to be minimized.

the area of the base is x^2 cm^2, and the area of each side panel is xy cm^2. The expression for surface area is therefore

$$S = x^2 + 4xy. \tag{1}$$

Find an auxiliary equation to eliminate one variable in (2).

To eliminate one of the variables from the right-hand side of this equation, we use the fact that volume equals 4000 cm^3. This gives the auxiliary equation

$$x^2y = 4000, \qquad \text{or} \qquad y = \frac{4000}{x^2}.$$

Substituting this expression for y into the principal equation (1) gives

$$\begin{aligned} S(x) &= x^2 + 4x\left(\frac{4000}{x^2}\right) \\ &= x^2 + \frac{16{,}000}{x}. \end{aligned}$$

The problem is to minimize $S(x)$. However, the independent variable x now lies within the *open* interval $(0, \infty)$ since, given any positive value for x, a corresponding value for y is determined by the equation $y = 4000/x^2$. Rather than minimizing $S(x)$ over a closed interval, we face the problem of finding the *absolute minimum* for $S(x)$. We proceed to do so by finding the critical numbers for $S(x)$.

To minimize $S(x)$ find critical numbers by setting $S'(x) = 0$.

Since

$$S'(x) = 2x - \frac{16{,}000}{x^2} = 0$$

implies $x^3 = 8000$, we obtain the critical number $x = 20$. (Note that $x = 0$ is also a critical number, but $x = 0$ is not physically possible.) To determine whether $x = 20$ is a relative extremum we examine the sign of $S'(x)$.

Inspect critical numbers to check for *relative* minimum.

Since we can write

$$S'(x) = \frac{2(x^3 - 8000)}{x^2}$$

Find a physical argument to show that the relative minimum is indeed the absolute minimum.

we conclude that $S'(x)$ is negative for all $x \in (0, 20)$, and $S'(x)$ is positive for all $x \in (20, \infty)$. This shows that $S(20)$ is a relative minimum. Since $x = 20$ corresponds to the *only* relative minimum, and since it seems clear that a box of minimum surface area must exist, we conclude that

$$S(20) = 20^2 + \frac{16{,}000}{20} = 1200 \text{ cm}^2$$

is the minimum surface area, which occurs for dimensions $x = 20$ cm and $y = 4000/20^2 = 10$ cm.
(See Figure 4.16.) ■

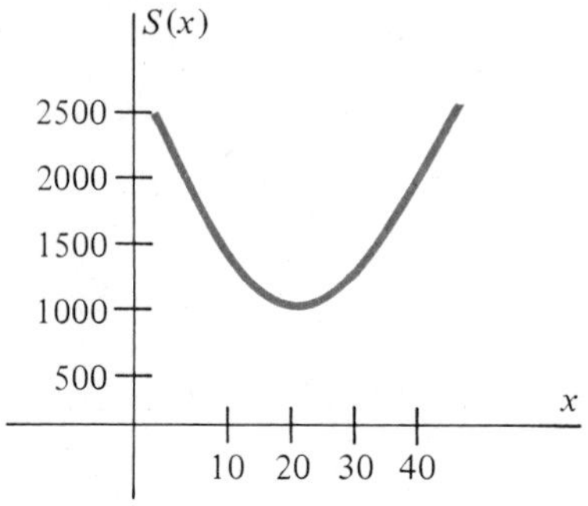

Figure 4.16 $S(x) = x^2 + \dfrac{1600}{x}$.

Exercise Set 5.4

In each of Exercises 1–22, find all critical numbers. Then classify each critical number as corresponding to a relative maximum, relative minimum, or neither.

1. $f(x) = x(x - 3)$

2. $f(x) = x^2 - x - 6$

3. $f(x) = 6x^{5/2} - 70x^{3/2} + 15$

4. $f(x) = \dfrac{x + 1}{x}$

5. $f(x) = x^3 - 3x^2 + 6$

6. $f(x) = 4x^3 + 9x^2 - 12x + 7$

7. $f(x) = x^3 + x^2 - 8x + 8$

8. $f(x) = x^4 + 4x^3 - 8x^2 - 48x + 9$

9. $f(x) = x + \sin x$

10. $f(x) = \tan^2 x$

11. $f(x) = \dfrac{x - 1}{x + 1}$

12. $f(x) = 9x - x^{-1}$

13. $y = \tan(1 - x^2)$

14. $f(x) = \csc x$

15. $y = |9 - x^2|$

16. $f(x) = \sqrt[3]{x}(x - 7)^2$

17. $f(x) = \sqrt[3]{x - 2} + \sqrt{x + 2}$

18. $y = \dfrac{x - 1}{x^2 + 2}$

19. $f(x) = x^{5/3} - 5x^{2/3} + 3$

20. $f(x) = \dfrac{\sqrt{x}}{4 - x}$

21. $f(x) = \dfrac{1 - x}{1 + x^2}$

22. $f(x) = x^{2/3}\sqrt[3]{1 - x}$

23. Discuss the relative extrema for functions of the form $y = ax + b$.

24. Explain why a function of the form $y = x^2 + bx + c$ always has precisely one relative minimum and no relative maximum.

25. What is the maximum number of local extrema that a polynomial of degree n can have? Could such a function have fewer?

26. The function $f(x) = x^2 - ax + b$ has a local minimum at $x = 2$. Find a.

27. The function $f(x) = x^3 + ax^2 + bx + 7$ has relative extrema at $x = 1$ and $x = -3$.
a. Find a and b.
b. Classify the extrema (as relative maxima or minima).

28. The function $f(x) = a(x^2 - bx + 16)$ has a local extrema at $x = 5$.
a. Find b.
b. Determine the sign of a if $(5, f(5))$ is a local minimum.

29. The function $y = x^3 + ax + 7$ has a critical number at $x = 0$. Find a.

30. Two sides of a triangle are of length a. Find the length of the third side so that area is a maximum.

31. A salt container is to be made in the shape of a right circular cylinder and is to contain 500 cm^3 of salt. Find the dimensions for the container that requires the least amount of material.

32. Find the dimensions of the box with square top and bottom and rectangular sides of capacity V that can be constructed from a minimum amount of material.

33. Prove that the function $f(x) = ax^2 + bx + c$, $a \neq 0$, always has precisely one relative extremum which occurs at the number $x = -\dfrac{b}{2a}$.

34. The rate at which an outbreak of influenza spreads through a college dormitory is proportional to the product of the number of students infected and the number of students not yet infected. Show that the influenza is spreading most rapidly when precisely one half of the students are infected. (Assume that infected students remain so for the duration of the epidemic.)

35. Find the point on the parabola $y = x^2$ nearest the point $(-3, 0)$.

36. A storage bunker is to contain 96 cubic meters and is to have a square base and vertical sides. If the material for the base costs \$10 per square meter and the material for the sides and top costs \$6 per square meter, find the dimensions that minimize cost.

37. Show that of all right circular cylinders of fixed volume V the one with minimum surface area has height equal to the diameter of its base.

38. Prove that the distance from a point $P = (x_0, y_0)$ in the plane to the line with equation $ax + by + c = 0$ is the number

$$d = \frac{|ax_0 + by_0 + c|}{\sqrt{a^2 + b^2}}.$$

39. Explain geometrically why there is no relative extremum at the critical number $x = 1$ in Example 4. Is there a tangent to the graph of $y = f(x)$ at $(1, 0)$? What is its slope?

40. Prove statements (ii) and (iii) of Theorem 5.

5.5 SIGNIFICANCE OF THE SECOND DERIVATIVE: CONCAVITY

Suppose that the function $f(x)$ is differentiable on the interval (a, b). If the derivative $f'(x)$ is an increasing function on (a, b), then the slopes of the lines tangent to the graph of $y = f(x)$ increase (i.e., rotate counterclockwise) as x increases. This means that the graph will have a "cupped up" shape as illustrated in Figure 5.1. We formalize this concept by saying that the graph of $y = f(x)$ is **concave up** (or that $y = f(x)$ has positive concavity) on (a, b) if $f'(x)$ is an increasing function on (a, b). Similarly, the graph of $y = f(x)$ is said to be **concave down** when $f'(x)$ is a decreasing function (negative concavity) (see Figure 5.2).

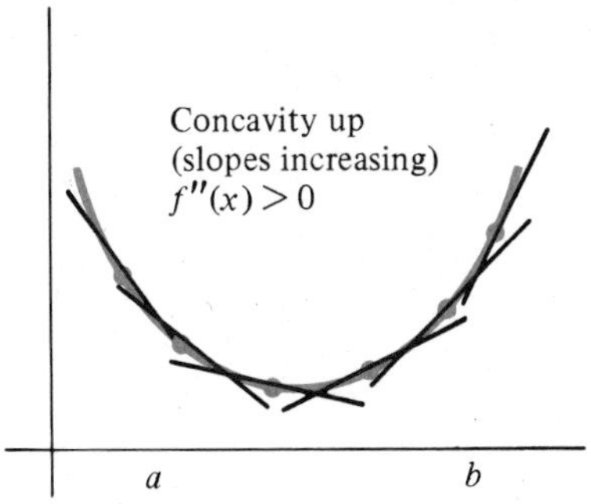

Figure 5.1 Graph of $y = f(x)$ is concave up if $f'(x)$ is increasing, that is, $f''(x) > 0$.

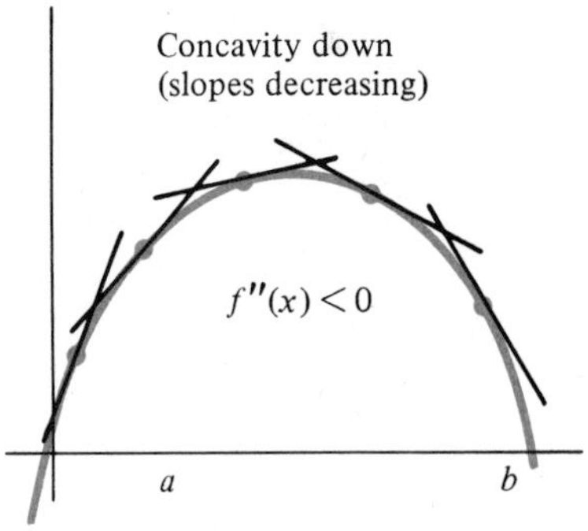

Figure 5.2 Graph of $y = f(x)$ is concave down if $f'(x)$ is decreasing, that is, $f''(x) < 0$.

REMARK: We are defining concavity in analytic terms (that is, by use of $f'(x)$) rather than geometrically, in order quickly to develop the relationship between concavity and the second derivative. A more geometric approach would be to say that the graph of $f(x)$ is concave up on (a, b) if, for each $c \in (a, b)$, the graph "lies above the tangent at $(c, f(c))$," as is the case in Figure 5.1. In Exercise 37 you are asked to show that this geometric condition implies that $f''(x) > 0$ on (a, b), the condition we have chosen to use. A similar geometric definition can be given for concave down.

The following theorem shows that concavity is determined by the sign of the second derivative.

THEOREM 6

If $f(x)$ is twice differentiable on (a, b), then

(i) $y = f(x)$ is concave up on (a, b) if $f''(x) > 0$ for all $x \in (a, b)$;
(ii) $y = f(x)$ is concave down on (a, b) if $f''(x) < 0$ for all $x \in (a, b)$.

Proof: Since $f''(x) = \frac{d}{dx}f'(x)$, $f'(x)$ is increasing when $f''(x) > 0$, by Theorem 3, Section 5.3. But if $f'(x)$ is increasing, the graph of $y = f(x)$ is concave up. Thus, the graph of $y = f(x)$ is concave up on (a, b) if $f''(x) > 0$ for all $x \in (a, b)$. This proves (i). The proof of (ii) is similar. ■

Theorem 6 gives a *procedure for determining the intervals on which $f(x)$ is concave up or concave down when $f'(x)$ is continuous:*

(i) Find all numbers of x for which $f''(x) = 0$ or $f''(x)$ fails to exist.
(ii) Determine the sign of $f''(x)$ on each of the resulting intervals, and apply Theorem 6.

Points on the graph of $y = f(x)$ which separate arcs of opposite concavity are called **inflection points.**

Example 1 Determine the concavity of the graph of the function $f(x) = x^4 - 3x^2 + 2$.

Strategy
Calculate $f''(x)$.

Solution
Since $f(x) = x^4 - 3x^2 + 2$ we have

$$f'(x) = 4x^3 - 6x$$

and

$$f''(x) = 12x^2 - 6.$$

Find all x with $f''(x) = 0$ or $f''(x)$ undefined.

Setting $f''(x) = 0$ gives $12x^2 = 6$, or $x^2 = \frac{1}{2}$.

The zeros of the second derivative $f''(x)$ are, therefore,

$$x_1 = -\frac{\sqrt{2}}{2}, \quad \text{and} \quad x_2 = \frac{\sqrt{2}}{2}.$$

Determine the sign of $f''(x)$ on each of the resulting intervals by checking $f''(t)$ at a single number. Then apply Theorem 6.

There are no numbers x for which $f''(x)$ is undefined. Since $f''(x)$ is a continuous function (it's a polynomial), it must have constant sign on each of the intervals determined by its zeros. We may therefore determine the sign of $f''(x)$ on each of these intervals by checking the sign of $f''(t)$ at a "test number" t. The results of doing so appear in Table 5.1.

Table 5.1

Interval I	Test number $t \in I$	Value of $f''(t)$	Sign of $f''(t)$
$\left(-\infty, -\frac{\sqrt{2}}{2}\right)$	$t = -1$	$f''(-1) = 6$	+
$\left(-\frac{\sqrt{2}}{2}, \frac{\sqrt{2}}{2}\right)$	$t = 0$	$f''(0) = -6$	−
$\left(\frac{\sqrt{2}}{2}, \infty\right)$	$t = 1$	$f''(1) = 6$	+

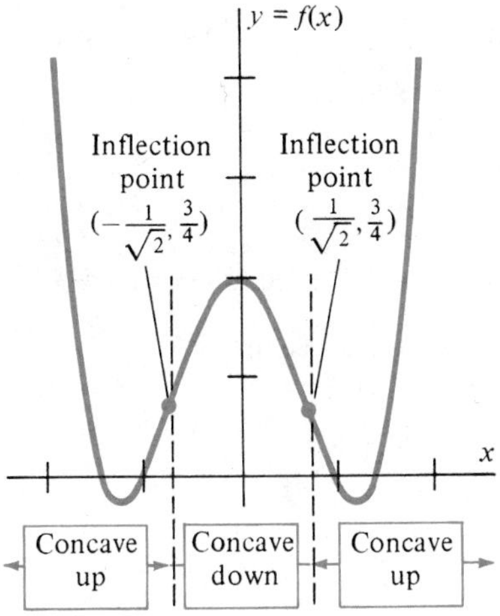

Figure 5.3 Concavity for graph of $f(x) = x^4 - 3x^2 + 2$.

From the results in Table 5.1 and by Theorem 6, we may conclude that the graph of $y = f(x)$ is

(i) concave up on $\left(-\infty, -\frac{\sqrt{2}}{2}\right)$,

(ii) concave down on $\left(-\frac{\sqrt{2}}{2}, \frac{\sqrt{2}}{2}\right)$,

(iii) concave up on $\left(\frac{\sqrt{2}}{2}, \infty\right)$.

Since the concavity of the graph of $y = f(x)$ changed both at $(\sqrt{2}/2, f(-\sqrt{2}/2)) = (-\sqrt{2}/2, 3/4)$ and at $(\sqrt{2}/2, f(\sqrt{2}/2)) = (\sqrt{2}/2, 3/4)$, both of these points are *inflection points* (see Figure 5.3). ■

It is important to note that the point $(x_0, f(x_0))$ need not be an inflection point for the graph of $y = f(x)$, even though $f''(x_0) = 0$ or $f''(x_0)$ fails to exist. The following example shows that the concavity can be the same on either side of such a point.

Example 2 Determine the concavity and find the inflection points for the graph of $f(x) = x^{2/3} - \frac{1}{5}x^{5/3}$.

Solution: We have

$$f'(x) = \frac{2}{3}x^{-1/3} - \frac{1}{3}x^{2/3},$$

so

$$\begin{aligned} f''(x) &= -\frac{2}{9}x^{-4/3} - \frac{2}{9}x^{-1/3} \\ &= -\frac{2}{9}x^{-4/3}(1 + x) \\ &= \frac{-2(1 + x)}{9x^{4/3}}. \end{aligned}$$

Thus $f''(x) = 0$ for $x = -1$, and $f''(x)$ is undefined for $x = 0$. The sign of $f''(x)$ must therefore be checked on the intervals $(-\infty, -1)$, $(-1, 0)$, and $(0, \infty)$. This time, rather than substituting particular test numbers and constructing a table, it is simpler just to examine the signs of the numerator and denominator on each interval, as illustrated in Figure 5.4.

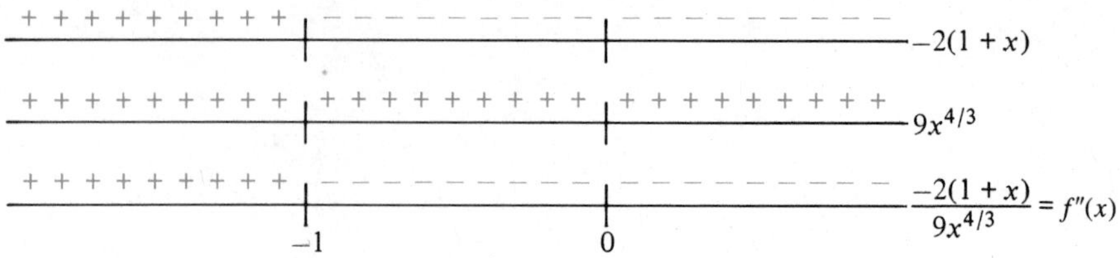

Figure 5.4 Sign analysis for $f''(x) = \dfrac{-2(1 + x)}{9x^{4/3}}$.

From Figure 5.4, we can see that $f''(x)$ is positive on $(-\infty, -1)$ but negative on both $(-1, 0)$ and $(0, \infty)$. Thus, by Theorem 6, the graph is

(i) concave up on $(-\infty, -1)$,
(ii) concave down on $(-1, 0)$ *and* on $(0, \infty)$.

This means that the point $(-1, f(-1)) = \left(-1, \dfrac{6}{5}\right)$ is an inflection point, but that the point $(0, f(0)) = (0, 0)$ is *not* an inflection point (see Figure 5.5). ■

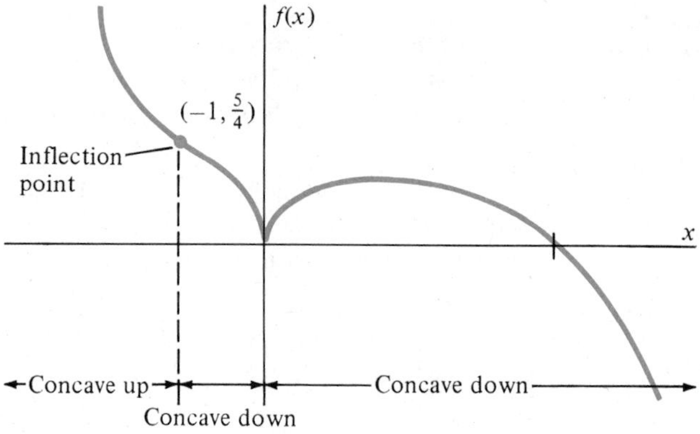

Figure 5.5 Graph of $f(x) = x^{2/3} - \dfrac{1}{5}x^{5/3}$.

The Second Derivative Test for Extrema

Notice that the determination of concavity divides the graph in Figure 5.3 into three distinct arcs, each of which contains one relative extremum. There is a very direct relationship between concavity and the nature of extrema (i.e., whether an extremum is a relative maximum or a relative minimum) as the following theorem shows.

THEOREM 7
Second Derivative Test

Let $f(x)$ be differentiable on an open interval containing the critical number $x = a$, with $f'(a) = 0$. Suppose also that $f''(a)$ exists.

(i) If $f''(a) < 0$, then $f(a)$ is a relative maximum.
(ii) If $f''(a) > 0$, then $f(a)$ is a relative minimum.
(iii) If $f''(a) = 0$, there is no conclusion.

Proof: A formal proof of Theorem 7 is outlined in Exercise 38. We prefer to give only an informal sketch of the argument here to avoid clouding the simple idea with details.

In case (i), the condition $f''(a) < 0$ may be interpreted as saying that the graph of $y = f(x)$ is concave down near $x = a$. This means that $f'(x)$ is a decreasing function near a. Since $f'(a) = 0$, it follows that $f'(x) > 0$ for $x < a$ and $f'(x) < 0$ for $x > a$. Thus, by the First Derivative Test, $f(a)$ is a relative maximum. A similar argument applies for statement (ii). ■

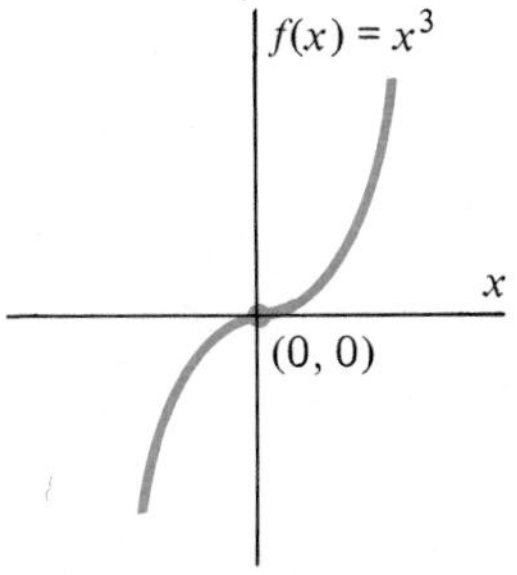

Figure 5.6 $f'(0) = f''(0) = 0$, but $(0, 0)$ is not an extremum.

In applying the Second Derivative Test be sure to note statement (iii)—if $f''(a) = 0$, we obtain no conclusion about whether $f(a)$ is a relative extrema. The following simple examples show why.

Example 3 For $f(x) = x^3, f'(x) = 3x^2$, and $f''(x) = 6x$. Thus $x = 0$ is a critical number, and $f''(0) = 0$. But $(0, f(0)) = (0, 0)$ is neither a relative maximum nor a relative minimum (see Figure 5.6). ■

Example 4 For $f(x) = x^4$, $f'(x) = 4x^3$ and $f''(x) = 12x^2$. Again, $x = 0$ is a critical number, and $f''(0) = 0$. For this function, $(0, f(0)) = (0, 0)$ is a relative minimum (see Figure 5.7). ■

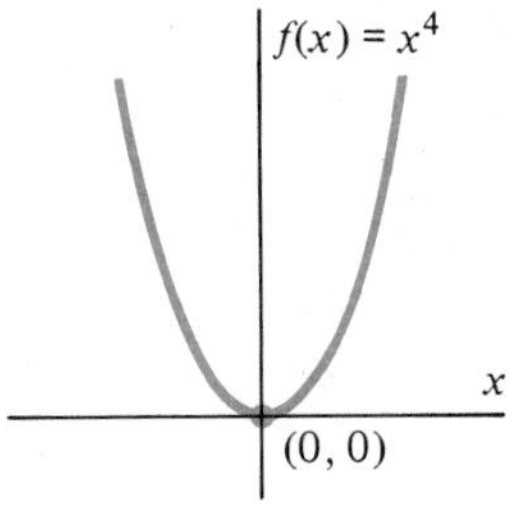

Figure 5.7 $f'(0) = f''(0) = 0$; $(0, 0)$ is a relative minimum.

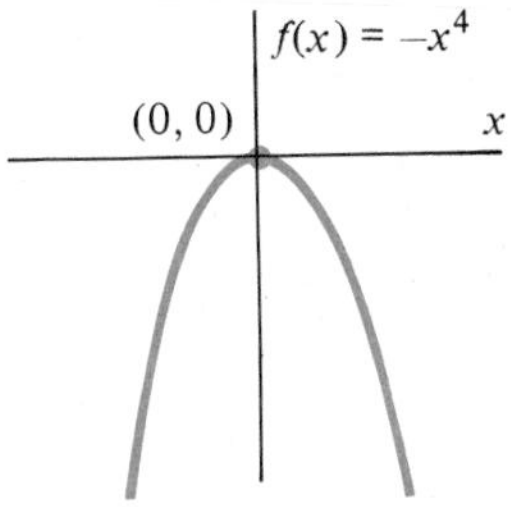

Figure 5.8 $f'(0) = f''(0) = 0$; $(0, 0)$ is a relative maximum.

Example 5 The function $f(x) = -x^4$ has derivatives $f'(x) = -4x^3$ and $f''(x) = -12x^2$. Here also, $x = 0$ is a critical number, and $f''(0) = 0$. This time $(0, f(0)) = (0, 0)$ is a relative maximum (see Figure 5.8). ■

Example 6 Find all relative extrema for the function $f(x) = x^4 - 3x^2 + 2$ in Example 1.

Strategy
Set $f'(x) = 0$ to find critical numbers.

Solution
Setting $f'(x) = 4x^3 - 6x = 0$, we obtain the equation

$$2x(2x^2 - 3) = 0$$

Classify critical numbers using the Second Derivative Test.

so $x = 0$, $x = -\sqrt{3/2}$, and $x = \sqrt{3/2}$ are critical numbers. We can classify these by the Second Derivative Test.

Since $f''(x) = 12x^2 - 6$ we find that

$f''(-\sqrt{3/2}) = 12 > 0$, so $(-\sqrt{3/2}, -1/4)$ is a relative minimum;
$f''(0) = -6 < 0$, so $(0, 2)$ is a relative maximum;
$f''(\sqrt{3/2}) = 12 > 0$, so $(\sqrt{3/2}, -1/4)$ is a relative minimum (see Figure 5.3). ■

Example 7 Discuss the relative extrema and concavity for the function $y = x + \sin x$.

Strategy
Find dy/dx and d^2y/dx^2.

Solution
Here $dy/dx = 1 + \cos x$, so $d^2y/dx^2 = -\sin x$.

Evaluate d^2y/dx^2 at critical numbers to check for extrema by Theorem 7. If Theorem 7 fails, use First Derivative Test.

The equation $dy/dx = 0$ yields the critical numbers $x = \pm\pi, \pm 3\pi, \pm 5\pi, \ldots$. However, at such points $d^2y/dx^2 = 0$, so the Second Derivative Test is inconclusive. However, since $dy/dx = 1 + \cos x \geq 0$ for all x, the First Derivative Test shows that there are no relative extrema.

Examine sign of d^2y/dx^2 between zeros of $d^2y/dx^2 = 0$ to determine concavity by Theorem 6.

The equation $d^2y/dx^2 = -\sin x = 0$ has solutions $x = 0, \pm\pi, \pm 2\pi, \ldots$. Inspection of the sign of d^2y/dx^2 on the resulting intervals shows that the graph of y is concave up on the intervals $(\pi, 2\pi), (3\pi, 4\pi), \ldots, ((2n-1)\pi, 2n\pi), \ldots$ and concave down on the intervals $(0, \pi), (2\pi, 3\pi), \ldots, (2n\pi, 2(n+1)\pi), \ldots$.

Thus all points $(n\pi, n\pi)$, $n = 0, \pm 1, \pm 2, \ldots$ are inflection points (see Figure 5.9). ■

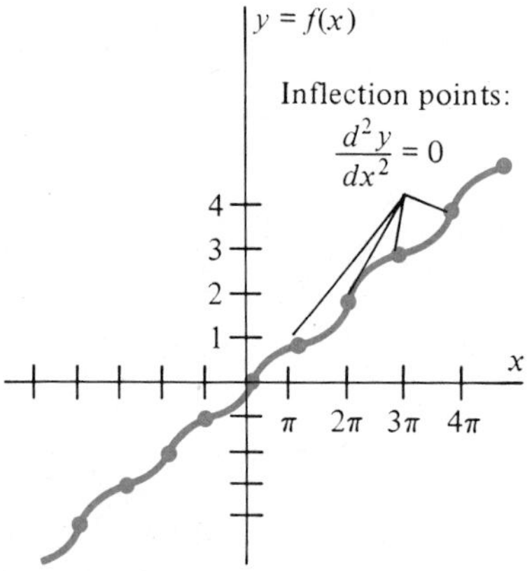

Figure 5.9 Graph of $f(x) = x + \sin x$.

Exercise Set 5.5

In each of Exercises 1–20, describe the concavity of the graph and find the inflection points (if any).

1. $f(x) = \dfrac{1}{x}$
2. $f(x) = \sin x$
3. $f(x) = \dfrac{1}{x-4}$
4. $f(x) = \dfrac{x}{x+1}$
5. $f(x) = (x+2)^{1/3}$
6. $f(x) = \cos x$
7. $f(x) = (2x+1)^3$
8. $f(x) = \sqrt{x+4}$
9. $f(x) = 2x^3 - 3x^2 + 18x - 12$
10. $f(x) = \dfrac{x+4}{x-4}$
11. $y = \tan x$

12. $y = 2x^3 + 12x^2 + 18x + 12$

13. $y = \sin^2 x$

14. $y = |4 - x^2|$

15. $y = x^4 + 2x^3 - 36x^2 + 24x - 6$

16. $y = \sec x$

17. $y = x^4 - 4x^3 - 18x^2 + 48x - 28$

18. $y = \dfrac{x}{x^2 + 1}$

19. $f(x) = \dfrac{\sqrt[3]{x}}{1 - x}$

20. $f(x) = x^{5/3} - 5x^{2/3} + 3$

In each of Exercises 21–28, use the Second Derivative Test to determine whether $f(x)$ has a relative extremum at the given value of x.

21. $f(x) = \sin^2 x, \quad x = \pi/2$

22. $f(x) = x^2 + \dfrac{2}{x}, \quad x = 1$

23. $f(x) = \cos(1 + x^2), \quad x = 0$

24. $f(x) = 2x^3 - 3x^2, \quad x = 1$

25. $f'(x) = \dfrac{x - 1}{x + 1}, \quad x = 1$

26. $f'(x) = \dfrac{1 - \sin 2x}{\cos x}, \quad x = \pi/4$

27. $f'(x) = (x - 1)(x + 2), \quad x = -2$

28. $f'(x) = \sqrt{x^3 - 4x}, \quad x = 2$

29. What conditions must hold for the constants a, b, and c in order that the general quadratic function $f(x) = ax^2 + bx + c$ be concave up for all x?

30. Find an example of a function $f(x)$ and a number a so that $f(a)$ is a relative maximum but $f''(a) = 0$.

31. Find an example of a function $f(x)$ and a number b so that $f(b)$ is a relative minimum but $f''(b) = 0$.

32. Find an example of a function $f(x)$ and a number c for which $f'(c) = f''(c) = 0$ but $f(c)$ is not a relative extremum.

33. How many inflection points can a polynomial of degree $n = 2$ possess? $n = 3$? $n = k$?

34. Find an example of a function $f(x)$ for which
a. $f(x)$ is concave up on $(-\infty, 2)$ and concave down on $(2, \infty)$,
b. $f(x)$ is increasing on $(-\infty, 1)$ and $(3, \infty)$,
c. $f(x)$ is decreasing on $(1, 3)$.

35. Find an example of a function which is concave down on $(-\infty, 0)$, concave up on $(0, \infty)$, and decreasing throughout its domain.

36. True or false? If $f(0) = 0$ and $y = f(x)$ is concave down for all x then $f(x) \leq 0$ for all x.

37. Let $f(x)$ be differentiable on an open interval I. Assume that throughout I "the graph of $f(x)$ is above the tangent." (That is, for each $c \in I$ if $y = g(x)$ is an equation for the tangent at $(c, f(c))$ then $g(x) < f(x)$ for all $x \in I$ with $x \neq c$.) Show that $f'(x)$ is increasing on I.

38. Prove statement (i) of the Second Derivative Test (Theorem 7) as follows:
a. Assume that $f''(a) = L < 0$
b. Show that $\lim_{x \to a} \dfrac{f'(x)}{x - a} = L < 0$
c. Conclude that there exists an interval $I = (a - h, a + h)$ so that $\dfrac{f'(x)}{x - a} < 0$ whenever $x \in I$.
d. Show that $f'(x) > 0$ for $x \in (a - h, a)$
e. Show that $f'(x) < 0$ for $x \in (a, a + h)$
f. Apply the First Derivative Test to conclude that $(a, f(a))$ is a relative maximum.

5.6 ASYMPTOTES

In the discussions of extrema and concavity of this chapter, you have probably noticed several examples of functions whose values become infinite as x approaches some constant or whose values approach a constant as x becomes large. The purpose of this section is to develop a terminology for describing such properties and a set of techniques by which these properties can be identified.

Limits at Infinity

The function $f(x) = -\dfrac{2x + 1}{x + 2}$ is an example of a function whose values approach

Table 6.1

x	$f(x) = -\dfrac{2x+1}{x+2}$
0	−0.50000
5	−1.57143
10	−1.75000
20	−1.86364
50	−1.94231
100	−1.97059
250	−1.98811
500	−1.99402
1,000	−1.99701
2,000	−1.99850
5,000	−1.99940
10,000	−1.99997

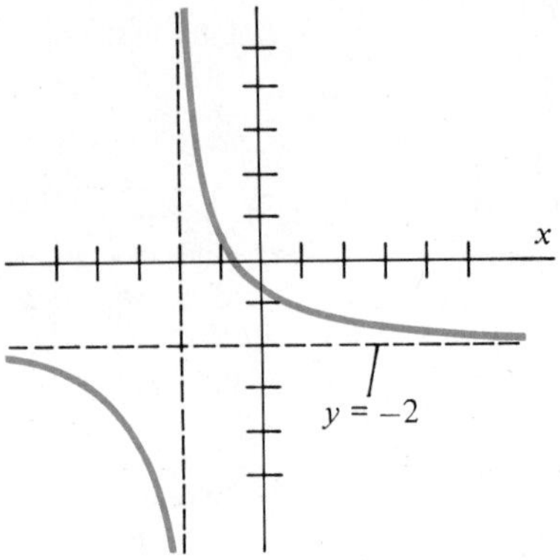

Figure 6.1 Graph of $f(x) = -\dfrac{2x+1}{x+2}$.

a constant as x becomes large. In fact, both the entries of Table 6.1 and the graph of $f(x)$ in Figure 6.1 suggest that the values of $f(x)$ approach the number $L = -2$ as x becomes very large.

In this case we write

$$\lim_{x\to\infty} -\frac{2x+1}{x+2} = -2$$

and we refer to the line $y = -2$ as a *horizontal asymptote*. More generally, we make the following informal definition.

DEFINITION 3

The expression $\lim_{x\to\infty} f(x) = L$ means that the values $f(x)$ approach the number L as x increases without bound. The expression $\lim_{x\to-\infty} f(x) = M$ means that the values $f(x)$ approach the number M as x decreases without bound.

As for earlier definitions of limit, Definition 3 is an informal working definition. A formal definition is given in Exercise 49.

Example 1

(a) $\lim_{x\to\infty} \dfrac{x+2}{x} = 1$ (Figure 6.2).

(b) $\lim_{x\to\infty} \dfrac{1-x}{1+x} = -1$ (Figure 6.3).

(c) $\lim_{x\to\infty} \sin x$ does not exist since $\sin x$ oscillates between 1 and -1, no matter how large x becomes. (Figure 6.4.)

(d) $\lim_{x\to\infty} \dfrac{\sin x}{x} = 0$, since $\left|\dfrac{1}{x}\sin x\right| = \left|\dfrac{1}{x}\right| \cdot |\sin x|$

$\leq \dfrac{1}{x} \to 0$ as $x \to \infty$. (Figure 6.5.) ■

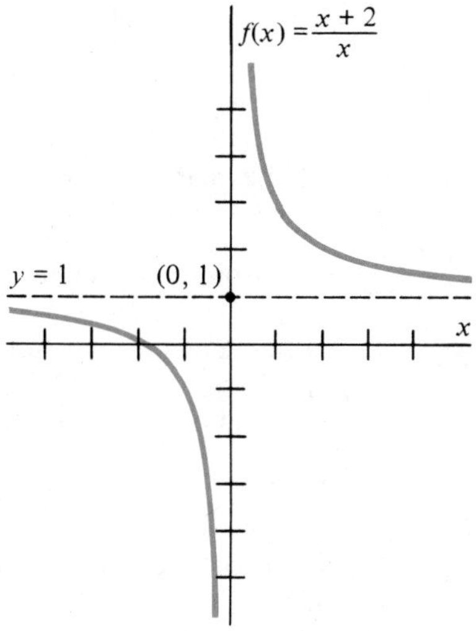

Figure 6.2 $\lim_{x\to\infty}\left(\frac{x+2}{x}\right) = 1$; $\lim_{x\to-\infty}\left(\frac{x+2}{x}\right) = 1$.

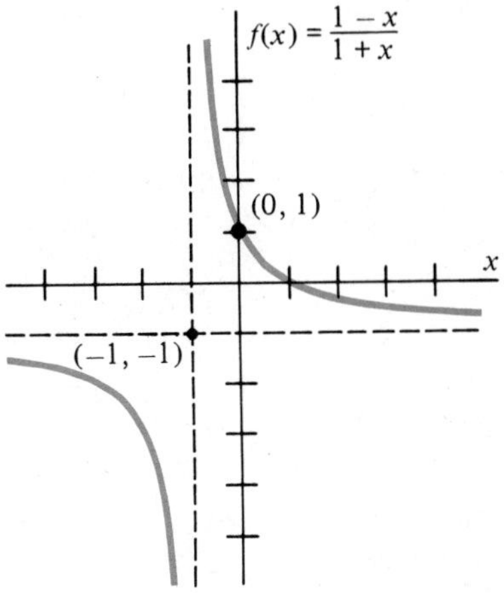

Figure 6.3 $\lim_{x\to\infty}\left(\frac{1-x}{1+x}\right) = -1$; $\lim_{x\to-\infty}\left(\frac{1-x}{1+x}\right) = -1$.

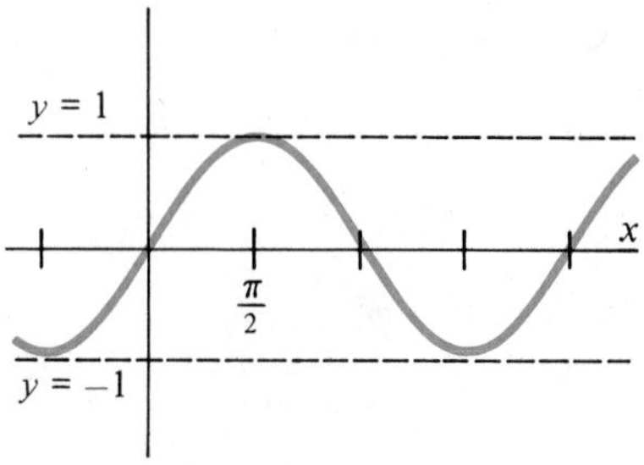

Figure 6.4 $\lim_{x\to\infty} \sin x$ does not exist.

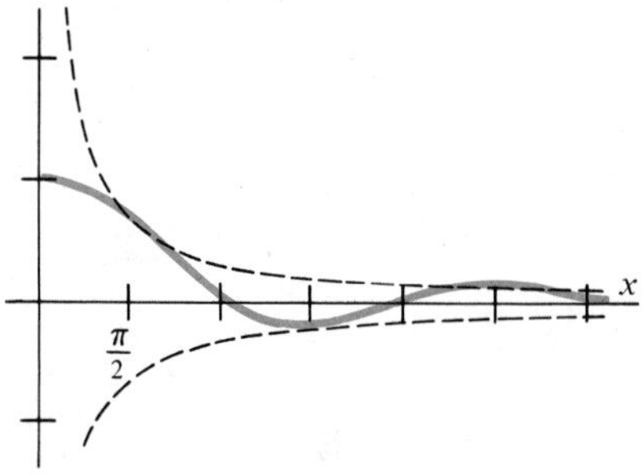

Figure 6.5 $\lim_{x\to\infty} \frac{1}{x} \sin x = 0$.

In general, a **horizontal asymptote** is a line of the form $y = L$ that the graph of $y = f(x)$ approaches as either $x \to \infty$ or $x \to -\infty$. In other words, the line $y = L$ is a horizontal asymptote if either

$$L = \lim_{x\to\infty} f(x) \qquad \text{or} \qquad L = \lim_{x\to-\infty} f(x).$$

(See Figure 6.6.)

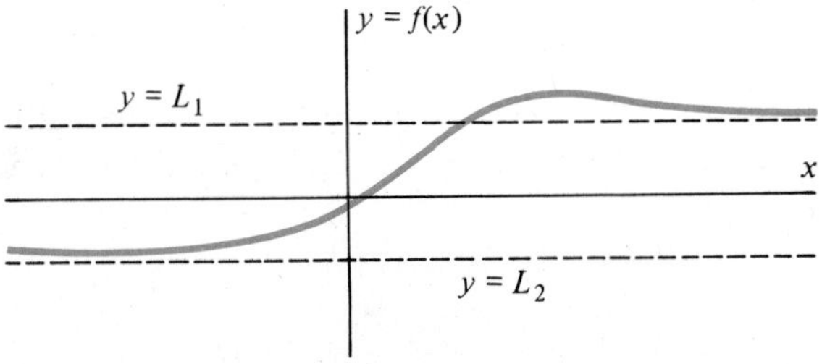

Figure 6.6 Horizontal asymptotes $y = L_1$ and $y = L_2$.

Example 2 Identify the horizontal asymptotes for the functions in Example 1.

Solution: From the information in Example 1 and Figures 6.2 through 6.5 we have

(a) $y = \dfrac{x+2}{x}$ has horizontal asymptote $y = 1$;

(b) $y = \dfrac{1-x}{1+x}$ has horizontal asymptote $y = -1$;

(c) $y = \sin x$ has no horizontal asymptote.

(d) $y = \dfrac{\sin x}{x}$ has horizontal asymptote $y = 0$. ■

Evaluating Limits at Infinity

To handle nontrivial examples of limits as $x \to \infty$ or as $x \to -\infty$, we need to know what properties such limits share and how these properties can be exploited. The first issue is addressed by the following theorem, which states that limits at infinity have the same algebraic properties as limits evaluated at real numbers. Its proof is left as Exercise 54.

THEOREM 8

Suppose that $\lim_{x\to\infty} f(x)$ and $\lim_{x\to\infty} g(x)$ both exist. Then

(i) $\lim_{x\to\infty} [f(x) + g(x)] = \lim_{x\to\infty} f(x) + \lim_{x\to\infty} g(x)$,

(ii) $\lim_{x\to\infty} [cf(x)] = c \lim_{x\to\infty} f(x)$, $\quad c$ constant,

(iii) $\lim_{x\to\infty} [f(x)g(x)] = [\lim_{x\to\infty} f(x)] \cdot [\lim_{x\to\infty} g(x)]$,

(iv) $\lim_{x\to\infty} \left[\dfrac{f(x)}{g(x)}\right] = \dfrac{\lim_{x\to\infty} f(x)}{\lim_{x\to\infty} g(x)}$, $\quad$ provided $\lim_{x\to\infty} g(x) \neq 0$.

Statements similar to (i)–(iv) in Theorem 8 hold for limits as $x \to -\infty$. The algebraic properties of Theorem 8 will be used together with the following Theorem, which we state without proof, in evaluating limits at infinity.

THEOREM 9

Let $r = p/q$ be any positive rational number. Then

(i) $\lim_{x\to\infty} \dfrac{1}{x^r} = 0$ $\quad$ if q is even, $\quad$ and

(ii) $\lim_{x\to\pm\infty} \dfrac{1}{x^r} = 0$ $\quad$ if q is odd.

Theorem 9 is just the observation that if r is a *positive* exponent, x^r increases without bound as x increases without bound. Thus, $1/x^r \to 0$ as $x \to \infty$.

TECHNIQUE (LIMITS AT INFINITY): To evaluate limits of the form

$$\lim_{x\to\infty} \frac{p(x)}{q(x)} \quad \text{or} \quad \lim_{x\to-\infty} \frac{p(x)}{q(x)}$$

where $p(x)$ and $q(x)$ are polynomials, divide both $p(x)$ and $q(x)$ by the highest power of x present. Then use the fact that $\lim_{x\to\pm\infty} 1/x^r = 0$ for $r > 0$ (Theorem 9).

(Although the technique is stated for rational functions only, it works equally as well when $p(x)$ and $q(x)$ involve fractional exponents, for obvious reasons. Also, note that there is no difficulty in assuring $x \neq 0$ in executing this technique, since we may assume that x is large.)

Example 3 Using the above technique we obtain

Strategy	*Examples*
Divide numerator and denominator by x^2.	(a) $\lim_{x\to\infty} \dfrac{3x^2+7x-4}{1-x^2} = \lim_{x\to\infty} \dfrac{3+\frac{7}{x}-\frac{4}{x^2}}{\frac{1}{x^2}-1}$
Apply Theorem 8 (algebra of limits) and then Theorem 9.	$= \dfrac{3+7\left(\lim_{x\to\infty}\frac{1}{x}\right)-4\left(\lim_{x\to\infty}\frac{1}{x^2}\right)}{\left(\lim_{x\to\infty}\frac{1}{x^2}\right)-1} = \dfrac{3+7(0)-4(0)}{0-1} = -3.$
Divide by x^3.	(b) $\lim_{x\to-\infty} \dfrac{x-3}{2x+x^3} = \lim_{x\to-\infty} \dfrac{\frac{1}{x^2}-\frac{3}{x^3}}{\frac{2}{x^2}+1}$
Apply Theorems 8 and 9.	$= \dfrac{\left(\lim_{x\to-\infty}\frac{1}{x^2}\right)-3\left(\lim_{x\to-\infty}\frac{1}{x^3}\right)}{2\left(\lim_{x\to-\infty}\frac{1}{x^2}\right)+1} = \dfrac{0-3(0)}{2(0)+1} = 0.$
Divide by $x^{2/3}$. Apply Theorems 8 and 9.	(c) $\lim_{x\to\infty} \dfrac{\sqrt{x}+\sqrt[3]{x}}{x^{2/3}} = \lim_{x\to\infty} \dfrac{\frac{1}{x^{1/6}}+\frac{1}{x^{1/3}}}{1} = \dfrac{\left(\lim_{x\to\infty}\frac{1}{x^{1/6}}\right)+\left(\lim_{x\to\infty}\frac{1}{x^{1/3}}\right)}{1} = \dfrac{0+0}{1} = 0.$ ■

Infinite Limits

You have undoubtedly noticed by now that many functions do not have limits as $x \to \pm\infty$. For example, in the case of the parabola $y = x^2$, neither $\lim_{x\to\infty} x^2$, nor $\lim_{x\to-\infty} x^2$ exist in the sense of Definition 3.

However, in this case we can write

$$\lim_{x\to\infty} x^2 = \infty \tag{1}$$

and

$$\lim_{x\to-\infty} x^2 = \infty, \tag{2}$$

which means that the values of $f(x) = x^2$ *increase without bound* as $x \to \infty$ or as $x \to -\infty$ (Figure 6.7).

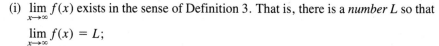

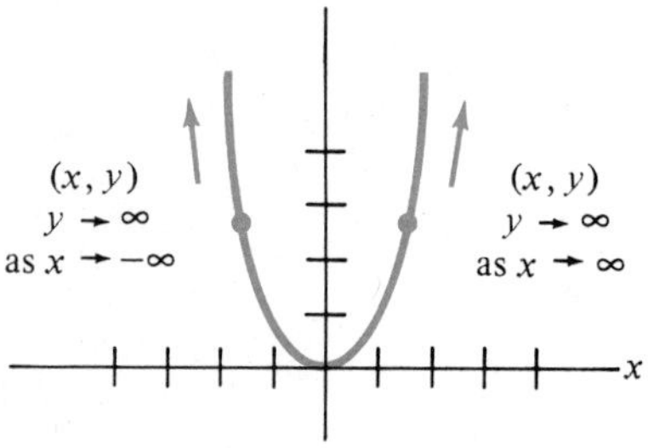

Figure 6.7 For $f(x) = x^2$, both $\lim_{x\to\infty} f(x) = \infty$ and $\lim_{x\to-\infty} f(x) = \infty$.

Since the symbol ∞ is not a number, but rather a concept, equations (1) and (2) do *not* say that the corresponding limits exist. Rather, they give us information about why the limit fails to exist: $f(x) = x^2$ becomes infinitely large as $x \to \pm\infty$.

This situation is to be contrasted with the situation which occurs for the function $f(x) = \sin x$ (Figure 6.4). In this case $\lim_{x\to\infty} f(x)$ fails to exist, but we cannot write $\lim_{x\to\infty} \sin x = \infty$ since the values of $f(x) = \sin x$ do not increase without bound. Rather, they simply fail to *converge* to a single number L as $x \to \infty$.

We summarize all this by saying that for the expression $\lim_{x\to\infty} f(x)$, one of three situations occurs:

(i) $\lim_{x\to\infty} f(x)$ exists in the sense of Definition 3. That is, there is a *number* L so that $\lim_{x\to\infty} f(x) = L$;

(ii) $\lim_{x\to\infty} f(x)$ fails to exist because $\lim_{x\to\infty} f(x) = \infty$ or $\lim_{x\to\infty} f(x) = -\infty$ (such as for $f(x) = x^2$);

(iii) $\lim_{x\to\infty} f(x)$ fails to exist, and neither $\lim_{x\to\infty} f(x) = \infty$ nor $\lim_{x\to\infty} f(x) = -\infty$ (such as for $f(x) = \sin x$).

The obvious corresponding statements hold for $\lim_{x\to-\infty} f(x)$.

Example 4 Find $\lim_{x\to\infty} \dfrac{4x^3 + 3x + 3}{2x^2 + x}$.

Strategy

Use technique of Example 3—divide both numerator and denominator by x^3.

Solution

$$\lim_{x\to\infty} \frac{4x^3 + 3x + 3}{2x^2 + x} = \lim_{x\to\infty} \left[\frac{4 + \dfrac{3}{x^2} + \dfrac{3}{x^3}}{\dfrac{2}{x} + \dfrac{1}{x^2}} \right].$$

Examine limits of numerator and denominator separately, using Theorems 8 and 9.

In the numerator, we have

$$\lim_{x\to\infty} \left(4 + \frac{3}{x^2} + \frac{3}{x^3}\right) = 4 + 0 + 0 = 4$$

while in the denominator,

$$\lim_{x\to\infty}\left(\frac{2}{x}+\frac{1}{x^2}\right)=0+0=0.$$

Since the denominator "shrinks" to zero while the numerator approaches 4, the quotient "blows up," i.e., increases without bound. Thus

$$\lim_{x\to\infty}\frac{4x^3+3x+3}{2x^2+x}=+\infty. \qquad \blacksquare$$

Vertical Asymptotes

The three situations summarized above for $\lim_{x\to\infty} f(x)$ can also occur for $\lim_{x\to a} f(x)$, $\lim_{x\to a^+} f(x)$, or $\lim_{x\to a^-} f(x)$, that is, for one- or two-sided limits at $x = a$. For example,

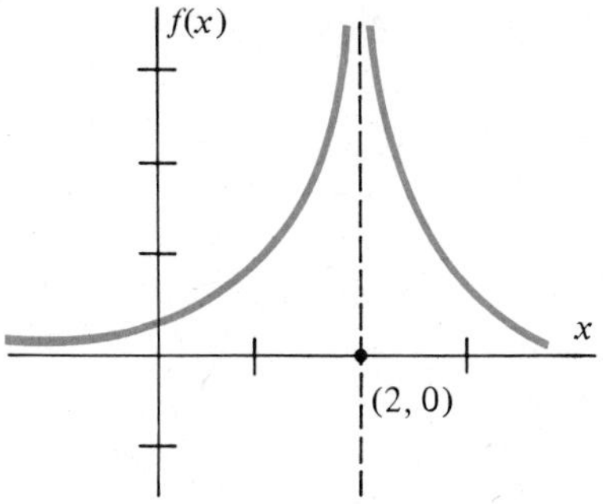

Figure 6.8 $\lim_{x\to 2}\frac{1}{(x-2)^2}=\infty.$

(a) for $f(x)=\sqrt{x}$, $\quad \lim_{x\to 2} f(x)=\sqrt{2}$ (Case (i)),

(b) for $f(x)=\dfrac{1}{(x-2)^2}$, $\quad \lim_{x\to 2} f(x)=+\infty$, $\quad$ since $(x-2)^2\to 0$ as $x\to 2$ (Figure 6.8, case (ii)),

(c) for $f(x)=\sin\left(\dfrac{1}{x}\right)$, $\lim_{x\to 0^+} f(x)$ does not exist since $\sin\left(\dfrac{1}{x}\right)$ oscillates between 1 and -1 as $x\to 0^+$ (Figure 3.7, Chapter 2).

We formalize statements like $\lim_{x\to 2} f(x)=+\infty$ in (b) above with the following working definition.

DEFINITION 4

The statement $\lim_{x\to a} f(x)=\infty$ means that the values $f(x)$ increase without bound as x approaches a from either direction. The statement $\lim_{x\to a^+} f(x)=\infty$ means that the values of $f(x)$ increase without bound as x approaches a from the right. Similar meanings are attached to the statements

$$\lim_{x\to a^-} f(x)=\infty,\qquad \lim_{x\to a} f(x)=-\infty,\qquad \lim_{x\to a^+} f(x)=-\infty,\qquad \text{and}$$
$$\lim_{x\to a^-} f(x)=-\infty.$$

Evaluating infinite limits at $x = a$ will require a fair amount of intuition on your part. The following theorem formalizes the central idea.

THEOREM 10

Let a be any constant and let $r=\dfrac{p}{q}$ be a positive rational number. Assume that p and q have no common factors except 1. Then

(i) If p is even, $\quad \lim_{x\to a}\dfrac{1}{(x-a)^r}=\infty.$

(ii) If p is odd, $\quad \lim_{x\to a^+}\dfrac{1}{(x-a)^r}=\infty,$

and, for q odd $\quad \lim_{x\to a^-}\dfrac{1}{(x-a)^r}=-\infty.$

The function $f(x) = \dfrac{1}{(x-2)^2}$ is an example of case (i) of Theorem 10. Since the exponent 2 is even, the denominator $(x-2)^2$ is small *and positive* when $x \neq 2$ is close to 2, regardless of whether $x > 2$ or $x < 2$. Since $\lim_{x\to 2} (x-2)^2 = 0$, as $x \to 2$ from either direction the denominator approaches zero *through positive values*. Thus, the quotient becomes infinitely large and positive as $x \to 2$. That is, $\lim_{x\to 2} \dfrac{1}{(x-2)^2} = \infty$ (Figure 6.8).

However, the situation for $g(x) = \dfrac{1}{(x-2)^3}$ is different. If $x - 2$ is positive, so is $(x-2)^3$. But if $(x-2)$ is negative, $(x-2)^3$ is also. Thus,

(a) if x approaches 2 from the right, $(x-2)^3$ approaches zero through positive values, so

$$\lim_{x\to 2^+} \frac{1}{(x-2)^3} = \infty;$$

(b) however, if x approaches 2 from the left, $(x-2)^3$ is *negative,* so $(x-2)^3$ approaches zero through negative values. Thus

$$\lim_{x\to 2^-} \frac{1}{(x-2)^3} = -\infty$$

(Figure 6.9).

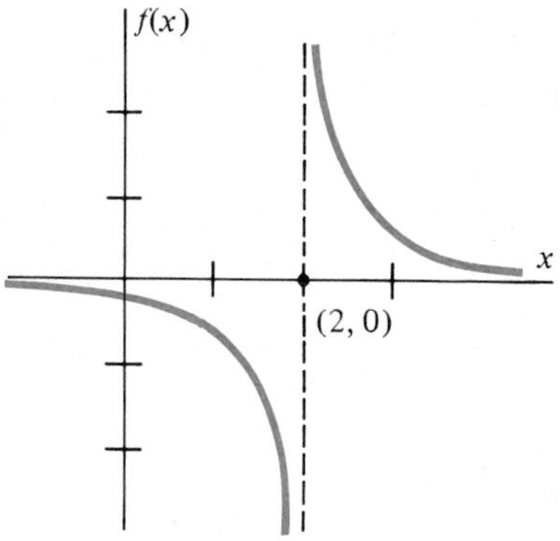

Figure 6.9 $\lim_{x\to 2^+} \dfrac{1}{(x-2)^3} = \infty$; $\lim_{x\to 2^-} \dfrac{1}{(x-2)^3} = -\infty$.

Rather than rely too much on the formal statement of Theorem 10, you should use this type of reasoning when evaluating a limit involving a denominator that is approaching zero.

Notice that in both Figures 6.8 and 6.9 the points $(x, f(x))$ on the graph of $f(x)$ approach the vertical line $x = 2$ as x approaches 2. For this reason we refer to the vertical line $x = 2$ as a *vertical asymptote,* both for the graph of $f(x) = \dfrac{1}{(x-2)^2}$ and for the graph of $g(x) = \dfrac{1}{(x-2)^3}$.

More generally, the vertical line $x = a$ is a **vertical asymptote** for the graph of $f(x)$ if one or more of the limits $\lim_{x\to a} f(x)$, $\lim_{x\to a^+} f(x)$, or $\lim_{x\to a^-} f(x)$ is infinite.

The technique for finding vertical asymptotes is rooted in Theorem 10: look for numbers for which a factor in the denominator of $f(x)$ becomes zero.

Example 5 Find all vertical asymptotes for the graph of $f(x) = \dfrac{x^2}{4 - x^2}$ and determine the corresponding limits.

Solution: The denominator of $f(x)$ can be factored as follows:

$$f(x) = \frac{x^2}{4 - x^2} = \frac{x^2}{(2-x)(2+x)}.$$

The denominator therefore equals zero for $x = -2$ and $x = 2$. These numbers are candidates for vertical asymptotes. To find the four corresponding one-sided limits, we must determine the signs of each of the factors of $f(x)$. This information is easily obtained by use of a marked number line such as Figure 6.10.

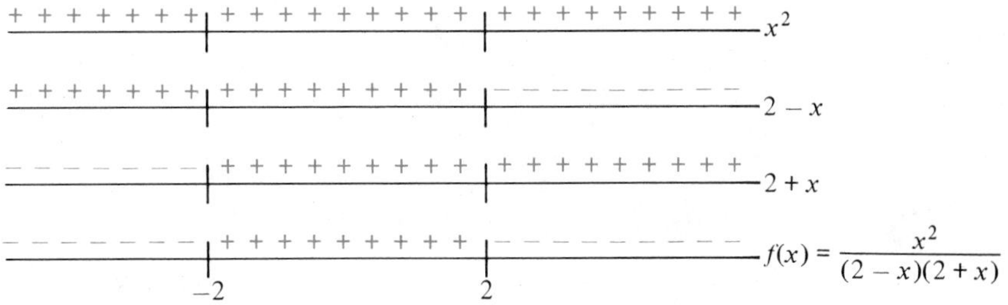

Figure 6.10 Sign analysis for $f(x)$.

Since the numerator of $f(x) = \dfrac{x^2}{4 - x^2}$ approaches 4 and the denominator approaches zero as $x \to \pm 2$, the one-sided limits at $x = \pm 2$ are all infinite. Figure 6.10 allows us to determine whether each limit is infinitely large and positive or infinitely large and negative. We have

$$\lim_{x \to -2^-} \frac{x^2}{4 - x^2} = -\infty,$$

$$\lim_{x \to -2^+} \frac{x^2}{4 - x^2} = \infty,$$

$$\lim_{x \to 2^-} \frac{x^2}{4 - x^2} = \infty, \quad \text{and}$$

$$\lim_{x \to 2^+} \frac{x^2}{4 - x^2} = -\infty.$$

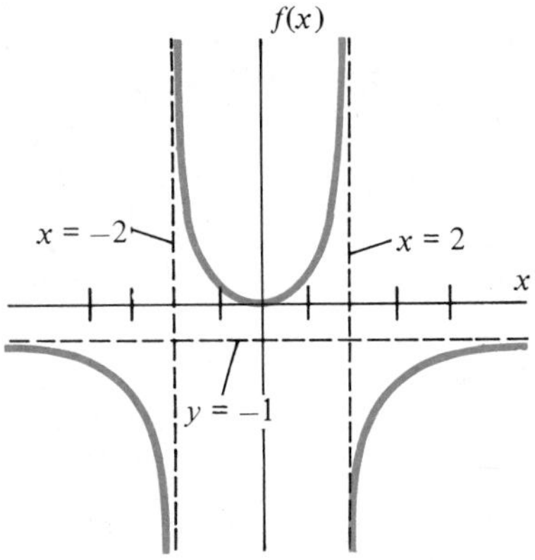

Figure 6.11 $f(x) = \dfrac{x^2}{5 - x^2}$.

The vertical asymptotes are, therefore, $x = -2$ and $x = 2$. (See Figure 6.11. Note that the graph has a *horizontal* asymptote $y = -1$.) ■

Exercise Set 5.6

In Problems 1–26, find the indicated limits.

1. $\displaystyle \lim_{x \to \infty} \frac{3x^2 + 2}{10x^2 - 3x}$

2. $\displaystyle \lim_{x \to \infty} \frac{6x^4 - 8x}{7 - 3x^4}$

3. $\displaystyle \lim_{x \to \infty} \frac{x(4 - x^3)}{3x^4 + 2x^2}$

4. $\displaystyle \lim_{x \to \infty} \frac{(x - 3)(x + 4)}{2x^2 + 2}$

5. $\displaystyle \lim_{x \to -\infty} \frac{x^2 - 7}{3 + 4x^2}$

6. $\displaystyle \lim_{x \to -\infty} \frac{2x + 6}{x^2 + 1}$

7. $\displaystyle \lim_{x \to \infty} \frac{3x^2 + 7x}{1 - x^4}$

8. $\displaystyle \lim_{x \to \infty} \frac{x^4 - 4}{x^3 + 7x^2}$

9. $\displaystyle \lim_{x \to \infty} \frac{\sin x}{x}$

10. $\displaystyle \lim_{x \to \infty} \frac{\cos(2x)}{1 - x^2}$

11. $\displaystyle \lim_{x \to -\infty} \frac{x^4 - 3\sin x}{3x + 5x^5}$

12. $\displaystyle \lim_{x \to \infty} \frac{x^2|2 + x^2|}{4 - x^4}$

13. $\displaystyle \lim_{x \to \infty} \frac{\sqrt{x - 1}}{x^2}$

14. $\displaystyle \lim_{x \to \infty} \frac{x^{2/3} + x^{4/3}}{x^2}$

15. $\displaystyle \lim_{x \to \infty} = \frac{x^{2/3} + x}{1 + x^{3/4}}$

16. $\displaystyle \lim_{x \to \infty} \frac{\sqrt{x} + 7}{1 - \sqrt[3]{x}}$

17. $\displaystyle \lim_{x \to 2^+} \frac{1}{x - 2}$

18. $\displaystyle \lim_{x \to -2^-} \frac{1}{(x + 2)^2}$

19. $\displaystyle \lim_{x \to \pi/2^-} \tan x$

20. $\displaystyle \lim_{x \to \pi/2^+} \tan x$

21. $\displaystyle \lim_{x \to 0^-} \frac{|x|}{x}$

22. $\displaystyle \lim_{x \to 0^+} \frac{|x|}{x}$

23. $\lim_{x\to 2^+} \dfrac{x^2+1}{x^2-4}$

24. $\lim_{x\to 2^-} \dfrac{x^2+1}{x^2-4}$

25. $\lim_{x\to 3^-} \dfrac{2}{x^2-9}$

26. $\lim_{x\to 0} x + \dfrac{1}{x}$

In Exercises 27–34, find the horizontal asymptotes for the given functions.

27. $y = \dfrac{1}{1+x}$

28. $y = \dfrac{1+x}{3-x}$

29. $y = \dfrac{2x^2}{x^2+1}$

30. $y = \dfrac{x^2}{1-x}$

31. $y = \dfrac{2x^2}{(x^2+1)^2}$

32. $y = \dfrac{x^2+3}{1-3x^2}$

33. $y = 3 + \dfrac{\sin x}{x}$

34. $y = \dfrac{4x - \sqrt{x}}{x^{2/3}+x}$

35. The function $y = \dfrac{ax+7}{4-x}$ has a horizontal asymptote of $y = 3$. Find a.

Find all vertical asymptotes for the function in

36. Exercise 17.

37. Exercise 18.

38. Exercise 23.

39. Exercise 24.

40. Exercise 25.

41. Exercise 26.

42. The function $y = \dfrac{x^r + 3x}{7 - x^{4/3}}$ has a horizontal asymptote of $y = -1$. Find r.

43. The function $y = \dfrac{\pi + ax^r}{1 - 3x^{2/3}}$ has a horizontal asymptote of $y = -2$. Find a and r.

44. The function $y = \dfrac{3x^r + 2x^3}{rx^3}$ has a horizontal asymptote of $y = 5/3$. Find r.

45. Show that $\lim_{x\to\infty} \dfrac{5x^3+3x^2-x+4}{x^3-x+1} = 5.$

46. Show that $\lim_{x\to\infty} \dfrac{x^3-6x^2+2x-1}{4x^2+4x+4} = +\infty.$

47. Show that $\lim_{x\to\infty} \dfrac{x^4+4x^3-x+7}{x^5+2x+2} = 0.$

48. Show that $\lim_{x\to\infty} \dfrac{a_nx^n + a_{n-1}x^{n-1} + \cdots + a_1x + a_0}{b_mx^m + b_{m-1}x^{m-1} + \cdots + b_1x + b_0}$ equals

a. 0 if $n < m$,
b. ∞ if $n > m$,
c. a_n/b_m if $n = m$.

49. A formal way of saying $\lim_{x\to\infty} f(x) = L$ is the following: "Given $\epsilon > 0$ there is an integer N so that $|f(x) - L| < \epsilon$ whenever $x > N$." Use this definition to prove that

a. $\lim_{x\to\infty} \dfrac{1}{x} = 0,$

b. $\lim_{x\to\infty} \dfrac{x}{2x^2+2} = 0,$

c. $\lim_{x\to\infty} \dfrac{x^2+6}{2x^2+5} = \dfrac{1}{2}.$

50. State a definition analogous to that of Exercise 49 for $\lim_{x\to-\infty} f(x) = L$ and use it to prove that:

a. $\lim_{x\to-\infty} \dfrac{1}{x} = 0$

b. $\lim_{x\to-\infty} \dfrac{x}{2x+1} = \dfrac{1}{2}$

c. $\lim_{x\to-\infty} \dfrac{6x-2}{x+1} = 6$

51. The formal way to say that $\lim_{x\to a} f(x) = \infty$ is: "Given $N > 0$ there exists a number $\delta > 0$ so that $f(x) > N$ whenever $|x - a| < \delta$." Use this definition to prove that

a. $\lim_{x\to 0} \dfrac{1}{|x|} = \infty,$

b. $\lim_{x\to 0} \dfrac{1}{x^2} = \infty,$

c. $\lim_{x\to \pi/2} |\tan x| = \infty.$

52. State a formal definition for the statement $\lim_{x\to a^-} f(x) = \infty$ and use it to prove that $\lim_{x\to \pi/2^-} \tan x = \infty.$

53. State a formal definition for the statement $\lim_{x\to a^-} f(x) = -\infty$ and use it to prove that $\lim_{x\to 0^-} \dfrac{1}{x} = -\infty.$

54. Use the definition in Exercise 49 to prove Theorem 8.

5.7 APPLICATIONS TO CURVE SKETCHING

Sections 3–6 of this chapter have addressed ways in which information about limits and derivatives can be used to identify certain properties of the graph of a function. Our purpose here is to summarize these and other ideas that one should utilize when sketching the graph of a given function. The goal in mastering these techniques is to become able to quickly determine a function's behavior without having to develop a precise graph.

To plot the graph of a given function $y = f(x)$:

(1) Determine the domain and, if possible, the range.
(2) If possible, locate the zeros by solving the equation $f(x) = 0$.
(3) Find all vertical asymptotes by finding the values a for which $\lim_{x\to a} f(x)$, $\lim_{x\to a^+} f(x)$, or $\lim_{x\to a^-} f(x)$ is infinite (Section 5.6).
(4) Locate all horizontal asymptotes by examining $\lim_{x\to\pm\infty} f(x)$ (Section 5.6).
(5) Find all critical numbers, classify the extrema, and determine whether $f(x)$ is increasing or decreasing on the resulting intervals (Sections 5.3 and 5.4).
(6) Determine the concavity and locate the inflection points (Section 5.5).
(7) Calculate the values of the function at a few convenient numbers and locate the corresponding points on the graph. Then sketch in the graph according to the above information.

Example 1 Sketch the graph of $f(x) = \dfrac{x}{1 - x^2}$.

Solution: Following the outline stated above, we find that:

(1) The *domain* of $f(x)$ is all numbers except $x = 1$ and $x = -1$. That is because the zeros of the denominator are $x = \pm 1$. The *range* is all real numbers. To see this note that $f(0) = 0$, but that $f(x)$ becomes infinitely large and positive as x increases toward 1. Similarly, as x decreases from 0 toward -1, $f(x)$ becomes infinitely large and negative.
(2) The equation $f(x) = 0$ has the single solution $x = 0$. Thus, $x = 0$ is the only zero for $f(x)$.
(3) Since the zeros of the denominator are $x = -1$ and $x = 1$, we check for vertical asymptotes at $x = \pm 1$. Since

$$\lim_{x\to -1^-}\left(\frac{x}{1-x^2}\right) = \infty, \quad \lim_{x\to -1^+}\left(\frac{x}{1-x^2}\right) = -\infty,$$

$$\lim_{x\to 1^-}\left(\frac{x}{1-x^2}\right) = \infty, \quad \text{and} \quad \lim_{x\to 1^+}\left(\frac{x}{1-x^2}\right) = -\infty,$$

both $x = -1$ and $x = 1$ are vertical asymptotes.
(4) The line $y = 0$ (i.e., the x-axis) is a horizontal asymptote, since

$$\lim_{x\to\infty}\left(\frac{x}{1-x^2}\right) = \lim_{x\to\infty}\left(\frac{\dfrac{1}{x}}{\dfrac{1}{x^2} - 1}\right) = \frac{0}{0-1} = 0, \quad \text{and}$$

$$\lim_{x\to -\infty}\left(\frac{x}{1-x^2}\right) = \lim_{x\to -\infty}\left(\frac{\dfrac{1}{x}}{\dfrac{1}{x^2} - 1}\right) = \frac{0}{0-1} = 0.$$

(5) To find the critical numbers we use the equation

$$f'(x) = \frac{d}{dx}\left(\frac{x}{1 - x^2}\right) = \frac{(1)(1 - x^2) - (x)(-2x)}{(1 - x^2)^2} = \frac{1 + x^2}{(1 - x^2)^2} = 0.$$

However, this equation has no solutions, since $1 + x^2 \neq 0$ for all x. Although $f'(x)$ is undefined for $x = \pm 1$, these numbers are not in the domain of $f(x)$. There are, therefore, no critical numbers; hence, no relative extrema. Since $1 + x^2 \geq 1$ for all x and $(1 - x^2)^2 > 0$ for $x \neq \pm 1$, we can see that $f'(x) > 0$ if $x \neq \pm 1$. Thus, $f(x)$ is increasing on all intervals on which $f(x)$ is defined.

(6) Since

$$f''(x) = \frac{d}{dx}\left[\frac{1 + x^2}{(1 - x^2)^2}\right] = \frac{2x(1 - x^2)^2 - (1 + x^2)(2)(1 - x^2)(-2x)}{(1 - x^2)^4}$$

$$= \frac{2x^3 + 6x}{(1 - x^2)^3}$$

$$= \frac{2x(x^2 + 3)}{(1 - x^2)^3},$$

the candidates for inflection points are $x = 0$ ($f''(0) = 0$) and $x = \pm 1$ ($f''(\pm 1)$ undefined). The sign of $f''(x)$ on each of the resulting intervals can be obtained from Figure 7.1, which records the sign of each of the factors $2x$, $x^2 + 3$, and $(1 - x^2)^3$.

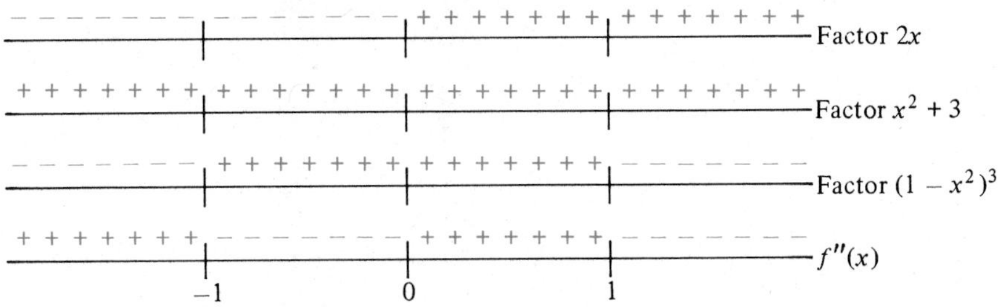

Figure 7.1 Sign analysis for $f''(x) = \dfrac{2x(x^2 + 3)}{(1 - x^2)^3}$.

From the sign of $f''(x)$ we conclude that the graph of $y = f(x)$ is

concave up on $(-\infty, -1)$,
concave down on $(-1, 0)$,
concave up on $(0, 1)$, and
concave down on $(1, \infty)$.

The point $(0, f(0)) = (0, 0)$ is an inflection point.

(7) Since $f(-2) = 2/3$, $f(0) = 0$, and $f(2) = -2/3$, the points $(-2, 2/3)$, $(0, 0)$, and $(2, -2/3)$ are on the graph. The graph is sketched in Figure 7.2. ■

Example 2 Sketch the graph of $f(x) = \dfrac{\sin x}{2 + \cos x}$ for $-\pi \leq x \leq \pi$.

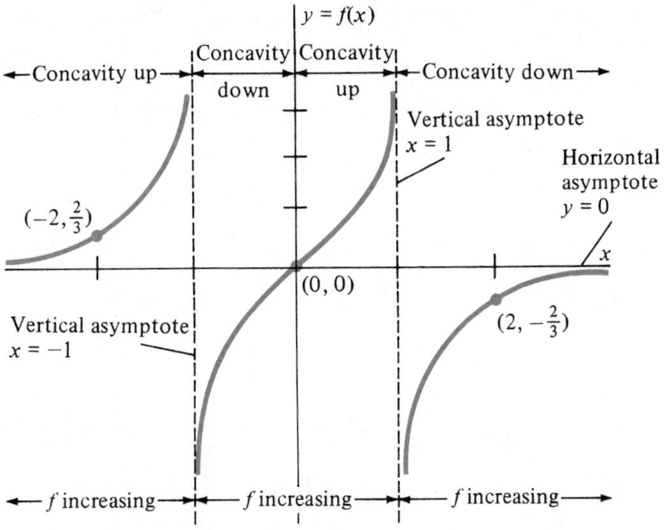

Figure 7.2 Graph of $f(x) = \dfrac{x}{1 - x^2}$.

Solution: Proceeding as in Example 1, we find that

(1) The domain of $f(x)$ is explicitly stated as the interval $[-\pi, \pi]$. Since $|\sin x| \le 1$ and $|\cos x| \le 1$ for all x, the denominator can never be smaller than $2 + (-1) = 1$, so

$$|f(x)| = \left|\frac{\sin x}{2 + \cos x}\right| \le |\sin x| \le 1.$$

Thus, the range must lie within the interval $[-1, 1]$.

(2) The only solution of the equation $f(x) = \dfrac{\sin x}{2 + \cos x} = 0$ in $[-\pi, \pi]$ occurs when $\sin x = 0$, that is, at $x = -\pi$, 0 or π. These are the zeros for $f(x)$.

(3) Since $f(x)$ is defined for all x, there are no vertical asymptotes.

(4) Since the domain of $f(x)$ is restricted to a finite interval there can be no horizontal asymptotes.

(5) The first derivative is

$$f'(x) = \frac{d}{dx}\left(\frac{\sin x}{2 + \cos x}\right) = \frac{(2 + \cos x)(\cos x) - (\sin x)(-\sin x)}{(2 + \cos x)^2}$$

$$= \frac{2\cos x + 1}{(2 + \cos x)^2}.$$

The equation $f'(x) = 0$ implies $2\cos x + 1 = 0$, or $\cos x = -1/2$. The solutions of this equation in $[-\pi, \pi]$ are $x = \pm\dfrac{2\pi}{3}$. Since $f'(x)$ is defined for all x, the critical numbers are $x = \pm\dfrac{2\pi}{3}$.

By checking the sign of $f'(x)$ at one "test number" t in each of the resulting intervals, we obtain the information in Table 7.1.

Table 7.1 Analysis of sign of $f'(x)$

Interval I	Test number $t \in I$	Value of $f'(t)$	Sign of $f'(t)$
$\left[-\pi, -\frac{2\pi}{3}\right)$	$t = -\frac{5\pi}{6}$	$\dfrac{-\sqrt{3}+1}{\left(2-\frac{\sqrt{3}}{2}\right)^2}$	$-$
$\left(-\frac{2\pi}{3}, \frac{2\pi}{3}\right)$	$t = 0$	$\dfrac{2+1}{(2+1)^2}$	$+$
$\left(\frac{2\pi}{3}, \pi\right)$	$t = \frac{5\pi}{6}$	$\dfrac{-\sqrt{3}+1}{\left(2+\frac{\sqrt{3}}{2}\right)^2}$	$-$

We conclude that $f(x)$ is

$$\text{decreasing on } \left[-\pi, -\frac{2\pi}{3}\right),$$

$$\text{increasing on } \left(-\frac{2\pi}{3}, \frac{2\pi}{3}\right), \text{ and}$$

$$\text{decreasing on } \left(\frac{2\pi}{3}, \pi\right).$$

Consequently, $f\left(-\frac{2\pi}{3}\right) = \frac{\sqrt{3}}{3} \approx -0.58$ is a relative minimum and $f\left(\frac{2\pi}{3}\right) = \frac{\sqrt{3}}{3}$ is a relative maximum.

(6) $$f''(x) = \frac{(2+\cos x)^2(-2\sin x) - (2\cos x + 1)\cdot 2(2+\cos x)(-\sin x)}{(2+\cos x)^4}$$

$$= \frac{2\sin x(\cos x - 1)}{(2+\cos x)^3}$$

so $f''(x) = 0$ implies $2\sin x(\cos x - 1) = 0$. The solutions of this equation are $x = 0$ and the endpoints $x = -\pi$ and $x = \pi$. Since $\cos x - 1 \le 0$ and $2 + \cos x \ge 0$ for all x, the sign of $f''(x)$ will be the opposite of the sign of the factor $\sin x$. Thus, $f''(x) > 0$ on $(-\pi, 0)$ and $f''(x) < 0$ on $(0, \pi)$. The graph is therefore concave up on $(-\pi, 0)$ and concave down on $(0, \pi)$, and the point $(0, 0)$ is an inflection point.

(7) The graph appears in Figure 7.3. ■

Example 3 Sketch the graph of $f(x) = \dfrac{\sqrt[3]{x}}{1-x}$.

Solution:

(1) The domain of $f(x)$ is all $x \neq 1$. Since $f(0) = 0$ and $f(x)$ becomes infinite and positive as x increases to 1, the range includes the interval $[0, \infty)$. Moreover, as x *decreases* toward 1, $f(x)$ becomes large and negative. We therefore expect the range to include all numbers.

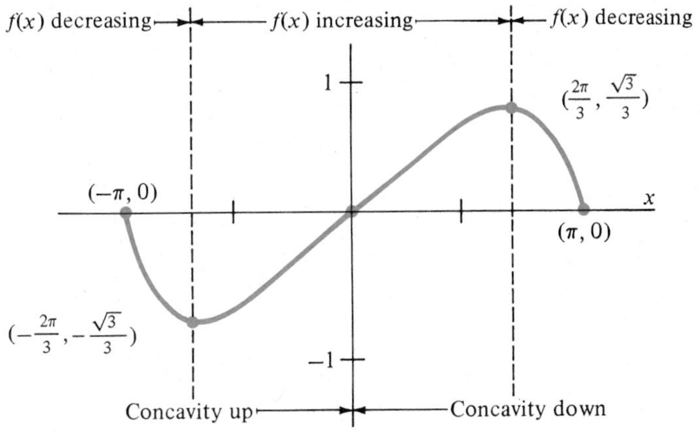

Figure 7.3 Graph of $f(x) = \dfrac{\sin x}{2 + \cos x}$.

(2) The equation $f(x) = 0$ has the single solution $x = 0$, so $(0, 0)$ is the only x-intercept.

(3) Since the denominator $1 - x$ equals zero for $x = 1$, we inspect the limits at $x = 1$ to check for vertical asymptotes. Since, as x approaches 1 from the right, the number $1 - x$ approaches zero *through negative values* while $\sqrt[3]{x}$ approaches 1, we have

$$\lim_{x \to 1^+} \frac{\sqrt[3]{x}}{1 - x} = -\infty.$$

Similarly, since $1 - x$ approaches zero *through positive* values as x approaches 1 from the left, we have

$$\lim_{x \to 1^-} \frac{\sqrt[3]{x}}{1 - x} = \infty.$$

Thus, $x = 1$ is a vertical asymptote.

(4) Since

$$\lim_{x \to \infty} \frac{\sqrt[3]{x}}{1 - x} = \lim_{x \to \infty} \frac{\left(\dfrac{1}{x^{2/3}}\right)}{\left(\dfrac{1}{x}\right) - 1} = \frac{0}{0 - 1} = 0 = \lim_{x \to -\infty} \frac{\sqrt[3]{x}}{1 - x},$$

the line $y = 0$ (that is, the x-axis) is a horizontal asymptote.

(5) The first derivative is

$$f'(x) = \frac{d}{dx}\left(\frac{x^{1/3}}{1 - x}\right) = \frac{\frac{1}{3}x^{-2/3}(1 - x) - (-1)x^{1/3}}{(1 - x)^2}$$

$$= \frac{1 + 2x}{3x^{2/3}(1 - x)^2}.$$

Thus, $f'(x) = 0$ when $x = -1/2$, and $f'(x)$ is undefined for $x = 0$ and $x = 1$. Since $x = 1$ is not in the domain of $f(x)$, the critical numbers are $x = -1/2$ and

$x = 0$. (However, we must still check for a sign change in $f'(x)$ at $x = 1$, since $f'(x)$ is discontinuous at $x = 1$.) The sign of $f'(x)$ is determined by the signs of its factors $1 + 2x$, $x^{2/3}$, and $(1 - x)^2$, as recorded in Figure 7.4.

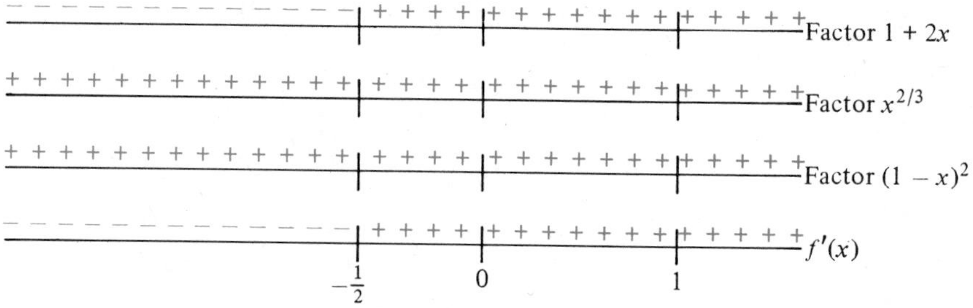

Figure 7.4 Sign analysis for $f'(x) = \dfrac{1 + 2x}{3x^{2/3}(1 - x)^2}$.

Thus, $f(x)$ is

$$\text{decreasing on } \left(-\infty, -\frac{1}{2}\right), \text{ and}$$

$$\text{increasing on } \left(-\frac{1}{2}, 0\right), (0, 1) \text{ and } (1, \infty).$$

The value $f\left(-\dfrac{1}{2}\right) = -\dfrac{2}{3\sqrt[3]{2}}$ is therefore a relative minimum. There are no other relative extrema.

(6) We leave it as a (messy) exercise for you to show that

$$\begin{aligned} f''(x) &= \frac{d}{dx}\left(\frac{1 + 2x}{3x^{2/3}(1 - x)^2}\right) \\ &= \frac{2(5x^2 + 5x - 1)}{9x^{5/3}(1 - x)^3}. \end{aligned}$$

To find the solutions of the equation $f''(x) = 0$, we must use the quadratic formula to solve $5x^2 + 5x - 1 = 0$. We obtain

$$x = \frac{-5 \pm \sqrt{(5)^2 - 4(5)(-1)}}{2(5)} = -\frac{1}{2} \pm \frac{3}{10}\sqrt{5}.$$

Since $f''(x)$ is undefined for $x = 0$ and $f''(x)$ is discontinuous at $x = 1$, to determine concavity we must check the sign of $f''(x)$ on the intervals determined by the numbers $x = -\dfrac{1}{2} - \dfrac{3\sqrt{5}}{10} \approx -1.17$, $x = 0$, $x = -\dfrac{1}{2} + \dfrac{3\sqrt{5}}{10} \approx 0.17$, and $x = 1$. The results of doing so appear in Figure 7.5.

Thus, the graph of $f(x)$ is

concave down on $(-\infty, -1.17)$,
concave up on $(-1.17, 0)$,
concave down on $(0, 0.17)$,

concave up on (0.17, 1), and
concave down on $(1, \infty)$.

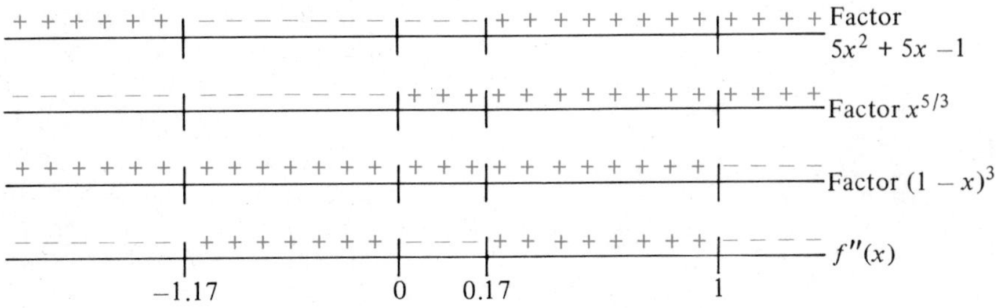

Figure 7.5 Sign analysis for $f''(x) = \dfrac{10x^2 + 10x - 2}{9x^{5/3}(1 - x)^3}$.

(7) The points (−1.17, −0.49), (0, 0), and (0.17, 0.67) are inflection points. The graph appears in Figure 7.6.

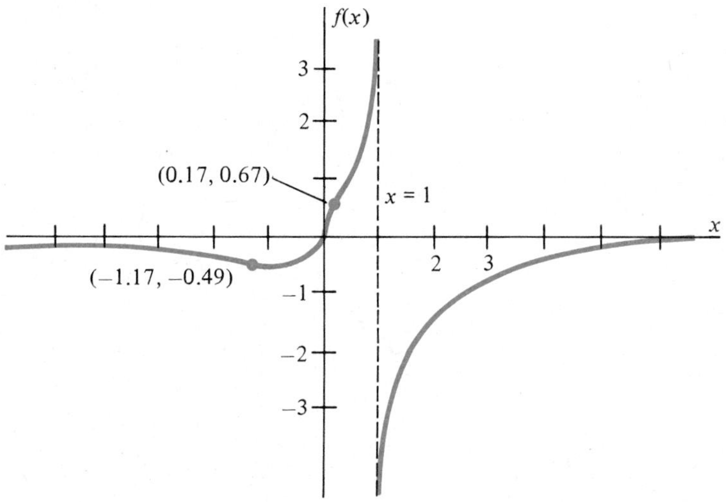

Figure 7.6 Graph of $f(x) = \dfrac{x^{1/3}}{1 - x}$.

Exercise Set 5.7

For each of Exercises 1–40, sketch the graph of $f(x)$ according to the points described in this section.

1. $f(x) = x^2 - 2x - 8$
2. $f(x) = x^2 - 9x$
3. $f(x) = 2x^3 - 3x^2$
4. $f(x) = 3x^4 - 4x^3$
5. $f(x) = x^3 + x^2 - 8x + 8$
6. $f(x) = x^3 + 2x^2 - x - 2$
7. $f(x) = x^3 - 7x + 6$
8. $f(x) = \dfrac{x - 1}{x + 1}$
9. $f(x) = \dfrac{x + 4}{x - 4}$
10. $f(x) = \cos(2x - 1)$
11. $f(x) = 9x - x^{-1}$
12. $f(x) = x + 3x^{2/3}$
13. $f(x) = \tan^2 x + 1$
14. $f(x) = \sqrt{x + 4}$
15. $f(x) = |4 - x^2|$
16. $f(x) = x^2 + \dfrac{2}{x}$
17. $f(x) = \sin x \cos x$
18. $f(x) = (x + 1)^{5/3} - (x + 1)^{2/3}$
19. $f(x) = \dfrac{1}{x(x - 4)}$
20. $f(x) = x^2 - \dfrac{9}{x^2}$

21. $f(x) = \dfrac{x}{(2x + 1)^2}$

22. $f(x) = \sqrt{2x - x^2}$

23. $f(x) = (x - 3)^{2/3} + 1$

24. $f(x) = \sqrt{6x - x^2 - 8}$

25. $f(x) = \dfrac{1}{3}x^3 - x^2 - 3x + 4$

26. $f(x) = \dfrac{x^2 + 1}{x}$

27. $f(x) = \dfrac{x^2 - 4x + 5}{x - 2}$

28. $f(x) = \dfrac{3}{2}x^{2/3} - x$

29. $f(x) = 4x^2(1 - x^2)$

30. $f(x) = x^4 - 3x^2 + 2$

31. $f(x) = x^{2/3} - \dfrac{1}{5}x^{5/3}$

32. $f(x) = \dfrac{2x + 1}{x + 2}$

33. $f(x) = \dfrac{x^2}{9 - x^2}$

34. $f(x) = \dfrac{x^2}{x^2 - 16}$

35. $f(x) = 5x^3 - x^5$

36. $f(x) = x^4 - 2x^2 + 1$

37. $f(x) = 16 - 20x^3 + 3x^5$

38. $f(x) = x^4 - 18x^2 + 32$

39. $f(x) = \dfrac{1 - x^2}{x^3}$

40. $f(x) = 3x^{5/3} - 15x^{2/3} + 5$

5.8 INDETERMINATE FORMS: l'HÔPITAL'S RULES

On several occasions we have encountered the indeterminate form 0/0 in attempting to evaluate a limit of the form

$$\lim_{x \to a} \frac{f(x)}{g(x)}. \tag{1}$$

Specifically, this difficulty arises in limit (1) when $f(a) = g(a) = 0$. For example, recall

$$\lim_{x \to 0} \frac{\sin x}{x}, \qquad \left(\frac{0}{0} \text{ form}\right)$$

which required an application of the Pinching Theorem to resolve. More generally, we have seen that this difficulty will always arise in applying the definition of the derivative:

$$f'(x) = \lim_{\Delta x \to 0} \frac{f(x + \Delta x) - f(x)}{\Delta x} \qquad \left(\frac{0}{0} \text{ form}\right).$$

The theory of the derivative may be applied to establish the following simple procedure for dealing with many limits of the type (1) that yield the indeterminate forms 0/0 or ∞/∞. Although the result is credited to the Frenchman, G.F.A. l'Hôpital (1661–1704), it was actually discovered by his teacher, Johann Bernoulli (1667–1748).

THEOREM 11
l'Hôpital's Rule

Let $f(x)$ and $g(x)$ be differentiable on an interval containing $x = a$, except possibly at $x = a$, and suppose that $f(a) = g(a) = 0$. If $g'(x) \neq 0$ for each $x \neq a$ in this interval,

$$\lim_{x \to a} \frac{f(x)}{g(x)} = \lim_{x \to a} \frac{f'(x)}{g'(x)}$$

provided the limit on right side of the equation exists.

In other words, l'Hôpital's Rule states that if, on attempting to evaluate limit (1) we obtain the indeterminate form 0/0, we may separately differentiate both numerator and denominator and try again. If we then succeed in obtaining a limit it will also be the limit of the original quotient. (Caution: differentiate $f(x)$ and $g(x)$ *separately*—do *not* apply the Quotient Rule.)

We will establish Theorem 11 after considering several examples and extensions.

Example 1 Since both $\sin x = 0$ and $x = 0$ at $x = 0$, l'Hôpital's Rule gives

$$\lim_{x\to 0} \frac{\sin x}{x} = \lim_{x\to 0} \frac{\frac{d}{dx}(\sin x)}{\frac{d}{dx}(x)} = \lim_{x\to 0} \frac{\cos x}{1} = 1.$$ ■

Example 2 Find $\lim_{x\to 8} \frac{\sqrt[3]{x} - 2}{x - 8}$.

Strategy
Check that both $\sqrt[3]{8} - 2 = 0$ and

$8 - 8 = 0$

in the numerator and denominator.

Solution
Since both the numerator and denominator become zero as $x \to 8$, we apply Theorem 11:

$$\lim_{x\to 8} \frac{\sqrt[3]{x} - 2}{x - 8} \quad \left(\frac{0}{0} \text{ form}\right)$$

Apply l'Hôpital's Rule.

$$= \lim_{x\to 8} \frac{\frac{1}{3}x^{-2/3}}{1}$$

$$= \frac{1}{3(8)^{2/3}} = \frac{1}{12}.$$ ■

REMARK 1: Frequently, on applying l'Hôpital's Rule to $\lim_{x\to a} \frac{f(x)}{g(x)}$, we find that $f'(a) = g'(a) = 0$. That is, we may again obtain the indeterminate form 0/0. In such cases we simply apply l'Hôpital's Rule again, provided the hypotheses of Theorem 11 are fulfilled for the functions $f'(x)$ and $g'(x)$.

Example 3 Find $\lim_{x\to 0} \frac{x - \tan x}{x - \sin x}$.

Strategy
Verify that 0/0 form results from setting $x = 0$.

Solution
Since both $x - \tan x = 0$ and $x - \sin x = 0$ for $x = 0$, we apply l'Hôpital's Rule to find that

$$\lim_{x\to 0} \frac{x - \tan x}{x - \sin x} \quad \left(\frac{0}{0} \text{ form}\right)$$

Apply l'Hôpital's Rule.

$$= \lim_{x\to 0} \frac{1 - \sec^2 x}{1 - \cos x} \quad \left(\frac{0}{0} \text{ form again}\right)$$

Apply l'Hôpital's Rule again!

$$= \lim_{x\to 0} \frac{-2\sec^2 x \tan x}{\sin x} \quad \left(\text{still } \frac{0}{0} \text{ form}\right)$$

Apply l'Hôpital's Rule again!

$$= \lim_{x\to 0} \frac{-4\sec^2 x \tan^2 x - 2\sec^4 x}{\cos x}$$

This limit can be evaluated by setting $x = 0$.

$$= \frac{-4(1)(0) - 2(1)}{1} = -2. \qquad \blacksquare$$

REMARK 2: Be careful not to apply l'Hôpital's Rule in situations where the hypotheses of Theorem 11 are not fulfilled. If the limit (1) is not of form 0/0, an application of l'Hôpital's Rule is often incorrect. For example,

$$\lim_{x\to 0} \frac{\sin x}{x - \sin x} \qquad \left(\frac{0}{0} \text{ form}\right)$$

$$= \lim_{x\to 0} \frac{\cos x}{1 - \cos x}$$

$$= +\infty.$$

since $(1 - \cos x) \to 0$ as $x \to 0$ and $\cos x \to 1$ as $x \to 0$. However, applying l'Hôpital's Rule to the quotient $\dfrac{\cos x}{1 - \cos x}$ would lead to the incorrect result

$$\lim_{x\to 0} \frac{-\sin x}{\sin x} = -1.$$

REMARK 3: If $\lim_{x\to a} f(x) = \lim_{x\to a} g(x) = \infty$, then the limit in (1) yields the indeterminate form ∞/∞. L'Hôpital's Rule can be shown to apply in such situations as well. That is, if the hypotheses of Theorem 11 are fulfilled except that the statement $f(a) = g(a) = 0$ is replaced by the statement $\lim_{x\to a} f(x) = \lim_{x\to a} g(x) = \infty$, then

$$\lim_{x\to a} \frac{f(x)}{g(x)} = \lim_{x\to a} \frac{f'(x)}{g'(x)} \qquad \left(\frac{\infty}{\infty} \text{ form}\right)$$

provided the limit on the right side exists.

Example 4

$$\lim_{x\to \pi/2} \frac{\tan^2 x}{\sec^2 x} \qquad \left(\frac{\infty}{\infty} \text{ form}\right)$$

$$= \lim_{x\to \pi/2} \frac{2\tan x \sec^2 x}{2\sec^2 x \tan x}$$

$$= \lim_{x\to \pi/2} (1) = 1. \qquad \blacksquare$$

REMARK 4: L'Hôpital's Rule also applies to one-sided limits and limits at infinity. In other words, if lim denotes any one of $\lim_{x\to a^+}$, $\lim_{x\to a^-}$, $\lim_{x\to\infty}$, or $\lim_{x\to-\infty}$ and if the hypotheses of Theorem 11 or Remark 3 are fulfilled on an interval of the appropriate type, then

$$\lim \frac{f(x)}{g(x)} = \lim \frac{f'(x)}{g'(x)}.$$

Example 5

$$\lim_{x\to\infty} \frac{5x^2 - 7x + 6}{3 - x + x^2} \qquad \left(\frac{\infty}{\infty} \text{ form}\right)$$

$$= \lim_{x\to\infty} \frac{10x - 7}{-1 + 2x} \qquad \left(\text{still } \frac{\infty}{\infty} \text{ form}\right)$$

$$= \lim_{x\to\infty} \frac{10}{2} = 5.$$

■

Example 6

$$\lim_{x\to 0^+} \frac{\sin x}{x^{3/2}} \qquad \left(\frac{0}{0} \text{ form}\right)$$

$$= \lim_{x\to 0^+} \frac{\cos x}{\frac{3}{2}x^{1/2}} = \infty$$

since $\cos x \to 1$ and $\frac{3}{2}x^{1/2} \to 0$ as $x \to 0^+$. ■

The justification for l'Hôpital's Rule depends upon a generalization of the Mean Value Theorem due to Augustin L. Cauchy (1789–1857). Recall, the Mean Value Theorem guarantees the existence of a number $c \in (a, b)$ so that

$$f'(c) = \frac{f(b) - f(a)}{b - a} \tag{2}$$

when $f(x)$ is differentiable on (a, b) and continuous on $[a, b]$. Cauchy's generalization arises by viewing the denominator on the right side of equation (2) as $g(b) - g(a)$, where g is the identity function $g(x) = x$, and then asking what results by allowing $g(x)$ to instead be *any* differentiable function of x. Since $g'(c) = 1$ when $g(x)$ is the identity function, the following result is a direct generalization of the Mean Value Theorem.

THEOREM 12
Cauchy's Mean Value Theorem

Let the functions $f(x)$ and $g(x)$ be continuous on the closed interval $[a, b]$ and differentiable for all x in the open interval (a, b). Also, assume $g'(x) \neq 0$ for all x in (a, b). Then there exists a number c in (a, b) so that

$$\frac{f'(c)}{g'(c)} = \frac{f(b) - f(a)}{g(b) - g(a)}.$$

Proof: The Mean Value Theorem was proved (Section 5.2) by considering the function

$$d(x) = f(x) - \left[f(a) + \frac{f(b) - f(a)}{b - a}(x - a)\right]. \tag{3}$$

Since Theorem 12 is obtained from the Mean Value Theorem by generalizing the identity function $g(x) = x$ to more general functions $g(x)$, we consider the function

$D(x)$ obtained from (3) by replacing $b - a$ with $g(b) - g(a)$, and $x - a$ with $g(x) - g(a)$:

$$D(x) = f(x) - \left[f(a) + \frac{f(b) - f(a)}{g(b) - g(a)}(g(x) - g(a))\right]. \tag{4}$$

Now for this function $D(x)$ to make sense, we must know that $g(b) - g(a) \neq 0$. However since $g(x)$ satisfies the hypotheses of the Mean Value Theorem,

$$\frac{g(b) - g(a)}{b - a} = g'(d)$$

for some number d in (a, b). Since we are assuming $g'(d) \neq 0$ for all d in (a, b) this shows that $g(b) - g(a) \neq 0$. Thus, there is no difficulty in defining $D(x)$ by equation (4).

Now, just as in the proof of the Mean Value Theorem, the function $D(x)$ in (4) is continuous on $[a, b]$ and differentiable on (a, b) by our assumptions on $f(x)$ and $g(x)$. Moreover, $D(a) = D(b) = 0$. We may therefore apply Rolle's Theorem to conclude that for some number $c \in (a, b)$

$$D'(c) = 0. \tag{5}$$

From equation (4) we can see that

$$D'(x) = f'(x) - g'(x)\left(\frac{f(b) - f(a)}{g(b) - g(a)}\right). \tag{6}$$

Letting $x = c$ and combining statements (5) and (6) then gives the desired result. ■

We are now equipped to prove Theorem 11. We begin by showing that, under the hypotheses of Theorem 11,

$$\lim_{x \to a^+} \frac{f(x)}{g(x)} = \lim_{x \to a^+} \frac{f'(x)}{g'(x)}. \tag{7}$$

To do this we let $x > a$. Then, when x is sufficiently close to a, the hypotheses of Theorem 11, including the assumption that $g(x) \neq 0$ in (a, x), guarantee that we can apply Theorem 12 on the interval $[a, x]$. This means that there is a number c between a and x so that

$$\frac{f(x) - f(a)}{g(x) - g(a)} = \frac{f'(c)}{g'(c)}. \tag{8}$$

Since we are assuming that $f(a) = g(a) = 0$, equation (8) is just the equality

$$\frac{f(x)}{g(x)} = \frac{f'(c)}{g'(c)}, \qquad a < c < x, \tag{9}$$

which holds for all x sufficiently close to a. Since the number c lies between a and x, we must have that $c \to a$ as $x \to a$. Thus, from (9) we conclude that

$$\lim_{x \to a^+} \frac{f(x)}{g(x)} = \lim_{c \to a^+} \frac{f'(c)}{g'(c)} = \lim_{x \to a^+} \frac{f'(x)}{g'(x)},$$

which proves statement (7). The proof for the case $\lim_{x\to a^-} \frac{f(x)}{g(x)}$ is handled in the same way, by applying Theorem 12 to the interval $[x, a]$. The two cases taken together establish Theorem 11.

Exercise Set 5.8

In Exercises 1–37, find the indicated limit.

1. $\lim_{x\to 3} \frac{x^2 - 9}{x - 3}$
2. $\lim_{x\to 0} \frac{\sin 5x}{x}$
3. $\lim_{x\to 0} \frac{\sin x^2}{x}$
4. $\lim_{x\to 2} \frac{x^3 - x^2 - x - 2}{x - 2}$
5. $\lim_{x\to 0} \frac{1 - \cos x}{x^2}$
6. $\lim_{x\to \pi/2} \frac{1 - \sin x}{\cos x}$
7. $\lim_{\theta\to 0} \frac{\tan \theta}{\theta}$
8. $\lim_{x\to 0^+} \frac{1 - \cos x}{x^3}$
9. $\lim_{\theta\to 0} \frac{\tan \theta - \theta}{\theta - \sin \theta}$
10. $\lim_{x\to \infty} \frac{x^3 + 2x + 1}{4x^3 + 1}$
11. $\lim_{x\to 0} \frac{1 + \sin \sqrt{x}}{\cos x}$
12. $\lim_{x\to 1} \frac{\sqrt{x} - \sqrt[4]{x}}{x - 1}$
13. $\lim_{x\to 0} \frac{1 + \cos \sqrt{x}}{\sin x}$
14. $\lim_{x\to 0^+} \frac{1 - \cos \sqrt{x}}{\sin x}$
15. $\lim_{x\to 0^+} \frac{x^2}{x - \sin x}$
16. $\lim_{x\to \infty} \frac{\sqrt{1 + x^2}}{x}$
17. $\lim_{x\to 0} \frac{x - \sin x}{x^3}$
18. $\lim_{x\to 0} \frac{\sin x - x \cos x}{x}$
19. $\lim_{x\to 0} \frac{\sin x}{\sqrt[3]{x}}$
20. $\lim_{x\to 0^+} \frac{\cos x - x}{\sqrt{x}}$
21. $\lim_{x\to 0^+} \frac{\cos x}{\sqrt{x}}$
22. $\lim_{x\to \pi^+} \frac{\sin x}{\sqrt{x - \pi}}$
23. $\lim_{x\to 0} \frac{\sin x - x}{x - \tan x}$
24. $\lim_{x\to 0^+} \frac{\cos x - x}{x - \tan x}$
25. $\lim_{x\to 0} \frac{\sin x - x}{x^{2/3}}$
26. $\lim_{x\to 1^-} \frac{x^{5/2} - 1 + \sqrt{1 - x}}{\sqrt{1 - x^2}}$
27. $\lim_{x\to 2} \frac{2\sqrt{x} - \sqrt{6 + x}}{x - \sqrt{6 - x}}$
28. $\lim_{x\to \pi/2} \frac{\cos 3x}{\cos x}$
29. $\lim_{x\to \pi/2} \frac{\tan(x/2) - 1}{x - \pi/2}$
30. $\lim_{x\to \infty} \frac{\tan(1/x)}{1/x}$
31. $\lim_{x\to \infty} \frac{\sqrt{x} - 3x^2}{x(6 - x)}$
32. $\lim_{x\to \infty} \frac{(ax + b)^3}{(x + c)^3}$
33. $\lim_{x\to \infty} \frac{\sqrt{x} - \sqrt{a}}{\sqrt{x} + \sqrt{a}}$
34. $\lim_{x\to \infty} \frac{x^3 - 7x^2 + 6x - 5}{(x - 3)(5 - x^2)}$
35. $\lim_{x\to \infty} \frac{3x^2 - 4}{4x^2 + 3}$
36. $\lim_{x\to \infty} \frac{x^2(x - 1)(x + 3)}{(x^3 - 6)(2x^2 + x + 1)}$
37. $\lim_{x\to \infty} \frac{\sin x}{x^2 + \pi}$

38. If $\lim_{x\to a} f(x) = 0$ and $\lim_{x\to a} g(x) = \infty$, the limit $\lim_{x\to a} f(x)g(x)$ yields the indeterminate form $0 \cdot \infty$. In this case we can often apply l'Hôpital's Rule by writing

a. $\lim_{x\to a} f(x)g(x) = \lim_{x\to a} \frac{f(x)}{(1/g(x))}$ $\left(\frac{0}{0} \text{ form}\right)$, or

b. $\lim_{x\to a} f(x)g(x) = \lim_{x\to a} \frac{g(x)}{(1/f(x))}$ $\left(\frac{\infty}{\infty} \text{ form}\right)$.

Use this idea to find $\lim_{x\to 0^+} x \cot x$.

In Exercises 39–46, use the idea of Exercise 38 to find the limit.

39. $\lim_{x\to \infty} x \sin(1/x)$
40. $\lim_{x\to 0^+} x^{-1/2} \sin x$
41. $\lim_{x\to \pi/4} (1 - \tan x)\sec 2x$
42. $\lim_{x\to \pi/2^-} (x - \pi/2) \sec x$
43. $\lim_{x\to \infty} x \tan (1/x)$
44. $\lim_{x\to \infty} x^2 \sin(1/x)$
45. $\lim_{x\to 0^+} x^2 \cot x$
46. $\lim_{x\to 0^+} x^2 \csc x$

47. If $\lim_{x\to a} f(x) = \lim_{x\to a} g(x) = \infty$, the limit $\lim_{x\to a} (f(x) - g(x))$ yields the indeterminate form $\infty - \infty$. In such cases we can often rewrite the difference $f(x) - g(x)$ as a quotient to which l'Hôpital's Rule is applicable. For example, $\lim_{x\to \pi/2} (\tan x - \sec x)$ yields the indeterminate form $\infty - \infty$. Show that $\lim_{x\to \pi/2} (\tan x - \sec x) = 0$ by writing

$$\tan x - \sec x = \frac{\sin x}{\cos x} - \frac{1}{\cos x} = \frac{\sin x - 1}{\cos x}$$

and applying l'Hôpital's Rule.

In Exercises 48–51, use the idea of Exercise 47 to find the limit.

48. $\lim_{x \to 0^+} \left(\frac{1}{x} - \frac{1}{\sin x} \right)$

49. $\lim_{x \to 0} (\csc x - \cot x)$

50. $\lim_{x \to 0} \left(\frac{1}{\sin x} - \frac{1}{x^2} \right)$

51. $\lim_{x \to 0^+} \left(\frac{1}{x} - \frac{1}{\sqrt{x}} \right)$

52. Prove l'Hôpital's Rule for the case $x \to \infty$ as follows:

a. Set $t = 1/x$. Then

$$\lim_{x \to \infty} \frac{f(x)}{g(x)} = \lim_{t \to 0^+} \frac{f(1/t)}{g(1/t)}.$$

b. Show that

$$\lim_{t \to 0^+} \frac{f(1/t)}{g(1/t)} = \lim_{t \to 0^+} \frac{(-1/t^2)f'(1/t)}{(-1/t^2)g'(1/t)} = \lim_{t \to 0^+} \frac{f'(1/t)}{g'(1/t)}$$

when $\lim_{x \to \infty} \frac{f(x)}{g(x)}$ has form $\left(\frac{0}{0}\right)$ or $\left(\frac{\infty}{\infty}\right)$.

c. Conclude that, in this case, $\lim_{x \to \infty} \frac{f(x)}{g(x)} = \lim_{x \to \infty} \frac{f'(x)}{g'(x)}$.

5.9 ANTIDERIVATIVES

We have previously seen that for a particle moving along a line the functions representing position, velocity, and acceleration are related by the process of differentiation. In particular, if $s(t)$ denotes the location of a particle on the number line at time t and if $v(t)$ and $a(t)$ denote its velocity and acceleration, respectively, then

$$v(t) = s'(t), \qquad \text{and} \qquad a(t) = v'(t) = s''(t) \tag{1}$$

provided these derivatives exist. In Leibniz's notation, statements in (1) become

$$v = \frac{ds}{dt}, \qquad \text{and} \qquad a = \frac{dv}{dt} = \frac{d^2s}{dt^2}. \tag{2}$$

We now wish to ask the reverse questions, namely,

(i) If the acceleration function of a particle is known, can we calculate its velocity function?

(ii) If the velocity function is known, can we calculate the particle's position?

Looking at equations (1) or (2), we can see that questions (i) and (ii) are both examples of the following more general problem:

> Antidifferentiation Problem: Given the function $f(x)$, find another function $F(x)$ so that
>
> $$F'(x) = f(x) \tag{3}$$
>
> or, in Leibniz' notation, so that
>
> $$\frac{dF}{dx} = f.$$

The function $F(x)$ in equation (3), if it can be found, is called an **antiderivative** of the function $f(x)$. The process of finding $F(x)$ is called **antidifferentiation.** Obviously, antidifferentiation requires an ability to "think backwards" through the process of differentiation—that is, to answer the question, "What function is it whose derivative is $f(x)$?"

Example 1 An antiderivative for $f(x) = 2x$ is the function $F(x) = x^2$, since $F'(x) = \dfrac{d}{dx}(x^2) = 2x = f(x)$. ■

Example 2 An antiderivative for $f(x) = \sin x$ is $F(x) = -\cos x$, since $F'(x) = \dfrac{d}{dx}(-\cos x) = \sin x = f(x)$. ■

Things are not as clear-cut as Examples 1 and 2 might suggest, however. The problem stems in part from the fact that antiderivatives are not unique. For example, the function $f(x) = 2x$ in Example 1 also has the following antiderivatives, among many others:

$$F_1(x) = x^2 + 1, \qquad F_2(x) = x^2 - 10, \qquad \text{and} \qquad F_3(x) = x^2 + \pi.$$

Similarly, $f(x) = \sin x$ in Example 2 has antiderivatives

$$G_1(x) = 1 - \cos x, \qquad G_2 = \pi - \cos x, \qquad \text{and} \qquad G_3(x) = C - \cos x,$$

where C is any constant.

Obviously, we need a clearer understanding of what to expect as the antiderivative of a given function. We therefore return to our geometric interpretation of the derivative.

Recall, starting with the differentiable function $F(x)$, the process of *differentiation* associates a *slope,* $F'(x)$, with each value of x in the domain of $F(x)$ (Figure 9.1).

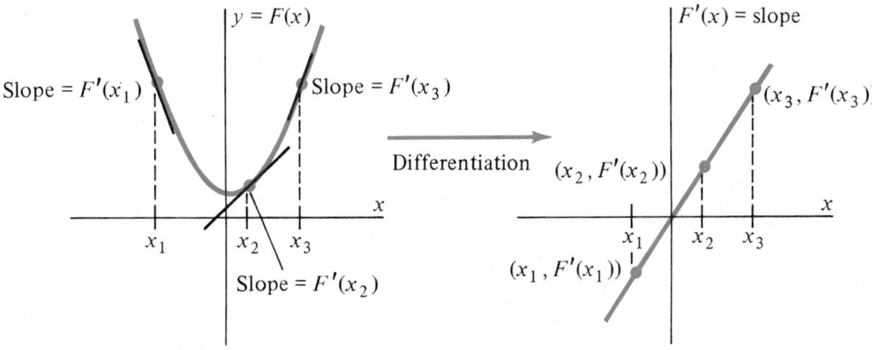

Figure 9.1 Differentiating $F(x)$ produces a slope, $F'(x)$, for each value of x.

Thus, if we begin with the function $f(x) = F'(x)$ and seek $F(x)$, we are beginning with a slope function and asking "what function $F(x)$ has slope $f(x) = F'(x)$ at each value of x?" Figure 9.2 illustrates the difficulty in asking such a question—since we do not know $F(x)$, we do not know the particular points $(x, F(x))$ on the graph of $y = F(x)$. We only know the slope at each point.

For example, suppose we ask for an antiderivative for the function $f(x) = \cos x$. That is, we seek $F(x)$ with $F'(x) = \cos x$. Figure 9.3 shows the required slopes $F'(x)$ for various values of x, and Figure 9.4 suggests several functions whose

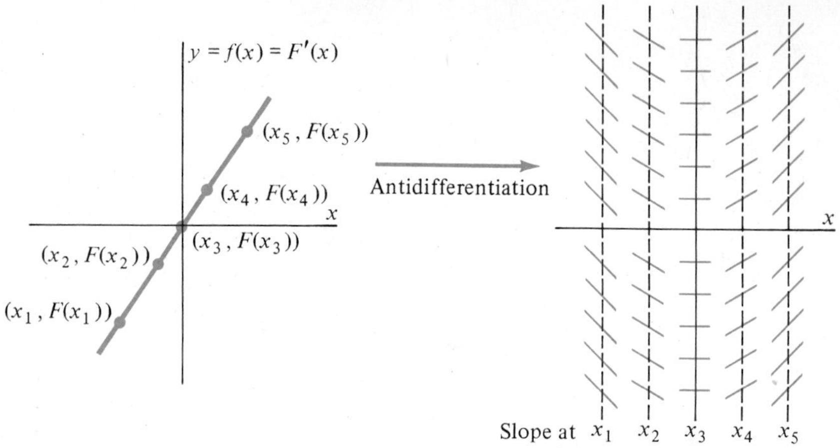

Figure 9.2 Geometric interpretation of antidifferentiation—values of $f(x) = F'(x)$ interpreted as slopes of $F(x)$.

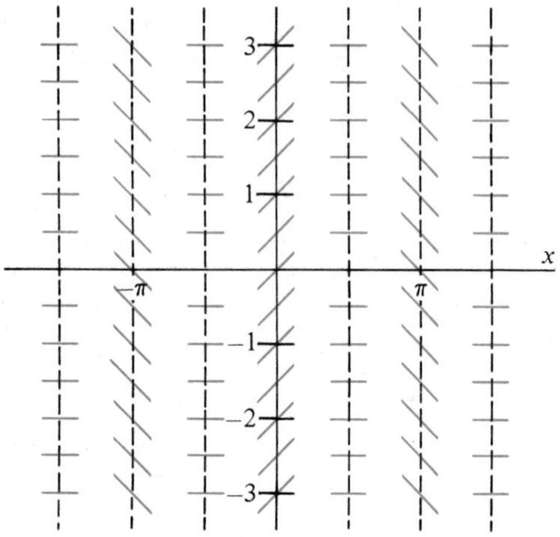

Figure 9.3 Slopes for $F(x)$ if $F'(x) = \cos x$.

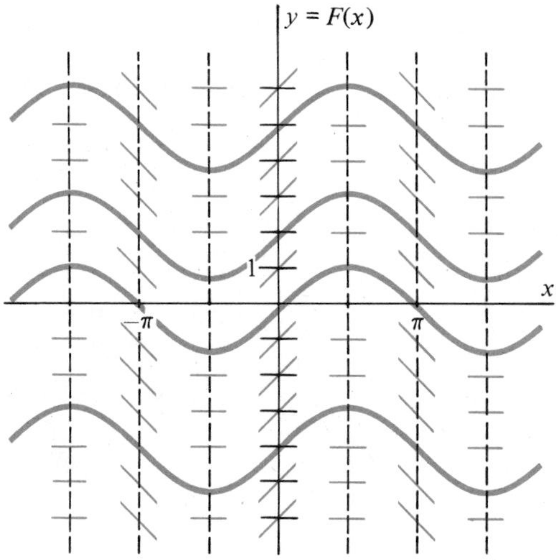

Figure 9.4 Antiderivatives for $F(x) = \cos x$.

graphs have the required slopes. As Figure 9.4 suggests, we will see that the *graphs of all antiderivatives $F(x)$ for $f(x) = \cos x$ have the same shape*. They differ only by the property that one is a "vertical shift" of the other. In other words, if $F_1(x)$ and $F_2(x)$ are both antiderivatives for $f(x) = \cos x$, then $F_2(x) = F_1(x) + C$ for some constant C.

Notice this difference between the process of differentiation and that of antidifferentiation: differentiating a particular function produces another (unique) function, while finding an antiderivative seems to result in an entire family of functions. The following theorem makes these observations precise.

THEOREM 13

Suppose the function $F(x)$ is an antiderivative for the function $f(x)$ on the interval I. That is, assume $F'(x) = f(x)$ for all $x \in I$. Then *any* antiderivative $G(x)$ for $f(x)$ on I must have the form

$$G(x) = F(x) + C \tag{4}$$

for some constant C. Moreover, every function $G(x)$ of this form is an antiderivative for $f(x)$.

In other words, *any two antiderivatives for $f(x)$ differ at most by a constant,* as suggested by our discussion concerning slopes. Before proving Theorem 13, we will establish a special case, using the Mean Value Theorem.

THEOREM 14 The only antiderivatives of the zero function are the constant functions. That is, if $F'(x) = 0$ for all x, then $F(x) \equiv C$* for some constant C.

Proof: Suppose $F(x)$ is an antiderivative for the zero function. This means that $F(x)$ is differentiable for all x, so $F(x)$ satisfies the hypotheses of the Mean Value Theorem. Let a be any constant and let $x \neq a$ be any other number. The Mean Value Theorem guarantees the existence of a number b between a and x so that

$$\frac{F(x) - F(a)}{x - a} = F'(b)$$

But $F'(b) = 0$ by our hypothesis, and $x - a \neq 0$. Thus, $F(x) - F(a) = 0$, so $F(x) = F(a)$ for all values of x. Setting $C = F(a)$ shows that $F(x) = C$ for all x, as claimed. ■

We can now use Theorem 14 to give a simple proof of Theorem 13. Suppose that $F(x)$ and $G(x)$ are two antiderivatives for $f(x)$. This means that $F'(x) = G'(x) = f(x)$. The function $h(x) = G(x) - F(x)$ is therefore differentiable, and

$$h'(x) = G'(x) - F'(x) = f(x) - f(x) = 0$$

for all $x \in I$. By Theorem 14 we conclude that $h(x) = C$ for some constant C and all $x \in I$. Thus, $h(x) = G(x) - F(x) = C$, or $G(x) = F(x) + C$, as desired.

Notation and Terminology

If $F'(x) = f(x)$, the most general expression for the antiderivative of $f(x)$ is $F(x) + C$, according to Theorem 13. The notation for this is

$$\int f(x)\,dx = F(x) + C \tag{5}$$

The symbols $\int$, called an **integral sign,** and dx enclose the function for which we are denoting the antiderivative. For now, you may think of the integral sign $\int$ as calling for the antiderivative and the symbol dx, not as a differential, but merely as a device for indicating the name of the independent variable. For example, $\int x^2\,dx$ is read "the antiderivative of the function x^2 with respect to x," while $\int \cos u\,du$ is "the antiderivative of the function $\cos u$ with respect to u."

Some authors use the terms *integral,* or *indefinite integral,* rather than the term *antiderivative*. The three are synonymous, as are the terms *antidifferentiation* and *integration*. We prefer the term antiderivative as more indicative of the process just

*Recall, the notation $F(x) \equiv C$ means $F(x) = C$ for all x.

described. (To be consistent, we should probably also refer to the symbol $\int$ as the *antiderivative sign,* but such terminology is quite rare for reasons we shall see later.)

Example 3 The following table shows a set of differentiation facts and the corresponding antidifferentiation facts (Table 9.1). ■

Table 9.1 Some differentiation and antidifferentiation results

$\frac{d}{dx} F(x) = f(x)$	$\int f(x)\,dx = F(x) + C$
$\frac{d}{dx}\left(\frac{1}{2}x^2\right) = x$	$\int x\,dx = \frac{1}{2}x^2 + C$
$\frac{d}{dx}\left(x^3 + \frac{1}{6}x^6\right) = 3x^2 + x^5$	$\int (3x^2 + x^5)\,dx = x^3 + \frac{1}{6}x^6 + C$
$\frac{d}{dt}(\sin t) = \cos t$	$\int \cos t\,dt = \sin t + C$
$\frac{d}{d\theta}(\tan\theta) = \sec^2\theta$	$\int \sec^2\theta\,d\theta = \tan\theta + C$
$\frac{d}{dx}(\sqrt{x}) = \frac{1}{2\sqrt{x}}$	$\int \frac{1}{2\sqrt{x}}\,dx = \sqrt{x} + C$
$\frac{d}{dt}\left(\frac{1}{2}\sin t^2\right) = t\cos t^2$	$\int t\cos t^2\,dt = \frac{1}{2}\sin t^2 + C$
$\frac{d}{dx}(\sqrt{ax^2 + bx + c}) = \frac{2ax + b}{2\sqrt{ax^2 + bx + c}}$	$\int \frac{2ax + b}{2\sqrt{ax^2 + bx + c}}\,dx = \sqrt{ax^2 + bx + c} + C$

Properties of Antiderivatives

Since antidifferentiation is, roughly speaking, the reverse process of differentiation, it is not surprising that the following properties hold.

THEOREM 15

Suppose both $f(x)$ and $g(x)$ have antiderivatives on a common interval I. Then, on I,

(a) $\int [f(x) + g(x)]\,dx = \int f(x)\,dx + \int g(x)\,dx,$ and

(b) $\int [cf(x)]\,dx = c \cdot \int f(x)\,dx$

Proof: Let $F(x)$ and $G(x)$ be particular antiderivatives for $f(x)$ and $g(x)$, respectively. That is, assume that $F'(x) = f(x)$ and $G'(x) = g(x)$ for all $x \in I$. Then, by Theorem 5, Chapter 3, we have

(i) $\frac{d}{dx}[F(x) + G(x)] = F'(x) + G'(x) = f(x) + g(x),$ and

(ii) $\frac{d}{dx}[cF(x)] = cF'(x) = cf(x).$

Equation (i) shows that $[F(x) + G(x)]$ is an antiderivative for $[f(x) + g(x)]$, and equation (ii) shows that $cF(x)$ is an antiderivative for $cf(x)$. ■

Theorem 15 states that, in finding antiderivatives for sums or multiples of functions, we may first find antiderivatives for the individual functions and then form the appropriate sums or multiples, incorporating all arbitrary constants into a single constant.*

Example 4 Find $\int [7x^2 - 2x^3]\, dx.$

Strategy

Apply Theorem 15.

$\int x^2\, dx = \frac{1}{3}x^3 + C_1$

$\int x^3\, dx = \frac{1}{4}x^4 + C_2$

$C = C_1 + C_2$

Check result by differentiating.

Solution

$$\begin{aligned}\int [7x^2 - 2x^3]\, dx &= 7\int x^2\, dx - 2\int x^3\, dx\\ &= \left[7\left(\frac{x^3}{3}\right) + C_1\right] - 2\left[\frac{1}{4}x^4 + C_2\right]\\ &= \frac{7}{3}x^3 - \frac{1}{2}x^4 + C.\end{aligned}$$

To verify this result, we check

$$\begin{aligned}\frac{d}{dx}\left(\frac{7}{3}x^3 - \frac{1}{2}x^4 + C\right) &= \frac{7}{3}(3)x^2 - \frac{1}{2}(4)x^3 + 0\\ &= 7x^2 - 2x^3.\end{aligned}$$ ■

In Example 4, we have made use of the formula

$$\int x^n\, dx = \frac{1}{n+1}x^{n+1} + C, \qquad n \neq -1, \tag{6}$$

which holds for all rational exponents n. Formula (6) may be verified by differentiation:

$$\frac{d}{dx}\left[\frac{1}{n+1}x^{n+1} + C\right] = \frac{1}{n+1}(n+1)x^n = x^n.$$

Obviously, any differentiation rule of the form $f'(x) = g(x)$ gives a corresponding antidifferentiation rule $\int g(x)\, dx = f(x) + C$. However, not every function has an antiderivative of the kind described in this section. For example, nothing we have seen up to this point enables us to find an antiderivative for the function $f(x) = \sin(x^2)$, as you will conclude after making a few attempts at writing down such an antiderivative. Indeed, many functions simply do not have elementary antiderivatives. We shall need to develop a more sophisticated theory (Chapter 6) to address this problem.

*Thus, if C_1 and C_2 denote arbitrary (and, therefore, unknown) constants, the sum $(C_1 + C_2)$ and the multiple kC_1 are again just unknown constants. We will therefore always combine arbitrary constants by equations of the form

$$C_1 + C_2 = C \qquad \text{and} \qquad kC_1 = C.$$

U-Substitutions

A helpful technique for finding antiderivatives is based on the Chain Rule formula for differentiation. If $F'(x) = f(x)$ we can write this formula as

$$[F(u(x))]' = f(u(x)) \cdot u'(x). \tag{7}$$

Forming the antiderivatives of both sides of equation (7) gives the formula

$$\boxed{\int f(u(x)) \cdot u'(x)\, dx = F(u(x)) + C} \tag{8}$$

The technique suggested by formula (8) is this: If the function $g(x)$ in $\int g(x)\, dx$ can be viewed as a product of two functions in the form $g(x) = f(u(x)) \cdot u'(x)$ then the antiderivative is given by (8).

Since this technique depends on the identification of an "inside" function $u(x)$, an effective procedure for implementing this idea is to use the notation by which we defined the differential du:

$$du = u'(x)\, dx \qquad \left(\text{since } \frac{du}{dx} = u'(x)\right).$$

Then, beginning with the left side of (8) we can write

$$\int f(u(x)) \cdot u'(x)\, dx = \int f(u)\, du = F(u) + C. \tag{9}$$

We say that we have made the **u-substitution**

$$u = u(x); \qquad du = u'(x)\, dx$$

in (9), since we have effectively changed independent variables from x to u. Of course, we must remember to substitute back for $u = u(x)$ in our final answer.

Example 5 Find $\int \sin^3 x \cos x\, dx$ by the method of u-substitution.

Strategy
Since $(\sin x)' = \cos x$, the inside function is $u = \sin x$.

Solution
We use the u-substitution

$$u = \sin x; \qquad du = \cos x\, dx.$$

We obtain

Substitute for u and du.

Find antiderivative with respect to u.

Substitute back in terms of x.

$$\begin{aligned}\int \sin^3 x \cos x\, dx &= \int u^3 du \\ &= \frac{1}{4}u^4 + C \\ &= \frac{1}{4}\sin^4 x + C\end{aligned}$$

■

Example 6 Find $\int x^2(x^3 + 7)^4\, dx$.

Strategy
Take $u = x^3 + 7$ to be the inside function, since $du = 3x^2\, dx$ has a factor of x^2.

Solution
We make the u-substitution

$$u = x^3 + 7, \qquad du = 3x^2\, dx.$$

Then $x^2\,dx = \frac{1}{3}\,du$, and we obtain

$$\int x^2 (x^3 + 7)^4\,dx = \int (x^3 + 7)^4 x^2\,dx$$

Substitute for u and for du.

$$= \int u^4 \cdot \left(\frac{1}{3}\,du\right)$$

$$= \frac{1}{3}\int u^4 du$$

Apply the Power Rule (equation 6).

$$= \frac{1}{3}\cdot\frac{1}{5}u^5 + C$$

Substitute back for x.

$$= \frac{1}{15}(x^3 + 7)^5 + C. \qquad \blacksquare$$

Example 7 Find $\displaystyle\int \frac{x^2\,dx}{\sqrt{1 + x^3}}$.

Strategy

The only hope is in setting $u = 1 + x^3$.

Solution

We make the u-substitution

$$u = 1 + x^3; \qquad du = 3x^2\,dx.$$

Solve for the factor $x^2\,dx$ in terms of du.

Then $x^2\,dx = \frac{1}{3}\,du$, so we obtain

Make the u-substitution.

$$\int \frac{x^2\,dx}{\sqrt{1 + x^3}} = \int \frac{\frac{1}{3}\,du}{\sqrt{u}}$$

Factor out the $\frac{1}{3}$, by Theorem 15.

$$= \frac{1}{3}\int u^{-1/2} du$$

Find the antiderivative with respect to u using (6).

$$= \left(\frac{1}{3}\right)[(2)u^{1/2} + C_1]$$

Substitute back in terms of x.

$$= \frac{2}{3}\sqrt{1 + x^3} + C, \qquad C = \frac{1}{3}C_1. \qquad \blacksquare$$

Example 8 Find $\displaystyle\int \frac{\sin x \cos x\,dx}{\sqrt{1 + \sin^2 x}}$.

Solution: A first attempt might be to try the u-substitution

$$u = \sin x; \qquad du = \cos x\,dx.$$

Indeed, this works, since we obtain

$$\int \frac{\sin x \cos x\,dx}{\sqrt{1 + \sin^2 x}} = \int \frac{u \cdot du}{\sqrt{1 + u^2}}.$$

To handle the antiderivative on the right, we can make a second substitution

$$w = 1 + u^2; \qquad dw = 2u\,du,$$

so $u\,du = \frac{1}{2}\,dw$. With this substitution we have

$$\int \frac{\sin x \cos x\,dx}{\sqrt{1 + \sin^2 x}} = \int \frac{u\,du}{\sqrt{1 + u^2}} = \int \frac{\frac{1}{2}\,dw}{\sqrt{w}} = \int \frac{1}{2} w^{-1/2}\,dw$$

$$= \sqrt{w} + C$$
$$= \sqrt{1 + u^2} + C$$
$$= \sqrt{1 + \sin^2 x} + C.$$

A shorter solution to this problem is obtained if we simply take

$$u = 1 + \sin^2 x; \qquad du = 2 \sin x \cos x\,dx.$$

Then $\sin x \cos x\,dx = \frac{1}{2}\,du$, and we obtain

$$\int \frac{\sin x \cos x\,dx}{\sqrt{1 + \sin^2 x}} = \int \frac{\frac{1}{2}\,du}{\sqrt{u}} = \int \frac{1}{2} u^{-1/2} = \sqrt{u} + C$$

$$= \sqrt{1 + \sin^2 x} + C.$$

Obviously, in some cases more than one choice for u will work. ■

Position and Velocity

Returning now to the issues of position, velocity, and acceleration we can use the notation developed here to write the corresponding antidifferentiation formulas for equation (1):

$$s(t) = \int v(t)\,dt + C \tag{10}$$

and

$$v(t) = \int a(t)\,dt + C \tag{11}$$

The following examples illustrate how the constants in equations (10) and (11) are determined in particular situations.

Example 9 When a hockey player strikes a puck with a certain force, the puck moves along the ice with velocity $v(t) = 64 - \sqrt{t}$ meters per second at time t, where $0 \le t \le 8$ seconds. Let $s(t)$ denote the distance of the puck from the player after t seconds. Find $s(t)$.

Solution: Using (10) we find the position function $s(t)$ to be

$$s(t) = \int (64 - t^{1/2})\,dt = 64t - \frac{2}{3} t^{3/2} + C.$$

To determine C we use the fact that $s(0) = 0$, which says that the distance between the player and the puck is zero at the initial time $t = 0$. This gives

$$0 = s(0) = 64(0) - \frac{2}{3}(0)^{3/2} + C = C,$$

so

$$C = 0.$$

Thus,

$$s(t) = 64t - \frac{2}{3}t^{3/2}.$$

■

The condition that $s(0) = 0$ in Example 9 is referred to as an **initial condition** since it provides the value of $s(t)$ at a particular value of t. (It is not necessary that this value be restricted at $t = 0$. Information about $s(t)$ at *any* value of $t \in [0, 8]$ will allow us to determine C.) In problems involving antidifferentiation, we will need one initial condition to determine the arbitrary constant C for each antidifferentiation performed.

Example 10 When an object is moving freely in the atmosphere it is pulled toward the earth by the force of gravity. Near the surface of the earth the acceleration due to gravity is approximately 9.8 m/sec^2. If a fireworks display rocket is fired vertically from ground level at an initial velocity of 45 m/sec and fails to explode, when does it strike the ground?

Strategy

Find $a(t)$ from the problem statement.

Solution

Since the only force acting on the rocket is acceleration due to gravity, the acceleration function is the constant

$$a(t) = -9.8 \text{ m/sec}^2.$$

(We take the positive direction to be upward, so the correct sign for $a(t)$ is $-$.)

Find $v(t)$ by using equation (11).

The velocity function is therefore

$$\begin{aligned} v(t) &= \int (-9.8)\, dt \\ &= -9.8t + C_1. \end{aligned}$$

Apply initial condition $v(0) = 45$ to find C_1.

To find the constant C_1, we use the initial condition $v(0) = 45$. Setting $t = 0$ gives

$$v(0) = 45 = (-9.8)(0) + C_1$$

so

$$C_1 = 45.$$

The explicit velocity function is therefore

$$v(t) = -9.8t + 45 \text{ m/sec}.$$

Find $s(t)$ from equation (10).

The position function is

$$\begin{aligned} s(t) &= \int (-9.8t + 45)\, dt \\ &= -4.9t^2 + 45t + C_2. \end{aligned}$$

Apply initial condition $s(0) = 0$ to find C_2.

Since the rocket is fired from ground level, the initial condition on $s(t)$ is $s(0) = 0$. Thus

$$s(0) = 0 = -4.9(0)^2 + 45(0) + C_2,$$

so

$$C_2 = 0.$$

The explicit position function is therefore

$$s(t) = -4.9t^2 + 45t \text{ m}.$$

Set $s(t) = 0$ to find desired time.

The rocket strikes the ground when $s(t) = 0$ so we set

$$s(t) = -4.9t^2 + 45t = 0$$

and obtain $t = 0$ (launch), or

$$-4.9t + 45 = 0,$$

which gives

$$t = \frac{45}{4.9} \approx 9.18 \text{ seconds (after launching).}$$ ■

The following diagram summarizes the notion that differentiation and antidifferentiation can "almost" be thought of as inverse processes. The difficulty with this observation is that the first of these two operations produces a unique outcome while the second does not. More formally put, we have

$$\frac{d}{dx}\left\{\int f(x)\,dx\right\} = f(x)$$

while

$$\int \left\{\frac{d}{dx} f(x)\right\} dx = f(x) + C.$$

This notion will be pursued more fully in Chapter 9.

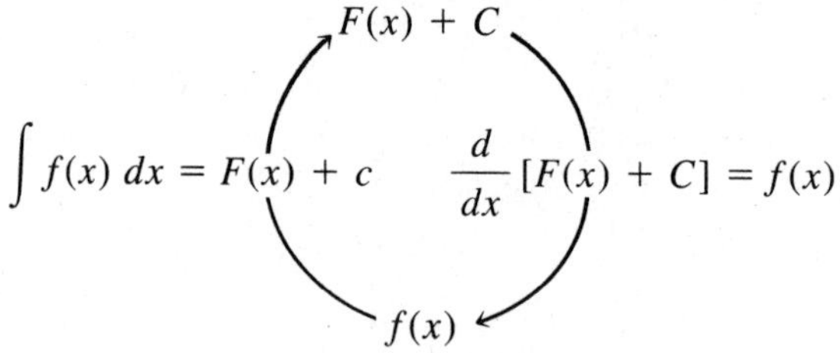

Exercise Set 5.9

1. Figures a–d illustrate the graphs of four particular functions. Figures i–iv represent the associated slopes for the antiderivatives. Match each function with the correct slope portrait for its antiderivative.

a.

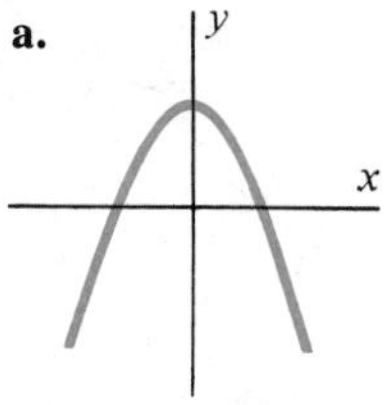

i.

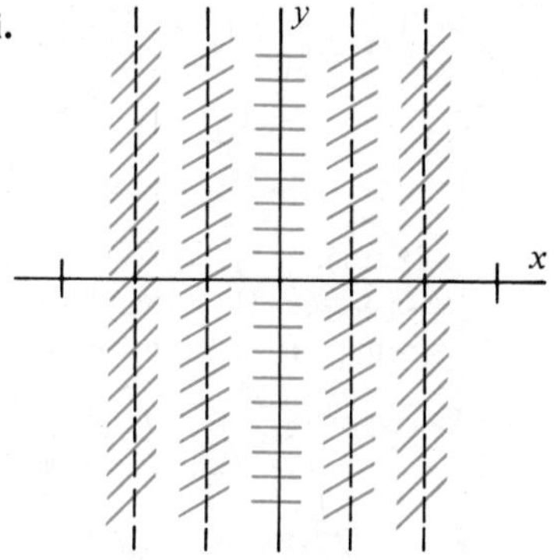

b.

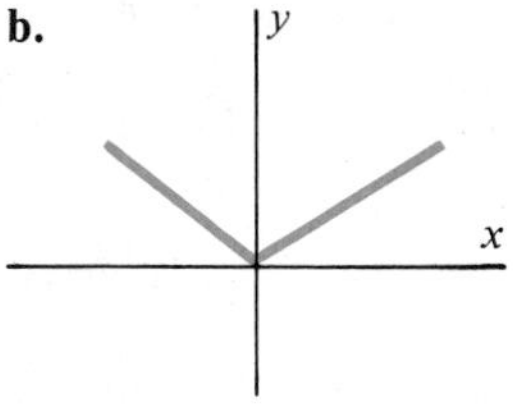

ii.

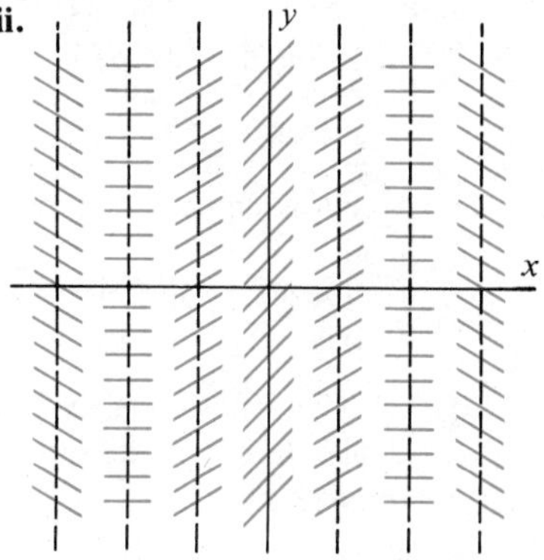

c.

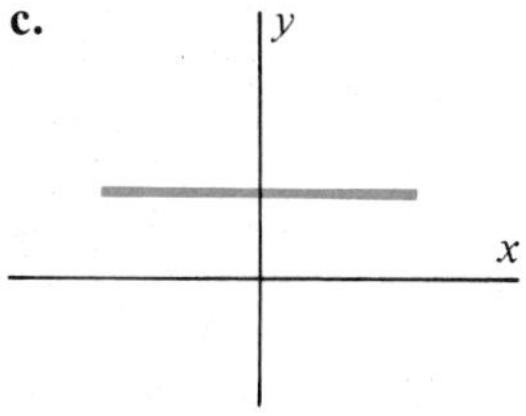

iii.

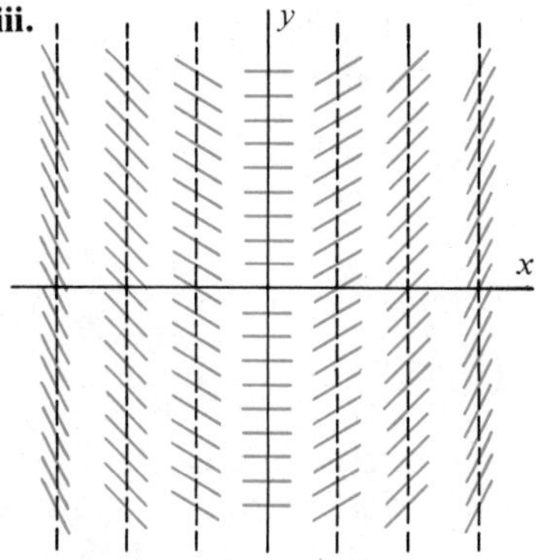

d.

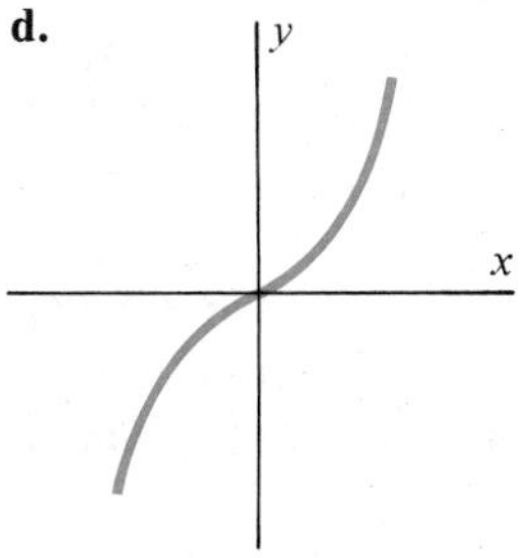

iv.

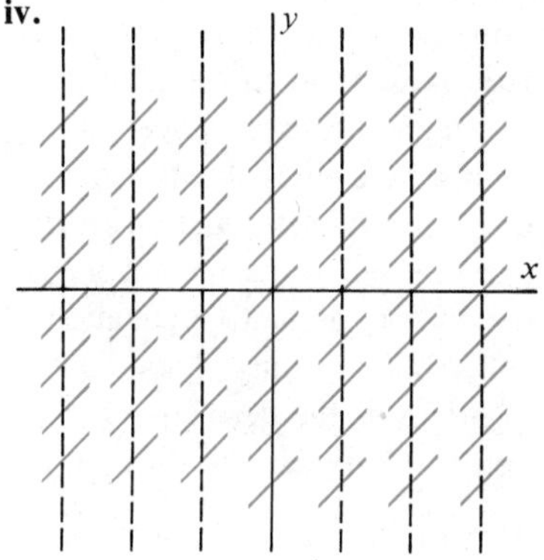

In each of Exercises 2–9, sketch the slope portrait for the antiderivative of the given function.

2. $f(x) = x$

3. $f(x) = 1 - x$

4. $f(x) = \sqrt{x}$

5. $f(x) = \sin x$

6. $y = 1 - x^2$

7. $y = (1 - x)^2$

8. $y = 1/x$

9. $y = x^3 - 2$

In Exercises 10–37, find the general form of the indicated antiderivative.

10. $\int (2x^2 + 1)\, dx$

11. $\int (1 - x^3)\, dx$

12. $\int (x^{2/3} + x^{5/2})\, dx$

13. $\int \sqrt{x + 2}\, dx$

14. $\int \sin 4x\, dx$

15. $\int (2 \cos t - \sin t)\, dt$

16. $\int (\sin x + \sec^2 x)\, dx$

17. $\int \sqrt{t}(t^2 + 1)\, dt$

18. $\int (t - 1)(t + 1)\, dt$

19. $\int (t^2 + 1)(t + 2)\, dt$

20. $\int \frac{1}{\sqrt{x + 1}}\, dx$

21. $\int x\sqrt{x^2 + 1}\, dx$

22. $\int x \cos x^2\, dx$

23. $\int \sec 2\theta \tan 2\theta\, d\theta$

24. $\int \csc(\pi x) \cot(\pi x)\, dx$

25. $\int x \csc^2(x^2)\, dx$

26. $\int \frac{\sin \sqrt{x} \cos \sqrt{x}}{\sqrt{x}}\, dx$

27. $\int (1 - t^2)\sqrt{3t^3 - 9t + 9}\, dt$

28. $\int (1 + \sqrt{x})(1 - \sqrt{x})\, dx$

29. $\int \sec^2(x - \pi)\, dx$

30. $\int x \sec^2 x^2\, dx$

31. $\int \frac{(2\sqrt{x} + 3)^2}{\sqrt{x}}\, dx$

32. $\int (t^3 - 6t + 7)^5 (2 - t^2)\, dt$

33. $\int x\sqrt{x - 1} \cdot \sqrt{x + 1}\, dx$

34. $\int \sqrt{1 - \sin x}\, \sqrt{1 + \sin x}\, dx$

35. $\int x \sec(\pi - x^2) \tan(\pi - x^2)\, dx$

36. $\int \dfrac{\sec^2 \sqrt{2x+1}}{\sqrt{2x+1}}\, dx$

37. $\int (x^5 - 2x^3)(x^6 - 3x^4)^{5/2}\, dx$

In Exercises 38–44, find the position function $s(t)$ corresponding to the given velocity function and initial condition.

38. $v(t) = \cos t, \quad s(0) = 2$

39. $v(t) = 1 + 2t, \quad s(0) = 0$

40. $v(t) = 2t^2, \quad s(0) = 4$

41. $v(t) = t(t+2), \quad s(0) = 2$

42. $v(t) = \sqrt{t}(t+4), \quad s(0) = 0$

43. $v(t) = \dfrac{(\sqrt{t}+3)^3}{\sqrt{t}}, \quad s(0) = 0$

44. $v(t) = (\sqrt{t}+4)(\sqrt{t}-4), \quad s(0) = 5$

In Exercises 45–51, find the velocity function $v(t)$ and position function $s(t)$ corresponding to the given acceleration function and initial conditions.

45. $a(t) = 2, \quad v(0) = 3, \quad s(0) = 0$

46. $a(t) = -2, \quad v(0) = 6, \quad s(0) = 10$

47. $a(t) = 3t, \quad v(0) = 0, \quad s(0) = 20$

48. $a(t) = 200, \quad v(0) = 100, \quad s(0) = 200$

49. $a(t) = 4t + 4, \quad v(0) = 8, \quad s(0) = 12$

50. $a(t) = \cos t, \quad v(0) = 1, \quad s(0) = 9$

51. $a(t) = \sin t, \quad v(0) = 0, \quad s(0) = 2$

52. A stone is thrown vertically upward with an initial velocity 15 m/sec. How long will it take for the stone to **(a)** stop rising, **(b)** strike the ground?

53. How long will it take for the rocket in Example 10 to reach the highest point of its trajectory?

54. A particle moves along a line with constant acceleration $a = 3$ m/sec^2.
- **a.** How fast is it moving after 6 seconds if its initial velocity is $v(0) = 10$ m/sec?
- **b.** What is its initial velocity $v(0)$ if its speed after 3 seconds is 15 m/sec?

55. A coin is dropped from the top of a building 150 meters high.
- **a.** How long will it take for the coin to strike the ground?
- **b.** With what velocity will the coin strike the ground?

56. What constant acceleration will enable the driver of an automobile to increase its speed from 20 m/sec to 25 m/sec in 10 seconds?

57. What constant negative acceleration (deceleration) is required to bring an automobile travelling at a rate of 72 km/hr to a full stop in 100 meters? (*Caution:* Be careful to work in common units.)

5.10 DIFFERENTIAL EQUATIONS (Optional)

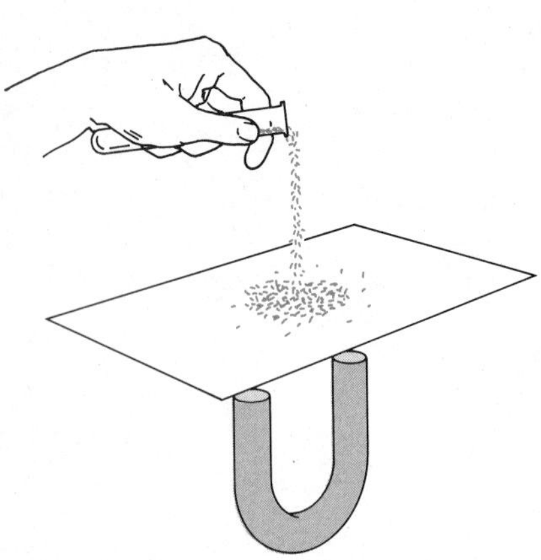

Figure 10.1 Iron filing-magnet experiment.

Calculating functions from their slope functions and finding velocity and position from acceleration are not the only types of problems in which one attempts to determine a function from information about its derivative. For example, a classic experiment in elementary physics is that of sprinkling iron filings on a sheet of paper suspended over a horseshoe magnet (Figure 10.1). The iron filings assume the directions of the magnetic force field surrounding the magnet (Figure 10.2). Here we obtain information about the direction (slopes) of the field lines, and we can then ask for the equations of the field lines themselves.

A general theory for determining a function from certain types of information about its derivative(s) is referred to as the theory of **differential equations.** A typical differential equation (or, derivative equation) is an equation involving an unknown function together with various of its derivatives.

For example, since velocity is the derivative of position, we have already seen that the position function $s(t)$ satisfies the differential equation

$$s'(t) = v(t). \tag{1}$$

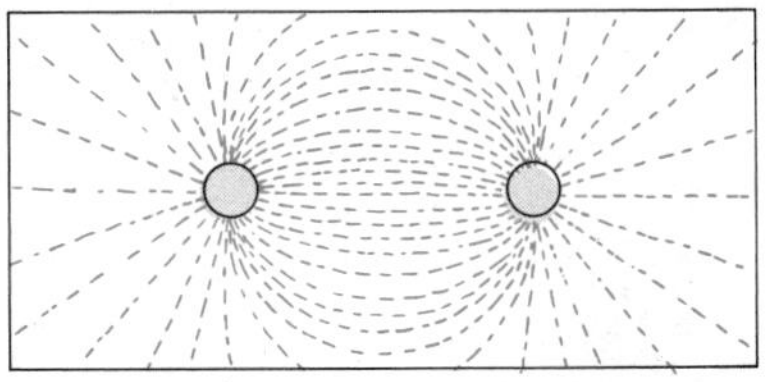

Figure 10.2 Filings point in direction of magnetic field lines.

Later, we shall see that the function $x(t)$ representing the displacement of the end of an oscillating spring satisfies the differential equation

$$x''(t) + bx'(t) + cx(t) = 0. \tag{2}$$

Equation (1) is an example of a **first order** differential equation since it involves only a first derivative. Equation (2) is an example of a **second order** differential equation.

A **solution** of a differential equation is a function that is differentiable as many times as the equation requires and that satisfies the equation. For example, we have already seen that the differential equation

$$\frac{dy}{dx} = \cos x$$

has the solution $y = \sin x$, while the differential equation

$$x\frac{d^2y}{dx^2} - \frac{dy}{dx} = 0$$

has solution $y = x^2$. (To see this note that $\frac{dy}{dx} = 2x$ and $\frac{d^2y}{dx^2} = 2$ for $y = x^2$. Thus, $x \cdot \frac{d^2y}{dx^2} - \frac{dy}{dx} = (x)(2) - 2 \cdot x = 0$.)

Probably the simplest form for a differential equation is

$$\frac{dy}{dx} = f(x). \tag{3}$$

Here, any antiderivative $y = F(x)$ for the function $f(x)$ is a solution, since $\frac{dy}{dx} = F'(x) = f(x)$. In fact, once we have identified a particular antiderivative $F(x)$, Theorem 13 guarantees that all solutions of (3) have the form

$$y = F(x) + C, \qquad C \text{ constant.} \tag{4}$$

We refer to (4) as the **complete solution** of differential equation (3) since all particular solutions of (3) may be obtained from (4) by appropriately specifying the constant C.

Frequently we will encounter differential equation (3) in the *differential* form

$$dy = f(x)\,dx. \tag{5}$$

To see that equations (3) and (5) are indeed equivalent, recall that the derivative $\frac{dy}{dx}$ may be viewed as the quotient of the differential $dy = \frac{dy}{dx} \cdot dx$ and the differential dx. Multiplying both sides of equation (3) by dx produces Equation (5), and vice versa.

Example 1 Find the most general solution of the differential equation

$$x\,dx - 3\,dy = 0.$$

Strategy

Solve for $\frac{dy}{dx}$.

Solution

Here $3\,dy = x\,dx$, so

$$\frac{dy}{dx} = \frac{x}{3}.$$

Find y by antidifferentiation.

Thus

$$y = \int \frac{x}{3}\,dx$$
$$= \frac{x^2}{6} + C$$

is the most general solution (see Figure 10.3). ■

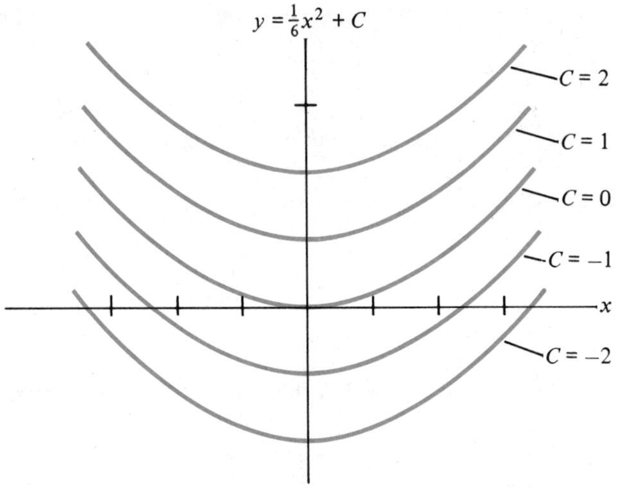

Figure 10.3 Solutions of $x\,dx - 3\,dy = 0$ are $y = \frac{1}{6}x^2 + C$.

As Example 1 illustrates, the most general solution of equation (3) is not a function but rather a *family* of functions of the form (4), one such function corresponding to each choice of the constant C. To obtain a *unique* solution to (3) we must also specify an *initial condition*. An **initial condition** is simply a point (x_0, y_0) through which the desired solution curve must pass. For example, if we specify the initial condition $y(1) = 3$ in Example 1, we obtain the equation

$$3 = y(1) = \frac{1}{6}(1)^2 + C = \frac{1}{6} + C,$$

so

$$C = 3 - \frac{1}{6} = \frac{17}{6}.$$

The particular solution satisfying this initial condition is therefore

$$y = \frac{1}{6}x^2 + \frac{17}{6}.$$

A problem consisting of a differential equation together with an initial condition is called an **initial value problem.**

Example 2 Solve the initial value problem.

$$dy = x\sqrt{x^2+1}\,dx, \qquad y(0) = 1$$

Strategy

Solve for $\frac{dy}{dx}$.

Find the antiderivative by the u-substitution method with $u = x^2 + 1$, $du = 2x\,dx$.

Apply initial condition to find C.

Solution

Here $\frac{dy}{dx} = x\sqrt{x^2 + 1}$,

so

$$\begin{aligned} y &= \int x\sqrt{x^2 + 1}\,dx \\ &= \int u^{1/2}\left(\frac{1}{2}\,du\right) \\ &= \frac{1}{2}\int u^{1/2}\,du \\ &= \frac{1}{2}\left(\frac{2}{3}u^{3/2} + C_1\right) \\ &= \frac{1}{3}u^{3/2} + C, \qquad C = \frac{1}{2}C_1 \\ &= \frac{1}{3}(x^2 + 1)^{3/2} + C. \end{aligned}$$

Then $y(0) = 1$ gives the equation

$$\begin{aligned} 1 = y(0) &= \frac{1}{3}(0^2 + 1)^{3/2} + C \\ &= \frac{1}{3} + C \end{aligned}$$

so

$$C = \frac{2}{3}.$$

The desired solution is therefore

$$y = \frac{1}{3}(x^2 + 1)^{3/2} + \frac{2}{3}.$$

■

Separation of Variables

Another type of differential equation for which we can often find a complete solution has the form

$$\frac{dy}{dx} = \frac{f(x)}{g(y)}, \qquad g(y) \neq 0. \tag{6}$$

Differential equations of this type are called **separable.** That is because if we multiply both sides of Equation (6) by the function $g(y)$ we obtain the equation

$$g(y) \cdot \frac{dy}{dx} = f(x). \tag{7}$$

If we can find antiderivatives G for g and F for f, we will have succeeded in **separating the variables** in equation (6), since the left side of (7) is the derivative

$$\frac{d}{dx}G(y) = G'(y)\frac{dy}{dx} = g(y)\frac{dy}{dx}$$

of the composite function $G(y) = G(y(x))$, and the right side of (7) is the derivative $F'(x) = f(x)$ of the function $F(x)$. It then follows, from Theorem 13, that

$$G(y) = F(x) + C, \qquad C = \text{constant} \tag{8}$$

on the domain common to $F(x)$ and $G(y(x))$. Equation (8), in general, defines the solution $y(x)$ of equation (6) implicitly, but equation (8) can often be solved to produce the solution $y(x)$ as an explicit function of x.

REMARK: The differential notation $dy = \left(\frac{dy}{dx}\right) dx$ is often used to abbreviate the above discussion by saying that

$$\frac{dy}{dx} = \frac{f(x)}{g(y)} \text{ implies that } g(y)\, dy = f(x)\, dx,$$

so

$$\int g(y)\, dy = \int f(x)\, dx,$$

and we obtain the solution y by "integrating both sides."

Example 3 Use the technique of separation of variables to find the complete solution of the differential equation

$$\frac{dy}{dx} = \frac{x}{y}, \qquad y \neq 0.$$

Strategy
Separate variables.

Solution
Multiplying both sides by y gives

$$y \cdot \frac{dy}{dx} = x$$

Find antiderivatives of both sides.

so

$$\int y\left(\frac{dy}{dx}\right) dx = \int x\, dx.$$

Thus

$$\frac{1}{2}y^2 + C_1 = \frac{1}{2}x^2 + C_2,$$

Solve for y^2, combining all arbitrary constants.

so

$$y^2 + 2C_1 = x^2 + 2C_2,$$

or

$$y^2 = x^2 + C$$

where $C = 2C_2 - 2C_1$.

Graphs of this equation, for various values of C, appear in Figure 10.4.

To verify this result, note that for $y > 0$,

$$y = (x^2 + C)^{1/2}$$

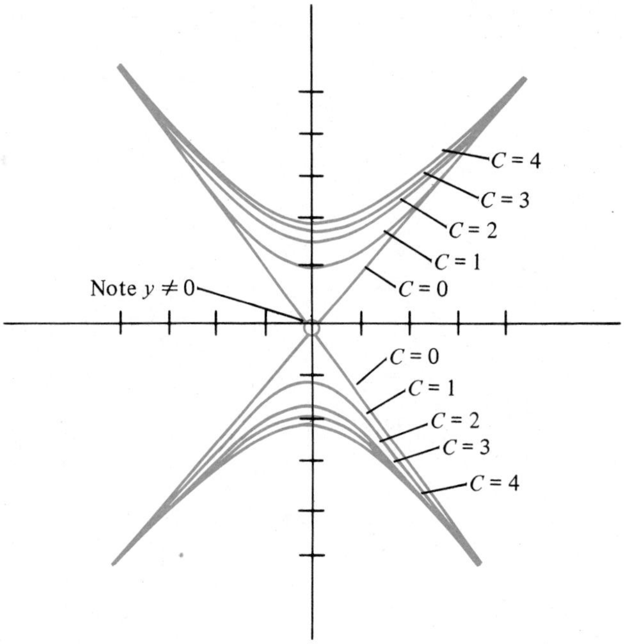

Figure 10.4 Solutions of the differential equation $\frac{dy}{dx} = \frac{x}{y}$.

and

$$\begin{aligned}\frac{dy}{dx} &= \frac{1}{2}(x^2 + C)^{-1/2}(2x)\\ &= x(x^2 + C)^{-1/2}\\ &= \frac{x}{y}\end{aligned}$$

and similarly for $y < 0$. ■

Example 4 Find the solution of the initial value problem

$$dy = 2xy^2\,dx, \qquad y(0) = 1/2.$$

Strategy

Solution

We have

Separate variables.

$$y^{-2} \cdot \frac{dy}{dx} = 2x,$$

so

Find antiderivatives of both sides.

$$\int y^{-2}\left(\frac{dy}{dx}\right) dx = \int 2x\,dx$$

and, therefore,

Solve for y.

$$-y^{-1} + C_1 = x^2 + C_2.$$

Thus

$$\frac{1}{y} = -x^2 - C, \qquad C = C_2 - C_1$$

so

$$y = \frac{-1}{x^2 + C}.$$

From the initial condition $y(0) = 1/2$ we obtain

Apply initial condition to solve for C.

$$\frac{1}{2} = y(0) = \frac{-1}{0^2 + C} = \frac{-1}{C}$$

so

$$C = -2.$$

The solution is therefore

$$y = \frac{-1}{x^2 - 2}.$$

■

Differential equations are an important part of applied mathematics, physics, engineering mathematics, mathematical biology, and physical chemistry. In later chapters we shall encounter other types of differential equations, techniques for their solutions, and applications involving such equations.

Exercise Set 5.10

In Exercises 1–10, find the most general solution of the differential equation.

1. $\frac{dy}{dx} = x - 1$

2. $\frac{dy}{dx} = 2 + \sec^2 x$

3. $\frac{dy}{dx} = x^2 - \frac{1}{x^2}$

4. $\frac{dy}{dx} = 4 \cos 2x$

5. $dy = (x - \sqrt{x})\, dx$

6. $dy + 3\, dx = 0$

7. $dy = \frac{x}{y}\, dx$

8. $\frac{dy}{dx} = \frac{x}{y^2}$

9. $dy = -4xy^2\, dx$

10. $dy = \frac{\sqrt{x}}{\sqrt{y}}\, dx$

In Exercises 11–18, find the solution of the initial value problem.

11. $\frac{dy}{dx} = \frac{x}{y}, \quad y(1) = 4$

12. $2y\, dy = \sqrt{x}\, dx, \quad y(0) = 2$

13. $dy = \frac{x^2}{(1 + x^3)^2}\, dx, \quad y(0) = \frac{1}{2}$

14. $dy = \frac{x}{y\sqrt{1 + x^2}}\, dx, \quad y(0) = 2$

15. $dy = y^2(1 + x^2)\, dx, \quad y(0) = 1$

16. $\frac{dy}{dx} = 2x^2y^2, \quad y(0) = -3$

17. $\frac{d^2y}{dx^2} = x^2, \quad y(0) = 2, \quad y'(0) = 4$

18. $\frac{d^2y}{dx^2} = \sin x, \quad y(0) = \pi, \quad y'(0) = 2$

19. The slope of the line tangent to the graph of $y = f(x)$ at $(x, f(x))$ is $x - \sqrt{x}$, $x > 0$. If the point $(1, 2)$ is on the curve, find $f(x)$.

20. Find a curve in the xy-plane which contains the point $(2, 1)$ and whose normal at (x, y) has slope y/x, $x > 0$.

SUMMARY OUTLINE OF CHAPTER 5

- ***Theorem*** (Mean Value Theorem): If $f(x)$ is continuous on $[a, b]$ and differentiable on (a, b), there exists $c \in (a, b)$ with

$$f'(c) = \frac{f(b) - f(a)}{b - a}.$$

- The function $f(x)$ is **increasing** on (a, b) if $f'(x) > 0$ for all $x \in (a, b)$, and **decreasing** on (a, b) if $f'(x) < 0$ for all $x \in (a, b)$.

- The number $f(c)$ is a relative **maximum** if $f(c) \geq f(x)$ for all x near c. The number $f(c)$ is a relative **minimum** if $f(c) \leq f(x)$ for all x near c. Relative maxima and minima are referred to as **relative extrema.**

- ***Theorem:*** If $f(c)$ is a relative extremum, then either $f'(c) = 0$ or $f'(c)$ fails to exist.

- ***Theorem:*** The graph of $y = f(x)$ is

  (a) **concave up** on $[a, b]$ if $f''(x) > 0$ for all $x \in (a, b)$,
  (b) **concave down** on $[a, b]$ if $f''(x) < 0$ for all $x \in (a, b)$.

- ***Theorem:*** If $f'(c) = 0$ and $f''(c)$ exists, then $f(c)$ is a

  (a) relative maximum if $f''(c) < 0$,
  (b) relative minimum if $f''(c) > 0$.

- ***Theorem*** (l'Hôpital's Rule): If $f(x)$ and $g(x)$ are differentiable and either $\lim_{x\to a} f(x) = \lim_{x\to a} g(x) = 0$, or $\lim_{x\to a} f(x) = \lim_{x\to a} g(x) = \infty$, then

$$\lim_{x\to a} \frac{f(x)}{g(x)} = \lim_{x\to a} \frac{f'(x)}{g'(x)}$$

provided this limit exists.

- The function $F(x)$ is called an **antiderivative** (or **indefinite integral**) for the function $f(x)$ if $F'(x) = f(x)$. The most general antiderivative for $f(x)$, if one exists, is $F(x) + C$ where $F'(x) = f(x)$ and C is an arbitrary constant. This is written

$$\int f(x)\, dx = F(x) + C.$$

- The **differential equation** of the form

$$\frac{dy}{dx} = \frac{f(x)}{g(y)}, \qquad \text{or} \qquad dy = \frac{f(x)}{g(y)}\, dy$$

may be solved by the technique of **separation of variables:**

$$\frac{dy}{dx} = \frac{f(x)}{g(y)} \Rightarrow g(y)\, dy = f(x)\, dx$$
$$\Rightarrow \int g(y)\, dy = \int f(x)\, dx$$
$$\Rightarrow G(y) = F(x) + C$$

where $G'(y) = g(y)$ and $F'(x) = f(x)$. The constant C is determined by an **initial condition** of the form $y(x_0) = y_0$.

REVIEW EXERCISES—CHAPTER 5

In Exercises 1–20, find the intervals on which $f(x)$ is increasing and decreasing, find all relative extrema and points of inflection, determine the concavity, and sketch the graph.

1. $y = 4x - x^2$

2. $y = x(x - 1)(x + 3)$

3. $f(x) = x^2 - 2x + 3$

4. $y = \sin 4x$

5. $f(x) = 2\sin(\pi x)$

6. $y = \dfrac{1 - x}{x}$

7. $f(t) = 2\cos^2(2t)$

8. $f(x) = \sqrt{1 - \sin^2 x}$

9. $y = \dfrac{x - 3}{x + 3}$

10. $f(t) = \dfrac{t^2}{t^2 - 1}$

11. $f(x) = x\sqrt{16 - x^2}$

12. $y = \dfrac{\sqrt{x}}{1 + \sqrt{x}}$

13. $y = \dfrac{t^2}{t^2 + 9}$

14. $y = \dfrac{2}{\sqrt{x}} + \dfrac{\sqrt{x}}{2}$

15. $f(x) = x^4 - 2x^2$

16. $y = (x + 1)(x - 1)^2$

17. $y = t^3 - 3t^2 + 2$

18. $f(x) = x + \cos x$

19. $f(x) = x^2 + \dfrac{2}{x}$

20. $y = \tan x + \cot x$

21. The sum of two numbers is 10. If one is increasing, what can you say about the other?

22. The sum of two positive numbers is 10. For what value(s) will the difference of their squares be a maximum?

23. Sketch a possible graph for a function that is continuous for all x and for which
a. $f(2) = 4$,
b. $f'(2) = 1$, and
c. $f''(x) < 0$ for all x.

24. Sketch a possible graph for a function that is continuous for all x and for which
a. $f(-2) = -f(2) = -3$,
b. $f'(-2) = f'(2) = 0$,
c. $f''(-2) > 0$, and
d. $f''(2) < 0$.

25. A particle moves along a line with velocity $v(t) = 2t - (t + 1)^{-2}$.
a. Find $s(t)$, its position at time t, if $s(0) = 0$.
b. Find $a(t)$, its acceleration at time t.

26. A particle moves along a line so that after t seconds it is $s(t) = t^3 - 6t^2 + 9t - 4$ units on the positive side of the origin.
a. Where does the particle lie at time $t = 0$?
b. In which direction is it moving at time $t = 0$?
c. When does it change direction?
d. How many times does it change direction?

27. A fence 3 meters tall is parallel to the wall of a building and 1 meter from the building. What is the length of the shortest ladder which can extend from the ground over the fence to the building wall?

28. Use the Intermediate Value Theorem and Rolle's Theorem to prove that if $f(x)$ has a continuous derivative for $x \in [a, b]$, if $f(a)$ and $f(b)$ have opposite signs, and if $f'(x) \neq 0$ for all $x \in [a, b]$, then the equation $f(x) = 0$ has *precisely* one root in $[a, b]$.

29. Prove that for any quadratic function $f(x) = ax^2 + bx + c$ and any interval $[\alpha, \beta]$, the midpoint $\gamma = \frac{1}{2}(\alpha + \beta)$ satisfies the conclusion of the Mean Value Theorem.

30. Sketch examples of graphs of functions defined on an interval $[a, b]$ for which the conclusion of the Mean Value Theorem is satisfied:
a. at precisely two points,
b. at precisely three points.

31. True or false? If $f(x)$ is increasing on $[a, b]$ and $f^{-1}(x)$ is defined on $[f(a), f(b)]$, then $f^{-1}(x)$ must be an increasing function on $[f(a), f(b)]$.

32. Show that the function $f(t) = t^3 - 6t^2 + 14t + 5$ is increasing for all t. (*Hint:* Complete the square on $f'(t)$.)

33. For $f(x) = x^2 + bx + c$ find a condition involving b and c which will ensure that $f(x) > 0$ for all x. (*Hint:* Ensure that the absolute minimum is greater than zero.)

In Exercises 34–43, find the limit using l'Hôpital's Rule where necessary.

34. $\lim_{x\to 0} \dfrac{x - \sin x}{\tan x}$

35. $\lim_{x\to 0} \dfrac{\pi - \csc x}{\pi + \cot x}$

36. $\lim_{x\to 0} \dfrac{\sin x}{\sqrt{x}}$

37. $\lim_{x\to 0} \dfrac{\tan 2x}{x}$

38. $\lim_{x\to 0^+} \dfrac{x - \tan x^2}{x - \sin x}$

39. $\lim_{x\to 3^+} \dfrac{9 - \sqrt{x}}{3 - \sqrt{x}}$

40. $\lim_{x\to 0} \left(\csc x - \dfrac{1}{x}\right)$

41. $\lim_{x\to 0^+} \left(\dfrac{1}{x} - \cot x\right)$

42. $\lim_{x\to 0^+} \left(2 \csc 2x - \dfrac{1}{x}\right)$

43. $\lim_{x\to 0} \dfrac{\tan 2x}{\tan x}$

In Exercises 44–57, find the most general antiderivative for the given function.

44. $f(x) = 6x^2 - 2x + 1$

45. $f(x) = (x^2 - 6x)^2$

46. $y = \dfrac{\sin \sqrt{t}}{\sqrt{t}}$

47. $f(x) = x \cdot \sqrt{9x^4}$

48. $f(x) = 3\sqrt{x} + 3/\sqrt{x}$

49. $y = (t + \sqrt{t})^3$

50. $y = x \sec^2 x^2$

51. $f(s) = s\sqrt{9 - s^2}$

52. $f(x) = \dfrac{x^3 - 7x^2 + 6x}{x}$

53. $y = \dfrac{x^3 + x^2 - x + 2}{x + 2}$

54. $f(x) = x \cos(1 + x^2)$

55. $f(x) = \dfrac{\sec x}{1 + \tan^2 x}$

56. $y = (2x - 1)(2x + 1)$

57. $y = \dfrac{x}{4x^4 + 4x^2 + 1}$

In Exercises 58–61, find the particular function satisfying the stated conditions.

58. $f'(x) = 1 + \cos x, \quad f(0) = 3$

59. $f'(x) = \dfrac{x}{\sqrt{1 + x^2}}, \quad f(0) = 4$

60. $f''(x) = 3, \quad f'(1) = 6, \quad f(0) = 4$

61. $f''(x) = \sin x - \cos x, \quad f'(0) = 3, \quad f(0) = 0$

62. Find an equation for the graph whose slope at any point is twice the x-coordinate of that point, and that contains the point (2, 9).

63. Find an equation for the graph containing the point (0, 1) whose slope at any point is the quotient of its x-coordinate divided by its y-coordinate.

64. From what height must a ball be dropped in order to strike the ground with a velocity of −49 m/sec? (The acceleration due to gravity is −9.8 m/sec.)

In Exercises 65–68, find the solution of the initial value problem.

65. $\dfrac{dy}{dx} = \sec^2 x, \quad y(0) = 1$

66. $\dfrac{dy}{dx} = \dfrac{x + 1}{y}, \quad y(1) = 2$

67. $dy = 2xy^2\,dx, \quad y(0) = -1/2$

68. $dy = \dfrac{\sqrt{x + 1}}{\sqrt{y}}\,dx, \quad y(0) = 1$

Archimedes

Bonaventura Cavalieri

John Wallis

Georg Friedrich Bernhard Riemann

Carl Friedrich Gauss

UNIT 3

INTEGRATION

The Origins of Integration

Historically, the problem of finding the area of a region under a curve, which leads to the integral, was studied earlier and more intensively than the problem of finding the tangent to a curve, which leads to the derivative.

More than two millenia ago the Greek mathematician Eudoxus of Cnidus (an island in the Mediterranean Sea near present-day Turkey) (c. 408–355 B.C.) stated what is called the Method of Exhaustion. This principle says that if one successively subtracts from a quantity at least half of it, and from the remainder at least its half, and so on, eventually there will remain something smaller than any preassigned quantity. (''Quantity'' may refer to a length, an area, or a volume.) As one example, Eudoxus applied this concept to achieve a difference between the area of a circle and that of an inscribed regular polygon that was arbitrarily small, and thus eventually to prove that the ratio of the areas of two circles equals the ratio of the squares of their diameters. He did not actually determine a formula for the area of a circle, however. He also proved a similar theorem concerning the volumes of spheres. Archimedes wrote that Eudoxus was the first to prove that the volume of a cone is one-third the volume of a cylinder having the same base and altitude. Eudoxus's work is contained in Book V of Euclid's *Elements,* and many applications are given in Book VI.

Archimedes of Syracuse (in Sicily) (287–212 B.C.) was the greatest mathematician of antiquity. He made much use of the Method of Exhaustion. However, it was not until 1906 that the discovery of a manuscript in a Constantinople library showed how Archimedes actually thought. This manuscript, copied in a tenth-century hand, contains several works of Archimedes; among them is a letter to his friend Eratosthenes, a mathematician and librarian at Alexandria. The Archimedes material had been washed away in the thirteenth century in order to re-use the parchment for religious texts, but most of the earlier writing can be read. (Such a re-used parchment is called a *palimpsest*.) Archimedes explained to his friend that he was not very certain of the validity of his method, although it did seem to give correct results. He would cut an area or volume to be measured into infinitely many parallel lines or sections, which (in his imagination) he would place at one end of a lever so as to balance an area or volume at the other end whose measure and center of gravity were known. Of course, Archimedes had no idea that he was so close to integral calculus, two thousand years before Newton!

In 1635, an Italian named Bonaventura Cavalieri (1598–1647) introduced, in *Geometria indivisibilibus,* his Method of Indivisibles. It was a precursor to integration, and an extension of Archimedes's work. Cavalieri did not define his terms well, but apparently an indivisible of a planar region is a line across that region, and an indivisible of a solid is a section by a plane. The region or solid is made up of infinitely many parallel chords or sections. If these parallel lines (or planes) are slid into other conformations, the total area (or volume) is unchanged. This principle can be used in ingenious ways to find areas and volumes of complex structures. Cavalieri wrote on optics and astronomy as well as mathematics, and is credited with introducing logarithms in Italy.

John Wallis (1616–1703) was Professor of Geometry (i.e., Mathematics) at Oxford University in England for an astonishing 54 years. He was one of the first to treat conics as equations of the second degree rather than as sections of a cone, and was the first to explain negative, fractional, and zero exponents in detail. He also created the symbol (∞) still used for infinity. In the direction of calculus, he performed many definite integrations, essentially equivalent to $\int_0^1 x^m\, dx$, and his work with exponents allowed him to let m take on many values (except -1).

After Newton and Leibniz published their new calculus, much controversy ensued as to who was first. Others indulged in a more dignified argument concerning the foundations of the mathematics itself. After about 150 years, the German mathematician Georg Friedrich Bernhard Riemann (1826–1866) put the integral onto a firm logical basis, treating it as the limit of upper and lower sums rather than as the inverse of differentiation. The modern presentation of the integral is mostly due to him. Riemann became a lecturer at the University of Göttingen at the age of 28. His probationary lecture was attended by his teacher, the great mathematician Carl Friedrich Gauss. For one of the few times in his life, Gauss applauded a mathematics lecture. Riemann's topic was the hypotheses that underlie the foundations of geometry; half a century later the concepts that he presented were used by Albert Einstein as the basis for the theory of general relativity.

Pronunciation Guide

Eudoxus (You-dox'us)

Archimedes (Ark'i-me'deez)

Bonaventura Cavalieri (Bone'a-ven-tur'a Kav'a-leer'ee)

Georg Friedrich Bernhard Riemann (Jorj Freed'rik Bern'hard Ree'mahn)

Carl Friedrich Gauss (Carl Freed'rik Gouse)

(Photographs: Archimedes, Culver Service; Bonaventura Cavalieri and Georg Friedrich Bernhard Riemann, David Eugene Papers, Rare Book and Manuscript Library, Columbia University; John Wallis and Carl Friedrich Gauss, the Library of Congress.)

CHAPTER 6

THE DEFINITE INTEGRAL

6.1 INTRODUCTION

The primary motivation for the discussions in this chapter is the problem of defining and calculating the area of a region in the plane, one of the oldest problems in mathematics. When the region consists of one or more figures from plane geometry (e.g., square, circle), the formulas developed in high school geometry quickly give the solution. But when the region is bounded by an arbitrary smooth curve, the answer is not so easily obtained.

For example, consider the region R represented in Figure 1.1. Rather than trying to determine the area of R directly, we will use a version of a procedure used by the ancient Greeks to *approximate* the area, A, of R.

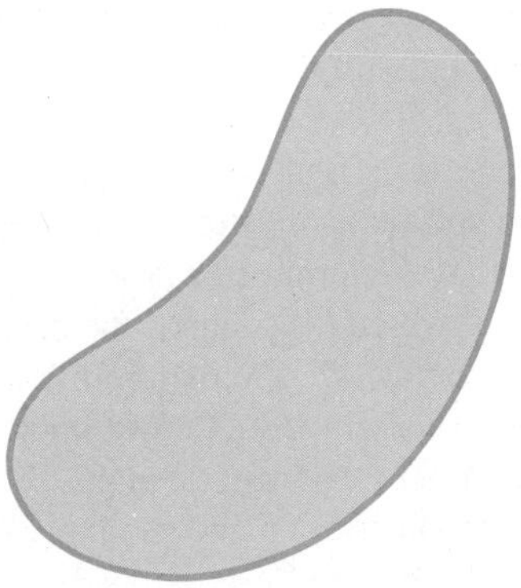

Figure 1.1 Region R.

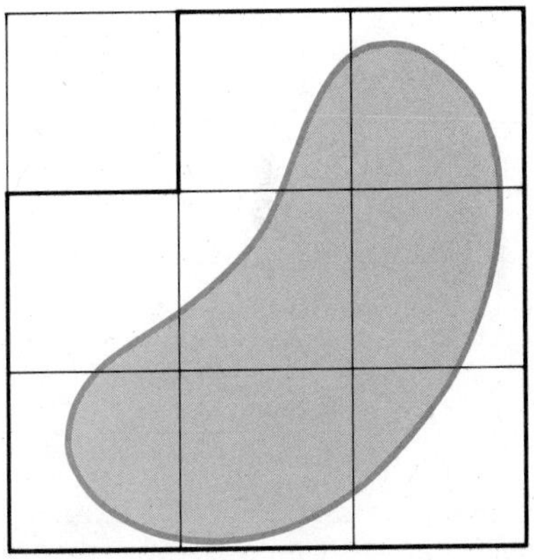

Figure 1.2 No square is contained in R; 8 squares together contain R.

If we impose a grid consisting of one-centimeter squares upon region R (see Figure 1.2) and count the number of squares that (a) lie entirely within R (in this case, none do) and (b) contain some portion of R (in this case, eight do), we obtain the estimate

$$0 \leq A \leq 8 \text{ cm}^2.$$

We can improve this first estimate by subdividing each square into four smaller squares, each of area $1/4$ cm^2 (see Figure 1.3). Doing so, we observe (a) that 7

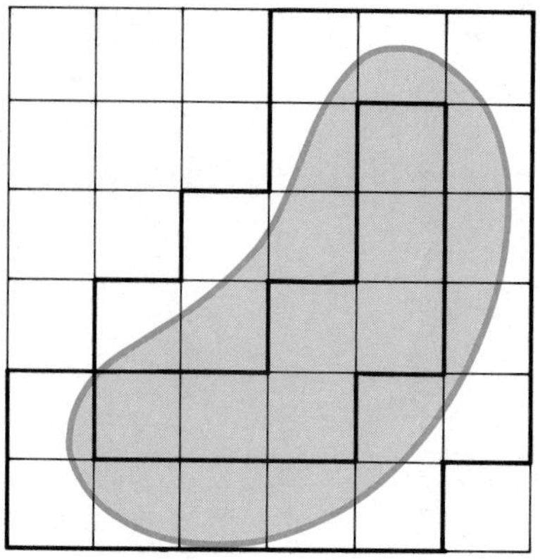

Figure 1.3 7 squares contained in R; 26 squares together contain R.

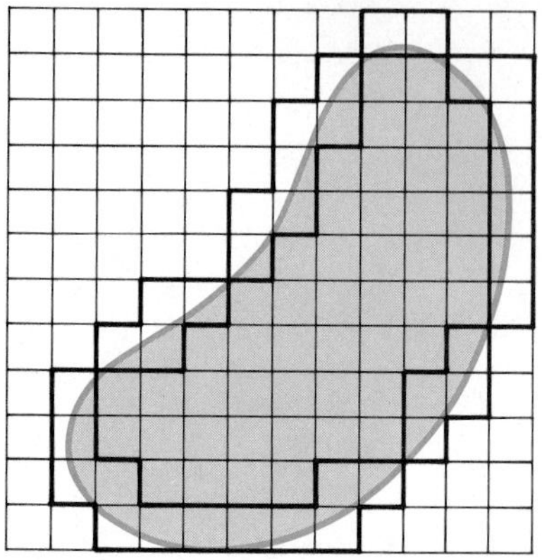

Figure 1.4 48 squares contained in R; 84 squares together contain R.

squares, each of area $1/4 \text{ cm}^2$, lie entirely within R and (b) that 26 squares contain some portion of R. Thus, we arrive at the improved estimate

$$\frac{7}{4} \text{ cm}^2 \leq A \leq \frac{26}{4} \text{ cm}^2.$$

A further subdivision of the grid into squares of area $1/16 \text{ cm}^2$ (see Figure 1.4) gives the estimate

$$\frac{48}{16} \text{ cm}^2 \leq A \leq \frac{84}{16} \text{ cm}^2.$$

Clearly, this procedure can be repeated many times, with each halving of the grid dimensions yielding a new estimate. (No wonder that the Greeks referred to this as the ''Method of Exhaustion''!)

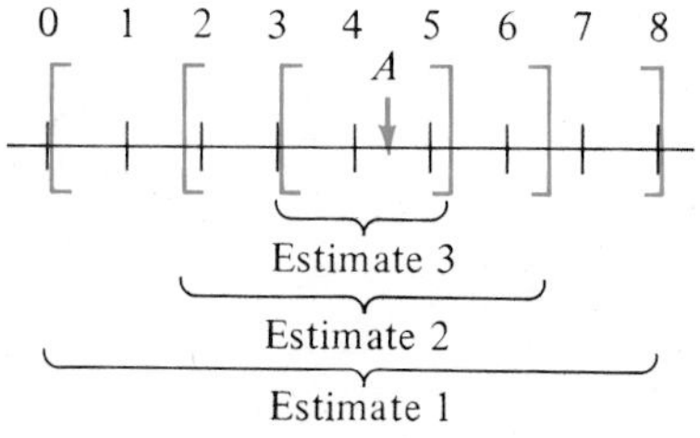

Figure 1.5 Nested intervals, each containing the desired number A.

To explain our interest in the preceding description, we consider the set of estimates obtained so far, as illustrated by use of the number line in Figure 1.5. The point is that we have determined a procedure for generating a sequence of ''nested'' intervals, each of which (a) contains the desired number, A, and (b) is smaller than the preceding interval. We are therefore led to conjecture that ''in the limit'' this sequence of intervals will shrink down upon a single number, which we take to be the desired area A. The exercise set in this section contains several examples of how familiar area formulas from plane geometry can be obtained by approximation procedures similar to the one described here.

Area as a Limit of a Sequence

Beginning in Section 6.2, we shall pursue the area problem in a more structured manner than we have done here. However, a characteristic that will continue throughout the chapter is that we shall define and calculate areas as limits of sequences of numbers. An infinite **sequence** of numbers is simply an infinite list

$$a_1, a_2, a_3, \ldots, a_n, \ldots.$$

We will use the bracket notation $\{a_1, a_2, a_3, \ldots\}$ to denote a sequence. Some sequences can be constructed by a formula that gives the value of the nth number in terms of n; in that case the entire sequence may be represented as $\{f(n)\}$. For instance, $\left\{\frac{1}{n}\right\}$ represents the sequence $\left\{1, \frac{1}{2}, \frac{1}{3}, \ldots\right\}$.

We say that a sequence $\{a_1, a_2, a_3, \ldots\}$ has a *limit* L if the numbers a_n approach the number L as n becomes increasingly large. That is,

$$L = \lim_{n\to\infty} a_n \quad \text{if} \quad a_n \to L \text{ as } n \to \infty.$$

For example, the sequence

$$\left\{\frac{1}{n}\right\} = \left\{1, \frac{1}{2}, \frac{1}{3}, \frac{1}{4}, \ldots, \frac{1}{n}, \ldots\right\}$$

has limit $0 = \lim_{n\to\infty} \frac{1}{n}$, while the sequence

$$\left\{\frac{n+3}{2n}\right\} = \left\{\frac{4}{2}, \frac{5}{4}, \frac{6}{6}, \frac{7}{8}, \ldots\right\}$$

has limit $\frac{1}{2} = \lim_{n\to\infty} \frac{n+3}{2n} = \lim_{n\to\infty} \frac{1 + \left(\frac{3}{n}\right)}{2} = \frac{1+0}{2}$.

We shall make a detailed study of infinite sequences in Chapter 13. At that point we will formally prove some of the statements that we justify only on an intuitive level in Chapter 6. However, we will be using little more than the straightforward notion that the statement $L = \lim_{n\to\infty} a_n$ for the sequence $\{a_1, a_2, a_3, \ldots\}$ represents just the same notion as $L = \lim_{x\to\infty} f(x)$ for functions, except that the **terms** of the sequence $\{a_n\} = \{f(n)\}$ occur only at integers.

Summation Notation

Throughout the first two sections of this chapter we will be adding up lists of numbers to develop approximations to certain areas. The summation notation involving the symbol Σ (sigma) is very useful in doing this. Recall its meaning:

$$\sum_{j=1}^{n} x_j \quad \text{means} \quad x_1 + x_2 + x_3 + \cdots + x_{n-1} + x_n.$$

Example 1

$$\sum_{j=1}^{4} (2j+1) = (2(1)+1) + (2(2)+1) + (2(3)+1) + (2(4)+1)$$
$$= 24.$$

■

Example 2

$$\sum_{j=1}^{3} (9 - j^2) = (9 - 1^2) + (9 - 2^2) + (9 - 3^2) = 13.$$

■

General Remarks

The theory we shall develop here is concerned with continuous functions. Although many of the results obtained in this chapter are true for more general types of functions, we restrict our considerations to continuous functions for three reasons:

(1) Much of the theory of this chapter can be established for the case of continuous functions by use of the Mean Value Theorem, and we use this approach. This allows us to emphasize the strong interrelationships between the differential and integral calculus.

(2) The ability to apply the definite integral to a variety of geometric and physical problems depends on a strong understanding of the integral as the limit of an approximating sum. By restricting our attention to continuous functions, we can develop the definite integral in precisely this way.

(3) The notions concerning the definite integral are both new and important. We prefer to keep the discussion as simple as possible, focusing on basic properties and techniques without striving too quickly for generalities.

Exercise Set 6.1

1. Figures 1.6 and 1.7 show the result of using n circumscribing rectangles of uniform thickness h/n to approximate the area of a triangle of base b and altitude h.

a. Show that the approximation to the area A illustrated by Figure 1.6 is

$$A \approx \frac{bh}{n^2}\{1 + 2 + \cdots + n\}.$$

b. Use the formula $1 + 2 + 3 + \cdots + k = \dfrac{k(k+1)}{2}$ to show that the approximation in (a) can be written

$$A \approx \frac{1}{2}bh + \frac{bh}{2n}.$$

c. Show that $\lim\limits_{n\to\infty}\left(\dfrac{1}{2}bh + \dfrac{bh}{2n}\right) = \dfrac{1}{2}bh$, and conclude that the area of the triangle is the limit of the sequence of approximations.

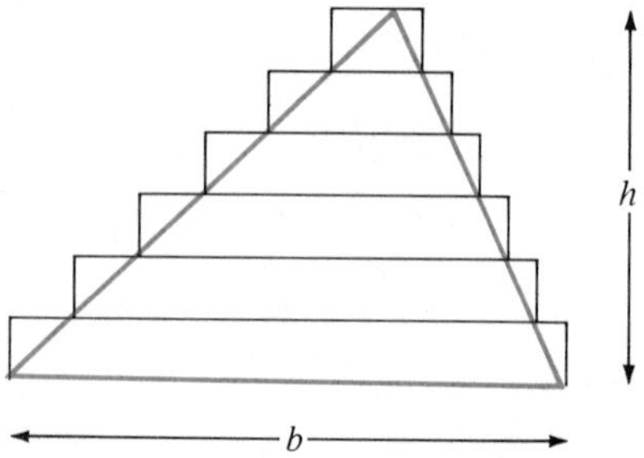

Figure 1.6 Area of triangle approximated by circumscribing rectangles.

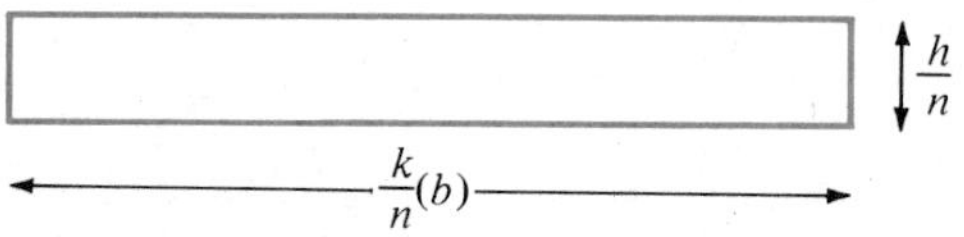

Figure 1.7 Area of kth rectangle from the top is $A_k = \left(\dfrac{h}{n}\right)\cdot\left(\dfrac{kb}{n}\right)$.

2. Figures 1.8 and 1.9 show how the idea of Exercise 1 can be used to approximate the area of a circle of radius r.

a. Show that the resulting approximation is

$$A \approx 2\sum_{k=1}^{n}\left(\frac{2r^2}{n}\right)\sqrt{1-\frac{k^2}{n^2}} = \frac{4r^2}{n}\sum_{k=1}^{n}\sqrt{1-\frac{k^2}{n^2}}$$

b. Use a hand calculator or a computer to calculate values of this approximation for various values of n and r, and compare your results with the actual value $A = \pi r^2$. (The results of having done so on a computer appear in Table 1.1).

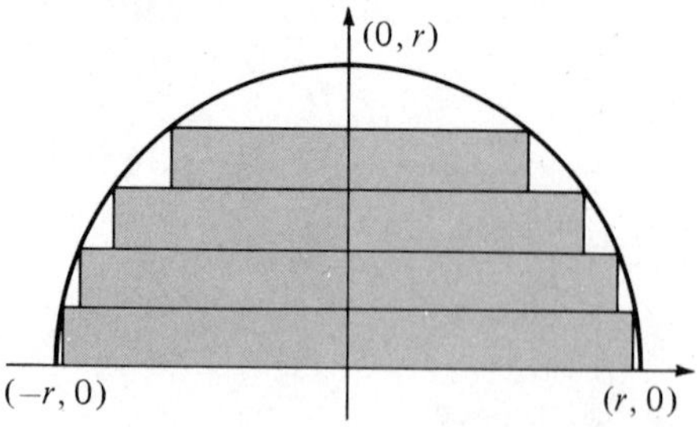

Figure 1.8 Semicircle approximated by inscribed rectangles.

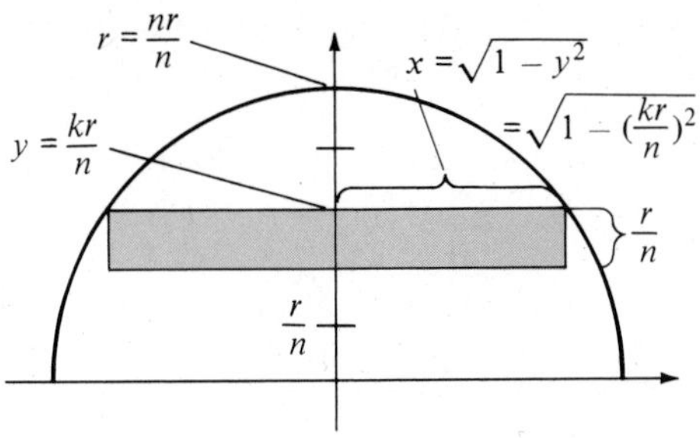

Figure 1.9 Approximating rectangle has area $A_k = 2xy = 2\left(\frac{kr}{n}\right)\sqrt{1 - \left(\frac{kr}{n}\right)^2}$.

Table 1.1 Approximations to the areas of circles of various radii using the approximation in Exercise 2.

r	n	Approximation	$A = \pi r^2$(actual)
1	5	2.63705	3.14159
1	10	2.90452	
1	20	3.02846	
1	50	3.09827	
1	100	3.12042	
1	500	3.13749	
3	5	23.73344	28.27433
3	10	26.14066	
3	20	27.25618	
3	50	27.88442	
3	100	28.08375	
3	500	28.23739	
8	5	168.77113	201.06193
8	10	185.88917	
8	20	193.82175	
8	50	198.28918	
8	100	199.70669	
8	500	200.79920	
8	1000	200.93155	

3. Using a ruler and compass, draw a circle circumscribing a square and circumscribed by another square as in Figure 1.10. By measuring the perimeters of the two squares, obtain an interval estimate for the circumference of the circle. State a procedure for generating an improved estimate, and use this procedure to obtain one or two better estimates. What should you observe about the relationship between the diameter of the circle and your estimates?

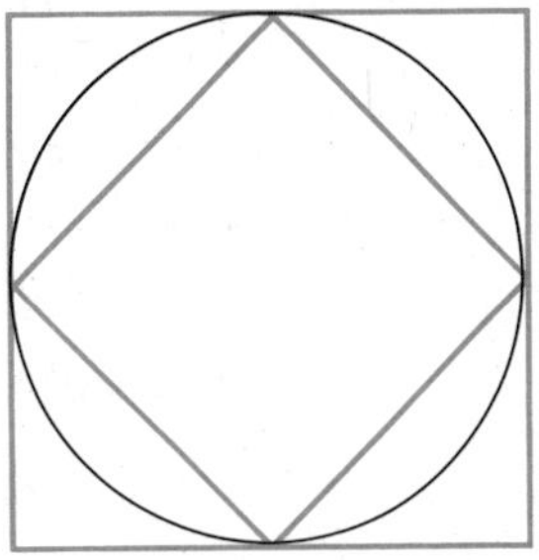

Figure 1.10

4. What can you say about the accuracy of the interval estimates in Figure 1.5? What is the geometric significance of the possible error in these estimates? (Refer to Figures 1.2–1.4.)

5. Here is a procedure for approximating the area of a circle different from that of Exercise 2. As in Exercise 1, we construct a regular polygon of n sides circumscribed by the circle of radius r (see Figure 1.11). If we construct radii from the center of the circle to each vertex of the polygon, we divide the polygon into $2n$ congruent right triangles.

a. Show that the base angle of each triangle is π/n.

b. Show that the area of each triangle is

$$\bar{A} = \frac{r^2}{4}\sin\left(\frac{2\pi}{n}\right)$$

(*Hint:* Use the identity $\sin 2\theta = 2\sin\theta\cos\theta$).

c. Show that the resulting approximation to the area of the circle is

$$A = \frac{nr^2}{2}\sin\left(\frac{2\pi}{n}\right).$$

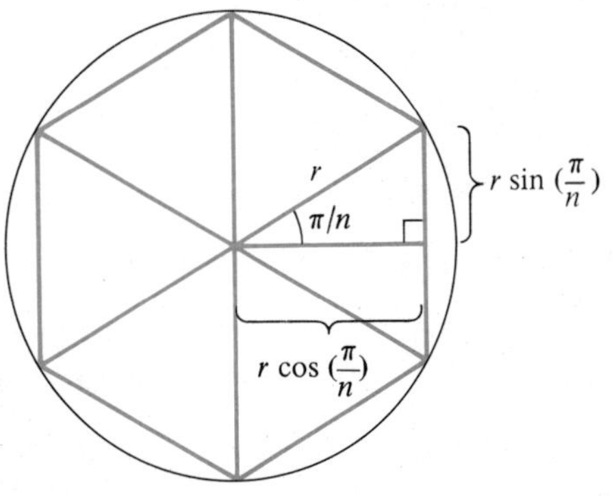

Figure 1.11 Circle of radius r.

d. If you have access to a hand calculator or computer, complete Table 1.2 comparing the approximation obtained in part c with the actual value $A = \pi r^2$ for various values of n and r.

Table 1.2

r	n	$A \approx \frac{nr^2}{2}\sin\left(\frac{2\pi}{n}\right)$	$A = \pi r^2$
1	5	2.37764	3.14159
1	20		3.14159
1	50	3.13333	3.14159
.4	5	30.04225	
.4	20	49.44268	
.4	50	50.13325	
.4	200		
.10	5	237.76405	
.10	20	309.06174	
.10	50	313.33282	
.10	200		
.10	500		
.10	1000	314.15693	

6. The approximations in Table 1.1 are smaller than the actual values they approximate. Why is this true?

7. Will the approximations to the area of a triangle obtained using the approximation in Exercise 1 be larger or smaller than the actual value? Why?

8. Here is another way to approximate the area of a circle. This time we construct the regular polygon of n sides circumscribing the circle of radius r (Figure 1.12). We then construct line segments from the center of the circle to each vertex of the polygon and also construct radii joining the center of the circle with the midpoints of each of the sides of the polygon. This divides the polygon into $2n$ congruent right triangles.

a. Show that the area of each triangle is

$$\overline{A} = \frac{r^2}{2}\tan\left(\frac{\pi}{n}\right).$$

b. Show that the resulting approximation to the area of the circle is

$$A = nr^2\tan\left(\frac{\pi}{n}\right).$$

c. If a calculator or computer is available to you, construct a table like Table 1.2, using the approximation in (b).

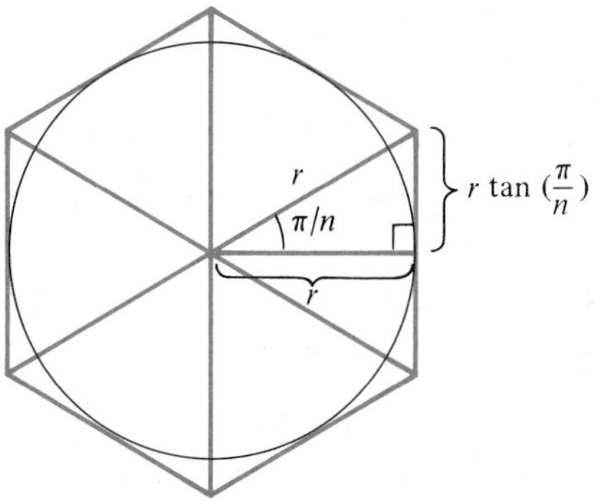

Figure 1.12

9. Use the method of Exercise 1 to obtain the formula $A = \frac{1}{2}h(B + b)$ for the area of the trapezoid with bases of lengths B and b and altitude h as a limit of an approximation scheme.

10. Use the method of Exercise 2 to find an approximation procedure for calculating the area of the region bounded by the graph of $y = 4 - x^2$ and the x-axis.

11. *(Computer)* Use the approximation procedure in Exercise 10 to construct a table of approximations to the area of the region described in Exercise 8.

6.2 THE AREA PROBLEM: APPROXIMATING SUMS

In order to make use of the language of functions, we shall pursue the following formulation of the Area Problem.

AREA PROBLEM: Let $f(x)$ be a continuous function for $x \in [a, b]$. Find the area of the region R bounded by the graph of $f(x)$, the x-axis, and the lines $x = a$ and $x = b$.

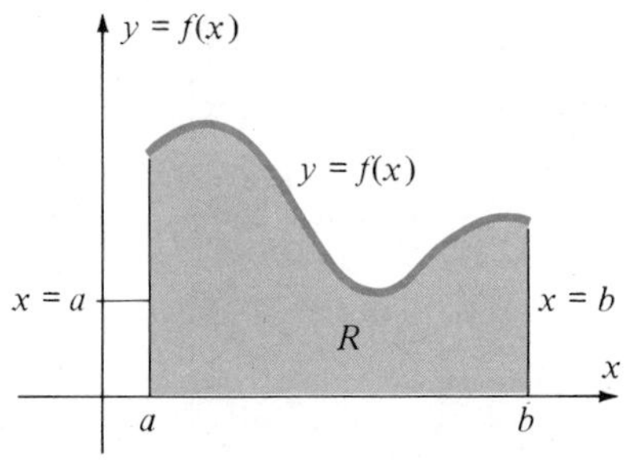

Figure 2.1 Region R bounded by graph of $f(x)$ and x-axis for $a \leq x \leq b$.

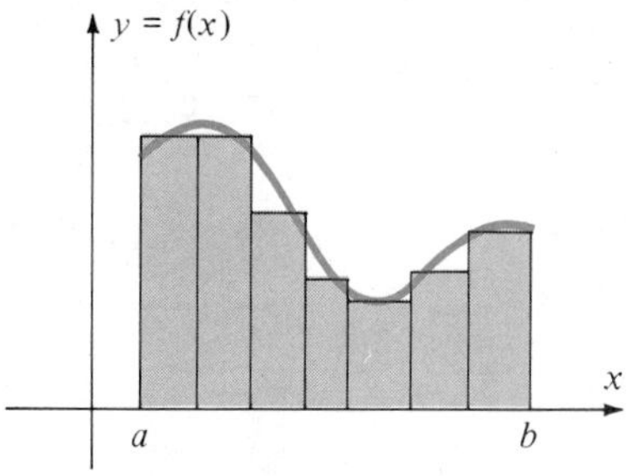

Figure 2.2 Approximating the region R by rectangles.

Figure 2.1 represents a typical region R for the case of a nonnegative function $f(x)$. Figure 2.2 shows our strategy for attacking the Area Problem. In order to define the area A, we use the idea of approximating the region R with figures of known area, as in Section 6.1. The shapes that we shall use are rectangles, and we will execute our approximation scheme in such a way that, ultimately, we will be able to calculate the limit of the sequence of approximations for a wide class of functions $f(x)$.

Figure 2.3 illustrates the basic idea behind the approximation of R by rectangles. As the number of rectangles increases, and the sizes of the individual rectangles decrease, the union of the set of approximating rectangles more accurately ''fits'' the region R. Our intention is to define the area A of R to be the limiting value of the approximations associated with these rectangles. Of course, we must first show that such a limit exists, and that is the purpose of this section.

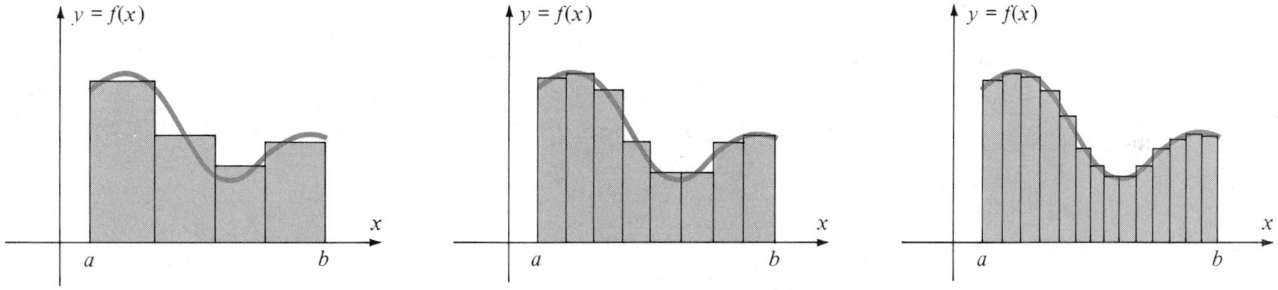

Figure 2.3 As the number n of subintervals increases, the approximation of R by rectangles becomes more precise.

Lower Approximating Sums

A **lower approximating sum** $\underline{S}_n$ for the region R is an approximation of R by n rectangles of equal width *each of which is entirely contained within R* (see Figures 2.4 and 2.5). Since we use n rectangles of equal width, the width of each is $\Delta x = \dfrac{b-a}{n}$, and the endpoints of the resulting subintervals are

$$x_0 = a,\ x_1 = a + \Delta x,\ x_2 = a + 2\,\Delta x,\ \ldots,\ x_n = a + n\,\Delta x = b. \tag{1}$$

Obviously, we want the height of the rectangle constructed over the interval $[x_{j-1}, x_j]$ to be the minimum value of $f(x)$ on $[x_{j-1}, x_j]$. We can state this more precisely by recalling that, since $f(x)$ is continuous on $[x_{j-1}, x_j]$, there is at least one number $c_j \in [x_{j-1}, x_j]$ with

$$f(c_j) = \min\{f(x) \mid x_{j-1} \leq x \leq x_j\}$$

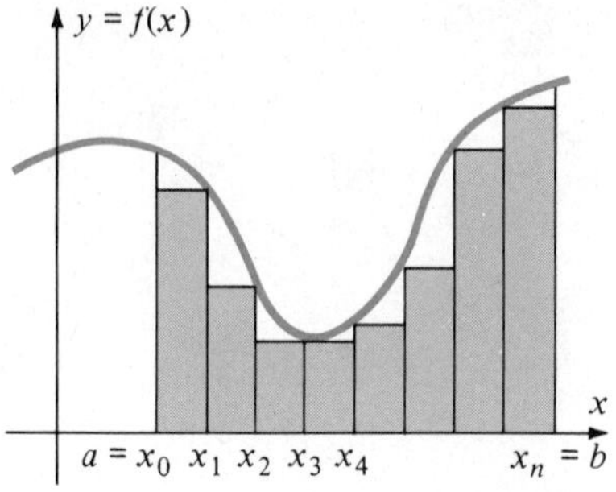

Figure 2.4 A lower approximating sum for R. Union of rectangles lies entirely within R.

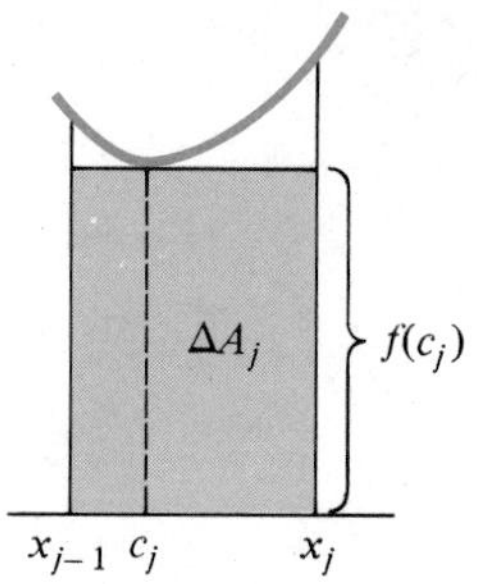

Figure 2.5 The jth approximating rectangle has area $\Delta A_j = f(c_j)\,\Delta x$.

(Theorem 1, Chapter 4). Using this notation we can write the area ΔA_j of the jth *inscribed* rectangle as

$$\Delta A_j = f(c_j)\Delta x.$$

The lower approximating sum $\underline{S}_n$ is therefore

$$\begin{aligned}\underline{S}_n &= \sum_{j=1}^{n} \Delta A_j \\ &= \Delta A_1 + \Delta A_2 + \cdots + \Delta A_n \\ &= f(c_1)\Delta x + f(c_2)\Delta x + \cdots + f(c_n)\Delta x \\ &= \sum_{j=1}^{n} f(c_j)\Delta x.\end{aligned}$$

Example 1 Find the lower approximating sum $\underline{S}_4$ for the area of the region R bounded by the graph of $f(x) = 4 - x^2$ and the x-axis between $x = 0$ and $x = 2$.

Solution: Since we use subintervals of equal length Δx, and since $n = 4$, the length of each subinterval is

$$\Delta x = \frac{2-0}{4} = \frac{1}{2}.$$

The endpoints of the subintervals are therefore

$$x_0 = 0, \quad x_1 = \frac{1}{2}, \quad x_2 = 1, \quad x_3 = \frac{3}{2}, \quad \text{and} \quad x_4 = 2.$$

Since $f(x) = 4 - x^2$ is decreasing on $[0, 2]$, the smallest value of $f(x)$ on each subinterval occurs at the right endpoint. Thus,

$$c_1 = \frac{1}{2}, \quad c_2 = 1, \quad c_3 = \frac{3}{2}, \quad \text{and} \quad c_4 = 2.$$

The lower approximating sum is, therefore,

$$\underline{S}_4 = f\left(\frac{1}{2}\right)\cdot\left(\frac{1}{2}\right) + f(1)\left(\frac{1}{2}\right) + f\left(\frac{3}{2}\right)\left(\frac{1}{2}\right) + f(2)\left(\frac{1}{2}\right)$$

$$= \left[4 - \left(\frac{1}{2}\right)^2\right]\left(\frac{1}{2}\right) + [4 - (1)^2]\left(\frac{1}{2}\right) + \left[4 - \left(\frac{3}{2}\right)^2\right]\left(\frac{1}{2}\right) + [4 - (2)^2]\left(\frac{1}{2}\right)$$

$$= \left[\frac{15}{4} + 3 + \frac{7}{4} + 0\right]\left(\frac{1}{2}\right)$$

$$= \frac{17}{4}.$$

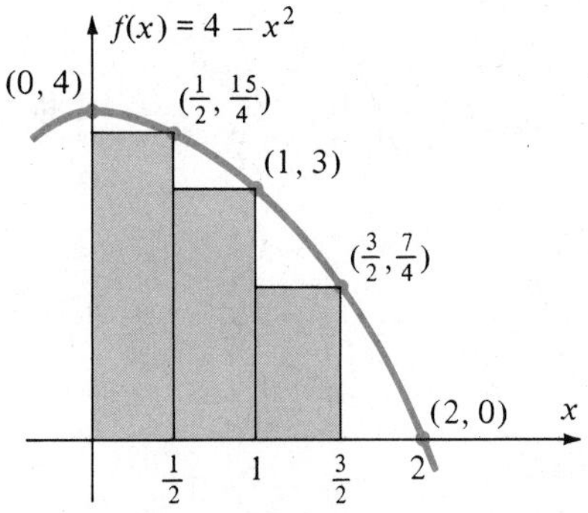

Figure 2.6 Lower approximating sum $\underline{S}_4$ in Example 1.

We will later show that the actual value of the area is $A = \dfrac{16}{3}$. Thus, $\underline{S}_4 < A$ (see Figure 2.6). ■

A simple but important observation concerning lower sums is that, since each of the associated rectangles lies entirely within the region R, our definition of area should provide that

$$\underline{S}_n \leq A \qquad (2)$$

for all lower approximating sums $\underline{S}_n$.

Upper Approximating Sums

If, instead of using the smallest value of $f(x)$ on each subinterval $[x_{j-1}, x_j]$, we use the *largest* value, we obtain what is called an **upper approximating sum,** $\overline{S}_n$. That is, we again divide $[a, b]$ into n subintervals of equal length $\Delta x = \dfrac{b-a}{n}$, with endpoints as in (1), but we take the height of the rectangle over the interval $[x_{j-1}, x_j]$ to be the value $f(d_j)$, where

$$f(d_j) = \max\{f(x) \,|\, x_{j-1} \leq x \leq x_j\}.$$

(Again, the existence of at least one such number $d_j \in [x_{j-1}, x_j]$ is guaranteed by the continuity of $f(x)$.) The area of the jth approximating rectangle is, therefore, $\Delta A_j = f(d_j)\Delta x$, and the upper approximating sum is

$$\overline{S}_n = \sum_{j=1}^{n} \Delta A_j = \sum_{j=1}^{n} f(d_j)\Delta x$$

(See Figures 2.7 and 2.8.)

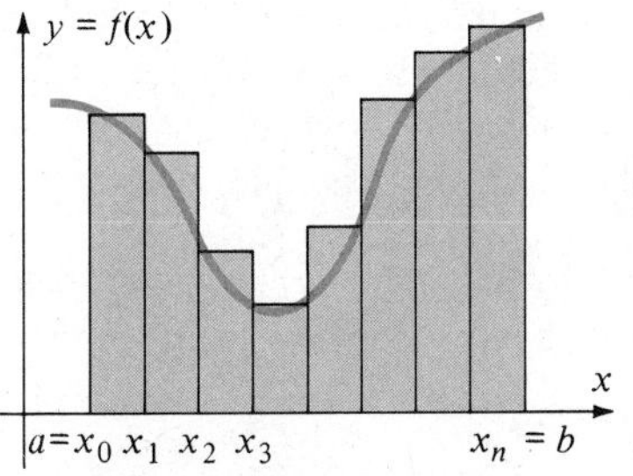

Figure 2.7 An upper approximating sum. The region R lies entirely within the union of the rectangles.

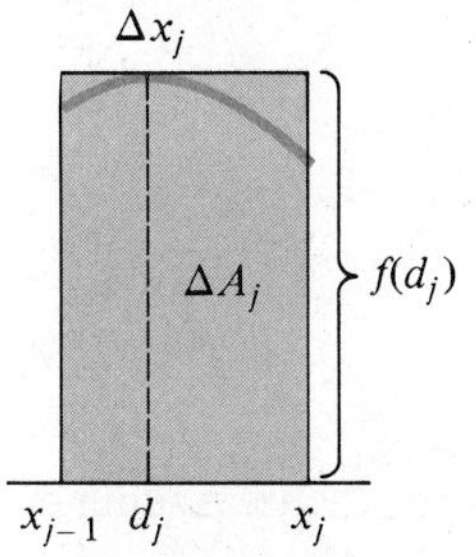

Figure 2.8 The jth approximating rectangle has area $\Delta A_j = f(d_j)\,\Delta x$.

Example 2 Find the upper approximating sum $\overline{S}_4$ for the region R in Example 1.

Solution: As in Example 1, we have subintervals of length $\Delta x = 1/2$ and with endpoints

$$x_0 = 0, \qquad x_1 = \frac{1}{2}, \qquad x_2 = 1, \qquad x_3 = \frac{3}{2}, \qquad \text{and} \qquad x_4 = 2.$$

However, since $f(x) = 4 - x^2$ is decreasing on $[0, 2]$, the maximum value of $f(x)$ will occur at the *left* endpoint of each subinterval. We will therefore have

$$c_1 = 0, \qquad c_2 = \frac{1}{2}, \qquad c_3 = 1, \qquad \text{and} \qquad c_4 = \frac{3}{2}.$$

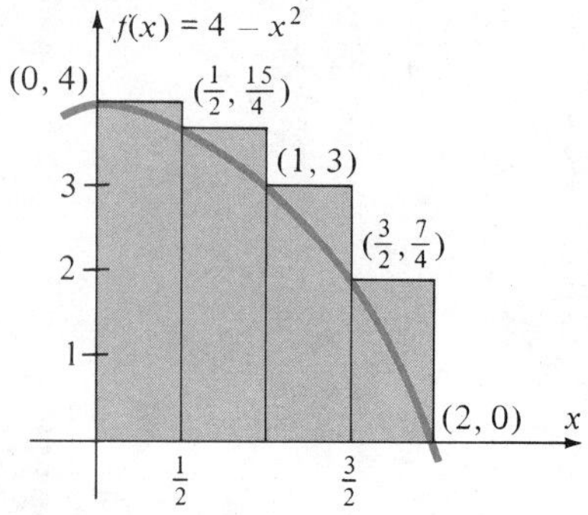

Figure 2.9 Upper approximating sum $\overline{S}_4$ in Example 2.

The upper approximating sum $\overline{S}_4$ is

$$\begin{aligned}\overline{S}_4 &= f(0)\left(\frac{1}{2}\right) + f\left(\frac{1}{2}\right)\left(\frac{1}{2}\right) + f(1)\left(\frac{1}{2}\right) + f\left(\frac{3}{2}\right)\left(\frac{1}{2}\right)\\ &= [4 - (0)^2]\left(\frac{1}{2}\right) + \left[4 - \left(\frac{1}{2}\right)^2\right]\left(\frac{1}{2}\right) + [4 - (1)^2]\left(\frac{1}{2}\right) + \left[4 - \left(\frac{3}{2}\right)^2\right]\left(\frac{1}{2}\right)\\ &= \left[4 + \frac{15}{4} + 3 + \frac{7}{4}\right]\left(\frac{1}{2}\right)\\ &= \frac{25}{4}.\end{aligned}$$

Since $A = 16/3$, we have $A < \overline{S}_4$. (See Figure 2.9.) ■

Since, for upper approximating sums, the union of the approximating rectangles entirely contains the region R, our definition of area should provide that

$$A \le \overline{S}_n \tag{3}$$

for all upper approximating sums $\overline{S}_n$, as in Example 2.

Combining inequalities (2) and (3), we see that a definition of area should provide that

$$\underline{S}_n \le A \le \overline{S}_n \tag{4}$$

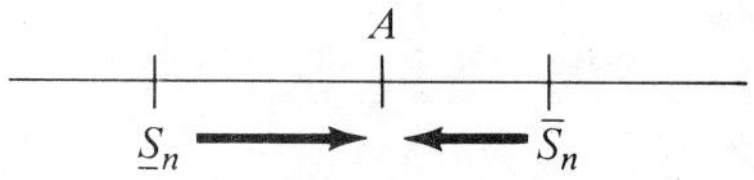

Figure 2.10 The area A of R is the unique number satisfying $\underline{S}_n \le A \le \overline{S}_n$ for all lower and upper approximating sums.

for all lower approximating sums $\underline{S}_n$ and upper approximating sums $\overline{S}_n$. Also, by the way $\underline{S}_n$ and $\overline{S}_n$ are defined, you can see that $\underline{S}_n$ increases and $\overline{S}_n$ decreases as $n \to \infty$. Now if it were the case that $\underline{S}_n$ and $\overline{S}_n$ had the same limit S as $n \to \infty$, we could both satisfy inequality (4) and obtain an unambiguous definition of the area A of R by taking $A = S$ (see Figure 2.10). Indeed, this is precisely how things work out. Before taking up the general case, we look at a particular example.

Example 3 Let R be the region bounded above by the graph of $f(x) = x^2$, below by the x-axis, on the left by $x = 0$, and on the right by $x = 1$. Show that the number 1/3 satisfies the inequality

$$\underline{S}_n \le \frac{1}{3} \le \overline{S}_n$$

for all lower and upper approximating sums, and that

$$\lim_{n\to\infty} \underline{S}_n = \frac{1}{3} = \lim_{n\to\infty} \overline{S}_n.$$

Solution: We shall make use of the formula

$$1 + 2^2 + 3^2 + \cdots + k^2 = \frac{k(k+1)(2k+1)}{6}. \tag{5}$$

(See Appendix II for a proof of this formula.)

To form the lower approximating sum for $f(x) = x^2$ on $[0, 1]$, we use n subintervals of equal size $\Delta x = \dfrac{1}{n}$, and endpoints

$$x_0 = 0, \qquad x_1 = \frac{1}{n}, \qquad x_2 = \frac{2}{n}, \ldots, x_n = \frac{n}{n} = 1.$$

Since $f(x) = x^2$ is *increasing* on $[0, 1]$, the minimum value of $f(x)$ on each subinterval $[x_{j-1}, x_j]$ will occur at the *left* endpoint, x_{j-1}. That is, $c_j = x_{j-1} = \dfrac{j-1}{n}$ for each $j = 1, 2, \ldots, n$. Thus,

$$\begin{aligned}
\underline{S}_n &= f(0)\cdot\frac{1}{n} + f\left(\frac{1}{n}\right)\cdot\frac{1}{n} + f\left(\frac{2}{n}\right)\cdot\frac{1}{n} + \cdots + f\left(\frac{n-1}{n}\right)\cdot\frac{1}{n} \\
&= \left[0^2 + \left(\frac{1}{n}\right)^2 + \left(\frac{2}{n}\right)^2 + \cdots + \left(\frac{n-1}{n}\right)^2\right]\left(\frac{1}{n}\right) \\
&= [1 + 2^2 + 3^2 + \cdots + (n-1)^2]\left(\frac{1}{n^3}\right).
\end{aligned}$$

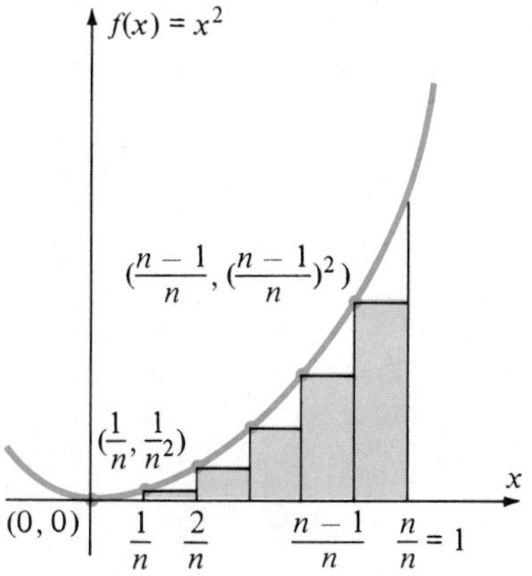

Figure 2.11 The lower approximating sum $\underline{S}_n$ for $f(x) = x^2$ on $[0, 1]$ is $\underline{S}_n = \frac{1}{3} - \left(\frac{3n-1}{6n^2}\right)$.

Using formula (5) with $k = n - 1$, we can write this sum as

$$\begin{aligned}
\underline{S}_n &= \left\{\frac{(n-1)[(n-1)+1][2(n-1)+1]}{6}\right\}\left(\frac{1}{n^3}\right) \\
&= \frac{(n-1)(n)(2n-1)}{6n^3}
\end{aligned}$$

so

$$\underline{S}_n = \frac{2n^2 - 3n + 1}{6n^2} = \frac{1}{3} - \left(\frac{3n-1}{6n^2}\right). \qquad \text{(Figure 2.11.)} \tag{6}$$

To obtain the upper approximating sum we use $d_j = x_j$ on the interval $[x_{j-1}, x_j]$, $j = 1, 2, \ldots, n$. That is, d_j is the *right* endpoint of the jth interval. Thus,

$$\begin{aligned}
\overline{S}_n &= f\left(\frac{1}{n}\right)\cdot\frac{1}{n} + f\left(\frac{2}{n}\right)\cdot\frac{1}{n} + \cdots + f\left(\frac{n}{n}\right)\cdot\frac{1}{n} \\
&= \left[\left(\frac{1}{n}\right)^2 + \left(\frac{2}{n}\right)^2 + \cdots + \left(\frac{n}{n}\right)^2\right]\left(\frac{1}{n}\right) \\
&= [1 + 2^2 + 3^2 + \cdots + n^2]\cdot\left(\frac{1}{n^3}\right) \\
&= \frac{n(n+1)(2n+1)}{n^3} \qquad \text{(using formula (5))} \\
&= \frac{2n^2 + 3n + 1}{6n^2},
\end{aligned}$$

so

$$\overline{S}_n = \frac{1}{3} + \left(\frac{3n+1}{6n^2}\right) \qquad \text{(Figure 2.12.)} \tag{7}$$

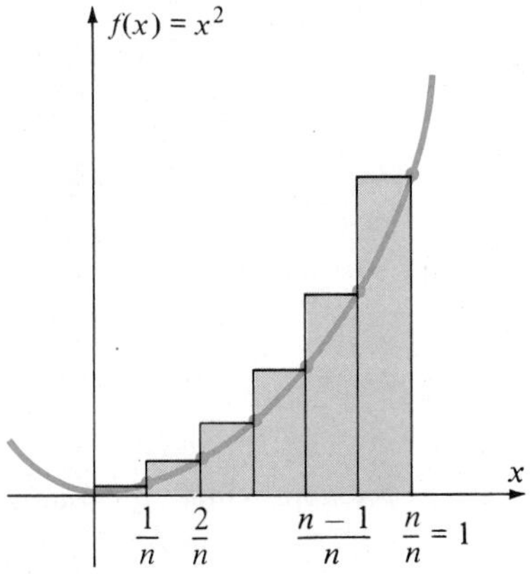

Figure 2.12 The upper approximating sum $\overline{S}_n$ for $f(x) = x^2$ on $[0, 1]$ is $\overline{S}_n = \frac{1}{3} + \left(\frac{3n+1}{6n^2}\right)$.

Now since the numbers $\dfrac{3n-1}{6n^2}$ and $\dfrac{3n+1}{6n^2}$ are positive for $n \geq 2$, we have from (6) and (7) that

$$\frac{1}{3} - \left(\frac{3n-1}{6n^2}\right) = \underline{S}_n < \frac{1}{3} < \overline{S}_n = \frac{1}{3} + \left(\frac{3n+1}{6n^2}\right) \tag{8}$$

for all $n \geq 2$. Moreover, the number $1/3$ is the *only* number that can occupy the middle position in inequality (8), since

$$\lim_{n\to\infty} \underline{S}_n = \lim_{n\to\infty} \left[\frac{1}{3} - \left(\frac{3n-1}{6n^2}\right)\right] = \lim_{n\to\infty} \left(\frac{1}{3} - 0\right) = \frac{1}{3},$$

and

$$\lim_{n\to\infty} \overline{S}_n = \lim_{n\to\infty} \left[\frac{1}{3} + \left(\frac{3n+1}{6n^2}\right)\right] = \lim_{n\to\infty} \left(\frac{1}{3} + 0\right) = \frac{1}{3}.$$

We therefore define the area of the region R to be the number $A = 1/3$. ■

The following theorem shows that the result of Example 3 is true for *any* continuous function.

THEOREM 1

Let $f(x)$ be continuous on the interval $[a, b]$. Let $\underline{S}_n$ and $\overline{S}_n$ denote the lower and upper approximating sums for $f(x)$ on $[a, b]$. Then $\lim_{n\to\infty} \underline{S}_n$ and $\lim_{n\to\infty} \overline{S}_n$ exist, and

$$\lim_{n\to\infty} \underline{S}_n = \lim_{n\to\infty} \overline{S}_n.$$

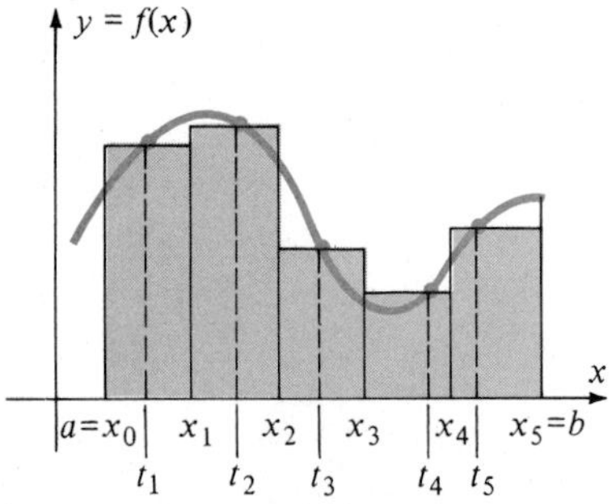

Figure 2.13 An arbitrary approximating sum S_5 for $f(x)$ on $[a, b]$. The numbers $t_j \in [x_{j-1}, x_j]$ are arbitrary.

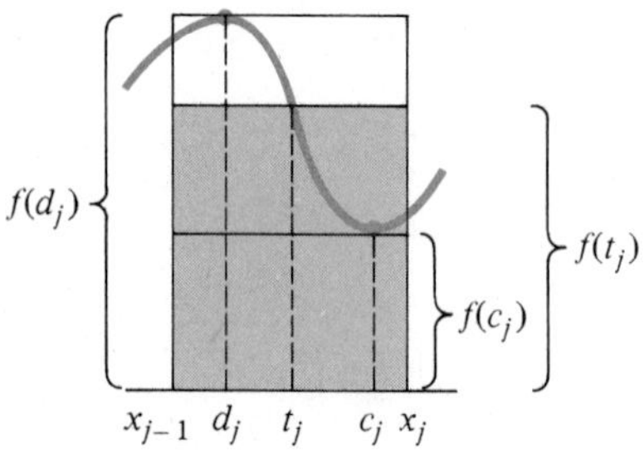

Figure 2.14 $f(c_j) \leq f(t_j) \leq f(d_j)$, so $\underline{S}_n \leq S_n \leq \overline{S}_n$.

We shall defer a proof of Theorem 1 to the end of this section and focus here on the significance of this result.

Combining inequality (4) with the statement of Theorem 1, we conclude that the caption on Figure 2.10 is indeed correct: as $n \to \infty$, $\underline{S}_n$ increases, $\overline{S}_n$ decreases, and the two approximations meet at a single number $S = \lim_{n\to\infty} \underline{S}_n = \lim_{n\to\infty} \overline{S}_n$ "in the limit." This is the number that we shall define to be the area A of the region R. In doing so, however, we want to make use of one additional notion: an *arbitrary* approximating sum (or, just an "approximating sum.")

Let's again divide the interval $[a, b]$ into n subintervals of equal length $\Delta x = \dfrac{b-a}{n}$, and with endpoints

$$a = x_0 < x_1 < x_2 < \cdots < x_n = b.$$

But this time, for the height of the rectangle over the jth interval $[x_{j-1}, x_j]$, we use $f(t_j)$, where t_j is *any* number in the interval $[x_{j-1}, x_j]$. (That is, $t_j \in [x_{j-1}, x_j]$ is **arbitrary.**) An (arbitrary) approximating sum is therefore

$$S_n = \sum_{j=1}^{n} f(t_j)\Delta x, \qquad t_j \in [x_{j-1}, x_j].$$

Since $f(c_j) \leq f(t_j) \leq f(d_j)$ in each interval $[x_{j-1}, x_j]$, any approximating sum S_n must satisfy the inequality

$$\underline{S}_n \leq S_n \leq \overline{S}_n \tag{9}$$

for each $n = 1, 2, \ldots$ (see Figures 2.13 and 2.14). Since $\lim_{n\to\infty} \underline{S}_n = \lim_{n\to\infty} \overline{S}_n$, it follows from an argument analogous to the Pinching Theorem that

$$\lim_{n\to\infty} \underline{S}_n = \lim_{n\to\infty} S_n = \lim_{n\to\infty} \overline{S}_n.$$

This observation is so important that we state it as a corollary to Theorem 1.

COROLLARY 1

Let $f(x)$ be continuous on $[a, b]$. There exists a unique number S so that

$$S = \lim_{n\to\infty} S_n = \lim_{n\to\infty} \sum_{j=1}^{n} f(t_j)\Delta x, \qquad t_j \in [x_{j-1}, x_j]$$

regardless of how the numbers t_j are chosen in the intervals $[x_{j-1}, x_j]$.

The number S in Corollary 1 is how we define the area A of the region R in the statement of the Area Problem.

DEFINITION 1

Let R be the region bounded above by the graph of the continuous function $f(x)$, below by the x-axis, on the left by $x = a$, and on the right by $x = b$. The **area** of R is the number A defined by the equation

$$A = \lim_{n\to\infty} \sum_{j=1}^{n} f(t_j)\,\Delta x, \qquad t_j \in [x_{j-1}, x_j]$$

where $x_0, x_1, x_2, \ldots, x_n$ are the endpoints obtained by dividing $[a, b]$ into n subintervals of equal length $\Delta x = \dfrac{b-a}{n}$.

We can summarize the work of this section by saying that if $f(x)$ is continuous on $[a, b]$, all approximating sums (lower, upper, or arbitrary) have the same limit S as $n \to \infty$, and the area A of the region R is this limit. To find A for a particular region R, we may therefore construct *any* approximating sum S_n and calculate the limit $A = \lim_{n\to\infty} S_n$. (Although this is a cumbersome task in the manner described here, the ideas developed in this section will allow us to establish a much more powerful and practical means for calculating areas in Section 6.4.)

The next example shows that Definition 1 agrees with the usual definition of area for a trapezoidal region. It makes use of the formula

$$1 + 2 + 3 + \cdots + k = \frac{k(k+1)}{2}, \qquad (10)$$

which is proved in Appendix II.

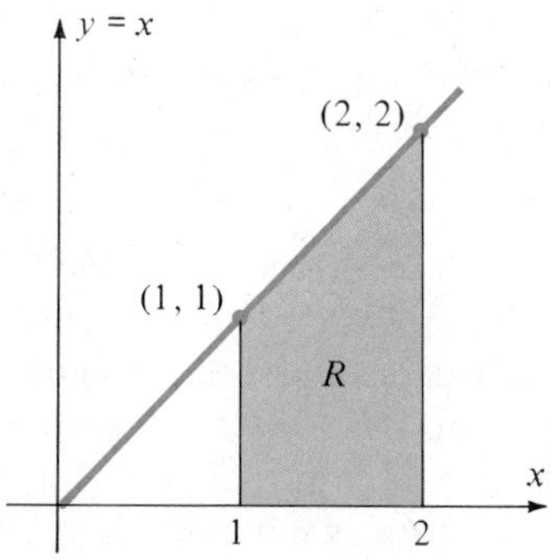

Figure 2.15 Area of R is 3/2.

Example 4 Consider the trapezoidal region R bounded above by the line $y = x$, below by the x-axis, on the left by the line $x = 1$, and on the right by the line $x = 2$. By plane geometry we know the area of R to be $A = 3/2$. Show that Definition 1 produces this same value for the area of R (see Figure 2.15).

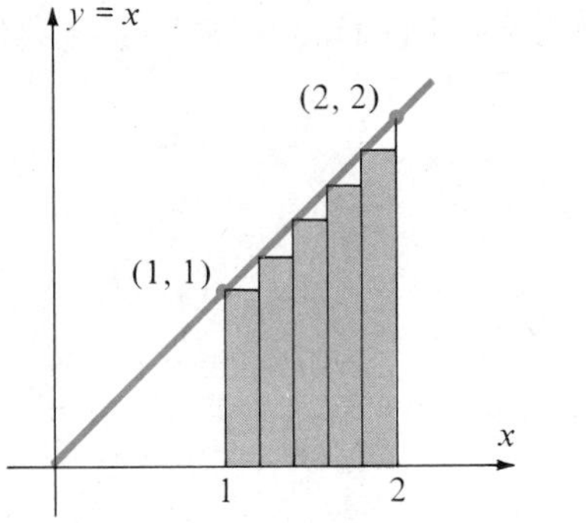

Figure 2.16 Approximating area of R by lower sums.

Solution: Since we are free to use *any* approximating sum, we will choose to use a *lower* approximating sum. Dividing [1, 2] into n subintervals of equal length $\Delta x = \frac{2-1}{n} = \frac{1}{n}$ produces endpoints

$$x_0 = 1, \qquad x_1 = 1 + \frac{1}{n}, \qquad x_2 = 1 + \frac{2}{n}, \ldots, \qquad x_n = 1 + \frac{n}{n} = 2.$$

Since $f(x) = x$ is *increasing* on [1, 2], we take the "test numbers" t_j to be the *left* endpoints of each interval. Thus,

$$\begin{aligned}
\underline{S}_n &= f(1)\left(\frac{1}{n}\right) + f\left(1 + \frac{1}{n}\right)\left(\frac{1}{n}\right) + f\left(1 + \frac{2}{n}\right)\left(\frac{1}{n}\right) + \cdots + f\left(1 + \frac{n-1}{n}\right)\left(\frac{1}{n}\right) \\
&= \left[1 + (1 + \frac{1}{n}) + \left(1 + \frac{2}{n}\right) + \cdots + \left(1 + \frac{n-1}{n}\right)\right]\left(\frac{1}{n}\right) \\
&= \left[\underbrace{(1 + 1 + \cdots + 1)}_{n \text{ terms}} + \left(\frac{1}{n} + \frac{2}{n} + \cdots + \frac{n-1}{n}\right)\right]\left(\frac{1}{n}\right) \\
&= \left[n + \frac{1}{n}(1 + 2 + \cdots + n - 1)\right]\left(\frac{1}{n}\right) \\
&= 1 + \frac{1}{n^2}(1 + 2 + \cdots + n - 1) \\
&= 1 + \frac{1}{n^2}\left[\frac{(n-1)n}{2}\right] \qquad \text{(Apply (10) with } k = n - 1\text{)} \\
&= 1 + \frac{n-1}{2n}.
\end{aligned}$$

Thus,

$$A = \lim_{n\to\infty}\left[1 + \frac{n-1}{2n}\right] = \lim_{n\to\infty}\left[1 + \frac{1 - \left(\frac{1}{n}\right)}{2}\right] = 1 + \frac{1-0}{2} = \frac{3}{2}. \quad ■$$

Although the idea behind approximating sums is straightforward, the calculations involved in executing a particular problem are both repetitious and time-consuming. This is precisely the kind of situation in which computers can be very helpful. Program 5 in Appendix I is a BASIC program which computes a lower approximating sum for the function $f(x) = 3x^2 + 7$ on the interval [2, 3]. Program 6 is a BASIC program which computes upper approximating sums for this same function on [2, 3]. Using this program on a microcomputer we obtained the results in Table 2.1.

Table 2.1 Lower and upper approximating sums for $f(x) = 3x^2 + 7$ on [2, 3] obtained on a computer.

n	$\underline{S}_n$	$\overline{S}_n$
2	22.3749	29.8750
5	24.5199	27.5200
10	25.2549	26.7550
25	25.7008	26.3008
100	25.9250	26.0750
500	25.9850	26.0150

REMARK: The results of Table 2.1 illustrate vividly both the power and the limitations of computers in studying and applying the ideas of the calculus. Implementing Program 5 or Program 6 on a computer makes the calculation of particular approximating sums relatively effortless. (If you disagree with this statement, try producing the results of Table 2.1 by hand!) However, while it seems clear from Table 2.1 that the values of the approximating sums obtained are approaching a limiting value, it is not at all clear what the precise value of that limit happens to be. By contrast, the method of approximating sums presented in Examples 3 and 4, tedious as it may be, will provide a convincing demonstration that the limit in question is precisely 26 (see Exercise 15).

Proof of Theorem 1 This should really be regarded as a sketch of the proof, rather than a rigorous argument, since several facts must be assumed that we cannot prove here. However, the basic notion is a straightforward calculation.

To show that $\lim_{n\to\infty} \underline{S}_n$ exists, we note that $\underline{S}_{n+1} \geq \underline{S}_n$, since the subinterval size decreases as n increases. (Thus, the minimum values $f(c_j)$ can only increase, not decrease.) Thus, $\{\underline{S}_n\}$ is an increasing sequence of numbers which is bounded above (by any upper sum). Such sequences always have limits, as we shall prove in Chapter 13. The proof that $\lim_{n\to\infty} \overline{S}_n$ exists is similar.

To prove $\lim_{n\to\infty} \underline{S}_n = \lim_{n\to\infty} \overline{S}_n$, it is sufficient to prove that

$$\lim_{n\to\infty} (\overline{S}_n - \underline{S}_n) = 0. \tag{11}$$

According to the definitions of $\underline{S}_n$ and $\overline{S}_n$ we have that

$$\begin{aligned}\lim_{n\to\infty} (\overline{S}_n - \underline{S}_n) &= \lim_{n\to\infty} \left\{\sum_{j=1}^{n} f(d_j)\,\Delta x - \sum_{j=1}^{n} f(c_j)\,\Delta x\right\} \\ &= \lim_{n\to\infty} \sum_{j=1}^{n} [\,f(d_j) - f(c_j)]\,\Delta x.\end{aligned} \tag{12}$$

Now recall that $\Delta x = (x_j - x_{j-1}) = \dfrac{b-a}{n}$, so as $n \to \infty$ the width of each of the subintervals approaches zero. Since $f(x)$ is a continuous function, the difference between the maximum value $f(d_j)$ and the minimum value $f(c_j)$ must approach zero as the width of the interval $[x_{j-1}, x_j]$ approaches zero.* Thus, if we let

$$M_n = \max\{\,f(d_j) - f(c_j)\,|\,1 \leq j \leq n\},$$

we can state in mathematical notation both that

$$f(d_j) - f(c_j) \leq M_n \qquad \text{for all } j = 1, 2, \ldots, n, \tag{13}$$

(that is, M_n is the largest difference) and that

$$\lim_{n\to\infty} M_n = 0 \tag{14}$$

(that is, the differences approach zero as the interval width approaches zero).

Now, returning to equation (12) and using inequality (13) we obtain the statement

$$\begin{aligned}\lim_{n\to\infty} (\overline{S}_n - \underline{S}_n) &= \lim_{n\to\infty} \sum_{j=1}^{n} [\,f(d_j) - f(c_j)]\,\Delta x \\ &\leq \lim_{n\to\infty} \sum_{j=1}^{n} M_n\,\Delta x \\ &= \lim_{n\to\infty} M_n \cdot \sum_{j=1}^{n} \Delta x.\end{aligned} \tag{15}$$

* An entirely rigorous proof of this statement involves the concept of *uniform continuity,* a topic for more advanced courses on analysis.

Since

$$\sum_{j=1}^{n} \Delta x = \sum_{j=1}^{n} \left(\frac{b-a}{n}\right) = \frac{n(b-a)}{n} = b - a,$$

equation (14) and inequality (15) now give

$$\lim_{n\to\infty} (\overline{S}_n - \underline{S}_n) = (b-a) \lim_{n\to\infty} M_n = 0,$$

which establishes the desired equation (11). ■

Exercise Set 6.2

In each of Exercises 1–6, find the lower approximating sum $\underline{S}_n$ for the area of the region bounded by the graph of $f(x)$, the x-axis, the line $x = a$, and the line $x = b$.

1. $f(x) = x, \quad a = 0, \quad b = 4, \quad n = 4$
2. $f(x) = 2x + 5, \quad a = 1, \quad b = 3, \quad n = 4$
3. $f(x) = 2x^2 + 3, \quad a = 0, \quad b = 4, \quad n = 4$
4. $f(x) = 9 - x^2, \quad a = -3, \quad b = 0, \quad n = 3$
5. $f(x) = \sin x, \quad a = 0, \quad b = \pi, \quad n = 4$
6. $f(x) = x^3 + x + 1, \quad a = 0, \quad b = 6, \quad n = 3$

In each of Exercises 7–12, find the upper approximating sum $\overline{S}_n$ for the area of the region bounded by the graph of $f(x)$, the x-axis, the line $x = a$, and the line $x = b$.

7. $f(x) = 2x, \quad a = 0, \quad b = 3, \quad n = 3$
8. $f(x) = 6 - x, \quad a = 0, \quad b = 3, \quad n = 6$
9. $f(x) = 3x^2 + 10, \quad a = -1, \quad b = 3, \quad n = 4$
10. $f(x) = 4 - x^2, \quad a = 0, \quad b = 2, \quad n = 4$
11. $f(x) = \cos x, \quad a = -\pi/2, \quad b = \pi/2, \quad n = 4$
12. $f(x) = x^3 + x + 1, \quad a = 0, \quad b = 6, \quad n = 3$

13. Let $f(x) = 3x + 2$.
    a. Find the lower approximating sum $\underline{S}_4$ for $f(x)$ on $[0, 2]$ with $n = 4$ subintervals.
    b. Find an expression for the lower approximation sum $\underline{S}_n$ for $f(x)$ on $[0, 2]$ with n subintervals.
    c. Find $\lim_{n\to\infty} \underline{S}_n$ with $\underline{S}_n$ as in part (b).
    d. Find an expression for the upper approximation sum $\overline{S}_n$ for $f(x)$ on $[0, 2]$ with n subintervals.
    e. Find $\lim_{n\to\infty} \overline{S}_n$ and $\lim_{n\to\infty} (\overline{S}_n - \underline{S}_n)$.
    f. What do you conclude about the area bounded by the graph of $y = f(x)$, the x-axis, the line $x = 0$, and the line $x = 2$?

14. Let $f(x) = x^2$.
    a. Find the lower approximating sum $\underline{S}_n$ for $f(x)$ on the interval $[0, a]$.
    b. Find $\lim_{n\to\infty} \underline{S}_n$ with $\underline{S}_n$ as in part (a).
    c. Find the lower approximating sum $\underline{S}_n$ for the function $f(x) = x^2$ on the interval $[a, b]$ where $0 \le a \le b$.
    d. Find $\lim_{n\to\infty} \underline{S}_n$ with $\underline{S}_n$ as in part (c).
    e. How could you have answered part (d) from the answer to part (b) and area considerations alone?
    f. What is the area of the region bounded by the graph of $y = x^2$ between $x = 1$ and $x = 5$?

15. Use the method of lower approximating sums to calculate the area of the region bounded above by the graph of $y = 3x^2 + 7$, on the left by $x = 2$, on the right by $x = 3$, and below by the x-axis. Compare your results with those of Table 2.1.

16. Rework Exercise 15 using upper approximating sums.

17. Rework Exercise 13 for the function $f(x) = 3x + 4$ and the interval $[1, 2]$.

18. Rework Exercise 13 for the function $f(x) = 8 - 2x$ and the interval $[0, 4]$.

19. Rework Exercise 15 for the function $f(x) = 3x^2 + 6$ and the interval $[0, 1]$.

20. Rework Exercise 15 for the function $f(x) = 2x^2 + x + 3$ and the interval $[0, 2]$.

21. Rework Exercise 15 for the function $f(x) = |1 - x^2|$ and the interval $[0, 2]$.

22. *(Calculator)* Use a hand calculator to approximate the area enclosed by the graph of the ellipse

$$\frac{x^2}{9} + \frac{y^2}{4} = 1$$

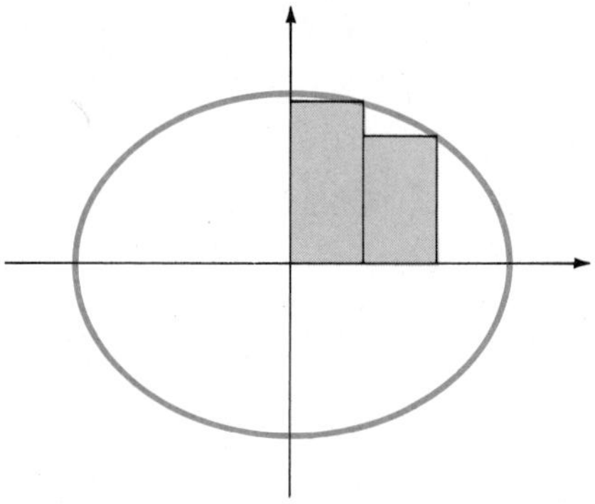

as follows.

a. Solve the equation for y to obtain

$$y = \pm 2 \cdot \sqrt{1 - x^2/9}.$$

Choose the + sign to obtain the equation for the upper half of the figure.

b. Observe that the total area will be four times the area of the portion of the ellipse lying in the first quadrant.

c. Use your calculator and the function $f(x) = 2\sqrt{1 - x^2/9}$ to obtain the lower approximating sum $\underline{S}_3$. Multiply by four to estimate the total area.

d. Obtain an improved estimate by using $n = 6$.

23. *(Calculator)* Consider the unit circle.

a. Write down the equation for the function $y = f(x)$ that describes the top half of the graph.

b. Use your calculator to obtain an estimate for the area of the portion of the circle lying in the first quadrant by calculating the lower approximating sum $\underline{S}_3$.

c. Calculate $\underline{S}_6$ for the same region.

d. Compare your estimates with what you know to be the exact value of that area.

24. *(Calculator)* Rework Exercise 22 using upper approximating sums.

25. *(Calculator)* Rework Exercise 23 using upper approximating sums.

26. *(Computer)*

a. Modify Program 5 to compute lower approximating sums for the function $f(x) = 3x^2 + 7$ on the interval $[A, B]$ instead of the interval $[2, 3]$.

b. Complete the following table using your modified program.

$[A, B]$	n	$\underline{S}_n$
[1, 5]	5	
[1, 5]	20	
[1, 5]	100	
[5, 7]	5	
[5, 7]	20	
[5, 7]	100	
[1, 7]	5	
[1, 7]	20	
[1, 7]	100	

c. What do you conjecture, based on your results in part (b) about the relationship between $\underline{S}_n$ on $[1, 7]$, $\underline{S}_n$ on $[1, 5]$, and $\underline{S}_n$ on $[5, 7]$?

27. *(Computer)* Make the same changes in Program 6 as indicated for Program 5 in Exercise 26. Then answer the same questions as in Exercise 26 for the corresponding upper sums $\overline{S}_n$.

28. *(Computer)* Modify Program 5 to compute lower approximating sums for the function $f(x) = \sqrt{1 + x^2}$ on the interval $[2, 3]$. Then use the program to approximate the area bounded by the graph of $f(x) = \sqrt{1 + x^2}$, the x-axis, the line $x = 2$, and the line $x = 3$.

29. *(Computer)* Modify the program you obtained in Exercise 28 so as to compute lower approximating sums for $f(x) = \sqrt{1 + x^2}$ on the interval $[a, b]$ (instead of $[2, 3]$).

30. *(Computer)* Write a program to approximate the area of the ellipse described in Exercise 22.

31. *(Computer)* Write a program to approximate the area bounded by the arc of $y = \sin x$ and the x-axis between $x = 0$ and $x = \pi/4$ using lower approximating sums.

6.3 RIEMANN SUMS: THE DEFINITE INTEGRAL

In this section, we generalize the notion of an approximating sum. We want to do this for two reasons. First, the approximating sum defined in Section 6.2 applies only to nonnegative functions and the problem of calculating areas of regions bounded by their graphs. We shall want to apply a similar approximation procedure in situations which can involve functions that have negative values or to issues other than area. Second, we want to free ourselves from the restriction of having to use subintervals of equal size, for reasons we shall see in this section.

Partitions and Norms

If $[a, b]$ is a closed interval, a **partition** P_n of $[a, b]$ is any set of $n + 1$ numbers $\{x_0, x_1, x_2, \ldots, x_n\}$ with

$$a = x_0 < x_1 < x_2 < \cdots < x_n = b.$$

As you can see in Figure 3.1, a partition P_n divides the interval $[a, b]$ into n nonoverlapping subintervals.

However, these subintervals are not necessarily of equal length. We define the **norm** of the partition P_n, denoted by $\|P_n\|$, to be the largest of these lengths. That is,

$$\|P_n\| = \max\{(x_1 - x_0), (x_2 - x_1), \ldots, (x_n - x_{n-1})\}.$$

Then

$$x_j - x_{j-1} \leq \|P_n\|, \qquad j = 1, 2, 3, \ldots, n. \tag{1}$$

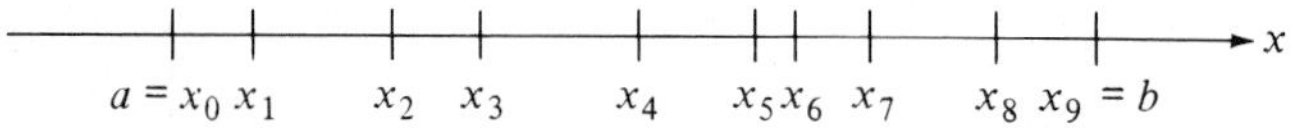

Figure 3.1 A partition P_9 for the interval $[a, b]$.

Example 1 A partition of $[0, 4]$ is $P_7 = \left\{0, 1, \frac{3}{2}, \frac{7}{4}, 2, \frac{10}{3}, \frac{11}{3}, 4\right\}$.

The norm of this partition is $\|P_7\| = \frac{10}{3} - 2 = \frac{4}{3}$. ■

If $f(x)$ is continuous on $[a, b]$, we define a *Riemann sum** for $f(x)$ on $[a, b]$ in much the same way as we defined arbitrary approximating sums in Section 6.2, except that we do not require the subintervals to be of equal length. Nor do we require $f(x) \geq 0$.

DEFINITION 2

Let $f(x)$ be continuous on $[a, b]$. A **Riemann sum** for $f(x)$ on $[a, b]$ is any sum of the form

$$R_n = \sum_{j=1}^{n} f(t_j)\Delta x_j, \qquad t_j \in [x_{j-1}, x_j]$$

where $\{x_0, x_1, x_2, \ldots, x_n\}$ is a partition of $[a, b]$, $\Delta x_j = x_j - x_{j-1}$, and t_j is an element of $[x_{j-1}, x_j]$, $j = 1, 2, \ldots, n$.

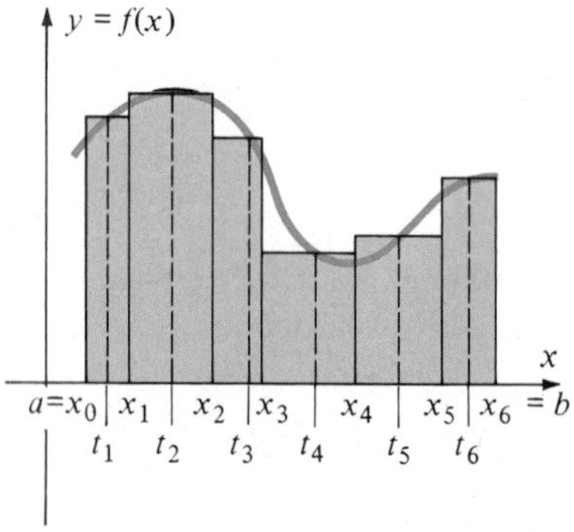

Figure 3.2 A Riemann sum R_6.

Thus, every approximating sum S_n is also a Riemann sum, but the latter concept is more general. Figure 3.2 shows a geometric interpretation for a typical Riemann sum.

As in Section 6.2, we shall want to use Riemann sums to calculate areas (at least in principle). But in generalizing the notion of approximating sum we have introduced a minor difficulty. We must now be more careful about speaking of the limit of a sequence $\{R_n\}$ of Riemann sums as $n \to \infty$. The reason is illustrated by Figure 3.3. Since the lengths of the subintervals in the partition P_n are not necessarily

*Riemann sums, and most of what is presented here as the theory of the definite integral, are originally the work of the German mathematician Bernhard Riemann, 1826–1866.

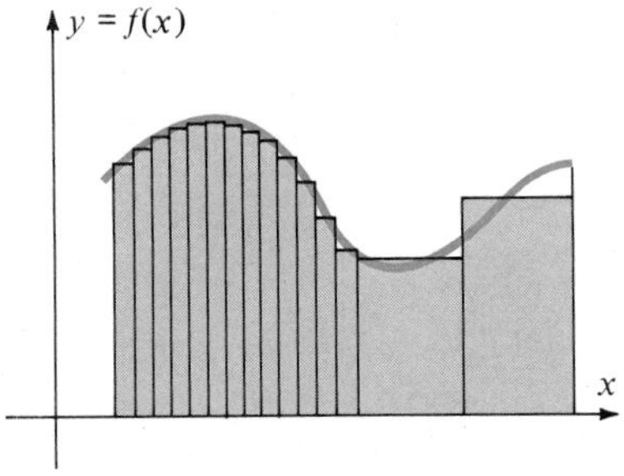

Figure 3.3 As $n \to \infty$, the widths of some rectangles may remain large.

equal, the lengths of all subintervals need not necessarily become small as n becomes large. If we want this to happen (as we do when approximating regions by rectangles) we must also specify that P_n is a partition for which $\|P_n\| \to 0$ as $n \to \infty$. By inequality (1) this guarantees that $\Delta x_j \to 0$ for all $j = 1, 2, \ldots, n$.

Riemann sums share an important property with approximating sums: If $f(x)$ is continuous on $[a, b]$, all Riemann sums for $f(x)$ on $[a, b]$ have the same limit, roughly speaking.

THEOREM 2

Let $f(x)$ be continuous on $[a, b]$. There exists a unique number S with

$$S = \lim_{n\to\infty} R_n = \lim_{n\to\infty} \sum_{j=1}^{n} f(t_j)\, \Delta x_j, \qquad t_j \in [x_{j-1}, x_j]$$

for all Riemann sums R_n with partition $P_n = \{x_0, x_1, x_2, \ldots, x_n\}$ on $[a, b]$ where $\Delta x_j = x_j - x_{j-1}$ and $\|P_n\| \to 0$ as $n \to \infty$. ■

A rigorous proof of Theorem 2 is beyond the scope of this text, although the basic idea in the proof is the same as that in the proof of Theorem 1. In fact, by rechecking the proof of Theorem 1 you will see that it did not make use of the fact that $f(x) \geq 0$. Thus, Theorem 1, as Theorem 2 above, remains true if $f(x)$ takes on negative values.

The Definite Integral

Theorem 2 states that a unique number S may be associated with the continuous function $f(x)$ on $[a, b]$ by the process of taking limits of Riemann sums. We use the richer notation $S = \int_a^b f(x)\, dx$ to denote both the function and the interval associated with this number, and we refer to this number as the *definite integral* of $f(x)$ on $[a, b]$.

DEFINITION 3

The **definite integral** of the continuous function $f(x)$ on the interval $[a, b]$ is the number

$$\int_a^b f(x)\, dx = \lim_{n\to\infty} \sum_{j=1}^{n} f(t_j)\, \Delta x_j, \qquad t_j \in [x_{j-1}, x_j] \tag{2}$$

where $P_n = \{x_0, x_1, \ldots, x_n\}$ is a partition of $[a, b]$, with $\Delta x_j = x_j - x_{j-1}$ and $\|P_n\| \to 0$ as $n \to \infty$.

The symbol $\int_a^b f(x)\, dx$ is read, "the definite integral from a to b of $f(x)$ with respect to x." As with the symbol $\int f(x)\, dx$ for antiderivatives, we refer to $\int$ as the **integral sign,** and we write the symbol dx following the **integrand** $f(x)$ to indicate that x is the independent variable for $f(x)$. (We shall later see that the symbol dx has a meaning associated with the differential $dx = \Delta x$, as suggested by the Riemann sum. But for now simply regard dx as part of the notation identifying the definite integral.) The endpoints a and b are referred to as the **limits** of integration, and their presence at the extremities of the integral sign distinguishes the definite integral $\int_a^b f(x)\,dx$ from the antiderivative $\int f(x)\,dx$.

REMARK 1: It is very important to note that the definite integral $\int_a^b f(x)\,dx$ is a **number,** while the antiderivative $\int f(x)\,dx$ is a family of **functions.** That is,

$$\int_a^b f(x)\,dx = S \in \mathbb{R}, \qquad \text{while} \qquad \int f(x)\,dx = F(x) + C$$

where $F'(x) = f(x)$. Although this is a point of possible confusion now, the results of Section 6.4 will reveal a strong relationship between definite integrals and antiderivatives, justifying this similarity of notation.

REMARK 2: At this point, the only means available to us for evaluating the definite integral $\int_a^b f(x)\,dx$ results from Theorem 1: Find any sequence of Riemann sums S_n for $f(x)$ on $[a, b]$, with $\|P_n\| \to 0$ as $n \to \infty$. As defined in Definition 3, $\int_a^b f(x)\,dx$ will be the limit of this sequence.

Example 2 Since every approximating sum is also a Riemann sum, we may use the definite integral to write the result of Example 3, Section 6.2, as

$$\int_0^1 x^2\,dx = \frac{1}{3},$$

and the result of Example 4, Section 6.2, is

$$\int_1^2 x\,dx = \frac{3}{2}.$$

■

Example 3 Use Definition 3 to find $\displaystyle\int_0^3 (2x - 4)\,dx$.

Solution: It is most convenient to use a partition P_n with subintervals of equal length $\Delta x_j = \dfrac{3-0}{n} = \dfrac{3}{n}$, $j = 1, 2, \ldots, n$ and endpoints

$$x_0 = 0, \quad x_1 = \frac{3}{n}, \quad x_2 = 2\left(\frac{3}{n}\right), \quad x_3 = 3\left(\frac{3}{n}\right), \ldots, \quad x_n = n\left(\frac{3}{n}\right) = 3.$$

We are free to choose any t_j in the jth interval $[x_{j-1}, x_j]$, so we arbitrarily choose t_j to be the right endpoint, $t_j = x_j$, $j = 1, 2, \ldots, n$. Then, for each n, a Riemann sum satisfying Definition 3 is

$$\begin{aligned}
R_n &= f\left(\frac{3}{n}\right)\left(\frac{3}{n}\right) + f\left(\frac{6}{n}\right)\left(\frac{3}{n}\right) + f\left(\frac{9}{n}\right)\left(\frac{3}{n}\right) + \cdots + f\left(\frac{3n}{n}\right)\left(\frac{3}{n}\right) \\
&= \left\{\left[2\left(\frac{3}{n}\right) - 4\right] + \left[2\left(\frac{6}{n}\right) - 4\right] + \left[2\left(\frac{9}{n}\right) - 4\right] + \cdots + \left[2\left(\frac{3n}{n}\right) - 4\right]\right\}\left(\frac{3}{n}\right) \\
&= \left[\left(\frac{2\cdot 3}{n}\right)(1 + 2 + 3 + \cdots + n) - 4n\right]\left(\frac{3}{n}\right)
\end{aligned}$$

$$= \left\{\left(\frac{6}{n}\right)\left[\frac{n(n+1)}{2}\right] - 4n\right\}\left(\frac{3}{n}\right) \qquad \text{(by formula (10), Section 6.2.)}$$

$$= \frac{9(n+1)}{n} - 12.$$

Thus,

$$\int_0^3 (2x-4)\,dx = \lim_{n\to\infty}\left[\frac{9n+9}{n} - 12\right] = \lim_{n\to\infty}\left[\frac{9+\left(\frac{9}{n}\right)}{1} - 12\right]$$

$$= \frac{9+0}{1} - 12 = -3. \qquad \blacksquare$$

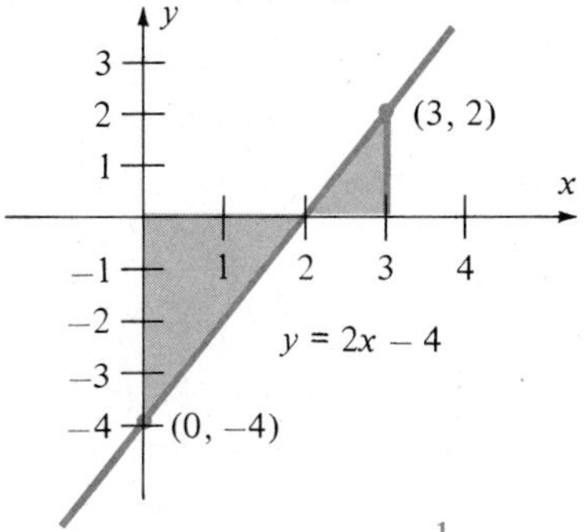

Figure 3.4 Area of R is $\frac{1}{2}(2)(4) + \frac{1}{2}(1)(2) = 5$, but $\int_0^3 (2x-4)\,dx = -3$.

It is important to note in Example 3 that the value of the integral $\int_0^3 (2x-4)\,dx = -3$ is *not* the area of a region bounded by the graph of $f(x) = 2x - 4$ (Figure 3.4). This is because the function $f(x)$ is negative for some numbers $x \in [0, 3]$. *It is only when $f(x) \geq 0$ that the area interpretation is valid for definite integrals,* although the definite integral exists whenever $f(x)$ is continuous.

The Definite Integral and Area

The preceding remark is so important that we emphasize it here in the form of a theorem.

THEOREM 3

Let $f(x)$ be continuous on $[a, b]$ with $f(x) \geq 0$ for all $x \in [a, b]$. Then the area A of the region R bounded above by the graph of $f(x)$, below by the x-axis, on the left by $x = a$, and on the right by $x = b$ is given by the definite integral

$$A = \int_a^b f(x)\,dx.$$

Proof: According to Definition 1, the area A is defined to be the limit

$$A = \lim_{n\to\infty} \sum_{j=1}^{n} f(t_j)\,\Delta x, \qquad t_j \in [x_{j-1}, x_j] \tag{3}$$

of a sequence of approximating sums for the region R. Since every approximating sum is also a Riemann sum, we may combine Definition 3 with equation (3) to conclude that

$$A = \lim_{n\to\infty} \sum_{j=1}^{n} f(t_j)\,\Delta x = \int_a^b f(x)\,dx. \qquad \blacksquare$$

Example 4 When the integrand $f(x)$ is nonnegative, we may sometimes evaluate $\int_a^b f(x)\,dx$ by identifying it with a region R for which we already know the area,

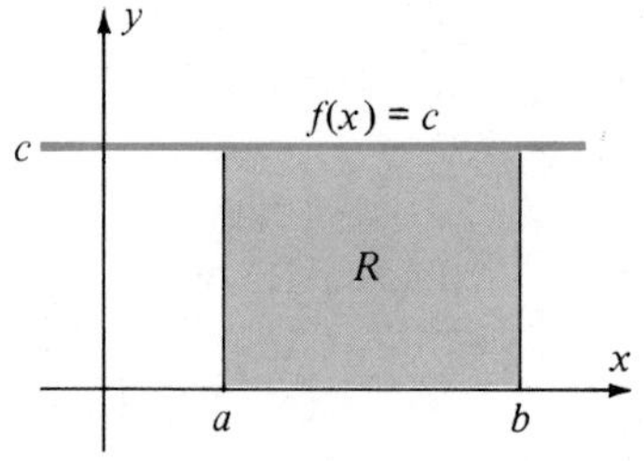

Figure 3.5 $\int_a^b c\,dx = c(b - a) =$ area of R.

usually by a formula from plane geometry:

(a) $\int_a^b c\,dx = c(b - a)$ for any constant $c > 0$, since the integrand and the x-axis, for $a \le x \le b$, bound a rectangle of width $b - a$ and height c (Figure 3.5).

(b) $\int_a^b (cx + d) = \frac{c}{2}(b^2 - a^2) + d(b - a)$ if $cx + d \ge 0$ for $x \in [a, b]$, since this integrand and the x-axis, for $a \le x \le b$, bound a trapezoid with bases $B_1 = f(a) = ca + d$ and $B_2 = f(b) = cb + d$, and with altitude $h = b - a$. The area of this trapezoid, from plane geometry, is

$$\begin{aligned} A &= \frac{1}{2}(B_1 + B_2)h = \frac{1}{2}[(ca + d) + (cb + d)](b - a) \\ &= \frac{1}{2}(ca + cb + 2d)(b - a) \\ &= \frac{c}{2}(b + a)(b - a) + d(b - a) \end{aligned}$$

(Figure 3.6.)

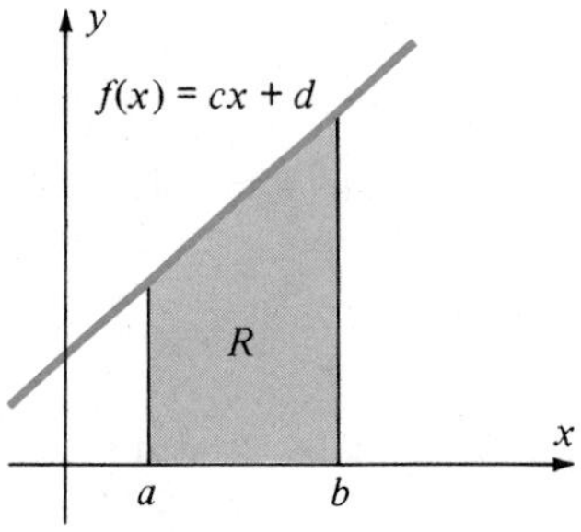

Figure 3.6 $\int_a^b (cx + d)\,dx = \frac{c}{2}(b^2 - a^2) + d(b - a) =$ area of R.

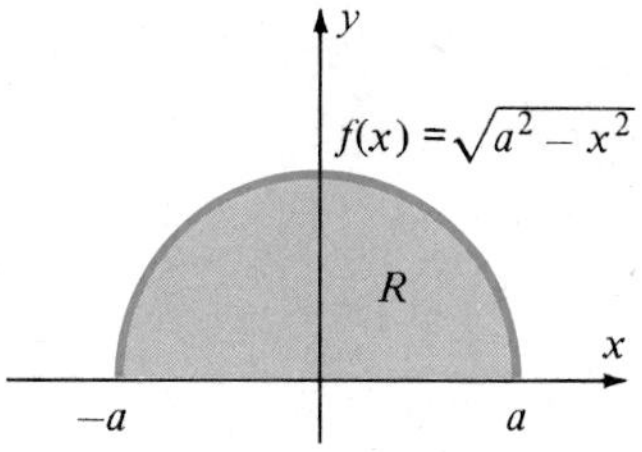

Figure 3.7 $\int_{-a}^a \sqrt{a^2 - x^2}\,dx = \frac{1}{2}\pi a^2 =$ area of R.

(c) $\int_{-a}^a \sqrt{a^2 - x^2}\,dx = \frac{1}{2}\pi a^2$, since the graph of $f(x) = \sqrt{a^2 - x^2}$ and the interval $[-a, a]$ bound a semicircle of radius a (Figure 3.7). ■

Properties of Definite Integrals

We will make frequent use of the properties of the definite integral established by the following theorem.

THEOREM 4
Properties of the Definite Integral

Let $f(x)$ and $g(x)$ be continuous on $[a, b]$ and let c be a number. Then

(a) $\int_a^b [f(x) + g(x)]\,dx = \int_a^b f(x)\,dx + \int_a^b g(x)\,dx$

(b) $\int_a^b [cf(x)]\,dx = c\int_a^b f(x)\,dx$

(c) $\int_a^b f(x)\,dx = \int_a^c f(x)\,dx + \int_c^b f(x)\,dx \qquad$ for any $c \in [a, b]$.

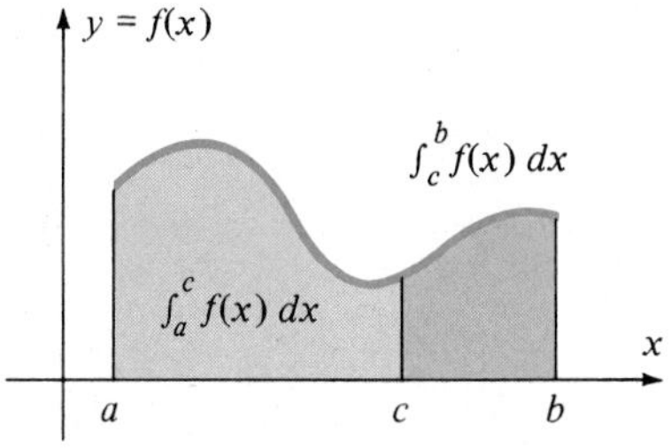

Figure 3.8 $\int_a^b f(x)\,dx = \int_a^c f(x)\,dx + \int_c^b f(x)\,dx$.

Statement (a) says that we can evaluate a definite integral of the sum of two (or more) functions as the sum of the definite integrals of the individual functions. (Recall that this *additive* property is also true of derivatives and antiderivatives.) Statement (b) says that we can "factor out" constants, bringing them past the integral sign. (Again, this property is also shared by derivatives and by antiderivatives.) Property (c) says that we can "break" the integral over $[a, b]$ into the sum of the integral over $[a, c]$ and the integral over $[c, b]$. When $f(x) \geq 0$ for all $x \in [a, b]$, this property is geometrically obvious from area considerations (see Figure 3.8), but statement (c) assures that it is true for *all* continuous integrands.

Example 5 Since

$$\int_0^1 x^2\,dx = \frac{1}{3}, \qquad \text{(Example 2)},$$

$$\int_0^1 x\,dx = \frac{1}{2}, \qquad \text{(Example 4, part (b))},$$

and

$$\int_0^1 c\,dx = c, \qquad c > 0 \qquad \text{(Example 4, part (a))},$$

we have, from Theorem 4, parts (a) and (b), that

(a) $\displaystyle\int_0^1 4x^2\,dx = 4\int_0^1 x^2\,dx = 4\left(\frac{1}{3}\right) = \frac{4}{3}$

(b)
$$\begin{aligned}\int_0^1 (2x^2 - 4x + 6)\,dx &= \int_0^1 2x^2\,dx + \int_0^1 (-4x)\,dx + \int_0^1 6\,dx\\ &= 2\int_0^1 x^2\,dx - 4\int_0^1 x\,dx + \int_0^1 6\,dx\\ &= 2\left(\frac{1}{3}\right) - 4\left(\frac{1}{2}\right) + 6\\ &= \frac{14}{3}.\end{aligned}$$

(c)
$$\begin{aligned}\int_0^1 (7 - 4x)^2\,dx &= \int_0^1 (49 - 56x + 16x^2)\,dx\\ &= \int_0^1 49\,dx - 56\int_0^1 x\,dx + 16\int_0^1 x^2\,dx\\ &= 49 - 56\left(\frac{1}{2}\right) + 16\left(\frac{1}{3}\right)\\ &= \frac{79}{3}.\end{aligned}$$

■

Proof of Theorem 4 A rigorous proof of Theorem 4 depends on the careful development of the theory of infinite sequences that lies ahead, but our working understanding of limits of sequences of Riemann sums allows us to expose the ideas behind the proof.

To prove statement (a), suppose that $R_n^f = \sum_{j=1}^n f(t_j)\,\Delta x_j$ and $R_n^g = \sum_{j=1}^n g(t_j)\,\Delta x_j$ are Riemann sums for $\int_a^b f(x)\,dx$ and for $\int_a^b g(x)\,dx$, respectively, over common partitions. (That is, assume the t_j's and the x_j's are the same in the two Riemann sums.) Since

$$\sum_{j=1}^{n} f(t_j)\,\Delta x_j + \sum_{j=1}^{n} g(t_j)\,\Delta x_j = \sum_{j=1}^{n} [f(t_j) + g(t_j)]\,\Delta x_j, \tag{4}$$

the Riemann sums for $\int_a^b f(x)\,dx$ and $\int_a^b g(x)\,dx$ add together to produce a Riemann sum for $\int_a^b [f(x) + g(x)]\,dx$. Since the limits of the sums on the left side of equation (4) exist, we may use the additive property of limits of sequences (which we will prove in Chapter 13) to conclude that

$$\begin{aligned}\int_a^b f(x)\,dx + \int_a^b g(x)\,dx &= \lim_{n\to\infty} \sum_{j=1}^{n} f(t_j)\,\Delta x_j + \lim_{n\to\infty} \sum_{j=1}^{n} g(t_j)\,\Delta x_j \\ &= \lim_{n\to\infty} \sum_{j=1}^{n} [f(t_j) + g(t_j)]\,\Delta x_j \\ &= \int_a^b [f(x) + g(x)]\,dx.\end{aligned}$$

The argument for statement (b) is similar.

The idea behind the proof of statement (c) is to require that c be an endpoint in all partitions in a sequence R_n of Riemann sums with limit $\int_a^b f(x)\,dx$. Write $n = m + p$ where m is the number of subintervals in $[a, c]$ and p is the number of subintervals in $[c, b]$. Use the notation $z_k = x_{m+k}$ for the x_j's to the right of c, and the notation $u_k = t_{m+k}$ for the t_j's to the right of c. Then

$$\begin{aligned}\int_a^b f(x)\,dx &= \lim_{n\to\infty} \sum_{j=1}^{n} f(t_j)\,\Delta x_j \\ &= \lim_{n\to\infty} \left[\sum_{j=1}^{m} f(t_j)\,\Delta x_j + \sum_{k=1}^{p} f(u_k)\,\Delta z_k\right] \qquad (n = m + p) \\ &= \lim_{m\to\infty} \sum_{j=1}^{m} f(t_j)\,\Delta x_j + \lim_{p\to\infty} \sum_{k=1}^{p} f(u_k)\,\Delta z_k \\ &= \int_a^c f(x)\,dx + \int_c^b f(x)\,dx.\end{aligned}$$

This is messy, but it is important that you at least appreciate that this is why we need to allow partitions with unequal subinterval lengths. If all subintervals were of equal length, we could not guarantee that c would correspond to an endpoint, so we could not "break the sum into two pieces at $x = c$" as we have done here. ■

Extending the Definition of $\int_a^b f(x)\,dx$

Since $\int_a^b f(x)\,dx$ has been defined for *intervals* $[a, b]$, it is implicit in Definition 3 that $a < b$. We will have need to work with definite integrals with $a \geq b$, which we now define.

DEFINITION 4

Let $f(x)$ be continuous on $[a, b]$. Then

(a) $\displaystyle\int_a^a f(x)\,dx = 0$

(b) $\int_b^a f(x)\, dx = -\int_a^b f(x)\, dx.$

These definitions have simple geometric motivation. Statement (a) says that a region of zero width and finite height has zero area, at least when $f(x) \geq 0$. The motivation for (b) is this. If $a < b$, the integral on the right is "from left to right." Since the integral on the left is then "from right to left," we are reversing the order in which we march along the number line adding up terms in the Riemann sums. The effect is to change the sign on each term $\Delta x_j = (x_j - x_{j-1})$, leaving $f(x)$ unchanged, and therefore, changing the sign on the Riemann sums associated with $\int_a^b f(x)\, dx$. We may paraphrase (b) by saying that "reversing the order of integration changes the sign of the integral."

Using Definition 4, we may extend property (c) of Theorem 4 to any possible ordering of the numbers a, b, and c.

COROLLARY 2

Let $f(x)$ be continuous on an interval containing the numbers a, b, and c. Then

$$\int_a^b f(x)\, dx = \int_a^c f(x)\, dx + \int_c^b f(x)\, dx$$

regardless of the ordering of the numbers a, b, and c.

Proof: We will consider the case $c < a < b$, leaving the other five possible orderings for you (Exercise 35).

Since $c < a < b$, we have by Theorem 4 that

$$\int_c^b f(x)\, dx = \int_c^a f(x)\, dx + \int_a^b f(x)\, dx. \tag{5}$$

But, $\int_c^a f(x)\, dx = -\int_a^c f(x)$ by Definition 4. Solving equation (5) for $\int_a^b f(x)\, dx$ and using this fact gives

$$\int_a^b f(x)\, dx = -\int_c^a f(x)\, dx + \int_c^b f(x)\, dx$$

$$= \int_a^c f(x)\, dx + \int_c^b f(x)\, dx,$$

as claimed. ■

Exercise Set 6.3

1. Given that $\int_0^2 f(x)\, dx = 3$, $\int_2^5 f(x)\, dx = -2$, and $\int_5^8 f(x)\, dx = 5$, find

   a. $\int_2^0 f(x)\, dx$
   b. $\int_0^5 f(x)\, dx$
   c. $\int_2^8 f(x)\, dx$
   d. $\int_0^8 f(x)\, dx$
   e. $\int_5^0 f(x)\, dx$
   f. $\int_8^2 f(x)\, dx$

2. Given that $\int_1^3 f(x)\, dx = 5$ and $\int_1^3 g(x)\, dx = -2$, find

   a. $\int_1^3 [f(x) + g(x)]\, dx$
   b. $\int_1^3 [6g(x) - 2f(x)]\, dx$
   c. $\int_3^1 2f(x)\, dx - \int_1^3 g(x)\, dx$
   d. $\int_1^3 [2f(x) + 3]\, dx$

e. $\int_{3}^{1} [g(x) - 4f(x) + 5]\, dx$

f. $\int_{1}^{3} [1 - 4g(x) + 3f(x)]\, dx$

In Exercises 3–16, use the facts that

$$\int_a^b x^2\, dx = \frac{b^3}{3} - \frac{a^3}{3}; \qquad \int_a^b x\, dx = \frac{b^2}{2} - \frac{a^2}{2};$$

$$\int_a^b c\, dx = c(b - a)$$

to find the value of the given integral (see Exercises 33, 34).

3. $\int_0^5 6\, dx$

4. $\int_0^3 (2x + 5)\, dx$

5. $\int_{-2}^4 (3x - 2)\, dx$

6. $\int_5^2 (5x + 7)\, dx$

7. $\int_4^0 (9 - 2x)\, dx$

8. $\int_2^5 2x^2\, dx$

9. $\int_3^1 (6 - x^2)\, dx$

10. $\int_{-4}^{-1} (3x^2 - 1)\, dx$

11. $\int_0^1 (x^2 - x + 2)\, dx$

12. $\int_{-2}^2 (2x^2 - 3x + 2)\, dx$

13. $\int_0^4 x(2 + x)\, dx$

14. $\int_3^{-1} (x - 2)(x + 2)\, dx$

15. $\int_{-4}^{-1} 2x(1 - 2x)\, dx$

16. $\int_{-1}^1 (ax^2 + bx + c)\, dx$

In Exercises 17–24, sketch a region in the plane whose area is given by the definite integral. Find the area of that region by geometry.

17. $\int_{-2}^3 (x + 1)\, dx$

18. $\int_0^3 \sqrt{9 - x^2}\, dx$

19. $\int_{-2}^2 |x - 1|\, dx$

20. $\int_{-1}^1 \sqrt{1 - x^2}\, dx$

21. $\int_{-4}^0 (\sqrt{4 - (x + 2)^2} + 1)\, dx$

22. $\int_1^4 |2x - 7|\, dx$

23. $\int_{-3}^3 (4\sqrt{9 - x^2} + 3)\, dx$

24. $\int_0^4 (5x - \sqrt{16 - x^2})\, dx$

(Calculator) The **midpoint rule** for approximating $\int_a^b f(x)\, dx$ uses a Riemann sum where $f(x)$ is evaluated in each subinterval at the midpoint, i.e., we take $\Delta x = (b - a)/n$, $x_0 = a$, $x_1 = a + \Delta x$, $x_2 = a + 2\,\Delta x$, $\ldots$, $x_n = a + n\,\Delta x = b$ and in each interval $[x_{j-1}, x_j]$ we choose $t_j = (x_{j-1} + x_j)/2$. The *midpoint rule* approximation is then

$$\int_a^b f(x)\, dx \approx \frac{b - a}{n}\{f(t_1) + f(t_2) + \cdots + f(t_n)\}.$$

(See Figure 3.9.)

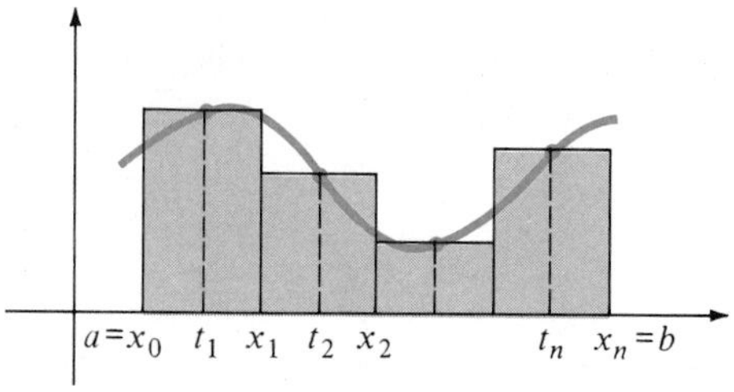

Figure 3.9 Approximation by midpoint rule: $t_j = \dfrac{x_{j-1} + x_j}{2}$.

In Exercises 25–30, use a calculator or computer to obtain the midpoint rule approximation for the given integral and the given value of n. Compare the results with those obtained using lower Riemann sums with the same values n to approximate the integrals.

25. $\int_0^3 (x^3 + 1)\, dx, \quad n = 6$

26. $\int_0^{\pi/2} \sin x\, dx, \quad n = 6$

27. $\int_0^{\pi/4} \tan x\, dx, \quad n = 4$

28. $\int_0^{\pi} \cos x\, dx, \quad n = 6$

29. $\int_0^4 (x^4 + x^3)\, dx, \quad n = 8$

30. $\int_0^2 \sqrt{4 - x^2}\, dx, \quad n = 8$

Program 7 in Appendix I is a BASIC program which implements the midpoint rule for the function $f(x) = x^3 + 1$ on an interval $[a, b]$ with n equal subdivisions. A, B, and N are inputs by the user.

31. *(Computer)* Use Program 7 to approximate the integral in Exercise 25 with $n = 10$, 50, and 100.

32. *(Computer)* Modify this program to compute approximations for the integral in Exercise 26. Approximate this integral with $n = 10$, 20, 50, and 100.

33. Use the definition of the integral to prove that $\int_a^b x\, dx = \dfrac{b^2}{2} - \dfrac{a^2}{2}$.

34. Use the definition of the integral to prove that $\int_a^b x^2\,dx = \frac{b^3}{3} - \frac{a^3}{3}$.

35. Prove the remaining cases of Corollary 2.

36. Give an argument for statement (b) of Theorem 4 similar to that given for statement (a).

37. Prove that if $f(x)$ and $g(x)$ are continuous on $[a, b]$, and if $f(x) \le g(x)$ for all $x \in [a, b]$, then

$$\int_a^b f(x)\,dx \le \int_a^b g(x)\,dx$$

(*Hint:* Write $g(x) = f(x) + h(x)$. What do you know about the sign of $h(x)$? Of $\int_a^b h(x)\,dx$?)

38. Prove that if $f(x)$ is continuous on $[a, b]$, then

$$\left|\int_a^b f(x)\,dx\right| \le \int_a^b |f(x)|\,dx.$$

6.4 THE FUNDAMENTAL THEOREM OF CALCULUS

In Section 6.3 we saw that the definite integral provides a precise answer to the Area Problem. Up to this point, however, we have been able to actually calculate $\int_a^b f(x)\,dx$ for a very limited class of functions. The next theorem shows how definite integrals may be calculated for a wide variety of integrands and reveals a remarkable relationship between antiderivatives and definite integrals. It is such an important result that it is called the Fundamental Theorem of Calculus.

THEOREM 5
Fundamental Theorem of Calculus

Let $f(x)$ be continuous on the interval $[a, b]$. If $F(x)$ is an antiderivative for $f(x)$ on $[a, b]$, then

$$\int_a^b f(x)\,dx = F(b) - F(a) \tag{1}$$

In other words, the value of a definite integral is entirely determined by an antiderivative $F(x)$ (*any* antiderivative) and the endpoints of the interval $[a, b]$.

In using the Fundamental Theorem to evaluate definite integrals, we shall use the notation

$$F(x)]_a^b = F(b) - F(a).$$

Before proving the Fundamental Theorem, we consider several examples.

Example 1

$$\int_1^2 (4x + 6)\,dx = 2x^2 + 6x]_1^2 = (2 \cdot 2^2 + 6 \cdot 2) - (2 \cdot 1^2 + 6 \cdot 1)$$
$$= (8 + 12) - (2 + 6)$$
$$= 12$$

since an antiderivative for $f(x) = 4x + 6$ is $F(x) = 2x^2 + 6x$. ■

Example 2

$$\int_{-2}^2 (x^3 - 3x^2 + 2x - 5)\,dx = \frac{x^4}{4} - x^3 + x^2 - 5x]_{-2}^2$$
$$= \left(\frac{2^4}{4} - 2^3 + 2^2 - 5 \cdot 2\right)$$

$$-\left(\frac{(-2)^4}{4} - (-2)^3 + (-2)^2 - 5(-2)\right)$$
$$= (4 - 8 + 4 - 10) - (4 + 8 + 4 + 10)$$
$$= -36$$

since an antiderivative for $f(x) = x^3 - 3x^2 + 2x - 5$ is

$$F(x) = \frac{x^4}{4} - x^3 + x^2 - 5x.$$ ■

Example 3

$$\int_0^{3\pi/2} \cos x\, dx = \sin x\Big]_0^{3\pi/2} = \sin\left(\frac{3\pi}{2}\right) - \sin(0)$$
$$= -1 - 0$$
$$= -1$$

since an antiderivative for $f(x) = \cos x$ is $F(x) = \sin x$. ■

Example 4

$$\int_1^4 \frac{x+1}{\sqrt{x}}\, dx = \int_1^4 \left(\sqrt{x} + \frac{1}{\sqrt{x}}\right) dx$$
$$= \int_1^4 (x^{1/2} + x^{-1/2})\, dx$$
$$= \frac{2}{3}x^{3/2} + 2x^{1/2}\Big]_1^4$$
$$= \left(\frac{2}{3}\cdot 4^{3/2} + 2\cdot 4^{1/2}\right) - \left(\frac{2}{3}\cdot 1^{3/2} + 2\cdot 1^{1/2}\right)$$
$$= \frac{2}{3}\cdot 8 + 2\cdot 2 - \frac{2}{3} - 2$$
$$= \frac{20}{3}$$

since an antiderivative for $f(x) = x^{1/2} + x^{-1/2}$ is $F(x) = \frac{2}{3}x^{3/2} + 2x^{1/2}$. ■

REMARK 1: In order to use the Fundamental Theorem, it is necessary to be able to find an antiderivative $F(x)$ for $f(x)$. Here is a summary list of the antiderivative formulas we have developed up to this point:

$$\int m\, dx = mx + C \qquad \int \sec^2 x\, dx = \tan x + C$$

$$\int x^n\, dx = \frac{x^{n+1}}{n+1} + C, \quad n \neq -1 \qquad \int \csc^2 x\, dx = -\cot x + C$$

$$\int \sin x\, dx = -\cos x + C \qquad \int \sec x \tan x\, dx = \sec x + C$$

$$\int \cos x\, dx = \sin x + C \qquad \int \csc x \cot x\, dx = -\csc x + C$$

REMARK 2: You may wonder why we are writing just $F(x)$ rather than $F(x) + C$ for an antiderivative of $f(x)$. The answer is that we need only find *any particular* antiderivative for $f(x)$, since all antiderivatives yield the same value for the expression $F(b) - F(a)$. To see this note that

$$\begin{aligned} F(x) + C\Big]_a^b &= [F(b) + C] - [F(a) + C] \\ &= F(b) - F(a) + (C - C) \\ &= F(b) - F(a). \end{aligned}$$

Example 5 Find $\int_{-1}^{5} |x - 2|\, dx$.

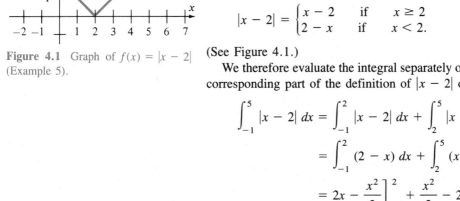

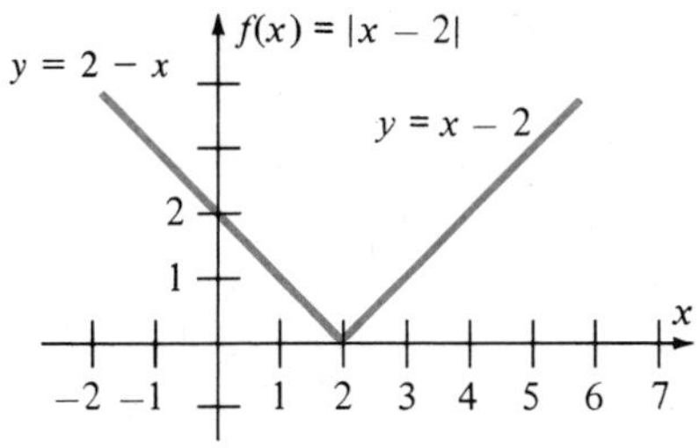

Figure 4.1 Graph of $f(x) = |x - 2|$ (Example 5).

Solution: Applying the definition of absolute value we see that

$$|x - 2| = \begin{cases} x - 2 & \text{if} \quad x - 2 \ge 0 \\ -(x - 2) & \text{if} \quad x - 2 < 0. \end{cases}$$

That is,

$$|x - 2| = \begin{cases} x - 2 & \text{if} \quad x \ge 2 \\ 2 - x & \text{if} \quad x < 2. \end{cases}$$

(See Figure 4.1.)

We therefore evaluate the integral separately on $[-1, 2]$ and on $[2, 5]$, using the corresponding part of the definition of $|x - 2|$ on each interval:

$$\begin{aligned} \int_{-1}^{5} |x - 2|\, dx &= \int_{-1}^{2} |x - 2|\, dx + \int_{2}^{5} |x - 2|\, dx \\ &= \int_{-1}^{2} (2 - x)\, dx + \int_{2}^{5} (x - 2)\, dx \\ &= 2x - \frac{x^2}{2}\Big]_{-1}^{2} + \frac{x^2}{2} - 2x\Big]_{2}^{5} \\ &= \left[\left(2\cdot 2 - \frac{2^2}{2}\right) - \left(2(-1) - \frac{(-1)^2}{2}\right)\right] \\ &\quad + \left[\left(\frac{5^2}{2} - 2\cdot 5\right) - \left(\frac{2^2}{2} - 2\cdot 2\right)\right] \\ &= \left(2 + \frac{5}{2}\right) + \left(\frac{5}{2} + 2\right) \\ &= 9. \end{aligned}$$

■

Proof of the Fundamental Theorem: In proving Theorem 5, we will make use of the Mean Value Theorem in the form

$$\frac{F(x_j) - F(x_{j-1})}{\Delta x_j} = F'(t_j) \tag{2}$$

where $\Delta x_j = x_j - x_{j-1}$ and t_j is some number in $[x_{j-1}, x_j]$. (We are assured that $F(x)$ satisfies the hypotheses of this theorem since $F'(x) = f(x)$ exists for each $x \in [a, b]$.)

Since the left side of equation (1) is a definite integral, we begin by writing down a Riemann sum for $f(x)$ on $[a, b]$, using partitions $P_n = \{a = x_0, x_1, x_2, \ldots, x_n = b\}$, where $\Delta x_j = x_j - x_{j-1}$ and $\|P_n\| \to 0$ as $n \to \infty$:

$$R_n = \sum_{j=1}^{n} f(t_j)\,\Delta x_j, \qquad t_j \in [x_{j-1}, x_j].$$

Recall that, by Theorem 2, the number t_j can be taken to be *any* number in $[x_{j-1}, x_j]$, $j = 1, 2, \ldots, n$.

We now specify each t_j in a very special way. In each interval $[x_{j-1}, x_j]$, we choose t_j to be that number that satisfies the conclusion of the Mean Value Theorem for the function $F(x)$ on $[x_{j-1}, x_j]$. That is, we choose each t_j so as to satisfy equation (2). Then, recalling that $f(t_j) = F'(t_j)$, we use equation (2) to proceed as follows:

$$\begin{aligned}\int_a^b f(x)\,dx &= \lim_{n\to\infty} \sum_{j=1}^{n} f(t_j)\,\Delta x_j \\ &= \lim_{n\to\infty} \sum_{j=1}^{n} F'(t_j)\,\Delta x_j \\ &= \lim_{n\to\infty} \sum_{j=1}^{n} \left[\frac{F(x_j) - F(x_{j-1})}{\Delta x_j}\right] \Delta x_j \\ &= \lim_{n\to\infty} \sum_{j=1}^{n} [F(x_j) - F(x_{j-1})].\end{aligned}$$

Now the last line above contains a **telescoping sum** in that every term drops out except $F(x_0) = F(a)$ and $F(x_n) = F(b)$. To see this, notice that

$$\begin{aligned}\sum_{j=1}^{n} [F(x_j) - F(x_{j-1})] &= [F(x_1) - F(x_0)] + [F(x_2) - F(x_1)] + [F(x_3) - F(x_2)] \\ &\quad + \cdots + [F(x_{n-1}) - F(x_{n-2})] + [F(x_n) - F(x_{n-1})] = F(x_n) - F(x_0).\end{aligned}$$

We have therefore shown that

$$\int_a^b f(x)\,dx = \lim_{n\to\infty} [F(x_n) - F(x_0)] = F(b) - F(a),$$

which completes the proof.* ■

u-Substitutions in Definite Integrals

In the next example, a u-substitution, first introduced in Section 5.9, is used to find the antiderivative. Note that we substitute back in terms of x's *before* the integral is evaluated.

Example 6 Find $\displaystyle\int_{-1}^{3} \frac{x}{\sqrt{7 + x^2}}\,dx$.

*The fact that the need to evaluate the limit dropped out as soon as the Mean Value Theorem was employed demonstrates the power of this theorem. While it is of little use for actual computations, its value in developing the relevant theory cannot be overemphasized.

Solution: To find the antiderivative $\int \frac{x\,dx}{\sqrt{7+x^2}}$, we use the substitution

$$u = 7 + x^2; \qquad du = 2x\,dx.$$

Then $x\,dx = \frac{1}{2}\,du$, and we have

$$\int \frac{x\,dx}{\sqrt{7+x^2}} = \int \frac{\frac{1}{2}\,du}{\sqrt{u}} = \frac{1}{2}\int u^{-1/2}\,du = \frac{1}{2}\cdot(2u^{1/2} + C_1)$$

$$= \sqrt{7+x^2} + C, \qquad C = \frac{1}{2}C_1.$$

We therefore use $F(x) = \sqrt{7+x^2}$ and the Fundamental Theorem to find that

$$\int_{-1}^{3} \frac{x\,dx}{\sqrt{7+x^2}} = \sqrt{7+x^2}\Big]_{-1}^{3} = \sqrt{7+3^2} - \sqrt{7+(-1)^2}$$

$$= \sqrt{16} - \sqrt{8}$$

$$= 4 - 2\sqrt{2}.$$

■

In evaluating definite integrals involving u-substitutions, it is often simpler to change the limits of integration, as well as the integrand, from x's to u's, and then to evaluate the antiderivative directly as a function of u. The following theorem justifies doing this.

THEOREM 6

Let $g'(x)$ be continuous on $[a, b]$ and let $f(x)$ be continuous on an interval I containing the values $u = g(x)$, $x \in [a, b]$. If f has an antiderivative on I, then

$$\int_a^b f(g(x))\cdot g'(x)\,dx = \int_{g(a)}^{g(b)} f(u)\,du.$$

Proof: If F is an antiderivative for f, then the composite function $F(g(x))$ is an antiderivative for $f(g(x))g'(x)$. Thus, by the Fundamental Theorem,

$$\int_a^b f(g(x))g'(x)\,dx = F(g(x))\Big]_a^b = F(g(b)) - F(g(a)).$$

But, again using the Fundamental Theorem, we can write this last expression as

$$F(g(b)) - F(g(a)) = F(u)\Big]_{g(a)}^{g(b)} = \int_{g(a)}^{g(b)} f(u)\,du.$$

■

Example 7 Using Theorem 6 we can evaluate the definite integral $\int_{-1}^{3} \frac{x\,dx}{\sqrt{7+x^2}}$ in Example 6 as follows:

As before, we use the substitution

$$u = 7 + x^2; \qquad du = 2x\,dx,$$

so $x\,dx = \frac{1}{2}\,du$. Also, we change limits by noting that

(i) when $x = -1$, $u = 7 + (-1)^2 = 8$, and
(ii) when $x = 3$, $u = 7 + (3)^2 = 16$.

Thus, by Theorem 6,

$$\int_{-1}^{3} \frac{x\,dx}{\sqrt{7 + x^2}} = \int_{8}^{16} \frac{\frac{1}{2}\,du}{\sqrt{u}} = \frac{1}{2}\int_{8}^{16} u^{-1/2}\,du$$
$$= \frac{1}{2}\cdot 2u^{1/2}\Big]_8^{16}$$
$$= \sqrt{16} - \sqrt{8}$$
$$= 4 - 2\sqrt{2}.$$

■

At this point you may choose between two approaches to definite integrals requiring u-substitutions. One approach is to use Theorem 6, "transforming" both the integrand *and* the limits of integration via the u-substitution, and evaluating the antiderivative directly as a function of u. The second is to use the u-substitution only to find the antiderivative, which is then rewritten as a function of x before it is evaluated.

With the first approach, you must do four evaluations (first transform both limits of integration, and finally evaluate the antiderivative at each limit), but the functions being evaluated are relatively simple. With the second approach, only two evaluations are required, but the antiderivative may be quite complicated.

We encourage you to become comfortable with both methods by working many problems; at the same time, you will develop the ability to decide which method will be easiest for each type of integrand.

Example 8 Find $\int_0^{\pi/4} \tan^2 x \sec^2 x\,dx$.

Solution: To find an antiderivative for $f(x) = \tan^2 x \sec^2 x$, we use the u-substitution

$$u = \tan x; \qquad du = \sec^2 x\,dx.$$

(The substitution $u = \sec x$ would not work, since we would have $du = \sec x \tan x\,dx$, leaving one factor of $\tan x$ not accounted for.)

If we use the u-substitution only to find the antiderivative, we proceed as follows:

$$\int \tan^2 x \sec^2 x\,dx = \int u^2\,du = \frac{1}{3}u^3 + C = \frac{1}{3}\tan^3 x + C$$

so

$$\int_0^{\pi/4} \tan^2 x \sec^2 x\,dx = \frac{1}{3}\tan^3 x\Big]_0^{\pi/4} = \frac{1}{3}\tan^3\left(\frac{\pi}{4}\right) - \frac{1}{3}\tan^3(0)$$
$$= \frac{1}{3}\cdot 1^3 - \frac{1}{3}\cdot 0^3$$
$$= \frac{1}{3}.$$

To use the approach of transforming the limits of integration via the u-substitution, we note that

$$\text{if } x = 0, \quad \text{then} \quad u = \tan(0) = 0;$$

$$\text{if } x = \frac{\pi}{4}, \quad \text{then} \quad u = \tan\left(\frac{\pi}{4}\right) = 1.$$

Thus,

$$\int_0^{\pi/4} \tan^2 x \sec^2 x\, dx = \int_0^1 u^2\, du = \frac{1}{3}u^3\Big]_0^1 = \frac{1}{3}\cdot 1^2 - \frac{1}{3}\cdot 0^2 = \frac{1}{3}. \quad ■$$

Example 9 Find $\displaystyle\int_{-1}^{2} \frac{x^2\, dx}{(x^3+4)^2}$.

Solution: Here we use the substitution

$$u = x^3 + 4; \qquad du = 3x^2\, dx.$$

Thus, $x^2\, dx = \frac{1}{3}\, du$. Also,

$$\text{if } x = -1, \quad u = (-1)^3 + 4 = -1 + 4 = 3;$$

$$\text{if } x = 2, \quad u = 2^3 + 4 = 8 + 4 = 12.$$

Thus,

$$\begin{aligned}\int_{-1}^{2} \frac{x^2\, dx}{(x^3+4)^2} = \int_3^{12} \frac{\frac{1}{3}\, du}{u^2} &= \frac{1}{3}\int_3^{12} u^{-2}\, du \\ &= -\frac{1}{3}u^{-1}\Big]_3^{12} \\ &= -\frac{1}{3}\left(\frac{1}{12}\right) - \left[-\frac{1}{3}\left(\frac{1}{3}\right)\right] \\ &= \frac{1}{12}. \quad ■\end{aligned}$$

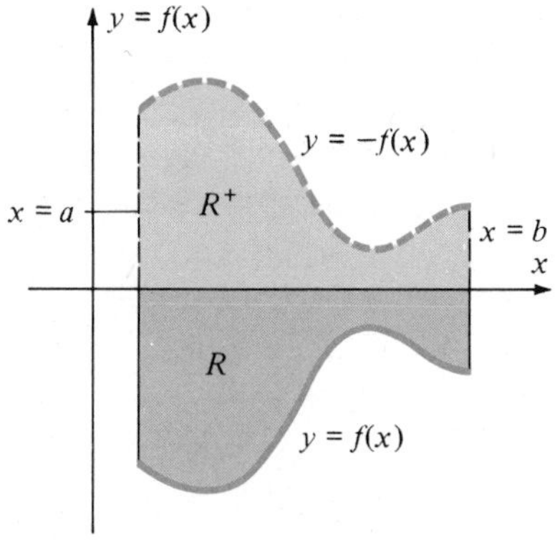

Figure 4.2 If $f(x) \le 0$, area of $R = -\int_a^b f(x)\, dx$.

More on Area

Having now discovered an efficient means of evaluating definite integrals, we wish to extend our notion of area so that the integral can be used to calculate areas of more general regions. The first step in this extension is to define the area of a region bounded by the graph of $f(x)$ and the x-axis, for $a \le x \le b$, when $f(x) \le 0$ for all $x \in [a, b]$. (See Figure 4.2.)

Let R be such a region. Then R is congruent to the region R^+ bounded *above* by the graph of $g(x) = -f(x)$, below by the x-axis, on the left by $x = a$, and on the right by $x = b$. Since the two regions R and R^+ are congruent, we want to define the area A of R to be the same as the area A^+ of R^+. That is, we want

$$A = \text{area of } R = \text{area of } R^+ = \int_a^b [-f(x)\, dx] = -\int_a^b f(x)\, dx.$$

DEFINITION 5

Let $f(x)$ be continuous on $[a, b]$ with $f(x) \le 0$ for all $x \in [a, b]$. The **area** of the region bounded below by the graph of $f(x)$, above by the x-axis, on the left by $x = a$, and on the right by $x = b$, is given by the integral

$$A = -\int_a^b f(x)\, dx.$$

Thus, area is the *negative* of the integral when the integrand $f(x)$ is negative. Remember, the integral itself measures area only when the integrand is positive.

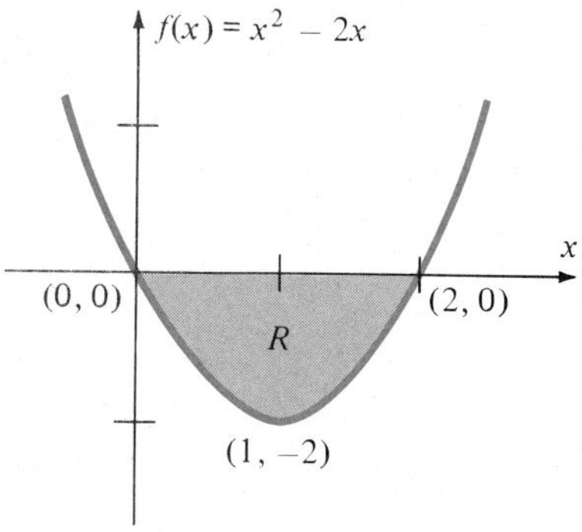

Figure 4.3 Region R in Example 10.

Example 10 Find the area of the region R bounded by the graph of $f(x) = x^2 - 2x$ and the x-axis.

Solution: In problems of this type, it is important first to sketch the graph of $f(x)$, so as to be able to identify the region R. The sketch in Figure 4.3 suggests that the only region bounded by the graph of $f(x)$ and the x-axis lies *below* the x-axis, for $0 \le x \le 2$. We verify this by noting that the zeros of $f(x) = x^2 - 2x = x(x-2)$ are $x = 0$ and $x = 2$, that $f(x) \le 0$ for $x \in [0, 2]$, and that $f(x) > 0$ if $x \notin [0, 2]$.

Thus, by Definition 5,

$$\begin{aligned} A &= -\int_0^2 (x^2 - 2x)\, dx \\ &= -\left(\frac{x^3}{3} - x^2\right]_0^2 \\ &= -\left[\left(\frac{2^3}{3} - 2^2\right) - \left(\frac{0^3}{3} - 0^2\right)\right] \\ &= -\left(\frac{8}{3} - 4\right) \\ &= \frac{4}{3}. \end{aligned}$$

■

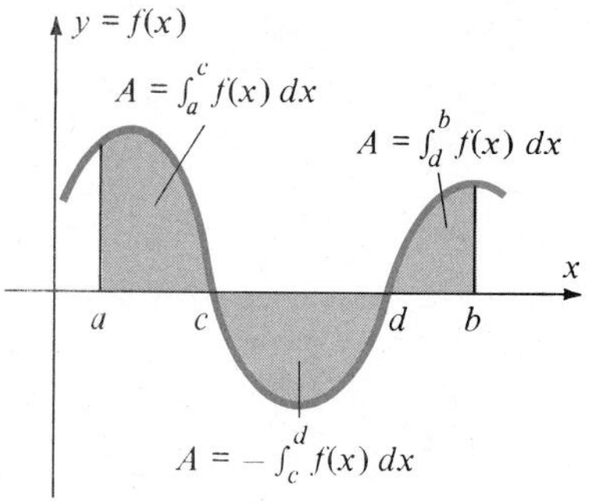

Figure 4.4 Rule for calculating area when $f(x)$ changes sign.

Figure 4.4 illustrates *the method we can use to find the area of a region bounded by the graph of the continuous function $f(x)$ and the x-axis for $a \le x \le b$ when $f(x)$ has both positive and negative values on $[a, b]$:*

(i) Find all zeros for $f(x)$ in $[a, b]$.

(ii) Determine the sign of $f(x)$ on each of the intervals determined by these zeros and the endpoints a and b.

(iii) Calculate the area of the part of the region corresponding to each interval according to the definitions

$$\text{Area} = \int_c^d f(x)\, dx \quad \text{if} \quad f(x) \ge 0 \text{ for } x \in [c, d]$$

$$\text{Area} = -\int_c^d f(x)\, dx \quad \text{if} \quad f(x) \le 0 \text{ for } x \in [c, d]$$

(iv) Sum the resulting areas.

Example 11 Find the area of the region bounded by the graph of $y = x^2 - 2x$, the x-axis, and the lines $x = 0$ and $x = 4$.

Strategy

Graph $f(x)$ for $0 \le x \le 4$.

Set $f(x) = 0$ to find zeros.

Determine sign of $f(x)$ on $[0, 2]$ and $[2, 4]$.

Apply definitions of area.

Use the Fundamental Theorem to evaluate the integrals.

Solution

The graph of $f(x) = x^2 - 2x$ appears in Figure 4.5. To find the zeros we set

$$f(x) = x^2 - 2x = x(x-2) = 0$$

and obtain $x = 0, 2$. Since $f(x) \le 0$ for $x \in [0, 2]$ and $f(x) \ge 0$ for $x \in [2, 4]$, we find that

$$\begin{aligned}\text{Area} &= -\int_0^2 (x^2 - 2x)\,dx + \int_2^4 (x^2 - 2x)\,dx \\ &= -\left[\frac{x^3}{3} - x^2\right]_0^2 + \left[\frac{x^3}{3} - x^2\right]_2^4 \\ &= -\left[\left(\frac{8}{3} - 4\right) - 0\right] + \left[\left(\frac{64}{3} - 16\right) - \left(\frac{8}{3} - 4\right)\right] \\ &= \frac{4}{3} + \frac{20}{3} = 8.\end{aligned}$$

■

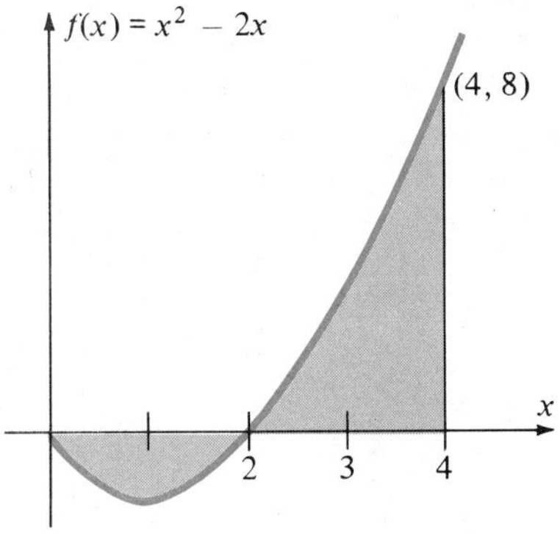

Figure 4.5 Region in Example 11.

Example 12 Find the area of the region R bounded by the graph of $f(x) = x(x^2 - 1)^{1/3}$ and the x-axis for $-2 \le x \le 2$.

Solution: The graph of $f(x)$ is sketched in Figure 4.6. As Figure 4.6 suggests, the part of R lying to the left of the y-axis is congruent with the part lying to the right of the y-axis. This is because $f(x)$ is an *odd* function, i.e., because

$$f(-x) = (-x)((-x)^2 - 1)^{1/3} = -x(x^2 - 1)^{1/3} = -f(x).$$

Thus, we need only calculate the area of the part of R lying in the right half-plane and double the result.

Since the equation

$$f(x) = x(x^2 - 1)^{1/3} = 0$$

has solutions $x = 0$, $x = -1$, and $x = 1$, the zeros of $f(x)$ in $[0, 2]$ are $x = 0$ and $x = 1$. Since $f(x) \le 0$ for $x \in [0, 1]$ $\left(\text{check } f\left(\frac{1}{2}\right), \text{ say, to verify this}\right)$ and $f(x) \ge 0$ for $x \in [1, 2]$, we have

$$\text{Area} = 2\left\{-\int_0^1 x(x^2 - 1)^{1/3}\,dx + \int_1^2 x(x^2 - 1)^{1/3}\,dx\right\}. \tag{3}$$

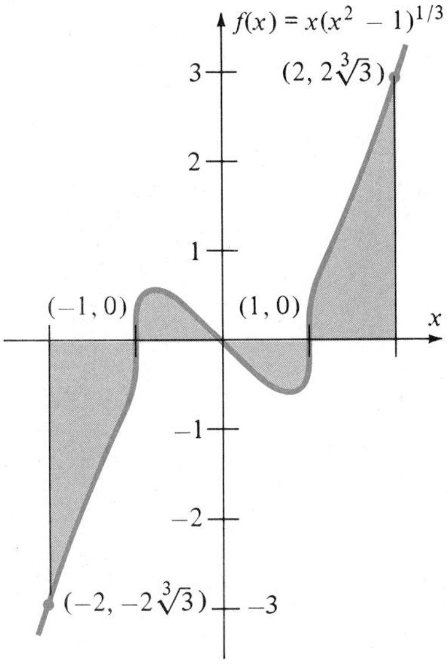

Figure 4.6 Region R in Example 12.

In the first integral, we use the substitution

$$u = x^2 - 1; \qquad du = 2x\,dx.$$

Then $x\,dx = \dfrac{1}{2}\,du$, and the new limits of integration are found as follows:

If $x = 0$, $\quad u = 0^2 - 1 = -1$.
If $x = 1$, $\quad u = 1^2 - 1 = 0$.

Thus,

$$\int_0^1 x(x^2-1)^{1/3}\,dx = \int_{-1}^{0} u^{1/3}\cdot\frac{1}{2}\,du = \frac{1}{2}\int_{-1}^{0} u^{1/3}\,du \tag{4}$$
$$= \frac{1}{2}\cdot\frac{3}{4}u^{4/3}\Big]_{-1}^{0}$$
$$= \frac{3}{8}(0^{4/3}-(-1)^{4/3})$$
$$= -\frac{3}{8}.$$

We use the same u-substitution in the second integral, with limits of integration determined as follows:

If $x = 1$, $\quad u = 1^2 - 1 = 0.$
If $x = 2$, $\quad u = 2^2 - 1 = 3.$

Thus,

$$\int_1^2 x(x^2-1)^{1/3}\,dx = \int_0^3 u^{1/3}\cdot\frac{1}{2}\,du = \frac{1}{2}\int_0^3 u^{1/3}\,du \tag{5}$$
$$= \frac{1}{2}\cdot\frac{3}{4}u^{4/3}\Big]_0^3$$
$$= \frac{3}{8}(3^{4/3}-0)$$
$$= \frac{3}{8}(3)^{4/3}$$

Using (4) and (5) we return to (3) and find that

$$\text{Area} = 2\left\{-\left(-\frac{3}{8}\right)+\frac{3}{8}(3)^{4/3}\right\}$$
$$= \frac{3}{4}(1+\sqrt[3]{81})$$

■

Exercise Set 6.4

In each of Exercises 1–38, evaluate the definite integral using the Fundamental Theorem of Calculus and, where necessary, a u-substitution.

1. $\int_{-2}^{2} (x^3 - 1)\,dx$

2. $\int_0^4 (x^2 - 2x + 3)\,dx$

3. $\int_2^0 x(\sqrt{x} - 1)\,dx$

4. $\int_9^4 \left(\sqrt{x} + \frac{1}{\sqrt{x}}\right) dx$

5. $\int_0^{2\pi/3} \sin(2x)\,dx$

6. $\int_{-1}^{2} \left(\frac{2}{x^3} + 5x\right) dx$

7. $\int_0^{\pi/4} (1 - \cos 2x)\,dx$

8. $\int_0^3 t(\sqrt[3]{t} - 2)\,dt$

9. $\int_0^1 2x(x^2 + 1)^2\,dx$

10. $\int_0^1 x\sqrt{1 - x^2}\,dx$

11. $\int_0^4 |x - 3|\,dx$

12. $\int_4^9 \left(\sqrt{x} - \frac{1}{\sqrt{x}}\right) dx$

13. $\int_1^2 x\sqrt{4 - x^2}\,dx$

14. $\int_0^4 |x - \sqrt{x}|\,dx$

15. $\int_1^3 (t^2 + 2)^2\,dt$

16. $\int_0^1 (x^{3/5} - x^{5/3})\,dx$

17. $\int_1^2 \frac{1 - x}{x^3}\,dx$

18. $\int_0^4 \frac{dt}{\sqrt{2t + 1}}$

19. $\int_0^{\sqrt{\pi}/2} t \sin(\pi - t^2)\, dt$

20. $\int_0^{\pi/2} \cos^3 t \sin t\, dt$

21. $\int_{-\pi/4}^{\pi/4} \frac{\sin x}{\cos^2 x}\, dx$

22. $\int_0^{\pi/3} \cos t\sqrt{1 - \sin^2 t}\, dt$

23. $\int_0^1 x\sqrt{ax^2 + b}\, dx$

24. $\int_0^2 \frac{x}{\sqrt{16 + x^2}}\, dx$

25. $\int_{-1}^1 \frac{x}{(1 + x^2)^3}\, dx$

26. $\int_0^3 x\sqrt{9 - x^2}\, dx$

27. $\int_{\pi^2/2}^{\pi^2} \frac{\sin\sqrt{x}}{\sqrt{x}}\, dx$

28. $\int_0^{\pi/4} \sin x\sqrt{\cos x}\, dx$

29. $\int_1^2 \frac{x}{(2x^2 - 1)^3}\, dx$

30. $\int_4^9 x^{1/2}(1 - x^{3/2})\, dx$

31. $\int_0^4 |9 - x^2|\, dx$

32. $\int_{-2}^2 |1 - x^2|\, dx$

33. $\int_0^1 \frac{1}{(4 - x)^2}\, dx$

34. $\int_{-1}^4 |x^2 - x - 2|\, dx$

35. $\int_0^{\pi/4} \sin^2 x\, dx$ (*Hint:* Use the identity

$$\sin^2 x = \frac{1}{2} - \frac{1}{2}\cos 2x.)$$

36. $\int_0^{\pi} \cos^2 x\, dx$ (*Hint:* Use the identity

$$\cos^2 x = \frac{1}{2} + \frac{1}{2}\cos 2x.)$$

37. $\int_{-2}^3 f(x)\, dx$, where $f(x) = \begin{cases} x^2 - 1 & \text{for} \quad x \le 1 \\ x - 1 & \text{for} \quad x > 1 \end{cases}$

38. $\int_1^5 f(x)\, dx$, where

$$f(x) = \begin{cases} \sqrt{x} & \text{for} \quad 0 \le x \le 4 \\ 2x^2 - 7x & \text{for} \quad x > 4 \end{cases}.$$

39. Let $f(x)$ be defined by

$$f(x) = \begin{cases} x & \text{if} \quad x \le 2 \\ 4 - x & \text{if} \quad x > 2 \end{cases}.$$

From area considerations, evaluate the following integrals (Figure 4.7).

a. $\int_0^2 f(x)\, dx$

b. $\int_0^4 f(x)\, dx$

c. $\int_{-2}^2 f(x)\, dx$

d. $\int_{-1}^2 f(x)\, dx$

e. $\int_4^0 f(x)\, dx$

f. $\int_2^5 f(x)\, dx$

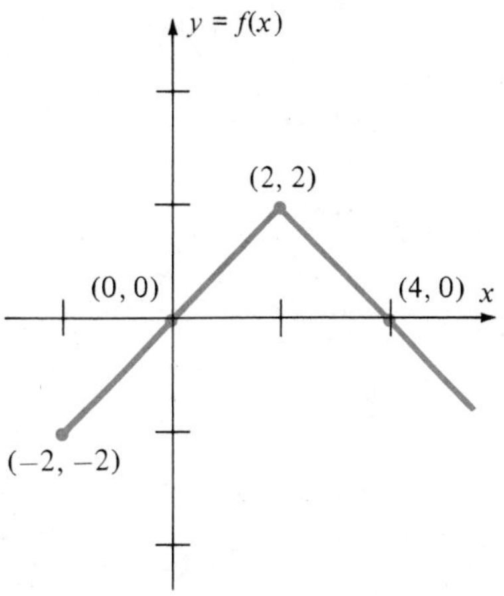

Figure 4.7

In Exercises 40–49, find **(a)** the definite integral of the given function over the given interval and **(b)** the area bounded by the graph of the given function and the x-axis over the given interval.

40. $f(x) = 2x + 1,\quad [0, 3]$

41. $f(x) = 3 - x,\quad [0, 2]$

42. $f(x) = 3 - x^2,\quad [0, \sqrt{3}]$

43. $f(x) = 2x^2 - 1,\quad [1, 4]$

44. $f(x) = 2x^2 - 6x + 1,\quad [1, 2]$

45. $f(x) = x^2 - 6,\quad [0, 3]$

46. $f(x) = x(x + 1),\quad [-1, 3]$

47. $f(x) = (2x + 1)(2x - 1),\quad [-2, 2]$

48. $f(x) = x,\quad [-3, 3]$

49. $f(x) = x - x^2,\quad [0, 3]$

In Exercises 50–57, find the area of region bounded by the graph of $f(x)$ and the x-axis between the lines $x = a$ and $x = b$.

50. $f(x) = 4 - x^2,\quad a = 0,\quad b = 3$

51. $f(x) = \sqrt{x} - 4,\quad a = 0,\quad b = 25$

52. $f(x) = \sin x,\quad a = 0,\quad b = 3\pi/2$

53. $f(x) = \sqrt{x} - x,\quad a = 0,\quad b = 4$

54. $f(x) = x\sqrt{9 - x^2},\quad a = 0,\quad b = 3$

55. $f(x) = x^2(2x^3 + 2),\quad a = -2,\quad b = 1$

56. $f(x) = x\cos(\pi + x^2),\quad a = 0,\quad b = \sqrt{\pi}/2$

57. $f(x) = \cos^3 x \sin x,\quad a = 0,\quad b = 3\pi/4$

58. Find the area bounded by the graph of $f(x) = \cos 2x$ and the x-axis for $0 \le x \le \pi$.

59. Show that

a. $\displaystyle\int_0^{\pi/2} \sin x\, dx = \int_0^{\pi/2} \cos x\, dx,$

b. $\displaystyle\int_0^{\pi/2} \sin^2 x\, dx = \int_0^{\pi/2} \cos^2 x\, dx.$

What do you conjecture about $\displaystyle\int_0^{\pi/2} \sin^n x\, dx$ and $\displaystyle\int_0^{\pi/2} \cos^n x\, dx$? Can you prove this?

60. Generalize the result of Example 12 by showing that if $f(-x) = -f(x)$ for all x and $f(x)$ is continuous, then

a. the area of the region bounded by the graph of $f(x)$ and the x-axis for $-a \le x \le a$ is twice that for $0 \le x \le a$, but

b. $\displaystyle\int_{-a}^{a} f(x)\, dx = 0.$

6.5 DISTANCE AND VELOCITY

In Chapter 3, we saw that the functions describing the position, velocity, and acceleration of an object moving along a line were related by the process of differentiation. (In particular, $v(t) = s'(t)$, $a(t) = v'(t)$.) In this section we shall show that the distance D travelled by such an object from time $t = a$ to time $t = b$ can be calculated as a definite integral involving the velocity function $v(t)$. In particular,

$$D = \int_a^b |v(t)|\, dt. \tag{1}$$

There are two reasons for addressing this topic at this point, midway through the development of the theory of the definite integral. First, we want to demonstrate, for a problem other than that of calculating area, how the definite integral arises as the solution to a problem when the approximations to the solution are recognized as Riemann sums. Second, as an application of the Fundamental Theorem, we will be able to show the relationship between distance as defined in (1) and distance determined by an antiderivative. (They are, of course, the same.)

We begin with the problem of an object moving along a line with velocity $v(t)$ at time t. The question we want to answer is this: Given that $v(t)$ is a known, continuous function, how can we calculate the distance travelled by the object from time $t = a$ to time $t = b$?

For the romantics among us, here is one centuries-old approach to solving this problem.

> As for distance sailed east or west, early mariners relied solely on dead reckoning, estimating their speed by peering over the side and noting the rate at which foam slipped past the planking. The results they came up with were little more than a compound of hunch, guesswork and intuition. Samuel Eliot Morison calculates that on Columbus' first Atlantic passage, he consistently overestimated the distance of his day's run by about 9 per cent, which scales up to a 90-mile error for every 1,000 miles traveled. However, the remarkable fact is not how far off these navigators were, but how amazingly close they came using such primitive methods.
>
> Progress in the following centuries came slowly. In the late 1500s, some ingenious mariner—name and nationality unknown—invented the common, or chip, log, a simple speed-measuring device. The chip log was a piece of wood shaped like a slice of pie. Attached to it was a light line, knotted at certain equal intervals. To check the ship's speed, a man threw the log overboard, turned a small 30-second sandglass, and began counting the knots that slipped through his fingers as the ship left the log astern.

The number of knots he counted in 30 seconds was translated into the nautical miles per hour his ship was traveling. With this device began usage of the term ''knots'' to mean ''nautical miles per hour.''
Life Science Library/SHIPS by Edward V. Lewis, Robert O'Brien and the Editors of LIFE. Time-Life Books, Inc., Publisher. © 1965 Time Inc.

In our more mathematical setting, we begin by partitioning the time interval $[a, b]$ into n subintervals of equal length, $\Delta t = \dfrac{b-a}{n}$, and with endpoints

$$t_0 = a, \qquad t_1 = a + \Delta t, \qquad t_2 = a + 2\,\Delta t, \ldots, t_n = a + n\,\Delta t = b.$$

Our idea is to assume that $v(t)$ is constant on each subinterval of time, to approximate the distance travelled during each subinterval, and then to sum the individual distances to obtain an approximation to D, the distance travelled by the object.

To do so we let c_j denote any number (time) in the jth interval $[t_{j-1}, t_j]$, and we assume that $v(t) \equiv v(c_j)$ for all $t \in [t_{j-1}, t_j]$. Using the definition

distance = speed × time

we approximate the distance ΔD_j travelled by the object during the time subinterval $[t_{j-1}, t_j]$ as

$$\Delta D_j = |v(c_j)|(t_j - t_{j-1}) = |v(c_j)|\,\Delta t, \qquad j = 1, 2, \ldots, n.$$

(Recall, if $v(t)$ is velocity, speed $= |v(t)|$. We care only about the rate of the motion, not its direction.) Summing these individual approximations gives an approximation to D as

$$D \approx \sum_{j=1}^{n} \Delta D_j = \sum_{j=1}^{n} |v(c_j)|\,\Delta t, \qquad c_j \in [t_{j-1}, t_j] \tag{2}$$

Now the sum on the right-hand side of approximation (2) is a Riemann sum for the function $|v(t)|$ on the interval $[a, b]$. If $v(t)$ is continuous on $[a, b]$, so is $|v(t)|$, so by Theorem 2 this Riemann sum will have as its limit the integral

$$\int_a^b |v(t)|\,dt = \lim_{n\to\infty} \sum_{j=1}^{n} |v(c_j)|\,\Delta t, \qquad c_j \in [t_{j-1}, t_j].$$

Finally, we argue that as $n \to \infty$, $\Delta t \to 0$, so the error involved in approximating $v(t)$ by $v(c_j)$ on the interval $[t_{j-1}, t_j]$ approaches zero. That is, the approximation (2) should become increasingly accurate as $n \to \infty$. We therefore *define* the number D to be the limit of this approximating sum. That is, we define distance by the equation

$$D = \lim_{n\to\infty} \sum_{j=1}^{n} |v(c_j)|\,\Delta t = \int_a^b |v(t)|\,dt \tag{3}$$

Example 1 Starting at time $t = 0$, a particle moves along a line with velocity $v(t) = 2 + \sqrt{t}$ m/sec. How far does it travel during the first 4 seconds?

Solution: Here $v(t) = 2 + \sqrt{t} \geq 0$ for all $t \in [0, 4]$, so $|v(t)| = v(t)$ in (3).

Thus, by (3),

$$D = \int_0^4 (2 + \sqrt{t})\, dt = 2t + \frac{2}{3}t^{3/2}\Big]_0^4 = \left(2 \cdot 4 + \frac{2}{3} \cdot 8\right) - (0)$$
$$= \frac{40}{3} \text{ meters.}$$ ■

Example 2 A particle moves along a line with velocity $v(t) = 5 - t^2$ m/sec. Find the distance travelled from time $t = 3$ sec to time $t = 5$ sec.

Solution: This time $v(t) = 5 - t^2 < 0$ for $t \in [3, 5]$, so $|v(t)| = -v(t)$ in equation (3). Thus,

$$D = \int_3^5 -(5 - t^2)\, dt = \frac{t^3}{3} - 5t\Big]_3^5 = \left(\frac{125}{3} - 25\right) - \left(\frac{27}{3} - 15\right)$$
$$= \frac{68}{3} \text{ meters.}$$ ■

Obviously, a difficulty arises in applying (3) when $v(t)$ has both positive and negative values on $[a, b]$. In such cases, we must first find the zeros of $v(t)$, and then apply formula (3) separately on each of the subintervals of $[a, b]$ determined by the zeros and the endpoints.

Example 3 If an object moves along a line with velocity $v(t) = t^3 - 3t^2 - t + 3$ m/sec, how far does it travel from time $t = 0$ to time $t = 4$ seconds?

Solution: In order to determine the sign of $v(t)$ on $[0, 4]$, we must find the solutions of the equation $v(t) = 0$. By performing a polynomial long division, if necessary, we find that $v(t)$ has factorization

$$v(t) = (t + 1)(t - 1)(t - 3).$$

Thus, $v(t) = 0$ on $[0, 4]$ if $t = 1$ or $t = 3$. By evaluating $v(t)$ at numbers in each of the resulting intervals you can verify that

$$v(t) \geq 0 \quad \text{for} \quad t \in [0, 1]$$
$$v(t) \leq 0 \quad \text{for} \quad t \in [1, 3]$$
$$v(t) \geq 0 \quad \text{for} \quad t \in [3, 4].$$

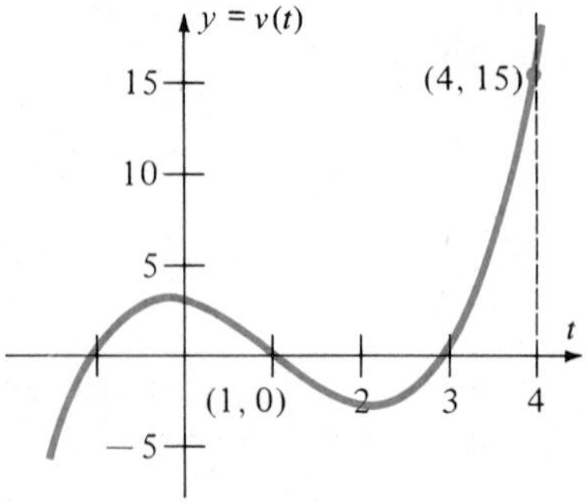

Figure 5.1 $v(t) = t^3 - 3t^2 - t + 3$ in Example 3.

(See Figure 5.1.)
Thus,

$$D = \int_0^1 (t^3 - 3t^2 - t + 3)\, dt + \int_1^3 -(t^3 - 3t^2 - t + 3)\, dt$$
$$+ \int_3^4 (t^3 - 3t^2 - t + 3)\, dt$$
$$= \left[\frac{t^4}{4} - t^3 - \frac{t^2}{2} + 3t\right]_0^1 + \left[-\frac{t^4}{4} + t^3 + \frac{t^2}{2} - 3t\right]_1^3$$
$$+ \left[\frac{t^4}{4} - t^3 - \frac{t^2}{2} + 3t\right]_3^4$$

$$= \left(\frac{7}{4} - 0\right) + \left(\frac{9}{4} - \left(-\frac{7}{4}\right)\right) + \left(4 - \left(-\frac{9}{4}\right)\right)$$

$$= 12 \text{ meters.}$$ ■

Example 4 The position of the end of a vibrating spring is given by the function $s(t) = 6 \cos \pi t$, where t is in seconds and s is in centimeters. How far does the end of the spring travel between times $t = 0$ and $t = 3$?

Strategy

Differentiate $s(t)$ to find $v(t)$.

Solution

Since $s(t) = 6 \cos \pi t$ is the *position* of the end of the spring, its velocity is

$$v(t) = \frac{d}{dt}(6 \cos \pi t) = -6\pi \sin \pi t.$$

Set $v(t) = 0$ to find zeros in $[0, 3]$.

Then $v(t) = 0$ when $\sin \pi t = 0$, which happens for

$$\pi t = n\pi, \qquad \text{or} \qquad t = n = 0, \pm 1, \pm 2, \ldots .$$

The zeros for $v(t)$ in $[0, 3]$ are therefore $t = 0, 1, 2,$ and 3. By checking particular numbers, we find that

Determine the sign of $v(t)$ on each of the resulting intervals.

$$\begin{aligned} v(t) &\le 0 \quad \text{for} \quad t \in [0, 1] \\ v(t) &\ge 0 \quad \text{for} \quad t \in [1, 2] \\ v(t) &\le 0 \quad \text{for} \quad t \in [2, 3]. \end{aligned}$$

(See Figure 5.2.)

Thus, by (3),

Apply the definition (3) on each interval according to the sign of $v(t)$.

$$\begin{aligned} D &= \int_0^1 -v(t)\, dt + \int_1^2 v(t)\, dt + \int_2^3 -v(t)\, dt \\ &= \int_0^1 6\pi \sin \pi t\, dt + \int_1^2 (-6\pi \sin \pi t)\, dt + \int_2^3 6\pi \sin \pi t\, dt \\ &= -6 \cos \pi t\Big]_0^1 + 6 \cos \pi t\Big]_1^2 - 6 \cos \pi t\Big]_2^3 \\ &= [(-6)(-1) - (-6)(1)] + [(6)(1) - (6)(-1)] + [(-6)(-1) - (-6)(1)] \\ &= 36 \text{ cm.} \end{aligned}$$ ■

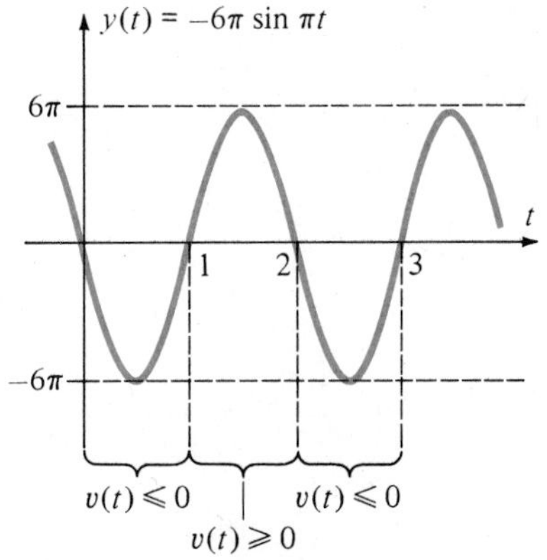

Figure 5.2 Velocity function in Example 4.

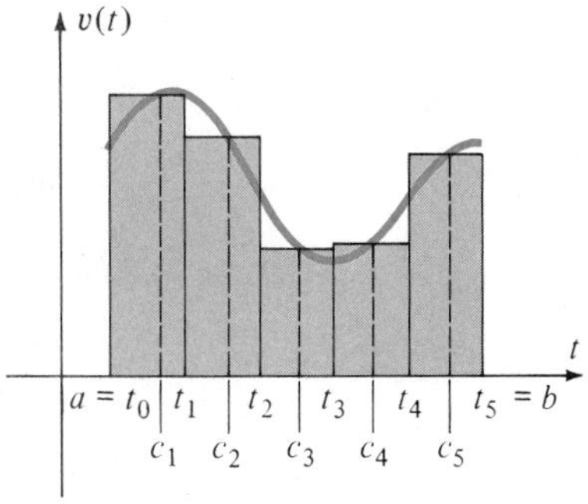

Figure 5.3 Riemann sum R_5 for $D = \int_a^b |v(t)|\, dt$.

You have undoubtedly noted the similarity between the calculation of area and the calculation of distance. Both are nonnegative quantities, and both are calculated by integrating functions which themselves are nonnegative. In fact, a sketch of a Riemann sum for the integral $D = \int_a^b |v(t)|\, dt$ (Figure 5.3) shows that the sum could just as well be interpreted as an approximation to the area of the region bounded by the graph of $y = |v(t)|$ and the t-axis for $a \le t \le b$. In fact, we can employ the absolute value notation used here to write the solution to the Area Problem succinctly as

$$A = \int_a^b |f(x)|\, dx.$$

(See Exercise 20.)

Distance as an Antiderivative

Now anyone can define anything, so whenever we define a quantity (such as distance), we must always check to make sure that our new definition agrees with other accepted definitions for the same quantity. For example, when we defined the area of the region R bounded by the graph of $f(x) \ge 0$ and the x-axis for $a \le x \le b$, by $A = \int_a^b f(x)\, dx$, it was important that

$$\int_a^b c\, dx = c(b - a), \qquad c > 0,$$

since this is the agreed upon area of a rectangle with width $b - a$ and height c.

For the case of a nonnegative velocity function $v(t)$, we know from Chapter 3 that the distance travelled from time $t = a$ to time $t = b$ by a particle travelling with velocity $v(t)$ is

$$D = s(b) - s(a) \tag{4}$$

where $s(t)$ is an antiderivative (a position function) for $v(t)$. To verify that equations (3) and (4) define the same quantity, we use the Fundamental Theorem of Calculus. If $s'(t) = v(t)$, we have

$$D = \int_a^b v(t)\, dt = s(t)\Big]_a^b = s(b) - s(a).$$

The cases $v(t) \le 0$ and $v(t)$ of mixed sign are similar and are left as Exercises 21 and 22.

Exercise Set 6.5

In Exercises 1–10, calculate the distance travelled by a particle moving along a line with the given velocity from time $t = a$ to time $t = b$.

1. $v(t) = 9 - t^2, \quad a = 0, \quad b = 3$
2. $v(t) = \sqrt{t + 1}, \quad a = 0, \quad b = 8$
3. $v(t) = \sin t, \quad a = 0, \quad b = \pi$
4. $v(t) = t - t^3, \quad a = 0, \quad b = 1$
5. $v(t) = 2 + t^2, \quad a = 0, \quad b = 5$
6. $v(t) = t^2 + t - 6, \quad a = 0, \quad b = 2$
7. $v(t) = \cos \pi t, \quad a = \frac{1}{2}, \quad b = \frac{3}{2}$
8. $v(t) = \dfrac{1 - t}{\sqrt{t}}, \quad a = 2, \quad b = 4$
9. $v(t) = \sin^2 t \cos t, \quad a = 0, \quad b = \frac{\pi}{2}$
10. $v(t) = \sqrt{t}(t + 2), \quad a = 0, \quad b = 1$

In each of Exercises 11–16, the motion of a particle along a line is described by its velocity function. Taking care to note where $v(t)$ changes sign, compute the distance travelled by the particle during the time interval specified (see Example 3).

11. $v(t) = t - 4, \quad t = 0 \quad \text{to} \quad t = 6$

12. $v(t) = 2 - t^2, \quad t = 0 \quad \text{to} \quad t = 4$

13. $v(t) = \sin(2t), \quad t = 0 \quad \text{to} \quad t = \pi/2.$

14. $v(t) = t^{1/3} - t^{2/3}, \quad t = 0 \quad \text{to} \quad t = 8$

15. $v(t) = (t - 1)^5, \quad t = 0 \quad \text{to} \quad t = 2$

16. $v(t) = \dfrac{t^2 - 9}{t + 3}, \quad t = 0 \quad \text{to} \quad t = 4$

17. *(Calculator)* A particle moves along a line so that its velocity at various times is as follows:

t	1	3	5	7	9
$v(t)$	1	2	5	18	39

Use the midpoint rule with $n = 5$ to approximate the distance travelled by the particle between times $t = 0$ and $t = 10$ seconds (see Exercise 25, Section 6.3).

18. *(Calculator)* Use a Riemann sum of your own choosing to determine the distance travelled by a particle between times $t = 0$ and $t = 30$ seconds whose motion is along a line and whose velocities are observed as follows:

t	0	2	4	6	8	10	12	14	16	18	20	22	24	26	28	30
$v(t)$	3	7	13	6	1	−3	−5	−9	−4	−1	5	8	9	10	11	10

19. *(Calculator* or *Computer)* A particle moves along a line with velocity given by $v(t) = \sin\sqrt{1 + t^2}$ meters per second.

a. Use the midpoint rule with $n = 6$ to approximate the distance travelled between $t = 0$ and $t = 3$ seconds.

b. Use a modification of Program 9 to approximate this distance with $n = 10$, 50, and 200.

20. Show, from the results of Section 6.3, that the area of the region bounded by the graph of $f(x)$ and the x-axis for $a \le x \le b$ can be written

$$A = \int_a^b |f(x)|\, dx$$

regardless of the sign of $f(x)$.

21. Verify that $\int_a^b |v(t)|\, dt = -[s(b) - s(a)] = s(a) - s(b)$ if $s'(t) = v(t)$ and $v(t) \le 0$ for all $t \in [a, b]$.

22. What can you say about the relationship between $\int_a^b |v(t)|\, dt$ and the number $s(b) - s(a)$ when $v(t)$ has both positive and negative values on $[a, b]$?

6.6 FUNCTIONS DEFINED BY INTEGRATION

The Fundamental Theorem of Calculus provides a powerful method for evaluating $\int_a^b f(x)\, dx$ when we can find an antiderivative $F(x)$ for $f(x)$. A question that naturally arises is whether or not a continuous function $f(x)$ need have an antiderivative. In this section we will show that the answer is yes, as long as we are willing to accept a new kind of function, called a function defined by integration.

Example 1 Consider the function $F(x)$ defined by

$$F(x) = \int_0^x t^2\, dt.$$

Notice that the independent variable x appears as the upper limit of integration in the definite integral that defines $F(x)$. (The variable t is sometimes called the "dummy" variable of integration.) For each $x > 0$, $F(x)$ can be interpreted as the area of the region bounded by the graph of the function $f(t) = t^2$ and the t-axis

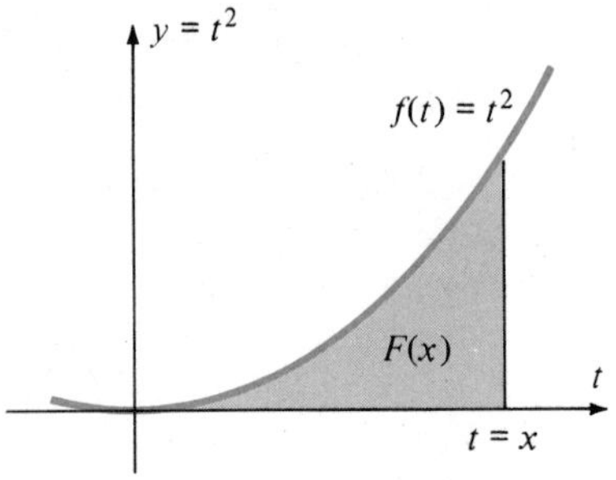

Figure 6.1 $F(x) = \int_0^x t^2\,dt$.

between $t = 0$ and $t = x$ (see Figure 6.1). In this case, we can simplify $F(x)$ using the Fundamental Theorem as follows:

$$F(x) = \int_0^x t^2\,dt = \frac{1}{3}t^3\Big]_0^x = \frac{1}{3}x^3.$$

It is therefore easy to see that

$$F'(x) = x^2.$$ ■

Notice that in Example 1 the function defined by integration, namely,

$$F(x) = \int_0^x t^2\,dt$$

turned out to be an antiderivative for the integrand. The following theorem shows that this will always be the case when the integrand is continuous. (This theorem is often also referred to as the Fundamental Theorem of Calculus since, in a certain sense, it is the logical equivalent of the preceding theorem. We shall not pursue this relationship here.)

THEOREM 7

Suppose that $f(x)$ is continuous for all $x \in [a, b]$. Then the function $F(x)$ defined by

$$F(x) = \int_a^x f(t)\,dt, \qquad x \in [a, b]$$

is an antiderivative for $f(x)$. That is, $F(x)$ is differentiable for $a < x < b$, and

$$\frac{d}{dx}\left\{\int_a^x f(t)\,dt\right\} = f(x), \qquad x \in (a, b). \tag{1}$$

Proof: Let $x \in (a, b)$. We will show that $F'(x)$ exists and equals $f(x)$. By the definition of the derivative, we must examine the limit of the difference quotient

$$\frac{F(x + \Delta x) - F(x)}{\Delta x} \qquad \text{as} \qquad \Delta x \to 0.$$

Now

$$\frac{F(x + \Delta x) - F(x)}{\Delta x} = \frac{\displaystyle\int_a^{x+\Delta x} f(t)\,dt - \int_a^x f(t)\,dt}{\Delta x}. \tag{2}$$

$$= \frac{1}{\Delta x}\int_x^{x+\Delta x} f(t)\,dt.$$

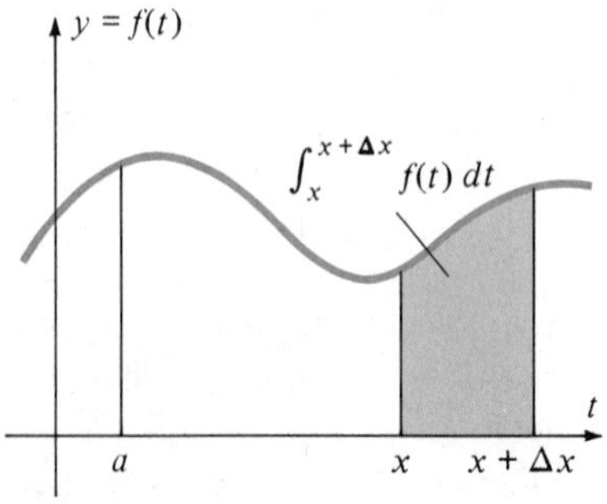

Figure 6.2

(See Figure 6.2.) For clarity, let us first assume that $\Delta x > 0$. On the interval $[x, x + \Delta x]$, let m denote the minimum value of $f(x)$ and let M denote the maximum value of $f(x)$. (These numbers exist since $f(x)$ is continuous on $[a, b]$.) Then, since $m \le f(t) \le M$, $t \in [a, b]$, it follows that

$$\int_x^{x+\Delta x} m\,dt \le \int_x^{x+\Delta x} f(t)\,dt \le \int_x^{x+\Delta x} M\,dt$$

(See Exercise 23.) Thus,

$$m\ \Delta x \le \int_x^{x+\Delta x} f(t)\ dt \le M\ \Delta x$$

since m and M are constants for any choice of Δx. Since we are assuming $\Delta x \neq 0$, we may divide by Δx to obtain the inequality

$$m \le \frac{1}{\Delta x} \int_x^{x+\Delta x} f(t)\ dt \le M. \tag{3}$$

(In the case $\Delta x \le 0$, inequality (3) is obtained by the same argument applied to the interval $[x + \Delta x, x]$.) Now since m and M are values of the function $f(t)$ on the interval $[x, x + \Delta x]$, we have

$$\lim_{\Delta x \to 0} m = \lim_{\Delta x \to 0} M = f(x), \tag{4}$$

since in this limit the interval shrinks to the single number $\{x\}$ and $f(x)$ is continuous at x. Thus, by the Pinching Theorem, we conclude from inequality (3) and equation (4) that

$$\lim_{\Delta x \to 0} \frac{1}{\Delta x} \int_x^{x+\Delta x} f(t)\ dt = f(x). \tag{5}$$

From equations (2) and (5) we now have

$$F'(x) = \lim_{\Delta x \to 0} \frac{F(x + \Delta x) - F(x)}{\Delta x} = f(x),$$

as claimed. ■

The role of Theorem 7 may seem somewhat obscure at this point. Why introduce the rather messy concept of a function defined by integration? The answer is that Theorem 7 guarantees that the function

$$F(x) = \int_a^x f(t)\ dt \tag{6}$$

is *always* available as an antiderivative for $f(x)$. Sometimes (happily, often!) a simpler antiderivative can be found—one that does not involve an integral sign. However, this is not always the case. (For example, try to find an antiderivative for $f(x) = \sin(x^2)$.) Theorem 7 guarantees that, at least in theory, every continuous function has an antiderivative.

This information is very important in the theory of ordinary and partial differential equations. Theorem 7 also gives us a very simple and explicit rule for differentiating functions defined by integration.

Example 2 For $F(x) = \int_2^x \sqrt{t} \sin t^2\ dt$, find $F'(x)$.

Solution: We apply equation (1) directly to obtain

$$F'(x) = \sqrt{x} \sin x^2.$$

(Note that the value of the constant 2 in the lower limit of integration is immaterial in determining $F'(x)$. This is because the rate of change of the value of the integral has to do only with the upper limit, representing the right boundary of the region with area $F(x)$.) ■

Example 3 For $F(x) = \int_{x^3}^{5} \sqrt{2 + \cos t}\, dt$, find $F'(x)$.

Solution: Here the variable is in the *lower* limit of integration. To bring the integral into the form (6), we interchange the order of integration to obtain

$$F(x) = -\int_{5}^{x^3} \sqrt{2 + \cos t}\, dt.$$

The function may now be viewed as a composite function $F(u(x))$ with

$$F(u) = -\int_{5}^{u} \sqrt{2 + \cos t}\, dt.$$

We can now apply Theorem 7 together with the Chain Rule to obtain

$$\begin{aligned} F'(x) &= -\sqrt{2 + \cos u} \cdot \frac{du}{dx} \\ &= -3x^2\sqrt{2 + \cos x^3}. \end{aligned}$$ ■

Theorem 7, together with the Mean Value Theorem for Derivatives, can be applied to prove the Mean Value Theorem for Integrals, as follows.

THEOREM 8
Mean Value Theorem for Integrals

Let $f(x)$ be continuous on the interval $[a, b]$. Then there exists a number $c \in (a, b)$ so that

$$\int_a^b f(x)\, dx = f(c) \cdot (b - a).$$

When $f(x) \geq 0$ for all $x \in [a, b]$, we may interpret Theorem 8 as saying that the area of the region bounded by the graph of $f(x)$ and the x-axis between $x = a$ and $x = b$ is the same as the area of the rectangle with height $f(c)$ and base $(b - a)$. For this reason the number

$$f(c) = \frac{1}{b - a} \int_a^b f(x)\, dx$$

is often called the **average value** of the function $f(x)$ on the interval $[a, b]$ (see Figure 6.3).

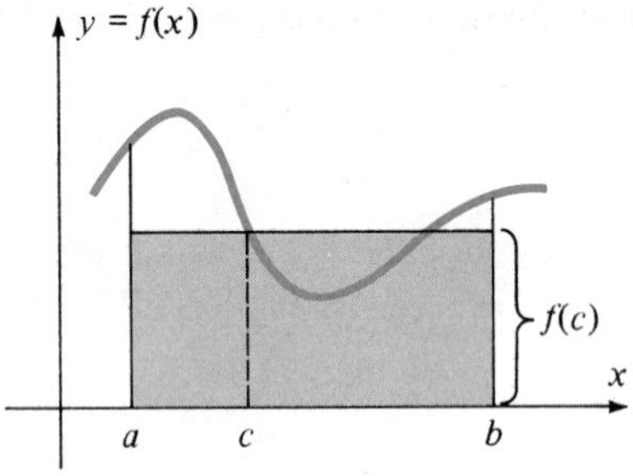

Figure 6.3 Mean Value Theorem for Integrals.

Proof of Theorem 8: By Theorem 7 the function

$$F(x) = \int_a^x f(t)\, dt \tag{7}$$

is differentiable on (a, b) and, since $f(x)$ is continuous on $[a, b]$, so is $F(x)$ (see Exercise 22). Thus, $F(x)$ satisfies the hypotheses of the Mean Value Theorem for Derivatives. By that theorem there exists a number $c \in [a, b]$ so that

$$F'(c) = \frac{F(b) - F(a)}{b - a}. \tag{8}$$

Using Theorem 7 and equation (7) we can rewrite equation (8) as

$$f(c) = \frac{\int_a^b f(t)\,dt - \int_a^a f(t)\,dt}{b - a}.$$

Now the second integral is zero, and the limits of integration in the first integral are constants. We may therefore replace the dummy variable t by x to obtain

$$f(c) = \frac{1}{b-a}\int_a^b f(x)\,dx. \qquad \blacksquare$$

This proof of the Mean Value Theorem for Integrals, like our proof of the Fundamental Theorem of Calculus, illustrates the intimate relationship between the differential and integral calculus. (An alternate proof of the Mean Value Theorem for Integrals, depending only on the Intermediate Value Theorem, is sketched in Exercise 25.) Although the Mean Value Theorem for Integrals may seem more elegant than practical, we can use it to establish a very important result concerning the error associated with the approximation of a definite integral by a Riemann sum. (Rules for approximating integrals by Riemann sums are called "rectangular rules," since they involve an approximation of a region by rectangles. One such rule, the "midpoint rule," was discussed in Exercises 25–30 of Section 6.3.)

THEOREM 9

Let $f(x)$ be differentiable on $[a, b]$ and assume that $|f'(x)| \leq M$ for all $x \in [a, b]$. Let $\int_a^b f(x)\,dx$ be approximated by a sum S_n of the form

$$S_n = \sum_{j=1}^{n} f(t_j)\,\Delta x$$

where $\Delta x = (b-a)/n$, $x_0 = a$, $x_1 = a + \Delta x$, $x_2 = a + 2\,\Delta x$, . . . , $x_n = a + n\,\Delta x = b$ and $t_j \in [x_{j-1}, x_j]$ for each $j = 1, 2, \ldots, n$. Then

$$\left| \int_a^b f(x)\,dx - S_n \right| \leq \frac{M(b-a)^2}{n}.$$

That is, the approximation of $\int_a^b f(x)\,dx$ by S_n produces an error no greater than $\dfrac{M(b-a)^2}{n}$.

Proof: We first make an observation needed in the midst of the proof. Let c_j and t_j be arbitrary elements of the interval $[x_{j-1}, x_j]$. Then, since $f(x)$ is differentiable on $[x_{j-1}, x_j]$, by the Mean Value Theorem for Derivatives there is a number e_j between c_j and t_j so that

$$\left| \frac{f(c_j) - f(t_j)}{c_j - t_j} \right| = |f'(e_j)|$$

or

$$|f(c_j) - f(t_j)| = |f'(e_j)| \cdot |c_j - t_j|. \tag{9}$$

(We use absolute value signs in stating this observation since we do not know the sign of $c_j - t_j$.)

The proof now proceeds as follows. By the Mean Value Theorem for Integrals, in each interval $[x_{j-1}, x_j]$ there exists a number c_j so that

$$\int_{x_{j-1}}^{x_j} f(x)\,dx = f(c_j)(x_j - x_{j-1}) = f(c_j)\,\Delta x. \tag{10}$$

We therefore obtain

$$\begin{aligned}\left|\int_a^b f(x)\,dx - S_n\right| &= \left|\sum_{j=1}^n \int_{x_{j-1}}^{x_j} f(x)\,dx - \sum_{j=1}^n f(t_j)\,\Delta x\right| \\ &= \left|\sum_{j=1}^n f(c_j)\,\Delta x - \sum_{j=1}^n f(t_j)\,\Delta x\right| \qquad \text{(using (10))} \\ &= \left|\sum_{j=1}^n [f(c_j) - f(t_j)]\,\Delta x\right| \\ &\le \sum_{j=1}^n |f(c_j) - f(t_j)|\,\Delta x.\end{aligned}$$

Applying equation (9) together with the fact that $|c_j - t_j| \le |x_j - x_{j-1}| = \Delta x$, we conclude that

$$\begin{aligned}\left|\int_a^b f(x)\,dx - S_n\right| &\le \sum_{j=1}^n |f'(e_j)|\,\Delta x^2 \\ &\le nM\,\Delta x^2 \\ &= \frac{M(b-a)^2}{n}.\end{aligned}$$

■

Example 4 Find a bound on the error involved in using a Riemann sum with 12 equal subintervals to approximate the integral

$$\int_0^\pi \sin x\,dx.$$

Solution: We have $|f'(x)| = |\cos x| \le 1$ for all x so we may take $M = 1$. By the statement of Theorem 9 with $a = 0$, $b = \pi$, and $n = 12$, we have

$$\left|\int_0^\pi \sin x\,dx - S_n\right| = \text{Error} \le \frac{(\pi - 0)^2}{12} = 0.8225.$$

This is the **absolute error.** Since we can actually calculate this integral via the Fundamental Theorem as

$$\int_0^\pi \sin x\,dx = -\cos x\Big]_0^\pi = 1 - (-1) = 2,$$

we can calculate the *relative error* as being no greater than $\dfrac{.8225}{2.0} = .4112$, and the **percentage error** as being no more than $\dfrac{.8225}{2} \times 100\% = 41.12\%$.* ■

*The **relative error** in a calculation is the absolute error divided by the actual value of what is being calculated. For example, if the number $x_0 = 8$ is *approximated* by number $x = 10$, the absolute error is $10 - 8 = 2$, and the relative error is $\dfrac{10-8}{8} = 0.25$.

Example 5 Determine how large n must be chosen so that the error involved in using a Riemann sum with n equal subintervals to approximate the integral

$$\int_1^2 \sqrt{1 + x^2}\, dx$$

is no greater than 0.05.

Solution: For $f(x) = \sqrt{1 + x^2}$, we have $f'(x) = \dfrac{x}{\sqrt{1 + x^2}}$. For $x \in [1, 2]$, the maximum value of the function $f'(x)$ is $f'(2) = \dfrac{2}{\sqrt{5}}$ (see Chapter 4, Section 3, if you forget how to do this calculation). According to Theorem 9, we need to choose n large enough to ensure that

$$\frac{M(b - a)^2}{n} < 0.05.$$

Since $M = \dfrac{2}{\sqrt{5}}$, $a = 1$, and $b = 2$, this requires that

$$\frac{2}{\sqrt{5}\, n} \leq \frac{1}{20}, \qquad \text{or} \qquad n \geq \frac{40}{\sqrt{5}}.$$

Since $\dfrac{40}{\sqrt{5}} = 17.89$, $n \geq 18$ will suffice. ■

Exercise Set 6.6

In each of Exercises 1–6, express $F(x)$ as a function of x not involving an integral sign, as in Example 1. Then verify equation (1) by differentiating your result.

1. $F(x) = \int_1^x t\, dt$

2. $F(x) = \int_{-2}^x (t^2 + 2t)\, dt$

3. $F(x) = \int_0^x \cos t\, dt$

4. $F(x) = \int_2^x \sqrt{1 + t}\, dt, \quad x \geq -1$

5. $F(x) = \int_0^x (at^2 + bt + c)\, dt$

6. $F(x) = \int_1^x \dfrac{t}{\sqrt{1 + t^2}}\, dt$

In Exercises 7–14, use Theorem 7 to find $F'(x)$.

7. $F(x) = \int_2^x t^2 \sin t^2\, dt$

8. $F(x) = \int_0^x \sqrt{t^4 + 1}\, dt$

9. $F(x) = \int_x^1 t^3 \cos^2 t\, dt$

10. $F(x) = \int_0^{3x} \sqrt{1 + \sin^4 t}\, dt$

11. $F(x) = \int_{x^2}^2 \sin(t^4 + 1)\, dt$

12. $F(x) = \int_{-x}^x \sqrt{t^2 + 1}\, dt$

13. $F(x) = x \int_3^x e^{-t^2}\, dt$

14. $F(x) = \int_3^x e^{-t^2}\,dt - \int_5^x e^{-t^2}\,dt$

15. Find the maximum error involved in using a Riemann sum with $n = 8$ equal subintervals to approximate the integral

$$\int_0^2 \sqrt{1 + x^3}\,dx.$$

16. Find the maximum error involved in using a rectangular approximation with $n = 6$ equal subintervals to approximate the integral

$$\int_0^{\pi/2} x \cos x\,dx.$$

17. Find the maximum error involved in the approximation obtained in Exercise 25, Section 6.3.

18. Find the maximum error involved in the approximation obtained in Exercise 26, Section 6.3.

19. Find how large n must be taken to ensure that the approximation of

$$\int_0^2 \sqrt{4 - x^2}\,dx$$

by the midpoint rule will be in error by less than 0.05.

20. Find how large n must be taken so that the error involved in using the midpoint rule to approximate

$$\int_0^{\pi/2} \sin x^2\,dx$$

will be less than 0.01.

21. Find a bound in the error associated with Exercise 19a., Section 6.5.

22. Let $f(t)$ be continuous on $[a, b]$ and define

$$F(x) = \int_a^x f(t)\,dt, \qquad x \in [a, b].$$

Prove that $F(x)$ is continuous on $[a, b]$ as follows:

a. Find an expression for $F(a + \Delta x)$ where $\Delta x > 0$.

b. Use the fact that $f(t)$ is bounded on $[a, b]$ to show that $\lim_{\Delta x \to 0^+} F(a + \Delta x) = F(a)$. This shows that $F(x)$ is continuous at $x = a$.

c. Find an expression for $F(b - \Delta x)$ where $\Delta x > 0$.

d. Use the expression in part **(c)** to show that $\lim_{\Delta x \to 0^+} F(b - \Delta x) = F(b)$. This shows that $F(x)$ is continuous at $x = b$.

e. Explain why $F(x)$ is continuous for $a < x < b$. (*Hint:* Theorem 7.)

23. Prove that if $m \le f(t) \le M$ for all $t \in [a, b]$, then

$$\int_x^{x+\Delta x} m\,dt \le \int_x^{x+\Delta x} f(t)\,dt \le \int_x^{x+\Delta x} M\,dt, \qquad [x, x + \Delta x] \subseteq [a, b].$$

24. Prove that $\left|\sum_{j=1}^{n} \int_{a_j}^{b_j} f(x)\,dx\right| \le \sum_{j=1}^{n} \int_{a_j}^{b_j} |f(x)|\,dx$ if $f(x)$ is continuous on each of the intervals $[a_j, b_j]$. (*Hint:* See Exercise 38, Section 6.3.)

25. Verify the details of the following proof of the Mean Value Theorem for Integrals.

a. Let m and M be the minimum and maximum values of $f(x)$ on $[a, b]$. From the definition of Riemann sums it follows that

$$m(b - a) \le \int_a^b f(x)\,dx \le M(b - a).$$

b. Dividing by $b - a$ gives

$$m \le \frac{1}{b - a}\int_a^b f(x)\,dx \le M.$$

c. Thus, by the Intermediate Value Theorem there exists $c \in [a, b]$ so that

$$f(c) = \frac{1}{b - a}\int_a^b f(x)\,dx.$$

d. Multiplying both sides by $b - a$ then gives

$$f(c) \cdot (b - a) = \int_a^b f(x)\,dx.$$

6.7 RULES FOR APPROXIMATING INTEGRALS

We have seen that many functions, such as $\sin x^2$ or $\sqrt{x^3 + 1}$, do not have antiderivatives that are easily found. In fact, it can be proved that many such functions simply do not have antiderivatives expressible as elementary functions. The Fundamental Theorem of Calculus is therefore not applicable to integrals involving such functions, and we can only approximate the values of these integrals. However, we have already seen how to approximate the integral

$$\int_a^b f(x)\,dx$$

using Riemann sums, including the midpoint rule. Moreover, Theorem 9 assures that the integral above can be approximated to within any desired degree of accuracy provided $f'(x)$ is bounded on $[a, b]$.

In this section we present two additional techniques for approximating definite integrals, the Trapezoidal Rule and Simpson's Rule. The advantage in these rules over the rectangular rules is that they achieve greater accuracy for a given value of n.

The Trapezoidal Rule

Figure 7.1 illustrates the fact that, in intervals over which the graph of $f(x)$ is changing rapidly (the slope has relatively large positive or negative values), an approximating rectangle does not fit the curve well. There are large areas under the curve that are not included in the rectangle and vice versa. We can try to improve the estimate by drawing a line from $(a, f(a))$ to $(b, f(b))$, which is a secant line that approximates the curve between those points, and using the trapezoidal area below this line in the same way that we previously used approximating rectangles. From elementary geometry, the area of a trapezoid with parallel sides $f(a)$ and $f(b)$ and altitude $(b - a)$ is $\frac{1}{2}(b - a)[f(a) + f(b)]$.

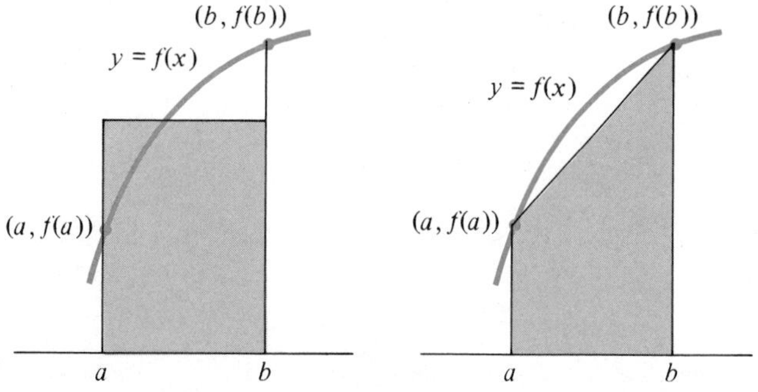

Figure 7.1 Approximations by trapezoids provide a "better fit," in general, than do approximations by rectangles.

To implement this idea, we divide the interval $[a, b]$ into n subintervals of equal length $\Delta x = \frac{b-a}{n}$, and we denote the endpoints of the subintervals by

$$a = x_0 < x_1 < x_2 < \cdots < x_n = b.$$

However, instead of using the area of an approximating rectangle in each subinterval we use the approximation

$$\int_{x_{j-1}}^{x_j} f(x)\,dx \approx \frac{1}{2}[f(x_{j-1}) + f(x_j)]\,\Delta x.$$

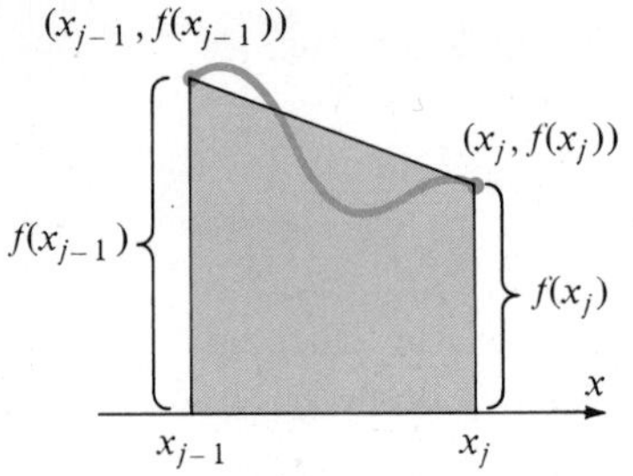

Figure 7.2 Trapezoidal approximation.

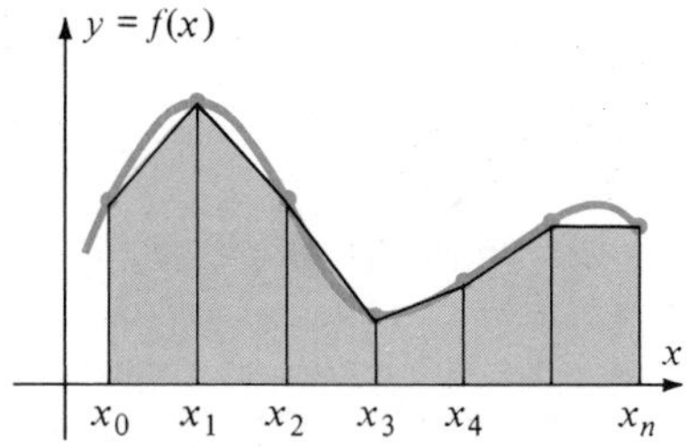

Figure 7.3 The Trapezoidal Rule.

As illustrated in Figures 7.2 and 7.3, this quantity corresponds to the area of the trapezoid with vertices $(x_{j-1}, 0)$, $(x_j, 0)$, $(x_{j-1}, f(x_{j-1}))$, and $(x_j, f(x_j))$ when $f(x) \geq 0$ on the interval $[a, b]$.

Summing these individual approximations gives

$$\int_a^b f(x)\,dx \approx \frac{\Delta x}{2}\{[f(x_0) + f(x_1)] + [f(x_1) + f(x_2)] + [f(x_2) + f(x_3)] + \cdots + [f(x_{n-2}) + f(x_{n-1})] + [f(x_{n-1}) + f(x_n)]\}.$$

Notice that each function value $f(x_j)$, except $f(x_0)$ and $f(x_n)$, occurs twice in the above expression. That is because each such value is involved as a dimension in two trapezoids, once as a left vertical side and once as a right vertical side. We may therefore state our approximation as follows.

> Trapezoidal Rule: Let $\Delta x = \dfrac{b-a}{n}$ and let
>
> $x_0 = a,\ x_1 = a + \Delta x,\ x_2 = a + 2\,\Delta x,\ \ldots,\ x_n = a + n\,\Delta x = b.$
>
> Then
>
> $$\int_a^b f(x)\,dx \approx \frac{b-a}{2n}\{f(x_0) + 2f(x_1) + 2f(x_2) + \cdots + 2f(x_{n-1}) + f(x_n)\}.$$

Example 1 Approximate $\displaystyle\int_1^4 \frac{1}{x}\,dx$ using the Trapezoidal Rule with $n = 6$.

Solution: With $a = 1$, $b = 4$, and $n = 6$, we have $\Delta x = \dfrac{4-1}{6} = \dfrac{1}{2}$, so

$$x_0 = 1,\ x_1 = \frac{3}{2},\ x_2 = 2,\ x_3 = \frac{5}{2},\ x_4 = 3,\ x_5 = \frac{7}{2},\ x_6 = 4.$$

(See Figure 7.3.)

The Trapezoidal Rule thus gives the approximation

$$\int_1^4 \frac{1}{x}\,dx \approx \frac{3}{12}\left\{1 + \frac{4}{3} + 1 + \frac{4}{5} + \frac{2}{3} + \frac{4}{7} + \frac{1}{4}\right\} = \frac{2361}{1680} = 1.4054.$$

■

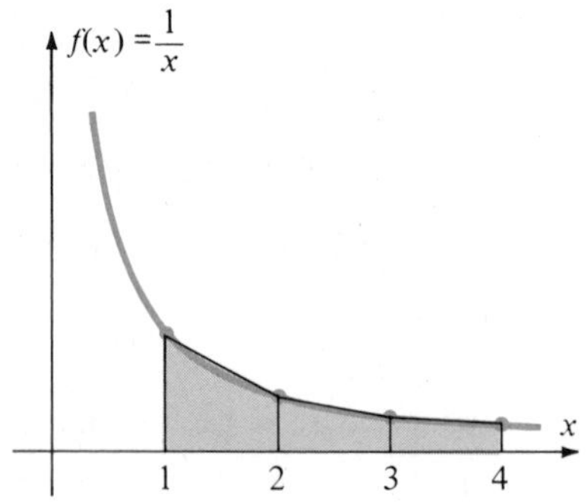

Figure 7.4 Approximation in Example 1.

From the geometry of Figure 7.4, we can see that the approximation in Example 1 is actually *larger* than the actual value of the integral. This is because the graph of $f(x)$ is concave up on $[1, 4]$.

ERROR IN TRAPEZOIDAL APPROXIMATIONS: In more advanced courses it is proved that the error in using the Trapezoidal Rule to approximate the integral $\int_a^b f(x)\,dx$ with n subdivisions has the following bound:

$$|\text{Error}| \leq \frac{(b-a)^3 M}{12n^2} \tag{1}$$

where M is the maximum value of $f''(x)$ on the interval $[a, b]$.

Example 2 Find a bound on the error for the approximation in Example 1.

Solution: For $f(x) = \dfrac{1}{x}$, $f''(x) = \dfrac{2}{x^3}$ which has a maximum value of $M = 2$ on the interval $[1, 4]$. Thus, by inequality (1) we have

$$|\text{Error}| \le \frac{3^3}{12(6^2)}(2) = \frac{1}{8} = 0.125. \qquad \blacksquare$$

Program 8 in Appendix I is a BASIC program implementing the Trapezoidal Rule. This program approximates the integral $\int_a^b \frac{1}{x}\,dx$.

Running this program on a computer, we obtained the results contained in Table 7.1. The student with access to a computer is encouraged to experiment with modifications of this program on integrals previously calculated using the Fundamental Theorem. Notice that the integral approximated in Table 7.1 *cannot* be evaluated by the Fundamental Theorem at this point as we do not yet have available an antiderivative for the integrand $1/x$.

Table 7.1

$[A, B]$	n	S
[1, 4]	5	1.41348
[1, 4]	20	1.38805
[1, 4]	100	1.38636
[1, 4]	250	1.38631
[1, 20]	5	3.76955
[1, 20]	20	3.06566
[1, 20]	100	2.99872
[1, 20]	250	2.99621

Simpson's Rule

This approximation procedure uses parabolic arcs, rather than line segments, to approximate portions of the graph of $y = f(x)$. It uses the fact that if x_0, x_1, and x_2 are three values of x with

$$x_1 - x_0 = x_2 - x_1 = \Delta x$$

then the definite integral of the parabola passing through the three points (x_0, y_0), (x_1, y_1), and (x_2, y_2) equals

$$\frac{\Delta x}{3}[y_0 + 4y_1 + y_2]. \tag{2}$$

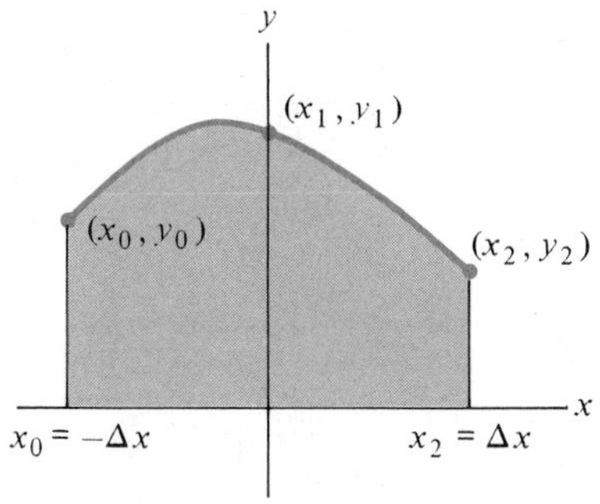

Figure 7.5

To see this let $x_0 = -\Delta x$, $x_1 = 0$, and $x_2 = \Delta x$ as in Figure 7.5. Let the parabola through (x_0, y_0), (x_1, y_1), and (x_2, y_2) be denoted by

$$y = ax^2 + bx + c$$

(Every parabola with a vertical axis has an equation of this form, as we shall prove in Chapter 15.) Then we have

$$y_0 = a(\Delta x)^2 - b(\Delta x) + c,$$
$$y_1 = c, \qquad \text{and}$$
$$y_2 = a(\Delta x)^2 + b(\Delta x) + c.$$

Using these values we find that

$$\begin{aligned}\int_{-\Delta x}^{\Delta x}(ax^2 + bx + c)\,dx &= \left[\frac{1}{3}ax^3 + \frac{1}{2}bx^2 + cx\right]_{-\Delta x}^{\Delta x}\\ &= \frac{\Delta x}{3}[2a(\Delta x)^2 + 6c]\\ &= \frac{\Delta x}{3}[y_0 + 4y_1 + y_2].\end{aligned}$$

We apply this result to obtain Simpson's Rule. As before, we divide $[a, b]$ into n subintervals, but now we must require that n be an *even* integer. We then approximate the integral

$$\int_{x_{2j-2}}^{x_{2j}} f(x)\, dx$$

over each *pair* of subintervals by expression (2). Thus we are approximating the actual value of the integral of $f(x)$ over $[x_{2j-2}, x_{2j}]$ by the integral of the approximating parabola. We obtain the approximation

$$\begin{aligned}\int_a^b f(x)\, dx \approx \frac{\Delta x}{3}\{&[f(x_0) + 4f(x_1) + f(x_2)] + [f(x_2) + 4f(x_3) + f(x_4)]\\ &+ [f(x_4) + 4f(x_5) + f(x_6)] + \cdots\\ &+ [f(x_{n-4}) + 4f(x_{n-3}) + f(x_{n-2})]\\ &+ [f(x_{n-2}) + 4f(x_{n-1}) + f(x_n)]\}.\end{aligned}$$

Again, notice that each $f(x_{2j})$ is counted twice, except for $f(x_0)$ and $f(x_n)$, for the same reason as in the Trapezoidal Rule. Letting $\Delta x = \dfrac{b-a}{n}$, we can state this approximation rule as follows:

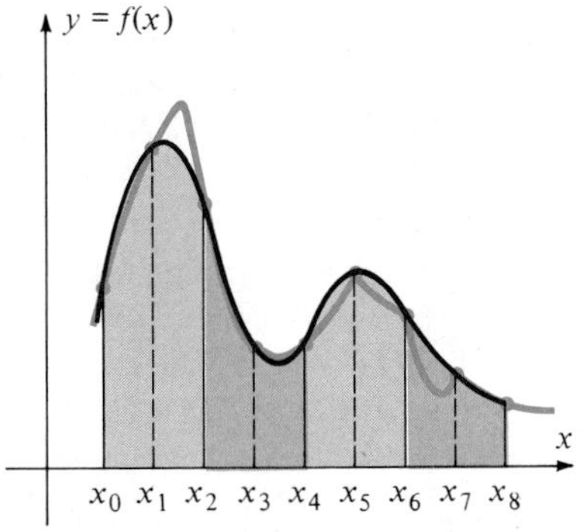

Figure 7.6 Approximation by Simpson's Rule.

> Simpson's Rule: Let n be an even integer and let $\Delta x = \dfrac{b-a}{n}$.
>
> Let
>
> $x_0 = a,\ x_1 = a + \Delta x,\ x_2 = a + 2\Delta x,\ \ldots,\ x_n = a + n\,\Delta x = b.$
>
> Then
>
> $$\begin{aligned}\int_a^b f(x)\, dx \approx \frac{b-a}{3n}\{&f(x_0) + 4f(x_1) + 2f(x_2) + 4f(x_3)\\ &+ 2f(x_4) + \cdots + 2f(x_{n-2}) + 4f(x_{n-1}) + f(x_n)\}.\end{aligned}$$

(See Figure 7.6.)

Example 3 Use Simpson's Rule with $n = 6$ to approximate $\displaystyle\int_1^4 \frac{1}{x}\, dx$.

Solution: With $x_0, x_1, \ldots, x_6$ as in Example 1, we obtain from the statement of Simpson's Rule the approximation

$$\begin{aligned}\int_1^4 \frac{1}{x}\, dx &\approx \frac{4-1}{3\cdot 6}\left\{1 + 4\left(\frac{2}{3}\right) + 2\left(\frac{1}{2}\right) + 4\left(\frac{2}{5}\right) + 2\left(\frac{1}{3}\right) + 4\left(\frac{2}{7}\right) + \frac{1}{4}\right\}\\ &= \frac{3497}{2520} = 1.3877.\end{aligned}$$

■

ERROR IN SIMPSON'S RULE APPROXIMATIONS: For the approximation provided by Simpson's Rule it is elsewhere proved that

$$|\text{Error}| \le \frac{(b-a)^5 M}{180n^4} \tag{3}$$

where M is the maximum value of $|f^{(iv)}(x)|$, the fourth derivative, on the interval $[a, b]$.

Example 4 Find how large n must be in order that the approximation of the integral in Example 3 by Simpson's Rule will be accurate to within 0.005.

Solution: For $f(x) = \dfrac{1}{x}$, $f^{(iv)}(x) = 24x^{-5}$. The maximum value of this fourth derivative on $[1, 4]$ is $M = 24$. According to inequality (3) we must find n sufficiently large that

$$\frac{(4-1)^5}{180n^4}(24) \le 0.005$$

Solving this inequality for n, we find that

$$n^4 \ge \frac{(24)(243)}{(.005)(180)} = 6480.$$

Here $n = 10$ will suffice since $10^4 = 10{,}000 > 6480$. ■

Table 7.2

$[A, B]$	n	S
[1, 4]	6	1.38770
[1, 4]	20	1.38631
[1, 4]	100	1.38629
[1, 4]	250	1.38629
[1, 20]	6	3.21696
[1, 20]	20	3.00677
[1, 20]	100	2.99577
[1, 20]	250	2.99573

Program 9 in Appendix I is a BASIC program which implements Simpson's Rule to approximate the integral $\int_a^b \frac{1}{x}\,dx$. Table 7.2 was obtained by use of this program. (Compare these results with those of Table 7.1 for the Trapezoidal Rule.)

In conclusion, several observations should be made about the two approximation procedures presented in this section.

(1) While the Trapezoidal Rule approximates curves by line segments, Simpson's Rule fits second degree curves to the given curve. Thus, we would expect Simpson's Rule to be more accurate for the same value of n. (Compare the results in Tables 7.1 and 7.2.)

(2) Since the error formula for the Trapezoidal Rule involves $f''(x)$, the Trapezoidal Rule gives exact information for first degree polynomials, for which $f''(x) \equiv 0$.

(3) However, since the error formula for Simpson's Rule involves the fourth derivative of $f(x)$, Simpson's Rule gives exact results for polynomials of degree three or less, for which $f^{(iv)}(x) \equiv 0$.

Exercise Set 6.7

In Exercises 1–5, use the Trapezoidal Rule with the given value of n to approximate the integral.

1. $\int_0^4 (x^3 - 7x + 4)\,dx; \quad n = 4$

2. $\int_0^2 \frac{1}{1+x^2}\,dx; \quad n = 4$

3. $\int_0^\pi \sin x\,dx; \quad n = 4$

4. $\int_0^\pi \sin x\,dx; \quad n = 6$

5. $\int_0^4 \frac{1}{1+x^2}\,dx; \quad n = 8$

In Exercises 6–10, use Simpson's Rule to approximate the integral.

6. In Exercise 1.

7. In Exercise 2.

8. In Exercise 3.

9. In Exercise 4.

10. In Exercise 5.

(Calculator) In Exercises 11–14, use the Trapezoidal Rule with $n = 6$ to approximate the given integral.

11. $\int_0^{\pi/4} \frac{1}{1+x^2}\,dx$

12. $\int_0^3 \sqrt{1+x^2}\,dx$

13. $\int_1^3 \sqrt{x^3 + 2}\,dx$

14. $\int_0^1 \sin x^2\,dx$

(Calculator) In Exercises 15–18, use Simpson's Rule with $n = 6$ to approximate the integral.

15. In Exercise 11.

16. In Exercise 12.

17. In Exercise 13.

18. In Exercise 14.

19. *(Computer)* Modify Program 8 in Appendix I to approximate the integral in Exercise 11 with $n = 5, 20, 100$, and 250.

20. *(Computer)* Modify Program 9 in Appendix I to approximate the integral in Exercise 11 with $n = 5, 20, 100$, and 250.

21. *(Calculator)* Use the Trapezoidal Rule with $n = 6$ to approximate

a. $\int_0^3 \sin\sqrt{x}\,dx$ **b.** $\int_0^{\pi/3} \frac{1}{\sqrt{\cos x}}\,dx.$

22. *(Calculator)* Repeat Exercise 21 using Simpson's Rule.

SUMMARY OUTLINE OF CHAPTER 6

- An **approximating sum** for the area of the region R bounded by the graph of $f(x) \geq 0$ and the x-axis for $a \leq x \leq b$ is an expression of the form $S_n = \sum_{j=1}^{n} f(t_j)\,\Delta x$ where $\Delta x = x_j - x_{j-1}$ and $t_j \in [x_{j-1}, x_j]$.
- The **area** of R is $A = \lim_{n\to\infty} S_n = \lim_{n\to\infty} \sum_{j=1}^{n} f(t_j)\,\Delta x.$
- If $f(x)$ is continuous on $[a, b]$, the **definite integral** of $f(x)$ on $[a, b]$ is defined to be the limit of the approximating Riemann sums, that is,

$$\int_a^b f(x)\,dx = \lim_{n\to\infty} \sum_{j=1}^{n} f(t_j)\,\Delta x_j, \qquad t_j \in [x_{j-1}, x_j].$$

- If $f(x) \geq 0$ on $[a, b]$ and $f(x)$ is continuous, then $\int_a^b f(x)\,dx$ is the **area** of the region bounded by the graph of $f(x)$ and the x-axis between $x = a$ and $x = b$.
- ***Properties of the Definite Integral***

$$\int_a^b [f(x) + g(x)]\,dx = \int_a^b f(x)\,dx + \int_a^b g(x)\,dx$$

$$\int_a^b [cf(x)]\,dx = c\int_a^b f(x)\,dx$$

$$\int_a^b f(x)\,dx = \int_a^c f(x)\,dx + \int_c^b f(x)\,dx$$

$$\int_a^a f(x)\,dx = 0; \qquad \int_b^a f(x)\,dx = -\int_a^b f(x)\,dx$$

- If $v(t)$ is the **velocity** of a particle moving along a line, then the **distance** traveled by the particle from time $t = a$ to time $t = b$ is given by the integral

$$D = \int_a^b |v(t)|\,dt$$

- The **Fundamental Theorem of Calculus** states that if $F'(x) = f(x)$ on $[a, b]$, then

$$\int_a^b f(x)\,dx = F(b) - F(a).$$

- Theorem: $\frac{d}{dx}\int_a^x f(t)\,dt = f(x).$
- Theorem: $\int_a^b f(x)\,dx = f(c)(b - a)$ for some number c in (a, b) if $f(x)$ is continuous.
- Trapezoidal Rule:

$$\int_a^b f(x)\,dx \approx \frac{b-a}{2n}\{f(x_0) + 2f(x_1) + 2f(x_2) + \cdots + 2f(x_{n-1}) + f(x_n)\}$$

- Simpson's Rule:

$$\int_a^b f(x)\,dx \approx \frac{b-a}{3n}\{f(x_0) + 4f(x_1) + 2f(x_2) + 4f(x_3) + \cdots + 4f(x_{n-1}) + f(x_n)\}, \qquad n \text{ even.}$$

REVIEW EXERCISES—CHAPTER 6

In Exercises 1–6, write **(a)** a lower approximating sum and **(b)** an upper approximating sum for the given function and interval, using n subintervals of equal size.

1. $f(x) = 3x - 1, \quad x \in [0, 3], \quad n = 6$

2. $f(x) = \frac{1}{x}, \quad x \in [1, 3], \quad n = 4$

3. $f(x) = \frac{1}{1 + x^2}, \quad x \in [-1, 1], \quad n = 6$

4. $f(x) = \sin \pi x, \quad x \in [0, 1], \quad n = 6$

5. $f(x) = \tan x, \quad x \in [-\pi/3, \pi/3], \quad n = 4$

6. $f(x) = \tan x, \quad x \in [0, 1], \quad n = 4$

7. Find the area of the region bounded above by the graph of $y = 2x - 2$ and below by the x-axis for $2 \le x \le 4$ by calculating the limit of lower approximating sums.

8. Rework Exercise 7 using upper approximating sums.

In Exercises 9–44, evaluate the definite integral using the Fundamental Theorem of Calculus.

9. $\int_0^9 \sqrt{x}\, dx$

10. $\int_0^3 3\sqrt{x+1}\, dx$

11. $\int_0^1 (x^{2/3} - x^{1/2})\, dx$

12. $\int_1^2 \frac{1-t}{t^3}\, dt$

13. $\int_0^1 x^3(x+1)\, dx$

14. $\int_0^\pi \sin^2 x\, dx \quad \left(\textit{Hint: } \sin^2\theta = \frac{1}{2} - \frac{1}{2}\cos 2\theta\right)$

15. $\int_0^1 (3x+2)\, dx$

16. $\int_1^3 (7+3x)\, dx$

17. $\int_{-3}^5 (x^2+2)\, dx$

18. $\int_1^5 (3x^2-2)\, dx$

19. $\int_1^4 \sqrt{x}\, dx$

20. $\int_0^8 \sqrt{x+1}\, dx$

21. $\int_1^4 \left(\sqrt{x} - \frac{1}{\sqrt{x}}\right) dx$

22. $\int_1^8 (x^{1/3} - 1)\, dx$

23. $\int_0^2 (x+7)((2x+2)\, dx$

24. $\int_1^4 (x^2-1)(x+2)\, dx$

25. $\int_0^{\pi/4} \sin x\, dx$

26. $\int_{-\pi/4}^{-\pi/4} \cos x\, dx$

27. $\int_0^{\pi/4} \sin(2x)\, dx$

28. $\int_{\pi/4}^{\pi/2} \cos(\pi - 2x)\, dx$

29. $\int_1^2 (x+4)^{10}\, dx$

30. $\int_0^4 (\sqrt{a} + \sqrt{x})^2\, dx$

31. $\int_2^4 \frac{t^2-2t}{t}\, dt$

32. $\int_1^2 \frac{1+t}{t^3}\, dt$

33. $\int_0^3 \frac{dt}{(t+1)^2}$

34. $\int_3^8 \frac{1}{\sqrt{x+1}}\, dx$

35. $\int_0^{\pi/4} \frac{1}{\cos^2 x}\, dx$

36. $\int_0^{\pi/3} \frac{\sin x}{\cos^2 x}\, dx$

37. $\int_0^1 (x - \sqrt{x})^2\, dx$

38. $\int_1^4 x^{1/2}(1 + x^{3/2})^5\, dx$

39. $\int_1^8 t(\sqrt[3]{t} - 2t)\, dt$

40. $\int_1^4 \frac{x^2+2x+4}{\sqrt{x}}\, dx$

41. $\int_1^2 \frac{1}{(1-2x)^3}\, dx$

42. $\int_{-\pi/4}^{\pi/4} \sec^2 t\, dt$

43. $\int_0^1 \frac{x^3+8}{x+2}\, dx$

44. $\int_0^2 x^2\sqrt{x^3+1}\, dx$

In Exercises 45–51, find the area of the region bounded by the graph of the given function, the x-axis, the line $x = a$, and the line $x = b$.

45. $f(x) = \sqrt{x-1}, \quad a = 1, \quad b = 5$

46. $f(x) = \frac{1}{\sqrt{x-1}}, \quad a = 2, \quad b = 10$

47. $f(x) = (x^2+2)^2, \quad a = 0, \quad b = 1$

48. $f(x) = (x-1)(x+2), \quad a = 0, \quad b = 2$

49. $f(x) = 4x - x^2, \quad a = 4, \quad b = 5$

50. $f(x) = \sin \pi x, \quad a = 0, \quad b = 2$

51. $f(x) = 3 + 2x - x^2, \quad a = 1, \quad b = 4$

In each of Exercises 52–57, find the area of the region bounded above by the graph of $y = f(x)$ and below by the x-axis for the specified intervals.

52. $f(x) = x^3 - x, \quad a = 1, \quad b = 3$

53. $f(x) = x^2 - x - 2, \quad a = 2, \quad b = 4$

54. $f(x) = \sin(2x), \quad a = 0, \quad b = \frac{\pi}{4}$

55. $f(x) = \sqrt{1-x}, \quad a = 1, \quad b = 1$

56. $f(x) = \frac{1}{(x-2)^2}, \quad a = -2, \quad b = 1$

57. $f(x) = (ax^2+3)^2, \quad a = 0, \quad b = 3$

In each of Exercises 58–62, a region is described. Sketch the region, noting where the bounding function $f(x)$ is positive and where it is negative. Then calculate the area of the region using one or more integrals.

58. The region bounded by the graph $f(x) = 9 - x^2$ and the x-axis between $x = -3$ and $x = 3$.

59. The region bounded by the graph of $f(x) = x^2 + 2x - 3$ between $x = -3$ and $x = 1$.

60. The region bounded by the graph of $f(x) = (x-1)^3$ between $x = 0$ and $x = 2$.

61. The region bounded by the graph of $f(x) = 2x - 4$ and the x-axis for $0 \le x \le 4$.

62. The region bounded by the graph of $f(x) = \sqrt{x} - 2$ and the x-axis for $0 \le x \le 9$.

63. An object moves along a line with velocity $v(t) = t(6 - 3t)$ meters per second. Initially it is located 5 units to the right of the origin.
a. Find the location of the object after 4 seconds.
b. Find the distance travelled by the object during the first 4 seconds.
c. In which direction is the object moving after 4 seconds?

64. An object moves along a line with velocity $v(t) = 2 + 4\sin t$. Originally it is located at the origin.
a. Find its location after 2π seconds.
b. Find the distance travelled by the object during the first 2π seconds.

65. An object moves along a line with acceleration $a(t) = 4 - 6t$. Initially its velocity is zero. Find the distance travelled by the object during the first 2 seconds.

66. True or false? If $f(x)$ is an even function, then $\int_{-a}^{a} f(x)\,dx = 2\int_{0}^{a} f(x)\,dx$. What if $f(x)$ is odd?

67. Let $F(x) = \int_{0}^{x} t^2\sqrt{1+t}\,dt$ for $x > -1$. Find

a. $F(0)$ **b.** $F'(x)$
c. $F'(3)$ **d.** $F'(2x)$

68. Let $F(x) = \int_{0}^{2x} \frac{1}{1+t^2}\,dt$. Find

a. $F'(x)$ **b.** $F'(1)$
c. $F(x^2)$ **d.** $F'(x^2)$

69. True or false? $\frac{d}{dx}\left\{\int_{a}^{x} f(t)\,dt\right\} = \int_{a}^{x} \frac{d}{dt}\left\{f(t)\right\}dt$

70. True or false? If $\int_{a}^{b} f(x)\,dx = 0$ then $f(x) \equiv 0$.

71. True or false? If $f(x)$ is continuous on $[a, b]$ and $\int_{a}^{b} f(x)\,dx = 0$, then $f(c) = 0$ for at least one $c \in [a, b]$.

72. True or false? If $\int_{a}^{b} |f(x)|\,dx = 0$ and $f(x)$ is continuous on $[a, b]$, then $f(x) = 0$ for all $x \in [a, b]$.

73. Find $\frac{d}{dx}\left\{\int_{2x}^{0} \sec t\,dt\right\}$

74. Find $\frac{d}{dx}\left\{\int_{x}^{x^2} \sqrt{1+t^2}\,dt\right\}$

In Exercises 75–80, sketch a region in the plane whose area corresponds to the given integral. Then find the value of the integral from area considerations.

75. $\int_{1}^{4} (2x+1)\,dx$

76. $\int_{-1}^{1} \sqrt{1-x^2}\,dx$

77. $\int_{0}^{3} (1+\sqrt{9-x^2})\,dx$

78. $\int_{-3}^{3} |3-x|\,dx$

79. $\int_{0}^{4} -\sqrt{16-x^2}\,dx$

80. $\int_{-1}^{1} (3+\sqrt{1-x^2})\,dx$

In Exercises 81–83, use the Trapezoidal Rule with n subdivisions to approximate the given integral.

81. $\int_{-2}^{2} \frac{1}{1+x^2}\,dx, \quad n = 4$

82. $\int_{0}^{4} \sqrt{1+x^2}\,dx, \quad n = 4$

83. $\int_{-1}^{2} \frac{1}{x+2}\,dx, \quad n = 6$

In Exercises 84–85, use Simpson's Rule with n subdivisions to approximate the given integral.

84. $\int_{0}^{2} \frac{1}{1+x}\,dx, \quad n = 6$

85. $\int_{0}^{\pi/3} \cos x\,dx, \quad n = 6$ (Use a cosine table or calculator.)

CHAPTER 7

MATHEMATICAL APPLICATIONS OF THE DEFINITE INTEGRAL

7.1 INTRODUCTION

In this chapter and the next, you will encounter a variety of physical and mathematical problems, all of which share the property that their solutions arise in the form of definite integrals. In each instance, we will be led to the solution through a sequence of steps similar to those by which the definite integral arose as the solution to the area problem in Chapter 6.

Because of the strong parallels that exist among the problems and solutions of this chapter, we summarize here the essential concepts involved in the definition of the integral. You should refer to this introductory summary from time to time, as an aid to understanding the seminal idea of this chapter.

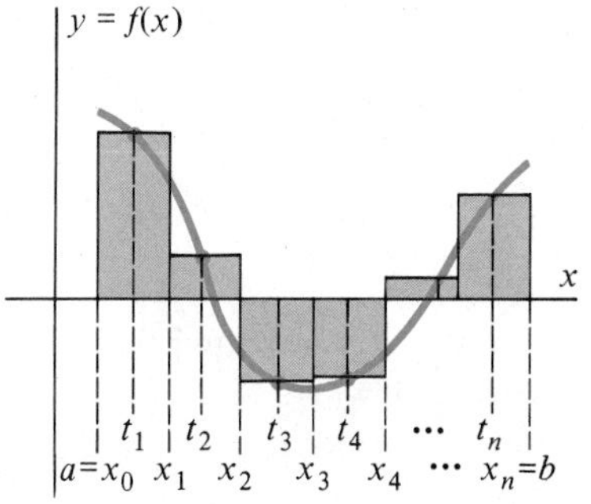

Figure 1.1 Approximating Riemann sum.

SUMMARY DEFINITION OF THE INTEGRAL $\int_a^b f(x)\,dx$: Let $f(x)$ be a continuous function for $x \in [a, b]$.

1. For each positive integer n, the **partition** of the interval $[a, b]$ into n equal* subintervals means the selection of $n + 1$ numbers

$$a = x_0 < x_1 < x_2 < \cdots < x_n = b$$

so that $(x_j - x_{j-1}) = \Delta x = \dfrac{b-a}{n}$ for $j = 1, 2, \ldots, n$.

2. A **Riemann sum** for $f(x)$ on $[a, b]$ is an expression of the form

$$\sum_{j=1}^{n} f(t_j)\,\Delta x$$

where t_j is an arbitrary element of the subinterval $[x_{j-1}, x_j]$ for each $j = 1, 2, \ldots, n$ and the numbers $x_0, x_1, \ldots, x_n$ constitute a partition of $[a, b]$ (see Figures 1.1 and 1.2).

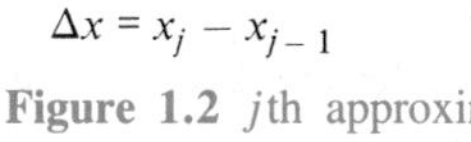

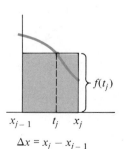

Figure 1.2 jth approximating rectangle.

*The most general definition of partition, as given in Chapter 6, allows for subintervals of varying length $\Delta x_j = (x_j - x_{j-1})$. Since all Riemann sums for $f(x)$ on $[a, b]$ have the limit $\int_a^b f(x)\,dx$ and since we will actually be constructing the partitions that are used here, we will work with **regular partitions**—those with subintervals of equal length $\Delta x = \dfrac{b-a}{n}$.

3. The limit of this Riemann sum as $n \to \infty$,

$$\int_a^b f(x)\,dx = \lim_{n\to\infty} \sum_{j=1}^{n} f(t_j)\,\Delta x,$$

exists independent of how the numbers $t_j \in [x_{j-1}, x_j]$ are chosen.

In each of the questions taken up in this chapter, the following problem-solving strategy will unfold:

(a) A continuous function will be identified that determines the quantity to be calculated.
(b) We will approximate the solution by assuming that this continuous function actually assumes only a finite number of distinct values. This will produce an approximation in the form of a Riemann sum.
(c) We will assume that as the number of distinct values of the function becomes infinite, the approximations approach the actual value of the quantity to be calculated.
(d) The desired quantity will be defined as the limiting value of the approximating Riemann sum as $n \to \infty$. That is, the desired quantity will be obtained as a definite integral (point 3 above).

You are urged, in each case, to study carefully the procedure by which the solution is obtained rather than simply accepting the formula that results. Only by striving to understand the ways in which integrals arise from approximation schemes can you gain the insight you will need when encountering new problems whose solutions involve definite integrals.

7.2 AREA BETWEEN TWO CURVES

We begin with the problem of calculating the area A of a region bounded by two curves. A typical situation is that illustrated in Figure 2.1 where the region R is bounded above and below by the graphs of the continuous functions $f(x)$ and $g(x)$, respectively, on the left by the line $x = a$, and on the right by the line $x = b$.

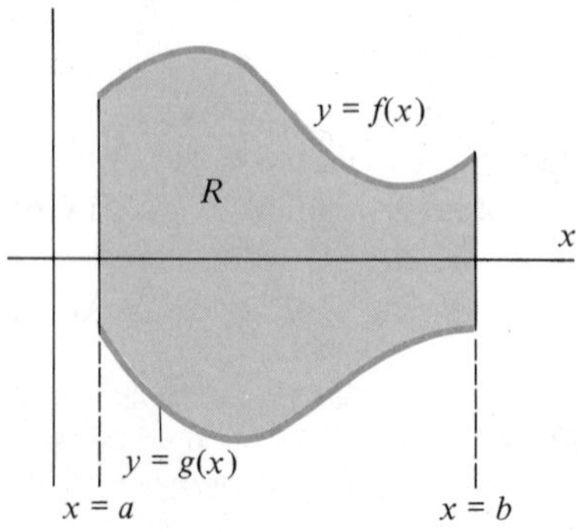

Figure 2.1 Region bounded by two curves.

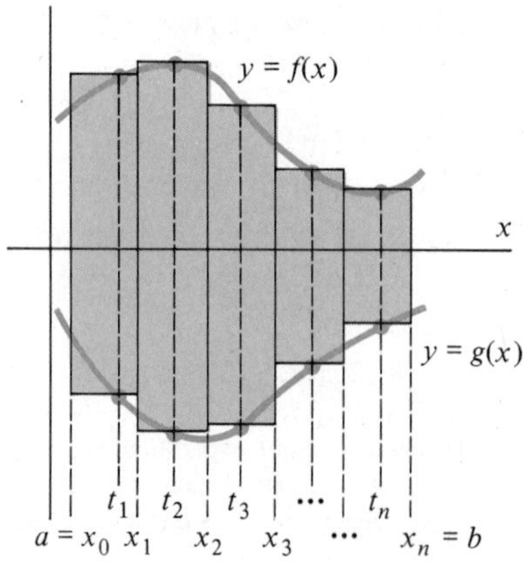

Figure 2.2 Approximating Riemann sum.

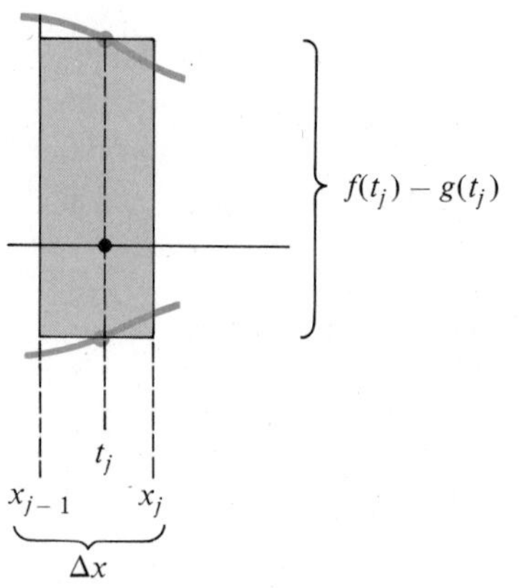

Figure 2.3 jth approximating rectangle has area $A_j = [f(t_j) - g(t_j)]\,\Delta x$.

To approximate A, we partition $[a, b]$ into n equal subintervals and choose t_j in $[x_{j-1}, x_j]$ arbitrarily for each $j = 1, 2, \ldots, n$. Then the area ΔA_j corresponding to the jth interval is approximated by the rectangle whose height is $f(t_j) - g(t_j)$ and whose width is $\Delta x = \dfrac{b-a}{n}$ (see Figures 2.2 and 2.3), that is,

$$\Delta A_j \approx [f(t_j) - g(t_j)]\,\Delta x. \tag{1}$$

Summing these approximations over all subintervals gives the approximation

$$A = \sum_{j=1}^{n} \Delta A_j \approx \sum_{j=1}^{n} [f(t_j) - g(t_j)]\,\Delta x. \tag{2}$$

Now the expression on the right-hand side of approximation (2) is a Riemann sum for the function $f(x) - g(x)$ on the interval $[a, b]$. Furthermore, we claim that approximation (2) becomes increasingly accurate as n becomes large. We therefore define the area to be the limiting value of the above approximation as $n \to \infty$. By the definition of the integral, this means that

$$A = \lim_{n\to\infty} \sum_{j=1}^{n} [f(t_j) - g(t_j)]\,\Delta x = \int_a^b [f(x) - g(x)]\,dx.$$

In other words,

> If $f(x)$ and $g(x)$ are continuous for $x \in [a, b]$ and $f(x) \geq g(x)$ for all $x \in [a, b]$, then the area of the region bounded by the graphs of $f(x)$ and $g(x)$ for $a \leq x \leq b$ is defined by the integral
>
> $$A = \int_a^b [f(x) - g(x)]\,dx. \tag{3}$$

(In Exercise 36 you are asked to show that this definition agrees with and extends our earlier definitions of area.)

REMARK: Notice that the integrand in equation (3) is nonnegative. This means that the integral will always be nonnegative, regardless of whether the region lies above or below the x-axis. In using equation (3), one must be careful to determine which curve is ''on top.''

Example 1 Find the area of the region bounded by the graphs of $y = \sqrt{x}$ and $y = -x - 1$ between $x = 1$ and $x = 4$.

Strategy

Sketch the region in question.

Determine which curve is on top.

Apply (3).

Solution

We first sketch the region showing one typical approximating rectangle (Figure 2.4). The upper boundary is the curve $f(x) = \sqrt{x}$ and the lower boundary is the curve $g(x) = -x - 1$.

Applying equation (3) we obtain

$$A = \int_1^4 [x^{1/2} - (-x - 1)]\,dx$$

$$= \frac{2}{3}x^{3/2} + \frac{1}{2}x^2 + x\Big]_1^4$$

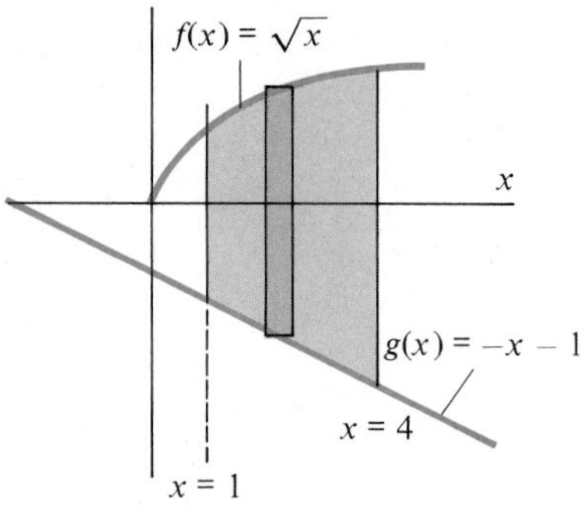

Figure 2.4

$$= \left[\frac{2}{3}(4)^{3/2} + \frac{1}{2}(4)^2 + 4\right] - \left[\frac{2}{3} + \frac{1}{2} + 1\right]$$

$$= \frac{91}{6}. \qquad \blacksquare$$

In some problems, the two curves bound a region by intersecting, and particular endpoint values for x are not specified. In such situations, we must find the x-coordinates of the points of intersection before equation (3) can be applied.

Example 2 Find the area of the region bounded by the graphs of the functions $y = x^4 + 1$ and $y = 2x^2$.

Strategy

Sketch region.

Solution

A rough sketch of the region shows that the graphs intersect at two points. To find these points we equate the two functions, obtaining

Find points of intersection.

$$x^4 + 1 = 2x^2$$

or

Solve by factoring. (If the equation cannot be easily factored, use the quadratic formula.)

$$x^4 - 2x^2 + 1 = (x^2 - 1)^2 = 0$$

so $x^2 - 1 = 0$ and $x = \pm 1$. The points of intersection are therefore $(-1, 2)$ and $(1, 2)$.

Determine which curve is on top (see Figure 2.5).

Checking any particular value of x in $(-1, 1)$ shows that $f(x) = x^4 + 1$ determines the upper boundary and the graph of $g(x) = 2x^2$ determines the lower boundary. Thus, by equation (3),

Apply (3).

$$A = \int_{-1}^{1} [(x^4 + 1) - 2x^2]\, dx$$

$$= \frac{1}{5}x^5 + x - \frac{2}{3}x^3\Big]_{x=-1}^{x=1}$$

$$= \left(\frac{1}{5} + 1 - \frac{2}{3}\right) - \left(-\frac{1}{5} - 1 + \frac{2}{3}\right)$$

$$= \frac{16}{15}. \qquad \blacksquare$$

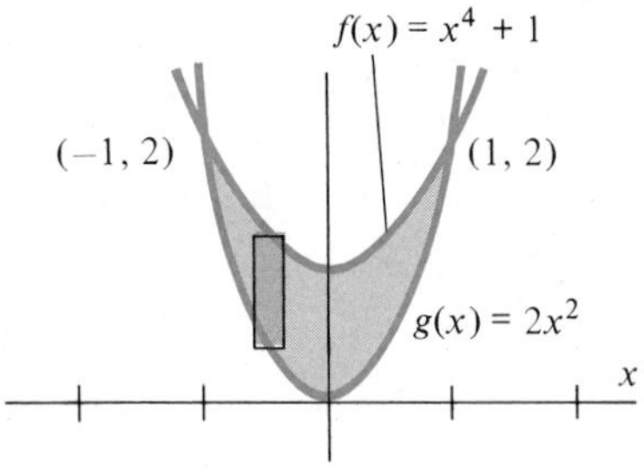

Figure 2.5

Often the graphs bounding the region in question cross several times. In such cases, we apply equation (3) in each of the resulting subregions, being careful to note for each region the upper and lower boundaries.

Example 3 Find the area of the region bounded by the graphs of $f(x) = \sin x$ and $g(x) = 1/2$ for $0 \le x \le 2\pi$.

Solution: The two graphs cross where $\sin x = 1/2$. The solutions of this equation in $[0, 2\pi]$ are $x = \pi/6$ and $x = 5\pi/6$. By checking particular values of x we can see that, on the resulting intervals,

$$\sin x \le \frac{1}{2} \qquad \text{for} \qquad x \in \left[0, \frac{\pi}{6}\right],$$

$$\sin x \geq \frac{1}{2} \qquad \text{for} \qquad x \in \left[\frac{\pi}{6}, \frac{5\pi}{6}\right],$$

$$\sin x \leq \frac{1}{2} \qquad \text{for} \qquad x \in \left[\frac{5\pi}{6}, 2\pi\right].$$

(See Figure 2.6.)

Thus, by (3)

$$\begin{aligned} A &= \int_0^{\pi/6} \left(\frac{1}{2} - \sin x\right) dx + \int_{\pi/6}^{5\pi/6} \left(\sin x - \frac{1}{2}\right) dx + \int_{5\pi/6}^{2\pi} \left(\frac{1}{2} - \sin x\right) dx \\ &= \left[\frac{x}{2} + \cos x\right]_0^{\pi/6} + \left[-\cos x - \frac{x}{2}\right]_{\pi/6}^{5\pi/6} + \left[\frac{x}{2} + \cos x\right]_{5\pi/6}^{2\pi} \\ &= \left[\left(\frac{\pi}{12} + \frac{\sqrt{3}}{2}\right) - (0 + 1)\right] + \left[\left(\frac{\sqrt{3}}{2} - \frac{5\pi}{12}\right) - \left(-\frac{\sqrt{3}}{2} - \frac{\pi}{12}\right)\right] \\ &\quad + \left[(\pi + 1) - \left(\frac{5\pi}{12} - \frac{\sqrt{3}}{2}\right)\right] \\ &= \frac{\pi}{3} + 2\sqrt{3} \approx 4.51. \end{aligned}$$

■

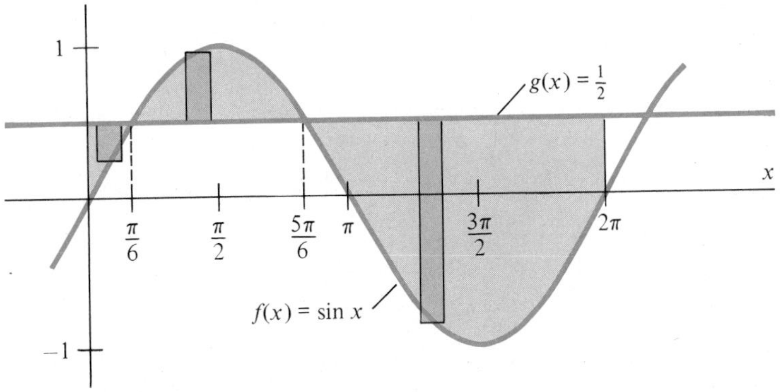

Figure 2.6 Area bounded by $f(x) = \sin x$ and $g(x) = 1/2$, $0 \leq x \leq 2\pi$.

Our next example illustrates that it may sometimes be easier to calculate the desired area by integrating with respect to y rather than with respect to x.

Example 4 Find the area bounded by the graphs of the equations $y = x$ and $x = y^2 - 2$

Solution: The area in question is illustrated in Figure 2.7. The points of intersection are found by setting

$$y = y^2 - 2$$

or

$$y^2 - y - 2 = (y + 1) \cdot (y - 2) = 0,$$

so

$$y = -1 \qquad \text{or} \qquad y = 2.$$

The points are therefore $(-1, -1)$ and $(2, 2)$.

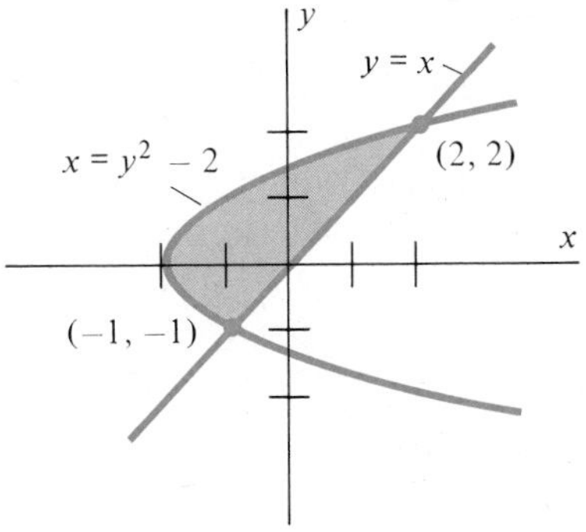

Figure 2.7

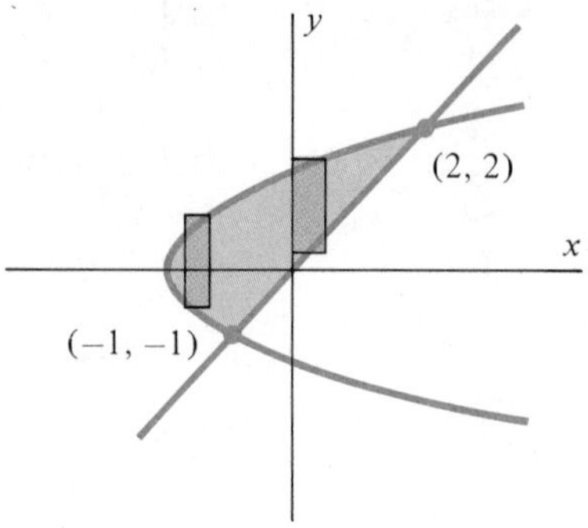

Figure 2.8

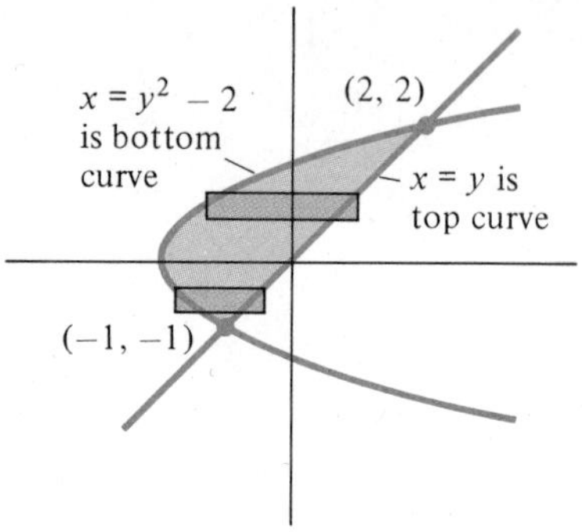

Figure 2.9

Figure 2.8 shows that by partitioning the x-axis (i.e., by using vertical rectangles) one encounters a difficulty—some approximating rectangles have both bottoms and tops on the graph of $x = y^2 - 2$, while others have tops on the graphs of $x = y^2 - 2$ and bottoms on the graph of $y = x$.

A simpler approach is to partition the y-axis, thus viewing y as the independent variable. Then all approximating rectangles run from the curve $x = y^2 - 2$ to the line $y = x$ (Figure 2.9). The area is thus calculated as

$$\begin{aligned}\text{Area} &= \int_{y=-1}^{y=2} [y - (y^2 - 2)]\, dy \\ &= -\frac{1}{3}y^3 + \frac{1}{2}y^2 + 2y\Big]_{-1}^{2} \\ &= 9/2.\end{aligned}$$

■

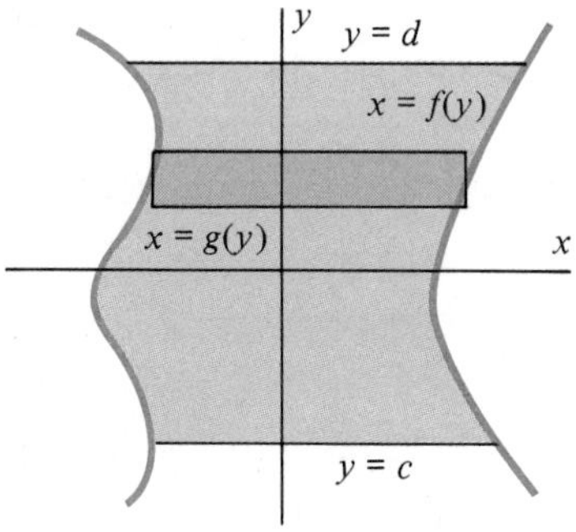

Figure 2.10 Region bounded by graphs of functions of y and horizontal lines.

In general, when the region in question is bounded by the graphs of the functions $x = f(y)$ and $x = g(y)$ (see Figure 2.10), we have the area formula

$$\text{Area} = \int_c^d [f(y) - g(y)]\, dy$$

if $f(y) \geq g(y)$.

Example 5 Find the area of the region bounded by the graphs of the equations $3y - x = 6$, $x + y = -2$, and $x + y^2 = 4$ illustrated in Figure 2.11.

Solution: The graphs of the first two equations are lines. The graph of the third is a parabola.

As Figure 2.11 suggests, the simplest solution involves partitioning the y-axis, so that the approximating rectangles lie parallel to the x-axis. In this case, the approximating rectangles all will have their right edge lying on the parabola. Their left edge will lie on one of the two lines.

To pursue this approach, we express each equation as a function of y, obtaining

$$f(y) = 3y - 6, \qquad g(y) = -y - 2, \qquad \text{and} \qquad h(y) = 4 - y^2.$$

Equating these functions two at a time and solving gives the three points of intersection among these graphs: $(-3, 1)$, $(0, 2)$, and $(0, -2)$.

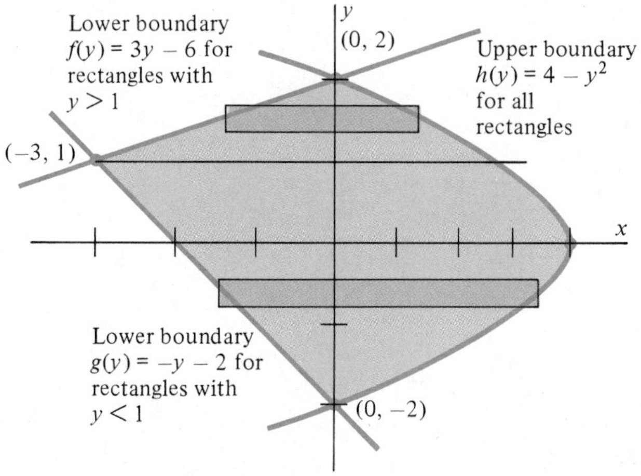

Figure 2.11 Region in Example 5: Partitioning the y-axis.

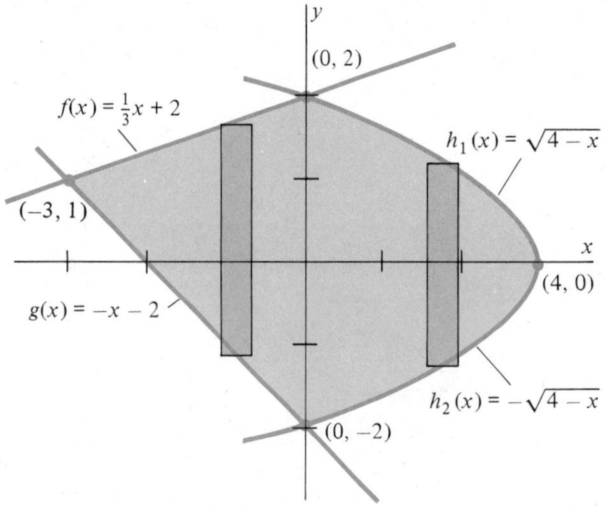

Figure 2.12 Region in Example 5: Partitioning the x-axis.

For $-2 \le y \le 1$, the left edges of the approximating rectangles are determined by the graph of $g(y) = -y - 2$, while for $1 \le y \le 2$ these edges intersect the graph of $f(y) = 3y - 6$. The area of the region is therefore

$$\begin{aligned}
A &= \int_{-2}^{1} [(4 - y^2) - (-y - 2)]\, dy + \int_{1}^{2} [(4 - y^2) - (3y - 6)]\, dy \\
&= \int_{-2}^{1} (6 + y - y^2)\, dy + \int_{1}^{2} (10 - 3y - y^2)\, dy \\
&= 6y + \frac{y^2}{2} - \frac{y^3}{3}\Big]_{-2}^{1} + 10y - \frac{3y^2}{2} - \frac{y^3}{3}\Big]_{1}^{2} \\
&= \left[\left(6 + \frac{1}{2} - \frac{1}{3}\right) - \left(-12 + 2 + \frac{8}{3}\right)\right] \\
&\quad + \left[\left(20 - 6 - \frac{8}{3}\right) - \left(10 - \frac{3}{2} - \frac{1}{3}\right)\right] \\
&= \frac{50}{3} \approx 16.67.
\end{aligned}$$

A different solution to this problem involves partitioning the x-axis rather than the y-axis. To do so we attempt to solve each of the given equations for y as a function of x. We obtain

$$f(x) = \frac{1}{3}x + 2; \qquad g(x) = -x - 2; \qquad h(x) = \pm\sqrt{4 - x}.$$

Of course, the equation $h(x) = \pm\sqrt{4 - x}$ determines not one, but *two* functions, $h_1(x) = \sqrt{4 - x}$ and $h_2(x) = -\sqrt{4 - x}$. The region may then be described as follows (see Figure 2.12):

$$\begin{aligned}
g(x) = -x - 2 \le y \le \frac{1}{3}x + 2 = f(x), &\qquad \text{for} \qquad -3 \le x \le 0 \\
h_2(x) = -\sqrt{4 - x} \le y \le \sqrt{4 - x} = h_1(x), &\qquad \text{for} \qquad 0 \le x \le 4.
\end{aligned}$$

The calculation of area is, therefore,

$$\begin{aligned} A &= \int_{-3}^{0}\left[\left(\frac{1}{3}x+2\right)-(-x-2)\right]dx+\int_{0}^{4}\left[\sqrt{4-x}-(-\sqrt{4-x})\right]dx \\ &= \int_{-3}^{0}\left(\frac{4}{3}x+4\right)dx+\int_{0}^{4}2\sqrt{4-x}\,dx \\ &= \left[\frac{2}{3}x^2+4x\right]_{-3}^{0}+\left[\left(-\frac{4}{3}\right)(4-x)^{3/2}\right]_{0}^{4} \\ &= \left[0-\left(\frac{18}{3}-12\right)\right]+\left[0-\left(-\frac{4}{3}\right)(8)\right] \\ &= \frac{50}{3}. \end{aligned}$$

■

In each of the preceding examples, we have emphasized the necessity of ensuring that integrand $[f(x) - g(x)]$ be nonnegative when calculating the area between the graphs of $f(x)$ and $g(x)$. A simple way to state the area formula so as to highlight this concern is

$$A = \int_a^b |f(x) - g(x)|\,dx$$

which is entirely equivalent to the statement involving equation (3).

Exercise Set 7.2

In Exercises 1–12, sketch the region bounded by the graphs of the given functions between the indicated values of x. Then calculate the area of the region.

1. $f(x) = x + 1, \quad g(x) = -2x + 1, \quad 0 \le x \le 2$

2. $f(x) = 2x + 3, \quad g(x) = x^2 - 4, \quad -1 \le x \le 1$

3. $f(x) = \sqrt{x}, \quad g(x) = -x^2, \quad 0 \le x \le 4$

4. $f(x) = \dfrac{1}{x^2}, \quad g(x) = x^{2/3}, \quad 1 \le x \le 8$

5. $f(x) = \sin x, \quad g(x) = \cos x, \quad 0 \le x \le 2\pi$

6. $f(x) = 1, \quad g(x) = \cos x, \quad 0 \le x \le 2\pi$

7. $f(y) = y - 3, \quad g(y) = y^2, \quad -2 \le y \le 2$

8. $f(x) = x\sqrt{9 - x^2}, \quad g(x) = -x, \quad -3 \le x \le 3$

9. $f(x) = |4 - x^2|, \quad g(x) = 5, \quad -3 \le x \le 3$

10. $f(x) = \sin x, \quad g(x) = x, \quad 0 \le x \le \pi$

11. $f(x) = \dfrac{x^2 - 1}{x^2}, \quad g(x) = \dfrac{1 - x^2}{x^2}, \quad 1 \le x \le 2$

12. $f(x) = x^{2/3}, \quad g(x) = x^{1/3}, \quad -1 \le x \le 1$

In Exercises 13–24, sketch the region bounded by the graphs of the given equations. Then calculate the area of the region.

13. $y = 4 - x^2, \quad y = x - 2$

14. $y = 9 - x^2, \quad 9y - x^2 + 9 = 0$

15. $y = x^2, \quad y = x^3$

16. $y = x^3, \quad y = x$

17. $x + y^2 = 4, \quad y = x + 2$

18. $y = 6x - x^2, \quad y - x = 6, \quad y = -2x$

19. $y = x^2 - 6, \quad y = x, \quad y = -x$
(*Hint:* The region consists of three pieces.)

20. $x = y^3, \quad y = 2, \quad y - x = 6$

21. $x = \sqrt{y}, \quad x = \sqrt[3]{y}$

22. $2x = y^2, \quad x + 2 = y^2$

23. $y = x^{2/3}, \quad y = 2 - x^2$

24. $y = x^{2/3}, \quad y = x^2$

25. Find the area of the region bounded by the graph of $y = x^{2/3}$ and the line $y = 1$ by partitioning the x-axis.

26. Rework Exercise 25, partitioning the y-axis.

27. Find the area of the region bounded by the graphs of $y = x^3$, $y = -x^3$, $y = 1$, and $y = -1$ by partitioning the x-axis.

28. Rework Exercise 27, partitioning the y-axis.

29. Find the area of the region bounded by the graphs of $x = y^2 - 4$, $y = 3x$, and $y = -\frac{1}{3}x + 2$, by partitioning the y-axis. (This will involve two integrals.)

30. Rework Exercise 29, partitioning the x-axis. (This will involve three integrals.)

31. Find a number a so that the line $y = a$ divides the region bounded by the x-axis and the graph of the equation $y = 4 - x^2$ into two regions of equal area.

32. Verify that the method of this section produces the same value for the area of the region bounded by the following lines as does the appropriate formula from plane geometry.

a. $y = x, \quad x = 3, \quad y = -1$
b. $y = 2x + 2, \quad y = -2x - 1, \quad x = 1, \quad x = 3.$

33. *(Calculator/Computer)* Use the Trapezoidal Rule with $n = 6$ to approximate the area bounded by the graphs of $y = \sqrt{9 - x^2}$ and $y = \sqrt{5}$.

34. *(Calculator/Computer)* Use Simpson's Rule with $n = 8$ to approximate the area of the region common to the two circles $x^2 + y^2 = 25$ and $(x - 6)^2 + y^2 = 25$. (*Hint:* The points $(3, \pm 4)$ lie on both circles.)

35. *(Calculator/Computer)* Use Simpson's Rule to approximate the area of the region enclosed by the ellipse $16x^2 + 9y^2 = 144$.

36. Verify that the definition of area in equation (3) agrees with the definition given in Chapter 6 for the area of the region bounded by the graph of $f(x)$ and the x-axis for $a \le x \le b$, when $f(x)$ is nonnegative on $[a, b]$.

7.3 CALCULATING VOLUMES BY SLICING

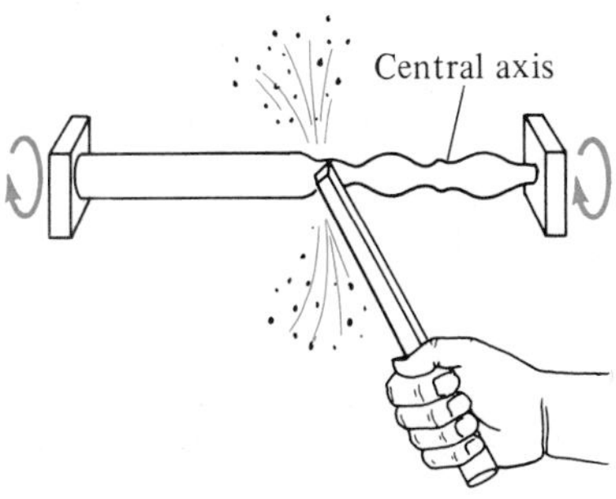

Figure 3.1 Table leg produced on a lathe by pressing chisel against rotating block of wood stock.

Even though the definite integral involves only two variables, there are certain types of three-dimensional objects whose volumes can be calculated as definite integrals of functions of a single variable. For example, it is common for solid objects to be produced by a process of milling a rotating piece of stock. In using a lathe to produce a wooden table leg, a craftsman presses a chisel against a rapidly rotating block of wood (Figure 3.1). Similarly, a potter works a ball of clay into a vase by using a potter's wheel, which allows the clay to be rotated at a uniform speed about a central axis (Figure 3.2).

The problem of calculating the volume of such solids is idealized mathematically as follows. Let $f(x)$ be a continuous nonnegative function for $a \le x \le b$. Let R denote the region bounded by the graph of $y = f(x)$, the x-axis, and the lines $x = a$ and $x = b$ (Figure 3.3). As the region R rotates about the x-axis (Figure 3.4), it sweeps out a **solid of revolution,** S. Just as for the lathe and pottery wheel illustra-

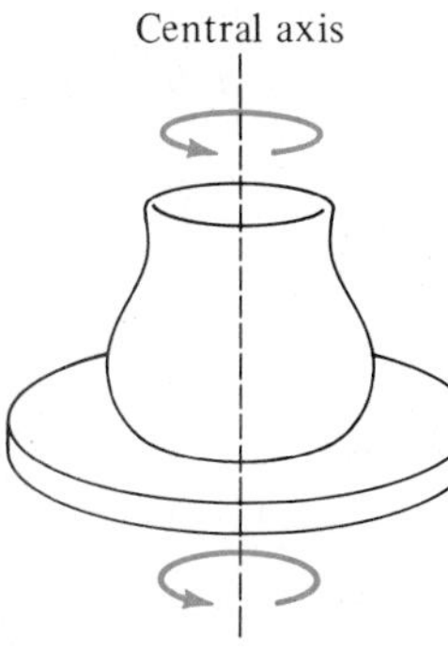

Figure 3.2 Pottery produced by shaping clay rotating on a wheel.

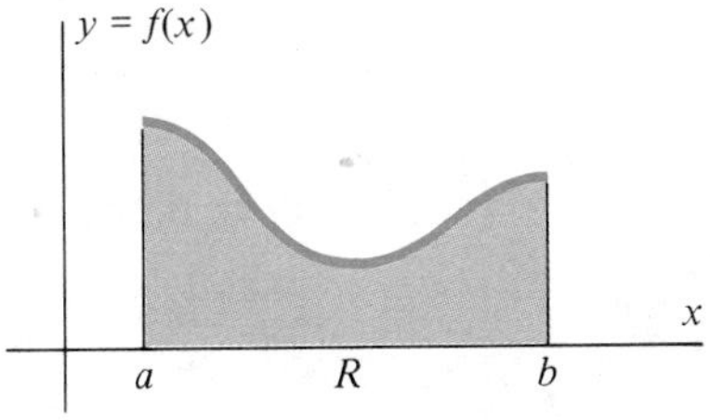

Figure 3.3 Region to be rotated.

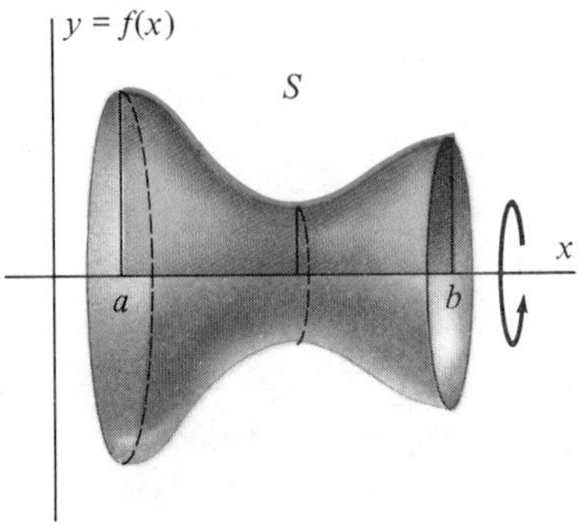

Figure 3.4 Solid obtained by rotating R about the x-axis.

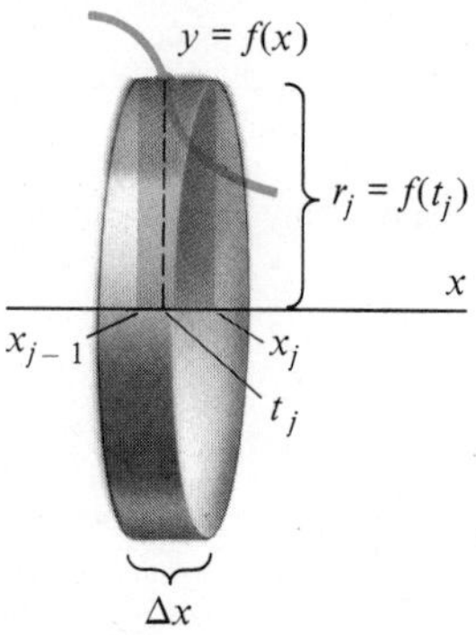

Figure 3.5 Rectangle of radius $r_j = f(t_j)$ generates disc of volume $V_j = \pi f^2(t_j)\,\Delta x$.

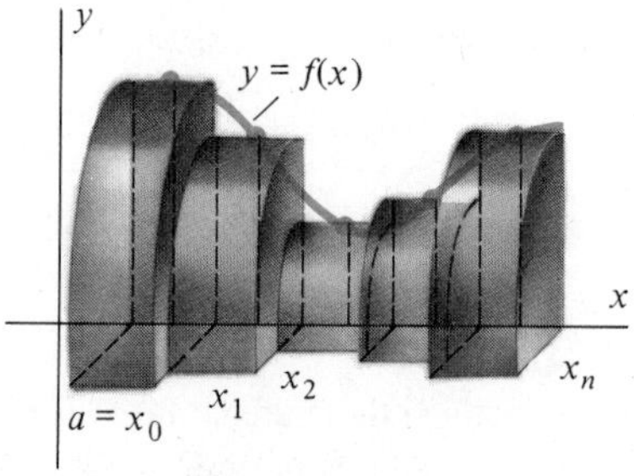

Figure 3.6 One quarter of the approximation to the volume of revolution S obtained by assuming $f(x)$ constant on subintervals.

tions, the cross sections for S taken perpendicular to the x-axis will be circles of radius $r = f(x)$. This is because the cross section taken at location x is described by rotating about the x-axis the line segment from $(x, 0)$ to $(x, f(x))$.

To find a formula for the volume of S, we begin by developing an approximation to the solid S. We do this by partitioning the interval $[a, b]$ into n equal subintervals of length $\Delta x = \dfrac{b-a}{n}$, and with endpoints $a = x_0, x_1, \ldots, x_n = b$. We arbitrarily select one "test number" t_j in each interval $[x_{j-1}, x_j]$, and we assume that the function $f(x)$ is constant throughout the interval $[x_{j-1}, x_j]$, with value $f(x) \equiv f(t_j)$.

In our approximation, corresponding to each interval $[x_{j-1}, x_j]$, we will be rotating a horizontal line segment $y = f(t_j)$ about the x-axis. This will generate a disc, of radius $r_j = f(t_j)$ and thickness Δx. The volume of this disc is, therefore,

$$\Delta V_j = \pi r_j^2\,\Delta x = \pi f^2(t_j)\,\Delta x \qquad \text{(here } f^2(t) \text{ means } [f(t)]^2\text{)}.$$

(See Figure 3.5.)

Summing the volumes of these individual discs from 1 to n gives the volume of our approximating solid as

$$\sum_{j=1}^{n} \Delta V_j = \sum_{j=1}^{n} \pi f^2(t_j)\,\Delta x.$$

(See Figure 3.6.) Next, we argue that as $n \to \infty$ and the size Δx of each individual disc becomes small, the volume of our approximating solid should approach the volume of S. That is, we want to *define* the volume V of S by the equation

$$V = \lim_{n\to\infty} \sum_{j=1}^{n} \Delta V_j = \lim_{n\to\infty} \sum_{j=1}^{n} \pi f^2(t_j)\,\Delta x.$$

Since the sum on the right is a Riemann sum for the function $\pi f^2(x)$, we have arrived at the following definition:

> Let $f(x)$ be continuous for $a \le x \le b$ and let R denote the region bounded by the graph of $y = f(x)$, the x-axis, and the lines $x = a$ and $x = b$. The volume of the solid obtained by rotating R about the x-axis is
>
> $$V = \int_a^b \pi[f(x)]^2\,dx. \qquad (1)$$

Example 1 Verify that equation (1) produces the formula $V = \frac{1}{3}\pi r^2 h$ for the volume of a right circular cone.

Strategy

Label variables.

Use a coordinate system to view cone as solid of revolution.

Find an equation for the line bounding the cross section from above.

Apply (1).

Solution

Let S be the cone with radius r and height h. We impose a coordinate system on S as illustrated in Figure 3.7. We can then view the cone as the solid obtained by rotating the triangle with vertices $(0, 0)$, $(h, 0)$, and (h, r) about the x-axis. Since the equation for the hypotenuse is $f(x) = \dfrac{r}{h}x$, we apply equation (1) to find that

$$V = \int_0^r \pi\left[\frac{r}{h}x\right]^2 dx = \frac{\pi r^2}{3h^2}x^3\Big]_0^h = \frac{1}{3}\pi r^2 h,$$

which is the desired formula. ■

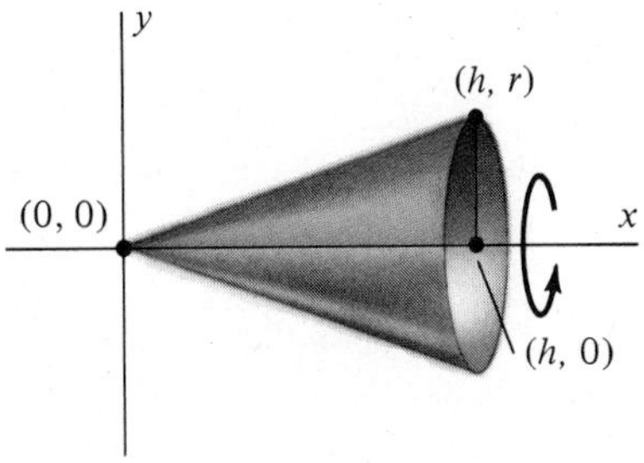

Figure 3.7 Cone as a volume of revolution.

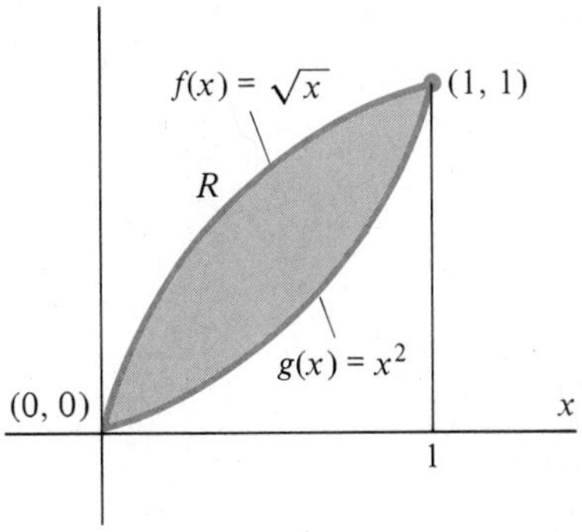

Figure 3.8 Region to be rotated about x-axis.

Example 2 Find the volume of the solid obtained by rotating the region bounded by the graphs of $f(x) = \sqrt{x}$ and $g(x) = x^2$ about the x-axis.

Strategy

Find points where the graphs cross.

View solid as difference of two solids of revolution.

Solution

The two graphs cross at (0, 0) and (1, 1) since the equation $\sqrt{x} = x^2$ implies $x = x^4$ or $x(1 - x^3) = 0$. Since $\sqrt{x} > x^2$ for $0 < x < 1$, the region is bounded above by the graph of $f(x) = \sqrt{x}$ and below by the graph of $g(x) = x^2$. As Figure 3.8 indicates, we may view the resulting solid as the solid obtained by rotation of $f(x) = \sqrt{x}$ from which the solid obtained by rotation of $g(x) = x^2$ is removed. The calculation for volume, by equation (1), is therefore

Apply (1) to each.

$$
\begin{aligned}
V &= \int_0^1 \pi(\sqrt{x})^2\,dx - \int_0^1 \pi(x^2)^2\,dx \\
&= \frac{\pi}{2}x^2\Big]_0^1 - \frac{\pi}{5}x^5\Big]_0^1 \\
&= \frac{3\pi}{10}.
\end{aligned}
$$

■

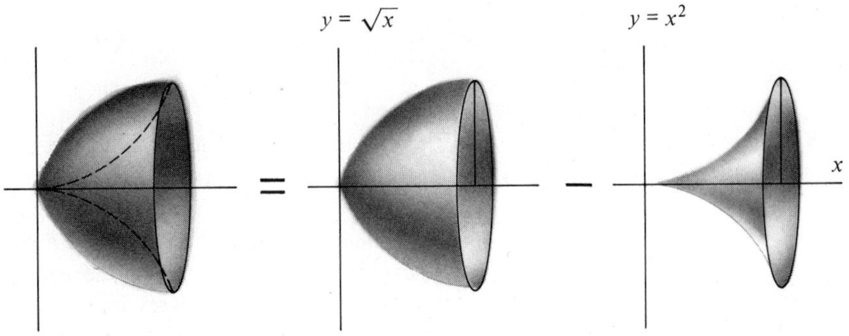

Figure 3.9 Volume obtained by expressing area between curves $\sqrt{x}$ and x^2 as the difference of volumes corresponding to upper curve $\sqrt{x}$ and lower curve x^2.

REMARK: As you may have observed, the volume in Example 2 could have been calculated by the single integral

$$V = \int_0^1 \pi[(\sqrt{x})^2 - (x^2)^2]\,dx.$$

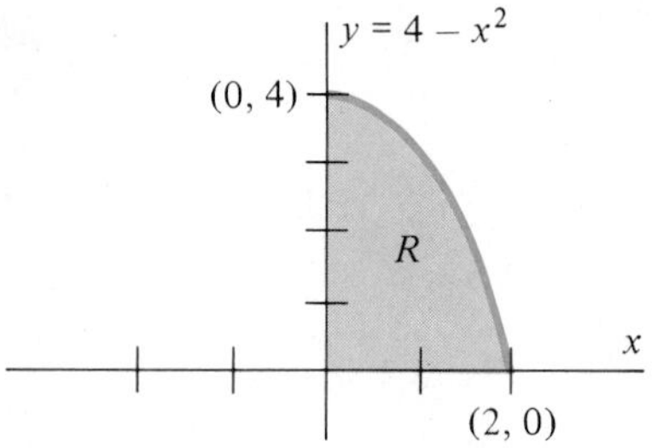

Figure 3.10 Region to be rotated about y-axis.

In general, if the region R is bounded above by $y = f(x)$ and below by $y = g(x) \geq 0$ for $a \leq x \leq b$, then the formula for volume is

$$V = \int_a^b \pi([f(x)]^2 - [g(x)]^2)\, dx \qquad (2)$$

However, caution must be taken not to misinterpret the integrand as $[f(x) - g(x)]^2$.

Example 3 The region in the first quadrant bounded by the graph of $y = 4 - x^2$ and the coordinate axes is rotated about the y-axis. Find the volume of the resulting solid.

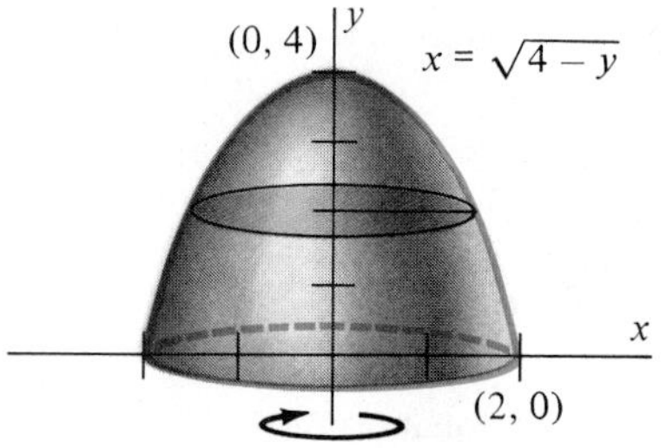

Figure 3.11 Cross sections perpendicular to y-axis are circles of radius $r = \sqrt{4 - y}$.

Solution: This problem is similar to that of Example 1 except that the roles of x and y are reversed. Solving the given equation for x as a function of y gives $f(y) = \sqrt{4 - y}$. Since the rotation is about the y-axis, the integration will be with respect to y, and the limits of integration are from $y = 0$ to $y = 4$ (see Figures 3.10 and 3.11).

We obtain

$$V = \int_0^4 \pi[\sqrt{4 - y}]^2\, dy$$

$$= \pi\left(4y - \frac{1}{2}y^2\right)\Big]_0^4$$

$$= 8\pi.$$

■

The result of Example 3 generalizes as follows:

> If the region R is bounded by the graph of the continuous function $x = f(y)$ and the y-axis from $y = a$ to $y = b$, then the volume of the solid obtained by rotating R about the y-axis is
>
> $$V = \int_a^b \pi[f(y)]^2\, dy.$$
>
> (See Figure 3.12.)

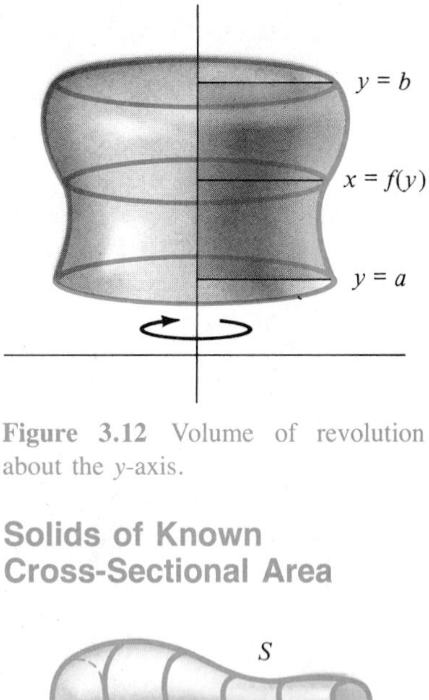

Figure 3.12 Volume of revolution about the y-axis.

Solids of Known Cross-Sectional Area

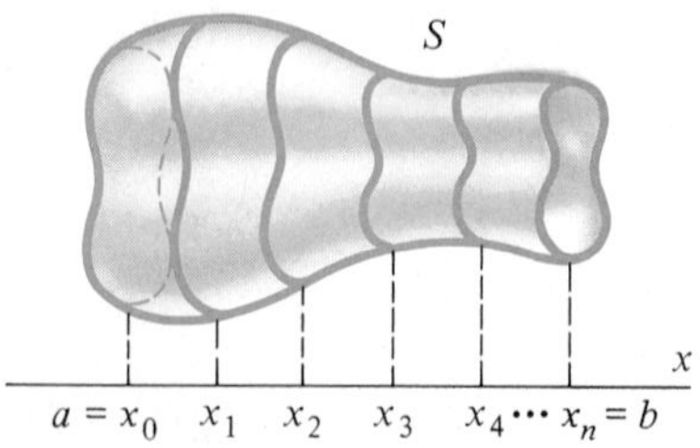

Figure 3.13 Cross sections of known area perpendicular to axis.

For the solid of revolution S, the area of a cross section taken perpendicular to the x-axis at $x = x_0$ will be $A(x_0) = \pi f^2(x_0)$, since the radius of the disc-shaped cross section is $r = f(x_0)$ (see Figure 3.4). We can interpret formula (1) for the volume of S as the integral of the cross-sectional area $A(x) = \pi f^2(x)$ from $x = a$ to $x = b$. Our next objective is to show that the volume of *any* solid of known cross-sectional area (possibly not a solid of revolution) can be calculated in this same manner.

In particular, suppose that S is a solid for which the area of each cross section perpendicular to a given axis is known (Figure 3.13). Let $A(x)$ denote the area of the cross section taken at location x, and assume that the solid extends from $x = a$ to $x = b$. If we partition the interval $[a, b]$ into equal subintervals of length $\Delta x = \dfrac{b - a}{n}$ with endpoints $a = x_0 < x_1 < x_2 < \cdots < x_n = b$, the solid is partitioned by the cross sections taken at these endpoints into n slices of equal thickness Δx (see Figure 3.14). By selecting one number t_j in each subinterval $[x_{j-1}, x_j]$, we approximate the volume of the slice over the interval $[x_{j-1}, x_j]$ by the volume ΔV_j of the cylinder with base area $A(t_j)$ and thickness Δx (Figure 3.16), that is,

$$\Delta V_j = A(t_j)\, \Delta x.$$

Summing these approximations for $j = 1, 2, \ldots, n$ gives the approximation to the volume V of S

$$V \approx \sum_{j=1}^{n} \Delta V_j = \sum_{j=1}^{n} A(t_j)\, \Delta x, \tag{3}$$

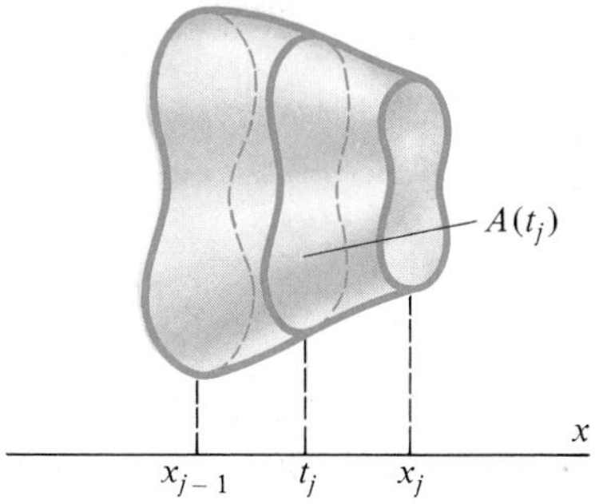

Figure 3.14 Slice over interval $[x_{j-1}, x_j]$.

which should approach the volume V as $n \to \infty$ and the lengths Δx approach zero (Figure 3.15). If the function $A(x)$ is continuous for $a \le x \le b$, the sum on the right-hand side of approximation (3) is a Riemann sum which therefore has a limit as $n \to \infty$. We define the volume V of S to be this limit.

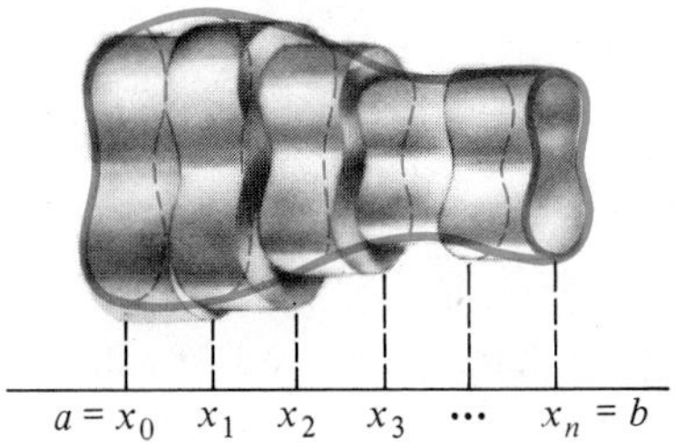

Figure 3.15 Approximation by volume elements.

If $A(x)$ denotes the area of the cross section of S for $a \le x \le b$ then, if $A(x)$ is continuous on $[a, b]$, the volume V of S is

$$V = \int_a^b A(x)\, dx \tag{4}$$

Notice that equation (4) generalizes the familiar formula for the volume of a cylinder with base area A and height h: $V = Ah$. It also contains formula (1) for the volume of a solid of revolution as a special case since, for such solids, $A(x) = \pi f^2(x)$.

Example 4 The base of a solid is a circle of radius 4 cm. All cross sections perpendicular to a particular axis are squares. Find the volume of this solid (see Figure 3.17).

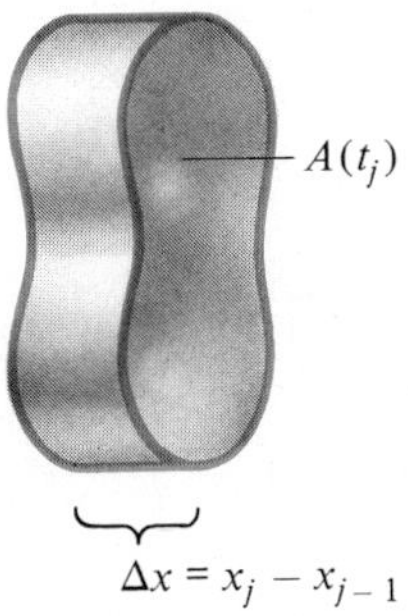

Figure 3.16 Volume element $A(t_j)\,\Delta x$ approximates volume of slice.

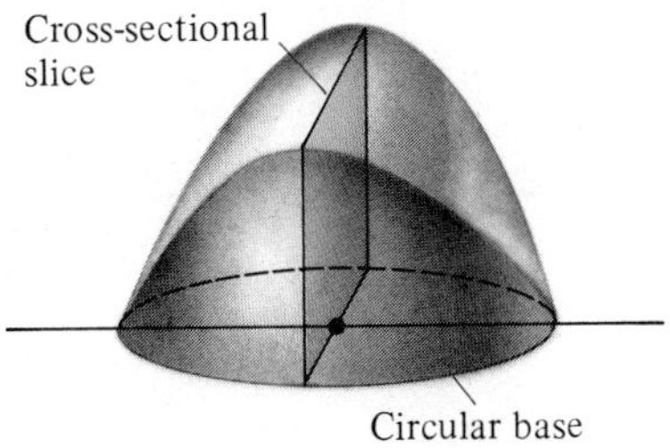

Figure 3.17 Cross sections perpendicular to the x-axis.

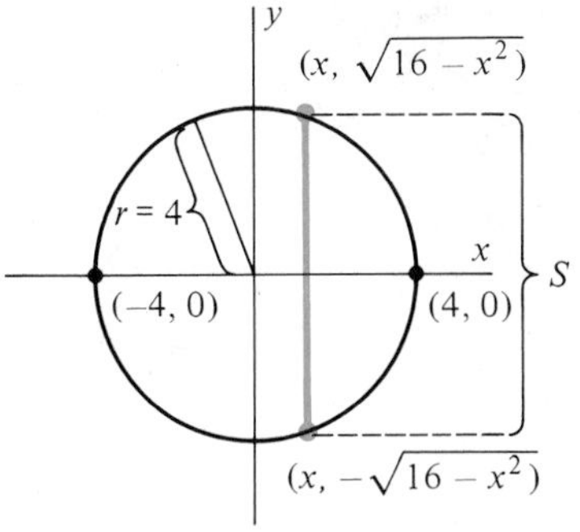

Figure 3.18 Side $s = 2\sqrt{16 - x^2}$.

Strategy

Find an expression for the area of a cross section. Begin by finding an equation for the boundary of the base.

Solution

If we impose an xy-coordinate system on the circular base so that the x-axis corresponds to the given axis, then the equation for the boundary of the base is

$$x^2 + y^2 = 16 \qquad \text{(Figure 3.18).}$$

The equation for the upper semicircle is $y = \sqrt{16 - x^2}$, and the equation for the lower semicircle is $y = -\sqrt{16 - x^2}$. A cross section perpendicular to the x-axis will therefore intersect this circular base in a chord of length $s = 2\sqrt{16 - x^2}$. Since this chord is one side of the square cross section, the area of the cross section is

$$\begin{aligned} A(x) &= s^2 \\ &= (2\sqrt{16 - x^2})^2 \\ &= 64 - 4x^2. \end{aligned}$$

Find the limits of integration.

The smallest and largest values of x are, respectively, -4 and 4. The volume, by equation (4), is therefore

Apply (4).

$$\begin{aligned} V &= \int_{-4}^{4} A(x)\,dx = \int_{-4}^{4} (64 - 4x^2)\,dx \\ &= 64x - \frac{4}{3}x^3\Big]_{-4}^{4} \\ &= \frac{1024}{3}\ \text{cm}^3. \end{aligned}$$ ■

Although the solid in Example 5 is a solid of revolution, the axis of rotation is neither the x- nor the y-axis. Note how using equation (4) helps clarify the issue.

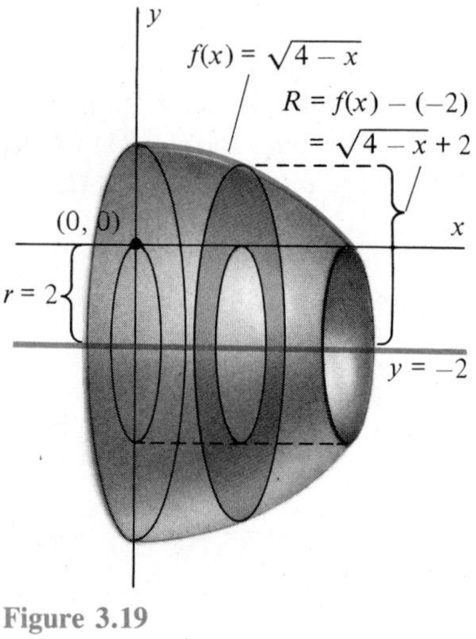

Figure 3.19

Example 5 Find the volume of the solid generated by rotating the region bounded by the graph of $f(x) = \sqrt{4 - x}$ and the x-axis for $0 \le x \le 4$ about the line $y = -2$.

Solution: Since the region in question is being rotated about a line other than the x-axis, we cannot use formula (1). As illustrated in Figure 3.19, a cross section taken at location x will consist of a circle of radius $R = [\sqrt{4 - x} - (-2)]$ from which a smaller circle of radius $r = 2$ has been removed. The cross-sectional area is therefore

$$\begin{aligned} A(x) &= \pi R^2 - \pi r^2 \\ &= \pi(\sqrt{4 - x} + 2)^2 - \pi \cdot 2^2 \\ &= \pi(4 - x + 4\sqrt{4 - x}). \end{aligned}$$

We can now apply equation (4) to find that

$$\begin{aligned} V &= \int_{0}^{4} \pi(4 - x + 4\sqrt{4 - x})\,dx \\ &= \pi\left(4x - \frac{1}{2}x^2 - \frac{8}{3}(4 - x)^{3/2}\right)\Big]_{0}^{4} \\ &= \frac{88\pi}{3}. \end{aligned}$$ ■

Exercise Set 7.3

In Exercises 1–7, find the volume of the solid obtained by revolving the region bounded by the given curve and the x-axis, for $a \le x \le b$, about the x-axis.

1. $f(x) = 2x + 1, \quad 1 \le x \le 4$

2. $f(x) = \sqrt{4x - 1}, \quad 1 \le x \le 5$

3. $f(x) = \sin x, \quad 0 \le x \le \pi$
$\left(Hint: \sin^2 x = \dfrac{1}{2} - \dfrac{1}{2}\cos 2x.\right)$

4. $f(x) = |x - 1|, \quad 0 \le x \le 3$

5. $f(x) = \tan x, \quad 0 \le x \le \pi/4$

6. $f(x) = \sqrt{4 - x^2}, \quad 1 \le x \le 2$

7. $f(x) = \dfrac{\sqrt{1 + x}}{x^{3/2}}, \quad 1 \le x \le 2$

In Exercises 8–11, find the volume of the solid obtained by revolving about the x-axis the region bounded by the given curves.

8. $f(x) = x^2, \quad g(x) = x^3$

9. $f(x) = \dfrac{1}{4}x^2, \quad g(x) = x$

10. $f(x) = \dfrac{1}{x}, \quad g(x) = \sqrt{x}, \quad 1 \le x \le 4$

11. $f(x) = \sin x, \quad g(x) = \cos x, \quad 0 \le x \le \pi$

In each of Exercises 12–17, find the volume of the solid obtained by rotating the region bounded by the given curves about the y-axis.

12. $x + y = 4, \quad x = 0, \quad 0 \le y \le 4$

13. $y = x^2, \quad y = 0, \quad 0 \le x \le 2$

14. $y = 4, \quad y = x^2$

15. $y = x^3, \quad x = 2, \quad y = 0$

16. $y = x^2, \quad y = x^3$

17. $y = \dfrac{1}{x}, \quad y = 0, \quad \dfrac{1}{4} \le x \le 1$

18. Find the volume of the right pyramid whose base is a square 10 cm on a side and whose altitude is 8 cm (Figure 3.20).

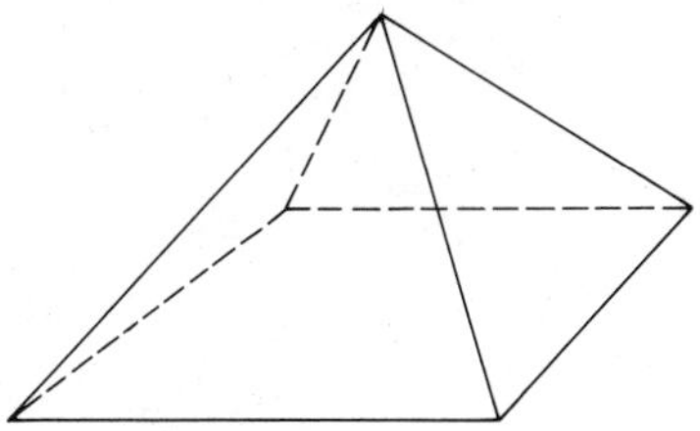

Figure 3.20

19. By integration, derive the formula for the volume of a sphere of radius r.

20. The base of a solid is a circle of radius 4. Find the volume of the solid if all cross sections perpendicular to a given axis are equilateral triangles (Figure 3.21).

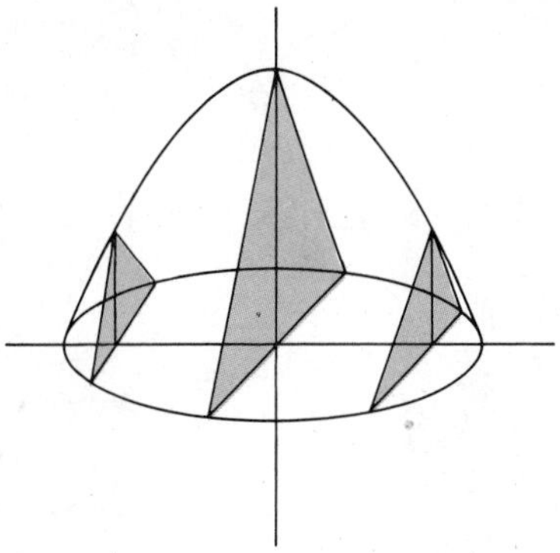

Figure 3.21

21. Find the volume of the solid in Exercise 20 if the cross sections perpendicular to the given axis are isosceles right triangles with bases lying along the base of the solid (Figure 3.22).

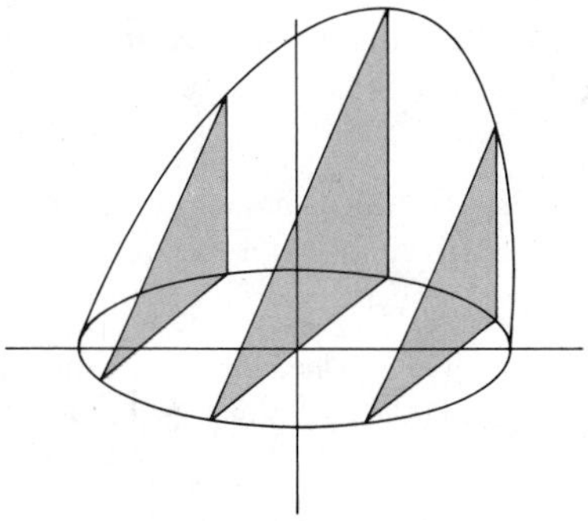

Figure 3.22

22. A hemispherical water tank of radius 10 meters contains water to a depth of 6 meters. How many cubic meters of water does the tank contain?

23. The base of a solid is the region bounded by the graphs of $f(x) = x^2$ and $g(x) = 8 - x^2$. Find the volume of the solid if all cross sections perpendicular to the x-axis are squares.

24. Find the volume of the solid in Exercise 23 if the cross sections perpendicular to the x-axis are semicircles.

25. Find the formula for the volume of the frustum of a cone with bases of radius r and R, and height h (see Figure 3.23).

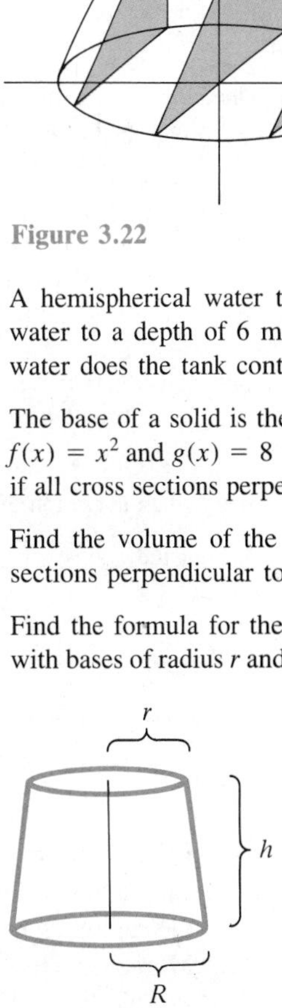

Figure 3.23

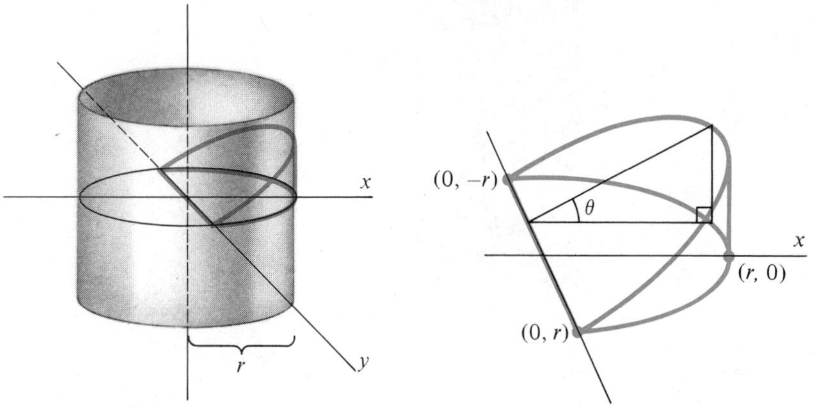

Figure 3.24

26. Find the volume of the solid generated by revolving the triangle with vertices (0, 0), (2, 5), and (5, 0) about the x-axis.

27. Water is running into the tank in Exercise 22 at a rate of 10 cubic meters per minute. How fast is the water level rising when its depth is 4 meters?

28. A hole of radius 2 cm is drilled through the center of a spherical ball of radius 4 cm. What is the volume of the remaining solid?

29. Find the volume of the solid obtained by rotating the region bounded by the x-axis and the graph of $y = 1 - x^2$ about the line $y = -3$.

30. Find the volume of the solid obtained by rotating the region bounded by the graphs of $y = \sqrt{x}$ and $y = \dfrac{1}{2}x$ about the line $y = 4$.

31. Find the volume of the solid obtained when the region bounded by the graphs of $y = \sqrt{x}$, $y = 0$, and $x = 9$ is rotated about the line $y = -2$.

32. Find the volume of the solid obtained when the graph of the ellipse $\dfrac{x^2}{a^2} + \dfrac{y^2}{b^2} = 1$ is rotated about the x-axis.

33. Find the volume of the solid obtained when the region in Exercise 32 is rotated about the y-axis.

34. What formula is obtained from Exercises 32 and 33 when $a = b = r$?

35. When a right circular cylinder of radius r is sliced by two planes, one perpendicular to the central axis of the cylinder and the other at an angle θ with the first, as in Figure 3.24, a wedge results. Find its volume.

36. *(Calculator/Computer)* Use the Trapezoidal Rule with $n = 8$ to approximate the volume of the solid obtained by rotating the region bounded by the graph of $f(x) = \dfrac{1}{\sqrt{1 + x^2}}$ and the line $y = 0$ for $0 \le x \le 4$ about the x-axis.

37. *(Calculator/Computer)* Use Simpson's rule with $n = 12$ to approximate the volume of the solid in Exercise 36.

38. *(Calculator/Computer)* Use an approximation scheme of your own choosing to approximate the volume of the solid obtained by rotating the region bounded by the graph of $y = x^{-1/2}$ and the x-axis for $1 \le x \le 4$ about the x-axis.

39. *(Calculator/Computer)* Approximate the volume of the solid obtained by revolving the region in Exercise 32 about the line $y = -2$ if $a = 3$ and $b = 1$.

7.4 CALCULATING VOLUMES BY THE METHOD OF CYLINDRICAL SHELLS

In addition to the slicing methods of Section 7.3, there is another approach to calculating volumes of solids of revolution. This method is particularly useful when the region to be rotated is not adjacent to the axis about which it is to be rotated. Moreover, this method will sometimes succeed when the method of slicing fails, and vice versa. While the slicing method is based on the idea of approximating cross sections taken perpendicular to the axis of rotation, the **method of cylindrical shells** uses approximating cylinders centered about the axis of rotation.

To describe this method, we let R be the region bounded by the graph of the continuous nonnegative function $f(x)$ and the x-axis for $a \le x \le b$ (Figure 4.1). If the region R is rotated about the y-axis, a solid S is generated (Figure 4.2).

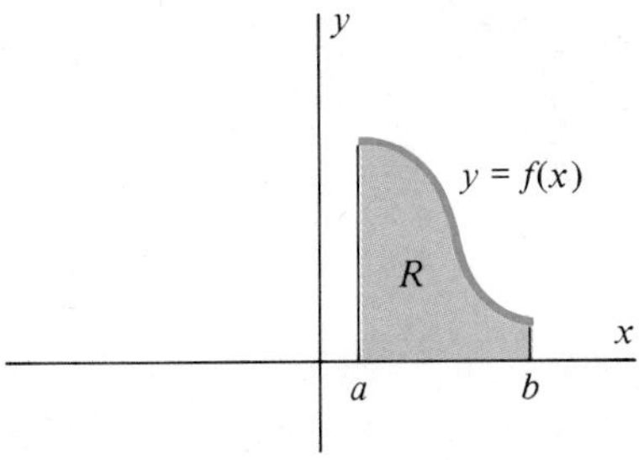

Figure 4.1 Region R to be rotated.

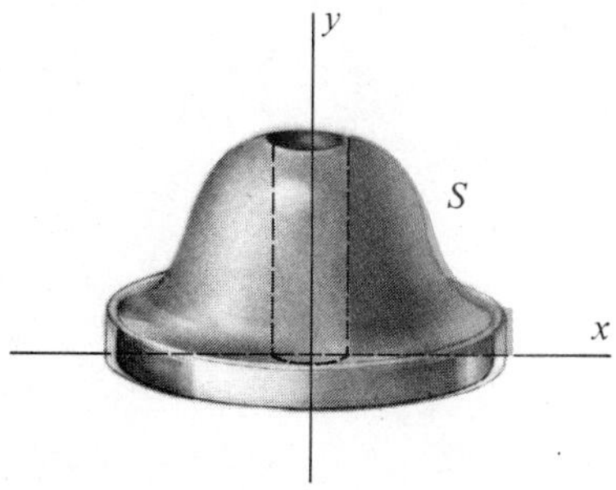

Figure 4.2 Solid obtained by rotating region R about the y-axis.

To approximate the volume of S, we partition the interval $[a, b]$ with numbers $a = x_0 < x_1 < x_2 < \cdots < x_n = b$ with equal differences $x_j - x_{j-1} = \Delta x = \dfrac{b-a}{n}$, $j = 1, 2, \ldots, n$. In each subinterval $[x_{j-1}, x_j]$, we arbitrarily select a number t_j. Then the area of the region R over the interval $[x_{j-1}, x_j]$ is approximated by $f(t_j)\,\Delta x$, the area of the rectangle with base $[x_{j-1}, x_j]$ and height $f(t_j)$ (see Figure 4.3). As each rectangle rotates about the y-axis, it generates a cylindrical shell. The totality of all such shells (Figure 4.4) constitutes a solid whose volume we take as the approximation to the volume of S.

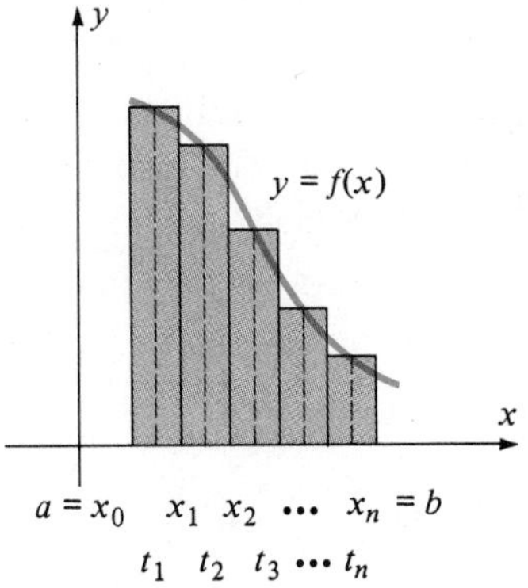

Figure 4.3 Approximating the area of R by rectangles.

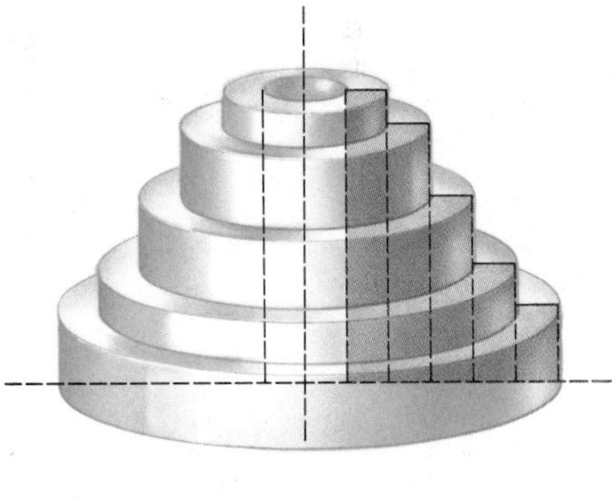

Figure 4.4 Rotating the approximating rectangles produces the approximation to the solid S.

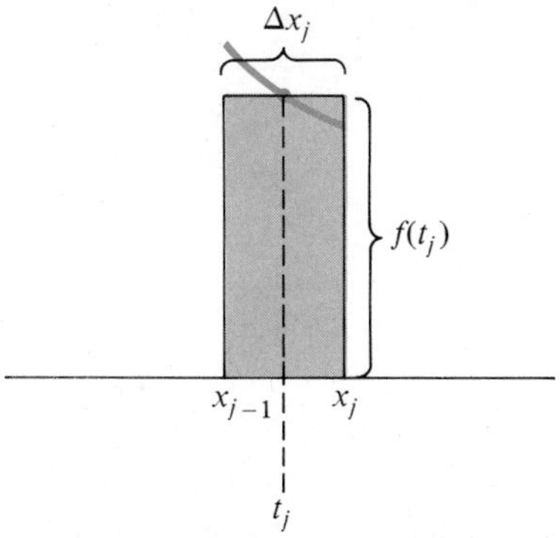

Figure 4.5 Approximating rectangle over interval $[x_{j-1}, x_j]$.

Since this approximation will be the sum of the volumes of the approximating cylinders, we note that when the rectangle over $[x_{j-1}, x_j]$ in Figure 4.5 is rotated about the y-axis, the volume ΔV_j of the resulting cylindrical "volume element" is

$$\begin{aligned} \Delta V_j &= \pi x_j^2 f(t_j) - \pi x_{j-1}^2 f(t_j) \\ &= \pi f(t_j)(x_j^2 - x_{j-1}^2). \end{aligned} \tag{1}$$

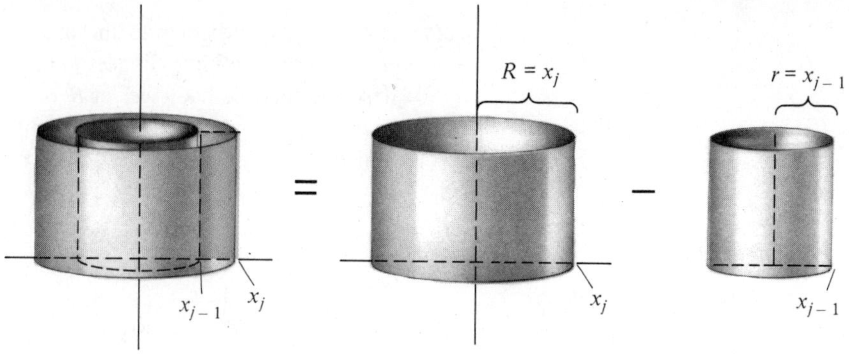

Figure 4.6 Volume element is a cylindrical shell—a cylinder minus a smaller cylinder.

The explanation for equation (1) is that the cylindrical shell whose volume is ΔV_j can be viewed as a cylinder of radius $R = x_j$ and height $f(t_j)$ from which a smaller cylinder of radius $r = x_{j-1}$ and height $f(t_j)$ has been removed (Figure 4.6).

Summing the approximations given by equation (1) over all $j = 1, 2, \ldots, n$ gives the approximation

$$V \approx \sum_{j=1}^{n} \Delta V_j = \sum_{j=1}^{n} \pi f(t_j)(x_j^2 - x_{j-1}^2). \tag{2}$$

We have now arrived at the step at which the desired quantity V should arise as the limit as $n \to \infty$ of the approximating sum on the right-hand side of equation (2). However, careful examination of this expression shows that it is in fact *not* a Riemann sum. We can therefore conclude nothing about the limiting value of approximation (2) at this point.

By comparing the sum in (2) with the standard form for a Riemann sum (Section 7.1), we see that approximation (2) has two difficulties:

(a) The factor $\Delta x = x_j - x_{j-1}$ is missing in each term.
(b) There are *three* points in each interval at which the expression being summed is evaluated: x_{j-1}, t_j, and x_j. A Riemann sum involves a function evaluated at only *one* point in each subinterval.

We overcome these difficulties in two steps.

Step 1: Notice that the terms $(x_j^2 - x_{j-1}^2)$ in approximation (2) may be factored as

$$\begin{aligned} x_j^2 - x_{j-1}^2 &= (x_j + x_{j-1})(x_j - x_{j-1}) \\ &= (x_{j-1} + x_j)\,\Delta x, \end{aligned} \tag{3}$$

producing the desired factor Δx.

Step 2: The term $(x_{j-1} + x_j)$ can be written as $2\left(\dfrac{x_{j-1} + x_j}{2}\right)$, which may be interpreted as twice the midpoint of the interval $[x_{j-1}, x_j]$. Since $t_j \in [x_{j-1}, x_j]$ is arbitrary, we *choose* t_j to be the midpoint $\dfrac{1}{2}(x_{j-1} + x_j)$. This will allow all independent variables appearing in the approximating sum to be written in terms of t_j.

Applying the ideas of steps 1 and 2 to approximation (2), we obtain

$$V \approx \sum_{j=1}^{n} \pi f(t_j)(x_j^2 - x_{j-1}^2), \qquad t_j \in [x_{j-1}, x_j] \text{ arbitrary} \tag{4}$$

$$= \sum_{j=1}^{n} \pi f(t_j)(x_j + x_{j-1})\,\Delta x \qquad \text{(step 1)}$$

$$= \sum_{j=1}^{n} 2\pi f(t_j)\left(\frac{x_{j-1} + x_j}{2}\right)\Delta x$$

$$= \sum_{j=1}^{n} 2\pi t_j f(t_j)\,\Delta x, \qquad t_j = \frac{x_{j-1} + x_j}{2} \qquad \text{(step 2).}$$

Now the final sum in approximation (4) *is* a Riemann sum for the function $g(x) = 2\pi x f(x)$. Therefore, by the theory of Chapter 6, as $n \to \infty$ the approximating sum approaches the integral of $g(x)$, that is, we obtain

$$\lim_{n\to\infty} \sum_{j=1}^{n} 2\pi t_j f(t_j)\,\Delta x = \int_a^b 2\pi x f(x)\,dx.$$

This is the number that we define to be the desired volume.

> If the region R bounded by the graph of the continuous nonnegative function $f(x)$ and the x-axis, for $0 \le a \le x \le b$, is rotated about the y-axis, the volume V of the resulting solid may be calculated as
>
> $$V = \int_a^b 2\pi x f(x)\,dx. \tag{5}$$

Example 1 The region bounded by the graph of $y = -2x^2 + 8x - 6$ and the x-axis is rotated about the y-axis. Find the volume V of the resulting solid.

Strategy
Factor equation for y to find limits of integration.

Solution
Since $y = -2x^2 + 8x - 6 = -2(x - 1)(x - 3)$, the region lies between the lines $x = 1$ and $x = 3$. Thus

Apply equation (5).

$$V = \int_1^3 2\pi x[-2x^2 + 8x - 6]\,dx$$

$$= 2\pi \cdot \int_1^3 (-2x^3 + 8x^2 - 6x)\,dx$$

$$= 2\pi\left[-\frac{2}{4}x^4 + \frac{8}{3}x^3 - \frac{6}{2}x^2\right]_1^3$$

$$= \frac{32\pi}{3}.$$

(See Figure 4.7.) ■

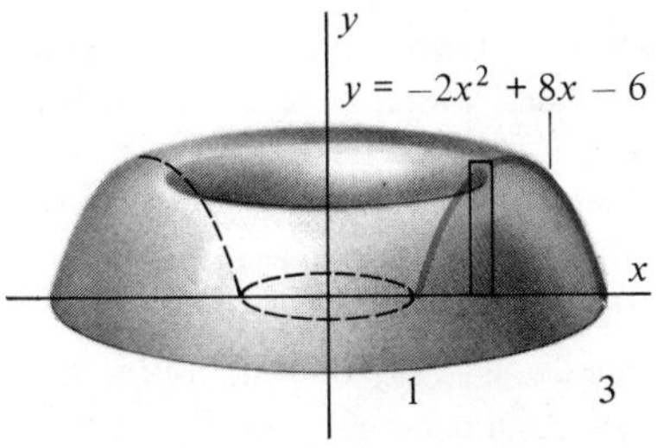

Figure 4.7 Volume of revolution about y-axis.

Equation (5) can be generalized to regions bounded below by curves other than the x-axis, as the following example shows.

Example 2 The region R is bounded by the graphs of $f(x) = -2x^2 + 8x - 6$ and $g(x) = 2x - 6$. Find the volume of the solid generated by revolving R about the y-axis.

Strategy
Set $f(x) = g(x)$ to find the points where the curves cross.

Solution
To find the points of intersection, we set $f(x) = g(x)$ and obtain the equation

$$-2x^2 + 8x - 6 = 2x - 6,$$

or

$$-2x^2 = -6x,$$

Note which curve bounds the top of the region and which curve bounds the bottom.

so $x = 0$ or $x = 3$. As Figure 4.8 illustrates, when the interval [0, 3] is partitioned, the approximating rectangles are bounded above by $f(x) = -2x^2 + 8x - 6$ and below by $g(x) = 2x - 6$. Since the factor $f(x)$ in equation (5) represents the height of the approximating rectangles, in this case equation (5) is modified to the following:

Apply equation (5) to both curves.

$$\begin{aligned} V &= \int_0^3 2\pi x[f(x) - g(x)]\,dx \\ &= \int_0^3 2\pi x[-2x^2 + 8x - 6 - (2x - 6)]\,dx \\ &= \int_0^3 2\pi x(-2x^2 + 6x)\,dx \\ &= 2\pi\left[-\frac{2}{4}x^4 + \frac{6}{3}x^3\right]_0^3 \\ &= 27\pi. \end{aligned}$$

■

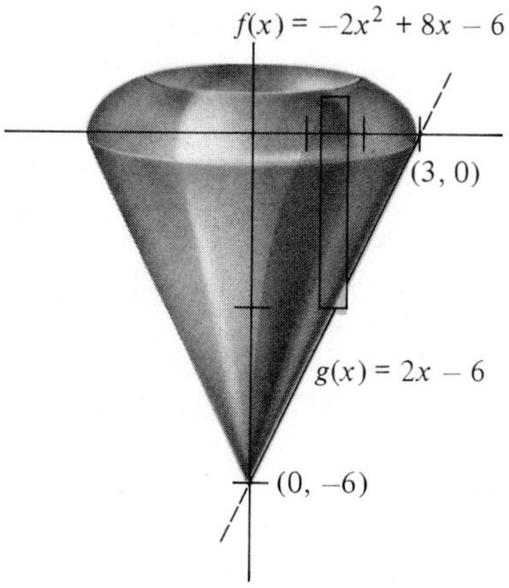

Figure 4.8 Region R bounded above by $f(x)$ and below by $g(x)$.

REMARK 1: We may state the generalization of equation (5) observed in Example 2 as follows: If the region R bounded above by the graph of $y = f(x)$ and below by the graph of $y = g(x)$, for $0 \le a \le x \le b$, is rotated about the y-axis, the volume of the resulting solid is

$$V = \int_a^b 2\pi x[f(x) - g(x)]\,dx. \qquad (6)$$

REMARK 2: You may have noticed that the integrand in equations (5) and (6) have simple geometric interpretations. If the region R to be rotated about the y-axis is sliced vertically at location x and the resulting line segment rotated about the y-axis, a band of height $[f(x) - g(x)]$ is generated (see Figure 4.9). Since the circumference of this band is $2\pi x$, the surface area of the band is precisely the integrand $2\pi x[f(x) - g(x)]$ in equation (6). We may therefore interpret equations (5) and (6) by saying that the volume V is found by integrating the "radial" cross section from $x = a$ to $x = b$.

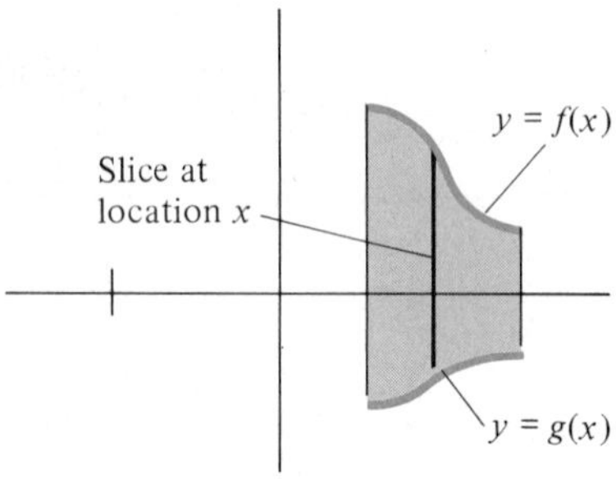

Figure 4.9 Vertical slice taken at location x.

REMARK 3: Many of the problems on volumes of revolution in this section and the preceding section may be solved by using either the method of slicing by perpendicular cross sections or the method of cylindrical shells. Usually one method will be simpler to use, depending on the geometry of the region being rotated.

Example 3 Find the volume of the solid obtained by revolving the region bounded by the graph of $y = 1 - x^2$ and the x-axis about the line $y = -1$.

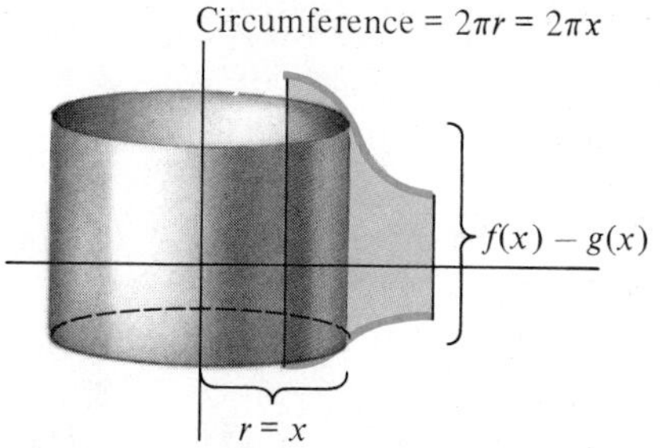

Figure 4.10 Slice generates band of surface area $2\pi x[f(x) - g(x)]$.

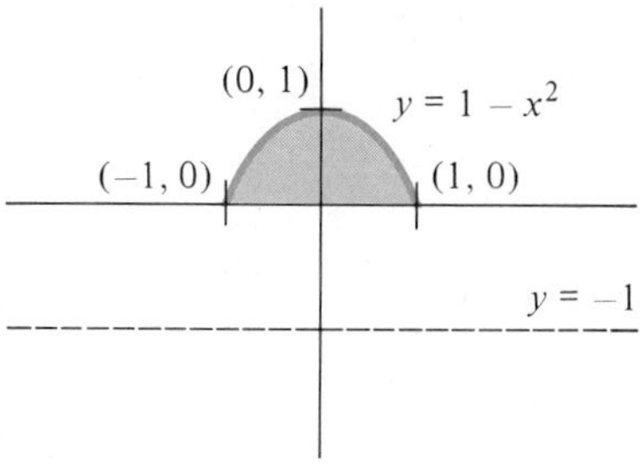

Figure 4.11 Region to be revolved about the line $y = -1$.

Solution: The graph of $y = 1 - x^2$ meets the x-axis at $x = \pm 1$. The relevant interval for x is therefore $-1 \le x \le 1$ (Figure 4.11).

Method 1: To use the method of cylindrical shells, we must replace x with y in equation (6), since the axis of revolution is the line $y = -1$. Solving the equation $y = 1 - x^2$ for x as a function of y gives $x = \pm\sqrt{1 - y}$. A line segment slicing the region parallel to the x-axis at height y therefore has length $\sqrt{1 - y} - (-\sqrt{1 - y}) = 2\sqrt{1 - y}$. Since the axis of revolution is $y = -1$, the radius of the shell swept out by this line segment is $r = y + 1$. The surface area of this shell is therefore $2\pi(y + 1) \cdot 2\sqrt{1 - y}$. Applying equation (6) then gives

$$V = \int_0^1 2\pi(y + 1) \cdot 2\sqrt{1 - y}\, dy.$$

Evaluating this integral requires a rather subtle u-substitution. The idea is to substitute u for $1 - y$ so that the term $\sqrt{1 - y}$ becomes just $\sqrt{u}$. We let

$$u = 1 - y; \qquad du = -dy.$$

Then $y = 1 - u$ and $dy = -du$, so the term $(y + 1)\, dy$ becomes $(y + 1)\, dy = [(1 - u) + 1](-du) = -(2 - u)\, du$. Finally, we obtain the new limits of integration by noting that

$$\begin{aligned} &\text{if } y = 0, \quad u = 1 - 0 = 1, \quad \text{and} \\ &\text{if } y = 1, \quad u = 1 - 1 = 0. \end{aligned}$$

With these substitutions the volume integral becomes

$$\begin{aligned} V = \int_0^1 2\pi(y + 1)2\sqrt{1 - y}\, dy &= 4\pi \int_1^0 -(2 - u)u^{1/2}\, du \\ &= -4\pi \int_1^0 (2u^{1/2} - u^{3/2})\, du \\ &= -4\pi\left[\frac{4}{3}u^{3/2} - \frac{2}{5}u^{5/2}\right]_1^0 \\ &= 4\pi\left(\frac{4}{3} - \frac{2}{5}\right) \\ &= \frac{56\pi}{15}. \end{aligned}$$

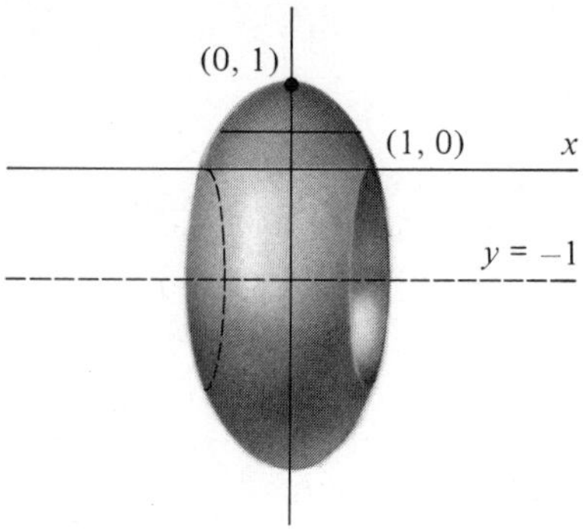

Figure 4.12 Solid of revolution obtained.

The method of slicing, discussed below, leads to a simpler solution.

Method 2: To calculate this volume by the method of slicing, we note that a rectangular slice taken perpendicular to the x-axis will extend from the line $y = 0$ to the curve $y = 1 - x^2$. Since the axis of revolution is $y = -1$, this slice will generate a disc of radius $R = (1 - x^2) + 1$ from which a disc of radius $r = 1$ is removed (Figures 4.14 and 4.15).

Using equation (4) of Section 7.3, we obtain

$$V = \int_{-1}^{1} \{\pi[(1 - x^2) + 1]^2 - \pi(1)^2\}\, dx$$

$$= \pi \int_{-1}^{1} (3 - 4x^2 + x^4)\, dx$$

$$= \pi\left[3x - \frac{4}{3}x^3 + \frac{1}{5}x^5\right]_{-1}^{1}$$

$$= \frac{56\pi}{15}.$$

■

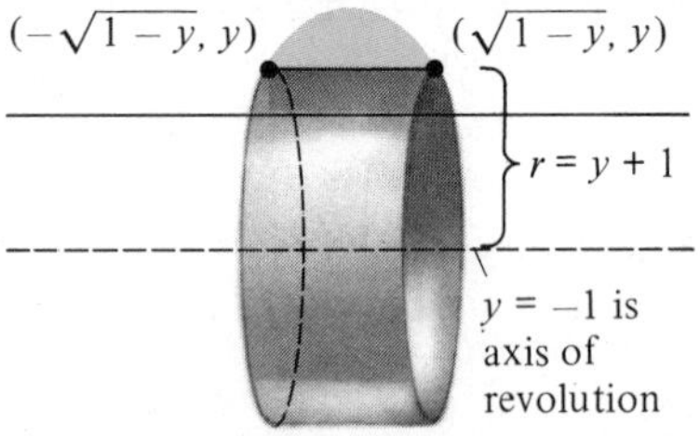

Figure 4.13 Approximating rectangle sweeps out shell of surface area $(y + 1) \cdot 2\sqrt{1 - y}$.

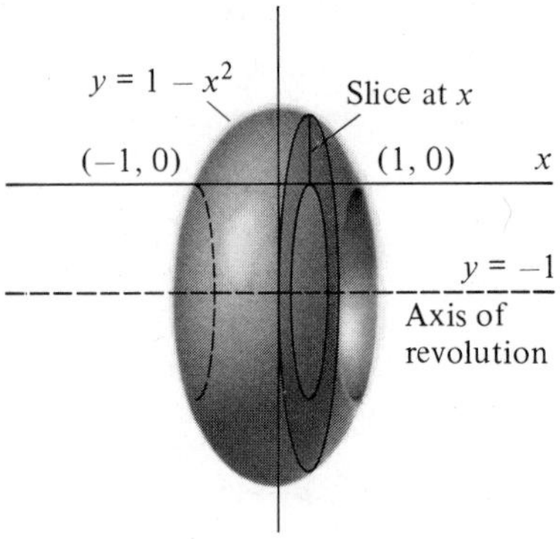

Figure 4.14 Slice perpendicular to x-axis.

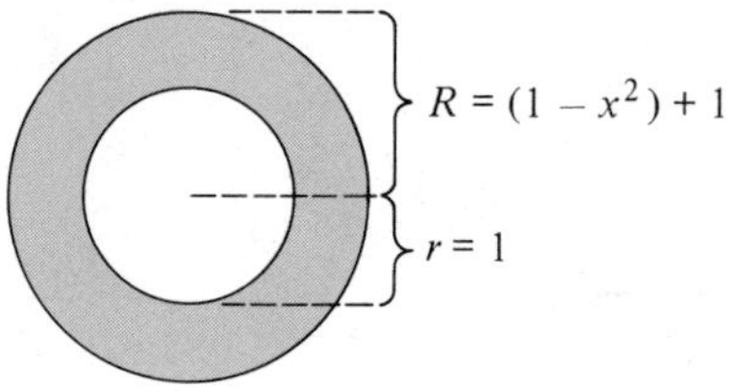

Figure 4.15 Slice generates annulus of area $\pi[(2 - x^2)^2x - 1]$.

Exercise Set 7.4

For each of Exercises 1–13, sketch the region bounded by the graphs of the given functions for the specified values of x or y. Then find the volume of the solid obtained by revolving the region about the y-axis.

1. $x + y = 1, \quad x = 0, \quad y = 0$
2. $y = \sqrt{x}, \quad y = 0, \quad 1 \le x \le 4$
3. $y = x^3, \quad y = 0, \quad 1 \le x \le 3$
4. $y = \sqrt{1 + x^2}, \quad y = 0, \quad 0 \le x \le 3$
5. $y = \sqrt{x}, \quad y = -x, \quad x = 4$
6. $y = 1 + x + x^2, \quad y = -2, \quad 1 \le x \le 3$
7. $y = \dfrac{1}{x}, \quad y = 1, \quad x = 4$
8. $y = 2, \quad y = \sqrt[3]{x}, \quad x = 0$
9. $y = \sin x^2, \quad y = 1, \quad 0 \le x \le \sqrt{\dfrac{\pi}{2}}$
10. $y = \dfrac{1}{\sqrt{4 - x^2}}, \quad y = 0, \quad 0 \le x \le 1$
11. $y = \sqrt{9 - x^2}, \quad y = 0, \quad 0 \le x \le 3$
12. $y = \dfrac{1}{\sqrt{x}} + \sqrt{x}, \quad y = 0, \quad 1 \le x \le 4$

Figure 4.16 Torus.

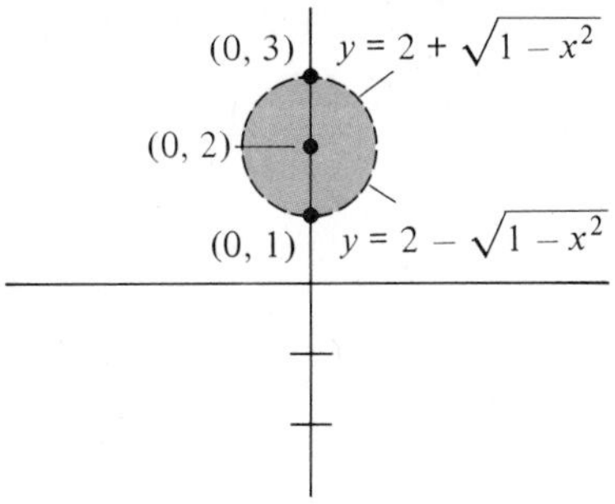

Figure 4.17 Torus generated by revolving this circle about the x-axis.

13. $y = \dfrac{\sqrt{1 + x^{3/2}}}{\sqrt{x}}$, $\quad y = 0, \quad 1 \le x \le 4$

In Exercises 14–21, use either the method of cylindrical shells or the method of slicing to find the volume of the solid described.

14. The region R, bounded by the graphs of $y = \sin x$ and $y = -x$, for $0 \le x \le \pi$ is rotated about the y-axis.

15. The region bounded by the graph of $x = y^2$ and the line $x = 4$ is revolved about the line $x = -1$.

16. The region bounded by the graph of $x = \sqrt{4 + y}$, the line $x = 0$, and the line $y = 0$ is revolved about the line $x = -2$.

17. The region bounded by the graphs of $y = x^2$ and $y = \sqrt{x}$ is revolved about the line $y = -2$.

18. The region contained within the triangle with vertices $(1, 0)$, $(2, 4)$, and $(4, 0)$ is rotated about the y-axis.

19. The region bounded by the graphs of $y = x$ and $y = x^3$ is revolved about the y-axis.

20. The region bounded by the graphs of $y = 4$ and $y = x^2$ is revolved about the line $y = -2$.

21. The triangle with vertices $(1, 3)$, $(1, 7)$, and $(4, 7)$ is revolved about the line $y = 1$.

22. Use the method of cylindrical shells to find the volume of the solid obtained by rotating the region bounded by the ellipse $\dfrac{x^2}{a^2} + \dfrac{y^2}{b^2} = 1$ about the y-axis.

23. Figure 4.16 shows a solid called a *torus*. A **torus** is generated by revolving a disc about an axis to which it is not adjacent. Figure 4.17 shows a region R that is interior to the circle of radius $r = 1$ and center $(0, 2)$. Let T be the torus obtained by revolving R about the x-axis.

a. Show that the volume of T can be expressed as

$$V = \pi \int_{-1}^{1} [\sqrt{1 - x^2} + 2]^2\, dx - \pi \int_{-1}^{1} [2 - \sqrt{1 - x^2}]^2\, dx.$$

b. Show that V simplifies to the integral

$$V = 8\pi \int_{-1}^{1} \sqrt{1 - x^2}\, dx.$$

c. Evaluate the integral in (b) by interpreting it as the area of a semicircle.

d. Find the volume V of T.

24. *(Calculator/Computer)* Use the method of cylindrical shells and the Trapezoidal Rule to approximate the volume of the solid obtained by revolving the region bounded by the graph of $y = \sin x$ and the x-axis, for $0 \le x \le \pi$ about the y-axis.

25. *(Calculator/Computer)* Use Simpson's Rule to approximate the volume described in Exercise 24.

26. *(Calculator/Computer)* Approximate the volume of the solid obtained by revolving the region bounded by the graph of $y = \sqrt{1 + x^4}$, the line $y = 0$, and the lines $x = 0$ and $x = 2$ about the axis $y = -1$.

7.5 ARC LENGTH AND SURFACE AREA

In addition to the area and volume problems discussed in Sections 7.2 through 7.4, two other types of geometric problems can be treated by use of the definite integral. These are the problems of computing the length of a curve in the plane and the surface area of a solid of revolution. We treat both problems in this section and then turn to different settings for the remaining applications of the definite integral.

Arc Length

The problem here is to define and calculate the length of the arc of the graph of the function $y = f(x)$ from the point $(a, f(a))$ to the point $(b, f(b))$ (Figure 5.1). In the analysis that follows we will discover what specific conditions $f(x)$ need satisfy, but at the outset it seems necessary to assume at least that $f(x)$ be continuous for $a \le x \le b$.

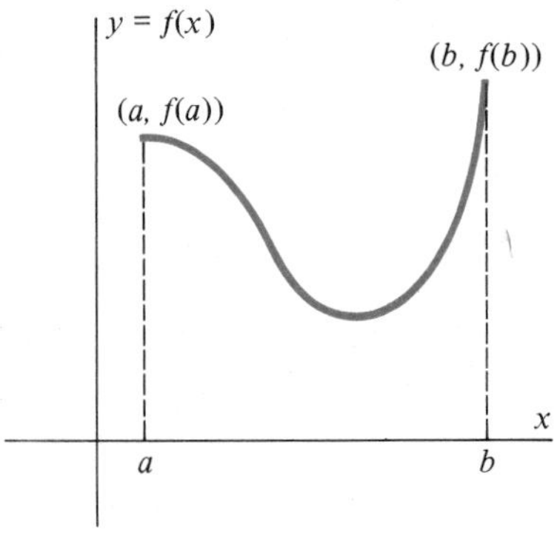

Figure 5.1 Arc whose length is to be calculated.

We approach this problem in a familiar way, by partitioning the interval $[a, b]$ into n subintervals of equal length. As before, we use partitions $a = x_0 < x_1 < x_2 < \cdots < x_n = b$, where $x_j - x_{j-1} = \Delta x = \dfrac{b-a}{n}$ for $j = 1, 2, \ldots, n$. On each subinterval $[x_{j-1}, x_j]$, we wish to approximate the length of the arc by an expression that we can easily calculate. We do this by connecting the endpoints of the arc, $P_{j-1} = (x_{j-1}, f(x_{j-1}))$ and $P_j = (x_j, f(x_j))$ with a line segment ℓ_j (see Figure 5.2).

We will approximate the length of the arc over the subinterval $[x_{j-1}, x_j]$ by the length of the line segment ℓ_j. The length of the entire arc is thus approximated by the length of the polygonal path through $P_0, P_1, P_2, \ldots, P_n$. If the limit of this approximation exists as $n \to \infty$, we will take this limit as the definition of the arc length in question.

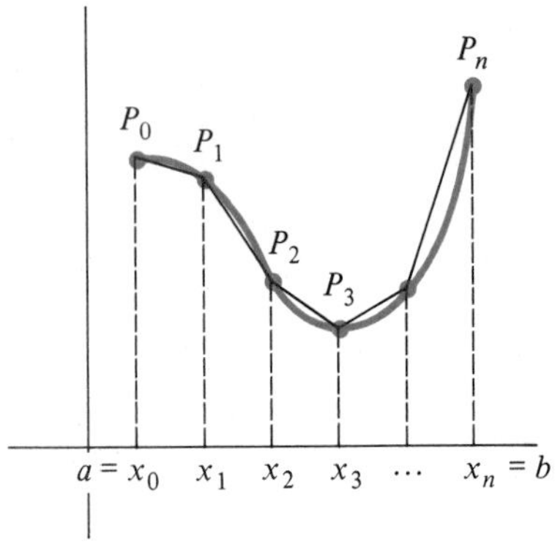

Figure 5.2 Approximating arc by polygonal path.

An application of the distance formula gives the length of ℓ_j as

$$\text{Length } \ell_j = \sqrt{(x_j - x_{j-1})^2 + [f(x_j) - f(x_{j-1})]^2} \tag{1}$$

$$= \sqrt{1 + \left[\frac{f(x_j) - f(x_{j-1})}{x_j - x_{j-1}}\right]^2} \cdot (x_j - x_{j-1}).$$

If $f(x)$ is differentiable on $[x_{j-1}, x_j]$, the Mean Value Theorem guarantees the existence of a number $t_j \in [x_{j-1}, x_j]$ so that

$$f'(t_j) = \frac{f(x_j) - f(x_{j-1})}{x_j - x_{j-1}}. \qquad \text{(See Figure 5.3.)}$$

Substituting this expression into equation (1) gives

$$\text{Length of } \ell_j = \sqrt{1 + [f'(t_j)]^2} \cdot \Delta x.$$

Summing these expressions over all $j = 1, 2, \ldots, n$ gives the approximation

$$\text{Arc length} \approx \sum_{j=1}^{n} \sqrt{1 + [f'(t_j)]^2}\, \Delta x. \tag{2}$$

Now the sum on the right side of approximation (2) will be a Riemann sum if the function $\sqrt{1 + [f'(t_j)]^2}$ is continuous. For this to be true, $f'(x)$ must be a continu-

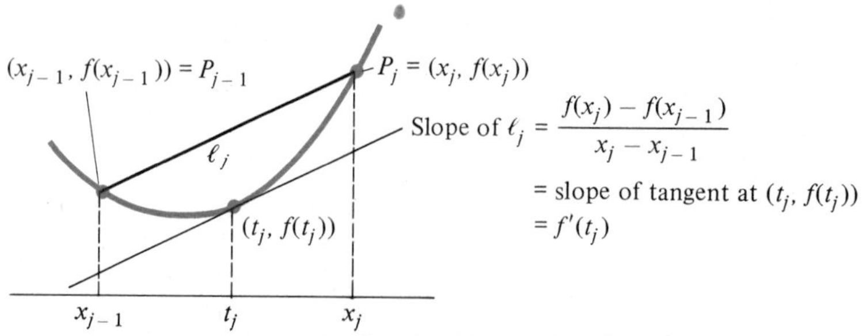

Figure 5.3 Slope of ℓ_j equals $f'(t_j)$ by Mean Value Theorem.

ous function on $[a, b]$. When this is the case, the Riemann sum converges to the corresponding integral as $n \to \infty$, which we take as the desired arc length L. That is, we *define*

$$L = \lim_{n\to\infty} \sum_{j=1}^{n} \sqrt{1 + [f'(t_j)]^2}\, \Delta x$$

$$= \int_a^b \sqrt{1 + [f'(x)]^2}\, dx.$$

We summarize our definition as follows:

If $f(x)$ has a continuous derivative $f'(x)$ on $[a, b]$, the length of the graph of $y = f(x)$ from $(a, f(a))$ to $(b, f(b))$ is given by

$$L = \int_a^b \sqrt{1 + [f'(x)]^2}\, dx \tag{3}$$

or, in Leibniz notation,

$$L = \int_a^b \sqrt{1 + \left[\frac{dy}{dx}\right]^2}\, dx.$$

Example 1 Verify that the expression for arc length in equation (3) agrees with the Euclidean distance formula for the case of a nonvertical line segment joining two points in the plane.

Strategy

First, determine the length according to the Euclidean distance formula.

Solution

Let $P = (x_1, y_1)$ and $Q = (x_2, y_2)$ be two such points. Then the Euclidean distance formula gives the length L of the line segment PQ as

$$L = \sqrt{(\Delta x)^2 + (\Delta y)^2} = \sqrt{(x_2 - x_1)^2 + (y_2 - y_1)^2}. \tag{4}$$

Find an equation for the line in question. Use form

$y - y_1 = m(x - x_1)$

where m = slope.

On the other hand, the equation for the line passing through P and Q is

$$y - y_1 = \frac{y_2 - y_1}{x_2 - x_1}(x - x_1)$$

so

Differentiate to find dy/dx.

$$\frac{dy}{dx} = \frac{y_2 - y_1}{x_2 - x_1}.$$

Equation (3) therefore gives the length of PQ as

Apply equation (3) to find length by integration.

$$L = \int_{x_1}^{x_2} \sqrt{1 + \left(\frac{y_2 - y_1}{x_2 - x_1}\right)^2}\, dx$$

$$= \left\{\sqrt{1 + \left(\frac{y_2 - y_1}{x_2 - x_1}\right)^2}\right\} \cdot x \Big]_{x_1}^{x_2}$$

$$= \sqrt{1 + \left(\frac{y_2 - y_1}{x_2 - x_1}\right)^2} \cdot (x_2 - x_1)$$

$$= \sqrt{(x_2 - x_1)^2 + (y_2 - y_1)^2},$$

Check that the two results agree.

which agrees with expression (4). ■

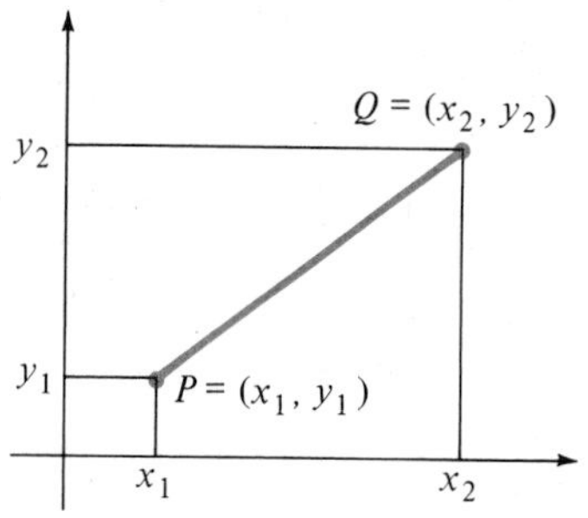

Figure 5.4 Euclidean distance agrees with arc length for line segments.

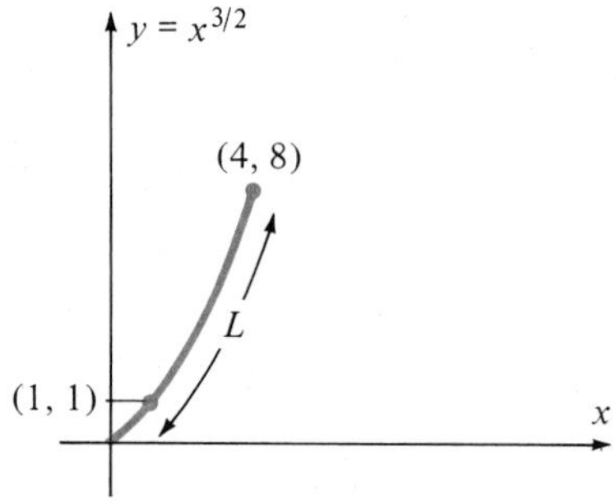

Figure 5.5

Example 2 Find the length of the arc of the graph of $y = x^{3/2}$ between (1, 1) and (4, 8).

Strategy
Check that dy/dx is continuous.

Apply equation (3).

Use the u-substitution $u = 1 + \frac{9}{4}x$; $du = \frac{9}{4}\,dx$. Then

$$\int \sqrt{1 + \frac{9}{4}x}\,dx = \int \sqrt{u}\cdot\frac{4}{9}\,du$$
$$= \frac{4}{9}\cdot\frac{2}{3}u^{3/2} + C$$
$$= \frac{4}{9}\cdot\frac{2}{3}\left(1 + \frac{9}{4}x\right)^{3/2} + C$$

is the antiderivative.

Solution
Here $\frac{dy}{dx} = \frac{3}{2}x^{1/2}$ is continuous on the interval [1, 4], so equation (3) applies. We obtain

$$L = \int_1^4 \sqrt{1 + \left[\frac{3}{2}x^{1/2}\right]^2}\,dx$$
$$= \int_1^4 \sqrt{1 + \frac{9}{4}x}\,dx$$
$$= \frac{4}{9}\cdot\frac{2}{3}\left(1 + \frac{9}{4}x\right)^{3/2}\Bigg]_1^4$$
$$= \frac{8}{27}\left[10^{3/2} - \left(\frac{13}{4}\right)^{3/2}\right] \approx 7.634.$$

(Figure 5.5.) ■

Example 3 Find the length of the arc of the graph of the equation

$$6xy - y^4 - 3 = 0 \tag{5}$$

from (19/12, 2) to (14/3, 3).

Strategy
Solve for x rather than y due to presence of y^4 term.

Solution
The form of the equation suggests that we solve for x as a function of y. Doing so we obtain

$$x = f(y) = \frac{1}{6}y^3 + \frac{1}{2y}.$$

Then

$$f'(y) = \frac{1}{2}y^2 - \frac{1}{2y^2}.$$

Rewrite equation (3) for functions of y.

The form of equation (3) for the arc of the graph of $x = f(y)$ connecting (a, c) and (b, d) is

$$L = \int_c^d \sqrt{1 + [f'(y)]^2}\, dy.$$

Apply equation obtained.

(By squaring the binomial term and factoring the result, the term under the radical can be brought into the form of a perfect square.)

We therefore obtain

$$\begin{aligned} L &= \int_2^3 \sqrt{1 + \left[\frac{1}{2}y^2 - \frac{1}{2y^2}\right]^2}\, dy \\ &= \int_2^3 \sqrt{1 + \left[\frac{1}{4}y^4 - \frac{1}{2} + \frac{1}{4y^4}\right]}\, dy \\ &= \int_2^3 \sqrt{\frac{1}{4}y^4 + \frac{1}{2} + \frac{1}{4y^4}}\, dy \\ &= \int_2^3 \sqrt{\left(\frac{1}{2}y^2 + \frac{1}{2y^2}\right)^2}\, dy \\ &= \int_2^3 \left(\frac{1}{2}y^2 + \frac{1}{2y^2}\right) dy \\ &= \frac{1}{6}y^3 - \frac{1}{2y}\Big]_2^3 \\ &= \frac{13}{3} - \frac{13}{12} = \frac{13}{4}. \end{aligned}$$

■

REMARK: You have no doubt observed that the integral in equation (3) is, in general, very difficult to evaluate by use of the Fundamental Theorem. Although we will study additional techniques of integration in Chapter 11, nevertheless we will remain unable to find convenient antiderivatives for most integrands arising from an application of equation (3). Thus, the procedures for approximating integrals will be needed in most practical applications of the formulas for arc length.

Surface Area

We saw in Section 7.3 that when the region bounded by the graph of $y = f(x)$ and the x-axis for $a \le x \le b$ is revolved about the x-axis, a solid of revolution results. We wish now to determine how to calculate the lateral surface area of such a solid.

As for the arc length problem, we partition $[a, b]$ with numbers $a = x_0 < x_1 < x_2 < \cdots < x_n = b$ where $x_j - x_{j-1} = \Delta x = \dfrac{b-a}{n}$, $j = 1, 2, \ldots, n$. We then connect successive points $P_j = (x_j, f(x_j))$ with line segments to form a polygonal approximation to the graph of $y = f(x)$ from $P_0 = (a, f(a))$ to $P_n = (b, f(b))$ (see Figure 5.6). As this polygonal path is revolved about the x-axis, it sweeps out a solid whose surface area we will take as an approximation to the surface area S of the solid in question (Figure 5.7).

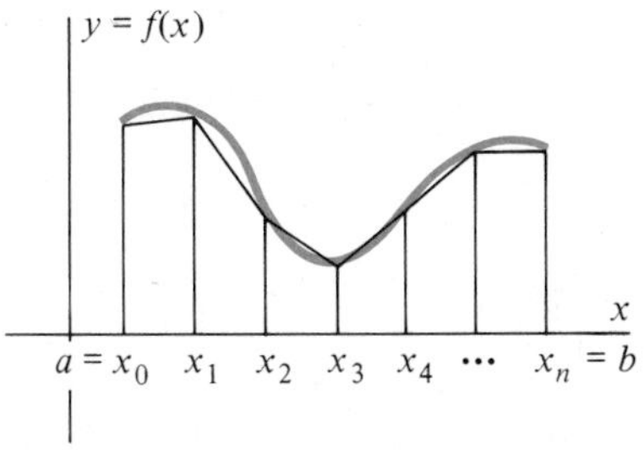

Figure 5.6 Polygonal approximation to graph of $y = f(x)$.

As the line segment joining P_{j-1} and P_j revolves about the x-axis (Figure 5.8), it sweeps out a frustum of a cone, with radii $f(x_{j-1})$ and $f(x_j)$, and a slant height

$$P_{j-1}P_j = \sqrt{(x_j - x_{j-1})^2 + [f(x_j) - f(x_{j-1})]^2}. \tag{6}$$

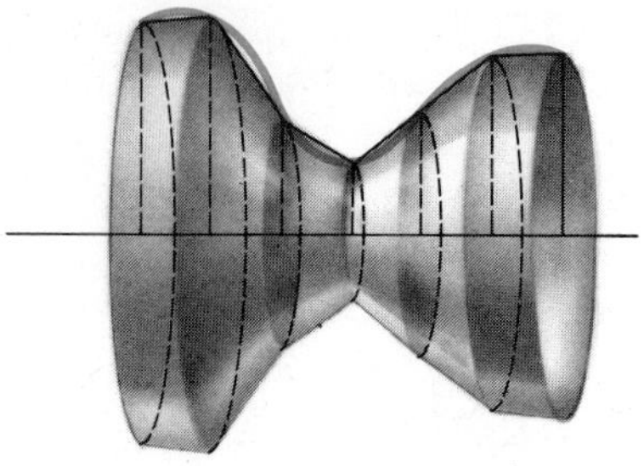

Figure 5.7 Figure approximating solid of revolution for surface area problem.

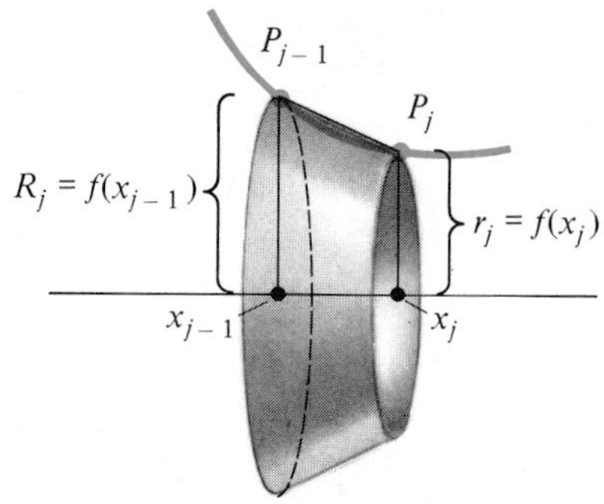

Figure 5.8 Element of surface area is frustum of cone with height Δx.

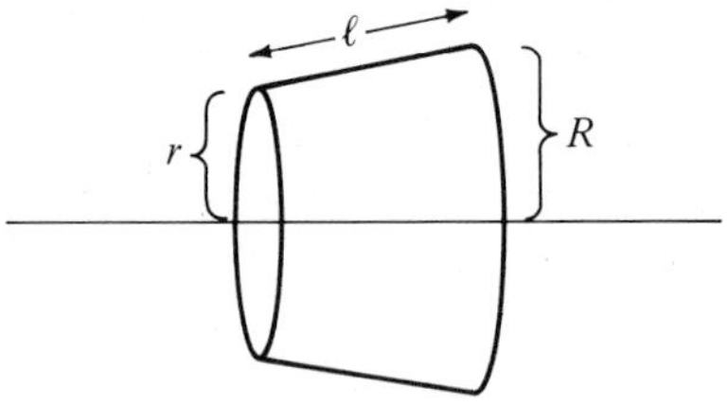

Figure 5.9 $s = \pi(r + R)\ell$.

Now the formula* for the lateral surface area of a frustum of a cone with radii r and R and slant height ℓ is $s = \pi(r + R)\ell$ (Figure 5.9), so the surface area of the frustum generated by $P_{j-1}P_j$ is

$$\Delta S_j = \pi(f(x_{j-1}) + f(x_j))\sqrt{(x_j - x_{j-1})^2 + [f(x_j) - f(x_{j-1})]^2} \tag{7}$$

$$= \pi(f(x_{j-1}) + f(x_j))\sqrt{1 + \left[\frac{f(x_j) - f(x_{j-1})}{x_j - x_{j-1}}\right]^2} \cdot \Delta x.$$

However, as was the case for the method of volumes by cylindrical shells, summing approximations of this type will *not* produce a Riemann sum. The difficulty again is that the function is being evaluated at more than one point in the interval $[x_{j-1}, x_j]$.

As in the derivation of the formula for arc length, we can use the Mean Value Theorem to write

$$\frac{f(x_j) - f(x_{j-1})}{x_j - x_{j-1}} = f'(t_j) \tag{8}$$

for some $t_j \in [x_{j-1}, x_j]$, $j = 1, 2, \ldots, n$ if $f(x)$ is continuous on $[a, b]$ and differentiable on (a, b). To simplify the expression $(f(x_{j-1}) + f(x_j))$, we observe that the number $y_j = \dfrac{f(x_{j-1}) + f(x_j)}{2}$ lies between the numbers $f(x_{j-1})$ and $f(x_j)$. If $f(x)$ is continuous on $[a, b]$, the Intermediate Value Theorem guarantees the existence of a number s_j in $[x_{j-1}, x_j]$ for which $f(x)$ assumes this intermediate value, i.e., for which

$$\frac{f(x_{j-1}) + f(x_j)}{2} = f(s_j). \tag{9}$$

Substituting the expressions in equations (8) and (9) for the corresponding factors in expression (7) gives

$$\Delta S_j = 2\pi f(s_j)\sqrt{1 + [f'(t_j)]^2} \cdot \Delta x. \tag{10}$$

The approximation for surface area is now obtained by summing expression (10) over all subintervals. We obtain

$$S \approx \sum_{j=1}^{n} \Delta S_j = \sum_{j=1}^{n} 2\pi f(s_j)\sqrt{1 + [f'(t_j)]^2}\,\Delta x. \tag{11}$$

*Exercise 25, Section 7.3.

Were the expression on the right side of approximation (11) a Riemann sum, we would now be in a position to take the limiting value of this approximation as our expression for surface area, that is, we would have

$$S = \lim_{n\to\infty} \sum_{j=1}^{n} 2\pi f(s_j)\sqrt{1 + [f'(t_j)]^2}\, \Delta x \tag{12}$$

$$= \int_a^b 2\pi f(x)\sqrt{1 + [f'(x)]^2}\, dx$$

when $f'(x)$ is continuous on $[a, b]$. Of course, there remains a difficulty, which is that the function in the approximating sum in (11) is being evaluated at two distinct points, s_j and t_j, in each interval $[x_{j-1}, x_j]$. Having come this close to establishing equation (12) we must, nevertheless, state that justifying equation (12) as the limiting value of approximation (11) requires a deeper treatment of the integral than we have developed here. However, under the assumption that $f'(x)$ exists and is continuous for all $x \in [a, b]$ it can be shown that equation (12) holds. (For a proof of this result the reader is referred to more advanced texts.) We therefore define surface area S according to equation (12).

> If the function $f(x)$ has a continuous derivative on the interval $[a, b]$, then the surface area S of the solid of revolution obtained by revolving about the x-axis the region bounded by the graph of $y = f(x) \geq 0$ and the x-axis, for $a \leq x \leq b$, is
>
> $$S = \int_a^b 2\pi f(x)\sqrt{1 + [f'(x)]^2}\, dx \tag{13}$$
>
> or, in Leibniz notation
>
> $$S = \int_a^b 2\pi y\sqrt{1 + \left[\frac{dy}{dx}\right]^2}\, dx. \tag{14}$$

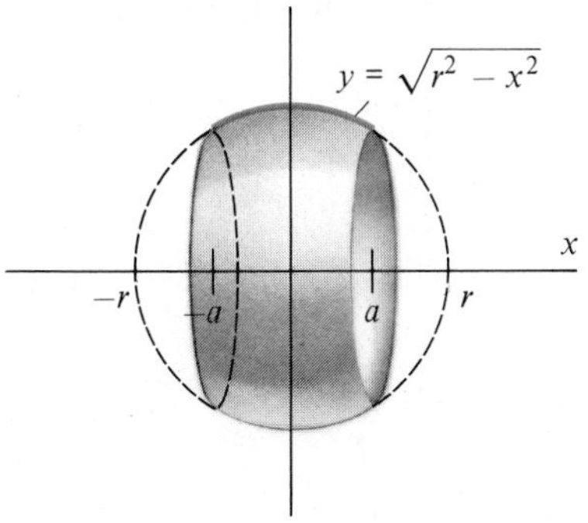

Figure 5.10

REMARK: It is important to observe that we require $y = f(x) \geq 0$ in equations (13) and (14). If $f(x)$ is negative for some $x \in [a, b]$, we must replace $f(x)$ by $|f(x)|$ in (13) and y by $|y|$ in (14).

Example 4 Find the surface area of the band of the sphere generated by revolving the arc of the circle $x^2 + y^2 = r^2$ lying above the interval $[-a, a]$, $a < r$, about the x-axis (Figure 5.10).

Solution: Here $y = \sqrt{r^2 - x^2}$, so

$$\frac{dy}{dx} = \frac{-x}{\sqrt{r^2 - x^2}}.$$

Using equation (14), we obtain

$$S = \int_{-a}^{a} 2\pi\sqrt{r^2 - x^2}\sqrt{1 + \left[\frac{-x}{\sqrt{r^2 - x^2}}\right]^2} \cdot dx$$

$$= \int_{-a}^{a} 2\pi\sqrt{r^2 - x^2}\sqrt{\frac{(r^2 - x^2) + x^2}{r^2 - x^2}} \cdot dx$$

$$= \int_{-a}^{a} 2\pi\sqrt{r^2}\, dx$$
$$= 2\pi rx]_{-a}^{a}$$
$$= 4\pi ar$$
■

Example 5 Why is it incorrect in Example 4 to set $a = r$ and conclude that the surface area of the sphere of radius r is $S = 4\pi r^2$?

Solution: Although the conclusion is correct, the reasoning is faulty. Applying formula (14) requires that the function $y = f(x)$ have a continuous derivative for all x in the interval in question. Since $\frac{dy}{dx} = \frac{-x}{\sqrt{r^2 - x^2}}$ is undefined for $x = r$, equation (14) cannot be applied on the interval $[-r, r]$. We will obtain a satisfactory solution to this problem in Chapter 16 when we study curves described by parametric equations. ■

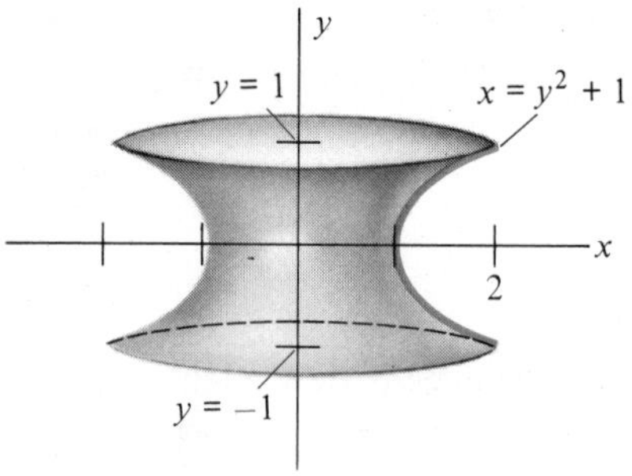

Figure 5.11

Example 6 Find an integral giving the surface area of the "spool" obtained by revolving the region bounded by the graph of $y^2 - x + 1 = 0$ and the y-axis, for $-1 \leq y \leq 1$, about the y-axis (Figure 5.11).

Solution: Solving for x as a function of y we obtain

$$x = y^2 + 1$$

so

$$\frac{dx}{dy} = 2y.$$

Since dx/dy is continuous for $-1 \leq y \leq 1$, we may use the analogue of equation (14) for rotations about the y-axis. We obtain

$$S = \int_{-1}^{1} 2\pi x\sqrt{1 + \left[\frac{dx}{dy}\right]^2}\, dy, \qquad \text{or}$$

$$S = \int_{-1}^{1} 2\pi(y^2 + 1)\sqrt{1 + 4y^2}\, dy. \tag{15}$$

As will be the case for many functions to which equation (13) is applied, we cannot find a convenient antiderivative for the integrand in the above integral for surface area. ■

Example 7 Use the Trapezoidal Rule to approximate the surface area of the solid described in Example 6.

Solution: The purpose of this example is to remind you that, even though the integral in Example 6 cannot be evaluated precisely via the Fundamental Theorem, nevertheless we have available several procedures (including Simpson's Rule) for approximating such integrals to any desired degree of accuracy.

We denote the integrand in integral (15) as

$$f(y) = 2\pi(y^2 + 1)\sqrt{1 + 4y^2}$$

and we subdivide the interval $[a, b] = [-1, 1]$ into n equal subintervals of length $\dfrac{1-(-1)}{n} = \dfrac{2}{n}$. Then the endpoints of these intervals are

$$y_0 = -1, \qquad y_1 = -1 + \frac{2}{n},$$
$$y_2 = -1 + (2)\cdot\frac{2}{n}, \; \ldots, \; y_n = -1 + (n)\left(\frac{2}{n}\right) = 1$$

and the formula for the Trapezoidal Rule (Section 6.7) becomes

$$A = \frac{1}{n}\left\{2\pi\cdot 2\cdot\sqrt{5} + \sum_{j=1}^{n-1} 2^2\pi\left[\left(-1+\frac{2j}{n}\right)^2 + 1\right]\cdot \sqrt{1 + 4\left(-1+\frac{2j}{n}\right)^2} + 2\pi\cdot 2\cdot\sqrt{5}\right\}.$$

From Section 6.7, we know that the error associated with this calculation will be no greater than

$$\frac{(b-a)^3\cdot M}{12n^2} = \frac{[1-(-1)]^3}{12n^2}\cdot M = \frac{2M}{3n^2} \tag{16}$$

where M is the maximum value of $|f''(y)|$ on $[-1, 1]$.

A calculation shows that

$$f''(y) = \frac{12\pi(1 + 6y^2 + 16y^4)}{(1+4y^2)^{3/2}}. \tag{17}$$

Rather than using the theory of Chapter 5 to obtain the precise value of M, we can do a ''worst case'' analysis by observing that the denominator in the expression for $f''(y)$ can never be smaller than 1, and for $-1 \le y \le 1$, the numerator cannot exceed $12\pi\cdot(1+6+16) = 276\pi$. Thus the error bound in equation (16) cannot exceed

$$E = \frac{2\cdot 276\pi}{3n^2} = \frac{184\pi}{n^2}.$$

Implementing these findings on a computer produced the results contained in Table 5.1.

Of course, care must be taken in using large values of n (such as in the last two rows of Table 5.1) that roundoff error is not affecting the accuracy of the approximation. This question is addressed in courses on numerical analysis. You may verify that for a given value of the error bound E, fewer subintervals (smaller n) are required by Simpson's Rule than by the Trapezoidal Rule (see Section 6.7).

Table 5.1

n	Trapezoidal approximation	Error bound
5	27.5483	23.1221
10	26.5413	5.7805
25	26.2583	0.9249
50	26.2179	0.2312
100	26.2078	0.0578
200	26.2052	0.0145

Exercise Set 7.5

In each of Exercises 1–11, find the length of the arc determined by the given equation.

1. $y = 2x + 3,$ from $(-3, -3)$ to $(2, 7)$
2. $y = x^{3/2},$ from $(0, 0)$ to $(4, 8)$
3. $y - 2 = (x + 2)^{3/2}$ from $(-2, 2)$ to $(2, 10)$
4. $3y = (x^2 + 2)^{3/2}$ for $0 \le x \le 2$
5. $3y = 2(x + 1)^{3/2}$ from $(0, 2/3)$ to $(3, 16/3)$
6. $y = \dfrac{x^3}{3} + \dfrac{1}{4x}$ for $1 \le x \le 3$
7. $(3y - 6)^2 = (x^2 + 2)^3$ for $0 \le x \le 3$
8. $y = \dfrac{x^4}{4} + \dfrac{1}{8x^2}$ for $1 \le x \le 2$
9. $y = \dfrac{1}{6}x^3 + \dfrac{1}{2}x^{-1},$ from $\left(1, \dfrac{2}{3}\right)$ to $\left(4, \dfrac{259}{24}\right)$
10. $y = \dfrac{2}{3}x^{3/2} - \dfrac{1}{2}x^{1/2},$ from $\left(1, \dfrac{1}{6}\right)$ to $\left(3, \dfrac{3\sqrt{3}}{2}\right)$
11. $y = \dfrac{x^3}{6} + \dfrac{1}{2x},$ from $\left(1, \dfrac{2}{3}\right)$ to $\left(3, \dfrac{14}{3}\right)$

In each of Exercises 12–20, find the area of the lateral surface of the solid generated by revolving the stated region about the given axis.

12. The region bounded by the graph of $y = 2x + 3$ and the x-axis for $1 \le x \le 3$ is revolved about the x-axis.
13. The region bounded by the graph of $y = x^3$ and the x-axis for $0 \le x \le 3$ is revolved about the x-axis.
14. The region bounded by the graph of $y = x^{1/3}$ and the y-axis for $0 \le y \le 2$ is revolved about the y-axis.
15. The region bounded by the graph of $x = 2\sqrt{y}$ and the y-axis for $1 \le y \le 4$ is revolved about the y-axis.
16. The region bounded by the graph of $y = \sqrt{x}$ and the x-axis for $1 \le x \le 2$ is revolved about the x-axis.
17. The region bounded by the graph of $y = \sqrt{2x - 1}$ and the x-axis for $2 \le x \le 8$ is revolved about the x-axis.
18. The region bounded by the graph of $12xy - 3y^4 = 4$ and the y-axis for $1 \le y \le 3$ is revolved about the y-axis.
19. The region bounded by the graph of $y = \dfrac{x^3}{4} + \dfrac{1}{3x}$ and the x-axis for $1 \le x \le 3$ is revolved about the x-axis.
20. *(Calculator/Computer)* The small arc of the circle with equation $x^2 + (y - 1)^2 = 25$ from $(3, 5)$ to $(4, 4)$ is revolved about the x-axis. Use the Trapezoidal Rule to approximate the lateral surface area of the resulting solid.
21. *(Calculator/Computer)* A circle has center $(1, 1)$ and radius 5. The small arc joining the points $(4, 5)$ and $(5, 4)$ on the circle is revolved about the x-axis. Use the Trapezoidal Rule to approximate the lateral surface area of the resulting solid.
22. *(Calculator/Computer)* Write down an integral giving the length of the arc of the graph of $f(x) = \sin x$ between $(0, 0)$ and $(\pi/2, 1)$. Use the Trapezoidal Rule with $n = 10$ to approximate the value of this integral.
23. *(Calculator/Computer)* Write down an integral giving the length of the graph of the function $f(x) = x^2$ lying between $(-1, 1)$ and $(1, 1)$. Approximate this integral using Simpson's Rule with $n = 10$.
24. *(Calculator/Computer)* Write down an integral giving the lateral surface area of the figure obtained by revolving the graph of the ellipse $\dfrac{x^2}{a^2} + \dfrac{y^2}{b^2} = 1$, for $-\dfrac{a}{2} \le x \le \dfrac{a}{2}$, about the x-axis. Approximate the value of this integral for $a = 4$ and $b = 3$ using Simpson's Rule with $n = 10$.

SUMMARY OUTLINE OF CHAPTER 7

- The area of the region bounded between the graphs of $f(x)$ and $g(x)$ for $a \le x \le b$ is

$$A = \int_a^b |f(x) - g(x)|\, dx.$$

- The volume of the solid whose cross-sectional area is $A(x)$ and that lies between $x = a$ and $x = b$ is

$$V = \int_a^b A(x)\, dx.$$

- The volume of the solid obtained by revolving about the x-axis the region bounded by the graph of $y = f(x) \ge 0$ and the x-axis for $a \le x \le b$ is

$$V = \int_a^b \pi[f(x)]^2\, dx.$$

- The volume of the solid obtained by rotating about the y-axis the region bounded by the graph of $y = f(x) \geq 0$ and the x-axis for $a \leq x \leq b$ is

$$V = \int_a^b 2\pi x f(x)\,dx.$$

- The length of the graph of $y = f(x)$ from $(a, f(a))$ to $(b, f(b))$ is

$$L = \int_a^b \sqrt{1 + [f'(x)]^2}\,dx.$$

- The lateral surface area of the solid obtained by rotating about the x-axis the region bounded by the graph of $y = f(x) \geq 0$, and the x-axis for $a \leq x \leq b$ is

$$S = \int_a^b 2\pi f(x)\sqrt{1 + [f'(x)]^2}\,dx.$$

REVIEW EXERCISES—CHAPTER 7

In Exercises 1–12, find the area of the region bounded by the graphs of the given equations.

1. $y = x^3, \quad y = 0, \quad x = -2, \quad x = 2$

2. $y = 1 - x^2, \quad y = -x - 1, \quad x = -2, \quad x = 1$

3. $y = \sqrt{x}, \quad y = -\sqrt{x}, \quad x = 0, \quad x = 4$

4. $y = x^2, \quad y = 2 - x, \quad x = -2, \quad x = 1$

5. $y = x^2 - 4x + 2, \quad x + y = 6$

6. $y = 2 - 2x - x^2, \quad x = -y$

7. $y = \dfrac{x - 2}{\sqrt{x^2 - 4x + 8}}, \quad x = 0, \quad y = 0$

8. $x - y^2 + 3 = 0, \quad x - 2y = 0$

9. $x + y^2 = 0, \quad x + y + 2 = 0$

10. $x = 2y^2 - 3, \quad x = y^2 + 1$

11. $x = y^{2/3}, \quad x = y^2$

12. $y = 4x^2, \quad 4x + y - 8 = 0$

13. Find the area of the region bounded by the parabola $y = x^2 + x - 2$ and the line through $(-1, -2)$ and $(1, 0)$.

In Exercises 14–25, use integration to find the volume of the solid S described.

14. The sphere of radius r.

15. The solid obtained by rotating about the x-axis the region bounded by the line $x = 4$ and the parabola $y^2 = x$.

16. The solid obtained by rotating about the x-axis the region bounded by the graph of $y = 1/x$ and the x-axis for $1 \leq x \leq 4$.

17. The solid obtained by rotating about the x-axis the region bounded by the graph of $y = \sec\left(\dfrac{\pi x}{2}\right)$ and the x-axis for $-\dfrac{1}{2} \leq x \leq \dfrac{1}{2}$.

18. The solid obtained by rotating about the x-axis the region bounded by the graph of $y = \sin x$ and the x-axis for $0 \leq x \leq 3\pi/2$.

19. The solid obtained by rotating about the x-axis the region bounded by the graphs of $f(x) = \sqrt{1 - x^2}$ and $g(x) = 1/2$.

20. The solid obtained by rotating about the y-axis the region bounded by the graph of $f(x) = x^{2/3}$ and the x-axis for $0 \leq x \leq 2$.

21. The solid obtained by rotating about the y-axis the region bounded by the graph of $y = \sin x^2$ and the x-axis for $0 \leq x \leq \sqrt{\pi}/2$.

22. The solid obtained by rotating about the y-axis the region bounded by the graphs of $y_1 = \sin x^2$ and $y_2 = \cos x^2$ for $0 \leq x \leq \sqrt{\pi}/2$.

23. The solid obtained by rotating about the line $y = -2$ the region bounded by the graph of $y = 1/x^2$ and the x-axis for $1 \leq x \leq 4$.

24. The solid obtained by rotating about the line $y = -3$ the region bounded by the graph of $y = x^{2/3} + 1$ and the x-axis for $0 \leq x \leq 2$.

25. The solid obtained by rotating about the line $x = 2$ the region bounded by the graph of $y = 3 + x^3$ and the lines $x = 0$ and $y = 0$.

In Exercises 26–29, find the length of the graph of the given equation between the indicated points.

26. $y = 1 + x^{3/2}$ from $(0, 1)$ to $(4, 9)$

27. $x^2 = y^3$ from $(0, 0)$ to $(1, 1)$

28. $y^2 = (x + 1)^3$ from $(0, 1)$ to $(1, \sqrt{8})$

29. $y = \dfrac{1}{8}\left(x^4 + \dfrac{2}{x^2}\right)$ from $(1, 3/8)$ to $(2, 33/16)$

30. Find the volume of a triangular pyramid with base area A and height h.

31. The base of a solid is the region bounded by the ellipse $x^2 + \dfrac{y^2}{4} = 1$. Find the volume given that the cross sections perpendicular to the x-axis are

a. squares,

b. equilateral triangles.

32. A water tank has the shape of a hemisphere of radius r. To what percent capacity is it filled when the water has depth $\frac{1}{2}r$?

33. Find the area of the region bounded by the parabola $y = 2 - x - x^2$ and the line through $(-1, 2)$ and $(1, 0)$.

34. Find the area of the region bounded by the graphs of the equations $3x + 5y = 23$, $5x - 2y = 28$, and $2x - 7y = -26$.

35. Find the area of the region bounded by the graphs of $y = \sin 2x$ and $y = \cos x$ for $0 \leq x \leq 2\pi$.

36. Find the area of the region bounded by the graphs of $y = x^3 - x$ and $y = x - x^2$.

37. *(Calculator/Computer)* Approximate the length of the graph of $f(x) = \cos^2 x$ from $x = 0$ to $x = \pi$.

38. *(Calculator/Computer)* Approximate the lateral surface area of the solid obtained when one arc ($0 \leq x \leq \pi$, say) of the graph of $f(x) = \sin x$ is revolved about the x-axis.

CHAPTER 8

FURTHER APPLICATIONS OF THE DEFINITE INTEGRAL

8.1 INTRODUCTION

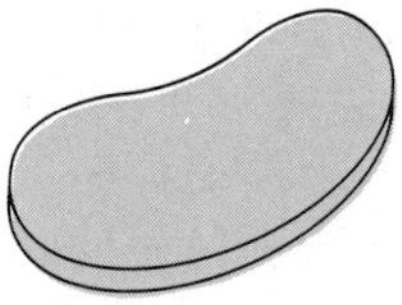

Figure 1.1 Plate P.

From the purely mathematical concerns of the preceding chapter, we turn now to several important questions in engineering and physics. We can give simple illustrations of these questions at the outset by imagining a plate P with surface area A and of uniform thickness and density (Figure 1.1). The three questions we will raise are the following:

(a) If P is submerged in water, what is the total force exerted on P by the water (Figure 1.2)? (This is a particularly important question if P is part of the wall of a dam or a nuclear reactor.)
(b) How much work is done on P by lifting it a distance d (Figure 1.3)? (This type of question is directly related to questions about the capacity of machines to perform certain tasks and the energy required to do so.)
(c) What is the location on the face of P of the point where P balances (Figure 1.4)? (This question is particularly important for engineers because calculations can often be greatly simplified by assuming that an object is concentrated at its "center of gravity.")

We shall make a rather careful study of each of these questions and then conclude the chapter with the unifying topic of the average value of a function.

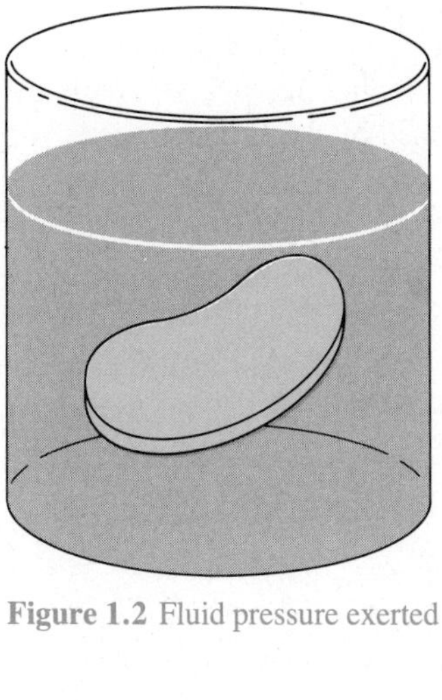

Figure 1.2 Fluid pressure exerted on P.

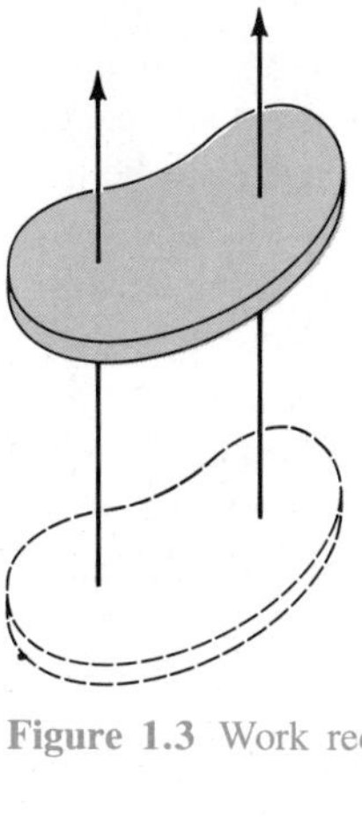

Figure 1.3 Work required to raise P.

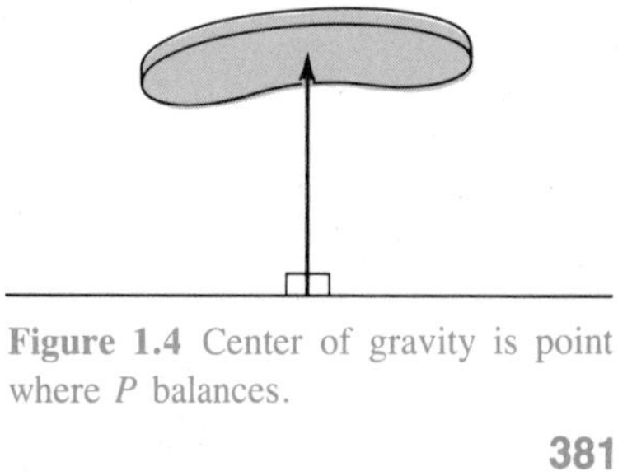

Figure 1.4 Center of gravity is point where P balances.

8.2 HYDROSTATIC PRESSURE

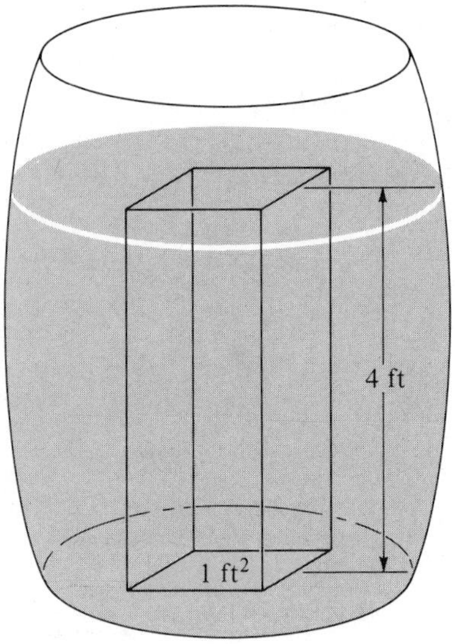

Figure 2.1

The word hydrostatic means a property of still fluids (as opposed to hydrodynamic, which refers to properties of moving fluids). Our interest here is in calculating pressures and forces exerted by water on the walls of containers.*

The **pressure** acting on an object is defined to be the force per unit area acting on that object. For example, if a rain barrel is filled to a depth $h = 4$ feet with water, then above each square foot of the bottom of the barrrel extends a rectangular column of water of volume 1 ft $\times$ 1 ft $\times$ 4 ft $=$ 4 ft^3. Since the density of water is 62.4 lb/ft^3, the pressure exerted on the bottom of the barrel by the column of water is

$$\begin{aligned} p &= \text{density} \times \text{height} \\ &= (62.4 \text{ lb/ft}^3) \times (4 \text{ ft}) \\ &= 249.6 \text{ lb/ft}^2. \end{aligned}$$

(Figure 2.1.)

The total force F exerted on the bottom of the barrel is found by multiplying the area A of the base by the pressure p. If, in this example, the base is a circle of radius 2 ft then

$$A = \pi r^2 = 4\pi \text{ ft}^2.$$

so

$$\begin{aligned} F = pA &= (249.6 \text{ lb/ft}^2) \cdot (4\pi \text{ ft}^2) \\ &= 998.4\pi \text{ lb}. \end{aligned}$$

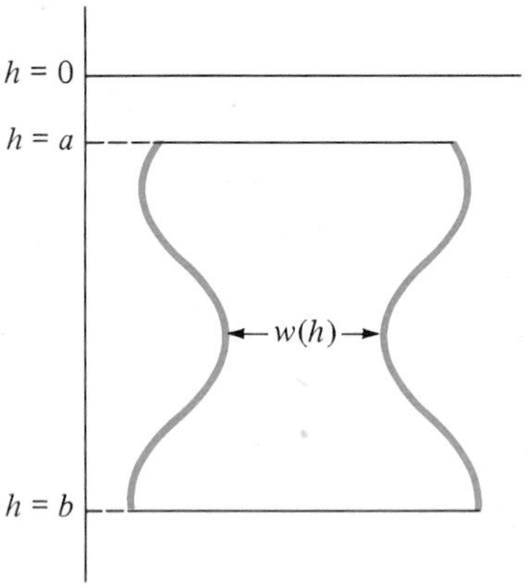

Figure 2.2 Face of submerged object; $w(h)$ is the width at depth h.

The problem of calculating the force acting on the side wall of the barrel is less straightforward. The preceding discussion shows that pressure depends on depth. However, **Pascal's Principle** states that the pressure exerted by a fluid on a submerged particle acts equally in all directions. This means, for example, that the hydrostatic pressure at a point Q within the barrel is independent of whether Q is located along a side wall, at the bottom, or even at a location entirely surrounded by water. In other words, the hydrostatic pressure at Q depends *only* on depth. However, the calculation of the total force acting on the side wall involves more than a simple multiplication, since the wall exists at various depths.

To pursue this question, we idealize it somewhat by considering a submerged object whose face consists of a region bounded above by the line $h = a$, below by the line $h = b$, and whose width at depth h is determined by the function $w(h)$. We take the surface of the water to lie at depth $h = 0$, we take the positive h-axis to extend vertically downward, and we assume that the face of the object is sitting vertically in the fluid (see Figure 2.2).

Since pressure varies continuously with the depth, we partition the interval $[a, b]$ and develop an *approximation* to the total force by assuming the pressure to be constant on each of the resulting rectangles.

To do this, we use a regular partition $a = h_0 < h_1 < h_2 < \cdots < h_n = b$, and we select one arbitrary number t_j in each interval $[h_{j-1}, h_j]$. We then make two types

*The same principles apply to all fluids. In order to reduce the number of constants (such as density), we will concentrate on the most common fluid, water.

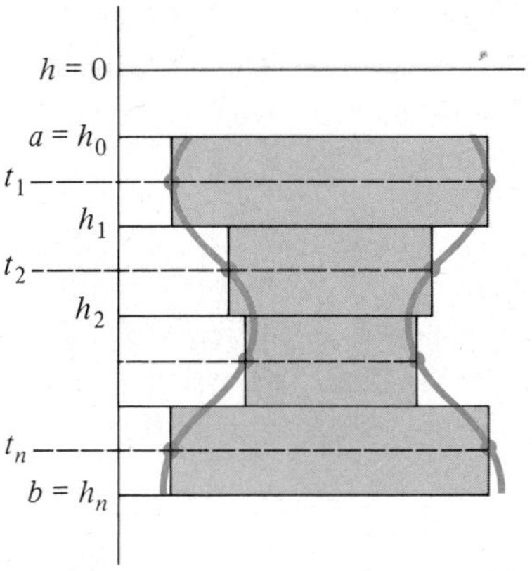

Figure 2.3 Approximating rectangles and test points t_j.

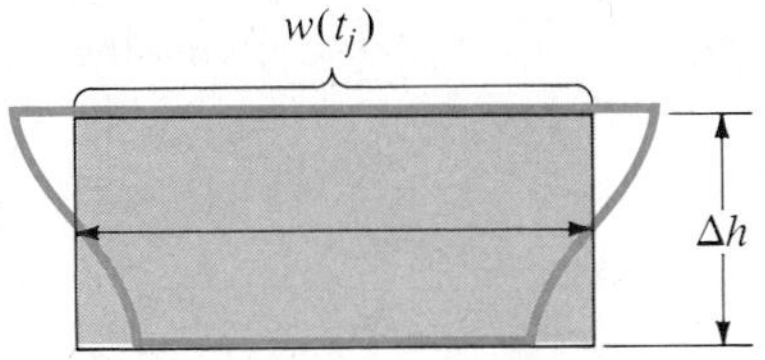

Figure 2.4 jth approximating rectangle.

of approximations:

1. We approximate the slice of the face lying between depth h_{j-1} and depth h_j by the rectangle whose width is $w(t_j)$ and whose height is $\Delta h = \dfrac{b-a}{n}$ (see Figures 2.3 and 2.4).
2. We approximate the depth of each point in the jth rectangle by the depth t_j.

That is, we assume the width $w(t)$ to be the constant $w(t_j)$ throughout the jth rectangle, and we assume the pressure throughout the jth rectangle to be constant and equal to $p(t_j) = (62.4)t_j$. (Recall that the factor 62.4 is the density of water in lb/ft^3.) Since the jth rectangle has area $\Delta A_j = w(t_j)\Delta h$, the force ΔF_j acting on the jth rectangle is

$$\Delta F_j = (62.4)t_j \Delta A_j = (62.4)t_j w(t_j)\Delta h, \qquad \Delta h = \frac{b-a}{n}.$$

Finally, we approximate the total force F acting on the face by the sum of these individual forces:

$$F \approx \sum_{j=1}^{n} \Delta F_j = \sum_{j=1}^{n} (62.4)t_j w(t_j)\Delta h. \tag{1}$$

If the width of the face is a continuous function of depth for $a \le h \le b$, then the function $(62.4)hw(h)$ in approximation (1) (which is being approximated by the constants $t_j w(t_j)$) is a continuous function of h. Consequently, the sum on the right-hand side of approximation (1) is a Riemann sum that, since the function $w(h)$ is continuous, must have a limit as $n \to \infty$. We *define* this limit as the total force acting on the face, that is, we take

$$\begin{aligned} F &= \lim_{n\to\infty} \sum_{j=1}^{n} (62.4)t_j w(t_j)\,\Delta h \\ &= \int_a^b (62.4)hw(h)\,dh. \end{aligned} \tag{2}$$

Equation (2) states that the total force in pounds acting on a submerged vertical face lying between depths $h = a$ and $h = b$ and whose width at depth h is $w(h)$ (all dimensions in feet) is found by integrating the product $(62.4)hw(h)$ over the interval $[a, b]$.

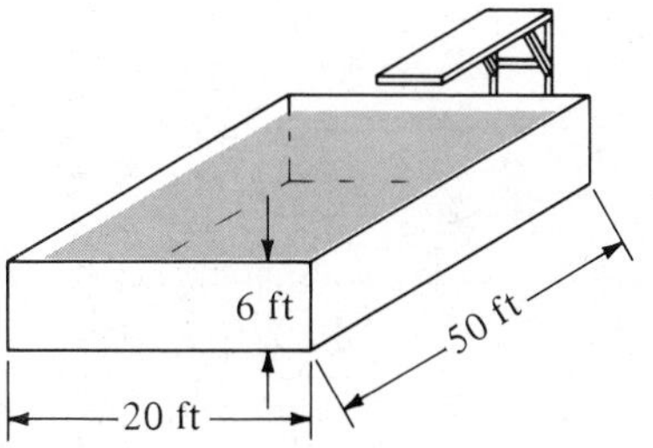

Figure 2.5 Swimming pool.

Example 1 An above-ground swimming pool has the shape of a rectangular box 50 feet long, 20 feet wide, and 6 feet high (Figure 2.5). Find the force exerted on an end wall when the pool is filled to a depth of 5 feet (Figure 2.6).

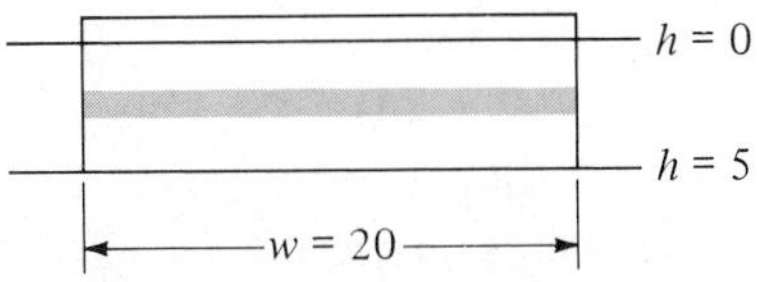

Figure 2.6 End wall of pool.

Solution: We take the water level to be depth $h = 0$. Then the submerged portion of the wall is a rectangle, extending from depth $h = 0$ to depth $h = 5$ feet. The width, $w = 20$ feet, is constant. An application of formula (2) gives

$$\begin{aligned} F &= \int_0^5 (62.4)h(20)\,dh \\ &= 624h^2\Big]_0^5 \\ &= 15{,}600 \text{ pounds}. \end{aligned}$$

■

Example 2 The lower portion of the vertical wall of a dam is a trapezoid of upper base 40 feet, lower base 20 feet, and height 5 feet. Find the force acting on this portion of the wall if the water level is 4 feet above the upper base (see Figure 2.7).

Strategy
Sketch the face.

Establish a coordinate axis for h.

Use similar triangles to find an equation for $w(h)$.

Solution
We sketch the trapezoid extending from depth $h = 4$ to depth $h = 9$ as in Figure 2.8. To determine the width $w(h)$ of the trapezoid as a function of depth, we construct a line segment from point A parallel to one side of the trapezoid. This results in two similar triangles with common vertex, A. The altitude of the smaller triangle is $(9 - h)$ and the altitude of the larger triangle is 5. If we let x denote the base of the smaller triangle, we have

$$\frac{x}{20} = \frac{9-h}{5}$$

so

$$\begin{aligned} x &= 4(9-h) \\ &= 36 - 4h. \end{aligned}$$

Since $w = 20 + x$, we have

$$\begin{aligned} w(h) &= 20 + (36 - 4h) \\ &= 56 - 4h. \end{aligned}$$

Apply equation (2).

This is indeed correct, since (a) $w(h)$ is linear, (b) $w(4) = 40$, and (c) $w(9) = 20$. We may now apply equation (2) to conclude that

$$\begin{aligned} F &= \int_4^9 (62.4)h(56 - 4h)\,dh \\ &= 62.4\int_4^9 (56h - 4h^2)\,dh \\ &= 62.4(28h^2 - \frac{4}{3}h^3\Big]_4^9 \\ &\approx 58{,}240 \text{ pounds.} \end{aligned}$$

■

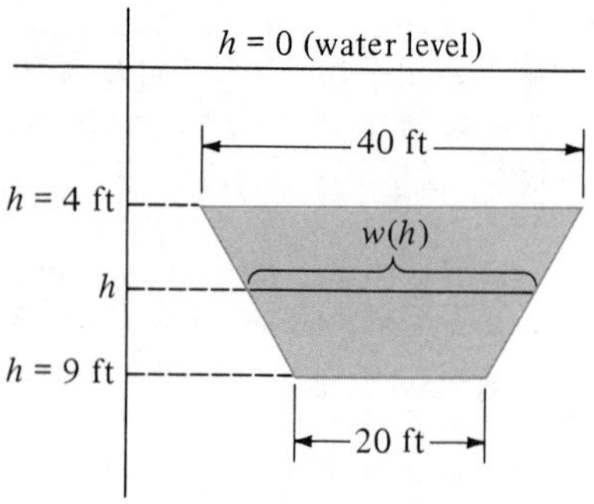

Figure 2.7 Trapezoid lies 4 ft below water level.

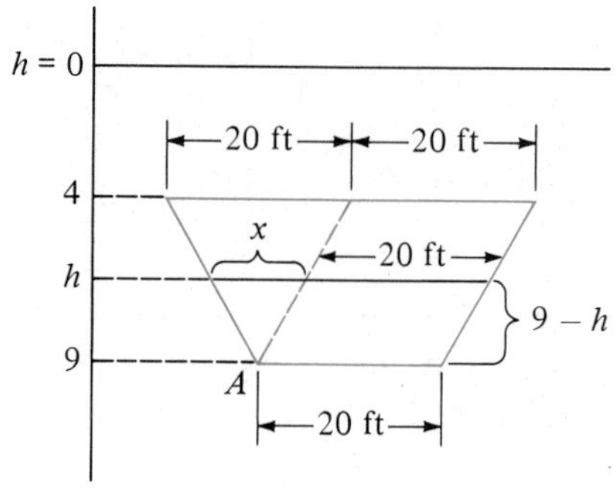

Figure 2.8 $w(h) = 20 + x$.

Example 3 A submarine has a circular window of radius 6 inches. Find the hydrostatic force on the window when the submarine is submerged so that the top of the window lies 20 feet below water level. (Assume the window to be mounted vertically.)

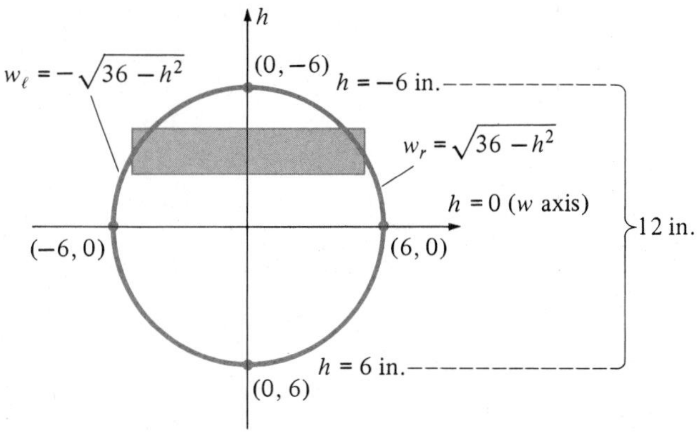

Figure 2.9 A *wh* coordinate system for submarine window.

Strategy

Use a two-dimensional coordinate system to describe width as a function of depth.

Write standard form for equation of circle.

Solve for w as a function of h.

Solution

If we impose a wh-coordinate system at the center of the window (Figure 2.9), we can write the equation for the edge of the window as

$$w^2 + h^2 = 36.$$

Solving for w as a function of h, we find that the right edge of the window is described by the function

$$w_r = \sqrt{36 - h^2}\,, \qquad -6 \le h \le 6$$

and the left by

$$w_\ell = -\sqrt{36 - h^2}\,, \qquad -6 \le h \le 6.$$

The width of the window is therefore

$$w(h) = w_r - w_\ell = 2\sqrt{36 - h^2}\,, \qquad -6 \le h \le 6. \tag{3}$$

To calculate the correct value of h in equation (2), note that the center of the coordinate system ($h = 0$) corresponds to a depth of 20 ft, 6 in = 246 in.

Apply equation (2).

Since the top of the window lies 20 feet below the water surface, the depth of a horizontal element of area varies from 20 ft to 21 ft as the variable h varies from -6 inches to 6 inches. We must therefore use the quantity $(246 + h)$ inches in place of the factor h in equation (2). Using this observation together with equations (2) and (3) and converting the weight of water from 62.4 pounds per cubic foot to $62.4/1728 = .0361$ pounds per cubic inch, we find that

$$F = \int_{-6}^{6} (.0361)(246 + h)(2)\sqrt{36 - h^2}\,dh \text{ pounds.} \tag{4}$$

$$= (2)(246)(.0361)\int_{-6}^{6} \sqrt{36 - h^2}\,dh + (2)(.0361)\int_{-6}^{6} h\sqrt{36 - h^2}\,dh.$$

Evaluate second integral using u-substitution.

The second integral may be evaluated by a u-substitution:

Let $u = 36 - h^2$; $\quad du = -2h\,dh$.
Then $u = 0$ when $h = 6$ and $u = 0$ when $h = -6$.

Thus,

$$\int_{-6}^{6} h\sqrt{36 - h^2}\,dh = \int_{0}^{0} \sqrt{u}\left(-\frac{1}{2}\right) du = 0. \tag{5}$$

(This result was to have been expected, since $h\sqrt{36 - h^2}$ is an *odd* function, and the integral of any odd function over an interval of the form $[-a, a]$ is zero.)

Evaluate first integral geometrically.

We do not yet have available a technique for finding an antiderivative for the first integral in (4). However, the definite integral itself may be interpreted as the area of the semicircle bounded above by the graph of $y = \sqrt{36 - x^2}$ and below by the x-axis (Figure 2.10). Since the radius of this semicircle is $r = 6$, we have

$$\int_{-6}^{6} \sqrt{36 - h^2}\, dh = \frac{1}{2}\pi \cdot 6^2 = 18\pi. \tag{6}$$

Combine results.

From equations (4), (5), and (6) we now have

$$F = (2)(246)(.0361)(18\pi) + 2(.0361)(0)$$
$$\approx 1004 \text{ pounds.}$$

■

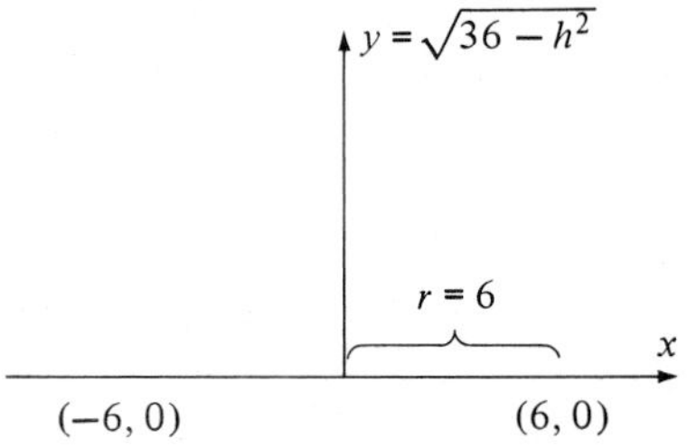

Figure 2.10

$$A = \frac{1}{2}\pi r^2 = \int_{-r}^{r} \sqrt{r^2 - h}\, dh.$$

REMARK 1: The first integral in equation (4) above can be evaluated either by a theorem due to Pappus (Section 8.5) or by the technique of trigonometric substitutions (Section 11.3). The geometric technique used here was simplest, but it depended heavily on the fact that the limits of integration correspond to endpoints of radii.

REMARK 2: When we are simply unable to find an antiderivative in problems such as Example 3, we always have the option of approximating the integral by one of our techniques for numerical integration. For example, the results of applying the Trapezoidal Rule and Simpson's Rule to approximate the integral in equation (4) are given in Table 2.1.

Table 2.1 Approximations of the integral in equation (4)

n	$\approx F$ (Trapezoidal Rule)	$\approx F$ (Simpson's Rule)
10	971.34 lb	991.13 lb
20	992.85 lb	997.71 lb
50	1001.67 lb	1003.20 lb
100	1003.60 lb	1003.96 lb

Exercise Set 8.2

1. The vertical ends of a water trough are rectangles 3 feet wide and 18 inches high. Find the hydrostatic force on the end panels when the trough is full of water.

2. A cardboard container for mineral water has a square base and rectangular sides 5 inches wide and 10 inches high. Find the force acting on a side when the carton is full of water.

3. The vertical end of a water trough is an isosceles right triangle with the right angle at the bottom. Find the hydrostatic force acting on the end when the trough is filled to a depth of 2 feet (Figure 2.11).

4. The vertical end of a water trough is an equilateral triangle whose sides are each 2 feet long. Find the hydrostatic force on an end panel when the trough is full of water.

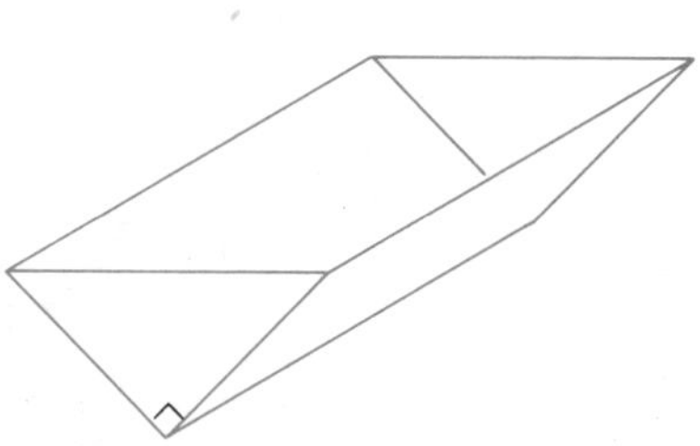

Figure 2.11

5. The wall of a dam has the shape of the parabola $y = \frac{1}{100}x^2$. Find the hydrostatic force acting on the wall if the wall is 80 feet wide across its top and the dam is full of

water. (*Hint:* The given equation assumes a coordinate system whose origin lies at the bottom of the wall. You have to find the depth h from the top of the wall.)

6. The end panels of a water trough are rectangles 2 feet wide and one foot high. To what depth must the trough be filled so that the force acting on an end panel is exactly half the force acting on the end panel when the trough is full?

7. Find the hydrostatic force on an end panel of a water trough in the shape of a semicircle of radius 2 feet when the trough is full of water (Figure 2.12).

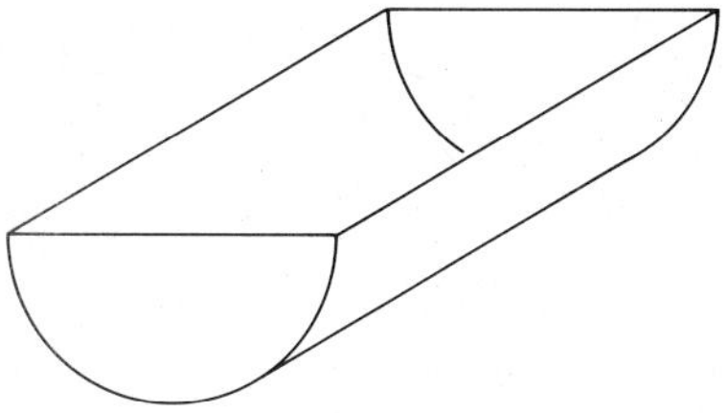

Figure 2.12

8. A submarine has rectangular windows 10 inches wide and 5 inches high. Find the hydrostatic force on such a window when the top of the window lies 30 feet below water level.

9. Find the hydrostatic force on the submarine window in Exercise 8 if it has the shape of an equilateral triangle with side $s = 10$ inches, bottom side horizontal, and the top of the window 30 feet below water level.

10. Suppose an object is submerged in a fluid weighing c pounds per cubic foot. How should equation (2) be modified so as to give the force acting on a face of this object?

11. Use the result of Exercise 10 to find the force acting on one wall of a carton of heavy cream weighing 72 pounds per cubic foot if the walls of the carton are rectangles of width 4 inches and height 6 inches and the carton is full of cream.

12. A fuel oil truck has a tank in the shape of a cylinder whose base is described by the ellipse $\frac{x^2}{64} + \frac{y^2}{36} = 1$. The tank is mounted horizontally on the truck (i.e., the elliptical end panels are vertical). Find the force acting on one end panel when the tank is half filled with fuel oil weighing 58 pounds per cubic foot (see Exercise 10).

13. If a rectangular plate is suspended in water, the theory of this section enables us to compute the hydrostatic force acting on one face of the plate. Why does the plate not move as a result of this force?

14. *(Calculator/Computer)* Use Simpson's Rule to approximate the force acting on an end panel of the oil tank in Exercise 12 when the tank is full of oil.

In each of Exercises 15–17, compute the hydrostatic force on a vertical plate of the given shape submerged to the given depth d below water level. In each case, compare your result with the product $F = (62.4)A \cdot h$ where A is the area of the figure and h is the depth of the center of the object below water level.

15. A rectangle of width 15 inches and height 10 inches lying $d = 2$ feet below water level.

16. A square of side 2 feet lying $d = 6$ feet below water level.

17. A rectangle of width 3 feet and height 2 feet lying 7 feet below water level.

18. What do you conjecture, based on the results of Exercises 15–17, about the relationship between the area, depth of center, and hydrostatic force acting on a submerged plate? Do the results of Exercise 14 support this conjecture? (These ideas are pursued in Section 8.4 on Centers of Gravity.)

8.3 WORK

When a constant force F moves an object a distance d in a direction parallel to the force, the product

$$W = Fd \tag{1}$$

is defined as the **work** done on that object. For example, in lifting a 2-pound textbook 3 feet you do an amount of work equal to $W = (3 \text{ feet}) \times (2 \text{ lb}) = 6$ foot-pounds (Figure 3.1).

The choice of the unit foot-pounds reflects the definition of work as the product of force times distance. If force and distance are measured in units other than feet and pounds, the units for work are modified appropriately (inch-pounds or newton-meters, for example).

In many practical situations, the calculation of work cannot be done by a simple multiplication because the force F required to move an object is not constant but,

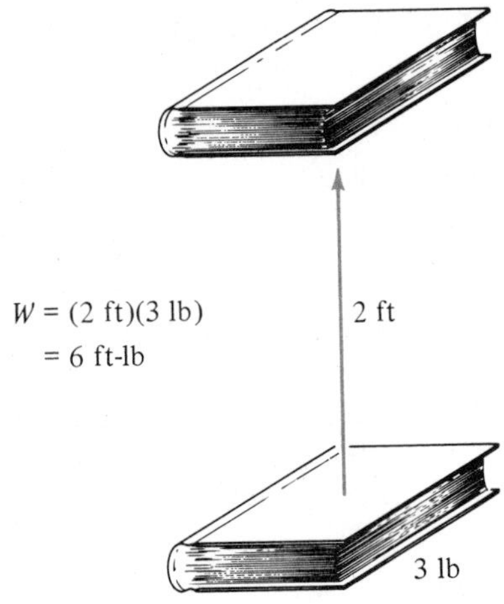

Figure 3.1

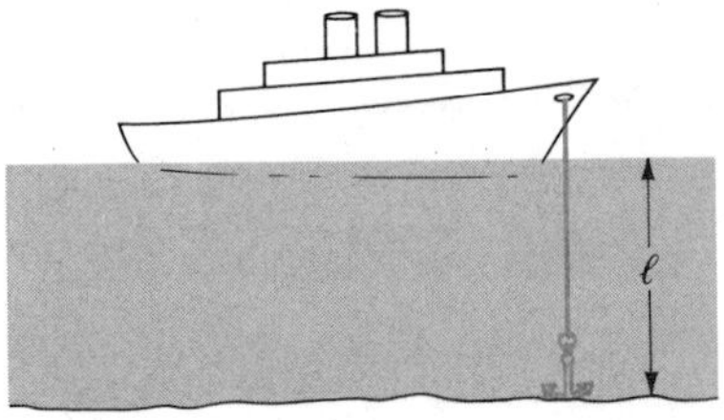

Figure 3.2 Force required to raise anchor depends upon ℓ.

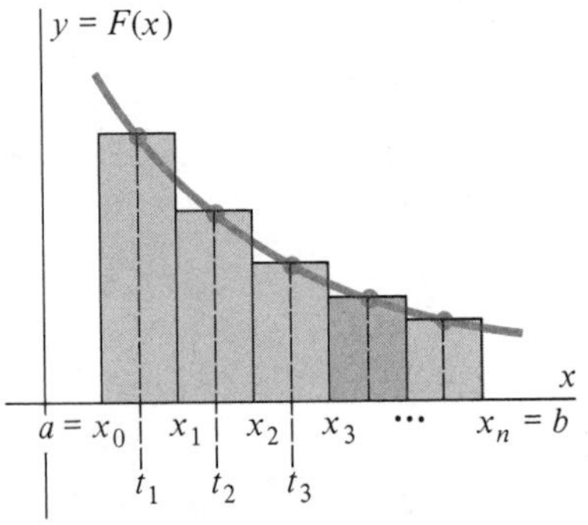

Figure 3.3 Approximating work done over $[a, b]$ by variable force $F(x)$.

rather, depends upon a one-dimensional variable. For example, the force required to raise a ship's anchor depends on the length of cable, ℓ, between the ship and the anchor (Figure 3.2). This is because the cable usually has substantial weight, and both the cable and the anchor must be raised. As the anchor is reeled in, the weight of the remaining cable is a decreasing function of ℓ.

When the motion of the object is along a line, the force required to move the object can sometimes be written as a function $F(x)$ of the object's position x along the line (as we will see in Example 1). In such cases we can generalize equation (1) by use of the definite integral, as follows.

Suppose that an object is to be moved from location $x = a$ to location $x = b$. We can approximate the work done on the object over the entire interval $[a, b]$ by partitioning $[a, b]$ into small subintervals, assuming $F(x)$ to be constant on each subinterval, and then summing the resulting values of work required on each subinterval (Figure 3.3). To do so we choose a partition $a = x_0 < x_1 < x_2 < \cdots < x_n = b$ so that $x_j - x_{j-1} = \Delta x = \dfrac{b-a}{n}$ for each $j = 1, 2, \ldots, n$, and we select one number t_j in each interval $[x_{j-1}, x_j]$. If we approximate the force function $F(x)$ throughout the interval $[x_{j-1}, x_j]$ by the constant $F(t_j)$, then by equation (1) the work ΔW_j required to move the object from x_{j-1} to x_j is the product $\Delta W_j = F(t_j)(x_j - x_{j-1}) = F(t_j)\,\Delta x$ (Figure 3.4). By summing these approximations, we obtain the approximation for the work done in moving the object from $x = a$ to $x = b$ as

$$W \approx \sum_{j=1}^{n} \Delta W_j = \sum_{j=1}^{n} F(t_j)\,\Delta x. \tag{2}$$

If the force function $F(x)$ is continuous for $x \in [a, b]$, then the sum on the right-hand side of approximation (2) is a Riemann sum that has a limit as $n \to \infty$. Since our approximation was designed to provide an increasingly accurate measurement of the actual value of the work as $n \to \infty$ (and, hence, as the size of the intervals approaches zero), we *define* work to be the limiting value of this approximating sum, that is,

$$\begin{aligned} W &= \lim_{n\to\infty} \sum_{j=1}^{n} F(t_j)\,\Delta x \qquad (3) \\ &= \int_a^b F(x)\,dx. \end{aligned}$$

Equation (3) states that the work done by a variable force in moving an object along a line from location $x = a$ to location $x = b$ is found by integrating the force function from $x = a$ to $x = b$. Figure 3.3 illustrates a typical force function and the geometric interpretation of our approximation scheme. Although the region bounded by the graph of $F(x)$ does not correspond to any physical aspect of the work problem, the area of that region is precisely the work being calculated.

Example 1 A principle from elementary physics known as **Hooke's Law** states that the force required to stretch or compress a spring a distance of x units from its natural length is proportional to that distance, that is,

$$F(x) = kx \qquad \text{(Hooke's Law).} \tag{4}$$

Here k is a constant, called the **spring constant,** which depends on the particular spring. Since the force varies with distance, calculating the work done in stretching

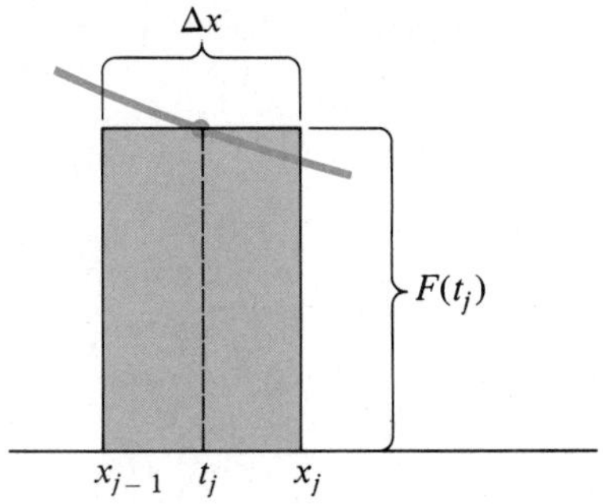

Figure 3.4 Approximation to work over jth subinterval $[x_{j-1}, x_j]$ is $\Delta W_j = F(t_j)\,\Delta x$.

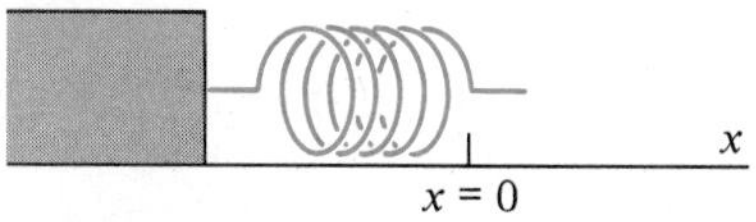

Figure 3.5 Spring in natural state.

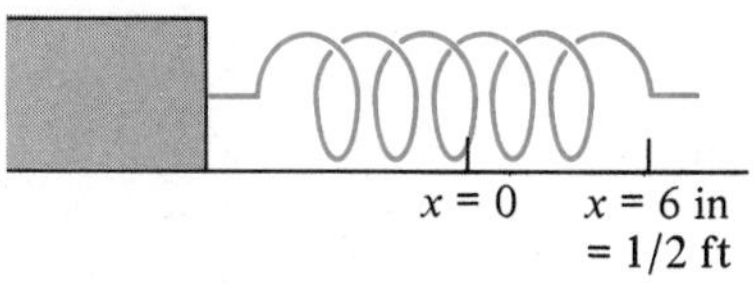

Figure 3.6 Spring stretched 6 in $= \dfrac{1}{2}$ ft.

or compressing a spring requires use of equation (3) as the following examples illustrate.

(a) A spring has spring constant $k = 20$ lb/ft. The work done in stretching the spring 6 inches beyond its natural length is found by use of equations (3) and (4):

$$\begin{aligned} W &= \int_0^{1/2} F(x)\,dx \\ &= \int_0^{1/2} 20 \cdot x \cdot dx \\ &= 10x^2\Big]_0^{1/2} \\ &= 10\left(\frac{1}{4} - 0\right) \\ &= 2.5 \text{ foot-pounds.} \end{aligned}$$

(Figures 3.5 and 3.6.)

(b) The work done in stretching the spring in part (a) from 3 inches beyond its natural length to 6 inches $= \frac{1}{2}$ ft beyond its natural length is

$$W = \int_{1/4}^{1/2} 20x\,dx = 10x^2\Big]_{1/4}^{1/2} = \frac{15}{8} \text{ foot-pounds.}$$

(c) Suppose that 4 foot-pounds of work are required to stretch a spring 1 foot beyond its natural length. We may calculate the spring constant from these data since

$$4 = \int_0^1 kx\,dx = \frac{k}{2}x^2\Big]_0^1 = k/2.$$

Thus, $4 = k/2$, so $k = 8$ lb/ft. ■

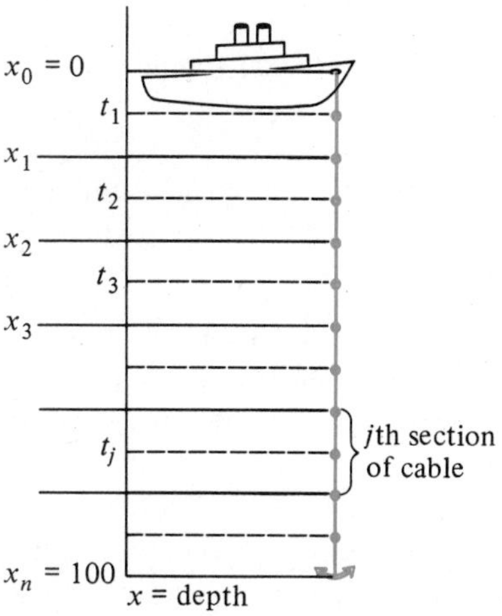

Figure 3.7 jth section of cable is assumed to lie uniformly at a depth of t_j feet.

Example 2 Suppose that a ship's anchor weighs 2 tons (4000 pounds) in water and that the anchor is hanging taut from 100 feet of cable. Find the work required to wind in the anchor if the cable weighs 20 pounds per foot in water.

Solution: This example does not quite fit the general framework within which equation (3) was developed. Although the anchor is being moved from depth $x = 100$ to depth $x = 0$, the cable itself is not being lifted 100 feet. Rather, each "slice" of cable lying at depth x feet is being lifted only x feet. To approximate the work done on the entire cable, we can partition the interval $[0, 100]$ as usual, with n equally spaced numbers $x_0 = 0 < x_1 < x_2 < \cdots < x_n = 100$ (Figure 3.7). Then the force required to lift each section of cable, ΔF_j, is its weight, that is,

$$\Delta F_j = (20 \text{ lb/ft})(\Delta x \text{ feet}).$$

By selecting one number t_j in each interval $[x_{j-1}, x_j]$, we approximate the work required to lift the jth section of cable to the surface by the product of the force ΔF_j and the distance t_j, that is,

$$\begin{aligned} \Delta W_j &\approx \Delta F_j \cdot t_j \\ &= 20t_j\,\Delta x \text{ ft-lb.} \end{aligned}$$

The work W_c required to raise the cable is therefore approximated by the sum

$$W_c \approx \sum_{j=1}^{n} \Delta W_j = \sum_{j=1}^{n} 20t_j \, \Delta x. \tag{5}$$

Since the expression on the right side of equation (5) is a Riemann sum for the function $f(x) = 20x$, we conclude that

$$\begin{aligned} W_c &= \lim_{n\to\infty} \sum_{j=1}^{n} 20t_j \, \Delta x \\ &= \int_0^{100} 20x \, dx. \end{aligned} \tag{6}$$

Since the work required to raise a 4000-lb anchor 100 feet is $W_a = (4000 \text{ lb}) \times (100 \text{ ft}) = 400{,}000$ ft-lb, we obtain the total work required as

$$\begin{aligned} W = W_c + W_a &= 400{,}000 \text{ ft-lb} + \int_0^{100} 20x \, dx \\ &= 400{,}000 \text{ ft-lb} + 10x^2\Big]_0^{100} \\ &= 500{,}000 \text{ ft-lb}. \end{aligned}$$

■

REMARK: Notice that the integral obtained for the work W_c required to raise the cable in equation (6) cannot be obtained by applying equation (3). The reason is that equation (3) applies to a variable force acting to move an object from location $x = a$ to location $x = b$. However, equation (6) refers to moving a one-dimensional object of uniform weight through a *variable* distance. We could attempt to generalize equation (6), but the wording would be clumsy and, necessarily, vague. Instead, you should note the following cautions:

(i) *Equation (3) may be applied only when a variable force $F(x)$ moves the point at which it is applied from location $x = a$ to location $x = b$.* (Thus, equation (3) was correctly applied in Example 1, since the point of application corresponded to the end of the spring. However, in Example 2 there was no such "point of application.")

(ii) In all other problems involving the calculation of work, you should
   (a) partition the relevant interval,
   (b) approximate the work required in each interval,
   (c) write the resulting approximation to the total work as a Riemann sum, and
   (d) obtain the appropriate integral as the limit of the approximating Riemann sum.

Thus, work problems not falling under point (i) leave us without a handy formula for their solution. However, such problems provide an excellent test of our ability to apply the theory of the definite integral by implementing the now familiar steps of point (ii).

Example 3 A swimming pool is 40 feet long and 20 feet wide. The floor of the pool has a constant slope from a depth of 2 feet at one end to a depth of 10 feet at the other. Find the work required to pump all the water out through a valve located at the top edge of the pool when the pool is full (Figure 3.8).

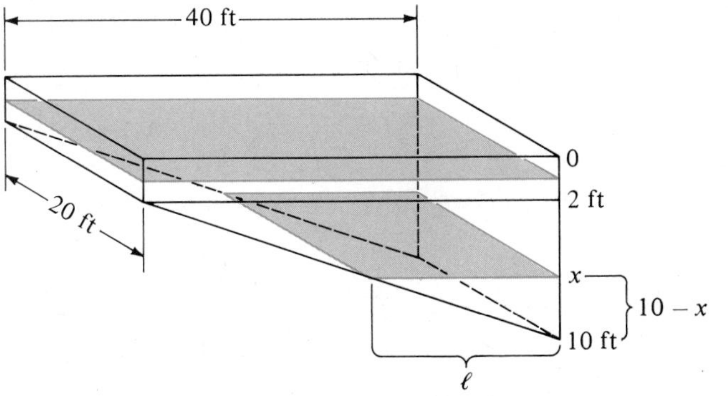

Figure 3.8 Swimming pool with sloping floor.

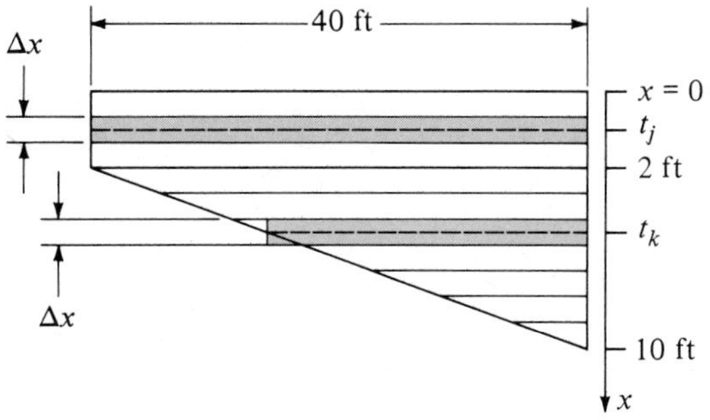

Figure 3.9 Partitioning the depth axis.

Strategy

Label the independent variable.

Treat the problem in two parts according to the geometry.

Solution

Let the variable x denote depth. We imagine the result of partitioning the interval $[0, 10]$ and slicing the water volume by horizontal planes into slabs of thickness Δx. As Figure 3.9 illustrates, the resulting slabs are of two types, those with fixed length 40 ft (lying above depth 2 ft) and those with variable length (lying below depth 2 ft). We treat the two types separately.

Partition the interval $[0, 2]$ and approximate the volume of each slab.

For depths t_j with $0 \leq t_j \leq 2$, the slab of thickness Δx has volume

$$\begin{aligned} \Delta V_j &= (20 \text{ ft})(40 \text{ ft})(\Delta x \text{ ft}) \\ &= 800 \ \Delta x \text{ ft}^3. \end{aligned}$$

The force (weight) required to lift this slab is therefore

$\Delta F = (62.4)\ \Delta V$

$$\begin{aligned} \Delta F_j &= (62.4 \text{ lb/ft}^3)(800 \ \Delta x \text{ ft}^3) \\ &= 49{,}920 \ \Delta x \text{ lb}. \end{aligned}$$

If we assume this slab to be concentrated at depth t_j, the work required to lift it to the top of the pool will be

$\Delta W = t \cdot \Delta F.$

$$\Delta W_j = t_j \Delta F_j = 49{,}920 t_j \ \Delta x \text{ ft-lb}.$$

Sum the individual approximations. The result is a Riemann sum.

Assuming the interval $[0, 2]$ to have been partitioned into n subintervals, the work W_T required to pump out the top 2 feet of water is approximated by

$$W_T \approx \sum_{j=1}^{n} \Delta W_j = \sum_{j=1}^{n} 49{,}920 t_j \ \Delta x$$

Obtain the integral as the limit of the Riemann sum.

so

$$W_T = \int_0^2 49{,}920 x \, dx.$$

Use similar triangles to find length as a function of depth.

For depths t_k with $2 \leq t_k \leq 10$, the length ℓ_k of the slab depends on its depth. By similar triangles (Figure 3.10), we have

$$\frac{\ell_k}{10 - t_k} = \frac{40}{8}$$

so

$$\ell_k = 5(10 - t_k).$$

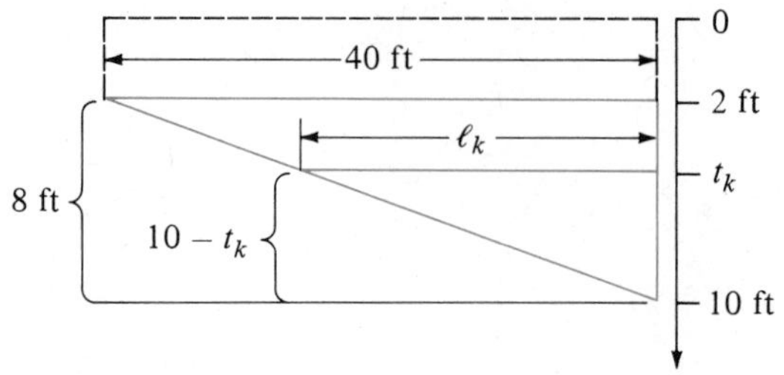

Figure 3.10 Similar triangles give length ℓ_k at depth t_k.

Proceed as in the first part.

A slab of water of thickness Δx assumed to be concentrated at depth t_k therefore has volume

$$\begin{aligned}\Delta V_k &= (20\text{ ft})[5(10 - t_k)\text{ft}](\Delta x\text{ ft})\\ &= 100(10 - t_k)\ \Delta x\text{ ft}^3\end{aligned}$$

and requires a lifting force to overcome its weight of

$$\begin{aligned}\Delta F_k &= (62.4\text{ lb/ft}^3)[100(10 - t_k)\ \Delta x\text{ ft}^3]\\ &= 6240(10 - t_k)\ \Delta x\text{ lb.}\end{aligned}$$

The work required to lift this slab to the top edge is therefore

$$\Delta W_k = 6240t_k(10 - t_k)\ \Delta x\text{ ft-lb.}$$

Summing over all slabs lying between depths 2 ft and 10 ft (assuming n such slabs) gives the approximation to W_B, the work required to pump out all water lying below 2 ft:

$$W_B \approx \sum_{k=1}^{n} \Delta W_k = \sum_{k=1}^{n} 6240t_k(10 - t_k)\ \Delta x.$$

Thus

$$W_B = \int_2^{10} 6240x(10 - x)\ dx.$$

Combine results.

Combining the expressions for W_T and W_B then gives the total work W required as

1 ft-ton = 2000 ft-lb.

$$\begin{aligned}W &= W_T + W_B\\ &= \int_0^2 49{,}920x\ dx + \int_2^{10} 6240x(10 - x)\ dx\\ &= 24{,}960x^2\Big]_0^2 + [31{,}200x^2 - 2080x^3]_2^{10}\\ &= 1{,}031{,}680\text{ ft-lb}\\ &= 515.84\text{ ft-tons.}\end{aligned}$$

■

Exercise Set 8.3

1. How much work is done in compressing a spring with spring constant $k = 40$ pound per foot a distance of 6 inches?

2. A spring has spring constant $k = 5$ newtons per meter. How much work is done in stretching the spring 80 cm beyond

its natural length? (Here work will be in units of newton · meters = joules.)

3. How much work is done in compressing the spring in Exercise 1 another 6 inches?

4. How much work is done in stretching the spring in Exercise 2 another 40 cm?

5. A force of 50 pounds stretches a spring 4 inches. Find the work required to stretch the spring an additional 4 inches.

6. If 40 foot-pounds of work are required to stretch a spring 6 inches beyond its natural length, what is its spring constant?

7. True or false? If W_1 is the work done on an object in moving it from point A to point B, and W_2 is the work done on the same object in moving it from point B to point C, then the work W done in moving the object from point A to point C satisfies the equation $W = W_1 + W_2$. What property of definite integrals is involved in this question?

8. The work required to stretch a spring 6 centimeters beyond its natural length is $W = 2$ ergs. What is the value of the spring constant for this spring? (The units for this spring constant will be dynes per centimeter, where 1 erg = 1 dyne · centimeter.)

9. A spring hangs from one end, which is attached to a supporting beam. When a 10-pound weight is attached to the free end of the spring it stretches a distance of 2 feet.
   a. Find the spring constant k.
   b. Find the additional weight that must be added so that the spring will be stretched 3 feet beyond its natural length.
   c. If both weights are removed from the end of the spring, what is the work required to again stretch the spring a distance of 3 feet beyond its natural length?

10. A children's wading pool is 6 feet wide, 10 feet long, and $\frac{1}{2}$ foot deep. How much work is required to pump all the water out through a valve at the top edge of the pool when the pool is full of water?

11. A conical water tank has radius $r = 10$ feet and height $h = 12$ feet and is mounted with its base horizontal and its tip pointing downward. How much work is required to pump the water out through a valve in the top of the tank when the tank is full?

12. A water tank has the shape of a right circular cylinder with radius $r = 10$ feet and depth $h = 20$ feet. It is mounted with its axis vertical. The tank contains 10 feet of water. Find the work done in pumping this water out through a valve in the top of the tank.

13. Find the work done in pumping the water out of the tank in Exercise 12 when the tank is full of water.

14. A water tank has the shape of a right circular cone of height $h = 20$ feet and radius $r = 6$ feet. It is mounted with its axis vertical and tip down. It contains 4 feet of water. Find the work done in pumping this water out through a valve in the top of the tank.

15. Find the work done in pumping the water out of the tank in Exercise 14 if the tank is full of water.

16. Find the work done in Exercise 12 if the water is pumped to a height 10 feet above the top of the tank.

17. Find the work done in Exercise 14 if the water is pumped to a height of 15 feet above the top of the tank.

18. A hemispherical water tank has radius $r = 5$ feet. The tank is mounted with its circular base on top, lying horizontally. How much work is required to pump the water out through a valve on the top edge of the tank when the tank is full?

19. A 20-foot section of rope weighing 2 pounds per foot is hanging from a windlass (crank). How much work is done in winding up the entire length of rope?

20. A chain weighs 40 newtons per meter. How much work is done in winding a 30-meter section of chain hanging from a windlass?

21. Coulomb's Law governing the force of attraction between two electrically charged bodies states that the force of attraction or repulsion between two point charges is directly proportional to the product of the charges and inversely proportional to the square of the distance between them, that is,

$$F = K\frac{q_1q_2}{r^2}$$

where q_1 and q_2 are the magnitudes of the charges at points P_1 and P_2, respectively, r is the distance between P_1 and P_2, and K is the constant of proportionality. Suppose that charges of equal magnitude q exist at points P_1 and P_2 which are 4 meters apart, and that the force with which the two charges repel each other is known to be 10 newtons.
   a. What is the value of the constant K in terms of q?
   b. How much work is required to bring one charge from point P_2 to a distance of 2 meters from point P_1?

22. Suppose that an electron with charge +1 is stationary and located at (5, 0). How much work is required to move a second electron with charge +1 from (0, 0) to (4, 0)? (See Exercise 21.)

23. What is the answer to Exercise 22 if a third electron with charge +1 is located at (−8, 0) and stationary?

24. The weight of an object varies inversely with the square of its distance from the center of the earth, that is,

$$w(r) = \frac{k}{r^2}$$

where $w(r)$ is the weight of the object when located at a distance r from the earth's center. A satellite weighs 5 tons at the earth's surface, when it is 4000 miles from the earth's center.

a. Find the value of k for this satellite.

b. Find the amount of work done in propelling this satellite to a height of 100 miles above the earth's surface.

8.4 MOMENTS AND CENTERS OF GRAVITY

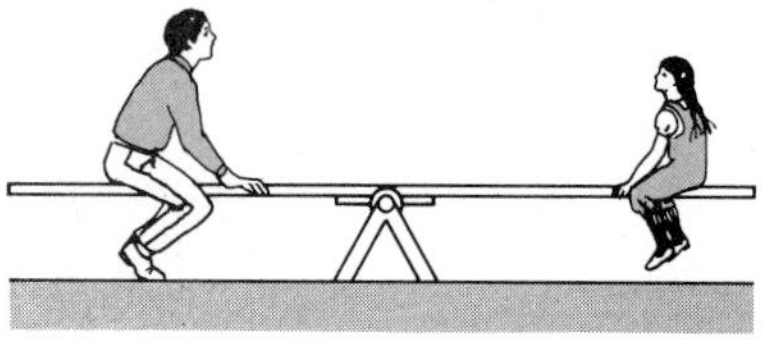

Figure 4.1 A larger weight at a smaller distance balances a smaller weight at a larger distance.

Most children can identify the two factors that determine whether they can ''balance'' another child seated opposite them on a seesaw: their weight and the distance at which they are seated from the pivot (Figure 4.1). Increasing either of these two quantities increases the tendency of the seesaw to rotate their end downward. We wish here to pursue the mathematical generalizations of these ideas. The results, combined with the Theorem of Pappus in the next section, enable us to extend and simplify the techniques of the preceding two sections.

We idealize the situation of the seesaw by imagining two objects of weight w_1 and w_2 placed on a flat weightless rod, which in turn is mounted on a fulcrum (pivot). The rod is assumed to be free to rotate about the fulcrum, and we assume the weights of the objects to be located at distances ℓ_1 and ℓ_2 from the fulcrum, respectively (Figure 4.2).

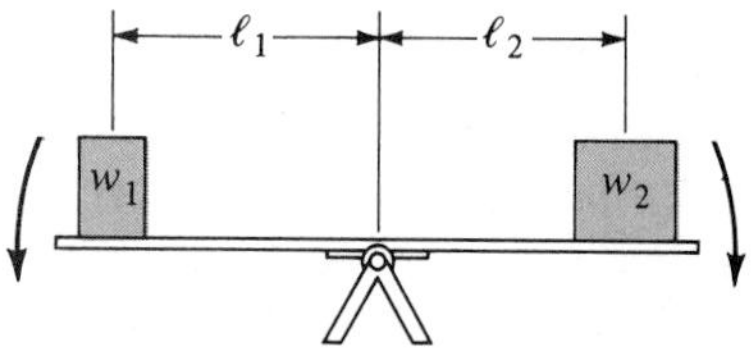

Figure 4.2 Weights balanced on a fulcrum.

The tendency of the rod to rotate in the counterclockwise direction is called the **torque** $\ell_1 w_1$. Similarly, the torque $\ell_2 w_2$ represents the tendency of the rod to rotate in the clockwise direction. A physical principle called the **law of the lever** states that the rod will balance (equilibrium will occur) when the magnitudes of the opposing torques are equal, that is, when

$$\ell_1 w_1 = \ell_2 w_2. \tag{1}$$

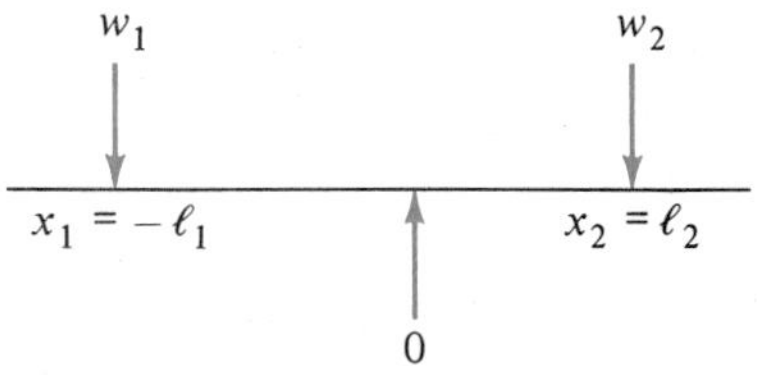

Figure 4.3 Assigning coordinates to the lever.

We can further simplify the model by introducing a one-dimensional coordinate system along the rod (Figure 4.3). If we denote the point corresponding to the fulcrum as $x = 0$ and take the positive x-axis as the direction in which weights produce clockwise rotation, then for $x_1 = -\ell_1$ and $x_2 = \ell_2$ the equilibrium equation (1) becomes

$$x_1 w_1 + x_2 w_2 = 0. \tag{2}$$

Equation (2) can now be easily generalized to the condition for equilibrium for n weights $w_1, w_2, \ldots, w_n$ located at positions $x_1, x_2, \ldots, x_n$, respectively (Figure 4.4). The condition is

$$x_1 w_1 + x_2 w_2 + \cdots + x_n w_n = 0$$

or

$$\sum_{j=1}^{n} x_j w_j = 0. \tag{3}$$

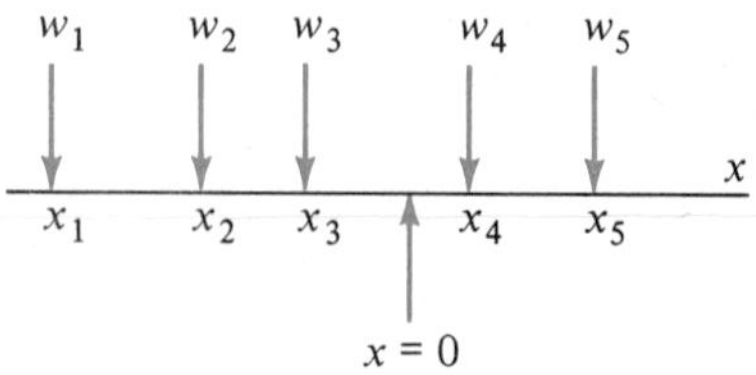

Figure 4.4 Equilibrium condition is $\Sigma x_j w_j = 0$ for fulcrum at $x = 0$.

Finally, we observe that if the fulcrum is located at $x = \bar{x}$ rather than at $x = 0$ (which corresponds to moving the origin of the coordinate system, but *does not* change the physical situation), the (signed) distance of the weight w_j away from the fulcrum becomes $(x_j - \bar{x})$ rather than x_j (Figure 4.5). In this case the equilibrium condition (3) becomes

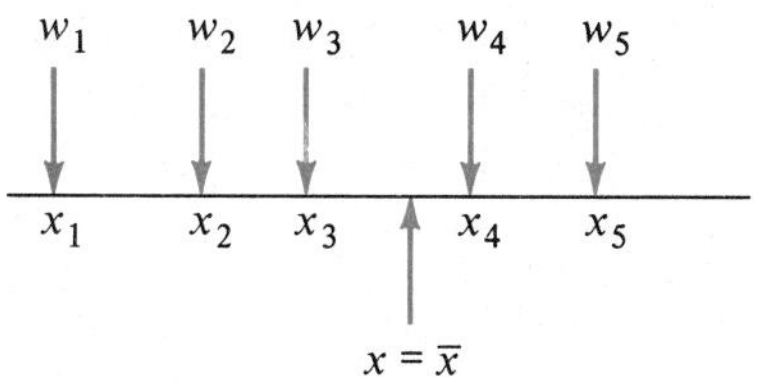

Figure 4.5 Fulcrum at $\overline{x}$. (Note that all we have done is change the label from 0 to $\overline{x}$.)

$$\sum_{j=1}^{n} (x_j - \overline{x})w_j = 0. \tag{4}$$

Since we can write equation (4) as

$$\sum_{j=1}^{n} x_j w_j - \overline{x} \sum_{j=1}^{n} w_j = 0$$

we may solve for $\overline{x}$ as

$$\overline{x} = \frac{\displaystyle\sum_{j=1}^{n} x_j w_j}{\displaystyle\sum_{j=1}^{n} w_j}. \tag{5}$$

Equation (5) specifies the location of the fulcrum in our idealized balance problem when equilibrium occurs. The advantage of this version of the equilibrium condition is that it enables us to calculate $\overline{x}$ from the weights $w_1, w_2, \ldots, w_n$ and the locations $x_1, x_2, \ldots, x_n$. Not surprisingly, $\overline{x}$ is referred to as the **center of gravity** of the system so described.

Example 1 A system of weights $w_1 = 10$ lb, $w_2 = 20$ lb, $w_3 = 10$ lb, $w_4 = 20$ lb, and $w_5 = 25$ lb is located along a line at points $x_1 = -6$ ft, $x_2 = -2$ ft, $x_3 = 1$ ft, $x_4 = 3$ ft, and $x_5 = 6$ ft, respectively. Find the center of gravity.

Solution:

$$\overline{x} = \frac{[(-6)(10) + (-2)(20) + (1)(10) + (3)(20) + (6)(25)] \text{ ft-lb}}{(10 + 20 + 10 + 20 + 25) \text{ lb}}$$

$$= \frac{24}{17} \text{ ft.}$$

■

REMARK 1: The physical definition of the weight w of an object is $w = mg$, where m is the mass of the object and g is the acceleration due to gravity. Since g is a constant, w and m are proportional. Thus, $\overline{x}$ in equation (5) is also the **center of mass** for the n particles whose masses are $m_1 = w_1/g$, $m_2 = w_2/g, \ldots, m_n = w_n/g$, since all factors of g in equation (5) will cancel. Henceforth we will refer exclusively to centers of mass.

REMARK 2: If we write the equilibrium equation (4) in terms of masses $m_j = w_j/g$, we obtain the equation

$$\sum_{j=1}^{n} (x_j - \overline{x})m_j = 0. \tag{6}$$

The sum on the left side of equation (6) is called the **first moment** of the mass system about the number $\overline{x}$. The first moment of a system about a number may be regarded as a measurement of the tendency of an idealized rod supported at $\overline{x}$ and with masses m_j at locations x_j to rotate, as in the earlier discussion involving torque. Since equation (5), written for masses $m_j = w_j/g$, becomes

$$\overline{x} = \frac{\sum_{j=1}^{n} x_j m_j}{\sum_{j=1}^{n} m_j}, \tag{7}$$

we may interpret $\overline{x}$ as *the first moment of the mass system about* $x = 0$ *divided by the total mass.*

The preceding discussion was only preliminary to our main interest in this section, which is *to calculate the center of mass of a three-dimensional object of uniform thickness and density*. To achieve this objective, we must pursue one additional mathematical abstraction, the rod of nonuniform density. In this discussion, we will pass from the setting of a *discrete system* involving finitely many masses $m_1, m_2, \ldots, m_n$, to the *continuous case,* as follows.

Imagine a cylindrical rod of length ℓ and uniform cross-sectional area A. Imagine also that the density ρ of the material is constant across any cross section of the rod, but that the density varies continuously along the rod. We impose a one-dimensional coordinate system along the rod, letting x denote displacement along the axis, and we position the rod to lie between $x = 0$ and $x = \ell$ (Figure 4.6). We assume that $\rho(x)$ is continuous for $0 \le x \le \ell$.

Since the density of the rod is not uniform, the center of mass will not necessarily lie at the midpoint. To approximate the center of mass $\overline{x}$, we partition the interval $[0, \ell]$ into n subintervals of equal length with endpoints $0 = x_0 < x_1 < x_2 < \cdots < x_n = \ell$. By slicing the rod with planes perpendicular to the axis through each of the points $x_0, x_1, x_2, \ldots, x_n$, we divide the rod into n cylinders, each of length $\Delta x = x_j - x_{j-1} = \dfrac{\ell}{n}$ and of cross-sectional area A. The volume of each small cylinder is therefore $V_j = A\ \Delta x$.

Now the mass m of an object of *uniform* density ρ and volume V is the product $m = \rho V$. Since the small cylinders are not of uniform density, we select one number t_j in each interval $[x_{j-1}, x_j]$ and approximate the density throughout the jth cylinder by the density at $x = t_j$, that is, we assume that $\rho \equiv \rho(t_j)$ throughout the interval $[x_{j-1}, x_j]$. This assumption allows us to approximate the mass m_j of the jth cylinder as

$$m_j \approx \rho(t_j)V_j = \rho(t_j)A\ \Delta x. \tag{8}$$

Finally, we assume the mass m_j of each cylinder to be concentrated at the point $t_j \in [x_{j-1}, x_j]$. With these assumptions, we have described a system consisting of a finite number of masses, so we may apply the equilibrium condition of equation (6)

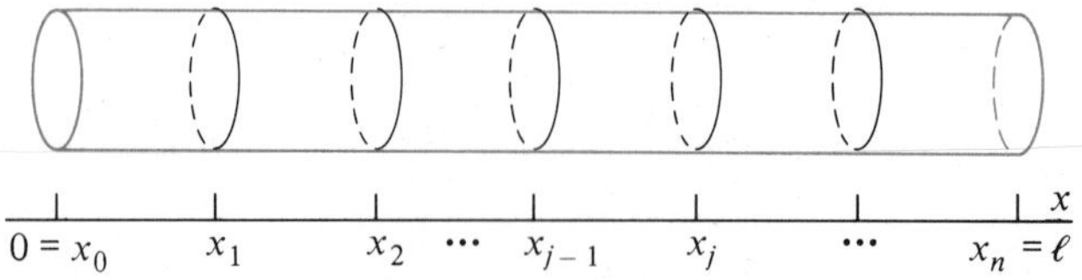

Figure 4.6 Rod of constant cross-sectional area and varying density.

for the center of mass $\overline{x}$. We obtain, from equation (6) and approximation (8), the condition

$$0 = \sum_{j=1}^{n} (t_j - \overline{x})m_j \approx \sum_{j=1}^{n} (t_j - \overline{x})\rho(t_j)A\ \Delta x. \tag{9}$$

Since $\rho(x)$ is assumed continuous on $[0, \ell]$, the sum in equation (9) is a Riemann sum for the function $f(x) = (x - \overline{x})\rho(x)A$. We assume that in the limit as $n \to \infty$ the error in the approximation approaches zero as the length of the individual small cylinders approaches zero. We therefore conclude that

$$\begin{aligned} 0 &= \lim_{n\to\infty} \sum_{j=1}^{n} (t_j - \overline{x})\rho(t_j)A\ \Delta x \\ &= \int_0^{\ell} (x - \overline{x})\rho(x)A\ dx \\ &= \int_0^{\ell} x\rho(x)A\ dx - \overline{x}\cdot \int_0^{\ell} \rho(x)A\ dx. \end{aligned}$$

We can now solve for $\overline{x}$. We obtain

$$\overline{x} = \frac{\displaystyle\int_0^{\ell} x\rho(x)A\ dx}{\displaystyle\int_0^{\ell} \rho(x)A\ dx}. \tag{10}$$

Equation (10) gives the *center of mass for a rod of constant cross-sectional area A and density $\rho(x)$*. By analogy with the discrete case described by equation (7), the integral in the numerator is called the first moment of the rod about the endpoint $x = 0$, and the integral in the denominator is the total mass.

Example 2 A rod 10 cm long has uniform cross-sectional area $A = 4\ \text{cm}^2$. Find the center of mass of the rod if

(a) the density ρ is uniform, or
(b) the density $\rho(x)$ varies linearly from 3 grams per cubic centimeter at one end to 6 grams per cubic centimeter at the other.

Solution: (a) We may apply equation (10) with $\rho(x) \equiv \rho$ (constant) to obtain

$$\overline{x} = \frac{\displaystyle\int_0^{10} \rho\cdot 4\cdot x\ dx}{\displaystyle\int_0^{10} \rho\cdot 4\cdot dx} = \frac{4\rho\left[\frac{1}{2}x^2\right]_0^{10}}{4\rho[x]_0^{10}} = \frac{4\rho\cdot 50}{4\rho\cdot 10} = 5,$$

which is the midpoint of the rod, as expected.

(b) For the variable density case, we must find an equation representing the density function. The statement that the density $\rho(x)$ varies linearly from $\rho(0) = 3$ to $\rho(10) = 6$ means that $\rho(x) = cx + d$ for some constants c and d.

Since $\rho(0) = c\cdot 0 + d = 3$, we have $d = 3$. Then, from $\rho(10) = c\cdot 10 + 3 = 6$, we obtain $c = \frac{3}{10}$. The density function is therefore $\rho(x) = \frac{3}{10}x + 3$. Thus, from (10),

$$\bar{x} = \frac{\int_0^{10} 4x\left[\frac{3}{10}x + 3\right] dx}{\int_0^{10} 4\left[\frac{3}{10}x + 3\right] dx} = \frac{\int_0^{10}\left(\frac{6}{5}x^2 + 12x\right) dx}{\int_0^{10}\left(\frac{6}{5}x + 12\right) dx}$$

$$= \frac{\frac{2}{5}x^3 + 6x^2\Big]_0^{10}}{\frac{3}{5}x^2 + 12x\Big]_0^{10}}$$

$$= \frac{50}{9} \approx 5.56.$$

Because of the variation of the density, the center of mass is closer to the high-density end of the rod, rather than at the midpoint. ■

Example 3 Find an equation for the center of mass $\bar{x}$ of a rod of length ℓ of constant density ρ whose cross-sectional area $A(x)$ is a continuous function for $0 \leq x \leq \ell$.

Solution: The differences between this question and the question answered by equation (10) are that (i) the density $\rho(x)$ is now assumed constant and (ii) the cross-sectional area A is now assumed to vary. We proceed with the same analysis as before, except that in approximation (8) we approximate the volume V_j by assuming the cross-sectional area $A(t_j)$ to be uniform throughout the jth cylinder. Thus, the analogue of approximation (8) is

$$m_j = \rho V_j \approx \rho A(t_j)\,\Delta x.$$

The analogue of equation (9) is

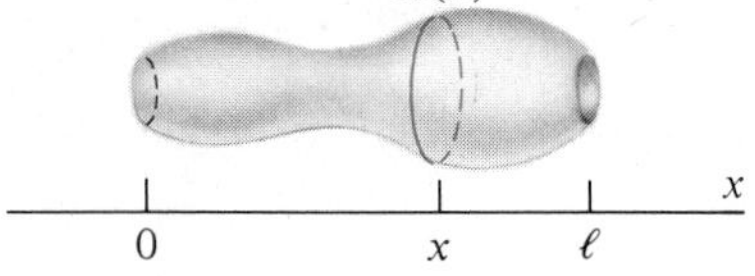

Figure 4.7 Rod of constant density and variable cross-sectional area.

$$0 = \sum_{j=1}^{n}(t_j - \bar{x})m_j \approx \sum_{j=1}^{n}(t_j - \bar{x})\rho A(t_j)\,\Delta x,$$

which, in the limit as $n \to \infty$, gives

$$\int_0^{\ell}(x - \bar{x})\rho A(x)\,dx = 0.$$

Solving this equation for $\bar{x}$ produces the desired equation

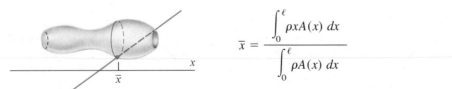

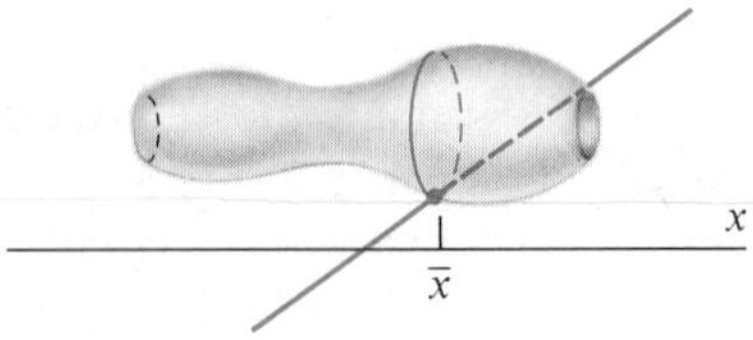

Figure 4.8 Rod "balances" at center of mass (gravity) $x = \bar{x}$.

$$\bar{x} = \frac{\int_0^{\ell}\rho x A(x)\,dx}{\int_0^{\ell}\rho A(x)\,dx} \tag{11}$$

for *the center of mass (or gravity) along the x-axis of a rod of uniform density and continuously varying cross-sectional area $A(x)$* (Figures 4.7 and 4.8). ■

Center of Mass of a Plate

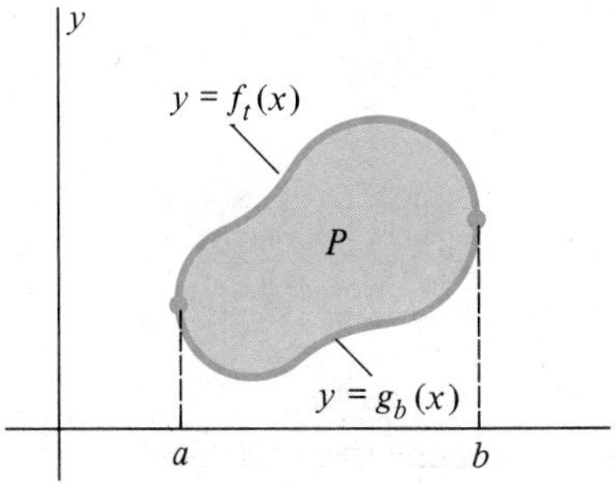

Figure 4.9 Face of thin plate bounded by graphs of $y = f_t(x)$ and $y = g_b(x)$.

We may apply the result of Example 3 to a thin plate of uniform density ρ and uniform thickness c. Suppose that the plate lies in the xy-plane and that the face is bounded above by the graph of $y = f_t(x)$ and below by the graph of $y = g_b(x)$ for $a \le x \le b$ (Figure 4.9), where the subscripts refer to "top" and "bottom." Then the area of a cross section taken perpendicular to the x-axis is $A(x) = c[f_t(x) - g_b(x)]$ (Figure 4.10). Regarding the plate as a rod of constant density and variable cross-sectional area, we apply equation (11) to find the x-coordinate of the center of mass $\bar{x}$:

$$\bar{x} = \frac{\displaystyle\int_a^b c\rho x[f_t(x) - g_b(x)]\,dx}{\displaystyle\int_a^b c\rho[f_t(x) - g_b(x)]\,dx}. \tag{12}$$

(See Figure 4.11.)

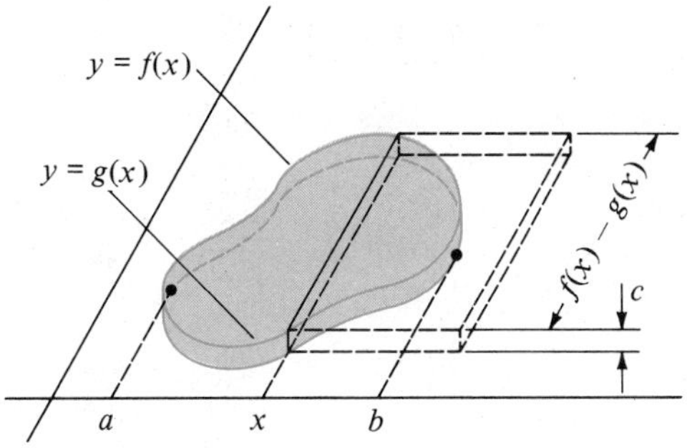

Figure 4.10 Cross section perpendicular to x-axis has area $A(x) = [f(x) - g(x)]c$.

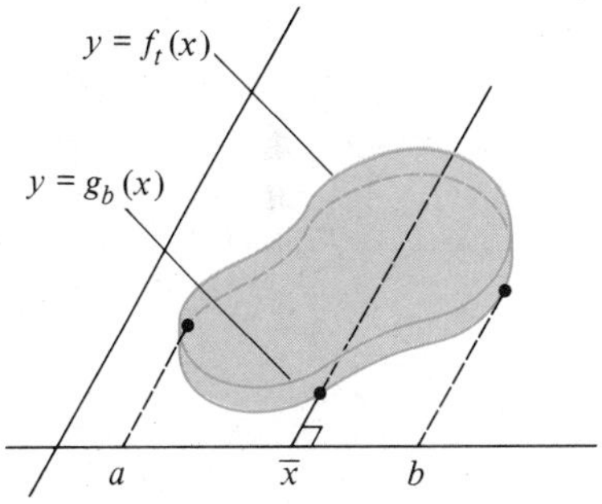

Figure 4.11 Plate "balances" at perpendicular to x-axis through center of mass $\bar{x}$.

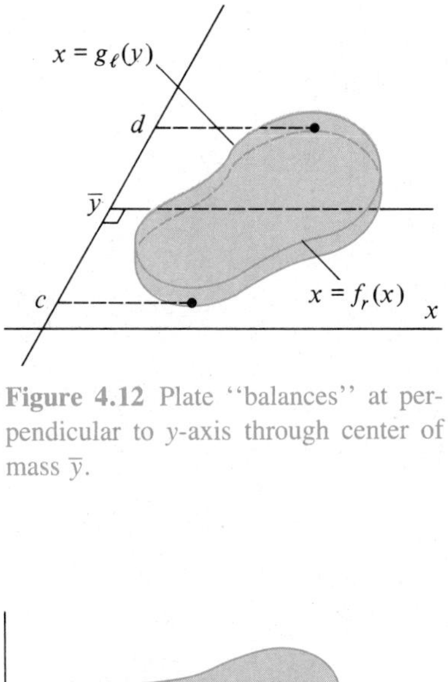

Figure 4.12 Plate "balances" at perpendicular to y-axis through center of mass $\bar{y}$.

Similarly, if the face of the plate is bounded on the right by the graph of $x = f_r(y)$ and on the left by the graph of $x = g_\ell(y)$, for $c \le x \le d$, then the y-coordinate of the center of mass is

$$\bar{y} = \frac{\displaystyle\int_c^d c\rho y[f_r(y) - g_\ell(y)]\,dy}{\displaystyle\int_c^d c\rho[f_r(y) - g_\ell(y)]\,dy}. \tag{13}$$

(See Figure 4.12.)

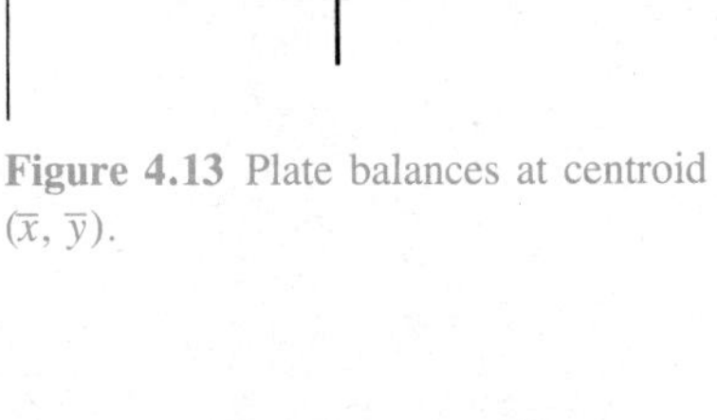

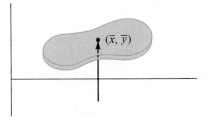

Figure 4.13 Plate balances at centroid $(\bar{x}, \bar{y})$.

The point $(\bar{x}, \bar{y})$ whose coordinates are given by equations (12) and (13) is called the center of mass, or **centroid,** of the plate. Since the centroid lies on the lines representing the centers of gravity with respect to both the x- and y-axes, it is properly referred to as the (two-dimensional) center of gravity of the plate. It is the point where the plate "balances" (see Figure 4.13).

We may simplify equations (12) and (13) in two ways. First, note that the constants c and ρ may be factored out of both equations completely. This shows that the centroid is a purely geometric property of the face R of the plate, independent of both the uniform thickness and uniform density. Second, notice that the denominators in both equations (12) and (13) are simply the area of the face multiplied by the constant $c\rho$. We may therefore restate our findings as follows:

> If the region R is bounded above and below by the continuous functions $f_t(x)$ and $g_b(x)$ for $a \le x \le b$, and on the right and left by the continuous functions $f_r(y)$ and $g_\ell(y)$ for $c \le y \le d$, then the coordinates $(\bar{x}, \bar{y})$ of the centroid of R are
>
> $$\bar{x} = \frac{1}{A}\int_a^b x[f_t(x) - g_b(x)]\,dx \tag{14}$$
>
> $$\bar{y} = \frac{1}{A}\int_c^d y[f_r(y) - g_\ell(y)]\,dy \tag{15}$$
>
> where A is the area of R.

REMARK 3: The integrals in equations (14) and (15) are referred to as the *moment of R about the y-axis* and the *moment of R about the x-axis*, respectively.

REMARK 4: An alternate formula for $\bar{y}$, given in terms of $f_t(x)$ and $g_b(x)$, is developed in Exercise 17.

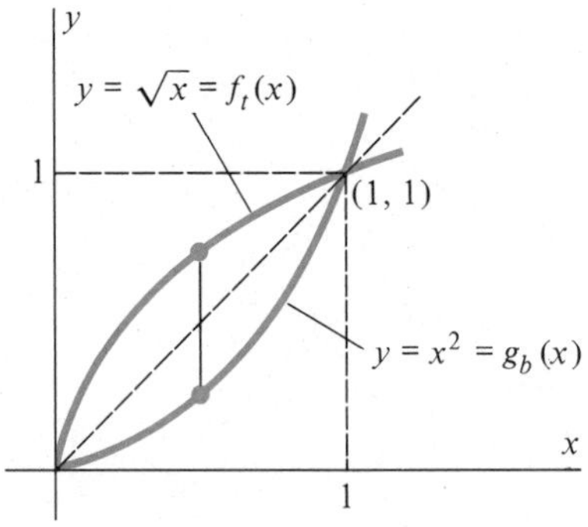

Figure 4.14

Example 4 Find the centroid of the region R bounded by the graphs of $y = \sqrt{x}$ and $y = x^2$.

Strategy

Find the upper and lower bounding functions.

Solution

For $0 \le x \le 1$, the region is bounded above by $f_t(x) = \sqrt{x}$ and below by $g_b(x) = x^2$ (Figure 4.14).

Find $A = \int_a^b [f_t - g_b]\,dx.$

$$A = \int_0^1 (\sqrt{x} - x^2)\,dx = \frac{2}{3}x^{3/2} - \frac{1}{3}x^3\Big]_0^1$$

$$= \frac{1}{3}.$$

Find $\bar{x}$ from (14).

We may apply equation (14) directly to obtain

$$\bar{x} = 3\int_0^1 x[\sqrt{x} - x^2]\,dx$$

$$= 3\left[\frac{2}{5}x^{5/2} - \frac{1}{4}x^4\right]_0^1$$

$$= 3\left(\frac{2}{5} - \frac{1}{4}\right) = \frac{9}{20}.$$

Find the right and left bounding functions.

Regarding y as the independent variable, we observe that for $0 \le y \le 1$ the region R is bounded on the right by $f_r(y) = \sqrt{y}$ and on the left by $g_\ell(y) = y^2$ (Figure 4.15).

Apply (15) to find $\bar{y}$.

Therefore, by equation (15)

$$\bar{y} = 3\int_0^1 y[\sqrt{y} - y^2]\,dy = 9/20.$$

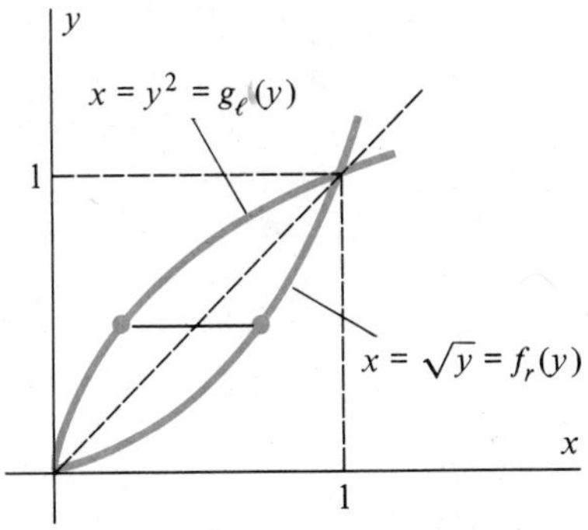

Figure 4.15

The centroid of R is therefore

$$(\bar{x}, \bar{y}) = (9/20, 9/20).$$

■

REMARK 5: The region R in Example 4 is symmetric with respect to the line $y = x$, meaning that the figure obtained by reflecting R in the line $y = x$ is congruent to R. Notice that the centroid (9/20, 9/20) lies on this line. In fact, *if R is any region which is symmetric with respect to a line ℓ the centroid of R must lie on ℓ.* If this were not true, reflecting R in the line of symmetry would result in two congruent figures with different "centers of gravity," a physical impossibility. Happily, many of the figures for which one needs to find centroids in practice contain one or more lines of symmetry. In such cases, this observation can simplify the calculation of the centroid greatly.

Example 5 Find the centroid of the quarter circle R with center (0, 0) and radius $r = 4$ lying in the first quadrant.

Strategy

Find A by geometry.

Solution

Since R is a quarter circle,

$$A = \frac{1}{4}\pi r^2 = \frac{16\pi}{4} = 4\pi.$$

Find upper and lower bounding functions and find $\bar{x}$ from (14). Centroid lies on line $x = \bar{x}$.

The region R is bounded above by $f_x(x) = \sqrt{16 - x^2}$ and below by $g_b(x) = 0$. Thus

$$\begin{aligned}\bar{x} &= \frac{1}{4\pi}\int_0^4 x\sqrt{16 - x^2}\, dx \\ &= \frac{1}{4\pi}\left(-\frac{1}{3}\right)(16 - x^2)^{3/2}\Big]_0^4 = \frac{16}{3\pi}.\end{aligned}$$

Use symmetry to obtain second line containing centroid. Centroid lies on the intersection.

Since R is symmetric about the line $y = x$, the centroid must lie on the line $y = x$. Thus $\bar{x} = \bar{y} = \dfrac{16}{3\pi}$, so the centroid is $\left(\dfrac{16}{3\pi}, \dfrac{16}{3\pi}\right)$ (see Figure 4.16).

■

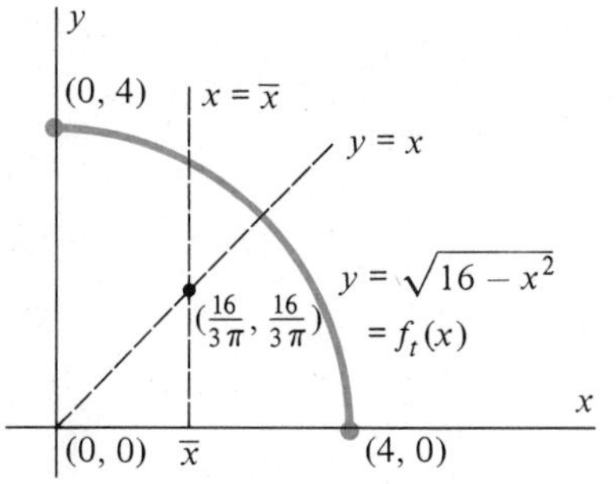

Figure 4.16 Centroid lies on intersection of $y = \bar{x}$ and $y = x$.

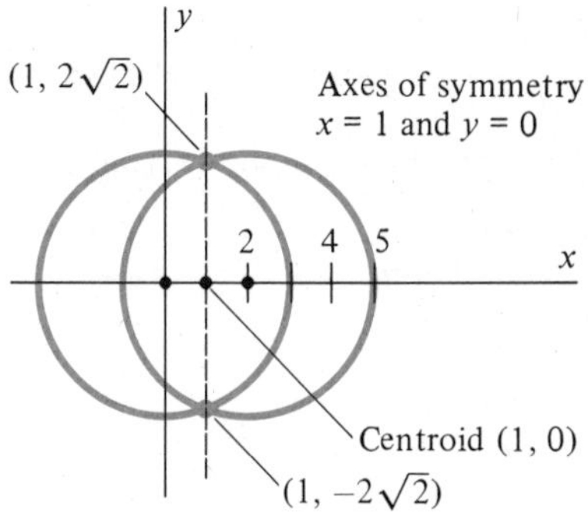

Figure 4.17 Centroid of region common to two circles.

Example 6 Find the centroid of the region R common to both the circle $x^2 + y^2 = 9$ and the circle $(x - 2)^2 + y^2 = 9$ (Figure 4.17).

Strategy

Find points of intersection of the circles.

One line of symmetry is through the centers.

The other is through the points of intersection.

Solution

The circles intersect where $x^2 = (x - 2)^2$. Solving this equation gives $-4x + 4 = 0$ or $x = 1$. The points of intersection are therefore $(1, 2\sqrt{2})$ and $(1, -2\sqrt{2})$. Since both circles have centers on the x-axis, the region R is symmetric with respect to the x-axis. Since the circles have the same radius $r = 3$, the region is also symmetric with respect to the vertical line $x = 1$. The centroid must therefore lie on both lines $x = 1$ and $y = 0$, that is, $(\bar{x}, \bar{y}) = (1, 0)$. ■

Exercise Set 8.4

1. Find the center of gravity for the system consisting of weights $w_1 = 1$ lb, $w_2 = 31$ lb, and $w_3 = 7$ lb located along a line at positions $x_1 = -4$, $x_2 = 2$, and $x_3 = 6$, respectively.

2. Find the center of mass for the system consisting of masses $m_1 = 2$ grams, $m_2 = 5$ grams, $m_3 = 1$ gram, and $m_4 = 10$ grams located along a line at positions $x_1 = -10$, $x_2 = -3$, $x_3 = 1$, and $x_4 = 5$, respectively.

3. An 80-pound child is sitting 3 feet from the pivot on a seesaw. How far from the pivot must a 50-pound child sit in order that the seesaw balance?

4. A manufacturing firm ships 100 items per year to city A and 300 items per year to city B. A straight section of interstate highway 200 miles long joins the two cities. The items are shipped by truck along this highway. Where along this highway should the manufacturing plant be located if shipping costs are to be equal? What is the relationship between this problem and equation (1)?

5. A system of masses $m_1, m_2, \ldots, m_n$ are located in an xy-plane at points $(x_1, y_1), (x_2, y_2), \ldots, (x_n, y_n)$, respectively. By analogy with the discussion leading up to equations (4) and (6), explain why the center of mass with respect to the x-axis, $\bar{x}$, satisfies the equation $\sum_{j=1}^{n} (x_j - \bar{x})m_j = 0$. (*Hint:* For the purposes of this calculation, we may regard the points as being distributed along a line parallel to the x-axis.)

6. Let $\bar{y}$ denote the center of mass for the system of Exercise 5. Explain why

$$\sum_{j=1}^{n} (y_j - \bar{y})m_j = 0.$$

7. Conclude from Exercise 5 and Exercise 6 that the center of mass $(\bar{x}, \bar{y})$ of the system of masses $m_1, m_2, \ldots, m_n$ located at points $(x_1, y_1), (x_2, y_2), \ldots, (x_n, y_n)$ is given by the coordinates

$$\bar{x} = \frac{\sum_{j=1}^{n} x_j m_j}{\sum_{j=1}^{n} m_j}, \qquad \bar{y} = \frac{\sum_{j=1}^{n} y_j m_j}{\sum_{j=1}^{n} m_j}$$

8. Use the result of Exercise 7 to find the center of mass of the system consisting of masses $m_1 = 10$ grams, $m_2 = 15$ grams, $m_3 = 2$ grams, and $m_4 = 10$ grams located at points $(x_1, y_1) = (0, 0)$, $(x_2, y_2) = (2, -2)$, $(x_3, y_3) = (4, -2)$, and $(x_4, y_4) = (1, 1)$.

9. Find (x_3, y_3) if the mass system consisting of mass $m_1 = 2$ grams at point $(1, 1)$, mass $m_2 = 4$ grams at point $(-2, 3)$, and mass $m_3 = 3$ grams at point (x_3, y_3) has center of mass $(\bar{x}, \bar{y}) = (0, 1)$.

In each of Exercises 10–16, find the centroid of the region bounded by the given curves.

10. $y = x^3$ and $y = x^{1/3}$, for $0 \le x \le 1$

11. $y = x^2$ and $y = x^3$

12. $x = 4 - y^2$ and $x = 0$ (*Hint:* Obtain $\bar{y}$ by symmetry considerations. Use the substitution $u = 4 - x$ to find $\bar{x}$.)

13. $y = 4 - x$, $x = 0$, and $y = 0$

14. $y = x$, $y = 4 - x$, and $y = 0$

15. $y = \sqrt{4 - x^2}$ and $y = 0$

16. $x = y^2$ and $x = 4 - y^2$ (*Hint:* This can be done entirely by symmetry considerations.)

17. Assuming $f(x)$ and $g(x)$ to be continuous, derive the formula

$$\bar{y} = \frac{\frac{1}{2}\int_a^b [f(x)^2 - g(x)^2]\,dx}{\int_a^b [f(x) - g(x)]\,dx}$$

for the y-coordinate of the centroid of the region R bounded above by the graph of $y = f(x)$ and below by the graph of $y = g(x)$, for $a \le x \le b$, as follows:

a. Imagine the region R to be the face of a plate of uniform density ρ and uniform thickness c.

b. Partition the interval $[a, b]$ into n subintervals of equal length $\Delta x = \dfrac{b-a}{n}$ and select one number t_j in each subinterval $[x_{j-1}, x_j]$.

c. Approximate the "slice" of the plate over the subinterval $[x_{j-1}, x_j]$ to be the rectangular "slab" of height $[f(t_j) - g(t_j)]$, width Δx, thickness c, and density ρ. The mass of this slice is therefore approximated by $m_j \approx c\rho[f(t_j) - g(t_j)]\,\Delta x$.

d. Observe that the center of mass of this rectangular slab, with respect to the y-axis, must lie on the line

$$y = y_j = \frac{1}{2}[f(t_j) + g(t_j)].$$

e. Write down the analogue of equilibrium equation (6) as

$$0 = \sum_{j=1}^{n} (y_j - \bar{y})m_j \approx \sum_{j=1}^{n} (y_j - \bar{y})c\rho[f(t_j) - g(t_j)]\,\Delta x.$$

Substitute for y_j from (d) and apply $\lim_{n\to\infty}$.

In Exercises 18–26, use the result of Exercise 17 together with equation (14) to find the centroid of the region bounded by the graphs of the given functions.

18. $f(x) = 1 - x^2, \quad g(x) = 2, \quad \text{for} \quad -1 \le x \le 1$

19. $f(x) = x^3, \quad g(x) = 0, \quad \text{for} \quad 0 \le x \le 1$

20. $f(x) = \sqrt{x}, \quad g(x) = 0, \quad \text{for} \quad 0 \le x \le 4$

21. $f(x) = x, \quad g(x) = -x, \quad \text{for} \quad 0 \le x \le 4$

22. $f(x) = \sin x, \quad g(x) = 0, \quad \text{for} \quad 0 \le x \le \pi$

23. $f(x) = \sqrt{9 - x^2}, \quad g(x) = 0, \quad -3 \le x \le 3$

24. $f(x) = 1, \quad g(x) = \cos x, \quad \text{for} \quad -\pi/2 \le x \le \pi/2$

25. $f(x) = x^2 + x + 1, \quad g(x) = 0 \quad \text{for} \quad 1 \le x \le 3$

26. $f(x) = 4, \quad g(x) = 4 - x^2$

27. The density of a rod of length $\ell = 10$ cm is given by $\rho(x) = 1 + x$ grams/cm^3, where x represents distance from the lighter end. Find the center of mass of the rod.

28. Find the center of mass of a 10-cm rod if the density of the rod varies uniformly from 2 grams/cm^3 at one end to 12 grams/cm^3 at the other.

29. A right circular cone has radius $r = 4$ cm and height $h = 10$ cm. The density of the material from which the cone is made is uniform. Use equation (11) to find the center of mass of the cone.

30. Find the center of mass of the solid obtained by revolving about the x-axis the region bounded by the graph of $y = x^2$ for $0 \le x \le 2$. (Assume the solid to have uniform density.)

31. Give a symmetry argument to show that the centroid of a circle is its center.

32. Use a symmetry argument to find the centroid of the region bounded above by $y = \sin x$ and below by $y = -\sin x$ for $0 \le x \le \pi$.

33. Find the centroid of the region bounded above by the graph of $f(x) = 4 - x^2$ and below by the graph of the function

$$f(x) = \begin{cases} -x - 2 & \text{for} \quad -2 \le x \le 0 \\ x - 2 & \text{for} \quad 0 \le x \le 2 \end{cases}$$

34. From symmetry considerations, find the centroid of the region enclosed by the ellipse with equation $x^2 + y^2 - xy = 6$. (*Hint:* Note that the equation is unchanged under the substitutions $y = x$ and $y = -x$.)

35. Show that the medians of a triangle intersect at the centroid.

8.5 THE THEOREM OF PAPPUS

You may have noticed a similarity between equation (14) in Section 8.4 and the formula for finding the volume of a solid of revolution by the method of cylindrical shells. To make this observation more precise, suppose that a region R in the xy-plane is bounded above by the graph of $y = f(x)$ and below by the graph of $g(x)$, for $a \le x \le b$ (Figure 5.1).

If $f(x)$ and $g(x)$ are continuous, the volume V of the solid obtained by revolving R about the y-axis is

$$V = \int_a^b 2\pi x[f(x) - g(x)]\,dx. \tag{1}$$

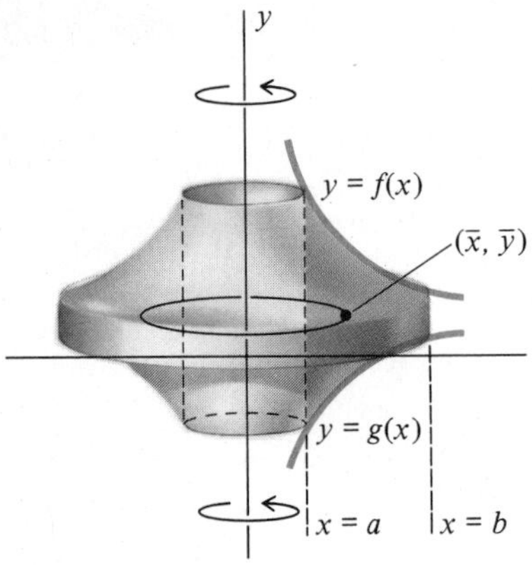

Figure 5.1 Centroid at $(\bar{x}, \bar{y})$ travels distance $2\pi\bar{x}$ as R is revolved about y-axis.

Also, the x-coordinate of the centroid $(\bar{x}, \bar{y})$ of R is given by equation (14) of the preceding section as

$$\bar{x} = \frac{1}{A} \cdot \int_a^b x[f(x) - g(x)]\, dx. \tag{2}$$

By comparing equations (1) and (2), we see that

$$V = 2\pi A\bar{x}. \tag{3}$$

Now the quantity $2\pi\bar{x}$ is simply the circumference of the circle swept out by the centroid $(\bar{x}, \bar{y})$ as R is revolved about the y-axis. We have therefore proved a particular case of the following theorem, due to the Greek mathematician Pappus, who lived approximately 300 A.D.

THEOREM 1
Pappus

Let R be a region lying in a plane and let ℓ be a line not intersecting R. The volume V of the solid obtained by revolving the region R about the line ℓ is given by the equation $V = cA$ where A is the area of R and c is the circumference of the circle swept out by the centroid of R as R revolves about the line ℓ.

By using this theorem, we can calculate certain volumes which the techniques of Chapter 7 cannot handle.

Example 1 Find the volume of the torus obtained by revolving the circle $(x - 2)^2 + y^2 = 1$ about the y-axis.

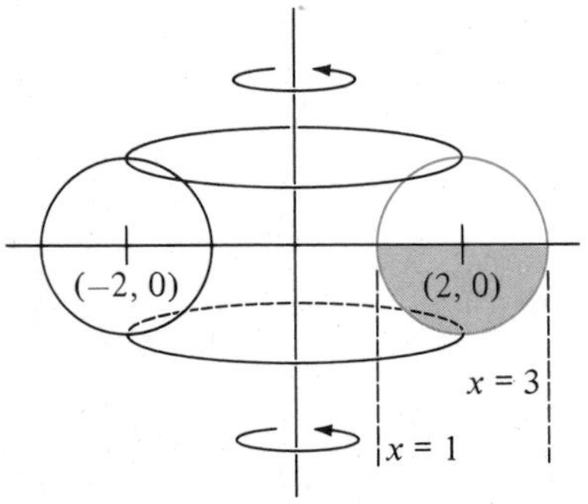

Figure 5.2 Torus obtained by revolving circle about y-axis.

Solution: Let's first note why the method of shells will not handle this problem, at least with the integration techniques developed thus far. We can write the equation for the upper semicircle as

$$y = \sqrt{1 - (x - 2)^2}$$

and the equation for the bottom semicircle as

$$y = -\sqrt{1 - (x - 2)^2}.$$

Applying the method of cylindrical shells gives the volume as

$$V = \int_1^3 2\pi x[2\sqrt{1 - (x - 2)^2}]\, dx,$$

and we have not yet developed a technique which will allow us to find an antiderivative for this integrand. (However, we *can* evaluate the definite integral by geometry—see Exercise 11.) But it is easy to see that

(i) the area of the circle is $A = \pi \cdot 1^2 = \pi$,
(ii) the centroid of the circle is $(\bar{x}, \bar{y}) = (2, 0)$, and
(iii) as the circle revolves around the y-axis, the centroid sweeps out a circle of radius $2\pi\bar{x} = 4\pi$.

Therefore, by Theorem 1, $V = (4\pi)\pi = 4\pi^2$. ∎

The Theorem of Pappus can also aid in certain calculations of hydrostatic pressure. Recall from Section 8.2 that if the flat vertical face of a submerged object lying between depths $h = a$ and $h = b$ has width $w(h)$, $a \le h \le b$ (all in feet), then the total hydrostatic force (in pounds) acting on that face is given by the integral

$$F = \int_a^b (62.4)h \cdot w(h)\, dh. \tag{4}$$

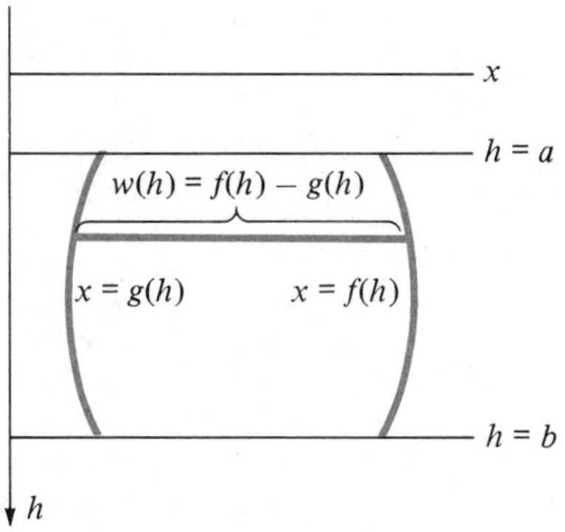

Figure 5.3 Submerged face.

If the right and left edges of the face are determined by the continuous functions $f(h)$ and $g(h)$ then we can write

$$w(h) = f(h) - g(h)$$

and equation (4) becomes

$$F = \int_a^b (62.4)h[f(h) - g(h)]\, dh. \tag{5}$$

(See Figure 5.3.)

Since the h-coordinate of the centroid of the face is given by

$$\bar{h} = \frac{1}{A}\int_a^b h[f(h) - g(h)]\, dh, \tag{6}$$

we conclude from equations (5) and (6) that

$$F = (62.4)A\bar{h}. \tag{7}$$

Equation (7) states that to calculate the hydrostatic force on a vertical face of a submerged object we may regard the face as lying horizontally at the depth of its centroid.

Example 2 Use equation (7) to calculate the hydrostatic force on the submarine window in Example 3 of Section 8.2.

Solution: Recall that the window is circular, of radius $r = 6$ inches, and that the window is positioned vertically with its top 20 feet below water level. The centroid therefore lies at a depth of $\bar{h} = 20$ ft $+ 6$ in $= 246$ in, and the area of the window is $\pi(1/2)^2 = \pi/4$ ft$^2 = 36\pi$ in^2. Since we are working in units of inches rather than feet, we must convert 62.4 lb/ft^3 to $62.4/1728 = .03611$ lb/inch^3. From equation (7) we obtain

$$\begin{aligned} F &= (.03611\ \text{lb/in}^3) \times 36\pi\ \text{in}^2 \times 246\ \text{in} \\ &\approx 1004\ \text{lb}. \end{aligned}$$

■

Exercise Set 8.5

1. Find the volume of the torus obtained by revolving the region enclosed by the circle with the equation $x^2 + (y - 5)^2 = 9$ about the x-axis.

2. Find the volume of the torus obtained by revolving the region enclosed by the circle $(x - 2)^2 + (y + 3)^2 = 4$ about the line $y = 3$.

3. Use the Theorem of Pappus to find the volume of the solid obtained by revolving the region bounded by the graphs of $y = x^2$ and $y = \sqrt{x}$ about the y-axis (see Example 4, Section 8.4).

4. Find the volume of the solid obtained by revolving the region in the first quadrant bounded by the graph of the equa-

tion $x^2 + y^2 = 16$ about the line $y = -x$ (see Example 5, Section 8.4).

5. The region R common to the circles $x^2 + y^2 = 9$ and $(x - 2)^2 + y^2 = 9$ is revolved about the line $x = -3$. Find the volume of the resulting solid in terms of A, the area of R (see Example 6, Section 8.4).

6. Find the volume of the solid obtained by revolving the region bounded by the graph of $y = 2 + \sqrt{4 - x^2}$ and the x-axis for $-2 \leq x \leq 2$ about the x-axis.

7. Find the force exerted on the end wall of a swimming pool if the wall is a rectangle 10 feet wide and 8 feet deep. Assume the pool to be completely filled.

8. Find the force on a vertical window of a submarine if the window is in the shape of a circle of radius 4 inches and the top of the window is located 30 feet below the surface of the water.

9. Suppose the window in Exercise 8 has the shape of a rectangle 8 inches high and 6 inches long with a semicircle of radius 4 inches attached at either end. Find the force on the window (Figure 5.4).

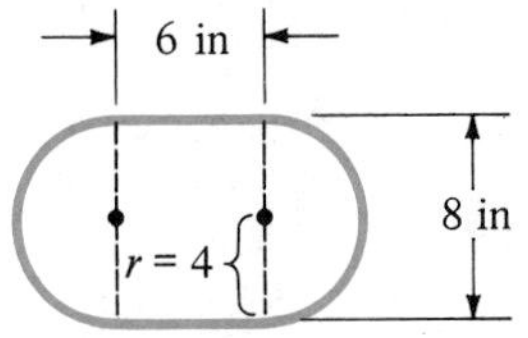

Figure 5.4

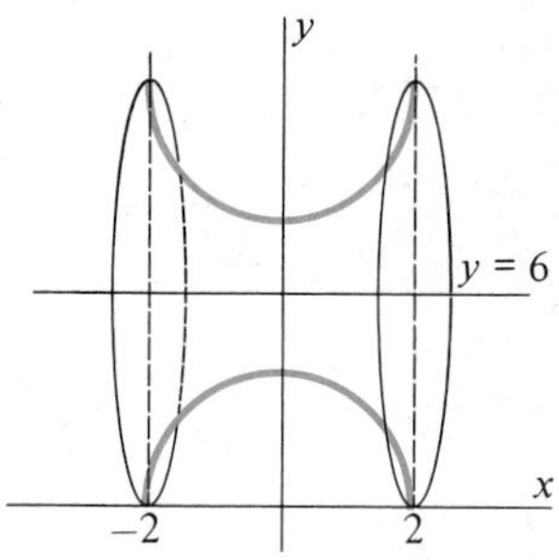

Figure 5.5

10. Find the volume of the solid obtained by revolving the graph of the semicircle $y = \sqrt{4 - x^2}$ about the line $y = 6$ (Figure 5.5).

11. Show that the integral in Example 1 can be evaluated using the u-substitution $u = x - 2$ together with the observation that

$$\int_{-1}^{1} \sqrt{1 - u^2}\, du = \pi/2.$$

(This integral gives the area of the semicircle of radius 1.)

8.6 AVERAGE VALUE OF A FUNCTION

Often a single number is sought that describes the "typical" or "average" value of a function $f(x)$ on an interval $[a, b]$. For example, consider the problem of determining average daily temperature, for the purpose of calculating energy consumption in an office building. By noting the two daily temperature functions $g(t)$ and $h(t)$ in Figures 6.1 and 6.2, we can see that such an average should not be computed simply from the difference between the maximum and minimum temperatures for the day in question. Indeed, intuition suggests that heating costs would be greater on the day whose temperature function $h(t)$ is given in Figure 6.2 since the tempera-

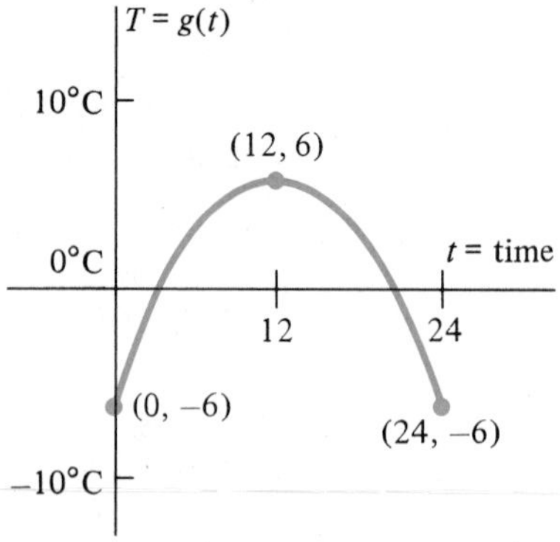

Figure 6.1 Temperature $g(t) = 6 - \frac{1}{12}(t - 12)^2$.

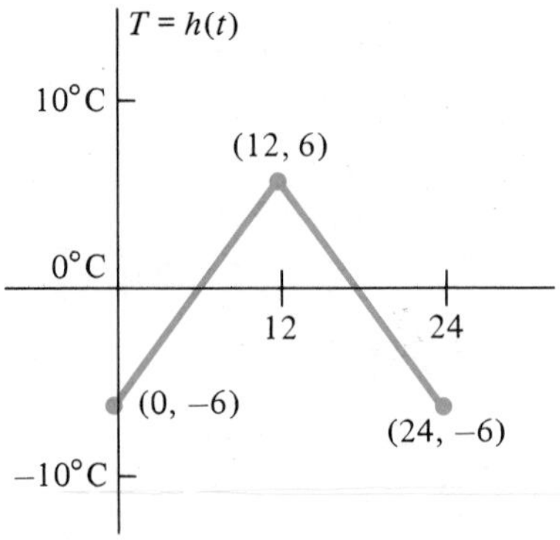

Figure 6.2 Temperature $h(t) = \begin{cases} t - 6, & 0 \leq t \leq 12 \\ 18 - t, & 12 \leq t \leq 24 \end{cases}$.

tures $h(t)$ are lower than the temperatures $g(t)$, except at times $t = 0$, $t = 12$, and $t = 24$ hours. Notice, however, that on both days the maximum temperatures (6°C) and minimum temperatures (−6°C) are the same.

The above remarks suggest that a measurement of average temperature for a temperature function $f(t)$, for $a \le t \le b$, should reflect not only the difference between maximum and minimum temperatures but also some measurement of how long the temperature function remained at various temperature levels in between. One way to do this is to partition the interval $[a, b]$ into n subintervals of equal length $\Delta t = \frac{b - a}{n}$ at points $a = t_0 < t_1 < t_2 < \cdots < t_n = b$. If we select one time s_j in each of the subintervals $[t_{j-1}, t_j]$, the n numbers $f(s_1), f(s_2), \ldots, f(s_n)$ represent temperature readings taken from among equally spaced time intervals. By averaging these values we arrive at a number

$$\bar{f}_n = \frac{f(s_1) + f(s_2) + \cdots + f(s_n)}{n}$$

reflecting an approximate value for $f(t)$ in each subinterval $[t_{j-1}, t_j]$ of the time interval $[a, b]$. We therefore argue that by letting $n \to \infty$ we should obtain an average $\bar{f}$, reflecting *each* individual function value $f(t)$, $t \in [a, b]$. That is, we take

$$\bar{f} = \lim_{n \to \infty} \bar{f}_n = \lim_{n \to \infty} \frac{1}{n} \sum_{j=1}^{n} f(s_j). \tag{1}$$

Since $\Delta t = \frac{b - a}{n}$, we can write the sum in equation (1) as a Riemann sum as follows:

$$\begin{aligned} \frac{1}{n} \sum_{j=1}^{n} f(s_j) &= \sum_{j=1}^{n} f(s_j) \cdot \frac{1}{n} \\ &= \left(\frac{b - a}{b - a}\right) \sum_{j=1}^{n} f(s_j) \cdot \frac{1}{n} \\ &= \frac{1}{b - a} \sum_{j=1}^{n} f(s_j) \left(\frac{b - a}{n}\right) \\ &= \frac{1}{b - a} \sum_{j=1}^{n} f(s_j)\, \Delta t. \end{aligned}$$

If $f(t)$ is continuous for $t \in [a, b]$, the limit as $n \to \infty$ of this Riemann sum is a definite integral:

$$\bar{f} = \lim_{n \to \infty} \left\{\frac{1}{b - a} \sum_{j=1}^{n} f(s_j)\, \Delta t\right\} = \frac{1}{b - a} \int_a^b f(t)\, dt.$$

Thus, if $f(x)$ is a continuous function for $x \in [a, b]$ we *define* the *average value A of $f(x)$ on $[a, b]$* by

$$A = \bar{f} = \frac{1}{b - a} \int_a^b f(x)\, dx. \tag{2}$$

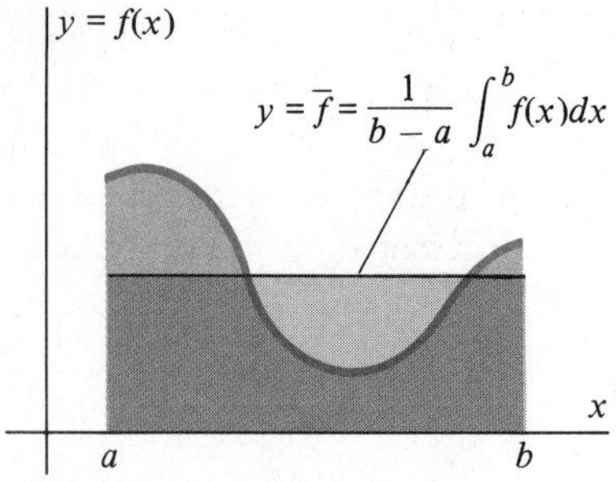

Figure 6.3 When $f(x) \geq 0$ for all $x \in [a, b]$, $\bar{f}$ is the height of the rectangle with base $(b - a)$ and area $\int_a^b f(x)\, dx$.

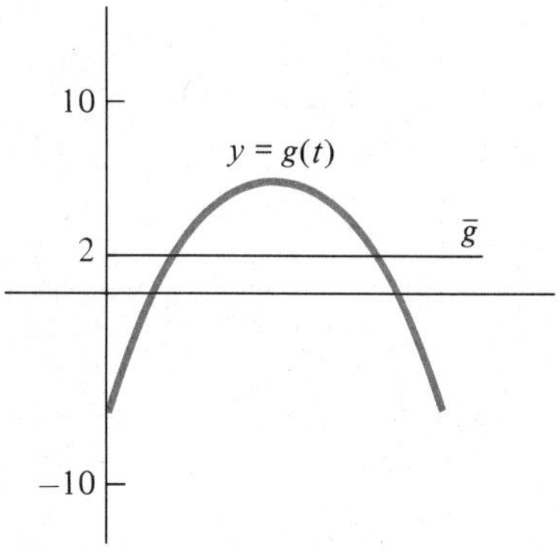

Figure 6.4 $\bar{g} = 2°C$.

Of course, we have seen equation (2) before. Recall from Chapter 6 that the Mean Value Theorem for Integrals guarantees that $\bar{f}$ exists and that $\bar{f} = f(c)$ for some $c \in [a, b]$. When $f(x)$ is nonnegative, we may interpret $\bar{f}$ as the height of a rectangle whose base has length $b - a$ and whose height is $\int_a^b f(x)\, dx$. Our purpose in reintroducing this concept is to extend its interpretation beyond the area considerations of Chapter 6 (see Figure 6.3).

Example 1 Find the average temperature for the temperature functions $g(t)$ and $h(t)$ in Figures 6.1 and 6.2.

Solution: Applying equation (2) to $g(t) = 6 - \frac{1}{12}(t - 12)^2$, we obtain

$$\begin{aligned}\bar{g} &= \frac{1}{24 - 0}\int_0^{24}\left[6 - \frac{1}{12}(t - 12)^2\right] dt\\ &= \frac{1}{24}\left[6t - \frac{1}{36}(t - 12)^3\right]_0^{24}\\ &= \frac{1}{24}\{144 - 2(48)\}\\ &= 2°\text{C} \qquad \text{(Figure 6.4).}\end{aligned}$$

For the temperature function $h(t) = \begin{cases} t - 6, & 0 \leq t \leq 12 \\ 18 - t, & 12 \leq t \leq 24 \end{cases}$, we obtain

$$\begin{aligned}\bar{h} &= \frac{1}{24 - 0}\left\{\int_0^{12}(t - 6)\, dt + \int_{12}^{24}(18 - t)\, dt\right\}\\ &= \frac{1}{24}\left\{\left[\frac{1}{2}t^2 - 6t\right]_0^{12} + \left[18t - \frac{1}{2}t^2\right]_{12}^{24}\right\}\\ &= \frac{1}{24}\{(72 - 72) + (144 - 144)\}\\ &= 0°\text{C}.\end{aligned}$$

■

Example 2 Suppose that the amount of heating oil required to heat a house for one day during the winter months is the product of the difference between 20°C and the average daily temperature multiplied by 0.6 gal. Find the levels of oil consumption associated with each of the temperature functions in Figures 6.1 and 6.2.

Solution: For temperature function g, the oil consumption is

$$\begin{aligned}C_g &= (20°\text{C} - A_g)\cdot(0.6 \text{ gal per °C})\\ &= (20 - 2)(.6) = 10.8 \text{ gal.}\end{aligned} \tag{3}$$

For the temperature function h, the oil consumption is

$$\begin{aligned}C_h &= (20°\text{C} - A_h)\cdot(0.6 \text{ gal per °C})\\ &= (20° - 0°)(.6) = 12 \text{ gal.}\end{aligned}$$

Thus, our remark about the cost of heating was correct: the function $h(t)$ does lead to a greater heating cost than does $g(t)$. ■

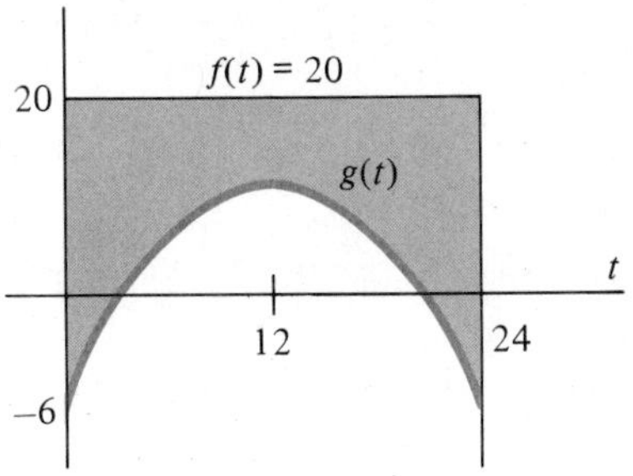

Figure 6.5 Consumption is average value of $(0.6)[20 - g(t)]$.

REMARK 1: In Example 2, the consumption for temperature function $g(t)$ can be associated with the area bounded above by the graph of $f(t) \equiv 20°\text{C}$ and below by the graph of $g(t)$ (see Figure 6.5). The calculation in equation (3) can therefore be viewed as the product of (0.6 gal per °C) times the average value of the difference $d(t) = 20° - g(t)$, since

$$C_g = (0.6)(20 - A_g) = (0.6)\left\{\frac{24 \cdot 20}{24} - \frac{1}{24} \cdot \int_0^{24} g(t)\,dt\right\}$$
$$= (0.6)\left\{\frac{1}{24} \cdot \int_0^{24} [20 - g(t)]\,dt\right\}.$$

REMARK 2: Each of the formulas developed in Chapters 7 and 8 for calculating mathematical and physical quantities as definite integrals can be restated in terms of average value. For example,

(a) The area, A, bounded by the graphs of $y = f(x)$ and $y = g(x)$ for $a \le x \le b$ can be written as

$$A = \int_a^b |f(x) - g(x)|\,dx = (b - a)\left\{\frac{1}{b - a}\int_a^b |f(x) - g(x)|\,dx\right\}$$
$$= (b - a) \cdot (\text{average value of } |f(x) - g(x)|)$$
$$= (b - a)\,\overline{|f(x) - g(x)|}.$$

(b) The volume, V, of the solid obtained by revolving the region bounded by the graph of $y = f(x)$ and the x-axis about the x-axis is

$$V = \int_a^b \pi[f(x)]^2\,dx = \pi(b - a)\left\{\frac{1}{b - a}\int_a^b [f(x)]^2\,dx\right\}$$
$$= \pi(b - a) \cdot (\text{average value of } f^2(x))$$
$$= \pi(b - a)\overline{f^2(x)}.$$

(c) The work, W, required to move an object requiring force $f(x)$ from $x = a$ to $x = b$ is

$$W = \int_a^b f(x)\,dx = (b - a)\left\{\frac{1}{b - a}\int_a^b f(x)\,dx\right\}$$
$$= (b - a) \cdot (\text{average value of force, } f(x))$$
$$= (b - a)\overline{f(x)}.$$

Clearly, we could have conducted the entire discussion on applications of the integral in the language of average value. Rather, we have opted to retain this topic as a final unifying commentary, noting for one last time the essential characteristic of the definite integral—the single number, arising as the limit of an approximation scheme, which best describes the behavior of a continuous function throughout an interval $[a, b]$.

Exercise Set 8.6

1. Find an estimate for the average temperature $\overline{T}$ for the function in Figure 6.6 using $n = 8$ time periods of equal size and the data in Table 6.1, using the left endpoint of each interval as the "test point" t_j.

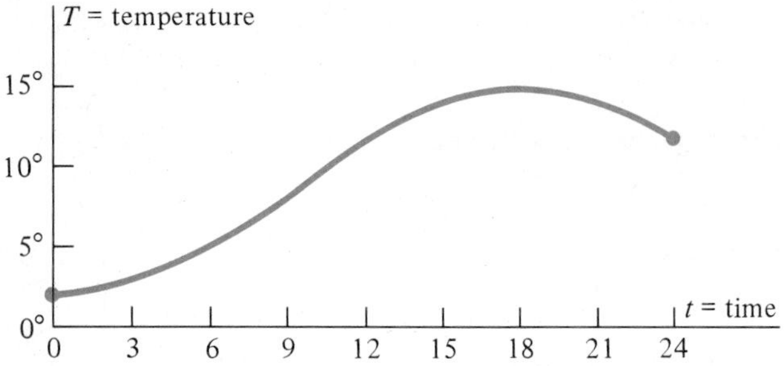

Figure 6.6 Temperature in degrees Celsius as a function of time during a 24-hour period.

Table 6.1

t	T
12 a.m.	2°
3 a.m.	3°
6 a.m.	5°
9 a.m.	8°
12 p.m.	12°
3 p.m.	14°
6 p.m.	15°
9 p.m.	14°
12 a.m.	12°

t = time
T = temperature

2. Repeat Exercise 1 for $n = 12$ using the data in Table 6.2. Again use the left endpoints as "test points."

Table 6.2

t	0	2	4	6	8	10	12	2	4	6	8	10	12
T	2	2	4	5	7	9	12	13	14	15	15	13	12

3. Find estimates for average temperature, $\overline{T}$, using (a) $n = 4$, (b) $n = 6$, and (c) $n = 12$ and the data in Table 6.3, and using left endpoints as test points.

Table 6.3

t	12	2	4	6	8	10	12.	2	4	6	8	10	12
T	−6	−8	−8	−7	−5	−3	0	2	1	−1	−3	−5	−8

4. Find the average value of the function $f(x) = x$ on the interval $[0, 3]$ by geometry.

5. Find the average value of the function $f(x) = 2x + 1$ on the interval $[1, 3]$ by geometry.

6. Find the average value of the function $f(x) = \sqrt{5 - x^2}$ by geometry.

In each of Exercises 7–14, find the average value of the given function on the given interval.

7. $f(x) = 2x - 1, \quad x \in [-2, 2]$

8. $f(x) = x^2 - 7, \quad x \in [-1, 3]$

9. $f(x) = \dfrac{x^3 + 1}{1 + x}, \quad x \in [2, 4]$

10. $f(x) = \sqrt{x + 2}, \quad x \in [-2, 2]$

11. $f(x) = x \sin(\pi - x^2), \quad x \in [0, \sqrt{\pi}/2]$

12. $f(x) = \sqrt{9 - x^2}, \quad x \in [-3, 3]$ (*Hint:* Use area considerations.)

13. $f(x) = \sin(x + x^3), \quad x \in [-\pi/4, \pi/4]$ (*Hint:* Use symmetry considerations.)

14. $f(x) = x \sec^2(\pi + x^2), \quad x \in \left[0, \dfrac{\sqrt{\pi}}{2}\right]$

15. During a 24-hour period the outside temperature rose from −4°C to 19°C at a constant rate. Find the average temperature.

16. Find the oil consumption for the house in Example 2 on a day when the temperature was given by the function

$$f(t) = \begin{cases} 2t - 10, & 0 \le t \le 8 \\ 6, & 8 \le t \le 16 \\ 22 - t, & 16 \le t \le 24. \end{cases}$$

17. Find the average value of the cost function $C(x) = 10 + 5 \sin \pi_x$ for $0 \le x \le 8$.

18. Write down the formula for the surface area of a volume of revolution in terms of average value.

19. Write down the formula for the centroid of a plane region R bounded above by $f_t(x)$, below by $g_b(x)$, on the right by $f_r(y)$, and on the left by $g_\ell(y)$ in terms of average value.

20. Show that the average value of the difference of two continuous functions on $[a, b]$ is the difference of their average values. What is the relationship for $cf(x)$? What about products $f(x)g(x)$?

21. Find a number a so that the average value of $f(x) = x^2 + 3$ on $[0, a]$ is $A = 6$.

22. Find a so that the average value of $f(x) = x^3 + \sqrt[3]{x}$ on $[-2, a]$ is $A = 0$. (*Hint:* Use the symmetry of $f(x)$.)

23. What is the relationship between the average value of $f(x)$ on $[a, b]$ and the area bounded by the graph of $y = f(x)$ and the x-axis if $f(x) \geq 0$ for all $x \in [a, b]$?

24. What is the relationship between average velocity as defined in Chapter 3 and the average value of a velocity function?

25. True or false: If A is the average value of $f(x)$ on $[a, b]$, then there is a number $c \in [a, b]$ so that $f(c) = A$, and $\int_a^c f(x)\,dx = \int_c^b f(x)\,dx$.

26. True or false: If A_1 is the average value of $f(x)$ on $[a, b]$, and A_2 is the average of $f(x)$ on $[b, c]$, then $A_1 + A_2$ is the average value of $f(x)$ on $[a, c]$.

27. True or false: The average value of the product of two continuous functions is the product of their average values.

SUMMARY OUTLINE OF CHAPTER 8

- The **hydrostatic force** acting on a submerged face lying between depths $h = a$ and $h = b$ whose width at depth h is $w(h)$ is given by the integral

$$F = \int_a^b (62.4) \cdot h \cdot w(h)\,dh.$$

- The **work** done by a variable force $F(x)$ in moving an object from point $x = a$ to point $x = b$ is given by the integral

$$W = \int_a^b F(x)\,dx.$$

- The **center of mass** of a rod of constant cross-sectional area A and continuously varying density $\rho(x)$ is

$$\bar{x} = \frac{\int_0^\ell x\rho(x)A\,dx}{\int_0^\ell \rho(x)A\,dx}.$$

- The **center of mass** of a rod of uniform density ρ and continuously varying cross-sectional area $A(x)$ is

$$\bar{x} = \frac{\int_0^\ell \rho x A(x)\,dx}{\int_0^\ell \rho A(x)\,dx}.$$

- The **centroid** of the region R bounded above by the graph of $y = f_t(x)$, below by the graph of $y = g_b(x)$, on the right by the graph of $x = f_r(y)$, and on the left by the graph of $x = g_\ell(y)$ is $(\bar{x}, \bar{y})$, where

$$\bar{x} = \frac{1}{A} \cdot \int_a^b x[f_t(x) - g_b(x)]\,dx \qquad (A = \text{area})$$

$$\bar{y} = \frac{1}{A} \cdot \int_c^d y[f_r(y) - g_\ell(y)]\,dy \qquad (A = \text{area})$$

- ***Theorem of Pappus:*** The volume V of the solid obtained by revolving the region R about the line ℓ (not intersecting R) is $V = cA$, where A is the area of R and c is the circumference of the circle swept out by the centroid of R.

- The **average value** of the continuous function $y = f(x)$ on the interval $[a, b]$ is

$$\bar{f} = \frac{1}{b - a} \int_a^b f(x)\,dx.$$

REVIEW EXERCISES—CHAPTER 8

1. A spring is stretched 10 inches by a force of 20 pounds. How much work is needed to stretch it 15 inches?

2. A cylindrical tank 10 feet in diameter and 20 feet high is full of water. How much work will be done in pumping all the water out through a valve in the top?

3. A tank in the shape of an inverted right circular cone has a radius of 8 feet and a depth of 12 feet. The tank is full of water. How much work will be done in pumping all the water to a height 10 feet above the top of the tank?

4. How much work is done in Problem 3 if only half the water is pumped to a height of 10 feet above the tank?

5. A 400-pound chain 20 feet long hangs from a windlass. How much work is done in winding in the chain?

6. How much work is done in Problem 5 if a 50-pound hook hangs at the end of the chain?

7. A water trough has end panels in the shape of trapezoids with lower bases 12 inches and upper bases 18 inches. The altitude of the trapezoid is 12 inches. Find the hydrostatic pressure on an end panel when the trough is full of water.

8. The vertical wall of a dam has the shape of the region enclosed by the parabola $y = \frac{1}{16}x^2$. Find the hydrostatic pressure on the wall when the water level behind it is 16 feet.

9. Show that the hydrostatic force against one face of an object submerged vertically in water is the product of the pressure at the centroid of the face times the area of the face.

10. A semicircular plate of radius 12 inches is submerged vertically in water with its diameter at the top edge and parallel to the surface of the water. Find the hydrostatic force on one face of the plate if the diameter lies 6 inches below water level.

11. A 20-cm rod has uniform cross-sectional area $A = 16\pi \text{ cm}^2$. Find its center of mass if its density varies uniformly from 2 grams per cm^3 at one end to 8 grams per cm^3 at the other.

12. A 10-cm rod has uniform density $\rho = 10$ grams per cm^3. Find the center of mass of the rod if its cross sectional area at a point x cm from one end is $(2 + .05x^2)\text{ cm}^2$.

13. Three particles of mass 2 grams, 5 grams, and 7 grams are located on the x-axis at points having coordinates -5, 1, and 4, respectively. Find the center of mass of the system.

14. Find the coordinates of the center of mass of the system consisting of four particles having coordinates $(-2, 4)$, $(-2, -2)$, $(0, 6)$, and $(1, -4)$ if the particles are of equal mass.

15. Find the center of mass of a rod of length 100 cm if the cross-sectional area is constant and the density is proportional to $1 + \sqrt{x}$ grams per cm^3 where x is the distance from one end.

16. Sketch an example of a region R in the plane whose centroid does not lie within R.

17. Find the centroid of the region bounded above by the graph of $y = x$, below by the x-axis, on the left by the line $x = 0$, and on the right by the line $x = 2$.

18. Find the centroid of the region bounded above by the graph of $y = 16 - x^2$ and below by the x-axis.

19. Find the centroid of the region bounded above by the graph of $y = x$ and below by the graph of $y = x^2$.

20. Find the centroid of the region bounded above by the graph of $y = 6x - x^2$ and below by the x-axis.

21. Find the centroid of the ellipse $9x^2 + 4y^2 = 36$.

22. Find the centroid of the region bounded by the graphs of $y = 6x - x^2$ and $y = 3 - |x - 3|$.

23. Use the Theorem of Pappus to find the volume of the cone obtained by rotating about the y-axis the triangle with vertices $(4, 0)$, $(0, 8)$, and $(0, 0)$.

24. Use the Theorem of Pappus to find the volume of the solid obtained by rotating the region bounded above by the graph of $y = \sqrt{4 - x^2}$ and below by the x-axis about the line

a. $y = 0$,
b. $y = -2$,
c. $y = 6$.

In Exercises 25–30, find the average value of the given function.

25. $y = x\sqrt{1 - x^2}$ on $[-1, 1]$

26. $y = \sin x \cos x$ on $[0, \pi/2]$

27. $y = x^3 \sin x^2$ on $[-\pi, \pi]$ (*Hint:* Use the fact that $y = f(x)$ is odd.)

28. $y = \dfrac{x}{\sqrt{1 + x^2}}$ on $[0, 2]$

29. $y = \sqrt{x} + \sqrt[3]{x}$ on $[0, 1]$

30. $y = \sin x$ on $[0, 10\pi]$

31. Show that the average value of $f(x) = \sin x$ on intervals of the form $[2n\pi, 2m\pi]$ is $\bar{f} = 0$, where n and m are integers. What about intervals of the form $[x_0 + 2n\pi, x_0 + 2m\pi]$?

32. Show that the average value of $f(x) = \sin^2 x$ on intervals of the form $[0, n\pi]$ is $\bar{f} = \frac{1}{2}$ where n is an integer.

33. Give an argument involving average values to show that $\lim_{L\to\infty} \frac{1}{L}\int_0^L \sin^2 x\, dx = \frac{1}{2}$. What about

$$\lim_{L\to\infty} \frac{1}{L}\int_0^L \cos^2 x\, dx?$$

Leonhard Euler

Johann Bernoulli

Jakob Bernoulli

UNIT 4

THE TRANSCENDENTAL FUNCTIONS

Discovery of the Transcendental Functions

At the beginning of the eighteenth century, the calculus was still in its infancy. Logically precise foundations of the subject were still to be developed, and only polynomials could be differentiated or integrated easily. The logarithmic and exponential functions were still to be invented, and the trigonometric functions were not yet really understood as functions.

One of the greatest innovators in mathematics then appeared: Leonhard Euler (1707–1783). Born in Switzerland, he spent many years in St. Petersburg (now Leningrad), and then taught in Berlin for 25 years. Catherine the Great then hired him back to Russia, where he spent the last 17 years of his life. He was perhaps the most prolific mathematician of all time—he published 520 books and papers and, after his death, the *Proceedings of the St. Petersburg Academy* continued to publish at least one new paper by Euler in every issue for the next 47 years! His output is all the more incredible for the fact that he was blind for virtually all of his second Russian period, dictating his work to one of his sons.

No one else did more to put collegiate mathematics in its present form. Among the symbols Euler initiated are the sigma (Σ) for summation, e to represent the constant 2.71828. . . , i for the imaginary $\sqrt{-1}$, and even a, b, and c for the sides of a triangle and A, B, and C for the opposite angles. Although one William Jones had first used the symbol π for 3.14159. . . in 1706, Euler made its use standard.

Euler wrote textbooks in differential and integral calculus, and the general form of these texts is still in use today. His greatest work was the *Introductio in analysin infinitorum* (1748), in which Euler did for the calculus what Euclid had done for geometry and the theory of numbers in ancient Greece. He systematized differentiation and the method of fluxions (integral calculus), creating a new system that is often called *analysis*—the study of infinite processes. The French physicist and astronomer Arago called Euler "analysis incarnate."

Euler transformed the trigonometric ratios into the trigonometric functions as we think of them today, and even first used the abbreviations sin, cos, and tan. He treated logarithms and exponents as functions, whereas their creators (John Napier and Henry Briggs) had thought of them merely as tools to aid calculation.

Leonhard Euler received both his bachelor's and his master's degrees at the age of 15, studying mathematics primarily under Johann Bernoulli. Bernoulli was one of a dozen mathematicians of that name, stretching over six generations, though the most famous are Johann (1667–1748) and his brother Jakob (1654–1705). The two brothers corresponded for many years with Leibniz about the calculus, and made many discoveries with which we are familiar today. Johann first introduced integration by partial fractions. He also discovered a relation between the trigonometric and logarithmic functions, thus paving the way to the realization that there are only two basic types of elementary functions: polynomial, rational, and algebraic functions on one hand, and the transcendental (trigonometric, logarithmic, exponential, and hyperbolic) functions on the other. Johann Bernoulli and his brother Jakob both studied such things as arc length, the curvature of curves, and points of inflection. Their ideas may seem somewhat naive today, however. One postulate states, "Each curved line consists of infinitely many straight lines, these themselves being infinitely small."

Jakob Bernoulli was the first to publish polar coordinates (although Newton had the general idea some years before). He studied the catenary or hanging chain curve, which appears in this unit as the hyperbolic cosine curve. He suggested for the first time the term *calculus integralis* as a name for what Leibniz had called *calculus summatorius,* and Leibniz agreed that this was a better term. This is why we use the terminology "integral calculus" rather than "summation calculus."

Many other mathematicians also contributed to the rapid growth of calculus in the eighteenth century. However, the greatest influence on the modern calculus classroom was that of the functions, symbols, and methods discovered and promoted by Leonhard Euler and the brothers Bernoulli.

Pronunciation Guide

Leonhard Euler (Len′ard Oy′ler)

Johann Bernoulli (Yo′hahn Burr-noo′lee)

(Photograph of Leonhard Euler from the Library of Congress. Photographs of Johann and Jakob Bernoulli from the David Eugene Smith Papers, Rare Book and Manuscript Library, Columbia University.)

CHAPTER 9

LOGARITHMIC AND EXPONENTIAL FUNCTIONS

9.1 INTRODUCTION

In high school mathematics, you encountered the **logarithm to the base b,** defined by the equation

$$\log_b y = x \qquad \text{if and only if} \qquad y = b^x. \tag{1}$$

In writing equation (1) we assume that b is a positive constant not equal to 1 and that $y > 0$. For example,

$$\begin{aligned} \log_{10} 100 &= 2, & &\text{since} & 10^2 &= 100, \\ \log_2 8 &= 3, & &\text{since} & 2^3 &= 8, \\ \log_{16} 4 &= \frac{1}{2}, & &\text{since} & \sqrt{16} &= 4, \end{aligned}$$

and

$$\log_2 \frac{1}{4} = -2, \qquad \text{since} \qquad 2^{-2} = \frac{1}{4}.$$

Logarithms, as defined in equation (1), have the following properties:

$(L1)$ $\quad \log_b(mn) = \log_b m + \log_b n,$

$(L2)$ $\quad \log_b\left(\dfrac{m}{n}\right) = \log_b m - \log_b n,$

$(L3)$ $\quad \log_b(n^r) = r \log_b n, \qquad r$ rational.

These properties follow from the laws of exponents (see Section 1.2) and the definition in equation (1) (see Exercises 19–21).

Notice that equation (1) defines the logarithm function as the *inverse* of the exponential function $y = b^x$, according to our definition of inverse function in Chapter 3. However, since b^x is defined only when x is a rational number, neither of these functions is either continuous or differentiable.

Natural Logarithm and Exponential Functions

The goal of this chapter is to use the theory of the calculus to develop more general logarithmic and exponential functions. These new functions, called the *natural* logarithm and exponential functions, will be both continuous and differentiable throughout their domains. Besides answering certain mathematical questions, these new functions are useful in developing mathematical models for the growth or decay of certain chemicals and biological populations.

There are two approaches to developing the natural logarithm and exponential functions, both of which are commonly used in calculus texts. Regardless of the approach taken, the natural logarithm and exponential functions will turn out to be inverses of each other. One may therefore begin by defining either function from first principles and then obtain the other function as the inverse of the first.

A yet unanswered question leads us to begin by defining the natural logarithm function. Recall, from Chapter 5, that the antidifferentiation formula

$$\int x^n\,dx = \frac{x^{n+1}}{n+1} + C, \qquad n \neq -1 \tag{2}$$

does not apply when $n = -1$. In fact, we have not yet found an "elementary" antiderivative for the reciprocal function $f(x) = 1/x$. However, by Theorem 6 of Chapter 6 we are assured that the function

$$L(x) = \int_1^x \frac{1}{t}\,dt, \qquad x > 0 \tag{3}$$

is differentiable, and that $\frac{d}{dx}L(x) = 1/x$. We therefore *define* the natural logarithm function to be $L(x)$ in equation (3). That is, we define the natural logarithm function as an antiderivative for the function $f(x) = 1/x$. We will then show that

(a) $L(x)$ in equation (3) actually represents a logarithm function, that is, it satisfies equations $(L1)$ through $(L3)$,
(b) $L(x)$ has an inverse function, which we will denote by $\exp(x)$,
(c) $\exp(x)$ actually is an exponential function, and
(d) $\exp(x)$ leads to solutions of models for quantities obeying the Law of Natural Growth.

Before proceeding with the details of this program, you should refresh yourself on the basic properties of logarithmic and exponential functions in the exercise set.

Exercise Set 9.1

In Exercises 1–10, find the logarithm.

1. $\log_{10} 1000$

2. $\log_7 343$

3. $\log_8 2$

4. $\log_4 (0.5)$

5. $\log_5\left(\frac{1}{125}\right)$

6. $\log_2 \sqrt{2}$

7. $\log_4 8$

8. $\log_{343} 7$

9. $\log_{10} 1$

10. $\log_b 1$

In Exercises 11–14, sketch the graph of each function by first plotting several points.

11. $f(x) = 2^x$
$g(x) = \log_2 x$

12. $f(x) = 10^x$
$g(x) = \log_{10} x$

13. $f(x) = \left(\frac{1}{2}\right)^x$
$g(x) = \log_{1/2} x$

14. $f(x) = 4^x$
$g(x) = \log_4 x$

15. Find a if $\log_a 16 = 4$.

16. Find a if $a^3 = 125$.

17. The exponential function occurs naturally in formulas for the periodic compounding of interest. For example, if an annual rate of interest of r percent is applied to a principal amount P_0 placed in a savings account, the amount $P(1)$ on deposit at the end of one year is

$$P(1) = P_0 + rP_0 = (1 + r)P_0.$$

In the second year the interest rate r is applied to the new principal, $(1 + r)P_0$. Thus, the amount $P(2)$ on deposit at the end of two years is

$$P(2) = (1 + r)P_0 + r[(1 + r)P_0] = (1 + r)^2 P_0.$$

Show that the amount $P(n)$ on deposit at the end of n years is given by the exponential function

$$P(n) = (1 + r)^n P_0.$$

18. Show that if interest is compounded k times per year at an annual percentage rate r, the amount $P(n)$ on deposit in a savings account after n years is

$$P(n) = \left(1 + \frac{r}{k}\right)^{nk} P_0$$

where P_0 is the amount of the original principal.

19. Prove property ($L1$) of logarithms as follows:

a. Let $x = \log_b m$ and $y = \log_b n$. Show that $m = b^x$ and $n = b^y$.

b. By the laws of exponents, show that $mn = b^{x+y}$.

c. Explain why $\log_b b^z = z$, z rational.

d. Conclude that $\log_b(mn) = \log_b[b^{x+y}] = x + y = \log_b m + \log_b n$.

20. Prove property ($L2$) (see Exercise 19).

21. Prove property ($L3$) (see Exercise 19).

22. The Beer-Lambert law relates the absorption of light travelling through a material to the concentration and the thickness of the material. If I_0 and I denote the intensities of light of a particular wavelength before and after passing through the material, respectively, and if x denotes the length of the path followed by the beam of light passing through the material, then

$$\log_{10}\left(\frac{I}{I_0}\right) = kx$$

where k is a constant depending on the material. Express I as a function of x.

23. True or false? The values of $y = \log_5 x$, as defined in this section, must be rational numbers.

24. True or false? If $0 < x < y$ then $\log_a x < \log_a y$. Why?

9.2 THE NATURAL LOGARITHM FUNCTION

We begin by recalling Theorem 6 of Chapter 6: If $f(t)$ is continuous for all t in an interval $[a, b]$, then for $c \in (a, b)$ the function

$$F(x) = \int_c^x f(t)\, dt, \qquad a \le x \le b$$

is a differentiable function of x, and $\frac{d}{dx} F(x) = f(x)$. This theorem is the key to answering the question raised in Section 9.1 about finding an antiderivative for the function $f(x) = 1/x$. In light of this theorem, we *define* the function $y = \ln x$, for $x > 0$ by

$$\ln x = \int_1^x \frac{1}{t}\, dt, \qquad x > 0. \tag{1}$$

Then $y = \ln x$ is a differentiable function for all x in the interval $(0, \infty)$, and

$$\frac{d}{dx} \ln x = \frac{1}{x}, \qquad x > 0. \tag{2}$$

Since a differentiable function is necessarily continuous, this shows that $f(x) = \ln x$ is continuous on $(0, \infty)$.

The geometric interpretation of the expression $y = \ln x$ is simple:

(a) If $x > 1$, $\ln x$ is the area of the region bounded by the graph of $f(t) = 1/t$ and the t-axis between $t = 1$ and $t = x$ (Figure 2.1).

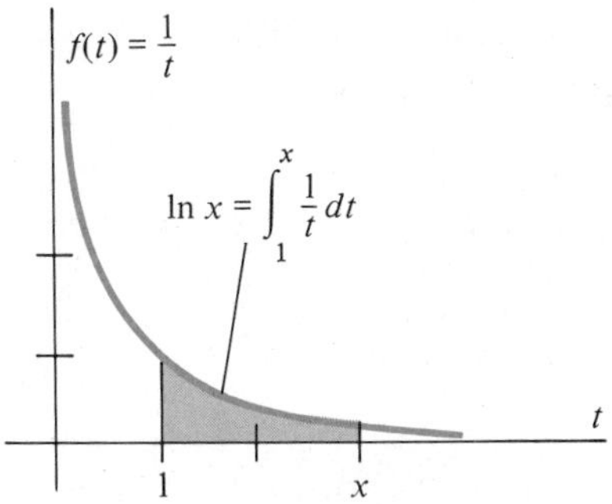

Figure 2.1 $\ln x > 0$ if $x > 1$.

(b) If $x < 1$, $\ln x$ is the *negative* of the area bounded by the graph of $f(t) = 1/t$ and the t-axis between $t = x$ and $t = 1$ (Figure 2.2).

(c) If $x = 1$, $\quad \ln x = \int_1^1 \frac{1}{t}\,dt = 0, \qquad$ so $\ln 1 = 0$.

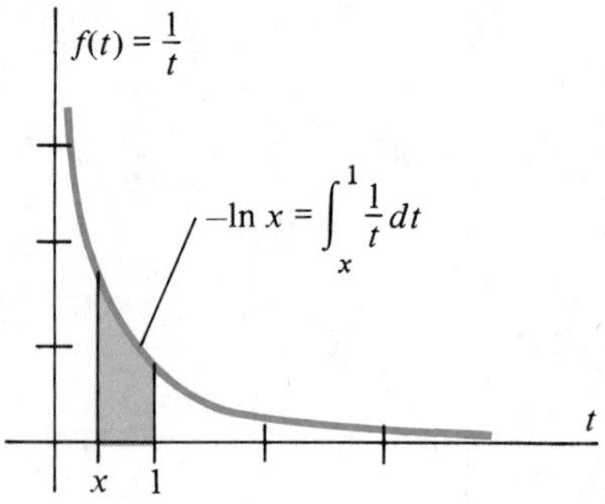

Figure 2.2 $\ln x < 0$ if $x < 1$.

REMARK: We could have chosen any number in $(0, \infty)$ for the fixed boundary c of the area function. The natural logarithm function is defined with $c = 1$ in order to have $\ln 1 = 0$. Since $\log_a 1 = 0$ for all logarithm functions (because $a^0 = 1$ by definition), this agrees with the behavior we expect for $\ln x$.

We refer to the function $y = \ln x$ defined by equation (1) as the **natural logarithm function.** For $x > 0$ the value of $\ln x$ can be approximated to any desired accuracy by applying one of the procedures for approximating integrals (such as Simpson's Rule) to the integral in equation (1). A table of values for $\ln x$ appears in the Appendix, and most hand calculators and computers are preprogrammed to provide values of $\ln x$. *Note that* $\ln x$ *is not defined for* $x \leq 0$. If $u(x)$ is a differentiable function of x, we may use the Chain Rule to generalize the differentiation formula of equation (2) to:

$$\frac{d}{dx}\ln u = \frac{1}{u}\cdot\frac{du}{dx}, \qquad u > 0. \tag{3}$$

Example 1 Find $f'(x)$ for (a) $f(x) = \ln 3x$, and (b) $f(x) = x\ln(1 + x^2)$.

Solution:

(a) $f'(x) = \frac{1}{3x}\cdot 3 = \frac{1}{x}$, by equation (3).

(b) By the Product Rule and equation (3) we obtain

$$\begin{aligned} f'(x) &= (1)\ln(1 + x^2) + x\cdot\frac{d}{dx}[\ln(1 + x^2)] \\ &= \ln(1 + x^2) + x\left(\frac{1}{1 + x^2}\right)\cdot\frac{d}{dx}(1 + x^2) \\ &= \ln(1 + x^2) + x\left(\frac{1}{1 + x^2}\right)(2x) \\ &= \ln(1 + x^2) + \frac{2x^2}{1 + x^2}. \end{aligned}$$

■

But what does the function defined by equation (1) have to do with logarithms? The answer is provided by the following theorem which shows that *the function* $y = \ln x$ *satisfies properties* (L1)–(L3) *of logarithms* (see Section 9.1).

THEOREM 1

The function $y = \ln x$ satisfies the following properties for all real numbers a, x, and r with $a > 0$ and $x > 0$:

$$\begin{aligned} &(L1) \quad \ln(ax) = \ln a + \ln x, \\ &(L2) \quad \ln(a/x) = \ln a - \ln x, \\ &(L3) \quad \ln(x^r) = r\ln x. \end{aligned}$$

The proof of Theorem 1 makes use of Theorem 13, Chapter 5, which states that *if two functions have the same derivative, they differ at most by a constant*. That is, if $f'(x) = g'(x)$, then $f(x) = g(x) + C$ for some constant C. We prove only property ($L1$), leaving ($L2$) and ($L3$) as exercises.

Strategy
Viewing x as the independent variable, name the functions $\ln ax$ and $\ln x$.

Proof of (L1)
Let $f(x)$ and $g(x)$ be the functions

$$f(x) = \ln ax \text{ and } g(x) = \ln x.$$

Then, by equation (3),

Show that $f'(x) = g'(x)$.

$$f'(x) = \frac{1}{ax} \cdot a = \frac{1}{x},$$

and

$$g'(x) = \frac{1}{x}.$$

Conclude that $f(x) = g(x) + C$.

Since $f'(x) = g'(x)$, Theorem 13 of Chapter 5 implies that

$$f(x) = g(x) + C$$

for some constant C, that is,

$$\ln ax = \ln x + C. \tag{4}$$

Find C by setting $x = 1$ and using fact that $\ln 1 = 0$.

To determine the constant C, we use the fact that $\ln 1 = 0$. Setting $x = 1$ in equation (4) gives

$$\ln a = 0 + C,$$

so $C = \ln a$. Equation (4) now gives the desired result. ■

The point of Theorem 1 is that the natural logarithm function defined by equation (1) satisfies all the properties characteristic of a logarithmic function. Moreover, $\ln x$ is defined for *all* real numbers $x > 0$, both rational and irrational. In Section 9.4 we will show that $\ln x$ can actually be written in the customary form, $\log_e x$, for a particular base number e, and we will justify the terminology *natural* logarithm. The remainder of this section concerns additional properties of $\ln x$.

Graph of $y = \ln x$

The graph of the natural logarithm function appears in Figure 2.3. It has the following properties:

(a) The domain of $y = \ln x$ is $(0, \infty)$.
(b) The function $y = \ln x$ is increasing for all x in its domain. (This follows from the fact that the derivative $1/x$ is positive for $x > 0$.)
(c) The graph of $y = \ln x$ is concave down on $(0, \infty)$. (The second derivative is

$$\frac{d^2}{dx^2} \ln x = \frac{d}{dx}\left(\frac{1}{x}\right) = -\frac{1}{x^2},$$

which is negative for all $x > 0$.)
(d) $\ln(1) = \int_1^1 \frac{1}{t}\, dt = 0.$

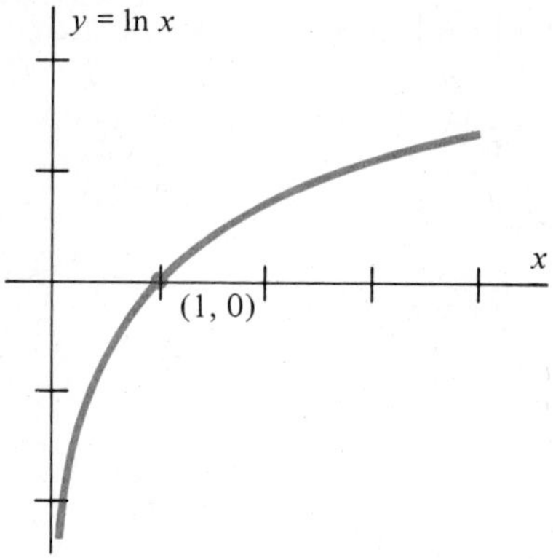

Figure 2.3 Graph of the natural logarithm function.

(e) $\lim_{x\to\infty} \ln x = +\infty$; $\lim_{x\to 0^+} \ln x = -\infty$. Thus, the *range* of $f(x) = \ln x$ is $(-\infty, \infty)$.

(To see the first part of statement (e), we use the fact that $\ln x$ is an increasing function for all x. Thus $\ln 2 > \ln 1 = 0$. Also, by property ($L3$), $\ln 2^n = n \ln 2$ for any integer $n > 0$. Since $\ln 2 > 0$,

$$\lim_{n\to\infty} \ln 2^n = \lim_{n\to\infty} n \cdot \ln 2 = +\infty.$$

Letting $x = 2^n$, it now follows that $\lim_{x\to\infty} \ln x = +\infty$.

The proof that $\lim_{x\to 0^+} \ln x = -\infty$ is similar, and is left as an exercise.)

The Number *e*

Since $y = \ln x$ is a continuous, increasing function whose range is the entire real line, there is precisely one number x for which $\ln x = 1$. We denote this number by the letter e, that is, we define the number e by the equation $\ln e = 1$. Since e satisfies the equation

$$1 = \int_1^e \frac{1}{t}\, dt,$$

we can obtain approximations for e by applying one of our procedures for approximating integrals (Simpson's Rule, for example) to the integral $\int_1^x \frac{1}{t}\, dt$ (see Exercise 66). The number e has been shown to be an irrational number, and the decimal expansion for e correct to 12 decimal places is known to be

$$e \approx 2.718281828459. \qquad (5)$$

We will see in the next two sections that the number e turns out to be the ''natural'' choice as a base for both the logarithmic and exponential functions. Note also that we can use approximation (5) together with property ($L3$) to obtain coordinates for points on the graph of $y = \ln x$ of the form

$$(e^n, \ln e^n) = (e^n, n \ln e) = (e^n, n).$$

(See Table 2.1 and Figure 2.4.)

Table 2.1

n	$x = e^n$	$y = n \cdot \ln e$
-2	$\frac{1}{e^2} \approx 0.13534$	-2
-1	$\frac{1}{e} \approx 0.36788$	-1
0	$e^0 = 1$	0
1	$e \approx 2.71828$	1
2	$e^2 \approx 7.38906$	2

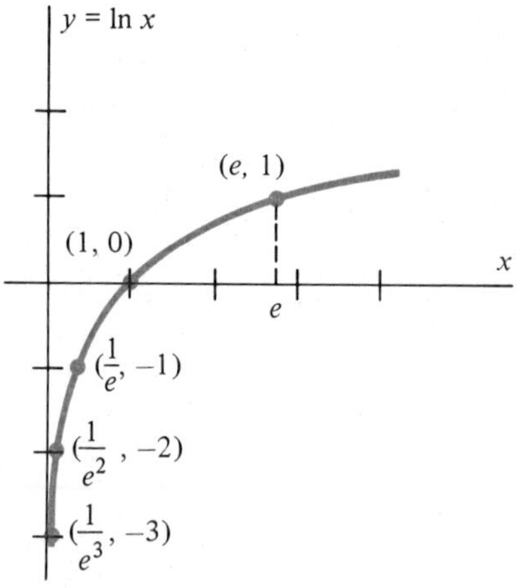

Figure 2.4 Points (e^n, n) on the graph of $y = \ln x$.

Natural Logarithms as Antiderivatives

From the differentiation formulas (2) and (3), we have the integration formulas

$$\int \frac{1}{x}\,dx = \ln|x| + C, \qquad x \neq 0 \tag{6}$$

and

$$\int \frac{u'(x)}{u(x)}\,dx = \int \frac{1}{u}\,du = \ln|u(x)| + C, \qquad u(x) \neq 0. \tag{7}$$

To see that equation (6) is valid for the case $x < 0$, recall that, in this case, $|x| = -x$. Thus

$$\frac{d}{dx}\ln|x| = \frac{d}{dx}\ln(-x) = \frac{1}{-x}(-1) = \frac{1}{x},$$

as required. The demonstration that (7) holds for $u(x) < 0$ is similar.

Example 2 Find $\displaystyle\int \frac{1 + \cos x}{x + \sin x}\,dx, \qquad x + \sin x \neq 0.$

Solution: The numerator of the integrand appears to be the derivative of the denominator, so we try the u-substitution

$$u = x + \sin x; \qquad du = (1 + \cos x)\,dx.$$

Then

$$\begin{aligned}\int \frac{1 + \cos x}{x + \sin x}\,dx &= \int \frac{1}{u}\,du \\ &= \ln|u| + C \\ &= \ln|x + \sin x| + C.\end{aligned}$$

■

Example 3 Find $\displaystyle\int \frac{x + 1}{x^2 + 2x}\,dx.$

Strategy

Make u-substitution to obtain form

$$\int \frac{u'(x)}{u(x)}\,dx = \int \frac{1}{u}\,du.$$

Remember,

$du = u'(x)\,dx.$

Apply equation (7).

Solution

Since the degree of the numerator is one less than the degree of the denominator, we attempt to bring the integral to the form of equation (7). Accordingly, we make the substitution

$$u = x^2 + 2x.$$

Then

$$du = (2x + 2)\,dx = 2(x + 1)\,dx,$$

so

$$(x + 1)\,dx = \frac{1}{2}\,du.$$

With these substitutions we obtain

$$\int \frac{(x + 1)\,dx}{x^2 + 2x} = \int \frac{1}{u}\cdot\left(\frac{1}{2}\,du\right)$$

$$= \frac{1}{2} \int \frac{1}{u}\, du$$

$$= \frac{1}{2} \ln|u| + C$$

$$= \frac{1}{2} \ln|x^2 + 2x| + C. \qquad \blacksquare$$

Example 4 Find $\displaystyle\int \frac{x^2 + 3x}{x + 1}\, dx, \qquad x \neq -1.$

Strategy

The integrand is an improper fraction. Perform a long division.

Solution

We first reduce the integrand to a proper fraction

$$\begin{array}{r@{}l} & \quad x + 2 \\ x + 1 \,\big)\!\! & \overline{x^2 + 3x + 0} \\ & \underline{x^2 + \;\; x} \\ & \qquad 2x \\ & \qquad \underline{2x + 2} \\ & \qquad\quad -\,2. \end{array}$$

Thus

$$\frac{x^2 + 3x}{x + 1} = x + 2 - \frac{2}{x + 1},$$

so

Apply formula (7).

$$\int \frac{x^2 + 3x}{x + 1}\, dx = \int \left[x + 2 - \frac{2}{x + 1}\right] dx$$

$$= \int x\, dx + 2 \int dx - 2 \int \frac{1}{x + 1}\, dx$$

$$= \frac{x^2}{2} + 2x - 2 \ln|x + 1| + C. \qquad \blacksquare$$

Example 5 Find the area of the region bounded by the graph of $y = x - \dfrac{1}{x}$ and the x-axis between $x = 1$ and $x = e$.

Strategy

Determine where $f(x) = x - \dfrac{1}{x}$ is nonnegative.

Solution

Since $x \geq \dfrac{1}{x}$ for $x \geq 1$, the function $y = x - \dfrac{1}{x}$ is nonnegative for $1 \leq x \leq e$.

The area A is therefore

Use formula

$$A = \int_a^b f(x)\, dx.$$

$$A = \int_1^e \left(x - \frac{1}{x}\right) dx$$

$$= \frac{1}{2} x^2 - \ln x \Big]_1^e$$

Use facts that $\ln e = 1$ and $\ln 1 = 0$.

$$= \left(\frac{1}{2} e^2 - \ln e\right) - \left(\frac{1}{2} - \ln 1\right)$$

$$= \left(\frac{1}{2} e^2 - 1\right) - \left(\frac{1}{2} - 0\right)$$

Use approximation $e \approx 2.71828$.

$$= \frac{1}{2} e^2 - 3/2 \approx 2.19.$$

(Figure 2.5.) ■

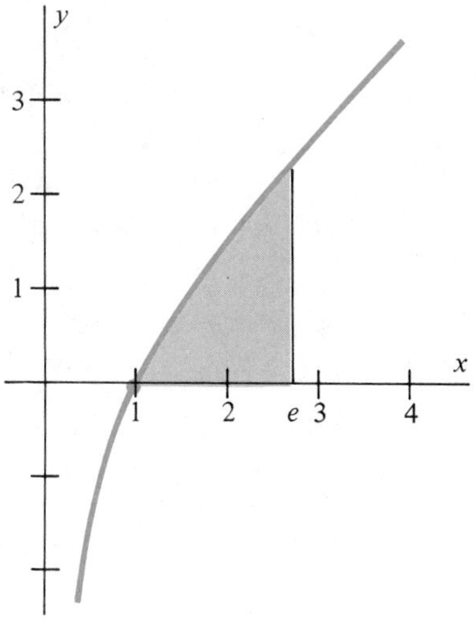

Figure 2.5 Graph of $y = x - 1/x$.

Notice in the next example how we must take care to observe the absolute value signs in equation (7). The value of the integral is negative, since the region in the plane associated with the integral is bounded below by the nonpositive integrand (see Figure 2.6).

Example 6 Find $\displaystyle\int_0^2 \frac{x}{x^2 - 9}\, dx$.

Solution: We use the u-substitution

$$u = x^2 - 9; \qquad du = 2x\, dx.$$

Then $x\, dx = \frac{1}{2}\, du$, and the limits of integration change as follows:

$$\begin{aligned} &\text{If } x = 0, && u = 0^2 - 9 = -9 \\ &\text{If } x = 2, && u = 2^2 - 9 = -5. \end{aligned}$$

Then

$$\begin{aligned} \int_0^2 \frac{x}{x^2 - 9}\, dx = \int_{-9}^{-5} \frac{1}{u} \cdot \frac{1}{2}\, du &= \frac{1}{2} \int_{-9}^{-5} \frac{1}{u}\, du \\ &= \frac{1}{2} \ln|u| \Big]_{-9}^{-5} \\ &= \frac{1}{2} [\ln|-5| - \ln|-9|] \\ &= \frac{1}{2} (\ln 5 - \ln 9) \\ &\approx -0.294. \end{aligned}$$

■

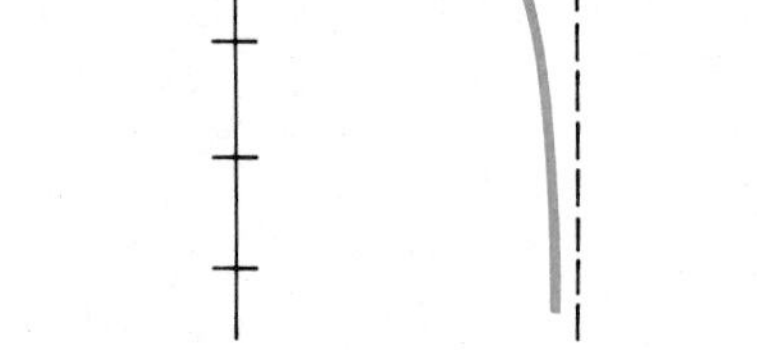

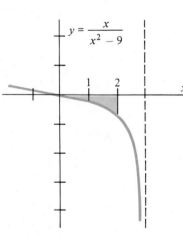

Figure 2.6

The last example involves a natural logarithm, although this may not be obvious on first glance.

Example 7 Find $\displaystyle\int \frac{dx}{\sqrt{x}(1 + \sqrt{x})}$.

Solution: We try the u-substitution

$$u = 1 + \sqrt{x}; \qquad du = \frac{1}{2\sqrt{x}}\, dx.$$

Then $\dfrac{dx}{\sqrt{x}} = 2\,du$, and we obtain

$$\begin{aligned}\int \frac{dx}{\sqrt{x}(1+\sqrt{x})} = \int \frac{1}{u}\cdot 2\,du &= 2\int \frac{1}{u}\,du \\ &= 2\ln|u| + C \\ &= 2\ln(1+\sqrt{x}) + C.\end{aligned}$$

($|1+\sqrt{x}| = 1+\sqrt{x}$ since $\sqrt{x} \geq 0$.) ■

Exercise Set 9.2

1. Use properties of ln x to solve for x. (You will need to use the table in the Appendix or a hand calculator.)
   - **a.** $\ln 2x = 0$
   - **b.** $2\ln x = \ln 2x$
   - **c.** $\ln(2/x) - \ln x = 0$
   - **d.** $\ln 3^{x^2} = 0$
   - **e.** $\sqrt{\ln \sqrt{x}} = 1$
   - **f.** $3\ln x + x = 2 + \ln x^3$
   - **g.** $\int_1^x \frac{3}{t}\,dt = 6$
   - **h.** $\int_2^x \frac{1}{t}\,dt = 0$
2. Why is the function $y = \ln x$ continuous?
3. True or false? Every number is a natural logarithm.

In Exercises 4–25, find the derivative.

4. $y = \ln 2x$
5. $y = x\ln x$
6. $y = \ln(6 - x^2)$
7. $f(x) = \ln\sqrt{x^3 - x}, \quad x > 1$
8. $f(t) = \ln(\ln t)), \quad t > 1$
9. $y = \sin(\ln x)$
10. $h(t) = \sqrt{1 + \ln t}$
11. $f(x) = \dfrac{x}{1 + \ln x}$
12. $f(x) = x^2 \ln^2 x$
13. $y = (3\ln\sqrt{x})^4$
14. $f(x) = (\sin x)\ln(1 + \sqrt{x})$
15. $f(t) = \dfrac{\ln(a + bt)}{\ln(c + dt)}$
16. $y = \ln(\sin t - t\cos t)$
17. $f(x) = \ln\cos x$
18. $y = x\ln(x^2 - x - 3), \quad x > 3$
19. $u(t) = \dfrac{t}{1 + \ln^2 t}$
20. $y = \ln(xy)$
21. $\ln(x + y) + \ln(x - y) = 1$
22. $x\ln y = 3$
23. $x\ln y + y\ln x = x$
24. $y = \displaystyle\int_x^1 \frac{1}{2t}\,dt, \quad x > 0$
25. $y = \displaystyle\int_1^{x^2} \ln 2t\,dt$

In each of Exercises 26–31, sketch the graph of the given function, noting all relative extrema.

26. $y = x\ln x, \quad x \geq 0$
27. $y = x - \ln x, \quad x > 0$
28. $y = \ln(2 + \sin x)$
29. $y = \ln(1 + x^2)$
30. $\ln(xy) = 1$
31. $y = \ln(\ln x), \quad x > 1$

Evaluate the following integrals.

32. $\displaystyle\int \frac{dx}{2x + 1}$
33. $\displaystyle\int \frac{dx}{1 - x}$
34. $\displaystyle\int \frac{x\,dx}{x^2 + 1}$
35. $\displaystyle\int \frac{x - 1}{x^2 - 2x}\,dx$
36. $\displaystyle\int \frac{x}{1 - 3x^2}\,dx$
37. $\displaystyle\int \frac{x^2 + 3}{x^3 + 9x}\,dx$
38. $\displaystyle\int \frac{\cos x}{\sin x}\,dx$
39. $\displaystyle\int \frac{dx}{x\ln x}$
40. $\displaystyle\int \frac{\sin t\,dt}{4 + 2\cos t}$
41. $\displaystyle\int \frac{\ln^2 x}{x}\,dx$
42. $\displaystyle\int \frac{x^2}{x + 1}\,dx$
43. $\displaystyle\int \frac{1}{\sqrt{x}\,(1 - \sqrt{x})}\,dx$
44. $\displaystyle\int \frac{1 - 2t^2}{1 - t}\,dt$
45. $\displaystyle\int \frac{x^4 + 3x^2 + x + 1}{x + 1}\,dx$
46. $\displaystyle\int_1^e \frac{1}{x}\,dx$
47. $\displaystyle\int_e^{e^2} \frac{1}{x\ln x}\,dx$
48. $\displaystyle\int_1^e \frac{\ln x}{x}\,dx$
49. $\displaystyle\int_e^{e^2} \frac{1}{x\ln(x^2)}\,dx$
50. $\displaystyle\int_2^3 \frac{x}{x^2 + 1}\,dx$
51. $\displaystyle\int_1^2 \left(\frac{1}{1 + x} - \frac{1}{2 + x}\right)dx$
52. Find the equation of the line tangent to the graph of the equation $y = x(\ln x)^2 + \dfrac{x}{\ln x}$ at the point $(e, 2e)$.

53. Find the area of the region bounded by the graph of $y = \dfrac{1}{x-2}$ and the x-axis for $3 \le x \le 4$.

54. Find the area of the region in the first quadrant bounded by the graphs of $x + y - 6 = 0$ and $xy = 8$.

55. Let ϵ be a small real number.

a. Use differentials to obtain a formula for approximating the quantity $\ln(1 + \epsilon)$.

b. Use the formula obtained in part (a) to approximate $\ln(0.8)$, $\ln(0.95)$, $\ln(1.05)$, and $\ln(1.2)$.

c. Compare results obtained in part (b) with those obtained from a hand calculator or the table in the Appendix.

56. True or false? For a given value of $x > 0$, the graphs of $y = \ln x$ and $y = \ln ax$ have the same slope.

57. Find the volume of the solid generated by rotating the region bounded by the graph of $y = \sqrt{\dfrac{2}{x-1}}$ and the x-axis, for $3 \le x \le 5$, about the x-axis.

58. Verify that the function $F(x) = x \ln x - x + C$ is an antiderivative for $f(x) = \ln x$, $x > 0$, that is, show that

$$\int \ln x \, dx = x \ln x - x + C, \qquad x > 0.$$

59. Use the result of Exercise 58 to find the average value of the natural logarithm function on the interval $[1, e]$.

60. A particle moves along a line with acceleration $a(t) = \dfrac{1}{t+2}$ m/sec^2. Find the distance travelled by the particle during the time interval $[0, 4]$ if $v(0) = 0$. (*Hint:* Use Exercise 58.)

61. Economists define the **growth of a function** $y = f(t)$ as the ratio

$$G = \frac{\frac{dy}{dt}}{y} = \frac{f'(t)}{f(t)} = \frac{y'}{y}.$$

For example, if $f(2) = 6$ and $f'(2) = 3$, then the growth of the function $y = f(t)$ at time $t = 2$ is $G = \dfrac{3}{6} = 0.5$.

a. Show that the growth of a function may be calculated as the derivative of the natural logarithm of that function, that is, $G = \dfrac{d}{dt} \ln(f(t))$.

b. Show that the growth of the product of two functions is the sum of the individual growths.

c. An oil company determines that the price it obtains for heating oil is increasing at a rate of 15% per year but that the number of gallons of heating oil sold is decreasing by 10% per year. Find the rate at which revenues obtained from the sale of heating oil are increasing.

62. The vapor pressure, in mm, of a certain fluid is related to its temperature by the equation

$$\ln P = \frac{-2000}{T} + 5.5.$$

a. Find $\dfrac{dP}{dT}$.

b. Find $\dfrac{dT}{dP}$.

63. (*Calculator*) Use the Trapezoidal Rule with $n = 10$ to approximate

a. $\ln 2$ **b.** $\ln 10$ **c.** $\ln\left(\dfrac{1}{2}\right)$

64. The approximations obtained in Exercise 63 are all *larger* than the actual values. Why?

65. (*Calculator*) Use Newton's Method to approximate the solution of the equation $\ln x = x - 4$.

66. (*Computer*) Use Simpson's Rule, including the error formula, to show that

$$\ln 2.7 < 1 < \ln 2.8.$$

Hence, $2.7 < e < 2.8$. How many subdivisions n must you use to insure that this result is correct?

67. (*Computer*) Use a program for Simpson's Rule, together with the definition of $\ln x$, to complete the following table (n is the number of subdivisions).

n	$\sim\ln 2$	$\sim\ln 4$	$\sim\ln 10$
10			
20			
50			
200			

68. (*Computer*) Compare the results obtained in Exercise 67 with those obtained from the following approximations to $\ln(1 + x)$:

a. $\ln(1 + x) \sim P_1 = x$

b. $\ln(1 + x) \sim P_2 = x - \dfrac{x^2}{2}$

c. $\ln(1 + x) \sim P_3 = x - \dfrac{x^2}{2} + \dfrac{x^3}{3}$

d. $\ln(1 + x) \sim P_4 = x - \dfrac{x^2}{2} + \dfrac{x^3}{3} - \dfrac{x^4}{4}$.

(We will see how such approximations are obtained in Chapter 12.)

69. Prove statement ($L2$) of Theorem 1.

70. Prove statement ($L3$) of Theorem 1 for the case r rational. (The case r irrational must await the extension of the Power Rule $\frac{d}{dx}x^r = rx^{r-1}$ to irrational exponents. This is done in Section 9.4.)

9.3 THE NATURAL EXPONENTIAL FUNCTION

In Section 9.2 we saw that the natural logarithm function is an increasing function for all $x \in (0, \infty)$ with range $(-\infty, \infty)$. These conditions are sufficient to guarantee the existence of the *inverse* of the natural logarithm function (see Section 3.8). We denote this function initially by $\exp(x)$. In other words, we *define* the exponential function as follows:

$$y = \exp(x) \quad \text{if and only if} \quad x = \ln y. \tag{1}$$

The definition embodied in statement (1) says that "the value of the exponential function $\exp(x)$ is the number y whose natural logarithm is the number x." Read the opposite way, the statement says that "the value of the natural logarithm function $\ln y$ is the number x for which the value of the exponential function is y."

Now recall from Section 3.8 that if the function $g(x)$ is the inverse of the function $f(x)$, then $f(x)$ is also the inverse of $g(x)$ for all $g(x)$ in the domain of $f(x)$. Thus, we have both

$$g(f(x)) = x \quad \text{and} \quad f(g(x)) = x$$

for appropriate values of x. Since $\exp(x)$ is defined to be the inverse of the natural logarithm function $\ln x$, we therefore have both the identities

$$\exp(\ln x) = x, \quad x > 0 \quad \text{and} \quad \ln(\exp(x)) = x, \quad -\infty < x < \infty.$$

The Graph of $y = \exp(x)$

Since the graph of $y = \exp(x)$ is the reflection in the line $y = x$ of the graph $y = \ln x$,* it has the following properties (Figure 3.1):

1. The domain of $y = \exp(x)$ is $(-\infty, \infty)$ since this is the range of $y = \ln x$, that is, the function $y = \exp(x)$ is defined for all numbers.
2. The range of $y = \exp(x)$ is $(0, \infty)$. This is because the domain of $y = \ln x$ is $(0, \infty)$.
3. Since all points of the form (e^n, n), $n = 0, \pm 1, \pm 2, \ldots$ lie on the graph of $y = \ln x$, the points of the form (n, e^n), $n = 0, \pm 1, \pm 2, \ldots$ lie on the graph of the exponential function.
4. Since $y = \ln x$ is increasing for all $x \in (0, \infty)$, the function $y = \exp(x)$ is increasing for all $x \in (-\infty, \infty)$ (see Exercise 76).

Figures 3.2 and 3.3 show graphs of various functions of the form $y = \exp(kx)$ with $k > 0$. Notice that these graphs rise (or fall) more quickly for increasing x as k increases.

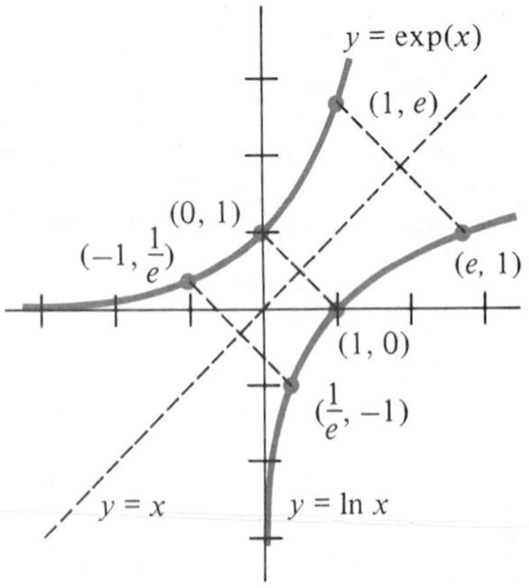

Figure 3.1 Function $y = \exp(x)$ is the inverse of the natural log function $y = \ln x$.

---

* It is a property of inverse functions that the graph of the inverse of $f(x)$ is the reflection in the line $y = x$ of the graph of $y = f(x)$. See Section 3.8 for a review of these concepts.

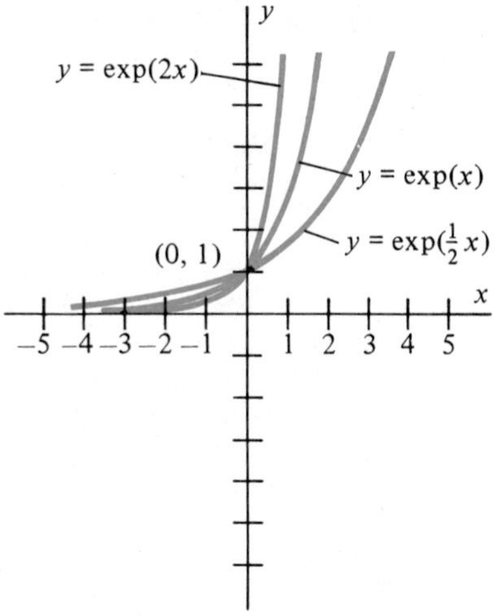

Figure 3.2 Graphs of $y = \exp(kx)$, $k > 0$.

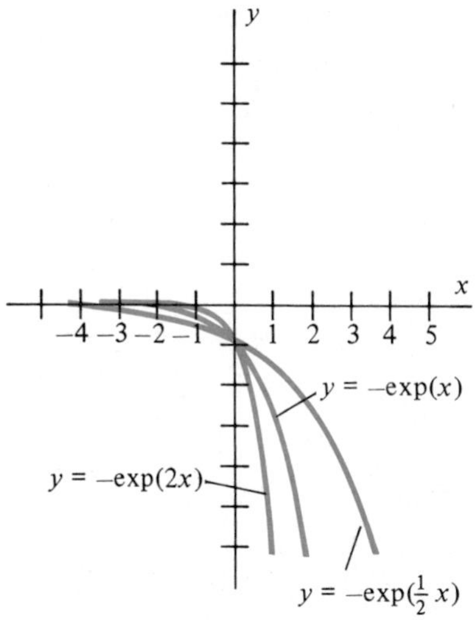

Figure 3.3 Graphs of $y = -\exp(kx)$, $k > 0$.

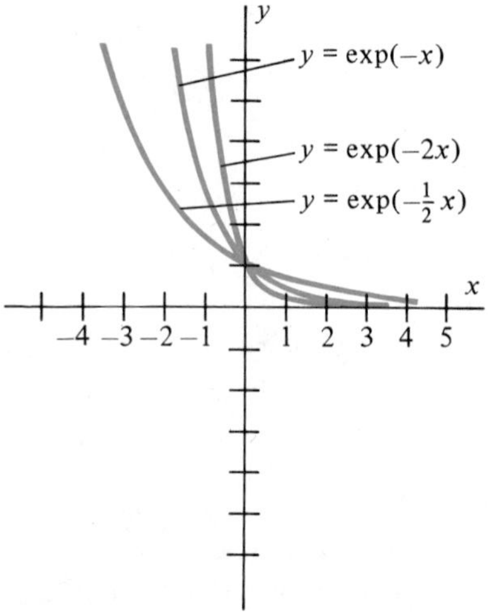

Figure 3.4 Graphs of $y = \exp(kx)$, $k < 0$.

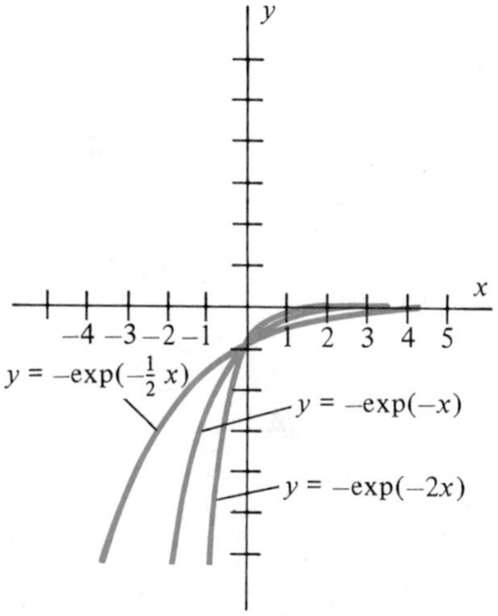

Figure 3.5 Graphs of $y = -\exp(kx)$, $k < 0$.

Figures 3.4 and 3.5 show graphs of various functions of the form $y = \exp(kx)$ with $k < 0$. These graphs fall (or rise) more quickly for increasing x as k decreases.

Exp(x) is an Exponential Function

At this point we can show that the function $\exp(x)$ is really an exponential function, in the usual sense. In particular, for any rational number x,

$$\boxed{\exp(x) = e^x} \tag{2}$$

where e is the number defined by the equation $\ln e = 1$. This follows from the fact that $\ln e^x = x \cdot \ln e = x \cdot 1 = x$ for any rational number x, according to Theorem 1. Since $\ln[e^x] = x$, $\exp(x) = e^x$ by statement (1).

From now on we will use the notation e^x rather than $\exp(x)$ to denote the natural exponential function. However, you should keep in mind the fact that statements (1) and (2) together define e^x for *all* real exponents x, both rational and irrational.

The following theorem shows that the function e^x satisfies the characteristic properties of exponents for *all* values of x.

THEOREM 2

Let x_1, x_2, and r be any real numbers. Then

(i) $e^{x_1} \cdot e^{x_2} = e^{x_1+x_2}$,

(ii) $\dfrac{e^{x_1}}{e^{x_2}} = e^{x_1-x_2}$,

(iii) $[e^{x_1}]^r = e^{rx_1}$.

Strategy

Convert exponentials to logarithms by taking natural logs of both sides of the equations.

Use property ($L1$):

$\ln y_1 + \ln y_2 = \ln(y_1y_2)$.

Convert back to exponential form.

Use identity $e^{\ln x} = x$.

Substitute back.

Proof

Let $y_1 = e^{x_1}$ and $y_2 = e^{x_2}$.

Then

$$\ln y_1 = x_1 \quad \text{and} \quad \ln y_2 = x_2.$$

By property ($L1$), Section 9.2,

$$x_1 + x_2 = \ln y_1 + \ln y_2 = \ln(y_1y_2).$$

Thus

$$\begin{aligned} e^{x_1+x_2} &= e^{\ln(y_1y_2)} \\ &= y_1y_2 \\ &= e^{x_1} \cdot e^{x_2} \end{aligned}$$

as required. This proves statement (i). The proofs of statements (ii) and (iii) are similar, and are left as Exercises 77 and 78. ■

The Derivative of $y = e^x$

In asking for the derivative of the exponential function, we uncover one of the most remarkable facts of the calculus. Recall from Section 8, Chapter 3 that if the function $y = f(x)$ has an inverse $x = g(y)$, and if $f(x)$ is differentiable in an interval containing x_0, then $g(y)$ is also differentiable in an interval containing $y_0 = f(x_0)$ if $f'(x_0) \neq 0$. In other words, the inverse of a differentiable function is differentiable, where defined. Thus, since $y = \ln x$ is differentiable, so is its inverse $y = e^x$. To obtain the derivative of $y = e^x$, we apply the Chain Rule to the identity

$$\ln e^x = x. \tag{3}$$

Differentiating both sides of equation (3) we obtain

$$\frac{1}{e^x} \cdot \frac{d}{dx} e^x = 1. \tag{4}$$

Multiplying both sides of equation (4) by e^x gives the desired result:

$$\boxed{\frac{d}{dx} e^x = e^x.} \tag{5}$$

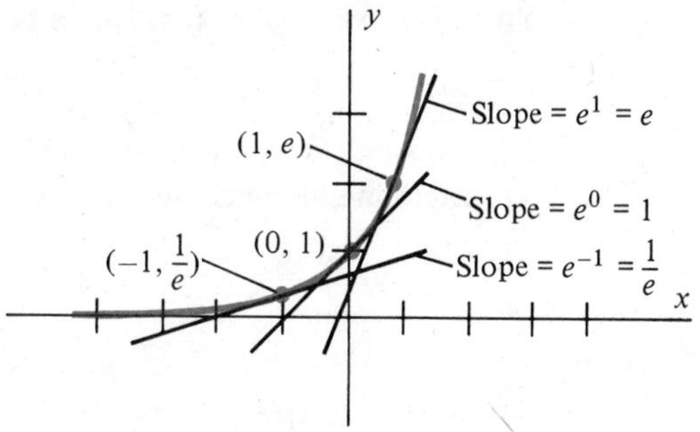

Figure 3.6 Slope of $y = e^x$ equals y-coordinate.

In other words, the exponential function $y = e^x$ is its own derivative! This function, and its multiples, are the only functions in the calculus with this property. (In fact, we could develop the entire theory of exponential and logarithm functions by seeking a function $f(x)$ for which $f'(x) = f(x)$.) Figure 3.6 illustrates the geometric interpretation of formula (5): The slope of the tangent to the graph of $y = e^x$ at (x, e^x) is precisely the y-coordinate e^x.

If $u(x)$ is a differentiable function of x, the Chain Rule gives the following generalization of formula (4):

$$\frac{d}{dx} e^u = e^u \cdot \frac{du}{dx}. \tag{6}$$

Example 1 Find $\frac{dy}{dx}$ for (a) $y = e^{6x}$, and (b) $y = e^{x \sin x}$

Strategy

(a) Let $u = 6x$; then

$$\frac{du}{dx} = 6.$$

(b) Let $u = x \sin x$. Then

$$\frac{du}{dx} = \sin x + x \cos x.$$

Solution

(a) Equation (6) gives

$$\frac{dy}{dx} = e^{6x} \cdot \frac{d}{dx}(6x) = 6e^{6x}.$$

(b) Equation (6) gives

$$\frac{dy}{dx} = e^{x \sin x} \cdot \frac{d}{dx}(x \sin x)$$
$$= e^{x \sin x}[\sin x + x \cos x].$$

■

Example 2 Sketch the graph of $y = xe^{1-x^2}$.

Strategy

Odd or even?

Solution

(a) For $f(x) = xe^{1-x^2}$ we find by substituting $-x$ that

$$f(-x) = (-x)e^{1-(-x)^2} = -xe^{1-x^2} = -f(x).$$

Thus, $f(x)$ is an *odd* function, and the graph of $y = f(x)$ is symmetric about the origin.

Find zeros.

(b) Since the exponential function is never zero, the equation

$$y = xe^{1-x^2} = 0$$

has the single solution $x = 0$. The only zero is $x = 0$.

Calculate dy/dx using the Product Rule and formula (5).

(c) To determine the intervals on which $f(x)$ is increasing we calculate dy/dx:

$$\frac{dy}{dx} = (1)e^{1-x^2} + x[-2xe^{1-x^2}]$$

$$\frac{dy}{dx} = (1 - 2x^2)e^{1-x^2}. \tag{7}$$

Critical numbers solutions of $\frac{dy}{dx} = 0$.

Thus, $\frac{dy}{dx} = 0$ for $x = -\frac{1}{\sqrt{2}}$ and $\frac{1}{\sqrt{2}}$. Inspection of the sign of $\frac{dy}{dx}$ in (7) gives the following information:

Interval	Sign of $f'(x)$	Nature of $f(x)$
$\left(-\infty, -\frac{1}{\sqrt{2}}\right)$	negative	decreasing
$\left(-\frac{1}{\sqrt{2}}, \frac{1}{\sqrt{2}}\right)$	positive	increasing
$\left(\frac{1}{\sqrt{2}}, \infty\right)$	negative	decreasing

Apply first derivative test to classify critical numbers.

(d) The information in equation (7) shows that

(i) $\left(-\frac{1}{\sqrt{2}}, -\sqrt{\frac{e}{2}}\right)$ is a relative minimum,

(ii) $\left(\frac{1}{\sqrt{2}}, \sqrt{\frac{e}{2}}\right)$ is a relative maximum.

Calculate d^2y/dx^2 using Product Rule.

(e) To determine concavity we calculate the second derivative. Using (7) we obtain

$$\frac{d^2y}{dx^2} = \frac{d}{dx}\{(1 - 2x^2)e^{1-x^2}\}$$
$$= (-4x)e^{1-x^2} + (1 - 2x^2)[(-2x)e^{1-x^2}]$$

$$\frac{d^2y}{dx^2} = 2x(2x^2 - 3)e^{1-x^2}. \tag{8}$$

Find zeros of d^2y/dx^2.

Thus

$$\frac{d^2y}{dx^2} = 0 \quad \text{if} \quad x = -\sqrt{\frac{3}{2}}, 0 \quad \text{or} \quad \sqrt{\frac{3}{2}}.$$

Examine sign of d^2y/dx^2 between zeros.

Examining the sign of d^2y/dx^2 in equation (8) gives the following information:

Interval	Sign of $\frac{d^2y}{dx^2}$	Concavity
$\left(-\infty, -\sqrt{\frac{3}{2}}\right)$	negative	down
$\left(-\sqrt{\frac{3}{2}}, 0\right)$	positive	up
$\left(0, \sqrt{\frac{3}{2}}\right)$	negative	down
$\left(\sqrt{\frac{3}{2}}, \infty\right)$	positive	up

The graph therefore has inflection points at

$$\left(-\sqrt{\frac{3}{2}}, -\sqrt{\frac{3}{2e}}\right), \ (0, 0), \text{ and } \left(\sqrt{\frac{3}{2}}, \sqrt{\frac{3}{2e}}\right).$$

The graph appears in Figure 3.7. ■

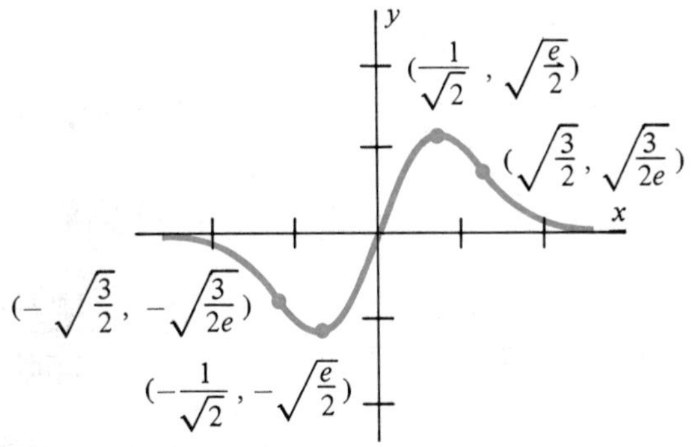

Figure 3.7 Graph of $y = xe^{1-x^2}$.

Integrals Involving e^x

The differentiation formula (5) gives the integration formula

$$\int e^x\,dx = e^x + C. \tag{9}$$

For a differentiable function $u(x)$, the integration formula corresponding to formula (6) is

$$\int e^{u(x)}\,u'(x)\,dx = \int e^u\,du = e^u + C. \tag{10}$$

Example 3 Find the following integrals:

(a) $\int e^{-3x}\, dx$

(b) $\int x^2 e^{x^3+1}\, dx.$

Strategy
Let $u = -3x$ and apply equation (9).

Solution
(a) For $u = -3x$, $du = -3\, dx$, so $dx = -\frac{1}{3}\, du$. We may therefore write

$$\begin{aligned}\int e^{-3x}\, dx &= \int e^{u}\left(-\frac{1}{3}\right) du \\ &= -\frac{1}{3}\int e^{u}\, du \\ &= -\frac{1}{3} e^{u} + C \\ &= -\frac{1}{3} e^{-3x} + C.\end{aligned}$$

Try to view the factor x^2 as a factor of du. This works for the substitution $u = x^3 + 1$.

(b) If we use the u-substitution $u = x^3 + 1$, we have $du = 3x^2\, dx$, so $x^2\, dx = \frac{1}{3}\, du$. Thus

$$\begin{aligned}\int x^2 e^{x^3+1}\, dx &= \int e^{u} \cdot \frac{1}{3} \cdot du \\ &= \frac{1}{3}\int e^{u}\, du \\ &= \frac{1}{3} e^{x^3+1} + C.\end{aligned}$$

Apply equation (10). ■

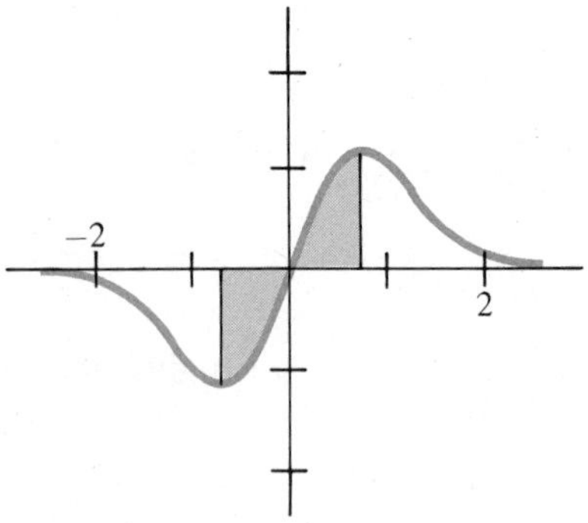

Figure 3.8

Example 4 Find the area bounded by the graph of $y = xe^{1-x^2}$ and the x-axis for $-2 \le x \le 2$ (see Figures 3.7 and 3.8).

Strategy
Recall, y was graphed in Example 2.

Solution
Since $y = xe^{1-x^2}$ is an odd function, its graph is symmetric about the origin. We

may therefore calculate the desired area A as

Area on $[-2, 2]$ equals twice the area on $[0, 2]$.

$$A = 2 \cdot \int_0^2 xe^{1-x^2}\, dx.$$

To evaluate this integral, we use the u-substitution

Find a substitution u so that x is a factor of du. The substitution $u = 1 - x^2$ works.

$$u = 1 - x^2, \qquad du = -2x\, dx.$$

Then

$$x\, dx = -\frac{1}{2}\, du$$

and the limits of integration become

Change limits of integration according to u-substitution.

$$\begin{aligned} u &= 1 - 0^2 = 1 \quad &\text{if} \quad x = 0 \\ u &= 1 - 2^2 = -3 \quad &\text{if} \quad x = 2. \end{aligned}$$

With these substitutions, we obtain

Apply formula (10).

Factor out $\left(-\frac{1}{2}\right)$.

Apply Fundamental Theorem of Calculus.

$$\begin{aligned} A &= 2\int_0^2 xe^{1-x^2}\, dx \\ &= 2\int_1^{-3} e^u\left(-\frac{1}{2}\right) du \\ &= 2\left(-\frac{1}{2}\right)\int_1^{-3} e^u\, du \\ &= -[e^u]_1^{-3} \\ &= -[e^{-3} - e] \\ &= e - \frac{1}{e^3} \\ &\approx 2.67. \end{aligned}$$

■

More on the Number e

The number e is difficult to approximate from the defining property $\ln e = \int_1^e \frac{1}{x}\, dx = 1$, since it appears as a limit of integration. This same number arises in a seemingly unrelated manner as the limit

$$e = \lim_{x\to\infty}\left(1 + \frac{1}{x}\right)^x. \tag{11}$$

It is much easier to approximate e as this limit, since we need only calculate values of the function $\left(1 + \frac{1}{x}\right)^x$ for x large.

In Section 9.5 we will see how this limit arises in a natural way in the calculation of compound interest. For now we shall simply show how the limit is verified and note several related examples.

To prove (11), we shall prove the more general statement

$$e^r = \lim_{x\to\infty}\left(1 + \frac{r}{x}\right)^x, \qquad r \text{ constant.} \tag{12}$$

Since $\lim_{x\to\infty}\left(1+\frac{r}{x}\right) = 1 + 0 = 1$, the limit in (12) has the indeterminate form 1^∞. Since the natural logarithm is a continuous function, we begin by considering the logarithm of the desired limit.

$$\begin{aligned}
\ln\left[\lim_{x\to\infty}\left(1+\frac{r}{x}\right)^x\right] &= \lim_{x\to\infty} \ln\left(1+\frac{r}{x}\right)^x && \text{(Theorem 11, Chapter 2)}\\
&= \lim_{x\to\infty} x\cdot\ln\left(1+\frac{r}{x}\right) && \text{(Theorem 1)}\\
&= \lim_{x\to\infty}\frac{\ln\left(1+\frac{r}{x}\right)}{\frac{1}{x}} && \left(\frac{0}{0}\text{ form}\right)\\
&= \lim_{x\to\infty}\frac{\left(1+\frac{r}{x}\right)^{-1}\left(-\frac{r}{x^2}\right)}{\frac{-1}{x^2}} && \text{(l'Hôpital's Rule)}\\
&= \lim_{x\to\infty}\frac{r}{1+\frac{r}{x}} && \text{(Simplifying by algebra)}\\
&= r,
\end{aligned}$$

since $\lim_{x\to\infty}\frac{r}{x} = 0$. Applying the exponential function to both sides of the equation $\ln\left[\lim_{x\to\infty}\left(1+\frac{r}{x}\right)^x\right] = r$ gives $\lim_{x\to\infty}\left(1+\frac{r}{x}\right)^x = e^r$, the desired result.

The results of approximating e by the expression $\left(1+\frac{1}{n}\right)^n$ for various values of n appear in Table 3.1.

Table 3.1 Approximating $e \approx 2.7182818, \ldots$ using $\left(1+\frac{1}{n}\right)^n$

n	$\left(1+\frac{1}{n}\right)^n$	n	$\left(1+\frac{1}{n}\right)^n$
1	2.000000	500	2.715569
5	2.488320	1,000	2.716925
20	2.653298	2,500	2.717742
50	2.691588	5,000	2.718016
100	2.704814	10,000	2.718160
250	2.712865	100,000	2.718442

Example 5 According to equation (12) we have

(a) $\lim_{x\to\infty}\left(1-\frac{1}{x}\right)^x = e^{-1} = \frac{1}{e}$

(b) $\lim_{x\to\infty}\left(1+\frac{3}{x}\right)^x = e^3$

(c) $\lim_{x\to\infty}\left(1+\frac{6}{x}\right)^{2x} = \lim_{x\to\infty}\left[\left(1+\frac{6}{x}\right)^x\right]^2 = (e^6)^2 = e^{12}$

(d)
$$\begin{aligned}\lim_{x\to\infty}\left(1-\frac{1}{x^2}\right)^x &= \lim_{x\to\infty}\left[\left(1-\frac{1}{x}\right)\left(1+\frac{1}{x}\right)\right]^x\\ &= \lim_{x\to\infty}\left[\left(1-\frac{1}{x}\right)^x\cdot\left(1+\frac{1}{x}\right)^x\right]\\ &= e^{-1}\cdot e^1\\ &= 1.\end{aligned}$$

■

The Indeterminate Forms 1^∞, 0^0, and ∞^0

The proof of the limit (12) involved a standard technique for dealing with limits yielding the exponential indeterminate forms 1^∞, 0^0, and ∞^0. The technique is to first calculate the limit of the *natural logarithm* of the given expression. By property (*L*3) of Theorem 1, this logarithm can then be brought into one of the indeterminate forms $0/0$ or ∞/∞, to which l'Hôpital's Rule applies. However, you must remember that the resulting limit L is the natural logarithm of the original limit, which is, therefore, e^L. (See Section 5.8 for a review of l'Hôpital's Rule and indeterminate forms.)

Example 6 Find $\lim_{x\to 0^+}(\sin x)^x$.

Strategy

Solution

Check to see if limit can be obtained by setting $x = 0$.

Setting $x = 0$ gives the indeterminate form 0^0. We therefore begin by considering the natural logarithm of the limit.

Apply natural logarithm. Move ln inside limit by Theorem 11, Chapter 2.

$$\ln\left[\lim_{x\to 0^+}(\sin x)^x\right] = \lim_{x\to 0^+}\ln(\sin x)^x$$

Bring exponent out in front of ln by property (*L*3).

$$= \lim_{x\to 0^+} x\cdot\ln(\sin x) \qquad (0\cdot\infty \text{ form})$$

Invert one factor and divide to obtain ∞/∞ form.

$$= \lim_{x\to 0^+}\frac{\ln(\sin x)}{\frac{1}{x}} \qquad \left(\frac{\infty}{\infty}\text{ form}\right)$$

Apply l'Hôpital's Rule.

$$= \lim_{x\to 0^+}\frac{\frac{\cos x}{\sin x}}{-\frac{1}{x^2}}$$

Simplify.

$$= \lim_{x\to 0^+}\frac{-x^2}{\tan x} \qquad \left(\frac{0}{0}\text{ form}\right)$$

Apply l'Hôpital's Rule again.

$$= \lim_{x\to 0^+}\frac{-2x}{\sec^2 x}$$

Obtain limit $L = 0$. $\qquad = \dfrac{-2 \cdot 0}{1} = 0.$

Original limit is $e^L = e^0 = 1$. $\qquad$ Thus, $\lim_{x \to 0^+} (\sin x)^x = e^0 = 1.$ ■

Exercise Set 9.3

1. Simplify the following expressions:

a. $e^{\ln 2}$ **b.** $e^{-\ln 4}$
c. $e^{(\ln x - \ln y)}$ **d.** $\ln e^{-x^2}$
e. $\ln xe^{\sqrt{x}} - \ln x$ **f.** $e^{x \ln 2}$
g. $e^{\ln(1/x)}$ **h.** $e^{4 \ln x}$
i. $\ln xe^{x^2}$ **j.** $e^{x - \ln x}$.

2. Solve the following equations for x.

a. $\ln x = 2$ **b.** $\ln x^2 = 9$
c. $e^{x^2} = 5$ **d.** $e^{2x} - 2e^x + 1 = 0$

3. Find y if $e^{x-y} = x + 3$.

4. Find y if $e^{(y-1)^2} = x^4 + 1, \quad y > 1$.

In Exercises 5–20, find the derivative of the given function.

5. $y = e^{3x}$ **6.** $f(x) = xe^{-x}$

7. $f(t) = e^{\sqrt{t}}$ **8.** $y = \dfrac{e^x + 1}{e^x}$

9. $y = e^{x^2 - x}$ **10.** $f(t) = e^{\sin t}$

11. $f(x) = e^x \sin x$ **12.** $y = xe^{-3 \ln x}$

13. $f(x) = \ln \dfrac{e^x + 1}{x + 1}$ **14.** $f(x) = \dfrac{e^x - 1}{e^x + 1}$

15. $f(x) = (2 - e^{x^2})^3$ **16.** $y = x \ln(e^x + x)$

17. $y = \dfrac{1}{2}(e^x + e^{-x})$

18. $f(x) = (x^2 + x - 1)e^{x^2+3}$

19. $y = e^{\sqrt{x}} \cdot \ln \sqrt{x}$ **20.** $f(t) = te^{(1/t)^2}$

In Exercises 21–24, find $\dfrac{dy}{dx}$ by implicit differentiation.

21. $e^{xy} = x$ **22.** $e^{x-y} = ye^x$

23. $\ln(x + 2y) = e^y$ **24.** $y^2e^x + y \ln x = 2$

25. True or false? For $y = e^{kx}$, y' is proportional to y.

26. True or false? The equation $y = e^x$ has a solution x for every $y \geq 0$.

27. True or false? You can *always* divide by e^x.

In Exercises 28–41, evaluate the given indefinite integrals.

28. $\int e^{2x}\, dx$ **29.** $\int e^{-x}\, dx$

30. $\int e^{2x+6}\, dx$ **31.** $\int xe^{x^2+3}\, dx$

32. $\int x^2e^{1-x^3}\, dx$ **33.** $\int e^{2x}(1 + e^{2x})^3\, dx$

34. $\int \dfrac{e^{\sqrt{x}}}{\sqrt{x}}\, dx$ **35.** $\int \dfrac{e^{1/x}}{x^2}\, dx$

36. $\int \dfrac{e^x}{\sqrt{e^x + 1}}\, dx$ **37.** $\int \dfrac{e^x}{1 + e^x}\, dx$

38. $\int \cos x \cdot e^{\sin x}\, dx$ **39.** $\int \dfrac{e^x}{(3 + e^x)^2}\, dx$

40. $\int \dfrac{1 + e^{-ax}}{1 - e^{-ax}}\, dx$ **41.** $\int \dfrac{(1 + e^{\sqrt{x}})e^{\sqrt{x}}}{\sqrt{x}}\, dx$

In Exercises 42–49, evaluate the definite integrals.

42. $\int_0^1 e^{2x}\, dx$ **43.** $\int_1^{\ln 4} e^{-x}\, dx$

44. $\int_0^1 xe^{1-x^2}\, dx$ **45.** $\int_0^{\pi/2} \cos xe^{\sin x}\, dx$

46. $\int_0^{\ln 2} \dfrac{e^x}{2 + e^x}\, dx$ **47.** $\int_1^{e^2} \dfrac{\ln x}{x}\, dx$

48. $\int_0^2 \dfrac{e^x + e^{-x}}{2}\, dx$ **49.** $\int_0^2 e^x\, dx$

In Exercises 50–59, find the limit.

50. $\lim_{x \to 0^+} x^x$ **51.** $\lim_{x \to \infty} x^x$

52. $\lim_{x \to 0^+} x^{\cos x}$ **53.** $\lim_{x \to \infty} \left(\dfrac{1}{x} \ln \dfrac{1}{x}\right)$

54. $\lim_{x \to 0^+} (x + 2)^x$ **55.** $\lim_{x \to \infty} \left(\dfrac{x + 1}{x}\right)^x$

56. $\lim_{x \to \infty} \left(\dfrac{x}{x + 1}\right)^{x+1}$ **57.** $\lim_{x \to 0^+} x^{\sin x}$

58. $\lim_{x\to 0^+} x^{\sqrt{x}}$

59. $\lim_{x\to\infty} (1 + x^2)^{e^{-x}}$

60. Find the maximum value of the function $y = (3 - x^2)e^x$.

61. Find all relative extrema for the function $y = x^2e^{1-x^2}$.

62. Sketch the graph of the function $y = xe^{1-x^3}$, noting intervals where y is increasing, relative extrema, points of inflection, and concavity.

63. The line $y = -\dfrac{1}{e}$ is tangent to the graph of $y = xe^x$ at point P. Find P.

64. Find the area of the region bounded by the graphs of $y = e^x$ and $y = e^{-x}$ for $-1 \le x \le 1$.

65. Find the volume of the solid obtained by revolving about the x-axis the region bounded by the graph of $y = e^{2x}$ and the x-axis for $0 \le x \le 2$.

66. Show that the function $y = Ae^{kt}$ satisfies the differential equation $y' = ky$ and the initial condition $y(0) = A$. Use this information to solve the following *initial value problems:*

a. $y' = y, \quad y(0) = 1$
b. $y' = \pi y, \quad y(0) = -2$
c. $y' + 3y = 0, \quad y(0) = 2.$

67. Find the average value of the function $y = xe^{1-x^2}$ on the interval $[-2, 2]$.

68. Show that $f(x) = \dfrac{e^x + e^{-x}}{2}$ is an even function, and that $g(x) = \dfrac{e^x - e^{-x}}{2}$ is an odd function (see Exercise 49, Section 1.5).

69. Lambert's law states that the intensity of light, after passing through a thickness l of absorbing liquid, is $I = I_0e^{-kl}$ where I_0 and k are constants. Find k if $\dfrac{dI}{dl} = 4I$.

70. The function $f(x) = \dfrac{1}{\sigma\sqrt{2\pi}}\, e^{-(x-\mu)^2/2\sigma^2}$ is called the *normal* probability density function with mean μ and variance σ^2.

a. Show that the graph of $f(x)$ is symmetric about the line $x = \mu$.
b. Show that inflection points exist for $x = \mu \pm \sigma$.
c. Graph $f(x)$, obtaining the familiar bell-shaped curve.

71. Human growth in height from age one year to adulthood typically looks like the graph in Figure 3.9. A function that has been used to model this kind of growth is the **double logistic**

$$y(t) = \frac{a}{1 + e^{-b_1(t-c_1)}} + \frac{f - a}{1 + e^{-b_2(t-c_2)}}$$

where the parameters a, b_1, b_2, c_1, c_2, and f are different for each individual.

a. Show that the parameter f is adult height, that is, show that

$$f = \lim_{t\to\infty} y(t).$$

b. If $b_1 = b_2$ and $c_1 = c_2 = c$, show that the time of most rapid growth is $t = c$.

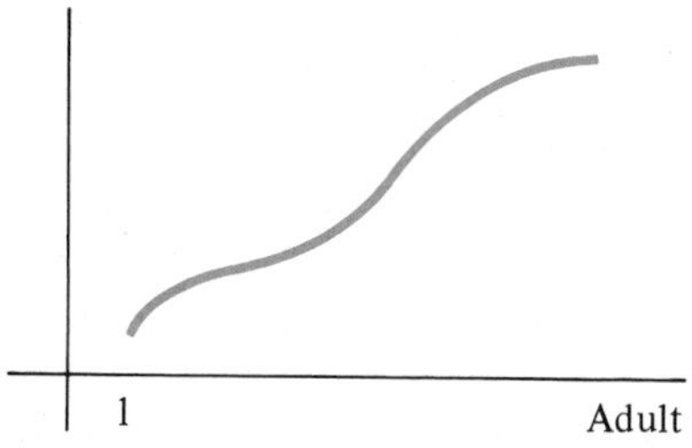

Figure 3.9 Typical double logistic.

72. *(Computer)* An ion with n more electrons than protons carries a negative charge of $-np$ where p is the charge on one electron. An ion with m fewer electrons than protons carries a charge of mp. The attraction potential energy between two such ions is

$$E_+ = \frac{-mnp^2}{r}$$

where r is the distance separating the nuclei. However, since the electrons of the two ions repel each other there is a repulsion potential energy of

$$E_- = ae^{-r/b}$$

where a and b are constants. Assume $m = n = p = b = 1$ and $a = 10$, and let $E = E_+ + E_-$ be the total potential energy.

a. Find the relative maximum and minimum values of the total potential energy on $(0, \infty)$ and the numbers r for which they occur. (*Hint:* You will need to use Newton's Method to solve the equation $dE/dr = 0$. There are two zeros of this equation.)
b. Graph the function E.

73. *(Computer)* Solve the equation $e^{2x} - 8x + 1 = 0$. (*Hint:* Use Newton's Method.)

74. *(Computer)* Solve the equation $e^x + x = 5, \quad x > 0.$

75. *(Computer)* Write a program to calculate entries for the following table, comparing actual values of e^x with those of the approximating polynomials

$$P_1(x) = 1 + x$$

$$P_2(x) = 1 + x + \frac{x^2}{2!}$$

$$P_3(x) = 1 + x + \frac{x^2}{2!} + \frac{x^3}{3!}$$

$$P_4(x) = 1 + x + \frac{x^2}{2!} + \frac{x^3}{3!} + \frac{x^4}{4!}$$

(Recall, $n! = n(n-1)(n-2) \cdot \cdots \cdot 2 \cdot 1$.)

x	e^x	$P_1(x)$	$P_2(x)$	$P_3(x)$	$P_4(x)$
1					
0.5					
2					
−2					

76. Prove that if the function $f(x)$ has an inverse $f^{-1}(x)$ defined for all x, then $f^{-1}(x)$ is increasing on the interval $[f(a), f(b)]$ if $f(x)$ is increasing on the interval $[a, b]$.

77. Prove statement (ii) of Theorem 2.

78. Prove statement (iii) of Theorem 2 for the case r rational.

79. Use l'Hôpital's Rule to show that $\lim_{x\to\infty} \frac{x^n}{e^x} = 0$.

80. Generalize the result of Exercise 79 by showing that $\lim_{x\to\infty} \frac{p(x)}{e^x} = 0$ where $p(x)$ is any polynomial.

9.4 EXPONENTIALS AND LOGS TO OTHER BASES

By use of the (natural) exponential function e^x, we can extend the domain of the exponential function $y = a^x$, $a > 0$, to *all* real numbers, both rational and irrational. To do so we use the identity

$$a = e^{\ln a}, \qquad a > 0$$

to write

$$a^x = [e^{\ln a}]^x. \tag{1}$$

Now if both a and x are rational numbers, we can apply the Laws of Exponents (specifically, property (c) of Theorem 3, Section 1.2) to conclude that

$$[e^{\ln a}]^x = e^{x \ln a}. \tag{2}$$

Combining equations (1) and (2) we conclude that

$$\boxed{a^x = e^{x \ln a}} \tag{3}$$

for a, x rational. However, the right side of equation (3) is defined for all real numbers x and all real numbers $a > 0$. We therefore *define* the function $y = a^x$ by equation (3). Our demonstration shows that this definition agrees with the algebraic definition of a^x when x is rational, so we have now succeeded in extending this definition to all real numbers x. The following theorem shows that we have succeeded in preserving the usual laws of exponents in making this extension.

THEOREM 3

If $a > 0$ is any real number, the equations

(E1) $\quad a^x a^y = a^{x+y}$
(E2) $\quad a^x/a^y = a^{x-y}$
(E3) $\quad (a^x)^y = a^{xy}$

hold for all real numbers x and y.

Strategy

Proof
We prove only ($E3$), leaving ($E1$) and ($E2$) as exercises. By equation (3)

Apply equation (3).
Use property ($L3$), Section 9.2.
Equation (3) again.

$$\begin{aligned}(a^x)^y &= e^{y \ln a^x}\\ &= e^{xy \ln a}\\ &= a^{xy}\end{aligned}$$

(See Exercises 30 and 31.) ■

Differentiating $y = a^x$

Since the exponential function $y = e^x$ and the linear function $y = (\ln a)x$ are both differentiable, equation (3) defines the function $y = a^x$ as the composition of two differentiable functions. As such, $y = a^x$ must be differentiable. To find its derivative we apply the Chain Rule to the right-hand side of equation (3):

$$\begin{aligned}\frac{d}{dx} a^x &= \frac{d}{dx} e^{x \ln a}\\ &= e^{x \ln a} \cdot \frac{d}{dx} \{x \ln a\}\\ &= e^{x \ln a} \cdot \ln a\\ &= a^x \ln a\end{aligned}$$

That is,

$$\boxed{\frac{d}{dx} a^x = a^x \ln a.} \tag{4}$$

If $u = u(x)$ is a differentiable function of x, the Chain Rule and equation (4) give

$$\frac{d}{dx} a^u = a^u \cdot \ln a \cdot \frac{du}{dx}. \tag{5}$$

Example 1 Find $\frac{dI}{dx}$ for the Beer-Lambert law of Exercise 22, Section 9.1.

Solution: Recall that

$$I = I_0 10^{kx}$$

where I_0 and k are constants. Equation (5) gives

$$\frac{dI}{dx} = I_0 10^{kx} \cdot \ln 10 \cdot \frac{d}{dx}(kx) = kI_0 10^{kx} \ln 10.$$ ■

Observe that equation (5) agrees with the differentiation formula for $y = e^u$ when $a = e$, since $\ln e = 1$. This fact explains why the number e is referred to as the *natural* choice for the base of an exponential function: among all bases, e is the one for which the derivative of the exponential function has the simplest form.

The integration formulas corresponding to the differentiation formulas (4) and (5) are

$$\boxed{\int a^x\, dx = \frac{a^x}{\ln a} + C, \qquad a > 0, \qquad a \neq 1} \tag{6}$$

and

$$\int a^{u(x)}u'(x)\,dx = \int a^u\,du = \frac{a^u}{\ln a} + C, \qquad a > 0, \qquad a \neq 1. \tag{7}$$

Example 2 Find

$$\int \sin x \cos x \cdot 10^{\sin^2 x}\,dx.$$

Solution: Since the exponent is $\sin^2 x$, we try the u-substitution

$$u = \sin^2 x; \qquad du = 2\sin x \cos x\,dx.$$

Then $\sin x \cos x\,dx = \frac{1}{2}\,du$, and we obtain

$$\begin{aligned}\int \sin x \cos x \cdot 10^{\sin^2 x}\,dx &= \int 10^u \cdot \frac{1}{2}\,du \\ &= \frac{1}{2}\int 10^u\,du \\ &= \frac{1}{2\ln 10}\,10^u + C \\ &= \frac{1}{2\ln 10}\,10^{\sin^2 x} + C.\end{aligned}$$

■

The Function $y = \log_a x$

For any positive real number $a \neq 1$, equation (3) together with the algebraic definition of logarithm (Section 9.1) allows us to view the equation $y = \log_a x$ as defining a differentiable function of $x > 0$. To see this we write

$$y = \log_a x \text{ if and only if } x = a^y = e^{y \ln a}. \tag{8}$$

Applying the natural logarithm to both sides of the equation

$$x = e^{y \ln a}$$

gives

$$\ln x = \ln[e^{y \ln a}] = y \ln a \cdot \ln e = y \ln a,$$

so

$$y = \frac{\ln x}{\ln a}, \qquad a \neq 1. \tag{9}$$

In other words, we have found that the function $\log_a x$ is simply a multiple of the function $\ln x$, and conversely, since statements (8) and (9) together give the equation

$$\log_a x = \frac{\ln x}{\ln a}, \qquad a \neq 1. \tag{10}$$

Example 3 Find formulas for converting common logs to natural logs and vice versa.

Strategy

Find ln 10 from a natural log table or calculator.

Solution

Common logs are logarithms to the base $a = 10$. Since

$$\ln(10) = 2.303,$$

equation (10) gives

Write equation (10) with $a = 10$.

$$\log_{10}x \approx \frac{1}{2.303} \ln x = .434 \ln x.$$

Solve equation (10) for ln x.

Solving for ln x we obtain

$$\ln x \approx 2.303 \log_{10}x.$$

■

The differentiation formula for $y = \log_a x$ can be found by differentiating both sides of equation (10). We obtain

$$\frac{d}{dx} \log_a x = \frac{1}{x \ln a}, \qquad x > 0, \qquad a \neq 1. \tag{11}$$

If $u = u(x)$ is a differentiable function of x, the Chain Rule and formula (11) give:

$$\frac{d}{dx} \log_a u = \frac{1}{u \ln a} \cdot \frac{du}{dx}, \qquad u > 0, \qquad a \neq 1. \tag{12}$$

REMARK: Equations (11) and (12) show why $\ln x = \log_e x$ is called the *natural* logarithm function. Among all logarithm functions, ln x is the one for which the derivative has simplest form, since the factor ln a is just the integer 1.

Example 4 Find the minimum value of the function

$$y = \log_{10}(1 + x^2), \qquad -\infty < x < \infty.$$

Strategy

Find dy/dx by (12) with $u = 1 + x^2$.

Find critical numbers by setting $dy/dx = 0$.

Solution

To find the critical numbers we set

$$\frac{dy}{dx} = \frac{2x}{(1 + x^2) \ln 10} = 0 \tag{13}$$

and obtain the single critical number $x = 0$.

Verify that the relative minimum is the absolute minimum.

From the form of dy/dx in equation (13) it is easy to see that $dy/dx < 0$ if $x < 0$, and $dy/dx > 0$ if $x > 0$. Thus, $y = \log_{10}(1 + x^2)$ decreases on $(-\infty, 0)$ and increases on $(0, \infty)$. The value $y = 0$ at the point (0, 0) is therefore a relative (and absolute) minimum value of y on $(-\infty, \infty)$. ■

Logarithmic Differentiation

There are two types of functions for which it is advantageous to calculate the derivative by differentiating the natural logarithm of the function rather than the function itself. The first concerns functions of the form

$$y = u(x)^{v(x)}, \qquad u(x) \geq 0. \tag{14}$$

Since the exponent $v(x)$ is not constant, the Power Rule is not applicable. Moreover, the base $u(x)$ is not constant, so the rule (5) for differentiating an exponential does not apply. However, by applying the natural logarithm function to both sides of equation (14) we obtain

$$\ln y = \ln[u(x)^{v(x)}] = v(x) \ln u(x). \tag{15}$$

We may then obtain dy/dx from equation (15) by the technique of implicit differentiation.

Example 5 Find $\dfrac{dy}{dx}$ for $y = x^{\sin x}, \quad x > 0.$

Strategy
Take natural logs of both sides.

Solution
Since $y = x^{\sin x}$ we obtain

$$\ln y = \ln[x^{\sin x}] = \sin x \cdot \ln x.$$

Differentiate both sides, using Chain Rule on left, Product Rule on right.

Differentiating implicitly we find that

$$\frac{1}{y} \cdot \frac{dy}{dx} = \cos x \cdot \ln x + \frac{\sin x}{x},$$

so

Solve for dy/dx and substitute for y.

$$\frac{dy}{dx} = y\left[\cos x \cdot \ln x + \frac{\sin x}{x}\right]$$

$$= x^{\sin x}\left[\cos x \cdot \ln x + \frac{\sin x}{x}\right].$$

■

The second type of function for which this technique is useful is one that is complicated by a large number of algebraic operations. Properties ($L1$)–($L3$) can often be applied to greatly simplify the form of $\ln y$. The derivative dy/dx is then calculated implicitly as above.

Example 6 Find $\dfrac{dy}{dx}$ for $y = \dfrac{\sqrt{1 + x^2}\,(3x + 2)^3}{\sqrt[3]{x^2(x + 1)}}.$

Strategy
Take natural logs of both sides.

Use properties ($L1$)–($L3$) of natural logs to simplify right-hand side.

Solution
We obtain

$$\ln y = \ln\left\{\frac{\sqrt{1 + x^2}\,(3x + 2)^3}{\sqrt[3]{x^2(x + 1)}}\right\}$$

$$= \ln[(1 + x^2)^{1/2}(3x + 2)^3] - \ln[x^2(x + 1)]^{1/3}$$

$$= \frac{1}{2}\ln(1 + x^2) + 3\ln(3x + 2)$$

$$- \frac{1}{3}[2 \cdot \ln x + \ln(x + 1)].$$

Differentiate both sides.

Differentiating both sides then gives

$$\frac{1}{y} \cdot \frac{dy}{dx} = \frac{1}{2} \cdot \frac{2x}{1 + x^2} + 3 \cdot \frac{3}{3x + 2} - \frac{1}{3}\left[\frac{2}{x} + \frac{1}{x + 1}\right]$$

$$= \frac{x}{1 + x^2} + \frac{9}{3x + 2} - \frac{2}{3x} - \frac{1}{3(x + 1)}$$

Solve for dy/dx.

so

$$\frac{dy}{dx} = y\left[\frac{x}{1+x^2} + \frac{9}{3x+2} - \frac{2}{3x} - \frac{1}{3(x+1)}\right].$$

If necessary, you could substitute for y in terms of x. ■

Exercise Set 9.4

1. Write each of the following in the form e^u.
a. $2x$ **b.** π^3 **c.** $7^{\ln x}$
d. $4^{\sqrt{2}}$ **e.** $3^{\sin x}$ **f.** $2^x \cdot 4^{1-x}$

2. Use a calculator with a natural exponential key or a table of values for the exponential function to find decimal expressions for the numbers in Exercise 1, parts (b) and (d).

3. Find a formula for converting powers of 10 to powers of e and vice versa.

4. True or false? $e^{x \ln a} = a^x$ for any a and x. Prove your answer.

In Exercises 5–20, find the derivative of the given function.

5. $y = 3^x$

6. $f(x) = \pi^{x^2-1}$

7. $f(x) = \log_{10}(2x - 1)$

8. $y = x10^{x-1}$

9. $y = \log_{10}(\ln x)$

10. $f(x) = 2^{1-x} \cdot \log_2\sqrt{x}$

11. $g(t) = t^2\log_2 t$

12. $y = \log_{10}e^{\sqrt{t^2+1}}$

13. $f(x) = \pi^x + x^\pi$

14. $y = \log_{10}2^{x^2}$

15. $y = x^x$

16. $y = x^{\cos x}$

17. $y = x^{\sqrt{x}}$

18. $y = x^{\ln x}$

19. $y = (\cos x)^{\sin x}$, $\cos x \geq 0$

20. $y = [\ln x]^x$

In Exercises 21–28, evaluate the given integral.

21. $\int 5^x \, dx$

22. $\int x2^{1-x^2} \, dx$

23. $\int \frac{\pi^{\sqrt{x}}}{\sqrt{x}} \, dx$

24. $\int 2^{1-x}2^{1+x} \, dx$

25. $\int a^{2 \ln x} \, dx$

26. $\int 7^{ax^2+bx}\left(ax + \frac{b}{2}\right) dx$

27. $\int_0^2 x3^{x^2} \, dx$

28. $\int_0^2 2^{3t-1} \, dt$

29. True or false? The number a^x is nonzero for all x if $a > 0$.

30. Prove statement ($E1$) in Theorem 3.

31. Prove statement ($E2$) in Theorem 3.

32. Sketch the graph of the function $y = x2^{1-x}$ by finding
a. the relative extrema,
b. the intervals on which y is increasing,
c. the inflection points,
d. the concavity.

33. Find the number $a > 0$ so that the function $y = a^x$ satisfies the differential equation $y' - 2y = 0$.

34. *(Calculator)* Use differentials to obtain a formula for approximating $\log_{10}(10^n + x)$ where n is an integer and x is a small real number. Use this formula to approximate
a. $\log_{10}(10.2)$ **b.** $\log_{10}(98)$
c. $\log_{10}(995)$ **d.** $\log_{10}(.09)$

35. Find the length of the graph $y = \frac{1}{2}(e^x + e^{-x})$ from $(0, 1)$ to $\left(\ln 2, \frac{5}{4}\right)$.

36. Find the lateral surface area of the solid obtained by revolving the arc of the graph in Exercise 35 about the x-axis.

37. If P_0 dollars are invested at 10% interest, compounded annually, the amount on deposit after n years is $P(n) = (1.10)^n P_0$ (see Exercise 17, Section 9.1). Find the number r so that

$$P(n) = (1.10)^n P_0 = e^{nr}P_0, \qquad n = 1, 2, \ldots.$$

38. The owner of a valuable oil painting estimates that the value of the painting will be approximately $V(t) = (5000)2^{\sqrt{t}}$ dollars over the next few years. Here t represents time in years, and the present value of the painting is $V(0) = 5000$ dollars. How fast will the value of the painting be increasing in 4 years?

In Exercises 39–44, find dy/dx by logarithmic differentiation.

39. $y = \frac{x(x+1)(x+2)}{(x+3)(x+4)(x+5)}$

40. $y = \sqrt[3]{\frac{x+2}{x+3}}$

41. $y = \frac{\sqrt[3]{x^2+1}}{(x+1)\cdot(x+2)^2}$

42. $y = \dfrac{(x^2 + 2)^3 \cdot (x - 1)^5}{x \cdot \sqrt{x + 1} \cdot \sqrt{x + 2}}$

43. $y = \sqrt{x\sqrt{x\sqrt{x}}}$

44. $y = x^{\sqrt{x+1}}$

45. Use the definition of a^x in equation (3) to show that property ($L3$) of Theorem 1 holds for all r, both rational and irrational (see Exercise 70, Section 9.2).

46. Use the definition of a^x in equation (3) to show that property (iii) of Theorem 2 holds for all r, both rational and irrational (see Exercise 78, Section 9.3).

47. Prove that the Power Rule, $\dfrac{d}{dx} x^r = rx^{r-1}$, is valid for all r, both rational and irrational.

9.5 EXPONENTIAL GROWTH AND DECAY

Imagine the following experiment associated with a course in biology. On a certain day, a number n of fruit flies are placed in an enclosed environment such as a large bell jar. If the environment is supportive (e.g., sufficient food supply and proper temperature), the number of fruit flies will increase as time passes. The experiment is to record the number of fruit flies present after t days, $N(t)$, and to find a mathematical relationship between time and population size. In experiments of this kind, data such as those presented in Table 5.1 are often obtained. (See Figure 5.1.)

Table 5.1 Typical data on growth of fruit flies

t (days)	0	4	8	12	16	20	24
$N(t)$ (fruit-flies)	10	18	35	72	107	208	361

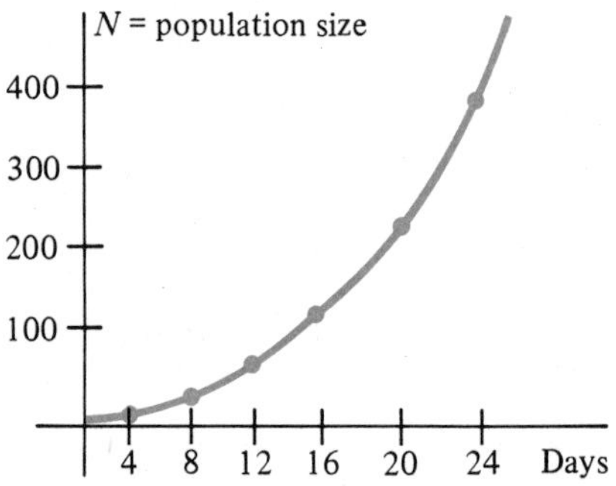

Figure 5.1 Plot of typical data on growth of fruit flies.

On examining these data, one observes that the population size, $N(t)$, increases more rapidly as this population size itself grows, i.e., the larger the population the more rapid the growth rate. Biologists refer to this phenomenon as the **Law of Natural Growth,** which is often stated as:

$$\{\text{Rate of growth of population}\} \propto \{\text{Present population size}\}. \quad (1)$$

The symbol "$\propto$" means "is proportional to." The statement of proportionality, $A \propto B$, is equivalent to the equation $A = kB$, where k is a constant. We can develop a mathematical model for populations that grow according to law (1) by assuming that the population size $N(t)$ may be approximated by a differentiable function of time. Since the derivative $N'(t) = dN/dt$ gives the rate of growth (increase or decrease) of the population size, the mathematical formulation of the growth law (1) is

$$\frac{dN}{dt} = kN. \tag{2}$$

The constant k in equation (2) is referred to as the **growth constant.** Since the mathematical model (2) consists of a *differential equation,* we must find a differentiable function $y = N(t)$ that satisfies the equation.

We have already determined (Section 9.3) that any exponential function of the form $y = Ce^{kt}$ satisfies equation (2), since

$$\begin{aligned}\frac{dy}{dt} &= \frac{d}{dt}(Ce^{kt}) \\ &= C\left(\frac{d}{dt}e^{kt}\right) \\ &= kCe^{kt} \\ &= ky.\end{aligned}$$

In Exercise 34 you will show that other than the trivial solution $y = 0$, there are no other possible solutions of equation (2). In other words, all solutions of the differential equation (2) must have the form $y = Ce^{kt}$. Finally, we can be more specific about the constant C. Setting $t = 0$ in the equation $y = Ce^{kt}$ gives

$$y(0) = Ce^{k\cdot 0} = C,$$

so $C = y(0)$. We refer to the constant $C = y(0) = y_0$ as the **initial condition** associated with the differential equation (2). We summarize our findings as follows.

THEOREM 4 The unique solution of the differential equation

$$\frac{dy}{dt} = ky, \tag{3}$$

satisfying the initial condition $y(0) = y_0$ is the exponential function

$$y = y_0e^{kt}. \tag{4}$$

Theorem 4 provides the strategy for solving problems involving populations or quantities that satisfy the Law of Natural Growth (1): If we can safely assume that the size or amount of the population or quantity y can be viewed as a differentiable function of t (which usually represents time), then y has the form given by equation (4). By substituting known data into equation (4) one obtains the growth constant k and the initial population size or amount, y_0. The function y is then completely determined.

Graphs of the growth function (4) are shown in Figure 5.2 for the two general cases $k > 0$ (growth) and $k < 0$ (decay).

Example 1 Assume that the rate of growth of a population of fruit flies is proportional to the size of the population at each instant of time. If 100 fruit flies are present initially and 300 are present after 10 days, how many will be present after 15 days?

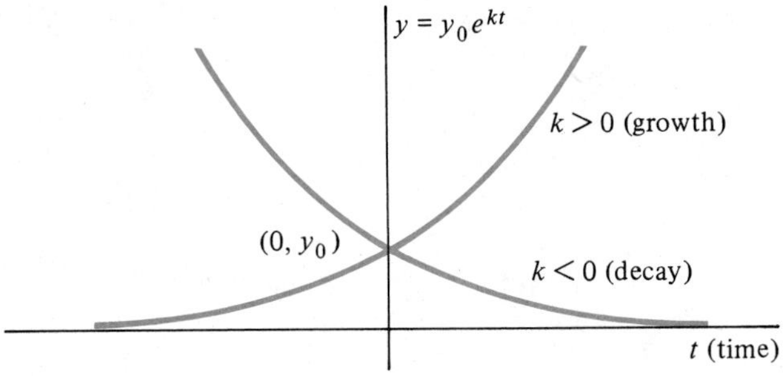

Figure 5.2 Graph of the natural growth function.

Strategy
Apply Theorem 4.

Solution
Since it is stated that the population size N satisfies the differential equation

$$\frac{dN}{dt} = kN,$$

the function $N(t)$ is given by equation (4):

$$N(t) = N_0 e^{kt}. \tag{5}$$

Find $N_0 = N(0)$ from given data.

We are given that $N = 100$ when $t = 0$, that is,

$$N_0 = 100.$$

Also, we are given that $N = 300$ when $t = 10$. Inserting these values in (5) gives

$$300 = 100e^{10k}$$

so

Take natural logs of both sides and solve for k.

$$e^{10k} = 3.$$

Thus

$$10k = \ln 3$$

so

$$k = \frac{\ln 3}{10}.$$

State explicit formulation of $N(t)$.

The function $N(t)$ is therefore

$$N(t) = 100e^{\left(\frac{\ln 3}{10}\right)t},$$

Find $N(15)$ using tables or calculator, rounding to nearest integer.

so

$$N(15) = 100e^{\frac{15 \ln 3}{10}} = 100e^{\frac{3}{2}\ln 3} \quad 100 \cdot 3^{3/2}$$
$$\approx 520 \text{ fruit flies.}$$

■

Example 2 A rule of thumb in the study of rates of certain chemical reactions is that the reaction rate R (the number of molecules reacting per second) will double with each 10°C increase in temperature.

(a) Show that an exponential function modelling the relationship between the reaction rate R and temperature satisfies this rule of thumb.
(b) According to the model obtained in part (a) if a chemical reaction proceeds at the rate of 3×10^{-4} moles per liter per second at 20°C, at what temperature will the rate be three times as fast?

Strategy

Assume an exponential form for R. Note that t stands for temperature, not time, in this application.

Solution

(a) We write the reaction rate R in the form

$$R(t) = R_0 e^{kt}, \qquad t = \text{temperature} \tag{6}$$

Apply the rule of thumb to (6) to find k. Any two temperatures that differ by 10°C will give the same value of k; but 0°C simplifies the calculation.

and show that this function satisfies the rule of thumb given above. To determine the constant k we apply the rule for two temperatures differing by 10°C, say 0°C and 10°C. The rule says that

$$R(10°) = 2R(0°)$$

so

$$R_0 e^{10k} = 2R_0 e^{0 \cdot k}$$

Cancel factors of R_0 and take ln's of both sides.

or

$$e^{10k} = 2.$$

Thus

$$k = \frac{\ln 2}{10}$$

and we can write $R(t)$ in (6) as

$$R(t) = R_0 e^{\left(\frac{\ln 2}{10}\right)t}. \tag{7}$$

Show that

$R(t + 10) = 2R(t)$

to verify that the preceding calculation was independent of the starting temperature.

To test (7) against the rule of thumb for the general case we calculate $R(t + 10)$ from (7):

$$\begin{aligned} R(t+10) &= R_0 e^{\frac{\ln 2}{10}(t+10)} \\ &= R_0 e^{\left(\frac{\ln 2}{10}\right)t} \cdot e^{\left(\frac{\ln 2}{10}\right)10} \\ &= e^{\ln 2} \cdot R_0 e^{\left(\frac{\ln 2}{10}\right)t} \\ &= 2 \cdot R(t), \end{aligned}$$

which verifies that (7) satisfies the rule of thumb.

(b) We are given that $R(20) = 3 \times 10^{-4}$. Substituting this information in (7) gives

$$3 \times 10^{-4} = R_0 e^{2 \ln 2} = R_0 e^{\ln 2^2} = 4R_0$$

so

$$R_0 = \frac{3}{4} \times 10^{-4} = 0.75 \times 10^{-4}. \tag{8}$$

Desired temperature T satisfies equation $R(T) = 3R(20)$. Solve this equation for T.

We seek a temperature T so that

$$R(T) = 3R(20) = 9 \times 10^{-4}. \tag{9}$$

Combining (7), (8), and (9) we have

$$\left(\frac{3}{4} \times 10^{-4}\right) e^{\left(\frac{\ln 2}{10}\right)T} = 9 \times 10^{-4}.$$

Take natural logs of both sides.

Thus

$$e^{\left(\frac{\ln 2}{10}\right)T} = \frac{9 \cdot 4}{3} = 12,$$

so

Solve for T.

$$\left(\frac{\ln 2}{10}\right)T = \ln 12,$$

or

$$T = 10\,\frac{\ln 12}{\ln 2}$$
$$\approx 35.85°\text{C}.$$

■

Example 3 Certain radioactive isotopes, such as uranium, are known to decay at a rate proportional to the amount present. If a block of 50 grams of such material decays to 40 grams in 5 days, find the half-life of this material. (The half-life is the amount of time required for the material to decay to half its original amount.)

Strategy

Label the variables.

Solution

Let $A(t)$ denote the amount present at time t. Since A is given as satisfying the differential equation (3), by Theorem 4, A has the form

State form of A from Theorem 4.

$$A(t) = A_0 e^{kt}, \qquad t = \text{time}. \tag{10}$$

We are given

Apply initial condition $A(0) = 50 = A_0$.

$$A_0 = A(0) = 50 \text{ grams}.$$

To find k we use the data $A(5) = 40$ together with equation (10). We obtain

Use equation $A(5) = 40$ to obtain equation for k.

$$40 = A_0 e^{5k} = 50e^{5k},$$

so

$$e^{5k} = \frac{40}{50} = \frac{4}{5}.$$

Solve for k by taking ln's of both sides.

Thus

$$k = \frac{1}{5}\ln\left(\frac{4}{5}\right). \tag{11}$$

Write down condition for T: $A(T) = \frac{1}{2}A(0)$.

We seek a time T so that

$$A(T) = \frac{1}{2}A_0. \tag{12}$$

Combine (10) and (12).

Using (10) we obtain

$$A_0 e^{kT} = \frac{1}{2}A_0$$

Solve for T, taking ln's of both sides and using (11).

so

$$e^{kT} = \frac{1}{2}.$$

Thus

$$T = \frac{\ln\left(\frac{1}{2}\right)}{k}.$$

Using the value for k in (11), we find

$$T = \frac{\ln\left(\frac{1}{2}\right)}{\frac{1}{5}\ln\left(\frac{4}{5}\right)} = \frac{5[\ln 1 - \ln 2]}{\ln 4 - \ln 5} \approx 15.53 \text{ days.}$$ ■

COMPOUND INTEREST: Recall from Exercise 17, Section 9.1, that if P_0 dollars are invested at r percent annual interest, compounded annually, the amount on deposit after n years, denoted by $P(n)$, is

$$P(n) = (1 + r)^n P_0 \text{ dollars.}$$

In Exercise 18, Section 9.1, it was shown that if interest is compounded k times per year, the amount on deposit after n years is

$$P(n) = \left(1 + \frac{r}{k}\right)^{nk} P_0 \text{ dollars.} \tag{13}$$

Many banks advertise ''continuous'' compounding of interest. The meaning of this concept is that the number of compoundings per year is regarded as having become infinitely large (i.e., interest is added to your account ''each instant'').

To obtain an expression for $P(n)$ under continuous compounding of interest we apply $\lim_{k\to\infty}$ to expression (13). We obtain

$$P(n) = \lim_{k\to\infty} \left\{\left(1 + \frac{r}{k}\right)^{nk} P_0\right\}. \tag{14}$$

Since $\lim_{x\to\infty} cf(x) = c \lim_{x\to\infty} f(x)$ and $\lim_{x\to\infty} [f(x)]^n = \left[\lim_{x\to\infty} f(x)\right]^n$ (see Theorem 8, Section 5.6), we may write expression (14) as

$$P(n) = P_0\left[\lim_{k\to\infty} \left(1 + \frac{r}{k}\right)^k\right]^n. \tag{15}$$

Now since Equation (12), Section 9.3 gives

$$\lim_{k\to\infty} \left(1 + \frac{r}{k}\right)^k = e^r,$$

equation (15) gives *the equation for continuous compounding of interest:*

$$P(n) = P_0 e^{rn}. \tag{16}$$

(Since the development leading to equation (16) did not require that n be an integer, equation (16) actually holds for all real values of n.)

Example 4 Find the value of an initial deposit of \$1,000 invested at 12% annual interest compounded continuously for 6 years.

Solution: We have $P_0 = 1000$, $r = 0.12$, and $n = 6$. From equation (16),

$$\begin{aligned} P(6) &= 1000e^{6(.12)} \\ &= 2504.43 \text{ dollars.} \end{aligned}$$

■

Example 5 Just as the amount of money available after compounding P_0 dollars continually for n years can be found using equation (16), the *initial* amount P_0 needed to produce a given amount $P_n = P(n)$ after n years can be calculated. To do this we multiply both sides of equation (16) by e^{-rn} to obtain

$$P_0 = P_n e^{-rn}. \tag{17}$$

In this case P_0 is called the **present value** of an investment that will yield P_n dollars in n years if compounded continuously at r percent. Find the size of the deposit which, when compounded continuously for 10 years at 12% interest, will yield $20,000.

Solution: From equation (17) we obtain

$$\begin{aligned} P_0 &= e^{-(0.12)\cdot 10}(20{,}000) \\ &= 6{,}023.88 \text{ dollars.} \end{aligned}$$

■

Example 6 The owner of a stand of timber estimates that his timber is presently worth $5,000 and that the timber increases in value with time according to the formula

$$V(t) = 5000e^{\sqrt{t}/2} \text{ dollars}$$

where t is time in years. If the prevailing interest rates over the foreseeable future are expected to average 12%, when should he sell his timber so as to maximize profit?

Strategy
Convert the revenue received at a future date to present value, using (17).

Solution
Since dollars received at different times have different present values, we must discount the changing value of the revenue received for the sale of the timber by converting it to its *present value*. Using equation (17) we find the present value of the revenue received in selling the timber after t years to be

$$\begin{aligned} R(t) &= V(t)e^{-.12t} \\ &= 5000e^{(\sqrt{t}/2 - .12t)} \end{aligned}$$

To maximize $R(t)$ we set $R'(t) = 0$:

Set $R'(t) = 0$.

$$R'(t) = 5000e^{(\sqrt{t}/2 - .12t)}\left[\frac{1}{4\sqrt{t}} - .12\right] = 0. \tag{18}$$

Since the exponential function is never zero we must have

$$\frac{1}{4\sqrt{t}} - .12 = 0$$

Solve for t.

or

$$\sqrt{t} = \left(\frac{1}{.48}\right)$$

so

$$t = \left(\frac{1}{.48}\right)^2 = 4.34 \text{ years},$$

Verify that the critical point obtained yields a maximum.

or approximately 4 years and 4 months. (We can verify that this value of t indeed produces a profit maximum by applying the first derivative test to $R'(t)$ in equation (18).) ■

The question of modelling natural growth of biological populations requires further discussion. Although the Law of Natural Growth provides an adequate model for certain populations during certain intervals of time, the exponential solution $P(t) = P_0 e^{kt}$ clearly cannot hold for all future values of time since all biological populations exist in limited environments. We will return to this issue in Chapter 11 where we will refine our model to reflect a limited environment.

Exercise Set 9.5

In Exercises 1–6, find an exponential function of the form $y = Ae^{kt}$ which satisfies the given conditions.

1. $y' = 2y$ and $y(0) = 1$

2. $y' = -4y$ and $y(0) = 2$

3. $y' - 5y = 0$ and $y(1) = 1$

4. $y(0) = 2$ and $y(2) = 6$

5. $y(0) = 1$ and $y(1) = e^{-2}$

6. $y(0) = y(1) = 0$

7. True or false? If the population $P(t)$ of a city is increasing at a rate of 5% per year, then the function $P(t)$, if differentiable, satisfies the equation $P'(t) = .05P(t)$. If this is false, what is the correct equation?

In Exercises 8–13, find the indicated limit.

8. $\lim\limits_{x \to \infty} \left(1 + \dfrac{2}{x}\right)^x$

9. $\lim\limits_{t \to \infty} \left(1 + \dfrac{1}{t}\right)^{t/2}$

10. $\lim\limits_{t \to \infty} \left(1 - \dfrac{3}{t}\right)^{2t}$

11. $\lim\limits_{x \to \infty} \left(1 - \dfrac{4}{x^2}\right)^x$

12. $\lim\limits_{t \to \infty} \left(4 - \dfrac{8}{t}\right)^{-t}$

13. $\lim\limits_{x \to 0^+} (1 - x)^{1/x}$

14. What rate of interest, compounded continuously, will produce an effective annual yield of 8%? (**Effective annual yield** is the rate required under *annual* compounding of interest to obtain the specified yield.)

15. An annuity pays 10% interest, compounded continuously. What amount of money, deposited today, will have grown to \$2,500 in 8 years?

16. Show that the effective annual rate of interest, i, for continuous compounding at rate r is $i = e^r - 1$.

17. The population of a certain city is increasing at a rate of 5% per year. If the population $P(t)$ is assumed differentiable, and if $P'(t) = kP(t)$, find k (see Exercise 7).

18. How long does it take for a deposit of P_0 dollars to double at 10% interest compounded continuously?

19. True or false? The half-life of an isotope depends upon the amount present. Explain.

20. A radioactive isotope I decays exponentially with a half-life of 20 days. If 50 mg of the isotope remain after 10 days,
a. How much of the isotope was present initially?
b. How much will remain after 30 days?
c. When will 90% of the initial amount have disintegrated?

21. Suppose an object travelling through a fluid is subject only to a force resisting its motion, and this resisting force is proportional to the velocity of the object. Let's write this resisting force as $f_r = cv$ where c is constant and v denotes velocity. By Newton's second law we have $f_r = ma$ where m is the mass of the object and a is its acceleration. Since $a = \dfrac{dv}{dt}$, we may combine our two equations to obtain

$$m\frac{dv}{dt} = -cv, \qquad \text{or} \qquad \frac{dv}{dt} = -\left(\frac{c}{m}\right)v.$$

a. Find a solution $v(t)$ of this differential equation for which $v(0) = v_0$.
b. Find a solution of the equation for $v_0 = 10$ m/sec, $c = 40$, and $m = 10$.
c. Does $v(t)$ approach a limit as $t \to \infty$? If so, what is this limit?

22. In a **first order chemical reaction** the rate of disappearance of reactant is proportional to the amount present. Chemists use the notation $[A]$ for the concentration of reactant A in moles per liter. Thus, in first order reactions, the concentration $[A]$ as a function of time is a solution of the differential equation

$$\frac{d[A]}{dt} = -k[A].$$

a. Find an expression for the concentration $[A(t)]$ after t seconds if $[A_0] = [A(0)]$.
b. If the reaction begins with a concentration $[A_0] = 10$ moles per liter, and 6 moles per liter of reactant A remain after 20 minutes, find the concentration of reactant A one hour after the reaction begins.

23. What is the present value of \$1000 five years from now if the prevailing rate of interest, when compounded continuously, is 10%? What is the present value if 10% is the prevailing *effective annual yield?* (See Exercise 14.)

24. The number of bacteria in a certain culture grows from 50 to 400 in 12 hours. Assuming that the rate of increase is proportional to the number of bacteria present
a. How long does it take for the number of bacteria present to double?
b. How many bacteria will be present after 16 hours?

25. If we assume temperature to be constant at all altitudes, the rate of decrease in atmospheric pressure p, as a function of altitude, h, is proportional to p. That is,

$$\frac{dp}{dh} = -kp$$

where the constant k is approximately $k = 0.116\ \text{km}^{-1}$. Find the pressure in atmospheres 8 km (5 mi) above sea level if the pressure at sea level is $p_0 = 1.2\ \text{kg/m}^3$. Use the fact that 1 atmosphere = 13.595 grams per cm^3.

26. Find the effective annual yield (see Exercise 14) on
a. an interest rate of 10% compounded semiannually,
b. an interest rate of 10% compounded quarterly,
c. an interest rate of 10% compounded daily,
d. an interest rate of 10% compounded continuously.

27. Land bought for speculation is expected to be worth $V(t) = 10{,}000(1.2)^{\sqrt{t}}$ dollars after t years. If the cost of money holds constant at 10%, when should the land be sold so as to maximize profits? Does the original price of the land affect this calculation?

28. In a simple model of the growth of a biological population, the rate of increase of N, the number of individuals, is proportional to the difference between the birth rate, b, and the death rate, d. That is, the simple birth-death model of population growth is

$$\frac{dN}{dt} = bN - dN.$$

a. Find an expression for population size, N, as a function of the birth rate r, death rate d, and initial population size N_0.
b. If in a population modelled by this process there are 20 births per year per 100 individuals and 12 deaths per year per 100 individuals, find the anticipated size of the population in 20 years if its present size is $N_0 = 200$ individuals.

29. When a foreign substance is introduced into the body, the body's defense mechanisms move to break down the substance and excrete it. The rate of excretion is usually proportional to the concentration present in the body, and the half-life of the resulting exponential decay is referred to as the **biological half-life** of the substance. If, after 12 hours, 30% of a massive dosage of a substance has been excreted by the body, what is the biological half-life of the substance?

30. All living matter contains two types of carbon, ^{14}C and ^{12}C, in its molecules. While the organism is alive the ratio of ^{14}C to ^{12}C is constant, as the ^{14}C is interchanged with fixed levels of $^{14}CO_2$ in the atmosphere. However, when a plant or animal dies this replenishment ceases, and the radioactive ^{14}C present decays exponentially with a half-life of 5760 years. By examining the $^{14}C/^{12}C$ ratio, archeologists can determine the percent of the original ^{14}C level remaining and therefore date a once living fossil.
a. If 80% of the original amount of ^{14}C remains, how old is the fossil?
b. A fossil contains 10% of its original amount of ^{14}C. What is its age?

31. The body concentrates iodine in the thyroid gland. This observation leads to the treatment of thyroid cancer by injecting radioactive iodine into the bloodstream. One isotope used has a half-life of approximately 8 days and decays exponentially in time. If 50 micrograms of this isotope are injected, what amount remains in the body after three weeks?

32. A chemical dissolves in water at a rate proportional to the amount still undissolved. If 20 grams of the chemical are placed in water and 10 grams remain undissolved 5 minutes later, when will 90% of the chemical be dissolved?

33. The value of a case of a certain French wine t years after being imported to the United States is $V(t) = 100(1.5)^{\sqrt{t}}$. If the cost of money under continuous compounding of interest is expected to remain constant at 12.5% per year, when should the wine be sold so as to maximize profits? (Assume storage costs to be negligible.)

34. Prove that $y = Ce^{kt}$ is the only nontrivial $(y \neq 0)$ solution of the differential equation $y' = ky$ as follows.

a. Assume that the differentiable function $y = f(t)$ is also a solution.

b. Show that the product $g(t) = f(t)e^{-kt}$ is differentiable, and that $g'(t) = 0$.

c. Conclude that $g(t) = C$ for some constant C, that is, necessarily $f(t) = Ce^{kt}$.

9.6 MORE GENERAL EQUATIONS FOR GROWTH AND DECAY (Optional)

The point of Section 9.5 was that the simple first order differential equation

$$\frac{dy}{dt} = ky \tag{1}$$

always has exponential solutions of the form $y = Ce^{kt}$. In this section we shall briefly consider some generalizations of equation (1) whose solutions also involve exponential functions.

For example, suppose we arbitrarily add a constant b to the right side of equation (1). We obtain the differential equation

$$\frac{dy}{dt} = ky + b. \tag{2}$$

By letting $a = -\dfrac{b}{k}$ we can write equation (2) as

$$\frac{dy}{dt} = k(y - a). \tag{3}$$

The interpretation of equation (3) is that the rate of increase or decrease of y is proportional to the *difference* between y and the constant a. Before turning to practical examples of this type, we observe that equations (2) and (3) may be solved by the method of *separation of variables,* as described in Section 5.10. The following example illustrates the use of this method.

Example 1 Find a solution of the differential equation

$$\frac{dy}{dt} = 2 - ay, \qquad a \neq 0. \tag{4}$$

Strategy

Separate variables.

Solution

Recall that **separating variables** means isolating y terms on one side of the equation and t terms on the other. To do so we must assume $2 - ay \neq 0$. We then divide both sides by $(2 - ay)$ and multiply both sides by dt to obtain

$$\frac{dy}{2 - ay} = dt.$$

Integrate both sides.

Integrating both sides then gives

$$\int \frac{dy}{2 - ay} = \int dt,$$

so

(Only one constant of integration need be displayed.)

$$-\frac{1}{a}\ln|2 - ay| = t + C \qquad (C \text{ arbitrary})$$

or

$$\ln|2 - ay| = -at - aC.$$

Thus

Solve for y by applying the exponential function to both sides.

$$\begin{aligned} |2 - ay| &= e^{(-at-aC)} \\ &= Ke^{-at}, \qquad K = e^{-aC}. \end{aligned}$$

Since the arbitrary constant K must be positive, we may drop the absolute value signs and solve for y as

$$y = \frac{2}{a} + Ae^{-at}, \qquad A = -\frac{K}{a} \tag{5}$$

A is determined by initial conditions.

where the constant A is to be determined from an initial condition. For example, if $y(0) = 1$ we obtain

$$1 = \frac{2}{a} + Ae^{0}, \qquad \text{so} \qquad A = \left(1 - \frac{2}{a}\right).$$

Verify that assumption on $2 - ay$ holds for solutions obtained.

To check that our assumption $2 - ay \neq 0$ holds, notice that from (5)

$$\begin{aligned} 2 - ay &= 2 - a\left[\frac{2}{a} + Ae^{-at}\right] \\ &= -aAe^{-at}, \end{aligned}$$

which is not zero unless $A = 0$. However, if $A = 0$ equation (5) still gives the valid solution $y = \frac{2}{a}$. Thus, in all cases equation (5) represents a solution of (4). ■

Examples 2 and 3 illustrate how differential equations of the above type arise as models in certain "mixing" problems.

Example 2 A tank in which chocolate milk is being mixed contains 460 liters of milk and 40 liters of chocolate syrup initially. Syrup and milk are then added to the tank at the rates of 2 liters per minute of syrup and 8 liters per minute of milk. Simultaneously, mixture is withdrawn at the rate of 10 liters per minute. Assuming perfect mixing of the milk and syrup, find the function $y(t)$ giving the amount of syrup in the tank at time t.

Strategy
Label variables.

Solution
If y represents the amount of syrup in the tank, dy/dt is the rate at which the amount of syrup is changing. This rate is the rate at which syrup enters the tank (2 liters per minute) minus the rate at which syrup leaves the tank $\left(\frac{y}{500} \times 10\right.$ liters per minute). The function y therefore satisfies the differential equation

Use fact that dy/dt is a *rate* to find a differential equation satisfied by y.

$$\frac{dy}{dt} = 2 - .02y.$$

Find y by solving this differential equation.

This is a particular form of equation (4) treated in Example 1, with $a = .02$. From (5) the solution is

$$y = Ae^{-.02t} + 100.$$

Determine the constant A by applying *initial conditions*.

To determine A we apply the initial condition that $y = 40$ when $t = 0$. We obtain the equation

$$40 = A + 100,$$

so

$$A = -60.$$

State the solution.

The desired solution is therefore

$$y = (100 - 60e^{-.02t}) \text{ liters.} \tag{6}$$

Verify that restrictions on y are met.

Note that the requirement of Example 1 that

$$y \neq \frac{2}{a} = \frac{2}{.02} = 100$$

is met since $60e^{-.02t} > 0$ for all t. The graph of the solution appears in Figure 6.1 and values for $y(t)$ are listed in Table 6.1. ■

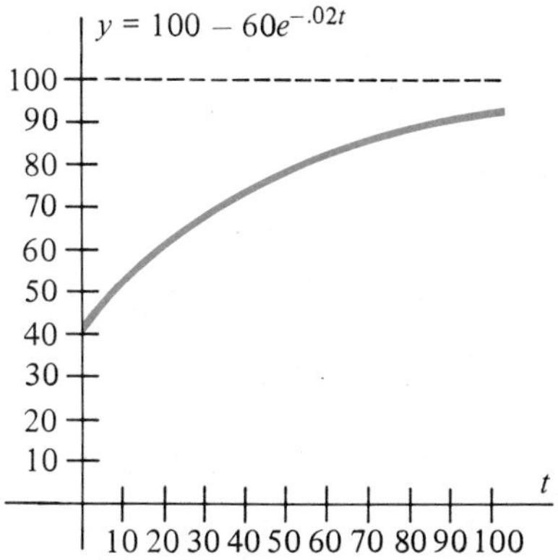

Figure 6.1 Solution of mixing problem in Example 2.

Table 6.1 Calculated values of $y(t)$ in Figure 6.1

t	0	1	2	3	4	5	6	10	15	20
$y = 100 - 60e^{-.02t}$	40	50.9	59.8	67.1	73.1	77.9	81.9	91.9	97.0	98.9

REMARK: Figure 6.1 illustrates an important property of solutions of differential equations of the form of equation (2), namely, that the solution approaches a horizontal asymptote (constant limit) as $t \to \infty$. For the solution y of Example 2 it is easy to see that

$$\lim_{t\to\infty} y = \lim_{t\to\infty} (100 - 60e^{-.02t}) = 100,$$

as Figure 6.1 suggests. This asymptote $y = 100$ is called the **steady state** part of the solution, while the term $60e^{-.02t}$ is referred to as the **transient** part of the solution. The result of Example 2 is precisely what intuition suggests—regardless of the initial concentration of syrup, the steady state concentration will be $100/500 = .2$, since 2 liters of syrup enter for every 8 liters of milk.

Example 3 Newton's law of cooling states that the rate at which an object gains or loses heat is proportional to the difference in temperature between the object and its surroundings. A bottle of soda at temperature 6°C is removed from a refrigerator and placed in a room at temperature 22°C. If, after 10 minutes, the temperature of the soda is 14°C, find its temperature after 20 minutes according to Newton's law.

Strategy

Solution

Label variables.

Let $y(t)$ denote the temperature of the soda t minutes after it is removed from the refrigerator. Newton's law of cooling states that

Interpret Newton's law as a differential equation for y.

$$\frac{dy}{dt} = k \cdot (22 - y).$$

Thus

Separate variables.

$$\frac{dy}{22 - y} = k\,dt$$

so

$$\int \frac{dy}{22 - y} = \int k\,dt.$$

Integrate.

Integrating on both sides we obtain

$$-\ln(22 - y) = kt + C,$$

so

Apply exponential function to both sides.

$$\begin{aligned} 22 - y &= e^{-(kt+C)} \\ &= Ae^{-kt}, \qquad A = e^{-C}. \end{aligned}$$

Thus

Solve for y.

$$y = 22 - Ae^{-kt}. \tag{7}$$

Use initial data to find A.

To find A we apply the initial data* that $y_0 = 6°$ when $t = 0$. Inserting this information in (7) gives

$$6 = 22 - Ae^0,$$

so

$$A = 16. \tag{8}$$

Use additional data to find k.

To find k we use (7), (8), and the data $y(10) = 14$. We obtain the equation

$$14 = 22 - 16e^{-10k}.$$

Thus

$$e^{-10k} = \frac{-8}{-16} = \frac{1}{2},$$

so

$$\ln(e^{-10k}) = -10k = \ln\left(\frac{1}{2}\right) = -\ln 2,$$

so

$$k = \frac{-\ln 2}{-10} = \frac{\ln 2}{10}.$$

$$k = \frac{\ln 2}{10}.$$

The function y is now completely determined as

$$y = 22 - 16e^{-(\ln 2/10)t}. \tag{9}$$

The temperature after 20 minutes is therefore

Find $y(20)$

$$e^{\ln 2^{-2}} = 2^{-2} = \frac{1}{4}.$$

$$\begin{aligned} y(20) &= 22 - 16e^{-\left(\frac{\ln 2}{10}\right)20} \\ &= 22 - 16e^{-2\ln 2} \\ &= 22 - 16e^{\ln 2^{-2}} \end{aligned}$$

*Initial data refer to information about a function at *any* value of the independent variable, not necessarily $t = 0$.

$$= 22 - 16 \cdot \frac{1}{4}$$

$$= 18°.$$

(See Table 6.2 and Figure 6.2.) ■

Table 6.2

t	y in (9)
0	6.00
4	9.87
8	12.81
12	15.04
16	16.72
20	18.00
24	18.97
28	19.70
32	20.26
36	20.68

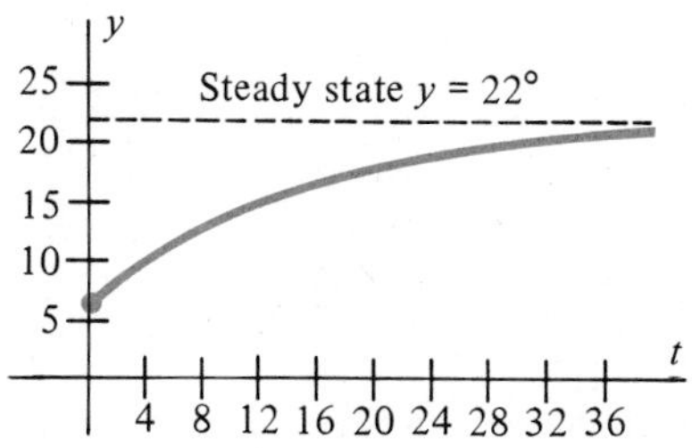

Figure 6.2 Graph of $y = 22 - 16e^{-\left(\frac{\ln 2}{10}\right)t}$.

Integrating Factors

A second generalization of equation (1) is to differential equations of the form

$$\frac{dy}{dt} + p(t)y = g(t) \tag{10}$$

where $p(t)$ and $g(t)$ are continuous functions of t. The differences between equation (10) and equation (1) involve these two functions. In equation (1) we have $p(t) \equiv -k$, and $g(t) \equiv 0$.

We may find solutions to equations of the form (10) if we let $u(t) = \int p(t)\,dt$ and note that *if we multiply both sides of* (10) *by the integrating factor*

$$e^{u(t)} = e^{\int p(t)\,dt} \tag{11}$$

it is easy to find the antiderivative of the left side of equation (10). (In (11) the expression $\int p(t)\,dt$ denotes any *particular* antiderivative for $p(t)$.) To demonstrate our remark we note that

$$\left[\frac{dy}{dt} + p(t)y\right] \cdot e^{u(t)} = \frac{dy}{dt}\,e^{\int p(t)\,dt} + yp(t)e^{\int p(t)\,dt} \tag{12}$$

$$= \frac{d}{dt}\left[ye^{\int p(t)\,dt}\right]$$

$$= \frac{d}{dt}\left[y \cdot e^{u(t)}\right].$$

To find a solution of the differential equation (10), we begin by multiplying both sides by an integrating factor $e^{u(t)}$ in (11) to obtain

$$\left[\frac{dy}{dt} + p(t)y\right]e^{u(t)} = g(t)e^{u(t)}.$$

Using (12) we can write this equation as

$$\frac{d}{dt}[ye^{u(t)}] = g(t)e^{u(t)}$$

so

$$ye^{u(t)} = \int g(t)e^{u(t)}\,dt + C$$

or

$$y = e^{-u(t)}\left\{\int g(t)e^{u(t)}\,dt + C\right\}. \tag{13}$$

The function y in (13) is called the **general solution** of equation (10) because, as is shown in more advanced courses, every solution of (10) has the form (13) for some choice of the constant C.

Rather than remembering formula (13), you should simply remember that a differential equation of the form (10) requires an integrating factor $e^{u(x)} = e^{\int p(t)\,dt}$.

Example 4 Find a solution of the differential equation

$$\frac{dy}{dt} + 2ty = t. \tag{14}$$

Solution: This equation has the form (10) with $p(t) = 2t$, $g(t) = t$. We therefore use the integrating factor

$$e^{u(t)} = e^{\int 2t\,dt} = e^{t^2}.$$

Multiplying both sides of (14) by e^{t^2} gives

$$e^{t^2} \cdot \frac{dy}{dt} + e^{t^2} \cdot 2t \cdot y = te^{t^2}.$$

We next rewrite the left side as a derivative

$$\frac{d}{dt}[e^{t^2} \cdot y] = e^{t^2} \cdot \frac{dy}{dt} + e^{t^2} \cdot 2t \cdot y$$

and find antiderivatives of both sides:

$$\begin{aligned} e^{t^2}y &= \int te^{t^2}\,dt \\ &= \frac{1}{2}e^{t^2} + C. \end{aligned}$$

Finally, we multiply both sides by e^{-t^2} to obtain

$$\begin{aligned} y &= e^{-t^2}\left[\frac{1}{2}e^{t^2} + C\right] \\ &= Ce^{-t^2} + \frac{1}{2}. \end{aligned}$$

■

Example 5 Find a solution of $\frac{dy}{dt} - \frac{2}{t}y = t^3$, $t > 0$ satisfying the initial condition $y(1) = 2$.

Strategy
Find an integrating factor.

Solution
We use the integrating factor

$$e^{u(t)} = e^{\int -2/t\,dt} = e^{-2\ln t} = e^{\ln t^{-2}} = \frac{1}{t^2}.$$

Multiply both sides of the equation by the integrating factor.

We then obtain

$$\frac{1}{t^2}\cdot\frac{dy}{dt} - \frac{2}{t^3}\cdot y = t^3\left(\frac{1}{t^2}\right).$$

We can write this equation as

Write the left side as a derivative.

$$\frac{d}{dt}\left[\frac{1}{t^2}y\right] = t$$

so

Integrate both sides.

$$\frac{1}{t^2}y = \int t\,dt = \frac{1}{2}t^2 + C.$$

Solve for y.

Thus

$$\begin{aligned} y &= t^2\left[\frac{1}{2}t^2 + C\right] \\ &= \frac{1}{2}t^4 + Ct^2. \end{aligned}$$

Apply initial conditions to solve for C.

The initial condition $y(1) = 2$ gives

$$2 = \frac{1}{2} + C$$

so

$$C = 2 - \frac{1}{2} = \frac{3}{2}.$$

The desired solution is

$$y(t) = \frac{1}{2}t^4 + \frac{3}{2}t^2.$$ ■

Second Order Linear Equations

A third generalization of differential equation (1) occurs when a second order derivative is added to the equation. One form for such equations is

$$\frac{d^2y}{dt^2} + a\frac{dy}{dt} + by = 0. \tag{15}$$

Equations of this form cannot be solved by the method of separation of variables because of the presence of the second order derivative. However, based on our experience with the preceding cases we might suspect an exponential solution of the form $y = e^{rt}$. For such functions we have $dy/dt = re^{rt}$ and $d^2y/dt^2 = r^2e^{rt}$. If we insert these expressions in equation (15), we find that $y = e^{rt}$ is a solution of (15) if and only if

$$r^2e^{rt} + are^{rt} + be^{rt} = 0,$$

or

$$(r^2 + ar + b)e^{rt} = 0. \tag{16}$$

Since the exponential function e^{rt} is never zero, the only way in which equation (16) can hold is if the **characteristic polynomial**

$$r^2 + ar + b \tag{17}$$

equals zero. In other words, the function $y = e^{rt}$ is a solution of the differential equation (15) if and only if the number r is a root of the characteristic polynomial (17). Since the polynomial (17) can have two, one, or zero real roots, this method can yield two, one, or no solutions of (15) of the form $y = e^{rt}$.

Example 6 Find all solutions of the differential equation

$$\frac{d^2y}{dt^2} - \frac{dy}{dt} - 2y = 0. \tag{18}$$

Strategy

Insert solution $y = e^{rt}$ in equation (18).

Solution

For $y = e^{rt}$ we obtain from equation (15) that

$$r^2e^{rt} - re^{rt} - 2e^{rt} = 0,$$

or

Factor e^{rt} to find characteristic polynomial.

$$(r^2 - r - 2)e^{rt} = 0.$$

Since the polynomial

Factor polynomial to find roots r_1 and r_2. (Or, solve by quadratic equation.)

$$(r^2 - r - 2) = (r - 2)(r + 1)$$

has roots $r = -1$ and $r = 2$, we obtain the solutions

Solutions are e^{r_1t} and e^{r_2t}.

$$y_1 = e^{-t}$$

and

$$y_2 = e^{2t},$$

which we can verify by substituting into equation (18). ■

In Exercise 29 you are asked to show that if $y_1(t)$ and $y_2(t)$ are solutions of the differential equation (15), then so is any function $y(t)$ of the form

$$y(t) = c_1y_1(t) + c_2y_2(t)$$

where c_1 and c_2 are constants. In courses on differential equations it is actually proved that, if $y_1(t)$ is not a multiple of $y_2(t)$, then any solution of equation (15) must have this form.

As noted above, a differential equation of the form (15) need not have any solutions of the form e^{rt}. For example, the differential equation

$$\frac{d^2y}{dt^2} + y = 0 \tag{19}$$

has characteristic polynomial $r^2 + 1$, which has only the imaginary roots $r = \pm i$, where $i^2 = -1$. The theory by which solutions of (19) are obtained involves extending our notion of exponential functions to include complex numbers as exponents. However, to do this we must first develop the notion of an infinite series representation for the exponential function. Although this task may sound hopelessly complicated, you should be encouraged to know that the result leads us back to two particularly simple solutions of equation (19), namely $y_1 = \sin x$ and $y_2 = \cos x$. We shall leave the topic of differential equations for now, having seen another glimpse of a subject whose elementary aspects will appear again from time to time throughout the remaining chapters of the text.

Exercise Set 9.6

In Exercises 1–4, find a particular solution of the initial value problem. Identify the steady-state and transient parts of the solution.

1. $\frac{dy}{dt} = y + 1 \quad y(0) = 1$

2. $\frac{dy}{dt} = 2y + 2 \quad y(0) = -1$

3. $\frac{dy}{dt} = -y + \pi \quad y(0) = \pi/2$

4. $\frac{dy}{dx} + y = 2 \quad y(0) = 2$

In Exercises 5–10, find the characteristic polynomial for the differential equation and any exponential solutions that may exist.

5. $\frac{d^2y}{dx^2} - y = 0$

6. $\frac{d^2y}{dx^2} + 3\frac{dy}{dx} - 4y = 0$

7. $\frac{d^2y}{dx^2} + 2\frac{dy}{dx} - 3y = 0$

8. $y'' + 4y' + 4y = 0$

9. $y'' + 4y' + y = 0$

10. $y'' + y' + 4y = 0$

11. The differential equation in Exercise 8 has solution $y_1 = e^{-2t}$. Show that $y_2 = te^{-2t}$ is also a solution.

In Exercises 12–21, find the general solution of the differential equation.

12. $y' + 2y = 4$

13. $y' + ay = b$

14. $y' - \frac{2}{t}y = t^4$

15. $y' = \frac{1}{t^2}y$

16. $y' - 2y = e^{2t}\sin t$

17. $y' = 2ty + 3t$

18. $y' + y = e^t$

19. $ty' - 2y = -2t$

20. $ty' + y = t$

21. $y' = y - 2e^{-t}$

22. Use the method of separation of variables to find a solution of the differential equation $dy/dx = axy$ where a is constant and $y > 0$.

23. A curve in the xy-plane has the property that at each point (x, y) the slope of the tangent equals xy, the product of the coordinates. Find an equation for such a curve containing the point $(0, 1)$ and sketch its graph. (*Hint:* Use the result of Exercise 22.)

24. A tank contains 200 liters of brine. The initial concentration of salt is 0.5 kilogram per liter of brine. Fresh water is then added at a rate of 3 liters per minute and the brine is drawn off from the bottom of the tank at the same rate. Assume perfect mixing.

a. Find a differential equation for the number of kilograms of salt, $p(t)$, in the solution at time t.

b. How much salt is present after 20 minutes?

c. What is the steady state concentration of salt?

25. A bottle of water whose temperature is 26°C is placed in a room at temperature 12°C. If the temperature of the water after 30 minutes is 20°C, find its temperature after 2 hours.

26. A person initially places \$100 in a savings account that pays interest at a rate of 10% per year compounded continuously. If the person makes additional deposits of \$100 per year continuously throughout each year, find:

a. a differential equation for $P(t)$, the amount on deposit after t years, and

b. the amount on deposit after 7 years.

27. An electrical circuit contains a battery supplying E volts, a resistance R, and an inductance L. When the battery is connected, the resulting current I satisfies the differential equation

$$L\frac{dI}{dt} + RI = E, \qquad I(0) = 0.$$

a. Find the solution I as a function of t.
b. Obtain Ohm's Law, $I = E/R$, as the steady state part of the solution.

28. If, in the circuit described in Exercise 27, the inductance is replaced by a capacitance C the equation becomes

$$R\frac{dI}{dt} + \frac{I}{C} = \frac{dE}{dt}.$$

Find the solution $I(t)$, if the battery is switched off (i.e., $dE/dt = 0$ at the time $t = 0$.)

29. Show that if $y_1(t)$ and $y_2(t)$ are solutions of the differential equation

$$\frac{d^2y}{dt^2} + a\frac{dy}{dt} + by = 0,$$

then so is any function of the form

$$f(t) = c_1y_1(t) + c_2y_2(t),$$

where c_1 and c_2 are constants.

SUMMARY OUTLINE OF CHAPTER 9

- ***Definition:*** $\log_b x = y$ if and only if $b^y = x$.
- ***Properties of logarithms:***

(L1) $\log_b(mn) = \log_b m + \log_b n.$
(L2) $\log_b(m/n) = \log_b m - \log_b n.$
(L3) $\log_b(n^r) = r \log_b n.$

- ***Definition:*** $\ln x = \int_1^x \frac{1}{t}\,dt, \quad x > 0.$
- ***Theorem:*** $\frac{d}{dx}\ln x = \frac{1}{x}; \quad \frac{d}{dx}\ln u = \frac{1}{u}\cdot\frac{du}{dx}.$
- ***Theorem:***

$$\int \frac{1}{x}\,dx = \ln|x| + C; \qquad \int \frac{u'(x)}{u(x)}\,dx = \ln|u| + C.$$

- ***Definition:*** e is defined by the equation $\ln e = 1$. $e \approx 2.718281828459$.
- ***Definition:*** $y = e^x$ if and only if $x = \ln y$.
- ***Theorem:*** $\frac{d}{dx}e^x = e^x; \quad \frac{d}{dx}e^u = e^u\cdot\frac{du}{dx}.$
- ***Theorem:***

$$\int e^x\,dx = e^x + C; \qquad \int e^{u(x)}\cdot u'(x)\,dx = e^u + C.$$

- ***Definition:*** $a^x = e^{x \ln a}, \quad a > 0.$
- ***Properties of a^x:***

(E1) $a^x \cdot a^y = a^{x+y}.$
(E2) $a^x/a^y = a^{x-y}.$
(E3) $(a^x)^y = a^{xy}.$

- ***Theorem:*** $\frac{d}{dx}a^x = a^x \ln a; \quad \frac{d}{dx}a^u = a^u\cdot\frac{du}{dx}\cdot\ln a.$
- ***Theorem:***

$$\int a^x\,dx = \frac{a^x}{\ln a} + C; \qquad \int a^{u(x)}u'(x)\,dx = \frac{a^u}{\ln a} + C.$$

- ***Definition:*** $y = \log_a x$ if and only if $x = a^y = e^{y \ln a}$.
- ***Conversion Formula:*** $\log_a x = \frac{\ln x}{\ln a}, \quad a \neq 1.$
- ***Theorem:***

$$\frac{d}{dx}\log_a x = \frac{1}{x \ln a}; \qquad \frac{d}{dx}\log_a u = \frac{1}{u \ln a}\cdot\frac{du}{dx}.$$

- ***Law of Natural Growth:*** $\frac{dN}{dt} \propto N.$
- ***Model for Natural Growth:*** $\frac{dy}{dt} = ky.$
- ***Solution of Model:*** $y = Ce^{kt}, \quad C = y(0).$
- ***Formula for continuous compounding of interest:*** $P(t) = e^{rt}P_0.$
- ***Present value of P_t dollars in t years:*** $P(0) = e^{-rt}P_t.$
- ***Alternate definition of e^r:***

$$e^r = \lim_{k\to\infty}\left(1 + \frac{r}{k}\right)^k,$$

$$e = \lim_{k\to\infty}\left(1 + \frac{1}{k}\right)^k.$$

- ***Differential equations*** of the form $\frac{dy}{dt} = ky + b$ are solved by the method of separation of variables.
- ***The differential equation*** $\frac{d^2y}{dt^2} + a\frac{dy}{dt} + by = 0$ has solution $y = e^{rt}$ if r is a root of the characteristic polynomial $r^2 + ar + b$ (0, 1, or 2 such solutions will exist).
- Differential Equations of the form $\frac{dy}{dt} + p(t)y = g(t)$ require an **integrating factor** $e^{u(t)} = e^{\int p(t)\,dt}$.

REVIEW EXERCISES—CHAPTER 9

1. Simplify the following.
a. $8^{2/3}$ **b.** $36^{-5/2}$
c. $16^{5/4}$ **d.** $4^{-3/4}$

2. Find the following logarithms:
a. $\log_2 16$ **b.** $\log_8 16$
c. $\ln e^2$ **d.** $\log_2 3\sqrt{2}$

3. Find the inverses of the following functions:
a. $f(x) = x + 2$ **b.** $g(x) = x^2, \quad x \geq 0$
c. $y = 2x + 1$ **d.** $y = \sqrt{x}$

4. Solve for x:
a. $3\ln x - \ln 3x = 0$ **b.** $\ln x^3 = 3$
c. $\int_1^x \frac{2}{t}\,dt = 4$ **d.** $\int_2^x \frac{1}{t}\,dt = 3 + \ln 2$

Find the derivative of each function with respect to its independent variable.

5. $y = x^2 \ln(x - a)$

6. $f(x) = xe^{1-x}$

7. $f(x) = \ln(\ln^2 x)$

8. $y = (2\ln\sqrt{x})^3$

9. $y = e^{x^2}\tan 2x$

10. $f(t) = \sin^2 t\cos(\ln t)$

11. $f(t) = \frac{te^t}{1 + e^t}$

12. $y = x\ln(\sqrt{x} - e^{-x})$

13. $y = \frac{\ln(t - a)}{\ln(t^2 + b)}$

14. $f(t) = \ln\cos t$

15. $f(t) = e^{\sqrt{t} - \ln t}$

16. $y = \frac{1}{a}\ln\left(\frac{a + bx}{x}\right)$

17. $y = \frac{1}{3x} + \frac{1}{4}\ln\left(\frac{1 - 2x}{\sqrt{x}}\right)$

18. $f(t) = \sin(\ln t + te^{\sqrt{t}})$

19. $x\ln y^2 + y\ln x = 1 \quad \left(\text{Find } \frac{dy}{dx}\right)$

20. $y = e^{4x}(e^{-x} + \ln x)^2$

21. $x^2 + e^{xy} - y^2 = 2 \quad \left(\text{Find } \frac{dy}{dx}\right)$

22. $y = xe^{1-\sqrt{x}\ln x}$

23. $e^{xy} = \sqrt{xy} \quad \left(\text{Find } \frac{dy}{dx}\right)$

24. $y = \sqrt{e^{-x} + e^x}$

25. $f(x) = \ln\sqrt{e^{2x} + \sin\sqrt{x}}$

26. $y = \int_1^{\sin x} \ln(t + 1)\,dt$

Find the integral.

27. $\int \frac{x\,dx}{1 - x^2}$

28. $\int xe^{1-x^2}\,dx$

29. $\int \frac{x + 1}{4x + 2x^2}\,dx$

30. $\int \frac{dx}{(x + 1)\ln^3(x + 1)}$

31. $\int \sqrt{e^x}\,dx$

32. $\int (e^x + 1)^2\,dx$

33. $\int_e^{e^2} \frac{1}{x\sqrt{\ln x}}\,dx$

34. $\int \frac{e^x - e^{-x}}{e^x + e^{-x}}\,dx$

35. $\int_0^1 \frac{x^3 - 1}{x + 1}\,dx$

36. $\int_1^4 \frac{e^{\sqrt{x}}}{\sqrt{x}}\,dx$

37. $\int_0^2 x(e^{x^2} + 1)\,dx$

38. $\int_4^9 \frac{1}{\sqrt{x}(1 + \sqrt{x})}\,dx$

39. $\int e^x(1 - e^{2x})^3\,dx$

40. $\int \frac{\cot\sqrt{x}}{\sqrt{x}}\,dx$

41. $\int \frac{2x + 3x^2}{x^3 + x^2 - 7}\,dx$

42. $\int_1^{\ln 2} (e^x + 1)(e^x - 1)\,dx$

43. $\int_{-3}^{-2} \frac{x}{x^2 + 1}\,dx$

44. $\int_0^1 \left(\frac{1}{x + 1} - \frac{1}{x + 2}\right)dx$

Use logarithmic differentiation to find $\frac{dy}{dx}$ for

45. $y = x^{\sqrt{x}}$

46. $y = \sqrt{\frac{(x - 1)^2(x + 1)}{x(x + 3)^3}}$

47. Find all relative extrema for the function $y = x^2\ln x$.

48. Find the maximum and minimum values of the function $f(x) = e^{x^2/3}$ for x in the interval $[-1, \sqrt{8}]$.

49. On what intervals is the graph of the function $y = \ln(1 + x^2)$ concave up?

50. Find the area of the region bounded by the graph of $y = \dfrac{x}{x^2 + 1}$ and the x-axis for $-2 \le x \le 2$.

51. Find the area of the region bounded by the graphs of $y = \dfrac{1}{x}$ and $y = \dfrac{10x - 21}{4x - 10}$. (*Hint:* Use long division.)

52. Find all relative extrema for the function $y = x^2 - \ln x^2$ and sketch the graph.

53. Find the volume of the solid generated by rotating the portion of the graph of $y = \sqrt{\dfrac{2}{x - 1}}$ from $x = 3$ to $x = 5$ about the x-axis.

54. The function

$$f(x) = \begin{cases} \lambda e^{-\lambda x}, & x > 0 \\ 0, & x \le 0 \end{cases}$$

is called the **exponential probability density function** with parameter λ. For any density function the probability that x will fall in the interval $[a, b]$ is

$$P\{a \le x \le b\} = \int_a^b f(x)\, dx.$$

For $\lambda = 1$, find the following probabilities associated with the exponential distribution:
a. $P\{0 \le x \le 1\}$,
b. $P\{0 \le x \le \ln 2\}$,
c. $P\{0 \le x \le 4\}$.

55. Graph the exponential probability density function for $\lambda = 1$ (Exercise 54).

56. *(Calculator/Computer)* The *standard* normal probability density function is $f(x) = \dfrac{1}{\sqrt{2\pi}} e^{-x^2/2}$. Use a program for numerical integration to approximate

a. $P\{-1 \le x \le 1\} = \displaystyle\int_{-1}^{1} \frac{1}{\sqrt{2\pi}} e^{-x^2/2}\, dx$

b. $P\{0 \le x \le 2\}$

57. Sketch the graph of the standard normal probability density function in Exercise 56.

58. Find a formula for approximating the function $y = x \ln x^2$ near $x = e$. Use this formula to approximate $y(2.70) = (2.70) \ln(2.70)^2$.

59. Approximate $e^{\sqrt{4.1}}$ by differentials.

60. Let $f(x) = \dfrac{e^x + e^{-x}}{2}$, $\quad g(x) = \dfrac{e^x - e^{-x}}{2}$.
a. Show that $f'(x) = g(x)$ and that $g'(x) = f(x)$.
b. Show that both functions satisfy the differential equation $y'' - y = 0$.
c. Find two solutions of the differential equation $y'' - k^2 y = 0$.

61. A particle, starting from rest, moves along a line with acceleration $a(t) = \dfrac{1}{(t + 2)^2}$ m/sec^2. Find
a. the velocity function $v(t)$,
b. the distance travelled after 10 seconds.

62. *(Calculator)* Approximate $\displaystyle\int_0^5 e^{x^2}\, dx$ using a rectangular rule with $n = 10$.

63. Find r if $2^x = e^{rx}$ for all x.

64. Find the equation of the line tangent to the graph of $y = xe^{2x}$ at the point $(\ln 2, 4 \cdot \ln 2)$.

65. A cylindrical jar 12 centimeters in diameter is leaking water at a rate of $\ln t^2$ cubic centimeters per minute. How fast is the water level in the jar falling?

66. *(Calculator)* Use Simpson's rule with $n = 10$ to approximate
a. $\ln 4$ **b.** $\ln 5$ **c.** $\ln(1/2)$

67. *(Calculator/Computer)* Use Newton's Method to approximate the solution of the equation $2 + \ln x = x$.

Find the solution of the initial value problem:

68. $\dfrac{dy}{dx} = 2y$, $\quad y(0) = 1$

69. $\dfrac{dy}{dx} + y = 0$, $\quad y(\ln 2) = 2$

70. $2y' + 4y = 0$, $\quad y(0) = \pi$

71. $y' + y = 2$, $\quad y(0) = 1$

72. $4y' - 4y + 4 = 0$, $\quad y(0) = 1$

73. $(y' + y)^2 = 0$, $\quad y(0) = 1$

74. Find the average value of the function $y = xe^{x^2+1}$ on the interval $[0, \sqrt{\ln 2}]$.

75. *(Calculator/Computer)* Approximate the solution of the equation $e^x + 2x = 0$.

76. The population of the United States in 1970 was 203 million. By 1980 the population had grown to 227 million. Assuming exponential growth, what will the population be in 1990? 2000?

77. In the chemical reaction called the inversion of raw sugar, the inversion rate is proportional to the amount of raw sugar remaining. If 100 kilograms of raw sugar is reduced to 75 kilograms in 6 hours, how long will it be until

a. half the raw sugar has been inverted?
b. 90% of the raw sugar has been inverted?

78. A population of fruit flies grows exponentially. If initially there were 100 flies and if after 10 days there were 500 flies, how many flies were present after 4 days?

Find the following limits:

79. $\lim_{t\to\infty}\left(1+\frac{3}{t}\right)^t$

80. $\lim_{k\to\infty}\left\{\left(1+\frac{1}{k}\right)\left(1-\frac{1}{k}\right)\right\}^k$

81. $\lim_{x\to 0^+}(1-3x)^{1/x}$ **82.** $\lim_{t\to 0^+}\sqrt[t]{\frac{(1-t)}{1+t}}$

83. A snowball melts at a rate proportional to its surface area. If the snowball originally had a radius of 10 cm and, after 20 minutes, its radius is 8 cm, find

a. a differential equation for the radius r,
b. a solution of this differential equation involving two constants,
c. an explicit solution of this equation, and
d. the radius of the snowball after one hour.

(*Hint:* In part (a) show that $\frac{dv}{dt} = s\frac{dr}{dt}$ where v = volume and s = surface area.)

84. 100 grams of a radioactive substance is reduced to 40 grams in 6 hours. If the decay is exponential, what is the half-life?

85. A pan of boiling water is removed from a burner and cools to 80°C in three minutes. Find its temperature after 10 minutes, according to Newton's law of cooling, if the surrounding temperature is 30°C.

86. Show that the exponential function $f(x) = Ce^{kx}$ satisfies the differential equation $f'(x+a) = kf(x+a)$ for every real number a.

CHAPTER 10

TRIGONOMETRIC AND INVERSE TRIGONOMETRIC FUNCTIONS

10.1 INTRODUCTION

We have already encountered the six trigonometric functions and their derivatives. The objectives of this chapter are to

(i) obtain antiderivatives for all six trigonometric functions,
(ii) develop techniques for integrating more complicated expressions involving trigonometric functions,
(iii) define the inverses of the trigonometric functions and find their derivatives and antiderivatives,
(iv) define the hyperbolic trigonometric functions, and
(v) study some models of physical phenomena involving these functions.

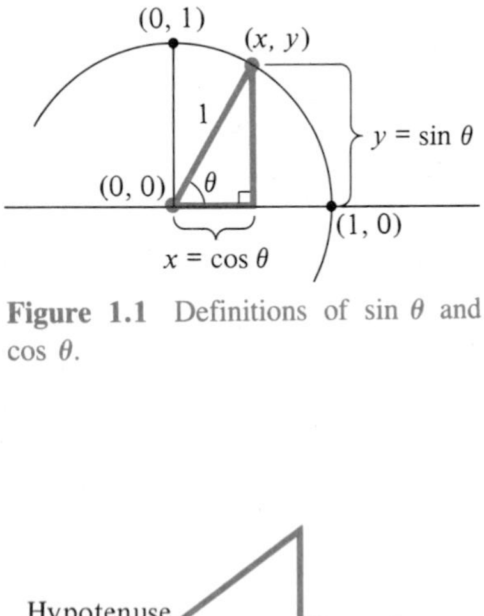

Figure 1.1 Definitions of $\sin\theta$ and $\cos\theta$.

Before proceeding, we should have in mind a clear summary of the basic information about trigonometric functions. Recall that the sine and cosine functions are defined by use of the unit circle (Figure 1.1). If θ denotes the angle formed between the radius from (0, 0) to (x, y) and the positive x-axis, measured counterclockwise, we define

$$\sin\theta = y\text{-coordinate of } (x, y) = y \tag{1}$$

and

$$\cos\theta = x\text{-coordinate of } (x, y) = x. \tag{2}$$

If θ is in the interval $0 < \theta < \pi/2$, an equivalent definition can be stated in terms of a right triangle of any size, such as the one in Figure 1.2. Because they both contain a right angle and the angle θ, the triangles in Figures 1.1 and 1.2 are similar; this requires that

$$\sin\theta = \frac{y}{1} = \frac{\text{opposite}}{\text{hypotenuse}} \quad \text{and} \quad \cos\theta = \frac{x}{1} = \frac{\text{adjacent}}{\text{hypotenuse}}. \tag{3}$$

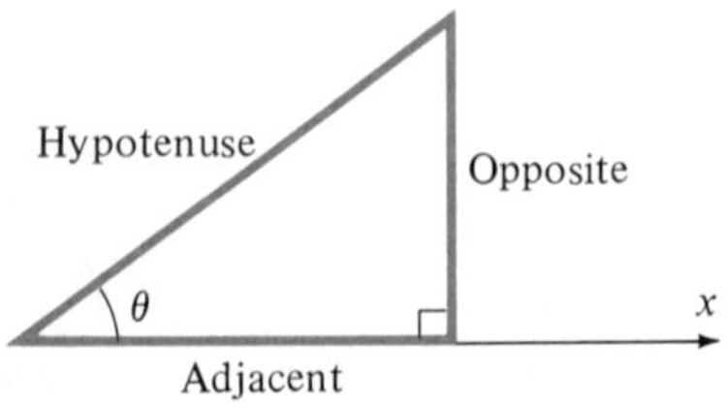

Figure 1.2 Labeling of sides for defining trig functions by ratios, $0 \le \theta \le \pi/2$.

For $\theta = 0$ and $\theta = \pi/2$, the triangle "degenerates" into a horizontal or vertical line segment, respectively. When $\theta = 0$, we have $x = 1$ and $y = 0$, so $\cos 0 = 1$ and $\sin 0 = 0$. When $\theta = \pi/2$, we have $x = 0$ and $y = 1$, so $\cos\pi/2 = 0$ and $\sin\pi/2 = 1$.

However, in extending this notion to angles outside the interval $[0, \pi/2]$, we must remember that the hypotenuse must always be viewed as a radius extending from the origin, and that the angle θ is formed between this radius and the positive x-axis. Thus, in these cases θ will not be an *interior* angle of the triangle, and the sign associated with the lengths of the adjacent and opposite sides will depend upon whether they extend in the positive or negative x or y directions. Figure 1.3 illustrates the sign conventions in the four possible cases. (Note that the sign of the hypotenuse is always positive.)

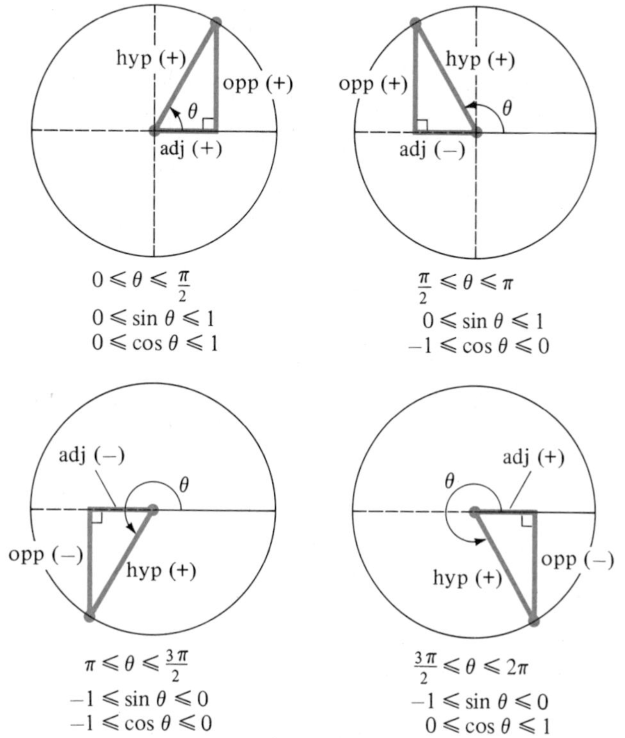

Figure 1.3 Sign conventions for opposite-adjacent-hypotenuse formulation of trigonometric functions.

Four additional trigonometric functions are defined in terms of the sine and cosine function as follows:

Function	Notation		Formal definition		Unit circle interpretation		Triangle interpretation
tangent	$\tan\theta$	=	$\frac{\sin\theta}{\cos\theta}$	=	$\frac{y}{x}$	=	$\frac{\text{opposite}}{\text{adjacent}}$
cotangent	$\cot\theta$	=	$\frac{\cos\theta}{\sin\theta}$	=	$\frac{x}{y}$	=	$\frac{\text{adjacent}}{\text{opposite}}$
secant	$\sec\theta$	=	$\frac{1}{\cos\theta}$	=	$\frac{1}{x}$	=	$\frac{\text{hypotenuse}}{\text{adjacent}}$
cosecant	$\csc\theta$	=	$\frac{1}{\sin\theta}$	=	$\frac{1}{y}$	=	$\frac{\text{hypotenuse}}{\text{opposite}}$

We have already determined the following differentiation and integration formulas:

1. $\dfrac{d}{dx}\sin x = \cos x$ 1′. $\int \cos x\, dx = \sin x + C$

2. $\dfrac{d}{dx}\cos x = -\sin x$ 2′. $\int \sin x\, dx = -\cos x + C$

3. $\dfrac{d}{dx}\tan x = \sec^2 x$ 3′. $\int \sec^2 x\, dx = \tan x + C$

4. $\dfrac{d}{dx}\cot x = -\csc^2 x$ 4′. $\int \csc^2 x\, dx = -\cot x + C$

5. $\dfrac{d}{dx}\sec x = \sec x \tan x$ 5′. $\int \sec x \cdot \tan x\, dx = \sec x + C$

6. $\dfrac{d}{dx}\csc x = -\csc x \cot x$ 6′. $\int \csc x \cdot \cot x\, dx = -\csc x + C$

(See Sections 3.5 and 5.9.)

Identities involving the six trigonometric functions, together with graphs of these six functions, appear in Chapter 1. These include:

$$\sin^2\theta + \cos^2\theta = 1$$
$$\tan^2\theta + 1 = \sec^2\theta$$
$$1 + \cot^2\theta = \csc^2\theta$$

$$\sin(\theta \pm \phi) = \sin\theta\cos\phi \pm \cos\theta\sin\phi$$
$$\cos(\theta \pm \phi) = \cos\theta\cos\phi \mp \sin\theta\sin\phi$$
$$\tan(\theta \pm \phi) = \frac{\tan\theta \pm \tan\phi}{1 \mp \tan\theta\tan\phi}, \qquad 1 \mp \tan\theta\tan\phi \neq 0$$

$$\sin^2\theta = \frac{1}{2}(1 - \cos 2\theta)$$
$$\cos^2\theta = \frac{1}{2}(1 + \cos 2\theta)$$

$$\sin 2\theta = 2\sin\theta\cos\theta$$
$$\cos 2\theta = \cos^2\theta - \sin^2\theta$$

$$\sin(-\theta) = -\sin\theta$$
$$\cos(-\theta) = \cos\theta$$

Remember, all angles are measured in *radians* when working with trigonometric functions in calculus.

10.2 INTEGRALS OF TAN x AND SEC x

The problem of finding antiderivatives for $\sin x$ and $\cos x$ was simple to solve, since these functions occur as derivatives of $-\cos x$ and $\sin x$, respectively. However, this is not the case for the remaining four trigonometric functions.

To find the antiderivative for $y = \tan x$ we make use of the formula

$$\int \frac{v'(x)}{v(x)}\,dx = \ln|v(x)| + C \tag{1}$$

from Chapter 9. Since $\tan x = \dfrac{\sin x}{\cos x}$, we let $v(x) = \cos x$. Then $v'(x)\,dx = -\sin x\,dx$. Using equation (1) we obtain

$$\begin{aligned}\int \tan x\,dx &= \int \frac{\sin x}{\cos x}\,dx \\ &= -\int \frac{v'(x)}{v(x)}\,dx \\ &= -\ln|v(x)| + C \\ &= -\ln|\cos x| + C.\end{aligned}$$

Since $-\ln|\cos x| = \ln|\cos x|^{-1} = \ln|\sec x|$, we have the formula

$$\boxed{\int \tan x\,dx = \ln|\sec x| + C, \qquad \cos x \neq 0.} \tag{2}$$

More generally, if u is a differentiable function of x we have

$$\boxed{\int \tan u\,du = \ln|\sec u| + C, \qquad \cos u \neq 0.} \tag{3}$$

To find the antiderivative for $y = \cot u$, we write $\cot u = \dfrac{\cos u}{\sin u}$ and apply equation (1) as above. We obtain

$$\boxed{\int \cot u\,du = \ln|\sin u| + C, \qquad \sin u \neq 0.} \tag{4}$$

Example 1 Find $\int x \cot x^2\,dx$.

Strategy

Find a u-substitution so that x is a factor of du.

Solution

We make the u-substitution

$$u = x^2; \qquad du = 2x\,dx.$$

Then $x\,dx = \frac{1}{2}\,du$, and we obtain

Make u-substitution.

$$\int x \cot x^2\,dx = \int \cot u \cdot \frac{1}{2}\,du$$

$$= \frac{1}{2} \int \cot u \, du$$

Apply equation (4).

$$= \frac{1}{2} \ln |\sin u| + C$$

Substitute back.

$$= \frac{1}{2} \ln |\sin x^2| + C.$$

■

Example 2 Find $\int \tan^3 x \, dx$.

Solution: We use the identity $\tan^2 x = \sec^2 x - 1$ to write

$$\begin{aligned}\int \tan^3 x \, dx &= \int \tan x[\sec^2 x - 1] \, dx \\ &= \int \tan x \sec^2 x \, dx - \int \tan x \, dx\end{aligned}$$

In the second step, we have distributed $\tan x$ and used the additive property of integrals. In the first integral we use the substitution

$$u = \tan x; \qquad du = \sec^2 x \, dx$$

to obtain

$$\begin{aligned}\int \tan x \sec^2 x \, dx &= \int u \, du \\ &= \frac{1}{2} u^2 + C \\ &= \frac{1}{2} \tan^2 x + C.\end{aligned}$$

The second integral is handled by equation (3). Thus,

$$\int \tan^3 x \, dx = \frac{1}{2} \tan^2 x - \ln |\sec x| + C.$$

■

The antiderivatives for $\sec x$ and $\csc x$ also involve the natural logarithm function, although the derivations are not so straightforward. To handle $\sec x$ we multiply by $\dfrac{\sec x + \tan x}{\sec x + \tan x}$ to obtain

$$\begin{aligned}\int \sec x \, dx &= \int \sec x \cdot \left(\frac{\sec x + \tan x}{\sec x + \tan x}\right) dx \\ &= \int \frac{\sec x \tan x + \sec^2 x}{\sec x + \tan x} \, dx \\ &= \int \frac{\frac{d}{dx}(\sec x + \tan x)}{\sec x + \tan x} \, dx \\ &= \ln |\sec x + \tan x| + C \qquad \text{(by equation (1)).}\end{aligned}$$

Thus, if u is a differentiable function of x,

$$\int \sec u \, du = \ln|\sec u + \tan u| + C. \tag{5}$$

Similarly, we can show that

$$\int \csc u \, du = \ln|\csc u - \cot u| + C. \tag{6}$$

(See Exercise 36.)

Example 3 Find the length of the graph of the function

$y = \ln \sec x$ from $(0, 1)$ to $(\pi/3, \ln 2)$.

Strategy

Find dy/dx.

Use formula for arc length

$$L = \int_a^b \sqrt{1 + \left[\frac{dy}{dx}\right]^2}\, dx.$$

Use identity

$1 + \tan^2 x = \sec^2 x$.

Use formula (5).

Solution

Since $dy/dx = \tan x$ we have

$$\begin{aligned} L &= \int_0^{\pi/3} \sqrt{1 + \tan^2 x}\, dx \\ &= \int_0^{\pi/3} \sqrt{\sec^2 x}\, dx \\ &= \int_0^{\pi/3} \sec x \, dx \\ &= \ln|\sec x + \tan x|\Big|_0^{\pi/3} \\ &= \ln|2 + \sqrt{3}| - \ln|1 + 0| = \ln(2 + \sqrt{3}). \\ &\approx 1.317. \end{aligned}$$

■

Example 4 Find $\displaystyle\int \frac{\csc\sqrt{x}}{\sqrt{x}}\, dx$.

Strategy

Find a u-substitution involving $x^{-1/2}$ as factor of du.

Make u-substitution.

Factor constant.

Apply formula (6).

Substitute back.

Solution

We make the u-substitution

$$u = \sqrt{x} = x^{1/2}; \qquad du = \frac{dx}{2\sqrt{x}} = \frac{1}{2}x^{-1/2}\, dx.$$

Then $x^{-1/2}\, dx = 2\, du$, and

$$\begin{aligned} \int \frac{\csc\sqrt{x}}{\sqrt{x}}\, dx &= \int \csc u \cdot 2\, du \\ &= 2\int \csc u \cdot du \\ &= 2\ln|\csc u - \cot u| + C \\ &= 2\ln|\csc\sqrt{x} - \cot\sqrt{x}| + C. \end{aligned}$$

■

The exercise set below provides a general review of integration of trigonometric functions as well as exercises concerning equations (3)–(6).

Exercise Set 10.2

In Exercises 1–30, find the integral.

1. $\int \cos 3x\, dx$

2. $\int \sec^2 2x\, dx$

3. $\int \sec 2x \tan 2x\, dx$

4. $\int x \csc x^2\, dx$

5. $\int (\tan^2 x + 1)\, dx$

6. $\int \frac{\tan \sqrt{x}}{\sqrt{x}}\, dx$

7. $\int \frac{x}{\cos x^2}\, dx$

8. $\int \frac{1}{\sqrt{x} \sec \sqrt{x}}\, dx$

9. $\int x \tan(3x^2 - 1)\, dx$

10. $\int \sin x \sqrt{1 + \cos x}\, dx$

11. $\int \tan^3 x \sec^2 x\, dx$

12. $\int \cot x \cdot \ln(\sin x)\, dx, \quad \sin x > 0$

13. $\int \sec^2 x e^{\tan x}\, dx$

14. $\int \frac{\cos^2(2x - 1)}{\sin(2x - 1)}\, dx$

15. $\int \frac{\sec^2 x + 1}{x + \tan x}\, dx$

16. $\int \frac{\csc^2(x + \pi/4)}{\cot^3(x + \pi/4)}\, dx$

17. $\int x \cot^2(1 - x^2)\, dx$

18. $\int \frac{\sec^2 x}{\sqrt{\tan x}}\, dx$

19. $\int (\sec x + \tan x)\, dx$

20. $\int \frac{\sin x + \cos x}{\sin x - \cos x}\, dx$

21. $\int (1 - x)\sec(x^2 - 2x)\, dx$

22. $\int \frac{\sec^5 \sqrt{x} \tan \sqrt{x}}{\sqrt{x}}\, dx$

23. $\int \sqrt{1 - \sin^2 x}\, e^{\sin x}\, dx$

24. $\int \frac{\sec \ln x}{x}\, dx$

25. $\int \frac{dx}{\sqrt{x}(1 + \cos \sqrt{x})}$

26. $\int \csc^4(2x)\, dx$

27. $\int \frac{\cos 4x}{\sqrt{1 + \sin 4x}}\, dx$

28. $\int \sin^3 2\theta\, d\theta$

29. $\int \frac{\sec^2 \theta\, d\theta}{\sqrt{\tan \theta + 1}}$

30. $\int \frac{\tan x}{1 + \tan^2 x}\, dx$

31. Find the area bounded by the graph of $y = x \sec x^2$ and the x-axis for $0 \le x \le \sqrt{\pi/3}$.

32. Find the area bounded by the graph of $y = \sec x \cdot \tan x$ and the x-axis for $-\pi/3 \le x \le \pi/3$.

33. Find the average value of the function $f(x) = \csc x$ for $\pi/6 \le x \le \pi/3$.

34. Find $\int \sin x \cdot \cos x\, dx$ in two ways. (*Hint:* $\sin 2x = 2 \sin x \cos x$.)

35. Find $\int \tan x \sec^2 x\, dx$ in two ways.

36. Derive formula (6).

37. Find the length of the curve $y = \ln \sin x$ from $(\pi/6, -\ln 2)$ to $(\pi/3, \ln(\sqrt{3}/2))$.

38. Find the volume of the solid obtained by revolving the graph of $y = \sec x \tan x$, $-\pi/4 \le x \le \pi/4$, about the x-axis.

39. Find the volume of the solid obtained by revolving about the x-axis the region bounded by the graphs of $y = \sec x \tan x$ and $y = \tan x$ for $-\pi/4 \le x \le \pi/4$.

40. Find the area of one of the regions bounded by the graph of $y = \sec x$ and the line $y = -2$.

41. Use Simpson's Rule with $n = 6$ to approximate the length of the graph of $y = \cos x$ for $0 \le x \le 2\pi$.

10.3 INTEGRALS INVOLVING PRODUCTS OF TRIGONOMETRIC FUNCTIONS

Integrals that involve products of the trigonometric functions can often be handled by use of identities and appropriate u-substitutions. For each of the various types of integrals discussed in this section we will present a typical example, followed by a general statement of strategy, followed by more challenging examples. You should be cautioned that integrating expressions of this kind is rarely straightforward. Indeed, the best recipe for success at this business is knowledge of general strategies coupled with the experience of working many such problems.

Integrals Involving Odd Powers of Sine and Cosine

Example 1 Find $\int \sin^3 x \cos x \, dx$.

Solution: We make the u-substitution

$$u = \sin x; \qquad du = \cos x \, dx$$

so that $\cos x$ is a factor of du. Then

$$\begin{aligned}\int \sin^3 x \cos x \, dx &= \int u^3 \, du \\ &= \frac{1}{4} u^4 + C \\ &= \frac{1}{4} \sin^4 x + C.\end{aligned}$$

■

> **General Strategy:** In an integral of the form
>
> $$\int \sin^n x \cdot \cos^m x \, dx, \qquad \text{with } n \text{ or } m \text{ odd:}$$
>
> (i) *If m is odd,* make the substitution $u = \sin x$. Then $du = \cos x \, dx$ involves one factor of $\cos x$. The remaining even factors of $\cos x$ may be converted to a function of $\sin x$ by the identity $\cos^2 x = 1 - \sin^2 x$. The integral then has the form
>
> $$\int \sin^n x[1 - \sin^2 x]^k \cdot \cos x \, dx = \int u^n(1 - u^2)^k \, du$$
>
> with $k = \dfrac{m - 1}{2}$.
>
> (ii) *If n is odd,* make the substitution $u = \cos x$. Then $du = -\sin x \, dx$ and the remaining even factors of $\sin x$ may be converted to a function of $\cos x$. Proceed as above.

Example 2 Find $\int \sin^2 x \cos^3 x \, dx$.

Strategy

Use one factor of $\cos x$ for du.

Convert remaining factors of $\cos x$ to functions of $\sin x$.

Solution

The exponent on $\cos x$ is odd, so we make the u-substitution

$$u = \sin x; \qquad du = \cos x \, dx.$$

Then, using the identity $\cos^2 x = 1 - \sin^2 x$, we obtain

$$\begin{aligned}\int \sin^2 x \cos^3 x \, dx &= \int \sin^2 x[1 - \sin^2 x] \cos x \, dx \\ &= \int u^2(1 - u^2) \, du \\ &= \frac{1}{3} u^3 - \frac{1}{5} u^5 + C \\ &= \frac{1}{3} \sin^3 x - \frac{1}{5} \sin^5 x + C.\end{aligned}$$

■

Example 3 Find $\int \sin^5 2x\, dx$.

Strategy
Use one factor of $\sin 2x$ for du.

Solution
Even though no factor of $\cos x$ is present, we may make the u-substitution

$$u = \cos 2x; \qquad du = -2 \sin 2x\, dx.$$

We obtain

Convert remaining $\sin^4 2x$ to $[1 - \cos^2 2x]^2$.

Make u-substitution.

$$\begin{aligned}\int \sin^5 2x\, dx &= \int \sin^4 2x \cdot \sin 2x\, dx \\ &= \int [1 - \cos^2 2x]^2 \cdot \sin 2x\, dx \\ &= \int [1 - u^2]^2 \left(-\frac{1}{2}\right) du \\ &= -\frac{1}{2} \int (1 - 2u^2 + u^4)\, du \\ &= -\frac{1}{2} u + \frac{1}{3} u^3 - \frac{1}{10} u^5 + C \\ &= -\frac{1}{2} \cos 2x + \frac{1}{3} \cos^3 2x - \frac{1}{10} \cos^5 2x + C.\end{aligned}$$

■

Integrals Involving Even Powers Of Sine and Cosine

Example 4 Find $\int \sin^2 x \cos^2 x\, dx$.

Strategy
Use half angle formulas to reduce exponents by half.

Solution
We use the identities

$$\sin^2 x = \frac{1}{2} - \frac{1}{2} \cos 2x$$

$$\cos^2 x = \frac{1}{2} + \frac{1}{2} \cos 2x$$

to obtain

Use identity

$$\cos^2 2x = \frac{1}{2} + \frac{1}{2} \cos 4x$$

(i.e., substitute for $\cos^2$ a second time).

$$\begin{aligned}\int \sin^2 x \cos^2 x\, dx &= \int \left(\frac{1}{2} - \frac{1}{2} \cos 2x\right)\left(\frac{1}{2} + \frac{1}{2} \cos 2x\right) dx \\ &= \int \left(\frac{1}{4} - \frac{1}{4} \cos^2 2x\right) dx \\ &= \int \left[\frac{1}{4} - \frac{1}{4}\left(\frac{1}{2} + \frac{1}{2} \cos 4x\right)\right] dx \\ &= \int \left(\frac{1}{8} - \frac{1}{8} \cos 4x\right) dx \\ &= \frac{x}{8} - \frac{1}{32} \sin 4x + C.\end{aligned}$$

■

General Strategy: In integrals of the form

$$\int \sin^n x \cos^m x \, dx$$

where both n and m are even, use the identities

$$\sin^2 \theta = \frac{1}{2} - \frac{1}{2}\cos 2\theta \qquad (1)$$

$$\cos^2 \theta = \frac{1}{2} + \frac{1}{2}\cos 2\theta \qquad (2)$$

repeatedly until an integrand involving only constants and cosine terms is obtained.

Example 5 Find $\int \cos^2\left(\frac{x}{2}\right)\sin^4\left(\frac{x}{2}\right) dx$.

Strategy

Both exponents are even so we use identities (1) and (2) with $\theta = x/2$.

Use identity (2) again on $\cos^2 x$.

Handle $\cos^3 x$ term as in Example 2.

Solution

$$\begin{aligned}
\int \cos^2\left(\frac{x}{2}\right)\sin^4\left(\frac{x}{2}\right) dx &= \int \cos^2\left(\frac{x}{2}\right)\left[\sin^2\left(\frac{x}{2}\right)\right]^2 dx \\
&= \int \left[\frac{1}{2} + \frac{1}{2}\cos x\right]\left[\frac{1}{2} - \frac{1}{2}\cos x\right]^2 dx \\
&= \frac{1}{8}\int [1 + \cos x][1 - 2\cos x + \cos^2 x]\, dx \\
&= \frac{1}{8}\int [1 - \cos x - \cos^2 x + \cos^3 x]\, dx \\
&= \frac{1}{8}\int \left[1 - \cos x - \left(\frac{1}{2} + \frac{1}{2}\cos 2x\right) + \cos^3 x\right] dx \\
&= \frac{1}{8}\int \left[\frac{1}{2} - \cos x - \frac{1}{2}\cos 2x + \cos^3 x\right] dx \\
&= \frac{1}{8}\int \left[\frac{1}{2} - \cos x - \frac{1}{2}\cos 2x + (1 - \sin^2 x)\cos x\right] dx \\
&= \frac{1}{8}\int \left[\frac{1}{2} - \frac{1}{2}\cos 2x - \sin^2 x \cos x\right] dx \\
&= \frac{x}{16} - \frac{1}{32}\sin 2x - \frac{1}{24}\sin^3 x + C.
\end{aligned}$$

■

Example 6 Find $\int \sin^4 x\, dx$.

Strategy

Use identity (1) with $\theta = x$.

Solution

$$\begin{aligned}
\int \sin^4 x\, dx &= \int [\sin^2 x]^2\, dx \\
&= \int \left[\frac{1}{2} - \frac{1}{2}\cos 2x\right]^2 dx
\end{aligned}$$

Use identity (2) on $\cos^2 2x$ with $\theta = 2x$.

$$= \int \left[\frac{1}{4} - \frac{1}{2}\cos 2x + \frac{1}{4}\cos^2 2x\right] dx$$

$$= \int \left[\frac{1}{4} - \frac{1}{2}\cos 2x + \frac{1}{4}\left(\frac{1}{2} + \frac{1}{2}\cos 4x\right)\right] dx$$

$$= \int \left(\frac{3}{8} - \frac{1}{2}\cos 2x + \frac{1}{8}\cos 4x\right) dx$$

$$= \frac{3x}{8} - \frac{1}{4}\sin 2x + \frac{1}{32}\sin 4x + C.$$

■

Integrals Involving Powers of Secant and Tangent

Example 7 Find $\int \tan^2 x \sec^2 x\, dx$.

Solution: We make the u-substitution

$$u = \tan x; \qquad du = \sec^2 x\, dx.$$

Then

$$\int \tan^2 x \sec^2 x\, dx = \int u^2\, du$$

$$= \frac{1}{3}u^3 + C$$

$$= \frac{1}{3}\tan^3 x + C.$$

■

Example 8 Find $\int \tan^3 x \sec x\, dx$.

Solution: We use the identity $\tan^2 x = \sec^2 x - 1$ to write

$$\int \tan^3 x \sec x\, dx = \int \tan^2 x \cdot \tan x \sec x\, dx$$

$$= \int (\sec^2 x - 1)\tan x \sec x\, dx.$$

We now make the u-substitution

$$u = \sec x; \qquad du = \tan x \sec x\, dx.$$

With this substitution we have

$$\int \tan^3 x \sec x\, dx = \int (u^2 - 1)\, du$$

$$= \frac{1}{3}u^3 - u + C$$

$$= \frac{1}{3}\sec^3 x - \sec x + C.$$

■

General Strategy: In integrals of the form

$$\int \sec^n x \tan^m x\, dx:$$

(i) *If n is even,* write $\sec^{n-2} x$ as a function of $\tan x$ using the identity $\sec^2 x = \tan^2 x + 1$. Then make the u-substitution $u = \tan x$. The remaining factor $\sec^2 x\, dx$ becomes du.
(ii) *If m is odd,* write $\tan^{m-1} x$ as a function of $\sec x$ using the identity above. Make the u-substitution $u = \sec x$ using one factor of $\sec x$, and using the remaining factor of $\tan x$ as $du = \sec x \tan x\, dx$.
(iii) *If n is odd and m is even,* the technique of integration by parts is needed. (This will be discussed in Chapter 11.)

Example 9 Find $\int \tan^4 x\, dx$.

Strategy

Use identity $\tan^2 x = \sec^2 x - 1$.
Use identity again.

Let $u = \tan x$

$du = \sec^2 x\, dx$

in first term.

Solution

$$\begin{aligned}\int \tan^4 x\, dx &= \int (\sec^2 x - 1) \tan^2 x\, dx\\ &= \int (\sec^2 x \tan^2 x - \tan^2 x)\, dx\\ &= \int (\sec^2 x \tan^2 x - \sec^2 x + 1)\, dx\\ &= \frac{1}{3}\tan^3 x - \tan x + x + C.\end{aligned}$$ ■

Integrals of the Forms

$\int \sin mx \cos nx\, dx$

$\int \sin mx \sin nx\, dx$

$\int \cos mx \cos nx\, dx$

In these integrals the general strategy is to use the identities

$$\sin mx \cdot \cos nx = \frac{1}{2}[\sin(m+n)x + \sin(m-n)x] \tag{3}$$

$$\sin mx \cdot \sin nx = \frac{1}{2}[\cos(m-n)x - \cos(m+n)x] \tag{4}$$

$$\cos mx \cdot \cos nx = \frac{1}{2}[\cos(m+n)x + \cos(m-n)x]. \tag{5}$$

(See Exercises 45–47.)

Example 10 Find $\int \sin 5x \cos 2x\, dx$.

Solution: We use identity (3) to write

$$\begin{aligned}\int \sin 2x \cdot \cos 5x\, dx &= \int \frac{1}{2}[\sin(2x+5x) + \sin(2x-5x)]\, dx\\ &= \frac{1}{2}\int (\sin 7x + \sin(-3x))\, dx\\ &= \frac{1}{2}\left[\left(\frac{1}{7}\right)\cdot(-\cos 7x) + \left(-\frac{1}{3}\right)(-\cos(-3x))\right] + C\\ &= -\frac{1}{14}\cos 7x + \frac{1}{6}\cos(-3x) + C\\ &= -\frac{1}{14}\cos 7x + \frac{1}{6}\cos 3x + C.\end{aligned}$$ ■

Exercise Set 10.3

In Exercises 1–36, find the integral.

1. $\int \sin^2 2x\,dx$

2. $\int_0^{\pi/4} \sec^2 x\,dx$

3. $\int \sin x \cos^2 x\,dx$

4. $\int \sin x \cos^3 x\,dx$

5. $\int \sin^5 x\,dx$

6. $\int \tan x \sec^2 x\,dx$

7. $\int_0^{\pi/2} \sin^2 x \cos^3 x\,dx$

8. $\int_0^{\pi/4} \sin^2 x \cos^2 x\,dx$

9. $\int \sec^3 x \tan^3 x\,dx$

10. $\int \sec^4(2x - 1)\,dx$

11. $\int_0^{\pi} \sin^6 x\,dx$

12. $\int \tan^3 x \sec^4 x\,dx$

13. $\int \cot^3 x\,dx$

14. $\int \sin^3 x\sqrt{\cos x}\,dx$

15. $\int_0^{\pi/4} \sin x \sin 3x\,dx$

16. $\int \sin 5x \cos 3x\,dx$

17. $\int (\sin x + \cos x)^2\,dx$

18. $\int \sin^2 x \cos^5 x\,dx$

19. $\int \tan^4 x \sec^4 x\,dx$

20. $\int \cot^5 x\,dx$

21. $\int \tan^5 x \sec^3 x\,dx$

22. $\int \frac{\sec^2 x}{1 + \tan x}\,dx$

23. $\int \sin x \sin 2x\,dx$

24. $\int \sin 4x \cos 3x\,dx$

25. $\int \frac{\cos^3 x}{\sqrt{\sin x}}\,dx$

26. $\int (\tan x + \cot x)^2\,dx$

27. $\int \frac{\sec^2 x}{(1 + \tan x)^4}\,dx$

28. $\int \csc x \cdot \cot x\,dx$

29. $\int \cos 5x \cdot \cos 3x\,dx$

30. $\int \cos 7x \sin 2x\,dx$

31. $\int \frac{\sin^3 \theta}{\cos \theta}\,d\theta$

32. $\int \frac{\tan^3 \theta}{\sec \theta}\,d\theta$

33. $\int \sec^2 x\sqrt{\tan x}\,dx$

34. $\int \sin ax \cdot \sin bx \cdot \cos cx\,dx$

35. $\int \sin ax \cdot \cos bx \cdot \cos cx\,dx$

36. $\int \sin(ax + b)\cos(cx + d)\,dx$

Evaluate the following definite integrals which arise often in applied mathematics (n and m are integers).

37. $\frac{1}{\pi} \cdot \int_{-\pi}^{\pi} \cos n\theta\,d\theta, \quad n \geq 1$

38. $\frac{2}{\pi} \cdot \int_0^{\pi} \sin n\theta\,d\theta, \quad n \geq 1$

39. $\int_{-L}^{L} \cos^2\left(\frac{n\pi x}{L}\right)dx$

40. $\int_0^{2\pi} \sin^2 nx\,dx$

41. $\int_0^{2\pi} \sin nx \cos mx\,dx$

42. $\int_0^{2\pi} \cos nx \cos mx\,dx$

Show how the following reduction formulas are obtained.

43. $\int \tan^n x\,dx = \frac{\tan^{n-1} x}{n - 1} - \int \tan^{n-2} x\,dx, \quad n \neq 1$

44. $\int \cot^n x\,dx = -\frac{\cot^{n-1} x}{n - 1} - \int \cot^{n-2} x\,dx, \quad n \neq 1$

45. Prove identity (3) by adding the identities

$$\sin(\theta_1 + \theta_2) = \sin\theta_1 \cos\theta_2 + \cos\theta_1 \sin\theta_2,$$
$$\sin(\theta_1 - \theta_2) = \sin\theta_1 \cos\theta_2 - \cos\theta_1 \sin\theta_2.$$

46. Prove identity (4) by subtracting the identities

$$\cos(\theta_1 - \theta_2) = \cos\theta_1 \cos\theta_2 + \sin\theta_1 \sin\theta_2,$$
$$\cos(\theta_1 + \theta_2) = \cos\theta_1 \cos\theta_2 - \sin\theta_1 \sin\theta_2.$$

47. Prove identity (5).

48. Find the average value of the function $y = \tan x \sec^2 x$ for $-\pi/4 \leq x \leq \pi/4$.

49. Find the volume of the solid obtained by revolving about the x-axis the region bounded by the graph of $y = \tan x \sec x$ for $0 \leq x \leq \pi/4$.

50. Find the area of the region bounded by the graphs of $f(x) = \tan^2 x$ and $g(x) = -\tan^2 x$ for $0 \leq x \leq \pi/4$.

51. Find the area of the region bounded above by the graph of $y = \sec x$ and below by the x-axis for $0 \leq x \leq \pi/4$.

52. Find the average value of the function $f(x) = \sin^2 x$ on $[0, \pi]$.

53. A particle moves along a line with acceleration function $a(t) = \cos^2 t$. Find
    - **a.** its velocity function $v(t)$, if $v(0) = 0$;
    - **b.** its position function $s(t)$, if $s(0) = 4$.

54. Find the volume of the solid obtained by revolving about the x-axis the region bounded by the graph of $f(x) = \sec^2 x \tan x$ and the x-axis for $0 \leq x \leq \pi/4$.

55. Find the length of the graph of $y = \ln \sec x$ from $(0, 0)$ to $(\pi/3, \ln 2)$.

10.4 THE INVERSE TRIGONOMETRIC FUNCTIONS

Consider the following problem. A television camera is located 3 kilometers from the launch pad for a rocket that will propel a new satellite into orbit. The problem is to determine the angle $\theta(t)$ at which the camera should be inclined at each instant t so as to track the rocket during the initial phase of its ascent, when its trajectory is nearly vertical.

If the acceleration $a(t)$ of the rocket is known, the height of the tip of the rocket above the ground, $s(t)$, may be calculated by integration. Thus, at each instant t we may compute $\tan \theta$ by the equation

$$\tan \theta = \frac{s}{3}. \tag{1}$$

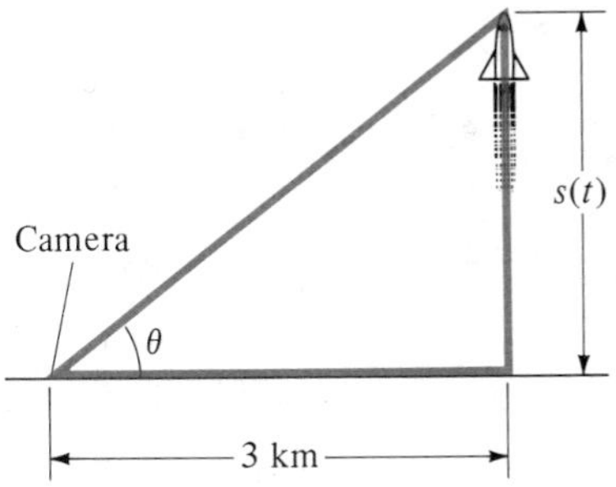

Figure 4.1 Elevation of camera determined by equation (1).

(See Figure 4.1.)

The difficulty with equation (1) is that it describes the *tangent* of the desired angle rather than the angle θ itself. The situation is therefore analogous to that encountered in Chapter 9 when we obtained equations of the form

$$\ln x = a. \tag{2}$$

Recall that equations of this type are solved by applying the *inverse* of the natural logarithm function to both sides of the equation. That is, by applying the exponential function e^x to both sides of equation (2) we obtain

$$e^{\ln x} = e^a, \qquad \text{or} \qquad x = e^a.$$

If we are to use this same idea in solving equation (1) for θ, we must find an *inverse* for the tangent function. Suppose for a moment that such a function, say $g(\theta)$, could be found. This would mean that $g(\tan \theta) = \theta$. Applying this function to both sides of equation (1) would then give the solution

$$g(\tan \theta) = g\left(\frac{s}{3}\right), \qquad \text{or} \qquad \theta = g\left(\frac{s}{3}\right). \tag{3}$$

With this simple motivation in mind, we therefore set out to find inverses for each of the six trigonometric functions. As we succeed at doing so, we will also obtain some very useful integration formulas.

The Inverse Tangent Function

To find the inverse function $g(x)$ for the tangent function $y = \tan x$, we review several important facts about inverse functions (see Section 3.8):

1. In order for the inverse of $y = \tan x$ to exist, the function $y = \tan x$ must be one-to-one. That is, for each y_0 in the range of $y = \tan x$ there must be one and only one x_0 in the domain of $y = \tan x$ with $y_0 = \tan x_0$.
2. If the inverse function $g(x)$ exists, then by definition,

   $$g(y) = x \qquad \text{if and only if} \qquad y = \tan x$$

   for all points (x_0, y_0) with x_0 in the domain of $\tan x$ and y_0 in the domain of $g(y)$.

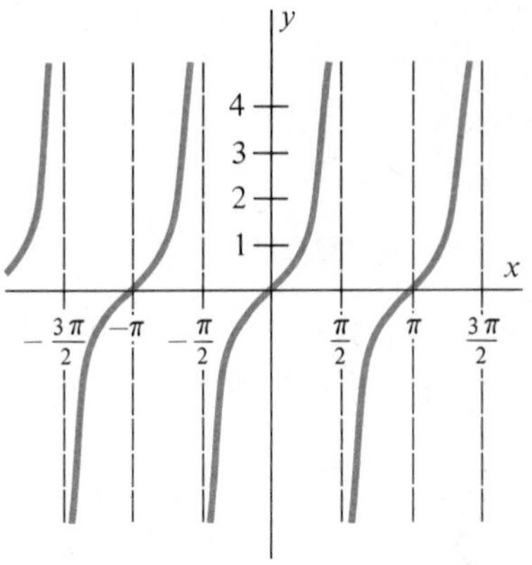

Figure 4.2 Graph of $y = \tan x$.

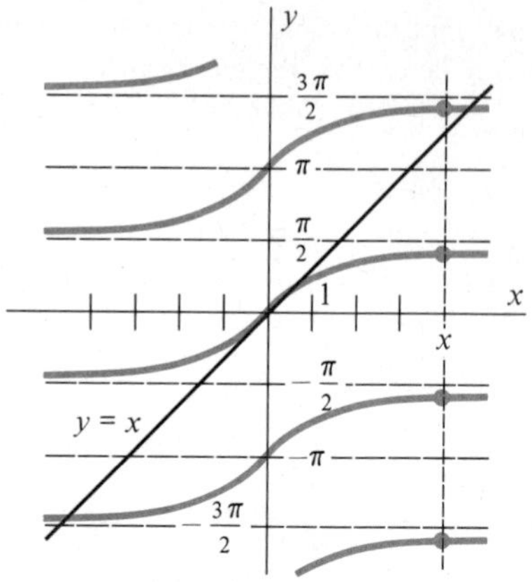

Figure 4.3 Reflection of $y = \tan x$ in $y = x$. (Many y's correspond to a single value of x.)

3. The graph of $y = g(x)$ is the reflection in the line $y = x$ of the graph of $y = \tan x$, that is, the graph of $y = g(x)$ is $\{(x, y) \mid x = \tan y\}$.

Now consider the graph of $y = \tan x$ (Figure 4.2) and its reflection in the line $y = x$ (Figure 4.3).

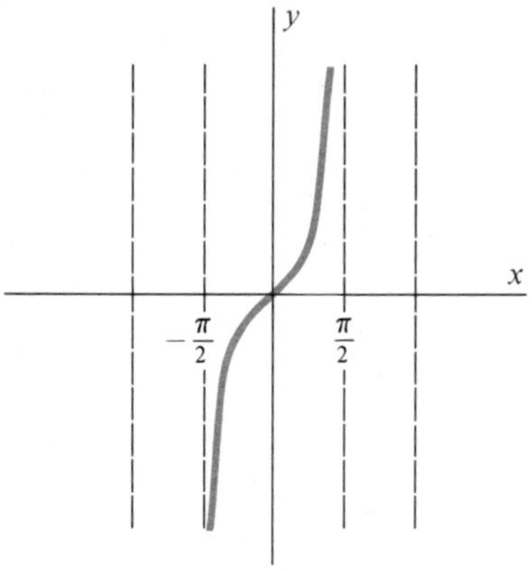

Figure 4.4 Graph of restricted function $y = \text{Tan } x$.

On observing these two graphs, one immediately concludes that there is no hope of defining an inverse for the function $y = \tan x$ for all x in the domain of this function. The difficulty is that $y = \tan x$ is *periodic*. In particular, $\tan(x + n\pi) = \tan x$ for all $n = \pm 1, +2, \ldots$, so each number y in the range of $y = \tan x$ corresponds to *an infinite number* of x's. For example, the value $y = 1$ corresponds to $x = \pi/4, 5\pi/4, 9\pi/4, \ldots$. In other words, the function $y = \tan x$ is not 1-1. Thus, the "definition" of the inverse function g in equation (3) makes no sense, and the graph of the reflection (Figure 4.3) is not the graph of a *function*.

However, all is not lost. Notice in Figure 4.2 that the arc of the graph of $y = \tan x$ corresponding to the interval $-\pi/2 < x < \pi/2$ has the property that each value of y corresponds to a *unique* value of x. This is because the function $y = \tan x$ is *increasing* throughout the interval $(-\pi/2, \pi/2)$ and, hence, one-to-one (see Exercise 43). Our strategy is therefore to restrict the domain of the function $y = \tan x$ to the interval $(-\pi/2, \pi/2)$ and then define the *inverse* of the tangent function, denoted for now by g, by the equation

$$g(y) = x \quad \text{if and only if} \quad y = \tan x, \quad -\frac{\pi}{2} < x < \frac{\pi}{2}.$$

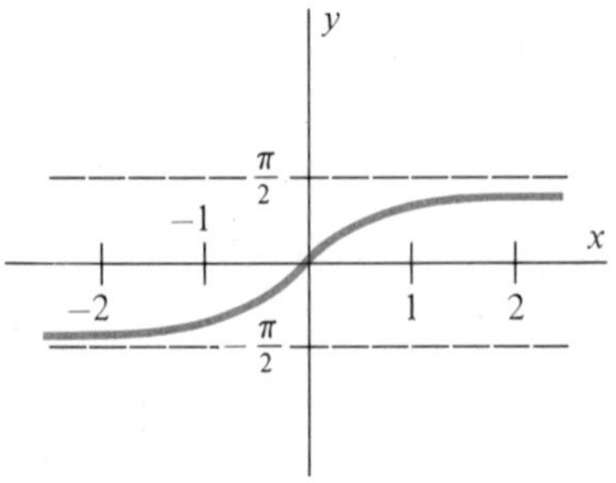

Figure 4.5 Reflection of graph of $y = \text{Tan } x$ is the graph of the inverse of $y = \text{Tan } x$.

We denote the restriction of the tangent function to the domain $(-\pi/2, \pi/2)$ by $y = \text{Tan } x$, and we refer to its graph as the **principal branch** of the graph of $y = \tan x$. Figures 4.4 and 4.5 show that the reflection of the graph of $y = \text{Tan } x$ in the line $y = x$ indeed represents the graph of a function. Figure 4.5 is therefore the graph of the *inverse* of the restricted tangent function $y = \text{Tan } x$.

Since the restricted tangent function $y = \text{Tan } x$ has domain $(-\pi/2, \pi/2)$ and range $(-\infty, \infty)$, the inverse will be defined on $(-\infty, \infty)$ with values in $(-\pi/2, \pi/2)$. The formal definition is the following.

DEFINITION 1

For $x \in (-\infty, \infty)$ and $y \in (-\pi/2, \pi/2)$ the **inverse tangent function,** $\tan^{-1} x$, is defined by

$$y = \tan^{-1} x \qquad \text{if and only if} \qquad x = \tan y. \tag{4}$$

Be careful to note that $\tan^{-1} x$ means the inverse of the function $y = \text{Tan } x$, *not* $(\tan x)^{-1}$. Other notation for the inverse tangent function common in textbooks is $y = \text{Tan}^{-1} x$, or $y = \text{Arc tan } x$.

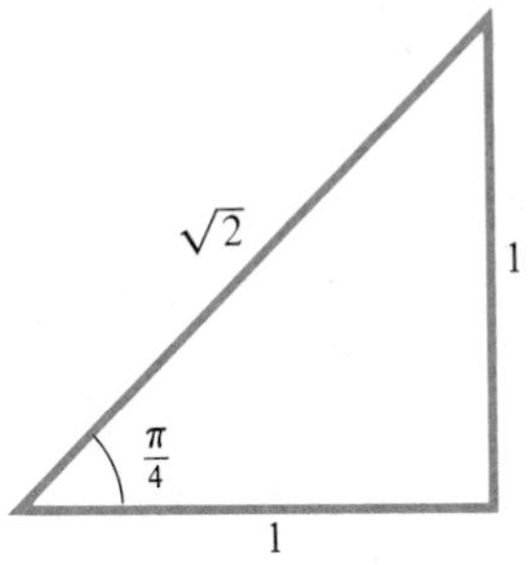

Figure 4.6 $\tan(\pi/4) = 1$; $\tan^{-1}(1) = \pi/4$.

Example 1 Find (a) $\tan^{-1}(1)$, (b) $\tan^{-1}(0)$, and (c) $\sin(\tan^{-1}(1/\sqrt{3}))$.

Solution:

(a) We seek an angle y so that $-\pi/2 < y < \pi/2$, and $\tan y = 1$. This angle is $y = \pi/4$. Thus,

$$\tan^{-1}(1) = \frac{\pi}{4}, \qquad \text{since } \tan\left(\frac{\pi}{4}\right) = 1 \qquad \text{(Figure 4.6).}$$

(b) $\tan^{-1}(0) = 0$, since $\tan(0) = 0$.

(c) Since $\tan(\pi/6) = 1/\sqrt{3}$, $\sin(\tan^{-1}(1/\sqrt{3})) = \sin(\pi/6) = 1/2$ (see Figure 4.7). ■

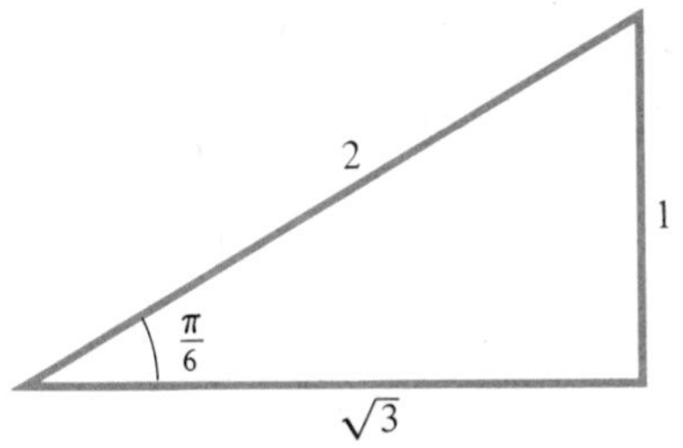

Figure 4.7 $\tan(\pi/6) = 1/\sqrt{3}$; $\tan^{-1}(1/\sqrt{3}) = \pi/6$; $\sin(\tan^{-1}(1/\sqrt{3})) = 1/2$.

Example 2 At what angle should the camera be inclined when the rocket in Figure 4.1 is 7.8 kilometers above the ground?

Solution: From equation (1), when $s = 7.8$ km we obtain

$$\tan\theta = \frac{7.8}{3} = 2.6.$$

If you have access to a hand calculator or computer with the ability to compute values of the inverse tangent, simply compute $\tan^{-1}(2.6)$. If not, use the table of trigonometric functions in the Appendix to find an angle whose tangent is approximately 2.6. By either method the answer is

$$\begin{aligned}\theta = \tan^{-1}(2.6) &\approx 1.20 \text{ radians}\\ &\approx 69^\circ.\end{aligned}$$

■

Other Inverse Trigonometric Functions

To find inverses for the remaining five trigonometric functions, we use the same method followed to obtain $\tan^{-1} x$, namely:

(i) Restrict the domain of the trigonometric function to an interval on which the function is one-to-one and on which the function values cover the entire range.

(ii) Define the inverse of the trigonometric function $f(x)$ by the usual definition

$$y = f^{-1}(x) \qquad \text{if and only if} \qquad x = f(y).$$

We obtain the following inverse functions.

DEFINITION 2

The remaining five inverse trigonometric functions are defined as follows:

(a) For $-1 \leq x \leq 1$ and $-\pi/2 \leq y \leq \pi/2$ the **inverse sine function** is defined by $y = \sin^{-1} x$ if and only if $x = \sin y$.
(b) For $-1 \leq x \leq 1$ and $0 \leq y \leq \pi$ the **inverse cosine function** is defined by $y = \cos^{-1} x$ if and only if $x = \cos y$.
(c) For $|x| \geq 1$ and $y \in [0, \pi/2) \cup [\pi, 3\pi/2)$ the **inverse secant function** is defined by $y = \sec^{-1} x$ if and only if $x = \sec y$.
(d) For $|x| \geq 1$ and $y \in (0, \pi/2] \cup (\pi, 3\pi/2]$ the **inverse cosecant function** is defined by $y = \csc^{-1} x$ if and only if $x = \csc y$.
(e) For $-\infty < x < \infty$ and $0 < y < \pi$ the **inverse cotangent function** is defined by $y = \cot^{-1} x$ if and only if $x = \cot y$.

(Note that the domains of both $\sec^{-1} x$ and $\csc^{-1} x$ are unions of disjoint intervals. This is because the *ranges* of $\sec x$ and $\csc x$ consist of disjoint intervals. It is possible to choose different domains for $\sec^{-1} x$ and $\csc^{-1} x$, such as $[0, \pi/2) \cup (\pi/2, \pi]$ for $\sec^{-1} x$.)

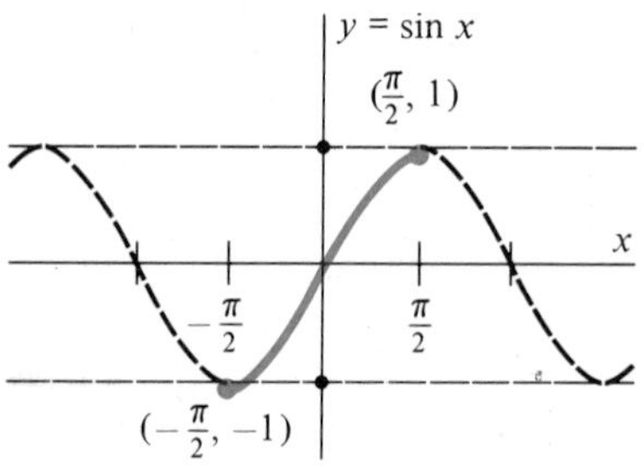

Figure 4.8 Principal branch of $y = \sin x$ is $-\pi/2 \leq x \leq \pi/2$.

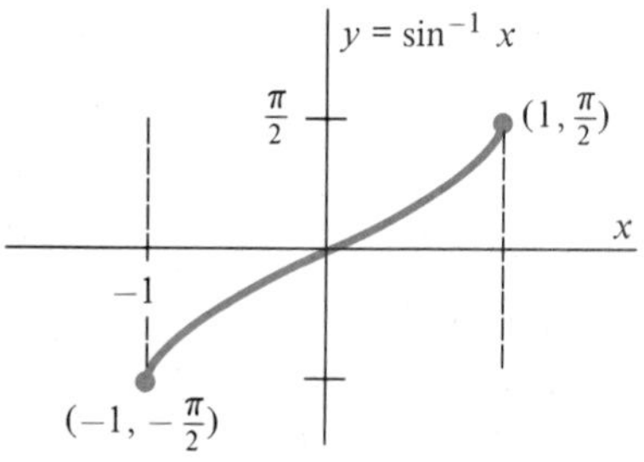

Figure 4.9 Graph of $y = \sin^{-1} x$ is reflection of graph of principal branch of $y = \sin x$ in line $y = x$.

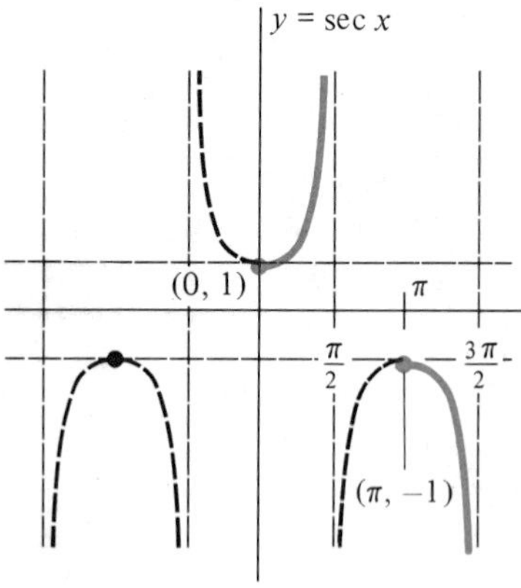

Figure 4.10 Principal branch of $y = \sec x$ is $x \in [0, \pi/2) \cup [\pi, 3\pi/2)$.

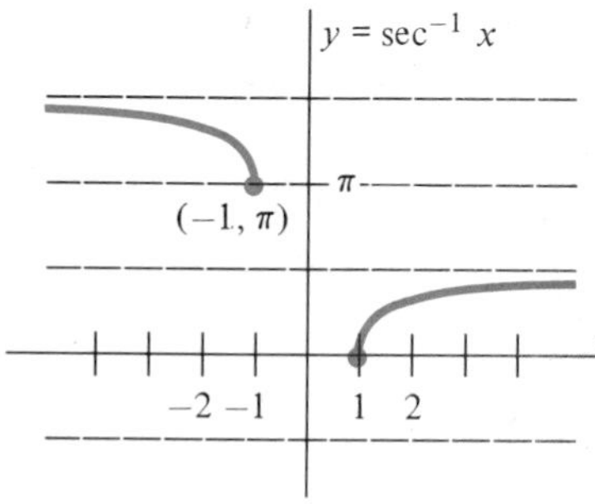

Figure 4.11 Graph of $y = \sec^{-1} x$ is reflection of principal branch of $y = \sec x$ in line $y = x$.

Figures 4.8 through 4.11 illustrate the principal branches of the sine and secant functions, together with the graphs of the corresponding inverse functions. (The graphs of the other inverse functions are constructed in a similar manner (see Exercise 42).)

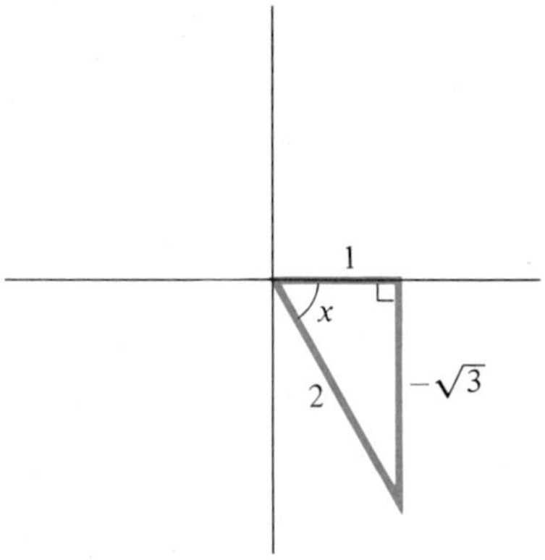

Figure 4.12 $\sin\left(-\frac{\pi}{3}\right) = -\frac{\sqrt{3}}{2}$ so $\sin^{-1}\left(-\frac{\sqrt{3}}{2}\right) = -\frac{\pi}{3}$.

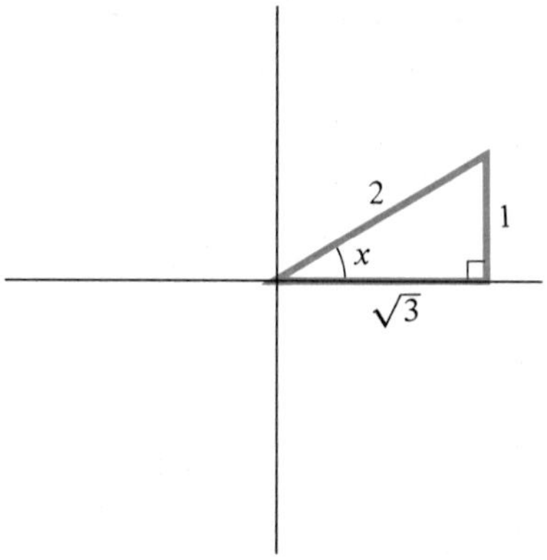

Figure 4.13 $\csc\left(\frac{\pi}{6}\right) = 2$ so $\csc^{-1}(2) = \frac{\pi}{6}$.

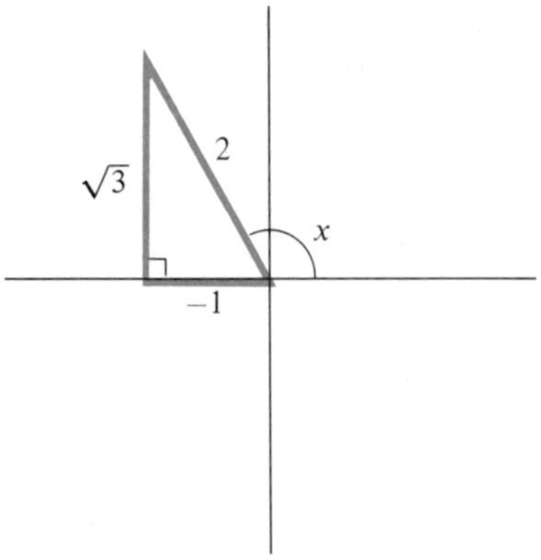

Figure 4.14 $\cos\left(\frac{2\pi}{3}\right) = -\frac{1}{2}$ so $\cos^{-1}\left(-\frac{1}{2}\right) = \frac{2\pi}{3}$, $\tan\left(\cos^{-1}\left(-\frac{1}{2}\right)\right) = -\sqrt{3}$.

Example 3 Find (a) $\sin^{-1}(-\sqrt{3}/2)$, (b) $\csc^{-1}(2)$, and (c) $\tan(\cos^{-1}(-1/2))$.

Solution:

(a) An angle x that satisfies $\sin x = -\sqrt{3}/2$ and $-\pi/2 \le x \le \pi/2$ is $x = -\pi/3$. Thus, $\sin^{-1}(-\sqrt{3}/2) = -\pi/3$ (Figure 4.12).
(b) An angle x with $\csc x = 2$ and $x \in (0, \pi/2] \cup (\pi, 3\pi/2]$ is $x = \pi/6$. Thus, $\csc^{-1}(2) = \pi/6$ (Figure 4.13).
(c) An angle x with $\cos x = -1/2$ and $0 \le x \le \pi$ is $x = 2\pi/3$. Thus, $\tan(\cos^{-1}(-1/2)) = \tan(2\pi/3) = -\sqrt{3}$ (Figure 4.14). ■

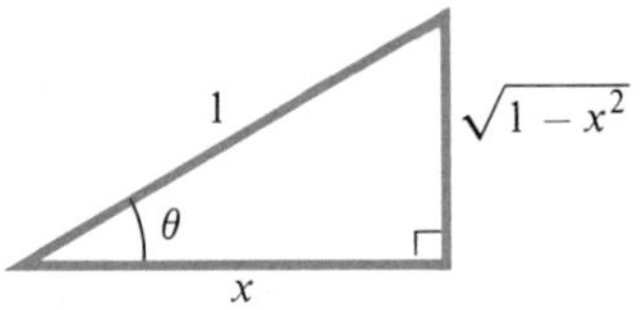

Figure 4.15 $\cos\theta = x$; $\sin\theta = \sqrt{1-x^2}$.

The following example demonstrates the rather remarkable fact that a composition of a trigonometric function with an inverse trigonometric function produces an algebraic function.

Example 4 Let $-1 \le x \le 1$. Express $y = \sin(\cos^{-1} x)$ as an algebraic function of x.

Strategy
Construct a triangle with an angle whose cosine is x, using Pythagorean Theorem.

Solution
A right triangle containing an angle θ whose cosine is x may be constructed with hypotenuse 1, adjacent side x, and opposite side $\sqrt{1-x^2}$ (see Figure 4.15). We see that

Read off desired sine.

$$\sin(\cos^{-1} x) = \sin\theta = \frac{\text{opp}}{\text{hyp}} = \sqrt{1-x^2}.$$

Alternatively, we could solve this problem by using the identity

$$\sin\theta = \sqrt{1-\cos^2\theta}.$$

Then

Use $\cos(\cos^{-1} x) = x$.

$$\begin{aligned}\sin(\cos^{-1} x) &= \sqrt{1-[\cos(\cos^{-1} x)]^2}\\ &= \sqrt{1-x^2}.\end{aligned}$$

(See Exercise 35.) ■

Certain identities among inverse trigonometric functions follow from the definitions of the trigonometric and inverse trigonometric functions.

Example 5 Prove the identity $\cos^{-1} x = \pi/2 - \sin^{-1} x$.

Strategy
Begin with an angle whose sine is x.

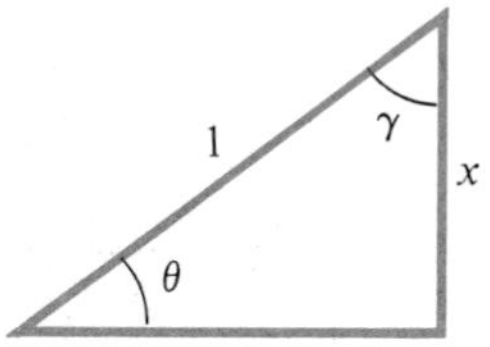

Figure 4.16 $\theta + \gamma = \pi/2$
$\gamma = \cos^{-1} x$
$\theta = \sin^{-1} x$.

Solution
Let θ be the base angle of a right triangle so that $\sin \theta = x$. In order that we may apply the inverse sine to x we must have $0 \le \theta \le \pi/2$.

If γ denotes the remaining acute angle of such a triangle, θ and γ are complementary. That is,

$$\theta + \gamma = \pi/2$$

so

$$\gamma = \pi/2 - \theta.$$

But Figure 4.16 shows that

$$\gamma = \cos^{-1} x \qquad \text{and} \qquad \theta = \sin^{-1} x.$$

so

$$\cos^{-1} x = \pi/2 - \sin^{-1} x.$$

■

Exercise Set 10.4

In Exercises 1–19, find the values indicated.

1. $\sin^{-1}(1/2)$

2. $\cos^{-1}(-\sqrt{3}/2)$

3. $\tan^{-1}(0)$

4. $\sec^{-1}(2)$

5. $\cos^{-1}(-1)$

6. $\cot^{-1}(-\sqrt{3})$

7. $\sin^{-1}(1) + \cos^{-1}(1)$

8. $\tan^{-1}\left(-\frac{1}{\sqrt{3}}\right)$

9. $\tan\left(\sin^{-1}\left(\frac{1}{2}\right)\right)$

10. $\cos^{-1}(\sin \pi/4)$

11. $\sec(\tan^{-1}(\sqrt{3}))$

12. $\cot^{-1}\left(\tan\left(\frac{2\pi}{3}\right)\right)$

13. $\cos[2 \tan^{-1}(\sqrt{3})]$

14. $\sin\left[2 \cot^{-1}\left(\frac{3}{4}\right)\right]$

15. $\tan^2\left(\pi - \sin^{-1}\left(\frac{\sqrt{2}}{2}\right)\right)$

16. $\sec^{-1}\left(2 \tan \frac{\pi}{4}\right)$

17. $\cos^{-1}\left(2 - \sqrt{2} \sin\left(\frac{\pi}{4}\right)\right)$

18. $\sec(2 \tan^{-1} 1)$

19. $\tan(\sec^{-1} 2)$

In Exercises 20–29, find an algebraic expression for the given function of x.

20. $y = \cos(\sin^{-1} x)$

21. $y = \tan(\sin^{-1} x)$

22. $y = \tan(\tan^{-1} x)$

23. $y = \sin[2 \sin^{-1} (x)]$

24. $y = \sin(\tan^{-1} x)$

25. $y = \sin(\sec^{-1} x)$

26. $y = \cos(\tan^{-1} x)$

27. $y = \tan(\cot^{-1} x)$

28. $y = \tan(\sec^{-1} x)$

29. $y = \cos(\csc^{-1} x)$

In Exercises 30–34, verify the stated identities.

30. $\sin^{-1} x = \frac{\pi}{2} - \cos^{-1} x$

31. $\sin^{-1}(-x) = -\sin^{-1} x$

32. $\tan^{-1}(-x) = -\tan^{-1} x$

33. $\cos^{-1}(-x) = \pi - \cos^{-1} x$

34. $\tan^{-1} x + \tan^{-1}\left(\frac{1}{x}\right) = \frac{\pi}{2}$

35. Explain why the identity $\sin(\sin^{-1} x) = x$ is valid for $-1 \le x \le 1$. Are similar identities true for the other inverse trigonometric functions? If so, for what values of x do they hold?

36. The graph of a function must have the **vertical line property:** any vertical line meets the graph in at most one point.
a. If a function has an inverse, what property must its graph possess concerning horizontal lines?
b. Show that each of the six trigonometric functions fails this test.
c. Show that the graphs of the principal branches of the six trigonometric functions do indeed satisfy this test.

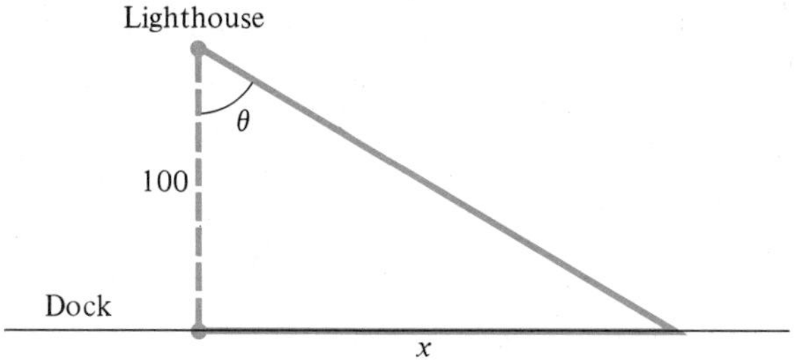

Figure 4.17 Lighthouse problem.

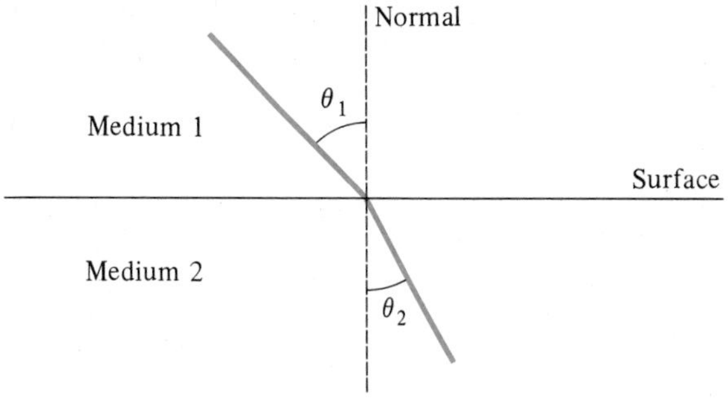

Figure 4.18 Snell's Law of Refraction.

37. True or false? $\sec^{-1}(\cos \pi/4) = \dfrac{2}{\sqrt{2}}$.

38. The selection of the principal branch of a trigonometric function is somewhat arbitrary.

a. Show that the selection of the principal branch of the sine function as $\pi/2 \leq x \leq 3\pi/2$ leads to a legitimate inverse function for $y = \sin x$.

b. What would be the disadvantage of choosing the principal branch of $y = \sin x$ to be $0 \leq x \leq \pi/2$?

c. What properties should the principal branch of a trigonometric function possess?

39. A lighthouse lies 100 meters off a straight shoreline. A boat dock is located on the shoreline at the point nearest the lighthouse. Write an equation describing θ, the angle between the light beam and the line joining the lighthouse and the dock, as a function of the distance x between the dock and the point where the beam of light strikes the shore (see Figure 4.17).

40. Snell's Law of Refraction (See Section 4.4) states that when a ray of light passes from one medium to another, the angles θ_1 and θ_2 formed between the rays and the normal (perpendicular line) to the boundary are related by the equation $\dfrac{\sin \theta_1}{n_1} = \dfrac{\sin \theta_2}{n_2}$ where n_1 and n_2 are constants. Express θ_2 as a function of θ_1 (see Figure 4.18).

41. A patrol car is parked on an overpass 20 meters above a roadway, and the officer is using radar to time the cars moving along the roadway. Ignoring the heights of the cars, express the angle formed between the radar beam and the vertical when a car is x meters away from the point on the roadway directly beneath the patrol car.

42. Graph the inverse functions $y = \cot^{-1} x$, $y = \cos^{-1} x$, and $y = \csc^{-1} x$.

43. Use the Mean Value Theorem to prove that an increasing differentiable function must be one-to-one.

10.5 DERIVATIVES OF THE INVERSE TRIGONOMETRIC FUNCTIONS

Section 10.4 began with the question of determining the angle θ at which a camera located 3 kilometers from a launch pad should be inclined so as to focus on a rocket s kilometers above the launch pad (see Figure 4.1). The equation

$$\tan \theta = \frac{s}{3}$$

together with the development of the inverse tangent function led to the answer

$$\theta = \tan^{-1}\left(\frac{s}{3}\right). \tag{1}$$

Since $s = s(t)$ is not constant, equation (1) actually describes $\theta = \theta(t)$ as a function of s, and, therefore, of t. The natural next question to ask is whether θ in (1) is a *differentiable* function of t. If the answer is yes, the derivative $d\theta/dt$ determines the *rate* at which the angle of the camera must be increased for the camera to track the rocket.

Obviously, we are raising the larger questions of whether the inverse trigonometric functions are differentiable, and, if so, how their derivatives may be computed. The answer to the first question is easily resolved by invoking Theorem 13 of Section 3.8: $f^{-1}(x)$ is differentiable whenever $f(x)$ is differentiable and $f'(f^{-1}(x)) \neq 0$. Since each of the six trigonometric functions is differentiable, so are the six inverse trigonometric functions.

To find the derivative of $y = \tan^{-1} x$ we begin with the identity

$$\tan(\tan^{-1} x) = x.$$

Differentiating both sides with respect to x, using the Chain Rule, gives

$$\sec^2(\tan^{-1} x) \cdot \frac{d}{dx} \tan^{-1} x = 1,$$

so

$$\frac{d}{dx} \tan^{-1} x = \frac{1}{\sec^2(\tan^{-1} x)}.$$

The identity $\sec^2 \theta = 1 + \tan^2 \theta$ then gives

$$\frac{d}{dx} \tan^{-1} x = \frac{1}{1 + [\tan(\tan^{-1} x)]^2} = \frac{1}{1 + x^2}. \tag{2}$$

Since $\sec^2(\tan^{-1} x) = 1 + x^2 \neq 0$ for all x, the derivative in equation (2) exists for all values of x. If u denotes a differentiable function of x, equation (2), together with the Chain Rule, gives

$$\boxed{\frac{d}{dx} \tan^{-1} u = \frac{1}{1 + u^2} \cdot \frac{du}{dx}.} \tag{3}$$

Example 1 The rocket described above is launched vertically and undergoes a constant acceleration of 600 m/sec^2. Find the time at which the angle of inclination of the camera is increasing most rapidly.

Strategy

Find equation for $s(t)$ using

$$v(t) = \int a(t)\, dt$$

$$s(t) = \int v(t)\, dt.$$

$v(0) = 0; \quad s(0) = 0.$

Solution

Since $a = 600 \text{ m/sec}^2 = 0.6 \text{ km/sec}^2$,

$$\text{velocity} = v(t) = .6t + v(0) = .6t \text{ km/sec, and}$$
$$\text{distance} = s(t) = .3t^2 + s(0) = .3t^2 \text{ km}.$$

Equation (1) is therefore

$$\theta(t) = \tan^{-1}\left(\frac{.3t^2}{3}\right) = \tan^{-1}(.1t^2).$$

The *rate* at which θ is changing is $d\theta/dt$. By formula (3):

Find $d\theta/dt$ by (3).

$$\frac{d\theta}{dt} = \frac{1}{1 + [.1t^2]^2} \cdot \frac{d}{dt}(.1t^2) = \frac{.2t}{1 + .01t^4}$$

To maximize $\frac{d\theta}{dt}$ set

$$\frac{d}{dt}\left(\frac{d\theta}{dt}\right) = \frac{d^2\theta}{dt^2} = 0$$

and solve to find critical numbers. (There are no critical numbers for which $d^2\theta/dt^2$ is undefined.)

To find the time at which $d\theta/dt$ is a maximum, we compute

$$\begin{aligned}\frac{d^2\theta}{dt^2} &= \frac{.2[1 + .01t^4] - (.2t)(.04t^3)}{[1 + .01t^4]^2}\\ &= \frac{.2 - .006t^4}{[1 + .01t^4]^2}.\end{aligned}$$

Setting $d^2\theta/dt^2 = 0$ gives $t^4 = 100/3$. The only meaningful critical number is therefore

$$t = \left(\frac{100}{3}\right)^{1/4} \approx 2.40 \text{ seconds.}$$

The first derivative test, applied to $d\theta/dt$, shows that this time indeed corresponds to the maximum value of $d\theta/dt$. ■

To find the derivative of $y = \sin^{-1} x$ we begin with the identity

$$\sin(\sin^{-1} x) = x, \qquad |x| \le 1.$$

Differentiating both sides with respect to x gives

$$\cos(\sin^{-1}(x)) \cdot \frac{d}{dx} \sin^{-1} x = 1.$$

Dividing both sides by $\cos(\sin^{-1} x)$ and using the identity $\cos\theta = \sqrt{1 - \sin^2\theta}$ gives

$$\begin{aligned}\frac{d}{dx}\sin^{-1} x &= \frac{1}{\sqrt{1 - [\sin(\sin^{-1} x)]^2}} \qquad (4)\\ &= \frac{1}{\sqrt{1 - x^2}}\end{aligned}$$

where we must require that $\cos(\sin^{-1} x) = \sqrt{1 - x^2} \neq 0$. If u is a differentiable function of x, we generalize equation (4) by the Chain Rule to the formula

$$\frac{d}{dx}\sin^{-1} u = \frac{1}{\sqrt{1 - u^2}} \cdot \frac{du}{dx}, \qquad |u| < 1. \qquad (5)$$

We may establish differentiation formulas for each of the four remaining inverse trigonometric functions by proceeding in exactly the same fashion. The results are the following:

$$\frac{d}{dx}\cos^{-1} u = \frac{-1}{\sqrt{1 - u^2}} \cdot \frac{du}{dx}, \qquad |u| < 1 \qquad (6)$$

$$\frac{d}{dx}\cot^{-1} u = \frac{-1}{1 + u^2} \cdot \frac{du}{dx}, \qquad (7)$$

$$\frac{d}{dx}\sec^{-1} u = \frac{1}{u\sqrt{u^2 - 1}} \cdot \frac{du}{dx}, \qquad |u| > 1 \qquad (8)$$

$$\frac{d}{dx}\csc^{-1} u = \frac{-1}{u\sqrt{u^2 - 1}} \cdot \frac{du}{dx}, \qquad |u| > 1 \qquad (9)$$

It is important to note the restrictions $|u| < 1$ in (6) and $|u| > 1$ in (8) and (9). If these inequalities do not hold the corresponding formulas are meaningless.

Example 2 Find $\dfrac{dy}{dx}$ for $y = x\tan^{-1} x + \cos^{-1} x$.

Strategy
Use Product Rule on first term, together with (2).

Solution

$$\frac{dy}{dx} = \tan^{-1} x + x \cdot \frac{d}{dx}\tan^{-1} x + \frac{d}{dx}\cos^{-1} x$$
$$= \tan^{-1} x + \frac{x}{1 + x^2} + \frac{-1}{\sqrt{1 - x^2}}.$$

Use (6) on second term.

This expression is defined only for $|x| < 1$. ■

Example 3 Prove the identity $\tan^{-1} x = \pi/2 - \tan^{-1}(1/x)$ by differentiation.

Strategy
Show

$$\frac{d}{dx}\tan^{-1} x = -\frac{d}{dx}\tan^{-1}\frac{1}{x}$$

by computing $\dfrac{d}{dx}\tan^{-1}\left(\dfrac{1}{x}\right)$.

Solution
By equation (3) we have

$$\frac{d}{dx}\tan^{-1}\left(\frac{1}{x}\right) = \frac{1}{1 + \left(\frac{1}{x}\right)^2} \cdot \left(-\frac{1}{x^2}\right)$$
$$= \frac{-1}{1 + x^2}$$
$$= -\frac{d}{dx}\tan^{-1} x.$$

Recall, if $f'(x) = g'(x)$ then $f(x) = g(x) + C$.

Thus

$$\tan^{-1} x = -\tan^{-1}\left(\frac{1}{x}\right) + C$$

To find C, substitute particular value of x.

for some constant C. To find C we substitute $x = 1$ to obtain

$$\tan^{-1} 1 = -\tan^{-1} 1 + C$$

Use fact that $\tan^{-1} 1 = \dfrac{\pi}{4}$.

or

$$\frac{\pi}{4} = -\frac{\pi}{4} + C$$

so

$$C = \frac{\pi}{2},$$

as required. ■

The differentiation formulas (3), (5), and (8) immediately give the following integration formulas:

$$\int \frac{du}{\sqrt{1 - u^2}} = \sin^{-1} u + C \qquad (10)$$

$$\int \frac{du}{1 + u^2} = \tan^{-1} u + C \qquad (11)$$

$$\int \frac{du}{u\sqrt{u^2 - 1}} = \sec^{-1} u + C \qquad (12)$$

(We do not state the integration formulas corresponding to differentiation formulas (6), (7), and (9), since they involve only the negatives of the above integrals.)

The next two examples show how we can make use of u-substitutions to bring integrals into the form of equations (10)–(12).

Example 4 Find $\int \frac{1}{\sqrt{9 - x^2}}\, dx$.

Strategy

Solution

Factor out 9 to obtain form of equation (10).

$$\int \frac{dx}{\sqrt{9 - x^2}} = \int \frac{dx}{\sqrt{9\left(1 - \frac{x^2}{9}\right)}} = \int \frac{dx}{3\sqrt{1 - \left(\frac{x}{3}\right)^2}}$$

Make u-substitution.

We make the u-substitution

$$u = \frac{x}{3}; \qquad du = \frac{1}{3}\, dx.$$

Then $dx = 3\, du$, so our integral becomes

Apply equation (10).

$$\frac{1}{3}\int \frac{3 \cdot du}{\sqrt{1 - u^2}} = \int \frac{du}{\sqrt{1 - u^2}}$$
$$= \sin^{-1}\left(\frac{x}{3}\right) + C. \quad \blacksquare$$

Example 5 Find $\int_1^3 \frac{dx}{\sqrt{x}(1 + x)}$.

Strategy

Make a u-substitution so that $1/\sqrt{x}$ is a factor of du. Here $u = \sqrt{x}$ works.

Solution

We make the u-substitution

$$u = \sqrt{x}; \qquad du = \frac{1}{2\sqrt{x}}.$$

Change limits of integration.

Thus $\frac{1}{\sqrt{x}}\, dx = 2\, du$. We must also change limits of integration:

If $x = 1$, $\quad u = \sqrt{1} = 1$.
If $x = 3$, $\quad u = \sqrt{3}$.

Use formula (11).

With these substitutions we obtain

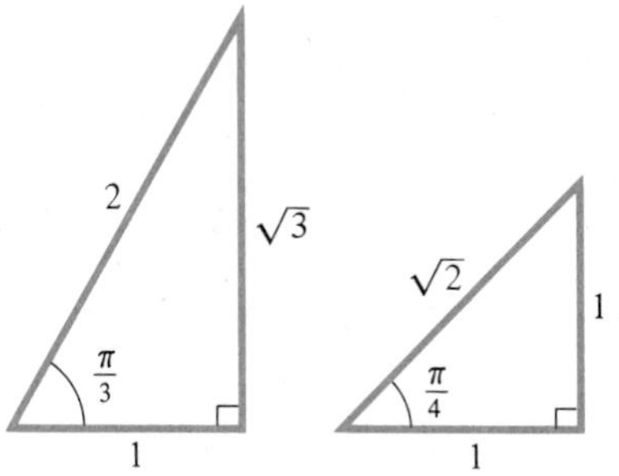

$$\int_1^3 \frac{dx}{\sqrt{x}(1 + x)} = \int_1^{\sqrt{3}} \frac{2\, du}{1 + u^2}$$
$$= 2\tan^{-1} u\Big]_1^{\sqrt{3}}$$
$$= 2(\tan^{-1}\sqrt{3} - \tan^{-1} 1)$$
$$= 2(\pi/3 - \pi/4)$$
$$= \pi/6. \quad \blacksquare$$

The derivatives of inverse trigonometric functions are algebraic functions. Among other things, this remarkable fact allows us to obtain a simple procedure for approximating the irrational number π, as the following example shows.

Example 6 Use integration formula (11) to approximate π.

Strategy

The idea here is to select limits so as to obtain π as the value of the definite integral. We then approximate this integral.

Solution

Since

$$\int \frac{dx}{1+x^2} = \tan^{-1} x + C$$

we have

$$\int_0^1 \frac{dx}{1+x^2} = \tan^{-1} x\Big]_0^1$$
$$= \tan^{-1}(1) - \tan^{-1}(0)$$
$$= \pi/4.$$

Thus

$$\pi = 4\int_0^1 \frac{dx}{1+x^2}.$$

We may therefore approximate π by approximating the definite integral. The results of using Simpson's Rule with various values of n to approximate this integral appear in Table 5.1. ■

Table 5.1

n	Approximation to π using Simpson's Rule
4	3.14156863
10	3.14159262
20	3.14159253
50	3.14159251

Exercise Set 10.5

In Exercises 1–14, find the derivative.

1. $y = \sin^{-1} 3x$
2. $f(x) = x \tan^{-1}(x-1)$
3. $f(t) = \sin^{-1}\sqrt{t}$
4. $y = x^3 \csc^{-1}(1+x)$
5. $f(x) = \sin^{-1} e^{-x}$
6. $y = \tan^{-1}\sqrt{x^2-1}$
7. $y = \sqrt{\cos^{-1} x}$
8. $f(x) = \sin^{-1} x^2 - x \sec^{-1}(x+3)$
9. $f(x) = \ln \tan^{-1} x$
10. $y = \tan^{-1}\left(\frac{1-x}{1+x}\right)$
11. $y = \sec^{-1}\left(\frac{1}{x}\right)$
12. $y = e^{\tan^{-1}\sqrt{x}}$
13. $f(x) = \frac{\tan^{-1} x}{1+x^2}$
14. $f(x) = x^2 \sin^{-1} 2x$

In Exercises 15–32, find the indicated integral.

15. $\int \frac{1}{4+x^2}\,dx$
16. $\int_0^1 \frac{dx}{\sqrt{1-x^2}}$
17. $\int \frac{1}{2x\sqrt{x^2-16}}\,dx$
18. $\int \frac{\cos x}{1+\sin^2 x}\,dx$
19. $\int \frac{x}{\sqrt{1-x^2}}\,dx$
20. $\int_1^2 \frac{1}{x\sqrt{x^2-1}}\,dx$
21. $\int \frac{1}{\sqrt{e^{2x}-1}}\,dx$
22. $\int \frac{x}{\sqrt{1-x^4}}\,dx$
23. $\int \frac{e^{\sqrt{x}}}{\sqrt{x}(1+e^{2\sqrt{x}})}\,dx$
24. $\int \frac{x^2}{\sqrt{1-x^6}}\,dx$
25. $\int \frac{x^2}{1+x^6}\,dx$
26. $\int \frac{\sin x}{1+\cos^2 x}\,dx$
27. $\int_0^1 \frac{\tan^{-1} x}{1+x^2}\,dx$
28. $\int \frac{\cos x}{\sqrt{4-\sin^2 x}}\,dx$
29. $\int_1^2 \frac{dx}{x^2+2}$
30. $\int_2^3 \frac{dx}{x\sqrt{16x^2-25}}$
31. $\int_{1/2}^{\sqrt{2}/2} \frac{dx}{2x\sqrt{4x^2-1}}$
32. $\int_1^e \frac{1}{x\sqrt{1-\ln^2 x}}\,dx$

33. Use separation of variables (see Section 5.10) to solve the initial value problem

$$dy/dx = 4 + y^2, \qquad y(0) = 1.$$

34. Solve the initial value problem

$$dy/dx = \sqrt{1 - y^2}, \qquad y(0) = 1.$$

35. In the radar problem of Exercise 41, Section 10.4, how fast must the radar gun rotate to track an automobile 40 meters away from the point directly under the radar gun if the automobile is moving at a rate of 120 km/hr?

36. A tapestry 3 meters high hangs on a wall so that its lower edge is 1 meter above an observer's eye level. How far from the tapestry should the observer stand so as to maximize the angle subtended in the observer's eye by the tapestry?

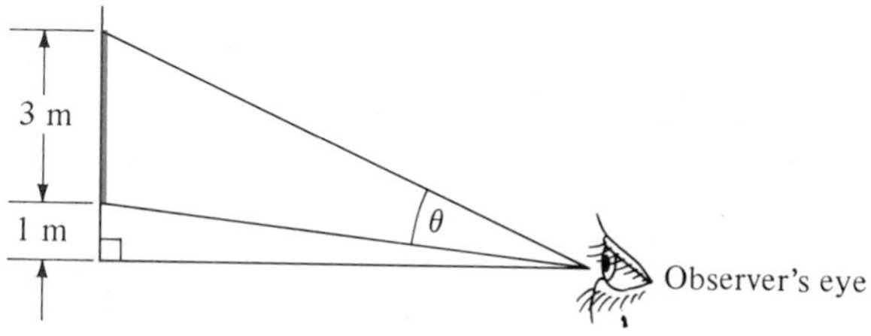

37. A ladder 10 meters long leans against a wall. If the base of the ladder is slipping away from the wall at a rate of 2 meters per second, how fast is the angle between the ladder and the wall increasing when the top of the ladder is 8 meters above the ground?

38. Analysis shows that the first minimum for the diffraction pattern of a circular aperture of diameter d, assuming certain conditions, is given by the equation

$$\sin \theta = 1.22\left(\frac{\lambda}{d}\right), \qquad 0 < \theta < \frac{\pi}{2}$$

where λ is the wavelength of light (a constant). Assuming that θ is a differentiable function of d, show that θ is a decreasing function of d.

39. Find an equation for the line tangent to the graph of $y = \sin^{-1} x^2$ at the point $(\sqrt{2}/2, \pi/6)$.

40. Find the area of the region bounded by the graph of $f(x) = \dfrac{x^2}{1 + x^6}$ and the x-axis for $0 \le x \le 1$.

41. Find the volume of the solid generated by revolving the region bounded by the graph of $f(x) = \dfrac{1}{\sqrt{1 - x^4}}$ and the x-axis, for $0 \le x \le \sqrt{2}/2$, about the y-axis.

42. Find the average value of the function $f(x) = \dfrac{xe^{-x^2}}{\sqrt{1 - e^{-2x^2}}}$ on the interval $[\sqrt{\ln 2}, 1]$.

43. Derive formula (6).

44. Derive formula (7).

45. Derive formula (8).

46. Derive formula (9).

10.6 THE HYPERBOLIC FUNCTIONS (Optional)

Recall from Chapter 9 that the function $y = Ce^x$ is the only nontrivial differentiable function satisfying the differential equation

$$\frac{dy}{dx} = y. \tag{1}$$

We have already seen that equation (1) has various practical interpretations corresponding to the various interpretations of the first derivative (e.g., velocity, rate of growth, and slope of tangent). Since the second derivative also has meaningful interpretations (acceleration, concavity, change in rate of growth) it is natural to ask which nonzero functions satisfy the second order differential equation.

$$\frac{d^2y}{dx^2} = y. \tag{2}$$

A quick check shows that both $y_1 = e^x$ and $y_2 = e^{-x}$ are solutions of equation (2). Moreover, certain combinations of these two functions also satisfy equation (2) and, in addition, satisfy properties quite similar to the properties of the trigonometric functions. We are referring to the following functions.

DEFINITION 3 The hyperbolic sine and cosine functions, $\sinh x$ and $\cosh x$, are defined as follows:

$$\sinh x = \frac{1}{2}(e^x - e^{-x}), \tag{3}$$

$$\cosh x = \frac{1}{2}(e^x + e^{-x}). \tag{4}$$

Roughly speaking, $\cosh x$ represents the average of exponential growth and exponential decay, while $\sinh x$ represents half the difference between these two phenomena.

Derivative Formulas

The differentiation formulas for the hyperbolic functions closely resemble those for the trigonometric functions.

For example, from equations (3) and (4) we obtain

$$\begin{aligned}\frac{d}{dx}\sinh x &= \frac{d}{dx}\left\{\frac{1}{2}(e^x - e^{-x})\right\}\\ &= \frac{1}{2}(e^x + e^{-x})\\ &= \cosh x,\end{aligned}$$

and

$$\begin{aligned}\frac{d}{dx}\cosh x &= \frac{d}{dx}\left\{\frac{1}{2}(e^x + e^{-x})\right\}\\ &= \frac{1}{2}(e^x - e^{-x})\\ &= \sinh x.\end{aligned}$$

(Note that the derivative of $\cosh x$ is $\sinh x$, not $-\sinh x$.)

As a consequence of equations (3) and (4), both $\sinh x$ and $\cosh x$ are their own second derivatives (just as e^x and e^{-x}):

$$\frac{d^2}{dx^2}(\sinh x) = \frac{d}{dx}(\cosh x) = \sinh x.$$

and

$$\frac{d^2}{dx^2}(\cosh x) = \frac{d}{dx}(\sinh x) = \cosh x.$$

Therefore, both $y_1 = \sinh x$ and $y_2 = \cosh x$ are solutions of differential equation (2).

REMARK: It is easy to show that, if $f(x)$ and $g(x)$ are solutions of differential equation (2), then $c_1 f(x) + c_2 g(x)$ is also a solution for any choice of constants c_1

and c_2. (This is called a **linear combination** of solutions.) The functions sinh x and cosh x are merely special linear combinations of the exponential solutions of (2).

In Chapter 15 we will see why the terminology "hyperbolic" is used in conjunction with these functions. In courses on differential equations and engineering mathematics these functions occur frequently. (The most common application of these functions is probably that of describing the tension at any point along a cable suspended from two points.) We will limit the discussion here to the questions of identities, derivatives, integrals, and inverses for these functions.

Graphs of Sinh *x* and Cosh *x*

From the definition of sinh x we can see that

$$\sinh(-x) = \frac{1}{2}[e^{-x} - e^{-(-x)}] = \frac{1}{2}[e^{-x} - e^{x}] = -\frac{1}{2}[e^{x} - e^{-x}] = -\sinh x.$$

Thus, sinh x is an *odd* function, and its graph is symmetric with respect to the origin. Moreover, since

$$\frac{d}{dx}\sinh x = \cosh x = \frac{1}{2}[e^{x} + e^{-x}] > 0$$

for all x, sinh x is an increasing function on all intervals. Finally, since

$$\frac{d^2}{dx^2}\sinh x = \sinh x \text{ is } \begin{cases}\text{positive, if } x > 0\\ \text{negative, if } x < 0\end{cases}$$

the graph of sinh x is concave down on $(-\infty, 0)$ and concave up on $(0, \infty)$ (see Figure 6.1).

A similar analysis shows that

(a) cosh x is an *even* function: $\cosh(-x) = \cosh x$ for all x. Its graph is therefore symmetric with respect to the y-axis.
(b) $\cosh x > 0$ for all x.

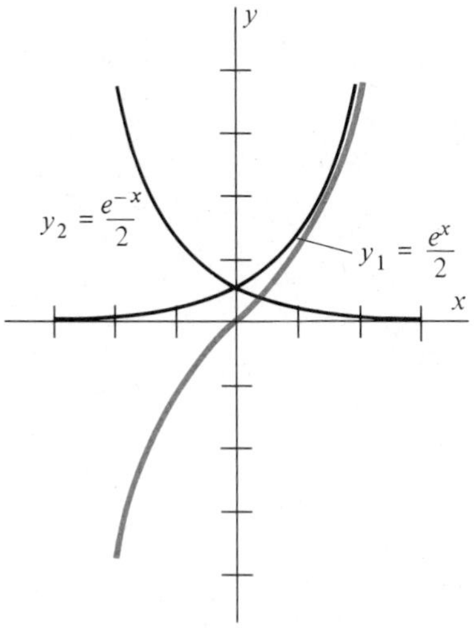

Figure 6.1 $\sinh x = \dfrac{e^x - e^{-x}}{2}$.

(c) $\cosh x$ is decreasing on $(-\infty, 0]$, increasing on $[0, \infty)$.
(d) The graph of $\cosh x$ is concave up on $(-\infty, \infty)$.

The graph of $\cosh x$ appears in Figure 6.2.

As with the trigonometric functions, we can define four other hyperbolic functions in terms of $\sinh x$ and $\cosh x$.

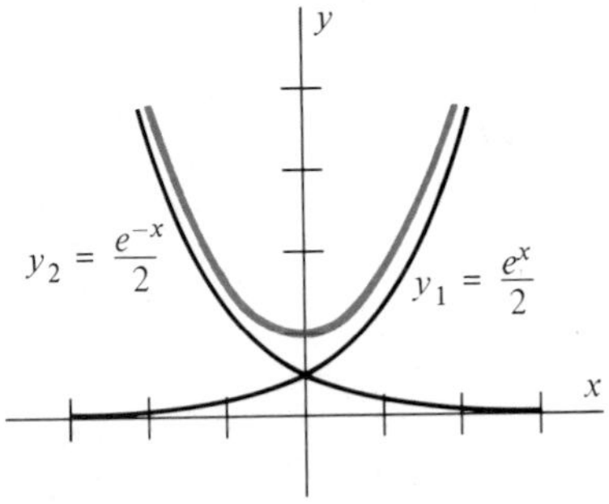

Figure 6.2 $\cosh x = \dfrac{e^x + e^{-x}}{2}$.

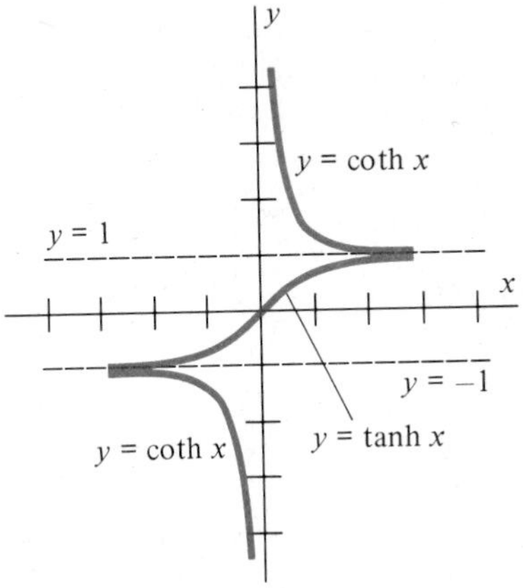

Figure 6.3 Graphs of $y = \tanh x$ and $y = \coth x$.

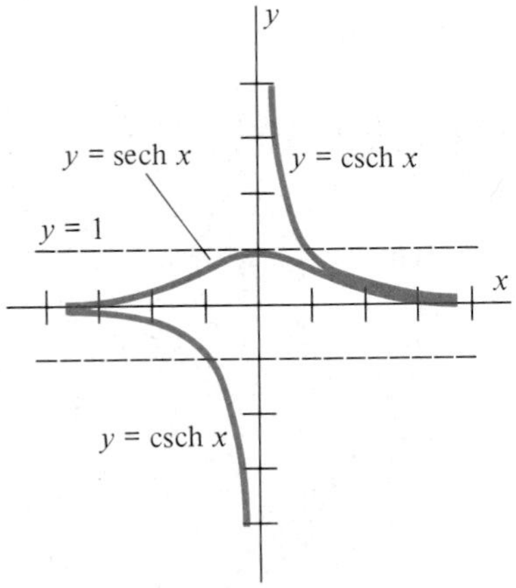

Figure 6.4 Graphs of $y = \text{sech}\, x$ and $y = \text{csch}\, x$.

DEFINITION 4

The hyperbolic tangent, cotangent, secant, and cosecant functions are defined by

$$\tanh x = \frac{\sinh x}{\cosh x} = \frac{e^x - e^{-x}}{e^x + e^{-x}}, \tag{5}$$

$$\coth x = \frac{\cosh x}{\sinh x} = \frac{e^x + e^{-x}}{e^x - e^{-x}}, \tag{6}$$

$$\text{sech}\, x = \frac{1}{\cosh x} = \frac{2}{e^x + e^{-x}}, \tag{7}$$

$$\text{csch}\, x = \frac{1}{\sinh x} = \frac{2}{e^x - e^{-x}}. \tag{8}$$

Graphs of these functions appear in Figures 6.3 and 6.4.

The definitions in equations (3)–(8), together with the Chain Rule, give the following differentiation formulas:

$$\frac{d}{dx} \sinh u = \cosh u \cdot \frac{du}{dx} \tag{9}$$

$$\frac{d}{dx} \cosh u = \sinh u \cdot \frac{du}{dx} \tag{10}$$

$$\frac{d}{dx} \tanh u = \text{sech}^2\, u \cdot \frac{du}{dx} \tag{11}$$

$$\frac{d}{dx}\coth u = -\operatorname{csch}^2 u \cdot \frac{du}{dx} \tag{12}$$

$$\frac{d}{dx}\operatorname{sech} u = -\operatorname{sech} u \cdot \tanh u \cdot \frac{du}{dx} \tag{13}$$

$$\frac{d}{dx}\operatorname{csch} u = -\operatorname{csch} u \cdot \coth u \cdot \frac{du}{dx} \tag{14}$$

Example 1 For $f(x) = x \operatorname{sech}^2 \sqrt{x}$,

$$\begin{aligned} f'(x) &= 1 \cdot \operatorname{sech}^2 \sqrt{x} + x \cdot \frac{d}{dx}(\operatorname{sech}^2 \sqrt{x}) \\ &= \operatorname{sech}^2 \sqrt{x} + x\left\{2 \operatorname{sech} \sqrt{x} \cdot (-\operatorname{sech} \sqrt{x} \cdot \tanh \sqrt{x}) \cdot \frac{1}{2\sqrt{x}}\right\} \\ &= \operatorname{sech}^2 \sqrt{x} - \sqrt{x} \operatorname{sech}^2 \sqrt{x} \tanh \sqrt{x}. \end{aligned}$$

(Note that the derivative of sech x is $-\operatorname{sech} x \tanh x$, not $\operatorname{sech} x \tanh x$ as the analogy with sec x would suggest.) ■

Integration Formulas

Differentiation formulas (9)–(14) give the following integration formulas:

$$\int \cosh u \cdot du = \sinh u + C \tag{15}$$

$$\int \sinh u \cdot du = \cosh u + C \tag{16}$$

$$\int \operatorname{sech}^2 u \cdot du = \tanh u + C \tag{17}$$

$$\int \operatorname{csch}^2 u \cdot du = -\coth u + C \tag{18}$$

$$\int \operatorname{sech} u \tanh u \, du = -\operatorname{sech} u + C \tag{19}$$

$$\int \operatorname{csch} u \cdot \coth u \, du = -\operatorname{csch} u + C \tag{20}$$

Identities

The hyperbolic functions satisfy identities similar to those for the trigonometric functions. For example,

$$\begin{aligned} \cosh^2 x &= \left[\frac{1}{2}(e^x + e^{-x})\right]^2 \\ &= \frac{1}{4}(e^{2x} + 2 - e^{-2x}) \\ &= 1 + \frac{1}{4}(e^{2x} - 2 - e^{-2x}) \end{aligned}$$

$$= 1 + \left[\frac{1}{2}(e^x - e^{-x})\right]^2$$
$$= 1 + \sinh^2 x.$$

Thus,

$$\cosh^2 x - \sinh^2 x = 1. \tag{21}$$

Similarly, we can show that

$$\operatorname{sech}^2 x = 1 - \tanh^2 x \tag{22}$$

and

$$\operatorname{csch}^2 x = \coth^2 x - 1. \tag{23}$$

(Other identities are presented in the exercise set.)

Example 2 Show that the function $y = \tanh(ax)$ is a solution of the differential equation

$$\frac{dy}{dx} = a(1 - y^2).$$

Solution: By formula (11), the derivative of the given function is

$$\frac{dy}{dx} = \frac{d}{dx}\tanh(ax) = a \operatorname{sech}^2(ax).$$

Thus, with identity (22) we obtain

$$\frac{dy}{dx} = a \operatorname{sech}^2(ax)$$
$$= a[1 - \tanh^2(ax)]$$
$$= a(1 - y^2).$$

■

Inverses

We have already verified that the hyperbolic sine function is increasing for all values of x. This condition guarantees that $y = \sinh x$ is one-to-one. That is, each number y corresponds to precisely one value of x via the equation $y = \sinh x$. We may therefore define the **inverse** of the hyperbolic sine function $\sinh^{-1} x$ to be the function that reverses this correspondence. In other words

$$y = \sinh^{-1} x \quad \text{if and only if} \quad x = \sinh y. \tag{24}$$

We can obtain an explicit formulation for the inverse hyperbolic sine function by combining equations (3) and (24): If

$$y = \sinh^{-1} x, \qquad \text{then} \qquad x = \sinh y = \frac{1}{2}(e^y - e^{-y})$$

so

$$e^y - 2x - e^{-y} = 0. \tag{25}$$

Multiplying both sides of (25) by e^y gives

$$e^{2y} - 2xe^y - 1 = 0. \tag{26}$$

Viewing (26) as a quadratic expression in the variable e^y and applying the quadratic formula gives

$$\begin{aligned} e^y &= \frac{2x \pm \sqrt{4x^2 + 4}}{2} \\ &= x \pm \sqrt{x^2 + 1}. \end{aligned}$$

Since e^y is never negative, and $\sqrt{x^2 + 1} > x$, the ambiguous sign $\pm$ must be $+$. Taking natural logs of both sides of the last equation now gives

$$y = \ln(x + \sqrt{x^2 + 1})$$

so

$$\sinh^{-1} x = \ln(x + \sqrt{x^2 + 1}), \qquad -\infty < x < \infty. \tag{27}$$

By suitably restricting domains where necessary, we may similarly define inverse functions for each of the remaining hyperbolic functions. By rewriting each hyperbolic function in terms of exponential functions, we can then obtain explicit formulas for these inverse functions.

The results are as follows:

$$\cosh^{-1} x = \ln(x + \sqrt{x^2 - 1}), \qquad x \geq 1 \tag{28}$$

$$\operatorname{sech}^{-1} x = \ln\left(\frac{1 + \sqrt{1 - x^2}}{x}\right), \qquad 0 < x \leq 1 \tag{29}$$

$$\operatorname{csch}^{-1} x = \ln\left(\frac{1}{x} + \frac{\sqrt{1 + x^2}}{|x|}\right), \qquad x \neq 0 \tag{30}$$

$$\tanh^{-1} x = \frac{1}{2} \ln\left(\frac{1 + x}{1 - x}\right), \qquad |x| < 1 \tag{31}$$

$$\coth^{-1} x = \frac{1}{2} \ln\left(\frac{x + 1}{x - 1}\right), \qquad |x| > 1 \tag{32}$$

Derivative and integral formulas follow from equations (27)–(32) by applications of the corresponding rules for differentiating natural logarithms. For example, from (27) we have

$$\begin{aligned} \frac{d}{dx} \sinh^{-1} x &= \frac{d}{dx} \ln(x + \sqrt{x^2 + 1}) \\ &= \frac{1 + \dfrac{1}{2\sqrt{x^2 + 1}}(2x)}{x + \sqrt{x^2 + 1}} \\ &= \frac{\sqrt{x^2 + 1} + x}{\sqrt{x^2 + 1}(x + \sqrt{x^2 + 1})} \\ &= \frac{1}{\sqrt{x^2 + 1}}. \end{aligned}$$

Thus

$$\frac{d}{dx} \sinh^{-1} u = \frac{1}{\sqrt{u^2 + 1}} \cdot \frac{du}{dx}. \tag{33}$$

We leave as exercises the formulas

$$\frac{d}{dx}\cosh^{-1} u = \frac{1}{\sqrt{u^2-1}}\cdot\frac{du}{dx}, \tag{34}$$

$$\frac{d}{dx}\tanh^{-1} u = \frac{1}{1-u^2}\cdot\frac{du}{dx}, \quad |u| < 1 \tag{35}$$

$$\frac{d}{dx}\coth^{-1} u = \frac{1}{1-u^2}\cdot\frac{du}{dx}, \quad |u| > 1 \tag{36}$$

$$\frac{d}{dx}\operatorname{sech}^{-1} u = \frac{-1}{u\sqrt{1-u^2}}\cdot\frac{du}{dx}, \quad 0 < u < 1 \tag{37}$$

$$\frac{d}{dx}\operatorname{csch}^{-1} u = \frac{-1}{u\sqrt{1+u^2}}\cdot\frac{du}{dx}, \quad u > 0 \tag{38}$$

By now you have no doubt realized that the hyperbolic functions and their inverses are really not "new" functions. Rather, they are simply combinations of exponential and logarithmic functions. However, these particular combinations occur often in certain kinds of applications (such as those in the next section). Because the differentiation and integration formulas for the hyperbolic functions are simpler than those for the corresponding exponential and logarithmic forms, it is convenient to use them as shorthand notation.

Exercise Set 10.6

In Exercises 1–10, find the value of the indicated expression.

1. $\sinh 0$
2. $\sinh 1$
3. $\cosh(\ln 2)$
4. $\sinh(\ln 4)$
5. $\tanh(\ln 2)$
6. $\operatorname{sech}(1)$
7. $\coth(\ln 4)$
8. $\operatorname{csch}(\ln \pi^3)$
9. $\sinh^{-1}(1)$
10. $\tanh^{-1}(1/2)$

In Exercises 11–32, find the derivative of the given function.

11. $y = \sinh 2x$
12. $f(x) = x \cosh x$
13. $f(t) = \sin x \tanh x$
14. $y = \sinh^3(1 - x^2)$
15. $f(x) = \sqrt{\cosh(4x)}$
16. $y = \dfrac{1 - \cosh x}{1 + \cosh x}$
17. $y = \dfrac{1}{\cosh x}$
18. $f(x) = \ln(\tanh x^2)$
19. $f(x) = e^x \operatorname{csch} x^2$
20. $x + \cosh xy = y$
21. $y = \sinh^{-1} 2x$
22. $f(x) = \cosh^{-1}(1 + x^2)$
23. $f(s) = \tanh^{-1} s^2$
24. $y = \sqrt{\sinh^{-1} x}$
25. $f(x) = \ln \cosh^{-1} \pi x$
26. $y = \cosh^{-1} e^{x^2}$
27. $y = \cosh x \cosh x^2$
28. $f(x) = \ln(1 + \cosh \pi x)$
29. $y = \sinh \ln x^2$
30. $y = \sqrt{1 + \cosh^2 3x}$
31. $y = \tanh \sqrt{1 + x^2}$
32. $y = x \sinh^{-1}(1 + x^2)$

In Exercises 33–48, find the integrals

33. $\displaystyle\int_1^2 \frac{1}{\sqrt{1 + x^2}}\,dx$
34. $\displaystyle\int_1^2 x \sinh x^2\,dx$
35. $\displaystyle\int \tanh^2(2x)\,dx$
36. $\displaystyle\int e^x \sinh x\,dx$
37. $\displaystyle\int_0^2 \sinh x \cosh x\,dx$
38. $\displaystyle\int_0^1 \frac{1}{4 - x^2}\,dx$
39. $\displaystyle\int_2^6 \frac{dx}{\sqrt{x^2 - 4}}$
40. $\displaystyle\int_2^4 \frac{dx}{2 - x^2}$
41. $\displaystyle\int_0^1 \frac{1}{\sqrt{9x^2 + 25}}\,dx$
42. $\displaystyle\int_1^2 \frac{dx}{x\sqrt{4 + x^2}}$
43. $\displaystyle\int \frac{\sinh x}{\cosh x}\,dx$
44. $\displaystyle\int \frac{1}{\sinh^2 x}\,dx$
45. $\displaystyle\int \frac{\cosh x}{\sqrt{\sinh x}}\,dx$
46. $\displaystyle\int \frac{\sinh x}{1 + \cosh^2 x}\,dx$

47. $\int \tanh^2 x \operatorname{sech}^2 x \, dx$

48. $\int \frac{1 - \tanh^2 x}{1 + \tanh^2 x} \, dx$

49. True or false? The hyperbolic sine function is a periodic function with period 2π.

Find the following limits.

50. $\lim_{x\to\infty} \frac{\sinh x}{x}$

51. $\lim_{x\to 0} \frac{\sinh x - x}{\tanh x - x}$

Verify the following identities.

52. $1 - \tanh^2 x = \operatorname{sech}^2 x$

53. $\cosh x \pm \sinh x = e^{\pm x}$

54. $\sinh 2x = 2 \sinh x \cosh x$

55. $\cosh 2x = 2 \sinh^2 x + 1$

Verify the addition formulas.

56. $\sinh(x + y) = \sinh x \cdot \cosh y + \sinh y \cdot \cosh x$.

57. $\cosh(x + y) = \cosh x \cdot \cosh y + \sinh x \cdot \sinh y$.

58. A flexible cable fastened at both ends hangs in the shape of the **catenary** $y = a \cosh(x/a)$. Find the length of the graph of the catenary $y = \cosh x$ from the point with x-coordinate 0 to the point with x-coordinate b.

59. Show that the functions $y_1 = \sinh(kx)$ and $y_2 = \cosh(kx)$ satisfy the differential equation $d^2y/dx^2 = k^2y$. Find two solutions of this equation that are not hyperbolic functions. How are these solutions related?

Find as many solutions as you can for the following differential equations.

60. $\frac{d^2y}{dx^2} - 4y = 0$

61. $y'' - k^2\pi^2 y = 0$

62. Verify that the function $y = A \sin x + B \cos x + C \sinh x + D \cosh x$ is a solution of the fourth order differential equation $\frac{d^4y}{dx^4} = y$.

63. Find the area of the region bounded by the graphs of $y = \coth x$, $y = \tanh x$, $x = \ln 2$ and $x = \ln 4$.

64. Find the volume of the solid obtained by rotating about the x-axis the region bounded above by the graph of $y = \cosh x$ and below by the x-axis, for $-\ln 2 \le x \le \ln 2$.

Find all relative extrema.

65. $f(x) = \cosh x - \sinh x$

66. $f(x) = 2 \cosh x + 5 \sinh x$

67. $f(x) = 5 \cosh x - 2 \sinh x$

In Exercises 68–70, find a second order differential equation satisfied by the function.

68. $f(x) = \cosh 2x$

69. $f(x) = \cosh 3x - 2 \sinh 3x$

70. $f(x) = A \cosh 6x + B \sinh 6x$.

71. Determine which of the hyperbolic trigonometric functions are odd and which are even.

72. Show that $e^x = \cosh x + \sinh x$. Find a similar expression for e^{-x}.

73. Verify differentiation formulas (10)–(14).

74. Verify identity (23).

75. According to the definition of $\cosh^{-1} x$ in equation (28), what is the *principal branch* of the graph of $y = \cosh x$?

76. Verify that the function $y = \cosh x$ is increasing on $[0, \infty)$ and therefore one-to-one and invertible on this interval. Sketch the graph of $y = \cosh^{-1} x$ by reflecting the principal branch of $y = \cosh x$ in the line $y = x$.

77. Find the principal branch for each of the functions $\tanh x$, $\coth x$, $\operatorname{sech} x$, and $\operatorname{csch} x$, according to equations (29)–(32), and verify that each function is invertible when restricted to its principal branch.

78. Sketch the graphs of the inverse functions defined by equations (29)–(32).

79. Verify differentiation formulas (34)–(38).

80. Consider the situation in which one edge of a rectangular metal plate is held at a constant high temperature, while the opposite edge is held at a constant low temperature; the remaining two edges are insulated. In determining an equation that gives the temperature at each point on the plate, part of the problem involves solving the *second order* differential equation

$$\frac{d^2g(y)}{dy^2} - k^2g(y) = 0$$

where k is a nonzero constant. Verify that

$$g(y) = c_1 \sinh(ky) + c_2 \cosh(ky)$$

satisfies the differential equation for any choice of constants c_1 and c_2.

10.7 MODELLING OSCILLATORY MOTION (Optional)

Nature provides many examples of the phenomenon of oscillation. For example, a mass attached to the end of a spring will oscillate back and forth when pulled away from its equilibrium position and released. A similar phenomenon occurs when a violin string is plucked or when an automobile strikes a pothole. Sizes of certain biological populations are known to exhibit oscillatory behavior when the organisms depend upon seasonal food supplies or compete with other populations.

The sine and cosine functions appear frequently in mathematical descriptions of such phenomena. This is not surprising, since these functions exhibit the oscillatory behavior required in such models. The purpose of this section is to observe how these functions arise from basic physical or biological principles governing situations in which oscillation occurs.

The Harmonic Oscillator

Imagine a mass m lying on a frictionless surface and attached to a spring (see Figure 7.1). Recall from Section 8.3 that when the mass is moved x units from its natural rest position the spring will exert a restoring force, F_s, proportional to the displacement x (Hooke's Law). Since the restoring force acts in the direction opposite to the motion of the mass, we write

$$F_s = -kx \tag{1}$$

where the constant k is called the **spring constant** and depends on the particular spring.

Newton's second law of motion states that the sum of all forces acting on the mass must equal the product of the mass m and acceleration a. Since the net force* acting on the mass after it is displaced and released is F_s, we can write Newton's law as

$$F_s = ma, \qquad a = \text{acceleration}.$$

Since the variable x represents the displacement of the mass from its rest position, the time derivative dx/dt is the velocity of the mass, and the second derivative d^2x/dt^2 is its acceleration. Assuming the displacement x to be a twice differentiable function of t, we can rewrite Newton's law as

$$F_s = m \cdot \frac{d^2x}{dt^2}. \tag{2}$$

Combining equations (1) and (2) now gives the differential equation

$$m \cdot \frac{d^2x}{dt^2} = -kx,$$

or

$$\boxed{\frac{d^2x}{dt^2} + \frac{k}{m} \cdot x = 0.} \tag{3}$$

*The force of gravity is exactly counteracted by the supporting force exerted by the surface, so the net *vertical* force is zero.

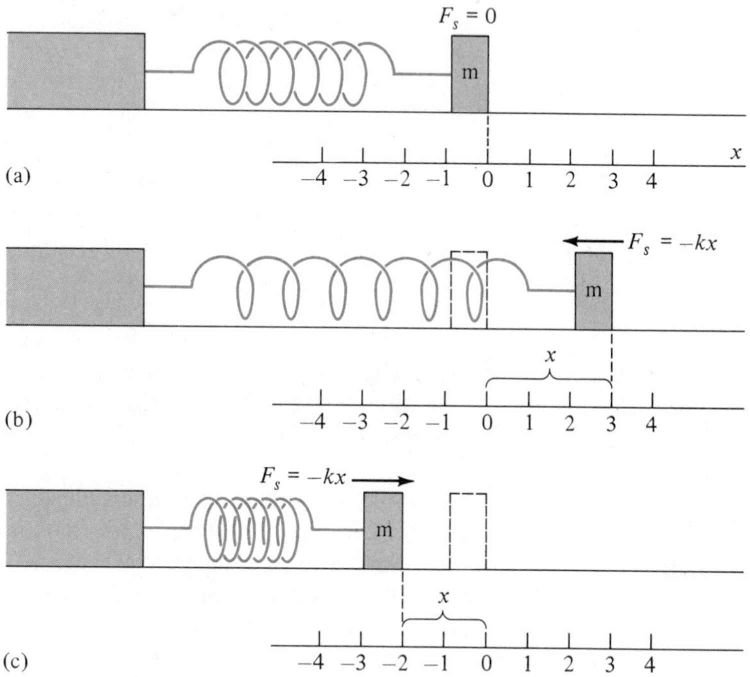

Figure 7.1 Mass-spring system (a) at equilibrium, (b) stretched, and (c) compressed. Displacement, x, is measured with respect to a particular point on the block.

Equation (3) is called the **harmonic oscillator equation,** and it provides the desired mathematical model for the motion of the mass for the situation described above. Of course, if equation (3) is to be a useful model we must be able to (a) find solutions of the model and (b) show that they exhibit the desired oscillatory behavior.

We observe that both $x_1 = \sin\sqrt{\frac{k}{m}}\,t$ and $x_2 = \cos\sqrt{\frac{k}{m}}\,t$ are solutions of (3), since

$$\frac{d^2x_1}{dt} = \frac{d}{dt}\left(\sqrt{\frac{k}{m}} \cdot \cos\sqrt{\frac{k}{m}}\,t\right) = -\frac{k}{m}\sin\sqrt{\frac{k}{m}}\,t = -\frac{k}{m}x_1$$

and

$$\frac{d^2x_2}{dt} = \frac{d}{dt}\left(-\sqrt{\frac{k}{m}}\sin\sqrt{\frac{k}{m}}\,t\right) = -\frac{k}{m}\cos\sqrt{\frac{k}{m}}\,t = -\frac{k}{m}x_2.$$

Thus

$$\frac{d^2x_1}{dt^2} + \frac{k}{m}x_1 = 0 \quad \text{and} \quad \frac{d^2x_2}{dt^2} + \frac{k}{m}x_2 = 0.$$

Moreover, any combination of these two solutions of the form

$$x(t) = A\cos\sqrt{\frac{k}{m}}\,t + B\sin\sqrt{\frac{k}{m}}\,t \tag{4}$$

is also a solution of (3), since

$$\frac{dx}{dt} = -A\sqrt{\frac{k}{m}}\sin\sqrt{\frac{k}{m}}\,t + B\sqrt{\frac{k}{m}}\cos\sqrt{\frac{k}{m}}\,t, \qquad \text{and}$$

$$\frac{d^2x}{dt^2} = -A\left(\frac{k}{m}\right)\cos\sqrt{\frac{k}{m}}\,t - B\left(\frac{k}{m}\right)\sin\sqrt{\frac{k}{m}}\,t$$

$$= -\left(\frac{k}{m}\right)x(t).$$

The solution (4) is referred to as the *general solution* of the model given by equation (3). Our remaining task is to find values of the constants A and B so that we can evaluate (4) for any particular time t. We do this by imposing two **initial conditions** on the problem (that is, we specify the values of the initial displacement $x(0)$ and the initial velocity $x'(0)$).

REMARK: The **order** of a differential equation is the order of the highest derivative appearing in it; equation (3) is a second-order equation. The number of arbitrary constants that occur in the general solution of a differential equation must equal the order of the equation. In order to solve an nth-order differential equation completely, n initial conditions must be specified.

Example 1 An 8-kg mass is attached to a spring as in Figure 7.1. The spring constant is $k = 2$ N/m. Find the resulting motion if the mass is displaced 10 cm from its equilibrium position and released from rest (zero initial velocity).

Strategy

Solution

Write differential equation (3), using given values of k and m.

Here $\frac{k}{m} = \frac{2}{8} = \frac{1}{4}$, so the differential equation (3) is

$$\frac{d^2x}{dt^2} + \frac{1}{4}x = 0.$$

Write general solution from (4).

From (4), with $\sqrt{k/m} = 1/2$, the general solution is

$$x(t) = A\cos\frac{t}{2} + B\sin\frac{t}{2}. \tag{5}$$

Set $x(0) = 10\text{ cm} = \frac{1}{10}$ m to find one equation in A and B.

To find the values of A and B, we apply the given *initial conditions*

(i) $x(0) = \frac{1}{10}$ $\left(\text{initial displacement 10 cm} = \frac{1}{10}\text{ m}\right)$, and

(ii) $x'(0) = 0$ (initial velocity zero).

From equation (5) and condition (i) we have

$$\frac{1}{10} = A\cos 0 + B\sin 0 = A,$$

so $A = \frac{1}{10}$. Since, from (5)

Differentiate $x(t)$ to obtain equation for $x'(t)$.

$$\frac{dx}{dt} = -\frac{A}{2}\sin\frac{t}{2} + \frac{B}{2}\cos\frac{t}{2} \tag{6}$$

condition (ii) gives

Set $x'(0) = 0$ to obtain second equation in A and B.

$$0 = -\frac{A}{2}\sin 0 + \frac{B}{2}\cos 0 = \frac{B}{2}.$$

Thus, $B = 0$. The solution is therefore

$$x(t) = \frac{1}{10}\cos\frac{t}{2}\ \text{(meters)}.$$

■

Example 2 Repeat Example 1, except that the mass now is released with an initial velocity of 3 m/sec.

Strategy

Same solution as for Example 1 except that the condition $x'(0) = 0$ is replaced by the condition $x'(0) = 3$.

Solution

As in Example 1 we have

$$x(t) = A\cos\frac{t}{2} + B\sin\frac{t}{2}.$$

Also, we still have $x(0) = \frac{1}{10}$, so $A = \frac{1}{10}$.

However, the condition that the initial velocity is 3 m/sec means that $x'(0) = 3$. Equation (6) therefore gives

Set $x'(0) = 3$ and solve for B.

$$3 = -\frac{A}{2}\sin 0 + \frac{B}{2}\cos 0 = \frac{B}{2}$$

so $B = 6$. In this case the solution is therefore

$$x(t) = \frac{1}{10}\cos\frac{t}{2} + 6\sin\frac{t}{2}.$$

■

The solution in Example 1 is clearly an oscillatory solution, since it is simply a multiple of $\cos\frac{t}{2}$. However, you may not so easily accept the claim that the solution in Example 2 is oscillatory. Is the sum of a sine function (oscillatory) and a cosine function (oscillatory) always oscillatory?

The answer to this question is provided by Exercise 9 at the end of this section: For any function of the form

$$x(t) = A\cos\omega t + B\sin\omega t$$

there exist constants R and c so that $x(t)$ can be written as

$$x(t) = R\cos(\omega t - c). \tag{7}$$

Thus, the solution to the *harmonic oscillator* equation

$$\frac{d^2x}{dt^2} + \omega^2 x = 0, \qquad \omega = \sqrt{\frac{k}{m}} \tag{8}$$

is indeed an oscillatory function, given by $x(t)$ in equation (7).

A Predator–Prey System

A seemingly quite different situation involving oscillatory phenomena occurs in a biological setting where a population of predators (foxes, say) depends upon a population of prey (rabbits) for its food supply. We assume that

(i) the rate of growth of the predator population depends upon the size of the prey population (i.e., more rabbits means more foxes);
(ii) the rate of decline of the prey population depends upon the size of the predator population (i.e., more foxes means *fewer* rabbits).

As the rabbit population grows, this increase in food supply for the foxes means that the foxes will increase in number, causing life to be more hazardous (and shorter!) for the rabbits. Thus the rabbit population dwindles and therefore, so does the fox population, and so on (see Figure 7.2).

In order to treat this situation mathematically, we will make several simplifying assumptions. First, we will denote the rabbit population by $x(t)$ and the fox population by $y(t)$, where t represents time, and we will assume that both of these are *continuous* functions that are at least twice differentiable for all t. Then we translate assumptions (i) and (ii) into the specific relations

$$\text{(i}'\text{)}\quad \frac{dy}{dt} = a(x - \overline{x}), \qquad a > 0, \qquad \overline{x} > 0$$

$$\text{(ii}'\text{)}\quad \frac{dx}{dt} = -b(y - \overline{y}), \qquad b > 0, \qquad \overline{y} > 0$$

Here $\overline{x}$ and $\overline{y}$ represent certain fixed "equilibrium" sizes for the two populations. Since $a > 0$, assumption (i′) states that the fox population will increase when the rabbit population is larger than $\overline{x}$ and decrease when the rabbit population falls below $\overline{x}$. Similarly, hypothesis (ii′) states that the rabbit population will decline when the fox population exceeds $\overline{y}$ and grow otherwise.

The variables x and y seem hopelessly intertwined in equations (i′) and (ii′), since dx/dt depends upon y and dy/dt depends upon x. (We say that the equations are **coupled** for this reason.) However, a simple "trick" allows us to find a solution.

Assuming $x(t)$ to be twice differentiable, we differentiate both sides of equation (ii′) to obtain the equation

$$\frac{d^2x}{dt^2} = -b\frac{dy}{dt}. \tag{9}$$

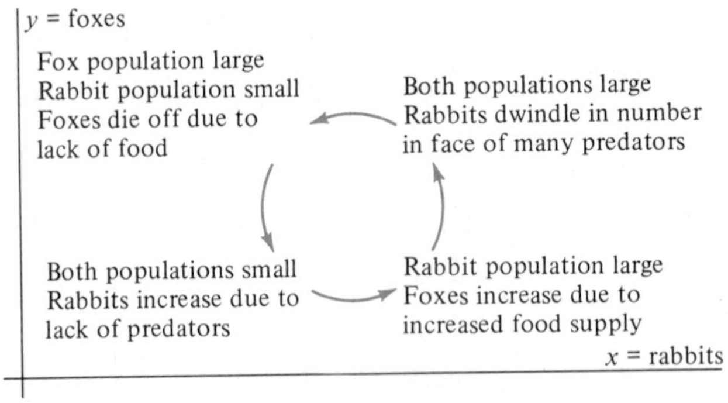

Figure 7.2

Combining equations (i′) and (9) now gives the equation

$$\frac{d^2x}{dt^2} + abx = ab\bar{x}. \tag{10}$$

Notice that equation (10) resembles the harmonic oscillator, equation (8), except that the right-hand side of (10) is the constant $ab\bar{x}$ rather than the constant zero. In other words, the difference between a solution x of (10) and a solution x_h of the harmonic oscillator is that when we form the sum of a multiple of x_h and its second derivative, the constant $ab\bar{x}$ remains. Since differentiation "kills" constants, we seek a solution of (10) in the form of the solution of the harmonic oscillator plus a constant, that is, we try

$$x = R\cos(\sqrt{ab}\,t - c) + A. \tag{11}$$

Then

$$\frac{dx}{dt} = -\sqrt{ab}\,R\sin(\sqrt{ab}\,t - c) \quad \text{and} \quad \frac{d^2x}{dt^2} = -abR\cos(\sqrt{ab}\,t - c).$$

If x is given by (11), we have

$$\begin{aligned}\frac{d^2x}{dt^2} + abx &= [-abR\cos(\sqrt{ab}\,t - c)] + ab[R\cos(\sqrt{ab}\,t - c) + A]\\ &= Aab.\end{aligned}$$

This shows that if we take $A = \bar{x}$ in equation (11), the resulting function

$$x(t) = R\cos(\sqrt{ab}\,t - c) + \bar{x} \tag{12}$$

is a solution of equation (10) for any constants R and c. We can now find y by differentiating equation (11) and using equation (ii′). We obtain

$$\begin{aligned}y(t) &= \bar{y} - \frac{1}{b}\frac{dx}{dt} && \text{(from equation (ii′))}\\ &= \bar{y} - \frac{1}{b}\{-\sqrt{ab}\,R\sin(\sqrt{ab}\,t - c)\} && \text{(from (12))}\\ &= \sqrt{\frac{a}{b}}\,R\sin(\sqrt{ab}\,t - c) + \bar{y}.\end{aligned}$$

We summarize our findings as follows.

The predator-prey model

$$\frac{dy}{dt} = a(x - \bar{x}), \qquad a > 0, \qquad \bar{x} > 0 \tag{13}$$

$$\frac{dx}{dt} = -b(y - \bar{y}), \qquad b > 0, \qquad \bar{y} > 0 \tag{14}$$

has solution

$$x(t) = \bar{x} + R\cos(\sqrt{ab}\,t - c) \tag{15}$$

$$y(t) = \bar{y} + \sqrt{\frac{a}{b}}\,R\sin(\sqrt{ab}\,t - c) \tag{16}$$

where R and c are arbitrary constants that can be determined by initial conditions.

The form of the solutions obtained in equations (15) and (16) makes it clear that the sizes of both populations will oscillate, since this is the nature of the sine and cosine functions. However, you must understand that the solutions obtained here depend heavily on the specific form assumed for the differential equations (13) and (14). That is, the success of the solutions in describing a real situation always depends on the validity of the equations used to model the phenomenon in question. In fact, more realistic models for predator–prey systems are usually more complicated than the model described here.

Example 3 Find a solution for the predator–prey model consisting of the differential equations

$$\frac{dy}{dt} = 2(x - 20)$$

$$\frac{dx}{dt} = -(y - 10)$$

if at time $t = 0$ there are $x(0) = 25$ rabbits and $y(0) = 12$ foxes present in an ecological niche.

Strategy
Determine the constants a, b, $\overline{x}$, and $\overline{y}$ from the given equations.

Solution
Comparing the equations above with (13) and (14) we see that

$$a = 2,\ b = 1,\ \overline{x} = 20,\ \text{and}\ \overline{y} = 10.$$

Obtain general form of solutions from (15) and (16).

From (15) and (16) the solutions are

$$x(t) = 20 + R\cos(\sqrt{2}\,t - c)$$
$$y(t) = 10 + \sqrt{2}\,R\sin(\sqrt{2}\,t - c).$$

Apply initial data to determine constants R and c.

To determine the constants R and c we can use the initial data:

$$25 = x(0) = 20 + R\cos(-c),$$

so

$$R\cos(-c) = 5. \tag{17}$$

Both (17) and (18) have factors of R. Dividing one equation by the other eliminates R. (Alternatively, solve both equations for R and equate.)

Also,

$$12 = y(0) = 10 + \sqrt{2}\,R\sin(-c),$$

so

$$R\sin(-c) = \frac{2}{\sqrt{2}}. \tag{18}$$

Dividing (18) by (17) then gives

$$\frac{R\sin(-c)}{R\cos(-c)} = \tan(-c) = \frac{2}{5\sqrt{2}}$$

so

$$c = -\tan^{-1}\left(\frac{2}{5\sqrt{2}}\right) \approx -.276 \text{ radians.}$$

Return known value of c to either (17) or (18) to solve for R.

Using this value for c in line (17) allows us to solve for R as

$$R = \frac{5}{\cos(.276)} \approx 5.2.$$

The solutions are therefore

$$x(t) = 20 + 5.2 \cos(\sqrt{2}\, t + .276)$$
$$y(t) = 10 + \sqrt{2}(5.2) \sin(\sqrt{2}\, t + .276).$$

These solutions are graphed in Figure 7.3. ■

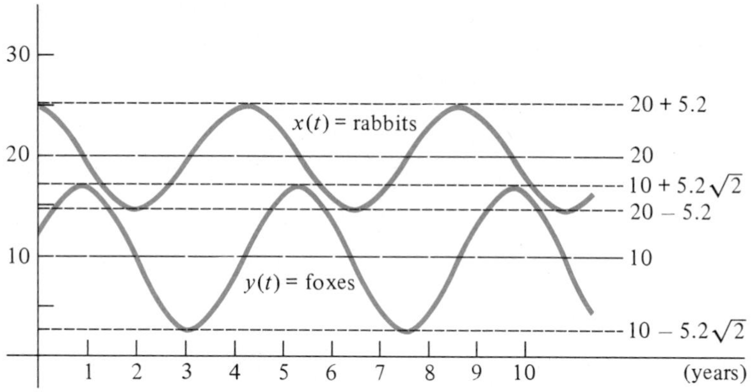

Figure 7.3 Oscillatory nature of predator-prey populations.

Linear Second Order Differential Equations

The general linear second order differential equation is an equation of the form

$$a\frac{dx^2}{dt^2} + b\frac{dx}{dt} + cx = f(t) \tag{19}$$

where a, b, c, and $f(t)$ may be either constants or functions of t. While a full treatment of equation (19) must await our development of the complex exponential function, we can summarize here the special cases we have handled thus far.

1. The equation

$$\frac{d^2x}{dt^2} + \omega^2 x = 0, \qquad \omega^2 > 0 \tag{20}$$

is the *harmonic oscillator,* with solutions of the form

$$x(t) = R\cos(\omega t - c), \qquad R,\ c \text{ constants.}$$

2. The equation

$$\frac{d^2x}{dt^2} + \omega^2 x = A \tag{21}$$

is the harmonic oscillator equation modified by a nonzero right-hand side. It has solutions

$$x(t) = \frac{A}{\omega^2} + R\cos(\omega t - c), \qquad A, R, c \text{ constants.}$$

3. The equation

$$\frac{d^2x}{dt^2} - \omega^2 x = 0 \tag{22}$$

is *not* the harmonic oscillator equation, since the sign of the term $\omega^2 x$ is negative. We have already seen (Section 10.6) that equations of this type have solutions $y_1 = Ae^{\omega t}$ and $y_2 = Be^{-\omega t}$. By combining such solutions we can also write solutions as $y_3 = C\sinh(\omega t)$ and $y_4 = D\cosh(\omega t)$.

Notice the dramatic difference between the solutions of equation (20) and those of equation (22): by simply changing a sign in the differential equation, one changes the solutions from bounded oscillating solutions to solutions that grow or decay exponentially with time.

Example 4 Find a solution of the differential equation

$$\frac{d^2x}{dt^2} + 9x = 3$$

satisfying the initial conditions $x(0) = 1$, $x'(0) = 2$.

Solution: The equation is of the type (21), so a solution will have the form

$$x(t) = \frac{1}{3} + R\cos(3t - c).$$

Applying the condition $x(0) = 1$ gives the equation

$$1 = \frac{1}{3} + R\cos(-c). \tag{23}$$

To apply the second condition, we first calculate

$$x'(t) = -3R\sin(3t - c).$$

Then the initial condition $x'(0) = 2$ gives

$$2 = -3R\sin(-c). \tag{24}$$

Solving both (23) and (24) for R and equating the results gives

$$\frac{2/3}{\cos(-c)} = \frac{2}{-3\sin(-c)}$$

so

$$\tan(-c) = -1,$$

or

$$c = -\tan^{-1}(-1) \approx \frac{\pi}{4}.$$

From (24) we can now solve for R:

$$R = \frac{2}{-3\sin(-\pi/4)} = \frac{2}{-3(-\sqrt{2}/2)} = \frac{2\sqrt{2}}{3}$$

The desired solution is therefore

$$x(t) = \frac{1}{3} + \frac{2\sqrt{2}}{3}\cos(3t - \pi/4).$$ ■

In this section we have seen how mathematical models can be developed in two "real world" applications—the spring-weight problem and the predator-prey problem. Although we do not have room for many examples of the kind of modelling discussed in this section, it is important to present these few examples and to summarize again the essential steps in using mathematics to solve applied problems:

(1) Identify the underlying physical or biological law(s) governing the problem. (Hooke's and Newton's laws for the spring problem, the relationships between growth rates and population sizes for the predator-prey problem.)
(2) Formulate a mathematical model, or problem, reflecting the physical or biological problem. (In both cases the mathematical model consisted of one or two differential equations.)
(3) Solve the mathematical model or problem. (In both cases, strictly calculus.)
(4) Interpret the mathematical solutions in terms of the original problem. (Masses attached to springs oscillate, and so do populations involved in predator-prey circumstances.)

Exercise Set 10.7

In each of Exercises 1–4, find a solution of the differential equation of the form $x(t) = A\cos\omega t + B\sin\omega t$.

1. $\frac{d^2x}{dt^2} + 16x = 0$
2. $3x''(t) + 12x(t) = 0$
3. $x''(t) = -9x(t)$
4. $\frac{d^2x}{dt^2} + 5x = 0$
5. Find a solution of Exercise 1 for which $x(0) = 1$ and $x'(0) = 0$.
6. Find a solution of Exercise 2 for which $x(0) = 2$ and $x'(0) = 1$.
7. Find a solution of Exercise 3 for which $x(0) = x'(0) = 0$.
8. Find a solution of Exercise 4 for which $x\left(\frac{\pi}{4\sqrt{5}}\right) = 1$ and $x'\left(\frac{\pi}{4\sqrt{5}}\right) = 0.$
9. Show that one can find constants R and c so that the function $x(t) = A\sin\omega t + B\cos\omega t$ can be written as $x(t) = R\cos(\omega t - c)$ as follows:
   a. Take $R = \sqrt{A^2 + B^2}$.
   b. Set $t = 0$ and equate the two expressions for $x(t)$ to obtain the equation $B = \sqrt{A^2 + B^2}\cos(-c)$. Since $\cos(-c) = \cos(c)$, conclude that
   $$\cos c = \frac{B}{\sqrt{A^2 + B^2}}. \tag{25}$$
   c. Differentiate the expressions for $x(t)$ to obtain the equation
   $$\omega A\cos\omega t - \omega B\sin\omega t = -\omega R\sin(\omega t - c).$$
   Again, set $t = 0$ to obtain the statement
   $$A = -\sqrt{A^2 + B^2}\sin(-c).$$
   Since $\sin(-c) = -\sin(c)$, conclude that
   $$\sin c = \frac{A}{\sqrt{A^2 + B^2}}. \tag{26}$$
   d. Explain why c is completely determined by equations (25) and (26), but not necessarily by either equation alone, and verify that the constants R and c indeed satisfy the identity $A\sin\omega t + B\cos\omega t = R\cos(\omega t - c)$.
10. Show that $\sin t + \cos t = \sqrt{2}\cos\left(t - \frac{\pi}{4}\right)$ for all t.

11. Show that $4 \sin 2t + 3 \cos 2t = 5 \cos(2t - .927)$ for all t.

In Exercises 12–15, express the given function in the form $x(t) = R \cos(\omega t - c)$.

12. $x(t) = \sin 3t + \cos 3t$

13. $x(t) = \sin 2t - \sqrt{3} \cos 2t$

14. $x(t) = \sqrt{3} \sin \pi t + \cos \pi t$

15. $x(t) = 6 \sin t - 6 \cos t$

16. Recall from our review of the trigonometric functions in Section 1.6 that, for the function $x(t) = R \cos(\omega t - c)$,

(i) the **amplitude** of the oscillation is $|R|$,

(ii) the **period** of the oscillation is $T = \left|\frac{2\pi}{\omega}\right|$, and

(iii) the **phase angle** for the oscillation is $\frac{c}{\omega}$.

a. Find the amplitude, period, and phase angle for the function in Exercise 10.
b. Same instructions as part (a) for Exercise 11.

17. A 0.5 kg mass is attached to a spring with spring constant $k = 2$ N/m, as in Figure 7.1. The mass is pulled 10 cm from the equilibrium position and released with zero velocity. Find an equation describing the resulting motion.

18. A 0.5 kg mass is attached to a spring with spring constant $k = 100$ N/m, as in Figure 7.1. The mass is tapped while in its equilibrium position so as to give it an initial velocity of 10 cm/sec.
a. Find a function describing the resulting motion.
b. When will the mass return to its equilibrium point for the first time?
c. What is the amplitude of the resulting motion?

19. The **frequency of oscillation** for a function of the form $x(t) = R \cos(\omega t - c)$ is defined to be $f = \frac{1}{T}$, where T is the period (see Exercise 16). Thus, $f = \frac{\omega}{2\pi}$. Find the frequency of oscillation for each of the functions in Exercises 12–15.

20. True or false? In the spring-mass problem the amplitude of the oscillation depends upon the initial conditions. Explain.

21. What are some of the details we have overlooked in describing the spring-mass problem? (For example, can a mass attached to a spring 20 cm long be given an initial displacement of 25 cm? What about the requirements of Hooke's Law concerning the elasticity of the spring?)

22. Find the solution of the predator-prey system

$$\frac{dy}{dt} = x - 25$$

$$\frac{dx}{dt} = -2(y - 16)$$

if $x(0) = y(0) = 20$.

23. What is the solution of the predator-prey system in Exercise 22 under the initial conditions $x(0) = 25$, $y(0) = 16$?

24. What restrictions must be placed on the initial conditions associated with the predator-prey system described by equations (13)–(16)?

25. Discuss the long-term behavior of the solutions of the predator-prey system

$$\frac{dy}{dt} = ax, \qquad a > 0$$

$$\frac{dx}{dt} = -by, \qquad b > 0.$$

26. Sketch the graph of the solution of the initial value problem

$$\frac{d^2x}{dt^2} + kx = 0$$

$$x(0) = 1$$

$$x'(0) = 0$$

for $k = 0, \pm 1, \pm 4, \pm 9$. What can you conclude about this solution as
a. $k \to \infty$?
b. $k \to -\infty$?
c. $k \to 0^+$?
d. $k \to 0^-$?

SUMMARY OUTLINE OF CHAPTER 10

- $\int \sin u \, du = -\cos u + C$

$\int \cos u \, du = \sin u + C$

$\int \sec^2 u \, du = \tan u + C$

$\int \csc^2 u \, du = -\cot u + C$

$\int \sec u \tan u \, du = \sec u + C$

$\int \csc u \cot u \, du = -\csc u + C$

$\int \tan u\,du = \ln|\sec u| + C, \qquad \cos u \neq 0$

$\int \cot u\,du = \ln|\sin u| + C, \qquad \sin u \neq 0$

$\int \sec u\,du = \ln|\sec u + \tan u| + C$

$\int \csc u\,du = \ln|\csc u - \cot u| + C$

- To integrate $\int \sin^m x \cos^n x\,dx$:

(i) If one of m or n is odd (say n), make u-substitution $u = \sin x$, so $du = \cos x\,dx$. Change remaining even factors of $\cos x$ to factors of $\sin x$ using the identity $\cos^2 x = 1 - \sin^2 x$.

(ii) If both n and m are even, use the identities $\sin^2 x = \frac{1}{2} - \frac{1}{2}\cos 2x$; $\cos^2 x = \frac{1}{2} + \frac{1}{2}\cos 2x$.

Use a similar strategy in integrals of the form

$$\int \sec^n x \tan^m x\,dx.$$

- The **hyperbolic trigonometric functions** are as follows:

$$\sinh x = \frac{1}{2}(e^x - e^{-x})$$

$$\cosh x = \frac{1}{2}(e^x + e^{-x})$$

$$\tanh x = \frac{\sinh x}{\cosh x}$$

$$\coth x = \frac{\cosh x}{\sinh x}$$

$$\operatorname{sech} x = \frac{1}{\cosh x}$$

$$\operatorname{csch} x = \frac{1}{\sinh x}$$

- The **inverse hyperbolic trigonometric** functions are as follows:

$\sinh^{-1} x = \ln(x + \sqrt{x^2 + 1}), \qquad -\infty < x < \infty$

$\cosh^{-1} x = \ln(x + \sqrt{x^2 - 1}), \qquad x \geq 1$

$\operatorname{sech}^{-1} x = \ln\left(\frac{1 + \sqrt{1 - x^2}}{x}\right), \qquad 0 < x \leq 1$

$\operatorname{csch}^{-1} x = \ln\left(\frac{1}{x} + \frac{\sqrt{1 + x^2}}{|x|}\right), \qquad x \neq 0$

$\tanh^{-1} x = \frac{1}{2}\ln\left(\frac{1 + x}{1 - x}\right), \qquad |x| < 1$

$\coth^{-1} x = \frac{1}{2}\ln\left(\frac{x + 1}{x - 1}\right), \qquad |x| > 1$

$\frac{d}{dx}\sinh^{-1} u = \frac{1}{\sqrt{u^2 + 1}} \cdot \frac{du}{dx}$

$\frac{d}{dx}\cosh^{-1} u = \frac{1}{\sqrt{u^2 - 1}} \cdot \frac{du}{dx}$

$\frac{d}{dx}\tanh^{-1} u = \frac{1}{1 - u^2} \cdot \frac{du}{dx}, \qquad |u| < 1$

$\frac{d}{dx}\coth^{-1} u = \frac{1}{1 - u^2} \cdot \frac{du}{dx}, \qquad |u| > 1$

$\frac{d}{dx}\operatorname{sech}^{-1} u = \frac{-1}{u\sqrt{1 - u^2}} \cdot \frac{du}{dx}, \qquad 0 < |u| < 1$

$\frac{d}{dx}\operatorname{csch}^{-1} u = \frac{-1}{|u|\sqrt{1 + u^2}} \cdot \frac{du}{dx}, \qquad u > 0$

- The differential equation $\frac{d^2x}{dx^2} + \omega^2 x = 0$ is referred to as the **harmonic oscillator** equation. Its solutions have the form

$x(t) = R\cos(\omega t - c), \qquad R, c$ constants.

- A system of differential equations for the **predator-prey** problem described in Section 10.7 is

$$\frac{dy}{dt} = a(x - \bar{x})$$

$$\frac{dx}{dt} = -b(y - \bar{y})$$

where a, b, $\bar{x}$, and $\bar{y}$ are positive constants. The solutions of this system have the form

$x(t) = \bar{x} + R\cos(\sqrt{ab}\cdot t - c),$

$y(t) = \bar{y} + \sqrt{\frac{a}{b}}\,R\sin(\sqrt{ab}\cdot t - c).$

- The **inverse trigonometric functions** are defined as follows:

$y = \tan^{-1} x$ if and only if $x = \tan y, \qquad -\infty < x < \infty, \qquad -\pi/2 < y < \pi/2$
$y = \sin^{-1} x$ if and only if $x = \sin y, \qquad -1 \leq x \leq 1, \qquad -\pi/2 \leq y \leq \pi/2$
$y = \cos^{-1} x$ if and only if $x = \cos y, \qquad -1 \leq x \leq 1, \qquad 0 \leq y \leq \pi$
$y = \sec^{-1} x$ if and only if $x = \sec y, \qquad |x| \geq 1, \qquad y \in [0, \pi/2) \cup [\pi, 3\pi/2)$
$y = \csc^{-1} x$ if and only if $x = \csc y, \qquad |x| > 1, \qquad y \in (0, \pi/2] \cup (\pi, 3\pi/2]$
$y = \cot^{-1} x$ if and only if $x = \cot y, \qquad -\infty < x < \infty, \qquad 0 < y < \pi.$

■ ***Differentiation and Integration Formulas for Inverse Trigonometric Functions:***

$$\frac{d}{dx}\tan^{-1} u = \frac{1}{1+u^2}\cdot\frac{du}{dx}; \qquad \int \frac{du}{1+u^2} = \tan^{-1} u + C$$

$$\frac{d}{dx}\sin^{-1} u = \frac{1}{\sqrt{1-u^2}}\cdot\frac{du}{dx}; \qquad \int \frac{du}{\sqrt{1-u^2}} = \sin^{-1} u + C$$

$$\frac{d}{dx}\sec^{-1} u = \frac{1}{u\sqrt{u^2-1}}\cdot\frac{du}{dx}; \qquad \int \frac{du}{u\sqrt{u^2-1}} = \sec^{-1} u + C, \qquad |u| \geq 1$$

$$\frac{d}{dx}\cot^{-1} u = \frac{-1}{1+u^2}\cdot\frac{du}{dx}$$

$$\frac{d}{dx}\cos^{-1} u = \frac{-1}{\sqrt{1-u^2}}\cdot\frac{du}{dx}$$

$$\frac{d}{dx}\csc^{-1} u = \frac{-1}{u\sqrt{u^2-1}}\cdot\frac{du}{dx}$$

■ ***Differentiation and Integration Formulas for Hyperbolic Trigonometric Functions:***

$$\frac{d}{dx}\sinh u = \cosh u\cdot\frac{du}{dx}; \qquad \int \cosh u\cdot du = \sinh u + C$$

$$\frac{d}{dx}\cosh u = \sinh u\cdot\frac{du}{dx}; \qquad \int \sinh u\cdot du = \cosh u + C$$

$$\frac{d}{dx}\tanh u = \operatorname{sech}^2 u\cdot\frac{du}{dx}; \qquad \int \operatorname{sech}^2 u\cdot du = \tanh u + C$$

$$\frac{d}{dx}\coth u = -\operatorname{csch}^2 u\cdot\frac{du}{dx}; \qquad \int \operatorname{csch}^2 u\cdot du = -\coth u + C$$

$$\frac{d}{dx}\operatorname{sech} u = -\operatorname{sech} u\cdot\tanh u\cdot\frac{du}{dx}; \qquad \int \operatorname{sech} u\cdot\tanh u\cdot du = -\operatorname{sech} u + C$$

$$\frac{d}{dx}\operatorname{csch} u = -\operatorname{csch} u\cdot\coth u\cdot\frac{du}{dx}; \qquad \int \operatorname{csch} u\cdot\coth u\cdot du = -\operatorname{csch} u + C$$

REVIEW EXERCISES—CHAPTER 10

In Exercises 1–9, evaluate the given expression.

1. $\sin^{-1}\left(\frac{\sqrt{3}}{2}\right)$

2. $\cos^{-1}(-1/2)$

3. $\tan^{-1}\sqrt{3}$

4. $\sin^{-1}(\sin \pi/4)$

5. $\cos^{-1}(\cos(-\pi/4))$

6. $\tan^{-1}(\sin \pi/2)$

7. $\cot\left(\sin^{-1}\frac{\sqrt{2}}{2}\right)$

8. $\tan(\sec^{-1}(-2))$

9. $\sinh(\ln 2)$

In Exercises 10–33, find the derivative of the given function.

10. $y = \tan^{-1} 3x$

11. $y = \sin^{-1}\sqrt{x}$

12. $f(x) = \tan^{-1}\left(\frac{2-x}{2+x}\right)$

13. $f(t) = \cot^{-1}(1-t^2)$

14. $y = \cosh(\ln x)$

15. $f(x) = x^2 \sinh(1-x)$

16. $y = \tan^{-1}\frac{\sqrt{x}}{2}$

17. $y = \csc(\cot 6x)$

18. $y = \ln^2(\cos^2 x^2)$

19. $f(x) = \sin^{-1}[\ln(2x+1)]$

20. $y = \sin^{-1}(\cos e^{-x})$

21. $f(x) = \frac{\tan^{-1} x}{1+x^2}$

22. $y = \dfrac{\sin^{-1} e^x}{1 + e^x}$

23. $f(x) = \sec^{-1} \sqrt{x^2 + 4}$

24. $y = \sqrt{\cosh x^2}$

25. $y = \dfrac{1}{\pi + \tanh x}$

26. $f(x) = \ln^2(\sinh x)$

27. $y = (\sinh x + \cos^{-1} 2x)^{1/5}$

28. $y = \ln|\tan x|$

29. $y = \ln(\sin^{-1} x)$

30. $f(x) = \dfrac{\tan 2x}{2 + \sec 2x}$

31. $f(x) = x \tanh^{-1}(\ln x)$

32. $y = x^2 \sinh^{-1}(e^x)$

33. $y = \ln \sqrt{\tanh^{-1}(x^2)}$

In Exercises 34–70, find the indicated integral.

34. $\displaystyle\int \frac{dx}{\sqrt{1 + 9x^2}}$

35. $\displaystyle\int \frac{dx}{\sqrt{4x^2 - 1}}$

36. $\displaystyle\int \frac{dx}{x\sqrt{4 + x^2}}$

37. $\displaystyle\int \sin^4 2x \cos 2x \, dx$

38. $\displaystyle\int \sec^5 x \tan x \, dx$

39. $\displaystyle\int_{\pi^2/4}^{\pi^2/16} \frac{\sin^2\sqrt{x}}{\sqrt{x}} \, dx$

40. $\displaystyle\int_0^{\pi/2} \sin^{5/2} x \cos x \, dx$

41. $\displaystyle\int_{-1}^{1} \frac{dx}{\sqrt{2 - x^2}}$

42. $\displaystyle\int x \csc^2 x^2 \, dx$

43. $\displaystyle\int_0^{\pi/4} \sqrt{\tan x} \sec^2 x \, dx$

44. $\displaystyle\int \frac{\sec^2 \sqrt{x} \tan \sqrt{x}}{\sqrt{x}} \, dx$

45. $\displaystyle\int \frac{1 + \sin^2 2x}{\cos^2 2x} \, dx$

46. $\displaystyle\int \tan(\sec x) \sec x \tan x \, dx$

47. $\displaystyle\int_0^{\pi/8} \tan^2(2x) \sec^2(2x) \, dx$

48. $\displaystyle\int \frac{x}{x^4 + 1} \, dx$

49. $\displaystyle\int \frac{dx}{\sqrt{4 - x^2}}$

50. $\displaystyle\int \frac{e^{-x}}{2 + e^{-2x}} \, dx$

51. $\displaystyle\int \frac{\cos^{-1} x}{\sqrt{1 - x^2}} \, dx$

52. $\displaystyle\int \frac{\sqrt{x}}{x^3 + 4} \, dx$

53. $\displaystyle\int \frac{e^{\sqrt{x}}}{\sqrt{x}(1 + e^{2\sqrt{x}})} \, dx$

54. $\displaystyle\int_1^2 \frac{x^2}{x^6 + 9} \, dx$

55. $\displaystyle\int \frac{\sqrt{\sin^{-1} x}}{\sqrt{1 - x^2}} \, dx$

56. $\displaystyle\int \frac{\sin x}{\sqrt{2 - \cos^2 x}} \, dx$

57. $\displaystyle\int_0^{\pi/9} \sec^3 2x \tan 2x \, dx$

58. $\displaystyle\int_0^{\pi/2} \tan^3\left(\frac{x}{2}\right) dx$

59. $\displaystyle\int_0^{\pi/3} \sec^4 x \, dx$

60. $\displaystyle\int \sec^3 x \tan^3 x \, dx$

61. $\displaystyle\int_0^{\pi/3} \cos x \cos 5x \, dx$

62. $\displaystyle\int_0^{\pi/4} \sin x \cos 2x \, dx$

63. $\displaystyle\int \sqrt{\cos x} \sin^3 x \, dx$

64. $\displaystyle\int \cot^4 2x \, dx$

65. $\displaystyle\int \frac{\sin^3 x}{\cos^2 x} \, dx$

66. $\displaystyle\int_0^{\pi/2} \sin^2 x \cos^4 x \, dx$

67. $\displaystyle\int \frac{\cosh x}{1 + \sinh x} \, dx$

68. $\displaystyle\int x \coth x^2 \, dx$

69. $\displaystyle\int \sinh^3 x \, dx$

70. $\displaystyle\int \frac{e^{3x}}{\sqrt{9 + e^{6x}}} \, dx$

71. Find the area of the region bounded by the graph of $y = \sin^2 x$ and the x-axis between $x = 0$ and $x = \pi$.

72. Find the area of the region bounded by the graphs of $y = \sin^2 x$ and $y = \cos^2 x$ for x between $x = 0$ and $x = \pi/2$.

73. Find the volume of the solid generated by rotating about the x-axis the region bounded by the graph of $y = \tan x$ and the x-axis for $0 \le x \le \pi/4$.

74. Find the volume of the solid generated by rotating about the x-axis the region bounded by the graph of $y = \dfrac{1}{\sqrt{1 + x^2}}$ and the x-axis for $0 \le x \le 1$.

75. Find the volume of the solid generated by revolving about the y-axis the region bounded by the graph of $y = \dfrac{1}{1 + x^4}$ and the x-axis between $x = 0$ and $x = 2$.

76. Find the average value of the function $y = \cosh x$ on the interval $[-\ln 2, \ln 2]$.

77. Find dy/dx if $y = \cosh^{-1} x^2 y$.

78. Find the area bounded by the graph of $y = \cosh x$ and the x-axis for $-1 \le x \le 1$.

79. For the function $y = \sinh x \cosh x$, find all relative extrema, determine the concavity, and sketch the graph.

80. A particle moves along a line with velocity $v(t) = \sin^2 \pi t$. Find the distance travelled by the particle between times $t = 0$ and $t = 4$.

81. Use differentials to approximate $\sin^{-1}(0.48)$.

82. Prove that $\cosh x > \sinh x$ for all x.

83. Find the equation of the line tangent to the graph of $y = \sin^{-1} x^2$ at the point with x-coordinate $\sqrt{2}/2$.

84. An airplane is flying directly away from a radar station at an altitude of 3 kilometers and a ground speed of 400 km/hr. How fast is the angle of elevation of the tracking antenna decreasing when the airplane is directly over a point 4 kilometers from the radar station? (Use inverse trigonometric functions.)

85. Find the points where the line tangent to the graph of $y = \cot^{-1} x$ is parallel to the line with equation $x + 5y - 10 = 0$.

86. A particle moves along a line with acceleration $a(t) = -1 - \dfrac{2t}{(1 + t^2)^2}$ m/sec^2. Find the function $s(t)$ giving the distance travelled after t seconds if the particle starts at the origin with initial velocity 4 m/sec.

87. A small boat is tied to a rope which is connected to a windlass (crank). The windlass is mounted on the edge of a dock 10 meters above the water level. The windlass is pulling the boat ashore by winding in the rope at the rate of 1 meter per second. Find the rate at which the angle between the rope and the horizontal is increasing at the instant when 20 meters of rope remain out.

88. Why is the hyperbolic sine function continuous throughout its domain?

89. A particle moves along a line with acceleration $a(t) = \sin t + \cos t$. Describe the motion.

90. Prove that when a body falls from rest and encounters air resistance proportional to the square of its speed, the resulting velocity at time t is $v(t) = A \tanh(at)$ where A and a are positive constants, as follows:

a. Observe that two forces act on the body:

F_g = force due to gravity = mg (downward),
F_r = force due to air resistance = kv^2 (upward).

b. Apply Newton's second law to conclude

$$ma = F_g - F_r, \qquad \text{or} \qquad ma = mg - kv^2.$$

c. Since $a = \dfrac{dv}{dt}$, the equation becomes

$$m\frac{dv}{dt} = mg - kv^2.$$

d. Separate variables and integrate.

91. Find two solutions of the differential equation $x'' + 25x = 0$.

92. Find a solution of the differential equation $d^2x/dt^2 + 16x = 0$ of the form $x(t) = A \cos \omega t + B \sin \omega t$ for which $x(0) = 4$ and $x'(0) = 2$.

93. Find a solution of the differential equation $y'' + 2y = 0$ of the form $y = R \cos(\omega t - c)$ for which $y(0) = 1$ and $y'(0) = 0$.

94. A 16-kg mass is attached to a spring and is sitting on a horizontal frictionless surface. The spring constant is $k = 4$ N/m. Find the resulting motion if the mass is displaced 10 cm from its rest position and released with zero initial velocity.

95. Find a solution of the differential equation $x'' + 16x = 0$.

CHAPTER 11

TECHNIQUES OF INTEGRATION

11.1 INTRODUCTION

Up to this point we have referred to the function $F(x)$ as an **antiderivative** for the function $f(x)$ if $F'(x) = f(x)$, and we have written the most general antiderivative for $f(x)$ as

$$\int f(x)\,dx = F(x) + C. \tag{1}$$

In Chapter 6 we learned, from the Fundamental Theorem of Calculus, that the **definite integral** $\int_a^b f(x)\,dx$ of a continuous function $f(x)$ could be evaluated using an antiderivative $F(x)$ by the equation

$$\int_a^b f(x)\,dx = F(b) - F(a). \tag{2}$$

Because of this strong relationship between antiderivatives and definite integrals, we will henceforth refer to antiderivatives as **indefinite integrals.** Thus, the term "integration" and the phrase "finding the integral of $f(x)$" refer either to the indefinite integral in (1) or the definite integral in (2) above. The context will always make clear which meaning is intended.

For many functions the associated integrals are not obvious. The goal of this chapter is to develop a "toolbox" of techniques by which many of these integrals may be found. Unfortunately, you will see that the techniques for finding integrals are neither as straightforward nor as all-encompassing as the techniques developed in Chapter 3 for calculating derivatives. Skill at finding integrals involves a good sense of which types of functions arise from differentiating which other (or similar) types of functions. The best way to develop such skills is simply to tackle many different types of integrals and look for patterns among the integrals that you succeed in solving. However, the techniques presented here will give you a good start in this program by outlining several broad classes of integrals with similar solutions.

Before beginning the discussion of these techniques, we summarize the integration formulas with which we are already familiar.

Constants and Powers:

1. $\int a\,dx = ax + C$
2. $\int x^n\,dx = \dfrac{x^{n+1}}{n+1} + C, \qquad n \neq -1$
3. $\int \dfrac{1}{x}\,dx = \ln|x| + C$

Logarithmic and Exponential Functions:

4. $\int e^x\,dx = e^x + C$
5. $\int a^x\,dx = \dfrac{a^x}{\ln a} + C, \qquad a \neq 1, \qquad a > 0$

Trigonometric Functions:

6. $\int \sin x\,dx = -\cos x + C$
7. $\int \cos x\,dx = \sin x + C$
8. $\int \sec^2 x\,dx = \tan x + C$
9. $\int \csc^2 x\,dx = -\cot x + C$
10. $\int \sec x \tan x\,dx = \sec x + C$
11. $\int \csc x \cdot \cot x\,dx = -\csc x + C$
12. $\int \tan x\,dx = \ln|\sec x| + C$
13. $\int \cot x\,dx = \ln|\sin x| + C$
14. $\int \sec x\,dx = \ln|\sec x + \tan x| + C$
15. $\int \csc x\,dx = \ln|\csc x - \cot x| + C$

Algebraic Functions:

16. $\int \dfrac{dx}{1+x^2} = \tan^{-1} x + C$
17. $\int \dfrac{dx}{\sqrt{1-x^2}} = \sin^{-1} x + C$
18. $\int \dfrac{dx}{x\sqrt{x^2-1}} = \sec^{-1} x + C$

Hyperbolic Functions:

19. $\int \sinh x \, dx = \cosh x + C$

20. $\int \cosh x \, dx = \sinh x + C$

21. $\int \operatorname{sech}^2 x \, dx = \tanh x + C$

22. $\int \operatorname{csch}^2 x \, dx = -\coth x + C$

23. $\int \operatorname{sech} x \tanh x \, dx = -\operatorname{sech} x + C$

24. $\int \operatorname{csch} x \coth x \, dx = -\operatorname{csch} x + C$

In the endpapers of this text and in books of scientific tables, you will find many more types of integration formulas than those listed above. We will address the use of integral tables in Section 11.7. While such tables of integrals are of great practical value, you are encouraged not to rely too heavily on such aids for handling the integrals presented in the first six sections of this chapter. The ability to evaluate an integral by identifying and executing the appropriate integration technique is important in the material that lies ahead, as well as in further course work in differential equations, mathematical physics, and engineering mathematics.

The exercise set of this section offers additional practice in the use of the above integration formulas together with the technique of integration by ***u*-substitution** (see Section 5.9). Recall that an integral of the form

$$\int f(g(x)) \cdot g'(x) \, dx$$

may be evaluated by means of the substitution

$$u = g(x); \qquad du = g'(x) \, dx.$$

We obtain

$$\int f(g(x)) \cdot g'(x) \, dx = \int f(u) \, du.$$

Example 1 Find $\displaystyle\int \frac{x}{\sqrt{x^2 + 2}} \, dx.$

Strategy
Recognize that the denominator is a composite function and that the required power of x is present in the numerator.

Solution
We use the substitution

$$u = x^2 + 2; \qquad du = 2x \, dx.$$

Then $x \, dx = \dfrac{1}{2} du$, so we obtain

Substitute into the integral.

$$\int \frac{x}{\sqrt{x^2 + 2}} \, dx = \int \frac{1/2 \, du}{\sqrt{u}}$$

Evaluate the integral using formula (2).

$$= \frac{1}{2} \int u^{-1/2}\, du$$

$$= \frac{1}{2}(2u^{1/2}) + C$$

Substitute back.

$$= \sqrt{x^2 + 2} + C. \quad \blacksquare$$

Example 2 Evaluate the integral $\displaystyle\int_0^{\sqrt{2}/2} \frac{x}{\sqrt{1 - x^4}}\, dx$.

Strategy

We *cannot* use $u = 1 - x^4$ since this requires

$du = -4x^3\, dx,$

and only one factor of x is present. The fact that

$\dfrac{d}{dx}(x^2) = 2x$

suggests that we try $u = x^2$.

Change limits of integration by the same substitution formula $u = x^2$.

Evaluate the integral using formula (17).

Solution

We recognize the denominator as $\sqrt{1 - (x^2)^2}$, which suggests the derivative of the inverse sine function. We therefore let

$$u = x^2; \qquad du = 2x\, dx.$$

Then $x\, dx = \dfrac{1}{2}\, du.$

To find the new limits of integration we note that

$$u = 0^2 = 0 \quad \text{when} \quad x = 0, \quad \text{and}$$

$$u = \left(\frac{\sqrt{2}}{2}\right)^2 = \frac{1}{2} \quad \text{when} \quad x = \frac{\sqrt{2}}{2}.$$

Thus,

$$\int_0^{\sqrt{2}/2} \frac{x}{\sqrt{1 - x^4}}\, dx = \int_0^{1/2} \frac{\frac{1}{2}\, du}{\sqrt{1 - u^2}}$$

$$= \frac{1}{2} \int_0^{1/2} \frac{du}{\sqrt{1 - u^2}}$$

$$= \frac{1}{2} \sin^{-1} u \Big]_0^{1/2}$$

$$= \frac{1}{2} \sin^{-1}\left(\frac{1}{2}\right)$$

$$= \frac{1}{2} \cdot \frac{\pi}{6}$$

$$= \frac{\pi}{12}. \quad \blacksquare$$

Exercise Set 11.1

Evaluate the following integrals:

1. $\displaystyle\int x\sqrt{1 + x^2}\, dx$

2. $\displaystyle\int (x^2 - 3)\sqrt{x^3 - 9x}\, dx$

3. $\displaystyle\int \frac{x}{1 + x}\, dx$

4. $\displaystyle\int x2^{x^2}\, dx$

5. $\displaystyle\int_0^1 \sin \pi x\, dx$

6. $\displaystyle\int_0^1 \frac{x^4}{1 + x^5}\, dx$

7. $\displaystyle\int \frac{x^3}{\sqrt{1 - x^4}}\, dx$

8. $\displaystyle\int_0^1 \frac{x}{1 + x^4}\, dx$

9. $\displaystyle\int_0^1 \frac{e^x}{1+e^x}\,dx$

10. $\displaystyle\int \tan^2 x \sec^2 x\,dx$

11. $\displaystyle\int \frac{x+2}{x^2+4x+7}\,dx$

12. $\displaystyle\int \frac{\cos \pi x}{4+\sin \pi x}\,dx$

13. $\displaystyle\int_0^{1/4} \sec^2 \pi x e^{\tan \pi x}\,dx$

14. $\displaystyle\int x^{1/3}\sqrt{1+x^{4/3}}\,dx$

15. $\displaystyle\int_0^1 \frac{e^{\sqrt{x}}}{\sqrt{x}}\,dx$

16. $\displaystyle\int \frac{\cos x}{1+\sin^2 x}\,dx$

17. $\displaystyle\int_0^{\pi/3} \sin^3 x \cos x\,dx$

18. $\displaystyle\int e^{-\pi x}\,dx$

19. $\displaystyle\int \frac{e^x\,dx}{1+e^{2x}}$

20. $\displaystyle\int \sqrt{1-\cos 2x}\,dx$

21. $\displaystyle\int_0^{1/4} \sin^4 \pi x\,dx$

22. $\displaystyle\int x \tan x^2\,dx$

23. $\displaystyle\int_0^{\pi/4} \frac{dx}{\cos^2 x}$

24. $\displaystyle\int \frac{\sec^2 x\,dx}{\sqrt{1+2\tan x}}$

25. $\displaystyle\int_1^2 \frac{dx}{x^2+2x+1}$

26. $\displaystyle\int 2^x e^{1+x}\,dx$

27. $\displaystyle\int x \ln(x^2+\pi)\,dx$ (*Hint:* $\displaystyle\int \ln u\,du = u \ln|u| - u + C$.)

28. $\displaystyle\int_0^3 \frac{3^{\sqrt{x+1}}}{\sqrt{x+1}}\,dx$

29. $\displaystyle\int \csc^3 x \cot x\,dx$

30. $\displaystyle\int e^x \tan(1+e^x)\,dx$

31. $\displaystyle\int x \sec(\pi + x^2)\,dx$

32. $\displaystyle\int \frac{\cos x\,dx}{\sqrt{1-\sin^2 x}}$

33. $\displaystyle\int (3^x+5)^2\,dx$

34. $\displaystyle\int_e^{e^2} \frac{1}{x \ln x}\,dx$

35. $\displaystyle\int \frac{(1+3\sqrt{x})^3}{\sqrt{x}}\,dx$

36. $\displaystyle\int \frac{1}{\sqrt{x}(1+\sqrt{x})^2}\,dx$

37. $\displaystyle\int \frac{\sqrt{x}+1-(1/2)x^{-1/2}}{(1+\sqrt{x})^2}\,dx$

38. $\displaystyle\int \frac{x^3+5x^2+3x-9}{x-1}\,dx$

39. $\displaystyle\int \frac{\tan x^2 - 2x^2 \sec^2 x^2}{\tan^2 x^2}\,dx$

40. $\displaystyle\int \frac{3x^2}{(9+x^3)\sqrt{\ln(9+x^3)}}\,dx$

41. $\displaystyle\int (\sin x - \cos x)^3(\cos x + \sin x)\,dx$

11.2 INTEGRATION BY PARTS

Just as the technique of u-substitution is obtained from the Chain Rule, the formula for the technique of *integration by parts* may be obtained from the Product Rule. Recall that if the functions $f(x)$ and $g(x)$ are differentiable, the Product Rule states that

$$[f(x)g(x)]' = f(x)g'(x) + f'(x)g(x).$$

Solving this equation for $f(x)g'(x)$ gives

$$f(x)g'(x) = [f(x)g(x)]' - f'(x)g(x). \tag{1}$$

Now the meaning of the equals sign in equation (1) is that the functions on either side are the same. Thus, they must have the same indefinite integrals. That is,

$$\begin{aligned}\int f(x)g'(x)\,dx &= \int \{[f(x)g(x)]' - f'(x)g(x)\}\,dx \qquad (2)\\ &= \int [f(x)g(x)]'\,dx - \int f'(x)g(x)\,dx.\end{aligned}$$

Since $\int [f(x)g(x)]'\,dx = f(x)g(x) + C$, we can write equation (2) as

$$\boxed{\int f(x)g'(x)\,dx = f(x)g(x) - \int f'(x)g(x)\,dx.} \tag{3}$$

Equation (3) is referred to as the **integration by parts formula.** It is useful when integrating products of the form $f(x)g'(x)$. In such cases equation (3) allows us to replace the original integrand with a new one, which is the product of the *derivative* of the function $f(x)$ and an *integral* of the function $g'(x)$. Of course, the formula is useful only when the resulting integral is simpler to evaluate than the original integral. (Note that since equation (3) still involves integrals on both sides, there is no need to display a constant of integration with the term $f(x)g(x)$. When the integral $\int f'(x)g(x)\,dx$ is finally evaluated, a constant of integration must appear.)

Example 1 Use the integration by parts formula to find the integral $\int x \cos x \, dx$.

Strategy

Choose one factor of $x \cos x$ as $f(x)$, the other as $g'(x)$.

Solution

We view $x \cos x$ as the product $f(x)g'(x)$ where

$$f(x) = x, \qquad g'(x) = \cos x.$$

Find $f'(x)$, $g(x)$.

Then

$$f'(x) = 1, \qquad g(x) = \sin x,$$

so by equation (3)

Apply equation (3).

$$\int x \cos x \, dx = x \sin x - \int \sin x \, dx$$

$$= x \sin x + \cos x + C. \qquad \blacksquare$$

REMARK 1: We could have chosen $f(x) = \cos x$ and $g'(x) = x$ in Example 1. However, this would not have been wise since we would then have had $f'(x) = -\sin x$ and $g(x) = x^2/2$. Integration by parts then would have given

$$\int x \cos x \, dx = \frac{x^2}{2} \cos x - \int \frac{x^2}{2} \sin x \, dx. \tag{4}$$

This is not incorrect, but the integral on the right side of equation (4) is no simpler to solve than the one on the left. The general strategy in using equation (3) is to pick the factors $f(x)$ and $g'(x)$ so that

(i) the integral $g(x)$ can be found, and
(ii) the resulting integral $\int f'(x)g(x)\,dx$ is as simple to evaluate as possible.

REMARK 2: There is no need to introduce an arbitrary constant when finding an integral $g(x)$ for $g'(x)$, since (3) holds for *any* integral of $g'(x)$. The simplest choice is to let the arbitrary constant in $g(x)$ be zero.

There is a somewhat simpler form by which the integration by parts formula can be remembered. If we make the substitutions

$$u = f(x), \qquad dv = g'(x)\,dx$$

then

$$du = f'(x)\,dx, \qquad v = g(x)$$

and equation (3) becomes

$$\int u\,dv = uv - \int v\,du. \qquad (5)$$

Equation (5) is the Leibniz notation form for the integration by parts formula.

Example 2 Find $\int x^2 \ln x\,dx$.

Strategy
The derivative of $\ln x$ is a power of x, so take $u = \ln x$, $dv = x^2\,dx$.

Solution
We let

$$u = \ln x, \qquad dv = x^2\,dx,$$

so

Find du, v.

$$du = \frac{1}{x}\,dx, \qquad v = \frac{1}{3}x^3.$$

Apply equation (5).

Integrating by parts then gives

$$\begin{aligned}\int x^2 \ln x\,dx &= \frac{1}{3}x^3 \ln x - \int \frac{1}{3}x^3 \cdot \frac{1}{x}\,dx \\ &= \frac{1}{3}x^3 \ln x - \frac{1}{3}\int x^2\,dx \\ &= \frac{1}{3}x^3 \ln x - \frac{1}{9}x^3 + C.\end{aligned}$$

■

Often more than one application of the integration by parts formula is required, as the following example shows.

Example 3 Find $\int x^2 e^x\,dx$.

Strategy
Take $u = x^2$ since differentiation reduces the exponent by one.

Solution
We take

$$u = x^2, \qquad dv = e^x\,dx.$$

Then

$$du = 2x\,dx, \qquad v = e^x,$$

and integration by parts gives

Apply equation (5).

$$\int x^2 e^x\,dx = x^2 e^x - 2\int x e^x\,dx. \qquad (6)$$

In $\int x e^x\,dx$ again take $u = x$, since differentiation will yield simply $du = 1 \cdot dx$.

The integral on the right requires a second application of the parts formula: In

$$\int x e^x\,dx,$$

we take

$$u = x, \qquad dv = e^x\,dx,$$

so

$$du = dx, \qquad v = e^x.$$

Apply equation (5) again.

Then, a second integration by parts gives

$$\begin{aligned}\int xe^x\,dx &= xe^x - \int e^x\,dx \\ &= xe^x - e^x + C_1.\end{aligned} \tag{7}$$

Combine results.

Combining lines (6) and (7) we now have

$$\begin{aligned}\int x^2e^x\,dx &= x^2e^x - 2\{xe^x - e^x + C_1\} \\ &= e^x(x^2 - 2x + 2) + C\end{aligned}$$

where $C = -2C_1$. (Since C_1 is an arbitrary constant, C is also an arbitrary constant; we choose to write the answer in the simpler form.) ■

In the exercise set you will encounter integrals requiring even more than two applications of the parts formula. A different type of problem requiring two applications of the parts formula is the following.

Example 4 Find $\int e^x \cos x\,dx$.

Solution: This time there is no clue as to which function to take as u. We arbitrarily choose

$$u = e^x, \qquad dv = \cos x\,dx.$$

Then

$$du = e^x\,dx, \qquad v = \sin x$$

and the integration by parts formula gives

$$\int e^x \cos x\,dx = e^x \sin x - \int e^x \sin x\,dx. \tag{8}$$

Now the integral on the right side of (8) seems no simpler than the integral on the left. However, let's try another application of the integration by parts formula in the integral $\int e^x \sin x\,dx$. Again taking $u = e^x$ we have

$$u = e^x, \qquad dv = \sin x\,dx$$

and

$$du = e^x\,dx, \qquad v = -\cos x.$$

Thus,

$$\begin{aligned}\int e^x \sin x\,dx &= -e^x \cos x - \int (-\cos x)e^x\,dx \\ &= -e^x \cos x + \int e^x \cos x\,dx.\end{aligned} \tag{9}$$

Note that the desired integral has reappeared on the right side of equation (9)! This happens because the second derivatives of e^x and $\cos x$ (as well as $\sin x$, $\sinh x$, and $\cosh x$) are simply multiples of the original functions. Combining equations (8) and (9) gives

$$\int e^x \cos x\,dx = e^x \sin x - \left\{-e^x \cos x + \int e^x \cos x\,dx\right\}$$
$$= e^x(\sin x + \cos x) - \int e^x \cos x\,dx.$$

We may now add $\int e^x \cos x\,dx$ to both sides to obtain

$$2\int e^x \cos x\,dx = e^x(\sin x + \cos x) + C_1$$

so

$$\int e^x \cos x\,dx = \frac{e^x}{2}(\sin x + \cos x) + C, \qquad C = \frac{C_1}{2}.$$ ■

We can apply the parts formula to any integral of the form $\int f(x)\,dx$ simply by taking $u = f(x)$ and $dv = dx$. The resulting integral may or may not be solvable.

Example 5 Find $\int \sin^{-1} x\,dx$.

Strategy
Here our only choice is to take

$u = \sin^{-1} x, \qquad dv = dx.$

Apply equation (5).

(Handle $\int \frac{x}{\sqrt{1-x^2}}\,dx$ by a u-substitution with $u = 1 - x^2$.)

Solution
With $u = \sin^{-1} x, \qquad dv = dx,$

$$du = \frac{dx}{\sqrt{1-x^2}}, \qquad \text{and} \qquad v = x.$$

The parts formula gives

$$\int \sin^{-1} x\,dx = x\sin^{-1} x - \int \frac{x}{\sqrt{1-x^2}}\,dx$$
$$= x\sin^{-1} x + \sqrt{1-x^2} + C.$$ ■

Example 6 Show that $\int \ln x\,dx = x\ln x - x + C$.

Solution: We do this by the method of Example 5. Since the only factor of the integrand is $\ln x$, we take

$$u = \ln x, \qquad dv = dx.$$

Then,

$$du = \frac{1}{x}\,dx, \qquad \text{and} \qquad v = x$$

so an application of the integration by parts formula gives

$$\int \ln x\,dx = x \cdot \ln x - \int x\left(\frac{1}{x}\right)dx$$

$$= x \cdot \ln x - \int 1\, dx$$
$$= x \ln x - x + C. \qquad \blacksquare$$

The following is one of the trickier integrals involving integration by parts. It should be noted for future reference.

Example 7 Find $\int \sec^3 x\, dx$.

Strategy
Break $\sec^3 x$ into two factors.

Solution
We let

$$u = \sec x, \qquad dv = \sec^2 x\, dx.$$

Then

$$du = \sec x \tan x\, dx, \qquad v = \tan x,$$

so

Apply equation (5).

Use identity $\tan^2 x = \sec^2 x - 1$.

$$\begin{aligned}\int \sec^3 x\, dx &= \sec x \tan x - \int \tan^2 x \sec x\, dx\\ &= \sec x \tan x - \int [\sec^2 x - 1] \sec x\, dx\\ &= \sec x \tan x - \int \sec^3 x\, dx + \int \sec x\, dx\\ &= \sec x \tan x - \int \sec^3 x\, dx + \ln|\sec x + \tan x|\end{aligned}$$

Thus

Add $\int \sec^3 x\, dx$ to both sides and divide by 2.

$$\int \sec^3 x\, dx = \frac{1}{2} \sec x \tan x + \frac{1}{2} \ln|\sec x + \tan x| + C. \qquad \blacksquare$$

Finally, we note that the integration by parts formula can be applied to definite integrals as well as to indefinite integrals. The corresponding formulation of equation (3) is

$$\int_a^b f(x)g'(x)\, dx = f(x)g(x)\Big]_a^b - \int_a^b f'(x)g(x)\, dx.$$

When using the uv formulation of the integration by parts formula to evaluate definite integrals, it is simplest to write all functions in terms of the original independent variable, thus avoiding confusion over the correct limits of integration.

Example 8 Find $\int_0^{\sqrt{\pi/2}} x^3 \cos x^2\, dx$.

Strategy
We take $u = x^2$ so the remaining factor $dv = x \cos x^2$ is a multiple of the derivative of $\sin x^2$.

Solution
Let

$$u = x^2, \qquad dv = x \cos x^2\, dx.$$

Then

$$du = 2x\,dx, \qquad v = \frac{1}{2}\sin x^2,$$

so the parts formula gives

$$\int_0^{\sqrt{\pi/2}} x^3 \cos x^2\,dx = \frac{1}{2}x^2 \sin x^2\Big]_0^{\sqrt{\pi/2}} - \int_0^{\sqrt{\pi/2}} x \sin x^2\,dx$$

Apply (5).

$$= \frac{1}{2}\left(\frac{\pi}{4}\right)\sin\left(\frac{\pi}{4}\right) - \left[-\frac{1}{2}\cos x^2\right]_0^{\sqrt{\pi/2}}$$

Evaluate resulting indefinite integral over original limits of integration.

$$= \frac{\sqrt{2}\,\pi}{16} + \frac{1}{2}\left(\frac{\sqrt{2}}{2} - 1\right)$$

$$\approx .1312.$$

■

Exercise Set 11.2

In Exercises 1–36, evaluate the given integral.

1. $\int x\,e^x\,dx$

2. $\int x \ln x\,dx$

3. $\int x \cos x\,dx$

4. $\int x^2 \ln x\,dx$

5. $\int \tan^{-1} x\,dx$

6. $\int x \sec x \tan x\,dx$

7. $\int x \sec^2 \pi x\,dx$

8. $\int e^{ax} \sin x\,dx$

9. $\int \sin(\ln x)\,dx$

10. $\int \sin x \sinh x\,dx$

11. $\int_0^1 x(x+2)^8\,dx$

12. $\int_0^4 x\sqrt{x+1}\,dx$

13. $\int (\ln x)^2\,dx$

14. $\int x \ln x^2\,dx$

15. $\int xe^{2x}\,dx$

16. $\int_1^e x^3 \ln x\,dx$

17. $\int_0^1 x^3 e^{2x}\,dx$

18. $\int \sec^5 x\,dx$

19. $\int x \sinh x\,dx$

20. $\int_0^1 \frac{x^3}{\sqrt{x^2+1}}\,dx$

21. $\int e^{2x} \cos x\,dx$

22. $\int \cos^{-1} 2x\,dx$

23. $\int x \sin^{-1} x\,dx$

24. $\int (\cos^{-1} x)^2\,dx$

25. $\int \sqrt{x} \ln x\,dx$

26. $\int (\ln x)^2\,dx$

27. $\int \frac{x^3}{\sqrt{1+x^2}}\,dx$

28. $\int x^3\sqrt{9-x^2}\,dx$

29. $\int \tanh^{-1} x\,dx$

30. $\int x^3 e^{ax^2}\,dx$

31. $\int \sinh^{-1} x\,dx$

32. $\int x \tanh^{-1} x\,dx$

33. $\int x \sinh^{-1} x\,dx$

34. $\int \sqrt{x}\,e^{-\sqrt{x}}\,dx$

35. $\int 2xe^{-\sqrt{x}}\,dx$

36. $\int \sec^3 x \tan^2 x\,dx$

37. Find the area of the region bounded by the graph of $y = x \sin \pi x$ and the x-axis between the lines $x = 0$ and $x = 1$.

38. Find the area of the region bounded by the graphs of $y = \ln(1 + x)$, $y = x$, and $x = 1$.

39. Find the volume of the solid obtained by revolving about the x-axis the region bounded by the graph of $y = \ln x$ and the x-axis for $1 \le x \le e$.

40. Find the volume of the solid obtained by revolving the region in Exercise 39 about the y-axis.

41. Find the average value of the function $f(x) = e^{-x} \sin \pi x$ for $x \in [0, 1]$.

42. Find the average value of the function $y = \sin^{-1} x$ for $x \in [0, 1]$.

43. A particle moves along a line with velocity $v(t) = t \sin \pi t$ meters per second. Find the distance travelled by the particle between times $t = 0$ and $t = 2$ seconds.

44. A water tank has the shape of the solid obtained by revolving the region bounded by the graph of $y = \ln x$, the x-axis, and the line $x = 5$ about the x-axis. The tank is full of water. Find the work done in pumping all the water to the top of the tank.

45. A **reduction formula** is one that reduces the order of the integrand; by repeated applications the order can be reduced to zero or one. Establish the reduction formula

$$\int \sin^n x\, dx = -\frac{1}{n}\sin^{n-1} x \cos x + \frac{n-1}{n}\int \sin^{n-2} x\, dx$$

using integration by parts.

46. Use the formula in Exercise 45 to find $\int \sin^2 x\, dx$.

47. Use the formula in Exercise 45 to find $\int \sin^4 x\, dx$.

48. Establish the reduction formula

$$\int x^n e^x\, dx = x^n e^x - n\int x^{n-1} e^x\, dx$$

using integration by parts.

49. Use the formula in Exercise 48 to find $\int x^4 e^x\, dx$.

50. Establish the reduction formula

$$\int x^n \sin ax\, dx = -\frac{x^n}{a}\cos ax + \frac{n}{a}\int x^{n-1}\cos ax\, dx$$

using integration by parts.

51. Use the formula in Exercise 50 to find $\int x \sin 3x\, dx$.

52. Show that the integral $\int p(x)e^x\, dx$ can be handled by n applications of the integration by parts formula when $p(x)$ is a polynomial of degree n.

11.3 TRIGONOMETRIC SUBSTITUTIONS

Certain types of integrals may be evaluated by means of a trigonometric substitution of the form $x = a \sin \theta$, $x = a \tan \theta$, or $x = a \sec \theta$. You will see that the idea is based on the simple Pythagorean Theorem for right triangles, and it is used to handle integrands involving factors of $\sqrt{x^2 + a^2}$, $\sqrt{a^2 - x^2}$, and $\sqrt{x^2 - a^2}$.

Integrals Involving $\sqrt{x^2 + a^2}$

The idea is to visualize a right triangle for which the quantity $\sqrt{x^2 + a^2}$ is the length of one of the three legs. The simplest way to do this is to draw a right triangle whose two legs are labelled x and a, respectively (Figure 3.1). The Pythagorean Theorem then states that the length of the hypotenuse is the quantity $\sqrt{x^2 + a^2}$.

Next, let θ denote one of the acute angles of the triangle constructed above. Since a is constant, θ is a function of x, and vice versa. In fact, you can see that in Figure 3.1 x and θ are related by the equation

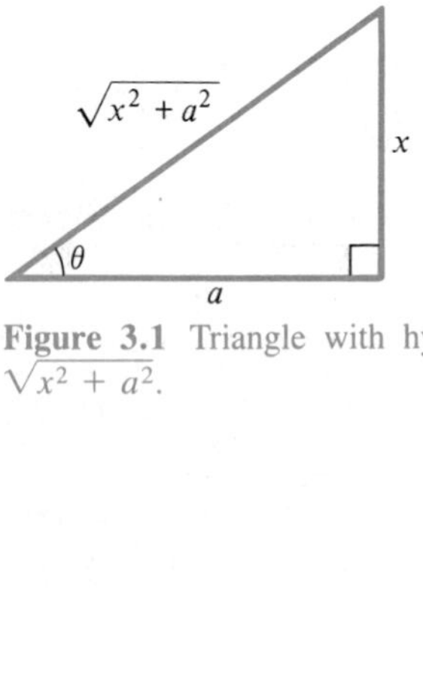

Figure 3.1 Triangle with hypotenuse $\sqrt{x^2 + a^2}$.

$$\frac{x}{a} = \tan \theta,$$

or

$$x = a \tan \theta. \tag{1}$$

If we make the substitution indicated by equation (1), the expression $\sqrt{x^2 + a^2}$ becomes

$$\sqrt{x^2 + a^2} = \sqrt{(a \tan \theta)^2 + a^2} = a\sqrt{\tan^2 \theta + 1} = a \sec \theta. \tag{2}$$

In other words, *the trigonometric substitution $x = a \tan \theta$ eliminates the radical.* This is the objective in using trigonometric substitutions. Of course, in using substitution (1) we must also substitute for the differential dx since we will now be integrating with respect to θ. From equation (1) and the definition of the differential we obtain the required equation

$$dx = \frac{d}{d\theta}(a \tan \theta)\, d\theta = a \sec^2 \theta\, d\theta. \tag{3}$$

Example 1 Find $\displaystyle\int \frac{dx}{\sqrt{x^2+1}}$.

Strategy

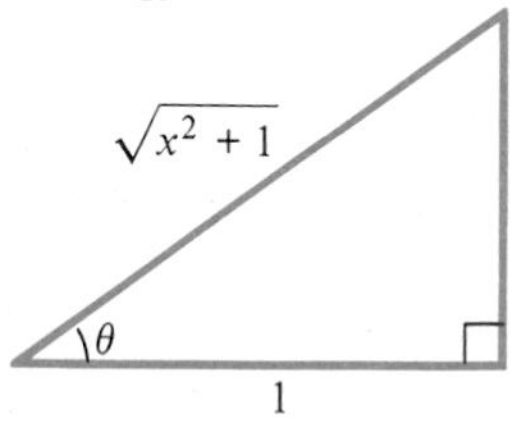

Figure 3.2

Solution

This integral has the form

$$\int \frac{dx}{\sqrt{x^2+a^2}} \qquad \text{with} \qquad a = 1.$$

We therefore construct a right triangle with legs of length x and 1 and hypotenuse of length $\sqrt{x^2+1}$ (Figure 3.2). From the triangle, or from equation (1), we have the substitution equation

Write the equation for x as a function of θ. Differentiate this equation to obtain the expression for dx.

$$x = \tan\theta,$$

so

$$dx = \sec^2\theta\, d\theta.$$

With these substitutions we obtain

Substitute for x and for dx in the original integral.

Find the resulting integral in terms of θ.

$$\begin{aligned}\int \frac{dx}{\sqrt{x^2+1}} &= \int \frac{\sec^2\theta\, d\theta}{\sqrt{\tan^2\theta+1}}\\ &= \int \frac{\sec^2\theta}{\sqrt{\sec^2\theta}}\, d\theta\\ &= \int \sec\theta\, d\theta\\ &= \ln|\sec\theta+\tan\theta| + C.\end{aligned}$$

Substitute back in terms of x using the triangle.

From the triangle in Figure 3.2 we can read directly that

$$\sec\theta = \sqrt{x^2+1}; \qquad \tan\theta = x,$$

as can be proved using the substitution equation. The integral is therefore

$$\int \frac{dx}{\sqrt{x^2+1}} = \ln|\sqrt{x^2+1}+x| + C.$$

(Since $\sqrt{x^2+1}+x>0$ for all x, the absolute value signs may be dropped.) ■

REMARK 1: You may wonder how we decided to label the leg opposite angle θ rather than the leg adjacent to θ as x in Figure 3.2. The answer is that it really makes no difference, except that the latter choice would have led to the substitution $x = \cot\theta$, $dx = -\csc^2\theta\, d\theta$. The result, of course, would have been the same.

REMARK 2: There is a certain amount of sloppiness going on here. (We wanted to present the basic idea in a simple way before discussing the following detail.) The problem is in equation (2) where we said that $\sqrt{\tan^2\theta+1} = \sec\theta$. The correct statement is that $\sqrt{\tan^2\theta+1} = |\sec\theta|$. To avoid the difficulty of dealing with absolute values (and to make the calculations of Example 1 correct) we will agree to work only with the *principal values* of the trigonometric functions as we did in Chapter 10 in defining the inverse trigonometric functions. To see how this avoids the above difficulty in equation (2), recall that the principal values of $y = \tan\theta$ are $-\pi/2 < \theta < \pi/2$. For this range of values of θ, $\sec\theta > 0$, so statement (2) is correct. This procedure will also help avoid confusion in changing limits of integration in definite integrals (see Example 4).

REMARK 3: We could also have written the integral in Example 1 simply as

$$\int \frac{dx}{\sqrt{x^2+1}} = \sinh^{-1} x + C.$$

(See Section 10.8.) The technique of trigonometric substitutions reduces the number of such formulas that you must remember.

Example 2 Find $\int \sqrt{x^2+4}\,dx$.

Strategy

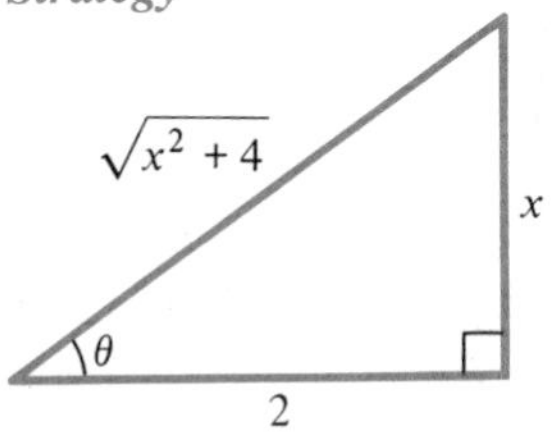

Figure 3.3

Solution
The integral has the form

$$\int \sqrt{x^2+a^2}\,dx \qquad \text{with} \qquad a = 2.$$

We therefore construct a right triangle with legs of length x and 2 and hypotenuse $\sqrt{x^2+4}$ (Figure 3.3).

The substitution is

Read off x as a function of θ from the triangle. Find dx by differentiation.

$$x = 2\tan\theta, \qquad -\frac{\pi}{2} < \theta < \frac{\pi}{2}$$

$$dx = 2\sec^2\theta\,d\theta.$$

and the integral becomes

Substitute, simplify.

$$\int \sqrt{x^2+4}\,dx = \int \sqrt{4\tan^2\theta+4}\cdot 2\sec^2\theta\,d\theta$$

$$= 4\int \sqrt{\tan^2\theta+1}\cdot\sec^2\theta\,d\theta$$

Use Example 7, Section 11.2.

$$= 4\int \sec^3\theta\,d\theta$$

$$= 4\left\{\frac{1}{2}\sec\theta\tan\theta + \frac{1}{2}\ln|\sec\theta+\tan\theta|\right\} + C$$

Substitute back in terms of x using the triangle.

$$= 2\cdot\frac{\sqrt{x^2+4}}{2}\cdot\frac{x}{2} + 2\ln\left|\frac{\sqrt{x^2+4}}{2}+\frac{x}{2}\right| + C$$

$$= \frac{1}{2}x\sqrt{x^2+4} + 2\ln\left|\frac{\sqrt{x^2+4}}{2}+\frac{x}{2}\right| + C. \quad ■$$

Integrals Involving $\sqrt{a^2-x^2}$

Here the idea is the same, except we must use a right triangle for which one of the sides has length $\sqrt{a^2-x^2}$.

According to the Pythagorean Theorem, this can be achieved by labelling the hypotenuse a and one of the legs x. The other leg then has length $\sqrt{a^2-x^2}$. As you can see from the triangle in Figure 3.4 this leads to the equation

$$\frac{x}{a} = \sin\theta, \qquad -\frac{\pi}{2} \le \theta \le \frac{\pi}{2}$$

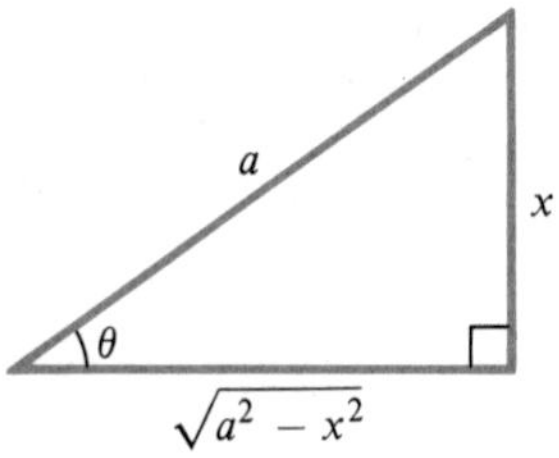

Figure 3.4 $x = a \sin \theta$.

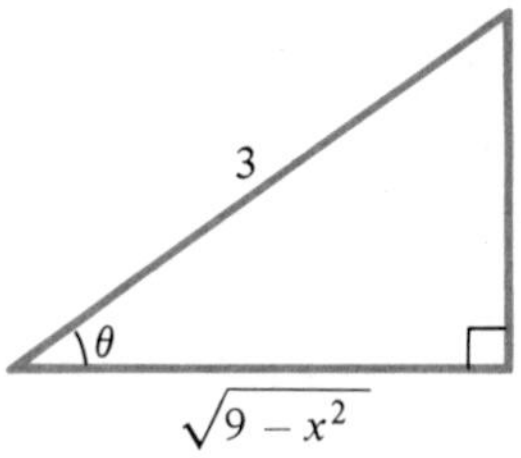

Figure 3.5

and, therefore, to the substitution

$$x = a \sin \theta,$$
$$dx = a \cos \theta \, d\theta.$$

This gives

$$\sqrt{a^2 - x^2} = \sqrt{a^2 - a^2 \sin^2 \theta} = a\sqrt{1 - \sin^2 \theta} = a \cos \theta,$$

since $\cos \theta \geq 0$ for $-\pi/2 \leq \theta \leq \pi/2$.

Example 3 Find $\displaystyle\int \frac{dx}{x^2\sqrt{9 - x^2}}$.

Solution: Here $a = 3$ (Figure 3.5), so we use the substitution

$$x = 3 \sin \theta, \qquad -\pi/2 \leq \theta \leq \pi/2$$
$$dx = 3 \cos \theta \, d\theta.$$

We obtain

$$\int \frac{dx}{x^2\sqrt{9 - x^2}} = \int \frac{3 \cos \theta \, d\theta}{9 \sin^2 \theta\sqrt{9 - 9 \sin^2 \theta}} = \int \frac{3 \cos \theta \, d\theta}{9 \sin^2 \theta \cdot 3 \cdot \sqrt{1 - \sin^2 \theta}}$$

$$= \frac{1}{9} \int \frac{\cos \theta \, d\theta}{\sin^2 \theta \cdot \cos \theta}$$

$$= \frac{1}{9} \int \csc^2 \theta \, d\theta$$

$$= -\frac{1}{9} \cot \theta + C$$

$$= -\frac{1}{9} \cdot \frac{\sqrt{9 - x^2}}{x} + C. \blacksquare$$

Example 4 Find the area of the region bounded above by the semicircle $y = \sqrt{4 - x^2}$, below by the x-axis, and on the left by the line $x = -1$ (Figure 3.7).

Strategy
Write the integral giving the area.

Solution
The area is given by the integral

$$A = \int_{-1}^{2} \sqrt{4 - x^2} \, dx.$$

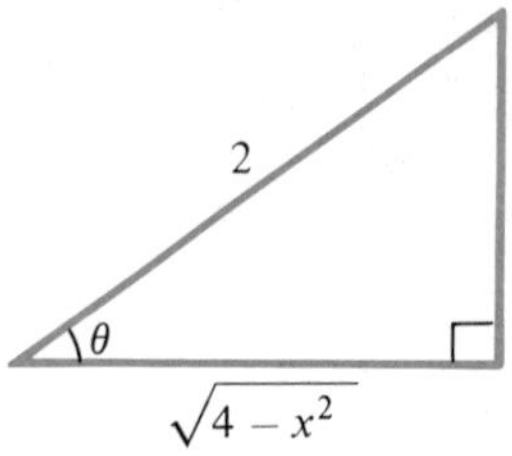

Figure 3.6

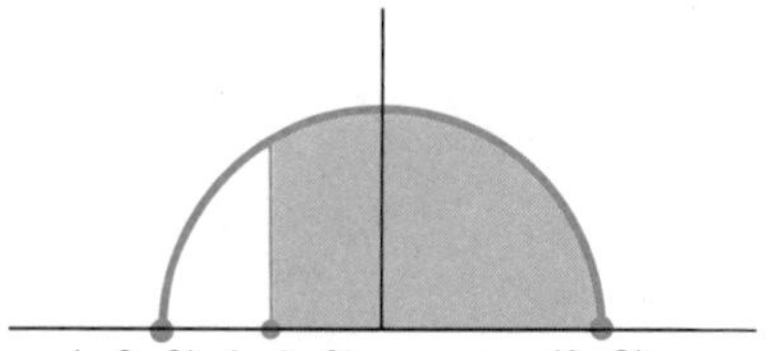

Figure 3.7 Region in Example 4.

Use the substitution for form $\sqrt{a^2 - x^2}$.

To evaluate the integral, we use the trigonometric substitution

$$x = 2 \sin \theta, \qquad -\pi/2 \le \theta \le \pi/2,$$
$$dx = 2 \cos \theta \, d\theta.$$

Change limits of integration, using the principal values of the function $\sin^{-1} \theta$.

To find the limits of integration with respect to θ we note that

$$\sin \theta = -\frac{1}{2} \quad \text{when} \quad x = -1, \quad \text{and}$$
$$\sin \theta = 1 \quad \text{when} \quad x = 2.$$

To invert these relationships, we must recall the domain of the function $\sin^{-1} \theta$: $-\pi/2 \le \theta \le \pi/2$. Thus

$$\theta = \sin^{-1}(-1/2) = -\pi/6 \quad \text{when} \quad x = -1$$

and

$$\theta = \sin^{-1}(1) = \pi/2 \quad \text{when} \quad x = 2.$$

Substitute for x and dx, and use new limits of integration.

Our integral therefore becomes

$$\begin{aligned} A = \int_{-1}^{2} \sqrt{4 - x^2}\, dx &= \int_{-\pi/6}^{\pi/2} \sqrt{4 - (2 \sin \theta)^2} \cdot 2 \cos \theta \, d\theta \\ &= \int_{-\pi/6}^{\pi/2} 2\sqrt{1 - \sin^2 \theta} \cdot 2 \cos \theta \, d\theta \\ &= 4 \int_{-\pi/6}^{\pi/2} \cos^2 \theta \, d\theta \\ &= 4 \int_{-\pi/6}^{\pi/2} \left[\frac{1}{2} + \frac{1}{2} \cos 2\theta\right] d\theta \\ &= 4\left[\frac{\theta}{2} + \frac{1}{4} \sin 2\theta\right]_{-\pi/6}^{\pi/2} \\ &= 4\left[\pi/3 + \frac{\sqrt{3}}{8}\right] \\ &\approx 5.055. \end{aligned}$$

Use identity

$$\cos^2 \theta = \frac{1}{2} + \frac{1}{2} \cos 2\theta.$$

■

REMARK: Notice that we did not have to substitute back in terms of x in the integral in Example 4. Once we have changed variables and limits of integration in a definite integral, we may proceed to evaluate the integral directly.

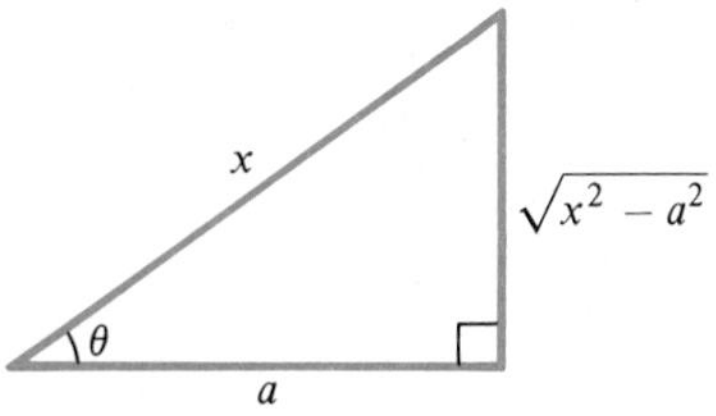

Figure 3.8 $x = a \sec \theta$.

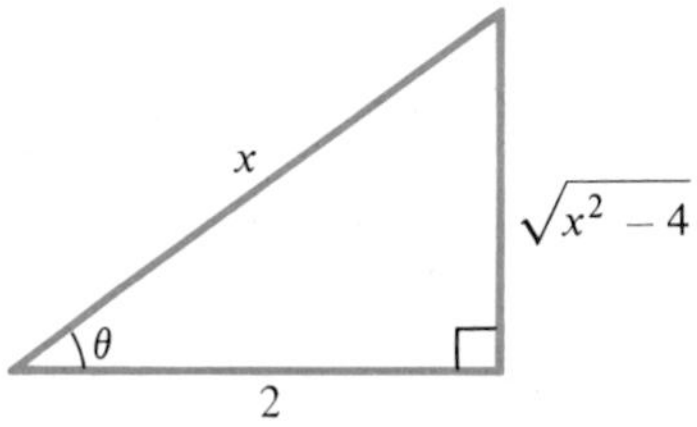

Figure 3.9 $x = 2 \sec \theta$.

Integrals Involving $\sqrt{x^2 - a^2}$

The situation for integrals involving $\sqrt{x^2 - a^2}$ is similar to those for the preceding two types except that we must label the hypotenuse x. Thus, one leg of the triangle is a, the other $\sqrt{x^2 - a^2}$. As suggested by Figure 3.8, we use the substitutions

$$x = a \sec \theta \qquad 0 \le \theta < \pi/2 \quad \text{or} \quad \pi \le \theta < 3\pi/2,$$
$$dx = a \sec \theta \tan \theta \, d\theta.$$

Then the radical $\sqrt{x^2 - a^2}$ simplifies to

$$\sqrt{x^2 - a^2} = \sqrt{a^2 \sec^2 \theta - a^2} = a\sqrt{\tan^2 \theta} = a \tan \theta,$$

since $\tan \theta \ge 0$ when $0 \le \theta < \pi/2$ or $\pi \le \theta < 3\pi/2$.

Example 5 Find $\displaystyle\int \frac{dx}{x^2\sqrt{x^2 - 4}}$.

Strategy

Solution

Because of the presence of the radical $\sqrt{x^2 - 4}$, we use the substitution (Figure 3.9)

$$x = 2 \sec \theta, \qquad 0 \le \theta < \pi/2 \quad \text{or} \quad \pi \le \theta < 3\pi/2,$$
$$dx = 2 \sec \theta \tan \theta \, d\theta.$$

The integral becomes

Make trigonometric substitution.

$$\int \frac{dx}{x^2\sqrt{x^2 - 4}} = \int \frac{2 \sec \theta \cdot \tan \theta \, d\theta}{(2 \sec \theta)^2 \sqrt{4 \sec^2 \theta - 4}}$$

Use $\sec^2 \theta - 1 = \tan^2 \theta$.

$$= \int \frac{2 \sec \theta \tan \theta \, d\theta}{4 \sec^2 \theta \cdot 2 \tan \theta}$$

$$= \frac{1}{4} \int \frac{1}{\sec \theta} \, d\theta$$

$\sec \theta = \dfrac{1}{\cos \theta}$.

$$= \frac{1}{4} \int \cos \theta \, d\theta$$

$$= \frac{1}{4} \cdot \sin \theta + C$$

From Figure 3.9,

$\sin \theta = \dfrac{\sqrt{x^2 - 4}}{x}$.

$$= \frac{1}{4} \cdot \frac{\sqrt{x^2 - 4}}{x} + C. \qquad \blacksquare$$

The following example shows how two different integration techniques can apply to the same integral.

Example 6 Find $\int \frac{1}{2x^2 + 8}\, dx$.

Strategy
Factor 8 from denominator to bring integrand to form

$$\frac{1}{u^2 + 1}$$

corresponding to $\tan^{-1} u$.

Solution
(Method i—a u-substitution.)

We write

$$\int \frac{1}{2x^2 + 8}\, dx = \frac{1}{8} \int \frac{1}{\left(\frac{x}{2}\right)^2 + 1}\, dx.$$

Then, using the u-substitution

$$u = \frac{x}{2}, \qquad du = \frac{1}{2}\, dx$$

Since $du = \frac{1}{2}\, dx$,

$dx = 2\, du$.

we have $dx = 2\, du$, so

$$\begin{aligned}\int \frac{1}{2x^2 + 8}\, dx &= \frac{1}{8} \int \frac{2\, du}{u^2 + 1}\\ &= \frac{1}{4} \int \frac{du}{u^2 + 1}\\ &= \frac{1}{4} \tan^{-1} u + C\\ &= \frac{1}{4} \tan^{-1}\left(\frac{x}{2}\right) + C.\end{aligned}$$

Apply formula (16), Section 11.1.

Write

$x^2 + 2^2 = (\sqrt{x^2 + 2^2})^2$.

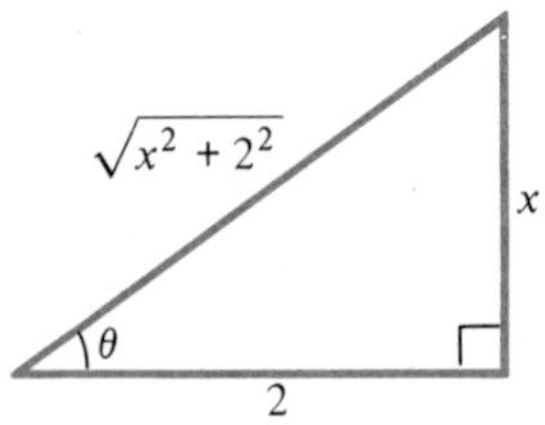

Figure 3.10 $x = 2 \tan \theta$.

(Method ii—a trigonometric substitution.)

Since

$$\int \frac{1}{2x^2 + 8}\, dx = \frac{1}{2} \int \frac{1}{(\sqrt{x^2 + 2^2})^2}\, dx$$

we use the trigonometric substitution (Figure 3.10)

$$x = 2 \tan \theta, \qquad -\pi/2 < \theta < \pi/2,$$
$$dx = 2 \sec^2 \theta\, d\theta.$$

We obtain

$$\begin{aligned}\int \frac{1}{2x^2 + 8}\, dx &= \frac{1}{2} \int \frac{2 \sec^2 \theta\, d\theta}{(2 \sec \theta)^2}\\ &= \frac{1}{4} \int d\theta\\ &= \frac{1}{4}\theta + C\\ &= \frac{1}{4} \tan^{-1}\left(\frac{x}{2}\right) + C.\end{aligned}$$

■

Read $\theta = \tan^{-1}\left(\frac{x}{2}\right)$ from Figure 3.10.

In summary, the following trigonometric substitutions, together with the principal values indicated, are used to handle integrals involving the indicated radicals.

$\sqrt{x^2 + a^2}$	requires	$x = a \tan \theta,$	$-\pi/2 < \theta < \pi/2,$
$\sqrt{a^2 - x^2}$	requires	$x = a \sin \theta,$	$-\pi/2 \le \theta \le \pi/2,$
$\sqrt{x^2 - a^2}$	requires	$x = a \sec \theta,$	$0 \le \theta < \pi/2$ or $\pi \le \theta < 3\pi/2.$

Exercise Set 11.3

In Exercises 1–33, evaluate the given integral.

1. $\int \frac{x}{\sqrt{1 - x^2}}\, dx$

2. $\int \frac{x}{\sqrt{4 + x^2}}\, dx$

3. $\int \frac{dx}{\sqrt{9 - x^2}}$

4. $\int_1^2 \sqrt{x^2 - 1}\, dx$

5. $\int \frac{x^2}{\sqrt{x^2 - 4}}\, dx$

6. $\int \frac{x}{\sqrt{9 - x^2}}\, dx$

7. $\int \frac{\sqrt{1 - x^2}}{x}\, dx$

8. $\int \frac{1}{x^2\sqrt{x^2 + 16}}\, dx$

9. $\int \frac{1}{x^3\sqrt{x^2 - 4}}\, dx$

10. $\int \frac{x}{x^2 + 4}\, dx$

11. $\int \frac{1}{(16 + x^2)^2}\, dx$

12. $\int \frac{dx}{(4 - x^2)^{3/2}}$

13. $\int \frac{x\, dx}{(1 - x^2)^{3/2}}$

14. $\int \frac{dx}{(x^2 - 4)^{3/2}}$

15. $\int x\sqrt{x^2 - 1}\, dx$

16. $\int \sqrt{4 - x^2}\, dx$

17. $\int \frac{x^2\, dx}{(x^2 + 3)^{3/2}}$

18. $\int \frac{\sqrt{1 - x^2}}{x^4}\, dx$

19. $\int \frac{x^2}{\sqrt{x^2 + a^2}}\, dx$

20. $\int \frac{\sqrt{x^2 - a^2}}{x}\, dx$

21. $\int \frac{\sqrt{x^2 - a^2}}{x^2}\, dx$

22. $\int \frac{x^2}{\sqrt{x^2 - a^2}}\, dx$

23. $\int \frac{x\, dx}{\sqrt{(x^2 + a^2)^3}}$

24. $\int \frac{\sqrt{x^2 + a^2}}{x}\, dx$

25. $\int \frac{dx}{x\sqrt{x^2 - a^2}}$

26. $\int \frac{dx}{x\sqrt{x^2 + a^2}}$

27. $\int \frac{\sqrt{x^2 + a^2}}{x^2}\, dx$

28. $\int \frac{x^2\, dx}{(x^2 + a^2)^{3/2}}$

29. $\int \frac{x^2 + 3}{\sqrt{x^2 + 9}}\, dx$

30. $\int \frac{x - 5}{x^2 - 16}\, dx$

31. $\int \frac{x^2 - 3}{x\sqrt{x^2 + 4}}\, dx$

32. $\int \frac{1 - x}{x\sqrt{1 + x^2}}\, dx$

33. $\int \frac{x^2 - 4x + 5}{x\sqrt{9 - x^2}}\, dx$

34. Use trigonometric substitutions to obtain formula 16, Section 11.1.

35. Find the area of the region bounded by the graph of $y = 5 - \sqrt{x^2 + 9}$ and the x-axis.

36. Find the volume of the solid generated by revolving about the y-axis the region bounded by the graph of $y = x\sqrt{9 - x^2}$ and the x-axis.

37. A water tank is in the shape of a cylinder with circular end panels of radius 5 meters. The tank is mounted with the circular ends vertical. To what percentage capacity is the tank filled when the depth of the water is 6 meters?

38. Find the area of the region enclosed by the ellipse $\frac{x^2}{4} + \frac{y^2}{3} = 1.$

39. Find the length of the graph of $y = x^2$ from (0, 0) to (1/2, 1/4).

40. The region bounded above by the graph of $y = x(16 - x^2)^{1/4}$ and below by the x-axis is rotated about the x-axis. Find the volume of the resulting solid.

41. Find the length of the graph of $y = \ln x$ from (1, 0) to (3, ln 3).

11.4 INTEGRALS INVOLVING QUADRATIC EXPRESSIONS

The method of u-substitutions may be used together with the method of trigonometric substitutions to handle integrals involving quadratic expressions more general than those encountered in Section 11.3, such as the following.

Example 1 Find $\int \sqrt{16 - \left(\frac{x-2}{\sqrt{3}}\right)^2}\, dx.$

Strategy
Make a u-substitution for $\frac{x-2}{\sqrt{3}}$.
(This simplifies the term involving x.)

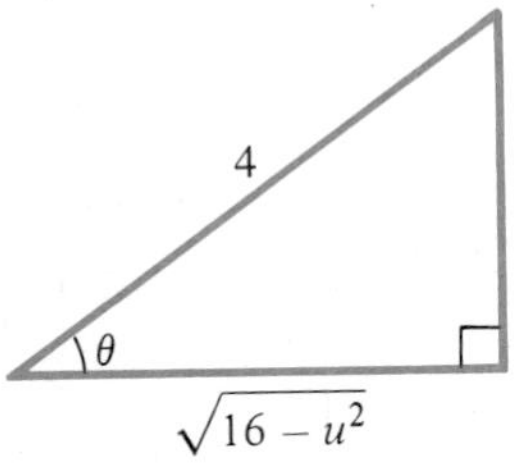

Figure 4.1 $u = 4 \sin \theta$
$du = 4 \cos \theta\, d\theta.$

Solution
We first let

$$u = \frac{x-2}{\sqrt{3}}, \qquad du = \frac{1}{\sqrt{3}}\, dx. \tag{1}$$

Thus $dx = \sqrt{3}\, du$, and the integral becomes

$$\int \sqrt{16 - \left(\frac{x-2}{\sqrt{3}}\right)^2}\, dx = \sqrt{3} \int \sqrt{16 - u^2}\, du.$$

The integral on the right now requires the trigonometric substitution (Figure 4.1)

$$u = 4 \sin \theta, \qquad -\pi/2 \le \theta \le \pi/2, \tag{2a}$$
$$du = 4 \cos \theta\, d\theta. \tag{2b}$$

This gives

$$\begin{aligned}\sqrt{3} \int \sqrt{16 - u^2}\, du &= \sqrt{3} \int 4 \cos \theta \cdot 4 \cos \theta\, d\theta \\ &= 16 \cdot \sqrt{3} \int \cos^2 \theta\, d\theta \\ &= 8\sqrt{3} \int [1 + \cos 2\theta]\, d\theta \\ &= 8\sqrt{3}\left(\theta + \frac{1}{2} \sin 2\theta\right) + C.\end{aligned}$$

Substitute back in terms of u by use of Figure 4.1. (You must first reduce the double angle.)

To express this integral in terms of the original variable x, we must undo both substitutions *in the order opposite to the order in which they were made.* Since

$$\begin{aligned}\sin 2\theta &= 2 \sin \theta \cos \theta \\ &= 2\left(\frac{u}{4}\right) \frac{\sqrt{16 - u^2}}{4} \\ &= \frac{1}{8} u\sqrt{16 - u^2},\end{aligned}$$

and

$$\theta = \sin^{-1}\left(\frac{u}{4}\right)$$

Substitute back in terms of the original variable x using equation (1).

(both follow from Figure 4.1), we reverse the substitutions (2a) and (2b) by

$$8\sqrt{3}\left(\theta + \frac{1}{2} \sin 2\theta\right) + C = 8\sqrt{3} \sin^{-1}\left(\frac{u}{4}\right) + \frac{\sqrt{3}}{2} u\sqrt{16 - u^2} + C.$$

We then reverse the substitution (1) by setting $u = \dfrac{x-2}{\sqrt{3}}$ to obtain

$$\int \sqrt{16 - \left(\frac{x-2}{\sqrt{3}}\right)^2}\, dx = 8\sqrt{3}\left\{\sin^{-1}\left(\frac{x-2}{4\sqrt{3}}\right) + \frac{1}{16}\left(\frac{x-2}{\sqrt{3}}\right)\sqrt{16 - \left(\frac{x-2}{\sqrt{3}}\right)^2}\right\} + C.$$

■

We are about to encounter a type of integral for which two or more successive substitutions will be required. It is important to keep in mind a clear idea of which substitutions have been carried out and in what order so that you can correctly state the solution in terms of the original variable. The type of integral we study here involves a square root of a general quadratic expression $ax^2 + bx + c$. Since the methods of trigonometric substitutions apply to many integrals involving perfect squares, we can handle integrals involving square roots of general quadratics by first completing the square and then applying a u-substitution (if necessary), followed by a trigonometric substitution. (Remember to keep the general strategy well in mind as there are several major steps involved in these solutions!)

Example 2 Find $\displaystyle\int \frac{dx}{\sqrt{x^2 - 2x + 5}}$.

Strategy

First complete the square under the radical.

Solution

By completing the square, we find

$$\sqrt{x^2 - 2x + 5} = \sqrt{(x^2 - 2x + 1) + 4} = \sqrt{(x-1)^2 + 2^2}.$$

Make the u-substitution

$u = x - 1$.

We therefore use the u-substitution

$$u = x - 1; \qquad du = dx$$

Write the original integral in terms of the u-substitution.

to write the integral as

$$\int \frac{dx}{\sqrt{x^2 - 2x + 5}} = \int \frac{dx}{\sqrt{(x-1)^2 + 2^2}} = \int \frac{du}{\sqrt{u^2 + 2^2}}.$$

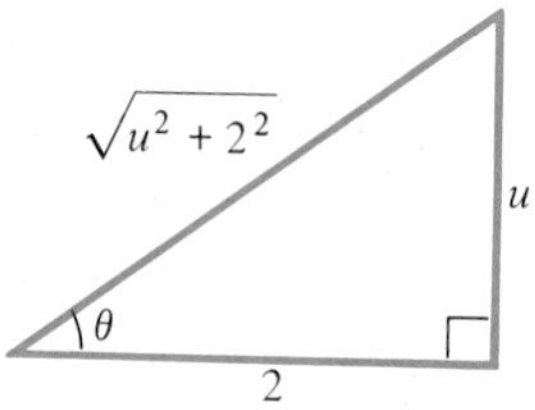

Figure 4.2 $u = 2\tan\theta$
$du = 2\sec^2\theta\, d\theta$.

The last integral requires the trigonometric substitution (Figure 4.2)

$$u = 2\tan\theta, \qquad -\pi/2 < \theta < \pi/2,$$
$$du = 2\sec^2\theta\, d\theta.$$

With this substitution $\sqrt{u^2 + 2^2} = 2\sec\theta$, so the integral becomes

$$\int \frac{du}{\sqrt{u^2 + 2^2}} = \int \frac{2\sec^2\theta \cdot d\theta}{2\sec\theta} = \int \sec\theta\, d\theta$$

Apply formula (14), Section 11.1.

$$= \ln|\sec\theta + \tan\theta| + C$$

Substitute back in terms of u.

$$= \ln\left|\frac{\sqrt{u^2 + 2^2}}{2} + \frac{u}{2}\right| + C$$

Substitute back in terms of x.

$$= \ln\left(\frac{\sqrt{(x-1)^2 + 2^2}}{2} + \frac{x-1}{2}\right) + C$$

Use property: $\ln(z/2) = \ln z - \ln 2$.

$$= \ln(\sqrt{x^2 - 2x + 5} + x - 1) - \ln 2 + C$$

Incorporate $-\ln 2$ in constant of integration.

$$= \ln(\sqrt{x^2 - 2x + 5} + x - 1) + C.$$ ■

Example 3 Find $\displaystyle\int \frac{x+3}{\sqrt{2x^2 - 8x}}\,dx$.

Strategy

Complete the square under the radical.

Solution

We have

$$\sqrt{2x^2 - 8x} = \sqrt{2(x^2 - 4x + 4) - 2(4)}$$
$$= \sqrt{2(x-2)^2 - 8}.$$

Make a u-substitution to simplify x · term.

We therefore make the u-substitution

$$u = x - 2; \qquad du = dx.$$

Then

$$x = u + 2, \qquad \text{so} \qquad x + 3 = u + 5.$$

With these substitutions the integral becomes

$$\int \frac{x+3}{\sqrt{2x^2 - 8x}}\,dx = \int \frac{x+3}{\sqrt{2(x-2)^2 - 8}}\,dx$$
$$= \int \frac{u+5}{\sqrt{2u^2 - 8}}\,du$$
$$= \frac{1}{\sqrt{2}}\int \frac{u+5}{\sqrt{u^2 - 4}}\,du$$

Split integrand into two terms.

$$= \frac{1}{\sqrt{2}}\int \frac{u}{\sqrt{u^2 - 4}}\,du + \frac{1}{\sqrt{2}}\int \frac{5}{\sqrt{u^2 - 4}}\,du.$$

First integrand has form $\displaystyle\int w^{-1/2}\,dw$ where $w = u^2 - 4$.

The first of these two integrals is evaluated using a simple substitution and the Power Rule:

$$\frac{1}{\sqrt{2}}\int \frac{u}{\sqrt{u^2 - 4}}\,du = \frac{1}{2\sqrt{2}}\int \frac{2u}{\sqrt{u^2 - 4}}\,du$$
$$= \frac{1}{\sqrt{2}} \cdot \sqrt{u^2 - 4} + C_1.$$

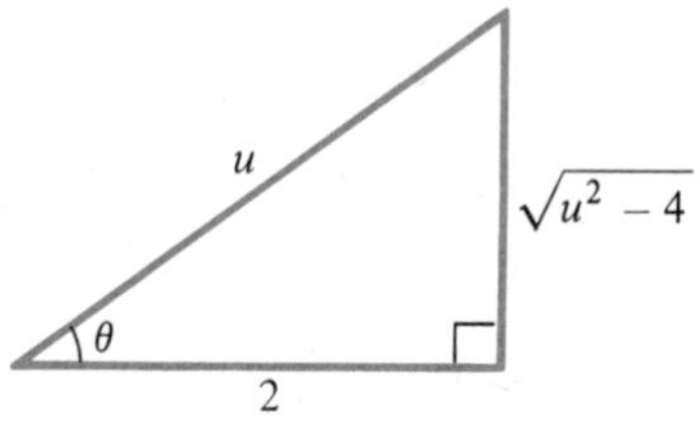

Figure 4.3 $u = 2\sec\theta$.

The second integral requires the trigonometric substitution

$$u = 2\sec\theta, \qquad 0 \le \theta < \pi/2 \qquad \text{or} \qquad \pi \le \theta < \frac{3\pi}{2},$$

$$du = 2\sec\theta\tan\theta \qquad \text{(Figure 4.3)}.$$

Apply trigonometric substitution.

It becomes

$$\frac{1}{\sqrt{2}}\int\frac{5}{\sqrt{u^2-4}}\,du = \frac{1}{\sqrt{2}}\int\frac{5\cdot 2\cdot\sec\theta\tan\theta\,d\theta}{2\tan\theta}$$
$$= \frac{5}{\sqrt{2}}\int\sec\theta\,d\theta$$
$$= \frac{5}{\sqrt{2}}\ln|\sec\theta+\tan\theta| + C_2$$
$$= \frac{5}{\sqrt{2}}\ln\left|\frac{u}{2}+\frac{\sqrt{u^2-4}}{2}\right| + C_2.$$

Combine results.

Let $C = C_1 + C_2$.

Incorporate factor $-\ln 2$ in C (see Example 2).

Combining these results and recalling that $u = x - 2$ we have

$$\int\frac{x+3}{\sqrt{2x^2+8x}}\,dx = \frac{1}{\sqrt{2}}\sqrt{(x-2)^2-4}$$
$$+\frac{5}{\sqrt{2}}\ln\left|\frac{x-2}{2}+\frac{\sqrt{(x-2)^2-4}}{2}\right| + C$$
$$= \frac{1}{\sqrt{2}}\{\sqrt{x^2-4x} + 5\ln|x-2+\sqrt{x^2-4x}|\} + C.$$

■

Exercise Set 11.4

Evaluate the following integrals.

1. $\int\frac{dx}{x^2-4x+4}$
2. $\int\frac{dx}{x^2-6x+12}$
3. $\int\frac{dx}{\sqrt{x^2+6x+13}}$
4. $\int\frac{dx}{\sqrt{5+4x-x^2}}$
5. $\int\frac{dx}{\sqrt{x^2-6x}}$
6. $\int\frac{x\,dx}{x^2+6x+15}$
7. $\int\frac{x\,dx}{\sqrt{6x-x^2}}$
8. $\int\frac{1+x}{\sqrt{x^2-6x+13}}\,dx$
9. $\int\frac{2x+1}{\sqrt{18+6x+x^2}}\,dx$
10. $\int\frac{3x-1}{\sqrt{x^2+8x}}\,dx$
11. $\int\frac{x+2}{x^2-8x-7}\,dx$
12. $\int\frac{2x-1}{6x^2+12x}\,dx$
13. $\int\frac{x}{(x^2+2x+2)^2}\,dx$
14. $\int(2x-3)\sqrt{16+12x-4x^2}\,dx$
15. $\int\frac{dx}{(9-x)(88-18x+x^2)^{3/2}}$
16. $\int\frac{\sqrt{x^2+10x+21}}{x^3+15x^2+75x+125}\,dx$
17. $\int\frac{4x^2+20x+25}{\sqrt{4x^2+20x+29}}\,dx$
18. $\int\frac{dx}{(3-x)\sqrt{18-6x+x^2}}$

11.5 THE METHOD OF PARTIAL FRACTIONS

In elementary algebra you learned the rule for adding two fractions with different denominators:

$$\frac{a}{b}+\frac{c}{d}=\frac{ad+bc}{bd}.$$

The idea was simply to find a common denominator. For the addition of rational functions of x, the rule is the same. For example,

$$\frac{1}{x+1} + \frac{3}{x-2} = \frac{1(x-2) + 3(x+1)}{(x+1)(x-2)} = \frac{4x+1}{x^2 - x - 2}. \tag{1}$$

By reversing this procedure, we obtain a technique for integrating rational functions of x such as that on the right side of equation (1). Using equation (1) we see that

$$\int \frac{4x+1}{x^2-x-2}\,dx = \int \left[\frac{1}{x+1} + \frac{3}{x-2}\right] dx \tag{2}$$

$$= \ln|x+1| + 3\ln|x-2| + C.$$

Clearly the second integral in equation (2) is easier to handle than the first, since the terms in the integrand involve lower degree polynomials. The goal of the method of partial fractions is to decompose rational functions into sums of simpler terms, each of which can be integrated by methods we have already developed.

Before describing the method in its full detail, we will try to gain some insight into how one goes about finding such a decomposition. Obviously we need to keep two points in mind:

POINT 1: In searching for a decomposition of the rational function $\frac{p(x)}{q(x)}$ we must work only with denominators *which are factors of* $q(x)$. Otherwise, the terms obtained cannot possibly add up to $\frac{p(x)}{q(x)}$.

POINT 2: We will work only with proper fractions. That is, in all rational expressions $\frac{p(x)}{q(x)}$ we will require that the degree of $p(x)$ be less than the degree of $q(x)$. (If this is not the case we will first simplify by a polynomial long division).

With these two points and our general goal in mind, we next take up several examples.

Example 1 Find a partial fraction decomposition for

$$f(x) = \frac{4x-7}{x^2-x-6}.$$

Strategy

Factor the denominator.

Solution

We first factor the denominator as

$$x^2 - x - 6 = (x+2)(x-3).$$

Allow one term for each factor of the denominator. (Assume constant numerators for first degree denominators.)

We therefore seek a decomposition of the form

$$\frac{4x-7}{x^2-x-6} = \frac{A}{x+2} + \frac{B}{x-3}. \tag{3}$$

where A and B are constants. (Neither A nor B can be polynomials of degree higher than zero, or we would be introducing improper fractions, violating Point 2.)

Determine A and B by recombining fractions and equating numerators.

To determine A and B, we simply recombine the terms on the right of equation (3) according to the rule for adding fractions:

$$\frac{4x-7}{x^2-x-6} = \frac{A}{x+2} + \frac{B}{x-3} \tag{4}$$

$$= \frac{A(x-3)+B(x+2)}{(x+2)\cdot(x-3)}$$

Collect like terms in x in numerator on right-hand side.

$$= \frac{(A+B)x+(-3A+2B)}{(x+2)\cdot(x-3)}.$$

Use fact that if two polynomials are equal, coefficients of like powers of x must be the same (see Exercise 35).

Now since the numerators of the two fractions on the left and right sides of equation (4) are equal, their coefficients of x must be the same and their constant terms must be the same. This gives the two equations

Equate coefficients of like powers of x.

$$\begin{aligned} A + B &= 4 \qquad \text{(coefficients of } x\text{)},\\ -3A + 2B &= -7 \qquad \text{(constants)}. \end{aligned}$$

Solve the resulting system either by substitution or by elimination.

The solution of this system of equations may be found by adding three times the top equation to the bottom equation. (Or, by substitution, if you prefer.) Doing so we obtain

$$5B = 5, \qquad \text{so} \qquad B = 1.$$

From either equation we then obtain $A = 3$.

We therefore have

$$\frac{4x-7}{x^2-x-6} = \frac{3}{x+2} + \frac{1}{x-3}.$$

■

Example 2 Find $\displaystyle\int \frac{4x-7}{x^2-x-6}\,dx.$

Solution: Using the results of Example 1 we find

$$\int \frac{4x-7}{x^2-x-6}\,dx = \int \frac{3}{x+2}\,dx + \int \frac{1}{x-3}\,dx$$

$$= 3\ln|x+2| + \ln|x-3| + C.$$

■

Example 3 Find a partial fraction decomposition for the function

$$f(x) = \frac{4x^2-3x+1}{x^3-x^2-x}.$$

Strategy

Factor the denominator.

Solution

The denominator factors as

$$x^3 - x^2 + x = x(x^2 - x + 1).$$

(This is the best we can do, since the term $x^2 - x + 1$ does not have linear factors.*) We therefore seek a partial fraction decomposition of the form

Allow one term for each factor of the denominator. (Degree of numerator in each term is one less than the degree of the denominator.)

$$\frac{4x^2 - 3x + 1}{x^3 - x^2 + x} = \frac{A}{x} + \frac{Bx + C}{x^2 - x + 1}$$

Here we must allow for a first degree numerator in the last term on the right since the denominator has degree two. To find the constants A, B, and C we recombine terms to find that

Recombine terms collecting like terms in x in numerator.

$$\begin{aligned}\frac{4x^2 - 3x + 1}{x^3 - x^2 + x} &= \frac{A}{x} + \frac{Bx + C}{x^2 - x + 1}\\ &= \frac{A(x^2 - x + 1) + (Bx + C)(x)}{x(x^2 - x + 1)}\\ &= \frac{(A + B)x^2 + (-A + C)x + A}{x(x^2 - x + 1)}.\end{aligned}$$

Equate coefficients of like powers of x.

Solve resulting system.

Equating coefficients of the various powers of x, we obtain the equations

$$\begin{cases} A + B = 4 & \text{(coefficients of } x^2) \\ -A + C = -3 & \text{(coefficients of } x) \\ A = 1 & \text{(constants)} \end{cases}$$

The system has solution $A = 1, B = 3, C = -2$. The partial fraction decomposition is therefore

$$\frac{4x^2 - 3x + 1}{x^3 - x^2 + x} = \frac{1}{x} + \frac{3x - 2}{x^2 - x + 1}.$$ ■

Example 4 Find $\displaystyle\int \frac{2x^3 - 8x^2 + 9x + 1}{x^2 - 4x + 4}\,dx.$

Strategy

Integrand is an improper fraction. Reduce it to a proper fraction, plus a polynomial, by performing a polynomial long division.

Solution

Before seeking a partial fraction decomposition we note that the integrand is an improper fraction. We therefore perform a polynomial long division:

$$\begin{array}{r} 2x \\ x^2 - 4x + 4\overline{)2x^3 - 8x^2 + 9x + 1} \\ \underline{2x^3 - 8x^2 + 8x} \\ x + 1 \end{array}$$

This calculation shows that

$$\frac{2x^3 - 8x + 9x + 1}{x^2 - 4x + 4} = 2x + \frac{x + 1}{x^2 - 4x + 4}.$$

*A quadratic polynomial $ax^2 + bx + c$ has linear factors $ax^2 + bx + c = a(x - A)(x - B)$ if and only if A and B are **roots** of the polynomial. Since the roots are given by the quadratic formula $x = \dfrac{-b \pm \sqrt{b^2 - 4ac}}{2a}$, the polynomial has linear factors only if the **discriminant** $b^2 - 4ac$ is nonnegative. (This statement is called the **Discriminant Test** for the existence of linear factors.) In this case $a = 1$, $b = -1$, and $c = 1$, so $b^2 - 4ac = (-1)^2 - (4)(1)(1) = 1 - 4 < 0$. Thus there can be no linear factors.

The first term presents no difficulties so we concentrate on finding a partial fraction decomposition for the second term. The denominator factors as

Factor denominator in second term.

$$x^2 - 4x + 4 = (x - 2)^2$$

so we seek a decomposition of the form

$$\frac{x + 1}{x^2 - 4x + 4} = \frac{A}{x - 2} + \frac{B}{(x - 2)^2}.$$

Allow one term for each factor $(x - 2)$ and $(x - 2)^2$. (Only a constant numerator in the second term is required–see Exercise 36). This situation can be viewed in yet another way: If the numerator of the second term were written as $B + Cx$, we would arrive at a system of two equations in three unknowns; this means that the value of C is arbitrary. We just choose $C = 0$.

Note that both $(x - 2)$ and $(x - 2)^2$ are factors of the denominator, so terms with both denominators must be included. (However, we need not include an x term in the numerator of the last fraction since an x term will appear when we combine the fractions as written.) Combining terms we obtain

Combine terms.

$$\begin{aligned}\frac{x + 1}{x^2 - 4x + 4} &= \frac{A}{x - 2} + \frac{B}{(x - 2)^2}\\ &= \frac{A(x - 2) + B}{(x - 2)^2}\\ &= \frac{Ax + (-2A + B)}{(x - 2)^2}.\end{aligned}$$

Equate coefficients of like powers of x.

Equating like powers of x we obtain the equations

$$\begin{aligned} A &= 1 \qquad \text{(coefficients of } x\text{)},\\ -2A + B &= 1 \qquad \text{(constants)}.\end{aligned}$$

This system has solution $A = 1$, $B = 3$, so

$$\frac{x + 1}{x^2 - 4x + 4} = \frac{1}{x - 2} + \frac{3}{(x - 2)^2}.$$

Combining our results we now have

$$\begin{aligned}\int \frac{2x^3 - 8x^2 + 9x + 1}{x^2 - 4x + 4}\, dx &= \int 2x\, dx + \int \frac{1}{x - 2}\, dx + \int \frac{3}{(x - 2)^2}\, dx\\ &= x^2 + \ln|x - 2| - \frac{3}{x - 2} + C. \qquad \blacksquare\end{aligned}$$

It is now time to state the formal procedure lurking behind Examples 1–4.

The obvious first step in obtaining a partial fraction decomposition for the rational function $\frac{p(x)}{q(x)}$ is to factor the denominator. A theorem about polynomials tells us what to expect as the outcome: *every polynomial $q(x)$ with real coefficients can be factored into the product of only two types of factors:*

(i) powers of linear terms, that is, terms of the form $(x - a)^n$, and/or
(ii) powers of irreducible quadratic terms, that is, terms of the form $(x^2 + bx + c)^m$.

(Thus we are assured, for example, that a cubic polynomial such as $x^3 - 6x^2 + 5x - 9$ can be factored at least into a quadratic term and a linear term, if not three linear terms. Unfortunately, however, this theorem provides no indication of what these factors actually might be.)

We shall not prove this theorem here, but it is central to understanding why the following procedure encompasses all rational integrands.

Procedure for Finding the Partial Fraction Decomposition of the Rational Function

$$f(x) = \frac{p(x)}{q(x)}$$

1. Check that the degree of $p(x)$ is smaller than the degree of $q(x)$. If not, perform a polynomial long division.
2. Factor the denominator $q(x)$ into the product of powers of linear and irreducible quadratic terms. That is, obtain the factorization

$$q(x) = d(x - a_1)^{n_1} \cdot \cdots \cdot (x - a_k)^{n_k} \cdot (x^2 + b_1x + c_1)^{m_1} \cdot \cdots \cdot (x^2 + b_jx + c_j)^{m_j}.$$

3. For each linear factor $(x - a)^n$ allow the terms

$$\frac{A_1}{x - a} + \frac{A_2}{(x - a)^2} + \cdots + \frac{A_n}{(x - a)^n}$$

in the partial fraction expansion. For each factor $(x^2 + bx + c)^m$ allow the terms

$$\frac{B_1x + C_1}{x^2 + bx + c} + \frac{B_2x + C_2}{(x^2 + bx + c)^2} + \cdots + \frac{B_mx + C_m}{(x^2 + bx + c)^m}.$$

4. Combine all terms in the partial fraction expansion and collect coefficients of like powers of x.
5. Equate coefficients of like powers of x between the partial fraction expansion and the original function $\dfrac{p(x)}{q(x)}$.
6. Solve the resulting system for the unknown constants.

The procedure is stated in its fullest generality. Do not be intimidated by the notation, as usually only a small number of factors appear. However, in deciding what types of terms to allow, step 3 provides a helpful reference.

Example 5 The following are several examples of rational functions and the forms of their associated partial fraction expansions.

Function	*Partial Fraction Expansion*
$\dfrac{1}{(x-1)(x+2)}$	$\dfrac{A}{x-1} + \dfrac{B}{x+2}$
$\dfrac{x+2}{(x-1)^2}$	$\dfrac{A}{x-1} + \dfrac{B}{(x-1)^2}$
$\dfrac{x^2-6x+1}{x(x^2-x-1)}$	$\dfrac{A}{x} + \dfrac{Bx+c}{x^2-x-1}$
$\dfrac{x+7}{x^2(x-2)^2}$	$\dfrac{A}{x} + \dfrac{B}{x^2} + \dfrac{C}{x-2} + \dfrac{D}{(x-2)^2}$
$\dfrac{x^3+6x+1}{x(x^2+x+1)^2}$	$\dfrac{A}{x} + \dfrac{Bx+C}{x^2+x+1} + \dfrac{Dx+E}{(x^2+x+1)^2}$
$\dfrac{x^4-x^2+1}{(x-1)^3(x^2+x+2)^2}$	$\dfrac{A}{x-1} + \dfrac{B}{(x-1)^2} + \dfrac{C}{(x-1)^3} + \dfrac{Dx+E}{x^2+x+2} + \dfrac{Fx+G}{(x^2+x+2)^2}$ ■

Example 6 Find $\displaystyle\int \frac{3x^4 + 9x^3 + 15x^2 + 10x + 4}{x(x^2 + 2x + 2)^2}\,dx.$

Solution: The numerator is fourth degree, and the denominator is fifth degree. Thus, long division is not needed. The denominator is in factored form so we proceed to seek the partial fraction decomposition:

$$\begin{aligned}
&\frac{3x^4 + 9x^3 + 15x^2 + 10x + 4}{x(x^2 + 2x + 2)^2} \\
&= \frac{A}{x} + \frac{Bx + C}{x^2 + 2x + 2} + \frac{Dx + E}{(x^2 + 2x + 2)^2} \\
&= \frac{A(x^2 + 2x + 2)^2 + x(Bx + C)(x^2 + 2x + 2) + x(Dx + E)}{x(x^2 + 2x + 2)^2} \\
&= \frac{(A + B)x^4 + (4A + 2B + C)x^3 + (8A + 2B + 2C + D)x^2 + (8A + 2C + E)x + 4A}{x(x^2 + 2x + 2)^2}
\end{aligned}$$

Equating coefficients of like powers of x gives the equations

$$\begin{aligned}
A + B &= 3 && \text{(coefficients of } x^4\text{)} \\
4A + 2B + C &= 9 && \text{(coefficients of } x^3\text{)} \\
8A + 2B + 2C + D &= 15 && \text{(coefficients of } x^2\text{)} \\
8A + 2C + E &= 10 && \text{(coefficients of } x\text{)} \\
4A &= 4 && \text{(constants)}
\end{aligned}$$

The solution of this system is $A = 1$, $B = 2$, $C = 1$, $D = 1$, and $E = 0$.
We therefore conclude that

$$\begin{aligned}
&\int \frac{3x^4 + 9x^3 + 15x^2 + 10x + 4}{x(x^2 + 2x + 2)^2}\,dx \\
&= \int \frac{1}{x}\,dx + \int \frac{2x + 1}{x^2 + 2x + 2}\,dx + \int \frac{x}{(x^2 + 2x + 2)^2}\,dx \\
&= \int \frac{1}{x}\,dx + \int \frac{2x + 2}{x^2 + 2x + 2}\,dx - \int \frac{1}{(x + 1)^2 + 1}\,dx + \int \frac{x}{((x + 1)^2 + 1)^2}\,dx.
\end{aligned}$$

The first two of these integrals produce logarithms and the third yields an inverse tangent function. The fourth is handled by a trigonometric substitution (see Exercise 13, Section 11.4). The result is

$$\begin{aligned}
&\int \frac{3x^4 + 9x^3 + 15x^2 + 10x + 4}{x(x^2 + 2x + 2)^2}\,dx \\
&= \ln|x| + \ln(x^2 + 2x + 2) - \frac{3}{2}\tan^{-1}(x + 1) - \frac{x + 2}{2[(x + 1)^2 + 1]} + C.
\end{aligned}$$

■

The Heaviside Method*

There is a procedure that often can speed the calculation of the unknown coefficients in an assumed partial fraction expansion. Suppose we wish to find constants A, B, and C so that

$$\frac{p(x)}{(x - a)(x - b)(x - c)} = \frac{A}{x - a} + \frac{B}{x - b} + \frac{C}{x - c}. \tag{5}$$

*This technique is due to Oliver Heaviside (1850–1925).

To find A, we (i) multiply both sides of the equation by $(x - a)$, obtaining

$$\frac{p(x)}{(x-b)(x-c)} = A + B\left(\frac{x-a}{x-b}\right) + C\left(\frac{x-a}{x-c}\right). \tag{6}$$

Then we (ii) set $x = a$, obtaining from equation (6) the statement

$$A = \frac{p(a)}{(a-b)(a-c)}.$$

Similarly, we can find B and C by repeating steps (i) and (ii) with the factors $(x - b)$ and $(x - c)$, respectively. We find that

$$B = \frac{p(b)}{(b-a)(b-c)}; \qquad C = \frac{p(c)}{(c-a)(c-b)}.$$

This procedure, valid only for the case of distinct linear factors, is sometimes described by saying that one ''strikes'' the factor on the left side of equation (5) corresponding to the desired constant and then evaluates this expression at the corresponding value of x.

Example 7 Find $\displaystyle\int \frac{6x^2 + 6x - 6}{x^3 + 2x^2 - x - 2}\,dx.$

Strategy
Find the factors of $x^3 + 2x^2 - x - 2$.

Solution
To factor the denominator, we recall that $(x - a)$ can be a factor of $x^3 + 2x^2 - x - 2$ only if a is a factor of the constant term 2. Thus, the only possible linear factors are $(x \pm 1)$ and $(x \pm 2)$. Trial and error with polynomial long division quickly shows that

$$x^3 + 2x^2 - x - 2 = (x - 1)(x + 1)(x + 2).$$

Write the form of the partial fraction decomposition.

We therefore seek the partial fraction decomposition

$$\frac{6x^2 + 6x - 6}{(x-1)(x+1)(x+2)} = \frac{A}{x-1} + \frac{B}{x+1} + \frac{C}{x+2}. \tag{7}$$

Use the Heaviside method to find A, B, and C.

Using the Heaviside method to find A, we strike the factor $(x - 1)$ in the left side of (7) and set $x = 1$ to obtain

$$A = \frac{6(1)^2 + 6(1) - 6}{(1+1)(1+2)} = \frac{6}{6} = 1.$$

Similarly we find that

$$B = \frac{6(-1)^2 + 6(-1) - 6}{(-1-1)(-1+2)} = \frac{-6}{-2} = 3$$

and

$$C = \frac{6(-2)^2 + 6(-2) - 6}{(-2-1)(-2+1)} = \frac{6}{3} = 2.$$

Thus

$$\int \frac{6x^2 + 6x - 6}{x^3 + 2x^2 - x - 2}\,dx = \int \frac{1}{x-1}\,dx + \int \frac{3}{x+1}\,dx + \int \frac{2}{x+2}\,dx$$
$$= \ln|x-1| + 3\ln|x+1| + 2\ln|x+2| + C.$$

■

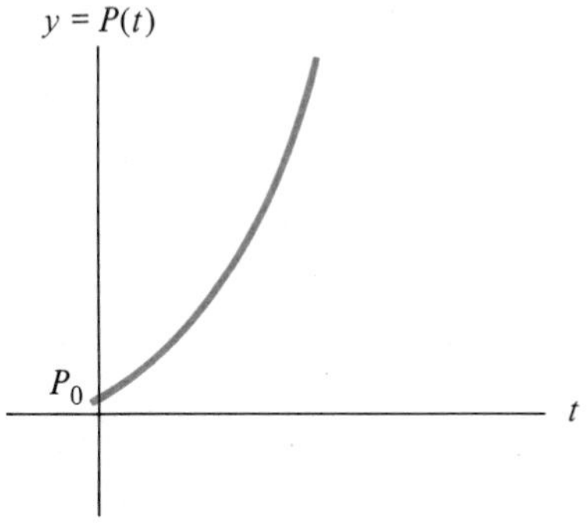

Figure 5.1 Model for natural growth assumes an unlimited environment.

Example 8 An important application of the method of partial fraction decomposition occurs in the logistic growth model. Recall from Chapter 9 that the Law of Natural Growth

$$\{\text{Rate of growth}\} \propto \{\text{Present size of population}\} \tag{8}$$

leads to the differential equation

$$\frac{dP}{dt} = kP, \tag{9}$$

which has solutions $P(t) = P_0 e^{kt}$. The constant k in equation (9) is referred to as the growth constant. A typical solution curve for equation (9) is depicted in Figure 5.1. The objection to this growth model raised in Chapter 9 is that it assumes an unlimited environment. That is, when $k > 0$ the model assumes that the population can increase at an exponential rate forever.

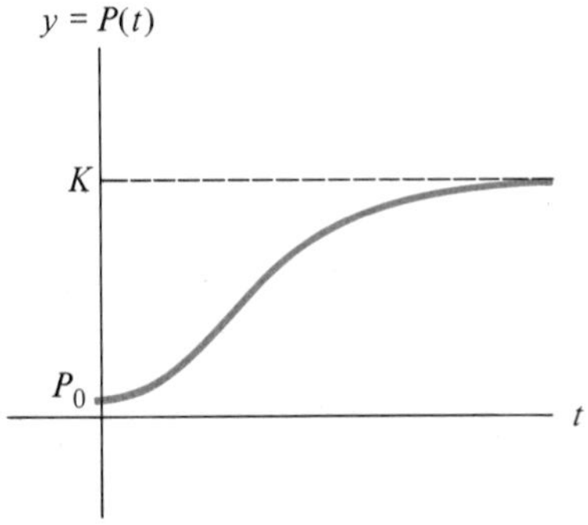

Figure 5.2 Logistic growth model assumes a maximum possible population size K.

The logistic growth model postulates a maximum possible population size K (sometimes called the **carrying capacity** of the system) by assuming that the growth rate is proportional to both present population size and the ''unutilized capacity for growth'' in the system (see Figure 5.2). That is,

$$\text{Rate of Growth} \propto (\text{Present size}) \cdot (\text{Unutilized capacity for growth}). \tag{10}$$

Since the fraction of the carrying capacity K unutilized at population size $P < K$ is $\left(\frac{K-P}{K}\right)$, the differential equation corresponding to the logistic growth law in (10) is

$$\frac{dP}{dt} = kP\left(\frac{K-P}{K}\right) \tag{11}$$

where k is again called the natural growth constant.

To solve (11), we separate variables obtaining

$$\frac{dP}{P(K-P)} = \frac{k}{K}\,dt,$$

so

$$\int \frac{dP}{P(K-P)} = \int \frac{k}{K}\,dt = \left(\frac{k}{K}\right)t + C. \tag{12}$$

To find the integral on the left side, we use the method of partial fractions: Setting

$$\frac{1}{P(K-P)} = \frac{A}{P} + \frac{B}{K-P}$$

$$= \frac{A(K - P) + BP}{P(K - P)}$$

$$= \frac{(-A + B)P + AK}{P(K - P)}$$

gives the equations

$$\begin{cases} -A + B = 0 & \text{(coefficients of } P\text{)}, \\ AK \quad\;\, = 1 & \text{(constants)}. \end{cases}$$

Thus, $A = B = \dfrac{1}{K}$. The integral on the left side of (12) is therefore

$$\begin{aligned} \int \frac{dP}{P(K - P)} &= \frac{1}{K} \cdot \int \frac{1}{P}\, dP + \frac{1}{K} \cdot \int \frac{1}{K - P}\, dP \\ &= \frac{1}{K} \ln P - \frac{1}{K} \ln(K - P) \\ &= \frac{1}{K} \ln\left(\frac{P}{K - P}\right). \end{aligned}$$

Thus equation (12) becomes

$$\frac{1}{K} \ln\left(\frac{P}{K - P}\right) = \frac{k}{K} t + C$$

or

$$\frac{P}{K - P} = e^{kt + KC}.$$

Solving for P now gives

$$\begin{aligned} Pe^{-kt}e^{-KC} &= K - P \\ P(1 + Me^{-kt}) &= K \qquad \text{where} \qquad M = e^{-KC} \\ P &= \frac{K}{1 + Me^{-kt}}. \end{aligned} \tag{13}$$

From the form of the solution in line (13) we can indeed see that

$$\lim_{t \to \infty} P(t) = \lim_{t \to \infty} \frac{K}{1 + Me^{-kt}} = K$$

for populations satisfying the logistic growth model. ■

Exercise Set 11.5

In Exercises 1–30, find the indicated integral.

1. $\displaystyle\int \frac{1}{x(x + 1)}\, dx$

2. $\displaystyle\int \frac{3x + 2}{x(x + 1)}\, dx$

3. $\displaystyle\int \frac{2x + 4}{1 - x^2}\, dx$

4. $\displaystyle\int \frac{7x + 2}{x^2 + x - 2}\, dx$

5. $\displaystyle\int \frac{1}{1 - x^2}\, dx$

6. $\displaystyle\int \frac{x^2 + 1}{x^2 - 1}\, dx$

7. $\displaystyle\int \frac{1}{x + x^3}\, dx$

8. $\displaystyle\int \frac{x^2 + 3x + 3}{x(1 + x)}\, dx$

9. $\displaystyle\int \frac{1}{(x + 1)(x^2 + 1)}\, dx$

10. $\displaystyle\int \frac{ax^2 + bx + a}{x + x^3}\, dx$

11. $\int \frac{4x^2 + x + 1}{(x + 2)(x^2 + 1)}\, dx$

12. $\int \frac{1}{x(x + 1)^2}\, dx$

13. $\int \frac{x^2 + 5x + 1}{x(x + 1)^2}\, dx$

14. $\int \frac{dx}{(x + 1)^2}$

15. $\int \frac{2x^2 + 3x + 2}{x^3 + 2x^2 + x}\, dx$

16. $\int \frac{x^3 + x^2 + 1}{x^4 + 2x^2 + 1}\, dx$

17. $\int \frac{dx}{x^3 + 3x^2 + 2x}\, dx$

18. $\int \frac{3x^2 + 6x + 2}{x(x + 1)(x + 2)}\, dx$

19. $\int \frac{5x^3 - 9x^2 + 3x - 3}{x(x - 3)(x^2 + 1)}\, dx$

20. $\int \frac{6x^3 - 4x^2 + 3x - 3}{x(x - 1)(1 + x^2)}\, dx$

21. $\int \frac{4}{x^3 + x}\, dx$

22. $\int \frac{dx}{x^3 - 6x^2 + 13x - 10}$

23. $\int \frac{dx}{(a^2 - x^2)^2}$

24. $\int \frac{2x^3 - 4x^2 - 7x - 18}{x^2 - 2x - 8}\, dx$

25. $\int \frac{10x^2 + 22x + 57}{x^3 + 7x^2 + 12x + 18}\, dx$

26. $\int \frac{2x^2 - 3x - 2}{x^3 + x^2 - 2x}\, dx$

27. $\int \frac{5x^2 + 2x + 2}{x^3 - 1}\, dx$

28. $\int \frac{6x^2 - 21x - 9}{x^3 - 6x^2 + 3x + 10}\, dx$

29. $\int \frac{9x^2 + 26x - 16}{x^3 + 2x^2 - 8x}\, dx$

30. $\int \frac{x^4 - x^3 + 3x^2 - 10x + 8}{x^3 - x^2 - 4}\, dx$

31. Find the area of the region bounded above by the graph of $y = \frac{7x + 3}{x^2 + x}$ and below by the x-axis for $1 \le x \le 2$.

32. The region bounded above by the graph of $y = \frac{x}{x^2 + 1}$ and below by the axis for $0 \le x \le 1$ is revolved about the x-axis. Find the volume of the resulting solid.

33. The region bounded by the graph of $y = \frac{1}{(x + 1)(x + 3)}$ and the x-axis for $0 \le x \le 2$ is revolved about the y-axis. Find the volume of the resulting solid.

34. Find the average value of the function $f(x) = \frac{4x^2}{(1 + x^2)^2}$ on the interval $[0, 1]$.

35. Prove that if

$p(x) = a_n x^n + a_{n-1}x^{n-1} + \cdots + a_1 x + a_0$ and $q(x) = b_n x^n + b_{n-1}x^{n-1} + \cdots + b_1 x + b_0$

are polynomials with $p(x) = q(x)$ for all x, then $a_n = b_n$, $a_{n-1} = b_{n-1}, \ldots, a_0 = b_0$. That is, if two polynomials are equal then corresponding coefficients of like powers of x must be equal. (*Hint:* Set $x = 0$ to conclude that $a_0 = b_0$. Differentiate.)

36. Show that the expression $\frac{A}{x - a} + \frac{B}{(x - a)^2}$ can always be brought to the form $\frac{Cx + D}{(x - a)^2}$ by correctly choosing A and B. (This shows that we need not allow for a first degree numerator in the term $\frac{B}{(x - a)^2}$ when seeking a partial fraction decomposition for $\frac{p(x)}{(x - a)^2}$.)

37. Find a solution of the logistic differential equation $\frac{dP}{dt} = 2P\left(\frac{100 - P}{100}\right)$. What is the carrying capacity?

38. Let $P(t)$ be a solution of the differential equation $\frac{dP}{dt} = P(1 - P)$ so that $P(0) > 0$. What is $\lim_{t \to \infty} P(t)$?

39. Verify the evaluation of the integrals in Example 6.

11.6 MISCELLANEOUS SUBSTITUTIONS

With the exception of the integration by parts formula, all integration techniques encountered up to this point may be viewed as substitution methods of one sort or another. In this section we present several additional types of substitution techniques.

Radicals of Polynomial Expressions

In integrands involving fractional powers of polynomial expressions, such as $\sqrt[n]{p(x)}$, it is often useful to make a substitution of the form $u^n = p(x)$. This substitution will eliminate the radical sign since $\sqrt[n]{p(x)} = \sqrt[n]{u^n} = u$.

Example 1 Find $\displaystyle\int \frac{dx}{x\sqrt{x+1}}$, $\qquad x > 0$.

Strategy

To obtain a square under the square root we use

$u^2 = x + 1$.

Then solve for x and differentiate to find dx.

Substitute.

Form the partial fraction decomposition.

Integrate.

Solve $u^2 = x + 1$ for $u = \sqrt{x+1}$ and substitute back.

Solution

We make the substitution $u^2 = x + 1$. Then $x = u^2 - 1$, so $dx = 2\,u\,du$. The integral becomes

$$\int \frac{dx}{x\sqrt{x+1}} = \int \frac{2\,u\,du}{(u^2-1)\cdot u}$$

$$= \int \frac{2\,du}{u^2-1}$$

$$= \int \left\{ \frac{1}{u-1} - \frac{1}{u+1} \right\} du$$

$$= \ln\left| \frac{u-1}{u+1} \right| + C$$

$$= \ln\left(\frac{\sqrt{x+1}-1}{\sqrt{x+1}+1} \right) + C. \qquad \blacksquare$$

Example 2 Find $\displaystyle\int \frac{dx}{\sqrt{x} - \sqrt[3]{x}}$, $\qquad x > 0$.

Strategy

We need $x = u^n$ where both $\dfrac{n}{2}$ and $\dfrac{n}{3}$ are integers. Thus, we use $n = 6$.

Substitute and simplify.

Reduce integrand by long division.

$$\begin{array}{r} u^2 + u + 1 \\ u - 1 \overline{)u^3 } \\ \underline{u^3 - u^2} \\ u^2 \\ \underline{u^2 - u} \\ u \\ \underline{u - 1} \\ 1 \end{array} + \frac{1}{u-1}$$

Substitute back

$u = x^{1/6}$.

Solution

We use the substitution $x = u^6$. Then $dx = 6u^5\,du$, so we obtain

$$\int \frac{dx}{\sqrt{x} - \sqrt[3]{x}} = \int \frac{6u^5\,du}{\sqrt{u^6} - \sqrt[3]{u^6}}$$

$$= \int \frac{6u^5\,du}{u^3 - u^2}$$

$$= 6 \int \frac{u^3\,du}{u-1}$$

$$= 6 \int \left\{ u^2 + u + 1 + \frac{1}{u-1} \right\} du$$

$$= 2u^3 + 3u^2 + 6u + 6\ln|u-1| + C$$

$$= 2x^{1/2} + 3x^{1/3} + 6x^{1/6} + 6\ln|x^{1/6}-1| + C. \qquad \blacksquare$$

Example 3 Find $\int \sqrt{1 + e^x}\,dx$.

Strategy
Substitute a square for the quantity $1 + e^x$.

Solution
We use the substitution $u^2 = 1 + e^x$. Then

$$e^x = u^2 - 1$$

so

Note that $u^2 > 1$ for all x, so these expressions are well defined.

$$x = \ln(u^2 - 1),$$

and

$$dx = \frac{2\,u\,du}{u^2 - 1}.$$

We obtain

Substitute and simplify.

$$\int \sqrt{1 + e^x}\,dx = \int \sqrt{u^2}\left(\frac{2u}{u^2 - 1}\right) du$$

$$= \int \frac{2u^2}{u^2 - 1}\,du$$

Reduce integrand by long division.

$$= \int \left(2 + \frac{2}{u^2 - 1}\right) du$$

Form the partial fraction decomposition.

$$= \int \left(2 + \frac{1}{u - 1} - \frac{1}{u + 1}\right) du$$

Integrate.

$$= 2u + \ln\left|\frac{u - 1}{u + 1}\right| + C$$

Substitute back $u = \sqrt{1 + e^x}$.

$$= 2\sqrt{1 + e^x} + \ln\left(\frac{\sqrt{1 + e^x} - 1}{\sqrt{1 + e^x} + 1}\right) + C. \qquad \blacksquare$$

Rational Expressions in Sine and Cosine

There is a special substitution that allows us to handle a number of integrals involving rational functions of sin x and cos x. While the technique is not elegant, it is the only technique available for integrating such expressions. The substitution is

$$u = \tan\left(\frac{x}{2}\right). \tag{1}$$

To find the values for sin x and cos x corresponding to substitution (1) we observe that

$$\cos\left(\frac{x}{2}\right) = \frac{1}{\sec\left(\frac{x}{2}\right)} = \frac{1}{\sqrt{1 + \tan^2\left(\frac{x}{2}\right)}} = \frac{1}{\sqrt{1 + u^2}}$$

and

$$\sin\left(\frac{x}{2}\right) = \tan\left(\frac{x}{2}\right)\cos\left(\frac{x}{2}\right) = \frac{u}{\sqrt{1 + u^2}}.$$

Thus

$$\sin x = 2\sin\left(\frac{x}{2}\right)\cos\left(\frac{x}{2}\right) = \frac{2u}{1+u^2} \tag{2}$$

and

$$\cos x = \sqrt{1-\sin^2 x} = \left\{1 - \left(\frac{2u}{1+u^2}\right)^2\right\}^{1/2} = \left\{\frac{1-2u^2+u^4}{1+2u^2+u^4}\right\}^{1/2} \tag{3}$$

$$= \frac{1-u^2}{1+u^2}.$$

Finally, from (1) we have $x = 2\tan^{-1} u$, so

$$dx = \frac{2}{1+u^2}\,du. \tag{4}$$

Example 4 Find $\displaystyle\int \frac{dx}{2+\cos x}$.

Strategy
Use substitution (1) in the form of equations (3) and (4).

Solution
Using equations (3) and (4) we obtain

$$\int \frac{dx}{2+\cos x} = \int \frac{\frac{2}{1+u^2}}{2+\left(\frac{1-u^2}{1+u^2}\right)}\,du$$

Simplify.

$$= \int \frac{\frac{2}{1+u^2}}{\frac{2(1+u^2)+(1-u^2)}{1+u^2}}\,du$$

Factor into form $\displaystyle\int \frac{dz}{1+z^2}$.

$$= \int \frac{2\,du}{3+u^2}$$

$$= \frac{2}{3}\int \frac{du}{1+\left(\frac{u}{\sqrt{3}}\right)^2}$$

$$= \frac{2}{\sqrt{3}}\int \frac{\frac{1}{\sqrt{3}}\,du}{1+\left(\frac{u}{\sqrt{3}}\right)^2}$$

$$= \frac{2}{\sqrt{3}}\tan^{-1}\left(\frac{u}{\sqrt{3}}\right) + C$$

Substitute back using (1).

$$= \frac{2}{\sqrt{3}}\tan^{-1}\left(\frac{1}{\sqrt{3}}\tan\frac{x}{2}\right) + C. \quad ■$$

Example 5 Find $\int \frac{1 + \sin x}{1 + \cos x}\, dx$.

Strategy

Use substitution (1).

Solution

Using equations (2)–(4) we obtain

$$\int \frac{1 + \sin x}{1 + \cos x}\, dx = \int \frac{\left(1 + \frac{2u}{1 + u^2}\right)}{\left(1 + \frac{1 - u^2}{1 + u^2}\right)} \cdot \left(\frac{2}{1 + u^2}\right) du$$

Simplify.

$$= \int \left(\frac{1 + 2u + u^2}{2}\right) \cdot \left(\frac{2}{1 + u^2}\right) du$$

Multiply factors in integrand.

$$= \int \frac{1 + 2u + u^2}{1 + u^2}\, du$$

Simplify by performing a long division:

$$\begin{array}{r} 1 \\ u^2 + 1 \overline{\smash{)}u^2 + 2u + 1} \\ \underline{u^2 \qquad + 1} \\ 2u \end{array} + \frac{2u}{u^2 + 1}$$

$$= \int \left(1 + \frac{2u}{u^2 + 1}\right) du$$

$$= u + \ln(u^2 + 1) + C$$

$$= \tan\left(\frac{x}{2}\right) + \ln\left(\tan^2\left(\frac{x}{2}\right) + 1\right) + C. \quad \blacksquare$$

Exercise Set 11.6

Find the following integrals.

1. $\int \frac{\sqrt{x}}{1 + \sqrt{x}}\, dx$

2. $\int \frac{\sqrt{x + 1}}{x}\, dx$

3. $\int \frac{dx}{\sqrt{x} + 2\sqrt[3]{x}}$

4. $\int \frac{\sqrt[3]{x} + 1}{\sqrt[3]{x} - 1}\, dx$

5. $\int \frac{x}{\sqrt[3]{x + 1}}\, dx$

6. $\int \frac{x + 2}{\sqrt{x - 2}}\, dx$

7. $\int \frac{dx}{2 + \sin x}$

8. $\int \frac{dx}{1 - \sin x}$

9. $\int \frac{dx}{1 - \cos x}$

10. $\int \frac{x^3}{\sqrt{1 + x^2}}\, dx$

11. $\int \frac{x^2}{\sqrt[3]{1 + x}}\, dx$

12. $\int \frac{1 + \sqrt{x}}{1 - \sqrt{x}}\, dx$

13. $\int \frac{dx}{\sin x + \cos x}$

14. $\int \frac{1 - \sin x}{1 + \cos x}\, dx$

15. $\int \frac{x^3}{\sqrt{1 - 2x}}\, dx$

16. $\int x^2(x^2 + 1)^{3/2}\, dx$

17. $\int \sqrt{\frac{1 + x}{1 - x}}\, dx$

18. $\int \frac{dx}{1 + e^x}$

19. $\int \frac{dx}{x\sqrt{a + bx}}$

20. $\int x\sqrt{1 + x}\, dx$

21. $\int \frac{x}{\sqrt{2 + x}}\, dx$

11.7 THE USE OF INTEGRAL TABLES

The endpapers of this text constitute what is referred to as a *table of integrals*—a list of antiderivatives for various types of functions. While the list presented here is relatively short, many tables of integrals have been published that contain hundreds of entries. Many of the formulas contained in such tables are established using the techniques of this chapter. Such tables are very useful when you are faced with a large number of integrals to evaluate or when you are unwilling to spend the time required to execute the techniques of this section.

Using a table of integrals is a relatively straightforward task. You must match the form of the integrand with those appearing in the table, select an integral matching yours in form, and identify the corresponding values of the constants. For example, the integral

$$\int \frac{dx}{x(3 + 5x^2)}$$

is evaluated using formula (64), which states that

$$\int \frac{du}{u(a + bu^2)} = \frac{1}{2a} \ln\left|\frac{u^2}{a + bu^2}\right| + C.$$

To do so we must take $a = 3$, $b = 5$, and $u = x$. We obtain

$$\int \frac{dx}{x(3 + 5x^2)} = \frac{1}{2 \cdot 3} \ln\left|\frac{x^2}{3 + 5x^2}\right| + C.$$

However, it is often necessary to first perform a u-substitution to bring an integrand into the form of one of the entries in a table. In doing so you must be especially careful to carry out the substitution for all variables, including the differential dx. The following examples demonstrate this technique.

Example 1 Use the table of integrals to evaluate

$$\int \frac{dx}{1 + e^{\pi x}}.$$

Solution: We use formula (103):

$$\int \frac{du}{1 + e^u} = \ln\left(\frac{e^u}{1 + e^u}\right) + C.$$

To do so we must use the u-substitution

$$u = \pi x; \qquad du = \pi\, dx$$

Then $dx = \dfrac{1}{\pi}\, du$, and we obtain

$$\int \frac{dx}{1 + e^{\pi x}} = \int \frac{\frac{1}{\pi}\, du}{1 + e^u} = \frac{1}{\pi} \int \frac{du}{1 + e^u} = \frac{1}{\pi} \ln\left(\frac{e^{\pi x}}{1 + e^{\pi x}}\right) + C. \quad \blacksquare$$

Example 2 Use the table of integrals to evaluate

$$\int x\sqrt{x^2 - 4x + 1}\, dx.$$

Solution: This time it is not so clear which formula might correspond to the given integral. Since the table contains no entries involving the general quadratic $ax^2 + bx + c$, we begin by completing the square on the quadratic term:

$$\begin{aligned} x^2 - 4x + 1 &= (x^2 - 4x + 4) - 3 \\ &= (x - 2)^2 - 3. \end{aligned} \tag{1}$$

Our next step is to use the u-substitution

$$u = x - 2; \qquad du = dx.$$

Then $x = u + 2$, and using equation (1), we obtain

$$\int x\sqrt{x^2 - 4x + 1}\, dx = \int x\sqrt{(x-2)^2 - 3}\, dx \tag{2}$$

$$= \int (u + 2)\sqrt{u^2 - 3}\, du$$

$$= \int u\sqrt{u^2 - 3}\, du + 2\int \sqrt{u^2 - 3}\, du.$$

To handle the first integral, we use formula 46:

$$\int u\sqrt{u^2 \pm a^2}\, du = \frac{1}{3}(u^2 \pm a^2)^{3/2} + C. \tag{3}$$

The second integral is evaluated using formula 45:

$$\int \sqrt{u^2 \pm a^2}\, du = \frac{1}{2}[u\sqrt{u^2 \pm a^2} \pm a^2 \ln(u + \sqrt{u^2 \pm a^2})] + C. \tag{4}$$

Taking $a = \sqrt{3}$ and the sign choice $\pm = -$, we combine (2), (3), and (4) to obtain

$$\int x\sqrt{x^2 - 4x + 1}\, dx = \frac{1}{3}[(x-2)^2 - 3]^{3/2} +$$

$$2 \cdot \frac{1}{2}[(x-2)\sqrt{(x-2)^2 - 3} - 3\ln(x - 2 + \sqrt{(x-2)^2 - 3})] + C. \quad \blacksquare$$

Exercise Set 11.7

In Exercises 1–30, use the table of integrals in the endpapers to evaluate the integral.

1. $\int \frac{dx}{1 + e^{5x}}$
2. $\int \frac{dx}{x(3 + x)}$
3. $\int \frac{dx}{x^2(9 + 2x)^2}$
4. $\int \frac{5\, dx}{\sqrt{14x - x^2}}$
5. $\int \frac{x + 3}{2 + 5x^2}\, dx$
6. $\int \frac{3\, dx}{\sqrt{6x + x^2}}$
7. $\int \tan^4 \pi x\, dx$
8. $\int x^2 \ln x\, dx$
9. $\int \sin 6x \sin 3x\, dx$
10. $\int \frac{dx}{8 + 3e^{5x}}$
11. $\int \frac{dx}{x^2(2 - 3x)}$
12. $\int \sqrt{12x - 4x^2}\, dx$
13. $\int \frac{9\, dx}{25 - 4x^2}$
14. $\int \frac{5\, dx}{\sqrt{10x + x^2}}$
15. $\int \tan^3 5x\, dx$
16. $\int e^x \sin 3x\, dx$
17. $\int \frac{6x^2\, dx}{\sqrt{6 + x}}$
18. $\int \frac{\sqrt{9 - x^2}}{x^2}\, dx$
19. $\int \frac{6\, dx}{(14 - x^2)^{3/2}}$
20. $\int x \cos^{-1} 3x\, dx$
21. $\int \frac{\pi}{4 + 4 \sin 2x}\, dx$
22. $\int \frac{4\, dx}{x\sqrt{9 - x^2}}$
23. $\int \frac{3\, dx}{x\sqrt{x^2 + 16}}$
24. $\int \sin \ln \pi x\, dx$
25. $\int \frac{dx}{9x^2 + 12x + 5}$
26. $\int \frac{dx}{\sqrt{7 - 6x - x^2}}$
27. $\int \frac{7\, dx}{(x + 4)\sqrt{x^2 + 8x + 20}}$
28. $\int \frac{4\, dx}{4x^2 + 20x + 16}$
29. $\int \frac{(2x^2 - 7)\, dx}{\sqrt{16 - x^2}}$
30. $\int \frac{dx}{(2x - x^2)^{3/2}}$

11.8 IMPROPER INTEGRALS

In Section 6.6 we encountered functions defined by integration of the form

$$F(t) = \int_a^t f(x)\,dx. \tag{1}$$

When $f(x) \geq 0$ for $x \geq a$, we identified the value of $F(t)$ with the area of the region bounded by the graph of $y = f(x)$ and the x-axis between $x = a$ and $x = t$ (Figure 8.1).

Now if $f(x)$ in equation (1) is continuous on $[a, \infty)$, the function $F(t)$ is a perfectly well defined function for all $t > a$. We may therefore ask about its limit as $t \to \infty$. To do so, we define the **improper integral** $\int_a^\infty f(x)\,dx$ as this limit, when it exists (Figure 8.2).

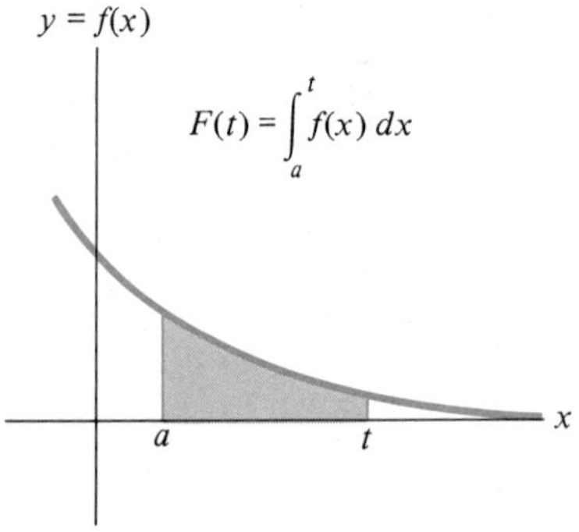

Figure 8.1 $F(t)$ defined by integration.

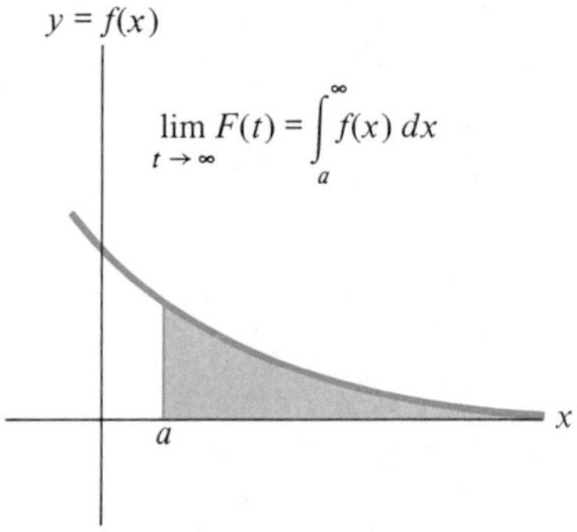

Figure 8.2 $\lim_{t\to\infty} F(t)$ defines an improper integral.

DEFINITION 1

$$\int_a^\infty f(x)\,dx = \lim_{t\to\infty} \int_a^t f(x)\,dx.$$

If the limit in Definition (1) exists, we say that the improper integral **converges.** Otherwise it is said to **diverge.**

There are several important reasons for studying improper integrals at this point, including the following:

(i) In our study of infinite series (Chapters 12–14), we shall need to be able to determine whether certain regions of infinite length might nevertheless contain only a finite amount of area. Since the definite integral determines area when $f(x) \geq 0$, Definition 1 gives us a way to extend the notion of area to regions of infinite length.

(ii) In Chapter 9 we determined that the **present value** P_0 of an asset assumed to be worth P dollars after T years was $P_0 = e^{-rT}P$, assuming continuous compounding at the annual rate r. When the asset itself is income-producing, such as a forest continually harvested for lumber, P is a function of t, and the calculation becomes

$$P_0 = \int_0^T e^{-rt}P(t)\,dt.$$

(See Example 6.) The calculation of present value for an asset with an infinite life expectancy then leads to the improper integral

$$P_0 = \int_0^{\infty} e^{-rt} P(t)\, dt.$$

Calculation of improper integrals of the type given in Definition 1 is simply a matter of carefully evaluating the indicated limit.

Example 1 Find $\displaystyle\int_1^{\infty} \frac{1}{x^2}\, dx$.

Strategy

Apply Definition (1).

Evaluate the definite integral from $x = 1$ to $x = t$.

Evaluate the limit as $t \to \infty$ of this integral.

Solution

$$\begin{aligned}\int_1^{\infty} \frac{1}{x^2}\, dx &= \lim_{t\to\infty} \int_1^{t} \frac{1}{x^2}\, dx \\ &= \lim_{t\to\infty} \left\{ -\frac{1}{x}\right]_{x=1}^{x=t} \Big\} \\ &= \lim_{t\to\infty} \left[\left(-\frac{1}{t}\right) - (-1)\right] \\ &= 0 - (-1) = 1.\end{aligned}$$

This improper integral converges. ■

Example 2 Find $\displaystyle\int_1^{\infty} \frac{1}{x}\, dx$.

Strategy

Apply Definition (1).

Evaluate the definite integral.

Evaluate the limit using the fact that $\lim_{t\to\infty} \ln t = +\infty$.

Solution

$$\begin{aligned}\int_1^{\infty} \frac{1}{x}\, dx &= \lim_{t\to\infty} \int_1^{t} \frac{1}{x}\, dx \\ &= \lim_{t\to\infty} \{\ln x]_{x=1}^{x=t}\} \\ &= \lim_{t\to\infty} [\ln t - \ln 1] \\ &= +\infty.\end{aligned}$$

This improper integral diverges. ■

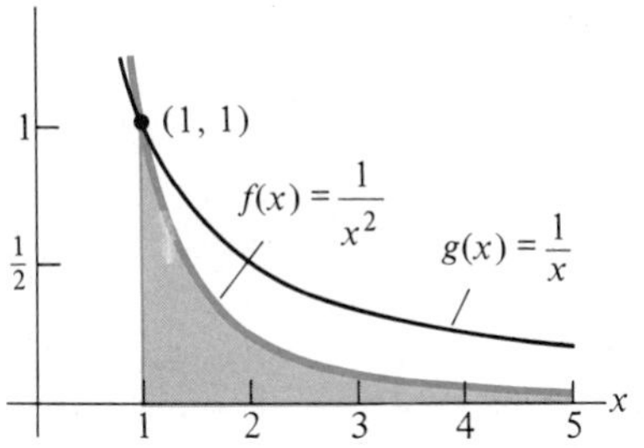

Figure 8.3 Graph of $f(x) = \dfrac{1}{x^2}$ approaches x-axis more quickly than graph of $g(x) = \dfrac{1}{x}$.

REMARK: It is instructive to note the difference between the improper integrals in Examples 1 and 2. Both are of the form $\int_1^{\infty} \frac{1}{x^p}\, dx$, but their behavior is quite different. As Figure 8.3 illustrates, the graph of $y = 1/x^2$ approaches the x-axis more quickly as x increases than does the graph of $y = 1/x$. In an intuitive sense, it is the rapidity with which the graph of such a function approaches the x-axis that determines whether its improper integral converges. However, this determination cannot be made simply by inspecting a graph. The next example gives a more precise criterion.

Example 3 For what values of p does the improper integral $\displaystyle\int_1^{\infty} \frac{1}{x^p}\, dx$ converge?

Strategy
Evaluate the improper integral, carrying along the constant p.

Solution
We have

$$\int_1^\infty \frac{1}{x^p}\,dx = \lim_{t\to\infty}\int_1^t x^{-p}\,dx$$

Assume $p \neq 1$, so the integral is obtained via the Power Rule.

$$= \lim_{t\to\infty}\left\{\left.\frac{x^{-p+1}}{-p+1}\right]_{x=1}^{x=t}\right\}, \qquad p \neq 1$$

$$= \lim_{t\to\infty}\left[\frac{t^{1-p}}{1-p} - \frac{1}{1-p}\right].$$

The existence of the limit depends upon the sign of the exponent $1 - p$. Examine each case.

Three cases arise:

(a) If $1 - p < 0$, then $t^{1-p} \to 0$ as $t \to \infty$ and the integral converges.
(b) If $1 - p > 0$, then $t^{1-p} \to \infty$ as $t \to \infty$ and the integral diverges.
(c) In the case $1 - p = 0$, excluded above, the integral is $\int_1^\infty \frac{1}{x}\,dx$, which diverges (Example 2).

We therefore conclude that

$$\int_1^\infty \frac{1}{x^p}\,dx \text{ converges if and only if } p > 1. \qquad \blacksquare$$

By use of the improper integral, we can extend the notion of area to regions of infinite length.

DEFINITION 2

Let $f(x) \geq 0$ for all $x \geq a$. The area of the region bounded above by the graph of $y = f(x)$ and below by the x-axis and extending to the right of the line $x = a$ is defined to be

$$A = \int_a^\infty f(x)\,dx.$$

if this integral converges.

REMARK: Our previous definition of area as

$$A = \int_a^b f(x)\,dx$$

involved A as an infinite limit of approximating sums. Definition 2 actually involves a "double limit," as both the number of approximating rectangles and the interval over which they are constructed become infinite. The notions of volume and surface area for solids of revolution may be extended to regions of infinite length in the same way.

Example 4 Find the area of the region bounded above by the graph of $y = 1/x^3$, below by the x-axis, and on the left by the line $x = 1$.

Solution: By Definition 2 and Example 3 we have

$$A = \int_1^\infty \frac{1}{x^3}\,dx = \lim_{t\to\infty}\left\{\int_1^t x^{-3}\,dx\right\} = \lim_{t\to\infty}\left\{\frac{t^{-2}}{-2} - \frac{1^{-2}}{-2}\right\}$$
$$= \frac{1}{2}.$$

■

Example 5 Find the volume of the solid obtained by revolving about the x-axis the region R bounded above by the graph of $y = \frac{1}{x}$ and below by the x-axis for $1 \le x < \infty$ (Figure 8.4).

Solution: This volume is $V = \int_1^\infty \pi\left[\frac{1}{x}\right]^2 dx = \pi$, by Example 1. Thus, although R has infinite area, it sweeps out a region of finite volume. (This peculiarity is a result only of the way in which we have extended these definitions to the infinite case. Such an intriguing object cannot be realized.) ■

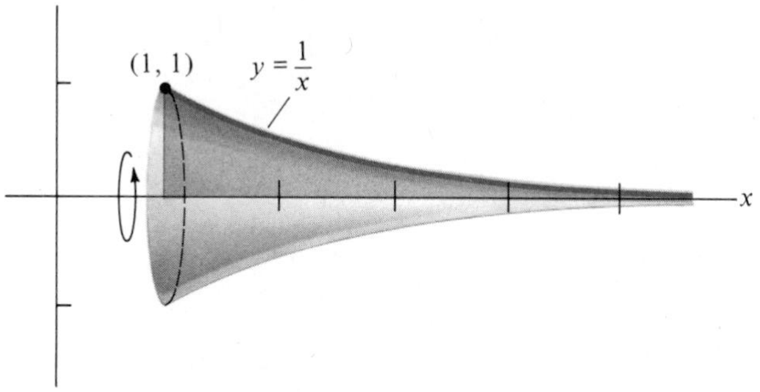

Figure 8.4 A region of infinite area generates a solid of finite volume.

Example 6 In Chapter 9 we saw that the **present value** of a sum P of money that will become available after t years is $P_0 = e^{-rt}P$, where r is the rate of continuous compounding of interest. Closely related to this notion is that of a **revenue stream,** which is simply an anticipated flow of money over time. For example, consider the question of defining the present value of the profits from a small business over the next five years if profits are expected at a rate of $P(t)$, where $P(t)$ is a function of time.

If we partition the time interval $[0, 5]$ into small intervals of length Δt, then the present value of the profits earned during the time interval $[t_{j-1}, t_j]$ is approximated by

$$\Delta P_j = e^{-rs_j}P(s_j)\,\Delta t \qquad 1 \le j \le n$$

where s_j is an arbitrary element of the interval $[t_{j-1}, t_j]$. An approximation to the total present value of these profits is then

$$P_0 \approx \sum_{j=1}^{n} e^{-rs_j}P(s_j)\,\Delta t.$$

As $n \to \infty$ this Riemann sum approaches the integral $\int_0^5 e^{-rt}P(t)\,dt$, so we define

$$P_0 = \int_0^5 e^{-rt}P(t)\,dt.$$

If we assume the business to have an infinite expected lifetime, the present value of all future profits is defined by the improper integral

$$P_0 = \int_0^\infty e^{-rt}P(t)\,dt.$$ ■

Obviously, we can define improper integrals of the form $\int_{-\infty}^a f(x)\,dx$ by analogy with Definition 1. We can extend the notion of improper integral to integrals with two infinite limits as follows.

DEFINITION 3

The improper integral $\int_{-\infty}^\infty f(x)\,dx$ is defined by

$$\int_{-\infty}^\infty f(x)\,dx = \int_{-\infty}^0 f(x)\,dx + \int_0^\infty f(x)\,dx.$$

The integral on the left is said to converge only when both integrals on the right converge. Otherwise it is said to diverge.

Example 7 Find $\int_{-\infty}^\infty \frac{1}{1+x^2}\,dx$.

Solution: By Definition 3,

$$\begin{aligned}\int_{-\infty}^\infty \frac{1}{1+x^2}\,dx &= \int_{-\infty}^0 \frac{1}{1+x^2}\,dx + \int_0^\infty \frac{1}{1+x^2}\,dx \\ &= \lim_{t\to-\infty} \{\tan^{-1} x]_{x=t}^{x=0}\} + \lim_{t\to\infty} \{\tan^{-1} x]_{x=0}^{x=t}\} \\ &= \lim_{t\to-\infty} \{-\tan^{-1}(t)\} + \lim_{t\to\infty} \{\tan^{-1}(t)\} \\ &= -\left(-\frac{\pi}{2}\right) + \frac{\pi}{2} = \pi.\end{aligned}$$ ■

A Comparison Test

We will sometimes need to determine whether an improper integral converges or diverges, even though we cannot find an elementary antiderivative for its integrand. In such cases we can often resolve the issue of convergence for $\int_a^\infty f(x)\,dx$ by comparing the integrand $f(x)$ with another function $g(x)$, and using information about $\int_a^\infty g(x)\,dx$.

> Comparison Test: Suppose $f(x)$ and $g(x)$ are continuous on $[a, \infty)$ with $0 \le f(x) \le g(x)$ for all $x \ge a$. Then
>
> (i) If $\int_a^\infty g(x)\,dx$ converges, so does $\int_a^\infty f(x)\,dx$.
>
> (ii) If $\int_a^\infty f(x)\,dx$ diverges, so does $\int_a^\infty g(x)\,dx$.

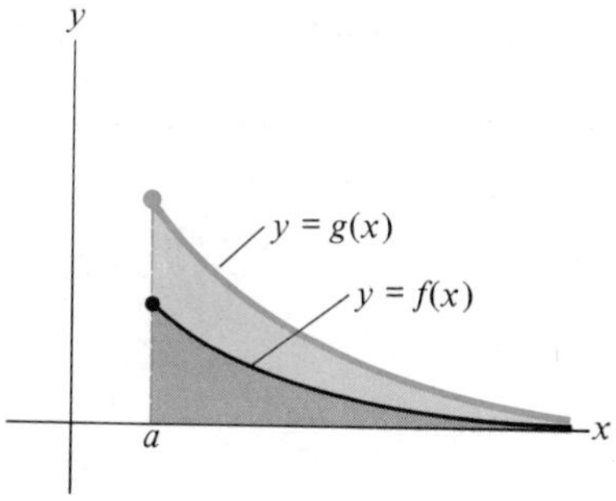

Figure 8.5 Comparison Test: If $0 \le f(x) \le g(x)$ and $\int_a^\infty g(x)\,dx$ converges, then $\int_a^\infty f(x)$ converges.

Figure 8.5 illustrates statement (i), which we justify intuitively by saying that if the area of the region R_1 bounded by $g(x)$ and the x-axis, $a \le x < \infty$ is finite, then the area of the region R_2 bounded by the graph of $f(x)$ and the x-axis, $a \le x < \infty$, must also be finite, since $R_2 \subseteq R_1$. A similar argument applies to statement (ii): If R_2 is not finite, then R_1 cannot be finite. A more formal proof is given in Chapter 13 for an analogous statement concerning infinite series.

Example 8 Use the Comparison Test to determine whether the improper integral

$$\int_1^\infty e^{-(1+x^2)}\,dx$$

converges.

Solution: Since $1 + x^2 > x$ for $x \ge 1$ and e^{-x} is a *decreasing* function, we have the comparison

$$0 < e^{-(1+x^2)} < e^{-x}, \qquad x \ge 1.$$

Since

$$\int_1^\infty e^{-x}\,dx = \lim_{t\to\infty} -e^{-x}\Big]_1^t = \lim_{t\to\infty}\left(\frac{1}{e} - e^{-t}\right) = \frac{1}{e}$$

converges, so must

$$\int_1^\infty e^{-(1+x^2)}\,dx.$$

(Note, however, that we cannot determine the *value* of this integral by this method—but only whether or not it converges.) ■

Improper Integrals—Infinite Integrands

A second general type of improper integral occurs when the integrand in a definite integral becomes infinite at one or both limits of integration. For example, suppose that $f(x)$ is defined and continuous on $[a, b)$ but that $f(b)$ fails to exist (Figure 8.6). In this case the Riemann integral $\int_a^b f(x)\,dx$ as defined in Chapter 6 does not exist. However, we define this **improper integral** as follows:

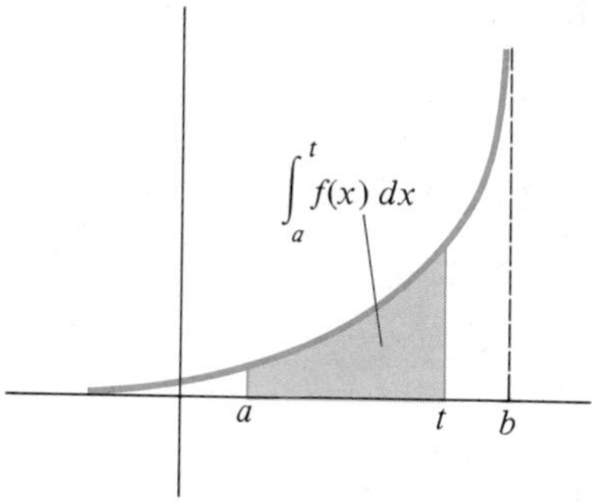

Figure 8.6 $f(b)$ undefined.

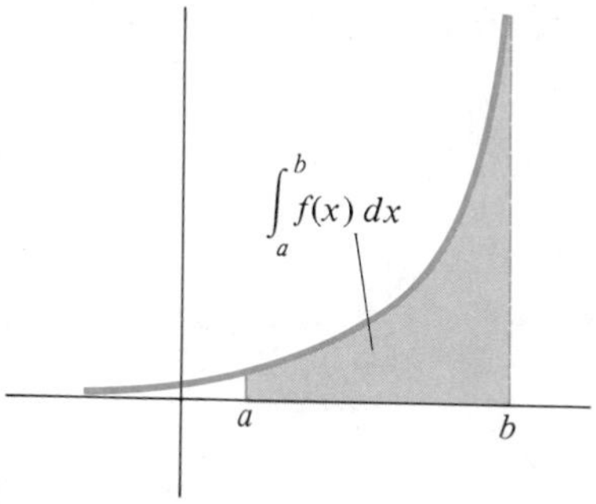

Figure 8.7 $\int_a^b f(x)\,dx = \lim_{t\to b^-} \int_a^t f(x)\,dx.$

DEFINITION 4

If $f(x)$ is continuous on $[a, b)$ but $f(b)$ fails to exist, we define

$$\int_a^b f(x)\,dx = \lim_{t\to b^-} \int_a^t f(x)\,dx.$$

If the limit on the right exists, the integral is said to converge. Otherwise it is said to diverge (Figure 8.7).

Example 9 Determine whether the improper integral

$$\int_0^1 \frac{x}{\sqrt{1-x^2}}\,dx$$

converges or diverges.

Strategy

Determine where the integrand fails to exist.

Apply Definition 4.

Use a u-substitution with

$u = 1 - x^2, \qquad du = -2x\,dx.$

$$\int \frac{x}{\sqrt{1-x^2}}\,dx = \int \frac{-1/2}{\sqrt{u}}\,du = -\sqrt{u} + C.$$

Solution

The integrand becomes infinite at $x = 1$. We have

$$\begin{aligned}\int_0^1 \frac{x}{\sqrt{1-x^2}}\,dx &= \lim_{t\to 1^-} \left\{\int_0^t \frac{x}{\sqrt{1-x^2}}\,dx\right\} \\ &= \lim_{t\to 1^-} \{-\sqrt{1-x^2}\,\big]_{x=0}^{x=t}\} \\ &= \lim_{t\to 1^-} \{-\sqrt{1-t^2}+1\} \\ &= 1.\end{aligned}$$

The integral converges. ■

When the integrand fails to exist at the left endpoint, the improper integral is defined by analogy with Definition 4 (see Figure 8.8). That is,

$$\int_a^b f(x)\,dx = \lim_{t\to a^+} \int_t^b f(x)\,dx.$$

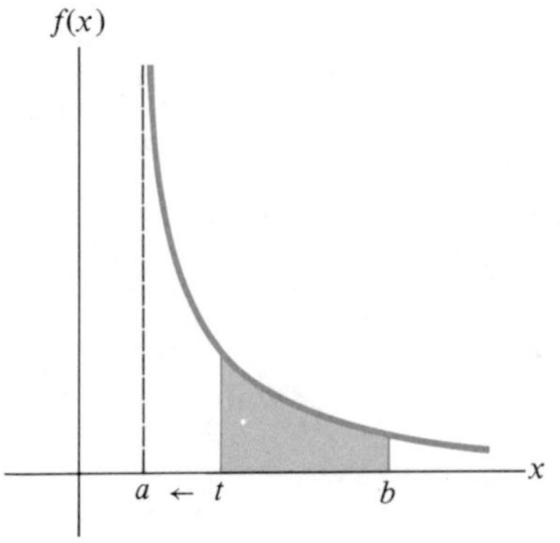

Figure 8.8 $\int_a^b f(x)\,dx = \lim_{t\to a^+} \int_t^b f(x)\,dx.$

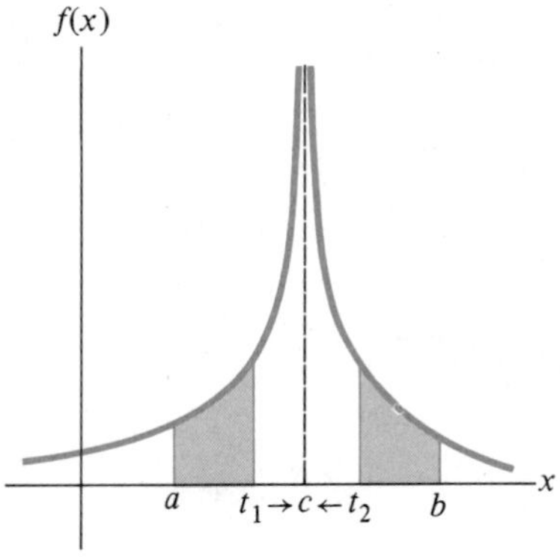

Figure 8.9 $\int_a^b f(x)\,dx$, $f(c)$ undefined.

When an integrand fails to exist at a point $c \in (a, b)$, we extend Definition 4 as follows:

$$\int_a^b f(x)\,dx = \int_a^c f(x)\,dx + \int_c^b f(x)\,dx \tag{2}$$
$$= \lim_{t_1 \to c^-} \int_a^{t_1} f(x)\,dx + \lim_{t_2 \to c^+} \int_{t_2}^b f(x)\,dx.$$

(See Figure 8.9.) The improper integral on the left in equation (2) is said to converge only if both improper integrals on the right converge. Otherwise it is said to diverge.

Example 10 Determine whether the improper integral

$$\int_0^3 (x-2)^{-4/3}\,dx$$

converges or diverges.

Solution: The integrand becomes unbounded (i.e., fails to exist) at $c = 2$. Applying (2) we find that

$$\begin{aligned}\int_0^3 (x-2)^{-4/3}\,dx &= \lim_{t_1\to 2^-}\int_0^{t_1}(x-2)^{-4/3}\,dx + \lim_{t_2\to 2^+}\int_{t_2}^3 (x-2)^{-4/3}\,dx\\ &= \lim_{t_1\to 2^-}\{-3(x-2)^{-1/3}]_0^{t_1}\} + \lim_{t_2\to 2^+}\{-3(x-2)^{-1/3}]_{t_2}^3\}\\ &= \lim_{t_1\to 2^-}\left[\frac{-3}{\sqrt[3]{t_1-2}} + \frac{3}{\sqrt[3]{-2}}\right] + \lim_{t_2\to 2^+}\left[-3 + \frac{3}{\sqrt[3]{t_2-2}}\right]\\ &= \infty + \infty = \infty.\end{aligned}$$

Since both limits fail to exist, the improper integral diverges. ■

Exercise Set 11.8

In Exercises 1–28, determine whether the improper integral converges. Evaluate those which converge.

1. $\int_3^\infty \frac{1}{x}\,dx$
2. $\int_2^\infty \frac{1}{(x+1)^2}\,dx$
3. $\int_1^\infty \frac{x}{\sqrt{1+x^2}}\,dx$
4. $\int_1^\infty \frac{dx}{4+x^2}$
5. $\int_0^\infty e^{-x}\,dx$
6. $\int_2^\infty \frac{x}{(4+x^2)^{3/2}}\,dx$
7. $\int_0^\infty e^{-x}\sin x\,dx$
8. $\int_0^\infty e^x\cos x\,dx$
9. $\int_{-\infty}^2 e^{2x}\,dx$
10. $\int_{-\infty}^0 \frac{1}{1+x^2}\,dx$
11. $\int_0^\infty xe^{-x}\,dx$
12. $\int_{-\infty}^\infty \frac{dx}{4+x^2}$
13. $\int_0^1 \frac{1}{x}\,dx$
14. $\int_0^1 \frac{1}{x^2}\,dx$
15. $\int_0^3 \sqrt{9-x^2}\,dx$
16. $\int_e^\infty \frac{1}{x\ln x}\,dx$
17. $\int_0^1 x\ln x\,dx$
18. $\int_0^1 \frac{dx}{\sqrt{1-x}}$
19. $\int_0^e x^2\ln x\,dx$
20. $\int_0^1 x^{-2/3}\,dx$
21. $\int_{-1}^1 \frac{1}{x}\,dx$
22. $\int_0^3 \frac{dx}{x^2+2x-3}$
23. $\int_0^\infty e^{-2x}\cos 2x\,dx$
24. $\int_0^\infty e^{-3x}\sin 2x\,dx$
25. $\int_0^\infty \frac{e^{-2x}}{\sqrt{x}}\,dx$
26. $\int_0^\infty \sqrt{x}\,e^{-x}\,dx$

27. $\int_0^1 x\ln(1+x)\,dx$ **28.** $\int_0^1 \ln^2 x\,dx$

29. Show that the integral

$$\int_0^\infty \sin x\,dx$$

diverges. Thus, it is not necessary that

$$\lim_{t\to\infty}\int_a^t f(x)\,dx = \pm\infty$$

for an improper integral to diverge.

30. Give a geometric argument that $\int_1^\infty \frac{dx}{x^2} = \int_0^1 \frac{dx}{\sqrt{x}} - 1$.

31. Find the values of p for which the integral $\int_1^\infty x^p\,dx$ converges.

32. Find the values of p for which the integral $\int_0^\infty x^p\,dx$ converges.

33. Give a geometric argument to show that if $0 \le f(x) \le g(x)$ for all $x > a$ then

$$\int_a^\infty g(x)\,dx \text{ diverges} \quad \text{if} \quad \int_a^\infty f(x)\,dx \text{ diverges.}$$

Use the Comparison Test to determine whether the following integrals converge or diverge.

34. $\int_1^\infty \frac{1}{1+x^3}\,dx$ **35.** $\int_1^\infty \frac{e^x}{\sqrt{1+x^2}}\,dx$

36. $\int_1^\infty \frac{1}{\sqrt{1+x^5}}\,dx$ **37.** $\int_1^\infty \frac{|\sin x|}{x^2}\,dx$

38. $\int_1^\infty e^{\sqrt{x}}\,dx$ **39.** $\int_1^\infty \sqrt{e^x + \sin x}\,dx$

40. $\int_1^\infty \frac{dx}{\sqrt{e^{x^2} + x + \cos x}}$ **41.** $\int_1^\infty e^{-x}\sin x\,dx$

42. Explain why

$$\int_{-\infty}^\infty f(x)\,dx = \lim_{t\to\infty}\int_{-t}^t f(x)\,dx$$

is wrong. (*Hint:* Consider the function $f(x) = x^3$, for example.)

43. Show that if $\int_{-\infty}^\infty f(x)\,dx$ converges then

$$\int_{-\infty}^\infty f(x)\,dx = \int_{-\infty}^a f(x)\,dx + \int_a^\infty f(x)\,dx$$

for all $a \in (-\infty, \infty)$.

44. Consider the region R bounded above by the graph of $y = 1/x$ and below by the x-axis for $1 \le x < \infty$.
a. Show that the surface area of the solid obtained by revolving R about the x-axis is infinite.
b. Compare the result of part (a) with that of Example 5. Comment on the observation that Figure 8.4 represents an infinite horn that (i) cannot be painted with a finite amount of paint but (ii) can hold a finite amount of paint.

45. Find the volume and the surface area of the solid obtained by revolving about the x-axis the region bounded above by the graph of $f(x) = e^{-x}$ and below by the x-axis for $1 \le x < \infty$.

46. Find the values of n for which the integral $\int_0^1 x^n \ln x\,dx$ converges.

47. A business expects to generate one million dollars profit per year indefinitely.
a. Find the present value of its anticipated profits over the next five-year period, assuming an annual rate of interest of $r = 8\%$.
b. Find the present value of all anticipated future profits.

48. What is the present value of an investment that will produce revenues at the rate of $R(t) = 1000 + 200t$ dollars per year, assuming an annual rate of continuous compounding of $r = 10\%$,
a. over the period of the next 5 years?
b. over an infinite future?

49. Show that

$$\int_0^1 [\ln x]^n\,dx = (-1)^n \cdot n!$$

$$(n! = n(n-1)(n-2)\cdot \cdots \cdot 2 \cdot 1).$$

50. Show that $\int_1^\infty \frac{dx}{1+e^x} = \ln(1+e) - 1$.

51. Use the notion of an improper integral to explain how you can conclude from the result of Example 4 in Section 7.5 that the surface area of the sphere is $4\pi r^2$. (*Hint:* See Example 5 of that section.)

SUMMARY OUTLINE OF CHAPTER 11

- Integration by parts formula: $\int u\,dv = uv - \int v\,du$.
- Trigonometric substitutions:

(a) $\sqrt{x^2 + a^2}$ requires $x = a\tan\theta$, $-\pi/2 < \theta < \pi/2$
$dx = a\sec^2\theta\,d\theta$

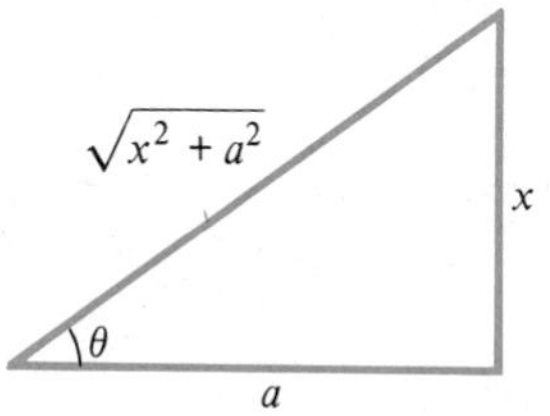

(b) $\sqrt{a^2 - x^2}$ requires $x = a\sin\theta$, $-\pi/2 \le \theta \le \pi/2$
$dx = a\cos\theta\,d\theta$

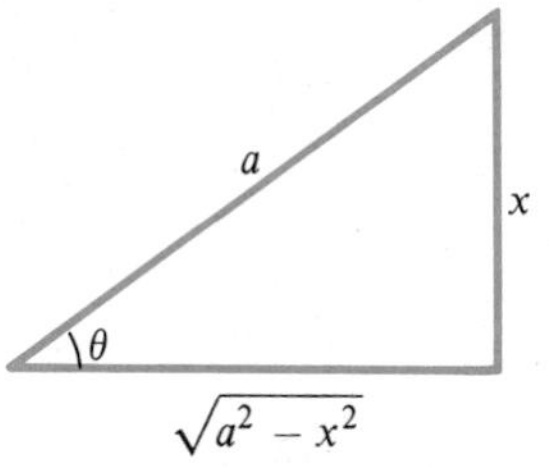

(c) $\sqrt{x^2 - a^2}$ requires $x = a\sec\theta$, $0 \le \theta < \pi/2$ or
$\pi \le \theta < 3\pi/2$
$dx = a\sec\theta\tan\theta\,d\theta$

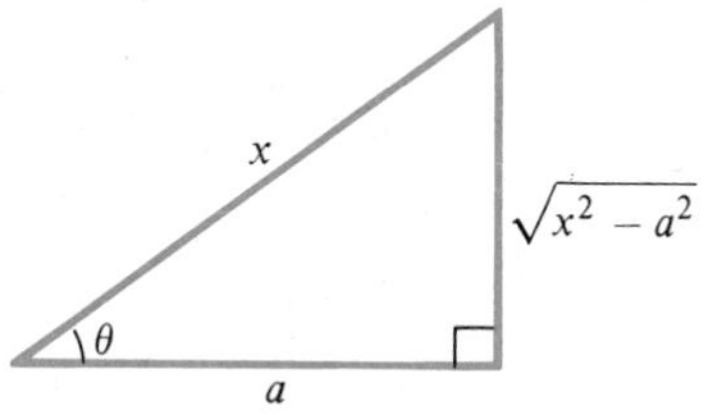

- To handle integrals involving $\sqrt{ax^2 + bx + c}$, complete the square and use the corresponding trigonometric substitution.
- The method of partial fractions decomposes rational functions into sums of simpler form:

$$\frac{p(x)}{(x-a)(x-b)} = \frac{A}{x-a} + \frac{B}{x-b}$$
$$\frac{p(x)}{(x-a)(x^2+bx+c)} = \frac{A}{x-a} + \frac{Bx+C}{x^2+bx+c}.$$

- The improper integral $\int_a^\infty f(x)\,dx$ is defined by

$$\int_a^\infty f(x)\,dx = \lim_{t\to\infty}\int_a^t f(x)\,dx.$$

- If $f(b)$ fails to exist, the improper integral $\int_a^b f(x)\,dx$ is defined by

$$\int_a^b f(x)\,dx = \lim_{t\to b^-}\int_a^t f(x)\,dx.$$

REVIEW EXERCISES—CHAPTER 11

In Exercises 1–86, evaluate the given integral.

1. $\int x\sqrt{x^2+9}\,dx$

2. $\int xe^{3x}\,dx$

3. $\int \sqrt{x^2+a^2}\,dx$

4. $\int \frac{dx}{x^2-6x+9}$

5. $\int \frac{2x+3}{x(x+3)}\,dx$

6. $\int \frac{\sqrt{x+1}}{\sqrt{x}}\,dx$

7. $\int_{-1}^{1} x^{-5/3}\,dx$

8. $\int x\tan^{-1}x\,dx$

9. $\int x\ln^2 x\,dx$

10. $\int \frac{dx}{\sqrt{x^2+a^2}}$

11. $\int \frac{dx}{x^2-4x+9}$

12. $\int \frac{4x+5}{x^2-x+2}\,dx$

13. $\int \frac{\sqrt{x+4}}{x}\,dx$

14. $\int \frac{1}{\sqrt[3]{x+1}}\,dx$

15. $\int_0^2 x\sqrt{x+2}\,dx$

16. $\int x(x+3)^6\,dx$

17. $\int \frac{dx}{\sqrt{a^2-x^2}}$

18. $\int \frac{dx}{\sqrt{x^2+8x+25}}$

19. $\int \frac{3x^2+x+1}{x^2(x+1)}\,dx$

20. $\int \frac{dx}{2\sqrt{x}+3\sqrt[3]{x}}$

21. $\int_0^{\pi/4} \tan^2 x\, dx$

22. $\int_{-\infty}^{\infty} xe^{-x^2}\, dx$

23. $\int_1^e x^2 \ln x\, dx$

24. $\int \sqrt{a^2 - x^2}\, dx$

25. $\int \frac{dx}{\sqrt{1 + 4x - x^2}}$

26. $\int \frac{4x^2 + x + 2}{(x - 2)(x^2 + 1)}\, dx$

27. $\int \frac{2 + \sqrt{x}}{2 - \sqrt{x}}\, dx$

28. $\int \frac{x + 1}{\sqrt[3]{x + 1}}\, dx$

29. $\int_0^{\infty} \frac{dx}{\sqrt[3]{x + 2}}$

30. $\int x \csc^2 \pi x\, dx$

31. $\int \sqrt{x^2 - a^2}\, dx$

32. $\int \frac{dx}{\sqrt{x^2 - 8x}}$

33. $\int \frac{x^2 + 2x + 3}{(x + 2)(x^2 + x + 1)}\, dx$

34. $\int \frac{x - 3}{\sqrt{x + 2}}\, dx$

35. $\int_{-\infty}^{0} xe^x\, dx$

36. $\int_1^2 x \sec^{-1} x\, dx$

37. $\int (x^2 - 4)^{3/2}\, dx$

38. $\int \frac{x\, dx}{x^2 + 4x + 8}$

39. $\int \frac{3x^3 + x + 2}{x(1 + x^3)}\, dx$

40. $\int \frac{dx}{1 + 2 \sin x}$

41. $\int_0^{\pi/2} \frac{dx}{1 - \sin x}$

42. $\int \csc^3 x\, dx$

43. $\int \frac{e^{2x}}{\sqrt{1 - e^x}}\, dx$

44. $\int \frac{dx}{x\sqrt{9 + x^2}}$

45. $\int \frac{2 + x}{\sqrt{x^2 - 6x + 13}}\, dx$

46. $\int \frac{3x^2 - x + 10}{(x^2 + 2)(4 - x)}\, dx$

47. $\int \frac{dx}{1 - 2 \sin x}$

48. $\int \frac{\cos x}{2 - \cos x}\, dx$

49. $\int \frac{x^2}{\sqrt{1 + x^2}}\, dx$

50. $\int \frac{dx}{e^x - e^{-x}}$

51. $\int_e^{\infty} \frac{dx}{x \ln^2 x}$

52. $\int \sec^5 x\, dx$

53. $\int \frac{x}{a^2 + x^2}\, dx$

54. $\int \frac{2x - 1}{\sqrt{17 + 8x + x^2}}\, dx$

55. $\int \frac{5x^2 - 2x + 12}{x^3 - x^2 + 4x - 4}\, dx$

56. $\int \frac{dx}{1 + x^3}$

57. $\int_{-\infty}^{\infty} \frac{dx}{9 + x^2}$

58. $\int \frac{dx}{1 + e^x}$

59. $\int \frac{x}{\sqrt{36 + x^2}}$

60. $\int \frac{3x + 1}{\sqrt{x^2 + 8x}}\, dx$

61. $\int \frac{6x^2 + 5x - 2}{x^3 + x^2 - 2x}\, dx$

62. $\int \frac{x^5\, dx}{\sqrt{1 + x^2}}$

63. $\int_{-\infty}^{\infty} e^{-|x|}\, dx$

64. $\int \frac{1}{x^2\sqrt{x^2 - 25}}\, dx$

65. $\int \frac{x^5}{(x^2 + 2)^2}\, dx$

66. $\int \sqrt{1 - \cos x}\, dx$

67. $\int_0^{\infty} \frac{e^{-\sqrt{x}}}{\sqrt{x}}\, dx$

68. $\int \frac{1}{\sqrt{4x^2 - 25}}\, dx$

69. $\int \sqrt{1 + \sin x}\, dx$

70. $\int_1^2 \frac{x\, dx}{1 - x}$

71. $\int_0^{\pi/2} \tan x\, dx$

72. $\int \ln^2 x\, dx$

73. $\int \frac{7\, dx}{13 + x^2}$

74. $\int \frac{3\, dx}{x^2 - 5}, \quad x > \sqrt{5}$

75. $\int \frac{\pi\, dx}{x\sqrt{36 - x^2}}$

76. $\int \frac{dx}{x\sqrt{9 - \pi x}}$

77. $\int \frac{4\, dx}{(7 - 3x)^2}$

78. $\int \frac{5\, dx}{x^2(7 - 2x)}$

79. $\int \frac{dx}{x(\pi + 4x)^2}$

80. $\int \frac{3x\, dx}{\sqrt{9 + 2x}}$

81. $\int \frac{4x^2\, dx}{\sqrt{49 - x^2}}$

82. $\int \frac{dx}{x^2\sqrt{8 - x^2}}$

83. $\int \frac{dx}{\sqrt{5x - x^2}}$

84. $\int \frac{dx}{9 + 4 \cos x}$

85. $\int x \sin^{-1} 2x\, dx$

86. $\int x^3 \ln x\, dx$

Brook Taylor

Colin Maclaurin

Joseph Fourier

Karl Weierstrass

UNIT 5

THE THEORY OF INFINITE SERIES

The Theory of Infinite Series

We noted in the introduction to Unit I that Newton and Leibniz both found themselves expressing functions as infinite sums. Later workers in analysis, particularly Leonhard Euler and Johann Bernoulli, also dealt with these infinite series. This early work was not on a firm mathematical footing, however, and operations were sometimes performed that just happened to work because of the circumstances in a particular problem. The development of a logical theory was yet to happen, a process that unfolded over a long period of time.

Brook Taylor (1685–1731) and Isaac Newton knew each other through their membership in the Royal Society of London. Newton was its president for many years, and Taylor served for several years as its secretary. Because he thought highly of the new Newtonian calculus, Taylor attempted to clarify the subject in his writings. However, he was a rather murky writer, and he dealt only with algebraic functions. (Trigonometric, logarithmic, and exponential functions were not well known until Euler's time.)

In this unit you will write functions in the form of a special type of infinite series called Taylor's series. Taylor first used this series in 1713 and published it two years later. Because his writing was so difficult to follow, it had little immediate impact on mathematics; it was some forty years before Euler made use of Taylor's series in differential calculus. Much of the rest of Taylor's work was never appreciated, and he failed to receive full credit for it.

Also introduced in this unit is Maclaurin's series, a special case of Taylor's series that was used by Taylor (as was acknowledged by Maclaurin in 1742). Colin Maclaurin (1698–1746), a Scotsman, was the most outstanding British mathematician in the generation following Newton. He was a prodigy, matriculating at the University of Glasgow at the age of 11. He received his M.A. degree at 15. By age 19 he was a college mathematics teacher, and at 21 he published his first mathematical work of importance.

In 1719 Maclaurin met Isaac Newton, and he quickly became a devoted disciple of Newtonian calculus. When Bishop George Berkeley wrote a tract attacking Newton's fluxions, Maclaurin responded in 1742 with his *Treatise on Fluxions,* the first complete and systematic presentation of Newton's calculus. Although it was still not the final answer to rigor in calculus, it stood as a standard for nearly a century. (Maclaurin was not comfortable with the limit concept, and tried to base the calculus on geometry instead.)

Among Maclaurin's mathematical discoveries was what is now known as Cramer's rule for evaluating determinants. There is some poetic justice here, for Maclaurin didn't discover Maclaurin's series, while he did originate Cramer's rule, which is known by another man's name. (It should be noted in fairness that Cramer's notation was better than Maclaurin's.) After Maclaurin's death, British mathematics declined in importance until the nineteenth century.

Joseph Fourier (1768–1830) was the only mathematician ever to serve as Governor of Lower Egypt. He had supported the French Revolution, and was rewarded with an appointment to the École Polytechnique. However, he had always wanted to be an army officer, a career denied him because he was the son of a tailor. When the opportunity came to accompany Napoleon on a military campaign in Egypt, Fourier resigned his teaching position and went along, and was appointed Governor in 1798. When the British took Egypt in 1801, Fourier returned to France.

Fourier studied the flow of heat in metallic plates and rods. The theory that he developed now has applications in industry and in the study of the temperature of the earth's interior. He discovered that many functions could be expressed as infinite sums of sine and cosine terms, now called a trigonometric series, or Fourier series. A paper that he submitted to the Academy of Science in Paris in 1807 was studied by several eminent mathematicians and rejected because he failed to prove his claims. They suggested that he reconsider and refine his paper, and even made heat flow the topic for a prize to be awarded in 1812. Fourier won the prize, but the Academy still declined to publish his paper because of its lack of rigor. (When Fourier became the secretary of the Academy in 1824, the 1812 paper was then published without change.)

As Fourier grew older, he developed at least one peculiar notion. Whether influenced by his stay in the heat of Egypt or by his own studies of the flow of heat in metals, he became obsessed with the idea that extreme heat was the natural condition for the human body. He was always heavily bundled in woolen clothing, and kept his rooms at high temperatures. He died in his sixty-third year, as expressed by Howard Eves in *An Introduction to the History of Mathematics,* "thoroughly cooked."

Many creative mathematicians have shown their genius at an early age. A notable exception was Karl Weierstrass (1816–1897), who did not really become a mathematician until he was 40 years old. He had studied law at the University of Bonn at his father's insistence, but he spent much of his time fencing and drinking. He did not complete his studies, and did not get his degree even after four years. Instead he turned to mathematics, but did not complete his degree in that subject, either. Eventually Weierstrass became a *gymnasium* (high school) teacher in subjects such as writing and gymnastics. The mathematician in him was struggling to break through, and he did some mathematical research, although he had no contact with other mathematicians. He did manage to publish a few minor papers that attracted some attention. After thirteen years of secondary level teaching, Weierstrass obtained a position at a technical school, and was then called to the University of Berlin. Here his mathematical and pedagogical abilities surfaced; he is considered the greatest teacher of higher mathematics of the nineteenth century, as judged by the number of his students who became significant researchers. Judged on his own mathematical creativity, he is sometimes called the leading analyst of his time, and the father of modern analysis.

Weierstrass worked in several areas, but his principal contribution was the study of functions of complex numbers through power series, which we will mention in Chapter 14. This was an extension of the work of such earlier mathematicians as Maclaurin, Taylor, and Euler, but with a new rigor that they had been unable to attain.

With all this creativity, Weierstrass did not publish many papers, and much of what he accomplished is known only through notes taken in class by his students. Indeed, he was uninterested in publicity or fame, and seemingly did not mind that some of his students used material from his lectures as their own. Hence, the record of the mathematical achievements of Karl Weierstrass has never been completely sorted out.

Pronunciation Guide

Joseph Fourier (For-yay)

Karl Weierstrass (Vy'er-shtrahs)

(Photographs from the David Eugene Smith Papers, Rare Book and Manuscript Library, Columbia University.)

CHAPTER 12

THE APPROXIMATION PROBLEM: TAYLOR POLYNOMIALS

12.1 INTRODUCTION: THE APPROXIMATION PROBLEM

We are about to embark on the third major topic of the calculus, the theory of infinite series. The purpose of this brief chapter is to pose a motivating problem that will carry us throughout this unit (Chapters 12, 13, and 14).

The approximation problem is simply this: *Given a function $f(x)$ and a number $x = a$, how can we approximate values of $f(x)$ for x near a using polynomials, and what is the accuracy of these approximations?*

Of course, if $f(x)$ is a polynomial the answer is easy. We simply use $f(x)$ to approximate itself, with complete accuracy. Our real interest is in knowing how to approximate transcendental functions, like $\sin x$ or e^x, or algebraic functions like $\sqrt[5]{x}$.

The reason for our interest in approximating such functions is that although we have developed a rich theory of differentiation and integration for these functions, we do not yet have available a means for calculating their values, except in special cases. For example, how does one calculate $\sin 53°$, $e^{0.05}$, or $\sqrt[5]{216}$? The theory of differentials (Chapter 3) is a first step in this direction, although it is quite limited. Even if you feel secure in the knowledge that your calculator is equipped with a special key for each of these functions or that a five-place table for such functions will always be available somewhere, the question remains as to how that calculator was programmed or how the entries in such tables are determined.

As stated above, our objective is to approximate arbitrary functions using *polynomials*. The reason for using polynomials is that values of polynomial functions can be easily computed with complete accuracy by simple multiplications and additions. In other words, we want to know how to approximate troublesome functions with convenient ones.

In the three sections that lie ahead, we shall find strong indications that for a large class of functions (i) we can approximate any of these functions using a particular type of polynomial, (ii) the accuracy of our approximation improves as the degree of the polynomial increases, and (iii) we can achieve as great an accuracy as desired for a particular value of x by using a polynomial of sufficiently high degree. The ideas developed in the course of this discussion will carry us naturally into the topic of infinite series (Chapter 13).

12.2 TAYLOR POLYNOMIALS

Given the function $f(x)$, suppose we ask for the polynomial of lowest degree that best approximates $f(x)$ near $x = a$. Since the polynomials of lowest degree are constants, the obvious answer to this question is the constant polynomial

$$P_0(x) \equiv f(a) \qquad \text{(Figure 2.1).} \tag{1}$$

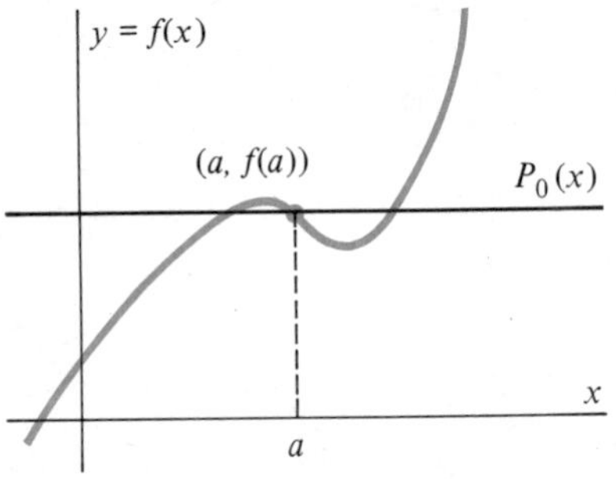

Figure 2.1 Constant polynomial $P_0 \equiv a$ approximating $f(x)$ near a.

Next, let us ask for the polynomial of degree one (that is, a linear function) that best approximates $f(x)$ near $x = a$. We shall write this polynomial in the form

$$P_1(x) = b + c(x - a) \tag{2}$$

and seek to determine the constants b and c. Since we again want $P_1(a) = f(a)$, we obtain $b = f(a)$. Now by the theory of the derivative, the line through $(a, f(a))$ that best approximates $f(x)$ has slope $f'(a)$, if $f'(a)$ exists. In other words, we also want $f'(a) = P_1'(a) = c$. $P_1(x)$ is therefore

$$P_1(x) = f(a) + f'(a)(x - a). \tag{3}$$

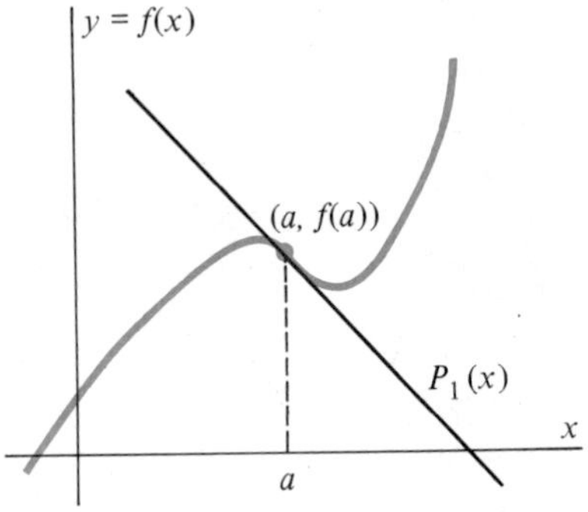

Figure 2.2 First degree polynomial $P_1(x) = f'(a)(x - a) + f(a)$.

Continuing in this way, we write the second degree polynomial approximating $f(x)$ as

$$P_2(x) = b + c(x - a) + d(x - a)^2.$$

The condition that $P_2(a) = f(a)$ gives $b = f(a)$ and the condition that the polynomial and the function $f(x)$ have the same slope at $x = a$ gives $P_2'(a) = c = f'(a)$. To determine the remaining constant d we require that $P_2(x)$ and $f(x)$ have the same measure of concavity at $x = a$; that is, we set $P_2''(a) = f''(a)$.

This gives the equation

$$P_2''(a) = 2d = f''(a).$$

Thus, $d = \dfrac{f''(a)}{2}$, and $P_2(x)$ is therefore (see Figure 2.3)

$$P_2(x) = f(a) + f'(a)(x - a) + \frac{f''(a)}{2}(x - a)^2. \tag{4}$$

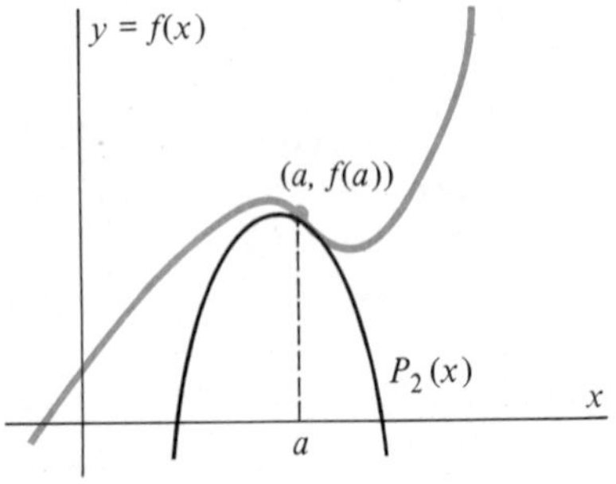

Figure 2.3 Second degree approximating polynomial $P_2(x)$.

We must carry this analysis one step further for the general form of these polynomials to become clear. Although we do not have available a geometric interpretation for derivatives of order higher than two, the idea is that for an approximating polynomial P_n of degree n, each derivative of order 1 through n must equal the corresponding derivative of $f(x)$ at $x = a$.

Write the general third-degree polynomial $P_3(x)$ in the form

$$P_3(x) = b + c(x - a) + d(x - a)^2 + e(x - a)^3.$$

Then, requiring that $P_3(x)$ and $f(x)$, together with their first three derivatives, agree at $x = a$ gives

$$\begin{aligned} P_3(a) &= b = f(a); & b &= f(a) \\ P_3'(a) &= c = f'(a); & c &= f'(a) \\ P_3''(a) &= 2d = f''(a); & d &= \frac{f''(a)}{2} \\ P_3'''(a) &= 3 \cdot 2 \cdot e = f'''(a); & e &= \frac{f'''(a)}{3!}. \end{aligned}$$

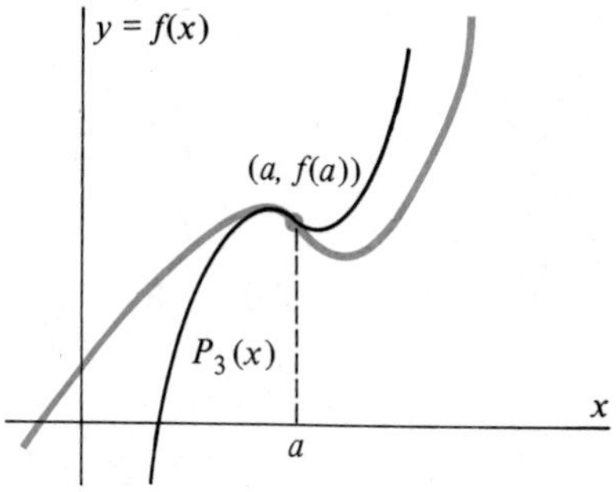

Figure 2.4 Third degree approximating polynomial $P_3(x)$.

The polynomial $P_3(x)$ is therefore (see Figure 2.4)

$$P_3(x) = f(a) + f'(a)(x-a) + \frac{f''(a)}{2}(x-a)^2 + \frac{f'''(a)}{3!}(x-a)^3. \tag{5}$$

(The notation $n!$ in equation (5) is the **factorial symbol,** defined by

$$n! = n(n-1)(n-2)\cdots 3\cdot 2\cdot 1.)$$

Figures 2.1 through 2.4 suggest that the polynomials $P_n(x)$ more closely approximate the function $f(x)$ near $x = a$ as the degree of $P_n(x)$ increases. We will verify this observation for particular choices of the function $f(x)$.

We can generalize equations (1), (3), and (4) by stating that the approximating polynomial of degree n for the function $f(x)$ near $x = a$ is

$$P_n(x) = f(a) + f'(a)(x-a) + \frac{f''(a)}{2}(x-a)^2 + \frac{f'''(a)}{3!}(x-a)^3 + \cdots + \frac{f^{(n)}(a)}{n!}(x-a)^n. \tag{6}$$

The polynomial $P_n(x)$ in (6) is referred to as the **Taylor polynomial** of degree n* for $f(x)$, expanded about $x = a$. The ideas we discuss here concerning $P_n(x)$ were developed by the English mathematician Brook Taylor (1686–1731) early in the eighteenth century. In Section 12.3 we will take up the question of how well the polynomial $P_n(x)$ approximates $f(x)$. The remaining part of this section concerns finding $P_n(x)$ for various choices of $f(x)$.

Example 1 Find $P_0(x)$, $P_1(x)$, $P_2(x)$, and $P_3(x)$ for the function $f(x) = e^x$ at $x = 0$.

Solution: The values of $f(x) = e^x$ and the first three derivatives at $a = 0$ are as follows:

$$\begin{aligned} f(x) &= e^x; & f(0) &= e^0 = 1 \\ f'(x) &= e^x; & f'(0) &= 1 \\ f''(x) &= e^x; & f''(0) &= 1 \\ f'''(x) &= e^x; & f'''(0) &= 1. \end{aligned}$$

From this information and equation (6) we obtain

$$\begin{aligned} P_0(x) &= 1 \\ P_1(x) &= 1 + x \\ P_2(x) &= 1 + x + \frac{x^2}{2} \\ P_3(x) &= 1 + x + \frac{x^2}{2} + \frac{x^3}{3!}. \end{aligned}$$

Figure 2.5 illustrates the Taylor polynomials of Example 1.

* There is a slight abuse of terminology here. Since it may be the case that $f^{(n)}(a) = 0$, a Taylor polynomial of degree n as defined above may actually be a polynomial of lower degree (see Example 2). We shall not concern ourselves further with this observation.

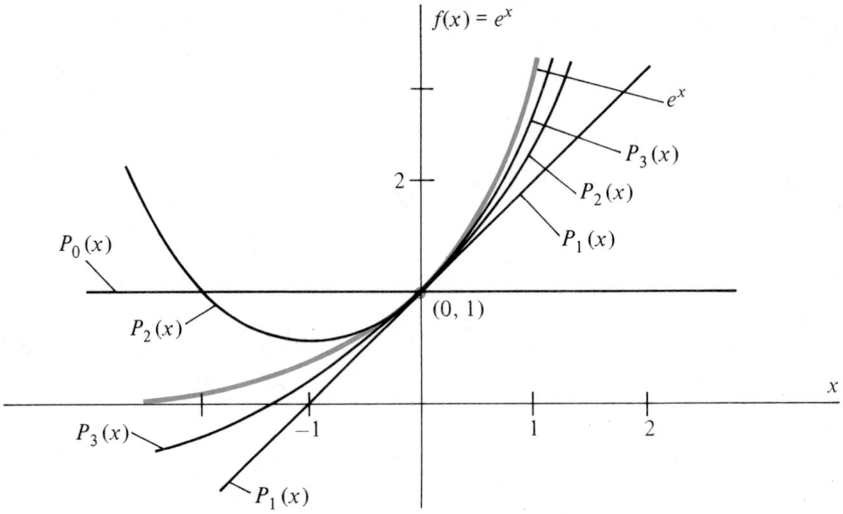

Figure 2.5 Taylor polynomials for $f(x) = e^x$ expanded about $a = 0$.

Table 2.1 indicates the accuracy with which these polynomials approximate $f(x) = e^x$ for various values of x.

Table 2.1 Values of Taylor approximations to e^x near $a = 0$

x	$P_3(x)$	$P_0(x)$	$P_1(x)$	$P_2(x)$	e^x
−1.5	0.0625	1.0	−0.5	0.6250	0.2231
−1.0	0.3333	1.0	0	0.5	0.3679
0	1.0	1.0	1.0	1.0	1.0
0.5	1.6458	1.0	1.5	1.625	1.6487
1.0	2.6667	1.0	2.0	2.5	2.7183
1.5	4.1875	1.0	2.5	3.6250	4.4817

■

Example 2 Find the Taylor polynomials of degrees 0 to 5 for $f(x) = \sin x$ expanded about $a = 0$.

Solution: Here

$$
\begin{aligned}
f(x) &= \sin x; & f(0) &= 0\\
f'(x) &= \cos x; & f'(0) &= 1\\
f''(x) &= -\sin x; & f''(0) &= 0\\
f'''(x) &= -\cos x; & f'''(0) &= -1\\
f^{(4)}(x) &= \sin x; & f^{(4)}(0) &= 0\\
f^{(5)}(x) &= \cos x; & f^{(5)}(0) &= 1.
\end{aligned}
$$

Thus, by (6),

$$
\begin{aligned}
P_0(x) &= 0\\
P_1(x) &= x\\
P_2(x) &= x
\end{aligned}
$$

$$P_3(x) = x - \frac{x^3}{3!}$$

$$P_4(x) = x - \frac{x^3}{3!}$$

$$P_5(x) = x - \frac{x^3}{3!} + \frac{x^5}{5!}$$

(See Figure 2.6.) ■

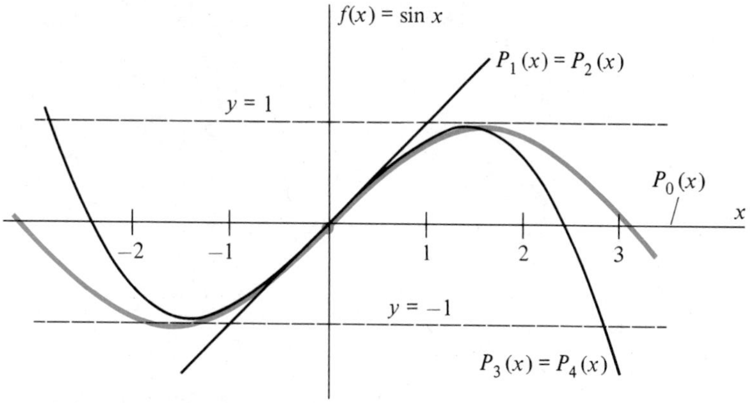

Figure 2.6 Five Taylor polynomials for $f(x) = \sin x$ at $x = 0$. Note that $P_1(x)$ and $P_2(x)$ coincide, as do $P_3(x)$ and $P_4(x)$.

Example 3 In Example 2 the even powers of x in the Taylor polynomial for $f(x) = \sin x$ at $a = 0$ dropped out, since even derivatives of $\sin x$ are zero at $a = 0$. For comparison, we next determine the Taylor polynomial of degree 3 for $f(x) = \sin x$ expanded about $a = \pi/6$.

$$f(x) = \sin x; \qquad f(\pi/6) = \frac{1}{2}$$

$$f'(x) = \cos x; \qquad f'(\pi/6) = \frac{\sqrt{3}}{2}$$

$$f''(x) = -\sin x; \qquad f''(\pi/6) = -\frac{1}{2}$$

$$f'''(x) = -\cos x; \qquad f'''(\pi/6) = -\frac{\sqrt{3}}{2}.$$

By (6),

$$P_3(x) = \frac{1}{2} + \frac{\sqrt{3}}{2}\left(x - \frac{\pi}{6}\right) + \left(\frac{1}{2}\right)\left(-\frac{1}{2}\right)\left(x - \frac{\pi}{6}\right)^2 + \left(\frac{1}{3!}\right)\left(-\frac{\sqrt{3}}{2}\right)\left(x - \frac{\pi}{6}\right)^3$$

$$= \frac{1}{2} + \frac{\sqrt{3}}{2}\left(x - \frac{\pi}{6}\right) - \frac{1}{4}\left(x - \frac{\pi}{6}\right)^2 - \frac{\sqrt{3}}{12}\left(x - \frac{\pi}{6}\right)^3.$$ ■

Example 4 Find the Taylor polynomials of degrees 0 to 4 for the function $f(x) = \cos x$ expanded about $a = 0$.

Solution: The required derivatives are as follows:

$$\begin{aligned}
f(x) &= \cos x; & f(0) &= 1\\
f'(x) &= -\sin x; & f'(0) &= 0\\
f''(x) &= -\cos x; & f''(0) &= -1\\
f'''(x) &= \sin x; & f'''(0) &= 0\\
f^{(4)}(x) &= \cos x; & f^{(4)}(0) &= 1.
\end{aligned}$$

The Taylor polynomials are

$$\begin{aligned}
P_0(x) &= 1\\
P_1(x) &= 1\\
P_2(x) &= 1 - \frac{x^2}{2}\\
P_3(x) &= 1 - \frac{x^2}{2}\\
P_4(x) &= 1 - \frac{x^2}{2} + \frac{x^4}{4!}.
\end{aligned}$$

■

Example 5 Find the Taylor polynomials of degrees 0 to 4 for the function $f(x) = \ln(1 + x)$ expanded about $a = 1$.

Solution: We have

$$\begin{aligned}
f(x) &= \ln(1 + x); & f(1) &= \ln(1 + 1) = \ln 2\\
f'(x) &= (1 + x)^{-1}; & f'(1) &= (1 + 1)^{-1} = \frac{1}{2}\\
f''(x) &= -(1 + x)^{-2}; & f''(1) &= -(1 + 1)^{-2} = -\frac{1}{4}\\
f'''(x) &= 2(1 + x)^{-3}; & f'''(1) &= 2(1 + 1)^{-3} = \frac{1}{4}\\
f^{(4)}(x) &= -6(1 + x)^{-4}; & f^{(4)}(1) &= -6(1 + 1)^{-4} = -\frac{3}{8}.
\end{aligned}$$

Thus, by (6),

$$\begin{aligned}
P_0(x) &= \ln 2\\
P_1(x) &= \ln 2 + \frac{1}{2}(x - 1)\\
P_2(x) &= \ln 2 + \frac{1}{2}(x - 1) + \left(\frac{1}{2}\right)\left(-\frac{1}{4}\right)(x - 1)^2\\
&= \ln 2 + \frac{1}{2}(x - 1) - \frac{1}{8}(x - 1)^2\\
P_3(x) &= \ln 2 + \frac{1}{2}(x - 1) - \frac{1}{8}(x - 1)^2 + \left(\frac{1}{3!}\right)\left(\frac{1}{4}\right)(x - 1)^3
\end{aligned}$$

$$= \ln 2 + \frac{1}{2}(x-1) - \frac{1}{8}(x-1)^2 + \frac{1}{24}(x-1)^3$$

$$P_4(x) = \ln 2 + \frac{1}{2}(x-1) - \frac{1}{8}(x-1)^2 + \frac{1}{24}(x-1)^3 + \left(\frac{1}{4!}\right)\left(-\frac{3}{8}\right)(x-1)^4$$

$$= \ln 2 + \frac{1}{2}(x-1) - \frac{1}{8}(x-1)^2 + \frac{1}{24}(x-1)^3 - \frac{1}{64}(x-1)^4.$$

Table 2.2 compares values of these approximating polynomials for various values of x.

Table 2.2 Approximations to $f(x) = \ln(1 + x)$

x	$\ln(1+x)$	$P_0(x)$	$P_1(x)$	$P_2(x)$	$P_3(x)$	$P_4(x)$
−.5	−.6931	.6931	−.0569	−.3381	−.4787	−.5578
0	0	.6931	.1931	.0681	.0265	.0109
1.0	.6931	.6931	.6931	.6931	.6931	.6931
1.5	.9163	.6931	.9431	.9119	.9171	.9161
2.0	1.0986	.6931	1.1931	1.0681	1.1098	1.0941
5.0	1.7918	.6931	2.6931	0.6931	3.3598	−0.6402

■

The entries of Tables 2.1 and 2.2 and the figures of this section emphasize that we will need to be quite careful about the accuracy of using Taylor polynomials to approximate functions. Although accuracy seems to improve as the degree of $P_n(x)$ increases, notice also that accuracy decreases as x moves away from the number a.

We can write formula (6) for $P_n(x)$ more compactly using sigma notation if we observe the following conventions:

$$0! = 1, \qquad \text{and} \qquad f^{(0)}(x) \text{ means } f(x).$$

Then

$$\boxed{P_n(x) = \sum_{j=0}^{n} \frac{f^{(j)}(a)}{j!}(x-a)^j.}$$

Finally, we remark that Taylor polynomials are defined only for functions $f(x)$ with the requisite number of derivatives at $x = a$. We will take a closer look at this property in Section 12.3.

Exercise Set 12.2

In Exercises 1–16, find the Taylor polynomial of degree n for the function $f(x)$ expanded about $x = a$.

1. $f(x) = e^{-x}$, $a = 0$, $n = 4$
2. $f(x) = e^x$, $a = 1$, $n = 4$
3. $f(x) = \cos x$, $a = \pi/4$, $n = 6$

4. $f(x) = \sin x, \quad a = \pi/3, \quad n = 5$

5. $f(x) = \ln(1 + x), \quad a = 0, \quad n = 4$

6. $f(x) = \dfrac{1}{1 + x}, \quad a = 0, \quad n = 4$

7. $f(x) = \tan x, \quad a = 0, \quad n = 3$

8. $f(x) = \dfrac{1}{1 + x^2}, \quad a = 0, \quad n = 2$

9. $f(x) = e^{x^2}, \quad a = 0, \quad n = 3$

10. $f(x) = \sqrt{x}, \quad a = 2, \quad n = 2$

11. $f(x) = \sqrt{1 - x}, \quad a = 0, \quad n = 3$

12. $f(x) = x \sin x, \quad a = 0, \quad n = 7$

13. $f(x) = \sec x, \quad a = \pi/4, \quad n = 3$

14. $f(x) = \tan^{-1} x, \quad a = 0, \quad n = 3$

15. $f(x) = \dfrac{1}{1 + e^x}, \quad a = 0, \quad n = 3$

16. $f(x) = x^4, \quad a = 2, \quad n = 4$

17. Compare the results of Exercises 5 and 6 and those of Exercises 8 and 14. Formulate a conjecture on the relationship between the Taylor polynomial for $f(x)$ and that for $f'(x)$ and test your conjecture on a few examples.

18. What is the relationship between the Taylor polynomial for $f(x)$ at $a = 0$ and that for $xf(x)$? Can you prove this?

19. True or false? If $f(x)$ is an even function, the Taylor polynomial for $f(x)$ expanded about $a = 0$ will contain only even powers of x. Explain.

20. The Taylor polynomial $P_n(x)$ is defined for $f(x)$ if each of the first n derivatives of $f(x)$ exists at $x = a$. Does this imply that $f(x)$ is defined for all x for which $P_n(x)$ is defined? (*Hint:* Consider this question for the function in Example 5.)

12.3 TAYLOR'S THEOREM

If $P_n(x)$ denotes the Taylor polynomial of degree n for the function $f(x)$ at $x = a$, our claim is that $P_n(x)$ may be used to approximate $f(x)$ for x near a. In order to determine the accuracy of this approximation we write

$$f(x) = P_n(x) + R_n(x). \tag{1}$$

Since $R_n(x) = f(x) - P_n(x)$, the term $R_n(x)$ represents the *error* made in approximating $f(x)$ by $P_n(x)$ (see Figure 3.1). It is referred to as the **remainder** in a Taylor approximation.

Taylor's Theorem provides a way to estimate the size of $R_n(x)$ when making a Taylor approximation.

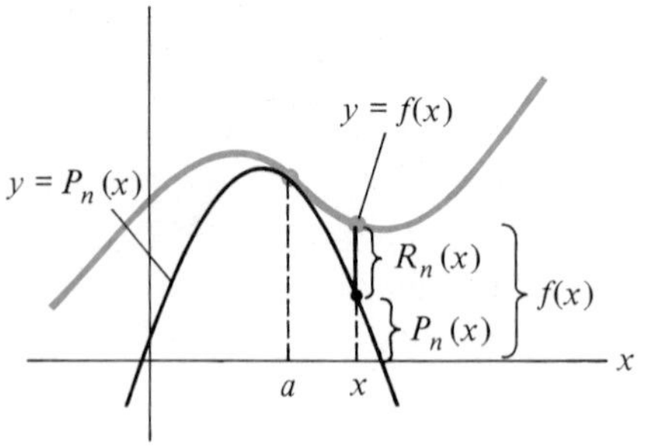

Figure 3.1 $R_n(x) = f(x) - P_n(x)$.

THEOREM 1
Taylor's Theorem

Let n be an integer and let $f(x)$ be a function for which $f^{(n+1)}(x)$ exists for each x in the closed interval $[\alpha, \beta]$. Let a be any number in the open interval (α, β). Then for each $x \in [\alpha, \beta]$ there exists a number c between a and x so that

$$f(x) = f(a) + f'(a)(x-a) + \frac{f''(a)}{2!}(x-a)^2 + \cdots + \frac{f^{(n)}(a)}{n!}(x-a)^n + \frac{f^{(n+1)}(c)}{(n+1)!}(x-a)^{n+1}.$$

(This is called Taylor's formula with remainder, or just Taylor's formula.) In other words,

$$f(x) = P_n(x) + R_n(x)$$

where $P_n(x)$ is the Taylor polynomial of degree n and

$$R_n(x) = \frac{f^{(n+1)}(c)}{(n+1)!}(x-a)^{n+1}. \tag{2}$$

We may paraphrase Taylor's Theorem by saying that the error involved in approximating an $(n+1)$ times differentiable function by its Taylor polynomial $P_n(x)$ is just the term we would add to $P_n(x)$ to obtain $P_{n+1}(x)$, except that the coefficient $\frac{f^{(n+1)}(c)}{(n+1)!}$ is evaluated at a specific number *between* a and x rather than at $x = a$.

We will see in the next section how to determine a bounding value for $R_n(x)$.

Example 1 Write Taylor's formula using the third degree Taylor polynomial for $f(x) = e^x$ expanded about $a = 0$.

Solution: From Example 1 of Section 12.2 we have

$$P_3(x) = 1 + x + \frac{x^2}{2!} + \frac{x^3}{3!}.$$

For $f(x) = e^x$, $f^{(4)}(x) = e^x$, so $f^{(4)}(c) = e^c$. Thus, by Taylor's Theorem,

$$R_3(x) = \frac{e^c}{4!}(x-0)^4$$

where c lies between 0 and x. Taylor's formula is therefore

$$e^x = 1 + x + \frac{x^2}{2} + \frac{x^3}{3!} + \frac{e^c}{4!}x^4.$$

■

Since Taylor's Theorem is both important and useful, it is unfortunate that no easily motivated proof has been found. We present one of the standard proofs in the remainder of this section.

Proof of Theorem 1: In addition to the number a, choose a number $b \in [\alpha, \beta]$. Then we can define a number C (which depends on a, b, and n) by the equation

$$f(b) = f(a) + f'(a)(b-a) + \cdots + \frac{f^{(n)}(a)}{n!}(b-a)^n + C. \tag{3}$$

That is, C is the difference $C = f(b) - P_n(b)$, where $P_n(b)$ is the Taylor polynomial of degree n expanded about a.

First note that if $b = a$, then (3) collapses to $f(a) = f(a)$, and $C = 0$. Thus, we consider $b \neq a$ (it makes no difference whether $a < b$ or $b < a$) and attempt to prove that C has the form of the remainder $R_n(x)$ given in the theorem.

In order to do this, we construct a rather odd-looking function of x, whose purpose is to allow us to apply Rolle's Theorem on the interval between a and b:

$$F(x) = f(b) - \left\{f(x) + f'(x)(b - x) + \cdots + \frac{f^{(n)}(x)}{n!}(b - x)^n + C\frac{(b - x)^{n+1}}{(b - a)^{n+1}}\right\}. \tag{4}$$

Now we show the following facts about $F(x)$:

(i) If $x = a$, then

$$\begin{aligned} F(a) &= f(b) - \left\{f(a) + f'(a)(b - a) + \cdots + \frac{f^{(n)}(a)}{n!}(b - a)^n + C\frac{(b - a)^{n+1}}{(b - a)^{n+1}}\right\} \\ &= 0 \qquad \text{(by (3))}. \end{aligned}$$

(ii) If $x = b$, then

$$\begin{aligned} F(b) &= f(b) - \left\{f(b) + f'(b)(b - b) + \cdots + \frac{f^{(n)}(b)}{n!}(b - b)^n + C\frac{(b - b)^{n+1}}{(b - a)^{n+1}}\right\} \\ &= f(b) - f(b) = 0 \end{aligned}$$

(iii) Because $f^{(n+1)}(x)$ exists for each $x \in [\alpha, \beta]$, $F(x)$ is differentiable on the entire interval, and

$$F'(x) = -\frac{f^{(n+1)}(x)}{n!}(b - x)^n + (n + 1)C\frac{(b - x)^n}{(b - a)^{n+1}}.$$

To see that this is true, we simply differentiate $F(x)$ in equation (4), using the Product Rule, and note that many terms cancel:

$$\begin{aligned} F'(x) = 0 - \Bigg\{ & f'(x) + f''(x)(b - x) - f'(x) \\ & + \frac{f'''(x)}{2}(b - x)^2 - f''(x)(b - x) \\ & + \frac{f^{(4)}(x)}{3!}(b - x)^3 - \frac{f'''(x)}{2}(b - x)^2 \\ & + \cdots + \frac{f^{(n)}(x)}{(n - 1)!}(b - x)^{n-1} - \frac{f^{(n-1)}(x)}{(n - 2)!}(b - x)^{n-2} \\ & + \frac{f^{(n+1)}(x)}{n!}(b - x)^n - \frac{f^{(n)}(x)}{(n - 1)!}(b - x)^{n-1} - (n + 1)C\frac{(b - x)^n}{(b - a)^{n+1}}\Bigg\} \\ = -&\frac{f^{(n+1)}(x)}{n!}(b - x)^n + (n + 1)C\frac{(b - x)^n}{(b - a)^{n+1}}. \end{aligned}$$

From statements (i) and (ii) and Rolle's Theorem, it follows that there is a number c between a and b such that $F'(c) = 0$. Using statement (iii), we may write this condition as

$$\frac{f^{(n+1)}(c)}{n!}(b-c)^n = (n+1)C\frac{(b-c)^n}{(b-a)^{n+1}}.$$

Solving for C now gives

$$C = \frac{f^{(n+1)}(c)}{(n+1)!}(b-a)^{n+1}. \tag{5}$$

Since we have placed no restriction on b other than $b \neq a$ (and we already know that $C = 0$ if $b = a$), we can substitute x for b in (5), which yields exactly $C = R_n(x)$. This completes the proof. ■

Exercise Set 12.3

In Exercises 1–14, write Taylor's formula (Theorem 1) for $f(x)$ using the Taylor polynomial of degree n expanded about $x = a$.

1. $f(x) = e^{-x}$, $n = 3$, $a = 0$
2. $f(x) = \sin x$, $n = 3$, $a = 0$
3. $f(x) = \cos x$, $n = 4$, $a = 0$
4. $f(x) = \sin x$, $n = 3$, $a = \pi/4$
5. $f(x) = \tan^{-1} x$, $n = 3$, $a = 0$
6. $f(x) = \sqrt{x}$, $n = 3$, $a = 4$
7. $f(x) = \dfrac{1}{1 + x^2}$, $n = 2$, $a = 1$
8. $f(x) = 3x^4 + 2x + 2$, $n = 3$, $a = 2$
9. $f(x) = \sec x$, $n = 2$, $a = \pi/4$
10. $f(x) = \ln(1 + x^2)$, $n = 3$, $a = 0$
11. $f(x) = \sinh x$, $n = 4$, $a = 0$
12. $f(x) = x \sinh x$, $n = 3$, $a = 0$
13. $f(x) = \cosh x$, $n = 3$, $a = \ln 2$
14. $f(x) = (1 + x)^{3/2}$, $n = 3$, $a = 0$
15. Let $f(x)$ be a polynomial of degree n. Use Theorem 1 to prove that $f(x) = P_n(x)$.
16. Refer to the precise hypotheses of Rolle's Theorem in Chapter 5. Then show that the hypothesis of Theorem 1 that $f^{(n+1)}(x)$ exists for each x in $[\alpha, \beta]$ can be relaxed to the condition that $f^{(n)}(x)$ is continuous on $[\alpha, \beta]$ and $f^{n+1}(x)$ exists for each x in (α, β).
17. Use Taylor's Theorem to justify Newton's Method for approximating the root of a function.
18. Let $P_n(x)$ be the Taylor polynomial of degree n for $f(x) = \sin x$ expanded about $a = 0$. Show that $\lim_{n\to\infty} R_n(x) = 0$ if $|x| < 1$.
19. By multiplying both sides of the equation by $1 - x$, show that

$$\frac{1}{1-x} = 1 + x + x^2 + x^3 + \cdots + x^n + \frac{x^{n+1}}{1-x}, \quad x \neq 1.$$

Conclude that $R_n(x) = \dfrac{x^{n+1}}{1-x}$. Does this contradict Taylor's Theorem? Use Taylor's Theorem to verify that, for $f(x) = \dfrac{1}{1-x}$ and $a = 0$, the Taylor polynomial of degree n for $f(x) = \dfrac{1}{1-x}$ is

$$P_n(x) = 1 + x + x^2 + \cdots + x^n.$$

20. By substituting $-x$ for x in the equation in Exercise 19 conclude that

$$\frac{1}{1+x} = 1 - x + x^2 - x^3 + \cdots + (-1)^n x^n + \frac{(-1)^{n+1}x^{n+1}}{1+x}, \quad x \neq -1.$$

Using Taylor's Theorem with $a = 0$, verify that the Taylor polynomial for $f(x) = \dfrac{1}{1+x}$ of degree n is

$$P_n(x) = 1 - x + x^2 - x^3 + \cdots + (-1)^n x^n.$$

21. Replace x by x^2 in Exercise 20 to conclude that

$$\frac{1}{1+x^2} = 1 - x^2 + x^4 - \cdots + (-1)^n x^{2n} + \frac{(-1)^{n+1}x^{2n+2}}{1+x^2}.$$

What is the Taylor Polynomial $P_n(x)$, of degree n expanded about $a = 0$, for $f(x) = \dfrac{1}{1+x^2}$?

22. Replace x by t in Exercise 20 and integrate between 0 and x to conclude that

$$\int_0^x \frac{1}{1+t}\,dt = \int_0^x \left(1 - t + t^2 - t^3 + \cdots + (-1)^{n-1}t^{n-1} + \frac{(-1)^n t^n}{1+t}\right) dt$$

$$= x - \frac{x^2}{2} + \frac{x^3}{3} - \frac{x^4}{4} + \cdots + \frac{(-1)^{n-1}x^n}{n} + (-1)^n \int_0^x \frac{t^n}{1+t}\,dt.$$

What is the Taylor polynomial of degree n, expanded about $x = 0$, for $f(x) = \ln(1+x)$?

23. Let $P_n(x)$ be as in Exercise 18. Show that $\lim_{n\to\infty} R_n(x) = 0$ for all x. $\left(\textit{Hint:}\text{ You must first show that } \lim_{n\to\infty} \dfrac{x^n}{n!} = 0.\right)$

24. Use Taylor's Theorem to prove the Binomial Theorem:

$$(x+a)^n = a^n + na^{n-1}x + \frac{n(n-1)}{2}a^{n-2}x^2 + \cdots + \binom{n}{r}a^{n-r}x^r + \cdots + nax^{n-1} + x^n$$

where n is a positive integer and $\binom{n}{r}$ is the **binomial coefficient**

$$\binom{n}{r} = \frac{n!}{r!(n-r)!}.$$

12.4 APPLICATIONS OF TAYLOR'S THEOREM

According to Taylor's Theorem, if we approximate $f(x)$ by the nth degree Taylor polynomial $P_n(x)$ expanded about $x = a$, the error (remainder) in the calculation is

$$R_n(x) = \frac{f^{(n+1)}(c)}{(n+1)!}(x-a)^{n+1} \tag{1}$$

where c lies between a and x. Since x is given, the choices that influence the size of the error involve the numbers a and n. Obviously, we will want to choose a close to x so that the factors $(x - a)$ in $R_n(x)$ are small. Also, we will want to choose a so that the function value $f(a)$ and the various derivatives $f^{(k)}(a)$ are easy to compute.

Once a convenient value for a is chosen, the magnitude of the error depends only on the integer n. Since we seldom know the actual value for c, we usually cannot determine $R_n(x)$ precisely. However, in most applications we are concerned only with knowing the approximate size of $R_n(x)$. For example, if we are to approximate $\ln(1.2)$ to within 0.01, we need only be able to demonstrate that $|R_n(1.2)| < 0.01$.

In general, if we wish to achieve a Taylor approximation for $f(x)$ to within ϵ, the objective will be to find a value of n sufficiently large to guarantee that the inequality

$$\left|\frac{f^{(n+1)}(c)}{(n+1)!}(x-a)^{n+1}\right| < \epsilon$$

holds. By (1) this will assure the desired accuracy.

Example 1 Use the Taylor polynomial of degree 3 for $f(x) = \ln x$ expanded about $a = 1$ to approximate $\ln(1.5)$ and find the accuracy of this approximation.

Strategy
Find $P_3(x)$ as in Section 12.2.

Solution
We have

$$f(x) = \ln x; \qquad f(1) = 0$$

$$f'(x) = \frac{1}{x}; \qquad f'(1) = 1$$

$$f''(x) = -\frac{1}{x^2}; \qquad f''(1) = -1$$

$$f'''(x) = \frac{2}{x^3}; \qquad f'''(1) = 2.$$

Thus

$$P_3(x) = 1(x-1) - \frac{1}{2}(x-1)^2 + \frac{2}{3!}(x-1)^3.$$

Evaluate $P_3(x)$ at $x = 1.5$ to obtain approximation $P_3(1.5)$.

The approximation is therefore

$$\ln(1.5) \approx P_3(1.5) = (.5) - \frac{1}{2}(.5)^2 + \frac{1}{3}(.5)^3$$

$$= .416\bar{6}.$$

Since $f^{(4)}(x) = -6x^{-4}$ for $f(x) = \ln x$, by (1) the error in the approximation is

Write expression for $|R_3(1.5)|$ using (1).

$$|R_3(1.5)| = \left|\frac{-6c^{-4}}{4!}(1.5-1)^4\right| = \left|\frac{-6c^{-4}}{24}(.5)^4\right|$$

$$= |c^{-4}| \cdot \frac{(.5)^4}{4}.$$

Since c is unknown, we must determine the largest possible value for $|f^{(4)}(c)|$ to obtain an upper bound on the error. We know that c is between $a = 1$ and $x = 1.5$.

Since $|c^{-4}| < 1$ for $1 < c < 1.5$, we have the inequality

$$|R_3(1.5)| < \frac{(.5)^4}{4} = .015625.$$

We may conclude only that the approximation is accurate to one decimal place. ■

REMARK: When we say that a number is accurate to k decimal places, we mean that the error is less than $5 \times 10^{-(k+1)}$. That is, accuracy to one decimal place means an error less than $5 \times 10^{-2} = .05$, accuracy to two decimal places means an error less than .005, and so forth. Thus we could claim that the approximation in Example 1 was accurate to one decimal place, and the approximation in Example 2 (below) is accurate to three decimal places.

Example 2 Suppose we wish to use the approximation

$$\sin x \approx x$$

for angles satisfying the inequality $|x| < \pi/45$. What is the maximum possible error associated with this approximation?

Strategy

Determine the Taylor polynomial $P_n(x)$ associated with the approximation.

Solution

For $f(x) = \sin x$, both $P_1(x) = x$ and $P_2(x) = x$ when $a = 0$. We may therefore take $n = 2$ as corresponding to the above approximation. Since $f'''(c) = -\cos(c)$, the error is

Once $f(x)$, n, and a are known, use equation (1).

$$|R_2(x)| = \left|\frac{-\cos(c)}{3!}(x - 0)^3\right| = \frac{|\cos(c)|}{3!}|x|^3.$$

Estimate the maximum possible size of $|f''(c)|$. Use to obtain bound on $|R_2(x)|$.

Since $|\cos(c)| \le 1$ for all values of c and since $|x| < \pi/45$, we have

$$|R_2(x)| \le \frac{1}{3!}|x|^3 < \frac{1}{3!}\left(\frac{\pi}{45}\right)^3 < .00006.$$

The maximum error in the approximation is $.00006 = 6 \times 10^{-5}$. Note that if we had used $n = 1$ in these calculations, we would have had $|f''(c)| = |-\sin(c)| \le 1$ and we could have claimed only that

$$|R_1(x)| \le \frac{1}{2!}|x|^2 < \frac{1}{2}\left(\frac{\pi}{45}\right)^2 < .0025.$$ ■

Example 3 How large must n be chosen so that $\cos 48°$ is approximated with four decimal place accuracy using a Taylor polynomial for $f(x) = \cos x$ expanded about $a = 45° = \pi/4$?

Strategy

Set up the expression for $|R_n(x)|$ using $x = 48° = \dfrac{4\pi}{15}$.

Find an upper bound for $|R_n(x)|$.

Solution

For $f(x) = \cos x$, $f^{(n+1)}(c)$ is either $\pm\sin c$ or $\pm\cos c$. In either case, $|f^{(n+1)}(c)| \le 1$. Using this inequality we obtain

$$\left|R_n\left(\frac{4\pi}{15}\right)\right| = \left|\frac{f^{(n+1)}(c)}{(n+1)!}\left(\frac{4\pi}{15} - \frac{\pi}{4}\right)^{n+1}\right| \tag{2}$$

$$< \frac{1}{(n+1)!}\cdot\left(\frac{\pi}{60}\right)^{n+1}$$

Require that the upper bound be less than the desired degree of accuracy.

To ensure that $\left|R_n\left(\frac{4\pi}{15}\right)\right| < 5 \times 10^{-5}$, we need to find n large enough so that

$$\frac{1}{(n+1)!}\cdot\left(\frac{\pi}{60}\right)^{n+1} < 5 \times 10^{-5}. \tag{3}$$

We now proceed simply by trial and error to find n sufficiently large that (3) holds. Values of the left side of inequality (3) for various values of n are as follows

n	$\dfrac{1}{(n+1)!}\left(\dfrac{\pi}{60}\right)^{n+1}$
1	.0013708
2	.0000240
3	.0000003

Thus $n = 2$ is sufficient to give the desired accuracy. ■

Example 4 Find a bound on the magnitude of $|x|$ so that the approximation

$$e^x \approx P_3(x) = 1 + x + \frac{x^2}{2!} + \frac{x^3}{3!}$$

is accurate to within .001.

Strategy

Solution

First note that the form of the given approximation is a Taylor polynomial expanded about $a = 0$, and that $f^{(n)}(a) = e^0 = 1$ for all n.

Set up the expression for $|R_3(x)|$ using (1).

Since $f^{(n)}(c) = e^c$ for all n, we have

$$|R_3(x)| = \left|\frac{e^c}{4!}x^4\right| = \frac{e^c}{24}x^4.$$

Find an expression for the maximum size of the factor e^c.

Now since e^c is an increasing function and c lies between 0 and x, the maximum value of e^c for $-|x| < c < |x|$ is $e^{|x|}$. Thus

$$|R_3(x)| \le \frac{e^{|x|}}{24}x^4. \tag{4}$$

Since e^x is what is being approximated, the factor $e^{|x|}$ must be replaced by a "safe" upper bound.

For $|R_3(x)|$ in (4) to be less than .001, we will clearly need to take $|x| < 1$. Thus we may safely bound the term $e^{|x|}$ by, say, 3, since $e^1 \approx 2.718 < 3$. We therefore need to find the maximum value of x for which

Solve the resulting inequality for x^n.

$$\frac{3}{24}x^4 < .001, \qquad \text{or} \qquad x^4 < .008.$$

By trial and error (or by extracting the fourth root) find a value of x satisfying the inequality.

Since $(.25)^4 < .0039 < .008$, the bound $|x| < .25$ will assure the desired accuracy. ■

The exercise set of this section deals with calculations similar to those of Examples 1–4. Before leaving this topic, however, we want to remark on two generalizations suggested by Example 3.

Since $\frac{\pi}{60} < 1$, we may write inequality (2) of Example 3 as

$$\left|R_n\left(\frac{4\pi}{15}\right)\right| < \frac{1}{(n+1)!}\left(\frac{\pi}{60}\right)^{n+1} < \frac{1}{(n+1)!}.$$

This shows that $R_n\left(\frac{4\pi}{15}\right) \to 0$ as $n \to \infty$. In other words, cos 48° can be approximated to any desired degree of accuracy by simply taking n sufficiently large. Since increasing n corresponds to taking polynomials of higher degree, this means that the *sequence* of estimates

$$P_0\left(\frac{4\pi}{15}\right) = \frac{\sqrt{2}}{2}$$

$$P_1\left(\frac{4\pi}{15}\right) = \frac{\sqrt{2}}{2} - \frac{\sqrt{2}}{2}\left(\frac{\pi}{60}\right)$$

$$P_2\left(\frac{4\pi}{15}\right) = \frac{\sqrt{2}}{2} - \frac{\sqrt{2}}{2}\left(\frac{\pi}{60}\right) - \frac{\sqrt{2}}{4}\left(\frac{\pi}{60}\right)^2$$

$$P_3\left(\frac{4\pi}{15}\right) = \frac{\sqrt{2}}{2} - \frac{\sqrt{2}}{2}\left(\frac{\pi}{60}\right) - \frac{\sqrt{2}}{4}\left(\frac{\pi}{60}\right)^2 + \frac{\sqrt{2}}{12}\left(\frac{\pi}{60}\right)^3$$
$$\vdots$$

approaches the number $\cos\left(\frac{4\pi}{15}\right)$. In the language of Chapter 13 we will say that

(i) the **sequence** of approximations $\left\{P_0\left(\frac{4\pi}{15}\right), P_1\left(\frac{4\pi}{15}\right), P_2\left(\frac{4\pi}{15}\right), \ldots\right\}$ approaches $\cos\left(\frac{4\pi}{15}\right)$ **in the limit,** that is,

$$\lim_{n\to\infty} P_n\left(\frac{4\pi}{15}\right) = \cos\left(\frac{4\pi}{15}\right);$$

(ii) the **series** of terms

$$\sum_{j=0}^{n} \frac{f^{(j)}(\pi/4)}{j!}\left(\frac{\pi}{60}\right)^j$$
$$= \frac{\sqrt{2}}{2} - \frac{\sqrt{2}}{2}\left(\frac{\pi}{60}\right) - \frac{\sqrt{2}}{4}\left(\frac{\pi}{60}\right)^2 + \cdots + \frac{f^{(n)}\left(\frac{\pi}{4}\right)}{n!}\left(\frac{\pi}{60}\right)^n$$

approaches $\cos\left(\frac{4\pi}{15}\right)$ **in the limit** as $n \to \infty$, that is,

$$\sum_{j=0}^{\infty} \frac{f^{(j)}\left(\frac{4\pi}{15}\right)}{j!}\left(\frac{\pi}{60}\right)^j = \cos\left(\frac{4\pi}{15}\right).$$

Obviously, things are getting a bit complicated. Before trying to achieve the objectives stated in the introduction to this chapter, we must first carefully develop the notions of infinite sequence and infinite series in a simpler setting. After having done so in Chapter 13 we will return to the topic of representing functions by polynomials in Chapter 14.

Exercise Set 12.4

In Exercises 1–8, use Taylor's Theorem to make the indicated approximation and estimate the accuracy using equation (1).

1. $\ln(1.5)$ $\quad f(x) = \ln(x+1), \quad a = 0, \quad n = 3$

2. $\cos 36°$ $\quad f(x) = \cos x, \quad a = \pi/4, \quad n = 2$

3. $\sqrt{3.91}$ $\quad f(x) = \sqrt{x}, \quad a = 4, \quad n = 2$

4. $e^{0.2}$ $\quad f(x) = e^x, \quad a = 0, \quad n = 3$

5. $\cos 1$ $\quad f(x) = \cos x, \quad a = \pi/3, \quad n = 2$

6. $\sin^{-1}(0.2)$ $\quad f(x) = \sin^{-1} x, \quad a = 0, \quad n = 1$

7. $\sqrt[3]{10}$ $\quad f(x) = \sqrt[3]{x}, \quad a = 8, \quad n = 2$

8. $\tan^{-1}\left(\frac{1}{2}\right)$ $\quad f(x) = \tan^{-1} x, \quad a = 0, \quad n = 2$

In Exercises 9–15, determine a bound on the accuracy of the given approximation for the indicated range of x.

9. $\sin x \approx x, \quad |x| < .05$

10. $\sin x \approx x - \frac{x^3}{3!}, \quad |x| < .15$

11. $\cos x \approx \frac{1}{2} - \frac{\sqrt{3}}{2}\left(x - \frac{\pi}{3}\right), \quad \left|x - \frac{\pi}{3}\right| < .05$

12. $\tan x \approx 1 + 2\left(x - \frac{\pi}{4}\right), \quad \left|x - \frac{\pi}{4}\right| < \frac{\pi}{36}$

13. $\sqrt[3]{1+x} \approx 1 + \frac{x}{3}, \quad |x| < .025$

14. $\ln x \approx (x-1) - \frac{1}{2}(x-1)^2 + \frac{1}{3}(x-1)^3$,

$|x-1| < 0.1$

15. $\sqrt{1+x} \approx 1 + \frac{x}{2}, \qquad 0 < x < .02$

In Exercises 16–20, determine how large n must be taken to ensure accuracy to four decimal places in approximating

16. $\sqrt{38}$ using $f(x) = \sqrt{x}, \qquad a = 36$

17. $\ln 1.3$ using $f(x) = \ln(x+1), \qquad a = 0$

18. $\sin 9°$ using $f(x) = \sin x, \qquad a = 0$

19. $\cos 42°$ using $f(x) = \cos x, \qquad a = \pi/4$

20. $e^{0.3}$ using $f(x) = e^x, \qquad a = 0.$

21. Approximate e correct to four decimal places.

22. (Another way to approximate $\ln x$.)

a. Find the Taylor polynomial with $a = 0$ of degree 3 for $f(x) = \ln\left(\frac{1+x}{1-x}\right)$ including the remainder term.

b. Find the value of x for which $\frac{1+x}{1-x} = 1.5$.

c. Find the accuracy in using the polynomial in part (a) to approximate $\ln(1.5)$.

d. Compare this accuracy with that obtained in Example 1 using the Taylor polynomial of degree 3 for $f(x) = \ln(1+x)$.

23. A scientist needing to calculate $\cos x$ for small angles, say $|x| < 6°$, wonders how much accuracy is lost in simply using the approximation $\cos x \approx 1$. What is the answer?

24. Suppose that you needed to make many hand calculations of the function $f(x) = x^5 + 3x^3 + 2x + 6$ for values of x between 0.8 and 1.0. Explain how you might obtain the approximation

$$f(x) \approx 12 + 16(x-1).$$

What is the accuracy to be expected from such approximations? What if, instead, you use the approximation

$$f(x) \approx 12 + 16(x-1) + 19(x-1)^2?$$

25. Use Taylor's Theorem to prove the second derivative test for relative extrema.

26. Show that $\sin x = x - \frac{x^3}{3!} + R_4(x)$ where $|R_4(x)| \le \frac{|x|^5}{5!}$. Use this to show, assuming that $R_4(x)$ is continuous, that

$$\int_0^1 \sin x\,dx = \int_0^1 \left(x - \frac{x^3}{3!} + R_4(x)\right) dx = \frac{11}{24} + \int_0^1 R_4(x)\,dx.$$

Conclude that the number $\frac{11}{24}$ provides an approximation to the integral

$$\int_0^1 \sin x\,dx$$

with an error E no greater than

$$E = \left|\int_0^1 \sin x\,dx - \frac{11}{24}\right| = \left|\int_0^1 R_4(x)\,dx\right| \le \int_0^1 |R_4(x)|\,dx = \int_0^1 \frac{x^5}{5!}\,dx \le \frac{1}{6!} \approx 0.0014.$$

In Exercises 27–30, use the method of Exercise 26 to obtain an approximation accurate to two decimal places for the given integral.

27. $\int_0^1 \sin x^2\,dx$

28. $\int_0^1 e^{x^2}\,dx$

29. $\int_0^1 \cos x^2\,dx$

30. $\int_0^1 \frac{\sin x}{x}\,dx$

SUMMARY OUTLINE OF CHAPTER 12

- The Taylor polynomial of degree n for $f(x)$, expanded about $x = a$, is

$$P_n(x) = f(a) + f'(a)(x-a) + \frac{f''(a)}{2!}(x-a)^2 + \cdots + \frac{f^{(n)}(a)}{n!}(x-a)^n = \sum_{j=0}^{n} \frac{f^{(j)}(a)}{j!}(x-a)^j.$$

- ***Theorem:*** If $f(x)$ is $(n+1)$ times differentiable and

$$f(x) = P_n(x) + R_n(x)$$

then

$$R_n(x) = \frac{f^{(n+1)}(c)}{(n+1)!}(x-a)^{n+1}$$

where c lies between a and x.

REVIEW EXERCISES—CHAPTER 12

In Exercises 1–10, find the Taylor polynomial of degree n for $f(x)$ expanded about $x = a$.

1. $f(x) = \sin 2x, \quad a = 0 \quad n = 5$

2. $f(x) = \ln(1 + x^2), \quad a = 0 \quad n = 2$

3. $f(x) = x \cos x, \quad a = \pi/4 \quad n = 3$

4. $f(x) = \sqrt{1 + x^2}, \quad a = 0 \quad n = 3$

5. $f(x) = e^{x^2}, \quad a = 0 \quad n = 3$

6. $f(x) = \sqrt{2x + 3}, \quad a = 11 \quad n = 3$

7. $f(x) = \tan^{-1} x, \quad a = 1 \quad n = 3$

8. $f(x) = x \ln(1 + x), \quad a = 0 \quad n = 4$

9. $f(x) = 2^x, \quad a = 0 \quad n = 3$

10. $f(x) = \dfrac{1}{1 + x^3}, \quad a = 0 \quad n = 2$

In Exercises 11–15, find the accuracy of the approximation of the given quantity by the Taylor polynomial of degree n for the function $f(x)$ expanded about $x = a$.

11. $\sin 34°, \quad f(x) = \sin x, \quad a = \pi/6, \quad n = 3$

12. $\tan(\pi/12), \quad f(x) = \tan x, \quad a = 0, \quad n = 2$

13. $\sqrt[5]{35}, \quad f(x) = \sqrt[5]{x}, \quad a = 32, \quad n = 2$

14. $\sec\left(\dfrac{3\pi}{16}\right), \quad f(x) = \sec x, \quad a = \dfrac{\pi}{4}, \quad n = 2$

15. $e^{.25}, \quad f(x) = e^x, \quad a = 0, \quad n = 3$

16. Show that the Taylor polynomial of degree 4 for $f(x) = \cosh x^2$ expanded about $a = 0$ is $P_4(x) = 1 + \dfrac{x^4}{2}$.

17. By integrating the Taylor polynomial of degree $n - 1$ for $f(x) = \dfrac{1}{1 - x}$, show that

$$\ln \frac{1}{1 - x} = x + \frac{x^2}{2} + \frac{x^3}{3} + \cdots + \frac{x^n}{n} + \int_0^x \frac{t^n\, dt}{1 - t}.$$

18. Show that, if $P_n(x)$ is the Taylor polynomial of degree n for $f(x) = \sinh x$ expanded about $a = 0$, and if $Q_{n+1}(x)$ is the Taylor polynomial of degree $n + 1$ for $\cosh x$ expanded about $a = 0$, then

$$\frac{d}{dx} Q_{n+1}(x) = P_n(x).$$

What if $a \neq 0$?

19. Let $P_n(x)$ be the Taylor polynomial of degree n, expanded about $a = 0$, for $f(x) = x^{7/2}$. Show that $P_0(0) = P_1(0) = P_2(0) = P_3(0) = 0$, but that $P_n(0)$ is undefined for $n \geq 4$.

20. Let $P_n(x)$ be the Taylor polynomial of degree n for $f(x)$ expanded about $x = a$. If $P_n(x) = 0$ for all n for which $P_n(x)$ is defined, must $f(x)$ be the zero function? (*Hint:* See Exercise 19.)

CHAPTER 13

THE THEORY OF INFINITE SERIES

13.1 INTRODUCTION

In Chapter 12 we saw that values of the transcendental function $f(x) = e^x$ could be approximated with increasing accuracy by the Taylor polynomials

$$P_0(x) \equiv 1$$
$$P_1(x) = 1 + x$$
$$P_2(x) = 1 + x + \frac{x^2}{2}$$
$$P_3(x) = 1 + x + \frac{x^2}{2} + \frac{x^3}{3!}$$
$$P_4(x) = 1 + x + \frac{x^2}{2} + \frac{x^3}{3!} + \frac{x^4}{4!}$$
$$\vdots$$
$$P_n(x) = 1 + x + \frac{x^2}{2} + \frac{x^3}{3!} + \frac{x^4}{4!} + \cdots + \frac{x^n}{n!}$$
$$\vdots$$

This observation, supported by Taylor's Theorem, suggests two types of conclusions:

(i) For each particular value of x, the *sequence* of numbers

$$P_0(x), P_1(x), P_2(x), P_3(x), \ldots, P_n(x), \ldots$$

leads to a *limiting value* as $n \to \infty$, which we expect to be precisely the number $f(x) = e^x$.

(ii) Although $P_n(x)$ is a sum of finitely many terms, its limiting value as $n \to \infty$ may be viewed as the sum of infinitely many terms

$$1 + x + \frac{x^2}{2} + \frac{x^3}{3!} + \cdots + \frac{x^n}{n!} + \cdots.$$

Indeed, we will succeed in verifying these conclusions for the function $f(x) = e^x$ and many others in Chapter 14. To do so, however, we must first develop a clear understanding and a sound theory for the ideas appearing in statements (i) and (ii). This is the purpose of Chapter 13.

Statement (i) raises the question of determining whether an unending list of numbers

$$a_1, a_2, a_3, a_4, \ldots, a_n, \ldots \tag{1}$$

can determine a single number L as its limiting value, written

$$L = \lim_{n\to\infty} a_n. \tag{2}$$

We shall refer to unending strings such as (1) as **infinite sequences,** and we develop the theory concerning limits of the form (2) in Sections 13.2 and 13.3.

Statement (ii) concerns expressions of the form

$$\sum_{n=1}^{\infty} a_n = a_1 + a_2 + a_3 + \cdots + a_n + \cdots, \tag{3}$$

which we shall refer to as **infinite series.** The theory of infinite series of the form (3) is developed in Sections 13.4 through 13.8. At the outset, you should carefully note the distinction between an infinite sequence and an infinite series: the first is simply an infinite *list* of numbers, while the second involves the *sum* of infinitely many numbers.

Because the theory of infinite sequences and series involves both new and somewhat technical ideas, we shall deal primarily with sequences and series of constants in this chapter. However, our ultimate objective is to apply this theory to infinite series of functions, which we shall do in Chapter 14.

13.2 INFINITE SEQUENCES

In Section 6.1 we rather casually defined an infinite sequence to be an unending string of numbers of the form

$$a_1, a_2, a_3, \ldots, a_n, \ldots. \tag{1}$$

It is important to note, however, that expression (1) indicates an *order* in which these numbers appear in the string (the subscripts) as well as the numbers themselves (the a_n's.) A more precise notion of an infinite sequence is therefore the following.

DEFINITION 1 An infinite sequence is a function whose domain is the positive integers.

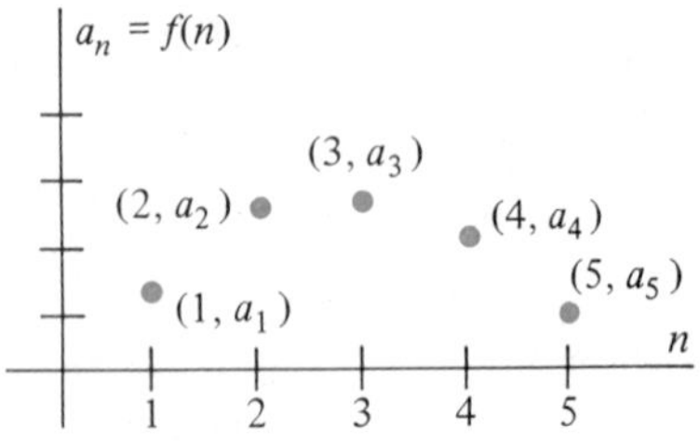

Figure 2.1 Plotting the sequence $\{a_1, a_2, a_3, \ldots\}$ where $a_n = f(n)$.

For example, the infinite sequence $\{1^2, 2^2, 3^2, 4^2, 5^2, \ldots, n^2, \ldots\}$ can be viewed as the set of values of the function $f(n) = n^2$, where $f(1) = 1^2$ is the first term, $f(2) = 2^2$ is the second term, and so on. Using the function concept, we can graph sequences on a coordinate plane. Figure 2.1 shows the graph of an arbitrary sequence $\{a_n\}$.

Usually, a sequence will be specified by a rule of the form $a_n = f(n)$ that determines the nth term of the sequence for each integer n, just as functions are usually specified by an equation of the form $y = f(x)$. For example, the rule $a_n = 2^n$ determines the sequence

$$\{2^n\} = \{2^1, 2^2, 2^3, 2^4, 2^5, \ldots\}$$
$$= \{2, 4, 8, 16, 32, \ldots\},$$

while the rule $a_n = (-1)^n$ determines the sequence

$$\{(-1)^n\} = \{-1, 1, -1, 1, -1, 1, -1, \ldots\}.$$

In such cases the term $a_n = f(n)$ is referred to as the **general term** of the sequence. Note that we use braces { } to denote the entire sequence.

Example 1 Write out the first few terms of the sequences whose general terms are

(a) $a_n = 2n + 1$,

(b) $a_n = 2 + \dfrac{(-1)^n}{n}$.

Solution: In part (a) we have

$$a_1 = 2(1) + 1 = 3, \qquad a_2 = 2(2) + 1 = 5, \qquad a_3 = 2(3) + 1 = 7,$$

and so on,

so

$$\{2n + 1\} = \{3, 5, 7, 9, 11, 13, 15, \ldots\},$$

while in (b) we have

$$a_1 = 2 + \frac{(-1)}{1} = 1, \qquad a_2 = 2 + \frac{(-1)^2}{2} = \frac{5}{2},$$

$$a_3 = 2 + \frac{(-1)^3}{3} = \frac{5}{3}, \qquad \text{and so on,}$$

so

$$\left\{2 + \frac{(-1)^n}{n}\right\} = \left\{1, \frac{5}{2}, \frac{5}{3}, \frac{9}{4}, \frac{9}{5}, \frac{13}{6}, \frac{13}{7}, \frac{17}{8}, \frac{17}{9}, \ldots\right\}.$$

Graphs of these two sequences appear in Figures 2.2 and 2.3. ■

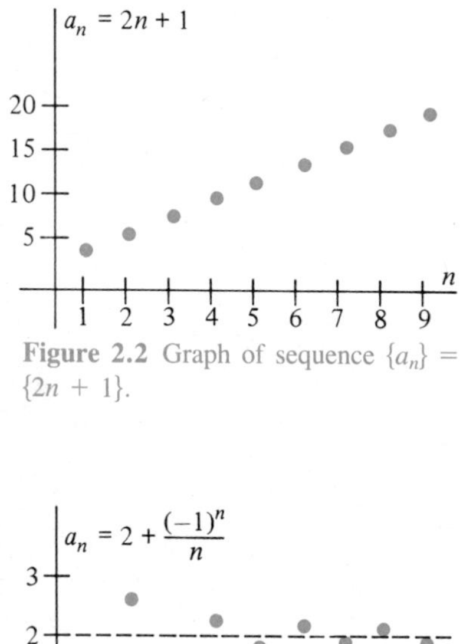

Figure 2.2 Graph of sequence $\{a_n\} = \{2n + 1\}$.

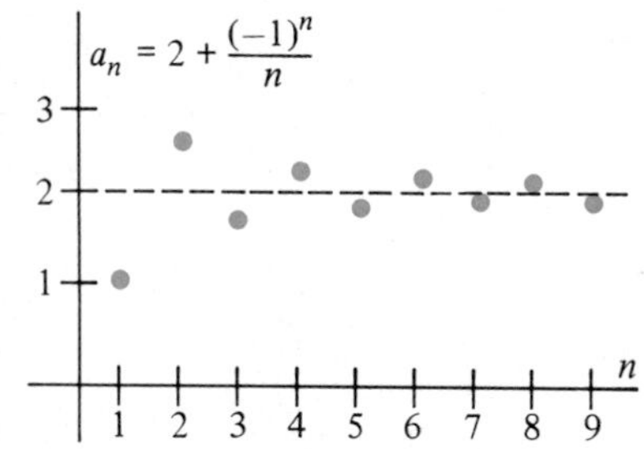

Figure 2.3 Graph of the sequence $\{a_n\} = \left\{2 + \dfrac{(-1)^n}{n}\right\}$.

There is an important difference between the sequence $\{2n + 1\}$ in Figure 2.2 and the sequence $\left\{2 + \dfrac{(-1)^n}{n}\right\}$ in Figure 2.3. In the first of these, the terms of the sequence increase uniformly, not approaching any particular number. In fact, in the language of Chapter 5, we would say that

$$\lim_{n\to\infty} \{2n + 1\} = +\infty$$

since the terms of the sequence increase without bound as $n \to \infty$.

However, the terms of the sequence $\left\{2 + \dfrac{(-1)^n}{n}\right\}$ "approach" the number $L = 2$ as $n \to \infty$, which we wish to write as

$$2 = \lim_{n\to\infty} \left\{2 + \frac{(-1)^n}{n}\right\}.$$

This notion of the *limit* of a sequence strongly parallels the notion of $\lim_{x\to\infty} f(x) = L$ developed for more general functions in Chapter 5. As we did in Chapter 5, we begin the discussion on limits by suggesting a *working definition.*

DEFINITION 2
Working Definition of Limit

$L = \lim_{n\to\infty} a_n$ means that the numbers a_n approach the number L as n increases without bound.

Using this intuitive definition, together with some simple algebra, we can evaluate many types of limits.

Example 2 Find $\lim_{n\to\infty} \dfrac{n+3}{n^2}$.

Strategy

Separate terms in the numerator and simplify.

Use fact that

$$\frac{c}{n^k} \to 0 \quad \text{as} \quad n \to \infty$$

if $k > 0$ (Chapter 5).

Solution

$$\lim_{n\to\infty} \frac{n+3}{n^2} = \lim_{n\to\infty}\left(\frac{1}{n} + \frac{3}{n^2}\right)$$
$$= 0.$$

■

Example 3 Find $\lim_{n\to\infty} \dfrac{6n^3 + 5n^2 + 7}{4n^3 - 2n + 2}$.

Strategy

Divide all terms by n^3 (the highest power of n in the denominator).

Use fact that

$$\frac{c}{n^k} \to 0 \quad \text{as} \quad n \to \infty$$

if $k > 0$.

Solution

$$\lim_{n\to\infty} \frac{6n^3 + 5n^2 + 7}{4n^3 - 2n + 2} = \lim_{n\to\infty} \frac{6 + \dfrac{5}{n} + \dfrac{7}{n^3}}{4 - \dfrac{2}{n^2} + \dfrac{2}{n^3}}$$
$$= \frac{6 + 0 + 0}{4 - 0 + 0} = \frac{3}{2}.$$

■

Example 4 Find $\lim_{n\to\infty} \left[\ln(n+4) - \dfrac{1}{2}\ln(n)\right]$.

Strategy

Use properties of $\ln x$ to reduce expression to the logarithm of a single number.

Solution

$$\lim_{n\to\infty} [\ln(n+4) - 1/2 \ln(n)] = \lim_{n\to\infty} [\ln(n+4) - \ln(n^{1/2})]$$
$$= \lim_{n\to\infty} \ln\left(\frac{n+4}{\sqrt{n}}\right)$$

Divide both terms in numerator by $\sqrt{n}$.

$$= \lim_{n\to\infty} \ln\left(\sqrt{n} + \frac{4}{\sqrt{n}}\right)$$

$$= \infty,$$

Use fact that $\ln x \to \infty$ as $x \to \infty$.

since $\sqrt{n} \to \infty$ and $\dfrac{4}{\sqrt{n}} \to 0$ as $n \to \infty$.

(Strictly speaking, we say that this limit does not exist since $+\infty$ is not a real number.) ■

Example 5 Find $\lim_{n\to\infty} (-1)^n \left(\frac{n+1}{n}\right)$.

Strategy

Because of the factor $(-1)^n$, examine even and odd terms separately.

Solution

For even values of n, $(-1)^n = 1$, so we find that

Divide through by n.

$$\lim_{n\to\infty} (-1)^n \left(\frac{n+1}{n}\right) \qquad \text{(even integers only)}$$

$$= \lim_{n\to\infty} (1)\left(1 + \frac{1}{n}\right) = 1.$$

However, for odd values of n, $(-1)^n = -1$, so for such values of n

Divide through by n.

$$\lim_{n\to\infty} (-1)^n \left(\frac{n+1}{n}\right) \qquad \text{(odd integers only)}$$

$$= \lim_{n\to\infty} (-1)\left(1 + \frac{1}{n}\right) = -1.$$

Since this shows that the terms do not approach a *single* number as $n \to \infty$, this limit does not exist (see Figure 2.4). ■

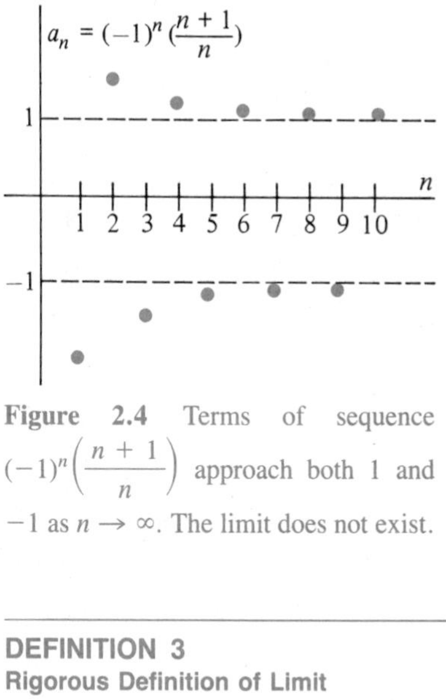

Figure 2.4 Terms of sequence $(-1)^n\left(\frac{n+1}{n}\right)$ approach both 1 and -1 as $n \to \infty$. The limit does not exist.

If $L = \lim_{n\to\infty} a_n$ exists, we say that the sequence $\{a_n\}$ **converges.** Otherwise the sequence is said to **diverge.** Note from Examples 4 and 5 that a sequence may diverge either because a_n becomes infinite as $n \to \infty$ or because a_n, remaining bounded, fails to approach a *single* number as $n \to \infty$.

We next turn to a more rigorous definition of the limit of a sequence.

DEFINITION 3
Rigorous Definition of Limit

We say that $L = \lim_{n\to\infty} a_n$ if and only if for each number $\epsilon > 0$ there exists an integer N so that

$$|a_n - L| < \epsilon \qquad \text{whenever } n \ge N.$$

Definition 3 says this: If $L = \lim_{n\to\infty} a_n$, we will find all terms of the sequence $\{a_n\}$, beyond the Nth term, lying within ϵ units of the number L. Since the integer N, in

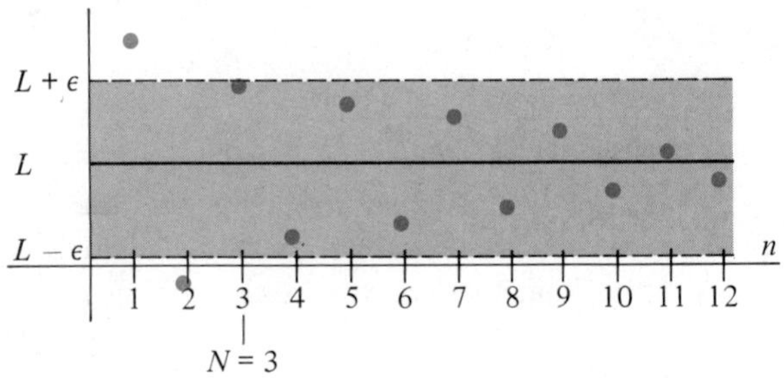

Figure 2.5 Large ϵ.

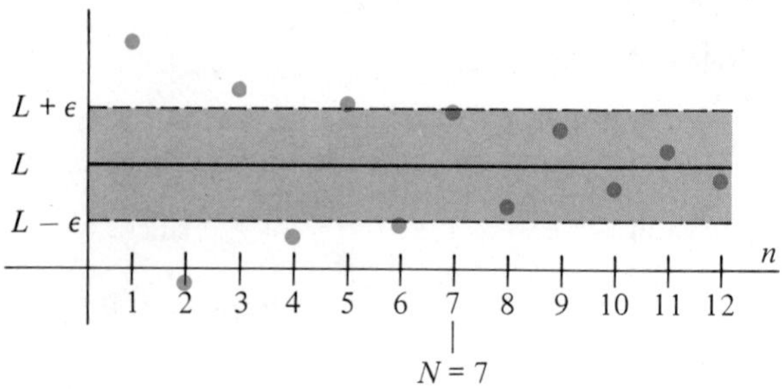

Figure 2.6 Medium ϵ.

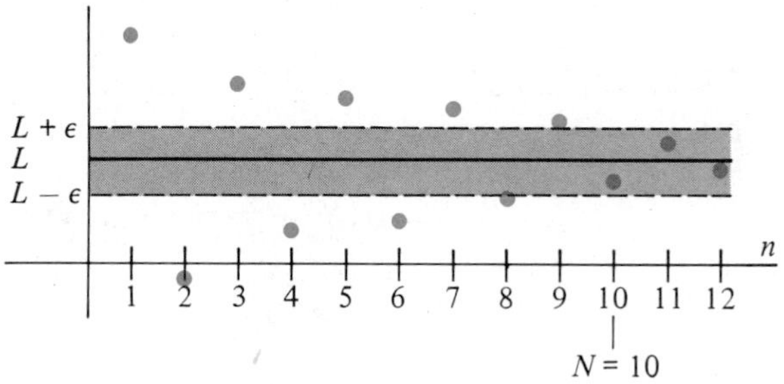

Figure 2.7 Small ϵ.

general, depends upon the number ϵ, we expect to have to look further along the sequence to observe this "closeness" as ϵ decreases in size. Figures 2.5 through 2.7 illustrate choices of N corresponding to three different values of ϵ for a typical sequence $\{a_n\}$.

Definition 3 allows us to rigorously prove statements of the form $L = \lim_{n\to\infty} a_n$ once we have found L. It does not, however, tell us how L is determined from the general term for the sequence $\{a_n\}$. For this, the intuitive Definition 2 and familiarity with examples such as Examples 2–5 above are essential.

The following examples illustrate how Definition 3 is used.

Example 6 Prove that $\lim_{n\to\infty} \frac{1}{n} = 0$.

Strategy
Set up the inequality

$|a_n - L| < \epsilon$

and solve for n to find a relationship between n and ϵ.

Solution
According to Definition 3, we allow $\epsilon > 0$ to be any fixed positive number. Since $a_n = 1/n$ and $L = 0$, we must determine how large to choose n to guarantee that

$$\left|\frac{1}{n} - 0\right| = \frac{1}{n} < \epsilon. \tag{2}$$

Solving inequality (2) for n we see that it is equivalent to the inequality

$$n > \frac{1}{\epsilon}. \tag{3}$$

Choose N large enough that the inequality holds whenever $n \geq N$.

We therefore take N to be any integer larger than $1/\epsilon$. Then, inequality (3) holds, whenever $n \geq N$, which guarantees that inequality (2) holds. In other words, if

$N > 1/\epsilon$ then

$$\left|\frac{1}{n} - 0\right| < \epsilon \qquad \text{whenever} \qquad n \geq N,$$

as required by Definition 3. ■

Example 7 Prove that $\lim_{n\to\infty} \frac{2n-1}{n+2} = 2.$

Solution

Strategy
Set up the inequality
$|a_n - L| < \epsilon.$

We assume $\epsilon > 0$ to be an arbitrary fixed number. Since $a_n = \frac{2n-1}{n+2}$ and $L = 2$, we must determine how large to choose n to guarantee that

$$\left|\frac{2n-1}{n+2} - 2\right| < \epsilon. \tag{4}$$

Solve this inequality for n.

Inequality (4) is equivalent to the inequality

$$\left|\frac{2n-1-2(n+2)}{n+2}\right| < \epsilon,$$

or

$$\frac{5}{n+2} < \epsilon. \tag{5}$$

$$\frac{5}{n+2} < \epsilon$$
$$\Leftrightarrow 5 < \epsilon(n+2)$$
$$\Leftrightarrow \frac{5}{\epsilon} < n+2$$
$$\Leftrightarrow n > \frac{5}{\epsilon} - 2.$$

Solving inequality (5) for n we find that it is equivalent to the inequality

$$n > \frac{5}{\epsilon} - 2. \tag{6}$$

We therefore take N to be any integer larger than $\frac{5}{\epsilon} - 2$. Then, inequality (6) holds whenever $n \geq N$. Since inequality (6) is equivalent to inequality (4), this shows that, for $N > \frac{5}{\epsilon} - 2$, we will have

$$\left|\frac{2n-1}{n+2} - 2\right| < \epsilon \qquad \text{whenever} \qquad n \geq N,$$

as required by Definition 3. ■

Definition 3 can be used to prove several theorems giving rules by which limits of sequences may be calculated. Since the proofs of these theorems are similar to the proofs given in Chapter 2 for the corresponding theorems on limits of functions, several will be left as exercises.

THEOREM 1
Properties of Limits of Sequences

If $\lim_{n\to\infty} a_n = L$, $\lim_{n\to\infty} b_n = M$ and c is any real number, then

(i) $\lim_{n\to\infty} (a_n + b_n) = L + M$,

(ii) $\lim_{n\to\infty} (ca_n) = cL$,

(iii) $\lim_{n\to\infty} (a_n b_n) = LM$,

(iv) $\lim_{n\to\infty} \left(\frac{a_n}{b_n}\right) = \frac{L}{M}$, $\quad b_n \neq 0, \quad M \neq 0$.

Proof: The proofs of (i) and (ii) are straightforward and are left as exercises (see Exercises 42 and 43).

To prove (iii), we assume $\epsilon > 0$ is given and seek to determine how to guarantee that

$$|a_n b_n - LM| < \epsilon. \tag{7}$$

We can now write

$$\begin{aligned} |a_n b_n - LM| &= |a_n b_n - a_n M + a_n M - LM| \\ &\le |a_n b_n - a_n M| + |a_n M - LM| \quad \text{(by the triangle inequality)} \\ &= |a_n||b_n - M| + |M||a_n - L|. \end{aligned} \tag{8}$$

Because $L = \lim_{n\to\infty} a_n$ and $M = \lim_{n\to\infty} b_n$, we can make $|b_n - M|$ and $|a_n - L|$ as small as we like by taking n large.

Let's first focus on the term $|a_n||b_n - M|$ in inequality (8). Since the sequence $\{a_n\}$ converges to L, it is a **bounded sequence** (see Exercise 37). This means that there is a number K so that

$$|a_n| \le K \qquad \text{for all} \qquad n = 1, 2, 3, \ldots. \tag{9}$$

Also, since $\lim_{n\to\infty} b_n = M$, we may apply the definition of limit to conclude that there is an integer N_1 so that

$$|b_n - M| < \frac{\epsilon}{2K} \qquad \text{whenever} \qquad n \ge N_1. \tag{10}$$

Combining inequalities (9) and (10) we may now conclude that

$$|a_n||b_n - M| \le K\left(\frac{\epsilon}{2K}\right) = \frac{\epsilon}{2} \qquad \text{whenever} \qquad n \ge N_1. \tag{11}$$

To get a similar expression for the term $|M||a_n - L|$, we use the fact that $\lim_{n\to\infty} a_n = L$ to conclude that there exists a number N_2 so that

$$|a_n - L| < \frac{\epsilon}{2|M|} \qquad \text{whenever} \qquad n \ge N_2.$$

Thus,

$$|M||a_n - L| < |M|\left(\frac{\epsilon}{2|M|}\right) = \frac{\epsilon}{2} \qquad \text{whenever} \qquad n \ge N_2. \tag{12}$$

Now let N be the larger of N_1 and N_2. Then whenever $n \geq N$, we have that both $n \geq N_1$ and $n \geq N_2$. We may therefore combine inequalities (8), (11), and (12) to conclude that

$$\begin{aligned} |a_n b_n - LM| &\leq |a_n||b_n - M| + |M||a_n - L| \\ &< \frac{\epsilon}{2} + \frac{\epsilon}{2} \\ &= \epsilon \end{aligned}$$

whenever $n \geq N$. This proves (iii).

Similar reasoning establishes (iv) (see Exercise 44). ■

Theorem 1 together with the proof in Example 6 makes legitimate the calculations in Examples 1–3. The next theorem addresses a situation that occurred in Example 4.

THEOREM 2

Suppose that $\lim_{n\to\infty} a_n = L$ and each number a_n lies in the domain of the function $f(x)$. If $f(x)$ is continuous at $x = L$ then

$$\lim_{n\to\infty} f(a_n) = f(L).$$

Proof: Let $\epsilon > 0$ be given. We need to demonstrate that

$$|f(a_n) - f(L)| < \epsilon \tag{13}$$

for sufficiently large n. Since $f(x)$ is a continuous function there exists a number $\delta > 0$ so that

$$|f(x) - f(L)| < \epsilon \tag{14}$$

whenever

$$|x - L| < \delta \qquad \text{(see Definition 9, Section 2.7).} \tag{15}$$

Now since $L = \lim_{n\to\infty} a_n$, there exists a number N so that

$$|a_n - L| < \delta \qquad \text{whenever} \qquad n \geq N. \tag{16}$$

Then, with $a_n = x$, and $n \geq N$, inequality (16) tells us that (15) is satisfied; therefore, (13) and (14) are identical and both are true. Thus, $\lim_{n\to\infty} f(a_n) = f(L)$. ■

Example 8 Since $f(x) = \ln x$ is continuous for $x > 0$,

$$\lim_{n\to\infty} \ln\left(\frac{n-1}{n+1}\right) = \ln\left(\lim_{n\to\infty} \frac{n-1}{n+1}\right) = \ln 1 = 0.$$ ■

Example 9 Since $f(x) = \tan x$ is continuous for $-\pi/2 < x < \pi/2$,

$$\lim_{n\to\infty} \tan\left(\frac{\pi n^2 + 1}{3 - 4n^2}\right) = \tan\left[\lim_{n\to\infty} \left(\frac{\pi n^2 + 1}{3 - 4n^2}\right)\right] = \tan\left(-\frac{\pi}{4}\right) = -1.$$ ■

Example 10 Since $f(x) = \sqrt{x}$ is continuous for $x \geq 0$,

$$\lim_{n\to\infty} \sqrt{\frac{4n+1}{n}} = \left\{\lim_{n\to\infty} \frac{4n+1}{n}\right\}^{1/2} = \sqrt{4} = 2. \qquad \blacksquare$$

The following is the analogue for sequences of the Pinching Theorem for functions (Theorem 5, Chapter 2).

THEOREM 3
Pinching Theorem

Let $\{a_n\}$, $\{b_n\}$, and $\{c_n\}$ be sequences and let P be a positive integer. Let $a_n \leq b_n \leq c_n$ for all integers $n \geq P$. If

$$\lim_{n\to\infty} a_n = L = \lim_{n\to\infty} c_n$$

then $\lim_{n\to\infty} b_n = L$ also.

Proof: Let $\epsilon > 0$ be given. We must show that there exists an integer N so that

$$|b_n - L| < \epsilon \qquad \text{whenever} \qquad n \geq N. \tag{17}$$

Since $\lim_{n\to\infty} a_n = L$ there exists an integer N_a so that

$$|a_n - L| < \epsilon \qquad \text{whenever} \qquad n \geq N_a. \tag{18}$$

We shall make use of inequality (18) in its equivalent form

$$-\epsilon < a_n - L < \epsilon, \qquad n \geq N_a,$$

or

$$L - \epsilon < a_n < L + \epsilon, \qquad n \geq N_a. \tag{19}$$

Also, since $\lim_{n\to\infty} c_n = L$, this same reasoning shows that there exists an integer N_c so that

$$L - \epsilon < c_n < L + \epsilon \qquad \text{whenever} \qquad n \geq N_c. \tag{20}$$

Finally, we restate our hypothesis as

$$a_n \leq b_n \leq c_n, \qquad n \geq P. \tag{21}$$

Now let N be the largest of the integers N_a, N_b, and P. Then whenever $n \geq N$ all three inequalities (19) through (21) hold. Combining the left side of (19) and the right side of (20) with (21) shows that

$$L - \epsilon < a_n \leq b_n \leq c_n < L + \epsilon, \qquad n \geq N$$

or

$$L - \epsilon < b_n < L + \epsilon, \qquad n \geq N.$$

Thus

$$-\epsilon < b_n - L < \epsilon \qquad \text{whenever} \qquad n \geq N. \tag{22}$$

Since inequality (22) is equivalent to inequality (17), the proof is complete. $\blacksquare$

Example 11 Show that $\lim_{n\to\infty} \frac{1}{n^p} = 0$ if $p \geq 1$.

Solution: If $p \geq 1$, $n^p \geq n$ for all $n = 1, 2, 3, \ldots$. Thus

$$0 \leq \frac{1}{n^p} \leq \frac{1}{n}, \qquad n \geq 1.$$

Since $\lim_{n\to\infty} \{0\} = 0 = \lim_{n\to\infty} \left\{\frac{1}{n}\right\}$, the conclusion follows by the Pinching Theorem. ■

Example 12 Find $\lim_{n\to\infty} \frac{\sin n}{n}$.

Strategy
Find bounds on $\sin n$.

Solution
We have $|\sin n| \leq 1$ for all n, that is,

$$-1 \leq \sin n \leq 1, \qquad n \geq 1.$$

Divide by n to find bounds on $\frac{\sin n}{n}$.

Thus

$$-\frac{1}{n} \leq \frac{\sin n}{n} \leq \frac{1}{n}.$$

Apply Theorem 3.

Since $\lim_{n\to\infty} \left(-\frac{1}{n}\right) = 0 = \lim_{n\to\infty} \left(\frac{1}{n}\right)$,

$$\lim_{n\to\infty} \frac{\sin n}{n} = 0$$

by the Pinching Theorem. ■

We summarize the ideas of this section by noting that finding the limit of sequence $\{a_n\}$ is similar to finding horizontal asymptotes for the function $f(x)$ with $f(n) = a_n$, $n = 1, 2, \ldots$. The principal difference is that $\{a_n\}$ is a function defined only for positive integers. Thus $L = \lim_{n\to\infty} a_n$ if $L = \lim_{x\to\infty} f(x)$, but the converse need not be true (see Exercise 35).

Exercise Set 13.2

In each of Exercises 1–28, write out the first four terms of the given sequence and determine whether the sequence converges or diverges. If the sequence converges, find its limit.

1. $\left\{\frac{n}{2n+1}\right\}$

2. $\left\{\frac{2n-1}{n+3}\right\}$

3. $\left\{\frac{n-4}{n^2+2}\right\}$

4. $\left\{\frac{n^2+1}{3n(n+2)}\right\}$

5. $\left\{\frac{1}{1+n^2}\right\}$

6. $\left\{\frac{1}{e^n}\right\}$

7. $\{\sqrt{5}\}$

8. $\left\{\frac{(n-1)(n+1)}{2n^2+2n+2}\right\}$

9. $\left\{\frac{20n}{1+\sqrt{n}}\right\}$

10. $\left\{\frac{6-n^{3/2}}{(\sqrt{n}+1)^2}\right\}$

11. $\left\{\dfrac{3 + (-1)^n\sqrt{n}}{n + 2}\right\}$

12. $\{(-1)^n \sin n\}$

13. $\left\{\sqrt{1 + \dfrac{1}{n}}\right\}$

14. $\left\{1 + \dfrac{(-1)^n}{2^n}\right\}$

15. $\left\{\cos\left(\dfrac{n - 1}{n^2}\right)\right\}$

16. $\left\{\dfrac{n + 1}{n}\right\}$

17. $\left\{\dfrac{n^{3/2} + 2}{2n^{3/2}}\right\}$

18. $\left\{\dfrac{e^n - e^{-n}}{e^n + e^{-n}}\right\}$

19. $\left\{\dfrac{1}{n} - \dfrac{1}{n + 1}\right\}$

20. $\left\{\dfrac{2^n}{5^{n+2}}\right\}$

21. $\{\sqrt{n + 1} - \sqrt{n}\}$

22. $\left\{\dfrac{\cos^2 n}{n}\right\}$

23. $\left\{\dfrac{\sqrt{2n^2 + 1}}{n}\right\}$

24. $\left\{\tan^{-1}\left(\dfrac{n + 2}{2}\right)\right\}$

25. $\left\{n \cdot \sin \dfrac{\pi}{2n}\right\}$

26. $\left\{\ln \dfrac{n^2 + 1}{(n + 2)(n + 3)}\right\}$

27. $\left\{\left(1 + \dfrac{1}{n}\right)^n\right\}$

28. $\left\{\left(1 - \dfrac{1}{n}\right)^n\right\}$

In Exercises 29–33, use the definition of limit to prove that $\lim_{n\to\infty} a_n = L$.

29. $a_n = \dfrac{3}{n}$; $L = 0$

30. $a_n = \dfrac{1}{2n + 1}$; $L = 0$

31. $a_n = \dfrac{n}{3n + 1}$; $L = \dfrac{1}{3}$

32. $a_n = \dfrac{3n - 1}{n + 1}$; $L = 3$

33. $a_n = \dfrac{n^2 + 2n + 3}{1 + n^2}$; $L = 1$

34. Use Theorem 2 to show that $\lim_{n\to\infty} \sqrt[n]{a} = \lim_{n\to\infty} e^{\ln a^{1/n}} = 1$ if $a > 0$.

35. Let $a_n = \sin \pi n$, $n = 1, 2, \ldots$, and let $f(x) = \sin \pi x$. Then $f(n) = a_n$. Show that $\lim_{n\to\infty} a_n = 0$ but that $\lim_{x\to\infty} f(x)$ does not exist.

36. Prove that if $\lim_{n\to\infty} a_n$ exists, this limit is unique. (*Hint:* Assume that both $\lim_{n\to\infty} a_n = L$ and $\lim_{n\to\infty} a_n = M$. Then show $L = M$ by examining the inequality
$|L - M| \le |L - a_n| + |a_n - M|$.)

37. A sequence is called **bounded** if there is a number M so that $|a_n| \le M$ for all terms a_n of the sequence. Prove that a convergent sequence must be bounded.

38. Give an example showing that a bounded sequence need *not* converge.

39. Prove that the sum of two bounded sequences is again bounded. What about the product of two bounded sequences? The quotient?

40. Prove that $\lim_{n\to\infty} a_n = L$ if and only if $\lim_{n\to\infty} |a_n - L| = 0$.

41. Prove that $\lim_{n\to\infty} \dfrac{1}{a^n} = 0$ if $a > 1$, using the Pinching Theorem.

42. Prove part (i) of Theorem 1.

43. Prove part (ii) of Theorem 1.

44. Prove part (iv) of Theorem 1.

13.3 MORE ON INFINITE SEQUENCES

In Section 13.2 we observed that if $\{a_n\}$ is a sequence and $f(x)$ is a function for which $f(n) = a_n$, $n = 1, 2, \ldots$, then we can conclude that $\lim_{n\to\infty} a_n = L$ whenever $\lim_{x\to\infty} f(x) = L$. Geometrically this can be explained by saying that the sequence $\{a_n\}$ is "sampling" the function $f(x)$ at positive integers. Thus, if the values $f(x)$ approach the asymptote $y = L$ as $x \to \infty$, so must the particular values $a_n = f(n)$ (Figure 3.1).

This observation allows us to reformulate the problem of finding $\lim_{n\to\infty} a_n$ as the problem of finding $\lim_{x\to\infty} f(x)$. In the latter problem, l'Hôpital's Rule may often be

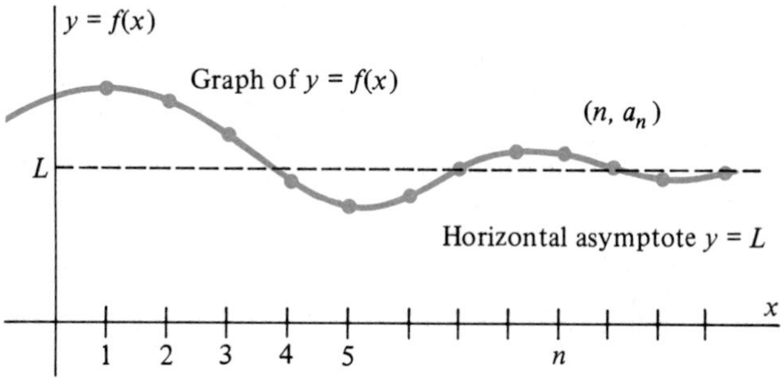

Figure 3.1 If $f(n) = a_n$, $a_n \to L$ as values of $f(x)$ approach the asymptote $y = L$.

successfully applied to calculate an otherwise ambiguous limit. (See Section 5.8 if you need to review l'Hôpital's Rule.)

Example 1 Find $\lim_{n\to\infty} \frac{n}{e^n}$.

Solution: The limit is ambiguous because both numerator and denominator become infinite as $n \to \infty$. Using the above notion together with l'Hôpital's Rule gives

$$\begin{aligned} \lim_{n\to\infty} \frac{n}{e^n} &= \lim_{x\to\infty} \frac{x}{e^x} \qquad \left(\frac{\infty}{\infty} \text{ form}\right) \\ &= \lim_{x\to\infty} \frac{\frac{d}{dx}(x)}{\frac{d}{dx}(e^x)} \qquad \text{(l'Hôpital's Rule)} \\ &= \lim_{x\to\infty} \frac{1}{e^x} \\ &= 0. \end{aligned}$$

■

Example 2 In Section 6.1 it was shown that when a circle of radius r is partitioned into n congruent sectors, the expression $A_n = \frac{nr^2}{2} \cdot \sin\left(\frac{2\pi}{n}\right)$ approximates the area A of the circle. Show that this sequence of approximations converges to the correct expression for the area of the circle. That is, show that

$$\pi r^2 = \lim_{n\to\infty} \frac{nr^2}{2} \cdot \sin\left(\frac{2\pi}{n}\right).$$

(Figure 3.2.)

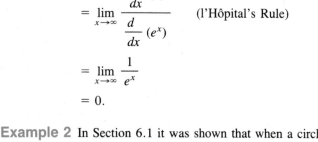

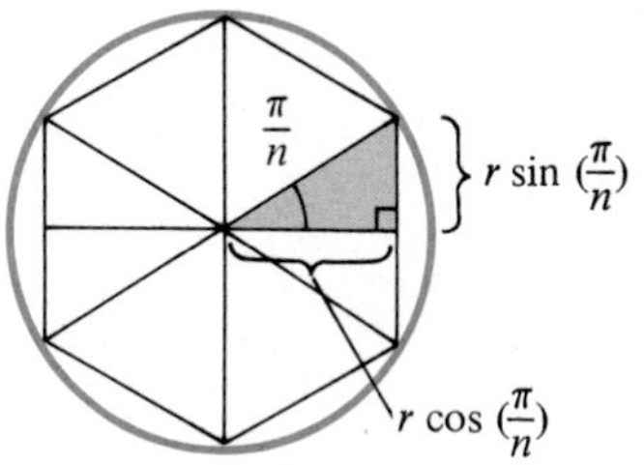

Figure 3.2 Approximating the area of a circle.

Strategy
Consider the limit of the corresponding function $f(x)$.

Solution
We have

$$\lim_{n\to\infty} \frac{nr^2}{2} \sin\left(\frac{2\pi}{n}\right) = \lim_{x\to\infty} \frac{xr^2}{2} \sin\left(\frac{2\pi}{x}\right) \qquad (\infty \cdot 0 \text{ form})$$

Bring indeterminate form $0 \cdot \infty$ to form $\frac{0}{0}$.

$$= \lim_{x\to\infty} \frac{\frac{r^2}{2}\sin\left(\frac{2\pi}{x}\right)}{\frac{1}{x}} \qquad \left(\frac{0}{0}\text{ form}\right)$$

Apply l'Hôpital's Rule.

$$= \lim_{x\to\infty} \frac{\frac{r^2}{2}\cos\left(\frac{2\pi}{x}\right)\cdot\left(\frac{-2\pi}{x^2}\right)}{\frac{-1}{x^2}}$$

Simplify by cancelling factors of $\left(-\frac{1}{x^2}\right)$ and 2.

$$= \lim_{x\to\infty} \pi r^2 \cos\left(\frac{2\pi}{x}\right)$$

Apply limit.

$$= \pi r^2 \cos(0) = \pi r^2. \qquad \blacksquare$$

When calculating limits involving exponents it is often helpful to first calculate $\lim_{n\to\infty} \ln(a_n)$, $a_n > 0$, using the property of logarithms that $\ln(x^n) = n \ln x$. (Such problems often also involve an application of l'Hôpital's Rule.) You must note that the limit $L = \lim_{n\to\infty} \ln a_n$ obtained is not the desired limit but, rather, its logarithm. Thus, $\lim_{n\to\infty} a_n = e^L$.

Example 3 Find $\lim_{n\to\infty} (2n + 1)^{1/n^2}$.

Strategy

Consider the *logarithm* of the corresponding function $f(x)$.

Solution

Because of the variable exponent we consider

Use property of logs: $\ln x^r = r \ln x$

$$\lim_{n\to\infty} \ln[(2n+1)^{1/n^2}] = \lim_{x\to\infty} \ln[(2x+1)^{1/x^2}]$$

$$= \lim_{x\to\infty} \frac{1}{x^2}\cdot\ln(2x+1) \qquad (0\cdot\infty\text{ form})$$

Bring to $\frac{\infty}{\infty}$ form.

$$= \lim_{x\to\infty} \frac{\ln(2x+1)}{x^2} \qquad \left(\frac{\infty}{\infty}\text{ form}\right)$$

Apply l'Hôpital's Rule.

$$= \lim_{x\to\infty} \frac{\frac{2}{2x+1}}{2x}$$

Simplify and obtain limit L.

$$= \lim_{x\to\infty} \frac{1}{2x^2 + x}$$

$$= 0.$$

Solution is e^L.

The original limit is therefore

$$\lim_{n\to\infty} (2n+1)^{1/n^2} = e^0 = 1. \qquad \blacksquare$$

We may now combine the preceding notions with those of Section 13.2 to calculate several limits that arise frequently.

$$\boxed{\lim_{n\to\infty} x^n = 0 \quad \text{if} \quad |x| < 1} \tag{1}$$

Proof: Let ϵ be an arbitrary positive number. Then, since

$$\lim_{n\to\infty} \frac{1}{n} = 0,$$

$\lim_{n\to\infty} \epsilon^{1/n} = \epsilon^0 = 1$. Thus there exists an integer N, by Definition 3, such that $\epsilon^{1/n} > |x|$ whenever $n \geq N$. Thus

$$|x|^n < (\epsilon^{1/n})^n = \epsilon, \qquad n \geq N.$$

Equivalently, noting that $|x|^n = |x^n|$, we have

$$|x^n - 0| < \epsilon \qquad \text{whenever} \qquad n \geq N.$$

Thus,

$$\lim_{n\to\infty} x^n = 0 \qquad \text{by Definition 3.}$$

■

$$\boxed{\lim_{n\to\infty} \frac{x^n}{n!} = 0} \tag{2}$$

Proof: Let x be given and let N be an integer such that $N > |x|$. Write

$$J = \left|\left(\frac{x}{1}\right)\left(\frac{x}{2}\right)\left(\frac{x}{3}\right) \cdots \left(\frac{x}{N-1}\right)\right|$$

and note that J is constant since N and x are fixed. Then for $n > N$ we have

$$\begin{aligned} \left|\frac{x^n}{n!}\right| &= \left|\left(\frac{x}{1}\right)\left(\frac{x}{2}\right)\left(\frac{x}{3}\right) \cdots \left(\frac{x}{N-1}\right)\left(\frac{x}{N}\right) \cdots \left(\frac{x}{n}\right)\right| \\ &= J\left|\left(\frac{x}{N}\right)\left(\frac{x}{N+1}\right) \cdots \left(\frac{x}{n}\right)\right| \qquad (n-N+1) \text{ factors} \\ &\leq J \cdot \left|\frac{x}{N}\right|^{(n-N+1)} \end{aligned}$$

Since $|x| < N$, we have $\left|\frac{x}{N}\right| < 1$, so

$$\lim_{n\to\infty} J \cdot \left|\frac{x}{N}\right|^{(n-N+1)} = J \cdot \left|\frac{x}{N}\right|^{(-N+1)} \cdot \lim_{n\to\infty} \left|\frac{x}{N}\right|^n = 0$$

by (1). This establishes (2) by the preceding inequality. ■

$$\boxed{\lim_{n\to\infty} \frac{\ln(n)}{n} = 0} \tag{3}$$

Proof: $\lim_{n\to\infty} \frac{\ln(n)}{n} = \lim_{x\to\infty} \frac{\ln x}{x} = \lim_{x\to\infty} \frac{\frac{1}{x}}{1} = 0$, by l'Hôpital's Rule. ■

$$\lim_{n\to\infty} \sqrt[n]{n} = 1 \tag{4}$$

Proof: $\lim_{n\to\infty} \ln(n^{1/n}) = \lim_{n\to\infty} \dfrac{\ln(n)}{n} = 0$, by (3).

Thus,

$$\lim_{n\to\infty} n^{1/n} = e^0 = 1.$$

■

$$\lim_{n\to\infty} \left(1 + \frac{t}{n}\right)^n = e^t \tag{5}$$

Proof:

$$\begin{aligned}
\lim_{n\to\infty} \ln\left[\left(1 + \frac{t}{n}\right)^n\right] &= \lim_{x\to\infty} \ln\left[\left(1 + \frac{t}{x}\right)^x\right] \\
&= \lim_{x\to\infty} x \ln\left(1 + \frac{t}{x}\right) && (\infty \cdot 0 \text{ form}) \\
&= \lim_{x\to\infty} \frac{\ln\left(1 + \frac{t}{x}\right)}{\frac{1}{x}} && \left(\frac{0}{0} \text{ form}\right) \\
&= \lim_{x\to\infty} \frac{\left(1 + \frac{t}{x}\right)^{-1} \cdot \left(-\frac{t}{x^2}\right)}{\frac{-1}{x^2}} && \text{(l'Hôpital's Rule)} \\
&= \lim_{x\to\infty} \frac{t}{1 + \frac{t}{x}} \\
&= t.
\end{aligned}$$

The limit is therefore e^t.

■

Example 4

$$\begin{aligned}
\lim_{n\to\infty} \left(\frac{5}{n}\right)^{1/n} &= \lim_{n\to\infty} \frac{\sqrt[n]{5}}{\sqrt[n]{n}} \\
&= \frac{\{\lim_{n\to\infty} \sqrt[n]{5}\}}{\{\lim_{n\to\infty} \sqrt[n]{n}\}} \\
&= \frac{1}{1} = 1
\end{aligned}$$

by statement (4).

■

Example 5

$$\lim_{n\to\infty}\left(\frac{n+4}{n}\right)^{3n} = \lim_{n\to\infty}\left(1+\frac{4}{n}\right)^{3n}$$
$$= \left[\lim_{n\to\infty}\left(1+\frac{4}{n}\right)^{n}\right]^3$$
$$= (e^4)^3 = e^{12}$$

by statement (5). ■

Theorems (1) through (3) of Section 13.2 and limits (1) through (5) above will enable you to calculate limits for most of the sequences to be encountered in what follows. However, we shall have need of one further theorem on sequences, which we develop here.

A set S of numbers is said to be *bounded* if there exists a number M so that $|x| \leq M$ for every number $x \in S$. A fundamental property of the real number system states that among all such bounds M there can always be found a smallest bound. This property is referred to as the **completeness axiom,** and it is formally stated as follows.

COMPLETENESS AXIOM FOR REAL NUMBERS: If S is any nonempty bounded set of real numbers there exists a least upper bound L for S. That is, there exists a number L for which

(i) $x \leq L$ for every $x \in S$, and
(ii) if M is any upper bound for S, then $L \leq M$.

The completeness axiom and its role in the definition of the real number system are topics for more advanced courses. We shall use this axiom to establish a theorem about *increasing* sequences. As for functions, we define the sequence $\{a_n\}$ to be **increasing** if $a_n > a_m$ whenever $n > m$.

THEOREM 4 Every bounded increasing sequence converges.

Proof: We use the definition of the limit of a sequence and the completeness axiom. Let $\epsilon > 0$ be given, and let $\{a_n\}$ denote a bounded increasing sequence. Then the set of numbers $\{a_1, a_2, a_3, \ldots\}$ is bounded, so it has a least upper bound L according to the completeness axiom. We will complete the proof by showing that $\lim_{n\to\infty} a_n = L$.

Since $\epsilon > 0$, we must have $L - \epsilon < L$, so $L - \epsilon$ cannot be an upper bound for the sequence. Thus, there must exist an integer N such that

$$a_N > L - \epsilon. \qquad \text{(Figure 3.3.)}$$

But since $\{a_n\}$ is increasing we must have $a_n > a_N$ for all $n > N$. Thus we have the following five numbers in increasing order:

$$L - \epsilon < a_N < a_n \leq L < L + \epsilon, \qquad n > N.$$

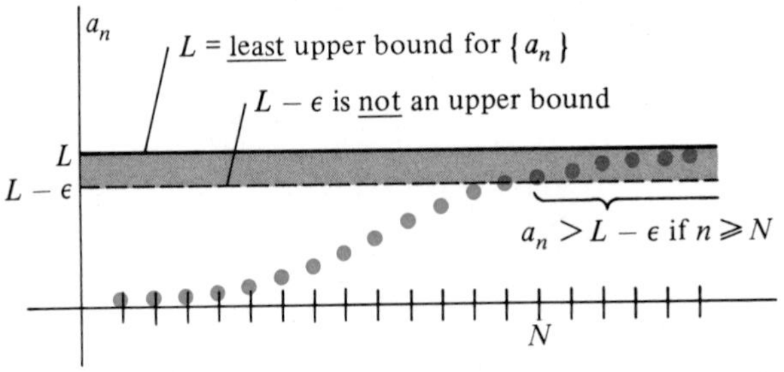

Figure 3.3 A bounded increasing sequence converges to its least upper bound.

It now follows from $L - \epsilon < a_n < L + \epsilon$ that

$$|a_n - L| < \epsilon \qquad \text{whenever} \qquad n > N.$$

Thus, $L = \lim_{n\to\infty} a_n$ according to Definition 3. ■

There is an obvious extension of Theorem 4 to bounded *decreasing* sequences. If $\{a_n\}$ is a bounded decreasing sequence, then the sequence $\{-a_n\}$ is a bounded *increasing* sequence, which, by Theorem 4, has a limit, say L. By Theorem 1 we then have

$$\lim_{n\to\infty} a_n = -\lim_{n\to\infty} \{-a_n\} = -L.$$

We summarize these remarks as follows.

COROLLARY 1 Every bounded decreasing sequence converges.

Before leaving this topic we note that Theorem 4 and Corollary 1 have already played a significant role in the development of the definite integral. Recall that if the function $f(x)$ is continuous and nonnegative on $[a, b]$ the lower Riemann sum $\underline{S}_n$ is defined to be the number

$$\underline{S}_n = \sum_{j=1}^{n} c_j\left(\frac{b-a}{n}\right)$$

where c_j is the minimum value of $f(x)$ on the jth subinterval of $[a, b]$. It was argued in Chapter 6 that since the *sequence* of lower Riemann sums, $\{\underline{S}_n\}$ was both increasing and bounded, the limit $S = \lim_{n\to\infty} \underline{S}_n$ exists. The existence of this limit allowed us to define the definite integral in terms of lower Riemann sums as

$$\int_a^b f(x)\,dx = \lim_{n\to\infty} \underline{S}_n = S.$$

Theorem 4 provides the justification for this reasoning. Similarly, Corollary 1 shows why the decreasing sequence of *upper* Riemann sums $\overline{S}_n$ must have a limit, which was shown also to be S. Finally, the Pinching Theorem (Theorem 3) justifies our earlier conclusion that since an *arbitrary* Riemann sum S_n must satisfy the inequality

$$\underline{S}_n \leq S_n \leq \overline{S}_n$$

for all n, the limit $S = \lim_{n\to\infty} S_n$ exists independent of the particular type of approximating sum. Thus the notion of infinite sequence provides a theory by which our earlier conclusions about definite integrals are made more precise.

Recursively Defined Sequences

In many applications, particularly in computer science, sequences are defined *recursively* rather than as functions of the integer n. One of the most famous recursively defined sequences is the sequence of **Fibonacci numbers**

$$1, 1, 2, 3, 5, 8, 13, 21, 34, \ldots . \tag{6}$$

This sequence was first discovered by the Italian mathematician Leonardo of Pisa (who also went by the name Fibonacci) around the year 1200 A.D. (Fibonacci is regarded by many as the most brilliant of the pre-Renaissance mathematicians).

The Fibonacci sequence $\{F_n\}$ in (6) is determined by the rules

$$F_1 = 1, \tag{7}$$

$$F_2 = 1, \quad \text{and} \tag{8}$$

$$F_{n+2} = F_n + F_{n+1}, \qquad n = 2, 3, 4, \ldots . \tag{9}$$

That is, every term in the sequence beyond the second is found by adding the two preceding terms. This is what we mean by saying that the sequence $\{a_n\}$ is **recursively defined:** the term a_n is a function of one or more preceding terms, such as a_{n-1}, a_{n-2}, etc. But a_n is *not* written as an explicit function of n.

It is often possible to rewrite a recursively defined sequence as an explicit function of n, and vice versa. For example, it has been shown that the nth term of the Fibonacci sequence can be written

$$F_n = \frac{1}{\sqrt{5}}\left(\frac{1+\sqrt{5}}{2}\right)^{n+1} - \frac{1}{\sqrt{5}}\left(\frac{1-\sqrt{5}}{2}\right)^{n+1}.$$

(See Exercise 22 for a biological interpretation of the Fibonacci sequence.) Another such example is the factorial sequence

$$\{a_n\} = \{n!\} = \{0!, 1!, 2!, 3!, \ldots\} = \{1, 1, 2, 6, 24, \ldots\}.$$

It can be defined recursively as the sequence $\{f_n\} = \{f_0, f_1, f_2, \ldots\}$ where

$$f_0 = 1$$
$$f_n = nf_{n-1}, \qquad n = 1, 2, 3, \ldots .$$

However, not every sequence $\{a_n\}$ can be defined recursively.

Recursively defined sequences occur frequently in the analysis of computer algorithms. In such situations one is concerned only with how to determine the next term in the sequence given the present term (and, possibly, several preceding terms), not with the correspondence between integers n and the terms a_n. Exercises 22–26 in this section concern recursively defined sequences.

Exercise Set 13.3

In each of Exercises 1–18, find the indicated limit, if it exists.

1. $\lim_{n\to\infty} n \sin\left(\frac{2}{n}\right)$

2. $\lim_{n\to\infty} \frac{\sin^3 n}{n}$

3. $\lim_{n\to\infty} \sqrt[n]{4n}$

4. $\lim_{n\to\infty} \frac{\ln(n)}{\sqrt{n}}$

5. $\lim_{n\to\infty} (n + 1)e^{-n}$

6. $\lim_{n\to\infty} \frac{n}{e^n}$

7. $\lim_{n\to\infty} \frac{n^n}{n!}$

8. $\lim_{n\to\infty} n^{3/n}$

9. $\lim_{n\to\infty} (n + \pi)^{1/n}$

10. $\lim_{n\to\infty} \left(1 - \frac{3}{n}\right)^n$

11. $\lim_{n\to\infty} \frac{3^n}{(n + 3)!}$

12. $\lim_{n\to\infty} \left(\frac{e}{n} \ln \frac{e}{n}\right)$

13. $\lim_{n\to\infty} \frac{n^2 \ln(n)}{2^n}$

14. $\lim_{n\to\infty} n^{\sin(\pi/n)}$

15. $\lim_{n\to\infty} \left(\frac{n + 3}{n}\right)^n$

16. $\lim_{n\to\infty} \left(1 + \frac{1}{n^2}\right)^n$

17. $\lim_{n\to\infty} \sqrt[n]{n^3}$

18. $\lim_{n\to\infty} \frac{n - \sin n}{n + \cos n}$

19. Refer to Exercise 5, Section 6.1. Develop the approximation $C_n = 2nr \sin\left(\frac{\pi}{n}\right)$ to the circumference of the circle of radius r. Show that this sequence converges to the correct value for the circumference.

20. Refer to Exercise 8, Section 6.1, where we developed the approximation $A_n = nr^2 \tan(\pi/n)$ to the area of the circle of radius r. Show that $\lim_{n\to\infty} A_n$ exists and is the correct value of the area.

21. Prove that $\lim_{n\to\infty} x^n$ does not exist if $|x| > 1$.

22. The Fibonacci sequence $\{1, 1, 2, 3, 5, 8, 13, 21, 34, \ldots\}$ arises as a mathematical model for the size of a population of rabbits under the following conditions. We assume that we begin with a single pair of rabbits, that this and each other pair of rabbits become fertile one month after birth, that each pair of fertile rabbits gives birth to one new pair of rabbits each month, and that no rabbits die. If F_n represents the number of pairs of rabbits in the population after n months, show that

a. $F_1 = 1$,

b. $F_2 = 1$, and

c. $F_{n+2} = 2F_n + (F_{n+1} - F_n)$. (*Hint:* The term $2F_n$ is explained as follows. Every pair of rabbits that was present two months ago is still present along with one pair of offspring. The second term accounts for the fact that those rabbits which were fertile two months ago have produced *two* pairs of offspring since then, one of which is not counted in the first term.)

d. Conclude from (c) that $F_{n+2} = F_n + F_{n+1}$.

23. Let $\{a_n\}$ be a sequence recursively defined by the equations

$$a_0 = 1$$

$$a_n = 2a_{n-1}, \qquad n = 1, 2, \ldots .$$

Show that the general term for this sequence is $a_n = 2^n$.

24. Find the general term for the sequence $\{a_n\}$ recursively defined by the equations

$$a_0 = 4$$

$$a_n = a_{n-1} + 1, \qquad n = 1, 2, \ldots .$$

25. Find the general term for the sequence $\{a_n\}$ recursively defined by the equations

$$a_0 = -5$$

$$a_n = a_{n-1} + 2, \qquad n = 1, 2, \ldots .$$

26. For the Fibonacci sequence (Exercise 22) show that

a. $F_{n+3} = 2F_{n+1} + F_n$

b. $F_{n+4} = 3F_{n+1} + 2F_n$

c. $F_{n+p} = F_pF_{n+1} + F_{p-1}F_n, \qquad p = 3, 4, \ldots .$

27. Give an example of a bounded sequence that does not converge.

28. Give an example of an increasing sequence that does not converge.

29. Must every convergent sequence be bounded and either increasing or decreasing?

13.4 INFINITE SERIES

Recall that if the continuous function $y = f(x)$ is defined for all $x \geq 1$, the improper integral $\int_1^\infty f(x)\,dx$ is defined as a limit:

$$\int_1^\infty f(x)\,dx = \lim_{n\to\infty} \int_1^n f(x)\,dx. \tag{1}$$

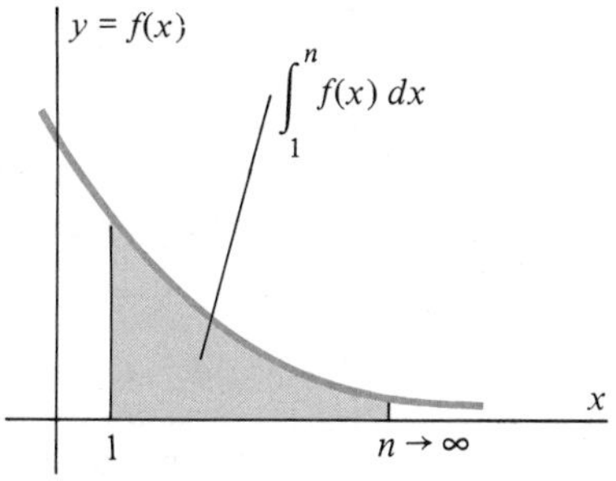

Figure 4.1 $\int_1^\infty f(x)\,dx$ is the limit of the partial integrals $\int_1^n f(x)\,dx$ as $n \to \infty$.

Here n denotes any positive real number. The integral on the right-hand side of (1) may be referred to as a **partial integral** for the integral on the left since it is evaluated over only part of the interval $[1, \infty)$. If $f(x) \geq 0$ for all $x \geq 1$, the integrals in equation (1) may be associated with areas of regions bounded by the graph of $y = f(x)$ (Figure 4.1).

Since an infinite sequence $\{a_n\}$ is a function defined for all integers $n \geq 1$, we may define the analogous operation for infinite series as follows. For each integer n we define the ***n*th partial sum,** S_n, by

$$S_n = \sum_{k=1}^{n} a_k = a_1 + a_2 + a_3 + \cdots + a_n.$$

That is,

$$\begin{aligned} S_1 &= a_1 \\ S_2 &= a_1 + a_2 \\ S_3 &= a_1 + a_2 + a_3 \\ &\vdots \\ S_n &= a_1 + a_2 + a_3 + a_4 + a_5 + \cdots + a_n. \end{aligned}$$

If the limit of the infinite sequence $S_1, S_2, S_3, \ldots$, of the partial sums exists we say that the *infinite series* $\Sigma_{k=1}^{\infty} a_k$ converges, and we refer to $S = \lim_{n\to\infty} S_n$ as the *sum* of the series. If this limit does not exist we say that the infinite series *diverges*.

If $a_k \geq 0$ for all $k \geq 0$, we may interpret these notions in terms of area. Notice that the area of the rectangle of height $h = a_1$, constructed over the interval $[1, 2]$ in Figure 4.2, has area a_1, since the length of its base is $2 - 1 = 1$. Similarly, the area of the rectangle over $[2, 3]$ is a_2, and so on. Figure 4.3 illustrates this notion for the general case of the kth term of the sequence $\{a_k\}$.

We may therefore identify the partial sum S_n with the area of the region contained by the first n rectangles in Figure 4.2. The question of whether the limit $S = \lim_{n\to\infty} S_n$ exists is geometrically equivalent to the question of whether the limit of the partial integrals in Figure 4.1 exists. In both cases we are asking about the area of a region unbounded in the positive x direction.

We summarize these notions in the following definition.

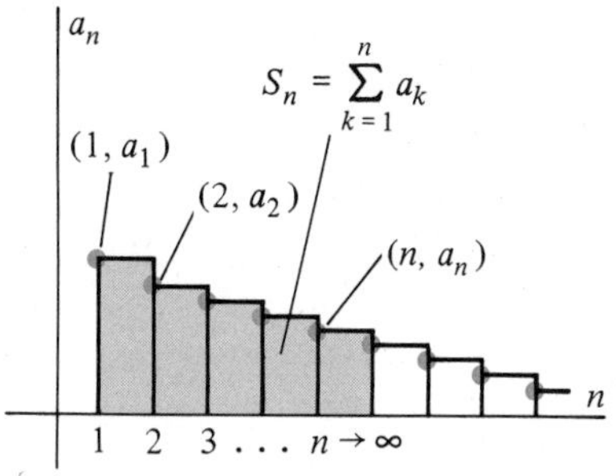

Figure 4.2 $\sum_{k=1}^{\infty} a_k$ is the limit of the partial sums $\sum_{k=1}^{n} a_k$ as $n \to \infty$.

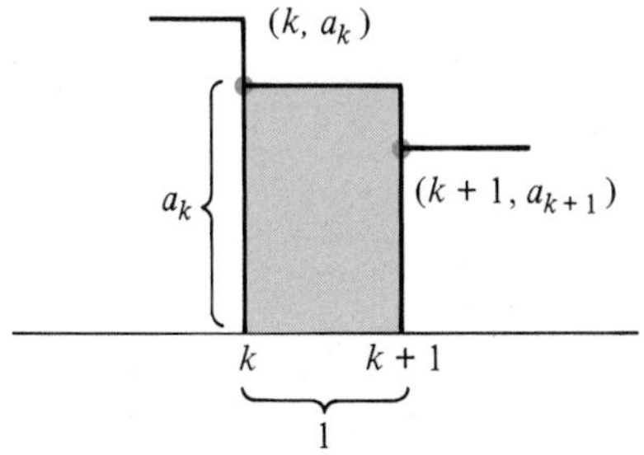

Figure 4.3 Area of rectangle with vertex (k, a_k) is also a_k.

DEFINITION 4

An **infinite series** is an expression of the form

$$\sum_{k=1}^{\infty} a_k = a_1 + a_2 + a_3 + \cdots. \tag{2}$$

The infinite series

$$\sum_{k=1}^{\infty} a_k$$

is said to **converge** to the **sum** S if

$$S = \lim_{n\to\infty} S_n, \tag{3}$$

where S_n denotes the nth **partial sum**

$$S_n = a_1 + a_2 + a_3 + \cdots + a_n. \tag{4}$$

If the limit in (3) does not exist, the series $\sum_{k=1}^{\infty} a_k$ is said to **diverge.**

Notice that the concept of infinite sequence is involved in Definition 4 in two distinct ways. First, we may interpret equation (2) as saying that the infinite series $\Sigma_{k=1}^{\infty} a_k$ is formed by summing the terms of the infinite sequence $\{a_1, a_2, a_3, \ldots\}$. (By "summing" we do not mean the actual performing of an infinite number of additions but, rather, the limit calculation illustrated in Figure 4.2.) Second, the value of the infinite series is the limit, if it exists, of the sequence of partial sums $\{S_1, S_2, S_3, \ldots\}$ defined by equation (4). Thus, while the kth terms of both the sequence $\{a_k\}$ and the series $\Sigma_{k=1}^{\infty} a_k$ are denoted by a_k, the **limit of the sequence** $\{a_k\}$ is simply $L = \lim_{k\to\infty} a_k$ while the **sum of the series** $\sum_{k=1}^{\infty} a_k$ is

$$S = \lim_{n\to\infty} \left\{\sum_{k=1}^{n} a_k\right\}.$$

Example 1 The repeating decimal $0.6666\overline{6}$ may be interpreted as the infinite series

$$\begin{aligned} 0.6666\overline{6} &= .6 + (.06) + (.006) + (.0006) + \cdots \\ &= \frac{6}{10} + \frac{6}{10^2} + \frac{6}{10^3} + \frac{6}{10^4} + \cdots \\ &= \sum_{k=1}^{\infty} \frac{6}{10^k}. \end{aligned}$$

Let us verify that this interpretation is consistent with the ordinary notion that $0.66\overline{6} = 2/3$. The nth partial sum of this series is

$$S_n = \frac{6}{10} + \frac{6}{10^2} + \frac{6}{10^3} + \cdots + \frac{6}{10^n}. \tag{5}$$

Now note that each term of the sum is 10 times the following term. Multiplying both sides of equation (5) by $\frac{1}{10}$ gives

$$\frac{1}{10} S_n = \frac{6}{10^2} + \frac{6}{10^3} + \frac{6}{10^4} + \cdots + \frac{6}{10^{n+1}}. \tag{6}$$

Subtracting equation (6) from equation (5) we conclude that

$$S_n - \frac{1}{10} S_n = \frac{6}{10} - \frac{6}{10^{n+1}},$$

so

$$S_n = \frac{10}{9}\left(\frac{6}{10} - \frac{6}{10^{n+1}}\right) = \frac{2}{3}\left(1 - \frac{1}{10^n}\right).$$

According to the definition of an infinite series, the sum of the series is therefore

$$S = \lim_{n\to\infty} S_n = \lim_{n\to\infty} \frac{2}{3}\left(1 - \frac{1}{10^n}\right) = 2/3.$$

This shows that

$$\frac{2}{3} = \sum_{k=1}^{\infty} \frac{6}{10^k} = .6666\bar{6}.$$

■

Example 2 The infinite series

$$\sum_{k=1}^{\infty} (-1)^k = -1 + 1 - 1 + 1 - 1 + \cdots$$

does not converge. To see why, observe that the partial sums are

$$\begin{aligned}
S_1 &= -1 \\
S_2 &= -1 + 1 = 0 \\
S_3 &= -1 + 1 - 1 = -1 \\
S_4 &= -1 + 1 - 1 + 1 = 0 \\
&\vdots \\
S_{2n-1} &= -1 + 1 - 1 + \cdots + 1 - 1 = -1 \qquad (2n - 1 \text{ terms}) \\
S_{2n} &= -1 + 1 - 1 + \cdots + 1 - 1 + 1 = 0 \qquad (2n \text{ terms}).
\end{aligned}$$

Thus, the terms of the sequence $\{S_n\}$ are alternately -1 or 0, so $\lim_{n\to\infty} S_n$ does not exist.

■

Example 3 The infinite series

$$\sum_{k=1}^{\infty} k^2 = 1^2 + 2^2 + 3^2 + 4^2 + \cdots$$

does not converge. To see this, recall the formula for the nth partial sum of this series (Section 6.2, equation (5)):

$$S_n = 1^2 + 2^2 + 3^2 + \cdots + n^2 = \frac{n(n + 1)(n + 2)}{6}.$$

Thus

$$\lim_{n\to\infty} S_n = \lim_{n\to\infty} \frac{n(n+1)(n+2)}{6} = +\infty$$

so $\lim_{n\to\infty} S_n$ does not exist. Intuitively, this result is obvious since the terms of the series are increasing in size. This means that the terms of the sequence $\{S_n\}$ grow at an increasing rate and, therefore, cannot converge. ■

Example 4 Determine whether the infinite series

$$\sum_{k=1}^{\infty} \frac{1}{k(k+1)}$$

converges. If it does, find its sum.

Solution: By the method of partial fractions we can show that

$$\frac{1}{k(k+1)} = \frac{1}{k} - \frac{1}{k+1}, \qquad k = 1, 2, 3, \ldots .$$

We can therefore write the partial sum S_n for this series as

$$\begin{aligned} S_n = \sum_{k=1}^{n} \frac{1}{k(k+1)} &= \frac{1}{1\cdot 2} + \frac{1}{2\cdot 3} + \frac{1}{3\cdot 4} + \cdots + \frac{1}{(n-1)n} + \frac{1}{n(n+1)} \\ &= \left[\frac{1}{1} - \frac{1}{2}\right] + \left[\frac{1}{2} - \frac{1}{3}\right] + \left[\frac{1}{3} - \frac{1}{4}\right] \\ &\quad + \cdots + \left[\frac{1}{n-1} - \frac{1}{n}\right] + \left[\frac{1}{n} - \frac{1}{n+1}\right] \\ &= 1 - \frac{1}{n+1} \end{aligned}$$

since all other terms in this "telescoping sum" cancel. This shows that

$$\sum_{k=1}^{\infty} \frac{1}{k(k+1)} = \lim_{n\to\infty} S_n = \lim_{n\to\infty} \left(1 - \frac{1}{n+1}\right) = 1.$$

Thus the series converges, and its sum is $S = 1$. ■

REMARK: Up to this point we have been indexing all infinite series so that the first term has index $k = 1$. This is not necessary. For example, we could rewrite

$$\sum_{k=1}^{\infty} a_k \quad \text{as} \quad \sum_{k=2}^{\infty} b_k$$

where $b_k = a_{k-1}$, since in this case we would have

$$\sum_{k=1}^{\infty} a_k = a_1 + a_2 + a_3 + a_4 + \cdots$$

and

$$\sum_{k=2}^{\infty} b_k = b_2 + b_3 + b_4 + b_5 + \cdots$$
$$= a_{(2-1)} + a_{(3-1)} + a_{(4-1)} + a_{(5-1)} + \cdots$$
$$= a_1 + a_2 + a_3 + a_4 + \cdots.$$

Similarly, we can write

$$\sum_{k=1}^{\infty} a_k \quad \text{as} \quad \sum_{k=0}^{\infty} c_k, \quad \text{where} \quad c_k = a_{k+1}.$$

In particular, *the convergence of an infinite series does not depend upon which index is associated with its first term*. To see this, let S_n denote the nth partial sum of the series $\Sigma_{k=1}^{\infty} a_k$. Then,

(i) The nth partial sum (sum of the first n terms) of the series $\Sigma_{k=2}^{\infty} a_k$ is

$$(a_2 + a_3 + \cdots + a_{n+1}) = (a_1 + a_2 + \cdots + a_{n+1}) - a_1$$
$$= S_{n+1} - a_1.$$

(ii) The nth partial sum of the series $\Sigma_{k=0}^{\infty} a_k$ is

$$(a_0 + a_1 + \cdots + a_{n-1}) = a_0 + (a_1 + a_2 + \cdots + a_{n-1})$$
$$= a_0 + S_{n-1}.$$

Thus the partial sums for the series in (i) and (ii) will converge precisely when $\lim_{n\to\infty} S_n$ exists. Another way to look at this is to say that *adding or subtracting a finite number of terms cannot affect the convergence or divergence of an infinite series*. But be careful—there are two distinct issues here:

(a) When we are concerned only with determining *whether an infinite series converges,* we need not be concerned with the particular value of the first index. In such cases we will often write just Σa_k instead of $\sum_{k=1}^{\infty} a_k$, for example.

(b) However, when working with an explicit formula *for the partial sum* of a series, as in Example 4, it is very important to note the leading index. For example, you can verify that if the series in Example 4 had been $\sum_{k=2}^{\infty} \frac{1}{k(k+1)}$, the sum would have been $S = 1/2$, rather than $S = 1$.

Both statements (a) and (b) are analogues of statements which hold for the improper integral $\int_a^{\infty} f(x)\,dx$ when $f(x)$ is continuous and bounded:

(a') Whether $\int_a^{\infty} f(x)\,dx$ *converges* does not depend on the value of a, but

(b') if $\int_a^{\infty} f(x)\,dx$ converges, the *value* of the improper integral *does* depend, in general, on the constant a.

Geometric Series

In high school algebra one encounters the formula for the sum of a **geometric progression:**

$$1 + x + x^2 + x^3 + \cdots + x^{n-1} = \frac{1 - x^n}{1 - x}, \qquad x \neq 1. \tag{7}$$

(Equation (7) may be verified by multiplying both sides by $1 - x$. On the left side all terms cancel except $1 - x^n$.)

In equation (7) the variable x, referred to as the **ratio term,** can represent any number except $x = 1$. This formula leads directly to the definition of the **geometric series** with ratio term x:

$$\sum_{k=0}^{\infty} x^k = x^0 + x^1 + x^2 + x^3 + \cdots + x^k + \cdots$$
$$= 1 + x + x^2 + x^3 + \cdots + x^k + \cdots.$$

This series may converge for some values of x but not for others. We now investigate its behavior.

The formula for the nth partial sum of this series is conveniently given by equation (7):

$$S_n = \frac{1 - x^n}{1 - x}, \qquad x \neq 1.$$

To determine the values of x for which the geometric series converges, we apply the properties of limits (Theorem 1, Section 13.2) to conclude that

$$\lim_{n\to\infty} S_n = \lim_{n\to\infty} \frac{1 - x^n}{1 - x} \tag{8}$$
$$= \frac{1 - \{\lim_{n\to\infty} x^n\}}{1 - x}.$$

Now by equation (1), Section 13.3, $\lim_{n\to\infty} x^n = 0$ if $|x| < 1$ and, by Exercise 21, Section 13.3, $\lim_{n\to\infty} x^n$ does not exist if $|x| > 1$. Thus (8) shows that the geometric series $\Sigma_{k=0}^{\infty} x^k$

(i) converges to $S = \lim_{n\to\infty} S_n = \dfrac{1}{1 - x}$ if $|x| < 1$, and

(ii) diverges if $|x| > 1$.

The two remaining cases correspond to $|x| = 1$, namely, $x = \pm 1$. If $x = 1$, the geometric series becomes the constant series

$$\sum_{k=0}^{\infty} 1^k = 1 + 1 + 1 + \cdots,$$

which diverges. If $x = -1$ the series is

$$\sum_{k=0}^{\infty} (-1)^k = 1 - 1 + 1 - 1 + \cdots,$$

which diverges, as in Example 2.

This completely determines the convergence properties of the geometric series, as summarized in the following theorem.

THEOREM 5
Convergence of Geometric Series

If $|x| < 1$, the geometric series converges to the sum

$$\sum_{k=0}^{\infty} x^k = \frac{1}{1-x}. \tag{9}$$

If $|x| \geq 1$, the geometric series $\Sigma_{k=0}^{\infty} x^k$ diverges.

Example 5 The series

$$1 + \frac{2}{3} + \frac{4}{9} + \frac{8}{27} + \cdots + \frac{2^k}{3^k} + \cdots$$

is a geometric series since the $(k + 1)$st term can be written $x^k = \dfrac{2^k}{3^k} = \left(\dfrac{2}{3}\right)^k$.

The ratio term is therefore $x = \dfrac{2}{3}$. Since $\left|\dfrac{2}{3}\right| < 1$, the series converges and its sum, by Theorem 5, is

$$S = \sum_{k=0}^{\infty} \left(\frac{2}{3}\right)^k = \frac{1}{1-\frac{2}{3}} = 3.$$

■

Example 6 The series

$$\frac{1}{2} + \frac{1}{4} + \frac{1}{8} + \frac{1}{16} + \cdots + \frac{1}{2^k} + \cdots$$

has the form of the geometric series with $x^k = \dfrac{1}{2^k} = \left(\dfrac{1}{2}\right)^k$ except that the term $1 = \left(\dfrac{1}{2}\right)^0$ is missing. In order to use Theorem 5, we write

$$\begin{aligned}
\sum_{k=1}^{\infty} \frac{1}{2^k} &= \frac{1}{2} + \frac{1}{4} + \frac{1}{8} + \frac{1}{16} + \cdots + \frac{1}{2^k} + \cdots \\
&= \left\{1 + \frac{1}{2} + \frac{1}{4} + \frac{1}{8} + \frac{1}{16} + \cdots + \frac{1}{2^k} + \cdots\right\} - 1 \\
&= \sum_{k=0}^{\infty} \frac{1}{2^k} - 1 \\
&= \left(\frac{1}{1-\frac{1}{2}}\right) - 1 \\
&= 1.
\end{aligned}$$

■

The Algebra of Convergent Infinite Series

The following theorem shows that a limited amount of algebra may be performed on convergent infinite series.

THEOREM 6

Suppose that the series Σa_k and Σb_k both converge, with sums

$$S = \sum a_k \quad \text{and} \quad T = \sum b_k.$$

Then

(i) the series $\Sigma(a_k + b_k)$ converges, with sum

$$\sum (a_k + b_k) = S + T, \qquad \text{and}$$

(ii) for any real number c the series Σca_k converges, with sum

$$\sum ca_k = cS.$$

Proof: To prove part (i) we let S_n denote the nth partial sum for Σa_k and we let T_n denote the nth partial sum for Σb_k. Then $(S_n + T_n)$ is a partial sum for the series $\Sigma(a_k + b_k)$, since

$$\begin{aligned}\sum_{k=1}^{n} (a_k + b_k) &= (a_1 + b_1) + (a_2 + b_2) + \cdots + (a_n + b_n)\\ &= (a_1 + a_2 + \cdots + a_n) + (b_1 + b_2 + \cdots + b_n)\\ &= S_n + T_n.\end{aligned}$$

Applying Definition 4 and Theorem 1, part (i), we see that

$$\begin{aligned}\sum_{k=1}^{\infty} (a_k + b_k) &= \lim_{n\to\infty} (S_n + T_n)\\ &= \lim_{n\to\infty} S_n + \lim_{n\to\infty} T_n\\ &= S + T.\end{aligned}$$

The proof of part (ii) is similar and is left as an exercise. ■

REMARK: We may paraphrase Theorem 6 by writing

$$\sum (a_k + b_k) = \sum a_k + \sum b_k \tag{10}$$

and

$$\sum ca_k = c \sum a_k. \tag{11}$$

Equations (10) and (11) must be read from right to left—if the sums on the right exist then so do the sums on the left, and equality holds.

Example 7 To find the sum of the series

$$\sum_{k=0}^{\infty} \frac{3 \cdot 2^k + 3^k}{5^k} = 4 + \frac{9}{5} + \frac{21}{25} + \frac{51}{125} + \cdots$$

we use equations (10) and (11) together with Theorem 5:

$$\sum_{k=0}^{\infty} \frac{3 \cdot 2^k + 3^k}{5^k} = \sum_{k=0}^{\infty} \left(\frac{3 \cdot 2^k}{5^k} + \frac{3^k}{5^k} \right)$$

$$= 3 \sum_{k=0}^{\infty} \frac{2^k}{5^k} + \sum_{k=0}^{\infty} \frac{3^k}{5^k}$$

$$= 3 \sum_{k=0}^{\infty} \left(\frac{2}{5} \right)^k + \sum_{k=0}^{\infty} \left(\frac{3}{5} \right)^k$$

$$= 3 \left(\frac{1}{1 - \frac{2}{5}} \right) + \left(\frac{1}{1 - \frac{3}{5}} \right)$$

$$= 3 \left(\frac{5}{3} \right) + \frac{5}{2}$$

$$= \frac{15}{2}. \qquad \blacksquare$$

Unfortunately, about the only types of infinite series yielding explicit formulas for their sums are the geometric series and those series whose partial sums telescope, as in Example 4. However, in most applications of the ideas developed in this chapter, we shall need only to determine whether a particular infinite series converges or diverges. The remaining sections of this chapter concern how this determination is made for various types of infinite series.

Exercise Set 13.4

In Exercises 1–6, write out the first four terms of the infinite series

1. $\displaystyle\sum_{k=1}^{\infty} \frac{\cos \pi k}{2^k}$

2. $\displaystyle\sum_{k=1}^{\infty} \frac{\sqrt{k} \sin \pi k}{k + 1}$

3. $\displaystyle\sum_{k=1}^{\infty} \frac{2^k + 1}{3^k + 2}$

4. $\displaystyle\sum_{k=0}^{\infty} \tan\left(\frac{\pi k}{3}\right)$

5. $\displaystyle\sum_{k=1}^{\infty} \ln\left(\frac{k}{k + 1}\right)$

6. $\displaystyle\sum_{k=1}^{\infty} k^k$

In Exercises 7–24, determine whether the given series converges or diverges. If it converges find its sum.

7. $\displaystyle\sum_{k=0}^{\infty} \frac{1}{7^k}$

8. $\displaystyle\sum_{k=1}^{\infty} \frac{1}{3^k}$

9. $\displaystyle\sum_{k=0}^{\infty} 4^k$

10. $\displaystyle\sum_{k=1}^{\infty} \frac{7^k + 3^k}{5^k}$

11. $\displaystyle\sum_{k=0}^{\infty} \frac{2^{2k}}{3^{3k}}$

12. $\displaystyle\sum_{k=2}^{\infty} \frac{-1}{3^k}$

13. $\displaystyle\sum_{k=0}^{\infty} \frac{1}{(2 + x)^k}, \quad |x| < 1$

14. $\displaystyle\sum_{k=2}^{\infty} \frac{1}{k(k + 1)}$

15. $\displaystyle\sum_{k=1}^{\infty} \left[\frac{1}{k + 2} - \frac{1}{k + 1}\right]$

16. $\displaystyle\sum_{k=2}^{\infty} \frac{2^{k+1} + 2 \cdot 7^k}{9^k}$

17. $\displaystyle\sum_{k=1}^{\infty} \cos \pi k$

18. $\displaystyle\sum_{k=2}^{\infty} \frac{1}{k^2 - 1}$

19. $\displaystyle\sum_{k=1}^{\infty} \frac{1}{k^2 + 5k + 6}$

20. $\displaystyle\sum_{k=4}^{\infty} \frac{1}{k^2 - 9}$

21. $\displaystyle\sum_{k=1}^{\infty} \ln\left(\frac{k}{k + 1}\right)$

22. $\displaystyle\sum_{k=1}^{\infty} \frac{2^{k-2} + 3^{k+1}}{5^k}$

23. $\displaystyle\sum_{k=0}^{\infty} \frac{2^{k/2}}{3^k}$

24. $\displaystyle\sum_{k=1}^{\infty} \frac{3^k}{3^{k/2}}$

In Exercises 25–30, write the given decimal fraction as **(a)** an infinite series, and **(b)** the quotient of two integers.

25. $0.333\overline{3}\ldots$

26. $0.777\overline{7}\ldots$

27. $0.9292\overline{92}\ldots$

28. $0.321515\overline{15}\ldots$

29. $0.412412\overline{412}\ldots$

30. $0.0213434\overline{34}\ldots$

In Exercises 31–34, use Theorem 5 on the convergence of geometric series to establish the stated fact

31. $\sum_{k=0}^{\infty} (-1)^k x^k = \dfrac{1}{1+x} \quad \text{if} \quad |x| < 1$

32. $\sum_{k=0}^{\infty} x^{2k} = \dfrac{1}{1-x^2} \quad \text{if} \quad |x| < 1$

33. $\sum_{k=0}^{\infty} \dfrac{x^k}{y^k} = \dfrac{y}{y-x} \quad \text{if} \quad |x| < |y|$

34. $\sum_{k=1}^{\infty} x^k = \dfrac{x}{1-x} \quad \text{if} \quad |x| < 1.$

35. When dropped from a height h, a ball rebounds to a height $\frac{2}{3}h$. Write an infinite series expressing the total distance travelled by the ball as it bounces an infinite number of times. What is this distance?

36. Let Σa_k be the series $\sum_{k=0}^{\infty} \left(1 + \dfrac{1}{2^k}\right)$ and let Σb_k be the series $\sum_{k=0}^{\infty} (-1)$. Show that the statement $\Sigma(a_k + b_k) = \Sigma a_k + \Sigma b_k$ is false, but that $\Sigma(a_k + b_k)$ converges.

37. If $\Sigma c a_k$ converges for a particular number c, must the series Σa_k converge? Why or why not?

38. Prove Theorem 6, part (ii).

39. Prove that if Σa_k diverges, so must $\Sigma c a_k$ for any $c \neq 0$.

40. Prove that if $\sum_{k=1}^{\infty} a_k$ converges, then $\sum_{k=p}^{\infty} a_k$ converges for any $p \geq 1$. (*Hint:* If S_n is the nth partial sum for $\sum_{k=1}^{\infty} a_k$, then $S_{p+n-1} - S_{p-1}$ is the nth partial sum for $\sum_{k=p}^{\infty} a_k$.)

41. Find a formula that defines the partial sum S_n for the series Σa_k *recursively*, in terms of the partial sum S_{n-1} and the nth term a_n.

42. *(Computer)* Program 10 in Appendix I is a BASIC program that computes partial sums for the geometric series $\sum_{k=p}^{\infty} ax^k$.

For example, partial sums for the series $\sum_{k=2}^{\infty} 7 \cdot \left(\dfrac{4}{5}\right)^k$ are obtained by specifying $p = 2$, $a = 7$, and $x = \dfrac{4}{5}$. Results appear in Table 4.1.

Table 4.1

n	$S_n = \sum_{k=2}^{n} 7\left(\frac{4}{5}\right)^k$
5	13.224959
10	19.393521
25	22.294217
50	22.399598
100	22.399997
200	22.399998
500	22.399998

Show that $\lim_{n\to\infty} S_n = \sum_{k=2}^{\infty} 7\left(\dfrac{4}{5}\right)^k = 22.4.$

In Exercises 43–46, use Program 10 to find the partial sums S_5, S_{10}, S_{20}, S_{50}, and S_{100}. Then find $\lim_{n\to\infty} S_n$.

43. $\sum_{k=0}^{\infty} \left(\dfrac{2}{3}\right)^k$

44. $\sum_{k=3}^{\infty} 4\left(\dfrac{9}{10}\right)^k$

45. $\sum_{k=5}^{\infty} -2\left(\dfrac{5}{7}\right)^k$

46. $\sum_{k=1}^{\infty} 6\left(\dfrac{3}{2}\right)^k.$

13.5 THE INTEGRAL AND COMPARISON TESTS

In this section we develop four tests for determining whether a given infinite series converges or diverges. The first is actually a test by which only divergent series may be detected.

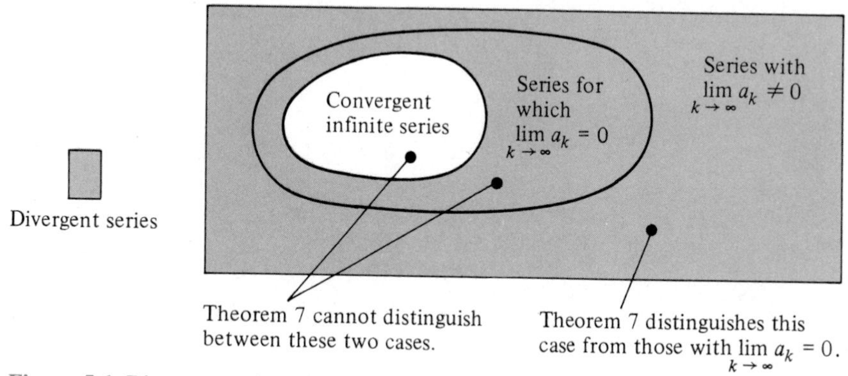

Figure 5.1 Diagram on the applicability of Theorem 7.

THEOREM 7
Necessary Condition for Convergence

If the infinite series Σa_k converges then $\lim_{k\to\infty} a_k = 0$.

Proof: Let S_k denote the kth partial sum for Σa_k. If Σa_k converges there exists a sum S for the series. That is, $\lim_{k\to\infty} S_k = S$. Then $\lim_{k\to\infty} S_{k-1} = S$ also. But $a_k = S_k - S_{k-1}$, so

$$\begin{aligned}\lim_{k\to\infty} a_k &= \lim_{k\to\infty} (S_k - S_{k-1})\\ &= \lim_{k\to\infty} S_k - \lim_{k\to\infty} S_{k-1}\\ &= S - S\\ &= 0.\end{aligned}$$

■

REMARK: It is important to note that Theorem 7 is *not* useful in demonstrating that the series Σa_k converges. Although the condition $\lim_{k\to\infty} a_k = 0$ is **necessary** for convergence (meaning all convergent series have this property), it is **not sufficient** to guarantee convergence (meaning some divergent series also have this property). However, Theorem 7 does establish the divergence of a series Σb_k for which $\lim_{k\to\infty} b_k \neq 0$. The Venn diagram in Figure 5.1 illustrates the type of series to which Theorem 7 applies.

The following equivalent formulation of Theorem 7 summarizes these remarks and is the one you will generally find useful in your work.

COROLLARY 2

If $\lim_{k\to\infty} a_k \neq 0$, the series Σa_k diverges.

Example 1

$$\sum_{k=1}^{\infty} \frac{k}{k+2}$$

diverges, by application of Corollary 2:

$$\lim_{k\to\infty} a_k = \lim_{k\to\infty} \frac{k}{k+2} = 1 \neq 0.$$

■

Example 2

$$\sum_{k=1}^{\infty} \cos(\pi k)$$

diverges, by Corollary 2, since $\lim_{k\to\infty} a_k = \lim_{k\to\infty} \cos(\pi k)$ does not exist. ■

Example 3 Whether the series

$$\sum_{k=1}^{\infty} \frac{1}{k}$$

converges cannot be determined using Theorem 7 or Corollary 2, since $\lim_{k\to\infty} \frac{1}{k} = 0$. However, the test discussed next will show that this series diverges. ■

The Integral Test

This test exploits the relationship between an infinite series of positive terms and the improper integral of a positive continuous function.

THEOREM 8
Integral Test

Let $a_k > 0$ for all $k = 1, 2, \ldots$ and let $f(x)$ be a continuous decreasing function defined on $[1, \infty)$ so that $f(k) = a_k$ for each $k = 1, 2, \ldots$. Then

$$\Sigma a_k \text{ converges} \qquad \text{if and only if} \qquad \int_1^{\infty} f(x)\,dx \text{ converges.}$$

(See Figures 5.2 and 5.3.) That is, the series and the improper integral either both converge or both diverge.

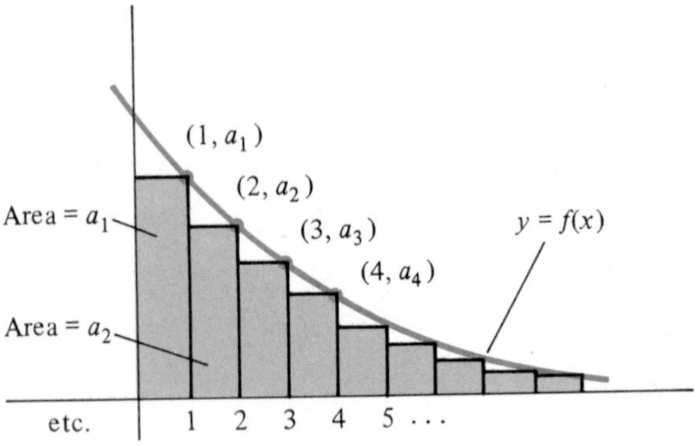

Figure 5.2 $\sum a_k$ converges if $\int_1^{\infty} f(x)\,dx < \infty$.

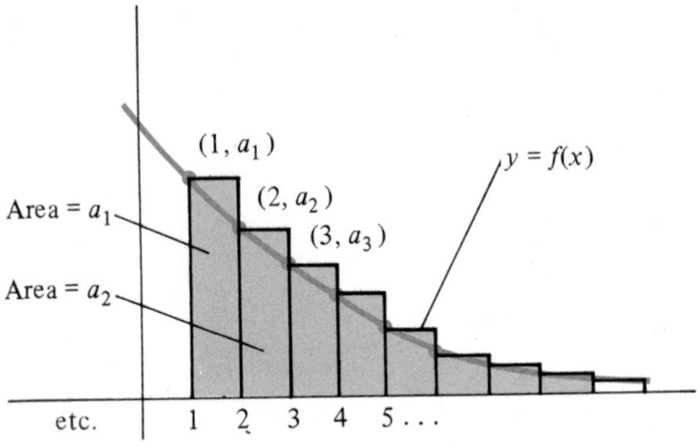

Figure 5.3 $\sum a_k$ diverges if $\int_1^{\infty} f(x)\,dx = \infty$.

Since $f(x) > 0$ for all $x \in [1, \infty)$, we may interpret these two conclusions geometrically by saying that

(i) If the area of the region R_f, bounded by the graph of $y = f(x)$ and the x-axis for $1 \leq x < \infty$, is finite, so is the area, R_s, of the region enclosed by the rectangles of area $a_1, a_2, a_3, \ldots$ (Figure 5.2).
(ii) If the area of the region R_f is infinite, so is the area of the region R_s (Figure 5.3).

Proof of Theorem 8: We focus on the relationship between the partial sums for Σa_k and the partial integrals for $f(x)$. Since $f(x)$ is decreasing on $[1, \infty)$, for any integer $k \geq 2$,

$$f(x) \geq f(k) = a_k \qquad \text{for all } x \in [k-1, k], \qquad \text{and} \tag{1}$$

$$f(x) \leq f(k) = a_k \qquad \text{for all } x \in [k, k+1]. \tag{2}$$

(See Figures 5.4 and 5.5.)

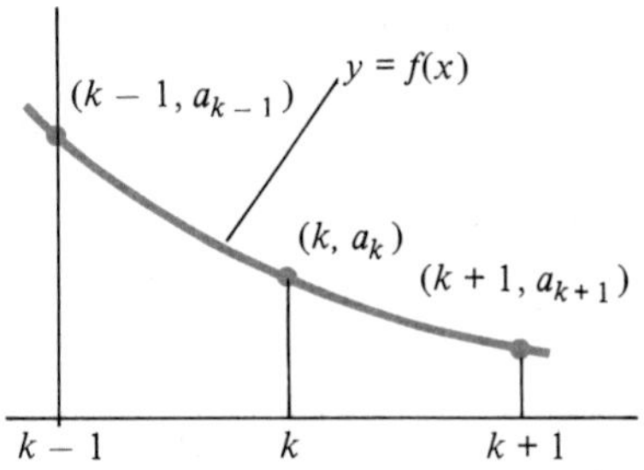

Figure 5.4 $f(k) = a_k$.

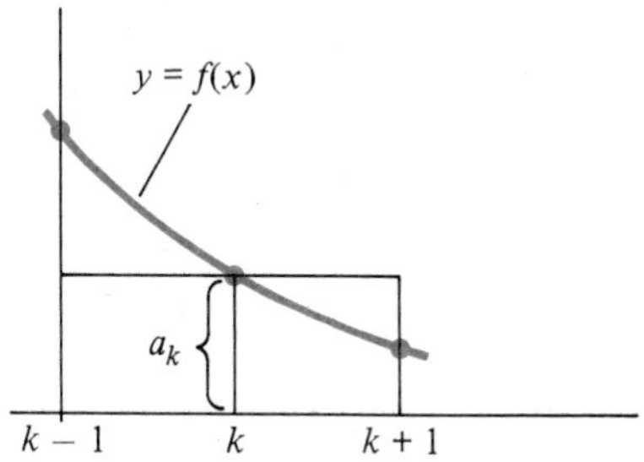

Figure 5.5 $f(x) \geq a_k,\ x \leq k;$ $f(x) \leq a_k,\ x \geq k.$

Inequalities (1) and (2), together with the Mean Value Theorem for Integrals imply that

$$\int_{k-1}^{k} f(x)\,dx \geq a_k, \qquad \text{and} \tag{3}$$

$$\int_{k}^{k+1} f(x)\,dx \leq a_k, \qquad k = 2, 3, 4, \ldots \qquad \text{(Figure 5.6)}. \tag{4}$$

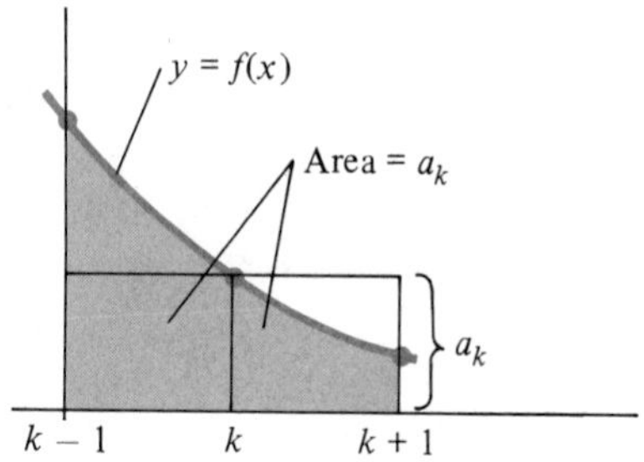

Figure 5.6
$\int_{k-1}^{k} f(x)\,dx \geq a_k \geq \int_{k}^{k+1} f(x)\,dx.$
The block of height a_k extending from $k - 1$ to k has area a_k; so does the block from k to $k + 1$.

For any integer n we may sum inequality (3) from $k = 2$ to n to conclude that

$$\sum_{k=2}^{n} \int_{k-1}^{k} f(x)\,dx = \int_{1}^{n} f(x)\,dx \geq \sum_{k=2}^{n} a_k = S_n - a_1 \tag{5}$$

where S_n denotes the nth partial sum of the series $\sum_{k=1}^{n} a_k$. Thus,

$$S_n \leq a_1 + \int_{1}^{n} f(x)\,dx < a_1 + \int_{1}^{\infty} f(x)\,dx. \tag{6}$$

(The last inequality holds since $f(x) > 0$ on $[1, \infty)$.) Inequality (6) shows that if the improper integral $\int_1^\infty f(x)\,dx$ converges, the sequence $\{S_n\}$ of partial sums for the series Σa_k is a *bounded* sequence. Since $a_k > 0$ for each k, $\{S_k\}$ is also an *increasing* sequence. Since every bounded increasing sequence converges, the sequence $\{S_n\}$ converges. That is, if the improper integral $\int_1^\infty f(x)\,dx$ converges, so does the infinite series Σa_k.

The converse conclusion is obtained from inequality (4). Summing both sides from $k = 1$ to $k = n$ shows that

$$\int_{1}^{n+1} f(x)\,dx \leq \sum_{k=1}^{n} a_k = S_n. \tag{7}$$

Now $\int_1^{n+1} f(x)\,dx$ is an increasing function of n since $f(x) > 0$ for $x \in [1, \infty)$.

Thus, if the improper integral $\int_1^\infty f(x)\,dx$ diverges, we may conclude from inequality (7) that

$$\lim_{n\to\infty} S_n \geq \lim_{n\to\infty} \int_1^{n+1} f(x)\,dx = +\infty,$$

so the sequence $\{S_n\}$ diverges. This shows that if the improper integral diverges, so does the series Σa_k. This completes the proof. ■

Example 4 The **harmonic series** is the series

$$\sum_{k=1}^{\infty} \frac{1}{k} = 1 + \frac{1}{2} + \frac{1}{3} + \frac{1}{4} + \cdots + \frac{1}{k} + \cdots.$$

To test this series for convergence, we use the function $f(x) = \dfrac{1}{x}$, which satisfies all hypotheses of the integral test.

Since

$$\int_1^\infty \frac{1}{x}\,dx = \lim_{t\to\infty} \int_1^t \frac{1}{x}\,dx = \lim_{t\to\infty} \ln t = +\infty$$

the harmonic series diverges. (Note that $\lim_{k\to\infty} a_k = \lim_{k\to\infty} \dfrac{1}{k} = 0$, although this series does not converge.) ■

Example 5 To test the series

$$\sum_{k=1}^{\infty} ke^{-k^2}$$

for convergence, we use the function $f(x) = xe^{-x^2}$. Since

$$f'(x) = e^{-x^2} - 2x^2e^{-x^2} = (1 - 2x^2)e^{-x^2} < 0 \qquad \text{if} \qquad x > \frac{\sqrt{2}}{2},$$

we are assured that $f(x)$ is decreasing on $[1, \infty)$. Since

$$\begin{aligned}\int_1^\infty xe^{-x^2}\,dx &= \lim_{t\to\infty} \int_1^t xe^{-x^2}\,dx \\ &= \lim_{t\to\infty} -\frac{1}{2}e^{-x^2}\Big]_1^t \\ &= \lim_{t\to\infty} \left(\frac{1}{2e} - \frac{1}{2e^{t^2}}\right) = \frac{1}{2e}\end{aligned}$$

the infinite series $\sum_{k=1}^{\infty} ke^{-k^2}$ converges, by the Integral Test. ■

Example 6 A ***p*-series** is a series of the form

$$\sum_{k=1}^{\infty} \frac{1}{k^p} = 1 + \frac{1}{2^p} + \frac{1}{3^p} + \frac{1}{4^p} + \cdots, \qquad p > 0.$$

Determine for which values of p a p-series converges.

Strategy

Solution

Identify a function $f(x)$ with which to apply the Integral Test.

We use the function $f(x) = \frac{1}{x^p} = x^{-p}$.

Since

Verify that $f(x)$ is decreasing on $[1, \infty)$.

$$f'(x) = -px^{-p-1} = \frac{-p}{x^{p+1}}$$

is negative for $x > 0$, the function $f(x)$ is decreasing on $[1, \infty)$.

Set up the improper integral and evaluate, carrying along the unknown constant p.

The improper integral for $f(x)$ is

$$\begin{aligned} \int_1^{\infty} x^{-p}\,dx &= \lim_{t\to\infty} \int_1^t x^{-p}\,dx \\ &= \lim_{t\to\infty} \frac{x^{-p+1}}{1-p}\bigg]_1^t \qquad (p \neq 1) \\ &= \left\{\lim_{t\to\infty} \frac{t^{-p+1}}{1-p}\right\} - \frac{1}{1-p}. \end{aligned}$$

Determine the values of p for which the improper integral converges.

This limit will exist only if the exponent of t is negative; that is,

$$-p + 1 < 0, \qquad \text{or} \qquad p > 1.$$

Handle the remaining case ($p = 1$) directly (see Example 4).

Thus, the improper integral converges if $p > 1$, and diverges if $0 < p < 1$. In the case $p = 1$, (excluded above), the p-series becomes simply the harmonic

Apply Theorem 8.

series, which diverges. We therefore conclude that *the p-series* $\sum_{k=1}^{\infty} \frac{1}{k^p}$ *converges if* $p > 1$ *and diverges if* $0 < p \leq 1$. ■

REMARK: We can paraphrase the statement of the Integral Test as follows: After carefully checking that all hypotheses are satisfied, integrate the general term for Σa_k from one to infinity. If this integral converges, so does the series. If the integral diverges, so does the series.

As with techniques of integration, you should develop a strategy for determining which of the tests for convergence apply to a given infinite series. The key to determining whether the Integral Test applies is whether you can determine if the improper integral of the general term converges. Of course, you must also verify that the terms of the series are positive and decreasing.

The Basic Comparison Test

Like the Integral Test, this test is motivated by area considerations. However, instead of comparing terms of a given series with certain integrals, we compare them with terms of another series.

THEOREM 9
Basic Comparison Test

Let Σa_k and Σb_k be infinite series with $0 < a_k \leq b_k$ for each $k = 1, 2, 3, \ldots$. Then

(i) If Σb_k converges, so does Σa_k.
(ii) If Σa_k diverges, so does Σb_k.

Proof: We let

$$S_n = \sum_{k=1}^{n} a_k \qquad \text{and} \qquad T_n = \sum_{k=1}^{n} b_k$$

be the corresponding partial sums. Since $a_k \leq b_k$ for all k, we have

$$S_n \leq T_n \qquad \text{for all} \qquad n = 1, 2, \ldots. \tag{8}$$

To prove (i), we note that if Σb_k converges, $\lim_{n\to\infty} T_n$ exists. Thus the sequence $\{T_n\}$ is bounded and, by inequality (8), so is the sequence $\{S_n\}$. Since $a_k > 0$ for all k, this shows that $\{S_n\}$ is a bounded increasing sequence, so by Theorem 4, $\lim_{n\to\infty} S_n$ exists. That is, Σa_k converges.

To prove (ii), we note that since $\{S_n\}$ is an increasing sequence, if Σa_k diverges, we must have $\lim_{n\to\infty} S_n = +\infty$. Thus, by inequality (8)

$$\lim_{n\to\infty} T_n \geq \lim_{n\to\infty} S_n = +\infty,$$

so Σb_k diverges. ■

Example 7 The series $\sum \frac{1}{1 + k^3}$ converges, by comparison with the convergent p-series $\sum \frac{1}{k^3}$, since $\frac{1}{1+k^3} < \frac{1}{k^3}$. ■

Example 8 To determine the convergence of the series $\sum \frac{\sqrt{k} - 1}{k^2 + 2}$, we make the comparison

$$\frac{\sqrt{k} - 1}{k^2 + 2} < \frac{\sqrt{k}}{k^2 + 2} < \frac{\sqrt{k}}{k^2} = \frac{1}{k^{3/2}}.$$

Since the series $\sum \frac{1}{k^{3/2}}$ is a p-series with $p = 3/2 > 1$, it converges. So, therefore, does the series $\sum \frac{\sqrt{k} - 1}{k^2 + 2}$. ■

Example 9 The series

$$\sum_{k=1}^{\infty} \frac{1 + \ln k}{k}$$

diverges by comparison with the harmonic series $\sum_{k=1}^{\infty} \frac{1}{k}$, since

$$\frac{1 + \ln k}{k} > \frac{1}{k} \quad \text{if} \quad k > 1. \qquad \blacksquare$$

REMARK 1: Successful application of the Comparison Test to the series Σa_k involves finding another series Σb_k with which you can make a useful comparison. If you suspect that Σa_k converges, you should look for a *convergent* series Σb_k with $a_k \le b_k$. But if you suspect that Σa_k diverges, look for a divergent series Σb_k with $a_k \ge b_k$.

REMARK 2: We say that the series Σa_k *is dominated by* the series Σb_k (or, that Σb_k dominates Σa_k) if $a_k \leqq b_k$ for all k. Using this terminology we may paraphrase Theorem 9 by saying that a positive series that is dominated by a convergent series must converge, while a series that dominates a positive divergent series must diverge.

It is important to note that two cases remain (dominated by a divergent series and dominating a convergent series) in which no valid conclusions can be drawn. The following table may help you remember which comparisons yield valid conclusions.

Table 5.1 Conclusions concerning Σa_k when compared with Σb_k.

	Σb_k converges	Σb_k diverges
$0 < a_k < b_k$	Σa_k converges	no valid conclusion
$a_k > b_k > 0$	no valid conclusion	Σa_k diverges

The Basic Comparison Test is used whenever an appropriate comparison series can be found. As the list of series with which you are familiar expands, your facility with the Basic Comparison Test will grow.

The Limit Comparison Test

The following variation on the Basic Comparison Test is often easier to apply.

THEOREM 10
Limit Comparison Test

Let Σa_k and Σb_k be series with $a_k > 0$, $b_k > 0$ for all $k = 1, 2, \ldots$.
If the limit

$$\rho = \lim_{k \to \infty} \frac{a_k}{b_k}$$

exists, and $\rho \neq 0$, then either both series converge or both series diverge.

The Limit Comparison Test states that if the ratio of the general terms of two series tends to a positive limit, then the two series have the same convergence property—either both converge or both diverge. This is most useful in determining the convergence of series whose general terms are rational functions or quotients of functions involving rational exponents.

Before proving this theorem, we consider several examples of its use.

Example 10 The series

$$\sum_{k=1}^{\infty} \frac{1}{2k + 1}$$

diverges, as can be shown by Theorem 10. Comparing this series with the harmonic series

$$\sum b_k = \sum_{k=1}^{\infty} \frac{1}{k},$$

we find that

$$\rho = \lim_{k\to\infty} \left(\frac{\dfrac{1}{2k+1}}{\dfrac{1}{k}} \right) = \lim_{k\to\infty} \frac{k}{2k + 1} = \frac{1}{2} > 0.$$

Since the limit $\rho > 0$ exists and since the harmonic series diverges, so must the given series. ■

Example 11 To apply the Limit Comparison Test to the series

$$\sum_{k=1}^{\infty} \frac{\sqrt{k} + 2}{k^2 + k + 1},$$

we observe that the general term is a quotient containing a highest exponent of 1/2 in its numerator and a highest exponent of 2 in its denominator. This suggests a comparison with the series whose general term is $b_k = \dfrac{k^{1/2}}{k^2} = \dfrac{1}{k^{3/2}}$. We therefore take Σb_k to be the p-series

$$\sum_{k=1}^{\infty} \frac{1}{k^{3/2}}.$$

We obtain

$$\rho = \lim_{k\to\infty} \left(\frac{\dfrac{\sqrt{k} + 2}{k^2 + k + 1}}{\dfrac{1}{k^{3/2}}} \right) = \lim_{k\to\infty} \left(\frac{k^2 + 2k^{3/2}}{k^2 + k + 1} \right) = 1 > 0.$$

We may now conclude that, since the p-series $\displaystyle\sum_{k=1}^{\infty} \frac{1}{k^{3/2}}$ converges, so does the series $\displaystyle\sum_{k=1}^{\infty} \frac{\sqrt{k} + 2}{k^2 + k + 1}$. ■

Example 12 Determine whether the series

$$\sum_{k=1}^{\infty} \sin\left(\frac{\pi}{k}\right)$$

converges.

Solution: Let's systematically try each of the techniques available thus far:

(a) Theorem 7 provides no conclusion, since

$$\lim_{k\to\infty} a_k = \lim_{k\to\infty} \sin\left(\frac{\pi}{k}\right) = \sin(0) = 0.$$

(b) Theorem 8 (Integral Test) is not appropriate, since the improper integral

$$\int_1^\infty \sin\left(\frac{\pi}{x}\right)\, dx$$

cannot easily be evaluated.

(c) Theorem 9 (Basic Comparison Test) depends on our ability to choose a valid comparison series. We do have the comparison

$$0 < \sin\left(\frac{\pi}{k}\right) < \frac{\pi}{k}, \qquad k = 1, 2, 3, \ldots$$

but the Comparison Test gives no information since the series $\Sigma\pi/k = \pi\Sigma 1/k$ diverges.

However, the Limit Comparison Test may be applied by recalling that we have previously established the limit

$$\rho = \lim_{k\to\infty} \frac{\sin\left(\frac{\pi}{k}\right)}{\left(\frac{\pi}{k}\right)} = \lim_{x\to 0^+} \frac{\sin x}{x} = 1.$$

Thus, by Theorem 10, the series $\sum_{k=1}^{\infty} \sin\left(\frac{\pi}{k}\right)$ has the same convergence property as does the series $\sum_{k=1}^{\infty} \frac{\pi}{k}$. That is, $\sum_{k=1}^{\infty} \sin\left(\frac{\pi}{k}\right)$ diverges. ■

Proof of Theorem 10: Since all terms of Σa_k and Σb_k are positive, and $\rho \neq 0$, we must have

$$\rho = \lim_{k\to\infty} \frac{a_k}{b_k} > 0. \tag{9}$$

Now let $\epsilon = \frac{\rho}{2} > 0$. Then by (9) and the definition of the limit of a sequence there exists an integer N such that

$$\left|\frac{a_k}{b_k} - \rho\right| < \epsilon \qquad \text{whenever} \qquad k \geq N. \tag{10}$$

Statement (10) is equivalent to the statement

$$\rho - \epsilon < \frac{a_k}{b_k} < \rho + \epsilon, \qquad k \geq N. \tag{11}$$

Substituting for $\epsilon = \frac{1}{2}\rho$ and multiplying through by $b_k > 0$ in (11) gives

$$\frac{1}{2}\rho \cdot b_k < a_k < \frac{3}{2}\rho \cdot b_k, \qquad k \geq N. \tag{12}$$

Using the left half of inequality (12), we conclude that Σa_k dominates a multiple of Σb_k, since $\Sigma \frac{1}{2}\rho b_k = \left(\frac{1}{2}\rho\right) \Sigma b_k$. Thus Σb_k converges if Σa_k converges (Theorem 9). The right half of inequality (12) shows that Σa_k is dominated by $\Sigma \frac{3}{2}\rho b_k = \left(\frac{3}{2}\rho\right) \Sigma b_k$, so Σb_k diverges if Σa_k diverges. These two conclusions taken together establish Theorem 10. ■

Before concluding this section, we need to make one further observation which applies to each of the tests embodied by Theorems 8, 9, and 10. We have previously noted that if the series $\Sigma_{k=N}^{\infty} a_k$ converges, then so does the series $\Sigma_{k=1}^{\infty} a_k$, since

$$\sum_{k=1}^{\infty} a_k = \{a_1 + a_2 + \cdots + a_{N-1}\} + \sum_{k=N}^{\infty} a_k. \tag{13}$$

Since the sum inside braces in (13) is finite, the two infinite series either both converge or both diverge.

This observation means that we need not require that the conditions of Theorems 8 through 10 hold on the full series $\Sigma_{k=1}^{\infty} a_k$ but rather only on any "tail" of the series of the form $\Sigma_{k=N}^{\infty} a_k$. In such cases the conclusions of Theorems 8 through 10 remain valid. Note that we used this idea in the proof of Theorem 10: the comparisons obtained from inequality (11) held only for $k \geq N$.

For example, the series $\displaystyle\sum_{k=1}^{\infty} \frac{7+k}{1+k^3}$ may be successfully compared to the series $\displaystyle\sum_{k=1}^{\infty} \frac{2}{k^2} = 2\sum_{k=1}^{\infty} \frac{1}{k^2}$ for $k > 7$ since

$$\frac{7+k}{1+k^3} < \frac{k+k}{1+k^3} = \frac{2k}{1+k^3} < \frac{2k}{k^3} = \frac{2}{k^2}$$

whenever $k > 7$. The Comparison Test then shows that the "tail" $\displaystyle\sum_{k=8}^{\infty} \frac{7+k}{1+k^3}$ converges, since the p-series $\displaystyle\sum_{k=1}^{\infty} \frac{1}{k^2}$ converges. Thus the original series $\displaystyle\sum_{k=1}^{\infty} \frac{7+k}{1+k^2}$ also converges.

THEOREM 11

Let N be a positive integer. The series $\Sigma_{k=1}^{\infty} a_k$ converges if and only if the series $\Sigma_{k=N}^{\infty} a_k$ converges for each integer $N > 1$.

Exercise Set 13.5

In Exercises 1–6, use the Integral Test to determine whether the given series converges or diverges.

1. $\sum \frac{1}{2k+1}$
2. $\sum \frac{1}{(3k+1)^2}$
3. $\sum \frac{1}{k \cdot \ln k}$
4. $\sum \frac{1}{k(\ln k)^2}$
5. $\sum \frac{1}{1+k^2}$
6. $\sum k^2 e^{-k^3}$

In Exercises 7–12, use the Basic Comparison Test to determine whether the given series converges or diverges.

7. $\sum \frac{1}{1+k^2}$
8. $\sum \frac{1}{k^{1/2}+k^{3/2}}$
9. $\sum \frac{\sqrt{k}}{1+k^3}$
10. $\sum \frac{\sqrt{k}}{1+k}$
11. $\sum \frac{3}{\sqrt{k}+2}$
12. $\sum (k-1)e^{-k}$

In Exercises 13–16, use the Limit Comparison Test to determine whether the given series converge or diverge.

13. $\sum \frac{k+3}{2k^2+1}$
14. $\sum \frac{k^2-4}{k^3+k+5}$
15. $\sum \frac{k+\sqrt{k}}{k+k^3}$
16. $\sum \frac{2k+2}{\sqrt{k^3+2}}$

In Exercises 17–46, determine whether the series converges or diverges and state which test you used.

17. $\sum \frac{1}{\sqrt{k+1}}$
18. $\sum \frac{\cos^2 \pi k}{k^2}$
19. $\sum \frac{k+1}{k}$
20. $\sum \frac{k(k+1)}{(k+2)(k^2+1)}$
21. $\sum k^2 e^{-k^3}$
22. $\sum \cos\left(\frac{\pi k}{4}\right)$
23. $\sum \frac{2k^2+3k-1}{k^4-6k+10}$
24. $\sum \frac{\sin(\pi k)}{k}$
25. $\sum \frac{2^k}{3^k+1}$
26. $\sum \frac{\tan^{-1} k}{1+k^2}$
27. $\sum \frac{k}{\sqrt{1+k^2}}$
28. $\sum \frac{k^2-2}{k^2+2}$
29. $\sum \frac{1}{\sqrt{4k(k+1)}}$
30. $\sum \frac{2k+2}{\sqrt{k^3+2}}$
31. $\sum \frac{1}{\sqrt[3]{k^2+2k}}$
32. $\sum \frac{k+1}{2 \ln k}$
33. $\sum \frac{\ln k}{k}$
34. $\sum \frac{\tan^{-1} k}{k^2}$
35. $\sum \frac{k+1}{\sqrt{k^{3/2}+1}}$
36. $\sum \frac{\ln k}{1+\ln k}$
37. $\sum \frac{\ln(k+1)}{k+2}$
38. $\sum \frac{k}{1+k^2}$
39. $\sum \frac{n^2}{n^3+1}$
40. $\sum \frac{n^2+3}{2n^4+n-6}$
41. $\sum \frac{\ln n}{n^3}$
42. $\sum \frac{3^k}{k+7}$
43. $\sum \frac{2k^2}{\sqrt{k^3+5}}$
44. $\sum \frac{k+3}{(k+2)2^k}$
45. $\sum \frac{\sqrt{k}}{\cos(2k-6)+k^2}$
46. $\sum \frac{\tan^{-1}\sqrt{k}}{\pi+6k^2}$

47. Prove that if Σa_k and Σb_k are positive term series and $\lim_{k\to\infty} \frac{a_k}{b_k} = 0$, then Σa_k converges if Σb_k converges.

48. Prove that if Σa_k and Σb_k are positive term series and $\lim_{k\to\infty} \frac{a_k}{b_k} = +\infty$ then Σa_k diverges if Σb_k diverges.

49. Prove that if Σa_k converges, then $\lim_{N\to\infty} \sum_{k=N}^{\infty} a_k = 0$. That is, if Σa_k converges its "tails" tend toward zero.

50. Let Σa_k be a convergent infinite series that satisfies the hypotheses of the integral test. Show, using inequalities (5) and (7), that

$$\int_{n+1}^{\infty} f(x)\,dx \le \sum_{k=n+1}^{\infty} a_k \le \int_{n}^{\infty} f(x)\,dx \tag{14}$$

where $f(k) = a_k$ and $f(x)$ is continuous and decreasing on $[1, \infty)$. Inequality (14) gives an estimate for the **truncation error** $R_n = \sum_{n+1}^{\infty} a_k$ which occurs in using the partial sum $S_n = \sum_{k=1}^{n} a_k$ to approximate the sum of a convergent series $S = \sum_{k=1}^{\infty} a_k$ when $\{a_k\}$ is a decreasing sequence with $\lim_{k\to\infty} a_k = 0$.

51. Use the result of Exercise 50 to estimate the magnitude of the error that occurs in approximating the sum of the p-series $\sum_{k=1}^{\infty} \frac{1}{k^2}$ by the partial sum $\sum_{k=1}^{10} \frac{1}{k^2}$.

52. Use the result of Exercise 50 to determine how large n must be taken so that the partial sum S_n for the series $\sum_{k=2}^{\infty} \frac{1}{k \ln^2 k}$ differs from its sum S by less than .01.

53. Use the result of Exercise 50 to determine how large n must be taken so that the partial sum S_n for the series $\sum_{k=1}^{\infty} ke^{-k^2}$ differs from its sum S by less than .02.

54. Prove that the harmonic series $\sum_{k=1}^{\infty} \frac{1}{k}$ diverges without using the Integral Test, as follows

a. Write the harmonic series as

$$\sum_{k=1}^{\infty} \frac{1}{k} = \{1\} + \left\{\frac{1}{2}\right\} + \underbrace{\left\{\frac{1}{3} + \frac{1}{4}\right\}}_{2^1 \text{ terms}} + \underbrace{\left\{\frac{1}{5} + \frac{1}{6} + \frac{1}{7} + \frac{1}{8}\right\}}_{2^2 \text{ terms}}$$

$$+ \left\{\frac{1}{9} + \cdots + \frac{1}{16}\right\} \quad (2^3 \text{ terms})$$

$$+ \left\{\frac{1}{17} + \cdots + \frac{1}{32}\right\} \quad (2^4 \text{ terms})$$

$$\cdots + \left\{\frac{1}{2^n + 1} + \cdots + \frac{1}{2^{n+1}}\right\} \quad (2^n \text{ terms}).$$

b. Show that the sums of the terms in each of the blocks above are greater than $\frac{1}{2}$.

c. Conclude that $S_{2^{n+1}} \geq 1 + \frac{(n+1)}{2}$.

d. From (c), conclude that $\lim_{n \to \infty} S_n = +\infty$.

55. Use the technique of Exercise 54 to prove the "rule of thumb" for the partial sum $S_{2^n} = \sum_{k=1}^{2^n} \frac{1}{k}$ of the harmonic series: $S_{2^n} \geq 1 + \frac{n}{2}$.

13.6 THE RATIO AND ROOT TESTS

In this section we discuss two additional tests for series with positive terms. The first involves the limit of the ratio of successive terms of the series.

THEOREM 12
Ratio Test

Let $a_k > 0$ for all $k = 1, 2, 3, \ldots$ and let

$$\rho = \lim_{k \to \infty} \frac{a_{k+1}}{a_k}. \quad (1)$$

Then

(i) If $\rho < 1$, the series $\sum_{k=1}^{\infty} a_k$ converges.

(ii) If $\rho > 1$, the series $\sum_{k=1}^{\infty} a_k$ diverges.

(iii) If $\rho = 1$, no conclusion may be drawn.

Before proving Theorem 12, we discuss several examples in which the Ratio Test applies.

Example 1 For the series $\sum \frac{k^2}{3^k}$, the kth term is $a_k = \frac{k^2}{3^k}$, so $a_{k+1} =$

$\dfrac{(k+1)^2}{3^{k+1}}$. The limit in (1) is therefore

$$\rho = \lim_{k\to\infty} \frac{\dfrac{(k+1)^2}{3^{k+1}}}{\dfrac{k^2}{3^k}} = \lim_{k\to\infty} \frac{3^k(k+1)^2}{3^{k+1}k^2} = \lim_{k\to\infty} \frac{1}{3}\left(\frac{k+1}{k}\right)^2 = \frac{1}{3}.$$

Since $\rho = \dfrac{1}{3} < 1$, the series $\sum \dfrac{k^2}{3^k}$ converges, by the Ratio Test. ■

REMARK: The Ratio Test is often useful in dealing with series involving factorial expressions. Recall that

$$k! = k(k-1)(k-2)\cdot\cdots\cdot 3\cdot 2\cdot 1.$$

Thus, ratios of factorials can be simplified as follows:

$$\frac{(k+1)!}{k!} = \frac{(k+1)k!}{k!} = k+1$$

$$\frac{(k+2)!}{k!} = \frac{(k+2)(k+1)k!}{k!} = (k+2)(k+1)$$

$$\frac{k!}{(k+p)!} = \frac{k!}{(k+p)(k+p-1)\cdot\cdots\cdot(k+1)\cdot k!}$$

$$= \frac{1}{(k+p)(k+p-1)\cdot\cdots\cdot(k+1)}.$$

Example 2 For the series $\sum \dfrac{k^2}{(k+1)!}$ the limit (1) is

$$\rho = \lim_{k\to\infty} \frac{\dfrac{(k+1)^2}{((k+1)+1)!}}{\dfrac{k^2}{(k+1)!}} = \lim_{k\to\infty} \frac{(k+1)!(k+1)^2}{(k+2)!(k^2)}$$

$$= \lim_{k\to\infty} \left(\frac{1}{k+2}\right)\left(\frac{k+1}{k}\right)^2$$

$$= 0\cdot 1^2 = 0.$$

Thus, the series $\sum \dfrac{k^2}{(k+1)!}$ converges, by the Ratio Test. ■

Example 3 For the series $\sum \dfrac{k!}{k^k}$ the ratio limit is

$$\rho = \lim_{k\to\infty} \frac{\dfrac{(k+1)!}{(k+1)^{k+1}}}{\dfrac{k!}{k^k}} = \lim_{k\to\infty} \frac{(k+1)!k^k}{k!(k+1)^{k+1}}$$

$$= \lim_{k\to\infty} (k+1)\left[\frac{k^k}{(k+1)^{k+1}}\right]$$

$$= \lim_{k\to\infty} \frac{k+1}{k+1}\left[\frac{k^k}{(k+1)^k}\right]$$

$$= \lim_{k\to\infty} \left(\frac{k}{k+1}\right)^k$$

$$= \lim_{k\to\infty} \frac{1}{\left(1+\frac{1}{k}\right)^k} = \frac{1}{e}.$$

The series $\sum \frac{k!}{k^k}$ therefore converges, by the Ratio Test. ■

Example 4 If we attempt to apply the Ratio Test to the series $\sum \frac{k}{k^2+1}$ we find that

$$\rho = \lim_{k\to\infty} \frac{\left(\frac{k+1}{(k+1)^2+1}\right)}{\left(\frac{k}{k^2+1}\right)}$$

$$= \lim_{k\to\infty} \frac{(k+1)(k^2+1)}{k[(k+1)^2+1]} = \lim_{k\to\infty} \frac{k^3+k^2+k+1}{k^3+2k^2+2k} = 1,$$

so the Ratio Test is inconclusive. However, we can handle this series by a limit comparison with the harmonic series $\Sigma b_k = \sum \left(\frac{1}{k}\right)$ (Theorem 10). We find

$$\lim_{k\to\infty} \left(\frac{a_k}{b_k}\right) = \lim_{k\to\infty} \frac{\left(\frac{k}{k^2+1}\right)}{\left(\frac{1}{k}\right)} = \lim_{k\to\infty} \frac{k^2}{k^2+1} = 1.$$

Thus, since the harmonic series diverges, so does the series $\sum \frac{k}{k^2+1}$. ■

Proof of Theorem 12: To prove statement (i), ($\rho < 1$ implies convergence), we will compare the series $\Sigma_{k=1}^{\infty} a_k$ with a geometric series. Since $\rho < 1$, we can find a number γ so that $\rho < \gamma < 1$.

Since

$$\lim_{k\to\infty} \frac{a_{k+1}}{a_k} = \rho,$$

there exists an integer N so that

$$\frac{a_{k+1}}{a_k} \le \gamma \qquad \text{whenever} \qquad k \ge N. \tag{2}$$

(To see this take $\epsilon = \gamma - \rho > 0$ in Definition 3.)

Inequality (2) shows that

$$a_{N+1} \le \gamma a_N$$

$$a_{N+2} \le \gamma a_{N+1} \le \gamma^2 a_N$$

and in general, that

$$a_{N+k} \le \gamma^k a_N, \qquad k = 1, 2, 3, \ldots . \tag{3}$$

Now let $b_k = \gamma^k a_N$ for $k = 1, 2, \ldots$. Then $\Sigma b_k = \Sigma \gamma^k a_N = a_N \Sigma \gamma^k$ is a convergent series since $\gamma < 1$ and the series $\Sigma \gamma^k$ is geometric. Thus, inequality (3) shows that the series $\Sigma_{k=1}^{\infty} a_{N+k}$ is dominated by the convergent series $\Sigma_{k=1}^{\infty} b_k$. The series $\Sigma_{k=1}^{\infty} a_{N+k} = \Sigma_{k=N+1}^{\infty} a_k$ therefore converges, by the Comparison Test. This shows, by Theorem 10, that the series Σa_k converges.

The proof of statement (ii) is similar (except that $\rho > 1$, so the comparison is with a divergent geometric series) and is left as an exercise.

To prove statement (iii) note that

(a) the harmonic series $\sum \frac{1}{k}$ diverges, although

$$\rho = \lim_{k\to\infty} \frac{\frac{1}{k+1}}{\frac{1}{k}} = \lim_{k\to\infty} \frac{k}{k+1} = 1, \qquad \text{and}$$

(b) the p-series $\sum \frac{1}{k^2}$ converges, although

$$\rho = \lim_{k\to\infty} \frac{\frac{1}{(k+1)^2}}{\frac{1}{k^2}} = \lim_{k\to\infty} \frac{k^2}{(k+1)^2} = 1.$$

■

In Chapter 14 we shall concern ourselves with series whose terms contain variables. For such series the Ratio Test is often helpful in determining those values of the variable for which the series converges. The following is one such example.

Example 5 For which values of $x > 0$ does the infinite series $\Sigma x^k/k$ converge?

Solution: Applying the Ratio Test we find that

$$\rho = \lim_{k\to\infty} \left(\frac{\frac{x^{k+1}}{k+1}}{\frac{x^k}{k}} \right) = \lim_{k\to\infty} \frac{x^{k+1} \cdot k}{x^k(k+1)}$$

$$= \lim_{k\to\infty} x\left(\frac{k}{k+1}\right)$$

$$= x.$$

Since $\rho = x$, the series will converge for $\rho = x < 1$, by the Ratio Test, and diverge for $x > 1$. The case $\rho = x = 1$ is inconclusive by the Ratio Test, but if $x = 1$ the series becomes simply the harmonic series, which diverges. Thus, the series $\Sigma x^k/k$, $x > 0$, converges only if $0 < x < 1$. ■

The Root Test

The following test for convergence uses information about the kth root of the general term for Σa_k.

THEOREM 13
Root Test

Let $a_k \geq 0$ for all $k = 1, 2, 3, \ldots$ and let $\rho = \lim_{k\to\infty} \sqrt[k]{a_k}$. Then

(i) If $\rho < 1$, the series $\sum_{k=1}^{\infty} a_k$ converges.

(ii) If $\rho > 1$, the series $\sum_{k=1}^{\infty} a_k$ diverges.

(iii) If $\rho = 1$, no conclusions may be drawn.

Proof: We proceed as in the proof of the Ratio Test. If $\rho < 1$, there exists a constant γ so that $\rho < \gamma < 1$. Then, since $\lim_{k\to\infty} \sqrt[k]{a_k} = \lim_{k\to\infty} a_k^{1/k} = \rho$, there exists a constant N so that

$$a_k^{1/k} < \gamma \qquad \text{whenever} \qquad k \geq N. \tag{4}$$

(To see this take $\epsilon = \gamma - \rho$ in Definition 3.) Inequality (4) may be written

$$a_k < \gamma^k, \qquad k \geq N,$$

which shows that the series $\Sigma_{k=N}^{\infty} a_k$ is dominated by the geometric series $\Sigma_{k=N}^{\infty} \gamma^k$. Since $\gamma < 1$, this geometric series converges. Thus the series $\Sigma_{k=N}^{\infty} a_k$ converges by comparison with a geometric series, so $\Sigma_{k=1}^{\infty} a_k$ converges by Theorem 11. This proves statement (i).

The proof of statement (ii) is similar to that of (i) and is left as an exercise.

The proof of statement (iii) proceeds as in the proof of Theorem 12. We note that

(a) $\rho = \lim_{k\to\infty} \left(\frac{1}{k}\right)^{1/k} = \lim_{k\to\infty} \frac{1}{\sqrt[k]{k}} = 1$ by statement (4), Section 13.3, and the series $\sum \frac{1}{k}$ diverges.

(b) $\rho = \lim_{k\to\infty} \left(\frac{1}{k^2}\right)^{1/k} = \lim_{k\to\infty} \left(\frac{1}{\sqrt[k]{k}}\right)^2 = 1$, and the series $\sum \frac{1}{k^2}$ converges. ■

The Root Test is less versatile than the Ratio Test. It is primarily used in series involving kth powers.

Example 6 The series $\Sigma e^k/k^k$ converges. This can be shown using the Root Test since

$$\rho = \lim_{k\to\infty} \left(\frac{e^k}{k^k}\right)^{1/k} = \lim_{k\to\infty} \frac{e}{k} = 0 < 1.$$ ■

Example 7 For the series $\Sigma k^2/2^k$ we find that

$$\rho = \lim_{k\to\infty}\left(\frac{k^2}{2^k}\right)^{1/k} = \lim_{k\to\infty}\frac{(k^{1/k})^2}{2} = \frac{1}{2} < 1$$

so the series converges. ■

Exercise Set 13.6

In each of Exercises 1–20, determine whether the given series converges or diverges and state the test used to verify your conclusion.

1. $\sum \frac{2^k}{k+2}$

2. $\sum \frac{k3^k}{(k+1)!}$

3. $\sum k^{10}e^{-k}$

4. $\sum \frac{k!}{2^{k+2}}$

5. $\sum \frac{\ln k}{e^k}$

6. $\sum \frac{(3k)!}{(k!)^3}$

7. $\sum \frac{k+2}{1+k^3}$

8. $\sum \left(\frac{k}{2k+1}\right)^k$

9. $\sum \frac{1}{(\ln k)^k}$

10. $\sum \frac{k!}{k^k}$

11. $\sum \left(1+\frac{2}{k}\right)^k$

12. $\sum \left(\frac{k}{k+1}\right)^k$

13. $\sum \frac{3^k}{k^3 2^{k+2}}$

14. $\sum \frac{k^3 2^{k+3}}{2^{2k}}$

15. $\sum \frac{(k!)^2}{(3k)!}$

16. $\sum \left(\frac{k}{1+k^3}\right)^k$

17. $\sum \frac{(k+2)!}{4!k!2^k}$

18. $\sum \frac{1}{(k+1)!}$

19. $\sum \frac{k!}{e^{3k}}$

20. $\sum \frac{1}{k\sqrt{\ln k}}$

In Exercises 21–24, find those values of $x > 0$ for which the given series converges.

21. $1 + x + \frac{x^2}{2!} + \frac{x^3}{3!} + \cdots = \sum_{k=0}^{\infty} \frac{x^k}{k!}$ $(0! = 1)$

22. $1 + x^2 + x^4 + \cdots = \sum_{k=0}^{\infty} x^{2k}$

23. $1 + \frac{x^2}{2} + \frac{x^4}{4} + \cdots = 1 + \sum_{k=1}^{\infty} \frac{x^{2k}}{2k}$

24. $1 + \frac{x^2}{2} + \frac{x^4}{3} + \cdots = \sum_{k=0}^{\infty} \frac{x^{2k}}{k+1}$

25. Prove Theorem 12, part (ii). (*Hint:* If $\rho > 1$, let γ be a constant so that $1 < \gamma < \rho$. Then, if $\frac{a_{k+1}}{a_k} > \gamma$, it follows that $a_k > \gamma^k$. Proceed as in the proof of part (i).).

26. Prove Theorem 13, part (ii). (*Hint:* If $\rho > 1$, let γ be a constant so that $1 < \gamma < \rho$. Then if $\sqrt[k]{a_k} > \gamma$, it follows that $a_k > \gamma^k$. Proceed as in the proof of part (i).).

27. For the *sequence* $\{a_k\}$ with $a_k > 0$, prove that if $\lim_{k\to\infty} \frac{a_{k+1}}{a_k} < 1$, then $\lim_{k\to\infty} a_k = 0$.

28. For the *sequence* $\{a_k\}$ with $a_k > 0$, if $\lim_{k\to\infty} \frac{a_{k+1}}{a_k} > 1$, what can you say about $\lim_{k\to\infty} a_k$?

13.7 ABSOLUTE AND CONDITIONAL CONVERGENCE

Each of the convergence tests discussed thus far applies only to series with strictly positive terms. In this section we discuss two criteria by which series containing both positive and negative terms may be tested for convergence.

Absolute Convergence

One straightforward procedure for testing the series Σa_k for convergence is simply to replace each term by its absolute value and then test the resulting series, $\Sigma|a_k|$, for convergence. This is the idea of *absolute convergence*.

DEFINITION 5

The series Σa_k is said to **converge absolutely** if the series of absolute values, $\Sigma|a_k|$, converges.

Example 1 The series

$$\sum_{k=1}^{\infty} \frac{(-1)^{n+1}}{k^2} = 1 - \frac{1}{4} + \frac{1}{9} - \frac{1}{16} + \frac{1}{25} - \cdots$$

converges absolutely since

$$\sum_{k=1}^{\infty} \left|\frac{(-1)^{n+1}}{k^2}\right| = \sum_{k=1}^{\infty} \frac{1}{k^2}$$

is a convergent p-series. ■

Example 2 The series

$$\sum_{k=1}^{\infty} (-1)^{k+1} = 1 - 1 + 1 - 1 + \cdots$$

does not converge absolutely since $\sum_{k=1}^{\infty} |(-1)^{n+1}| = 1 + 1 + 1 + \cdots$ diverges. ■

The following theorem shows that the terminology of Definition 5 is well chosen—absolutely convergent series indeed converge.

THEOREM 14

If the series $\Sigma|a_k|$ converges then so does the series Σa_k. That is, every absolutely convergent series converges.

Proof: Assume that the series $\Sigma|a_k|$ converges. Since the inequality

$$-|a_k| \le a_k \le |a_k|$$

holds for all terms a_k, we may add $|a_k|$ to all terms in this inequality, to get

$$0 \le a_k + |a_k| \le 2|a_k|, \qquad k = 1, 2, 3, \ldots \tag{1}$$

Since the series $\Sigma|a_k|$ converges, so does the series $2\Sigma|a_k| = \Sigma 2|a_k|$. Thus inequality (1) shows that the series $\Sigma(a_k + |a_k|)$ is a series with positive terms that is dominated by a convergent series. The Comparison Test therefore guarantees that the series $\Sigma(a_k + |a_k|)$ converges. We may now apply Theorem 6 (Section 13.4) to conclude that the series

$$\Sigma a_k = \Sigma(a_k + |a_k|) - \Sigma|a_k|$$

converges. This completes the proof. ■

The information supplied by Theorem 14 is simple to apply. If the series Σa_k contains negative terms, we test the positive series $\Sigma|a_k|$ for convergence. If $\Sigma|a_k|$ converges, so does Σa_k. However, if $\Sigma|a_k|$ diverges, we can draw no conclusions about the original series Σa_k.

Example 3 To test the series

$$\sum_{k=0}^{\infty} \frac{(-1)^k k^2}{(k+1)!}$$

for convergence, we apply the Ratio Test to the series

$$\sum_{k=0}^{\infty} \left| \frac{(-1)^k k^2}{(k+1)!} \right| = \sum_{k=0}^{\infty} \frac{k^2}{(k+1)!}.$$

We find that

$$\rho = \lim_{k\to\infty} \frac{a_{k+1}}{a_k} = \lim_{k\to\infty} \frac{\dfrac{(k+1)^2}{(k+2)!}}{\dfrac{k^2}{(k+1)!}} = \lim_{k\to\infty} \frac{(k+1)^2}{k^2(k+2)} = 0.$$

The series $\sum_{k=0}^{\infty} \frac{(-1)^k k^2}{(k+1)!}$ therefore converges absolutely. ■

Example 4 The series

$$\sum_{k=0}^{\infty} \frac{\cos \pi k}{2^k} = 1 - \frac{1}{2} + \frac{1}{4} - \frac{1}{8} + \cdots$$

contains both positive and negative terms. To test for absolute convergence we consider the series (which turns out to be a geometric series)

$$\sum_{k=0}^{\infty} \left| \frac{\cos \pi k}{2^k} \right| = \sum_{k=0}^{\infty} \frac{|\cos \pi k|}{2^k} = \sum_{k=0}^{\infty} \frac{1}{2^k}$$

$$= \sum_{k=0}^{\infty} \left(\frac{1}{2}\right)^k = \frac{1}{1 - 1/2} = 2 \qquad \text{(geometric).}$$

The series $\sum_{k=0}^{\infty} \frac{\cos \pi k}{2^k}$ converges absolutely. ■

Example 5 For the series

$$\sum_{k=1}^{\infty} \frac{(-1)^{k+1}}{k} = 1 - \frac{1}{2} + \frac{1}{3} - \frac{1}{4} + \cdots$$

the series of absolute values is the harmonic series $\sum_{k=1}^{\infty} 1/k$, which diverges. Theorem 14 therefore gives no conclusion concerning the convergence of the series $\sum_{k=1}^{\infty} \frac{(-1)^{k+1}}{k}$. ■

Example 6 If the variable x is unrestricted, the series

$$\sum_{k=1}^{\infty} \frac{x^k}{k} = x + \frac{x^2}{2} + \frac{x^3}{3} + \cdots$$

contains negative terms for $x < 0$. To determine values of x for which this series converges we consider the series of absolute values

$$\sum_{k=1}^{\infty} \left|\frac{x^k}{k}\right| = \sum_{k=1}^{\infty} \frac{|x|^k}{k}.$$

Applying the Ratio Test to this series, we find that

$$\rho = \lim_{k\to\infty} \frac{|a_{k+1}|}{|a_k|} = \lim_{k\to\infty} \frac{\dfrac{|x|^{k+1}}{k+1}}{\dfrac{|x|^k}{k}} = \lim_{k\to\infty} |x|\left(\frac{k}{k+1}\right) = |x|.$$

Thus, $\rho < 1$ only if $|x| < 1$. This shows that $\sum_{k=1}^{\infty} x^k/k$ converges if $|x| < 1$. However, we obtain no conclusions for $|x| \geq 1$ by this procedure. ■

Alternating Series

Several types of infinite series containing negative terms can be shown to converge even though they do not converge absolutely. Among these are certain of the *alternating series*.

DEFINITION 6

An **alternating series** is a series of the form

$$\sum_{k=0}^{\infty} (-1)^k a_k = a_0 - a_1 + a_2 - a_3 + a_4 - \cdots$$

where $a_k > 0$ for each $k = 0, 1, 2, 3, \ldots$.

The following theorem indicates the conditions under which an alternating series converges. We will demonstrate its use before we give a proof.

THEOREM 15
Alternating Series Test

Let $a_k > 0$ for each $k = 0, 1, 2, \ldots$. The alternating series

$$\sum_{k=0}^{\infty} (-1)^k a_k$$

converges if both the following hold:

(i) The terms of the sequence $\{a_k\}_{k=0}^{\infty}$ are decreasing, and
(ii) $\lim_{k\to\infty} a_k = 0$.

Requirement (ii) in Theorem 15 is just a restatement of the necessary condition for convergence given by Theorem 7. To verify condition (i) in particular cases it is often useful to consider a function $f(x)$ with $f(k) = a_k$, as in the statement of the Integral Test. If $f'(x) < 0$ for $x \geq 0$, the function $f(x)$ is decreasing on $[0, \infty)$, which assures that condition (i) holds. (If $f'(x) < 0$ and the integral converges only for $x \geq N$ where N is some positive integer, then the "tail" $\Sigma_{k=N}^{\infty} (-1)^k a_k$ is convergent, and so is the original series.)

Example 7 For the series

$$\sum_{k=1}^{\infty} \frac{(-1)^{k+1}}{k} = 1 - \frac{1}{2} + \frac{1}{3} - \frac{1}{4} + \cdots$$

the general term is $(-1)^{k+1}a_k$, where $a_k = \dfrac{1}{k} > 0$. Since the sequence $\left\{\dfrac{1}{k}\right\}$ is decreasing and $\lim_{k\to\infty} \dfrac{1}{k} = 0$, the series converges by Theorem 15. ■

Combining the results of Examples 5 and 7, we see that the **alternating harmonic series** $\sum_{k=1}^{\infty} \dfrac{(-1)^{k+1}}{k} = 1 - \dfrac{1}{2} + \dfrac{1}{3} - \dfrac{1}{4} + \cdots$ converges, but does not converge absolutely. (This shows that the converse of Theorem 14 is not true.) Such series are said to *converge conditionally*.

DEFINITION 7

The series Σa_k is said to **converge conditionally** if Σa_k converges and $\Sigma|a_k|$ diverges.

Example 8 Determine whether the series

$$\sum_{k=1}^{\infty} \frac{(-1)^{k+1}\, k}{1 + k^2}$$

converges absolutely, converges conditionally, or diverges.

Strategy
First, test for absolute convergence.

Solution
The series of absolute values is

$$\sum_{k=1}^{\infty} \frac{k}{1 + k^2}.$$

Use the Integral Test on the series of absolute values.

We may test this series by the Integral Test. Since

$$\int_1^{\infty} \frac{x}{1 + x^2}\, dx = \lim_{t\to\infty} \int_1^{t} \frac{x}{1 + x^2}\, dx$$

$$= \lim_{t\to\infty} \frac{1}{2} \ln(1 + t^2) = +\infty$$

the series of absolute values diverges. Thus, the original series does not converge absolutely.

Since the series is alternating but does not converge absolutely, apply the alternating series test.

We next test for conditional convergence using Theorem 15. Since

$$\frac{d}{dx}\left(\frac{x}{1+x^2}\right) = \frac{1+x^2-x(2x)}{(1+x^2)^2} = \frac{1-x^2}{(1+x^2)^2}$$

is negative for $x > 1$, the sequence $\left\{\frac{k}{1+k^2}\right\}$ is decreasing. Also, $\lim_{k\to\infty} \frac{k}{1+k^2} = 0$. Thus, both conditions of Theorem 15 are met, so the series converges conditionally. ■

Example 9 For which values of x does the series

$$\sum_{k=1}^{\infty} \frac{x^k}{k}$$

converge?

Solution: We have already determined that the series converges absolutely for $|x| < 1$ (Example 6). If $|x| > 1$, we may apply l'Hôpital's Rule (differentiating with respect to k) to conclude that

$$\lim_{k\to\infty} \frac{|x|^k}{k} \quad \left(\frac{\infty}{\infty} \text{ form}\right)$$

$$= \lim_{k\to\infty} \frac{|x|^k \ln|x|}{1} = +\infty.$$

Thus, $\lim_{k\to\infty} \frac{x^k}{k}$ does not exist if $|x| > 1$, so $\sum_{k=1}^{\infty} \frac{x^k}{k}$ diverges for these values of x, by Theorem 7. The remaining cases are $x = 1$ and $x = -1$.

If $x = 1$, the series is the harmonic series $\Sigma_{k=1}^{\infty} 1/k$, which diverges. If $x = -1$, the series is

$$\sum_{k=1}^{\infty} \frac{(-1)^k}{k} = -1 + \frac{1}{2} - \frac{1}{3} + \frac{1}{4} - \cdots.$$

This is the negative of the alternating harmonic series, which converges conditionally (Example 7).

We have therefore shown that the series $\Sigma_{k=1}^{\infty} x^k/k$ converges for $x \in [-1, 1)$ and diverges for all other values of x. In particular, the series converges absolutely for $x \in (-1, 1)$ and converges conditionally for $x = -1$. ■

Proof of Theorem 15: For the alternating series

$$\sum_{k=0}^{\infty} (-1)^k a_k,$$

the even partial sums have the form

$$S_{2n} = (a_0 - a_1) + (a_2 - a_3) + \cdots + (a_{2n-2} - a_{2n-1}) + a_{2n}.$$

Since the sequence $\{a_k\}$ is decreasing, each of the terms in parentheses is positive.

This shows that all even partial sums are positive. Also, since $a_{2n+1} > a_{2n+2}$

$$\begin{aligned} S_{2n+2} &= S_{2n} - a_{2n+1} + a_{2n+2} \\ &= S_{2n} - (a_{2n+1} - a_{2n-2}) \\ &< S_{2n}. \end{aligned} \tag{2}$$

Thus, the sequence of even partial sums $\{S_{2n}\}$ is decreasing. In particular, this shows that the sequence of positive terms $\{S_{2n}\}$ is bounded above by $S_0 = a_0$ and below by zero. Thus $\{S_{2n}\}$ is a bounded decreasing sequence which, by Corollary 1, must converge. We denote the limit of this sequence by $S = \lim_{n\to\infty} S_{2n}$.

Next, since $\lim_{n\to\infty} a_{2n+1} = 0$, we see that

$$\begin{aligned} \lim_{n\to\infty} S_{2n+1} &= \lim_{n\to\infty} (S_{2n} - a_{2n+1}) \\ &= S - 0 \\ &= S, \end{aligned}$$

so both even and odd partial sums converge to S. Thus $S = \lim_{n\to\infty} S_n$ and the series $\Sigma_{k=0}^{\infty} (-1)^k a_k$ converges. ■

Estimating Remainders for Alternating Series

We remarked at the end of Section 13.4 that very few types of series have a sum that can be expressed as an explicit formula. Indeed, alternating series generally cannot be summed this way. For this reason, it is important to be able to *estimate* the sum of a given alternating series. Fortunately, there is a simple way to obtain both an estimate of the sum and a bound on the error associated with that estimate.

As in line (2) in the proof of Theorem 15, we may show that the odd partial sums of an alternating series $\Sigma_{k=0}^{\infty} (-1)^k a_k$ form an *increasing* sequence:

$$\begin{aligned} S_{2n+1} &= S_{2n-1} + (a_{2n} - a_{2n+1}) \\ &> S_{2n-1}, \qquad n = 1, 2, \ldots \end{aligned} \tag{3}$$

since $a_{2n} > a_{2n+1}$.

If we write

$$S = \sum_{k=0}^{\infty} (-1)^k a_k,$$

inequalities (2) and (3) together imply that

$$S_{2p-1} < S < S_{2m} \tag{4}$$

for any positive integers p and m. That is, *any* even partial sum is larger than *any* odd partial sum, and the sum of the series lies between them (see Figure 7.1).

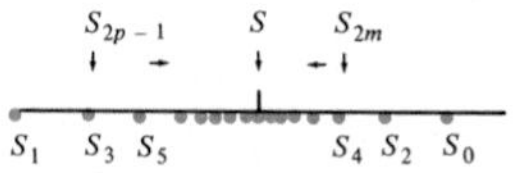

Partial sums (S_{2p-1}) increase to S.
Partial sums (S_{2m}) decrease to S.

Figure 7.1

Now let

$$S = S_n + R_n \tag{5}$$

where

$$R_n = \sum_{k=n+1}^{\infty} (-1)^k a_k.$$

Then R_n is the **truncation error,** or **remainder,** associated with the approximation of S by the partial sum S_n. From (4) and (5) it follows, by setting $m = n$ and $p = n + 1$, that

$$|R_{2n}| = |S - S_{2n}| < |S_{2n+1} - S_{2n}| = a_{2n+1}. \tag{6}$$

Similarly, by setting $m = p = n$ we find

$$|R_{2n-1}| = |S - S_{2n-1}| < |S_{2n} - S_{2n-1}| = a_{2n}. \tag{7}$$

Thus, regardless of whether we truncate an alternating series after an even or an odd number of terms, *the absolute value of the remainder (or tail) is less than the absolute value of the next term in the series*. This result is sufficiently important that we carefully summarize what we have proved in the form of a theorem.

THEOREM 16
Truncation Error

Let $a_k > 0$ for each $k = 0, 1, 2, \ldots$ and assume that

(i) $a_k > a_{k+1}$ for each $k = 0, 1, 2, \ldots$, and
(ii) $\lim_{k\to\infty} a_k = 0$.

The error associated with approximating the sum of the alternating series

$$\sum_{k=0}^{\infty} (-1)^k a_k$$

by the partial sum

$$S_n = \sum_{k=0}^{n} (-1)^k a_k$$

is less than the first truncated term, a_{n+1}.

Example 10 Find an approximation to the sum of the alternating harmonic series

$$\sum_{k=0}^{\infty} (-1)^k \left(\frac{1}{k+1}\right) = 1 - \frac{1}{2} + \frac{1}{3} - \frac{1}{4} + \cdots$$

accurate to within .05.

Solution: If we approximate this sum by the partial sum

$$S_n = \sum_{k=0}^{n} (-1)^k \left(\frac{1}{k+1}\right)$$

the error in this approximation is less than $a_{n+1} = \dfrac{1}{n+2}$. We therefore choose n sufficiently large to satisfy the inequality

$$\frac{1}{n+2} \le .05 = \frac{1}{20}.$$

This gives

$$n + 2 \geq 20, \qquad \text{or} \qquad n \geq 18.$$

The partial sum S_{18} therefore has the desired accuracy. It is

$$S_{18} = 1 - \frac{1}{2} + \frac{1}{3} - \frac{1}{4} + \cdots + \frac{1}{19} \approx 0.72.$$ ■

REMARK: The series in Example 10 is identical to the series in Example 7, as can be seen by making the substitution $k = j - 1$ and noting that $(-1)^{j-1} = (-1)^{j+1}$.

Example 11 How many terms must be included in the partial sum for the alternating series

$$\sum_{k=0}^{\infty} \frac{(-1)^k}{1 + k^2} = 1 - \frac{1}{2} + \frac{1}{5} - \frac{1}{10} + \cdots$$

to ensure that the partial sum approximates the sum of the series accurately to within 0.01?

Solution: It is easier to work with S_{n-1}, so that the absolute value of the first truncated term is $a_n = \dfrac{1}{1 + n^2}$. We therefore need

$$\frac{1}{1 + n^2} \leq \frac{1}{100}, \qquad \text{or} \qquad n^2 \geq 99.$$

The choice $n = 10$ therefore will suffice, and the partial sum S_9 will have the desired accuracy. ■

Exercise Set 13.7

In each of Exercises 1–30, test the given series for **(a)** absolute convergence, **(b)** conditional convergence.

1. $\sum_{k=1}^{\infty} \frac{(-1)^k}{k^3}$

2. $\sum_{k=1}^{\infty} \frac{(-1)^k}{2k + 1}$

3. $\sum_{k=1}^{\infty} \frac{(-1)^{k+1}k^2}{(k + 2)!}$

4. $\sum_{k=2}^{\infty} (-1)^{k+1} \frac{k}{\ln k}$

5. $\sum_{k=1}^{\infty} (-1)^k \frac{k!}{(2k + 1)!}$

6. $\sum_{k=1}^{\infty} \frac{(-k)^3}{3^k}$

7. $\sum_{k=1}^{\infty} \frac{(-1)^k}{1 + \sqrt{k}}$

8. $\sum_{k=1}^{\infty} \frac{\cos \pi k}{k}$

9. $\sum_{k=1}^{\infty} \frac{\sin\left(\frac{\pi k}{2}\right)}{k}$

10. $\sum_{k=1}^{\infty} \frac{(-1)^k}{\sqrt{k(k + 2)}}$

11. $\sum_{k=1}^{\infty} \frac{(-1)^k \cdot k^3}{2^{k+2}}$

12. $\sum_{k=2}^{\infty} \frac{k^2(-1)^{k+1}}{\ln k}$

13. $\sum_{k=2}^{\infty} \frac{k^2(-1)^{k+1}}{\ln k}$

14. $\sum_{k=1}^{\infty} \frac{(2k + 1)(-1)^k}{6k + 2}$

15. $\sum_{k=1}^{\infty} \frac{(-1)^k\sqrt{k}}{k + 1}$

16. $\sum_{k=1}^{\infty} \frac{(-1)^{k+1}2^k}{k^3 \cdot 3^{k+2}}$

17. $\sum_{k=2}^{\infty} \frac{(-1)^k}{k \ln^2 k}$

18. $\sum_{k=1}^{\infty} \frac{(-1)^{k+1}k!}{6^k}$

19. $\sum_{k=1}^{\infty} \frac{(-1)^k}{1 + k^2}$

20. $\sum_{k=1}^{\infty} \sin\left(\frac{k\pi}{4}\right)$

21. $\sum_{k=2}^{\infty} \frac{(-1)^k k}{\ln \sqrt{k}}$

22. $\sum_{k=1}^{\infty} \frac{(-1)^k k^2}{(2k + 1)(k + 3)}$

23. $\sum_{k=1}^{\infty} \frac{(-1)^k \sinh k}{3e^{2k}}$

24. $\sum_{k=1}^{\infty} \frac{(-1)^k \tan^{-1} k}{\sqrt{k}}$

25. $\sum_{k=1}^{\infty} \frac{\cos \pi k}{k+2}$

26. $\sum_{k=2}^{\infty} \frac{(-1)^k}{\sqrt[k]{\ln k}}$

27. $\sum_{k=2}^{\infty} \frac{k \sin\left[\frac{(2k+1)\pi}{2}\right]}{\sqrt{1+e^{\sqrt{k}}}}$

28. $\sum_{k=2}^{\infty} \frac{(-1)^k(k^{3/2}+3k)}{7-k^2+2k^{5/2}}$

29. $\sum_{k=1}^{\infty} \frac{\sinh k - \cosh k}{k}$

30. $\sum_{k=2}^{\infty} \frac{(-1)^k \cosh 2k}{ke^k}$

In Exercises 31–34, determine the values of x for which the given series converges absolutely.

31. $\sum_{k=0}^{\infty} \frac{x^k}{k!} = 1 + x + \frac{x^2}{2!} + \frac{x^3}{3!} + \cdots$

32. $\sum_{k=0}^{\infty} \frac{x^k}{2k+1} = 1 + \frac{x}{3} + \frac{x^2}{5} + \cdots$

33. $\sum_{k=0}^{\infty} (-1)^k \frac{x^{2k+1}}{(2k+1)!} = x - \frac{x^3}{3!} + \frac{x^5}{5!} - \frac{x^7}{7!} + \cdots$

34. $\sum_{k=0}^{\infty} (-1)^k \frac{x^{2k}}{(2k)!} = 1 - \frac{x^2}{2!} + \frac{x^4}{4!} - \frac{x^6}{6!} + \cdots$

35. How large must n be taken so that the partial sum $S_n = \sum_{k=0}^{n} (-1)^k \left(\frac{1}{1+k}\right)$ approximates the sum of the alternating harmonic series accurate to within .01?

36. With what accuracy does the sum

$$\sum_{k=0}^{10} \frac{(-1)^k}{(k+1)^3}$$

approximate the sum of the series

$$\sum_{k=0}^{\infty} \frac{(-1)^k}{(k+1)^3}?$$

37. Prove that if Σa_k and Σb_k converge absolutely then so does $\Sigma(a_k + b_k)$.

38. Prove that if Σa_k converges absolutely and c is any real number the series Σca_k converges absolutely.

39. Prove for the (not necessarily positive) series Σa_k that if

$$\rho = \lim_{k\to\infty} \left|\frac{a_{k+1}}{a_k}\right|,$$

(i) Σa_k converges absolutely if $0 \le \rho < 1$;
(ii) Σa_k diverges if $\rho > 1$;
(iii) no conclusions may be drawn if $\rho = 1$.

40. Prove that Σa_k^2 converges if Σa_k converges absolutely. Is the converse true?

SUMMARY OUTLINE OF CHAPTER 13

- An **infinite sequence** $\{a_n\}$ is a function whose domain is the set of positive integers.
- $L = \lim_{n\to\infty} a_n$ if, given $\epsilon > 0$, there exists an integer N so that $|a_n - L| < \epsilon$ whenever $n \ge N$.
- The sequence $\{a_n\}$ is said to **converge** if $L = \lim_{n\to\infty} a_n$ exists.
- The nth **partial sum** of the infinite series $\sum_{k=1}^{\infty} a_k$ is the sum of the first n terms: $S_n = \sum_{k=1}^{n} a_k$.
- S is called the **sum** of the infinite series if $S = \lim_{n\to\infty} S_n = \lim_{n\to\infty} \sum_{k=1}^{n} a_k$.
- The series Σa_k is said to **converge** if the sum S exists.
- The **geometric series** $\sum_{k=0}^{\infty} x^k = 1 + x + x^2 + x^3 + \cdots$ converges, if $|x| < 1$, to the sum $S = \frac{1}{1-x}$, and diverges otherwise.
- The **harmonic series** $\sum_{k=1}^{\infty} \frac{1}{k} = 1 + \frac{1}{2} + \frac{1}{3} + \frac{1}{4} + \cdots$ diverges.
- The ***p*-series** $\sum_{k=1}^{\infty} \frac{1}{k^p} = 1 + \frac{1}{2^p} + \frac{1}{3^p} + \frac{1}{4^p} + \cdots$ converges if $p > 1$ and diverges otherwise.
- A **necessary condition** for Σa_k to converge: $\lim_{k\to\infty} a_k = 0$ (however, this condition does not guarantee convergence).
- ***Tests for convergence of Σa_k when $a_k \ge 0$ for all k:***

  (1) (Comparison Test) Let $0 \le a_k \le b_k$ for all k.
  a. If Σb_k converges, Σa_k converges.
  b. If Σa_k diverges, Σb_k diverges.

(2) (Integral Test) If $f(k) = a_k$ for all k and $f(x)$ is decreasing on $[1, \infty)$ then Σa_k converges if and only if $\int_1^\infty f(x)\,dx$ converges.

(3) (Ratio Test) If $\rho = \lim_{k\to\infty} \frac{a_{k+1}}{a_k}$ exists, then

a. Σa_k converges if $\rho < 1$.
b. Σa_k diverges if $\rho > 1$.
c. No conclusion may be drawn if $\rho = 1$.

(4) (Root Test) If $\rho = \lim_{k\to\infty} \sqrt[k]{a_k}$ exists, same conclusions as for the Ratio Test.

(5) (Limit Comparison Test) If $b_k > 0$ for all k and $\rho = \lim_{k\to\infty} \frac{a_k}{b_k} > 0$ exists, then Σa_k and Σb_k either both converge or both diverge.

- The series Σa_k **converges absolutely** if $\Sigma|a_k|$ converges.
- ***Theorem:*** Σa_k converges if $\Sigma|a_k|$ converges.
- An **alternating series** is a series of the form $\sum_{k=0}^{\infty} (-1)^k a_k$ with all $a_k > 0$.
- ***Theorem:*** An alternating series converges if
  a. $a_k > a_{k+1}$ for all k, and
  b. $\lim_{k\to\infty} a_k = 0$.
- The **error** associated with the approximation of the sum of the alternating series $\sum_{k=0}^{\infty} (-1)^k a_k$ by the partial sum $S_n = \sum_{k=0}^{n} (-1)^k a_k$ is no greater than the first discarded term, a_{n+1}.

REVIEW EXERCISES—CHAPTER 13

In Exercises 1–10, determine whether the given sequence converges. If it does, find its limit.

1. $\left\{\sin \frac{n\pi}{2}\right\}$

2. $\left\{\frac{1}{\sqrt{n+1}}\right\}$

3. $\left\{\frac{\sqrt{n}}{\ln(\sqrt{n}+1)}\right\}$

4. $\left\{\left(1 - \frac{2}{n}\right)^{2n}\right\}$

5. $\left\{\frac{3^n}{n!}\right\}$

6. $\left\{\frac{(-1)^{n+1}}{\sqrt{n^2+1}}\right\}$

7. $\left\{\frac{(-1)^{n+1}\cdot n^3}{n(1-n+n^2)}\right\}$

8. $\left\{\frac{n^2+2n+2}{n^{3/2}}\right\}$

9. $\{e^{2\ln(n)}\}$

10. $\{\ln(n+1) - \ln(n-1)\}$

In Exercises 11–14, find the sum of the given geometric series.

11. $1 + \frac{1}{6} + \frac{1}{36} + \frac{1}{216} + \cdots$

12. $10 + \frac{10}{2} + \frac{10}{4} + \frac{10}{8} + \cdots$

13. $\frac{2}{3} + \frac{4}{9} + \frac{8}{27} + \cdots$

14. $1 - \frac{1}{5} + \frac{1}{25} - \frac{1}{125} + \cdots$

In Exercises 15 and 16, express the repeating decimal as an infinite series and find its rational form.

15. $.3737\overline{37}\ldots$

16. $.0263263\overline{263}\ldots$

In Exercises 17–20, find the sum of the series.

17. $\sum_{k=3}^{\infty} \left(\frac{1}{4}\right)^k$

18. $\sum_{k=1}^{\infty} \frac{3^k + 2^k}{5^k}$

19. $\sum_{k=2}^{\infty} \frac{20}{3^{k+1}}$

20. $\sum_{k=1}^{\infty} \left[\frac{3}{2^k} + \frac{1}{(k+1)(k+2)}\right]$

In Exercises 21–58, determine whether the given series converges or diverges and state which test you used.

21. $\sum_{k=1}^{\infty} \frac{1}{k(k+1)}$

22. $\sum_{k=1}^{\infty} \frac{3^k}{k^3}$

23. $\sum_{k=3}^{\infty} \frac{2^k}{k^5}$

24. $\sum_{k=1}^{\infty} \frac{(-1)^k}{2k+1}$

25. $\sum_{k=1}^{\infty} \frac{k^3}{k!}$

26. $\sum_{k=0}^{\infty} \frac{1}{10k+2}$

27. $\sum_{k=2}^{\infty} \frac{\cos \pi k}{\sqrt{k}}$

28. $\sum_{k=2}^{\infty} \frac{k^k}{k!}$

29. $\sum_{k=3}^{\infty} \frac{\ln k}{k^2}$

30. $\sum_{k=0}^{\infty} \frac{2^k + k}{(k+1)!}$

31. $\sum_{k=1}^{\infty} \frac{1}{9+k^2}$

32. $\sum_{k=2}^{\infty} \frac{(-1)^k k}{\ln k}$

33. $\sum_{k=2}^{\infty} \frac{2}{k \ln(k+1)}$

34. $\sum_{k=1}^{\infty} \frac{\sqrt{k}}{2^k}$

35. $\sum_{k=0}^{\infty} \frac{k^4 \cdot 2^k}{(k+2)!}$

36. $\sum_{k=1}^{\infty} (-1)^k \frac{\ln k}{k}$

37. $\sum_{k=0}^{\infty} \frac{1}{\sqrt{4+k^2}}$

38. $\sum_{k=1}^{\infty} \left(\frac{k-1}{k+1}\right)^k$

39. $\sum_{k=1}^{\infty} (-1)^{k+1} \frac{\sqrt{k}}{2k+1}$

40. $\sum_{k=1}^{\infty} k^2 e^{-k^3}$

41. $\sum_{k=1}^{\infty} \left(\frac{k!}{k^k}\right)^k$

42. $\sum_{k=2}^{\infty} (-1)^k \frac{k^2+2}{1+k+k^2}$

43. $\sum_{k=1}^{\infty} 2ke^{-k}$

44. $\sum_{k=0}^{\infty} \left(\frac{k}{1+k}\right)^k$

45. $\sum_{k=0}^{\infty} (-1)^k \sin \pi k$

46. $\sum_{k=2}^{\infty} \frac{(2k+1)!}{k^2(k+1)!}$

47. $\sum_{k=1}^{\infty} \frac{k!}{e^{2k}}$

48. $\sum_{k=2}^{\infty} \frac{1}{(k+6)^{3/2}}$

49. $\sum_{k=1}^{\infty} (-1)^k \frac{3}{\sqrt{k}}$

50. $\sum_{k=2}^{\infty} \frac{1}{(\ln k)^k}$

51. $\sum_{k=1}^{\infty} \frac{2^{2k-1}}{(2k-1)!}$

52. $\sum_{k=2}^{\infty} \frac{1}{\sqrt{k}(\ln k)^3}$

53. $\sum_{k=0}^{\infty} \frac{(-1)^k}{1+k^2}$

54. $\sum_{k=1}^{\infty} \tan^{-1} k$

55. $\sum_{k=1}^{\infty} k\left(\frac{\pi}{k}\right)^k$

56. $\sum_{k=0}^{\infty} \frac{(-1)^k k(k+2)}{(k+1)!}$

57. $\sum_{k=1}^{\infty} \left(\frac{k}{2k+1}\right)^k$

58. $\sum_{k=0}^{\infty} \frac{(-1)^k \sin^{-1}\left(\frac{1}{k+1}\right)}{k+1}$

In Exercises 59–68, determine whether the given series converges absolutely, converges conditionally, or diverges.

59. $\sum_{k=0}^{\infty} \frac{(-1)^{k+1}}{3k+2}$

60. $\sum_{k=2}^{\infty} \frac{(-1)^k}{k \ln k}$

61. $\sum_{k=1}^{\infty} (-1)^k \frac{k^k}{k!}$

62. $\sum_{k=0}^{\infty} \frac{\cos \pi k}{1+k^2}$

63. $\sum_{k=1}^{\infty} \frac{(-1)^{k+1}}{k(k+1)}$

64. $\sum_{k=0}^{\infty} (-1)^k \left(\frac{2}{3}\right)^k$

65. $\sum_{k=1}^{\infty} \frac{k!}{(-2)^k}$

66. $\sum_{k=1}^{\infty} \frac{\cos \frac{\pi k}{3}}{1+k^2}$

67. $\sum_{k=1}^{\infty} \frac{(-1)^k k^2}{(k+1)!}$

68. $\sum_{k=1}^{\infty} \frac{(-1)^{k+1}}{\sqrt{k(k+1)}}$

In Exercises 69–72, find a bound on the error associated with the approximation of the sum of the given series by the partial sum S_4 consisting of the sum of the first four terms.

69. $\sum_{k=1}^{\infty} \frac{2(-1)^k}{1+k^2} = -1 + \frac{2}{5} - \frac{2}{10} + \cdots$

70. $\sum_{k=1}^{\infty} \frac{(-1)^k}{k^4} = -1 + \frac{1}{16} - \frac{1}{81} + \cdots$

71. $\sum_{k=1}^{\infty} \frac{(-1)^{k+1}}{2k+1} = \frac{1}{3} - \frac{1}{5} + \frac{1}{7} - \frac{1}{9} + \cdots$

72. $\sum_{k=1}^{\infty} \frac{(-1)^{k+1} \sin^2\left(\frac{\pi}{2k}\right)}{k+1} = \frac{1}{2} - \frac{1}{6} + \frac{1}{16} - \cdots$

73. Define the sequence $\{a_k\}$ by $a_1 = c$, $a_{k+1} = a_1 \cdot a_k$, $k \geq 1$. For what values of c does $\{a_k\}$ converge?

74. Prove that if the sequence $\{a_n\}$ is not bounded it cannot converge.

75. If Σa_k converges but $\Sigma(a_k + b_k)$ diverges, what can you conclude about Σb_k?

76. Explain how one could use an infinite series to determine at what time between 1 P.M. and 2 P.M. is the minute hand on a clock directly over the hour hand. (*Hint:* The minute hand moves 12 times as fast as the hour hand.)

77. A ball is dropped from a height of 10 meters. Each time it strikes the ground, it rebounds to 5/6 of the height from which it fell. Find the total distance travelled by the ball
a. when it has struck the ground 10 times.
b. when it has come to rest.

78. Prove that if Σa_k diverges then $\Sigma c a_k$ diverges for any number $c \neq 0$.

79. Let Σa_k and Σb_k be series with positive terms. If $\Sigma(a_k + b_k)^2$ converges, what can you conclude about
a. Σa_k^2?
b. Σb_k^2?
c. $\Sigma a_k b_k$?

80. Let $p(k)$ and $q(k)$ be polynomials in k with positive coefficients. Let n be the degree of $p(k)$ and let m be the degree of $q(k)$. Show that
a. $\sum \frac{p(k)}{q(k)}$ converges if $m > n + 1$
b. $\sum \frac{p(k)}{q(k)}$ diverges if $m \leq n + 1$.

CHAPTER 14

POWER SERIES

14.1 INTRODUCTION

The purpose of this chapter is to complete the discussion of the *approximation problem*. Recall that we determined in Chapter 12 that if the function $f(x)$ has $n + 1$ derivatives at $x = a$, we can obtain the polynomial approximation

$$f(x) \approx f(a) + f'(a)(x - a) + \frac{f''(a)}{2!}(x - a)^2 + \cdots + \frac{f^{(n)}(a)}{n!}(x - a)^n. \tag{1}$$

The error in this approximation is given by

$$R_n(x) = \frac{f^{(n+1)}(c)}{(n + 1)!}(x - a)^{n+1} \tag{2}$$

for some number c between a and x (Taylor's Theorem).

In this chapter we will use the theory of infinite series to show that if the function $f(x)$ has derivatives of all orders, then for certain values of x we can actually extend the approximation in line (1) to an *equation* of the form

$$f(x) = f(a) + f'(a)(x - a) + \frac{f''(a)}{2!}(x - a)^2 + \cdots = \sum_{k=0}^{\infty} \frac{f^{(k)}(a)}{k!}(x - a)^k. \tag{3}$$

Since, for each particular value of x, the expression on the right-hand side of equation (3) is an infinite series of constants, the theory developed in Chapter 13 will enable us to determine for which values of x such series converge and how they can be manipulated to solve certain types of problems. In particular, we will be able to complete the discussion begun in Chapters 9 and 10 concerning possible types of solutions for second order differential equations of the form

$$y'' + ay' + by = 0.$$

Before focusing on the idea of *Taylor series,* which are series of the form in equation (3), we take up the more general notion of a *power series* in the variable x.

Such series have the form

$$a_0 + a_1(x - a) + a_2(x - a)^2 + \cdots = \sum_{k=0}^{\infty} a_k(x - a)^k. \tag{4}$$

Thus, a Taylor series is a particular type of power series with

$$a_0 = f(a), \qquad a_1 = f'(a), \qquad \ldots, \qquad a_k = \frac{f^{(k)}(a)}{k!}, \ldots .$$

14.2 GENERAL POWER SERIES

A power series in powers of $x - c$ is an expression of the form

$$\sum_{k=0}^{\infty} a_k(x - c)^k = a_0 + a_1(x - c) + a_2(x - c)^2 + \cdots + a_k(x - c)^k + \cdots \tag{1}$$

where the coefficients $a_0, a_1, a_2, \ldots$ are constants and x is regarded as an independent variable.*

Since expression (1) may be simplified by the change of variable $u = x - c$, we will work almost exclusively with power series of the form

$$\sum_{k=0}^{\infty} a_k x^k = a_0 + a_1 x + a_2 x^2 + \cdots + a_k x^k + \cdots. \tag{2}$$

Here are several examples of power series:

$$\sum_{k=0}^{\infty} x^k = 1 + x + x^2 + x^3 + \cdots + x^k \cdots. \tag{3}$$

$$\sum_{k=0}^{\infty} \frac{x^k}{k!} = 1 + x + \frac{x^2}{2!} + \frac{x^3}{3!} + \cdots + \frac{x^k}{k!} + \cdots \qquad (0! = 1). \tag{4}$$

$$\sum_{k=0}^{\infty} \frac{(-1)^k}{1 + k} x^k = 1 - \frac{x}{2} + \frac{x^2}{3} + \cdots + \frac{(-1)^k}{1 + k} x^k + \cdots. \tag{5}$$

From our work in Chapter 13 we know that, for any particular value of x, the series in equation (2) may or may not converge. For example, the series in equation (3) is the geometric series, which we have previously shown to converge only for

---

*Also it should be noted at the outset that any infinite series $\Sigma a_k(bx - c)^k$ involving powers of a *linear* function $f(x) = bx - c$ is a power series, since the general term may be written

$$a_k(bx - c)^k = a_k\left[b\left(x - \frac{c}{b}\right)\right]^k = b^k a_k\left(x - \frac{c}{b}\right)^k = A_k(x - C)^k$$

with $A_k = b^k a_k$ and $C = c/b$.

$|x| < 1$. On the other hand, if we test the series in equation (4) for absolute convergence using the Ratio Test, we find that

$$\rho = \lim_{k\to\infty} \frac{\left|\dfrac{x^{k+1}}{(k+1)!}\right|}{\left|\dfrac{x^k}{k!}\right|} = \lim_{k\to\infty} \frac{|x|}{k+1} = 0$$

for all x, so the series in equation (4) converges for all values of x.

To determine the values of x for which the series in equation (5) converges we again apply the Ratio Test to test for absolute convergence: Since

$$\rho = \lim_{k\to\infty} \frac{\left|\dfrac{(-1)^{k+1}x^{k+1}}{k+2}\right|}{\left|\dfrac{(-1)^k x^k}{k+1}\right|} = \lim_{k\to\infty} \left(\frac{k+1}{k+2}\right)|x| = |x|, \tag{6}$$

this power series converges absolutely if $|x| < 1$. Also, we can see that if $x = 1$ we obtain the alternating series $\sum_{k=0}^{\infty} \frac{(-1)^k}{1+k}$, which converges, and when $x = -1$ we obtain the harmonic series $\sum_{k=0}^{\infty} \frac{1}{1+k}$, which diverges. However, this leaves the convergence for $|x| > 1$ yet to be determined. Rather than pursue specific examples in this way, we now establish two theorems that tell us the convergence properties of power series in general.

THEOREM 1

(i) If the power series

$$\sum_{k=0}^{\infty} a_k x^k$$

converges for $x = c \neq 0$, then it converges absolutely for all x with $|x| < |c|$.

(ii) If the power series

$$\sum_{k=0}^{\infty} a_k x^k$$

diverges for $x = d$, then it diverges for all x with $|x| > |d|$.

Proof: To prove part (i) we assume that $\Sigma a_k c^k$ converges. It is a necessary condition (Theorem 7, Section 13.5) that $\lim_{k\to\infty} a_k c^k = 0$. Letting $\epsilon = 1$ in Definition 3, Chapter 13, we conclude that there exists an integer N so that $|a_k c^k| < 1$ whenever $k \geq N$. Now let x be any number such that $|x| < |c|$, and let $\gamma = \dfrac{|x|}{|c|} < 1$. Then whenever $k \geq N$ we have

$$|a_k x^k| = \frac{|a_k c^k x^k|}{|c^k|} = |a_k c^k| \left(\frac{|x|}{|c|}\right)^k < \gamma^k.$$

This shows that the series $\sum_{k=N}^{\infty} |a_k x^k|$ is dominated by the convergent geometric series $\sum_{k=N}^{\infty} \gamma^k$. Thus, by the Comparison Test, the series $\sum_{k=N}^{\infty} |a_k x^k|$ converges. The series $\Sigma a_k x^k$ therefore converges absolutely if $|x| < |c|$.

To prove part (ii), we assume that the series $\Sigma a_k d^k$ diverges and let x be any number so that $|x| > |d|$. Then $\Sigma a_k x^k$ cannot converge, since the convergence of $\Sigma a_k x^k$ would imply the convergence of $\Sigma a_k d^k$, by part (i). Thus $\Sigma a_k x^k$ diverges whenever $|x| > |d|$. This completes the proof. ■

Theorem 1 gives insight into the geometry of the set of values of x for which a given power series converges. Given the power series $\Sigma a_k x^k$, let S be this set.

Now suppose that there exists a number $d \notin S$. Then $\Sigma a_k d^k$ diverges, so by Theorem 1, $|x| \le |d|$ for every $x \in S$. This shows that the set S is bounded if it is not the entire real line, $(-\infty, \infty)$. By the Completeness Axiom (Section 13.3) there then exists a least upper bound r for the set S. Let us look at two cases:

(i) If $|x| < r$, then $|x|$ is not an upper bound for S, so there exists an element $c \in S$ with $|x| < c$. Since $c \in S$, $\Sigma a_k c^k$ converges. Thus $\Sigma a_k x^k$ converges absolutely, by Theorem 1.
(ii) If $|x| > r$, then $x \notin S$ so $\Sigma a_k x^k$ diverges.

Case (i) shows that S contains the interval $(-r, r)$, and that $\Sigma a_k x^k$ converges absolutely for every $x \in (-r, r)$. Case (ii) shows that the only other possible elements of S are the endpoints r and $-r$. For this reason the number r is called the **radius of convergence** of the power series. The interval $(-r, r)$, $[-r, r)$, $(-r, r]$, or $[-r, r]$ on which $\Sigma a_k x^k$ converges is called the **interval of convergence.** We summarize these findings as follows:

THEOREM 2

Given the power series $\Sigma a_k x^k$, precisely one of the following holds:

(a) The power series converges only for $x = 0$.
(b) There exists a positive number r so that the power series converges absolutely for $|x| < r$ and diverges for $|x| > r$. (The series may or may not converge for $x = \pm r$.)
(c) The power series converges for all values of x.

(See Figure 2.1.)

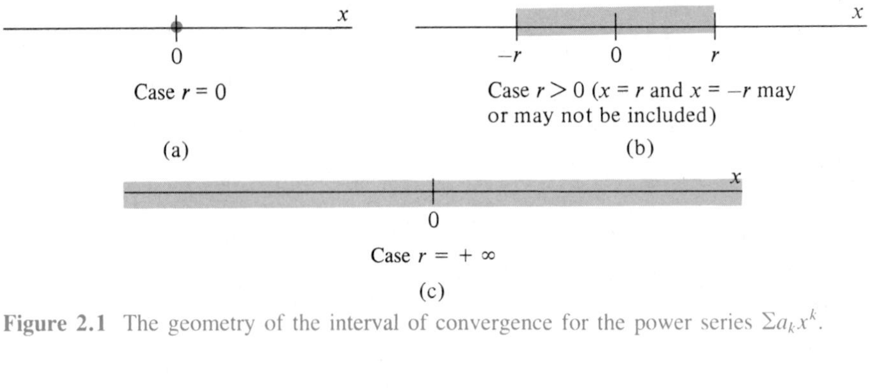

Figure 2.1 The geometry of the interval of convergence for the power series $\Sigma a_k x^k$.

Example 1

(a) The radius of convergence for the geometric power series Σx^k in equation (3) is $r = 1$, and the interval of convergence is $(-1, 1)$.

(b) The radius of convergence for the power series $\Sigma \dfrac{x^k}{k!}$ in equation (4) is $r = \infty$, and the interval of convergence is $(-\infty, \infty)$.

(c) To determine the radius of convergence of the power series $\Sigma \dfrac{(-1)^k}{1+k} x^k$ in equation (5) we test for absolute convergence using the Ratio Test. We found in equation (6) that $\rho = |x|$.

Thus, $\rho < 1$ if and only if $|x| < 1$, so $r = 1$ is the radius of convergence, and the power series converges absolutely for $|x| < 1$. We also checked the two values $x = r = 1$ and $x = -r = -1$, finding that the power series converges for $x = 1$ and diverges for $x = -1$. The interval of convergence for the power series $\displaystyle\sum_{k=0}^{\infty} \frac{(-1)^k}{1+k} x^k$ is therefore $(-1, 1]$. ■

Example 2 Find the radius and interval of convergence for the power series $\displaystyle\sum_{k=1}^{\infty} \frac{1}{k2^k} x^k$.

Strategy

First, find the radius of convergence, using the Ratio Test. (The Root Test will also work well in this example.)

Solution

We test for absolute convergence, using the Ratio Test:

$$\rho = \lim_{k\to\infty} \frac{\left|\dfrac{1}{(k+1)2^{k+1}} x^{k+1}\right|}{\left|\dfrac{1}{k2^k} \cdot x^k\right|}$$

$$= \lim_{k\to\infty} \left(\frac{k}{k+1}\right) \cdot \left(\frac{1}{2}\right) \cdot |x| = \frac{1}{2}|x|.$$

Thus $\rho < 1$ if $|x| < 2$. The radius of convergence is therefore $r = 2$.

To determine the interval of convergence check the endpoints $x = r = 2$ and $x = -r = -2$.

If $x = 2$ we obtain the harmonic series

$$\sum_{k=1}^{\infty} \frac{1}{k \cdot 2^k} (2^k) = \sum_{k=1}^{\infty} \frac{1}{k} = 1 + \frac{1}{2} + \frac{1}{3} + \cdots,$$

which diverges. If $x = -2$ we obtain the *alternating* harmonic series

$$\sum_{k=1}^{\infty} \frac{1}{k \cdot 2^k} (-2)^k = \sum_{k=1}^{\infty} \frac{(-1)^k}{k} = -1 + \frac{1}{2} - \frac{1}{3} + \cdots,$$

which converges. The interval of convergence is therefore $[-2, 2)$. ■

Example 3 Find the interval of convergence for the power series

$$\sum_{k=0}^{\infty} \frac{x^k}{3^k}.$$

Solution: Testing for absolute convergence using the Root Test we find that

$$\rho = \lim_{k\to\infty}\left\{\left|\frac{x^k}{3^k}\right|\right\}^{1/k} = \lim_{k\to\infty}\frac{|x|}{3} = \frac{|x|}{3}.$$

Thus, $\rho < 1$ if $|x| < 3$. The radius of convergence is $r = 3$. (The Ratio Test will also work well in obtaining this conclusion.)

If $x = 3$, the series becomes

$$\sum_{k=0}^{\infty}\frac{3^k}{3^k} = \sum_{k=0}^{\infty} 1 = 1 + 1 + 1 + \cdots,$$

which diverges. When $x = -3$, we obtain the divergent series

$$\sum_{k=0}^{\infty}\frac{(-3)^k}{3^k} = \sum_{k=0}^{\infty}(-1)^k = 1 - 1 + 1 - 1 + \cdots.$$

The interval of convergence is therefore $(-3, 3)$. ■

Example 4 The radius of convergence for the power series

$$\sum_{k=0}^{\infty} k!x^k$$

is $r = 0$ since, by the Ratio Test,

$$\rho = \lim_{k\to\infty}\frac{|(k+1)!x^{k+1}|}{|k!x^k|} = \lim_{k\to\infty}(k+1)|x| = \begin{cases}0, & x = 0\\ \infty, & x \neq 0\end{cases}$$

The "interval" of convergence is simply $\{0\}$. ■

The next example involves a power series of the form $\Sigma a_k(x - c)^k$. As we stated at the outset, we will use the change of variable $u = x - c$ to bring the power series into the form $\Sigma a_k u^k$. Note, however, that the interval of convergence will be centered at $x = c$ rather than at $x = 0$.

Example 5 Find the interval of convergence for the power series

$$\sum_{k=1}^{\infty}\frac{3^k}{k}(2x - 1)^k.$$

Strategy

Use the substitution $u = 2x - 1$ to bring power series to the form of equation (2). (The interval of convergence will therefore be centered about $x = 1/2$.) Apply Ratio Test to find radius of convergence in the variable u.

Solution

Letting $u = 2x - 1$ we obtain the series

$$\sum_{k=1}^{\infty}\frac{3^k}{k}u^k.$$

Testing this series for absolute convergence using the Ratio Test we find that

$$\rho = \lim_{k\to\infty}\frac{\left|\dfrac{3^{k+1}\cdot u^{k+1}}{(k+1)}\right|}{\left|\dfrac{3^k\cdot u^k}{k}\right|}$$

$$= \lim_{k\to\infty} 3\left(\frac{k}{k+1}\right)|u| = 3|u|.$$

Rewrite the inequality for $|u|$ in terms of the original variable x.

The series converges absolutely for $|u| < 1/3$ or for $|2x - 1| < 1/3$. This last inequality is equivalent to

$$-\frac{1}{3} < 2x - 1 < \frac{1}{3},$$

or

$$\frac{1}{3} < x < \frac{2}{3}.$$

Test the endpoints individually.

The radius of convergence (about $x = 1/2$) is therefore $r = 1/6$. At the endpoint $x = 2/3$ we obtain the series

$$\sum_{k=1}^{\infty} \frac{3^k}{k}\left(\frac{1}{3}\right)^k = \sum_{k=1}^{\infty} \frac{1}{k},$$

which diverges. At the endpoint $x = 1/3$ we obtain

$$\sum_{k=1}^{\infty} \frac{3^k}{k}\left(-\frac{1}{3}\right)^k = \sum_{k=1}^{\infty} \frac{(-1)^k}{k},$$

which converges. The interval of convergence is therefore $[1/3, 2/3)$. ■

Exercise Set 14.2

In Exercises 1–35, find the interval of convergence of the given power series.

1. $\sum_{k=0}^{\infty} \frac{x^k}{k+2}$
2. $\sum_{k=1}^{\infty} \frac{x^k}{2k}$
3. $\sum_{k=0}^{\infty} \frac{(-1)^{k+1}}{k!}x^k$
4. $\sum_{k=0}^{\infty} \frac{2^k x^k}{(k+1)!}$
5. $\sum_{k=1}^{\infty} \frac{(k^2+1)}{k!}x^k$
6. $\sum_{k=0}^{\infty} \frac{k \cdot x^k}{2^k}$
7. $\sum_{k=2}^{\infty} \frac{x^k}{\ln k}$
8. $\sum_{k=1}^{\infty} \frac{(-1)^k e^k}{k^2}x^k$
9. $\sum_{k=1}^{\infty} \frac{\cos \pi k}{1+k}x^k$
10. $\sum_{k=1}^{\infty} \frac{(2k+1)!}{2k!}x^{2k}$
11. $\sum_{k=1}^{\infty} \frac{(-1)^k}{k(k+1)}x^k$
12. $\sum_{k=1}^{\infty} k^2 c^k x^k$
13. $\sum_{k=1}^{\infty} \frac{(-1)^k}{k!}(x-3)^k$
14. $\sum_{k=1}^{\infty} \frac{k}{3^k}(x-\pi)^k$
15. $\sum_{k=0}^{\infty} k!(x-1)^k$
16. $\sum_{k=1}^{\infty} \frac{3^k}{k^2}(2x-1)^k$
17. $\sum_{k=2}^{\infty} \frac{(-1)^k}{\ln k}(3x-2)^k$
18. $\sum_{k=2}^{\infty} \frac{1}{(\ln k)^k}(x-1)^k$
19. $\sum_{k=1}^{\infty} \frac{1}{k+2}x^{2k+1}$
20. $\sum_{k=1}^{\infty} \frac{(x+2)^k}{(k+1)3^k}$
21. $\sum_{k=1}^{\infty} \frac{(x-3)^k}{k(k+1)}$
22. $\sum_{k=0}^{\infty} \frac{x^{2k+1}}{\pi^k}$
23. $\sum_{k=1}^{\infty} \frac{k(x-2)^k}{e^k}$
24. $\sum_{k=0}^{\infty} \frac{(2x+5)^k}{\sqrt{2k+8}}$
25. $\sum_{k=2}^{\infty} \frac{k(7x+1)^k}{2^k}$
26. $\sum_{k=1}^{\infty} \frac{(x-2)^k}{3^k \cdot k^2}$

27. $\sum_{k=3}^{\infty} kx^k$

28. $\sum_{k=1}^{\infty} \frac{x^k}{\ln(k+1)}$

29. $\sum_{k=1}^{\infty} \frac{(x-1)^k}{3^k\sqrt{k+1}}$

30. $\sum_{k=0}^{\infty} \frac{(2x-1)^k}{5^k}$

31. $\sum_{k=1}^{\infty} \frac{(k+4)x^k}{(k+1)(k+2)e^k}$

32. $\sum_{k=0}^{\infty} \frac{k^3(x-2)^k}{3^k}$

33. $\sum_{k=0}^{\infty} \frac{(-1)^k x^{2k+1}}{2(k+1)!}$

34. $\sum_{k=0}^{\infty} \frac{(-1)^k x^{2k}}{(2k)!}$

35. $\sum_{k=1}^{\infty} \frac{x^k}{k(k+1)}$

36. Show that the radius of convergence of the power series $\sum_{k=1}^{\infty} \frac{1}{k}(2x-3)^k$ is $r = 1/2$.

37. Prove that if the interval of convergence of the power series $\Sigma a_k x^k$ is $[-r, r)$, then the power series is conditionally convergent, but not absolutely convergent, for $x = -r$.

38. Prove that if the power series $\Sigma a_k x^k$ has radius of convergence r, then the power series $\Sigma a_k x^{ck}$ has radius of convergence $r^{1/c}$, $c > 0$.

39. Prove that if the power series $\Sigma a_k x^k$ has radius of convergence r_a and if the power series $\Sigma b_k x^k$ has radius of convergence r_b, then the series $\Sigma(a_k + b_k)x^k$ converges absolutely for $|x| < c$ where c is the smaller of r_a and r_b.

40. Show that the power series $\Sigma a_k(x-c)^k$ always converges at $x = c$.

14.3 DIFFERENTIATION AND INTEGRATION OF POWER SERIES

Within its interval of convergence, a power series represents a perfectly legitimate function of x. Thus, if the power series $\Sigma a_k x^k$ converges on the interval I_a, we may write

$$f(x) = \sum a_k x^k, \qquad x \in I_a$$

If $\Sigma b_k x^k$ is a second power series with interval of convergence I_b, then $g(x) = \Sigma b_k x^k$ is again a function of x, but $g(x)$ is defined on a (possibly) different interval than is $f(x)$. However, if the intervals I_a and I_b overlap (that is, if $I = I_a \cap I_b$ is not empty), we may form sums and differences of $f(x)$ and $g(x)$ as follows:

$$\begin{aligned}(f+g)(x) &= f(x) + g(x)\\ &= \sum a_k x^k + \sum b_k x^k = \sum (a_k + b_k)x^k, \qquad x \in I_a \cap I_b\end{aligned}$$

$$\begin{aligned}(f-g)(x) &= f(x) - g(x)\\ &= \sum a_k x^k - \sum b_k x^k = \sum (a_k - b_k)x^k, \qquad x \in I_a \cap I_b.\end{aligned}$$

(The concepts of multiplication and division for power series are more complicated and will not be pursued here.)

Example 1 The formula for the sum of a geometric series shows that the function $f(x) = \frac{1}{1-x}$ may be represented as a power series for $|x| < 1$:

$$\frac{1}{1-x} = \sum_{k=0}^{\infty} x^k = 1 + x + x^2 + x^3 + \cdots. \tag{1}$$

Multiplying both sides of equation (1) by x shows that

$$\frac{x}{1-x} = \sum_{k=0}^{\infty} x^{k+1} = x + x^2 + x^3 + x^4 + \cdots, \qquad |x| < 1. \tag{2}$$

Replacing x by $-x$ in (1) gives

$$\frac{1}{1+x} = \sum_{k=0}^{\infty} (-x)^k = 1 - x + x^2 - x^3 + \cdots, \qquad |x| < 1. \tag{3}$$

Adding equations (1) and (2) (which is allowed because their intervals of convergence are identical) shows that

$$\begin{aligned} \frac{1}{1-x} + \frac{1}{1+x} &= \frac{2}{(1-x)(1+x)} \\ &= \sum_{k=0}^{\infty} x^k + \sum_{k=0}^{\infty} (-x)^k = 2 + 2x^2 + 2x^4 + 2x^6 + \cdots \\ &= 2\sum_{k=0}^{\infty} x^{2k}, \qquad |x| < 1. \end{aligned}$$

However, the following calculation is *not* legitimate:

$$\frac{1}{1-\left(\frac{1}{x}\right)} = \frac{x}{x-1} = \sum_{k=0}^{\infty} \left(\frac{1}{x}\right)^k = 1 + \frac{1}{x} + \frac{1}{x^2} + \frac{1}{x^3} + \cdots, \tag{4}$$

so add (2) and (4) to obtain

$$\begin{aligned} \frac{x}{1-x} + \frac{x}{x-1} &= 0 \\ &= \cdots + x^{-3} + x^{-2} + x^{-1} + 1 + x + x^2 + x^3 + \cdots. \end{aligned} \tag{5}$$

Of course, equation (5) cannot hold for *any* values of x. The error in this calculation lies in the fact that the series in equation (4) converges only if $\left|\frac{1}{x}\right| < 1$, that is, if $|x| > 1$, while the series in equation (2) converges only if $|x| < 1$. There are no values of x for which both series converge.* ■

If you are beginning to suspect that within its interval of convergence the power series

$$f(x) = a_0 + a_1x + a_2x^2 + \cdots \tag{6}$$

behaves just like an "infinitely long polynomial," that is precisely the point we are trying to make. In fact, the following theorem shows that we may even differentiate expression (6) term by term, just as for polynomials, obtaining

$$f'(x) = a_1 + 2a_2x + 3a_3x^2 + \cdots,$$

which is the power series representation for the derivative of the function in equation (6). Its proof is given in more advanced courses.

*This absurdity is credited to Euler. See *An Introduction to the History of Mathematics,* 5th ed., by Howard Eves. Philadelphia: Saunders College Publishing, 1983, page 347.

THEOREM 3
Differentiation of Power Series

Suppose that the power series $\Sigma a_k x^k$ has a radius of convergence $r \neq 0$ and that the function $f(x)$ is defined to be its sum:

$$f(x) = \sum_{k=0}^{\infty} a_k x^k = a_0 + a_1 x + a_2 x^2 + a_3 x^3 + \cdots, \qquad |x| < r.$$

Then

(i) the function $f(x)$ is differentiable for $x \in (-r, r)$,

(ii) the power series $\sum_{k=0}^{\infty} k a_k x^{k-1}$ converges absolutely for each $x \in (-r, r)$, and

(iii) $f'(x) = \sum_{k=0}^{\infty} k a_k x^{k-1} = a_1 + 2a_2 x + 3a_3 x^2 + 4a_4 x^3 + \cdots, \qquad |x| < r.$

Example 2 Applying Theorem 3 to the geometric series

$$\frac{1}{1-x} = \sum_{k=0}^{\infty} x^k = 1 + x + x^2 + x^3 + \cdots, \qquad |x| < 1,$$

we conclude that

$$\frac{d}{dx}\left\{\frac{1}{1-x}\right\} = \frac{1}{(1-x)^2} = \sum_{k=0}^{\infty} kx^{k-1}$$
$$= 1 + 2x + 3x^2 + 4x^3 + \cdots, \qquad |x| < 1.$$

■

Example 3 Find a power series representation for the function

$$f(x) = \frac{x}{(1-x^2)^2}.$$

Solution: Since $f(x)$ does not resemble any of the functions for which we already know the power series representations, we look for another way to attack the problem. Since the numerator is a multiple of the derivative of the denominator, $f(x)$ is easily integrated, leading to the result $f(x) = \frac{1}{2}\frac{d}{dx}\left(\frac{1}{1-x^2}\right)$. Now we can recognize the expression in parentheses as the sum of the geometric series in x^2, which converges for $|x^2| < 1$. Then we have

$$\frac{1}{1-x^2} = \sum_{k=0}^{\infty} (x^2)^k = \sum_{k=0}^{\infty} x^{2k} = 1 + x^2 + x^4 + x^6 + \cdots, \qquad |x| < 1.$$

We conclude from Theorem 3 that

$$\frac{x}{(1-x^2)^2} = \frac{1}{2}\frac{d}{dx}\left\{\sum_{k=0}^{\infty} x^{2k}\right\} = \frac{1}{2}\sum_{k=0}^{\infty} 2kx^{2k-1} = x + 2x^3 + 3x^5 + \cdots.$$

This series converges absolutely for $|x| < 1$. ■

Example 4 The power series

$$\sum_{k=0}^{\infty} \frac{x^k}{k!} = 1 + x + \frac{x^2}{2!} + \frac{x^3}{3!} + \cdots$$

converges for all values of x, as noted in Example 1, Section 14.2. On differentiating this series we find that

$$\begin{aligned}\frac{d}{dx}\left\{\sum_{k=0}^{\infty} \frac{x^k}{k!}\right\} &= \frac{d}{dx}\left\{1 + x + \frac{x^2}{2!} + \frac{x^3}{3!} + \frac{x^4}{4!} + \cdots\right\}\\ &= \left\{1 + \frac{2x}{2!} + \frac{3x^2}{3!} + \frac{4x^3}{4!} + \cdots\right\}\\ &= \left\{1 + x + \frac{x^2}{2!} + \frac{x^3}{3!} + \cdots\right\}\\ &= \sum_{k=0}^{\infty} \frac{x^k}{k!}.\end{aligned}$$

That is, if $f(x) = \sum_{k=0}^{\infty} \frac{x^k}{k!}$, then $f'(x) = f(x)$. Recall that we have previously shown that the only function other than $f(x) \equiv 0$ that satisfies this differential equation is the function $f(x) = Ce^x$ (Exercise 34, Section 9.5). This shows that, for some constant C,

$$Ce^x = \sum_{k=0}^{\infty} \frac{x^k}{k!} = 1 + x + \frac{x^2}{2!} + \frac{x^3}{3!} + \cdots.$$

Setting $x = 0$ shows that $C = 1$, so we conclude that the power series representation for the function e^x is

$$e^x = \sum_{k=0}^{\infty} \frac{x^k}{k!} = 1 + x + \frac{x^2}{2!} + \frac{x^3}{3!} + \cdots$$

and that this representation is valid for all values of x. ■

The statement of Theorem 3 raises an obvious question about integrals of functions represented by power series. The answer is just what you might suspect.

THEOREM 4
Integration of Power Series

Suppose that the power series $\Sigma a_k x^k$ has radius of convergence $r \neq 0$ and that the function $f(x)$ is defined as its sum:

$$f(x) = \sum_{k=0}^{\infty} a_k x^k = a_0 + a_1 x + a_2 x^2 + a_3 x^3 + \cdots, \qquad |x| < r.$$

Then

(i) $\int f(x)\, dx$ exists for $x \in (-r, r)$,

(ii) the power series $\sum_{k=0}^{\infty}\left(\frac{a_k}{k+1}\right)x^{k+1}$ converges absolutely for each $x \in (-r, r)$, and

(iii) $$\int f(x)\,dx = \sum_{k=0}^{\infty}\left(\frac{a_k}{k+1}\right)x^{k+1} + C$$
$$= \left\{a_0x + \frac{a_1}{2}x^2 + \frac{a_2}{3}x^3 + \frac{a_3}{4}x^4 + \cdots\right\} + C, \qquad |x| < r.$$

In other words, a power series may be integrated term by term within its radius of convergence. The proof of Theorem 4 is left for more advanced courses. Note that, rather than writing a constant of integration for each term, we collect all constants into a single number C.

Example 5 Since $\ln(1 + x) = \int \frac{1}{1+x}\,dx$, we integrate equation (3) according to Theorem 4.

$$\ln(1 + x) = \sum_{k=0}^{\infty}\left\{\int (-x)^k\,dx\right\}$$
$$= \sum_{k=0}^{\infty} -\frac{(-x)^{k+1}}{k+1} + C$$
$$= \sum_{k=0}^{\infty} \frac{(-1)^k x^{k+1}}{k+1} + C$$
$$= \left\{x - \frac{x^2}{2} + \frac{x^3}{3} - \frac{x^4}{4} + \cdots\right\} + C, \qquad |x| < 1.$$

Setting $x = 0$ gives $\ln 1 = 0 = C$, so

$$\ln(1 + x) = \sum_{k=0}^{\infty} -\frac{(-x)^{k+1}}{k+1}$$
$$= x - \frac{x^2}{2} + \frac{x^3}{3} - \frac{x^4}{4} + \cdots, \qquad |x| < 1. \tag{7}$$

Equation (7) provides a practical means for calculating values of $\ln a$ for $0 < a < 2$. For example, to approximate $\ln(1.2)$ we set $x = 0.2$ and apply (7) to obtain

$$\ln(1.2) \approx .2 - \frac{(.2)^2}{2} + \frac{(.2)^3}{3} - \frac{(.2)^4}{4} = .182266$$

with an error of less than $\frac{.2^5}{5} = .000064$ (Theorem 16, Section 13.7). ■

Example 6 Replacing x by x^2 in equation (3) shows that

$$\frac{1}{1+x^2} = \sum_{k=0}^{\infty}(-x^2)^k = 1 - x^2 + x^4 - x^6 + \cdots, \qquad |x| < 1.$$

Since $\int \frac{1}{1+x^2}\,dx = \tan^{-1} x + C$, Theorem 4 gives

$$\tan^{-1} x = \sum_{k=0}^{\infty} \left\{ \int (-x^2)^k\,dx \right\} + C$$

$$= \sum_{k=0}^{\infty} \frac{(-1)^k}{2k+1} \cdot x^{2k+1} + C$$

$$= \left\{ x - \frac{x^3}{3} + \frac{x^5}{5} - \cdots \right\} + C, \qquad |x| < 1.$$

Setting $x = 0$ gives $\tan^{-1}(0) = 0 = C$. Thus

$$\tan^{-1} x = \sum_{k=0}^{\infty} \frac{(-1)^k}{2k+1} \cdot x^{2k+1} = x - \frac{x^3}{3} + \frac{x^5}{5} - \cdots.$$

■

Exercise Set 14.3

In Exercises 1–10, find a power series representation for the given function using equation (1). State the radius of convergence for the power series obtained.

1. $\frac{1}{1-2x}$
2. $\frac{x^2}{1-x}$
3. $\frac{1}{1+4x^2}$
4. $\frac{1}{1-9x^2}$
5. $\frac{x}{1+x^2}$
6. $\frac{x}{1-x^2}$
7. $\frac{x-1}{x+1}$
8. $\frac{x-1}{1+x^2}$
9. $\frac{1}{1-x^4}$
10. $\frac{x}{4+x^2}$

In Exercises 11–19, find a power series representation for the given function using Theorem 3. State the radius of convergence.

11. $f(x) = \frac{2}{(1+x)^2}$ $\left(\textit{Hint: } f(x) = -2\frac{d}{dx}\left(\frac{1}{1+x}\right).\right)$
12. $f(x) = \frac{2}{(1-x)^3}$ $\left(\textit{Hint: } f(x) = \frac{d^2}{dx^2}\left(\frac{1}{1-x}\right).\right)$
13. $f(x) = \frac{x}{(1+x^2)^2}$
14. $f(x) = \frac{1-x^2}{(1+x^2)^2}$ $\left(\textit{Hint: } f(x) = \frac{d}{dx}\left(\frac{x}{1+x^2}\right).\right)$
15. $f(x) = \frac{1+x^2}{(1-x^2)^2}$
16. $f(x) = \frac{1}{(1+4x)^2}$
17. $f(x) = \frac{8x}{(1+4x^2)^2}$ (*Hint:* Use Exercise 3.)
18. $f(x) = \frac{1+2x-x^2}{(1+x^2)^2}$ (*Hint:* Use Exercise 8.)
19. $f(x) = \frac{2}{(x+1)^2}$ (*Hint:* Use Exercise 7.)

In Exercises 20–27, find a power series representation for the given function using Theorem 4. State the radius of convergence.

20. $f(x) = \ln(1+x)$
21. $f(x) = \ln(1-x)$
22. $f(x) = x\ln(1+x)$
23. $f(x) = \tan^{-1}(2x)$
24. $x\tan^{-1} x$
25. $\ln(4+x)$
26. $\int \frac{dx}{1+x^4}$
27. $\ln(1+x^2)$

28. Use the results of Exercises 20 and 21 to show that

$$\ln\left(\frac{1+x}{1-x}\right) = 2\left(x + \frac{x^3}{3} + \frac{x^5}{5} + \cdots\right) \qquad |x| < 1.$$

29. For the power series

$$\sum_{k=0}^{\infty} (-1)^k \frac{x^{2k}}{(2k)!} = 1 - \frac{x^2}{2!} + \frac{x^4}{4!} - \frac{x^6}{6!} + \cdots$$

a. Use the Ratio Test to show that the series converges absolutely for all values of x.
b. Using Theorem 3, show that the function

$$f(x) = \sum_{k=0}^{\infty} (-1)^k \frac{x^{2k}}{(2k)!}$$

satisfies the differential equation $f''(x) = -f(x)$.
c. Show that $f(0) = 1$.
d. What function have you seen previously which satisfies both properties (b) and (c)?

30. For the power series

$$\sum_{k=0}^{\infty} (-1)^k \cdot \frac{x^{2k+1}}{(2k+1)!} = x - \frac{x^3}{3!} + \frac{x^5}{5!} - \frac{x^7}{7!} + \cdots :$$

a. Use the Ratio Test to show that the series converges absolutely for all values of x.
b. Using Theorem 3, show that the function $g(x) = \sum_{k=0}^{\infty} (-1)^k \frac{x^{2k+1}}{(2k+1)!}$ satisfies the differential equation $g''(x) = -g(x)$.
c. Show that $g(0) = 0$.
d. Show that $g'(x) = f(x)$ and $f'(x) = -g(x)$ where $f(x)$ is the function in Exercise 29. For what common function is $g(x)$ a power series representation?

31. Use the Chain Rule to show that if the power series

$$f(x) = \sum_{k=0}^{\infty} a_k(bx + c)^k$$

converges for $|bx + c| < r$, then

$$f'(x) = \sum_{k=0}^{\infty} kba_k(bx + c)^{k-1}, \qquad |bx + c| < r.$$

14.4 TAYLOR AND MACLAURIN SERIES

We are about to uncover a remarkable relationship between Taylor polynomials and power series, one that will provide a precise solution to the approximation problem posed in Chapter 12.

Recall that if the function $f(x)$ has $n + 1$ derivatives in an interval containing $x = c$, then the Taylor polynomial of degree n for $f(x)$ is

$$\begin{aligned} P_n(x) &= f(c) + f'(c)(x - c) \\ &\quad + \frac{f''(c)}{2!}(x - c)^2 + \cdots + \frac{f^{(n)}(c)}{n!}(x - c)^n \\ &= \sum_{k=0}^{n} \frac{f^{(k)}(c)}{k!}(x - c)^k. \end{aligned} \tag{1}$$

We refer to the coefficient $\frac{f^{(k)}(c)}{k!}$ of $(x - c)^k$ as the ***k*th Taylor coefficient.** Also, if we write

$$f(x) = P_n(x) + R_n(x), \tag{2}$$

the remainder term $R_n(x)$ is bounded as follows (Taylor's Theorem):

$$|R_n(x)| \le \left| \frac{f^{(n+1)}(d)}{(n+1)!}(x - c)^{n+1} \right| \tag{3}$$

where d lies between c and x.

Now note that the right-hand side of equation (1) is precisely the nth partial sum for the power series

$$\sum_{k=0}^{\infty} \frac{f^{(k)}(c)}{k!}(x-c)^k = f(c) + f'(c)(x-c) + \frac{f''(c)}{2!}(x-c)^2 + \frac{f'''(c)}{3!}(x-c)^3 + \cdots. \tag{4}$$

Series (4) is referred to as the **Taylor series** for $f(x)$, expanded about $x = c$. It is simply the result of allowing the Taylor polynomial $P_n(x)$ to become "infinitely long." In other words

$$\sum_{k=0}^{\infty} \frac{f^{(k)}(c)}{k!}(x-c)^k = \lim_{n\to\infty} \sum_{k=0}^{n} \frac{f^{(k)}(c)}{k!}(x-c)^k = \lim_{n\to\infty} P_n(x). \tag{5}$$

Of course, the expression in equation (4) makes sense only if the function $f(x)$ is infinitely differentiable. That is, the derivative $f^{(n)}(c)$ must exist for all integers $n = 1, 2, \ldots$ in order that all Taylor coefficients be defined.

The question of determining the values of x for which a Taylor series converges may be handled by use of the remainder term, $R_n(x)$. Applying $\lim_{n\to\infty}$ to both sides of equation (2) and using (5) we see that

$$\begin{aligned} f(x) &= \lim_{n\to\infty} P_n(x) + \lim_{n\to\infty} R_n(x) \\ &= \sum_{k=0}^{\infty} \frac{f^{(k)}(c)}{k!}(x-c)^k + \lim_{n\to\infty} R_n(x). \end{aligned}$$

This shows that *the Taylor series (4) converges to $f(x)$ if and only if* $\lim_{n\to\infty} R_n(x) = 0$.

Example 1 Find the Taylor series for the function $f(x) = \sin x$ expanded about $c = \pi/4$ and determine the values of x for which the series converges.

Strategy

Find the Taylor coefficients

$$\frac{f^{(k)}(c)}{k!}$$

with $c = \dfrac{\pi}{4}$.

Solution

We have

$$\begin{aligned} f(x) &= \sin x; & f\left(\frac{\pi}{4}\right) &= \frac{\sqrt{2}}{2} \\ f'(x) &= \cos x; & f'\left(\frac{\pi}{4}\right) &= \frac{\sqrt{2}}{2} \\ f''(x) &= -\sin x; & f''\left(\frac{\pi}{4}\right) &= -\frac{\sqrt{2}}{2} \\ f'''(x) &= -\cos x; & f'''\left(\frac{\pi}{4}\right) &= -\frac{\sqrt{2}}{2} \\ f^4(x) &= \sin x; & f^4\left(\frac{\pi}{4}\right) &= \frac{\sqrt{2}}{2} \\ &\vdots \end{aligned}$$

$$f^{(2k)}(x) = (-1)^k \sin x; \qquad f^{(2k)}\left(\frac{\pi}{4}\right) = (-1)^k \frac{\sqrt{2}}{2}$$

$$f^{(2k+1)}(x) = (-1)^k \cos x; \qquad f^{(2k+1)}\left(\frac{\pi}{4}\right) = (-1)^k \frac{\sqrt{2}}{2}$$

$$\vdots$$

Insert the Taylor coefficients and $c = \pi/4$ in (4) to obtain the Taylor series.

The Taylor series is therefore

$$\sin x = \frac{\sqrt{2}}{2} + \frac{\sqrt{2}}{2}\left(x - \frac{\pi}{4}\right) - \frac{\sqrt{2}}{2 \cdot 2!}\left(x - \frac{\pi}{4}\right)^2 - \frac{\sqrt{2}}{2 \cdot 3!}\left(x - \frac{\pi}{4}\right)^3 + \frac{\sqrt{2}}{2 \cdot 4!}\left(x - \frac{\pi}{4}\right)^4 + \cdots.$$

Determine where $\lim_{n\to\infty} |R_n(x)| = 0$ to find the values of x for which the series converges.

Use fact that $\lim_{n\to\infty} \frac{x^n}{n!} = 0$ (limit (2), Section 13.3).

Since $f^{(n+1)}(d)$ is either $\pm \sin d$ or $\pm \cos d$, we know that $|f^{(n+1)}(d)| \le 1$ for all values of n. Thus

$$\begin{aligned}\lim_{n\to\infty} |R_n(x)| &= \lim_{n\to\infty} \left| \frac{f^{(n+1)}(d)}{(n+1)!}\left(x - \frac{\pi}{4}\right)^{n+1} \right| \\ &\le \lim_{n\to\infty} \frac{\left| x - \frac{\pi}{4} \right|^{n+1}}{(n+1)!} \\ &= 0\end{aligned}$$

for all values of x. Thus the series converges for all values of x. ■

The special case of a Taylor series expanded about $x = 0$ is referred to as a **Maclaurin series:**

$$\sum_{k=0}^{\infty} \frac{f^{(k)}(0)}{k!} x^k = f(0) + f'(0)x + \frac{f''(0)}{2!}x^2 + \cdots + \frac{f^{(k)}(0)}{k!}x^k + \cdots. \quad (6)$$

Example 2 Find a Maclaurin series for $f(x) = e^x$ and determine the values of x for which it converges.

Solution: Every derivative of $f(x) = e^x$ is the same:

$$\begin{aligned} f(x) &= e^x; \qquad f(0) = 1 \\ f'(x) &= e^x; \qquad f'(0) = 1 \\ &\vdots \\ f^{(n)}(x) &= e^x; \qquad f^{(n)}(0) = 1 \\ &\vdots \end{aligned}$$

Thus

$$\begin{aligned} e^x &= 1 + x + \frac{x^2}{2!} + \frac{x^3}{3!} + \cdots + \frac{x^k}{k!} + \cdots \\ &= \sum_{k=0}^{\infty} \frac{x^k}{k!}. \end{aligned}$$

Here, for all values of n,

$$|f^{n+1}(d)| = |e^d| = e^d \le e^{|d|} \le e^{|x|}$$

since $|d| \le |x|$, so

$$\begin{aligned}\lim_{n\to\infty} |R_n(x)| &= \lim_{n\to\infty} \left| \frac{e^d}{(n+1)!} x^{n+1} \right| \\ &\le \lim_{n\to\infty} e^{|x|} \frac{|x|^{n+1}}{(n+1)!} \\ &= 0\end{aligned}$$

for all values of x. The Maclaurin series for e^x therefore converges to e^x for all values of x. (Of course, we found in Section 14.2 that the power series converges for all x.) ■

Example 3 Find the Maclaurin series for $f(x) = \ln(1 + x)$.

Solution: Here

$$\begin{aligned}
f(x) &= \ln(1+x); & f(0) &= 0\\
f'(x) &= (1+x)^{-1}; & f'(0) &= 1\\
f''(x) &= -(1+x)^{-2}; & f''(0) &= -1\\
f'''(x) &= 2(1+x)^{-3}; & f'''(0) &= 2\\
&\vdots\\
f^{(k)}(x) &= (-1)^{k+1}(k-1)!(1+x)^{-k}; & f^{(k)}(0) &= (-1)^{k+1}(k-1)!\\
&\vdots
\end{aligned}$$

Using (6) we obtain

$$\begin{aligned}\ln(1+x) &= 0 + 1 \cdot x - \frac{1}{2!}x^2 + \frac{2}{3!}x^3 - \cdots \\ &\quad + \frac{(-1)^{k+1}(k-1)!}{k!}x^k + \cdots \\ &= x - \frac{x^2}{2} + \frac{x^3}{3} - \frac{x^4}{4} + \cdots + \frac{(-1)^{k+1}}{k}x^k + \cdots \\ &= \sum_{k=1}^{\infty} \frac{(-1)^{k+1}}{k} x^k.\end{aligned}$$

This is the power series that we showed to converge to $\ln(1 + x)$ for $|x| < 1$ in Example 5 of Section 14.3. ■

Up to this point, we have seen two ways to represent functions by infinite series. In Section 14.3 we were able to obtain power series expansions for certain functions by using the formula for the sum of a geometric series, together with term by term differentiation or integration. In this section, we have obtained an explicit formula for the Taylor (or Maclaurin) series of a given function. Obviously, every Taylor or Maclaurin series is also a power series. We will now show that the converse is also true. To do so we assume that $f(x)$ is represented as a power series with radius of convergence $r > 0$:

$$f(x) = a_0 + a_1 + a_2x^2 + \cdots + a_kx^k + \cdots. \tag{7}$$

Using Theorem 3, we may repeatedly differentiate both sides of (7), obtaining

$$\begin{aligned}
f'(x) &= a_1 + 2a_2x + 3a_3x^2 + \cdots + ka_kx^{k-1} + \cdots \\
f''(x) &= 2a_2 + 2\cdot 3a_3x + \cdots + (k-1)ka_kx^{k-2} + \cdots \\
&\vdots \\
f^{(k)}(x) &= k!a_k + (k+1)!a_{k+1}x + \frac{(k+2)!}{2!}a_{k+2}x^2 + \cdots.
\end{aligned} \tag{8}$$

Setting $x = 0$ in each equation causes all terms containing powers of x to vanish, so

$$\begin{aligned}
f(0) &= a_0, & a_0 &= \frac{f(0)}{0!} \\
f'(0) &= a_1, & a_1 &= \frac{f'(0)}{1!} \\
f''(0) &= 2a_2, & a_2 &= \frac{f''(0)}{2!} \\
&\vdots & &\vdots \\
f^{(k)}(0) &= k!a_k, & a_k &= \frac{f^{(k)}(0)}{k!}.
\end{aligned}$$

In other words, if $f(x)$ is represented by a convergent power series, as in equation (7), then the coefficients a_k must be precisely the Maclaurin coefficients

$$a_k = \frac{f^{(k)}(0)}{k!}.$$

If we replace x by $x - c$ in equation (7) and evaluate the derivatives at $x = c$, we prove the following theorem for Taylor series.

THEOREM 5 If the function $f(x)$ can be represented by a power series of the form

$$f(x) = a_0 + a_1(x-c) + a_2(x-c)^2 + \cdots + a_k(x-c)^k + \cdots$$

$$= \sum_{k=0}^{\infty} a_k(x-c)^k$$

with a radius of convergence $r > 0$, then the coefficients a_k in the power series expansion must be the Taylor coefficients

$$a_k = \frac{f^{(k)}(c)}{k!}, \qquad k = 0, 1, 2, \ldots.$$

Example 4 In Exercise 2 you are asked to obtain the Maclaurin series expansion

$$\sin x = x - \frac{x^3}{3!} + \frac{x^5}{5!} - \frac{x^7}{7!} + \cdots + \frac{(-1)^k x^{2k+1}}{(2k+1)!} + \cdots$$

and to show that it is valid for all values of x. We may therefore differentiate this series term by term, to obtain a power series for $\cos x = \dfrac{d}{dx} \sin x$:

$$\cos x = 1 - \frac{3x^2}{3!} + \frac{5x^4}{5!} - \frac{7x^6}{7!} + \cdots + \frac{(-1)^k(2k+1)x^{2k}}{(2k+1)!} + \cdots$$

$$= 1 - \frac{x^2}{2!} + \frac{x^4}{4!} - \frac{x^6}{6!} + \cdots + \frac{(-1)^k x^{2k}}{(2k)!} + \cdots,$$

which is also valid for all x. Theorem 5 then assures that this series is, in fact, the Maclaurin series for $f(x) = \cos x$. ■

Example 5 Find a Maclaurin series for $x^3 e^{x^2}$.

Strategy

Do *not* calculate Taylor coefficients for $f(x) = x^3 e^{x^2}$. Work directly with the Maclaurin series for e^x and substitute x^2 for x.

Solution

Replacing x by x^2 in the series for e^x in Example 2 shows that

$$e^{x^2} = 1 + x^2 + \frac{x^4}{2!} + \frac{x^6}{3!} + \cdots = \sum_{k=0}^{\infty} \frac{x^{2k}}{k!},$$

which converges for all x. Thus

Multiply by x^3.

$$x^3 e^{x^2} = x^3\left\{1 + x^2 + \frac{x^4}{2!} + \frac{x^6}{3!} + \cdots\right\}$$

$$= x^3 + x^5 + \frac{x^7}{2!} + \frac{x^9}{3!} + \cdots$$

$$= \sum_{k=0}^{\infty} \frac{x^{2k+3}}{k!}$$

converges for all values of x. By Theorem 5 this must be the Maclaurin series for $x^3 e^{x^2}$. ■

The following examples show how Taylor and Maclaurin series may be used to calculate certain constants and integrals.

Example 6 Since the Maclaurin series

$$e^x = \sum_{k=0}^{\infty} \frac{x^k}{k!} = 1 + x + \frac{x^2}{2!} + \frac{x^3}{3!} + \cdots$$

converges for all x, we may use it to approximate e^a for any value of a. Setting $x = 1$ gives

$$e = 1 + 1 + \frac{1}{2!} + \frac{1}{3!} + \cdots + \frac{1}{k!} + \cdots$$

$$= \sum_{k=0}^{\infty} \frac{1}{k!},$$

and setting $x = -1$, we obtain

$$\frac{1}{e} = 1 - 1 + \frac{1}{2!} - \frac{1}{3!} + \frac{1}{4!} - \cdots + \frac{(-1)^k}{k!} + \cdots$$

$$= \sum_{k=0}^{\infty} \frac{(-1)^k}{k!}.$$

Since this last series is an alternating series, we may approximate $\frac{1}{e}$ by, say,

$$\frac{1}{e} \approx \frac{1}{2!} - \frac{1}{3!} + \frac{1}{4!} - \frac{1}{5!} = .3667$$

with accuracy no worse than $\frac{1}{6!} = .0014$. ■

Example 7 We may approximate the integral $\int_0^1 e^{-x^2}\,dx$ using Maclaurin series as follows. Since

$$e^x = 1 + x + \frac{x^2}{2!} + \frac{x^3}{3!} + \cdots + \frac{x^k}{k!} + \cdots$$

converges for all x, the series

$$e^{-x^2} = 1 - x^2 + \frac{x^4}{2!} - \frac{x^6}{3!} + \cdots + \frac{(-x^2)^k}{k!} + \cdots$$

also converges for all x. By Theorem 4, an antiderivative for e^{-x^2} is represented as

$$\int e^{-x^2}\,dx = \int 1\,dx - \int x^2\,dx + \int \frac{x^4}{2!}\,dx - \int \frac{x^6}{3!}\,dx + \cdots + C$$

$$= x - \frac{x^3}{3} + \frac{x^5}{5 \cdot 2!} - \frac{x^7}{7 \cdot 3!} + \cdots + \frac{(-1)^k x^{2k+1}}{(2k+1)k!} + \cdots + C.$$

Thus,

$$\int_0^1 e^{-x^2}\,dx = 1 - \frac{1}{3} + \frac{1}{5 \cdot 2!} - \frac{1}{7 \cdot 3!} + \cdots + \frac{(-1)^k}{(2k+1)k!} + \cdots.$$

Since this is an alternating series, we may approximate this series to any desired degree of accuracy by taking sufficiently many terms. For example, to obtain accuracy of .01, we find by trial and error that for $k = 4$,

$$\frac{1}{(2 \cdot 4 + 1)4!} = \frac{1}{9 \cdot 24} = \frac{1}{216} < \frac{1}{100}.$$

Thus, we use terms up through $k = 3$ to obtain

$$\int_0^1 e^{-x^2}\,dx \approx 1 - \frac{1}{3} + \frac{1}{5 \cdot 2!} - \frac{1}{7 \cdot 3!} = .74,$$

accurate to within .01. ■

The Binomial Series

According to the Binomial Theorem, we know that

$$(1 + x)^n = 1 + nx + \frac{n(n-1)}{2}x^2 + \cdots + \binom{n}{k}x^k + \cdots + nx^{n-1} + x^n \tag{9}$$

when n is a positive integer. In writing (9) we have used the notation $\binom{n}{k}$ for the **binomial coefficient**

$$\binom{n}{k} = \frac{n!}{k!(n-k)!} = \frac{n(n-1)(n-2)\cdot\cdots\cdot(n-k+1)}{k!}.$$

(See Exercise 24, Section 12.3.)

We can extend equation (9) to the case where n is not a positive integer. However, the result involves an infinite series, rather than a polynomial in x. The result is called the **Binomial Series:**

$$\begin{aligned}(1 + x)^r &= 1 + rx + \frac{r(r-1)}{2}x^2 + \frac{r(r-1)(r-2)}{3!}x^3 + \cdots \\ &= 1 + \sum_{k=1}^{\infty} \frac{r(r-1)(r-2)\cdot\cdots\cdot(r-k+1)}{k!}x^k, \qquad |x| < 1.\end{aligned} \tag{10}$$

(Note that the infinite series in (10) reduces to the polynomial in (9) when $r = n$ is a positive integer, since the term $(r - k + 1)$ becomes zero when k reaches $r + 1$.)

Proving the validity of (10) involves showing that the right-hand side is precisely the Maclaurin series for $f(x) = (1 + x)^r$, and that the radius of convergence for this series is one. To do so we note that

$$\begin{aligned}
f(x) &= (1 + x)^r \Rightarrow f(0) = 1 \\
f'(x) &= r(1 + x)^{r-1} \Rightarrow f'(0) = r \\
f''(x) &= r(r-1)(1 + x)^{r-2} \Rightarrow f''(0) = r(r-1) \\
&\vdots \\
f^{(k)}(x) &= r(r-1)\cdot\cdots\cdot(r-k+1)(1 + x)^{r-k} \\
&\Rightarrow f^{(k)}(0) = r(r-1)\cdot\cdots\cdot(r-k+1)
\end{aligned}$$

Thus, the kth Maclaurin coefficient for $f(x) = (1 + x)^r$ is indeed

$$\frac{f^{(k)}(0)}{k!} = \frac{r(r-1)(r-2)\cdot\cdots\cdot(r-k+1)}{k!}$$

as in the Binomial Series (10).

To verify that the series converges for $|x| < 1$, we let a_k denote the kth term

$$a_k = \frac{r(r-1)\cdot\cdots\cdot(r-k+1)}{k!}x^k.$$

Then

$$\rho = \lim_{k\to\infty}\left|\frac{a_{k+1}}{a_k}\right| = \lim_{k\to\infty}\left|\frac{r(r-1)\cdot\cdots\cdot(r-k)}{r(r-1)\cdot\cdots\cdot(r-k+1)}\cdot\frac{k!}{(k+1)!}\cdot\frac{x^{k+1}}{x^k}\right|$$

$$= \lim_{k\to\infty} \left| \frac{(r-k)}{1} \cdot \frac{1}{(k+1)} \cdot x \right|$$
$$= \lim_{k\to\infty} \left| \frac{r-k}{k+1} \right| \cdot |x|$$
$$= |x|.$$

Thus, if $|x| < 1$ we have $\rho < 1$, and the series converges absolutely by the Ratio Test.

Example 8 The Binomial Series, with $r = \frac{1}{2}$, gives

$$\sqrt{1+x} = (1+x)^{1/2} = 1 + \frac{1}{2}x + \frac{\frac{1}{2}\left(\frac{1}{2}-1\right)}{2}x^2 + \frac{\frac{1}{2}\left(\frac{1}{2}-1\right)\left(\frac{1}{2}-2\right)}{3!}x^3$$
$$+ \frac{\frac{1}{2}\left(\frac{1}{2}-1\right)\left(\frac{1}{2}-2\right)\left(\frac{1}{2}-3\right)}{4!}x^4 + \cdots$$
$$= 1 + \frac{1}{2}x - \frac{1}{8}x^2 + \frac{1}{16}x^3 - \frac{5}{128}x^4 + \cdots, \qquad |x| < 1. \qquad \blacksquare$$

Example 9 Replacing x by x^3 in Example 8 gives

$$\sqrt{1+x^3} = (1+x^3)^{1/2}$$
$$= 1 + \frac{1}{2}x^3 - \frac{1}{8}x^6 + \frac{1}{16}x^9 - \frac{5}{128}x^{12} + \cdots, \quad |x| < 1. \qquad \blacksquare$$

Example 10 With $r = -\frac{1}{3}$ and $-x$ in place of x, (10) gives

$$\frac{1}{\sqrt[3]{1-x}} = [1+(-x)]^{-1/3} = 1 + \left(-\frac{1}{3}\right)(-x) + \frac{\left(-\frac{1}{3}\right)\left(-\frac{1}{3}-1\right)}{2}(-x)^2$$
$$+ \frac{\left(-\frac{1}{3}\right)\left(-\frac{1}{3}-1\right)\left(-\frac{1}{3}-2\right)}{3!}(-x)^3$$
$$+ \frac{\left(-\frac{1}{3}\right)\left(-\frac{1}{3}-1\right)\left(-\frac{1}{3}-2\right)\left(-\frac{1}{3}-3\right)}{4!}(-x)^4 + \cdots$$
$$= 1 + \frac{1}{3}x + \frac{2}{9}x^2 + \frac{14}{81}x^3 + \frac{35}{243}x^4 + \cdots, \qquad |x| < 1. \qquad \blacksquare$$

Summary

In answer to the approximation problem posed in Chapter 12, we have found that

(1) If the function $f(x)$ has derivatives of all orders in a neighborhood of $x = c$, we may approximate $f(x)$ (for a particular value of x) by using the Taylor polynomials

$$P_n(x) = \sum_{k=0}^{n} \frac{f^{(k)}(c)}{k!}(x - c).$$

(2) The accuracy in this approximation is given by the remainder

$$R_n(x) = f(x) - P_n(x) = \frac{f^{n+1}(d)}{(n+1)!}(x - c)^{n+1}$$

where d lies between c and x.

(3) If $\lim_{n\to\infty} R_n(x) = 0$, the Taylor series converges to the function $f(x)$.

(4) If $f(x)$ is represented by any convergent power series with a radius of convergence $r > 0$, then this power series is, in fact, the Taylor series for $f(x)$.

(5) Within its radius of convergence, a Taylor (power) series for $f(x)$ may be differentiated or integrated term by term to obtain the Taylor series expansion for $f'(x)$ or $\int f(x)\,dx$, respectively.

Exercise Set 14.4

In Exercises 1–18, find the Taylor or Maclaurin series for the given function expanded about the given point and determine the values of x for which the series converges.

1. $f(x) = e^{2x}$, $c = 0$

2. $f(x) = \sin x$, $c = 0$

3. $f(x) = \cos x$, $c = \pi/4$

4. $f(x) = \sin x$, $c = \pi/6$

5. $f(x) = 1 + x^2$, $c = 2$

6. $f(x) = \dfrac{1}{1 + x}$, $c = 0$

7. $f(x) = \ln(3 + x)$, $c = 0$

8. $f(x) = x \sin 2x$, $c = 0$

9. $f(x) = 2^x$, $c = 0$

10. $f(x) = (1 + x)^n$, $c = 0$

11. $f(x) = (1 + x)^{3/2}$, $c = 0$

12. $f(x) = \sqrt{x}$, $c = 4$

13. $f(x) = \sqrt{x + 1}$, $c = 0$

14. $f(x) = \dfrac{1}{x}$, $c = 2$

15. $f(x) = \dfrac{\sin x}{x}$, $c = 0$

16. $f(x) = x^2 e^{-x}$, $c = 0$

17. $f(x) = x \sin x$, $c = \pi/4$

18. $f(x) = x^2 \ln(1 + x)$, $c = 0$

19. Find a Maclaurin series for $f(x) = \cos^2 x$ by use of the identity $\cos^2 x = \dfrac{1}{2}[1 + \cos 2x]$.

20. Find the first four terms of the Maclaurin series for $f(x) = \tan x$.

21. Find the first three terms of the Maclaurin series for $f(x) = \sec^2 x$, using the result of Exercise 20.

22. Find the Maclaurin series for $x \sin x^2$ using the Maclaurin series for $\sin x$.

23. Find a Maclaurin series for $\cos x^2$ using the technique of Exercise 22.

24. Find the Maclaurin series for $f(x) = \sinh x$ from the Maclaurin series for e^x and e^{-x}.

25. Find the Maclaurin series for $x \cosh x$.

26. Find the Maclaurin series for $f(x) = \sin^2 x$ (see Exercise 19).

In Exercises 27–32, use Taylor series to approximate the given quantity accurately to three decimal places.

27. $\sin 2°$

28. e^2

29. $\ln(1.1)$

30. $\int_0^{\pi/4} \sin x^2\, dx$

31. $\int_0^{1/2} \frac{1}{1+x^3}\, dx$

32. $\int_0^1 e^{-x^2}\, dx$

In Exercises 33–36, use the Binomial Series to find a power series for the given function and the radius of convergence.

33. $f(x) = \sqrt{1+2x}$

34. $f(x) = \sqrt[3]{27+x}$

35. $f(x) = (9+3x)^{3/2}$

36. $f(x) = \frac{1}{\sqrt[3]{8-x^2}}$

37. Define the function $f(x)$ by

$$f(x) = \begin{cases} e^{-(1/x^2)}, & x \neq 0 \\ 0, & x = 0 \end{cases}.$$

a. Show that $f^{(n)}(0)$ exists and equals zero for every integer $n \geq 1$.

b. Show that the Maclaurin series for $f(x)$ is identically equal to zero for all values of x. Thus the function $f(x)$ does not equal its Maclaurin series in any interval containing zero. (*Remark:* This example shows that the existance of all derivatives is not a sufficient condition for a function to equal its Taylor series. The condition that $\lim_{n\to\infty} R_n(x) = 0$ is essential.)

38. Prove that power series representations are unique. That is, prove that if

$$f(x) = \sum_{k=0}^{\infty} a_k x^k \quad \text{and} \quad f(x) = \sum_{k=0}^{\infty} b_k x^k,$$

then $\quad a_0 = b_0,\ a_1 = b_1,\ \ldots,\ a_k = b_k,\ \ldots.$

(*Hint:* Set $x = 0$ to obtain $a_0 = b_0$. Then differentiate and set $x = 0$ again, and so on.)

14.5 POWER SERIES SOLUTIONS OF DIFFERENTIAL EQUATIONS (Optional)

The uniqueness of power series representations (see Exercise 38, Section 14.4) provides a method for solving certain types of differential equations. For differential equations of the form

$$f(x, y, y') = 0 \tag{1}$$

the idea is to express the (unknown) solution y as a power series in the independent variable x:

$$y = \sum_{k=0}^{\infty} a_k x^k. \tag{2}$$

Then, assuming (2) has a nonzero radius of convergence, we apply Theorem 3 to conclude that

$$y' = \sum_{k=0}^{\infty} k a_k x^{k-1}. \tag{3}$$

We next insert representations (2) and (3) into the differential equation (1). For each integer k, coefficients of x^k will appear in various places in equation (1). By equating coefficients of x^k on either side of equation (1) we can often find a set of equations which allow the coefficients $a_0, a_1, a_2, a_3, \ldots$ to be determined completely. In such cases the power series representation for the solution is obtained.

Example 1 Use the power series method to solve the differential equation

$$y' = y. \tag{4}$$

Solution: We assume that $y = \sum_{k=0}^{\infty} a_k x^k$, so that $y' = \sum_{k=0}^{\infty} k a_k x^{k-1}$, and we assume that both of these series have (the same) nonzero radius of convergence. Inserting these expansions in the differential equation (4) gives

$$\sum_{k=0}^{\infty} ka_k x^{k-1} = \sum_{k=0}^{\infty} a_k x^k$$

or

$$\begin{aligned} a_1 + 2a_2x + 3a_3x^2 + \cdots + ka_kx^{k-1} + \cdots \\ = a_0 + a_1x + a_2x^2 + \cdots + a_{k-1}x^{k-1} + \cdots. \end{aligned} \tag{5}$$

Since the two series in equation (5) are equal, the coefficients of like powers of x must be the same (Exercise 38, Section 14.4). Thus,

$$\begin{aligned} a_1 &= a_0 && \text{(constant terms)} \\ 2a_2 &= a_1 && (x \text{ terms}) \\ 3a_3 &= a_2 && (x^2 \text{ terms}) \\ &\vdots \\ ka_k &= a_{k-1} && (x^{k-1} \text{ terms}) \\ &\vdots \end{aligned}$$

Using these equations, we may solve for each coefficient $a_2, a_3, a_4, \ldots$ in terms of the one preceding and, therefore, in terms of a_0:

$$\begin{aligned} a_1 &= a_0 \\ a_2 &= \frac{1}{2}a_1 = \frac{1}{2}a_0 = \frac{1}{2!}a_0 \\ a_3 &= \frac{1}{3}a_2 = \frac{1}{3\cdot 2}a_0 = \frac{1}{3!}a_0 \\ &\vdots \\ a_k &= \frac{1}{k}a_{k-1} = \frac{1}{k(k-1)\cdot\cdots\cdot 3\cdot 2}a_0 = \frac{1}{k!}a_0. \end{aligned} \tag{6}$$

Returning to the expansion for y, we can now write

$$\begin{aligned} y &= a_0 + a_0x + \frac{1}{2!}a_0x^2 + \frac{1}{3!}a_0x^3 + \cdots + \frac{1}{k!}a_0x^k + \cdots \\ &= a_0\left\{1 + x + \frac{x^2}{2!} + \frac{x^3}{3!} + \cdots + \frac{x^k}{k!} + \cdots\right\}, \end{aligned} \tag{7}$$

which we recognize as the power series representation for the function $y = a_0e^x$. Since we have previously verified that this power series converges for all values of x, the solution $y = a_0e^x$ is valid for all x. The constant a_0 is determined by an initial condition for the equation (4). ■

REMARK: The result of Example 1 should not have been unexpected, since we had determined in Chapter 9 that the solution of the differential equation $y' = y$ is $y = Ce^x$. However, we see here that the use of power series provides an entirely different approach to solving differential equations.

The equation

$$a_k = \frac{1}{k}a_{k-1}$$

on the left side of equation (6) warrants special attention. It is referred to as the **recurrence relation** for the differential equation because it gives the general formula by which each coefficient in the series expansion for the solution (except the first) may be determined from its predecessor. If you can succeed in finding a recurrence relation, you will be able to generate all coefficients in the expansion for the solution of a differential equation. However, two issues then remain to be addressed:

(1) Does the power series obtained from the recurrence relation actually converge? If so, for which values of x? You will have succeeded in finding a legitimate solution of the differential equation only if the power series has a nonzero radius of convergence.
(2) Can a closed form expression* for the power series solution be recognized, as in Example 1 where the solution was recognized as the power series for a_0e^x? The answer to this question is not critical if you are willing to accept solutions in the form of power series. In fact, many important functions in mathematical physics arise as series solutions of differential equations and do not have closed form expressions (see Example 3).

Example 2 Use the power method to solve the initial value problem

$$\begin{cases} y' = xy \\ y(0) = 1. \end{cases}$$

Strategy

Assume a power series form for the solution y. Differentiate to find series form for y'.

Solution

We assume that

$$y = \sum_{k=0}^{\infty} a_k x^k, \qquad \text{so} \qquad y' = \sum_{k=0}^{\infty} k a_k x^{k-1}.$$

Insert the expansions for y and y' into the differential equation.

This gives

$$\sum_{k=0}^{\infty} k a_k x^{k-1} = x \sum_{k=0}^{\infty} a_k x^k$$

$$= \sum_{k=0}^{\infty} a_k x^{k+1}$$

Write out the first few terms, including the general terms on both sides *for the same power of* x. (Here we arbitrarily picked x^{k-1}.)

or,

$$a_1 + 2a_2x + 3a_3x^2 + \cdots + ka_kx^{k-1} + \cdots$$
$$= a_0x + a_1x^2 + a_2x^3 + \cdots + a_{k-2}x^{k-1} + \cdots$$

Equate coefficients of like powers of x.

so

$$a_1 = 0 \qquad \text{(constant terms)}$$

$$a_2 = \frac{1}{2}a_0 \qquad (x \text{ terms})$$

*A closed form expression is one which does not involve an infinite process, such as summation. $f(x) = \dfrac{1}{1-x}$ is expressed in closed form, while $f(x) = \sum_{k=0}^{\infty} x^k$ is not.

$$a_3 = \frac{1}{3}a_1 = 0 \qquad (x^2 \text{ terms})$$

$$\vdots$$

The recurrence relation is obtained by equating coefficients of x^{k-1}.

$$a_k = \frac{1}{k}a_{k-2} \qquad (x^{k-1} \text{ terms})$$

$$\vdots$$

Determine the form of all coefficients using the recurrence relation.

From the recurrence relation $a_k = \dfrac{1}{k}a_{k-2}$ and the fact that $a_1 = 0$, we see that all odd coefficients are zero: $0 = a_1 = a_3 = a_5 = \cdots$.

Also we can see that the even coefficients are

$$a_2 = \frac{1}{2}a_0$$

$$a_4 = \frac{1}{4}a_2 = \frac{1}{4\cdot 2}a_0 = \frac{1}{2^2}\left(\frac{1}{2\cdot 1}\right)a_0 = \frac{1}{2^2}\cdot\frac{1}{2!}a_0$$

$$a_6 = \frac{1}{6}a_4 = \frac{1}{6\cdot 4\cdot 2}a_0 = \frac{1}{2^3}\left(\frac{1}{3\cdot 2\cdot 1}\right)a_0 = \frac{1}{2^3}\cdot\frac{1}{3!}a_0$$

$$\vdots$$

$$a_{2k} = \frac{1}{2k}a_{2k-2} = \frac{1}{2k(2k-2)\cdot\cdots\cdot 2}a_0 = \frac{1}{2^k}\cdot\frac{1}{k!}a_0.$$

Write the series for y using coefficients found above.

The power series for the solution y is therefore

Try to bring the series for y into the form of a known power series. (Here we use the fact that

$$e^u = \sum_{k=0}^{\infty}\frac{u^k}{k!}$$

for all values of u.)

$$y = a_0 + a_1x + a_2x^2 + a_3x^3 + a_4x^4 + \cdots$$

$$= a_0 + \frac{1}{2}a_0x^2 + \frac{1}{2^2}\cdot\frac{1}{2!}a_0x^4 + \frac{1}{2^3}\cdot\frac{1}{3!}a_0x^6 + \cdots + \frac{1}{2^k}\cdot\frac{1}{k!}a_0x^{2k} + \cdots$$

$$= a_0\left\{1 + \frac{x^2}{2} + \frac{\left(\frac{x^2}{2}\right)^2}{2!} + \frac{\left(\frac{x^2}{2}\right)^3}{3!} + \cdots + \frac{\left(\frac{x^2}{2}\right)^k}{k!} + \cdots\right\}$$

$$= a_0\sum_{k=0}^{\infty}\frac{(x^2/2)^k}{k!}.$$

Apply initial condition to determine a_0.

This is the power series expansion for $y = a_0e^{x^2/2}$, which converges for all values of x. The general solution for the differential equation $y' - xy = 0$ is therefore $y = a_0e^{x^2/2}$. The constant a_0 is determined from the initial condition:

$$1 = y(0) = a_0e^0 = a_0.$$

The solution of the initial value problem is $y = e^{x^2/2}$. ■

Power series methods may be used in higher order differential equations as well. Note in the following example that the power series for y must be differentiated twice since the differential equation is of order 2.

Example 3 Find a power series solution for the differential equation

$$xy'' - y = 0.$$

Solution: We shall work with the equation in the form

$$xy'' = y. \tag{8}$$

We assume the solution y to have the form

$$y = \sum_{k=0}^{\infty} a_k x^k.$$

Then

$$y' = \sum_{k=0}^{\infty} k a_k x^{k-1}, \qquad \text{and} \qquad y'' = \sum_{k=0}^{\infty} k(k-1) a_k x^{k-2}.$$

Equation (8) becomes

$$x \sum_{k=0}^{\infty} k(k-1) a_k x^{k-2} = \sum_{k=0}^{\infty} a_k x^k,$$

or

$$\begin{aligned} 2a_2 x + 3\cdot 2a_3 x^2 + 4\cdot 3\cdot a_4 x^3 + \cdots + k(k-1)a_k x^{k-1} + \cdots \\ = a_0 + a_1 x + a_2 x^2 + a_3 x^3 + \cdots + a_{k-1} x^{k-1} + \cdots. \end{aligned}$$

Thus

$$\begin{aligned} a_0 &= 0 \\ 2a_2 &= a_1 \Rightarrow a_2 = \frac{1}{2} a_1 \\ 3\cdot 2\cdot a_3 &= a_2 \Rightarrow a_3 = \frac{1}{3\cdot 2} a_2 = \frac{1}{3\cdot 2^2} a_1 = \frac{1}{3(2!)^2} a_1 \\ 4\cdot 3\cdot a_4 &= a_3 \Rightarrow a_4 = \frac{1}{4\cdot 3} a_3 = \frac{1}{4\cdot 3^2\cdot 2^2} a_1 = \frac{1}{4(3!)^2}\cdot a_1 \\ &\vdots \\ k(k-1)a_k &= a_{k-1} \Rightarrow a_k = \frac{1}{k(k-1)} a_{k-1} = \cdots = \frac{1}{k[(k-1)!]^2}\cdot a_1 \end{aligned}$$

The solution y is therefore

$$y = a_1\left[x + \frac{1}{2}x^2 + \frac{1}{3\cdot 2^2}x^3 + \frac{1}{4\cdot (3!)^2}x^4 + \cdots + \frac{1}{k[(k-1)!]^2}x^k + \cdots\right]. \tag{9}$$

To determine the radius of convergence for (9) we apply the Ratio Test.

$$\rho = \lim_{k\to\infty} \left| \frac{\dfrac{a_1}{(k+1)(k!)^2} x^{k+1}}{\dfrac{a_1}{k[(k-1)!]^2} x^k} \right| = \lim_{k\to\infty} \left(\frac{k}{k+1}\right)\left(\frac{1}{k^2}\right)|x| = 0$$

for all values of x, so the series in (9) converges for all x. We have therefore obtained a legitimate solution for the differential equation (8), although the solution is expressed in power series form and is not immediately recognizable as the power series for a known function in closed form. ■

As you might suspect at this point, the theory associated with the use of power series in solving differential equations is not simple. In fact, the examples we have presented here were carefully chosen to convey the basic idea while avoiding the difficulties surrounding this method. These difficulties include the following:

(i) It is often possible only to obtain the first few terms of the series solution rather than the general term as we have succeeded in doing in these examples. This may leave the question of convergence unanswerable, although in many cases one can obtain at least an approximation to the solution.

(ii) The assumption that the solution y has a power series representation contains the implicit assumption that y has derivatives of all orders. Since the solution to a differential equation of degree n need have only n derivatives, the power series approach will necessarily fail to identify solutions to certain equations since it assumes too much.

Further work on power series methods in differential equations is left to more specialized courses. Should you take such a course you will find that the material in Chapters 12 through 14 constitutes an important foundation on which much of the work of that course depends.

Exercise Set 14.5

In Exercises 1–8, use the method of this section to find a power series form of the solution of the differential equation or initial value problem. Check your work by solving the equation by the method of separation of variables.

1. $y' + y = 0$
2. $y' + 2y = 0$
3. $y' - 6y = 0$
4. $y' + xy = 0$
5. $y' - 2xy = 0$
6. $y' + ax = 0$
7. $\begin{cases} y' + 4y = 0 \\ y(0) = 1 \end{cases}$
8. $\begin{cases} y' + xy = 0 \\ y(0) = 2 \end{cases}$
9. Find a power series solution for the second order differential equation $xy'' + y = 0$.
10. Find a power series solution for the initial value problem

$$\begin{cases} y'' + y = 0 \\ y(0) = 0 \\ y'(0) = 1. \end{cases}$$

(*Hint:* Let $y = \sum_{k=0}^{\infty} a_k x^k$. Since $y(0) = 0$, $a_0 = 0$. This observation simplifies the recurrence relation.)

11. Find a power series solution for the initial value problem

$$\begin{cases} y'' + y = 0 \\ y(0) = 1 \\ y'(0) = 0. \end{cases}$$

12. Conclude from Exercises 10 and 11 that the function $y = A \sin x + B \cos x$ is a solution of the differential equation $y'' + y = 0$.

14.6 THE COMPLEX FUNCTION $e^{i\theta}$ (Optional)

In this section we see how ideas about power series may be used to complete the discussion concerning solutions of the second order linear differential equation

$$y'' + Ay' + By = 0 \tag{1}$$

where A and B are constants. Recall from Section 9.7 that if a solution of (1) is assumed to have the form $y = e^{rx}$ we obtain, on inserting this function in (1), the equation

$$r^2e^{rx} + Are^{rx} + Be^{rx} = e^{rx}[r^2 + Ar + B] = 0. \tag{2}$$

Since e^{rx} is never zero, equation (2), hence equation (1), can be true only if r is a root of the characteristic equation

$$r^2 + Ar + B = 0. \tag{3}$$

We have previously determined that

(i) If (3) has two distinct roots, r_1 and r_2, equation (1) has solutions $y_1 = e^{r_1x}$ and $y_2 = e^{r_2x}$.
(ii) If (3) has a single repeated real root, r_1, equation (1) has solutions $y_1 = e^{r_1x}$ and $y_2 = xe^{r_1x}$.

But what happens when (3) has a complex* root $r = a + ib$? A theorem (which we shall not prove) assures us that if $r_1 = a + ib$ is a root of (3), so is its conjugate $r_2 = a - ib$. That is, either (3) has two distinct complex roots or else all roots of (3) are real.

Because solutions of (1) have the form e^{rx} when r is real, we are led to ask whether it makes sense to try to consider the function $f(x) = e^{(a+ib)x}$ as a solution of (1) when the characteristic equation (3) has the complex root $r = a + ib$. If we can succeed in doing this in any reasonable way, the resulting laws of exponents should give

$$e^{rx} = e^{(a+ib)x} = e^{ax+ibx} = e^{ax}e^{ibx}. \tag{4}$$

Since a and b are real numbers, the function e^{ax} on the right side of (4) is well understood. It is the second factor e^{ibx} that we need to define. Also, since $e^{ibx} = (e^{ix})^b$, we will deal first only with the function $g(x) = e^{ix}$, where $i^2 = -1$.

To define the function $g(x) = e^{ix}$ we recall the power series expansion for the function e^t:

$$e^t = 1 + t + \frac{t^2}{2} + \frac{t^3}{3!} + \frac{t^4}{4!} + \cdots + \frac{t^k}{k!} + \cdots = \sum_{k=0}^{\infty} \frac{t^k}{k!}. \tag{5}$$

We formally *define* e^{ix} by this same power series:

$$e^{ix} = 1 + ix + \frac{(ix)^2}{2} + \frac{(ix)^3}{3!} + \frac{(ix)^4}{4!} + \cdots + \frac{(ix)^k}{k!} + \cdots \tag{6}$$

$$= \sum_{k=0}^{\infty} \frac{(ix)^k}{k!}.$$

However, equation (6) really has no meaning yet. We have defined power series only for real variables, so equation (5) is defined only when t is a real number. This is not to say that equation (6) should be discarded. Rather, we will try to define what is meant by equation (6) in a way that extends equation (5) to the complex case. We do so as follows.

Notice that since $i^2 = -1$, $i^4 = (i^2)^2 = (-1)^2 = 1$, $i^6 = (-1)^3 = -1$, $i^8 = (-1)^4 = 1$, and so on, the coefficients of even powers of x in equation (6) are actually real numbers. In fact,

---

* An introduction to complex numbers is presented in Appendix III. The ideas contained in that appendix are assumed in this discussion.

$$\begin{aligned} 1 &+ \frac{(ix)^2}{2} + \frac{(ix)^4}{4!} + \frac{(ix)^6}{6!} + \cdots + \frac{(ix)^{2k}}{(2k)!} + \cdots \\ &= 1 + \frac{i^2x^2}{2} + \frac{i^4x^4}{4!} + \frac{i^6x^6}{6!} + \cdots + \frac{i^{2k}x^{2k}}{(2k)!} + \cdots \\ &= 1 - \frac{x^2}{2} + \frac{x^4}{4!} - \frac{x^6}{6!} + \cdots + \frac{(-1)^k x^{2k}}{(2k)!} + \cdots. \end{aligned}$$

Since this is precisely the Maclaurin series for $\cos x$, we see that *the even terms in (6) sum to $\cos x$ for all x.*

Similarly, since $i^3 = i(i^2) = -i$, $i^5 = i(i^4) = i$, $i^7 = i(i^6) = -i$, and so on, the terms involving the odd powers of x in equation (6) are

$$\begin{aligned} ix &+ \frac{i^3x^3}{3!} + \frac{i^5x^5}{5!} + \cdots + \frac{i^{2k+1}x^{2k+1}}{(2k+1)!} + \cdots \\ &= ix - \frac{ix^3}{3!} + \frac{ix^5}{5!} - \cdots + \frac{(-1)^k i x^{2k+1}}{(2k+1)!} + \cdots \\ &= i\left(x - \frac{x^3}{3!} + \frac{x^5}{5!} - \cdots + \frac{(-1)^k x^{2k+1}}{(2k+1)!} + \cdots\right). \end{aligned}$$

Now the expression in parentheses above is just the Maclaurin series for $\sin x$, so *the odd terms in (6) sum to $i \cdot \sin x$ for all x.*

To summarize, our observations suggest that

$$\begin{aligned} e^{ix} &= \sum_{k=0}^{\infty} \frac{(ix)^k}{k!} = \sum_{k=0}^{\infty} \frac{(-1)^k x^{2k}}{(2k)!} + i \sum_{k=0}^{\infty} \frac{(-1)^k x^{2k+1}}{(2k+1)!} \\ &= \cos x + i \sin x. \end{aligned}$$

We therefore *define* the complex function e^{ix} by

$$e^{ix} = \cos x + i \sin x, \qquad i^2 = -1, \qquad -\infty < x < \infty. \tag{7}$$

This definition is usually referred to as Euler's formula. We next make a further definition, one which extends Euler's formula to include the desired equation (4).

DEFINITION 1

Let a and b be real numbers. Let $i^2 = -1$ and let x be a real variable. Let r be the complex number $r = a + ib$. The **complex exponential function** e^{rx} is defined by

$$e^{rx} = e^{(a+ib)x} = e^{ax}e^{ibx} = e^{ax}(\cos bx + i \sin bx).$$

With this definition it can be shown (Exercise 23) that all the laws of algebra hold for the complex exponential function:

(a) $e^{r_1x} \cdot e^{r_2x} = e^{(r_1+r_2)x}$

(b) $\dfrac{e^{r_1x}}{e^{r_2x}} = e^{(r_1-r_2)x}$

(c) $[e^{rx}]^n = e^{nrx}$.

Finally, we need to extend the idea of differentiation to complex functions. We do this as follows.

DEFINITION 2

Let $u(x)$ and $v(x)$ be differentiable functions of the real variable x. The derivative of the complex function

$$f(x) = u(x) + iv(x)$$

is

$$f'(x) = u'(x) + iv'(x).$$

With this definition, we can differentiate the complex exponential function $e^{rx} = e^{(a+ib)x}$:

$$\begin{aligned}
\frac{d}{dx}e^{rx} = \frac{d}{dx}e^{(a+ib)x} &= \frac{d}{dx}[e^{ax}(\cos bx + i\sin bx)] \\
&= \frac{d}{dx}[e^{ax}\cos bx + ie^{ax}\sin bx] \\
&= \frac{d}{dx}(e^{ax}\cos bx) + i\frac{d}{dx}(e^{ax}\sin bx) \\
&= (ae^{ax}\cos bx - be^{ax}\sin bx) \\
&\quad + i(ae^{ax}\sin bx + be^{ax}\cos bx) \\
&= (a + ib)(e^{ax}\cos bx + ie^{ax}\sin bx) \\
&= re^{rx}.
\end{aligned}$$

That is,

$$\frac{d}{dx}e^{rx} = re^{rx}, \qquad r = a + ib. \tag{8}$$

Equation (8) directly answers the question raised at the beginning of this discussion. If $r_1 = a + ib$ is a complex root of the characteristic polynomial (3), then so is $r_2 = a - ib$, and both $y_1 = e^{r_1x} = e^{(a+ib)x}$ and $y_2 = e^{r_2x} = e^{(a-ib)x}$ are solutions of the differential equation (1). To verify this, simply note that

$$y_1'' + ay_1' + by_1 = (r_1^2 + ar_1 + b)e^{r_1x} = 0,$$

and similarly for y_2, as before.

However, $y_1 = e^{r_1x} = e^{(a+ib)x}$ and $y_2 = e^{r_2x} = e^{(a-ib)x}$ are not very convenient solutions to (1) since both are complex-valued. Happily, we can generate two real-valued solutions from y_1 and y_2 as follows. Notice that

$$\begin{aligned}
\frac{y_1 + y_2}{2} &= \frac{e^{ax}[\cos bx + i\sin bx] + e^{ax}[\cos(-bx) + i\sin(-bx)]}{2} \\
&= \frac{e^{ax}[\cos bx + i\sin bx] + e^{ax}[\cos bx - i\sin bx]}{2} \\
&= e^{ax}\cos bx.
\end{aligned}$$

Similarly,

$$\begin{aligned}
\frac{y_1 - y_2}{2i} &= \frac{e^{ax}[\cos bx + i\sin bx] - e^{ax}[\cos bx - i\sin bx]}{2i} \\
&= e^{ax}\sin bx.
\end{aligned}$$

Since we have previously seen that sums and multiples of solutions of equation

(1) are again solutions, the functions

$$z_1 = e^{ax}\cos bx, \qquad z_2 = e^{ax}\sin bx \tag{9}$$

must be solutions of (1) when $r = a + ib$ is a complex root of the characteristic equation (3). You can verify this by substituting z_1 and z_2 directly into equation (1).

Example 1 Find two solutions of the differential equation

$$y'' + 4y' + 8y = 0.$$

Solution: The characteristic equation (3) is

$$r^2 + 4r + 8 = 0,$$

which, by the quadratic formula, has roots

$$r = \frac{-4 \pm \sqrt{16 - 4\cdot 8}}{2} = -2 \pm 2i.$$

The complex roots are therefore $r_1 = -2 + 2i$ and $r_2 = -2 - 2i$. By (9) the two solutions are

$$y_1 = e^{-2x}\cos(2x) \qquad \text{and} \qquad y_2 = e^{-2x}\sin(2x).$$ ■

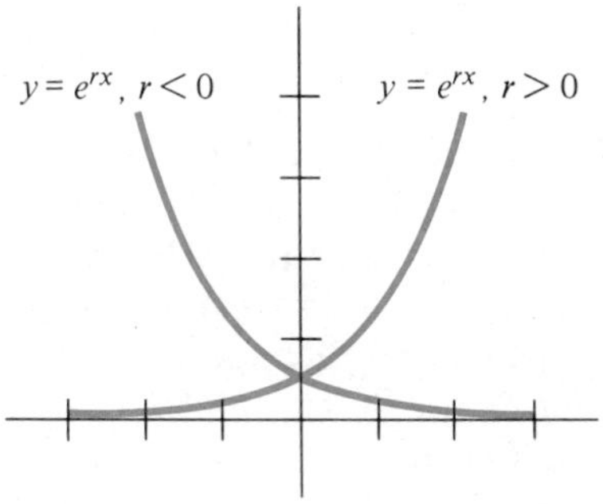

Figure 6.1 Case i.

Combining the findings of Section 9.7 with those above allows us to give a complete description of the real solutions of the differential equation

$$y'' + Ay' + By = 0.$$

Case i: If the characteristic equation (3) has distinct real roots r_1 and r_2, two solutions are

$$y_1 = e^{r_1x}, \qquad y_2 = e^{r_2x}$$

and, in general,

$$y = C_1e^{r_1x} + C_2e^{r_2x}, \qquad C_1, C_2 \text{ constants} \tag{10}$$

is a solution (Figure 6.1).

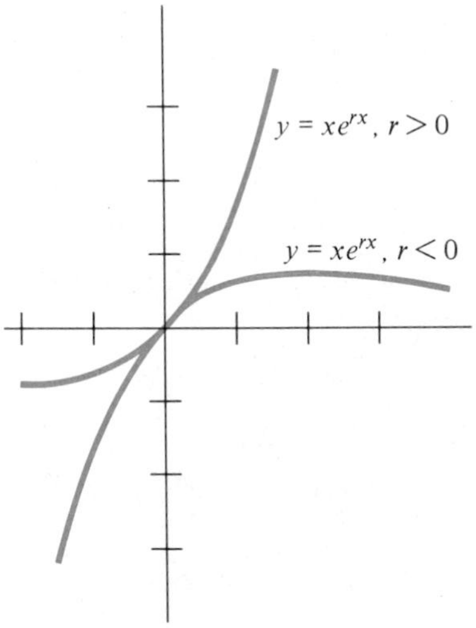

Figure 6.2 Case ii.

Case ii: If the characteristic equation has the repeated real root r, two solutions are

$$y_1 = e^{rx} \qquad \text{and} \qquad y_2 = xe^{rx}$$

and, in general,

$$y = C_1e^{rx} + C_2xe^{rx} \tag{11}$$

is a solution (Figure 6.2).

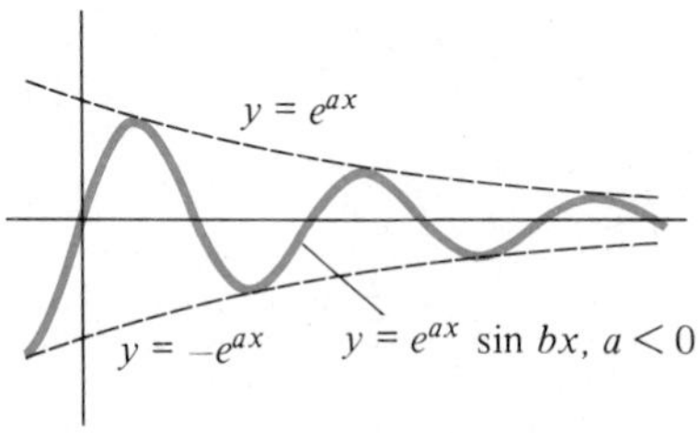

Figure 6.3 Case iii.

Case iii: If the characteristic equation has the complex root $a + ib$, then $a - ib$ is also a root, two solutions are

$$y_1 = e^{ax}\cos bx \qquad \text{and} \qquad y_2 = e^{ax}\sin bx,$$

and, in general,

$$y = e^{ax}[C_1\cos bx + C_2\sin bx] \tag{12}$$

is a solution (Figure 6.3).

Example 2 Find the general form of solutions of the differential equation

$$y'' + y = 0.$$

Solution: Here the characteristic equation is

$$r^2 + 1 = 0$$

so the roots are $r = \pm\sqrt{-1} = \pm i$. By (12) the general form of solution is

$$\begin{aligned} y &= e^0[A \cos x + B \sin x] \\ &= A \cos x + B \sin x \end{aligned}$$

as we have previously determined. ■

We could have simply stated at the outset that the general form for solutions of the differential equation (1) is that given by expression (12) when the characteristic equation has complex roots. However, the whole point of this section is to see how the notion of power series representations for the functions e^x, $\sin x$ and $\cos x$ led to the important Definition 1. In fact, this notion of power series representations was the key idea in the extension of the exponential function to the case of the complex exponent $(a + ib)x$. Functions that have power series representations, called *analytic* functions, play a very important role in higher mathematics. In fact, much of the theory of functions of a complex variable is developed by use of the power series concept as we have done here in a very limited setting. Power series, therefore, are important in the development of mathematical theory as well as in dealing with the more applied ideas presented in Chapters 12 and 14.

Exercise Set 14.6

In Exercises 1–6, find all real or complex roots of the given polynomial.

1. $r^2 + 4r + 7 = 0$

2. $r^2 - 2r + 2 = 0$

3. $r^2 - 6r + 13 = 0$

4. $r^2 + 9 = 0$

5. $r^3 - 4r^2 - 5r = 0$

6. $r^4 - 16 = 0$

In Exercises 7–20, find the general form of solutions to the differential equation.

7. $y'' + 2y' - 3y = 0$

8. $y'' - y' = 0$

9. $y'' + 4y' + 7y = 0$

10. $y'' - 2y' + 2y = 0$

11. $y'' + 3y' - 10y = 0$

12. $y'' - 6y' + 13y = 0$

13. $y'' + 9y = 0$

14. $y'' + 4y' + 4y = 0$

15. $y'' - 6y' + 9y = 0$

16. $y'' + 6y' + 14y = 0$

17. $y'' - 4y' + 4y = 0$

18. $y'' - 9y = 0$

19. $y'' - 2y' + 5y = 0$

20. $y'' = 0$

21. For which values of a will all solutions of the differential equation $y'' + ay = 0$ remain bounded as $x \to \infty$?

22. Show that all solutions of the differential equation $y'' + 2y' + y = 0$ approach zero as $x \to \infty$.

23. For the complex exponential function $e^{rx} = e^{(a+ib)x}$ prove that

a. $e^{r_1x} \cdot e^{r_2x} = e^{(r_1+r_2)x}$.

b. $\dfrac{e^{r_1x}}{e^{r_2x}} = e^{(r_1-r_2)x}$.

c. $[e^{rx}]^n = e^{nrx}$.

d. $\dfrac{d}{dx}[e^{r_1x} + e^{r_2x}] = \dfrac{d}{dx}e^{r_1x} + \dfrac{d}{dx}e^{r_2x}$.

e. $\dfrac{d}{dx}[ce^{r_1x}] = c\dfrac{d}{dx}e^{r_1x}$.

24. Use $e^{ix} = \cos x + i \sin x$ to find a relation between e, π, and i.

SUMMARY OUTLINE OF CHAPTER 14

- A **power series** has the form $\sum_{k=0}^{\infty} a_k(x-c)^k$.
- ***Theorem:*** If the power series $\Sigma a_k x^k$ converges for $x = c$, it does so for all x with $|x| < |c|$.
- ***Theorem:*** The set of all x for which a power series converges is either (i) $\{c\}$, (ii) an interval with midpoint $x = c$ and radius $r \neq 0$, or (iii) $(-\infty, \infty)$. (The radius r is called the **radius of convergence.**)
- ***Theorem:*** If $f(x) = \Sigma a_k x^k$ with radius of convergence r, then $f'(x)$ exists, and $f'(x) = \Sigma k a_k x^{k-1}$ with radius of convergence r.
- ***Theorem:*** If $f(x) = \Sigma a_k x^k$ with radius of convergence r, then $\int f(x)\,dx = \Sigma\left(\frac{a_k}{k+1}\right)x^{k+1} + C$ with radius of convergence r.
- A **Taylor series** is a power series of the form

$$\sum_{k=0}^{\infty} \frac{f^{(k)}(c)}{k!}(x-c)^k = f(c) + f'(c)(x-c) + \frac{f''(c)}{2!}(x-c)^2 + \cdots + \frac{f^{(k)}(c)}{k!}(x-c)^k + \cdots.$$

- If $c = 0$, this series is called a **Maclaurin series.**
- The **binomial series** is

$$(1+x)^r = 1 + \sum_{k=1}^{\infty} \frac{r(r-1)(r-2)\cdot\cdots\cdot(r-k+1)}{k!}x^k,$$

$|x| < 1$.

- A differential equation can sometimes be solved by assuming a power series representation for the solution, inserting it into the equation, and solving for successive coefficients of x^k.
- The complex exponential function e^{rx} is

$$e^{rx} = e^{(a+ib)x} = e^{ax}(\cos bx + i \sin bx), \qquad r = a + ib.$$

It is a solution of the differential equation

$$y'' + Ay' + By = 0$$

when r is a root of the characteristic polynomial $r^2 + Ar + B = 0$.

REVIEW EXERCISES—CHAPTER 14

In Exercises 1–6, find the interval of convergence for the given series.

1. $\sum k(x-3)^k$

2. $\sum (-1)^k k^2 (x-1)^k$

3. $\sum \frac{\sqrt{k}(x-2)^k}{k+3}$

4. $\sum \frac{(x+2)^k}{2^k}$

5. $\sum \frac{(-1)^k x^k}{k \ln k}$

6. $\sum \frac{\ln k(x-1)^k}{e^k}$

7. Show that $\sum_{k=0}^{\infty} a_k x^{k+1} = \sum_{k=1}^{\infty} a_{k-1}x^k$.

8. Show that $\sum_{k=1}^{\infty} k(k+1)x^k = \sum_{k=2}^{\infty} (k-1)kx^{k-1}$.

9. Find numbers $a_1, a_2, \ldots$ so that

$$\sum na_n x^{n-1} + c\sum a_n x^n = 0, \qquad c \neq 0.$$

What is the function $f(x)$ whose power series is $\sum a_k x^k$?

10. Find the numbers $a_2, a_3, a_4, \ldots$ so that

$$\sum_{n=2}^{\infty} n(n-1)a_n x^{n-2} + 4\sum_{n=0}^{\infty} a_n x^n = 0.$$

11. Find a power series representation for the function

$$f(x) = \frac{1}{1+x^4}.$$

12. Find a Maclaurin series for the function $f(x) = \sin x \cos x$.

13. Find a Maclaurin series for $f(x) = \sqrt[3]{1+x^2}$.

14. Find a power series for the function $f(x) = \frac{1}{(1-x)(2-x)}$ using the geometric series. What is the radius of convergence?

15. Find the first 3 terms of the Taylor series for $f(x) = \sin\sqrt{x}$ expanded about $x = \pi^2/4$.

16. What is the radius of convergence of a power series for $e^{\sqrt{x}}$ expanded about $a = 0$?

17. Find a Maclaurin series for $f(x) = x^2 e^{x^2}$ using the series for e^x.

18. Find the Taylor series for $f(x) = \sqrt{x + 4}$ expanded about $a = 2$. For what values of x does it converge?

19. Use a Taylor series to approximate $\sqrt{e}$ accurate to three decimal places.

20. Use a power series to find a solution to the initial value problem

$$\begin{cases} y' - xy = 0 \\ y(0) = 2. \end{cases}$$

21. Find a power series solution for the second order differential equation $xy'' - y = 0$.

22. Find a power series solution for the initial value problem

$$\begin{cases} xy'' + y = 0 \\ y(0) = 0 \\ y'(0) = 1 \end{cases}$$

Why is the condition $y(0) = 0$ *necessary?*

23. Find a Maclaurin series for $f(x) = x - x^3$.

24. Find the general form of solutions to the differential equation

$$y'' - 4y' + 6y = 0.$$

25. Find the solution of the initial value problem

$$\begin{cases} y'' - 4y' + 8y = 0 \\ y(0) = 0 \\ y'(0) = 2. \end{cases}$$

René Descartes

Pierre de Fermat

UNIT 6

GEOMETRY IN THE PLANE AND IN SPACE

Geometry in the Plane and in Space

René Descartes (1596–1650) and Pierre de Fermat (1601–1665) had at least two things in common: both were French, and neither was a professional mathematician. They shared something else of far greater importance—each independently discovered analytic geometry.

Descartes left school at the age of 16 and went to Paris, where he only occasionally studied mathematics. At 21 he became a soldier-for-pay, hiring out to the armies of various nations and minor political subdivisions. He alternated military service with study and travel throughout Europe, and met a number of mathematicians. His soldiering was apparently successful; he was offered a commission as a lieutenant general, but declined it. After nine years of this life, his growing interest in matters of philosophy led him to the relative peace of the Netherlands, where he studied and wrote for twenty years.

On the night of November 10, 1619, Descartes had three dreams that profoundly affected his life. He claimed that, in one of these dreams, he was given the key to understanding nature. Although he never revealed precisely what this was, it is widely believed that he was referring to the relationship between algebra and geometry. This may therefore be said to be the founding date of analytic geometry, though publication was not to occur for 18 years.

In 1637 Descartes published *Discours de la méthode,* or *Discourse on the Method of Rightly Conducting the Reason and Seeking Truth in the Sciences*. This was a program for conducting philosophical research—an attempt to reach valid conclusions in many fields by systematic reasoning. The *Discours* included three appendices. Each was a brilliant illustration of the application of the method to a different field. The first, *La dioptrique*, included the first published statement of the law of refraction [although the Dutch mathematician Willebrord Snell (1591–1626) had discovered it earlier]. The second appendix, on meteorology, included an explanation of the colors of the rainbow. The third, and most famous, was *La géométrie*. Starting with an ancient problem from the Greek mathematician Pappus (third century A.D.), Descartes applied algebra to geometry and vice versa. This was not quite today's analytic geometry, however. Coordinate axes were implied but not used explicitly, and were thought of as oblique rather than the perpendicular coordinates that we now label Cartesian. The notions of distance, angle between lines, and slope were not included. In no case was a curve plotted from an equation. Descartes did not consider negative numbers or negative coordinates. Only one equation was considered in detail, the general second degree equation discussed in Chapter 15. Descartes gave the conditions on the coefficients of the equation required for the curve to be an ellipse, parabola, or hyperbola. The details were omitted, and the entire work was difficult to read. Indeed, readers often found it difficult to see how the appendices were related to the philosophy set forth in the main text. In spite of these inadequacies, the concepts of analytic geometry were introduced in this little work, the only book on mathematics Descartes ever produced. It was a striking achievement.

Descartes was lured away from the Netherlands to a position with Queen Christina of Sweden, who was interested in philosophy and wanted him to establish an academy of sciences. Unfortunately, she wanted her tutoring in philosophy at 5:00 a.m., which conflicted with Descartes' lifelong practice of late rising. He was unable to maintain his health in the cold, damp Swedish climate, and died of pneumonia at 54 years of age.

Pierre de Fermat was a lawyer, but he loved mathematics and pursued a variety of mathematical areas as a hobby. His work in analytic geometry was done several years before that of Descartes, but it was not published until 1679, fourteen years after Fermat's death and fifty years after its completion. Descartes wrote in more modern notation than Fermat, but the latter's ideas more closely resemble modern mathematics. In his rather short work, *Isogoge ad locus planos et solidos,* or *Introduction to Plane and Solid Loci,* Fermat presented in archaic terminology an astonishing amount of material that is familiar to us today. His linear equation written "*D in A aequitur B in E*" we would write as $Dx = By$, and Fermat sketched it as a line—or, rather, as a half-line, since he did not accept negative coordinates. He dealt with hyperbolas, parabolas, circles, and ellipses. He even extended his analytic geometry to space, a generalization that had not occurred to Descartes.

Fermat considered the family of equations of the form $y = x^n$, where n is a positive or negative integer. In an attempt to find the maxima and minima of the associated curves, he found the derivatives of these special types of function. He also used what was essentially the differentiation process (but without the idea of the limit) for finding tangents to these same curves. Later Fermat rectified the functions, or found the area of the region between such a curve and a line, which corresponds to the integral of x^n. It is curious that he did not seem to have recognized the inverse nature of these two processes. Although Fermat is sometimes called a discoverer of the calculus, he worked with only a few polynomial functions, he lacked the concept of the limit, and he failed to note the fundamental theorem of calculus. Thus, he cannot be said to have reached the intellectual summit that might have allowed him to be ranked with Newton and Leibniz.

Fermat's chief interest was in such things as prime numbers, perfect numbers, and magic squares; he made notable discoveries about these topics, which fall under the heading of the theory of numbers. He was also one of the earliest to develop the theory of probability. But while he did more significant work than many professional mathematicians, he published virtually nothing during his lifetime, and his brilliance was not recognized until long after his death.

Pronunciation Guide

René Descartes (Reh-nay′ Day-cart′)

Pierre de Fermat (Pee-air′ day Fair-mah)

(Photographs from the David Eugene Smith Papers, Rare Book and Manuscript Library, Columbia University.)

CHAPTER 15

THE CONIC SECTIONS

15.1 INTRODUCTION

This is the first of three chapters that serve as a bridge from the calculus of a single variable (Chapters 2–14) to the calculus for vector-valued functions and functions of several variables (Chapters 18–21). In making this transition, it is important to develop a broader familiarity with the types of planar curves that arise as graphs of equations of the form

$$Ax^2 + Bxy + Cy^2 + Dx + Ey + F = 0. \qquad (1)$$

Such curves will occur as graphs of vector-valued functions, as boundaries for domains of functions of two variables, and as intersections of planes with surfaces in three dimensions. Moreover, graphs of certain equations of the form (1) have important and useful geometric properties, as we shall see in this chapter.

Equation (1) is referred to as the general second degree equation in the two variables x and y. We have previously examined two special cases of this equation rather carefully:

(i) If $A = B = C = 0$, the equation is

$$Dx + Ey + F = 0.$$

Its graph is a straight line in the plane (Section 1.4).

(ii) If $A = C \neq 0$ and $B = 0$ the equation can be written

$$x^2 + y^2 + dx + ey + f = 0,$$

where $d = \dfrac{D}{A}$, $e = \dfrac{E}{A}$, $f = \dfrac{F}{A}$. By completing the squares in x and in y this equation can be brought to the standard form for a circle:

$$(x - h)^2 + (y - k)^2 = r^2.$$

(See Section 1.3.)

In the following three sections we shall see that if $B = 0$, the remaining possibilities for the graph of equation (1) are limited to the parabola, the ellipse, or the hyperbola. The case $B \neq 0$ is handled in Section 15.5, where we show that by

Figure 1.1 Circle.

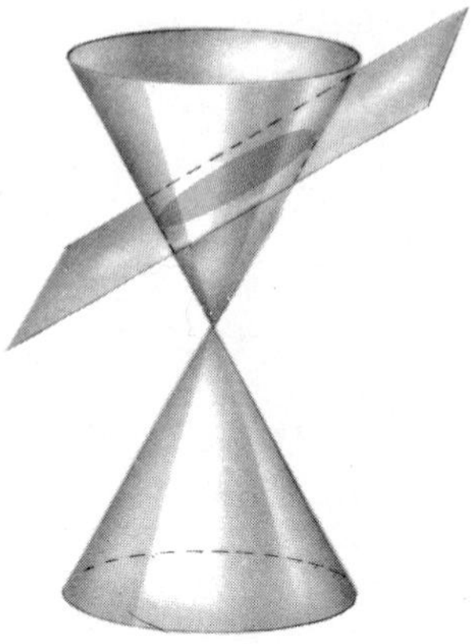
Figure 1.2 Ellipse.

Figure 1.3 Parabola.

Figure 1.4 Hyperbola.

rotating the xy-axes we may find new coordinates for the plane in which the xy term in equation (1) is not present. Thus, we will show that the graph of any second degree equation of the form (1) must correspond to one of the five figures mentioned above.

The circle, parabola, ellipse, and hyperbola are referred to as **conic sections** because each arises as the curve of intersection of an infinite cone and a plane, as illustrated in Figures 1.1 through 1.4. (A line is referred to as a **degenerate** conic section.)

We shall not pursue this three-dimensional interpretation of the conic sections further. Rather, we shall use the techniques of analytic geometry in the plane to analyze equation (1) in the cases corresponding to each of these figures.

Translation of Axes

In constructing a number line, we are free to locate the point $x = 0$ anywhere along the line. We are also free to choose the location for $x = 1$, but these two choices then determine the locations of the points corresponding to all other real numbers. Similarly, in assigning coordinates to the plane we are free to locate the axes wherever we like (as long as they are perpendicular). In graphing certain equations in two variables it is sometimes convenient to relocate the x- and y-axes. We describe one such technique here, called **translation of axes.**

Figure 1.5 shows a typical xy-coordinate plane. Suppose we wished to move the y-axis so that it remained vertical, but so that it passed through the point $(h, 0)$, $h > 0$. (This is called **translating** the y-axis.) How would this change the coordinates of the points in this plane?

Let's refer to the vertical axis in its new position as the Y-axis. Also, we write the xY-coordinates of a point with pointed brackets, $\langle a, b\rangle$, while (a, b) denotes the xy-coordinates of a point. Since the y-coordinate of a point denotes its distance above or below the x-axis, translating the y-axis causes no change in y-coordinates. However, since the x-coordinate of a point is its distance to the right or left of the y-axis, translating the y-axis h units to the right *subtracts* h units from every x-

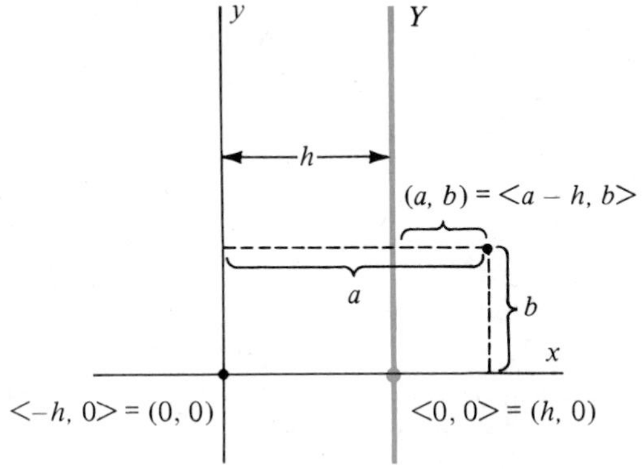

Figure 1.5 Translating the y-axis h units subtracts h from each x-coordinate.

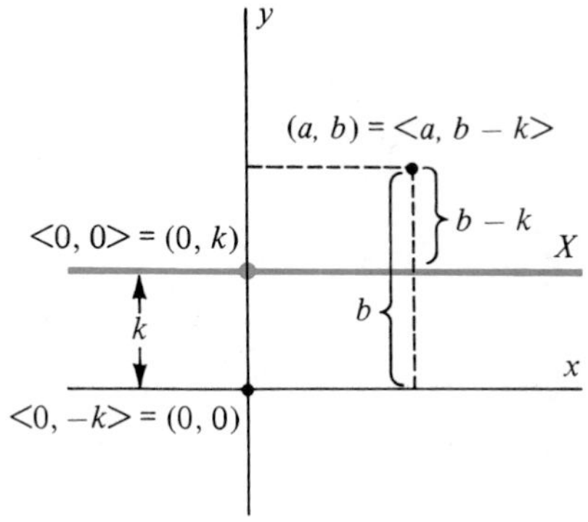

Figure 1.6 Translating the x-axis k units subtracts k from each y-coordinate.

coordinate. That is, if $(x, y) = \langle X, Y\rangle$, then

$$X = x - h \qquad \text{and} \qquad Y = y. \tag{2}$$

(See Figure 1.5.)

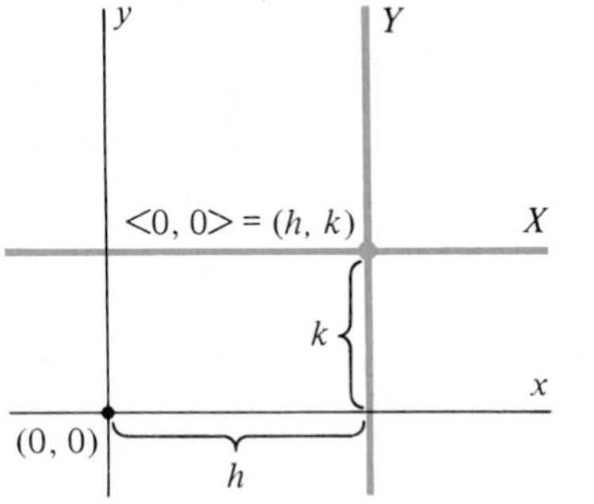

Figure 1.7 Translating both x and y axes: $X = x - h$, $Y = y - k$.

Similarly, translating the x-axis so that it remains horizontal but passes through the point $(0, k)$ rather than $(0, 0)$, we obtain an Xy-coordinate system where $\langle X, Y\rangle = (x, y)$ if and only if

$$X = x \qquad \text{and} \qquad Y = y - k. \tag{3}$$

This is because our translation has the effect of subtracting k units from every y-coordinate (see Figure 1.6).

In many applications, we will want to translate both axes. We do this by combining the two translations described above. In particular, if we wish to translate axes so that the origin $\langle 0, 0\rangle$ in the new XY-coordinate system lies at the point with xy-coordinates (h, k), we use both the first equation in (2) and the second equation in (3). That is, $\langle X, Y\rangle = (x, y)$ if and only if

$$X = x - h, \tag{4}$$

$$Y = y - k. \tag{5}$$

(See Figure 1.7.) It is important to note that equations (4) and (5) are valid for all h and k, although we have illustrated the case $h > 0$, $k > 0$ in Figure 1.7.

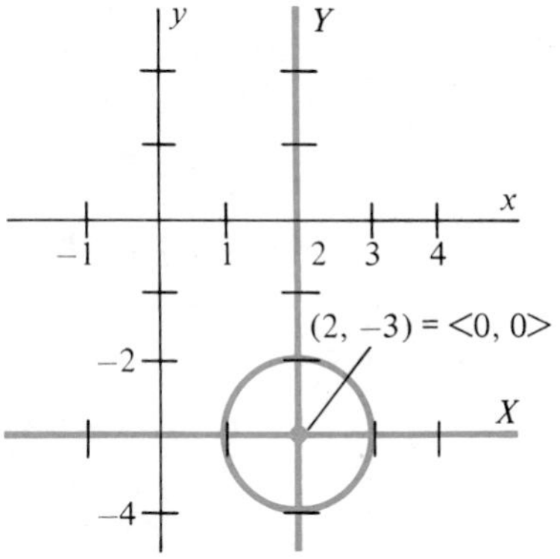

Figure 1.8 The graph of the equation $x^2 + y^2 - 4x + 6y + 12 = 0$ is the unit circle in the XY-plane.

Figure 1.8 illustrates why we are interested in translations of axes. On completion of the squares in x and y, the equation

$$x^2 + y^2 - 4x + 6y + 12 = 0$$

becomes

$$(x - 2)^2 + (y + 3)^2 = 1.$$

With the substitutions $X = x - 2$, $Y = y + 3$ (i.e., with $h = 2$, $k = -3$), this equation becomes

$$X^2 + Y^2 = 1,$$

easily recognized as the unit circle in the XY-coordinates. While this example contains nothing really new, it illustrates the way in which we shall use translations of axes throughout this chapter.

15.2 PARABOLAS

We begin with the purely geometric definition of the parabola.

DEFINITION 1 A parabola is the set of all points P in the plane lying equidistant from a fixed line ℓ and a fixed point F.

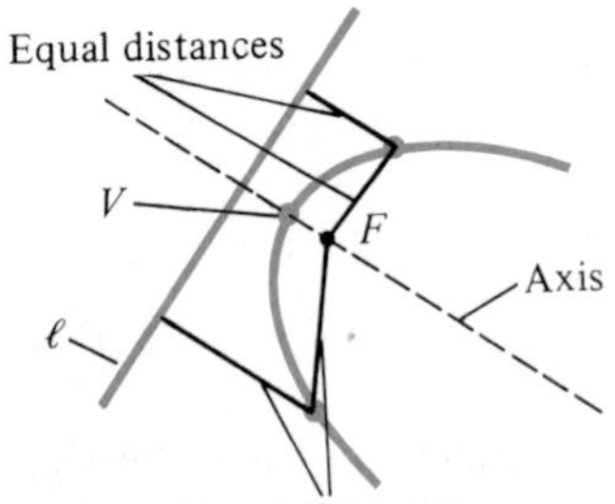

Figure 2.1 Parabola (ℓ = directrix, F = focus, V = vertex).

In Definition 1 the fixed point F is called the **focus** of the parabola, and the fixed line ℓ is called the **directrix.** The line through the focus perpendicular to the directrix is called the **axis** of the parabola.

It is easy to see from Definition 1 that the point V on the axis midway between the focus and directrix lies on the parabola. This point is called the **vertex** of the parabola. A typical parabola, illustrating the requirement of Definition 1, is sketched in Figure 2.1.

To determine the equation satisfied by the coordinates of a point $P = (x, y)$ on the graph of a parabola, we position the parabola in the xy-plane so that the vertex V lies at the origin and so that the focus F lies at the point $(0, c)$ (see Figure 2.2). The directrix then has equation ℓ: $y = -c$. (The number c, representing the (signed) distance from the vertex to the focus, is referred to as the **focal length** of the parabola.)

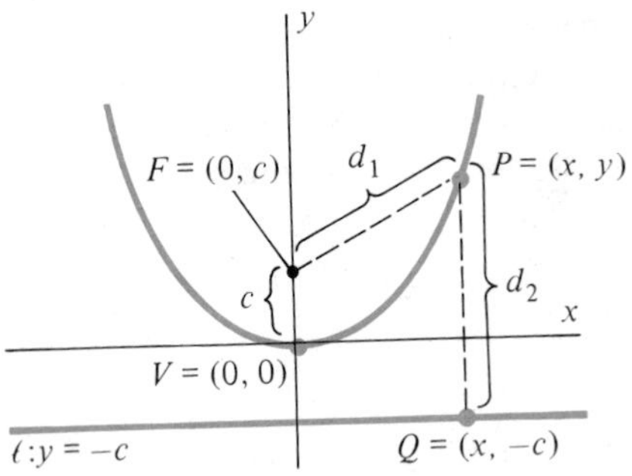

Figure 2.2 Parabola in standard position.

Now let $P = (x, y)$ be any point on the parabola, and let Q be the point on ℓ nearest P. Then $Q = (x, -c)$, and the distance from P to ℓ is the same as the distance from P to Q. Using the distance formula we can state the requirement $d_1 = d_2$ of Definition 1 as

$$\sqrt{(x-0)^2 + (y-c)^2} = \sqrt{(x-x)^2 + (y-(-c))^2},$$

or

$$\sqrt{x^2 + (y-c)^2} = |y + c|.$$

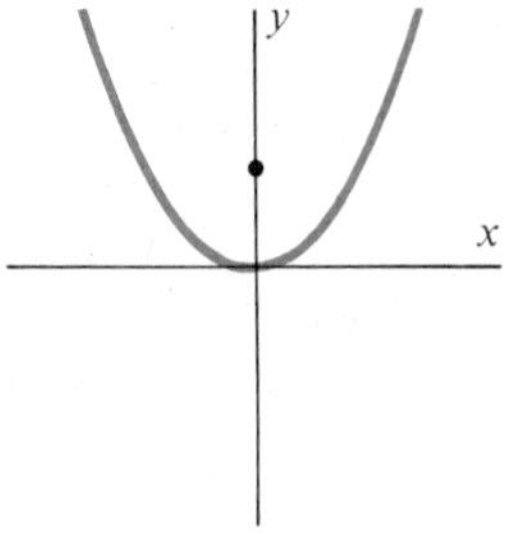

Figure 2.3 $x^2 = 4cy$, $c > 0$.

Squaring both sides gives

$$x^2 + (y - c)^2 = (y + c)^2$$

$$x^2 + (y^2 - 2cy + c^2) = y^2 + 2cy + c^2,$$

$$\boxed{x^2 = 4cy.} \qquad (1)$$

Equation (1) is the **standard form** for the parabola with vertex at the origin and focus on the y-axis. In Figure 2.2 we have assumed $c > 0$. However, the calculations above did not depend on this assumption, so equation (1) is also valid if $c < 0$. In the case $c < 0$ the parabola opens in the negative y-direction (see Figures 2.3 and 2.4).

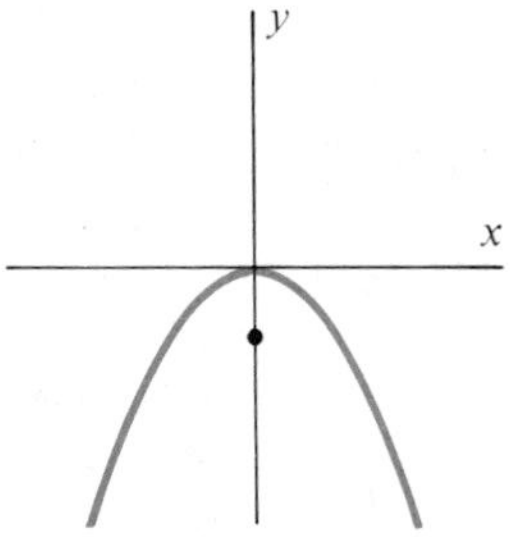

Figure 2.4 $x^2 = 4cy$, $c < 0$.

REMARK: The above demonstration shows that any point (x, y) satisfying the *geometric* condition of Definition 1 satisfies the *algebraic* condition of equation (1). In Exercise 39 you are asked to show the converse—that any point (x, y) satisfying the algebraic condition (1) also satisfies the geometric condition of Definition 1.

For a parabola with vertex at (0, 0), the axis may lie along the x-axis rather than the y-axis. In this case the focus is labelled $(c, 0)$, the directrix is $x = -c$, and the requirement of Definition 1 is that

$$\sqrt{(x - c)^2 + (y - 0)^2} = \sqrt{(x - (-c))^2 + (y - y)^2},$$

which simplifies to the equation

$$\boxed{y^2 = 4cx.} \qquad (2)$$

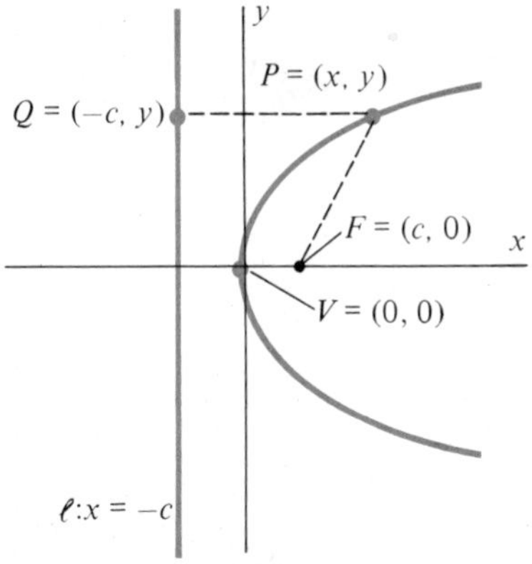

Figure 2.5 $y^2 = 4cx$, $c > 0$.

(See Figure 2.5.) Equation (2) is the standard form for the parabola with vertex at (0, 0) and focus on the x-axis. When $c > 0$, the parabola opens in the positive x-direction. When $c < 0$ it opens in the negative x-direction (Figure 2.6).

Example 1 The equation $x^2 + 8y = 0$ may be written in the form

$$x^2 = -8y$$

corresponding to equation (1). Its graph is therefore a parabola with vertex (0, 0), with focal length $c = -2$, and opening in the negative y direction. Its focus is at $(0, -2)$ (see Figure 2.7). ■

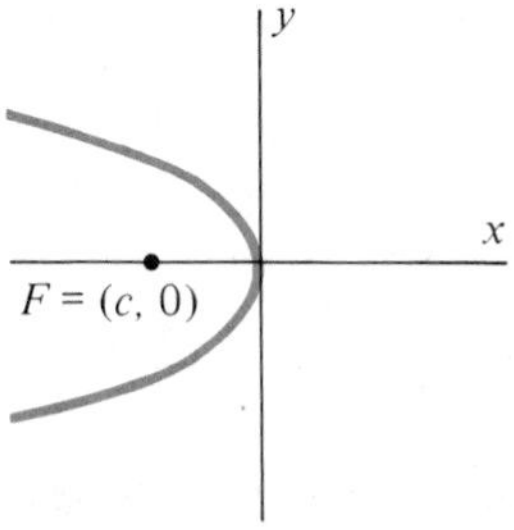

Figure 2.6 $y^2 = 4cx$, $c < 0$.

Example 2 The graph of the equation $\frac{1}{2}y^2 - x = 0$ can be written in the form of equation (2) as

$$y^2 = 2x.$$

Its graph is a parabola opening in the positive x direction with vertex (0, 0), focal length $c = 1/2$, and focus $\left(\frac{1}{2}, 0\right)$ (Figure 2.8). ■

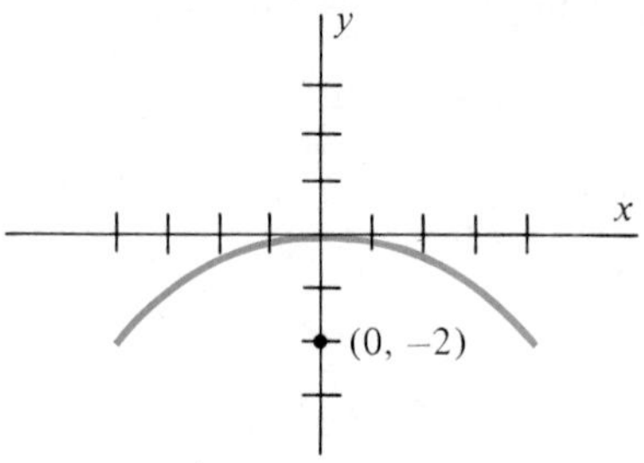

Figure 2.7 $x^2 = -8y$.

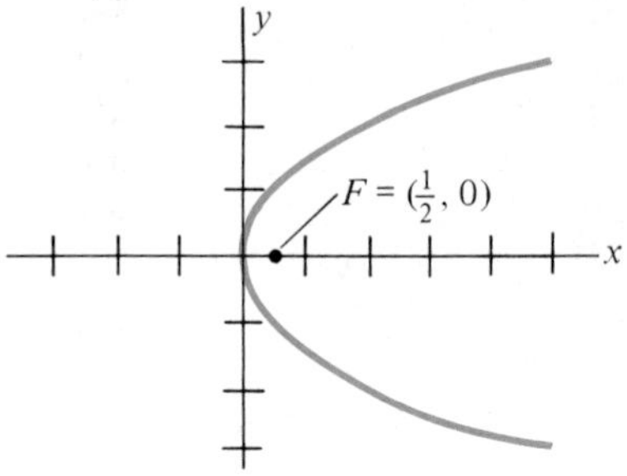

Figure 2.8 $y^2 = 2x$.

Translating the Parabola

When the vertex of a parabola lies at a point (h, k) rather than at $(0, 0)$, we may apply the technique of **translation of axes,** to analyze its graph. To do so we introduce new XY-coordinates, via the substitutions

$$X = x - h \qquad \text{and} \qquad Y = y - k. \tag{3}$$

These substitutions determine new X- and Y-axes, parallel to the x- and y-axes, but with the origin in XY-coordinates located at the vertex of the parabola (see Figure 2.9).

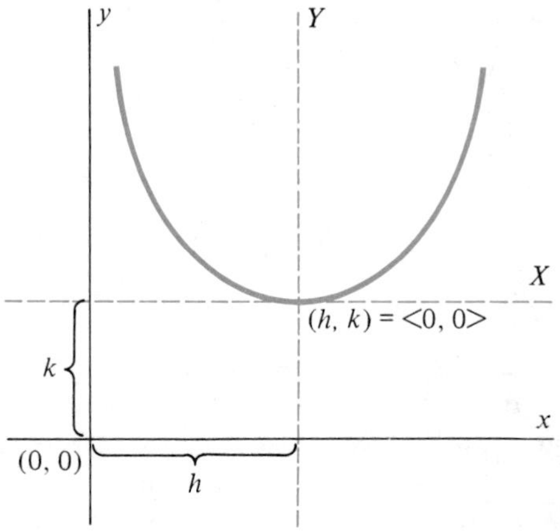

Figure 2.9 $(\ ,\ )$ = xy-coordinates $\langle\ ,\ \rangle$ = XY-coordinates

If, after translating axes, the parabola opens along the Y-axis with focal length c, the equation for the parabola in XY-coordinates is

$$X^2 = 4cY, \tag{4}$$

according to equation (1). Combining equation (4) with substitutions (3) we conclude that

$$\boxed{(x - h)^2 = 4c(y - k)} \tag{5}$$

is the *standard form for the equation for the parabola with vertex at (h, k), focal length c, and axis parallel to the y-axis*. Similarly, from equation (2) and substitutions (3) it follows that

$$\boxed{(y - k)^2 = 4c(x - h)} \tag{6}$$

is the *standard form for the equation of the parabola with vertex at (h, k), focal length c, and axis parallel to the x-axis.*

As was the case for equations of circles in Section 1.4, the usual technique for analyzing the equation of a parabola is to first complete the square, if necessary, and then to bring the resulting equation into the form of equation (5) or equation (6).

Example 3 Describe the graph of the equation

$$(y - 1)^2 = 6(x + 2).$$

Strategy

Identify the form of the equation as that of equation (6).

Solution

We note that the equation has the form of equation (6) with

$$h = -2 \qquad \text{and} \qquad k = 1.$$

Since the parabola opens in the x direction we have

$$\begin{aligned}&\text{vertex} = (h, k),\\&\text{focal length} = c,\\&\text{axis: } y = k,\\&\text{focus: } (h + c, k).\end{aligned}$$

so the graph is a parabola with axis parallel to the x-axis. Since $6 = 4c$, we have $c = 3/2$.

The vertex is $(-2, 1)$, the focal length is $3/2$, the axis is $y = 1$, and the focus is $\left(-\frac{1}{2}, 1\right)$ (see Figure 2.10). ■

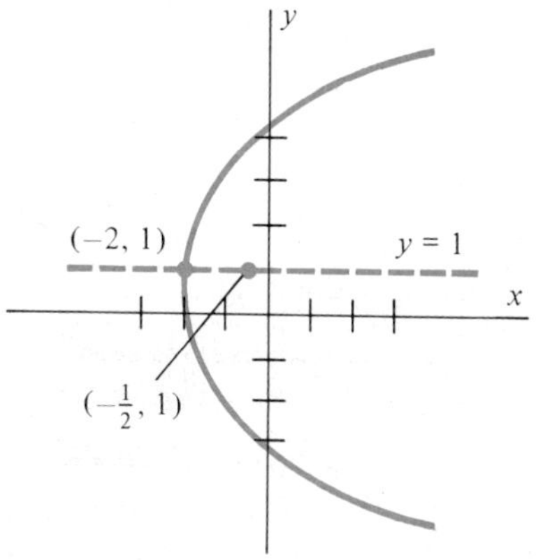

Figure 2.10 $(y - 1)^2 = 6(x + 2)$.

Example 4 Find the vertex, focus, axis, and focal length for the parabola with equation $x^2 - 6x + 6y - 3 = 0$.

Strategy

Solution

We first observe that

$$x^2 - 6x + 6y - 3 = 0$$

gives

Complete the square in x by adding 9 to both sides.

$$[x^2 - 6x + 9] + 6y - 3 = 9,$$
$$(x - 3)^2 + 6y = 12,$$

Move all terms except the squared term to the right side.

$$\begin{aligned}(x - 3)^2 &= -6y + 12\\&= -6(y - 2)\\&= 4\left(-\frac{3}{2}\right)(y - 2).\end{aligned}$$

Identify the form of the resulting equation.

This equation has the form of equation (5) with

$$h = 3, \qquad k = 2, \qquad c = -3/2.$$

Thus the vertex is (3, 2), the focal length is $c = -3/2$, the axis is $x = h = 3$, and the focus is $(h, k + c) = (3, 1/2)$. The graph appears in Figure 2.11. ■

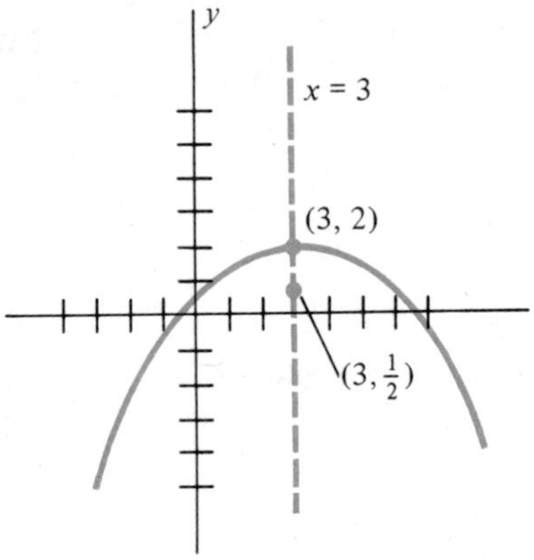

Figure 2.11 $x^2 - 6x + 6y - 3 = 0$.

Example 5 Find the equation of the parabola with directrix $y = -2$ and focus $F = (2, 5)$.

Solution: Since the axis lies perpendicular to the directrix, the axis is parallel to the y-axis. The standard equation for the parabola is therefore that of equation (5). Since the vertex lies midway between the directrix and the focus, the vertex has coordinates $\left(2, -2 + \frac{1}{2}(5 - (-2))\right) = (2, 3/2)$. Thus, $h = 2$ and $k = 3/2$. Since the distance from the focus to the vertex is $5 - 3/2 = 7/2$, the focal length is $c = 7/2$. The desired equation is therefore

$$(x - 2)^2 = 4\left(\frac{7}{2}\right)(y - 3/2),$$

or

$$(x - 2)^2 = 14(y - 3/2).$$

■

Reflecting Properties of Parabolas

An important application of parabolic arcs follows from the physical law that when a ray of light strikes a reflecting surface the angle of incidence equals the angle of reflection. The relationship between this law and parabolas lies in the fact that the line tangent to the graph of a parabola at a point P makes equal angles with the ray from P to the focus and the line through P parallel to the axis of the parabola (Figure 2.12 and Exercise 31).

Thus, rays of light parallel to the axis of a parabolic reflector will all be reflected precisely to the focus of the parabola (Figure 2.13). This principle is used in reflecting telescopes, where the parallel rays of light from the stars are reflected off a parabolic mirror to the eyepiece located at the focus. The same principle applies to radar signals, which are gathered by means of parabolic reflecting "dishes."

The same principle in reverse explains the design of automobile headlamps. By placing the bulb at the focus of a parabolic mirror, one ensures that the beams of light will leave the headlamp parallel to the axis of the parabolic mirror.

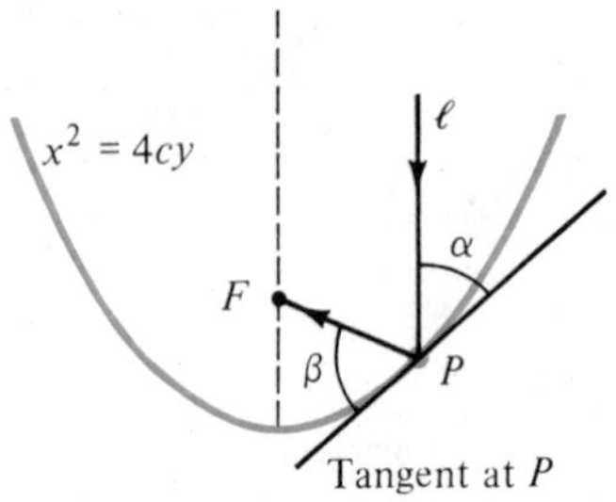

Figure 2.12 Reflecting property: $\alpha = \beta$ if ℓ is parallel to axis.

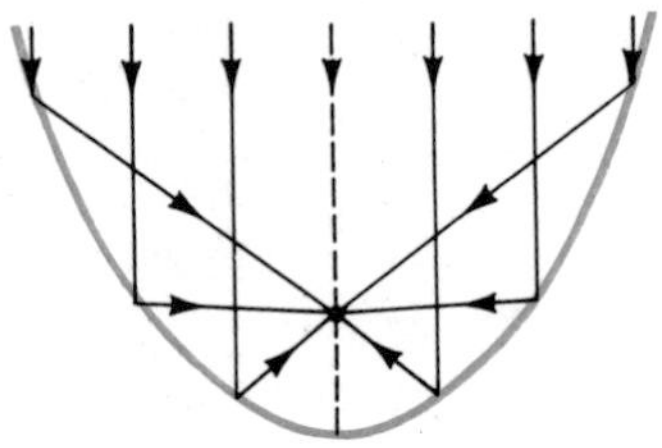

Figure 2.13 Rays parallel to the axis are all reflected through the focus of a parabola, and conversely.

Exercise Set 15.2

In Exercises 1–20, sketch the parabola whose equation is given, noting the vertex, focal length, axis, and focus.

1. $y = 2x^2$

2. $y = 4x^2 + 8$

3. $y = -2x^2 + 2$

4. $x = -y^2$

5. $x = 3y^2 + 1$

6. $x + 2y^2 = 4$

7. $8y + x^2 = 2$

8. $y - x^2 = 4$

9. $(x - 2)^2 = 8(y - 1)$

10. $x^2 = 12(y + 2)$

11. $(y - 2)^2 = 6(x + 3)$

12. $(y - 1)^2 = -7x$

13. $y^2 - 2y - 4x - 7 = 0$

14. $y^2 + 4y + x + 7 = 0$

15. $x^2 + 2x - 6y - 11 = 0$

16. $x^2 - 4x - 4y = 0$

17. $x^2 + 6x - 10y + 19 = 0$

18. $y^2 - 4y - 8x + 20 = 0$

19. $y^2 + 4y - 2x + 6 = 0$

20. $y^2 + 6y - 3x + 6 = 0.$

In Exercises 21–27, find an equation for the parabola with the stated properties.

21. Vertex (0, 0) and focus (0, 3).

22. Vertex (−1, 2) and focus (−1, −2).

23. Vertex (1, 3) and directrix $y = -5$.

24. Focus (−1, −1) and directrix $x = 3$.

25. Focus (3, −6) and vertex (−1, −6).

26. Vertex (0, 4) and directrix $x = 2$.

27. Axis $x = 0$, directrix $y = -3$, and containing the point (0, 1).

28. State a procedure for sketching the graph of a parabola according to Definition 1, assuming that you are given the equation for the directrix and the location of the focus and assuming that you can use only a ruler and compass.

29. The chord through the focus of a parabola perpendicular to its axis is called the **latus rectum** of the parabola (See Figure 2.14). Prove that the length of the latus rectum of a parabola is $|4c|$ where $|c|$ is the focal length.

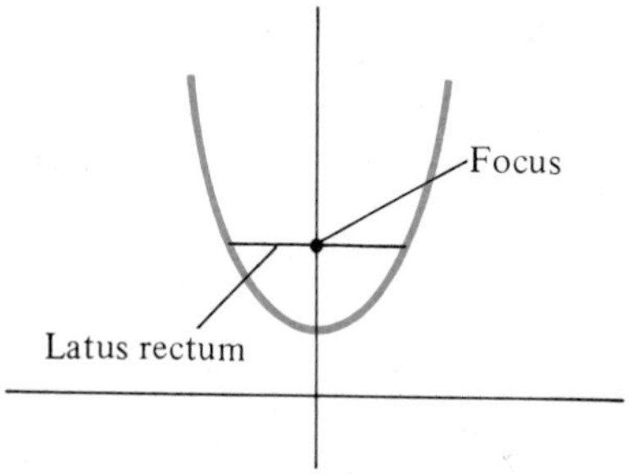

Figure 2.14 Latus rectum.

30. Prove that a circle that has a diameter corresponding to the latus rectum of a parabola must be tangent to the directrix of that parabola.

31. In Figure 2.12, α is the angle formed between the tangent at P and the line ℓ through P parallel to the axis of the parabola. β is the angle formed between the tangent at P and the line segment from P to the focus F. Prove that $\alpha = \beta$.

(*Hint:* Construct a coordinate system with origin at the vertex of the parabola and extend ℓ through P toward the directrix. Also draw a line through F, perpendicular to the tangent at P, until it intersects ℓ. What can you show about the triangle formed in this way?)

32. Prove that the graph of the second degree equation

$$Ax^2 + Bxy + Cy^2 + Dx + Ey + F = 0$$

is a parabola with axis parallel to either the x- or y-axes whenever $B = 0$ and either $A = 0$ or $C = 0$.

33. Find the area of the region bounded by the graph of $y = p(x)$ and the x-axis if $p(x)$ is a parabola with focus at the origin and directrix $y = -2$.

34. Find the volume of the solid obtained by revolving about the y-axis the area of the region bounded by the y-axis and the parabola with vertex (−1, 0) and directrix $x = -3/2$.

35. Find the area of the region bounded by the x-axis and the parabola with vertex (4, 2) and directrix $y = 4$.

36. Find the volume of the solid obtained by revolving about the y-axis the region bounded by the y-axis and the parabola with focus (0, 4) and directrix $x = 4$.

37. Find the average value, on the interval [0, 5], of the function whose graph is a parabola with focus (2, 3) and vertex (2, −1).

38. Find the area of the region bounded by the graph of the function in Exercise 34 and the line $x = 5$.

39. Show that if a point (x, y) satisfies the equation $x^2 = 4cy$ then it has the property that its distance from the directrix $y = -c$ equals its distance from the focus $(0, c)$. Thus, any point satisfying the algebraic condition $x^2 = 4cy$ for a parabola also satisfies the geometric condition of Definition 1.

15.3 THE ELLIPSE

The geometric definition of the ellipse is the following.

DEFINITION 2 An ellipse is the set of all points in the plane the sum of whose distances from two fixed points is a constant.

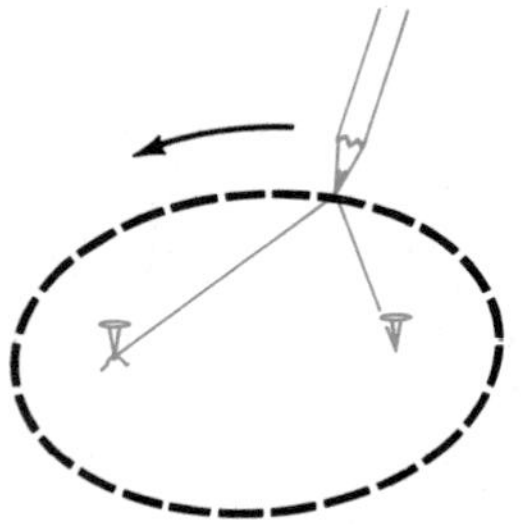

Figure 3.1 Tracing out an ellipse.

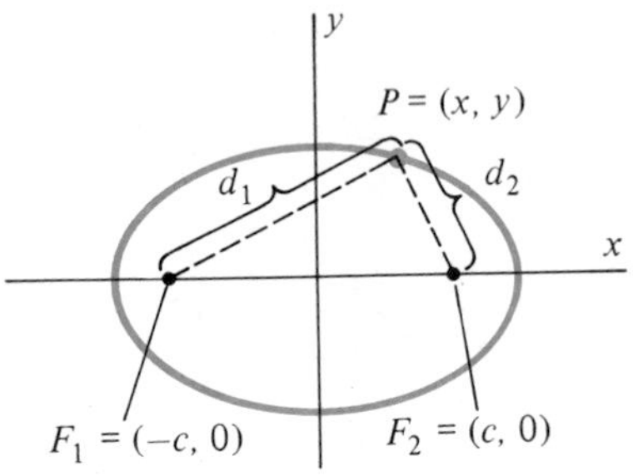

Figure 3.2 Ellipse in standard position.

The two fixed points in Definition 2 are referred to as the **foci** of the ellipse. We may interpret Definition 2 by imagining a length of string tied at each end to one of two tacks driven into a flat surface. The tacks represent the foci. If a pencil is held against the surface as in Figure 3.1, so that the string is kept taut, the figure traced out by the possible locations for the pencil tip is an ellipse. This is because the sum of the distances of the pencil tip from the foci remains constant.

To determine the equation for the ellipse, we construct a coordinate system so that the foci for the ellipse are located at points $F_1 = (-c, 0)$ and $F_2 = (c, 0)$ (Figure 3.2). We let $P = (x, y)$ be an arbitrary point on the ellipse, and we let d_1 and d_2 be the distances from P to F_1 and from P to F_2, respectively. The condition that P be on the ellipse is that $d_1 + d_2$ equal a constant, which we take to be $2a$. (Note that we must have $a > c$, since the distance between the foci is $2c$.)

Using the distance formula, we can now write the requirement of Definition 2 as

$$d_1 + d_2 = 2a,$$

or

$$\sqrt{(x + c)^2 + y^2} + \sqrt{(x - c)^2 + y^2} = 2a.$$

To simplify this equation we move the second radical to the right-hand side and square both sides. We obtain

$$(x + c)^2 + y^2 = 4a^2 - 4a\sqrt{(x - c)^2 + y^2} + (x - c)^2 + y^2.$$

After the squared terms are expanded, this simplifies to

$$2cx = 4a^2 - 4a\sqrt{(x - c)^2 + y^2} - 2cx,$$

so

$$a\sqrt{(x - c)^2 + y^2} = a^2 - cx.$$

Squaring both sides again gives

$$a^2[(x - c)^2 + y^2] = c^2x^2 - 2a^2cx + a^4,$$

or

$$(a^2 - c^2)x^2 + a^2y^2 = a^2(a^2 - c^2). \tag{1}$$

To simplify (1), we define the constant b by

$$b = \sqrt{a^2 - c^2}. \tag{2}$$

Substituting (2) into (1) then gives

$$b^2x^2 + a^2y^2 = a^2b^2,$$

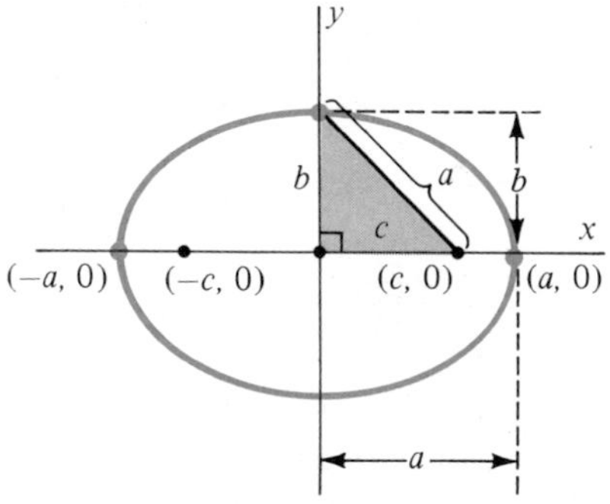

Figure 3.3 $\frac{x^2}{a^2} + \frac{y^2}{b^2} = 1.$

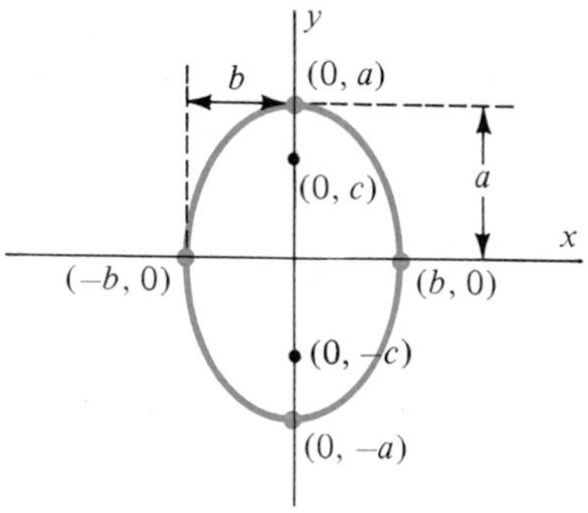

Figure 3.4 $\frac{x^2}{b^2} + \frac{y^2}{a^2} = 1.$

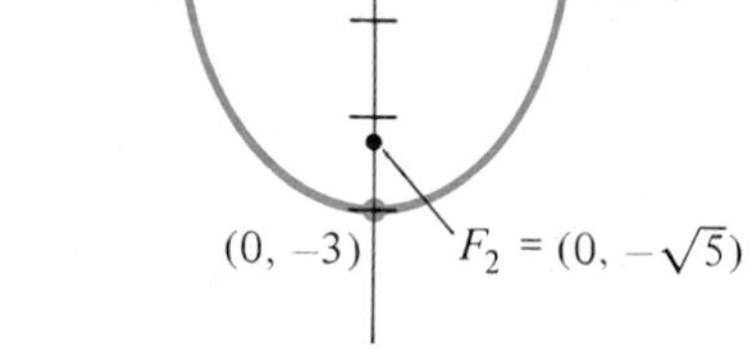

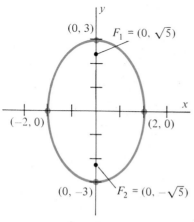

Figure 3.5 $\frac{x^2}{4} + \frac{y^2}{9} = 1.$

or

$$\frac{x^2}{a^2} + \frac{y^2}{b^2} = 1, \qquad a > b. \tag{3}$$

Equation (3) is the *standard form for the equation of the ellipse with center at (0, 0) and foci on the x-axis*. (The **center** of an ellipse is the midpoint of the line segment joining the foci.) In Exercise 39 you are asked to show the converse—that any point satisfying equation (3) also satisfies the geometric condition of Definition 2.

The geometry associated with equation (3) is worth noting. Setting $y = 0$ in (3) shows that the x-intercepts of the ellipse are $x = \pm a$ (see Figure 3.3). The line segment joining the points $(-a, 0)$ and $(a, 0)$ is referred to as the *major* axis of the ellipse. Its length is $2a$. In general, the **major axis** of an ellipse is the chord passing through both foci.

Setting $x = 0$ in (3) shows that the y-intercepts are $y = \pm b$. The line segment joining $(0, b)$ and $(0, -b)$ is called the *minor* axis of the ellipse. Its length is $2b$. More generally, the **minor axis** of an ellipse is the chord through the center and perpendicular to the major axis.

Figure 3.4 illustrates the ellipse that results when the foci are located along the y-axis at $(0, c)$ and at $(0, -c)$. The equation describing this ellipse may be developed just as for the previous case. However, it is much simpler to note that in passing from Figure 3.3 to Figure 3.4 we have merely interchanged the roles of x and y. Thus,

$$\frac{x^2}{b^2} + \frac{y^2}{a^2} = 1, \qquad a > b \tag{4}$$

is *the standard form for the equation of the ellipse with center at (0, 0) and foci along the y-axis*.

Note in equation (4) that the constant a still represents half the length of the major axis and that b is again half the length of the minor axis. For either equation (3) or equation (4), the distance between the origin and a focus is found by solving equation (2) for c:

$$c = \sqrt{a^2 - b^2}. \tag{5}$$

Example 1 Describe the graph of the equation $\frac{x^2}{4} + \frac{y^2}{9} = 1$.

Solution: The equation has the form of equation (4) with $b^2 = 4$, $a^2 = 9$. Thus, $b = 2$ and $a = 3$. The figure is an ellipse with center at $(0, 0)$, with major axis vertical and of length $2a = 6$, and with minor axis of length $2b = 4$. From equation (5) we find that $c = \sqrt{9 - 4} = \sqrt{5}$, so the foci are located at $(0, \sqrt{5})$ and $(0, -\sqrt{5})$ (see Figure 3.5). ■

Example 2 Describe the graph of the equation $4x^2 + 16y^2 - 64 = 0$.

Solution: To bring the equation into standard form, we move the constant term to the right and divide both sides by this constant. We obtain

$$4x^2 + 16y^2 = 64,$$

so

$$\frac{x^2}{4^2} + \frac{y^2}{2^2} = 1.$$

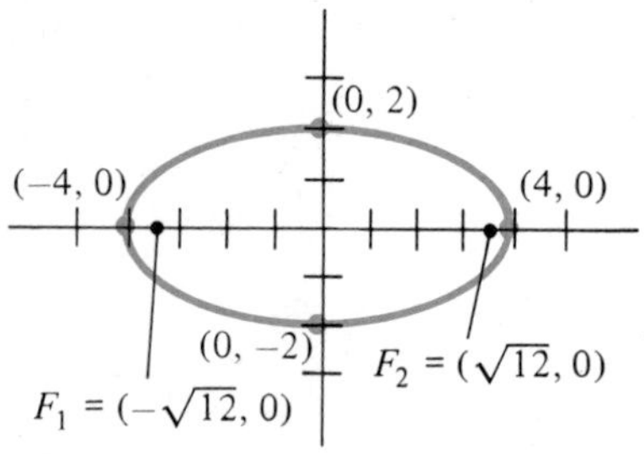

Figure 3.6 $4x^2 + 16y^2 - 64 = 0$.

By setting $y = 0$, we obtain the x-intercepts, $x = \pm 4$. The endpoints of the major axis are therefore $(-4, 0)$ and $(4, 0)$. Setting $x = 0$ gives the y-intercepts, $y = \pm 2$. The endpoints of the minor axis are $(0, 2)$ and $(0, -2)$. Since the equation is of the form of equation (3), with $a = 4$ and $b = 2$, we have from equation (5) that $c = \sqrt{4^2 - 2^2} = \sqrt{12}$. The foci are therefore $(\pm\sqrt{12}, 0)$ (Figure 3.6). ■

Translating the Ellipse

When the center of an ellipse lies at the point (h, k) rather than at the origin, we can analyze its graph by first translating the coordinate axes by the substitutions

$$X = x - h \quad \text{and} \quad Y = y - k \quad \text{(Figure 3.7.)} \tag{6}$$

In XY-coordinates, the center of the ellipse lies at the origin. If the major axis of the ellipse is parallel to the x-axis, the equation of the ellipse in XY-coordinates is, by equation (3),

$$\frac{X^2}{a^2} + \frac{Y^2}{b^2} = 1. \tag{7}$$

Using equation (6), we conclude that

$$\boxed{\frac{(x-h)^2}{a^2} + \frac{(y-k)^2}{b^2} = 1, \qquad a > b} \tag{8}$$

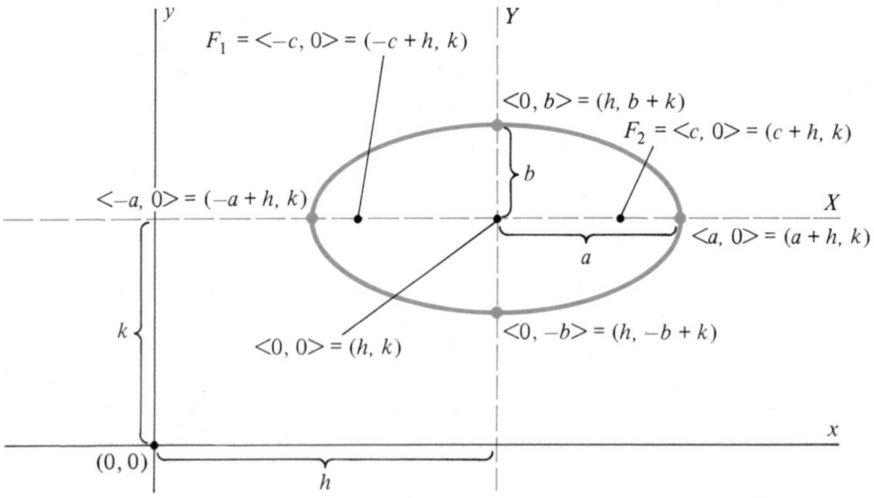

Figure 3.7 Ellipse $\frac{(x-h)^2}{a^2} + \frac{(y-k)^2}{b^2} = 1$, $a > b$. $\langle\ ,\ \rangle$ = XY-coordinates, $(\ ,\ )$ = xy-coordinates.

is *the standard form of the equation for the ellipse with center at (h, k) and major axis parallel to the x-axis*. In (8), the constant a represents half the length of the major axis while b is half the length of the minor axis. Similarly,

$$\frac{(x-h)^2}{b^2} + \frac{(y-k)^2}{a^2} = 1, \qquad a > b \tag{9}$$

is *the standard form of the equation for the ellipse with center at (h, k), with major axis parallel to the y-axis and of length 2a, and with minor axis of length 2b*.

For either equation (8) or equation (9), the distance between the center of the ellipse and either focus is given by c in equation (5).

Example 3 Describe the graph of the equation

$$4x^2 + 9y^2 - 16x - 54y + 61 = 0.$$

Strategy

Complete the square in both variables.

Solution

$$\begin{aligned} &4x^2 + 9y^2 - 16x - 54y + 61 \\ &\quad = 4(x^2 - 4x + 4) + 9(y^2 - 6y + 9) + 61 - 16 - 81 \\ &\quad = 4(x-2)^2 + 9(y-3)^2 - 36. \end{aligned}$$

Write the original equation in squared form.

The equation is therefore equivalent to

$$4(x-2)^2 + 9(y-3)^2 = 36,$$

Bring equation to form (8) or (9). Identify all constants.

or

$$\frac{(x-2)^2}{3^2} + \frac{(y-3)^2}{2^2} = 1.$$

Center is (h, k).
Major axis has length $2a$.
Minor axis has length $2b$.
Foci are at $(h \pm c, k)$ where $c = \sqrt{a^2 - b^2}$.

This equation has the form of equation (8) with $h = 2$, $k = 3$, $a = 3$, $b = 2$, and $c = \sqrt{3^2 - 2^2} = \sqrt{5}$. The graph is an ellipse with center (2, 3), major axis parallel to the x-axis and of length $2a = 6$, and minor axis of length $2b = 4$. Foci are at $(2 - \sqrt{5}, 3)$ and $(2 + \sqrt{5}, 3)$. The graph appears in Figure 3.8. ■

Example 4 Find the equation for the ellipse with center $(-4, 3)$, with minor axis of length 6, and with foci at $(-4, 3 \pm 4)$.

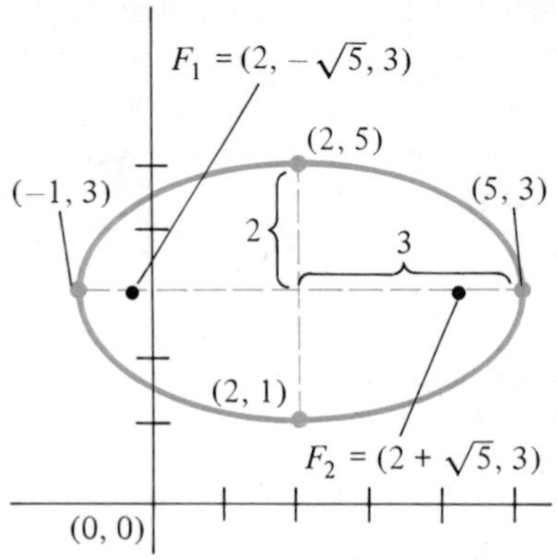

Figure 3.8 Graph of equation $4x^2 + 9y^2 - 16x + 54y - 61 = 0$.

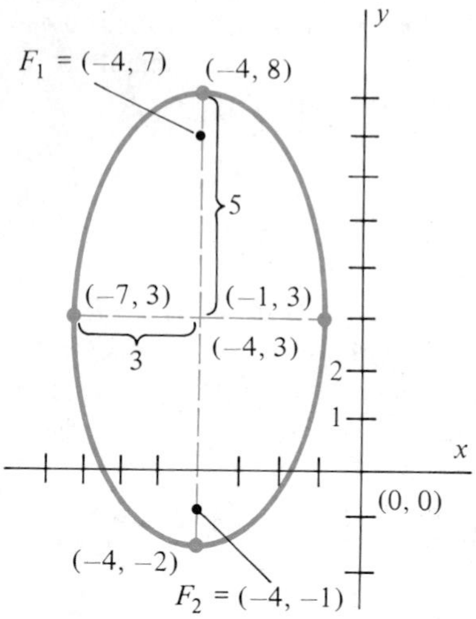

Figure 3.9

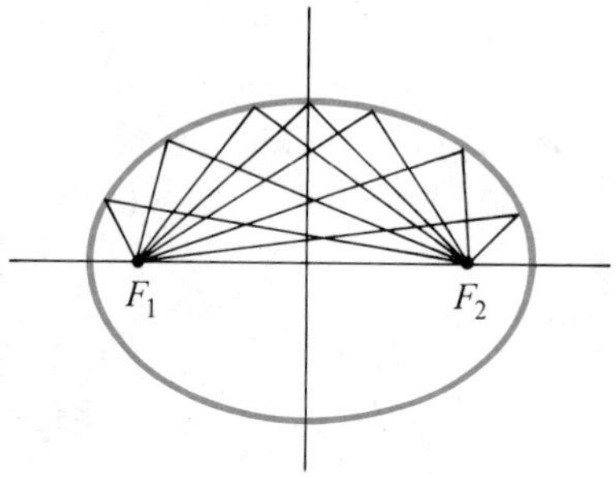

Figure 3.10 Reflecting property of ellipses.

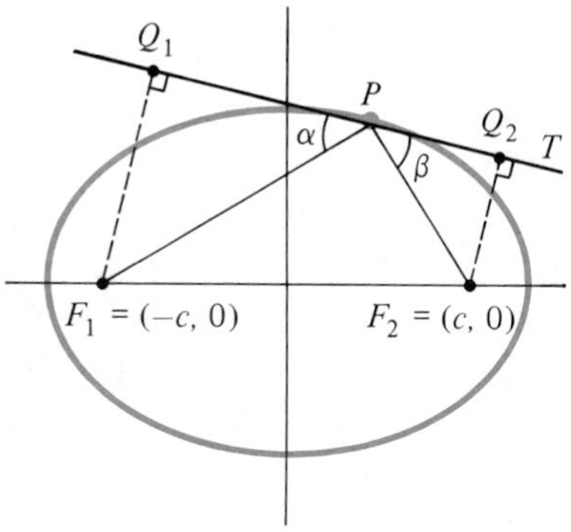

Figure 3.11 $\alpha = \beta$ if T is the tangent to the ellipse at P.

Solution: The distance from the center of the ellipse to one of the foci is $c = 4$. The length of the minor axis is $2b = 6$, so $b = 3$. Thus, by equation (5),

$$a^2 = b^2 + c^2 = 3^2 + 4^2 = 5^2,$$

so $a = 5$. Also, $h = -4$ and $k = 3$ are the coordinates of the center. Since the major axis is vertical, we use equation (9) to write

$$\frac{(x+4)^2}{3^2} + \frac{(y-3)^2}{5^2} = 1.$$

(See Figure 3.9.) ■

Reflecting Property of Ellipses

In Exercise 32 you are asked to show that the rays $\overline{PF_1}$ and $\overline{PF_2}$, from a point P on an ellipse to the foci, make equal angles with the line tangent to the ellipse at P. Applying the physical law that angle of incidence equals angle of reflection, we conclude that if a ray of light or a sound wave emanates from the focus F_1 of an elliptical reflecting surface, it will be reflected through the second focus F_2. Visitors to the United States Capitol Building hear tales of just such a phenomenon. In the days when congressmen were seated on the floor beneath the beautiful elliptical dome, the unfortunate politician seated at a focus risked losing whispered secrets to the statesman seated at the other focus (see Figures 3.10 and 3.11).

Exercise Set 15.3

In Exercises 1–16, find the center and foci for the given ellipse. State the length of the major and minor axes and sketch the graph.

1. $\dfrac{x^2}{9} + \dfrac{y^2}{1} = 1$

2. $\dfrac{x^2}{16} + \dfrac{y^2}{49} = 1$

3. $4x^2 + 16y^2 = 64$

4. $9x^2 + 4y^2 = 36$

5. $50x^2 + 18y^2 - 450 = 0$

6. $8x^2 + 18y^2 - 72 = 0$

7. $\dfrac{(x-2)^2}{3} + \dfrac{(y-1)^2}{4} = 1$

8. $\dfrac{(x-2)^2}{3} + \dfrac{(y+2)^2}{4} = 25$

9. $(x+3)^2 + 4(y-1)^2 = 16$

10. $5(x + 2)^2 + 4(y - 3)^2 = 25$

11. $9x^2 + 4y^2 - 18x + 8y = 23$

12. $x^2 + 4y^2 + 16y + 12 = 0$

13. $9x^2 + 25y^2 + 54x - 50y - 119 = 0$

14. $9x^2 + 4y^2 - 36x + 48y + 144 = 0$

15. $2x^2 + 3y^2 + 6y - 4x - 1 = 0$

16. $4x^2 + 2y^2 + 24x + 12y + 46 = 0$

In Exercises 17–23, find the equation for the ellipse with the stated properties.

17. Center at (0, 0), major axis of length 8, minor axis of length 6, foci on x-axis.

18. Foci at $(\pm 4, 0)$, major axis of length 10.

19. Foci at $(\pm 4, 0)$, minor axis of length 6.

20. Foci at $(0, \pm\sqrt{3})$, minor axis of length 2.

21. Foci at (4, 5) and (4, −1), minor axis of length 8.

22. Center at (2, −6), foci at (2, −3), and (2, −9), major axis of length 10.

23. Foci at (−2, 3) and (−2, 9), minor axis of length 8.

24. Find the area of the region enclosed by the ellipse

$$\frac{x^2}{4} + \frac{y^2}{9} = 1.$$

25. Find the volume of the solid obtained by revolving about the x-axis the region in the first quadrant bounded by the graph of the ellipse $25x^2 + 9y^2 = 225$.

26. Find the volume of the solid obtained by revolving the region described in Exercise 25 about the y-axis.

27. A chord through a focus of an ellipse and parallel to the minor axis is called a **latus rectum.** Find the length of a latus rectum for the ellipse

$$\frac{x^2}{a^2} + \frac{y^2}{b^2} = 1, \qquad a^2 > b^2.$$

28. The **eccentricity** of the ellipse $\frac{x^2}{a^2} + \frac{y^2}{b^2} = 1$ is defined to be the ratio

$$e = \frac{c}{a} = \frac{\sqrt{a^2 - b^2}}{a}.$$

a. Show that $0 \le e < 1$ for every ellipse.
b. Discuss the change in the nature of the ellipse as e approaches zero.
c. Discuss the change in the nature of the ellipse as e approaches one.

29. True or false? A circle is a special case of an ellipse. Explain.

30. Show that for the general second degree equation

$$Ax^2 + Bxy + Cy^2 + Dx + Ey + F = 0$$

if $B = 0$, $A \neq 0$, and $C \neq 0$, the graph is an ellipse if A and B have the same sign.

31. Consider the equation for an ellipse: $\frac{x^2}{a^2} + \frac{y^2}{b^2} = 1.$

a. Differentiate implicitly with respect to x to show that

$$\frac{dy}{dx} = \frac{-b^2x}{a^2y}, \qquad y \neq 0.$$

b. Use part (a) to show that the equation for the line tangent to the graph of this ellipse at (x_0, y_0) can be written

$$(b^2x_0)x + (a^2y_0)y - a^2b^2 = 0.$$

32. Use the result of Exercise 31 to prove that triangle F_1Q_1P is similar to triangle F_2Q_2P in Figure 3.11. Conclude that $\alpha = \beta$.

33. Find the equation for the set of all points the sum of whose distances from (2, 3) and (−4, 3) is 12.

34. Find the equation for the set of all points whose distances from the point (4, 0) is half the distance from the line $y = 3$.

35. Find the volume of the solid obtained by revolving the region bounded by the ellipse $16x^2 + 9y^2 = 144$ about the y-axis.

36. A water tank has the shape of a cylinder with elliptical end panels. The end panels have major axes horizontal, of length 32 feet, and minor axes vertical, of length 18 feet. The tank is full. Find the work done in pumping the water out through a valve in the top of the tank.

37. Find the answer to Exercise 36 if the tank is only half full.

38. State the domain and range of the function $f(x) = \sqrt{b^2\left(1 - \frac{x^2}{a^2}\right)}$. What is the relationship between the graph of this function and the ellipse with equation $b^2x^2 + a^2y^2 = a^2b^2$?

39. Show that if (x, y) is a point whose coordinates satisfy the equation

$$\frac{x^2}{a^2} + \frac{y^2}{b^2} = 1,$$

then (x, y) satisfies the geometric condition of Definition 2, that the sum of its distances from the foci $(-c, 0)$ and $(c, 0)$ is $2a$, where $c^2 = a^2 - b^2$.

15.4 THE HYPERBOLA

As for the parabola and the ellipse, we begin with the geometric definition of the hyperbola and then develop the corresponding equations.

DEFINITION 3

A hyperbola is the set of all points in the plane the difference of whose distances from two fixed points is a constant.

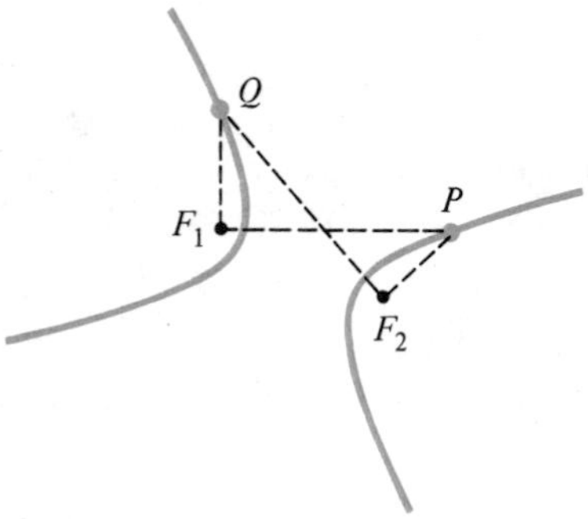

Figure 4.1 $d(P, F_1) - d(P, F_2) = d(Q, F_2) - d(Q, F_1)$.

To interpret Definition 3, we locate two fixed points, F_1 and F_2 (called **foci**) in the plane. Let P be a point on the hyperbola, and let $k = d(P, F_1) - d(P, F_2)$. The hyperbola with foci F_1 and F_2 containing the point P is the set of all points Q such that either $d(Q, F_1) - d(Q, F_2) = k$ or else $d(Q, F_2) - d(Q, F_1) = k$ (see Figure 4.1).

To determine the equations for the hyperbola, we construct a pair of xy-coordinate axes so that the foci are located at $F_1 = (-c, 0)$ and $F_2 = (c, 0)$ (Figure 4.2). (We refer to the midpoint of the line segment F_1F_2 as the **center** of the hyperbola, so the center in this case is the origin.) Let $P = (x, y)$ be a point on the hyperbola, and let $2a$ be the constant referred to in Definition 3. Then P is on the hyperbola if and only if

$$|d(P, F_2) - d(P, F_1)| = 2a.$$

Using the distance formula and the coordinates for P, F_1, and F_2, we can rewrite this equation as

$$|\sqrt{(x - c)^2 + y^2} - \sqrt{(x + c)^2 + y^2}| = 2a, \qquad c > a. \tag{1}$$

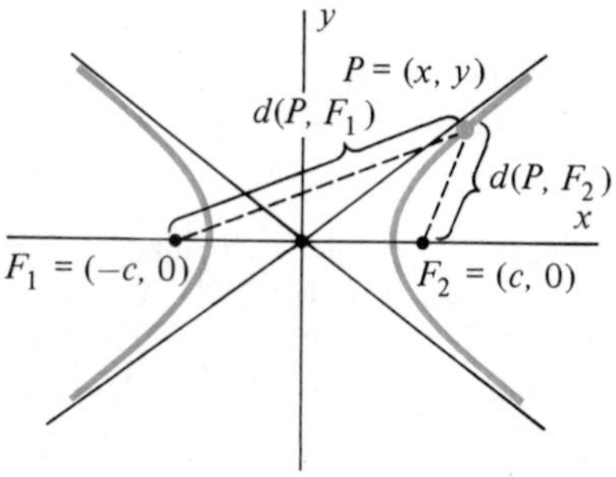

Figure 4.2 Hyperbola with center at (0, 0).

By squaring both sides of this equation and simplifying, just as we did in Section 16.3 for the ellipse, we can obtain the *standard form of the equation for the hyperbola with center at (0, 0) and foci on the x-axis:*

$$\frac{x^2}{a^2} - \frac{y^2}{b^2} = 1, \qquad b = \sqrt{c^2 - a^2}. \tag{2}$$

By setting $y = 0$ in equation (2), we obtain the x-intercepts of the hyperbola $x = \pm a$. The points $V_1 = (-a, 0)$ and $V_2 = (a, 0)$ are called the **vertices** of the hyperbola. (By setting $x = 0$ you will observe that the graph of equation (2) has no y-intercepts. This observation will help you in determining whether the foci of a given hyperbola lie along the x-axis or along the y-axis.)

In graphing equation (2) it is helpful to note that *the lines* $y = \pm\frac{b}{a}x$ *are asymptotes for the graph of equation (2).* This means that the points on the graph of the hyperbola approach the lines $y = \pm\frac{b}{a}x$ as $|x|$ becomes large. To see this we solve (2) for y:

$$\frac{y^2}{b^2} = \frac{x^2}{a^2} - 1,$$

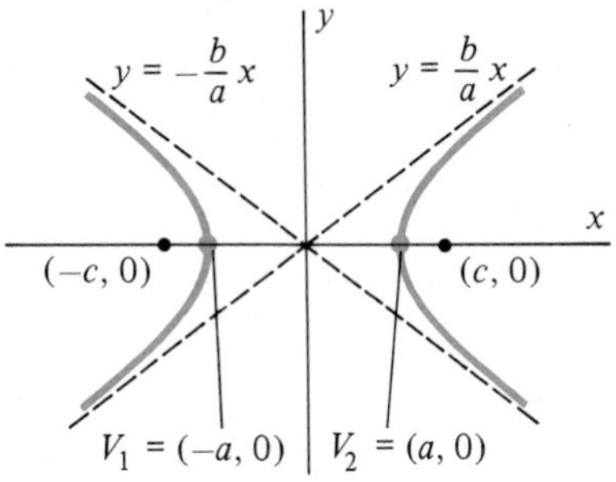

Figure 4.3 The hyperbola $\frac{x^2}{a^2} - \frac{y^2}{b^2} = 1$.

so

$$y = \pm b\sqrt{\frac{x^2}{a^2} - 1} = \pm\frac{b}{a}x\sqrt{1 - \frac{a^2}{x^2}}. \tag{3}$$

Since $\lim\limits_{|x|\to\infty} \sqrt{1 - \frac{a^2}{x^2}} = \sqrt{1 - 0} = 1$, equation (3) shows that

$$\boxed{y = \pm\frac{b}{a}x} \tag{4}$$

are the equations for asymptotes to the graph of equation (2) (see Figure 4.3).

It is worth noting the geometric significance of b in equation (2). If a rectangle is constructed with vertical sides passing through the vertices $(-a, 0)$ and $(0, a)$ and with horizontal sides passing through the points $(0, b)$ and $(-b, 0)$, the diagonals of this rectangle pass through $(0, 0)$ and have slope $\pm\frac{b}{a}x$ (Figure 4.4).

Thus, the asymptotes (4) may be sketched by extending the diagonals of this rectangle. Also, the second equation in line (2) may be solved for c to give $c^2 = a^2 + b^2$. Thus, c may be interpreted as the length of the hypotenuse of one of the eight triangles determined by the diagonals of this rectangle.

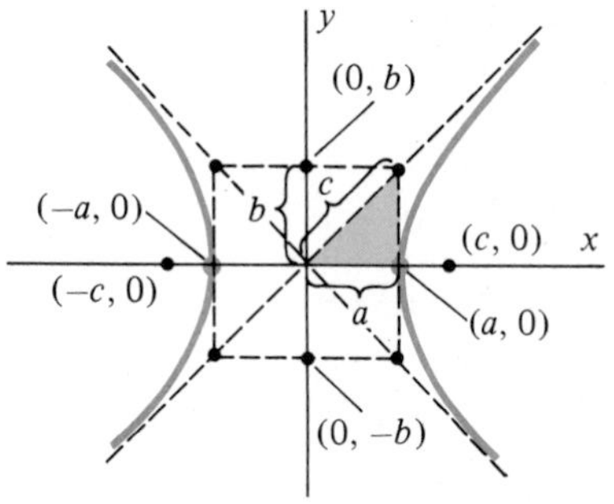

Figure 4.4 Relationships among the constants a, b and c.

Example 1 Sketch the hyperbola $\frac{x^2}{9} - \frac{y^2}{4} = 1$.

Solution: This equation has the form of equation (2) with $a = 3$ and $b = 2$. The vertices are therefore $V_1 = (-3, 0)$ and $V_2 = (3, 0)$, and the asymptotes are $y = \pm\frac{2}{3}x$. The graph appears in Figure 4.5. ■

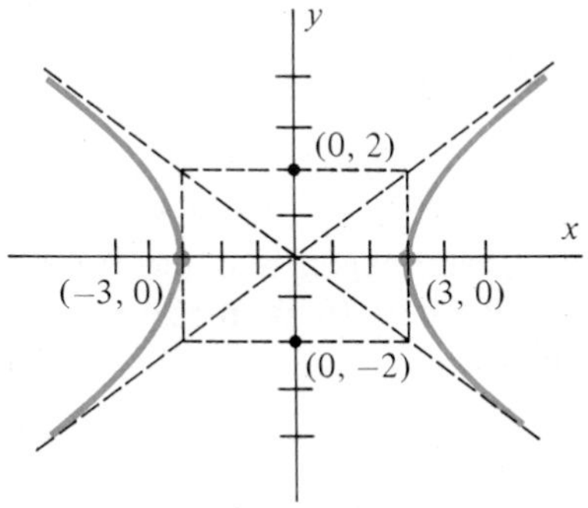

Figure 4.5 $\frac{x^2}{9} - \frac{y^2}{4} = 1$.

If the foci of a hyperbola lie along the y-axis at $F_1 = (0, c)$ and $F_2 = (0, -c)$, respectively, the condition

$$|d(P, F_2) - d(P, F_1)| = 2a$$

becomes

$$|\sqrt{x^2 + (y + c)^2} - \sqrt{x^2 + (y - c)^2}| = 2a, \tag{5}$$

which simplifies to the *standard form of the equation for the hyperbola with center at (0, 0) and foci along the y-axis*

$$\boxed{\frac{y^2}{a^2} - \frac{x^2}{b^2} = 1, \qquad b = \sqrt{c^2 - a^2}, \qquad c > a.} \tag{6}$$

As in the previous case, we may solve for y to determine the equations for the asymptotes:

$$y^2 = a^2\left(\frac{x^2}{b^2} + 1\right),$$

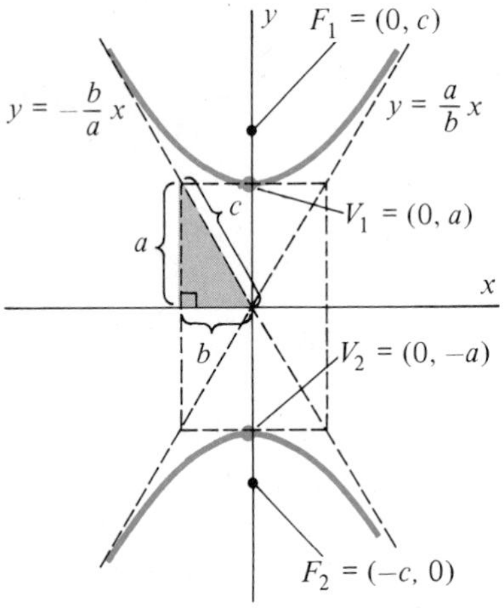

Figure 4.6 Hyperbola $\frac{y^2}{a^2} - \frac{x^2}{b^2} = 1$.

so

$$y = \pm a\sqrt{\frac{x^2}{b^2} + 1} = \pm\frac{a}{b}x\sqrt{1 + \frac{b^2}{x^2}}. \tag{7}$$

Since $\lim_{x\to\infty} \sqrt{1 + \frac{b^2}{x^2}} = 1$, equation (7) shows that *the asymptotes for the graph of equation (6) are given by*

$$\boxed{y = \pm\frac{a}{b}x.} \tag{8}$$

(See Figure 4.6.)

REMARK: A simple way to avoid confusion in working with equations (4) and (8) for the asymptotes associated with the standard form for a hyperbola is to note that in either case the equations are

$$y = \pm\left(\frac{\text{constant in denominator of } y\text{-term}}{\text{constant in denominator of } x \text{ term}}\right)^{1/2} \cdot x.$$

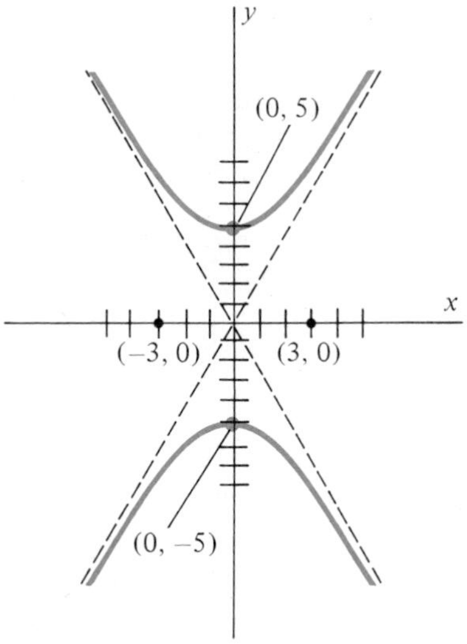

Figure 4.7 Graph of $25x^2 - 9y^2 + 225 = 0$.

Example 2 Graph the hyperbola $25x^2 - 9y^2 + 225 = 0$.

Solution: Moving the constant term to the right-hand side and dividing by -225 gives

$$\frac{y^2}{25} - \frac{x^2}{9} = 1, \qquad \text{or} \qquad \frac{y^2}{5^2} - \frac{x^2}{3^2} = 1.$$

This equation has the form of equation (6) with $a = 5$ and $b = 3$. By (8) the asymptotes are $y = \pm\frac{5}{3}x$. The vertices are $(0, a) = (0, 5)$ and $(0, -a) = (0, -5)$. The graph appears as Figure 4.7. ■

Translating the Hyperbola

If the center of a hyperbola is located at (h, k) rather than at $(0, 0)$, we translate axes using the equations

$$X = x - h \qquad \text{and} \qquad Y = y - k. \tag{9}$$

If the foci of the hyperbola lie on the X-axis, the equation in XY-coordinates, from (2), is

$$\frac{X^2}{a^2} - \frac{Y^2}{b^2} = 1, \tag{10}$$

and the equations for the asymptotes are

$$Y = \pm\frac{b}{a}X. \tag{11}$$

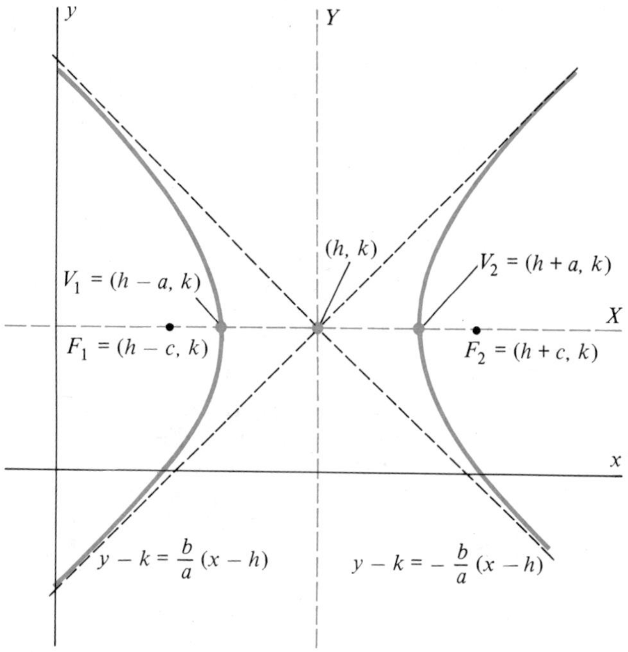

Figure 4.8 Hyperbola with center (h, k).

Using (9), we may write these equations in xy-coordinates. The equation

$$\frac{(x-h)^2}{a^2} - \frac{(y-k)^2}{b^2} = 1, \qquad b = \sqrt{c^2 - a^2}, \qquad a < c \tag{12}$$

is *the standard form of the equation for the hyperbola with center at (h, k) and foci on a line parallel to the x-axis*. The asymptotes for this hyperbola are

$$y - k = \pm\frac{b}{a}(x - h) \qquad \text{(see Figure 4.8).} \tag{13}$$

Similarly, *the standard form of the equation for the hyperbola with center at (h, k) and foci on a line parallel to the y-axis is*

$$\frac{(y-k)^2}{a^2} - \frac{(x-h)^2}{b^2} = 1, \qquad b = \sqrt{c^2 - a^2}, \qquad a < c \tag{14}$$

and the asymptotes for this hyperbola are

$$y - k = \pm\frac{a}{b}(x - h). \tag{15}$$

Example 3 Describe the graph of the hyperbola

$$16(x + 3)^2 - 4(y - 2)^2 = 64.$$

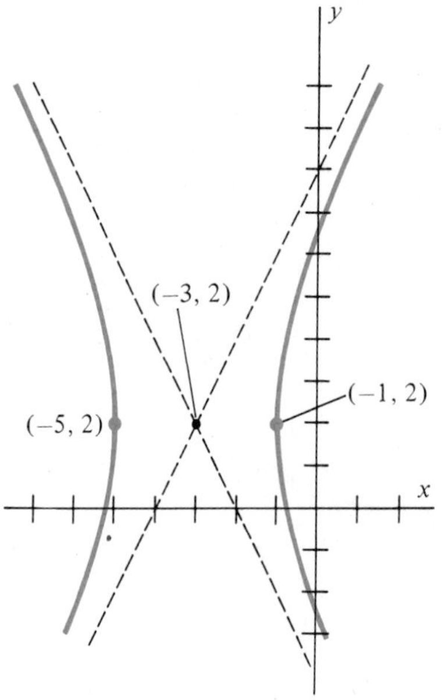

Figure 4.9 Graph of $16(x + 3)^2 - 4(y - 2)^2 = 64$.

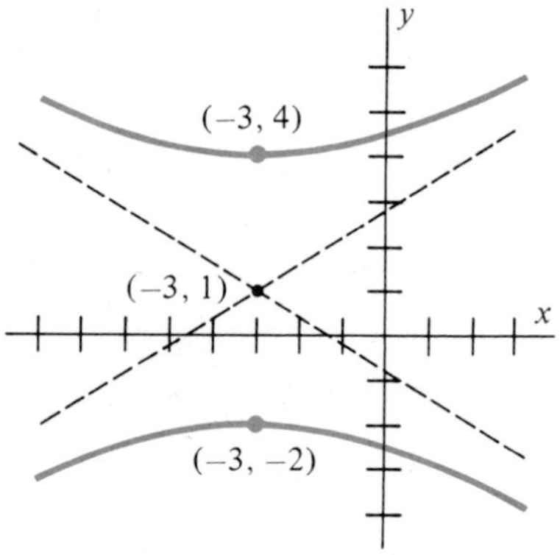

Figure 4.10 Graph of equation $25y^2 - 9x^2 - 50y - 54x - 281 = 0$.

Solution: Dividing both sides by 64 shows that the standard form for the hyperbola is

$$\frac{(x + 3)^2}{2^2} - \frac{(y - 2)^2}{4^2} = 1.$$

This equation has the form of equation (12) with $h = -3$, $k = 2$, $a = 2$, $b = 4$, and $c = \sqrt{a^2 + b^2} = \sqrt{20} = 2\sqrt{5}$. The hyperbola therefore has center $(-3, 2)$, vertices $(-5, 2)$ and $(-1, 2)$, foci $(-3 - 2\sqrt{5}, 2)$ and $(-3 + 2\sqrt{5}, 2)$, and asymptotes $y - 2 = \pm 2(x + 3)$. The graph appears as Figure 4.9. ■

Example 4 Describe the graph of the equation

$$25y^2 - 9x^2 - 50y - 54x - 281 = 0.$$

Solution: We complete the square on both variables on the left-hand side to find that

$$25(y^2 - 2y + 1) - 9(x^2 + 6x + 9) - 281 = 25 - 81,$$

so

$$25(y - 1)^2 - 9(x + 3)^2 = 225,$$

or

$$\frac{(y - 1)^2}{3^2} - \frac{(x + 3)^2}{5^2} = 1.$$

This equation has the form of equation (14) with $h = -3$, $k = 1$, $a = 3$, $b = 5$, and $c = \sqrt{a^2 + b^2} = \sqrt{34}$. Its graph is a hyperbola with center $(-3, 1)$, vertices $(-3, -2)$ and $(-3, 4)$, foci $(-3, 1 - \sqrt{34})$ and $(-3, 1 + \sqrt{34})$, and asymptotes $y - 1 = \pm\frac{3}{5}(x + 3)$ (see Figure 4.10). ■

Exercise Set 15.4

In Exercises 1–20, graph the hyperbola corresponding to the given equation. State the center and the equations for the asymptotes.

1. $\frac{x^2}{2^2} - \frac{y^2}{3^2} = 1$
2. $\frac{y^2}{4^2} - \frac{x^2}{2^2} = 1$
3. $\frac{x^2}{5} - \frac{y^2}{9} = 1$
4. $\frac{x^2}{9} - \frac{y^2}{2} = 1$
5. $16x^2 - 4y^2 - 64 = 0$
6. $x^2 - 9y^2 = 9$
7. $4y^2 - 7x^2 - 28 = 0$
8. $5x^2 - 9y^2 - 10x = 25$
9. $9x^2 - 4y^2 + 54x + 32y + 119 = 0$
10. $16y^2 - 4x^2 - 32y - 8x - 52 = 0$
11. $9x^2 - 4y^2 - 36x - 24y - 36 = 0$
12. $4y^2 - x^2 - 2x = 2$
13. $x^2 - 4y^2 = 4$
14. $4x^2 - y^2 + 24x + 4y + 28 = 0$
15. $2x^2 - y^2 - 4\sqrt{2}\,x + 2\sqrt{2}\,y = 6$
16. $3x^2 - 5y^2 + 6x - 20y = -8$.
17. $16x^2 - 25y^2 + 64x + 150y = 561$
18. $9x^2 - y^2 + 36x + 4y + 23 = 0$
19. $9y^2 - 16x^2 - 126y - 160x = 184$
20. $4y^2 - 16y - x^2 - 4x + 11 = 0$

In Exercises 21–27, find the equation for the hyperbola with the stated properties.

21. Foci at $(-2, 0)$ and $(2, 0)$, vertices at $(-1, 0)$ and $(1, 0)$.
22. Foci at $(2, 5)$ and $(2, -3)$, vertices at $(2, 3)$ and $(2, -1)$.
23. Vertices at $(0, 4)$ and $(0, 0)$, asymptotes $y = \pm 2x$.
24. Foci at $(-3, 0)$ and $(3, 0)$, asymptotes $y = \pm\frac{3}{2}x$.
25. Foci at $(-5, 2)$ and $(3, 2)$, asymptotes $y - 2 = \pm 2(x + 1)$.
26. Center at the origin, one vertex $(0, 4)$, and containing the point $(2, 8)$.
27. Center at the origin, one vertex $(0, 4)$, and one focus $(0, -5)$.
28. Find an equation for the line tangent to the graph of the hyperbola $4x^2 - 16y^2 = 64$ at the point $(4\sqrt{2}, 2)$.
29. What properties do the graph of the hyperbola

$$\frac{x^2}{a^2} - \frac{y^2}{b^2} = 1$$

and its **conjugate** $\frac{y^2}{b^2} - \frac{x^2}{a^2} = 1$ have in common?

30. Show that the line tangent to the graph of the hyperbola $b^2x^2 - a^2y^2 = a^2b^2$ at the point (x_0, y_0) has slope $m = \frac{b^2x_0}{a^2y_0}$.
31. Using the result of Exercise 30, find the equation for the line tangent to the graph of $b^2x^2 - a^2y^2 = a^2b^2$ at the point (x_0, y_0).
32. Show that for the general second degree equation

$$Ax^2 + Bxy + Cy^2 + Dx + Ey + F = 0$$

if $B = 0$ and both $A \neq 0$ and $C \neq 0$, the graph is a hyperbola if A and B have opposite signs.

33. A chord of a hyperbola through a focus perpendicular to the principal axis is called a **latus rectum.** Show that the length of a latus rectum is $\frac{2b^2}{a}$.
34. Find an equation for the hyperbola with center at the origin, latus rectum of length 20, and one vertex at $(0, 4)$.
35. Show that any point (x, y) satisfying equation (2) satisfies the geometric condition of Definition 3.

15.5 ROTATION OF AXES

Up to this point, we have discussed all possible second degree equations of the form

$$Ax^2 + Bxy + Cy^2 + Dx + Ey + F = 0 \tag{1}$$

except those involving an xy term. In this section we complete that discussion by considering the case $B \neq 0$. On encountering such an expression, our objective will be to find a suitable change of variables from (x, y)-coordinates to (x', y')-coordinates so that in the new coordinates equation (1) contains no $x'y'$ term.

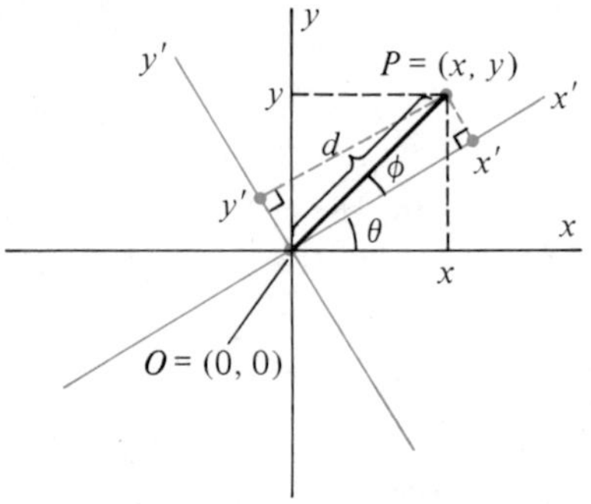

Figure 5.1 Rotation of axes through θ radians.

The simplest change of variables that accomplishes this objective is a rotation of the xy-axes through an angle θ. After having constructed new x'- and y'-axes by rotating the x- and y-axes counterclockwise through an angle θ, we observe that a point P with xy-coordinates (x, y) will have $x'y'$-coordinates given by the equations

$$x' = d \cos \phi, \tag{2a}$$

$$y' = d \sin \phi, \tag{2b}$$

where d is the distance from the origin, 0, to P and ϕ is the angle between $\overline{OP}$ and the x'-axis (Figure 5.1).

Similarly, we can express the xy-coordinates of P as

$$x = d \cos(\theta + \phi), \tag{3a}$$

$$y = d \sin(\theta + \phi). \tag{3b}$$

Using the addition laws for sine and cosine we can combine equations (2) and (3) to obtain the equations

$$x = d(\cos \theta \cos \phi - \sin \theta \sin \phi) = x' \cos \theta - y' \sin \theta, \tag{4a}$$

$$y = d(\sin \theta \cos \phi + \cos \theta \sin \phi) = x' \sin \theta + y' \cos \theta. \tag{4b}$$

Solving this system for x' and y', we obtain

$$x' = x \cos \theta + y \sin \theta, \tag{5a}$$

$$y' = -x \sin \theta + y \cos \theta. \tag{5b}$$

By using equations (4) and (5), we can pass back and forth from xy- to $x'y'$-coordinates. In particular, if we substitute the right-hand sides of equations (4) for x and y into equation (1), we find that the coefficient of the $x'y'$ term is

$$\begin{aligned} &2(C - A)\sin \theta \cos \theta + B(\cos^2 \theta - \sin^2 \theta) \\ &= (C - A)\sin 2\theta + B \cos 2\theta. \end{aligned}$$

If we wish this coefficient to be zero, we need only require that

$$\frac{\cos 2\theta}{\sin 2\theta} = \frac{A - C}{B}$$

or that*

$$\cot 2\theta = \frac{A - C}{B}, \qquad 0 < \theta < \frac{\pi}{2}. \tag{6}$$

Since θ is not yet specified, this shows that by choosing θ to satisfy equation (6) we can ensure that equation (1), when written in $x'y'$-coordinates, will contain no $x'y'$ term. Thus the graph of equation (1), in the rotated $x'y'$-coordinate system, will be a circle, a parabola, an ellipse, or a hyperbola.

Example 1 For the equation $5x^2 + 6xy + 5y^2 - 8 = 0$,

(a) find a rotation that eliminates the xy term, and
(b) sketch the graph.

---

*Notice that equation (6) is defined for all B except $B = 0$. In this case the original equation (1) contains no xy term, so a rotation is not needed.

Solution: We have $A = 5$, $B = 6$, and $C = 5$, so equation (6) becomes $\cot 2\theta = \dfrac{5-5}{6} = 0$. The equation $\cot 2\theta = 0$ has solution $2\theta = \pi/2$, or $\theta = \pi/4$. With this value of θ equations (4) become

$$x = \frac{\sqrt{2}}{2}x' - \frac{\sqrt{2}}{2}y',$$
$$y = \frac{\sqrt{2}}{2}x' + \frac{\sqrt{2}}{2}y'.$$

Inserting these expressions for x and y in the original equation gives

$$5\left(\frac{\sqrt{2}}{2}x' - \frac{\sqrt{2}}{2}y'\right)^2 + 6\left(\frac{\sqrt{2}}{2}x' - \frac{\sqrt{2}}{2}y'\right)\left(\frac{\sqrt{2}}{2}x' + \frac{\sqrt{2}}{2}y'\right) + 5\left(\frac{\sqrt{2}}{2}x' + \frac{\sqrt{2}}{2}y'\right)^2 - 8 = 0$$

or, on simplifying,

$$x'^2 + \frac{y'^2}{4} = 1.$$

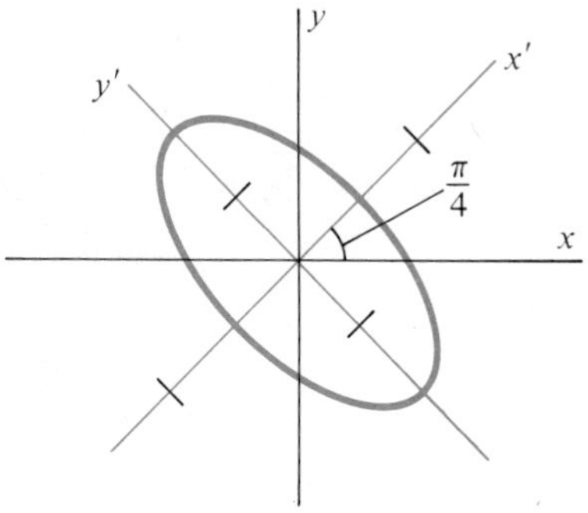

Figure 5.2 Graph of $5x^2 + 6xy + 5y^2 - 8 = 0$ is a rotated ellipse.

The graph of this ellipse is shown in Figure 5.2. ■

Example 2 Graph the equation $xy = 1$.

Strategy

Solution

Identify form of the equation.

This is an equation of the form of the general second degree equation (1) with $B = 1$ and $A = C = D = E = 0$. Although we are already familiar with the general nature of its graph, we may use the method of rotation of axes to determine more precise information.

Rotate axes to eliminate xy term.

Find the rotation angle from (6).

From equation (6) we find that

$$\cot 2\theta = \frac{A - C}{B} = 0$$

Express x and y as functions of x' and y' and insert these expressions into the given equation.

so $\theta = \pi/4$, as in Example 1. From equations (4) we then obtain

$$x = \frac{\sqrt{2}}{2}x' - \frac{\sqrt{2}}{2}y',$$
$$y = \frac{\sqrt{2}}{2}x' + \frac{\sqrt{2}}{2}y'.$$

Inserting these expressions in $xy = 1$ gives

$$\left(\frac{\sqrt{2}}{2}x' - \frac{\sqrt{2}}{2}y'\right)\left(\frac{\sqrt{2}}{2}x' + \frac{\sqrt{2}}{2}y'\right) = 1,$$

or

Bring resulting equation to standard form.

$$\frac{x'^2}{2} - \frac{y'^2}{2} = 1.$$

The graph is therefore a hyperbola in $x'y'$-coordinates, with center at (0, 0) and asymptotes $y' = \pm x'$ (see Figure 5.3). ■

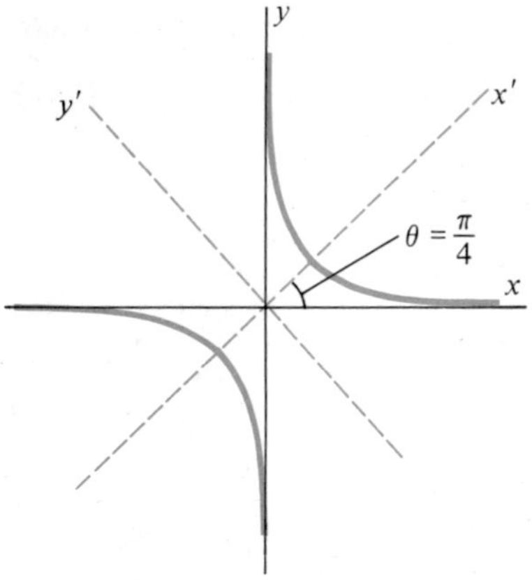

Figure 5.3 Graph of $xy = 1$ is a rotated hyperbola.

We have now completed our analysis of the general second degree equation

$$Ax^2 + Bxy + Cy^2 + Dx + Ey + F = 0. \tag{7}$$

If $B \neq 0$, we can employ a rotation of axes, as described above, to obtain an equation in $x'y'$-coordinates of the form

$$A'x'^2 + C'y'^2 + D'x' + E'y' + F' = 0 \tag{8}$$

The graph of equation (8) must then fall under one of the following cases:

(i) If $A' = C' \neq 0$, the graph of (8) is a circle (Section 15.1).
(ii) If $A' = 0$ or $C' = 0$ (but not both), the graph of (8) is a parabola (Exercise 32, Section 15.2).
(iii) If $A' \neq 0$ and $C' \neq 0$ and $A' \neq C'$, the graph of (8) is
  (a) an ellipse, if A' and C' have the same sign (Exercise 30, Section 15.3),
  (b) a hyperbola, if A' and C' have opposite signs (Exercise 32, Section 15.4).
(iv) If $A' = C' = 0$, the graph of (8) is a line (Section 1.7).

Since this list exhausts all possibilities for the coefficients in equation (7), the graph of any equation of the form (7) must be a circle, a parabola, an ellipse, a hyperbola, or a line, or a degenerate case of one of these.

Exercise Set 15.5

In each of Exercises 1–6, find a rotation θ that eliminates the xy term.

1. $3x^2 + 2xy + y^2 + 6x + 9 = 0$
2. $6x^2 + 2xy + 2y^2 + 8y + 8 = 0$
3. $x^2 - 3xy + y^2 - x + y + 10 = 0$
4. $4x^2 + \sqrt{3}\,xy + 3y^2 + 2x + 7y - 8 = 0$
5. $5x^2 + 4xy + 2y^2 = 1$
6. $9x^2 - 24xy + 16y^2 - 40x - 30y + 100 = 0$

In each of Exercises 7–16, find a rotation θ that eliminates the xy term. Then use equations (4) to obtain an equation in $x'y'$-coordinates. Finally, graph the equation.

7. $5x^2 + 6xy + 5y^2 - 32 = 0$
8. $3x^2 + 10xy + 3y^2 - 32 = 0$
9. $3x^2 + 2\sqrt{3}\,xy + y^2 + 2x - 2\sqrt{3}\,y + 16 = 0$
10. $5x^2 + 4xy + 5y^2 = 21$
11. $9x^2 - 24xy + 16y^2 - 40x - 30y + 100 = 0$
12. $x^2 - 6xy + y^2 + 16 = 0$
13. $7x^2 - 2\sqrt{3}xy + 5y^2 - 16 = 0$
14. $23x^2 - 26\sqrt{3}xy - 3y^2 + 144 = 0$
15. $x^2 + 2xy + y^2 + \sqrt{2}x - \sqrt{2}\,y = 0$
16. $31x^2 + 10\sqrt{3}xy + 21y^2 = 144.$
17. Find the endpoints of the major axis of the ellipse in Exercise 16.
18. Find the coordinates of the focus of the parabola in Exercise 15.
19. Find the vertices of the hyperbola in Exercise 14.
20. Find the endpoints of the minor axis of the ellipse in Exercise 13.

SUMMARY OUTLINE OF CHAPTER 15

- A **translation of axes** that relocates the origin (0, 0) at point (h, k) is accomplished by the substitutions

  $X = x - h$ and $Y = y - k$

  from xy- to XY-coordinates.

- The standard forms of the equations for the **parabola** with vertex at (h, k) and focal length c are

  $(x - h)^2 = 4c(y - k)$ (axis vertical),
  $(y - k)^2 = 4c(x - h)$ (axis horizontal).

- The standard form of the equation for the **ellipse** with center at (h, k) is

$$\frac{(x-h)^2}{a^2} + \frac{(y-k)^2}{b^2} = 1.$$

- The standard forms of the equations for the **hyperbola** with center at (h, k) are

$$\frac{(x-h)^2}{a^2} - \frac{(y-k)^2}{b^2} = 1 \text{ (foci on line parallel to } x\text{-axis)},$$

$$\frac{(y-k)^2}{a^2} - \frac{(x-h)^2}{b^2} = 1 \text{ (foci on line parallel to } y\text{-axis)}.$$

- A **rotation of axes** to eliminate the xy term in

$$Ax^2 + Bxy + Cy^2 + Dx + Ey + F = 0$$

is accomplished by the substitutions

$$x = x' \cos\theta - y' \sin\theta$$
$$y = x' \sin\theta + y' \cos\theta$$

where θ is determined by the equation

$$\cot 2\theta = \frac{A-C}{B}, \qquad 0 < \theta < \frac{\pi}{2}.$$

REVIEW EXERCISES—CHAPTER 15

1. Write an equation for the circle with center $(3, -4)$ and radius 6.

2. Write an equation for the circle with center $(-2, -3)$ and radius 5.

3. Write an equation for the circle with center $(2, 3)$ that contains the point $(2, -1)$.

4. Write equations for the circles with radius 5 containing the points $(-1, -4)$ and $(4, 1)$. How many such circles exist?

5. Write an equation for the parabola with vertex $(0, 4)$ and directrix $x = -2$.

6. Write an equation for the parabola with axis horizontal, vertex on the y-axis, and containing the points $(2, 1)$ and $(18, 3)$.

7. Write an equation for the parabola with axis vertical and containing the points $(0, 5)$, $(2, 5)$, and $(3, 14)$.

8. Find an equation for the circle that passes through the focus and vertex of the parabola $y = 8x^2$ and that is tangent to the parabola at $(0, 0)$.

9. Find an equation of the ellipse with center $(-1, 1)$, focus $(1, 1)$, and vertex $(2, 1)$.

10. Find an equation for the ellipse with focus $(0, 1)$ and vertices $(0, 0)$ and $(0, 8)$.

11. Does the origin lie inside or outside the ellipse with equation $(x-4)^2 + 4(y-3)^2 = 4$?

Find the points of intersection of each of the following pairs of graphs.

12. The line $x + y - 4 = 0$ and the circle $x^2 + y^2 = 16$.

13. The line $x - y - 2 = 0$ and the parabola $y = 4 - x^2$.

14. The line $4y - 3x + 12 = 0$ and the ellipse $9x^2 + 16y^2 = 144$.

15. The parabola $y = 6 - x^2$ and the parabola $y = x^2 - 2$.

In Exercises 16–45, write the equation in standard form, identify the graph, and sketch.

16. $3y^2 - x - 4 = 0$

17. $x^2 - 2x + y^2 + 6y + 2 = 0$

18. $16x^2 + 7y^2 - 112 = 0$

19. $2x^2 - y^2 + 4 = 0$

20. $x^2 - 3y^2 + 3 = 0$

21. $16x^2 + 12y^2 - 192 = 0$

22. $3x^2 + 3y^2 - 3 = 0$

23. $9x^2 + 4y^2 - 36 = 0$

24. $16x^2 - 7y^2 + 112 = 0$

25. $4x^2 - 9y^2 - 36 = 0$

26. $4x^2 + 9y^2 - 36 = 0$

27. $y^2 + x - 2 = 0$

28. $7y^2 - 2x + 6 = 0$

29. $x^2 + 14x + y^2 - 10y + 70 = 0$

30. $9x^2 + 8y^2 - 72 = 0$

31. $9x^2 - 8y^2 - 72 = 0$

32. $4x^2 + y + 8 = 0$

33. $4x^2 + 25y^2 - 100 = 0$

34. $x^2 + 3y^2 - 3 = 0$

35. $3x^2 - 3y^2 + 12 = 0$

36. $x^2 - 6x + y^2 - 4y + 8 = 0$

37. $x^2 - 2ax + y^2 + 4ay + 5a^2 - 1 = 0$

38. $x^2 - 2x + y^2 - 8 = 0$

39. $y^2 - x^2 - 2 = 0$

40. $3x + 2y^2 - 3 = 0$

41. $x^2 + y^2 - 2x - 6y + 3 = 0$

42. $4x^2 - 25y^2 + 100 = 0$

43. $9y^2 - 25x^2 - 225 = 0$

44. $4x^2 - 9y^2 - 36 = 0$

45. $2x - 16y^2 - 12 = 0.$

CHAPTER 16

POLAR COORDINATES AND PARAMETRIC EQUATIONS

16.1 INTRODUCTION

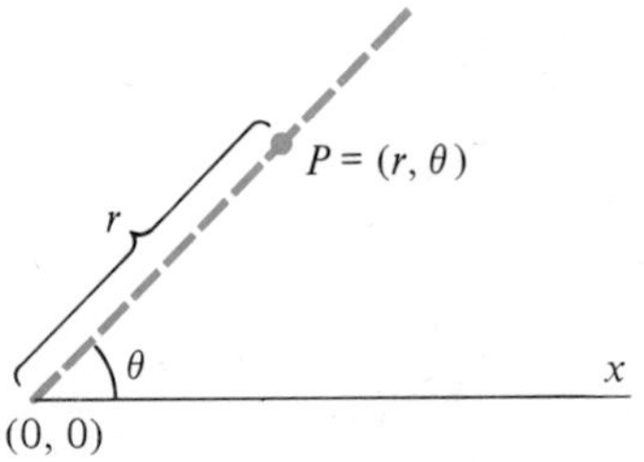

Figure 1.1 The polar coordinates of a point P.

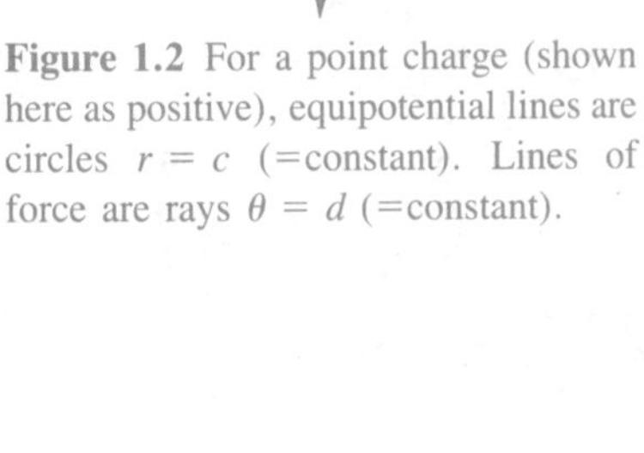

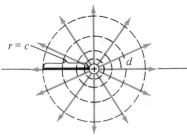

Figure 1.2 For a point charge (shown here as positive), equipotential lines are circles $r = c$ (=constant). Lines of force are rays $\theta = d$ (=constant).

Up to this point, we have been using Cartesian coordinates for the plane, by which we mean that the rectangular or xy-coordinates of a point $P(x, y)$ denote its distances from two perpendicular axes. This is a convenient scheme to use when graphing functions whose variables represent naturally perpendicular quantities (length versus height, for example) or when attempting to approximate areas by rectangles.

However, many plane curves can be more easily described in a coordinate system in which the coordinates of a point $P(r, \theta)$ measure its (radial) distance, r, along a ray from the origin that makes an angle θ with some fixed reference ray (usually the positive x-axis; see Figure 1.1).

These coordinates are especially convenient for describing curves that are symmetric with respect to the origin. An example of this situation occurs in the theory of electric potential fields. For a point charge, the lines representing equal electric potentials (called equipotentials) in any plane containing the point charge are circles with the point as center (Figure 1.2). These equipotential circles can most easily be described simply by stating their radius. Similarly, the lines of force emanating from such a point charge are rays, which are described completely by the angle they form with the positive x-axis.

The system in which points in the plane are coordinatized by their radial and angular variables, r and θ, is called the **polar coordinate system.** The goal of Sections 16.2 through 16.4 is to develop this system and its relationship to the familiar rectangular coordinate system and to determine how to graph equations and calculate areas in polar coordinates.

In Sections 16.5 and 16.6, we introduce the idea of *parametric equations* to bring a new viewpoint to certain curves in the plane. We do this because we want to interpret these curves as graphs of functions where we have previously been unable to do so. For example, the graph of an ellipse in xy-coordinates has up until now been regarded as not corresponding to a function $y = f(x)$, since some values of x are associated with more than one value of y (i.e., the vertical line test fails for this curve).

This same difficulty can arise in polar coordinates. For example, consider the shadow cast by the bob of a conical pendulum on a plane perpendicular to its axis by

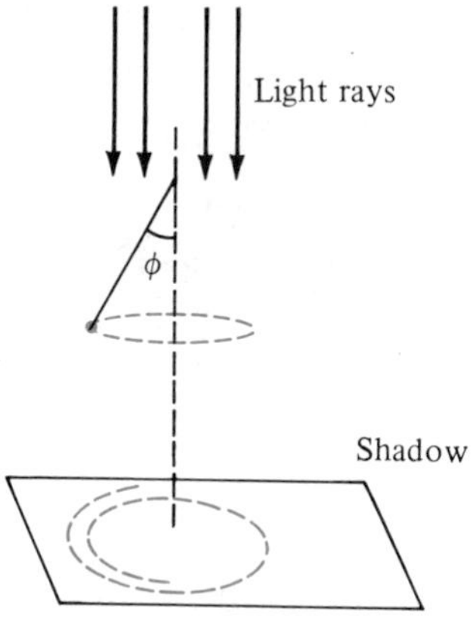

Figure 1.3 Conical pendulum.

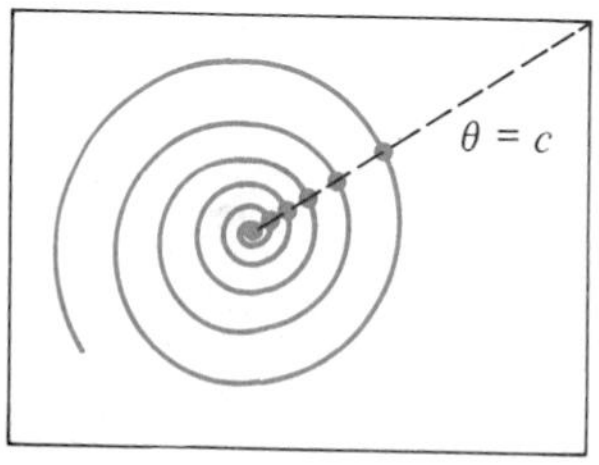

Figure 1.4 Shadow of bob traces out a spiral in the plane.

light rays parallel to the axis (Figure 1.3). As the bob simultaneously rotates and approaches the axis, the shadow traces out the spiral shown in Figure 1.4. Even though points on this spiral may be described conveniently by their polar coordinates, r and θ, the spiral cannot be interpreted as the graph of a function $r = f(\theta)$ since many points correspond to a given value of θ.

This difficulty may be overcome by viewing the curve in question as the path traced out by the point $P(t) = (x(t), y(t))$ as the independent variable t ranges through some interval, usually $[0, \infty)$ or $(-\infty, \infty)$. Here $x(t)$ and $y(t)$, the rectangular coordinates of P, are themselves functions of the single independent variable t. (Similarly, we can obtain the graph of a curve in polar coordinates as the path traced out by the point $P(t) = (r(t), \theta(t))$.) In this way we associate the desired curve with a *function* of t, since only one point corresponds to a particular value of t. We refer to the coordinate functions $x(t)$ and $y(t)$ (or, $r(t)$ and $\theta(t)$) as **parametric equations** for the curve, and we call the independent variable t the **parameter.**

Thus, the present chapter serves as the bridge linking the calculus of a single variable with the multidimensional calculus. Although the ideas and techniques that appear here are still those of the univariate calculus, they nevertheless point the way to developing a theory for functions whose values are *vectors* in the plane or in space. The study of functions whose dependent or independent variables are vectors occupies the remaining chapters of this text.

16.2 THE POLAR COORDINATE SYSTEM

Let O be any point in the plane, called the **pole,** and let ℓ be any half infinite ray emanating from O. The ray ℓ is called the **polar axis** for the plane. The choice of a pole and a polar axis determines a **polar coordinate system** for the plane in which any point P may be assigned **polar coordinates** $P = (r, \theta)$, where

(i) r, called the **radial variable,** is the distance from O to P, and
(ii) θ, called the **angular variable,** is the angle formed between the polar axis and the ray $\overline{OP}$, measured in the counter clockwise direction (Figure 2.1).

Figure 2.2 illustrates various points plotted in polar coordinates. As in most applications of trigonometry in the calculus, we prefer to work in radians.

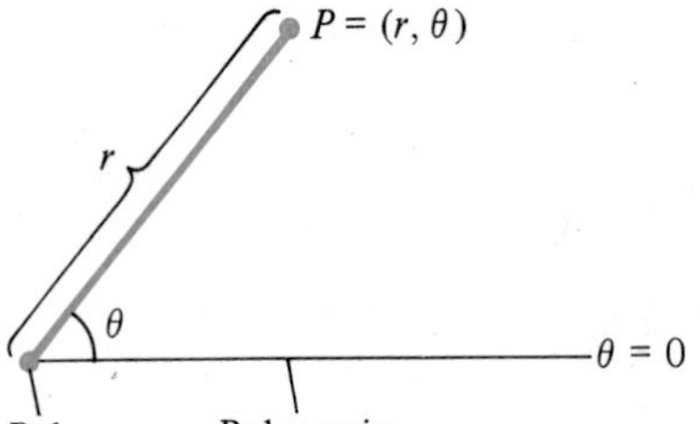

Figure 2.1 Polar coordinates for point P.

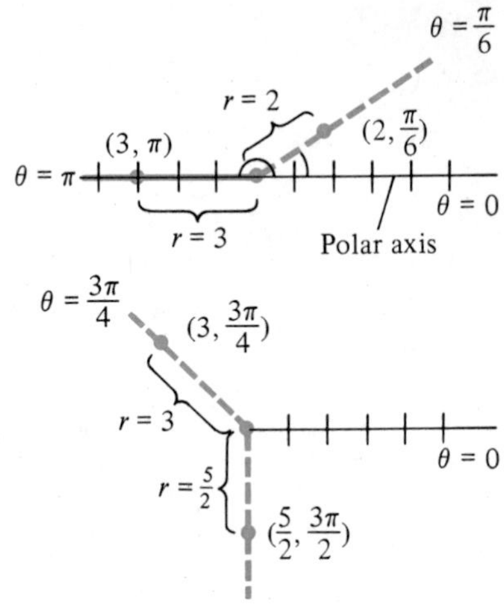

Figure 2.2 Polar coordinates for several points.

In the xy-coordinate plane, the graphs of equations of the form x = constant or y = constant are lines. In the polar coordinate plane, graphs of equations of the form r = constant are circles, while graphs of equations of the form θ = constant are rays (see Figure 2.3).

You may have already noticed that polar coordinates are not unique. Indeed, since the angles $\theta = \theta_0 + 2n\pi$, $n = 0, \pm 1, \pm 2, \ldots$ all determine the same ray, we have

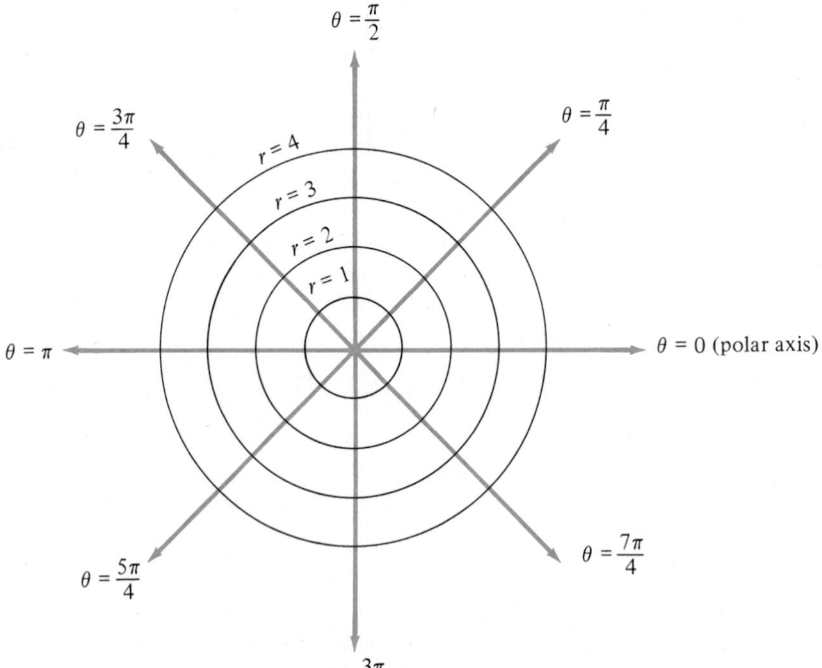

Figure 2.3 The polar coordinate plane.

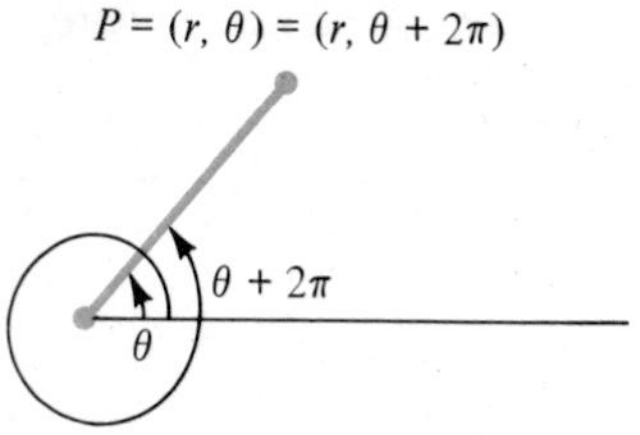

Figure 2.4 $(r, \theta) = (r, \theta + 2n\pi)$.

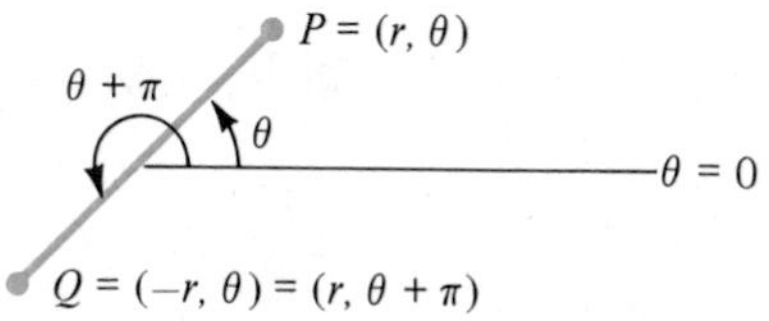

Figure 2.5 $(-r, \theta) = (r, \theta + \pi)$.

$$(2, \pi) = (2, 3\pi) = (2, 5\pi) = (2, -\pi) = \cdots,$$

$$\left(1, \frac{\pi}{4}\right) = \left(1, \frac{9\pi}{4}\right) = \left(1, \frac{17\pi}{4}\right) = \left(1, -\frac{7\pi}{4}\right) = \cdots,$$

and, in general,

$$(r, \theta) = (r, \theta + 2n\pi), \qquad n = \pm 1, \pm 2, \pm 3, \ldots \tag{1}$$

Moreover, it is customary to extend the range of values for the radial variable r to include negative numbers via the identity

$$(-r, \theta) = (r, \theta + \pi) \tag{2}$$

In other words, the point with polar coordinates $(-r, \theta)$ lies r units along the ray pointing in the direction opposite that specified by the angular variable θ.

Example 1 By identities (1) and (2) we have

(a) $\left(3, \frac{\pi}{6}\right) = \left(3, \frac{13\pi}{6}\right) = \left(3, -\frac{11\pi}{6}\right)$,

(b) $\left(-4, \frac{\pi}{4}\right) = \left(4, \frac{5\pi}{4}\right) = \left(4, -\frac{3\pi}{4}\right)$,

(c) $\left(7, -\frac{\pi}{2}\right) = \left(-7, \frac{\pi}{2}\right) = \left(-7, \frac{5\pi}{2}\right) = \left(-7, -\frac{3\pi}{2}\right)$. ■

The *pole* O in the polar coordinate plane has coordinates $(0,\theta)$ for every real number θ.

Polar versus Rectangular Coordinates

The relationships between polar and xy-coordinates are easily determined when the origin in xy-coordinates is identified with the pole, and the positive x-axis is taken to be the polar axis (see Figure 2.6). If the point P has rectangular coordinates (x, y) and polar coordinates (r, θ), and if $r \geq 0$, the definitions of $\sin\theta$ and $\cos\theta$ give immediately

$$x = r\cos\theta \qquad \text{and} \qquad y = r\sin\theta \tag{3}$$

(Figure 2.6). In Exercise 54, you are asked to show that equations (3) hold for $r < 0$ as well.

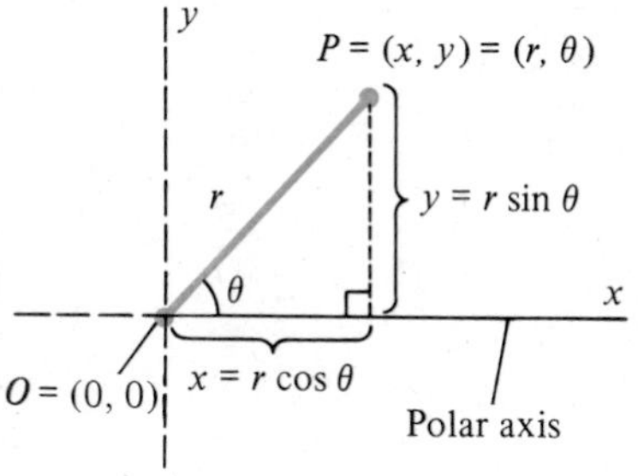

Figure 2.6 Relationships between rectangular and polar coordinates.

It follows from the equations in (3) (or from Figure 2.6) that

$$r = \sqrt{x^2 + y^2}; \qquad \tan\theta = \frac{y}{x}, \qquad x \neq 0. \tag{4}$$

Equations in line (3) are used in changing from polar to rectangular coordinates while the equations in line (4) allow us to change from rectangular to polar coordinates.

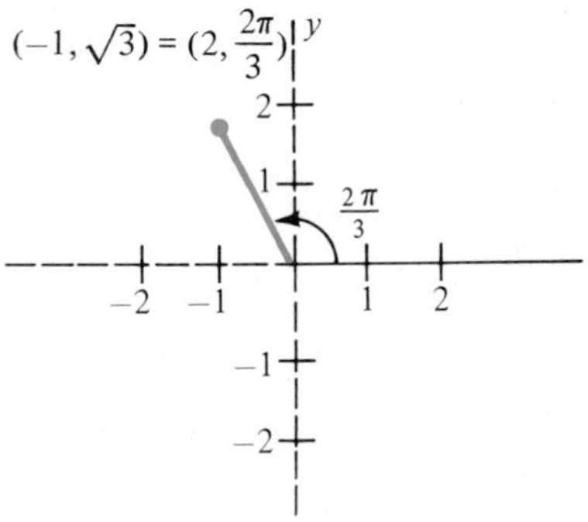

Figure 2.7

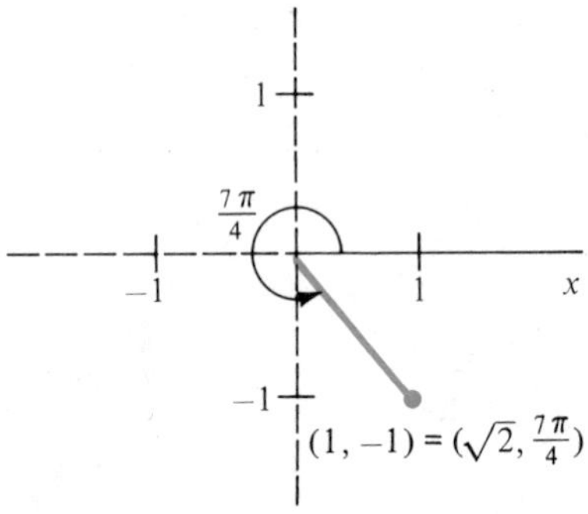

Figure 2.8

Example 2 To find the rectangular coordinates corresponding to the polar coordinates $(2, 2\pi/3)$, we use equations (3) with $r = 2$, $\theta = 2\pi/3$:

$$x = 2\cos\frac{2\pi}{3} = 2\left(-\frac{1}{2}\right) = -1,$$

$$y = 2\sin\frac{2\pi}{3} = 2\left(\frac{\sqrt{3}}{2}\right) = \sqrt{3}.$$

(See Figure 2.7.) ■

Example 3 To find the polar coordinates of the point with rectangular coordinates $(1, -1)$, we use the equations in (4):

$$r = \sqrt{x^2 + y^2} = \sqrt{1^2 + 1^2} = \sqrt{2}, \tag{5}$$

$$\tan\theta = \frac{y}{x} = \frac{-1}{1} = -1. \tag{6}$$

Now r is clearly determined by equation (5), but θ is *not* uniquely determined by equation (6), since both $\theta = 3\pi/4$ and $\theta = 7\pi/4$ satisfy $\tan\theta = -1$. The correct value, $\theta = 7\pi/4$, is identified by noting that the point $(1, -1)$ lies in the fourth quadrant (Figure 2.8). ■

REMARK: It is important to note, as in Example 3, that equations (4) do not uniquely determine θ for given values of x and y. This is because each value of $\tan\theta$ occurs twice for $\theta \in [0, 2\pi)$. This difficulty is overcome most simply by noting the quadrant in which the point (x, y) lies, since $\tan\theta$ is one-to-one in each quadrant. An alternative approach is to first determine r from equation (4), and then determine θ as the (unique) simultaneous solution of equations (3): $x = r\cos\theta$; $y = r\sin\theta$.

Example 4 Graph the polar equation $r = 1 + \cos\theta$ and find the rectangular form for the equation.

Solution: The simplest approach to graphing polar equations is to select several (convenient) values of θ (beginning with $\theta = 0$), calculate the corresponding value of r, and plot the points (r, θ) obtained. The graph is then completed by sketching a curve connecting these points. For the equation $r = 1 + \cos\theta$, we obtain the following points:

θ	0	$\frac{\pi}{6}$	$\frac{\pi}{4}$	$\frac{\pi}{3}$	$\frac{\pi}{2}$	$\frac{2\pi}{3}$	$\frac{3\pi}{4}$	$\frac{5\pi}{6}$	π
r	2	$\frac{2+\sqrt{3}}{2}$	$\frac{2+\sqrt{2}}{2}$	$\frac{3}{2}$	1	$\frac{1}{2}$	$\frac{2-\sqrt{2}}{2}$	$\frac{2-\sqrt{3}}{2}$	0

θ	$\frac{7\pi}{6}$	$\frac{5\pi}{4}$	$\frac{4\pi}{3}$	$\frac{3\pi}{2}$	$\frac{5\pi}{3}$	$\frac{7\pi}{4}$	$\frac{11\pi}{6}$	2π
r	$\frac{2-\sqrt{3}}{2}$	$\frac{2-\sqrt{2}}{2}$	$\frac{1}{2}$	1	$\frac{3}{2}$	$\frac{2+\sqrt{2}}{2}$	$\frac{2+\sqrt{3}}{2}$	2

The graph is the **cardioid** in Figure 2.9. ■

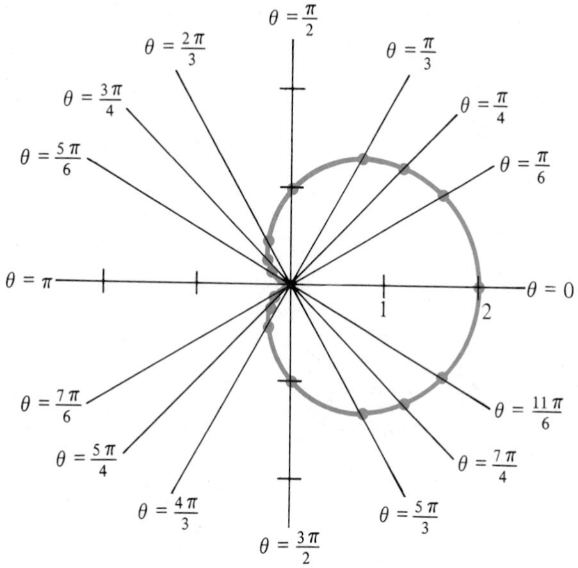

Figure 2.9 Cardioid $r = 1 + \cos\theta$.

To find the rectangular form of the equation $r = 1 + \cos\theta$, we multiply both sides by r to obtain

$$r^2 = r + r\cos\theta.$$

Using the equations in (3) and (4), we obtain

$$x^2 + y^2 = \sqrt{x^2 + y^2} + x.$$

Clearly the original equation is simpler to graph than its rectangular counterpart.

Symmetry in Polar Coordinates

Perhaps you noticed in Example 4 that the graph of the equation $r = 1 + \cos\theta$ is symmetric about the x-axis. This is due to the identity $\cos(-\theta) = \cos\theta$, or $\cos(2\pi - \theta) = \cos\theta$, $0 \le \theta \le \pi$. More generally, graphs of polar equations will be

(i) *symmetric about the x-axis* if $(r, -\theta)$ lies on the graph whenever (r, θ) does (Figure 2.10);

(ii) *symmetric about the y-axis* if $(r, \pi - \theta)$ lies on the graph whenever (r, θ) does (Figure 2.11);

(iii) *symmetric about the origin* if $(-r, \theta) = (r, \theta + \pi)$ lies on the graph whenever (r, θ) does (Figure 2.12).

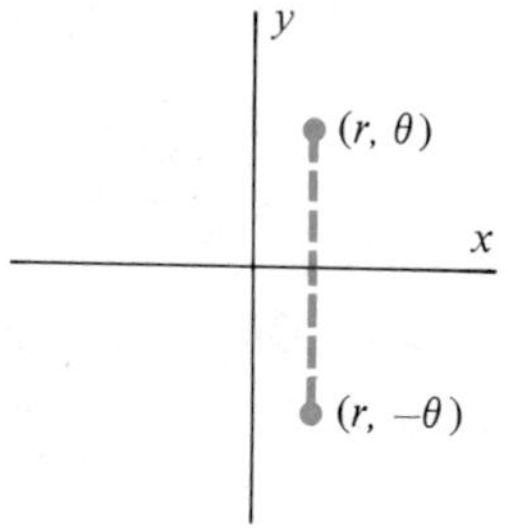

Figure 2.10 Symmetry about the x-axis.

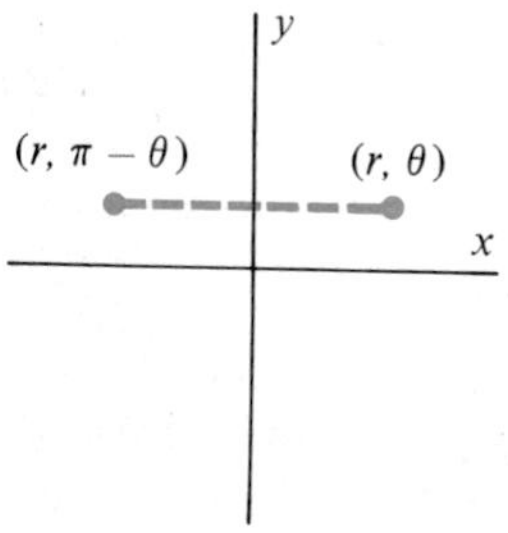

Figure 2.11 Symmetry about the y-axis.

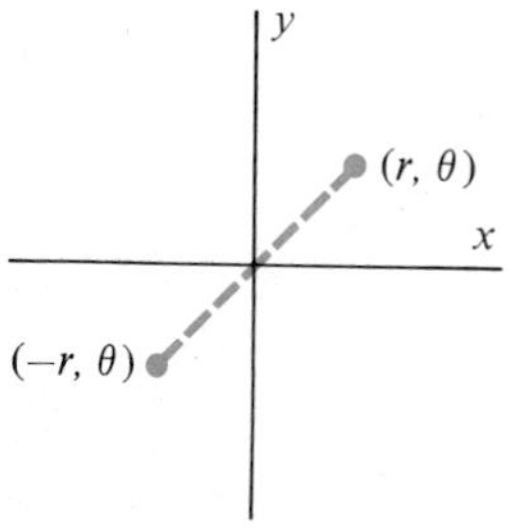

Figure 2.12 Symmetry about the origin.

Example 5 Graph the polar equation $r = \sin 2\theta$.

Solution: For values of θ with $0 \le \theta \le \pi$, we obtain the following pairs of points:

θ	0	$\frac{\pi}{6}$	$\frac{\pi}{4}$	$\frac{\pi}{3}$	$\frac{\pi}{2}$	$\frac{2\pi}{3}$	$\frac{3\pi}{4}$	$\frac{5\pi}{6}$	π
r	0	$\frac{\sqrt{3}}{2}$	1	$\frac{\sqrt{3}}{2}$	0	$\frac{-\sqrt{3}}{2}$	-1	$\frac{-\sqrt{3}}{2}$	0

Plotting these points gives the two leaves shown in Figure 2.13 (note that values of r are negative for $\pi/2 < \theta < \pi$). ■

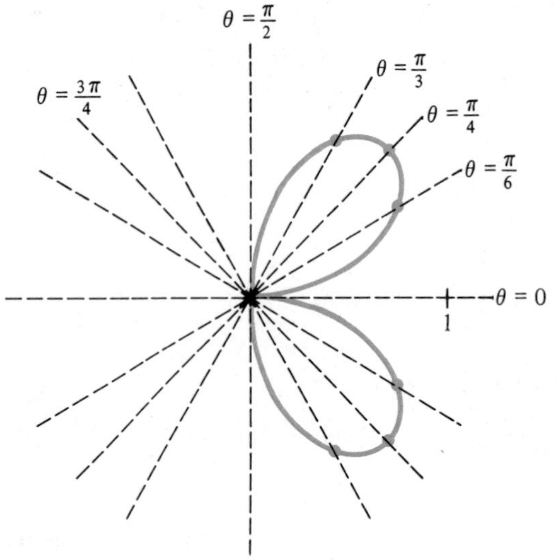

Figure 2.13 Graph of $r = \sin 2\theta$ for $0 \le \theta \le \pi$.

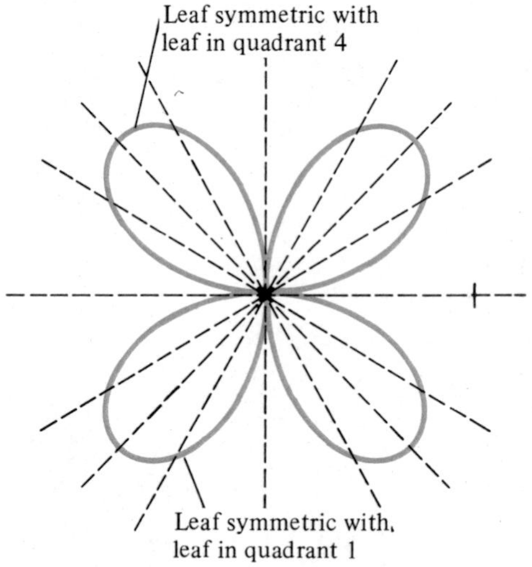

Figure 2.14 Graph of $r = \sin 2\theta$ for $0 \le \theta \le 2\pi$.

We can use symmetry considerations to obtain the remaining portion of the curve. To do so, we use the double angle identity to write

$$r(\theta) = \sin 2\theta = 2 \sin \theta \cos \theta.$$

Since $\sin(\theta + \pi) = -\sin \theta$ and $\cos(\theta + \pi) = -\cos \theta$, we obtain

$$r(\theta + \pi) = 2 \cdot \sin(\theta + \pi) \cos(\theta + \pi) = 2 \cdot \sin \theta \cos \theta = r(\theta).$$

This shows that $(r, \theta + \pi) = (-r, \theta)$ is on the graph whenever (r, θ) is. That is, we may sketch a leaf in the third quadrant symmetric with the leaf in the first quadrant, and a leaf in the second quadrant symmetric with the leaf in the fourth quadrant (Figure 2.14). (Actually we could obtain the entire graph by applying symmetry arguments just to the leaf obtained for $0 \le \theta \le \pi/2$. Can you see how this could be done?) The resulting figure is called a four-leaved rose.

Example 6 For the polar equation $r = 2 \cos \theta$,

(a) sketch the graph;
(b) note any symmetries of the graph;
(c) find the rectangular form of the equation.

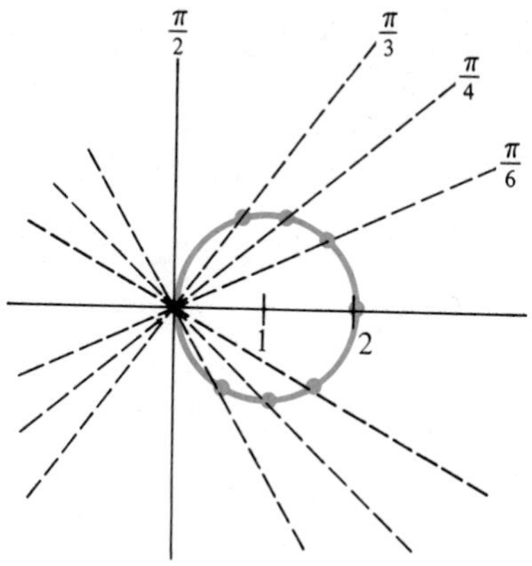

Figure 2.15 Graph of $r = 2\cos\theta$.

Solution:

(a) For $0 \le \theta \le \pi$, we obtain the following points:

θ	0	$\frac{\pi}{6}$	$\frac{\pi}{4}$	$\frac{\pi}{3}$	$\frac{\pi}{2}$	$\frac{2\pi}{3}$	$\frac{3\pi}{4}$	$\frac{5\pi}{6}$	π
r	2	$\sqrt{3}$	$\sqrt{2}$	1	0	-1	$-\sqrt{2}$	$-\sqrt{3}$	-2

These points produce the curve shown in Figure 2.15.

For $\pi < \theta < 2\pi$, we note that $r(\theta_0 + \pi) = 2\cos(\theta_0 + \pi) = -2\cos(\theta_0) = -r(\theta_0)$ where $0 < \theta_0 < \pi$. Thus, these values simply trace over the same points a second time.

(b) Figure 2.15 suggests symmetry about the x-axis. This is indeed the case, since, for $r(\theta) = 2\cos\theta$,

$$r(-\theta) = 2\cos(-\theta) = 2\cos\theta = r(\theta).$$

Thus, $(r, -\theta)$ is on the graph whenever (r, θ) is.

(c) To find the rectangular form for $r = 2\cos\theta$, we multiply by r to obtain

$$r^2 = 2r\cos\theta.$$

Using equations (3) and (4) we obtain

$$x^2 + y^2 = 2x, \qquad \text{or} \qquad x^2 - 2x + y^2 = 0.$$

Completing the square in x gives

$$(x^2 - 2x + 1) + y^2 = 1$$

or

$$(x - 1)^2 + y^2 = 1.$$

The graph is therefore a circle of radius $r = 1$ with center $(1, 0)$. ■

Exercise Set 16.2

In Exercises 1–8, find rectangular coordinates for the point given in polar coordinates.

1. $(1, \pi/2)$

2. $(3, \pi/6)$

3. $(0, \pi)$

4. $(-2, \pi/4)$

5. $(\sqrt{2}, -\pi/4)$

6. $\left(-1, -\frac{3\pi}{2}\right)$

7. $(-3, \pi)$

8. $(\pi, -\pi)$

In Exercises 9–16, find polar coordinates, subjected to the stated restrictions on θ, for the point given in rectangular coordinates.

9. $(1, 1)$, $\quad 0 < \theta < \pi$

10. $(1, 1)$, $\quad \pi < \theta < 2\pi$

11. $(-3, 0)$, $\quad -\pi/2 < \theta < \pi/2$

12. $(1, -\sqrt{3})$, $\quad \pi < \theta < 2\pi$

13. $(1, -\sqrt{3})$, $\quad 0 < \theta < \pi$

14. $(-2, 2)$, $\quad 0 < \theta < \pi$

15. $(-2, 2)$, $\quad 3\pi < \theta < 4\pi$

16. $(-3, 4)$, $\quad \pi < \theta < 2\pi$

In Exercises 17–24, identify all symmetries (about the x-axis, the y-axis, or the origin) possessed by the graph of the given equation.

17. $r = 4\cos\theta$

18. $r = 2\sin\theta$

19. $r = 1 + \sin\theta$

20. $r^2 = \cos\theta$

21. $r^2 = \cos 2\theta$

22. $r = \sin 3\theta$

23. $r = 2$

24. $r = 4 \sin 2\theta$

In Exercises 25–32, find an equation in polar coordinates for the given equation.

25. $x^2 + y^2 = 4$

26. $y^2 = 16x$

27. $4x^2 + y^2 = 4$

28. $xy = 2$

29. $x = 6$

30. $y = 4$

31. $x^2 + y^2 + 2y = 0$

32. $y = x$

In Exercises 33–40, find an equation in rectangular coordinates for the given polar equation.

33. $r = 4 \sin \theta$

34. $r = 6$

35. $r = \tan \theta$

36. $r = \csc \theta$

37. $r = 4 \sec \theta$

38. $r = 1 + \sin \theta$

39. $r = \dfrac{1}{1 - \cos \theta}$

40. $r^2 = 4 \sec \theta$

In Exercises 41–50, sketch the graph of the given polar equation.

41. $r = \dfrac{1}{2}\theta$ (spiral of Archimedes)

42. $r = 2 \sin \theta$ (circle)

43. $r = 1 + \sin \theta$ (cardioid)

44. $r = \cos 2\theta$ (4-leaved rose)

45. $r = 1 + 2 \cos \theta$ (limaçon)

46. $r^2 = \cos 2\theta$ (lemniscate)

47. $r = e^{\theta}$, $\theta \geq 0$ (spiral)

48. $r = \dfrac{1}{e^{\theta}}$, $\theta \geq 0$

49. $r = 1 - 2 \cos \theta$ (limaçon)

50. $r = \sin 4\theta$ (8-leaved rose)

51. Prove that the graph of $r = 2 \sin \theta - 2 \cos \theta$ is a circle. Find its center and radius.

52. Prove that the graph of $r = a \sin \theta$ is a circle. Sketch the graph.

53. Prove that the graph of $r = a \cos \theta$ is a circle. Sketch the graph.

54. Show that equations (3) are valid for the case $r < 0$.

16.3 GRAPHING TECHNIQUES FOR POLAR EQUATIONS

In Section 16.2 we graphed polar equations by plotting a few points (r_j, θ_j) and sketching in a curve, a technique also used to graph equations in rectangular coordinates. A slightly different approach is often simpler to use. That is simply to put the polar equation in the form $r = f(\theta)$, if possible, and then to "think dynamically," noting the behavior of r as θ sweeps out one or more revolutions about the pole.

For example, to graph the polar equation $r = a(1 + \cos \theta)$ we first note that the maximum value of $r = a(1 + 1) = 2a$ will occur when $\cos \theta = 1$, that is, when $\theta = 0$. The curve is therefore contained within the circle $r = 2a$ (Figure 3.1). To

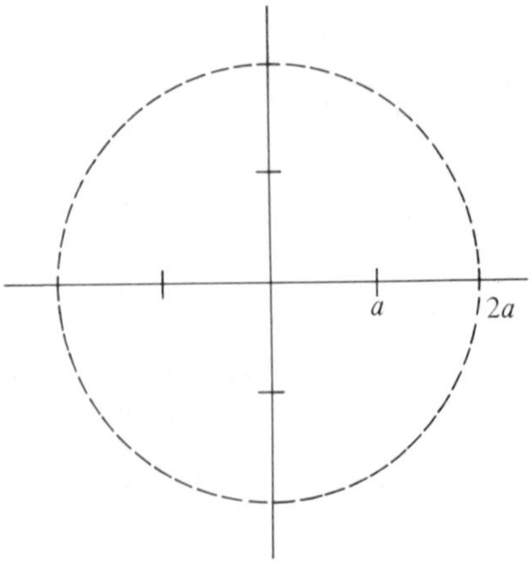

Figure 3.1 Circle $r = 2a$.

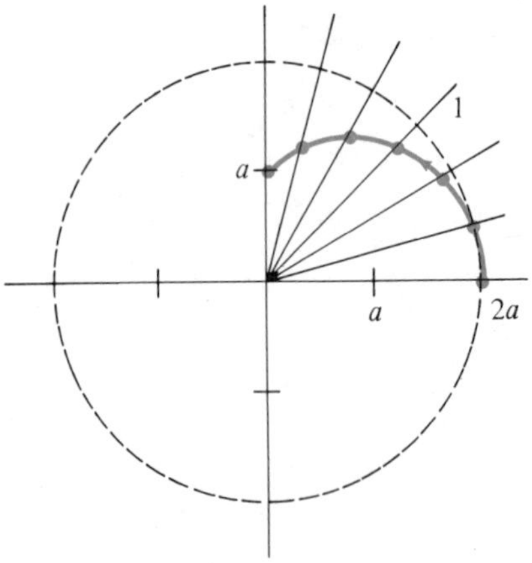

Figure 3.2 As θ increases toward $\pi/2$, r decreases from $2a$ to a.

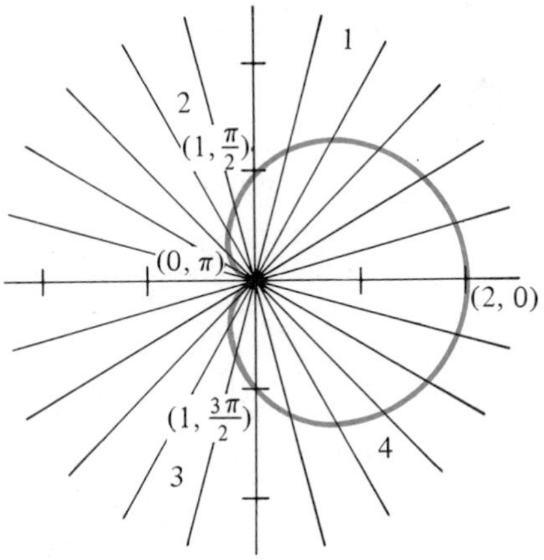

Figure 3.3 Graph of $r = a(1 + \cos \theta)$.

plot the curve we begin at the point $(2a, 0)$ and note that as θ increases from 0 to $\pi/2$, r *decreases* from length $r(0) = 2a$ to length $r(\pi/2) = a\left(1 + \cos\frac{\pi}{2}\right) = a$. This observation enables us to sketch in the arc labelled 1 in Figure 3.2. We next note the behavior of r as θ increases from $\pi/2$ to π: r decreases from $r(\pi/2) = a$ to $r(\pi) = a(1 + (-1)) = 0$. This produces the arc labelled 2 in Figure 3.3. Similarly, arcs 3 and 4 are "swept out" as θ increases from π to 2π. Allowing θ to increase beyond 2π simply retraces the arcs already obtained, so the complete curve corresponding to $r = a(1 + \cos\theta)$ is obtained for $0 \le \theta \le 2\pi$ (Figure 3.3).

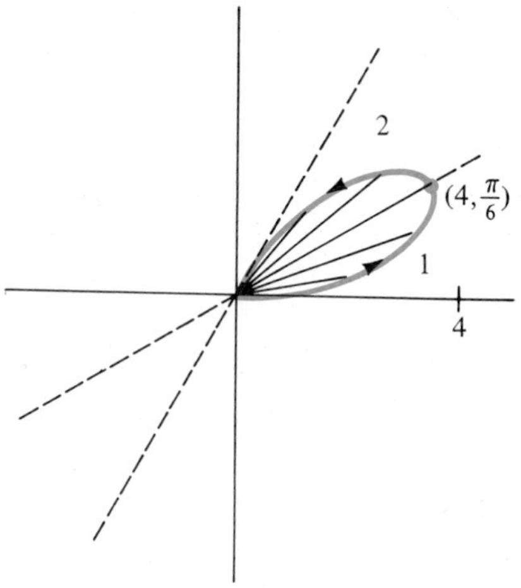

Figure 3.4 $r = 4\sin 3\theta,\ 0 \le \theta \le \pi/3$.

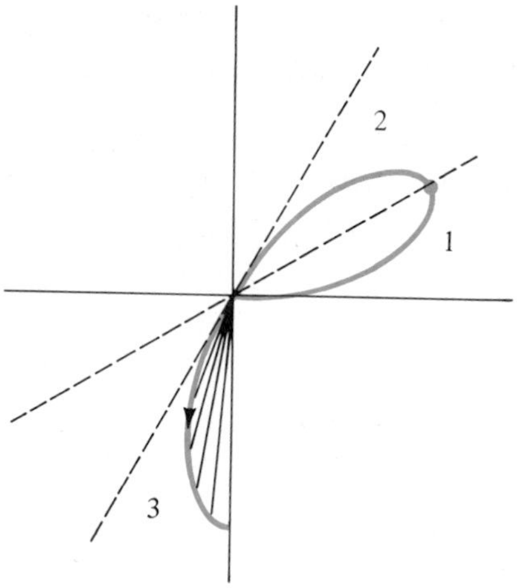

Figure 3.5 $r = 4\sin 3\theta,\ 0 \le \theta \le \pi/2$.

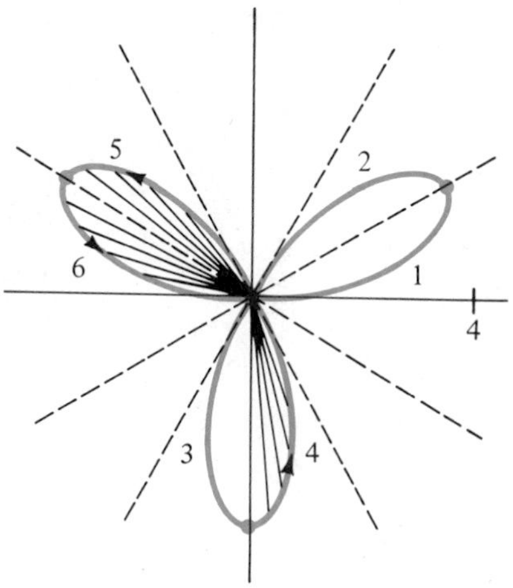

Figure 3.6 Three-leaved rose $r = 4\sin 3\theta,\ 0 \le \theta \le \pi$.

Example 1 Sketch the graph of the polar equation $r = 4\sin 3\theta$.

Solution: Since $|\sin 3\theta| \le 1$ for all θ, the curve must lie within the circle $r = 4$. Beginning at the point with $\theta = 0$, we note the following:

(a) As θ turns from 0 to $\pi/6$, 3θ turns from 0 to $\pi/2$, so r *increases* from $r(0) = 4\sin 0 = 0$ to $r(\pi/6) = 4\sin 3(\pi/6) = 4$. This produces arc 1 in Figure 3.4.

(b) As θ turns from $\pi/6$ to $\pi/3$, 3θ turns from $\pi/2$ to π, so r *decreases* from $r(\pi/6) = 4$ to $r(\pi/3) = 0$ (arc 2 in Figure 3.4).

(c) As θ turns from $\pi/3$ to $\pi/2$, 3θ turns from π to $3\pi/2$, so r *decreases* from $r(\pi/3) = 0$ to $r(\pi/2) = 4\sin 3(\pi/2) = -4$, producing arc 3 in Figure 3.5.

(d) As θ increases through the next quarter turn, from $\pi/2$ to π, 3θ increases from $3\pi/2$ to 3π. Thus, r increases from -4 to 4 for $\pi/2 \le \theta \le 5\pi/6$, producing arcs 4 and 5, and decreases from 4 to 0 for $5\pi/6 \le \theta \le \pi$, producing arc 6 (Figure 3.6).

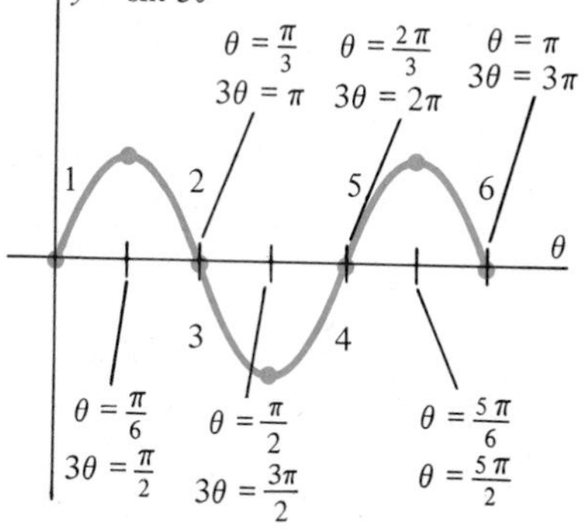

Figure 3.7 Graph of $f(\theta) = \sin 3\theta$.

By noting the behavior of r for $\theta > \pi$, we can see that the entire curve (a three-leaved rose) has been obtained with only $0 \le \theta \le \pi$. (Figure 3.7 recalls the graph of the function $f(\theta) = \sin 3\theta$. Note how the arcs numbered 1–6 correspond to the arcs in Figures 3.4–3.6.) ■

Example 2 Graph the lemniscate $r^2 = a \cos 2\theta$, $a > 0$.

Solution: Taking square roots of both sides shows that the given equation corresponds to the *pair* of equations

$$r_1 = \sqrt{a \cos 2\theta} \tag{1}$$

and

$$r_2 = -\sqrt{a \cos 2\theta}. \tag{2}$$

Thus, some values of θ will correspond to two values of r, while others will not correspond to any real values of r.

Beginning with $\theta = 0$, we observe that

(i) As θ increases from 0 to $\pi/4$, 2θ increases from 0 to $\pi/2$. Thus,

(a) r_1 decreases from $r_1(0) = \sqrt{a}$ to $r_1(\pi/4) = 0$ (arc 1 in Figure 3.8),
(b) r_2 increases from $r_2(0) = -\sqrt{a}$ to $r_2(\pi/4) = 0$ (arc 2).

(ii) As θ increases from $\pi/4$ to $3\pi/4$, 2θ increases from $\pi/2$ to $3\pi/2$. Since $\cos 2\theta < 0$ for $\pi/2 < 2\theta < 3\pi/2$, equations (1) and (2) have no solutions in this interval.

(iii) As θ increases from $3\pi/4$ to π, 2θ increases from $3\pi/2$ to 2π. Thus,

(a) r_1 increases from $r_1(3\pi/2) = 0$ to $r_1(2\pi) = \sqrt{a}$ (arc 3 in Figure 3.9),
(b) r_2 decreases from $r_2(3\pi/2) = 0$ to $r_2(2\pi) = -\sqrt{a}$ (arc 4).

Checking equations (1) and (2) for values of $\theta > \pi$ shows that all points on the graph are obtained by using only $0 \le \theta \le \pi/4$ and $3\pi/4 \le \theta \le \pi$. ■

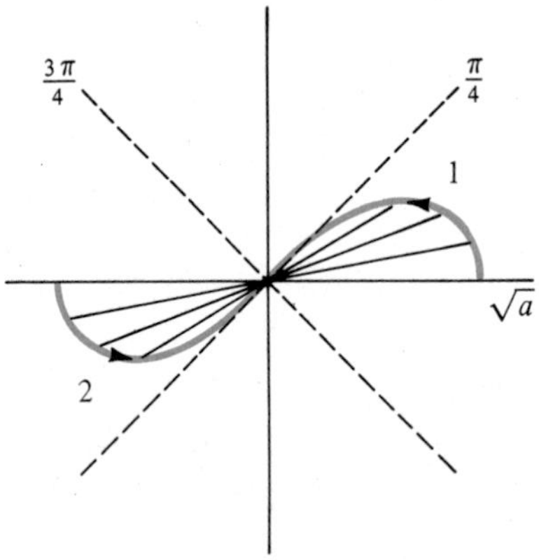

Figure 3.8 $r^2 = a \cos 2\theta$, $0 \le \theta \le \pi/4$.

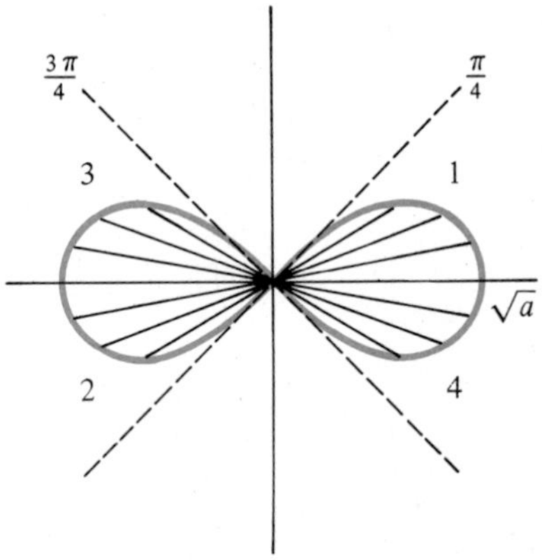

Figure 3.9 $r^2 = a \cos 2\theta$.

Intersections of Graphs in Polar Coordinates

In Section 16.4, we will need to be able to determine points of intersection for graphs of pairs of equations of the form

$$r_1 = f(\theta), \tag{3}$$

$$r_2 = g(\theta). \tag{4}$$

As for equations in rectangular coordinates, points of intersection may be found by equating the right-hand sides of equations (3) and (4) and solving the resulting equation for θ. However, unlike the situation for rectangular equations, *this method will not necessarily produce all points of intersection,* as the following examples show.

Example 3 Find all points of intersection for the graphs of the equations $r_1 = 1 + \cos\theta$ and $r_2 = 3\cos\theta$.

Solution: Setting $r_1 = r_2$ gives the equation

$$1 + \cos\theta = 3\cos\theta,$$

so

$$\cos\theta = 1/2.$$

The solutions of this equation for $-\pi < \theta < \pi$ are $\theta_1 = \pi/3$ and $\theta_2 = -\pi/3$. The corresponding points of intersection are $(3/2, \pi/3)$ and $(3/2, -\pi/3)$. However, as Figure 3.10 illustrates, the *pole* $(0, \theta)$ is also common to both graphs. Setting $r_1(\theta) = r_2(\theta)$ misses this point, since the pole corresponds to $\theta = \pi$ in the graph of r_1, but it corresponds to $\theta = \pi/2$ in the graph of r_2. ■

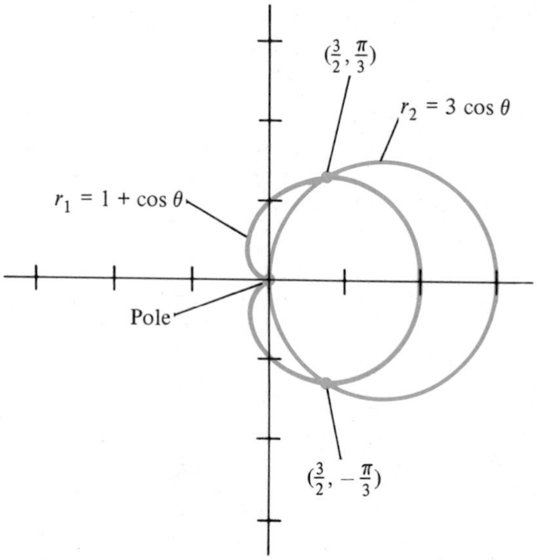

Figure 3.10 Graphs of r_1 and r_2 intersect in 3 points.

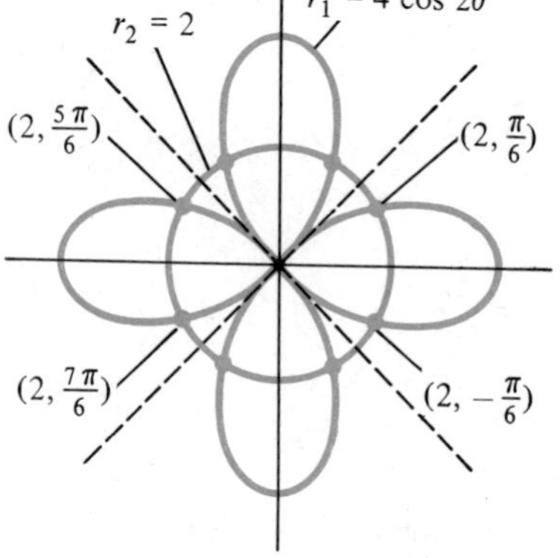

Figure 3.11 Graphs of r_1 and r_2 intersect in 8 points.

Example 4 Find all points of intersection of the four-leaved rose $r_1 = 4\cos 2\theta$ and the circle $r_2 = 2$.

Solution: Setting $r_1 = r_2$ gives $4\cos 2\theta = 2$, or $\cos 2\theta = 1/2$. Thus $2\theta = \pm\pi/3$ or $2\theta = \pm 7\pi/3$. The four distinct solutions of this equation in the interval $(0, 2\pi)$ are $\theta = \pi/6$, $5\pi/6$, $7\pi/6$, and $11\pi/6$. These give the four points of intersection $(2, \pi/6)$, $(2, 5\pi/6)$, $(2, 7\pi/6)$, and $(2, 11\pi/6)$. However, the method of setting $r_1 = r_2$ has failed to detect *four* additional points of intersection $(2, \pi/3)$,

$(2, 2\pi/3)$, $(2, 4\pi/3)$, and $(2, 5\pi/3)$ (see Figure 3.10). The reason for this is that each of these last four points corresponds to *different* values of θ on the graph of r_1 than on the graph of r_2. ■

Examples 3 and 4 show that simply setting $r_1 = r_2$ and solving the resulting equation $f(\theta) = g(\theta)$ for θ will not necessarily yield all points common to the graphs of equations (3) and (4). The reason for this difficulty is the nonuniqueness of polar coordinates. For example, the point $(2, \pi/3)$, common to both graphs $r_1 = 4 \cos 2\theta$ and $r_2 = 2$ in Example 4, actually occurs on the graph of r_1 in its equivalent form $(-2, 4\pi/3)$. Since these coordinates do not satisfy $r_2 = 2$, the point does not result from setting $r_1 = r_2$. Rather than trying to work out an algorithm allowing for all possible ways in which points of intersection of two polar equations can arise, *you should simply develop the habit of always sketching the two curves to ensure that all such points are found.*

Exercise Set 16.3

In Exercises 1–10, sketch the graph of the given polar equation using the method of Examples 1 and 2.

1. $r = a \cos \theta$

2. $r = a \sin \theta$

3. $r = a \sin 2\theta$

4. $r = a(1 + \cos \theta)$

5. $r = a(1 - \sin \theta)$

6. $r = 1 + 2 \sin \theta$

7. $r = 2\theta$

8. $r^2 = 4 \cos 2\theta$

9. $r = \sin \theta + \cos \theta$

10. $r = a(1 - \cos \theta)$

In Exercises 11–20, find all points of intersection of the graphs of the two given equations.

11. $r_1 = 2 \cos \theta$, $r_2 = 2 \sin \theta$

12. $r_1 = 2 \cos \theta$, $r_2 = 1$

13. $r_1 = 1 + \sin \theta$, $r_2 = 3 \sin \theta$

14. $r_1 = a \sin 3\theta$, $r_2 = a$

15. $r_1 = a(1 + \cos \theta)$, $r_2 = 3a \cos \theta$

16. $r_1 = 2a \cos 2\theta$, $r_2 = a$

17. $r_1 = a(1 + \cos \theta)$, $r_2 = a(1 - \cos \theta)$

18. $r_1^2 = \cos 2\theta$, $r_2 = \dfrac{\sqrt{2}}{2}$

19. $r = \dfrac{1}{\theta}$, $\theta = \pi/4$

20. $r_1 = a \sin 4\theta$, $r_2 = a$

21. Show that the graphs of the polar equations

$$r_1 = \frac{a}{1 \pm \cos \theta}; \qquad r_2 = \frac{a}{1 \pm \sin \theta}$$

are parabolas. (*Hint:* Convert to rectangular equations.)

22. Show that the graphs of the polar equations

$$r_1 = \frac{ab}{1 \pm b \cos \theta}; \qquad r_2 = \frac{ab}{1 \pm b \sin \theta}$$

are ellipses if $0 < b < 1$. (*Hint:* Convert to rectangular equations.)

In Exercises 23–28, find a rectangular form for the given polar equation, identify the graph as either a parabola or an ellipse, and sketch the graph. (See Exercises 21 and 22.)

23. $r = \dfrac{1}{1 + \cos \theta}$

24. $r = \dfrac{4}{1 - \sin \theta}$

25. $r = \dfrac{2}{1 + \frac{1}{2} \cos \theta}$

26. $2r = \dfrac{4}{2 - \sin \theta}$

27. $r - r \cos \theta = 2$

28. $3r + r \cos \theta - 6 = 0$

16.4 CALCULATING AREA IN POLAR COORDINATES

When a region R in the plane is determined by the graph of an equation written in the *polar* form $r = f(\theta)$, the area of R cannot be calculated by a direct application of the equations developed in Chapters 7 and 8. The difficulty lies in the fact that these formulas were obtained for functions $y = g(x)$ written in *rectangular* coordinates.

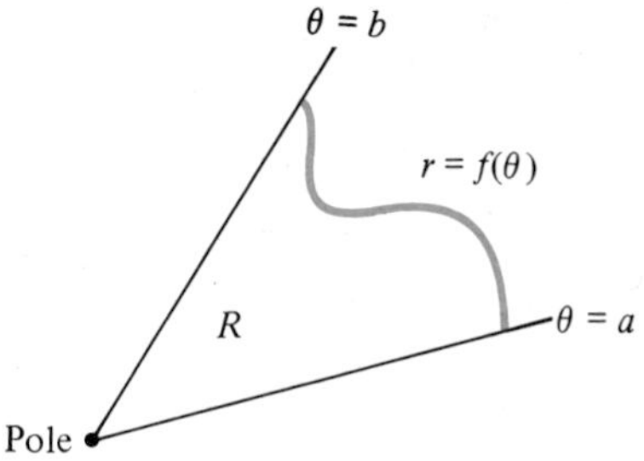

Figure 4.1 Region determined by $r = f(\theta)$, $a \leq \theta \leq b$.

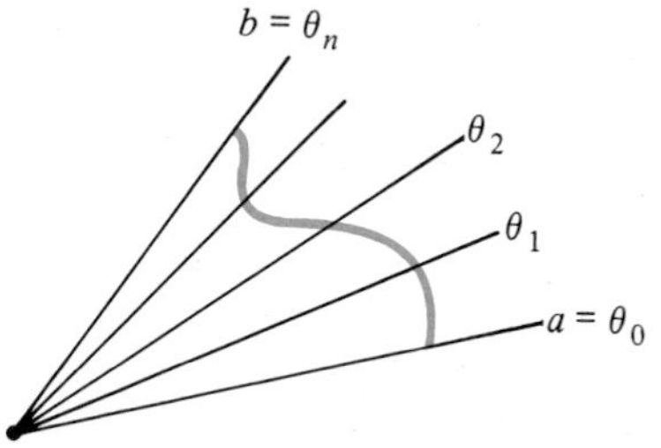

Figure 4.2 Partition $\theta_0 < \theta_1 < \cdots < \theta_n$ slices R into wedge-shaped regions.

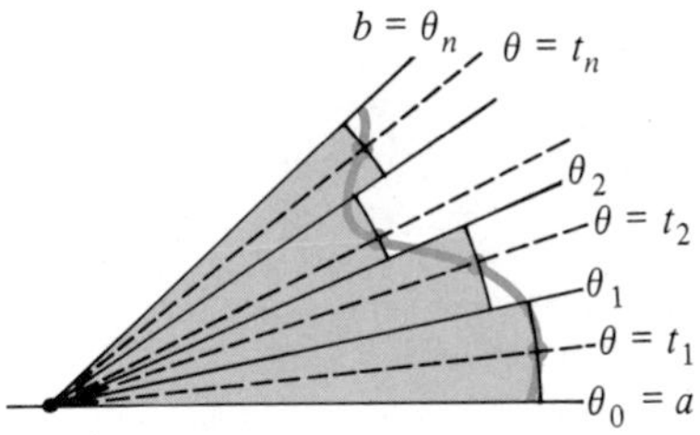

Figure 4.3 Area of R approximated by areas of sectors of circles.

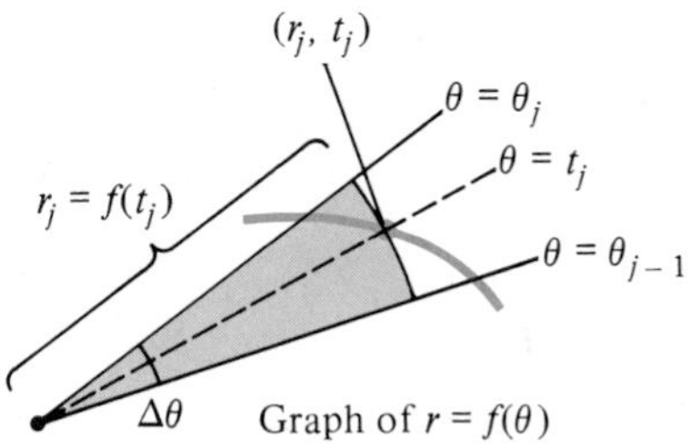

Figure 4.4 $\Delta A_j = \dfrac{1}{2} r_j^2 \, \Delta\theta$.

To calculate areas of regions determined by the graphs of polar equations, we might try rewriting such equations in rectangular coordinates, using the substitutions $x = r\cos\theta$ and $y = r\sin\theta$, as in Section 16.2. However, this strategy often leads to one of two undesirable consequences. Either the resulting rectangular equations are much more complicated than the original polar equations, or, even worse, the resulting equations might not describe either variable as a *function* of the other.

A more satisfactory approach is to return to the basic theory of the definite integral and determine the formulas appropriate for calculating area in polar coordinates, just as the formulas for the various applications of the definite integral were developed in Chapters 7 and 8. The small amount of time spent in this effort will allow us to avoid the step of converting to rectangular coordinates for each area problem in polar coordinates.

Let us suppose that a region R is bounded by the graph of the function $r = f(\theta)$ for values of θ lying between the numbers $\theta = a$ and $\theta = b$ (Figure 4.1). Since θ is the independent variable, we partition the interval $[a, b]$ into n equal subintervals* by using the increment $\Delta\theta = \dfrac{b - a}{n}$ and the endpoints

$$\theta_0 = a, \qquad \theta_1 = a + \Delta\theta, \qquad \theta_2 = a + 2\Delta\theta, \ldots, \theta_n = a + n\Delta\theta = b.$$

The rays $\theta = \theta_j$, $j = 0, 1, 2, \ldots, n$ then "slice" the region R into n wedge-shaped subregions (Figure 4.2).

We now approximate the area of each of the subregions. To do so we choose one number t_j in each interval $[\theta_{j-1}, \theta_j]$, $j = 1, 2, \ldots, n$ (Figure 4.3). We then approximate the area of the jth subregion by the area of the circular **sector** of constant radius $r_j = f(t_j)$ and angle $\Delta\theta$ (see Figure 4.4).

Since the area of the circle of radius r_j is πr_j^2, the area of the circular sector comprising the fractional part $\dfrac{\Delta\theta}{2\pi}$ of the entire circle is

$$\Delta A_j = \pi r_j^2\left(\frac{\Delta\theta}{2\pi}\right) = \frac{1}{2} r_j^2 \, \Delta\theta = \frac{1}{2}[f(t_j)]^2 \, \Delta\theta.$$

Summing these approximations, one for each subregion, gives an approximation to the area A of the whole region R of the form

$$A \approx \sum_{j=1}^{n} \Delta A_j = \sum_{j=1}^{n} \frac{1}{2}[f(t_j)]^2 \, \Delta\theta. \tag{1}$$

If the function $r = f(\theta)$ is continuous for $a \leq \theta \leq b$, the sum on the right-hand side of approximation (1) is a *Riemann* sum for the function $\dfrac{1}{2}[f(\theta)]^2$. By the theory of the integral developed in Chapter 6, this approximation "converges" to the definite integral

$$\int_a^b \frac{1}{2} \cdot [f(\theta)]^2 \, d\theta = \lim_{n\to\infty} \sum_{j=1}^{n} \frac{1}{2}[f(t_j)]^2 \, \Delta\theta \tag{2}$$

*A more general development would involve allowing subintervals of arbitrary (that is, not necessarily equal) size $\Delta\theta_j = \theta_j - \theta_{j-1}$. It is unnecessary to accommodate this generality here, since we have already seen that all Riemann sums converge to the same integral when f is continuous.

as $n \to \infty$, if the limit on the right exists. We therefore argue that as $n \to \infty$, the circular sectors provide an increasingly accurate approximation to the region R and that the area of R should be defined as follows.

The area A of the region bounded by the graph of the continuous function $r = f(\theta)$ between the rays $\theta = a$ and $\theta = b$, $a < b$, is given by the definite integral

$$A = \int_a^b \frac{1}{2}[f(\theta)]^2\, d\theta. \tag{3}$$

The same cautions should be observed in using equation (3) as in calculating areas in rectangular coordinates. The region should first be carefully sketched, so that the proper limits of integration may be determined.

Example 1 Find the area of the region bounded by the rays $\theta = -\pi/4$ and $\theta = \pi/4$, and the graph of the equation $r = 1 + \sin\theta$.

Solution: The region is the portion of the cardioid swept out by the radius of length $r = 1 + \sin\theta$ as θ increases from $\theta = -\pi/4$ to $\theta = \pi/4$ (Figure 4.5). By equation (3) the area is

$$\begin{aligned} A &= \int_{-\pi/4}^{\pi/4} \frac{1}{2}[1 + \sin\theta]^2\, d\theta \\ &= \int_{-\pi/4}^{\pi/4} \frac{1}{2}[1 + 2\sin\theta + \sin^2\theta]\, d\theta \\ &= \int_{-\pi/4}^{\pi/4} \frac{1}{2}\left[1 + 2\sin\theta + \left(\frac{1}{2} - \frac{1}{2}\cos 2\theta\right)\right] d\theta \quad \left(\text{using the identity } \sin^2\theta = \frac{1}{2} - \frac{1}{2}\cos 2\theta\right. \\ &= \int_{-\pi/4}^{\pi/4} \frac{1}{2}\left[\frac{3}{2} + 2\sin\theta - \frac{1}{2}\cos 2\theta\right] d\theta \end{aligned}$$

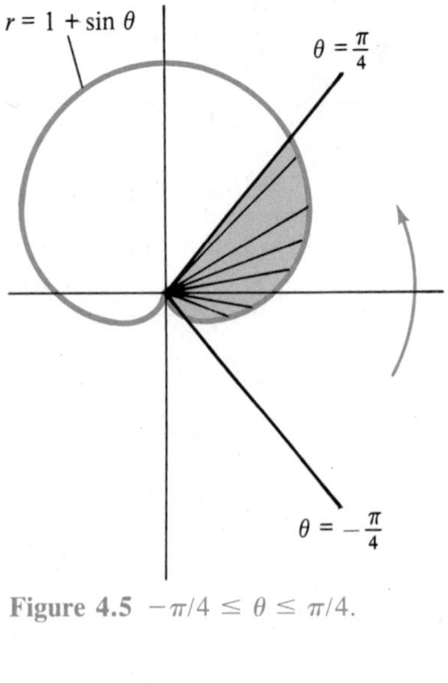

Figure 4.5 $-\pi/4 \le \theta \le \pi/4$.

$$= \frac{1}{2}\left[\frac{3\theta}{2} - 2\cos\theta - \frac{1}{4}\sin 2\theta\right]_{-\pi/4}^{\pi/4}$$

$$= \frac{1}{2}\left\{\left(\frac{3\pi}{8} - \sqrt{2} - \frac{1}{4}\right) - \left(-\frac{3\pi}{8} - \sqrt{2} + \frac{1}{4}\right)\right\}$$

$$= \frac{3\pi}{8} - \frac{1}{4} \approx 0.928.$$

■

Example 2 Find the area of the region enclosed by the graph of the cardioid $r = 1 + \cos\theta$.

Strategy

Sketch the region.

Find a pair of smallest and largest values of θ required to sweep out the region. These are the limits of integration. (Note that $0 \le \theta \le 2\pi$ works just as well as $-\pi \le \theta \le \pi$.)

Set up the integral given by equation (3).

Square $f(\theta)$ in the integrand and simplify, using the identity

$\cos^2\theta = \frac{1}{2} + \frac{1}{2}\cos 2\theta$.

Integrate.

Solution

The cardioid is sketched in Figure 4.6. Since the cardioid is obtained as the graph of the function $f(\theta) = 1 + \cos\theta$ as θ increases from $\theta = -\pi$ to $\theta = \pi$, the limits of integration are $a = -\pi$ and $b = \pi$. By equation (3) the area is

$$A = \int_{-\pi}^{\pi} \frac{1}{2}[1 + \cos\theta]^2\, d\theta$$

$$= \frac{1}{2}\int_{-\pi}^{\pi} [1 + 2\cos\theta + \cos^2\theta]\, d\theta$$

$$= \frac{1}{2}\int_{-\pi}^{\pi} \left[1 + 2\cos\theta + \left(\frac{1}{2} + \frac{1}{2}\cos 2\theta\right)\right] d\theta$$

$$= \frac{1}{2}\int_{-\pi}^{\pi} \left[\frac{3}{2} + 2\cos\theta + \frac{1}{2}\cos 2\theta\right] d\theta$$

$$= \frac{1}{2}\left[\frac{3\theta}{2} + 2\sin\theta + \frac{1}{4}\sin 2\theta\right]_{-\pi}^{\pi}$$

$$= \frac{1}{2}\left\{\left(\frac{3\pi}{2} + 2\cdot 0 + \frac{1}{4}\cdot 0\right) - \left(\frac{3}{2}(-\pi) + 2\cdot 0 + \frac{1}{4}\cdot 0\right)\right\}$$

$$= \frac{3\pi}{2}.$$

■

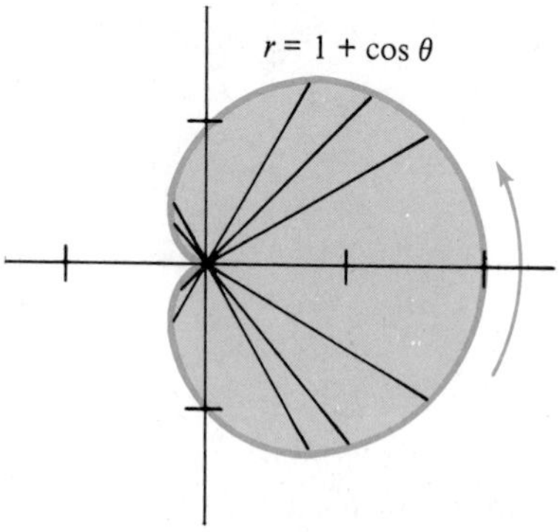

Figure 4.6 $-\pi \le \theta \le \pi$.

Often, symmetry considerations may be used to simplify the calculation of area, as the following example illustrates.

Example 3 Find the area of the region enclosed by the graph of the polar equation $r = a \sin 3\theta$.

Strategy

Sketch the figure. (Recall Example 1, Section 16.3.)

Look for symmetries to simplify the calculation.

Determine the limits of integration for the region whose area is to be calculated. Multiply the integral by the constant determined by the use of symmetries.

Apply equation (3)

$\sin^2 3\theta = \frac{1}{2} - \frac{1}{2}\cos 6\theta.$

Solution

The region is a three-leaved rose, as illustrated in Figure 4.7. Since the 3 leaves are congruent, and since the leaf determined by $0 \le \theta \le \pi/3$ is symmetric about the ray $\theta = \pi/6$, we may obtain the area of the entire figure by calculating the area of the half leaf determined by $0 \le \theta \le \pi/6$ and multiplying the result by 6. We obtain

$$\begin{aligned} A &= 6 \cdot \int_0^{\pi/6} \frac{1}{2}[a \sin 3\theta]^2 \, d\theta \\ &= 3a^2 \int_0^{\pi/6} \sin^2 3\theta \, d\theta \\ &= 3a^2 \int_0^{\pi/6} \left[\frac{1}{2} - \frac{1}{2}\cos 6\theta\right] d\theta \\ &= 3a^2 \left[\frac{\theta}{2} - \frac{1}{12}\sin 6\theta\right]_0^{\pi/6} \\ &= \frac{a^2\pi}{4}. \end{aligned}$$

■

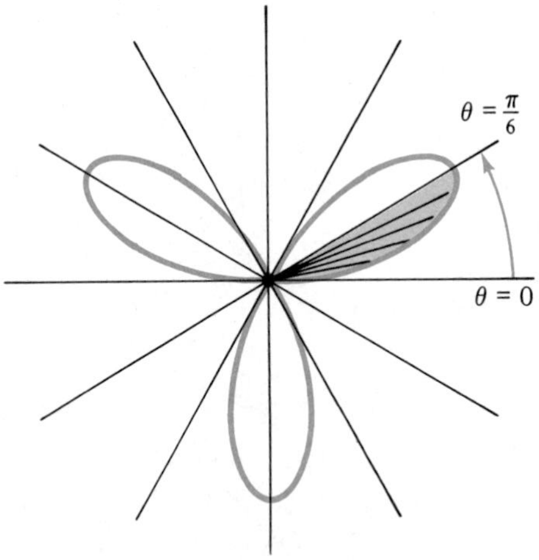

Figure 4.7 Area of three-leaved rose is 6 times area of region determined by $0 \le \theta \le \pi/6$.

As happens with equations in rectangular coordinates, an equation in polar coordinates may fail to determine r as a function of θ. The following example shows how symmetry may be used to overcome this difficulty.

Example 4 Find the area of the region enclosed by the lemniscate $r^2 = \cos 2\theta$.

Strategy

Sketch the figure. (Recall Example 2, Section 16.3.)

Take square roots to solve for r. Obtain a *function* $r = f(\theta)$.

Compare region determined by the function with that of the original equation.

Use symmetry to simplify calculation.

Apply equation (3).

Solution

The lemniscate is sketched in Figure 4.8. By extracting square roots on both sides of the equation $r^2 = \cos 2\theta$ we obtain the equation $r = \pm\sqrt{\cos 2\theta}$. By choosing the positive sign we obtain the function

$$f(\theta) = \sqrt{\cos 2\theta}.$$

This function is defined only for

$$-\frac{\pi}{4} \le \theta \le \frac{\pi}{4} \qquad \text{and} \qquad \frac{3\pi}{4} \le \theta \le \frac{5\pi}{4}.$$

However, as θ ranges through these two intervals the two lobes of the lemniscate are traced out. We may therefore integrate over $-\pi/4 \le \theta \le \pi/4$ to obtain the area of one lobe and double the result. We obtain

$$\begin{aligned} A &= 2\int_{-\pi/4}^{\pi/4} \frac{1}{2}[\sqrt{\cos 2\theta}]^2\, d\theta \\ &= \int_{-\pi/4}^{\pi/4} \cos 2\theta\, d\theta \\ &= \frac{1}{2}\sin 2\theta\Big]_{-\pi/4}^{\pi/4} \\ &= 1. \end{aligned}$$

■

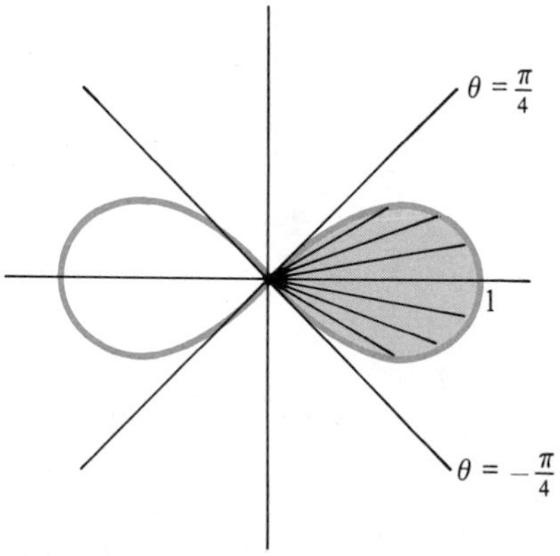

Figure 4.8 $-\pi/4 \le \theta \le \pi/4$ determines half the area of the lemniscate $r^2 = \cos 2\theta$.

Area Between Two Curves

When a region R lies between the graphs of two polar equations, as in Figure 4.9, we may calculate the area of R by subtracting the area enclosed by the inner curve from area enclosed by the outer curve. That is, if R lies inside the graph of $r = f(\theta)$ and outside the graph of $r = g(\theta)$, for $a \le \theta \le b$, the area A of R is given by the integral

$$A = \int_a^b \frac{1}{2}[f(\theta)]^2\, d\theta - \int_a^b \frac{1}{2}[g(\theta)]^2\, d\theta. \tag{4}$$

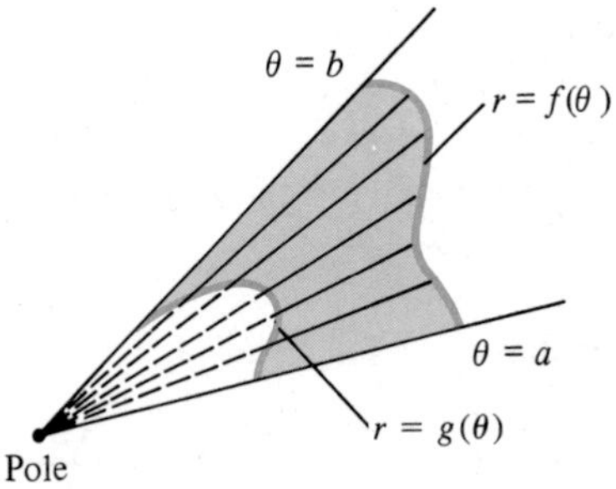

Figure 4.9 Region between graphs of $r = f(\theta)$ and $r = g(\theta)$ for $a \le \theta \le b$.

Example 5 Find the area of the region lying inside the circle $r = 3\cos\theta$ and outside the cardioid $r = 1 + \cos\theta$.

Strategy

Sketch the region.

Determine the points of intersection. This gives the limits of integration.

Determine which is the outer curve.

Apply equation (4).

Simplify integrand.

Solution

The graphs of the two equations are sketched in Figure 4.10. By solving the two equations simultaneously, we find the points of intersection to be $(3/2, \pm\pi/3)$ (see Example 3, Section 16.3). The limits of integration are therefore $-\pi/3 \le \theta \le \pi/3$.

In this interval, the outer (greater) function is $f(\theta) = 3\cos\theta$. The inner function is $g(\theta) = 1 + \cos\theta$. By equation (4) the area is

$$A = \int_{-\pi/3}^{\pi/3} \frac{1}{2}[3\cos\theta]^2\,d\theta - \int_{-\pi/3}^{\pi/3} \frac{1}{2}[1 + \cos\theta]^2\,d\theta$$

$$= \int_{-\pi/3}^{\pi/3} \frac{1}{2}\{9\cos^2\theta - (1 + 2\cos\theta + \cos^2\theta)\}\,d\theta$$

$$= \int_{-\pi/3}^{\pi/3} \frac{1}{2}[8\cos^2\theta - 1 - 2\cos\theta]\,d\theta$$

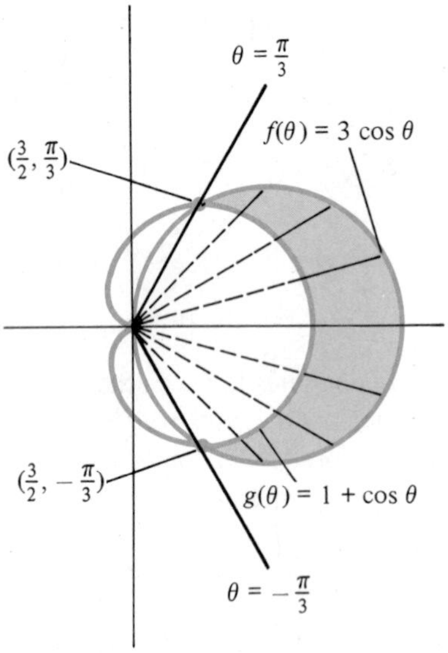

Figure 4.10

$\cos^2\theta = \frac{1}{2} + \frac{1}{2}\cos 2\theta.$

$$= \int_{-\pi/3}^{\pi/3} \frac{1}{2}\left[8\left(\frac{1}{2} + \frac{1}{2}\cos 2\theta\right) - 1 - 2\cos\theta\right] d\theta$$

$$= \int_{-\pi/3}^{\pi/3} \left(\frac{3}{2} + 2\cos 2\theta - \cos\theta\right) d\theta$$

$$= \frac{3\theta}{2} + \sin 2\theta - \sin\theta\Bigg]_{-\pi/3}^{\pi/3}$$

$$= \pi.$$

■

Exercise Set 16.4

In Exercises 1–8, find the area of the region determined by the given equations and inequalities.

1. $r = 1 + \sin\theta, \quad -\pi/2 < \theta < \pi/2$

2. $r = a\sin 2\theta, \quad 0 \le \theta \le \pi/2$

3. $r = 2\sin\theta, \quad 0 \le \theta \le \pi$

4. $r = \theta, \quad 0 \le \theta \le \pi$

5. $r = 2 + \sin\theta, \quad 0 \le \theta \le \pi$

6. $r = a\cos 3\theta, \quad \pi/2 < \theta < \frac{2\pi}{3}$

7. $r = 4 + \sin\theta, \quad \pi/4 \le \theta \le \frac{3\pi}{4}$

8. $r = 3\cos 2\theta, \quad -\frac{\pi}{4} \le \theta \le \frac{\pi}{4}$

In Exercises 9–14, find the area of the region enclosed by the graph of the given equation.

9. $r = 2\cos\theta$

10. $r = 1 + \cos\theta$

11. $r = a\sin 2\theta$

12. $r = a\sin 4\theta$

13. $r = a\cos 3\theta$

14. $r^2 = \cos\theta$

In Exercises 15–20, find the area of the region described.

15. The region inside the cardioid $r = 1 + \sin\theta$ and outside the circle $r = 2\sin\theta$.

16. The region inside the cardioid $r = 1 + \cos\theta$ and outside the circle $r = 3\cos\theta$. (*Hint:* You will need to treat the intervals $\left[\frac{\pi}{3}, \frac{\pi}{2}\right]$ and $\left[\frac{\pi}{2}, \pi\right]$ separately.)

17. The regions common to both cardioids $r = a(1 + \cos\theta)$ and $r = a(1 - \cos\theta)$.

18. The region outside the three-leaved rose $r = 4\sin 3\theta$ and inside the circle $r = 4$.

19. The region common to the circles $r = 2\cos\theta$ and $r = 2\sin\theta$.

20. The region common to the circle $r = \frac{\sqrt{2}}{2}$ and the lemniscate $r^2 = \cos 2\theta$.

21. Find the area of the region bounded by the graph of the spiral $r_1 = \theta$ and $r_2 = e^{\theta}$ for $0 \le \theta \le \pi$.

22. Find the area of the region inside the graph of $r = 3 + \sin\theta$ and outside the graph of $r = 4\sin\theta$.

23. Find the area of the region outside the graph of $r = 1 + \cos\theta$ and inside the graph of $r = 2 - \cos\theta$.

24. Find the area of the region bounded by the graph of the equation $r - r\cos\theta = 2$ and the line $\theta = \frac{\pi}{2}$.

25. Show that the result of calculating the area of the region enclosed by the graph of $r = 2\cos\theta$ in polar coordinates agrees with the result of calculating the area of the same region in rectangular coordinates.

26. Let $f(\theta) \ge g(\theta)$ for all θ. Sketch the region whose area is given by the integral $\int_a^b \frac{1}{2}[f(\theta) - g(\theta)]^2\, d\theta$. Compare this region with the region whose area is given by the integral in equation (4). Show that, in general, the two integrals are not equal.

27. *(Computer)* Program 11 in Appendix I is a BASIC program that approximates $\int_a^b \frac{1}{2}[f(\theta)]^2\, d\theta$ for the function $f(\theta) = 1 + \sqrt{\sin\theta}$.

a. Use Program 11 to approximate the integral $\int_0^{\pi/2} \frac{1}{2}[1 + \sqrt{\sin\theta}]^2\, d\theta$ with $n = 10$, 100, and 200.

b. Use Program 11 to approximate the area of the region enclosed by the graph of $f(\theta) = 1 + \sqrt{\sin\theta}$ for $0 \le \theta \le \pi$.

c. Modify Program 11 to approximate the area of the region enclosed by the graph of $f(\theta) = 1 - \sqrt{\cos\theta}$, $0 \le \theta \le \pi/2$.

16.5 PARAMETRIC EQUATIONS

Parametric equations provide a means of describing a curve in the plane by functions, even though the curve may not represent the graph of a function of the form $y = f(x)$. Rather than describing either coordinate as a function of the other, we allow both coordinates to be functions of a third (independent) variable, called a **parameter***, which is usually denoted by t. That is, we describe the curve as the collection of points $P(t) = (x(t), y(t))$ where t ranges through some specified domain.

For example, the graph of the equation

$$x^2 + y^2 = 1 \tag{1}$$

is the familiar unit circle in the plane. However, equation (1) does not determine y as a function of x, since two distinct values of y correspond to each x with $-1 < x < 1$. We may use our knowledge of trigonometry to obtain parametric equations, or a **parameterization,** for the unit circle as follows. Let t denote the angle formed between the positive x-axis and the radius from $(0, 0)$ to (x, y), measured in the counterclockwise direction. Then

$$x = \cos t$$

$$y = \sin t$$

are the coordinate functions of the point (x, y). By inspection we can see that as t increases from $t = 0$ to $t = 2\pi$, the point $(x, y) = (\cos t, \sin t)$ traverses the unit circle once in the counterclockwise direction (Figure 5.1). We therefore say that a parameterization of the unit circle is given by the equations

$$\left.\begin{aligned} x(t) &= \cos t \\ y(t) &= \sin t \end{aligned}\right\} \quad 0 \le t < 2\pi. \qquad \begin{matrix}(2a)\\(2b)\end{matrix}$$

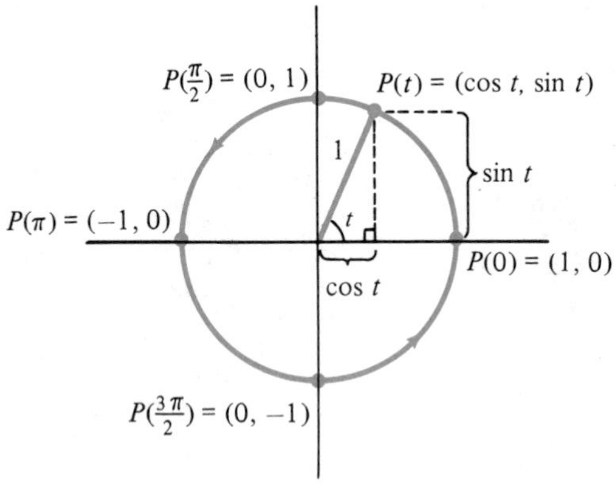

Figure 5.1 Parameterization of unit circle: $(x(t)), \;\; y(t)) = (\cos t, \sin t)$ $0 \le t < 2\pi$.

---

*The term *parameter* denotes a variable that often has no geometric or physical interpretation. Here the parameter t has no geometric interpretation with respect to the curve it helps describe. In some applications the parameter t represents time, while the variables x and y represent coordinates of points in the plane.

Note that equations (2a) and (2b) describe each of the coordinates x and y as a *function* of the parameter t. This observation will allow us to apply a certain amount of the theory of the calculus to curves that do not represent y as a function of x or vice versa. However, it is important to note that parameterizations for curves are not unique. For example, each of the following pairs of equations also represents a parameterization of the unit circle

$$\left.\begin{aligned} x(t) &= \cos 2t \\ y(t) &= \sin 2t \end{aligned}\right\} \quad 0 \le t < \pi,$$

$$\left.\begin{aligned} x(t) &= \cos t \\ y(t) &= \sin(-t) \end{aligned}\right\} \quad 0 \le t < 2\pi,$$

$$\left.\begin{aligned} x(t) &= \cos(2\pi - t) \\ y(t) &= \sin(2\pi - t) \end{aligned}\right\} \quad 0 \le t < 2\pi.$$

In the first parameterization, the point $P(t)$ traces out the circle counterclockwise, but the parameter t increases only from 0 to π. If we interpret t as time, then $P(t)$ moves twice as fast as in the parameterization (2). For the second parameterization, $P(t)$ traverses the circle in a *clockwise* direction, as shown by the arrows in Figure 5.2. Using the identities $\cos(2\pi - t) = \cos(-t) = \cos t$ and $\sin(2\pi - t) = \sin(-t)$, you can show that the behavior of $P(t)$ for the third parameterization is the same as that shown in Figure 5.2.

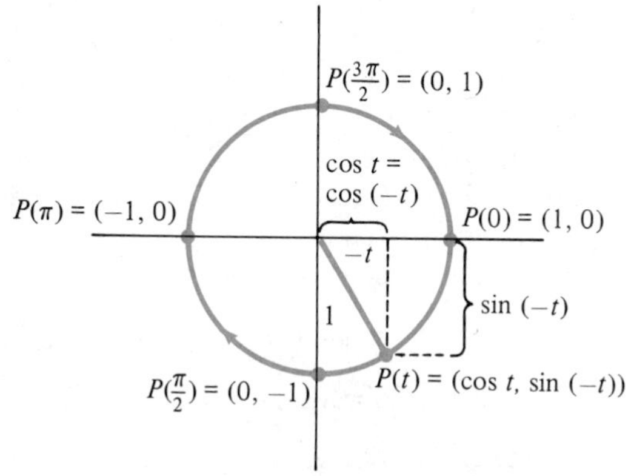

Figure 5.2 Alternate parameterization of unit circle: $(x(t),\ y(t)) = (\cos t, \sin(-t))$, $0 \le t < 2\pi$.

Example 1 Find parametric equations for the line ℓ containing the point $P_0 = (2, 4)$ and with slope 3.

Solution: We first find the equation for ℓ in xy-coordinates, using the point-slope form. We obtain

$$y - 4 = 3(x - 2), \tag{3}$$

so

$$y = 3x - 2. \tag{4}$$

If we set $x = t$ in equation (4), we find that $y = 3t - 2$, and we obtain the parametric equations

$$\left.\begin{aligned} x(t) &= t \\ y(t) &= 3t - 2 \end{aligned}\right\} -\infty < t < \infty.$$

A different parameterization for ℓ may be obtained by setting $t = x - 2$ in equation (3). Then, $y - 4 = 3t$, so $y = 3t + 4$, and the parameterization is

$$\left.\begin{aligned} x(t) &= t + 2 \\ y(t) &= 3t + 4 \end{aligned}\right\} -\infty < t < \infty.$$

There is no reason why we could not set $x = t^3$ in equation (4). In this case we obtain $y = 3t^3 - 2$ and the parameterization

$$\left.\begin{aligned} x(t) &= t^3 \\ y(t) &= 3t^3 - 2 \end{aligned}\right\} -\infty < t < \infty.$$

However, we could *not* use $x = t^2$, $y = 3t^2 - 2$ as our parameterization, since t^2 is never negative. The parameterization

$$\left.\begin{aligned} x(t) &= t^2 \\ y(t) &= 3t^2 - 2 \end{aligned}\right\} -\infty < t < \infty$$

would describe only the half line with nonnegative x-coordinates. ■

Up to this point, we have considered only the problem of finding parameterizations for curves described by equations in xy-coordinates. Just as important is the problem of describing a curve that is presented in parametric form. One technique that is always available is to simply sketch the curve by selecting various values for t, calculating the corresponding coordinates of the point $P(t) = (x(t), y(t))$, and sketching in a curve passing through the various points obtained (in order!). However, through a familiarity with the xy-coordinate forms of various curves you can often convert the parametric equations directly into their xy counterparts.

Example 2 A point moves along a path with coordinates

$$(x(t), y(t)) = (a \cos t, b \sin t)$$

at time t. Describe the path.

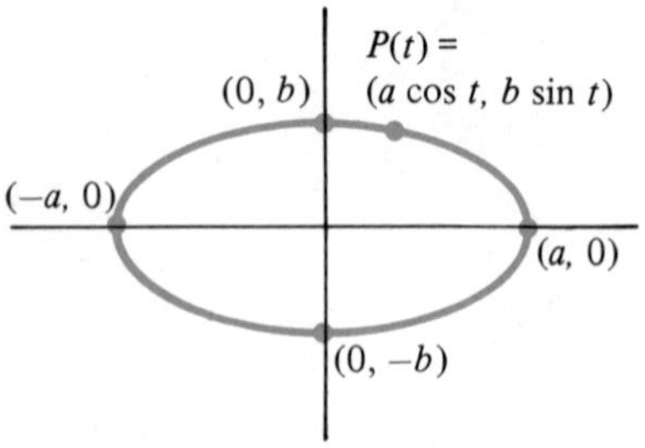

Figure 5.3 Parameterization for the ellipse $\dfrac{x^2}{a^2} + \dfrac{y^2}{b^2} = 1$.

Solution: By squaring both coordinate functions we observe that

$$[x(t)]^2 = a^2 \cos^2 t; \qquad [y(t)]^2 = b^2 \sin^2 t.$$

We obtain

$$\frac{[x(t)]^2}{a^2} + \frac{[y(t)]^2}{b^2} = \cos^2 t + \sin^2 t = 1.$$

The curve is, therefore, the ellipse $\dfrac{x^2}{a^2} + \dfrac{y^2}{b^2} = 1$ seen in Figure 5.3. ■

Example 3 Convert the parametric equations

$$x(t) = t + 1$$

$$y(t) = 2t^2 - 3$$

to a single equation in the variables x and y, and identify the curve.

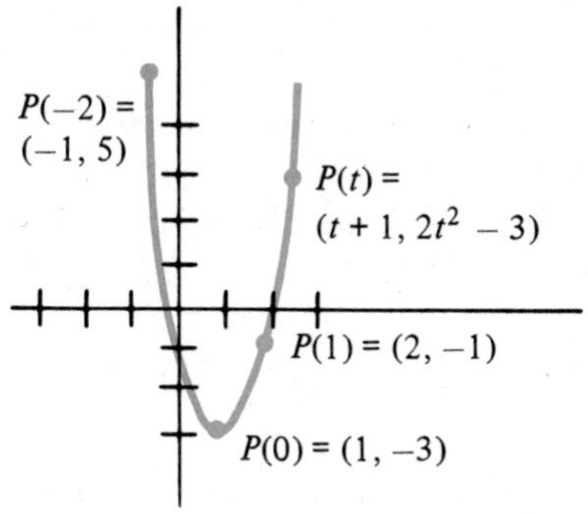

Figure 5.4 Parameterization for the parabola $y = 2(x - 1)^2 - 3$.

Solution: This time we use the more general strategy of solving one equation for t and inserting this expression in the other. Solving the first equation for t gives

$$t = x - 1.$$

Inserting this expression for t in the second equation then gives

$$y = 2(x - 1)^2 - 3.$$

This curve is the parabola sketched in Figure 5.4. ■

Example 4 To convert the parametric equations

$$x(t) = t^2 - 2$$
$$y(t) = t^2 + 1$$

to an equation in x and y alone, we begin by noting that

$$y = t^2 + 1 = (t^2 - 2) + 3 = x + 3.$$

But the resulting equation is not simply $y = x + 3$. This is because the range of the function $x(t) = t^2 - 2$ is $[-2, \infty)$. (That is, we cannot have points (x, y) with $x < -2$, since $t^2 \geq 0$ for all t.) We must therefore restrict the domain of x by writing

$$y = x + 3, \qquad x \geq -2.$$

(See Figure 5.5.) ■

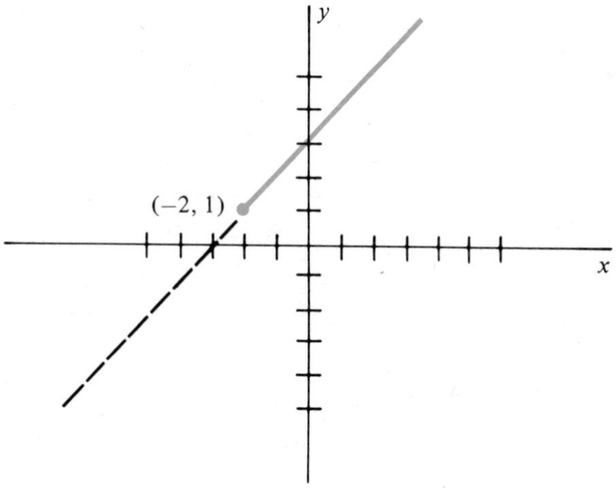

Figure 5.5 Graph of parametric equations $x(t) = t^2 - 2$, $y(t) = t^2 + 1$.

Having conveyed the general notion of parametric equations, we should pause at this point to sharpen our use of the terminology **curve.** In general, a curve C in the plane is the set of all points of the form

$$C = \{(x(t), y(t)) | t \in I\} \tag{5}$$

where I is some interval, and where the coordinate functions $x(t)$ and $y(t)$ are **continuous** functions of $t \in I$.

The requirement that $x(t)$ and $y(t)$ be continuous assures that the curve will itself be continuous (unbroken) in the sense of the graph of a continuous function $y =$

$f(x)$. The curve C in (5) is called **smooth** if the derivatives $x'(t)$ and $y'(t)$ are continuous functions of $t \in I$.

The preceding examples illustrate several situations in which one can pass from equations in xy-coordinates to parametric equations or vice versa. However, this is not always possible, as you may have guessed. The remaining discussions in this section concern only curves described in parametric form.

Equations of Tangents

Suppose that the curve C is given in the parametric form of equation (5) and that we wish to find the slope of the line tangent to the curve at the point $P(t_0) = (x(t_0), y(t_0))$. (For example, the curve C might be the trajectory of a rocket, and we might wish to describe its angle of elevation as a function of time.) To do so, we begin by letting Δt be a small real number. Then $P(t_0 + \Delta t) = (x(t_0 + \Delta t), y(t_0 + \Delta t))$ is, in general, a second point on C (see Figure 5.6). If $x(t_0) \neq x(t_0 + \Delta t)$, we can write the slope of the secant through $P(t_0)$ and $P(t_0 + \Delta t)$ as

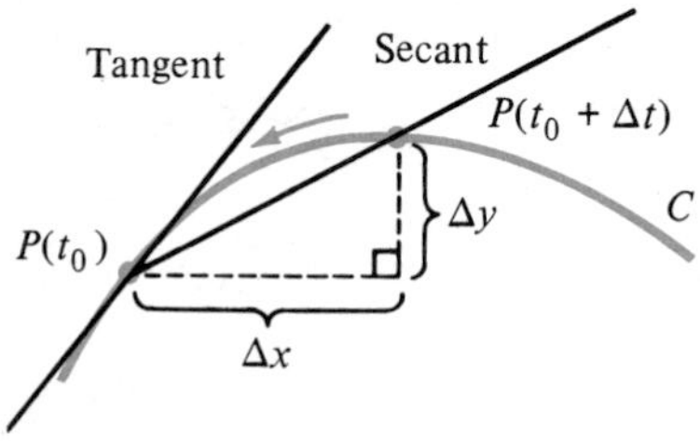

Figure 5.6 Slope of tangent is limit of slopes of secants.

$$m_{\text{sec}} = \frac{\Delta y}{\Delta x} = \frac{y(t_0 + \Delta t) - y(t_0)}{x(t_0 + \Delta t) - x(t_0)}. \tag{6}$$

The limit of this slope as Δt tends to zero is the number m, the desired slope of the tangent at $P(t_0)$. By dividing both Δy and Δx in line (6) by $\Delta t \neq 0$, we find that

$$m = \lim_{\Delta t \to 0} m_{\text{sec}} = \lim_{\Delta t \to 0} \frac{\left(\dfrac{y(t_0 + \Delta t) - y(t_0)}{\Delta t}\right)}{\left(\dfrac{x(t_0 + \Delta t) - x(t_0)}{\Delta t}\right)} = \frac{y'(t_0)}{x'(t_0)}. \tag{7}$$

Obviously, m in equation (7) makes sense only if both $y'(t_0)$ and $x'(t_0)$ exist and $x'(t_0) \neq 0$. Also, recall that we needed to assume that $x(t_0 + \Delta t) \neq x(t_0)$. However, this assumption will be valid for all Δt near zero whenever $x'(t_0)$ exists and is not zero (see Exercise 42). Thus, we have shown that *if $x'(t_0)$ and $y'(t_0)$ exist, and if $x'(t_0) \neq 0$, the slope of the line tangent to the graph of the curve $C = \{(x(t), y(t)) | t \in I\}$ at $P(t_0) = (x(t_0), y(t_0))$ is*

$$m = \frac{y'(t_0)}{x'(t_0)}, \qquad x'(t_0) \neq 0 \tag{8}$$

or, in Leibniz notation,

$$m = \frac{\dfrac{dy}{dt}}{\dfrac{dx}{dt}}, \qquad \frac{dx}{dt} \neq 0$$

The equation for this tangent line is therefore

$$y - y(t_0) = \frac{y'(t_0)}{x'(t_0)}(x - x(t_0))$$

or

$$x'(t_0)[y - y(t_0)] - y'(t_0)[x - x(t_0)] = 0 \tag{9}$$

If $x'(t_0) = 0$ and $y'(t_0) \neq 0$, equation (9) reduces to the equation $x = x(t_0)$ so the tangent is a vertical line through $(x(t_0), y(t_0))$. If both $x'(t_0) = 0$ and $y'(t_0) = 0$, no conclusion can be drawn about the tangent at $(x(t_0), y(t_0))$.

Example 5 Find the equation of the line tangent to the curve given by the parametric equations

$$\left.\begin{aligned} x(t) &= 2\cos t \\ y(t) &= 4\sin t \end{aligned}\right\} 0 \le t < 2\pi$$

at the point $(\sqrt{2}, 2\sqrt{2})$.

Strategy
Find t_0.

Solution
To find the value of t_0 for which $P(t_0) = (\sqrt{2}, 2\sqrt{2})$, we set

$$x(t_0) = 2\cos t_0 = \sqrt{2},$$

$$y(t_0) = 4\sin t_0 = 2\sqrt{2}.$$

This gives $\cos t_0 = \sqrt{2}/2$, $\sin t_0 = \sqrt{2}/2$, so $t_0 = \pi/4$.

Find $x'(t_0)$, $y'(t_0)$.

Then

$$x'(t_0) = -2\sin(\pi/4) = -\sqrt{2}$$

and

$$y'(t_0) = 4\cos(\pi/4) = 2\sqrt{2}.$$

Use equation (9).

Substituting into equation (9) then gives

$$-\sqrt{2}(y - 2\sqrt{2}) - 2\sqrt{2}(x - \sqrt{2}) = 0$$

or

$$y = -2x + 4\sqrt{2}.$$ ■

We can verify the result of Example 5 by converting to rectangular coordinates. As in Example 2, we have

$$\frac{[x(t)]^2}{4} + \frac{[y(t)]^2}{16} = \cos^2 t + \sin^2 t = 1,$$

so C is the graph of the ellipse

$$\frac{x^2}{4} + \frac{y^2}{16} = 1. \tag{10}$$

(See Figure 5.7.)

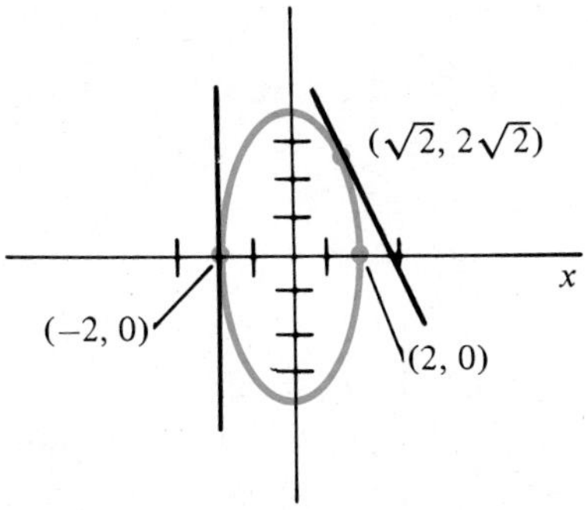

Figure 5.7 Tangents to curve in Example 5.

Differentiating implicitly with respect to x in equation (10) gives

$$\frac{x}{2} + \frac{y}{8} \cdot \frac{dy}{dx} = 0$$

so

$$\frac{dy}{dx} = -\frac{4x}{y}.$$

With $x = \sqrt{2}$ and $y = 2\sqrt{2}$, we obtain $m = dy/dx = -2$, which agrees with the result of Example 5.

Example 6 Determine the values of t and the corresponding points for which the curve in Example 5 has vertical tangents.

Strategy
Find t_0 so that $x'(t_0) = 0$.

Determine if $y'(t_0) \neq 0$ for each such t_0.

If so, the points

$$P(t_0) = (x(t_0), y(t_0))$$

yield vertical tangents.

Solution
Since $x'(t) = -2 \sin t$, the equation

$$x'(t) = -2 \sin t = 0$$

has solutions $t_0 = 0, \pi$ for $0 \leq t < 2\pi$.

Since

$$y'(0) = 4 \cos 0 = 4 \neq 0 \quad \text{and}$$
$$y'(\pi) = 4 \cos \pi = -4 \neq 0,$$

the points

$$P(0) = (2 \cos 0, 4 \sin 0) = (2, 0)$$

and

$$P(\pi) = (2 \cos \pi, 4 \sin \pi) = (-2, 0)$$

yield vertical tangents (see Figure 5.7). ■

Tangents to Polar Curves

If a curve in the plane is the graph of a polar equation of the form $r = f(\theta)$, we can use the equations

$$x = r \cos \theta \tag{11a}$$

$$y = r \sin \theta \tag{11b}$$

to obtain parametric equations for the curve. Multiplying the equation $r = f(\theta)$ on both sides by $\cos \theta$ gives the equation $r \cos \theta = f(\theta) \cos \theta$. Using (11a), we obtain

$$x(\theta) = f(\theta) \cos \theta. \tag{12a}$$

Similarly, multiplying both sides of $r = f(\theta)$ by $\sin \theta$ and using (11b), we obtain

$$y(\theta) = f(\theta) \sin \theta. \tag{12b}$$

If the derivative $f'(\theta) = dr/d\theta$ exists, we may differentiate both sides of equation (12a) and use equation (11b) to conclude that

$$\begin{aligned} x'(\theta) &= f'(\theta) \cos \theta - f(\theta) \sin \theta \\ &= \frac{dr}{d\theta} \cos \theta - r \sin \theta. \end{aligned} \tag{13a}$$

Similarly, differentiating both sides of (12b) and using (11a), we obtain

$$\begin{aligned} y'(\theta) &= f'(\theta) \sin \theta + f(\theta) \cos \theta \\ &= \frac{dr}{d\theta} \sin \theta + r \cos \theta. \end{aligned} \tag{13b}$$

Finally, we may apply equation (8) (with parameter θ rather than t) and equations (13a) and (13b) to conclude that *the slope of the line tangent to the graph of the polar equation $r = f(\theta)$ at the point (r, θ) is*

$$m = \frac{\dfrac{dr}{d\theta}\sin\theta + r\cos\theta}{\dfrac{dr}{d\theta}\cos\theta - r\sin\theta}. \tag{14}$$

As noted for equation (8), m in equation (14) is defined only if $x'(\theta) = \dfrac{dr}{d\theta}\cos\theta - r\sin\theta \neq 0$. If $x'(\theta) = 0$ and $y'(\theta) \neq 0$, the graph has a vertical tangent at (r, θ). If both $x'(\theta) = 0$ and $y'(\theta) = 0$, no conclusions may be drawn.

Example 7 Find the slope of the line tangent to the cardioid with equation $r = 1 + \sin\theta$ at the point where $\theta = \pi/3$.

Strategy

Find r corresponding to $\theta = \pi/3$.

Solution

If $\theta = \pi/3$, then

$$r = 1 + \sin\pi/3 = 1 + \frac{\sqrt{3}}{2}.$$

Differentiate to find $dr/d\theta$. Evaluate.

Also, for $r = 1 + \sin\theta$, we find $dr/d\theta = \cos\theta$. Thus, if $\theta = \pi/3$,

$$\sin\theta = \frac{\sqrt{3}}{2}, \qquad \cos\theta = \frac{1}{2}, \qquad \frac{dr}{d\theta} = \frac{1}{2}.$$

Substitute into equation (14).

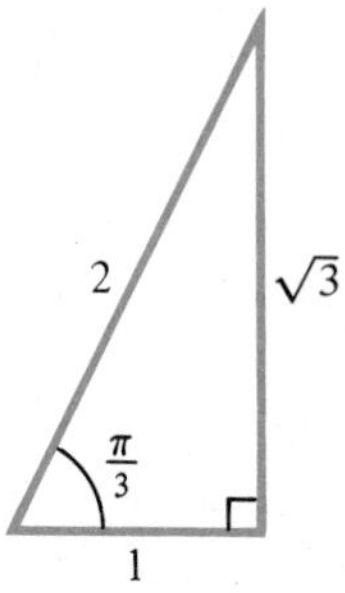

Thus, by equation (14),

$$m = \frac{\frac{1}{2}\cdot\frac{\sqrt{3}}{2} + \left(1 + \frac{\sqrt{3}}{2}\right)\cdot\frac{1}{2}}{\frac{1}{2}\cdot\frac{1}{2} - \left(1 + \frac{\sqrt{3}}{2}\right)\frac{\sqrt{3}}{2}}$$

$$= -1.$$

■

Example 8 Find the points where the four-leaved rose

$$r = \sin 2\theta, \qquad 0 \le \theta < 2\pi$$

has vertical tangents.

Strategy

Find a parameterization for the curve.

Solution

By equations (12a) and (12b), a parameterization for $f(\theta) = \sin 2\theta = 2\sin\theta\cos\theta$ is

$$x(\theta) = 2\sin\theta\cos^2\theta$$

$$y(\theta) = 2\sin^2\theta\cos\theta.$$

Find the values of θ for which $x'(\theta) = 0$.

Thus,

$$x'(\theta) = 2\cos^3\theta - 4\sin^2\theta\cos\theta = 0$$

implies

$$\cos\theta(\cos^2\theta - 2\sin^2\theta) = 0.$$

Thus, $x'(\theta) = 0$ whenever

$$\cos\theta = 0 \qquad \text{or} \qquad \cos^2\theta = 2\sin^2\theta.$$

The equation $\cos\theta = 0$ has solutions $\theta = \pi/2$ and $\theta = 3\pi/2$ in the interval $[0, 2\pi]$. The equation $\cos^2\theta = 2\sin^2\theta$ gives $\tan^2\theta = \frac{1}{2}$ or $\tan\theta = \pm\frac{\sqrt{2}}{2}$, which has 4 solutions in the interval $[0, 2\pi)$, $\theta = \pm\tan^{-1}\left(\pm\frac{\sqrt{2}}{2}\right)$. The 6 solutions of $x'(\theta) = 0$ are therefore $\theta = \frac{\pi}{2}$, $\theta = \frac{3\pi}{2}$, and $\theta = \pm\tan^{-1}\left(\pm\frac{\sqrt{2}}{2}\right)$.

Find $y'(\theta)$.

To determine which of these yield vertical tangents we must examine

$$\begin{aligned} y'(\theta) &= 4\sin\theta\cos^2\theta - 2\sin^3\theta \\ &= 2\sin\theta[2\cos^2\theta - \sin^2\theta]. \end{aligned}$$

Determine if $y'(\theta) \neq 0$ for each θ with $x'(\theta) = 0$.

If $\theta = \pi/2$ or $3\pi/2$, then

$$y'(\theta) = -2\sin^3\theta \neq 0.$$

Similarly, if $\theta = \pm\tan^{-1}\left(\pm\frac{\sqrt{2}}{2}\right)$, then

$$y'(\theta) = 2\left(\pm\sqrt{\frac{1}{3}}\right)\left[2\cdot\frac{2}{3} - \frac{1}{3}\right] \neq 0.$$

If so, (r, θ) is a point that yields a vertical tangent. (See Figure 5.8.)

Thus, all 6 values of θ yield vertical tangents (note in Figure 5.8 that both $\theta = \pi/2$ and $\theta = 3\pi/2$ correspond to the origin). ■

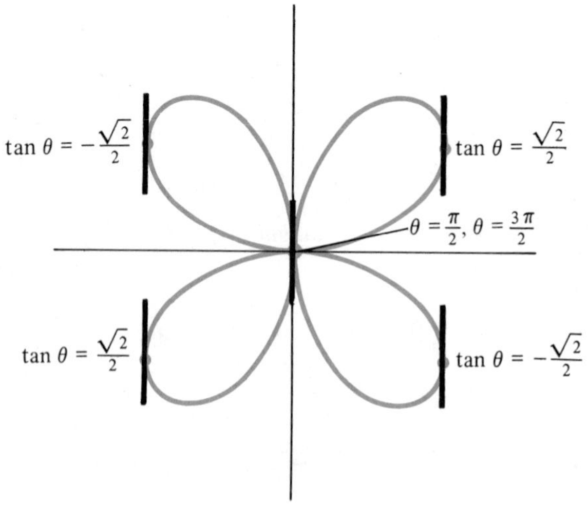

Figure 5.8 Points where graph of $r = 2\sin\theta$ has vertical tangents.

Exercise Set 16.5

In Exercises 1–12, sketch the curve described by the given parametric equations and find an equation in *xy*-coordinates whose graph contains the given curve.

1. $x(t) = t$, $y(t) = t + 6$

2. $x(t) = t + 2$, $y(t) = 3 - t$

3. $x(t) = 1 + 3t$, $y(t) = 2t + 2$

4. $x(t) = t$, $y(t) = t^2 + 2$

5. $x(t) = t^2 + 1$, $y(t) = t^2 - 1$

6. $x(t) = \sin t$, $y(t) = \cos^2 t$

7. $x(t) = e^t$, $y(t) = e^{3t}$

8. $x(t) = \sqrt{t}$, $y(t) = 3t - 2$

9. $x(t) = \sec t$, $y(t) = \tan t$

10. $x(t) = 3 \sin t$, $y(t) = 5 \cos t$

11. $x(t) = \sin t$, $y(t) = \sin 2t$

12. $x(t) = \cos t$, $y(t) = \sec t$

In Exercises 13–20, find the slope of the line tangent to the curve at the indicated point and the equation of the tangent.

13. $x(t) = t$, $y(t) = t^2 + 1$, $t = 2$

14. $x(t) = t^2$, $y(t) = 1 - t$, $t = 2$

15. $x(t) = \sin t$, $y(t) = \cos t$, $t = \pi/3$

16. $x(t) = \cos t$, $y(t) = \sin t$, $t = \pi/4$

17. $x(t) = \dfrac{1}{t}$, $y(t) = 3t^2 - 7$, $t = 2$

18. $x(t) = 3t^3$, $y(t) = \sin \pi t$, $t = 1$

19. $x(t) = 1 + \sqrt{t}$, $y(t) = 1 - \sqrt{t}$, $t = 4$

20. $x(t) = \sec t$, $y(t) = \tan t$, $t = \pi/4$

In Exercises 21–26, find the slope, if defined, of the line tangent to the graph of the polar equation at the point corresponding to the given value of θ.

21. $r = \cos \theta$, $\theta = \pi/4$

22. $r = 1 + \cos \theta$, $\theta = \pi/6$

23. $r = a \sin 2\theta$, $\theta = \pi/6$

24. $r = 2 \sin \theta$, $\theta = \pi/4$

25. $r = \theta$, $\theta = \pi/2$

26. $r = a \sin 3\theta$, $\theta = \pi/3$

In Exercises 27–32, find all points at which the curve described by the parametric equations has **(a)** a vertical tangent, **(b)** a horizontal tangent.

27. $x(t) = \cos t$, $y(t) = \sin t$

28. $x(t) = \sin 2t$, $y(t) = \sin t$

29. $x(t) = t^2 + 4$, $y(t) = 3t^2 - 6t + 2$

30. $x(t) = 5 + 2 \sin t$, $y(t) = 3 - \cos t$

31. $x(t) = 3t^2 + 6$, $y(t) = t - t^2$

32. $x(t) = t^{3/2}$, $y(t) = t + \cos t$

33. Find the point on the curve

$$C_1: \begin{array}{l} x(t) = t \\ y(t) = 2t^2 + 3 \end{array}$$

where the tangent is parallel to the line

$$C_2: \begin{array}{l} x(t) = t + 3 \\ y(t) = 4t - 10. \end{array}$$

34. Find parametric equations for the cardioid $r = 1 + \cos \theta$.

35. Find parametric equations for the three-leaved rose $r = a \sin 3\theta$.

36. Find parametric equations for the spiral $r = \theta$.

37. Find the points at which the cardioid $r = 1 + \sin \theta$ has vertical tangents.

38. Find the points at which the cardioid $r = 1 + \cos \theta$ has horizontal tangents.

39. Find the points at which the cardioid $r = 1 + \cos \theta$ has vertical tangents.

40. Show that the curve determined by the parametric equations

$$x(t) = a \cos t + h$$

$$y(t) = b \sin t + k$$

is an ellipse with center at (h, k).

41. A particle moves in the plane so that at time t it is at the point with coordinates $x(t) = t + 4$, $y(t) = 8 - t^2$. A second particle moves in the plane so that it is at the point with coordinates $x(t) = t + 4$, $y(t) = t + 6$, at time t.

a. Find equations in *xy*-coordinates for each of the curves.
b. Do the paths cross? If so, where?
c. Do the particles collide? If so, where and at what time?

42. Prove that if $x(t)$ is differentiable at t_0 and if $x'(t_0) \neq 0$, then $x(t_0 + \Delta t) \neq x(t_0)$ for all Δt in some neighborhood of $\Delta t = 0$. (*Hint:* Use the differential approximation $x(t_0 + \Delta t) \approx x(t_0) + x'(t_0)\Delta t$. See Chapter 3.)

43. Show that the notion of curve is a more general concept than that of the graph of a continuous function $y = f(x)$. That is, show that every graph of a continuous function $y = f(x)$ can be expressed as a curve in parametric form, but that the converse is not true.

44. Prove that the graph of the function $y = f(x)$ is a *smooth* curve if $f'(x)$ is continuous for all x in the domain of f.

16.6 ARC LENGTH AND SURFACE AREA REVISITED

In Section 7.5 we developed formulas for arc length and surface area calculations associated with the graph of a differentiable function $y = f(x)$. Since we have now seen that a curve is a more general notion than the graph of a function $y = f(x)$, we should ask whether the concepts of arc length and surface area can be extended to more general kinds of curves. In both cases the answer is yes, and the development here parallels that of Section 7.5 very closely. You should review that section carefully before reading further.

Arc Length

Suppose that C is a curve in the plane that is determined by the parametric equations

$$C: \left.\begin{aligned} x &= x(t) \\ y &= y(t) \end{aligned}\right\} \; t \in I,$$

where I is some interval. Suppose further that a and b are numbers in I, with $a < b$, so that the arc of the curve from $P = (x(a), y(a))$ to $Q = (x(b), y(b))$ does not intersect itself, except possibly if $P = Q$. The problem is to define and calculate the length L of the arc of C connecting P and Q (see Figure 6.1).

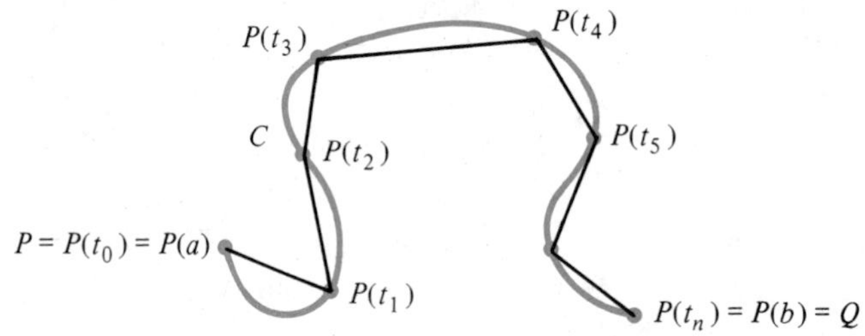

Figure 6.1 Polygonal path joining points $P(t_j) = (x(t_j), y(t_j))$ approximates C.

As in Section 7.5, we begin by partitioning the interval $[a, b]$ for the independent variable (parameter) t into n subintervals of equal length $\Delta t = \dfrac{b - a}{n}$ with endpoints

$$a = t_0 < t_1 = t_0 + \Delta t < t_2 = t_0 + 2\Delta t < \cdots < t_n = t_0 + n\Delta t = b.$$

For each integer $j = 0, 1, 2, \ldots, n$, we let $P(t_j) = (x(t_j), y(t_j))$. Then each $P(t_j)$ is a point on the arc of C joining P and Q. In each interval $[t_{j-1}, t_j]$, we use the length of the line segment $\overline{P(t_{j-1})P(t_j)}$ to approximate the length of the arc C_j connecting these two points. Using the distance formula, we can write this distance as

$$\begin{aligned} \Delta L_j &= \sqrt{\Delta x_j^2 + \Delta y_j^2} \qquad (1) \\ &= \sqrt{[x(t_j) - x(t_{j-1})]^2 + [y(t_j) - y(t_{j-1})]^2}, \qquad j = 1, 2, \ldots, n. \end{aligned}$$

(See Figure 6.2.)

If $x(t)$ is a differentiable function of t in each interval $[t_{j-1}, t_j]$, we may apply the Mean Value Theorem to conclude that there exists a number $c_j \in [t_{j-1}, t_j]$ so that

$$x'(c_j) = \frac{x(t_j) - x(t_{j-1})}{t_j - t_{j-1}} = \frac{x(t_j) - x(t_{j-1})}{\Delta t}.$$

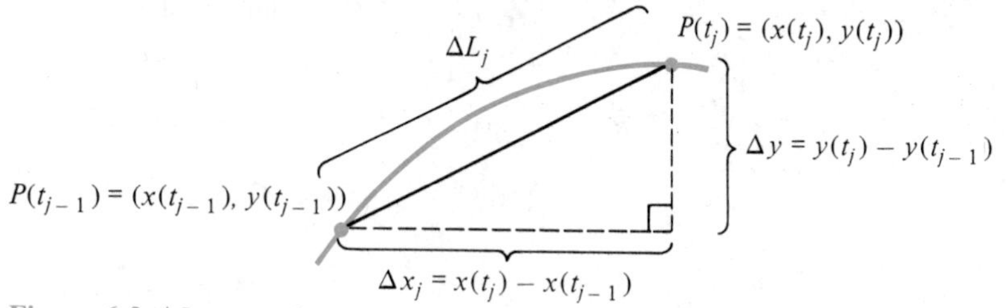

Figure 6.2 ΔL_j approximates length of arc from $P(t_{j-1})$ to $P(t_j)$.

Thus, we can write

$$x(t_j) - x(t_{j-1}) = x'(c_j)\,\Delta t, \qquad j = 1, 2, \ldots, n. \tag{2}$$

Similarly, if $y(t)$ is a differentiable function of t, there exist numbers $d_j \in [t_{j-1}, t_j]$ so that

$$y(t_j) - y(t_{j-1}) = y'(d_j)\,\Delta t, \qquad j = 1, 2, \ldots, n. \tag{3}$$

Combining equations (1)–(3), we obtain the length of the approximating line segment as

$$\begin{aligned} \Delta L_j &= \sqrt{[x'(c_j)\,\Delta t]^2 + [y'(d_j)\,\Delta t]^2} \\ &= \sqrt{[x'(c_j)]^2 + [y'(d_j)]^2}\,\Delta t, \qquad j = 1, 2, \ldots, n. \end{aligned} \tag{4}$$

Finally, we approximate the length L of the arc of C from P to Q by the length of the polygonal path connecting the points $P(t_0), P(t_1), \ldots, P(t_n)$. The latter is simply the sum of the lengths ΔL_j in line (4). We obtain the approximation

$$L \approx \sum_{j=1}^{n} \Delta L_j = \sum_{j=1}^{n} \sqrt{[x'(c_j)]^2 + [y'(d_j)]^2}\,\Delta t. \tag{5}$$

The expression on the right-hand side of equation (5) is "almost" a Riemann sum. The difficulty is that the functions $x'(t)$ and $y'(t)$ are being evaluated at two (possibly) different points in each subinterval. However, in more advanced courses, it is shown that if both $x'(t)$ and $y'(t)$ are continuous on $[a, b]$ this difficulty can be overcome, and that as $n \to \infty$ the approximating sum converges to the integral

$$L = \int_a^b \sqrt{[x'(t)]^2 + [y'(t)]^2}\,dt \tag{6}$$

or, in Leibniz notation

$$L = \int_a^b \sqrt{\left(\frac{dx}{dt}\right)^2 + \left(\frac{dy}{dt}\right)^2}\,dt. \tag{7}$$

As in Section 7.5, we argue that as $n \to \infty$ the polygonal paths more closely approximate the curve C, so that the limiting value of the length of the polygonal paths provides a reasonable definition of the length of the arc. It is important to keep in mind that *both $x'(t)$ and $y'(t)$ must be continuous on $[a, b]$ for formulas (6) and (7) to be valid.*

Example 1 Find the length of the arc of the curve

$$C: \begin{array}{l} x(t) = \cos t + t \sin t \\ y(t) = \sin t - t \cos t \end{array}$$

connecting the points $P = (x(0), y(0)) = (1, 0)$ and $Q = \left(x\left(\frac{\pi}{2}\right), y\left(\frac{\pi}{2}\right)\right) = \left(\frac{\pi}{2}, 1\right)$.

Solution: Here

$$x'(t) = -\sin t + \sin t + t \cos t = t \cos t$$

and

$$y'(t) = \cos t - \cos t + t \sin t = t \sin t.$$

By formula (6)

$$\begin{aligned} L &= \int_0^{\pi/2} \sqrt{(t \cos t)^2 + (t \sin t)^2}\, dt \\ &= \int_0^{\pi/2} \sqrt{t^2(\cos^2 t + \sin^2 t)}\, dt \\ &= \int_0^{\pi/2} t\, dt \\ &= \frac{t^2}{2}\bigg]_0^{\pi/2} \\ &= \frac{\pi^2}{8}. \end{aligned}$$

■

Arc Length in Polar Coordinates

In Section 16.5 we showed that the graph of the polar equation $r = f(\theta)$ can be parameterized, using θ as the parameter, as

$$x(\theta) = f(\theta) \cos \theta,$$

$$y(\theta) = f(\theta) \sin \theta.$$

(See equations (12a) and (12b) in Section 16.5.)

If $f'(\theta)$ is continuous, so are the functions

$$x'(\theta) = f'(\theta) \cos \theta - f(\theta) \sin \theta$$

and

$$y'(\theta) = f'(\theta) \sin \theta + f(\theta) \cos \theta,$$

so we may apply equation (6) to determine a formula for arc length in polar coordinates. Before doing so we note that

$$[x'(\theta)]^2 = [f'(\theta)]^2 \cos^2 \theta - 2f'(\theta)f(\theta) \sin \theta \cos \theta + [f(\theta)]^2 \sin^2 \theta,$$

and

$$[y'(\theta)]^2 = [f'(\theta)]^2 \sin^2 \theta + 2f'(\theta)f(\theta) \sin \theta \cos \theta + [f(\theta)]^2 \cos^2 \theta,$$

so

$$[x'(\theta)]^2 + [y'(\theta)]^2 = [f'(\theta)]^2 + [f(\theta)]^2. \tag{8}$$

From equations (6) and (8) it follows that

$$L = \int_a^b \sqrt{[f'(\theta)]^2 + [f(\theta)]^2}\, d\theta \tag{9}$$

gives the length of the arc of the graph of $r = f(\theta)$ from $\theta = a$ to $\theta = b$. In Leibniz notation equation (9) is

$$L = \int_a^b \sqrt{r^2 + \left(\frac{dr}{d\theta}\right)^2}\, d\theta. \tag{10}$$

Example 2 Find the length of the cardioid $r = 1 + \cos\theta$.

Strategy

Calculate $dr/d\theta$.

Determine the limits of integration.

Verify that equation (10) applies.

Solution

Here $dr/d\theta = -\sin\theta$, and the cardioid is swept out as θ makes one complete revolution. Appropriate limits of integration are therefore $\theta = 0$ to $\theta = 2\pi$.

Since $dr/d\theta$ is continuous on $[0, 2\pi]$, we may apply formula (10):

Apply (10).

Use identity $\cos^2\phi = \frac{1}{2} + \frac{1}{2}\cos 2\phi$ in reverse to simplify integrand.

Use the symmetry of the cosine function to handle the absolute value signs. (Alternatively, we could have used the symmetry of the cardioid and integrated over $[0, \pi]$.)

$$\begin{aligned}
L &= \int_0^{2\pi} \sqrt{(1+\cos\theta)^2 + (-\sin\theta)^2}\, d\theta \\
&= \int_0^{2\pi} \sqrt{(1 + 2\cos\theta + \cos^2\theta) + \sin^2\theta}\, d\theta \\
&= \int_0^{2\pi} \sqrt{2 + 2\cos\theta}\, d\theta \\
&= \int_0^{2\pi} \sqrt{4\left(\frac{1}{2} + \frac{1}{2}\cos\theta\right)}\, d\theta \\
&= \int_0^{2\pi} \sqrt{4\cos^2\left(\frac{\theta}{2}\right)}\, d\theta \\
&= \int_0^{2\pi} \left|2\cos\left(\frac{\theta}{2}\right)\right| d\theta \\
&= 2\int_0^{\pi} 2\cos\left(\frac{\theta}{2}\right) d\theta \\
&= 8\sin\left(\frac{\theta}{2}\right)\Big]_0^{\pi} = 8.
\end{aligned}$$

■

Differential for Arc Length

Using the Fundamental Theorem of Calculus, we may write L in equation (6) as a function of t,

$$L(t) = \int_a^t \sqrt{[x'(s)]^2 + [y'(s)]^2}\, ds$$

and differentiate with respect to t to conclude that

$$\frac{dL}{dt} = \sqrt{[x'(t)]^2 + [y'(t)]^2},$$

or, in Leibniz notation, that

$$\frac{dL}{dt} = \sqrt{\left(\frac{dx}{dt}\right)^2 + \left(\frac{dy}{dt}\right)^2}. \tag{11}$$

Multiplying both sides of equation (11) by the differential dt gives the *differential for arc length,*

$$dL = \sqrt{(dx)^2 + (dy)^2}. \tag{12}$$

Using (11) and (12), we may write the indefinite integral corresponding to equation (7) as

$$L = \int \left(\frac{dL}{dt}\right) dt = \int dL = \int \sqrt{(dx)^2 + (dy)^2}. \tag{13}$$

In other words, arc length is computed by integrating the differential for arc length between the corresponding limits of integration. The geometric interpretation of equation (12) is illustrated by Figure 6.3.

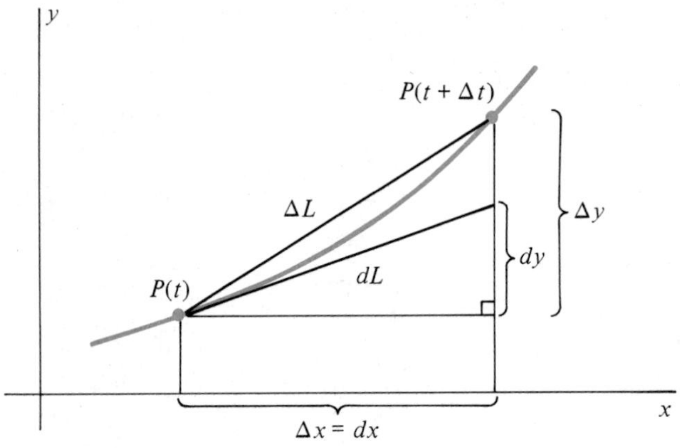

Figure 6.3 $dL = \sqrt{(dx)^2 + (dy)^2}$

Surface Area

In Section 7.5 we developed a formula for the lateral surface area of a figure that is obtained by revolving the graph of the continuously differentiable function $y = f(x)$, $a \le x \le b$, about the x-axis. Here we shall develop the corresponding formula for the case in which the curve C to be revolved about the x-axis is given by the parametric equations

$$C: \quad \left.\begin{array}{l} x = x(t) \\ y = y(t) \end{array}\right\} \; a \le t \le b.$$

As in the case of arc length, we will need to require that both $x'(t)$ and $y'(t)$ exist and are continuous for $t \in [a, b]$. Also, we shall require that $x'(t) \neq 0$ for all $t \in [a, b]$. (This last condition insures that the curve will have at most one point with any given x-coordinate (see Exercise 35).)

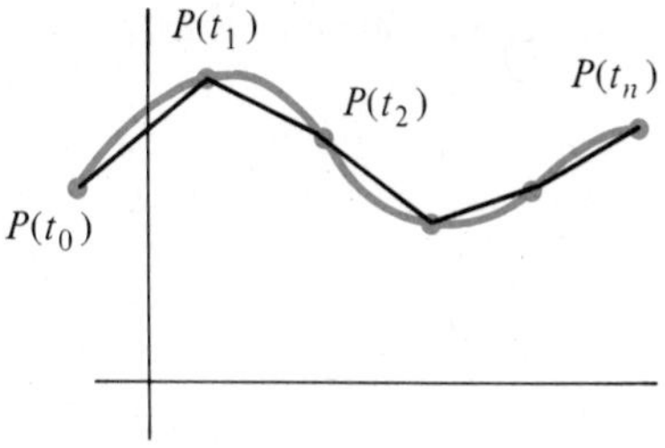

Figure 6.4 Curve to be revolved about the x-axis.

We begin with the usual step of partitioning the interval $[a, b]$ into n equal subintervals of length $\Delta t = \dfrac{b-a}{n}$, and with endpoints

$$a = t_0 < t_1 = t_0 + \Delta t < t_2 = t_0 + 2\Delta t < \cdots < t_n = t_0 + n\Delta t = b.$$

Let $P(t_j) = (x(t_j), y(t_j))$. We shall approximate C by the polygonal path consisting of the line segments $\overline{P(t_{j-1})P(t_j)}$ joining these consecutive points on C (see Figure 6.4). As each of these line segments is revolved about the x-axis it generates the frustum of a cone, as in Figure 6.5. The radii of the jth frustum are $y(t_{j-1})$ and $y(t_j)$, so the lateral surface area of the jth frustum is

$$\Delta S_j = \pi[y(t_{j-1}) + y(t_j)]\ell_j, \qquad j = 1, 2, \ldots, n. \tag{14}$$

where ℓ_j is the slant height. (See Section 7.5 for an explanation of this formula, if necessary.) Since the altitude of the jth frustum is $\Delta x_j = x(t_j) - x(t_{j-1})$, the slant height is

$$\begin{aligned}\ell_j &= \sqrt{(\Delta x_j)^2 + (\Delta y_j)^2} \\ &= \sqrt{[x(t_j) - x(t_{j-1})]^2 + [y(t_j) - y(t_{j-1})]^2}.\end{aligned}$$

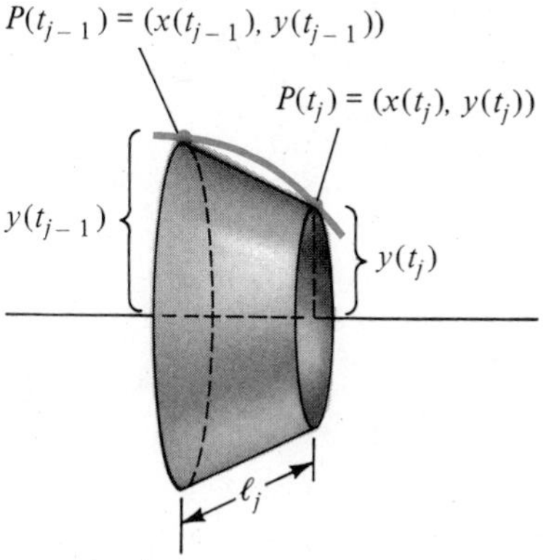

Figure 6.5 Frustum generated by $\overline{P(t_{j-1})P(t_j)}$.

As in our development of the formula for arc length, we now apply the Mean Value Theorem twice, once to the function $x(t)$ and once to the function $y(t)$, to conclude that there exist numbers c_j and d_j in $[t_{j-1}, t_j]$ so that

$$\ell_j = \sqrt{[x'(c_j)]^2 + [y'(d_j)]^2}\,\Delta t, \qquad j = 1, 2, \ldots, n. \tag{15}$$

Combining equations (14) and (15), and summing over all line segments $\overline{P(t_{j-1})P(t_j)}$, we obtain the following approximation to the lateral surface area S of the volume of revolution.

$$S \approx \sum_{j=1}^{n} \Delta S_j = \sum_{j=1}^{n} \pi[y(t_{j-1}) + y(t_j)]\sqrt{[x'(c_j)]^2 + [y'(d_j)]^2}\,\Delta t \tag{16}$$

A fact that we shall not prove is that if the derivatives $x'(t)$ and $y'(t)$ are continuous for $a \le t \le b$, the approximating sum on the right-hand side of approximation (16) converges to the definite integral

$$S = \int_a^b 2\pi y(t)\sqrt{[x'(t)]^2 + [y'(t)]^2}\, dt \tag{17}$$

or, in Leibniz notation,

$$S = \int_a^b 2\pi y \sqrt{\left(\frac{dx}{dt}\right)^2 + \left(\frac{dy}{dt}\right)^2}\, dt. \tag{18}$$

The usual argument supports formulas (17) and (18): as $n \to \infty$ the polygonal path more closely approximates the curve C, so the limiting value of the surface area generated by the approximating frustums must be the desired surface area.

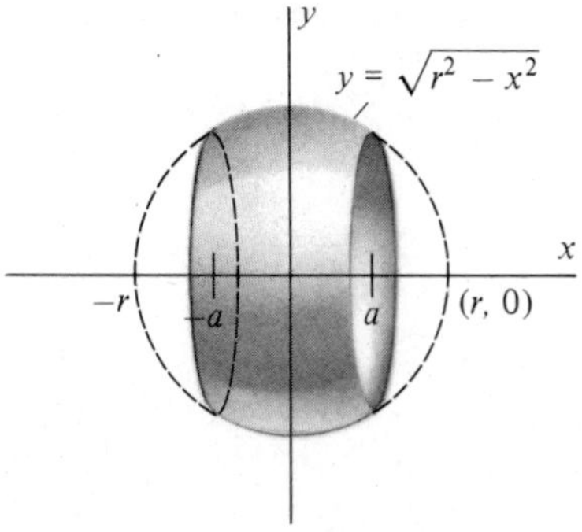

Figure 6.6 $S = 4\pi ar$.

Example 3 In Example 4, Section 7.5, it was shown that the surface area of the band of the sphere obtained by revolving the arc of the circle $x^2 + y^2 = r^2$ lying above the interval $[-a, a]$, $a < r$, about the x-axis is $S = 4\pi ar$ (Figure 6.6).

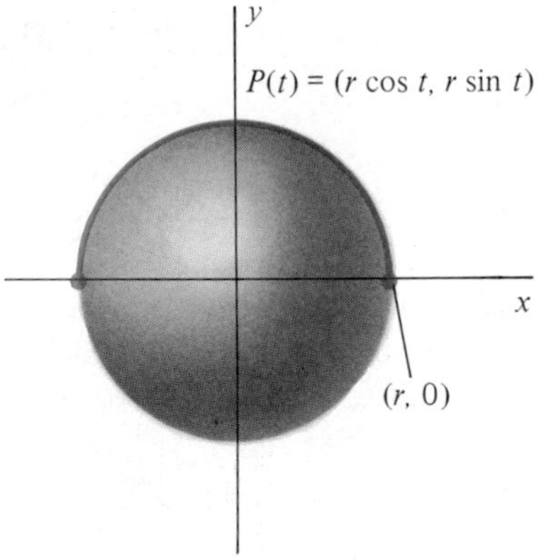

Figure 6.7 $S = 4\pi r^2$.

Moreover, we noted in Example 5 of that section that one could not conclude from this calculation that the surface area of a sphere is $S = 4\pi r^2$.

We may now use equation (17) to demonstrate that the surface area of the sphere of radius r is indeed $S = 4\pi r^2$ (Figure 6.7). To do so we use the parameterization of the upper semicircle

$$\left.\begin{array}{l} x(t) = r\cos t \\ y(t) = r\sin t \end{array}\right\} \quad 0 \le t \le \pi.$$

Then

$$x'(t) = -r\sin t; \qquad y'(t) = r\cos t.$$

Both of these derivatives are continuous for all t, in contrast to dy/dx (which is discontinuous at $x = r$ and $x = -r$). Thus, by equation (17),

$$\begin{aligned} S &= \int_0^{\pi} 2\pi(r\sin t)\sqrt{(-r\sin t)^2 + (r\cos^2 t)}\, dt \\ &= \int_0^{\pi} 2\pi r^2 \sin t\, dt \\ &= -2\pi r^2 \cos t\big]_0^{\pi} \\ &= -2\pi r^2(-1 - 1) \\ &= 4\pi r^2. \end{aligned}$$

■

Exercise Set 16.6

In Exercises 1–11, a curve is described by a pair of parametric equations. Find the length of the given curve.

1. $x(t) = t$, $y(t) = t^{3/2}$, $0 \le t \le \frac{4}{9}$
2. $x(t) = t + 1$, $y(t) = \frac{2}{3}t^{3/2}$, $0 \le t \le 3$
3. $x(t) = 3t^2$, $y(t) = 2t^3$, $0 \le t \le \sqrt{2}$
4. $x(t) = \sin t$, $y(t) = \cos t$, $0 \le t \le 2\pi$
5. $x(t) = 2(2t + 3)^{3/2}$, $y(t) = 3(t + 1)^2$, $0 \le t \le 2$
6. $x(t) = 2(1 - \cos t)$, $y(t) = 2\sin t$, $0 \le t \le \pi$
7. $x(t) = t^3 - 3t^2$, $y(t) = 3t^2$, $0 \le t \le 1$
8. $x(t) = \ln\cos t$, $y(t) = t$, $0 \le t \le \pi/3$
9. $x(t) = \cos^3 t$, $y(t) = \sin^3 t$, $0 \le t \le \pi$
10. $x(t) = 3t^3$, $y(t) = 3t^2$, $0 \le t \le \sqrt{5}$
11. $x(t) = e^t\cos t$, $y(t) = e^t\sin t$, $0 \le t \le \pi$

In Exercises 12–18, a curve is described by a pair of parametric equations. Find the surface area of the solid generated by revolving this curve about the x-axis.

12. $x(t) = t$, $y(t) = t^2$, $0 \le t \le 1$
13. $x(t) = t^2$, $y(t) = 4t$, $0 \le t \le 2$
14. $x(t) = t$, $y(t) = \frac{t^4}{4} + \frac{1}{8t^2}$, $0 \le t \le 2$
15. $x(t) = t$, $y(t) = \sqrt{t}$, $1 \le t \le 2$

16. $x(t) = t$, $y(t) = \dfrac{t^2 - 1}{2}$ $\quad 0 \le t \le 1$

17. $x(t) = 3t^2$, $y(t) = 2t^3$ $\quad 1 \le t \le 2$

18. $x(t) = \cos^3 t$, $y(t) = \sin^3 t$ $\quad 0 \le t \le \pi$

19. Find the length of the cardioid $r = \cos^2\left(\dfrac{\theta}{2}\right)$.

20. Find the length of the cardioid $r = 1 + \cos\theta$.

21. Find the length of the spiral $r = e^{2\theta}$, $\quad 0 \le \theta \le \pi$.

22. Find the area of the surface generated when one arch of the cycloid $x(t) = t - \sin t$, $y(t) = 1 - \cos t$ is revolved about the x-axis.

23. Find the length of the circle $x(t) = 2\cos t$, $y(t) = 2\sin t$, $0 \le t \le 2\pi$.

24. Find the length of the spiral $r = 2\theta$, $\quad 0 \le \theta \le \pi$.

25. Show, using equation (9), that the circumference of the circle $r = 3\cos\theta$ is 3π.

26. Find the length of the cardioid $r = a(1 + \sin\theta)$.

27. Find the length of the graph of $r = e^{\theta}$ for $0 \le \theta \le 2\pi$.

28. Find the length of the graph of the equation $r = 4\sec\theta$ for $0 \le \theta \le \dfrac{\pi}{4}$.

29. Find the surface area of the solid obtained by revolving the region bounded by the graph of $r = e^{\theta}$, $0 \le \theta \le \pi$, and the x-axis about the x-axis.

30. Find the area of the surface of the solid obtained by revolving the region bounded by the graph of $r = 4\sin\theta$ about the line $\theta = 0$.

31. Determine the formula for the surface area of the volume of the solid generated when an arc of the curve determined by the parametric equations $x = x(t)$, $y = y(t)$ is revolved about the y-axis.

32. Use the result of Exercise 31 to find the surface area of the solid generated when the curve given in Exercise 13 is revolved about the y-axis.

33. Use the result of Exercise 31 to find the surface area of the solid obtained by revolving the curve in Exercise 15 about the y-axis.

34. *(Calculator/Computer)* Use Simpson's Rule to approximate the length of the ellipse

$$x(t) = 4\cos t,$$
$$y(t) = 6\sin t.$$

35. Show that the condition $x'(t) \neq 0$ is sufficient to guarantee that a differentiable curve C have at most one point with any particular x-coordinate.

SUMMARY OUTLINE OF CHAPTER 16

- The **polar coordinates** (r, θ) for the point whose rectangular coordinates are (x, y) are determined by the equations

$$r = \sqrt{x^2 + y^2}$$

$$\tan\theta = y/x$$

while

$$x = r\cos\theta$$

$$y = r\sin\theta.$$

- The following identities hold for polar coordinates:

$$(-r, \theta) = (r, \theta + \pi)$$

$$(r, \theta + 2n\pi) = (r, \theta), \qquad n = \pm 1, \pm 2, \ldots.$$

- The graph of the polar equation $r = f(\theta)$ is
  1. symmetric with respect to the pole if $f(\theta + \pi) = f(\theta)$;
  2. symmetric with respect to the x-axis if $f(-\theta) = f(\theta)$;
  3. symmetric with respect to the y-axis if $f(\theta) = f(\pi - \theta)$.

- The area A of the region bounded by the graph of the function $r = f(\theta)$ and the rays $\theta = a$ and $\theta = b$ is

$$A = \int_a^b \frac{1}{2}[f(\theta)]^2\, d\theta.$$

- The slope of the line tangent to the curve determined by the **parametric equations** $x = x(t)$, $y = y(t)$, at the point $(x(t_0), y(t_0))$ is

$$m = \frac{y'(t_0)}{x'(t_0)}$$

provided $x'(t_0)$ and $y'(t_0)$ exist and $x'(t_0) \neq 0$. The equation of this tangent line is

$$x'(t_0)[y - y(t_0)] - y'(t_0)[x - x(t_0)] = 0.$$

■ The slope of the line tangent to the graph of the polar equation $r = f(\theta)$ at the point (r, θ) is

$$m = \frac{\frac{dr}{d\theta}\sin\theta + r\cos\theta}{\frac{dr}{d\theta}\cos\theta - r\sin\theta}.$$

■ For the arc of the curve C determined by the polar equations $x = x(t)$, $y = y(t)$, between the points $(x(a), y(a))$ and $(x(b), y(b))$:

1. The length of the arc is

$$L = \int_a^b \sqrt{[x'(t)]^2 + [y'(t)]^2}\, dt.$$

2. The surface area of the volume obtained by revolving the arc about the x-axis is

$$S = \int_a^b 2\pi y(t)\sqrt{[x'(t)]^2 + [y'(t)]^2}\, dt.$$

■ The length of the arc of the graph of the polar equation $r = f(\theta)$, from $\theta = a$ to $\theta = b$, is

$$L = \int_a^b \sqrt{[f'(\theta)]^2 + [f(\theta)]^2}\, d\theta.$$

REVIEW EXERCISES—CHAPTER 16

In Exercises 1–6, sketch the curve described by the given parametric equations.

1. $x = 5\cos\theta$, $y = 5\sin\theta$, $0 \le \theta \le 2\pi$

2. $x = 3 + 2t$, $y = 8 - 6t$, $-\infty < t < \infty$

3. $x = t\cos \pi t$, $y = t\sin \pi t$, $0 \le t \le 6$

4. $x = t^2 + 1$, $y = t^4 - 4$, $-\infty < t < \infty$

5. $x = \cos t$, $y = \sin 2t$, $0 \le t \le \pi$

6. $x = 3\sqrt{t} + 1$, $y = 1 - \sqrt{t}$, $0 \le t$

7. Eliminate the parameter in Exercise 3 to find an equation in x and y that represents the given curve.

8. Repeat Exercise 7 for the curve given in Exercise 4.

9. Repeat Exercise 7 for the curve given in Exercise 5.

In each of Exercises 10–13, find parametric equations for the given curves.

10. $x^2 + y^2 = 9$ **11.** $4x^2 + 9y^2 = 36$

12. $x + y = 7$ **13.** $x = 2y^2 + 4y + 5$

14. Show that the parametric equations $x(t) = t$, $y(t) = t^2$ describe the graph of the equation $y = x^2$ as well as the graph of the equation $y^3 = x^6$, but *not* the graph of $y^2 = x^4$.

15. Sketch the curve given parametrically in polar coordinates by the equations $r = 2t$, $\theta = \pi t^2$.

16. Sketch the curve given parametrically by the equations $\theta = t$, $r = e^t$.

17. Show that the points on the curve determined by the parametric equations $x(t) = t^2 - 1$, $y(t) = 3t$ lie on a parabola. Find the equation for this parabola in xy-coordinates.

18. Find an equation in polar coordinates for the circle with center at the origin and radius 5.

In Exercises 19–28, sketch the graph of the given polar equation.

19. $r = \cos 2\theta$ **20.** $r = 3\theta$

21. $r = 5\cos 3\theta$ **22.** $r = |\sin 2\theta|$

23. $r = 6\sin\theta$ **24.** $r = a\tan\theta$

25. $r = 2 + \sin 2\theta$ **26.** $r = \dfrac{1}{2 - \cos\theta}$

27. $\theta = \pi/2$ **28.** $r = 1 - 2\sin\theta$

In Exercises 29–34, find the area of the region enclosed by the given curve.

29. $r = 2(1 + \sin 2\theta)$ **30.** $r = 2 - \cos\theta$

31. $r = a(1 + \sin\theta)$ **32.** $r = \sqrt{1 - \sin\theta}$

33. $r = 6\cos 3\theta$ **34.** $r = 2 + 4\sin\theta\cos\theta$

In Exercises 36–40, find the length of the curve given by the parametric equations.

36. $x = 2t + 1$, $y = t^2$, $0 \le t \le 1$

37. $x = 3t$, $y = t^3$, $0 \le t \le 1$

38. $x = \sin t - t$, $y = \cos t$, $0 \le t \le \pi$

39. $x = 1 - \cos t$, $y = \sin t$, $0 \le t \le \pi$

40. $x = 9t^2, \quad y = 3t^3 - 9t^2, \quad 0 \le t \le 1$

In Exercises 41–44, find the length of the curve given by the polar equation.

41. $r = \cos\theta, \quad -\pi/4 \le \theta \le \pi/4$

42. $r = \theta^2, \quad 1 \le \theta \le 2$

43. $r = 1 + \sin\theta, \quad 0 \le \theta \le \pi/2$

44. $r = 1 - \cos\theta, \quad 0 \le \theta \le \pi$

45. Find the area of the region common to the circles $r = \sin\theta$ and $r = \cos\theta$.

46. Find the area of the region enclosed by the lemniscate $r^2 = 4\cos 2\theta$.

47. Find the area of the region inside the circle $r = 6\cos\theta$ and outside the cardioid $r = 2(1 + \cos\theta)$.

48. Find the area of the region inside the circle $r = 4\cos\theta$ and outside the circle $r = 2$.

49. Find the area of the region common to the circle $r = 4\sin\theta$ and the circle $r = 2$.

50. Find the slope of the line tangent to the curve given by the parametric equations $x(t) = \sin t$, $y(t) = \sin 2t$ at the point where $t = \pi/4$.

CHAPTER 17

VECTORS AND SPACE COORDINATES

17.1 INTRODUCTION

In Chapter 16, we saw that by use of parametric equations of the form $x = x(t)$, $y = y(t)$ we can define functions $f(t) = (x(t), y(t))$ whose values are pairs of real numbers rather than single numbers. Symbolically, we write

$$f: \mathbb{R} \to \mathbb{R}^2 \quad \text{if} \quad f(t) = (x(t), y(t)). \tag{1}$$

We may interpret (1) by saying that the domain of f is a subset of $\mathbb{R}$ (the real line) and that the range of f is a subset of $\mathbb{R}^2$, the real plane. Our next major goal is to develop the calculus of functions of this type. However, in attempting to do so we face several immediate difficulties due to the fact that the values of such functions are points in the plane rather than points on the real number line. For example, to speak of the *sum* of two functions of the form

$$f_1(t) = (x_1(t), y_1(t)), \qquad f_2(t) = (x_2(t), y_2(t)) \tag{2}$$

we will want to have the equation

$$f_1(t) + f_2(t) = (x_1(t), y_1(t)) + (x_2(t), y_2(t)). \tag{3}$$

But the right side of equation (3) calls for something we have not yet defined—the *sum* of two points in the plane. Similarly, the equation

$$cf_1(t) = c(x_1(t), y_1(t)), \tag{4}$$

referring to the product of c times the function $f_1(t)$, involves on its right-hand side the *product* of a number times a point in the plane.

What is needed is an **algebraic structure** for the xy-plane. By this we mean a way to add two points in the plane to produce another point in the plane (their sum) and a way to multiply points in the plane by real numbers. We achieve this additional structure for the plane through the concept of *vectors,* developed in Sections 17.2 through 17.4. We next introduce the Cartesian coordinate system for three-dimensional space in Section 17.5, and extend the concept of vectors to three-dimensional space in Sections 17.6 and 17.7. The chapter concludes with a brief look at two other means of coordinatizing three-dimensional space (Section 17.8) and a discussion of equations for cylinders and quadric surfaces in space (Section 17.9).

There is virtually no calculus contained in Chapter 17. However, the geometry of the plane and of space developed here provides the framework, language, and tools by which the concepts of the calculus will be extended to the multivariable settings addressed throughout the remainder of the text. It is important to master the concepts and techniques presented in this chapter before moving on, so that you do not become bogged down in the technical details of what unfolds in later chapters.

17.2 VECTORS IN THE PLANE: A GEOMETRIC DESCRIPTION

Figure 2.1 The line segment $\overline{PQ}$.

Figure 2.2 The vector $\overrightarrow{PQ}$.

Figure 2.3 The vector $\overrightarrow{QP}$.

If P and Q are points in the plane, we denote by $\overline{PQ}$ the line segment joining P and Q (Figure 2.1). As such, $\overline{PQ}$ has both a **length** and a **slope.** We can also make $\overline{PQ}$ indicate a **displacement,** meaning a motion either from P to Q or from Q to P, by specifying which of the two possible directions is intended. We do this using the concept of *vector,* as follows.

If P and Q are points in the plane, the **vector** $\overrightarrow{PQ}$ is the displacement from P to Q. We say that the vector $\overrightarrow{PQ}$ **originates** at point P, and **terminates** at point Q. In drawings, the vector $\overrightarrow{PQ}$ is represented by an arrow consisting of the line segment $\overline{PQ}$ to which a tip is added at point Q. The vector $\overrightarrow{QP}$ is the same line segment with the tip at point P (Figures 2.2 and 2.3). The length of either vector is the same as the length of the line segment $\overline{PQ}$, calculated by the distance formula.

In viewing a vector as a displacement, the two essential attributes are (i) the *length* of the vector and (ii) the *direction* of the vector. This suggests that a vector should be thought of as independent of any particular initial point. For example, if $P = (0, 0)$, $Q = (2, 4)$, $R = (-3, 1)$, and $S = (-1, 5)$, the vectors $\overrightarrow{PQ}$ and $\overrightarrow{RS}$ have the same lengths and directions (Figure 2.4). In recognition of the fact that a vector is not fastened to any particular point, we will use boldface notation ($\mathbf{v}$) for vectors from now on. We will denote the length (also called the **magnitude**) of $\mathbf{v}$ by $|\mathbf{v}|$.

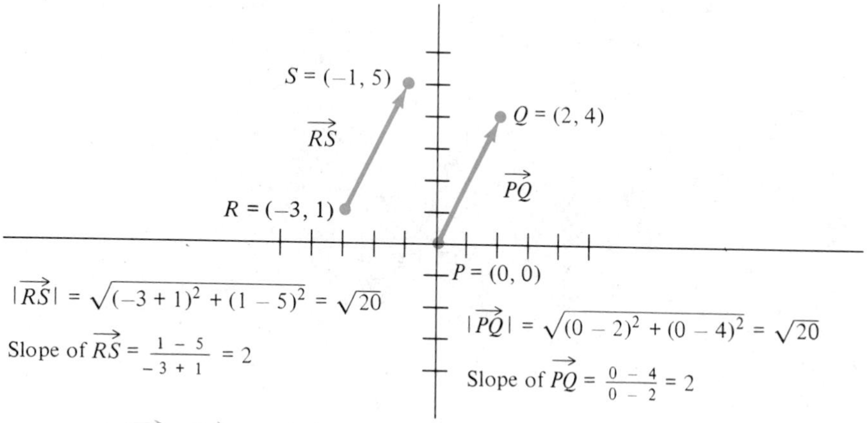

Figure 2.4 $\overrightarrow{PQ} = \overrightarrow{RS}$.

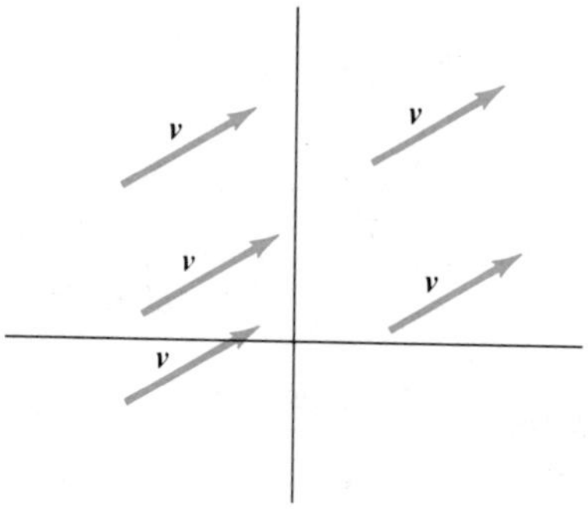

Figure 2.5 Various representations of the vector $\mathbf{v}$.

Thus, we arrive at the more general definition of a **vector** $\mathbf{v}$ as an object that has a **fixed length** and a **fixed direction,** and we say that two vectors, $\mathbf{v}$ and $\mathbf{w}$, are **equal** if they have the same length and the same direction. Thus $\overrightarrow{PQ} = \overrightarrow{RS}$ in Figure 2.4, as you can verify. Moreover, the vector $\mathbf{v}$ may be represented by any arrow in the plane whose length and direction are the same as those of $\mathbf{v}$ (Figure 2.5).

The relationship between the notation $\overrightarrow{PQ}$ and $\boldsymbol{v}$ for a particular vector is this: If the vector $\boldsymbol{v}$ is positioned in the plane with *initial point P* and *terminal point Q*, then $\boldsymbol{v} = \overrightarrow{PQ}$. However, $\overrightarrow{PQ}$ is only one of many possible representations (locations) for the vector $\boldsymbol{v}$. In other words, a vector $\boldsymbol{v}$ may be "translated" so as to originate at any point in the plane. The only thing that matters is that its length and direction are not changed.

Operations on Vectors

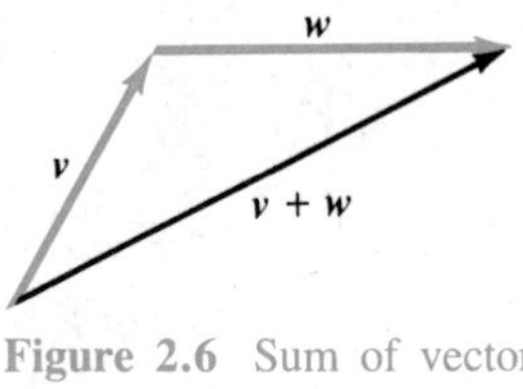

Figure 2.6 Sum of vectors $\boldsymbol{v}$ and $\boldsymbol{w}$.

Let $\boldsymbol{v}$ and $\boldsymbol{w}$ be vectors positioned so that the terminal point of $\boldsymbol{v}$ lies at the initial point of $\boldsymbol{w}$ (Figure 2.6). The vector $\boldsymbol{v} + \boldsymbol{w}$ is defined to be the vector that originates at the initial point of $\boldsymbol{v}$ and terminates at the terminal point of $\boldsymbol{w}$. This definition agrees with our concept of vectors as displacements—the result of the displacement $\boldsymbol{v}$ followed by the displacement $\boldsymbol{w}$ is precisely the displacement $\boldsymbol{v} + \boldsymbol{w}$.

You should note that this definition of addition applies only when the vectors $\boldsymbol{v}$ and $\boldsymbol{w}$ are positioned with the initial point of $\boldsymbol{w}$ at the terminal point of $\boldsymbol{v}$. If the vectors are not originally in such position, we must first translate $\boldsymbol{w}$ to the terminal point of $\boldsymbol{v}$ (Figure 2.7). By positioning $\boldsymbol{v}$ and $\boldsymbol{w}$ to originate at the same point, we can see that $\boldsymbol{v} + \boldsymbol{w}$ may also be interpreted as a diagonal of the parallelogram determined by $\boldsymbol{v}$ and $\boldsymbol{w}$ (Figure 2.8). Several other vector sums are illustrated in Figures 2.9 through 2.11.

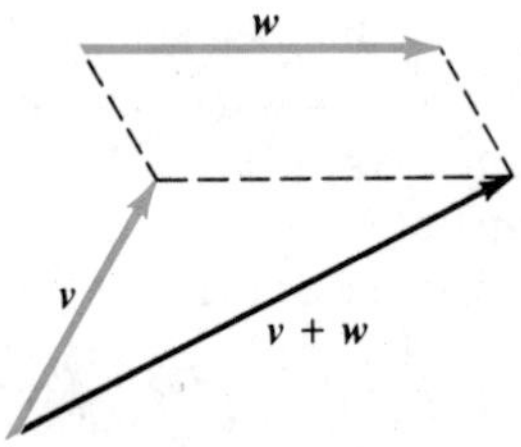

Figure 2.7 Translating $\boldsymbol{w}$ into position to form $\boldsymbol{v} + \boldsymbol{w}$.

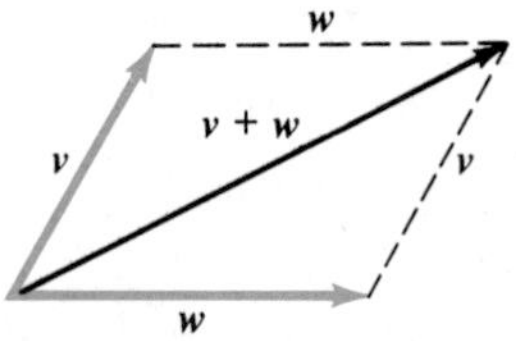

Figure 2.8 $\boldsymbol{v} + \boldsymbol{w}$ is the diagonal of the parallelogram whose sides are $\boldsymbol{v}$ and $\boldsymbol{w}$.

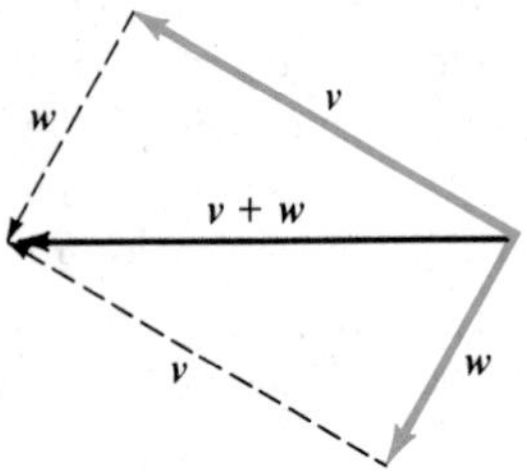

Figure 2.9

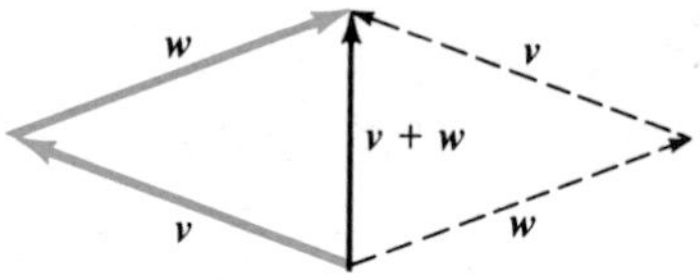

Figure 2.10

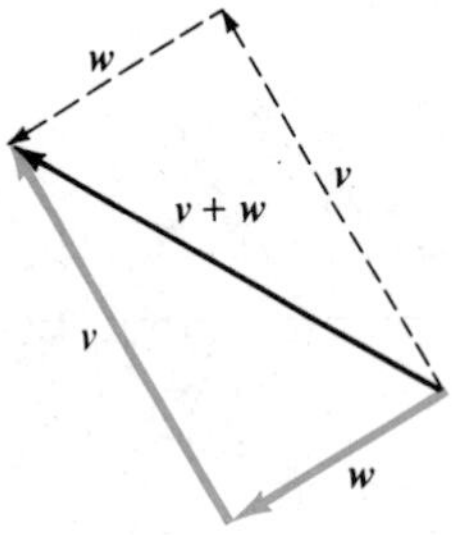

Figure 2.11

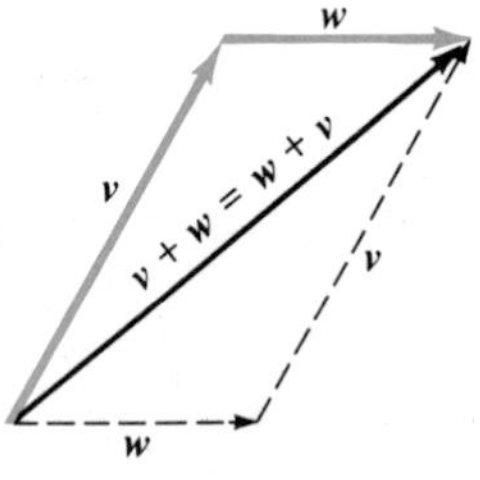

Figure 2.12

Figure 2.12 illustrates the fact that $\boldsymbol{v} + \boldsymbol{w} = \boldsymbol{w} + \boldsymbol{v}$, since the same diagonal is obtained regardless of the order in which the two adjacent sides of the parallelogram are listed.

For any real number c, we define the multiple $c\boldsymbol{v}$ to be the vector whose length is $|c|$ times the length of $\boldsymbol{v}$, and whose direction is

(i) the same as that of $\boldsymbol{v}$ if $c > 0$, or
(ii) opposite that of $\boldsymbol{v}$ if $c < 0$ (Figure 2.13).

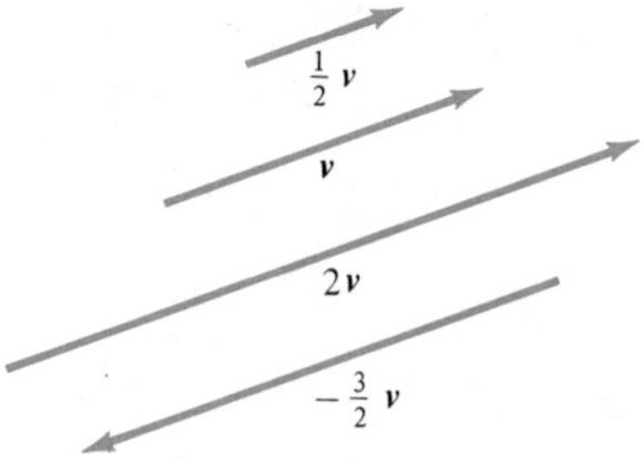

Figure 2.13

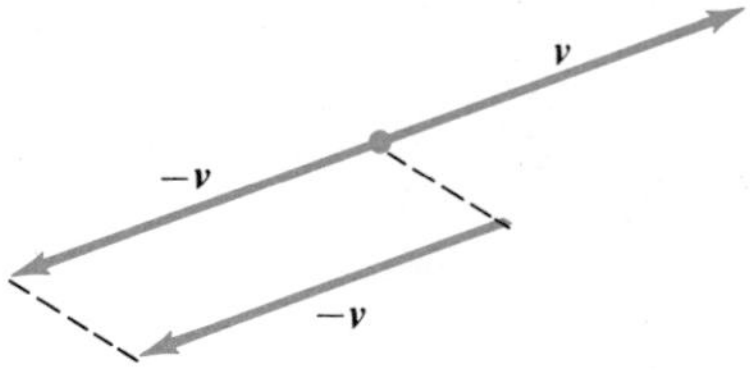

Figure 2.14

Two special cases are important: If $c = -1$, the result is $cv = -v$, called the negative of v (Figure 2.14); it has the same length as v, but the opposite direction. If $c = 0$, the result is a vector of zero length, called the **zero vector** and denoted by **0**. The zero vector has many properties similar to those of the real number zero, and it is the only vector for which a direction is not defined.

Finally, we may define the operation of **subtraction for vectors** by the equation

$$v - w = v + (-w). \tag{1}$$

That is, the difference $v - w$ is defined to be the *sum* of v and the negative of w. Figure 2.15 shows how $v - w$ may be obtained using equation (1) and the head-to-tail rule for vector addition. Figure 2.16 shows that the vector $v - w$ is the diagonal opposite $v + w$ in the parallelogram whose adjacent sides are v and w (compare Figure 2.12). Figure 2.17 illustrates the simplest method for constructing the vector $v - w$: if v and w originate at the same point, $v - w$ is the vector originating at the terminal point of w and terminating at the terminal point of v. This follows from the equation $w + (v - w) = v$.

Note that $v - v = \mathbf{0}$; that is, the result is the **zero vector**, not the real number zero.

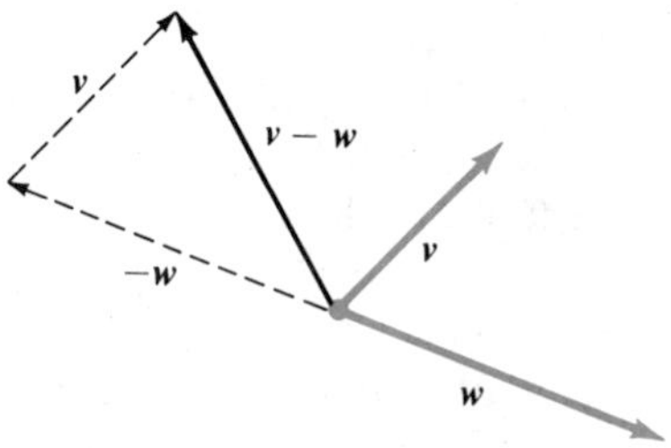

Figure 2.15 $v - w = v + (-w)$.

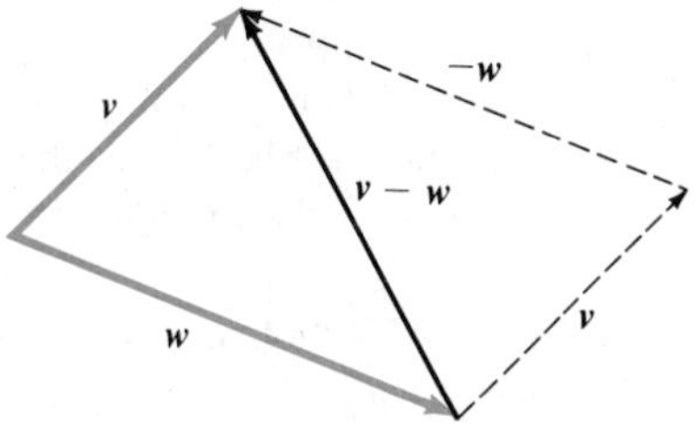

Figure 2.16 $v - w$.

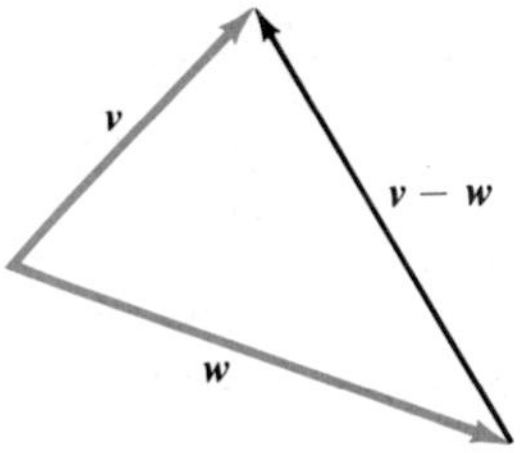

Figure 2.17 $v - w$.

Example 1 Let $P = (1, 2)$, $Q = (3, 6)$, and $R = (3, 0)$. Let $v = \overrightarrow{PQ}$ and $w = \overrightarrow{PR}$. Using the distance formula, we can see that

$$|v| = \text{length of } v = \text{length of } \overline{PQ} = \sqrt{(3-1)^2 + (6-2)^2} = 2\sqrt{5},$$

$$|w| = \text{length of } w = \text{length of } \overline{PR} = \sqrt{(3-1)^2 + (0-2)^2} = 2\sqrt{2}.$$

Also, the slope of v is $\dfrac{6-2}{3-1} = 2$, and the slope of w is $\dfrac{0-2}{3-1} = -1$. From this information it follows that

(a) v, when originating at $(3, 0)$, terminates at $(5, 4)$,
(b) w, when originating at $(3, 6)$, terminates at $(5, 4)$,
(c) $v + w$, when originating at $(1, 2)$, terminates at $(5, 4)$,
(d) $v - w$, when originating at $(3, 0)$, terminates at $(3, 6)$,
(e) $-w$, when originating at $(1, 2)$, terminates at $(-1, 4)$,
(f) $-2v$, when originating at $(1, 2)$, terminates at $(-3, -6)$,
(g) $-2v + w$, when originating at $(1, 2)$, terminates at $(-1, -8)$.

(See Figures 2.18 through 2.20.) ■

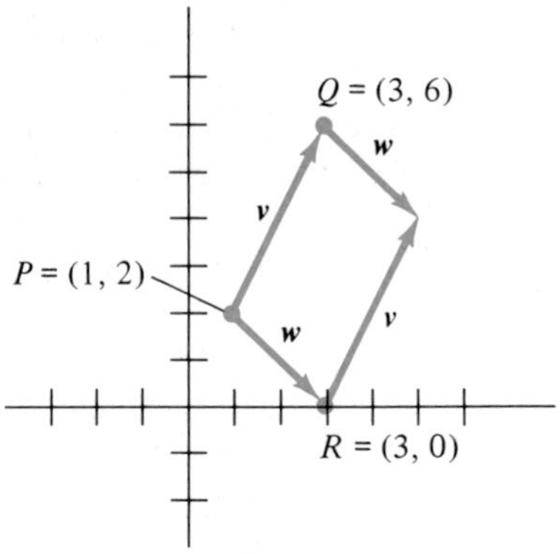

Figure 2.18

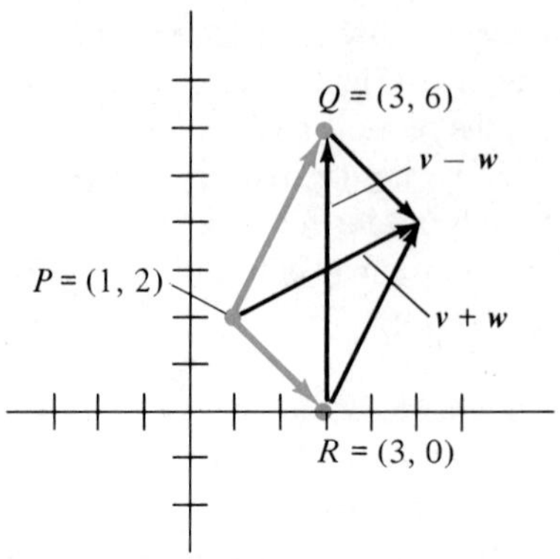

Figure 2.19

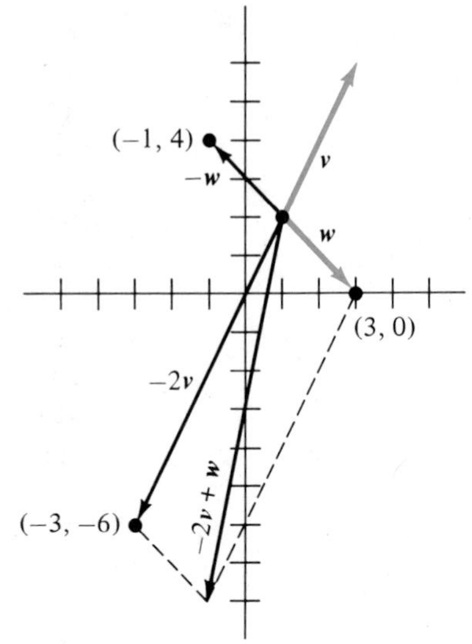

Figure 2.20

Elementary Uses of Vectors

Vectors are useful in determining the path of an object moving through a medium (usually air or water) that itself is in motion. The motion of the object relative to the earth can often be obtained as the vector sum of the motion of the object relative to the medium and the motion of the medium relative to the earth.

Example 2 A woman rows a boat at a rate of 20 meters per minute across a river 100 meters wide. The river is flowing with a speed of 10 meters per minute. The woman rows so that her boat is always perpendicular to the banks of the river. At what point will she arrive at the opposite bank? (Figure 2.21.)

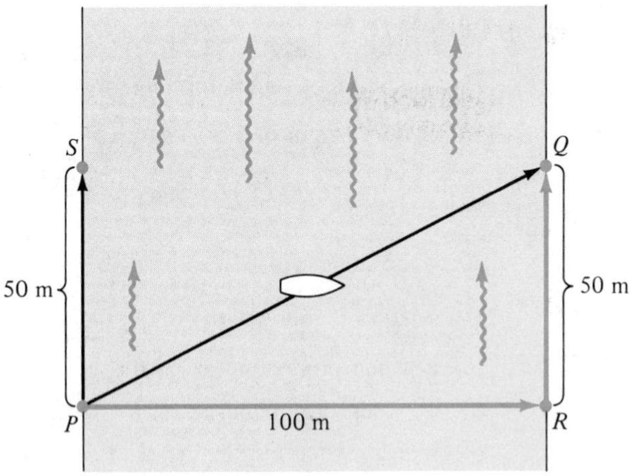

Figure 2.21 Motion of boat is sum of motion of boat in water and motion of water.

Solution: Since it takes the woman $\dfrac{100 \text{ m}}{20 \text{ m/min}} = 5$ minutes to row across the river, the water in the river moves 5×10 m/min $= 50$ meters downstream during

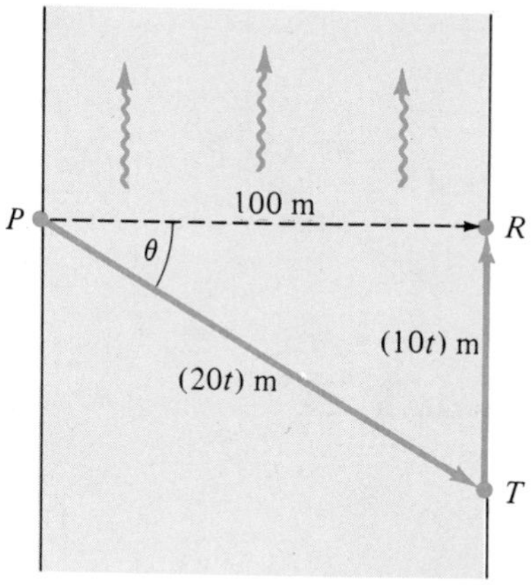

Figure 2.22 $\sin\theta = \frac{(10t)\text{m}}{(20t)\text{m}} = \frac{1}{2}$, $\theta = 30°$.

her journey. Her motion is therefore the sum of the vector $\overrightarrow{PR}$, perpendicular to the bank and of length 100 m, and the vector $\overrightarrow{PS}$, parallel to the bank and of length 50 m. The resulting vector terminates at Q, the point on the opposite bank 50 meters downstream from the starting point P. ■

Example 3 In what direction should the woman in Example 2 row so as to arrive at point R, directly opposite the starting point P?

Solution: Let θ be the positive angle at which the woman should row upstream, and let t denote the time required for the passage. The woman's motion is then the sum of the vector $\overrightarrow{PT}$, of length $20t$ meters, and the vector $\overrightarrow{TR}$, of length $10t$ meters (Figure 2.22). Since we require $\overrightarrow{PR} \perp \overrightarrow{TR}$, we can use the properties of the right triangle to calculate $\sin\theta = \frac{10t}{20t} = 0.5$. Thus, $\theta = 30°$. ■

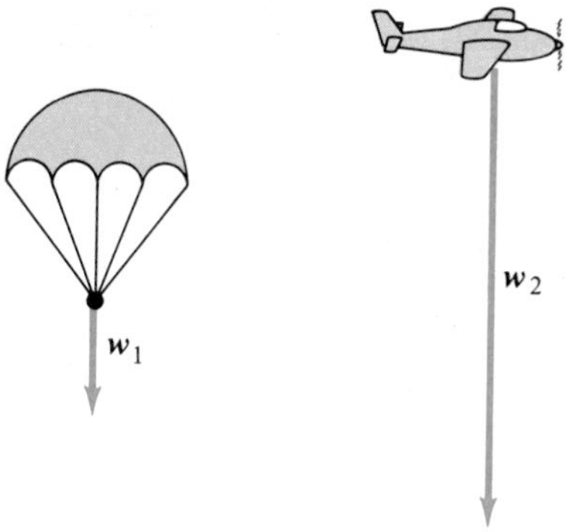

Figure 2.23 Force vectors, due to gravity, acting on a parachutist and an airplane.

Another important application of vectors occurs in physics problems involving forces. Since a force is described by its *magnitude* (usually in pounds, dynes, or newtons) and the *direction* in which it is applied, a force may be conveniently described by a vector whose length is the magnitude of the force it represents and whose direction is that of the force. For example, if g denotes acceleration due to gravity and m denotes the mass of an object, Newton's second law ($F = ma$ or, in this case $|\boldsymbol{w}_g| = ma$) tells us that the force vector $\boldsymbol{w}_g$, of magnitude $|\boldsymbol{w}_g| = mg$, acts vertically downward on each object near the surface of the earth (Figure 2.23).

In problems in which an object is constrained to move along an inclined plane, the force due to gravity causing the object to tend to slide along the plane must be "resolved" into the component forces acting parallel and perpendicular to the plane. (Figure 2.24.) That is, we must find two vectors with specified directions, whose sum is the original vector. We can do this because we know the trigonometric relationship among the three vectors.

Example 4 A block weighing 10 lb is placed on a plane inclined 30° from the horizontal. Find

(a) the magnitude of the force $\boldsymbol{f}_1$, acting parallel to the plane, and
(b) the magnitude of the force $\boldsymbol{f}_2$, acting perpendicular to the plane.

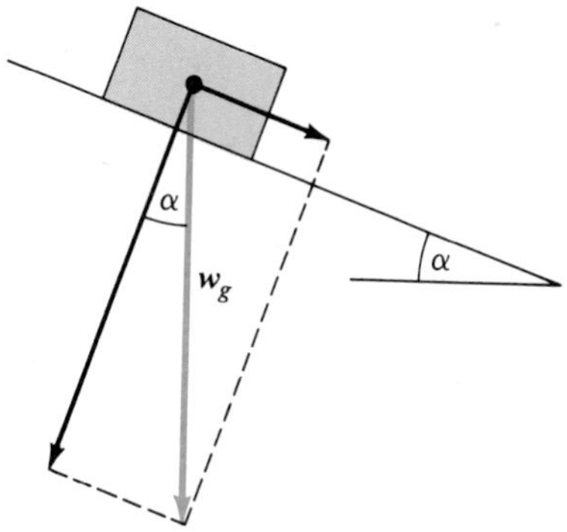

Figure 2.24 Force due to gravity is resolved into forces parallel and perpendicular to the inclined surface.

Solution: Since the desired vectors $\boldsymbol{f}_1$ and $\boldsymbol{f}_2$ are perpendicular, they form the adjacent sides of a rectangle for which $\boldsymbol{f} = \boldsymbol{f}_1 + \boldsymbol{f}_2$ is the given force due to gravity. Since the weight of the block is precisely this force, the magnitude of $\boldsymbol{f}$ is $|\boldsymbol{f}| = 10$ lb. From the two right triangles (see Figure 2.25), we can write the relations

$$|\boldsymbol{f}_1| = |\boldsymbol{f}|\cos 60° = 10(1/2) = 5 \text{ lb},$$

$$|\boldsymbol{f}_2| = |\boldsymbol{f}|\cos 30° = 10\left(\frac{\sqrt{3}}{2}\right) = 5\sqrt{3} \text{ lb}.$$

The significance of $\boldsymbol{f}_1$ and $\boldsymbol{f}_2$ in the physical situation is this: The vector $\boldsymbol{f}_2$, being perpendicular to the surface of the plane, is exactly counteracted by a "reaction force" exerted by the plane and causes no acceleration of the block. The vector $\boldsymbol{f}_1$, if it is not opposed by some external force, will cause acceleration down the plane, according to Newton's second law. If an external force equal to $-\boldsymbol{f}_1$ is applied to the block, the system will be in equilibrium (no acceleration). ■

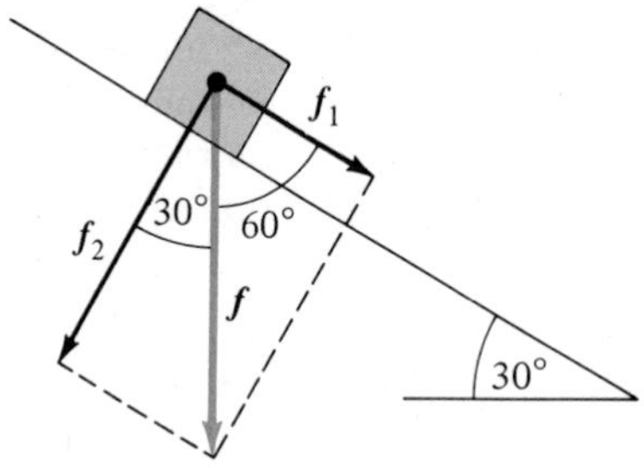

Figure 2.25 $|f_1| = |f| \cos 60°$, $|f_2| = |f| \cos 30°$.

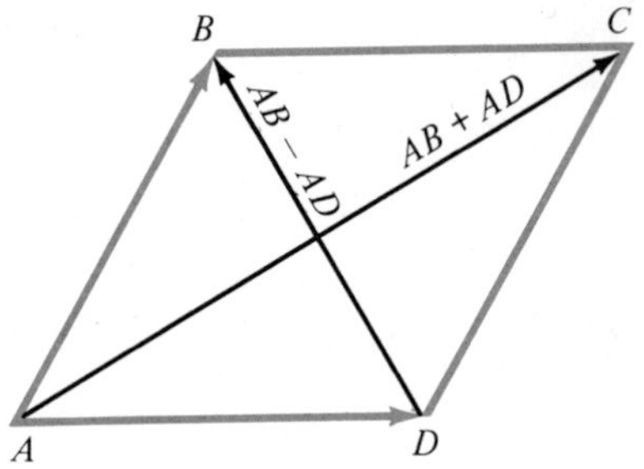

Figure 2.26 The diagonals of a parallelogram bisect each other.

Our final example shows how vector methods can be used in plane geometry.

Example 5 Use vectors to prove that the diagonals of a parallelogram bisect each other.

Solution: Let the vertices of the parallelogram be A, B, C, and D, as in Figure 2.26.

Since $\overrightarrow{AC} = \overrightarrow{AB} + \overrightarrow{AD}$, the vector from A to the midpoint of diagonal $\overrightarrow{AC}$ is

$$v = \frac{1}{2}(\overrightarrow{AB} + \overrightarrow{AD}).$$

Since $\overrightarrow{DB} = \overrightarrow{AB} - \overrightarrow{AD}$, the vector from A to the midpoint of diagonal $\overrightarrow{DB}$ is

$$\begin{aligned} w &= \overrightarrow{AD} + \frac{1}{2}\left(\overrightarrow{AB} - \overrightarrow{AD}\right) \\ &= \overrightarrow{AD} + \frac{1}{2}\overrightarrow{AB} - \frac{1}{2}\overrightarrow{AD} \\ &= \frac{1}{2}(\overrightarrow{AB} + \overrightarrow{AD}) \\ &= v. \end{aligned}$$

Since $v = w$, these midpoints are the same. ■

Exercise Set 17.2

1. Let $O = (0, 0)$, $A = (-2, -2)$, $B = (1, 1)$. Sketch the vectors $\overrightarrow{OA}$, $\overrightarrow{OB}$, and each of the following vectors.

a. $\overrightarrow{OA} + \overrightarrow{OB}$
b. $2\overrightarrow{OA}$
c. $-\overrightarrow{OB}$
d. $2\overrightarrow{OA} - \overrightarrow{OB}$
e. $\overrightarrow{OA} + 2\overrightarrow{OB}$
f. $\overrightarrow{AB} - \overrightarrow{OB}$
g. $\overrightarrow{AO} + \overrightarrow{AB}$
h. $2\overrightarrow{AB} - \overrightarrow{AO}$

2. Let $A = (2, 3)$, $B = (4, 5)$, $C = (4, 0)$. Sketch the following vectors.

a. $\overrightarrow{AB} + \overrightarrow{AC}$
b. $\overrightarrow{AB} - \overrightarrow{AC}$
c. $\overrightarrow{CA} + \overrightarrow{CB}$
d. $\overrightarrow{CA} - \overrightarrow{CB}$
e. $2\overrightarrow{AC} + \overrightarrow{AB}$
f. $\overrightarrow{AB} - 2\overrightarrow{AC}$
g. $\overrightarrow{BA} - \overrightarrow{BC}$
h. $\overrightarrow{BA} - \overrightarrow{CB}$

3. The points $A = (1, 1)$, $B = (5, 1)$, and $C = (6, 3)$ form 3 vertices of a parallelogram. Find the fourth vertex D if
a. A and C lie on a diagonal,
b. A and C lie on a common side.

4. Let $A = (0, 2)$, $B = (b, 5)$, $C = (5, 2)$, $D = (7, 5)$. Find b if $\overrightarrow{AB} + \overrightarrow{AC} = \overrightarrow{AD}$.

5. Find the length of the vector $\overrightarrow{AB}$ in Exercise 2.

6. Find the length of the vector $\overrightarrow{BC}$ in Exercise 2.

7. A man rows across a stream 150 meters wide so that his motion is always perpendicular to the bank. If the man rows at a rate of 50 meters per minute and the stream is flowing at a rate of 30 meters per minute, how far downstream from his starting point does the man reach the opposite bank?

8. How long will it take the woman in Example 3 to reach the opposite bank?

9. At what angle should the man in Exercise 7 row if he wishes to reach the opposite bank at the point directly opposite his starting point?

10. How long will it take the man in Exercise 9 to reach the opposite bank?

11. An airplane flies at a heading of due north at an air speed of 200 km/hr in a wind blowing due east at 30 km/hr. (The heading is the direction in which the airplane is pointed, and its air speed is its speed with respect to the air, rather than the ground.)

a. What is the location of the airplane after 2 hours?

b. How far does the airplane travel in 2 hours?

c. At which heading should the airplane fly so as to be moving due north?

12. What force must be overcome in pushing a 200-lb motorbike up a 30° incline, assuming that the only force to be overcome is that due to gravity? (See Example 4.)

13. A force of 50 lb is required to hold a block on a 45° frictionless incline. How much does the block weigh?

14. In the triangle with vertices A, B, and C, let D be the midpoint of side $\overline{AB}$ and let E be the midpoint of side $\overline{BC}$. Prove that $\overrightarrow{DE}$ is parallel to $\overrightarrow{AC}$ and half as long.

15. Prove that the midpoints of the sides of any quadrilateral form the vertices of a parallelogram.

17.3 THE ANALYTIC DEFINITION OF VECTORS IN THE PLANE

The preceding section developed the geometric notion of vectors as displacements, represented by arrows in the plane, and indicated several uses of this concept. Unfortunately, this intuitive idea of vectors does not provide a satisfactory framework in which to extend the concepts of the calculus. The goal of this section is therefore to develop a useful mathematical formulation of vectors in the plane.

To develop an analytic concept of vectors, we begin by associating with each point (a, b) in the xy-plane the (geometric) vector $\langle a, b\rangle$ originating at $(0, 0)$ and terminating at (a, b). We refer to the numbers a and b as the **components** of the vector $\langle a, b\rangle$. If we are to preserve the notion of head-to-tail addition and scalar multiplication as developed in Section 17.2, we must define

$$\langle a_1, b_1\rangle + \langle a_2, b_2\rangle = \langle a_1 + a_2, b_1 + b_2\rangle \qquad \text{(Figure 3.1)} \qquad (1)$$

and

$$c\langle a, b\rangle = \langle ca, cb\rangle \qquad \text{(Figure 3.2)} \qquad (2)$$

for any vectors $\langle a_1, b_1\rangle$ and $\langle a_2, b_2\rangle$ and any real number c.

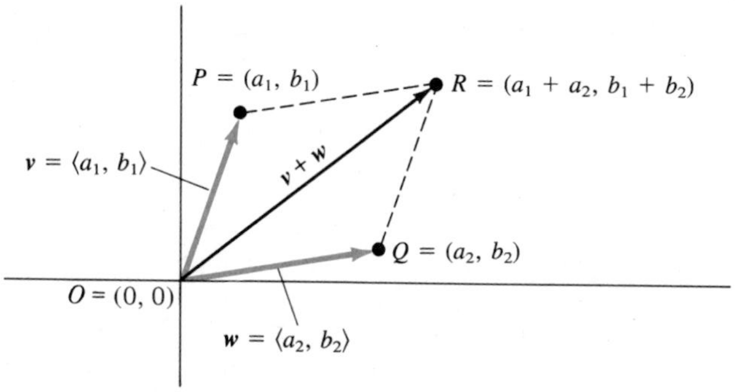

Figure 3.1 $v + w = \langle a_1 + a_2, b_1 + b_2\rangle$.

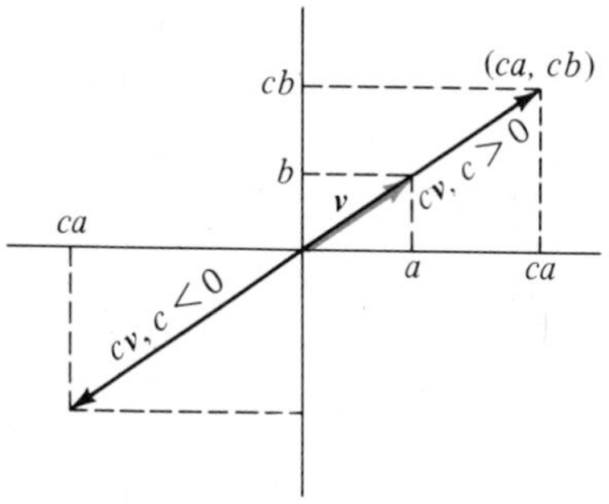

Figure 3.2 $c\langle a, b\rangle = \langle ca, cb\rangle$.

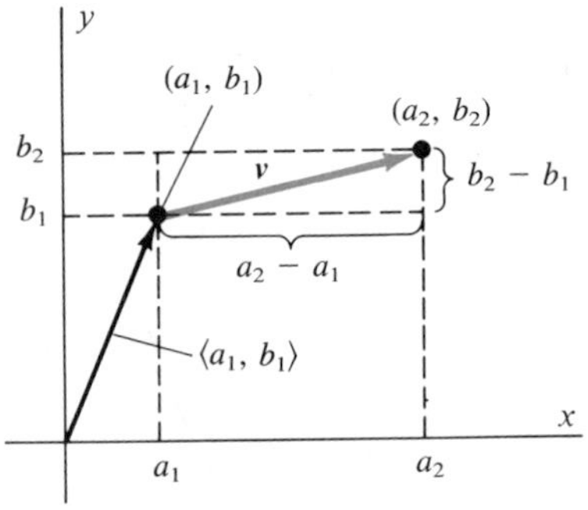

Figure 3.3 $v = \langle a_2 - a_1, b_2 - b_1\rangle$.

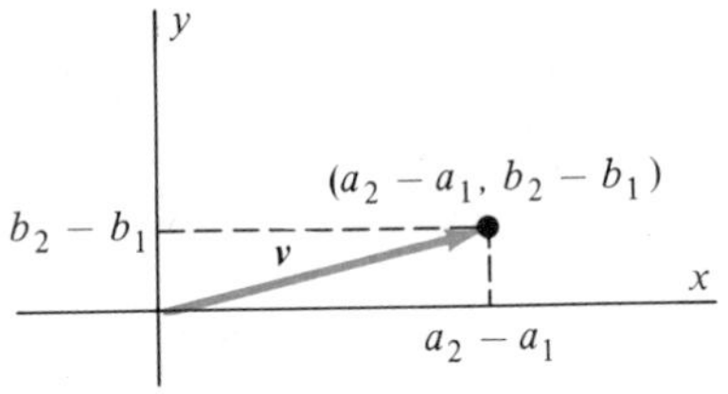

Figure 3.4 $v = \langle a_2 - a_1, b_2 - b_1\rangle$ is the same vector as v in Figure 3.3.

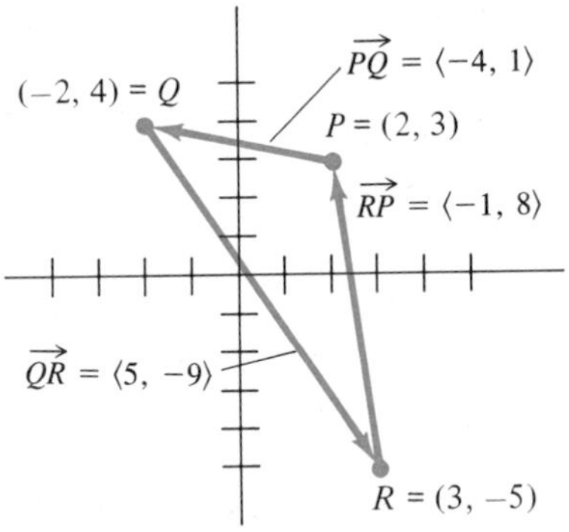

Figure 3.5

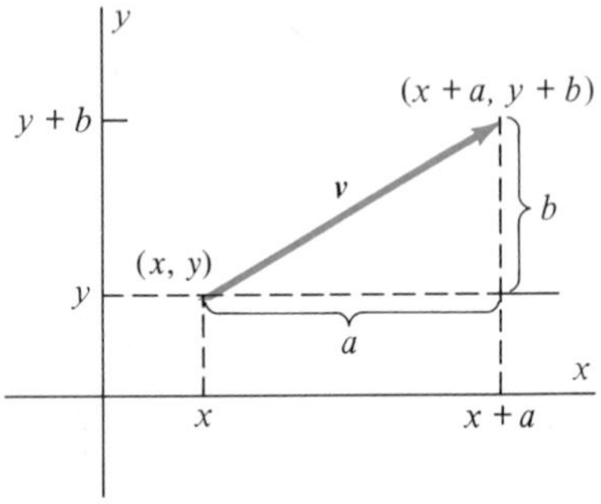

Figure 3.6 The vector $v = \langle a, b\rangle$.

Figure 3.1 shows why equation (1) preserves the notion of head-to-tail addition. If $\boldsymbol{v} = \langle a_1, b_1\rangle$ and $\boldsymbol{w} = \langle a_2, b_2\rangle$, then $\boldsymbol{v}$ terminates at point $P = (a_1, b_1)$ and $\boldsymbol{w}$ terminates at point $Q = (a_2, b_2)$. According to equation (1), the vector $\boldsymbol{v} + \boldsymbol{w}$ terminates at point $R = (a_1 + a_2, b_1 + b_2)$. Now this vector, $\boldsymbol{v} + \boldsymbol{w}$, satisfies the geometric rule of head-to-tail addition if and only if the quadrilateral $OPRQ$ is a parallelogram. Comparing the slopes of opposite sides we see that

$$\text{slope } \overline{PR} = \frac{(b_1 + b_2) - b_1}{(a_1 + a_2) - a_1} = \frac{b_2}{a_2} = \text{slope } \overline{OQ}, \qquad a_2 \neq 0.$$

Thus, either $\overline{PR}$ and $\overline{OQ}$ have equal slopes, or else both are vertical. Similarly

$$\text{slope } \overline{QR} = \frac{(b_1 + b_2) - b_2}{(a_1 + a_2) - a_2} = \frac{b_1}{a_1} = \text{slope } \overline{OP}, \qquad a_1 \neq 0,$$

and the same conclusions follow. The opposite sides of $OPRQ$ are therefore parallel, and $OPRQ$ is a parallelogram.

Since geometric vectors are independent of their initial points, we shall want our analytic definition of vectors to have this same property. To see how this is accomplished suppose that $\boldsymbol{v}$ is the (geometric) vector $\overrightarrow{PQ}$ originating at the point $P = (a_1, b_1)$ and terminating at the point $Q = (a_2, b_2)$. Equations (1) and (2) suggest that

$$\boldsymbol{v} = \overrightarrow{PQ} = \langle a_2 - a_1, b_2 - b_1\rangle, \tag{3}$$

since $\langle a_1, b_1\rangle + \boldsymbol{v} = \langle a_1, b_1\rangle + \langle a_2 - a_1, b_2 - b_1\rangle = \langle a_2, b_2\rangle$. Since geometric vectors may be translated anywhere in the plane, we shall interpret $\boldsymbol{v}$ in (3) as originating either at the point (a_1, b_1) or at the origin (or, for that matter, at *any* point in the plane—see Figures 3.3 and 3.4). Note, however, that *the components of $\overrightarrow{PQ}$ are the coordinates of the terminal point, Q, minus the coordinates of the initial point, P.*

Example 1 Let $P = (2, 3)$, $Q = (-2, 4)$, and $R = (3, -5)$. Then, as seen in Figure 3.5:

(a) $\overrightarrow{PQ} = \langle -2 - 2, 4 - 3\rangle = \langle -4, 1\rangle$,

(b) $\overrightarrow{QR} = \langle 3 - (-2), -5 - 4\rangle = \langle 5, -9\rangle$,

(c) $\overrightarrow{RP} = \langle 2 - 3, 3 - (-5)\rangle = \langle -1, 8\rangle$,

(d) $\overrightarrow{PQ} + \overrightarrow{QR} = \langle -4 + 5, 1 + (-9)\rangle = \langle 1, -8\rangle$,

(e) $3\overrightarrow{RP} = \langle 3(-1), 3(8)\rangle = \langle -3, 24\rangle$. ■

The upshot of the preceding discussion is that a vector in the plane may be regarded as an ordered pair $\langle a, b\rangle$ of numbers, subject to the laws of addition and scalar multiplication given by equations (1) and (2). Moreover, the vector $\langle a, b\rangle$ may be *represented* geometrically by an arrow originating at any point (x, y) and terminating at the point $(x + a, y + b)$ (Figure 3.6).

We therefore have arrived at the desired *analytic* definition of vectors in the plane.

DEFINITION 1

The set of all ordered pairs $\langle a, b\rangle$ of real numbers, subject to the laws

(i) $\langle a_1, b_1\rangle + \langle a_2, b_2\rangle = \langle a_1 + a_2, b_1 + b_2\rangle$
(ii) $c\langle a, b\rangle = \langle ca, cb\rangle$

for addition and multiplication by scalars, is called the **set of vectors in the plane.**

Example 2 For $\mathbf{v} = \langle 2, 5\rangle$ and $\mathbf{w} = \langle -3, 2\rangle$, we have

(a) $\mathbf{v} + \mathbf{w} = \langle 2 + (-3), 5 + 2\rangle = \langle -1, 7\rangle$,
(b) $3\mathbf{v} + \mathbf{w} = \langle 3\cdot 2, 3\cdot 5\rangle + \langle -3, 2\rangle = \langle 3, 17\rangle$,
(c) $4\mathbf{v} + 2\mathbf{w} = \langle 4\cdot 2, 4\cdot 5\rangle + \langle 2(-3), 2\cdot 2\rangle = \langle 2, 24\rangle$. ■

We define the vector $-\mathbf{v}$ as $(-1)\mathbf{v}$. That is,

$$\text{if } \mathbf{v} = \langle a, b\rangle, \qquad \text{then } -\mathbf{v} = \langle -a, -b\rangle. \tag{4}$$

As in Section 17.2, we define $\mathbf{v} - \mathbf{w}$ as $\mathbf{v} + (-\mathbf{w})$. That is, if

$$\mathbf{v} = \langle a_1, b_1\rangle, \qquad \mathbf{w} = \langle a_2, b_2\rangle,$$

then

$$\mathbf{v} - \mathbf{w} = \mathbf{v} + (-\mathbf{w}) = \langle a_1 - a_2, b_1 - b_2\rangle. \tag{5}$$

The **zero vector** is defined to be the vector $\mathbf{0} = \langle 0, 0\rangle$. Finally, we define two vectors to be **equal** if and only if they have the same components. That is,

$$\langle a_1, b_1\rangle = \langle a_2, b_2\rangle \qquad \text{if and only if} \qquad a_1 = a_2 \text{ and } b_1 = b_2.$$

Example 3 For $\mathbf{v} = \langle 1, -3\rangle$ and $\mathbf{w} = \langle 2, 5\rangle$,

(a) $\mathbf{v} - \mathbf{w} = \langle 1 - 2, -3 - 5\rangle = \langle -1, -8\rangle$.
(b) $2\mathbf{v} - 3\mathbf{w} = \langle 2, -6\rangle - \langle 6, 15\rangle = \langle -4, -21\rangle$.
(c) $\mathbf{v} - \mathbf{v} = \langle 1 - 1, -3 + 3\rangle = \langle 0, 0\rangle = \mathbf{0}$. ■

Properties of Vectors

Vectors in the plane satisfy many of the same properties as do real numbers. For example, vector addition is **commutative.** This means that $\mathbf{v} + \mathbf{w} = \mathbf{w} + \mathbf{v}$ for any two vectors $\mathbf{v}, \mathbf{w} \in \mathbb{R}^2$. However, not all properties of real numbers are shared by vectors. For example, we shall not encounter a useful way to define the *product* of two vectors as another vector in the plane.

The following theorem specifies the legitimate properties of vector addition and scalar multiplication for vectors in the plane.

THEOREM 1

Let $\mathbf{u}$, $\mathbf{v}$, and $\mathbf{w}$ be vectors in $\mathbb{R}^2$ and let a and b be real numbers. Then

(i) $\mathbf{v} + \mathbf{w} = \mathbf{w} + \mathbf{v}$ (commutativity),
(ii) $(\mathbf{u} + \mathbf{v}) + \mathbf{w} = \mathbf{u} + (\mathbf{v} + \mathbf{w})$ (associativity),
(iii) $\mathbf{0} + \mathbf{v} = \mathbf{v}$ (identity for vector addition),
(iv) $1\cdot\mathbf{v} = \mathbf{v}$ (identity for scalar multiplication),
(v) $a(\mathbf{v} + \mathbf{w}) = a\mathbf{v} + a\mathbf{w}$ (scalar multiplication distributes over vector addition).
(vi) $(a + b)\mathbf{v} = a\mathbf{v} + b\mathbf{v}$

To prove part (i), we write the vectors $\boldsymbol{v}$ and $\boldsymbol{w}$ in component form

$$\boldsymbol{v} = \langle x_1, y_1 \rangle, \qquad \boldsymbol{w} = \langle x_2, y_2 \rangle$$

and apply equation (1):

$$\boldsymbol{v} + \boldsymbol{w} = \langle x_1, y_1 \rangle + \langle x_2, y_2 \rangle = \langle x_1 + x_2, y_1 + y_2 \rangle, \tag{6}$$

$$\boldsymbol{w} + \boldsymbol{v} = \langle x_2, y_2 \rangle + \langle x_1, y_1 \rangle = \langle x_2 + x_1, y_2 + y_1 \rangle. \tag{7}$$

Since addition for real numbers is commutative, $x_1 + x_2 = x_2 + x_1$, and $y_1 + y_2 = y_2 + y_1$. Thus, the right-hand sides of equations (6) and (7) represent the same components, so $\boldsymbol{v} + \boldsymbol{w} = \boldsymbol{w} + \boldsymbol{v}$.

You are asked to prove statements (ii) through (vi) in the exercise set. Each statement is proved by writing the vector(s) in component form and using the corresponding properties of real numbers.

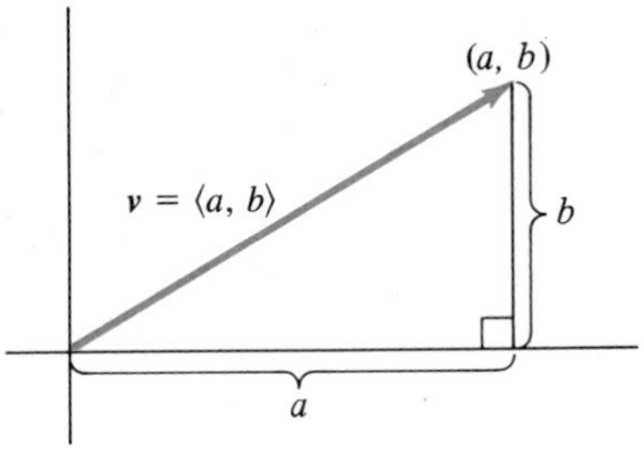

Figure 3.7 $|\boldsymbol{v}| = \sqrt{a^2 + b^2}$.

Length of Vectors

When represented by an arrow originating at (0, 0), the vector $\boldsymbol{v} = \langle a, b \rangle$ terminates at the point (a, b). The obvious way to define the length of $\boldsymbol{v}$ is to use the distance formula in the plane (see Figure 3.7).

DEFINITION 2

The **length of the vector** $\boldsymbol{v} = \langle a, b \rangle$ is $|\boldsymbol{v}| = \sqrt{a^2 + b^2}$.

Example 4 For $\boldsymbol{v} = \langle 2, 1 \rangle$ and $\boldsymbol{w} = \langle -1, 4 \rangle$,

(a) $|\boldsymbol{v}| = \sqrt{2^2 + 1^2} = \sqrt{5}$,
(b) $|\boldsymbol{w}| = \sqrt{(-1)^2 + 4^2} = \sqrt{17}$,
(c) $|2\boldsymbol{v} - \boldsymbol{w}| = \sqrt{(2 \cdot 2 - (-1))^2 + (2 \cdot 1 - 4)^2} = \sqrt{5^2 + 2^2} = \sqrt{29}$. ■

The following theorem summarizes the properties of length for vectors.

THEOREM 2

Let $\boldsymbol{v}$ and $\boldsymbol{w}$ be vectors and let c be a real number. Then

(i) $|\boldsymbol{v}| \geq 0$; $\quad |\boldsymbol{v}| = 0 \quad$ if and only if $\quad \boldsymbol{v} = \boldsymbol{0}$,
(ii) $|c\boldsymbol{v}| = |c| \cdot |\boldsymbol{v}|$,
(iii) $|\boldsymbol{v} + \boldsymbol{w}| \leq |\boldsymbol{v}| + |\boldsymbol{w}|$.

Properties (i) and (ii) are obvious for our geometric concept of vectors. To prove them according to Definition 1, we write $\boldsymbol{v}$ in the component form $\boldsymbol{v} = \langle a, b \rangle$. Then

$$|\boldsymbol{v}| = \sqrt{a^2 + b^2} \geq 0 \qquad \text{for all} \qquad a, b,$$

which proves the first part of (i). The second part of (i) is proved by noting that

$$|\boldsymbol{v}| = \sqrt{a^2 + b^2} = 0 \qquad \text{if and only if} \qquad a = b = 0,$$

in which case $\boldsymbol{v} = \langle a, b \rangle = \langle 0, 0 \rangle = \boldsymbol{0}$. Statement (ii) is just as easy:

$$|c\boldsymbol{v}| = |\langle ca, cb \rangle| = \sqrt{(ca)^2 + (cb)^2} = |c|\sqrt{a^2 + b^2} = |c| \cdot |\boldsymbol{v}|.$$

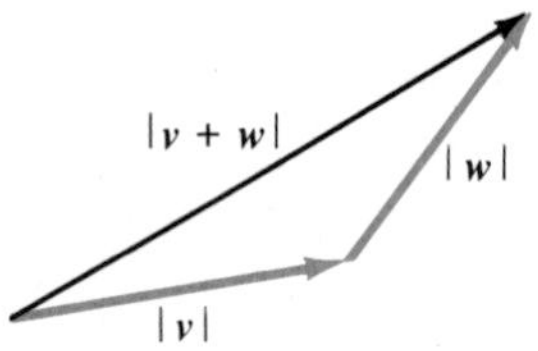

Figure 3.8 $|\boldsymbol{v} + \boldsymbol{w}| \leq |\boldsymbol{v}| + |\boldsymbol{w}|$.

Statement (iii) is referred to as the **triangle inequality.** It may be interpreted geometrically as saying that the length of one leg of a triangle cannot exceed the sum of the lengths of the other two legs (Figure 3.8). An analytic proof of statement (iii) is given in Section 17.4.

Unit Vectors

A vector u with $|u| = 1$ is called a **unit vector.** In what follows we shall frequently need to find a unit vector u in the same direction as a given vector v. We may do so using property (ii) of Theorem 2. Since

$$\left|\left(\frac{1}{|v|}\right)v\right| = \frac{1}{|v|}\cdot|v| = 1,$$

the vector

$$u = \frac{1}{|v|}v \tag{8}$$

is a unit vector pointing in the same direction as the vector v.

Example 5 Find a unit vector in the same direction as the vector $v = \langle 6, -4\rangle$.

Solution:

$$|v| = \sqrt{6^2 + (-4)^2} = \sqrt{52} = 2\sqrt{13}.$$

Thus, by equation (8), the desired unit vector is

$$u = \frac{1}{2\sqrt{13}}\langle 6, -4\rangle = \left\langle \frac{3}{\sqrt{13}}, \frac{-2}{\sqrt{13}}\right\rangle.$$

(Use Definition 2 to verify that $|u| = 1$.) ■

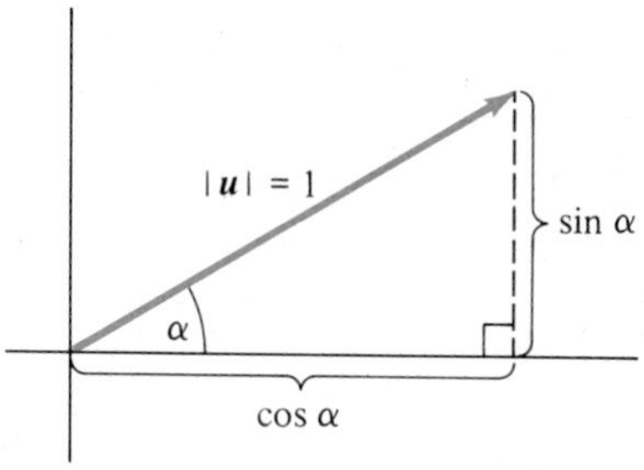

Figure 3.9 $u = \langle\cos\alpha, \sin\alpha\rangle$ is a unit vector.

A similar problem concerning unit vectors is that of finding a unit vector making an angle α with a given line or vector. For example, Figure 3.9 illustrates that *the unit vector forming an angle α with the positive x-axis is*

$$u = \langle\cos\alpha, \sin\alpha\rangle. \tag{9}$$

Equation (9) follows from the definition of the trigonometric functions $\sin\theta$ and $\cos\theta$.

Example 6 Find

(a) a unit vector u making an angle of 60° with the positive x-axis.
(b) a vector of length 5 making an angle of 45° with the positive x-axis.

Strategy

(a) Use equation (9) and the facts

$$\cos 60° = \cos\left(\frac{\pi}{3}\right) = \frac{1}{2},$$

$$\sin 60° = \sin\left(\frac{\pi}{3}\right) = \frac{\sqrt{3}}{2}.$$

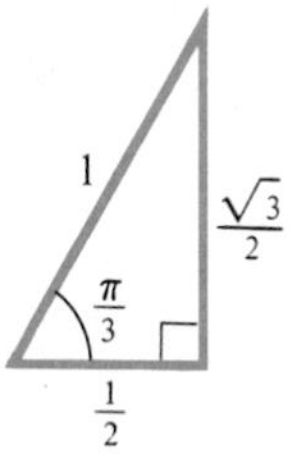

Solution

(a) Using equation (9) we obtain

$$u = \langle\cos 60°, \sin 60°\rangle = \left\langle\frac{1}{2}, \frac{\sqrt{3}}{2}\right\rangle.$$

(b) First, use equation (9) to find a unit vector $\boldsymbol{u}$ in the given direction.

(b) By equation (9) a *unit* vector making an angle of 45° with the positive x-axis is

$$\boldsymbol{u} = \langle \cos 45°, \sin 45° \rangle = \left\langle \frac{\sqrt{2}}{2}, \frac{\sqrt{2}}{2} \right\rangle.$$

The solution is then

$\boldsymbol{v} = 5\boldsymbol{u}.$

The desired vector is therefore

$$\boldsymbol{v} = 5\boldsymbol{u} = \left\langle \frac{5\sqrt{2}}{2}, \frac{5\sqrt{2}}{2} \right\rangle.$$

■

Unit Coordinate Vectors

Two special vectors enable us to develop yet another notation for representing vectors in the plane. They are the **unit coordinate vectors**

$$\boldsymbol{i} = \langle 1, 0 \rangle \qquad \text{and} \qquad \boldsymbol{j} = \langle 0, 1 \rangle.$$

(See Figure 3.10.) Using these two vectors, we may represent the vector $\langle a, b \rangle$ as

$$\langle a, b \rangle = a\langle 1, 0 \rangle + b\langle 0, 1 \rangle = a\boldsymbol{i} + b\boldsymbol{j}.$$

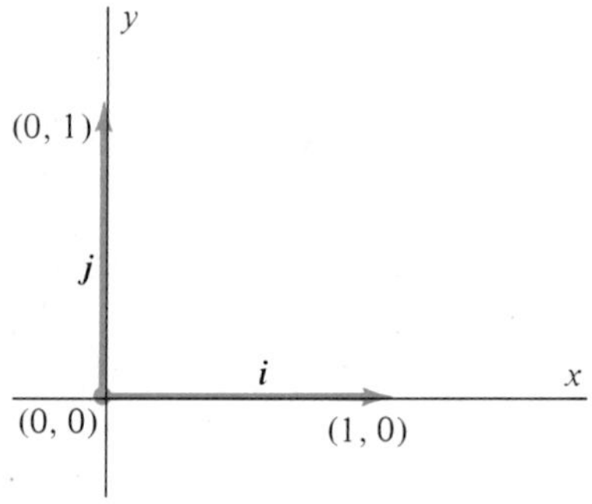

Figure 3.10 Unit coordinate vectors.

Geometrically, $a\boldsymbol{i}$ and $b\boldsymbol{j}$ represent the adjacent sides of the rectangle whose diagonal is $\langle a, b \rangle$ (Figure 3.11). When using this notation, we refer to the numbers a and b as the $\boldsymbol{i}$ and $\boldsymbol{j}$ components of the vector $a\boldsymbol{i} + b\boldsymbol{j}$, respectively.

The principal advantage of the coordinate vector notation is that it helps avoid confusion between points and vectors. We shall make frequent use of this notation in what follows.

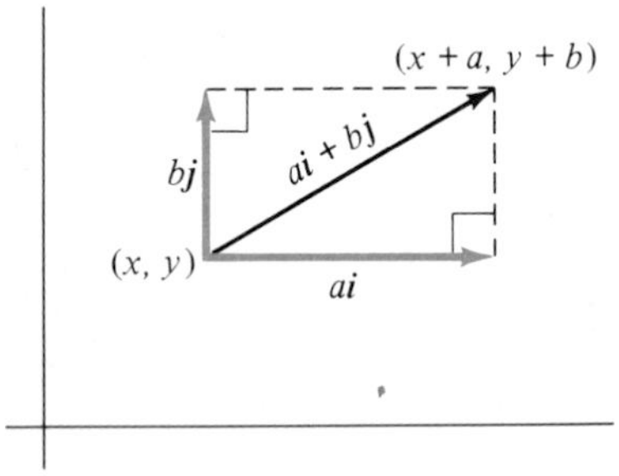

Figure 3.11 $\langle a, b \rangle = a\boldsymbol{i} + b\boldsymbol{j}$.

Example 7 For $\boldsymbol{v} = \langle -2, 1 \rangle$ and $\boldsymbol{w} = \langle 3, 4 \rangle$,

(a) $\boldsymbol{v} = -2\boldsymbol{i} + \boldsymbol{j}; \quad \boldsymbol{w} = 3\boldsymbol{i} + 4\boldsymbol{j},$
(b) $\boldsymbol{v} + \boldsymbol{w} = (-2\boldsymbol{i} + \boldsymbol{j}) + (3\boldsymbol{i} + 4\boldsymbol{j}) = (-2 + 3)\boldsymbol{i} + (1 + 4)\boldsymbol{j} = \boldsymbol{i} + 5\boldsymbol{j},$
(c) $3\boldsymbol{v} - 2\boldsymbol{w} = (-6\boldsymbol{i} + 3\boldsymbol{j}) - (6\boldsymbol{i} + 8\boldsymbol{j}) = (-6 - 6)\boldsymbol{i} + (3 - 8)\boldsymbol{j} = -12\boldsymbol{i} - 5\boldsymbol{j},$
(d) $|\boldsymbol{v}| = |-2\boldsymbol{i} + \boldsymbol{j}| = \sqrt{(-2)^2 + 1^2} = \sqrt{5}.$ ■

Exercise Set 17.3

In Exercises 1–10, sketch the given vector originating at the given point P.

1. $\langle 1, 3 \rangle$ at point $P = (3, 2)$

2. $\langle -6, 2 \rangle$ at point $P = (-1, 1)$

3. $\langle -4, -2 \rangle$ at point $P = (2, 6)$

4. $\langle -2, 4 \rangle$ at point $P = (1, 5)$

5. $\langle 3, -5 \rangle$ at point $P = (1, 1)$

6. $\boldsymbol{i} + \boldsymbol{j}$ at point $P = (3, 5)$

7. $-\boldsymbol{i} + 2\boldsymbol{j}$ at point $P = (6, 0)$

8. $5\boldsymbol{i} + 2\boldsymbol{j}$ at point $P = (-3, 4)$

9. $-3\boldsymbol{i} + 2\boldsymbol{j}$ at point $P = (3, -2)$

10. $4\boldsymbol{j}$ at point $P = (0, 2)$

In Exercises 11–16, let $P = (2, 5)$, $Q = (-1, 2)$, and $R = (2, -6)$. Find

11. $v = \overrightarrow{PQ}$

12. $v = \overrightarrow{PQ} + \overrightarrow{QR}$

13. $v = 2\overrightarrow{PQ} - \overrightarrow{RQ}$

14. $v = \overrightarrow{PR} + 4\overrightarrow{RQ}$

15. $v = 2\overrightarrow{RQ} - 2\overrightarrow{RP}$

16. $v = \overrightarrow{PQ} + 2\overrightarrow{QR} - 3\overrightarrow{RP}$

In Exercises 17–22, let $\boldsymbol{u} = \langle 3, 1\rangle$, $\boldsymbol{v} = \langle -2, 4\rangle$, and $\boldsymbol{w} = \langle -4, -2\rangle$. Find

17. $\boldsymbol{u} + 2\boldsymbol{v}$

18. $\boldsymbol{v} - 2\boldsymbol{u}$

19. $2\boldsymbol{u} + 2\boldsymbol{v} - 2\boldsymbol{w}$

20. $-3\boldsymbol{u} + 2\boldsymbol{v}$

21. $\boldsymbol{v} - \boldsymbol{u} - \boldsymbol{w}$

22. $7\boldsymbol{u} + 3\boldsymbol{w} - 6\boldsymbol{v}$

In Exercises 23–30, let $\boldsymbol{u} = 3\boldsymbol{i} - \boldsymbol{j}$, $\boldsymbol{v} = 2\boldsymbol{i} + 6\boldsymbol{j}$, and $\boldsymbol{w} = -\boldsymbol{i} + \boldsymbol{j}$. Find the indicated vector.

23. $\boldsymbol{u} + 2\boldsymbol{v}$

24. $\boldsymbol{u} - 4\boldsymbol{w}$

25. $\boldsymbol{u} + \boldsymbol{v} + \boldsymbol{w}$

26. $\boldsymbol{u} - \boldsymbol{v} - \boldsymbol{w}$

27. $3\boldsymbol{u} + 4\boldsymbol{v} - 2\boldsymbol{w}$

28. $6\boldsymbol{u} - 5\boldsymbol{w} + \boldsymbol{v}$

29. $3\boldsymbol{u} + 3\boldsymbol{v} + 3\boldsymbol{w}$

30. $-\boldsymbol{u} + \boldsymbol{v} + 2\boldsymbol{w}$

In Exercises 31–36, find the angle θ formed between the vector $\boldsymbol{v}$ and the positive x-axis.

31. $\boldsymbol{j}$

32. $\langle \sqrt{3}, 1\rangle$

33. $\langle -\sqrt{3}, 1\rangle$

34. $3\boldsymbol{i} - 3\boldsymbol{j}$

35. $\boldsymbol{i} - \sqrt{3}\boldsymbol{j}$

36. $-\boldsymbol{i} - \boldsymbol{j}$

In Exercises 37–44, find the length of the given vector.

37. $\langle 6, -1\rangle$

38. $\langle -3, 4\rangle$

39. $\langle a, a^2\rangle$

40. $\boldsymbol{i}$

41. $\boldsymbol{i} + \boldsymbol{j}$

42. $3\boldsymbol{i} + 4\boldsymbol{j}$

43. $6\boldsymbol{i} - 3\boldsymbol{j}$

44. $4(\boldsymbol{i} - 3\boldsymbol{j})$

45. Find a unit vector pointing in the direction opposite the positive x-axis.

46. Find a unit vector in the direction of $\boldsymbol{v} = 3\boldsymbol{i} + 4\boldsymbol{j}$.

47. Find a unit vector making an angle of 120° with the positive x-axis.

48. Find a vector of length 3 in the direction opposite of $\boldsymbol{v} = 2\boldsymbol{i} - 3\boldsymbol{j}$.

49. Find two unit vectors tangent to the graph of $y = x^3$ when positioned to originate at the point (1, 1).

50. Show that the length of the vector originating at $P = (x, y)$ and terminating at $Q = (x + a, y + b)$ is $\sqrt{a^2 + b^2}$.

51. Show that the analytic definition of scalar multiplication given by equation (2) agrees with the geometric definition of scalar multiplication given in Section 17.2.

52. Prove statements (ii) through (vi) of Theorem 1.

53. Let $\boldsymbol{v}_1 = \boldsymbol{i} + \boldsymbol{j}$ and $\boldsymbol{v}_2 = -\boldsymbol{i} + \boldsymbol{j}$. Show that for any vector $\boldsymbol{w}$ one can find constants c_1 and c_2 so that

$$\boldsymbol{w} = c_1\boldsymbol{v}_1 + c_2\boldsymbol{v}_2.$$

(*Hint:* Express $\boldsymbol{w}$ in component form and obtain two linear equations for the unknowns c_1 and c_2.)

54. Generalize Exercise 53 by showing that if $\boldsymbol{v}_1$ and $\boldsymbol{v}_2$ are any nonzero and nonparallel vectors, then any vector $\boldsymbol{w}$ can be expressed as

$$\boldsymbol{w} = c_1\boldsymbol{v}_1 + c_2\boldsymbol{v}_2$$

for appropriate c_1 and c_2.

55. Let P be a point in the plane and let $\boldsymbol{a}$ be the *position* vector $\boldsymbol{a} = \overrightarrow{OP}$, where O is the origin. Let $\boldsymbol{b}$ be any nonzero vector in the plane. Show that the set of all terminal points of the vectors

$$\boldsymbol{r}(t) = \boldsymbol{a} + t\boldsymbol{b}, \qquad -\infty < t < \infty$$

is a line, as follows.

a. Show that the vector $\boldsymbol{r}(t)$ originates at O for each value of t. For each t, let $P(t)$ be the terminal point of the vector $\boldsymbol{r}(t)$.

b. For any $t_1 \neq t_2$, show that the vector $\overrightarrow{P(t_1)P(t_2)}$ is parallel to $\boldsymbol{b}$. This shows that all points $P(t)$ lie on the same line ℓ.

c. Show that for any point Q on ℓ there is a number t_0 so that $Q = P(t_0)$. This shows that every point on the line corresponds to a vector $\boldsymbol{r}(t)$.

56. In Exercise 55, let $\boldsymbol{r}(t) = x(t)\boldsymbol{i} + y(t)\boldsymbol{j}$, $\boldsymbol{a} = a_1\boldsymbol{i} + a_2\boldsymbol{j}$, and $\boldsymbol{b} = b_1\boldsymbol{i} + b_2\boldsymbol{j}$. Find parametric equations for the components $x(t)$ and $y(t)$. Use these equations to provide an alternate solution to Exercise 55.

57. Use Exercise 55 to show that if P and Q are distinct points in the plane and ℓ is the line through P and Q, the point R lies on ℓ if and only if

$$\overrightarrow{OR} = \overrightarrow{OP} + t\overrightarrow{PQ}$$

for some number t. (O denotes the origin.)

58. Under what geometric conditions does equality hold in the triangle inequality (statement (iii) of Theorem 2), that is, when does $|\boldsymbol{v} + \boldsymbol{w}| = |\boldsymbol{v}| + |\boldsymbol{w}|$? How would you express this condition analytically?

17.4 THE DOT PRODUCT

Having defined addition and scalar multiplication for vectors in the plane, it is natural for us to ask whether one can find a useful way to define the product of two vectors. In this section we develop one such product, the *dot* product. After extending our discussion to vectors in space (Section 17.5), we shall encounter a second type of product, the *cross* product.

DEFINITION 3

The **dot product** of the vectors $\boldsymbol{v} = \langle x_1, y_1 \rangle$ and $\boldsymbol{w} = \langle x_2, y_2 \rangle$ is the number

$$\boldsymbol{v} \cdot \boldsymbol{w} = x_x x_2 + y_1 y_2. \tag{1}$$

In other words, the dot product of two vectors is found by multiplying their corresponding components and adding the resulting products. It is important to note that the dot product $\boldsymbol{v} \cdot \boldsymbol{w}$ is a *number,* although each of the factors is a vector. For vectors written in unit coordinate notation, equation (1) becomes

$$(x_1\boldsymbol{i} + y_1\boldsymbol{j}) \cdot (x_2\boldsymbol{i} + y_2\boldsymbol{j}) = x_1x_2 + y_1y_2.$$

The dot product is also referred to as the **scalar product** or the **inner product.**

Example 1 For $\boldsymbol{v} = \langle 2, 3 \rangle$ and $\boldsymbol{w} = \langle -4, 5 \rangle$,

$$\boldsymbol{v} \cdot \boldsymbol{w} = 2(-4) + 3 \cdot 5 = 7.$$ ■

Example 2 For $\boldsymbol{v} = \boldsymbol{i} - 3\boldsymbol{j}$ and $\boldsymbol{w} = 3\boldsymbol{i} + 3\boldsymbol{j}$,

$$\boldsymbol{v} \cdot \boldsymbol{w} = 1 \cdot 3 + (-3)3 = -6.$$ ■

Example 3 $\boldsymbol{i} \cdot \boldsymbol{j} = \langle 1, 0 \rangle \cdot \langle 0, 1 \rangle = 1 \cdot 0 + 0 \cdot 1 = 0.$ ■

Example 4 $(\boldsymbol{i} + \boldsymbol{j}) \cdot (\boldsymbol{i} - \boldsymbol{j}) = 1 \cdot 1 + 1(-1) = 0.$ ■

We shall make frequent use of the properties of the dot product given by the following theorem.

THEOREM 3

Let $\boldsymbol{u}$, $\boldsymbol{v}$, and $\boldsymbol{w}$ be vectors in the plane and let c be a real number. Then

(i) $\boldsymbol{v} \cdot \boldsymbol{v} = |\boldsymbol{v}|^2$,
(ii) $\boldsymbol{v} \cdot \boldsymbol{w} = \boldsymbol{w} \cdot \boldsymbol{v}$ (· is commutative),
(iii) $\boldsymbol{u} \cdot (\boldsymbol{v} + \boldsymbol{w}) = \boldsymbol{u} \cdot \boldsymbol{v} + \boldsymbol{u} \cdot \boldsymbol{w}$ (· distributes over vector addition),
(iv) $(c\boldsymbol{v}) \cdot \boldsymbol{w} = c(\boldsymbol{v} \cdot \boldsymbol{w})$ (scalars may be factored),
(v) $\boldsymbol{0} \cdot \boldsymbol{v} = 0$,
(vi) $|\boldsymbol{v} \cdot \boldsymbol{w}| \le |\boldsymbol{v}|\,|\boldsymbol{w}|$ (Schwarz inequality).

Proof: The proof of the Schwarz inequality will be given later, using Theorem 4. The proofs of statements (i) through (v) follow directly from equation (1). For example, to prove statement (i) we write $\boldsymbol{v}$ in component form as $\boldsymbol{v} = \langle x, y \rangle$. Then

$$\boldsymbol{v} \cdot \boldsymbol{v} = \langle x, y \rangle \cdot \langle x, y \rangle = x^2 + y^2 = (\sqrt{x^2 + y^2})^2 = |\boldsymbol{v}|^2.$$

To prove (ii), we let $v = \langle x_1, y_1\rangle$ and $w = \langle x_2, y_2\rangle$. Then

$$v \cdot w = x_1x_2 + y_1y_2 = x_2x_1 + y_2y_1 = w \cdot v.$$

The proofs of statements (iii) through (v) are similar and are left as exercises. ■

Angles Between Vectors

Examples 3 and 4 suggest that the dot product might have something to say about the angle between two vectors. In both cases the vectors are represented by perpendicular arrows, and in both cases the dot product is zero. In order to pursue this observation we need to define the concept of the angle between two vectors.

DEFINITION 4

Let the vectors $v = \langle x_1, y_1\rangle$ and $w = \langle x_2, y_2\rangle$ be represented by directed line segments in the plane, originating at the origin and terminating at the points (x_1, y_1) and (x_2, y_2), respectively. The **angle between the vectors** v and w is the smaller of the two angles formed between these line segments (Figure 4.1).

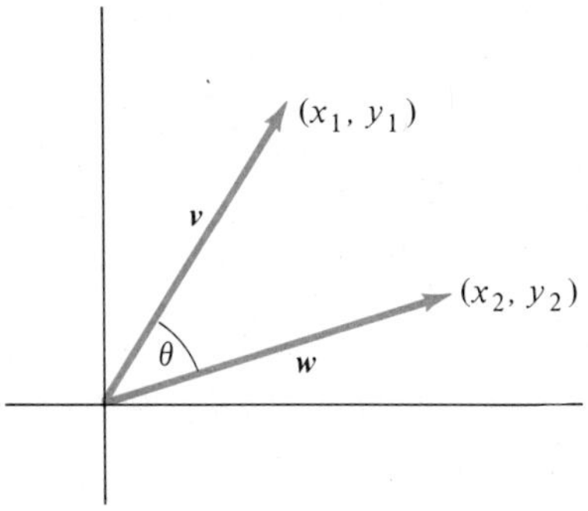

Figure 4.1 Angle between two vectors.

In other words, the angle between two vectors is simply the angle determined by their representations as arrows. We shall not concern ourselves with whether θ represents a clockwise or counterclockwise motion. That is, the angle between v and w will be taken to be the same as the angle between w and v.

We say that two vectors are **perpendicular** if the angle between them is $\pi/2$. Perpendicular vectors are also called **orthogonal vectors.** We say that two vectors are **parallel** if the angle between them is 0 or π.

The relationship between the dot product $v \cdot w$ and the angle θ formed between v and w is given by the following theorem.

THEOREM 4

Let v and w be nonzero vectors in the plane, and let θ be the angle between v and w. Then

$$v \cdot w = |v||w| \cos \theta. \tag{2}$$

Proof: According to our geometric concept of vectors, the vectors v and w determine a triangle in the plane for which the side opposite angle θ has length $|v - w|$

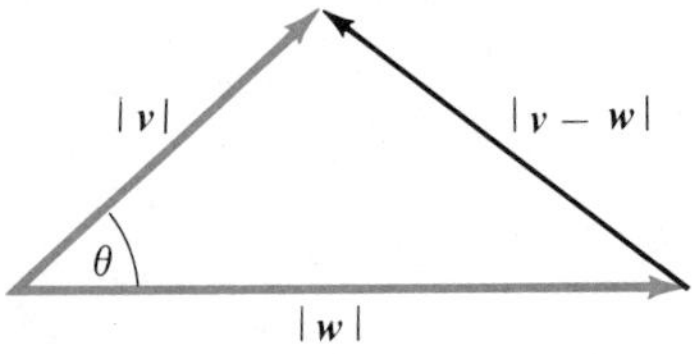

Figure 4.2 Law of Cosines: $|v - w|^2 = |v|^2 + |w|^2 - 2|v|\,|w| \cos\theta$.

(Figure 4.2). Thus, according to the law of cosines,

$$|\boldsymbol{v} - \boldsymbol{w}|^2 = |\boldsymbol{v}|^2 + |\boldsymbol{w}|^2 - 2|\boldsymbol{v}||\boldsymbol{w}| \cos\theta. \tag{3}$$

Let's write $\boldsymbol{v}$ and $\boldsymbol{w}$ in component form as

$$\boldsymbol{v} = \langle x_1, y_1\rangle, \qquad \boldsymbol{w} = \langle x_2, y_2\rangle.$$

Then $\boldsymbol{v} - \boldsymbol{w} = \langle x_1 - x_2, y_1 - y_2\rangle$, so equation (3) may be written in the form

$$(x_1 - x_2)^2 + (y_1 - y_2)^2 = (x_1^2 + y_1^2) + (x_2^2 + y_2^2) - 2|\boldsymbol{v}||\boldsymbol{w}| \cos\theta$$

which simplifies to the equation

$$-2x_1x_2 - 2y_1y_2 = -2|\boldsymbol{v}||\boldsymbol{w}| \cos\theta,$$

or

$$x_1x_2 + y_1y_2 = |\boldsymbol{v}||\boldsymbol{w}| \cos\theta.$$

Since the left-hand side of this equation is $\boldsymbol{v}\cdot\boldsymbol{w}$, the proof is complete. ■

When both $\boldsymbol{v} \neq \boldsymbol{0}$ and $\boldsymbol{w} \neq \boldsymbol{0}$, we may write equation (2) in the equivalent form

$$\cos\theta = \frac{\boldsymbol{v}\cdot\boldsymbol{w}}{|\boldsymbol{v}||\boldsymbol{w}|}. \tag{4}$$

Equation (4) is useful in calculating angles between vectors, as the following example shows.

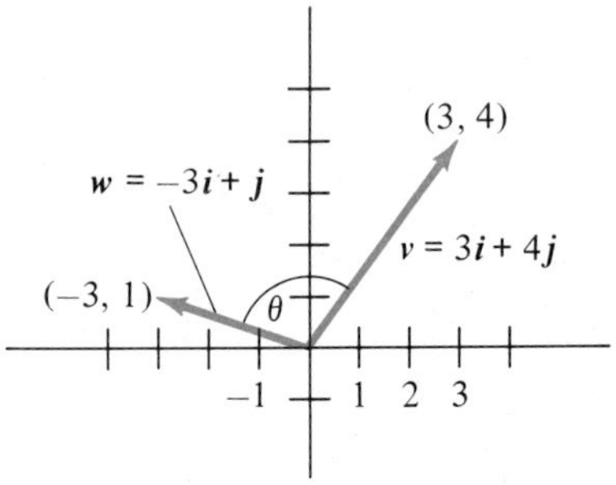

Figure 4.3

Example 5 Find the angle between the vectors $\boldsymbol{v} = 3\boldsymbol{i} + 4\boldsymbol{j}$ and $\boldsymbol{w} = -3\boldsymbol{i} + \boldsymbol{j}$ (Figure 4.3).

Solution: We use equation (4). Since

$$\boldsymbol{v}\cdot\boldsymbol{w} = 3(-3) + 4\cdot 1 = -5,$$

$$|\boldsymbol{v}| = \sqrt{3^2 + 4^2} = \sqrt{25} = 5,$$

$$|\boldsymbol{w}| = \sqrt{3^2 + 1^2} = \sqrt{10},$$

we have

$$\cos\theta = \frac{-5}{5\sqrt{10}} \approx -0.3162.$$

Since we have implied in Definition 4 that $0 \le \theta \le \pi$, we have

$$\theta = \cos^{-1}\left(\frac{-5}{5\sqrt{10}}\right) \approx 0.602\pi \ (\approx 108.4°).$$ ■

Theorem 4 provides a useful criterion for determining whether two nonzero vectors are orthogonal. If $|\boldsymbol{v}| \neq 0$ and $|\boldsymbol{w}| \neq 0$, equation (2) shows that $\boldsymbol{v}\cdot\boldsymbol{w} = 0$ if and only if $\cos\theta = 0$. Since θ is the angle between $\boldsymbol{v}$ and $\boldsymbol{w}$, $0 \le \theta \le \pi$. Thus, $\cos\theta = 0$ if and only if $\theta = \pi/2$. Combining these observations gives the following result.

COROLLARY 1

The nonzero vectors $\boldsymbol{v}$ and $\boldsymbol{w}$ are orthogonal if and only if $\boldsymbol{v} \cdot \boldsymbol{w} = 0$.

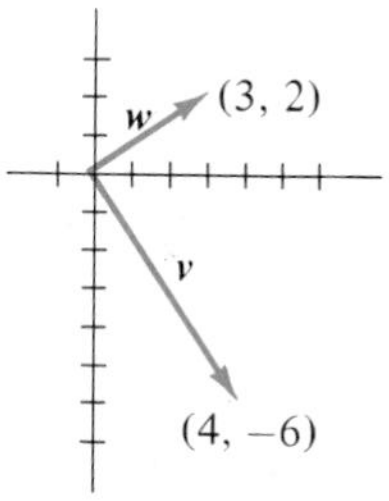

Figure 4.4

Example 6 Show that the vectors $\boldsymbol{v} = 4\boldsymbol{i} - 6\boldsymbol{j}$ and $\boldsymbol{w} = 3\boldsymbol{i} + 2\boldsymbol{j}$ are orthogonal (Figure 4.4).

Solution: We use Corollary 1. Since

$$\boldsymbol{v} \cdot \boldsymbol{w} = 4 \cdot 3 + (-6) \cdot 2 = 0,$$

the vectors are orthogonal. ■

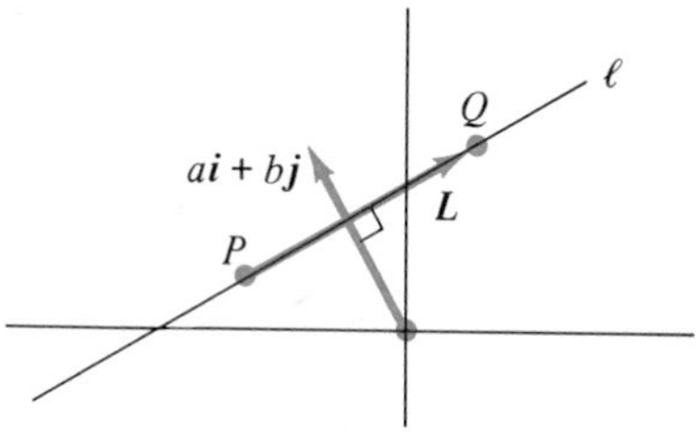

Figure 4.5

Example 7 Show that the vector $a\boldsymbol{i} + b\boldsymbol{j}$ is orthogonal to the line ℓ with equation $ax + by + c = 0$ (see Figure 4.5).

Solution: Our strategy is first to find a vector $\boldsymbol{L}$ parallel to the line ℓ and then to show that $(a\boldsymbol{i} + b\boldsymbol{j}) \cdot \boldsymbol{L} = 0$. We begin by letting $P = (x_1, y_1)$ and $Q = (x_2, y_2)$ be distinct points on ℓ. Then $\overrightarrow{PQ}$ is a segment of ℓ, so $\boldsymbol{L} = \overrightarrow{PQ}$ represents a vector parallel to ℓ, which has the component form

$$\boldsymbol{L} = (x_2 - x_1)\boldsymbol{i} + (y_2 - y_1)\boldsymbol{j}.$$

Now, since P and Q lie on ℓ, both

$$ax_1 + by_1 + c = 0 \qquad \text{and} \qquad ax_2 + by_2 + c = 0.$$

Subtracting corresponding sides of these equations gives the equation

$$a(x_2 - x_1) + b(y_2 - y_1) = 0,$$

which we can rewrite as $(a\boldsymbol{i} + b\boldsymbol{j}) \cdot \boldsymbol{L} = 0$. This shows that $a\boldsymbol{i} + b\boldsymbol{j}$ is orthogonal to $\boldsymbol{L}$, and, hence, to ℓ. ■

We next use Theorem 4 to prove the Schwarz inequality of Theorem 3. Since this inequality is so important, we restate it here as a corollary of Theorem 4.

COROLLARY 2
(Schwarz Inequality)

For any vectors $\boldsymbol{v}$ and $\boldsymbol{w}$,

$$|\boldsymbol{v} \cdot \boldsymbol{w}| \le |\boldsymbol{v}|\,|\boldsymbol{w}|.$$

Proof: If either $\boldsymbol{v} = \boldsymbol{0}$ or $\boldsymbol{w} = \boldsymbol{0}$, the inequality holds, since both sides are zero. Otherwise, an angle θ is determined between $\boldsymbol{v}$ and $\boldsymbol{w}$. Since $|\cos\theta| \le 1$ for all θ, we may apply Theorem 4 to conclude that

$$|\boldsymbol{v} \cdot \boldsymbol{w}| = |\boldsymbol{v}|\,|\boldsymbol{w}|\,|\cos\theta| \le |\boldsymbol{v}|\,|\boldsymbol{w}|.$$ ■

We can use the Schwarz inequality to prove the triangle inequality (Theorem 2, part (iii), Section 17.3):

$$|\boldsymbol{v} + \boldsymbol{w}| \le |\boldsymbol{v}| + |\boldsymbol{w}|. \tag{5}$$

The strategy will be to prove the inequality

$$|\boldsymbol{v} + \boldsymbol{w}|^2 \le (|\boldsymbol{v}| + |\boldsymbol{w}|)^2 \tag{6}$$

from which inequality (5) is obtained by taking square roots.

Applying Theorem 3, parts (i) through (iii), we find that

$$|v + w|^2 = (v + w)\cdot(v + w) \tag{7}$$
$$= v\cdot v + 2v\cdot w + w\cdot w$$
$$= |v|^2 + 2v\cdot w + |w|^2.$$

Now the Schwarz inequality shows that

$$2v\cdot w \le 2|v\cdot w| \le 2|v|\,|w|. \tag{8}$$

Combining (7) and (8), we obtain the inequality

$$|v + w|^2 \le |v|^2 + 2|v|\,|w| + |w|^2$$
$$= (|v| + |w|)^2,$$

which proves inequality (6).

Components

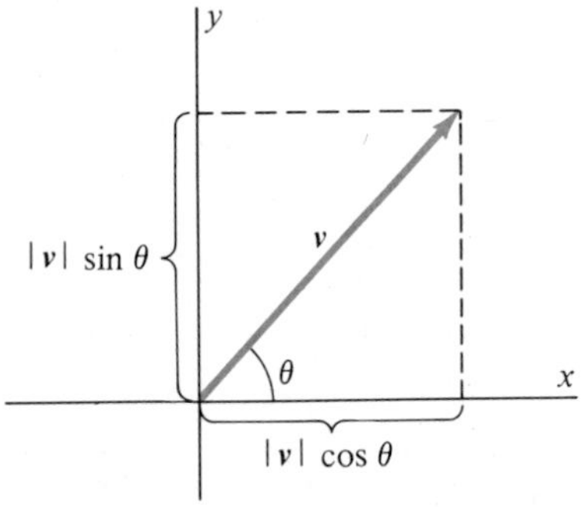

Figure 4.6 Component of v in x direction is $|v| \cos \theta = v \cdot i$.

If v is a nonzero vector and θ is the angle formed between v and the unit coordinate vector i, then v can be written in the component form

$$v = |v|(\cos\theta)i + |v|(\sin\theta)\,j. \tag{9}$$

(See Figure 4.6.) Notice that we can write the component of v in the direction of the vector i as

$$|v|\cos\theta = |v|\,|i|\cos\theta = v\cdot i \qquad \text{(Theorem 4)}. \tag{10}$$

Now suppose we are given a second vector w, with $w \neq 0$. Let θ be the angle formed between v and w. The number $|v| \cos \theta$ is called **the component of v with respect to w,** written $\text{comp}_w\, v = |v| \cos\theta$ (Figure 4.7). The number $\text{comp}_w\, v$ may be interpreted as the change, in the direction of w, represented by the vector v. Using the dot product we may write

$$|v|\cos\theta = \frac{|v|\,|w|\cos\theta}{|w|} = \frac{v\cdot w}{|w|}$$

so

$$\boxed{\text{comp}_w\, v = \frac{v\cdot w}{|w|}.} \tag{11}$$

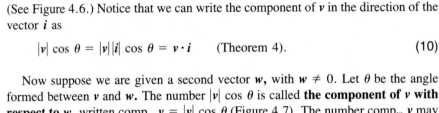

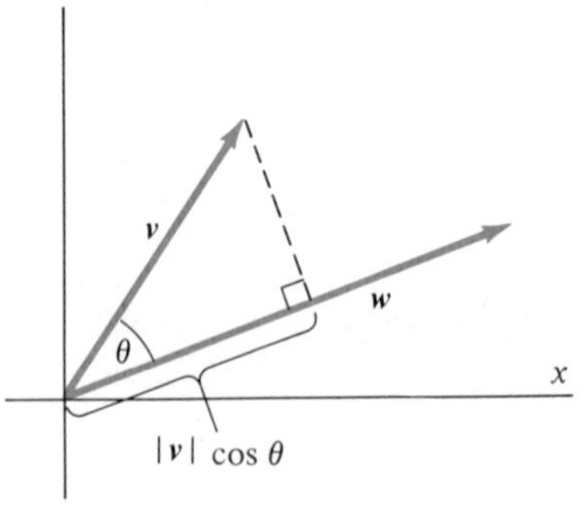

Figure 4.7 Component of v in the direction of w is $|v| \cos \theta = \dfrac{v \cdot w}{|w|}$.

Notice that $\text{comp}_i\, v = \dfrac{v\cdot i}{|i|} = v\cdot i$, so $\text{comp}_w\, v$ is a generalization of the concept of i-component in equation (10) (see Figure 4.7).

Example 8 For $v = i + 2j$ and $w = 3i + j$,

$$\text{comp}_w\, v = \frac{1\cdot 3 + 2\cdot 1}{\sqrt{3^2 + 1^2}} = \frac{5}{\sqrt{10}} \qquad \text{(Figure 4.8)}.$$ ■

Example 9 For $v = -4i + 3j$ and $w = i + j$

$$\text{comp}_w\, v = \frac{(-4)(1) + 3\cdot 1}{\sqrt{1^2 + 1^2}} = \frac{-1}{\sqrt{2}} \qquad \text{(Figure 4.9)}.$$ ■

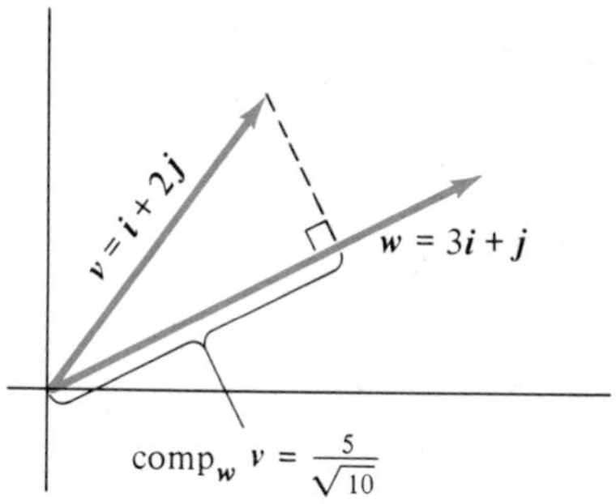

Figure 4.8

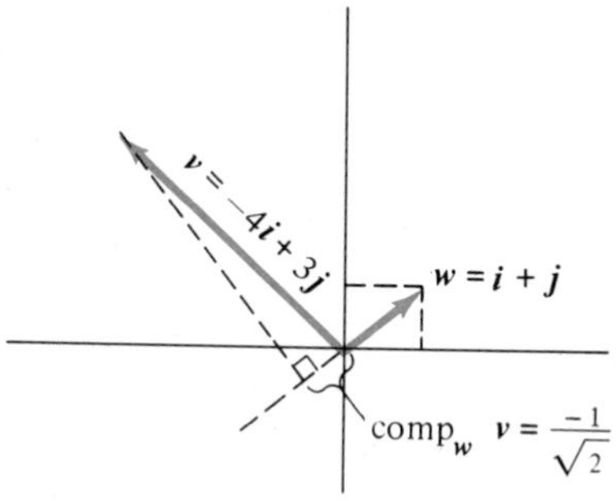

Figure 4.9

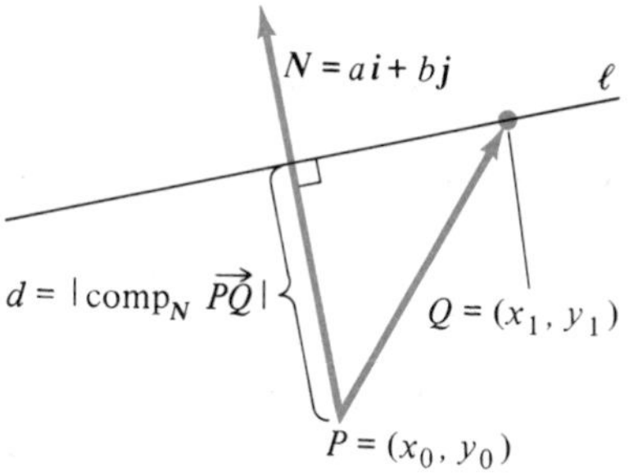

Figure 4.10 Distance from (x_0, y_0) to ℓ: $ax + by + c = 0$ is $|\text{comp}_N\, \overrightarrow{PQ}|$.

The notion of component can be used to develop a formula for the distance from the point $P = (x_0, y_0)$ to the line ℓ with equation ℓ: $ax + by + c = 0$. The idea is summarized by Figure 4.10: If N is a vector orthogonal to ℓ, and if $Q = (x_1, y_1)$ is any point on ℓ, then $|\text{comp}_N\, \overrightarrow{PQ}|$ is the required distance.

The details go like this: If $Q = (x_1, y_1)$ is a point on ℓ, then

$$\overrightarrow{PQ} = (x_1 - x_0)\boldsymbol{i} + (y_1 - y_0)\,\boldsymbol{j}.$$

By Example 7, we know that a vector orthogonal to ℓ is $\boldsymbol{N} = a\boldsymbol{i} + b\boldsymbol{j}$. The required distance is therefore

$$\begin{aligned} d = |\text{comp}_N\, \overrightarrow{PQ}| &= \frac{|\overrightarrow{PQ} \cdot \boldsymbol{N}|}{|\boldsymbol{N}|} \\ &= \frac{|a(x_1 - x_0) + b(y_1 - y_0)|}{\sqrt{a^2 + b^2}} \\ &= \frac{|(ax_1 + by_1) - (ax_0 + by_0)|}{\sqrt{a^2 + b^2}}. \end{aligned}$$

Since $Q = (x_1, y_1)$ is on ℓ, we have $ax_1 + by_1 + c = 0$, so $ax_1 + by_1 = -c$. We therefore find that

$$\boxed{d = \frac{|ax_0 + by_0 + c|}{\sqrt{a^2 + b^2}}} \tag{12}$$

is the distance from the point $P = (x_0, y_0)$ to the line ℓ with equation $ax + by + c = 0$.

Example 10 The distance from the point $(4, -2)$ to the line with equation $3x - y + 4 = 0$ is, by equation (12),

$$d = \frac{|3(4) + (-1)(-2) + 4|}{\sqrt{3^2 + 1^2}} = \frac{18}{\sqrt{10}}.$$

■

Projections

Let $\boldsymbol{v}$ and $\boldsymbol{w}$ be nonzero vectors in the plane. The vector

$$\text{proj}_{\boldsymbol{w}}\, \boldsymbol{v} = \left(\frac{\boldsymbol{v} \cdot \boldsymbol{w}}{|\boldsymbol{w}|^2}\right)\boldsymbol{w} \tag{13}$$

is called the **orthogonal projection** of $\boldsymbol{v}$ onto $\boldsymbol{w}$. Since we can write

$$\text{proj}_{\boldsymbol{w}}\, \boldsymbol{v} = \left(\frac{\boldsymbol{v} \cdot \boldsymbol{w}}{|\boldsymbol{w}|^2}\right)\boldsymbol{w} = \left(\frac{\boldsymbol{v} \cdot \boldsymbol{w}}{|\boldsymbol{w}|}\right)\frac{\boldsymbol{w}}{|\boldsymbol{w}|}, \tag{14}$$

we see that *$\text{proj}_{\boldsymbol{w}}\, \boldsymbol{v}$ is the product of the scalar $\text{comp}_{\boldsymbol{w}}\, \boldsymbol{v}$ times the unit vector $\frac{\boldsymbol{w}}{|\boldsymbol{w}|}$ in the direction of $\boldsymbol{w}$* (see Figure 4.11).

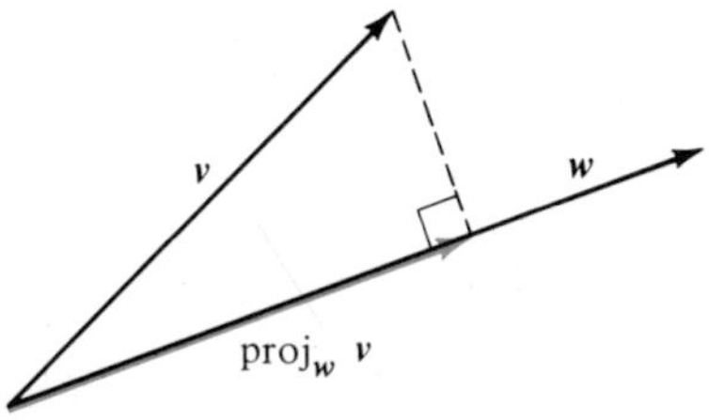

Figure 4.11 Projection of v onto w.

Since $\dfrac{\boldsymbol{w}}{|\boldsymbol{w}|}$ is a unit vector, equation (14) shows that

$$|\text{proj}_{\boldsymbol{w}}\,\boldsymbol{v}| = \left|\frac{\boldsymbol{w}\cdot\boldsymbol{v}}{|\boldsymbol{w}|}\right| = |\text{comp}_{\boldsymbol{w}}\,\boldsymbol{v}|. \tag{15}$$

That is, the length of the *vector* $\text{proj}_{\boldsymbol{w}}\,\boldsymbol{v}$ equals the absolute value of the *number* $\text{comp}_{\boldsymbol{w}}\,\boldsymbol{v}$.

The vector

$$\text{proj}_{\perp\boldsymbol{w}}\,\boldsymbol{v} = \boldsymbol{v} - \text{proj}_{\boldsymbol{w}}\,\boldsymbol{v} \tag{16}$$

is called the **orthogonal projection of $\boldsymbol{v}$ perpendicular to $\boldsymbol{w}$.**

It is obvious from equation (16) that

$$\boldsymbol{v} = \text{proj}_{\boldsymbol{w}}\,\boldsymbol{v} + \text{proj}_{\perp\boldsymbol{w}}\,\boldsymbol{v} \qquad \text{(Figure 4.12).} \tag{17}$$

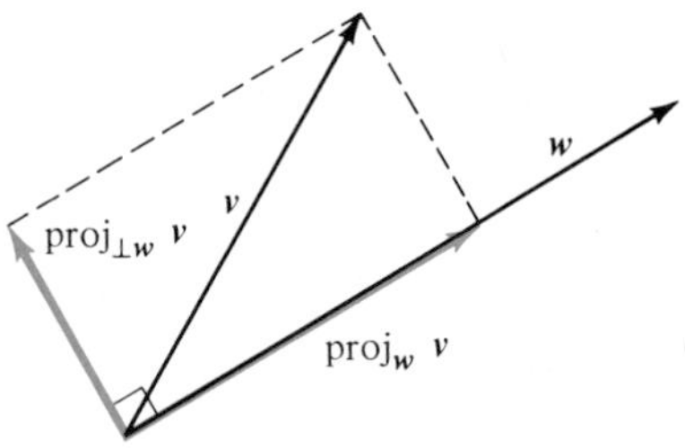

Figure 4.12 $\boldsymbol{v} = \text{proj}_{\boldsymbol{w}}\boldsymbol{v} + \text{proj}_{\perp\boldsymbol{w}}\,\boldsymbol{v}$.

In other words, $\text{proj}_{\boldsymbol{w}}\,\boldsymbol{v}$ and $\text{proj}_{\perp\boldsymbol{w}}\,\boldsymbol{v}$ may be interpreted as adjacent sides of a parallelogram for which $\boldsymbol{v}$ is the diagonal. Indeed, we can say more: the parallelogram is actually a rectangle. That is,

$$(\text{proj}_{\boldsymbol{w}}\,\boldsymbol{v})\cdot(\text{proj}_{\perp\boldsymbol{w}}\,\boldsymbol{v}) = 0. \tag{18}$$

The proof of (18) is simply a calculation:

$$\begin{aligned}(\text{proj}_{\boldsymbol{w}}\,\boldsymbol{v})\cdot(\text{proj}_{\perp\boldsymbol{w}}\,\boldsymbol{v}) &= \left[\left(\frac{\boldsymbol{v}\cdot\boldsymbol{w}}{|\boldsymbol{w}|^2}\right)\boldsymbol{w}\right]\cdot\left[\boldsymbol{v} - \left(\frac{\boldsymbol{v}\cdot\boldsymbol{w}}{|\boldsymbol{w}|^2}\right)\boldsymbol{w}\right]\\ &= \left(\frac{\boldsymbol{v}\cdot\boldsymbol{w}}{|\boldsymbol{w}|^2}\right)\boldsymbol{w}\cdot\boldsymbol{v} - \left(\frac{\boldsymbol{v}\cdot\boldsymbol{w}}{|\boldsymbol{w}|^2}\right)^2\boldsymbol{w}\cdot\boldsymbol{w}\\ &= \left(\frac{\boldsymbol{v}\cdot\boldsymbol{w}}{|\boldsymbol{w}|}\right)^2 - \left(\frac{\boldsymbol{v}\cdot\boldsymbol{w}}{|\boldsymbol{w}|^2}\right)^2|\boldsymbol{w}|^2\\ &= 0.\end{aligned}$$

Equation (18) explains the use of the terminology "orthogonal" projection in defining $\text{proj}_{\boldsymbol{w}}\,\boldsymbol{v}$. The projections $\text{proj}_{\boldsymbol{w}}\,\boldsymbol{v}$ and $\text{proj}_{\perp\boldsymbol{w}}\,\boldsymbol{v}$ allow us to express a given vector $\boldsymbol{v}$ as the sum of a vector parallel to a given vector or line, and a vector orthogonal to the given vector or line. This is a particularly useful concept in physics, where one wishes to "resolve" a force or velocity vector into **component vectors** parallel and perpendicular to a given force or velocity vector.

Example 11 For $\boldsymbol{v} = \boldsymbol{i} + 3\boldsymbol{j}$ and $\boldsymbol{w} = 4\boldsymbol{i} + 2\boldsymbol{j}$,

(a) $$\begin{aligned}\text{proj}_{\boldsymbol{w}}\,\boldsymbol{v} &= \left[\frac{(\boldsymbol{i}+3\boldsymbol{j})\cdot(4\boldsymbol{i}+2\boldsymbol{j})}{|4\boldsymbol{i}+2\boldsymbol{j}|^2}\right](4\boldsymbol{i}+2\boldsymbol{j})\\ &= \left[\frac{1\cdot 4 + 3\cdot 2}{4^2+2^2}\right](4\boldsymbol{i}+2\boldsymbol{j})\\ &= 2\boldsymbol{i}+\boldsymbol{j},\end{aligned}$$

and

$$\text{proj}_{\perp\boldsymbol{w}}\,\boldsymbol{v} = (\boldsymbol{i}+3\boldsymbol{j}) - (2\boldsymbol{i}+\boldsymbol{j}) = -\boldsymbol{i}+2\boldsymbol{j} \qquad \text{(Figure 4.13).}$$

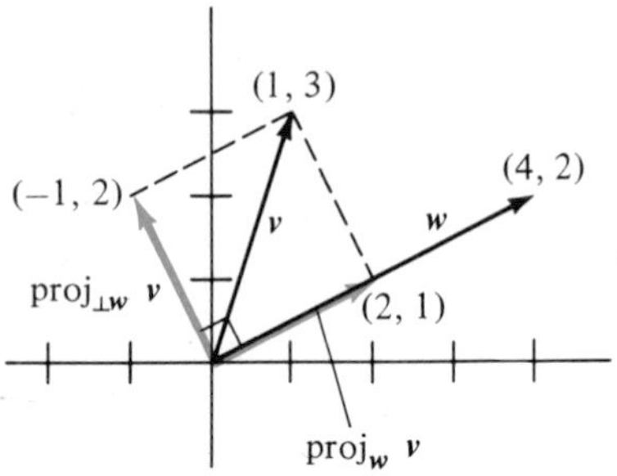

Figure 4.13 Projections of $\boldsymbol{v}$ onto and orthogonal to $\boldsymbol{w}$.

(b) $$\text{proj}_{\boldsymbol{v}}\,\boldsymbol{w} = \left[\frac{(4\boldsymbol{i}+2\boldsymbol{j})\cdot(\boldsymbol{i}+3\boldsymbol{j})}{|\boldsymbol{i}+3\boldsymbol{j}|^2}\right](\boldsymbol{i}+3\boldsymbol{j})$$

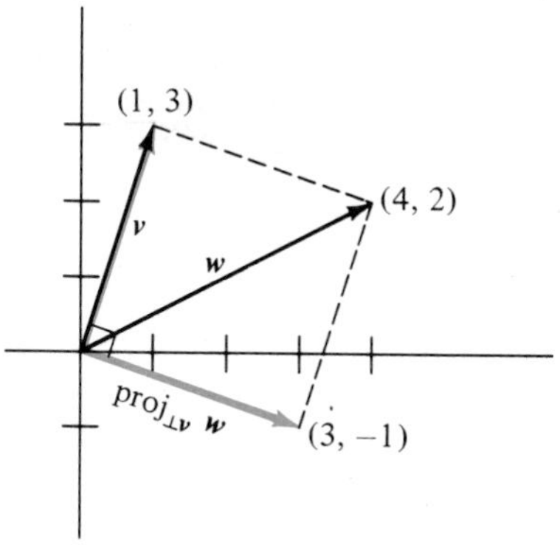

Figure 4.14 $v = \text{proj}_v\, w$.

$$= \left[\frac{4\cdot 1 + 2\cdot 3}{1^2 + 3^2}\right](\boldsymbol{i} + 3\boldsymbol{j})$$

$$= \boldsymbol{i} + 3\boldsymbol{j} \qquad (=\boldsymbol{v}),$$

and

$$\text{proj}_{\perp \boldsymbol{v}}\, \boldsymbol{w} = (4\boldsymbol{i} + 2\boldsymbol{j}) - (\boldsymbol{i} + 3\boldsymbol{j}) = 3\boldsymbol{i} - \boldsymbol{j} \qquad \text{(Figure 4.14).}$$ ■

Example 12 Find the force required to hold a 5000 kilogram automobile motionless on a 30° incline, assuming that the only force that must be overcome is that due to gravity.

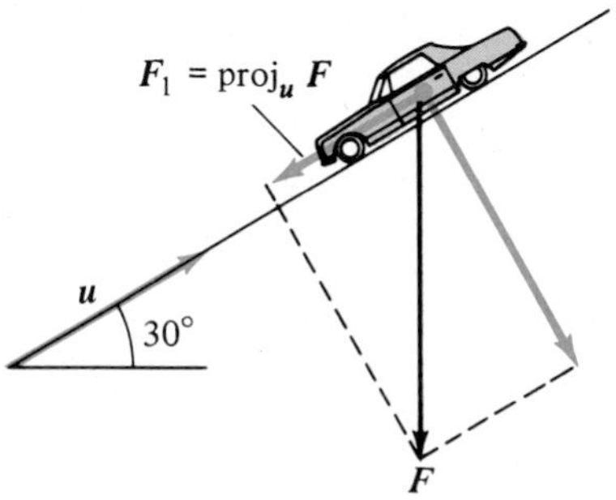

Figure 4.15 $|F_1| = |\text{proj}_u F|$ is the force required to prevent auto from rolling down the incline.

Solution: We may represent the force due to gravity as a vector $\boldsymbol{F}$ pointing vertically downward. According to Newton's second law, the magnitude of this force vector is $|\boldsymbol{F}| = mg = 5000\,g$, where $g = 9.8\ \text{m/sec}^2$ is the acceleration due to gravity. Thus

$$\boldsymbol{F} = (-5000\,g)\,\boldsymbol{j}.$$

By equation (9), Section 17.3, a unit vector $\boldsymbol{u}$ in the direction of the 30° incline is represented in Figure 4.15 and by the equation

$$\boldsymbol{u} = \cos 30°\boldsymbol{i} + \sin 30°\boldsymbol{j} = \frac{\sqrt{3}}{2}\,\boldsymbol{i} + \frac{1}{2}\boldsymbol{j}.$$

The force due to $\boldsymbol{F}$ acting in the direction of the incline is therefore represented by the vector

$$\boldsymbol{F}_1 = \text{proj}_{\boldsymbol{u}}\, \boldsymbol{F} = \frac{\boldsymbol{F}\cdot\boldsymbol{u}}{|\boldsymbol{u}|^2}\boldsymbol{u} = \frac{(-5000\,g)\boldsymbol{j}\cdot\left(\frac{\sqrt{3}}{2}\boldsymbol{i} + \frac{1}{2}\boldsymbol{j}\right)}{1^2}\left(\frac{\sqrt{3}}{2}\boldsymbol{i} + \frac{1}{2}\boldsymbol{j}\right)$$

$$= -2500\,g\left(\frac{\sqrt{3}}{2}\boldsymbol{i} + \frac{1}{2}\boldsymbol{j}\right).$$

The magnitude of this force is

$$|\boldsymbol{F}_1| = |\text{comp}_{\boldsymbol{u}}\, \boldsymbol{F}| = 2500\,g \qquad (=2500 \times 9\cdot 8 = 24{,}500 \text{ newtons}).$$

This is the force due to gravity which must be overcome to hold the automobile in place. ■

Example 13 When a constant force of magnitude F moves the point at which it is applied a distance d *in the direction in which the force is applied*, the work done by the force is defined to be the product

$$W = Fd. \tag{19}$$

However, if the point of application is confined to move along a line in a direction different from that of the applied force, equation (19) is not valid. If the vector $\boldsymbol{F}$ represents the magnitude and direction of the applied force, and the vector $\boldsymbol{d}$ represents the magnitude and direction of the resulting displacement, the work done is defined to be the dot product

$$W = \boldsymbol{F}\cdot\boldsymbol{d}. \tag{20}$$

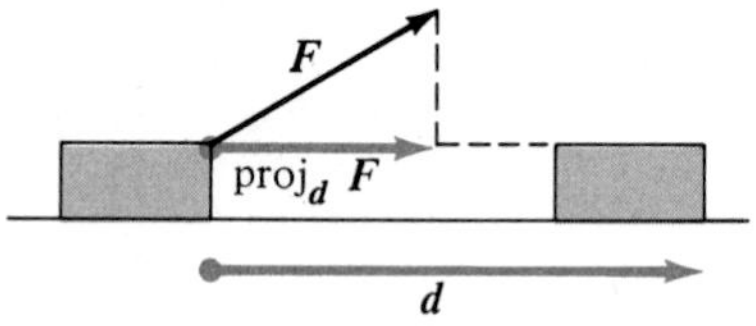

Figure 4.16 Work done by force **F** over the displacement d is $|\text{proj}_d \mathbf{F}||d| = F \cdot d$.

Figure 4.16 illustrates this definition. The vector component of F acting in the direction of d is $\text{proj}_d F$, whose magnitude is $\text{comp}_d F = \dfrac{F \cdot d}{|d|}$. Since this is the force acting in the direction of motion, equation (19) gives

$$W = (\text{comp}_d F)(|d|) = \frac{F \cdot d}{|d|}(|d|) = F \cdot d. \qquad \blacksquare$$

Exercise Set 17.4

In Exercises 1–6, find the dot product $v \cdot w$.

1. $v = i + 2j, \quad w = 3i - j$

2. $v = \langle -3, 5\rangle, \quad w = \langle 1, 4\rangle$

3. $v = \langle -3, 5\rangle, \quad w = \langle 6, -2\rangle$

4. $v = ai + bj, \quad w = ci + dj$

5. $v = i + 4j, \quad w = i - 4j$

6. $v = 2j, \quad w = i - 3j$

7. Find the cosine of the angle between the vectors $v = i - 3j$ and $w = -4i + j$.

8. Find the cosine of the angle between the vectors $v = \langle 3, 2\rangle$ and $w = \langle -2, 2\rangle$.

9. For the vectors $u = -2i + 4j$, $v = 3i + 5j$, and $w = 6i - 4j$ find

a. $u \cdot v$ **b.** $u \cdot w$
c. $u \cdot (v + w)$ **d.** $u \cdot (v - w)$
e. $(u + v) \cdot (v - w)$ **f.** $u \cdot (2v + 3w)$
g. $\text{comp}_v\, u$ **h.** $\text{comp}_w\, v$
i. $\text{proj}_u\, v$ **j.** $\text{proj}_{\perp w}\, u$

10. Let $v = ai + j$ and $w = 3i + 4j$. Find a so that

a. $v \cdot w = 0$ **b.** $v \cdot w = 10$
c. $\text{proj}_w\, v = v$ **d.** $\text{proj}_w\, v = \mathbf{0}$

11. For $v = 2i - 3j$ and $w = 5i + j$, find

a. $\text{comp}_w\, v$ **b.** $\text{comp}_i\, v$
c. $\text{comp}_v\, w$ **d.** $\text{proj}_w\, v$
e. $\text{proj}_{\perp w}\, v$ **f.** $\text{proj}_{\perp v}\, w$

12. Find the interior angles of the triangle with vertices (0, 0), (3, 0), and (3, 4).

13. Determine which of the following pairs of vectors are orthogonal

a. $v = 2i + 4j$, $w = 5i - j$ **b.** $v = 3i - 2j$, $w = 10i + 15j$

$v = i - 3j$, $w = 6i + 2j$ **d.** $v = i + j$, $w = -i - j$

14. Find the number a so that the vectors $v = 3i - 5j$ and $w = ai + 3j$ are orthogonal.

15. Find two unit vectors orthogonal to $v = 2i + j$.

16. Find an example of three nonzero vectors u, v, and w for which $u \cdot v = u \cdot w$, but $v \neq w$. This shows that a ''cancellation law'' for dot products cannot hold.

17. Show that $v + w$ and $v - w$ are perpendicular if $|v| = |w| \neq 0$. Conclude that the diagonals of a rhombus are perpendicular. (A rhombus is a parallelogram with all sides of equal length.)

18. Express the vector $v = 3i + 4j$ as the sum of a vector parallel to $w = 3i + j$ and a vector orthogonal to w.

19. Express the unit coordinate vector i as the sum of a vector parallel to the vector $v = 2i - j$ and a vector perpendicular to v.

20. Find the distance from the point (−4, 3) to the line with equation $2x - y - 3 = 0$.

21. Find the distance from the point (3, 6) to the line with equation $x + y - 1 = 0$.

22. Show that the points (1, 2), (3, 4), and (5, 2) are vertices of a right triangle. Which vertex corresponds to the right angle?

23. A child pulls a sled by exerting a 20-pound force on a rope which makes an angle of 45° with the horizontal. Find the work done by the child in pulling the sled a distance of 100 feet.

24. Prove statements (iii) through (v) of Theorem 3.

25. Assume $v \neq \mathbf{0}$ and $w \neq \mathbf{0}$. Under what conditions is $v \cdot w = |v| \cdot |w|$?

26. Find, to the nearest degree, the angle between $v = i + 4j$ and $w = 3i + 5j$.

27. Use the dot product to show that if $P = (a_1, b_1)$ and $Q = (a_2, b_2)$ are endpoints of the diameter of a circle and if $X = (x, y)$ is a point on the circle, then the vectors $\overrightarrow{XP}$ and $\overrightarrow{XQ}$ are orthogonal.

28. Use the result of Exercise 27 and dot product to find an equation for the circle having (a_1, b_1) and (a_2, b_2) as endpoints of a diameter.

29. Prove the converse of Exercise 17: In a parallelogram, if the diagonals are perpendicular, then the parallelogram is a rhombus.

17.5 SPACE COORDINATES; VECTORS IN SPACE

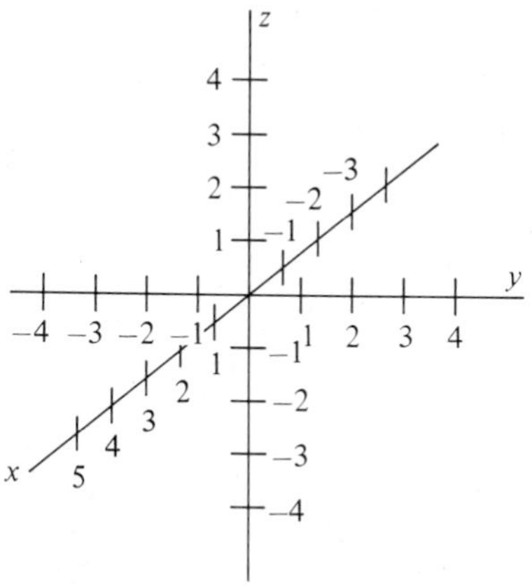

Figure 5.1 Coordinate axes for Euclidean space.

Euclidean three-dimensional space, $\mathbb{R}^3$, provides a framework in which to discuss space curves, surfaces, and graphs of functions of two variables. By Euclidean space, we mean the coordinatizing of space by three mutually perpendicular coordinate axes, labelled as in Figure 5.1.

Scales for each of the axes are chosen so that the point of intersection of the three axes corresponds to zero on each axis. The usual convention in sketching the coordinate axes is to display the positive half of the x-axis as pointing toward the reader, the positive half of the y-axis as pointing to the right, and the positive half of the z-axis as pointing upward.

Using the scales on the three coordinate axes, we may identify each point P in space with a unique triple of numbers $P = (x_0, y_0, z_0)$ as follows: the plane through P perpendicular to the x-axis intersects the x-axis at x_0; the plane through P perpendicular to the y-axis intersects the y-axis at y_0; the plane through P parallel to the z-axis intersects the z-axis at z_0 (see Figure 5.2). Examples of several points in space are shown in Figure 5.3.

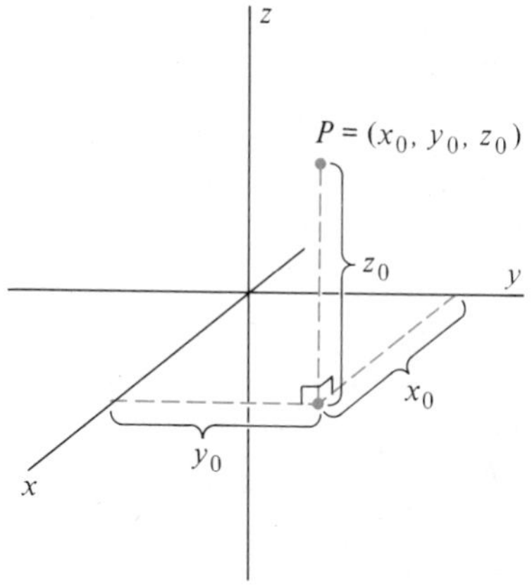

Figure 5.2 The coordinates of $P = (x_0, y_0, z_0)$.

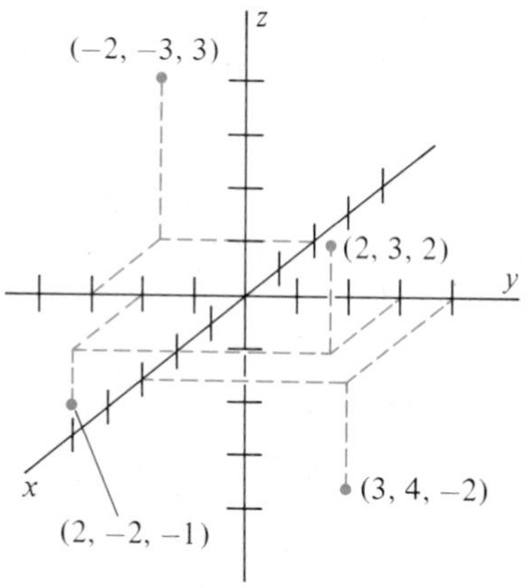

Figure 5.3 Points in space and their coordinates.

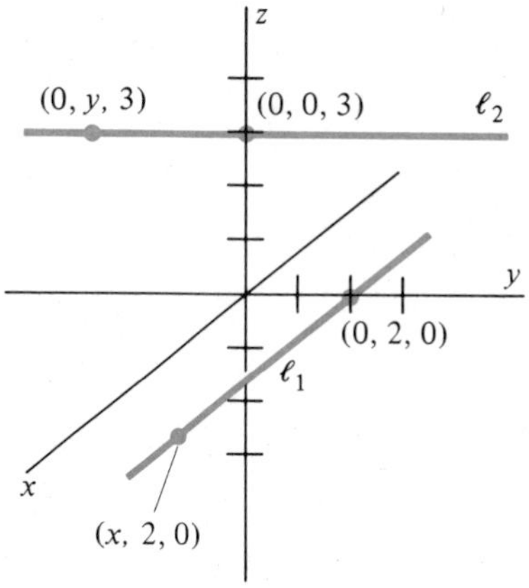

Figure 5.4 ℓ_1: The line $y = 2$, $z = 0$. ℓ_2: The line $x = 0$, $z = 3$.

Figure 5.4 illustrates that the set of all points obtained by specifying two of the three coordinates (and allowing the third to range unrestricted) is a line parallel to the axis of the unrestricted variable. If one specifies only one variable, allowing all values of the other two, the figure obtained is a plane perpendicular to the axis of the restricted variable (Figure 5.5). Equations for lines and planes not parallel or perpendicular to coordinate axes are more complicated and are the subject of Section 17.7.

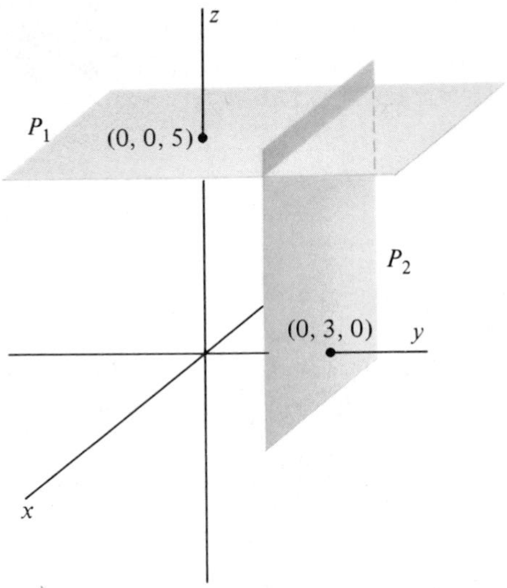

Figure 5.5 P_1: The plane $z = 5$.
P_2: The plane $y = 3$.

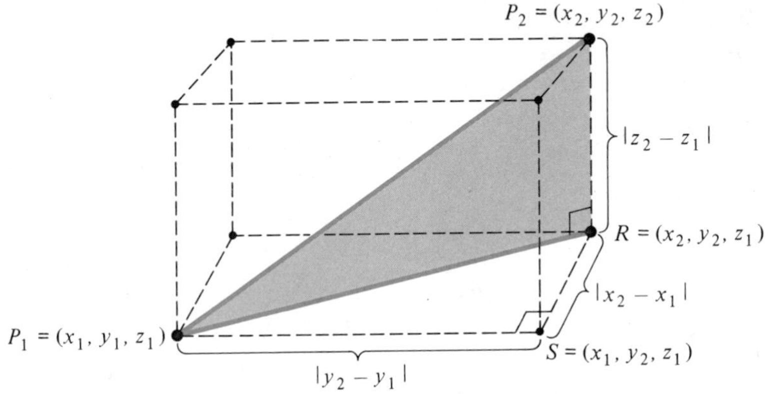

Figure 5.6 $d(P_1, P_2) = \sqrt{(x_2 - x_1)^2 + (y_2 - y_1)^2 + (z_2 - z_1)^2}$.

The Distance Formula

Let $P_1 = (x_1, y_1, z_1)$ and $P_2 = (x_2, y_2, z_2)$ be two points in space. To find the distance $d(P_1, P_2)$ between these points, we use the Pythagorean theorem twice.

Let $S = (x_1, y_2, z_1)$ and $R = (x_2, y_2, z_1)$, as in Figure 5.6. Then, since the segment $\overline{P_1R}$ is the hypotenuse of the right triangle ΔP_1RS, the Pythagorean theorem gives

$$\begin{aligned}[d(P_1, R)]^2 &= [d(S, R)]^2 + [d(P_1, S)]^2 \\ &= (x_2 - x_1)^2 + (y_2 - y_1)^2.\end{aligned}$$

Similarly, the segment $\overline{P_1P_2}$ is the hypotenuse of the right triangle ΔP_1RP_2. Thus, a second application of the Pythagorean theorem, together with the above equation, gives

$$\begin{aligned}[d(P_1, P_2)]^2 &= [d(P_1, R)]^2 + [d(R, P_2)]^2 \\ &= (x_2 - x_1)^2 + (y_2 - y_1)^2 + (z_2 - z_1)^2.\end{aligned}$$

Taking square roots of both sides of this equation gives the desired formula:

> The distance between $P_1 = (x_1, y_1, z_1)$ and $P_2 = (x_2, y_2, z_2)$ is
>
> $$d(P_1, P_2) = \sqrt{(x_2 - x_1)^2 + (y_2 - y_1)^2 + (z_2 - z_1)^2}.$$

Example 1 The distance between $P = (-3, 0, 4)$ and $Q = (2, 5, -2)$ is

$$\begin{aligned}d &= \sqrt{(2 - (-3))^2 + (5 - 0)^2 + (-2 - 4)^2} \\ &= \sqrt{5^2 + 5^2 + (-6)^2} = \sqrt{86}.\end{aligned}$$

■

Equations for Spheres

The sphere with center $C = (a, b, c)$ and radius $r > 0$ is the set of points $P = (x, y, z)$ for which

$$d(P, C) = r.$$

Using the distance formula, we may rewrite this equation as

$$\sqrt{(x-a)^2+(y-b)^2+(z-c)^2}=r,$$

or, in standard form,

$$(x-a)^2+(y-b)^2+(z-c)^2=r^2. \tag{1}$$

Example 2 The sphere with center $C=(2,-3,1)$ and radius $r=3$ has equation

$$(x-2)^2+(y+3)^2+(z-1)^2=9 \qquad \text{(Figure 5.7)}.$$ ■

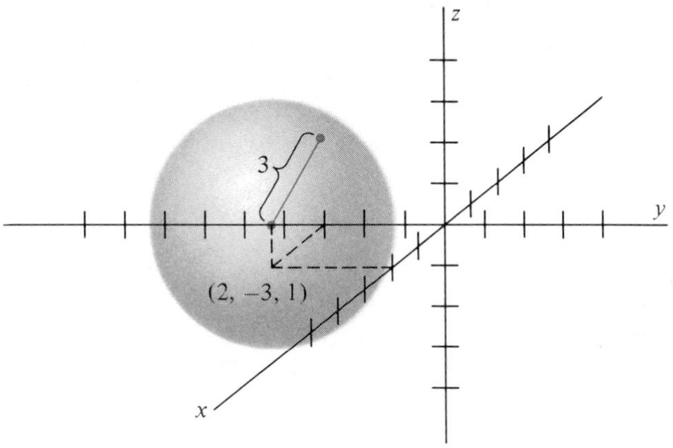

Figure 5.7 Sphere with center $(2, -3, 1)$ and radius 3.

Example 3 To determine whether the equation

$$x^2+y^2+z^2-4x+6z-3=0$$

is that of a sphere, we complete the square in each variable:

$$\begin{aligned}0 &= x^2+y^2+z^2-4x+6z-3\\ &= (x^2-4x)+y^2+(z^2+6z)-3\\ &= (x^2-4x+4)+y^2+(z^2+6z+9)-3-4-9.\end{aligned}$$

Thus,

$$(x-2)^2+y^2+(z+3)^2=16.$$

The graph is a sphere with center at $(2, 0, -3)$ and radius $r=4$. ■

Vectors in Space

The geometric concept of vectors carries over directly from the plane to three-dimensional space. The arrow representing $\boldsymbol{v}$ in space may be interpreted as a displacement, and the sum of the vectors $\boldsymbol{v}$ and $\boldsymbol{w}$ is again taken to be the diagonal of the parallelogram determined by $\boldsymbol{v}$ and $\boldsymbol{w}$. However, to give an analytic definition of vectors in space we must work with **ordered triples,** since points in $\mathbb{R}^3$ are determined by three coordinates.

DEFINITION 5 The set of all ordered triples $\langle x, y, z\rangle$ together with the laws

(i) $\langle x_1, y_1, z_1\rangle + \langle x_2, y_2, z_2\rangle = \langle x_1 + x_2, y_1 + y_2, z_1 + z_2\rangle$
(ii) $c\langle x, y, z\rangle = \langle cx, cy, cz\rangle, \qquad c \in \mathbb{R}$

for addition and scalar multiplication is the **set of vectors in space.**

For the vector $\boldsymbol{v} = \langle x_1, y_1, z_1\rangle$, the numbers x_1, y_1, and z_1 are again referred to as the **components** of $\boldsymbol{v}$. You have probably already noticed that Definition 5 prescribes that vector addition and scalar multiplication are to be carried out for vectors in space just as for vectors in the plane, except that now there is a third component to deal with. Accordingly, we define

(a) the **negative** of the vector $\boldsymbol{v} = \langle x, y, z\rangle$ as

$$-\boldsymbol{v} = \langle -x, -y, -z\rangle,$$

(b) the **difference** of two vectors $\boldsymbol{v} = \langle x_1, y_1, z_1\rangle$ and $\boldsymbol{w} = \langle x_2, y_2, z_2\rangle$ as

$$\boldsymbol{v} - \boldsymbol{w} = \boldsymbol{v} + (-\boldsymbol{w}) = \langle x_1 - x_2, y_1 - y_2, z_1 - z_2\rangle,$$

(c) the **zero vector** as

$$\boldsymbol{0} = \langle 0, 0, 0\rangle,$$

(d) **equality** of two vectors $\boldsymbol{v} = \langle x_1, y_1, z_1\rangle$ and $\boldsymbol{w} = \langle x_2, y_2, z_2\rangle$ by

$$\boldsymbol{v} = \boldsymbol{w} \quad \text{if and only if} \quad x_1 = x_2,\ y_1 = y_2 \text{ and } z_1 = z_2.$$

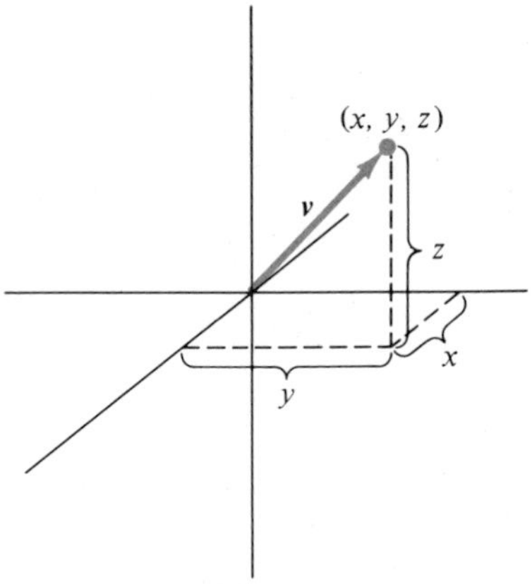

Figure 5.8 The vector $\boldsymbol{v} = \langle x, y, z\rangle$.

With these definitions, all properties of Theorem 1 for vectors in the plane carry over to vectors in space. The techniques of proof are the same as those used in the proof of Theorem 1, except that the third coordinate is included in all calculations (see Exercise 44).

The vector $\boldsymbol{v} = \langle x_1, y_1, z_1\rangle$ may be represented by an arrow originating at the origin and terminating at the point (x_1, y_1, z_1). More generally, if $P = (a, b, c)$ is any point in $\mathbb{R}^3$, the vector $\boldsymbol{v} = \langle x, y, z\rangle$ may be represented by an arrow originating at the point (a, b, c) and terminating at the point $(a + x, b + y, c + z)$ (see Figures 5.8 and 5.9).

As with vectors in the plane, we shall not concern ourselves with the distinction between vectors defined as ordered triples and their representations as arrows in space. For example, we say that the vector originating at the point $P = (x_1, y_1, z_1)$ and terminating at the point $Q = (x_2, y_2, z_2)$ may be written in component form as

$$\overrightarrow{PQ} = \langle x_2 - x_1, y_2 - y_1, z_2 - z_1\rangle.$$

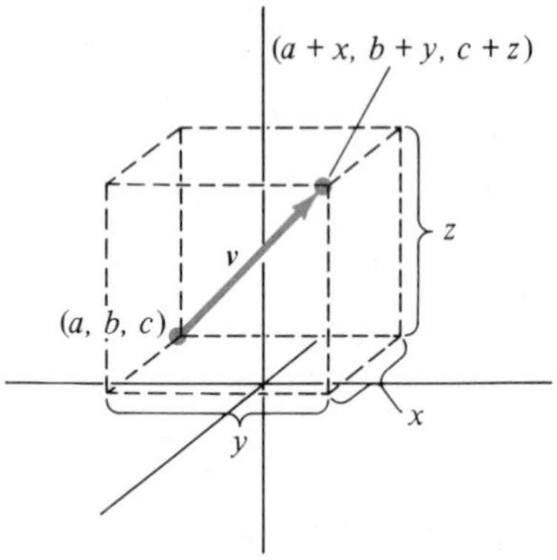

Figure 5.9 The vector $\boldsymbol{v} = \langle x, y, z\rangle$.

Finally, we define the **length of the vector** $\boldsymbol{v} = \langle x, y, z\rangle$ by applying the distance formula in Figure 5.6:

For $\boldsymbol{v} = \langle x, y, z\rangle$,
$$|\boldsymbol{v}| = \sqrt{x^2 + y^2 + z^2}. \tag{2}$$

With this definition of length, all properties of Theorem 2 carry over to vectors in space. For example, to prove that

$$|c\boldsymbol{v}| = |c|\,|\boldsymbol{v}|, \tag{3}$$

we express $\boldsymbol{v}$ in component form as $\boldsymbol{v} = \langle x, y, z \rangle$. Then

$$\begin{aligned} |c\boldsymbol{v}| &= \sqrt{(cx)^2 + (cy)^2 + (cz)^2} \\ &= |c|\sqrt{x^2 + y^2 + z^2} \\ &= |c|\,|\boldsymbol{v}|. \end{aligned}$$

You are asked to verify the remaining properties in Exercise 45.

Example 4 For the vectors $\boldsymbol{v} = \langle 2, -1, 4 \rangle$ and $\boldsymbol{w} = \langle 0, 3, -5 \rangle$,

(a) $\boldsymbol{v} + \boldsymbol{w} = \langle 2 + 0, -1 + 3, 4 + (-5) \rangle = \langle 2, 2, -1 \rangle$
(b) $3\boldsymbol{v} = \langle 3 \cdot 2, 3(-1), 3 \cdot 4 \rangle = \langle 6, -3, 12 \rangle$
(c) $\boldsymbol{v} - \boldsymbol{w} = \langle 2 - 0, -1 - 3, 4 - (-5) \rangle = \langle 2, -4, 9 \rangle$
(d) $|2\boldsymbol{v} + \boldsymbol{w}| = |\langle 2 \cdot 2 + 0, 2(-1) + 3, 2 \cdot 4 + (-5) \rangle|$
$\qquad = |\langle 4, 1, 3 \rangle| = \sqrt{4^2 + 1^2 + 3^2} = \sqrt{26}.$ ■

Unit Coordinate Vectors

We define the three unit coordinate vectors in space by

$$\boldsymbol{i} = \langle 1, 0, 0 \rangle, \qquad \boldsymbol{j} = \langle 0, 1, 0 \rangle, \qquad \boldsymbol{k} = \langle 0, 0, 1 \rangle.$$

(See Figure 5.10.) We may write the vector $\boldsymbol{v} = \langle x, y, z \rangle$ in terms of $\boldsymbol{i}, \boldsymbol{j}$, and $\boldsymbol{k}$ as

$$\begin{aligned} \boldsymbol{v} = \langle x, y, z \rangle &= x\langle 1, 0, 0 \rangle + y\langle 0, 1, 0 \rangle + z\langle 0, 0, 1 \rangle \\ &= x\boldsymbol{i} + y\boldsymbol{j} + z\boldsymbol{k} \qquad \text{(Figure 5.11).} \end{aligned}$$

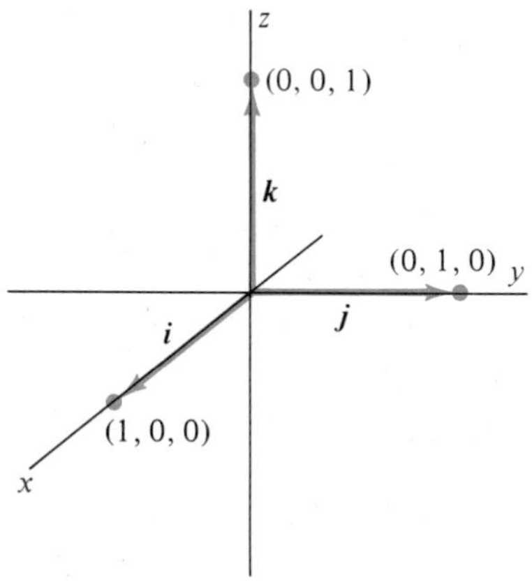

Figure 5.10 Unit coordinate vectors.

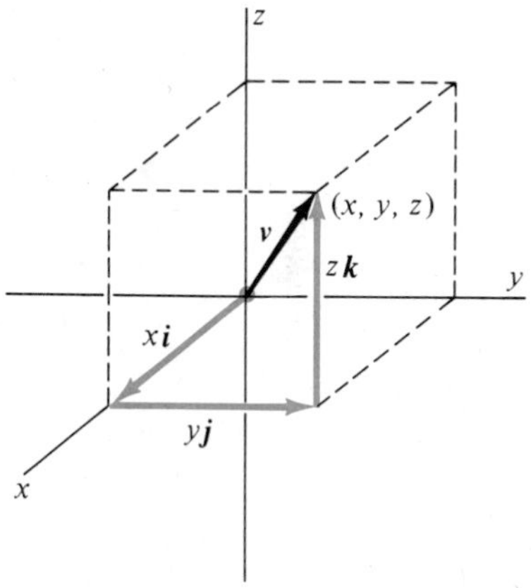

Figure 5.11 $\boldsymbol{v} = x\boldsymbol{i} + y\boldsymbol{j} + z\boldsymbol{k}$.

The Dot Product

The idea of dot product carries over to vectors in space by the obvious generalization of Definition 3:

DEFINITION 6

The **dot product** of the vectors $\boldsymbol{v} = \langle x_1, y_1, z_1 \rangle$ and $\boldsymbol{w} = \langle x_2, y_2, z_2 \rangle$ is the number

$$\boldsymbol{v} \cdot \boldsymbol{w} = x_1x_2 + y_1y_2 + z_1z_2.$$

All properties of Theorems 3 and 4 concerning the dot product remain true for vectors in space, including the Schwarz inequality

$$|\boldsymbol{v} \cdot \boldsymbol{w}| \le |\boldsymbol{v}|\,|\boldsymbol{w}| \tag{4}$$

and the representation of the dot product in terms of the lengths of v and w and the angle θ between them:

$$v \cdot w = |v||w| \cos \theta. \tag{5}$$

The proof of equation (5) is by direct analogy with that of Theorem 4, and the Schwarz inequality (4) follows from (5) as before since

$$|v \cdot w| = |v| \cdot |w||\cos \theta| \le |v||w|.$$

(See Exercises 46 and 47.)

Example 5 For the vectors $v = 3i + 2j - k$ and $w = i + 4j + k$,

(a) $$\begin{aligned} v + 2w &= (3i + 2j - k) + 2(i + 4j + k) \\ &= (3 + 2 \cdot 1)i + (2 + 2 \cdot 4)j + (-1 + 2 \cdot 1)k \\ &= 5i + 10j + k. \end{aligned}$$

(b) $v \cdot w = 3 \cdot 1 + 2 \cdot 4 + (-1)(1) = 3 + 8 - 1 = 10.$

(c) The cosine of the angle between v and w is

$$\cos \theta = \frac{v \cdot w}{|v| \cdot |w|} = \frac{10}{\sqrt{14} \cdot \sqrt{18}} = \frac{5}{3\sqrt{7}}.$$

(d) The **component** of v in the direction of w is the number

$$\text{comp}_w\, v = \frac{v \cdot w}{|w|} = \frac{10}{\sqrt{18}} = \frac{10}{3\sqrt{2}}.$$

(e) The **orthogonal projection** of v onto w is the vector

$$\begin{aligned} \text{proj}_{\perp w}\, v = \frac{v \cdot w}{|w|^2} w &= \frac{10}{(\sqrt{18})^2}(i + 4j + k) \\ &= \frac{5}{9}i + \frac{20}{9}j + \frac{5}{9}k. \end{aligned}$$

(f) A **unit vector** in the direction of w is

$$\frac{1}{|w|}w = \frac{1}{\sqrt{18}}(i + 4j + k) = \frac{1}{\sqrt{18}}i + \frac{4}{\sqrt{18}}j + \frac{1}{\sqrt{18}}k.$$ ■

Direction Cosines

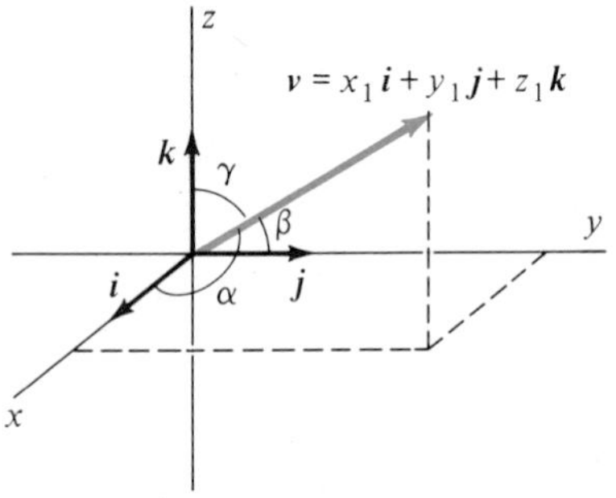

Figure 5.12 Direction angles for v.

Let $v = x_1 i + y_1 j + z_1 k$ be a nonzero vector, and let α, β, and γ denote the angles formed between v and the unit coordinate vectors i, j, and k, respectively (Figure 5.12).

We refer to α, β, and γ as the **direction angles** for v. Applying equation (5) we conclude that

$$\cos \alpha = \frac{v \cdot i}{|v||i|} = \frac{v \cdot i}{|v|} = \frac{x_1}{|v|}; \qquad x_1 = |v| \cos \alpha \tag{6}$$

$$\cos \beta = \frac{v \cdot j}{|v|\,|j|} = \frac{v \cdot j}{|v|} = \frac{y_1}{|v|}; \qquad y_1 = |v| \cos \beta \tag{7}$$

and

$$\cos \gamma = \frac{\boldsymbol{v} \cdot \boldsymbol{k}}{|\boldsymbol{v}|\,|\boldsymbol{k}|} = \frac{\boldsymbol{v} \cdot \boldsymbol{k}}{|\boldsymbol{v}|} = \frac{z_1}{|\boldsymbol{v}|}; \qquad z_1 = |\boldsymbol{v}| \cos \gamma. \tag{8}$$

Using (6) through (8), we may write the vector $\boldsymbol{v} = x_1\boldsymbol{i} + y_1\boldsymbol{j} + z_1\boldsymbol{k}$ as

$$\boldsymbol{v} = |\boldsymbol{v}|(\cos \alpha \boldsymbol{i} + \cos \beta \boldsymbol{j} + \cos \gamma \boldsymbol{k}). \tag{9}$$

Equation (9) yields two important conclusions. First, calculating lengths on both sides of (9) shows that

$$\begin{aligned} |\boldsymbol{v}| &= |\boldsymbol{v}|\,|\cos \alpha \boldsymbol{i} + \cos \beta \boldsymbol{j} + \cos \gamma \boldsymbol{k}| \\ &= |\boldsymbol{v}| \sqrt{\cos^2 \alpha + \cos^2 \beta + \cos^2 \gamma}. \end{aligned}$$

Cancelling factors of $|\boldsymbol{v}| \neq 0$ and squaring both sides gives

$$\cos^2 \alpha + \cos^2 \beta + \cos^2 \gamma = 1. \tag{10}$$

Second, for a unit vector $\boldsymbol{u}$, equation (9) becomes

$$\boldsymbol{u} = \cos \alpha \boldsymbol{i} + \cos \beta \boldsymbol{j} + \cos \gamma \boldsymbol{k}. \tag{11}$$

We refer to the cosines given in (6) through (8) as the **direction cosines** for the vector $\boldsymbol{v}$. Note that these numbers are defined independent of the particular representation for $\boldsymbol{v}$ as an arrow in space. Equation (10) states that *the sum of the squares of the direction cosines is always one*. Equation (11) states that *the direction cosines of a unit vector are precisely its components*.

Example 6 Let $\boldsymbol{v} = 2\boldsymbol{i} + \sqrt{5}\boldsymbol{j} + 4\boldsymbol{k}$. The direction cosines for $\boldsymbol{v}$ are, by equations (6) through (8),

$$\cos \alpha = \frac{2}{|\boldsymbol{v}|} = \frac{2}{\sqrt{4 + 5 + 16}} = \frac{2}{5},$$

$$\cos \beta = \frac{\sqrt{5}}{5}; \qquad \cos \gamma = \frac{4}{5}.$$ ■

Example 7 A vector $\boldsymbol{v}$ makes angles of 45° with the vector $\boldsymbol{i}$ and 60° with the vector $\boldsymbol{j}$.

(a) Find the angle between $\boldsymbol{v}$ and $\boldsymbol{k}$.
(b) Find a unit vector in the direction of $\boldsymbol{v}$.

Solution: We are given that

$$\cos \alpha = \cos 45° = \frac{\sqrt{2}}{2}; \qquad \cos \beta = \cos 60° = \frac{1}{2}.$$

We may therefore solve for $\cos \gamma$ using equation (10):

$$\begin{aligned} \cos^2 \gamma &= 1 - (\cos^2 \alpha + \cos^2 \beta) \\ &= 1 - \left(\frac{1}{2} + \frac{1}{4}\right) \\ &= \frac{1}{4}. \end{aligned}$$

Thus $\cos\gamma = \pm\frac{1}{2}$, so either $\gamma = 60°$ or $\gamma = 120°$.

If $\gamma = 60°$, a unit vector $\boldsymbol{u}$ in the direction of $\boldsymbol{v}$ is, by equation (11),

$$\boldsymbol{u} = \frac{\sqrt{2}}{2}\boldsymbol{i} + \frac{1}{2}\boldsymbol{j} + \frac{1}{2}\boldsymbol{k}.$$

If $\gamma = 120°$, then $\cos\gamma = -\frac{1}{2}$, so

$$\boldsymbol{u} = \frac{\sqrt{2}}{2}\boldsymbol{i} + \frac{1}{2}\boldsymbol{j} - \frac{1}{2}\boldsymbol{k}.$$

■

Exercise Set 17.5

1. Plot the following points: $P_1 = (-2, 1, 5)$, $P_2 = (3, 0, 2)$, $P_3 = (1, 1, 5)$, $P_4 = (2, -3, 6)$, $P_5 = (-3, -4, -5)$.

2. Sketch the following planes in coordinate space.
a. The xy-plane **b.** The yz-plane
c. The xz-plane **d.** The plane $x = 2$
e. The plane $y = 4$ **f.** The plane $z = -3$

3. Find the distance between the given pair of points.
a. $P = (1, 0, 1)$, $Q = (3, 2, 1)$
b. $P = (1, 2, -3)$, $Q = (-1, 4, 5)$
c. $P = (-3, 6, 2)$, $Q = (0, 3, 0)$
d. $P = (a, b, c)$, $Q = (2a, 2b, 2c)$

4. Find the midpoint of the line segment joining the points $P = (-4, 6, 1)$ and $Q = (-1, -3, -11)$.

5. Find the point one third of the distance from $P = (-4, 6, 1)$ to $Q = (-1, -3, -11)$.

6. Write an equation for the sphere with
a. center $(0, 0, 0)$ and radius $r = 2$,
b. center $(1, -1, 0)$ and radius $r = 1$,
c. center $(-2, 3, -5)$ and radius $r = 3$,
d. center $(4, 6, -2)$ and radius $r = 10$.

In Exercises 7–11, find the center and radius for the sphere with the given equation.

7. $x^2 + y^2 + z^2 - 4z = 5$

8. $x^2 + y^2 + z^2 - 2y - 4z = 4$

9. $x^2 + y^2 + z^2 - 4x + 2y - 6z = 2$

10. $x^2 + y^2 + z^2 - 4x + 4z = -4$

11. $x^2 + y^2 + z^2 - 6x + 2y + 4z = 11$

In Exercises 12–19, let $\boldsymbol{u} = \langle 2, -1, 5\rangle$, $\boldsymbol{v} = \langle -3, 5, 0\rangle$, and $\boldsymbol{w} = \langle 3, 3, 1\rangle$. Find the indicated vector or number.

12. $\boldsymbol{v} + \boldsymbol{w}$ **13.** $\boldsymbol{u} + 6\boldsymbol{v}$

14. $\boldsymbol{v} - \boldsymbol{w}$ **15.** $3\boldsymbol{u} - 2\boldsymbol{v}$

16. $\dfrac{\boldsymbol{v}}{|\boldsymbol{v}|}$ **17.** $|\boldsymbol{u} + 4\boldsymbol{v}|$

18. $\boldsymbol{v}\cdot\boldsymbol{w}$ **19.** $\boldsymbol{v}\cdot(\boldsymbol{u} + 2\boldsymbol{w})$

In Exercises 20–29, let $\boldsymbol{u} = \boldsymbol{i} + 2\boldsymbol{j} - \boldsymbol{k}$, $\boldsymbol{v} = 3\boldsymbol{i} - 2\boldsymbol{j} + 2\boldsymbol{k}$, and $\boldsymbol{w} = 5\boldsymbol{i} - \boldsymbol{j} + 3\boldsymbol{k}$. Find the indicated vector or number.

20. $\boldsymbol{u} + 2\boldsymbol{v} + \boldsymbol{w}$ **21.** $\boldsymbol{v} - 3\boldsymbol{w}$

22. $|3\boldsymbol{v} + \boldsymbol{w}|$ **23.** $\boldsymbol{u}\cdot(2\boldsymbol{v} + 3\boldsymbol{w})$

24. $|\boldsymbol{v}\cdot\boldsymbol{w} + \boldsymbol{w}\cdot\boldsymbol{u}|$ **25.** $\dfrac{\boldsymbol{v}}{|\boldsymbol{v}|} + \dfrac{\boldsymbol{w}}{|\boldsymbol{w}|}$

26. $\boldsymbol{v}\cdot\boldsymbol{w} - |\boldsymbol{v}||\boldsymbol{w}|$ **27.** $\text{proj}_{\boldsymbol{w}}\,\boldsymbol{v}$

28. $\text{comp}_{\boldsymbol{u}}\,\boldsymbol{w}$ **29.** $\text{comp}_{\boldsymbol{u}}\,\boldsymbol{v}$

30. Find a unit vector in the direction of $\boldsymbol{w} = \boldsymbol{i} + 4\boldsymbol{j} + 3\boldsymbol{k}$.

31. Find the vector $\boldsymbol{v} = \overrightarrow{PQ}$ for
a. $P = (1, 0, 1)$, $Q = (3, 1, 3)$,
b. $P = (-3, 2, 6)$, $Q = (1, 1, 1)$,
c. $P = (1, -5, 3)$, $Q = (7, -2, 2)$.

32. Find the point D so that $\overrightarrow{AB} = \overrightarrow{CD}$ if $A = (3, 1, 6)$, $B = (-2, 1, 5)$, and $C = (6, -2, 2)$.

33. Find the direction cosines for the given vector.
a. $\boldsymbol{v} = 3\boldsymbol{i} - \boldsymbol{j} + 2\boldsymbol{k}$
b. $\boldsymbol{v} = 6\boldsymbol{i} - 2\boldsymbol{j} + \boldsymbol{k}$
c. $\overrightarrow{PQ}$, where $P = (2, 1, 5)$ and $Q = (1, 3, 1)$

34. Which of the following sets of points are vertices of right triangles?
a. $P = (1, 3, 2)$, $Q = (4, 1, 4)$, $R = (6, 5, 5)$
b. $P = (0, 2, 5)$, $Q = (1, 3, 1)$, $R = (1, 4, 5)$
c. $P = (-3, 1, 2)$, $Q = (1, -3, 2)$, $R = (2, -2, 2)$

35. We say that two vectors $\boldsymbol{v}$ and $\boldsymbol{w}$ in space are **parallel** if there is a constant c so that $\boldsymbol{v} = c\boldsymbol{w}$. The vectors are said to point in the same direction if $c > 0$, and they are said to point in opposite directions if $c < 0$.
   a. Find two vectors of length 2 parallel to the vector $\boldsymbol{v} = 2\boldsymbol{i} - 4\boldsymbol{j} + 4\boldsymbol{k}$.
   b. Find the constant a so that the vectors $\boldsymbol{v} = \boldsymbol{i} - 3\boldsymbol{j} + 4\boldsymbol{k}$ and $\boldsymbol{w} = a\boldsymbol{i} + 9\boldsymbol{j} - 12\boldsymbol{k}$ are parallel.
   c. Find a vector of length 5 in the direction opposite that of $\boldsymbol{v} = \boldsymbol{i} - 2\boldsymbol{j} + 3\boldsymbol{k}$.
   d. Find a and b so that the vectors $3\boldsymbol{i} - \boldsymbol{j} + 4\boldsymbol{k}$ and $a\boldsymbol{i} + b\boldsymbol{j} - 2\boldsymbol{k}$ are parallel.

36. Describe the set of all points determined by the vector $\boldsymbol{r}(t) = (\boldsymbol{i} + 2\boldsymbol{j} + \boldsymbol{k}) + t(\boldsymbol{i} + \boldsymbol{j} + \boldsymbol{k})$ for $-\infty < t < \infty$, where $\boldsymbol{i} + 2\boldsymbol{j} + \boldsymbol{k}$ originates at $(0, 0, 0)$.

37. Let $\boldsymbol{u} = \boldsymbol{i} - 2\boldsymbol{j} + 3\boldsymbol{k}$, $\boldsymbol{v} = \boldsymbol{i} + 2\boldsymbol{j} + \boldsymbol{k}$, and $\boldsymbol{w} = \boldsymbol{i} - 4\boldsymbol{j} + \boldsymbol{k}$. Find t so that the vector $\boldsymbol{u} + t\boldsymbol{v}$ is orthogonal to $\boldsymbol{w}$.

38. Show that a nonzero vector is completely determined by its direction cosines and its length.

39. Which of the following triples can be direction angles of a single vector?
   a. 45°, 45°, 60°
   b. 30°, 45°, 60°
   c. 45°, 60°, 60°

40. Use the dot product to prove the following statements.
   a. $|\boldsymbol{v} + \boldsymbol{w}| = |\boldsymbol{v} - \boldsymbol{w}|$ if and only if $\boldsymbol{v}$ and $\boldsymbol{w}$ are orthogonal.
   b. $|\boldsymbol{v} + \boldsymbol{w}|^2 = |\boldsymbol{v}|^2 + |\boldsymbol{w}|^2$ if and only if $\boldsymbol{v}$ and $\boldsymbol{w}$ are orthogonal.

41. Let $\boldsymbol{v}_1$, $\boldsymbol{v}_2$, and $\boldsymbol{v}_3$ be mutually orthogonal vectors in space. Use the dot product to show that if c_1, c_2, and c_3 are scalars so that $c_1\boldsymbol{v}_1 + c_2\boldsymbol{v}_2 + c_3\boldsymbol{v}_3 = \boldsymbol{0}$, then $c_1 = c_2 = c_3 = 0$.

42. Give a geometric argument to show that three vectors $\boldsymbol{v}_1$, $\boldsymbol{v}_2$, and $\boldsymbol{v}_3$ are coplanar if and only if there exist constants c_1 and c_2 so that $\boldsymbol{v}_3 = c_1\boldsymbol{v}_1 + c_2\boldsymbol{v}_2$.

43. Let $\boldsymbol{v}_1 = \boldsymbol{i} + \boldsymbol{j}$, $\boldsymbol{v}_2 = \boldsymbol{j} + \boldsymbol{k}$, $\boldsymbol{v}_3 = \boldsymbol{i} + 2\boldsymbol{j} + 3\boldsymbol{k}$. Show that given any vector $\boldsymbol{w}$ there exist constants c_1, c_2, and c_3 so that $\boldsymbol{w} = c_1\boldsymbol{v}_1 + c_2\boldsymbol{v}_2 + c_3\boldsymbol{v}_3$. (*Hint:* Express $\boldsymbol{w}$ in component form and obtain 3 linear equations for the constants c_1, c_2, and c_3.)

44. State and prove the analogue of Theorem 1 for vectors in space.

45. State and prove the analogue of Theorem 2 for vectors in space.

46. Prove the Schwarz inequality for vectors in space.

47. Prove equality (5) for vectors in space.

17.6 THE CROSS PRODUCT

As we noted when we introduced the dot product, there is a second type of multiplication involving vectors. While the dot product of two vectors is a scalar, the *cross product* (or *vector* product) of two vectors is another vector. It is defined as follows.

DEFINITION 7

Let $\boldsymbol{v} = x_1\boldsymbol{i} + y_1\boldsymbol{j} + z_1\boldsymbol{k}$ and $\boldsymbol{w} = x_2\boldsymbol{i} + y_2\boldsymbol{j} + z_2\boldsymbol{k}$ be vectors in space. The **cross product, $\boldsymbol{v} \times \boldsymbol{w}$,** is the **vector**

$$\boldsymbol{v} \times \boldsymbol{w} = (y_1z_2 - z_1y_2)\boldsymbol{i} + (z_1x_2 - x_1z_2)\boldsymbol{j} + (x_1y_2 - y_1x_2)\boldsymbol{k}. \tag{1}$$

There is a geometric interpretation of expression (1) similar to that developed for the dot product. Before taking up that issue, we look to several examples for insights.

Example 1 If $\boldsymbol{v} = \boldsymbol{i}$ and $\boldsymbol{w} = \boldsymbol{j}$, then $x_1 = y_2 = 1$ and $y_1 = z_1 = x_2 = z_2 = 0$ in Definition 7. Thus, by equation (1)

$$\boldsymbol{i} \times \boldsymbol{j} = (0\cdot 0 - 0\cdot 1)\boldsymbol{i} + (0\cdot 0 - 1\cdot 0)\boldsymbol{j} + (1\cdot 1 - 0\cdot 0)\boldsymbol{k}.$$

That is,

$$\boldsymbol{i} \times \boldsymbol{j} = \boldsymbol{k} \quad \text{(Figure 6.1)}. \tag{2}$$

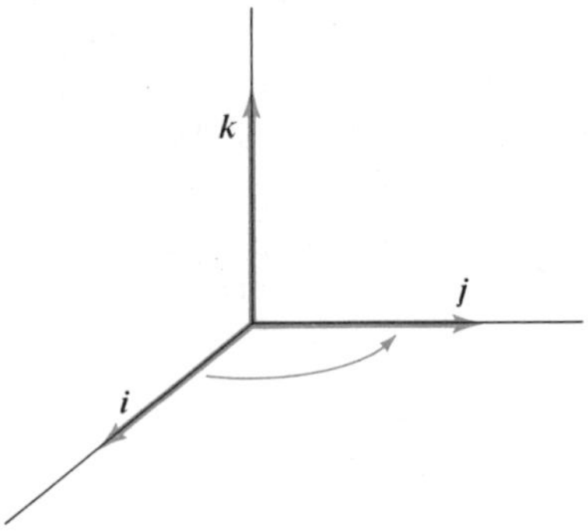

Figure 6.1 $i \times j = k$.

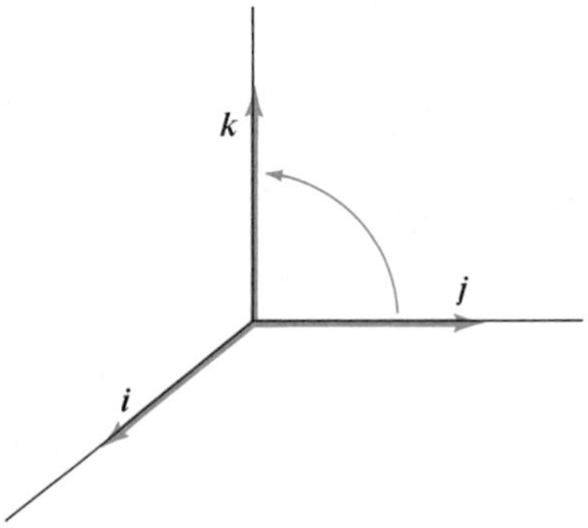

Figure 6.2 $j \times k = i$.

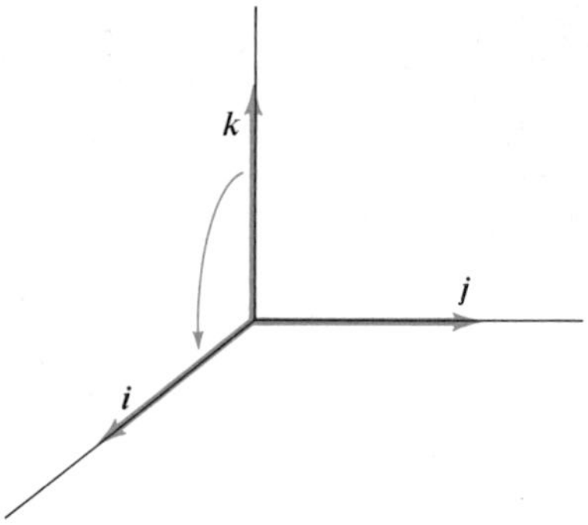

Figure 6.3 $k \times i = j$.

Similarly, you can verify that

$$\boldsymbol{j} \times \boldsymbol{k} = \boldsymbol{i} \qquad \text{(Figure 6.2)} \tag{3}$$

and

$$\boldsymbol{k} \times \boldsymbol{i} = \boldsymbol{j} \qquad \text{(Figure 6.3).} \tag{4}$$

The curved arrows in Figures 6.1 through 6.3 indicate the order of the vectors in the cross products.

However, you can also verify that the cross product is not commutative, since

$$\boldsymbol{j} \times \boldsymbol{i} = -\boldsymbol{k}, \qquad \boldsymbol{k} \times \boldsymbol{j} = -\boldsymbol{i}, \qquad \text{and} \qquad \boldsymbol{i} \times \boldsymbol{k} = -\boldsymbol{j}. \tag{5}$$

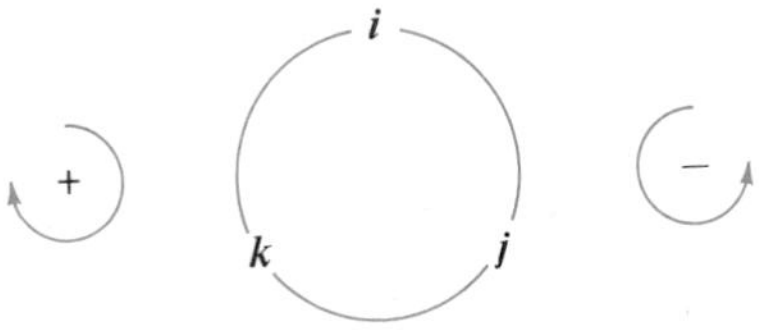

Figure 6.4 Device for remembering cross products of unit coordinate vectors.

Figure 6.4 presents a device for remembering these results. By beginning at the first factor and traversing the circle in the direction leading directly to the second factor, one determines the sign of the cross product: positive if the direction is clockwise (equations (2)–(4)); negative if the direction is counterclockwise (equation (5)). ■

Example 2 For $\boldsymbol{v} = 2\boldsymbol{i} + \boldsymbol{j} + 3\boldsymbol{k}$ and $\boldsymbol{w} = 4\boldsymbol{i} - \boldsymbol{j} + 2\boldsymbol{k}$,

$$\begin{aligned} \boldsymbol{v} \times \boldsymbol{w} &= [1 \cdot 2 - 3(-1)]\boldsymbol{i} + [3 \cdot 4 - 2 \cdot 2]\boldsymbol{j} + [2(-1) - 1 \cdot 4]\boldsymbol{k} \\ &= 5\boldsymbol{i} + 8\boldsymbol{j} - 6\boldsymbol{k}. \end{aligned}$$ ■

Properties of the Cross Product

It is easy to see that in each of the cross products of Example 1 the vector $\boldsymbol{v} \times \boldsymbol{w}$ is orthogonal to both factors, $\boldsymbol{v}$ and $\boldsymbol{w}$. The same is true for the vectors $\boldsymbol{v}$ and $\boldsymbol{w}$ of Example 2: $\boldsymbol{v} \times \boldsymbol{w} = 5\boldsymbol{i} + 8\boldsymbol{j} - 6\boldsymbol{k}$ is orthogonal to $\boldsymbol{v} = 2\boldsymbol{i} + \boldsymbol{j} + 3\boldsymbol{k}$, since $(\boldsymbol{v} \times \boldsymbol{w}) \cdot \boldsymbol{v} = 5 \cdot 2 + 8 \cdot 1 + (-6)(3) = 0$; $\boldsymbol{v} \times \boldsymbol{w}$ is orthogonal to $\boldsymbol{w}$, since $(\boldsymbol{v} \times \boldsymbol{w}) \cdot \boldsymbol{w} = 5 \cdot 4 + 8(-1) + 2(-6) = 0$.

These observations lead to the conjecture that $\boldsymbol{v} \times \boldsymbol{w}$ is *always* orthogonal both to $\boldsymbol{v}$ and to $\boldsymbol{w}$. To prove this conjecture, we write $\boldsymbol{v}$ and $\boldsymbol{w}$ in the component forms $\boldsymbol{v} = x_1\boldsymbol{i} + y_1\boldsymbol{j} + z_1\boldsymbol{k}$ and $\boldsymbol{w} = x_2\boldsymbol{i} + y_2\boldsymbol{j} + z_2\boldsymbol{k}$. Then, using Definition 7, we find that

$$\begin{aligned} (\boldsymbol{v} \times \boldsymbol{w}) \cdot \boldsymbol{v} &= [(y_1z_2 - z_1y_2)\boldsymbol{i} + (z_1x_2 - x_1z_2)\boldsymbol{j} \\ &\quad + (x_1y_2 - y_1x_2)\boldsymbol{k}] \cdot [x_1\boldsymbol{i} + y_1\boldsymbol{j} + z_1\boldsymbol{k}] \\ &= x_1(y_1z_2 - z_1y_2) + y_1(z_1x_2 - x_1z_2) + z_1(x_1y_2 - y_1x_2) \\ &= 0. \end{aligned}$$

Similarly, you can show that $(\boldsymbol{v} \times \boldsymbol{w}) \cdot \boldsymbol{w} = 0$. Thus, it follows from Corollary 1 (Section 17.4) that $\boldsymbol{v} \times \boldsymbol{w}$ is orthogonal to both $\boldsymbol{v}$ and $\boldsymbol{w}$.

This proves one part of the following theorem. The proofs of the other statements are carried out in the same way. Simply express all vectors in component form and apply Definition 7.

THEOREM 5
Properties of the Cross Product

Let $\boldsymbol{u}$, $\boldsymbol{v}$, and $\boldsymbol{w}$ be vectors and let c be a scalar. Then

(i) $\boldsymbol{v} \times \boldsymbol{w} = -\boldsymbol{w} \times \boldsymbol{v}$ (anticommutative)
(ii) $\boldsymbol{u} \times (\boldsymbol{v} + \boldsymbol{w}) = \boldsymbol{u} \times \boldsymbol{v} + \boldsymbol{u} \times \boldsymbol{w}$ (multiplication distributes over addition)
(iii) $c(\boldsymbol{v} \times \boldsymbol{w}) = (c\boldsymbol{v}) \times \boldsymbol{w} = \boldsymbol{v} \times (c\boldsymbol{w})$
(iv) $(\boldsymbol{v} \times \boldsymbol{w}) \perp \boldsymbol{v}$; $(\boldsymbol{v} \times \boldsymbol{w}) \perp \boldsymbol{w}$
(v) $\boldsymbol{v} \times \boldsymbol{v} = \boldsymbol{0}$ (self-annihilating)

The Determinant Notation

There is a useful formula for remembering equation (1) that involves the concept of **determinants.** The determinant of the 2×2 matrix

$$\begin{bmatrix} a_1 & b_1 \\ a_2 & b_2 \end{bmatrix}$$

is the number

$$\det\begin{bmatrix} a_1 & b_1 \\ a_2 & b_2 \end{bmatrix} = a_1b_2 - b_1a_2. \tag{6}$$

For example,

$$\det\begin{bmatrix} 3 & 2 \\ 4 & 1 \end{bmatrix} = 3\cdot 1 - 2\cdot 4 = 3 - 8 = -5.$$

The determinant of the 3×3 matrix

$$\begin{bmatrix} a_1 & b_1 & c_1 \\ a_2 & b_2 & c_2 \\ a_3 & b_3 & c_3 \end{bmatrix}$$

is the number

$$\begin{aligned} \det\begin{bmatrix} a_1 & b_1 & c_1 \\ a_2 & b_2 & c_2 \\ a_3 & b_3 & c_3 \end{bmatrix} &= a_1 \cdot \det\begin{bmatrix} b_2 & c_2 \\ b_3 & c_3 \end{bmatrix} - b_1 \cdot \det\begin{bmatrix} a_2 & c_2 \\ a_3 & c_3 \end{bmatrix} + c_1 \cdot \det\begin{bmatrix} a_2 & b_2 \\ a_3 & b_3 \end{bmatrix} \qquad (7) \\ &= a_1(b_2c_3 - c_2b_3) - b_1(a_2c_3 - c_2a_3) + c_1(a_2b_3 - b_2a_3). \end{aligned}$$

For example,

$$\begin{aligned} \det\begin{bmatrix} 1 & 2 & 3 \\ 4 & 5 & 6 \\ 7 & 8 & 9 \end{bmatrix} &= 1 \cdot \det\begin{bmatrix} 5 & 6 \\ 8 & 9 \end{bmatrix} - 2 \cdot \det\begin{bmatrix} 4 & 6 \\ 7 & 9 \end{bmatrix} + 3 \cdot \det\begin{bmatrix} 4 & 5 \\ 7 & 8 \end{bmatrix} \qquad (8) \\ &= (5\cdot 9 - 6\cdot 8) - 2(4\cdot 9 - 6\cdot 7) + 3(4\cdot 8 - 5\cdot 7) \\ &= 0. \end{aligned}$$

The reason for bringing up the topic of determinants is that by comparing equations (1) and (7) you can see that the *mnemonic* device

$$\boldsymbol{v} \times \boldsymbol{w} = \det \begin{bmatrix} \boldsymbol{i} & \boldsymbol{j} & \boldsymbol{k} \\ x_1 & y_1 & z_1 \\ x_2 & y_2 & z_2 \end{bmatrix}$$

$$= (y_1z_2 - z_1y_2)\boldsymbol{i} + (z_1x_2 - x_1z_2)\boldsymbol{j} + (x_1y_2 - y_1x_2)\boldsymbol{k} \tag{9}$$

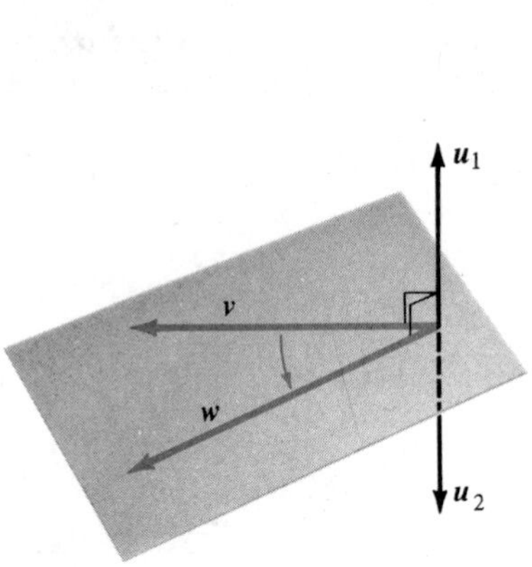

Figure 6.5 Both u_1 and u_2 are orthogonal to plane determined by v and w. Which one corresponds to $v \times w$?

provides a convenient way to remember equation (1). (Of course, equation (9) is not really a proper use of the determinant defined in equation (7), since the top row of the matrix consists of vectors rather than numbers.)

Determinants play an important role in the theory of systems of linear equations. Here we are making use of only a very limited aspect of determinants.

Example 3 For $\boldsymbol{v} = 3\boldsymbol{i} - \boldsymbol{j} - 4\boldsymbol{k}$ and $\boldsymbol{w} = -2\boldsymbol{i} + 2\boldsymbol{j} + \boldsymbol{k}$,

$$\boldsymbol{v} \times \boldsymbol{w} = \det \begin{bmatrix} \boldsymbol{i} & \boldsymbol{j} & \boldsymbol{k} \\ 3 & -1 & -4 \\ -2 & 2 & 1 \end{bmatrix} = ((-1)(1) - (-4)(2))\boldsymbol{i} + ((-4)(-2) - (3 \cdot 1))\boldsymbol{j} + (3 \cdot 2 - (-1)(-2))\boldsymbol{k}$$

$$= 7\boldsymbol{i} + 5\boldsymbol{j} + 4\boldsymbol{k}. \quad \blacksquare$$

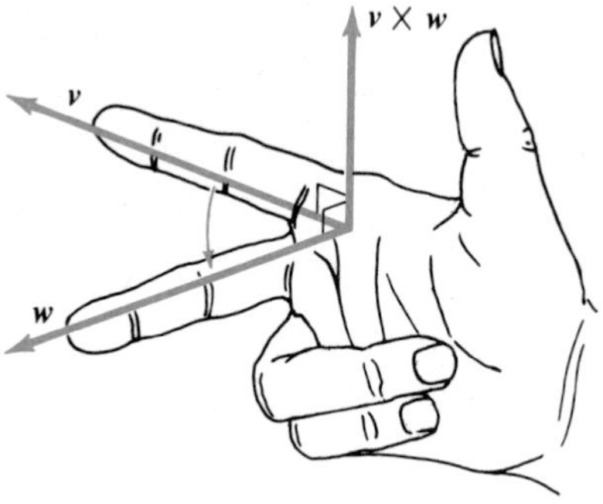

Figure 6.6 Right-hand rule determines direction of $v \times w$.

The next item on our agenda is a theorem that allows us to interpret the cross product geometrically. Before stating this theorem we need to clarify an earlier point. We have already shown that $\boldsymbol{v} \times \boldsymbol{w}$ is perpendicular both to $\boldsymbol{v}$ and to $\boldsymbol{w}$, so $\boldsymbol{v} \times \boldsymbol{w}$ is a vector perpendicular to the *plane* determined by $\boldsymbol{v}$ and $\boldsymbol{w}$. This leaves two possible directions for $\boldsymbol{v} \times \boldsymbol{w}$, as Figure 6.5 illustrates.

However, from Figures 6.1 through 6.3 you can see that each of the cross products $\boldsymbol{i} \times \boldsymbol{j}$, $\boldsymbol{j} \times \boldsymbol{k}$, and $\boldsymbol{k} \times \boldsymbol{i}$ satisfies the **right-hand rule:** *the thumb of a right hand points in the direction of* $\boldsymbol{v} \times \boldsymbol{w}$ *when the index finger points in the direction of* $\boldsymbol{v}$ *and the second finger points in the direction of* $\boldsymbol{w}$ (Figure 6.6).

In checking additional examples, you will observe that this rule holds in each case.

THEOREM 6

Let $\boldsymbol{v}$ and $\boldsymbol{w}$ be nonzero vectors forming an angle θ between them. Then

$$\boldsymbol{v} \times \boldsymbol{w} = (|\boldsymbol{v}||\boldsymbol{w}| \sin \theta)\boldsymbol{N} \tag{10}$$

where $\boldsymbol{N}$ is a unit vector orthogonal to both $\boldsymbol{v}$ and $\boldsymbol{w}$, in the direction determined by the right-hand rule.

Proof: We have already noted the direction of $\boldsymbol{v} \times \boldsymbol{w}$. What remains to be shown is that

$$|\boldsymbol{v} \times \boldsymbol{w}| = |(|\boldsymbol{v}||\boldsymbol{w}| \sin \theta)\boldsymbol{N}| = |\boldsymbol{v}||\boldsymbol{w}||\sin \theta|. \tag{11}$$

To prove equation (11), we write $\boldsymbol{v}$ and $\boldsymbol{w}$ in component form as

$$\boldsymbol{v} = x_1\boldsymbol{i} + y_1\boldsymbol{j} + z_1\boldsymbol{k}; \qquad \boldsymbol{w} = x_2\boldsymbol{i} + y_2\boldsymbol{j} + z_2\boldsymbol{k}.$$

Then, using equation (1) and the definition of the dot product we find that

$$|\boldsymbol{v} \times \boldsymbol{w}|^2 = |(y_1z_2 - z_1y_2)\boldsymbol{i} + (z_1x_2 - x_1z_2)\boldsymbol{j} + (x_1y_2 - y_1x_2)\boldsymbol{k}|^2 \tag{12}$$

$$\begin{aligned}&= (y_1z_2 - z_1y_2)^2 + (z_1x_2 - x_1z_2)^2 + (x_1y_2 - y_1x_2)^2\\ &= (x_1^2 + y_1^2 + z_1^2)(x_2^2 + y_2^2 + z_2^2) - (x_1x_2 + y_1y_2 + z_1z_2)^2\\ &= |\boldsymbol{v}|^2|\boldsymbol{w}|^2 - (\boldsymbol{v}\cdot\boldsymbol{w})^2.\end{aligned}$$

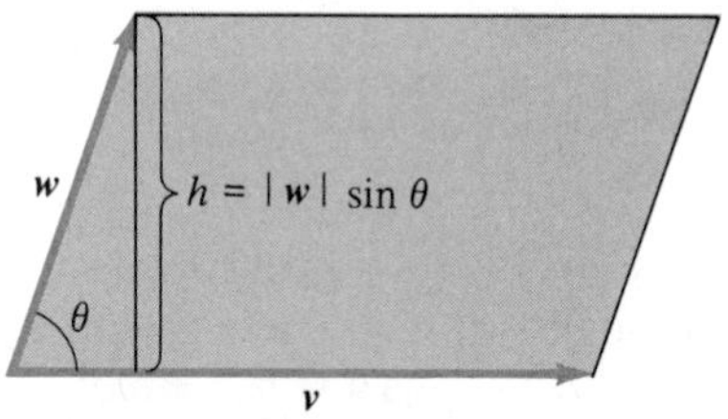

Figure 6.7 Area of parallelogram with base $|\boldsymbol{v}|$ and adjacent side $\boldsymbol{w}$ is $|\boldsymbol{v} \times \boldsymbol{w}|$.

Now by Theorem 4,

$$(\boldsymbol{v}\cdot\boldsymbol{w})^2 = |\boldsymbol{v}|^2|\boldsymbol{w}|^2\cos^2\theta. \tag{13}$$

Combining equations (12) and (13) we have

$$\begin{aligned}|\boldsymbol{v}\times\boldsymbol{w}|^2 &= |\boldsymbol{v}|^2|\boldsymbol{w}|^2 - |\boldsymbol{v}|^2|\boldsymbol{w}|^2\cos^2\theta\\ &= |\boldsymbol{v}|^2|\boldsymbol{w}|^2(1-\cos^2\theta)\\ &= |\boldsymbol{v}|^2|\boldsymbol{w}|^2\sin^2\theta.\end{aligned}$$

Taking square roots of both sides then gives equation (11). ■

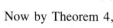

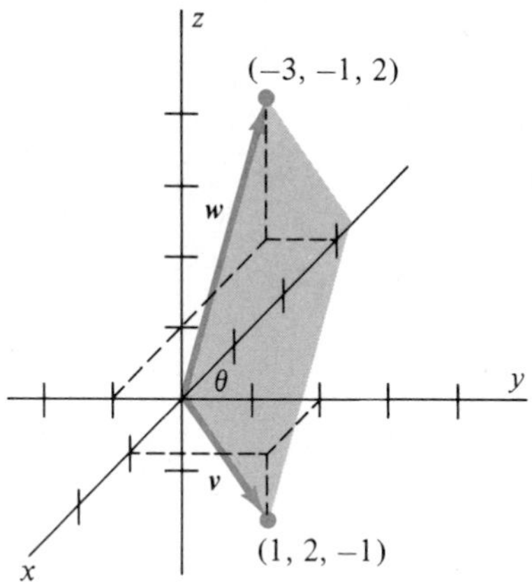

Figure 6.8

Figure 6.7 illustrates an immediate consequence of Theorem 6. If θ is the angle between $\boldsymbol{v}$ and $\boldsymbol{w}$, the altitude of the parallelogram determined by the vectors $\boldsymbol{v}$ and $\boldsymbol{w}$ is $h = |\boldsymbol{w}|\sin\theta$. Since the base has length $b = |\boldsymbol{v}|$, the area A must be

$$\boxed{A = bh = |\boldsymbol{v}||\boldsymbol{w}|\sin\theta = |\boldsymbol{v}\times\boldsymbol{w}|.} \tag{14}$$

Example 4 For $\boldsymbol{v} = \boldsymbol{i} + 2\boldsymbol{j} - \boldsymbol{k}$ and $\boldsymbol{w} = -3\boldsymbol{i} - \boldsymbol{j} + 2\boldsymbol{k}$,

$$\begin{aligned}\boldsymbol{v}\times\boldsymbol{w} = \det\begin{bmatrix}\boldsymbol{i} & \boldsymbol{j} & \boldsymbol{k}\\ 1 & 2 & -1\\ -3 & -1 & 2\end{bmatrix} &= (4-1)\boldsymbol{i} + (3-2)\boldsymbol{j} + (-1+6)\boldsymbol{k}\\ &= 3\boldsymbol{i} + \boldsymbol{j} + 5\boldsymbol{k}.\end{aligned}$$

The area of the parallelogram determined by $\boldsymbol{v}$ and $\boldsymbol{w}$ is, by equation (14),

$$A = |3\boldsymbol{i} + \boldsymbol{j} + 5\boldsymbol{k}| = \sqrt{9+1+25} = \sqrt{35} \qquad \text{(Figure 6.8).}$$ ■

Example 5 Find the sine of the angle between the vectors $\boldsymbol{v}$ and $\boldsymbol{w}$ in Example 4.

Solution: Solving equation (14) for $\sin\theta$ gives

$$\sin\theta = \frac{|\boldsymbol{v}\times\boldsymbol{w}|}{|\boldsymbol{v}||\boldsymbol{w}|}.$$

Since $|\boldsymbol{v}\times\boldsymbol{w}| = \sqrt{35}$, $|\boldsymbol{v}| = \sqrt{1^2+2^2+1^2} = \sqrt{6}$, and $|\boldsymbol{w}| = \sqrt{3^2+1^2+2^2} = \sqrt{14}$, we have

$$\sin\theta = \frac{\sqrt{35}}{\sqrt{6}\cdot\sqrt{14}} = \frac{1}{2}\sqrt{\frac{5}{3}}.$$ ■

We can use equation (14) to learn something about 2×2 determinants as follows. Let $\boldsymbol{v} = x_1\boldsymbol{i} + y_2\boldsymbol{j}$ and $\boldsymbol{w} = x_2\boldsymbol{i} + y_2\boldsymbol{j}$. Then $\boldsymbol{v}$ and $\boldsymbol{w}$ are vectors in the plane $z = 0$, and the area of the parallelogram they determine is

$$\begin{aligned}A = |\boldsymbol{v}\times\boldsymbol{w}| &= \det\begin{bmatrix}\boldsymbol{i} & \boldsymbol{j} & \boldsymbol{k}\\ x_1 & y_1 & 0\\ x_2 & y_2 & 0\end{bmatrix}\\ &= |(x_1y_2 - y_1x_2)\boldsymbol{k}|\end{aligned}$$

$$= |x_1y_2 - y_1x_2|$$

$$= \left| \det \begin{bmatrix} x_1 & y_1 \\ x_2 & y_2 \end{bmatrix} \right|$$

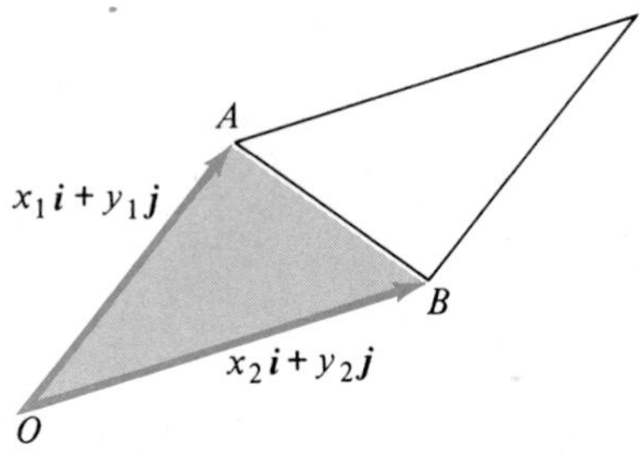

Figure 6.9 Area of ΔOAB is $\frac{1}{2}\left|\det\begin{bmatrix} x_1 & y_1 \\ x_2 & y_2 \end{bmatrix}\right|$

That is, the absolute value of the determinant of the matrix

$$\begin{bmatrix} x_1 & y_1 \\ x_2 & y_2 \end{bmatrix}$$

is the area of the parallelogram determined by the vectors $x_1\boldsymbol{i} + y_1\boldsymbol{j}$ and $x_2\boldsymbol{i} + y_2\boldsymbol{j}$. Since the diagonals of a parallelogram bisect the parallelogram into two congruent triangles, an even simpler interpretation of the 2×2 determinant is that *the absolute value of the determinant*

$$\det \begin{bmatrix} x_1 & y_1 \\ x_2 & y_2 \end{bmatrix}$$

is twice the area of the triangle determined by the vectors $x_1\boldsymbol{i} + y_1\boldsymbol{j}$ *and* $x_2\boldsymbol{i} + y_2\boldsymbol{j}$ (Figure 6.9).

Example 6 Find the area of the triangle with vertices (2, 3), (4, 7), and (5, 1).

Solution: Let $P = (2, 3)$, $Q = (4, 7)$, and $R = (5, 1)$. Two of the sides may be interpreted as the vectors

$$\boldsymbol{v} = \overrightarrow{PQ} = 2\boldsymbol{i} + 4\boldsymbol{j}$$

and

$$\boldsymbol{w} = \overrightarrow{PR} = 3\boldsymbol{i} - 2\boldsymbol{j} \qquad \text{(Figure 6.10).}$$

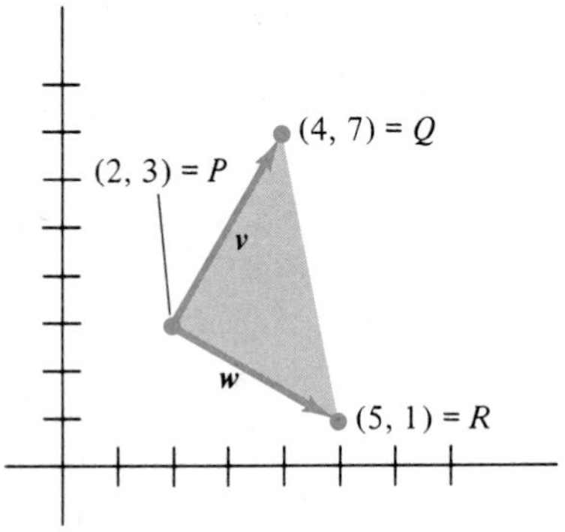

Figure 6.10

By the preceding observation, the area of the triangle is

$$A = \frac{1}{2}\left| \det \begin{bmatrix} 2 & 4 \\ 3 & -2 \end{bmatrix} \right| = \frac{1}{2} \cdot \left| 2(-2) - 4(3) \right| = \frac{16}{2} = 8.$$

■

Clearly, the geometry associated with the dot and cross products is rich. We shall use these ideas in Section 17.7 to develop equations for lines and planes in space.

Exercise Set 17.6

In Exercises 1–6, let $\boldsymbol{u} = \boldsymbol{i} + \boldsymbol{j} + \boldsymbol{k}$, $\boldsymbol{v} = 2\boldsymbol{i} - \boldsymbol{j} + 2\boldsymbol{k}$, and $\boldsymbol{w} = 3\boldsymbol{i} + 4\boldsymbol{j} - \boldsymbol{k}$. Find the indicated vector.

1. $\boldsymbol{v} \times \boldsymbol{w}$

2. $\boldsymbol{u} \times \boldsymbol{w}$

3. $\boldsymbol{w} \times \boldsymbol{u}$

4. $\boldsymbol{u} \times \boldsymbol{v}$

5. $\boldsymbol{u} \cdot (\boldsymbol{v} \times \boldsymbol{w})$

6. $\boldsymbol{v} \cdot (\boldsymbol{w} \times \boldsymbol{u})$

In Exercises 7–14, let $\boldsymbol{u} = 2\boldsymbol{i} - \boldsymbol{j} + 4\boldsymbol{k}$, $\boldsymbol{v} = \boldsymbol{i} - 3\boldsymbol{k}$, and $\boldsymbol{w} = 2\boldsymbol{i} + 3\boldsymbol{j} + 4\boldsymbol{k}$. Find the indicated vectors.

7. $\boldsymbol{u} \times \boldsymbol{v}$

8. $\boldsymbol{v} \times \boldsymbol{w}$

9. $\boldsymbol{u} \times \boldsymbol{w}$

10. $3\boldsymbol{u} \times 2\boldsymbol{w}$

11. $\boldsymbol{u} \times (\boldsymbol{v} \times \boldsymbol{w})$

12. $(\boldsymbol{u} \times \boldsymbol{v}) \times \boldsymbol{w}$

13. $\boldsymbol{u} \cdot (\boldsymbol{v} \times \boldsymbol{w})$

14. $\boldsymbol{w} \cdot (\boldsymbol{v} \times \boldsymbol{u})$

In Exercises 15–18, find the area of the parallelogram determined by the given vectors.

15. $\boldsymbol{v} = 2\boldsymbol{i} + \boldsymbol{j} + \boldsymbol{k}$, $\quad \boldsymbol{w} = 3\boldsymbol{i} - \boldsymbol{j} - 2\boldsymbol{k}$

16. $\boldsymbol{v} = -3\boldsymbol{i} + \boldsymbol{k}$, $\quad \boldsymbol{w} = 2\boldsymbol{i} + 2\boldsymbol{j} + \boldsymbol{k}$

17. $v = 2i - j, \quad w = 4i + 2j$

18. $v = -6i + 4j - 3k, \quad w = i - j - k$

In Exercises 19–21, find the area of the triangle in space with given vertices.

19. $A = (3, 0, 1), \quad B = (2, -1, 2), \quad C = (1, 3, -2)$

20. $A = (5, 2, -2), \quad B = (-1, 4, 2), \quad C = (-4, 5, 3)$

21. $A = (0, 2, 1), \quad B = (-4, 1, -2), \quad C = (1, 1, -2)$

22. Find the area of the triangle with vertices $(2, 4)$, $(-3, 8)$, and $(-5, -2)$.

23. Find the area of the triangle with vertices $(1, 5)$, $(-3, -4)$, and $(4, 6)$.

24. Find a unit vector orthogonal to both $v = 2i + j - k$ and $w = i + j + 4k$.

25. Find a unit vector orthogonal to both $v = i + 3j$ and $w = 4i - j$.

26. Show that the cross product is not associative. That is, find vectors u, v, and w so that $u \times (v \times w) \neq (u \times v) \times w$.

27. Find two unit vectors orthogonal to $v = i + j$ and $w = 2i - j + 3k$.

28. Show that $|v \times w| = |v||w|$ if v and w are orthogonal.

29. Show that if $u = x_1i + y_1j + z_1k$, $v = x_2i + y_2j + z_2k$, and $w = x_3i + y_3j + z_3k$, then

$$u \cdot (v \times w) = \det \begin{bmatrix} x_1 & y_1 & z_1 \\ x_2 & y_2 & z_2 \\ x_3 & y_3 & z_3 \end{bmatrix}.$$

30. Show that the volume of the parallelepiped formed by u, v, and w is $|u \cdot (v \times w)|$. (*Hint:* $|v \times w|$ is the area of the base. Thus, $|u \cdot (v \times w)| = |v \times w| \cdot (|u||\cos\theta|)$ is the area of the base times the altitude. Why?)

31. Use Exercises 29 and 30 to find the volume of the parallelepiped determined by the vectors $u = 2i + j - 4k$, $v = -i + 3j - 2k$, and $w = 4i - j + 3k$.

32. Find the volume of the parallelepiped determined by the vectors $u = -3i + j - k$, $v = 2i + 3j + 2k$, and $w = i + 4j + 2k$.

33. Prove that $u \times (v \times w) = (u \cdot w)v - (u \cdot v)w$.

34. Prove that $u \cdot (v \times w) = v \cdot (w \times u)$.

35. Prove that $u \cdot (v \times w) = w \cdot (u \times v)$.

36. Prove that $(u \times v) \times w = u \times (v \times w)$ only if $(u \times w) \times v = 0$.

37. Suppose that $u + v + w = 0$. Show that $u \times v = v \times w = w \times u$. What is the geometric interpretation of this result?

38. Why did we not try to define the cross product in $\mathbb{R}^2$?

17.7 EQUATIONS FOR LINES AND PLANES

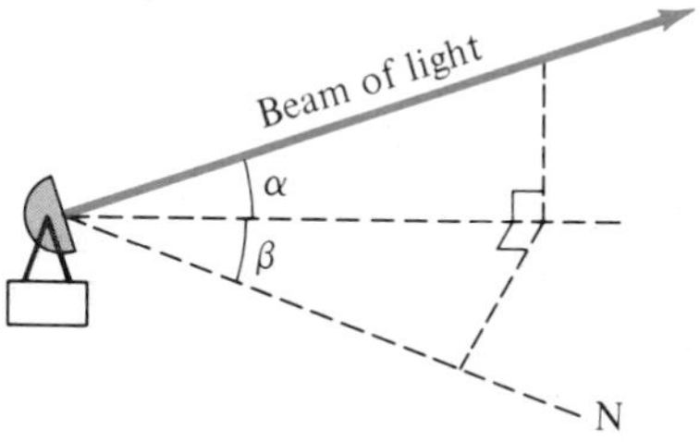

Figure 7.1 Searchlight positioned with elevation angle α and rotation angle β.

In the plane, a line ℓ is determined by its slope and one point on the line. However, a point and a slope do not provide sufficient information to determine a line in space. Indeed, even the term "slope" becomes ambiguous in space. For example, think about the problem of writing down an equation for the line determined by the beam from a searchlight as it sweeps across the sky. The beam is determined not only by a *point* (the location of the searchlight) and an *elevation* (perhaps the analogue of slope that you had in mind) but also by the angle at which the searchlight is *rotated* away from a fixed direction, say due north (Figure 7.1).

One of the simplest ways to think about how to find the general form for a line in space is to remember that two points determine a line. If P and Q are two points in space, then the vector $b = \overrightarrow{PQ}$ is parallel to the line ℓ containing P and Q. Moreover, any multiple tb is also parallel to ℓ (Figure 7.2).

We can use the point P and the vector b to determine points on the line ℓ as follows. Let $a = \overrightarrow{OP}$ be the **position vector** originating at the origin O and terminating at the point P on ℓ. Then, since tb is parallel to ℓ, the vector $r(t) = a + tb$ also terminates on the line ℓ (Figure 7.3). As the values of t increase from 0 to ∞, the vector $a + tb$ terminates successively at all points on ℓ lying on one side of P, while as t decreases from 0 to $-\infty$ the vector $a + tb$ determines all points on ℓ lying on the opposite side of P (Figure 7.4). (See Exercises 55–57 of Section 17.3 for a more explicit demonstration of this concept.)

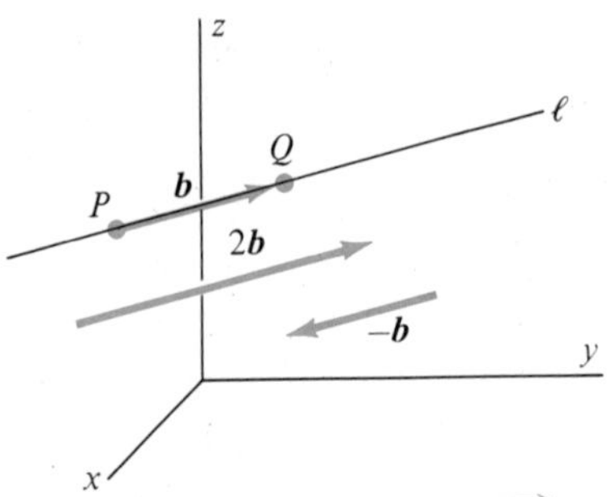

Figure 7.2 Multiples of $\boldsymbol{b} = \overrightarrow{PQ}$ are parallel to ℓ.

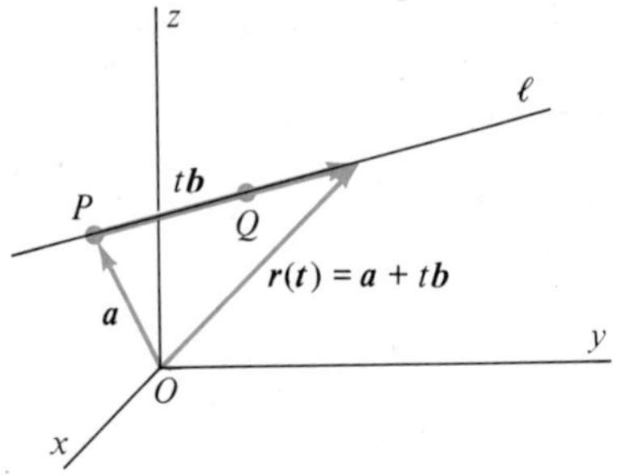

Figure 7.3 If $\boldsymbol{a} = \overrightarrow{OP}$, the vector $\boldsymbol{r}(t) = \boldsymbol{a} + t\boldsymbol{b}$ terminates on ℓ.

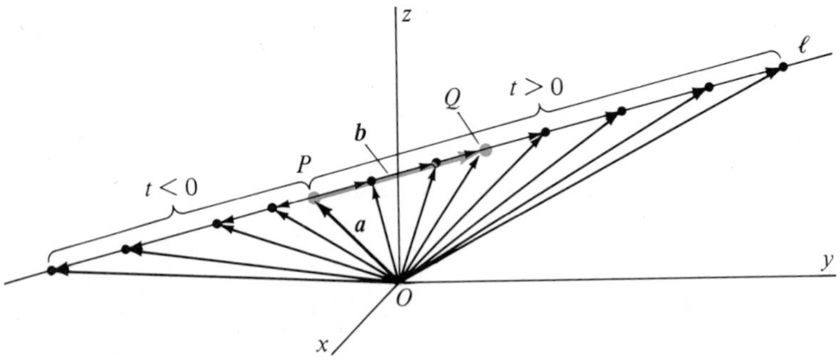

Figure 7.4 ℓ: $\{\boldsymbol{r}(t) = \boldsymbol{a} + t\boldsymbol{b} | -\infty < t < \infty\}$.

We can summarize this discussion by saying that a line ℓ in space is determined by a *position vector* ***a***, originating at the origin, and a *direction vector* ***b*** as

$$\ell\text{: } \boldsymbol{r}(t) = \boldsymbol{a} + t\boldsymbol{b}, \qquad -\infty < t < \infty. \tag{1}$$

Equation (1) is called the **vector form** of the equation for ℓ. If P and Q are distinct points on ℓ we can always take

$$\boldsymbol{a} = \overrightarrow{OP} \tag{2}$$

and

$$\boldsymbol{b} = \overrightarrow{PQ} \tag{3}$$

as in the preceding discussion. In this case equation (1) becomes

$$\ell\text{: } \boldsymbol{r}(t) = \overrightarrow{OP} + t\overrightarrow{PQ}, \qquad -\infty < t < \infty.$$

More generally, equation (1) allows us to interpret the line ℓ as an infinite collection of vectors, each terminating at a point on ℓ.

Example 1 Find an equation for the line containing the point $P = (2, -1, 4)$ and parallel to the vector $\boldsymbol{v} = \boldsymbol{i} - \boldsymbol{j} + 2\boldsymbol{k}$.

Solution: The position vector is

$$\boldsymbol{a} = \overrightarrow{OP} = 2\boldsymbol{i} - \boldsymbol{j} + 4\boldsymbol{k}$$

and the direction vector is given as

$$\boldsymbol{b} = \boldsymbol{v} = \boldsymbol{i} - \boldsymbol{j} + 2\boldsymbol{k}.$$

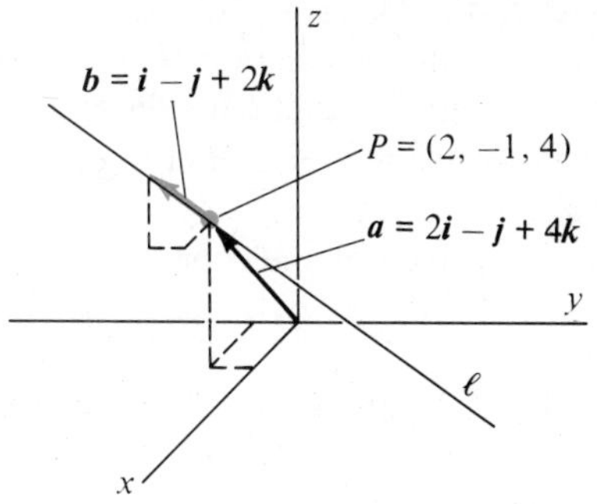

Figure 7.5

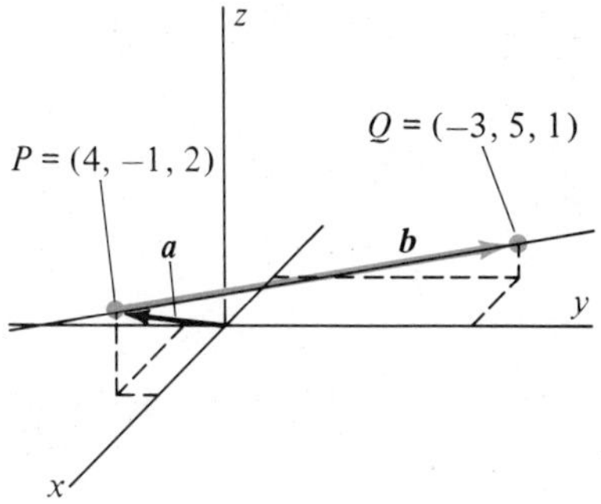

Figure 7.6

The line is, by equation (1), the set of all vectors of the form

$$\boldsymbol{r}(t) = (2\boldsymbol{i} - \boldsymbol{j} + 4\boldsymbol{k}) + t(\boldsymbol{i} - \boldsymbol{j} + 2\boldsymbol{k}) \qquad \text{(Figure 7.5).}$$ ■

Example 2 Find an equation for the line containing the points $P = (4, -1, 2)$ and $Q = (-3, 5, 1)$.

Solution: By equation (2) a position vector is

$$\boldsymbol{a} = \overrightarrow{OP} = 4\boldsymbol{i} - \boldsymbol{j} + 2\boldsymbol{k}$$

and, by (3), a direction vector is

$$\boldsymbol{b} = \overrightarrow{PQ} = (-3 - 4)\boldsymbol{i} + (5 - (-1))\boldsymbol{j} + (1 - 2)\boldsymbol{k} = -7\boldsymbol{i} + 6\boldsymbol{j} - \boldsymbol{k}.$$

The line has vector equation

$$\boldsymbol{r}(t) = (4\boldsymbol{i} - \boldsymbol{j} + 2\boldsymbol{k}) + t(-7\boldsymbol{i} + 6\boldsymbol{j} - \boldsymbol{k}) \qquad \text{(Figure 7.6).}$$ ■

Example 3 Find the point, if it exists, where the lines

$$\ell_1: \boldsymbol{r}_1(t) = (3\boldsymbol{i} + 2\boldsymbol{j} - \boldsymbol{k}) + t(-6\boldsymbol{i} + 4\boldsymbol{j} + 3\boldsymbol{k})$$

$$\ell_2: \boldsymbol{r}_2(t) = (5\boldsymbol{i} + 4\boldsymbol{j} + 7\boldsymbol{k}) + t(14\boldsymbol{i} - 6\boldsymbol{j} + 2\boldsymbol{k})$$

intersect.

Solution: If ℓ_1 and ℓ_2 intersect at point P, it is not necessary that they do so for the same value of t. The condition of intersection is therefore that

$$\boldsymbol{r}_1(t_1) = \boldsymbol{r}_2(t_2) \tag{4}$$

for some times t_1 and t_2. Equating $\boldsymbol{i}$-components in equation (4) gives

$$3 - 6t_1 = 5 + 14t_2,$$

or

$$6t_1 + 14t_2 = -2. \tag{5}$$

Equating $\boldsymbol{j}$-components in (4) gives the equation

$$2 + 4t_1 = 4 - 6t_2$$

or

$$4t_1 + 6t_2 = 2. \tag{6}$$

Solving equations (5) and (6) simultaneously (either by substitution or by elimination) gives $t_1 = 2$ and $t_2 = -1$. To ensure that equation (4) holds for these values of t we must write out $\boldsymbol{r}_1(t_1)$ and $\boldsymbol{r}_2(t_2)$ and verify that the $\boldsymbol{k}$-components are equal.

Since

$$\boldsymbol{r}_1(t_1) = (3\boldsymbol{i} + 2\boldsymbol{j} - \boldsymbol{k}) + 2(-6\boldsymbol{i} + 4\boldsymbol{j} + 3\boldsymbol{k}) = -9\boldsymbol{i} + 10\boldsymbol{j} + 5\boldsymbol{k}$$

and

$$\boldsymbol{r}_2(t_2) = (5\boldsymbol{i} + 4\boldsymbol{j} + 7\boldsymbol{k}) + (-1)(14\boldsymbol{i} - 6\boldsymbol{j} + 2\boldsymbol{k}) = -9\boldsymbol{i} + 10\boldsymbol{j} + 5\boldsymbol{k},$$

we can see that the lines indeed intersect at the point $(-9, 10, 5)$. ■

The Distance from a Point to a Line in Space

In space, a line does not determine a unique direction for vectors orthogonal to that line. (In fact, the family of vectors orthogonal to a given line determines an entire plane, as we shall see later in this section.) The problem of finding the distance d from a point $P = (x_0, y_0, z_0)$ to a line with equation $\boldsymbol{r}(t) = \boldsymbol{a} + t\boldsymbol{b}$, must be handled differently from the corresponding problem in the plane.

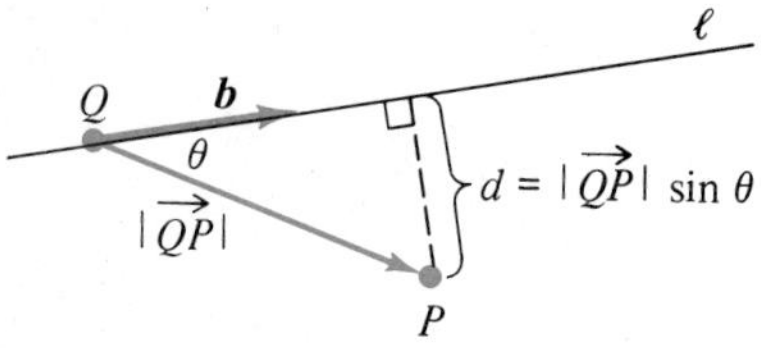

Figure 7.7 $d = |\overrightarrow{QP}| \sin\theta = \dfrac{|\overrightarrow{PQ} \times \boldsymbol{b}|}{|\boldsymbol{b}|}$.

Figure 7.7 illustrates that the desired distance d is given by the expression

$$d = |\overrightarrow{QP}||\sin\theta| \tag{7}$$

where Q is any point on ℓ and θ is the angle formed between the $\overrightarrow{QP}$ and the direction vector $\boldsymbol{b}$.

The right side of equation (7) is suggestive of the cross product. In fact, we may use Theorem 6 to rewrite equation (7) as

$$d = |\overrightarrow{QP}||\sin\theta| = \frac{|\overrightarrow{QP}||\boldsymbol{b}||\sin\theta|}{|\boldsymbol{b}|} = \frac{|\overrightarrow{PQ} \times \boldsymbol{b}|}{|\boldsymbol{b}|}.$$

Since $|\boldsymbol{b}| \neq 0$ for all lines ℓ, we have the formula

$$\boxed{d = \frac{|\overrightarrow{PQ} \times \boldsymbol{b}|}{|\boldsymbol{b}|}} \tag{8}$$

for the distance from a point P to a line with direction vector $\boldsymbol{b}$ and containing the point Q.

Example 4 Find the distance from the point $P = (1, -1, 2)$ to the line with vector equation

$$\ell: \boldsymbol{r}(t) = 3\boldsymbol{i} - 2\boldsymbol{j} + 4\boldsymbol{k} + t(\boldsymbol{i} - 2\boldsymbol{j} + 2\boldsymbol{k}).$$

Strategy

Solution

Find the direction vector $\boldsymbol{b}$.

Since the equation for ℓ is in the form of equation (1) the direction vector is

$$\boldsymbol{b} = \boldsymbol{i} - 2\boldsymbol{j} + 2\boldsymbol{k}.$$

Find a point Q on the line ℓ.

Setting $t = 0$ shows that the point

$$Q = (3, -2, 4)$$

lies on ℓ. Thus,

Form the vector $\overrightarrow{QP}$.

$$\begin{aligned}\overrightarrow{QP} &= (1-3)\boldsymbol{i} + (-1-(-2))\boldsymbol{j} + (2-4)\boldsymbol{k}\\ &= -2\boldsymbol{i} + \boldsymbol{j} - 2\boldsymbol{k}.\end{aligned}$$

Apply equation (8).

The distance from P to ℓ is therefore

$$\begin{aligned} d &= \frac{1}{\sqrt{1^2 + (-2)^2 + 2^2}} \cdot \left| \det \begin{bmatrix} \boldsymbol{i} & \boldsymbol{j} & \boldsymbol{k} \\ -2 & 1 & -2 \\ 1 & -2 & 2 \end{bmatrix} \right| \\ &= \frac{1}{3}\left| -2\boldsymbol{i} + 2\boldsymbol{j} + 3\boldsymbol{k} \right| \\ &= \frac{\sqrt{17}}{3}. \end{aligned}$$

■

Parametric Equations for Lines

Suppose that the vectors $\boldsymbol{a}$ and $\boldsymbol{b}$ have components $\boldsymbol{a} = a_1\boldsymbol{i} + a_2\boldsymbol{j} + a_3\boldsymbol{k}$, and $\boldsymbol{b} = b_1\boldsymbol{i} + b_2\boldsymbol{j} + b_3\boldsymbol{k}$. Then if we write $\boldsymbol{r}(t) = x(t)\boldsymbol{i} + y(t)\boldsymbol{j} + z(t)\boldsymbol{k}$, the vector equation $\boldsymbol{r}(t) = \boldsymbol{a} + t\boldsymbol{b}$ for a line has component form

$$\begin{aligned} x(t)\boldsymbol{i} + y(t)\boldsymbol{j} + z(t)\boldsymbol{k} &= (a_1\boldsymbol{i} + a_2\boldsymbol{j} + a_3\boldsymbol{k}) + t(b_1\boldsymbol{i} + b_2\boldsymbol{j} + b_3\boldsymbol{k}). \\ &= (a_1 + tb_1)\boldsymbol{i} + (a_2 + tb_2)\boldsymbol{j} + (a_3 + tb_3)\boldsymbol{k}. \end{aligned}$$

Equating components in this equation gives the **parametric form** for the line ℓ:

$$\begin{cases} x(t) = a_1 + tb_1 \\ y(t) = a_2 + tb_2 \\ z(t) = a_3 + tb_3. \end{cases} \tag{9}$$

The point (x, y, z) is on ℓ if and only if the coordinates satisfy the equations in (9) for some single value of t.

Example 5 Find parametric equations for the coordinates of the points on the line in Example 1.

Solution: From Example 1 we have

$$a_1 = 2, \quad a_2 = -1, \quad a_3 = 4, \quad b_1 = 1, \quad b_2 = -1, \quad b_3 = 2.$$

Thus, by (9)

$$\begin{cases} x(t) = 2 + t \\ y(t) = -1 - t \\ z(t) = 4 + 2t \end{cases}$$

are the parametric equations for ℓ. ■

Example 6 Find a vector parallel to the line with parametric equations

$$x = -4 + t, \qquad y = 3 - 6t, \qquad \text{and} \qquad z = 2 + 2t.$$

Solution: Comparing these equations with those in (9), we see that a direction vector for the line is

$$\boldsymbol{b} = b_1\boldsymbol{i} + b_2\boldsymbol{j} + b_3\boldsymbol{k} = \boldsymbol{i} - 6\boldsymbol{j} + 2\boldsymbol{k}.$$

This vector is parallel to the line. ■

Symmetric Equations for Lines

Solving each of the equations in (9) for t gives the equations

$$t = \frac{x(t) - a_1}{b_1}; \qquad t = \frac{y(t) - a_2}{b_2}; \qquad t = \frac{z(t) - a_3}{b_3}.$$

Equating the right-hand sides of each of these equations gives the **symmetric** (or, **rectangular**) form of the equations for the line ℓ with position vector $\boldsymbol{a} = a_1\boldsymbol{i} + a_2\boldsymbol{j} + a_3\boldsymbol{k}$ and direction vector $\boldsymbol{b} = b_1\boldsymbol{i} + b_2\boldsymbol{j} + b_3\boldsymbol{k}$:

$$\boxed{\frac{x - a_1}{b_1} = \frac{y - a_2}{b_2} = \frac{z - a_3}{b_3}.} \tag{10}$$

Since the position vector $\boldsymbol{a}$ terminates at the point (a_1, a_2, a_3) on ℓ, we may also refer to equation (10) as the symmetric form for the equations of the line containing the point (a_1, a_2, a_3) and having direction vector $\boldsymbol{b} = b_1\boldsymbol{i} + b_2\boldsymbol{j} + b_3\boldsymbol{k}$.

Example 7 Find symmetric equations for the line determined by parametric equations in Example 5.

Solution: Using the values of a_j and b_j, $j = 1, 2, 3$, in Example 5, together with equations in (10), we obtain

$$\frac{x-2}{1} = \frac{y+1}{-1} = \frac{z-4}{2}.$$

■

Example 8 Find a unit vector orthogonal to the line ℓ with symmetric equations

$$\frac{x-2}{3} = \frac{y-4}{-2} = \frac{z+1}{5}.$$

Strategy

Find a vector parallel to ℓ – its direction vector $\boldsymbol{b}$.

Solution

The direction vector for ℓ is, by equation (10),

$$\boldsymbol{b} = 3\boldsymbol{i} - 2\boldsymbol{j} + 5\boldsymbol{k}.$$

Find any vector $\boldsymbol{c}$ with $\boldsymbol{c} \cdot \boldsymbol{b} = 0$.

A vector $\boldsymbol{c} = c_1\boldsymbol{i} + c_2\boldsymbol{j} + c_3\boldsymbol{k}$ is orthogonal to $\boldsymbol{b}$ if and only if

$$\boldsymbol{b} \cdot \boldsymbol{c} = 3c_1 - 2c_2 + 5c_3 = 0.$$

One solution (among many) of this equation is

$$c_1 = 2, \qquad c_2 = 3, \qquad c_3 = 0.$$

The desired unit vector is

$$\boldsymbol{u} = \left(\frac{1}{|\boldsymbol{c}|}\right)\boldsymbol{c}.$$

The vector $\boldsymbol{c} = 2\boldsymbol{i} + 3\boldsymbol{j}$ is therefore orthogonal to ℓ. A *unit* vector in the same direction is

$$\boldsymbol{u} = \frac{\boldsymbol{c}}{|\boldsymbol{c}|} = \frac{\boldsymbol{c}}{\sqrt{13}} = \frac{2}{\sqrt{13}}\boldsymbol{i} + \frac{3}{\sqrt{13}}\boldsymbol{j}.$$

■

Example 9 Find the symmetric equations for the line containing the points $P = (7, 3, -1)$ and $Q = (4, -5, -2)$.

Strategy

Find the direction vector $\boldsymbol{b} = \overrightarrow{PQ}$.

Solution

The direction vector for the line is

$$\begin{aligned}\boldsymbol{b} = \overrightarrow{PQ} &= (4-7)\boldsymbol{i} + (-5-3)\boldsymbol{j} + (-2-(-1))\boldsymbol{k}\\ &= -3\boldsymbol{i} - 8\boldsymbol{j} - \boldsymbol{k}\end{aligned}$$

Use $P = (a_1, a_2, a_3)$ and substitute into equations in (10).

Using the point $P = (7, 3, -1)$ on the line and the direction vector $\boldsymbol{b}$ we obtain from the equations in (10) that

$$\frac{x-7}{-3} = \frac{y-3}{-8} = \frac{z+1}{-1}.$$

■

Equations for Planes

In solid geometry a plane is characterized by the following two facts:

(i) Through any point P on a line ℓ, there is precisely one plane σ perpendicular to ℓ.

(ii) The point $Q \neq P$ lies on the plane σ in (i) if and only if the line through P and Q is perpendicular to ℓ.

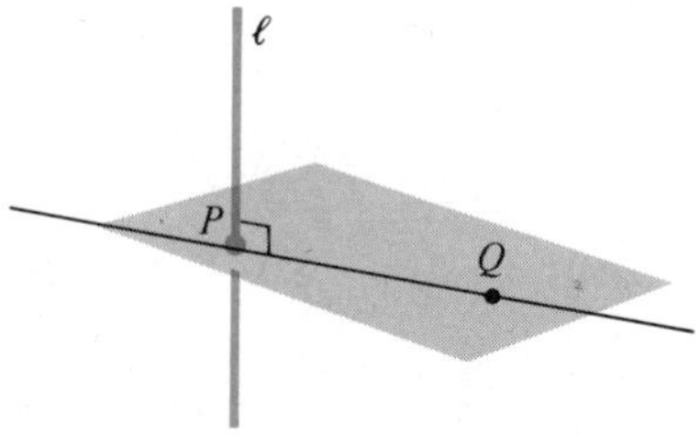

Figure 7.8 Plane σ determined by a point P and a line ℓ.

(Figure 7.8.)

We can use vectors to develop an analytic description of the plane σ as follows. If $\boldsymbol{n}$ is a direction vector for the line ℓ, the condition that the line through P and Q be perpendicular to the line ℓ is equivalent to the condition that the vectors $\overrightarrow{PQ}$ and $\boldsymbol{n}$ be orthogonal. By Corollary 1 of Section 17.4, this last condition is equivalent to the condition that $\overrightarrow{PQ}\cdot\boldsymbol{n} = 0$. We may therefore write *the plane σ determined by the point P and the vector $\boldsymbol{n}$* as

$$\sigma = \{Q \mid \overrightarrow{PQ}\cdot\boldsymbol{n} = 0\}. \tag{11}$$

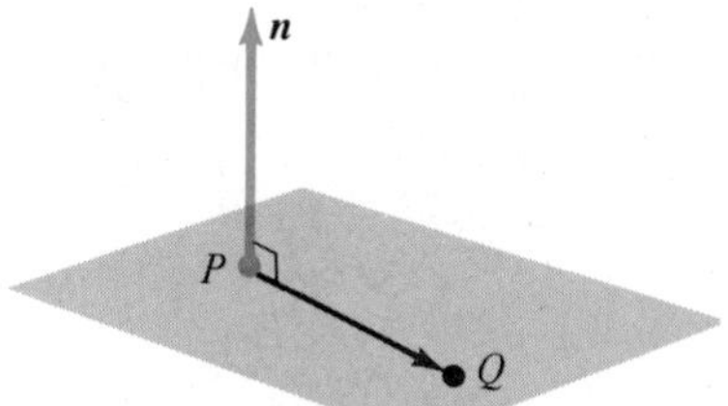

Figure 7.9 Plane σ determined by a point P and a normal vector $\boldsymbol{n}$.

The vector $\boldsymbol{n}$ in equation (11) is referred to as a **normal vector** for the plane σ (Figure 7.9).

To develop an equation in xyz-coordinates for the plane σ, we express $\boldsymbol{n}$ in component form as $\boldsymbol{n} = a\boldsymbol{i} + b\boldsymbol{j} + c\boldsymbol{k}$, and we assume the point P to have coordinates $P = (x_0, y_0, z_0)$. Let the point Q have coordinates $Q = (x, y, z)$. Then the vector $\overrightarrow{PQ}$ is

$$\overrightarrow{PQ} = (x - x_0)\boldsymbol{i} + (y - y_0)\boldsymbol{j} + (z - z_0)\boldsymbol{k}, \tag{12}$$

and the condition in line (11) is that

$$[(x - x_0)\boldsymbol{i} + (y - y_0)\boldsymbol{j} + (z - z_0)\boldsymbol{k}]\cdot(a\boldsymbol{i} + b\boldsymbol{j} + c\boldsymbol{k}) = 0. \tag{13}$$

Equation (13) simplifies to the equation

$$a(x - x_0) + b(y - y_0) + c(z - z_0) = 0,$$

or

$$\boxed{ax + by + cz = ax_0 + by_0 + cz_0.} \tag{14}$$

Equation (14) is *the equation for the plane with normal vector $\boldsymbol{n} = a\boldsymbol{i} + b\boldsymbol{j} + c\boldsymbol{k}$ containing the point* (x_0, y_0, z_0). Notice that the coefficients of x, y, and z are just the respective components of the normal vector $\boldsymbol{n}$, while the right-hand side is a *constant* determined from $\boldsymbol{n}$ and P.

Example 10 Find a vector normal to the plane with equation $3x + 5y - z = 9$.

Solution: The components of a normal vector are simply the coefficients of x, y, and z:

$$\boldsymbol{n} = 3\boldsymbol{i} + 5\boldsymbol{j} - \boldsymbol{k}.$$

■

Example 11 Find an equation for the plane with normal vector $\boldsymbol{n} = \boldsymbol{i} + 4\boldsymbol{j} + 2\boldsymbol{k}$, and containing the point $P = (6, -3, 4)$.

Solution: Let $Q = (x, y, z)$ be a point in the desired plane. We are given

$$a = 1, \qquad b = 4, \qquad c = 2$$

and

$$x_0 = 6, \qquad y_0 = -3, \qquad z_0 = 4.$$

Substituting into equation (14) gives

$$x + 4y + 2z = 1 \cdot 6 + 4(-3) + 2 \cdot 4 = 2.$$

The desired equation is therefore

$$x + 4y + 2z = 2.$$

■

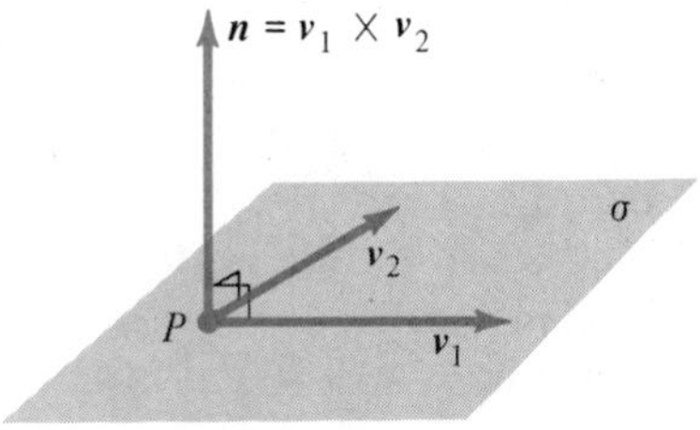

Figure 7.10 $\boldsymbol{n} = \boldsymbol{v}_1 \times \boldsymbol{v}_2$ is orthogonal to the plane σ if $\boldsymbol{v}_1$ and $\boldsymbol{v}_2$ lie in σ.

If a normal vector is not specified, we can sometimes determine a normal vector for a plane from other information about the plane. In particular, if we can find two vectors, $\boldsymbol{v}_1$ and $\boldsymbol{v}_2$, lying in the plane, then the cross product $\boldsymbol{n} = \boldsymbol{v}_1 \times \boldsymbol{v}_2$ is a normal to the plane, since it is orthogonal both to $\boldsymbol{v}_1$ and to $\boldsymbol{v}_2$ (Figure 7.10).

Example 12 Find an equation for the plane determined by the three points $P = (-4, 0, 2)$, $Q = (1, -3, 1)$, and $R = (2, -2, 6)$.

Strategy

Find two vectors.

$\boldsymbol{v}_1 = \overrightarrow{PQ}, \qquad \boldsymbol{v}_2 = \overrightarrow{PR}$

in the plane.

Solution

Two vectors in the plane are

$$\boldsymbol{v}_1 = \overrightarrow{PQ} = 5\boldsymbol{i} - 3\boldsymbol{j} - \boldsymbol{k}$$

and

$$\boldsymbol{v}_2 = \overrightarrow{PR} = 6\boldsymbol{i} - 2\boldsymbol{j} + 4\boldsymbol{k}.$$

Find a normal

$\boldsymbol{n} = \boldsymbol{v}_1 \times \boldsymbol{v}_2.$

A normal vector is therefore

$$\boldsymbol{n} = \boldsymbol{v}_1 \times \boldsymbol{v}_2 = \det\begin{bmatrix} \boldsymbol{i} & \boldsymbol{j} & \boldsymbol{k} \\ 5 & -3 & -1 \\ 6 & -2 & 4 \end{bmatrix} = -14\boldsymbol{i} - 26\boldsymbol{j} + 8\boldsymbol{k}.$$

Substitute into equation (14), using $\boldsymbol{n}$ and P.

Using $\boldsymbol{n}$ and the point P, we obtain from equation (14) that

$$\begin{aligned} -14x - 26y + 8z &= -14(-4) + (-26)(0) + 8 \cdot 2 \\ &= 72. \end{aligned}$$

The desired equation is

$$-14x - 26y + 8z = 72.$$

■

Example 13 Find the distance from the point $P = (3, -1, 2)$ to the plane with equation $2x - y + z = 4$.

Strategy

Find a normal $\boldsymbol{n}$ for the plane.

Solution

A normal vector for the plane is

$$\boldsymbol{n} = 2\boldsymbol{i} - \boldsymbol{j} + \boldsymbol{k}.$$

Set $x = y = 0$ and solve for z to find any convenient point Q in the plane.

A point Q in the plane is $Q = (0, 0, 4)$. A vector parallel to $\boldsymbol{n}$ originating at P and terminating in the plane is

$$\boldsymbol{v} = \text{proj}_{\boldsymbol{n}}\, \overrightarrow{PQ} \qquad \text{(see Figure 7.10).}$$

Project $\overrightarrow{PQ}$ onto $\boldsymbol{n}$. Recall that $|\text{proj}_{\boldsymbol{n}}\, \overrightarrow{PQ}| = |\text{comp}_{\boldsymbol{n}}\, \overrightarrow{PQ}|$ (Section 17.4).

The desired distance is the length of this vector

$$|\boldsymbol{v}| = |\text{comp}_{\boldsymbol{n}}\, \overrightarrow{PQ}| = \frac{|\overrightarrow{PQ} \cdot \boldsymbol{n}|}{|\boldsymbol{n}|}$$
$$= \frac{|(-3\boldsymbol{i} + \boldsymbol{j} + 2\boldsymbol{k}) \cdot (2\boldsymbol{i} - \boldsymbol{j} + \boldsymbol{k})|}{|2\boldsymbol{i} - \boldsymbol{j} + \boldsymbol{k}|}$$
$$= \frac{5}{\sqrt{6}} \qquad \text{(Figure 7.11).}$$

You may wish to try another convenient point Q, say (2, 0, 0), and verify that the result is the same. ■

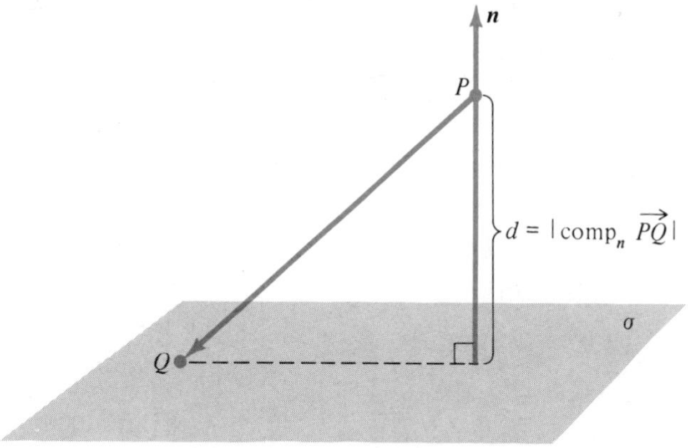

Figure 7.11 Distance from a point P to a plane σ.

Exercise Set 17.7

In Exercises 1–7, find a parametric form of the equations for the line described.

1. The line with direction vector $\boldsymbol{b} = \boldsymbol{i} + \boldsymbol{j} - \boldsymbol{k}$ and containing the point $P = (1, 2, 3)$.

2. The line with direction vector $\boldsymbol{b} = 2\boldsymbol{i} - \boldsymbol{j} + 3\boldsymbol{k}$ and containing the point $P = (3, -6, 2)$.

3. The line with position and direction vectors $\boldsymbol{a} = \boldsymbol{b} = 3\boldsymbol{i} - \boldsymbol{j} + 5\boldsymbol{k}$.

4. The line containing the points $P = (1, 3, -1)$ and $Q = (7, -2, 5)$.

5. The line containing the points $P = (-4, 2, 1)$ and $Q = (-3, 5, 3)$.

6. The line with position vector $\boldsymbol{a} = \boldsymbol{i} + 2\boldsymbol{j} - 6\boldsymbol{k}$ and direction vector $\boldsymbol{b} = 2\boldsymbol{i} - \boldsymbol{j} + 4\boldsymbol{k}$.

7. The line through (1, 2, 0) and orthogonal to *both* vectors $\boldsymbol{v} = \boldsymbol{i} + 3\boldsymbol{j} + \boldsymbol{k}$ and $\boldsymbol{w} = \boldsymbol{j} - 3\boldsymbol{k}$.

In Exercises 8–11, find rectangular (symmetric) equations for the line determined by the given parametric equations.

8. $x(t) = 2t$, $y(t) = 3 - t$, $z(t) = t$

9. $x(t) = t + 7$, $y(t) = 4t - 6$, $z(t) = 3 - 2t$

10. $x(t) = 2t + 5$, $y(t) = t - 6$, $z(t) = 5t + 2$

11. $x(t) = t$, $y(t) = 8t - 6$, $z(t) = 4t + 4$

Find symmetric equations for

12. The line in Exercise 2.

13. The line in Exercise 3.

14. The line in Exercise 5.

In Exercises 15–18, find parametric equations for the line determined by the given symmetric equations.

15. $\dfrac{x}{3} = \dfrac{y}{2} = \dfrac{z}{5}$

16. $\dfrac{x - 2}{3} = \dfrac{y + 1}{-4} = \dfrac{z}{3}$

17. $\dfrac{x+4}{4}=\dfrac{y-2}{-2}=\dfrac{z+3}{3}$

18. $x=\dfrac{y-3}{2}=z+1$

19. Find a vector parallel to the line in Exercise 9.

20. Find a vector parallel to the line in Exercise 16.

21. Find a direction vector for the line in Exercise 17.

22. Find a direction vector for the line in Exercise 18.

23. For the line ℓ with vector form $\boldsymbol{r}(t)=\boldsymbol{a}+t\boldsymbol{b}$, $\boldsymbol{b}\neq\boldsymbol{0}$, find the number t_0 for which $\boldsymbol{r}(t_0)\perp\boldsymbol{b}$. Conclude that $\boldsymbol{r}(t_0)\perp\ell$.

24. Use Exercise 23 to show that the distance from the origin to the line ℓ with equation $\boldsymbol{r}(t)=\boldsymbol{a}+t\boldsymbol{b}$ is $d=|\boldsymbol{r}(t_0)|$, where t_0 is as in Exercise 23.

25. Use Exercise 24 to find the distance from the origin to the line with vector equation $\boldsymbol{r}(t)=\boldsymbol{i}+2\boldsymbol{j}+t(\boldsymbol{j}-\boldsymbol{k})$.

26. Use Exercise 24 to find the distance from the origin to the line with symmetric equations

$$x-1=\frac{2-y}{2}=\frac{z}{-2}.$$

27. Find the distance from the point $P=(2,-1,4)$ to the line with equation $\boldsymbol{r}(t)=(3\boldsymbol{i}-\boldsymbol{j}-\boldsymbol{k})+t(\boldsymbol{i}+2\boldsymbol{j}+\boldsymbol{k})$.

28. Find the distance from the point $P=(1,3,-2)$ to the line with equations $x(t)=1+t$, $y(t)=3-2t$, $z(t)=2t-2$.

29. Find the distance from the point $P=(0,-2,1)$ to the line with equations $\dfrac{x-1}{4}=\dfrac{y+3}{-2}=\dfrac{z+1}{5}$.

30. For the line ℓ with vector form $\boldsymbol{r}(t)=\boldsymbol{a}+t\boldsymbol{b}$, $\boldsymbol{b}=b_1\boldsymbol{i}+b_2\boldsymbol{j}+b_3\boldsymbol{k}$, the numbers b_1, b_2, and b_3 are called the **direction numbers.** What condition must be satisfied by the direction numbers in order that the symmetric equations for ℓ exist?

31. Find the distance d between the lines

$$\ell_1:\ \frac{x-2}{3}=\frac{y+1}{2}=\frac{z-3}{5}$$

and

$$\ell_2:\ \frac{x+4}{1}=\frac{y-3}{2}=\frac{z+1}{2}$$

as follows:

a. Find direction vectors $\boldsymbol{b}_1$ and $\boldsymbol{b}_2$ for lines ℓ_1 and ℓ_2, respectively.

b. Observe that a vector perpendicular to both lines ℓ_1 and ℓ_2 is the cross product $\boldsymbol{n}=\boldsymbol{b}_1\times\boldsymbol{b}_2$.

c. Find a point P on ℓ_1 and a point Q on ℓ_2.

d. Obtain the distance d as

$$d=|\text{comp}_{\boldsymbol{n}}\,\overrightarrow{PQ}|.$$

32. Use Exercise 31 to find the distance between the lines ℓ_1 and ℓ_2:

$$\ell_1:\ x(t)=1+t,\quad y(t)=5t,\quad z(t)=1-t,$$
$$\ell_2:\ x(t)=2+t,\quad y(t)=2-3t,\quad z(t)=1+5t.$$

Two lines with vector equations ℓ_1: $\boldsymbol{r}_1(t)=\boldsymbol{a}_1+t\boldsymbol{b}_1$ and ℓ_2: $\boldsymbol{r}_2(t)=\boldsymbol{a}_2+t\boldsymbol{b}_2$ **intersect** if $\boldsymbol{r}_1(t_1)=\boldsymbol{r}_2(t_2)$ for some numbers t_1 and t_2. Find the point at which each of the following pairs of lines intersect.

33. $\boldsymbol{r}_1(t)=(2\boldsymbol{i}+\boldsymbol{j}+2\boldsymbol{k})+t(5\boldsymbol{i}+\boldsymbol{j}+3\boldsymbol{k})$
$\boldsymbol{r}_2(t)=(-4\boldsymbol{i}+7\boldsymbol{j}+10\boldsymbol{k})+t(3\boldsymbol{i}-3\boldsymbol{j}-4\boldsymbol{k})$

34. ℓ_1: $x(t)=1+t$, $y(t)=2-2t$, $z(t)=t+5$
ℓ_2: $x(t)=2+2t$, $y(t)=5-9t$, $z(t)=2+6t$

In each of Exercises 35–39, find an equation for the plane with the given normal containing the given point.

35. $\boldsymbol{n}=\boldsymbol{i}+2\boldsymbol{j}-\boldsymbol{k}$, $P=(1,2,-3)$

36. $\boldsymbol{n}=4\boldsymbol{i}-\boldsymbol{j}+\boldsymbol{k}$, $P=(-1,3,5)$

37. $\boldsymbol{n}=2\boldsymbol{i}-2\boldsymbol{j}+3\boldsymbol{k}$, $P=(0,-2,5)$

38. $\boldsymbol{n}=\boldsymbol{i}+\boldsymbol{j}$ $P=(4,2,-3)$

39. $\boldsymbol{n}=\boldsymbol{i}+2\boldsymbol{j}+\boldsymbol{k}$ $P=(0,-1,3)$

In Exercises 40–44, find an equation for the plane containing the given points.

40. $P=(0,1,-2)$, $Q=(1,1,1)$, $R=(3,5,1)$

41. $P=(-2,3,-4)$, $Q=(1,5,1)$, $R=(-2,-2,-2)$

42. $P=(7,0,-2)$, $Q=(1,-3,4)$, $R=(5,2,-3)$

43. $P=(-1,1,1)$, $Q=(0,2,-1)$, $R=(3,5,-2)$

44. $P=(-2,1,5)$, $Q=(2,0,-2)$, $R=(-1,-1,3)$

45. Find a vector normal to the plane with equation $2x-3y+z=5$.

46. Find a vector normal to the plane containing $(-2,1,3)$, $(5,1,5)$, and $(0,0,2)$.

47. Find an equation for the plane perpendicular to the line with symmetric equation

$$\frac{x-2}{3}=\frac{1-y}{6}=\frac{z+2}{2}$$

containing the point $(3,-2,5)$.

48. Find an equation for the plane containing the point $(3,-1,-1)$ and perpendicular to the line with parametric equations $x=3+t$, $y=-1+4t$, $z=5+3t$.

49. Find parametric equations for the line of intersection of the planes with equations $x+y-2z=6$ and $2x-4y+z=3$.

50. The angle between two planes is equal to the angle between vectors normal to the two planes. Find the angle between the planes with equations $x + y + z = 6$ and $3x - y + 2z = 5$.

51. Find the angle between the planes with equations $x - y + z = 2$ and $x + 3y + 3z = -4$ (see Exercise 50).

52. Show that the distance from the point $P_0 = (x_0, y_0, z_0)$ to the plane with equation $ax + by + cz = d$ is

$$d = \frac{|ax_0 + by_0 + cz_0 - d|}{\sqrt{a^2 + b^2 + c^2}}.$$

(*Hint:* See Example 13.)

53. Find the distance from the point $P = (3, -6, 5)$ to the plane with equation $6x - y + z = 2$.

54. Show that the plane with equation $ax + by + cz = 1$ intersects the coordinate axes at the respective values $x = \frac{1}{a}$, $y = \frac{1}{b}$, $z = \frac{1}{c}$ if $a \neq 0$, $b \neq 0$, $c \neq 0$.

55. Find parametric equations for the line containing the point $(2, 4, -3)$ and perpendicular to the plane with equation $2x + 3y - 7z = 9$.

56. Determine which of the following sets of points are coplanar.

a. $(2, 1, 0)$, $(3, 0, 5)$, $(1, 1, 1)$, $(2, 3, -12)$
b. $(1, -6, 2)$, $(3, -5, 11)$, $(4, 0, 4)$, $(1, 5, -2)$
c. $(1, 7, 2)$, $(3, 4, -2)$, $(1, 5, 1)$, $(1, 9, 3)$

57. Find an equation for the plane containing the point $(3, -1, 3)$ and the line with symmetric equations

$$\frac{x-2}{3} = \frac{y+1}{5} = \frac{z}{4}.$$

17.8 QUADRIC SURFACES AND CYLINDERS

In Chapter 15 we showed that the graph of the general second degree equation in two variables

$$Ax^2 + Bxy + Cy^2 + Dx + Ey + F = 0$$

if not degenerate, is either a line, a parabola, a circle, an ellipse, or a hyperbola in the plane. The general second degree equation in three variables has the form

$$Ax^2 + By^2 + Cz^2 + Dxy + Exz + Fyz + Gx + Hy + Iz + J = 0. \qquad (1)$$

Equation (1) is called a **quadric equation.** Graphs of quadric equations are surfaces in space, called **quadric surfaces.** In this section we shall examine six general types of quadric surfaces and a seventh special case that we refer to as the **quadric cylinder.**

The six general quadric surfaces are

1. the ellipsoid,
2. the elliptic paraboloid,
3. the elliptic cone,
4. the hyperboloid of one sheet,
5. the hyperboloid of two sheets,
6. the hyperbolic paraboloid.

We will study each of these figures by finding its *traces* in the coordinate planes and in planes parallel to the coordinate planes. To understand what this means, let us first recall the equations for the three coordinate planes:

$$z = 0 \text{ is the equation for the } xy\text{-plane,} \qquad \text{(2a)}$$

$$y = 0 \text{ is the equation for the } xz\text{-plane,} \qquad \text{(2b)}$$

$$x = 0 \text{ is the equation for the } yz\text{-plane.} \qquad \text{(2c)}$$

(See Figure 8.1).

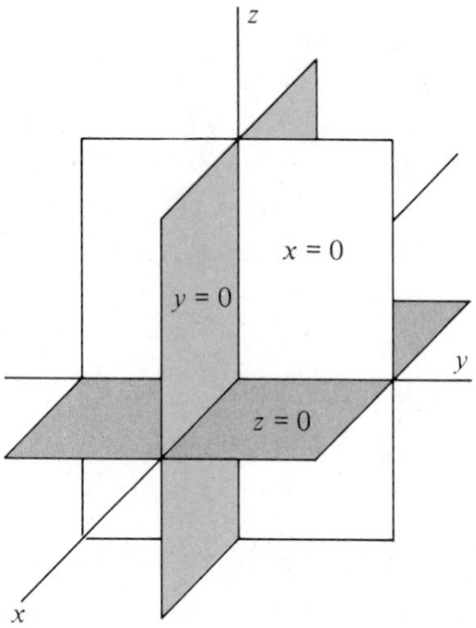

Figure 8.1 The three coordinate planes.

The **trace** of a surface in any plane is simply the intersection of the surface and the plane. The equation for this trace is obtained by substituting the constant value determining the plane for the corresponding variable in the equation for the surface. For example, we have already seen (Section 17.5) that the equation for the sphere with center (0, 0, 0) and radius r is

$$x^2 + y^2 + z^2 = r^2. \tag{3}$$

Combining equations (2a) and (3), we find that the equation for the trace of the sphere in the xy-plane is $x^2 + y^2 = r^2$. The trace is therefore the circle with center (0, 0) and radius r in the xy-plane (Figure 8.2).

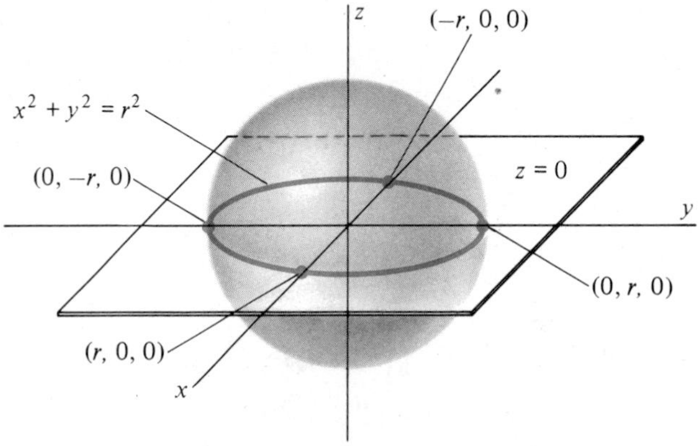

Figure 8.2 Trace of sphere $x^2 + y^2 + z^2 = r^2$ in plane $z = 0$ is circle $x^2 + y^2 = r^2$.

More generally, equations for planes *parallel* to the coordinate planes are

$$z = d, \quad \text{a plane parallel to the } xy\text{-plane,} \tag{4a}$$

$$y = d, \quad \text{a plane parallel to the } xz\text{-plane,} \tag{4b}$$

$$x = d, \quad \text{a plane parallel to the } yz\text{-plane.} \tag{4c}$$

Thus, to find the equation of the trace of the sphere in, say, the plane $z = d$, we combine equations (3) and (4a) to get $x^2 + y^2 + d^2 = r^2$, or $x^2 + y^2 = r^2 - d^2$. If $d^2 < r^2$, then the trace is again a circle (although smaller than the one in the xy-plane); if $d^2 > r^2$, there are no points of intersection (the plane is above or below the sphere).

The Ellipsoid: $\dfrac{x^2}{a^2} + \dfrac{y^2}{b^2} + \dfrac{z^2}{c^2} = 1$

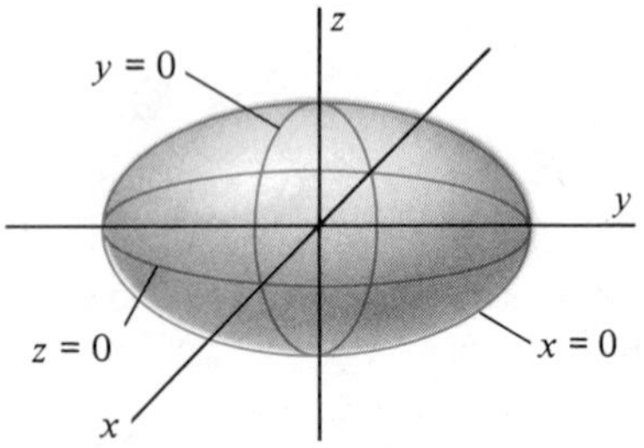

Figure 8.3 Traces of the ellipsoid $\dfrac{x^2}{a^2} + \dfrac{y^2}{b^2} + \dfrac{c^2}{d^2} = 1$ in the coordinate planes.

Setting $z = 0$ shows that the trace in the xy-plane is the ellipse

$$\frac{x^2}{a^2} + \frac{y^2}{b^2} = 1.$$

Similarly, setting $y = 0$ and then $x = 0$ shows that the traces in the xz- and yz-plane are also ellipses. These three traces are shown in Figure 8.3.

Using equations (4a) through (4c), we find that the traces in planes that are parallel to the coordinate planes and that intersect the ellipsoid are again ellipses.

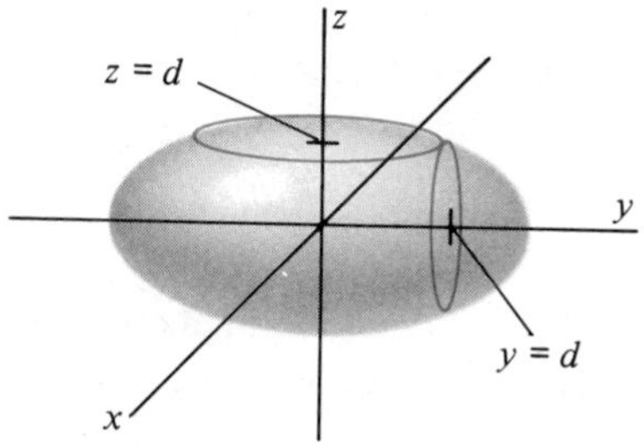

Figure 8.4 Traces of the ellipsoid in the planes $y = d$ and $z = d$.

For example, setting $z = d$ gives the trace

$$\frac{x^2}{a^2} + \frac{y^2}{b^2} = 1 - \frac{d^2}{c^2}. \tag{5}$$

The graph of equation (5) is an ellipse if $0 \le |d| < |c|$, it is a pair of points $(0, 0, \pm d)$ if $|d| = |c|$, and the equation has no solutions if $|d| > |c|$ (Figure 8.4). Figure 8.5 shows traces of the ellipsoid for various planes determined by equations (4a) through (4c).

The ellipsoid is symmetric with respect to each of the coordinate planes, since replacing x by $-x$, y by $-y$, or z by $-z$ does not change the equation. If $a = b = c$, the ellipsoid is, of course, a sphere.

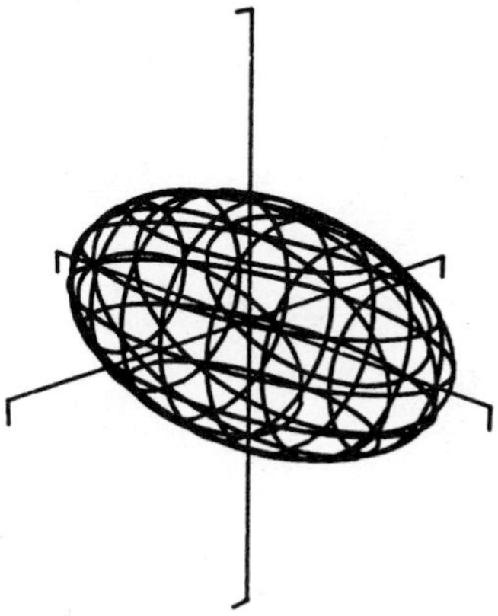

Figure 8.5 Various traces of an ellipsoid.

The Elliptic Paraboloid: $z = \dfrac{x^2}{a^2} + \dfrac{y^2}{b^2}$

Setting $x = 0$ shows that the trace of this figure in the yz-plane is the parabola $z = \dfrac{y^2}{b^2}$. Similarly, the trace in the xz-plane is the parabola $z = \dfrac{x^2}{a^2}$. The trace in the xy-plane ($z = 0$) is simply the origin $(0, 0)$ (Figure 8.6).

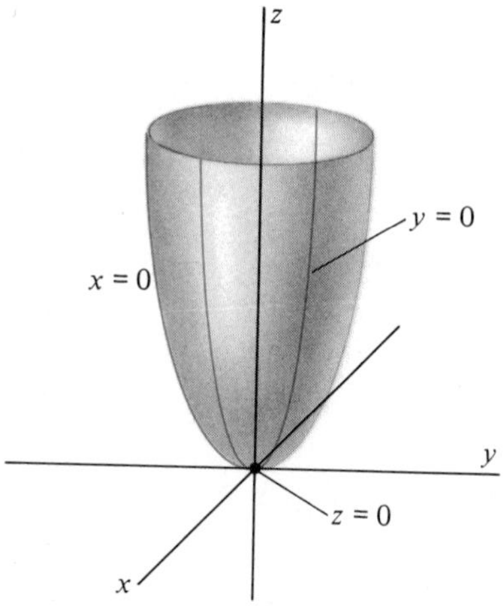

Figure 8.6 Traces of the elliptic paraboloid $z = \dfrac{x^2}{a^2} + \dfrac{y^2}{b^2}$ in the coordinate planes.

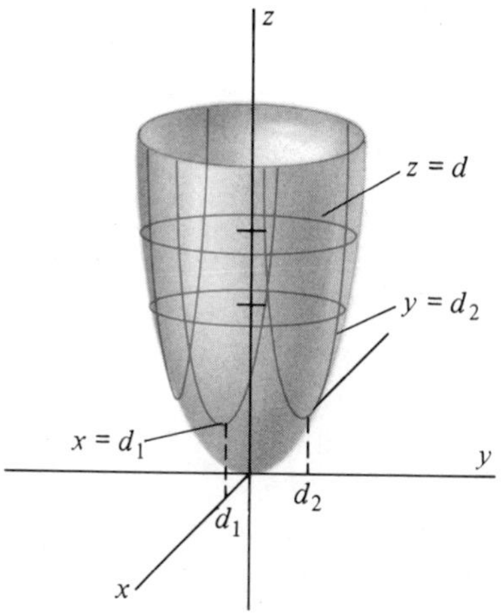

Figure 8.7 Traces of the elliptic paraboloid in planes $x = d_1$, $y = d_2$, $z = d$.

The traces in the planes $x = d_1$ and $y = d_2$ are again parabolas:

$$z = \frac{y^2}{b^2} + \frac{d_1^2}{a^2}; \qquad z = \frac{x^2}{a^2} + \frac{d_2^2}{b^2}.$$

However, traces in the plane $z = d$ are ellipses with equations

$$\frac{x^2}{a^2 d} + \frac{y^2}{b^2 d} = 1$$

if $d > 0$. If $d < 0$ there are no traces (Figure 8.7).

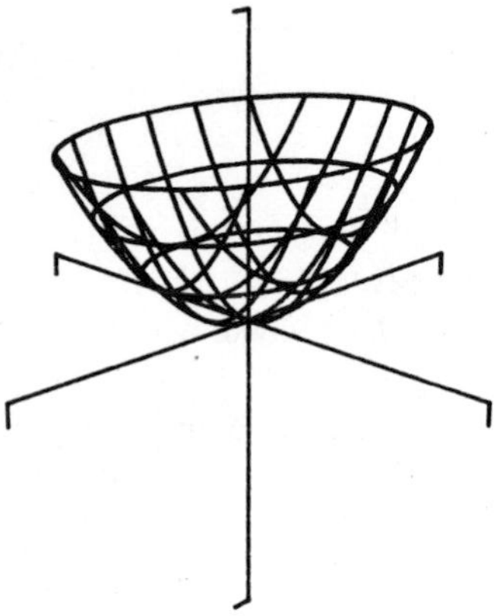

Figure 8.8 Various traces of an elliptic paraboloid.

The elliptic paraboloid is symmetric with respect to the xz- and yz-planes, as you can see by replacing x by $-x$, or y by $-y$ in its equation (Figure 8.8).

The Elliptic Cone: $z^2 = \dfrac{x^2}{a^2} + \dfrac{y^2}{b^2}$.

The difference between the equation of the elliptic cone and that of the elliptic paraboloid is the presence of z^2 rather than z. As a result, the traces in the xz-plane ($y = 0$) are the lines $z = \pm\dfrac{x}{a}$, and the traces in the yz-plane ($x = 0$) are the lines $z = \pm\dfrac{y}{b}$. The trace in the xy-plane ($z = 0$) is simply the origin (0, 0, 0) (Figure 8.9).

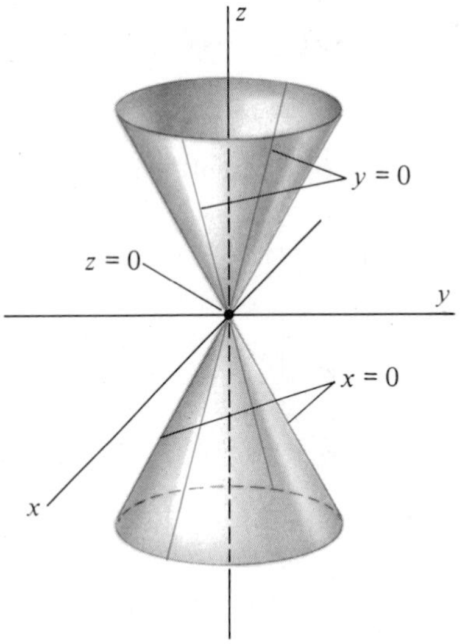

Figure 8.9 Traces of elliptic cone $z^2 = \dfrac{x^2}{a^2} + \dfrac{y^2}{b^2}$ in the coordinate planes.

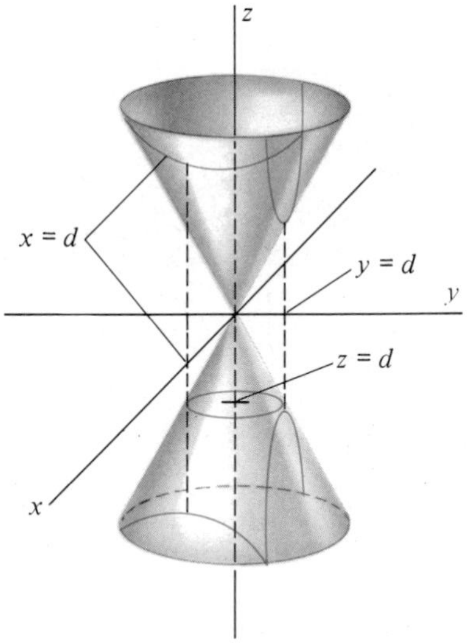

Figure 8.10 Traces of elliptic cone in planes $x = d$, $y = d$, and $z = d$.

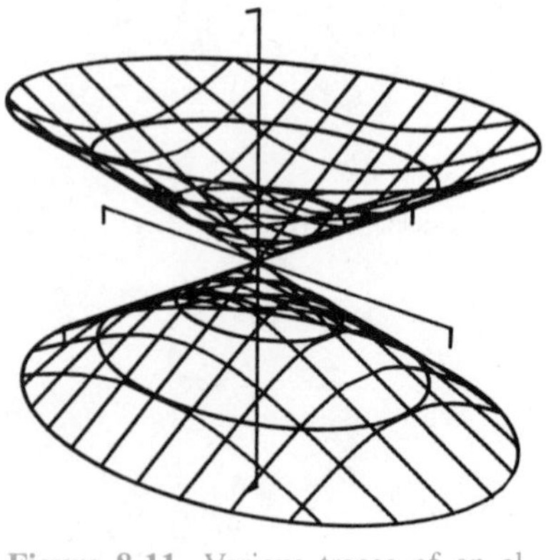

Figure 8.11 Various traces of an elliptic cone.

In the planes $x = d$, the traces are the hyperbolas

$$\frac{a^2z^2}{d^2} - \frac{a^2y^2}{b^2d^2} = 1.$$

Similarly, the traces in the planes $y = d$ are hyperbolas (Figure 8.10). However, in the planes $z = d$ and the traces are the ellipses with equations

$$\frac{x^2}{a^2d^2} + \frac{y^2}{b^2d^2} = 1.$$

The elliptic cone is symmetric with respect to each of the three coordinate planes. Also, it differs from the elliptic paraboloid in that it is unbounded in both the positive and negative z directions (Figure 8.11). If $a = b$, the cone is called a *right* or *circular* cone.

The Hyperboloid of One Sheet: $\dfrac{x^2}{a^2} + \dfrac{y^2}{b^2} - \dfrac{z^2}{c^2} = 1$

The trace in the xy-plane ($z = 0$) is the ellipse

$$\frac{x^2}{a^2} + \frac{y^2}{b^2} = 1,$$

while traces in the other two coordinate planes are the hyperbolas

$$\frac{x^2}{a^2} - \frac{z^2}{c^2} = 1 \quad \text{and} \quad \frac{y^2}{b^2} - \frac{z^2}{c^2} = 1$$

(Figure 8.12). Similarly, setting $z = d$ shows that traces parallel to the xy-plane are ellipses, while the traces in planes parallel to the xz- or yz-planes are hyperbolas. The hyperboloid of one sheet is symmetric with respect to each of the coordinate axes (Figures 8.13 and 8.14).

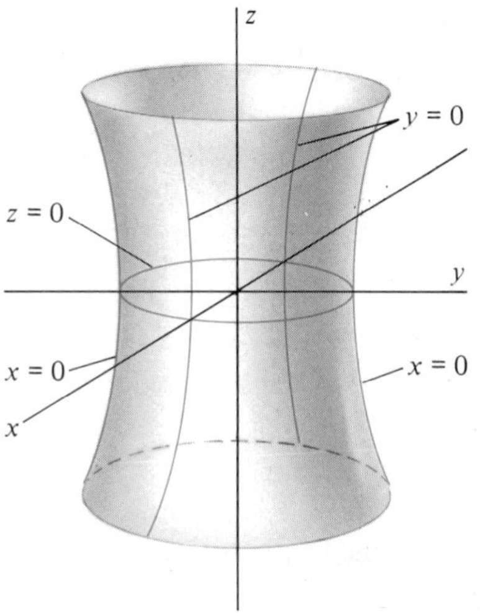

Figure 8.12 Traces of the hyperboloid of one sheet $\dfrac{x^2}{a^2} + \dfrac{y^2}{b^2} - \dfrac{z^2}{c^2} = 1$ in the coordinate planes.

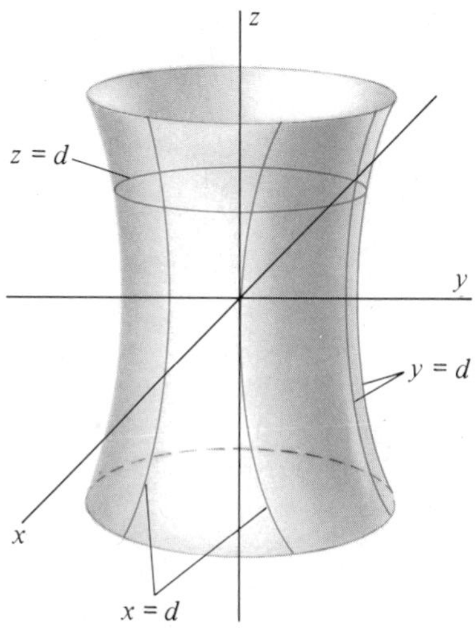

Figure 8.13 Traces of the hyperboloid of one sheet in the planes $x = d$, $y = d$, and $z = d$.

The Hyperboloid of Two Sheets: $\dfrac{x^2}{a^2} + \dfrac{y^2}{b^2} - \dfrac{z^2}{c^2} = -1$

Setting $z = 0$ shows that this surface has no trace in the xy-plane. The traces in the xz- and yz-planes are the hyperbolas

$$\frac{z^2}{c^2} - \frac{x^2}{a^2} = 1 \quad \text{and} \quad \frac{z^2}{c^2} - \frac{y^2}{b^2} = 1,$$

respectively. The traces in planes parallel to the xz- and yz-planes are also hyperbolas. However, traces in planes parallel to the xy-plane ($z = d$) are ellipses

$$\frac{x^2}{a^2} + \frac{y^2}{b^2} = \frac{d^2}{c^2} - 1$$

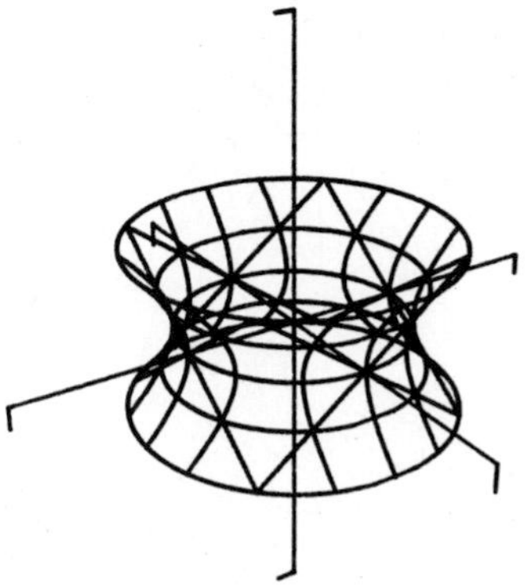

Figure 8.14 Various traces of a hyperboloid of one sheet.

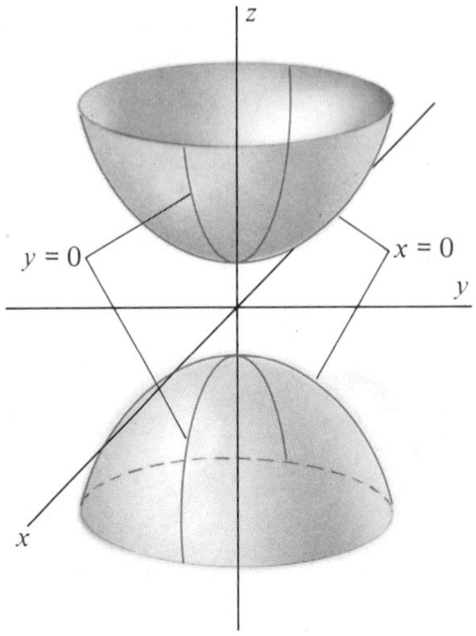

Figure 8.15 Traces of the hyperboloid of two sheets $\frac{x^2}{a^2} + \frac{y^2}{b^2} - \frac{z^2}{c^2} = -1$.

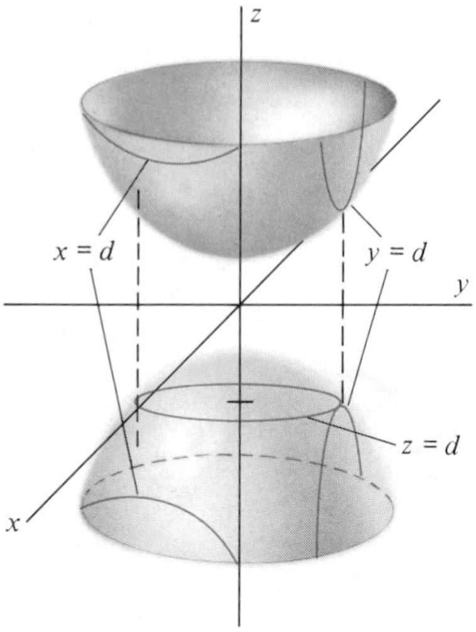

Figure 8.16 Traces of the hyperboloid of two sheets in planes parallel to the coordinate planes.

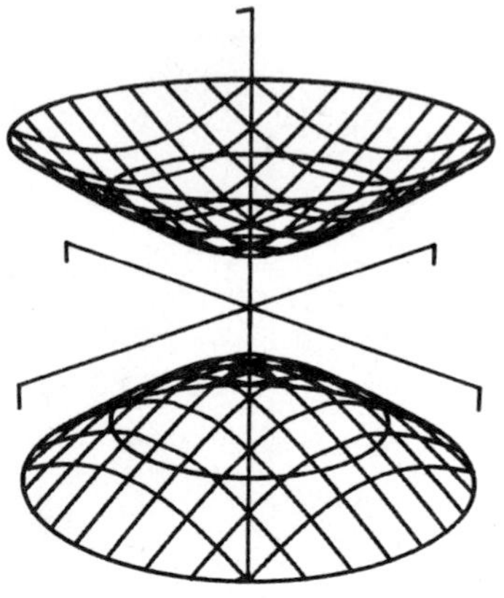

Figure 8.17 Traces of a hyperboloid of two sheets.

provided $|d| > |c|$. The hyperboloid of two sheets is symmetric with respect to each of the coordinate planes (Figures 8.15–8.17).

The Hyperbolic Paraboloid: $z = \dfrac{y^2}{b^2} - \dfrac{x^2}{a^2}$

Setting $z = 0$ shows that the trace in the xy-plane is the pair of intersecting straight lines $y = \pm \left|\dfrac{b}{a}\right| x$. In the xz-plane ($y = 0$) the trace is the parabola $z = -\dfrac{x^2}{b^2}$ (Figure 8.18). The traces in planes parallel to the xz- and yz-planes are also parabolas. However, the traces in planes parallel to the xy-plane ($z = d$) are hyperbolas. The hyperbolic paraboloid is symmetric with respect to the xz- and yz- planes (Figure 8.19).

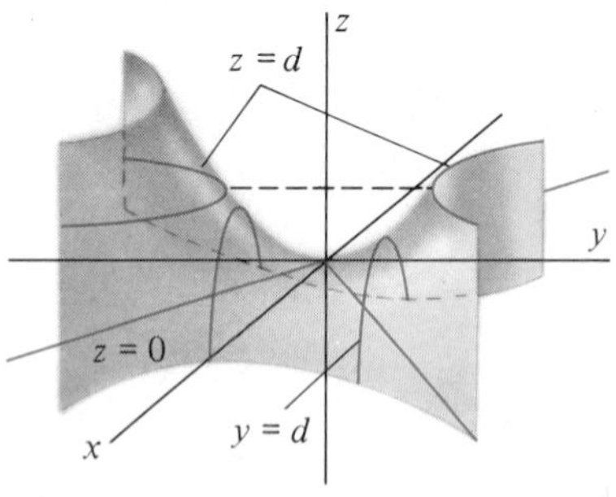

Figure 8.18 Traces of the hyperbolic paraboloid $z = \frac{y^2}{b^2} - \frac{x^2}{a^2}$.

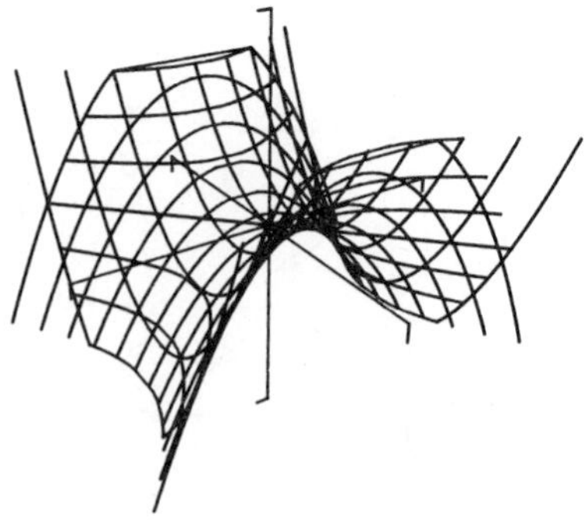

Figure 8.19 Various traces of a hyperbolic paraboloid.

The six figures described here do not exhaust all possibilities for the graph of the general second degree equation (1). We shall not attempt to present an all encompassing discussion as we were able to do for the two variable case in Chapter 15.

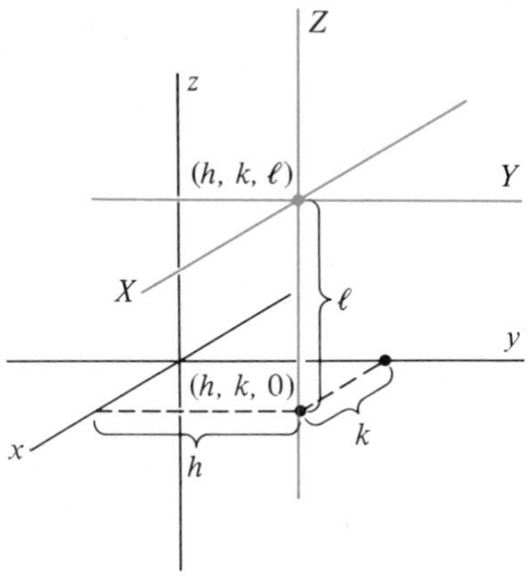

Figure 8.20 Translation of axes $X = x - h$, $Y = y - k$, $Z = z - \ell$.

However, we do wish to point out that the technique of completing the square may often be applied to reduce a given second degree equation to one of the six types discussed above. In problems of this type it is often useful to use the **translation of axes** defined by the following equations

$$X = x - h, \tag{6a}$$

$$Y = y - k, \tag{6b}$$

$$Z = z - \ell. \tag{6c}$$

As in the two variable case, these substitutions amount to a relocation of the coordinate axes so that the origin lies at (h, k, ℓ) and so that each of the coordinate axes lies parallel to its original position (see Figure 8.20).

Example 1 Describe the graph of the equation

$$x^2 + y^2 + 3z^2 - 2x + 4y - 4 = 0.$$

Strategy

Complete the square in x and in y.

Solution

$$x^2 - 2x + y^2 + 4y + 3z^2 - 4 = 0$$

$$(x^2 - 2x + 1) + (y^2 + 4y + 4) + 3z^2 - 4 - 1 - 4 = 0$$

$$(x - 1)^2 + (y + 2)^2 + 3z^2 - 9 = 0$$

Use translated variables (6a) through (6c) to simplify equation.

The given equation therefore has the form

$$X^2 + Y^2 + 3Z = 9,$$

or

Identify the form of the equation obtained.

$$\frac{X^2}{3^2} + \frac{Y^2}{3^2} + \frac{Z^2}{(\sqrt{3})^2} = 1$$

with $X = x - 1$, $Y = y + 2$, and $Z = z$. The graph of this equation is an ellipsoid with center at $(1, -2, 0)$ (Figure 8.21). ■

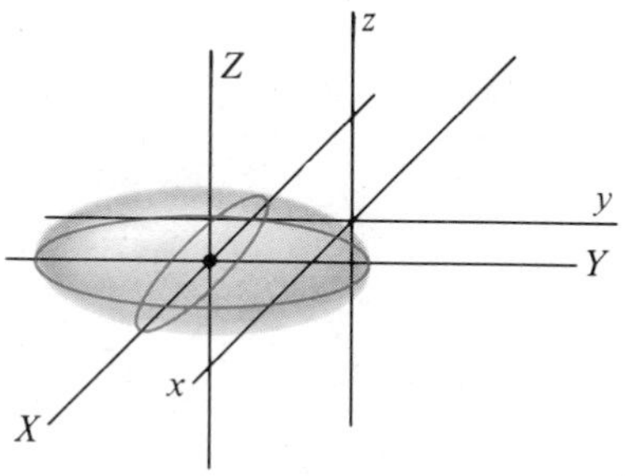

Figure 8.21 Ellipsoid $x^2 + y^2 + 3z^2 - 2x + 4y - 4 = 0$.

Cylinders

When an equation involves fewer variables than the number of axes on which it is graphed, the traces obtained by selecting various values of the missing variable must all be the same. The simplest example of this occurs when the equation $x = a$ is graphed in the xy-coordinate plane. Any trace obtained by setting $y = d$ is simply a point whose x-coordinate is a (Figure 8.22).

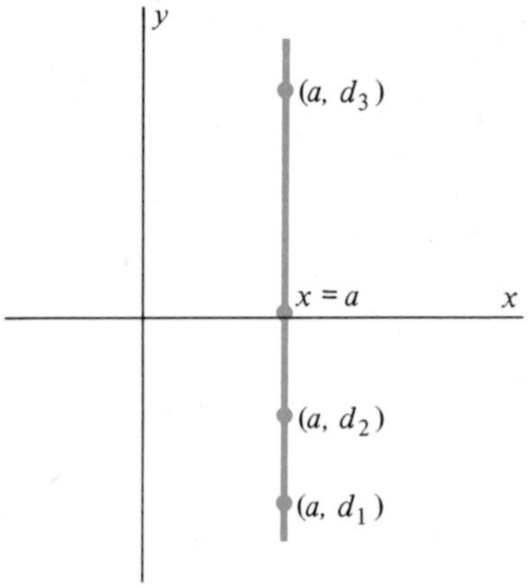

Figure 8.22 Graph of equation $x = a$ is a *cylinder* in the xy-plane.

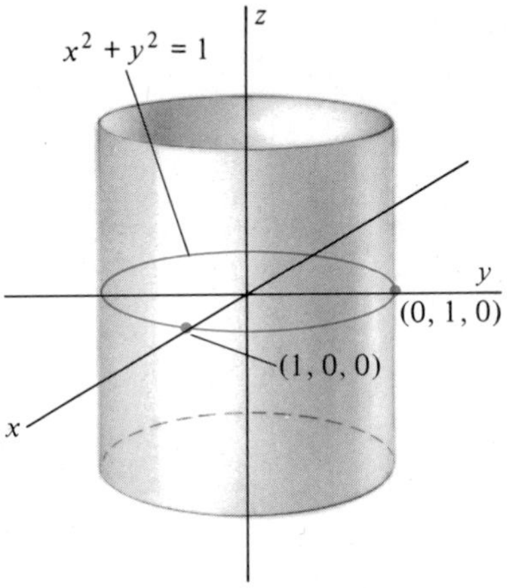

Figure 8.23 The equation $x^2 + y^2 = 1$ is a cylinder in xyz space.

A similar phenomenon occurs when an equation involving only two variables, say x and y, is graphed in xyz-space. Any trace corresponding to a fixed value of the missing variable, $z = d$, will be a "copy" of the graph of the equation in the xy-plane. The graph in xyz-space will therefore be a surface which can be thought of as an "infinite stack" of copies of the two dimensional figure. Graphs in xyz-space of equations with one or two variables missing are therefore called **cylinders.**

A typical cylinder is the graph of $x^2 + y^2 = 1$ in xyz-space (Figure 8.23). This is a special case of the **elliptic cylinder**

$$\frac{x^2}{a^2} + \frac{y^2}{b^2} = 1$$

in space. Three other examples of cylinders are

(a) a plane whose equation contains only two variables, such as $x + y = 2$ (Figure 8.24);

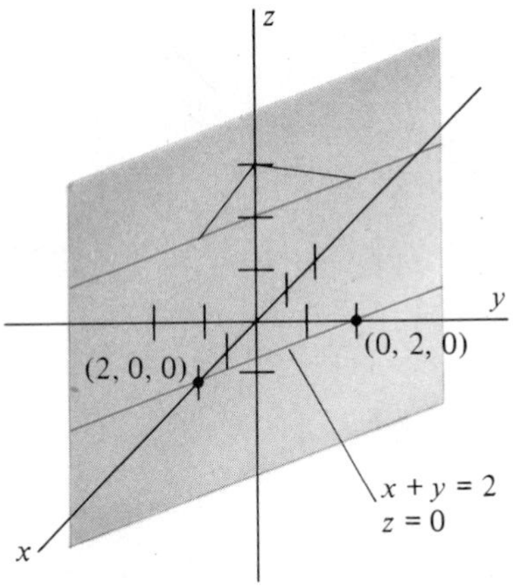

Figure 8.24 The plane $x + y = 2$ is a cylinder.

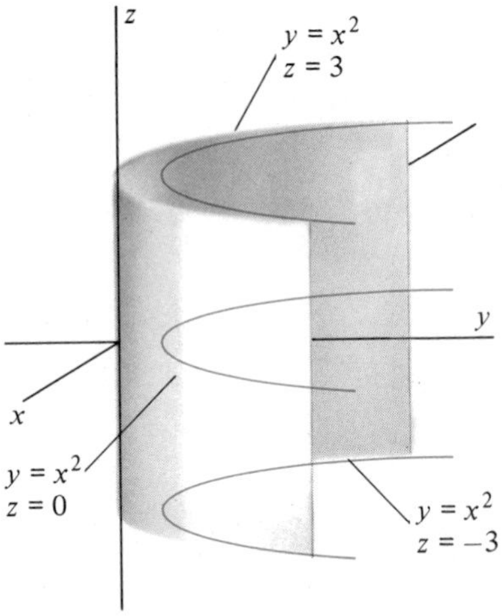

Figure 8.25 Parabolic cylinder $y = x^2$.

(b) the parabolic cylinder $y = x^2$ (Figure 8.25);

(c) the hyperbolic cylinder $\dfrac{y^2}{a^2} - \dfrac{x^2}{b^2} = 1$ (Figure 8.26).

Example 2 Sketch the cylinder $9x^2 + 4z^2 - 18x - 16z = 11$.

Solution: The first task is to complete the square in x and z:

$$\begin{aligned}
&9x^2 - 18x + 4z^2 - 16z = 11\\
&9(x^2 - 2x) + 4(z^2 - 4z) = 11\\
&9(x^2 - 2x + 1) + 4(z^2 - 4z + 4) = 11 + 9\cdot 1 + 4\cdot 4\\
&9(x-1)^2 + 4(z-2)^2 = 36\\
&\frac{(x-1)^2}{2^2} + \frac{(z-2)^2}{3^2} = 1.
\end{aligned}$$

This is the equation for an ellipse with center (1, 2) in the xz-plane. Since the variable y is missing, the figure is an elliptic cylinder with central axis parallel to the y-axis (Figure 8.27). ■

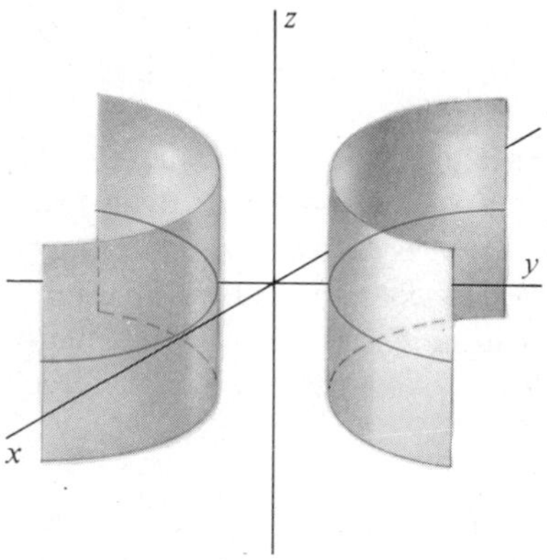

Figure 8.26 The hyperbolic cylinder $\dfrac{y^2}{a^2} - \dfrac{x^2}{b^2} = 1$.

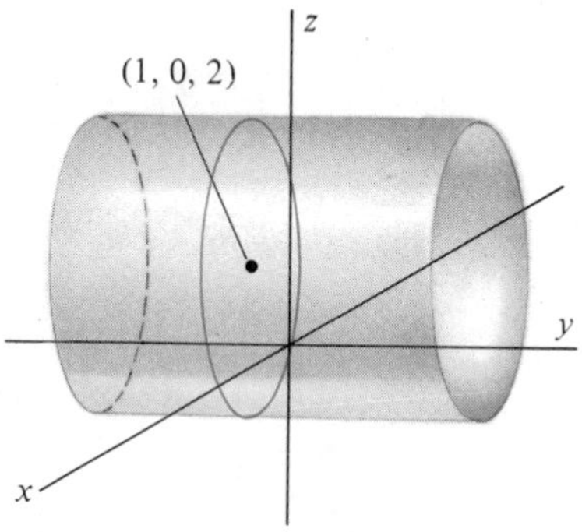

Figure 8.27 The elliptic cylinder $9x^2 - 18x + 4z^2 - 16z = 11$.

Exercise Set 17.8

In Exercises 1–20, describe the quadric surface and draw a rough sketch of its graph.

1. $x^2 + y^2 + z^2 = 10$
2. $\dfrac{x^2}{4} + \dfrac{y^2}{9} + \dfrac{z^2}{16} = 1$
3. $\dfrac{x^2}{4} + \dfrac{y^2}{9} = z^2$
4. $6x^2 + 12y^2 - 8z^2 - 24 = 0$
5. $4x^2 + 9y^2 - 36z^2 + 36 = 0$
6. $9y^2 - 6x^2 = 54$
7. $2x^2 - 3y^2 + 8x + 6y - 6z + 5 = 0$
8. $8x^2 + 4y^2 - 2z^2 + 24y - 4z + 44 = 0$
9. $x^2 + 2y^2 - 2z^2 + 6x - 4y + 7 = 0$
10. $x^2 + 4y^2 - 4z^2 + 6x - 8y - 8z + 9 = 0$
11. $6x^2 + 9y^2 + 36y - 54z + 36 = 0$
12. $4x^2 + 3y^2 + 3z^2 - 12y + 18z + 27 = 0$
13. $9z^2 + 4y^2 - 36x = 0$

14. $x^2 - 4y^2 + 4z^2 = 0$

15. $z^2 + y^2 - x^2 - 1 = 0$

16. $6x^2 - 12y^2 + 8z^2 + 24 = 0$

17. $x^2 + y^2 + z^2 - 6x + 2y + 6z + 18 = 0$

18. $9x^2 + 4y^2 + 9z^2 - 36x + 8y - 18z + 13 = 0$

19. $x^2 + y^2 + 4x - 2y - 36z + 175 = 0$

20. $y^2 - x^2 - 6x - 2y - 4z - 16 = 0$

In Exercises 21–36, sketch the given cylinder in *xyz*-space.

21. $x = 3$

22. $y = 6$

23. $x + y = 6$

24. $x^2 + y^2 = 5$

25. $\dfrac{x^2}{4} - \dfrac{y^2}{9} = 1$

26. $\dfrac{x^2}{2} + \dfrac{y^2}{4} = 1$

27. $xy = 1$

28. $y = \sin x$

29. $y = 4z^2$

30. $x = \sqrt{9 - z^2}$

31. $x^2 - z^2 = 1$

32. $z^2 - y^2 = 1$

33. $z = 1 - y^2$

34. $(x - 1)^2 + (z + 3)^2 = 9$

35. $3x^2 + 12x - z + 16 = 0$

36. $9x^2 + 4z^2 + 18x - 16z - 11 = 0$

37. Write the equation for the surface obtained by revolving the graph of $y = 2x$ about the y-axis.

38. Describe the set of all points P for which the distance from P to the y-axis is twice the distance from P to the xz-plane.

39. Find an equation for the solid obtained by revolving the graph of $y = z^2$ about the y-axis.

17.9 CYLINDRICAL AND SPHERICAL COORDINATES

We have seen that certain equations involving two variables are more appropriately graphed using polar coordinates than rectangular (Cartesian) coordinates for the plane. Similarly, certain surfaces in space are more easily described using *cylindrical* or *spherical* coordinates than by use of rectangular coordinates. In fact, some surfaces that are easily described in one of these coordinate systems are almost impossible to determine in rectangular coordinates.

Cylindrical Coordinates

In this coordinate system, the xy-plane is coordinatized by polar coordinates, and the third coordinate denotes the usual rectangular z-coordinate. That is, if the point P has cylindrical coordinates $P = (r, \theta, z)$ and rectangular coordinates $P = (x, y, z)$, we have the equations

$$x = r\cos\theta, \qquad r \geq 0 \tag{1a}$$
$$y = r\sin\theta, \qquad r \geq 0 \tag{1b}$$
$$z = z \tag{1c}$$

giving the rectangular coordinates in terms of the cylindrical coordinates, and the equations

$$r = \sqrt{x^2 + y^2} \tag{2a}$$
$$\tan\theta = \frac{y}{x} \tag{2b}$$
$$z = z \tag{2c}$$

for the cylindrical coordinates in terms of the rectangular coordinates (Figure 9.1). (Note that we now require $r \geq 0$, unlike the case for polar coordinates in the plane.)

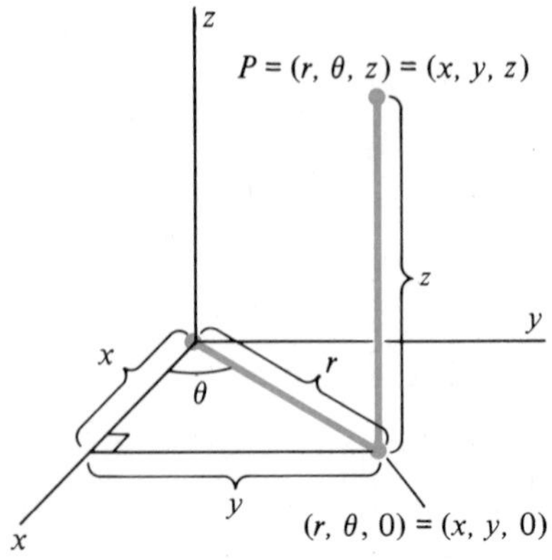

Figure 9.1 The cylindrical coordinate system.

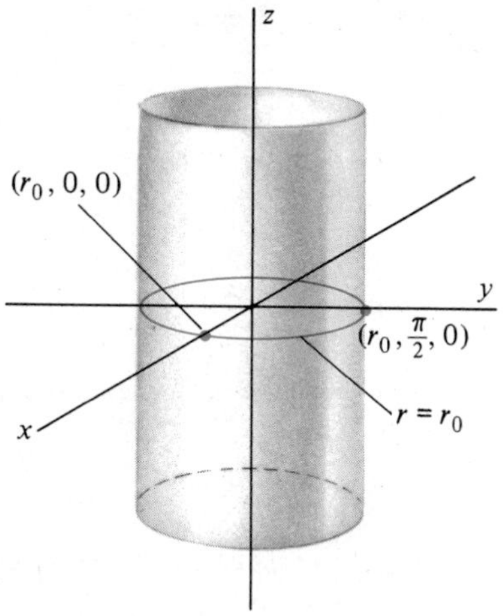

Figure 9.2 Graph of $r = r_0$ is a circular cylinder.

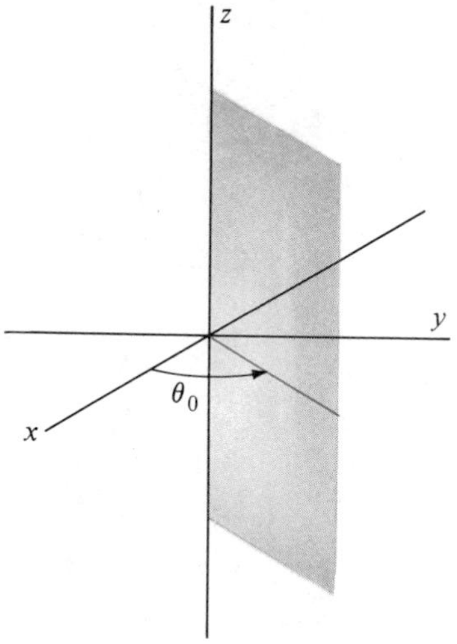

Figure 9.3 Graph of $\theta = \theta_0$ is a half plane.

Figure 9.2 shows why cylindrical coordinates are named as they are. The graph of the equation $r = r_0$, r_0 constant, is a cylinder in space. The graph of $\theta = \theta_0$, θ_0 constant, is a half plane, as in Figure 9.3. The graph of $z = z_0$, z_0 constant, is a horizontal plane, as in rectangular coordinates. Several points, plotted in both rectangular and cylindrical coordinates, appear in Figure 9.4.

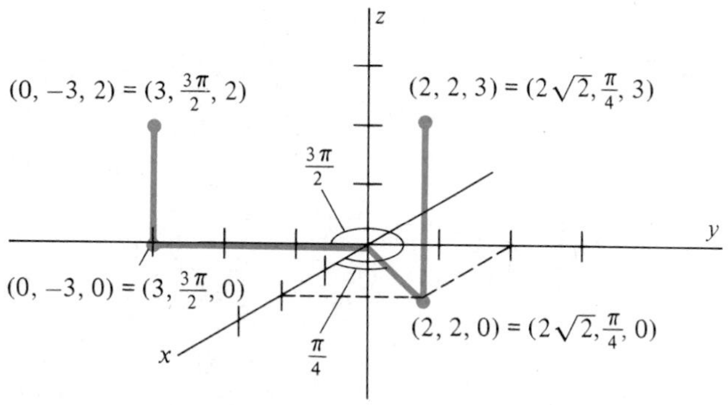

Figure 9.4 Four points in polar coordinates.

Example 1 Find cylindrical coordinates for the point with rectangular coordinates $(1, \sqrt{3}, 4)$.

Solution: Here $x = 1$, $y = \sqrt{3}$, $z = 4$. By equations (2a) through (2c),

$$r = \sqrt{1^2 + (\sqrt{3})^2} = 2,$$

$$\tan\theta = \sqrt{3} \Leftrightarrow \theta = \pi/3 \quad \text{(since } x \text{ and } y \text{ place the point in the first quadrant)},$$

$$z = 4.$$

The cylindrical coordinates are $(2, \pi/3, 4)$. ■

Example 2 Express the equation $z^2 = x^2 + y^2$ in cylindrical coordinates.

Solution: Using (1a) and (1b) the equation becomes

$$\begin{aligned} z^2 &= (r \cos \theta)^2 + (r \sin \theta)^2 \\ &= r^2 \cdot (\cos^2 \theta + \sin^2 \theta) = r^2. \end{aligned}$$

The equation therefore becomes

$$z = \pm r.$$

The graph of this equation is the cone in Figure 9.5. ■

Example 3 Graph the equation $r^2 = a^2 \sin \theta$ in cylindrical coordinates.

Solution: Since the variable z is missing, the graph is a cylinder in the z direction whose trace in the xy-plane is the graph of the lemniscate $r^2 = a^2 \sin \theta$ (see Figure 9.6). ■

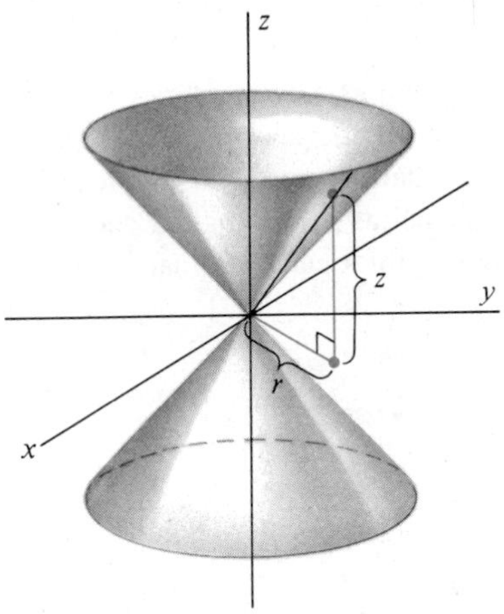

Figure 9.5 Graph of $z = \pm r$ is a cone.

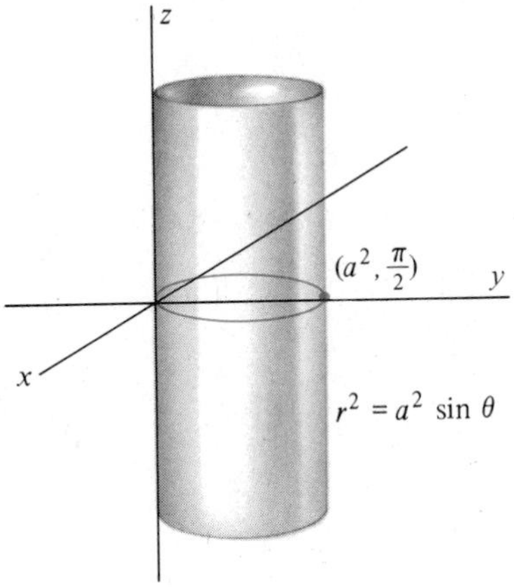

Figure 9.6 Graph of $r^2 = a^2 \sin \theta$ is a cylinder in cylindrical coordinates.

Spherical Coordinates

In spherical coordinates, a point is determined by the ordered triple $P = (\rho, \theta, \phi)$ where $\rho = |\overrightarrow{OP}|$ is the distance of the point P from the origin, θ is the polar angle associated with the vertical line through P, and ϕ is the (tilt) angle between the vector $\overrightarrow{OP}$ and the positive z-axis (Figure 9.7). By convention, we require $\rho \geq 0$, $0 \leq \theta < 2\pi$, and $0 \leq \phi \leq \pi$.

Figure 9.8 illustrates the reason for the terminology "spherical coordinates." The graph of the equation $\rho = \rho_0$, ρ_0 constant, is a sphere of radius ρ_0 centered at the origin. The graph of $\theta = \theta_0$, θ_0 constant, is a half plane, as in cylindrical coordinates. The graph of $\phi = \phi_0$, ϕ_0 constant, is a cone (Figure 9.9).

Figure 9.10 illustrates the relationships between the spherical coordinates (ρ, θ, ϕ) and the rectangular coordinates (x, y, z) for a point P. Since the vector $\overrightarrow{OP}$ is a diagonal of the rectangle $OQPR$,

$$|\overrightarrow{OQ}| = \rho \sin \phi, \qquad \text{and} \qquad z = \rho \cos \phi. \tag{3}$$

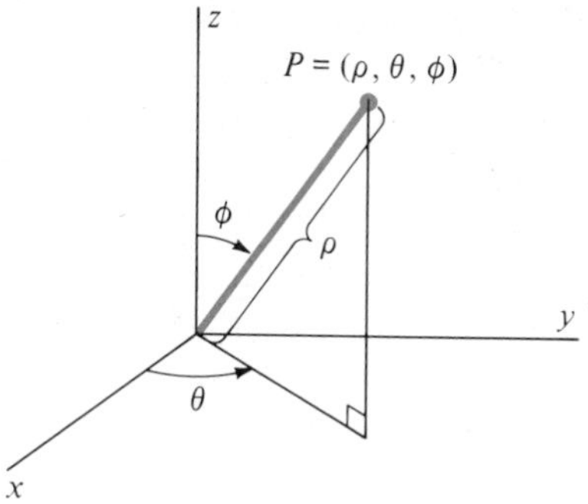

Figure 9.7 Spherical coordinates.

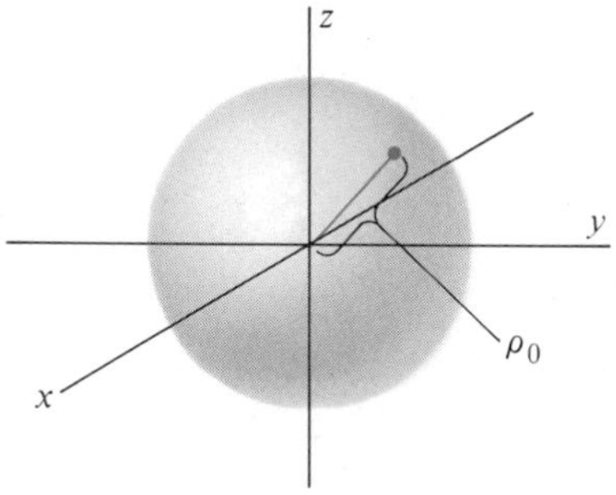

Figure 9.8 Graph of $\rho = \rho_0$ is a sphere.

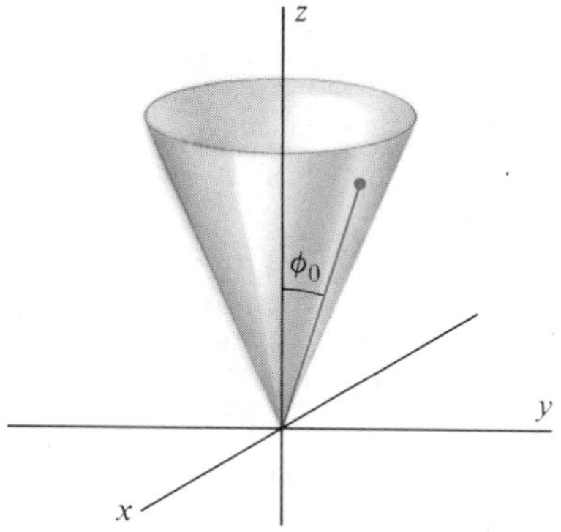

Figure 9.9 Graph of $\phi = \phi_0$ is a cone.

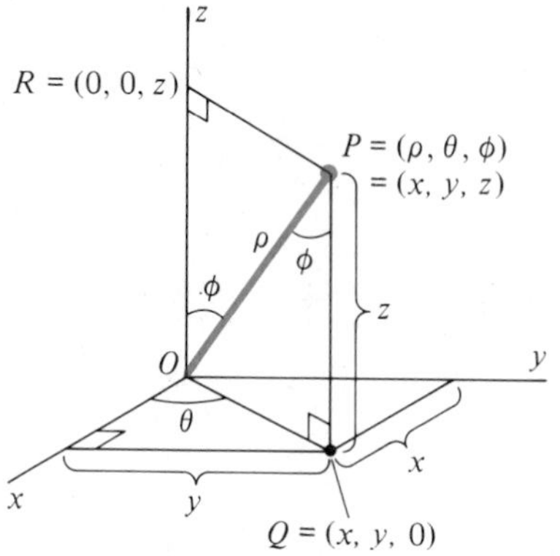

Figure 9.10 Relationships between rectangular and spherical coordinates.

Also,

$$x = |\overrightarrow{OQ}| \cos \theta, \qquad \text{and} \qquad y = |\overrightarrow{OQ}| \sin \theta. \tag{4}$$

Combining equations (3) and (4) gives

$$x = \rho \sin \phi \cos \theta, \qquad \rho \geq 0 \tag{5a}$$
$$y = \rho \sin \phi \sin \theta, \qquad \rho \geq 0 \tag{5b}$$
$$z = \rho \cos \phi, \qquad \rho \geq 0. \tag{5c}$$

Also, from the distance formula we find

$$\rho = \sqrt{x^2 + y^2 + z^2}. \tag{6}$$

Equations for θ and ϕ may be easily derived from equations (3) and (4).

Example 4 The point P has spherical coordinates $P = (3, \pi/3, \pi/4)$. Find rectangular coordinates for P.

Solution: Here $\rho = 3$, $\theta = \pi/3$, and $\phi = \pi/4$. By equations (5a) through (5c),

$$x = 3 \sin\left(\frac{\pi}{4}\right) \cdot \cos\left(\frac{\pi}{3}\right) = 3\left(\frac{\sqrt{2}}{2}\right)\left(\frac{1}{2}\right) = \frac{3\sqrt{2}}{4},$$

$$y = 3\sin\left(\frac{\pi}{4}\right)\cdot\sin\left(\frac{\pi}{3}\right) = 3\left(\frac{\sqrt{2}}{2}\right)\left(\frac{\sqrt{3}}{2}\right) = \frac{3\sqrt{6}}{4},$$

$$z = 3\cos\left(\frac{\pi}{4}\right) = \frac{3\sqrt{2}}{2}.$$

The rectangular coordinates are $P = \left(\frac{3\sqrt{2}}{4}, \frac{3\sqrt{6}}{4}, \frac{3\sqrt{2}}{2}\right)$. ■

Example 5 Express the equation $x^2 - y^2 + z^2 = 4$ in spherical coordinates.

Solution: Using equations (5a) through (5c) the equation becomes

$$(\rho \sin\phi \cos\theta)^2 - (\rho \sin\phi \sin\theta)^2 + (\rho \cos\phi)^2 = 4,$$

$$\rho^2 \sin^2\phi[\cos^2\theta - \sin^2\theta] + \rho^2\cos^2\phi = 4,$$

$$\rho^2 \sin^2\phi[1 - 2\sin^2\theta] + \rho^2\cos^2\phi = 4,$$

$$\rho^2 - 2\rho^2\sin^2\phi\sin^2\theta = 4.$$ ■

Exercise Set 17.9

1. The following points are given in rectangular coordinates. Find their cylindrical coordinates.
a. $(1, 1, 0)$ **b.** $(\sqrt{3}, 1, 3)$
c. $(-1, 1, -2)$ **d.** $(-1, \sqrt{3}, 4)$
e. $(0, 3, -5)$ **f.** $(-\sqrt{2}, \sqrt{2}, \sqrt{2})$

2. The following points are given in cylindrical coordinates. Find their rectangular coordinates.
a. $(2, \pi/4, -3)$ **b.** $(1, \pi/6, 4)$
c. $(4, \pi/3, -5)$ **d.** $(2, 4\pi/3, 2)$
e. $(1, \pi, 1)$ **f.** $(5, 5\pi/3, 5)$

3. The following points are given in rectangular coordinates. Find their spherical coordinates.
a. $(1, 0, 0)$ **b.** $(1, 1, \sqrt{2})$
c. $(1, -1, \sqrt{2})$ **d.** $(1, \sqrt{3}, 2)$
e. $(-\sqrt{3}, 1, -2)$ **f.** $(-2, 2, 2\sqrt{2})$

4. The following points are given in spherical coordinates. Find their rectangular coordinates.
a. $(2, \pi/4, \pi/3)$ **b.** $(1, \pi/2, \pi)$
c. $(2, 3\pi/4, \pi/4)$ **d.** $(3, 3\pi/2, 2\pi/3)$
e. $(5, \pi/6, 5\pi/6)$ **f.** $(2, 5\pi/3, 3\pi/4)$

5. The following points are given in cylindrical coordinates. Find their spherical coordinates.
a. $(1, 0, 0)$ **b.** $(\sqrt{2}, -\pi/4, \sqrt{2})$
c. $(2, \pi/3, 2)$ **d.** $(2, \pi/4, 2)$
e. $(2, 5\pi/3, 0)$ **f.** $(2, \pi/6, 2)$

6. The following points are given in spherical coordinates. Find their cylindrical coordinates.
a. $(2, \pi/4, \pi/2)$ **b.** $(2, \pi/2, \pi/4)$
c. $(3, 2\pi/3, \pi/2)$ **d.** $(1, \pi/2, 2\pi/3)$
e. $(4, \pi/4, 0)$ **f.** $(2, \pi/3, \pi/2)$

In Exercises 7–14, an equation in cylindrical coordinates is given. Write the equation in rectangular coordinates and sketch the graph.

7. $z = 3$ **8.** $r = 2$
9. $z = 2r$ **10.** $z = r\sin\theta$
11. $r^2 + z^2 = 4$ **12.** $z^2 = r^2\cos^2\theta - 1$
13. $\cos^2\theta - \sin^2\theta = \dfrac{a^2}{r^2}$ **14.** $z = r^2\cos^2\theta$

In Exercises 15–22, an equation in rectangular coordinates is given. Write the equation in cylindrical coordinates.

15. $x^2 + y^2 = 9$ **16.** $x^2 + y^2 + z^2 = 9$
17. $x^2 + y^2 = 9z$ **18.** $y^2 + z^2 = 1$
19. $x^2 + z^2 = 4$ **20.** $x + y + z = 4$
21. $xy - ax = 4$ **22.** $x^2 - y^2 = 4$

In Exercises 23–30, write the equation in the exercise referred to in spherical coordinates.

23. Exercise 15 **24.** Exercise 16
25. Exercise 17 **26.** Exercise 18
27. Exercise 19 **28.** Exercise 20
29. Exercise 21 **30.** Exercise 22

SUMMARY OUTLINE OF CHAPTER 17

- A **vector** in the plane is an ordered pair of numbers $v = \langle a, b\rangle$ subject to the laws

$$\langle a_1, b_1\rangle + \langle a_2, b_2\rangle = \langle a_1 + a_2, b_1 + b_2\rangle,$$

$$c\langle a, b\rangle = \langle ca, cb\rangle.$$

- A vector in space is an ordered triple $v = \langle a, b, c\rangle$ subject to the laws

$$\langle a_1, b_1, c_1\rangle + \langle a_2, b_2, c_2\rangle = \langle a_1 + a_2, b_1 + b_2, c_1 + c_2\rangle,$$

$$d\langle a, b, c\rangle = \langle da, db, dc\rangle.$$

- The **unit coordinate vectors** are $\boldsymbol{i} = \langle 1, 0, 0\rangle, \boldsymbol{j} = \langle 0, 1, 0\rangle, \boldsymbol{k} = \langle 0, 0, 1\rangle$.
- ***Notation:*** $\boldsymbol{v} = \langle a, b, c\rangle = a\boldsymbol{i} + b\boldsymbol{j} + c\boldsymbol{k}$.
- Vectors, in the plane or in space, satisfy the following properties:

(i) $\boldsymbol{v} + \boldsymbol{w} = \boldsymbol{w} + \boldsymbol{v}$
(ii) $(\boldsymbol{u} + \boldsymbol{v}) + \boldsymbol{w} = \boldsymbol{u} + (\boldsymbol{v} + \boldsymbol{w})$
(iii) $\boldsymbol{0} + \boldsymbol{v} = \boldsymbol{v}$
(iv) $1 \cdot \boldsymbol{v} = \boldsymbol{v}$
(v) $a(\boldsymbol{v} + \boldsymbol{w}) = a\boldsymbol{v} + a\boldsymbol{w}$
(vi) $(a + b)\boldsymbol{v} = a\boldsymbol{v} + b\boldsymbol{v}$.

- The **length** of the vector $\boldsymbol{v} = a\boldsymbol{i} + b\boldsymbol{j} + c\boldsymbol{k}$ is $|\boldsymbol{v}| = \sqrt{a^2 + b^2 + c^2}$.
- Vector length has the following properties:

(i) $|\boldsymbol{v}| \geq 0$; $|\boldsymbol{v}| = 0$ if and only if $\boldsymbol{v} = \boldsymbol{0}$,
(ii) $|c\boldsymbol{v}| = |c||\boldsymbol{v}|$,
(iii) $|\boldsymbol{v} + \boldsymbol{w}| \leq |\boldsymbol{v}| + |\boldsymbol{w}|$.

- A **unit vector** in the direction of $\boldsymbol{v}$ is $\boldsymbol{u} = \dfrac{\boldsymbol{v}}{|\boldsymbol{v}|}$.
- The **dot product** of the vectors $\boldsymbol{v} = x_1\boldsymbol{i} + y_1\boldsymbol{j} + z_1\boldsymbol{k}$ and $\boldsymbol{w} = x_2\boldsymbol{i} + y_2\boldsymbol{j} + z_2\boldsymbol{k}$ is the number

$$\boldsymbol{v} \cdot \boldsymbol{w} = x_1x_2 + y_1y_2 + z_1z_2.$$

- ***Theorem:*** $\boldsymbol{v} \cdot \boldsymbol{w} = |\boldsymbol{v}||\boldsymbol{w}| \cos\theta$, where θ is the angle between $\boldsymbol{v}$ and $\boldsymbol{w}$.
- Properties of the dot product:

(i) $\boldsymbol{v} \cdot \boldsymbol{v} = |\boldsymbol{v}|^2$
(ii) $\boldsymbol{v} \cdot \boldsymbol{w} = \boldsymbol{w} \cdot \boldsymbol{v}$
(iii) $\boldsymbol{u} \cdot (\boldsymbol{v} + \boldsymbol{w}) = \boldsymbol{u} \cdot \boldsymbol{v} + \boldsymbol{u} \cdot \boldsymbol{w}$
(iv) $(c\boldsymbol{v}) \cdot \boldsymbol{w} = c(\boldsymbol{v} \cdot \boldsymbol{w})$
(v) $\boldsymbol{0} \cdot \boldsymbol{v} = 0$
(vi) $|\boldsymbol{v} \cdot \boldsymbol{w}| \leq |\boldsymbol{v}||\boldsymbol{w}|$.

- The vectors $\boldsymbol{v}$ and $\boldsymbol{w}$ are **orthogonal** (perpendicular) if and only if $\boldsymbol{v} \cdot \boldsymbol{w} = 0$.
- The **component** of the vector $\boldsymbol{v}$ in the direction of the vector $\boldsymbol{w}$ is the **number**

$$\text{comp}_{\boldsymbol{w}}\, \boldsymbol{v} = \frac{\boldsymbol{v} \cdot \boldsymbol{w}}{|\boldsymbol{w}|}.$$

- The **projection** of the vector $\boldsymbol{v}$ along the vector $\boldsymbol{w}$ is the **vector**

$$\text{proj}_{\boldsymbol{w}}\, \boldsymbol{v} = \left(\frac{\boldsymbol{v} \cdot \boldsymbol{w}}{|\boldsymbol{w}|^2}\right)\boldsymbol{w}.$$

- The distance from the point $P = (x_0, y_0)$ in the plane to the line with equation $ax + by + c = 0$ is

$$d = \frac{|ax_0 + by_0 + c|}{\sqrt{a^2 + b^2}}.$$

- The distance d between two points $P = (x_1, y_1, z_1)$ and $Q = (x_2, y_2, z_2)$ in space is

$$d = \sqrt{(x_2 - x_1)^2 + (y_2 - y_1)^2 + (z_2 - z_1)^2}.$$

- The **cross product** of the vectors $\boldsymbol{v} = x_1\boldsymbol{i} + y_1\boldsymbol{j} + z_1\boldsymbol{k}$ and $\boldsymbol{w} = x_2\boldsymbol{i} + y_2\boldsymbol{j} + z_2\boldsymbol{k}$ in space is the **vector**

$$\boldsymbol{v} \times \boldsymbol{w} = \det\begin{bmatrix} \boldsymbol{i} & \boldsymbol{j} & \boldsymbol{k} \\ x_1 & y_1 & z_1 \\ x_2 & y_2 & z_2 \end{bmatrix}$$

$$= (y_1z_2 - z_1y_2)\boldsymbol{i} + (z_1x_2 - x_1z_2)\boldsymbol{j} + (x_1y_2 - y_1x_2)\boldsymbol{k}.$$

- Properties of the cross product:

(i) $\boldsymbol{v} \times \boldsymbol{w} = -(\boldsymbol{w} \times \boldsymbol{v})$
(ii) $\boldsymbol{u} \times (\boldsymbol{v} + \boldsymbol{w}) = \boldsymbol{u} \times \boldsymbol{v} + \boldsymbol{u} \times \boldsymbol{w}$
(iii) $c(\boldsymbol{v} \times \boldsymbol{w}) = (c\boldsymbol{v}) \times \boldsymbol{w} = \boldsymbol{v} \times (c\boldsymbol{w})$
(iv) $(\boldsymbol{v} \times \boldsymbol{w}) \perp \boldsymbol{v}$; $(\boldsymbol{v} \times \boldsymbol{w}) \perp \boldsymbol{w}$
(v) $\boldsymbol{v} \times \boldsymbol{v} = \boldsymbol{0}$.

- ***Theorem:*** $\boldsymbol{v} \times \boldsymbol{w} = (|\boldsymbol{v}||\boldsymbol{w}| \sin\theta)\boldsymbol{n}$, where $\boldsymbol{n}$ is a unit vector orthogonal to both $\boldsymbol{v}$ and $\boldsymbol{w}$ in the direction determined by the right-hand rule.
- $|\boldsymbol{v} \times \boldsymbol{w}|$ is the area of the parallelogram determined by $\boldsymbol{v}$ and $\boldsymbol{w}$.
- The **line** with position vector $\boldsymbol{a} = a_1\boldsymbol{i} + a_2\boldsymbol{j} + a_3\boldsymbol{k}$ and direction vector $\boldsymbol{b} = b_1\boldsymbol{i} + b_2\boldsymbol{j} + b_3\boldsymbol{k}$ has

(i) vector equation $\boldsymbol{r}(t) = \boldsymbol{a} + t\boldsymbol{b}$, $-\infty < t < \infty$.
(ii) parametric equations $x(t) = a_1 + tb_1$, $y(t) = a_2 + tb_2$, $z(t) = a_3 + tb_3$.
(iii) symmetric equations $\dfrac{x - a_1}{b_1} = \dfrac{y - a_2}{b_2} = \dfrac{z - a_3}{b_3}$.

- The **distance** d from the point P to the line with direction vector $\boldsymbol{b}$ containing the point Q is

$$d = \frac{|\overrightarrow{PQ} \times \boldsymbol{b}|}{|\boldsymbol{b}|}.$$

- The **plane** with normal vector $\boldsymbol{n}$ containing the point P is the set of all points Q for which

$$\overrightarrow{PQ} \cdot \boldsymbol{n} = 0.$$

- If $n = a\boldsymbol{i} + b\boldsymbol{j} + c\boldsymbol{k}$, $P = (x_0, y_0, z_0)$ and $Q = (x, y, z)$ the equation of the plane is

$$ax + by + cz = ax_0 + by_0 + cz_0.$$

- The cylindrical coordinates (r, θ, z) and the rectangular coordinates (x, y, z) for a point P are related via the equations

$$x = r\cos\theta, \quad r \geq 0, \qquad r = \sqrt{x^2 + y^2}$$
$$y = r\sin\theta, \quad r \geq 0, \quad \text{and} \quad \tan\theta = y/x,$$
$$z = z \qquad\qquad z = z$$

- The spherical coordinates (ρ, θ, ϕ) and the rectangular coordinates (x, y, z) for a point P are related via the equations

$$x = \rho\sin\phi\cos\theta, \quad \rho \geq 0 \qquad \rho = \sqrt{x^2 + y^2 + z^2}$$
$$y = \rho\sin\phi\sin\theta, \quad \rho \geq 0 \quad \text{and} \quad \tan\theta = y/x, \quad x \neq 0$$
$$z = \rho\cos\phi, \quad \rho \geq 0 \qquad \tan\phi = \frac{\sqrt{x^2 + y^2}}{z}, \quad z \neq 0.$$

REVIEW EXERCISES—CHAPTER 17

1. Find the distance between the points $P = (1, 2, -4)$ and $Q = (2, -5, 2)$.

2. Find an equation for the set of all points $P = (x, y, z)$ equidistant from the fixed points $P_1 = (1, -2, 1)$ and $Q = (3, 4, -3)$.

3. Describe the graph of the equation $x^2 + y^2 + z^2 - 6x + 4y - 2z + 10 = 0$.

4. Let $\boldsymbol{u} = \boldsymbol{i} + 2\boldsymbol{j} - 3\boldsymbol{k}$, $\boldsymbol{v} = 2\boldsymbol{i} + 2\boldsymbol{j} + 6\boldsymbol{k}$, $\boldsymbol{w} = \boldsymbol{i} - 4\boldsymbol{j} + 3\boldsymbol{k}$. Find
   a. $\boldsymbol{u} + 2\boldsymbol{v}$
   b. $\boldsymbol{u} - \boldsymbol{v}$
   c. $|3\boldsymbol{u} + \boldsymbol{v}|$
   d. $\boldsymbol{u} \cdot \boldsymbol{v}$
   e. $\boldsymbol{u} \cdot (\boldsymbol{v} \times \boldsymbol{w})$
   f. $|\boldsymbol{u} + \boldsymbol{v} - \boldsymbol{w}|$

5. Find the cosine of the angle between the vectors $\boldsymbol{v} = \boldsymbol{i} - 3\boldsymbol{j} + 2\boldsymbol{k}$ and $\boldsymbol{w} = 3\boldsymbol{i} + 3\boldsymbol{j} + 2\boldsymbol{k}$.

6. Determine whether the following pairs of lines intersect and, if so, at what point.
   a. $x = t$, $y = t + 2$, $z = 2t - 4$; $\quad x = 1 - t$, $y = 3 + t$, $z = 4t$
   b. $x = 1 + 2t$, $y = t - 2$, $z = 1 + t$; $\quad x = 2t + 1$, $y = 4 - 2t$, $z = 5 - t$

7. Find parametric equations for the line containing the point $(2, 1, -3)$ that is perpendicular to both of these lines.

   ℓ_1: $x = 2t$, $y = 3 + t$, $z = 5t$

   ℓ_2: $x = t + 5$, $y = 6 - 4t$, $z = 3t + 4$

8. Find the direction cosines for the vector $\boldsymbol{v} = 3\boldsymbol{i} + 4\boldsymbol{j} + 5\boldsymbol{k}$.

9. Find an equation for the plane containing the points $(-1, 2, 1)$, $(2, 5, 3)$, and $(-4, 0, 2)$.

10. Find an equation for the plane containing the line

    ℓ_1: $\boldsymbol{r}(t) = \boldsymbol{i} - 3\boldsymbol{j} + \boldsymbol{k} + t(4\boldsymbol{i} + 2\boldsymbol{j} - \boldsymbol{k})$

    and the point $(1, 2, 1)$.

11. Find a vector normal to the plane with equation $8x + y - 2z = 5$.

12. Find the distance from the point $P = (1, 2, -4)$ to the plane with equation $3x + 2y - 5z = 5$.

13. Find a vector equation for the line of intersection of the planes with equations $2x + 3y - z = 4$ and $x - 3y + 5z = 2$.

14. Show that if $\boldsymbol{v} \times \boldsymbol{w} = 0$ and $\boldsymbol{w} \neq 0$ then $\boldsymbol{v} = c\boldsymbol{w}$ for some constant c.

15. Show that $\boldsymbol{u} \times (\boldsymbol{v} \times \boldsymbol{w}) = (\boldsymbol{u} \cdot \boldsymbol{w})\boldsymbol{v} - (\boldsymbol{u} \cdot \boldsymbol{v})\boldsymbol{w}$.

16. Show that $(\boldsymbol{u} + \boldsymbol{v}) \times (\boldsymbol{u} - \boldsymbol{v}) = 2\boldsymbol{v} \times \boldsymbol{u}$.

17. Sketch the graph of the equation $y^2 = 9 + z^2$.

18. Sketch the graph of the equation $x^2 + z^2 = 1 + y$.

19. Find rectangular coordinates for the point with spherical coordinates $(2, \pi/4, \pi/3)$.

20. Find parametric equations for the line containing the points $(3, 5, -6)$ and $(-2, 3, -1)$.

21. Find an equation for the plane containing the point $(2, -5, 3)$ and perpendicular to the line with parametric equations $x = 4 + 2t$, $y = 3 - t$, $z = 5 + 6t$.

22. Find an equation for the sphere with center on the z-axis and containing the points $(5, 0, 0)$ and $(0, 0, 4)$.

23. For $\boldsymbol{v} = \boldsymbol{i} + 2\boldsymbol{j} + 4\boldsymbol{k}$ and $\boldsymbol{w} = 3\boldsymbol{j} + 4\boldsymbol{k}$, find
    a. $\text{comp}_{\boldsymbol{w}}\, \boldsymbol{v}$
    b. $\text{proj}_{\boldsymbol{w}}\, \boldsymbol{v}$

24. Find the distance from the point $(-4, 2, 5)$ to the line with equation $x = 3 + t$, $y = 4 - 2t$, $z = 5 + 5t$.

25. Find the area of the triangle with vertices $(1, 3, -2)$, $(1, 1, 1)$, and $(4, 0, 3)$.

26. Find a unit vector in the same direction as $\mathbf{v} = \mathbf{i} - 3\mathbf{j} + 4\mathbf{k}$.

27. Find an equation for the plane containing the points $(3, -2, 6)$ and $(4, -2, 2)$ that is perpendicular to the xz-plane.

28. Find cylindrical coordinates for the point with rectangular coordinates $(2, 2, 5)$.

29. Find symmetric equations for the line with parametric equations $x = 1 + 3t$, $y = 2 + 7t$, $z = 3 - t$.

In Exercises 30–33, find a rectangular equation for the given equation and sketch the graph.

30. $r = \cos\theta$

31. $\phi = \pi/4$

32. $\rho = 2\sec\phi$

33. $\rho\sin\phi = 2\cos\theta$

34. Find the vector of length 3 with direction cosines $\dfrac{\sqrt{2}}{2}$, 0, and $\dfrac{\sqrt{2}}{2}$.

35. For $\mathbf{v} = 2\mathbf{i} + 3\mathbf{j} - \mathbf{k}$ and $\mathbf{w} = \mathbf{i} + 3\mathbf{j} + \mathbf{k}$ find

a. $\text{comp}_{\mathbf{w}}\,\mathbf{v}$

b. $\text{proj}_{\mathbf{v}}\,\mathbf{w}$

36. Find the distance from the point $(1, 2, 1)$ to the plane containing the points $(1, 1, 1)$, $(2, -1, 5)$, and $(3, 1, -2)$.

37. Find the equation for the plane with intercepts $x = 3$, $y = 4$, $z = 5$.

38. Find an equation for the plane perpendicular to the line with symmetric equations

$$\frac{x-2}{3} = \frac{y+1}{2} = \frac{z-3}{-2}$$

and containing the point $(3, -2, 3)$.

39. Find parametric equations for the line parallel to the line of intersection of the two planes $x + y + 2z = 6$ and $x - y - 2z = 4$ and containing the point $(5, 2, -3)$.

In Exercises 40–46, identify and sketch the graph.

40. $9x^2 - 4y^2 + 36z^2 = -36$

41. $6x^2 + y^2 - 2z^2 = 6$

42. $36(x-1)^2 + 18(y+3)^2 + 8(z+1)^2 = 72$

43. $-x^2 + y^2 = 9z$

44. $y^2 = 4x^2 + 2z$

45. $9x^2 + 9y^2 - 4z^2 + 18x - 16z - 43 = 0$

46. $(x+1)(y-3) = 1$

47. Find all vectors of length 2 orthogonal to both $\mathbf{v} = 3\mathbf{i} + 2\mathbf{j} + \mathbf{k}$ and $\mathbf{w} = \mathbf{i} + \mathbf{j} - 2\mathbf{k}$.

48. Find the area of the parallelogram determined by the vectors $\mathbf{v} = 2\mathbf{i} + 3\mathbf{j} + \mathbf{k}$ and $\mathbf{w} = \mathbf{i} - \mathbf{j} + 2\mathbf{k}$.

49. Prove that the planes $a_1x + b_1y + c_1z = d_1$ and $a_2x + b_2y + c_2z = d_2$ are perpendicular if and only if $a_1a_2 + b_1b_2 + c_1c_2 = 0$.

50. Show that the points $(1, 1, -1)$, $(0, 1, -1/2)$, $(-1, 1, 0)$, and $(0, 0, 1/4)$ all lie in the same plane.

Josiah Willard Gibbs

George Gabriel Stokes

UNIT 7

CALCULUS IN HIGHER DIMENSIONS

Calculus in Higher Dimensions

A host of mathematicians contributed to the several topics included in this unit: partial differentiation, multiple integration, and vector analysis. Several of these people have been discussed in other units.

Newton differentiated functions of two variables by means of formulas that we now obtain by partial differentiation, a topic presented in Chapter 19. This work is recorded in his personal papers, but was not published. Leibniz also differentiated functions of two variables, but did not make much use of them. Jakob Bernoulli and his nephew Nicolaus used what were essentially partial derivatives around 1720, although they did not recognize that there is a difference between differentiation for a function of one variable and that for a function of two or more variables. The same symbol was used by many early writers for regular and partial derivatives, which led to much confusion. The "rounded d" symbol, ∂, was first used by Euler in 1776, but not in the way we use it today. The first to use $\partial y/\partial x$ was the Frenchman Adrien-Marie Legendre in 1786, but it was more than a century until that notation came into general use.

Multiple integration was introduced first by Newton, but his arguments were geometrical and somewhat unclear. In the first half of the eighteenth century Euler used repeated integrations in order to integrate over a bounded domain. Joseph Louis Lagrange (1736–1813) used a triple integral in a work on gravitation involving ellipsoids, around 1775. By the nineteenth century the use of multiple integrals became fairly common.

The concept of a vector was known in antiquity. Aristotle represented forces by vectors, and knew that two forces acting in different directions could be summed by what we call the parallelogram law. Representation of complex numbers in the plane was done by Wessel in Denmark, Argand in France, and Gauss in Germany between 1798 and 1806. It was fairly well known by 1830 that vectors could be neatly expressed as complex numbers. However, many physical situations involve more than two factors that are not all in the same plane. For many years mathematicians searched for a three-dimensional vector system that preserved all of the properties of real-number algebra. These efforts were unsuccessful, and the Irish mathematician William Rowan Hamilton (1805–1865) demonstrated in 1843 that it was not possible to construct such a system. He postulated a system with four mutually independent unit vectors, in which the basic objects are ordered quadruples of real numbers, called *quaternions*. Quaternions do not obey the commutative law (that is, $A \cdot B \neq B \cdot A$); this was the first algebra in which such behavior was studied. The use of quaternions was advocated throughout the nineteenth century by Hamilton and others, and courses in the subject were taught in British and American graduate schools well into the twentieth century. Eventually, however, it was superceded by the vector analysis pioneered by J. W. Gibbs (and presented here in Chapter 21).

Hamilton was a child prodigy, and largely self-taught. He read Greek, Latin, and Hebrew by the age of 5, and by age 14 had added nearly a dozen more languages. At age 22, still an undergraduate, he was appointed Professor of Astronomy at an Irish university and Royal Astronomer of Ireland. His life was not happy, however; his marriage was unpleasant and his quaternions were not well received on the Continent. He was knighted early, in 1835, but became an alcoholic in the last twenty years of his life.

Josiah Willard Gibbs (1839–1903) was an American mathematician and scientist in a day when American science was held in little esteem. Born in New Haven, Connecticut, he attended Yale College. He received a Ph.D. in physics in 1863, and then went to Europe for further study. Returning in 1871, he became Professor of Mathematical Physics at Yale and remained there the rest of his life. In 1881 he printed, at his own expense, a little pamphlet called *Elements of Vector Analysis* for the use of his students. In it, he created the subject much as it is known today. The pamphlet became well known, and E. B. Wilson published a book called *Vector Analysis,* based on Gibbs' lectures, in 1901.

Another self-taught mathematical physicist, Oliver Heaviside (1850–1925), independently created vector analysis. He had only an elementary school education in his native England, but studied by himself and, though he used unorthodox methods, accomplished a great deal. His work in electromagnetic theory made brilliant advances, but it was disdained by most mathematicians of the time because of his lack of formal education. Heaviside took ideas from Hamilton's theory of quaternions, and developed scalar and vector multiplication. His notions were well received by engineers and physicists.

Two applications of vectors are Green's Theorem and Stokes' Theorem, both presented in this unit. George Green (1793–1841) was also a self-taught mathematical physicist, and one of the first to treat static electricity and magnetism in mathematical terms. His 1828 pamphlet, privately published and entitled *An Essay on the Application of Mathematical Analysis to the Theories of Electricity and Magnetism,* was little read until discovered by Lord Kelvin, who printed it in a German journal between 1854 and 1858. Green's Theorem appears in this work.

George Gabriel Stokes (1819–1903) was an Irishman who was Lucasian Professor of Mathematics at Cambridge, the same post held by Newton a century and a half earlier. Though prestigious, the position did not pay well, and Stokes had to teach at a School of Mines in order to make ends meet. He studied various physical phenomena from a mathematical point of view, and through a brilliant paper established the foundations of hydrodynamics. The theorem named after him is a vector version of Green's Theorem.

Pronunciation Guide

Adrien-Marie Legendre	(Luh-zhahn′dre)
Joseph Louis Lagrange	(La-grahnj)

(Photograph of J. W. Gibbs from the Library of Congress. Photograph of George G. Stokes from the David Eugene Smith Papers, Rare Book and Manuscript Library, Columbia University.)

CHAPTER 18

VECTOR-VALUED FUNCTIONS

18.1 INTRODUCTION

A **vector-valued function** is a function of the form

$$\boldsymbol{f}(t) = x(t)\boldsymbol{i} + y(t)\boldsymbol{j} + z(t)\boldsymbol{k} \tag{1}$$

where $x(t)$, $y(t)$, and $z(t)$ are functions of the independent variable t.

REMARK: In discussing vector-valued functions, we will call the three *vectors* on the right-hand side of equation (1) "vector components," and we will refer to the *numbers* $x(t)$, $y(t)$, and $z(t)$ as "scalar components" or "component functions." When we use the term "components" alone, it will always be clear from the context which type we mean.

We have already seen two examples of functions of this type:

(i) The **line** with vector equation

$$\begin{aligned}\boldsymbol{r}(t) &= (a_1\boldsymbol{i} + a_2\boldsymbol{j} + a_3\boldsymbol{k}) + t(b_1\boldsymbol{i} + b_2\boldsymbol{j} + b_3\boldsymbol{k})\\ &= (a_1 + tb_1)\boldsymbol{i} + (a_2 + tb_2)\boldsymbol{j} + (a_3 + tb_3)\boldsymbol{k}\end{aligned}$$

has the form of equation (1) with $x(t) = (a_1 + tb_1)$, $y(t) = (a_2 + tb_2)$, and $z(t) = (a_3 + tb_3)$. The vector $\boldsymbol{r}(t)$ is interpreted as extending from the origin to the point $P(t)$ whose coordinates are the components of the vector $\boldsymbol{r}(t)$. If t denotes time, the vector function represents the motion of a particle moving in space.

(ii) The **curve** C in the plane determined by the parametric equations

$$C: \quad \begin{cases} x = x(t) \\ y = y(t) \end{cases} \qquad a \le t \le b$$

may be interpreted as the graph of the vector function

$$\boldsymbol{f}(t) = x(t)\boldsymbol{i} + y(t)\boldsymbol{j}, \qquad a \le t \le b$$

where the vector $\boldsymbol{f}(t)$ extends from the origin to the point $(x(t), y(t))$ on C.

The concept of vector function allows us to pursue generalizations of both of these examples in a unified way. By interpreting $\boldsymbol{f}(t)$ as a vector from the origin to a particle in space, we may study the motion of that particle, addressing the usual issues of position, speed, velocity, acceleration, distance, and elapsed time. These are the **dynamic aspects** associated with the vector function $\boldsymbol{f}(t)$. The **static consid-**

erations associated with vector functions concern the nature of the curve C traced out by the tip of the vector $\boldsymbol{f}(t)$—arc length, tangent and normal vectors, and curvature.

In dealing with both types of issues we shall make heavy use of the fact that the vector function $\boldsymbol{f}(t)$ is the sum of three vector components, each of which is a function of a single independent variable (Figure 1.1). The laws of vector algebra allow us to perform algebraic operations with vector functions, while the theory of the calculus (Chapters 2–11) will enable us to address many of the issues associated with derivatives and integrals of vector functions.

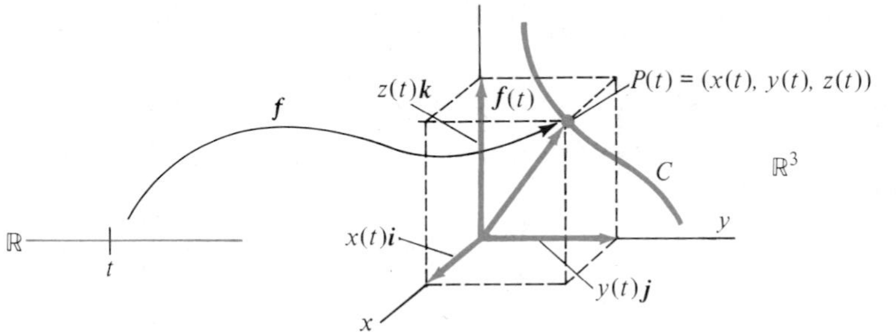

Figure 1.1 The vector function $\boldsymbol{f}(t) = x(t)\boldsymbol{i} + y(t)\boldsymbol{j} + z(t)\boldsymbol{k}$ from $\mathbb{R}$ to $\mathbb{R}^3$.

Example 1 Sketch the graph of the vector function

$$\boldsymbol{f}(t) = a \cos t\boldsymbol{i} + a \sin t\boldsymbol{j} + bt\boldsymbol{k}.$$

where a and b are constants.

Solution: We note that in any plane parallel to the xy-plane the distance from the point $(a \cos t, a \sin t, bt)$ to the point $(0, 0, bt)$ on the z-axis is

$$d = \sqrt{a^2 \cos^2 t + a^2 \sin^2 t} = |a|.$$

Thus, all points lie on the graph of the circular cylinder $x^2 + y^2 = a^2$. As t increases, the z-coordinate of the point determined by $\boldsymbol{f}(t)$ increases uniformly. For every point $(x(t), y(t), z(t))$ on the graph, infinitely many points $(x(t), y(t), z(t) + 2\pi nb)$, $n = \pm 1, \pm 2, \ldots$ also lie on the graph. The graph is the **circular helix** in Figure 1.2. ■

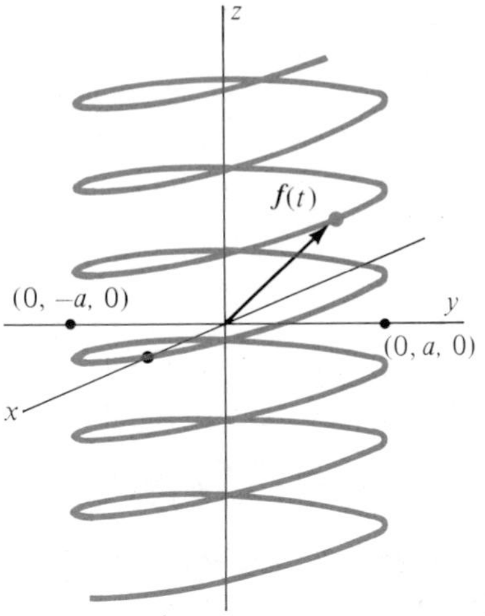

Figure 1.2 The circular helix $\boldsymbol{f}(t) = a \cos t\boldsymbol{i} + a \sin t\boldsymbol{j} + bt\boldsymbol{k}$.

Exercise Set 18.1

In Exercises 1–8, sketch the graph of the given vector function.

1. $\boldsymbol{f}(t) = \boldsymbol{i} + \boldsymbol{j} + t\boldsymbol{k}$

2. $\boldsymbol{f}(t) = \cos t\boldsymbol{i} + \sin t\boldsymbol{j} + t\boldsymbol{k}$

3. $\boldsymbol{f}(t) = t\boldsymbol{i} + \cos t\boldsymbol{j} + \sin t\boldsymbol{k}$

4. $\boldsymbol{f}(t) = e^{-t}\boldsymbol{i} + e^{-t}\boldsymbol{j} + t\boldsymbol{k}$

5. $\boldsymbol{f}(t) = \cos t\boldsymbol{i} + \sin t\boldsymbol{j} + \cos t\boldsymbol{k}$

6. $\boldsymbol{f}(t) = t\boldsymbol{i} + t^2\boldsymbol{j}$

7. $\boldsymbol{f}(t) = \sin t\boldsymbol{i} + \boldsymbol{j} + \boldsymbol{k}$

8. $\boldsymbol{f}(t) = \sin t\boldsymbol{i} + \sin t\boldsymbol{j} + t\boldsymbol{k}$

The **implicit domain** of a vector function $\boldsymbol{f}(t)$ is the largest set of values of t for which each of the component functions is defined. State the implicit domain for each of the following vector functions.

9. $\boldsymbol{f}(t) = t\boldsymbol{i} + \sqrt{t}\,\boldsymbol{j} + \sin t\boldsymbol{k}$

10. $\boldsymbol{f}(t) = t^2\boldsymbol{i} - t\boldsymbol{j} + \tan t\boldsymbol{k}$

11. $f(t) = \frac{1}{1+t}i + \frac{1}{t-1}j + \cos tk$

12. $f(t) = \frac{1}{\sqrt{1-t}}i + 2tj + \sin^2 tk$

13. $f(t) = \frac{1}{9-t^2}i + \frac{2}{1+t}j + \sqrt{t}k$

14. $f(t) = (t^2 - 1)i + \sec tj + t^3k$

15. $f(t) = t^2i + t^{3/2}j + t^{2/3}k$

16. $f(t) = \ln ti + \cos tj + \sqrt{9 - t^2}k$

17. An automobile assembly robot turns a machine screw at a constant rate of 10π radians per second. The **pitch** of the screw (the number of threads per millimeter of length) is such that the screw advances 0.5 mm for each complete revolution.

a. Write a vector function for the motion of a paint spot on one thread of the screw.

b. At what rate, in mm/sec, must the robot arm advance?

18. The **polarization** of a light wave is determined by the motion of the tip of the associated "electric vector" $E(t)$. If the motion follows a circular helix, the light is said to be circularly polarized. If a particular light wave is circularly polarized according to $E(t) = \cos(10^3t)i + \sin(10^3t)j + (3 \times 10^8)tk$, how far does the wave advance along the z-axis during one complete revolution of the electric vector?

19. Describe the graph of the vector function $f(t) = 3 \cos ti + 4 \sin tj + tk$.

18.2 ALGEBRA, LIMITS, AND CONTINUITY

Given two vector-valued functions

$$f(t) = x_1(t)i + y_1(t)j + z_1(t)k$$

and

$$g(t) = x_2(t)i + y_2(t)j + z_2(t)k,$$

we may form the following functions, as indicated.

(i) The **sum** of $f(t)$ and $g(t)$ is the vector-valued function

$$\begin{aligned}(f + g)(t) &= f(t) + g(t)\\ &= [x_1(t) + x_2(t)]i + [y_1(t) + y_2(t)]j + [z_1(t) + z_2(t)]k.\end{aligned}$$

In other words, we add vector functions component by component.

(ii) The **scalar multiple** of $f(t)$ by the real number c is the vector-valued function

$$(cf)(t) = cf(t) = [cx_1(t)]i + [cy_1(t)]j + [cz_1(t)]k.$$

That is, multiplication of vector functions by scalars is done component by component.

(iii) The **dot product** of $f(t)$ and $g(t)$ is the **real-valued function**

$$(f \cdot g)(t) = f(t) \cdot g(t) = x_1(t)x_2(t) + y_1(t)y_2(t) + z_1(t)z_2(t).$$

The values of the dot product function are *numbers* rather than vectors.

(iv) The **cross product** of $f(t)$ and $g(t)$ is the vector-valued function

$$\begin{aligned}(f \times g)(t) &= f(t) \times g(t)\\ &= [y_1(t)z_2(t) - z_1(t)y_2(t)]i + [z_1(t)x_2(t) - x_1(t)z_2(t)]j + [x_1(t)y_2(t) - y_1(t)x_2(t)]k.\end{aligned}$$

The values of the cross product function are *vectors*.

Example 1 From the vector functions

$$f(t) = ti + t^2j + \sqrt{t}k, \qquad t \geq 0$$

and

$$g(t) = \cos t\,\boldsymbol{i} + \sin t\,\boldsymbol{j} + t\boldsymbol{k},$$

the following functions may be formed:

(a) $(\boldsymbol{f} + \boldsymbol{g})(t) = (t + \cos t)\boldsymbol{i} + (t^2 + \sin t)\boldsymbol{j} + (\sqrt{t} + t)\boldsymbol{k}, \qquad t \geq 0.$
(b) $3\boldsymbol{f}(t) = 3t\boldsymbol{i} + 3t^2\boldsymbol{j} + 3\sqrt{t}\,\boldsymbol{k}, \qquad t \geq 0.$
(c) $(\boldsymbol{f} - 2\boldsymbol{g})(t) = (t - 2\cos t)\boldsymbol{i} + (t^2 - 2\sin t)\boldsymbol{j} + (\sqrt{t} - 2t)\boldsymbol{k}, \qquad t \geq 0.$
(d)

$$(\boldsymbol{f} \times \boldsymbol{g})(t) = \det \begin{bmatrix} \boldsymbol{i} & \boldsymbol{j} & \boldsymbol{k} \\ t & t^2 & \sqrt{t} \\ \cos t & \sin t & t \end{bmatrix}$$

$$= (t^3 - \sqrt{t}\sin t)\boldsymbol{i} + (\sqrt{t}\cos t - t^2)\boldsymbol{j} + (t\sin t - t^2\cos t)\boldsymbol{k}. \blacksquare$$

Example 2 For $\boldsymbol{f}(t) = \sqrt{t}\,\boldsymbol{i} + \sin t\boldsymbol{j} + e^t\boldsymbol{k}$ and $\Delta t \neq 0$, find the difference quotient

$$\boldsymbol{Q} = \frac{1}{\Delta t}[\boldsymbol{f}(t + \Delta t) - \boldsymbol{f}(t)].$$

Solution: $\boldsymbol{f}(t + \Delta t) = \sqrt{t + \Delta t}\,\boldsymbol{i} + \sin(t + \Delta t)\boldsymbol{j} + e^{(t+\Delta t)}\boldsymbol{k}$, so

$$\frac{1}{\Delta t}[\boldsymbol{f}(t + \Delta t) - \boldsymbol{f}(t)]$$

$$= \frac{1}{\Delta t}\{[\sqrt{t + \Delta t} - \sqrt{t}]\boldsymbol{i} + [\sin(t + \Delta t) - \sin t]\boldsymbol{j} + [e^{t+\Delta t} - e^t]\boldsymbol{k}\}$$

$$= \left[\frac{\sqrt{t + \Delta t} - \sqrt{t}}{\Delta t}\right]\boldsymbol{i} + \left[\frac{\sin(t + \Delta t) - \sin t}{\Delta t}\right]\boldsymbol{j} + \left[\frac{e^{t+\Delta t} - e^t}{\Delta t}\right]\boldsymbol{k}. \blacksquare$$

Limits of Vector Functions

Since addition and scalar multiplication for vector-valued functions are defined as the usual operations performed within each of the components, we define the *limit* of a vector function in the same way.

DEFINITION 1

Let $\boldsymbol{f}(t)$ be the vector function with component functions

$$\boldsymbol{f}(t) = x(t)\boldsymbol{i} + y(t)\boldsymbol{j} + z(t)\boldsymbol{k}.$$

If $\lim_{t\to t_0} x(t) = L_1, \quad \lim_{t\to t_0} y(t) = L_2, \quad$ and $\quad \lim_{t\to t_0} z(t) = L_3$, we define the **limit of the vector function** $\boldsymbol{f}(t)$ as t approaches t_0 to be

$$\lim_{t\to t_0} \boldsymbol{f}(t) = [\lim_{t\to t_0} x(t)]\boldsymbol{i} + [\lim_{t\to t_0} y(t)]\boldsymbol{j} + [\lim_{t\to t_0} z(t)]\boldsymbol{k}$$

$$= L_1\boldsymbol{i} + L_2\boldsymbol{j} + L_3\boldsymbol{k}.$$

Definition 1 states that the limit of a vector function is the vector whose components are the limits of the individual components. If one or more of these component limits fails to exist we say that the limit of the vector function fails to exist.

Example 3 Find $\lim_{t\to\pi/3} (t^2\boldsymbol{i} + \sin t\boldsymbol{j} + \cos t\boldsymbol{k})$.

Solution: The limits of the components are

$$L_1 = \lim_{t \to \pi/3} x(t) = \lim_{t \to \pi/3} t^2 = \frac{\pi^2}{9},$$

$$L_2 = \lim_{t \to \pi/3} y(t) = \lim_{t \to \pi/3} \sin t = \frac{\sqrt{3}}{2} \qquad \text{(Figure 2.1)},$$

$$L_3 = \lim_{t \to \pi/3} z(t) = \lim_{t \to \pi/3} \cos t = \frac{1}{2}.$$

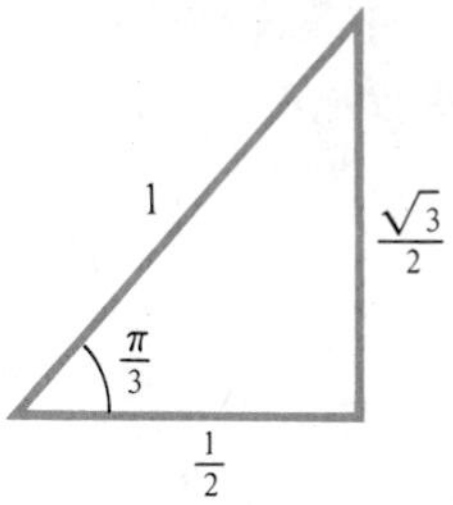

Figure 2.1

Thus, by Definition 1,

$$\lim_{t \to \pi/3} (t^2\boldsymbol{i} + \sin t\boldsymbol{j} + \cos t\boldsymbol{k}) = \frac{\pi^2}{9}\boldsymbol{i} + \frac{\sqrt{3}}{2}\boldsymbol{j} + \frac{1}{2}\boldsymbol{k}. \qquad \blacksquare$$

Example 4 Find $\lim_{\Delta t \to 0} \left\{ \frac{1}{\Delta t} [\boldsymbol{f}(t + \Delta t) - \boldsymbol{f}(t)] \right\}$ for the vector function

$$\boldsymbol{f}(t) = \sqrt{t}\,\boldsymbol{i} + \sin t\boldsymbol{j} + e^t\boldsymbol{k}, \qquad t > 0.$$

Strategy
Simplify the given function, as in Example 2.

Solution
From Example 2, we have

$$\lim_{\Delta t \to 0} \frac{1}{\Delta t} [\boldsymbol{f}(t + \Delta t) - \boldsymbol{f}(t)]$$
$$= \lim_{\Delta t \to 0} \left\{ \left[\frac{\sqrt{t + \Delta t} - \sqrt{t}}{\Delta t} \right] \boldsymbol{i} + \left[\frac{\sin(t + \Delta t) - \sin t}{\Delta t} \right] \boldsymbol{j} + \left[\frac{e^{t+\Delta t} - e^t}{\Delta t} \right] \boldsymbol{k} \right\}$$

Evaluate the limit of each component separately.

Use the fact that

$$\lim_{\Delta t \to 0} \frac{g(t + \Delta t) - g(t)}{\Delta t} = \frac{d}{dt}(g(t)).$$

Using the definition of the derivative of a function of one variable we find the following:

$$L_1 = \lim_{\Delta t \to 0} \left[\frac{\sqrt{t + \Delta t} - \sqrt{t}}{\Delta t} \right] = \frac{d}{dt}(\sqrt{t}) = \frac{1}{2\sqrt{t}}, \qquad t > 0.$$

$$L_2 = \lim_{\Delta t \to 0} \left[\frac{\sin(t + \Delta t) - \sin t}{\Delta t} \right] = \frac{d}{dt}(\sin t) = \cos t.$$

$$L_3 = \lim_{\Delta t \to 0} \left[\frac{e^{t+\Delta t} - e^t}{\Delta t} \right] = \frac{d}{dt}(e^t) = e^t.$$

Apply Definition 1.

Thus, by Definition 1,

$$\lim_{\Delta t \to 0} \left\{ \frac{1}{\Delta t} [\boldsymbol{f}(t + \Delta t) - \boldsymbol{f}(t)] \right\} = \frac{1}{2\sqrt{t}}\boldsymbol{i} + \cos t\boldsymbol{j} + e^t\boldsymbol{k}. \qquad \blacksquare$$

Continuity

Continuity for vector functions is defined in terms of the continuity of each of the individual component functions.

DEFINITION 2

The vector function $\boldsymbol{f}(t) = x(t)\boldsymbol{i} + y(t)\boldsymbol{j} + z(t)\boldsymbol{k}$ is **continuous** at $t = t_0$ if each of the component functions $x(t)$, $y(t)$, and $z(t)$ is continuous at t_0.

We say that a vector function is continuous on an interval I if it is continuous at each $t \in I$. Note that each component of $f(t)$ must be continuous at t_0 if $f(t)$ is continuous at t_0. Thus, a vector function is **discontinuous** at t_0 if one or more of its component functions is discontinuous at t_0.

Example 5 Find the intervals on which the vector function

$$f(t) = \frac{1}{t}\boldsymbol{i} + t^2\boldsymbol{j} + \frac{2}{t^2 - 4}\boldsymbol{k}$$

is continuous.

Strategy
Determine the values of t where one or more of the component functions is discontinuous.

Solution
The $\boldsymbol{i}$-component $x(t) = 1/t$ is discontinuous at $t = 0$.

The $\boldsymbol{j}$-component $y(t) = t^2$ is continuous for all t.

The $\boldsymbol{k}$-component $z(t) = \dfrac{2}{t^2 - 4} = \dfrac{2}{(t-2)(t+2)}$ is discontinuous for $t = -2, 2$.

The vector function is continuous for all other values of t.

The vector function $f(t)$ is therefore discontinuous at $t = -2$, 0, and 2, so $f(t)$ is continuous on the intervals

$$(-\infty, -2),\ (-2, 0),\ (0, 2), \text{ and } (2, \infty).$$

■

Recall that the real valued function $g(t)$ is defined to be continuous at $t = t_0$ if (a) $g(t_0)$ exists, (b) $\lim_{t \to t_0} g(t)$ exists, and (c) $g(t_0) = \lim_{t \to t_0} g(t)$ (Definition 4, Section 2.6). We may combine Definitions 1 and 2 to obtain a corresponding equivalent definition of continuity for vector functions: The vector function $f(t)$ is continuous at $t = t_0$ if

(a) $f(t_0)$ exists,
(b) $\lim_{t \to t_0} f(t)$ exists, and
(c) $f(t_0) = \lim_{t \to t_0} f(t)$.

In defining the limit $L = \lim_{t \to t_0} g(t)$ for a function of a single variable (Chapter 2), we worked with the intuitive notion of $g(t)$ "approaching" L as t approaches t_0. More formally, we characterized $L = \lim_{t \to t_0} g(t)$ as meaning that $|g(t) - L|$ becomes small as $|t - t_0|$ becomes small. At this point you may wonder whether Definition 1 preserves this notion for vector functions. That is, does the statement $\boldsymbol{L} = \lim_{t \to t_0} f(t)$ guarantee that the vector $f(t)$ is "close" to the vector $\boldsymbol{L}$ when $|t - t_0|$ is small? The following theorem shows that this is indeed the case. Note that the number $|f(t) - \boldsymbol{L}|$ used to measure the "closeness" of $f(t)$ and $\boldsymbol{L}$ is the (nonnegative) length of the vector $f(t) - \boldsymbol{L}$.

THEOREM 1

Let $f(t)$ be the vector function with components

$$f(t) = x(t)\boldsymbol{i} + y(t)\boldsymbol{j} + z(t)\boldsymbol{k}$$

and let $\boldsymbol{L}$ be the vector $\boldsymbol{L} = L_1\boldsymbol{i} + L_2\boldsymbol{j} + L_3\boldsymbol{k}$. Then $\boldsymbol{L} = \lim_{t\to t_0} \boldsymbol{f}(t)$ if and only if given $\epsilon > 0$ there exists a number $\delta > 0$ so that

$$|\boldsymbol{f}(t) - \boldsymbol{L}| < \epsilon \text{ whenever } 0 < |t - t_0| < \delta. \tag{1}$$

That is, $\boldsymbol{L} = \lim_{t\to t_0} \boldsymbol{f}(t)$ if and only if $|\boldsymbol{f}(t) - \boldsymbol{L}|$ is small whenever $|t - t_0|$ is small (Figure 2.2).

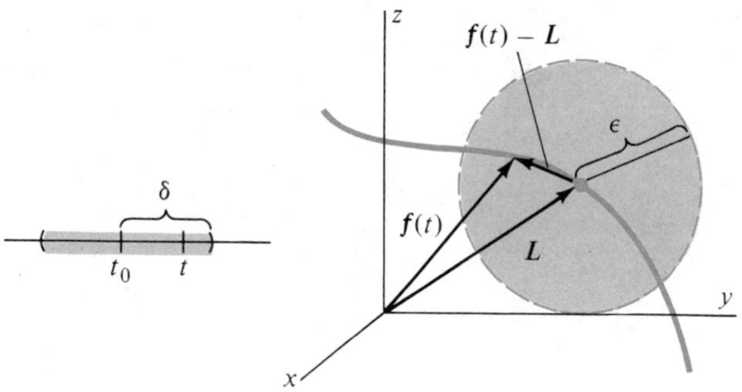

Figure 2.2 $|\boldsymbol{f}(t) - \boldsymbol{L}| < \epsilon$ if $0 < |t - t_0| < \delta$.

Proof: First, let us assume that $\boldsymbol{L} = \lim_{t\to t_0} \boldsymbol{f}(t)$. Let $\epsilon > 0$ be given. We must show that there exists a number $\delta > 0$ for which statement (1) holds. By Definition 1, we know that

$$L_1 = \lim_{t\to t_0} x(t), \tag{2}$$

$$L_2 = \lim_{t\to t_0} y(t), \quad \text{and} \tag{3}$$

$$L_3 = \lim_{t\to t_0} z(t). \tag{4}$$

According to the definition of the limit of a function of a single variable (Definition 7, Chapter 2) and equation (2), there exists a number δ_1 so that

$$|x(t) - L_1| < \epsilon/3 \qquad \text{whenever} \qquad 0 < |t - t_0| < \delta_1. \tag{5}$$

(You will see in a few lines why we use $\epsilon/3$ rather than just ϵ.) Similarly, it follows from equations (3) and (4) that there exist positive numbers δ_2 and δ_3 so that

$$|y(t) - L_2| < \epsilon/3 \qquad \text{whenever} \qquad 0 < |t - t_0| < \delta_2 \tag{6}$$

and

$$|z(t) - L_3| < \epsilon/3 \qquad \text{whenever} \qquad 0 < |t - t_0| < \delta_3. \tag{7}$$

If we take δ to be the smallest of the numbers δ_1, δ_2, and δ_3, then each of the inequalities on the right-hand sides of (5) through (7) is fulfilled when $0 < |t - t_0| < \delta$. In this case, using inequalities (5) through (7) and the triangle inequality, we find that

$$\begin{aligned} |\boldsymbol{f}(t) - \boldsymbol{L}| &= |(x(t)\boldsymbol{i} + y(t)\boldsymbol{j} + z(t)\boldsymbol{k}) - (L_1\boldsymbol{i} + L_2\boldsymbol{j} + L_3\boldsymbol{k})| \\ &= |(x(t) - L_1)\boldsymbol{i} + (y(t) - L_2)\boldsymbol{j} + (z(t) - L_3)\boldsymbol{k}| \\ &\le |x(t) - L_1| + |y(t) - L_2| + |z(t) - L_3| \end{aligned}$$

$$< \frac{\epsilon}{3} + \frac{\epsilon}{3} + \frac{\epsilon}{3}$$
$$= \epsilon.$$

That is,

$$|f(t) - L| < \epsilon \quad \text{whenever} \quad 0 < |t - t_0| < \delta. \tag{8}$$

To prove the converse, we assume that whenever $\epsilon > 0$ is given we can find a number $\delta > 0$ so that statement (1) holds, and we show that this guarantees that $L = \lim_{t \to t_0} f(t)$. To do so we observe that if t is any number for which $0 < |t - t_0| < \delta$ (with δ as in equation (8)) we have

$$\begin{aligned} |x(t) - L_1| &= \sqrt{|x(t) - L_1|^2} \\ &\leq \sqrt{|x(t) - L_1|^2 + |y(t) - L_2|^2 + |z(t) - L_3|^2} \\ &= |f(t) - L| \\ &< \epsilon. \end{aligned}$$

That is, $|x(t) - L_1| < \epsilon$ whenever $0 < |t - t_0| < \delta$. This shows that $\lim_{t \to t_0} x(t) = L_1$.

Similar comparisons show that $\lim_{t \to t_0} y(t) = L_2$ and that $\lim_{t \to t_0} z(t) = L_3$. This shows that $L = \lim_{t \to t_0} f(t)$, according to Definition 1. This completes the proof. ■

The following corollary of Theorem 1 provides the analogous characterization of continuity for vector functions. The proof is similar and is left as an exercise.

COROLLARY 1 The vector function $f(t)$ is continuous at $t = t_0$ if and only if for any $\epsilon > 0$ there exists a corresponding number $\delta > 0$ so that

$$|f(t) - f(t_0)| < \epsilon \quad \text{whenever} \quad |t - t_0| < \delta. \tag{9}$$

REMARK: Be sure to note, in both statement (1) and statement (9), that the inequality on the left-hand side concerns the length of a vector, while the inequality on the right concerns the absolute value of a number.

Exercise Set 18.2

In Exercises 1–6, let $f(t) = t\boldsymbol{i} + 3t\boldsymbol{j} + t^2\boldsymbol{k}$ and $g(t) = \sin t\boldsymbol{i} + \cos t\boldsymbol{j} + \boldsymbol{k}$. Find the indicated functions.

1. $(f + g)(t)$

2. $(f - g)(t)$

3. $(3f + 2g)(t)$

4. $(4f - 3g)(t)$

5. $(f \cdot g)(t)$

6. $(f \times g)(t)$

In Exercises 7–12, let $f(t) = (1 + t)\boldsymbol{i} + (1 - t^2)\boldsymbol{j} + t\boldsymbol{k}$, $g(t) = \sqrt{t}\,\boldsymbol{i} + (1 - t)\boldsymbol{j} + e^t\boldsymbol{k}$, and $h(t) = \sin t$. Find the indicated functions.

7. $(f + 3g)(t)$

8. $h(t)f(t)$

9. $h(t)(f - g)(t)$

10. $2f(t) - 3h(t)g(t)$

11. $(f \cdot g)(t)$

12. $(3f \times 2g)(t)$

In Exercises 13–18, find the indicated limit.

13. $\lim_{t \to \pi/4} (\sin t\boldsymbol{i} + \tan t\boldsymbol{j} + \cos t\boldsymbol{k})$

14. $\lim_{t \to 2} (t^2\boldsymbol{i} + (3t - 2)\boldsymbol{j} + 6t\boldsymbol{k})$

15. $\lim_{t \to 0} \left\{ \left(\frac{\sin t}{t} \right)\boldsymbol{i} + \left(\frac{1 - \cos t}{t} \right)\boldsymbol{j} + e^{2t}\boldsymbol{k} \right\}$

16. $\lim_{t \to 0} \left\{ \frac{\sin 3t}{t}\boldsymbol{i} + \frac{\tan 2t}{3t}\boldsymbol{j} + \ln(1 + t)\boldsymbol{k} \right\}$

17. $\lim_{t\to 3}\left\{\left(\frac{9-t^2}{3-t}\right)\boldsymbol{i}+\left(\frac{t^2+t-12}{t-3}\right)\boldsymbol{j}+\left(\frac{t^3-13t-12}{t-3}\right)\boldsymbol{k}\right\}$

18. $\lim_{t\to 2}\left\{(t+3)\boldsymbol{i}+\left(\frac{t^2-4}{t-2}\right)\boldsymbol{j}+\left(\frac{t^3+t^2-4t-4}{t-2}\right)\boldsymbol{k}\right\}$

In Exercises 19–25, determine the intervals on which the given vector function is continuous.

19. $\boldsymbol{f}(t)=2t\boldsymbol{i}+\cos t\boldsymbol{j}+\tan t\boldsymbol{k}$

20. $\boldsymbol{f}(t)=\sqrt{9-t^2}\,\boldsymbol{i}+\frac{1}{t}\boldsymbol{j}+\sin t\boldsymbol{k}$

21. $\boldsymbol{f}(t)=\frac{1}{1+\sqrt{t}}\boldsymbol{i}+\frac{1}{1-\sqrt{t}}\boldsymbol{j}+t^{3/2}\boldsymbol{k}$

22. $\boldsymbol{f}(t)=\left(\frac{1}{t^2-t+12}\right)\boldsymbol{i}+\ln(2t)\boldsymbol{j}+e^{-t}\boldsymbol{k}$

23. $\boldsymbol{f}(t)=\ln(1+t^2)\boldsymbol{i}+\ln(1-t^2)\boldsymbol{j}+\sqrt{t}\boldsymbol{k}$

24. $\boldsymbol{f}(t)=\begin{cases} t\boldsymbol{i}+(3t-2)\boldsymbol{j}+(3-t)\boldsymbol{k}, & -\infty<t<2\\ 2\boldsymbol{i}+(2+t)\boldsymbol{j}+t\boldsymbol{k}, & 2\le t<\infty\end{cases}$

25. $\boldsymbol{f}(t)=\begin{cases} t^2\boldsymbol{i}+(3-t)\boldsymbol{j}+4t\boldsymbol{k}, & -\infty<t<1\\ t\boldsymbol{i}+(4+t)\boldsymbol{j}+(1-t^2)\boldsymbol{k}, & 1\le t<\infty\end{cases}$

26. Prove that if $\lim_{t\to t_0}\boldsymbol{f}(t)=\boldsymbol{L}$ and $\lim_{t\to t_0}\boldsymbol{g}(t)=\boldsymbol{M}$, then $\lim_{t\to t_0}(\boldsymbol{f}+\boldsymbol{g})(t)=\boldsymbol{L}+\boldsymbol{M}$ as follows.
   a. Write $\boldsymbol{f}(t)$ and $\boldsymbol{g}(t)$ in component form.
   b. Obtain $(\boldsymbol{f}+\boldsymbol{g})(t)$ in component form.
   c. Apply Definition 1, and then Theorem 1 of Chapter 2.

27. Prove that if $\boldsymbol{f}(t)$ and $\boldsymbol{g}(t)$ are continuous at $t=t_0$ then so is
   a. $(\boldsymbol{f}+\boldsymbol{g})(t)$,
   b. $(c\boldsymbol{f})(t)$,
   c. $(\boldsymbol{f}\times\boldsymbol{g})(t)$.

28. Prove that if $\boldsymbol{f}(t)$ and $\boldsymbol{g}(t)$ are continuous vector-valued functions then so are the real-valued functions
   a. $(\boldsymbol{f}\cdot\boldsymbol{g})(t)$,
   b. $|\boldsymbol{f}(t)|$.

29. For each set of vector functions, find the intervals on which $\boldsymbol{f}(t)$, $\boldsymbol{g}(t)$, $\boldsymbol{h}(t)$, and $d(t)$ are continuous.
   a. $\boldsymbol{f}(t)=\sqrt{t}\,\boldsymbol{i}-\frac{1}{t}\boldsymbol{j}+t\boldsymbol{k},\ \boldsymbol{g}(t)=-\sqrt{t}\,\boldsymbol{i}+\frac{1}{t}\boldsymbol{j}+3t\boldsymbol{k},$
   $\boldsymbol{h}(t)=(\boldsymbol{f}+\boldsymbol{g})(t)$
   b. $\boldsymbol{f}(t)=\frac{1}{t}\boldsymbol{i}+\ln(1+t)\boldsymbol{j}+\sqrt{t}\,\boldsymbol{k},\quad \boldsymbol{g}(t)=t\boldsymbol{i}+\sqrt{t}\,\boldsymbol{k},$
   $d(t)=(\boldsymbol{f}\cdot\boldsymbol{g})(t)$

30. If $(\boldsymbol{f}+\boldsymbol{g})(t)$ is continuous, what can you say about the continuity of $\boldsymbol{f}(t)$ and $\boldsymbol{g}(t)$? What conclusions can you draw from the continuity of the dot product $(\boldsymbol{f}\cdot\boldsymbol{g})(t)$?

31. Prove Corollary 1.

18.3 DERIVATIVES AND INTEGRALS OF VECTOR FUNCTIONS

Let $\boldsymbol{f}(t)$ be a vector function and let t be a number in the domain of $\boldsymbol{f}(t)$. Also, assume that the numbers $t+\Delta t$ lie in the domain of $\boldsymbol{f}(t)$ for $|\Delta t|$ sufficiently small. The vector

$$\frac{\boldsymbol{f}(t+\Delta t)-\boldsymbol{f}(t)}{\Delta t}=\left(\frac{1}{\Delta t}\right)\left\{\boldsymbol{f}(t+\Delta t)-\boldsymbol{f}(t)\right\},\qquad \Delta t\neq 0 \tag{1}$$

is referred to as a **difference quotient** for the vector function $\boldsymbol{f}(t)$ (recall Example 2 of the preceding section). The vector in equation (1) may be interpreted as the average change in the vector $\boldsymbol{f}(t)$ per unit change in t over an interval of length Δt. As for functions of a single variable, we are interested in knowing about the limiting value of this vector as $\Delta t\to 0$, which we refer to as the *derivative* of the vector function $\boldsymbol{f}(t)$. The next three sections in part concern the geometric and physical interpretations of this derivative. Our interest here focuses on defining and calculating this particular limit.

DEFINITION 3

The vector function $\boldsymbol{f}(t)$ is said to be **differentiable** at the number t if the limit

$$\boldsymbol{f}'(t)=\lim_{\Delta t\to 0}\frac{\boldsymbol{f}(t+\Delta t)-\boldsymbol{f}(t)}{\Delta t} \tag{2}$$

exists. The vector $\boldsymbol{f}'(t)$ is called the **derivative** of the vector function $\boldsymbol{f}(t)$ at t.

As for functions of a single variable, the vector function $\boldsymbol{f}(t)$ is said to be differentiable on an open interval I if $\boldsymbol{f}'(t)$ in (2) exists for every $t \in I$. Thus, the derivative of a vector function is again a vector function, and the domain of $\boldsymbol{f}'(t)$ is the subset of the domain of $\boldsymbol{f}(t)$ for which the limit in line (2) exists.

The following theorem shows that the derivative $\boldsymbol{f}'(t)$ may be calculated directly from the components of $\boldsymbol{f}(t)$.

THEOREM 2

Let $\boldsymbol{f}(t) = x(t)\boldsymbol{i} + y(t)\boldsymbol{j} + z(t)\boldsymbol{k}$. The vector function $\boldsymbol{f}(t)$ is differentiable if and only if each of the component functions $x(t)$, $y(t)$, and $z(t)$ is differentiable. In this case

$$\boldsymbol{f}'(t) = x'(t)\boldsymbol{i} + y'(t)\boldsymbol{j} + z'(t)\boldsymbol{k} \tag{3}$$

or, in Leibniz notation

$$\frac{d}{dt}(\boldsymbol{f}(t)) = \frac{dx}{dt}\boldsymbol{i} + \frac{dy}{dt}\boldsymbol{j} + \frac{dz}{dt}\boldsymbol{k}.$$

Theorem 2 is proved by examining the difference quotient in Definition 3. According to Definition 1,

$$\begin{aligned}
&\lim_{\Delta t\to 0} \frac{\boldsymbol{f}(t+\Delta t) - \boldsymbol{f}(t)}{\Delta t} \\
&= \lim_{\Delta t\to 0} \frac{[x(t+\Delta t)\boldsymbol{i} + y(t+\Delta t)\boldsymbol{j} + z(t+\Delta t)\boldsymbol{k}] - [x(t)\boldsymbol{i} + y(t)\boldsymbol{j} + z(t)\boldsymbol{k}]}{\Delta t} \\
&= \lim_{\Delta t\to 0} \left\{\left(\frac{x(t+\Delta t) - x(t)}{\Delta t}\right)\boldsymbol{i} + \left(\frac{y(t+\Delta t) - y(t)}{\Delta t}\right)\boldsymbol{j} + \left(\frac{z(t+\Delta t) - z(t)}{\Delta t}\right)\boldsymbol{k}\right\} \\
&= \left[\lim_{\Delta t\to 0}\left(\frac{x(t+\Delta t) - x(t)}{\Delta t}\right)\right]\boldsymbol{i} + \left[\lim_{\Delta t\to 0}\left(\frac{y(t+\Delta t) - y(t)}{\Delta t}\right)\right]\boldsymbol{j} \\
&\quad + \left[\lim_{\Delta t\to 0}\left(\frac{z(t+\Delta t) - z(t)}{\Delta t}\right)\right]\boldsymbol{k}.
\end{aligned}$$

This shows that $\boldsymbol{f}(t)$ is differentiable if and only if each of the component functions is differentiable. Applying each of the indicated limits and using Definition 3 establishes equation (3).

Example 1 Let $\boldsymbol{f}(t) = t^2\boldsymbol{i} + \cos t\boldsymbol{j} + e^{3t}\boldsymbol{k}$. Find $\boldsymbol{f}'(t)$.

Solution:
$$\begin{aligned}
\boldsymbol{f}'(t) &= \left[\frac{d}{dt}(t^2)\right]\boldsymbol{i} + \left[\frac{d}{dt}(\cos t)\right]\boldsymbol{j} + \left[\frac{d}{dt}(e^{3t})\right]\boldsymbol{k} \\
&= 2t\boldsymbol{i} - \sin t\boldsymbol{j} + 3e^{3t}\boldsymbol{k}.
\end{aligned}$$
■

If the derivative $\boldsymbol{f}'(t)$ is itself differentiable, we define its derivative to be the *second* derivative of $\boldsymbol{f}(x)$. That is, $\boldsymbol{f}''(x) = [\boldsymbol{f}'(x)]'$. The second derivative of $\boldsymbol{f}(t)$ is again a vector function. Just as for real-valued functions, some or all derivatives of order higher than second may exist as well.

Example 2 Let $\boldsymbol{f}(t) = \sin 2t\boldsymbol{i} + \cos 2t\boldsymbol{j} + \sqrt{t}\boldsymbol{k}$. Find both $\boldsymbol{f}'(t)$ and $\boldsymbol{f}''(t)$.

Solution: For $t > 0$ we have

$$f'(t) = 2\cos 2t\boldsymbol{i} - 2\sin 2t\boldsymbol{j} + \frac{1}{2}t^{-1/2}\boldsymbol{k},$$

and

$$f''(t) = -4\sin 2t\boldsymbol{i} - 4\cos 2t\boldsymbol{j} - \frac{1}{4}t^{-3/2}\boldsymbol{k}.$$ ■

Example 3 Show that the function $\boldsymbol{f}(t) = A\sin\omega t\boldsymbol{i} + B\cos\omega t\boldsymbol{j}$ satisfies the **vector differential equation**

$$\boldsymbol{f}''(t) + \omega^2\boldsymbol{f}(t) = \boldsymbol{0}. \tag{4}$$

Strategy
Calculate $\boldsymbol{f}'(t)$ and $\boldsymbol{f}''(t)$.

Solution
Here

$$\boldsymbol{f}'(t) = \omega A\cos\omega t\boldsymbol{i} - \omega B\sin\omega t\boldsymbol{j}$$

and

$$\boldsymbol{f}''(t) = -\omega^2 A\sin\omega t\boldsymbol{i} - \omega^2 B\cos\omega t\boldsymbol{j}.$$

Substitute $\boldsymbol{f}$ and $\boldsymbol{f}''$ into the left-hand side of (4) and verify that $\boldsymbol{0}$ is obtained.

Substituting into equation (4) shows that

$$\begin{aligned}\boldsymbol{f}''(t) + \omega^2\boldsymbol{f}(t) &= [-\omega^2 A\sin\omega t\boldsymbol{i} - \omega^2 B\cos\omega t\boldsymbol{j}] \\ &\quad + \omega^2[A\sin\omega t\boldsymbol{i} + B\cos\omega t\boldsymbol{j}] \\ &= \boldsymbol{0}.\end{aligned}$$ ■

The following theorem shows how derivatives of various other vector functions may be calculated.

THEOREM 3

Let the vector functions $\boldsymbol{f}(t)$ and $\boldsymbol{g}(t)$ and the scalar-valued function $h(t)$ be differentiable on appropriate intervals and let c be a number. Then the vector functions $(\boldsymbol{f} + \boldsymbol{g})(t)$, $(c\boldsymbol{f})(t)$, $(h\boldsymbol{f})(t)$, $(\boldsymbol{f} \times \boldsymbol{g})(t)$, and $\boldsymbol{f}(h(t))$, and the scalar-valued function $(\boldsymbol{f}\cdot\boldsymbol{g})(t)$ are differentiable, and

(i) $(\boldsymbol{f} + \boldsymbol{g})'(t) = \boldsymbol{f}'(t) + \boldsymbol{g}'(t)$,
(ii) $(c\boldsymbol{f})'(t) = c\boldsymbol{f}'(t)$,
(iii) $(h\boldsymbol{f})'(t) = h(t)\boldsymbol{f}'(t) + h'(t)\boldsymbol{f}(t)$,
(iv) $(\boldsymbol{f} \times \boldsymbol{g})'(t) = [\boldsymbol{f}(t) \times \boldsymbol{g}'(t)] + [\boldsymbol{f}'(t) \times \boldsymbol{g}(t)]$,
(v) $[\boldsymbol{f}(h(t))]' = h'(t)\boldsymbol{f}'(h(t))$ (Chain Rule),
(vi) $(\boldsymbol{f}\cdot\boldsymbol{g})'(t) = \boldsymbol{f}(t)\cdot\boldsymbol{g}'(t) + \boldsymbol{f}'(t)\cdot\boldsymbol{g}(t)$.

Before commenting on the proof of Theorem 3, we consider two examples of its use.

Example 4 Let $\boldsymbol{f}(t) = t^2\boldsymbol{i} + e^t\boldsymbol{k}$ and $\boldsymbol{g}(t) = \sin t\boldsymbol{i} + \cos t\boldsymbol{j} + \boldsymbol{k}$. Find the derivative of the vector function $\boldsymbol{r}(t) = (\boldsymbol{f} \times \boldsymbol{g})(t)$.

Solution: The derivatives of the given functions are

$$\boldsymbol{f}'(t) = 2t\boldsymbol{i} + e^t\boldsymbol{k} \qquad \text{and} \qquad \boldsymbol{g}'(t) = \cos t\boldsymbol{i} - \sin t\boldsymbol{j}.$$

According to part (iv) of Theorem 3 we have

$$\begin{aligned}
\mathbf{r}'(t) &= [\mathbf{f}(t) \times \mathbf{g}'(t)] + [\mathbf{f}'(t) \times \mathbf{g}(t)] \\
&= \det\begin{bmatrix} \mathbf{i} & \mathbf{j} & \mathbf{k} \\ t^2 & 0 & e^t \\ \cos t & -\sin t & 0 \end{bmatrix} + \det\begin{bmatrix} \mathbf{i} & \mathbf{j} & \mathbf{k} \\ 2t & 0 & e^t \\ \sin t & \cos t & 1 \end{bmatrix} \\
&= [e^t \sin t\mathbf{i} + e^t \cos t\mathbf{j} - t^2 \sin t\mathbf{k}] \\
&\quad + [-e^t \cos t\mathbf{i} + (e^t \sin t - 2t)\mathbf{j} + 2t \cos t\mathbf{k}] \\
&= e^t(\sin t - \cos t)\mathbf{i} + [e^t(\sin t + \cos t) - 2t]\mathbf{j} \\
&\quad + (2t \cos t - t^2 \sin t)\mathbf{k}.
\end{aligned}$$

This same result may be obtained directly by noting that

$$\begin{aligned}
\mathbf{r}(t) &= \mathbf{f}(t) \times \mathbf{g}(t) \\
&= \det\begin{bmatrix} \mathbf{i} & \mathbf{j} & \mathbf{k} \\ t^2 & 0 & e^t \\ \sin t & \cos t & 1 \end{bmatrix} \\
&= -e^t \cos t\mathbf{i} + (e^t \sin t - t^2)\mathbf{j} + t^2 \cos t\mathbf{k}
\end{aligned}$$

and differentiating component by component, according to Theorem 2. ■

Example 5 Let $\mathbf{f}(t)$ and $\mathbf{g}(t)$ be as in Example 4. Find the derivative of the scalar-valued function $s(t) = (\mathbf{f} \cdot \mathbf{g})(t)$.

Solution: By Theorem 3, part (vi),

$$\begin{aligned}
s'(t) &= \mathbf{f}(t) \cdot \mathbf{g}'(t) + \mathbf{f}'(t) \cdot \mathbf{g}(t) \\
&= [(t^2\mathbf{i} + e^t\mathbf{k}) \cdot (\cos t\mathbf{i} - \sin t\mathbf{j})] + [(2t\mathbf{i} + e^t\mathbf{k}) \cdot (\sin t\mathbf{i} + \cos t\mathbf{j} + \mathbf{k})] \\
&= t^2 \cos t + 2t \sin t + e^t.
\end{aligned}$$

To calculate this derivative directly, we first compute

$$\begin{aligned}
s(t) = \mathbf{f}(t) \cdot \mathbf{g}(t) &= t^2 \sin t + (0)(\cos t) + e^t \\
&= t^2 \sin t + e^t.
\end{aligned}$$

Differentiating then gives the above result. ■

Examples 4 and 5 suggest that Theorem 3 may be proved by writing the vector functions $\mathbf{f}(t)$ and $\mathbf{g}(t)$ in component form and applying Theorem 2. This is indeed the case. We prove part (v) here and leave the other statements as exercises. To do so we write $\mathbf{f}(t)$ in component form as

$$\mathbf{f}(t) = x(t)\mathbf{i} + y(t)\mathbf{j} + z(t)\mathbf{k}.$$

Then

$$\mathbf{f}(h(t)) = x(h(t))\mathbf{i} + y(h(t))\mathbf{j} + z(h(t))\mathbf{k}.$$

According to Theorem 2 and the Chain Rule of Chapter 4,

$$\begin{aligned}
[\mathbf{f}(h(t))]' &= [x(h(t))]'\mathbf{i} + [y(h(t))]'\mathbf{j} + [z(h(t))]'\mathbf{k} \\
&= [x'(h(t)) \cdot h'(t)]\mathbf{i} + [y'(h(t)) \cdot h'(t)]\mathbf{j} + [z'(h(t)) \cdot h'(t)]\mathbf{k} \\
&= h'(t)[x'(h(t))\mathbf{i} + y'(h(t))\mathbf{j} + z'(h(t))\mathbf{k}] \\
&= h'(t) \cdot \mathbf{f}'(h(t)).
\end{aligned}$$

■

The result of the following example will be useful in later calculations.

Example 6 Let $\boldsymbol{f}(t)$ be a vector function and let c be a constant. Prove that if $\boldsymbol{f}(t)$ is differentiable on some interval I and if $|\boldsymbol{f}(t)| = c$ for all $t \in I$, then $\boldsymbol{f}(t)$ and $\boldsymbol{f}'(t)$ are orthogonal for all $t \in I$.

Strategy

Express $|\boldsymbol{f}(t)|^2$ as a dot product using $|\boldsymbol{v}|^2 = \boldsymbol{v} \cdot \boldsymbol{v}$.

The resulting expression equals c^2.

Differentiate both sides of this equation using part (vi) of Theorem 3.

Use fact that $\boldsymbol{v}$ is orthogonal to $\boldsymbol{w}$ if and only if $\boldsymbol{v} \cdot \boldsymbol{w} = 0$.

Solution

By Theorem 3, Chapter 17,

$$|\boldsymbol{f}(t)|^2 = \boldsymbol{f}(t) \cdot \boldsymbol{f}(t).$$

Thus, since $|\boldsymbol{f}(t)| = c$, we have

$$\boldsymbol{f}(t) \cdot \boldsymbol{f}(t) = c^2, \qquad t \in I.$$

Differentiating both sides of this equation with respect to t gives

$$\boldsymbol{f}(t) \cdot \boldsymbol{f}'(t) + \boldsymbol{f}'(t) \cdot \boldsymbol{f}(t) = 0$$

or,

$$2\boldsymbol{f}(t) \cdot \boldsymbol{f}'(t) = 0.$$

Thus $\boldsymbol{f}'(t) \cdot \boldsymbol{f}(t) = 0, \quad t \in I$, so $\boldsymbol{f}(t)$ and $\boldsymbol{f}'(t)$ are orthogonal for all $t \in I$. ■

Integrals of Vector Functions

Since derivatives of vector functions are obtained by simply differentiating the individual components, we define integrals for such functions in the same way.

DEFINITION 4

The **indefinite integral** (or **antiderivative**) of the vector function $\boldsymbol{f}(t) = x(t)\boldsymbol{i} + y(t)\boldsymbol{j} + z(t)\boldsymbol{k}$ is

$$\int \boldsymbol{f}(t)\,dt = \left[\int x(t)\,dt\right]\boldsymbol{i} + \left[\int y(t)\,dt\right]\boldsymbol{j} + \left[\int z(t)\,dt\right]\boldsymbol{k} \tag{5}$$

provided that the integrals of the component functions exist.

REMARK: It is important to note in equation (5) that *three* constants of integration will arise—one for each of the antiderivatives of the respective components. That is, if $X'(t) = x(t)$, $Y'(t) = y(t)$, and $Z'(t) = z(t)$, we can write equation (5) as either

$$\int \boldsymbol{f}(t)\,dt = [X(t) + c_1]\boldsymbol{i} + [Y(t) + c_2]\boldsymbol{j} + [Z(t) + c_3]\boldsymbol{k}, \tag{6}$$

or,

$$\int \boldsymbol{f}(t)\,dt = X(t)\boldsymbol{i} + Y(t)\boldsymbol{j} + Z(t)\boldsymbol{k} + \boldsymbol{C} \tag{7}$$

where

$$\boldsymbol{C} = c_1\boldsymbol{i} + c_2\boldsymbol{j} + c_3\boldsymbol{k}.$$

A common mistake is to write the constant $\boldsymbol{C}$ in equation (7) as a number rather than as a vector.

Example 7 Find the antiderivative $\int f(t)\,dt$ for

$$f(t) = 2t\boldsymbol{i} + e^{-3t}\boldsymbol{j} + \sec^2 t\boldsymbol{k}.$$

Solution: By Definition (4)

$$\begin{aligned}\int f(t)\,dt &= \left[\int 2t\,dt\right]\boldsymbol{i} + \left[\int e^{-3t}\,dt\right]\boldsymbol{j} + \left[\int \sec^2 t\,dt\right]\boldsymbol{k}\\ &= (t^2 + c_1)\boldsymbol{i} + \left(-\frac{1}{3}e^{-3t} + c_2\right)\boldsymbol{j} + (\tan t + c_3)\boldsymbol{k}\\ &= t^2\boldsymbol{i} - \frac{1}{3}e^{-3t}\boldsymbol{j} + \tan t\boldsymbol{k} + \boldsymbol{C}, \qquad \boldsymbol{C} = c_1\boldsymbol{i} + c_2\boldsymbol{j} + c_3\boldsymbol{k}.\end{aligned}$$

■

Example 8 Find the vector function $f(t)$ for which

$$f'(t) = \boldsymbol{i} + t^3\boldsymbol{j} + \left(\frac{1}{1+t^2}\right)\boldsymbol{k},$$

and $f(0) = 2\boldsymbol{i} - \boldsymbol{j} + 3\boldsymbol{k}$.

Solution: By antidifferentiation we obtain

$$\begin{aligned}f(t) &= \int f'(t)\,dt\\ &= \int\left[\boldsymbol{i} + t^3\boldsymbol{j} + \left(\frac{1}{1+t^2}\right)\boldsymbol{k}\right]dt\\ &= (t + c_1)\boldsymbol{i} + \left(\frac{1}{4}t^4 + c_2\right)\boldsymbol{j} + (\tan^{-1} t + c_3)\boldsymbol{k}.\end{aligned}$$

Setting $t = 0$ and using the given *initial condition* $f(0) = 2\boldsymbol{i} - \boldsymbol{j} + 3\boldsymbol{k}$ gives the equation

$$f(0) = c_1\boldsymbol{i} + c_2\boldsymbol{j} + c_3\boldsymbol{k} = 2\boldsymbol{i} - \boldsymbol{j} + 3\boldsymbol{k}.$$

Thus, $c_1 = 2$, $c_2 = -1$, and $c_3 = 3$. Thus,

$$f(t) = (t + 2)\boldsymbol{i} + \left(\frac{1}{4}t^4 - 1\right)\boldsymbol{j} + (\tan^{-1} t + 3)\boldsymbol{k}$$

is the desired solution. ■

Finally, the definite integral of the vector function is defined as the vector whose components are the respective definite integrals of the components of $f(t)$.

DEFINITION 5

Let $f(t) = x(t)\boldsymbol{i} + y(t)\boldsymbol{j} + z(t)\boldsymbol{k}$. If the definite integral of each of the components exists on the interval $[a, b]$, we define

$$\int_a^b f(t)\,dt = \left[\int_a^b x(t)\,dt\right]\boldsymbol{i} + \left[\int_a^b y(t)\,dt\right]\boldsymbol{j} + \left[\int_a^b z(t)\,dt\right]\boldsymbol{k}.$$

Example 9 For $\boldsymbol{f}(t) = \sin 3t\boldsymbol{i} + \cos t\boldsymbol{j} + \boldsymbol{k}$, find

$$\int_0^{\pi/2} \boldsymbol{f}(t)\, dt.$$

Solution: By Definition 5,

$$\begin{aligned}\int_0^{\pi/2} \boldsymbol{f}(t)\, dt &= \left[\int_0^{\pi/2} \sin 3t\, dt\right]\boldsymbol{i} + \left[\int_0^{\pi/2} \cos t\, dt\right]\boldsymbol{j} + \left[\int_0^{\pi/2} dt\right]\boldsymbol{k}\\ &= \left[-\frac{1}{3}\cos 3t\Big|_0^{\pi/2}\right]\boldsymbol{i} + \left[\sin t\Big|_0^{\pi/2}\right]\boldsymbol{j} + \left[t\Big|_0^{\pi/2}\right]\boldsymbol{k}\\ &= \left[0 - \left(-\frac{1}{3}\right)\right]\boldsymbol{i} + [1 - 0]\boldsymbol{j} + \left[\frac{\pi}{2} - 0\right]\boldsymbol{k}\\ &= \frac{1}{3}\boldsymbol{i} + \boldsymbol{j} + \frac{\pi}{2}\boldsymbol{k}.\end{aligned}$$

■

We may summarize the results of this section by saying that for the purposes of calculating limits, derivatives, or integrals, the vector function $\boldsymbol{f}(t) = x(t)\boldsymbol{i} + y(t)\boldsymbol{j} + z(t)\boldsymbol{k}$ is regarded as a sum of three component functions of a single variable. In the remaining sections of this chapter, we shall see how these results may be combined with a broader interpretation of $\boldsymbol{f}(t)$ as a vector in space to develop the concepts of tangent, normal, velocity, acceleration, and curvature associated with curves in space.

Exercise Set 18.3

In Exercises 1–4, state the intervals on which the given vector function is differentiable.

1. $\boldsymbol{f}(t) = t\boldsymbol{i} + \cos t\boldsymbol{j} + \sqrt{t}\,\boldsymbol{k}$

2. $\boldsymbol{f}(t) = t^{-2}\boldsymbol{i} + |t - 3|\boldsymbol{j} + t^3\boldsymbol{k}$

3. $\boldsymbol{f}(t) = \ln(3 + t)\boldsymbol{i} + \left(\dfrac{1}{t + 2}\right)\boldsymbol{j} + e^t\boldsymbol{k}$

4. $\boldsymbol{f}(t) = \sqrt{1 - t^2}\boldsymbol{i} + \ln t\boldsymbol{j} + \dfrac{1}{1 - t}\boldsymbol{k}$

In Exercises 5–10, find the derivative of the given vector function.

5. $\boldsymbol{f}(t) = t\boldsymbol{i} + \sqrt{t}\,\boldsymbol{j}$

6. $\boldsymbol{f}(t) = \boldsymbol{i} + \sin t\boldsymbol{j} + \cos 2t\boldsymbol{k}$

7. $\boldsymbol{f}(t) = \sqrt{t}\,\boldsymbol{i} + t^{-3/2}\boldsymbol{j} + \ln(2t - 1)\boldsymbol{k}$

8. $\boldsymbol{f}(t) = \cos^2 t\boldsymbol{i} + \sec t\boldsymbol{j} + \tan^{-1} t\boldsymbol{k}$

9. $\boldsymbol{f}(t) = \sin^{-1} t\boldsymbol{i} + \sqrt{1 + t^2}\,\boldsymbol{j} + e^{-t^3}\boldsymbol{k}$

10. $\boldsymbol{f}(t) = e^{\sqrt{t}}\boldsymbol{i} + 3\boldsymbol{j} - \cos^{-1} 2t\boldsymbol{k}$

11. Find $\boldsymbol{f}''(t)$ for the function $\boldsymbol{f}(t)$ in Exercise 7.

12. Find $\boldsymbol{f}''(t)$ for the function $\boldsymbol{f}(t)$ in Exercise 10.

13. Show that the vector function $\boldsymbol{f}(t) = Ae^{\omega t}\boldsymbol{i} + Be^{-\omega t}\boldsymbol{j}$ satisfies the vector differential equation

$$\boldsymbol{f}''(t) - \omega^2\boldsymbol{f}(t) = 0.$$

14. Show that the vector function

$$\boldsymbol{f}(t) = C \sinh \omega t\boldsymbol{i} + D \cosh \omega t\boldsymbol{j}$$

satisfies the vector differential equation in Exercise 13.

In Exercises 15–20, let

$$\begin{aligned}\boldsymbol{f}(t) &= \sin t\boldsymbol{i} + \boldsymbol{j} + t^2\boldsymbol{k},\\ \boldsymbol{g}(t) &= t\boldsymbol{i} + \cos t\boldsymbol{k},\\ h(t) &= e^{3t}.\end{aligned}$$

Find the derivative of the given function.

15. $\boldsymbol{r}(t) = (3\boldsymbol{f} - 2\boldsymbol{g})(t)$

16. $\boldsymbol{r}(t) = h(t)\boldsymbol{f}(t)$

17. $\boldsymbol{r}(t) = (\boldsymbol{f} \times \boldsymbol{g})(t)$

18. $\boldsymbol{r}(t) = (\boldsymbol{f} \cdot \boldsymbol{g})(t)$

19. $\boldsymbol{r}(t) = \boldsymbol{f}(h(t))$

20. $\boldsymbol{r}(t) = |\boldsymbol{g}(t)|^2$

21. Verify that $\boldsymbol{f}(t)$ and $\boldsymbol{f}'(t)$ are orthogonal if $\boldsymbol{f}(t)$ is the vector function $\boldsymbol{f}(t) = \cos 2t\boldsymbol{i} + \sin 2t\boldsymbol{j} + 3\boldsymbol{k}$.

In Exercises 22–26, find the antiderivative for the given vector function.

22. $f(t) = t\boldsymbol{i} + \cos t\boldsymbol{j} + \sin t\boldsymbol{k}$

23. $f(t) = \sqrt{t}\,\boldsymbol{i} + e^{2t}\boldsymbol{j} + \dfrac{1}{t}\boldsymbol{k}$

24. $f(t) = t \sin t\boldsymbol{i} + \cos^2 t\boldsymbol{j} + (1 - t)\boldsymbol{k}$

25. $f(t) = \ln t\boldsymbol{i} + \dfrac{1}{t \ln t}\boldsymbol{j}$

26. $f(t) = t \sec^2(1 - t^2)\boldsymbol{i} + \sqrt{1 - t}\,\boldsymbol{j} + \sqrt{t}\,\boldsymbol{k}$

27. Find $f(t)$ if $f'(t) = \boldsymbol{i} + t^2\boldsymbol{j}$ and $f(0) = 3\boldsymbol{i} + 5\boldsymbol{j}$

28. Find $f(t)$ if $f'(t) = \cos 2t\boldsymbol{i} + \left(\dfrac{1}{1 + t^2}\right)\boldsymbol{j} + \dfrac{t}{\sqrt{1 + t^2}}\boldsymbol{k}$ and $f(0) = -4\boldsymbol{i} + \boldsymbol{k}$.

29. Find $f(t)$ if $f'(t) = \cos^2 t\boldsymbol{i} + \sin^2 t\boldsymbol{j} + e^{-t}\boldsymbol{k}$ and $f(0) = 3\boldsymbol{i} - 6\boldsymbol{j} + 2\boldsymbol{k}$.

30. Find a solution of the vector differential equation $f'(t) = f(t)$ for which

$$f(0) = \boldsymbol{i} + 2\boldsymbol{j} - \boldsymbol{k}.$$

31. Find a solution of the vector differential equation

$$f'(t) + 4f(t) = 0$$

satisfying the initial condition $f(0) = \boldsymbol{i} + 4\boldsymbol{k}$.

32. Find a family of solutions to the vector differential equation

$$f''(t) - 2f(t) = 0.$$

33. Find $\displaystyle\int_a^b (t\boldsymbol{i} + t^2\boldsymbol{j} - t^3\boldsymbol{k})\,dt$.

34. Find $\displaystyle\int_0^1 f(t)\,dt$ if $f(t) = t^2\boldsymbol{i} - 2t\boldsymbol{j} + \sqrt{t}\,\boldsymbol{k}$.

35. Find $\displaystyle\int_0^{\pi/4} f(t)\,dt$ if $f(t) = \cos t\boldsymbol{i} + \sin 2t\boldsymbol{j} + \cos^2 t\boldsymbol{k}$.

36. Find $\displaystyle\int_1^4 f(t)\,dt$ if $f(t) = e^t\boldsymbol{i} + t^2\boldsymbol{j} + \ln 2t\boldsymbol{k}$.

37. Show that if $f'(t) = \boldsymbol{0}$ for all $t \in I$, then $f(t)$ is constant on the interval I.

38. Show that if $f'(t) = g'(t)$ for all $t \in I$ then $f(t) = g(t) + \boldsymbol{C}$ for some vector $\boldsymbol{C}$ and all $t \in I$.

39. True or false? If $f(t)$ is differentiable on $[a, b]$, there exists a constant $c \in [a, b]$ so that

$$f'(c) = \frac{f(b) - f(a)}{b - a}.$$

40. Prove statements (i) through (iv) and (vi) of Theorem 3 by first writing the vector functions in component form.

41. Complete the following alternate proof of differentiation formula (iii) of Theorem 3:

$$(hf)'(t) = h'(t)f(t) + h(t)f'(t).$$

a. Show that we can write

$$\begin{aligned}(hf)'(t) &= \lim_{\Delta t \to 0} \frac{1}{\Delta t}[h(t + \Delta t)f(t + \Delta t) - h(t)f(t)]\\ &= \lim_{\Delta t \to 0} \frac{1}{\Delta t}[h(t + \Delta t) - h(t)]f(t + \Delta t)\\ &\quad + \lim_{\Delta t \to 0} \frac{1}{\Delta t}[f(t + \Delta t) - f(t)]h(t)\end{aligned}$$

b. Show that the limit in part (a) is

$$(hf)'(t) = h'(t)f(t) + h(t)f'(t).$$

42. Use a method similar to that of Exercise 41 to prove part (vi) of Theorem 3:

$$(f \cdot g)'(t) = f'(t) \cdot g(t) + f(t) \cdot g'(t).$$

18.4 TANGENT VECTORS AND ARC LENGTH

We have already seen (Section 16.5) that a curve C in the plane can be described by a pair of parametric equations. For example, the unit circle is parameterized by the equations

$$C: \begin{cases} x(t) = \cos t \\ y(t) = \sin t \end{cases} \qquad 0 \le t < 2\pi.$$

In a similar way, we may interpret the vector function

$$C: \quad \boldsymbol{r}(t) = x(t)\boldsymbol{i} + y(t)\boldsymbol{j} + z(t)\boldsymbol{k} \qquad t \in I \tag{1}$$

as giving a parameterization of a curve C in space. That is, we interpret C as the set of all points $P(t) = (x(t), y(t), z(t))$ lying at the tips of the position vectors $\boldsymbol{r}(t)$.

(Recall that interpreting $\boldsymbol{r}(t)$ as a **position vector** means that $\boldsymbol{r}(t)$ originates at the origin.) As for curves in the plane, we shall say that the curve C is **differentiable** if the vector function $\boldsymbol{r}(t)$ is differentiable on I, which in turn requires that each of the component functions $x(t)$, $y(t)$, and $z(t)$ is differentiable (Theorem 2). The curve C is called **smooth** if $\boldsymbol{r}'(t)$ is continuous on I.

When the derivative $\boldsymbol{r}'(t_0)$ exists, we refer to it as a *tangent vector*.

DEFINITION 6

For the differentiable curve C in equation (1), if $\boldsymbol{r}'(t_0) \neq \boldsymbol{0}$, the vector

$$\boldsymbol{r}'(t_0) = x'(t_0)\boldsymbol{i} + y'(t_0)\boldsymbol{j} + z'(t_0)\boldsymbol{k}, \qquad t_0 \in I$$

is called a **tangent vector** for C at the tip of the vector $\boldsymbol{r}(t)$ (Figure 4.1).

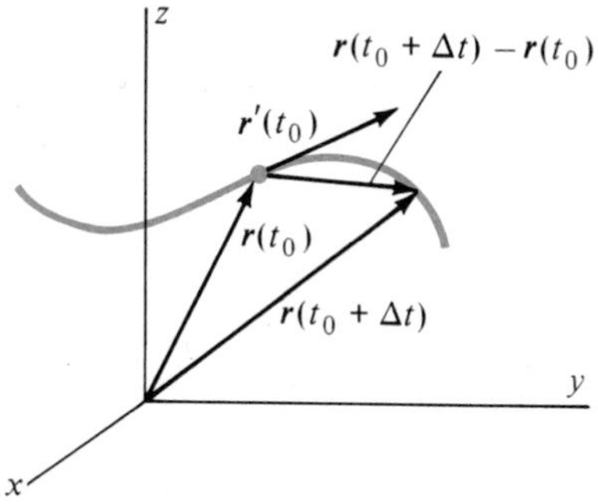

Figure 4.1 $\boldsymbol{r}'(t_0)$ is tangent to C at tip of $\boldsymbol{r}(t_0)$.

To see that Definition 6 agrees with our earlier concept of tangent, observe that the vector

$$\boldsymbol{r}(t_0 + \Delta t) - \boldsymbol{r}(t_0) \tag{2}$$

originates at the tip of $\boldsymbol{r}(t_0)$ and terminates at the tip of $\boldsymbol{r}(t_0 + \Delta t)$ (see Figure 4.1). This vector may therefore be regarded as a secant, joining the points $\boldsymbol{P}(t_0)$ and $\boldsymbol{P}(t_0 + \Delta t)$ on C. As $\Delta t \to 0$ the secant line through these same points approaches the line we shall call the *tangent* to C at $\boldsymbol{r}(t_0)$. However, we cannot obtain a direction vector for this line by simply applying $\lim_{\Delta t \to 0}$ to the vector in equation (1) since the length of this vector will approach zero as $\Delta t \to 0$. We therefore multiply the vector in (2) by the scalar $\frac{1}{\Delta t}$ (changing its length, but not its direction) before applying the limit. Thus, the direction of the tangent to C at $\boldsymbol{r}(t_0)$ is given by the vector

$$\boldsymbol{r}'(t_0) = \lim_{\Delta t \to 0} \frac{\boldsymbol{r}(t_0 + \Delta t) - \boldsymbol{r}(t_0)}{\Delta t}. \tag{3}$$

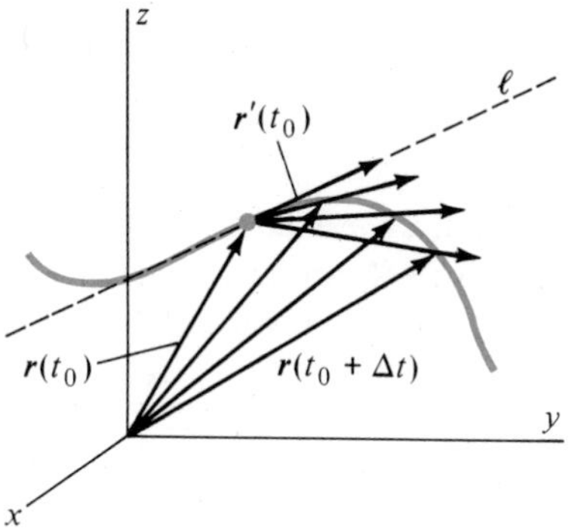

Figure 4.2 Line ℓ tangent to C at tip of $\boldsymbol{r}(t_0)$ has direction vector $\boldsymbol{r}'(t_0)$.

(See Figure 4.2.) Since we assume $\boldsymbol{r}'(t_0) \neq \boldsymbol{0}$, the tangent vector has nonzero length.

Example 1 Show that a vector tangent to the curve

$$C: \quad \boldsymbol{r}(t) = a \cos t\boldsymbol{i} + a \sin t\boldsymbol{j} + b\boldsymbol{k}, \qquad 0 \le t < 2\pi$$

at $\boldsymbol{r}(t)$ is orthogonal to $\boldsymbol{r}(t)$.

Solution: A vector tangent to C is, by Definition 6,

$$\boldsymbol{r}'(t) = -a \sin t\boldsymbol{i} + a \cos t\boldsymbol{j}.$$

Since $\boldsymbol{r}(t) \cdot \boldsymbol{r}'(t) = -a^2 \cos t \sin t + a^2 \sin t \cos t = 0$, the vectors $\boldsymbol{r}(t)$ and $\boldsymbol{r}'(t)$ are orthogonal. (Note that since $|\boldsymbol{r}(t)| = \sqrt{a^2 \cos^2 t + a^2 \sin^2 t + b^2} = \sqrt{a^2 + b^2}$ is constant, this result also follows by Example 6, Section 18.3.) ■

If $\boldsymbol{r}'(t_0) \neq \boldsymbol{0}$, a unit vector $\boldsymbol{T}$ tangent to the curve C at $\boldsymbol{r}(t_0)$ is given by

$$\boldsymbol{T} = \frac{\boldsymbol{r}'(t_0)}{|\boldsymbol{r}'(t_0)|} \quad \text{(unit tangent).} \tag{4}$$

Example 2 Find a unit vector tangent to the circular helix $\boldsymbol{r}(t) = a \cos t\boldsymbol{i} + a \sin t\boldsymbol{j} + bt\boldsymbol{k}$ at the point $(0, a, b\pi/2)$.

Strategy

Solution

Find t_0 for which $\boldsymbol{r}(t_0)$ terminates at the point $\left(0, a, \frac{b\pi}{2}\right)$.

Setting $\boldsymbol{r}(t_0) = 0\boldsymbol{i} + a\boldsymbol{j} + \frac{b\pi}{2}\boldsymbol{k}$ gives the equations

$$a \cos t_0 = 0, \qquad a \sin t_0 = a, \qquad bt_0 = \frac{b\pi}{2},$$

to which the solution is $t_0 = \pi/2$.

Find $\boldsymbol{r}'(t_0)$.

According to Definition 6 a tangent vector at $\boldsymbol{r}(t_0) = \boldsymbol{r}(\pi/2)$ is

$$\begin{aligned} \boldsymbol{r}'\left(\frac{\pi}{2}\right) &= -a \sin\left(\frac{\pi}{2}\right)\boldsymbol{i} + a \cos\left(\frac{\pi}{2}\right)\boldsymbol{j} + b\boldsymbol{k} \\ &= -a\boldsymbol{i} + b\boldsymbol{k}. \end{aligned}$$

Find $|\boldsymbol{r}'(t_0)|$.

This vector has length

$$\left|\boldsymbol{r}'\left(\frac{\pi}{2}\right)\right| = \sqrt{a^2 + b^2}.$$

Apply (4).

The desired unit tangent is

$$\boldsymbol{T} = \frac{\boldsymbol{r}'\left(\frac{\pi}{2}\right)}{\left|\boldsymbol{r}'\left(\frac{\pi}{2}\right)\right|} = \frac{-a\boldsymbol{i} + b\boldsymbol{k}}{\sqrt{a^2 + b^2}}.$$ ■

To find an equation for the *line* ℓ tangent to C at $\boldsymbol{r}(t_0)$, we use the fact that $\boldsymbol{r}(t_0)$ is a position vector for ℓ and $\boldsymbol{r}'(t_0)$ is a direction vector for ℓ. Since t is the parameter for C, we should use a different letter to parameterize ℓ, say ω. We may then use equation (1), Section 17.7, to write a parameterization for ℓ as

$$\ell\colon\ \boldsymbol{R}(\omega) = \boldsymbol{r}(t_0) + \omega\boldsymbol{r}'(t_0). \tag{5}$$

(Be careful when using equation (5): $\boldsymbol{R}(\omega)$ is the position vector of a point on ℓ, while $\boldsymbol{r}(t_0)$ and $\boldsymbol{r}'(t_0)$ are fixed vectors determined by the curve C.)

Example 3 Find vector, parametric, and symmetric equations for the line ℓ tangent to the curve

$$C: \quad r(t) = t^2 i + (3 - t)j + t^3 k$$

at the point $(4, 1, 8)$.

Strategy

Find t_0 for which $r(t_0)$ terminates at the point $(4, 1, 8)$. $r(t_0)$ is the position vector for ℓ.

Find $r'(t_0)$. This is the direction vector for ℓ.

Use equation (5) to write the vector form of ℓ.

The parametric equations for x, y, and z are the components of $R(\omega)$.

The symmetric equations are found from the parametric equations as in Section 17.7.

Solution

Setting $r(t_0) = 4i + j + 8k$ gives the equations

$$t^2 = 4, \qquad 3 - t = 1, \qquad \text{and} \qquad t^3 = 8$$

for which the solution is $t_0 = 2$.

Since $r'(t) = 2ti - j + 3t^2 k$,

$$r'(t_0) = r'(2) = 4i - j + 12k.$$

The vector form of ℓ is therefore

$$\ell: R(\omega) = (4i + j + 8k) + \omega(4i - j + 12k).$$

To find parametric equations for ℓ, we let $P(\omega) = (x(\omega), y(\omega), z(\omega))$ be a point on ℓ. Then

$$
\begin{aligned}
x(\omega) &= 4 + 4\omega && (\textbf{\textit{i}}\text{-components}) \\
y(\omega) &= 1 - \omega && (\textbf{\textit{j}}\text{-components}) \\
z(\omega) &= 8 + 12\omega && (\textbf{\textit{k}}\text{-components})
\end{aligned}
$$

are the parametric equations for the coordinates of P. Solving each of these equations for ω and equating the results gives the symmetric equations

$$\frac{x - 4}{4} = 1 - y = \frac{z - 8}{12}$$

for ℓ. $\blacksquare$

Arc Length

If C is a smooth curve in space parameterized by the vector function

$$r(t) = x(t)i + y(t)j + z(t)k$$

an expression for the length L of the arc of C from $r(a)$ to $r(b)$ is obtained in a manner similar to that for plane curves (see Section 7.5). First, the interval $[a, b]$ is partitioned into n subintervals of equal length $\Delta t = \dfrac{b - a}{n}$ and with endpoints $a = t_0 < t_1 < t_2 < \cdots < t_n = b$. This determines a polygonal path connecting the tips of the vectors $r(t_0)$, $r(t_1)$, . . . , $r(t_n)$ of total length

$$\sum_{j=1}^{n} \Delta L_j = \sum_{j=1}^{n} \sqrt{\Delta x_j^2 + \Delta y_j^2 + \Delta z_j^2} \tag{6}$$

where $\Delta x_j = x(t_j) - x(t_{j-1})$, $\Delta y_j = y(t_j) - y(t_{j-1})$, and $\Delta z_j = z(t_j) - z(t_{j-1})$. As before, applications of the Mean Value Theorem (once for each component) lead to an approximation of the form

$$L \approx \sum_{j=1}^{n} \sqrt{[x'(u_j)]^2 + [y'(v_j)]^2 + [z'(w_j)]^2} \, \Delta t_j. \tag{7}$$

As $n \to \infty$, $\Delta t \to 0$ and we obtain

$$L = \int_a^b \sqrt{[x'(t)]^2 + [y'(t)]^2 + [z'(t)]^2}\, dt \tag{8}$$

which we take as the definition of the arc length L. (Since this is familiar ground, we do not pursue the details of this development here.) Since the expression under the radical in equation (8) is equal to $\boldsymbol{r}'(t) \cdot \boldsymbol{r}'(t) = |\boldsymbol{r}'(t)|^2$, the arc length may be expressed more compactly as

$$L = \int_a^b |\boldsymbol{r}'(t)|\, dt. \tag{9}$$

Example 4 Find the length of the curve

$$C: \quad \boldsymbol{r}(t) = \cos \pi t\boldsymbol{i} + \sin \pi t\boldsymbol{j} + t^{3/2}\boldsymbol{k}$$

from $\boldsymbol{r}(0) = \boldsymbol{i}$ to $\boldsymbol{r}(4) = \boldsymbol{i} + 8\boldsymbol{k}$.

Solution: Here

$$\boldsymbol{r}'(t) = -\pi \sin \pi t\boldsymbol{i} + \pi \cos \pi t\boldsymbol{j} + \frac{3}{2}\sqrt{t}\,\boldsymbol{k}$$

so

$$|\boldsymbol{r}'(t)| = \sqrt{\pi^2 \sin^2 \pi t + \pi^2 \cos^2 \pi t + \frac{9}{4}t}$$
$$= \sqrt{\pi^2 + \frac{9}{4}t}.$$

By formula (9),

$$L = \int_0^4 \sqrt{\pi^2 + \frac{9}{4}t}\, dt = \frac{8}{27}\left(\pi^2 + \frac{9}{4}t\right)^{3/2}\Bigg]_0^4$$
$$= \frac{8}{27}[(\pi^2 + 9)^{3/2} - \pi^3] \approx 15.1$$

(Figure 4.3). ■

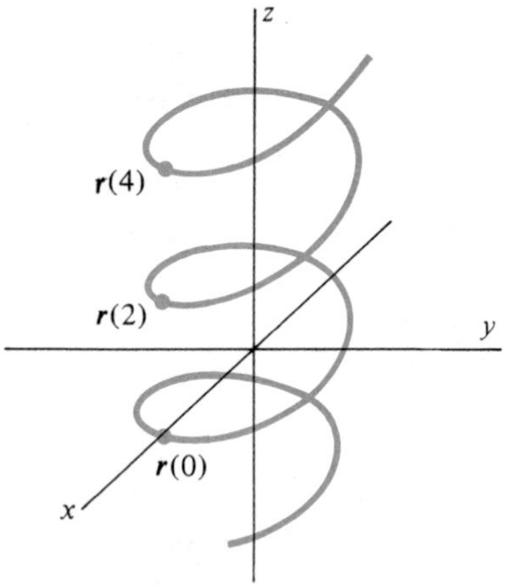

Figure 4.3

Differential for Arc Length

As for plane curves, we refer to the arc length function, $s(t)$, as

$$s(t) = \int_a^t |\boldsymbol{r}'(\omega)|\, d\omega. \tag{10}$$

The function $s(t)$ gives the length of the arc of the curve C: $\boldsymbol{r}(t) = x(t)\boldsymbol{i} + y(t)\boldsymbol{j} + z(t)\boldsymbol{k}$ from $\boldsymbol{r}(a)$ to $\boldsymbol{r}(t)$. By Theorem 7, Chapter 6, we have

$$s'(t) = |\boldsymbol{r}'(t)| = \sqrt{[x'(t)]^2 + [y'(t)]^2 + [z'(t)]^2} \tag{11}$$

or, in Leibniz notation,

$$\frac{ds}{dt} = \sqrt{\left(\frac{dx}{dt}\right)^2 + \left(\frac{dy}{dt}\right)^2 + \left(\frac{dz}{dt}\right)^2} \tag{12}$$

Multiplying both sides of equation (12) by dt gives the differential for arc length (of space curves)

$$ds = \sqrt{(dx)^2 + (dy)^2 + (dz)^2}. \tag{13}$$

Parameterization by Arc Length

Up to this point, we have not concerned ourselves with the *rate* at which the vector function $\boldsymbol{f}(t)$ traces out a curve in space. In order to proceed further with our analysis of curves, we will now have to do so. A simple example of this type of concern is the observation that the vector functions $\boldsymbol{f}(t) = \boldsymbol{a} + t\boldsymbol{b}$ and $\boldsymbol{g}(t) = \boldsymbol{a} + 2t\boldsymbol{b}$ both trace out the same line in space, although a particle located at the tip of $\boldsymbol{g}(t)$ moves along the line at a speed twice that of a particle located at the tip of $\boldsymbol{f}(t)$.

A baseline for comparing rates at which the vector function $\boldsymbol{f}(t)$ traces out a curve in space is provided by the notion of *parameterization by arc length*. Intuitively, this notion says that as the parameter t varies through an interval of length t_0, an arc of this same length $s = t_0$ is traced out by the vector function $\boldsymbol{f}(t)$ (Figure 4.4). A more precise formulation of this concept is the following.

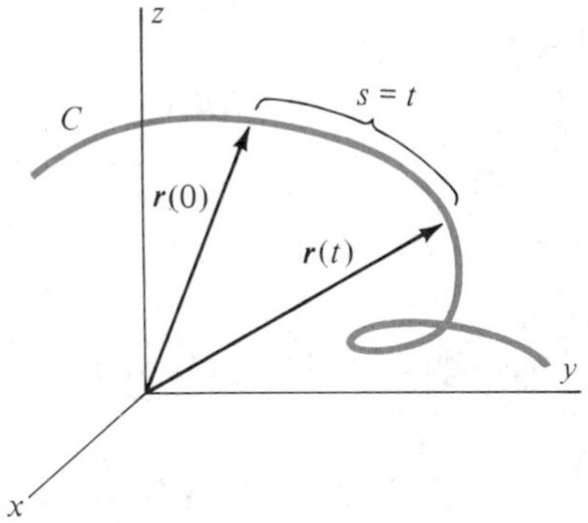

Figure 4.4 $\boldsymbol{r}(t)$ is parameterized by arc length if $\Delta s = \Delta t$.

DEFINITION 7

Let the curve C be the graph of the differentiable vector function $\boldsymbol{r}(t) = x(t)\boldsymbol{i} + y(t)\boldsymbol{j} + z(t)\boldsymbol{k}$ defined on an interval I containing the number zero. The curve C (and the function $\boldsymbol{r}(t)$) is said to be **parameterized by arc length** if

$$\int_0^t |\boldsymbol{r}'(\omega)|\, d\omega = t \tag{14}$$

for all t in the interval I.

Since the integral in equation (14) is the length of the arc of C from $\boldsymbol{r}(0)$ to $\boldsymbol{r}(t)$, Definition 7 agrees with the intuitive description of parameterization by arc length given above. Differentiating both sides of equation (14) with respect to t shows that the condition

$$\boxed{|\boldsymbol{r}'(t)| = 1 \qquad \text{for all} \qquad t \in I} \tag{15}$$

is equivalent to C being parameterized by arc length. Thus, a curve is *parameterized by arc length if and only if its derivative has constant length equal to one.*

Example 5 The unit circle

$$r(t) = \cos t\mathbf{i} + \sin t\mathbf{j}, \qquad t \in [0, 2\pi)$$

is parameterized by arc length, since

$$\begin{aligned} |r'(t)| &= |-\sin t\mathbf{i} + \cos t\mathbf{j}| \\ &= \sqrt{\sin^2 t + \cos^2 t} \\ &= 1 \end{aligned}$$

for all $t \in [0, 2\pi)$. (This should not be surprising. Both the circumference of the unit circle and the length of the interval $[0, 2\pi)$ are 2π.) ■

Example 6 The circle of radius ρ centered at the origin is traced out by the vector function

$$r(t) = \rho \cos \alpha t\mathbf{i} + \rho \sin \alpha t\mathbf{j}, \qquad t \in \left[0, \frac{2\pi}{\alpha}\right), \qquad \alpha > 0.$$

Find the value of α for which this circle is parameterized by arc length.

Solution: We first find $|r'(t)|$:

$$r'(t) = -\alpha\rho \sin \alpha t\mathbf{i} + \alpha\rho \cos \alpha t\mathbf{j}.$$

Thus

$$\begin{aligned} |r'(t)| &= \sqrt{(-\alpha\rho \sin \alpha t)^2 + (\alpha\rho \cos \alpha t)^2} \\ &= |\alpha\rho|. \end{aligned}$$

In order that equation (14) hold, we must have

$$|r'(t)| = |\alpha\rho| = 1.$$

We therefore take $\alpha = 1/\rho$. The circle of radius ρ is parameterized by arc length by the function

$$r(t) = \rho \cos\left(\frac{t}{\rho}\right)\mathbf{i} + \rho \sin\left(\frac{t}{\rho}\right)\mathbf{j}.$$ ■

We conclude by highlighting the comment following equation (15). If the curve C determined by the vector function $r(s)$ is parameterized by arc length, the **unit tangent** $T(s)$ at $r(s)$ is

$$\boxed{T(s) = r'(s).} \tag{16}$$

Exercise Set 18.4

In Exercises 1–6, find a vector tangent to the given curve at the given point.

1. $r(t) = t^3\mathbf{i} + \sqrt{t}\,\mathbf{j} + 2\mathbf{k}, \quad t = 4$
2. $r(t) = \sin t\mathbf{i} + \cos 2t\mathbf{j} + e^{3t}\mathbf{k}, \quad t = 0$
3. $r(t) = \sec t\mathbf{i} + \tan t\mathbf{j}, \quad t = \pi/4$
4. $r(t) = \sqrt{1 + t^2}\,\mathbf{i} + \tan^{-1} t\mathbf{j} + \ln(1 + t)\mathbf{k}, \quad t = 0$
5. $r(t) = (1 + t^2)\mathbf{i} + (t - 4)\mathbf{j} + 6t\mathbf{k}$, at the point $(2, -5, -6)$
6. $r(t) = a\mathbf{i} + bt\mathbf{j} + ct^2\mathbf{k}$ at the point (a, b, c).

7. Find a unit vector tangent to the curve $\boldsymbol{r}(t) = a\cos\omega t\boldsymbol{i} + b\sin\omega t\boldsymbol{j}$ at the point where $t = \pi/4$.

In Exercises 8–10, let $\boldsymbol{r}(t) = (t^2 + 2)\boldsymbol{i} + 3t\boldsymbol{j}$.

8. Find the value(s) of t for which $\boldsymbol{r}(t)$ and $\boldsymbol{r}'(t)$ are orthogonal.

9. Find the value(s) of t for which $\boldsymbol{r}(t)$ and $\boldsymbol{r}'(t)$ have the same direction.

10. Find the value(s) of t for which $\boldsymbol{r}(t)$ and $\boldsymbol{r}'(t)$ have opposite direction.

11. Find an equation in vector form for the line tangent to the curve in Exercise 1 at the given point.

12. Find an equation in parametric form for the line tangent to the curve in Exercise 2 at the given point.

13. Find an equation in symmetric form for the line tangent to the curve in Exercise 5 at the given point.

14. Show that every tangent to the unit circle is orthogonal to the radius vector through the point of tangency.

15. Let $\boldsymbol{r}(t) = e^t\cos t\boldsymbol{i} + e^t\sin t\boldsymbol{j}$. Find the angle between $\boldsymbol{r}(t)$ and the tangent $\boldsymbol{r}'(t)$ for

a. $t = 0$,
b. $t = \pi/4$.

In Exercises 16–20, find the length of the indicated arc.

16. $\boldsymbol{r}(t) = \cos t\boldsymbol{i} + \sin t\boldsymbol{j}, \quad 0 \le t \le \pi/4$

17. $\boldsymbol{r}(t) = \cos t\boldsymbol{i} + \frac{\sqrt{2}}{2}\sin t\boldsymbol{j} + \frac{\sqrt{2}}{2}\sin t\boldsymbol{k}, \quad 0 \le t \le \pi/2$

18. $\boldsymbol{r}(t) = e^t\cos t\boldsymbol{i} + e^t\sin t\boldsymbol{j}, \quad 0 \le t \le \pi/2$

19. $\boldsymbol{r}(t) = t\boldsymbol{i} + (t^2 + 1)\boldsymbol{j} + t\boldsymbol{k}, \quad 0 \le t \le 1$

20. $\boldsymbol{r}(t) = t^3\boldsymbol{i} + 3t^2\boldsymbol{j} + 3t\boldsymbol{k}, \quad 0 \le t \le 3$

21. Find a unit tangent $\boldsymbol{T}(t)$ to the graph of $y = x^2 + 3$ at the point (2, 7). (*Hint:* Let $x = t$. Then $y = t^2 + 3$, and a vector parameterization for the curve is $\boldsymbol{r}(t) = t\boldsymbol{i} + (t^2 + 3)\boldsymbol{j}$.)

22. Use the method of Exercise 21 to find a unit tangent $\boldsymbol{T}(t)$ to the graph of $y^3 = 1 - x^2$ at the point $(3, -2)$.

23. Determine the constant α so that the curve

$$\boldsymbol{r}(t) = \cos\alpha t\boldsymbol{i} + \sin\alpha t\boldsymbol{j} + \alpha t\boldsymbol{k}, \quad 0 \le t \le \frac{2\pi}{\alpha}, \quad \alpha > 0.$$

is parameterized by arc length.

24. Find a parameterization by arc length for the curve $y = 3x - 2$ so that $x = 0$ and $y = -2$ when $s = 0$.

25. Show that if $\boldsymbol{r}(t) = x(t)\boldsymbol{i} + y(t)\boldsymbol{j}$ and $x'(t_0) \ne 0$, Definition 6 for the *tangent* at $\boldsymbol{r}(t_0)$ agrees with our earlier definition of tangent for the curve C determined by the parametric equations

$$C: \begin{cases} x = x(t) \\ y = y(t). \end{cases}$$

26. Use Example 6, Section 18.3, to show that if $\boldsymbol{T}(t) = \frac{\boldsymbol{r}'(t)}{|\boldsymbol{r}'(t)|}$ is differentiable on $\boldsymbol{T}$, then $\boldsymbol{T}'(t)$ is orthogonal to $\boldsymbol{T}(t)$ for $t \in I$. If $\boldsymbol{T}'(t) \ne 0$, we define $\boldsymbol{N}(t) = \frac{\boldsymbol{T}'(t)}{|\boldsymbol{T}'(t)|}$ to be the **principal unit normal** to the curve $C = \{\boldsymbol{r}(t) | t \in I\}$.

27. Find the principal unit normals to the curve with parameterization

$$\boldsymbol{r}(t) = \sqrt{2}\cos t\boldsymbol{i} + \sqrt{2}\sin t\boldsymbol{j}, \qquad 0 \le t \le 2\pi.$$

(See Exercise 26.)

28. Show that the line with vector equation

$$\boldsymbol{R}(\omega) = \boldsymbol{r}(t_0) + \omega\boldsymbol{r}''(t_0)$$

is orthogonal to the tangent vector $\boldsymbol{T}(t_0)$ if $\boldsymbol{r}(t)$ is a parameterization by arc length, $\boldsymbol{r}(t)$ is twice differentiable at $\boldsymbol{r}(t_0)$, $\boldsymbol{r}'(t_0) \ne \boldsymbol{0}$, and $\boldsymbol{r}''(t_0) \ne \boldsymbol{0}$. (We refer to this line as the line **normal** to the curve at $\boldsymbol{r}(t_0)$.)

29. Find an equation for the line normal to the curve in Example 3 at the point (4, 1, 8) (see Exercise 28).

30. Give an example to show that $\boldsymbol{r}(t)$ is not always orthogonal to $\boldsymbol{r}'(t)$.

18.5 VELOCITY AND ACCELERATION

If a particle moves in space so that its location at time t is the point $P(t) = (x(t), y(t), z(t))$, its motion may be described by the **position vector** function

$$\boldsymbol{r}(t) = x(t)\boldsymbol{i} + y(t)\boldsymbol{j} + z(t)\boldsymbol{k}. \tag{1}$$

That is, we think of the position of the particle as the tip of the vector $\boldsymbol{r}(t)$ originating at the origin and terminating at the point $P(t)$.

In such cases the change in the particle's position from time t to time $t + \Delta t$ is

$$\Delta \boldsymbol{r} = \boldsymbol{r}(t + \Delta t) - \boldsymbol{r}(t).$$

Since the corresponding change in time is $\Delta t = (t + \Delta t) - t$, the **average velocity** during such a time interval is

$$\begin{aligned}\text{average velocity} &= \frac{\text{change in position}}{\text{change in time}}\\ &= \frac{\Delta \boldsymbol{r}}{\Delta t}\\ &= \frac{\boldsymbol{r}(t+\Delta t) - \boldsymbol{r}(t)}{\Delta t}.\end{aligned}$$

(See Figure 5.1.)

If our earlier concept of (instantaneous) velocity as the limit of average velocity as $\Delta t \to 0$ is to carry over to motion in space, we should define

$$\text{velocity at time } t = \lim_{\Delta t \to 0} \frac{\boldsymbol{r}(t+\Delta t) - \boldsymbol{r}(t)}{\Delta t}. \tag{2}$$

(See Figure 5.2.)

We have already seen that the limit on the right-hand side of equation (2), when it exists, is the derivative $\boldsymbol{r}'(t)$ of the vector function $\boldsymbol{r}(t)$. According to Theorem 2, this limit exists precisely when each of the component functions $x(t)$, $y(t)$, and $z(t)$ is differentiable. The definition of the *velocity* of a particle moving in space is therefore the following.

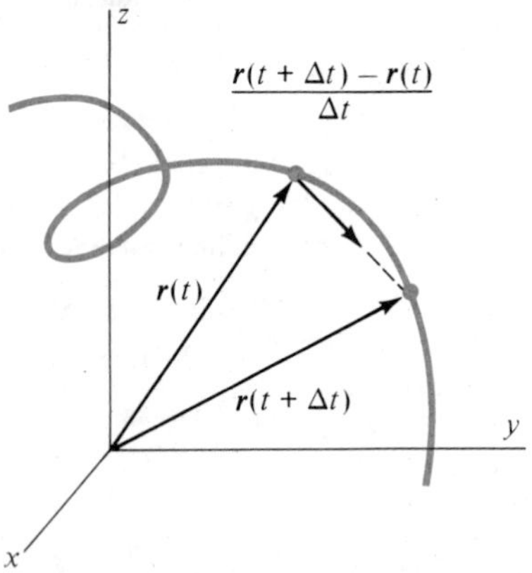

Figure 5.1 Average velocity is the vector $\frac{\boldsymbol{r}(t+\Delta t) - \boldsymbol{r}(t)}{\Delta t}$.

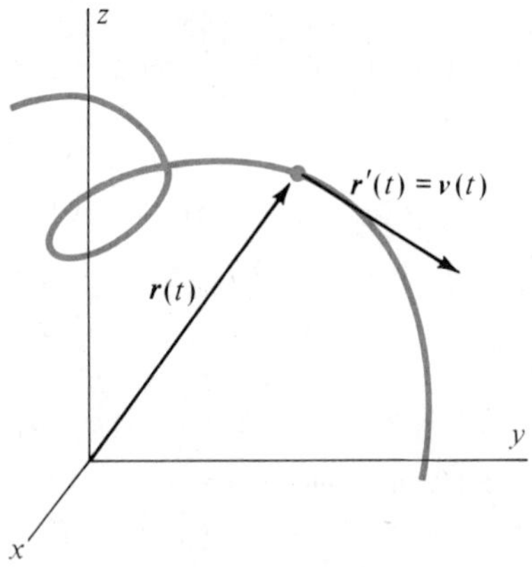

Figure 5.2 Velocity is the vector $\boldsymbol{v}(t) = \boldsymbol{r}'(t) = \lim_{\Delta t \to 0} \frac{\boldsymbol{r}(t+\Delta t) - \boldsymbol{r}(t)}{\Delta t}$, and $|\boldsymbol{v}(t_0)|$ = speed.

DEFINITION 8

Let the vector function

$$\boldsymbol{r}(t) = x(t)\boldsymbol{i} + y(t)\boldsymbol{j} + z(t)\boldsymbol{k}$$

be the position function for a particle moving in space and let the parameter t denote time. If $\boldsymbol{r}(t)$ is differentiable at time $t = t_0$, the **velocity** of the particle at $t = t_0$ is the vector

$$\boldsymbol{v}(t_0) = \boldsymbol{r}'(t_0) = x'(t_0)\boldsymbol{i} + y'(t_0)\boldsymbol{j} + z'(t_0)\boldsymbol{k}. \tag{3}$$

If $\boldsymbol{r}(t)$ is differentiable on an interval I, equation (3) determines the **velocity function** for $\boldsymbol{r}(t)$ on the interval I.

Since velocity is just the derivative of the position function $\boldsymbol{r}(t)$, we have already determined the two essential attributes of the velocity vector $\boldsymbol{v}(t_0)$:

(i) $\boldsymbol{v}(t_0)$ is *tangent* to the graph of $\boldsymbol{r}(t)$ at $\boldsymbol{r}(t_0)$, and points in the direction of increasing t, provided $\boldsymbol{v}(t_0) \neq \boldsymbol{0}$ (Section 18.4).

(ii) $|\boldsymbol{v}(t_0)| = |\boldsymbol{r}'(t_0)| = \left|\frac{ds}{dt}\right|$, where $s = s(t)$ denotes the arc length function (equation (11), Section 18.4). Thus, the length of the velocity vector is the **speed** (rate of change of distance along the path) of the particle at time t_0.

Example 1 Show that a particle moving about the unit circle in the xy-plane with position function

$$\boldsymbol{r}(t) = \cos \alpha t\boldsymbol{i} + \sin \alpha t\boldsymbol{j}$$

moves at a constant speed.

Solution: By Definition 8, the velocity function is

$$\boldsymbol{v}(t) = -\alpha \sin \alpha t\boldsymbol{i} + \alpha \cos \alpha t\boldsymbol{j}.$$

The speed at time t is therefore

$$|\boldsymbol{v}(t)| = \sqrt{(\alpha \sin \alpha t)^2 + (\alpha \cos \alpha t)^2} = |\alpha|,$$

which is constant. For this reason, physicists call this **uniform circular motion.** ■

As for motion along a line, we define the *acceleration* of a particle moving in space to be the rate of change of velocity with respect to time.

DEFINITION 9

Let the vector function

$$\boldsymbol{r}(t) = x(t)\boldsymbol{i} + y(t)\boldsymbol{j} + z(t)\boldsymbol{k}$$

be the position function of a particle moving in space and let the parameter t denote time. If $\boldsymbol{r}(t)$ is twice differentiable at time $t = t_0$, the **acceleration** of the particle at $t = t_0$ is the vector

$$\boldsymbol{a}(t_0) = \boldsymbol{r}''(t_0) = x''(t_0)\boldsymbol{i} + y''(t_0)\boldsymbol{j} + z''(t_0)\boldsymbol{k}.$$

Thus,

$$\boldsymbol{a}(t_0) = \boldsymbol{v}'(t_0). \tag{4}$$

Example 2 Show that for the particle moving about the unit circle in Example 1, the acceleration vector always points toward the center of the circle and is of constant magnitude.

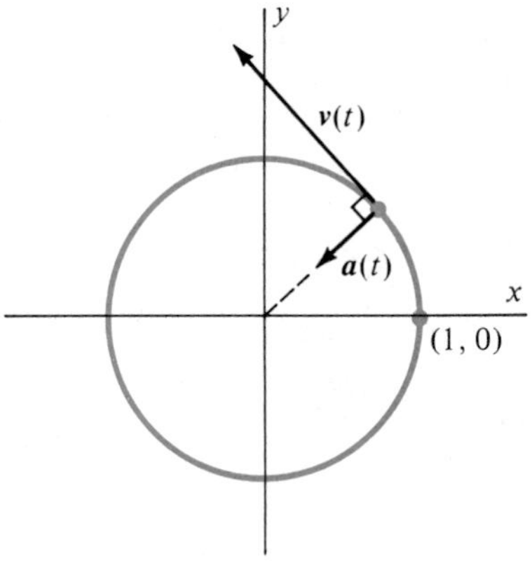

Figure 5.3 Velocity and acceleration vectors associated with uniform circular motion.

Solution: From the solution to Example 1 and equation (4) we have

$$\begin{aligned} \boldsymbol{a}(t) = \boldsymbol{v}'(t) &= -\alpha^2 \cos \alpha t\boldsymbol{i} - \alpha^2 \sin \alpha t\boldsymbol{j} \\ &= -\alpha^2(\cos \alpha t\boldsymbol{i} + \sin \alpha t\boldsymbol{j}) \\ &= -\alpha^2\boldsymbol{r}(t). \end{aligned}$$

Thus, $\boldsymbol{a}(t)$ points in the direction opposite $\boldsymbol{r}(t)$ and has length

$$|\boldsymbol{a}(t)| = \sqrt{(\alpha^2 \cos \alpha t)^2 + (\alpha^2 \sin \alpha t)^2} = \alpha^2.$$

(See Figure 5.3.) Note that $\boldsymbol{v}(t)$ and $\boldsymbol{a}(t)$ are orthogonal, an observation that follows from Example 6, Section 18.3, since $|\boldsymbol{v}(t)| = |\alpha| =$ constant. ■

Example 3 Find the velocity and acceleration functions associated with the position function

$$\boldsymbol{r}(t) = \cos t\boldsymbol{i} + \sin t\boldsymbol{j} + t^2\boldsymbol{k}$$

and show that $|\boldsymbol{a}(t)|$ is constant.

Solution:

$$\begin{aligned} \boldsymbol{v}(t) &= \boldsymbol{r}'(t) = -\sin t\boldsymbol{i} + \cos t\boldsymbol{j} + 2t\boldsymbol{k} \\ \boldsymbol{a}(t) &= \boldsymbol{v}'(t) = -\cos t\boldsymbol{i} - \sin t\boldsymbol{j} + 2\boldsymbol{k}. \end{aligned}$$

Since $|\boldsymbol{a}(t)| = \sqrt{\cos^2 t + \sin^2 t + 4} = \sqrt{5}$, $|\boldsymbol{a}(t)|$ is constant. The vectors $\boldsymbol{r}(t)$, $\boldsymbol{v}(t)$, and $\boldsymbol{a}(t)$ are sketched in Figure 5.4. ■

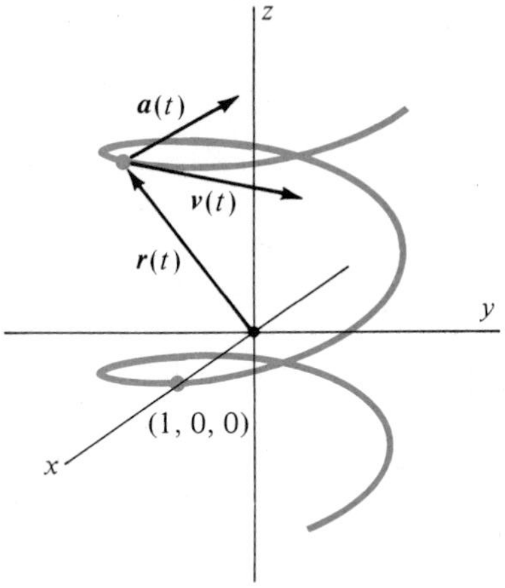

Figure 5.4 Velocity and acceleration vectors for the helix $\boldsymbol{r}(t) = \cos t\boldsymbol{i} + \sin t\boldsymbol{j} + t^2\boldsymbol{k}$.

Just as velocity and acceleration can be determined by differentiating the position function $\boldsymbol{r}(t)$, the functions for position and velocity can be obtained by integrating the acceleration function, provided antiderivatives of each component function can be found. More precisely, equation (4) gives

$$\boldsymbol{v}(t) = \int \boldsymbol{a}(t)\, dt, \tag{5}$$

and equation (3) shows that

$$\boldsymbol{r}(t) = \int \boldsymbol{v}(t)\, dt. \tag{6}$$

In using equations (5) and (6), you must remember that one constant of integration will appear in each component in each integration. These constants may be determined when *initial conditions* are specified for the functions $\boldsymbol{v}(t)$ and $\boldsymbol{r}(t)$.

Example 4 A particle starts from rest at the point $P_0 = (2, 0, -3)$ and moves with an acceleration of $\boldsymbol{a}(t) = 2\boldsymbol{i} + 6t\boldsymbol{j}$ m/sec^2. Find

(a) the velocity function for the particle,

(b) the position function for the particle,

(c) the location of the particle after 3 seconds.

Strategy
Identify the initial conditions.

Solution
Since the particle starts at $P_0 = (2, 0, -3)$, we have

$$\boldsymbol{r}_0 = \boldsymbol{r}(0) = 2\boldsymbol{i} - 3\boldsymbol{k}.$$

Since the particle starts from rest, we have

$$\boldsymbol{v}_0 = \boldsymbol{v}(0) = \boldsymbol{0}.$$

Obtain $\boldsymbol{v}(t)$ by integrating $\boldsymbol{a}(t)$. (Don't forget that

$$\left[\int 0\, dt\right]\boldsymbol{k} = (0 + C_3)\boldsymbol{k}.)$$

Using equation (5), we have

$$\begin{aligned}\boldsymbol{v}(t) &= \int (2\boldsymbol{i} + 6t\boldsymbol{j})\, dt \\ &= (2t + c_1)\boldsymbol{i} + (3t^2 + c_2)\boldsymbol{j} + c_3\boldsymbol{k}.\end{aligned}$$

Apply the initial condition $\boldsymbol{v}_0 = 0$ to determine c_1, c_2, and c_3.

Setting $t = 0$ and applying the initial condition $\boldsymbol{v}(0) = \boldsymbol{0}$ gives

$$\boldsymbol{0} = c_1\boldsymbol{i} + c_2\boldsymbol{j} + c_3\boldsymbol{k},$$

so

$$c_1 = c_2 = c_3 = 0.$$

The velocity function is therefore

$$\boldsymbol{v}(t) = 2t\boldsymbol{i} + 3t^2\boldsymbol{j}.$$

Obtain $\boldsymbol{r}(t)$ by integrating $\boldsymbol{v}(t)$.

Using equation (6), we find that

$$\begin{aligned}\boldsymbol{r}(t) &= \int (2t\boldsymbol{i} + 3t^2\boldsymbol{j})\, dt \\ &= (t^2 + d_1)\boldsymbol{i} + (t^3 + d_2)\boldsymbol{j} + d_3\boldsymbol{k}.\end{aligned}$$

Determine the constants d_1, d_2, and d_3 by applying the initial condition $\boldsymbol{r}(0) = 2\boldsymbol{i} - 3\boldsymbol{k}$.

Setting $t = 0$ and applying the initial condition $\boldsymbol{r}(0) = 2\boldsymbol{i} - 3\boldsymbol{k}$ gives

$$2\boldsymbol{i} - 3\boldsymbol{k} = d_1\boldsymbol{i} + d_2\boldsymbol{j} + d_3\boldsymbol{k}$$

so

$$d_1 = 2, \qquad d_2 = 0, \qquad \text{and} \qquad d_3 = -3.$$

Thus,

$$\boldsymbol{r}(t) = (t^2 + 2)\boldsymbol{i} + t^3\boldsymbol{j} - 3\boldsymbol{k}.$$

The location of the particle after 3 seconds is $\boldsymbol{r}(3)$.

The location of the particle after 3 seconds is

$$\begin{aligned}\boldsymbol{r}(3) &= (3^2 + 2)\boldsymbol{i} + 3^3\boldsymbol{j} - 3\boldsymbol{k} \\ &= 11\boldsymbol{i} + 27\boldsymbol{j} - 3\boldsymbol{k},\end{aligned}$$

or $(11, 27, -3)$. Note that, because there is no component of $\boldsymbol{a}(t)$ in the z direction, the entire motion takes place in the plane $z = -3$. ■

Projectile Motion

A frequent application of equations (5) and (6) concerns the trajectory followed by a projectile launched with a prescribed initial velocity, $\boldsymbol{v}_0$. The term "projectile" refers to any object, such as a missile, a flare, or a baseball, which is not self-propelling. Thus, the trajectory followed by a projectile is completely determined by the speed and direction (i.e., the velocity) at which it is launched and by the force of gravity. In the discussion that follows, we shall assume that air resistance is negligible, so that the only force acting on the projectile after launch is the force due to gravity. Also, we shall take the y-axis as vertical and the x-axis as horizontal, with the positive x-axis pointing in the direction of the horizontal component of the initial velocity. The motion is then restricted entirely to the xy-plane.

The motion of projectiles (and, for that matter, of all moving objects) is governed by the vector form of Newton's second law of motion:

$$\boldsymbol{F} = m\boldsymbol{a}. \tag{7}$$

Here $\boldsymbol{F}$ is the vector sum of all forces acting on the projectile, m is its mass, and $\boldsymbol{a}$ is its acceleration.

According to our use of the term projectile, the only force acting on the object in flight is the force due to gravity,

$$\boldsymbol{F} = -mg\boldsymbol{j} \tag{8}$$

where $g = 9.81$ m/sec^2 is the gravitational constant. (We use a negative sign since this force acts downward.)

Combining equations (7) and (8) we find that a projectile in flight experiences the acceleration

$$\boldsymbol{a} = -g\boldsymbol{j}. \tag{9}$$

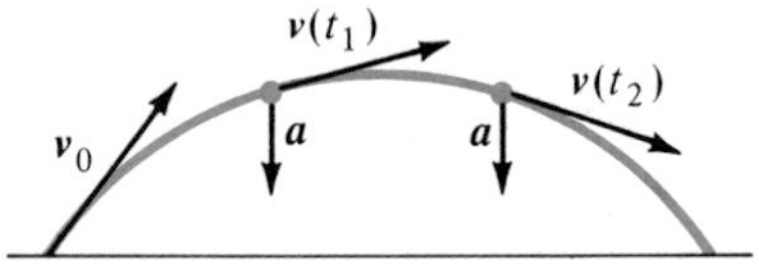

Figure 5.5 Motion of a projectile with initial velocity $\boldsymbol{v}_0$.

Figure 5.5 illustrates a typical trajectory for a projectile fired with initial velocity $\boldsymbol{v}_0$.

Now consider the problem of determining the trajectory of a projectile fired from the origin at an angle θ with the horizontal and with an initial speed s_0. This means that the initial velocity $\boldsymbol{v}_0$ has length $|\boldsymbol{v}_0| = s_0$. We may therefore write $\boldsymbol{v}_0$ in component form as

$$\boldsymbol{v}_0 = s_0 \cos \theta\boldsymbol{i} + s_0 \sin \theta\boldsymbol{j}. \tag{10}$$

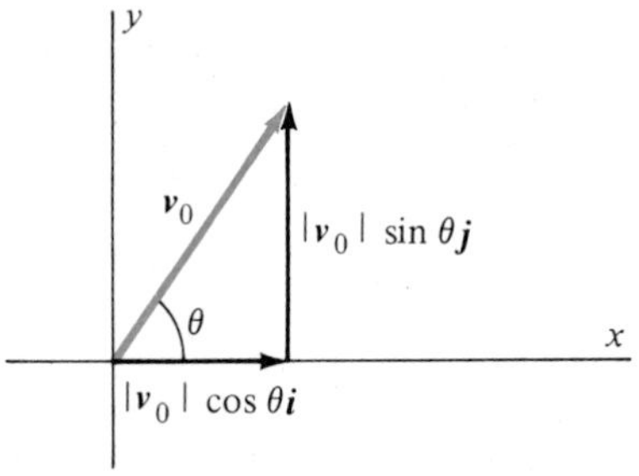

Figure 5.6 Components of initial velocity are

$\boldsymbol{v}_0 = |\boldsymbol{v}_0|\cos \theta\boldsymbol{i} + |\boldsymbol{v}_0|\sin \theta\boldsymbol{j}.$

(See Figure 5.6.)

Applying equation (5) with $\boldsymbol{a}(t)$ as in equation (9), we find $\boldsymbol{v}(t)$ to be

$$\boldsymbol{v}(t) = \int \boldsymbol{a}\, dt = \int -g\boldsymbol{j}\, dt = c_1\boldsymbol{i} + (c_2 - gt)\boldsymbol{j}.$$

Setting $t = 0$ and applying the initial condition given by equation (10), we obtain

$$\boldsymbol{v}(0) = c_1\boldsymbol{i} + c_2\boldsymbol{j} = s_0 \cos \theta\boldsymbol{i} + s_0 \sin \theta\boldsymbol{j}.$$

Thus,

$$c_1 = s_0 \cos \theta; \qquad c_2 = s_0 \sin \theta,$$

and therefore

$$\boldsymbol{v}(t) = s_0 \cos \theta\boldsymbol{i} + (s_0 \sin \theta - gt)\boldsymbol{j}.$$

According to equation (6), the position function $\boldsymbol{r}(t)$ is

$$\begin{aligned}\boldsymbol{r}(t) &= \int \boldsymbol{v}(t)\, dt = \int [s_0(\cos \theta)\boldsymbol{i} + (s_0 \sin \theta - gt)\boldsymbol{j}]\, dt \\ &= [s_0(\cos \theta)t + d_1]\boldsymbol{i} + \left[s_0(\sin \theta)t - \frac{1}{2}gt^2 + d_2\right]\boldsymbol{j}.\end{aligned}$$

Since the projectile is fired from the origin, we have the initial condition $\boldsymbol{r}(0) = \boldsymbol{0}$. Setting $t = 0$ in the above expression for $\boldsymbol{r}(t)$ and applying this initial condition gives

$$\boldsymbol{r}(0) = d_1\boldsymbol{i} + d_2\boldsymbol{j} = 0,$$

so

$$d_1 = d_2 = 0.$$

The position function for the projectile motion is therefore

$$r(t) = [s_0(\cos\theta)t]i + \left[s_0(\sin\theta)t - \frac{1}{2}gt^2\right]j.$$

We summarize our findings as follows

> The trajectory of a projectile fired from the origin with initial speed s_0 and angle of elevation θ is parameterized by the vector function
>
> $$r(t) = [s_0(\cos\theta)t]i + \left[s_0(\sin\theta)t - \frac{1}{2}gt^2\right]j. \quad (11)$$

Example 5 A flare is fired from ground level at an elevation of 60° with an initial speed of 50 m/sec. Find

(a) the maximum height of the trajectory,
(b) the length of time during which the flare is airborne, and
(c) the distance from the point of launch to the point of impact.

Strategy

Write the expression for $r(t) = x(t)i + y(t)j$ using (11).

Solution

With $\theta = 60° = \pi/3$ and $s_0 = 50$ m/sec, $r(t)$ is

$$r(t) = 50\left(\cos\frac{\pi}{3}\right)ti + \left(50\left(\sin\frac{\pi}{3}\right)t - \frac{1}{2}gt^2\right)j$$
$$= 25ti + \left(25\sqrt{3}t - \frac{1}{2}gt^2\right)j.$$

Find the time, t_{max}, of maximum height by maximizing the j-component of $r(t)$.

The height of the flare is given by the j-component

$$y(t) = 25\sqrt{3}t - \frac{1}{2}gt^2.$$

The maximum height occurs when

$$y'(t) = 25\sqrt{3} - gt = 0,$$

or

$$t_{max} = \frac{25\sqrt{3}}{g} = \frac{25\sqrt{3}}{9.81} \approx 4.4 \text{ seconds.}$$

$y(t_{max})$ is the maximum height.

The maximum height is

$$y(t_{max}) = 25\sqrt{3}\left(\frac{25\sqrt{3}}{9.81}\right) - \frac{1}{2}(9.81)\left(\frac{25\sqrt{3}}{9.81}\right)^2$$
$$= \frac{\left(\frac{1}{2}\right)25^2\cdot 3}{9.81} \approx 95.6 \text{ meters.}$$

Find the time, t_{end}, at which the projectile lands by setting $y(t) = 0$.

The projectile lands when the j-component reaches zero, that is, when

$$y(t) = 25\sqrt{3}t - \frac{1}{2}gt^2 = 0.$$

This equation has solutions $t = 0$ and $t = \dfrac{50\sqrt{3}}{g}$.

The time $t = 0$ corresponds to launch. The time at which the projectile returns to ground level is therefore

$$t_{\text{end}} = \frac{50\sqrt{3}}{g} = \frac{50\sqrt{3}}{9.81} \approx 8.8 \text{ seconds.}$$

Distance from launch point to impact point is $x(t_{\text{end}})$.

The distance from the point of launch to the point of impact is the $\boldsymbol{i}$-component of $\boldsymbol{r}(t_{\text{end}})$.

$$x(t_{\text{end}}) = 25(8.8) = 220 \text{ meters.}$$

■

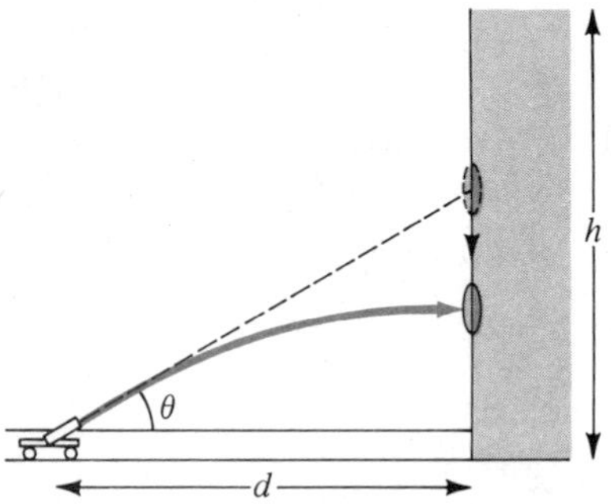

Figure 5.7 The cannon-target experiment.

Example 6 The following experiment is performed in an elementary physics class. A target is hung on a wall of a gymnasium at a height of h meters above the floor. A small cannon, which fires a lead slug, is located on the floor d meters from the wall. The muzzle velocity of the cannon is adjustable. The cannon is aimed at the bull's eye on the target. By means of a trip mechanism, the target is released from its hanger at the precise instant at which the cannon is fired and falls toward the floor. The result of repeated trials of the experiment is that no matter what muzzle velocity is chosen, the slug always strikes the bull's eye (see Figure 5.7). How can this be explained?

Solution: Let θ be the angle of elevation of the barrel of the cannon when aimed at the bull's eye, and let s_0 be the muzzle velocity of the cannon. We assume the cannon to be located at the origin of an xy-coordinate system. The trajectory of the slug is then given by $\boldsymbol{r}(t)$ in equation (11).

The slug reaches the wall when the $\boldsymbol{i}$-component of $\boldsymbol{r}(t)$ satisfies the equation

$$x(t) = s_0(\cos\theta)\, t = d$$

which occurs at time $t_f = \dfrac{d}{s_0 \cos\theta}$. At this time the $\boldsymbol{j}$-component of the slug is

$$\begin{aligned} y(t_f) &= s_0 \sin\theta \left(\frac{d}{s_0\cos\theta}\right) - \frac{1}{2} g \left(\frac{d}{s_0 \cos\theta}\right)^2 \\ &= d\tan\theta - \frac{1}{2} g t_f^2. \end{aligned}$$

Since $\dfrac{h}{d} = \tan\theta$, $h = d\tan\theta$. Thus

$$y(t_f) = h - \frac{1}{2} g t_f^2$$

is the $\boldsymbol{j}$-component of $\boldsymbol{r}(t)$ when the slug reaches the wall.

Since the target is a freely falling body, after t_f seconds it has fallen a distance $\dfrac{1}{2} g t_f^2$, so the $\boldsymbol{j}$-component of the bull's eye's position vector after t_0 seconds is also $h - \dfrac{1}{2} g t_f^2$. Since the slug and the bull's eye have the same position vector, $\boldsymbol{r}(t_f) = d\boldsymbol{i} + (h - \dfrac{1}{2} g t_f^2)\boldsymbol{j}$, at time $t = t_f$, the slug must always score a direct hit on the target.

■

Exercise Set 18.5

In Exercises 1–7, find the velocity and acceleration functions corresponding to the given position functions.

1. $\boldsymbol{r}(t) = \boldsymbol{i} + 2t\boldsymbol{j}$

2. $\boldsymbol{r}(t) = 3t\boldsymbol{i} + t^3\boldsymbol{j} + (2t + 3)\boldsymbol{k}$

3. $\boldsymbol{r}(t) = \cos 3t\boldsymbol{i} + \sin 3t\boldsymbol{j}$

4. $\boldsymbol{r}(t) = e^t\boldsymbol{i} + e^{-t}\boldsymbol{j} + \sqrt{t}\,\boldsymbol{k}, \quad t \geq 1$

5. $\boldsymbol{r}(t) = \ln t^2\boldsymbol{i} + \cos 2t\boldsymbol{j} + \dfrac{1}{t}\boldsymbol{k}, \quad t \geq 1$

6. $\boldsymbol{r}(t) = t(t - 1)\boldsymbol{i} + \tan^{-1} t\boldsymbol{j} + te^t\boldsymbol{k}$

7. $\boldsymbol{r}(t) = e^t \cos t\boldsymbol{i} + e^t \sin t\boldsymbol{j} + \cos t\boldsymbol{k}$

8. A particle moves with position function $\boldsymbol{r}(t) = \cos \pi t\boldsymbol{i} + \sin \pi t\boldsymbol{j}$. Find its speed at time $t = 1/4$.

9. A particle moves with position vector $\boldsymbol{r}(t) = t^2\boldsymbol{i} + 2t\boldsymbol{j} + t^3\boldsymbol{k}$. Find its speed at time $t = 2$.

In Exercises 10–13, find the position function $\boldsymbol{r}(t)$ from the given velocity function and the initial condition $\boldsymbol{r}(0) = \boldsymbol{r}_0$.

10. $\boldsymbol{v}(t) = 2t\boldsymbol{i} + t^2\boldsymbol{j}, \quad \boldsymbol{r}_0 = \boldsymbol{i} + 4\boldsymbol{j}$

11. $\boldsymbol{v}(t) = \cos t\boldsymbol{i} + \sin t\boldsymbol{j}, \quad \boldsymbol{r}_0 = 3\boldsymbol{i} + 2\boldsymbol{j}$

12. $\boldsymbol{v}(t) = e^t\boldsymbol{i} + \sqrt{t}\boldsymbol{j} + 2t\boldsymbol{k}, \quad \boldsymbol{r}_0 = 6\boldsymbol{i} + 4\boldsymbol{k}$

13. $\boldsymbol{v}(t) = \left(\dfrac{1}{1 + t^2}\right)\boldsymbol{i} + te^{t^2}\boldsymbol{j}, \quad \boldsymbol{r}_0 = -2\boldsymbol{i} + \boldsymbol{j} + 4\boldsymbol{k}$

In Exercises 14–18, find the velocity function $\boldsymbol{v}(t)$ and the position function $\boldsymbol{r}(t)$ given the acceleration function $\boldsymbol{a}(t)$ and the initial conditions $\boldsymbol{v}_0 = \boldsymbol{v}(0)$ and $\boldsymbol{r}_0 = \boldsymbol{r}(0)$.

14. $\boldsymbol{a} = \boldsymbol{i} + \boldsymbol{j}, \quad \boldsymbol{v}_0 = \boldsymbol{i} + \boldsymbol{j}, \quad \boldsymbol{r}_0 = \boldsymbol{0}$

15. $\boldsymbol{a} = 2\boldsymbol{i} + \boldsymbol{k}, \quad \boldsymbol{v}_0 = \boldsymbol{0}, \quad \boldsymbol{r}_0 = 3\boldsymbol{i} - \boldsymbol{j} + 4\boldsymbol{k}$

16. $\boldsymbol{a} = 6t\boldsymbol{i} + e^t\boldsymbol{j}, \quad \boldsymbol{v}_0 = \boldsymbol{0}, \quad \boldsymbol{r}_0 = 4\boldsymbol{i} + \boldsymbol{j} + 2\boldsymbol{k}$

17. $\boldsymbol{a} = \cos t\boldsymbol{i} + \sin t\boldsymbol{j}, \quad \boldsymbol{v}_0 = \boldsymbol{k}, \quad \boldsymbol{r}_0 = 3\boldsymbol{i} - \boldsymbol{j} + 2\boldsymbol{k}$

18. $\boldsymbol{a} = e^t\boldsymbol{i} + t\boldsymbol{j} + e^{-t}\boldsymbol{k}, \quad \boldsymbol{v}_0 = \boldsymbol{i} + \boldsymbol{k}, \quad \boldsymbol{r}_0 = \boldsymbol{i} + 4\boldsymbol{j} + \boldsymbol{k}$

19. Verify that the vectors $\boldsymbol{v}(t)$ and $\boldsymbol{a}(t)$ of Example 2 are orthogonal for each t.

20. Prove that if a particle moves with constant speed its velocity and acceleration vectors are orthogonal.

21. A particle moves about a circle in the xy-plane with position function $\boldsymbol{r}(t) = \rho \cos \alpha t\boldsymbol{i} + \rho \sin \alpha t\boldsymbol{j}$. Find a relationship between the radius of the circle and the speed of the particle.

22. A projectile is fired from ground level at an elevation angle of $\theta = 45°$ with an initial speed of 60 meters per second. Find
a. the position function $\boldsymbol{r}(t)$ of the projectile,
b. the maximum altitude of the projectile,
c. the flight time,
d. the distance from the launch point to the impact point,
e. the speed on impact.

23. Find each of the quantities in Exercise 22 for a projectile fired at an elevation angle of $\theta = 30°$ with an initial speed of 100 meters per second.

24. A bale of hay is dropped from an airplane flying at a ground speed of 200 km/hr and an elevation of 1000 meters. How far from the drop point does the bale strike the ground?

25. A handgun used in a physics experiment simultaneously fires one slug through the barrel and drops a second slug mounted on the side of the barrel. Independent of the muzzle velocity, the two slugs are observed to strike the ground at the same instant. Why?

26. How far along the path $\boldsymbol{r}(t)$ does the particle of Example 4 travel during the first 3 seconds of its motion? What is its distance from the starting point P_0 at $t = 3$ seconds?

27. a. Write the integral that expresses the arc length of the motion of the flare in Example 5.
b. *(Calculator/Computer)* Use Simpson's Rule with $n = 8$ to approximate the value of the integral in (a).

18.6 CURVATURE

Curvature is a measure of the rate at which a curve in space turns or twists. To make this idea more precise, we consider a curve C determined parametrically by the vector function

$$C: \quad \boldsymbol{r}(s) = x(s)\boldsymbol{i} + y(s)\boldsymbol{j} + z(s)\boldsymbol{k}, \qquad a \leq s \leq b. \tag{1}$$

We assume that C is parameterized by arc length, so that the tip of $\boldsymbol{r}(s)$ moves along C uniformly. That is, we assume that $|\boldsymbol{r}'(s)| = 1$, $a \leq s \leq b$.

Since $\boldsymbol{r}'(s)$ is a unit tangent at each point $\boldsymbol{r}(s)$, the derivative $(\boldsymbol{r}'(s))' = \boldsymbol{r}''(s)$ measures the rate at which the unit tangent changes (turns or twists) as s increases. This is analogous to the one variable case in which the second derivative $f''(t)$ measures the rate at which the slope function $f'(t)$ is changing. However, the derivatives here are vectors rather than numbers. We therefore make the following definition.

DEFINITION 10

Let the curve C in (1) be parameterized by arc length, and let $\boldsymbol{r}(s)$ in (1) be twice differentiable for $a \le s \le b$. The vector $\boldsymbol{r}''(s)$ is called the **curvature vector** at $\boldsymbol{r}(s)$. Its length, $|\boldsymbol{r}''(s)|$, is called the **curvature** of C at $\boldsymbol{r}(s)$.

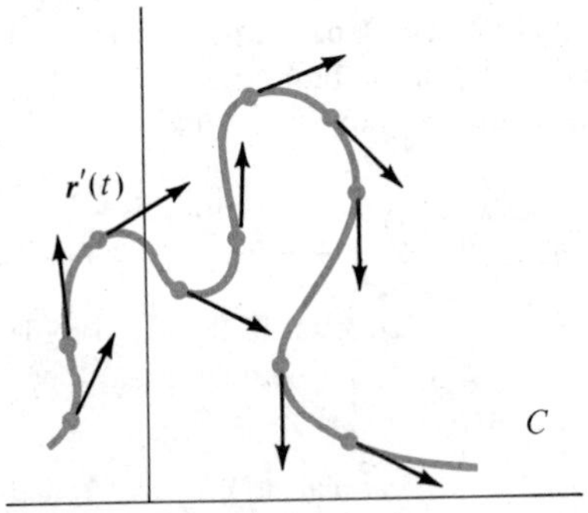

Figure 6.1 $\kappa(s)$ is large when C turns sharply.

REMARK: If C were not parameterized by arc length, then $|\boldsymbol{r}''(s)|$ would depend upon both the direction and the magnitude of $\boldsymbol{r}'(s)$. By insisting on parameterization by arc length, we have removed the dependence of the curvature on the "speed" of the curve, $|\boldsymbol{r}'(s)|$.

The Greek letter κ (kappa) is usually used to denote curvature. Thus, Definition 10 states that

$$\kappa(s) = |\boldsymbol{r}''(s)|, \qquad a \le s \le b. \tag{2}$$

Figures 6.1 and 6.2 illustrate the essence of curvature: $\kappa(s)$ is large when C turns sharply (rapid change in $\boldsymbol{r}'(s)$), and $\kappa(s)$ is small when C turns slowly (small change in $\boldsymbol{r}'(s)$).

Example 1 In Example 6, Section 18.4, we determined that the circle of radius ρ centered at the origin is parameterized by arc length by the vector function

$$\boldsymbol{r}(s) = \rho \cos\left(\frac{s}{\rho}\right)\boldsymbol{i} + \rho \sin\left(\frac{s}{\rho}\right)\boldsymbol{j}, \qquad 0 \le s \le 2\rho\pi.$$

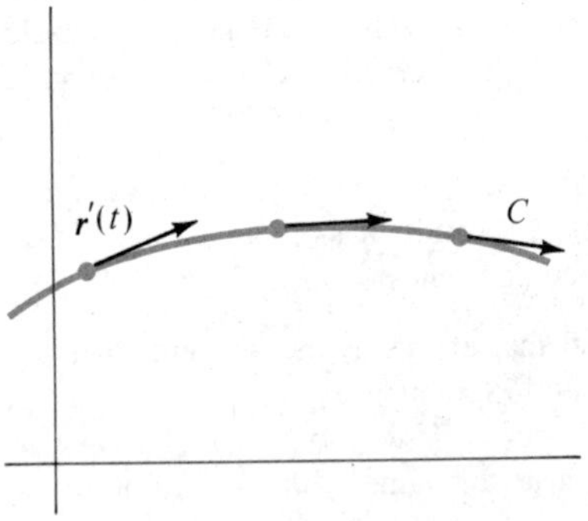

Figure 6.2 $\kappa(s)$ is small when C turns slowly.

The curvature vector is therefore

$$\boldsymbol{r}''(s) = -\frac{1}{\rho}\cos\left(\frac{s}{\rho}\right)\boldsymbol{i} - \frac{1}{\rho}\sin\left(\frac{s}{\rho}\right)\boldsymbol{j} = -\frac{1}{\rho^2}\boldsymbol{r}(s)$$

and the curvature at $\boldsymbol{r}(s)$ is

$$\kappa(s) = |\boldsymbol{r}''(s)| = \frac{1}{\rho}.$$

When the circle is allowed to have any center, a slight generalization of this argument shows that *the curvature of a circle of radius* ρ *is* $\kappa = 1/\rho$ *at each point on the circle* (see Exercise 33). ■

Figure 6.3 shows the curvature vectors for several circles of various radii.

For more general plane curves C parameterized by arc length, the expression $\rho(s) = \dfrac{1}{\kappa(s)}$ is called the **radius of curvature** at $\boldsymbol{r}(s)$. Since $\rho(s)$ is the radius of the circle whose curvature is also $\kappa(s)$, the number $\rho(s)$ determines a circle that

(i) is tangent to C at $\boldsymbol{r}(s)$,

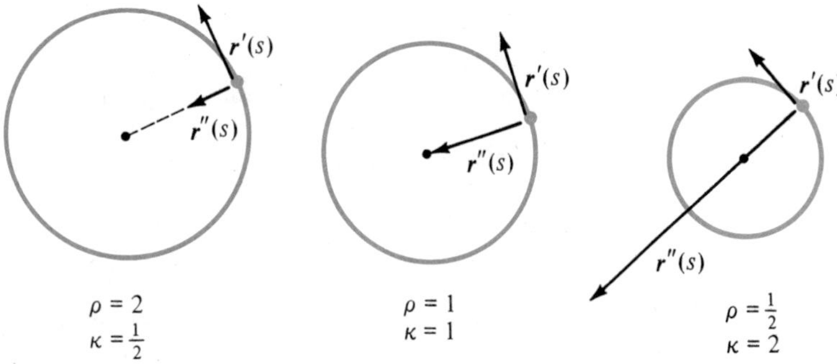

Figure 6.3 Tangent and curvature vectors for various circles.

(ii) has radius $\rho(s) = \dfrac{1}{\kappa(s)}$,

(iii) has the same curvature vector as C at $\boldsymbol{r}(s)$.

This circle is called the **osculating circle,** or **circle of curvature,** at $\boldsymbol{r}(s)$. The center of this circle is called the **center of curvature.** Figures 6.4 and 6.5 show the osculating circles at various points on a curve C.

For plane curves that are not parameterized by arc length, the following two theorems indicate how curvature may be computed.

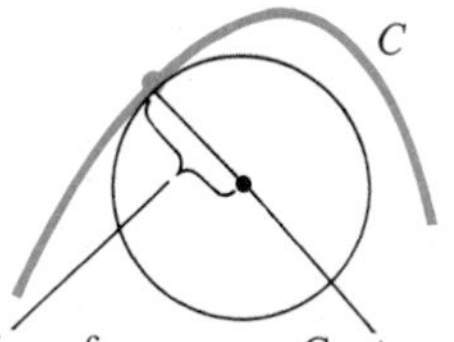

Figure 6.4 Circle of curvature for a plane curve C.

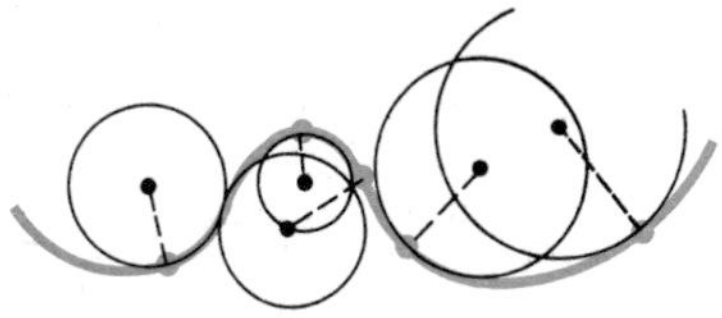

Figure 6.5 Osculating circles at various points on C.

THEOREM 4

Let the curve C be determined parametrically by the twice differentiable vector function

$$\boldsymbol{r}(t) = x(t)\boldsymbol{i} + y(t)\boldsymbol{j}, \qquad a \le t \le b. \tag{3}$$

The curvature $\kappa(t)$ of C at the point $(x(t), y(t))$ is

$$\kappa(t) = \frac{|x'(t)y''(t) - y'(t)x''(t)|}{([x'(t)]^2 + [y'(t)]^2)^{3/2}}. \tag{4}$$

Before indicating the proof of Theorem 4, we consider two examples.

Example 2 For the curve C given parametrically by the vector function

$$\boldsymbol{r}(t) = (2t^2 + 1)\boldsymbol{i} + (t^3 - 3)\boldsymbol{j}, \qquad 1 \le t \le 3$$

find

(a) the curvature function $\kappa(t)$, and
(b) the radius of curvature at the point $(3, -2)$.

Solution: Here $x(t) = 2t^2 + 1$ and $y(t) = t^3 - 3$. Thus

$$x'(t) = 4t \qquad y'(t) = 3t^2,$$

$$x''(t) = 4 \qquad y''(t) = 6t.$$

According to Theorem 4, the curvature is

$$\kappa(t) = \frac{|4t \cdot 6t - 3t^2 \cdot 4|}{([4t]^2 + [3t^2]^2)^{3/2}} = \frac{12t^2}{(16t^2 + 9t^4)^{3/2}}.$$

The point $(3, -2)$ corresponds to $t = 1$. The curvature at this point is therefore

$$\kappa(1) = \frac{12}{(16 + 9)^{3/2}} = \frac{12}{125}.$$

The radius of curvature at this point is

$$\rho(1) = \frac{1}{\kappa(1)} = \frac{125}{12}.$$ ■

Example 3 Find the curvature function $\kappa(t)$ for the graph of the function $f(t) = \sin t$.

Solution: We can write the function $f(t) = \sin t$ in parametric (vector) form as

$$\boldsymbol{r}(t) = t\boldsymbol{i} + \sin t\boldsymbol{j}.$$

Thus, $x(t) = t$ and $y(t) = \sin t$, so

$$x'(t) = 1 \qquad y'(t) = \cos t,$$

$$x''(t) = 0 \qquad y''(t) = -\sin t.$$

By Theorem 4,

$$\kappa(t) = \frac{|1 \cdot (-\sin t) - 0 \cdot \cos t|}{(1^2 + [\cos t]^2)^{3/2}} = \frac{|\sin t|}{(1 + \cos^2 t)^{3/2}}.$$

Notice that $\kappa(t)$ is a minimum (zero) at $t = 0, \pm\pi, \pm 2\pi, \ldots$ and that $\kappa(t)$ is a maximum (one) at $t = \pi/2 \pm n\pi$, $n = 1, 2, \ldots$. ■

Example 3 shows how Theorem 4 may be used to calculate the curvature for the graph of a twice differentiable function $y = f(x)$. Such functions can always be written in vector form as

$$\boldsymbol{r}(x) = x\boldsymbol{i} + f(x)\boldsymbol{j}.$$

Then $x' = 1$, $x'' = 0$, $y' = f'(x)$, and $y'' = f''(x)$, so the curvature $\kappa(x)$, according to Theorem 4, is

$$\kappa(x) = \frac{|0 \cdot f'(x) - f''(x) \cdot 1|}{(1 + [f'(x)]^2)^{3/2}}.$$

We state this result as a corollary of Theorem 4.

COROLLARY 1

Let $y = f(x)$ be a twice differentiable function of x. The curvature $\kappa(x)$ at the point $(x, f(x))$ on the graph of $y = f(x)$ is

$$\kappa(x) = \frac{|f''(x)|}{(1 + [f'(x)]^2)^{3/2}}.$$

Example 4 Show that the curvature of a nonvertical line in the plane is zero.

Solution: The equation for a nonvertical line in the plane can be written in the form $f(x) = mx + b$ where m and b are constants. Thus $f'(x) = m$ and $f''(x) = 0$. Thus, $\kappa(x) = 0$, according to Corollary 1. ■

Proof of Theorem 4: We will demonstrate the proof up to a point at which you can complete the argument with a calculation (Exercise 34).

The difficulty to be overcome is that the parameterization given by $\boldsymbol{r}(t)$ in equation (3) is not necessarily a parameterization by arc length. We address this problem by letting ℓ denote the length of the arc of C from $\boldsymbol{r}(a)$ to $\boldsymbol{r}(b)$. Then the arc length function

$$s(t) = \int_a^t |\boldsymbol{r}'(\omega)|d\omega \qquad \text{(equation (10), Section 18.4)} \tag{5}$$

has the properties that

$$s(a) = 0,$$
$$s(b) = \ell,$$

and

$$s'(t) = |\boldsymbol{r}'(t)|, \qquad a \le t \le b \qquad \text{(equation (11), Section 18.4).} \tag{6}$$

Now, suppose that the vector function

$$\boldsymbol{R}(s) = X(s)\boldsymbol{i} + Y(s)\boldsymbol{j}, \qquad 0 \le s \le \ell$$

is an arc length parameterization for C. Then the composition of $\boldsymbol{R}(s)$ with $s(t)$ in (5) is

$$\boldsymbol{R}(s(t)) = X(s(t))\boldsymbol{i} + Y(s(t))\boldsymbol{j}, \qquad a \le t \le b.$$

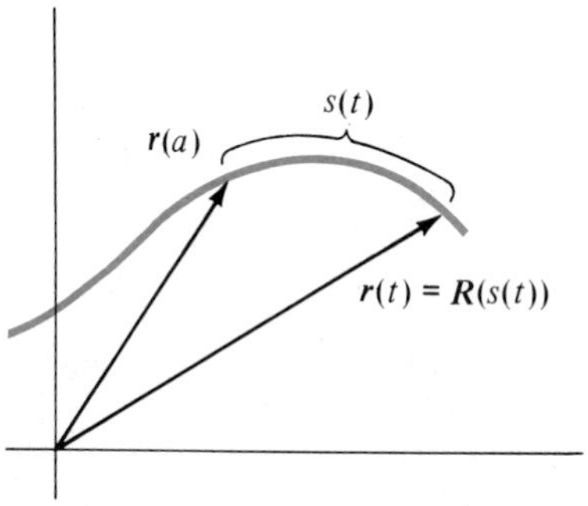

Figure 6.6 $\boldsymbol{R}(s(t)) = \boldsymbol{r}(t)$, $a \le t \le b$.

It is important to note that $\boldsymbol{R}(s(t)) = \boldsymbol{r}(t)$ for each $t \in [a, b]$, since both vectors terminate at a point $s(t)$ units along C from $\boldsymbol{r}(a)$ (see Figure 6.6).

Since $\boldsymbol{R}(s)$ is a parameterization by arc length,

$$\boldsymbol{T}(t) = \boldsymbol{R}'(s) = \frac{d}{ds}\boldsymbol{R}(s(t)) \tag{7}$$

is a unit tangent for C at $\boldsymbol{r}(t) = \boldsymbol{R}(s(t))$ (equation (16), Section 18.4). Differentiating both sides of equation (7) with respect to t, gives

$$\boldsymbol{T}'(t) = \frac{d}{dt}\left[\frac{d}{ds}\boldsymbol{R}(s)\right]$$
$$= \frac{d}{ds}\left[\frac{d}{ds}\boldsymbol{R}(s)\right]\frac{ds}{dt} \qquad \text{(by Chain Rule)}$$

$$= \boldsymbol{R}''(s)\,\frac{ds}{dt}$$

$$= \boldsymbol{R}''(s)\,|\boldsymbol{r}'(t)|, \qquad \text{(by equation (6))}$$

so

$$|\boldsymbol{R}''(s)| = \frac{|\boldsymbol{T}'(t)|}{|\boldsymbol{r}'(t)|}. \tag{8}$$

The expression in equation (8) is the curvature of C at $\boldsymbol{r}(t) = \boldsymbol{R}(s(t))$. It is left for you to show that

$$\frac{|\boldsymbol{T}'(t)|}{|\boldsymbol{r}'(t)|} = \frac{|x''(t)y'(t) - x'(t)y''(t)|}{([x'(t)]^2 + [y'(t)]^2)^{3/2}}, \tag{9}$$

which will complete the proof. ■

Curvature for Space Curves

If C is a space curve parameterized by arc length by the vector function

$$C:\ \ \boldsymbol{r}(s) = x(s)\boldsymbol{i} + y(s)\boldsymbol{j} + z(s)\boldsymbol{k}, \qquad a \le s \le b, \tag{10}$$

the vector $\boldsymbol{T}(s) = \boldsymbol{r}'(s)$ is a unit tangent to the curve C at $\boldsymbol{r}(s)$. Since $|\boldsymbol{r}'(s)| = 1$, the curvature vector $\boldsymbol{r}''(s)$ is orthogonal to the tangent $\boldsymbol{T}(s)$ (Example 6, Section 18.3). We therefore refer to the curvature vector $\boldsymbol{r}''(s)$ as being **normal** to C at $\boldsymbol{r}(s)$, and the **principal unit normal** to C at $\boldsymbol{r}(s)$ is the vector

$$\boldsymbol{N}(s) = \frac{\boldsymbol{r}''(s)}{|\boldsymbol{r}''(s)|}. \tag{11}$$

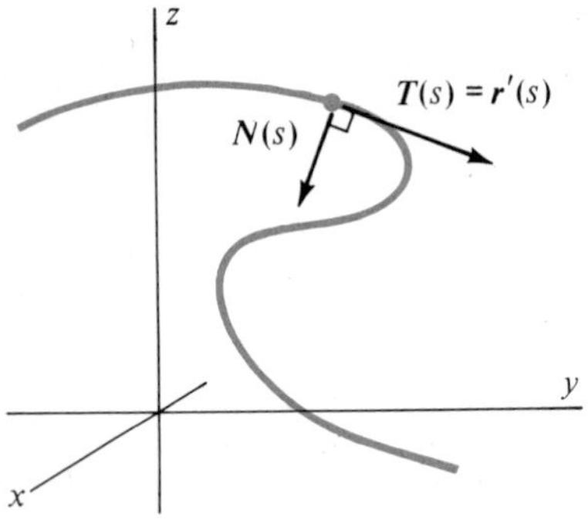

Figure 6.7 Principal unit normal is $N(s) = \dfrac{r''(s)}{|r''(s)|}$.

Of course, the principal unit normal in (11) is defined only when $\boldsymbol{r}(s)$ is twice differentiable and $|\boldsymbol{r}''(s)| \neq 0$ (see Figure 6.7). If C in (10) is a plane curve, the vector $\boldsymbol{N}(s)$ points toward the center of curvature. Multiplying (11) through by $|\boldsymbol{r}''(s)| = \kappa(s)$ and writing $\boldsymbol{r}'(s) = \boldsymbol{T}(s)$ gives the equation

$$\boldsymbol{T}'(s) = \kappa(s)\boldsymbol{N}(s), \tag{12}$$

an equation by which we could alternatively have defined curvature for space curves.

Normal and Tangential Components of Acceleration

We may apply these ideas to obtain an informative expression for the velocity of a particle moving in space with position function given by

$$\boldsymbol{r}(t) = x(t)\boldsymbol{i} + y(t)\boldsymbol{j} + z(t)\boldsymbol{k}. \tag{13}$$

We assume that $\boldsymbol{r}(t)$ is twice differentiable (although not necessarily an arc length parameterization for the path).

Letting s denote arc length and $\boldsymbol{T} = \dfrac{d\boldsymbol{r}}{ds}$ the unit tangent, we apply the Chain Rule to obtain the velocity function $\boldsymbol{v}$ as

$$\boldsymbol{v}(t) = \frac{d\boldsymbol{r}}{dt} = \frac{d\boldsymbol{r}}{ds}\frac{ds}{dt} = \frac{ds}{dt}\boldsymbol{T}.$$

Similarly, the acceleration function $\boldsymbol{a}(t)$ is

$$\begin{aligned}\boldsymbol{a}(t) &= \frac{d\boldsymbol{v}}{dt} = \frac{d}{dt}\left(\frac{d\boldsymbol{r}}{ds}\frac{ds}{dt}\right) \qquad (14)\\ &= \frac{d^2s}{dt^2}\frac{d\boldsymbol{r}}{ds} + \frac{ds}{dt}\left(\frac{d^2\boldsymbol{r}}{ds^2}\frac{ds}{dt}\right)\\ &= \frac{d^2s}{dt^2}\frac{d\boldsymbol{r}}{ds} + \left(\frac{ds}{dt}\right)^2\left[\frac{d}{ds}\left(\frac{d\boldsymbol{r}}{ds}\right)\right]\\ &= \frac{d^2s}{dt^2}\boldsymbol{T}(s) + \left(\frac{ds}{dt}\right)^2\kappa(s)\boldsymbol{N}(s) \qquad \text{(equation 12).}\end{aligned}$$

That is, we can express acceleration as the sum of a vector parallel to the unit tangent $\boldsymbol{T}(s)$ and a vector parallel to the unit normal $\boldsymbol{N}(s)$. It is customary to refer to the **tangential component of acceleration** as the coefficient $a_T = \dfrac{d^2s}{dt^2}$, and to the **normal component of acceleration** as the coefficient $a_N = \left(\dfrac{ds}{dt}\right)^2\kappa(s)$. We may then write

$$\boldsymbol{a} = a_T\boldsymbol{T} + a_N\boldsymbol{N}. \qquad (15)$$

The components a_T and a_N have straightforward interpretations. The tangential component is

$$a_T = \frac{d^2s}{dt^2} = \frac{d}{dt}\left(\frac{ds}{dt}\right) = \frac{d}{dt}(|\boldsymbol{v}(t)|),$$

so that a_T is just the rate at which speed is changing. For example, if the speed of the particle is constant, then $a_T = 0$. The normal component of acceleration can be written $a_N = \left(\dfrac{ds}{dt}\right)^2\kappa(s) = |\boldsymbol{v}|^2\kappa$. Thus, a_N is influenced both by speed and by curvature. Since $\boldsymbol{N}$ is orthogonal to $\boldsymbol{T}$, this explains why the force tending to pull an object (such as an automobile) from its path is influenced both by the speed of the object and by the curvature of the path.

Since $a_T = \dfrac{d}{dt}(|\boldsymbol{v}(t)|)$, computing a_T from $\boldsymbol{r}(t)$ is usually simple. However, calculating a_N can be considerably more difficult. An aid to this calculation is obtained by using equation (15) and the fact that $\boldsymbol{T}$ and $\boldsymbol{N}$ are orthogonal unit vectors:

$$\begin{aligned}|\boldsymbol{a}|^2 = \boldsymbol{a}\cdot\boldsymbol{a} &= (a_T\boldsymbol{T} + a_N\boldsymbol{N})\cdot(a_T\boldsymbol{T} + a_N\boldsymbol{N})\\ &= (a_T)^2\,|\boldsymbol{T}|^2 + a_N|\boldsymbol{N}|^2\\ &= a_T^2 + a_N^2.\end{aligned}$$

Thus

$$a_N = \sqrt{|\boldsymbol{a}|^2 - a_T^2}. \qquad (16)$$

Example 5 Find the tangential and normal components of acceleration for a particle moving along the helical path

$$\boldsymbol{r}(t) = \cos t^2\boldsymbol{i} + \sin t^2\boldsymbol{j} + t\boldsymbol{k}.$$

Strategy
Find $\boldsymbol{v}(t) = \boldsymbol{r}'(t)$.

Solution
Here

$$\boldsymbol{v}(t) = -2t \sin t^2\boldsymbol{i} + 2t \cos t^2\boldsymbol{j} + \boldsymbol{k}$$

so

Find $|\boldsymbol{v}(t)|$.

$$|\boldsymbol{v}(t)| = \sqrt{4t^2 \sin^2 t^2 + 4t^2 \cos^2 t^2 + 1} = \sqrt{4t^2 + 1}.$$

Thus

$a_T = \dfrac{d}{dt}(|\boldsymbol{v}(t)|)$.

$$a_T = \frac{d}{dt}(\sqrt{4t^2 + 1}) = \frac{4t}{\sqrt{4t^2 + 1}}.$$

Find $\boldsymbol{a}(t)$.

To find $\boldsymbol{a}(t)$ we differentiate $\boldsymbol{v}(t)$:

$$\boldsymbol{a}(t) = (-2 \sin t^2 - 4t^2 \cos t^2)\boldsymbol{i} + (2 \cos t^2 - 4t^2 \sin t^2)\boldsymbol{j},$$

so

Find $|\boldsymbol{a}(t)|$.

$$|\boldsymbol{a}(t)| = \{4 \sin^2 t^2 + 16t^2 \sin t^2 \cos t^2 + 16t^4 \cos^2 t^2 + 4 \cos^2 t^2 - 16t^2 \sin t^2 \cos t^2 + 16t^4 \sin^2 t^2\}^{1/2} = \sqrt{4 + 16t^4}.$$

Apply equation (16).

Thus

$$a_N = \sqrt{|\boldsymbol{a}|^2 - a_T^2} = \sqrt{(4 + 16t^4) - \frac{16t^2}{4t^2 + 1}} = \sqrt{\frac{64t^6 + 16t^4 + 4}{4t^2 + 1}}.$$

Notice that $\lim_{t\to\infty} a_T = 2$, while $\lim_{t\to\infty} a_N = +\infty$. Can you explain why this is so?

■

Vector Form of Curvature

A drawback of our work on curvature up to this point is that $\kappa(s)$ is difficult to compute, using Definition 10, when the curve C determined by $\boldsymbol{r}(t)$ is not parameterized by arc length. We shall now develop a formula for κ that is easier to apply. We imagine the parameterization $\boldsymbol{r}(t)$ for C as the position function of a particle moving in space. Then, writing

$$\boldsymbol{v}(t) = \frac{d\boldsymbol{r}}{dt} = \frac{d\boldsymbol{r}}{ds}\cdot\frac{ds}{dt} = \left(\frac{ds}{dt}\right)\boldsymbol{T}$$

and using equation (14) we find that

$$\boldsymbol{v} \times \boldsymbol{a} = \left(\frac{ds}{dt}\right)\boldsymbol{T} \times \left[\left(\frac{d^2s}{dt^2}\right)\boldsymbol{T} + \kappa\left(\frac{ds}{dt}\right)^2\boldsymbol{N}\right] = \kappa\left(\frac{ds}{dt}\right)^3(\boldsymbol{T} \times \boldsymbol{N})$$

since $\boldsymbol{T} \times \boldsymbol{T} = \boldsymbol{0}$. Also, since $\boldsymbol{T}$ and $\boldsymbol{N}$ are orthogonal unit vectors,

$$|\boldsymbol{T} \times \boldsymbol{N}| = |\boldsymbol{T}||\boldsymbol{N}||\sin \theta| = 1.$$

Thus,

$$|\boldsymbol{v} \times \boldsymbol{a}| = \kappa \left(\frac{ds}{dt}\right)^3 = \kappa |\boldsymbol{v}|^3,$$

so

$$\boxed{\kappa = \frac{|\boldsymbol{v} \times \boldsymbol{a}|}{|\boldsymbol{v}|^3}.} \tag{17}$$

Example 6 Find the curvature of the helix

$$C: \quad \boldsymbol{r}(t) = \cos t\boldsymbol{i} + \sin t\boldsymbol{j} + t\boldsymbol{k}.$$

Solution: Here

$$\boldsymbol{v}(t) = -\sin t\boldsymbol{i} + \cos t\boldsymbol{j} + \boldsymbol{k}$$

and

$$\boldsymbol{a}(t) = -\cos t\boldsymbol{i} - \sin t\boldsymbol{j}.$$

Thus

$$\begin{aligned}\boldsymbol{v} \times \boldsymbol{a} &= \det \begin{bmatrix} \boldsymbol{i} & \boldsymbol{j} & \boldsymbol{k} \\ -\sin t & \cos t & 1 \\ -\cos t & -\sin t & 0 \end{bmatrix} \\ &= \sin t\boldsymbol{i} - \cos t\boldsymbol{j} + (\sin^2 t + \cos^2 t)\boldsymbol{k},\end{aligned}$$

so

$$|\boldsymbol{v} \times \boldsymbol{a}| = \sqrt{\sin^2 t + \cos^2 t + 1} = \sqrt{2}.$$

Also,

$$|\boldsymbol{v}| = \sqrt{\sin^2 t + \cos^2 t + 1} = \sqrt{2}.$$

Thus, by (17),

$$\kappa = \frac{\sqrt{2}}{(\sqrt{2})^3} = \frac{1}{2}.$$

■

Exercise Set 18.6

In Exercises 1–4, verify that the given curve is parameterized by arc length. Then find (a) the unit tangent $\boldsymbol{T}(s)$, (b) the curvature vector $\boldsymbol{r}''(s)$, and (c) the curvature $\kappa(s)$.

1. $\boldsymbol{r}(s) = \frac{1}{2}s\boldsymbol{i} + \frac{\sqrt{3}}{2}s\boldsymbol{j}$

2. $\boldsymbol{r}(s) = (4 + \cos s)\boldsymbol{i} + (2 + \sin s)\boldsymbol{j}$

3. $\boldsymbol{r}(s) = \frac{\sqrt{2}}{2}\cos s\boldsymbol{i} + \frac{\sqrt{2}}{2}\sin s\boldsymbol{j} + \frac{\sqrt{2}}{2}s\boldsymbol{k}$

4. $\boldsymbol{r}(s) = \sin\left(\frac{s}{2}\right)\boldsymbol{i} + \frac{\sqrt{3}}{2}s\boldsymbol{j} + \cos\left(\frac{s}{2}\right)\boldsymbol{k}$

In Exercises 5–10, find the curvature function $\kappa(t)$ for the given plane curves.

5. $\boldsymbol{r}(t) = 3\boldsymbol{i} + t^2\boldsymbol{j}$

6. $\boldsymbol{r}(t) = t\boldsymbol{i} + (t^2 + 3)\boldsymbol{j}$

7. $\boldsymbol{r}(t) = t\boldsymbol{i} + e^t\boldsymbol{j}$

8. $\boldsymbol{r}(t) = e^t\boldsymbol{i} + e^{-t}\boldsymbol{j}$

9. $\boldsymbol{r}(t) = (3 + 2\sin t)\boldsymbol{i} + (5 + 2\cos t)\boldsymbol{j}$

10. $\boldsymbol{r}(t) = (t - \sin t)\boldsymbol{i} + (1 - \cos t)\boldsymbol{j}$

In Exercises 11–16, find the curvature of the graph of the given function.

11. $f(x) = 3 + x^2$

12. $f(x) = \sqrt{x + 4}$

13. $f(x) = x^3$

14. $f(x) = \ln x$

15. $f(x) = \cos x$

16. $f(x) = e^{x^2}$

In Exercises 17–20, find the curvature $\kappa(t)$ for the space curve determined by $\boldsymbol{r}(t)$.

17. $\boldsymbol{r}(t) = \boldsymbol{i} + t\boldsymbol{j} + t^2\boldsymbol{k}$

18. $\boldsymbol{r}(t) = \cos t\boldsymbol{i} + \sin t\boldsymbol{j} + \boldsymbol{k}$

19. $\boldsymbol{r}(t) = e^t \cos t\boldsymbol{i} + e^t \sin t\boldsymbol{j} + t\boldsymbol{k}$

20. $\boldsymbol{r}(t) = \ln t\boldsymbol{i} + t\boldsymbol{j} + t\boldsymbol{k}, \quad t > 0$

21. Find the center of curvature for the graph of $y = \sin x$ at the point $(\pi/2, 1)$.

22. Find the point on the graph of $y = \ln x$ where curvature is a maximum.

23. For the graph of $y = \sqrt{x}$, find
a. the curvature at $(1, 1)$,
b. the center of curvature at $(1, 1)$,
c. $\lim_{x\to 0^+} \kappa(x)$.

24. Find the equation for the osculating circle (circle of curvature) for the graph of $y = e^{-x}$ at the point where $x = 1$.

25. Show that the curvature of a line in space is zero.

26. Show that if the curve C is the graph of the polar equation $r = f(\theta)$ and if $f''(\theta)$ exists, then the curvature $\kappa(\theta)$ is given by

$$\kappa(\theta) = \frac{|f(\theta)f''(\theta) - 2[f'(\theta)]^2 - [f(\theta)]^2|}{([f'(\theta)]^2 + [f(\theta)]^2)^{3/2}}.$$

(*Hint:* C can be written as $\boldsymbol{r}(\theta) = x(\theta)\boldsymbol{i} + y(\theta)\boldsymbol{j}$, with $x(\theta) = f(\theta)\cos\theta$, $y(\theta) = f(\theta)\sin\theta$.)

In Exercises 27–30, use the result of Exercise 26 to find the curvature function $\kappa(\theta)$.

27. $r = 1 + \sin\theta$

28. $r = \theta$

29. $r = 1 - \cos\theta$

30. $r = e^{\theta}$

31. For a curve C in the plane, explain why the curvature vector $\boldsymbol{r}''(s)$ always points in the direction of the concave side of the curve.

32. Find the curvature of the ellipse $x^2 + 2y^2 = 4$ at the point $(2, 0)$.

33. Prove that the curvature of a circle of radius ρ is $\kappa = \dfrac{1}{\rho}$.

34. Complete the proof of Theorem 4.

SUMMARY OUTLINE OF CHAPTER 18

- A **vector function** $\boldsymbol{f}(t)$ has the form

$$\boldsymbol{f}(t) = x(t)\boldsymbol{i} + y(t)\boldsymbol{j}, \qquad \text{or} \qquad \boldsymbol{f}(t) = x(t)\boldsymbol{i} + y(t)\boldsymbol{j} + z(t)\boldsymbol{k}$$

where $x(t)$, $y(t)$, and $z(t)$ are real-valued functions.

- For vector functions $\boldsymbol{f}(t)$ and $\boldsymbol{g}(t)$, and scalar c,

(i) $(\boldsymbol{f} + \boldsymbol{g})(t) = \boldsymbol{f}(t) + \boldsymbol{g}(t)$
(ii) $(c\boldsymbol{f})(t) = c\boldsymbol{f}(t)$
(iii) $(\boldsymbol{f}\cdot\boldsymbol{g})(t) = \boldsymbol{f}(t)\cdot\boldsymbol{g}(t)$
(iv) $(\boldsymbol{f} \times \boldsymbol{g})(t) = \boldsymbol{f}(t) \times \boldsymbol{g}(t)$

- For $\boldsymbol{f}(t) = x(t)\boldsymbol{i} + y(t)\boldsymbol{j} + z(t)\boldsymbol{k}$,

(i) $\lim_{t\to t_0} \boldsymbol{f}(t) = \left[\lim_{t\to t_0} x(t)\right]\boldsymbol{i} + \left[\lim_{t\to t_0} y(t)\right]\boldsymbol{j} + \left[\lim_{t\to t_0} z(t)\right]\boldsymbol{k}$

(ii) $\boldsymbol{f}'(t) = x'(t)\boldsymbol{i} + y'(t)\boldsymbol{j} + z'(t)\boldsymbol{k}$

(iii) $\int \boldsymbol{f}(t)\,dt = \left[\int x(t)\,dt\right]\boldsymbol{i} + \left[\int y(t)\,dt\right]\boldsymbol{j} + \left[\int z(t)\,dt\right]\boldsymbol{k}$

(iv) $\int_a^b \boldsymbol{f}(t)\,dt = \left[\int_a^b x(t)\,dt\right]\boldsymbol{i} + \left[\int_a^b y(t)\,dt\right]\boldsymbol{j} + \left[\int_a^b z(t)\,dt\right]\boldsymbol{k}.$

(v) $\boldsymbol{f}(t)$ is continuous at t_0 if and only if $x(t)$, $y(t)$, and $z(t)$ are all continuous at t_0.

(vi) $\boldsymbol{L} = \lim_{t\to t_0} \boldsymbol{f}(t)$ if and only if $|\boldsymbol{f}(t) - \boldsymbol{L}| \to 0$ as $|t - t_0| \to 0$ (Theorem 1).

- Derivatives of vector functions are calculated as follows:

(i) $(\boldsymbol{f} + \boldsymbol{g})'(t) = \boldsymbol{f}'(t) + \boldsymbol{g}'(t)$
(ii) $(c\boldsymbol{f})'(t) = c\boldsymbol{f}'(t)$
(iii) $(h\boldsymbol{f})'(t) = h(t)\boldsymbol{f}'(t) + h'(t)\boldsymbol{f}(t)$
(iv) $(\boldsymbol{f} \times \boldsymbol{g})'(t) = [\boldsymbol{f}(t) \times \boldsymbol{g}'(t)] + [\boldsymbol{f}'(t) \times \boldsymbol{g}(t)]$
(v) $[\boldsymbol{f}(h(t))]' = h'(t)\boldsymbol{f}'(h(t))$
(vi) $(\boldsymbol{f}\cdot\boldsymbol{g})'(t) = \boldsymbol{f}(t)\cdot\boldsymbol{g}'(t) + \boldsymbol{f}'(t)\cdot\boldsymbol{g}(t)$.

- If $\boldsymbol{r}'(t_0) \neq 0$, the **tangent** to the graph of $\boldsymbol{r}(t)$ at $\boldsymbol{r}(t_0)$ is

$$\boldsymbol{r}'(t_0) = x'(t_0)\boldsymbol{i} + y'(t_0)\boldsymbol{j} + z'(t_0)\boldsymbol{k}.$$

- The **unit tangent** to the graph of $\boldsymbol{r}(t)$ at $\boldsymbol{r}(t_0)$ is

$$\boldsymbol{T} = \frac{\boldsymbol{r}'(t_0)}{|\boldsymbol{r}'(t_0)|}, \qquad \boldsymbol{r}'(t_0) \neq 0.$$

- The **length,** L, of the differentiable curve C; $\boldsymbol{r}(t) = x(t)\boldsymbol{i} + y(t)\boldsymbol{j} + z(t)\boldsymbol{k}$ from $\boldsymbol{r}(a)$ to $\boldsymbol{r}(b)$ is

$$L = \int_a^b \sqrt{[x'(t)]^2 + [y'(t)]^2 + [z'(t)]^2}\, dt$$

$$= \int_a^b |\boldsymbol{r}'(t)|\, dt.$$

- The curve C above is **parameterized by arc length** on the interval I if

$$\int_0^t |\boldsymbol{r}'(\omega)| d\omega = t$$

for all t in I. Equivalently, $|\boldsymbol{r}'(t)| = 1, \qquad t \in I.$

- For a particle moving in space with the twice differentiable **position function** $\boldsymbol{r}(t)$,

  (i) the **velocity** function is $\boldsymbol{v}(t) = \boldsymbol{r}'(t)$
  (ii) the **speed** is $|\boldsymbol{v}(t)| = |\boldsymbol{r}'(t)|$
  (iii) the **acceleration** function is $\boldsymbol{a}(t) = \boldsymbol{v}'(t) = \boldsymbol{r}''(t)$.

  Thus,

  (iv) $\boldsymbol{v}(t) = \int \boldsymbol{a}(t)\, dt$

  (v) $\boldsymbol{r}(t) = \int \boldsymbol{v}(t)\, dt.$

- The **trajectory** of a projectile fired from the origin with initial speed s_0 and angle of elevation θ is

$$\boldsymbol{r}(t) = s_0 \cos\theta\, t\boldsymbol{i} + \left(s_0 \sin\theta t - \frac{1}{2}gt^2\right)\boldsymbol{j}.$$

- If C: $\boldsymbol{r}(s)$ is parameterized by arc length, the **curvature** at $\boldsymbol{r}(s)$ is

$$\kappa(s) = |\boldsymbol{r}''(s)|.$$

- The **radius of curvature** at $r(s)$ is

$$\rho(s) = \frac{1}{\kappa(s)}, \qquad \kappa(s) \neq 0.$$

- If the plane curve C is parameterized by the equations $x = x(t)$, $y = y(t)$, the **curvature** at $(x(t), y(t))$ is

$$\kappa(t) = \frac{|x'(t)y''(t) - y'(t)x''(t)|}{([x'(t)]^2 + [y'(t)]^2)^{3/2}}.$$

- If the curve C is the graph of $y = f(x)$,

$$\kappa(x) = \frac{|f''(x)|}{(1 + [f'(x)]^2)^{3/2}}.$$

- The **principal unit normal** to $\boldsymbol{r}(s)$ is $\boldsymbol{N}(s) = \dfrac{\boldsymbol{r}''(s)}{|\boldsymbol{r}''(s)|}$ when $\boldsymbol{r}(s)$ is parameterized by arc length.

- If $\boldsymbol{r}(t)$ is a twice differentiable position function, its acceleration function can be written

$$\boldsymbol{a}(t) = a_T\boldsymbol{T} + a_N\boldsymbol{N}$$

where the **tangential component of acceleration** is

$$a_T = \frac{d^2s}{dt^2}$$

and the **normal component of acceleration** is

$$a_N = \left(\frac{ds}{dt}\right)^2 \kappa(s) = |\boldsymbol{v}|^2 \kappa.$$

- If the curve C is determined by the twice differentiable **position function** $\boldsymbol{r}(t)$, then

$$\kappa = \frac{|\boldsymbol{v} \times \boldsymbol{a}|}{|\boldsymbol{v}|^3}$$

at each point on C.

REVIEW EXERCISES—CHAPTER 18

1. Find an equation in rectangular coordinates for the graph of the vector function $\boldsymbol{r}(t) = a\cos t\boldsymbol{i} + b\sin t\boldsymbol{j}$.

2. Sketch the graph of the vector function $\boldsymbol{r}(t) = \boldsymbol{i} + \cos t\boldsymbol{j} + \boldsymbol{k}$.

3. Find the implicit domain of the vector function

$$\boldsymbol{r}(t) = t\boldsymbol{i} + \sqrt{1 - t^2}\boldsymbol{j} + \frac{1}{t}\boldsymbol{k}.$$

4. Show that the graph of the vector function $\boldsymbol{r}(t) = \tan t\boldsymbol{i} + \sec t\boldsymbol{j}$, $-\pi/2 < t < \pi/2$, is one branch of the hyperbola with rectangular equation $y^2 - x^2 = 1$. What are the asymptotes for this curve?

5. Sketch the graph of the **cycloid** given by the vector function $\boldsymbol{r}(t) = a(t - \sin t)\boldsymbol{i} + a(1 - \cos t)\boldsymbol{j}$, $t \geq 0$. Find the values of t for which the tangent vector is zero. (The cycloid gives the path of a point on the rim of a wheel of radius a as the wheel rolls along a horizontal surface.)

In Exercises 6–8, find the indicated limit.

6. $\lim_{t\to 0^+} \left\{\cos 2t\boldsymbol{i} + t\cos t\boldsymbol{j} + \dfrac{t}{\sqrt{t+2}}\boldsymbol{k}\right\}$

7. $\lim_{t\to 0} \left\{\left(\dfrac{\sin 3t}{t}\right)\boldsymbol{i} + \sqrt{t+2}\boldsymbol{j} + \dfrac{\tan 2t}{t}\boldsymbol{k}\right\}$

8. $\lim_{t\to 2^+} \left\{\sqrt{2+t}\boldsymbol{i} + \ln\sqrt{t}\boldsymbol{j} + \dfrac{t^3 - 8}{t-2}\boldsymbol{k}\right\}$

In Exercises 9–11, determine the values of t for which the function is discontinuous.

9. $\boldsymbol{r}(t) = \sin t\boldsymbol{i} + \sec t\boldsymbol{j}, \quad 0 \le t \le 2\pi$

10. $\boldsymbol{r}(t) = \dfrac{1}{\sqrt{t}}\boldsymbol{i} + \tan t\boldsymbol{j} + \ln(1 + t)\boldsymbol{k}, \quad 0 \le t \le \pi$

11. $\boldsymbol{r}(t) = \dfrac{1}{t^2 - 9}\boldsymbol{i} + \dfrac{t + 2}{t^2 - 4}\boldsymbol{j} + t\boldsymbol{k}, \quad -4 \le t \le 4$

In Exercises 12–15, find $\boldsymbol{r}'(t)$ and $\boldsymbol{r}''(t)$.

12. $\boldsymbol{r}(t) = e^{-2t}\boldsymbol{i} + \cos\left(\dfrac{\pi t}{4}\right)\boldsymbol{j} + \boldsymbol{k}$

13. $\boldsymbol{r}(t) = te^t\boldsymbol{i} + te^{-t}\boldsymbol{j}$

14. $\boldsymbol{r}(t) = \tan^{-1}2t\boldsymbol{i} + \sqrt{t}\boldsymbol{j} + t\boldsymbol{k}$

15. $\boldsymbol{r}(t) = \ln 3t\boldsymbol{i} + e^{\sqrt{t}}\boldsymbol{j}$

16. Find $\displaystyle\int_0^1 \boldsymbol{r}(t)\,dt$ for $\boldsymbol{r}(t)$ in Exercise 12.

17. Find $\displaystyle\int_0^1 \boldsymbol{r}(t)\,dt$ for $\boldsymbol{r}(t)$ in Exercise 13.

18. Find a vector function $\boldsymbol{r}(t)$ for which $\boldsymbol{r}'(t) = 4\boldsymbol{r}(t)$ and $\boldsymbol{r}(0) = 4\boldsymbol{i} - \boldsymbol{j} + \pi\boldsymbol{k}$.

19. Find a vector function $\boldsymbol{r}(t)$ for which $\boldsymbol{r}''(t) = 9\boldsymbol{r}(t)$ and $\boldsymbol{r}(0) = 4\boldsymbol{i} + 3\boldsymbol{j}$.

20. Find the values of α for which the graph of the vector function

$$\boldsymbol{r}(t) = \alpha t\boldsymbol{i} + \cos(2\alpha t)\boldsymbol{j} + \sin(2\alpha t)\boldsymbol{k}, \quad 0 \le t \le \pi$$

is parameterized by arc length.

21. Find an arc length parameterization for the circle $x^2 + y^2 = 25$.

22. Find the unit tangent $\boldsymbol{T}(t)$ for the ellipse $\boldsymbol{r}(t) = a\cos t\boldsymbol{i} + b\sin t\boldsymbol{j}$, $0 \le t \le 2\pi$.

23. For the graph of the vector function $\boldsymbol{r}(t) = t\boldsymbol{i} + \cos t\boldsymbol{j} + \sin t\boldsymbol{k}$, find
- **a.** the curvature, $\kappa(t)$,
- **b.** the unit tangent $\boldsymbol{T}(t)$.

24. A particle moves in space with position function $\boldsymbol{r}(t) = a\cos t\boldsymbol{i} + tb\boldsymbol{j} + a\sin t\boldsymbol{k}$. Show that
- **a.** the velocity vector $\boldsymbol{v}(t)$ and the acceleration vector $\boldsymbol{a}(t)$ have constant lengths,
- **b.** $\boldsymbol{v}(t)$ and $\boldsymbol{a}(t)$ are orthogonal for each t.

25. Find the length of the helix $\boldsymbol{r}(t) = 3t\boldsymbol{i} + \cos 2t\boldsymbol{j} + \sin 2t\boldsymbol{k}$ between the points $\boldsymbol{r}(0)$ and $\boldsymbol{r}(3)$.

26. Find the length of the curve determined by the position function $\boldsymbol{r}(t) = \ln t\boldsymbol{i} + t\boldsymbol{j}$ between $\boldsymbol{r}(1)$ and $\boldsymbol{r}(2)$.

27. Find the curvature of the cubic curve $y = x^3$ at the point $(2, 8)$.

28. Find the curvature of the curve determined by the vector function $\boldsymbol{r}(t) = 3\boldsymbol{i} + \sqrt{t}\,\boldsymbol{j} + t^2\boldsymbol{k}$ at the point $\boldsymbol{r}(4)$.

29. Let C be a curve in space determined by the differentiable vector function $\boldsymbol{r}(t)$, $a \le t \le b$. Show that if $t_0 \ne a$, $t_0 \ne b$, and $\boldsymbol{r}(t_0)$ is a point on C either nearest or farthest from the origin, then $\boldsymbol{r}(t_0)$ and $\boldsymbol{r}'(t_0)$ are orthogonal. (*Hint:* Consider $|\boldsymbol{r}(t)|^2 = \boldsymbol{r}(t)\cdot\boldsymbol{r}(t)$.)

30. Use Newton's law $\boldsymbol{F} = m\boldsymbol{a}$ to show that if a particle moves in space subject to zero net force, then
- **a.** the motion is along a line, and
- **b.** both the tangential and normal components of acceleration are zero.

31. Prove that for a particle moving in space, the speed of the particle is constant if and only if its velocity and acceleration vectors are orthogonal at all points along its path.

32. Find the radius of curvature and an equation for the circle of curvature at the point $(0,1)$ on the graph of $y = \cos x$.

33. Find the point on the graph of $y = e^x$ at which the radius of curvature is a minimum.

34. A particle moves along a curve with position function $\boldsymbol{r}(t) = t\cos t\boldsymbol{i} + t\sin t\boldsymbol{j}$. Find its speed as a function of t.

35. A projectile is fired with an angle of elevation $\alpha = 30°$ and an initial speed $|\boldsymbol{v}(0)| = 20$ m/sec at a vertical wall 10 meters away. At what height does it strike the wall?

CHAPTER 19

DIFFERENTIATION FOR FUNCTIONS OF SEVERAL VARIABLES

19.1 INTRODUCTION

Up to this point, we have discussed only functions of a single independent variable, which we have usually written in the form $y = f(x)$. Recall the meaning: Specifying a value for the independent variable x (input) determines a *unique* value of the dependent variable y (output). This concept is sometimes represented symbolically as

$$f\colon\ \mathbb{R} \to \mathbb{R}, \qquad \text{or} \qquad x \xrightarrow{f} y \tag{1}$$

when $f(x)$ is defined for all x in $\mathbb{R}$, the set of real numbers. If the domain of f is only a subset D of $\mathbb{R}$, the first expression in (1) is written

$$f\colon\ D \to \mathbb{R}, \qquad D \subset \mathbb{R} \tag{2}$$

The goal of this chapter is to extend the theory of differentiation to functions that involve more than one independent variable. Examples of such functions abound in nature and in the social sciences. Here are two.

(i) In chemistry, the ideal gas law relates the pressure, volume, and temperature of an ideal gas by the equation

$$PV = nRT \tag{3}$$

where n is the number of moles of the gas present and R is a constant. When written in the form

$$P = (nR)\left(\frac{T}{V}\right) = f(T, V),$$

equation (3) determines P as a function of both T and V. That is, to calculate pressure we must know both temperature and volume. Similarly, equation (3) can be put in the form

$$V = (nR)\left(\frac{T}{P}\right) = g(T, P)$$

or in the form

$$T = \left(\frac{1}{nR}\right)PV = h(P, V).$$

(ii) In economics, one encounters the simple equation

$$R = Px \tag{4}$$

for the revenue obtained by selling x items at a price of P dollars per item. Equation (4) has the form $R = f(P, x)$, meaning that one must know both the selling price and the number of sales in order to calculate revenue.

By a **function of two variables,** we shall mean a rule that assigns a unique number to each *pair* (x, y) of values of the independent variables. Functions of two variables will generally be written in the form

$$z = f(x, y). \tag{5}$$

In input-output terminology, equation (5) says that two inputs (x and y) are required to determine the output z. If the function $f(x, y)$ is defined for all ordered pairs (x, y), we may represent this situation symbolically as

$$f\colon\ \mathbb{R}^2 \to \mathbb{R}, \qquad \text{or} \qquad (x, y) \xrightarrow{f} z. \tag{6}$$

where $\mathbb{R}^2$ represents two "copies" of the real numbers, from which x and y are chosen independently. If the function $f(x, y)$ is defined only for (x, y) in a subset D of $\mathbb{R}^2$ (called the **domain** of the function), we write

$$f\colon\ D \to \mathbb{R}, \qquad D \subset \mathbb{R}^2 \tag{7}$$

As for functions of a single variable, we will refer to the set of values $f(x, y)$, for $(x, y) \in D$, as the **range** of the function f.

Figure 1.1 shows how we may represent functions of the form (5) or (7) geometrically. We interpret the ordered pairs (x, y) in the domain of $f(x, y)$ as points in the xy-plane. We indicate the value of the function, $z = f(x, y)$, by plotting the point (x, y, z) in space. Then, the height of the point (x, y, z) above or below the point $(x, y, 0)$ represents the number (or **value**) assigned by the function to the ordered pair (x, y). The set of all such points is called the **graph of the function** $z = f(x, y)$.

Notice that while graphs of functions of a single variable are *curves in the plane* (in general), graphs of functions of two variables are *surfaces in space.*

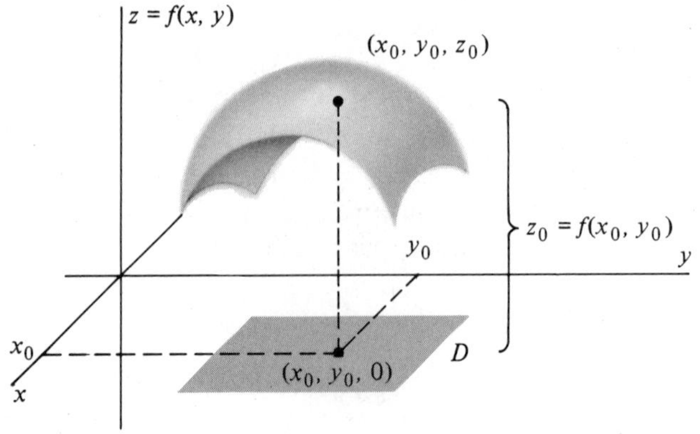

Figure 1.1 Graph of a function $z = f(x, y)$ of two variables with domain D.

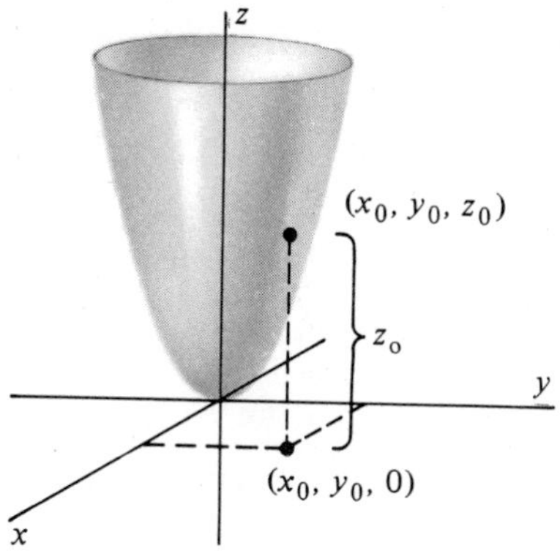

Figure 1.2 Paraboloid $z = x^2 + y^2$ is the graph of a function of two variables.

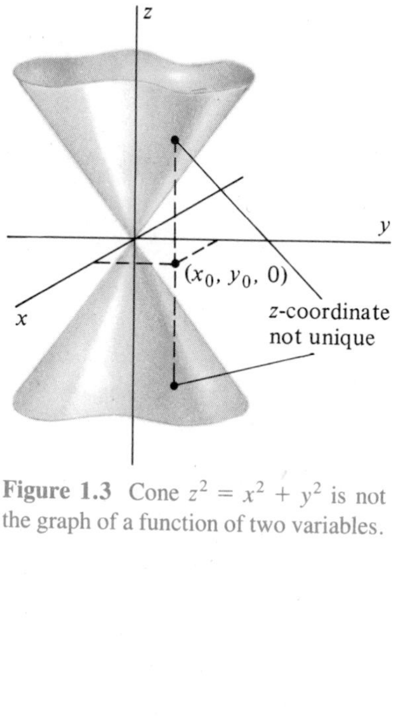

Figure 1.3 Cone $z^2 = x^2 + y^2$ is not the graph of a function of two variables.

An example of a function of two variables is given by the equation

$$z = x^2 + y^2.$$

As described in Section 17.8, the graph of this function is the paraboloid in Figure 1.2. However, the equation

$$z^2 = x^2 + y^2 \tag{8}$$

does *not* describe a function $z = f(x, y)$. The reason is that each pair of independent variables $(x, y) \neq (0, 0)$ corresponds to *two* values of z. This can be seen by writing equation (8) in the form

$$z = \pm \sqrt{x^2 + y^2}.$$

The graph of equation (8) is the cone in Figure 1.3.

Functions of three or more independent variables are defined in the analogous way. A function w of the three independent variables x, y, and z takes the form

$$w = f(x, y, z),$$

and we write

$$f\colon\ \mathbb{R}^3 \to \mathbb{R}, \qquad \text{or} \qquad (x, y, z) \xrightarrow{f} w.$$

An example of a function of three variables is the temperature function $T(x, y, z)$ for the room in which you are sitting. If you use a rectangular coordinate system to describe the space within the room, the number $w_0 = T(x_0, y_0, z_0)$ is the temperature at the point in the room with coordinates (x_0, y_0, z_0). Unfortunately, we cannot sketch a graph for a function of three variables, since all three axes are required just to plot the points in its domain.

More generally, a function of the n independent variables $x_1, x_2, \ldots, x_n$ has the form

$$w = f(x_1, x_2, \ldots, x_n).$$

Symbolically, we describe such functions by writing

$$f\colon\ \mathbb{R}^n \to \mathbb{R}, \qquad \text{or} \qquad (x_1, x_2, \ldots, x_n) \xrightarrow{f} w.$$

In this chapter, we will pursue the theory of differentiation for functions of two or more independent variables. The basic concepts of limit, continuity, rate of change, extreme values, mean values, and differentials will again appear. Moreover, the geometric insights provided by the theory of vectors in space will play an important role in what follows.

Exercise Set 19.1

Which of the following equations determine z as a function of the variables x and y?

1. $z = xy$

2. $z = \sin xy$

3. $z^2 = x^2 - y^2$

4. $z^3 = x + y$

5. $x - z = y + z$

6. $xyz = 1$

7. $x^2yz^2 = 1$

8. $x^3y^2z^5 = 1$

State the implicit domain for each of the following functions.

9. $f(x, y) = \dfrac{1}{x^2 + y^2}$

10. $f(x, y) = \dfrac{x - y}{x + y}$

11. $f(x, y) = \sqrt{y - x}$

12. $f(x, y) = \sqrt{1 - xy^2}$

13. $f(x, y) = \sin xy^2$

14. $f(x, y) = \dfrac{x + y}{4x - x^2y}$

15. $f(x, y, z) = \sqrt{4 - x^2 - y^2 - z^2}$

16. $f(x, y, z) = xyz$

17. $f(x, y) = \dfrac{1}{x - 3} + \dfrac{2}{x - y}$

18. $f(x, y, z) = \dfrac{y}{xyz}$

19. $f(x, y, z) = \ln\left(\dfrac{zx^2}{xy + yx}\right)$

20. Let $f(x, y) = x^2 + y^2$, and $g(z) = 2z + 3$. Write the composite function $h(x, y) = g(f(x, y))$ as an explicit function of x and y.

21. Let $f(x, y) = x + y^2$, and $g(z) = \sqrt{z}$.
a. Write the composite function $h(x, y) = g(f(x, y))$ as an explicit function of x and y.
b. Find the domain of the function $h(x, y)$.

22. Let $f(x, y) = xe^y$ and $g(z) = \dfrac{z + 1}{z - 1}$. Write the composite function $h(x, y) = g(f(x, y))$ as an explicit function of x and y.

23. Let $f(u, v) = u^2 + v^2$, $u(x, y) = x - \cos y$, and $v(x, y) = y + \sin x$. Write the composite function $h(x, y) = f(u(x, y), v(x, y))$ as an explicit function of x and y.

24. Let $f(x, y) = \dfrac{xy}{x^2 + y^2}$, $x(r, s) = r\sqrt{s}$, and $y(r, s) = s - r$. Write the composite function $h(r, s) = f(x(r, s), y(r, s))$ as an explicit function of r and s.

25. Let V be the volume of a right circular cone. Write $V = V(r, h)$ as a function of the radius r and height h.

26. Let S be the total exterior surface area of a right circular cylinder with top and bottom. Express S as a function $S(r, h)$ of the radius and the height of the cylinder.

27. An object is dropped from rest h meters above the ground. Express the time required for it to fall to the ground, $T(g, h)$ as a function of h and g, where g is the acceleration due to gravity (ignore air resistance).

28. A missile is fired from ground level at an angle θ with the horizontal and with an initial speed of s_0 meters per second. Express its altitude as a function $a(t, s_0, \theta, g)$ of t = time, s_0, θ, and g, where g is the acceleration due to gravity.

29. Let $f(x, y) = x(1 + 2y)$. Find
a. $f(2, 3)$
b. $f(2x, -y)$
c. $f(x^2, y^3)$

30. Let $f(x, y) = \sqrt{xy}$. Let $g(x) = 1 + x^3$, and $h(y) = 1 - y$. Find
a. $f(1, 5)$
b. $f(3x, -y)$
c. $f(g(x), 5)$
d. $f(g(x), h(y))$

31. What is the range of the function in Exercise 9?

32. What is the range of the function in Exercise 10?

19.2 LIMITS, CONTINUITY, AND GRAPHING TECHNIQUES

If $f(x_1, x_2, \ldots, x_n)$ is a function of the n independent variables $x_1, x_2, \ldots, x_n$, we may use the idea of a position vector to identify the point $(x_1, x_2, \ldots, x_n) \in \mathbb{R}^n$ with the vector

$$\boldsymbol{x} = \langle x_1, x_2, \ldots, x_n \rangle.$$

The equation

$$z = f(x_1, x_2, \ldots, x_n)$$

can then be written in the form

$$z = f(\boldsymbol{x}), \tag{1}$$

which emphasizes the fact that the input for f is the n-tuple $(x_1, x_2, \ldots, x_n) \in \mathbb{R}^n$, while the output is a number. We shall restrict our considerations almost entirely to the cases $n = 2$ and $n = 3$.

Neighborhoods and Open Sets

By a **neighborhood** of a point $P_0 = (x_0, y_0)$ in the plane, we shall mean an open **disc** N with center (x_0, y_0). That is,

$$N = \{(x, y) | \sqrt{(x - x_0)^2 + (y - y_0)^2} < r\} \tag{2}$$

where r is the radius of the disc N. (By analogy with open intervals on the real number line, the term ''open'' means that the boundary points are not included in the set.) Similarly, a neighborhood of a point $Q = (x_0, y_0, z_0)$ in space is an open **ball** N with center (x_0, y_0, z_0),

$$N = \{(x, y, z) | \sqrt{(x - x_0)^2 + (y - y_0)^2 + (z - z_0)^2} < r\}. \tag{3}$$

Using vector notation, we may generalize both (2) and (3) by saying that a neighborhood of the vector $\boldsymbol{x}_0$ is a set of vectors

$$N = \{\boldsymbol{x} | |\boldsymbol{x} - \boldsymbol{x}_0| < r\}. \tag{4}$$

In equation (4), r is called the **radius** of the neighborhood.

We shall make use of the concept of neighborhood in defining limits for functions of two and three variables. Before doing this, we define another concept that will be used later.

DEFINITION 1

A subset S of $\mathbb{R}^2$ or of $\mathbb{R}^3$ is said to be **open** if every element $\boldsymbol{x}$ of S has a neighborhood that is contained entirely within S.

According to Definition 1, a subset S of $\mathbb{R}^2$ is open if, for every point (x_0, y_0) in S, some small open disc with center (x_0, y_0) lies entirely within S. For example, the half plane $H = \{(x, y)|x < 2\}$ is an open set in $\mathbb{R}^2$, while the square region $S = \{(x, y)|0 \le x \le 1, 0 \le y \le 1\}$ is not (Figures 2.1 and 2.2).

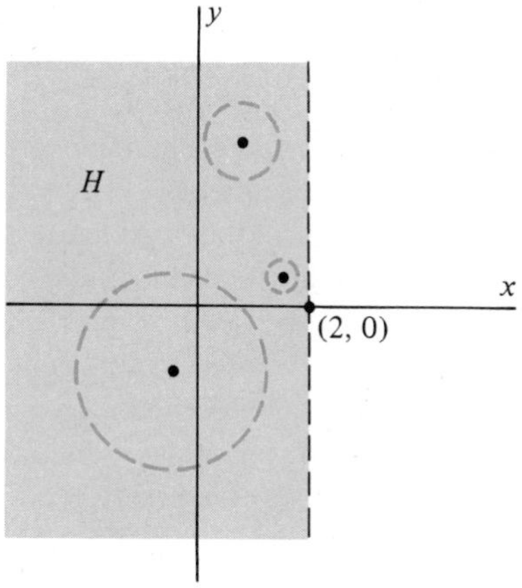

Figure 2.1 Half plane $x < 2$ is open in $\mathbb{R}^2$.

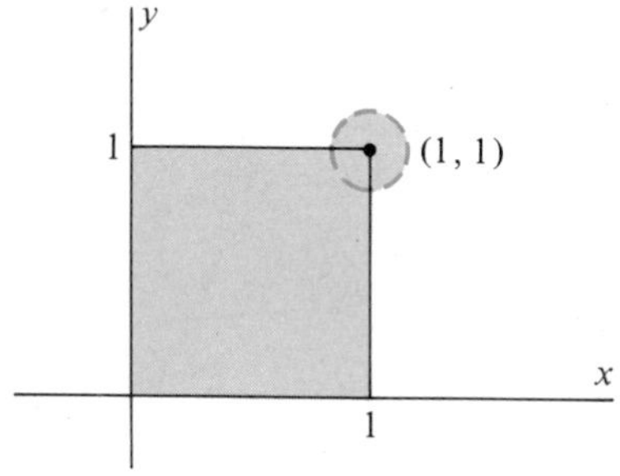

Figure 2.2 Square $0 \le x \le 1$, $0 \le y \le 1$ is not open. No disc with center (1, 1) lies within S.

In $\mathbb{R}^3$, a subset S is open if, for any point (x_0, y_0, z_0) in S, some small open ball with center (x_0, y_0, z_0) lies entirely within S. Thus, the region within (but not on) the ellipsoid $4x^2 + y^2 + 4z^2 = 4$, that is, the set $\{(x, y, z) | 4x^2 + y^2 + 4z^2 < 4\}$, is an open set in $\mathbb{R}^3$, while the solid cone $\{(x, y, z) | z^2 \geq x^2 + y^2\}$ is not (Figures 2.3 and 2.4).

Our interest in open sets is this: If $\boldsymbol{x}$ is a point in an open set S, then points lying "close to" $\boldsymbol{x}$ will also belong to S. We will need this concept in developing the theory of differentiation for functions of several variables.

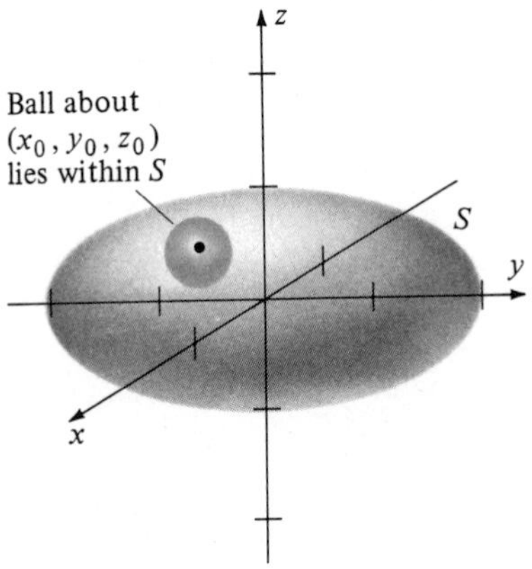

Figure 2.3 Interior of the ellipsoid $4x^2 + y^2 + 4z^2 = 4$ is an open set in $\mathbb{R}^3$.

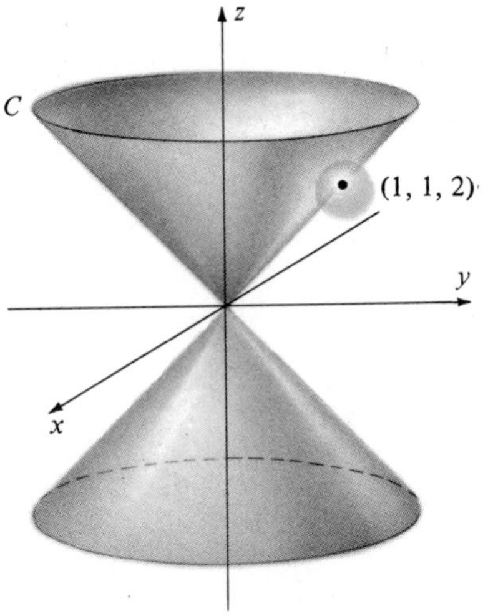

Figure 2.4 Solid cone $z^2 \geq x^2 + y^2$ is not an open set in $\mathbb{R}^3$. No ball with center (1, 1, 2) lies within C.

Limits

Just as we did for functions of a single variable, we want the statement

$$\lim_{\boldsymbol{x} \to \boldsymbol{x}_0} f(\boldsymbol{x}) = L \tag{5}$$

to mean that the values $f(\boldsymbol{x})$ of the function f "approach" the number L as the vector (point) $\boldsymbol{x}$ approaches the fixed vector $\boldsymbol{x}_0$. However, in formulating this definition we must realize that $\boldsymbol{x}$ may approach $\boldsymbol{x}_0$ along many different paths. This observation points out a significant difference between functions of one variable and functions of several variables. In the one-variable case, a variable can approach a constant only from one of two directions along the number line. In writing statement (5) we shall mean that $f(\boldsymbol{x}) \to L$ as $\boldsymbol{x} \to \boldsymbol{x}_0$ *regardless of the path* along which $\boldsymbol{x}$ approaches $\boldsymbol{x}_0$. Figure 2.5 illustrates this concept. Figure 2.6 suggests a situation in which the limit in (5) fails to exist.

Before formulating a precise definition for the limit of a function of several variables, we examine several examples.

Example 1 Consider the function $f(x, y) = 4 - x - y$. We claim that

$$\lim_{(x, y) \to (1, 1)} f(x, y) = 2. \tag{6}$$

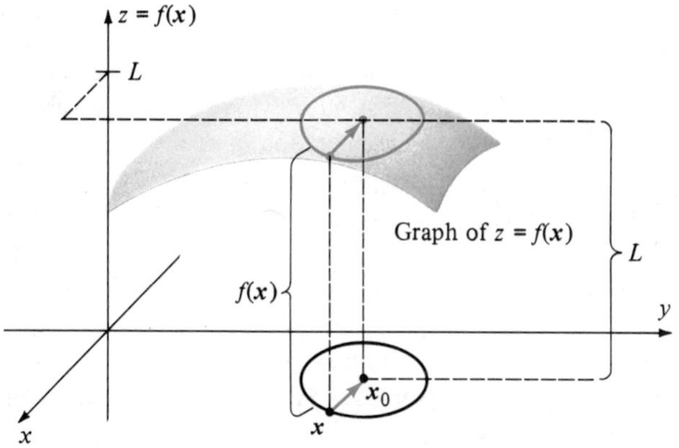

Figure 2.5 $\lim_{\mathbf{x}\to\mathbf{x}_0} f(\mathbf{x}) = L$ if $f(\mathbf{x}) \to L$ as $\mathbf{x}$ approaches $\mathbf{x}_0$ from any direction.

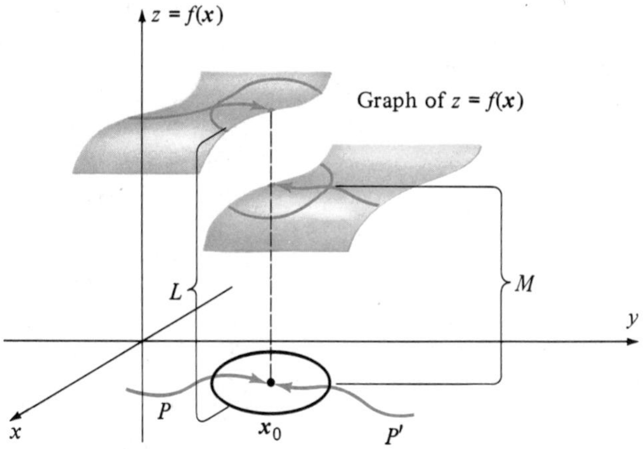

Figure 2.6 $f(\mathbf{x}) \to L$ as $\mathbf{x} \to \mathbf{x}_0$ along path P, but $f(\mathbf{x}) \to M$ as $\mathbf{x} \to \mathbf{x}_0$ along path P', so $\lim_{\mathbf{x}\to\mathbf{x}_0} f(\mathbf{x})$ does not exist.

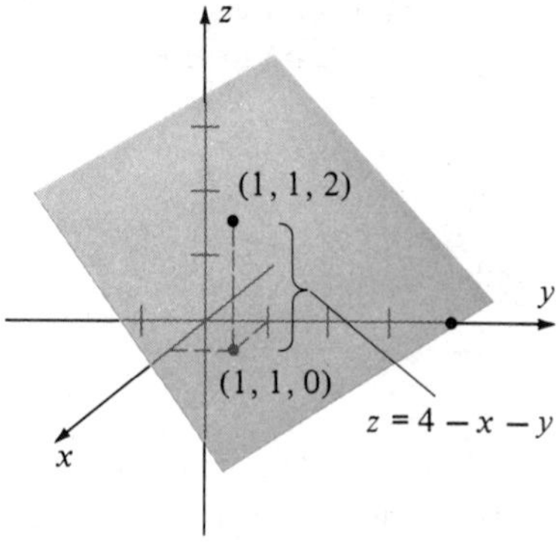

Figure 2.7 $\lim_{(x,y)\to(1,1)} (4 - x - y) = 2.$

Notice first that we have not used the vector notation of statement (5) in writing statement (6), since the number and names of the independent variables are known explicitly. From the form of $f(x, y)$ it is easy to see that $f(x, y) \to (4 - 1 - 1) = 2$ as $x \to 1$ and $y \to 1$. Furthermore, we can see that this result is independent of the path along which (x, y) approaches $(1, 1)$ by setting $z = f(x, y)$ and recalling (Chapter 17) that the graph of the equation $z = 4 - x - y$ is a plane (see Figure 2.7). Regardless of how (x, y) approaches $(1, 1)$, the point (x, y, z) on the plane approaches the point $(1, 1, 2)$. ■

Example 2 Consider the function $f(x, y) = \dfrac{xy}{x^2 + y^2}$. We claim that

$$\lim_{(x,y)\to(0,0)} \frac{xy}{x^2 + y^2} \tag{7}$$

does not exist. On first glance this conclusion may not be obvious. For example, we may allow (x, y) to approach $(0, 0)$ along the x-axis by setting $y = 0$. We obtain

$$\lim_{(x,0)\to(0,0)} \frac{xy}{x^2 + y^2} = \lim_{x\to 0} \frac{x \cdot 0}{x^2 + 0^2} = 0 \qquad \text{(see Figure 2.8).} \tag{8}$$

Similarly, allowing (x, y) to approach $(0, 0)$ along the y-axis $(x = 0)$ gives

$$\lim_{(0,y)\to(0,0)} \frac{xy}{x^2 + y^2} = \lim_{y\to 0} \frac{0 \cdot y}{0^2 + y^2} = 0. \tag{9}$$

However, if we allow (x, y) to approach $(0, 0)$ along the line $y = x$, we find that

$$\lim_{(x,x)\to(0,0)} \frac{xy}{x^2 + y^2} = \lim_{x\to 0} \frac{x^2}{x^2 + x^2} = \frac{1}{2}. \tag{10}$$

Since the limit in (10) does not agree with that in (8) or in (9), we conclude that the limit in (7) cannot exist. (See Exercise 48 for a different look at this function.) ■

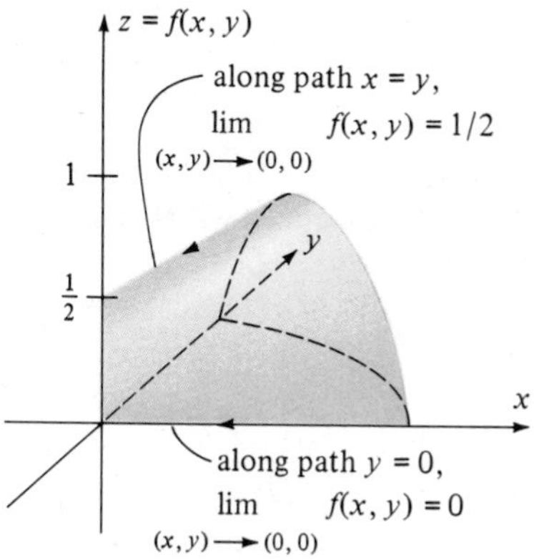

Figure 2.8 Portion of the graph of $f(x, y) = \dfrac{xy}{x^2 + y^2}$. $\lim_{(x,y)\to(0,0)} f(x, y)$ does not exist. (xy-plane has been rotated 90° for clarity.)

Example 3 Find $\lim_{(x, y, z) \to (2, \pi/3, 3)} x \cos yz$.

Solution: Since $f(x, y, z) = x \cos yz$ is a function of three variables, we cannot rely on a graph to determine this limit. We must, instead, rely on our intuition about the continuity of both the cosine function and the operation of multiplication. As $y \to \pi/3$ and $z \to 3$, $yz \to (\pi/3)(3) = \pi$, so $\cos yz \to \cos \pi = -1$. Thus, as $x \to 2$, $x \cos yz \to 2(-1) = -2$. That is,

$$\lim_{(x, y, z) \to (2, \pi/3, 3)} x \cos yz = -2. \qquad \blacksquare$$

Next, we give a formal definition for the limit in line (5). As was the case for the formal definition of $\lim_{x \to a} f(x)$, this definition is not easy to apply. However, when coupled with the geometric insights afforded by the vector notation developed in Chapter 17, it provides a useful and precise statement of this concept.

DEFINITION 2

Let $\boldsymbol{x}$ be the position vector for the point $(x_1, x_2, \ldots, x_n)$ in $\mathbb{R}^n$. Let $f(x_1, x_2, \ldots, x_n) = f(\boldsymbol{x})$ be a function of n variables defined for all $\boldsymbol{x}$ in a neighborhood of $\boldsymbol{x}_0$. Let L be a real number. We say that L is the **limit of the function** $f(\boldsymbol{x})$ as $\boldsymbol{x}$ approaches $\boldsymbol{x}_0$, written

$$\lim_{\boldsymbol{x} \to \boldsymbol{x}_0} f(\boldsymbol{x}) = L,$$

if and only if, for every $\epsilon > 0$, there exists a number $\delta > 0$ so that

$$|f(\boldsymbol{x}) - L| < \epsilon \qquad \text{whenever} \qquad 0 < |\boldsymbol{x} - \boldsymbol{x}_0| < \delta. \tag{11}$$

Note in statement (11) that the first inequality involves the *absolute value* of the *number* $f(\boldsymbol{x}) - L$, while the second concerns the *length* of the *vector* $\boldsymbol{x} - \boldsymbol{x}_0$. When applying Definition 2 in a particular situation, you should rewrite the second inequality in (11) in the appropriate component form.

Example 4 Use Definition 2 to prove that

$$\lim_{(x, y) \to (0, 0)} \sqrt{9 - x^2 - y^2} = 3.$$

Solution: Here $L = 3$, and $\boldsymbol{x}_0 = \boldsymbol{0}$ is the position vector associated with the origin. We begin by rewriting the second inequality in (11) as

$$0 < \sqrt{(x - 0)^2 + (y - 0)^2} = \sqrt{x^2 + y^2} < \delta. \tag{12}$$

Also, the first inequality in (11) becomes

$$\left|\sqrt{9 - x^2 - y^2} - 3\right| < \epsilon, \tag{13}$$

or

$$3 - \sqrt{9 - x^2 - y^2} < \epsilon$$

since $\sqrt{9 - x^2 - y^2} \le 3$. Solving this inequality for the radical term, we obtain

$$\sqrt{9 - x^2 - y^2} > 3 - \epsilon,$$

which holds if and only if the following chain of equivalent inequalities holds:

$$9 - (x^2 + y^2) > (3 - \epsilon)^2 \qquad \text{(assuming } \epsilon < 3\text{)},$$

$$x^2 + y^2 < 6\epsilon - \epsilon^2,$$

$$\sqrt{x^2 + y^2} < \sqrt{6\epsilon - \epsilon^2}. \tag{14}$$

Now let's assume that $\epsilon > 0$ is a small ($\epsilon < 3$) positive given number. If we define the number δ to be $\delta = \sqrt{6\epsilon - \epsilon^2}$, then (14) holds whenever inequality (12) holds. Since inequality (14) is equivalent to inequality (13), this shows that

$$\left|\sqrt{9 - x^2 - y^2} - 3\right| < \epsilon \qquad \text{whenever} \qquad 0 < \sqrt{x^2 + y^2} < \delta.$$

This proves the stated limit (see Figure 2.9). ■

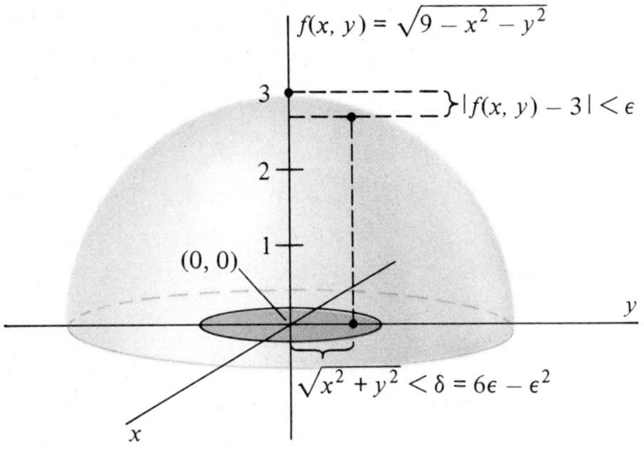

Figure 2.9 $\lim_{(x,y)\to(0,0)} \sqrt{9 - x^2 - y^2} = 3.$

The algebra of limits for functions of several variables is analogous to that for functions of a single variable. The following theorem may be proved using the same ideas used to prove Theorem 1, Section 2.4 (see Exercises 50 and 51).

THEOREM 1

Let $f(\mathbf{x})$ and $g(\mathbf{x})$ be functions of two or three variables defined in a neighborhood of $\mathbf{x}_0$. Suppose that both $\lim_{\mathbf{x}\to\mathbf{x}_0} f(\mathbf{x})$ and $\lim_{\mathbf{x}\to\mathbf{x}_0} g(\mathbf{x})$ exist. Let c be any constant. Then

(i) $\lim_{\mathbf{x}\to\mathbf{x}_0} [f(\mathbf{x}) + g(\mathbf{x})] = \lim_{\mathbf{x}\to\mathbf{x}_0} f(\mathbf{x}) + \lim_{\mathbf{x}\to\mathbf{x}_0} g(\mathbf{x})$,

(ii) $\lim_{\mathbf{x}\to\mathbf{x}_0} cf(\mathbf{x}) = c \cdot \lim_{\mathbf{x}\to\mathbf{x}_0} f(\mathbf{x})$,

(iii) $\lim_{\mathbf{x}\to\mathbf{x}_0} f(\mathbf{x})g(\mathbf{x}) = \left(\lim_{\mathbf{x}\to\mathbf{x}_0} f(\mathbf{x})\right)\left(\lim_{\mathbf{x}\to\mathbf{x}_0} g(\mathbf{x})\right)$,

(iv) $\lim_{\mathbf{x}\to\mathbf{x}_0} \dfrac{f(\mathbf{x})}{g(\mathbf{x})} = \dfrac{\lim_{\mathbf{x}\to\mathbf{x}_0} f(\mathbf{x})}{\lim_{\mathbf{x}\to\mathbf{x}_0} g(\mathbf{x})}, \qquad \lim_{\mathbf{x}\to\mathbf{x}_0} g(\mathbf{x}) \neq 0.$

Continuity

Continuity for functions of several variables is defined just as for functions of a single variable.

DEFINITION 3

The function $f(\boldsymbol{x})$, $\boldsymbol{x} \in \mathbb{R}^n$, is **continuous** at $\boldsymbol{x}_0$ if

(i) $f(\boldsymbol{x}_0)$ is defined,
(ii) $\lim_{\boldsymbol{x}\to\boldsymbol{x}_0} f(\boldsymbol{x})$ exists, and
(iii) $\lim_{\boldsymbol{x}\to\boldsymbol{x}_0} f(\boldsymbol{x}) = f(\boldsymbol{x}_0)$.

In other words, $f(\boldsymbol{x})$ is continuous at $\boldsymbol{x}_0$ if $f(\boldsymbol{x}) \to f(\boldsymbol{x}_0)$ as $\boldsymbol{x}\to\boldsymbol{x}_0$, regardless of the direction in which $\boldsymbol{x}$ approaches $\boldsymbol{x}_0$. Here are some examples.

1. The function $f(x, y) = 4 - x - y$ in Example 1 is continuous at (1, 1) since

$$\lim_{(x, y)\to(1, 1)} (4 - x - y) = 2 = f(1, 1).$$

2. The function in Figure 2.6 is discontinuous at $\boldsymbol{x}_0$, since $\lim_{\boldsymbol{x}\to\boldsymbol{x}_0} f(\boldsymbol{x})$ does not exist.

3. The function $f(x, y) = \dfrac{xy}{x^2 + y^2}$ in Example 2 is discontinuous at (0, 0), since $f(0, 0)$ is undefined.

4. The function $f(x, y, z) = x \cos yz$ in Example 3 is continuous at $(2, \pi/3, 3)$ since

$$\lim_{(x, y, z)\to(2, \pi/3, 3)} x \cos yz = -2 = f(2, \pi/3, 3).$$

5. The function $f(x, y) = \sqrt{9 - x^2 - y^2}$ in Example 4 is continuous at (0, 0) since

$$\lim_{(x, y)\to(0, 0)} \sqrt{9 - x^2 - y^2} = 3 = f(0, 0).$$

As is the case with limits of functions of several variables, continuity is a property that is often difficult to prove, however it can usually be determined by knowing how continuous functions combine to form other continuous functions. For example, since powers, products, and sums of continuous functions of a single variable are again continuous, we would expect a polynomial function such as

$$p(x, y) = x^3y^2 + 6xy + xy^4$$

to be a continuous function of two variables. Indeed, using Definition 3 and the ideas used to prove Theorems 7–10 of Chapter 3, we can prove that sums, multiples, products, powers, and quotients (where denominators are not zero) of continuous functions of several variables are continuous (see Exercises 52 and 53). Moreover, the composition of a continuous function of a single variable with a continuous function of n variables is a continuous function of n variables (see Exercise 54).

Points of discontinuity for functions of several variables occur for the usual reasons: a denominator becomes zero, an expression underneath a radical sign of even order becomes negative, and so forth.

Graphing Techniques: Traces and Level Curves

Graphing a function of two variables is difficult, at best. Two techniques frequently give a general idea of the appearance of the graph of the function $z = f(x, y)$. The first is simply setting one of the two independent variables equal to a constant, so that we obtain a function of a single variable, whose graph can be sketched in the appropriate plane in $\mathbb{R}^3$. Setting $x = c$ in $z = f(x, y)$ gives the equation $z = f(c, y) = g(y)$, whose graph is the intersection of the desired graph with the plane $x = c$; setting $y = c$ in $z = f(x, y)$ gives the equation $z = f(x, c) = h(x)$, whose graph is the intersection of the desired graph with the plane $y = c$.

We refer to the intersections of the graph of $z = f(x, y)$ with the planes $x = c$ or $y = c$ as the **traces** of $z = f(x, y)$ in the respective planes. By sketching traces of $z = f(x, y)$ in several planes, we can sometimes gain a fairly accurate picture of the graph of $z = f(x, y)$.

Example 5 Sketch several traces for the graph of $z = x^2 + y^2$.

Solution: We have previously sketched this paraboloid in Chapter 17, so it provides a familiar setting in which to try out the idea of sketching traces in planes. Tables 2.1 and 2.2 show the results of setting one of the independent variables equal to one of several constants.

Table 2.1

Plane $y = c$	Function $z = f(x, c)$
0	$z = x^2$
1	$z = x^2 + 1$
2	$z = x^2 + 4$
−1	$z = x^2 + 1$
−2	$z = x^2 + 4$

Table 2.2

Plane $x = c$	Function $z = f(c, y)$
0	$z = y^2$
1	$z = y^2 + 1$
2	$z = y^2 + 4$
−1	$z = y^2 + 1$
−2	$z = y^2 + 4$

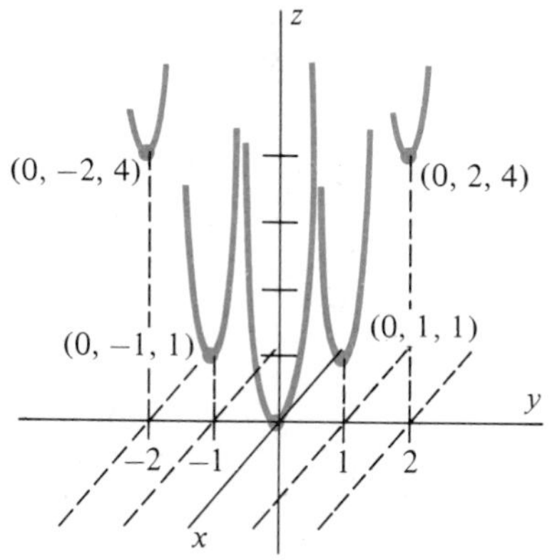

Figure 2.10 Traces of $z = x^2 + y^2$ in planes $y = c$.

Traces in each of these planes are indicated in Figures 2.10 and 2.11. They are combined in Figure 2.12. Figure 2.13 is a sketch of the corresponding paraboloid. ■

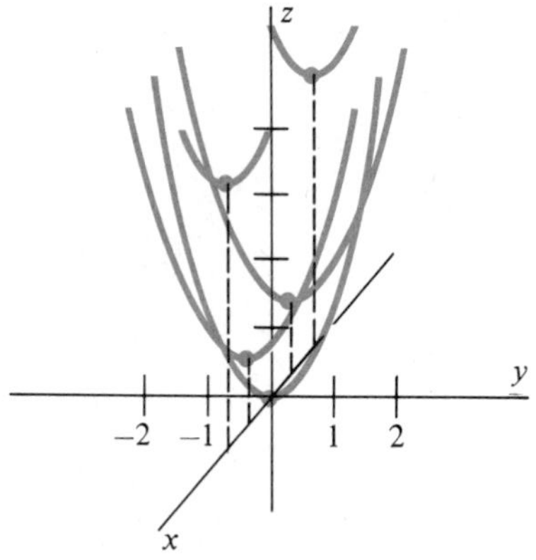

Figure 2.11 Traces of $z = x^2 + y^2$ in planes $x = c$.

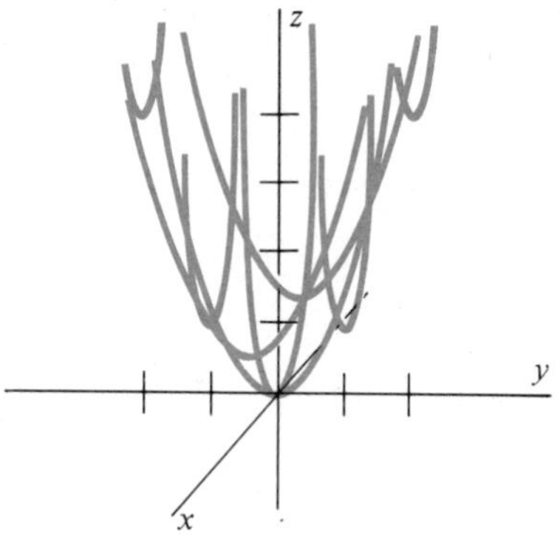

Figure 2.12 Traces of $z = x^2 + y^2$ in planes of both types.

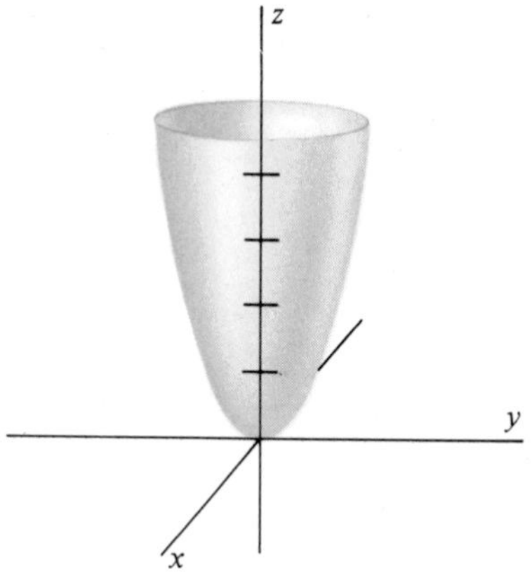

Figure 2.13 Graph of $z = x^2 + y^2$.

Example 6 For the function $f(x, y) = xy$, the equations of various traces are given in Tables 2.3 and 2.4.

Table 2.3

Plane $y = c$	Equation $z = f(x, c)$
0	$z = 0$
1	$z = x$
2	$z = 2x$
−1	$z = -x$
−2	$z = -2x$

Table 2.4

Plane $x = c$	Equation $z = f(c, y)$
0	$z = 0$
1	$z = y$
2	$z = 2y$
−1	$z = -y$
−2	$z = -2y$

These traces, and several others, are sketched in Figures 2.14 and 2.15. Figure 2.16 shows a portion of the graph of $f(x, y) = xy$. ■

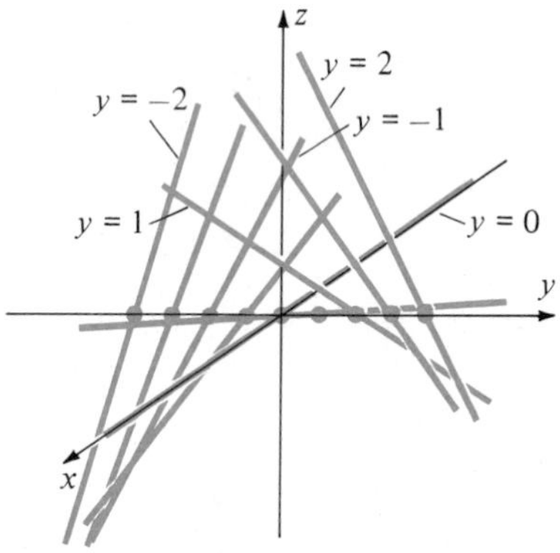

Figure 2.14 Traces of $z = xy$ in planes $y = c$.

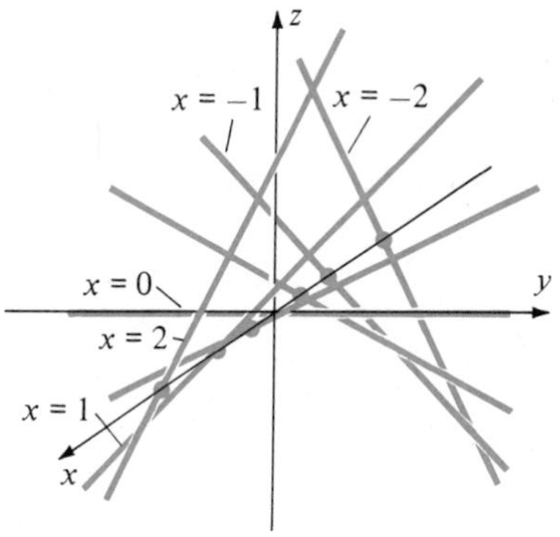

Figure 2.15 Traces of $z = xy$ in planes $x = c$.

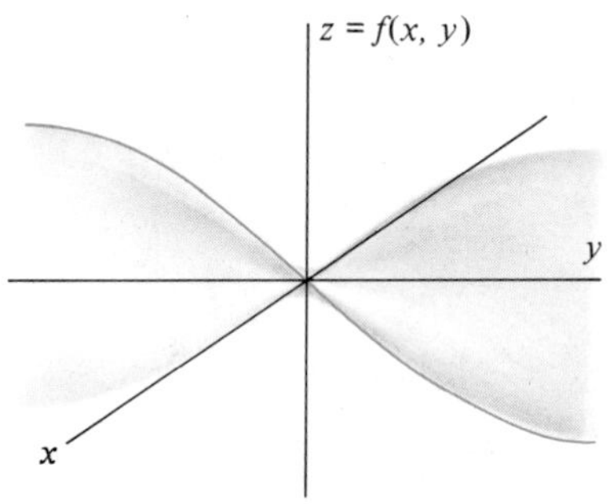

Figure 2.16 Graph of $f(x, y) = xy$.

Example 7 For the function $f(x, y) = y^2 - x^2$, the equations of various traces are given in Tables 2.5 and 2.6. They are all parabolas.

Table 2.5

Plane $y = c$	Equation $z = f(x, c)$
0	$z = -x^2$
1	$z = 1 - x^2$
2	$z = 4 - x^2$
−1	$z = 1 - x^2$
−2	$z = 4 - x^2$

Table 2.6

Plane $x = c$	Equation $z = f(c, y)$
0	$z = y^2$
1	$z = y^2 - 1$
2	$z = y^2 - 4$
−1	$z = y^2 - 1$
−2	$z = y^2 - 4$

Figure 2.17 shows both types of traces. Figure 2.18 shows the graph of $f(x, y) = y^2 - x^2$. ■

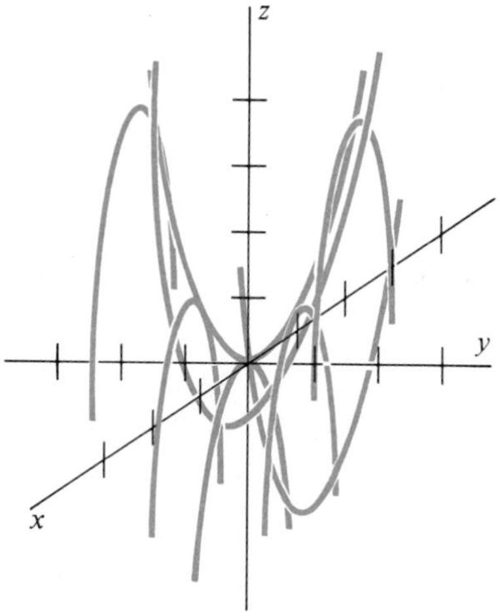

Figure 2.17 Traces of $z = y^2 - x^2$ in planes $y = c$ and $x = c$.

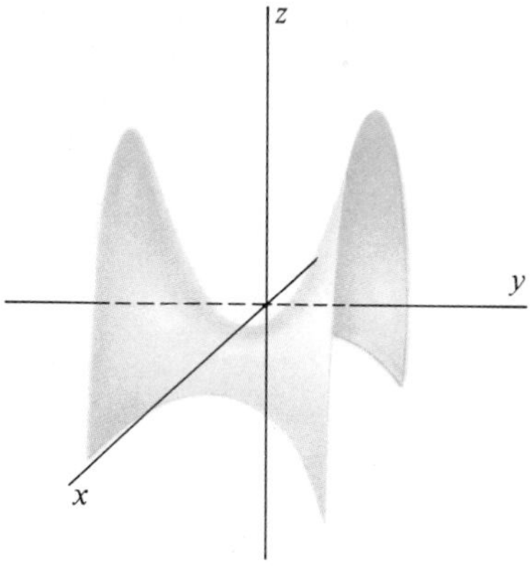

Figure 2.18 Graph of $z = y^2 - x^2$.

Level Curves and Surfaces

The idea behind using traces to describe the graph of $z = f(x, y)$ is that we actually try to sketch a three-dimensional graph. Another approach to representing a surface is to draw a purely two-dimensional sketch that conveys certain information about the surface. Here we want to know which points of the graph lie at various altitudes—that is, which points satisfy the equation $z = c$ for various values of c. This is the concept of **level curves.**

You have probably encountered level curves before. They are commonly used in topographical maps where one indicates the "lay of the land" by plotting curves of constant altitude (see Figures 2.19 and 2.20).

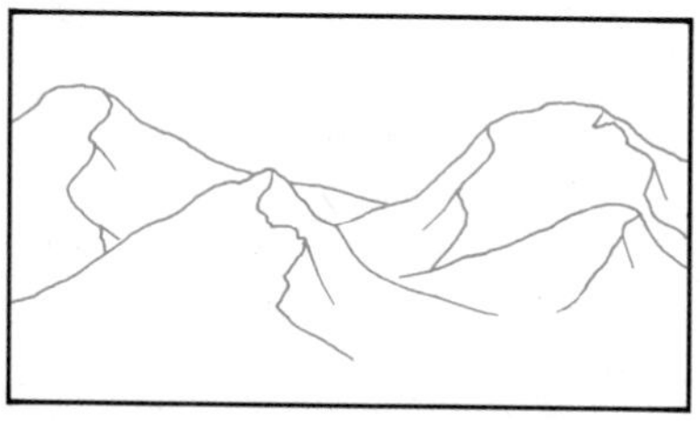

Figure 2.19 A picture of a mountainous region.

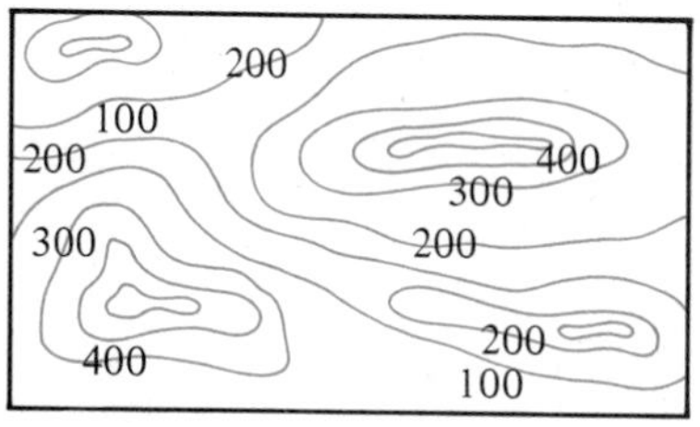

Figure 2.20 A topographical map of the region.

To use the notion of level curves to describe the graph of $z = f(x, y)$, we set $z = c$ and graph the resulting equation $f(x, y) = c$ in the xy-plane. Doing so for several values of c conveys the same type of information about the graph of $z = f(x, y)$ as does the topographical map for a geographic region. Note that although setting $z = c$ produces a trace of $z = f(x, y)$, we make no attempt to produce a three-dimensional sketch by this method. Also, you should be sure to label the value of z corresponding to each level curve sketched.

Example 8 Consider the function $z = x^2 + y^2$ in Example 5. Setting $z = c$ gives the equation $x^2 + y^2 = c$, valid for $c \geq 0$. The level curves are concentric circles, as sketched in Figure 2.21. ■

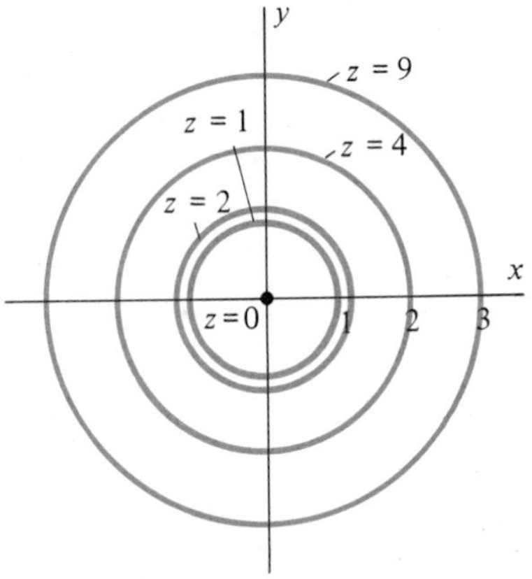

Figure 2.21 Level curves for $z = x^2 + y^2$ are circles $x^2 + y^2 = c$, $c \geq 0$. (See Figure 2.13.)

Example 9 For the function $z = xy$ in Example 6, the level curves are $xy = c$, or $y = \dfrac{c}{x}$. These are the hyperbolas in Figure 2.22. ■

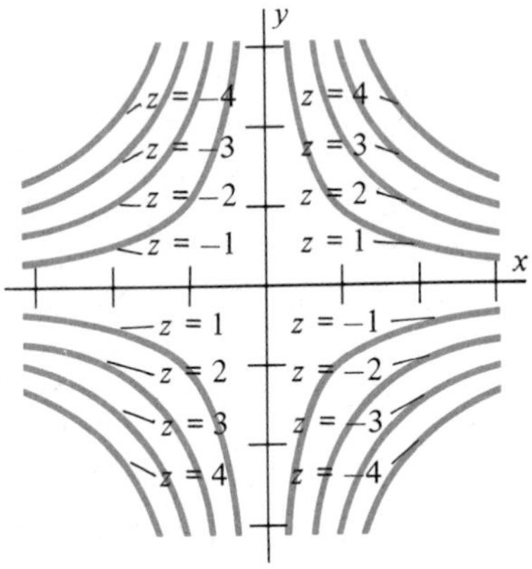

Figure 2.22 Level curves for $z = xy$ are curves $y = c/x$, $x \neq 0$. (See Figure 2.16.)

Example 10 For the function $f(x, y) = y^2 - x^2$ in Example 7, the level curves have equation $y^2 - x^2 = c$ or $y = \pm\sqrt{x^2 + c}$. Graphs of several such curves appear in Figure 2.23. ■

Although we cannot sketch graphs of functions of three variables, we can sketch **level surfaces** for functions of the form $w = f(x, y, z)$ by setting $w = c$ and sketching the graph of the equation $f(x, y, z) = c$. For example, level surfaces for the function $w = x^2 + y^2 + z^2$ are spheres $x^2 + y^2 + z^2 = c$, $c \geq 0$. More generally, any function of two variables $z = f(x, y)$ may be regarded as a level surface for a function of three variables

$$g(x, y, z) = z - f(x, y) = 0.$$

We will exploit this observation later in finding planes tangent to graphs of functions of two variables.

A simple example of a level surface is the set of all points in your classroom at which the temperature $T(x, y, z)$ has a particular value. Such surfaces of constant temperature are called **isothermal surfaces.** Another example occurs in the theory of electricity and magnetism, where surfaces on which an electric potential is constant are called **equipotential surfaces.**

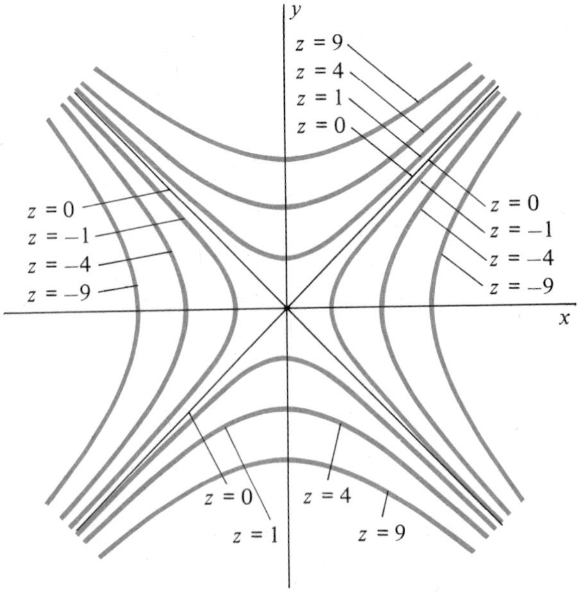

Figure 2.23 Level curves for $f(x, y) = y^2 - x^2$ are curves $y^2 - x^2 = c$. (See Figure 2.18.)

Exercise Set 19.2

In Exercises 1–12, evaluate the given limit.

1. $\lim_{(x,y)\to(1,3)} (x^2 - 2y)$

2. $\lim_{(x,y)\to(1,-1)} \sqrt{x - y}$

3. $\lim_{(x,y)\to(2,4)} \dfrac{x + y}{x - y}$

4. $\lim_{(x,y)\to(3,-1)} \dfrac{1}{\sqrt{x + y}}$

5. $\lim_{(x,y)\to(1,-2)} \dfrac{x + y^3}{x^2 + 2xy + y^2}$

6. $\lim_{(x,y)\to(\pi/2,1)} x \cos xy$

7. $\lim_{(x,y)\to(1,0)} xe^{y-x}$

8. $\lim_{(x,y)\to(\pi/2,1)} \ln \sin xy$

9. $\lim_{(x,y)\to(2,2)} \dfrac{\tan^{-1}(y/x)}{1 + xy}$

10. $\lim_{(x,y,z)\to(1,0,1)} \dfrac{x^2}{\sqrt{x^2 + y^2}}$

11. $\lim_{(x,y,z)\to(3,\pi/2,1)} x \cos yz$

12. $\lim_{(x,y,z)\to(1,0,1)} \dfrac{x^2 - 3yz}{\sec(xyz)}$

In Exercises 13–22, sketch several level curves for the given functions.

13. $f(x, y) = y - x^2$

14. $f(x, y) = x^2 + 4y^2$

15. $f(x, y) = 2xy$

16. $f(x, y) = x^2 - y^2$

17. $f(x, y) = xe^y$

18. $f(x, y) = x \cos y$

19. $f(x, y) = \sqrt{x + y}$

20. $f(x, y) = \dfrac{y}{x + y}$

21. $f(x, y) = \dfrac{x}{x^2 + y^2}$

22. $f(x, y) = \sqrt{y^2 - x^2}$

In Exercises 23–32, sketch the graph of $z = f(x, y)$ by first sketching several traces of $z = f(x, y)$ in the planes $x = c$ and $y = c$.

23. $f(x, y) = 3$

24. $f(x, y) = x$

25. $f(x, y) = y^2$

26. $f(x, y) = x + y$

27. $f(x, y) = x \cos y$

28. $f(x, y) = x^2 + 4y^2$

29. $f(x, y) = y - x$

30. $f(x, y) = y \sin x$

31. $f(x, y) = y/x$

32. $f(x, y) = xy^2$

In Exercises 33–38, determine whether the given set S is open in $\mathbb{R}^2$ or in $\mathbb{R}^3$, as stated.

33. $S = \{(x, y) | x^2 + y^2 < 2\}$, in $\mathbb{R}^2$

34. $S = \{(x, y) | x^2 + y^2 < 2\}$, in $\mathbb{R}^3$

35. $S = \{(x, y) | y < x\}$, in $\mathbb{R}^2$

36. $S = \{(x, y) | 1 \le x^2 + y^2 \le 3\}$, in $\mathbb{R}^2$

37. $S = \{(x, y, z) | x > 0, y > 0, z \ge 0\}$, in $\mathbb{R}^3$

38. $S = \{(x, y, z) | x \le y\}$, in $\mathbb{R}^3$

39. Find $\lim_{(x,y)\to(1,0)} \dfrac{xy - y}{x^2 + y^2 - 2x + 1}$, if it exists.

40. Is $f(x, y) = \dfrac{4xy}{x^2 + y^2}$ continuous at (0, 0)? What about $g(x, y) = \dfrac{4xy}{\sqrt{x^2 + y^2}}$?

41. Is the function

$$f(x, y) = \begin{cases} \dfrac{x^2y}{x^3 + y^3}, & (x, y) \ne (0, 0) \\ 0, & (x, y) = 0 \end{cases}$$

continuous at (0, 0)? (*Hint:* Consider the path $y = x$.)

42. Show that the function

$$f(x, y) = \begin{cases} \dfrac{x^3y}{x^3 + y^5}, & (x, y) \ne (0, 0) \\ 0, & (x, y) = 0 \end{cases}$$

is not continuous at (0, 0).

43. Is the function

$$f(x, y) = \begin{cases} \dfrac{x^3}{x^2 + y^2}, & (x, y) \ne (0, 0) \\ 0, & (x, y) = 0 \end{cases}$$

continuous at (0, 0)?

44. Let

$$f(x, y) = \begin{cases} \dfrac{\sin x \cos y}{x}, & (x, y) \ne (0, 0) \\ 0, & (x, y) = (0, 0). \end{cases}$$

Find $\lim_{(x,y)\to(0,0)} f(x, y)$. (*Hint:* Write $f(x, y) = g(x, y)h(x, y)$, $(x, y) \ne (0, 0)$, with $g(x, y) = \cos y$, $h(x, y) = \dfrac{\sin x}{x}$, and apply Theorem 1.)

45. Let

$$f(x, y) = \begin{cases} x \csc x \tan y, & (x, y) \ne (0, 0) \\ 0, & (x, y) = (0, 0). \end{cases}$$

Find $\lim_{(x,y)\to(0,0)} f(x, y)$ (see Exercise 44).

46. True or false? If $\lim_{(x,y)\to(a,b)} f(x, y) = L$, then $\lim_{x\to a} f(x, b) = L$. Explain.

47. True or false? If $\lim_{x\to a} f(x, b) = L$ and $\lim_{y\to b} f(a, y) = L$, then $\lim_{(x,y)\to(a,b)} f(x, y) = L$.

48. Consider the function $f(x, y) = \dfrac{xy}{x^2 + y^2}$ whose graph appears in Figure 2.8.

a. Letting $x = r \cos \theta$ and $y = r \sin \theta$, show that f may be expressed in polar coordinates as

$$f(r, \theta) = \frac{1}{2} \sin 2\theta.$$

b. Conclude from part (a) that the values $f(r, \theta)$ are independent of r, depending only on θ. Explain why this shows that f is not continuous at the origin.

49. Prove that $\lim_{(x, y) \to (0, 0)} (x^2 + y^2) = 0$ using Definition 2.

50. Prove part (i) of Theorem 1.

51. Prove part (ii) of Theorem 1.

52. Prove that sums and constant multiples of continuous functions of several variables are continuous.

53. Prove that the product of two continuous functions of several variables is continuous.

54. Prove that if $g(y)$ is a continuous function of a single variable and if $y = f(\mathbf{x})$ is a continuous function of several variables, then the composite function $R(\mathbf{x}) = g(f(\mathbf{x}))$ is a continuous function of several variables.

19.3 PARTIAL DIFFERENTIATION

The principal goal of this chapter is to develop a theory of differentiation for functions of several variables. Recall the one-variable case: The *derivative* of the function $y = f(x)$ is the limit

$$f'(x) = \lim_{\Delta x \to 0} \frac{f(x + \Delta x) - f(x)}{\Delta x}, \tag{1}$$

and this limit measures the *rate of change* of $f(x)$ with respect to change in x. In the case of two or more independent variables, the situation is less straightforward. As we have already seen, the question of calculating a rate of change for $z = f(x, y)$ at (x_0, y_0) by limits is a complicated issue, since the "nearby" point (x, y) may approach (x_0, y_0) along an infinite number of distinct paths.

We begin by examining rates at which $f(x, y)$ changes along paths parallel to the coordinate axes. This is the concept of **partial differentiation.**

DEFINITION 4

Let $z = f(x, y)$ be defined in a neighborhood of (x_0, y_0). The **partial derivative of $f(x, y)$ with respect to x** at (x_0, y_0) is the number

$$\frac{\partial f}{\partial x}(x_0, y_0) = \lim_{\Delta x \to 0} \frac{f(x_0 + \Delta x, y_0) - f(x_0, y_0)}{\Delta x}, \tag{2}$$

if this limit exists. Similarly, the **partial derivative of $f(x, y)$ with respect to y** at (x_0, y_0) is the number

$$\frac{\partial f}{\partial y}(x_0, y_0) = \lim_{\Delta y \to 0} \frac{f(x_0, y_0 + \Delta y) - f(x_0, y_0)}{\Delta y}, \tag{3}$$

provided this limit exists.

Comparing equations (1) and (2), we see that the partial derivative $\dfrac{\partial f}{\partial x}(x_0, y_0)$ is simply the result of holding the variable y constant and differentiating the function $z = f(x, y_0)$ as a function of x alone. Similarly, the partial derivative $\dfrac{\partial f}{\partial y}(x_0, y_0)$

results from treating the variable x as the constant $x = x_0$ and differentiating $z = f(x_0, y)$ as a function of y alone. Thus, partial derivatives may be calculated by the rules developed earlier for differentiating functions of a single variable.

Example 1 Calculate the partial derivatives with respect to x and y for the function $f(x, y) = x^2y^3 + e^x + \ln y$ when $x = 1$ and $y = 4$.

Solution: In computing the partial derivative with respect to x, we consider y (and any expression containing y alone) to be a constant. Thus,

$$\frac{\partial f}{\partial x}(x, y) = \left(\frac{d}{dx}x^2\right)y^3 + \frac{d}{dx}e^x + \frac{d}{dx}\ln y$$
$$= 2xy^3 + e^x + 0.$$

Similarly, when we consider x constant,

$$\frac{\partial f}{\partial y}(x, y) = x^2\left(\frac{d}{dy}y^3\right) + \frac{d}{dy}e^x + \frac{d}{dy}\ln y$$
$$= 3x^2y^2 + 0 + \frac{1}{y}, \qquad y \neq 0.$$

Thus,

$$\frac{\partial f}{\partial x}(1, 4) = 2\cdot 1\cdot 4^3 + e^1 = 128 + e$$

and

$$\frac{\partial f}{\partial y}(1, 4) = 3\cdot 1^2\cdot 4^2 + \frac{1}{4} = 48 + \frac{1}{4} = \frac{193}{4}.$$ ■

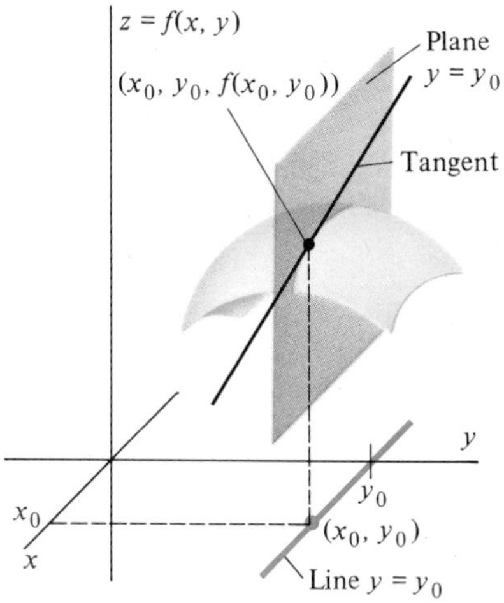

Figure 3.1 Partial derivative $\frac{\partial f}{\partial x}(x_0, y_0)$ is the slope of the line tangent to the trace of $z = f(x, y)$ in the plane $y = y_0$.

Example 2 For $f(x, y) = \sin xy$, we apply the rule $\frac{d}{dt}\sin at = a\cos at$, with one variable playing the role of a and the other playing the role of t:

$$\frac{\partial f}{\partial x}(x, y) = y\cos xy \qquad \text{and} \qquad \frac{\partial f}{\partial y}(x, y) = x\cos xy.$$ ■

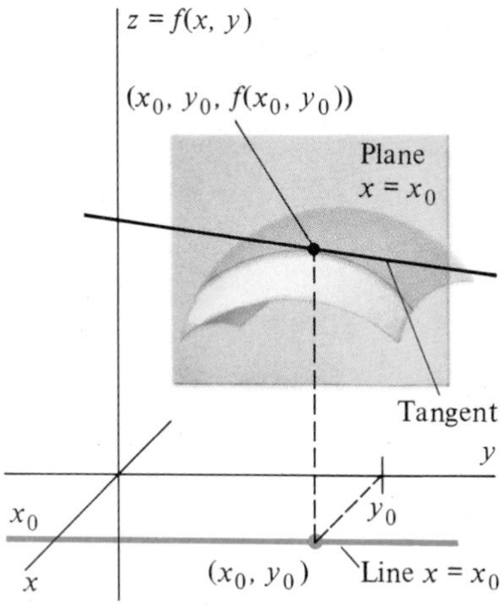

Figure 3.2 Partial derivative $\frac{\partial f}{\partial y}(x_0, y_0)$ is the slope of the line tangent to the trace of $z = f(x, y)$ in the plane $x = x_0$.

Figures 3.1 and 3.2 illustrate the geometric interpretations of the partial derivatives in Definition 4. To interpret $\frac{\partial f}{\partial x}(x_0, y_0)$, we note that the set of points (x, y, z) in $\mathbb{R}^3$ with $y = y_0$ is a plane. The intersection of this plane with the graph of $z = f(x, y)$ is the *trace* of $z = f(x, y)$ in the plane $y = y_0$. This curve may be viewed as the graph of the function of one variable

$$h(x) = f(x, y_0).$$

Since

$$h'(x_0) = \frac{\partial f}{\partial x}(x_0, y_0),$$

the partial derivative $\frac{\partial f}{\partial x}(x_0, y_0)$ gives the slope of the line tangent to this trace at the point $(x_0, y_0, f(x_0, y_0))$. A similar interpretation is valid for the partial derivative $\dfrac{\partial f}{\partial y}(x_0, y_0)$.

Example 3 For the function $f(x, y) = \sqrt{1 - x^2 - y^2}$ show that

(i) $\dfrac{\partial f}{\partial x}(x, y) = \dfrac{\partial f}{\partial y}(x, y)$ if and only if $y = x$, and

(ii) $\dfrac{\partial f}{\partial x}(x, y) = -\dfrac{\partial f}{\partial y}(x, y)$ if and only if $y = -x$

and interpret this result geometrically.

Solution: The given function can be considered as the composition of a function of a single variable with a function of two variables,

$$f(x, y) = f(u(x, y))$$

where $f(u) = \sqrt{u}$ and $u(x, y) = 1 - x^2 - y^2$. We therefore use the following generalization of the Chain Rule:

$$\frac{\partial f}{\partial x}(u(x, y)) = \frac{d}{du}f(u) \cdot \frac{\partial u}{\partial x}(x, y)$$

$$\frac{\partial f}{\partial y}(u(x, y)) = \frac{d}{du}f(u) \cdot \frac{\partial u}{\partial y}(x, y).$$

Thus,

$$\frac{\partial f}{\partial x}(x, y) = \frac{1}{2\sqrt{u}}(-2x) = \frac{-x}{\sqrt{1 - x^2 - y^2}},$$

$$\frac{\partial f}{\partial y}(x, y) = \frac{1}{2\sqrt{u}}(-2y) = \frac{-y}{\sqrt{1 - x^2 - y^2}}.$$

Thus

$$\frac{\partial f}{\partial x} = \frac{-2x}{\sqrt{1 - x^2 - y^2}} = \frac{-2y}{\sqrt{1 - x^2 - y^2}} = \frac{\partial f}{\partial y} \quad \text{if and only if } y = x.$$

Similarly,

$$\frac{\partial f}{\partial x} = -\frac{\partial f}{\partial y} \quad \text{if and only if} \quad y = -x.$$

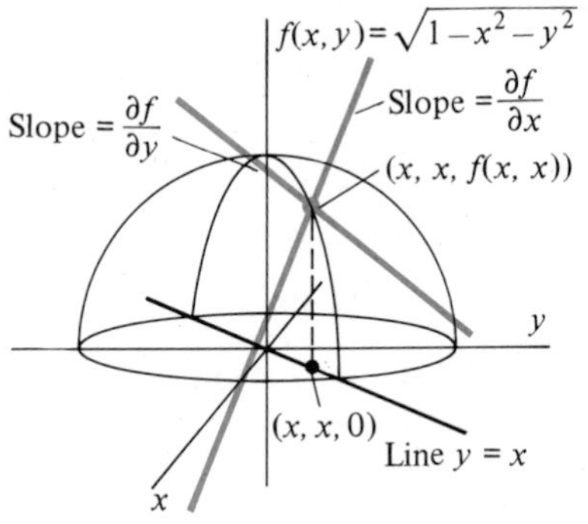

Figure 3.3 $\dfrac{\partial f}{\partial x} = \dfrac{\partial f}{\partial y}$ if $y = x$.

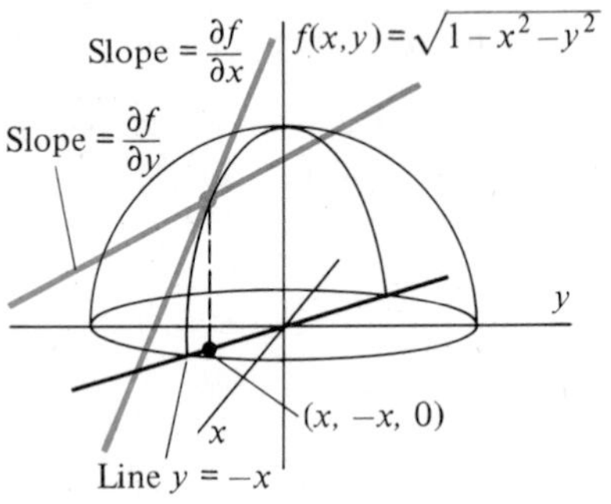

Figure 3.4 $\dfrac{\partial f}{\partial x} = -\dfrac{\partial f}{\partial y}$ if $y = -x$.

Figure 3.3 shows the geometric interpretation of this result. The graph of $z = f(x, y) = \sqrt{1 - x^2 - y^2}$ is a hemisphere. The only points on the surface of this hemisphere for which the tangent in the plane parallel to the x-axis has the same slope as the tangent in the plane parallel to the y-axis are the points lying above the line $y = x$. Similarly, the points at which these tangents have opposite slope lie above the line $y = -x$ in the xy-plane (see Figure 3.4). ■

Notation for Partial Derivatives

For a function of two variables $z = f(x, y)$, the symbol $\dfrac{\partial}{\partial x}$ denotes the partial derivative with respect to x just as the Leibniz notation $\dfrac{d}{dx}$ denotes the derivative of the function of a single variable $y = g(x)$. However, the prime notation $g'(x)$ does not have a convenient analogue for functions of more than one variable. Instead, we use subscripts to denote partial derivatives, such as $f_x(x, y)$ or z_x for $\dfrac{\partial f}{\partial x}(x, y)$.

We may summarize the various types of notation for partial derivatives as follows: If $z = f(x, y)$, then

$$\frac{\partial f}{\partial x}(x, y) = \frac{\partial}{\partial x} f(x, y) = z_x(x, y) = z_x,$$

and

$$\frac{\partial f}{\partial y}(x, y) = \frac{\partial}{\partial y} f(x, y) = z_y(x, y) = z_y.$$

Functions of More than Two Variables

Partial derivatives are defined for functions of more than two variables by the same idea used in Definition 4: Hold all variables constant except one, and differentiate the resulting function of a single variable as before. For example, if $w = f(x, y, z)$ is a function of the three independent variables x, y, and z, the three partial derivatives are defined as follows:

$$\frac{\partial f}{\partial x}(x, y, z) = \lim_{\Delta x \to 0} \frac{f(x + \Delta x, y, z) - f(x, y, z)}{\Delta x} \tag{4}$$

$$\frac{\partial f}{\partial y}(x, y, z) = \lim_{\Delta y \to 0} \frac{f(x, y + \Delta y, z) - f(x, y, z)}{\Delta y} \tag{5}$$

$$\frac{\partial f}{\partial z}(x, y, z) = \lim_{\Delta z \to 0} \frac{f(x, y, z + \Delta z) - f(x, y, z)}{\Delta z} \tag{6}$$

More generally, if $z = f(x_1, x_2, \ldots, x_n)$ is a function of n variables, the partial derivative with respect to the jth variable, x_j, is defined to be

$$\frac{\partial f}{\partial x_j}(x_1, x_2, \ldots, x_n) = \lim_{\Delta x_j \to 0} \frac{f(x_1, \ldots, x_{j-1}, x_j + \Delta x_j, x_{j+1}, \ldots, x_n) - f(x_1, x_2, \ldots, x_n)}{\Delta x_j}.$$

These partial derivatives are calculated just as partial derivatives for functions of two variables: Treat all variables except one as constants. Unfortunately, we lose our geometric interpretation in the presence of more than two independent variables. However, we do have the basic interpretation of derivative as rate of change. For example, the partial derivative $\dfrac{\partial f}{\partial x}$ in (4) is the rate at which the function $f(x, y, z)$ changes with respect to change in x. Also, we make use of subscript notation for partial derivatives for functions of more than two variables just as in the two-variable case. That is, if $w = f(x, y, z)$

$$\frac{\partial f}{\partial z}(x, y, z) = \frac{\partial}{\partial z}f(x, y, z) = f_z(x, y, z) = w_z.$$

Example 4 Let $f(x, y, z) = \sqrt{x}e^{y/z}$, $\quad z \neq 0, \quad x \geq 0$. Then

$$\frac{\partial f}{\partial x}(x, y, z) = \left\{\frac{d}{dx}\sqrt{x}\right\}e^{y/z} = \frac{1}{2\sqrt{x}} \cdot e^{y/z},$$

$$\frac{\partial f}{\partial y}(x, y, z) = \sqrt{x}\left\{\frac{\partial}{\partial y}(e^{y/z})\right\} = \sqrt{x}e^{y/z} \cdot \frac{1}{z} = \frac{\sqrt{x}}{z}e^{y/z},$$

$$\frac{\partial f}{\partial z}(x, y, z) = \sqrt{x}\left\{\frac{\partial}{\partial z}e^{y/z}\right\} = \sqrt{x}e^{y/z}\left(\frac{-y}{z^2}\right) = \frac{-y\sqrt{x}}{z^2}e^{y/z}. \quad ■$$

Example 5 The radial rate of heat flow in a substance between two concentric spheres is given by the function

$$H(r, R, t, T) = \frac{(t - T)4\pi kRr}{R - r}$$

where k is a constant, and where the inner sphere has radius r and temperature t and the outer sphere has radius R and temperature T. Thus (using the Quotient Rule to differentiate with respect to r and R),

$$H_r = \frac{\partial}{\partial r}H(r, R, t, T) = \frac{(R - r)[(t - T)4\pi kR] - (-1)[(t - T)4\pi kRr]}{(R - r)^2}$$

$$= \frac{(t - T)4\pi kR^2}{(R - r)^2},$$

$$H_R = \frac{\partial}{\partial R}H(r, R, t, T) = \frac{(R - r)[(t - T)4\pi kr] - (1)[(t - T)4\pi kRr]}{(R - r)^2}$$

$$= \frac{-(t - T)4\pi kr^2}{(R - r)^2},$$

$$H_t = \frac{\partial}{\partial t}H(r, R, t, T) = \frac{\partial}{\partial t}\left[t\frac{4\pi kRr}{R - r} - T\frac{4\pi kRr}{R - r}\right] = \frac{4\pi kRr}{R - r},$$

$$H_T = \frac{\partial}{\partial T}H(r, R, t, T) = \frac{\partial}{\partial T}\left[t\frac{4\pi kRr}{R - r} - T\frac{4\pi kRr}{R - r}\right] = -\frac{4\pi kRr}{R - r}. \quad ■$$

Example 6 For the vectors $\boldsymbol{x} = x_1\boldsymbol{i} + x_2\boldsymbol{j} + x_3\boldsymbol{k}$ and $\boldsymbol{y} = y_1\boldsymbol{i} + y_2\boldsymbol{j} + y_3\boldsymbol{k}$, the dot product

$$\boldsymbol{x} \cdot \boldsymbol{y} = x_1y_1 + x_2y_2 + x_3y_3$$

may be viewed as a function of the six independent variables (components) $x_1, x_2, x_3, y_1, y_2, y_3$. Thus,

$$\frac{\partial}{\partial x_1}(\boldsymbol{x} \cdot \boldsymbol{y}) = y_1; \qquad \frac{\partial}{\partial y_1}(\boldsymbol{x} \cdot \boldsymbol{y}) = x_1,$$

$$\frac{\partial}{\partial x_2}(\mathbf{x}\cdot\mathbf{y}) = y_2; \qquad \frac{\partial}{\partial y_2}(\mathbf{x}\cdot\mathbf{y}) = x_2,$$

$$\frac{\partial}{\partial x_3}(\mathbf{x}\cdot\mathbf{y}) = y_3; \qquad \frac{\partial}{\partial y_3}(\mathbf{x}\cdot\mathbf{y}) = x_3.$$

Thus, the rate at which $\mathbf{x}\cdot\mathbf{y}$ changes with respect to change in the component x_1 is just the component y_1, and so forth. ■

Limitations of Partial Derivatives

Before proceeding further, we should dispel a notion that students sometimes develop at this point. Although knowledge of the derivative $f'(x_0)$ completely determines the rate at which the function $y = f(x_0)$ changes at $x = x_0$, *knowledge of the partial derivatives $f_x(x_0, y_0)$ and $f_y(x_0, y_0)$ is not always sufficient to determine the rate at which the function of two variables $z = f(x, y)$ is changing at (x_0, y_0)*. The analogous statement for functions of more than two variables is also true.

To see this, let us return to the function

$$f(x, y) = \begin{cases} \dfrac{xy}{x^2 + y^2}, & (x, y) \neq (0, 0) \\ 0, & (x, y) = (0, 0) \end{cases}$$

introduced in Section 19.2. Using Definition 4, we find that

$$\frac{\partial f}{\partial x}(0, 0) = \lim_{\Delta x \to 0} \frac{f(0 + \Delta x, 0) - f(0, 0)}{\Delta x} = \lim_{\Delta x \to 0}\left(\frac{0}{\Delta x}\right) = 0$$

and

$$\frac{\partial f}{\partial y}(0, 0) = \lim_{\Delta y \to 0} \frac{f(0, 0 + \Delta y) - f(0, 0)}{\Delta y} = \lim_{\Delta y \to 0}\left(\frac{0}{\Delta y}\right) = 0.$$

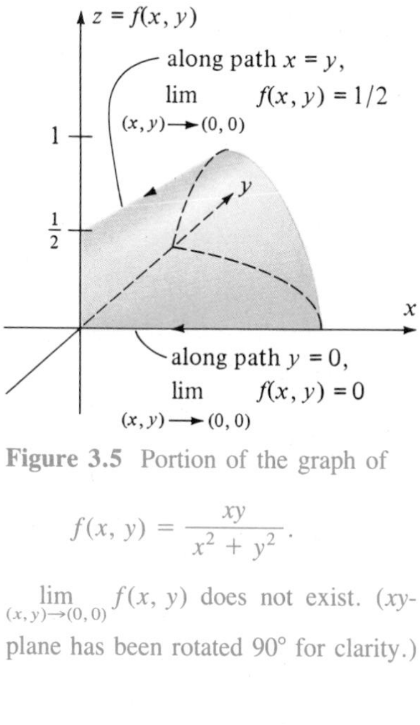

Figure 3.5 Portion of the graph of $f(x, y) = \dfrac{xy}{x^2 + y^2}$. $\lim_{(x,y)\to(0,0)} f(x, y)$ does not exist. (xy-plane has been rotated 90° for clarity.)

Thus, both partial derivatives at (0, 0) are zero. But it is false to conclude that the rate of change of $f(x, y)$ at (0, 0) in *every direction* is zero, as can be seen from Figure 3.5. For example, a small change from the point (0, 0) to the point $(\Delta x, \Delta x)$ causes an abrupt change in the function $f(x, y)$ from $f(0, 0) = 0$ to $f(\Delta x, \Delta x) = \dfrac{1}{2}$, no matter how small Δx may be.

It is important to understand that partial derivatives give information about functions *only* in the directions of the coordinate axes. We must develop a richer theory of differentiation in order to obtain general conclusions about the rate of change of a function of several variables near a particular point.

Higher Order Partial Derivatives

Repeated applications of partial differentiation lead to **higher order partial derivatives.** There is nothing terribly complicated about this concept, except that we must be very careful about notation since we encounter **mixed partial derivatives,** in which one differentiation is performed with respect to a particular variable, followed by another differentiation with respect to a different variable.

We will use the following notation:

$$\frac{\partial^2 f}{\partial x^2}(x, y) = \frac{\partial^2}{\partial x^2} f(x, y) \quad \text{means} \quad \frac{\partial}{\partial x}\left(\frac{\partial f}{\partial x}(x, y)\right). \tag{7}$$

$$\frac{\partial^2 f}{\partial y \partial x}(x, y) = \frac{\partial^2}{\partial y \partial x} f(x, y) \quad \text{means} \quad \frac{\partial}{\partial y}\left(\frac{\partial f}{\partial x}(x, y)\right). \tag{8}$$

$$\frac{\partial^2 f}{\partial x \partial y}(x, y) = \frac{\partial^2}{\partial x \partial y} f(x, y) \quad \text{means} \quad \frac{\partial}{\partial x}\left(\frac{\partial f}{\partial y}(x, y)\right). \tag{9}$$

$$\frac{\partial^2 f}{\partial y^2}(x, y) = \frac{\partial^2}{\partial y^2} f(x, y) \quad \text{means} \quad \frac{\partial}{\partial y}\left(\frac{\partial f}{\partial y}(x, y)\right). \tag{10}$$

Note that the order in which the differentiations are performed is indicated by reading the "denominator" of the derivative notation from *right* to *left*. Similar definitions hold for third and higher order partial derivatives.

Example 7 For the function $f(x, y) = x^2y^3 + \cos x \sin y$,

$$\frac{\partial^2 f}{\partial x^2}(x, y) = \frac{\partial}{\partial x}\left(\frac{\partial f}{\partial x}(x, y)\right) = \frac{\partial}{\partial x}(2xy^3 - \sin x \sin y) = 2y^3 - \cos x \sin y,$$

$$\frac{\partial^2 f}{\partial y \partial x}(x, y) = \frac{\partial}{\partial y}\left(\frac{\partial f}{\partial x}(x, y)\right) = \frac{\partial}{\partial y}(2xy^3 - \sin x \sin y) = 6xy^2 - \sin x \cos y,$$

$$\frac{\partial^2 f}{\partial x \partial y}(x, y) = \frac{\partial}{\partial x}\left(\frac{\partial f}{\partial y}(x, y)\right) = \frac{\partial}{\partial x}(3x^2y^2 + \cos x \cos y) = 6xy^2 - \sin x \cos y,$$

$$\frac{\partial^2 f}{\partial y^2}(x, y) = \frac{\partial}{\partial y}\left(\frac{\partial f}{\partial y}(x, y)\right) = \frac{\partial}{\partial y}(3x^2y^2 + \cos x \cos y) = 6x^2y - \cos x \sin y,$$

$$\frac{\partial^3 f}{\partial y^3}(x, y) = \frac{\partial}{\partial y}\left(\frac{\partial^2 f}{\partial y^2}(x, y)\right) = \frac{\partial}{\partial y}\left(\frac{\partial^2 f}{\partial y^2}\right) = \frac{\partial}{\partial y}(6x^2y - \cos x \sin y)$$
$$= 6x^2 - \cos x \cos y. \quad \blacksquare$$

We may also use subscript notation to indicate higher order partial derivatives as follows: If $z = f(x, y)$,

$$f_{xx}(x, y) \text{ or } z_{xx} \quad \text{means} \quad \frac{\partial^2 f}{\partial x^2}(x, y) = \frac{\partial}{\partial x}\left(\frac{\partial f}{\partial x}(x, y)\right) = (z_x)_x, \tag{11}$$

$$f_{xy}(x, y) \text{ or } z_{xy} \quad \text{means} \quad \frac{\partial^2 f}{\partial y \partial x}(x, y) = \frac{\partial}{\partial y}\left(\frac{\partial f}{\partial x}(x, y)\right) = (z_x)_y, \tag{12}$$

$$f_{yx}(x, y) \text{ or } z_{yx} \quad \text{means} \quad \frac{\partial^2 f}{\partial x \partial y}(x, y) = \frac{\partial}{\partial x}\left(\frac{\partial f}{\partial y}(x, y)\right) = (z_y)_x, \tag{13}$$

$$f_{yy}(x, y) \text{ or } z_{yy} \quad \text{means} \quad \frac{\partial^2 f}{\partial y^2}(x, y) = \frac{\partial}{\partial y}\left(\frac{\partial f}{\partial y}(x, y)\right) = (z_y)_y, \tag{14}$$

with similar statements holding for functions of more than two variables and for higher order derivatives.

It is important to note that, when subscripts are used, the differentiations are performed *in the order indicated by the subscripts, read from left to right*. This is just the opposite of the order in which the differentiations are indicated in the Leibniz notation on the right-hand sides of equations (11) through (14).

Example 8 For the function $f(x, y, z) = x^2y^3z^4$,

$$f_x = 2xy^3z^4; \qquad f_y = 3x^2y^2z^4; \qquad f_z = 4x^2y^3z^3$$

$$f_{xy} = \frac{\partial}{\partial y}(2xy^3z^4) = 6xy^2z^4; \qquad f_{yx} = \frac{\partial}{\partial x}(3x^2y^2z^4) = 6xy^2z^4$$

$$f_{yz} = \frac{\partial}{\partial z}(3x^2y^2z^4) = 12x^2y^2z^3; \qquad f_{zy} = \frac{\partial}{\partial y}(4x^2y^3z^3) = 12x^2y^2z^3$$

$$f_{xz} = \frac{\partial}{\partial z}(2xy^3z^4) = 8xy^3z^3; \qquad f_{zx} = \frac{\partial}{\partial x}(4x^2y^3z^3) = 8xy^3z^3$$

$$f_{xx} = 2y^3z^4; \qquad f_{yy} = 6x^2yz^4; \qquad f_{zz} = 12x^2y^3z^2$$

$$f_{xyz} = \frac{\partial}{\partial z}(6xy^2z^4) = 24xy^2z^3; \qquad f_{yzx} = \frac{\partial}{\partial x}(12x^2y^2z^2) = 24xy^2z^3$$

$$f_{xxx} = 0; \qquad f_{yyy} = 6x^2z^4; \qquad f_{zzz} = 24x^2y^3z.$$ ■

Equality of Mixed Partials

You have no doubt observed, in Example 7, that $f_{yx} = f_{xy}$ and, in Example 8, that $f_{xy} = f_{yx}$, $f_{yz} = f_{zy}$, and $f_{xz} = f_{zx}$. This is not true for all functions. However, when the function $f(x, y)$ and various of its partial derivatives are continuous, these mixed partials will be equal. The following theorem makes this precise. It is typically proved in courses on advanced calculus.

THEOREM 2
Equality of Mixed Partials

If the function $z = f(x, y)$ and the partial derivatives

$$\frac{\partial f}{\partial x}, \quad \frac{\partial f}{\partial y}, \quad \frac{\partial^2 f}{\partial x \partial y}, \text{ and } \frac{\partial^2 f}{\partial y \partial x}$$

are all continuous in a neighborhood of the point (x_0, y_0), then

$$\frac{\partial^2 f}{\partial x \partial y}(x_0, y_0) = \frac{\partial^2 f}{\partial y \partial x}(x_0, y_0). \tag{15}$$

Equation (15) plays an important role in the theory of differentiation that follows.

Exercise Set 19.3

In Exercises 1–20, find all first order partial derivatives.

1. $f(x, y) = xy$
2. $z = \sqrt{x + y^2}$
3. $z = x \tan y^2$
4. $f(x, y) = xy^3 + \sqrt{y}$
5. $f(x, y) = e^{x^2+y^2}$
6. $z = \tan^{-1}(y/x)$
7. $f(r, \theta) = r \sin(\pi/2 - \theta)$
8. $f(s, t) = \dfrac{s - t}{s + t}$
9. $z = \ln(xy^2 + x - y)$
10. $h(u, v) = e^{u-v} + e^{v-u}$
11. $f(x, y) = x^y$
12. $z = 2^x y^2$

13. $f(r, \theta) = r^2 \cos \theta$

14. $f(x, y, z) = x^3 e^y \ln z$

15. $f(x, y, z) = xy^3 - yz^2$

16. $w = \ln(x^2 + y^2 + z^2)$

17. $f(x, y, z) = \left(\dfrac{x - y}{x + y}\right)^z$

18. $f(u, v, w) = \dfrac{\sin u}{v^3 \tan^{-1} w}$

19. $f(r, s, t) = \dfrac{\sqrt{r}\, s \ln t}{\sqrt{s^2 - 2r + t}}$

20. $f(u, v, w) = \dfrac{ue^{vw}}{\sin^2 u + \tan^2 w}$

21. Find $\dfrac{\partial f}{\partial x}(2, 5)$ for $f(x, y) = xy^3 - y$.

22. Find $\dfrac{\partial f}{\partial y}(1, 0)$ for $f(x, y) = e^{x-y^2}$.

23. Find $z_x(2, 1)$ for $z = \sqrt{x + y^2}$.

24. Find f_{xx}, f_{xy}, f_{yx}, and f_{yy} for $f(x, y)$ in Exercise 4.

25. Find $\dfrac{\partial^2 f}{\partial r^2}$, $\dfrac{\partial^2 f}{\partial r \partial \theta}$, and $\dfrac{\partial^2 f}{\partial \theta^2}$ for $f(r, \theta)$ in Exercise 13.

26. Find f_{xx}, f_{xy}, f_{xz}, f_{yz}, f_{yy}, and f_{zz} for $f(x, y, z)$ in Exercise 15.

27. Find $w_{xx} + w_{yy} + w_{zz}$ for $w(x, y, z)$ in Exercise 16.

28. For a particle traveling in a circular orbit of radius r, the relation between angular speed ω and velocity v is $v = \omega r$. Find $\dfrac{\partial v}{\partial r}$ and $\dfrac{\partial v}{\partial \omega}$.

29. For the constant volume flow of an incompressible fluid through a tube of varying cross-sectional area, the equation $A_1 v_1 = A_2 v_2$ expresses the relationship between the respective cross-sectional areas and velocities at two points in the tube.

a. Express v_2 as a function of A_2, A_1, and v_1.

b. Find $\dfrac{\partial v_2}{\partial A_2}$, the rate at which v_2 changes with respect to change in A_2 alone.

c. Suppose that $A_1 = 5$ cm, $A_2 = 3$ cm, and $v_1 = 20$ cm/sec. Find the rate of change of v_2 with respect to A_2 if A_1 and v_1 are held constant.

d. With A_1, A_2, and v_1 as in (c), find the rate of change of A_2 with respect to change in v_1 if A_1 and v_2 are held constant.

In Exercises 30–36, use the equations

$$x = \rho \sin \phi \cos \theta$$

$$y = \rho \sin \phi \sin \theta$$

$$z = \rho \cos \phi$$

for changing from rectangular to spherical coordinates to find the indicated partial derivative.

30. $\dfrac{\partial x}{\partial \phi}$

31. $\dfrac{\partial y}{\partial \theta}$

32. $\dfrac{\partial z}{\partial \rho}$

33. $\dfrac{\partial x}{\partial \rho}$

34. $\dfrac{\partial y}{\partial \phi}$

35. $\dfrac{\partial z}{\partial \phi}$

36. $\dfrac{\partial \phi}{\partial z}$

37. For $f(x, y) = \displaystyle\int_x^{x+y} \cos t^2\, dt$ find

a. $\dfrac{\partial f}{\partial x}$ **b.** $\dfrac{\partial f}{\partial y}$

38. For $f(x, y, z) = \displaystyle\int_z^x \sqrt{t^3 + 1}\, dt - \int_y^z \sqrt{t^3 + 1}\, dt$ find

a. $\dfrac{\partial f}{\partial x}$ **b.** $\dfrac{\partial f}{\partial y}$ **c.** $\dfrac{\partial f}{\partial z}$

39. Show that the function $f(x, y) = \sin^{-1}\left(\dfrac{x - y}{x + y}\right)$ is a solution of the differential equation

$$x \frac{\partial z}{\partial x} + y \frac{\partial z}{\partial y} = 0.$$

40. Show that the function $w = \ln(e^x + e^y + e^z)$ satisfies the partial differential equation

$$\frac{\partial w}{\partial x} + \frac{\partial w}{\partial y} + \frac{\partial w}{\partial z} = 1.$$

41. The **electric potential** at an axial point for a charged disc is given by

$$v = \frac{\sigma}{2\epsilon_0}(\sqrt{a^2 + r^2} - r)$$

where σ and ϵ_0 are constants, a is the radius of the disc, and r is the distance from the point to the disc.

a. Find the rate of change of v with respect to a if r is held constant.

b. Find the rate of change of v with respect to r if a is held constant.

42. The national unemployment rate u may be viewed as the dot product $u = \mathbf{v} \cdot \mathbf{w}$ where the components of the vector $\mathbf{v} = \langle v_1, v_2, \ldots, v_n \rangle$ are the percentages of employable citizens in each of the n job categories and the components of the vector $\mathbf{w} = \langle w_1, w_2, \ldots, w_n \rangle$ are the unemployment rates within each category.

a. Find $\dfrac{\partial u}{\partial v_j}$ and interpret this rate.

b. Find $\dfrac{\partial u}{\partial w_j}$ and interpret this rate.

43. Let $f(x, y, z) = e^{x+y+z}$. Show that all partial derivatives of all orders equal $f(x, y, z)$.

44. Show that a function of the form $f(x, y) = e^{kx}g(y)$ satisfies the differential equation $\frac{\partial f}{\partial x}(x, y) = kf(x, y)$.

45. Show that the function $f(x, y) = \sin xy$ satisfies the differential equation

$$x\frac{\partial f}{\partial x}(x, y) - y\frac{\partial f}{\partial y}(x, y) = 0.$$

46. Show that the function $f(x, y) = \sin xy$ satisfies the differential equation

$$x^2\frac{\partial^2 f}{\partial x^2}(x, y) - y^2\frac{\partial^2 f}{\partial y^2}(x, y) = 0.$$

47. Find a solution of the differential equation

$$x^n\frac{\partial^n f}{\partial x^n} = y^n\frac{\partial^n f}{\partial y^n}, \qquad n = 1, 2, 3 \ldots.$$

(See Exercises 45 and 46.)

48. Show that the function $y(x, t) = g(t - cx)$ satisfies the **one-dimensional wave equation**

$$\frac{\partial^2 y}{\partial t^2} = \frac{1}{c^2}\frac{\partial^2 y}{\partial x^2}.$$

49. **Laplace's equation** for the function $f(x, y)$ is

$$\frac{\partial^2 f}{\partial x^2} + \frac{\partial^2 f}{\partial y^2} = 0.$$

Show that the following functions satisfy Laplace's equation.

a. $f(x, y) = e^x \sin y$
b. $f(x, y) = e^{-x}\cos y$
c. $f(x, y) = \ln\sqrt{x^2 + y^2}$

50. Show that a polynomial $P(x, y)$ in x and y (like $P(x, y) = 3x^2y^3 + xy^6 - 5x^3y$) has the property that $\frac{\partial^2 P}{\partial x \partial y} = \frac{\partial^2 P}{\partial y \partial x}$.

19.4 TANGENT PLANES

For the function of a single variable $y = f(x)$, knowledge of $f(a)$ and the derivative $f'(a)$ enables us to write an equation for the line tangent to the graph of $y = f(x)$ at the point $(a, f(a))$. The purpose of this brief section is to show how we can obtain an equation for a plane tangent to the graph of a function of two variables $z = f(x, y)$ from knowledge of its partial derivatives. We shall leave a careful discussion of the meaning of the term "tangent plane" for later. The issue here is simply how one goes about using partial derivatives to find such a plane, assuming that it exists. (See Exercise 32 for an explanation of why we seem to be hedging on the use of the term "tangent plane.")

Suppose that $z = f(x, y)$ is a function of two variables that is defined in a neighborhood of the point (x_0, y_0). Assume also that the partial derivatives $\frac{\partial f}{\partial x}(x_0, y_0)$ and $\frac{\partial f}{\partial y}(x_0, y_0)$ both exist. From the discussion of the geometric interpretation of partial derivatives in Section 19.3, we may conclude the following (see Figure 4.1).

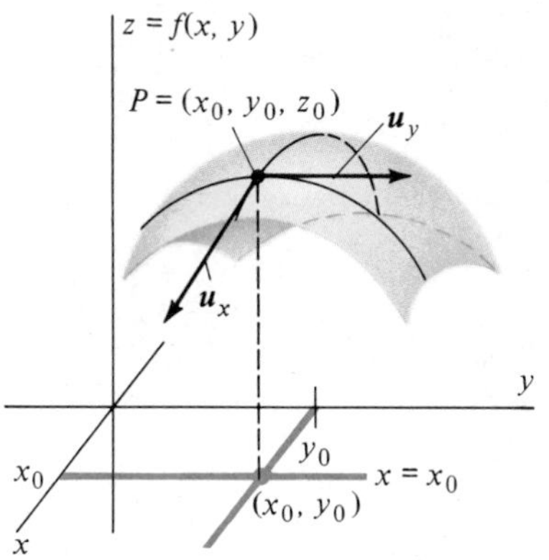

Figure 4.1 Partial derivatives determine vectors $u_x = i + f_x(a, b)k$ and $u_y = j + f_y(a, b)k$ tangent to graph of $z = f(x, y)$ at point P.

(i) The partial derivative $\frac{\partial f}{\partial x}(x_0, y_0)$ gives the slope of the line tangent to the trace of $z = f(x, y)$ in the plane $y = y_0$. Since the plane $y = y_0$ is parallel to the xz-coordinate plane, a vector parallel to this tangent line is

$$\boldsymbol{u}_x = \boldsymbol{i} + \frac{\partial f}{\partial x}(x_0, y_0)\boldsymbol{k}.$$

(To check this statement, note that the slope of the vector, thought of as a line segment in the plane $y = y_0$, is $\frac{\Delta z}{\Delta x} = \frac{\partial f}{\partial x}(x_0, y_0)$, as required.)

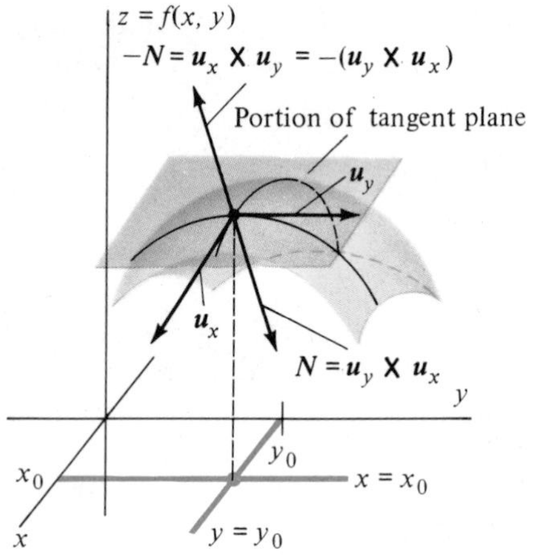

Figure 4.2 Vectors u_x and u_y determine normal $N = u_y \times u_x$ for plane tangent to graph of $z = f(x, y)$ at point P.

(ii) The partial derivative $\dfrac{\partial f}{\partial y}(x_0, y_0)$ gives the slope of the line tangent to the trace of $z = f(x, y)$ in the plane $x = x_0$. Since the plane $x = x_0$ is parallel to the yz-coordinate plane, a vector parallel to this tangent line is

$$\boldsymbol{u}_y = \boldsymbol{j} + \frac{\partial f}{\partial y}(x_0, y_0)\boldsymbol{k}.$$

Since both vectors $\boldsymbol{u}_x$ and $\boldsymbol{u}_y$ must lie in the tangent plane, they together determine a normal to the plane as

$$\boldsymbol{N} = \boldsymbol{u}_y \times \boldsymbol{u}_x = \det\begin{bmatrix} \boldsymbol{i} & \boldsymbol{j} & \boldsymbol{k} \\ 0 & 1 & \dfrac{\partial f}{\partial y}(x_0, y_0) \\ 1 & 0 & \dfrac{\partial f}{\partial x}(x_0, y_0) \end{bmatrix} \tag{1}$$

$$= \frac{\partial f}{\partial x}(x_0, y_0)\boldsymbol{i} + \frac{\partial f}{\partial y}(x_0, y_0)\boldsymbol{j} - \boldsymbol{k}.$$

According to equation (11), Section 17.7, an equation for the plane with normal vector $\boldsymbol{N}$, as above, and containing the point $(x_0, y_0, z_0) = (x_0, y_0, f(x_0, y_0))$ is

$$\frac{\partial f}{\partial x}(x_0, y_0)(x - x_0) + \frac{\partial f}{\partial y}(x_0, y_0)(y - y_0) - (z - z_0) = 0 \tag{2}$$

or

$$z = \frac{\partial f}{\partial x}(x_0, y_0)(x - x_0) + \frac{\partial f}{\partial y}(x_0, y_0)(y - y_0) + z_0. \tag{3}$$

Either equation (2) or equation (3) may be used to write the equation of the plane tangent to the graph of $z = f(x, y)$ at the point (x_0, y_0, z_0), where $z_0 = f(x_0, y_0)$. More generally, you may simply remember the idea used to develop these equations: the partial derivatives $\dfrac{\partial f}{\partial x}(x_0, y_0)$ and $\dfrac{\partial f}{\partial y}(x_0, y_0)$ determine **direction vectors,** $\boldsymbol{u}_x$ and $\boldsymbol{u}_y$, whose cross product $\boldsymbol{N} = \boldsymbol{u}_y \times \boldsymbol{u}_x$ is a normal to the desired plane. Equations (2) and (3) then follow by the theory of Section 17.7 concerning equations for lines and planes.

Example 1 Find an equation for the plane tangent to the graph of $f(x, y) = x^2 + 4y^2$ at the point $(2, 1, 8)$.

Solution: The required partial derivatives are

$$\frac{\partial f}{\partial x}(2, 1) = 2x\Big|_{\substack{x=2\\y=1}} = 4,$$

$$\frac{\partial f}{\partial y}(2, 1) = 8y\Big|_{\substack{x=2\\y=1}} = 8,$$

and $x_0 = 2$, $y_0 = 1$, $z_0 = 8$. Thus, by equation (3), the equation is

$$z = 4(x - 2) + 8(y - 1) + 8$$

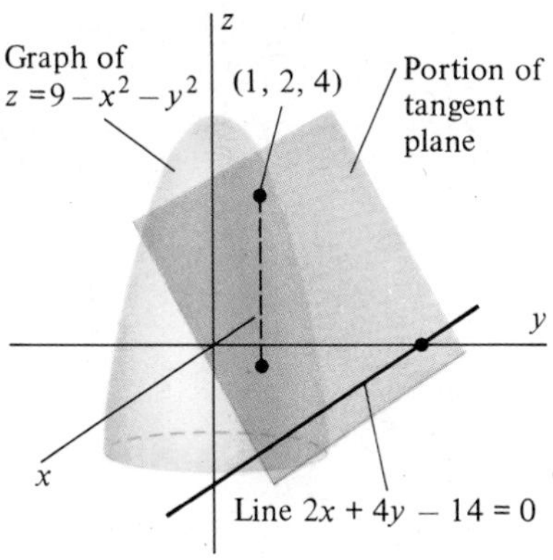

Figure 4.3 Plane tangent to graph of $y = 9 - x^2 - y^2$ at (2, 1, 4) intersects xy-plane along line $2x + 4y - 14 = 0$.

or

$$z = 4x + 8y - 8.$$ ■

Example 2 Find an equation for the line of intersection of the plane tangent to the graph of $z = 9 - x^2 - y^2$ at (1, 2, 4) and the xy-coordinate plane.

Solution: Since

$$z_x(1, 2) = -2x \Big|_{\substack{x=1\\y=2}} = -2$$

and

$$z_y(1, 2) = -2y \Big|_{\substack{x=1\\y=2}} = -4,$$

the equation of the tangent plane, according to (3), is

$$z = -2(x - 1) - 4(y - 2) + 4$$

or

$$z = -2x - 4y + 14.$$

Setting $z = 0$ gives the line of intersection of this plane with the xy-plane as $2x + 4y - 14 = 0$ (see Figure 4.3). ■

Normal Lines

If a surface has a tangent plane at point P, we say that a line through P is **normal** to the surface if it is normal to the tangent plane at P. Using N in (1) as a direction vector normal to the tangent plane for $z = f(x, y)$ at (x_0, y_0, z_0), we may write the normal line in vector form as

$$\ell: \boldsymbol{r}(t) = \boldsymbol{x}_0 + t\boldsymbol{N} \tag{4}$$

where $\boldsymbol{x}_0$ is the position vector $\boldsymbol{x}_0 = x_0\boldsymbol{i} + y_0\boldsymbol{j} + z_0\boldsymbol{k}$ (see Section 17.7). Writing $\boldsymbol{r}(t)$ in component form as $\boldsymbol{r}(t) = x(t)\boldsymbol{i} + y(t)\boldsymbol{j} + z(t)\boldsymbol{k}$ and using (1), we may write (4) in the component form

$$\begin{aligned} &x(t)\boldsymbol{i} + y(t)\boldsymbol{j} + z(t)\boldsymbol{k} \\ &= (x_0\boldsymbol{i} + y_0\boldsymbol{j} + z_0\boldsymbol{k}) + t\left(\frac{\partial f}{\partial x}(x_0, y_0)\,\boldsymbol{i} + \frac{\partial f}{\partial y}(x_0, y_0)\boldsymbol{j} - \boldsymbol{k}\right), \end{aligned} \tag{5}$$

in parametric form as

$$\begin{aligned} x(t) &= x_0 + t\,\frac{\partial f}{\partial x}(x_0, y_0), \\ y(t) &= y_0 + t\,\frac{\partial f}{\partial y}(x_0, y_0), \\ z(t) &= z_0 - t, \end{aligned} \tag{6}$$

or in symmetric form as

$$\frac{x - x_0}{\dfrac{\partial f}{\partial x}(x_0, y_0)} = \frac{y - y_0}{\dfrac{\partial f}{\partial y}(x_0, y_0)} = z_0 - z$$

$$\left(\frac{\partial f}{\partial x}(x_0, y_0) \neq 0,\ \frac{\partial f}{\partial y}(x_0, y_0) \neq 0\right). \tag{7}$$

Any of equations (5), (6), or (7) may be used to write an equation of the line normal to the graph of $z = f(x, y)$ at the point (x_0, y_0, z_0).

Example 3 Find equations for the line normal to the graph of $z = e^{y-x^2}$ at the point $(1, \ln 2, 2/e)$.

Solution: For the function $f(x, y) = e^{y-x^2}$,

$$\frac{\partial f}{\partial x}(1, \ln 2) = -2xe^{y-x^2}\bigg|_{\substack{x=1\\y=\ln 2}} = -2e^{\ln 2-1} = \frac{-4}{e}$$

and

$$\frac{\partial f}{\partial y}(1, \ln 2) = e^{y-x^2}\bigg|_{\substack{x=1\\y=\ln 2}} = e^{\ln 2-1} = \frac{2}{e}.$$

Also, $x_0 = 1$, $y_0 = \ln 2$, and $z_0 = \dfrac{2}{e}$. Using (6), we write parametric equations for the line as

$$x(t) = 1 - \frac{4t}{e},$$

$$y(t) = \ln 2 + \frac{2t}{e},$$

$$z(t) = \frac{2}{e} - t.$$

Symmetric equations for this line are

$$\frac{-e(x-1)}{4} = \frac{e(y - \ln 2)}{2} = \left(\frac{2}{e} - z\right).$$

Exercise Set 19.4

In each of Exercises 1–14, find an equation for the plane tangent to the graph of the given function at point P.

1. $f(x, y) = x^2 + y^2$; $P = (1, 3, 10)$
2. $f(x, y) = x^2 + 2xy + y^2$; $P = (3, -1, 4)$
3. $z = x^2 + y^2 - xy - 4x - 2y$; $P = (1, -1, 1)$
4. $f(x, y) = 2x^2 - y^2$; $P = (2, \sqrt{5}, 3)$
5. $f(x, y) = \dfrac{x-2}{y+2}$, $P = (4, -1, 2)$
6. $f(x, y) = \sqrt{9 - x^2 - y^2}$; $P = (1, -2, 2)$
7. $f(x, y) = \ln xy$; $P = (1, 1, 0)$
8. $f(x, y) = e^{-x} \sin y$; $P = (0, \pi/6, 1/2)$
9. $z = \dfrac{x}{x^2 + y^2}$; $P = (1, 1, 1/2)$
10. $f(x, y) = \tan^{-1}(y/x)$; $P = (2, 2, \pi/4)$
11. $z = \ln y^x$, $P = (1, 1, 0)$
12. $z = \dfrac{y-6}{3x^2-1}$, $P = (1, 4, -1)$
13. $f(x, y) = \ln\left(\dfrac{y-x}{y+x}\right)$, $P = (0, e, 0)$
14. $f(s, t) = \dfrac{1-\sqrt{s}}{t(s-\sqrt{t})}$, $P = \left(4, 1, -\dfrac{1}{3}\right)$

In Exercises 15–20, find a vector equation for the line normal to the graph of the given function at the point P, as described in the stated exercise.

15. Exercise 2.

16. Exercise 3.

17. Exercise 6.

18. Exercise 8.

19. Exercise 11.

20. Exercise 14.

21. Find the point on the graph of $z = -x^2 + xy + 2y^2$ where the tangent plane is parallel to the plane with equation $x - 14y + z = 4$.

22. Find a vector equation for the line of intersection of the plane tangent to the graph of $z = x^2 + 2y^2 - 4y + 2$ at $(2, 1, 4)$ and the xy-coordinate plane.

23. Show that all lines normal to the graph of $f(x, y) = x \sin y$ at points $(x, \pi/2)$ are parallel.

24. Show that at all points $(x, \pi/2)$, the planes tangent to the graph of $f(x, y) = x \sin y$ are the same. Find an equation for this plane.

25. Find an equation for the plane tangent to the paraboloid $z = 9x^2 + 4y^2$ at the point $(1, 2, 25)$.

26. Show that the volume of the tetrahedron formed by the planes $x = 0$, $y = 0$, and $z = 0$, and any tangent to the graph of the function $f(x, y) = \dfrac{c}{xy}$, $c > 0$ is $V = \dfrac{9c}{2}$.

27. Find an equation for the plane tangent to the sphere $x^2 + y^2 + z^2 = r^2$ at the point (x_0, y_0, z_0), $z_0 \neq 0$.

28. Find an equation for the plane tangent to the graph of the function $y = f(x, z)$ at the point $(x_0, y_0, z_0) = (x_0, f(x_0, z_0), z_0)$.

29. Use the result of Exercise 28 to find the equation of the plane tangent to the graph of $y = \tan^{-1}(z/x)$ at the point where $x = z = 1$.

30. Use the result of Exercise 28 to find a vector equation for the line of intersection of the plane tangent to the graph of $y = x^2 + 2xz - z^2$ at the point $(2, -8, -2)$ and the xz-coordinate plane.

31. Prove that every normal to a sphere passes through the center.

32. Consider (again) the function

$$f(x, y) = \begin{cases} \dfrac{xy}{x^2 + y^2}, & (x, y) \neq (0, 0) \\ 0, & (x, y) = (0, 0). \end{cases}$$

a. Use Definition 4 to show that $\dfrac{\partial f}{\partial x}(0, 0) = \dfrac{\partial f}{\partial y}(0, 0) = 0$ (see Section 19.3).

b. Use equation (3) to write the equation of the plane containing the point $(0, 0, 0)$ and having the normal $\mathbf{N}$ given by equation (1).

c. Explain why this plane should *not* be called a "tangent" plane for the graph of $z = f(x, y)$ at $(0, 0, 0)$ (see Figure 2.8).

19.5 RELATIVE AND ABSOLUTE EXTREMA

One of the principal applications of the derivative for functions of a single variable is in finding relative and absolute extrema. The purpose of this section is to show how partial derivatives may be used to find relative extrema for functions of several variables. Although the ideas discussed here are not confined to functions of just two independent variables, we will deal mainly with this case because of the opportunity to interpret the results geometrically.

We begin by defining what we mean by relative extrema.

DEFINITION 5

The number $z_0 = f(x_0, y_0)$ is called a **relative maximum** for the function $z = f(x, y)$ if there exists a neighborhood N of (x_0, y_0) so that $f(x, y)$ is defined for all $(x, y) \in N$, and so that

$$f(x_0, y_0) \geq f(x, y)$$

for all $(x, y) \in N$.

The number $z_0 = f(x_0, y_0)$ is called a **relative minimum** for $z = f(x, y)$ if there exists a neighborhood N of (x_0, y_0) so that $f(x, y)$ is defined for all $(x, y) \in N$ and so that

$$f(x_0, y_0) \leq f(x, y)$$

for all $(x, y) \in N$.

The number $z_0 = f(x_0, y_0)$ is a **relative extremum** for the function $z = f(x, y)$ if it is either a relative maximum or a relative minimum.

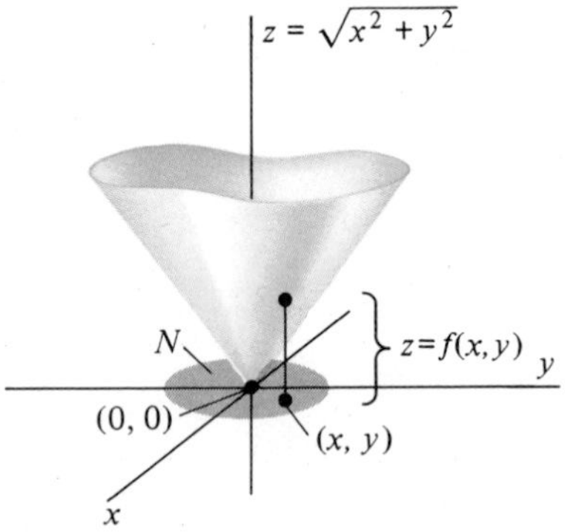

Figure 5.1 $f(0, 0) \leq f(x, y)$ for all (x, y) near $(0, 0)$; $f(0, 0)$ is a relative minimum.

Thus, a relative maximum for $z = f(x, y)$ is the precise analogue of a relative maximum for the function of a single variable $y = f(x)$: $z_0 = f(x_0, y_0)$ is the largest value $z = f(x, y)$ for all (x, y) "near" (x_0, y_0). A similar interpretation holds for relative minima.

Example 1 The function $f(x, y) = \sqrt{x^2 + y^2}$ has a relative minimum of $z_0 = 0$ at the point $(x_0, y_0) = (0, 0)$. This is easy to see, since

$$f(x, y) = \sqrt{x^2 + y^2} \geq 0 = f(0, 0)$$

for all points (x, y). Thus, Definition 5 is satisfied for any neighborhood (open disc) with center $(0, 0)$ (see Figure 5.1). ■

One way to verify that the number $z_0 = f(x_0, y_0)$ is a relative extremum is to simply compare the number $f(x_0, y_0)$ with values of the function $z = f(x, y)$ for points (x, y) near (x_0, y_0). While such comparisons are sometimes difficult, if not impossible, to make, this method does handle a variety of polynomial functions in two variables. The idea is to write $x = x_0 + h$ and $y = y_0 + k$ and then to examine the sign of the difference:

$$f(x_0, y_0) - f(x_0 + h, y_0 + k). \tag{1}$$

If this difference is nonnegative for all small values of h and k, we conclude that $f(x_0, y_0)$ is a relative maximum, since $f(x_0, y_0) \geq f(x_0 + h, y_0 + k)$ for all $(x_0 + h, y_0 + k)$ near (x_0, y_0). Similarly, if the difference in line (1) is nonpositive for all small values of h and k, we conclude that $f(x_0, y_0)$ is a relative minimum.

Example 2 Verify that $f(1, 2) = 4$ is a relative maximum for the function $f(x, y) = 2x + 4y - x^2 - y^2 - 1$.

Solution: Here $x_0 = 1$ and $y_0 = 2$, so the nearby point (x, y) is written $(x, y) = (x_0 + h, y_0 + k) = (1 + h, 2 + k)$. The difference in (1) is

$$\begin{aligned} f(1, 2) - f(1 + h, 2 + k) &= 4 - \{2(1 + h) + 4(2 + k) - (1 + h)^2 - (2 + k)^2 - 1\} \\ &= 4 - 2 - 2h - 8 - 4k + 1 + 2h + h^2 + 4 + 4k + k^2 + 1 \\ &= h^2 + k^2. \end{aligned} \tag{2}$$

Since $h^2 + k^2 \geq 0$ for all values of h and k, the above calculation shows that the difference in (2) is positive for all small values of h and k. Thus, $f(1, 2) = 4$ is a relative maximum for the function $f(x, y)$ (see Figure 5.2). ■

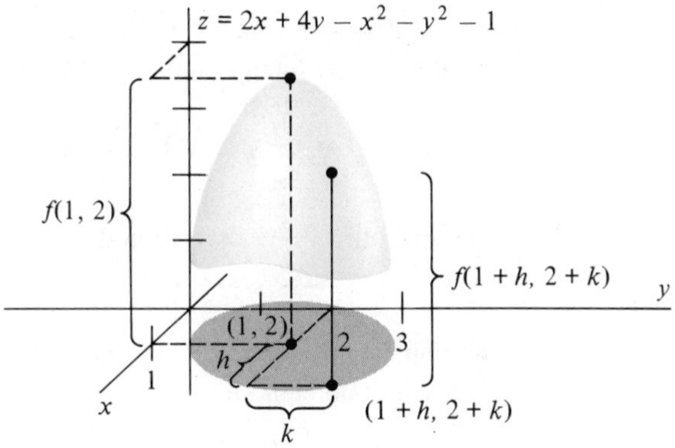

Figure 5.2 $f(1, 2) \geq f(1 + h, 2 + k)$ for small values of h and k; $f(1, 2)$ is a relative maximum.

Finding Relative Extrema

The preceding discussion addressed the issue of verifying that $f(x_0, y_0)$ is a relative extremum. But how do we find (x_0, y_0) to begin with? Recall the one-variable case again. If the function $y = f(x)$ has a relative extremum at $x = x_0$, then either $f'(x_0) = 0$ or else $f'(x_0)$ fails to exist. In the former case, the line tangent to the graph of $y = f(x)$ at $(x_0, f(x_0))$ is horizontal (see Figure 5.3). In the latter case, no tangent exists at $(x_0, f(x_0))$.

From a strictly geometric viewpoint, it seems that the same relationship should hold between the points (x_0, y_0), at which the function $z = f(x, y)$ has a relative extremum, and the *plane* tangent to the graph of $z = f(x, y)$ at $(x_0, y_0, f(x_0, y_0))$: At such points the tangent plane should either be horizontal or fail to exist. (Remember that our use of the term tangent plane here refers only to the plane determined by the partial derivatives $\frac{\partial f}{\partial x}(x_0, y_0)$ and $\frac{\partial f}{\partial y}(x_0, y_0)$—see Figure 5.4.)

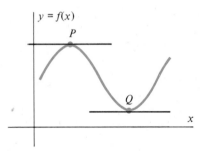

Figure 5.3 At relative extrema P or Q, $f'(x_0) = 0$ and tangent *line* is horizontal.

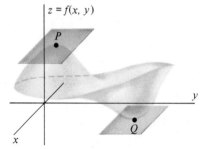

Figure 5.4 At relative extrema P or Q, $\frac{\partial f}{\partial x}(x_0, y_0) = \frac{\partial f}{\partial y}(x_0, y_0) = 0$ and tangent *plane* is horizontal.

Since a tangent plane is determined at the point $(x_0, y_0, f(x_0, y_0))$ by the partial derivatives $\frac{\partial f}{\partial x}(x_0, y_0)$ and $\frac{\partial f}{\partial y}(x_0, y_0)$, these geometric observations lead to the following theorem.

THEOREM 3 If the number $z_0 = f(x_0, y_0)$ is a relative extremum for the function $z = f(x, y)$ at the point (x_0, y_0), one of the following two conditions must hold:

(i) $\dfrac{\partial f}{\partial x}(x_0, y_0) = \dfrac{\partial f}{\partial y}(x_0, y_0) = 0$, or

(ii) one or both of $\dfrac{\partial f}{\partial x}(x_0, y_0)$ and $\dfrac{\partial f}{\partial y}(x_0, y_0)$ fails to exist.

Proof: First consider the case where $f(x_0, y_0)$ is a relative maximum. We will show that if statement (ii) is not true, then statement (i) must be true. Thus, we assume that both $\dfrac{\partial f}{\partial x}(x_0, y_0)$ and $\dfrac{\partial f}{\partial y}(x_0, y_0)$ exist, and we attempt to show that $\dfrac{\partial f}{\partial x}(x_0, y_0) = \dfrac{\partial f}{\partial y}(x_0, y_0) = 0$.

Holding $y = y_0$ fixed, we define the function (of a single variable) $g(x)$ by

$$g(x) = f(x, y_0).$$

Since $f(x_0, y_0)$ is a relative maximum, $f(x_0, y_0) \geq f(x, y)$ for all (x, y) near (x_0, y_0). Thus,

$$g(x_0) = f(x_0, y_0) \geq f(x, y_0) = g(x)$$

for all x near x_0. This shows that $g(x_0)$ is a relative maximum for the function $g(x)$. Since the derivative of $g(x)$ is

$$\begin{aligned} g'(x) &= \lim_{\Delta x \to 0} \frac{g(x + \Delta x) - g(x)}{\Delta x} \\ &= \lim_{\Delta x \to 0} \frac{f(x + \Delta x, y_0) - f(x, y_0)}{\Delta x} \\ &= \frac{\partial f}{\partial x}(x, y_0), \end{aligned}$$

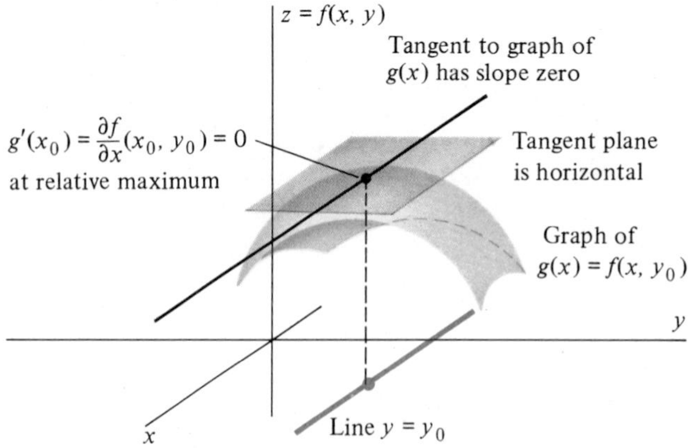

Figure 5.5 A relative maximum: If $\dfrac{\partial f}{\partial x}(x_0, y_0)$ exists, then $\dfrac{\partial f}{\partial x}(x_0, y_0) = 0$.

we know that $g'(x_0) = \dfrac{\partial f}{\partial x}(x_0, y_0)$ exists (Figure 5.5). But, according to Theorem 4, Section 5.4, if $g(x_0)$ is a relative maximum and $g'(x_0)$ exists, then $g'(x_0) = 0$. This shows that

$$\frac{\partial f}{\partial x}(x_0, y_0) = g'(x_0) = 0.$$

By repeating this argument with $x = x_0$ fixed, we can show that $\dfrac{\partial f}{\partial y}(x_0, y_0) = 0$ as well. Finally, we note that we have made no special use of the assumption that the extremum $f(x_0, y_0)$ was a relative maximum—the argument applies for relative minima as well. ■

Theorem 3 determines a procedure for finding relative extrema for $z = f(x, y)$: Find all points (x_0, y_0) where $\dfrac{\partial f}{\partial x}(x_0, y_0) = \dfrac{\partial f}{\partial y}(x_0, y_0) = 0$ or where either $\dfrac{\partial f}{\partial x}(x_0, y_0)$ or $\dfrac{\partial f}{\partial y}(x_0, y_0)$ fails to exist. (We call these points **critical points.**) Then test each critical point to determine whether it yields a relative extremum.

Example 3 For the function $f(x, y) = 2x + 4y - x^2 - y^2 - 1$, setting both partial derivatives equal to zero gives the equations

$$\frac{\partial f}{\partial x}(x, y) = 2 - 2x = 0, \qquad \frac{\partial f}{\partial y}(x, y) = 4 - 2y = 0.$$

The (simultaneous) solution of this pair of equations is $x = 1$, $y = 2$. Thus, condition (i) in Theorem 3 yields the single critical point $(1, 2)$. We have verified that $f(1, 2)$ is a relative maximum in Example 2. Since the partial derivatives are defined for all (x, y), there are no points satisfying condition (ii) of Theorem 3. Thus, the only relative extremum for this function is the relative maximum at $(1, 2)$ (see Figure 5.2). ■

Example 4 Find all relative extrema for the function

$$f(x, y) = \begin{cases} \sqrt{x^2 + y^2}; & (x, y) \neq (0, 0) \\ 0, & (x, y) = (0, 0). \end{cases}$$

Solution: The partial derivatives are

$$\frac{\partial f}{\partial x}(x, y) = \frac{x}{\sqrt{x^2 + y^2}}; \qquad \frac{\partial f}{\partial y}(x, y) = \frac{y}{\sqrt{x^2 + y^2}}.$$

Both partial derivatives are undefined at $(x, y) = (0, 0)$. For all other points at least one of the partial derivatives is nonzero. The graph of $f(x, y)$ in Figure 5.1 shows that $f(0, 0) = 0$ is a relative minimum. ■

It is important to understand that the conditions of Theorem 3 do not *guarantee* that $f(x_0, y_0)$ is a relative extremum. Theorem 3 merely provides *necessary* conditions for an extremum. Without satisfying condition (i) or (ii) of Theorem 3, $f(x_0, y_0)$ cannot be a relative extremum. However, some critical points do not yield extrema. The following is a typical such example.

Example 5 For the function $f(x, y) = y^2 - x^2$, the partial derivatives are

$$\frac{\partial f}{\partial x} = -2x; \qquad \frac{\partial f}{\partial y} = 2y.$$

Thus, $\frac{\partial f}{\partial x}(0, 0) = \frac{\partial f}{\partial y}(0, 0) = 0$. However, the number $f(0, 0) = 0$ is neither a relative maximum nor a relative minimum. To see this, we compare $f(0, 0)$ with $f(h, k)$ where (h, k) is a point near $(0, 0)$. Since

$$f(0, 0) - f(h, k) = 0 - (k^2 - h^2) = h^2 - k^2,$$

we see that the sign of this difference depends only on the relative sizes of $|h|$ and $|k|$. Since this difference is not of constant sign, $f(0, 0)$ is neither a maximum nor a minimum.

Figure 5.6 illustrates why $f(x, y) = y^2 - x^2$ does not have an extremum at $(0, 0)$. Although both partial derivatives are zero, the function $g(x) = f(x, 0) = -x^2$ reaches a relative *maximum* while the function $h(y) = f(0, y) = y^2$ reaches a relative *minimum* at $(0, 0)$. ■

In Example 5, the point $(0, 0)$ is called a *saddle point* for rather obvious reasons. The surface bows upward along one axis and downward along another. More generally, a point (x_0, y_0) in the domain of a function of two variables is called a **saddle point** if (x_0, y_0) is a critical point and if $f(x_0, y_0)$ is neither a relative maximum nor a relative minimum. Figure 5.7 shows the graph of another function that has a saddle point at $(0, 0)$ (see Exercise 20).

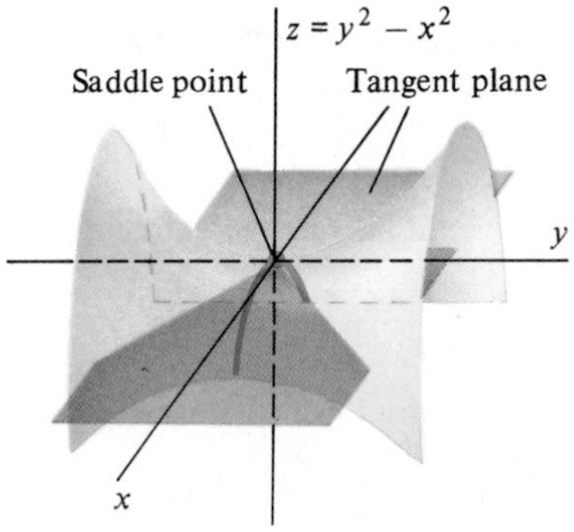

Figure 5.6 Graph of $z = y^2 - x^2$ has a *saddle* point, but no extremum, at $(0, 0)$.

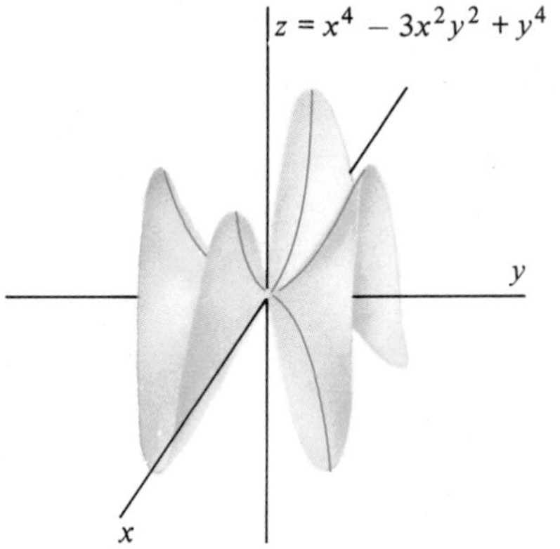

Figure 5.7 Graph of $f(x, y) = x^4 - 3x^2y^2 + y^4$ has a saddle point at $(0, 0)$.

Second Derivative Test

There is a theorem that helps sort out actual extrema from saddle points. It may be regarded as the analogue of the Second Derivative Test for functions of a single variable. A proof of this theorem may be found in texts on advanced calculus.

THEOREM 4
Second Derivative Test

Suppose that all second order partial derivatives of $f(x, y)$ are continuous in a neighborhood of (x_0, y_0) and that $\frac{\partial f}{\partial x}(x_0, y_0) = \frac{\partial f}{\partial y}(x_0, y_0) = 0$. Let

$$A = \frac{\partial^2 f}{\partial x^2}(x_0, y_0), \qquad B = \frac{\partial^2 f}{\partial y \partial x}(x_0, y_0), \qquad C = \frac{\partial^2 f}{\partial y^2}(x_0, y_0)$$

and

$$D = B^2 - AC.$$

Then

(i) If $D < 0$ and $A < 0$, $f(x_0, y_0)$ is a relative maximum.
(ii) If $D < 0$ and $A > 0$, $f(x_0, y_0)$ is a relative minimum.
(iii) If $D > 0$, (x_0, y_0) is a saddle point.
(iv) If $D = 0$, no conclusions may be drawn.

Theorem 4 verifies that the critical point (0, 0) in Example 5 is a saddle point, since

$$A = \frac{\partial^2 f}{\partial x^2}(0, 0) = -2, \qquad B = \frac{\partial^2 f}{\partial y \partial x}(0, 0) = 0, \qquad C = \frac{\partial^2 f}{\partial y^2}(0, 0) = 2$$

and

$$D = B^2 - AC = 4 > 0.$$

Similarly, for the function $f(x, y)$ in Example 3 and the critical point (1, 2), we have

$$A = \frac{\partial^2 f}{\partial x^2}(1, 2) = -2, \qquad B = \frac{\partial^2 f}{\partial y \partial x}(1, 2) = 0, \qquad C = \frac{\partial^2 f}{\partial y^2}(1, 2) = -2$$

and

$$D = B^2 - AC = -4.$$

Thus, since $D < 0$ and $A < 0$, the Second Derivative Test agrees with the conclusion of Example 3, that $f(1, 2)$ is a relative maximum.

Example 6 Find and classify all relative extrema for the function $f(x, y) = x^4 + y^4 - 4xy$.

Solution: The partial derivatives are

$$\frac{\partial f}{\partial x} = 4x^3 - 4y; \qquad \frac{\partial f}{\partial y} = 4y^3 - 4x.$$

Since both partial derivatives are defined for all (x, y), the extrema can occur only at points where

$$\frac{\partial f}{\partial x} = 4x^3 - 4y = 0, \qquad \text{or} \qquad y = x^3 \tag{3}$$

and

$$\frac{\partial f}{\partial y} = 4y^3 - 4x = 0, \qquad \text{or} \qquad x = y^3. \tag{4}$$

Substituting for x in (3) using (4) gives the equation $y = y^9$. Thus, either $y = 0$, or else $y^8 = 1$, which gives $y = \pm 1$. Using (4) we find that $x = 0$ if $y = 0$, $x = 1$ if $y = 1$, and $x = -1$ if $y = -1$. The three critical points are therefore $(0, 0)$, $(1, 1)$, and $(-1, -1)$.

Next, we calculate the second order partials:

$$\frac{\partial^2 f}{\partial x^2} = 12x^2; \qquad \frac{\partial^2 f}{\partial y \partial x} = -4; \qquad \frac{\partial^2 f}{\partial y^2} = 12y^2.$$

At the critical point $(1, 1)$ we have

$$A = \frac{\partial^2 f}{\partial x^2}(1, 1) = 12, \qquad B = \frac{\partial^2 f}{\partial y \partial x}(1, 1) = -4, \qquad C = \frac{\partial^2 f}{\partial y^2}(1, 1) = 12$$

and

$$D = B^2 - AC = 16 - 12 \cdot 12 = -128 < 0.$$

Thus, since $D < 0$ and $A > 0$, $f(1, 1) = -2$ is a relative minimum, according to Theorem 4.

At the critical point $(-1, -1)$ the values of A, B, C, and D are the same as for $(1, 1)$ so $f(-1, -1) = -2$ is also a relative minimum.

At the critical point $(0, 0)$, $A = C = 0$ and $B = -4$, so $D = B^2 - AC = 16 > 0$. Thus, $(0, 0)$ is a saddle point. ■

In each of Examples 7, 8, and 9, we obtain $D = 0$ at the critical point. These three examples show that any of the three possible outcomes (relative maximum, relative minimum, saddle point) may result when $D = 0$.

Example 7 For the function $f(x, y) = e^{-(x^4+y^4)}$, we have

$$\frac{\partial f}{\partial x} = -4x^3 e^{-(x^4+y^4)} \qquad \frac{\partial f}{\partial y} = -4y^3 e^{-(x^4+y^4)}$$

$$\frac{\partial^2 f}{\partial x^2} = (16x^6 - 12x^2)e^{-(x^4+y^4)} \qquad \frac{\partial^2 f}{\partial y^2} = (16y^6 - 12y^2)e^{-(x^4+y^4)}$$

$$\frac{\partial^2 f}{\partial x \partial y} = 16x^3 y^3 e^{-(x^4+y^4)}.$$

The point $(0, 0)$ is a critical point since $\frac{\partial f}{\partial x}(0, 0) = \frac{\partial f}{\partial y}(0, 0) = 0$. At this critical point

$$A = \frac{\partial^2 f}{\partial x^2}(0, 0) = 0; \qquad B = \frac{\partial^2 f}{\partial y \partial x}(0, 0) = 0; \qquad C = \frac{\partial^2 f}{\partial y^2}(0, 0) = 0.$$

Thus, the Second Derivative Test yields no conclusion about the critical point $(0, 0)$. However, it is easy to see that the expression $x^4 + y^4$ has a minimum at $(0, 0)$, so

$$f(x, y) = e^{-(x^4+y^4)} = \frac{1}{e^{x^4+y^4}}$$

has a relative maximum at (0, 0) (see Figure 5.8). ■

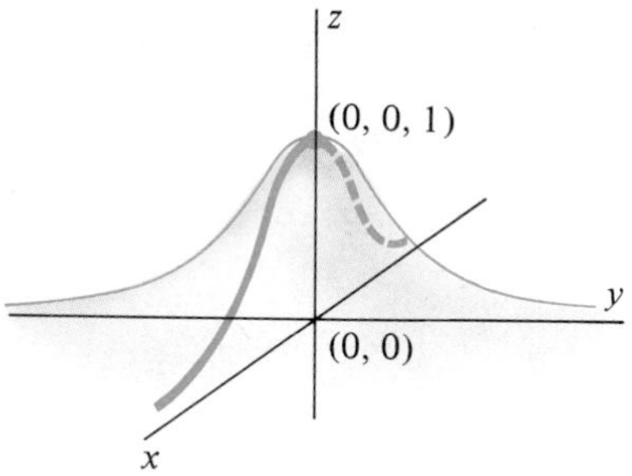

Figure 5.8 $f(x, y) = e^{-(x^4+y^4)}$ has a relative maximum at (0, 0).

Example 8 The function $f(x, y) = x^4 + y^4$ obviously has a relative minimum of $f(0, 0) = 0$, since $f(x, y) > 0$ at all other points. However, we find

$$A = \frac{\partial^2 f}{\partial x^2}(0, 0) = 12x^2\Big|_{x=0} = 0, \qquad B = \frac{\partial^2 f}{\partial y \partial x}(0, 0) = 0,$$

$$C = \frac{\partial^2 f}{\partial y^2}(0, 0) = 12y^2\Big|_{y=0} = 0, \qquad D = B^2 - AC = 0.$$

Thus, the Second Derivative Test fails to classify this critical point. ■

Example 9 For the function $f(x, y) = x^3 - y^3$, the only simultaneous solution of the two equations

$$\frac{\partial f}{\partial x}(x, y) = 3x^2 = 0 \qquad \text{and} \qquad \frac{\partial f}{\partial y}(x, y) = -3y^2 = 0$$

is $x = y = 0$, so (0, 0) is the only critical point. Since

$$\frac{\partial^2 f}{\partial x^2} = 6x, \qquad \frac{\partial^2 f}{\partial y \partial x} = 0, \qquad \frac{\partial^2 f}{\partial y^2} = -6y,$$

we have $A = B = C = D = 0$ at the critical point (0, 0).

Thus, the test gives no information as to the nature of this critical point. However, since $f(x, y) = x^3 - y^3$ takes on both positive and negative values in every neighborhood of (0, 0), the critical point (0, 0) is a saddle point. ■

Example 10 A builder wishes to design a rectangular house containing V cubic meters of heated space, so as to minimize heating costs. One wall of the building is to face south. The annual heating costs are estimated to be \$4 per square meter of floor space, \$3 per square meter for all exterior wall not facing south, and \$2 per square meter for exterior wall space facing south. What dimensions will produce the most energy-efficient building?

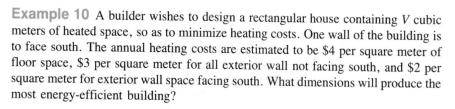

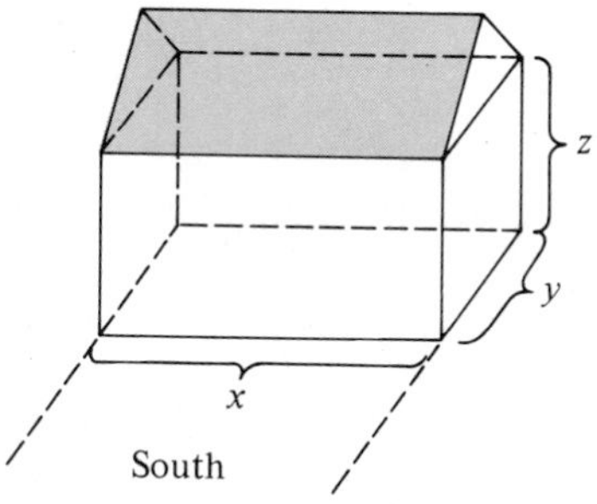

Figure 5.9

Solution: As in Figure 5.9, we let x denote the length of the wall facing south, and we refer to this dimension as the width of the house. We let y be the depth, and we let z be the height of the heated portion of the house. Since the area of the floor is xy, the area of each side wall is yz, and the area of each front and rear wall is xz, the annual heating cost is

$$C(x, y, z) = 4xy + 3(2yz) + 3xz + 2xz$$

- $2xz$: Cost for south wall
- $3xz$: Cost for rear (north) wall
- $3(2yz)$: Costs for side walls (both)
- $4xy$: Costs for roof (floor space)

$$= 4xy + 6yz + 5xz.$$

Since $V = xyz$, we have $z = \dfrac{V}{xy}$, and we may eliminate one of the three variables in $C(x, y, z)$ by this substitution:

$$\begin{aligned} C(x, y) &= 4xy + 6y\left(\frac{V}{xy}\right) + 5x\left(\frac{V}{xy}\right) \\ &= 4xy + \frac{6V}{x} + \frac{5V}{y}. \end{aligned}$$

To obtain the minimum value of this function of two variables, we set each of its partial derivatives equal to zero. We obtain

$$\frac{\partial C}{\partial x}(x, y) = 4y - \frac{6V}{x^2} = 0$$

or

$$y = \frac{3V}{2x^2} \tag{5}$$

and

$$\frac{\partial C}{\partial y}(x, y) = 4x - \frac{5V}{y^2} = 0$$

so

$$x = \frac{5V}{4y^2}. \tag{6}$$

Substituting the expression for y in (5) into equation (6) gives the equation

$$\begin{aligned} x &= \frac{5V}{4\left(\dfrac{3V}{2x^2}\right)^2} \\ &= \frac{5x^4}{9V}. \end{aligned}$$

Thus, $9Vx = 5x^4$, so $x = \sqrt[3]{\dfrac{9V}{5}}$. Returning to equation (5) we find that

$$y = \frac{3V}{2x^2} = \frac{3V}{2\left(\dfrac{9V}{5}\right)^{2/3}} = \frac{3}{2}\cdot\frac{\sqrt[3]{\dfrac{9V}{5}}}{\dfrac{9}{5}} = \frac{3}{2}\left(\frac{5x}{9}\right) = \frac{5}{6}x$$

and

$$z = \frac{V}{xy} = \frac{V}{x\left(\dfrac{5}{6}x\right)} = \frac{V}{\dfrac{5}{6}\left(\dfrac{9}{5}\right)^{2/3} V^{2/3}} = \frac{6}{5}\left(\frac{\sqrt[3]{\dfrac{9V}{5}}}{\dfrac{9}{5}}\right) = \frac{2}{3}x.$$

We leave it to you (Exercise 42) to verify that these dimensions actually correspond to the minimum value of $C(x, y, z)$, and that the optimal design requires a depth equal to five sixths the width and a height equal to two thirds the width. ■

Functions of More Than Two Variables

The discussion of this section has focussed on the question of finding relative extrema for functions of two variables. For functions of more than two variables, the statement of Theorem 3 generalizes directly:

(a) If the function of three variables $w = f(x, y, z)$ has a relative extremum at (x_0, y_0, z_0), then either

$$\frac{\partial f}{\partial x} = \frac{\partial f}{\partial y} = \frac{\partial f}{\partial z} = 0$$

at this point, or else one or more of these partial derivatives fails to exist at (x_0, y_0, z_0).

(b) More generally, the function $y = f(x_1, x_2, \ldots, x_n)$ of n variables can have a relative extremum at the point $(a_1, a_2, \ldots, a_n)$ only if one or more of the partial derivatives fails to exist, or if

$$\frac{\partial f}{\partial x_1} = \frac{\partial f}{\partial x_2} = \cdots = \frac{\partial f}{\partial x_n} = 0$$

at this point.

However, the Second Derivative Test does not generalize quite so easily. For functions of more than two variables, the classification of critical points is considerably more difficult and will not be pursued here.

Extrema on Closed Sets

Finally, we note that we have not addressed the question of *absolute* extrema, as we did for functions of a single variable in Chapter 4. In a manner analogous to the behavior of a function of one variable on a closed interval, a continuous function $z = f(x, y)$ of two variables will assume both an absolute maximum and an absolute minimum on any **closed bounded set** S in the plane. By a closed set in the plane we mean (roughly speaking) a set that includes its boundary, such as the disc $D = \{(x, y) | x^2 + y^2 \le 1\}$ or the square $S = \{(x, y) | 0 \le x \le 1, 0 \le y \le 1\}$. As you might suspect, such absolute extrema will occur either at critical points or at points on the boundary of S. Checking for extrema among boundary points can be a quite complicated task, so we have chosen to defer this topic to more advanced courses where appropriate theory and techniques are developed. (However, this issue is addressed in simple settings in Exercises 29–32.)

Exercise Set 19.5

In Exercises 1–16, find all critical points. Classify each as a relative maximum, relative minimum, or saddle point.

1. $f(x, y) = x^2 + y^2 + 4y + 4$
2. $f(x, y) = x^2 + y^2 + 4x - 2y + 11$
3. $f(x, y) = x^2 - y^2 + 6x + 4y + 5$
4. $f(x, y) = 7 - 2x + 2y - x^2 - y^2$
5. $f(x, y) = xy + 9$
6. $f(x, y) = x^2 + y^4 - 2x - 4y^2 + 5$
7. $f(x, y) = 5x^2 + y^2 - 10x - 6y + 15$
8. $f(x, y) = x^2 + y^3 - 3y$
9. $f(x, y) = x^3 - y^3$
10. $f(x, y) = e^{x^2-2x+y^2+4}$
11. $f(x, y) = x^2 - xy$
12. $f(x, y) = e^x \cos y$
13. $f(x, y) = x^3 + y^3 + 4xy$

14. $f(x, y) = x^4 + y^4 - 4xy$

15. $f(x, y) = x \cos y$

16. $f(x, y) = \dfrac{y}{x} - \dfrac{x}{y}$

17. Show that the function $z = 4 - \sqrt{x^2 + y^2}$ has a relative maximum at $(0, 0)$. (Note that none of the partial derivatives exists at this point.)

18. Show that the function $z = x^2 + y^2 - 4x - 2y + 9$ has a relative minimum at $(2, 1)$ using the method of Example 2.

19. Find the highest point on the graph of $z = 2y^3 - 3x^2 - 3xy + 9x$ if the positive z-axis is upward.

20. Show that the function $f(x, y) = x^4 - 3x^2y^2 + y^4$ has a saddle point at $(0, 0)$ (see Figure 5.7).

21. Show that the function $f(x, y) = x^2 - y^2 + 2x + 4y - 3$ has a saddle point at $(-1, 2)$ by the method of Example 2.

22. Consider the function $f(x, y) = \dfrac{x^2 + y^2}{(x + y)^2}$.

a. Show that $\dfrac{\partial f}{\partial x}(x, y) = \dfrac{\partial f}{\partial y}(x, y) = 0$ if and only if $y = x$ and $x \neq 0$.

b. Show that $f(x, y)$ has a relative minimum at each point (x, x) with $x \neq 0$.

c. Show that the function $g(\lambda) = f(x, \lambda x) = \dfrac{1 + \lambda^2}{(1 + \lambda)^2}$, $\lambda \neq -1$, has a relative minimum at $\lambda = 1$.

d. Explain the geometric and algebraic relationships between the functions $g(\lambda)$ and $f(x, y)$.

23. A rectangular box, with a top, is to hold 16 cubic meters. Find the dimensions that produce the least expensive box if the material for the side walls is half as expensive as the material for the top and the bottom.

24. Find the point on the plane $2x + 3y + z - 14 = 0$ nearest the origin.

25. Find the point on the plane with equation $ax + by + cz = d$ nearest the origin.

26. Find the dimensions of the closed rectangular box of volume $V = 8000$ cm^3 and of minimum surface area.

27. Find the dimensions of the rectangular package of largest volume that can be mailed under the restrictions that length plus girth cannot exceed 84 inches. (Girth is the perimeter of the cross section taken perpendicular to the length.)

28. A manufacturer of tape recorders intends to market x recorders through retail outlets and y recorders through wholesale outlets. The manufacturer estimates that the revenue per recorder sold retail will be $r(x, y) = 250 - \dfrac{x}{20} - \dfrac{y}{5}$, while the revenue per recorder on the wholesale market is estimated to be $w(x, y) = 200 - \dfrac{x}{50} - \dfrac{y}{20}$. The cost of producing the recorders is known to be \$100 each. Using the model Profit = Revenue − Cost, find the number of recorders that should be marketed retail and the number that should be marketed wholesale so as to maximize profit.

29. Find the *absolute* maximum and minimum values of the function $f(x, y) = 2x + 2y + 4$ on the *closed* disc $D = \{(x, y) | x^2 + y^2 \le 1\}$. (*Hint:* To check for absolute extrema along the boundary $C = \{(x, y) | x^2 + y^2 = 1\}$ use the parameterization $x = \cos t$, $y = \sin t$, and optimize $f(x(t), y(t))$ as a function of t.)

30. Find the absolute maximum and minimum values of the function $f(x, y) = 3x - 2y^3 + 2$ on the (closed) square $S = \{(x, y) | 0 \le x \le 1, 0 \le y \le 1\}$ (see Exercise 29).

31. Find the absolute maximum and minimum values of the function $z = x + 2y + 6$ on the closed set consisting of the ellipse $4x^2 + 2y^2 = 4$ and its interior (see Exercise 29).

32. Find the absolute maximum and minimum values for the function $f(x, y) = x^2 + 2y^2 - 2xy - 6x + 4y$ on the rectangle $0 \le x \le 8$, $-2 \le y \le 2$.

33. Given a set of points $(x_1, y_1), (x_2, y_2), \ldots, (x_n, y_n)$, one way to define a line $y = mx + b$ that "best fits" these points is given by the Method of Least Squares. As illustrated by Figure 5.10, the distance from the data point (x_j, y_j) to the point on the line $y = mx + b$ with x-coordinate x_j is

$$|y_j - (mx_j + b)| = |y_j - mx_j - b|.$$

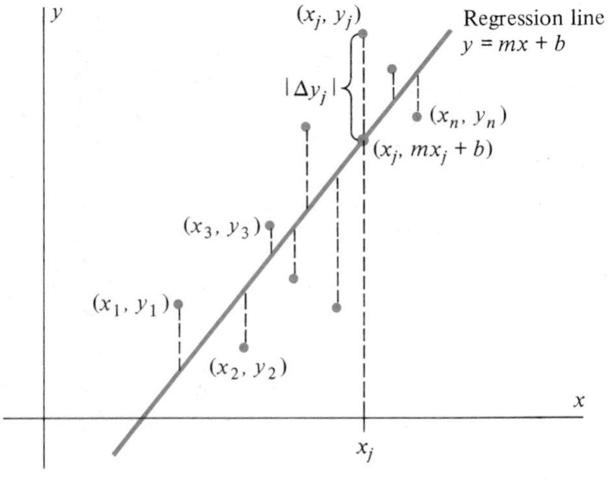

Figure 5.10 The Method of Least Squares determines the line that "best fits" the data points (x_1, y_1), $(x_2, y_2), \ldots, (x_n, y_n)$.

The Method of Least Squares defines the best fitting line $y = mx + b$ (called the **regression line**) to be the line that minimizes the sum of the squares of these individual distances:

$$S(m, b) = \sum_{j=1}^{n} (y_j - mx_j - b)^2.$$

Viewing $x_1, x_2, \ldots, x_n$ and $y_1, y_2, \ldots, y_n$ as fixed constants and m and b as independent variables, show that the values of m and b for which $S(m, b)$ is a minimum are given by the formulas

$$m = \frac{n\sum_{j=1}^{n} x_jy_j - \left(\sum_{j=1}^{n} x_j\right)\left(\sum_{j=1}^{n} y_j\right)}{n\sum_{j=1}^{n} x_j^2 - \left(\sum_{j=1}^{n} x_j\right)^2}$$

$$b = \frac{\left(\sum_{j=1}^{n} x_j^2\right)\left(\sum_{j=1}^{n} y_j\right) - \left(\sum_{j=1}^{n} x_j\right)\left(\sum_{j=1}^{n} x_jy_j\right)}{n\sum_{j=1}^{n} x_j^2 - \left(\sum_{j=1}^{n} x_j\right)^2}$$

(*Hint:* Set $\frac{\partial S}{\partial m}(m, b) = 0$ and $\frac{\partial S}{\partial b}(m, b) = 0$ and solve the resulting system of equations.)

34. For a group of $n = 6$ calculus students, achievement test scores (x_j) and final course averages in calculus y_j were as follows:

j	1	2	3	4	5	6
x_j	52	46	69	54	61	48
y_j	74	66	94	91	84	80

The regression line for these data appears in Figure 5.11. Use the result of Exercise 33 to show that this regression line has equation $y = 0.95x + 29$.

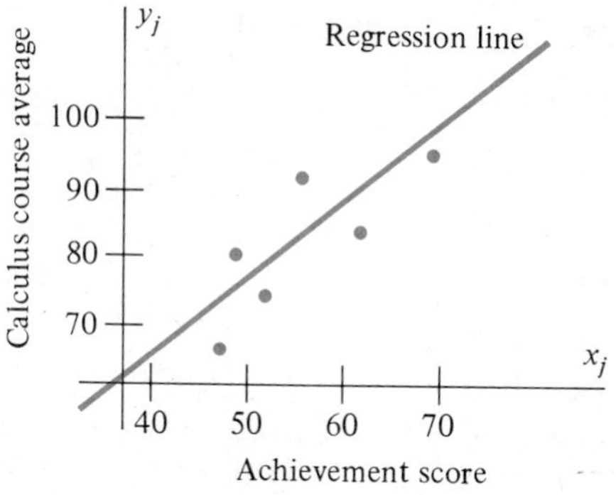

Figure 5.11 Achievement scores versus calculus course averages for six students.

35. A regression line corresponding to a set of data points $(x_1, y_1), (x_2, y_2), \ldots, (x_n, y_n)$ provides a model for *predicting* a value of y corresponding to any particular value of x. Use the regression line obtained in Exercise 34 to predict the course average for a student whose achievement score is $x = 60$.

36. Given the five points whose coordinates are:

x	0	1	1	2	3
y	2	4	3	6	6

a. Find the regression line for the data using the Method of Least Squares.
b. Plot the data points and the regression line.
c. Find the predicted value for $x = 4$.

37. Repeat Exercise 36 for the data below.

x	6	8	9	10	12
y	2	5	5	7	9

38. For data consisting of ordered triples $(x_1, y_1, z_1), \ldots, (x_n, y_n, z_n)$, the analogue of the regression line is the (regression) **plane** $z = ax + by + c$. This plane is defined as the plane that minimizes the sum of squares

$$S = \sum_{j=1}^{n} [z_j - (ax_j + by_j + c)]^2.$$

Show that the coefficients a, b, and c in this equation satisfy the equations

$$\text{(i) } a\sum_{j=1}^{n} x_j + b\sum_{j=1}^{n} y_j + nc = \sum_{j=1}^{n} z_j,$$

$$\text{(ii) } a\sum_{j=1}^{n} x_j^2 + b\sum_{j=1}^{n} x_jy_j + c\sum_{j=1}^{n} x_j = \sum_{j=1}^{n} x_jy_j,$$

$$\text{(iii) } a\sum_{j=1}^{n} x_jy_j + b\sum_{j=1}^{n} y_j^2 + c\sum_{j=1}^{n} y_j = \sum_{j=1}^{n} y_jz_j.$$

39. Recall, from Section 8.4, that the *centroid* of the n data points $(x_1, y_1), (x_2, y_2), \ldots, (x_n, y_n)$ is the point $(\bar{x}, \bar{y})$ where

$$\bar{x} = \frac{1}{n}\sum_{j=1}^{n} x_j, \qquad \bar{y} = \frac{1}{n}\sum_{j=1}^{n} y_j.$$

Show that the centroid lies on the regression line determined by the Method of Least Squares.

40. Suppose that you suspected that the n data points $(x_1, y_1), (x_2, y_2), \ldots, (x_n, y_n)$ could be "fitted" well by a curve of the form $y = m \ln x + b$ (a common situation in biology and chemistry). Explain how the Method of Least Squares could be used to find the constants m and b.

41. Prove Theorem 4 in the case of the function $f(x, y) = ax^2 + byx + cy^2$.

42. Verify that the critical point in Example 10 actually corresponds to an absolute minimum.

19.6 APPROXIMATIONS AND DIFFERENTIALS

For the function of a single variable $y = f(x)$, the existence of the derivative $f'(x)$ leads to the approximation formula

$$f(x + \Delta x) \approx f(x) + f'(x)\Delta x \tag{1}$$

and to the definition of the differential

$$df = f'(x)\, dx = f'(x)\Delta x \tag{2}$$

where $dx = \Delta x$. In this section, we take up the analogous issues for functions of several variables. The key to the entire discussion (and to the discussions of the sections that follow) is the Approximation Theorem. Again, note the important role played by the Mean Value Theorem in the proof of this theorem.

THEOREM 5
Approximation Theorem

Let the function $f(x, y)$ and its first partial derivatives $\frac{\partial f}{\partial x}(x, y)$ and $\frac{\partial f}{\partial y}(x, y)$ be continuous in an open rectangle $R = \{(x, y)|a_1 < x < a_2,\ b_1 < y < b_2\}$ in the xy-plane. Let the point (x_0, y_0) and the point $(x_0 + \Delta x, y_0 + \Delta y)$ both lie in R. Then

$$f(x_0 + \Delta x, y_0 + \Delta y) = f(x_0, y_0) + \frac{\partial f}{\partial x}(x_0, y_0)\Delta x + \frac{\partial f}{\partial y}(x_0, y_0)\Delta y + \epsilon_1\Delta x + \epsilon_2\Delta y \tag{3}$$

where $\lim_{\substack{\Delta x \to 0 \\ \Delta y \to 0}} \epsilon_1 = 0$ and $\lim_{\substack{\Delta x \to 0 \\ \Delta y \to 0}} \epsilon_2 = 0$.

Theorem 5 provides the relationship between the value of the function $f(x, y)$ at (x_0, y_0) and the value of this function at the "nearby" point $(x_0 + \Delta x, y_0 + \Delta y)$. It is the analogue of Theorem 14, Chapter 3, which showed that

$$f(x_0 + \Delta x) = f(x_0) + f'(x_0)\Delta x + \epsilon \cdot \Delta x \tag{4}$$

with $\lim_{\Delta x \to 0} \epsilon = 0$ when the function $y = f(x)$ is differentiable at $x = x_0$. By ignoring the "error term" $\epsilon \cdot \Delta x$ in (4), we obtained the approximation (1) for functions of a single variable. Similarly, ignoring the error term $(\epsilon_1\, \Delta x + \epsilon_2\, \Delta y)$ in equation (3) leads to the approximation

$$\boxed{f(x_0 + \Delta x, y_0 + \Delta y) \approx f(x_0, y_0) + \frac{\partial f}{\partial x}(x_0, y_0)\, \Delta x + \frac{\partial f}{\partial y}(x_0, y_0)\, \Delta y} \tag{5}$$

for functions of two variables.

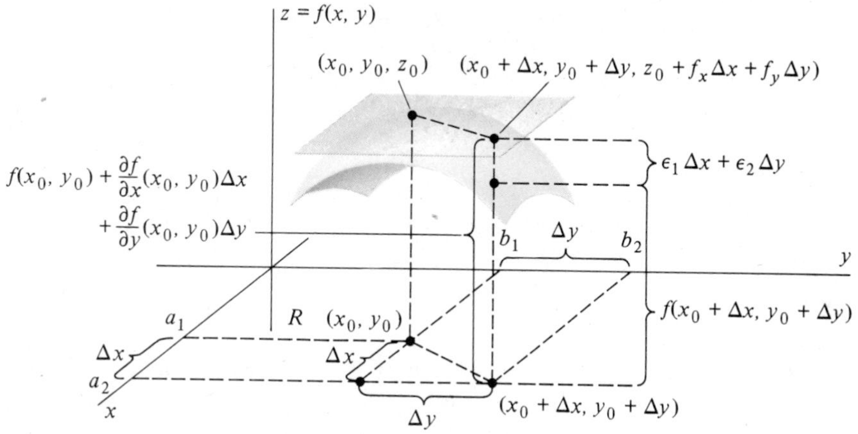

Figure 6.1 Point on tangent plane approximates point on graph of $z = f(x, y)$ above point $(x_0 + \Delta x, y_0 + \Delta y)$.

Figure 6.1 provides a geometric interpretation of approximation (5). Since both $\frac{\partial f}{\partial x}(x_0, y_0)$ and $\frac{\partial f}{\partial y}(x_0, y_0)$ exist, they determine a "tangent" plane with equation

$$z = \frac{\partial f}{\partial x}(x_0, y_0)(x - x_0) + \frac{\partial f}{\partial y}(x_0, y_0)(y - y_0) + f(x_0, y_0) \tag{6}$$

(equation (3), Section 19.4). To find the point on this plane above $(x, y) = (x_0 + \Delta x, y_0 + \Delta y)$, we set $x = x_0 + \Delta x$ and $y = y_0 + \Delta y$. Then $x - x_0 = \Delta x$ and $y - y_0 = \Delta y$, so (6) gives the z-coordinate as

$$z_T = \frac{\partial f}{\partial x}(x_0, y_0)\Delta x + \frac{\partial f}{\partial y}(x_0, y_0)\Delta y + f(x_0, y_0),$$

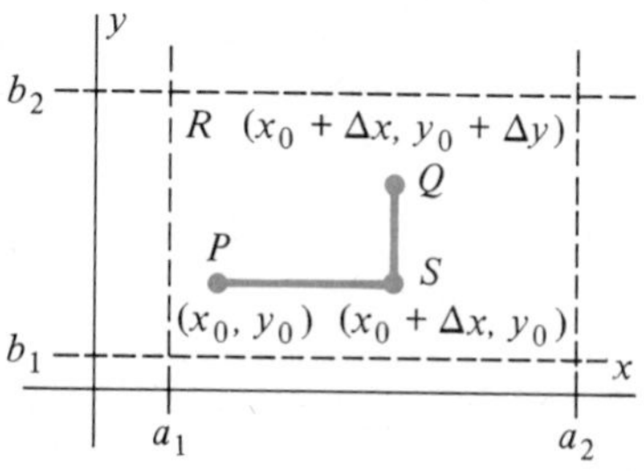

Figure 6.2 Diagram for proof of Theorem 5.

which is precisely the right-hand side of approximation (5). Thus, the left-hand side of approximation (5) is the z-coordinate of the point above $(x_0 + \Delta x, y_0 + \Delta y)$ *on the graph of* $z = f(x, y)$, the right-hand side of (5) is the z-coordinate of the point above $(x_0 + \Delta x, y_0 + \Delta y)$ *on the tangent plane*, and the difference between these two numbers is the "error term," $\epsilon_1 \Delta x + \epsilon_2 \Delta y$.

To help you better understand this observation, we rewrite equation (3) in Theorem 5 as follows:

$$\underbrace{f(x_0 + \Delta x, y_0 + \Delta y)}_{\substack{z\text{-coordinate of}\\ \text{point on graph of}\\ z = f(x,y) \text{ above}\\ (x_0+\Delta x,\, y_0+\Delta y).}} = \underbrace{f(x_0, y_0) + \frac{\partial f}{\partial x}(x_0, y_0)\,\Delta x + \frac{\partial f}{\partial y}(x_0, y_0)\,\Delta y}_{\substack{z\text{-coordinate of point on tangent plane}\\ \text{determined by } \frac{\partial f}{\partial x}(x_0,y_0) \text{ and } \frac{\partial f}{\partial y}(x_0,y_0)\\ \text{above } (x_0+\Delta x,\, y_0+\Delta y).}} + \underbrace{\epsilon_1\,\Delta x + \epsilon_2\,\Delta y}_{\substack{\text{Error term,}\\ \text{which tends}\\ \text{to zero as}\\ \Delta x \to 0 \text{ and}\\ \Delta y \to 0.}}$$

Proof of Theorem 5: Let $P = (x_0, y_0)$ and $Q = (x_0 + \Delta x, y_0 + \Delta y)$. Since R is a rectangle, the point $S = (x_0 + \Delta x, y_0)$ also lies in R, as do the line segments PS and SQ (see Figure 6.2).

Along the line segment PS the variable $y = y_0$ is fixed, so $g(x) = f(x, y_0)$ is a function of x alone, and $g'(x) = \dfrac{\partial f}{\partial x}(x, y_0)$ exists for all $x \in [x_0, x_0 + \Delta x]$. Thus, by the Mean Value Theorem, there exists a number $c \in [x_0, x_0 + \Delta x]$ so that

$$\frac{f(x_0 + \Delta x, y_0) - f(x_0, y_0)}{\Delta x} = \frac{\partial f}{\partial x}(c, y_0). \tag{7}$$

Similarly, along the line segment SQ, the function $h(y) = f(x_0 + \Delta x, y)$ is a function of y alone, and $h'(y) = \dfrac{\partial f}{\partial y}(x_0 + \Delta x, y)$ for each $y \in [y_0, y_0 + \Delta y]$. Again by the Mean Value Theorem, there exists a number $d \in [y_0, y_0 + \Delta y]$ so that

$$\frac{f(x_0 + \Delta x, y_0 + \Delta y) - f(x_0 + \Delta x, y_0)}{\Delta y} = \frac{\partial f}{\partial y}(x_0 + \Delta x, d). \tag{8}$$

Next we multiply both sides of equation (7) by Δx and both sides of equation (8) by Δy. Adding the resulting equations gives

$$\begin{aligned} f(x_0 + \Delta x, y_0 + \Delta y) - f(x_0, y_0) &= \frac{\partial f}{\partial x}(c, y_0)\Delta x \\ &\quad + \frac{\partial f}{\partial y}(x_0 + \Delta x, d)\Delta y. \end{aligned} \tag{9}$$

We now invoke the continuity of the partial derivatives. Since c lies between x_0 and $x_0 + \Delta x$, we have $c \to x_0$ as $\Delta x \to 0$. Thus, since $\dfrac{\partial f}{\partial x}(x, y)$ is continuous,

$$\lim_{\Delta x \to 0} \frac{\partial f}{\partial x}(c, y_0) = \frac{\partial f}{\partial x}(x_0, y_0). \tag{10}$$

Another way to write equation (10) is simply

$$\frac{\partial f}{\partial x}(c, y_0) = \frac{\partial f}{\partial x}(x_0, y_0) + \epsilon_1 \tag{11}$$

where $\epsilon_1 \to 0$ as $\Delta x \to 0$. Similarly, since d lies between y_0 and $y_0 + \Delta y$, $d \to y_0$ as $\Delta y \to 0$. Thus, since $\dfrac{\partial f}{\partial y}(x, y)$ is continuous, we conclude that

$$\lim_{\substack{\Delta x \to 0 \\ \Delta y \to 0}} \frac{\partial f}{\partial y}(x_0 + \Delta x, d) = \frac{\partial f}{\partial y}(x_0, y_0),$$

or, that

$$\frac{\partial f}{\partial y}(x_0 + \Delta x, d) = \frac{\partial f}{\partial y}(x_0, y_0) + \epsilon_2 \tag{12}$$

where $\epsilon_2 \to 0$ as $\Delta x \to 0$ and $\Delta y \to 0$.

Combining statements (9), (11), and (12) now gives

$$f(x_0 + \Delta x, y_0 + \Delta y) = f(x_0, y_0) + \frac{\partial f}{\partial x}(x_0, y_0)\Delta x + \frac{\partial f}{\partial y}(x_0, y_0)\Delta y + \epsilon_1 \Delta x + \epsilon_2 \Delta y$$

where $\epsilon_1 \to 0$ and $\epsilon_2 \to 0$ as $\Delta x \to 0$ and $\Delta y \to 0$ as desired. ■

Example 1 To see how this all works out in a particular example let us consider the specific function $f(x, y) = 2x^2 + 4y^2$ and the problem of calculating $f(1 + \Delta x, 2 + \Delta y)$. Here

$$(x_0, y_0) = (1, 2)$$

$$f(x_0, y_0) = f(1, 2) = 2 \cdot 1^2 + 4 \cdot 2^2 = 18$$

$$\frac{\partial f}{\partial x}(x_0, y_0) = 4x \Big|_{x=1} = 4$$

$$\frac{\partial f}{\partial y}(x_0, y_0) = 8y \Big|_{y=2} = 16$$

and

$$\begin{aligned} f(1 + \Delta x, 2 + \Delta y) &= 2(1 + \Delta x)^2 + 4(2 + \Delta y)^2 \\ &= 2[1 + 2\Delta x + \Delta x^2] + 4[4 + 4\Delta y + \Delta y^2] \\ &= 18 + 4\Delta x + 16\Delta y + 2\Delta x^2 + 4\Delta y^2 \\ &= \underbrace{18}_{f(1,2)} + \underbrace{4}_{\frac{\partial f}{\partial x}(1,2)}\Delta x + \underbrace{16}_{\frac{\partial f}{\partial y}(1,2)}\Delta y + \underbrace{(2\Delta x)}_{\epsilon_1}\Delta x + \underbrace{(4\Delta y)}_{\epsilon_2}\Delta y. \end{aligned} \tag{13}$$

Also, the approximation (5) is

$$f(1 + \Delta x, 2 + \Delta y) \approx 18 + 4\Delta x + 16\Delta y. \tag{14}$$

By comparing (13) and (14), we can see that the error in this approximation is

$$\epsilon_1 \Delta x + \epsilon_2 \Delta y = 2\Delta x^2 + 4\,\Delta y^2.$$

■

Example 2 Use approximation (5) to estimate the value of the expression $\sqrt{(3.04)^2 + (3.95)^2}$.

Solution: If we let $f(x, y) = \sqrt{x^2 + y^2}$, the problem is then to approximate $f(3.04, 3.95)$. To do so we note that for $x_0 = 3$ and $y_0 = 4$ it is easy to compute

$$f(x_0, y_0) = \sqrt{3^2 + 4^2} = \sqrt{25} = 5.$$

We therefore write

$$3.04 = x_0 + \Delta x \qquad \text{and} \qquad 3.95 = y_0 + \Delta y,$$

where

$$x_0 = 3, \qquad \Delta x = 0.04, \qquad y_0 = 4, \qquad \text{and} \qquad \Delta y = -0.05.$$

Also,

$$\frac{\partial f}{\partial x}(x_0, y_0) = \frac{\partial}{\partial x}(\sqrt{x^2 + y^2})\bigg|_{(3,4)} = \frac{3}{\sqrt{3^2 + 4^2}} = 0.6$$

and

$$\frac{\partial f}{\partial y}(x_0, y_0) = \frac{\partial}{\partial y}(\sqrt{x^2 + y^2}\bigg|_{(3,4)} = \frac{4}{\sqrt{3^2 + 4^2}} = 0.8.$$

Thus, by (5),

$$\begin{aligned}\sqrt{(3.04)^2 + (3.95)^2} &\approx 5 + (0.6)(0.04) + (0.8)(-0.05)\\ &= 4.984.\end{aligned}$$

The actual value of this expression, to four decimal places, is 4.9844. The *relative error* in the approximation is therefore

$$\frac{4.9844 - 4.984}{4.9844} \approx .00008,$$

a *percentage error* of less than 0.001%. ■

Approximation (5) is written in a form that is useful in approximating a particular value $f(x_0 + \Delta x, y_0 + \Delta y)$. By using the notation

$$\Delta f = f(x_0 + \Delta x, y_0 + \Delta y) - f(x_0, y_0),$$

we may rewrite approximation (5) as

$$\Delta f \approx \frac{\partial f}{\partial x}(x_0, y_0)\Delta x + \frac{\partial f}{\partial y}(x_0, y_0)\Delta y. \tag{15}$$

Approximation (15) is useful in approximating the *change* in the value of the function $f(x, y)$ due to changes Δx and Δy in the independent variables.

Example 3 The ideal gas law is $PV = nRT$, where P is pressure, V is volume, n is the number of moles of gas present, R is constant, and T is temperature. Find, according to the ideal gas law, the change in the volume of 1000 cc of gas at temperature 300°K and pressure 780 mm of mercury if the gas is heated by 10°K and the pressure is increased by 5 mm of mercury.

Solution: We first simplify the gas law by solving for the constant nR and using the given data. We find that

$$nR = \frac{PV}{T} = \frac{780 \cdot 1000}{300} = 2600.$$

Using this constant, we may solve the gas law for V as a function of T and P:

$$V(T, P) = \frac{nRT}{P} = \frac{2600T}{P}.$$

According to approximation (15),

$$\Delta V \approx \frac{\partial}{\partial T}\left(\frac{2600T}{P}\right)\Delta T + \frac{\partial}{\partial P}\left(\frac{2600T}{P}\right)\Delta P$$
$$= \frac{2600\Delta T}{P} - \frac{2600T\Delta P}{P^2}.$$

Using the data $T = 300$, $P = 780$, $\Delta T = 10$, and $\Delta P = 5$, we obtain

$$\Delta V \approx \frac{2600(10)}{780} - \frac{2600(300)(5)}{780^2} \tag{16}$$
$$= 26.92 \text{ cc}.$$

The actual value of V corresponding to the temperature $T = 310°$ and pressure $P = 785$ mm of mercury is

$$V = \frac{2600(310)}{785} = 1026.75$$

so the actual value of ΔV is 26.75. The approximation in (16) is therefore in error by $26.92 - 26.75 = 0.17$ cc, a percentage error of only $\left(\frac{0.17}{26.75}\right) \times 100\% = 0.64\%$. ■

Differentials

The preceding ideas may be written a bit more compactly using the terminology of **differentials.** Recall that in the one-variable case the differential df for the differentiable function $f(x)$ is defined to be

$$df = f'(x)\,dx \tag{17}$$

where $dx = \Delta x$ is an increment in the independent variable x. Note that the differential df in (17) is just the right-hand side of the approximation (1) when written in the form

$$f(x + \Delta x) - f(x) = \Delta f \approx f'(x)\Delta x.$$

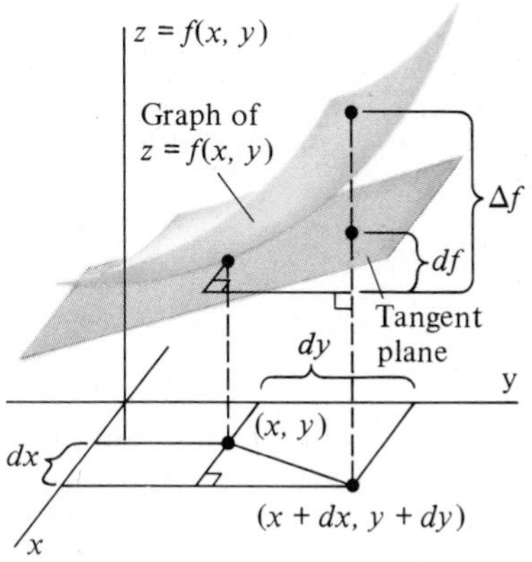

Figure 6.3 $df = f_x\,dx + f_y\,dy$ approximates change Δf by use of tangent plane.

In other words, the differential in (17) provides an approximation to the change Δf in $f(x)$, corresponding to the change Δx in x, obtained by following the tangent at $(x, f(x))$ rather than the graph of $y = f(x)$.

For precisely the same reason, we define the differential of the function $f(x, y)$ of two variables to be

$$df = \frac{\partial f}{\partial x}(x, y)\,dx + \frac{\partial f}{\partial y}(x, y)\,dy \tag{18}$$

where $dx = \Delta x$ and $dy = \Delta y$ are increments in the two independent variables x and y. Thus, the differential df in (18) is just the right-hand side of approximation (15). That is, $\Delta f \approx df$. As Figure 6.3 illustrates, the differential df provides an approximation to the change Δf corresponding to the changes $dx = \Delta x$ in x and $dy = \Delta y$ in y, obtained by following the tangent plane at $(x, y, f(x, y))$ rather than the graph of

$z = f(x, y)$. Of course, df is defined only when the partial derivatives $\frac{\partial f}{\partial x}(x, y)$ and $\frac{\partial f}{\partial y}(x, y)$ exist.

Other notation for the differential in (18) includes the statements

$$df = f_x\,dx + f_y\,dy, \tag{19}$$

$$dz = \frac{\partial f}{\partial x}(x, y)\,dx + \frac{\partial f}{\partial y}(x, y)\,dy \tag{20}$$

and

$$dz = f_x\,dx + f_y\,dy \tag{21}$$

where $z = f(x, y)$. The differential expressed by equations (18) through (21) is sometimes referred to as the **total differential** for the function $z = f(x, y)$.

Example 4 A cylindrical bearing of mass density λ gm/cc is measured with a caliper and found to have diameter 2.2 cm and height 5.8 cm. How large might one expect the error in the calculation of the mass of the bearing to be if the caliper is known to be accurate only to within 0.05 cm?

Solution: For the given measurements, the volume of the bearing is calculated to be

$$V = \pi\left(\frac{D}{2}\right)^2 h = \pi(1.1)^2(5.8) = 22.0477 \text{ cm}^3$$

so the calculated mass is $\lambda V = 22.0477\lambda$ gm.

To estimate the magnitude of the error in this calculation, we use the differential approximation dV:

$$\begin{aligned} dV &= \frac{\partial}{\partial D}\left[\pi\left(\frac{D}{2}\right)^2 h\right] dD + \frac{\partial}{\partial h}\left[\pi\left(\frac{D}{2}\right)^2 h\right] dh \\ &= \frac{\pi D h}{2}\,dD + \frac{\pi D^2}{4}\,dh. \end{aligned}$$

The value of dV corresponding to $D = 2.2$, $h = 5.8$, $dD = \Delta D = 0.05$, and $dh = \Delta h = 0.05$ is

$$\begin{aligned} dV &= \frac{\pi(2.2)(5.8)}{2}(0.05) + \frac{\pi(2.2)^2}{4}(0.05) \\ &= 1.1922. \end{aligned}$$

We conclude that the magnitude of the error in the calculation of mass is therefore no greater than 1.1922λ gm, a relative error of

$$\frac{1.1922\lambda}{22.0477\lambda} = .0541,$$

or a percentage error of 5.41%. ■

Example 5 In economic theory, the Cobb-Douglas production function relating output production y to the input of labor L and capital K has the form

$$y = \lambda L^{\alpha} K^{1-\alpha}$$

where λ and α are positive constants. The change Δy in production resulting from a change dL in labor input and a change dK in capital input is therefore approximated by the total differential

$$\begin{aligned} dy &= \frac{\partial}{\partial L}(\lambda L^{\alpha} K^{1-\alpha})dL + \frac{\partial}{\partial K}(\lambda L^{\alpha} K^{1-\alpha})dK \\ &= \alpha \lambda L^{\alpha-1} K^{1-\alpha} dL + (1-\alpha)\lambda L^{\alpha} K^{-\alpha} dK. \end{aligned}$$

■

Functions of Three Variables

The ideas of this section generalize directly to functions of three variables. The corresponding version of the Approximation Theorem (Theorem 5) is the following.

THEOREM 5′

Let $f(x, y, z)$ and its partial derivatives be continuous in the open rectangular box $B = \{(x, y, z) \mid a_1 < x < a_2,\ b_1 < y < b_2,\ c_1 < z < c_2\}$ in $\mathbb{R}^3$. If (x_0, y_0, z_0) and $(x_0 + \Delta x, y_0 + \Delta y, z_0 + \Delta z)$ both lie in B, then

$$\begin{aligned} f(x_0 + \Delta x, y_0 + \Delta y, z_0 + \Delta z) &= f(x_0, y_0, z_0) + \frac{\partial f}{\partial x}(x_0, y_0, z_0)\Delta x \\ &\quad + \frac{\partial f}{\partial y}(x_0, y_0, z_0)\Delta y + \frac{\partial f}{\partial z}(x_0, y_0, z_0)\Delta z \\ &\quad + \epsilon_1 \Delta x + \epsilon_2 \Delta y + \epsilon_3 \Delta z \end{aligned}$$

where $\epsilon_1 \to 0$, $\epsilon_2 \to 0$, and $\epsilon_3 \to 0$ as $\Delta x \to 0$, $\Delta y \to 0$, and $\Delta z \to 0$.

The proof of Theorem 5′ is similar to that of Theorem 5, involving three applications of the Mean Value Theorem (see Exercise 36).

Theorem 5′ leads directly to the approximations

$$f(x + \Delta x, y + \Delta y, z + \Delta z) \approx f(x, y, z) + \frac{\partial f}{\partial x}\Delta x + \frac{\partial f}{\partial y}\Delta y + \frac{\partial f}{\partial z}\Delta z$$

and

$$\Delta f \approx \frac{\partial f}{\partial x}\Delta x + \frac{\partial f}{\partial y}\Delta y + \frac{\partial f}{\partial z}\Delta z, \tag{22}$$

where $\Delta f = f(x + \Delta x, y + \Delta y, z + \Delta z) - f(x, y, z)$. Finally, the definition of the differential, suggested by approximation (22), is

$$\boxed{df = \frac{\partial f}{\partial x}(x, y, z)\,dx + \frac{\partial f}{\partial y}(x, y, z)\,dy + \frac{\partial f}{\partial z}(x, y, z)\,dz}$$

or

$$dz = f_x\,dx + f_y\,dy + f_z\,dz$$

where $z = f(x, y, z)$.

Example 6 According to Newton's Law of Universal Gravitation, the force of attraction between two bodies of mass m_1 and m_2 is

$$F(m_1, m_2, r) = \frac{Gm_1m_2}{r^2},$$

where r is the distance between the bodies and G is the universal gravitation constant. The change ΔF in this force caused by changes dm_1 and dm_2 in the masses and dr in the distance is approximated by the differential

$$dF = \frac{\partial}{\partial m_1}\left(\frac{Gm_1m_2}{r^2}\right) dm_1 + \frac{\partial}{\partial m_2}\left(\frac{Gm_1m_2}{r^2}\right) dm_2 + \frac{\partial}{\partial r}\left(\frac{Gm_1m_2}{r^2}\right) dr$$

$$= \frac{Gm_2}{r^2} dm_1 + \frac{Gm_1}{r^2} dm_2 - \frac{2Gm_1m_2}{r^3} dr. \quad ■$$

Exercise Set 19.6

In Exercises 1–12, find the total differential df.

1. $f(x, y) = x^2y^4$
2. $f(x, y) = x^{2/3}y^{5/2}$
3. $f(x, y) = \sqrt{x^2 + y^4}$
4. $f(x, y) = x \sin y^2$
5. $f(x, y) = e^{\sqrt{x}} \cos y$
6. $f(x, y) = \tan^{-1}(y/x)$
7. $f(x, y, z) = x^2yz^3$
8. $f(x, y, z) = \ln(x^2 + 3y + z^3)$
9. $f(x, y, z) = \dfrac{x}{y + 3^z}$
10. $f(x, y, z) = \sqrt{\dfrac{x}{y^2 + z^2}}$
11. $f(x, y, z) = \dfrac{x - y}{x^2 + y^2 + z^2}$
12. $f(x, y, z) = ze^{x^2 - y^3}$
13. Let $f(x, y) = 2xy^2 + x^2y$, $\quad x_0 = 2$, $\quad y_0 = 3$, $\Delta x = 0.1$, and $\Delta y = 0.2$.
    a. Calculate $f(x_0, y_0)$ and $f(x_0 + \Delta x, y_0 + \Delta y)$.
    b. Calculate $\Delta f = f(x_0 + \Delta x, y_0 + \Delta y) - f(x_0, y_0)$.
    c. Calculate $df(x_0, y_0)$.
    d. Find the difference $\Delta f - df$.
14. *(Calculator)* Let $f(x, y) = \sin x \cos y$, $x_0 = \pi/2$, $y_0 = \pi/4$, $\Delta x = \pi/16$, and $\Delta y = -\pi/16$. Answer the questions in Exercise 13.

In Exercises 15–22, use the total differential to approximate the given number.

15. $\sqrt{(3.02)^2 + (4.08)^2}$
16. $\sqrt{9.4} + \sqrt{15.6}$
17. $\sin 88° \cos 42°$
18. $(2.02)^3 + (3.96)^3$
19. $(5.03)^2(1.02)^3$
20. $(5.94)\ln(1.15)$
21. $(6.04)(3.1)(2.96)$
22. $\sqrt{(1.1)^2 + (4.05)^2 + (2.96)^2}$
23. *(Calculator)* Find the relative and percentage errors in the approximation in Exercise 15.
24. *(Calculator)* Find the relative and percentage errors in the approximation in Exercise 16.
25. *(Calculator)* Find the relative and percentage errors in the approximation in Exercise 17.
26. A rectangle is measured and found to have dimensions of 4 cm and 7 cm. If the measurements are known to be accurate only to within 0.2 cm, find the maximum error and the maximum percentage error in the calculated value $A = (4 \text{ cm})(7 \text{ cm}) = 28$ cm for the area of the rectangle.
27. A cone is measured and found to have diameter 10 cm and altitude 12 cm. There is a maximum error of 0.3 cm in the measurement of the diameter and a maximum error of 0.2 cm in the measurement of the altitude. Find the maximum error and the maximum percentage error in the calculation of the volume.
28. A cylindrical bearing of radius 2 cm and length 6 cm is dipped in a molten brass solution to produce a brass coating. If the thickness of the coating varies from 0.02 cm to 0.06 cm, find an approximation for the volume of the coated bearing.
29. The money supply in the economy is defined by the equation $M = C + D$ where C is the currency outstanding in banks and D is the amount of money in demand and term deposits. Use differentials to approximate the percentage change in the money supply caused by a 10% increase in the outstanding currency and a 5% decrease in total deposits.

30. For the Cobb-Douglas production function $P = \lambda L^3 K^{3/2}$, use differentials to approximate the percentage change in P caused by a 4% increase in labor (L) and a 10% decrease in capital K.

31. For the Cobb-Douglas production function $P = \lambda L^{\alpha} K^{\beta}$, what can you say about α and β if a percentage decrease in labor of $\gamma\%$, combined with a percentage increase in capital of $\gamma\%$, produces no change in P?

32. For small oscillations, the period of a simple pendulum is given by the equation

$$T = 2\pi \sqrt{\frac{s}{g}}$$

where s is the length of the pendulum, g is the acceleration due to gravity, and T is time. (If the units for s and g involve meters and seconds, the units for T are seconds.) What is the approximate error in the calculation of T resulting from its calculation using $s = 10$ m and $g = 9.8$ m/sec^2 if the actual values of these quantities are $s = 10.2$ and $g = 9.75$?

33. Approximate $f(x_0, y_0, z_0)$ for $f(x, y, z) = \dfrac{\sqrt{x^2 - y^2}}{\tan z}$ and $(x_0, y_0, z_0) = (5.04, 3.97, 0.78)$. (*Hint:* $\frac{\pi}{4} \approx 0.7853982$.)

34. When three resistors with resistances R_1, R_2, and R_3 are connected in parallel, the resistance of the resulting circuit is

$$R = \frac{1}{R_1 + R_2 + R_3}$$

What is the maximum percentage change in R resulting from a change of no more than 10% in each of R_1, R_2, and R_3?

35. The work involved in an adiabatic (no heat lost to the surroundings) expansion of an ideal gas is given by the equation

$$w = nC_v(T_2 - T_1)$$

where C_v is the average heat capacity of the gas, and T_1 and T_2 are the initial and final temperatures. Show that if both T_1 and T_2 are increased by λ percent, the amount of work involved is also increased by λ percent.

36. Prove Theorem 5′, using the Mean Value Theorem three times.

19.7 CHAIN RULES

In this section, we use the Approximation Theorem to establish two types of Chain Rule. Both are generalizations of the Chain Rule for functions of a single variable:

$$[f(u(x))]' = f'(u(x))u'(x). \tag{1}$$

Our first generalization of equation (1) concerns the composition of a function of several variables with a vector-valued function of a single variable. For example, let

$$\mathbf{r}(t) = x(t)\mathbf{i} + y(t)\mathbf{j} + z(t)\mathbf{k}, \qquad t \in (a, b) \tag{2}$$

be a vector function, and let

$$w = f(x, y, z), \qquad (x, y, z) \in Q \tag{3}$$

be a function of three variables. As in Chapter 18, we identify the *point* $(x(t), y(t), z(t))$ with the tip of the *position vector* $\mathbf{r}(t)$ in (2) (see Figure 7.1). By taking $x = x(t)$, $y = y(t)$, and $z = z(t)$, we may combine equations (2) and (3) to form the *composite* function

$$w(t) = f(x(t)), y(t), z(t)), \qquad t \in (a, b). \tag{4}$$

It is important to note that this composite function is just a function of a single variable, as discussed in Chapters 1 through 16. Our interest is in knowing when the derivative $w'(t)$ exists and, if it does, how it may be calculated.

Before stating the theorem that answers this question, we describe a simple setting in which a composite function of the type (4) occurs. Think of a region of airspace, such as that over the state of California, through which an airplane is about to fly. The region of airspace may be coordinatized with xyz-coordinates (which you

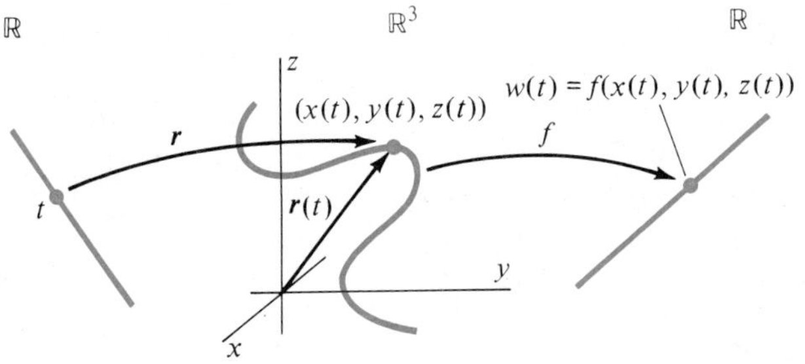

Figure 7.1 If $r: \mathbb{R} \to \mathbb{R}^3$ and $f: \mathbb{R}^3 \to \mathbb{R}$, the composite function $w(t) = f(r(t)) = f(x(t), y(t), z(t))$ maps $\mathbb{R}$ into $\mathbb{R}$.

may think of as representing latitude, longitude, and altitude, if you prefer), and the path of the airplane may be thought of as a curve parameterized by the vector function $\boldsymbol{r}(t) = x(t)\boldsymbol{i} + y(t)\boldsymbol{j} + z(t)\boldsymbol{k}$, $a \le t \le b$ giving the location of the airplane at time t. Assuming a steady state temperature distribution throughout the region, we let $T(x, y, z)$ be the temperature at location (x, y, z) in the airspace. *The composite function*

$$w(t) = T(x(t), y(t), z(t))$$

gives the temperature outside the airplane at time t.

What factors determine the rate, $\frac{dw}{dt}$, at which the temperature outside the airplane changes? Obviously $\frac{dx}{dt}$, $\frac{dy}{dt}$, and $\frac{dz}{dt}$ are involved, since they are the components of the velocity of the airplane. But $\frac{dw}{dt}$ will also depend upon the partial derivatives $\frac{\partial T}{\partial x}$, $\frac{\partial T}{\partial y}$, and $\frac{\partial T}{\partial z}$, since these determine the rate of change of temperature with respect to change in position. The precise relationship between these rates is given by the following theorem.

THEOREM 6
Chain Rule

Let $w = f(x, y, z)$ be a continuous function with continuous partial derivatives for all (x, y, z) in an open region Q in $\mathbb{R}^3$. Let

$$\boldsymbol{r}(t) = x(t)\boldsymbol{i} + y(t)\boldsymbol{j} + z(t)\boldsymbol{k}, \qquad a \le t \le b,$$

lie in Q for each $t \in [a, b]$ and assume that $\frac{dx}{dt}$, $\frac{dy}{dt}$, and $\frac{dz}{dt}$ exist for each $t \in (a, b)$. Then the composite function

$$w(t) = f(x(t), y(t), z(t))$$

is a differentiable function of $t \in (a, b)$, and

$$w'(t) = \frac{\partial f}{\partial x}(x, y, z)x'(t) + \frac{\partial f}{\partial y}(x, y, z)y'(t) + \frac{\partial f}{\partial z}(x, y, z)z'(t) \tag{5}$$

or, more compactly

$$\frac{df}{dt} = \frac{\partial f}{\partial x}\cdot\frac{dx}{dt} + \frac{\partial f}{\partial y}\cdot\frac{dy}{dt} + \frac{\partial f}{\partial z}\cdot\frac{dz}{dt}. \tag{6}$$

Proof: Let $t_0 \in (a, b)$ and let $\Delta t \neq 0$ be sufficiently small that $(t_0 + \Delta t)$ also lies in (a, b). Let

$$\begin{aligned} x_0 &= x(t_0) & \Delta x &= x(t_0 + \Delta t) - x(t_0) \\ y_0 &= y(t_0) & \Delta y &= y(t_0 + \Delta t) - y(t_0) \\ z_0 &= z(t_0), \quad \text{and} & \Delta z &= z(t_0 + \Delta t) - z(t_0) \end{aligned}$$

Then, according to the Approximation Theorem (Theorem 5′),

$$\begin{aligned} w(t_0 + \Delta t) = w(t_0) &+ \frac{\partial f}{\partial x}(x_0, y_0, z_0)\Delta x + \frac{\partial f}{\partial y}(x_0, y_0, z_0)\Delta y \\ &+ \frac{\partial f}{\partial z}(x_0, y_0, z_0)\Delta z + \epsilon_1 \Delta x + \epsilon_2 \Delta y + \epsilon_3 \Delta z \end{aligned} \tag{7}$$

where $\epsilon_1 \to 0$, $\epsilon_2 \to 0$, and $\epsilon_3 \to 0$ as $\Delta x \to 0$, $\Delta y \to 0$, and $\Delta z \to 0$.

Subtracting $w(t_0)$ from both sides of equation (7) and dividing by Δt gives the equation

$$\begin{aligned} \frac{w(t_0 + \Delta t) - w(t_0)}{\Delta t} &= \frac{\partial f}{\partial x}(x_0, y_0, z_0)\frac{\Delta x}{\Delta t} + \frac{\partial f}{\partial y}(x_0, y_0, z_0)\frac{\Delta y}{\Delta t} \\ &+ \frac{\partial f}{\partial z}(x_0, y_0, z_0)\frac{\Delta z}{\Delta t} + \epsilon_1 \frac{\Delta x}{\Delta t} + \epsilon_2 \frac{\Delta y}{\Delta t} + \epsilon_3 \frac{\Delta z}{\Delta t}. \end{aligned} \tag{8}$$

Now according to our definitions, we have

$$\lim_{\Delta t \to 0} \frac{\Delta x}{\Delta t} = \lim_{\Delta t \to 0} \frac{x(t_0 + \Delta t) - x(t_0)}{\Delta t} = x'(t_0).$$

Thus,

$$\begin{aligned} &\lim_{\Delta t \to 0} \left[\frac{\partial f}{\partial x}(x_0, y_0, z_0)\frac{\Delta x}{\Delta t}\right] \\ &= \frac{\partial f}{\partial x}(x_0, y_0, z_0) \cdot \lim_{\Delta t \to 0} \frac{\Delta x}{\Delta t} = \frac{\partial f}{\partial x}(x_0, y_0, z_0)x'(t_0) \end{aligned} \tag{9}$$

and, since $\Delta x \to 0$ as $\Delta t \to 0$,

$$\lim_{\Delta t \to 0} \epsilon_1 \cdot \frac{\Delta x}{\Delta t} = \left(\lim_{\Delta t \to 0} \epsilon_1\right)\left(\lim_{\Delta t \to 0} \frac{\Delta x}{\Delta t}\right) = 0 \cdot x'(t_0) = 0. \tag{10}$$

Equations analogous to (9) and (10) hold for the variables y and z.

Finally, from equation (8) and the equations of the form (9) and (10) for each of the variables x, y, and z, we conclude that

$$\begin{aligned} w'(t_0) &= \lim_{\Delta t \to 0} \frac{w(t_0 + \Delta t) - w(t_0)}{\Delta t} \\ &= \frac{\partial f}{\partial x}(x_0, y_0, z_0)x'(t_0) + \frac{\partial f}{\partial y}(x_0, y_0, z_0)y'(t_0) \\ &+ \frac{\partial f}{\partial z}(x_0, y_0, z_0)z'(t_0), \end{aligned}$$

as required. ■

Example 1 The function giving the temperature $T(x, y, z)$ at a point (x, y, z) in space is

$$T(x, y, z) = \lambda\sqrt{x^2 + y^2 + z^2}$$

where λ is constant.

Find the rate of change of temperature with respect to t along the elliptical helix

$$\mathbf{r}(t) = a\cos t\mathbf{i} + b\sin t\mathbf{j} + ct\mathbf{k}.$$

Solution: The temperature function along the helix is the composite function $T(x(t), y(t), z(t))$ where

$$x(t) = a\cos t, \qquad y(t) = b\sin t, \qquad z(t) = ct.$$

Using equation (6), we obtain the desired rate as

$$\begin{aligned}\frac{dT}{dt} &= \frac{\partial}{\partial x}(\lambda\sqrt{x^2 + y^2 + z^2})\cdot\frac{d}{dt}(a\cos t)\\ &\quad + \frac{\partial}{\partial y}(\lambda\sqrt{x^2 + y^2 + z^2})\cdot\frac{d}{dt}(b\sin t) + \frac{\partial}{\partial z}(\lambda\sqrt{x^2 + y^2 + z^2})\cdot\frac{d}{dt}(ct)\\ &= \frac{\lambda x(-a\sin t)}{\sqrt{x^2 + y^2 + z^2}} + \frac{\lambda yb\cos t}{\sqrt{x^2 + y^2 + z^2}} + \frac{\lambda cz}{\sqrt{x^2 + y^2 + z^2}}\\ &= \frac{\lambda(-a^2 + b^2)\sin t\cos t + \lambda c^2 t}{\sqrt{a^2\cos^2 t + b^2\sin^2 t + c^2t^2}}.\end{aligned}$$

■

Example 2 Let $f(x, y, z) = \sqrt{x}\,y^2e^{2z}$, $x(t) = 3t^2 + 2$, $y(t) = 6t$, $z(t) = 1 - t^3$, and $w(t) = f(x(t), y(t), z(t))$. Find $w'(t)$.

Solution: We have

$$\frac{\partial f}{\partial x}(x, y, z) = \frac{\partial}{\partial x}(\sqrt{x}\,y^2e^{2z}) = \frac{y^2e^{2z}}{2\sqrt{x}}$$

$$\frac{\partial f}{\partial y}(x, y, z) = \frac{\partial}{\partial y}(\sqrt{x}\,y^2e^{2z}) = 2\sqrt{x}\,ye^{2z}$$

and

$$\frac{\partial f}{\partial z}(x, y, z) = \frac{\partial}{\partial z}(\sqrt{x}\,y^2e^{2z}) = 2\sqrt{x}\,y^2e^{2z}.$$

Also,

$$x'(t) = 6t, \qquad y'(t) = 6, \qquad \text{and} \qquad z'(t) = -3t^2.$$

According to Theorem 6 we must have

$$\begin{aligned}w'(t) &= \frac{y^2e^{2z}}{2\sqrt{x}}(6t) + 2\sqrt{x}\,ye^{2z}(6) + 2\sqrt{x}\,y^2e^{2z}(-3t^2)\\ &= \frac{108t^3e^{2(1-t^3)}}{\sqrt{3t^2 + 2}} + 72t\sqrt{3t^2 + 2}\,e^{2(1-t^3)} - 216t^4\sqrt{3t^2 + 2}\,e^{2(1-t^3)}\\ &= \sqrt{3t^2 + 2}\,e^{2(1-t^3)}\left[\frac{108t^3}{3t^2 + 2} + 72t - 216t^4\right].\end{aligned}$$

■

Example 3 The **electric potential** at an axial point for a charged disc is given by the equation

$$V = \frac{\sigma}{2\epsilon_0}(\sqrt{a^2 + r^2} - r)$$

where a is the radius of the disc, r is the distance from the point to the disc, and σ and ϵ_0 are constants. If the radius of the disc is increasing at a rate of 2 cm/sec and the point is moving away from the disc at a rate of 5 cm/sec, find the rate at which the electric potential is changing when $a = 3$ cm and $r = 4$ cm.

Solution: Since V is a function of only two variables, a and r, we simply consider the third variable in the Chain Rule (6) to be $z = 0$, so $\frac{dz}{dt} = 0$. Applying the Chain Rule we find

$$\begin{aligned}\frac{dV}{dt} &= \frac{\partial}{\partial a}\left[\frac{\sigma}{2\epsilon_0}(\sqrt{a^2 + r^2} - r)\right]\frac{da}{dt} \\ &\quad + \frac{\partial}{\partial r}\left[\frac{\sigma}{2\epsilon_0}(\sqrt{a^2 + r^2} - r)\right]\frac{dr}{dt} \\ &= \frac{\sigma}{2\epsilon_0}\left(\frac{a}{\sqrt{a^2 + r^2}}\right)\frac{da}{dt} + \frac{\sigma}{2\epsilon_0}\left(\frac{r}{\sqrt{a^2 + r^2}} - 1\right)\frac{dr}{dt}.\end{aligned} \tag{11}$$

We are given

$$\frac{da}{dt} = 2 \text{ cm/sec}, \qquad \frac{dr}{dt} = 5 \text{ cm/sec}, \qquad a = 3 \text{ cm, and } r = 4 \text{ cm}.$$

Substituting these values into (11) gives

$$\begin{aligned}\frac{dV}{dt} &= \frac{\sigma}{2\epsilon_0}\left(\frac{3}{\sqrt{3^2 + 4^2}}\right)(2) + \frac{\sigma}{2\epsilon_0}\left(\frac{4}{\sqrt{3^2 + 4^2}} - 1\right)(5) \\ &= \frac{\sigma}{10\epsilon_0}.\end{aligned}$$

■

Other Types of Chain Rule

Various other sorts of composite functions can be formed using functions of several variables, each of which yields a corresponding Chain Rule. For example, if

$$x = x(s, t), \qquad y = y(s, t) \qquad \text{and} \qquad z = z(s, t)$$

are functions of two variables, and if

$$w = f(x, y, z)$$

is a function of three variables, the composite function

$$w(s, t) = f(x(s, t), y(s, t), z(s, t))$$

is again a function of two variables. To obtain $\frac{\partial w}{\partial s}$, we hold t constant and differentiate the resulting function of a single variable $g(s) = w(s, t)$ according to Theorem

6. Similarly, the partial derivative $\frac{\partial w}{\partial t}$ is obtained by holding s constant and differentiating with respect to t. The results are the following:

$$\frac{\partial f}{\partial s} = \frac{\partial f}{\partial x}\cdot\frac{\partial x}{\partial s} + \frac{\partial f}{\partial y}\cdot\frac{\partial y}{\partial s} + \frac{\partial f}{\partial z}\cdot\frac{\partial z}{\partial s}, \tag{12}$$

$$\frac{\partial f}{\partial t} = \frac{\partial f}{\partial x}\cdot\frac{\partial x}{\partial t} + \frac{\partial f}{\partial y}\cdot\frac{\partial y}{\partial t} + \frac{\partial f}{\partial z}\cdot\frac{\partial z}{\partial t}. \tag{13}$$

Another type of Chain Rule occurs when

$$x = x(r, s, t) \qquad \text{and} \qquad y = y(r, s, t)$$

are functions of three variables, and the function

$$w = f(x, y)$$

is a function of two variables. The resulting composite function

$$w(r, s, t) = f(x(r, s, t), y(r, s, t))$$

is a function of three variables. The various partial derivatives of $w(r, s, t)$ are found by holding two of the independent variables constant and differentiating with respect to the third. Doing so, we obtain the Chain Rule equations

$$\frac{\partial f}{\partial r} = \frac{\partial f}{\partial x}\cdot\frac{\partial x}{\partial r} + \frac{\partial f}{\partial y}\cdot\frac{\partial y}{\partial r}, \tag{14}$$

$$\frac{\partial f}{\partial s} = \frac{\partial f}{\partial x}\cdot\frac{\partial x}{\partial s} + \frac{\partial f}{\partial y}\cdot\frac{\partial y}{\partial s}, \tag{15}$$

$$\frac{\partial f}{\partial t} = \frac{\partial f}{\partial x}\cdot\frac{\partial x}{\partial t} + \frac{\partial f}{\partial y}\cdot\frac{\partial y}{\partial t}. \tag{16}$$

Because of the statement of Theorem 6, each of these Chain Rules provides that the partial derivative of a composite function with respect to any of its independent variables is calculated by differentiating "through" each of the intermediate variables involved in the composition.

The following theorem includes each of the Chain Rules (12)–(13), and (14)–(16) as a special case.

THEOREM 7
General Chain Rule

Suppose that $f(x_1, x_2, \ldots, x_n)$ is a function of the n variables $x_1, x_2, \ldots, x_n$, each of which in turn is a function $x_j(t_1, t_2, \ldots, t_m)$ of the m independent variables $t_1, t_2, \ldots, t_m$. If each of the first order partial derivatives exists, then

$$\frac{\partial f}{\partial t_k} = \frac{\partial f}{\partial x_1}\cdot\frac{\partial x_1}{\partial t_k} + \frac{\partial f}{\partial x_2}\cdot\frac{\partial x_2}{\partial t_k} + \cdots + \frac{\partial f}{\partial x_n}\cdot\frac{\partial x_n}{\partial t_k}$$

for $1 \le k \le m$.

Example 4 Let $f(x, y) = x^2y^3$ where the variables x and y are functions of the polar variables, r, θ:

$$x(r, \theta) = r^2 \cos \theta, \qquad y(r, \theta) = r(1 - \sin \theta).$$

According to Theorem 7 (with $n = m = 2$),

$$\begin{aligned}\frac{\partial f}{\partial r} &= \frac{\partial f}{\partial x}\cdot\frac{\partial x}{\partial r} + \frac{\partial f}{\partial y}\cdot\frac{\partial y}{\partial r}\\ &= \left[\frac{\partial}{\partial x}(x^2y^3)\right]\left[\frac{\partial}{\partial r}(r^2\cos\theta)\right] + \left[\frac{\partial}{\partial y}(x^2y^3)\right]\left[\frac{\partial}{\partial r}(r(1-\sin\theta))\right]\\ &= 2xy^3\cdot 2r\cos\theta + 3x^2y^2(1-\sin\theta)\\ &= 7r^6\cos^2\theta(1-\sin\theta)^3\end{aligned}$$

and

$$\begin{aligned}\frac{\partial f}{\partial \theta} &= \frac{\partial f}{\partial x}\cdot\frac{\partial x}{\partial \theta} + \frac{\partial f}{\partial y}\cdot\frac{\partial y}{\partial \theta}\\ &= 2xy^3(-r^2\sin\theta) + (3x^2y^2)(-r\cos\theta)\\ &= r^7\cos\theta(1-\sin\theta)^2[2\sin^2\theta - 2\sin\theta - 3\cos^2\theta].\end{aligned}$$

■

Example 5 The graph of the function $y = \sin(bx - c)$ is a sine curve with period $T = \dfrac{2\pi}{b}$ and "phase angle" $\dfrac{c}{b}$. The phase angle is distance along the x-axis between the origin and the nearest zero of $y = \sin(bx - c)$ with $\dfrac{dy}{dx} > 0$ (see Figure 7.2). If we replace the constant phase angle $\dfrac{c}{b}$ by the variable t, we obtain the function of two variables

$$u(x, t) = \sin(bx - t). \tag{17}$$

If we graph this function in the ux-coordinate plane for various values of t, we find that (17) represents a "travelling" wave that moves from left to right along the x-axis for increasing values of t (Figure 7.3). ■

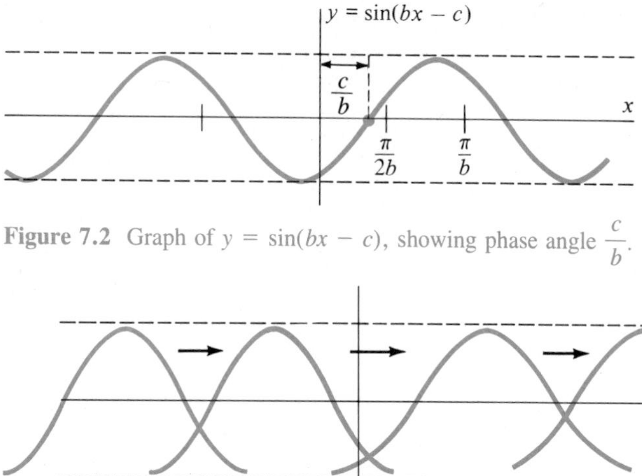

Figure 7.2 Graph of $y = \sin(bx - c)$, showing phase angle $\dfrac{c}{b}$.

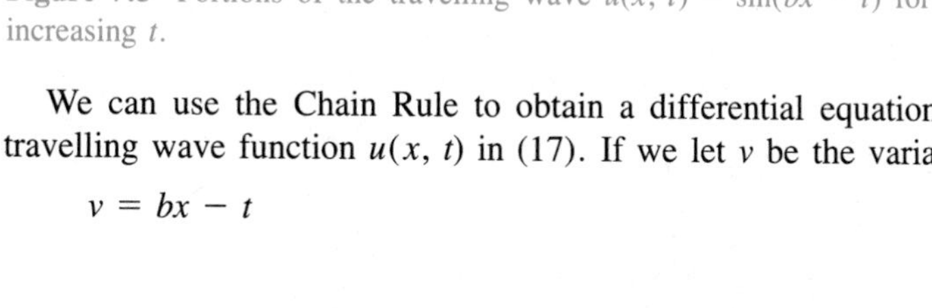

Figure 7.3 Portions of the travelling wave $u(x, t) = \sin(bx - t)$ for increasing t.

We can use the Chain Rule to obtain a differential equation satisfied by the travelling wave function $u(x, t)$ in (17). If we let v be the variable

$$v = bx - t$$

then $u(x, t)$ may be written simply as

$$u(x, t) = \sin(v).$$

Using the general Chain Rule with $n = 1$ and $m = 2$, we find that

$$\frac{\partial u}{\partial x}(x, t) = \frac{d}{dv}\sin(v)\cdot\frac{\partial v}{\partial x} = \cos(v)\cdot b = b\cdot\cos(bx - t) \tag{18}$$

and

$$\frac{\partial u}{\partial t}(x, t) = \frac{d}{dv}\sin(v)\cdot\frac{\partial v}{\partial t} = \cos(v)(-1) = -\cos(bx - t). \tag{19}$$

Comparing equations (18) and (19), we find that

$$\frac{\partial u}{\partial x} = -b\,\frac{\partial u}{\partial t}. \tag{20}$$

Equation (20) is referred to as a **partial differential equation** since it involves partial derivatives of the function $u(x, t)$. In general, solutions of partial differential equations are more numerous and more difficult to fully describe than are solutions of the types of ordinary differential equations that we have previously encountered (see Exercises 37–41).

Higher Order Derivatives

The various Chain Rules developed here may be applied repeatedly to calculate higher order partial derivatives of composite functions. For example, suppose that

$$z = f(x, y)$$

where

$$x = x(s, t), \qquad y = y(s, t)$$

and where each of these functions and the various first order partial derivatives satisfy the hypotheses of Theorem 7. The function $f(x, y) = f(x(s, t), y(s, t))$ is then a composite function of the independent variables s and t. According to Theorem 7 ($n = m = 2$),

$$\frac{\partial f}{\partial s} = \frac{\partial f}{\partial x}\cdot\frac{\partial x}{\partial s} + \frac{\partial f}{\partial y}\cdot\frac{\partial y}{\partial s} \tag{21}$$

and

$$\frac{\partial f}{\partial t} = \frac{\partial f}{\partial x}\cdot\frac{\partial x}{\partial t} + \frac{\partial f}{\partial y}\cdot\frac{\partial y}{\partial t}. \tag{22}$$

To calculate $\dfrac{\partial^2 f}{\partial s^2}$, we begin by differentiating $\dfrac{\partial f}{\partial s}$ in (21) according to Theorem 7:

$$\begin{aligned}\frac{\partial^2 f}{\partial s^2} &= \frac{\partial}{\partial s}\left(\frac{\partial f}{\partial s}\right) = \frac{\partial}{\partial s}\left(\frac{\partial f}{\partial x}\cdot\frac{\partial x}{\partial s}\right) + \frac{\partial}{\partial s}\left(\frac{\partial f}{\partial y}\cdot\frac{\partial y}{\partial s}\right) \\ &= \left[\frac{\partial}{\partial s}\left(\frac{\partial f}{\partial x}\right)\right]\frac{\partial x}{\partial s} + \frac{\partial f}{\partial x}\cdot\frac{\partial^2 x}{\partial s^2} \\ &\quad + \left[\frac{\partial}{\partial s}\left(\frac{\partial f}{\partial y}\right)\right]\frac{\partial y}{\partial s} + \frac{\partial f}{\partial y}\cdot\frac{\partial^2 y}{\partial s^2}\end{aligned} \tag{23}$$

To evaluate the first and third terms on the right side of equation (23), we must apply the Chain Rule again:

$$\frac{\partial}{\partial s}\left(\frac{\partial f}{\partial x}\right) = \frac{\partial}{\partial x}\left(\frac{\partial f}{\partial x}\right)\cdot\frac{\partial x}{\partial s} + \frac{\partial}{\partial y}\left(\frac{\partial f}{\partial x}\right)\cdot\frac{\partial y}{\partial s} \tag{24}$$
$$= \frac{\partial^2 f}{\partial x^2}\cdot\frac{\partial x}{\partial s} + \frac{\partial^2 f}{\partial y\partial x}\cdot\frac{\partial y}{\partial s},$$

$$\frac{\partial}{\partial s}\left(\frac{\partial f}{\partial y}\right) = \frac{\partial}{\partial x}\left(\frac{\partial f}{\partial y}\right)\cdot\frac{\partial x}{\partial s} + \frac{\partial}{\partial y}\left(\frac{\partial f}{\partial y}\right)\cdot\frac{\partial y}{\partial s} \tag{25}$$
$$= \frac{\partial^2 f}{\partial x\partial y}\cdot\frac{\partial x}{\partial s} + \frac{\partial^2 f}{\partial y^2}\cdot\frac{\partial y}{\partial s}.$$

Combining equations (24) and (25) with equation (23) gives

$$\frac{\partial^2 f}{\partial s^2} = \left[\frac{\partial^2 f}{\partial x^2}\cdot\frac{\partial x}{\partial s} + \frac{\partial^2 f}{\partial y\partial x}\cdot\frac{\partial y}{\partial s}\right]\frac{\partial x}{\partial s} + \frac{\partial f}{\partial x}\cdot\frac{\partial^2 x}{\partial s^2} \tag{26}$$
$$+ \left[\frac{\partial^2 f}{\partial x\partial y}\cdot\frac{\partial x}{\partial s} + \frac{\partial^2 f}{\partial y^2}\cdot\frac{\partial y}{\partial s}\right]\frac{\partial y}{\partial s} + \frac{\partial f}{\partial y}\cdot\frac{\partial^2 y}{\partial s^2}$$
$$= \frac{\partial^2 f}{\partial x^2}\left(\frac{\partial x}{\partial s}\right)^2 + 2\,\frac{\partial^2 f}{\partial y\partial x}\cdot\frac{\partial x}{\partial s}\cdot\frac{\partial y}{\partial s} + \frac{\partial^2 f}{\partial y^2}\left(\frac{\partial y}{\partial s}\right)^2$$
$$+ \frac{\partial f}{\partial x}\cdot\frac{\partial^2 x}{\partial s^2} + \frac{\partial f}{\partial y}\cdot\frac{\partial^2 y}{\partial s^2}.$$

The second order partial derivatives $\dfrac{\partial^2 f}{\partial t^2}$ and $\dfrac{\partial^2 f}{\partial s\partial t}$ are calculated in a similar way (see Exercises 32–34).

Example 6 Let $f(x, y) = x^2 + 2xy$, $x = r\cos\theta$ and $y = r\sin\theta$. Find (a) $\dfrac{\partial^2 f}{\partial r^2}$ and (b) $\dfrac{\partial^2 f}{\partial \theta^2}$.

Solution: For f, x, and y as given, we have

$$\frac{\partial f}{\partial x} = 2x + 2y \qquad \frac{\partial x}{\partial r} = \cos\theta \qquad \frac{\partial x}{\partial \theta} = -r\sin\theta$$

$$\frac{\partial^2 f}{\partial x^2} = 2 \qquad \frac{\partial^2 x}{\partial r^2} = 0 \qquad \frac{\partial^2 x}{\partial \theta^2} = -r\cos\theta$$

$$\frac{\partial f}{\partial y} = 2x \qquad \frac{\partial y}{\partial r} = \sin\theta \qquad \frac{\partial y}{\partial \theta} = r\cos\theta$$

$$\frac{\partial^2 f}{\partial y^2} = 0 \qquad \frac{\partial^2 y}{\partial r^2} = 0 \qquad \frac{\partial y}{\partial \theta^2} = -r\sin\theta$$

$$\frac{\partial^2 f}{\partial y\partial x} = 2$$

According to equation (26), with $s = r$, we have

$$\frac{\partial^2 f}{\partial r^2} = 2(\cos\theta)^2 + 2\cdot 2\cdot \cos\theta\sin\theta + (0)(\sin\theta)^2 + (2x+2y)(0) + (2x)(0) = 2\cos^2\theta + 4\sin\theta\cos\theta.$$

The formula for $\frac{\partial^2 f}{\partial \theta^2}$ is obtained from (25) by letting $s = \theta$:

$$\begin{aligned}\frac{\partial^2 f}{\partial \theta^2} &= 2(-r\sin\theta)^2 + 2\cdot 2(-r\sin\theta)(r\cos\theta) + (0)(r\cos\theta)^2 \\ &\quad + (2x+2y)(-r\cos\theta) + (2x)(-r\sin\theta) \\ &= 2r^2\sin^2\theta - 4r^2\sin\theta\cos\theta + [2r\cos\theta + 2r\sin\theta](-r\cos\theta) \\ &\quad + (2r\cos\theta)(-r\sin\theta) \\ &= 2r^2[\sin^2\theta - 4\sin\theta\cos\theta - \cos^2\theta].\end{aligned}$$

Exercise Set 19.7

In Exercises 1–10, find the rate of change of f with respect to t, $\frac{df}{dt}$, along the given curves.

1. $f(x, y) = x^2 + y^2$, $x(t) = 2t$, $y(t) = 6 - t^2$

2. $f(x, y) = x - y^2$, $x(t) = 3t^2 + 2$, $y(t) = 4 + t$

3. $f(x, y) = xy^2$, $\mathbf{r}(t) = \cos t\mathbf{i} + \sin t\mathbf{j}$

4. $f(x, y) = x^2 - y^2$, $\mathbf{r}(t) = e^t\mathbf{i} + e^{-t}\mathbf{j}$

5. $f(x, y) = \sqrt{x + y}$, $x(t) = \sqrt{t}$, $y(t) = e^{t^2}$

6. $f(x, y) = y^2 - x^2$, $x(t) = a\cos t$, $y(t) = a\sin t$

7. $f(x, y, z) = xy - x + z^2$, $\mathbf{r}(t) = t^2\mathbf{i} - 2t\mathbf{j} + \sin t\mathbf{k}$

8. $f(x, y, z) = x^2 + y^2 + z^2$, $\mathbf{r}(t) = a\cos \pi t\mathbf{i} + b\sin \pi t\mathbf{j} - t^2\mathbf{k}$

9. $f(x, y, z) = z(y^2 - x^2)$, $x(t) = a\cosh t$, $y(t) = b\sinh t$, $z(t) = e^{-2t}$

10. $f(x, y, z) = e^{xyz}$, $x(t) = \ln\sqrt{t}$, $y(t) = t\sin t$, $z(t) = 2^t$.

In Exercises 11–22, find the indicated derivative(s).

11. $f(x, y) = x^2y^3$, $x(t) = \cos t$, $y(t) = t\sin t$.
Find $\frac{df}{dt}$.

12. $f(x, y) = x^2y^3$, $x(s, t) = st$, $y(s, t) = s^2 - t^2$.
Find
a. $\frac{\partial f}{\partial s}$ **b.** $\frac{\partial f}{\partial t}$

13. $f(x, y, z) = x^2 + y^2 - z^2$, $x(s, t) = e^{st}$, $y(s, t) = st$, $z(s, t) = s - t$. Find
a. $\frac{\partial f}{\partial s}$ **b.** $\frac{\partial f}{\partial t}$

14. $f(x, y) = xy - y^2$, $x(r, s, t) = rst$, $y(r, s, t) = e^{rst}$.
Find
a. $\frac{\partial f}{\partial r}$ **b.** $\frac{\partial f}{\partial s}$ **c.** $\frac{\partial f}{\partial t}$

15. $f(x, y) = \sin(xy^2) - x^2y$, $x(s, t) = s^2 - st$, $y(s, t) = t^2s^2$. Find
a. $\frac{\partial f}{\partial s}$ **b.** $\frac{\partial f}{\partial t}$

16. $f(x, y) = x^2e^{y-x}$, $x(s, t) = s - t$, $y(s, t) = \sqrt{s + t}$.
Find
a. $\frac{\partial f}{\partial s}$ **b.** $\frac{\partial f}{\partial t}$

17. $f(x, y, z) = xy^3z^2$, $x(s, t) = s\cdot\sin t$, $y(s, t) = t\cos(s)$, $z(s, t) = t^2 - s^2$. Find
a. $\frac{\partial f}{\partial s}$ **b.** $\frac{\partial f}{\partial t}$

18. $f(x, y) = e^{xy}$, $x(r, s) = \sqrt{s^2 - r^2}$, $y(r, s) = \tan^{-1}\left(\frac{r}{s}\right)$. Find

a. $\frac{\partial f}{\partial r}$ **b.** $\frac{\partial f}{\partial s}$

19. $f(u, v) = e^{u^2} - e^{2v}$, $u(x, y) = xy^2$, $v(x, y) = x^2y$. Find

a. $\frac{\partial f}{\partial x}$ **b.** $\frac{\partial f}{\partial y}$

20. $f(x, y, z) = x^2 - xy + yz$, $x(r, \theta) = r\cos\theta$, $y(r, \theta) = 2r$, $z(r, \theta) = r\sin\theta$. Find

a. $\frac{\partial f}{\partial r}$ **b.** $\frac{\partial f}{\partial \theta}$

21. $f(r, \theta) = r^2(1 - \cos\theta)$, $r(t) = 1 + t^3$, $\theta(t) = \sqrt{1 + t^2}$. Find $\frac{df}{dt}$.

22. $f(x, y, z) = \ln(x^2 + y^2 + z^2)$, $x(u, v, w) = u\cos v$, $y(u, v, w) = v\sin u$, $z(u, v, w) = uvw$. Find

a. $\frac{\partial f}{\partial u}$ **b.** $\frac{\partial f}{\partial v}$ **c.** $\frac{\partial f}{\partial w}$

23. The radius of the base of a cone is 6 cm and it is increasing at a rate of 2 cm/sec. The height of the cone is 10 cm and it is increasing at a rate of 3 cm/sec. At what rate is the volume increasing?

24. The radius of a cylinder is 8 cm and it is increasing at a rate of 2 cm/sec. The height of the cylinder is 20 cm and it is increasing at a rate of 4 cm/sec.

a. Find the rate at which the volume is increasing.

b. Find the rate at which the lateral surface area is increasing.

25. A particle moves along a helix with position function $\mathbf{r}(t) = \cos t\,\mathbf{i} + \sin t\,\mathbf{j} + t\mathbf{k}$.

a. Find the function $D(x, y, z)$ giving the distance from the particle to the origin.

b. Find $\frac{dD}{dt}$, the rate at which the distance from the particle to the origin changes as a function of time.

c. Show that $\lim_{t\to\infty} D'(t) = 1$; $\lim_{t\to-\infty} D'(t) = -1$.

26. In economic theory, the **rate of growth** (as opposed to rate of change) of a function $f(t)$ is defined to be the ratio

$$\frac{f'(t)}{f(t)} = \frac{d}{dt}[\ln(f(t))] = \frac{\text{marginal function}}{\text{total function}}.$$

Show, according to this definition, that if the rate of growth of consumption C is a, and if the rate of growth of population P is b, then the rate of growth of per capita consumption, $R = \frac{C}{P}$, is $a - b$.

27. The money supply in the economy is defined by the equation $M = C + D$ where C is the cash on deposit in banks and D is the total of all time and demand deposits. If m, c, and d are the respective rates of growth of M, C, and D, show that

$$m = \frac{cC}{C + D} + \frac{dD}{C + D}.$$

(See Exercise 26 for the definition of rate of growth.)

28. Let $f(x, y) = xe^{y^2}$, $x(r, \theta) = r\cos\theta$, $y(r, \theta) = r\sin\theta$. Find

a. $\frac{\partial^2 f}{\partial r^2}$ **b.** $\frac{\partial^2 f}{\partial \theta^2}$

29. Let $f(x, y) = x^2 + y^4$, $x(s, t) = s^2t$, $y(s, t) = t^2 - s^2$. Find

a. $\frac{\partial^2 f}{\partial s^2}$ **b.** $\frac{\partial^2 f}{\partial t^2}$

30. Find $\frac{d^2f}{dt^2}$ for the funtion $f(x, y)$ in Exercise 11.

31. Find $\frac{\partial^2 f}{\partial r^2}$ for the function $f(x, y)$ in Exercise 14.

32. Find an expression for the mixed second order partial derivative $\frac{\partial^2 f}{\partial s \partial t}$ for the composite function

$$f(x, y) = f(x(s, t), y(s, t)).$$

33. Use the result of Exercise 32 to find $\frac{\partial^2 f}{\partial r \partial \theta}$ for the function f in Exercise 28.

34. Use the result of Exercise 32 to find $\frac{\partial^2 f}{\partial s \partial t}$ for the function f in Exercise 29.

35. By use of the Chain Rule, show that Laplace's equation

$$\frac{\partial^2 v}{\partial x^2} + \frac{\partial^2 v}{\partial y^2} + \frac{\partial^2 v}{\partial z^2} = 0$$

in cylindrical coordinates is

$$\frac{\partial^2 v}{\partial r^2} + \frac{1}{r}\frac{\partial v}{\partial r} + \frac{1}{r^2}\frac{\partial^2 v}{\partial \theta^2} + \frac{\partial^2 v}{\partial z^2} = 0.$$

36. Use the Chain Rule to show that Laplace's equation

$$\frac{\partial^2 v}{\partial x^2} + \frac{\partial^2 v}{\partial y^2} + \frac{\partial^2 v}{\partial z^2} = 0$$

in spherical coordinates is

$$\frac{\partial^2 v}{\partial \rho^2} + \frac{2}{\rho}\cdot\frac{\partial v}{\partial \rho} + \frac{1}{\rho^2}\cdot\frac{\partial^2 v}{\partial \theta^2} + \frac{\cot\theta}{\rho^2}\cdot\frac{\partial v}{\partial \theta} + \frac{\csc^2\theta}{\rho^2}\cdot\frac{\partial^2 v}{\partial \phi^2} = 0.$$

37. Show that $u = f(xy)$ satisfies the partial differential equation

$$x\frac{\partial u}{\partial x} = y\frac{\partial u}{\partial y}.$$

38. Show that $u = f(x + 2y^2)$ satisfies the partial differential equation

$$\frac{\partial u}{\partial y} = 4y\frac{\partial u}{\partial x}.$$

39. Show that $u = f(t - x)$ is a solution of the partial differential equation

$$\frac{\partial^2 u}{\partial x^2} + 2\frac{\partial^2 u}{\partial x \partial t} + \frac{\partial^2 u}{\partial t^2} = 0.$$

40. Show that $u = f(x^2 + y^2)$ satisfies the partial differential equation

$$\frac{\partial u}{\partial y} = \frac{y}{x}\cdot\frac{\partial u}{\partial x}.$$

41. Show that if $f(v)$ is any differentiable function of v, the composite function $u(x, t) = f(bx - t)$ is a solution of the partial differential equation

$$\frac{\partial u}{\partial x} = -b\frac{\partial u}{\partial t}.$$

19.8 DIRECTIONAL DERIVATIVES; THE GRADIENT

Let $z = f(x, y)$ be a function of two variables defined in a neighborhood of a point (x_0, y_0). The partial derivatives $\frac{\partial f}{\partial x}(x_0, y_0)$ and $\frac{\partial f}{\partial y}(x_0, y_0)$ measure the rate of change of $f(x, y)$ in the directions of the x- and y-coordinate axes, respectively. But how might we calculate the rate of change of $f(x, y)$ in an arbitrary direction? The answer to this question is provided by the *directional derivative*, the topic of this section.

Suppose that we wish to calculate the rate of change of $f(x, y)$ at (x_0, y_0) in the direction determined by the unit vector $\boldsymbol{u} = u_1\boldsymbol{i} + u_2\boldsymbol{j}$ (Figure 8.1). If t is a number for which $0 < t < 1$, the point $(x_0 + tu_1, y_0 + tu_2)$ lies on the line between the two points (x_0, y_0) and $(x_0 + u_1, y_0 + u_2)$, so the difference

$$f(x_0 + tu_1, y_0 + tu_2) - f(x_0, y_0)$$

represents the change in $f(x, y)$ in the direction of $\boldsymbol{u}$ over an interval of length $\sqrt{(tu_1)^2 + (tu_2)^2} = t\sqrt{u_1^2 + u_2^2} = t$, since $|\boldsymbol{u}| = \sqrt{u_1^2 + u_2^2} = 1$ (see Figure 8.2). The definition of the directional derivative is therefore the following.

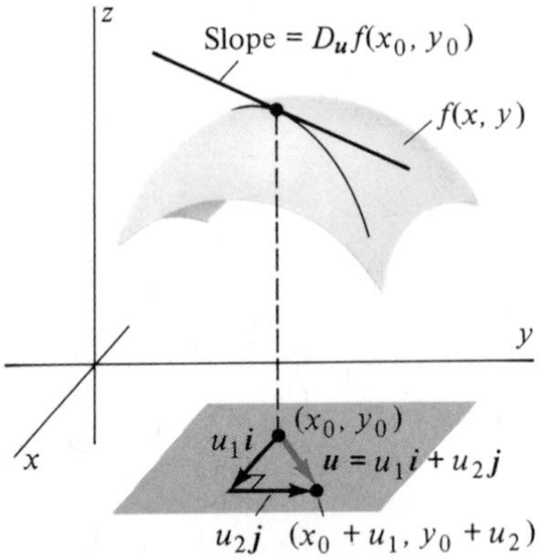

Figure 8.1 Directional derivative $D_{\boldsymbol{u}}f(x_0, y_0)$ is rate of change of f in the direction of the vector $\boldsymbol{u}$.

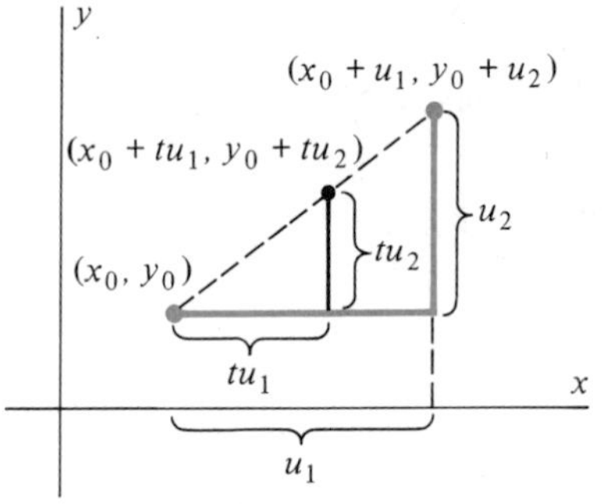

Figure 8.2 If $0 < t < 1$, the point $(x_0 + tu_1, y_0 + tu_2)$ lies between (x_0, y_0) and $(x_0 + u_1, y_0 + u_2)$.

DEFINITION 6

Let $f(x, y)$ be defined in a neighborhood of (x_0, y_0). Let $\boldsymbol{u} = u_1\boldsymbol{i} + u_2\boldsymbol{j}$ be a unit vector. The **directional derivative, $D_{\boldsymbol{u}}f(x_0, y_0)$**, in the direction of the unit vector $\boldsymbol{u}$, is the limit

$$D_{\boldsymbol{u}}f(x_0, y_0) = \lim_{t\to 0^+} \frac{f(x_0 + tu_1, y_0 + tu_2) - f(x_0, y_0)}{t}. \tag{1}$$

REMARK 1: Definition 6 has an obvious generalization to functions of more than two variables. If $\boldsymbol{u} = \langle u_1, u_2, \ldots, u_n\rangle$ is a unit vector in $\mathbb{R}^n$ and $f(x_1, x_2, \ldots, x_n)$ is a function of n variables, the directional derivative of f at $(x_1, x_2, \ldots, x_n)$ in the direction of $\boldsymbol{u}$ is the limit

$$\begin{aligned} &D_{\boldsymbol{u}}f(x_1, x_2, \ldots, x_n) \\ &= \lim_{t\to 0^+} \frac{f(x_1 + tu_1, x_2 + tu_2, \ldots, x_n + tu_n) - f(x_1, x_2, \ldots, x_n)}{t}. \end{aligned} \tag{2}$$

Unfortunately, equations (1) and (2) fail to provide a simple procedure for actually calculating the directional derivative when it exists. When $f(x, y)$ satisfies the hypotheses of the Approximation Theorem (Theorem 5), we can use this result to establish such a procedure. First, we write the numerator in (1) as

$$\begin{aligned} f(x_0 + tu_1, y_0 + tu_2) - f(x_0, y_0) &= \frac{\partial f}{\partial x}(x_0, y_0)(tu_1) + \frac{\partial f}{\partial y}(x_0, y_0)(tu_2) \\ &\quad + \epsilon_1(tu_1) + \epsilon_2(tu_2) \end{aligned}$$

where $\epsilon_1 \to 0$ and $\epsilon_2 \to 0$ as $t \to 0$. Dividing both sides by $t \neq 0$ and applying $\lim_{t\to 0^+}$ shows that

$$\begin{aligned} D_{\boldsymbol{u}}f(x_0, y_0) &= \lim_{t\to 0^+} \frac{\frac{\partial f}{\partial x}(x_0, y_0)(tu_1) + \frac{\partial f}{\partial y}(x_0, y_0)(tu_2) + \epsilon_1(tu_1) + \epsilon_2(tu_2)}{t} \\ &= \lim_{t\to 0^+} \left\{\frac{\partial f}{\partial x}(x_0, y_0)u_1 + \frac{\partial f}{\partial y}(x_0, y_0)u_2 + \epsilon_1 u_1 + \epsilon_2 u_2\right\} \\ &= \frac{\partial f}{\partial x}(x_0, y_0)u_1 + \frac{\partial f}{\partial y}(x_0, y_0)u_2 + \lim_{t\to 0^+} \epsilon_1 u_1 + \lim_{t\to 0^+} \epsilon_2 u_2. \end{aligned}$$

Since $\epsilon_1 \to 0$ and $\epsilon_2 \to 0$ as $t \to 0$, the last two terms on the right side of the above equation are zero. We have therefore established the following theorem.

THEOREM 8

If the function $f(x, y)$ and its first partial derivatives are continuous in some open rectangle containing (x_0, y_0), the directional derivative $D_{\boldsymbol{u}}f(x_0, y_0)$ in the direction of the unit vector $\boldsymbol{u} = u_1\boldsymbol{i} + u_2\boldsymbol{j}$ is given by

$$D_{\boldsymbol{u}}f(x_0, y_0) = \frac{\partial f}{\partial x}(x_0, y_0)u_1 + \frac{\partial f}{\partial y}(x_0, y_0)u_2. \tag{3}$$

REMARK 2: The corresponding result for functions of n variables, $n \geq 3$, is that the directional derivative in (2) may be expressed as

$$D_{\boldsymbol{u}}f(x_1, x_2, \ldots, x_n) = \frac{\partial f}{\partial x_1}(x_1, \ldots, x_n)u_1 + \frac{\partial f}{\partial x_2}(x_1, x_2, \ldots, x_n)u_2 + \cdots + \frac{\partial f}{\partial x_n}(x_1, x_2, \ldots, x_n)u_n \tag{4}$$

when $f(x_1, x_2, \ldots, x_n)$ and its first order partial derivatives are continuous in a neighborhood of $(x_1, x_2, \ldots, x_n)$ in $\mathbb{R}^n$.

REMARK 3: It is important to note that the vector $\boldsymbol{u}$ in expressions (1) through (4) is a *unit* vector. If you wish to calculate the directional derivative of f in the direction of a vector $\boldsymbol{w}$ with $|\boldsymbol{w}| \neq 1$, you must first obtain the unit vector $\boldsymbol{u} = \frac{\boldsymbol{w}}{|\boldsymbol{w}|}$ in the direction of $\boldsymbol{w}$.

REMARK 4: Note that if $\boldsymbol{u} = \boldsymbol{i}$ (so that $u_1 = 1$ and $u_2 = 0$), and if the partial derivative $\frac{\partial f}{\partial x}(x, y)$ exists, then the directional derivative $D_{\boldsymbol{i}}f(x, y)$ is equal to $\frac{\partial f}{\partial x}(x, y)$. Similarly, if $\boldsymbol{u} = \boldsymbol{j}$ and $\frac{\partial f}{\partial y}(x, y)$ exists, then $D_{\boldsymbol{j}}f(x, y)$ is equal to $\frac{\partial f}{\partial y}(x, y)$. This is most easily seen from the form for $D_{\boldsymbol{u}}f(x, y)$ in Theorem 8. However, $D_{\boldsymbol{u}}f(x, y)$ may exist, as defined in Definition 6, for all $\boldsymbol{u}$ even when one or both of the partial derivatives do not exist (see Exercise 41). Similarly, the directional derivative of the function $f(x_1, x_2, \ldots, x_n)$ in the direction of one of the coordinate axes in $\mathbb{R}^n$ agrees with the corresponding partial derivative when the partial derivative exists.

Example 1 Find the directional derivative $D_{\boldsymbol{u}}f(2, 1)$ where $f(x, y) = x^2e^{3y}$ and $\boldsymbol{u} = \frac{1}{\sqrt{5}}\boldsymbol{i} + \frac{2}{\sqrt{5}}\boldsymbol{j}$.

Solution: The partial derivatives are

$$\frac{\partial f}{\partial x}(2, 1) = 2xe^{3y}\bigg|_{(2,1)} = 4e^3$$

and

$$\frac{\partial f}{\partial y}(2, 1) = 3x^2e^{3y}\bigg|_{(2,1)} = 12e^3.$$

Since

$$|\boldsymbol{u}| = \sqrt{\frac{1}{5} + \frac{4}{5}} = 1,$$

we have from (3) that

$$D_{\boldsymbol{u}}f(2, 1) = (4e^3)\left(\frac{1}{\sqrt{5}}\right) + 12e^3\left(\frac{2}{\sqrt{5}}\right)$$
$$= \frac{28e^3}{\sqrt{5}} \approx 251.5.$$

■

Example 2 Find the directional derivative of the function $f(x, y, z) = e^x \cos y + xz$ at the point $(1, \pi, -1)$ in the direction of the vector $\boldsymbol{w} = \boldsymbol{i} - 3\boldsymbol{j} + 4\boldsymbol{k}$.

Solution: The partial derivatives of $f(x, y, z)$ are

$$\frac{\partial f}{\partial x}(1, \pi, -1) = e^x \cos y + z\Big|_{(1,\pi,-1)} = -e - 1,$$

$$\frac{\partial f}{\partial y}(1, \pi, -1) = -e^x \sin y\Big|_{(1,\pi,-1)} = 0,$$

and

$$\frac{\partial f}{\partial z}(1, \pi, -1) = x\Big|_{(1,\pi,-1)} = 1.$$

Since $|\boldsymbol{w}| = \sqrt{1^2 + 3^2 + 4^2} = \sqrt{26}$, a *unit* vector in the direction of $\boldsymbol{w}$ is $\boldsymbol{u} = \frac{1}{\sqrt{26}}(\boldsymbol{i} - 3\boldsymbol{j} + 4\boldsymbol{k})$, so $u_1 = \frac{1}{\sqrt{26}}$, $u_2 = \frac{-3}{\sqrt{26}}$, and $u_3 = \frac{4}{\sqrt{26}}$. According to (4),

$$D_{\boldsymbol{u}}f(1, \pi, -1) = (-e - 1)\left(\frac{1}{\sqrt{26}}\right) + (0)\left(\frac{-3}{\sqrt{26}}\right) + (1)\left(\frac{4}{\sqrt{26}}\right)$$
$$= \frac{-e + 3}{\sqrt{26}}$$
$$\approx 0.055.$$

■

If $\boldsymbol{u} = u_1\boldsymbol{i} + u_2\boldsymbol{j}$ is a unit vector and θ is the angle formed between $\boldsymbol{u}$ and the positive x-axis, then

$$u_1 = \frac{u_1}{|\boldsymbol{u}|} = \cos\theta, \quad \text{and} \quad u_2 = \frac{u_2}{|\boldsymbol{u}|} = \sin\theta.$$

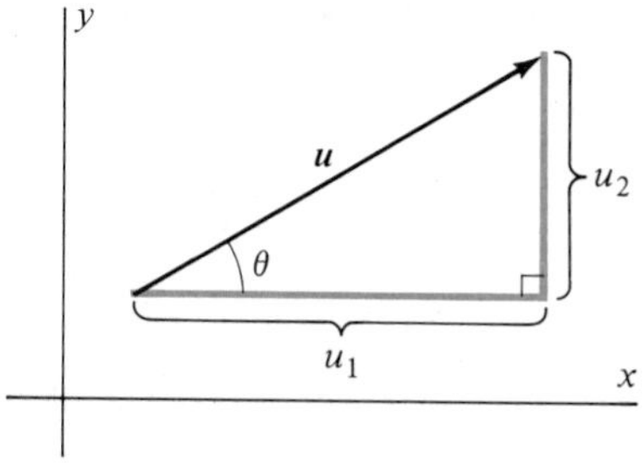

Figure 8.3 If $\boldsymbol{u} = u_1\boldsymbol{i} + u_2\boldsymbol{j}$ with $|\boldsymbol{u}| = 1$, then $u_1 = \cos\theta$, $u_2 = \sin\theta$.

(See Figure 8.3.) Using these equations, we may write the directional derivative $D_{\boldsymbol{u}}f(x_0, y_0)$ in (3) in the form

$$D_{\boldsymbol{u}}f(x_0, y_0) = \frac{\partial f}{\partial x}(x_0, y_0)\cos\theta + \frac{\partial f}{\partial y}(x_0, y_0)\sin\theta. \tag{5}$$

Equation (5) makes explicit the fact that when $f(x, y)$ and the first partials $\frac{\partial f}{\partial x}(x, y)$ and $\frac{\partial f}{\partial y}(x, y)$ are continuous, the directional derivative depends only on the partial derivatives and the direction of the unit vector $\boldsymbol{u}$.

Example 3 Let $f(x, y) = xy - y^3$. Find the unit vector $\boldsymbol{u}$ for which the directional derivative $D_{\boldsymbol{u}}f(2, 1)$ is a maximum.

Solution: We begin by calculating the partial derivatives for f:

$$\frac{\partial f}{\partial x}(2, 1) = y \Big|_{(2,1)} = 1,$$

$$\frac{\partial f}{\partial y}(2, 1) = x - 3y^2 \Big|_{(2,1)} = -1.$$

According to (5), the directional derivative $D_{\boldsymbol{u}}f(2, 1)$ is

$$\begin{aligned} D_{\boldsymbol{u}}f(2, 1) &= (1) \cos \theta + (-1) \sin \theta \\ &= \cos \theta - \sin \theta. \end{aligned}$$

We must therefore find the value of θ for which the function

$$f(\theta) = \cos \theta - \sin \theta$$

is a maximum. To do so we set

$$f'(\theta) = -\sin \theta - \cos \theta = 0$$

and obtain the equation

$$\sin \theta = -\cos \theta, \qquad \text{or} \qquad \tan \theta = -1,$$

which has solutions $\theta = \dfrac{3\pi}{4}$ and $\theta = \dfrac{7\pi}{4}$ for $0 \le \theta \le 2\pi$.

Since

$$f''\left(\frac{3\pi}{4}\right) = -\cos\left(\frac{3\pi}{4}\right) + \sin\left(\frac{3\pi}{4}\right) = \sqrt{2} > 0$$

and

$$f''\left(\frac{7\pi}{4}\right) = -\cos\left(\frac{7\pi}{4}\right) + \sin\left(\frac{7\pi}{4}\right) = -\sqrt{2} < 0,$$

the angle $\theta = \dfrac{7\pi}{4}$ corresponds to the maximum. For this angle, $\boldsymbol{u} = (\cos \theta)\boldsymbol{i} + (\sin \theta)\boldsymbol{j} = \dfrac{\sqrt{2}}{2}\boldsymbol{i} - \dfrac{\sqrt{2}}{2}\boldsymbol{j}$ is the required unit vector. ■

The Gradient

The form of the directional derivative given by Theorem 8 can be written as a dot product:

$$\begin{aligned} D_{\boldsymbol{u}}f(x_0, y_0) &= \frac{\partial f}{\partial x}(x_0, y_0)u_1 + \frac{\partial f}{\partial y}(x_0, y_0)u_2 \\ &= \left[\frac{\partial f}{\partial x}(x_0, y_0)\boldsymbol{i} + \frac{\partial f}{\partial y}(x_0, y_0)\boldsymbol{j}\right] \cdot [u_1\boldsymbol{i} + u_2\boldsymbol{j}]. \end{aligned} \tag{6}$$

The second factor in this dot product is just the unit vector $\boldsymbol{u} = u_1\boldsymbol{i} + u_2\boldsymbol{j}$. The first factor is called the *gradient* of $f(x, y)$ at (x_0, y_0). It is usually written as

$$\boldsymbol{\nabla} f(x_0, y_0) = \frac{\partial f}{\partial x}(x_0, y_0)\boldsymbol{i} + \frac{\partial f}{\partial y}(x_0, y_0)\boldsymbol{j} \tag{7}$$

or just

$$\nabla f = \frac{\partial f}{\partial x}\boldsymbol{i} + \frac{\partial f}{\partial y}\boldsymbol{j}.$$

For functions of more than two variables, the gradient is defined in a similar way.

DEFINITION 7

Let $f(x_1, x_2, \ldots, x_n)$ be a function of n variables for which each of the partial derivatives

$$\frac{\partial f}{\partial x_1}(x_1, x_2, \ldots, x_n), \quad \frac{\partial f}{\partial x_2}(x_1, x_2, \ldots, x_n), \ldots, \quad \frac{\partial f}{\partial x_n}(x_1, x_2, \ldots, x_n) \text{ exist.}$$

The **gradient** of f at $(x_1, x_2, \ldots, x_n)$ is the vector

$$\nabla f(x_1, x_2, \ldots, x_n) = \left\langle \frac{\partial f}{\partial x_1}(x_1, x_2, \ldots, x_n), \frac{\partial f}{\partial x_2}(x_1, x_2, \ldots, x_n), \ldots, \frac{\partial f}{\partial x_n}(x_1, x_2, \ldots, x_n) \right\rangle \tag{8}$$

or, more simply

$$\nabla f = \left\langle \frac{\partial f}{\partial x_1}, \frac{\partial f}{\partial x_2}, \ldots, \frac{\partial f}{\partial x_n} \right\rangle. \tag{9}$$

Thus, the gradient ∇f is just a vector whose components are the partial derivatives of f. In the next few pages, we shall see that the gradient has interesting geometric interpretations in the cases of functions of two and three variables.

Example 4 For $f(x, y) = x \cos y$, the gradient is

$$\begin{aligned}\nabla f(x, y) &= \left[\frac{\partial}{\partial x}(x \cos y)\right]\boldsymbol{i} + \left[\frac{\partial}{\partial y}(x \cos y)\right]\boldsymbol{j} \\ &= \cos y\boldsymbol{i} - x \sin x\boldsymbol{j}\end{aligned}$$

and the vector given by the gradient at $\left(2, \frac{\pi}{4}\right)$ is

$$\nabla f(2, \pi/4) = \frac{\sqrt{2}}{2}\boldsymbol{i} - \sqrt{2}\boldsymbol{j}.$$

■

Example 5 For $f(x, y, z) = \sqrt{x}\, e^y \tan^{-1} z$, the gradient is

$$\nabla f(x, y, z) = \frac{e^y \tan^{-1} z}{2\sqrt{x}}\boldsymbol{i} + \sqrt{x}\, e^y \tan^{-1} z\boldsymbol{j} + \frac{\sqrt{x}\, e^y}{1 + z^2}\boldsymbol{k},$$

and the vector given by the gradient at (4, 0, 1) is

$$\nabla f(4, 0, 1) = \frac{\pi}{16}\boldsymbol{i} + \frac{\pi}{2}\boldsymbol{j} + \boldsymbol{k}.$$

■

Vector Notation and Gradients

In discussing properties of the gradient, it is useful to represent points in $\mathbb{R}^n$ by their position vectors, as in Chapters 17 and 18. Thus, we write

$$(x_1, x_2) \in \mathbb{R}^2 \quad \text{as} \quad \mathbf{x} = x_1\mathbf{i} + x_2\mathbf{j} = \langle x_1, x_2 \rangle$$

$$(x_1, x_2, x_3) \in \mathbb{R}^3 \quad \text{as} \quad \mathbf{x} = x_1\mathbf{i} + x_2\mathbf{j} + x_3\mathbf{k} = \langle x_1, x_2, x_3 \rangle$$

and, in general

$$(x_1, x_2, \ldots, x_n) \in \mathbb{R}^n \text{ as } \mathbf{x} = \langle x_1, x_2, \ldots, x_n \rangle.$$

With this notation, $\nabla f(\mathbf{x})$ refers to the gradient of f evaluated at the point represented by the position vector $\mathbf{x}$.

Using vector notation, we may express the directional derivative of a function of n variables (4) as simply

$$D_{\mathbf{u}}f(\mathbf{x}) = \nabla f(\mathbf{x}) \cdot \mathbf{u} \tag{10}$$

where $\mathbf{x} = \langle x_1, x_2, \ldots, x_n \rangle$ and $\mathbf{u} = \langle u_1, u_2, \ldots, u_n \rangle$ are vectors in $\mathbb{R}^n$. In the cases $n = 2$ and $n = 3$ equation (10) provides insight into the geometry associated with $\nabla f(\mathbf{x})$. From Theorem 4, Chapter 17, we know that

$$\begin{aligned} \nabla f(\mathbf{x}) \cdot \mathbf{u} &= |\nabla f(\mathbf{x})| \cdot |\mathbf{u}| \cdot \cos\theta \\ &= |\nabla f(\mathbf{x})| \cos\theta \qquad (|\mathbf{u}| = 1) \end{aligned} \tag{11}$$

where θ is the angle between the vectors $\nabla f(\mathbf{x})$ and $\mathbf{u}$. Combining equations (10) and (11) gives

$$D_{\mathbf{u}}f(\mathbf{x}) = |\nabla f(\mathbf{x})| \cos\theta. \tag{12}$$

Thus, since $-1 \le \cos\theta \le 1$ for all θ, equation (12) shows that

(i) $-|\nabla f(\mathbf{x})| \le D_{\mathbf{u}}f(\mathbf{x}) \le |\nabla f(\mathbf{x})|$, and
(ii) $D_{\mathbf{u}}f(\mathbf{x})$ assumes its maximum value when $\cos\theta = 1$, that is, when $\theta = 0$.

Since the case $\theta = 0$ occurs precisely when $\nabla f(\mathbf{x})$ and the direction vector $\mathbf{u}$ point in the same direction, conclusion (ii) says that $\nabla f(\mathbf{x})$ points in the direction of the maximum value of $D_{\mathbf{u}}f(\mathbf{x})$. That is,

$$\nabla f(\mathbf{x}) \text{ points in the direction of most rapid increase for } f(\mathbf{x}). \tag{13}$$

(Assuming, of course, that the hypotheses of Theorem 8 hold.)

This observation has several important applications. If $\mathbf{x} = (x, y)$ is a point in the domain of the function $z = f(x, y)$, then $\nabla f(x, y)$ points in the direction in which a path through $(x, y, f(x, y))$ on the graph will rise most rapidly (see Figure 8.4). A path on a surface having the property that the tangent at each point of the path is parallel to the gradient of the function defining the surface is called a **path of steepest ascent.** Of obvious relevance to mountain climbers, this concept is also used to develop procedures for approximating extrema for complicated functions of several variables. (It should be geometrically obvious that proceeding in the opposite direction, $-\nabla f(x, y)$, leads to the **path of steepest descent.** In Exercise 33, you are asked to prove this analytically.)

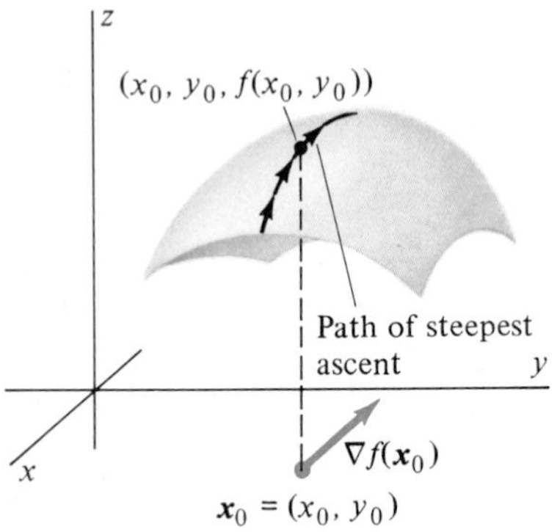

Figure 8.4 $\nabla f(x)$ points in the direction of most rapid increase of the function $z = f(x, y)$.

A second application of observation (13) concerns particles moving so as to maximize an attribute of the medium through which they are moving. Examples of this are insects flying toward a light source (maximizing the intensity of light), heat-seeking missiles (maximizing temperature), and sharks in water (maximizing blood concentration). In such situations, if $A(\boldsymbol{x})$ is a function giving the value of the attribute at location $\boldsymbol{x}$, the particle will move so that its tangent vector points in the direction of $\nabla A(\boldsymbol{x})$ (see Example 6).

Example 6 For the function $f(x, y) = 9 - \dfrac{x^2 + y^2}{4}$, the gradient at $\boldsymbol{x} = (x, y)$ is

$$\begin{aligned}\nabla f(\boldsymbol{x}) &= -\frac{x}{2}\boldsymbol{i} - \frac{y}{2}\boldsymbol{j}\\ &= -\frac{1}{2}(x\boldsymbol{i} + y\boldsymbol{j})\\ &= -\frac{1}{2}\boldsymbol{x}.\end{aligned}$$

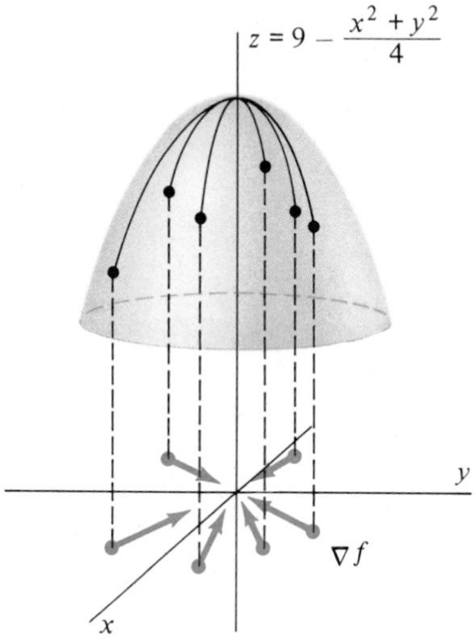

Figure 8.5 $\nabla f(\boldsymbol{x}) = -\dfrac{1}{2}\boldsymbol{x}$ for $f(x, y) = 9 - \dfrac{x^2 + y^2}{4}$.

Since $\boldsymbol{x} = x\boldsymbol{i} + y\boldsymbol{j}$ is the position vector of the point (x, y), the vector $\nabla f(\boldsymbol{x}) = -\dfrac{1}{2}\boldsymbol{x}$ points toward the origin for all $(x, y) \neq (0, 0)$. In particular,

if $\boldsymbol{x} = (2, 2)$, $\quad \nabla f = -\boldsymbol{i} - \boldsymbol{j}$,

if $\boldsymbol{x} = (1, 4)$, $\quad \nabla f = -\dfrac{1}{2}\boldsymbol{i} - 2\boldsymbol{j}$,

if $\boldsymbol{x} = (2, -1)$, $\quad \nabla f = -\boldsymbol{i} + \dfrac{1}{2}\boldsymbol{j}$.

This should not be surprising, since the graph of $z = f(x, y)$ is a paraboloid. At any point on this surface, the z-coordinate is increased most rapidly by moving directly toward the z-axis (see Figure 8.5). ■

Example 7 The temperature distribution across the surface of a rectangular plate is given by the function $T(x, y) = 100 - x^2 - 2y^2$. Find the path followed by a heat-seeking particle placed on the plate at the point (4, 2).

Solution: Let the path followed by the particle have parameterization

$$\boldsymbol{r}(t) = x(t)\boldsymbol{i} + y(t)\boldsymbol{j}.$$

Assuming the components $x(t)$ and $y(t)$ to be differentiable functions of t, we recall from Chapter 18 that the tangent vector $\boldsymbol{T}(t)$ at each point along the path is

$$\boldsymbol{T}(t) = x'(t)\boldsymbol{i} + y'(t)\boldsymbol{j}.$$

Since the particle is heat seeking, this tangent should point in the same direction as the gradient

$$\nabla T(x, y) = -2x\boldsymbol{i} - 4y\boldsymbol{j}.$$

This will be the case when

$$x'(t) = -2x(t) \qquad \text{and} \qquad y'(t) = -4y(t).$$

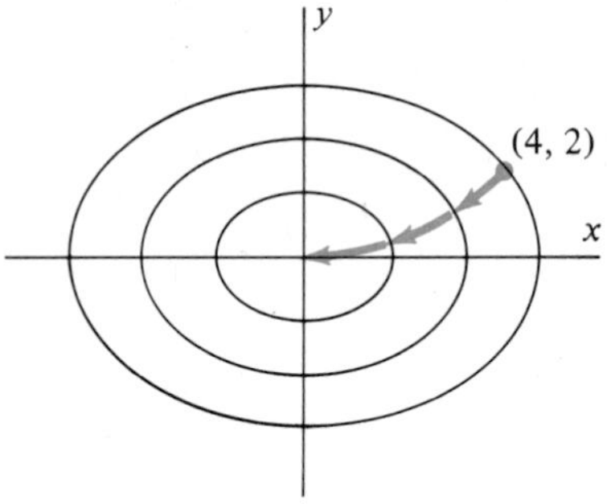

Figure 8.6 Heat-seeking particle approaches origin along parabolic path.

These are the familiar differential equations for exponential decay. Their solutions are

$$x(t) = x(0)e^{-2t}, \qquad \text{and} \qquad y(t) = y(0)e^{-4t}.$$

Since the particle begins at (4, 2), we have $x(0) = 4$ and $y(0) = 2$. Thus

$$x(t) = 4e^{-2t}, \qquad y(t) = 2e^{-4t}$$

are the components of parameterization of the path. We may eliminate the parameter t by noting that

$$y(t) = 2e^{-4t} = \frac{1}{8}(4e^{-2t})^2 = \frac{1}{8}[x(t)]^2.$$

The particle therefore approaches the origin along the parabola $y = \frac{1}{8}x^2$ (Figure 8.6). ■

Level Curves and Surfaces

Let $\boldsymbol{r}(t) = x(t)\boldsymbol{i} + y(t)\boldsymbol{j} + z(t)\boldsymbol{k}$ be a parameterization for a curve in $\mathbb{R}^3$, and let $w = f(x, y, z)$ be a function of three variables. Using the gradient and assuming that all necessary derivatives exist, we may write the Chain Rule (Theorem 6)

$$\frac{d}{dt}f(x, y, z) = \frac{\partial f}{\partial x}\cdot\frac{dx}{dt} + \frac{\partial f}{\partial y}\cdot\frac{dy}{dt} + \frac{\partial f}{\partial z}\cdot\frac{dz}{dt}$$

as

$$\boxed{\frac{d}{dt}f(\boldsymbol{r}(t)) = \boldsymbol{\nabla} f(\boldsymbol{r}(t))\cdot\boldsymbol{r}'(t).} \tag{14}$$

Equation (14) says that the derivative of the composite function $f(\boldsymbol{r}(t))$ is the dot product of the gradient $\boldsymbol{\nabla} f(\boldsymbol{r}(t))$ with the tangent vector $\boldsymbol{r}'(t) = x'(t)\boldsymbol{i} + y'(t)\boldsymbol{j} + z'(t)\boldsymbol{k}$ for each value of t. Equation (14) also holds for curves $\boldsymbol{r}(t) = x(t)\boldsymbol{i} + y(t)\boldsymbol{j}$ in the plane composed with functions $z = f(x, y)$ of two variables, as does Theorem 6. (Of course, the beauty of the vector notation is that we need not distinguish between these two cases in writing equation (14).)

Equation (14) reveals an important relationship between gradients and level curves. To see this, suppose that $z = f(x, y)$ is a function of two variables for which the equation

$$f(x, y) = k, \qquad k \text{ constant} \tag{15}$$

determines a curve C in the xy-plane with parameterization

$$\boldsymbol{r}(t) = x(t)\boldsymbol{i} + y(t)\boldsymbol{j}.$$

Then along this level curve the composite function

$$f(\boldsymbol{r}(t)) = f(x(t), y(t)) \tag{16}$$

is constant, according to equation (15).

Now let $\boldsymbol{x}_0 = \boldsymbol{r}(t_0) = (x(t_0), y(t_0))$ be a point on C. If the functions $f(x, y)$, $x(t)$ and $y(t)$ satisfy the hypotheses of Theorem 6 in a neighborhood of $\boldsymbol{x}_0$, we may

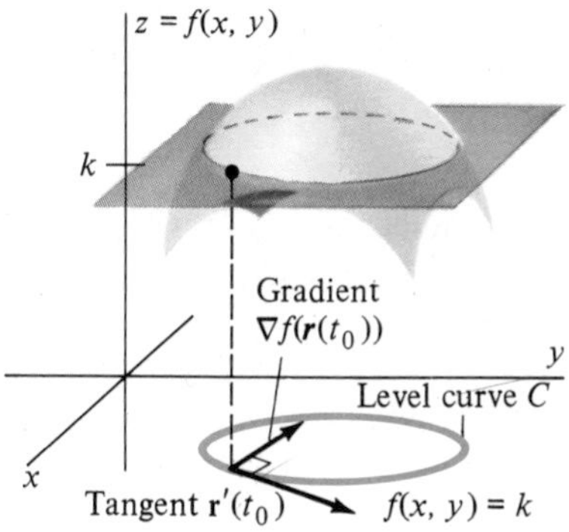

Figure 8.7 $\nabla f(x_0)$ is orthogonal to the level curve through x_0.

differentiate both sides of equation (16), using equation (14), to conclude that

$$\frac{d}{dt}f(\boldsymbol{r}(t))\bigg|_{t=t_0} = \nabla f(\boldsymbol{r}(t_0))\cdot\boldsymbol{r}'(t_0) = 0. \tag{17}$$

Since $\boldsymbol{r}'(t_0)$ is a vector tangent to the level curve C, equation (17) shows that *the gradient* $\nabla f(\boldsymbol{r}(t_0))$ *is orthogonal to the level curve* C *at* $\boldsymbol{x}_0$ (see Figure 8.7). That is,

> If $f(x, y)$ and its first partial derivatives are continuous, then at each point in the domain of $f(x, y)$ the gradient vector, if nonzero, is orthogonal to the level curve through that point. (18)

Example 8 The graph of the function $z = \sqrt{x^2 + y^2}$ is a cone. The level curves associated with this graph are the circles

$$\sqrt{x^2 + y^2} = k, \quad \text{or} \quad x^2 + y^2 = k^2.$$

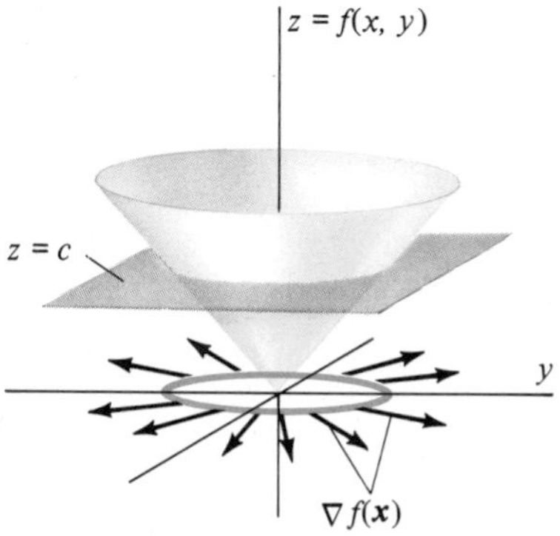

Figure 8.8 For $f(x, y) = \sqrt{x^2 + y^2}$ level curves are circles; $\nabla f(x)$ is parallel to a radius for the circle.

The gradient for this function is

$$\nabla f(x, y) = \frac{x}{\sqrt{x^2 + y^2}}\boldsymbol{i} + \frac{y}{\sqrt{x^2 + y^2}}\boldsymbol{j} = \frac{x\boldsymbol{i} + y\boldsymbol{j}}{\sqrt{x^2 + y^2}}.$$

In vector notation, the gradient may be written as

$$\nabla f(\boldsymbol{x}) = \frac{\boldsymbol{x}}{|\boldsymbol{x}|}.$$

This shows that the gradient vectors are parallel to radius vectors for the circular level curves and, therefore, orthogonal to the level curves (Figure 8.8). ■

A similar relationship exists between gradients for functions $f(x, y, z)$ of three variables and level surfaces. Suppose that

$$f(x, y, z) = k$$

is the equation of a level surface S for the function $w = f(x, y, z)$, and suppose that the point $\boldsymbol{x}_0$ lies on S. Choose a curve C on the surface S with parameterization

$$\boldsymbol{r}(t) = x(t)\boldsymbol{i} + y(t)\boldsymbol{j} + z(t)\boldsymbol{k}$$

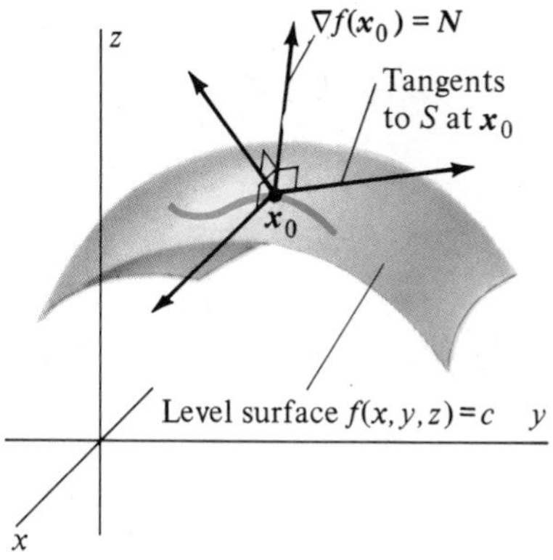

Figure 8.9 $\nabla f(x_0)$ is orthogonal to the level surface for $w = f(x, y, z)$ containing x_0.

for which $\boldsymbol{r}(t_0) = \boldsymbol{x}_0$ (see Figure 8.9). Then, along this curve

$$f(\boldsymbol{r}(t)) = f(x(t), y(t), z(t)) = k, \tag{19}$$

that is, the composite function $f(\boldsymbol{r}(t))$ is constant. As before, if the function $f(x, y, z)$ and the component functions $x(t)$, $y(t)$, and $z(t)$ satisfy the hypotheses of Theorem 6 in a neighborhood of $\boldsymbol{x}_0 = \boldsymbol{r}(t_0)$, we may differentiate both sides of equation (19), using the Chain Rule, to conclude that

$$\frac{d}{dt}f(\boldsymbol{r}(t))\bigg|_{t=t_0} = \nabla f(\boldsymbol{r}(t_0))\cdot\boldsymbol{r}'(t_0) = 0. \tag{20}$$

Equation (20) shows that the gradient $\nabla f(\boldsymbol{x}_0)$ is orthogonal to the tangent $\boldsymbol{r}'(t_0)$ to the curve C through $\boldsymbol{x}_0$ on the surface S. Since C was an *arbitrary* curve on S, it follows that $\nabla f(\boldsymbol{x}_0)$ is orthogonal to *every* tangent to S at $\boldsymbol{x}_0$. In particular, $\nabla f(\boldsymbol{x}_0)$ is orthogonal to the tangent plane determined by $\frac{\partial f}{\partial x}(x_0)$ and $\frac{\partial f}{\partial y}(x_0)$ and therefore to the level surface S itself.

> If $f(x, y, z)$ and its first partial derivatives are continuous, then at each point in the domain of $f(x, y, z)$ the gradient vector, if nonzero, is orthogonal to the level surface containing that point. (21)

Statement (21) will be useful in the discussion of tangent planes in Section 19.9.

Exercise Set 19.8

In Exercises 1–9, find the gradient of the given function at the point P.

1. $f(x, y) = x^2y, \quad P = (3, 1)$

2. $f(x, y) = xy^2 - ye^x, \quad P = (0, 2)$

3. $f(x, y) = x\cos(y - x), \quad P = (\pi/2, \pi/4)$

4. $f(x, y) = \dfrac{2x}{y - x}, \quad P = (2, 1)$

5. $f(x, y, z) = x^2y + xz^2, \quad P = (1, 1, 2)$

6. $f(x, y, z) = xe^{yz}, \quad P = (2, 0, 1)$

7. $f(x, y, z) = x^2y + xz^3 - y^2z, \quad P = (1, 2, 1)$

8. $f(x, y, z) = e^x\cos y - e^y\sin z, \quad P = (0, \pi/4, \pi/3)$

9. $f(x, y, z) = x\sqrt{y}\cosh z, \quad P = (2, 4, 1)$

In Exercises 10–19, find the directional derivative of the given function at the given point in the direction of the given vector.

10. $f(x, y) = 3x^2 - y^2, \quad P = (1, 2), \quad \boldsymbol{u} = \boldsymbol{i} + \boldsymbol{j}$

11. $f(x, y) = x^2y^2, \quad P = (-2, 3), \quad \boldsymbol{u} = \boldsymbol{i} - \boldsymbol{j}$

12. $f(x, y) = \sin(y - x), \quad P = (\pi/2, \pi/4), \quad \boldsymbol{u} = \boldsymbol{i} + 2\boldsymbol{j}$

13. $f(x, y) = \dfrac{x}{x + y}, \quad P = (1, 2), \quad y = \sqrt{3}\,\boldsymbol{i} + \boldsymbol{j}$

14. $f(x, y) = xe^{y^2} - y, \quad P = (1, 3), \quad \boldsymbol{u} = \sqrt{2}\,\boldsymbol{i} + \sqrt{2}\,\boldsymbol{j}$

15. $f(x, y, z) = xy + xz + yz, \quad P = (1, 2, 1),$
$\boldsymbol{u} = \boldsymbol{i} + \boldsymbol{j} - \boldsymbol{k}$

16. $f(x, y, z) = \cosh x - \sinh y + \tan^{-1} z,$
$P = (-1, 1, 3), \quad \boldsymbol{u} = \boldsymbol{i} - \boldsymbol{j} - 2\boldsymbol{k}$

17. $f(x, y, z) = xe^{yz}, \quad P = (2, 0, 1),$
$\boldsymbol{u} = \sqrt{3}\,\boldsymbol{i} + \sqrt{3}\,\boldsymbol{j} - \sqrt{5}\,\boldsymbol{k}$

18. $f(x, y, z) = xy^2 - x^2z + (yz)^3, \quad P = (1, 0, 1),$
$\boldsymbol{u} = \boldsymbol{i} + 2\boldsymbol{j} + \boldsymbol{k}$

19. $f(x, y, z) = x\cos y - y\sinh z, \quad P = (6, \pi/4, -1),$
$\boldsymbol{u} = \boldsymbol{i} - 2\boldsymbol{j} - 2\boldsymbol{k}$

In Exercises 20–24, write vector equations for the normal and tangent lines to the given curves at the given points.

20. $2x^2 + 3y^2 = 11, \quad P = (2, 1)$

21. $4x^2 - y^2 = 7, \quad P = (2, 3)$

22. $3x^2 - y^4 + x = 13, \quad P = (2, 1)$

23. $\sqrt{x} + \sqrt{y} = 4, \quad P = (4, 4)$

24. $6x^2 - 4y^2 = 18, \quad P = (3, -3)$

25. For $f(x, y) = xy^2 + ye^x$, find the directional derivative at $(0, 1)$ in the direction of most rapid increase of f.

26. For $f(x, y) = x^2 + 2y^3$, find the directional derivative at $(2, 3)$ in the direction toward the origin.

27. For $f(x, y)$ as in Exercise 25, find $D_{\boldsymbol{u}}f(0, 1)$ in the direction of the origin.

28. For $f(x, y) = \dfrac{x}{x + y}$, find $D_{\boldsymbol{u}}(1, 1)$ in the direction of the point $(2, 3)$.

29. For $f(x, y)$ as in Exercise 28, find $D_{\boldsymbol{u}}f(1, 1)$ in the direction of most rapid increase of $f(x, y)$.

30. Use the gradient to find the point(s) on the graph of the hyperbola $3x^2 - 2y^2 - 6x + 8y = 5$ where the tangent is horizontal.

31. Use the gradient to find the point(s) on the graph of the ellipse $x^2 - 6x + 2y^2 - 4y = -7$ where the tangent is
a. horizontal, **b.** vertical.

32. Use equation (10) to show that the directional derivative of f in the direction of the unit vector $\boldsymbol{u}$ is the component of f in the direction of $\boldsymbol{u}$.

33. Use equation (10) to show that the negative gradient, $-\nabla f(\boldsymbol{x})$ points in the direction of most rapid decrease of the function $f(x, y)$.

34. The temperature distribution in a room is given by the function $T(x, y, z) = 30 - (x^2 + 2y^2 + 3x^2)$. An insect flies so as to experience the most rapid decrease in temperature. In what direction does it move when located at point $(2, 1, 1)$?

35. Show that $|\nabla f(\boldsymbol{x}_0)|$ is the maximum value of $D_{\boldsymbol{u}}f(\boldsymbol{x}_0)$ when f and its first partials are continuous at $\boldsymbol{x}_0$.

36. Show that if $\cos\alpha$, $\cos\beta$, and $\cos\gamma$ are the direction cosines for the unit vector $\boldsymbol{u}$, then $D_{\boldsymbol{u}}f = \dfrac{\partial f}{\partial x}\cos\alpha\boldsymbol{i} + \dfrac{\partial f}{\partial y}\cos\beta\boldsymbol{j} + \dfrac{\partial f}{\partial z}\cos\gamma\boldsymbol{k}.$

37. Find the direction of steepest ascent at the point above $(4, 2)$ on the graph of $z = e^{x^2+y^2}$.

38. Find the direction of steepest ascent at the point above $(1, 1)$ on the graph of $z = xe^y + ye^x$.

39. Find the direction of steepest *descent* at the point above $(1, 2)$ on the graph of $z = 4x^2 - 2y^2$.

40. The temperature distribution on a metal plate is $T(x, y) = 200 - (x^2 + 4y^2)$. Find the path followed by a heat-seeking particle placed at the point $(0, 2)$.

41. Show that if the partial derivative $\dfrac{\partial f}{\partial x}(x_0, y_0)$ exists, then $D_i(x_0, y_0)$ exists and $D_i(x_0, y_0) = \dfrac{\partial f}{\partial x}(x_0, y_0)$. Give an example to show that $D_{\boldsymbol{u}}f(x_0, y_0)$ can exist for all $\boldsymbol{u}$ while $\dfrac{\partial f}{\partial x}(x_0, y_0)$ (and, therefore, $\nabla f(x_0, y_0)$) fails to exist.

19.9 DIFFERENTIABILITY: MORE ON TANGENT PLANES AND GRADIENTS

The purpose of this section is to tie up some loose ends concerning the notions of differentiability, tangent planes, and gradients. Having now seen many of the techniques and applications associated with partial and directional derivatives, we need to address the broader concept of an appropriate analogue of differentiability for functions of more than one variable. The failure of both the partial derivative and the directional derivative to provide a satisfactory generalization for the one-variable concept of differentiability can be seen by examining the function

$$f(x, y) = \begin{cases} \dfrac{xy^2}{x^2 + y^4}, & (x, y) \neq (0, 0) \\ 0, & (x, y) = (0, 0). \end{cases}$$

Although every directional derivative exists (Exercise 16), the function $f(x, y)$ is not continuous at $(0, 0)$. This last statement can be seen by noting that $f(x, y) = \dfrac{1}{2}$ for all (x, y) on the parabola $y^2 = x$ except $(0, 0)$. Since differentiable functions are necessarily continuous in the one-variable case, we need to develop a more powerful concept than just the directional derivative.

The key to developing the idea of differentiability for functions of several variables is to recall an equivalent formulation of differentiability in the one-variable case. If

$$f'(x_0) = \lim_{\Delta x \to 0} \frac{f(x_0 + \Delta x) - f(x_0)}{\Delta x} \tag{1}$$

exists, then for Δx small, we have

$$f'(x_0) = \frac{f(x_0 + \Delta x) - f(x_0)}{\Delta x} + \epsilon$$

where $\epsilon \to 0$ as $\Delta x \to 0$. Multiplying through by Δx and rearranging terms results in the equation

$$f(x_0 + \Delta x) = f(x_0) + f'(x_0)\Delta x + \epsilon \cdot \Delta x, \qquad \epsilon \to 0 \quad \text{as} \quad \Delta x \to 0. \tag{2}$$

Of course, equation (2) results from knowing that $f'(x_0)$ in equation (1) exists. However, if we know only that a number D exists for which

$$f(x_0 + \Delta x) = f(x_0) + D\Delta x + \epsilon\, \Delta x, \qquad \epsilon \to 0 \quad \text{as} \quad \Delta x \to 0 \tag{3}$$

then we can reverse the steps leading from equation (1) to equation (2) and show that $D = \lim_{\Delta x \to 0} \dfrac{f(x_0 + \Delta x) - f(x_0)}{\Delta x} = f'(x_0)$. That is, equation (3) provides an equivalent definition of differentiability for $f(x)$: If a number D exists for which equation (3) holds, then $f(x)$ is differentiable, and $f'(x_0) = D$ (see Section 3.9).

Equation (3) provides the notion of differentiability that we wish to generalize to functions of more than one variable. Before doing so, let's also review the geometry associated with equation (3). Since $f(x_0)$ is fixed once x_0 is chosen, the first two terms on the right-hand side of equation (3) constitute a linear function of Δx:

$$g(x_0 + \Delta x) = f(x_0) + D\Delta x. \tag{4}$$

This linear function approximates the actual value of $f(x_0 + \Delta x)$, with the error in the approximation being provided by the remaining term $\epsilon \cdot \Delta x$. Finally, note that since $\epsilon \to 0$ as $\Delta x \to 0$, this error term has the property that

$$\lim_{\Delta x \to 0} \frac{\text{error term}}{\Delta x} = \lim_{\Delta x \to 0} \frac{\epsilon \cdot \Delta x}{\Delta x} = \lim_{\Delta x \to 0} \epsilon = 0. \tag{5}$$

(See Figure 9.1.)

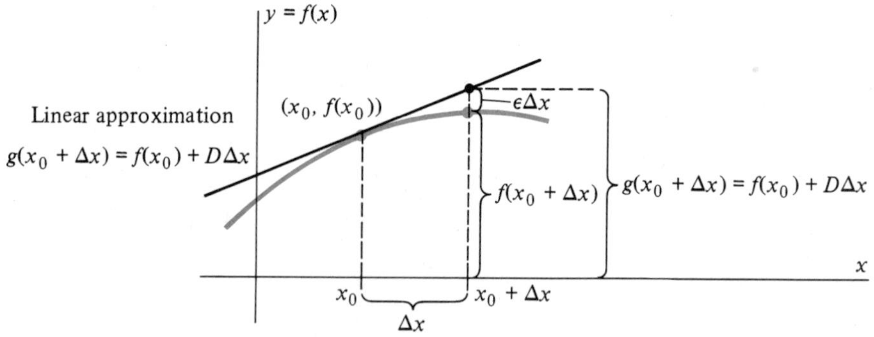

Figure 9.1 A differentiable function $f(x)$ is approximated by a linear function of Δx: the tangent at $(x_0, f(x_0))$.

Now let (x_0, y_0) be a point in the domain of the function $f(x, y)$ of two variables. If we wish to approximate $f(x_0 + \Delta x, y_0 + \Delta y)$ by a linear (meaning flat) object, that object is a plane. In order that the plane contain the point $(x_0, y_0, f(x_0, y_0))$, it must have an equation of the form

$$g(x_0 + \Delta x, y_0 + \Delta y) = f(x_0, y_0) + D_1\Delta x + D_2\Delta y \tag{6}$$

where D_1 and D_2 are constants. (This is the analogue of equation (4)). Also, by analogy with equation (5), we will require that if $\epsilon(\Delta x, \Delta y)$ denotes the error term for this approximation, then

$$\lim_{\substack{\Delta x\to 0\\ \Delta y\to 0}} \frac{\text{error term}}{|\Delta x\,\boldsymbol{i} + \Delta y\,\boldsymbol{j}|} = \lim_{\substack{\Delta x\to 0\\ \Delta y\to 0}} \frac{\epsilon(\Delta x,\Delta y)}{\sqrt{\Delta x^2 + \Delta y^2}} = 0. \tag{7}$$

Putting these ideas together, we arrive at our definition of differentiability for a function of two variables.

DEFINITION 8

The function $f(x, y)$ is said to be **differentiable** at (x_0, y_0) if there exist numbers D_1 and D_2 and a function $\epsilon(\Delta x, \Delta y)$ so that

$$f(x_0 + \Delta x, y_0 + \Delta y) = f(x_0, y_0) + D_1\Delta x + D_2\Delta y + \epsilon(\Delta x, \Delta y) \tag{8}$$

where

$$\lim_{\substack{\Delta x\to 0\\ \Delta y\to 0}} \frac{\epsilon(\Delta x, \Delta y)}{\sqrt{\Delta x^2 + \Delta y^2}} = 0. \tag{9}$$

Equation (8) is the precise analogue of equation (3), so we have succeeded in generalizing the concept of differentiability in the desired way. Moreover, just as equation (3) determines what is meant by the line tangent to the graph of $y = f(x)$ at $(x_0, f(x_0))$, Definition 8 provides the definition of what we mean by the plane tangent to the graph of $z = f(x_0, y_0)$ at $(x_0, y_0, f(x_0, y_0))$.

DEFINITION 9

The plane tangent to the graph of $z = f(x, y)$ at the point (x_0, y_0, z_0), if it exists, has the equation $z = g(x, y)$ where, for $x = x_0 + \Delta x$, $y = y_0 + \Delta y$,

$$g(x_0 + \Delta x, y_0 + \Delta y) = f(x_0, y_0) + D_1\Delta x + D_2\Delta y, \tag{10}$$

with D_1 and D_2 as in Definition 8.

We may paraphrase Definition 8 by saying that the function $f(x, y)$ is differentiable at (x_0, y_0) if it has a tangent plane that is a ''good approximation'' in the sense of equation (9) (Figure 9.2). It is a fact, which we shall not prove, that the tangent plane determined by Definition 8, if it exists, is unique.

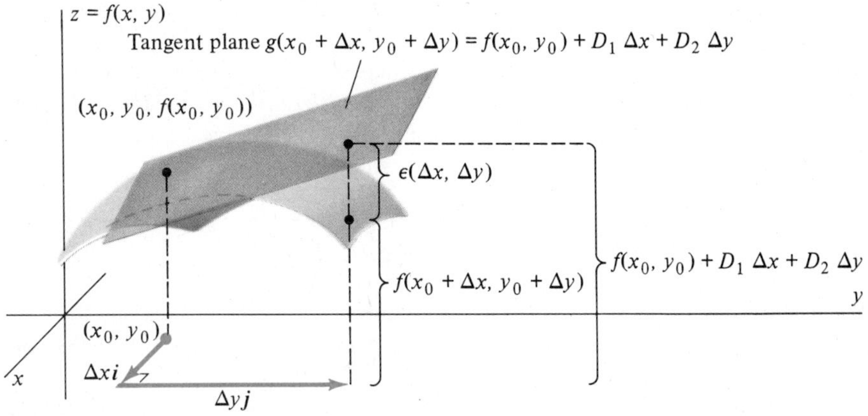

Figure 9.2 A differentiable function $f(x, y)$ is approximated by a tangent plane.

Let's now move toward bringing the discussion back to familiar ground. Recall from Theorem 5, Section 19.6, that if the function $f(x, y)$ and its first partial derivatives are continuous at (x_0, y_0) then

$$f(x_0 + \Delta x, y_0 + \Delta y) = f(x_0, y_0) + \frac{\partial f}{\partial x}(x_0, y_0)\Delta x + \frac{\partial f}{\partial y}(x_0, y_0)\Delta y + \epsilon_1 \Delta x + \epsilon_2 \Delta y \tag{11}$$

where $\epsilon_1 \to 0$ and $\epsilon_2 \to 0$ as $\Delta x \to 0$ and $\Delta y \to 0$. Now the error term in equation (11) is $\epsilon(\Delta x, \Delta y) = \epsilon_1 \Delta x + \epsilon_2 \Delta y$ and the result of Exercise 17 of this section shows that

$$\lim_{\substack{\Delta x \to 0 \\ \Delta y \to 0}} \frac{\epsilon_1 \Delta x + \epsilon_2 \Delta y}{\sqrt{\Delta x^2 + \Delta y^2}} = 0 \tag{12}$$

if $\epsilon_1 \to 0$ and $\epsilon_2 \to 0$ as $\Delta x \to 0$ and $\Delta y \to 0$. Thus, if we simply take $D_1 = \frac{\partial f}{\partial x}(x_0, y_0)$ and $D_2 = \frac{\partial f}{\partial y}(x_0, y_0)$, we can conclude that $f(x, y)$ is differentiable at (x_0, y_0). Moreover, using the vector notation $\boldsymbol{x}_0 = (x_0, y_0)$ and $\Delta \boldsymbol{x} = \Delta x\, \boldsymbol{i} + \Delta y\, \boldsymbol{j}$ we can write equations (8) and (9) as simply

$$f(\boldsymbol{x}_0 + \Delta \boldsymbol{x}) = f(\boldsymbol{x}_0) + \nabla f(\boldsymbol{x}_0) \cdot \Delta \boldsymbol{x} + \epsilon(\Delta \boldsymbol{x}) \tag{13}$$

where

$$\lim_{|\Delta \boldsymbol{x}| \to 0} \frac{\epsilon(\Delta \boldsymbol{x})}{|\Delta \boldsymbol{x}|} = 0. \tag{14}$$

Equations (13) and (14) indicate how differentiability is defined for functions of more than two variables: The function $f(\boldsymbol{x})$ is differentiable at $\boldsymbol{x}_0 \in \mathbb{R}^n$ if there exists a vector $\nabla f(\boldsymbol{x}_0)$ so that equations (13) and (14) hold.

Finally, with $z = g(x_0 + \Delta x, y_0 + \Delta y)$, $z_0 = f(x_0, y_0)$, $\Delta x = x - x_0$, and $\Delta y = y - y_0$, we may write the equation (10) for the tangent plane as

$$z = z_0 + D_1(x - x_0) + D_2(y - y_0).$$

With $D_1 = \frac{\partial f}{\partial x}(x_0, y_0)$ and $D_2 = \frac{\partial f}{\partial y}(x_0, y_0)$ this becomes

$$z = z_0 + \frac{\partial f}{\partial x}(x_0, y_0)(x - x_0) + \frac{\partial f}{\partial y}(x_0, y_0)(y - y_0),$$

which agrees with equation (3) of Section 19.4. That is, *the plane tangent to the graph of a differentiable function is the plane determined by the partial derivatives when they are continuous.*

We summarize these findings in the form of a theorem.

THEOREM 9 Let $f(x, y)$ and the first partial derivatives $\frac{\partial f}{\partial x}$ and $\frac{\partial f}{\partial y}$ be continuous in a neighborhood of the point (x_0, y_0). Then $f(x, y)$ is differentiable at (x_0, y_0) and the tangent plane at $(x_0, y_0, f(x_0, y_0))$ has equation

$$z = \frac{\partial f}{\partial x}(x_0, y_0)(x - x_0) + \frac{\partial f}{\partial y}(x_0, y_0)(y - y_0) + z_0 \tag{15}$$

and normal vector

$$N = \frac{\partial f}{\partial x}(x_0, y_0)\boldsymbol{i} + \frac{\partial f}{\partial y}(x_0, y_0)\boldsymbol{j} - \boldsymbol{k}, \tag{16}$$

where $z_0 = f(x_0, y_0)$.

Unfortunately, the converse of Theorem 9 is false. That is, there are examples of differentiable functions that do *not* have continuous partial derivatives. However, we shall not dwell on this distinction. In most instances we will simply require our functions to have continuous first partial derivatives at the point (x_0, y_0) in question. In such situations:

1. The tangent plane at (x_0, y_0, z_0) exists. Its normal vector and equation are as in Theorem 9.
2. Both partial derivatives and all directional derivatives exist, and the directional derivative $D_{\boldsymbol{u}}f(\boldsymbol{x}_0)$ may be calculated by the formula

   $D_{\boldsymbol{u}}f(\boldsymbol{x}_0) = \nabla f(\boldsymbol{x}_0) \cdot \boldsymbol{u}.$
3. The Chain Rule for the composite function $f(\boldsymbol{r}(t))$ is

   $\frac{d}{dt}f(\boldsymbol{r}(t)) = \boldsymbol{\nabla} f(\boldsymbol{r}(t)) \cdot \boldsymbol{r}'(t).$
4. All other Chain Rules in Section 19.7 apply.

Planes Tangent to Level Surfaces

Planes tangent to the graphs of functions of two variables $z = f(x, y)$ were discussed in Section 19.4. Additional exercises concerning such planes appear at the end of this section. However, there is another situation in which we will need to be able to find an equation for a plane tangent to a surface. That is the situation in which the surface is a *level surface* $f(x, y, z) = k$ for a function of three variables $w = f(x, y, z)$ (see Figure 9.3).

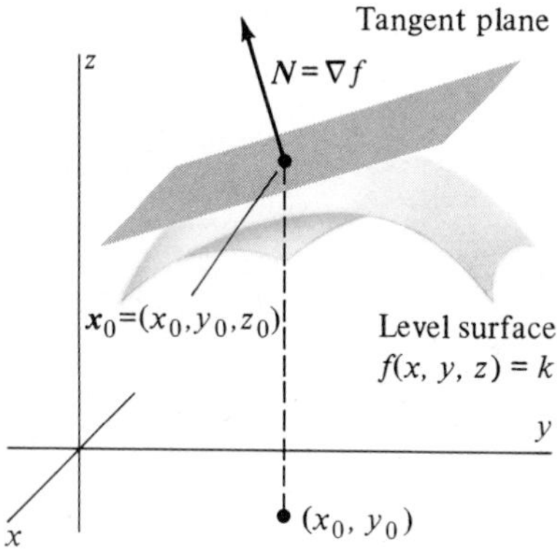

Figure 9.3

$N = \frac{\partial f}{\partial x}\boldsymbol{i} + \frac{\partial f}{\partial y}\boldsymbol{j} + \frac{\partial f}{\partial z}\boldsymbol{k} = \nabla f$ is normal to the level surface $f(x, y, z) = k$.

In such cases we may not be able to express the surface explicitly as the graph of a function $z = g(x, y)$ of two variables. However, it may be clear from the form of the function $w = f(x, y, z)$ that the equation $f(x, y, z) = k$ does define implicitly a function of two variables for which the first partial derivatives are continuous. (In such cases we say that the surface is *smooth,* which means just what you think it does—no jags, tears, or points.) Theorem 9 then guarantees that the tangent plane exists—the issue is only how to find its equation.

This is where statement (23) of Section 19.8 comes in. *Since* $\boldsymbol{\nabla} f(\boldsymbol{x}_0)$ *is orthogonal to the level surface* $f(x, y, z) = k$ *at* $\boldsymbol{x}_0$, $\nabla f(\boldsymbol{x}_0)$ *is a normal to the desired tangent plane.* In particular, if $\boldsymbol{x}_0 = (x_0, y_0, z_0)$, the normal vector is

$$N = \boldsymbol{\nabla} f(\boldsymbol{x}_0) = \frac{\partial f}{\partial x}(x_0, y_0, z_0)\boldsymbol{i} + \frac{\partial f}{\partial y}(x_0, y_0, z_0)\boldsymbol{j} + \frac{\partial f}{\partial z}(x_0, y_0, z_0)\boldsymbol{k} \tag{17}$$

and the plane equation $N \cdot (x - x_0) = 0$ becomes

$$\frac{\partial f}{\partial x}(x_0, y_0, z_0)(x - x_0) + \frac{\partial f}{\partial y}(x_0, y_0, z_0)(y - y_0) + \frac{\partial f}{\partial z}(x_0, y_0, z_0)(z - z_0) = 0. \tag{18}$$

Example 1 Find equations for a vector normal to the graph of the ellipsoid $2x^2 + 4y^2 + z^2 = 21$ at the point (2, 1, 3), and find an equation for the plane tangent to the graph at that point.

Solution: We regard the graph of the equation $2x^2 + 4y^2 + z^2 = 21$ as the level surface $f(x, y, z) = 21$ for the function

$$f(x, y, z) = 2x^2 + 4y^2 + z^2.$$

Since

$$\frac{\partial f}{\partial x}(2, 1, 3) = 4 \cdot 2 = 8, \qquad \frac{\partial f}{\partial y}(2, 1, 3) = 8 \cdot 1 = 8$$

and

$$\frac{\partial f}{\partial z}(2, 1, 3) = 2 \cdot 3 = 6,$$

a normal vector, according to (17) is

$$N = 8\boldsymbol{i} + 8\boldsymbol{j} + 6\boldsymbol{k}$$

and the equation of the tangent plane, from (18) is

$$8(x - 2) + 8(y - 1) + 6(z - 3) = 0$$

or

$$8x + 8y + 6z - 42 = 0.$$

■

REMARK 1: The result of Example 1 may also be obtained by solving the equation $2x^2 + 4y^2 + z^2 = 21$ for z as $z = \pm\sqrt{21 - 2x^2 - 4y^2}$ and working with the function $f(x, y) = \sqrt{21 - 2x^2 - 4y^2}$, which corresponds to the upper half of the ellipsoid.

REMARK 2: We now have two distinct methods for finding the plane tangent to a given surface:

(i) If the surface can be expressed as the graph of a function of two variables $z = f(x, y)$, the tangent plane has normal vector and equation as given by Theorem 9.

(ii) If the surface can be expressed as a level surface $f(x, y, z) = k$ for a function $w = f(x, y, z)$ of three variables, then ∇f is normal to the tangent plane, and equations (17) and (18) are used to find explicit formulations of the normal vector and tangent plane.

We conclude the section by showing that we have indeed developed a concept of differentiability for functions of two variables that is sufficient to guarantee continuity.

THEOREM 10 If the function $z = f(x, y)$ is differentiable at (x_0, y_0), then it is continuous at (x_0, y_0).

Proof Write $\boldsymbol{x}_0 = (x_0, y_0)$ and $\Delta \boldsymbol{x} = \Delta x\boldsymbol{i} + \Delta y\boldsymbol{j}$. Since $f(\boldsymbol{x})$ is differentiable at $\boldsymbol{x}_0$, we have

$$f(\boldsymbol{x}_0 + \Delta\boldsymbol{x}) = f(\boldsymbol{x}_0) + \nabla f(x_0)\cdot\Delta\boldsymbol{x} + \epsilon(\Delta\boldsymbol{x})$$

where

$$\lim_{|\Delta x|\to 0} \frac{\epsilon(\Delta \boldsymbol{x})}{|\Delta \boldsymbol{x}|} = 0.$$

Thus,

$$\begin{aligned}\lim_{\Delta x\to 0} (f(\boldsymbol{x}_0 + \Delta\boldsymbol{x}) - f(\boldsymbol{x}_0)) &= \lim_{\Delta x\to 0} \nabla f(x_0)\cdot\Delta\boldsymbol{x} + \lim_{\Delta x\to 0} \epsilon(\Delta\boldsymbol{x}),\\ &= \mathbf{0}.\end{aligned}$$

It follows that $f(\boldsymbol{x})$ is continuous at $\boldsymbol{x}_0$. ■

Exercise Set 19.9

In Exercises 1–8, find a normal vector and an equation for the tangent plane for the graph of the given equation at the given point. Where appropriate, you may use either of the two methods discussed in this section.

1. $x^2 + 4y^2 + 2z^2 = 9, \quad P = (-1, 0, 2)$
2. $xyz = 6, \quad P = (2, 1, 3)$
3. $x^2 + y^2 - 3yz = 8, \quad P = (1, 1, -2)$
4. $x^2 + y^2 = 9, \quad P = (1, \sqrt{8}, 2)$
5. $z = x^2 - y^3 + xy, \quad P = (2, 1, 5)$
6. $xy - zx + xz^2 = 1, \quad P = (1, -1, 2)$
7. $y = \sin x, \quad P = (\pi/2, 1, 5)$
8. $\sqrt{x} + \sqrt{y} + \sqrt{z} = 10, \quad P = (1, 16, 25)$
9. Find a unit normal to the graph of the equation $xy^2 + x^2y - xz = 4$ at the point $(1, 2, 3)$.
10. Find a vector equation for the normal to the graph of the equation $xyz = 6$ at the point $(1, 3, 2)$.
11. Show that the cylinder $x^2 + y^2 = 4$ and the sphere $x^2 + y^2 + z^2 - 8y - 6z + 21 = 0$ are tangent at the point $(0, 2, 3)$.
12. Show that the ellipsoid $9x^2 + 4y^2 + 9z^2 = 36$ and the sphere $x^2 + y^2 + z^2 - 10z + 16 = 0$ are tangent at $(0, 0, 2)$.
13. Show that the sphere $x^2 + y^2 + z^2 - 2\sqrt{2}\, z + 1 = 0$ and the cone $z = \sqrt{x^2 + y^2}$ share common tangent planes along a circle. Find an equation for that circle.
14. Show that if $f(x, y)$ has continuous partial derivatives in a neighborhood of (x_0, y_0) then $f(x, y)$ itself is continuous at (x_0, y_0).
15. Show that Theorem 9 gives, as a special case, the equation for the plane tangent to the graph of $z = f(x, y)$. (*Hint:* Regard the graph as a level surface for the function $g(x, y, z) = f(x, y) - z$.)
16. Use the definition of the directional derivative to show that the function

$$f(x, y) = \begin{cases} \dfrac{xy^2}{x^2 + y^4}, & (x, y) \neq (0, 0) \\ 0, & (x, y) = 0 \end{cases}$$

has directional derivatives in all directions at $(0, 0)$.

17. Show that $\lim_{\substack{\Delta x \to 0 \\ \Delta y \to 0}} \dfrac{\epsilon_1 \Delta x + \epsilon_2 \Delta y}{\sqrt{\Delta x^2 + \Delta y^2}} = 0$ if $\lim_{\substack{\Delta x \to 0 \\ \Delta y \to 0}} \epsilon_1 = 0$ and $\lim_{\substack{\Delta x \to 0 \\ \Delta y \to 0}} \epsilon_2 = 0$. (*Hint:* Write this as two separate fractions and use the fact that $\left| \dfrac{\Delta x}{\sqrt{\Delta x^2 + \Delta y^2}} \right| \le 1$.)

18. Prove the Mean Value Theorem for functions of two variables: If $f(x, y)$ has continuous partial derivatives in a domain containing the points $\boldsymbol{a} = (a_1, a_2)$ and $\boldsymbol{b} = (b_1, b_2)$, and the line segment joining $\boldsymbol{a}$ and $\boldsymbol{b}$, then there is a point $\boldsymbol{c} = (c_1, c_2)$ on this line segment for which

$$f(\boldsymbol{b}) - f(\boldsymbol{a}) = \nabla f(\boldsymbol{c}) \cdot (\boldsymbol{b} - \boldsymbol{a}).$$

(*Hint:* (a) Parameterize this line segment by $\boldsymbol{r}(t) = \boldsymbol{a} + t(\boldsymbol{b} - \boldsymbol{a})$, $0 \le t \le 1$. (b) Apply the Chain Rule to the composite function $g(t) = f(\boldsymbol{a} + t(\boldsymbol{b} - \boldsymbol{a}))$. (c) Apply the Mean Value Theorem for functions of a single variable.)

19. Let U be an open connected set in the domain of $f(x, y)$. (**Connected** means that any two points in U may be joined by a polygonal path lying entirely within U.) Prove that if $f(x, y)$ has continuous partial derivatives and if $\nabla f(x, y) = \mathbf{0}$ for all (x, y) in U, then $f(x, y) \equiv c$ for some constant c and all (x, y) in U. (*Hint:* Apply the Mean Value Theorem, Exercise 18.)

20. Let U and $f(x, y)$ be as in Exercise 19. Let $g(x, y)$ be defined for all $(x, y) \in U$ and assume that g has continuous partial derivatives. Prove that if $\nabla f(x, y) = \nabla g(x, y)$ for all $(x, y) \in U$ then $f(x, y)$ and $g(x, y)$ differ only by a constant. That is, $f(x, y) = g(x, y) + C$ for some constant C and all $(x, y) \in U$.

21. Let U be a rectangle in the domain of $f(x, y)$. Use the Mean Value Theorem for functions of two variables to prove that if $f(x, y)$ has continuous partial derivatives on U and if $|\nabla f(x, y)| \le K$ for all $(x, y) \in U$, then

$$|f(x_2, y_2) - f(x_1, y_1)| \le K\sqrt{(x_2 - x_1)^2 + (y_2 - y_1)^2}$$

for all (x_1, y_1) and (x_2, y_2) in U. Conclude that if $\nabla f(x, y)$ is bounded on U, then so is the function $f(x, y)$.

19.10 CONSTRAINED EXTREMA: THE METHOD OF LAGRANGE MULTIPLIERS

As an application of the theory of the gradient, we discuss a method due to the French mathematician Joseph L. Lagrange (1736–1813) for finding extrema of functions of several variables subject to constraints. Examples of this kind of problem are the following:

Example 1 Find the maximum and minimum values of the function $f(x, y) = 2x^2 + 4y^2$, given that $x^2 + y^2 = 1$.

Example 2 Find the point(s) on the hyperboloid $z = (y + 1)^2 - (x - 2)^2 + 1$ nearest the point $(2, -1, 2)$.

Example 3 A cylindrical tin can, with a top and a bottom, is to be manufactured using 100 cm^2 of tin, ignoring waste. What dimensions produce the can of maximum volume?

Each of these examples involves finding the extreme values of a function

$$w = f(\boldsymbol{x}), \qquad \boldsymbol{x} \in \mathbb{R}^2 \text{ or } \mathbb{R}^3 \qquad \text{(function to be optimized)} \tag{1}$$

subject to a **constraint** of the form

$$g(\boldsymbol{x}) = 0 \qquad \text{(constraint equation).} \tag{2}$$

In Example 1 the function to be optimized is $f(x, y) = 2x^2 + 4y^2$, and the constraint is $x^2 + y^2 = 1$. The latter equation can be put in the form of equation (2) by writing it as

$$g(x, y) = x^2 + y^2 - 1 = 0.$$

In Example 2, we are to minimize the distance function $D(x, y, z) = \sqrt{(x-2)^2 + (y+1)^2 + (z-2)^2}$ subject to the constraint that (x, y, z) lie on the hyperboloid. We can write this constraint in the form of equation (2) as

$$g(x, y, z) = z - (y+1)^2 + (x-2)^2 - 1 = 0.$$

In Example 3, the dimensions of the tin can are r(radius) and h(height), so we need to maximize the volume $V = \pi r^2 h$ subject to the constraint that total surface area equal 100 cm^2. This constraint may be expressed as the equation

$$g(r, h) = 2\pi r^2 + 2\pi rh - 100 = 0.$$

Each of these examples illustrates the important difference between finding *relative* extrema for functions $f(\boldsymbol{x})$ of several variables (the topic of Section 19.5) and finding *extrema in the presence of constraints,* or *constrained extrema*. The constraint equation $g(\boldsymbol{x}) = 0$ restricts the values of $\boldsymbol{x}$ that may be considered to just those satisfying this constraint. Thus, just as extreme values for $y = f(\boldsymbol{x})$ on a closed interval $[a, b]$ can occur at endpoints that do not correspond to relative extrema, constrained extrema in general will not correspond to relative extrema for the function $f(\boldsymbol{x})$.

Figure 10.1 illustrates the geometry associated with the constrained extrema problem of Example 1. Notice how the constraint $x^2 + y^2 = 1$ restricts the graph of $f(\boldsymbol{x})$ to just the curve C above the unit circle. Note also that neither the constrained maxima nor the constrained minima correspond to relative extrema for the function $f(\boldsymbol{x})$.

As we have seen in Section 19.5, it is sometimes possible to solve a problem involving a constraint by solving the constraint equation (2) for one of the independent variables and then substituting for this variable in equation (1). The resulting equation (1) may or may not yield a solution. The method due to Lagrange is more general, in that it does not depend on our ability to solve the constraint equation for any particular variable. It is based on the following theorem.

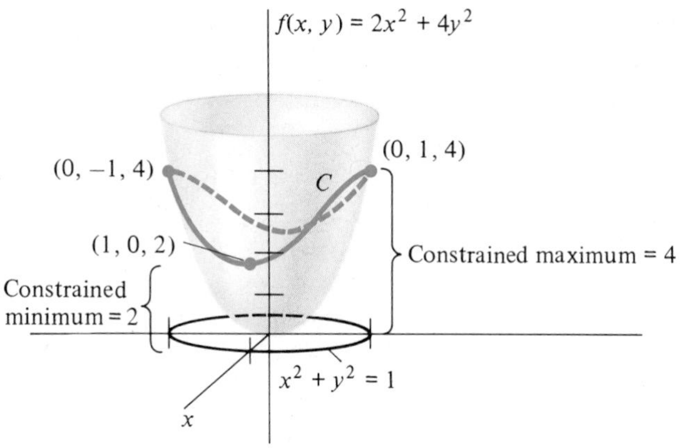

Figure 10.1 Constraint $g(x, y) = x^2 + y^2 - 1 = 0$ restricts graph of $f(x, y) = 2x^2 + 4y^2$ to the curve C.

THEOREM 11

Let $\boldsymbol{x}$ denote a point (vector) in either $\mathbb{R}^2$ or $\mathbb{R}^3$, and let $f(\boldsymbol{x})$ and $g(\boldsymbol{x})$ be functions of either two or three variables. Assume that both $f(\boldsymbol{x})$ and $g(\boldsymbol{x})$ have continuous partial derivatives in a neighborhood of the point $\boldsymbol{x}_0$. If $\boldsymbol{x}_0$ maximizes or minimizes the function

$$w = f(\boldsymbol{x})$$

subject to the constraint

$$g(\boldsymbol{x}) = 0,$$

and if $\nabla g(\boldsymbol{x}_0) \neq \mathbf{0}$, then

$$\nabla f(\boldsymbol{x}_0) = \lambda \nabla g(\boldsymbol{x}_0) \tag{3}$$

for some constant λ. In other words, $\nabla f(\boldsymbol{x}_0)$ and $\nabla g(\boldsymbol{x}_0)$ are parallel.

Theorem 11 is the analogue, for constrained extrema, of Theorem 3 for relative extrema. It says that constrained extrema can only occur when the gradient vectors $\nabla f(\boldsymbol{x}_0)$ and $\nabla g(\boldsymbol{x}_0)$ are parallel. Points $\boldsymbol{x}_0$ where this condition holds play the role of critical points in relative extrema problems. The method of Lagrange for finding relative extrema is now obvious—we simply check all points satisfying equation (3). Among the values $f(\boldsymbol{x}_0)$ must lie the constrained extrema, if such extrema exist. Before stating this procedure more precisely, we sketch a proof of Theorem 11.

Proof of Theorem 11 (Sketch): Let's first consider the planar case $\boldsymbol{x} = (x, y) \in \mathbb{R}^2$. Then the constraint equation $g(\boldsymbol{x}) = g(x, y) = 0$ determines a curve C in the xy-plane on which the point $\boldsymbol{x}_0$ must lie. This curve C is a *level curve* for the function $z = g(x, y)$. Thus, by statement (18), Section 19.8, if $\nabla g(\boldsymbol{x}_0) \neq \mathbf{0}$ then $\nabla g(\boldsymbol{x}_0)$ is orthogonal to C at $\boldsymbol{x}_0$. We will complete the proof for the planar case by showing that $\nabla f(\boldsymbol{x}_0)$ is also orthogonal to C at $\boldsymbol{x}_0$.

Let $\boldsymbol{r}(t) = x(t)\boldsymbol{i} + y(t)\boldsymbol{j}$ be a parameterization for C so that $\boldsymbol{r}(t_0) = \boldsymbol{x}_0$ and so that $\boldsymbol{r}'(t_0) \neq \mathbf{0}$.* Since $\boldsymbol{x}_0 = \boldsymbol{r}(t_0)$ maximizes $f(\boldsymbol{x})$ on C, the scalar function $f(\boldsymbol{r}(t))$ has a relative extremum at $t = t_0$. Thus, by the Chain Rule,

$$\left.\frac{d}{dt} f(\boldsymbol{r}(t))\right|_{t=t_0} = \nabla f(\boldsymbol{r}(t_0)) \cdot \boldsymbol{r}'(t_0) = 0. \tag{4}$$

This shows that $\nabla f(\boldsymbol{r}(t_0)) = \nabla f(\boldsymbol{x}_0)$ is orthogonal to $\boldsymbol{r}'(t_0)$. Since $\boldsymbol{r}'(t_0)$ is *tangent* to C at $\boldsymbol{x}_0$, it follows that $\nabla f(\boldsymbol{x}_0)$ is orthogonal to C at $\boldsymbol{x}_0$.

Since $\nabla f(\boldsymbol{x}_0)$, $\nabla g(\boldsymbol{x}_0)$, and C all lie in the xy-plane we may now conclude, since $\nabla f(\boldsymbol{x}_0)$ and $\nabla g(\boldsymbol{x}_0)$ are both orthogonal to C at $\boldsymbol{x}_0$, that $\nabla f(\boldsymbol{x}_0)$ and $\nabla g(\boldsymbol{x}_0)$ are parallel. That is, $\nabla f(\boldsymbol{x}_0) = \lambda \nabla g(\boldsymbol{x}_0)$ for some constant λ.

The space case $\boldsymbol{x} \in \mathbb{R}^3$ is similar to the planar case. However, the constraint equation $g(\boldsymbol{x}) = 0$ now determines a surface S in $\mathbb{R}^3$ that is a level surface for the function $u = g(\boldsymbol{x})$. By statement (21), Section 19.8, if $\nabla g(\boldsymbol{x}_0) \neq 0$ then $\nabla g(\boldsymbol{x}_0)$ is orthogonal to S at $\boldsymbol{x}_0$.

---

*$\boldsymbol{r}(t)$ can always be chosen so that $\boldsymbol{r}'(t_0) \neq \mathbf{0}$ if $\nabla g(\boldsymbol{x}_0) \neq \mathbf{0}$. This follows from the Implicit Function Theorem, a result usually discussed in courses on advanced calculus. Actually, the Implicit Function Theorem underlies several of the seemingly obvious statements being made here, which is why we refer to this argument only as a sketch of a proof.

As before, we will show that $\nabla f(\boldsymbol{x}_0)$ is also orthogonal to S at $\boldsymbol{x}_0$. We begin by letting C be an arbitrary curve on S with parameterization $\boldsymbol{r}(t) = x(t)\boldsymbol{i} + y(t)\boldsymbol{j} + z(t)\boldsymbol{k}$ for which $\boldsymbol{r}(t_0) = \boldsymbol{x}_0$ and $\boldsymbol{r}'(t_0) \neq \boldsymbol{0}$. Just as in the plane case, the scalar function $f(\boldsymbol{r}(t))$ has a relative extremum at t_0, so equation (4) again holds. Since $\boldsymbol{r}'(t_0)$ is tangent to C at S, it is also tangent to S at $\boldsymbol{x}_0$. This shows that $\nabla f(\boldsymbol{x}_0)$ is orthogonal to *a* tangent at S. But since the curve C was arbitrary, so is its tangent $\boldsymbol{r}'(t_0)$. It follows that $\nabla f(\boldsymbol{x}_0)$ is orthogonal to *all* tangents to S at $\boldsymbol{x}_0$, and therefore that $\nabla f(\boldsymbol{x}_0)$ is orthogonal to S. Since two vectors orthogonal to a (smooth) surface at a common point must be parallel, it follows that $\nabla f(\boldsymbol{x}_0) = \lambda \nabla g(\boldsymbol{x}_0)$ for some constant λ. ■

Theorem 11 establishes the **method of Lagrange multipliers** (for finding the extreme values of $f(\boldsymbol{x})$ subject to the constraint $g(\boldsymbol{x}) = 0$):

1. Find all simultaneous solutions of the equations

$$\nabla f(\boldsymbol{x}_0) = \lambda \nabla g(\boldsymbol{x}_0) \tag{5}$$

and

$$g(\boldsymbol{x}_0) = 0. \tag{6}$$

a. If $\boldsymbol{x} = (x, y) \in \mathbb{R}^2$, equations (5) and (6) are equivalent to the three equations

$$\begin{cases} \dfrac{\partial f}{\partial x}(x_0, y_0) = \lambda \dfrac{\partial g}{\partial x}(x_0, y_0), \\ \dfrac{\partial f}{\partial y}(x_0, y_0) = \lambda \dfrac{\partial g}{\partial y}(x_0, y_0), \\ g(x_0, y_0) = 0. \end{cases} \tag{7}$$

b. If $\boldsymbol{x} = (x, y, z) \in \mathbb{R}^3$, equations (5) and (6) are equivalent to the four equations

$$\begin{cases} \dfrac{\partial f}{\partial x}(x_0, y_0, z_0) = \lambda \dfrac{\partial g}{\partial x}(x_0, y_0, z_0), \\ \dfrac{\partial f}{\partial y}(x_0, y_0, z_0) = \lambda \dfrac{\partial g}{\partial y}(x_0, y_0, z_0), \\ \dfrac{\partial f}{\partial z}(x_0, y_0, z_0) = \lambda \dfrac{\partial g}{\partial z}(x_0, y_0, z_0), \\ g(x_0, y_0, z_0) = 0. \end{cases} \tag{8}$$

2. Calculate $f(\boldsymbol{x}_0)$ for all $\boldsymbol{x}_0$ obtained in step 1.
3. On geometric, analytic, or physical grounds, determine which of the numbers obtained in step 2 correspond to constrained extrema.

Solution to Example 1: We wish to find the extreme values of

$$f(x, y) = 2x^2 + 4y^2$$

subject to

$$g(x, y) = x^2 + y^2 - 1 = 0.$$

The three equations corresponding to equations (7) are:

$$4x = 2\lambda x \qquad (f_x = \lambda g_x) \tag{9}$$

$$8y = 2\lambda y \qquad (f_y = \lambda g_y) \tag{10}$$

and

$$x^2 + y^2 - 1 = 0 \qquad (g = 0) \tag{11}$$

To solve this system of three equations, we begin with equation (9). If $x = 0$ equation (9) is satisfied, equation (11) then becomes simply $y^2 - 1 = 0$, so $y = \pm 1$. We therefore obtain the two points $(0, 1)$ and $(0, -1)$ that must be checked for extrema.

On the other hand, if $x \neq 0$ in (9), we may divide both sides of (9) by x to obtain $4 = 2\lambda$, so $\lambda = 2$. Substituting this value of λ into (10) then gives $8y = 4y$, so y must equal zero. With $y = 0$, equation (11) becomes $x^2 - 1 = 0$, so $x = \pm 1$. We have therefore obtained two additional points, $(1, 0)$ and $(-1, 0)$.

Finally, we simply calculate the value of $f(x, y)$ for each of the four points $(0, 1)$, $(0, -1)$, $(1, 0)$, and $(-1, 0)$. We find

$$f(0, 1) = 4 \qquad \text{(maximum)},$$

$$f(0, -1) = 4 \qquad \text{(maximum)},$$

$$f(1, 0) = 2 \qquad \text{(minimum)},$$

$$f(-1, 0) = 2 \qquad \text{(minimum)}.$$

As Figure 10.1 illustrates, the constrained maximum is 4, occurring at $(0, 1)$ and $(0, -1)$, and the constrained minimum is $2 = f(1, 0) = f(-1, 0)$. ■

REMARK 1: In reading these examples, you will notice that the method by which we solve the systems of equations (7) and (8) varies from problem to problem. The great versatility of the method of Lagrange multipliers is that it applies to a wide variety of functions and constraints. However, the resulting systems of equations will therefore be of various types, and often most or all of the resulting equations will be nonlinear. Your success in using this method will depend on your ability to solve these various sorts of systems of equations.

REMARK 2: Note that although we obtained the value $\lambda = 2$ at one point in the solution of Example 1, the value of λ did not appear in the solution. This will always be the case, since the Lagrange multiplier λ is an "artificial" variable that is introduced merely as a device by which we can solve for the "real" variables in the problem.

The solution of Example 2 is similar to that of Example 1, and we leave it to you as Exercise 15.

Solution to Example 3: Since the volume of the cylinder is $V(r, h) = \pi r^2 h$, the areas of the top and bottom are each πr^2, and the lateral surface area is $2\pi rh$, we must maximize the function

$$V(r, h) = \pi r^2 h$$

subject to the side condition

$$g(r, h) = 2\pi r^2 + 2\pi rh - 100 = 0.$$

The equations corresponding to equations (7) are therefore

$$2\pi rh = \lambda(4\pi r + 2\pi h) \qquad (V_r = \lambda g_r), \tag{12}$$

$$\pi r^2 = \lambda(2\pi r) \qquad (V_h = \lambda g_h), \tag{13}$$

and

$$2\pi r^2 + 2\pi rh - 100 = 0 \qquad (g = 0). \tag{14}$$

Solving equation (12) for λ gives

$$\lambda = \frac{rh}{2r + h}, \tag{15}$$

while solving equation (13) for λ gives

$$\lambda = \frac{r}{2}. \tag{16}$$

Equating the right sides of (15) and (16) we find

$$\frac{rh}{2r + h} = \frac{r}{2}, \qquad \text{so} \qquad 2r + h = 2h,$$

or $h = 2r$. Substituting $h = 2r$ for h in (14) then gives

$$2\pi r^2 + 2\pi r(2r) - 100 = 0,$$

which has solution $r = \dfrac{10}{\sqrt{6\pi}}$. Since it is clear from the geometry of the situation that at least one pair (r, h) must produce a maximum volume, we conclude that the maximum volume of

$$V = \pi\left(\frac{10}{\sqrt{6\pi}}\right)^2\left(\frac{20}{\sqrt{6\pi}}\right) = \frac{1000}{3\sqrt{6\pi}}\ \text{cm}^3$$

corresponds to the dimensions $r = \dfrac{10}{\sqrt{6\pi}}$ cm and $h = 2r = \dfrac{20}{\sqrt{6\pi}}$ cm. ■

Example 4 As an example of an application of the method of Lagrange multipliers requiring a somewhat different technique of solution, we return to Example 10 of Section 19.5. The problem in that example was to find the dimensions x, y, and z for a rectangular building that would minimize the heating costs given by the function

$$C(x, y, z) = 4xy + 5xz + 6yz.$$

The constraint is that the volume of the building is to equal the constant V. That is,

$$g(x, y, z) = xyz - V = 0.$$

Applying equations (8), we find

$$4y + 5z = \lambda yz \qquad (C_x = \lambda g_x), \tag{17}$$

$$4x + 6z = \lambda xz \qquad (C_y = \lambda g_y), \tag{18}$$

$$5x + 6y = \lambda xy \qquad (C_z = \lambda g_z), \tag{19}$$

and

$$xyz - V = 0 \qquad (g = 0). \tag{20}$$

This time it is helpful to begin by multiplying both sides of equation (17) by x, both sides of (18) by y, and both sides of (19) by z. The result is the equations

$$4xy + 5xz = \lambda xyz, \tag{21}$$

$$4xy + 6yz = \lambda xyz, \tag{22}$$

and

$$5xz + 6yz = \lambda xyz. \tag{23}$$

Equating the left sides of (21) and (22) then gives $4xy + 5xz = 4xy + 6yz$, so $y = \frac{5}{6}x$. Similarly, equating the left sides of (22) and (23) gives $4xy + 6yz = 5xz + 6yz$, or $z = \frac{4y}{5} = \frac{2x}{3}$. With these substitutions (20) becomes

$$x\left(\frac{5}{6}x\right)\left(\frac{2}{3}x\right) = \frac{5}{9}x^3 = V,$$

so

$$x = \sqrt[3]{\frac{9V}{5}}, \qquad y = \frac{5}{6}\sqrt[3]{\frac{9V}{5}}, \qquad \text{and} \qquad z = \frac{2}{3}\sqrt[3]{\frac{9V}{5}},$$

as found in Example 10 of Section 19.5. ■

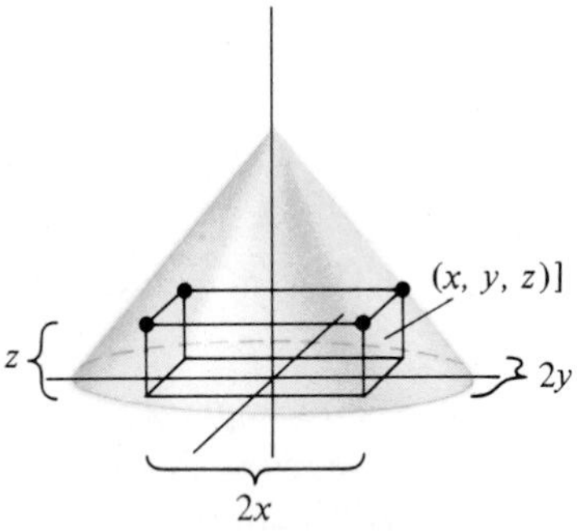

Figure 10.2 Box inscribed within a cone.

Example 5 A rectangular box is to be inscribed in the cone $z = 9 - \sqrt{x^2 + y^2}$, $z \geq 0$ (see Figure 10.2). Find the dimensions for the box that maximize its volume.

Solution: It is clear that to achieve maximum volume we should position the box with one face lying in the xy-plane. If (x, y, z) denotes one corner of the box lying in the first octant, $x, y, z > 0$, the dimensions of the box are

$$\text{length} = 2x, \qquad \text{width} = 2y, \qquad \text{and} \qquad \text{height} = z.$$

The problem is therefore to maximize the function

$$V(x, y, z) = 4xyz$$

subject to the constraint

$$g(x, y, z) = z + \sqrt{x^2 + y^2} - 9 = 0.$$

Applying the Lagrange criterion (8), we find

$$4yz = \frac{\lambda x}{\sqrt{x^2 + y^2}} \qquad (f_x = \lambda g_x), \tag{24}$$

$$4xz = \frac{\lambda y}{\sqrt{x^2 + y^2}} \qquad (f_y = \lambda g_y), \tag{25}$$

$$4xy = \lambda, \tag{26}$$

and

$$z + \sqrt{x^2 + y^2} - 9 = 0. \tag{27}$$

Multiplying (24) through by x, (25) by y, and (26) by z gives

$$4xyz = \frac{\lambda x^2}{\sqrt{x^2 + y^2}}; \qquad 4xyz = \frac{\lambda y^2}{\sqrt{x^2 + y^2}}; \qquad 4xyz = \lambda z.$$

Equating the right-hand sides of these equations gives

$$\frac{\lambda x^2}{\sqrt{x^2 + y^2}} = \frac{\lambda y^2}{\sqrt{x^2 + y^2}} \tag{28}$$

and

$$\lambda z = \frac{\lambda x^2}{\sqrt{x^2 + y^2}}. \tag{29}$$

From (28) it follows that $y = x$, since neither x nor y can be negative. With $y = x$, (29) gives $z = \dfrac{x^2}{\sqrt{2x^2}} = \dfrac{x}{\sqrt{2}}$. Substituting these expressions for y and z in (27) then gives

$$\frac{x}{\sqrt{2}} + \sqrt{x^2 + x^2} - 9 = \left(\frac{1}{\sqrt{2}} + \sqrt{2}\right)x - 9 = 0,$$

so

$$x = \frac{9}{\dfrac{1}{\sqrt{2}} + \sqrt{2}} = \frac{9\sqrt{2}}{1 + 2} = 3\sqrt{2}.$$

Thus, $y = 3\sqrt{2}$ and $z = 3$. The maximum volume is therefore

$$V = 4(3\sqrt{2})(3\sqrt{2})(3) = 216.$$

■

Exercise Set 19.10

In Exercises 1–14, find the maximum and minimum values of the given function subject to the given constraint.

1. $f(x, y) = 2x^2 + 4y^2$ subject to $x^2 + y^2 = 1$
2. $f(x, y) = x^2 + y$ subject to $x^2 + y^2 = 9$
3. $f(x, y) = x^3 - y^3$ subject to $x - y = 2$
4. $f(x, y) = xy$ subject to $x^2 + y^2 - 4y = 5$
5. $f(x, y) = xy$ subject to $x^2 + y^2 = 1$
6. $f(x, y) = y - x$ subject to $x^2 + y^2 = 2$
7. $f(x, y) = x^2 + 4x + 4y^2$ subject to $x^2 + 2y^2 = 4$
8. $f(x, y, z) = x + y + z$ subject to $x^2 + y^2 + z^2 = 4$
9. $f(x, y) = x^2 - 4x + 4y^2$ subject to $x^2 + y^2 = 1$
10. $f(x, y, z) = xyz$ subject to $2x^2 + y^2 + 4z^2 = 9$
11. $f(x, y, z) = x + 2y - z$ subject to $x^2 + y^2 + z^2 = 1$
12. $f(x, y, z) = \sqrt{xyz}$ subject to $x + y + z = 4$
13. $f(x, y, z) = x + y + z$ subject to $x^2 + y^2 + z^2 = 12$
14. $f(x, y, z) = x^2 + 2y^2 + 4z^2$ subject to $x^2 + y^2 + z^2 = 1$
15. Find the solution of Example 2.

16. Find the point on the ellipsoid $9x^2 + 36y^2 + 4z^2 = 36$ nearest the origin.

17. Find the point on the circle $(x - 3)^2 + (y + 2)^2 = 9$ nearest the origin.

18. Find the point on the sphere $x^2 + y^2 + z^2 = 1$ furthest from the point $(3, 2, 1)$.

19. Find the point on the ellipsoid $(x - 1)^2 + \dfrac{(y-2)^2}{4} + \dfrac{(z-1)^2}{9} = 1$ nearest the point $(1, -1, 1)$.

20. Find the point on the cone $f(x, y) = \sqrt{x^2 + y^2}$ nearest the point $(3, 1, 0)$.

21. Find the point(s) on the hyperboloid $z = (y + 1)^2 - (x - 2)^2 + 1$ nearest the point $(2, -1, 1)$.

22. The plane $2y - 3z = 8$ intersects the cone $z^2 = 4x^2 + 4y^2$ in an ellipse. Find the highest and lowest points of intersection.

23. Find the dimensions of the rectangular box of maximum volume that can be inscribed in a sphere of radius r.

24. Find the dimensions of the rectangular box of maximum volume which can be inscribed in the ellipsoid $x^2 + 4y^2 + 2z^2 = 8$.

25. Find the dimensions for the cylindrical jar, with volume 2 liters, that has minimum exterior surface area. (Assume that the jar has a lid and that the thickness of its walls is negligible.)

26–31. Rework Exercises 23–28 of Section 19.5 using the method of Lagrange multipliers.

19.11 RECONSTRUCTING A FUNCTION FROM ITS GRADIENT; EXACT DIFFERENTIAL EQUATIONS

In several ways, the gradient has emerged from the discussions of Sections 19.7 through 19.9 as the analogue of the derivative for functions of a single variable. It is therefore reasonable to ask whether the concept of antidifferentiation makes sense for gradients and, if so, whether this concept is useful. The goal of this section is to define what we mean by the ''antiderivative'' of a gradient (we don't really use this particular terminology), to show how such functions can be found, to prove a theorem indicating when such functions exist, and to show how all this can be applied to solve certain types of differential equations. We will restrict our discussion to functions of two variables, although analogous results may be developed in more general settings.

Potential Functions

We begin by recalling that a gradient is a vector-valued function of two variables

$$\nabla f(x, y) = M(x, y)\boldsymbol{i} + N(x, y)\boldsymbol{j} \tag{1}$$

obtained by partial differentiation from a function of two variables $f(x, y)$ by the rule

$$\nabla f(x, y) = \frac{\partial f}{\partial x}(x, y)\boldsymbol{i} + \frac{\partial f}{\partial y}(x, y)\boldsymbol{j}. \tag{2}$$

The antidifferentiation question for gradients is therefore the following: Given a vector-valued function of the form

$$\boldsymbol{F}(x, y) = M(x, y)\boldsymbol{i} + N(x, y)\boldsymbol{j} \tag{3}$$

is there a function of two variables $f(x, y)$ for which

$$\boldsymbol{F}(x, y) = \nabla f(x, y)? \tag{4}$$

Since the existence of the gradient alone is not a strong enough condition to guarantee that a function $f(x, y)$ is differentiable (in the sense of Definition 8), we prefer not to refer to $f(x, y)$ in (4) as the antiderivative for the vector-valued function $\boldsymbol{F}(x, y)$. Instead, we call it a *potential function,* or simply a *potential*. (The reason for this particular terminology has to do with the roles played by such functions in mechanics and in the theory of electricity and magnetism.)

DEFINITION 10

The function of two variables $f(x, y)$ is called a **potential** for the vector-valued function

$$\boldsymbol{F}(x, y) = M(x, y)\boldsymbol{i} + N(x, y)\boldsymbol{j}$$

in a region $Q \subseteq \mathbb{R}^2$ if $\dfrac{\partial f}{\partial x}(x, y)$ and $\dfrac{\partial f}{\partial y}(x, y)$ exist and if

$$\boldsymbol{F}(x, y) = \boldsymbol{\nabla} f(x, y)$$

for all (x, y) in Q.

Given the vector function $\boldsymbol{F}(x, y)$ in (3), the task of finding a function $f(x, y)$ for which $\boldsymbol{F}(x, y) = \boldsymbol{\nabla} f(x, y)$ is therefore referred to as finding a potential for $\boldsymbol{F}(x, y)$. This is also what we mean by "reconstructing the function $f(x, y)$ from its gradient." As the following example shows, potentials are not unique.

Example 1 If $f(x, y) = xe^{2y}$, then

$$\boldsymbol{\nabla} f(x, y) = e^{2y}\boldsymbol{i} + 2xe^{2y}\boldsymbol{j}.$$

Thus, the vector function

$$\boldsymbol{F}(x, y) = e^{2y}\boldsymbol{i} + 2xe^{2y}\boldsymbol{j}$$

has the function $f(x, y) = xe^{2y}$ as a potential. However, the function $g(x, y) = xe^{2y} + C$ is also a potential for $\boldsymbol{F}(x, y)$, since

$$\boldsymbol{\nabla} g(x, y) = e^{2y}\boldsymbol{i} + 2xe^{2y}\boldsymbol{j} = \boldsymbol{F}(x, y).$$

Thus, if $f(x, y)$ is a potential for $\boldsymbol{F}(x, y)$, so is $f(x, y) + C$ for any constant C. ■

Finding Potentials

Now, suppose we are faced with a vector function $\boldsymbol{F}(x, y)$ of the form (3). If a potential $f(x, y)$ exists, how is it found? By comparing the right-hand sides of equations (1) and (2), we obtain the following conclusion:

If $\boldsymbol{F}(x, y) = M(x, y)\boldsymbol{i} + N(x, y)\boldsymbol{j} = \boldsymbol{\nabla} f(x, y)$, then

$$M(x, y) = \frac{\partial f}{\partial x}(x, y) \tag{5}$$

and

$$N(x, y) = \frac{\partial f}{\partial y}(x, y). \tag{6}$$

Equations (5) and (6) are the keys to finding $f(x, y)$. Beginning with equation (5) and integrating "partially with respect to x" we find

$$f(x, y) = \int \left(\frac{\partial f}{\partial x}(x, y)\right) dx = \int M(x, y)\, dx = G_1(x, y) + h_1(y) + C_1. \tag{7}$$

By integrating "partially with respect to x" we mean treating y as a constant and integrating $M(x, y)$ as a function of x alone—just the reverse of partial differentiation. We must remember, however, that functions of y alone vanish entirely when differentiated partially with respect to x, so we must allow for the appearance of an entire function $h_1(y)$ of y alone when integrating partially with respect to x.

Applying the same idea to equation (6), we integrate partially with respect to y to obtain an equation of the form

$$f(x, y) = \int \left(\frac{\partial f}{\partial y}(x, y)\right) dy = \int N(x, y)\, dy = G_2(x, y) + h_2(x) + C_2, \tag{8}$$

where $h_2(x)$ is a function of x alone.

If we are fortunate, we can next equate the right-hand sides of equations (7) and (8) and determine the functions G_1, G_2, h_1, and h_2. The potential $f(x, y)$ can then be obtained from either equation (7) or equation (8). However, there are two reasons for which this procedure may fail. First, the function $\boldsymbol{F}(x, y)$ may simply fail to have a potential. Example 4 is one such function. (Theorem 12 of this section clarifies this existence question.) Second, the function $\boldsymbol{F}(x, y)$ might well have a potential, but we may not be clever enough to find it. This fact should not surprise you, since precisely the same difficulty surrounds the issue of finding antiderivatives for functions of a single variable. The same sorts of skills are required for finding potentials as are necessary for finding antiderivatives.

Example 2 Let $\boldsymbol{F}(x, y)$ be the vector-valued function

$$\boldsymbol{F}(x, y) = (3x^2 + 2y^2)\boldsymbol{i} + 4xy\boldsymbol{j}.$$

Find a function $f(x, y)$ for which $\boldsymbol{F}(x, y) = \nabla f(x, y)$.

Strategy

Identify $M(x, y)$ and $N(x, y)$.

Solution

Here $\boldsymbol{F}(x, y)$ has the form (3) with

$$M(x, y) = 3x^2 + 2y^2; \qquad N(x, y) = 4xy.$$

Integrate $M(x, y)$ partially with respect to x to find an expression for $f(x, y)$.

Partially integrating $M(x, y)$ with respect to x according to equation (7) gives

$$\begin{aligned} f(x, y) &= \int M(x, y)\, dx \\ &= \int (3x^2 + 2y^2)\, dx \\ &= x^3 + 2xy^2 + h_1(y) + C_1 \end{aligned} \tag{9}$$

Integrate $N(x, y)$ partially with respect to y to obtain a second expression for $f(x, y)$.

where $h_1(y)$ is a function of y alone. Then, partially integrating $N(x, y)$ with respect to y, according to equation (8), we find

$$\begin{aligned} f(x, y) &= \int N(x, y)\, dy \\ &= \int 4xy\, dy \\ &= 2xy^2 + h_2(x) + C_2 \end{aligned} \tag{10}$$

where $h_2(x)$ is a function of x alone.

Equate the two expressions for $f(x, y)$ and attempt to identify the unknown functions $h_1(y)$ and $h_2(x)$.

Equating these two expressions for $f(x, y)$ gives

$$x^3 + 2xy^2 + h_1(y) + C_1 = 2xy^2 + h_2(x) + C_2.$$

This equation is true if $h_1(y) = 0$, $h_2(x) = x^3$, and $C_1 = C_2 = C$. Both (9) and (10) then give the desired potential as

Read $f(x, y)$ from either expression.

$$f(x, y) = x^3 + 2xy^2 + C.$$

Partial differentiation verifies that

$$\nabla f(x, y) = \boldsymbol{F}(x, y).$$

■

Example 3 Find a potential for the function

$$\boldsymbol{F}(x, y) = (ye^{xy} - 2x \sin x^2)\boldsymbol{i} + \left(\frac{1}{\sqrt{y}} + xe^{xy}\right)\boldsymbol{j}.$$

Solution: Here $\boldsymbol{F}(x, y) = M(x, y)\boldsymbol{i} + N(x, y)\boldsymbol{j}$ with

$$M(x, y) = ye^{xy} - 2x \sin x^2; \qquad N(x, y) = \frac{1}{\sqrt{y}} + xe^{xy}.$$

Integrating $M(x, y)$ partially with respect to x gives

$$\begin{aligned} f(x, y) = \int M(x, y)\, dx &= \int (ye^{xy} - 2x \sin x^2)\, dx \\ &= e^{xy} + \cos x^2 + h_1(y) + C_1 \end{aligned} \tag{11}$$

where $h_1(y)$ is a function of y alone. Integrating $N(x, y)$ partially with respect to y gives

$$\begin{aligned} f(x, y) = \int N(x, y)\, dy &= \int \left(\frac{1}{\sqrt{y}} + xe^{xy}\right) dy \\ &= 2\sqrt{y} + e^{xy} + h_2(x) + C_2 \end{aligned} \tag{12}$$

where $h_2(x)$ is a function of x alone. Equating the right-hand sides of equations (11) and (12) then shows that

$$h_1(y) = 2\sqrt{y}, \qquad h_2(x) = \cos x^2, \qquad \text{and} \qquad C_1 = C_2 = C. \tag{13}$$

With the information in line (13), either of equations (11) and (12) gives that

$$f(x, y) = e^{xy} + \cos x^2 + 2\sqrt{y} + C.$$

Partial differentiation then verifies that $\nabla f(x, y) = \boldsymbol{F}(x, y)$. ■

Example 4 Show that the function

$$\boldsymbol{F}(x, y) = y\boldsymbol{i} - x\boldsymbol{j}$$

has no potential. That is, $\boldsymbol{F}(x, y)$ cannot be the gradient of a function of two variables $f(x, y)$.

Solution: On the contrary, let us assume that there does exist a function $f(x, y)$ with $\boldsymbol{F}(x, y) = \nabla f(x, y)$. Since

$$M(x, y) = y; \qquad N(x, y) = -x,$$

we have from equation (7) that

$$f(x, y) = \int M(x, y)\, dx = \int y\, dx = xy + h_1(y) + C_1 \tag{14}$$

where $h_1(y)$ is a function of y alone. Furthermore, since we are assuming that $\frac{\partial f}{\partial y}(x, y)$ exists, $h_1(y)$ must be a differentiable function of y. We may therefore differentiate both sides of (14) with respect to y. Doing so, and using equation (6), gives

$$\frac{\partial f}{\partial y}(x, y) = x + h'(y) = -x = N(x, y).$$

Thus,

$$h'(y) = -2x. \tag{15}$$

But since $h(y)$ must be a function of y alone, so is $h'(y)$. Thus, since $h'(y)$ cannot involve x, $\frac{\partial}{\partial x}(h'(y)) = 0$. This shows that the result of differentiating both sides of equation (15) with respect to x is the contradictory statement $0 = -2$. The assumption that $\boldsymbol{F}(x, y) = y\boldsymbol{i} - x\boldsymbol{j}$ has a potential therefore leads to a contradiction, and is therefore false. $\boldsymbol{F}(x, y) = y\boldsymbol{i} - x\boldsymbol{j}$ has no potential. ■

The result of Example 4 leaves us in need of information as to when a given vector function $\boldsymbol{F}(x, y)$ has a potential. The information is supplied by the following theorem, whose proof employs the ideas of Examples 2 through 4.

THEOREM 12

Let $M(x, y)$ and $N(x, y)$ have continuous first partial derivatives in an open rectangle R in the xy-plane. Then the vector-valued function

$$\boldsymbol{F}(x, y) = M(x, y)\boldsymbol{i} + N(x, y)\boldsymbol{j}$$

has a potential in R if and only if

$$\frac{\partial M}{\partial y}(x, y) = \frac{\partial N}{\partial x}(x, y) \tag{16}$$

for all points (x, y) in R.

Before giving the proof, we note that condition (16) is satisfied for $\boldsymbol{F}(x, y)$ in Example 2, since

$$\frac{\partial M}{\partial y}(x, y) = \frac{\partial}{\partial y}(3x^2 + 2y^2) = 4y = \frac{\partial N}{\partial x}(x, y).$$

Also, equation (16) holds for $\boldsymbol{F}(x, y)$ in Example (3):

$$\begin{aligned}\frac{\partial M}{\partial y}(x, y) &= \frac{\partial}{\partial y}(ye^{xy} - 2x \sin x^2)\\ &= e^{xy} + xye^{xy}\\ &= \frac{\partial}{\partial x}\left(\frac{1}{\sqrt{y}} + xe^{xy}\right)\\ &= \frac{\partial N}{\partial x}(x, y).\end{aligned}$$

However, as required by Theorem 12, condition (16) fails for $\boldsymbol{F}(x, y)$ in Example 4 since

$$\frac{\partial M}{\partial y} = \frac{\partial}{\partial y}(y) = 1 \neq -1 = \frac{\partial}{\partial x}(-x) = \frac{\partial N}{\partial x}.$$

Proof of Theorem 12: We first assume that $\boldsymbol{F}(x, y)$ has a potential $f(x, y)$ and show that equation (16) holds. According to equations (5) and (6) this means that

$$\boldsymbol{F}(x, y) = \nabla f(x, y)$$

where $M(x, y) = \dfrac{\partial f}{\partial x}(x, y)$ and $N(x, y) = \dfrac{\partial f}{\partial y}(x, y)$. Since both $M(x, y)$ and $N(x, y)$ are assumed to have continuous first partial derivatives in R, it follows that

$$\frac{\partial^2 f}{\partial y \partial x} = \frac{\partial}{\partial y}\left(\frac{\partial f}{\partial x}\right) = \frac{\partial M}{\partial y}(x, y) \tag{17}$$

and

$$\frac{\partial^2 f}{\partial x \partial y} = \frac{\partial}{\partial x}\left(\frac{\partial f}{\partial y}\right) = \frac{\partial N}{\partial x}(x, y). \tag{18}$$

Moreover, by Theorem 2 the mixed partial derivatives on the left-hand sides of equations (17) and (18) are equal. Thus, the right-hand sides of these equations are equal, and condition (16) is obtained.

The remaining part of the proof involves assuming that condition (16) holds and actually finding (at least formally) a potential for $\boldsymbol{F}(x, y)$. Taking a cue from Examples 2 through 4 we begin by fixing a point (x_0, y_0) in R and defining the function $f(x, y)$ as

$$f(x, y) = \int_{x_0}^{x} M(t, y_0)\, dt + \int_{y_0}^{y} N(x, s)\, ds. \tag{19}$$

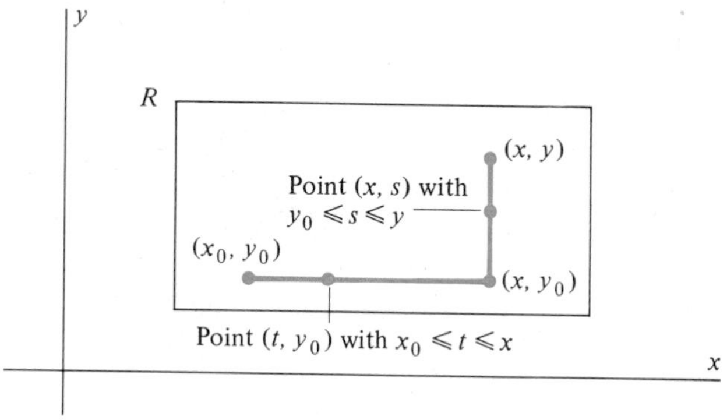

Figure 11.1 Domain of $M(x, y)$ and $N(x, y)$ in Theorem 12.

Figure 11.1 may help you understand how $f(x, y)$ has been "pulled out of the hat." Since we want to construct $f(x, y)$ in such a way that $\dfrac{\partial f}{\partial x}(x, y) = M(x, y)$ and $\dfrac{\partial f}{\partial y}(x, y) = N(x, y)$, the idea is to integrate M along a path involving only change in x and to integrate N along a path involving only change in y. The reason we require R to be a rectangle is simply to ensure that such a path lies entirely within R.

Since $M(x, y)$ and $N(x, y)$ are continuous in R, and since R is a rectangle, $f(x, y)$ is defined for all $(x, y) \in R$. Moreover, by the Fundamental Theorem of Calculus it follows that

$$\begin{aligned}\frac{\partial f}{\partial y}(x, y) &= \frac{\partial}{\partial y}\left\{\int_{x_0}^{x} M(t, y_0)\, dt + \int_{y_0}^{y} N(x, s)\, ds\right\} \\ &= \frac{\partial}{\partial y}\int_{y_0}^{y} N(x, s)\, ds \\ &= N(x, y)\end{aligned} \tag{20}$$

since the first integral inside the braces does not involve y. We also want to calculate $\frac{\partial f}{\partial x}(x, y)$, but this is a bit harder. We find that

$$\begin{aligned}\frac{\partial f}{\partial x}(x, y) &= \frac{\partial}{\partial x}\left\{\int_{x_0}^{x} M(t, y_0)\, dt + \int_{y_0}^{y} N(x, s)\, ds\right\} \\ &= M(x, y_0) + \frac{\partial}{\partial x}\int_{y_0}^{y} N(x, s)\, ds.\end{aligned} \tag{21}$$

In courses on advanced calculus it is proved that

$$\frac{\partial}{\partial x}\int_{y_0}^{y} N(x, s)\, ds = \int_{y_0}^{y} \frac{\partial}{\partial x} N(x, s)\, ds. \tag{22}$$

That is, we may pass the partial differentiation with respect to x under the y-integral sign in the integral on the right-hand side of equation (21). Using (22) we return to (21) and use condition (16) to find

$$\begin{aligned}\frac{\partial f}{\partial x}(x, y) &= M(x, y_0) + \int_{y_0}^{y} \frac{\partial}{\partial x} N(x, s)\, ds \\ &= M(x, y_0) + \int_{y_0}^{y} \frac{\partial}{\partial s} M(x, s)\, ds \qquad \left(\frac{\partial N}{\partial x}(x, s) = \frac{\partial M}{\partial s}(x, s)\right) \\ &= M(x, y_0) + M(x, s)\Big]_{s=y_0}^{s=y} \\ &= M(x, y_0) + [M(x, y) - M(x, y_0)] \\ &= M(x, y).\end{aligned} \tag{23}$$

Reviewing equations (20) and (23), we find that the function $f(x, y)$ defined in line (19) has the properties

$$\frac{\partial f}{\partial x}(x, y) = M(x, y); \qquad \frac{\partial f}{\partial y}(x, y) = N(x, y).$$

That is, $\nabla f(x, y) = \boldsymbol{F}(x, y)$, so $\boldsymbol{F}(x, y)$ indeed has a potential. ■

Exact Differential Equations

The ideas of this section may be applied to solve a certain type of ordinary differential equation. We begin this discussion by recalling that if $f(x, y)$ is a function of two variables with continuous first partial derivatives, the equation

$$f(x, y) = C \tag{24}$$

determines, in general, a level curve for the graph of the function $z = f(x, y)$. As Figures 11.2 and 11.3 illustrate, if (x_0, y_0) is a point on this level curve at which the tangent is not parallel to the y-axis, then equation (24) determines a function of a single variable,* $y = y(x)$, in some neighborhood N of the number x_0.

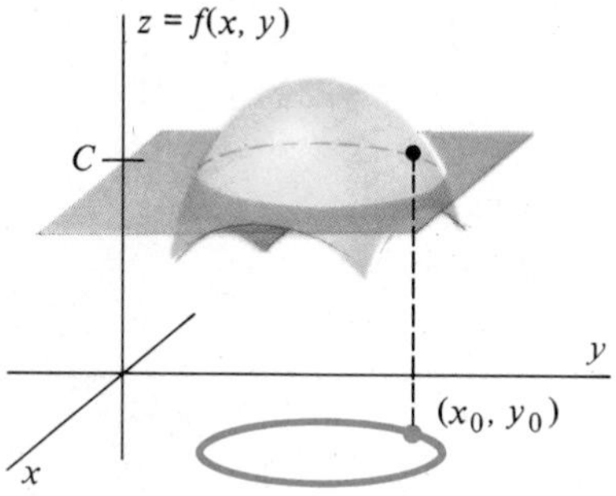

Figure 11.2 Level curve $f(x, y) = C$.

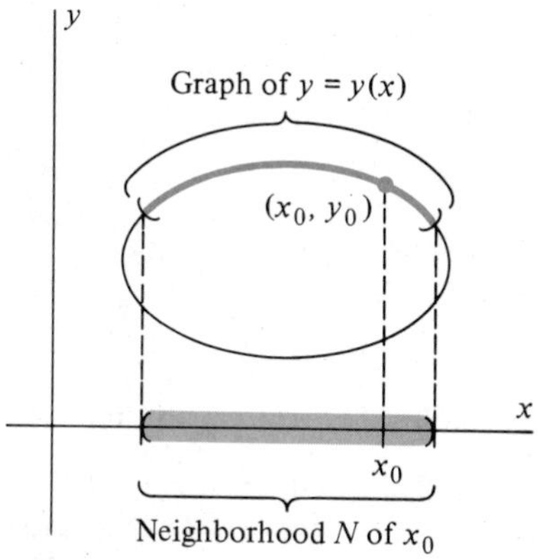

Figure 11.3 Level curve $f(x, y) = C$ determines a function $y = y(x)$ in a neighborhood N of x_0.

For $x \in N$, we may write equation (24) as

$$f(x, y(x)) = C. \tag{25}$$

Finally, differentiating both sides of (25), using the Chain Rule for functions of two variables, gives

$$\frac{\partial f}{\partial x}(x, y(x)) + \frac{\partial f}{\partial y}(x, y(x)) \cdot \frac{dy}{dx} = 0 \tag{26}$$

or, in the language of differentials

$$\frac{\partial f}{\partial x}(x, y(x))\, dx + \frac{\partial f}{\partial y}(x, y(x))\, dy = 0. \tag{27}$$

Now equation (26) is a differential equation of the form

$$M(x, y) + N(x, y)\frac{dy}{dx} = 0. \tag{28}$$

The general form for the equivalent equation (27) is

$$M(x, y)\, dx + N(x, y)\, dy = 0. \tag{29}$$

The question we wish to pursue is this: Given a differential equation of the form (28) or (29), and a point (x_0, y_0), when does there exist a function $y = y(x)$ that satisfies the differential equation in a neighborhood N of x_0, and for which $y(x_0) = y_0$? Obviously, we would also like to know how to find such a solution when it exists.

Comparing equations (26) and (28), we can see one situation in which equation (28) will have a solution: if $M(x, y) = \frac{\partial f}{\partial x}(x, y)$ and $N(x, y) = \frac{\partial f}{\partial y}(x, y)$, then equation (28) has the form of equation (26), and a solution is determined implicitly by equation (25). The constant C in (25) is determined by using the fact that the "initial point" (x_0, y_0) lies on the solution curve: $f(x_0, y_0) = C$.

Such differential equations are called *exact*. They are defined more precisely as follows.

*This, again, is because of the Implicit Function Theorem. However, the geometric explanation is simple—we need to know that the tangent at $(x_0, y(x_0))$ is not vertical if $y = y(x)$ is to be a function of x near $x = x_0$.

DEFINITION 11 The differential equation

$$M(x, y) + N(x, y)\frac{dy}{dx} = 0 \tag{30}$$

or

$$M(x, y)\,dx + N(x, y)\,dy = 0, \tag{31}$$

is called **exact** if there exists a function $f(x, y)$ for which

$$M(x, y) = \frac{\partial f}{\partial x}(x, y); \qquad N(x, y) = \frac{\partial f}{\partial y}(x, y). \tag{32}$$

Example 5 The differential equation

$$e^{2y} + 2xe^{2y}\frac{dy}{dx} = 0$$

is exact since, by Example 1,

$$M(x, y) = e^{2y} = \frac{\partial}{\partial x}(xe^{2y})$$

and

$$N(x, y) = 2xe^{2y} = \frac{\partial}{\partial y}(xe^{2y}).$$

A family of solutions is therefore determined by the level curves for the function $f(x, y) = xe^{2y}$:

$$xe^{2y} = C. \tag{33}$$

If $x > 0$, equation (33) may be solved explicitly for y as follows:

$$e^{2y} = \frac{c}{x};$$

$$2y = \ln\left(\frac{c}{x}\right);$$

$$y = \frac{1}{2}\ln\left(\frac{c}{x}\right)$$

$$= \ln\sqrt{\frac{c}{x}}.$$

A particular solution passing through the point (2, 0) may be found by substituting $x = x_0 = 2$ and $y = y_0 = 0$ in equation (33): $2e^0 = 2 = C$. Thus $y = \ln\sqrt{\dfrac{2}{x}}$, $x > 0$. ■

Example 6 From Example 2 we know that the differential equation

$$(3x^2 + 2y^2)\,dx + 4xy\,dy = 0$$

is exact, since the function $f(x, y) = x^3 + 2xy^2$ has the property that

$$M(x, y) = 3x^2 + 2y^2 = \frac{\partial}{\partial x}(x^3 + 2xy^2) = \frac{\partial f}{\partial x}(x, y)$$

and

$$N(x, y) = 4xy = \frac{\partial}{\partial y}(x^3 + 2xy^2) = \frac{\partial f}{\partial y}(x, y).$$

A family of solution curves is therefore

$$x^3 + 2xy^2 = C, \tag{34}$$

or

$$y = \pm\left(\frac{C - x^3}{2x}\right)^{1/2}. \tag{35}$$

A particular solution passing through the point (1, 2) is found by setting $x = 1$ and $y = 2$ in (34). This determines C as $C = 1^3 + 2(1)2^2 = 9$. Choosing the + sign in (35), we conclude that

$$y = \left(\frac{9 - x^3}{2x}\right)^{1/2}.$$

■

A Criterion for Exactness

The strong similarity between Examples 5 and 6 concerning differential equations and the earlier examples concerning the existence of potentials is no accident. In fact, by comparing Definitions 10 and 11 (with the aid of equations (5) and (6)) we conclude the following:

> The differential equation
>
> $$M(x, y)\,dx + N(x, y)\,dy = 0 \tag{36}$$
>
> is exact if and only if there exists a function $f(x, y)$ with
>
> $$M(x, y) = \frac{\partial f}{\partial x}(x, y)$$
>
> and
>
> $$N(x, y) = \frac{\partial f}{\partial y}(x, y)$$
>
> if and only if the function $f(x, y)$ is a potential for the vector function
>
> $$\mathbf{F}(x, y) = M(x, y)\mathbf{i} + N(x, y)\mathbf{j}. \tag{37}$$

The key to knowing when the differential equation (36) is exact is knowing when the vector function (37) has a potential. Since Theorem 12 provides just this sort of information, it may be restated in the following form relevant to the differential equation (36).

THEOREM 13 Let $M(x, y)$ and $N(x, y)$ have continuous partial derivatives in an open rectangle R. The differential equation

$$M(x, y)\,dx + N(x, y)\,dy = 0$$

is exact in R if and only if

$$\frac{\partial M}{\partial y}(x, y) = \frac{\partial N}{\partial x}(x, y) \tag{38}$$

for all (x, y) in R.

When faced with a differential equation of the form (36), with M and N as in Theorem 13, we may test for exactness by applying the now familiar criterion that $\frac{\partial M}{\partial y} = \frac{\partial N}{\partial x}$. If this criterion holds, a family of solution curves is given by the equation

$$f(x, y) = C$$

where $f(x, y)$ is a potential for the vector function $\boldsymbol{F}(x, y) = M(x, y)\boldsymbol{i} + N(x, y)\boldsymbol{j}$. In such cases $f(x, y)$ is found by the method of Examples 2 and 3.

Example 7 The differential equation

$$xy^3\,dx + x^3y\,dy = 0$$

is not exact, since

$$\frac{\partial M}{\partial y}(x, y) = \frac{\partial}{\partial y}(xy^3) = 3xy^2 \neq 3x^2y = \frac{\partial}{\partial x}(x^3y) = \frac{\partial N}{\partial x}(x, y).$$ ■

Example 8 Test the differential equation

$$(3x^2y + \cos x)\,dx + x^3\,dy = 0$$

for exactness. If it is exact, find a solution passing through the point $(1, \pi)$.

Solution: Here

$$M(x, y) = 3x^2y + \cos x; \qquad N(x, y) = x^3.$$

Since both M and N have continuous partial derivatives throughout $\mathbb{R}^2$, we may apply Theorem 13 to test for exactness. Since

$$\frac{\partial M}{\partial y}(x, y) = 3x^2 = \frac{\partial N}{\partial x}(x, y),$$

the equation is exact.

To solve the differential equation, we must find a potential $f(x, y)$ for the vector function (37). Proceeding as in Examples 2 and 3, we integrate $M(x, y)$ partially with respect to x to find

$$\begin{aligned} f(x, y) = \int M(x, y)\,dx &= \int (3x^2y + \cos x)\,dx \\ &= x^3y + \sin x + h_1(y) + K_1. \end{aligned}$$

Also,

$$f(x, y) = \int N(x, y)\, dy = \int x^3\, dy = x^3y + h_2(x) + K_2.$$

Equating these two expressions for $f(x, y)$ then gives

$$h_1(y) \equiv 0, \qquad h_2(x) = \sin x, \qquad \text{and} \qquad K_1 = K_2 = K.$$

Thus

$$f(x, y) = x^3y + \sin x + K$$

is the desired potential for any constant K. Arbitrarily taking $K = 0$ we obtain the family of level curves

$$x^3y + \sin x = C \tag{39}$$

as solution curves for the differential equation. To find the particular curve passing through the point $(x_0, y_0) = (1, \pi)$, we set $x = 1$ and $y = \pi$ in (39) to find

$$1^3\,(\pi) + \sin(\pi) = \pi = C,$$

so $C = \pi$. A solution passing through $(1, \pi)$ is therefore determined by the equation

$$x^3y + \sin x = \pi$$

as

$$y = \frac{\pi - \sin x}{x^3}, \qquad x^3 \neq 0.$$

■

Exercise Set 19.11

In Exercises 1–12, determine whether the given vector function has a potential. If so, find a potential for $\boldsymbol{F}(x, y)$.

1. $\boldsymbol{F}(x, y) = \boldsymbol{i} - \boldsymbol{j}$
2. $\boldsymbol{F}(x, y) = 6y\boldsymbol{i} + 6x\boldsymbol{j}$
3. $\boldsymbol{F}(x, y) = \pi y\boldsymbol{i} - \pi x\boldsymbol{j}$
4. $\boldsymbol{F}(x, y) = 3x\boldsymbol{i} + 4y\boldsymbol{j}$
5. $\boldsymbol{F}(x, y) = (x^2 - y)\boldsymbol{i} + (x + y^2)\boldsymbol{j}$
6. $\boldsymbol{F}(x, y) = \sin y\boldsymbol{i} + x\cos y\boldsymbol{j}$
7. $\boldsymbol{F}(x, y) = (3x^2\cos y - 1)\boldsymbol{i} + (2y - x^3\sin y)\boldsymbol{j}$
8. $\boldsymbol{F}(x, y) = (2xy^3 - 2x)\boldsymbol{i} + (3x^2y^2 + 1)\boldsymbol{j}$
9. $\boldsymbol{F}(x, y) = \left(2xe^{xy} + x^2ye^{xy}\right)\boldsymbol{i} + \left(x^3e^{xy} - \dfrac{1}{2\sqrt{y}}\right)\boldsymbol{j}$
10. $\boldsymbol{F}(x, y) = (x^3 - y\sin x)\boldsymbol{i} + (y^3 - \cos x)\boldsymbol{j}$
11. $\boldsymbol{F}(x, y) = (2xy^2 - y\sin x)\boldsymbol{i} + (2x^2y + \cos x)\boldsymbol{j}$
12. $\boldsymbol{F}(x, y) = 2x\tan^{-1} y\boldsymbol{i} + \dfrac{x^2}{1 + y^2}\boldsymbol{j}$

In Exercises 13–22, determine whether the given equation is exact. If it is, find a family of solution (level) curves of the form $f(x, y) = C$.

13. $2xy\, dx + x^2\, dy = 0$
14. $y^2 + xy\dfrac{dy}{dx} = 0$
15. $\dfrac{x\, dx}{\sqrt{x^2 + y^2}} + \dfrac{y\, dy}{\sqrt{x^2 + y^2}} = 0$
16. $\cos y^2 - 2xy\sin y^2\dfrac{dy}{dx} = 0$
17. $\dfrac{x}{y^2}\, dx + \dfrac{y}{x^2}\, dy = 0$
18. $\dfrac{y}{1 + x^2y^2} + \dfrac{x}{1 + x^2y^2}\dfrac{dy}{dx} = 0$
19. $(x - 2xy^2)\, dx + (y + x^2y)\, dy = 0$
20. $(y^2 - 2xy)\, dx + (2xy - x^2)\, dy = 0$
21. $\dfrac{x\, dx}{\sqrt{x + y^2}} + \dfrac{y\, dy}{\sqrt{x + y^2}} = 0$

22. $\dfrac{1}{\sqrt{x - y^2}} - \dfrac{2y}{\sqrt{x - y^2}}\dfrac{dy}{dx} = 0$

23. Find a solution of the differential equation in Exercise 13 passing through the point (1, 3).

24. Find a solution of the differential equation in Exercise 16 passing through the point (2, 0).

25. Find a solution of the differential equation in Exercise 20 passing through the point (1, 2).

26. Show that a separable differential equation $f(x)\,dx + g(y)\,dy = 0$ is always exact. What is the "solution" function $f(x, y)$?

27. Show that the differential equation $(1 + x^2y^2 + y)\,dx + x\,dy = 0$ is not exact. Then show that if this equation is multiplied through by the integrating factor $p(x, y) = \dfrac{1}{1 + x^2y^2}$ an exact differential equation results. Is a solution of this exact equation also a solution of the original equation?

SUMMARY OUTLINE OF CHAPTER 19

- If $\boldsymbol{x} \in \mathbb{R}^n$ and $f\colon \mathbb{R}^n \to \mathbb{R}$,

$$\lim_{\boldsymbol{x}\to\boldsymbol{x}_0} f(\boldsymbol{x}) = L \text{ means } |f(\boldsymbol{x}) - L| \to 0 \text{ as } \boldsymbol{x} \to \boldsymbol{x}_0.$$

- The function $f\colon \mathbb{R}^n \to \mathbb{R}$ is **continuous** at $\boldsymbol{x}_0$ if $\lim_{\boldsymbol{x}\to\boldsymbol{x}_0} f(\boldsymbol{x}) = f(\boldsymbol{x}_0)$.

- A **level curve** for the function $z = f(x, y)$ is the graph of the equation $f(x, y) = c$ in the xy-plane. A **level surface** for the function $w = f(x, y, z)$ is the graph of an equation $f(x, y, z) = c$ in space.

- The **partial derivatives** of $f(x, y, z)$ are the limits

$$\frac{\partial f}{\partial x}(x, y, z) = \lim_{\Delta x\to 0} \frac{f(x + \Delta x, y, z) - f(x, y, z)}{\Delta x},$$

$$\frac{\partial f}{\partial y}(x, y, z) = \lim_{\Delta y\to 0} \frac{f(x, y + \Delta y, z) - f(x, y, z)}{\Delta y},$$

and

$$\frac{\partial f}{\partial z}(x, y, z) = \lim_{\Delta z\to 0} \frac{f(x, y, z + \Delta z) - f(x, y, z)}{\Delta z}.$$

- ***Theorem:*** $\dfrac{\partial^2 f}{\partial y\partial x}(x, y) = \dfrac{\partial^2 f}{\partial x\partial y}(x, y)$ if f, $\dfrac{\partial f}{\partial x}$, $\dfrac{\partial f}{\partial y}$, $\dfrac{\partial^2 f}{\partial x\partial y}$, and $\dfrac{\partial^2 f}{\partial y\partial x}$ are continuous.

- The plane tangent to the graph of $z = f(x, y)$ at (x_0, y_0, z_0) has equation

$$\frac{\partial f}{\partial x}(x_0, y_0)(x - x_0) + \frac{\partial f}{\partial y}(x_0, y_0)(y - y_0) - (z - z_0) = 0.$$

- ***Theorem:*** If $z = f(x, y)$ has a relative extremum at (x_0, y_0) then either

  (i) $\dfrac{\partial f}{\partial x}(x_0, y_0) = \dfrac{\partial f}{\partial y}(x_0, y_0) = 0$, or

  (ii) one or both of $\dfrac{\partial f}{\partial x}(x_0, y_0)$ and $\dfrac{\partial f}{\partial y}(x_0, y_0)$ fail to exist.

- ***Theorem:*** If $\dfrac{\partial f}{\partial x}(x_0, y_0) = \dfrac{\partial f}{\partial y}(x_0, y_0) = 0$ and if $A = \dfrac{\partial^2 f}{\partial x^2}(x_0, y_0)$, $B = \dfrac{\partial^2 f}{\partial x\partial y}(x_0, y_0)$, $C = \dfrac{\partial^2 f}{\partial y^2}(x_0, y_0)$, and $D = B^2 - AC$, then

  (i) If $D < 0$, and $A < 0$, $f(x_0, y_0)$ is a relative maximum.
  (ii) If $D < 0$ and $A > 0$, $f(x_0, y_0)$ is a relative minimum.
  (iii) If $D > 0$, (x_0, y_0) is a saddle point.
  (iv) If $D = 0$ there is no conclusion.

- ***Theorem:*** (Linear Approximation) If $f(x, y)$ and its first partials are continuous, then

$$f(x_0 + \Delta x, y_0 + \Delta y) = f(x_0, y_0) + \frac{\partial f}{\partial x}(x_0, y_0)\Delta x + \frac{\partial f}{\partial y}(x_0, y_0)\Delta y + \epsilon_1\Delta x + \epsilon_2\Delta y$$

where $\epsilon_1 \to 0$ and $\epsilon_2 \to 0$ as $\Delta x \to 0$ and $\Delta y \to 0$.

- ***Chain Rules:***

  (i) $\dfrac{d}{dt}f(x(t), y(t)) = \dfrac{\partial f}{\partial x}\cdot\dfrac{dx}{dt} + \dfrac{\partial f}{\partial y}\cdot\dfrac{dy}{dt}$,

  (ii) $\dfrac{\partial}{\partial s}f(x(s, t), y(s, t)) = \dfrac{\partial f}{\partial x}\cdot\dfrac{\partial x}{\partial s} + \dfrac{\partial f}{\partial y}\cdot\dfrac{\partial y}{\partial s}$,

  (iii) $\dfrac{\partial}{\partial t}f(x(s, t), y(s, t)) = \dfrac{\partial f}{\partial x}\cdot\dfrac{\partial x}{\partial t} + \dfrac{\partial f}{\partial y}\cdot\dfrac{\partial y}{\partial t}$,

  ⋮

- The **directional derivative** $D_{\boldsymbol{u}}f(x_0, y_0)$ in the direction of the unit vector $\boldsymbol{u} = u_1\boldsymbol{i} + u_2\boldsymbol{j}$ is

$$D_{\boldsymbol{u}}f(x_0, y_0) = \lim_{t\to 0^+} \frac{f(x_0 + tu_1, y_0 + tu_2) - f(x_0, y_0)}{t}.$$

- The **gradient** $\nabla f(x_0, y_0)$, of the function $f(x, y)$ at (x_0, y_0) is the **vector**

$$\nabla f(x_0, y_0) = \frac{\partial f}{\partial x}(x_0, y_0)\boldsymbol{i} + \frac{\partial f}{\partial y}(x_0, y_0)\boldsymbol{j}.$$

- For $f(x, y, z)$

$$\nabla f(x_0, y_0, z_0) = \frac{\partial f}{\partial x}(x_0, y_0, z_0)\boldsymbol{i} + \frac{\partial f}{\partial y}(x_0, y_0, z_0)\boldsymbol{j} + \frac{\partial f}{\partial z}(x_0, y_0, z_0)\boldsymbol{k}.$$

- If $\frac{\partial f}{\partial x}(x, y)$ and $\frac{\partial f}{\partial y}(x, y)$ are continuous, then

$$D_{\boldsymbol{u}}f(x_0, y_0) = \frac{\partial f}{\partial x}(x_0, y_0)u_1 + \frac{\partial f}{\partial x}(x_0, y_0)u_2 = \nabla f(x_0, y_0)\cdot\boldsymbol{u}.$$

- $\nabla f(\boldsymbol{x}_0)$ points in the direction of most rapid increase of the function f at $\boldsymbol{x}_0$.
- $\nabla f(\boldsymbol{x}_0)$ is orthogonal to the level curve for $z = f(x, y)$ at $\boldsymbol{x}_0 = (x_0, y_0)$.
- $\nabla f(\boldsymbol{x}_0)$ is orthogonal to the level surface for $z = f(x, y, z)$ at $\boldsymbol{x}_0 = (x_0, y_0, z_0)$.
- The plane tangent to the surface $f(x, y, z) = c$ at (x_0, y_0, z_0) has equation

$$\frac{\partial f}{\partial x}(x_0, y_0, z_0)(x - x_0) + \frac{\partial f}{\partial y}(x_0, y_0, z_0)(y - y_0) + \frac{\partial f}{\partial z}(x_0, y_0, z_0)(z - z_0) = 0.$$

- The extreme values of the function $w = f(\boldsymbol{x})$ subject to the constraint $g(\boldsymbol{x}) = 0$ occur at points $\boldsymbol{x}_0$ where

$$\nabla f(\boldsymbol{x}_0) = \lambda g(\boldsymbol{x}_0)$$

where λ is constant.

- The vector function $\boldsymbol{F}(x, y) = M(x, y)\boldsymbol{i} + N(x, y)\boldsymbol{j}$ is said to have the **potential** $f(x, y)$ if $\boldsymbol{F}(x, y) = \nabla f(x, y)$.
- ***Theorem:*** If $M(x, y)$, $N(x, y)$ and their first partial derivatives are continuous in a rectangle R, then

$$\boldsymbol{F}(x, y) = M(x, y)\boldsymbol{i} + N(x, y)\boldsymbol{j}$$

has a potential in R if and only if $\frac{\partial M}{\partial y} = \frac{\partial N}{\partial x}$ for all $(x, y) \in R$.

- The differential equation

$$M(x, y)\,dx + N(x, y)\,dy = 0 \tag{1}$$

is **exact** if there exists a function f with $M = \frac{\partial f}{\partial x}$ and $N = \frac{\partial f}{\partial y}$. In this case a family of solution (level) curves is $f(x, y) = c$.

- ***Theorem:*** The differential equation (1) is exact if and only if $\frac{\partial M}{\partial y} = \frac{\partial N}{\partial x}$.

REVIEW EXERCISES—CHAPTER 19

1. Does $\lim_{(x,y)\to(0,0)} \frac{x^2 - y^2}{x^2 + y^2}$ exist? If so, find it.

2. Find $\lim_{(x,y)\to(0,0)} \frac{2x^2y^3}{(x^2 + y^2)^2}$.

3. Sketch level curves for the function $f(x, y) = \frac{x}{x^2 + y^2}$ corresponding to levels $z = -2, -1, 1, 2, 4$.

4. Find $\frac{df}{dt}$ if $f(x, y) = x\tan^{-1}\left(\frac{y}{x}\right)$, $x = 1 + t^2$, and $y = 1 - t$.

5. Find a vector normal to the curve $x^3 - 2xy^2 + y + 5 = 0$ at the point $(1, 2)$.

6. Find an equation for the plane tangent to the graph of the equation $x^3 + y^3 - 6xy + z = 0$ at the point $(2, 2, 8)$.

7. Find a vector normal to the surface $e^x \cos y - z = 4$ at the point $(0, \pi, -5)$.

8. What is the z-coordinate of the point $P = (1, 3, z)$ if P lies on the plane tangent to the ellipsoid $4x^2 + y^2 + 9z^2 = 17$ at the point $(1, 2, 1)$?

9. Find both first order partial derivatives for the function $f(x, y) = x\sin\sqrt{x^2 + y^2}$.

10. Find an equation for the plane tangent to the graph of $x^3 + y^3 + xz^2 + z^3 - 9 = 0$ at the point $(2, 1, -2)$.

11. Let $f(x, y, z) = xy^3 + x^2\sqrt{y^2 + z^2}$. Find the directional derivative of f in the direction of the vector $2\boldsymbol{i} + \boldsymbol{j} + 2\boldsymbol{k}$ at the point $(2, 3, 4)$.

12. Show that $f(x, y, z) = (ax + by + cz)^3$ satisfies the partial differential equation

$$x\frac{\partial f}{\partial x} + y\frac{\partial f}{\partial y} + z\frac{\partial f}{\partial z} = 3f.$$

13. Find $\frac{\partial f}{\partial r}$ and $\frac{\partial f}{\partial s}$ if $f(x, y) = 3x^3 + 2xy^2 - y^2$, $x = 2r + 5s$, and $y = r - 2s^2$.

14. Suppose that $f(x, y)$ has partial derivatives with respect to both variables. Show that the function $z = f(x - y, y - x)$ satisfies the partial differential equation

$$\frac{\partial z}{\partial x} + \frac{\partial z}{\partial y} = 0.$$

15. Find an equation for the line tangent to the graph of $x \cos \pi y + x^2 e^y = 6$ at the point $(2, 0)$.

16. Find the directional derivative $D_{\boldsymbol{u}} f$ of the function $f(x, y) = y^2 \sin^{-1} x + ye^x$ at the point $(0, 1)$ in the direction of the vector $\boldsymbol{u} = \boldsymbol{i} + \sqrt{e}\,\boldsymbol{j}$.

17. Let $f(x, y)$ and $g(x, y)$ have partial derivatives with respect to both variables. Show that $\nabla(\alpha f + \beta g) = \alpha \nabla f + \beta \nabla g$ where α and β are constants.

18. Find the maximum value of the directional derivative for the function $w = xy^3 + z \cos y - \ln(x^2 + y)$ at the point $(1, 0, 4)$.

19. Find a vector pointing in the direction of most rapid decrease for the function $f(x, y, z) = xyz - z \tan^{-1}(y/x)$ at the point $(1, 2, 3)$.

20. Find an equation for the plane tangent to the surface $z^2x - 2zy + e^{xy} = 9$ at the point $(2, 0, 2)$.

21. Show that the function $f(x, t) = e^{x+ct}$ is a solution of the wave equation

$$\frac{\partial^2 f}{\partial t^2} = c^2 \frac{\partial^2 f}{\partial x^2}.$$

22. Find the maximum and minimum values of the function $f(x, y) = 4x^2 + xy + 2y^2$ on the square $S = \{(x, y) | -1 \le x \le 1, -1 \le y \le 1\}$.

23. Use the method of Lagrange multipliers to find the rectangular box of largest volume that can be inscribed in the ellipsoid $x^2 + \frac{y^2}{9} + \frac{z^2}{4} = 1$.

Find and classify all relative extrema.

24. $f(x, y) = 4y^2 - 2x^2$

25. $f(x, y) = 3x^2 + xy - 6y^2$

26. $f(x, y) = x^2y + xy^2 + 4x + 4y$

27. $f(x, y) = e^{x^2-4xy}$

28. $f(x, y) = \ln(1 + x^2 + y^2)$

29. $f(x, y) = e^{1+x^2-y^2}$

30. $f(x, y) = 6x^2 - 2x - 3xy + y^2 + 5y + 5$

31. The directional derivative of $f(x, y)$ at P in the direction of the vector $\boldsymbol{u}_1 = \boldsymbol{j}$ is $D_{\boldsymbol{u}_1} f(P) = 3$, and the directional derivative of f at P in the direction of $\boldsymbol{u}_2 = 3\boldsymbol{i} + 4\boldsymbol{j}$ is $D_{\boldsymbol{u}_2} f(P) = 3$ also. Find

a. $\frac{\partial f}{\partial x}(P)$, **b.** $\frac{\partial f}{\partial y}(P)$, and **c.** $\nabla f(P)$.

32. The directional derivative of $f(x, y)$ at (x_0, y_0) in the direction of the unit vector $\boldsymbol{u}$ is 6. What is

a. $D_{2\boldsymbol{u}} f(x_0, y_0)$?

b. $D_{-\boldsymbol{u}} f(x_0, y_0)$?

33. Find a function $f(x, y)$ for which $\nabla f(x, y) = 2xe^y\boldsymbol{i} + (x^2e^y - \sin y)\boldsymbol{j}$.

34. Find a potential for the vector function $F(x, y) = (ye^x + e^y)\boldsymbol{i} + (1 + e^x + xe^y)\boldsymbol{j}$.

35. When three resistors r_1, r_2, and r_3 are connected in parallel, the net resistance R is determined by the equation

$$\frac{1}{R} = \frac{1}{r_1} + \frac{1}{r_2} + \frac{1}{r_3}.$$

Suppose $r_1 = 10$ ohms, $r_2 = 20$ ohms, and $r_3 = 25$ ohms. Approximate the change in the value of R that results from each of r_1, r_2, and r_3 being increased by 10%.

36. Is the function

$$f(x, y) = \begin{cases} \dfrac{6xy}{x^2 + y^2}, & (x, y) \ne (0, 0) \\ 0, & (x, y) = (0, 0) \end{cases}$$

differentiable at $(0, 0)$? Why or why not?

37. Show that the function

$$w = e^{x-y} + \cos(y - z) + \sqrt{z - x}$$

satisfies the partial differential equation

$$\frac{\partial w}{\partial x} + \frac{\partial w}{\partial y} + \frac{\partial w}{\partial z} = 0.$$

38. An airplane flying due north at a speed of 200 km/hr and an altitude of 2 km passes directly over an automobile travelling due east along a straight highway at a speed of 100 km/hr. At what rate are the plane and the automobile moving apart after 6 minutes?

39. Find the maximum and minimum values of the function

$$f(x, y) = 3x - 2y + 5$$

on or inside the ellipse $\frac{x^2}{4} + \frac{y^2}{9} = 1$.

40. A closed rectangular box is to contain 1000 cm^3. If the material for the top and bottom costs 2¢/cm^2 and the material for the sides costs 3¢/cm^2, find the dimensions which minimize cost.

41. Let $\boldsymbol{u}$ and $\boldsymbol{v}$ be distinct unit vectors. Is $D_{\boldsymbol{u}+\boldsymbol{v}}f(x_0, y_0) = D_{\boldsymbol{u}}f(x_0, y_0) + D_{\boldsymbol{v}}f(x_0, y_0)$? Why or why not?

42. Find a family of solution curves for the differential equation

$$(3x^2y + y^2 - \sin y)\, dx + (x^3 + 2xy - x\cos y)\, dy = 0.$$

43. Find an equation for the plane through the point $(1, 2, 1)$ with positive x-, y-, and z-intercepts that bounds the solid of least volume in the first octant.

44. Find the point on the surface

$$x^2 + y^2 + z^2 = 16$$

where the function $f(x, y, z) = x + 2y - 3z + 1$ is a maximum.

45. Suppose that the level surfaces $f(x, y, z) = K$ and $g(x, y, z) = L$ intersect in a curve C. Let $\boldsymbol{x}_0$ maximize $h(x, y, z)$ on C, and assume that f, g, and h have continuous partial derivatives. Show that $\nabla f(\boldsymbol{x}_0)$, $\nabla g(\boldsymbol{x}_0)$, and $\nabla h(\boldsymbol{x}_0)$ lie in a common plane when based at x_0. Conclude that $\nabla f(\boldsymbol{x}_0) = \lambda\nabla g(\boldsymbol{x}_0) + \mu\nabla h(\boldsymbol{x}_0)$ for some constants λ and μ.

46. Prove that if $f(x, y)$ is differentiable for all (x, y) in a closed and bounded disc D, then $f(x, y)$ is bounded on D. (This means that there exists a number M with $|f(x, y)| \le M$ for all $(x, y) \in D$).

47. The function $f(x, y, z)$ is called homogeneous of degree n if $f(tx, ty, tz) = t^n f(x, y, z)$ for all x, y, z, and t. Show that if n is an integer and f has continuous partial derivatives, then

$$x\frac{\partial f}{\partial x} + y\frac{\partial f}{\partial y} + z\frac{\partial f}{\partial z} = nf$$

when f is homogeneous of degree n.

48. For the function $f(x, y) = \sqrt{x^2 + y^2}$, show that all directional derivatives exist at $(0, 0)$, but that neither partial derivative exists at $(0, 0)$.

49. Show that the spheres with equations $x^2 + (y - 1)^2 + z^2 = 2$ and $x^2 + (y + 1)^2 + z^2 = 2$ intersect in a circle. Then show that at each point on the circle of intersection the tangent planes to the two circles are orthogonal.

50. Use the method of Lagrange multipliers to find the point on the plane $x - 3y + 2z = 6$ nearest the origin.

51. Find a vector in the direction of most rapid increase of the function $f(x, y, z) = e^{x-z}\tan^{-1}(y + z)$ at the point $(3, 1, 2)$.

52. Find an equation for the plane tangent to the surface $x^2 + 4y^2 + z^2 = 12$ at the point $(2, 1, 2)$.

53. Find $\dfrac{d^2f}{dt^2}$ if $f(x, y) = e^{y^2-x^2}$, $x(t) = 4 - t^2$, $y(t) = t\cos \pi t$.

54. Find the minimum value of the function $f(x, y) = 3y^2 - xy + x^2$ subject to the constraint that $x^2 + 3y^2 = 3$.

55. Find the maximum volume for a rectangular solid inscribed in the ellipsoid

$$\frac{x^2}{4} + \frac{y^2}{9} + \frac{z^2}{4} = 1.$$

CHAPTER 20

DOUBLE AND TRIPLE INTEGRALS

20.1 INTRODUCTION

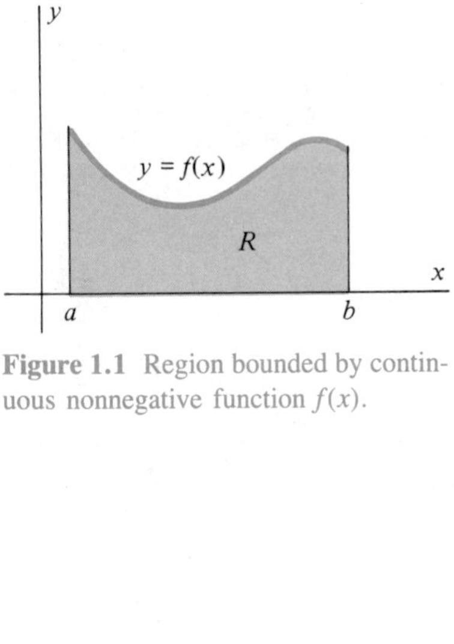

Figure 1.1 Region bounded by continuous nonnegative function $f(x)$.

The goal of this chapter is to extend the theory of the Riemann integral to functions of two and three variables. You will find that most of what we do here is a straightforward generalization of the theory of integration for functions of a single variable developed in Chapters 6 through 8. It is, therefore, very important that we begin with a clear recollection of the motivation and details associated with the integral $\int_a^b f(x)\,dx$.

The original motivation for the definite integral was the problem of calculating the area of a region R bounded by the graph of the continuous nonnegative function $y = f(x)$ and the x-axis for $a \leq x \leq b$ (Figure 1.1). We did so by using a partition P_n to divide the interval $[a, b]$ into subintervals with endpoints $a = x_0 < x_1 < x_2 < \cdots < x_n = b$ of length $\Delta x_j = x_j - x_{j-1}$. After choosing one number t_j arbitrarily in each interval, we formed the approximating Riemann sum

$$S_n = \sum_{j=1}^{n} f(t_j)\,\Delta x_j$$

representing the sum of the areas of the rectangles illustrated in Figure 1.2. We then proved that the limit of this Riemann sum, as $n \to \infty$ and as the norm of the partition $\|P_n\| \to 0$, is the desired area. This led to the definition of the definite integral

$$\int_a^b f(x)\,dx = \lim_{n\to\infty} \sum_{j=1}^{n} f(t_j)\,\Delta x_j$$

independent of whether or not $f(x)$ is nonnegative.

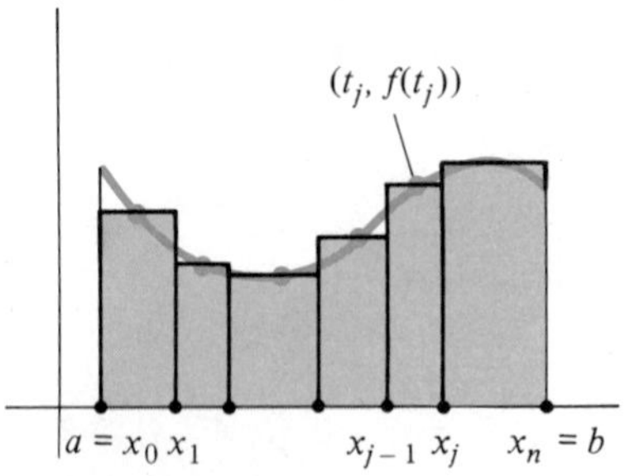

Figure 1.2 Riemann sum approximates region R by rectangles.

For functions of two variables, the primary motivation leading to the corresponding integral (called the *double* integral) will be the calculation of volume rather than area. This is because a region Q in the domain of $z = f(x, y)$ and the graph of $z = f(x, y)$ over Q bound a solid in space (see Figure 1.3). We will approximate the volume of this solid by rectangular prisms (Figure 1.4), and the entire development will be very much analogous to the one-variable case. Additionally, since the domain of $f(x, y)$ can be coordinatized by polar coordinates as well as rectangular coordinates, we will also carry out the development of double integrals in polar

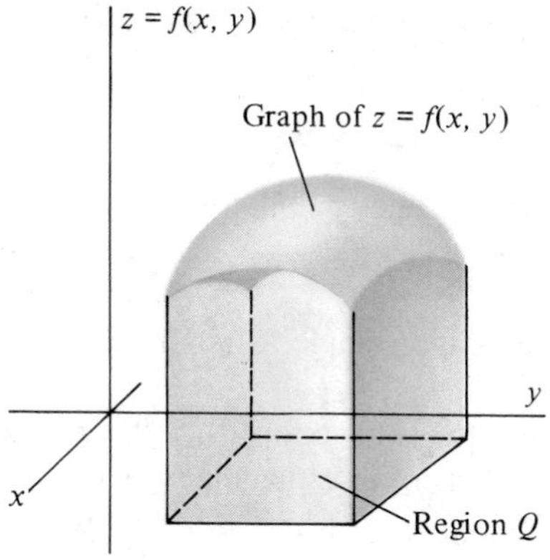

Figure 1.3 Graph of $z = f(x,y)$ over the region Q bounds a solid in xyz space.

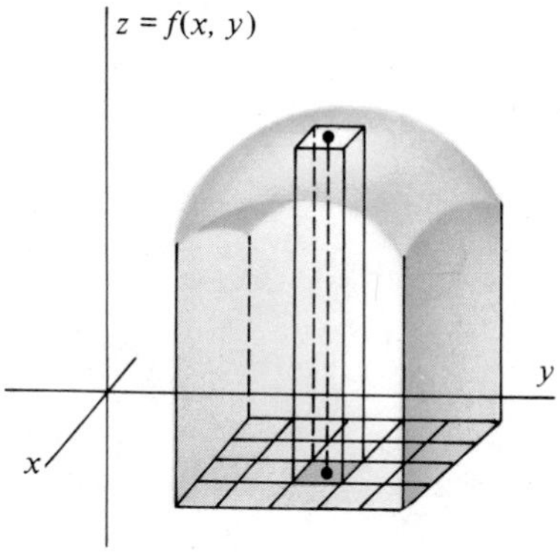

Figure 1.4 Volume of solid is approximated by using rectangular prisms.

coordinates. As for integrals of functions of a single variable, we will develop various physical applications of double integrals.

The chapter concludes with the development of the triple integral for functions of three variables. Although the geometry is more specialized, the issues of calculating volume and mass for three-dimensional objects still serve as primary motivation, and the triple integral is discussed in cylindrical and spherical coordinates as well as in rectangular coordinates.

20.2 THE DOUBLE INTEGRAL OVER A RECTANGLE

We begin with the problem of calculating the volume of a solid V bounded above by the graph of the continuous function $z = f(x, y)$, below by the rectangle $R = \{(x, y) \mid a \le x \le b,\ c \le y \le d\}$ in the xy-plane, and on four sides by the vertical planes $x = a$, $x = b$, $y = c$, and $y = d$ (see Figures 2.1 and 2.2). We let

$$\Delta x = \frac{b-a}{n}, \qquad \Delta y = \frac{d-c}{n}$$

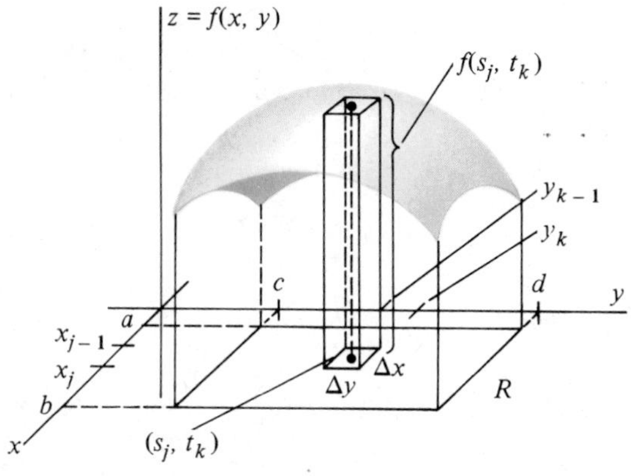

Figure 2.1 Volume of solid bounded by graph of $f(x,y)$ over R is approximated using rectangular prisms of volume $\Delta V_{jk} = f(s_j,t_k)\Delta x\Delta y$.

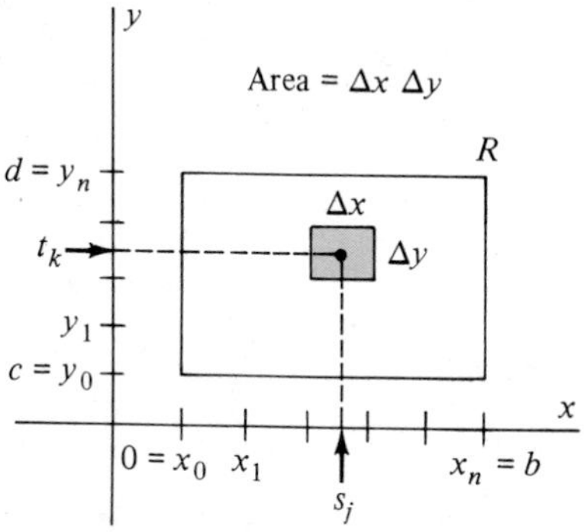

Figure 2.2 Rectangle R in xy-plane. Area of rectangle R_{jk} is $\Delta A = \Delta x\Delta y$.

where n is a positive integer. By drawing lines parallel to the y-axis with x-intercepts

$$x_0 = a, \qquad x_1 = a + \Delta x, \qquad x_2 = a + 2\Delta x, \ldots, x_n = a + n\Delta x = b,$$

and by drawing lines parallel to the x-axis with y-intercepts

$$y_0 = c, \qquad y_1 = c + \Delta y, \qquad y_2 = c + 2\Delta y, \ldots, y_n = c + n\Delta y = d,$$

we partition the rectangle R into n^2 rectangles R_{jk}, each of area $\Delta A = \Delta x \Delta y$.

The idea now is to approximate the volume of the region above the rectangle R_{jk} with the volume of a rectangular prism whose base has area $\Delta A = \Delta x \Delta y$. For the height of the prism we have several choices. If we were to use

$$m_{jk} = \text{minimum value of } f(x, y) \text{ on rectangle } R_{jk}$$

as the height, the volume of the prism would be no greater than the actual volume of the solid region over the rectangle R_{jk}. Summing over all rectangles R_{jk} would then yield a **lower approximating sum** satisfying the inequality

$$\underline{S}_n = \sum_{j=1}^{n} \sum_{k=1}^{n} m_{jk} \Delta A \le \text{volume of } V. \tag{1}$$

Similarly, if we were to take

$$M_{jk} = \text{maximum value of } f(x, y) \text{ on rectangle } R_{jk}$$

as the height, the volume of the prism would equal or exceed the volume of the solid region over the rectangle R_{jk}. Summing over all rectangles R_{jk} would produce an **upper approximating sum**

$$\overline{S}_n = \sum_{j=1}^{n} \sum_{k=1}^{n} M_{jk} \Delta A \ge \text{volume of } V. \tag{2}$$

By combining inequalities (1) and (2), we conclude

$$\underline{S}_n = \sum_{j=1}^{n} \sum_{k=1}^{n} m_{jk} \Delta A \le \text{volume of } V \le \sum_{j=1}^{n} \sum_{k=1}^{n} M_{jk} \Delta A = \overline{S}_n. \tag{3}$$

Just as in the one-variable case, it follows from the continuity of $f(x, y)$ that $\lim_{n\to\infty} \underline{S}_n = \lim_{n\to\infty} \overline{S}_n$. To see this note that, since $\underline{S}_n \le \overline{S}_n$ for all n, we have

$$\begin{aligned} 0 \le \overline{S}_n - \underline{S}_n &= \sum_{j=1}^{n} \sum_{k=1}^{n} (M_{jk} - m_{jk}) \Delta A \\ &\le \left(\max_{\substack{1 \le j \le n \\ 1 \le k \le n}} [M_{jk} - m_{jk}] \right) \sum_{j=1}^{n} \sum_{k=1}^{n} \Delta A \\ &= \left(\max_{\substack{1 \le j \le n \\ 1 \le k \le n}} [M_{jk} - m_{jk}] \right) (\text{Area of } R). \end{aligned} \tag{4}$$

The continuity of $f(x, y)$ assures that the first factor tends to zero as $n \to \infty$, since the dimensions of the rectangle R_{jk} tend to zero. (This fact depends on the *uniform*

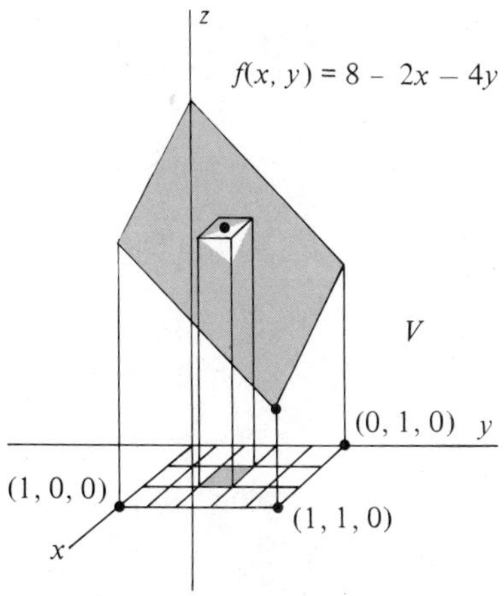

Figure 2.3 Solid bounded by $f(x) = 8 - 2x - 4y$ and xy-plane over unit square.

continuity of $f(x, y)$ and is proved in more advanced courses.) Since the second factor is a constant, the right-hand side of inequality (4) approaches zero as $n \to \infty$. The conclusion that $\lim_{n\to\infty} (\overline{S}_n - \underline{S}_n) = 0$ then follows from the Pinching Theorem.

Since both the upper and lower approximating sums converge to the same limit, this limit must be the desired volume, according to inequality (3). The volume of V may therefore be computed using either lower or upper sums, as the following example illustrates.

Example 1 Find the volume of the solid bounded above by the graph of $z = 8 - 2x - 4y$ and below by the xy-plane for (x, y) in the unit square $R = \{(x, y)|0 \le x \le 1, 0 \le y \le 1\}$ (see Figure 2.3).

Solution: A partition of the interval $[0, 1]$ into n equal subdivisions gives $\Delta x = \dfrac{1-0}{n} = \dfrac{1}{n}$, and endpoints

$$x_0 = 0, \quad x_1 = \frac{1}{n}, \quad x_2 = \frac{2}{n}, \ldots, x_n = \frac{n}{n} = 1.$$

Similarly, $\Delta y = \dfrac{1-0}{n} = \dfrac{1}{n}$, and

$$y_0 = 0, \quad y_1 = \frac{1}{n}, \quad y_2 = \frac{2}{n}, \ldots, y_n = \frac{n}{n} = 1.$$

Since the function $f(x, y) = 8 - 2x - 4y$ is decreasing in both variables, the minimum value of $f(x, y)$ on the square $x_{j-1} \le x \le x_j$, $y_{k-1} \le y \le y_k$ is

$$m_{jk} = f(x_j, y_k) = f\left(\frac{j}{n}, \frac{k}{n}\right) = 8 - 2\left(\frac{j}{n}\right) - 4\left(\frac{k}{n}\right). \tag{5}$$

In calculating the lower approximating sum, we will make use of the formula

$$1 + 2 + 3 + \cdots + n = \sum_{m=1}^{n} m = \frac{n(n+1)}{2}. \tag{6}$$

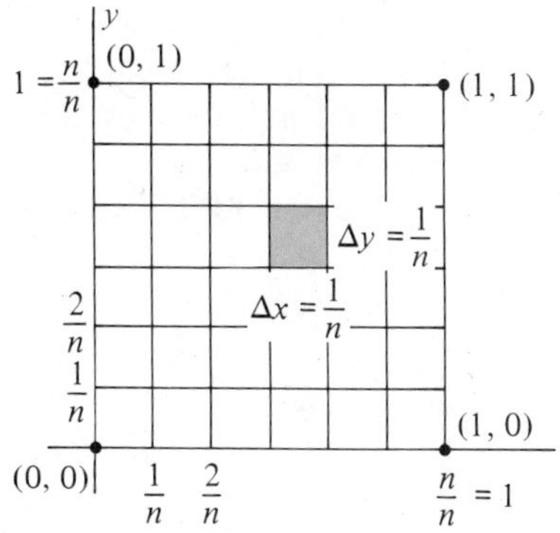

Figure 2.4 $\Delta x = \dfrac{1}{n}$ and $\Delta y = \dfrac{1}{n}$, so $\Delta A = \Delta x \Delta y = \dfrac{1}{n^2}$.

With m_{jk} as in (5) and $\Delta A = \Delta x \Delta y = \left(\dfrac{1}{n}\right)\left(\dfrac{1}{n}\right) = \dfrac{1}{n^2}$ (see Figure 2.4), the lower approximating double sum in (1) becomes

$$\underline{S}_n = \sum_{j=1}^{n} \sum_{k=1}^{n} \left\{\left[8 - 2\left(\frac{j}{n}\right) - 4\left(\frac{k}{n}\right)\right]\left(\frac{1}{n^2}\right)\right\} \tag{7}$$

$$= \sum_{j=1}^{n} \left\{\left[\sum_{k=1}^{n} 8 - 2\sum_{k=1}^{n}\left(\frac{j}{n}\right) - 4\sum_{k=1}^{n}\left(\frac{k}{n}\right)\right]\left(\frac{1}{n^2}\right)\right\}$$

$$= \left(\frac{1}{n^2}\right) \sum_{j=1}^{n} \left\{8n - 2n\left(\frac{j}{n}\right) - \left(\frac{4}{n}\right)\left[\frac{n(n+1)}{2}\right]\right\}$$

by equation (6)

$$= \left(\frac{1}{n^2}\right) \sum_{j=1}^{n} \{8n - 2j - 2(n+1)\}$$

$$= \left(\frac{1}{n^2}\right) \left\{\sum_{j=1}^{n} (8n) - 2\sum_{j=1}^{n} j - 2\sum_{j=1}^{n} (n+1)\right\}$$

$$= \left(\frac{1}{n^2}\right) \left\{8n^2 - 2\left[\frac{n(n+1)}{2}\right] - 2n(n+1)\right\} \quad \text{(by equation (5))}$$

$$= 8 - 3\left[\frac{n(n+1)}{n^2}\right].$$

Evaluating the limit as $n \to \infty$ of this approximating sum, we conclude that

$$\text{Volume of } V = \lim_{n\to\infty} \underline{S}_n = \lim_{n\to\infty} \left\{8 - 3\left[\frac{n(n+1)}{n^2}\right]\right\} = 5.$$

If we wish to calculate this volume by means of an upper approximating double sum, we must use the fact that the maximum value of the function $f(x, y)$ on the square $x_{j-1} \le x \le x_j$, $y_{k-1} \le y \le y_k$ occurs at the point (x_{j-1}, y_{k-1}), since $f(x, y)$ is decreasing in both variables x and y. Thus,

$$M_{jk} = f(x_{j-1}, y_{k-1}) = 8 - 2\left(\frac{j-1}{n}\right) - 4\left(\frac{k-1}{n}\right).$$

In Exercise 32 you are asked to show that, with this expression for M_{jk}, the corresponding upper approximating sum is

$$\overline{S}_n = 8 - 3\left[\frac{n(n-1)}{n^2}\right].$$

Thus, $\lim_{n\to\infty} \overline{S}_n = \lim_{n\to\infty} \left\{8 - 3\left[\frac{n(n-1)}{n^2}\right]\right\} = 5$, as above. ■

The success of applying the notions of upper and lower approximating sums to calculate the volume of the solid in Example 1 depended critically on formula (6) for evaluating the sums that occurred. However, such formulas are known only for a very limited number of sums, so the method of Example 1 is not very general. With the aid of computers we can treat a much broader class of functions, at least in terms of generating upper and lower approximating sums for various fixed values of n.

Program 12 in Appendix I is a BASIC program that calculates lower approximating sums over rectangles for functions that are decreasing in both variables, such as the function $f(x, y) = 8 - 2x - 4y$ in Example 1. Table 2.1 shows the results of using this program to obtain values of the lower approximating sum in Example 1. (You may verify these results by using the formula for $\underline{S}_n$ in line (7).)

Table 2.2 shows the results of using this same program to calculate lower approximating sums for the function $f(x, y) = 8 - 2x^2 - 2y^3$, again over the rectangle $R = \{(x, y) | 0 \le x \le 1, 0 \le y \le 1\}$. In Exercise 33 you are asked to show that the limit of these approximating sums as $n \to \infty$ is actually $S = \frac{41}{6} \approx 6.8333$.

Of course, we would like a more effective way to calculate volumes than simply working with approximations. The first step in developing such a theory is to define

Table 2.1 Results of using Program 12 to calculate lower approximating sums for $f(x) = 8 - 2x - 4y$ on the rectangle with $a = c = 0$ and $b = d = 1$

n	$\underline{S}_n$
5	4.40
10	4.70
20	4.85
50	4.94
100	4.97
200	4.985

Table 2.2 Results of using Program 12 to calculate lower approximating sums for $f(x) = 8 - 2x^2 - 2y^3$ on the rectangle with $a = c = 0$ and $b = d = 1$

n	$\underline{S}_n$
5	6.4
10	6.625
20	6.7312
50	6.7930
100	6.8132
200	6.8204

a more general type of approximating double sum. The key to doing this is to observe that if (s_j, t_k) is any point in the rectangle R_{jk}, and if m_{jk} and M_{jk} are the minimum and maximum values of $f(x, y)$ on R_{jk}, respectively, then

$$m_{jk} \leq f(s_j, t_k) \leq M_{jk}.$$

Thus, the approximating double sum

$$S_n = \sum_{j=1}^{n} \sum_{k=1}^{n} f(s_j, t_k)\Delta A, \qquad \Delta A = \Delta x \Delta y \tag{8}$$

has the property that

$$\underline{S}_n \leq S_n \leq \overline{S}_n$$

for all n. Since $\lim_{n\to\infty} \underline{S}_n = \lim_{n\to\infty} \overline{S}_n$ when $f(x, y)$ is continuous, the Pinching Theorem guarantees that the limit

$$S = \lim_{n\to\infty} S_n = \lim_{n\to\infty} \sum_{j=1}^{n} \sum_{k=1}^{n} f(s_j, t_k)\Delta A \tag{9}$$

will be the same, regardless of how the points (s_j, t_k) are chosen in the rectangles R_{jk}.

Just as we did for functions of a single variable, we refer to the approximating sum in line (8) as an arbitrary **Riemann (double) sum,** and we refer to its limit in line (9) as an integral. Since the summation is taken over increments in both variables x and y, we refer to the resulting integral as a **double integral.**

DEFINITION 1

Let $f(x, y)$ be continuous on the rectangle $R = \{(x, y) | a \leq x \leq b,\ c \leq y \leq d\}$. The **double integral of $f(x, y)$ over the rectangle R** is the number

$$\iint_R f(x,y)\, dA = \lim_{n\to\infty} \sum_{j=1}^{n} \sum_{k=1}^{n} f(s_j, t_k)\Delta A \tag{10}$$

where $\Delta A = \Delta x \Delta y = \left(\frac{b-a}{n}\right)\left(\frac{d-c}{n}\right)$ and (s_j, t_k) is an arbitrary point in the rectangle $R_{jk} = \{(x, y) | x_{j-1} \le x \le x_j,\ y_{k-1} \le y \le y_k\}$.*

In the symbol

$$\iint_R f(x, y)\, dA,$$

we use two integral signs to indicate that it represents the result of a double limit process—the x-interval $[a, b]$ and the y-interval $[c, d]$ have both been partitioned into increasingly many increasingly small subintervals. The subscripted R denotes the rectangle over which the integral is evaluated. For now, the symbol dA (which may also be written $dx\, dy$) indicates that the Riemann sum has been obtained by partitioning R into rectangles of area $\Delta A = \Delta x \Delta y$. As before, the function $f(x, y)$ is referred to as the **integrand.**

According to the development that led to the approximating Riemann sum, we define

$$\iint_R f(x, y)\, dA = \text{volume of V} \tag{11}$$

where V is the solid bounded above by the graph of the continuous nonnegative function $f(x, y)$ and below by the rectangle R in the xy-plane. But how do we evaluate the integral in (11)? Ideally, there would be a result analogous to the Fundamental Theorem of Calculus that would enable us to do so.

Actually, the answer to this question is straightforward. Recall from Section 7.3, formula (4), that the volume of V is given by the definite integral

$$\text{Volume of } V = \int_a^b A(x)\, dx, \qquad a < b \tag{12}$$

when $A(x)$ is the area of the cross section of V taken perpendicular to the x-axis. Now if $x_0 \in [a, b]$ is fixed, the area of this cross section is just

$$A(x_0) = \int_c^d f(x_0, y)\, dy, \qquad c < d, \tag{13}$$

since the cross section is bounded above by the continuous function $g(y) = f(x_0, y)$ (see Figure 2.5).

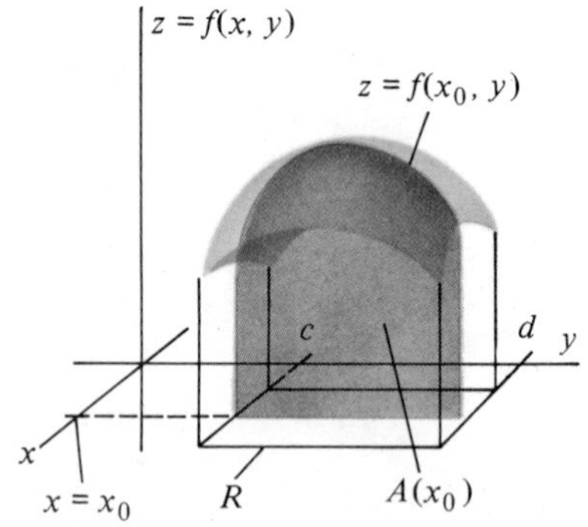

Figure 2.5 Area of cross section at x_0 is

$$A(x_0) = \int_c^d f(x_0, y)dy.$$

Combining equations (12) and (13) we conclude that

$$\text{Volume of } V = \int_a^b \left\{ \int_c^d f(x, y)\, dy \right\} dx, \qquad a < b, \qquad c < d. \tag{14}$$

The meaning of equation (14) is that the volume of V is calculated by first integrating $f(x, y)$ with respect to y (treating x as a constant) from c to d, and then integrat-

*A Riemann double sum is often defined more generally by dropping the requirement that the rectangles R_{jk} be of equal size. We will not be able to exploit this greater generality here, so we choose to simplify the discussion by working with a **regular grid,** that is, with rectangles R_{jk} of equal size.

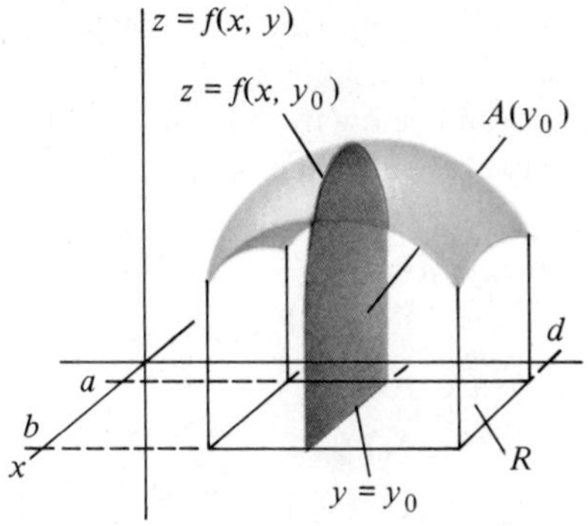

Figure 2.6 Area of cross section at y_0 is

$$A(y_0) = \int_a^b f(x, y_0)dx.$$

ing the resulting function of x from a to b. As Figure 2.6 illustrates, we could have begun by fixing y_0 and obtaining the area of the cross section perpendicular to the y-axis at $y = y_0$ as

$$A(y_0) = \int_a^b f(x, y_0)\, dx, \qquad a < b.$$

The resulting calculation for volume is

$$\text{Volume of } V = \int_c^d \left\{ \int_a^b f(x, y)\, dx \right\} dy, \qquad a < b,\ c < d. \tag{15}$$

The integrals in (14) and (15) are called **iterated integrals,** because they involve the composition of two integrations, each with respect to a single variable. We usually omit the braces and simply write

$$\int_a^b \int_c^d f(x, y)\, dy\, dx = \int_a^b \left\{ \int_c^d f(x, y)\, dy \right\} dx \tag{16}$$

and

$$\int_c^d \int_a^b f(x, y)\, dx\, dy = \int_c^d \left\{ \int_a^b f(x, y)\, dx \right\} dy. \tag{17}$$

It is important to note that iterated integrals are interpreted ''from inside out,'' meaning just what is specified by equations (16) and (17).

Example 2 Evaluate the iterated integral

$$\int_1^2 \int_1^3 (x^2 + 2xy)\, dy\, dx$$

and interpret the result geometrically.

Solution: The integral is evaluated as in equation (16):

$$\begin{aligned}
\int_1^2 \int_1^3 (x^2 + 2xy)\, dy\, dx &= \int_1^2 \left\{ \int_1^3 (x^2 + 2xy)\, dy \right\} dx \\
&= \int_1^2 \left\{ x^2y + xy^2 \Big]_{y=1}^{y=3} \right\} dx \\
&= \int_1^2 [(3x^2 + 9x) - (x^2 + x)]\, dx \\
&= \int_1^2 (2x^2 + 8x)\, dx \\
&= \frac{2}{3}x^3 + 4x^2 \Big]_1^2 \\
&= \left[\frac{2}{3}(8) + 4(4)\right] - \left[\frac{2}{3}(1) + 4(1)\right] \\
&= \frac{50}{3}.
\end{aligned}$$

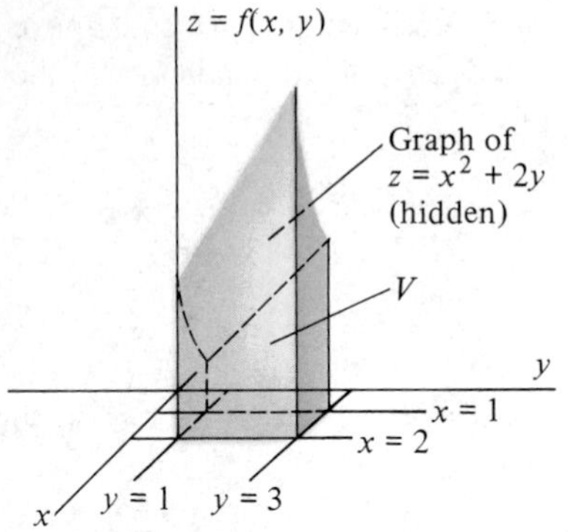

Figure 2.7 Volume of V is $\int_1^2 \int_1^3 (x^2 + 2y)dy\,dx$.

Since $f(x, y) = x^2 + 2xy$ is nonnegative for (x, y) in the rectangle $R = \{(x, y)|1 \le x \le 2,\ 1 \le y \le 3\}$, the number $\frac{50}{3}$ is the volume of the solid bounded above by the graph of $f(x, y) = x^2 + 2xy$ and below by the xy-plane over the rectangle R (Figure 2.7). This volume may also be calculated using (17) as

$$\begin{aligned}\int_1^3 \int_1^2 (x^2 + 2xy)\,dx\,dy &= \int_1^3 \left\{\int_1^2 (x^2 + 2xy)\,dx\right\} dy \\ &= \int_1^3 \left\{\frac{x^3}{3} + x^2y\Big]_{x=1}^{x=2}\right\} dy \\ &= \int_1^3 \left[\left(\frac{8}{3} + 4y\right) - \left(\frac{1}{3} + y\right)\right] dy \\ &= \int_1^3 \left(\frac{7}{3} + 3y\right) dy \\ &= \frac{7y}{3} + \frac{3y^2}{2}\Big]_1^3 \\ &= \frac{50}{3}.\end{aligned}$$

■

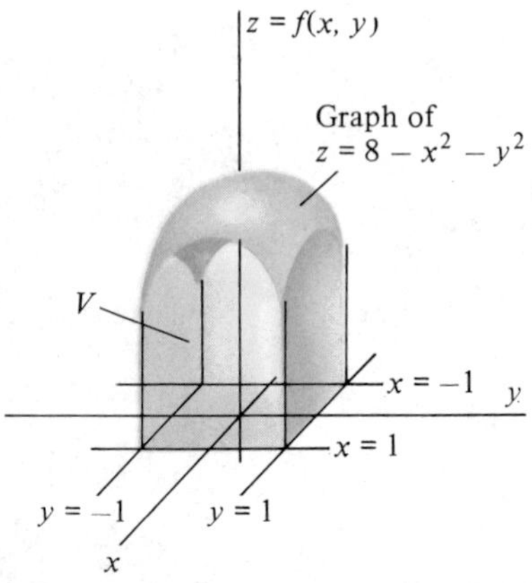

Figure 2.8 Volume of V is $\int_{-1}^1 \int_{-1}^1 (8 - x^2 - y^2)dx\,dy$.

Example 3 Calculate the volume of the solid over the square $R = \{(x, y)|-1 \le x \le 1,\ -1 \le y \le 1\}$ bounded above by the graph of $f(x, y) = 8 - x^2 - y^2$ and below by the xy-plane.

Solution: The desired volume may be calculated as

$$\begin{aligned}\iint_R (8 - x^2 - y^2)\,dA &= \int_{-1}^1 \left\{\int_{-1}^1 (8 - x^2 - y^2)\,dx\right\} dy \\ &= \int_{-1}^1 \left\{8x - \frac{1}{3}x^3 - xy^2\Big]_{x=-1}^{x=1}\right\} dy \\ &= \int_{-1}^1 \left(\frac{46}{3} - 2y^2\right) dy \\ &= \frac{46y}{3} - \frac{2y^3}{3}\Big]_{y=-1}^{y=1} \\ &= \frac{88}{3}.\end{aligned}$$

(See Figure 2.8.) ■

Finally, we note that the definition of the double integral in Definition 1 did not depend on the function $f(x, y)$ being nonnegative. The double integral $\iint_R f(x, y)\,dA$ is therefore defined for any continuous function $f(x, y)$, regardless of its sign of the rectangle R.

When $f(x, y) \geq 0$ for all (x, y) in R, we may combine statement (11) with statements (14) through (17) to obtain the equations

$$\iint_R f(x, y)\, dA = \int_a^b \int_c^d f(x, y)\, dy\, dx = \int_a^b \left\{ \int_c^d f(x, y)\, dy \right\} dx \tag{18}$$

and

$$\iint_R f(x, y)\, dA = \int_c^d \int_a^b f(x, y)\, dA = \int_c^d \left\{ \int_a^b f(x, y)\, dx \right\} dy. \tag{19}$$

Equations (18) and (19) are true regardless of the sign of $f(x, y)$, although we shall omit the details involved in proving the more general case (see Exercise 39 for a sketch of a proof of this statement). However, you must keep in mind that the integrals in (18) and (19) correspond to the volume of the region V between the rectangle R and the graph of $z = f(x, y)$ *only when* $f(x, y) \geq 0$ for all (x, y) in R, $a < b$, and $c < d$. We shall have more to say about properties of the double integral in the next section.

Exercise Set 20.2

In Exercises 1–14, evaluate the iterated integral.

1. $\int_0^1 \int_0^2 xy\, dx\, dy$

2. $\int_{-1}^1 \int_1^3 (y - x)\, dy\, dx$

3. $\int_1^3 \int_1^2 (4 + x - y)\, dx\, dy$

4. $\int_0^2 \int_{-1}^1 (x + y)^2\, dx\, dy$

5. $\int_0^1 \int_0^{\pi/2} x \sin y\, dy\, dx$

6. $\int_0^1 \int_0^1 ye^{x-y^2}\, dy\, dx$

7. $\int_0^2 \int_1^e y^2 \ln x\, dx\, dy$

8. $\int_0^9 \int_1^4 \sqrt{\frac{y}{x}}\, dx\, dy$

9. $\int_0^1 \int_0^1 x \cosh y\, dx\, dy$

10. $\int_0^2 \int_0^2 xye^{xy^2}\, dy\, dx$

11. $\int_0^1 \int_0^{\pi/2} xy \sin x\, dx\, dy$

12. $\int_0^{\pi} \int_0^1 \sinh x \cosh(\pi - y)\, dx\, dy$

13. $\int_0^1 \int_0^4 \frac{\sqrt{y}}{1 + x^2}\, dy\, dx$

14. $\int_0^{\pi/4} \int_0^{\pi/4} \tan x \sec^2 y\, dy\, dx$

In Exercises 15–21, evaluate the double integral over the rectangle R.

15. $\iint_R (x + y^2)\, dA,\quad R = \{(x, y) | 0 \leq x \leq 1, 0 \leq y \leq 1\}$

16. $\iint_R (x^2 + y^2)\, dA,\quad R = \{(x, y) | 0 \leq x \leq a, 0 \leq y \leq b\}$

17. $\iint_R x \cos y\, dA,\quad R = \{(x, y) | 0 \leq x \leq 4, 0 \leq y \leq \pi/2\}$

18. $\iint_R \frac{xy}{\sqrt{x^2 + y^2}}\, dA,\quad R = \{(x, y) | 1 \leq x \leq 2, 1 \leq y \leq 2\}$

19. $\iint_R xy \sec^2(xy^2)\, dA,\quad R = \{(x, y) | 0 \leq x \leq \pi/4, 0 \leq y \leq 1\}$

20. $\iint_R \frac{1}{\sqrt{x + y}}\, dA,\quad R = \{(x, y) | 4 \leq x \leq 8, 0 \leq y \leq 4\}$

21. $\iint_R y \cos(x + y)\, dA,\quad R = \{(x, y) | 0 \leq x \leq \pi/4, 0 \leq y \leq \pi/4\}$

In Exercises 22–27, use a double integral to calculate the volume of the solid bounded above by the graph of $z = f(x, y)$ and below by the rectangle R in the xy-plane.

22. $f(x, y) = 16 - 4x - 2y, \quad R = \{(x, y)|0 \le x \le 2, 0 \le y \le 1\}$

23. $f(x, y) = x, \quad R = \{(x, y)|0 \le x \le 2, 0 \le y \le 3\}$

24. $f(x, y) = 9 - x^2 - y^2, \quad R = \{(x, y)| -1 \le x \le 1, -1 \le y \le 1\}$

25. $f(x, y) = x \sin y, \quad R = \{(x, y)|0 \le x \le 1, 0 \le y \le \pi/2\}$

26. $f(x, y) = \dfrac{\sqrt{x}}{1 + y^2}, \quad R = \{(x, y)|0 \le x \le 4, 0 \le y \le 1\}$

27. $f(x, y) = \dfrac{e^{\sqrt{x}}}{\sqrt{xy}}, \quad R = \{(x, y)|1 \le x \le 4, 1 \le y \le 9\}$

28. Show that if $f(x, y)$ is continuous on the rectangle R, then

$$\iint_R cf(x, y)\, dA = c \iint_R f(x, y)\, dA$$

for any constant c.

29. Show that if $f(x, y)$ and $g(x, y)$ are continuous on the rectangle R then

$$\iint_R [f(x, y) + g(x, y)]\, dA = \iint_R f(x, y)\, dA + \iint_R g(x, y)\, dA.$$

30. Show that if $f(x, y) \le 0$ for all $(x, y) \in R = \{(x, y)|a \le x \le b, c \le y \le d\}$ and if V is the solid bounded by R and the graph of $z = f(x, y)$, then

$$\text{Volume of } V = -\iint_R f(x, y)\, dA.$$

31. Show that if $f(x, y)$ is continuous for $(x, y) \in R = \{(x, y)|a \le x \le b, c \le y \le d\}$ then

$$\int_a^b \int_c^d f(x, y)\, dy\, dx = -\int_b^a \int_c^d f(x, y)\, dy\, dx$$

$$= -\int_a^b \int_d^c f(x, y)\, dy\, dx$$

$$= \int_b^a \int_d^c f(x, y)\, dy\, dx.$$

32. Verify the formula $\overline{S}_n = 8 - 3\left[\dfrac{n(n-1)}{n^2}\right]$ for the upper approximating sum in Example 1.

33. Use an iterated integral to show that the limit of the approximating sums for $f(x, y) = 8 - 2x^2 - 2y^3$ in Table 2.2 is actually $\lim_{n\to\infty} \underline{S}_n = \dfrac{41}{6}$.

34. Use a lower approximating sum with $n = 4$ to approximate the volume of the solid bounded above by the graph of $f(x, y) = 2 + x + 2y$ and below by the rectangle

$$R = \{(x, y)|0 \le x \le 2, 0 \le y \le 2\}.$$

35. *(Computer)* Use Program 12 to approximate the volume of the solid bounded above by the graph of $z = \sin \sqrt{xy}$ and below by the rectangle

$$R = \{(x, y)|0 \le x \le 1, 0 \le y \le 1\}.$$

36. *(Computer)* Use Program 12 to approximate the double integral

$$\iint_R e^{x^2+y^2}\, dA \text{ where } R \text{ is the rectangle}$$

$$R = \{(x, y)|0 \le x \le 1, 0 \le y \le 2\}.$$

37. *(Computer)* Use Program 12 to approximate the iterated integral

$$\int_0^1 \int_1^2 \sin(x^2 + y^2)\, dx\, dy.$$

38. *(Computer)* Use Program 12 to approximate the iterated integral

$$\int_1^3 \int_2^4 \sqrt{x^3 + y^3}\, dx\, dy.$$

39. The purpose of this exercise is to enable you to sketch a proof of the fact that

$$\iint_R f(x, y)\, dA = \int_a^b \int_c^d f(x, y)\, dy\, dx$$

where $f(x, y)$ is continuous and $R = \{(x, y)|a \le x \le b, c \le y \le d\}$.

a. Begin with the partitions $a = x_0 < x_1 < \cdots < x_n = b$ and $c = y_0 < y_1 < \cdots < y_n = d$ and the arbitrary Riemann sum

$$S_n = \sum_{j=1}^n \sum_{k=1}^n f(s_j, t_k)\Delta y \Delta x$$

$$= \sum_{j=1}^n \left\{ \sum_{k=1}^n f(s_j, t_k)\Delta y \right\} \Delta x.$$

b. With j fixed, pick t_k in each interval $[y_{k-1}, y_k]$ to be the number for which

$$f(s_j, t_k)\Delta y = \int_{y_{k-1}}^{y_k} f(s_j, t)\, dt.$$

Why can this be done?

c. Conclude that

$$\sum_{k=1}^{n} f(s_j, t_k)\Delta y = \sum_{k=1}^{n} \int_{y_{k-1}}^{y_k} f(s_j, y)\, dy = \int_c^d f(s_j, y)\, dy.$$

Why is this true?

d. From (a) and (c) conclude that

$$S_n = \sum_{j=1}^{n} \left\{ \int_c^d f(s_j, y)\, dy \right\} \Delta x.$$

e. Note that S_n in (d) has the form $S_n = \sum_{j=1}^{n} G(s_j)\Delta x$ where

$G(s_j) = \int_c^d f(s_j, t)\, dt$. Conclude that

$$\lim_{n\to\infty} S_n = \lim_{n\to\infty} \sum_{j=1}^{n} G(s_j)\Delta x = \int_a^b G(x)\, dx$$
$$= \int_a^b \left\{ \int_c^d f(x, y)\, dy \right\} dx.$$

40. Show that if $f(x)$ is continuous on the interval $[a, b]$, and if $g(y)$ is continuous on the interval $[c, d]$, then

$$\iint_R f(x)g(y)\, dA = \left[\int_a^b f(x)\, dx \right] \cdot \left[\int_c^d g(y)\, dy \right]$$

where $R = \{(x, y) | a \le x \le b,\ c \le y \le d\}$.

20.3 DOUBLE INTEGRALS OVER MORE GENERAL REGIONS

Double integrals may be defined over regions in the plane more general than just rectangles. In particular, let Q be a region in the plane that is bounded (meaning it is contained in some finite rectangle) and whose boundary is a piecewise smooth curve. A rectangle is certainly an example of such a region. But so is a pentagon, a triangle, an ellipse, or even a more general region such as that in Figure 3.1.

Now suppose $f(x, y)$ is continuous on Q and that Q is entirely contained within a rectangle $R = \{(x, y) | a \le x \le b,\ c \le y \le d\}$. For each integer $n > 0$, we may construct a grid over Q as follows. Let $\Delta x = \dfrac{b - a}{n}$, and draw lines parallel to the y-axis with x-intercepts

$$x_0 = a, \quad x_1 = a + \Delta x, \quad x_2 = a + 2\Delta x, \ldots, x_n = a + n\Delta x = b.$$

Similarly, let $\Delta y = \dfrac{d - c}{n}$ and draw lines parallel to the x-axis with y-intercepts

$$y_0 = c, \quad y_1 = c + \Delta y, \quad y_2 = c + 2\Delta y, \ldots, y_n = c + n\Delta y = d.$$

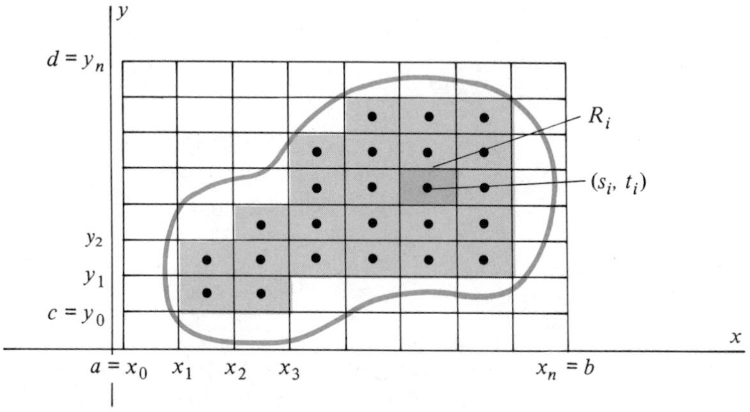

Figure 3.1 Partitioning a more general region Q.

This grid divides the rectangle R into n^2 rectangles, each of area $\Delta A = \Delta x \Delta y = \dfrac{(b - a)(d - c)}{n^2}$ (see Figure 3.1). (Again, we are using a *regular* grid for simplicity, although this is not necessary.)

Now some of these smaller rectangles will lie entirely within Q and some will not. Let $R_1, R_2, \ldots, R_m$ be a list of all such rectangles lying entirely within Q. For each such rectangle R_i, let (s_i, t_i) be a point in R_i. Finally, form the sum

$$S_n = \sum_{i=1}^{m} f(s_i, t_i)\Delta A. \tag{1}$$

Under the conditions stated on Q and on $f(x, y)$, as $n \to \infty$, the approximating sum in (1) will converge to the number we define to be the double integral of $f(x, y)$ over the region Q:

$$\iint_Q f(x, y)\, dA = \lim_{n\to\infty} \sum_{i=1}^{m} f(s_i, t_i)\Delta A. \tag{2}$$

Statement (2) is actually a definition of the double integral in this more general setting, although we prefer not to state it formally as such since this would require repeating nearly all the preceding statements. Before looking at ways in which such integrals may be evaluated, we wish to emphasize three points concerning this definition.

(i) Although only one summation sign appears in (2), this definition agrees with Definition 1 when Q is a rectangle. The difference is only that we have used a two-dimensional counting procedure (rows by columns) in stating Definition 1. In statement (2) we have simply listed all included rectangles $R_1, R_2, \ldots, R_m$ using a single index.

(ii) As $n \to \infty$, the approximation S_n in (1) becomes increasingly accurate for two reasons. First, since the dimensions of the individual rectangles R_i are decreasing, the union of these smaller rectangles more nearly "fills up" the region Q. Second, again since the dimensions of the R_i are decreasing, the number $f(s_i, t_i)$ becomes an increasingly accurate approximation to $f(x, y)$ on the rectangle R_i.

(iii) When $f(x, y) \geq 0$ for all $(x, y) \in Q$, the double integral (2) gives the volume of the solid V bounded by the graph of $z = f(x, y)$ and the region Q. The reason for this is the same as for the simpler case of the double integral over a rectangle: We may interpret S_n in (1) as the sum of a set of rectangular prisms, of base area ΔA and height $f(s_i, t_i)$, whose union "fills" the solid V in the limit as $n \to \infty$.

Regular Regions

There are two special types of regions for which the double integral in (2) can be evaluated as an iterated integral: x-simple and y-simple regions.

DEFINITION 2

A region Q in the xy-plane is called **y-simple** if there exist continuous functions $g_1(x)$ and $g_2(x)$ so that

$$Q = \{(x, y) | a \le x \le b, \ g_1(x) \le y \le g_2(x)\}.$$

The region Q is called **x-simple** if there exist continuous functions $h_1(y)$ and $h_2(y)$ so that

$$Q = \{(x, y) | c \le y \le d, \ h_1(y) \le x \le h_2(y)\}.$$

The region Q is called **regular** if it is both x-simple and y-simple.

Figure 3.2 gives two illustrations of y-simple regions. The condition that $g_1(x) \le y \le g_2(x)$ for all $x \in [a, b]$ simply means that the vertical line segment connecting $(x, g_1(x))$ and $(x, g_2(x))$ lies entirely within the region Q. Another way to say this is that lines parallel to the y-axis intersect the boundary of Q at most twice.

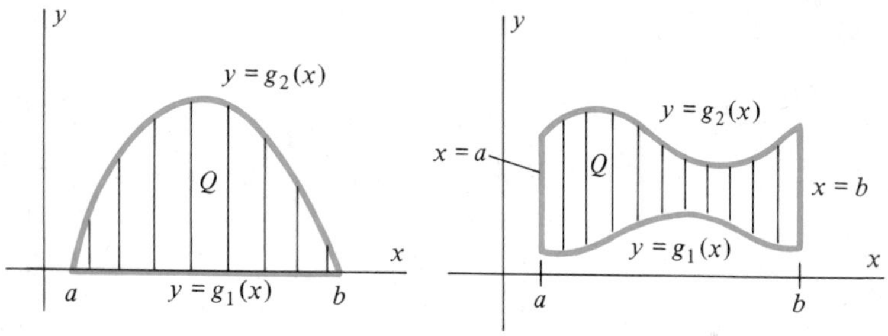

Figure 3.2 Two y-simple regions: Vertical lines intersect the boundary of Q at most twice.

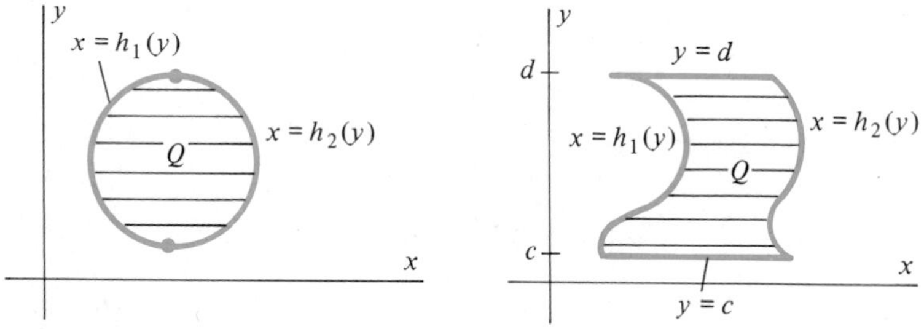

Figure 3.3 Two x-simple regions: Horizontal lines intersect the boundary of Q at most twice.

Figure 3.3 shows two x-simple regions. This condition means that lines parallel to the x-axis intersect the boundary of Q at most twice. Note also that Figures 3.2(a) and 3.3(a) are *regular* (both x-simple and y-simple). However, Figure 3.2(b) is not x-simple, and Figure 3.3(b) is not y-simple.

The following theorem shows how double integrals over x-simple or y-simple regions may be evaluated as iterated integrals.

THEOREM 1

Let $f(x, y)$ be continuous on the region Q.

(i) If $Q = \{(x, y) | a \le x \le b, g_1(x) \le y \le g_2(x)\}$ is y-simple, then

$$\iint_Q f(x, y)\, dA = \int_a^b \int_{g_1(x)}^{g_2(x)} f(x, y)\, dy\, dx = \int_a^b \left\{ \int_{g_1(x)}^{g_2(x)} f(x, y)\, dy \right\} dx. \tag{3}$$

(ii) If $Q = \{(x, y) | c \le y \le d, h_1(y) \le x \le h_2(y)\}$ is x-simple, then

$$\iint_Q f(x, y)\, dA = \int_c^d \int_{h_1(y)}^{h_2(y)} f(x, y)\, dx\, dy = \int_c^d \left\{ \int_{h_1(y)}^{h_2(y)} f(x, y)\, dx \right\} dy. \tag{4}$$

Equation (3) says that if Q is y-simple, we first integrate $f(x, y)$ as a function of y alone, between the limits $g_1(x)$ and $g_2(x)$. The resulting function of x alone is then integrated over the constant limits $a \le x \le b$. Equation (4) says that if Q is x-simple, we first integrate $f(x, y)$ as a function of x alone, between the limits $h_1(y)$ and $h_2(y)$.

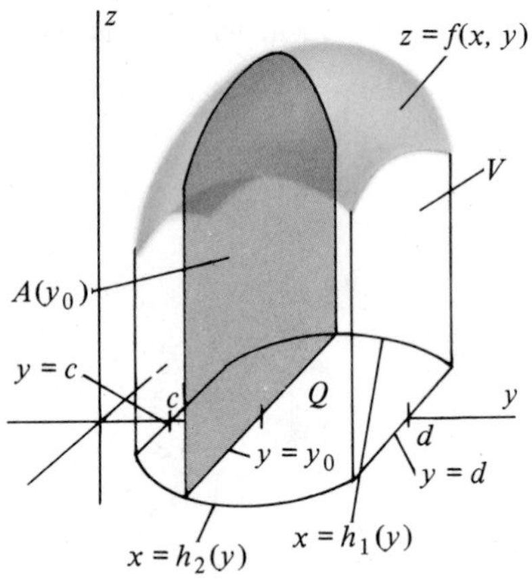

Figure 3.4 If Q is x-simple, area of cross section perpendicular to the y-axis at y_0 is

$$A(y_0) = \int_{h_1(y_0)}^{h_2(y_0)} f(x, y_0)\, dx.$$

We shall not prove Theorem 1. However, Figure 3.4 embodies an argument for the validity of equation (4) when $f(x, y)$ is nonnegative on Q and Q is x-simple: For y-fixed, the integral $\int_{h_1(y)}^{h_2(y)} f(x, y)\, dx = A(y)$ gives the area of the cross section of the solid bounded by the graph of $z = f(x, y)$ and the region Q. The familiar formula for the volume of a solid with known cross section then gives

$$\text{Volume of } V = \int_c^d A(y)\, dy = \int_c^d \int_{h_1(y)}^{h_2(y)} f(x, y)\, dx\, dy. \tag{5}$$

Since the left side of equation (5) corresponds to the double integral $\iint_Q f(x, y)\, dA$, we obtain equation (4).

In using Theorem 1 to evaluate double integrals, it is very important to first sketch the region Q to determine whether it is x-simple or y-simple. This determination will often dictate the order of integration in the iterated integral. (Of course, if Q is a regular region, you may proceed in either order. Just be careful that the limits of integration correspond to the chosen order of integration.)

Example 1 Evaluate the double integral $\iint_Q (2xy + y^2)\, dA$ where Q is the triangle with vertices (0, 0), (1, 0), and (1, 2).

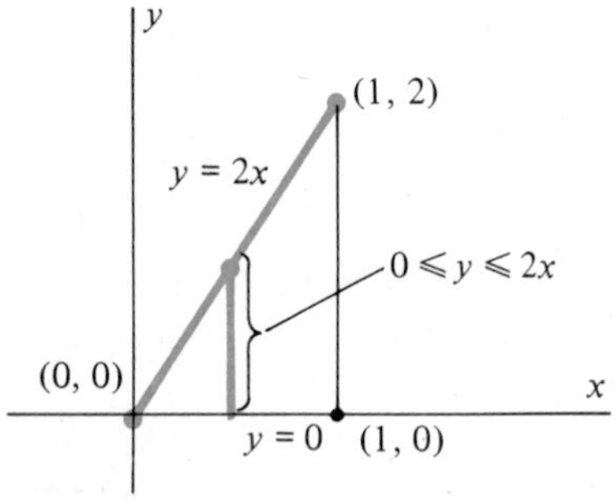

Figure 3.5 Treating the triangle Q as y-simple.

Solution: The triangular region Q is sketched in Figure 3.5. Since Q is y-simple, we find the equation of the line segment joining (0, 0) and (1, 2). It is simply $y = 2x$. Thus, a vertical line segment through Q extends from $y = g_1(x) = 0$ to $y = g_2(x) = 2x$. That is, the limits of integration are determined by the inequalities

$$0 \le x \le 1 \quad \text{and} \quad 0 \le y \le 2x. \tag{6}$$

According to equation (3), the double integral is evaluated as

$$\iint_Q (2xy + y^2)\, dA = \int_0^1 \int_0^{2x} (2xy + y^2)\, dy\, dx$$

$$= \int_0^1 \left\{ xy^2 + \frac{1}{3}y^3 \Big]_{y=0}^{y=2x} \right\} dx$$

$$= \int_0^1 \frac{20x^3}{3}\, dx$$

$$= \frac{5}{3}x^4 \Big]_0^1$$

$$= \frac{5}{3}.$$

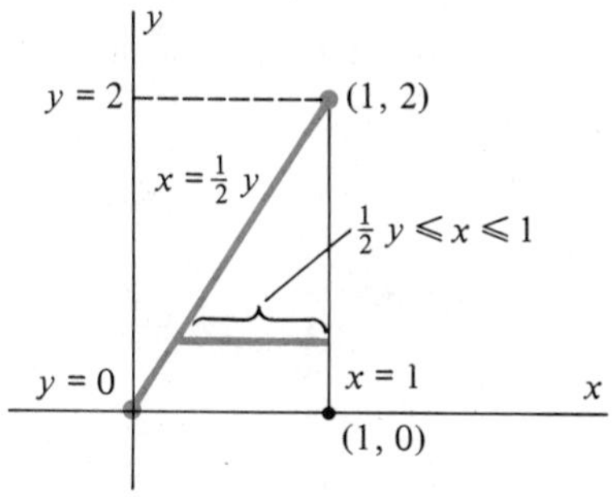

Figure 3.6 Treating the triangle Q as x-simple.

Now the region Q is also x-simple. However, to use equation (4) we must first write the equation for the line segment joining (0, 0) and (1, 2) as a function of y. Solving $y = 2x$ for x gives $x = h_1(y) = y/2$ as the left boundary. As Figure 3.6 illustrates, $h_2(y) = 1$ is the right boundary of Q. The limits of integration are therefore

$$0 \leq y \leq 2 \qquad \text{and} \qquad \frac{y}{2} \leq x \leq 1. \tag{7}$$

Equation (4) then gives

$$\iint_Q (2xy + y^2)\, dA = \int_0^2 \int_{y/2}^1 (2xy + y^2)\, dx\, dy$$

$$= \int_0^2 \left\{ x^2y + xy^2 \Big]_{x=y/2}^{x=1} \right\} dy$$

$$= \int_0^2 \left(-\frac{3}{4}y^3 + y^2 + y \right) dy$$

$$= -\frac{3}{16}y^4 + \frac{1}{3}y^3 + \frac{1}{2}y^2 \Big]_0^2$$

$$= \frac{5}{3}.$$

■

REMARK: Notice that the order of integration is entirely determined by our choice of whether to describe Q as y-simple, using inequalities (6), or as x-simple, using inequalities (7). Whichever integration is performed last must involve only constant limits of integration, since your answer cannot contain variables.

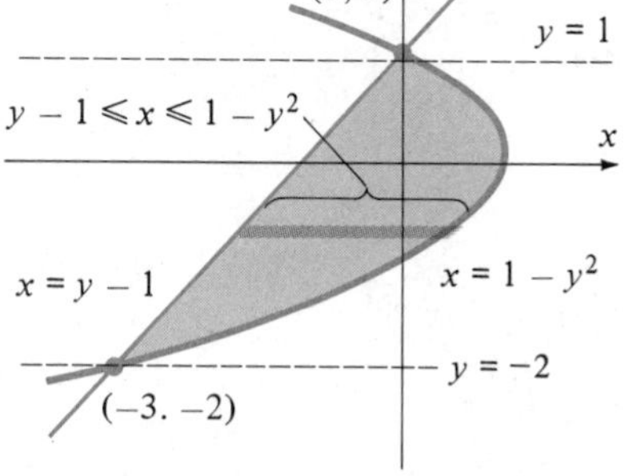

Figure 3.7 Region bounded by graphs of $y = x + 1$ and $x = 1 - y^2$ is x-simple.

Example 2 Evaluate the double integral

$$\iint_Q 4xy\, dA$$

where Q is the region bounded by the graphs of the equations $y = x + 1$ and $x = 1 - y^2$.

Solution: The region Q is sketched in Figure 3.7. This region, like that in Example 1, is both x-simple and y-simple. However, we shall choose to work with Q as an x-simple region, since every horizontal line through Q originates on the line $y = x + 1$ and terminates on the parabola $x = 1 - y^2$. (Viewing Q as y-simple would be much messier—some vertical lines terminate on the line and others terminate on the parabola.)

Solving the equation $y = x + 1$ for x gives $x = h_1(y) = y - 1$ as the left boundary of Q. The right boundary is $h_2(y) = 1 - y^2$. The limits of integration are therefore

$$-2 \le y \le 1 \quad \text{and} \quad y - 1 \le x \le 1 - y^2.$$

The double integral is evaluated as

$$\begin{aligned}
\iint_Q 4xy\, dA &= \int_{-2}^{1} \int_{y-1}^{1-y^2} 4xy\, dx\, dy \\
&= \int_{-2}^{1} \left\{ 2x^2y \Big]_{x=y-1}^{x=1-y^2} \right\} dy \\
&= \int_{-2}^{1} (2y^5 - 6y^3 + 4y^2)\, dy \\
&= \frac{1}{3}y^6 - \frac{3}{2}y^4 + \frac{4}{3}y^3 \Big]_{-2}^{1} \\
&= \frac{27}{2}.
\end{aligned}$$

■

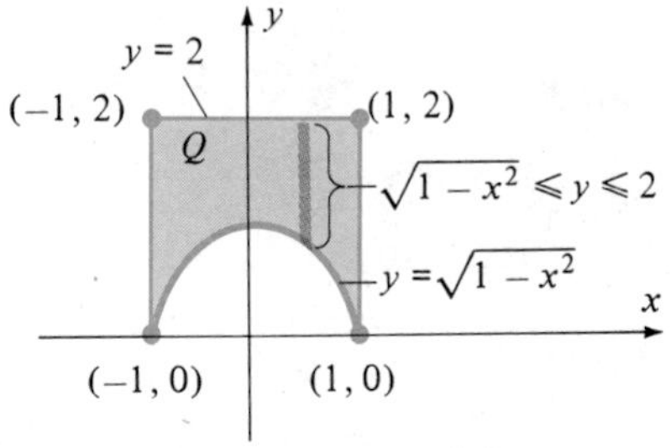

Figure 3.8 Region bounded by $y = \sqrt{1 - x^2}$ and $y = 2$ is y-simple, but not x-simple.

Example 3 Find the volume of the solid V bounded above by the graph of $f(x, y) = 2x^2y$ and below by the region Q bounded by the semicircle $y = \sqrt{1 - x^2}$ and the lines $y = 2$, $x = -1$, and $x = 1$.

Solution: The region Q is sketched in Figure 3.8. Since $f(x, y) \ge 0$ for all $(x, y) \in Q$, the volume of V is given by the double integral $\iint_Q 2x^2y\, dA$. As to the order of integration, we have no choice: Q is y-simple, but not x-simple. We must, therefore, integrate first with respect to y. The limits of integration are

$$-1 \le x \le 1 \quad \text{and} \quad \sqrt{1 - x^2} \le y \le 2.$$

The volume is

$$\begin{aligned}
\text{Volume of } V = \iint_Q 2x^2y\, dA &= \int_{-1}^{1} \int_{\sqrt{1-x^2}}^{2} 2x^2y\, dy\, dx \\
&= \int_{-1}^{1} \left\{ x^2y^2 \Big]_{y=\sqrt{1-x^2}}^{y=2} \right\} dx \\
&= \int_{-1}^{1} [4x^2 - x^2(1 - x^2)]\, dx \\
&= \int_{-1}^{1} (3x^2 + x^4)\, dx \\
&= x^3 + \frac{1}{5}x^5 \Big]_{-1}^{1} \\
&= \frac{12}{5}.
\end{aligned}$$

■

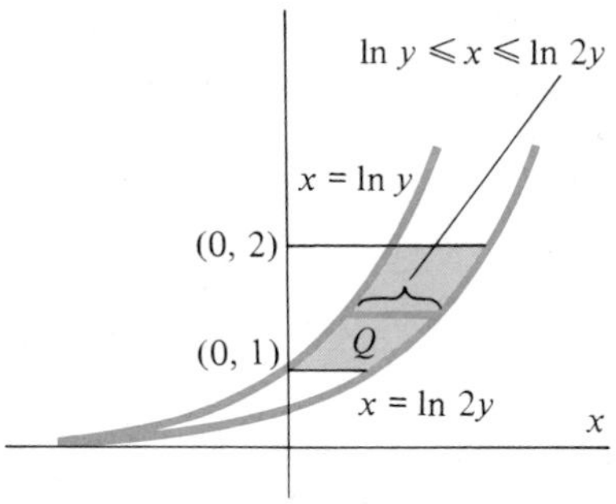

Figure 3.9 Region Q in Example 4.

Example 4 Find the volume of the solid V bounded by the graph of $f(x, y) = e^{x+y^2}$ and the region

$$Q = \{(x, y) | \ln y \le x \le \ln 2y, \quad 1 \le y \le 2\}.$$

Solution: The region Q is sketched in Figure 3.9. Although Q is both x-simple and y-simple, this time the order of integration is determined by the integrand $f(x, y) = e^{x+y^2}$. We must attempt to integrate first with respect to x, since there is no hope of finding an antiderivative with respect to y. Treating Q as an x-simple region, we use the limits

$$\ln y \le x \le \ln 2y \quad \text{and} \quad 1 \le y \le 2$$

and obtain

$$\begin{aligned}
\text{Volume of V} = \iint_Q e^{x+y^2}\, dA &= \int_1^2 \int_{\ln y}^{\ln 2y} e^{x+y^2}\, dx\, dy \\
&= \int_1^2 \left\{ \left[e^{x+y^2} \right]_{x=\ln y}^{x=\ln 2y} \right\} dy \\
&= \int_1^2 [(e^{\ln 2y})e^{y^2} - (e^{\ln y})e^{y^2}]\, dy \\
&= \int_1^2 (2ye^{y^2} - ye^{y^2})\, dy \\
&= \int_1^2 ye^{y^2}\, dy \\
&= \frac{1}{2} e^{y^2} \Big]_1^2 \\
&= \frac{e}{2}(e^3 - 1) \\
&\approx 25.94.
\end{aligned}$$

■

Finding Areas by Double Integration

The fact that volumes of solids may be calculated by use of double integrals means that areas of regions in the plane may also be calculated in this way. As Figure 3.10 illustrates, the area of Q is the same as the volume of the cylinder with base Q and uniform height $h = 1$. To calculate the area of Q, we simply evaluate the integral

$$\iint_Q 1\, dA = \text{Area of } Q. \tag{8}$$

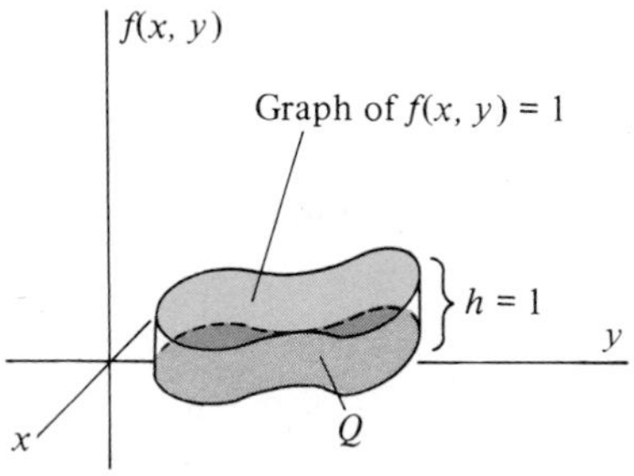

Figure 3.10 Area of $Q = \iint_Q 1\, dA$.

Example 5 Use a double integral to find the area of the region Q lying inside the circle $x^2 + y^2 = 4$ and above the line $y = 1$.

Solution: The points on the circle $x^2 + y^2 = 4$ with y-coordinate equal to 1 have x-coordinates $x = \pm\sqrt{4 - 1} = \pm\sqrt{3}$. The region may therefore be described by the inequalities

$$-\sqrt{3} \le x \le \sqrt{3}, \qquad 1 \le y \le \sqrt{4 - x^2}.$$

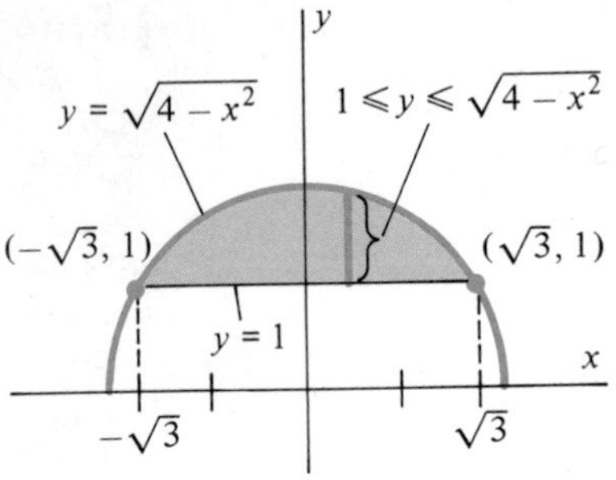

Figure 3.11 Region inside circle $x^2 + y^2 = 4$ and above $y = 1$.

(See Figure 3.11.) According to (8),

$$\begin{aligned}\text{Area of } Q &= \iint_Q 1\, dA \\ &= \int_{-\sqrt{3}}^{\sqrt{3}} \int_1^{\sqrt{4-x^2}} 1 \cdot dy\, dx \\ &= \int_{-\sqrt{3}}^{\sqrt{3}} \left\{ y \Big]_{y=1}^{y=\sqrt{4-x^2}} \right\} dx \\ &= \int_{-\sqrt{3}}^{\sqrt{3}} (\sqrt{4 - x^2} - 1)\, dx \\ &= 2\int_0^{\sqrt{3}} \sqrt{4 - x^2}\, dx - 2\sqrt{3}.\end{aligned}$$

A straightforward application of the method of trigonometric substitutions shows that

$$\int \sqrt{4 - x^2}\, dx = 2 \sin^{-1}\left(\frac{x}{2}\right) + \frac{x\sqrt{4 - x^2}}{2} + C.$$

The desired area is therefore

$$2\left[2 \sin^{-1}\left(\frac{\sqrt{3}}{2}\right) + \frac{\sqrt{3}}{2}\right] - 2\sqrt{3} \approx 2.457. \qquad \blacksquare$$

Interchanging the Order of Integration

There are occasions on which we need to interchange the order of integration in a given double integral. This usually occurs when we apply a particular formula involving a double integral (such as those we will encounter in Section 20.5, or one of the many such formulas that arise in mathematical statistics) with the result that an antiderivative cannot be found with respect to the ''inside'' variable. For example, in the double integral

$$\int_0^1 \int_{y^2}^1 ye^{x^2}\, dx\, dy \tag{9}$$

we are in deep trouble, since we cannot find a (formal) antiderivative for the integrand ye^{x^2} with respect to x.

However, such integrals can often be evaluated by reversing the order of integration. In particular, the antiderivative of ye^{x^2} with respect to y is easily seen to be the function $\frac{1}{2}y^2e^{x^2}$. But if we are going to reverse the order in which the antidifferentiations are performed, we must also determine correct limits of integration corresponding to this new order of integration. We do so as follows:

To reverse the order of integration in the iterated integral

$$\int_c^d \int_{h_1(y)}^{h_2(y)} f(x, y)\, dx\, dy:$$

(i) Identify the region Q for which the iterated integral can be written as the double integral

$$\int_c^d \int_{h_1(y)}^{h_2(y)} f(x, y)\, dx\, dy = \iint_Q f(x, y)\, dA.$$

(ii) Find constants a and b, and continuous functions $g_1(x)$ and $g_2(x)$, so that the region Q can be expressed as

$$Q = \{(x, y) | a \le x \le b,\ g_1(x) \le y \le g_2(x)\}.$$

(iii) Rewrite the iterated integral as

$$\int_c^d \int_{h_1(y)}^{h_2(y)} f(x, y)\, dx\, dy = \iint_Q f(x, y)\, dA$$

$$= \int_a^b \int_{g_1(x)}^{g_2(x)} f(x, y)\, dy\, dx.$$

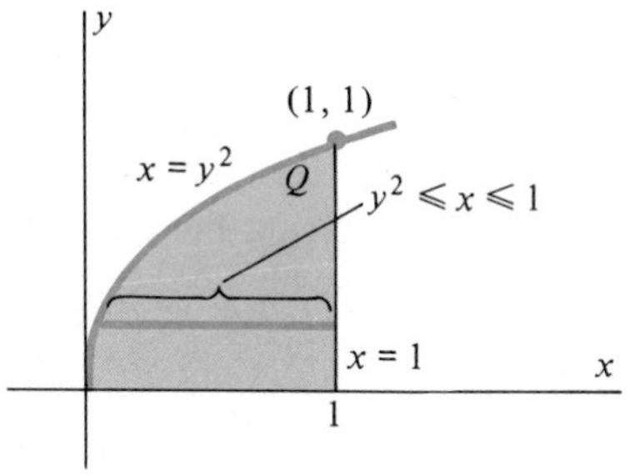

Figure 3.12 $Q = \{(x,y) | y^2 \le x \le 1, 0 \le y \le 1\}$.

Obviously, this procedure can be applied only when Q is a regular region. Also, though the procedure is stated for reversing the order of integration from $dx\, dy$ to $dy\, dx$, the procedure for changing from order $dy\, dx$ to order $dx\, dy$ is analogous. Finally, there is no guarantee that the resulting integral can be more easily evaluated than the original one.

Example 6 Use the procedure for reversing order of integration to evaluate the iterated integral

$$\int_0^1 \int_{y^2}^1 ye^{x^2}\, dx\, dy.$$

Solution: From the given limits of integration the region Q is described by the inequalities

$$y^2 \le x \le 1, \qquad 0 \le y \le 1.$$

That is, Q is the region bounded between the graphs of $x = y^2$ and $x = 1$ for $0 \le y \le 1$. From Figure 3.12 (which must always be sketched when applying this method), we can see that Q is regular and can also be described by the inequalities

$$0 \le x \le 1, \qquad 0 \le y \le \sqrt{x} \qquad \text{(Figure 3.13)}.$$

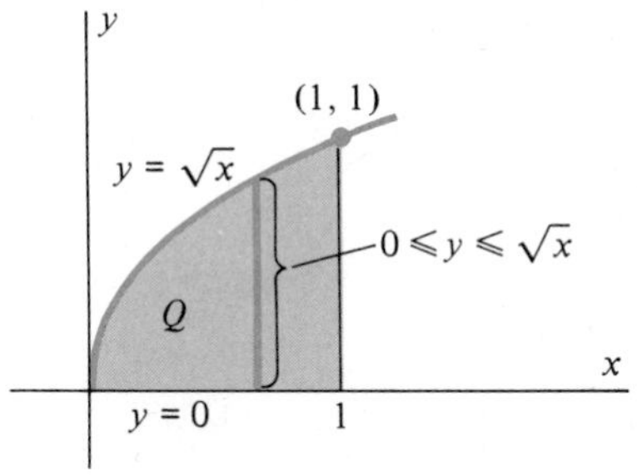

Figure 3.13 $Q = \{(x,y) | 0 \le x \le 1, 0 \le y \le \sqrt{x}\}$.

Beginning with the given integral, we therefore reverse the order of integration as follows:

$$\int_0^1 \int_{y^2}^1 ye^{x^2}\, dx\, dy = \iint_Q ye^{x^2}\, dA = \int_0^1 \int_0^{\sqrt{x}} ye^{x^2}\, dy\, dx$$

$$= \int_0^1 \left\{ \frac{1}{2} y^2 e^{x^2} \Big]_{y=0}^{y=\sqrt{x}} \right\} dx$$

$$= \int_0^1 \frac{1}{2} x e^{x^2} \, dx$$

$$= \frac{1}{4} e^{x^2} \Big]_0^1$$

$$= \frac{1}{4}(e - 1)$$

$$\approx 0.43.$$

■

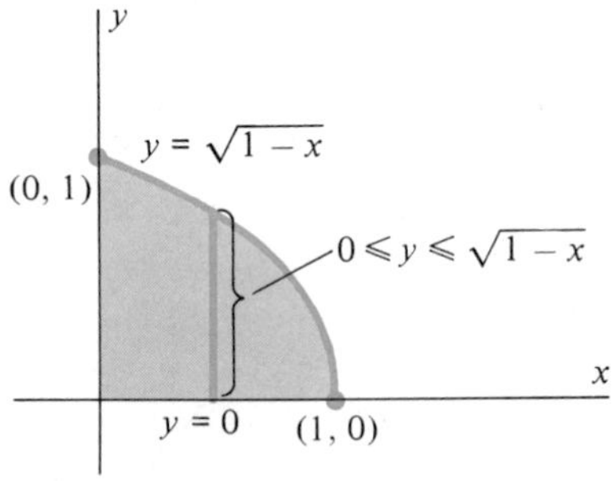

Figure 3.14 $Q = \{(x,y) | 0 \le x \le 1, 0 \le y \le \sqrt{1-x}\}$.

REMARK: It is important to note that we cannot simply interchange limits of integration when we interchange the order of integration. There is no alternative to sketching the region Q and working out the new limits of integration from knowledge of the boundary of Q.

Example 7 Evaluate the iterated integral

$$\int_0^1 \int_0^{\sqrt{1-x}} xy^2 \, dy \, dx$$

by first reversing the order of integration.

Solution: The given limits of integration are

$$0 \le x \le 1, \qquad 0 \le y \le \sqrt{1-x},$$

which describe the region Q bounded above by the graph of $y = \sqrt{1-x}$ and below by the x-axis for $0 \le x \le 1$ (see Figure 3.14). By solving the equation $y = \sqrt{1-x}$ for x, we find that this region may also be described by the inequalities

$$0 \le x \le 1 - y^2, \qquad 0 \le y \le 1.$$

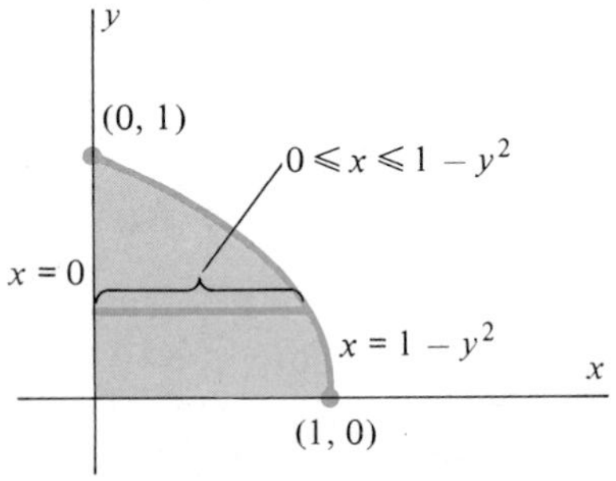

Figure 3.15 $Q = \{(x,y) | 0 \le x \le 1 - y^2, 0 \le y \le 1\}$.

(See Figure 3.15.) We may therefore evaluate the iterated integral as

$$\int_0^1 \int_0^{\sqrt{1-x}} xy^2 \, dy \, dx = \iint_Q xy^2 \, dA = \int_0^1 \int_0^{1-y^2} xy^2 \, dx \, dy$$

$$= \int_0^1 \left\{ \frac{1}{2} x^2 y^2 \Big]_{x=0}^{x=1-y^2} \right\} dy$$

$$= \int_0^1 \frac{1}{2}(1 - y^2)^2 y^2 \, dy$$

$$= \int_0^1 \left(\frac{1}{2} y^2 - y^4 + \frac{1}{2} y^6 \right) dy$$

$$= \frac{1}{6} y^3 - \frac{1}{5} y^5 + \frac{1}{14} y^7 \Big]_0^1$$

$$= \frac{4}{105}.$$

■

REMARK: If we attempt to evaluate the integral of Example 7 without first interchanging the order of integration, the following intermediate step will appear:

$$\int_0^1 \frac{1}{3} x(1 - x)^{3/2}\, dx.$$

Evaluation of this integral requires the technique of integration by parts, which involves somewhat more work than does the method of the example. As you work more problems, you will learn to recognize which order of integration is likely to be easier.

Properties of Double Integrals

As defined in this section, the double integral satisfies the following properties:

(i) $$\iint_Q [f(x, y) + g(x, y)]\, dA = \iint_Q f(x, y)\, dA + \iint_Q g(x, y)\, dA,$$

(ii) $$\iint_Q cf(x, y)\, dA = c \iint_Q f(x, y)\, dA, \qquad c = \text{constant},$$

(iii) $$\iint_{Q_1 \cup Q_2} f(x, y)\, dA = \iint_{Q_1} f(x, y)\, dA + \iint_{Q_2} f(x, y)\, dA.$$

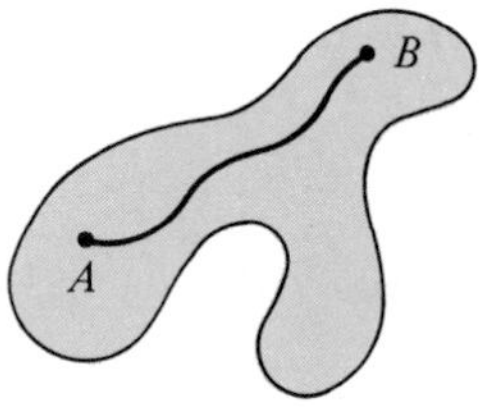

Figure 3.16 A connected region.

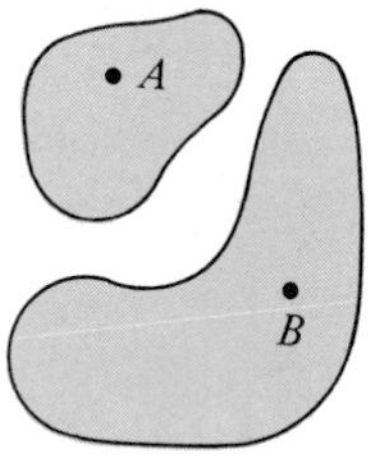

Figure 3.17 A disconnected region.

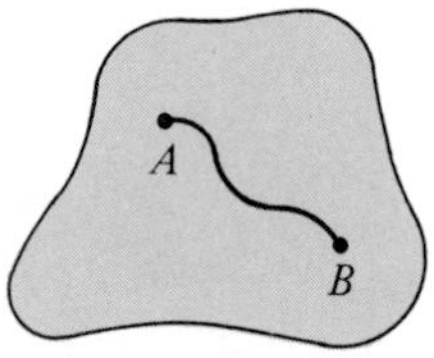

Figure 3.18 All regular regions are connected.

In (iii) we mean that $Q = Q_1 \cup Q_2$ is the union of two nonoverlapping regions Q_1 and Q_2, each satisfying the properties required of Q in this section. Property (iii) extends to unions of finitely many nonoverlapping regions $Q_1, Q_2, \ldots, Q_n$ of this type.

We shall not prove properties (i) through (iii), although they should not be surprising to you since analogous properties hold for definite integrals of functions of a single variable.

Just as the Mean Value Theorem for functions of a single variable is an important tool in developing the theory of the calculus of a single variable, we shall have need for a mean value property for double integrals. The statement of this property (Theorem 2 below) involves the notion of a *connected* region. We say that the region Q in the plane is **connected** if any two points in Q can be joined by a path that lies entirely within Q. Figure 3.16 illustrates this idea. Figure 3.17 shows a region that is not connected. (Such regions are called **disconnected.**) Figure 3.18 illustrates the fact, which we shall not prove, that all regular regions are connected.

THEOREM 2
Mean Value Property for Double Integrals

Let Q be a closed regular region in the plane and let $f(x, y)$ be continuous on Q. Then there exists a point (x_0, y_0) in Q so that

$$\iint_Q f(x, y)\, dA = f(x_0, y_0) \cdot (\text{Area of } Q). \tag{10}$$

REMARK: This theorem is the analogue of the mean value property for integrals of functions of a single variable (Theorem 7, Chapter 6). The number

$$f(x_0, y_0) = \frac{\iint\limits_Q f(x, y)\, dA}{\text{Area of } Q}$$

is called the **mean value** of $f(x, y)$ on Q.

Proof of Theorem 2: Let M and m be the maximum and minimum values of $f(x, y)$ on Q, respectively, and let (x_1, y_1) and (x_2, y_2) be points in Q for which

$$f(x_1, y_1) = M \qquad \text{and} \qquad f(x_2, y_2) = m.$$

Let A be the area of Q.

Since Q is a regular region, there exists a smooth curve connecting (x_1, y_1) and (x_2, y_2) that lies entirely within Q. Let

$$\boldsymbol{r}(t) = x(t)\boldsymbol{i} + y(t)\boldsymbol{j}, \qquad t_1 \le t \le t_2$$

be a parameterization for this curve, with $\boldsymbol{r}(t_1) = (x_1, y_1)$ and $\boldsymbol{r}(t_2) = (x_2, y_2)$. Since

$$m \le f(x, y) \le M \tag{11}$$

for all $(x, y) \in Q$, we have

$$m \le f(\boldsymbol{r}(t)) \le M \tag{12}$$

for all $t \in [t_1, t_2]$, where $f(\boldsymbol{r}(t)) = f(x(t), y(t))$.

From equation (11) it follows that

$$m \cdot A = \iint\limits_Q m\, dA \le \iint\limits_Q f(x, y)\, dA \le \iint\limits_Q M\, dA = M \cdot A.$$

Dividing through by A then gives

$$m \le \frac{\iint\limits_Q f(x, y)\, dA}{A} \le M. \tag{13}$$

Since $f(\boldsymbol{r}(t))$ is a continuous function of a single variable, it now follows from equations (12) and (13) and the Intermediate Value Theorem that there exists a number $t_0 \in [t_1, t_2]$ for which

$$f(\boldsymbol{r}(t_0)) = \frac{\iint\limits_Q f(x, y)\, dA}{A}. \tag{14}$$

Letting $(x_0, y_0) = \mathbf{r}(t_0)$, we may write equation (14) as

$$f(x_0, y_0) = \frac{\displaystyle\iint_Q f(x, y)\, dA}{\text{Area of } Q},$$

which is equation (10). ■

Exercise Set 20.3

In Exercises 1–12, sketch the region Q determined by the limits of integration and evaluate the iterated integral.

1. $\displaystyle\int_0^1 \int_0^x (y^2 - x^2)\, dy\, dx$

2. $\displaystyle\int_0^1 \int_0^{1-x} 2xy\, dy\, dx$

3. $\displaystyle\int_{-1}^0 \int_{-1}^{y+1} (xy - x)\, dx\, dy$

4. $\displaystyle\int_0^1 \int_0^y xye^{x^2}\, dx\, dy$

5. $\displaystyle\int_{-1}^1 \int_0^{1-x^2} xy\, dy\, dx$

6. $\displaystyle\int_0^1 \int_{-x}^{\sqrt{x}} \frac{y}{1+x}\, dy\, dx$

7. $\displaystyle\int_0^1 \int_{x^3}^{x^2} x\, dy\, dx$

8. $\displaystyle\int_0^{\sqrt{\pi/2}} \int_0^{\sqrt{y}} x \sin y^2\, dx\, dy$

9. $\displaystyle\int_0^{\pi/2} \int_0^{\sin x} 2y \cos x\, dy\, dx$

10. $\displaystyle\int_1^e \int_0^{\ln y} e^{x+y}\, dx\, dy$

11. $\displaystyle\int_0^1 \int_0^{y^2} e^{x/y}\, dx\, dy$

12. $\displaystyle\int_0^{\pi/2} \int_0^{\cos y} x \sin y\, dx\, dy$

In Exercises 13–22, evaluate the double integral.

13. $\displaystyle\iint_Q \sqrt{x}\, y\, dA, \quad Q = \{(x, y) | 0 \le x \le y^2,\ 0 \le y \le 1\}$

14. $\displaystyle\iint_Q x \cos \pi y\, dA,$
$Q = \{(x, y) | 0 \le x \le 1,\ 0 \le y \le x\}$

15. $\displaystyle\iint_Q (x + 2)\sqrt{1 + e^y}\, dA,$
$Q = \{(x, y) | 0 \le x \le e^{y/2},\ 0 \le y \le 1\}$

16. $\displaystyle\iint_Q \frac{x^2}{1 + y}\, dA,$
$Q = \{(x, y) | 0 \le x \le 1,\ 0 \le y \le e^x - 1\}$

17. $\displaystyle\iint_Q xy\, dA, \quad Q = \{(x, y) | y \le x \le \sqrt{y},\ 0 \le y \le 1\}$

18. $\displaystyle\iint_Q (x^2 + y^2)\, dA,$
$Q = \{(x, y) | -2 \le x \le 2,\ -3 \le y \le 1 - x^2\}$

19. $\displaystyle\iint_Q y\, dA, \quad Q = \{(x, y) | -1 \le x \le 1,\ e^x \le y \le e\}$

20. $\displaystyle\iint_Q ye^x\, dA,$
$Q = \{(x, y) | -\ln y \le x \le \ln y,\ 1 \le y \le 2\}$

21. $\displaystyle\iint_Q x\, dA,$
$Q = \{(x, y) | 0 \le x \le \sqrt{1 - y^2},\ 0 \le y \le 1\}$

22. $\displaystyle\iint_Q xy\, dA, \quad Q = \{(x, y) | 0 \le x^2 + y^2 \le 1\}$

In Exercises 23–28, sketch the region Q determined by the limits of integration, interchange the order of integration, and evaluate the given integral, where possible.

23. $\displaystyle\int_{-1}^1 \int_0^{x+1} (x + y)\, dy\, dx$

24. $\displaystyle\int_0^1 \int_{x^2}^1 xe^{y^2}\, dy\, dx$

25. $\displaystyle\int_0^1 \int_0^y xy^2\, dx\, dy$

26. $\displaystyle\int_{-2}^0 \int_{x^2}^4 xe^{y^2}\, dy\, dx$

27. $\displaystyle\int_1^e \int_0^{\ln x} f(x, y)\, dy\, dx$

28. $\displaystyle\int_0^1 \int_0^{\sin^{-1} x} f(x, y)\, dy\, dx$

In Exercises 29–33, use a double integral to find the area of Q.

29. Q is the region bounded by the graphs of $y = 4 - x^2$ and the line $y = x + 2$.

30. Q is the region bounded by the graphs of $y = x^2$ and $y = x^3$.

31. Q is the region bounded by the graphs of $y = \sin x$ and $y = \cos x$ for $0 \le x \le \pi/4$.

32. Q is the region bounded by the graphs of $y = x^3$ and $y = 4x$.

33. Q is the region bounded by the graphs of $y = \sqrt{x}$ and $y = x^2$.

34. Use a double integral to find the volume of the tetrahedron with vertices (1, 0, 0), (0, 1, 0), (0, 0, 0), and (0, 0, 1).

35. Find the area of the ellipse $\dfrac{x^2}{4} + \dfrac{y^2}{9} = 1$ by use of a double integral.

36. Use a double integral to establish the formula for the volume of a right circular cone of radius r and height h.

37. Find the volume of the solid bounded by the paraboloid $z = \dfrac{1}{4}(x^2 + y^2)$ and the paraboloid $z = 5 - x^2 - y^2$.

38. Use a double integral to find the volume of the sphere $x^2 + y^2 + z^2 = 4$ lying above the plane $z = 1$.

39. A hole of diameter 2 cm is drilled through the center of a spherical bearing of radius 4 cm. Find the volume of the remaining part of the bearing.

40. Find the volume of the portion of the solid bounded by the plane $z = x$, the cylinder $x^2 + y^2 = 4$, and the plane $z = 0$.

41. Find the volume of the solid bounded by the coordinate planes and the plane $6x + 3y + 2z = 6$.

42. Sketch the region in the first octant common to the two cylinders $z^2 = 1 - x^2$ and $z^2 = 1 - y^2$. Find its volume.

20.4 DOUBLE INTEGRALS IN POLAR COORDINATES

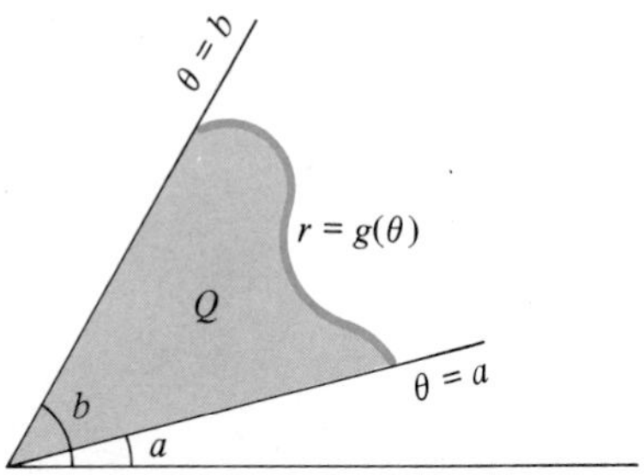

Figure 4.1 The region Q, bounded by the graph of $r = g(\theta)$ and the rays $\theta = a$ and $\theta = b$.

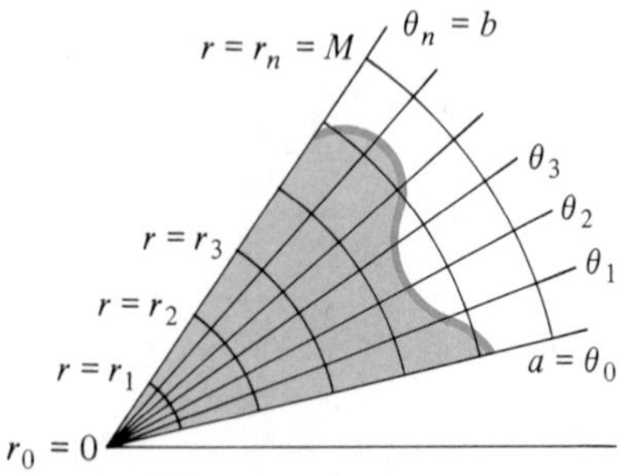

Figure 4.2 Q is partitioned into subregions by arcs $r = r_j$ and rays $\theta = \theta_k$.

So far, we have defined double integrals only in the case where Q is a region in the xy- (Cartesian) coordinate plane and $f(x, y)$ is a function written in terms of Cartesian coordinates. We can also define the double integral when the region Q and the function $f(r, \theta)$ are described in polar coordinates.

Figure 4.1 shows a region Q bounded by the rays $\theta = a$ and $\theta = b$ and the graph of the function $r = g(\theta)$. Let n be a positive integer and let $\Delta\theta = \dfrac{b - a}{n}$. Then the rays corresponding to the angles

$$a = \theta_0, \qquad \theta_1 = a + \Delta\theta, \qquad \theta_2 = a + 2\,\Delta\theta, \; . \; . \; . \; , \; \theta_n = a + n\,\Delta\theta = b$$

divide the region Q into n wedge-shaped regions (see Figure 4.2). Also, if M is any positive number with $|g(\theta)| \le M$ for all $\theta \in [a, b]$ and if $\Delta r = \dfrac{M}{n}$, then the concentric circles with radii

$$r_0 = 0, \qquad r_1 = \Delta r, \qquad r_2 = 2\,\Delta r, \; . \; . \; . \; , \; r_n = n\,\Delta r = M$$

divide the region Q into annular regions.

As Figure 4.2 illustrates, the circular arcs $r = r_j$ and the rays $\theta = \theta_k$ partition the region Q into subregions R_{jk} that are nearly rectangular in shape. To determine the area of R_{jk} we use two facts (see Figure 4.3):

(i) The region R_{jk} lies between the concentric circles $r = r_{j-1}$ and $r = r_j$. The area of the entire annulus lying between these circles is therefore

$$\text{Area of annulus} = \pi r_j^2 - \pi r_{j-1}^2.$$

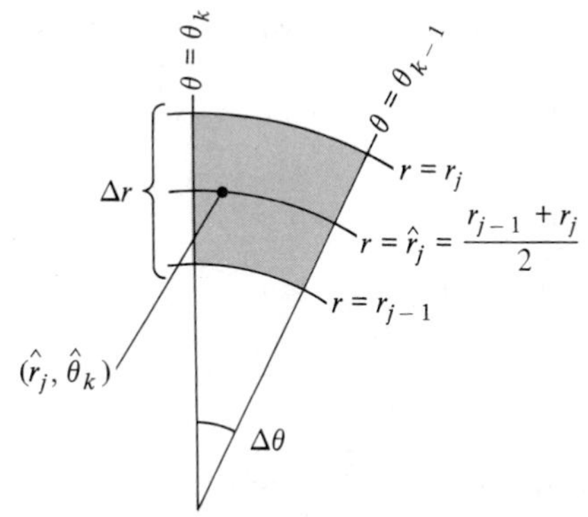

Figure 4.3 Area of region R_{jk} is $\Delta A_{jk} = \hat{r}_j \, \Delta r \, \Delta\theta$.

(ii) The angle $\Delta\theta$, measured in radians, determines the proportion of the entire annulus lying in the region R_{jk}. That is,

$$\text{Proportion of annulus in } R_{jk} = \frac{\Delta\theta}{2\pi}.$$

From statements (i) and (ii), you can see that the area of the region R_{jk} is

$$\begin{aligned} \Delta A_{jk} &= (\pi r_j^2 - \pi r_{j-1}^2)\left(\frac{\Delta\theta}{2\pi}\right) \qquad (1) \\ &= \frac{1}{2}(r_j^2 - r_{j-1}^2)\,\Delta\theta. \end{aligned}$$

Since

$$\begin{aligned} r_j^2 - r_{j-1}^2 &= (r_j + r_{j-1})(r_j - r_{j-1}) \\ &= 2\left(\frac{r_j + r_{j-1}}{2}\right)(r_j - r_{j-1}), \end{aligned}$$

we may write

$$r_j^2 - r_{j-1}^2 = 2\hat{r}_j \, \Delta r \qquad (2)$$

where $\hat{r}_j = \dfrac{r_j + r_{j-1}}{2}$ and $\Delta r = r_j - r_{j-1}$. It is important to note that the number $\hat{r}_j$ is simply the average of the radii r_{j-1} and r_j (see Figure 4.3). Combining (1) and (2), we may now write the area of the region R_{jk} as

$$\Delta A_{jk} = \hat{r}_j \, \Delta r \, \Delta\theta. \qquad (3)$$

Now let $f(r, \theta)$ be a continuous function defined on the region Q, and for each integer $k = 1, 2, \ldots, n$ let $\hat{\theta}_k$ be any number with $\theta_{k-1} \le \hat{\theta}_k \le \theta_k$. Then the number $f(\hat{r}_j, \hat{\theta}_k)$ is the value of the function $f(r, \theta)$ at the point $(\hat{r}_j, \hat{\theta}_k)$ in the region R_{jk}. If $f(r, \theta) \ge 0$ for all $(r, \theta) \in Q$ then the product $f(\hat{r}_j, \hat{\theta}_k)\,\Delta A_{jk}$ approximates the volume of the solid bounded by the graph of $f(r, \theta)$ over R_{jk}, and the sum

$$\sum_{j=1}^{n} \sum_{k=1}^{n} f(\hat{r}_j, \hat{\theta}_k)\,\Delta A_{jk} = \sum_{j=1}^{n} \sum_{k=1}^{n} f(\hat{r}_j, \hat{\theta}_k)\hat{r}_j \, \Delta r \, \Delta\theta \qquad (4)$$

approximates the volume of the solid bounded by the graph of $f(r, \theta)$ over the entire region Q. (In writing the sums in line (4) we mean that the sums are taken only over those regions R_{jk} that lie entirely within Q.) We therefore define the double integral of $f(r, \theta)$ over Q as

$$\iint_Q f(r, \theta)\, dA = \lim_{n\to\infty} \sum_{j=1}^{n} \sum_{k=1}^{n} f(\hat{r}_j, \hat{\theta}_k)\, \Delta A_{jk} \tag{5}$$

where $\Delta A_{jk} = \hat{r}_j\, \Delta r\, \Delta \theta$.

The integral in (5) is defined whenever $f(r, \theta)$ is continuous on Q, regardless of the sign of $f(r, \theta)$. However, as for the double integral in Cartesian coordinates, if $f(r, \theta) \geq 0$ for all $(r, \theta) \in Q$, we have

$$\iint_Q f(r, \theta)\, dA = \text{volume of solid bounded above by the graph of } f(r, \theta) \text{ and below by } Q, \text{ and} \tag{6}$$

$$\iint_Q 1 \cdot dA = \text{area of the region } Q. \tag{7}$$

Of course, the integral defined by equation (5) is of little value unless we know how to evaluate the limit of the approximating sums. The following theorem, analogous to Theorem 1, shows how this may be done. The idea behind its proof is contained in equation (4).

THEOREM 3

Let Q be the region bounded by the graph of the continuous function $r = g(\theta)$ and the rays $\theta = a$ and $\theta = b$, as in Figure 4.1. Let $f(r, \theta)$ be a continuous function defined on the region Q. Then

$$\iint_Q f(r, \theta)\, dA = \int_a^b \int_0^{g(\theta)} f(r, \theta) r\, dr\, d\theta. \tag{8}$$

Theorem 3 says that the double integral in (5) may be evaluated as an iterated integral, where we integrate first with respect to r alone, for $0 \leq r \leq g(\theta)$, and then integrate the resulting function of θ over the limits $a \leq \theta \leq b$. However, *it is important to note that the function to be integrated is the product $f(r, \theta)r$, not just $f(r, \theta)$.* The reason for this can be seen from equation (3): The area of the region R_{jk} is $\Delta A_{jk} = \hat{r}_j\, \Delta r\, \Delta \theta$. In the Cartesian-coordinate case the rectangles R_{jk} used in approximating the area of Q had area $\Delta A = \Delta x\, \Delta y$. The extra factor of r appearing in the iterated polar double integral appears because of this formula for the area of the approximating region R_{jk}.

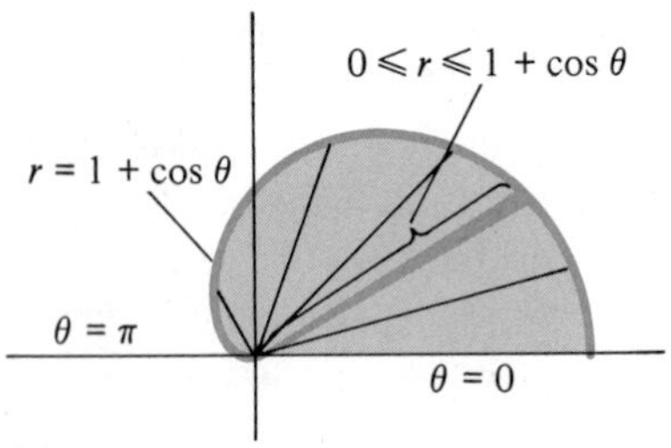

Figure 4.4 Region bounded by graph of $r = 1 + \cos\theta$ for $0 \leq \theta \leq \pi$.

Example 1 Evaluate the double integral $\iint_Q r \sin\theta\, dA$ where Q is the region inside the upper half of the cardioid $r = 1 + \cos\theta$ (see Figure 4.4).

Solution: Here $f(r, \theta) = r \sin \theta$, and the region Q may be described by the inequalities

$$0 \le r \le 1 + \cos \theta, \qquad 0 \le \theta \le \pi.$$

Thus, $g(\theta) = 1 + \cos \theta$ in equation (8), and

$$\iint_Q r \sin \theta \, dA = \int_0^{\pi} \int_0^{1+\cos \theta} r \sin \theta \cdot \underset{\uparrow}{r} \, dr \, d\theta \quad \text{note extra factor of } r$$

$$= \int_0^{\pi} \int_0^{1+\cos \theta} r^2 \sin \theta \, dr \, d\theta$$

$$= \int_0^{\pi} \left\{ \frac{1}{3} r^3 \sin \theta \Big]_{r=0}^{r=1+\cos \theta} \right\} d\theta$$

$$= \int_0^{\pi} \frac{1}{3} (1 + \cos \theta)^3 \sin \theta \, d\theta$$

$$= -\frac{1}{12} (1 + \cos \theta)^4 \Big]_{\theta=0}^{\theta=\pi}$$

$$= -\frac{1}{12} (0 - 2^4) = \frac{4}{3}.$$

■

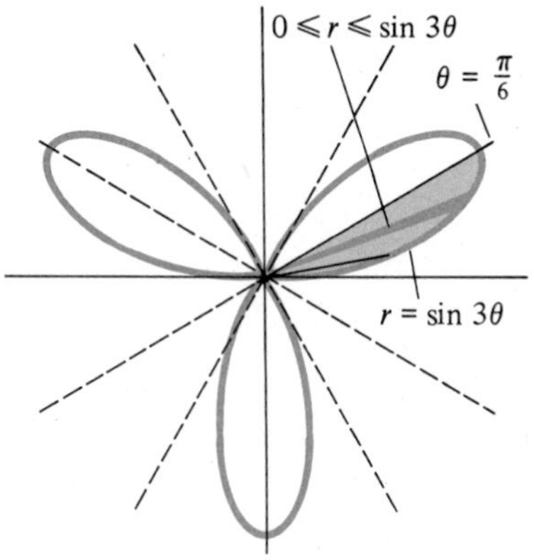

Figure 4.5 Region enclosed by three-leaved rose $r = \sin 3\theta$.

Example 2 Find the area of the region enclosed by the three-leaved rose $r = \sin 3\theta$ (see Figure 4.5).

Solution: As illustrated in Figure 4.5, one sixth of the entire region Q lies between the rays $\theta = 0$ and $\theta = \pi/6$. Using the symmetry of the region, equation (7), and equation (8) we find that

$$\text{Area of } Q = \iint_Q 1 \, dA$$

$$= 6 \int_0^{\pi/6} \int_0^{\sin 3\theta} 1 \cdot \underset{\uparrow}{r} \, dr \, d\theta \quad \text{note extra factor of } r$$

$$= 6 \int_0^{\pi/6} \left\{ \frac{r^2}{2} \Big]_{r=0}^{r=\sin 3\theta} \right\} d\theta = 3 \int_0^{\pi/6} \sin^2 3\theta \, d\theta$$

$$= \frac{3}{2} \int_0^{\pi/6} (1 - \cos 6\theta) \, d\theta \qquad \left(\sin^2 3\theta = \frac{1}{2}(1 - \cos 6\theta) \right)$$

$$= \frac{3}{2} \left[\theta - \frac{1}{6} \sin 6\theta \right]_{\theta=0}^{\theta=\pi/6}$$

$$= \frac{\pi}{4}.$$

■

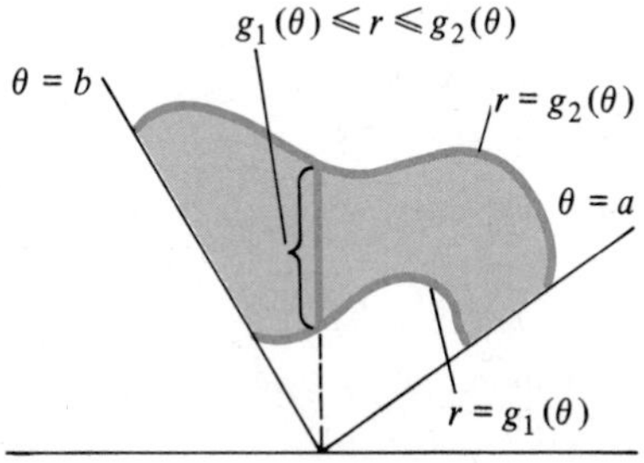

Figure 4.6 Region Q bounded by two polar curves, $r = g_1(\theta)$ and $r = g_2(\theta)$.

Figure 4.6 shows a region Q that lies *between* the graphs of the polar equations $r = g_1(\theta)$ and $r = g_2(\theta)$ for $a \le \theta \le b$. That is,

$$Q = \{(r, \theta) | g_1(\theta) \le r \le g_2(\theta), \qquad a \le \theta \le b\}.$$

If we let

$$Q_1 = \{(r, \theta) | 0 \le r \le g_1(\theta), \qquad a \le \theta \le b\}$$

and

$$Q_2 = \{(r, \theta) | 0 \le r \le g_2(\theta), \qquad a \le \theta \le b\}$$

then $Q = Q_2 - Q_1$. Accordingly, we have

$$\begin{aligned}\iint_Q f(r, \theta)\, dA &= \iint_{Q_2} f(r, \theta)\, dA - \iint_{Q_1} f(r, \theta)\, dA \\ &= \int_a^b \int_0^{g_2(\theta)} f(r, \theta)\, r\, dr\, d\theta - \int_a^b \int_0^{g_1(\theta)} f(r, \theta) r\, dr\, d\theta \\ &= \int_a^b \left\{ \int_0^{g_2(\theta)} f(r, \theta) r\, dr - \int_0^{g_1(\theta)} f(r, \theta) r\, dr \right\} d\theta \\ &= \int_a^b \int_{g_1(\theta)}^{g_2(\theta)} f(r, \theta) r\, dr\, d\theta.\end{aligned}$$

Thus, if Q is described by the polar inequalities

$$g_1(\theta) \le r \le g_2(\theta), \qquad a \le \theta \le b,$$

then

$$\iint_Q f(r, \theta)\, dA = \int_a^b \int_{g_1(\theta)}^{g_2(\theta)} f(r, \theta) r\, dr\, d\theta. \tag{9}$$

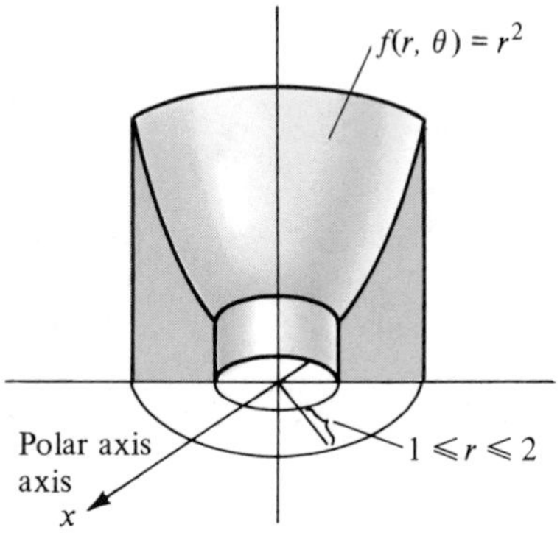

Figure 4.7 One half of the bearing sleeve of Example 3.

Example 3 A bearing sleeve has the shape of the solid bounded above by the graph of $f(r, \theta) = r^2$ and below by the xy-plane for $1 \le r \le 2$. Find the volume of the sleeve.

Solution: The solid is illustrated in Figure 4.7. The region Q over which the solid is defined is described by the inequalities

$$1 \le r \le 2, \qquad 0 \le \theta \le 2\pi.$$

The volume is therefore

$$\begin{aligned}\iint_Q f(r, \theta)\, dA &= \int_0^{2\pi} \int_1^2 r^2 \cdot r\, dr\, d\theta \\ &= \int_0^{2\pi} \int_1^2 r^3\, dr\, d\theta \\ &= \int_0^{2\pi} \left\{ \frac{1}{4} r^4 \Big]_{r=1}^{r=2} \right\} d\theta \\ &= \int_0^{2\pi} \frac{15}{4}\, d\theta \\ &= \frac{15\pi}{2}.\end{aligned}$$

■

Changing From Cartesian to Polar Coordinates

Often, a double integral in Cartesian coordinates is more easily evaluated by first changing to polar coordinates. For example, in Exercise 38 in the last section you were asked to find the volume of the sphere $x^2 + y^2 + z^2 = 4$ lying above the plane $z = 1$. This led to a double integral equivalent to

$$V = \int_{-\sqrt{3}}^{\sqrt{3}} \int_{-\sqrt{3-x^2}}^{\sqrt{3-x^2}} (\sqrt{4 - (x^2 + y^2)} - 1)\, dy\, dx. \tag{10}$$

(See Figure 4.8. It pays to read ahead!)

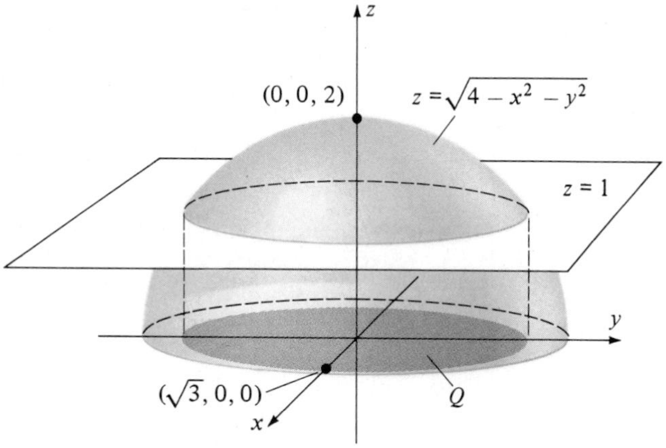

Figure 4.8 $Q = \{(x, y) | x^2 + y^2 \le 3\}$ may be described in polar coordinates as
$Q = \{(r, \theta) | 0 \le r \le \sqrt{3},\ 0 \le \theta \le 2\pi\}$.

We can calculate this volume much more simply using polar coordinates, together with the equations

$$x = r \cos \theta \tag{11a}$$

$$y = r \sin \theta \tag{11b}$$

for changing from Cartesian coordinates to polar coordinates.

The region

$$Q = \{(x, y) | -\sqrt{3} \le x \le \sqrt{3},\ -\sqrt{3 - x^2} \le y \le \sqrt{3 - x^2}\}$$

can be more easily described in polar coordinates as

$$Q = \{(r, \theta) | 0 \le r \le \sqrt{3},\ 0 \le \theta \le 2\pi\}$$

and the function $f(x, y) = \sqrt{4 - (x^2 + y^2)}$ in polar coordinates, obtained by using equations (11a) and (11b), is simply

$$f(r, \theta) = \sqrt{4 - (r^2 \cos^2 \theta + r^2 \sin^2 \theta)} = \sqrt{4 - r^2}.$$

The desired volume is then obtained using equations (6) and (8):

$$V = \iint_Q f(r \cos \theta, r \sin \theta)\, dA = \int_0^{2\pi} \int_0^{\sqrt{3}} (\sqrt{4 - r^2} - 1) r\, dr\, d\theta \tag{11}$$

note extra r

$$= \int_0^{2\pi} \left\{ -\frac{1}{3}(4 - r^2)^{3/2} - \frac{r^2}{2} \Big]_{r=0}^{r=\sqrt{3}} \right\} d\theta$$

$$= \int_0^{2\pi} \left[-\frac{1}{3}(1 - 4^{3/2}) - \frac{3}{2} \right] d\theta$$

$$= \int_0^{2\pi} \frac{5}{6}\, d\theta = \frac{5\pi}{3}.$$

This example is typical of the following more general problem: Given an iterated integral in Cartesian coordinates, how can we change from Cartesian coordinates to polar coordinates (without, of course, changing the value of the integral)? The answer may be seen by comparing equations (10) and (11) above:

To express the iterated integral

$$\int_c^d \int_{h_1(y)}^{h_2(y)} f(x, y)\, dx\, dy \qquad \text{(or equivalent)} \tag{12}$$

in polar coordinates,

(i) express the region $Q = \{(x, y) | h_1(y) \le x \le h_2(y),\ c \le y \le d\}$ in polar coordinates as

$$Q = \{(r, \theta) | g_1(\theta) \le r \le g_2(\theta), \qquad a \le \theta \le b\}, \text{ and}$$

(ii) using the substitutions $x = r \cos \theta$ and $y = \sin \theta$, replace the integrand

$$f(x, y) \qquad \text{by} \qquad f(r \cos \theta, r \sin \theta) \cdot r.$$

The result of steps (i) and (ii) is the iterated integral

$$\int_a^b \int_{g_1(\theta)}^{g_2(\theta)} f(r, \theta) r\, dr\, d\theta \tag{13}$$

which is equivalent to the integral (12).

Another way to write this is simply that

$$\iint_Q f(x, y)\, dx\, dy = \iint_Q f(r \cos \theta, r \sin \theta) r\, dr\, d\theta. \tag{14}$$

Equation (14) is referred to as a **change of variables** formula.

We shall not prove statement (14). However, it may be justified by comparing Theorem 1, which expresses $\iint_Q f\, dV$ as an iterated integral in Cartesian coordinates, with equation (9), which gives $\iint_Q f\, dV$ as an iterated integral in polar coordinates. We may paraphrase equation (14) by saying that in changing from Cartesian to polar coordinates we replace the *element of area*

$$dA = dx\, dy$$

in rectangular coordinates with the element of area

$$dA = r\, dr\, d\theta \tag{15}$$

in polar coordinates. It is very important to note the extra factor r in (15).

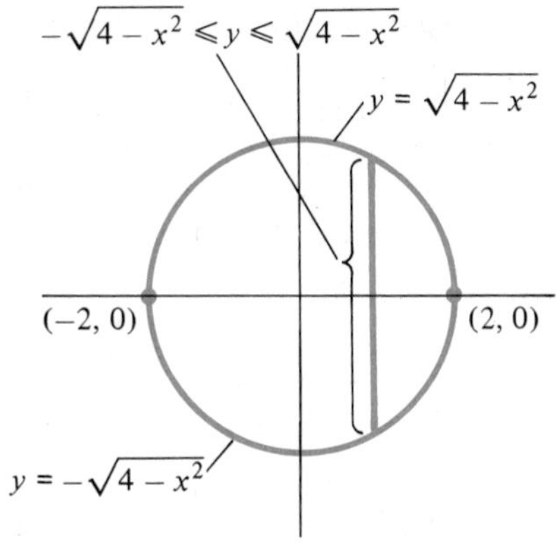

Figure 4.9 Q: $-2 \le x \le 2$
$-\sqrt{4-x^2} \le y \le \sqrt{4-x^2}$.

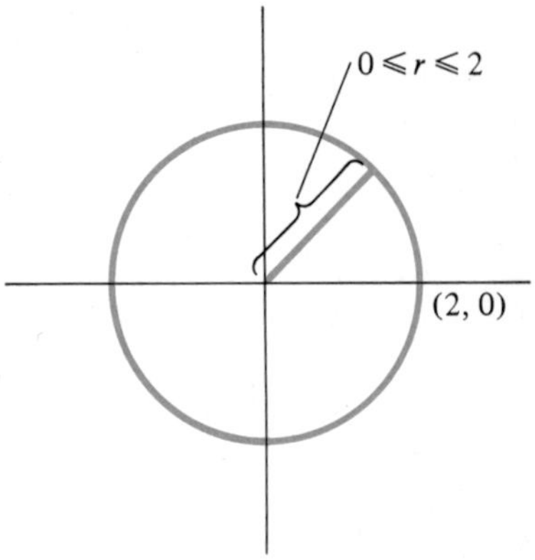

Figure 4.10 Q: $0 \le r \le 2$
$0 \le \theta \le 2\pi$.

Example 4 Evaluate the integral

$$\int_{-2}^{2} \int_{-\sqrt{4-x^2}}^{\sqrt{4-x^2}} e^{x^2+y^2}\, dy\, dx$$

by first changing to polar coordinates.

Solution: From the limits of integration, we can see that the integral is being evaluated over the disc-shaped region

$$Q = \{(x, y) | -2 \le x \le 2, -\sqrt{4-x^2} \le y \le \sqrt{4-x^2}\}.$$

(See Figure 4.9.) This region can be described in polar coordinates by the inequalities

$$0 \le r \le 2, \qquad 0 \le \theta \le 2\pi \qquad \text{(Figure 4.10).}$$

With $x = r\cos\theta$ and $y = r\sin\theta$, the function $f(x, y) = e^{x^2+y^2}$ becomes

$$\begin{aligned} f(r\cos\theta, r\sin\theta) &= e^{r^2\cos^2\theta + r^2\sin^2\theta} \\ &= e^{r^2}. \end{aligned}$$

Using (14), we obtain

$$\begin{aligned} \int_{-2}^{2} \int_{-\sqrt{4-x^2}}^{\sqrt{4-x^2}} e^{x^2+y^2}\, dy\, dx &= \int_0^{2\pi} \int_0^2 e^{r^2}\, \underset{\uparrow}{r\, dr\, d\theta} \quad \text{note the extra factor of } r \\ &= \int_0^{2\pi} \left\{ \frac{1}{2} e^{r^2} \Big]_{r=0}^{r=2} \right\} d\theta \\ &= \int_0^{2\pi} \frac{1}{2}(e^4 - 1)\, d\theta \\ &= \pi(e^4 - 1). \end{aligned}$$

■

REMARK: It is important to note that the double integral in Example 4 could not be evaluated in rectangular coordinates, since we would not be able to find an antiderivative for $e^{x^2+y^2}$ with respect to either x or y. Thus, it is sometimes essential that we change to polar coordinates, if possible. The next example involves a volume that could be calculated using Cartesian coordinates, but the calculation in polar coordinates is much easier. It is typical of the kind of problem we will encounter in Chapter 21.

Example 5 Interpret the iterated integral

$$\int_0^2 \int_0^{\sqrt{4-x^2}} \sqrt{5 - x^2 - y^2}\, dy\, dx$$

geometrically, and evaluate the integral by first changing to polar coordinates.

Solution: Since the integrand is nonnegative, the integral may be interpreted as the volume of the solid bounded above by the graph of $f(x, y) = \sqrt{5 - x^2 - y^2}$ and below by the quarter disc $Q = \{(x, y) | 0 \le x \le 2, 0 \le y \le \sqrt{4 - x^2}\}$ of radius 2 (Figure 4.11). In polar coordinates, Q is determined by the inequalities

$$0 \le r \le 2, \qquad 0 \le \theta \le \frac{\pi}{2}.$$

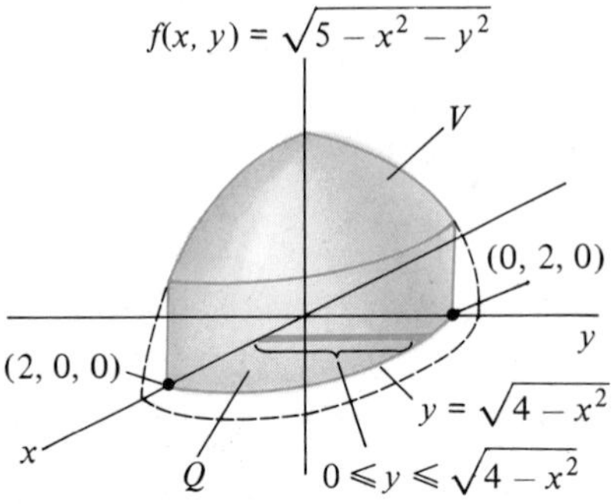

Figure 4.11 Solid V over region Q in Example 5.

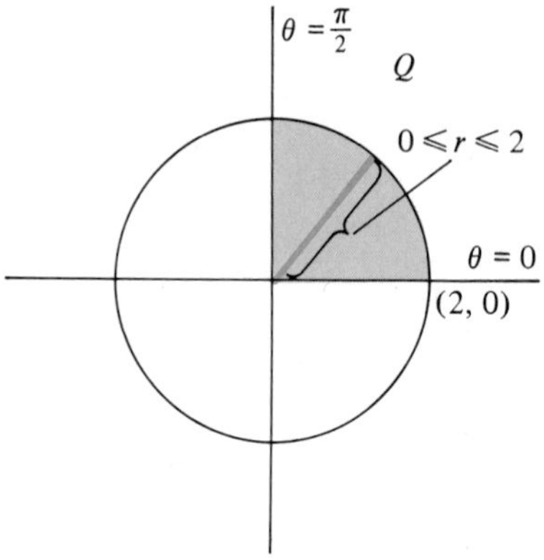

Figure 4.12 Region Q in Example 5: $0 \le r \le 2$, $0 \le \theta \le \pi/2$.

(See Figure 4.12.) Thus

$$\int_0^2 \int_0^{\sqrt{4-x^2}} \sqrt{5 - x^2 - y^2}\, dy\, dx$$

$$= \int_0^{\pi/2} \int_0^2 \sqrt{5 - (r\cos\theta)^2 - (r\sin\theta)^2} \cdot r\, dr\, d\theta$$ ← note the extra factor of r

$$= \int_0^{\pi/2} \int_0^2 \sqrt{5 - r^2}\, r\, dr\, d\theta$$

$$= \int_0^{\pi/2} \left\{ -\frac{1}{3}(5 - r^2)^{3/2} \Big]_{r=0}^{r=2} \right\} d\theta$$

$$= \int_0^{\pi/2} -\frac{1}{3}(1 - 5^{3/2})\, d\theta$$

$$= \frac{\pi}{6}(5^{3/2} - 1).$$

■

Usually, we will be interested in evaluating double integrals in polar coordinates for one of two reasons. One is that the integrand of a double integral in Cartesian coordinates may not yield easily to antidifferentiation. If the substitutions $x = r\cos\theta$ and $y = r\sin\theta$, plus the additional factor of r, produce a function $f(r\cos\theta, r\sin\theta)\cdot r$ for which antidifferentiations with respect to r and θ can be performed, then the integral can be evaluated in polar coordinates. The other reason for using polar coordinates has to do with the geometry of the region Q. When Q is more simply described in polar coordinates than in Cartesian coordinates, you should think about evaluating a double integral over Q in polar coordinates rather than in Cartesian coordinates.

Finally, we note that we have in this section considered only regions Q that are **r-simple,** in that they can be described by inequalities of the type

$$a \le \theta \le b, \qquad g_1(\theta) \le r \le g_2(\theta).$$

Regions that are **$\boldsymbol{\theta}$-simple** are described by inequalities of the type

$$c \le r \le d, \qquad h_1(r) \le \theta \le h_2(r).$$

Double integrals over θ-simple regions can also be evaluated as iterated integrals, with the integrations being performed first with respect to θ and then with respect to r. However, such integrals occur only infrequently and will not be discussed further.

Exercise Set 20.4

In Exercises 1–7, sketch the region Q and evaluate the double integral $\iint_Q f(r, \theta)\, dA$ for the given function over the given region Q.

1. $f(r, \theta) = r, \quad Q = \{(r, \theta) | 0 \le r \le 1,\ 0 \le \theta \le 2\pi\}$
2. $f(r, \theta) = 1 - r, \quad Q = \{(r, \theta) | 0 \le r \le 2,\ 0 \le \theta \le \pi\}$
3. $f(x, y) = e^{x^2+y^2}, \quad Q = \{(r, \theta) | r \le a\}$
4. $f(x, y) = e^{x^2+y^2}, \quad Q = \{(r, \theta) | 1 \le r \le 2,\ 0 \le \theta \le \pi/4\}$
5. $f(r, \theta) = 3r^2, \quad Q = \{(r, \theta) | 1 \le r \le 2,\ 0 \le \theta \le \pi/4\}$
6. $f(r, \theta) = 2 - r, \quad Q = \{(r, \theta) | 0 \le r \le (1 + \cos\theta),\ 0 \le \theta \le \pi/2\}$
7. $f(r, \theta) = r, \quad Q = \{(r, \theta) | 0 \le r \le \sin 3\theta,\ 0 \le \theta \le \pi/3\}$

In Exercises 8–14, find the area of the region Q by use of a double integral.

8. Q is the region enclosed by the circle $r = 2\cos\theta$.
9. Q is the region enclosed by the three-leaved rose $r = \sin 3\theta$.
10. Q is the region enclosed by the cardioid $r = a(1 + \cos\theta)$.
11. Q is the region inside the cardioid $r = 1 + \cos\theta$ and outside the circle $r = 1$.
12. Q is the region bounded by the graph of $r^2 = a^2 \sin 2\theta$.
13. Q is the region enclosed by the graph of $r = 2 - \sin\theta$.
14. Q is the region enclosed by the graph of the lemniscate $r^2 = 4\cos^2\theta$.

In Exercises 15–22, change the iterated integral from Cartesian to polar coordinates. Then evaluate the resulting integral.

15. $\displaystyle\int_0^1 \int_0^{\sqrt{1-x^2}} 2\, dy\, dx$
16. $\displaystyle\int_{-2}^{\sqrt{2}} \int_x^{\sqrt{4-x^2}} 1\, dy\, dx$
17. $\displaystyle\int_0^1 \int_0^{\sqrt{1-y^2}} e^{x^2+y^2}\, dx\, dy$
18. $\displaystyle\int_0^2 \int_{-\sqrt{2y-y^2}}^{\sqrt{2y-y^2}} \sqrt{x^2 + y^2}\, dx\, dy$
19. $\displaystyle\int_0^2 \int_0^{\sqrt{2x-x^2}} \frac{1}{\sqrt{x^2 + y^2}}\, dy\, dx$
20. $\displaystyle\int_0^1 \int_0^{\sqrt{4-x^2}} (x^2 + y^2)^{3/2}\, dy\, dx$
21. $\displaystyle\int_{-2}^2 \int_0^{\sqrt{4-x^2}} \frac{x}{\sqrt{x^2 + y^2}}\, dy\, dx$

22. $\int_0^1 \int_0^{\sqrt{1-y^2}} (x^2 + y^2)^{3/2}\, dx\, dy$

23. Find the volume of the portion of the cylinder $x^2 + (y - 1)^2 = 4$ bounded above by the plane $z = x + 4$ and below by the xy-plane.

24. Find the area of the region outside the spiral $r = \theta$ and inside the spiral $r = 2\theta$ for $0 \le \theta \le 2\pi$.

25. Find the volume of the region lying inside the sphere $x^2 + y^2 + z^2 = 4$ and outside the cylinder $x^2 + y^2 = 1$.

26. Find the volume of the region lying inside both the cone $z^2 = x^2 + y^2$ and the sphere $x^2 + y^2 + z^2 = 2$.

27. Find the volume of the solid bounded above by the plane $z = y + 2$ and below by the region inside the cardioid $r = 1 + \cos\theta$.

28. Use a double integral in polar coordinates to obtain the formula for the volume of a right circular cylinder of radius r and height h.

29. Find the volume of the solid bounded above by the graph of $z = 9 - x^2 - y^2$ and below by the graph of $z = 1 + x^2 + y^2$.

30. A hole 2 cm in diameter is drilled through the center of a spherical bearing of radius 3 cm. Find the volume of the remaining solid.

31. Change the order of integration in

$$\int_0^{\pi/2} \int_0^{\sin\theta} \sin\theta\, dr\, d\theta$$

and evaluate the integral.

32. Find the volume of the solid inside both the ellipsoid $z^2 + 4r^2 = 4$ and the cylinder $r = \sin\theta$.

33. Find the volume of the solid bounded by the cone $z = \sqrt{x^2 + y^2}$, the cylinder $x^2 + y^2 = 4$, and the xy-plane.

20.5 CALCULATING MASS AND CENTERS OF MASS USING DOUBLE INTEGRALS

In this section we return to the notions of mass, density, and centroid (or center of gravity) first discussed in Section 8.4. We will be concerned with using the double integral to calculate mass and centroids for lamina (thin flat objects) lying in the plane. The difference between the discussion of Section 8.4 and what we do here is that we previously had assumed the density ρ of the material to be constant. Here we will treat the more general case of a variable density $\rho(x, y)$. We assume the density function $\rho(x, y)$ to be continuous throughout the planar region Q that describes the lamina.

More specifically, let Q be a region in the plane (which we think of as the base of the lamina) and let $\rho(x, y)$ be the **mass per unit area** of the lamina at point (x, y). This is what we mean by density. (Thus, $\rho(x, y)$ is affected both by the thickness of the material and by its mass per unit volume. We will not be concerned about these two factors individually.) If R_j is a rectangle in Q of area ΔA, and if (s_j, t_j) is a point in R_j, then the product

$$\Delta M_j = \rho(s_j, t_j)\,\Delta A \qquad \text{(mass = mass per unit area times area)}$$

is an approximation to the mass of the lamina over the rectangle R_j. If, as in Section 20.2, the region Q is partitioned by a rectangular grid and $R_1, R_2, \ldots, R_n$ is a list of all rectangles lying within Q, then

$$M \approx \sum_{j=1}^{n} \Delta M_j = \sum_{j=1}^{n} \rho(s_j, t_j)\,\Delta A \tag{1}$$

is an approximation to the mass of the lamina over Q. Since, as $n \to \infty$, the union of the rectangles R_j provides an increasingly accurate approximation to the region Q, we use approximation (1) and equation (2) of Section 20.3 to define M, the mass of Q, by

$$M = \iint_Q \rho(x, y)\, dA. \tag{2}$$

Equation (2) generalizes the calculation of mass for the case of constant density $\rho(x, y) \equiv \rho$, since in that case

$$\begin{aligned} \text{Mass} &= (\text{density per unit area}) \cdot (\text{area}) \\ &= \rho \cdot \iint_Q 1\, dA \\ &= \iint_Q \rho\, dA. \end{aligned}$$

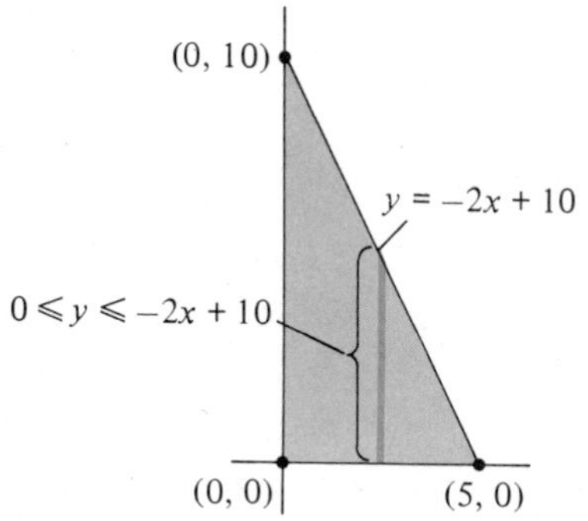

Figure 5.1 The triangular region in Example 1.

Example 1 A thin plate has the shape of a right triangle with legs of length 5 cm and 10 cm. The density, in terms of mass per unit area, at each point on the plate is proportional to the square of the distance from the vertex corresponding to the right angle. What is the mass of the plate?

Solution: With the triangle positioned as in Figure 5.1, the region it occupies may be described by the inequalities

$$0 \le x \le 5, \qquad 0 \le y \le -2x + 10.$$

Since the right angle is at the origin, the density function is

$$\rho(x, y) = \lambda(x^2 + y^2)$$

where λ is constant. By (2), the mass is

$$\begin{aligned} M = \iint_Q \lambda(x^2 + y^2)\, dA &= \int_0^5 \int_0^{-2x+10} \lambda(x^2 + y^2)\, dy\, dx \\ &= \int_0^5 \lambda \left\{ x^2y + \frac{1}{3}y^3 \right]_{y=0}^{y=-2x+10} \Bigg\} dx \\ &= \int_0^5 \lambda \left(-\frac{14}{3}x^3 + 50x^2 - 200x + \frac{1000}{3} \right) dx \\ &= \lambda \left[-\frac{7}{6}x^4 + \frac{50}{3}x^3 - 100x^2 + \frac{1000}{3}x \right]_0^5 \\ &= \frac{3125\lambda}{6}. \end{aligned}$$

The constant λ determines the appropriate units for mass. ■

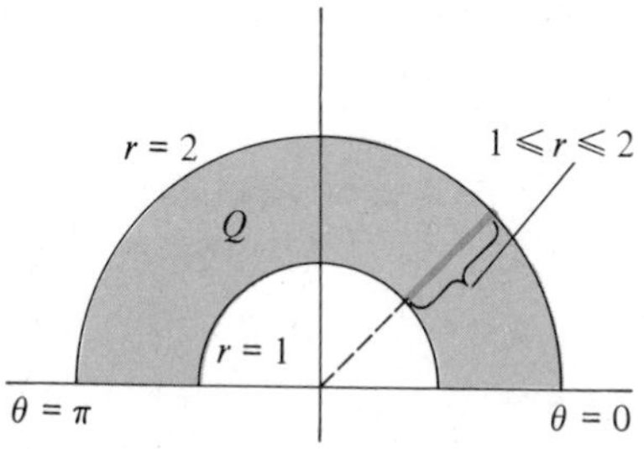

Figure 5.2 Annular region of Example 2.

Example 2 A machine part has the shape of a half annulus—the region between two concentric semicircles of radius $r = 1$ and $r = 2$ (see Figure 5.2). Find the mass of the part if the density at each point is proportional to the distance from that point to the common center of the two semicircles.

Solution: The density function can be written as

$$\rho(x, y) = \lambda\sqrt{x^2 + y^2}$$

where λ is constant. However, it is most convenient to describe the region Q in polar coordinates using the inequalities

$$0 \le \theta \le \pi, \qquad 1 \le r \le 2$$

in which case the density function is $\rho(r, \theta) = \lambda r$.

The mass is therefore

$$\begin{aligned} M = \iint_Q \lambda r\, dA &= \int_0^\pi \int_1^2 \lambda r \cdot r\, dr\, d\theta \\ &= \int_0^\pi \int_1^2 \lambda r^2\, dr\, d\theta = \int_0^\pi \left\{ \frac{\lambda}{3} r^3 \Big]_{r=1}^{r=2} \right\} d\theta \\ &= \int_0^\pi \frac{7\lambda}{3}\, d\theta = \frac{7\lambda\pi}{3}. \end{aligned}$$

■

Centers of Mass

In Section 8.4 we determined that the coordinates $(\overline{x}, \overline{y})$ of the *centroid* of the lamina with shape Q and constant density ρ are

$$\overline{x} = \frac{\int_a^b \rho x[f_t(x) - g_b(x)]\, dx}{\int_a^b \rho[f_t(x) - g_b(x)]\, dx} \tag{3}$$

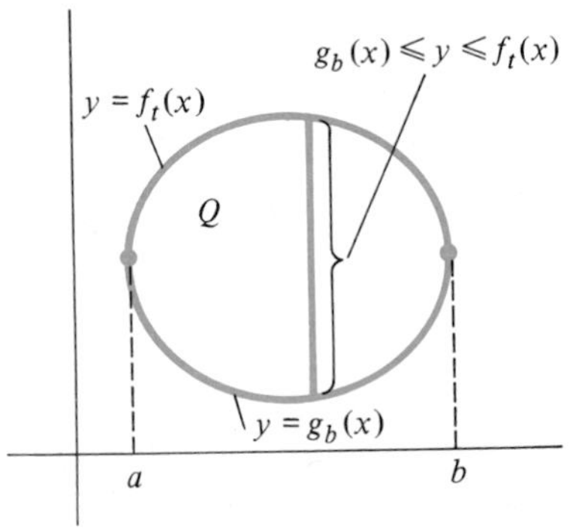

Figure 5.3

and

$$\overline{y} = \frac{\int_c^d \rho y[f_r(y) - g_\ell(y)]\, dy}{\int_c^d \rho[f_r(y) - g_\ell(y)]\, dy}. \tag{4}$$

In writing equations (3) and (4), we assumed that Q is a region that can be described both by the inequalities

$$a \le x \le b, \qquad g_b(x) \le y \le f_t(x) \qquad \text{(Figure 5.3)}$$

and also by the inequalities

$$g_\ell(y) \le x \le f_r(y), \qquad c \le y \le d \qquad \text{(Figure 5.4).}$$

In other words, Q is both x-simple and y-simple.

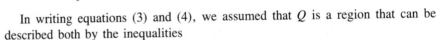

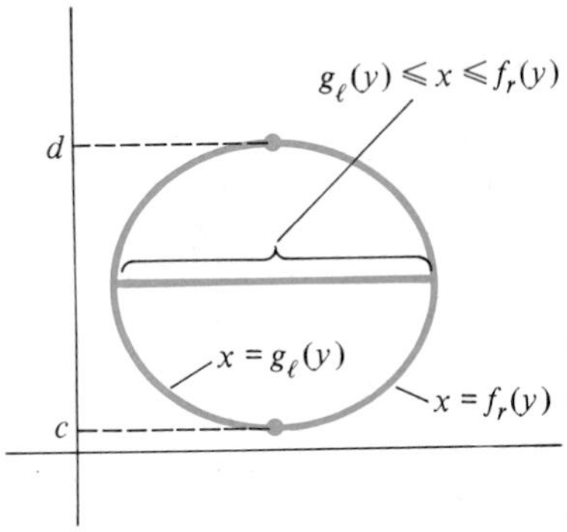

Figure 5.4

To generalize these formulas to the case of a variable density function $\rho(x, y)$, we begin with the integral in the numerator of $\overline{x}$ in (3), which we write as

$$\begin{aligned} \int_a^b \rho x[f_t(x) - g_b(x)]\, dx &= \int_a^b \left\{ \rho xy \Big]_{y=g_b(x)}^{y=f_t(x)} \right\} dx \\ &= \int_a^b \int_{g_b(x)}^{f_t(x)} \rho x\, dy\, dx. \end{aligned} \tag{5}$$

Similarly, the denominator of $\bar{x}$ can be written

$$\int_a^b \rho[f_t(x) - g_b(x)]\,dx = \int_a^b \left\{\rho y\Big]_{y=g_b(x)}^{y=f_t(x)}\right\} dx \tag{6}$$

$$= \int_a^b \int_{g_b(x)}^{f_t(x)} \rho\,dy\,dx.$$

Recall that the integral on the left side of (5) is called the **first moment** of Q about the y-axis. Its integrand may be interpreted as the mass of a thin vertical strip ($\rho[f_t(x) - g_b(x)]\,dx$) multiplied by the distance x of that strip from the y-axis. When summed (integrated) over all $x \in [a, b]$, this product provides a measure of the tendency of the region Q to rotate about the y-axis.

The right-hand side of (5) is interpreted in a similar way, except that now Q is thought of as having been partitioned into small rectangles, and the sum is taken over all such rectangles. However, the integrand may still be interpreted as the mass of a small rectangle ($\rho\,dx\,dy$) multiplied by x, the distance to the y-axis.

The right-hand side of (5) allows us to define the first moment of mass for Q about the y-axis when $\rho = \rho(x, y)$ is nonconstant. We simply replace ρ by $\rho(x, y)$ and define

$$M_y = \int_a^b \int_{g_b(x)}^{f_t(x)} x\rho(x, y)\,dy\,dx$$

or, more generally,

$$\boxed{M_y = \iint_Q x\rho(x, y)\,dA.} \tag{7}$$

Similarly, the first moment of mass for Q about the x-axis is defined to be

$$\boxed{M_x = \iint_Q y\rho(x, y)\,dA.} \tag{8}$$

To generalize the denominator of $\bar{x}$ in (3), we note that

$$\int_a^b \rho[f_t(x) - g_b(x)]\,dx = \int_a^b \left\{\rho y\Big]_{y=g_b(x)}^{y=f_t(x)}\right\} dx$$

$$= \int_a^b \int_{g_b(x)}^{f_t(x)} \rho\,dy\,dx$$

$$= \iint_Q \rho\,dy\,dx$$

which is just the mass (area times density) of Q. This same conclusion results for the denominator of $\bar{y}$ in (4). We are therefore ready to generalize the concept of centroid to that of *center of mass* (for variable density $\rho(x, y)$).

DEFINITION 3

Let Q be a region in the xy-plane that is either x-simple or y-simple, and let $\rho(x, y)$ be a continuous density function defined on Q. Then

(i) The **mass** of Q is the number

$$M = \iint_Q \rho(x, y)\, dA.$$

(ii) The **first moment of mass for Q about the y-axis** is the number

$$M_y = \iint_Q x\rho(x, y)\, dA.$$

(iii) The **first moment of mass for Q about the x-axis** is the number

$$M_x = \iint_Q y\rho(x, y)\, dA.$$

(iv) The **center of mass** of Q is the point $(\overline{x}, \overline{y})$, where

$$\overline{x} = \frac{M_y}{M}, \qquad \overline{y} = \frac{M_x}{M}.$$

Because of the way the concept of centroid is generalized by Definition 3, the physical interpretation of the center of mass is the same as before—the lamina associated with Q will "balance" at the point $(\overline{x}, \overline{y})$.

Example 3 A thin lamina has the shape of a triangle with vertices at (0, 0), (2, 0), and (2, 4). The density function associated with the lamina has equation $\rho(x, y) = 4x + 2y + 2$. Find the mass and center of mass of the lamina (see Figure 5.5).

Solution: The region Q associated with the lamina may be described by the inequalities

$$0 \le x \le 2, \qquad 0 \le y \le 2x.$$

Thus, by Definition 3,

$$\begin{aligned} M = \iint_Q (4x + 2y + 2)\, dA &= \int_0^2 \int_0^{2x} (4x + 2y + 2)\, dy\, dx \\ &= \int_0^2 \left\{ 4xy + y^2 + 2y \Big]_{y=0}^{y=2x} \right\} dx \\ &= \int_0^2 (12x^2 + 4x)\, dx \\ &= 4x^3 + 2x^2 \Big]_0^2 \\ &= 40, \end{aligned}$$

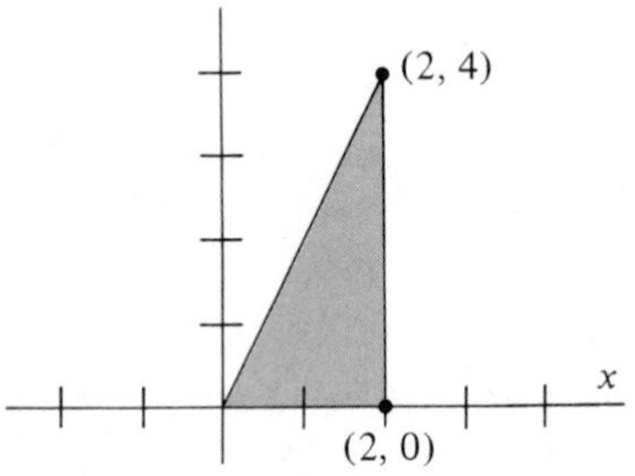

Figure 5.5

$$M_y = \iint_Q x(4x + 2y + 2)\,dA = \int_0^2 \int_0^{2x} (4x^2 + 2xy + 2x)\,dy\,dx$$

$$= \int_0^2 \left\{4x^2y + xy^2 + 2xy\Big]_{y=0}^{y=2x}\right\} dx$$

$$= \int_0^2 (12x^3 + 4x^2)\,dx = 3x^4 + \frac{4}{3}x^3\Big]_0^2$$

$$= \frac{176}{3},$$

and

$$M_x = \iint_Q y(4x + 2y + 2)\,dA = \int_0^2 \int_0^{2x} (4xy + 2y^2 + 2y)\,dy\,dx$$

$$= \int_0^2 \left\{2xy^2 + \frac{2}{3}y^3 + y^2\Big]_{y=0}^{y=2x}\right\} dx$$

$$= \int_0^2 \left(\frac{40}{3}x^3 + 4x^2\right) dx = \frac{10}{3}x^4 + \frac{4}{3}x^3\Big]_0^2$$

$$= \frac{192}{3}.$$

Thus,

$$\overline{x} = \frac{M_y}{M} = \frac{\left(\frac{176}{3}\right)}{40} = \frac{22}{15}$$

and

$$\overline{y} = \frac{M_x}{M} = \frac{\left(\frac{192}{3}\right)}{40} = \frac{8}{5}.$$

The center of mass is therefore $(\overline{x}, \overline{y}) = \left(\frac{22}{15}, \frac{8}{5}\right)$. ■

Example 4 Find the center of mass for the lamina described in Example 2.

Solution: The density function in Example 2 is

$$\rho(x, y) = \lambda\sqrt{x^2 + y^2}$$

and the region Q is described in polar coordinates as

$$0 \le \theta \le \pi, \qquad 1 \le r \le 2.$$

We evaluate the integral for M_y in polar coordinates as

$$M_y = \iint_Q \lambda x\rho(x, y)\,dA = \lambda \int_0^{\pi}\int_1^2 \underbrace{(r\cos\theta)}_{x = r\cos\theta}\underbrace{r}_{\rho = \sqrt{x^2+y^2} = r} \cdot \underbrace{r\,dr\,d\theta}_{dA}$$

$$= \lambda \int_0^{\pi}\int_1^2 r^3 \cos\theta\,dr\,d\theta$$

$$= \lambda \int_0^{\pi} \left\{\frac{1}{4}r^4 \cos\theta\right]_{r=1}^{r=2}\right\} d\theta$$

$$= \lambda \int_0^{\pi} \frac{15}{4}\cos\theta\,d\theta$$

$$= \lambda \left(\frac{15}{4}\sin\theta\right)\Bigg]_{\theta=0}^{\theta=\pi}$$

$$= 0.$$

(This should not surprise you since both Q and $\rho(x, y)$ are symmetric with respect to the y-axis.) The integral for M_x is

$$M_x = \iint_Q \lambda y\rho(x, y)\,dA = \lambda \int_0^{\pi}\int_1^2 \underbrace{(r\sin\theta)}_{y = r\sin\theta} \cdot \underbrace{r}_{\rho = \sqrt{x^2+y^2} = r} \cdot \underbrace{r\,dr\,d\theta}_{dA}$$

$$= \lambda \int_0^{\pi}\int_1^2 r^3 \sin\theta\,dr\,d\theta$$

$$= \lambda \int_0^{\pi} \left\{\frac{1}{4}r^4 \sin\theta\right]_{r=1}^{r=2}\right\} d\theta$$

$$= \lambda \int_0^{\pi} \frac{15}{4}\sin\theta\,d\theta$$

$$= \lambda \left(-\frac{15}{4}\cos\theta\right)\Bigg]_{\theta=0}^{\theta=\pi}$$

$$= \frac{15\lambda}{2}.$$

Since we found in Example 2 that $M = \dfrac{7\lambda\pi}{3}$, we have

$$\bar{x} = \frac{M_y}{M} = 0$$

and

$$\bar{y} = \frac{M_x}{M} = \frac{\left(\dfrac{15\lambda}{2}\right)}{\left(\dfrac{7\lambda\pi}{3}\right)} = \frac{45}{14\pi} \approx 1.02.$$

The center of mass is therefore $\left(0, \dfrac{45}{14\pi}\right)$. ■

REMARK: The center of mass is, in a sense, the "average" position of the mass of the lamina. The shape of the lamina and the form of the density function can sometimes combine in such a way that the center of mass actually lies *outside* the region Q! This is explored in Exercise 29.

Exercise Set 20.5

In Exercises 1–10, find the mass of a lamina with shape given by the region Q and density function $\rho(x, y)$.

1. $\rho(x, y) = x + y$,
$Q = \{(x, y) | 0 \le x \le 2,\ 0 \le y \le 1\}$

2. $\rho(x, y) = x^2 + y$,
$Q = \{(x, y) | 0 \le x \le 2,\ 0 \le y \le x\}$

3. $\rho(x, y) = 6 + x$,
$Q = \{(x, y) | -1 \le x \le 1,\ 0 \le y \le 1 - x^2\}$

4. $\rho(x, y) = xy$,
$Q = \{(x, y) | y^2 - 1 \le x \le 1 - y^2,\ -1 \le y \le 1\}$

5. $\rho(x, y) = \sin(x + y)$,
$Q = \{(x, y) | 0 \le x \le \pi/4,\ 0 \le y \le \pi/4\}$

6. $\rho(x, y) = xy$, $Q = \{(x, y) | 0 \le x \le 1,\ 0 \le y \le 1\}$

7. $\rho(x, y) = x^2 + y^2$,
$Q = \{(x, y) | -1 \le x \le 1,\ 0 \le y \le \sqrt{1 - x^2}\}$

8. $\rho(x, y) = \sqrt{x^2 + y^2}$,
$Q = \{(r, \theta) | 0 \le r \le 2,\ 0 \le \theta \le \pi/2\}$

9. $\rho(x, y) = \dfrac{1}{\sqrt{x^2 + y^2}}$, Q is the annulus $1 \le r \le 2$

10. $\rho(x, y) = xy$,
$\theta = \{(r, \theta) | 0 \le r \le 1,\ 0 \le \theta \le \pi/2\}$

11. Find the center of mass of the lamina in Exercise 1.

12. Find the center of mass of the lamina in Exercise 3.

13. Find the center of mass of the lamina in Exercise 8.

14. A lamina has the shape of a triangle with vertices (0, 0), (0, 4), and (1, 0). The density at each point (x, y) is $\rho(x, y) = y - x + 8$. Find the mass of the lamina.

15. Find the center of mass of the lamina in Exercise 14.

16. Find the centroid of the region bounded by the graph of $r = \sin 2\theta$ for $0 \le \theta \le \pi/2$. (Assume $\rho(x, y) = 1$ throughout the region.)

17. Find the centroid of the planar region bounded by the parabola $y = 4 - x^2$ and the x-axis.

18. Find the centroid of the region bounded by the graph of the function $f(x) = 1/x$ and the line $2x + 2y = 5$.

19. Find the centroid of the region obtained by connecting the points (0, 0), (4, 0), (4, 4), (2, 4), (2, 1), (0, 1), and (0, 0) in order by line segments.

20. Find the centroid of the region bounded by the graph of $y = x^2$ and $y = x^3$.

21. Show that the centroid of the region

$$R = \{(x, y) | 0 \le x \le a, \quad 0 \le y \le \sqrt{a^2 - x^2}\}$$

is $(\bar{x}, \bar{y}) = \left(\dfrac{4a}{3\pi}, \dfrac{4a}{3\pi}\right)$.

22. Find the centroid of the region bounded by the graph of $f(x) = \sinh x$ and the x-axis for $0 \le x \le 1$.

23. Find the centroid of the region bounded by the graph of $x = y(4 - y)$ and the y-axis.

24. A lamina has the shape of a circle of radius R. The density at any point P is proportional to the distance from the center. What is the mass?

25. A lamina has the shape of a right triangle with legs of length 2 and 6. The density at any point is proportional to the distance from the longer leg. What is the mass?

26. Find the distance from the vertex at the right angle to the center of mass for the triangle in Exercise 25.

27. Show that the denominator of $\bar{y}$ in (4) is the mass of the region over Q.

28. Assume that the region in Example 1 is oriented with the 10 cm leg of the triangle along the x-axis, again with the right angle at the origin. Show that the mass calculated in this way is the same as that found in Example 1.

29. Find the center of mass of a lamina with density function $\rho(r, \theta) = \lambda(1 + \cos^2 \theta)$ over the half annulus $Q = \{(r, \theta) | 1 \le r \le 2,\ 0 \le \theta \le \pi\}$. Locate the center of mass on a sketch of the region Q. State (in words) how the mass of the lamina is distributed, and what relation this has to the location of the center of mass.

20.6 SURFACE AREA

As our final application of double integrals, we consider the problem of calculating the area of a surface in space. We shall restrict our considerations to surfaces that are graphs of functions of two variables, although the ideas discussed here can be extended to more general types of surfaces.

Specifically, let Q be a region in the plane which is either x-simple or y-simple, and let the surface S be the graph of the continuous function $z = f(x, y)$ on Q. We wish to define the surface area of S in such a way that

(i) the surface area can actually be calculated, and
(ii) the surface area agrees with the usual notion of area when S is flat.

We begin in the usual way—we partition the region Q with a rectangular grid, and we denote $R_1, R_2, \ldots, R_n$ the rectangles lying entirely within Q. As before, the grid is constructed so that each of the rectangles R_m has dimensions Δx and Δy (see Figure 6.1). The grid on Q partitions the surface S into patches $S_1, S_2, \ldots, S_n$. Specifically, by the patch S_m we mean the portion of the surface S lying above the rectangle R_m.

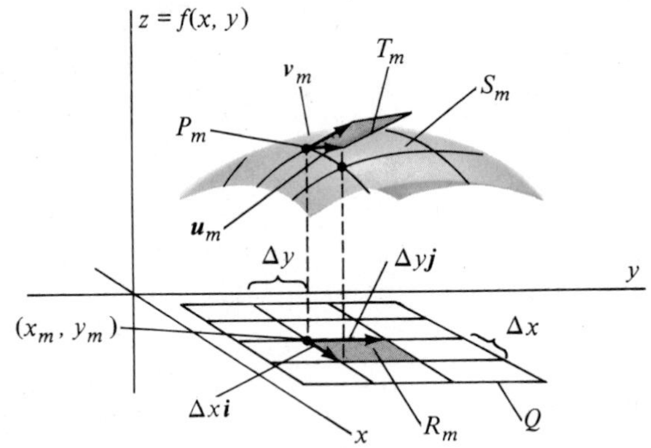

Figure 6.1 Surface is approximated by parallelograms pasted on above corners of approximating rectangles.

As Figure 6.1 suggests, the patch S_m is nearly a parallelogram, but not quite, since S_m is part of the (generally) curved surface S. We therefore construct a parallelogram T_m which approximates the patch S_m in the following way. Let P_m be the point on the corner of S_m nearest the z-axis and let (x_m, y_m) be the vertex of the rectangle R_m directly beneath the point P_m. Then the vector

$$\boldsymbol{u}_m = \Delta x\, \boldsymbol{i} + \frac{\partial f}{\partial x}(x_m, y_m)\,\Delta x\, \boldsymbol{k}, \tag{1}$$

when originating at point P_m, terminates at some point above the vertex $(x_m + \Delta x, y_m)$ of R_m. Moreover, $\boldsymbol{u}_m$ is tangent to S at P_m, by definition of the partial derivative $\frac{\partial f}{\partial x}(x_m, y_m)$. Similarly, the vector

$$\boldsymbol{v}_m = \Delta y\, \boldsymbol{j} + \frac{\partial f}{\partial y}(x_m, y_m)\,\Delta y\, \boldsymbol{k}, \tag{2}$$

when originating at P_m, terminates above the vertex $(x_m, y_m + \Delta y)$ of R_m and is tangent to S at P_m (see Figure 6.2).

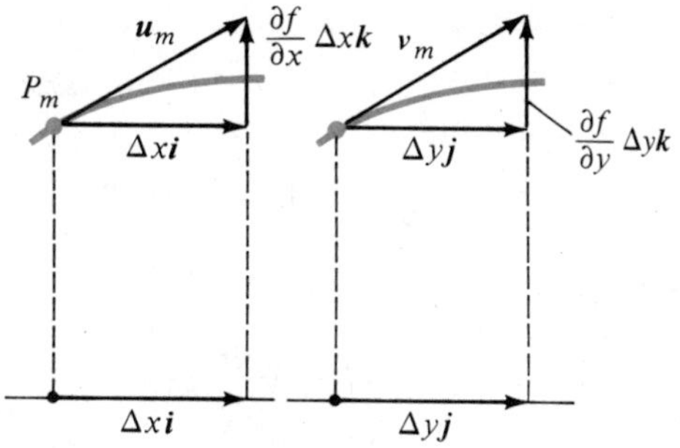

Figure 6.2 $u_m = \Delta x i + \dfrac{\partial f}{\partial x}\Delta x k$

$v_m = \Delta y j + \dfrac{\partial f}{\partial y}\Delta y k.$

From the properties noted for the vectors $\boldsymbol{u}_m$ and $\boldsymbol{v}_m$, we conclude that $\boldsymbol{u}_m$ and $\boldsymbol{v}_m$ determine a parallelogram T_m that

(i) lies directly above the rectangle R_m (and, hence, the patch S_m), and
(ii) lies tangent to the surface S at point P_m.

The idea is therefore to use the area of the parallelogram T_m as an approximation to the area of the patch S_m. From Section 6, Chapter 17, recall that the area of the parallelogram T_m is equal to the length of the cross product of the vectors $\boldsymbol{u}_m$ and $\boldsymbol{v}_m$, or

$$\begin{aligned}
\text{Area of } T_m &= |\boldsymbol{u}_m \times \boldsymbol{v}_m| \\
&= \left|\left(\Delta x\, \boldsymbol{i} + \frac{\partial f}{\partial x}(x_m, y_m)\Delta x\, \boldsymbol{k}\right) \times \left(\Delta y\, \boldsymbol{j} + \frac{\partial f}{\partial y}(x_m, y_m)\Delta y\, \boldsymbol{k}\right)\right| \\
&= (\Delta x \Delta y)\left|\det\begin{bmatrix} \boldsymbol{i} & \boldsymbol{j} & \boldsymbol{k} \\ 1 & 0 & \dfrac{\partial f}{\partial x}(x_m, y_m) \\ 0 & 1 & \dfrac{\partial f}{\partial y}(x_m, y_m)\end{bmatrix}\right| \\
&= (\Delta x \Delta y)\left| -\frac{\partial f}{\partial x}(x_m, y_m)\boldsymbol{i} - \frac{\partial f}{\partial y}(x_m, y_m)\boldsymbol{j} + \boldsymbol{k}\right| \\
&= \sqrt{\left[\frac{\partial f}{\partial x}(x_m, y_m)\right]^2 + \left[\frac{\partial f}{\partial y}(x_m, y_m)\right]^2 + 1}\; \Delta x \Delta y.
\end{aligned}$$

Summing these approximations over all patches $S_1, S_2, \ldots, S_n$ gives the approximation to the area of S as

$$\text{Area of } S \approx \sum_{m=1}^{n} \sqrt{\left[\frac{\partial f}{\partial x}(x_m, y_m)\right]^2 + \left[\frac{\partial f}{\partial y}(x_m, y_m)\right]^2 + 1}\; \Delta x\, \Delta y.$$

If the partial derivatives $\frac{\partial f}{\partial x}$ and $\frac{\partial f}{\partial y}$ are continuous on Q, this Riemann sum converges to an integral as $n \to \infty$ and as $\Delta x \to 0$ and $\Delta y \to 0$. This integral provides our definition of surface area.

DEFINITION 4

Let Q be a region in the plane that is either x-simple or y-simple, and let S be the graph of the function $z = f(x, y)$ for $(x, y) \in Q$. If $\frac{\partial f}{\partial x}(x, y)$ and $\frac{\partial f}{\partial y}(x, y)$ are continuous on Q, the **area of the surface S** is defined to be

$$A_s = \iint_Q \sqrt{\left[\frac{\partial f}{\partial x}(x, y)\right]^2 + \left[\frac{\partial f}{\partial y}(x, y)\right]^2 + 1}\, dA. \tag{3}$$

REMARK 1: Notice the similarity between the integral in (3) and the integral

$$L = \int_a^b \sqrt{1 + [f'(x)]^2}\, dx$$

for arc length (Section 7.5). Just as the differential

$$dL = \sqrt{1 + [f'(x)]^2}\, dx$$

is called the differential for arc length, we refer to the differential

$$dA_s = \sqrt{\left[\frac{\partial f}{\partial x}(x, y)\right]^2 + \left[\frac{\partial f}{\partial y}(x, y)\right]^2 + 1}\, dA$$

as the **differential for surface area.**

REMARK 2: Note that the integral in (3) can be written more compactly as

$$A_S = \iint_Q \sqrt{f_x^2 + f_y^2 + 1}\, dA. \tag{4}$$

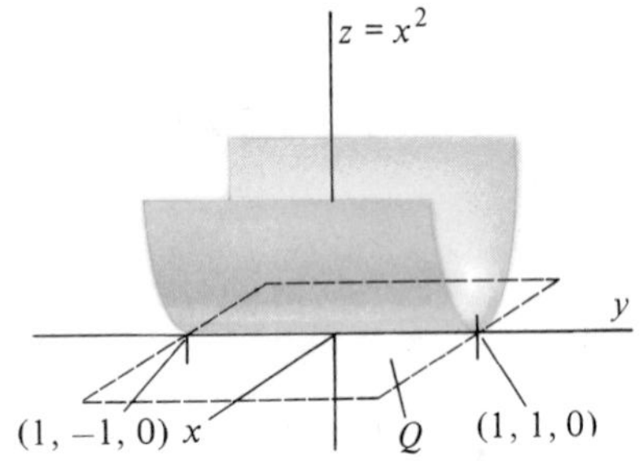

Figure 6.3 Cylinder $f(x,y) = x^2$ over the square Q: $-1 \le x \le 1$, $-1 \le y \le 1$.

Example 1 Find the area of the surface that is the graph of the equation $f(x, y) = x^2$ lying above the rectangle $Q = \{(x, y) \mid -1 \le x \le 1, -1 \le y \le 1\}$.

Solution: The surface is the "cylinder" sketched in Figure 6.3. Using the method of integration by trigonometric substitutions, we may evaluate the surface area integral (3) as

$$\begin{aligned}
A_s = \iint_Q \sqrt{(2x)^2 + 1}\, dA &= \int_{-1}^{1}\int_{-1}^{1} \sqrt{4x^2 + 1}\, dx\, dy \\
&= \int_{-1}^{1} \left\{ \frac{1}{2}x\sqrt{4x^2 + 1} + \frac{1}{4}\ln\left|2x + \sqrt{4x^2 + 1}\right| \Bigg]_{x=-1}^{x=1} \right\} dy \\
&= \int_{-1}^{1} \left\{ \sqrt{5} + \frac{1}{4}[\ln(2 + \sqrt{5}) - \ln(\sqrt{5} - 2)] \right\} dy \\
&= 2\sqrt{5} + \frac{1}{2}[\ln(2 + \sqrt{5}) - \ln(\sqrt{5} - 2)] \\
&\approx 5.92.
\end{aligned}$$

■

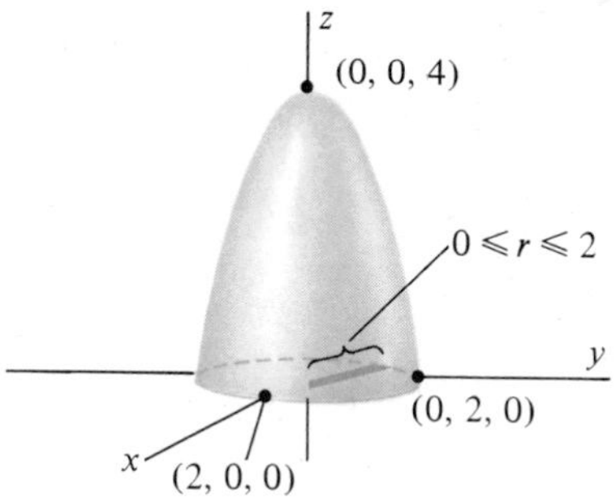

Figure 6.4 Paraboloid $z = 4 - x^2 - y^2$.

Example 2 Find the surface area of the portion of the paraboloid

$$z = 4 - x^2 - y^2$$

lying above the xy-plane.

Solution: The paraboloid intersects the xy-plane in the circle $4 - x^2 - y^2 = 0$, or $x^2 + y^2 = 4$. In xy-coordinates, this region may be described by the inequalities

$$-2 \le x \le 2, \qquad -\sqrt{4 - x^2} \le y \le \sqrt{4 - x^2}.$$

With $f(x, y) = 4 - x^2 - y^2$, we have

$$\frac{\partial f}{\partial x}(x, y) = -2x, \qquad \text{and} \qquad \frac{\partial f}{\partial y}(x, y) = -2y.$$

By (3) the integral giving the surface area is

$$A_s = \iint_Q \sqrt{4x^2 + 4y^2 + 1}\, dA = \int_{-2}^{2} \int_{-\sqrt{4-x^2}}^{\sqrt{4-x^2}} \sqrt{4x^2 + 4y^2 + 1}\, dy\, dx.$$

This integral is most easily evaluated in polar coordinates. With $x = r \cos \theta$ and $y = r \sin \theta$, and the region Q described as

$$0 \le \theta \le 2\pi, \qquad 0 \le r \le 2,$$

we have

$$\begin{aligned} A_s &= \int_0^{2\pi} \int_0^2 \sqrt{4(r \cos \theta)^2 + 4(r \sin \theta)^2 + 1}\, r\, dr\, d\theta \\ &= \int_0^{2\pi} \int_0^2 \sqrt{4r^2 + 1}\, r\, dr\, d\theta \\ &= \int_0^{2\pi} \frac{1}{12}(4r^2 + 1)^{3/2}\Big]_{r=0}^{r=2} d\theta \\ &= \frac{\pi}{6}(17^{3/2} - 1) \approx 11.5\,\pi. \end{aligned}$$

■

Example 3 Show that Definition 4 of surface area agrees with the usual definition of area when S is a flat surface over a rectangle in the plane.

Solution: If S is a flat surface in space, then S is a portion of a plane. Thus, we let

$$z = f(x, y) = Ax + By + C$$

be the equation for the plane, and we let

$$Q = \{(x, y) | \quad a \le x \le b, \quad c \le x \le d\}$$

be a rectangle in the plane.

As in Figure 6.1, the portion of the plane $Ax + By + C$ lying over the rectangle Q is a parallelogram. From (1) and (2) we see that the vectors

$$\boldsymbol{u} = \Delta x\, \boldsymbol{i} + \frac{\partial f}{\partial x}\Delta x\, \boldsymbol{k} = (b - a)\boldsymbol{i} + A(b - a)\boldsymbol{k}$$

and

$$v = \Delta y\, j + \frac{\partial f}{\partial y}\Delta y\, k = (d - c)j + B(d - c)k,$$

when positioned at the vertex $(a, b, f(a, b))$, form two adjacent sides of the parallelogram. By a standard calculation involving the crossproduct, the area of the parallelogram is

$$\begin{aligned} A_s &= |u \times v| \\ &= \left|\det \begin{bmatrix} i & j & k \\ b - a & 0 & A(b - a) \\ 0 & d - c & B(d - c) \end{bmatrix}\right| \\ &= |-A(b - a)(d - c)i - B(b - a)(d - c)j + (b - a)(d - c)k| \\ &= (b - a)(d - c)\sqrt{A^2 + B^2 + 1}. \end{aligned}$$

Using Definition (3) with $\frac{\partial f}{\partial x} = A$ and $\frac{\partial f}{\partial y} = B$ we find

$$A_s = \int_c^d \int_a^b \sqrt{A^2 + B^2 + 1}\, dx\, dy = (b - a)(d - c)\sqrt{A^2 + B^2 + 1}.$$

Thus, Definition 3 agrees with our usual concept of area for flat surfaces. ■

Recall that the calculation of arc length and surface area for volumes of revolution directly from the definition is generally difficult. The same is true for calculating surface area from Definition 4.

The difficulty has to do with finding an antiderivative for the integrand in the double integral in (3). The Exercise Set contains several examples of surfaces for which this calculation can be carried out without difficulty. Moreover, the general notion of calculating surface area by double integration will be put to important use in the next chapter.

Exercise Set 20.6

In Exercises 1–10, find the surface area of the graph of $z = f(x, y)$ above the region Q in the plane.

1. $f(x, y) = x + y + 6, \quad Q = \{(x, y)|0 \le x \le 1, 0 \le y \le 1\}$
2. $f(x, y) = 9 - x + 2y, \quad Q = \{(x, y)|0 \le x^2 + y^2 \le 1\}$
3. $f(x, y) = 9 - x^2 - y^2, \quad Q = \{(x, y)|0 \le x^2 + y^2 \le 3\}$
4. $f(x, y) = 4 + y^2, \quad Q = \{(x, y)|0 \le x \le 1, 0 \le y \le 2\}$
5. $f(x, y) = 2 - x - y, \quad Q = \{(x, y)|0 \le x \le 2, 0 \le y \le 2 - x\}$
6. $f(x, y) = 3 + y^2, \quad Q = \{(x, y)|0 \le x \le 2, 0 \le y \le 2\}$
7. $f(x, y) = \sqrt{x^2 + y^2}, \quad Q = \{(x, y)|1 \le x^2 + y^2 \le 4\}$
8. $f(x, y) = x + y^2, \quad Q = \{(x, y)|0 \le x \le 1, 0 \le y \le 2\}$
9. $f(x, y) = \sqrt{3}\, y - x^2, \quad Q = \{(x, y)|0 \le x \le 1, 0 \le y \le 1\}$
10. $f(x, y) = x^2 + y, \quad Q = \{(x, y)|0 \le x \le 1, 0 \le y \le x\}$
11. Find the surface area of the portion of the graph of $z = y + 2x^2$ over the triangular region with vertices (0, 0), (0, 1), and (1, 1).
12. Find the area of the part of the plane $x + y + z = 4$ bounded by the cylinder $x^2 + y^2 = 4$.
13. Find the surface area of the portion of the paraboloid $z = 16 - x^2 - y^2$ lying between the planes $z = 4$ and $z = 9$.
14. Find the surface area of the part of the sphere $x^2 + y^2 + z^2 = 4$ lying above the plane $z = 1$.

15. Find the surface area of the part of the hemisphere $z = \sqrt{4 - x^2 - y^2}$ lying inside the cylinder $x^2 + y^2 = 1$.

16. Find the surface area of the part of the paraboloid $z = 4 - x^2 - y^2$ lying outside the cylinder $x^2 + y^2 = 1$.

17. Use Simpson's Rule to approximate the surface area of the portion of the paraboloid $z = 3 - x^2 - y^2 + 2y$ lying above the plane $z = 2y + 2$.

18. Develop the formula for the surface area of a sphere using a double integral.

19. What relationship exists between the formula for the surface area of a solid of revolution and Definition 4?

20. (A coordinate-free formula for surface area) Let $\boldsymbol{u}_m$ and $\boldsymbol{v}_m$ be the vectors in (1) and (2). Let ΔT_m be the area of the parallelogram tangent to S at P_m, over the rectangle R_m of area ΔA, as before.

a. Show that $\boldsymbol{N}_m = \boldsymbol{u}_m \times \boldsymbol{v}_m$ is normal to S at P_m.

b. Show that $\boldsymbol{N} \cdot \boldsymbol{k} = \Delta x \Delta y = \Delta A$

c. Show that, also, $\boldsymbol{N} \cdot \boldsymbol{k} = |\boldsymbol{u}_m \times \boldsymbol{v}_m| \cos \theta$, where θ is the angle between $\boldsymbol{N}$ and $\boldsymbol{k}$.

d. Conclude from (b) and (c) that $\Delta T_m = |\boldsymbol{u}_m \times \boldsymbol{v}_m| = \dfrac{\Delta A}{\cos \theta} = \sec \theta \Delta A$.

e. Conclude from (d) that $A_s = \displaystyle\iint_Q \sec \theta \, dA$.

20.7 TRIPLE INTEGRALS

In this section, we define the triple integral for a continuous function $f(x, y, z)$ of three independent variables. Since all of $\mathbb{R}^3$ is required just to indicate a region Q in the domain of $f(x, y, z)$, we will not be able to identify the triple integral with a spatial entity, except in very special cases. However, by analogy with the way in which we defined the double integral, we may proceed to

(i) say how to partition the region Q into approximating rectangular blocks,
(ii) form an approximating (Riemann) sum for $f(x, y, z)$ over the region Q, and
(iii) obtain the limit of the approximating sum as the dimensions of the approximating blocks tend to zero and the number of blocks tends to infinity.

The result of these three steps is what we shall define to be the triple integral of $f(x, y, z)$ over the region Q, written $\displaystyle\iiint_Q f(x, y, z)\, dV$. As we did for double integrals, we shall first carry out this development for special types of regions Q (namely, boxes) and then indicate how the concept extends to more general regions. Also we shall continue to work exclusively with *regular* grids, meaning here that all rectangular blocks in a particular approximation will be of the same size.

The Triple Integral Over a Box

Let Q be the box-shaped region in $\mathbb{R}^3$ defined by the inequalities

$$a \le x \le b, \qquad c \le y \le d, \qquad p \le z \le q.$$

(See Figure 7.1.) Let $f(x, y, z)$ be a continuous function defined on Q, and let n be a positive integer. Taking $\Delta x = \dfrac{b - a}{n}$, we partition the x-interval $[a, b]$ into n subintervals of length Δx and with endpoints

$$x_0 = a, \qquad x_1 = a + \Delta x, \qquad x_2 = a + 2\Delta x, \ . \ . \ . \ , x_n = a + n\Delta x = b.$$

Similarly, with $\Delta y = \dfrac{d - c}{n}$ we partition the y-interval $[c, d]$ into n subintervals with endpoints

$$y_0 = c, \ y_1 = c + \Delta y, \qquad y_2 = c + 2\Delta y, \ . \ . \ . \ , y_n = c + n\Delta y = d.$$

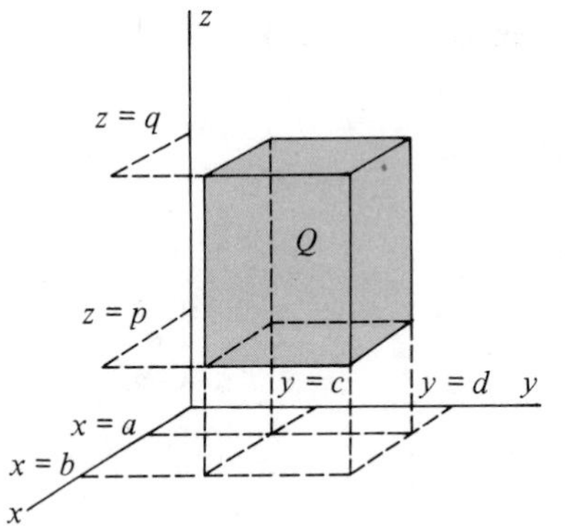

Figure 7.1 The rectangular box Q: $a \le x \le b$, $c \le y \le d$, $p \le z \le q$.

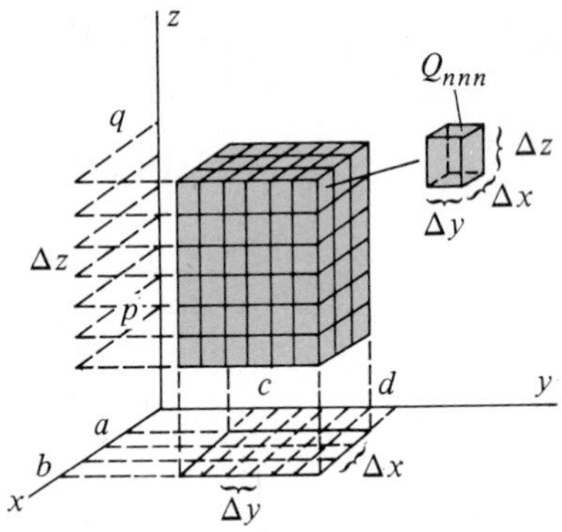

Figure 7.2 Partitioning the box into smaller boxes Q_{ijk} of volume $\Delta V = \Delta x \Delta y \Delta z$.

Also, using $\Delta z = \dfrac{q - p}{n}$, we partition $[p, q]$ into n subintervals with endpoints

$$z_0 = p, \qquad z_1 = p + \Delta z, \qquad z_2 = p + 2\Delta z, \ . \ . \ . \ , z_n = p + n\Delta z = q.$$

By constructing planes perpendicular to the x-axis at $x_0, x_1, \ldots, x_n$, planes perpendicular to the y-axis at $y_0, y_1, \ldots, y_n$, and planes perpendicular to the z-axis at $z_0, z_1, \ldots, z_n$, we partition the box Q into smaller rectangular boxes Q_{ijk}, each of which has volume $\Delta V = \Delta x \Delta y \Delta z$ (Figure 7.2).

Next, we select one point (s_i, t_j, u_k) in each box Q_{ijk}, and we form the approximating sum

$$S_n = \sum_{i=1}^{n} \sum_{j=1}^{n} \sum_{k=1}^{n} f(s_i, t_j, u_k)\Delta V. \tag{1}$$

By analogy with the one- and two-variable cases, this approximating sum is called a Riemann sum for $f(x, y, z)$ on Q. As in the one- and two-variable cases, if $f(x, y, z)$ is continuous on Q, this sum approaches a limit as $n \to \infty$ (and, hence, as $\Delta x \to 0$, $\Delta y \to 0$, and $\Delta z \to 0$). This limit is defined to be the **triple integral** of $f(x, y, z)$ on the box Q:

$$\iiint_Q f(x, y, z)\, dV = \lim_{n\to\infty} \sum_{i=1}^{n} \sum_{j=1}^{n} \sum_{k=1}^{n} f(s_i, t_j, u_k)\Delta V. \tag{2}$$

In the special case $f(x, y, z) \equiv 1$, we can give a geometric interpretation of the triple integral in (1). The terms in the approximating Riemann sum (1) are just

$$f(s_i, t_j, u_k)\Delta V = 1 \cdot \Delta V = \text{volume of } Q_{ijk}.$$

Since the union of the boxes Q_{ijk} is the box Q, it follows that

$$\iiint_Q 1\, dV = \text{Volume of } Q. \tag{3}$$

That is, the triple integral of the function $f(x, y, z) \equiv 1$ over the box Q is just the volume of Q.

As for double integrals, triple integrals over boxes can be evaluated as iterated integrals. In particular, with Q as above, we have

$$\iiint_Q f(x, y, z)\, dV = \int_p^q \int_c^d \int_a^b f(x, y, z)\, dx\, dy\, dz. \tag{4}$$

There is no reason why the first integration in the iterated integral in (4) must be with respect to x. Since each of the variables ranges between constant limits, the order of integration can be any of the six possible orders xyz, xzy, yxz, yzx, zxy, or zyx.

We shall not prove equation (4), although it is easy to explain in the case $f(x, y, z) \equiv 1$. If v is any number in the z-interval $[p, q]$, the plane $z = v$ determines a rectangular cross section of Q whose area is

$$A(z) = \int_c^d \int_a^b 1 \cdot dx\, dy.$$

Thus,

$$\begin{aligned} \text{Volume of } Q &= \int_p^q A(z)\, dz \\ &= \int_p^q \left\{ \int_c^d \int_a^b 1 \cdot dx\, dy \right\} dz \\ &= \int_p^q \int_c^d \int_a^b 1 \cdot dx\, dy\, dz. \end{aligned} \tag{5}$$

Combining equations (3) and (5) results in equation (4) in this special case.

Example 1 Evaluate the triple integral $\iiint_Q xe^y \cos z\, dV$ where Q is the box $\{(x, y, z) | 0 \le x \le 2,\ 0 \le y \le \ln 2,\ 0 \le z \le \pi/2\}$.

Solution: Using equation (4), we find

$$\begin{aligned} \iiint_Q xe^y \cos z\, dV &= \int_0^{\pi/2} \int_0^{\ln 2} \int_0^2 xe^y \cos z\, dx\, dy\, dz \\ &= \int_0^{\pi/2} \int_0^{\ln 2} \left\{ \frac{x^2}{2} e^y \cos z \Big]_{x=0}^{x=2} \right\} dy\, dz \\ &= \int_0^{\pi/2} \int_0^{\ln 2} 2e^y \cos z\, dy\, dz \\ &= \int_0^{\pi/2} \left\{ 2e^y \cos z \Big]_{y=0}^{y=\ln 2} \right\} dz \\ &= \int_0^{\pi/2} 2 \cos z\, dz \\ &= 2 \sin z \Big]_{z=0}^{z=\pi/2} \\ &= 2. \end{aligned}$$

■

Triple Integrals Over More General Regions

If Q is a bounded region in space (not necessarily a box), we define the triple integral of $f(x, y, z)$ over Q as follows. First, we find a box B containing the region Q. Next, we partition the box B into n^3 smaller rectangular boxes, just as we did when Q itself was a box, and we let $Q_1, Q_2, \ldots, Q_n$ be a complete list of all such rectangular boxes *lying entirely within the region* Q. For each such box we select a point $(s_j, t_j, u_j) \in Q_j$. (As before, the dimensions of the small rectangular boxes Q_j are Δx, Δy, and Δz—see Figure 7.3.) Finally, we form the approximating sum

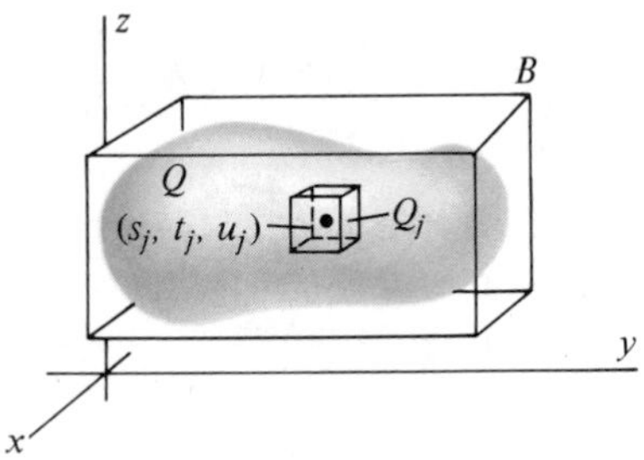

Figure 7.3 More general region Q enclosed by the box B.

$$S_n = \sum_{j=1}^{m} f(s_j, t_j, u_j)\Delta x \Delta y \Delta z. \tag{6}$$

The limit of this sequence of approximating sums, as $n \to \infty$ and Δx, Δy, and Δz approach zero, when it exists, is called the **triple integral** of $f(x, y, z)$ over Q:

$$\iiint_Q f(x, y, z)\, dA = \lim_{n\to\infty} \sum_{j=1}^{m} f(s_j, t_j, u_j)\Delta x \Delta y \Delta z. \tag{7}$$

The triple integral in (7) will exist whenever $f(x, y, z)$ is continuous on Q and Q is a "sufficiently nice" region in space. Rather than worry too much about the precise meaning of this last phrase, we shall state a theorem showing how triple integrals may be evaluated for regions of the type encountered in this text and in most applications. Before doing so, however, we need to make one observation concerning the approximating sum S_n in equation (6). In the special case $f(x, y, z) \equiv 1$, the terms in the approximating sum S_n are, as before, the volumes of the approximating boxes. The sum S_n therefore approximates the volume of Q, and in the limit we obtain

$$\iiint_Q 1 \cdot dV = \text{Volume of } Q \tag{8}$$

just as in the case when Q itself is a box.

Evaluating Triple Integrals

We say that a region Q in $\mathbb{R}^3$ is ***z*-simple** if every vertical line (that is, a line parallel to the z-axis) intersects the boundary of the region at most twice. The notions of ***x*-simple** and ***y*-simple** regions in $\mathbb{R}^3$ are defined accordingly. Figure 7.4 illustrates the fact that a torus is an example of a region that is z-simple, but neither x-simple nor y-simple. We shall describe a certain type of z-simple region Q for which the triple integral $\iiint_Q f(x, y, z)\, dV$ can be evaluated as an iterated integral. The same result applies to regions that are x-simple or y-simple by interchanging the roles of x and z or y and z.

Specifically, let Q be a z-simple region in $\mathbb{R}^3$ that can be described by inequalities of the form

$$g_1(x, y) \le z \le g_2(x, y), \qquad h_1(x) \le y \le h_2(x), \qquad a \le x \le b. \tag{9}$$

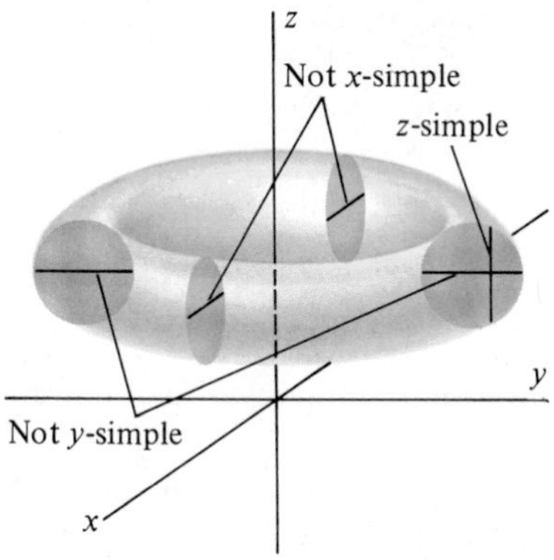

Figure 7.4 The torus is z-simple, but neither x-simple nor y-simple.

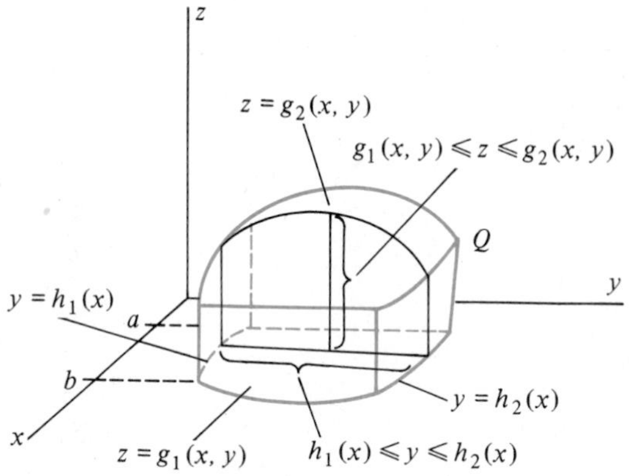

Figure 7.5 Region defined by inequalities (9).

In the inequalities in (9), we assume that g_1, g_2, h_1, and h_2 are continuous functions of their arguments. Figure 7.5 shows such a region. The inequality $g_1(x, y) \leq z \leq g_2(x, y)$ says that Q is bounded above by the graph of $z = g_2(x, y)$ and below by the graph of $z = g_1(x, y)$, and guarantees that the region Q is z-simple. The inequality $h_1(x) \leq y \leq h_2(x)$ says that Q is bounded on the right by the cylinder $y = h_2(x)$ and on the left by the cylinder $y = h_1(x)$. Finally, the inequality $a \leq x \leq b$ says that the region Q is bounded by the planes $x = a$ and $x = b$. The following theorem shows how triple integrals over such regions may be evaluated.

THEOREM 4

Let Q be a region in $\mathbb{R}^3$ described by the inequalities

$$a \leq x \leq b, \qquad h_1(x) \leq y \leq h_2(x), \qquad g_1(x, y) \leq z \leq g_2(x, y)$$

where h_1, h_2, g_1, and g_2 are continuous functions. Let $f(x, y, z)$ be continuous on Q. Then

$$\iiint_Q f(x, y, z)\, dV = \int_a^b \int_{h_1(x)}^{h_2(x)} \int_{g_1(x, y)}^{g_2(x, y)} f(x, y, z)\, dz\, dy\, dx. \qquad (10)$$

A proof of Theorem 4 requires a deeper treatment of multiple integrals than we have given here and is left to more advanced courses. Before proceeding to apply this result, we need to emphasize two points.

REMARK 1: Theorem 4 remains true with the roles of the independent variables interchanged. For example, if the region Q is described by the inequalities

$$c \leq y \leq d, \qquad h_1(y) \leq z \leq h_2(y), \qquad g_1(y, z) \leq x \leq g_2(y, z),$$

then the iterated integral formula is

$$\iiint_Q f(x, y, z)\, dV = \int_c^d \int_{h_1(y)}^{h_2(y)} \int_{g_1(y, z)}^{g_2(y, z)} f(x, y, z)\, dx\, dz\, dy.$$

REMARK 2: Theorem 4 may be paraphrased this way. To evaluate the triple integral

$$\iiint_Q f(x, y, z)\, dV:$$

(i) Find the constants and/or functions that bound the region Q in each of the three directions corresponding to the coordinate axes. (Write these down!)

(ii) Evaluate $\iiint_Q f(x, y, z)\, dV$ as an interated integral, integrating first with respect to a variable whose bounds depend on the other one or two variables, integrating second with respect to a variable whose limits involve the remaining variable, and integrating last with respect to a variable whose limits involve only constants.

It is important to note that we cannot evaluate an antiderivative over limits involving a variable for which an integration has already been performed. For example, the expressions

$$\int_a^b \int_{g_1(x,y)}^{g_2(x,y)} \int_{h_1(y)}^{h_2(y)} f(x, y, z)\, dz\, dy\, dx$$

and

$$\int_{h_1(x)}^{h_2(x)} \int_{g_1(x,z)}^{g_2(x,z)} \int_a^b f(x, y, z)\, dx\, dy\, dz$$

are nonsense.

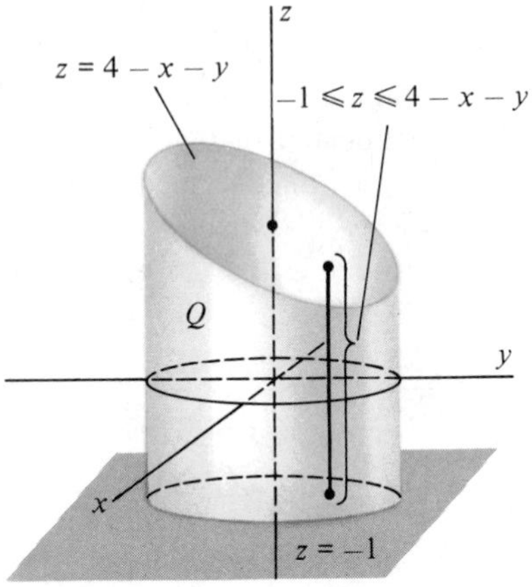

Figure 7.6 Region of Example 2.

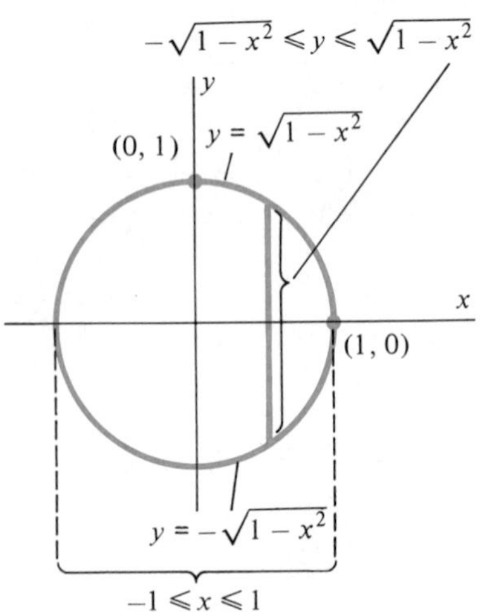

Figure 7.7 Projection of Q into the xy-plane.

Example 2 Evaluate the triple integral

$$\iiint_Q 2xy\, dV,$$

where Q is the region inside the cylinder $x^2 + y^2 = 1$ bounded by the planes $x + y + z = 4$ and $z = -1$.

Solution: The region is sketched in Figure 7.6. Figure 7.7 shows the projection of Q into the xy-plane and illustrates how the inequalities involving x and y are obtained. Solving the equation $x + y + z = 4$ gives $z = 4 - x - y$, so $-1 \le z \le 4 - x - y$. The region is therefore completely described by the inequalities

$$-1 \le x \le 1, \qquad -\sqrt{1 - x^2} \le y \le \sqrt{1 - x^2}, \qquad -1 \le z \le 4 - x - y.$$

Thus, according to Theorem 4,

$$\iiint_Q xyz\, dV = \int_{-1}^{1} \int_{-\sqrt{1-x^2}}^{\sqrt{1-x^2}} \int_{-1}^{4-x-y} 2xy\, dz\, dy\, dx$$

$$= \int_{-1}^{1} \int_{-\sqrt{4-x^2}}^{\sqrt{4-x^2}} \left\{ 2xyz \Big]_{z=-1}^{z=4-x-y} \right\} dy\, dx$$

$$= \int_{-1}^{1} \int_{-\sqrt{4-x^2}}^{\sqrt{4-x^2}} (10xy - 2x^2y - 2xy^2)\, dy\, dx$$

$$= \int_{-1}^{1} \left\{ 5xy^2 - x^2y^2 - \frac{2}{3}xy^3 \right]_{y=-\sqrt{4-x^2}}^{y=\sqrt{4-x^2}} \Bigg\} \, dx$$

$$= \int_{-1}^{1} -\frac{4}{3}x(4 - x^2)^{3/2} \, dx$$

$$= \frac{4}{15}(4 - x^2)^{5/2} \Bigg]_{x=-1}^{x=1}$$

$$= 0.$$

(This result is explained by the fact that both the integrand and the region Q are symmetric with respect to the plane $y = x$.) ■

REMARK: The integral in Example 2 could also have been evaluated as

$$\iiint_Q 2xy \, dV = \int_{-1}^{1} \int_{-\sqrt{1-y^2}}^{\sqrt{1-y^2}} \int_{-1}^{4-x-y} 2xy \, dz \, dx \, dy.$$

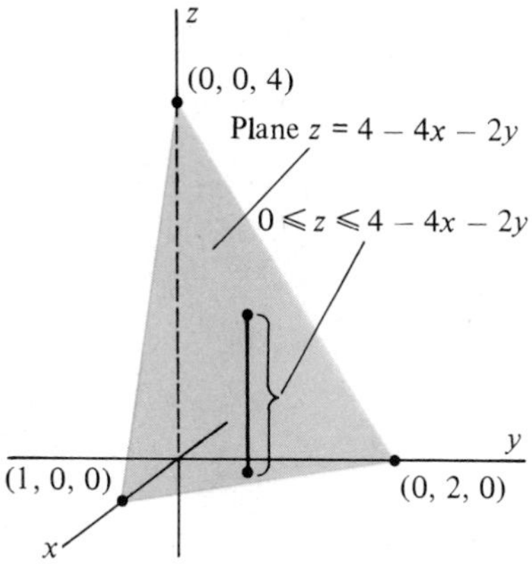

Figure 7.8 Tetrahedron of Example 3.

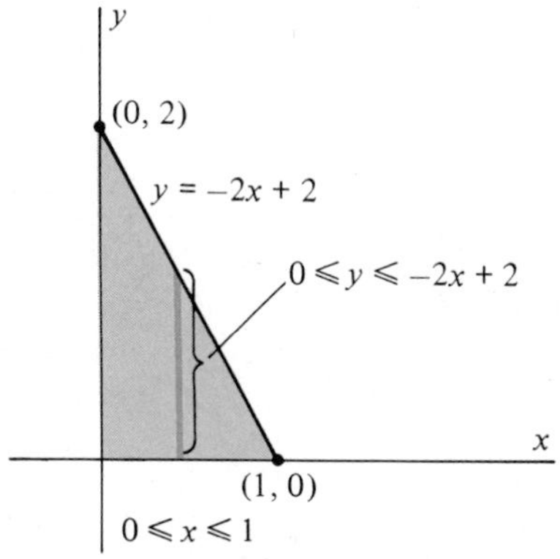

Figure 7.9 Projection of tetrahedron into the xy-plane.

Example 3 Evaluate the triple integral

$$\iiint_Q (x - y + z) \, dV$$

where Q is the tetrahedron with vertices $(0, 0, 0)$, $(1, 0, 0)$, $(0, 2, 0)$, and $(0, 0, 4)$.

Solution: The tetrahedron is sketched in Figure 7.8. Substituting the given points into the equation $z = Ax + By + C$ shows that the plane bounding Q above has equation $z = 4 - 4x - 2y$. As Figure 7.9 illustrates, the base of Q is a triangle bounded by the lines $x = 0$, $y = 0$, and $y = -2x + 2$. The region Q may therefore be described by the inequalities

$$0 \le x \le 1, \qquad 0 \le y \le -2x + 2, \qquad 0 \le z \le 4 - 4x - 2y.$$

The integral is therefore

$$\iiint_Q (x - y + z) \, dV = \int_0^1 \int_0^{2-2x} \int_0^{4-4x-2y} (x - y + z) \, dz \, dy \, dx$$

$$= \int_0^1 \int_0^{2-2x} \left\{ (x - y)z + \frac{1}{2}z^2 \right]_{z=0}^{z=4-4x-2y} \Bigg\} \, dy \, dx$$

$$= \int_0^1 \int_0^{2-2x} (4x^2 - 12x + 10xy - 12y + 4y^2 + 8) \, dy \, dx$$

$$= \int_0^1 \left\{ (4x^2 - 12x + 8)y + 5\,xy^2 - 6y^2 + \frac{4}{3}y^3 \right]_{y=0}^{y=2-2x} \Bigg\} \, dx$$

$$= \int_0^1 \left(\frac{4}{3}x^3 - 4x + \frac{8}{3} \right) dx$$

$$= \frac{1}{3}x^4 - 2x^2 + \frac{8}{3}x \Bigg]_0^1$$

$$= 1.$$

■

REMARK: The integral in Example 3 could also have been evaluated as

$$\int_0^2 \int_0^{1-y/2} \int_0^{4-4x-2y} (x - y + z)\, dz\, dx\, dy,$$

or

$$\int_0^1 \int_0^{4-4x} \int_0^{2-2x-1/2z} (x - y + z)\, dy\, dz\, dx$$

or as one of three other iterated integrals (see Exercise 29).

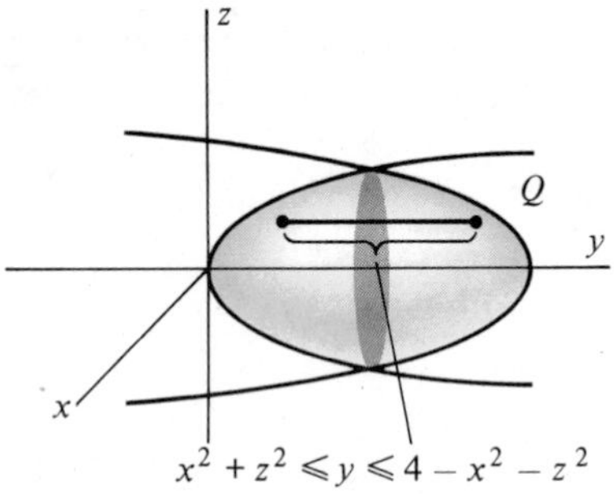

Figure 7.10 Intersecting paraboloids of Example 4.

Example 4 Find the volume of the solid bounded by the paraboloids $y = 4 - x^2 - z^2$ and $y = x^2 + z^2$.

Solution: The region is sketched in Figure 7.10. From the description of the region we can see that

$$x^2 + z^2 \le y \le 4 - x^2 - z^2. \tag{11}$$

We therefore seek inequalities for the variables x and z. Equating the two expressions for y gives

$$4 - x^2 - z^2 = x^2 + z^2$$

or

$$x^2 + z^2 = 2.$$

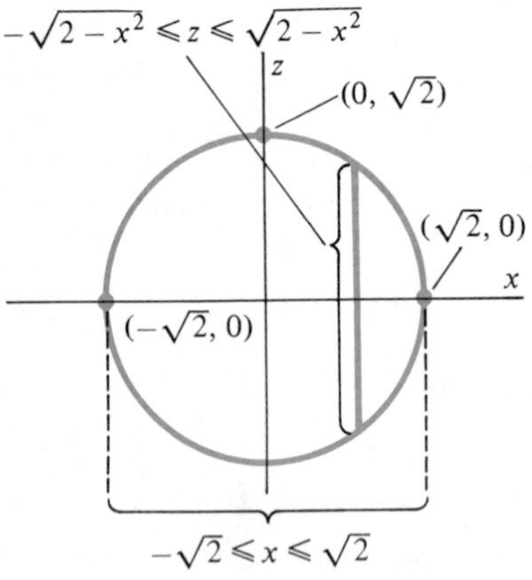

Figure 7.11 Projection of Q into xz-plane.

This is the equation of the circle (in the plane $y = 2$) where the paraboloids intersect. The projection of this circle into the xz-plane is shown in Figure 7.11. From this sketch we see that the inequalities on x and on z are

$$-\sqrt{2} \le x \le \sqrt{2}, \qquad -\sqrt{2 - x^2} \le z \le \sqrt{2 - x^2}. \tag{12}$$

Using statement (8) and inequalities (11) and (12), we obtain the desired volume as

$$\begin{aligned}
V = \iiint_Q 1\, dV &= \int_{-\sqrt{2}}^{\sqrt{2}} \int_{-\sqrt{2-x^2}}^{\sqrt{2-x^2}} \int_{x^2+z^2}^{4-x^2-z^2} 1\, dy\, dz\, dx \\
&= \int_{-\sqrt{2}}^{\sqrt{2}} \int_{-\sqrt{2-x^2}}^{\sqrt{2-x^2}} \left\{ y \Big]_{y=x^2+z^2}^{y=4-x^2-z^2} \right\} dz\, dx \\
&= \int_{-\sqrt{2}}^{\sqrt{2}} \int_{-\sqrt{2-x^2}}^{\sqrt{2-x^2}} (4 - 2x^2 - 2z^2)\, dz\, dx \\
&= \int_{-\sqrt{2}}^{\sqrt{2}} \left\{ (4 - 2x^2)z - \frac{2}{3}z^3 \Big]_{z=-\sqrt{2-x^2}}^{z=\sqrt{2-x^2}} \right\} dx \\
&= \int_{-\sqrt{2}}^{\sqrt{2}} \frac{8}{3}(2 - x^2)^{3/2}\, dx \qquad (13) \\
&= 4\pi.
\end{aligned}$$

(The integral in line (13) is evaluated by means of a trigonometric substitution.) ■

REMARK: The triple integral in Example 4 could also have been evaluated as

$$\int_{-\sqrt{2}}^{\sqrt{2}} \int_{-\sqrt{2-z^2}}^{\sqrt{2-z^2}} \int_{x^2+z^2}^{4-x^2-z^2} 1\, dy\, dx\, dz.$$

Density

The triple integral may also be used to calculate the mass of a solid object, if we know the *density* of the material in units of mass per unit volume (such as gram/cm^3) as a continuous *density function* $\rho(x, y, z)$. If Q is an object with a density function $\rho(x, y, z)$ of this type, we may approximate the mass of Q by partitioning Q into approximating boxes $Q_1, Q_2, \ldots, Q_n$, as before. If (s_j, t_j, u_j) is a point in Q_j, and if the dimensions of Q_j are small, then the quantity

$$M_j = \rho(s_j, t_j, u_j)\Delta x \Delta y \Delta z \qquad \text{(mass = density} \times \text{volume)}$$

provides an approximation to the mass of the jth box Q_j. Summing these approximations over all boxes contained within Q gives the approximating sum

$$M \approx \sum_{j=1}^{n} M_j = \sum_{j=1}^{n} \rho(s_j, t_j, u_j)\Delta x \Delta y \Delta z.$$

Thus, the **mass** of Q is defined by the triple integral

$$M = \iiint_Q \rho(x, y, z)\, dV. \tag{14}$$

That is, mass is the integral of the density function over the region Q, just as in the one- and two-variable cases.

Example 5 A small wedge has the shape of the region

$$Q = \{(x, y, z) | -1 \le x \le 1,\ 0 \le y \le 2,\ 0 \le z \le y\}.$$

Find its mass if the density at any point (x, y, z) is given by the density function $\rho(x, y, z) = 1 + x^2 + y^2$ gram/cm^3 and the dimensions for Q are in cm (Figure 7.12).

Solution: According to equation (14) and Theorem 4 the mass is

$$M = \iiint_Q (1 + x^2 + y^2)\, dV = \int_{-1}^{1} \int_0^2 \int_0^y (1 + x^2 + y^2)\, dz\, dy\, dx$$

$$= \int_{-1}^{1} \int_0^2 \left\{ (1 + x^2 + y^2)z \Big]_{z=0}^{z=y} \right\} dy\, dx$$

$$= \int_{-1}^{1} \int_0^2 (y + x^2y + y^3)\, dy\, dx$$

$$= \int_{-1}^{1} \left\{ \frac{1}{2}y^2 + \frac{1}{2}x^2y^2 + \frac{1}{4}y^4 \Big]_{y=0}^{y=2} \right\} dx$$

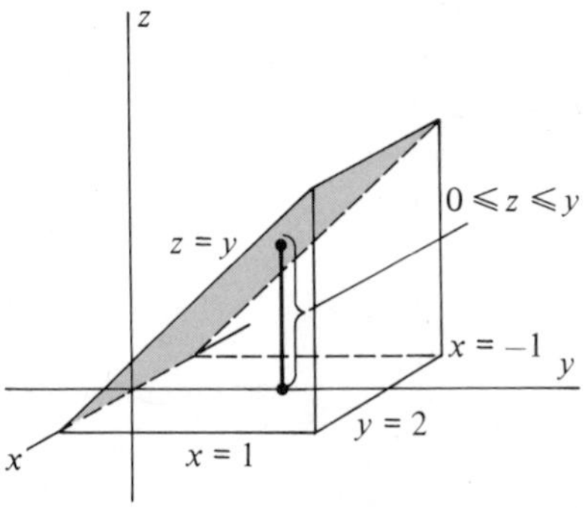

Figure 7.12 The wedge of Example 5.

$$= \int_{-1}^{1} (6 + 2x^2)\, dx$$

$$= 6x + \frac{2}{3}x^3 \Big]_{-1}^{1}$$

$$= \frac{40}{3} \text{ grams.} \qquad ■$$

Moments and Center of Mass

Let Q denote a solid in $\mathbb{R}^3$, and let $\rho(x, y, z)$ be a continuous function giving the density of Q at each point (x, y, z). As before, let Q be partitioned into rectangular boxes $Q_1, Q_2, \ldots, Q_n$ and let (s_j, t_j, u_j) be a point in Q_j, $j = 1, 2, \ldots, n$. Since s_j is the distance from the point (s_j, t_j, u_j) to the yz-plane, the product $s_j\rho(s_j, t_j, u_j)\Delta x\Delta y\Delta z$ may be interpreted as an approximation to the product of the mass of Q_j and the length of its "lever arm" extending from the yz-plane. By analogy with our previous discussions on moments (see Section 20.5) we define *the first moment, M_{yz}, of the solid Q with respect to the yz-plane* to be

$$M_{yz} = \iiint_Q x\rho(x, y, z)\, dV = \lim_{n\to\infty} \sum_{j=1}^{n} s_j\rho(s_j, t_j, u_j)\Delta x\Delta y\Delta z. \qquad (15)$$

Similarly, *the moments of Q with respect to the xz- and xy-planes* are

$$M_{xz} = \iiint_Q y\rho(x, y, z)\, dV \qquad (16)$$

and

$$M_{xy} = \iiint_Q z\rho(x, y, z)\, dV. \qquad (17)$$

Finally, if M denotes the mass of Q, the **center of mass** of Q is the point $(\bar{x}, \bar{y}, \bar{z})$ where

$$\bar{x} = \frac{M_{yz}}{M}; \qquad \bar{y} = \frac{M_{xz}}{M}; \qquad \bar{z} = \frac{M_{xy}}{M}. \qquad (18)$$

Example 6 Find the three first moments and the center of mass for the solid in Example 5.

Solution: With $\rho(x, y, z) = 1 + x^2 + y^2$ and Q as given in Example 5, we have

$$M_{yz} = \iiint_Q x(1 + x^2 + y^2)\, dV = \int_{-1}^{1}\int_0^2\int_0^y (x + x^3 + xy^2)\, dz\, dy\, dx$$

$$= \int_{-1}^{1}\int_0^2 \left\{(x + x^3 + xy^2)z\Big]_{z=0}^{z=y}\right\} dy\, dx$$

$$= \int_{-1}^{1}\int_0^2 [(x + x^3)y + xy^3]\, dy\, dx$$

$$= \int_{-1}^{1} \left\{(x + x^3)\frac{y^2}{2} + \frac{1}{4}xy^4\Big]_{y=0}^{y=2}\right\} dx$$

$$= \int_{-1}^{1} (6x + 2x^3)\, dx$$

$$= 3x^2 + \frac{1}{2}x^4 \Big]_{-1}^{1}$$

$$= 0.$$

Similar calculations show that

$$M_{xz} = \iiint_Q y(1 + x^2 + y^2)\, dV$$

$$= \int_{-1}^{1} \int_0^2 \int_0^y (y + x^2y + y^3)\, dz\, dy\, dx = \frac{896}{45}$$

and that

$$M_{xy} = \iiint_Q z(1 + x^2 + y^2)\, dV$$

$$= \int_{-1}^{1} \int_0^2 \int_0^y (z + x^2z + y^2z)\, dz\, dy\, dx = \frac{448}{45}.$$

Since we have previously calculated the mass to be $M = \frac{40}{3}$, the coordinates of the center of mass $(\bar{x}, \bar{y}, \bar{z})$, from equations (18), are

$$\bar{x} = \frac{M_{yz}}{M} = 0 \cdot \frac{3}{40} = 0,$$

$$\bar{y} = \frac{M_{xz}}{M} = \frac{896}{45} \cdot \frac{3}{40} = \frac{112}{75} \approx 1.49,$$

and

$$\bar{z} = \frac{M_{xy}}{M} = \frac{448}{45} \cdot \frac{3}{40} = \frac{56}{75} \approx 0.75.$$

■

Exercise Set 20.7

In Exercises 1–8, evaluate the iterated integral.

1. $\int_0^1 \int_0^1 \int_0^1 xyz\, dx\, dy\, dz$

2. $\int_0^2 \int_{-\pi/2}^{\pi/2} \int_1^2 x \cos ye^z\, dx\, dy\, dz$

3. $\int_0^1 \int_0^y \int_0^x 3\, dz\, dx\, dy$

4. $\int_1^3 \int_0^{\pi/4} \int_0^x \cos(x + y)\, dy\, dx\, dz$

5. $\int_0^2 \int_0^x \int_0^{x+y} z\, dz\, dy\, dx$

6. $\int_{-1}^1 \int_{-\sqrt{1-x^2}}^{\sqrt{1-x^2}} \int_0^{\sqrt{1-x^2-y^2}} 1\, dz\, dy\, dx$ (*Hint:* Use geometry.)

7. $\int_{-1}^1 \int_0^y \int_0^x ye^{x^2+y^2}\, dz\, dx\, dy$

8. $\int_0^2 \int_0^x \int_{x+y}^{x^2+y^2} 1\, dz\, dy\, dx$

9. Interchange the order of integration in

$$\int_0^1 \int_0^{2x} \int_0^{x+y} f(x, y, z)\, dz\, dy\, dx$$

from order $dz\, dy\, dx$ to order $dz\, dx\, dy$.

10. Use a triple integral to find the volume of the tetrahedron with vertices (0, 0, 0), (1, 0, 0), (1, 1, 0), and (1, 1, 1).

11. Sketch the solid whose volume is given by the iterated integral

$$\int_0^2 \int_0^{2x} \int_0^{x+y} dz\, dy\, dx.$$

12. Sketch the solid whose volume is given by the iterated integral

$$\int_{-1}^1 \int_{-\sqrt{1-x^2}}^{\sqrt{1-x^2}} \int_{\sqrt{x^2+y^2}}^{2-\sqrt{x^2+y^2}} dz\, dy\, dx.$$

13. Find the volume of the region in Exercise 12.

14. Find the volume of the tetrahedron bounded by the plane $x + y + z = 1$ and the coordinate planes $x = 0$, $y = 0$, and $z = 0$.

15. Find the volume of the region lying above the xy-plane, inside the cylinder $x^2 + y^2 = 9$, and below the plane $z = y + 3$.

16. Sketch the solid whose volume is given by the integral

$$V = \int_0^2 \int_0^{\sqrt{2x-x^2}} \int_0^{2-x} dz\, dy\, dx$$

and find the volume.

17. Evaluate the integral $\iiint_Q (3x + xz)\, dV$ where Q is the region bounded by the cylinder $x^2 + z^2 = 9$, the plane $y + z = 3$, and the plane $y = 0$.

18. Interchange the order of integration in the integral

$$\int_{-2}^2 \int_0^{\sqrt{4-x^2}} \int_{-\sqrt{4-x^2-y^2}}^{\sqrt{4-x^2-y^2}} f(x, y, z)\, dz\, dy\, dx$$

from $dz\, dy\, dx$ to $dy\, dz\, dx$.

19. Find the volume of the region bounded by the paraboloids $x = y^2 + z^2$ and $x = 2 - y^2 - z^2$.

20. Find the volume of the solid bounded above by the paraboloid $z = 2 - x^2 - y^2$ and below by the plane $z = 2 - 2x$.

21. Let Q be the solid bounded by the cylinder $x^2 + y^2 = 9$ and the planes $z = 0$ and $x + z = 3$. Find the mass of Q if the density at each point (x, y, z) is given by the function $\rho(x, y, z) = z$.

22. Find the volume of the region common to the cylinders $x^2 + z^2 = 1$ and $y^2 + z^2 = 1$.

23. The density at each point of the box $Q = \{(x, y, z) | 0 \le x \le 1, 0 \le y \le 2, 0 \le z \le 2\}$ is proportional to the square of the distance from the origin. Find the mass.

24. Find the center of mass of the region in Exercise 14 if the density is constant.

25. Find the center of mass of the solid in Exercise 15 if the density is constant.

26. Find the center of mass of the solid in Exercise 21.

27. Find the center of mass of the part of the region enclosed by the sphere $x^2 + y^2 + z^2 = 4$ lying in the first octant if the density is constant.

28. Find the center of mass of the cube $Q = \{(x, y, z) | 0 \le x \le 1, 0 \le y \le 1, 0 \le z \le 1\}$ if the density function is $\rho(x, y, z) = xyz$.

29. Find five other iterated integrals by which the triple integral of Example 3 may be evaluated.

30. Find the volume, mass, and center of gravity for the solid in Example 5 if the density is uniform $\rho(x, y, z) = 1$ gram/cm^3. Compare your answers with those in Examples 5 and 6, and account for the differences in physical terms.

20.8 TRIPLE INTEGRALS IN CYLINDRICAL AND SPHERICAL COORDINATES

In Section 17.9, we saw that certain solids can be described more naturally in cylindrical or spherical coordinates than in the usual Cartesian coordinate system. The purpose of this section is to define the triple integral for functions written in cylindrical or spherical coordinates. In part, we want to know how to calculate volumes and masses for solids described in these coordinate systems. Another reason for studying these topics is that certain triple integrals, originally expressed in Cartesian coordinates, are more easily evaluated by changing to either cylindrical or spherical coordinates.

Triple Integrals in Cylindrical Coordinates

Recall the relationship between cylindrical and Cartesian coordinates: If a point P has cylindrical coordinates $P = (r, \theta, z)$ and Cartesian coordinates $P = (x, y, z)$, then (r, θ) are the polar coordinates for the point (x, y) in the xy-plane (Figure 8.1).

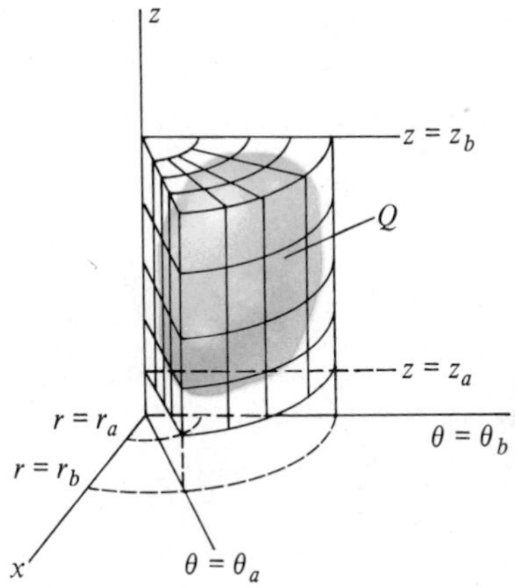

Figure 8.1 Partitioning the region Q using a grid specified in cylindrical coordinates.

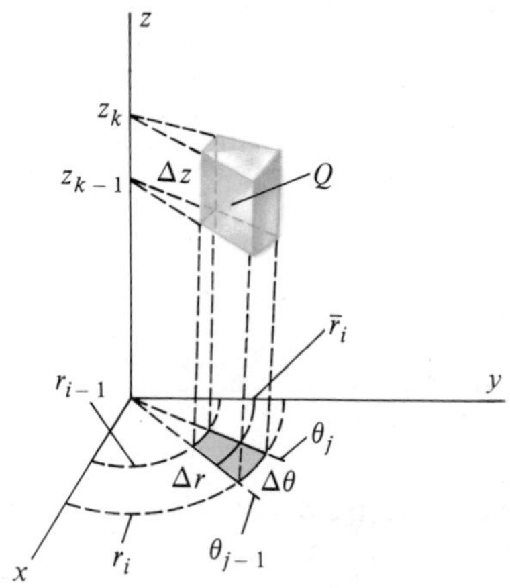

Figure 8.2 Volume of circular wedge is

$$\Delta V = \overline{r}\,\Delta r\,\Delta\theta\,\Delta z$$

where $\overline{r} = \dfrac{r_{i-1} + r_i}{2}$.

Now suppose that $\mathbb{R}^3$ is coordinatized by cylindrical coordinates, that Q is a region in $\mathbb{R}^3$, and that $f(r, \theta, z)$ is a continuous function defined on Q. We shall define the triple integral of $f(r, \theta, z)$ over Q in the usual way—by partitioning the region Q, forming an approximating sum, and obtaining the limit of the approximating sum.

Suppose that the region Q lies within the "cylindrical box" determined by the inequalities

$$r_a \le r \le r_b, \qquad \theta_a \le \theta \le \theta_b, \qquad z_a \le z \le z_b.$$

(See Figure 8.1.) Let n be any integer, and let $\Delta r = \dfrac{r_b - r_a}{n}$, $\Delta\theta = \dfrac{\theta_b - \theta_a}{n}$, and $\Delta z = \dfrac{z_b - z_a}{n}$. As illustrated in Figure 8.1, we partition the box containing the region Q by cylinders $r = r_i$, $i = 0, 1, 2, \ldots, n$, where

$$r_0 = r_a, \qquad r_1 = r_a + \Delta r, \qquad r_2 = r_a + 2\Delta r, \ldots, r_n = r_a + n\Delta r = r_b,$$

by half planes $\theta = \theta_j$, $j = 0, 1, 2, \ldots, n$, where

$$\theta_0 = \theta_a, \qquad \theta_1 = \theta_a + \Delta\theta, \qquad \theta_2 = \theta_a + 2\Delta\theta, \ldots, \theta_n = \theta_a + n\Delta\theta = \theta_b,$$

and by planes $z = z_k$, $k = 0, 1, 2, \ldots, n$, where

$$z_0 = z_a, \qquad z_1 = z_a + \Delta z, \qquad z_2 = z_a + 2\Delta z, \ldots, z_n = z_a + n\Delta z = z_b.$$

The effect of this partition is to divide the region containing Q into small cylindrical blocks

$$Q_{ijk} = \{(r, \theta, z) | r_{i-1} \le r \le r_i,\ \theta_{j-1} \le \theta \le \theta_j,\ z_{k-1} \le z \le z_k\}$$

as illustrated in Figure 8.2. According to equation (3), Section 20.4, the area of the base of block Q_{ijk} is

$$\Delta A_{ijk} = \hat{r}_i \Delta r \Delta\theta$$

where $\hat{r}_i = \dfrac{1}{2}(r_{i-1} + r_i)$. Since each block Q_{ijk} has height Δz, the volume of the block Q_{ijk} is

$$\Delta V_{ijk} = \hat{r}_i \Delta r \Delta\theta \Delta z. \tag{1}$$

Now let θ_j^* be any number in the interval $[\theta_{j-1}, \theta_j]$ and let z_k^* be any number in the interval $[z_{k-1}, z_k]$. Then, since $\hat{r}_i \in [r_{i-1}, r_i]$, the point $(\hat{r}_i, \theta_j^*, z_k^*)$ lies in the block Q_{ijk}. The sum

$$\sum_i \sum_j \sum_k f(\hat{r}_i, \theta_j^*, z_k^*)\Delta V_{ijk} = \sum_i \sum_j \sum_k f(\hat{r}_i, \theta_j^*, z_k^*)\hat{r}_i \Delta r \Delta\theta \Delta z \tag{2}$$

is the approximating Riemann sum for $f(r, \theta, z)$ over Q, where the sum is taken over all boxes Q_{ijk} lying entirely within Q. Its limit as $n \to \infty$, which exists when $f(r, \theta, z)$ is continuous and Q is as described in the following theorem, is the triple integral of $f(r, \theta, z)$ over Q:

$$\iiint_Q f(r, \theta, z)\, dV = \lim_{n\to\infty} \sum_i \sum_j \sum_k f(\hat{r}_i, \theta_j^*, z_k^*)\hat{r}_i \Delta r \Delta\theta \Delta z. \tag{3}$$

The following theorem shows how this triple integral may be evaluated for the types of regions encountered in this section and in most applications.

THEOREM 5 Let Q be a region in $\mathbb{R}^3$ of the form

$$Q = \{(r, \theta, z) | a \leq \theta \leq b,\ h_1(\theta) \leq r \leq h_2(\theta),\ g_1(r, \theta) \leq z \leq g_2(r, \theta)\}$$

where g_1, g_2, h_1, and h_2 are continuous functions. Let $f(r, \theta, z)$ be continuous on Q. Then

$$\iiint_Q f(r, \theta, z)\, dV = \int_a^b \int_{h_1(\theta)}^{h_2(\theta)} \int_{g_1(r,\theta)}^{g_2(r,\theta)} f(r, \theta, z)\, r\, dz\, dr\, d\theta. \tag{4}$$

REMARK 1: Note the extra factor r appearing in the iterated integral in (4). The reason for its appearance is the same as in Section 20.4: it can be considered a result of converting dV to cylindrical coordinates.

REMARK 2: When $f(r, \theta, z) \equiv 1$, it follows from equation (1) that the Riemann sum in (2) approximates the volume of Q. Thus

$$\boxed{\text{Volume of } Q = \iiint_Q dV} \tag{5}$$

holds in cylindrical coordinates as well as in rectangular coordinates.

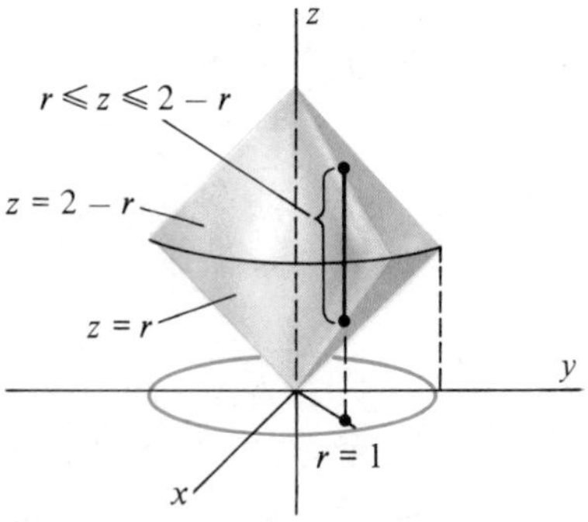

Figure 8.3 Solid Q in Example 1.

Example 1 A solid "top" has the shape of the region

$$Q = \{(r, \theta, z) | 0 \leq r \leq 1,\ 0 \leq \theta \leq 2\pi,\ r \leq z \leq 2 - r\}$$

as illustrated in Figure 8.3. Find the mass of the object if the density at any point is proportional to the distance from the z-axis.

Solution: The density function described here is $\rho(r, \theta, z) = \lambda r$, where λ is the constant of proportionality. By equation (14), Section 20.7, and Theorem 5, we have

$$\begin{aligned}
\text{Mass} = \iiint_Q \rho(r, \theta, z)\, dV &= \iiint_Q \lambda r\, dV \\
&= \int_0^{2\pi} \int_0^1 \int_r^{2-r} \lambda r^2\, dz\, dr\, d\theta \quad \text{note extra factor of } r \\
&= \int_0^{2\pi} \int_0^1 \left\{ \lambda r^2 z \Big]_{z=r}^{z=2-r} \right\} dr\, d\theta \\
&= \int_0^{2\pi} \int_0^1 (2\lambda r^2 - 2\lambda r^3)\, dr\, d\theta \\
&= \int_0^{2\pi} \left\{ \frac{2\lambda}{3} r^3 - \frac{\lambda}{2} r^4 \Big]_{r=0}^{r=1} \right\} d\theta
\end{aligned}$$

$$= \int_0^{2\pi} \left(\frac{\lambda}{6}\right) d\theta$$

$$= \frac{\pi\lambda}{3}. \qquad \blacksquare$$

Changing to Cylindrical Coordinates

You will sometimes find that a triple integral written in Cartesian coordinates is more easily evaluated by first changing to cylindrical coordinates. Doing so is analogous to changing a double integral from Cartesian to polar coordinates. Specifically, to write the iterated integral

$$\int_a^b \int_{h_1(x)}^{h_2(x)} \int_{g_1(x,y)}^{g_2(x,y)} f(x, y, z)\, dz\, dy\, dx$$

in cylindrical coordinates, we do the following:

(i) Express the region

$$Q = \{(x, y, z) | a \le x \le b,\ h_1(x) \le y \le h_2(x),\ g_1(x, y) \le z \le g_2(x, y)\}$$

in cylindrical coordinates as

$$Q = \{(r, \theta, z) | c \le \theta \le d,\ h_3(\theta) \le r \le h_4(\theta),\ g_3(r, \theta) \le z \le g_4(r, \theta)\}.$$

(ii) Using the substitutions $x = r\cos\theta$ and $y = r\sin\theta$, replace the integrand $f(x, y, z)$ by $f(r\cos\theta, r\ sin\ \theta, z)r$. (Do not forget the extra factor of r!)

(iii) Obtain the equation

$$\int_a^b \int_{h_1(x)}^{h_2(x)} \int_{g_1(x,y)}^{g_2(x,y)} f(x, y, z)\, dz\, dy\, dx \tag{6}$$
$$= \int_c^d \int_{h_3(\theta)}^{h_4(\theta)} \int_{g_3(r,\theta)}^{g_4(r,\theta)} f(r\cos\theta, r\sin\theta, z)\, r\, dz\, dr\, d\theta.$$

Equation (6) is verified by comparing Theorems 4 and 5, each of which expresses the triple integral $\iiint_Q f\, dV$ as one of the two iterated integrals in (6). One way to paraphrase equation (6) is to say that *in changing from Cartesian coordinates to cylindrical coordinates the* **volume element**

$$dV = dz\, dy\, dx$$

in Cartesian coordinates is replaced by the volume element

$$dV = r\, dz\, dr\, d\theta \tag{7}$$

in cylindrical coordinates. It is very important to note the extra factor of r that appears in the integrand on the right-hand side of (6) and in equation (7).

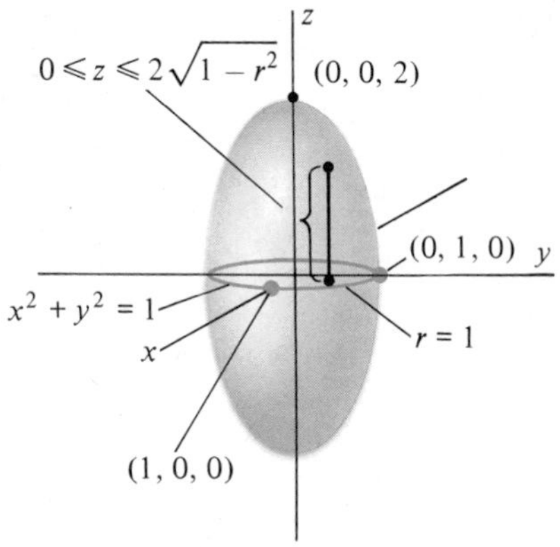

Figure 8.4 Ellipsoid $4x^2 + 4y^2 + z^2 = 4$ in Example 2.

Example 2 Calculate the volume of the ellipsoid

$$4x^2 + 4y^2 + z^2 = 4.$$

Solution: The ellipsoid is sketched in Figure 8.4. Since the ellipsoid is symmetric with respect to the xy-plane, we may calculate the volume as twice the volume of the region Q lying above the xy-plane. Since this region is described by the inequalities

$$0 \le z \le 2\sqrt{1 - x^2 - y^2}, \qquad -\sqrt{1 - x^2} \le y \le \sqrt{1 - x^2}, \qquad -1 \le x \le 1,$$

the volume is given by the iterated integral

$$V = 2\int_{-1}^{1}\int_{-\sqrt{1-x^2}}^{\sqrt{1-x^2}}\int_{0}^{2\sqrt{1-x^2-y^2}} 1\, dz\, dy\, dx. \tag{8}$$

Clearly, this integral will be difficult to evaluate, so we try switching to cylindrical coordinates. In cylindrical coordinates the region Q is described by the inequalities

$$0 \le r \le 1, \qquad 0 \le \theta \le 2\pi, \qquad 0 \le z \le 2\sqrt{1 - r^2}$$

since $2\sqrt{1 - x^2 - y^2} = 2\sqrt{1 - (x^2 + y^2)} = 2\sqrt{1 - r^2}$. Using equation (6) we may rewrite the integral (8) as

$$V = 2\int_{0}^{2\pi}\int_{0}^{1}\int_{0}^{2\sqrt{1-r^2}} r\, dz\, dr\, d\theta$$

note the extra factor of r.

$$= 2\int_{0}^{2\pi}\int_{0}^{1} \left\{ zr \Big]_{z=0}^{z=2\sqrt{1-r^2}} \right\} dr\, d\theta$$

$$= 2\int_{0}^{2\pi}\int_{0}^{1} 2r\sqrt{1 - r^2}\, dr\, d\theta$$

$$= 2\int_{0}^{2\pi} \left\{ -\frac{2}{3}(1 - r^2)^{3/2} \Big]_{r=0}^{r=1} \right\} d\theta$$

$$= \frac{8\pi}{3}.$$

■

Spherical Coordinates

Recall that the point P has spherical coordinates $P = (\rho, \theta, \phi)$ if $\rho \ge 0$ is its distance from the origin, θ is its rotation angle (as in cylindrical coordinates), and ϕ is its angle of inclination from the vertical (see Figure 8.5). Figure 8.6 illustrates the

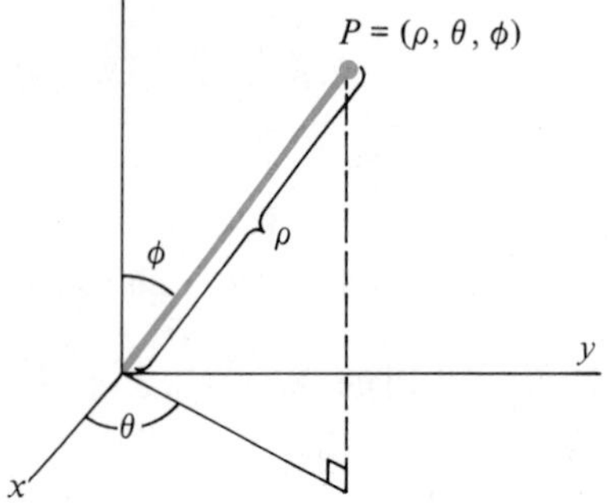

Figure 8.5 Spherical coordinates (ρ, θ, ϕ) for P.

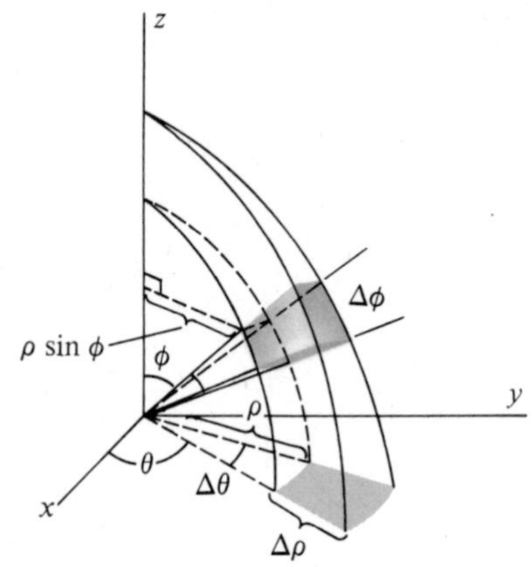

Figure 8.6 The spherical block.

"spherical block" determined by the inequalities

$$\rho_{i-1} \le \rho \le \rho_i, \qquad \theta_{j-1} \le \theta \le \theta_j, \qquad \phi_{k-1} \le \phi \le \phi_k \tag{9}$$

where $\Delta\rho = \rho_i - \rho_{i-1}$, $\Delta\theta = \theta_j - \theta_{j-1}$, and $\Delta\phi = \phi_k - \phi_{k-1}$. A fact, which we shall not prove, is that the volume of this spherical block is given by

$$\Delta V_{ijk} = \rho_i^{*2} \sin \phi_k^* \Delta\rho\Delta\theta\Delta\phi$$

for some $\rho_i^* \in [\rho_{i-1}, \rho_i]$ and $\phi_k^* \in [\phi_{k-1}, \phi_k]$ (see Exercise 31).

Thus, if a region Q in $\mathbb{R}^3$ lies within the spherical block determined by the inequalities $\rho_a \le \rho \le \rho_b$, $\theta_a \le \theta \le \theta_b$, $\phi_a \le \phi \le \phi_b$, a "spherical partition" of Q may be formed in the usual way: divide each of the intervals $[\rho_a, \rho_b]$, $[\theta_a, \theta_b]$, and $[\phi_a, \phi_b]$ into n equal subintervals, thus determining n^3 spherical blocks Q_{ijk} corresponding to inequalities (9). In order to define the integral of a continuous function $f(\rho, \theta, \phi)$ defined on Q, we select one point $(\rho_i^*, \theta_j^*, \phi_k^*)$ in each block Q_{ijk} and form the sum

$$\sum_i \sum_j \sum_k f(\rho_i^*, \theta_j^*, \phi_k^*)\Delta V_{ijk} \tag{10}$$

$$= \sum_i \sum_j \sum_k f(\rho_i^*, \theta_j^*, \phi_k^*)\rho_i^{*2} \sin \phi_k^* \Delta\rho\Delta\theta\Delta\phi.$$

As before, this sum is taken over only those spherical blocks Q_{ijk} lying entirely within Q. The limit of this sum as $n \to \infty$, when it exists, is the triple integral of $f(\rho, \theta, \phi)$ over Q:

$$\iiint_Q f(\rho, \theta, \phi)\, dV = \lim_{n\to\infty} \sum_i \sum_j \sum_k f(\rho_i^*, \theta_j^*, \phi_k^*)\rho_i^{*2} \sin \phi_k^* \Delta\rho\Delta\theta\Delta\phi. \tag{11}$$

As with other types of triple integrals, the triple integral in spherical coordinates may be evaluated as an iterated integral in certain cases.

THEOREM 6

Let Q be a region in $\mathbb{R}^3$ of the form

$$Q = \{\rho, \theta, \phi) | a \le \theta \le b,\ h_1(\theta) \le \phi \le h_2(\theta),\ g_1(\theta, \phi) \le \rho \le g_2(\theta, \phi)\}$$

where h_1, h_2, g_1, and g_2 are continuous functions. Let $f(\rho, \theta, \phi)$ be continuous on Q. Then,

$$\iiint_Q f(\rho, \theta, \phi)\, dV = \int_a^b \int_{h_1(\theta)}^{h_2(\theta)} \int_{g_1(\theta,\phi)}^{g_2(\theta,\phi)} f(\rho, \theta, \phi)\, \rho^2 \sin \phi\, d\rho\, d\phi\, d\theta. \tag{12}$$

REMARK: Note the extra factor $\rho^2 \sin \phi$ in the integrand in the iterated integral in line (12).

Example 3 Find the mass of the solid occupying the region

$$Q = \{(\rho, \theta, \phi) | 0 \le \rho \le 2,\ 0 \le \theta \le 2\pi,\ 0 \le \phi \le \pi/6\}$$

if the density at any point is proportional to its distance from the origin.

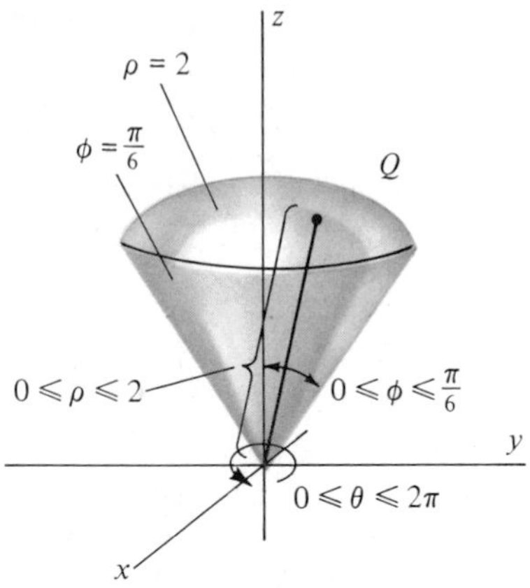

Figure 8.7 Region Q of Example 3.

Solution: The region Q is the "ice cream cone" region sketched in Figure 8.7. As before, we obtain the mass of the solid by integrating the density function $D(\rho, \theta, \phi) = \lambda\rho$. (Note that we are now using ρ for radial distance, not density.) Thus, by Theorem 6

$$\text{Mass} = \iiint_Q \lambda\rho\, dV = \int_0^{2\pi}\int_0^{\pi/6}\int_0^{2} (\lambda\rho)(\rho^2 \sin\phi)\, d\rho\, d\phi\, d\theta$$

note extra factor of $\rho^2 \sin\phi$.

$$= \int_0^{2\pi}\int_0^{\pi/6}\int_0^{2} \lambda\rho^3 \sin\phi\, d\rho\, d\phi\, d\theta$$

$$= \int_0^{2\pi}\int_0^{\pi/6} \left\{\frac{\lambda}{4}\rho^4 \sin\phi\Big]_{\rho=0}^{\rho=2}\right\} d\phi\, d\theta$$

$$= \int_0^{2\pi}\int_0^{\pi/6} 4\lambda \sin\phi\, d\phi\, d\theta$$

$$= \int_0^{2\pi} \left\{-4\lambda\cos\phi\Big]_{\phi=0}^{\phi=\pi/6}\right\} d\theta$$

$$= \int_0^{2\pi} 4\lambda\left(1 - \frac{\sqrt{3}}{2}\right) d\theta$$

$$= 8\lambda\pi\left(1 - \frac{\sqrt{3}}{2}\right).$$

■

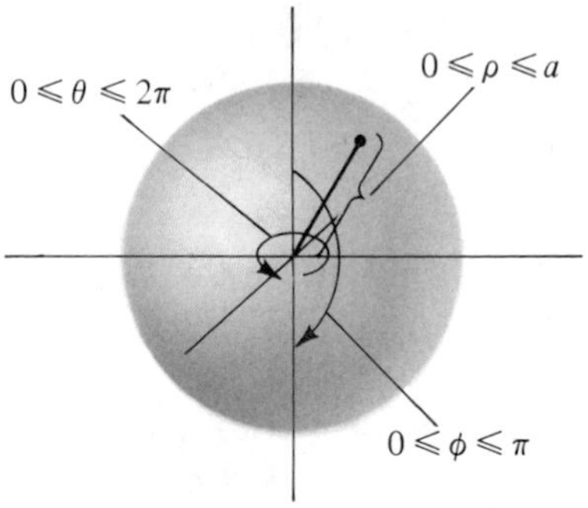

Figure 8.8 Sphere is determined by the inequalities

$0 \le \rho \le a,$
$0 \le \phi \le \pi,\ 0 \le \theta \le 2\pi.$

Example 4 Obtain the formula for the volume of a sphere of radius a using spherical coordinates.

Solution: The sphere of radius a is described by the spherical inequalities

$$0 \le \rho \le a, \qquad 0 \le \phi \le \pi, \qquad 0 \le \theta \le 2\pi \qquad \text{(Figure 8.8).}$$

Thus

$$\text{Volume} = \iiint_Q 1\, dV = \int_0^{2\pi}\int_0^{\pi}\int_0^{a} \rho^2 \sin\phi\, d\rho\, d\phi\, d\theta$$

note extra factor of $\rho^2 \sin\phi$.

$$= \int_0^{2\pi}\int_0^{\pi} \left\{\frac{1}{3}\rho^3 \sin\phi\Big]_{\rho=0}^{\rho=a}\right\} d\phi\, d\theta$$

$$= \int_0^{2\pi}\int_0^{\pi} \frac{a^3}{3} \sin\phi\, d\phi\, d\theta$$

$$= \int_0^{2\pi} \left\{-\frac{a^3}{3}\cos\phi\Big]_{\phi=0}^{\phi=\pi}\right\} d\theta$$

$$= \int_0^{2\pi} \frac{2a^3}{3}\, d\theta$$

$$= \frac{4}{3}\cdot\pi a^3.$$

■

Changing From Cartesian to Spherical Coordinates

If we encounter a function $f(x, y, z)$ written in Cartesian coordinates for which we wish to calculate a triple integral over a spherical region, it is often useful to rewrite the function in spherical coordinates using the substitutions

$$x = \rho\sin\phi\cos\theta, \qquad y = \rho\sin\phi\sin\theta, \qquad z = \rho\cos\phi.$$

(See Section 17.9 for a review of how these equations are obtained.) We shall not write out a general formula for converting triple integrals from Cartesian coordinates to spherical coordinates, since the situations in which such changes are possible are almost exclusively limited to functions involving the expression

$$\begin{aligned} x^2 + y^2 + z^2 &= \rho \sin^2 \phi \cos^2 \theta + \rho^2 \sin^2 \phi \sin^2 \theta + \rho^2 \cos^2 \phi \\ &= \rho^2 \sin^2 \phi(\cos^2 \theta + \sin^2 \theta) + \rho^2 \cos^2 \phi \\ &= \rho^2. \end{aligned}$$

Example 5 A hemispherical solid is described in Cartesian coordinates as

$$Q = \{(x, y, z)|0 \le x^2 + y^2 \le 1, 0 \le z \le \sqrt{1 - x^2 - y^2}\}.$$

Find the mass of the object if the density at the point (x, y, z) is given by the function

$$D(x, y, z) = e^{-(x^2+y^2+z^2)^{3/2}}.$$

Solution: The hemisphere may be described by the inequalities

$$0 \le \rho \le 1, \qquad 0 \le \phi \le \pi/2, \qquad 0 \le \theta \le 2\pi,$$

and the density function, in spherical coordinates, is

$$D(\rho, \theta, \phi) = e^{-\rho^3}.$$

Thus,

$$\begin{aligned} \text{Mass} &= \int_0^{2\pi} \int_0^{\pi/2} \int_0^1 e^{-\rho^3} \rho^2 \sin \phi \, d\rho \, d\phi \, d\theta \\ &= \int_0^{2\pi} \int_0^{\pi/2} \left\{ -\frac{1}{3} e^{-\rho^3} \sin \phi \Big]_{\rho=0}^{\rho=1} \right\} d\phi \, d\theta \\ &= \int_0^{2\pi} \int_0^{\pi/2} \frac{1}{3} \left(1 - \frac{1}{e}\right) \sin \phi \, d\phi \, d\theta \\ &= \int_0^{2\pi} \left\{ \left(-\frac{e-1}{3e}\right) \cos \phi \Big]_{\phi=0}^{\phi=\pi/2} \right\} d\theta \\ &= \int_0^{2\pi} \left(\frac{e-1}{3e}\right) d\theta \\ &= \frac{2\pi(e-1)}{3e}. \end{aligned}$$

note extra factor of $\rho^2 \sin \phi$.

■

Exercise Set 20.8

In Exercises 1–8, use cylindrical coordinates.

1. Find the volume of the solid bounded by the graphs of $z = x^2 + y^2$ and $z = 9$.

2. Find the volume of the solid bounded by the graphs of the equations $z = 9 - x^2 - y^2$, $x^2 + y^2 = 3$, and $z = 9$.

3. Find the center of mass of a solid having the shape of the solid in Exercise 1 if the density is constant.

4. Find the volume of the region lying inside both the cylinder $x^2 - 2x + y^2 = 0$ and the sphere $x^2 + y^2 + z^2 = 4$.

5. Find the volume of the solid remaining when the region inside the cylinder $r = 3 \sin \theta$ is removed from the solid region bounded by the sphere $x^2 + y^2 + z^2 = 9$.

6. A solid is bounded by the cylinder $y^2 + z^2 = 4$ and the planes $x = 0$ and $x = 4$. Find its mass if the density at each point is proportional to the distance of that point from the central axis of the cylinder.

7. Find the volume of the solid bounded above and below by the cone $z^2 = 2x^2 + 2y^2$ and on the sides by the cylinder $x^2 + y^2 - 4y = 0$.

8. Find the volume of the solid bounded above by the plane $z = x$ and below by the paraboloid $z = x^2 + y^2$.

In Exercises 9–12, evaluate the triple integral by first changing to cylindrical coordinates.

9. $\displaystyle\int_0^2 \int_{-\sqrt{2x-x^2}}^{\sqrt{2x-x^2}} \int_0^{\sqrt{x^2+y^2}} 1\, dz\, dy\, dx$

10. $\displaystyle\int_{-1}^1 \int_{-\sqrt{1-x^2}}^{\sqrt{1-x^2}} \int_{\sqrt{x^2+y^2}}^{1} x^2\, dz\, dy\, dx$

11. $\displaystyle\int_{-2}^2 \int_{-\sqrt{4-x^2}}^{\sqrt{4-x^2}} \int_0^{y+2} xy\, dz\, dy\, dx$

12. $\displaystyle\int_{-1}^1 \int_0^{\sqrt{1-x^2}} \int_{x^2+y^2}^{1} z\, dz\, dy\, dx$

In Exercises 13–18, use spherical coordinates.

13. Find the volume of the solid bounded above by the sphere $x^2 + y^2 + z^2 - 2z = 0$ and below by the cone $z = \sqrt{x^2 + y^2}$.

14. Find the mass of a sphere of radius $r = 2$ if the density at each point is proportional to the square of the distance from the center.

15. Find the center of mass of the quarter sphere

$$Q = \{(\rho, \theta, \phi) | 0 \le \rho \le 1,\ \ 0 \le \phi \le \pi/2,\ 0 \le \theta \le \pi\}$$

if the density is constant.

16. Find the mass of the solid lying between the spheres $x^2 + y^2 + z^2 = 1$ and $x^2 + y^2 + z^2 = 4$ if the density at each point is proportional to the reciprocal of the distance from the center of the spheres.

17. Find the volume of the solid remaining when the region lying inside the cone $z^2 = x^2 + y^2$ is removed from the region bounded by the sphere $x^2 + y^2 + z^2 = 4$.

18. Use spherical coordinates to obtain the formula for the volume of a right circular cone of radius r and height h.

In Exercises 19 and 20, evaluate the triple integral by first changing to spherical coordinates.

19. $\displaystyle\int_{-1}^1 \int_{-\sqrt{1-x^2}}^{\sqrt{1-x^2}} \int_0^{\sqrt{1-x^2-y^2}} 3\, dz\, dy\, dx$

20. $\displaystyle\int_{-2}^2 \int_{-\sqrt{4-x^2}}^{\sqrt{4-x^2}} \int_{-\sqrt{4-x^2-y^2}}^{\sqrt{4-x^2-y^2}} (x^2 + y^2)\, dz\, dy\, dx$

21. Two circular cylinders of radius R meet at right angles. Use triple integration to find the volume of the solid common to both cylinders.

22. Find the volume of the region lying inside the sphere $x^2 + y^2 + z^2 = 4$ and outside the cylinder $x^2 + z^2 = 1$.

23. Evaluate the triple integral

$$\iiint_Q \frac{z}{(x^2 + y^2)^{3/2}}\, dV$$

where Q is the region

$$Q = \{(x, y, z) | 1 \le x^2 + y^2 \le 3,\ 0 \le z \le 3\}.$$

24. Evaluate the triple integral $\displaystyle\iiint_Q \cos \pi y \cdot \sqrt{x^2 + z^2}\, dV$ where Q is the region bounded by the cylinders $x^2 + z^2 = 1$ and $x^2 + z^2 = 4$, and the planes $y = -1$ and $y = 2$.

25. Evaluate the triple integral $\displaystyle\iiint_Q \sqrt{\frac{x}{y^2 + z^2}}\, dV$ where Q is the region bounded by the cone $y^2 + z^2 = x^2$, the cylinder $y^2 + z^2 = 4$, and the planes $x = 0$ and $x = 2$.

26. Find the volume of the solid that remains when the cone $3y^2 = x^2 + z^2$ is removed from the sphere $x^2 + y^2 + z^2 = 4$.

27. Find the volume of the smaller of the two parts of the sphere $\rho = 4$ determined by the plane $y = 2$.

28. Evaluate the integral $\displaystyle\iiint_Q (x^2 + y^2)\, dV$ where Q is the sphere $x^2 + y^2 + z^2 \le 4$.

29. Find the volume of the solid lying between the spheres $x^2 + y^2 + z^2 = 1$ and $x^2 + y^2 + z^2 = 9$ and inside the cone $y^2 = x^2 + z^2$.

30. Find the volume of the solid bounded below by the graph of $x^2 + y^2 + z^2 + 2z = 0$ and above by the graph of $z^2 = x^4 + 2x^2y^2 + y^4$.

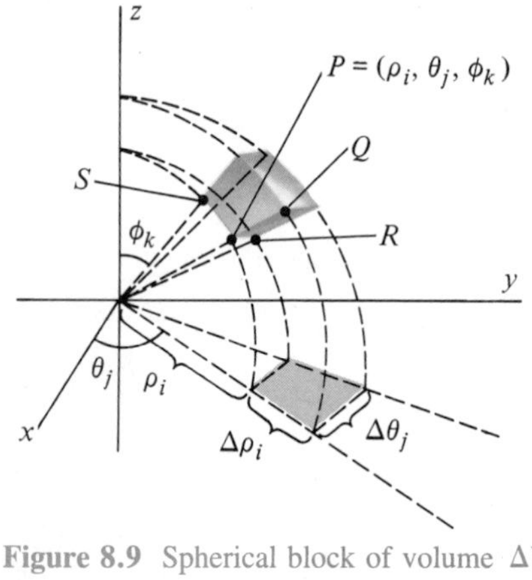

Figure 8.9 Spherical block of volume ΔV_{ijk}.

31. Obtain the approximation $\Delta V_{ijk} \approx \rho_i^2 \sin \phi_k \Delta\rho_i \Delta\theta_j \Delta\phi_k$ for the spherical block in Figure 8.9 as follows.

a. Observe that the spherical block is nearly a parallelepiped. We shall therefore approximate its volume by the product (length of PQ) · (length of PS) · (length of PR).

b. Show that length of $PQ = \Delta\rho_i$.

c. Show that the length of the arc PS is $\rho_i \Delta\phi_k$. (*Hint:* It lies on a circle of radius ρ_i.)

d. Show that the length of the arc PR is $\rho_i \sin \phi_k \Delta\theta_j$. (*Hint:* It lies on a horizontal circle of radius $\rho_i \sin \phi_k$. Why?)

SUMMARY OUTLINE OF CHAPTER 20

- If $f(x, y)$ is continuous on the planar region $Q = \{(x, y) | a \le x \le b,\ g_1(x) \le y \le g_2(x)\}$ then

  (i) $\displaystyle\iint_Q f(x, y)\, dA = \int_a^b \int_{g_1(x)}^{g_2(x)} f(x, y)\, dy\, dx.$

  (ii) Area of $Q = \displaystyle\iint_Q 1\, dA.$

  (iii) If the region Q can be described as

  $$Q = \{(x, y) | h_1(y) \le x \le h_2(y),\ c \le y \le d\},$$

  then also

  $$\iint_Q f(x, y)\, dA = \int_c^d \int_{h_1(y)}^{h_2(y)} f(x, y)\, dx\, dy.$$

- If $f(r, \theta)$ is continuous on the region Q described in polar coordinates as

  $$Q = \{(r, \theta) | g_1(\theta) \le r \le g_2(\theta),\ a \le \theta \le b\},$$

  then

  (i) $\displaystyle\iint_Q f(r, \theta)\, dA = \int_a^b \int_{g_1(\theta)}^{g_2(\theta)} f(r, \theta)\, r\, dr\, d\theta.$

  (ii) Area of $Q = \displaystyle\int_a^b \int_{g_1(\theta)}^{g_2(\theta)} r\, dr\, d\theta.$

- In changing a double integral from Cartesian to polar coordinates, the differential for area $dA = dx\, dy$ is replaced by the polar differential for area $dA = r\, dr\, d\theta$.

- If a thin lamina with density $\rho(x, y)$ has the shape of the planar region Q, then

  (i) Mass $= M = \displaystyle\iint_Q \rho(x, y)\, dA.$

  (ii) Center of Mass $= (\overline{x}, \overline{y})$ where $\overline{x} = \dfrac{M_y}{M}$, $\overline{y} = \dfrac{M_x}{M}$,

  $$M_y = \iint_Q x\rho(x, y)\, dA$$

  and

  $$M_x = \iint_Q y\rho(x, y)\, dA.$$

- The **surface area** of the graph of $z = f(x, y)$ over the region Q is

  $$A_s = \iint_Q \sqrt{\left[\frac{\partial f}{\partial x}(x, y)\right]^2 + \left[\frac{\partial f}{\partial y}(x, y)\right]^2 + 1}\, dA.$$

- If the region Q in $\mathbb{R}^3$ is described by the inequalities

  $$a \le x \le b,\quad h_1(x) \le y \le h_2(x),\quad g_1(x, y) \le z \le g_2(x, y),$$

  then

  (i) $\displaystyle\iiint_Q f(x, y, z)\, dV = \int_a^b \int_{h_1(x)}^{h_2(x)} \int_{g_1(x,y)}^{g_2(x,y)} f(x, y, z)\, dz\, dy\, dx.$

  (ii) Volume of $Q = \displaystyle\iiint_Q 1\, dV.$

- If a solid has the shape of the region Q in $\mathbb{R}^3$ and density function $\rho(x, y, z)$, then

  (i) Mass $= M = \displaystyle\iiint_Q \rho(x, y, z)\, dV.$

  (ii) Center of mass $= (\overline{x}, \overline{y}, \overline{z})$ where $\overline{x} = \dfrac{M_{yz}}{M}$, $\overline{y} = \dfrac{M_{xz}}{M}$, $\overline{z} = \dfrac{M_{xy}}{M}$,

  $$M_{yz} = \iiint_Q x\rho(x, y, z)\, dV,$$

  $$M_{xz} = \iiint_Q y\rho(x, y, z)\, dV,$$

  $$M_{xy} = \iiint_Q z\rho(x, y, z)\, dV.$$

- If the region Q is described in cylindrical coordinates by the inequalities

  $$a \le \theta \le b,\quad h_1(\theta) \le r \le h_2(\theta),\quad g_1(r, \theta) \le z \le g_2(r, \theta),$$

  then

  $$\iiint_Q f(r, \theta, z)\, dV = \int_a^b \int_{h_1(\theta)}^{h_2(\theta)} \int_{g_1(r,\theta)}^{g_2(r,\theta)} f(r, \theta, z) r\, dz\, dr\, d\theta.$$

- In changing a triple integral from Cartesian coordinates to cylindrical coordinates, the Cartesian differential for volume $dV = dx\,dy\,dz$ is replaced by the cylindrical differential for volume $dV = r\,dz\,dr\,d\theta$.

- If the region Q in $\mathbb{R}^3$ is described in spherical coordinates by the inequalities

$$a \le \theta \le b,\ \ h_1(\theta) \le \phi \le h_2(\theta),\ \ g_1(\theta, \phi) \le \rho \le g_2(\theta, \phi),$$

then

$$\iiint_Q f(\rho, \theta, \phi)\,dV = \int_a^b \int_{h_1(\theta)}^{h_2(\theta)} \int_{g_1(\theta,\phi)}^{g_2(\theta,\phi)} f(\rho, \theta, \phi)\rho^2 \sin\phi\,d\rho\,d\phi\,d\theta.$$

- In changing a triple integral from Cartesian to spherical coordinates, the Cartesian differential for volume $dV = dx\,dy\,dz$ is replaced by the spherical differential for volume $dV = \rho^2 \sin\phi\,d\rho\,d\phi\,d\theta$.

REVIEW EXERCISES—CHAPTER 20

1. Find the volume of the solid bounded above by the graph of $z = xy^2$ and below by the triangle with vertices $(0, 0)$, $(2, 0)$, and $(0, 1)$.

2. Evaluate the double integral $\iint_Q \frac{x}{xy + 2}\,dA$ where Q is the rectangle

$$Q = \{(x, y) | 0 \le x \le 1,\ 0 \le y \le 1\}.$$

3. Evaluate $\iint_Q e^{-x^2/2}\,dy\,dx$ where Q is the region

$$Q = \{(x, y) | 0 \le x \le 1,\ 0 \le y \le 2x\}.$$

4. Evaluate $\iint_Q (x - 4y)\,dA$ where Q is the rectangle

$$Q = \{(x, y) | -1 \le x \le 1,\ 0 \le y \le 2\}.$$

5. Find the volume of the solid in the first octant bounded by the graphs of $z = y^2$, $y = x$, $z = 0$, and $y = 4$.

6. Change the order of integration:

$$\int_{-1}^{2} \int_{x^2-2}^{x} f(x, y)\,dy\,dx.$$

7. Find the surface area of that part of the sphere $x^2 + y^2 + z^2 = 4$ lying inside the cylinder $x^2 + y^2 - 2x = 0$.

8. Find the center of mass of a solid having the shape of the region bounded above by the paraboloid $4z = 4 - (x^2 + y^2)$ and below by the xy-plane if the density of the material is uniform.

9. Find the volume of the solid bounded by the cylinder $r = 2\cos\theta$, the paraboloid $z = 2r^2$ and the plane $z = 0$ (use cylindrical coordinates).

10. Rewrite the integral

$$\int_{-\pi/2}^{\pi/2} \int_0^{2\cos\theta} \int_{-\sqrt{4-r^2}}^{\sqrt{4-r^2}} r\,dz\,dr\,d\theta$$

in rectangular coordinates.

11. Find the volume of the solid bounded by the graph of $\frac{x}{a} + \frac{y}{b} + \frac{z}{c} = 1$ $(a > 0,\ b > 0,\ c > 0)$ and the three coordinate planes.

12. Sketch the region Q corresponding to the iterated integral, reverse the order of integration, and evaluate the resulting integral:

$$\int_{-2}^{0} \int_{-\sqrt{x+2}}^{\sqrt{x+2}} y^2\,dy\,dx.$$

13. Calculate the volume of the solid lying inside the cylinder $x^2 + y^2 = 4$ and between the planes $y + z = 9$ and $z = 0$.

14. Use a double integral to calculate the area of the region lying between the graph of $\sqrt{x} + \sqrt{y} = 1$ and the graph of $x + y = 1$.

15. Find the center of mass of a solid bounded by the cylinder $r = 2$, the cone $z = r$, and the plane $z = 0$ if the density of the material is uniform.

16. Calculate the volume of the ellipse

$$\frac{x^2}{4} + \frac{y^2}{9} + \frac{z^2}{4} = 1.$$

17. Find the volume of the region common to the sphere $r^2 + z^2 = a^2$ and the cylinder $r = a\cos\theta$.

18. Find the surface area of the paraboloid $z = x^2 + y^2$ lying between the planes $z = 1$ and $z = 9$.

19. Find the area of the region cut from the plane $z = 4y$ by the cylinder $x^2 + y^2 = 4$.

20. Find the volumes of the two regions cut from the sphere $\rho = 4$ by the plane $x = 2$.

21. Find the volume of the region common to the sphere $x^2 + y^2 + z^2 = 16$ and the cylinder $x^2 + z^2 = 4$.

22. Use a double integral to find the area enclosed by the lemniscate $r^2 = 2\cos 2\theta$.

23. A solid is bounded below by the region bounded by the graphs of $y = x$ and $y = x^2 - 2$. It is bounded above by the plane $z - x + 2y = 10$. Find its volume.

24. Find $\iint_Q \frac{\sin x}{x}\, dA$ where Q is the triangle with vertices

(0, 0), (2, 0), and (2, 2).

25. Find the centroid of the half disc $r = 4$, $0 \le \theta \le \pi$.

26. Find the mass of a right circular cone of radius r and height h if the density at each point is proportional to the distance of that point from the vertex.

27. Find $\iiint_Q e^{(x^2+y^2+z^2)^{3/2}}\, dV$ where Q is the half sphere

$$Q = \{(r, \theta, \phi) | 0 \le r \le 1, 0 \le \phi \le \pi/2, 0 \le \theta \le 2\pi\}.$$

28. Show that if a thin lamina of constant density has the shape of the region Q described in polar coordinates, then the coordinates $(\bar{x}, \bar{y})$ of the center of mass may be calculated by the formulas

$$\bar{x} = \frac{1}{\text{area}} \iint_Q r^2 \cos\theta\, dr\, d\theta;$$

$$\bar{y} = \frac{1}{\text{area}} \iint_Q r^2 \sin\theta\, dr\, d\theta.$$

29. Use the result of Exercise 28 to calculate the center of mass for a thin lamina of constant density whose shape is the cardioid $r = 1 + \cos\theta$.

30. Find the volume of the solid bounded above by the plane $z - x = 2$ and below by the paraboloid $z = x^2 + y^2$.

31. Find the volume of the region bounded by the paraboloids

$$z = 4 - x^2 + 2x - y^2 - 4y$$

and

$$z = x^2 - 2x + y^2 + 4y + 5.$$

32. Find the volume of the solid bounded above by the cylinder $z = 4 - x^2$ and below by the paraboloid $3x^2 + y^2 = z$.

33. A solid corresponds to the region bounded by the cylinder $x^2 + y^2 = 4$ and the planes $z = 0$ and $z = 4$. Calculate the mass of the solid if the density at each point is proportional to the distance from the xy-plane.

34. Evaluate the integral $\iint_Q \cos\sqrt{x^2 + y^2}\, dA$ where Q is the disc

$$Q = \{(x, y) | 0 \le x^2 + y^2 \le 4\}.$$

35. Evaluate the integral $\iint_Q \frac{1}{\sqrt{1 + x^2 + y^2}}\, dA$ where Q is the quarter circle

$$Q = \{(x, y) | 0 \le x \le 1, 0 \le y \le \sqrt{1 - x^2}\}.$$

36. Find the volume of the solid bounded above by the graph of $z = 4 - r$ and below by the region in the $r\theta$ plane bounded by the graph of $r = 3\sin\theta$.

37. Find the volume of the solid bounded above by the cone $z^2 = x^2 + y^2$, on the sides by the cylinder $x^2 + y^2 - 4y = 0$, and below by the xy-plane.

38. Find the volume of the solid inside the cylinder $x^2 - 4x + y^2 = 0$ lying above the xy-plane and below the plane $z - x = 4$.

39. *(Computer)* Use Program 12 to approximate

$$\int_0^2 \int_1^3 \sin^2\sqrt{x - y^2}\, dy\, dx.$$

40. Find the area of the region bounded by the graphs of $y = \sqrt{x + 2}$ and $x - 3y + 2 = 0$ by double integration.

41. Find the volume of the solid bounded by the cone $z^2 = x^2 + y^2$ and the cylinder $x^2 + y^2 = 9$.

42. Find the centroid of the region bounded by the graph of $r = \cos 2\theta$.

43. Find the surface area of the part of the sphere $x^2 + y^2 + z^2 = 4$ lying outside the cylinder $x^2 + y^2 = 1$.

CHAPTER 21

VECTOR ANALYSIS

21.1 INTRODUCTION

The goal of this final chapter is to develop a theory of integral calculus for vector-valued functions defined in the plane or in space. In so doing we shall make use of many of the concepts developed in Chapters 17 through 21. The primary difference between our earlier work and what we do here is that we shall now be concerned with functions of the form $\boldsymbol{w} = \boldsymbol{F}(\boldsymbol{r})$ where both $\boldsymbol{w}$ and $\boldsymbol{r}$ are vectors. To avoid confusion between vector functions of this type and vector functions of a single real variable (the topic of Chapter 18), we shall use the term *vector fields*.

DEFINITION 1

Let Q be a subset of $\mathbb{R}^3$. A **vector field** on Q is a function

$$\boldsymbol{F}(x, y, z) = M(x, y, z)\boldsymbol{i} + N(x, y, z)\boldsymbol{j} + P(x, y, z)\boldsymbol{k} \tag{1}$$

that assigns a vector $\boldsymbol{w} = \boldsymbol{F}(x, y, z)$ to each point (x, y, z) in Q.

If Q is a subset of $\mathbb{R}^2$, a vector field on Q is a function of the form

$$\boldsymbol{F}(x, y) = M(x, y)\boldsymbol{i} + N(x, y)\boldsymbol{j}. \tag{2}$$

The real-valued functions $M(x, y, z)$, $N(x, y, z)$, and $P(x, y, z)$ in equation (1) are referred to as the **component** functions of the vector field $\boldsymbol{F}$. We say that a vector field is **continuous** if each of its component functions is continuous. Using the position vector

$$\boldsymbol{r} = x\boldsymbol{i} + y\boldsymbol{j} + z\boldsymbol{k}, \qquad (x, y, z) \in Q,$$

we may write the vector field in equation (1) as simply $\boldsymbol{w} = \boldsymbol{F}(\boldsymbol{r})$. Functions of the form $w = f(x, y, z)$, which assign real numbers to vectors, will now be referred to as **scalar functions** or scalar fields.

The following examples indicate the types of vector fields and scalar functions that we shall study in this chapter.

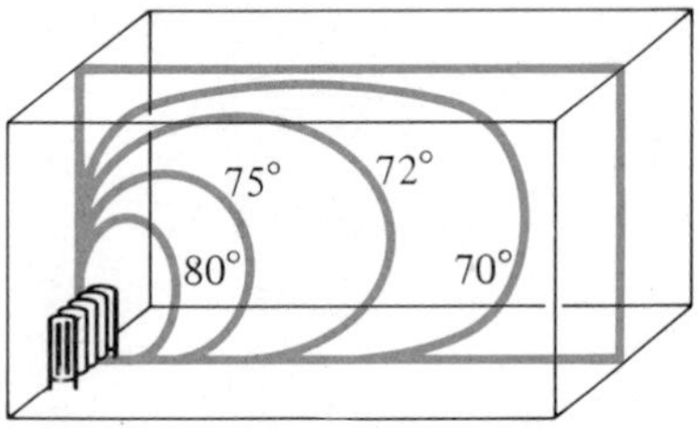

Figure 1.1 Isothermal curves in a cross section of a room with a single heat source.

Example 1 Figure 1.1 represents a room with a single radiator. The function $T(x, y, z)$, which gives the temperature at the point in the room with coordinates (x, y, z), is an example of a *scalar* function. The curves drawn in a cross section of the room are locations of points of equal temperature and are called **isothermal curves.**

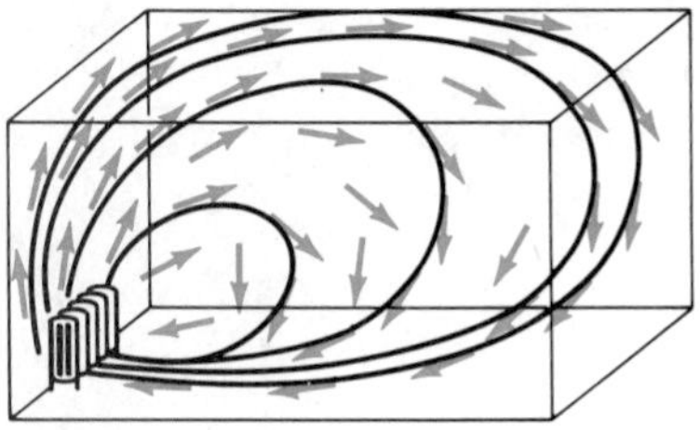

Figure 1.2 Tangents $F(x, y, z)$ to convection currents determine a vector field in the room.

Since warmer air rises and cooler air falls, the presence of a single heat source in a room causes air to flow about the room in **convection currents.** If the vector $\boldsymbol{F}(x, y, z)$ represents the velocity vector of the convection current passing through the point (x, y, z), the function $\boldsymbol{w} = \boldsymbol{F}(x, y, z)$ is an example of a vector field defined throughout the room (Figure 1.2). ■

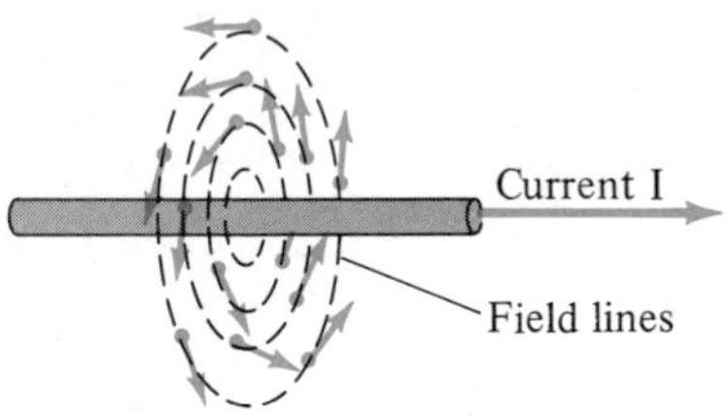

Figure 1.3 Tangents to magnetic field lines form a (magnetic) vector field in space.

Example 2 An electric current of magnitude I flowing through a thin wire induces a **magnetic field** around the wire. (The statement of Ampère's Law, giving the relationship between the electric current I and its magnetic effect, involves the idea of a *line integral,* introduced in Section 21.2.) The set of vectors $\boldsymbol{F}(x, y, z)$, giving the direction and intensity of the magnetic field at location (x, y, z), determines a vector field in the space surrounding the wire (Figure 1.3). ■

Example 3 An example of a force field is a **gravitational field.** According to Newton's Law of Gravitation, the magnitude of the force of attraction exerted on a particle P_2 of mass m_2 by a particle P_1 of mass m_1 is given by the equation

$$F = \frac{Gm_1m_2}{r^2} \tag{3}$$

where r is the distance between the particles and G is a constant. If we assume the particle P_1 to be located at the origin in xyz-space, then the force in (3) is a function of the coordinates (x, y, z) giving the location of the particle P_2. That is,

$$F(x, y, z) = \frac{Gm_1m_2}{x^2 + y^2 + z^2}. \tag{4}$$

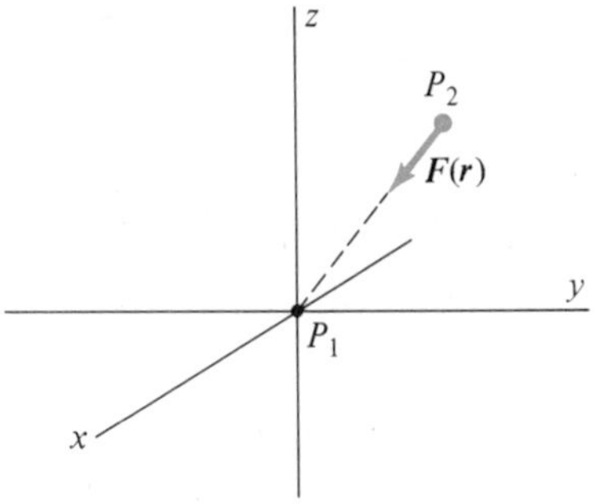

Figure 1.4 Gravitational force vector $\boldsymbol{F}(\boldsymbol{r})$ acting on particle P_2.

Since the force whose magnitude is given by (4) acts toward the origin, it acts in the direction opposite that of the position vector $\boldsymbol{r} = x\boldsymbol{i} + y\boldsymbol{j} + z\boldsymbol{k}$ of P_2. We may therefore write the force vector $\boldsymbol{F}(x, y, z)$ as

$$\boldsymbol{F}(x, y, z) = \left(\frac{Gm_1m_2}{(x^2 + y^2 + z^2)}\right)\left(-\frac{x\boldsymbol{i} + y\boldsymbol{j} + z\boldsymbol{k}}{\sqrt{x^2 + y^2 + z^2}}\right) \tag{5}$$

$$= \left(\frac{-Gm_1m_2}{(x^2 + y^2 + z^2)^{3/2}}\right)(x\boldsymbol{i} + y\boldsymbol{j} + z\boldsymbol{k}).$$

Using the position vector $\boldsymbol{r} = x\boldsymbol{i} + y\boldsymbol{j} + z\boldsymbol{k}$**,** we may write the force field in (5) as

$$\boldsymbol{F}(\boldsymbol{r}) = -\frac{Gm_1m_2}{|\boldsymbol{r}|^3}\boldsymbol{r}, \qquad \boldsymbol{r} \neq \boldsymbol{0} \qquad \text{(Figure 1.4).} \tag{6}$$

Since the vector $\boldsymbol{F}$ in (5) and (6) is defined for all position vectors $\boldsymbol{r} \neq \boldsymbol{0}$**,** either equation defines a vector field at all points of $\mathbb{R}^3$ except the origin (Figure 1.5). ■

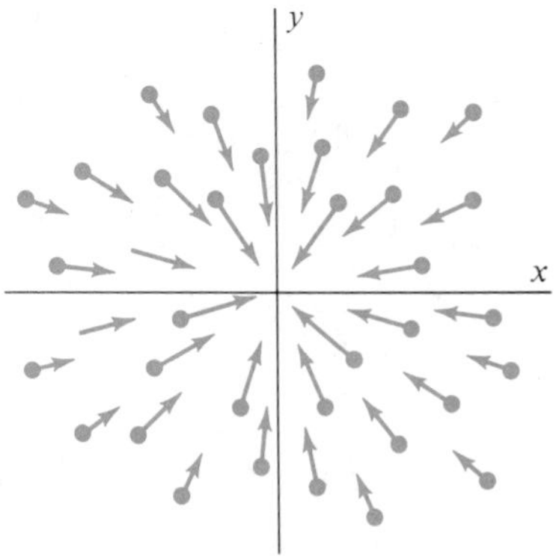

Figure 1.5 Gravitational force field (drawing represents xy-plane only).

The gravitational field of Example 3 is an example of a **central force field,** since the force vector at each point points toward the origin. More generally, a central force field in space has the form

$$\boldsymbol{F}(x, y, z) = f(x, y, z)(x\boldsymbol{i} + y\boldsymbol{j} + z\boldsymbol{k})$$

where $f(x, y, z)$ is a real-valued function of three variables.

Example 4 Another example of a central force field is obtained by letting $\boldsymbol{F}(x, y, z)$ be the vector representing the force exerted on a particle P_2 with electric charge q_2 at location (x, y, z) by a particle P_1 with electric charge q_1 located at the origin. According to **Coulomb's Law,** this force vector is

$$\boldsymbol{F}(\boldsymbol{r}) = \frac{kq_1q_2}{|\boldsymbol{r}|^3}\boldsymbol{r}, \qquad r \neq \boldsymbol{0} \tag{7}$$

where $\boldsymbol{r}$ is the position vector $\boldsymbol{r} = x\boldsymbol{i} + y\boldsymbol{j} + z\boldsymbol{k}$ of P_2, and k is a constant that depends on the choice of units for $\boldsymbol{r}$, q_1, and q_2. Figure 1.6 represents the **electric force field** $\boldsymbol{F}(\boldsymbol{r})$ (for positive charges P_2) due to a positive charge at the origin. Figure 1.7 represents the force field $\boldsymbol{F}(\boldsymbol{r})$ (for positive charges P_2) due to a negative charge at the origin. (Although the force field is defined in space, we show only the force vectors lying in the xy-plane in these figures.) ■

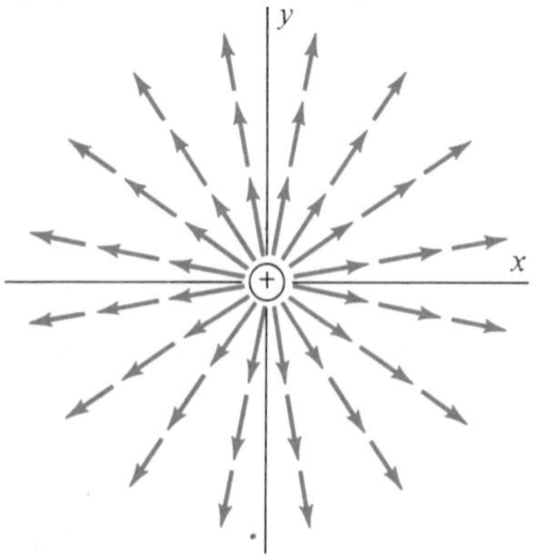

Figure 1.6 Electric force field due to a positive charge at the origin.

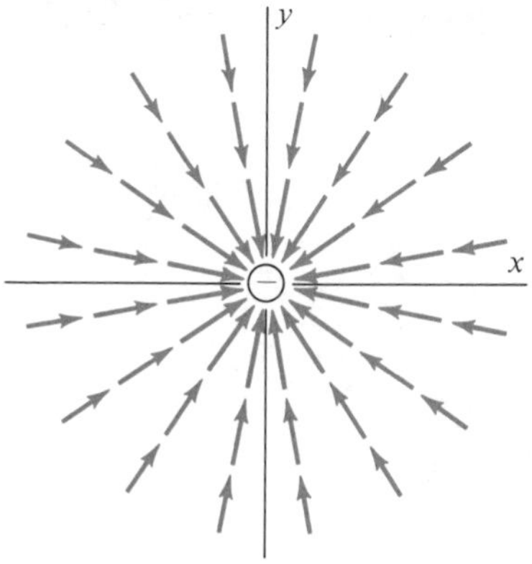

Figure 1.7 Electric field due to a negative charge at the origin.

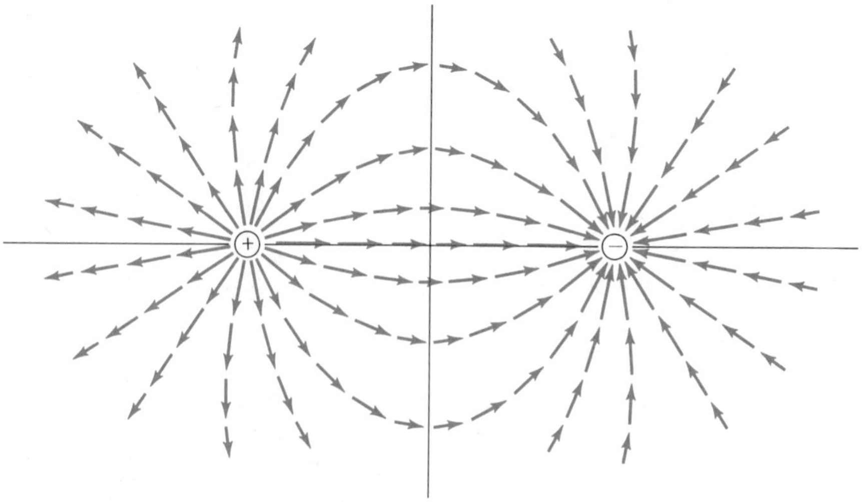

Figure 1.8 Electric force field due to two opposite charges is the vector sum of the individual central force fields.

If charge is considered to be analogous to mass, it is clear from comparison of equations (6) and (7) that the gravitational and electric force fields have the same form; physicists say that both forces obey an *inverse square law*. The major difference is that the direction of $\boldsymbol{F}(\boldsymbol{r})$ may be either toward or away from the origin for the electric field, but it can only be toward the origin for the gravitational field (since all masses are positive).

Electric force fields satisfy the **principle of superposition,** which states that the force acting on a charge at a point P due to two separate charges is the vector sum of the individual forces acting on the charge at P. Figure 1.8 represents the electric field, acting on a positive charge, that results from a positive charge at $(-1, 0)$ and a negative charge of equal magnitude at $(1, 0)$. Note that while the individual force fields in Figures 1.5 and 1.6 are central, their sum is not.

Example 5 If $f(x, y, z)$ is a differentiable scalar function of three variables, the gradient function

$$\boldsymbol{F}(x, y, z) = \nabla f(x, y, z) = f_x(x, y, z)\boldsymbol{i} + f_y(x, y, z)\boldsymbol{j} + f_z(x, y, z)\boldsymbol{k} \tag{8}$$

defines a vector field on $\mathbb{R}^3$. If $f(x, y)$ is a differentiable scalar function of two variables, the corresponding gradient field on $\mathbb{R}^2$ is

$$\boldsymbol{F}(x, y) = \nabla f(x, y) = f_x(x, y)\boldsymbol{i} + f_y(x, y)\boldsymbol{j}. \tag{9}$$

For example, the scalar function $f(x, y, z) = xe^{2y} \sin z$ yields the gradient vector field

$$\boldsymbol{F}(x, y, z) = e^{2y} \sin z\boldsymbol{i} + 2xe^{2y} \sin z\boldsymbol{j} + xe^{2y} \cos z\boldsymbol{k}. \qquad \blacksquare$$

As in Section 19.11, we will refer to the function $f(x, y, z)$ in equation (8) as a **scalar potential** for the vector field $\boldsymbol{F}(x, y, z)$. Although every differentiable scalar function yields a gradient vector field, a given vector field $\boldsymbol{F}(x, y, z)$ need not be the gradient of a scalar potential. Those vector fields that can be written as gradients of scalar potentials are called **conservative vector fields.**

There are many other familiar examples of vector fields. The flow of air currents around a moving automobile or airplane determines a **velocity field.** The study of such vector fields is called *aerodynamics*. Similarly, the velocity field determined by the flow of water through a container, such as a pipe or a dam, is the subject of *hydrodynamics*.

In this chapter we shall be answering two kinds of questions. The first is how to calculate the integral of a scalar function along a curve in the plane or in space. The principal motivation for this question is the problem of calculating the work done on an object moving along such a curve by a (force) vector field. The solution of this problem involves the notion of a *line integral*. Green's Theorem, the subject of Section 21.4, gives conditions under which the line integral of a given function over a closed plane curve is the same as the double integral of an associated function over the region interior to the curve.

The second major theme of the chapter is motivated by the question of calculating the rate at which a fluid (air or water, for example) is flowing in or out of a particular region, such as that determined by two cross sections in a water pipe. The idea of *surface integral* must be developed to pursue this issue (Section 21.5). The theorem due to Stokes (Section 21.6) relates the value of certain surface integrals over such regions to line integrals calculated over their bounding curves, while the

Divergence Theorem (Section 21.7) relates other types of surface integrals to triple integrals over the enclosed regions.

As these discussions develop, you are encouraged to keep in mind the simple examples of vector fields presented here as models by which you can interpret the ideas and results that lie ahead.

Exercise Set 21.1

In Exercises 1–10, sketch enough vectors in the given vector field to get a sense of the nature of the vector field.

1. $\boldsymbol{F}(x, y) = x\boldsymbol{i} - y\boldsymbol{j}$

2. $\boldsymbol{F}(x, y) = -y\boldsymbol{i} + x\boldsymbol{j}$

3. $\boldsymbol{F}(x, y) = 3\boldsymbol{i}$

4. $\boldsymbol{F}(x, y) = -x\boldsymbol{i} - y\boldsymbol{j}$

5. $\boldsymbol{F}(x, y) = \dfrac{x}{\sqrt{x^2 + y^2}}\boldsymbol{i} + \dfrac{y}{\sqrt{x^2 + y^2}}\boldsymbol{j}$

6. $\boldsymbol{F}(x, y) = \dfrac{x}{\sqrt{x^2 + y^2}}\boldsymbol{i} - \dfrac{y}{\sqrt{x^2 + y^2}}\boldsymbol{j}$

7. $\boldsymbol{F}(x, y, z) = \boldsymbol{j} + \boldsymbol{k}$

8. $\boldsymbol{F}(x, y, z) = x\boldsymbol{i} + y\boldsymbol{j}$

9. $\boldsymbol{F}(x, y, z) = 2x\boldsymbol{i} + 2y\boldsymbol{j} + 2z\boldsymbol{k}$

10. $\boldsymbol{F}(x, y, z) = \boldsymbol{i} - \boldsymbol{j} + \boldsymbol{k}$

11. Which of the vector fields in Exercises 1–10 are central?

Find the gradient vector field $\boldsymbol{F} = \nabla f$ for each of the functions in Exercises 12–17.

12. $f(x, y) = xy$

13. $f(x, y) = x^2 - y^2$

14. $f(x, y) = x \tan xy$

15. $f(x, y, z) = \sqrt{x^2 + y^2 + z^2}$

16. $f(x, y, z) = x \ln(y^2 + z^2)$

17. $f(x, y, z) = ze^{x-y}$

In Exercises 18–23, determine whether the given vector field is a gradient vector field. If so, find a potential ϕ with $\boldsymbol{F} = \nabla\phi$ (see Section 19.11).

18. $\boldsymbol{F}(x, y) = \sin y\boldsymbol{i} + \cos y\boldsymbol{j}$

19. $\boldsymbol{F}(x, y) = (\tan xy - xy \sec^2 xy)\boldsymbol{i} + x^2 \sec^2 xy\boldsymbol{j}$

20. $\boldsymbol{F}(x, y) = e^{\sqrt{xy}}\boldsymbol{i} - \sqrt{x}e^{\sqrt{xy}}\boldsymbol{j}$

21. $\boldsymbol{F}(x, y) = \left[\ln(y - x) + \dfrac{x}{y - x}\right]\boldsymbol{i} + \left(\dfrac{x}{y - x}\right)\boldsymbol{j}$

22. $\boldsymbol{F}(x, y, z) = x^2\boldsymbol{i} + x^2 \sin z\boldsymbol{j} - \cos yz\boldsymbol{k}$

23. $\boldsymbol{F}(x, y, z) = yze^{xyz}\boldsymbol{i} + xze^{xyz}\boldsymbol{j} + xye^{xyz}\boldsymbol{k}$

24. The **flow lines** for a vector field $\boldsymbol{F}(\boldsymbol{r})$ are curves $\boldsymbol{\alpha}(t)$ tangent to the vector field. That is, $\boldsymbol{\alpha}'(t) = \boldsymbol{F}(\boldsymbol{\alpha}(t))$ for every t and every flow line. (An example of flow lines associated with a vector field in the plane is shown in Figure 1.9.) Sketch several flow lines for each of the vector fields in Exercises 1–10.

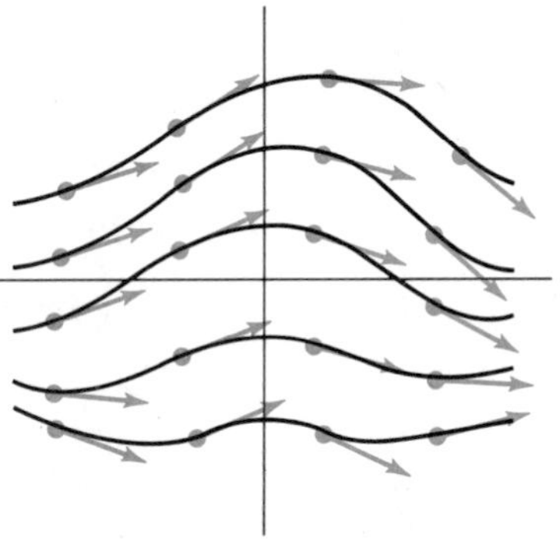

Figure 1.9 Flow lines for a vector field.

25. Imagine water flowing through a pipe as in Figure 1.10. At each point (x, y, z) within the pipe let $\boldsymbol{F}(x, y, z)$ be the velocity vector for the water at that point. This determines a vector field within the pipe. Why are the vectors in the narrow part of the pipe larger in magnitude?

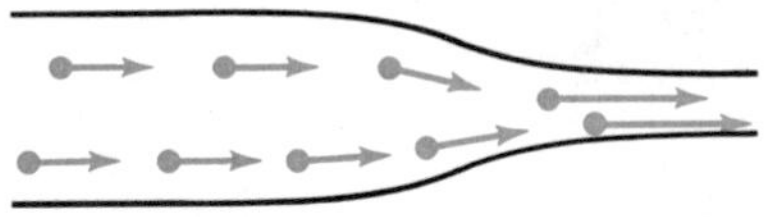

Figure 1.10 Velocity field in a water pipe.

26. Show that the central force field $\boldsymbol{F}(\boldsymbol{r}) = \left(\dfrac{k}{|\boldsymbol{r}|^3}\right)\boldsymbol{r}$ is the gradient of the scalar field $f(\boldsymbol{r}) = \dfrac{k}{|\boldsymbol{r}|}$. ($\boldsymbol{r}$ denotes the position vector $\boldsymbol{r} = x\boldsymbol{i} + y\boldsymbol{j} + z\boldsymbol{k}$ associated with the point (x, y, z).)

27. Let $\boldsymbol{F}(x, y, z)$ be the electric field describing the force on a particle with charge $q = +1$ at location (x, y, z) due to the combined effect of a charge $q_1 = +3$ at $(-1, 0, 0)$ and a charge $q_2 = -1$ at $(1, 0, 0)$.

a. Find the force vector $\boldsymbol{F}(0, 0, 0)$.

b. Find the force vector $\boldsymbol{F}(-2, 0, 0)$.

c. Sketch enough vectors $\boldsymbol{F}(x, y, 0)$ to get an idea of the force field in the xy-plane.

d. Sketch several of the flow lines for the vector field (see Exercise 24).

21.2 WORK AND LINE INTEGRALS

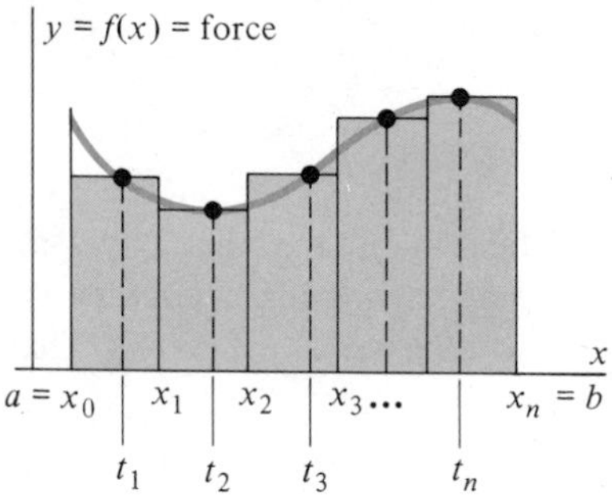

Figure 2.1 $W \approx \sum_{j=1}^{n} f(t_j)\,\Delta x$.

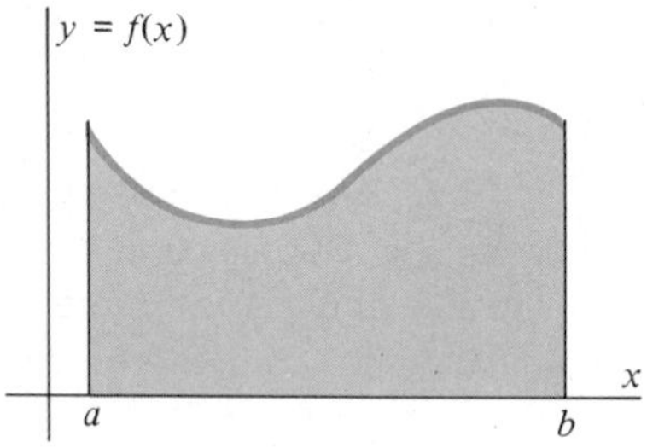

Figure 2.2 $W = \int_a^b f(x)\,dx$.

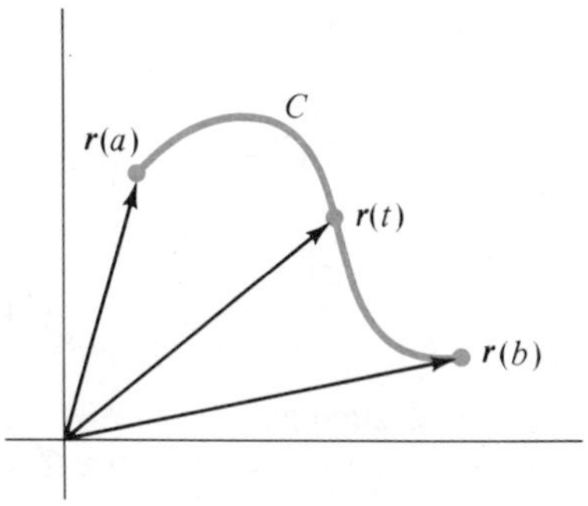

Figure 2.3 C: $\boldsymbol{r}(t) = x(t)\boldsymbol{i} + y(t)\boldsymbol{j}$.

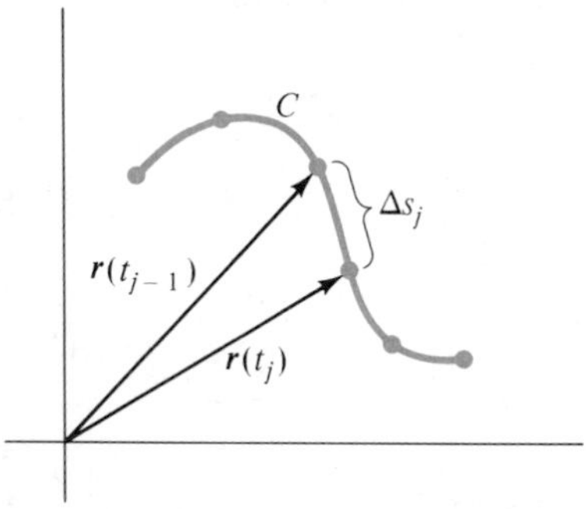

Figure 2.4 $\Delta s_j = \int_{t_{j-1}}^{t_j} |\boldsymbol{r}'(t)|\,dt$.

When a force of magnitude F moves an object d units along a line, the physicist defines the **work** done on the object by the force as the product $W = Fd$. That is,

$$\text{Work} = (\text{force}) \times (\text{distance}) \tag{1}$$

for constant forces applied along a line.

In Section 8.3 we applied the theory of the definite integral to calculate the work done by a continuously varying force $f(x)$ in moving an object from location $x = a$ to location $x = b$ along a line. The idea was simply to partition the interval $[a, b]$ into n subintervals of equal length $\Delta x = \dfrac{b-a}{n}$, with endpoints $a = x_0 < x_1 < x_2 < \cdots < x_n = b$, and to choose one "test number" t_j in each interval $[x_{j-1}, x_j]$. We then assumed the function $f(x)$ to have the constant value $f(t_j)$ throughout the interval $[x_{j-1}, x_j]$, and obtained the approximation $\Delta W_j \approx f(t_j)\,\Delta x$ to the work done by $f(x)$ on the jth interval $[x_{j-1}, x_j]$. This gave the approximation to the work done by $f(x)$ over the entire interval $[a, b]$ as

$$W = \sum_{j=1}^{n} \Delta W_j \approx \sum_{j=1}^{n} f(t_j)\,\Delta x. \tag{2}$$

Since $f(x)$ was assumed continuous on $[a, b]$, we concluded that the sum on the right-hand side of approximation (2) is a Riemann sum, whose limit as $n \to \infty$ is therefore a definite integral. This motivated the definition of the work done by $f(x)$ over the interval $[a, b]$ as

$$W = \lim_{n\to\infty} \sum_{j=1}^{n} f(t_j)\,\Delta x = \int_a^b f(x)\,dx, \tag{3}$$

which agrees with (1) when $f(x)$ is a constant force (Figures 2.1 and 2.2).

In this section we wish to generalize these concepts to calculate the work done on a particle by a force that moves the particle along a curve C lying in the plane or in space. Examples of this situation include small charged particles moving in electric fields, objects moving in gravitational fields, and others suggested by the discussion of Section 21.1.

To formulate the problem more precisely, assume that $\boldsymbol{F}$ is a vector field defined on an open subset D of $\mathbb{R}^3$ that determines a force vector $\boldsymbol{F}(\boldsymbol{r})$ for each position vector $\boldsymbol{r} = x\boldsymbol{i} + y\boldsymbol{j} + z\boldsymbol{k}$ in D. Also, let C be a curve lying within D that is parameterized by the vector function

$$C\colon\ \boldsymbol{r}(t) = x(t)\boldsymbol{i} + y(t)\boldsymbol{j} + z(t)\boldsymbol{k}, \qquad a \le t \le b \qquad \text{(see Figure 2.3)}. \tag{4}$$

If we partition the parameter interval $[a, b]$ into n subintervals of equal length $\Delta t = \dfrac{b-a}{n}$ with endpoints $a = t_0 < t_1 < t_2 < \cdots < t_n = b$, the position vectors $\boldsymbol{r}(t_0), \boldsymbol{r}(t_1), \ldots, \boldsymbol{r}(t_n)$ divide the curve C into n arcs. If the curve C is smooth (meaning that $\boldsymbol{r}'(t)$ is continuous for $t \in [a, b]$), equation (9) of Section 18.4 gives the length of the jth arc as

$$\Delta s_j = \int_{t_{j-1}}^{t_j} |\boldsymbol{r}'(t)|\,dt \qquad \text{(see Figure 2.4)}.$$

Under this assumption, the Mean Value Theorem for Integrals guarantees the existence of a number c_j in the interval $[t_{j-1}, t_j]$ for which

$$\Delta s_j = \int_{t_{j-1}}^{t_j} |\mathbf{r}'(t)|\, dt = |\mathbf{r}'(c_j)| \cdot (t_j - t_{j-1}) \tag{5}$$
$$= |\mathbf{r}'(c_j)|\, \Delta t.$$

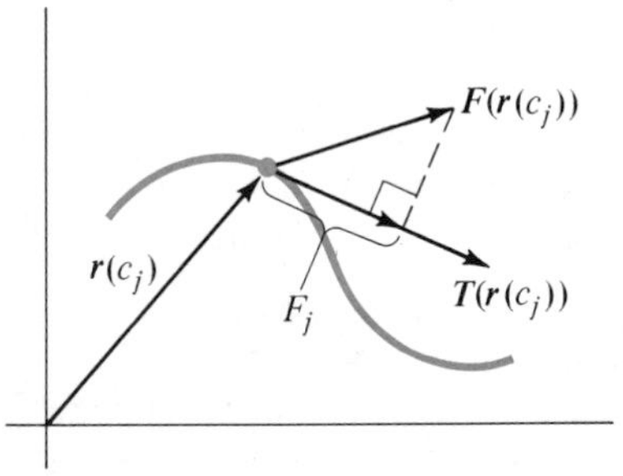

Figure 2.5 F_j is the component of the force field in the direction of the unit tangent at $\mathbf{r}(c_j)$.

To approximate the work ΔW_j done by the force field in moving a particle along the jth arc, we assume that the magnitude F_j of the force acting throughout the jth arc is the tangential component of $\mathbf{F}(\mathbf{r})$ at the point with position vector $\mathbf{r}(c_j)$. That is, we use

$$F_j = \text{comp}_{\mathbf{T}(\mathbf{r}(c_j))}\, \mathbf{F}(\mathbf{r}(c_j)) = \mathbf{F}(\mathbf{r}(c_j)) \cdot \mathbf{T}(\mathbf{r}(c_j)) \tag{6}$$

where $\mathbf{T}(\mathbf{r}(c_j))$ is the unit tangent to C at $\mathbf{r}(c_j)$ (see Figure 2.5).

Multiplying (5) and (6) and summing over all n arcs, we obtain the approximation

$$W \approx \sum_{j=1}^{n} F_j\, \Delta s_j = \sum_{j=1}^{n} \mathbf{F}(\mathbf{r}(c_j)) \cdot \mathbf{T}(\mathbf{r}(c_j)) |\mathbf{r}'(c_j)|\, \Delta t \tag{7}$$

for the work done by the force field $\mathbf{F}$ in moving the particle over the curve C. If $\mathbf{r}'(t)$ is continuous on $[a, b]$, the sum on the right-hand side of (7) is, as before, a Riemann sum. We therefore define the work W as the limit as $n \to \infty$ of this approximating sum:

$$W = \lim_{n\to\infty} \sum_{j=1}^{n} \mathbf{F}(\mathbf{r}(c_j)) \cdot \mathbf{T}(\mathbf{r}(c_j)) |\mathbf{r}'(c_j)|\, \Delta t \tag{8}$$
$$= \int_a^b \mathbf{F}(\mathbf{r}(t)) \cdot \mathbf{T}(\mathbf{r}(t)) |\mathbf{r}'(t)|\, dt.$$

Equation (8) may be simplified somewhat by recalling that, if $|\mathbf{r}'(t)| \neq 0$*, the unit tangent $\mathbf{T}(\mathbf{r}(t))$ to C at $\mathbf{r}(t)$ is given by

$$\mathbf{T}(\mathbf{r}(t)) = \frac{\mathbf{r}'(t)}{|\mathbf{r}'(t)|}. \tag{9}$$

Combining equations (8) and (9) leads, finally, to the desired definition of work done by a (continuous) force field in moving a particle along a smooth curve C.

DEFINITION 2

Let $\mathbf{F}$ be a continuous force field defined in some open set containing the smooth curve

$$C:\ \mathbf{r}(t) = x(t)\mathbf{i} + y(t)\mathbf{j} + z(t)\mathbf{k}, \qquad a \le t \le b, \qquad |\mathbf{r}'(t)| \neq 0.$$

The **work done by the force field F** in moving a particle along C from $\mathbf{r}(a)$ to $\mathbf{r}(b)$ is given by the definite integral

$$W = \int_a^b \mathbf{F}(\mathbf{r}(t)) \cdot \mathbf{r}'(t)dt \tag{10}$$

*The requirement that $|\mathbf{r}'(t)| \neq 0$ is not as restrictive as it might seem, since in most applications we are free to choose the particular parameterization used to describe the curve C.

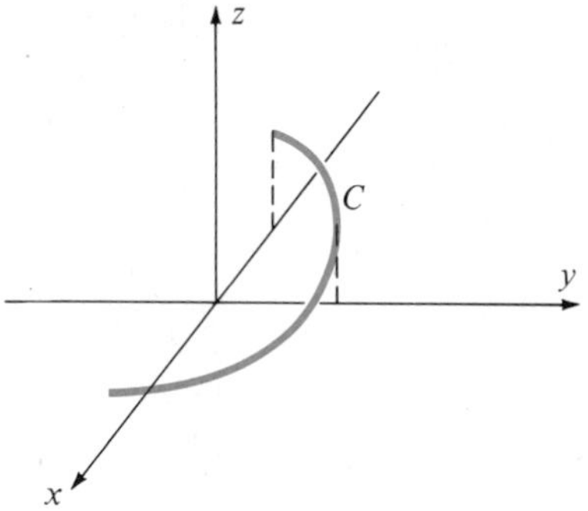

Figure 2.6 The helix C: $\boldsymbol{r}(t) = \cos t\boldsymbol{i} + \sin t\boldsymbol{j} + t\boldsymbol{k}$.

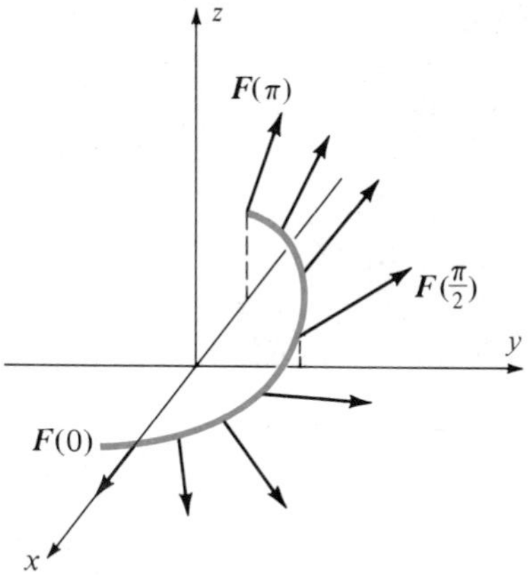

Figure 2.7 Selected vectors in the force field $\boldsymbol{F}(x, y, z) = 2x\boldsymbol{i} + 3y\boldsymbol{j} + z\boldsymbol{k}$ for which (x, y, z) lies on C.

REMARK: For the curve C as in Definition 2,

$$\boldsymbol{r}'(t) = \frac{d}{dt}(\boldsymbol{r}(t)) = \frac{dx}{dt}\boldsymbol{i} + \frac{dy}{dt}\boldsymbol{j} + \frac{dz}{dt}\boldsymbol{k},$$

which suggests the notation

$$d\boldsymbol{r} = \boldsymbol{r}'(t)\, dt. \tag{11}$$

Using (11), we sometimes abbreviate the integral in (10) as

$$\boxed{W = \int_C \boldsymbol{F} \cdot d\boldsymbol{r}.} \tag{12}$$

In using equation (12), it is important to remember that one must first find a smooth parameterization $\boldsymbol{r}(t)$ for C and then proceed as in equation (10). It is a fact, which we shall not prove, that the value of the integral in (12) does not depend on the particular parameterization chosen for C.

Example 1 Find the work done by the force field

$$\boldsymbol{F}(x, y, z) = 2x\boldsymbol{i} + 3y\boldsymbol{j} + z\boldsymbol{k} \qquad \text{(see Figures 2.6 and 2.7)}$$

in moving a particle along the circular helix

$$C: \quad \boldsymbol{r}(t) = \cos t\boldsymbol{i} + \sin t\boldsymbol{j} + t\boldsymbol{k}$$

from point $\boldsymbol{r}(0) = \boldsymbol{i}$ to point $\boldsymbol{r}(\pi) = -\boldsymbol{i} + \pi\boldsymbol{k}$.

Solution: The parameterization given for C is

$$x(t) = \cos t; \qquad y(t) = \sin t; \qquad z(t) = t,$$

so

$$\boldsymbol{F}(\boldsymbol{r}(t)) = 2\cos t\boldsymbol{i} + 3\sin t\boldsymbol{j} + t\boldsymbol{k}.$$

Also,

$$\boldsymbol{r}'(t) = -\sin t\boldsymbol{i} + \cos t\boldsymbol{j} + \boldsymbol{k}.$$

Thus, by Definition 2,

$$\begin{aligned} W &= \int_0^\pi [2\cos t\boldsymbol{i} + 3\sin t\boldsymbol{j} + t\boldsymbol{k}] \cdot [-\sin t\boldsymbol{i} + \cos t\boldsymbol{j} + \boldsymbol{k}]\, dt \\ &= \int_0^\pi (-2\cos t\sin t + 3\sin t\cos t + t)\, dt = \int_0^\pi (\sin t\cos t + t)\, dt \\ &= \frac{1}{2}\sin^2 t + \frac{1}{2}t^2\Big]_0^\pi = \frac{\pi^2}{2}. \end{aligned}$$

■

Example 2 Find the work done by the force field

$$\boldsymbol{F}(x, y) = 3y\boldsymbol{i} - x^2\boldsymbol{j}$$

in moving a particle along the plane curve $y = \sqrt{x}$ from the point (1, 1) to the point (4, 2).

Solution: Here we must first find a parameterization for the arc C of the graph of $y = \sqrt{x}$ from (1, 1) to (4, 2). We do so by setting $x(t) = t$, $y(t) = \sqrt{t}$. Then

$$\boldsymbol{r}(t) = t\boldsymbol{i} + \sqrt{t}\,\boldsymbol{j}, \qquad 1 \le t \le 4,$$

$$\boldsymbol{r}'(t) = \boldsymbol{i} + \frac{1}{2\sqrt{t}}\boldsymbol{j},$$

and

$$\boldsymbol{F}(\boldsymbol{r}(t)) = 3\sqrt{t}\,\boldsymbol{i} - t^2\boldsymbol{j}.$$

Thus, by (10)

$$\begin{aligned} W &= \int_1^4 [3\sqrt{t}\,\boldsymbol{i} - t^2\boldsymbol{j}] \cdot \left[\boldsymbol{i} + \frac{1}{2\sqrt{t}}\boldsymbol{j}\right] dt \\ &= \int_1^4 \left(3t^{1/2} - \frac{1}{2}t^{3/2}\right) dt \\ &= 2t^{3/2} - \frac{1}{5}t^{5/2}\Big]_1^4 \\ &= \frac{39}{5}. \end{aligned}$$

■

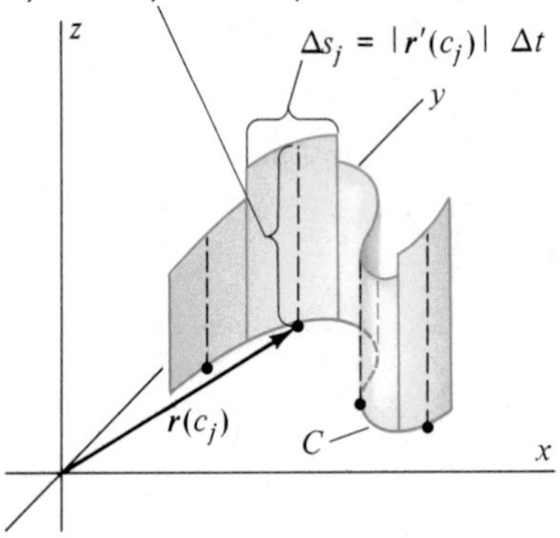

Figure 2.8 $\boldsymbol{F}(\boldsymbol{r}(c_j)) \cdot \boldsymbol{T}(\boldsymbol{r}(c_j))|\boldsymbol{r}'(c_j)|\,\Delta t$ in (8) is surface area of a curved slab over C.

Example 3 Show that the result of Example 2 remains unchanged if the parameterization $x(t) = t^2$, $y(t) = t$, $1 \le t \le 2$, is used for C.

Solution: Here $\boldsymbol{r}(t) = t^2\boldsymbol{i} + t\boldsymbol{j}$, $\quad \boldsymbol{r}'(t) = 2t\boldsymbol{i} + \boldsymbol{j}$, and $\boldsymbol{F}(\boldsymbol{r}(t)) = 3t\boldsymbol{i} - t^4\boldsymbol{j}$. Thus,

$$\begin{aligned} W &= \int_1^2 [3t\boldsymbol{i} - t^4\boldsymbol{j}] \cdot [2t\boldsymbol{i} + \boldsymbol{j}]\, dt \\ &= \int_1^2 (6t^2 - t^4)\, dt = 2t^3 - \frac{1}{5}t^5\Big]_1^2 = \frac{39}{5}. \end{aligned}$$

This result illustrates our earlier remark that the value of the integral in (10) depends only on $\boldsymbol{F}$ and on C, not on the particular parameterization chosen for C. ■

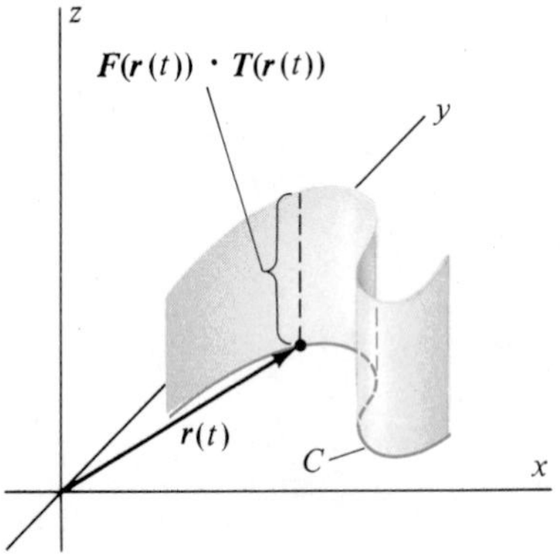

Figure 2.9 $\int_C \boldsymbol{F} \cdot d\boldsymbol{r}$ is a surface area.

Figure 2.8 gives a geometric interpretation of the approximating sum in equation (7). As Figure 2.9 illustrates, the integrals in equations (8) and (10) may be interpreted as surface areas of curved vertical "slabs" of variable height $\boldsymbol{F}(\boldsymbol{r}(t)) \cdot \boldsymbol{T}(\boldsymbol{r}(t))$.

Line Integrals

The existence of the integral in Definition 2 depends only on the mathematical properties of the force field $\boldsymbol{F}(x, y, z)$ and the curve C. We may therefore generalize Definition 2 to apply to any vector field $\boldsymbol{F}$, not necessarily a force field. The result is called a *line integral*.

DEFINITION 3

Let $\boldsymbol{F}$ be a continuous vector field in some open region D and let C be a smooth curve lying within D with parameterization $C = \{\boldsymbol{r}(t) | a \leq t \leq b\}$. The **line integral** of $\boldsymbol{F}$ over C is defined as

$$\int_C \boldsymbol{F} \cdot d\boldsymbol{r} = \int_a^b \boldsymbol{F}(\boldsymbol{r}(t)) \cdot \boldsymbol{r}'(t)\, dt. \tag{13}$$

Definitions 2 and 3 are the same except that Definition 2 refers to the more specific situation of calculating work when $\boldsymbol{F}$ is a force field.

The line integral in (13) may be written in various other ways. When the vector field $\boldsymbol{F}$ has the form

$$\boldsymbol{F}(x, y, z) = M(x, y, z)\boldsymbol{i} + N(x, y, z)\boldsymbol{j} + P(x, y, z)\boldsymbol{k}, \tag{14}$$

we use the notation

$$d\boldsymbol{r} = dx\, \boldsymbol{i} + dy\, \boldsymbol{j} + dz\, \boldsymbol{k}$$

to write $\boldsymbol{F} \cdot d\boldsymbol{r}$ as

$$\boldsymbol{F} \cdot d\boldsymbol{r} = M(x, y, z)\, dx + N(x, y, z)\, dy + P(x, y, z)\, dz. \tag{15}$$

If the parameterization for C has the form

$$\boldsymbol{r}(t) = x(t)\boldsymbol{i} + y(t)\boldsymbol{j} + z(t)\boldsymbol{k}, \tag{16}$$

then

$$\boldsymbol{r}'(t) = x'(t)\boldsymbol{i} + y'(t)\boldsymbol{j} + z'(t)\boldsymbol{k}. \tag{17}$$

With notation as in (15) and (17), the line integral in (13) can be written

$$\begin{aligned}\int_C \boldsymbol{F} \cdot d\boldsymbol{r} = \int_a^b \{&M(x(t), y(t), z(t))x'(t)\\ &+ N(x(t), y(t), z(t))y'(t)\\ &+ P(x(t), y(t), z(t))z'(t)\}\, dt.\end{aligned} \tag{18}$$

Equation (18) may be abbreviated simply as

$$\int_C \boldsymbol{F} \cdot d\boldsymbol{r} = \int_C M(x, y, z)\, dx + N(x, y, z)\, dy + P(x, y, z)\, dz. \tag{19}$$

Equation (18) is the most explicit of the three expressions for $\int_C \boldsymbol{F} \cdot d\boldsymbol{r}$ and, therefore, the one you will probably find most useful. In using either equation (13) or equation (19) you must keep in mind that an explicit parameterization of the form (16) must be found before the line integral may be evaluated (except in certain situations that we shall discuss in the next section).

When the vector field $\boldsymbol{F}$ and the curve C are defined in the plane by

$$\boldsymbol{F}(x, y) = M(x, y)\boldsymbol{i} + N(x, y)\boldsymbol{j}$$

and

$$\boldsymbol{r}(t) = x(t)\boldsymbol{i} + y(t)\boldsymbol{j}, \qquad a \leq t \leq b,$$

the line integral in Definition 3 takes the form

$$\int_C \boldsymbol{F} \cdot d\boldsymbol{r} = \int_a^b [M(x(t), y(t))x'(t) + N(x(t), y(t))y'(t)]\, dt \tag{20}$$

or

$$\int_C \boldsymbol{F} \cdot d\boldsymbol{r} = \int_C M(x, y)\, dx + N(x, y)\, dy. \tag{21}$$

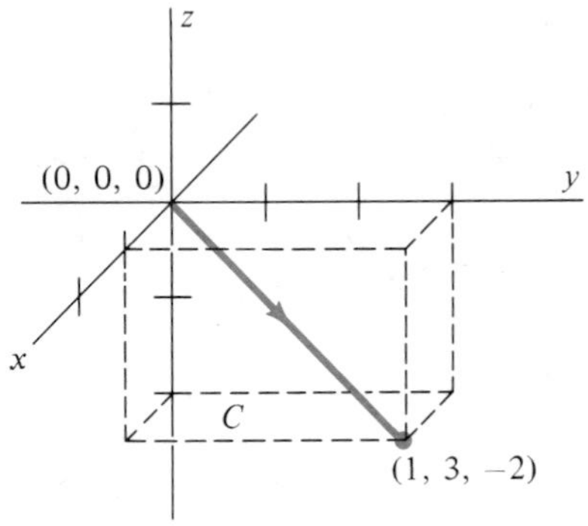

Figure 2.10 Path C from (0, 0, 0) to (1, 3, −2).

Example 4 Evaluate the line integral $\int_C \boldsymbol{F} \cdot d\boldsymbol{r}$ where $\boldsymbol{F}$ is the vector field

$$\boldsymbol{F}(x, y, z) = x^2\boldsymbol{i} + xy\boldsymbol{j} + xz\boldsymbol{k}$$

and C is the line segment from (0, 0, 0) to (1, 3, −2) (see Figure 2.10).

Solution: Here $\boldsymbol{F}$ has the form (14) with

$$M(x, y, z) = x^2, \qquad N(x, y, z) = xy, \qquad P(x, y, z) = xz. \tag{22}$$

The line segment C from (0, 0, 0) to (1, 3, −2) may be parameterized in the form (16) with

$$x(t) = t, \qquad y(t) = 3t, \qquad z(t) = -2t \tag{23}$$
$$\text{for} \quad 0 \le t \le 1 \quad \text{(Figure 2.10).}$$

Thus,

$$x'(t) = 1, \qquad y'(t) = 3, \qquad z'(t) = -2. \tag{24}$$

Combining equations (22) and (23) shows that

$$M(x(t), y(t), z(t)) = t^2; \qquad N(x(t), y(t), z(t)) = 3t^2; \qquad P(x(t), y(t), z(t)) = -2t^2. \tag{25}$$

Using equations (18), (24), and (25) we find

$$\int_C \boldsymbol{F} \cdot d\boldsymbol{r} = \int_0^1 [(t^2)(1) + (3t^2)(3) + (-2t^2)(-2)]\, dt$$

$$= \int_0^1 14t^2\, dt = \frac{14}{3} t^3\Big]_0^1 = \frac{14}{3}.$$

■

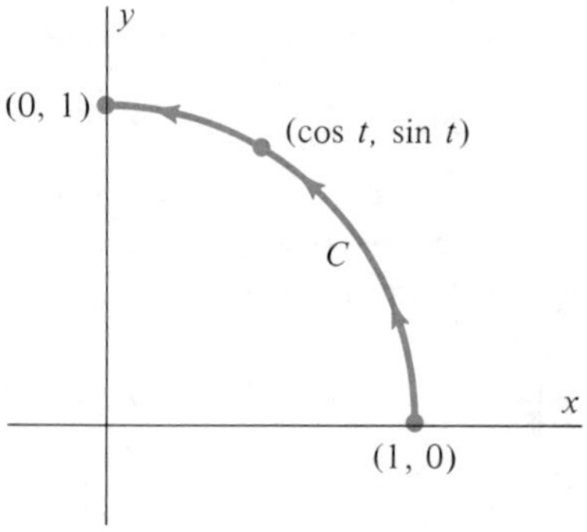

Figure 2.11 Path C from (1, 0) to (0, 1).

Example 5 Evaluate the line integral

$$\int_C xy\, dx - 2y^2\, dy$$

where C is the arc of the unit circle from (1, 0) to (0, 1) traversed counterclockwise (Figure 2.11).

Solution: The line integral has the form (21) with

$$M(x, y) = xy, \qquad N(x, y) = -2y^2.$$

A parameterization for the curve C is given by the equations

$$x(t) = \cos t, \qquad y(t) = \sin t, \qquad 0 \le t \le \pi/2 \qquad \text{(Figure 2.11).}$$

Thus

$$x'(t) = -\sin t, \qquad y'(t) = \cos t.$$

Equating the right sides of equations (20) and (21) and using the above functions we obtain

$$\begin{aligned}\int_C xy\,dx - 2y^2\,dy &= \int_0^{\pi/2} [(\cos t)(\sin t)(-\sin t) \\ &\qquad - 2(\sin^2 t)(\cos t)]\,dt. \\ &= \int_0^{\pi/2} -3\sin^2 t\cos t\,dt \\ &= -\sin^3 t\Big]_0^{\pi/2} \\ &= -1.\end{aligned}$$

■

REMARK: It is important to note that the value of the line integral in Definition 3 depends on the *direction* in which the curve C is traced out by the parameterization $\mathbf{r}(t)$. *Reversing the direction along C changes the sign of the line integral.* The reason for this can be most easily seen in line (8): Reversing the direction in which C is traversed changes the sign of the unit tangent vector $\mathbf{T}(\mathbf{r}(c_j))$.

To illustrate this remark, we evaluate the line integral in Example 5 over the same curve C, but traversed in the opposite direction, from (0, 1) to (1, 0). (We will refer to the curve C with its new orientation as $-C$). A parameterization for $-C$ is

$$x(t) = \cos\left(\frac{\pi}{2} - t\right), \qquad y(t) = \sin\left(\frac{\pi}{2} - t\right), \qquad 0 \le t \le \frac{\pi}{2},$$

so

$$x'(t) = \sin\left(\frac{\pi}{2} - t\right), \qquad y'(t) = -\cos\left(\frac{\pi}{2} - t\right),$$

and

$$\begin{aligned}\int_{-C} xy\,dx - 2y^2\,dy &= \int_0^{\pi/2} \left[\cos\left(\frac{\pi}{2} - t\right)\sin^2\left(\frac{\pi}{2} - t\right)\right. \\ &\qquad \left. + 2\sin^2\left(\frac{\pi}{2} - t\right)\cos\left(\frac{\pi}{2} - t\right)\right] dt \\ &= \int_0^{\pi/2} 3\sin^2\left(\frac{\pi}{2} - t\right)\cos\left(\frac{\pi}{2} - t\right) dt \\ &= -\sin^3\left(\frac{\pi}{2} - t\right)\Big]_0^{\pi/2} \\ &= 1.\end{aligned}$$

This is the negative of the result obtained in Example 5. We summarize this remark by writing

$$\boxed{\int_{-C} \mathbf{F}\cdot d\mathbf{r} = -\int_C \mathbf{F}\cdot d\mathbf{r}.}$$

When the curve C is a line segment parallel to the x-axis, the line integral in (18) and (19) reduces to

$$\int_C \boldsymbol{F} \cdot d\boldsymbol{r} = \int_C M(x, y, z)\, dx = \int_a^b M(x(t), y(t), z(t))x'(t)\, dt.$$

Such integrals are called **line integrals with respect to x.** Line integrals with respect to y and with respect to z are defined similarly. Using these definitions and Theorem 4, Chapter 6, we may write the line integral in (19) as

$$\int_C \boldsymbol{F} \cdot d\boldsymbol{r} = \int_C M(x, y, z)\, dx + \int_C N(x, y, z)\, dy + \int_C P(x, y, z)\, dz.$$

Paths Versus Curves

Up to this point we have defined the line integral only for smooth curves C and continuous vector fields $\boldsymbol{F}$. (Recall that the curve $C = \{\boldsymbol{r}(t) | a \le t \le b\}$ is *smooth* if $\boldsymbol{r}'(t)$ exists and is continuous for all t in the interval $[a, b]$.) The reason why C has been taken to be smooth is that the integrand in the definite integral in line (13) involves the derivative, $\boldsymbol{r}'(t)$. Thus, when $\boldsymbol{F}$ and $\boldsymbol{r}'(t)$ are continuous, the theory developed in Chapter 6 guarantees that the definite integral in equation (13) exists.

However, this integral exists under slightly less restrictive conditions. In particular, it was stated in Chapter 6 that the definite integral exists when the integrand is merely piecewise continuous. Thus, we need only require that $\boldsymbol{r}'(t)$ be piecewise continuous for the line integral in Definition 3 to exist. Curves of this type are called *piecewise smooth*. We shall refer to such curves as *paths* (see Figures 2.12 and 2.13).

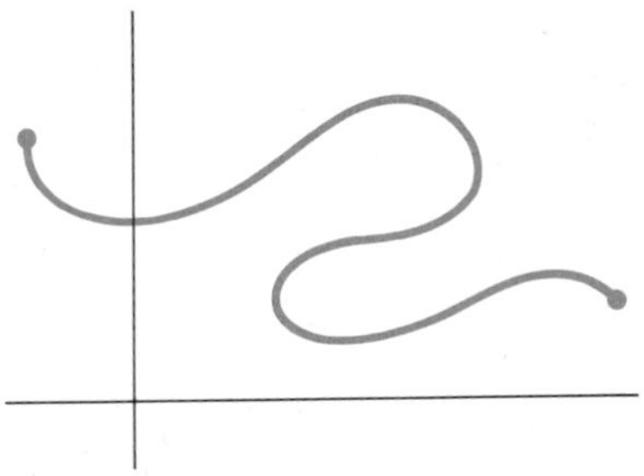

Figure 2.12 A smooth curve ($\boldsymbol{r}'(t)$ continuous).

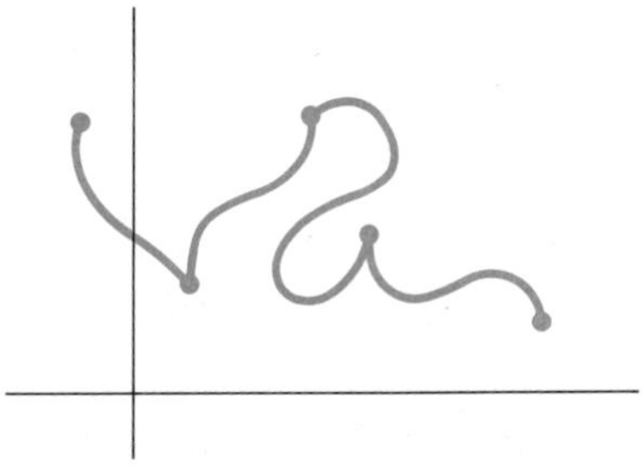

Figure 2.13 A piecewise smooth curve, or *path* ($\boldsymbol{r}'(t)$ piecewise continuous).

DEFINITION 4

A curve $C = \{\boldsymbol{r}(t) | a \le t \le b\}$ is **piecewise smooth** if there exist numbers $a = t_0 < t_1 < t_2 < \cdots < t_n = b$ so that $\boldsymbol{r}(t)$ is continuous for all $t \in [a, b]$ and so that $\boldsymbol{r}'(t)$ exists and is continuous on each of the subintervals $(t_0, t_1), (t_1, t_2), \ldots, (t_{n-1}, t_n)$. A piecewise smooth curve is called a **path.**

REMARK: For piecewise smooth arcs, we evaluate the line integral by first integrating over each of the subarcs on which $\boldsymbol{r}'(t)$ is continuous and then summing the results. This is justified by Theorem 4, Chapter 6, since

$$\int_C \boldsymbol{F} \cdot d\boldsymbol{r} = \int_a^b \boldsymbol{F}(\boldsymbol{r}(t)) \cdot \boldsymbol{r}'(t)\, dt \tag{26}$$

$$= \int_{t_0}^{t_1} \boldsymbol{F}(\boldsymbol{r}(t)) \cdot \boldsymbol{r}'(t)\, dt + \cdots + \int_{t_{n-1}}^{t_n} \boldsymbol{F}(\boldsymbol{r}(t)) \cdot \boldsymbol{r}'(t)\, dt$$

$$= \int_{C_1} \boldsymbol{F} \cdot d\boldsymbol{r} + \int_{C_2} \boldsymbol{F} \cdot d\boldsymbol{r} + \cdots + \int_{C_n} \boldsymbol{F} \cdot d\boldsymbol{r}$$

where C_j denotes the arc $C_j = \{\boldsymbol{r}(t) | t_{j-1} \le t \le t_j\}$.

Example 6 Evaluate the line integral

$$\int_C (x + y)\, dx + xy\, dy$$

over the path C consisting of the line segments

C_1: from (0, 0) to (1, 0),

C_2: from (1, 0) to (1, 2),

C_3: from (1, 2) to (0, 0).

(See Figure 2.14.)

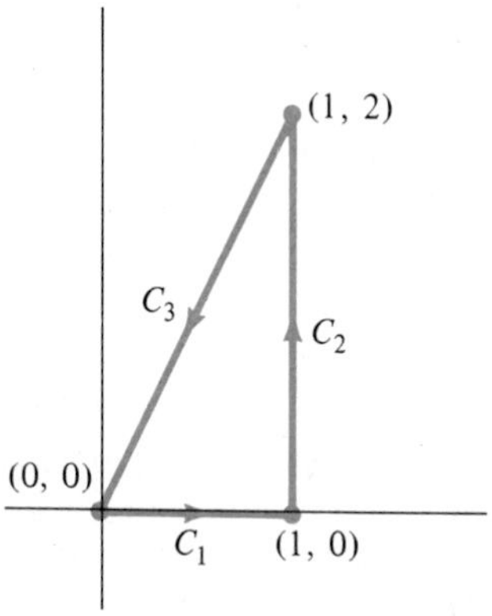

Figure 2.14

Solution: Here the vector field is $\boldsymbol{F}(x, y) = M(x, y)\boldsymbol{i} + N(x, y)\boldsymbol{j}$ with

$$M(x, y) = x + y, \qquad N(x, y) = xy.$$

A parameterization for C_1 is

$$C_1: \left.\begin{matrix} x(t) = t \\ y(t) = 0 \end{matrix}\right\} 0 \le t \le 1.$$

On this arc, $x'(t) = 1$ and $y'(t) = 0$. The line integral over C_1 is therefore

$$\int_{C_1} (x + y)\, dx + xy\, dy = \int_0^1 [(t + 0)(1) + (t \cdot 0)(0)]\, dt$$

$$= \int_0^1 t\, dt = \frac{1}{2}.$$

A parameterization for C_2 is

$$C_2: \left.\begin{matrix} x(t) = 1 \\ y(t) = t \end{matrix}\right\} 0 \le t \le 2.$$

On this arc, $x'(t) = 0$ and $y'(t) = 1$. Thus,

$$\int_{C_2} (x + y)\, dx + xy\, dy = \int_0^2 [(1 + t)(0) + (1 \cdot t)(1)]\, dt$$

$$= \int_0^2 t\, dt = 2.$$

Finally, a parameterization for C_3 is

$$C_3: \left.\begin{matrix} x(t) = 1 - t \\ y(t) = 2 - 2t \end{matrix}\right\} 0 \le t \le 1.$$

On this arc, $x'(t) = -1$ and $y'(t) = -2$. Thus, by (20),

$$\begin{aligned}\int_{C_3} (x+y)\,dx + xy\,dy &= \int_0^1 \{[(1-t)+(2-2t)](-1) \\ &\qquad + (1-t)(2-2t)(-2)\}\,dt \\ &= \int_0^1 (-4t^2 + 11t - 7)\,dt = -\frac{17}{6}.\end{aligned}$$

Applying equation (26), we conclude

$$\begin{aligned}\int_C (x+y)\,dx + xy\,dy &= \int_{C_1} (x+y)\,dx + xy\,dy \\ &\quad + \int_{C_2} (x+y)\,dx + xy\,dy \\ &\quad + \int_{C_3} (x+y)\,dx + xy\,dy \\ &= \frac{1}{2} + 2 - \frac{17}{6} \\ &= -\frac{1}{3}.\end{aligned}$$

■

Line Integrals with Respect to Arc Length

Thus far we have defined line integrals only for vector fields. Another type of line integral, called a **line integral with respect to arc length,** concerns a scalar valued function, say $f(x, y)$ on $\mathbb{R}^2$. Let C be a path with parameterization

$$C = \{\boldsymbol{r}(t) = x(t)\boldsymbol{i} + y(t)\boldsymbol{j} \mid a \le t \le b\}.$$

The usual partitioning of the interval $[a, b]$ produces arcs $C_1, C_2, \ldots, C_n$ whose lengths Δs_j, as before, are

$$\Delta s_j = \sqrt{[x'(c_j)]^2 + [y'(c_j)]^2}\,\Delta t, \qquad j = 1, 2, \ldots, n,$$

where c_j is a number in the jth subinterval of $[a, b]$. If the function $f(x, y)$ is continuous on an open set containing the path C, the Riemann sum

$$\sum_{j=1}^{n} f(x(c_j), y(c_j))\,\Delta s_j \tag{27}$$

$$= \sum_{j=1}^{n} f(x(c_j), y(c_j))\sqrt{[x'(c_j)]^2 + [y'(c_j)]^2}\,\Delta t$$

converges to the definite integral

$$\boxed{\int_C f(x, y)\,ds = \int_a^b f(x(t), y(t))\sqrt{[x'(t)]^2 + [y'(t)]^2}\,dt.} \tag{28}$$

The integral in equation (28) is called the *line integral* of $f(x, y)$ over C *with respect to arc length*. The difference between the integral in (28) and the line integral in Definition 3 is that (28) is an integral of a scalar function $f(x, y)$ with respect to change in arc length, while the integral in (13) is an integral of the dot product of a vector field and the unit tangent with respect to change in the parameter t. By recalling the definition

$$ds = \sqrt{[x'(t)]^2 + [y'(t)]^2}\, dt = |\mathbf{r}'(t)|\, dt$$

of the differential for arc length, we may interpret the line integral in Definition 3 as a special case of the integral in (28) with $f(x(t), y(t)) = \mathbf{F}(\mathbf{r}(t)) \cdot \mathbf{T}(\mathbf{r}(t))$. That is,

$$\int_C \mathbf{F} \cdot d\mathbf{r} = \int_C \mathbf{F} \cdot \mathbf{T}\, ds. \tag{29}$$

This can be seen by comparing the approximating sums in lines (8) and (27).

Although we shall work almost exclusively in what follows with line integrals as defined in Definition 3, equation (29) will be used in Section 21.4 to interpret certain statements about line integrals.

Example 7 Find $\int_C xy^3\, ds$ where C is the quarter circle

$$C = \left\{\cos t\mathbf{i} + \sin t\mathbf{j} \;\middle|\; 0 \le t \le \frac{\pi}{2}\right\}.$$

Solution: Here $x(t) = \cos t$, $y(t) = \sin t$, $x'(t) = -\sin t$, and $y'(t) = \cos t$. Thus, by (28)

$$\begin{aligned}\int_C xy^3\, ds &= \int_0^{\pi/2} (\cos t)(\sin^3 t)\sqrt{[-\sin t]^2 + [\cos t]^2}\, dt \\ &= \int_0^{\pi/2} \sin^3 t \cos t\, dt \\ &= \frac{1}{4}\sin^4 t\Big]_0^{\pi/2} \\ &= \frac{1}{4}.\end{aligned}$$

■

It is important to note that, unlike line integrals of the form $\int_C \mathbf{F} \cdot d\mathbf{r}$, line integrals with respect to arc length are independent of the direction along which the curve C is traversed (see Exercise 28).

An application of line integrals with respect to arc length concerns finding the mass M of a thin wire of variable density whose shape is given by the curve $C = \{\mathbf{r}(t) = x(t)\mathbf{i} + y(t)\mathbf{j} | a \le t \le b\}$. If the mass density (in units of mass per unit length) is given by the continuous function $f(x, y)$, the expression

$$\Delta m_j = f(x(c_j), y(c_j))\, \Delta s_j$$

approximates the mass of a section of the wire of length Δs_j, one point of which has coordinates $(x(c_j), y(c_j))$. The sum in equation (27) therefore approximates the total mass of the wire, which is given precisely by the line integral with respect to arc length in equation (28).

For curves in space, line integrals with respect to arc length are defined in an entirely analogous manner. That is, if $f(x, y, z)$ is continuous in an open set containing the piecewise smooth curve

$$C = \{\boldsymbol{r}(t) = x(t)\boldsymbol{i} + y(t)\boldsymbol{j} + z(t)\boldsymbol{k} \mid a \le t \le b\},$$

then

$$\int_C f(x, y, z)\, ds = \int_a^b f(x(t), y(t), z(t))\sqrt{[x'(t)]^2 + [y'(t)]^2 + [z'(t)]^2}\, dt.$$

Exercise Set 21.2

In Exercises 1–8, calculate the work done by the force field $\boldsymbol{F}$ in moving a particle along the specified path C.

1. $\boldsymbol{F}(x, y) = x\boldsymbol{i} + y\boldsymbol{j}$, C: $\boldsymbol{r}(t) = t^2\boldsymbol{i} + (3 + t)\boldsymbol{j}$, $0 \le t \le 2$
2. $\boldsymbol{F}(x, y) = xy\boldsymbol{i} + x^2\boldsymbol{j}$,
   C: $\boldsymbol{r}(t) = \cos t\boldsymbol{i} + \sin t\boldsymbol{j}$, $0 \le t \le \pi$
3. $\boldsymbol{F}(x, y) = x^2\boldsymbol{i} + y^2\boldsymbol{j}$,
   C: $\boldsymbol{r}(t) = \sqrt{t}\,\boldsymbol{i} + 3t\boldsymbol{j}$, $1 \le t \le 4$
4. $\boldsymbol{F}(x, y) = -y\boldsymbol{i} + x\boldsymbol{j}$,
   C: $\boldsymbol{r}(t) = (t + 1)^2\boldsymbol{i} + (t - 1)^2\boldsymbol{j}$, $0 \le t \le 2$
5. $\boldsymbol{F}(x, y) = (x^2 + y^2)\boldsymbol{i} + xy\boldsymbol{j}$,
   C: $\boldsymbol{r}(t) + t\boldsymbol{i} + t^2\boldsymbol{j}$, $0 \le t \le 2$
6. $\boldsymbol{F}(x, y) = x^2y\boldsymbol{i} + xy^2\boldsymbol{j}$,
   C: $\boldsymbol{r}(t) = t\boldsymbol{i} + 2t^2\boldsymbol{j}$, $0 \le t \le 1$
7. $\boldsymbol{F}(x, y, z) = xy\boldsymbol{i} + xz\boldsymbol{j} + yz\boldsymbol{k}$,
   C: $\boldsymbol{r}(t) = \cos t\boldsymbol{i} + \sin t\boldsymbol{j} + t\boldsymbol{k}$, $0 \le t \le \pi$
8. $\boldsymbol{F}(x, y, z) = x^2\boldsymbol{i} + y^2\boldsymbol{j} + z^2\boldsymbol{k}$,
   C: $\boldsymbol{r}(t) = t\boldsymbol{i} + 3t^2\boldsymbol{j} - 2t\boldsymbol{k}$; $0 \le t \le 2$

In Exercises 9–15, evaluate the line integral of the vector field $\boldsymbol{F}$ over the indicated path C.

9. $\boldsymbol{F}(x, y) = y\boldsymbol{i} + 2x\boldsymbol{j}$, C is the line segment from $(0, 0)$ to $(4, 2)$.
10. $\boldsymbol{F}(x, y) = -x^2\boldsymbol{i} + y^2\boldsymbol{j}$, C is the upper unit semicircle from $(1, 0)$ to $(-1, 0)$.
11. $\boldsymbol{F}(x, y) = (y - x)\boldsymbol{i} + xy\boldsymbol{j}$, C is the unit circle traversed counterclockwise.
12. $\boldsymbol{F}(x, y) = xy^2\boldsymbol{i} + (x + y)\boldsymbol{j}$, C is the triangular path from $(0, 0)$ to $(4, 0)$ to $(4, 2)$ to $(0, 0)$.
13. $\boldsymbol{F}(x, y) = (x + y)\boldsymbol{i} + (y^2 - x^2)\boldsymbol{j}$, C is the triangle with vertices $(-1, 0)$, $(1, -4)$, $(0, 2)$ traversed counterclockwise.
14. $\boldsymbol{F}(x, y, z) = xy\boldsymbol{i} + y\boldsymbol{j} + xz\boldsymbol{k}$, C is the line segment from $(0, 0, 0)$ to $(2, 4, -6)$.
15. $\boldsymbol{F}(x, y, z) = y\boldsymbol{i} - x\boldsymbol{j} + 2z\boldsymbol{k}$, C is the arc of the circular helix $\boldsymbol{r}(t) = \cos t\boldsymbol{i} + \sin t\boldsymbol{j} + t\boldsymbol{k}$ from $(1, 0, 0)$ to $(0, 1, \pi/2)$.
16. Evaluate $\int_C xy\, dx$ where C is the arc of the unit circle from $(0, 1)$ to $(-1, 0)$ traversed counterclockwise.
17. Evaluate $\int_C (x^2 + y^2)\, dy$ where C is the path in Exercise 16.
18. Evaluate $\int_C \boldsymbol{F} \cdot d\boldsymbol{r}$ where $\boldsymbol{F}(x, y) = (x^2 + y^2)\boldsymbol{i} + 2xy\boldsymbol{j}$ and C is the arc of the circle $x^2 + y^2 = 1$ from $(1, 0)$ to $(0, 1)$, followed by the line segment from $(0, 1)$ to $(-1, 1)$ (Figure 2.15).

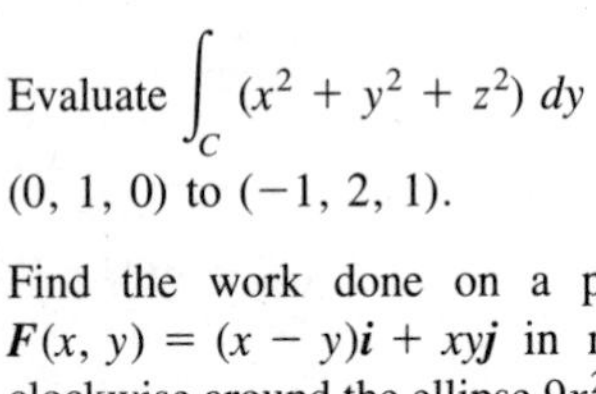

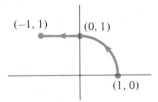

Figure 2.15

19. Evaluate $\int_C (x^2 + y^2 + z^2)\, dy$ along the line segment from $(0, 1, 0)$ to $(-1, 2, 1)$.
20. Find the work done on a particle by the force field $\boldsymbol{F}(x, y) = (x - y)\boldsymbol{i} + xy\boldsymbol{j}$ in moving a particle counterclockwise around the ellipse $9x^2 + 4y^2 = 36$ from $(2, 0)$ to $(-2, 0)$.

21. Evaluate $\int_C xy\,dx + xy^2\,dy$ where C is the curve in Exercise 18.

22. Find the work done by the force field $\boldsymbol{F}(x, y) = (x\boldsymbol{i} - y\boldsymbol{j})$ in moving a particle along the path in Figure 2.16.

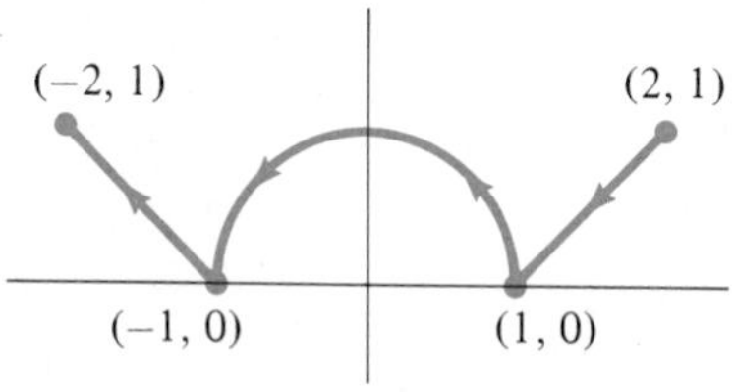

Figure 2.16

23. Evaluate $\int_C (x + y)\,dx + (x^2 - y^2)\,dy$ where C is the arc of the parabola $y = 2x^2$ from $(0, 0)$ to $(2, 8)$.

24. Evaluate $\int_C (x - y)\,dx + xy\,dy$ where C is the path consisting of the line segments from $(0, 1)$ to $(2, -1)$, from $(2, -1)$ to $(2, 5)$, and from $(2, 5)$ to $(4, 1)$.

25. Evaluate $\int_C x^2y\,dx + xy^2\,dy$ where C is the path from $(0, 2)$ to $(4, 1)$ along the three line segments parallel to the x- and y-axes.

26. Let $\boldsymbol{F}$ be the central force field $\boldsymbol{F}(\boldsymbol{r}) = \dfrac{2}{|\boldsymbol{r}|^3}\boldsymbol{r}, \boldsymbol{r} \neq \boldsymbol{0}$. Find the work done by $\boldsymbol{F}$ in moving a particle along the upper half of the unit circle from $(1, 0)$ to $(-1, 0)$.

27. Show that Definition 2 agrees with the definition of work when a force $f(x)$ is applied to a particle moving along the x-axis.

28. Show that the line integral with respect to arc length in line (28) is independent of the direction along which C is traversed.

Evaluate each of the following line integrals with respect to arc length.

29. $\int_C (x + 2y^2)\,ds$ where $C = \{\boldsymbol{r}(t) = 2t\boldsymbol{i} - 4t\boldsymbol{j} \mid 0 \le t \le 1\}$

30. $\int_C xe^y\,ds$ where $C = \{\boldsymbol{r}(t) = \cos t\boldsymbol{i} + \sin t\boldsymbol{j} \mid 0 \le t \le \pi\}$

31. $\int_C (8x + y - 3)\,ds$
where $C = \{\boldsymbol{r}(t) = t\boldsymbol{i} + 3\boldsymbol{j} + t^2\boldsymbol{k} \mid 0 \le t \le 1\}$

32. Give an argument to support the conclusion that if the curve C describes the shape of a thin wire of mass density $\rho(x, y, z)$ then the total mass of the wire is

$$M = \int_C \rho\,ds = \int_a^b \rho(x(t), y(t), z(t))\sqrt{[x'(t)]^2 + [y'(t)]^2 + [z'(t)]^2}\,dt.$$

33. Refer to Exercise 32. Find the total mass of a wire with shape

C: $\boldsymbol{r}(t) = \cos t\boldsymbol{i} - \sin t\boldsymbol{j} + 2\boldsymbol{k}, \quad 0 \le t \le \pi$

if the density is $\rho(x, y, z) = x^2 + y^2 + z^2$.

34. Refer to Exercise 32. Find the mass of a wire of constant density $\rho(x, y, z) = k$ shaped like the helix $\boldsymbol{r}(t) = 4\cos t\boldsymbol{i} + 4\sin t\boldsymbol{j} + 3t\boldsymbol{k}$ for $0 \le t \le 2\pi$.

21.3 LINE INTEGRALS: INDEPENDENCE OF PATH

The Fundamental Theorem of Calculus allows us to evaluate definite integrals of functions of a single variable by means of an antiderivative. In this section, we develop a similar theorem for line integrals. To do so requires the concept of an antiderivative for a vector field $\boldsymbol{F}$, which we will take to be a *scalar potential* ϕ for $\boldsymbol{F}$. Recall from Section 19.11 that the differentiable scalar function $\phi(x, y)$ is a scalar potential for the vector function $\boldsymbol{F}(x, y) = M(x, y)\boldsymbol{i} + N(x, y)\boldsymbol{j}$ if

$$\nabla\phi(x, y) = \boldsymbol{F}(x, y). \tag{1}$$

That is, $\boldsymbol{F}$ is conservative if $\boldsymbol{F}$ is a *gradient* field.

Using the position vector notation $\boldsymbol{r} = x\boldsymbol{i} + y\boldsymbol{j}$, we can rewrite equation (1) as

$$\nabla\phi(\boldsymbol{r}) = \boldsymbol{F}(\boldsymbol{r}). \tag{2}$$

We will also need to make use of the Chain Rule for vector functions: If $x(t)$ and $y(t)$ are differentiable component functions, then

$$\frac{d}{dt}\phi(x(t),\ y(t)) = \phi_x(x(t),\ y(t))x'(t) + \phi_y(x(t),\ y(t))y'(t) \tag{3}$$

$$= \nabla\phi(x(t),\ y(t)) \cdot [x'(t)\boldsymbol{i} + y'(t)\boldsymbol{j}].$$

Using the position vector notation, equation (3) can be written

$$\frac{d}{dt}\phi(\boldsymbol{r}(t)) = \nabla\phi(\boldsymbol{r}(t)) \cdot \boldsymbol{r}'(t). \tag{4}$$

Our Fundamental Theorem for line integrals is the following.

THEOREM 1

Let $C = \{\boldsymbol{r}(t) \mid a \le t \le b\}$ be a piecewise smooth curve in an open set D. Let $\boldsymbol{F}(\boldsymbol{r})$ be a vector field on D and let $\phi(\boldsymbol{r})$ be a differentiable scalar function for which $\boldsymbol{F}(\boldsymbol{r}) = \nabla\phi(\boldsymbol{r})$ for all r in D. Then

$$\int_C \boldsymbol{F} \cdot d\boldsymbol{r} = \phi(\boldsymbol{r}(b)) - \phi(\boldsymbol{r}(a)). \tag{5}$$

Theorem 1 says this. If the vector field $\boldsymbol{F}$ is the gradient of some scalar function ϕ, then the line integral in (5) over C from $A = \boldsymbol{r}(a)$ to $B = \boldsymbol{r}(b)$ depends only on the values of the scalar potential ϕ at A and at B. Thus, if the vector field $\boldsymbol{F}$ is *conservative,* we can determine the value of the line integral by finding a scalar potential ϕ for $\boldsymbol{F}$ (see Section 19.11).

Example 1 Find $\int_C \boldsymbol{F} \cdot d\boldsymbol{r}$ where $\boldsymbol{F}$ is the vector field $\boldsymbol{F}(x, y) = 2xy\boldsymbol{i} + x^2\boldsymbol{j}$ and C is the line segment from $(-1, 2)$ to $(4, -3)$.

Solution: Here $\phi(x, y) = x^2y$ is a potential for $\boldsymbol{F}$ since

$$\nabla\phi(x, y) = 2xy\boldsymbol{i} + x^2\boldsymbol{j} = \boldsymbol{F}(x, y).$$

Thus, by Theorem 1,

$$\int_C \boldsymbol{F}(\boldsymbol{r}) \cdot d\boldsymbol{r} = \phi(4, -3) - \phi(-1, 2) = 4^2(-3) - (-1)^2 2 = -50. \qquad \blacksquare$$

Example 2 Evaluate the line integral

$$\int_C 2xye^{x^2}\,dx + e^{x^2}\,dy$$

over a path C from $(0, 1)$ to $(1, 3)$.

Solution: This integral has the form $\int_C \boldsymbol{F} \cdot d\boldsymbol{r}$ where

$$\boldsymbol{F}(x, y) = 2xye^{x^2}\boldsymbol{i} + e^{x^2}\boldsymbol{j}$$

and $d\boldsymbol{r} = dx\,\boldsymbol{i} + dy\,\boldsymbol{j}$. A potential for $\boldsymbol{F}$ is

$$\phi(x, y) = ye^{x^2}$$

since

$$\nabla\phi(x, y) = 2xye^{x^2}\boldsymbol{i} + e^{x^2}\boldsymbol{j} = \boldsymbol{F}(x, y).$$

Thus, by Theorem 1

$$\begin{aligned}\int_C 2xye^{x^2}\,dx + e^{x^2}\,dy &= \phi(1, 3) - \phi(0, 1)\\ &= 3e - e^0\\ &= 3e - 1.\end{aligned}$$

■

Proof of Theorem 1: Assume first that C is a smooth curve. Then, under the hypotheses of Theorem 1, both equations (2) and (4) hold. Thus,

$$\begin{aligned}\int_C \boldsymbol{F}\cdot d\boldsymbol{r} &= \int_a^b \boldsymbol{F}(\boldsymbol{r}(t))\cdot\boldsymbol{r}'(t)\,dt\\ &= \int_a^b \nabla\phi(\boldsymbol{r}(t))\cdot\boldsymbol{r}'(t)\,dt \qquad \text{(equation (2))}\\ &= \int_a^b \frac{d}{dt}[\phi(\boldsymbol{r}(t))]\,dt \qquad \text{(equation (4))}\\ &= \phi(\boldsymbol{r}(t))\Big]_a^b\\ &= \phi(\boldsymbol{r}(b)) - \phi(\boldsymbol{r}(a)).\end{aligned}$$

If C is only piecewise smooth, let $C = C_1 \cup C_2 \cup \cdots \cup C_n$ where each C_j is smooth and has endpoints $\boldsymbol{r}(t_{j-1})$ and $\boldsymbol{r}(t_j)$. Then, using the first part of the proof on each C_j,

$$\begin{aligned}\int_C \boldsymbol{F}\cdot d\boldsymbol{r} &= \int_{C_1} \boldsymbol{F}\cdot d\boldsymbol{r} + \int_{C_2} \boldsymbol{F}\cdot d\boldsymbol{r} + \cdots + \int_{C_n} \boldsymbol{F}\cdot d\boldsymbol{r}\\ &= [\phi(\boldsymbol{r}(t_1)) - \phi(\boldsymbol{r}(t_0))] + [\phi(\boldsymbol{r}(t_2)) - \phi(\boldsymbol{r}(t_1))]\\ &\quad + \cdots + [\phi(\boldsymbol{r}(t_n)) - \phi(\boldsymbol{r}(t_{n-1}))]\\ &= \phi(\boldsymbol{r}(t_n)) - \phi(\boldsymbol{r}(t_0))\\ &= \phi(\boldsymbol{r}(b)) - \phi(\boldsymbol{r}(a)) \qquad (t_n = b,\ t_0 = a),\end{aligned}$$

since the sum telescopes. ■

In trying to determine whether the vector field $\boldsymbol{F}$ is conservative (i.e., whether it has a scalar potential ϕ), it is helpful to recall Theorem 12, Section 19.11: If the functions $M(x, y)$ and $N(x, y)$ are continuously differentiable in an open rectangle D, *the vector function*

$$\boldsymbol{F}(x, y) = M(x, y)\boldsymbol{i} + N(x, y)\boldsymbol{j} \qquad (6)$$

is conservative if and only if

$$\boxed{M_y(x, y) = N_x(x, y)} \qquad (7)$$

for all (x, y) *in* D. Equation (7) provides a quick check of whether Theorem 1 applies for a vector field $\boldsymbol{F}(x, y)$ in the plane. When equation (7) holds, the method of Section 19.11 can sometimes be applied to reconstruct the potential ϕ.

Example 3 Find the work done by the vector field

$$\boldsymbol{F}(x, y) = (e^y + 3x^2y^2)\boldsymbol{i} + (xe^y + 2x^3y + 2)\boldsymbol{j}$$

in moving a particle from the point (0, 0) to the point (3, 2) in the plane.

Solution: Here $\boldsymbol{F}(x, y) = M(x, y)\boldsymbol{i} + N(x, y)\boldsymbol{j}$ where

$$M(x, y) = e^y + 3x^2y^2; \qquad N(x, y) = xe^y + 2x^3y + 2.$$

Then

$$M_y(x, y) = e^y + 6x^2y = N_x(x, y),$$

so the force field $\boldsymbol{F}(x, y)$ is conservative. To find a potential $\phi(x, y)$ with

$$\begin{aligned}\nabla\phi(x, y) &= \phi_x(x, y)\boldsymbol{i} + \phi_y(x, y)\boldsymbol{j} \\ &= M(x, y)\boldsymbol{i} + N(x, y)\boldsymbol{j} = \boldsymbol{F}(x, y),\end{aligned} \tag{8}$$

we first equate $\boldsymbol{i}$-components and integrate partially with respect to x to find that

$$\phi(x, y) = \int (e^y + 3x^2y^2)\, dx = xe^y + x^3y^2 + f(y) + C \tag{9}$$

where $f(y)$ is a function of y alone. Equating y-components in (8) and integrating partially with respect to y shows that

$$\phi(x, y) = \int (xe^y + 2x^3y + 2)\, dy = xe^y + x^3y^2 + 2y + g(x) + C \tag{10}$$

where $g(x)$ is a function of x alone.

Equating the expressions for $\phi(x, y)$ in lines (9) and (10) shows that a scalar potential for $\boldsymbol{F}$ is

$$\phi(x, y) = xe^y + x^3y^2 + 2y.$$

We may therefore apply Theorem 1. The desired work is

$$\begin{aligned}W = \int_C \boldsymbol{F}\cdot d\boldsymbol{r} &= \phi(3, 2) - \phi(0, 0) \\ &= (3e^2 + 3^3\cdot 2^2 + 2\cdot 2) - 0 \\ &= 3e^2 + 112.\end{aligned}$$

■

Whether or not a potential ϕ can actually be found, Theorem 1 guarantees that the value of a line integral for a conservative vector field depends only on the endpoints of the path C and not on the path itself. (Physicists paraphrase this statement by saying that "work is a function of position, not of path.") The terminology we choose to use is that the line integral of a conservative force field is **independent of path.** The following theorem highlights this notion.

THEOREM 2

Let C be a piecewise smooth curve (i.e., a path) in an open set D. The line integral

$$\int_C \boldsymbol{F} \cdot d\boldsymbol{r}$$

is independent of path if and only if the vector field $\boldsymbol{F}$ is conservative in D.

The ''if'' part of this theorem follows directly from Theorem 1. The ''only if'' part is an important observation in more advanced courses in mathematics but will not be used here in a direct way. The proof is omitted.

The following corollary is a useful formulation of Theorem 2 and the statement concerning equation (7) for vector fields in the plane.

COROLLARY 1

Let D be an open rectangle in the plane and let C be any path from point A to point B lying within D. Let $M(x, y)$ and $N(x, y)$ be continuously differentiable in D. The line integral

$$\int_C M(x, y)\, dx + N(x, y)\, dy \tag{11}$$

is independent of path if and only if

$$M_y(x, y) = N_x(x, y) \tag{12}$$

for all (x, y) in D.

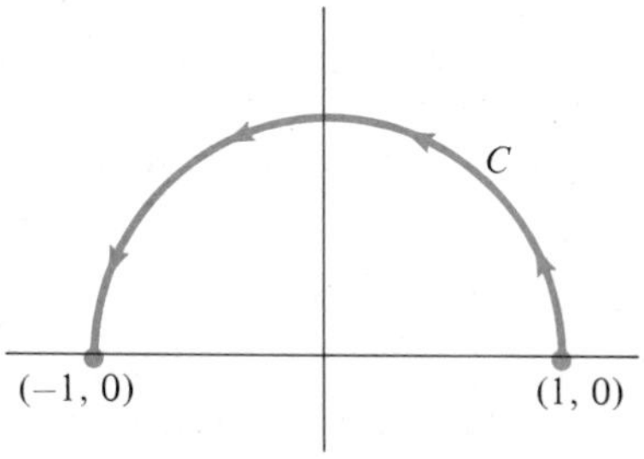

Figure 3.1 Path C from $(1, 0)$ to $(-1, 0)$.

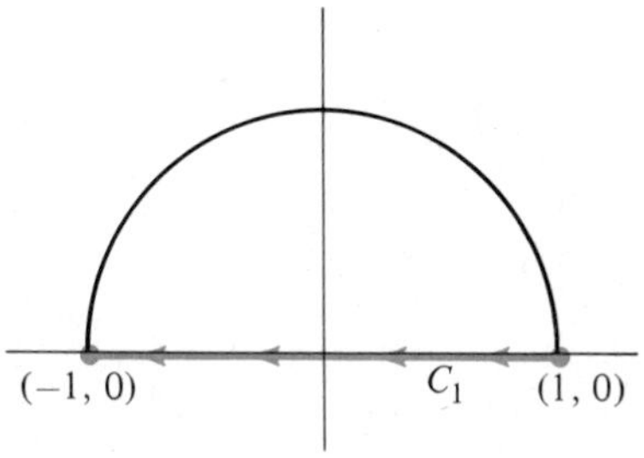

Figure 3.2 Path C_1 from $(1, 0)$ to $(-1, 0)$.

Example 4 Evaluate the line integral

$$\int_C (\cos xy - xy \sin xy)\, dx - x^2 \sin xy\, dy$$

where C is the upper unit semicircle from $(1, 0)$ to $(-1, 0)$.

Solution: This integral has the form (11) with

$$M(x, y) = \cos xy - xy \sin xy, \qquad N(x, y) = -x^2 \sin xy,$$

so

$$M_y(x, y) = -2x \sin xy - x^2 y \cos xy = N_x(x, y).$$

Thus, equation (12) holds, so the integral is independent of path. However, instead of applying the method of Example 3 to actually find the potential $\phi(x, y)$, we demonstrate a different way to apply Corollary 1 in evaluating this line integral.

Since the integral is independent of path, we may choose a path that makes the line integral easier to compute than does C. In this case, we choose the path C_1 consisting of the (straight) line segment from $(1, 0)$ to $(-1, 0)$ (see Figures 3.1 and 3.2). A parameterization for C_1 is

$$C_1: \left.\begin{aligned} x(t) &= 1 - t \\ y(t) &= 0 \end{aligned}\right\} \qquad 0 \le t \le 2.$$

Thus, $x'(t) = -1$ and $y'(t) = 0$. Calculating the line integral according to Definition 2 gives

$$\begin{aligned}
&\int_C (\cos xy - xy \sin xy)\, dx - x^2 \sin xy\, dy \\
&= \int_0^2 \{[\cos(0) - 0 \cdot \sin(0)](-1) - (1 - t)^2 \cdot \sin(0)(0)\}\, dt \\
&= \int_0^2 - 1\, dt \\
&= -2.
\end{aligned}$$

Now, with apologies for not revealing this information sooner, we note that a scalar potential for this vector field is $\phi(x, y) = x \cos xy$. With this information the line integral could have been evaluated using Theorem 1 as

$$\phi(-1, 0) - \phi(1, 0) = (-1)\cos(0) - (1)\cos(0) = -2.$$ ■

REMARK: Since a line integral over a path from point A to point B for a conservative force field $\boldsymbol{F}$ depends only on the points A and B, the notation

$$\int_C \boldsymbol{F} \cdot d\boldsymbol{r} = \int_A^B \boldsymbol{F} \cdot d\boldsymbol{r}$$

is sometimes used. If $\boldsymbol{F} = \nabla\phi$, Theorem 1 is stated in this notation as

$$\int_A^B \boldsymbol{F} \cdot d\boldsymbol{r} = \phi(B) - \phi(A).$$

In using this notation, one must exercise caution to ensure that the hypotheses of Theorem 1 are true. Otherwise, this notation is meaningless.

Example 5 In the above notation, the result of Example 4 can be stated as

$$\int_{(1,0)}^{(-1,0)} (\cos xy - xy \sin xy)\, dx - x^2 \sin xy\, dy = -2.$$ ■

Example 6 Theorems 1 and 2 apply also to paths and vector fields in space, although all of the preceding examples have dealt with paths in the plane. The line integral

$$\int_C yz\, dx + xz\, dy + xy\, dz$$

over the helical path

$$C: \quad \boldsymbol{r}(t) = \cos t\boldsymbol{i} + \sin t\boldsymbol{j} + t\boldsymbol{k}, \qquad 0 \le t \le \pi/4$$

may be handled by means of Theorem 1 by noting that the vector field

$$\boldsymbol{F}(x, y, z) = yz\boldsymbol{i} + xz\boldsymbol{j} + xy\boldsymbol{k}$$

is the gradient of the scalar potential

$$\phi(x, y, z) = xyz.$$

Since the endpoints of C are $A = \boldsymbol{r}(0) = (1, 0, 0)$ and $B = \boldsymbol{r}(\pi/4) = (\sqrt{2}/2, \sqrt{2}/2, \pi/4)$, we have

$$\begin{aligned}\int_C yz\,dx + xz\,dy + xy\,dz &= \int_{(1,0,0)}^{(\sqrt{2}/2,\ \sqrt{2}/2,\ \pi/4)} yz\,dx + xz\,dy + xy\,dz \\ &= \phi\left(\frac{\sqrt{2}}{2}, \frac{\sqrt{2}}{2}, \frac{\pi}{4}\right) - \phi(1, 0, 0) \\ &= \frac{\pi}{8}.\end{aligned}$$

■

Line Integrals Over Closed Paths

The path $C = \{\boldsymbol{r}(t)|a \le t \le b\}$ is called **closed** if $\boldsymbol{r}(a) = \boldsymbol{r}(b)$. Thus, circles, ellipses, triangles, squares, and rectangles are all examples of closed paths in the plane. There is a very simple situation governing line integrals over closed paths. *If the vector field* $\boldsymbol{F}$ *is conservative, the line integral* $\int_C \boldsymbol{F} \cdot d\boldsymbol{r}$ *over a closed path is zero.* This statement follows directly from Theorem 1: Since a closed path $C = \{\boldsymbol{r}(t)|a \le t \le b\}$ satisfies the equation $\boldsymbol{r}(a) = \boldsymbol{r}(b) = 0$,

$$\int_C \boldsymbol{F} \cdot d\boldsymbol{r} = \phi(\boldsymbol{r}(b)) - \phi(\boldsymbol{r}(a)) = 0.$$

This fact is well known to the physicist. In terms of work, it says that the work done by a conservative force field (meaning that no energy is lost in the process) in moving a particle around a closed path is zero.

Conservation of Energy

Let's take the physics one step further. Let $\boldsymbol{F}$ be a conservative force field. If a particle of mass m moves in the vector field $\boldsymbol{F}$ with velocity $\boldsymbol{v}(t) = \boldsymbol{r}'(t)$, physicists define the **kinetic energy** of the particle at time t as

$$K(t) = \frac{1}{2}m|\boldsymbol{v}(t)|^2. \tag{13}$$

Also, the change in **potential energy** U for the particle as it is moved from point A to point B is defined as the negative of the work done by the force field $\boldsymbol{F}$ in moving it from A to B:

$$U(B) - U(A) = -\int_A^B \boldsymbol{F} \cdot d\boldsymbol{r}. \tag{14}$$

Now if $\boldsymbol{a}(t)$ denotes the acceleration of the particle at time t, Newton's second law of motion states that

$$\boldsymbol{F}(\boldsymbol{r}(t)) = m\boldsymbol{a}(t) = m\boldsymbol{r}''(t),$$

so we can rewrite the integral of line (10) in Definition 2 as

$$\begin{aligned}\boldsymbol{F}(\boldsymbol{r}(t)) \cdot \boldsymbol{r}'(t) &= m\boldsymbol{r}''(t) \cdot \boldsymbol{r}'(t) \\ &= \frac{1}{2}m[2\boldsymbol{r}''(t) \cdot \boldsymbol{r}'(t)]\end{aligned} \tag{15}$$

$$= \frac{1}{2} m \frac{d}{dt} [\boldsymbol{r}'(t) \cdot \boldsymbol{r}'(t)]$$
$$= \frac{d}{dt} \left[\frac{1}{2} m |\boldsymbol{v}(t)|^2 \right].$$

Using (13) and (15) and the fact that $\boldsymbol{F}$ is conservative, we find that

$$\text{Work} = \int_A^B \boldsymbol{F} \cdot d\boldsymbol{r} = \int_a^b \boldsymbol{F}(\boldsymbol{r}(t)) \cdot \boldsymbol{r}'(t)\, dt \tag{16}$$
$$= \int_a^b \frac{d}{dt} \left[\frac{1}{2} m |\boldsymbol{v}(t)|^2 \right] dt$$
$$= \frac{1}{2} m |\boldsymbol{v}(b)|^2 - \frac{1}{2} m |\boldsymbol{v}(a)|^2$$
$$= K(B) - K(A)$$

where $A = \boldsymbol{r}(a)$, $B = \boldsymbol{r}(b)$. That is,

$$\boxed{\text{Work} = K(B) - K(A).} \tag{17}$$

Equation (17) may be interpreted as the principle that *in a conservative system, the work done on an object (by a force) equals the change in its kinetic energy.* Finally, combining equations (14) and (16) gives the equation

$$U(A) - U(B) = K(B) - K(A)$$

or

$$\boxed{K(A) + U(A) = K(B) + U(B).} \tag{18}$$

Equation (18) is the famous **principle of conservation of energy:** In a conservative system, the sum of kinetic and potential energy of a moving particle remains constant from point to point.

Exercise Set 21.3

In Exercises 1–10, evaluate the line integral by verifying that the vector field is conservative and applying Theorem 1.

1. $\int_C \boldsymbol{F} \cdot d\boldsymbol{r}$, $\boldsymbol{F}(x, y) = y\boldsymbol{i} + x\boldsymbol{j}$, C is a path from (0, 0) to (3, 1).

2. $\int_C \boldsymbol{F} \cdot d\boldsymbol{r}$, $\boldsymbol{F}(x, y) = 3x^2y^2\boldsymbol{i} + 2x^3y\boldsymbol{j}$, C is a path from (−3, 1) to (2, 2).

3. $\int_C \boldsymbol{F} \cdot d\boldsymbol{r}$, $\boldsymbol{F}(x, y) = ye^{xy}\boldsymbol{i} + xe^{xy}\boldsymbol{j}$, C is a path from (0, 0) to (1, 2).

4. $\int_C \boldsymbol{F} \cdot d\boldsymbol{r}$, $\boldsymbol{F}(x, y, z) = yze^{xyz}\boldsymbol{i} + xze^{xyz}\boldsymbol{j} + xye^{xyz}\boldsymbol{k}$, C is a path from (0, 0, 0) to (1, 0, 1).

5. $\int_C \boldsymbol{F} \cdot d\boldsymbol{r}$, $\boldsymbol{F}(x, y) = 2x\boldsymbol{i} + 2y\boldsymbol{j}$, C is the upper unit semicircle traversed counterclockwise.

6. $\int_{(1,1)}^{(2,3)} (1 + 2xy^2)\, dx + 2x^2y\, dy$

7. $\int_{(0,0)}^{(2,1)} e^{y^2}\, dx + 2xye^{y^2}\, dy$

8. $\int_{(0,0)}^{(\pi/2,1)} y \sin xy\, dx + x \sin xy\, dy$

9. $\int_{(0,0)}^{(1,\pi/4)} e^x \sin y\, dx + e^x \cos y\, dy$

10. $\int_{(0,0,0)}^{(0,\pi/4,\pi/4)} e^x \sin y \cos z\, dx$
$+ e^x \cos y \cos z\, dy - e^x \sin y \sin z\, dz$

11. Let $F(x, y)$ be a conservative vector field defined throughout the plane. Let C_1 be the path along the upper semicircle from $(1, 0)$ to $(-1, 0)$. Let C_2 be the path along the x-axis from $(-1, 0)$ to $(1, 0)$. If $\int_{C_1} \boldsymbol{F} \cdot d\boldsymbol{r} = a$, what is $\int_{C_2} \boldsymbol{F} \cdot d\boldsymbol{r}$? Why?

12. Evaluate $\int_C \sin y\, dx - x \cos y\, dy$ where C is the curve with parameterization C: $\boldsymbol{r}(t) = \cos t\boldsymbol{i} + \sin t\boldsymbol{j}$, $0 \le t \le \pi$.

13. Use Theorem 2 to prove that the vector field $\boldsymbol{F}$ is conservative in the open set D if and only if $\int_C \boldsymbol{F} \cdot d\boldsymbol{r} = 0$ for every closed path in D.

14. Use Theorem 2 and Exercise 13 to prove that the line integral $\int_C \boldsymbol{F} \cdot d\boldsymbol{r}$ is independent of path in the open set D if and only if $\int_C \boldsymbol{F} \cdot d\boldsymbol{r} = 0$ for every closed path in D.

15. Let $F(x, y) = 2xye^{x^2y}\boldsymbol{i} + x^2e^{x^2y}\boldsymbol{j}$. Let C_1 be the path from $A = (-1, 0)$ to $B = (1, 0)$ clockwise along the unit circle, and let C_2 be the path consisting of the line segment from A to B. Use the method of Section 21.2 to evaluate each of the line integrals $\int_{C_1} \boldsymbol{F} \cdot d\boldsymbol{r}$ and $\int_{C_2} \boldsymbol{F} \cdot d\boldsymbol{r}$. Are they equal? Why?

16. Repeat Exercise 15 for the vector field $F(x, y) = x^3y^2\boldsymbol{i} + x^3y\boldsymbol{j}$.

17. Show that, "in a conservative force field, the force is equal to the negative gradient of the potential."

18. Find a force field $\boldsymbol{F}(x, y, z)$ so that $\int_{(0,0,0)}^{(x,y,z)} \boldsymbol{F} \cdot d\boldsymbol{r} = xyz$ for all $(x, y, z) \in \mathbb{R}^3$, independent of path.

19. Are line integrals of $\boldsymbol{F}(x, y) = (x - y)\boldsymbol{i} + (x + y)\boldsymbol{j}$ independent of path in $\mathbb{R}^2$? Why or why not?

20. **a.** Show that

$$\int_C \frac{1}{x^2 + y^2}(x\, dy - y\, dx) = 2\pi$$

where C is the unit circle oriented counterclockwise.

b. Show that the vector field

$$\boldsymbol{F}(x, y) = \frac{x}{x^2 + y^2}\boldsymbol{i} - \frac{y}{x^2 + y^2}\boldsymbol{j}$$

is not conservative.

c. Show that, for

$$M(x, y) = \frac{-y}{x^2 + y^2} \quad \text{and} \quad N(x, y) = \frac{x}{x^2 + y^2},$$

$M_y(x, y) = N_x(x, y)$, with $x \neq 0$ and $y \neq 0$.

d. Explain why parts (a) through (c) do not contradict the statement involving equation (7).

21.4 GREEN'S THEOREM

The Fundamental Theorem of Calculus states that if $F'(x) = f(x)$ for all x in $[a, b]$, the definite integral $\int_a^b f(x)\, dx$ may be evaluated by means of the formula

$$\int_a^b f(x)\, dx = F(b) - F(a). \tag{1}$$

In other words, the value of the integral in (1) is completely determined by the associated antiderivative at the endpoints of the interval $[a, b]$. In this section, we establish a similar result for certain functions of two variables: The value of a double integral of such a function over a region R in the plane is the same as an associated line integral taken around the boundary of R. This remarkable result is due to George Green, an English mathematician and physicist (1793–1841).

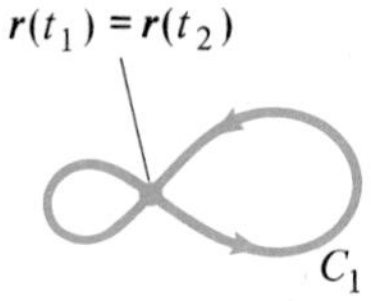

Figure 4.1 C_1 is not a simple closed curve.

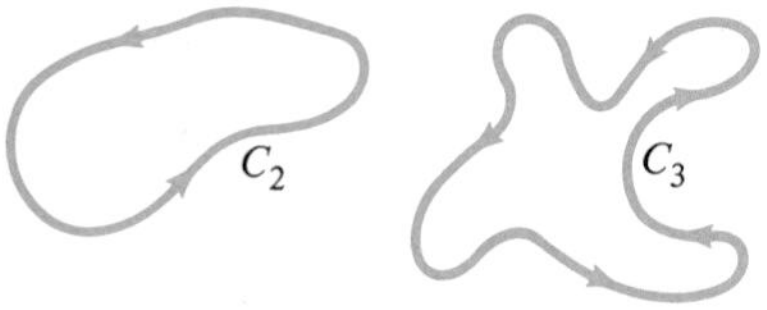

Figure 4.2 C_2 and C_3 are simple closed curves.

The statement of Green's Theorem involves the concept of **simple closed curves.** A closed curve $C = \{\boldsymbol{r}(t) | a \leq t \leq b\}$ is called simple if $\boldsymbol{r}(t_1) \neq \boldsymbol{r}(t_2)$ for all numbers t_1 and t_2 in (a, b) with $t_1 \neq t_2$. (Remember, we must have $\boldsymbol{r}(a) = \boldsymbol{r}(b)$ if C is closed.) In other words, *a simple closed curve cannot cross itself* (see Figures 4.1 and 4.2). Be careful not to confuse the notion of simple closed *curves* with our earlier concepts of vertically simple or horizontally simple *regions*.

THEOREM 3
Green's Theorem

Let C be a piecewise smooth simple closed curve in the plane, oriented counterclockwise, and let Q be the region enclosed by C. If the functions $M(x, y)$ and $N(x, y)$ have continuous first partial derivatives in an open region containing Q, then

$$\int_C M(x, y)\, dx + N(x, y)\, dy = \iint_Q \left(\frac{\partial N}{\partial x} - \frac{\partial M}{\partial y}\right) dA \qquad (2)$$

Before discussing the proof of Green's Theorem, let's try to get a feel for what is going on by means of a few remarks and examples. We begin by expressing the line integral in (2) in vector form. If we write

$$\boldsymbol{F}(x, y) = M(x, y)\boldsymbol{i} + N(x, y)\boldsymbol{j}$$

and

$$C = \{\boldsymbol{r}(t) | a \leq t \leq \text{b}\},$$

then the hypotheses of Green's Theorem are that C is a simple closed path and $\boldsymbol{F}(x, y)$ is a continuously differentiable vector field. The conclusion (2) may then be written as

$$\int_C \boldsymbol{F}(\boldsymbol{r}) \cdot d\boldsymbol{r} = \iint_Q \left(\frac{\partial N}{\partial x} - \frac{\partial M}{\partial y}\right) dA \qquad (3)$$

Now if $\boldsymbol{F}(x, y)$ is a conservative vector field, equation (3) is not surprising, since both integrals are zero. The line integral is zero since C is closed (Section 21.3), and the double integral is zero since $\dfrac{\partial N}{\partial x} - \dfrac{\partial M}{\partial y} = 0$ if $\boldsymbol{F}(x, y)$ is conservative. Thus, Green's Theorem is telling us something about evaluating line integrals for *nonconservative* vector fields. Figure 4.3 illustrates the result geometrically: the surface area associated with the line integral is numerically equal to the volume associated with the double integral (see Figure 2.8, Section 21.2).

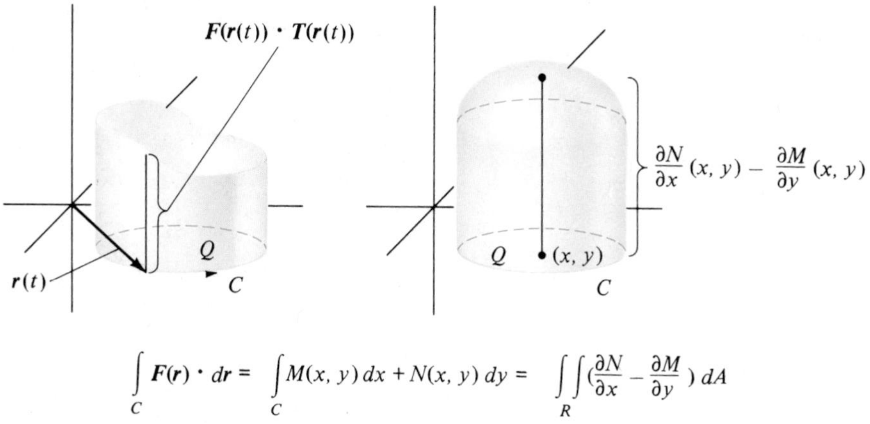
F(r(t)) · T(r(t))
r(t)
Q
C
Q
(x, y)
C
∂N/∂x (x, y) − ∂M/∂y (x, y)
∫_C F(r) · dr = ∫_C M(x, y) dx + N(x, y) dy = ∬_R (∂N/∂x − ∂M/∂y) dA

Figure 4.3 Statement of Green's Theorem.

Example 1 Use Green's Theorem to evaluate the line integral

$$\int_C xy^2\,dx + 2x^3y\,dy$$

where C is the path consisting of the upper unit semicircle traversed counterclockwise and the directed line segment from $(-1, 0)$ to $(1, 0)$ (see Figure 4.4).

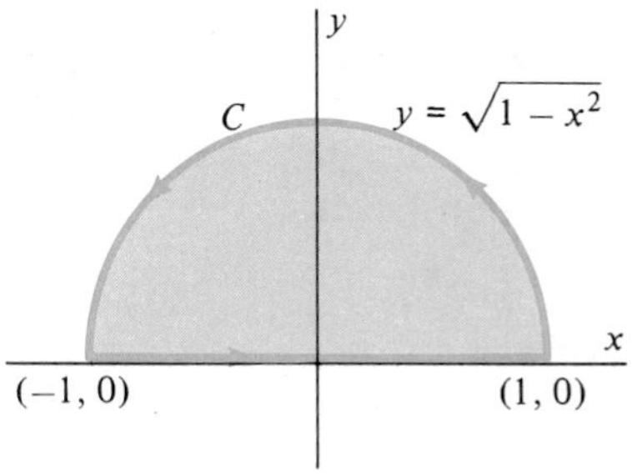

Figure 4.4 Region Q in Example 1.

Solution: Here $M(x, y) = xy^2$, $N(x, y) = 2x^3y$. The region Q enclosed by the path C may be described by the inequalities

$$Q: \quad -1 \le x \le 1; \qquad 0 \le y \le \sqrt{1 - x^2}.$$

With $dA = dx\,dy$, Green's Theorem says that

$$\begin{aligned}
\int_C xy^2\,dx + 2x^3y\,dy &= \int_{-1}^{1} \int_0^{\sqrt{1-x^2}} \left[\frac{\partial}{\partial x}(2x^3y) - \frac{\partial}{\partial y}(xy^2)\right] dy\,dx \\
&= \int_{-1}^{1} \int_0^{\sqrt{1-x^2}} (6x^2y - 2xy)\,dy\,dx \\
&= \int_{-1}^{1} \{[(3x^2 - x)y^2]_0^{\sqrt{1-x^2}}\}\,dx \\
&= \int_{-1}^{1} (3x^2 - x)(1 - x^2)\,dx \\
&= \int_{-1}^{1} (-3x^4 + x^3 + 3x^2 - x)\,dx \\
&= \frac{4}{5}.
\end{aligned}$$

■

Example 2 Use Green's Theorem to evaluate the line integral

$$\int_C xy\,dx + e^y\,dy$$

where C is the path from $(0, 0)$ to $(1, 1)$ along the graph of $y = x^2$ and from $(1, 1)$ to $(0, 0)$ along the graph of $y = \sqrt{x}$.

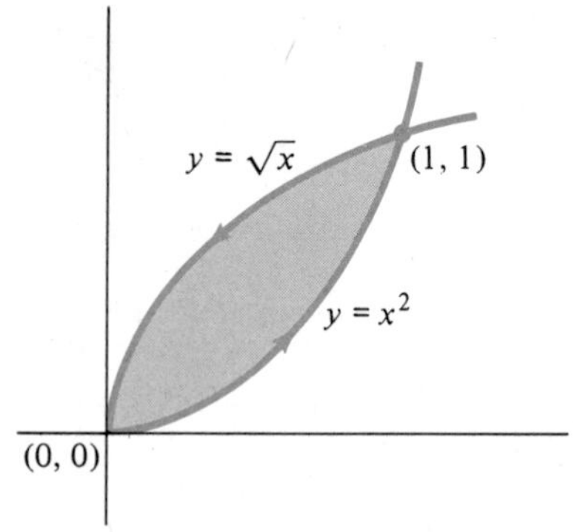

Figure 4.5 Region Q in Example 2.

Solution: Note in Figure 4.5 that C is oriented counterclockwise, as required. The region Q enclosed by C may be described by the inequalities

$$Q:\quad 0 \le x \le 1; \qquad x^2 \le y \le \sqrt{x}.$$

Since $M(x, y) = xy$ and $N(x, y) = e^y$, an application of Green's Theorem gives

$$\begin{aligned}\int_C xy\,dx + e^y\,dy &= \int_0^1 \int_{x^2}^{\sqrt{x}} \left[\frac{\partial}{\partial x}(e^y) - \frac{\partial}{\partial y}(xy)\right] dy\,dx \\ &= \int_0^1 \int_{x^2}^{\sqrt{x}} -x\,dy\,dx \\ &= \int_0^1 \{-xy]_{x^2}^{\sqrt{x}}\}\,dx \\ &= \int_0^1 (-x^{3/2} + x^3)\,dx \\ &= -\frac{3}{20}.\end{aligned}$$

■

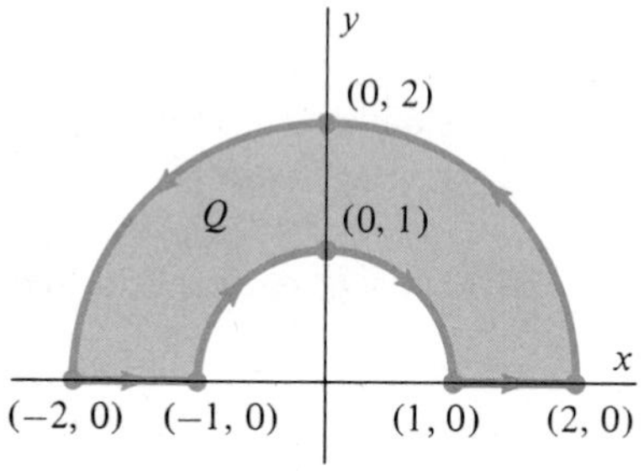

Figure 4.6

Example 3 Use Green's Theorem to evaluate the line integral

$$\int_C xy\,dx + (x + y)\,dy$$

where C is the path indicated in Figure 4.6.

Solution: The region Q enclosed by C may be described in polar coordinates by the inequalities

$$Q:\quad 1 \le r \le 2; \qquad 0 \le \theta \le \pi.$$

Since $M(x, y) = xy$ and $N(x, y) = x + y$, we have

$$\frac{\partial N}{\partial x} - \frac{\partial M}{\partial y} = \frac{\partial}{\partial x}(x + y) - \frac{\partial}{\partial y}(xy) = 1 - x. \tag{4}$$

Since we are using polar coordinates, we set

$$x = r\cos\theta, \qquad y = r\sin\theta, \qquad \text{and} \qquad dA = r\,dr\,d\theta. \tag{5}$$

Combining (4) and (5) with the statement of Green's Theorem, we obtain

$$\begin{aligned}\int_C xy\,dx + (x + y)\,dy &= \int_0^\pi \int_1^2 [1 - r\cos\theta] r\,dr\,d\theta \\ &= \int_0^\pi \left[\frac{r^2}{2} - \frac{r^3}{3}\cos\theta\right]_{r=1}^{r=2} d\theta \\ &= \int_0^\pi \left(\frac{3}{2} - \frac{7}{3}\cos\theta\right) d\theta \\ &= \frac{3\theta}{2} - \frac{7}{3}\sin\theta\Big]_0^\pi \\ &= \frac{3\pi}{2}.\end{aligned}$$

■

There should be no doubt in your mind that the Green's Theorem solution to this problem is much easier than a direct evaluation of the line integral, since the latter would require four separate parameterizations, one for each arc of the curve C.

Calculating Area by Line Integrals

So far we have seen examples in which the double integral over Q has been easier to evaluate than the line integral over C in Green's Theorem. Sometimes it is the other way around. Since areas of certain regions in the plane may be calculated by means of a double integral, Green's Theorem enables us to find expressions for the area of Q (in Theorem 3) in terms of line integrals over C. For example, if we use the functions $M(x, y) = 0$ and $N(x, y) = x$, we obtain

$$\int_C x\,dy = \iint_Q \left[\frac{\partial}{\partial x}(x) - \frac{\partial}{\partial y}(0)\right] dA = \iint_Q 1\,dA = \text{Area of } Q. \tag{6}$$

Similarly, by choosing $M(x, y) = y$ and $N(x, y) = 0$, we obtain

$$\int_C y\,dx = \iint_Q \left[\frac{\partial}{\partial x}(0) - \frac{\partial}{\partial y}(y)\right] dA = \iint_Q (-1)\,dA = -(\text{Area of } Q). \tag{7}$$

Subtracting the corresponding sides of equations (6) and (7) and dividing by 2 gives

$$\boxed{\text{Area of } Q = \frac{1}{2}\int_C x\,dy - y\,dx} \tag{8}$$

Any of equations (6), (7), or (8) may be used in calculating the area of Q. In the following example, equation (8) provides an easy solution to a problem that otherwise requires the technique of integration by trigonometric substitution.

Example 4 Calculate the area A of the region enclosed by the ellipse

$$\frac{x^2}{a^2} + \frac{y^2}{b^2} = 1.$$

Solution: A parameterization for the ellipse with counterclockwise orientation is

$$x(t) = a\cos t, \qquad y(t) = b\sin t, \qquad 0 \le t \le 2\pi.$$

Thus

$$x'(t) = -a\sin t, \qquad \text{and} \qquad y'(t) = b\cos t.$$

By equation (8),

$$\begin{aligned} A &= \frac{1}{2}\int_C x\,dy - y\,dx \\ &= \frac{1}{2}\int_0^{2\pi} [(a\cos t)(b\cos t) - (b\sin t)(-a\sin t)]\,dt \\ &= \frac{1}{2}\int_0^{2\pi} ab(\cos^2 t + \sin^2 t)\,dt \\ &= \pi ab. \end{aligned}$$

■

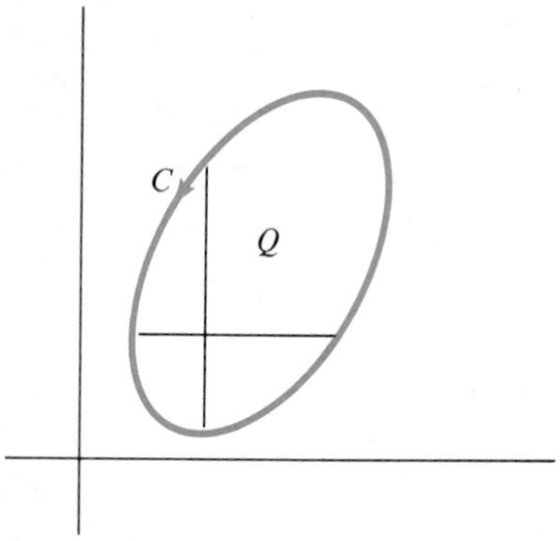

Figure 4.7 Q is both x-simple and y-simple.

Proof of Theorem 3: We will prove Theorem 3 only for regions which are both x-simple and y-simple and then comment on the idea by which the proof can be extended to more general regions.

Specifically, let us assume that Q is a region such as that in Figure 4.7 which can be described both as

$$Q = \{(x, y)|a \le x \le b, \quad f_1(x) \le y \le f_2(x)\} \quad \text{(Figure 4.7)} \tag{9}$$

or as

$$Q = \{(x, y)|g_1(y) \le x \le g_2(y), \quad c \le y \le d\} \quad \text{(Figure 4.8).} \tag{10}$$

According to the hypotheses of Theorem 3, each of the functions $f_1(x)$, $f_2(x)$, $g_1(x)$, and $g_2(x)$ is continuous and piecewise differentiable since its graph constitutes an arc of the curve C.

We will prove Theorem 3 by showing both that

$$\int_C M(x, y)\, dx = -\iint_Q \frac{\partial M}{\partial y}\, dA \tag{11}$$

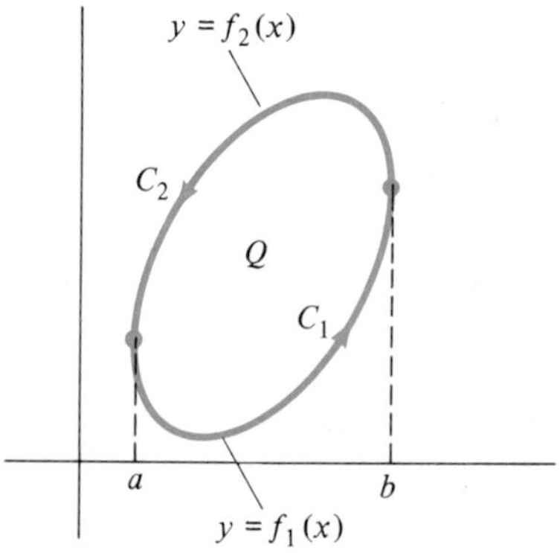

Figure 4.8 C_j is graph of $y = f_j(x)$, $j = 1, 2$.

and that

$$\int_C N(x, y)\, dy = \iint_Q \frac{\partial N}{\partial x}\, dA. \tag{12}$$

Adding the corresponding sides of equations (11) and (12) then establishes equation (2), which is what we are trying to prove.

To prove (11), we first use Figure 4.8 and equation (9) to find that

$$\begin{aligned}\int_C M(x, y)\, dx &= \int_{C_1} M(x, y)\, dx + \int_{C_2} M(x, y)\, dx\\ &= \int_a^b M(x, f_1(x))\, dx + \int_b^a M(x, f_2(x))\, dx\\ &= \int_a^b M(x, f_1(x))\, dx - \int_a^b M(x, f_2(x))\, dx.\end{aligned}$$

Thus,

$$\int_C M(x, y)\, dx = \int_a^b [M(x, f_1(x)) - M(x, f_2(x))]\, dx. \tag{13}$$

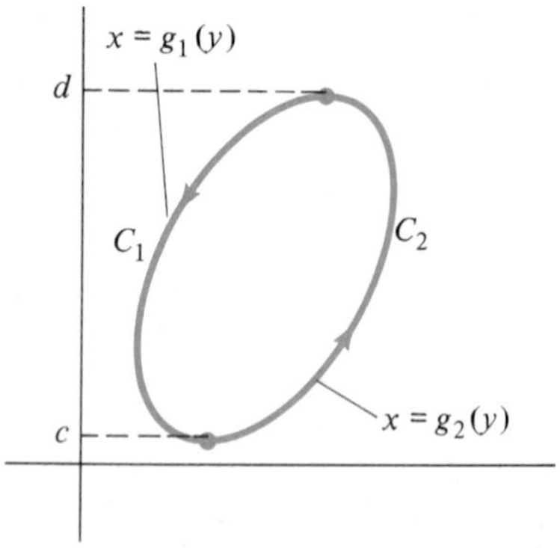

Figure 4.9 C_j is graph of $x = g_j(x)$, $j = 1, 2$.

On the other hand, we also have

$$\begin{aligned}\iint_Q \frac{\partial M}{\partial y}\, dA &= \int_a^b \int_{f_1(x)}^{f_2(x)} \frac{\partial}{\partial y} M(x, y)\, dy\, dx \\ &= \int_a^b \{M(x, y)]_{y=f_1(x)}^{y=f_2(x)}\}\, dx\\ &= \int_a^b [M(x, f_2(x)) - M(x, f_1(x))]\, dx.\end{aligned} \tag{14}$$

Comparing equations (13) and (14) shows that equation (11) holds. Similarly, using Figure 4.9 and equation (10) we can show that equation (12) holds. (This step is left to you as Exercise 23.) This completes the proof for regions Q of the stated type. ■

Extending Green's Theorem to More General Regions

Figure 4.10 shows a region that is neither x-simple nor y-simple, although it does satisfy the hypotheses of Theorem 3. To extend the preceding proof to this region, we make a "cut" (that is, we draw a line) from point P_1 to point P_2, chosen in such a way that the resulting subregions Q_1 and Q_2 are both x-simple and y-simple. Thus, if C_1 denotes the boundary of Q_1, and C_2 the boundary of Q_2, Green's Theorem on the individual subregions shows that

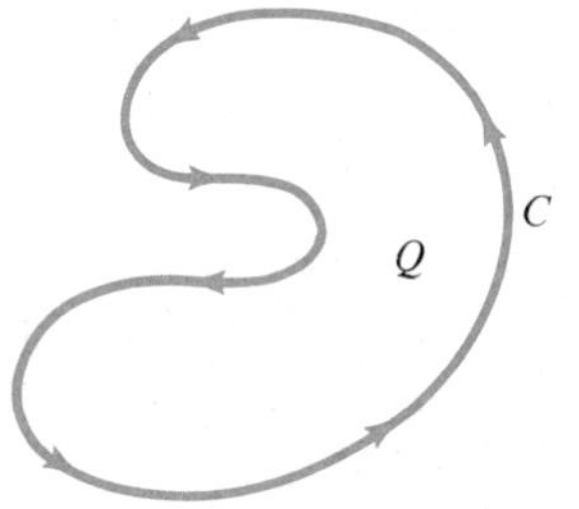

Figure 4.10 Q is neither x-simple nor y-simple.

$$\int_{C_1} M(x, y)\,dx + N(x, y)\,dy = \iint_{Q_1} \left[\frac{\partial N}{\partial x} - \frac{\partial M}{\partial y}\right] dA \tag{15}$$

and

$$\int_{C_2} M(x, y)\,dx + N(x, y)\,dy = \iint_{Q_2} \left[\frac{\partial N}{\partial x} - \frac{\partial M}{\partial y}\right] dA. \tag{16}$$

Adding the corresponding sides of these two equations then gives equation (2). The reason for this is that the contribution to the line integral in (15) resulting from the path from P_1 to P_2 is precisely the negative of the contribution of this same path to the line integral in (16), since the path is traversed in opposite directions by the two line integrals (see Figure 4.11). Thus, the two line integrals taken over the path P_1P_2 "cancel."

Green's Theorem may be extended to regions even more general than these in our statement of Theorem 3. In particular, Green's Theorem holds for regions Q containing "holes," such as that in Figure 4.12. For such regions, the curve C is the entire boundary—both the "inner" and "outer" curves. The meaning of "counterclockwise" for such regions is that each arc of C is traversed so as to keep the region Q on the left-hand side, as indicated in Figure 4.12. The idea by which Green's Theorem is extended to include such regions is again to make "cuts" (see Figure 4.13) dividing Q into subregions without holes, to which Theorem 3 applies. As before, the contributions to the line integral along each cut path sum to zero, so equation (2) results from adding the corresponding sides of the equations holding in each region.

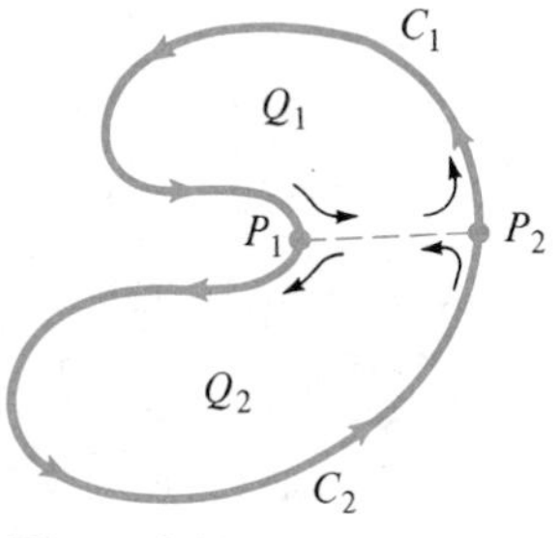

Figure 4.11 Q_1 and Q_2 are both x-simple and y-simple.

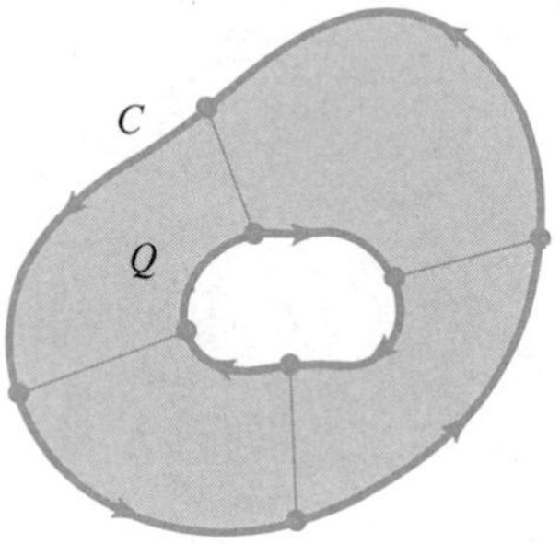

Figure 4.12 C consists of two arcs.

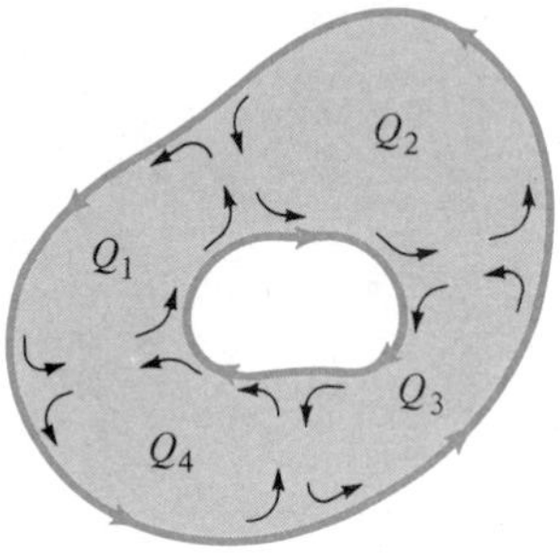

Figure 4.13 Cuts divide Q into regions without holes.

Other Formulations of Green's Theorem

There are two special forms in which Green's Theorem may be stated, each of which is a two-dimensional version of more general results yet to come. We will interpret each of these statements in terms of the flow of a thin layer of fluid, such as that very near the surface of the water in a swimming pool or in a cross section of a water pipe.

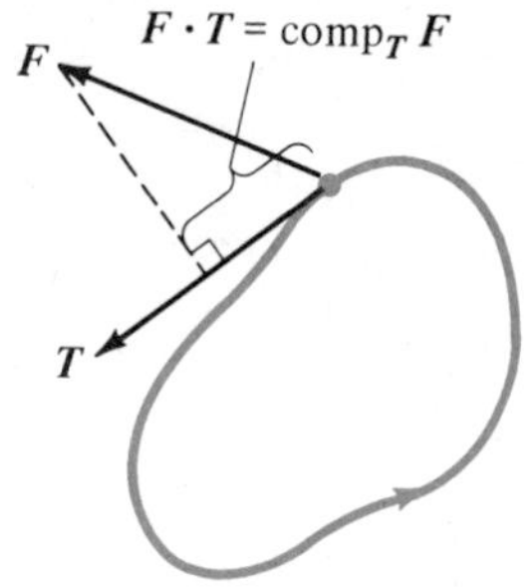

Figure 4.14 $F \cdot T$ is component of F in direction of T.

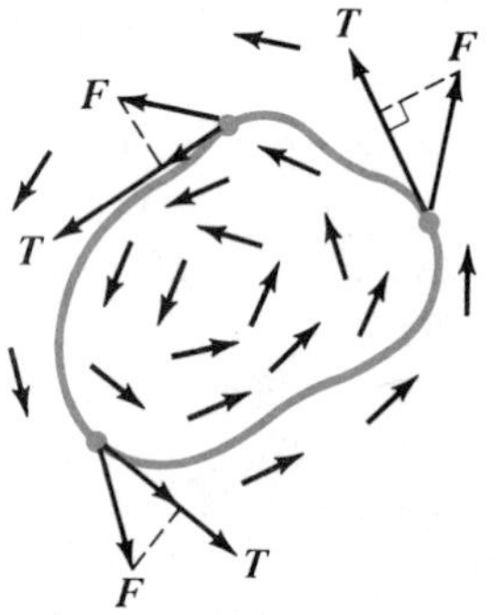

Figure 4.15 In a rotating fluid, $\left|\int_C F \cdot T\, ds\right|$ is large.

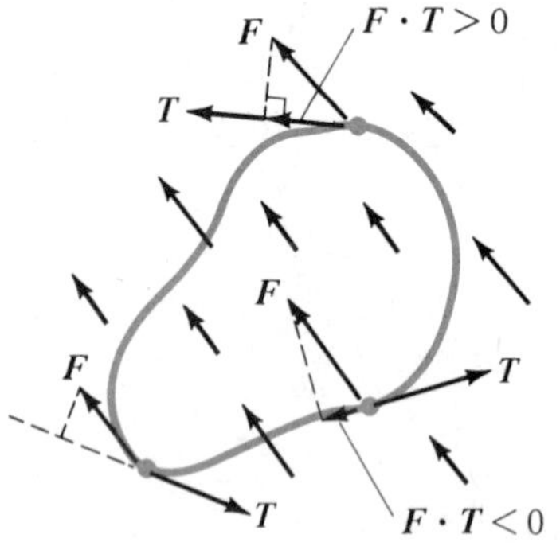

Figure 4.16 In a fluid with little rotation, $\left|\int_C F \cdot T\, ds\right|$ is small.

We begin by letting $\boldsymbol{F}(x, y) = M(x, y)\boldsymbol{i} + N(x, y)\boldsymbol{j}$ be the vector field giving the velocity (speed and direction) of the fluid flow at each point (x, y) in the thin layer. Using equation (29), Section 21.2, we may write the vector form (3) of Green's Theorem as

$$\int_C \boldsymbol{F} \cdot \boldsymbol{T}\, ds = \iint_Q \left(\frac{\partial N}{\partial x} - \frac{\partial M}{\partial y}\right) dA \tag{17}$$

where C is any smooth closed curve in the fluid layer. The integrand $\boldsymbol{F} \cdot \boldsymbol{T}$ in the line integral in (17) is just the component of the vector field (i.e., the flow) in the direction of the unit tangent $\boldsymbol{T}$ to the curve C at each point (see Figure 4.14). If the fluid has a tendency to circulate around the curve C (such as what you see when you pull the plug in your bathtub), we would intuitively expect the line integral in (17) to be large, since the velocity vector $\boldsymbol{F}$ and the tangent vector $\boldsymbol{T}$ should point rather consistently either in approximately the same directions (when the fluid rotates counterclockwise) or in opposite directions (when the fluid rotates clockwise), as illustrated in Figure 4.15. When the fluid has little tendency to rotate, such as occurs in a constant flow field, we expect the line integral in (17) to be small since the component of $\boldsymbol{F}$ on $\boldsymbol{T}$ will be of opposite sign on opposite sides of C (Figure 4.16). For these reasons we say that the line integral in (17) is a measure of the tendency of the fluid layer to circulate or to rotate. Vector fields in the plane for which this line integral is always zero are called **irrotational.**

Since the double integral in equation (17) has the same value as the line integral, it too measures the tendency of the fluid to rotate. However, it does so by integrating the scalar function $\dfrac{\partial N}{\partial x}(x, y) - \dfrac{\partial M}{\partial y}(x, y)$ over the region Q enclosed by C. For this reason we refer to this function as the **scalar curl** of the vector field $\boldsymbol{F}$. That is,

$$\text{curl } \boldsymbol{F}(x, y) = \frac{\partial N}{\partial x}(x, y) - \frac{\partial M}{\partial y}(x, y). \tag{18}$$

Using this definition, we may rewrite the formulation of Green's Theorem given in line (17) as

$$\boxed{\int_C \boldsymbol{F} \cdot \boldsymbol{T}\, ds = \iint_Q \text{curl } \boldsymbol{F}\, dA} \qquad \text{(Circulation of } \boldsymbol{F} \text{ around } C). \tag{19}$$

Equation (19) is referred to as **Stokes' Theorem in the plane.** In Section 21.6 we shall study the generalizations of curl and of Stokes' Theorem to three dimensions.

A second vector formulation of Green's Theorem concerns the outward unit normal to the closed smooth simple curve C. To obtain this vector we let $\boldsymbol{T}$ be the unit tangent at a point on the curve C (oriented counterclockwise) and we let θ be the angle formed between $\boldsymbol{T}$ and the unit vector $\boldsymbol{i}$ (Figure 4.17). Then

$$\boldsymbol{T} = \cos\theta \boldsymbol{i} + \sin\theta \boldsymbol{j}. \tag{20}$$

The unit vector $\boldsymbol{n}$ obtained by rotating $\boldsymbol{T}$ 90° in the counterclockwise direction is the **unit normal**

$$\begin{aligned} \boldsymbol{n} &= \cos(\theta + \pi/2)\boldsymbol{i} + \sin(\theta + \pi/2)\boldsymbol{j} \\ &= -\sin\theta \boldsymbol{i} + \cos\theta \boldsymbol{j}. \end{aligned}$$

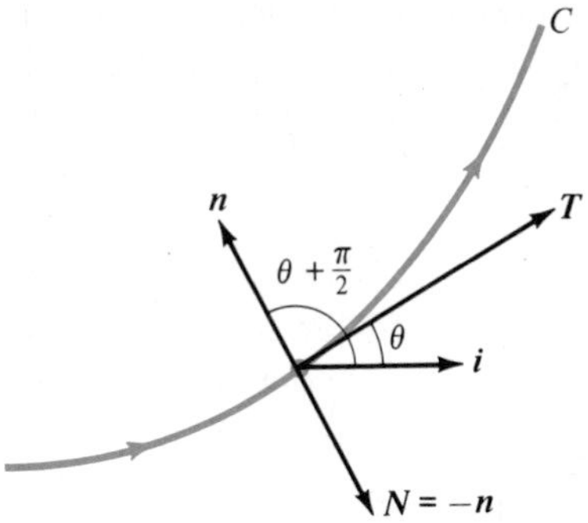

Figure 4.17 N is the outward unit normal.

However, this normal points inward for a closed curve C oriented counterclockwise, so the desired **outward unit normal** $\boldsymbol{N}$ is the vector

$$\boldsymbol{N} = -\boldsymbol{n} = \sin\theta\boldsymbol{i} - \cos\theta\boldsymbol{j}. \tag{21}$$

Now recall from Section 18.4 that the unit tangent vector $\boldsymbol{T}$ may be written

$$\boldsymbol{T} = \frac{dx}{ds}\boldsymbol{i} + \frac{dy}{ds}\boldsymbol{j} \tag{22}$$

where s is the arc length parameter and $\boldsymbol{r}(s) = x(s)\boldsymbol{i} + y(s)\boldsymbol{j}$ is a parameterization for C by arc length. From equations (20), (21), and (22) it follows that the outward unit normal $\boldsymbol{N}$ may be written

$$\boldsymbol{N} = \frac{dy}{ds}\boldsymbol{i} - \frac{dx}{ds}\boldsymbol{j}. \tag{23}$$

For the vector field $\boldsymbol{F}(x, y) = M(x, y)\boldsymbol{i} + N(x, y)\boldsymbol{j}$, the expression (23) for $\boldsymbol{N}$ suggests that

$$\int_C \boldsymbol{F}\cdot\boldsymbol{N}\,ds = \int_C M(x, y)\,dy - N(x, y)\,dx, \tag{24}$$

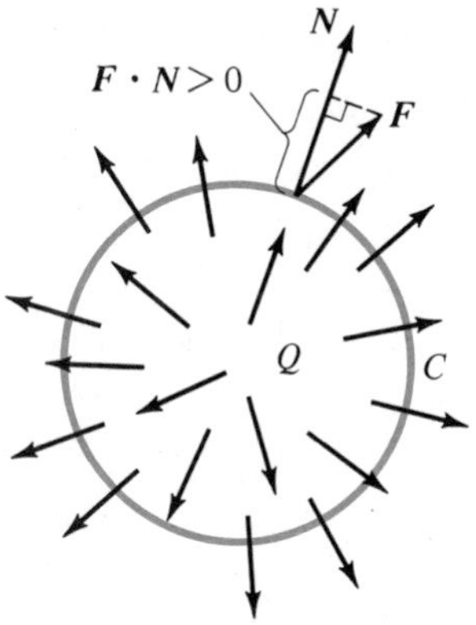

Figure 4.18 $\int_C \boldsymbol{F}\cdot\boldsymbol{N}\,ds > 0$ if net flow across C is outward.

and Green's Theorem, applied to the right-hand side of equation (24) gives

$$\int_C M(x, y)\,dy - N(x, y)\,dx = \iint_Q \left(\frac{\partial M}{\partial x} + \frac{\partial N}{\partial y}\right) dA. \tag{25}$$

Finally, combining (24) and (25), we obtain

$$\boxed{\int_C \boldsymbol{F}\cdot\boldsymbol{N}\,ds = \iint_Q \left(\frac{\partial M}{\partial x} + \frac{\partial N}{\partial y}\right) dA.} \tag{26}$$

Equation (26) is referred to as the *Divergence Theorem in the plane* for reasons that we will now explain. Since the scalar quantity $\boldsymbol{F}\cdot\boldsymbol{N}$ in the line integral is the component of the vector field (fluid flow) in the direction of the outward normal, the line integral in (26) measures the net flow of fluid outward across the boundary C of the region Q. The term **flux** is used to denote this net rate at which fluid flows across a boundary, and the line integral in (26) is called a **flux integral.** When the line integral is *positive,* more fluid is leaving Q than is entering. (Think of an open faucet somewhere inside Q.) When the line integral in (26) is *negative,* more fluid is flowing into Q than is leaving. (Think of an open drain somewhere inside Q.) When the line integral is small or zero, there is little or no net gain or loss of fluid in Q. (Think of fluid moving uniformly, or motionless fluid. See Figures 4.18 through 4.20.)

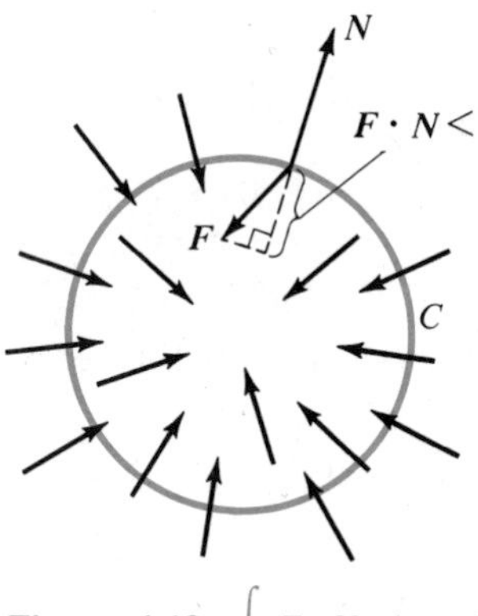

Figure 4.19 $\int_C \boldsymbol{F}\cdot\boldsymbol{N}\,ds < 0$ if net flow across C is inward.

Since the double integral in (26) gives the same measure of net flow as does the line integral, the integrand $\dfrac{\partial M}{\partial x} + \dfrac{\partial N}{\partial y}$ is referred to as the **divergence** of the vector field $\boldsymbol{F}$, written div $\boldsymbol{F}$. That is

$$\text{div}\,\boldsymbol{F}(x, y) = \frac{\partial}{\partial x}M(x, y) + \frac{\partial}{\partial y}N(x, y). \tag{27}$$

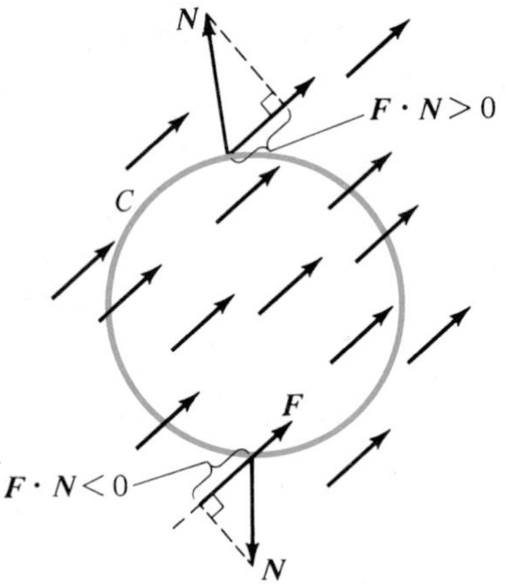

Figure 4.20 $\int_C F \cdot N \, ds = 0$ if net flow across C is zero.

Using (27), we may rewrite the Divergence Theorem in the plane as

$$\int_C \boldsymbol{F} \cdot \boldsymbol{N} \, ds = \iint_Q \text{div } \boldsymbol{F} \, dA. \qquad \text{(Flux of } F \text{ across } C\text{).} \tag{28}$$

In Section 21.7 we will encounter generalizations of both the concept of divergence of a vector field and the Divergence Theorem to three dimensions, the primary motivation being fluid flow in space.

Exercise Set 21.4

(In all exercises the closed curves C are oriented counterclockwise.) In Exercises 1–10, use Green's Theorem to evaluate the given line integral.

1. $\int_C xy\,dx + (x + y)\,dy$, C is the square with vertices $(0, 0)$, $(0, 1)$, $(1, 1)$, and $(1, 0)$.
2. $\int_C xy\,dx + (y - x)\,dy$, C is the unit circle.
3. $\int_C x^2y^3\,dx + (x^2 + 1)\,dy$, C is the square in Exercise 1.
4. $\int_C \sqrt{y}\,dx + \sqrt{x}\,dy$, C is the rectangle with vertices $(1, 1)$, $(1, 2)$, $(4, 1)$, and $(4, 2)$.
5. $\int_C xy^2\,dy - x^2y\,dx$, C is the quarter circular path from $(0, 0)$ to $(1, 0)$ to $(0, 1)$ to $(0, 0)$.
6. $\int_C e^x \tan y\,dx + e^x \sec^2 y\,dy$, C is the triangle with vertices $(-1, 2)$, $(3, 0)$, and $(1, 6)$.
7. $\int_C (x^3 + y)\,dx + (y - x^2)\,dy$, C is the boundary of the region bounded by the graphs of $y = x^2$ and $y = x^3$.
8. $\int_C (\tan^{-1} x + y^2)\,dx + (\ln^2 y - x^2)\,dy$, C is the circle $x^2 + y^2 = 4$.
9. $\int_C (\cos^5 x + \sqrt{x})\,dx + \tan^{-1} y\,dy$, C is the ellipse $4x^2 + y^2 = 1$.
10. $\int_C \sqrt{x}\,dx + \ln(x^2 + y^2)\,dy$, C is the curve in Figure 4.6.

In Exercises 11–14, use Green's Theorem to find $\int_C \boldsymbol{F} \cdot d\boldsymbol{r}$.

11. $\boldsymbol{F}(x, y) = 2y\boldsymbol{i} - 3x\boldsymbol{j}$, C is the unit circle.
12. $\boldsymbol{F}(x, y) = xy\boldsymbol{i} + x^2\boldsymbol{j}$, C is the square with vertices $(0, 0)$, $(0, 2)$, $(2, 2)$, and $(2, 0)$.
13. $\boldsymbol{F}(x, y) = x^2(y^2 - x^2)\boldsymbol{i} + \frac{2}{3}x^3y\boldsymbol{j}$, C is the triangle with vertices $(1, 1)$, $(4, 1)$, and $(3, 5)$.
14. $\boldsymbol{F}(x, y) = e^x \sin y\boldsymbol{i} + e^x \cos y\boldsymbol{j}$, C is the ellipse $x^2 + 9y^2 = 1$.
15. Find the work done by the force field $\boldsymbol{F}(x, y) = (e^{x^2} + 2y)\boldsymbol{i} + (ye^y - x)\boldsymbol{j}$ in moving a particle once around the path indicated in Figure 4.6.
16. Use equation (8) to find the area of the region bounded by the graphs of $y = x$ and $y = x^2$.
17. Use equation (8) to find the area of the region bounded by the graphs of $y = x$ and $y = \sqrt{x}$.
18. Find the area of the region enclosed by the graph of the parametric equations $x(t) = \sin t \cos t$, $y(t) = \sin t$ for $0 \le t \le \pi$.

19. Find the area of the region bounded by the graph of the parametric equations $x(t) = a\cos^3 t$, $y(t) = a\sin^3 t$, $0 \le t \le 2\pi$.

20. Let Q be a region in the plane whose boundary is a simple closed piecewise smooth curve C. Let A be the area of Q. Show that the coordinates of the centroid of Q are

$$\bar{x} = \frac{1}{2A}\int_C x^2\,dy, \qquad \bar{y} = -\frac{1}{2A}\int_C y^2\,dx.$$

21. Use Green's Theorem to prove that if $\boldsymbol{F}(x, y)$ is a conservative vector field in the plane, then any line integral $\int_C \boldsymbol{F}\cdot d\boldsymbol{r}$ over a piecewise smooth curve C is independent of path.

22. Use Green's Theorem to prove that if $\boldsymbol{F}(x, y)$ is a vector field in the plane for which $\int_A^B \boldsymbol{F}\cdot d\boldsymbol{r}$ is independent of path for all A and B, then $\int_C \boldsymbol{F}\cdot d\boldsymbol{r} = 0$ for every simple closed path C.

23. Complete the proof of Theorem 3 by showing that equation (12) holds.

24. Let $\boldsymbol{F}(x, y) = M(x, y)\boldsymbol{i} + N(x, y)\boldsymbol{j}$ where M and N have continuous first partial derivatives in some open rectangle D. Show that $\boldsymbol{F}$ is conservative in D if and only if $\text{curl}\,\boldsymbol{F} = 0$ for all $(x, y) \in D$.

21.5 SURFACE INTEGRALS

The primary motivation for the remainder of this chapter will be to extend the concept of flux, introduced in Section 21.4, to the flow of fluids in and out of regions in space. Since regions in space are bounded by surfaces rather than curves, we must begin by developing the concept of integrals over surfaces, or *surface integrals*. Although we are primarily interested in defining the integral of a *vector field* $\boldsymbol{F}(x, y, z)$ over a surface S (Figure 5.1), we shall begin by defining the integral of a scalar function $f(x, y, z)$ over a surface S (Figure 5.2). This is because the development of the surface integral for scalar functions very closely parallels the discussion of surface area in Section 20.6.

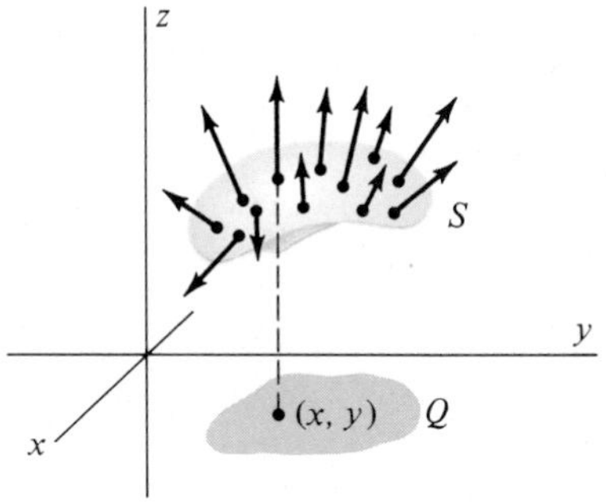

Figure 5.1 A vector field $F(x, y, z)$ assigns a vector to each point on the surface S.

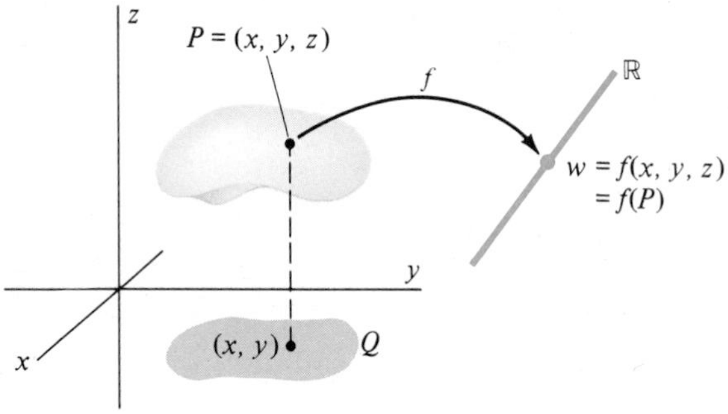

Figure 5.2 A scalar function $f(x, y, z)$ assigns a number to each point P on S.

By a surface S, we shall mean the graph of a continuous function of two variables, $z = g(x, y)$, over some regular region Q in the xy-plane. (Recall that the region Q is regular if it is both x-simple and y-simple.) Such a surface is illustrated in Figure 5.3. This is not the most general concept of surface, but techniques developed here will enable you to handle most surfaces encountered in practice. (For an excellent discussion of the more general notion of *parameterized surfaces*, see the text *Vector Calculus*, 2nd ed., Jerrold E. Marsden and Anthony J. Tromba, W. H. Freeman, 1981.)

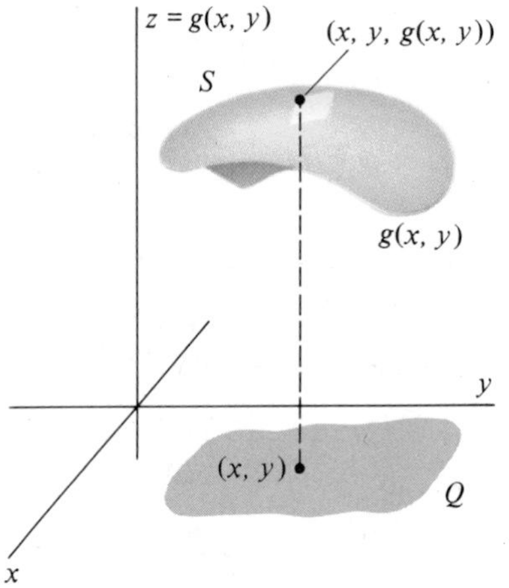

Figure 5.3 Surface S is the graph of a continuous function $z = g(x, y)$ over Q.

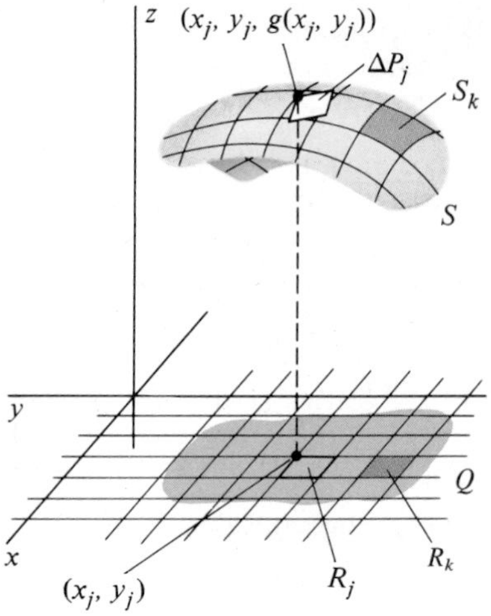

Figure 5.4 Parallelogram of area ΔP_j over rectangle R_j approximates the area ΔS_j of the "patch" S_j.

Suppose that the surface S is the graph of the continuous function $z = g(x, y)$ over the regular region Q, and suppose that the continuous scalar function $f(x, y, z)$ is defined for all points $(x, y, g(x, y))$ on the surface S. Recall from Section 20.6 that an inner partition of the region Q into n rectangles of area $\Delta A_j = \Delta x\,\Delta y$ partitions "most" of the surface S into curvilinear "patches" S_j of area ΔS_j (Figure 5.4.) Proceeding in a purely mathematical way, we form the product of the surface area ΔS_j of the patch S_j and the value of the scalar function $f(x_j, y_j, z_j)$ at an arbitrary point (x_j, y_j, z_j) on the patch. Summing these products over all n patches gives the approximating sum

$$\sum_{j=1}^{n} f(x_j, y_j, z_j)\,\Delta S_j, \qquad (x_j, y_j, z_j) \in S_j.$$

The limit as $n \to \infty$ (and as Δx and Δy approach zero) of this approximating sum, when it exists, is defined as the surface integral of $f(x, y, z)$ over S:

$$\iint_S f(x, y, z)\,dS = \lim_{n\to\infty} \sum_{j=1}^{n} f(x_j, y_j, z_j)\,\Delta S_j \tag{1}$$

We shall determine several ways in which this surface integral may be calculated. The first of these involves recalling that the area ΔS_j of the patch S_j may be approximated by the area ΔP_j of the parallelogram over R_j tangent to the surface S at one corner (Figure 5.4). That is, we have the approximation (see Section 20.6)

$$\Delta S_j \approx \Delta P_j = \sqrt{\left(\frac{\partial g}{\partial x}\right)^2 + \left(\frac{\partial g}{\partial y}\right)^2 + 1}\;\Delta x\,\Delta y \tag{2}$$

where the partial derivatives $\dfrac{\partial g}{\partial x}$ and $\dfrac{\partial g}{\partial y}$ are evaluated at one corner, (x_j, y_j), of the rectangle R_j. Using approximation (2) in equation (1) leads to the following definition.

DEFINITION 5

Let Q be a regular region in the plane and let $z = g(x, y)$ be a continuous function defined on some open set containing Q on which both $\frac{\partial g}{\partial x}$ and $\frac{\partial g}{\partial y}$ are continuous. Let S be the graph of $g(x, y)$ over Q, and let $f(x, y, z)$ be a continuous scalar function defined on S. The **surface integral** of $f(x, y, z)$ over S is defined by

$$\iint_S f(x, y, z)\, dS = \iint_Q f(x, y, g(x, y))\sqrt{\left(\frac{\partial g}{\partial x}\right)^2 + \left(\frac{\partial g}{\partial y}\right)^2 + 1}\, dx\, dy. \quad (3)$$

Example 1 Evaluate the surface integral

$$\iint_S (x^2 + y + z)\, dS$$

where S is the graph of the function $g(x, y) = \sqrt{3}\, y - x^2$ over the unit square

$$Q = \{(x, y) | 0 \le x \le 1,\ 0 \le y \le 1\}.$$

Solution: The integrand is the function

$$f(x, y, z) = x^2 + y + z.$$

Also,

$$\frac{\partial g}{\partial x} = -2x, \qquad \text{and} \qquad \frac{\partial g}{\partial y} = \sqrt{3}.$$

Substituting into equation (3) gives

$$\begin{aligned}\iint_S (x^2 + y + z)\, dS &= \iint_Q [x^2 + y + (\sqrt{3}\, y - x^2)]\sqrt{(-2x)^2 + (\sqrt{3})^2 + 1}\, dx\, dy\\ &= \int_0^1 \int_0^1 (1 + \sqrt{3})y\sqrt{4x^2 + 4}\, dy\, dx\\ &= (1 + \sqrt{3}) \int_0^1 \{y^2\sqrt{x^2 + 1}]_{y=0}^{y=1}\}\, dx\\ &= (1 + \sqrt{3}) \int_0^1 \sqrt{x^2 + 1}\, dx\\ &= (1 + \sqrt{3})\left(\frac{\sqrt{2}}{2} + \frac{1}{2}\ln(1 + \sqrt{2})\right)\\ &\approx 3.14.\end{aligned}$$

■

Example 2 Evaluate the surface integral

$$\iint_S (x^2 + y^2 + z)\, dS$$

where S is the portion of the graph of the function $z = 4 - x^2 - y^2$ bounded below by the xy-plane.

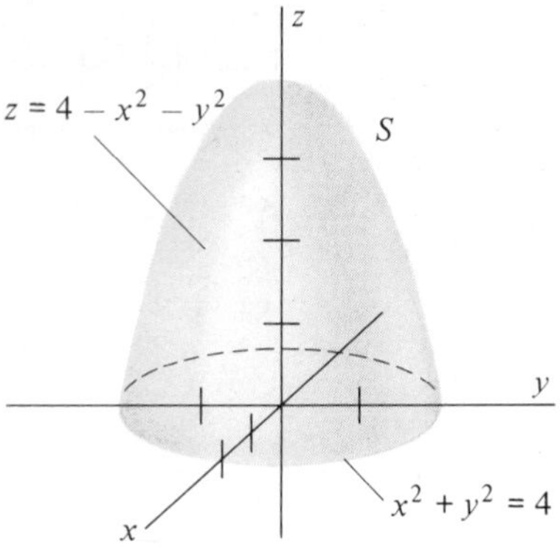

Figure 5.5 Surface S is graph of $z = 4 - x^2 - y^2$.

Solution: The integrand is $f(x, y, z) = x^2 + y^2 + z$, and the surface is the graph of the function $g(x, y) = 4 - x^2 - y^2$. Thus

$$\frac{\partial g}{\partial x} = -2x, \quad \text{and} \quad \frac{\partial g}{\partial y} = -2y.$$

The region Q in the xy-plane over which S is defined is

$$Q = \{(x, y) | x^2 + y^2 \leq 4\} \qquad \text{(Figure 5.5)}.$$

According to Definition 5, the surface integral is

$$\iint_S (x^2 + y^2 + z)\, dS = \iint_Q [x^2 + y^2 + (4 - x^2 - y^2)]\sqrt{(-2x)^2 + (-2y)^2 + 1}\, dx$$

$$= \iint_Q 4\sqrt{4x^2 + 4y^2 + 1}\, dx\, dy.$$

Switching to polar coordinates allows us to evaluate this integral as

$$\iint_S (x^2 + y^2 + z)\, dS = \int_0^{2\pi} \int_0^2 4\sqrt{4r^2 + 1}\, r\, dr\, d\theta$$

$$= \int_0^{2\pi} \left\{ \left[\frac{1}{3}(4r^2 + 1)^{3/2} \right]_{r=0}^{r=2} \right\} d\theta$$

$$= \frac{2\pi}{3}(17^{3/2} - 1)$$

$$\approx 144.71.$$ ■

REMARK: Definition 5 defines the surface integral only for surfaces that are graphs of functions of the form $z = g(x, y)$. Notice that by setting $f(x, y, z) = 1$ in Definition 5 we obtain the formula for surface area given by Definition 4, Chapter 20. As in our discussion of surface area in Chapter 20, there is no essential reason why z must be taken to be the dependent variable. For surfaces that are graphs of functions of the form $y = g(x, z)$ or $x = g(y, z)$, we may simply interchange roles among the variables x, y, and z as required. The following example presents one such situation. (Notice that the given surface cannot be described as the graph of a function of the form $z = g(x, y)$.)

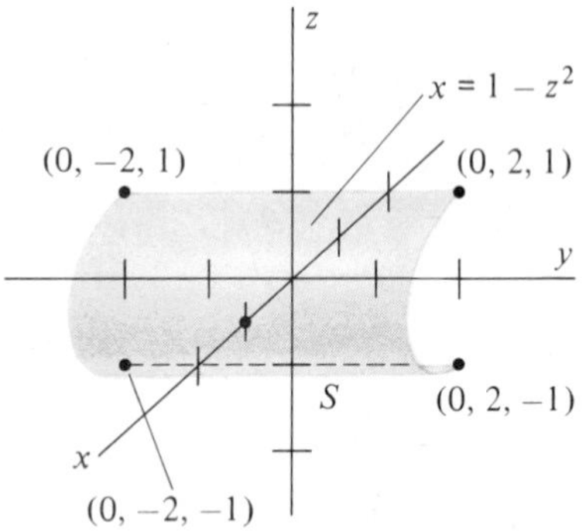

Figure 5.6 Surface S is graph of $x = 1 - z^2$.

Example 3 Evaluate the surface integral

$$\iint_S yz\, dS$$

where S is the portion of the graph of $x = 1 - z^2$ bounded by the yz-plane and the planes $y = -2$ and $y = 2$ (Figure 5.6).

Solution: In order to describe S as the graph of a function, we must take x as the dependent variable. Interchanging roles of x and z in Definition 5, we find that

$$f(x, y, z) = yz \quad \text{and} \quad g(y, z) = 1 - z^2.$$

Thus,

$$\frac{\partial g}{\partial z} = -2z \qquad \text{and} \qquad \frac{\partial g}{\partial y} = 0.$$

The region Q in the yz-plane over which S is defined is the rectangle

$$Q = \{(y, z) | -2 \le y \le 2, \ -1 \le z \le 1\}.$$

Definition 5 now gives

$$\begin{aligned}\iint_S yz\, dS &= \iint_Q yz\sqrt{(-2z)^2 + 1}\, dz\, dy \\ &= \int_{-2}^{2} \int_{-1}^{1} yz\sqrt{4z^2 + 1}\, dz\, dy \\ &= \int_{-2}^{2} \left\{ \frac{y}{12}(4z^2 + 1)^{3/2} \Big]_{z=-1}^{z=1} \right\} dy \\ &= \int_{-2}^{2} \frac{y}{12}(5^{3/2} - 5^{3/2})\, dy \\ &= 0.\end{aligned}$$

The result should not be surprising, since both $f(x, y, z)$ and the surface are symmetric with respect to both y and z. ■

Calculating Mass

A straightforward application of surface integrals of the form (1) occurs in the calculation of the total mass of a hollow object (such as a basketball or a vase) whose mass density per unit surface area, $\rho(x, y, z)$, varies continuously over the surface of the object. That is, we assume that $\rho(x, y, z)$ gives the mass (in grams/cm^2, say) of the object at location (x, y, z). Then if S_j is a small patch of area ΔS_j on the surface of the object, the total mass Δm_j of S_j may be approximated by assuming the mass density $\rho(x, y, z)$ to be constant throughout S_j. In other words, if (x_j, y_j, z_j) is a point on S_j, we have the approximation

$$\Delta M_j \approx \rho(x_j, y_j, z_j)\, \Delta S_j \qquad \left(\text{mass} = \frac{\text{mass}}{\text{unit area}} \times \text{area}\right). \tag{4}$$

Summing (4) over all patches in a partition covering S, we arrive at the approximation to total mass:

$$M \approx \sum_{j=1}^{n} \rho(x_j, y_j, z_j)\, \Delta S_j. \tag{5}$$

Comparing approximation (5) with equation (1), we conclude that in the limit as $n \to \infty$ (and $\Delta S_j \to 0$), assuming appropriate conditions on the shape of the object and on the density function $\rho(x, y, z)$,

$$\boxed{M = \int_S \rho(x, y, z)\, dS} \tag{6}$$

In other words, **the total mass of the object** is simply the integral, over the surface of the object, of the function giving the mass per unit area.

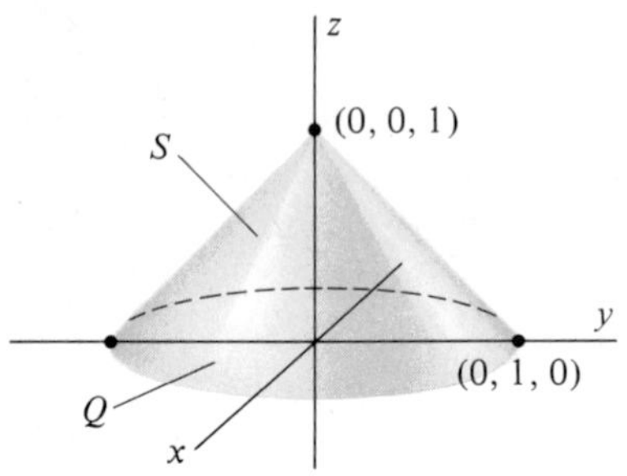

Figure 5.7

Example 4 Let S be the idealized cone described in cylindrical coordinates by the equation $z = 1 - r$, $0 \le r \le 1$. Find the mass of S if the density at any point is proportional to the square of the distance between the point and the origin (see Figure 5.7).

Solution: Here $\rho(x, y, z) = \lambda(x^2 + y^2 + z^2)$ is the mass density, where λ is a constant. We will work with ρ in cylindrical coordinates:

$$\rho(r, \theta, z) = \lambda(r^2 + z^2).$$

Also, the surface S is described by the function

$$z = g(r, \theta) = 1 - r.$$

Using the result of Exercise 25 of this section (for evaluating a surface integral in cylindrical coordinates) together with equation (6), we find that

$$M = \iint_S \rho(r, \theta, z)\, dS = \iint_Q \rho(r, \theta, g(r, \theta)) \sqrt{r^2 + r^2\left(\frac{\partial g}{\partial r}\right)^2 + \left(\frac{\partial g}{\partial \theta}\right)^2}\, dr\, d\theta$$

$$= \int_0^{2\pi}\int_0^1 \lambda(r^2 + (1 - r)^2) \cdot \sqrt{r^2 + r^2(-1)^2 + 0^2}\, dr\, d\theta$$

$$= \int_0^{2\pi}\int_0^1 \sqrt{2}\lambda \cdot (2r^3 - 2r^2 + r)\, dr\, d\theta$$

$$= \frac{2\sqrt{2}\pi\lambda}{3}.$$ ■

The concept of surface integral may be extended to various types of surfaces that do not fulfill the hypotheses of Definition 5. If a surface S is a union of a finite number of surfaces $S = S_1 \cup S_2 \cup \cdots \cup S_n$, each of which satisfies the hypotheses of Definition 5, we define the surface integral of $f(x, y, z)$ over S to be the sum of the individual surface integrals

$$\iint_S f(x, y, z)\, dS = \sum_{j=1}^{n} \iint_{S_j} f(x, y, z)\, dS \tag{7}$$

each of which is evaluated by use of Definition 5.

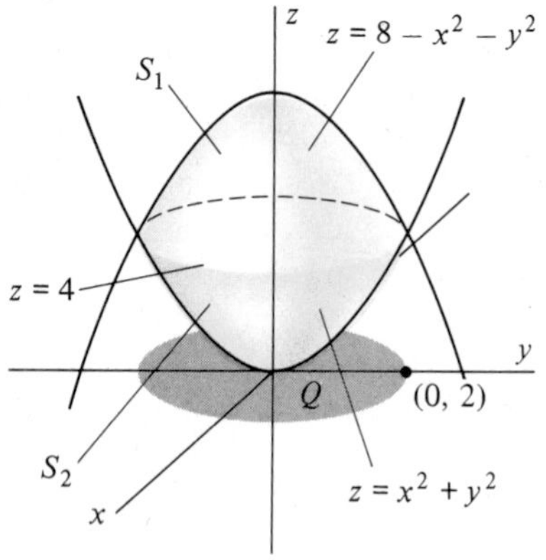

Figure 5.8 Surface S is not smooth along the curve $z = 4$.

Example 5 Evaluate $\iint_S (x^2 + y^2 + z)\, dS$ where the surface S is the boundary of the region bounded by the graphs of $g_1(x, y) = 8 - (x^2 + y^2)$ and $g_2(x, y) = x^2 + y^2$ (Figure 5.8).

Solution: The two graphs intersect along the circle $x^2 + y^2 = 4$, the interior of which we shall refer to as the region Q. We may therefore describe the surface S as $S = S_1 \cup S_2$ where

$$S_1 = \{(x, y, 8 - (x^2 + y^2)) | (x, y) \in Q\},$$

$$S_2 = \{(x, y, x^2 + y^2) | (x, y) \in Q\}.$$

On the surface S_1, with $f(x, y, z) = x^2 + y^2 + z$ and $g_1(x, y) = 8 - (x^2 + y^2)$, Definition 5 gives

$$\iint_{S_1} (x^2 + y^2 + z)\, dS = \iint_Q [x^2 + y^2 + (8 - (x^2 + y^2))]\sqrt{(-2x)^2 + (-2y)^2 + 1}\, dx\, dy$$

$$= \int_0^{2\pi}\int_0^2 8\sqrt{4r^2 + 1} \cdot r\, dr\, d\theta \qquad \text{(Switching to polar coordinates)}$$

$$= \frac{4\pi}{3}(17^{3/2} - 1).$$

On the surface S_2, with $f(x, y, z) = x^2 + y^2 + z$ and $g_2(x, y) = x^2 + y^2$, we obtain

$$\iint_{S_2} (x^2 + y^2 + z)\, dS = \iint_Q [x^2 + y^2 + (x^2 + y^2)]\sqrt{(2x)^2 + (2y)^2 + 1}\, dx\, dy$$

$$= \int_0^{2\pi}\int_0^2 2r^2\sqrt{4r^2 + 1}\, r\, dr\, d\theta$$

$$= \frac{\pi}{20}(17^{5/2} - 1) - \frac{\pi}{12}(17^{3/2} - 1).$$

Combining these results according to equation (7) gives

$$\iint_S (x^2 + y^2 + z)\, dS = \frac{\pi}{20}(17^{5/2} - 1) + \frac{5\pi}{4}(17^{3/2} - 1)$$

$$\approx 458.3.$$ ■

Definition 5 could not be applied directly to the entire surface S in Example 5 for two reasons. First, S is not the graph of a *function* $z = g(x, y)$, since there are two distinct points corresponding to some (x, y)-coordinates. Second, the surface S is not differentiable $\left(\text{meaning that } \dfrac{\partial g}{\partial x} \text{ and } \dfrac{\partial g}{\partial y} \text{ are not continuous}\right)$ along the "seam" where S_1 meets S_2.

Another difficulty that can arise in attempting to apply Definition 5 is that the integrand in the double integral has a singularity (i.e., becomes infinite) along the boundary of the region Q. In such cases we define $\iint_S f(x, y, z)\, dS$ to be the corresponding improper integral if this integral exists. The following example is typical. When combined with the notion of equation (7), this idea allows us to evaluate surface integrals over spheres, an important aspect of many applications of surface integrals.

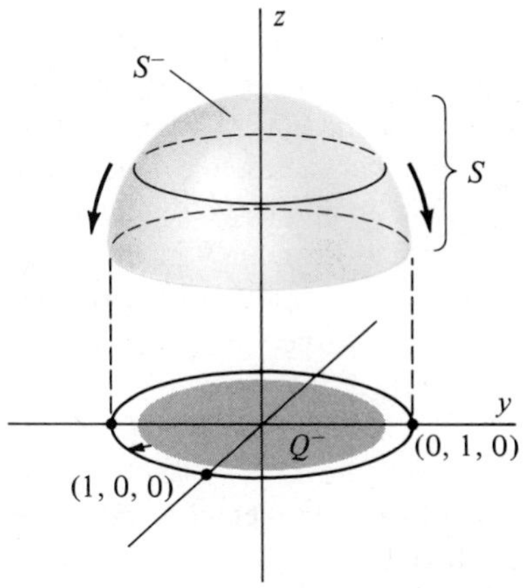

Figure 5.9 Surface integral over hemisphere $z = 1 + \sqrt{1 - x^2 - y^2}$ is evaluated as an improper integral.

Example 6 Evaluate the surface integral $\iint_S z\, dS$ where S is the hemisphere $z = 1 + \sqrt{1 - x^2 - y^2}$ (Figure 5.9).

Solution: Here $g(x, y) = 1 + \sqrt{1 - x^2 - y^2}$ and $f(x, y, z) = z = g(x, y)$. Equation (3) becomes

$$\iint_S z\, dS = \iint_Q [1 + \sqrt{1 - x^2 - y^2}] \cdot \sqrt{\left(\frac{-x}{\sqrt{1 - x^2 - y^2}}\right)^2 + \left(\frac{-y}{\sqrt{1 - x^2 - y^2}}\right)^2 + 1}\, dx\, dy$$

$$= \iint_Q \left[\frac{1}{\sqrt{1 - x^2 - y^2}} + 1\right] dx\, dy$$

where Q is the unit circle $x^2 + y^2 = 1$. However, this double integral is improper, since the denominator of the first term in the integrand vanishes as (x, y) approaches the boundary of Q. We evaluate this improper integral as suggested in Figure 5.9—by evaluating it over a disc Q^- of radius $\hat{r} < 1$ and then calculating the limit as $\hat{r}$ approaches 1. To do so we switch to polar coordinates and find that

$$\iint_S z\, dS = \iint_Q \left[\frac{1}{\sqrt{1 - r^2}} + 1\right] r\, dr\, d\theta$$

$$= \lim_{\hat{r} \to 1^-} \int_0^{\hat{r}} \int_0^{2\pi} \left[\frac{r}{\sqrt{1 - r^2}} + r\right] d\theta\, dr$$

$$= \lim_{\hat{r} \to 1^-} \int_0^{\hat{r}} 2\pi \left[\frac{r}{\sqrt{1 - r^2}} + r\right] dr$$

$$= \lim_{\hat{r} \to 1^-} 2\pi \left\{\left[-\sqrt{1 - r^2} + \frac{1}{2} r^2\right]_{r=0}^{r=\hat{r}}\right\}$$

$$= \lim_{\hat{r} \to 1^-} 2\pi \left\{-\sqrt{1 - \hat{r}^2} + \frac{1}{2}\hat{r}^2 - (-\sqrt{1})\right\}$$

$$= 3\pi.$$

■

REMARK: Before turning to the next topic, we make one final observation concerning Definition 5—it is very hard to work with, in general. The trouble is that the double integral in equation (3) will often be impossible to evaluate by antidifferentiation. We have chosen the examples here very carefully, so that this difficulty would not arise. This was done so that you could focus on becoming familiar with the new concept of surface integral with minimum distraction. (In other words, if you've been thinking that these integrals have been rather messy to evaluate, the news is discouraging. These are the *nice* ones!) In most situations, you will simply have to resort to a numerical integration procedure (such as Program 9 in Appendix I) to evaluate a given surface integral. In other cases, the symmetry of the surface and the function $f(x, y, z)$, a physical law, or some other aspect of the particular problem will allow you to deduce the value of the surface integral.

Flux Integrals

Returning now to our primary motivation concerning fluid flow, let us consider the motion of a fluid, such as water or a gas, through a region R in space. At each point (x, y, z) in the region, let $\mathbf{v}(x, y, z)$ be the velocity vector for the motion of the fluid. That is, $\mathbf{v}(x, y, z)$ is a vector pointing in the direction of motion whose length is the speed of the fluid at (x, y, z). We shall only consider the case of **steady state**

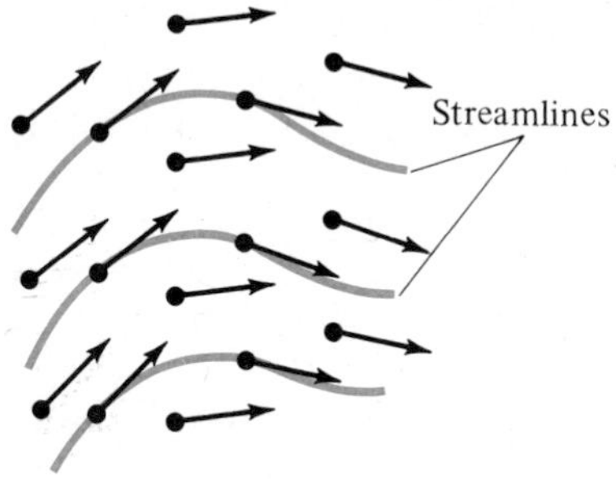

Figure 5.10 A velocity field and streamlines.

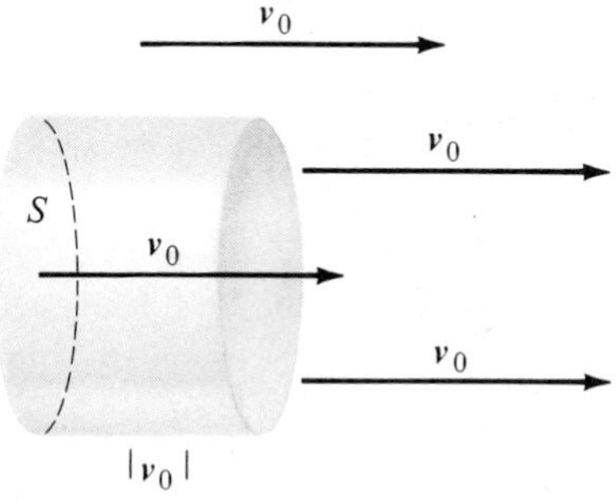

Figure 5.11 Volume of fluid flowing across S in unit time is $\Delta V = |v_0|\,\Delta S$.

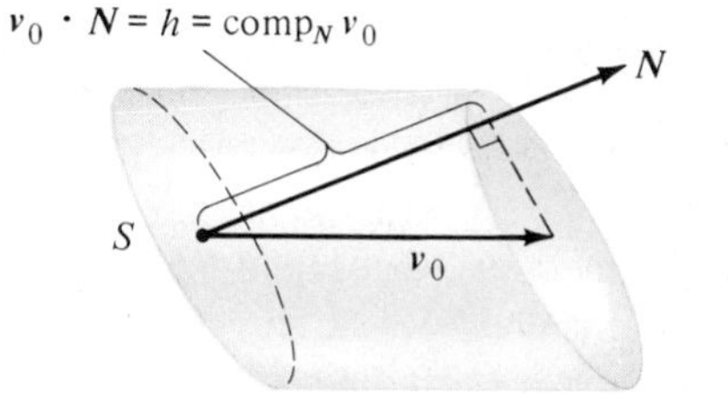

Figure 5.12 Volume of fluid crossing S in a unit of time is $\Delta V = (v_0 \cdot N)\,\Delta S$.

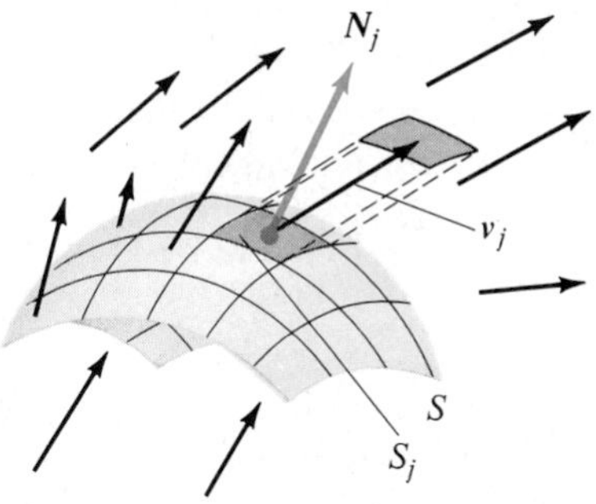

Figure 5.13 Volume of fluid crossing patch S_j is approximated by $\Delta V \approx (v_j \cdot N_j)\,\Delta S_j$.

fluid motion. This means that the velocity vector $\boldsymbol{v}(x, y, z)$ does not change as time changes. Thus, $\boldsymbol{v}(x, y, z)$ is a *vector field* defined on the region R. (The curves having the property of being tangent to the velocity field at each point are called the **streamlines** of the velocity field. See Figure 5.10.)

Now suppose that S is a permeable surface suspended in the fluid. (This means that the fluid can flow through the surface S, as coffee flows through a paper filter.) For a steady state velocity field $\boldsymbol{v}$, physicists define the **flux** of $\boldsymbol{v}$ across S as the rate at which mass is flowing across the surface S. Our interest here is in developing a formula by which flux can be calculated for a given surface and a given vector field.

We can simplify this problem somewhat by assuming that the density δ (mass per unit volume) is constant throughout the fluid. Since

$$\text{Mass} = (\text{density per unit volume}) \times (\text{volume}) \tag{8}$$

we need only find the rate at which volume (i.e., fluid) is flowing across S. Multiplying by δ then gives the desired flux, according to equation (8).

The simplest case of this problem occurs when $\boldsymbol{v}$ is a constant vector field, $\boldsymbol{v}(x, y, z) \equiv \boldsymbol{v}_0$, and S is a flat surface perpendicular to $\boldsymbol{v}_0$ (Figure 5.11). Since the speed at which fluid crosses S is $|\boldsymbol{v}_0|$, in one unit of time the quantity

$$\Delta V = |\boldsymbol{v}_0|\,\Delta S \tag{9}$$

crosses S, where ΔS is the area of S. If $\boldsymbol{v}(x, y, z) = \boldsymbol{v}_0$ is constant and S is flat, but not orthogonal to $\boldsymbol{v}_0$, the situation is only slightly more complicated. Let $\boldsymbol{N}$ be a unit vector normal to S. (There are two possible choices for $\boldsymbol{N}$. Take the one for which $0 \le \theta \le 90°$, where θ is the angle between $\boldsymbol{N}$ and $\boldsymbol{v}_0$.) The volume of fluid that crosses S in a unit of time can be thought of as a prism with base of area ΔS and height $h = \text{comp}_{\boldsymbol{N}}\boldsymbol{v}_0 = \boldsymbol{v}_0 \cdot \boldsymbol{N}$, as illustrated in Figure 5.12. The desired volume is thus

$$\Delta V = (\boldsymbol{v}_0 \cdot \boldsymbol{N})\,\Delta S \tag{10}$$

The more general case of a curved surface S and a nonconstant velocity field $\boldsymbol{v}(x, y, z)$ is handled in a familiar way. First, we assume that S has a normal vector $\boldsymbol{N}(x, y, z)$ at each point and that $\boldsymbol{N}(x, y, z)$ is continuous on S. Also, we assume that S can be recognized as having two sides and that $\boldsymbol{N}(x, y, z)$ always remains on the same side of S. (We will say more about this later.) All these assumptions are met if S is the graph of a differentiable function $z = g(x, y)$ over a regular region Q in the plane.

By partitioning Q, as in Figure 5.4, we divide S into nonoverlapping patches $S_1, S_2, \ldots, S_n$. For each patch we choose one point $(x_j, y_j, z_j) \in S_j$, and we let ΔS_j denote the surface area of S_j. Also, we approximate $\boldsymbol{v}(x, y, z)$ for every $(x, y, z) \in S_j$ by the vector $\boldsymbol{v}_j = \boldsymbol{v}(x_j, y_j, z_j)$ and we let $\boldsymbol{N}_j = \boldsymbol{N}(x_j, y_j, z_j)$. Using equation (10), we approximate the volume ΔV_j of fluid crossing the patch S_j in a unit of time as

$$\Delta V_j \approx (\boldsymbol{v}_j \cdot \boldsymbol{N}_j)\,\Delta S_j \qquad \text{(Figure 5.13).}$$

Summing these approximations over all patches gives the approximation

$$\Delta V \approx \sum_{j=1}^{n} (\boldsymbol{v}_j \cdot \boldsymbol{N}_j)\,\Delta S_j. \tag{11}$$

Finally, we argue that as the number of patches becomes large (and their individual sizes become small) the approximation in (11) should converge to the desired rate. Since the right-hand side of approximation (11) has the form of the approximating sum in equation (1), we conclude from equations (8) and (11) that the rate at which the mass of the fluid is flowing across S (that is, the flux) is given by the surface integral

$$\frac{dM}{dt} = \iint_S \delta \boldsymbol{v}(x, y, z) \cdot \boldsymbol{N}(x, y, z)\, dS. \tag{12}$$

The integral in (12) is sometimes called the **rate of mass transport** across S in the direction of $\boldsymbol{N}$.

Example 7 A fluid with mass density δ is emanating from the origin and flowing according to the central velocity field

$$\boldsymbol{v}(x, y, z) = \lambda(x\boldsymbol{i} + y\boldsymbol{j} + z\boldsymbol{k})$$

where λ is constant. Find the rate of mass transport, $\frac{dM}{dt}$, across the square region $S = \{(x, y, z) | z = 2, -1 \le x \le 1, -1 \le y \le 1\}$.

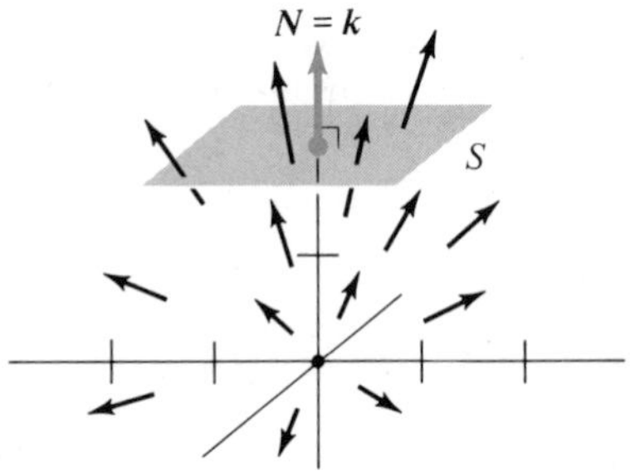

Figure 5.14 Flow across S in direction of $N = k$.

Solution: The square S is horizontal, so we use the unit normal $\boldsymbol{N} = \boldsymbol{k}$ at each point. Then $\boldsymbol{v}(x, y, z) \cdot \boldsymbol{N} = \lambda z$ so, by (12),

$$\frac{dM}{dt} = \iint_S \delta\lambda z\, dS = \int_{-1}^{1}\int_{-1}^{1} 2\delta\lambda\, dx\, dy = 8\delta\lambda.$$

Note that the choice of $\boldsymbol{N} = \boldsymbol{k}$ was arbitrary. If we had instead used $\boldsymbol{N} = -\boldsymbol{k}$, the resulting integral would have yielded $\frac{dM}{dt} = -8\delta\lambda$. This simply says that the flow is in the direction *opposite* the vector $\boldsymbol{N} = -\boldsymbol{k}$ (see Figures 5.14 and 5.15). ■

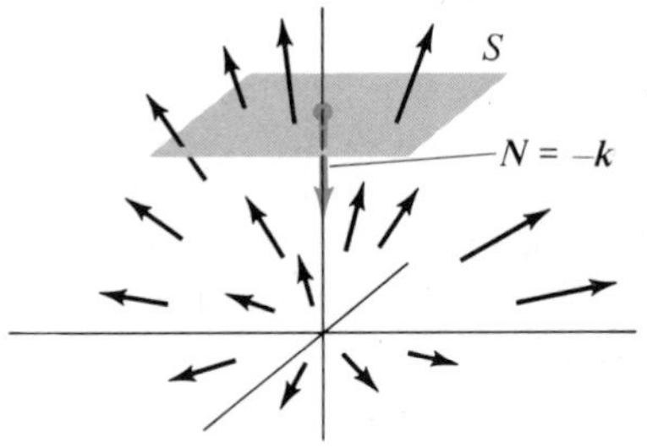

Figure 5.15 Flow in direction opposite to $N = -k$.

A more general concept than that of equation (11) is obtained by letting $\boldsymbol{F}(x, y, z)$ denote any vector field and $\boldsymbol{N} = \boldsymbol{N}(x, y, z)$ be a unit normal defined at each point on the surface S as described above. The flux of the vector field $\boldsymbol{F}$ across S in the direction of $\boldsymbol{N}$ is defined to be the surface integral

$$\text{Flux of } \boldsymbol{F} \text{ across } S = \iint_S \boldsymbol{F} \cdot \boldsymbol{N}\, dS \tag{13}$$

Thus (12) is a special case of (13) with $\boldsymbol{F} = \delta\boldsymbol{v}$. However, the vector field in (13) need not be a velocity field for a fluid. It may, for example, be an electric field induced by one or more point charges, or a gradient field for a temperature function.

In working with either equation (12) or equation (13) you will need to recall that a normal $\boldsymbol{N}(x, y, z)$ to the surface S that is the graph of $z = g(x, y)$ can be found by writing S as a level surface

$$f(x, y, z) = z - g(x, y) = 0$$

and forming

$$\boldsymbol{N}_1 = \frac{\boldsymbol{\nabla} f}{|\boldsymbol{\nabla} f|} = \frac{-\frac{\partial g}{\partial x}\boldsymbol{i} - \frac{\partial g}{\partial y}\boldsymbol{j} + \boldsymbol{k}}{\sqrt{\left(\frac{\partial g}{\partial x}\right)^2 + \left(\frac{\partial g}{\partial y}\right)^2 + 1}} \tag{14}$$

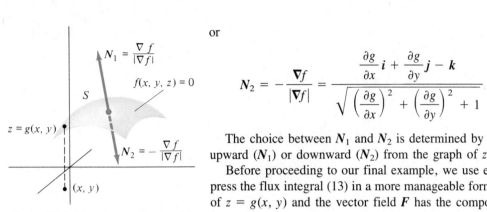

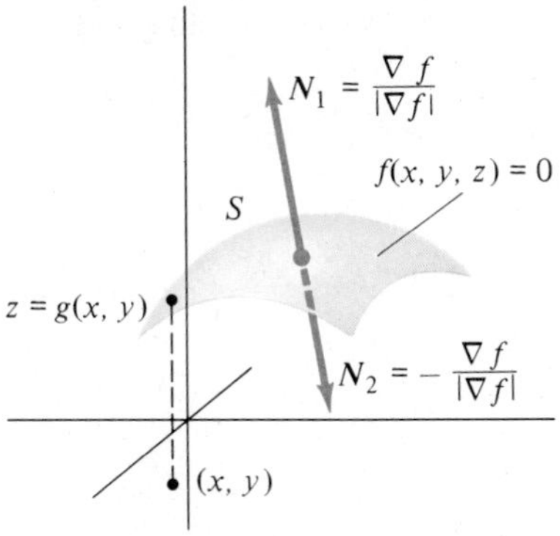

Figure 5.16 Choosing a unit normal to graph of $f(x, y, z) = z - g(x, y) = 0$.

or

$$N_2 = -\frac{\nabla f}{|\nabla f|} = \frac{\frac{\partial g}{\partial x}\boldsymbol{i} + \frac{\partial g}{\partial y}\boldsymbol{j} - \boldsymbol{k}}{\sqrt{\left(\frac{\partial g}{\partial x}\right)^2 + \left(\frac{\partial g}{\partial y}\right)^2 + 1}} \tag{15}$$

The choice between N_1 and N_2 is determined by whether you wish N to point upward (N_1) or downward (N_2) from the graph of $z = g(x, y)$ (Figure 5.16).

Before proceeding to our final example, we use equations (14) and (15) to express the flux integral (13) in a more manageable form. If the surface S is the graph of $z = g(x, y)$ and the vector field F has the component form

$$\boldsymbol{F}(x, y, z) = M(x, y, z)\boldsymbol{i} + N(x, y, z)\boldsymbol{j} + P(x, y, z)\boldsymbol{k},$$

then the flux integral (13) corresponding to the upward unit normal N_1 in (14) is, according to Definition 5,

$$\iint_S \boldsymbol{F}\cdot\boldsymbol{N}\,dS = \iint_Q (M\boldsymbol{i} + N\boldsymbol{j} + P\boldsymbol{k}) \cdot \left(\frac{-\frac{\partial g}{\partial x}\boldsymbol{i} - \frac{\partial g}{\partial y}\boldsymbol{j} + \boldsymbol{k}}{\sqrt{\left(\frac{\partial g}{\partial x}\right)^2 + \left(\frac{\partial g}{\partial y}\right)^2 + 1}}\right)\left(\sqrt{\left(\frac{\partial g}{\partial x}\right)^2 + \left(\frac{\partial g}{\partial y}\right)^2 + 1}\right) dx\,dy$$

where Q is the region in the xy-plane over which S is defined. This simplifies to

$$\iint_S \boldsymbol{F}\cdot\boldsymbol{N}\,dS = \iint_Q \left[-M(x, y, g(x, y))\frac{\partial g}{\partial x} - N(x, y, g(x, y))\frac{\partial g}{\partial y} + P(x, y, g(x, y))\right] dx\,dy \tag{16}$$

Similarly, the flux integral (13) corresponding to the downward unit normal N_2 in (15) is

$$\iint_S \boldsymbol{F}\cdot\boldsymbol{N}\,dS = \iint_Q \left[M(x, y, g(x, y))\frac{\partial g}{\partial x} + N(x, y, g(x, y))\frac{\partial g}{\partial y} - P(x, y, g(x, y))\right] dx\,dy \tag{17}$$

Example 8 Calculate the flux over the sphere S with center $(0, 0, 0)$ and radius R associated with the (central) inverse square field

$$\boldsymbol{F}(\boldsymbol{r}) = \frac{\lambda \boldsymbol{r}}{|\boldsymbol{r}|^3}, \qquad \boldsymbol{r} \neq 0, \qquad \lambda \text{ constant.}$$

That is,

$$\boldsymbol{F}(x, y, z) = \lambda \cdot \frac{x\boldsymbol{i} + y\boldsymbol{j} + z\boldsymbol{k}}{(x^2 + y^2 + z^2)^{3/2}}, \qquad x^2 + y^2 + z^2 \neq 0.$$

Solution: Since the force field F has the same symmetry with respect to the origin as does the sphere S, we will calculate the flux over the upper hemisphere only and

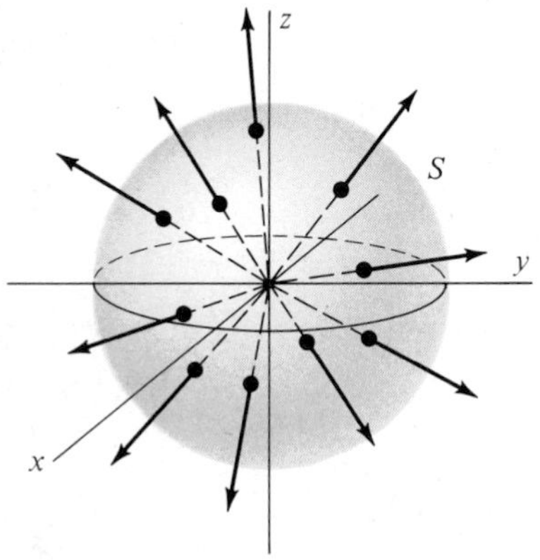

Figure 5.17 Flux of a central force field over a sphere is twice the flux over the upper hemisphere $z = \sqrt{R^2 - x^2}$.

then double the result (see Figure 5.17), realizing that we are dealing with an improper integral as in Example 6. Using equation (16) with

$$M(x, y, z) = \frac{\lambda x}{(x^2 + y^2 + z^2)^{3/2}} = \frac{\lambda x}{R^3},$$

$$N(x, y, z) = \frac{\lambda y}{(x^2 + y^2 + z^2)^{3/2}} = \frac{\lambda y}{R^3},$$

$$P(x, y, z) = \frac{\lambda z}{(x^2 + y^2 + z^2)^{3/2}} = \frac{\lambda z}{R^3},$$

and

$$g(x, y) = \sqrt{R^2 - x^2 - y^2} = z$$

gives

$$\begin{aligned}
\iint_S \boldsymbol{F} \cdot \boldsymbol{N}\, dS &= 2 \iint_Q \left[-\frac{\lambda x}{R^3} \left(\frac{-x}{\sqrt{R^2 - x^2 - y^2}} \right) \right. \\
&\quad \left. - \frac{\lambda y}{R^3} \left(\frac{-y}{\sqrt{R^2 - x^2 - y^2}} \right) + \frac{\lambda \sqrt{R^2 - x^2 - y^2}}{R^3} \right] dx\, dy \\
&= 2 \iint_Q \frac{\lambda}{R\sqrt{R^2 - x^2 - y^2}}\, dx\, dy \\
&= 2 \int_0^{2\pi} \int_0^R \frac{\lambda}{R\sqrt{R^2 - r^2}}\, r\, dr\, d\theta \qquad \text{(Switch to polar coordinates.)} \\
&= 2 \int_0^{2\pi} - \left[\frac{\lambda}{R} \sqrt{R^2 - r^2} \right]_{r=0}^{r=R} d\theta \\
&= 4\pi\lambda.
\end{aligned}$$

■

The surprising result of Example 8 is that *the flux of the inverse square field F over the sphere S is independent of the radius of S!* This result is true for any inverse square field, such as a gravitational field or an electric field. In the latter case the result of Example 8, together with another property of electric fields (Coulomb's Law), gives **Gauss's Law:**

> The flux of the electric field $\boldsymbol{E}$ through any closed surface S is
>
> $$\iint_S \boldsymbol{E} \cdot \boldsymbol{N}\, dS = 4\pi \sum_{j=1}^{n} q_j$$
>
> where $q_1, q_2, \ldots, q_n$ are the point changes contained within S.

The Orientation of a Surface

When the surface S is the graph of a single function $z = g(x, y)$, there is no ambiguity about what is meant by the *upward* unit normal (as opposed to the downward unit normal), so a flux integral is easily determined to be of the form (16) or of the form (17). Also, in the cases where S is the graph of the function $y = g(x, z)$ or $x = g(y, z)$, the corresponding meaning of upward normal and the modifications of equations (16) and (17) should be clear. However, the situation becomes less clear as we think about surfaces of the form $S = S_1 \cup S_2 \cup \cdots \cup S_n$ where each S_j is as above. We need to devote just a few more lines to the meaning of $\mathbf{N}$ in the flux integral (13) for these more general cases.

According to our original motivation, flux is a measure of the *net* (positive minus negative) rate of flow across S. If more fluid is flowing upward across S: $z = g(x, y)$ than is flowing downward, flux is positive, although fluid may actually be flowing upward across some portions of S and downward across others. It is therefore crucial that $\mathbf{N}$ always point on the same "side" of S, since $\mathbf{N}$ provides the gauge with which the rate of flow is measured at each point. Thus, in order for the flux integral in (13) to make sense, the surface S must be **orientable,** that is, we must be able to identify two distinct sides of S and to assign a unique unit normal $\mathbf{N}(x, y, z)$ that remains on one side or the other as (x, y, z) varies across S.

We have already seen that this is simple to do for surfaces of the form

$$S = \{(x, y, z) | z = g(x, y),\ (x, y) \in Q\}.$$

We just choose the unit normal with positive z-component (upward) or negative z-component (downward). For closed surfaces (such as a sphere or an ellipsoid) the two sides are the inside and the outside, and we must choose between the *inner* unit normal or the *outer* unit normal. (Figure 5.17 shows outer unit normals for a sphere.)

These two general situations (upper versus lower, or inner versus outer) encompass most of the surfaces with which you will need to deal, and we refer to any such surface as *orientable*. We shall not attempt to give a precise definition for this term (as is done in more advanced courses). However, you should know that not all surfaces are orientable. A classic example is the Möbius band, which is obtained by twisting one end of a long rectangular strip 180° and "gluing" it to the other end. As Figure 5.18 shows, a unit normal $\mathbf{N}$ beginning at point P and moving continuously once "around" the Möbius band arrives back at P pointing in the opposite direction.

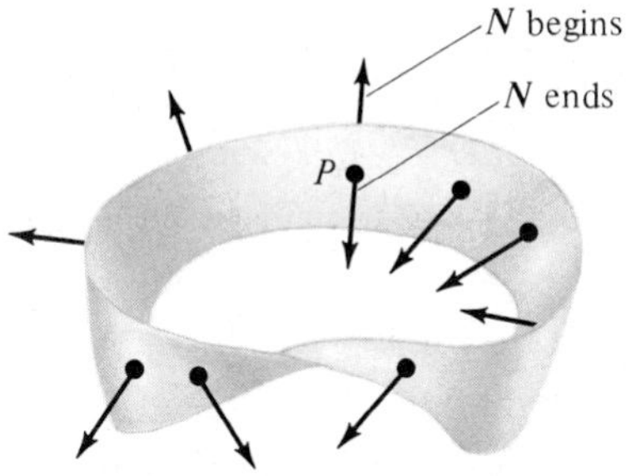

Figure 5.18 Möbius band is not orientable.

In summary, we caution you to note that the surface integral in Definition 5 does not involve the notion of orientation for S. However, the flux integral in (13) does depend on an orientation for S. Reversing the orientation for S (and, hence, replacing $\mathbf{N}$ by $-\mathbf{N}$) changes the sign of the integral in (13).

Exercise Set 21.5

1. Evaluate $\iint_S (x + 3y + z)\, dS$ over the portion of the plane $x - y + 2z = 4$ lying above the rectangle $Q = \{(x, y) | 0 \le x \le 1, 0 \le y \le 2\}$.

2. Evaluate the surface integral in Exercise 1 where S is the portion of the plane $x - y + 2z = 4$ lying above the unit circle.

3. Evaluate $\iint_S x^2 z\, dS$ where S is the portion of the plane $2x + 3y + z = 6$ lying in the first octant ($x \ge 0$, $y \ge 0$, $z \ge 0$).

4. Evaluate $\iint_S y^2 z\, dS$ where S is the portion of the cone $z^2 = x^2 + y^2$ lying between the planes $z = 1$ and $z = 4$.

5. Evaluate the surface integral $\iint_S (x^2 + 5y - z)\, dS$ where S is the graph of the cylinder $z = x^2$ over the square $-1 \le x \le 1$, $-1 \le y \le 1$.

6. Evaluate $\iint_S (x^2 - y^2 + 1)\, dS$ where S is the part of the plane $y = x + 2$ inside the cylinder $x^2 + y^2 = 4$.

7. Find the mass of the part of a sphere $x^2 + y^2 + z^2 = 9$ lying inside the cylinder $x^2 + y^2 = 4$ if the density per unit area δ is constant.

8. Find $\iint_S \boldsymbol{F} \cdot \boldsymbol{N}\, dS$ where $\boldsymbol{F} = 2x\boldsymbol{i} - y\boldsymbol{j} + 3z\boldsymbol{k}$, S is the part of the plane $x - y + 2z = 4$ bounded by the planes $y = 0$, $y = -4$, $x = 0$, and $x = 4$, and $\boldsymbol{N}$ is the upward unit normal.

9. Find $\iint_S (x + y + z^2)\, dS$ where S is the hemisphere $z = \sqrt{4 - x^2 - y^2}$.

10. Evaluate $\iint_S (y - x)\, dS$ where S is the part of the plane $6x + 4y + 2z = 8$ lying inside the cylinder $x^2 + y^2 = 4$.

11. Find the flux $\iint_S \boldsymbol{F} \cdot \boldsymbol{N}\, dS$ where $\boldsymbol{F} = x\boldsymbol{i} + y\boldsymbol{j} + z\boldsymbol{k}$, S is the upper half sphere $z = \sqrt{1 - x^2 - y^2}$, and $\boldsymbol{N}$ is the upward unit normal.

12. Use the result of Exercise 11 to find $\iint_S \boldsymbol{F} \cdot \boldsymbol{N}\, dS$ where $\boldsymbol{F} = x\boldsymbol{i} + y\boldsymbol{j} + z\boldsymbol{k}$, S is the entire unit sphere, and $\boldsymbol{N}$ is the outward unit normal.

13. Calculate the flux integral $\iint_S \boldsymbol{F} \cdot \boldsymbol{N}\, dS$ where $\boldsymbol{F} = 2x\boldsymbol{i} + y\boldsymbol{j} + z\boldsymbol{k}$, S is the surface of the paraboloid $z = x^2 + y^2$ bounded by the planes $z = 0$ and $z = 4$, and $\boldsymbol{N}$ is the outward unit normal.

14. Let $\boldsymbol{F} = 2\boldsymbol{i} - x\boldsymbol{j} + y\boldsymbol{k}$. Find the flux integral $\iint_S \boldsymbol{F} \cdot \boldsymbol{N}\, dS$ where S is the portion of the cylinder $x^2 + y^2 = 1$ lying between the planes $z = 0$ and $z = 1$, and $\boldsymbol{N}$ is the outward unit normal.

15. Find the flux outward through the sphere $x^2 + y^2 + z^2 = a^2$ for the vector field $\boldsymbol{F}(x, y, z) = z\boldsymbol{k}$. What is the flux if $\boldsymbol{F}(x, y, z) = y\boldsymbol{j}$? Can you give a geometric argument for these results?

16. Find $\iint_S \boldsymbol{F} \cdot \boldsymbol{N}\, dS$ where S is the ellipsoid $4x^2 + y^2 + z^2 = 4$, $\boldsymbol{F} = x^2\boldsymbol{i} + y^2\boldsymbol{j} + z^2\boldsymbol{k}$, and $\boldsymbol{N}$ is the outward unit normal. (This will require a geometric argument.)

17. Find the flux $\iint_S \boldsymbol{F} \cdot \boldsymbol{N}\, dS$ where $\boldsymbol{F} = x\boldsymbol{i} + y\boldsymbol{j} + z\boldsymbol{k}$, $\boldsymbol{N}$ is the outward unit normal, and S is the surface of the unit cube $\{(x, y, z) | 0 \le x \le 1,\ 0 \le y \le 1,\ 0 \le z \le 1\}$.

18. Suppose that $F(x, y, z) = ax\boldsymbol{i} + by\boldsymbol{j} + cz\boldsymbol{k}$ and that S is part of the plane $ax + by + cz = d$. Show that the flux integral $\iint_S \boldsymbol{F} \cdot \boldsymbol{N}\, dS$ is $(a^2 + b^2 + c^2)A$ where A is the area of S, if $\boldsymbol{N}$ is the unit normal with $\boldsymbol{N} \cdot c\boldsymbol{k} \ge 0$.

19. Is the result of Example 8 the same if $\boldsymbol{F}$ is an inverse *cube* field $\boldsymbol{F}(\boldsymbol{r}) = \dfrac{\boldsymbol{r}}{|\boldsymbol{r}|^4}$? What is the result in this case?

20. Show that the *flux* of the vector field $\boldsymbol{F} = M\boldsymbol{i} + N\boldsymbol{j} + P\boldsymbol{k}$ across a surface S can be written

$$\iint_S \boldsymbol{F} \cdot \boldsymbol{N}\, dS = \iint_S (P \cos \alpha + Q \cos \beta + R \cos \gamma)\, dS$$

where α, β, and γ are the direction numbers for the unit normal $\boldsymbol{N}$.

For a surface S, the coordinates $(\overline{x}, \overline{y}, \overline{z})$ of the *centroid* are defined to be

$$\overline{x} = \frac{1}{A} \iint_S x\, dS, \quad \overline{y} = \frac{1}{A} \iint_S y\, dS, \quad \overline{z} = \frac{1}{A} \iint_S z\, dS$$

where A is the surface area of S. Use these formulas in Exercises 21–23.

21. Find the centroid of the hemisphere $z = \sqrt{a^2 - x^2 - y^2}$, $z \ge 0$. (*Hint:* Use symmetry as much as possible.)

22. Find the centroid of the cylinder $z = 1 - x^2$, $-2 \le y \le 2$, $-1 \le x \le 1$.

23. Find the centroid of the part of the spherical surface $x^2 + y^2 + z^2 = 1$ lying in the first octant.

24. Show that the surface integral, as defined by equation (1), can be written in the form

$$\iint_S f(x, y, z)\, dS = \iint_Q f(x, y, g(x, y)) \sec \gamma\, dx\, dy$$

where $\gamma = \gamma(x, y)$ is the angle between the upward normal to the graph of $z = g(x, y)$ at (x, y, z) and the vector $\boldsymbol{k}$. (*Hint:* Refer to Section 20.6, Exercise 20.)

25. Show that if the function $f(r, \theta, z)$ is written using cylindrical coordinates for $\mathbb{R}^3$ and if the surface S is the graph of the function $z = g(r, \theta)$ in polar coordinates, then the formula in Definition 5 becomes

$$\iint_S f(r, \theta, z)\, dS = \iint_Q f(r, \theta, g(r, \theta)) \cdot \sqrt{r^2 + r^2\left(\frac{\partial g}{\partial r}\right)^2 + \left(\frac{\partial g}{\partial \theta}\right)^2}\, dr\, d\theta.$$

21.6 STOKES' THEOREM

In this section we encounter a generalization of Green's Theorem, which we have earlier stated as

$$\int_C \boldsymbol{F} \cdot d\boldsymbol{r} = \iint_Q \left(\frac{\partial N}{\partial x} - \frac{\partial M}{dy}\right) dA = \iint_Q (\text{curl } \boldsymbol{F})\, dA. \tag{1}$$

Recall the setting of equation (1): $\boldsymbol{F}(x, y) = M(x, y)\boldsymbol{i} + N(x, y)\boldsymbol{j}$ is a vector field in the plane (with continuously differentiable components), C is a piecewise smooth simple closed curve enclosing the region Q, and curl $\boldsymbol{F}$ is the scalar function

$$\text{curl } \boldsymbol{F} = \frac{\partial N}{\partial x}(x, y) - \frac{\partial M}{\partial y}(x, y). \tag{2}$$

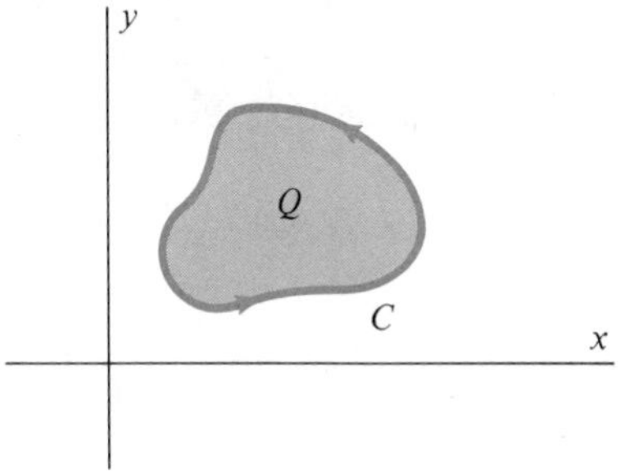

Figure 6.1 Curve C and region Q in Green's Theorem.

All this takes place in the plane, $\mathbb{R}^2$.

The generalization referred to above is Stokes' Theorem, which differs from Green's Theorem in two ways. First, the setting for Stokes' Theorem is $\mathbb{R}^3$. C will now be a closed space curve, and the vector field $\boldsymbol{F}$ will have the form

$$\boldsymbol{F}(x, y, z) = M(x, y, z)\boldsymbol{i} + N(x, y, z)\boldsymbol{j} + P(x, y, z)\boldsymbol{k}. \tag{3}$$

Second, whereas the integral on the right-hand side of equation (1) is a double integral over the plane region Q, the right-hand side of Stokes' Theorem will involve a surface integral evaluated over a smooth surface S bounded by the closed space curve C. Figures 6.1 and 6.2 illustrate the geometry associated with these two theorems.

Stokes' Theorem (which we shall state in more precise terms later) says this:

$$\boxed{\int_C \boldsymbol{F} \cdot d\boldsymbol{r} = \iint_S (\text{curl } \boldsymbol{F}) \cdot \boldsymbol{N}\, dS} \tag{4}$$

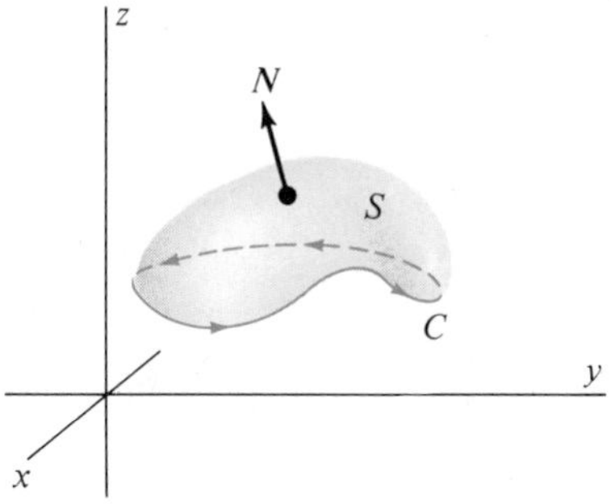

Figure 6.2 The surface S bounded by the space curve C in Stokes' Theorem.

where $\boldsymbol{N}$ is a unit normal for the surface S.

The strong similarity between equations (1) and (4) is obvious. However, the integrand in the surface integral in (4) has not yet been defined. This brings us to the second major difference between these two theorems.

Recall that the line integral in (1) was interpreted in Section 21.4 as the *circulation* around C of a fluid flowing with velocity field $\boldsymbol{F}(x, y)$. This interpretation issued from the equation

$$\int_C \boldsymbol{F} \cdot d\boldsymbol{r} = \int_C \boldsymbol{F} \cdot \boldsymbol{T}\, ds \tag{5}$$

discussed earlier. Now equation (5) holds for vector fields of the form (3) defined on space curves as well as in the planar case. Retaining the notion of fluid flow as our primary motivation, we would like to define curl $\boldsymbol{F}$ for $\boldsymbol{F}$ in (3) in such a way that equation (4) can be interpreted as a statement about fluid flow (and, of course, so that equation (4) is true!). The following definition of curl $\boldsymbol{F}$ does this. We shall defer further physical interpretation to the end of this section.

DEFINITION 6

Let $F(x, y, z)$ be a vector field of the form

$$F(x, y, z) = M(x, y, z)i + N(x, y, z)j + P(x, y, z)k$$

If all first order partial derivatives for each component function M, N, and P exist at (x, y, z), **curl** $F(x, y, z)$ is defined to be the vector

$$\text{curl } F(x, y, z) = \left(\frac{\partial P}{\partial y} - \frac{\partial N}{\partial z}\right)i + \left(\frac{\partial M}{\partial z} - \frac{\partial P}{\partial x}\right)j + \left(\frac{\partial N}{\partial x} - \frac{\partial M}{\partial y}\right)k \quad (6)$$

where each partial derivative is evaluated at (x, y, z).

It is important to note that curl $F(x, y, z)$ is a *vector* in $\mathbb{R}^3$ while curl $F(x, y)$ in (2) is a *number*. Now curl $F(x, y, z) \cdot N$ (the integrand in the surface integral in Stokes' Theorem) is a generalization of curl $F(x, y)$ (the integrand in the double integral in Green's Theorem) in the following sense. If we restrict $F(x, y, z)$ in (3) to the xy-plane by setting $z \equiv 0$ and $P(x, y, z) \equiv 0$, equation (6) gives

$$\text{curl } F(x, y, 0) = \left(\frac{\partial N}{\partial x} - \frac{\partial M}{\partial y}\right)k. \quad (7)$$

Since a unit normal to the xy-plane is $N = k$, we obtain

$$[\text{curl } F(x, y, 0)] \cdot N = \left[\left(\frac{\partial N}{\partial x} - \frac{\partial M}{\partial y}\right)k\right] \cdot k = \frac{\partial N}{\partial x} - \frac{\partial M}{\partial y}$$
$$= \text{curl } F(x, y), \text{ as claimed.}$$

You will often see the notation

$$\text{curl } F = \nabla \times F. \quad (8)$$

This results from thinking of the gradient ∇ as an "operator"

$$\nabla = \frac{\partial}{\partial x} i + \frac{\partial}{\partial y} j + \frac{\partial}{\partial z} k$$

and forming the cross product*

$$\nabla \times F = \det \begin{bmatrix} i & j & k \\ \frac{\partial}{\partial x} & \frac{\partial}{\partial y} & \frac{\partial}{\partial z} \\ M & N & P \end{bmatrix} \quad (9)$$

$$= \left(\frac{\partial P}{\partial y} - \frac{\partial N}{\partial z}\right)i + \left(\frac{\partial M}{\partial z} - \frac{\partial P}{\partial x}\right)j + \left(\frac{\partial N}{\partial x} - \frac{\partial M}{\partial y}\right)k.$$

Example 1 For the vector field $F(x, y, z) = yi - z^2 j + 2xk$, $M = y$, $N = -z^2$, and $P = 2x$. From equations (8) and (9) we find that

$$\text{curl } F = \det \begin{bmatrix} i & j & k \\ \frac{\partial}{\partial x} & \frac{\partial}{\partial y} & \frac{\partial}{\partial z} \\ y & -z^2 & 2x \end{bmatrix} = 2zi - 2j - k. \quad ■$$

*As in the mnemonic for remembering how to compute cross products, the determinant here is only formal notation, since only the third row of the matrix actually contains numbers.

One final detail needs to be cleaned up before we state Stokes' Theorem. We shall need to require that the surface S be **simply connected.** Basically, this means that S contains no holes. Another way to say this is that, given any point P_0 on S, the boundary curve C can be "continuously contracted" across S to an arbitrarily small closed curve containing P_0. (If S had a hole you could not do this since the contraction of C would get "hung up" at the hole (see Figure 6.3).

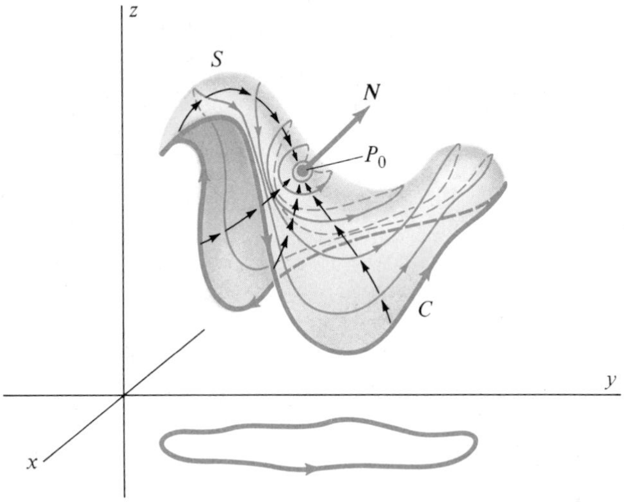

Figure 6.3 An oriented simply connected surface S. Contracting C continuously across S reveals the orientation of C induced by the orientation of S.

This notion of contracting C across S toward a point also allows us to clarify the issue of orientation. We shall require that the surface S be oriented. As defined in Section 21.5, this means that S comes equipped with a unit normal N that varies continuously across S, always remaining on the same side of S. In the statement of Stokes' Theorem we must require that the orientation of the boundary curve C be the orientation *induced* on C by the orientation of S. Here's what we mean. Let N be the unit normal to S at the point P_0. If S is smooth, then very near P_0 the surface of S may be approximated by a plane tangent to S at P_0. Think of a very small circle C_S on this tangent plane with center P_0. This small circle approximates a small closed curve on S, and we can continuously contract C across S so that it takes the shape of this small closed curve. The point is that the orientation of all three curves (C, the small curve on S, and the small circle C_S on the tangent plane) must be the same. N determines an orientation for C_S by a variation of the right-hand rule: Think of wrapping the index finger of your right hand around C_S, keeping your thumb parallel to N. Your index finger points out the orientation of C_S (see Figure 6.4). This is the orientation induced by N on C.

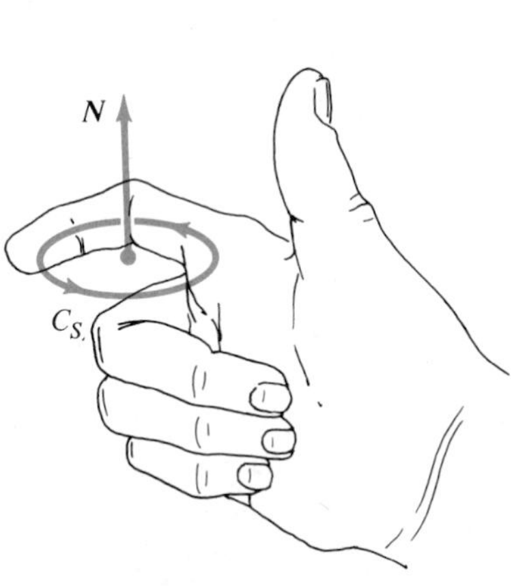

Figure 6.4 Right-hand rule determines orientation of C_s from direction of N.

We are now ready to state Stokes' Theorem.

THEOREM 4
Stokes' Theorem

Let S be a smooth, simply connected, oriented surface bounded by a piecewise smooth simple closed curve C with orientation induced by the orientation of S. Let

$$\boldsymbol{F}(x, y, z) = M(x, y, z)\boldsymbol{i} + N(x, y, z)\boldsymbol{j} + P(x, y, z)\boldsymbol{k}$$

be a vector field defined on an open set D containing S and C so that each of the component functions M, N, and P is continuous and has continuous partial derivatives on D. Then

$$\int_C \boldsymbol{F} \cdot d\boldsymbol{r} = \iint_S (\text{curl } \boldsymbol{F}) \cdot \boldsymbol{N}\, dS \tag{10}$$

where $\boldsymbol{N}$ is the unit normal to the oriented surface S.

Using equation (3), we may write equation (10) in the alternate form

$$\int_C M\, dx + N\, dy + P\, dz = \iint_S (\text{curl } \boldsymbol{F}) \cdot \boldsymbol{N}\, dS.$$

We may paraphrase Stokes' Theorem by saying that ''the line integral of the vector field around the boundary of S equals the integral of the normal component of curl $\boldsymbol{F}$ over the surface S.''

It is interesting to note that the particular shape of S is unimportant—the surface integral over S is entirely determined by the line integral along its boundary. For example, the surface integrals of (curl $\boldsymbol{F}$) $\cdot$ $\boldsymbol{N}$ over all of the surfaces in Figure 6.5 are the same, since each has the same boundary C.

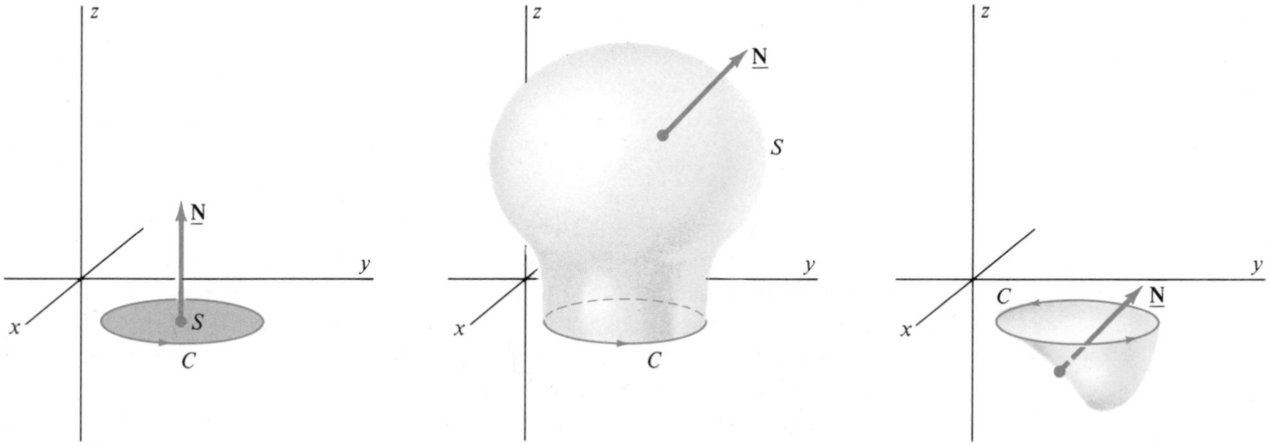

Figure 6.5 The integrals over all of these surfaces of (curl $\boldsymbol{F}$) $\cdot$ $\boldsymbol{N}$ are the same: $\iint_S (\text{curl } \boldsymbol{F}) \cdot \boldsymbol{N}\, dS = \int_C \boldsymbol{F} \cdot d\boldsymbol{r}$.

A proof of Stokes' Theorem is beyond the scope of this text. (If you've noticed that the concepts of this chapter sometimes appear to be less precise than in earlier chapters, you are correct. We are now dabbling in ideas that require considerably more advanced mathematics to treat rigorously than we can hope to achieve in a first calculus course.) However, we can discuss why Stokes' Theorem should hold.

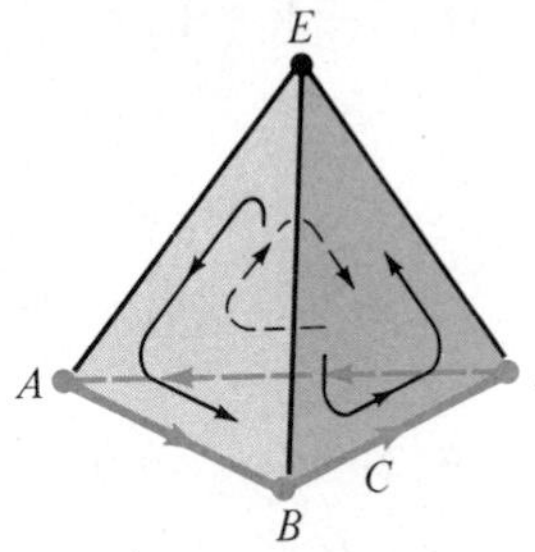

Figure 6.6 Stokes' Theorem holds for a tetrahedral surface.

First consider the case of the tetrahedron sketched in Figure 6.6. We shall take the surface S to be the union of the three faces ABE, BDE, and DAE. Then S is bounded by the triangle ABD, which is the curve C. We orient S by taking the outward unit normal $\boldsymbol{N}$ at each point, which induces the orientation ABD on C.

Since each face of the surface is a plane region, we apply Green's Theorem to each face and obtain the three equations

$$\int_{ABE} \boldsymbol{F}\cdot d\boldsymbol{r} = \iint_{ABE} (\text{curl } \boldsymbol{F})\cdot \boldsymbol{N}\, dS, \tag{11}$$

$$\int_{BDE} \boldsymbol{F}\cdot d\boldsymbol{r} = \iint_{BDE} (\text{curl } \boldsymbol{F})\cdot \boldsymbol{N}\, dS, \tag{12}$$

$$\int_{DAE} \boldsymbol{F}\cdot d\boldsymbol{r} = \iint_{DAE} (\text{curl } \boldsymbol{F})\cdot \boldsymbol{N}\, dS. \tag{13}$$

(In each of equations (11) through (13), we have used the remark concerning equation (7) to write the scalar curl in vector form.)

Next, we wish to add the corresponding sides in equations (11) through (13). When we do so, the contributions to the resulting line integral along the edges AE, BE, and DE will be zero, since each of these edges is traversed twice, once in each direction. Also, the three surface integrals should sum to the surface integral over S. The result, then, is

$$\int_{ABD} \boldsymbol{F}\cdot d\boldsymbol{r} = \iint_{S} (\text{curl } \boldsymbol{F})\cdot \boldsymbol{N}\, dS$$

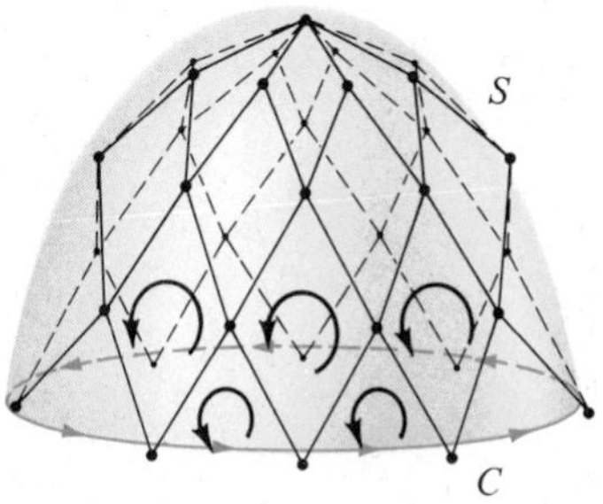

Figure 6.7 A polyhedral approximation to the surface S.

which is equation (10).

Now consider a more complex polyhedron S, such as that in Figure 6.7. The idea is the same. When we apply Green's Theorem to each face and sum the corresponding sides of the resulting equations, the only nonzero contributions to the line integral occur along the bottom edges. Along all other edges, the line integral will have been evaluated twice, once in each direction. Referring to the bottom edge as C, we again obtain equation (10).

The argument for a smooth surface S is to first approximate S by a sequence of polyhedra, each of whose vertices lies on the surface S. As the number of vertices in the polyhedra becomes large, the polyhedra become increasingly accurate approximations to the surface. Since equation (10) holds for each polyhedron, it holds for the smooth surface ''in the limit.''

Example 2 Find $\int_C \boldsymbol{F}\cdot d\boldsymbol{r}$ where $\boldsymbol{F}$ is the vector field $\boldsymbol{F}(x, y, z) = y\boldsymbol{i} - z^2\boldsymbol{j} + 2x\boldsymbol{k}$ and C is the unit circle $C = \{(x, y, 0) \mid x^2 + y^2 = 1\}$ oriented in the counterclockwise direction.

Solution: We could evaluate the line integral directly (and, painfully) by parameterizing C by $x(t) = \cos t$, $y(t) = \sin t$, $0 \le t \le 2\pi$. Instead, we will apply

Stokes' Theorem using any smooth surface S bounded by C. The simplest choice, of course, is the unit disc

$$S = \{(x, y, 0) \mid 0 \leq x^2 + y^2 \leq 1\}.$$

From Example 1 we know that

$$\text{curl } \boldsymbol{F} = 2z\boldsymbol{i} - 2\boldsymbol{j} - \boldsymbol{k}$$

and a unit normal giving S the required orientation is $\boldsymbol{N} = \boldsymbol{k}$. Thus

$$(\text{curl } \boldsymbol{F}) \cdot \boldsymbol{N} = (2z\boldsymbol{i} - 2\boldsymbol{j} - \boldsymbol{k}) \cdot \boldsymbol{k} = -1.$$

Thus, by Stokes' Theorem,

$$\int_C \boldsymbol{F} \cdot d\boldsymbol{r} = \iint_S (-1)\, dS = -\pi,$$

since π is the area of the unit disc. ■

Example 3 Verify Stokes' Theorem if $\boldsymbol{F}$ is the vector field

$$\boldsymbol{F}(x, y, z) = y\boldsymbol{i} - x^2\boldsymbol{j} + 2z^2\boldsymbol{k}$$

and S is the portion of the paraboloid $z = 4 - x^2 - y^2$ lying above the xy-plane. Take $\boldsymbol{N}$ to be the upward unit normal to S.

Solution: The curve C bounding S is the circle

$$C = \{(x, y, 0) \mid x^2 + y^2 = 4\}$$

which has parameterization $x(t) = 2 \cos t$, $y(t) = 2 \sin t$, $z(t) = 0$, $0 \leq t \leq 2\pi$. The line integral on the left side of equation (10) is therefore

$$\begin{aligned} \int_C \boldsymbol{F} \cdot d\boldsymbol{r} &= \int_C y\, dx - x^2\, dy + 2z^2\, dz \\ &= \int_0^{2\pi} (-4 \sin^2 t - 8 \cos^3 t)\, dt \\ &= -4\pi. \end{aligned}$$

To calculate the surface integral in (10), we first use (9) to find

$$\text{curl } \boldsymbol{F} = \nabla \times \boldsymbol{F} = \det \begin{bmatrix} \boldsymbol{i} & \boldsymbol{j} & \boldsymbol{k} \\ \dfrac{\partial}{\partial x} & \dfrac{\partial}{\partial y} & \dfrac{\partial}{\partial z} \\ y & -x^2 & 2z^2 \end{bmatrix} = -(2x + 1)\boldsymbol{k}.$$

Then, using equation (16), Section 21.5, we evaluate the surface integral as

$$\begin{aligned} \iint_S (\text{curl } \boldsymbol{F}) \cdot \boldsymbol{N}\, dS &= \iint_S [-(2x + 1)\boldsymbol{k}] \cdot \boldsymbol{N}\, dS \\ &= \iint_Q -(2x + 1)\, dx\, dy \end{aligned}$$

$$= \int_0^{2\pi}\int_0^2 -(2r\cos\theta + 1)\, r\, dr\, d\theta \qquad \text{(Switch to polar coordinates.)}$$

$$= -4\pi$$

which agrees with the value of the line integral. ■

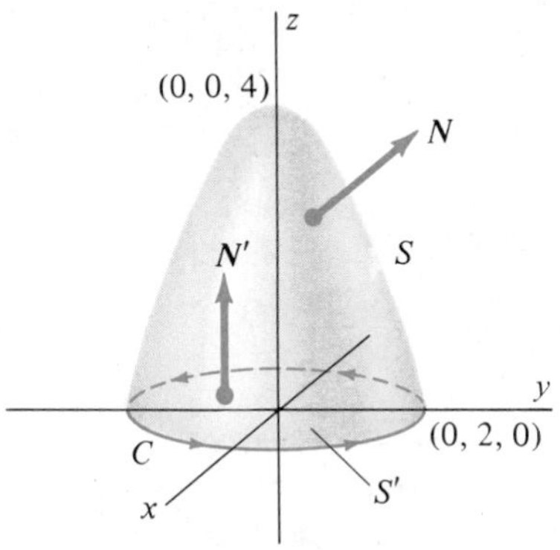

Figure 6.8 $\iint_S$ (curl $\boldsymbol{F}$) · $\boldsymbol{N}\, dS$ is the same as $\iint_{S'}$ (curl $\boldsymbol{F}$) · $\boldsymbol{N}'\, dS$, according to Stokes' Theorem.

REMARK: Notice that the calculation of the surface integral involved the same double integral as we would have obtained had the problem simply been to calculate $\iint_{S'}$ curl $\boldsymbol{F}\cdot\boldsymbol{N}'\, dS$ over the planar disc $S' = \{(x, y, 0) \mid 0 \le x^2 + y^2 \le 4\}$ (Figure 6.8). This is a direct illustration of the statement of Stokes Theorem—the value of the surface integral depends only on the line integral around the boundary.

Example 4 Find $\int_C \boldsymbol{F}\cdot d\boldsymbol{r}$ where the space curve C is the intersection of the plane $x - 2y + z = 5$ with the cylinder $x^2 + y^2 = 9$, oriented as in Figure 6.9, and $\boldsymbol{F}$ is the vector field $\boldsymbol{F}(x, y, z) = (x^2 - 3y^2)\boldsymbol{i} + (z^2 + y)\boldsymbol{j} + (x + 2z^2)\boldsymbol{k}$.

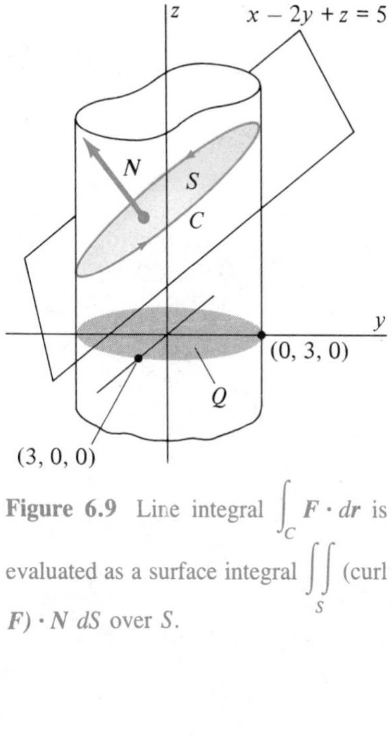

Figure 6.9 Line integral $\int_C \boldsymbol{F}\cdot d\boldsymbol{r}$ is evaluated as a surface integral $\iint_S$ (curl $\boldsymbol{F}$) · $\boldsymbol{N}\, dS$ over S.

Solution: Attempting to evaluate the line integral directly would be difficult at best, so we resort to Stokes' Theorem, taking S to be the portion of the plane enclosed by C. From the equation of the plane and the orientation of C, we see that a unit normal to S is

$$\boldsymbol{N} = \frac{1}{\sqrt{6}}(\boldsymbol{i} - 2\boldsymbol{j} + \boldsymbol{k})$$

which is upward. Also,

$$\text{curl } \boldsymbol{F} = \det\begin{bmatrix} \boldsymbol{i} & \boldsymbol{j} & \boldsymbol{k} \\ \dfrac{\partial}{\partial x} & \dfrac{\partial}{\partial y} & \dfrac{\partial}{\partial z} \\ (x^2 - 3y^2) & (z^2 + y) & (x + 2z^2) \end{bmatrix} = -2z\boldsymbol{i} - \boldsymbol{j} + 6y\boldsymbol{k}.$$

The equation for S becomes $z = g(x, y) = 5 - x + 2y$. By Stokes' Theorem and equation (16), Section 21.5, we have

$$\int_C \boldsymbol{F}\cdot d\boldsymbol{r} = \iint_S (\text{curl } \boldsymbol{F})\cdot\boldsymbol{N}\, dS$$

$$= \iint_Q \{-[-2(5 - x + 2y)](-1) - (-1)(2) + 6y\}\, dx\, dy$$

$$= \iint_Q (2x + 2y - 8)\, dx\, dy$$

$$= \int_0^{2\pi}\int_0^3 (2r\cos\theta + 2r\sin\theta - 8) r\, dr\, d\theta$$

$$= -72\pi.$$

(Note that we did not make use of the components of the unit normal $\boldsymbol{N}$—only the fact that it is oriented upward.) ■

Example 5 Calculate $\iint_S (\text{curl } \boldsymbol{F}) \cdot \boldsymbol{N}\, dS$ where S is the closed surface shown in Figure 6.10 and $\boldsymbol{F}$ is a vector field satisfying the hypotheses of Stokes' Theorem.

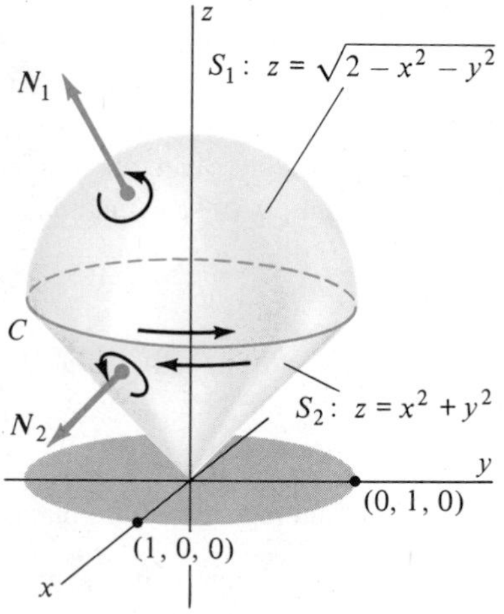

Figure 6.10 $\iint_S (\text{curl } F) \cdot N\, dS = 0$ for $S = S_1 \cup S_2$.

Solution: The closed surface S may be written $S = S_1 \cup S_2$ where S_1 is the hemisphere $z = \sqrt{2 - x^2 - y^2}$ and S_2 is the graph of the cone $z = x^2 + y^2$ over the unit disc. These two surfaces share the common boundary curve

$$C = \{(x, y, 1) \mid x^2 + y^2 = 1\}.$$

However, the two surfaces induce opposite orientations on C. This is because the unit normal to S_1 is upward, inducing a counterclockwise orientation on C, while the unit normal to S_2 is downward, inducing a clockwise orientation on C. Thus, by Stokes' Theorem,

$$\begin{aligned}\iint_S (\text{curl } \boldsymbol{F}) \cdot \boldsymbol{N}\, dS &= \iint_{S_1} (\text{curl } \boldsymbol{F}) \cdot \boldsymbol{N}_1\, dS + \iint_{S_2} (\text{curl } \boldsymbol{F}) \cdot \boldsymbol{N}_2\, dS \\ &= \int_C \boldsymbol{F} \cdot d\boldsymbol{r} + \int_{-C} \boldsymbol{F} \cdot d\boldsymbol{r} \\ &= \int_C \boldsymbol{F} \cdot d\boldsymbol{r} - \int_C \boldsymbol{F} \cdot d\boldsymbol{r} \\ &= 0.\end{aligned}$$

■

Curl *F* and Conservative Force Fields

For conservative planar vector fields $\boldsymbol{F}(x, y) = M\boldsymbol{i} + N\boldsymbol{j}$, we have previously seen that the condition that $\boldsymbol{F}$ be *conservative* in an open rectangle D (i.e., that $\boldsymbol{F} = \nabla\phi$ for some scalar function ϕ) is equivalent to the condition that curl $\boldsymbol{F}(x, y) = 0$ for all $(x, y) \in D$ (see Exercise 24, Section 21.4). We now sketch an argument for the same result in three dimensions.

First, suppose that curl $\boldsymbol{F}(x, y, z) = \boldsymbol{0}$ for all (x, y, z) in some simply connected region D in $\mathbb{R}^3$. Then if C is any closed path in D,

$$\int_C \boldsymbol{F} \cdot d\boldsymbol{r} = \iint_S \text{curl } \boldsymbol{F} \cdot \boldsymbol{N}\, dS = \iint_S \boldsymbol{O} \cdot \boldsymbol{N}\, dS = 0$$

where S is any smooth surface in D bounded by C. Thus, $\int_C \boldsymbol{F} \cdot d\boldsymbol{r} = 0$ for any closed path in D, so $\boldsymbol{F}(x, y, z)$ is conservative (Exercise 13, Section 21.3).

Conversely, suppose that $\boldsymbol{F}(x, y, z)$ is conservative in D. Then (again by Exercise 13, Section 21.3) $\int_C \boldsymbol{F} \cdot d\boldsymbol{r} = 0$ for any closed curve C in D. Now let (x_0, y_0, z_0) be any point in D, let C_r be a small circle of radius r and center (x_0, y_0, z_0), and let $\boldsymbol{N}$ be any unit normal to the plane containing C_r. Let S_r be the disc enclosed by C_r. Then, by Stokes' Theorem

$$\int_{C_r} \boldsymbol{F} \cdot d\boldsymbol{r} = \iint_{S_r} (\text{curl } \boldsymbol{F}) \cdot \boldsymbol{N}\, dS. \tag{14}$$

Now the line integral in (14) is zero, since $\boldsymbol{F}$ is conservative. Thus, $\iint_{S_r} (\text{curl } \boldsymbol{F})\cdot\boldsymbol{N}\, dS = 0$, no matter how small r is chosen. From this statement we conclude that

$$(\text{curl } \boldsymbol{F}(x_0, y_0, z_0))\cdot\boldsymbol{N} = 0. \tag{15}$$

(If the expression in (15) were either positive or negative, we could invoke the continuity of $\boldsymbol{F}(x, y, z)$ to contract C_r to a small circle on which $(\text{curl } \boldsymbol{F})\cdot\boldsymbol{N}$ is either always positive or always negative. The resulting surface integral in (14) would then be nonzero, a contradiction of equation (14).)

Since $\boldsymbol{N}$ was arbitrary, it follows from (15) that curl $\boldsymbol{F}(x_0, y_0, z_0) = \boldsymbol{0}$. Finally, since (x_0, y_0, z_0) was chosen arbitrarily in D, we conclude that curl $\boldsymbol{F}(x, y, z) = \boldsymbol{0}$ for all $(x, y, z) \in D$ when F is conservative.

The preceding argument is not a rigorous proof. However, it does convey the general flavor of the arguments linking the ideas of conservative vector fields, independence of path for line integrals, and the condition curl $\boldsymbol{F} = \boldsymbol{0}$. These relationships are summarized by the following theorem.

THEOREM 5

Let $\boldsymbol{F}(x, y, z) = M(x, y, z)\boldsymbol{i} + N(x, y, z)\boldsymbol{j} + P(x, y, z)\boldsymbol{k}$ be a vector field for which all of the component functions, together with all their first partial derivatives, are continuous in some open spherical region D in $\mathbb{R}^3$. The following conditions are all equivalent:

(i) $\boldsymbol{F}$ is conservative in D.

(ii) $\int_C \boldsymbol{F}\cdot d\boldsymbol{r}$ is independent of path C in D.

(iii) $\int_C \boldsymbol{F}\cdot d\boldsymbol{r} = 0$ for every closed path C in D.

(iv) curl $\boldsymbol{F}(x, y, z) = \boldsymbol{0}$ for all $(x, y, z) \in D$.

Theorem 5 gives the following condition for determining when a vector field $\boldsymbol{F} = M\boldsymbol{i} + N\boldsymbol{j} + P\boldsymbol{k}$ is conservative.

COROLLARY 2

Let $\boldsymbol{F}(x, y, z)$ and D be as in Theorem 5. Then $\boldsymbol{F}$ is conservative in D if and only if each of the following equations holds for all (x, y, z) in D:

(i) $\dfrac{\partial P}{\partial y}(x, y, z) = \dfrac{\partial N}{\partial z}(x, y, z)$,

(ii) $\dfrac{\partial M}{\partial z}(x, y, z) = \dfrac{\partial P}{\partial x}(x, y, z)$,

(iii) $\dfrac{\partial N}{\partial x}(x, y, z) = \dfrac{\partial M}{\partial y}(x, y, z)$.

The proof of Corollary 2 consists of applying Theorem 5, parts (i) and (iv) together with the definition of curl $\boldsymbol{F}$ (Definition 6).

A Physical Interpretation of Curl

As a last look at Stokes' Theorem, we use equation (5) to write it in the form

$$\int_C \boldsymbol{F}\cdot\boldsymbol{T}\,ds = \iint_S (\text{curl } \boldsymbol{F})\cdot\boldsymbol{N}\,dS. \tag{16}$$

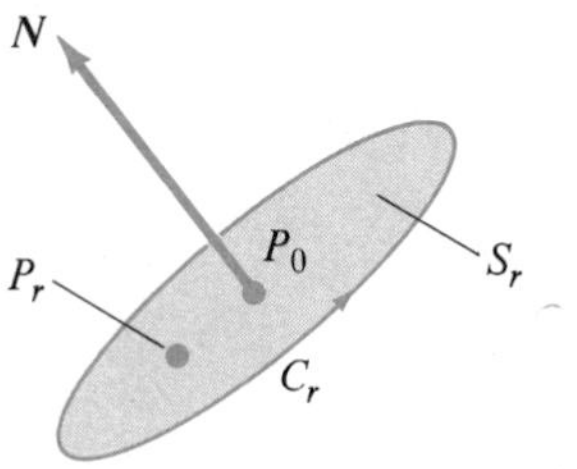

Figure 6.11

Returning to our primary motivation of fluid flow, we recall from Section 21.4 that if $\boldsymbol{F}$ is the velocity field of a moving fluid in which the curve C is submerged, the line integral in (16), called the **circulation** of $\boldsymbol{F}$ around C, is a measure of the tendency of the fluid to rotate, or circulate, around the curve C. Now let $P_0(x_0, y_0, z_0)$ be a point in the fluid, and let C_r be a circle with radius r and center P_0 (see Figure 6.11). Applying the Mean Value Theorem for double integrals (Theorem 2, Section 20.3) to equation (16), we find that for some point $P_r = (x_r, y_r, z_r)$ in S_r (the disc enclosed by C_r)

$$\begin{aligned}\int_{C_r} \boldsymbol{F}\cdot\boldsymbol{T}\,ds &= \iint_{S_r} (\text{curl } \boldsymbol{F})\cdot\boldsymbol{N}\,dS\\ &= \pi r^2[\text{curl } \boldsymbol{F}(x_r, y_r, z_r)]\cdot\boldsymbol{N}.\end{aligned}$$

Letting $r \to 0$, we conclude that

$$[\text{curl } \boldsymbol{F}(x_0, y_0, z_0)]\cdot\boldsymbol{N} = \lim_{r\to 0}\frac{1}{\pi r^2}\int_{C_r} \boldsymbol{F}\cdot\boldsymbol{T}\,ds. \tag{17}$$

The interpretation of equation (17) is this. The right-hand side, as described earlier, is a measure of the tendency of the fluid to circulate, or rotate, about the small circle C_r. Since the normal $\boldsymbol{N}$ to the disc enclosed by this circle appears on the left-hand side of (17), this tendency to rotate will be the largest when C_r is positioned so that $\boldsymbol{N}$ is parallel to curl $\boldsymbol{F}(x_0, y_0, z_0)$. Thus, curl $\boldsymbol{F}(x_0, y_0, z_0)$ is a vector whose *magnitude* is a measure of the tendency of the fluid at (x_0, y_0, z_0) to rotate (as about the drain hole in a bathtub), and whose *direction* is along the axis about which the fluid has the maximal tendency to rotate.

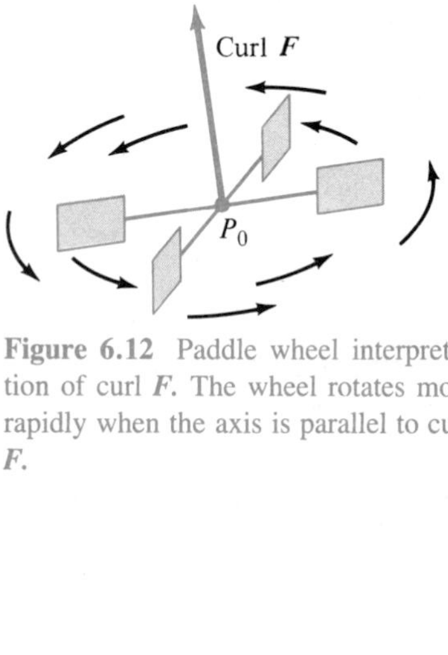

Figure 6.12 Paddle wheel interpretation of curl $\boldsymbol{F}$. The wheel rotates most rapidly when the axis is parallel to curl $\boldsymbol{F}$.

Figure 6.12 shows an interpretation of equation (17) in terms of a paddle wheel. The motion of the fluid will cause the paddle wheel, based at P_0, to rotate most quickly when the axis points parallel to curl $\boldsymbol{F}(x_0, y_0, z_0)$. For obvious reasons, fluid fields for which curl $\boldsymbol{F} = \boldsymbol{0}$ (such as conservative fields) are called *irrotational*.

It is important to note that what we have achieved in these last few observations is an interpretation of curl $\boldsymbol{F}(x, y, z)$ as a *local* property of the vector field—the tendency of the fluid *at location* (x, y, z) to rotate. In general, this tendency will vary from point to point within the fluid. What Stokes' Theorem says in this context is that the collective measure of this tendency taken over the entire surface S (that is, the value of the surface integral) is completely determined by, and equals, the tendency of the fluid to circulate around the boundary C (the value of the line integral).

Exercise Set 21.6

In Exercises 1–4, calculate curl $\boldsymbol{F}$.

1. $\boldsymbol{F}(x, y, z) = x\boldsymbol{i} - y\boldsymbol{j} + z^2\boldsymbol{k}$

2. $\boldsymbol{F}(x, y, z) = xy^2\boldsymbol{i} + xz^2\boldsymbol{j} + yz^2\boldsymbol{k}$

3. $\boldsymbol{F}(x, y, z) = xyz\boldsymbol{i} - \cos(xy)\boldsymbol{j} + \sin(yz)\boldsymbol{k}$

4. $\boldsymbol{F}(x, y, z) = e^{xy}\boldsymbol{i} - x^2z^2\boldsymbol{j} + \sqrt{xy}\,\boldsymbol{k}$

In Exercises 5–8, determine whether the given vector field is conservative.

5. $F(x, y, z) = yz\mathbf{i} + xz\mathbf{j} + xy\mathbf{k}$

6. $F(x, y, z) = 2xyz^2\mathbf{i} + x^2z^2\mathbf{j} + 2x^2yz\mathbf{k}$

7. $F(x, y, z) = \sin y\mathbf{i} + \cos x\mathbf{j} + \sqrt{xy}\,\mathbf{k}$

8. $F(x, y, z) = yze^{xyz}\mathbf{i} + xze^{xyz}\mathbf{j} + xye^{xyz}\mathbf{k}$

9. Verify Stokes' Theorem for the vector field $F(x, y, z) = z\mathbf{i} + x\mathbf{j} + y\mathbf{k}$ and the surface of the hemisphere $z = \sqrt{1 - x^2 - y^2}$.

10. Use Stokes' Theorem to calculate the line integral $\int_C F \cdot dr$ where $F(x, y, z) = x^2y^2\mathbf{i} + x^2z^2\mathbf{j} + y^2z^2\mathbf{k}$ and C is the perimeter of the rectangle with vertices (1, 1, 0), (1, 5, 0), (3, 1, 0), and (3, 5, 0) traversed in this order.

11. Verify Stokes' Theorem for $F(x, y, z) = x^2y\mathbf{i} + y^2z\mathbf{j} + xz\mathbf{k}$ where C is the boundary of the rectangle $Q = \{(x, y, 0) | 0 \leq x \leq 1, 0 \leq y \leq 2\}$ oriented counterclockwise.

12. Use Stokes' Theorem to calculate the line integral $\int_C F \cdot dr$ where $F(x, y, z) = x^2y\mathbf{i} + y^2z\mathbf{j} + xz\mathbf{k}$ and C is the intersection of the plane $x + 3y + z = 4$ with the cylinder $x^2 + y^2 = 1$ with orientation induced by the upward unit normal to the plane.

13. Verify Stokes' Theorem for the portion of the plane $x + 2y + z = 2$ in the first octant, the vector field $F(x, y, z) = z\mathbf{i} + x\mathbf{j} + y\mathbf{k}$, and the upward unit normal.

14. Calculate $\int_C F \cdot dr$ where $F(x, y, z) = xz\mathbf{i} + 2z\mathbf{j} - xy\mathbf{k}$ and C is the intersection of the plane $y = z + 2$ and the cylinder $x^2 + y^2 = 4$. The orientation on C is that induced by the upward unit normal to the plane.

15. Evaluate $\int_C F \cdot dr$ around the unit circle in the xy-plane, counterclockwise, where

$F(x, y, z) = (\sqrt{x} + y)\mathbf{i} + (e^y - x)\mathbf{j} + (\sin z + y)\mathbf{k}.$

16. Calculate $\int_C F \cdot dr$ where $F(x, y, z) = xyz\mathbf{i} + xz\mathbf{k}$ and where C is the intersection of the paraboloid $z = x^2 + y^2$ and the plane $z = 4$. The orientation on C is that induced by the upward unit normal on the paraboloid.

17. Verify Stokes' Theorem for the surface $z = 9 - x^2 - y^2$, $z \geq 0$ and the vector field $F(x, y, z) = x^2\mathbf{i} + y^2\mathbf{j} + z^2\mathbf{k}$.

18. Show that curl $F = \mathbf{0}$ for $F(x, y, z) = x^2\mathbf{i} + y^2\mathbf{j} + z^2\mathbf{k}$. Can you give a geometric interpretation for this result?

19. Let $F(x, y) = M(x, y)\mathbf{i} + N(x, y)\mathbf{j}$ be a conservative vector field. Let $G(x, y, z) = F(x, y) + \lambda z\mathbf{k}$ where λ is constant. Show that $G(x, y, z)$ is conservative. Is the result true if $z\mathbf{k}$ is replaced by $g(z)\mathbf{k}$?

20. Let $F(x, y, z)$ be a vector field satisfying the hypotheses of Stokes' Theorem. Show that $\iint_S (\text{curl } F) \cdot N\, dS = 0$ where S is the unit sphere.

21. Use Stokes' Theorem to show that a line integral is independent of parameterization if orientation is preserved.

21.7 THE DIVERGENCE THEOREM

In this last section, we generalize equation (28) of Section 21.4,

$$\int_C F \cdot N\, ds = \iint_Q \text{div } F(x, y)\, dA \qquad \text{(Flux of } F \text{ across } C\text{)}, \tag{1}$$

to an appropriate equation for vector fields $F(x, y, z)$ in $\mathbb{R}^3$. However, this time we begin with the physical interpretation of what we are after and see where the mathematics leads us.

Recall the interpretation for equation (1) in terms of fluid flow: $F(x, y)$ is a vector field in the plane giving the velocity vector $v(x, y) = F(x, y)$ for the steady state motion of a thin layer of fluid; C is a closed curve in the fluid layer; and $N(x, y)$ is the outward unit normal to C. In this setting the line integral $\int_C F \cdot N\, ds$

in (1) is interpreted as the **flux** of the fluid outward across C, that is, the net rate at which fluid is crossing the boundary C of Q (Figure 7.1). In Section 21.5 we have already encountered the notion of the flux of a vector field $\boldsymbol{F}(x, y, z)$ across a closed surface S in $\mathbb{R}^3$:

$$\text{Flux} = \iint_S \boldsymbol{F} \cdot \boldsymbol{N}\, dS, \tag{2}$$

where $\boldsymbol{N}(x, y, z)$ is the outward unit normal to S at (x, y, z). That is, the natural generalization of the line integral in (1), at least in terms of fluid flow, is the surface integral in (2) (see Figure 7.2).

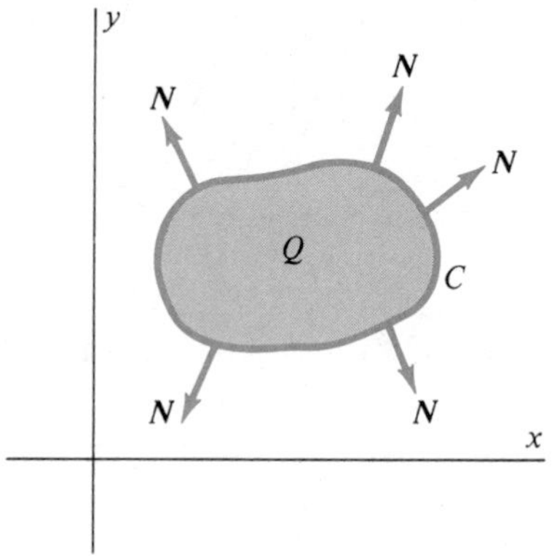

Figure 7.1 $\int_C \boldsymbol{F} \cdot \boldsymbol{N}\, ds$ is the flux of $\boldsymbol{F}(x, y)$ across C in $\mathbb{R}^2$.

Now the right-hand side of equation (1) is a double integral evaluated over Q, the region bounded by C, of the scalar function div $\boldsymbol{F}(x, y)$, and we can paraphrase equation (1) by saying that "div $\boldsymbol{F}(x, y)$ is a measure of the local behavior of $\boldsymbol{F}(x, y)$ that, when integrated over the enclosed region Q, determines the flux of $\boldsymbol{F}(x, y)$ across the boundary C."

Thus, to properly generalize equation (1) to a closed surface S in $\mathbb{R}^3$, we need to find a scalar function div $\boldsymbol{F}(x, y, z)$ that is a measure of the local behavior of $\boldsymbol{F}(x, y, z)$ and that, when integrated over the region R enclosed by S, gives the flux of $\boldsymbol{F}(x, y, z)$ across the surface S. That is, div $\boldsymbol{F}(x, y, z)$ must satisfy the equation

$$\iint_S \boldsymbol{F} \cdot \boldsymbol{N}\, dS = \iiint_R \text{div}\, \boldsymbol{F}\, dV. \tag{3}$$

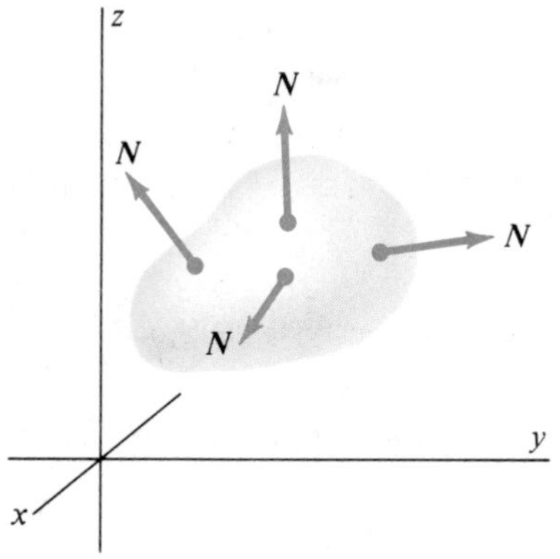

Figure 7.2 $\iint_S \boldsymbol{F} \cdot \boldsymbol{N}\, dS$ is the flux of $\boldsymbol{F}(x, y, z)$ across S in $\mathbb{R}^3$.

Although you can probably guess how we're going to define div $\boldsymbol{F}(x, y, z)$, let's proceed "experimentally" by investigating the surface integral in (3). For starters, we express $\boldsymbol{F}(x, y, z)$ in component form as

$$\boldsymbol{F}(x, y, z) = M(x, y, z)\boldsymbol{i} + N(x, y, z)\boldsymbol{j} + P(x, y, z)\boldsymbol{k}. \tag{4}$$

Also, we assume that S is a closed surface that is the union $S = S_1 \cup S_2$ of two smooth surfaces, which are graphs of the functions

$$S_1: \quad z = g_1(x, y), \qquad (x, y) \in Q$$

$$S_2: \quad z = g_2(x, y), \qquad (x, y) \in Q$$

with $g_1(x, y) \le g_2(x, y)$ for all $(x, y) \in Q$. Let C be the boundary common to S_1 and S_2 (see Figure 7.3).

Using (4), we write the left side of equation (3) as

$$\iint_S \boldsymbol{F} \cdot \boldsymbol{N}\, dS = \iint_S (M\boldsymbol{i}) \cdot \boldsymbol{N}\, dS + \iint_S (N\boldsymbol{j}) \cdot \boldsymbol{N}\, dS + \iint_S (P\boldsymbol{k}) \cdot \boldsymbol{N}\, dS. \tag{5}$$

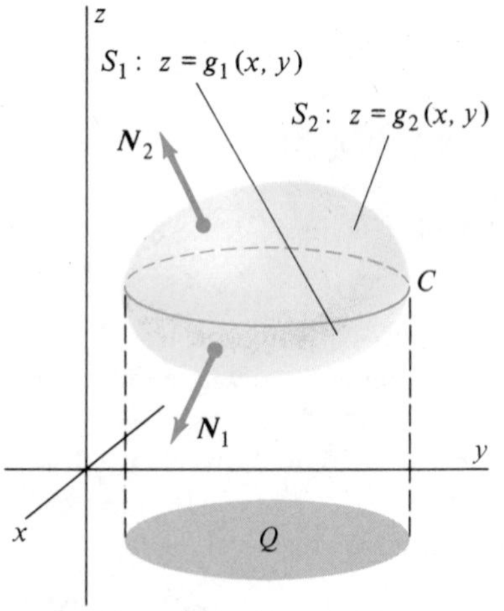

Figure 7.3 $S = S_1 \cup S_2$.

Let's now focus on the last integral on the right side of (5). Since the unit normal to S_2 is upward and the unit normal to S_1 is downward, we use both equation (16) and equation (17) of Section 21.5 to find that

$$\iint_S [P(x, y, z)\boldsymbol{k}] \cdot \boldsymbol{N}\, dS \tag{6}$$

$$= \iint_{S_2} [P(x, y, z)\boldsymbol{k}] \cdot \boldsymbol{N}_2\, dS + \iint_{S_1} [P(x, y, z)\boldsymbol{k}] \cdot \boldsymbol{N}_1\, dS$$

$$= \iint_Q P(x, y, g_2(x, y))\, dx\, dy + \iint_Q -P(x, y, g_1(x, y))\, dx\, dy$$

$$= \iint_Q \{P(x, y, z)]_{z=g_1(x,y)}^{z=g_2(x,y)}\}\, dx\, dy$$

$$= \iint_Q \left\{\int_{g_1(x,y)}^{g_2(x,y)} \frac{\partial}{\partial z} P(x, y, z)\, dz\right\} dx\, dy$$

$$= \iiint_R \frac{\partial}{\partial z} P(x, y, z)\, dx\, dy\, dz.$$

(Note that we have changed the order of integration from $dz\, dx\, dy$ to $dx\, dy\, dz$, as justified in Section 20.7.) That is,

$$\int_S (P\boldsymbol{k}) \cdot \boldsymbol{N}\, dS = \iiint_R \frac{\partial P}{\partial z}\, dx\, dy\, dz. \tag{7}$$

Similarly, if S can be written as in Figure 7.4, we can show that

$$\int_S (N\boldsymbol{j}) \cdot \boldsymbol{N}\, dS = \iiint_R \frac{\partial N}{\partial y}\, dx\, dy\, dz, \tag{8}$$

and if S can be written as in Figure 7.5, we obtain

$$\int_S (M\boldsymbol{i}) \cdot \boldsymbol{N}\, dS = \iiint_R \frac{\partial M}{\partial x}\, dx\, dy\, dz. \tag{9}$$

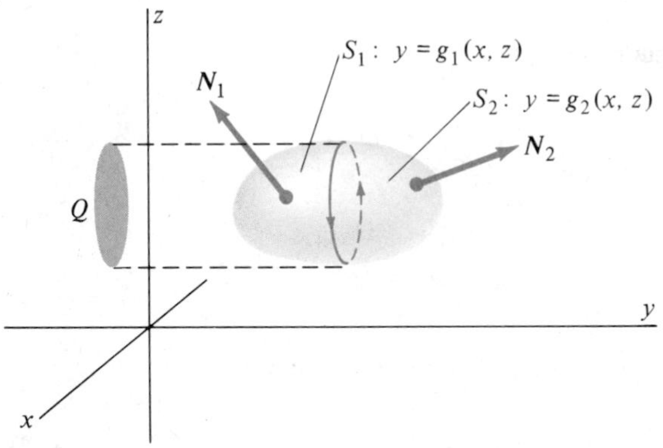

Figure 7.4 $S = S_1 \cup S_2$.

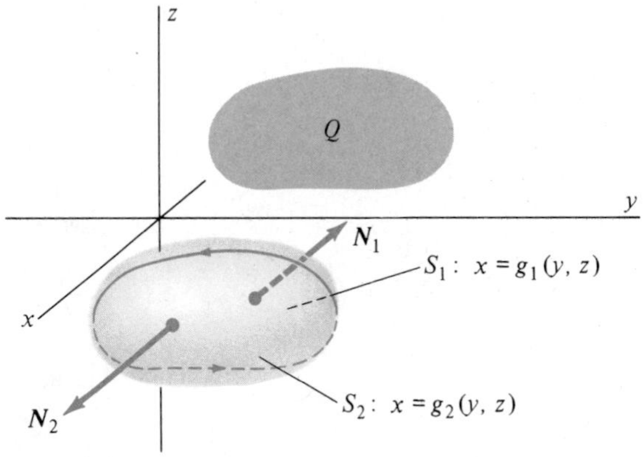

Figure 7.5 $S = S_1 \cup S_2$.

Adding the corresponding sides of equations (7), (8), and (9), using equation (4), we obtain the equation

$$\int_S \boldsymbol{F} \cdot \boldsymbol{N}\, dS = \iiint_R \left(\frac{\partial M}{\partial x} + \frac{\partial N}{\partial y} + \frac{\partial P}{\partial z}\right) dx\, dy\, dz. \tag{10}$$

Comparing equations (3) and (10) leads directly to the following definition.

DEFINITION 7

The **divergence of the vector field**

$$F(x, y, z) = M(x, y, z)\mathbf{i} + N(x, y, z)\mathbf{j} + P(x, y, z)\mathbf{k}, \tag{11}$$

written div $\boldsymbol{F}$, is the scalar function

$$\text{div } \boldsymbol{F}(x, y, z) = \frac{\partial}{\partial x}M(x, y, z) + \frac{\partial}{\partial y}N(x, y, z) + \frac{\partial}{\partial z}P(x, y, z). \tag{12}$$

With this definition of div $\boldsymbol{F}$, the preceding discussion shows that equation (3) holds for smooth closed surfaces that are simultaneously x-simple, y-simple, and z-simple. This is our desired generalization of the planar equation (1). A more general and precisely stated version of equation (3) is our final result.

THEOREM 6
Divergence Theorem

Let S be a closed piecewise smooth surface enclosing the region R. Let $\boldsymbol{F}(x, y, z) = M(x, y, z)\mathbf{i} + N(x, y, z)\mathbf{j} + P(x, y, z)\mathbf{k}$ be a vector field for which all of the components, together with all their first partial derivatives, are continuous throughout R. Then

$$\iint_S \boldsymbol{F} \cdot \boldsymbol{N}\, dS = \iiint_R \text{div } \boldsymbol{F}(x, y, z)\, dx\, dy\, dz \tag{13}$$

where $\boldsymbol{N} = \boldsymbol{N}(x, y, z)$ is the outward unit normal to S.

REMARK 1: The Divergence Theorem was first discovered by the German mathematician Carl Friedrich Gauss. The result is often referred to as Gauss' Theorem.

REMARK 2: Using the operator notation

$$\nabla = \frac{\partial}{\partial x}\mathbf{i} + \frac{\partial}{\partial y}\mathbf{j} + \frac{\partial}{\partial z}\mathbf{k}$$

we can write

$$\text{div } \boldsymbol{F} = \nabla \cdot \boldsymbol{F} = \frac{\partial M}{\partial x} + \frac{\partial N}{\partial y} + \frac{\partial P}{\partial z}.$$

With this notation, the Divergence Theorem may be written either as

$$\iint_S \boldsymbol{F} \cdot \boldsymbol{N}\, dS = \iiint_R \nabla \cdot \boldsymbol{F}\, dV, \qquad dV = dx\, dy\, dz \tag{14}$$

or as

$$\iint_S \boldsymbol{F} \cdot \boldsymbol{N}\, dS = \iiint_R \left(\frac{\partial M}{\partial x} + \frac{\partial N}{\partial y} + \frac{\partial P}{\partial z}\right) dx\, dy\, dz. \tag{15}$$

The appeal of equation (14) is that it expresses Theorem 6 in what is referred to as "coordinate free" form. On the other hand, equation (15) is the most explicit form for use when working in Cartesian (rectangular) coordinates.

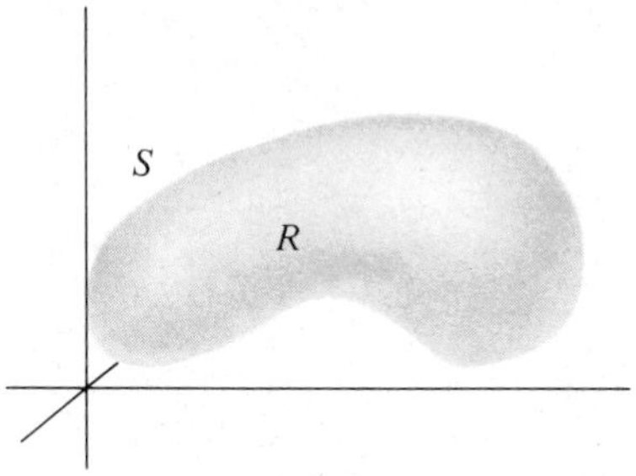

Figure 7.6 A region R that is not y-simple.

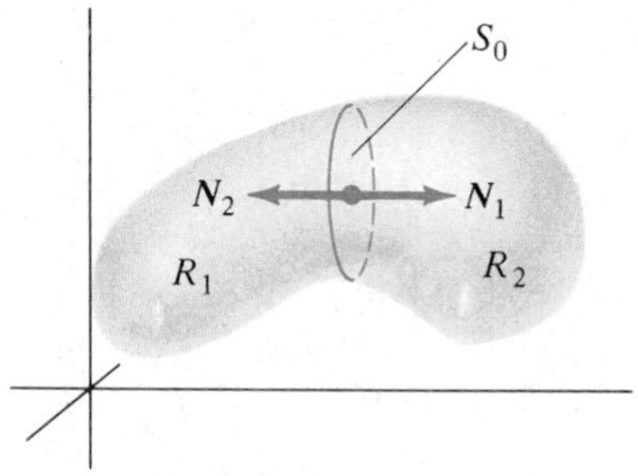

Figure 7.7 R_1 and R_2 are both y-simple. N_j is the outward unit normal for R_j, $j = 1, 2$.

REMARK 3: Our proof of the Divergence Theorem addressed only the case of a surface enclosing a region which was simultaneously x-simple, y-simple, and z-simple. Figures 7.6 and 7.7 illustrate how this proof may be extended to include more general regions. By constructing the surface S_0, we partition the region R into two regions, R_1 and R_2, both of which are simultaneously x-simple, y-simple, and z-simple. Letting S_1 denote the boundary of R_1 and S_2 the boundary of R_2, the previous proof shows that

$$\iint_{S_1} \boldsymbol{F} \cdot \boldsymbol{N}\, dS = \iiint_{R_1} \operatorname{div} \boldsymbol{F}\, dx\, dy\, dz \tag{16}$$

and

$$\iint_{S_2} \boldsymbol{F} \cdot \boldsymbol{N}\, dS = \iiint_{R_2} \operatorname{div} \boldsymbol{F}\, dx\, dy\, dz. \tag{17}$$

On the face S_0, the outward unit normal to S_1 is the *opposite* of the outward unit normal to S_2. Thus, the contributions over the face S_0 to the surface integrals in (16) and (17) cancel. Adding the corresponding sides of (16) and (17) then gives

$$\begin{aligned}\iint_{S} \boldsymbol{F} \cdot \boldsymbol{N}\, dS &= \iint_{S_1} \boldsymbol{F} \cdot \boldsymbol{N}\, dS + \iint_{S_2} \boldsymbol{F} \cdot \boldsymbol{N}\, dS \\ &= \iiint_{R_1} \operatorname{div} \boldsymbol{F}\, dx\, dy\, dz + \iiint_{R_2} \operatorname{div} \boldsymbol{F}\, dx\, dy\, dz \\ &= \iiint_{R} \operatorname{div} \boldsymbol{F}\, dx\, dy\, dz.\end{aligned}$$

The same argument can be applied to more complicated surfaces.

Before proceeding to examples, let's nail down a bit more precisely what the Divergence Theorem says about fluid flow. Of course, the left-hand side of equation (13) is the **net flux** (rate out minus rate in) of the fluid outward across the surface S. However, the volume integral on the right-hand side of (13) can be thought of as a sum taken throughout R of the product div $\boldsymbol{F}(x, y, z)\, dV$, where dV is an infinitesimal element of volume containing the point (x, y, z). It is this local property of div $\boldsymbol{F}(x, y, z)$ that we wish to understand better.

To do so, let $\Delta V_\epsilon = \dfrac{4}{3}\pi\epsilon^3$ be the volume of a small sphere S_ϵ of radius ϵ and center (x_0, y_0, z_0) contained within S. According to Theorem 6

$$\text{Flux of } \boldsymbol{F} \text{ out of } S_\epsilon = \iiint_{R_\epsilon} \operatorname{div} \boldsymbol{F}\, dx\, dy\, dz \tag{18}$$

where R_ϵ is the region enclosed by the sphere S_ϵ. We now proceed with an argument analogous to that of Section 21.6 for the interpretation of curl $\boldsymbol{F}$. The mean value property for triple integrals guarantees the existence of a point $(x_\epsilon, y_\epsilon, z_\epsilon)$ in R_ϵ for which

$$\iiint_{R_\epsilon} \operatorname{div} \boldsymbol{F}\, dx\, dy\, dz = \operatorname{div} \boldsymbol{F}(x_\epsilon, y_\epsilon, z_\epsilon)\, \Delta V_\epsilon. \tag{19}$$

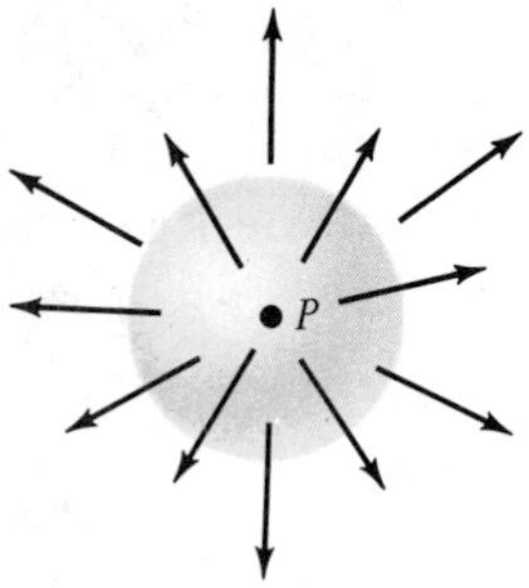

Figure 7.8 div $F > 0$; P is a source.

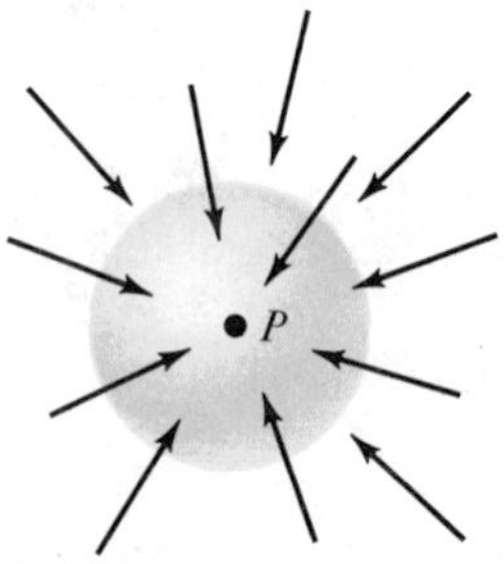

Figure 7.9 div $F < 0$; P is a sink.

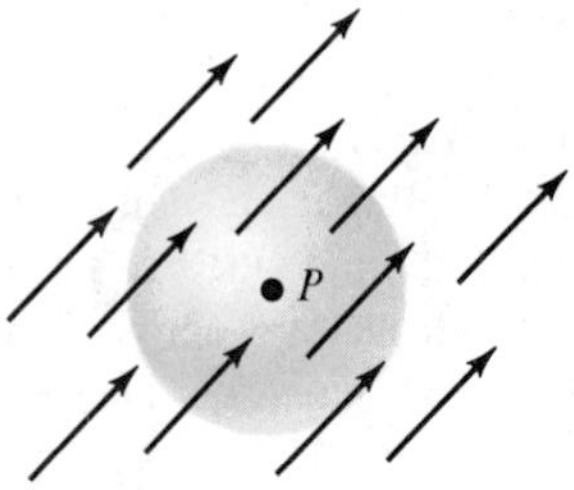

Figure 7.10 div $F = 0$; F is *incompressible* at P.

Combining (18) and (19) we conclude that

$$\text{div } \boldsymbol{F}(x_\epsilon, y_\epsilon, z_\epsilon) = \frac{\text{Flux of } \boldsymbol{F} \text{ out of } S_\epsilon}{\Delta V_\epsilon}. \tag{20}$$

Since $(x_\epsilon, y_\epsilon, z_\epsilon) \to (x_0, y_0, z_0)$ as $\epsilon \to 0$, we conclude that

$$\boxed{\text{div } \boldsymbol{F}(x_0, y_0, z_0) = \lim_{\epsilon \to 0} \frac{\text{Flux of } \boldsymbol{F} \text{ out of } S_\epsilon}{\Delta V_\epsilon}.} \tag{21}$$

In other words, div $\boldsymbol{F}(x_0, y_0, z_0)$ is the **flux per unit volume** at the point (x_0, y_0, z_0). Thus,

(i) If div $\boldsymbol{F}(x_0, y_0, z_0) > 0$, more fluid is flowing out across a small sphere centered at (x_0, y_0, z_0) than is flowing in. In this case we say that the fluid is **expanding** at (x_0, y_0, z_0), or that (x_0, y_0, z_0) is a **source** for the vector field $\boldsymbol{F}$ (see Figure 7.8).
(ii) If div $\boldsymbol{F}(x_0, y_0, z_0) < 0$, more fluid is flowing into a small sphere centered at (x_0, y_0, z_0) than is flowing out. The fluid is said, therefore, to be **contracting** at (x_0, y_0, z_0), and (x_0, y_0, z_0) is called a **sink** for $\boldsymbol{F}$ (see Figure 7.9).
(iii) If div $\boldsymbol{F}(x_0, y_0, z_0) = 0$, the amount of fluid flowing out of the small sphere centered at (x_0, y_0, z_0) equals the amount flowing in. In this case the fluid is said to be **incompressible** at (x_0, y_0, z_0) (see Figure 7.10).

The powerful statement of the Divergence Theorem is that knowledge of the local property div $\boldsymbol{F}(x, y, z)$, summed throughout the interior of S, determines the flux of $\boldsymbol{F}$ across the surface S, and conversely.

Example 1 Calculate the flux of the vector field

$$\boldsymbol{F}(x, y, z) = (x - y^2)\boldsymbol{i} + 3y\boldsymbol{j} + (x^3 - z)\boldsymbol{k}$$

outward over the sphere $S = \{(x, y, z) | x^2 + y^2 + z^2 = 9\}$.

Solution: Here

$$\begin{aligned} \text{div } \boldsymbol{F} &= \frac{\partial}{\partial x}(x - y^2) + \frac{\partial}{\partial y}(3y) + \frac{\partial}{\partial z}(x^3 - z) \\ &= 1 + 3 - 1 \\ &= 3. \end{aligned}$$

By the Divergence Theorem,

$$\text{Flux across } S = \iint_S \boldsymbol{F} \cdot \boldsymbol{N}\, dS = \iiint_R 3 \cdot dx\, dy\, dz = 3 \cdot \frac{4}{3}\pi \cdot 3^3 = 108\pi,$$

since the volume of the sphere is $\frac{4}{3}\pi r^3$. ■

Example 2 Evaluate the surface integral

$$\iint_S \boldsymbol{F} \cdot \boldsymbol{N}\, dS$$

where $\boldsymbol{F}(x, y, z) = xy\boldsymbol{i} + xz\boldsymbol{j} + yz\boldsymbol{k}$ and $\boldsymbol{N}$ is the outward unit normal to the ellipsoid $S = \{(x, y, z) \mid x^2 + 4y^2 + z^2 = 1\}$.

Solution: We use the Divergence Theorem rather than trying to evaluate the surface integral directly. We have

$$\text{div } \boldsymbol{F} = \frac{\partial}{\partial x}(xy) + \frac{\partial}{\partial y}(xz) + \frac{\partial}{\partial z}(yz) = 2y.$$

Letting R denote the region enclosed by the ellipsoid, we obtain

$$\iint_S \boldsymbol{F} \cdot \boldsymbol{N}\, dS = \iiint_R 2y\, dx\, dy\, dz$$
$$= \int_{-1}^{1} \int_{-(1/2)\sqrt{1-x^2}}^{(1/2)\sqrt{1-x^2}} \int_{-\sqrt{1-x^2-4y^2}}^{\sqrt{1-x^2-4y^2}} 2y\, dz\, dy\, dx.$$

Now this iterated integral can be evaluated directly. However, since the integrand satisfies the equation

$$\text{div } \boldsymbol{F}(x, -y, z) = -2y = -\text{div } \boldsymbol{F}(x, y, z),$$

and since the ellipsoid S is symmetric about the plane $y = 0$, the value of this integral is zero. ■

Example 3 Calculate the value of the surface integral $\iint_S \boldsymbol{F} \cdot \boldsymbol{N}\, dS$ where $\boldsymbol{F}(x, y, z) = x^2y\boldsymbol{i} + xy\boldsymbol{j} + y^2z^3\boldsymbol{k}$, S is the cube formed by the planes $x = \pm 1$, $y = \pm 1$, and $z = \pm 1$, and $\boldsymbol{N}$ is the outward unit normal on S.

Solution: Rather than calculate the surface integral directly for each of the six faces, we use the Divergence Theorem. Since

$$\text{div } \boldsymbol{F} = \frac{\partial}{\partial x}(x^2y) + \frac{\partial}{\partial y}(xy) + \frac{\partial}{\partial z}(y^2z^3) = 2xy + x + 3y^2z^2$$

we have, from Theorem 6,

$$\iint_S \boldsymbol{F} \cdot \boldsymbol{N}\, dS = \int_{-1}^{1} \int_{-1}^{1} \int_{-1}^{1} (2xy + x + 3y^2z^2)\, dx\, dy\, dz$$
$$= \int_{-1}^{1} \int_{-1}^{1} \left\{ \left[x^2y + \frac{1}{2}x^2 + 3y^2z^2x \right]_{x=-1}^{x=1} \right\} dy\, dz$$
$$= \int_{-1}^{1} \int_{-1}^{1} 6y^2z^2\, dy\, dz$$
$$= \int_{-1}^{1} 2y^3z^2 \Big]_{y=-1}^{y=1} dz$$
$$= \int_{-1}^{1} 4z^2\, dz$$
$$= \frac{8}{3}.$$
■

Example 4 Find the flux of the vector field

$$F(x, y, z) = y^2z^3\mathbf{i} + 4x^2yz^2\mathbf{j} + x^2z^3\mathbf{k}$$

outward across the unit sphere $S = \{(x, y, z) | x^2 + y^2 + z^2 = 1\}$.

Solution: Let R be the unit ball $R = \{(x, y, z) | x^2 + y^2 + z^2 \le 1\}$. We have

$$\text{div}\,\mathbf{F} = \frac{\partial}{\partial x}(y^2z^3) + \frac{\partial}{\partial y}(4x^2yz^2) + \frac{\partial}{\partial z}(x^2z^3) = 7x^2z^2.$$

According to the Divergence Theorem,

$$\text{Flux across } S = \iint_S \mathbf{F} \cdot \mathbf{N}\, dS = \iiint_R 7x^2z^2\, dx\, dy\, dz.$$

To evaluate the triple integral, we use spherical coordinates:

$$\begin{aligned} x &= \rho \cos\theta \sin\phi & 0 &\le \rho \le 1, \\ y &= \rho \sin\theta \sin\phi & 0 &\le \theta \le 2\pi, \\ z &= \rho \cos\phi & 0 &\le \phi \le \pi. \end{aligned}$$

We obtain

$$\begin{aligned} \iiint_R 7x^2z^2\, dx\, dy\, dz &= 7\int_0^1 \int_0^{2\pi} \int_0^{\pi} [\rho \cos\theta \sin\phi]^2[\rho\cos\phi]^2\rho^2 \sin\phi\, d\phi\, d\theta\, d\rho \\ &= 7\int_0^1 \int_0^{2\pi} \int_0^{\pi} \rho^6 \cos^2\theta \sin^3\phi \cos^2\phi\, d\phi\, d\theta\, d\rho \\ &= 7\int_0^1 \int_0^{2\pi} \int_0^{\pi} \rho^6 \cos^2\theta[1 - \cos^2\phi]\cos^2\phi \sin\phi\, d\phi\, d\theta\, d\rho \\ &= 7\int_0^1 \int_0^{2\pi} \left\{\rho^6 \cos^2\theta\left[-\frac{1}{3}\cos^3\phi + \frac{1}{5}\cos^5\phi\right]_0^{\pi}\right\} d\theta\, d\rho \\ &= 7 \cdot \frac{4}{15}\int_0^1 \int_0^{2\pi} \rho^6 \cos^2\theta\, d\theta\, d\rho \\ &= 7 \cdot \frac{4}{15} \cdot \pi \int_0^1 \rho^6\, d\rho \\ &= \frac{4\pi}{15}. \end{aligned}$$

■

The Divergence Theorem finds wide application in physics, engineering, and applied mathematics. We have restricted our discussion to the example of fluid flow to keep the discussion straightforward and to allow you to focus primarily on the mathematics. The subject area of electricity and magnetism makes considerable use of each of the major theorems presented in this chapter (and involves another theorem due to Gauss). An excellent discussion of these ideas may be found in the *Berkeley Physics Course, Volume 2: Electricity and Magnetism*.

Exercise Set 21.7

In Exercises 1–6, find div $\boldsymbol{F}$.

1. $\boldsymbol{F}(x, y, z) = x^2\boldsymbol{i} + y^2\boldsymbol{j} + z^2\boldsymbol{k}$

2. $\boldsymbol{F}(x, y, z) = x^2z\boldsymbol{i} + y^2x\boldsymbol{j} + xz^2\boldsymbol{k}$

3. $\boldsymbol{F}(x, y, z) = x\boldsymbol{i} + xy\boldsymbol{j} + xyz\boldsymbol{k}$

4. $\boldsymbol{F}(x, y, z) = (y - x)\boldsymbol{i} + (y - z)\boldsymbol{j} + (x - y)\boldsymbol{k}$

5. $\boldsymbol{F}(x, y, z) = \cos xy\boldsymbol{i} + e^{xyz}\boldsymbol{j} + y \sin(xz)\boldsymbol{k}$

6. $\boldsymbol{F}(x, y, z) = x \tan^{-1} y\boldsymbol{i} - \sqrt{yz}\,\boldsymbol{j} - z \sec y\boldsymbol{k}$

7. Find the flux of the vector field $\boldsymbol{F}(x, y, z) = z\boldsymbol{i} + x\boldsymbol{j} + y\boldsymbol{k}$ out of the ellipsoid $x^2 + 9y^2 + 4z^2 = 1$.

8. Find the value of the surface integral $\iint_S \boldsymbol{F} \cdot \boldsymbol{N}\, dS$ where $\boldsymbol{F}$ is the vector field $\boldsymbol{F}(x, y, z) = -y^2\boldsymbol{i} + x\boldsymbol{j} + z\boldsymbol{k}$, S is the sphere $x^2 + y^2 + z^2 = 5$ and $\boldsymbol{N}$ is the outward unit normal.

9. Find the flux of the vector field $\boldsymbol{F}(x, y, z) = 3x\boldsymbol{i} - z^2y\boldsymbol{j} + 2z\boldsymbol{k}$ outward across the rectangular box with vertices $(1, 0, 0)$, $(1, 3, 0)$, $(-2, 0, 0)$, $(-2, 3, 0)$, $(1, 0, 5)$, $(1, 3, 5)$, $(-2, 0, 5)$, and $(-2, 3, 5)$.

10. Find the value of the surface integral $\iint_S \boldsymbol{F} \cdot \boldsymbol{N}\, dS$ where $\boldsymbol{F}$ is the vector field $\boldsymbol{F}(x, y, z) = 3x\boldsymbol{i} - 4y\boldsymbol{j} + 5z\boldsymbol{k}$, and V is the volume of the solid enclosed by the smooth surface S.

11. Find the flux of the vector field $\boldsymbol{F}(x, y, z) = 2xy\boldsymbol{i} + z^2y\boldsymbol{j} + xz\boldsymbol{k}$ over the cube formed by the coordinate planes and the planes $x = 1$, $y = 1$, and $z = 1$.

12. Find the value of the surface integral $\iint_S \boldsymbol{F} \cdot \boldsymbol{N}\, dS$ where $\boldsymbol{F}(x, y, z) = (x + e^y)\boldsymbol{i} + (e^{xz} - y)\boldsymbol{j} + (xy + z)\boldsymbol{k}$, S is the cylinder $\{(x, y, z)|x^2 + y^2 = 4, 0 \le z \le 2\}$ and $\boldsymbol{N}$ is the outward unit normal.

13. Find the flux of the vector field $\boldsymbol{F}(x, y, z) = x^2y^2\boldsymbol{i} + xy^3\boldsymbol{j} + xy\boldsymbol{k}$ outward across the cylinder $\{(x, y, z)|x^2 + y^2 = 4, 0 \le z \le 2\}$. (*Hint:* Use cylindrical coordinates.)

14. Find the value of the surface integral $\iint_S \boldsymbol{F} \cdot \boldsymbol{N}\, dS$ where $\boldsymbol{F}$ is the vector field $\boldsymbol{F}(x, y, z) = x^2y\boldsymbol{i} + xy^2\boldsymbol{j} + xyz\boldsymbol{k}$, S is the surface of the quarter cylinder $C = \{(r, \theta, z)|0 \le r \le 1, 0 \le \theta \le \pi/2, 0 \le z \le 1\}$.

15. Find the flux of the vector field $\boldsymbol{F}(x, y, z) = x^3\boldsymbol{i} + xz^2\boldsymbol{j} + x^2z\boldsymbol{k}$ outward across the sphere $S = \{(x, y, z)|x^2 + y^2 + z^2 = 4\}$. (Use spherical coordinates.)

16. Find the value of the surface integral $\iint_S \boldsymbol{F} \cdot \boldsymbol{N}\, dS$ where $\boldsymbol{F}$ is the vector field $\boldsymbol{F}(x, y, z) = x^2y\boldsymbol{i} + xy^2\boldsymbol{j} + xyz\boldsymbol{k}$, S is the unit sphere $x^2 + y^2 + z^2 = 1$ and $\boldsymbol{N}$ is the outward unit normal. (Use spherical coordinates.)

17. Find the flux of $\boldsymbol{F}(x, y, z) = 6x\boldsymbol{i} - y\boldsymbol{j} + 4\boldsymbol{k}$ across the ellipsoid $x^2 + 4y^2 + z^2 = 4$. (Calculate the volume of the ellipsoid as a volume of revolution.)

18. Verify the Divergence Theorem for the vector field $\boldsymbol{F}(x, y, z) = x\boldsymbol{i} + y\boldsymbol{j} + z\boldsymbol{k}$ and the closed surface S of the cylinder $\{(x, y, z)|x^2 + y^2 \le 1, 0 \le z \le 2\}$.

In Exercises 19–21, verify the stated identities for differentiable vector fields $\boldsymbol{F}(x, y, z)$ and $\boldsymbol{G}(x, y, z)$.

19. $\nabla \times (\boldsymbol{F} + \boldsymbol{G}) = \nabla \times \boldsymbol{F} + \nabla \times \boldsymbol{G}$

20. $\nabla \cdot (\boldsymbol{F} + \boldsymbol{G}) = \nabla \cdot \boldsymbol{F} + \nabla \cdot \boldsymbol{G}$

21. $\nabla \cdot (\boldsymbol{F} \times \boldsymbol{G}) = (\nabla \times \boldsymbol{F}) \cdot \boldsymbol{G} - (\nabla \times \boldsymbol{G}) \cdot \boldsymbol{F}$

The **Laplacian of the scalar field** $\phi = \phi(x, y, z)$ is defined by the equation

$$\nabla^2\phi = \nabla \cdot (\nabla\phi) = \frac{\partial^2\phi}{\partial x^2} + \frac{\partial^2\phi}{\partial y^2} + \frac{\partial^2\phi}{\partial z^2}.$$

The equation $\dfrac{\partial^2\phi}{\partial x^2} + \dfrac{\partial^2\phi}{\partial y^2} + \dfrac{\partial^2\phi}{\partial z^2} = 0$ is called **Laplace's equation.** Functions which satisfy Laplace's equation are called **harmonic functions.**

22. Show that if $\boldsymbol{F} = \nabla\phi$, then ϕ is harmonic if and only if div $\boldsymbol{F} = 0$.

23. Determine which of the following vector fields are gradients of harmonic scalar functions.
    a. $\boldsymbol{F}(x, y, z) = yz\boldsymbol{i} + xz\boldsymbol{j} + xy\boldsymbol{k}$
    b. $\boldsymbol{F}(x, y, z) = 2xye^z\boldsymbol{i} + x^2e^z\boldsymbol{j} + x^2ye^z\boldsymbol{k}$
    c. $\boldsymbol{F}(x, y, z) = x \sin(yz)\boldsymbol{i} + z \cos yz\boldsymbol{j} + y \cos yz\boldsymbol{k}$

24. For the inverse square field $\boldsymbol{F}(\boldsymbol{r}) = \dfrac{\boldsymbol{r}}{|\boldsymbol{r}|^3}$, $|\boldsymbol{r}| \ne 0$, show that div $\boldsymbol{F} = 0$.

25. For the inverse square field of Exercise 24 show that $\iint_S \boldsymbol{F} \cdot \boldsymbol{N}\, dS = 0$ if S is a surface containing a region R, as in the statement of the Divergence Theorem, so that $(0, 0, 0) \notin S \cup R$.

26. Let S be a sphere with center $(0, 0, 0)$. Show that $\iint_S \boldsymbol{F} \cdot \boldsymbol{N}\, dS = 4\pi$ where $\boldsymbol{F}$ is the inverse square field of Exercise 24, and S is the outward unit normal. Why does this not contradict Theorem 6?

27. Show that $\iint_S \boldsymbol{F} \cdot \boldsymbol{N}\, dS = 0$ if the vector field $\boldsymbol{F}$ is incompressible for all $(x, y, z) \in S$. (S, $\boldsymbol{F}$, and $\boldsymbol{N}$ are as in Theorem 6.)

28. Show that if $\iint_S \boldsymbol{F} \cdot \boldsymbol{N}\, dS > 0$, then S must contain at least one source in its interior. (S, $\boldsymbol{F}$, and $\boldsymbol{N}$ are as in Theorem 6.) What if $\iint_S \boldsymbol{F} \cdot \boldsymbol{N}\, dS < 0$?

29. True or false? If $\iint_S \boldsymbol{F} \cdot \boldsymbol{N}\, dS = 0$, S may contain neither sources nor sinks. Explain.

30. Let $\boldsymbol{F}$, S, $\boldsymbol{N}$, and R satisfy the hypotheses of Theorem 6 in a region Ω. If $\iint_S \boldsymbol{F} \cdot \boldsymbol{N}\, dS = 0$ for all closed surfaces S within Ω must $F \equiv 0$ for all $(x, y, z) \in \Omega$? Why? or why not?

SUMMARY OUTLINE OF CHAPTER 21

- A **vector field** on $\mathbb{R}^3$ is a function

$$\boldsymbol{F}(x, y, z) = M(x, y, z)\boldsymbol{i} + N(x, y, z)\boldsymbol{j} + P(x, y, z)\boldsymbol{k}.$$

- The **work** done by the force field $\boldsymbol{F}(x, y, z)$ in moving an object along a curve $C = \{\boldsymbol{r}(t) | a \le t \le b\}$ is

$$W = \int_C \boldsymbol{F} \cdot d\boldsymbol{r} = \int_a^b \boldsymbol{F}(\boldsymbol{r}(t)) \cdot \boldsymbol{r}'(t)\, dt.$$

- The **line integral** of $\boldsymbol{F}(x, y, z)$ over $C = \{\boldsymbol{r}(t) = x(t)\boldsymbol{i} + y(t)\boldsymbol{j} + z(t)\boldsymbol{k} | a \le t \le b\}$

$$\begin{aligned}\int_C \boldsymbol{F} \cdot d\boldsymbol{r} &= \int_a^b \boldsymbol{F}(\boldsymbol{r}(t)) \cdot \boldsymbol{r}'(t)\, dt \\ &= \int_a^b [M(x(t), y(t), z(t))x'(t) \\ &\qquad + N(x(t), y(t), z(t))y'(t) \\ &\qquad + P(x(t), y(t), z(t))z'(t)]\, dt \\ &= \int_C M\, dx + N\, dy + P\, dz.\end{aligned}$$

- The **line integral** of the scalar function $f(x, y, z)$ **with respect to arc length** over the path C (above) is

$$\int_C f(x, y, z)\, ds = \int_a^b f(x(t), y(t), z(t))\sqrt{[x'(t)]^2 + [y'(t)]^2 + [z'(t)]^2}\, dt.$$

- ***Theorem 1:*** Under appropriate hypotheses, if $\boldsymbol{F} = \nabla\phi$ and $C = \{\boldsymbol{r}(t) | a \le t \le b\}$, then

$$\int_C \boldsymbol{F} \cdot d\boldsymbol{r} = \phi(\boldsymbol{r}(b)) - \phi(\boldsymbol{r}(a)).$$

- ***Theorem 2:*** $\int_C \boldsymbol{F} \cdot d\boldsymbol{r}$ is independent of path if and only if $\boldsymbol{F}$ is **conservative** (i.e., $\boldsymbol{F} = \nabla\phi$ for some scalar potential ϕ.)

- ***Green's Theorem:*** Let C be a simple closed path, oriented counterclockwise, that encloses Q. Then

$$\int_C M\, dx + N\, dy = \iint_Q \left(\frac{\partial N}{\partial x} - \frac{\partial M}{\partial y}\right) dA.$$

- The ***surface integral*** of the scalar function $f(x, y, z)$ over the surface $S = \{(x, y, g(x, y)) | (x, y) \in Q\}$ is

$$\iint_S f(x, y, z)\, dS = \iint_Q f(x, y, g(x, y))\sqrt{\left(\frac{\partial g}{\partial x}\right)^2 + \left(\frac{\partial g}{\partial y}\right)^2 + 1}\, dx\, dy.$$

- The **flux** of the vector field $\boldsymbol{F}$ across the surface S in the direction indicated by the unit normal $\boldsymbol{N}$ is

$$\text{Flux} = \iint_S \boldsymbol{F} \cdot \boldsymbol{N}\, dS.$$

- The **curl** of the vector field $\boldsymbol{F} = M\boldsymbol{i} + N\boldsymbol{j} + P\boldsymbol{k}$ is the **vector**

$$\text{curl } \boldsymbol{F} = \nabla \times \boldsymbol{F} = \left(\frac{\partial P}{\partial y} - \frac{\partial N}{\partial z}\right)\boldsymbol{i} + \left(\frac{\partial M}{\partial z} - \frac{\partial P}{\partial x}\right)\boldsymbol{j} + \left(\frac{\partial N}{\partial x} - \frac{\partial M}{\partial y}\right)\boldsymbol{k}.$$

- ***Stokes' Theorem:*** If the surface S is bounded by the curve C, then

$$\int_C \boldsymbol{F} \cdot d\boldsymbol{r} = \iint_S (\text{curl } \boldsymbol{F}) \cdot \boldsymbol{N}\, dS.$$

- The **divergence** of the vector field $F = Mi + Nj + Pk$ is the **scalar**

$$\text{div } F = \frac{\partial M}{\partial x} + \frac{\partial N}{\partial y} + \frac{\partial P}{\partial z}.$$

- ***Divergence Theorem (Gauss):*** If the surface S encloses the region R,

$$\iint_S F \cdot N\, dS = \iiint_R \text{div } F\, dV.$$

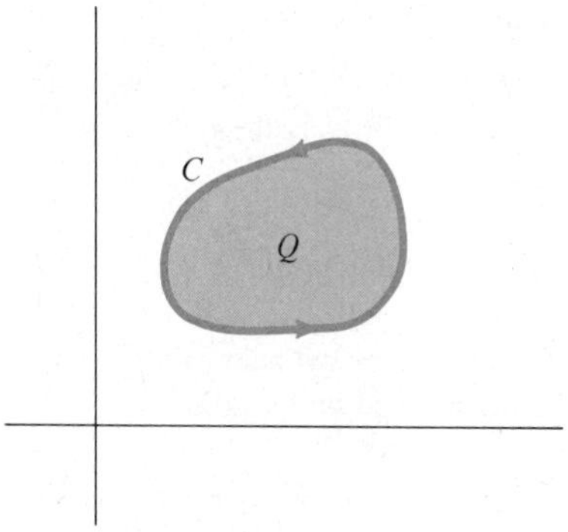

Green's Theorem

$$\int_C M\,dx + N\,dy = \iint_Q \left(\frac{\partial N}{\partial x} - \frac{\partial M}{\partial y}\right) dA$$

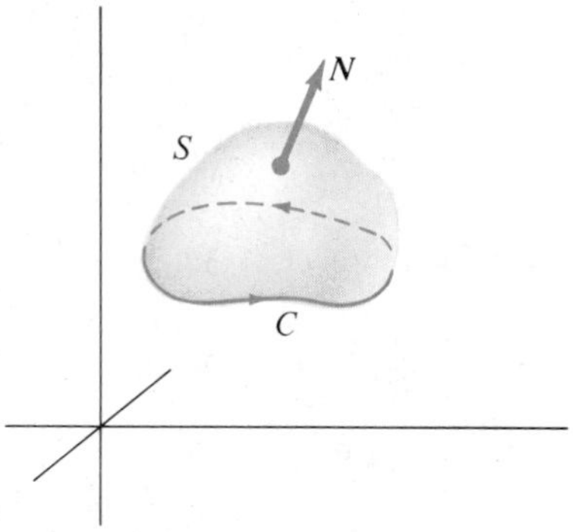

Stokes' Theorem

$$\int_C F \cdot dr = \iint_S (\text{curl } F) \cdot N\, dS$$

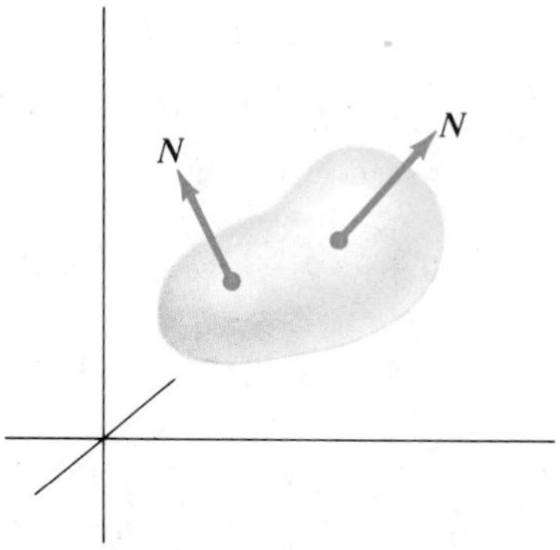

Divergence Theorem

$$\iint_S F \cdot N\, dS = \iiint_R \text{div } F\, dV$$

REVIEW EXERCISES—CHAPTER 21

1. Compute the gradient vector field of the given function.
a. $f(x, y, z) = y(\cos x)\ln z$
b. $f(x, y, z) = x^3yz^2 + \sin xy$

2. Calculate the work done by the force field $F(x, y) = x^2i + 2xyj$ when a particle is moved once around the triangle with vertices $(-1, 2)$, $(1, 0)$, and $(3, 2)$.

3. Show that the force field

$$F(x, y) = \frac{y}{x^2 + y^2}i - \frac{x}{x^2 + y^2}j \text{ is conservative.}$$

4. Show that curl $F = 0$ where F is the force field in Exercise 3.

5. Let C be the unit circle. Show that $\iint_C F \cdot dr \neq 0$ where F is the vector field in Exercise 3. Does this result together with the result of Exercise 4 contradict Stokes' Theorem? Why or why not?

6. For $F(x, y, z) = xyi + (y^2 - x^2)j + x^2z^2k$ find
a. curl F **b.** div F

7. Is the vector field $F(x, y, z) = y^2zi + 2xyzj + xy^2k$ conservative? Why or why not?

8. Show that $\iint_S F \cdot N\, dS = 0$ if S is the unit sphere and F is the vector field $F(r) = r_0$ where r_0 is constant.

9. Find the flux of the vector field $F(x, y, z) = zxi + xyj - x^2zk$ outward over the tetrahedron with vertices $(0, 0, 0)$, $(0, 1, 0)$, $(1, 0, 0)$, and $(0, 0, 1)$.

10. Show that the vector field $F(x, y, z) = xi + y^2j + z^3k$ is irrotational.

11. Let $f(x, y, z)$ be a differentiable scalar function. Show that curl $(\nabla f) = 0$.

12. Let $F(x, y, z)$ be a differentiable vector field. Show that div(curl F) $= 0$.

13. True or false? If curl $F = \mathbf{0}$ then $F = \nabla\phi$ for some differentiable scalar function $\phi(x, y, z)$. Explain.

14. Find the value of the surface integral $\iint_S F \cdot N\, dS$ where $F(x, y, z) = xyi + x^2zj + xzk$, S is the unit sphere and N is the outward unit normal.

15. Let $\boldsymbol{F}(x, y, z) = ax\boldsymbol{i} + by\boldsymbol{j} + cz\boldsymbol{k}$. Find the flux of $\boldsymbol{F}$ over the closed surface S in terms of the volume V of the region enclosed by S.

16. Find $\int_C \boldsymbol{F} \cdot d\boldsymbol{r}$ where $\boldsymbol{F}(x, y) = x\boldsymbol{i} - 3xy\boldsymbol{j}$ and C is the square with vertices (0, 0), (0, 2), (2, 2), (0, 2) oriented counterclockwise.

17. Find $\int_C \boldsymbol{F} \cdot d\boldsymbol{r}$ where $\boldsymbol{F}(x, y) = x^2\boldsymbol{i} - xy\boldsymbol{j}$ and C is the upper unit semicircle from (1, 0) to (−1, 0).

18. Find $\int_{(1,2)}^{(3,2)} \boldsymbol{F} \cdot d\boldsymbol{r}$ where $\boldsymbol{F}(x, y) = 4x^3y^2\boldsymbol{i} + 2x^4y\boldsymbol{j}$.

In Exercises 19–23, evaluate the given line integral.

19. $\int_C xz\,dx + yz\,dy + z\,dz$ where C is parameterized by $\boldsymbol{r}(t) = t\boldsymbol{i} + \boldsymbol{j} + \sin t\boldsymbol{k}$, $0 \le t \le \pi$.

20. $\int_C y\,dx + x\,dy$, C is the triangle with vertices (0, 0), (4, 0), and (2, 4), taken in that order.

21. $\int_C \boldsymbol{F} \cdot d\boldsymbol{r}$, $\boldsymbol{F}(x, y, z) = 2y\boldsymbol{i} - x\boldsymbol{j} + 3z^2\boldsymbol{k}$, $\boldsymbol{r}(t) = t\boldsymbol{i} - 3t\boldsymbol{j} + \boldsymbol{k}$, $0 \le t \le 1$.

22. $\int_C \boldsymbol{F} \cdot d\boldsymbol{r}$, $\boldsymbol{F}(x, y) = e^{2y}\boldsymbol{i} + \cos \pi x\boldsymbol{j}$, C is the triangle with vertices (0, 0), (1, 0), and (0, 1) taken in that order.

23. $\int_C \boldsymbol{F} \cdot d\boldsymbol{r}$, $\boldsymbol{F}(x, y) = xy^2\boldsymbol{i} + x^3y\boldsymbol{j}$, C is the upper unit semicircle taken counterclockwise.

In Exercises 24–28, evaluate the given surface integral.

24. $\iint_S (x^2 + 2)\,dS$, S is the portion of the plane $x + y + z = 6$ lying inside the cylinder $x^2 + y^2 = 9$.

25. $\iint_S (x + y)\,dS$, S is the portion of the plane $x - 2y + z = 8$ lying above the square with vertices (0, 0, 0), (1, 0, 0), (1, 1, 0), (0, 1, 0).

26. $\iint_S (x^2 + y^2 + 1)\,dS$, S is the part of the plane $z = x + 4$ inside the cylinder $x^2 + y^2 = 4$.

27. $\iint_S x^2z\,dS$, S is the portion of the cone $z^2 = x^2 + y^2$ lying between the planes $z = 1$ and $z = 4$.

28. $\iint_S \sqrt{x^2 + y^2}\,dS$, S is the part of the graph of $z = 2xy$ lying inside the cylinder $x^2 + y^2 = 1$.

In Exercises 29–33, find the flux $\iint \boldsymbol{F} \cdot \boldsymbol{N}\,dS$ of the given vector field outward over the given surface.

29. $\boldsymbol{F}(x, y, z) = xy\boldsymbol{i} + y\boldsymbol{j} + 2z\boldsymbol{k}$, S is the graph of the paraboloid $z = x^2 + y^2$ for $1 \le z \le 4$, $\boldsymbol{N}$ is the downward unit normal.

30. $\boldsymbol{F}(x, y, z) = y\boldsymbol{i} + x\boldsymbol{j}$, S is the unit sphere $x^2 + y^2 + z^2 = 1$.

31. $\boldsymbol{F}(x, y, z) = y\boldsymbol{i} + x\boldsymbol{j}$, S is the cylinder $x^2 + y^2 = 1$, $0 \le z \le 1$, including the top and bottom discs.

32. $\boldsymbol{F}(x, y, z) = (x + y^3)\boldsymbol{i} + (z^2 - y)\boldsymbol{j} + (x^3 - y^2)\boldsymbol{k}$, S is the unit sphere $x^2 + y^2 + z^2 = 1$.

33. $\boldsymbol{F}(x, y, z) = ax\boldsymbol{i} + by\boldsymbol{j} + cz\boldsymbol{k}$, S is the graph of $z = 4 - x^2 - y^2$, $z \ge 0$, $\boldsymbol{N}$ is the upward unit normal.

In Exercises 34–37, evaluate the given line integral.

34. $\int_C \boldsymbol{F} \cdot d\boldsymbol{r}$, $\boldsymbol{F}(x, y, z) = (x + y)\boldsymbol{i} + (z - y)\boldsymbol{j} + z\boldsymbol{k}$, C is the circle $x^2 + y^2 = 4$ oriented counterclockwise.

35. $\int_C \boldsymbol{F} \cdot d\boldsymbol{r}$, $\boldsymbol{F}(x, y, z) = 2y^2\boldsymbol{i} - 3z\boldsymbol{j} + 2x\boldsymbol{k}$, C is the triangle with vertices (0, 0, 0), (2, 4, 2), and (0, 4, 0), taken in that order.

36. $\int_C \boldsymbol{F} \cdot d\boldsymbol{r}$, $\boldsymbol{F}(x, y, z) = 3y\boldsymbol{i} - z\boldsymbol{j} + 4x\boldsymbol{k}$, C is the boundary of the part of the unit sphere $x^2 + y^2 + z^2 = 1$ lying in the first quadrant with orientation induced by the outward unit normal to the sphere.

37. $\int_C \boldsymbol{F} \cdot d\boldsymbol{r}$, $\boldsymbol{F}(x, y, z) = (x^3 - y)\boldsymbol{i} + (\cos y - x)\boldsymbol{j} + \sin z\boldsymbol{k}$, C is the triangle with vertices (0, 0, 0), (4, 2, 0), and (2, 4, 2), taken in that order.

38. Let $\boldsymbol{F}(x, y, z) = x^3\boldsymbol{i} + z\boldsymbol{j} + y\boldsymbol{k}$. Find the work done by $\boldsymbol{F}$ on an object that moves from (1, 0, 0) to (−1, 0, π) along a straight line.

39. Find the work done by $\boldsymbol{F}$ in Exercise 38 if the motion is along the helix $\boldsymbol{r}(t) = \cos t\boldsymbol{i} + \sin t\boldsymbol{j} + t\boldsymbol{k}$.

APPENDIX I

BASIC COMPUTER PROGRAMS

Included here are twelve BASIC computer programs that are referred to in various examples and exercises throughout the text. They are presented as ''bare-bones'' prototypes, which those with access to computing facilities (personal computers, programmable calculators, or large computers) can use in designing programs that actually operate on particular machines. Notation appearing in these programs includes the following:

1. $a*b$ means the product ab.
2. $a**b$ means the exponentiation a^b.
3. a/b means the quotient $\frac{a}{b}$.

Proper development of computer software requires full documentation, as well as the inclusion of checks to insure that the user does not attempt to supply inappropriate values to the program. (For example, in asking the user to specify the endpoints of an interval $[a, b]$, one should check to insure that $a < b$.) We have made no attempt to do either, since we wish to highlight only the algorithm involved in the program.

Program 1: Values of the Function $f(x) = \sin x$

```
 10 DEF FNF(T) = SIN(T)
 20 PRINT "enter a,b"
 30 INPUT A,B
 40 PRINT "enter the number of subdivisions"
 50 INPUT N
 60 LET D = (B - A)/N
 70 LET X = A
 80 FOR I = 0 TO N
 90  LET X = A + I*D
100  LET Y = FNF(X)
110  PRINT X,Y
120 NEXT I
130 END
```

Comment: This program prints values of the function $f(x) = \sin x$ for $n + 1$ values of x. The values of x are the endpoints obtained by dividing the interval $[a, b]$ into n equal subintervals. The parameters a, b, and n are supplied by the user.

Program 2: Difference Quotients for $f(x) = \sin x$

```
10 DEF FNF(T) = SIN(T)
20 PRINT "enter x,n"
30 INPUT X,N
40 LET Y1 = FNF(X)
50 FOR M = 1 TO N
60  LET Y2 = FNF(X + 1/M)
70  LET D = M*(Y2 - Y1)
80  PRINT D
90 NEXT M
100 END
```

Comment: This program prints values of the difference quotient

$$D = \frac{\sin\left(x + \dfrac{1}{m}\right) - \sin x}{\dfrac{1}{m}}$$

for $m = 1, 2, 3, \ldots, n$ where the value of x and the integer n are supplied by the user.

Program 3: Difference Quotient Approximations—Chain Rule

```
10 DEF FNF(T) = SQR(T)
20 DEF FNU(T) = 1 + T**2
30 PRINT "enter x"
40 INPUT X
50 FOR N = 1 TO 10
60  LET D1 = 0.5**N
70  LET U1 = FNU(X)
80  LET U2 = FNU(X + D1)
90  LET D2 = (U2 - U1)/D1
100  LET F1 = FNF(U1)
110  LET F2 = FNF(U2)
120  LET D3 = (F2 - F1)/(U2 - U1)
130  LET Z = D2*D3
140  LET W = X/SQR(1 + X**2)
150  PRINT Z,W
160 NEXT N
170 END
```

Comment: This program prints values of the product

$$\frac{f(u(x + \Delta x)) - f(u(x))}{u(x + \Delta x) - u(x)} \cdot \frac{u(x + \Delta x) - u(x)}{\Delta x}$$

for the functions $f(u) = \sqrt{u}$ and $u(x) = 1 + x^2$. The values of Δx are $\dfrac{1}{2^n}$ for $n = 1, 2, \ldots, 10$ and the value of x is supplied by the user. The precise value of the derivative

$$\frac{d}{dx}\sqrt{1 + x^2} = \frac{x}{\sqrt{1 + x^2}}$$

is also printed for comparison.

Program 4: Newton's Method for $f(x) = x^3 - 7$

```
 10 DEF FNF(T) = T**3 - 7
 20 DEF FND(T) = 3*(T**2)
 30 PRINT "how many iterations?"
 40 INPUT N
 50 PRINT "what is your first guess?"
 60 INPUT X
 70 FOR I = 1 TO N
 80  LET Z1 = FNF(X)
 90  LET Z2 = FND(X)
100  LET W = X - (Z1/Z2)
110  PRINT I,W
120  LET X = W
130 NEXT I
140 END
```

Comment: This program implements Newton's Method to locate a zero of the function $f(x) = x^3 - 7$. The value of the function is computed in line 10, the derivative $f'(x) = 3x^2$ is computed in line 20, and the formula for Newton's Method is implemented in line 100. The user supplies the number of iterations and a first guess at the root.

Program 5: Lower Riemann Sums for $f(x) = 3x + 2$

```
 10 DEF FNF(T) = 3*T + 2
 20 PRINT "enter interval endpoints a,b"
 30 INPUT A,B
 40 PRINT "how many subintervals?"
 50 INPUT N
 60 LET D = (B - A)/N
 70 LET S = 0
 80 FOR I = 1 TO N
 90  LET X = A + (I - 1)*D
100  LET S = S + FNF(X)*D
110 NEXT I
120 PRINT S
130 END
```

Comment: This program computes a lower Riemann sum for the function $f(x) = 3x + 2$ on the interval $[a, b]$ using n equal subintervals. The numbers a, b, and n

are supplied by the user. The program computes the function value at left endpoints of the resulting subintervals, since $f(x)$ is an increasing function.

Program 6: Upper Riemann Sums for $f(x) = 3x + 2$

```
10 DEF FNF(T)  = 3*T + 2
20 PRINT "enter interval endpoints a,b"
30 INPUT A,B
40 PRINT "how many subintervals?"
50 INPUT N
60 LET D = (B - A)/N
70 LET S = 0
80 FOR I = 1 TO N
90  LET X = A + I*D
100  LET S = S + FNF(X)*D
110 NEXT I
120 PRINT S
130 END
```

Comment: This program computes an upper Riemann sum for the function $f(x) = 3x + 2$ on the interval $[a, b]$ using n equal subintervals. The numbers a, b, and n are supplied by the user. The program computes the function values at right endpoints of the resulting subintervals, since $f(x)$ is an increasing function.

Program 7: The Midpoint Rule applied to $f(x) = x^3 + 1$

```
10 DEF FNF(T)  = T**3 + 1
20 PRINT "enter interval endpoints a,b"
30 INPUT A,B
40 PRINT "how many subintervals?"
50 INPUT N
60 LET D = (B - A)/N
70 LET S = 0
80 LET E = A - D/2
90 FOR I = 1 TO N
100  LET X = E + I*D
110  LET S = S + FNF(X)*D
120 NEXT I
130 PRINT S
140 END
```

Comment: This program implements the midpoint rule for approximate integration for the function $f(x) = x^3 + 1$ on the interval $[a, b]$ using n subintervals of equal length. The user supplies the values of a, b, and n.

Program 8: Trapezoidal Rule applied to $f(x) = x^2$

```
10 DEF FNF(T)  = T**2
20 PRINT "enter interval endpoints a,b"
30 INPUT A,B
```

```
 40 PRINT "how many subintervals?"
 50 INPUT N
 60 LET D = (B - A)/N
 70 LET S = FNF(A)
 80 FOR I = 1 TO (N - 1)
 90  LET X = A + I*D
100  LET S = S + 2*FNF(X)
110 NEXT I
120 LET S = S + FNF(B)
130 LET S = S*(D/2)
140 PRINT S
150 END
```

Comment: This program implements the Trapezoidal Rule for approximate integration for the function $f(x) = x^2$ on the interval $[a, b]$. The numbers a, b, and n (the number of subintervals) are supplied by the user. Values of the function are computed on line 300 in the subroutine at the end of the program.

Program 9: Simpson's Rule applied to the function $f(x) = x^2$

```
 10 DEF FNF(T) = T**2
 20 PRINT "enter interval endpoints a,b"
 30 INPUT A,B
 40 PRINT "how many subintervals (an even number)?"
 50 INPUT N
 60 LET C = (B - A)/N
 70 LET D = C/3
 80 LET S = FNF(A)
 90 FOR I = 1 TO (N - 1) STEP 2
100  LET S = S + 4*FNF(A + I*C)
110 NEXT I
120 FOR I = 2 TO (N - 2) STEP 2
130  LET S = S + 2*FNF(A + I*C)
140 NEXT I
150 LET S = S + FNF(B)
160 LET S = S*D
170 PRINT S
180 END
```

Comment: This program implements Simpson's Rule for approximate integration for the function $f(x) = x^2$ on the interval $[a, b]$. The numbers a, b, and n (the number of subintervals, *an even number*) are supplied by the user.

Program 10: Partial Sums of the Geometric Series

```
 10 PRINT "enter p,a,x,n"
 20 INPUT P,A,X,N
 30 LET S = 0
 40 FOR K = P TO N
 50  LET T = A*(X**K)
```

```
60   LET S = S + T
70 NEXT K
80 PRINT S
90 END
```

Comment: This program computes the partial sum

$$\sum_{k=p}^{n} ax^k$$

where the constants p, a, x, and n are supplied by the user.

Program 11: A Riemann Sum in Polar Coordinates for $f(\theta) = 1 + \sqrt{\sin\theta}$

```
10 DEF FNF(T) = 1 + SQR(SIN(T))
20 DEF FNG(T) = 0.5 * (T**2)
30 PRINT "enter interval endpoints a,b"
40 INPUT A,B
50 PRINT "how many subintervals?"
60 INPUT N
70 LET D = (B - A)/N
80 LET S = 0
90 FOR I = 1 TO N
100  LET Y = FNF(A + I*D)
110  LET S = S + FNG(Y)*D
120 NEXT I
130 PRINT S
140 END
```

Comment: This program computes a Riemann sum, using n subintervals of equal length, for the integral

$$\int_a^b \frac{1}{2}(1 + \sqrt{\sin\theta})^2\, d\theta$$

which approximates the area bounded by $f(\theta) = 1 + \sqrt{\sin\theta}$. The parameters a, b, and n are supplied by the user.

Program 12: Riemann Sum for a Double Integral

```
10 PRINT "enter a,b,c,d"
20 INPUT A,B,C,D
30 PRINT "how many subintervals?"
40 INPUT N
50 LET D1 = (B - A)/N
60 LET D2 = (D - C)/N
70 LET D3 = D1*D2
80 LET S = 0
90 FOR J = 1 TO N
100  FOR K = 1 TO N
```

```
110    LET S = S + (8 - 2*(A + J*D1) - 4*(C + K*D2))
120   NEXT K
130 NEXT J
140 LET S = S*D3
150 PRINT S
160 END
```

Comment: This program computes a Riemann sum for the double integral

$$\int_c^d \int_a^b (8 - 2x - 4y)\, dx\, dy$$

using a grid consisting of n^2 rectangles of dimension $(b - a)(d - c)$. The parameters a, b, c, d, and n are supplied by the user.

APPENDIX II

SOME ADDITIONAL PROOFS

Mathematical Induction

The *principle of mathematical induction* is a property of the real number system (we will think of it as an axiom) that allows us to prove statements about positive integers. Represented generally as $P(n)$, such statements are usually formulas involving some or all of the integers $1, 2, 3, \ldots, n$. Two particular formulas that we wish to prove here are equations (5) and (10) of Section 6.2:

$$1 + 2 + 3 + \cdots + n = \frac{n(n+1)}{2} \tag{1a}$$

and

$$1^2 + 2^2 + 3^2 + \cdots + n^2 = \frac{n(n+1)(2n+1)}{6}. \tag{1b}$$

The axiom that we shall use to prove these formulas is the following.

PRINCIPLE OF MATHEMATICAL INDUCTION (PMI)

If $P(n)$ represents a mathematical statement involving the positive integer n, then $P(n)$ is true for all positive integers n if both of the following conditions hold:

(i) $P(1)$ is true.
(ii) Whenever $P(n)$ is true, $P(n+1)$ is also true.

The PMI is not surprising. If a statement $P(1)$ is true, then setting $n = 1$ shows that $P(2) = P(1+1)$ is true by statement (ii). Since $P(2)$ is then known to be true, we next set $n = 2$ and conclude, again by (ii), that $P(3) = P(2+1)$ is true. Continuing in this way we verify that $P(n)$ is true for each of the positive integers $1, 2, 3, \ldots$.

Proof of Formula (1a): We prove formula (1a) by applying the PMI to the formula

$$P(n)\text{: } 1 + 2 + 3 + \cdots + n = \frac{n(n+1)}{2}. \tag{2}$$

First, setting $n = 1$ gives the statement

$$P(1)\text{: } 1 = \frac{1(1+1)}{2},$$

which is true, since $\dfrac{1(1+1)}{2} = \dfrac{1 \cdot 2}{2} = 1$. Thus, statement (i) of the PMI holds.

To show that statement (ii) of the PMI holds, we assume that $P(n)$ in (2) is true, and we attempt to show that

$$P(n+1)\colon 1 + 2 + 3 + \cdots + (n+1) = \frac{(n+1)[(n+1)+1]}{2} \tag{3}$$

is true. To do this we use statement (2) to write

$$\begin{aligned} 1 + 2 + 3 + \cdots + n + (n+1) &= \{1 + 2 + 3 + \cdots + n\} + n + 1 \\ &= \frac{n(n+1)}{2} + n + 1 \\ &= \frac{n(n+1) + 2(n+1)}{2} \\ &= \frac{(n+1)(n+2)}{2} \\ &= \frac{(n+1)[(n+1)+1]}{2}, \end{aligned}$$

which establishes (3). Thus, by the PMI, statement (2) holds for all positive integers. This proves formula (1a). ■

Proof of Formula (1b): This is done just as the proof of formula (1a), using the PMI. The statement to be proved is

$$P(n)\colon 1^2 + 2^2 + 3^2 + \cdots + n^2 = \frac{n(n+1)(2n+1)}{6} \tag{4}$$

The statement corresponding to $n = 1$ is

$$P(1)\colon 1^2 = \frac{1(1+1)(2 \cdot 1 + 1)}{6}$$

which is true, so (i) of the PMI holds.

Next, we assume that $P(n)$ in (4) holds, and we attempt to prove the statement

$$P(n+1)\colon \tag{5}$$

$$1^2 + 2^2 + 3^2 + \cdots + (n+1)^2 = \frac{(n+1)[(n+1)+1][2(n+1)+1]}{6}.$$

We do this using statement (4):

$$\begin{aligned} 1^2 + 2^2 + 3^2 + \cdots + (n+1)^2 &= \{1^2 + 2^2 + 3^2 + \cdots + n^2\} + (n+1)^2 \\ &= \frac{n(n+1)(2n+1)}{6} + (n+1)^2 \\ &= \frac{n(n+1)(2n+1) + 6(n+1)^2}{6} \\ &= \frac{(n+1)[n(2n+1) + 6(n+1)]}{6} \end{aligned}$$

$$= \frac{(n+1)[2n^2+7n+6]}{6}$$

$$= \frac{(n+1)[(n+2)(2n+3)]}{6}$$

$$= \frac{(n+1)[(n+1)+1][2(n+1)+1]}{6}.$$

Thus, $P(n+1)$ in (5) is true whenever $P(n)$ is true. Thus, statement (ii) of the PMI holds, so formula (1b) is proved by the PMI. ■

THEOREM 4, CHAPTER 2

Let n be a positive integer.

(i) If n is even, $\lim_{x \to a} \sqrt[n]{x} = \sqrt[n]{a}$ whenever $0 < a < \infty$.

(ii) If n is odd, $\lim_{x \to a} \sqrt[n]{x} = \sqrt[n]{a}$ for all $-\infty < a < \infty$.

Proof: We first prove case (i). According to Definition 7, Chapter 2, we must show that if $\epsilon > 0$ is given, there can be found a corresponding number δ so that

$$|\sqrt[n]{x} - \sqrt[n]{a}| < \epsilon \qquad \text{whenever} \qquad 0 < |x - a| < \delta. \tag{6}$$

The first of these two inequalities holds the key to determining how to choose δ, so we begin by examining the inequality

$$|\sqrt[n]{x} - \sqrt[n]{a}| < \epsilon \tag{7}$$

that we can rewrite as

$$-\epsilon < \sqrt[n]{x} - \sqrt[n]{a} < \epsilon.$$

The strategy is now to isolate the expression $x - a$ in the middle position, which we do as follows: From the above we have

$$\sqrt[n]{a} - \epsilon < \sqrt[n]{x} < \sqrt[n]{a} + \epsilon, \tag{8}$$

so

$$(\sqrt[n]{a} - \epsilon)^n < x < (\sqrt[n]{a} + \epsilon)^n, \tag{9}$$

so

$$(\sqrt[n]{a} - \epsilon)^n - a < x - a < (\sqrt[n]{a} + \epsilon)^n - a.$$

(There is a potential difficulty here. Going from step (8) to step (9) is valid only if $\sqrt[n]{a} - \epsilon$ is nonnegative. We can insure that this is true by requiring that $\epsilon < \sqrt[n]{a}$ to begin with. There is no problem in making this assumption, since if statement (6) holds for a smaller value of ϵ than that originally given it also must hold for the given value.)

Finally, we rewrite (9) as

$$-[a - (\sqrt[n]{a} - \epsilon)^n] < x - a < [(\sqrt[n]{a} + \epsilon)^n - a] \tag{10}$$

in which both expressions inside brackets are positive. Then, subject to the remark that $\epsilon < \sqrt[n]{a}$, inequality (10) is equivalent to inequality (7). Now suppose we take $\delta > 0$ to be so small that both

$$\delta < [a - (\sqrt[n]{a} - \epsilon)^n] \qquad \text{and} \qquad \delta < [(\sqrt[n]{a} + \epsilon)^n - a]. \tag{11}$$

Then the inequality

$$0 < |x - a| < \delta \tag{12}$$

can be rewritten as

$$-\delta < x - a < \delta. \tag{13}$$

From (11) it follows that if (13) holds, so does (10). Since (13) is equivalent to (12), and (10) is equivalent to (7), this shows that

$$|\sqrt[n]{x} - \sqrt[n]{a}| < \epsilon \qquad \text{whenever} \qquad 0 < |x - a| < \delta,$$

which proves statement (i).

To prove (ii), we assume that n is odd. If $a > 0$ the proof for part (i) applies. Thus, we must only concern ourselves with the case $a < 0$. But in this case $-a > 0$, so we may apply the proof of statement (i) to conclude that

$$\lim_{x \to a} \sqrt[n]{-x} = \sqrt[n]{-a}. \tag{14}$$

Of course, we now wish to simply "factor out" the -1's in (14), but we must do it formally. By Definition 7, Chapter 2, equation (14) means that, given $\epsilon > 0$, there exists a $\delta > 0$ so that

$$|\sqrt[n]{-x} - \sqrt[n]{-a}| < \epsilon \qquad \text{whenever} \qquad 0 < |-x - (-a)| < \delta. \tag{15}$$

Since, for odd n,

$$\begin{aligned} |\sqrt[n]{-x} - \sqrt[n]{-a}| &= |(-1)^n\sqrt[n]{x} - (-1)^n\sqrt[n]{a}| = |(-1)(\sqrt[n]{x} - \sqrt[n]{a})| \\ &= |\sqrt[n]{x} - \sqrt[n]{a}| \end{aligned}$$

and

$$|-x - (-a)| = |(-1)(x - a)| = |x - a|,$$

inequality (15) is equivalent to the statement

$$|\sqrt[n]{x} - \sqrt[n]{a}| < \epsilon \qquad \text{whenever} \qquad 0 < |x - a| < \delta,$$

which shows that $\lim_{x \to a} \sqrt[n]{x} = \sqrt[n]{a}$ in the present case. This completes the proof. ■

THEOREM 1, CHAPTER 2, (parts (ii) and (iii))

If $\lim_{x \to a} f(x) = L$ and $\lim_{x \to a} g(x) = M$, then

(ii) $\lim_{x \to a} f(x)g(x) = LM$, and

(iii) $\lim_{x \to a} \dfrac{f(x)}{g(x)} = \dfrac{L}{M}$, $M \neq 0$.

Proof: To prove (ii), we must show that, given $\epsilon > 0$, there exists a $\delta > 0$ so that

$$|f(x)g(x) - LM| < \epsilon \qquad \text{whenever} \qquad 0 \le |x - a| \le \delta.$$

We do this by using the triangle inequality to write

$$\begin{aligned} |f(x)g(x) - LM| &= |f(x)g(x) - f(x)M + f(x)M - LM| \\ &\le |f(x)g(x) - f(x)M| + |f(x)M - LM| \\ &= |f(x)||g(x) - M| + |M||f(x) - L|. \end{aligned} \tag{16}$$

We need to make three observations about the terms on the right side of inequality (16).

(a) Since $\lim_{x \to a} f(x) = L$, there exists a number $\delta_1 > 0$ so that

$$|f(x) - L| < 1 \qquad \text{whenever} \qquad 0 < |x - a| < \delta_1.$$

Thus,

$$|f(x)| < |L| + 1 \qquad \text{whenever} \qquad 0 < |x - a| < \delta_1. \tag{17}$$

(b) Also, since $\lim_{x \to a} f(x) = L$ there exists a number $\delta_2 > 0$ so that

$$|f(x) - L| < \frac{\epsilon}{2(|M| + 1)} \qquad \text{whenever} \qquad 0 < |x - a| < \delta_2. \tag{18}$$

(c) Finally, since $\lim_{x \to a} g(x) = M$, there exists a number δ_3 so that

$$|g(x) - M| < \frac{\epsilon}{2(|L| + 1)} \qquad \text{whenever} \qquad 0 < |x - a| < \delta_3. \tag{19}$$

Now, we take δ to be the smallest of the numbers δ_1, δ_2, and δ_3. Then each of the conditions in statements (17), (18), and (19) is fulfilled whenever $0 < |x - a| < \delta$. Combining inequality (16) with (17) through (19) then shows that

$$\begin{aligned} |f(x)g(x) - LM| &\le |f(x)| \cdot |g(x) - M| + |M| \cdot |f(x) - L| \\ &< (|L| + 1)\left[\frac{\epsilon}{2(|L| + 1)}\right] + |M|\left[\frac{\epsilon}{2(|M| + 1)}\right] \\ &< \frac{\epsilon}{2} + \frac{\epsilon}{2} \\ &= \epsilon \end{aligned}$$

whenever $0 < |x - a| < \delta$. This proves part (ii).

To prove (iii), we first show that

$$\lim_{x \to a} \frac{1}{g(x)} = \frac{1}{M}, \qquad M \neq 0. \tag{20}$$

Statement (iii) then follows from part (ii) and statement (20). To prove (20), we must show that, given $\epsilon > 0$, there exists a $\delta > 0$ so that

$$\left|\frac{1}{g(x)} - \frac{1}{M}\right| < \epsilon \qquad \text{whenever} \qquad 0 < |x - a| < \delta.$$

Since

$$\left|\frac{1}{g(x)} - \frac{1}{M}\right| = \left|\frac{M - g(x)}{Mg(x)}\right| = \left(\frac{1}{|M||g(x)|}\right)|M - g(x)|, \tag{21}$$

we note first that since $\lim_{x \to a} g(x) = M$, there exists a number δ_1 so that

$$|g(x) - M| < \frac{|M|}{2} \qquad \text{whenever} \qquad 0 < |x - a| < \delta_1. \tag{22}$$

From (22) it follows that $|g(x)| > \dfrac{|M|}{2}$, or

$$\frac{1}{|g(x)|} < \frac{2}{|M|} \qquad \text{whenever} \qquad 0 < |x - a| < \delta_1. \tag{23}$$

Next, again since $\lim_{x \to a} g(x) = M$, there exists a number δ_2 so that

$$|g(x) - M| < \frac{|M|^2 \epsilon}{2} \qquad \text{whenever} \qquad 0 < |x - a| < \delta_2. \tag{24}$$

Finally, we take δ to be the smaller of δ_1 and δ_2. Then from statements (21), (23), and (24) it follows that

$$\begin{aligned} \left| \frac{1}{g(x)} - \frac{1}{M} \right| &= \left(\frac{1}{|M||g(x)|} \right) |M - g(x)| \\ &< \left(\frac{1}{|M|} \right) \left(\frac{2}{|M|} \right) \left(\frac{|M|^2 \epsilon}{2} \right) \\ &= \epsilon \end{aligned}$$

whenever $0 < |x - a| < \delta$. This proves statement (iii). ■

Proof of Theorem 9, Chapter 3 (Chain Rule): Let x_0 be in the domain of $u(x)$. To prove that

$$[f(u(x_0))]' = f'(u(x_0))u'(x_0),$$

we begin by defining a new function $F(t)$ for t in the domain of f:

$$F(t) = \begin{cases} \dfrac{f(t) - f(u(x_0))}{t - u(x_0)} & \text{if} \quad t \neq u(x_0) \\ f'(u(x_0)) & \text{if} \quad t = u(x_0) \end{cases}$$

The function $F(t)$ is continuous at $t_0 = u(x_0)$, since

$$\lim_{t \to u(x_0)} F(t) = \lim_{t \to u(x_0)} \frac{f(t) - f(u(x_0))}{t - u(x_0)} = f'(u(x_0)) = F(u(x_0)).$$

The function $F(t)$ is also continuous at numbers $t = u(x) \neq u(x_0)$ since we have, for $s \neq u(x_0)$,

$$\lim_{s \to t} F(s) = \lim_{s \to t} \frac{f(s) - f(u(x_0))}{s - u(x_0)} = \frac{f(t) - f(u(x_0))}{t - u(x_0)} = F(t).$$

We wish now to work with the composite function $F(u(x))$. Since F is continuous and u is differentiable (hence, continuous), the composite function $F(u(x))$ is continuous on the domain of u. Thus, according to the definition of F,

$$\lim_{x \to x_0} F(u(x)) = F(u(x_0)) = f'(u(x_0)). \tag{25}$$

Next, we note that, if $x \neq x_0$, we can write

$$\frac{f(u(x)) - f(u(x_0))}{x - x_0} = F(u(x)) \left[\frac{u(x) - u(x_0)}{x - x_0} \right]. \tag{26}$$

To see this, note that if $t = u(x) \neq u(x_0)$, then

$$F(u(x)) = \frac{f(u(x)) - f(u(x_0))}{u(x) - u(x_0)}$$

$$F(u(x))[u(x) - u(x_0)] = f(u(x)) - f(u(x_0))$$

$$F(u(x))\left[\frac{u(x) - u(x_0)}{x - x_0}\right] = \frac{f(u(x)) - f(u(x_0))}{x - x_0}$$

while if $u(x) = u(x_0)$ for $x \neq x_0$, both sides of (26) are zero. Using equations (25) and (26) together with the continuity of F and u, we may now conclude that

$$\begin{aligned}[f(u(x_0))]' &= \lim_{x \to x_0} \frac{f(u(x)) - f(u(x_0))}{x - x_0} \\ &= \lim_{x \to x_0} F(u(x))\left[\frac{u(x) - u(x_0)}{x - x_0}\right] \\ &= f'(u(x_0))u'(x_0).\end{aligned}$$

■

APPENDIX III

COMPLEX NUMBERS

In the set of real numbers, the quadratic equation

$$x^2 = -1 \tag{1}$$

has no solution, since the square of a real number is never negative. The *complex numbers* constitute a number system containing the real number system (just as the real numbers contain the integers) and in this system equations such as equation (1) have solutions.

To form the complex number system, we first define the complex number i by the equation

$$i^2 = -1. \tag{2}$$

Another way to write (2) is to say that $i = \sqrt{-1}$. Thus, the complex number i is a solution to equation (1), as is the number $-i = -\sqrt{-1}$. However, i and $-i$ are not the only complex numbers. The following definition fully determines the complex number system.

DEFINITION 1

The **complex numbers** are all numbers of the form

$$z = a + bi,$$

where a and b are real numbers, together with the following operations:

(i) Scalar multiplication: $\lambda(a + bi) = \lambda a + (\lambda b)i, \qquad \lambda \in \mathbb{R}$
(ii) Addition: $(a_1 + b_1 i) + (a_2 + b_2 i) = (a_1 + a_2) + (b_1 + b_2)i$
(iii) Multiplication: $(a_1 + b_1 i) \cdot (a_2 + b_2 i) = (a_1 a_2 - b_1 b_2) + (a_1 b_2 + a_2 b_1)i$

Notice the following about Definition 1:

(a) When $b = 0$, the complex number $z = a + 0i = a$ is a real number. Moreover, the definitions of addition and multiplication agree with the definitions of addition and multiplication of real numbers when $b_1 = b_2 = 0$. Thus, the set of complex numbers of the form $\{a + bi \mid a \in \mathbb{R}, b = 0\}$ is just the set $\mathbb{R}$ of real numbers.

(b) The definition of multiplication is obtained by assuming that the usual distributive and commutative laws hold for all expressions involving a_1, a_2, b_1, b_2, and i:

$$\begin{aligned}(a_1 + b_1 i) \cdot (a_2 + b_2 i) &= a_1(a_2 + b_2 i) + (b_1 i)(a_2 + b_2 i)\\ &= a_1 a_2 + a_1 b_2 i + a_2 b_1 i + b_1 b_2 i^2\\ &= a_1 a_2 - b_1 b_2 + (a_1 b_2 + a_2 b_1)i.\end{aligned}$$

For $z = a + bi$, we say that z is **pure imaginary** if $a = 0$.

Example 1 If $z_1 = 3 + 2i$ and $z_2 = 4 - i$, then

(i) $4z_1 = 4(3 + 2i) = 4 \cdot 3 + (4 \cdot 2)i = 12 + 8i.$

(ii) $z_1 + z_2 = (3 + 2i) + (4 - i) = (3 + 4) + (2 + (-1))i = 7 + i.$

(iii) $$\begin{aligned}z_1 - z_2 = z_1 + (-z_2) &= (3 + 2i) + (-1)(4 - i)\\ &= (3 + 2i) + (-4 + i)\\ &= (3 + (-4)) + (2 + 1)i\\ &= -1 + 3i.\end{aligned}$$

(iv) $$\begin{aligned}z_1 z_2 = (3 + 2i)(4 - i) &= 3 \cdot 4 - (2)(-1) + [3(-1) + 2 \cdot 4]i\\ &= 14 + 5i.\end{aligned}$$ ■

Geometric and Vector Interpretations

Since the complex number $z = a + bi$ is determined by a pair of real numbers, we refer to a as the **real part** (or real component) of z, and we refer to b as the **imaginary part** (imaginary component) of z. (Remember that b is a real number.) Plotting the real part of z on the x-axis (called the real axis) and the imaginary part of z on the y-axis (called the imaginary axis), we can plot the complex number $z = a + bi$ as the ordered pair (a, b) in the xy-plane (see Figure III.1). Note that the real numbers (considered as a subset of the complex numbers) correspond to points on the real axis, while pure imaginary numbers correspond to points on the imaginary axis.

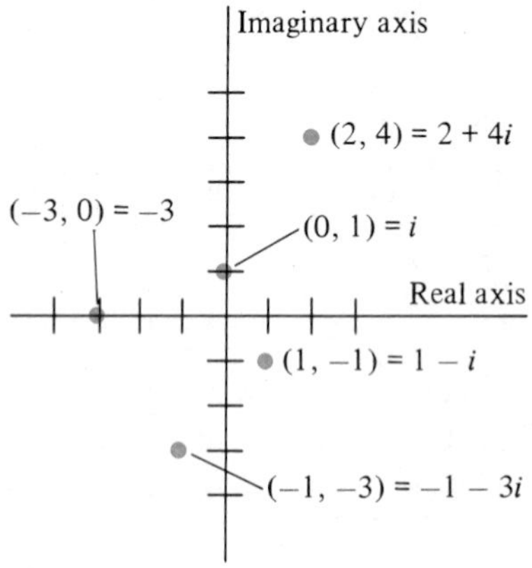

Figure III.1 Complex numbers $a + bi$ may be represented as points (a,b) in the complex plane.

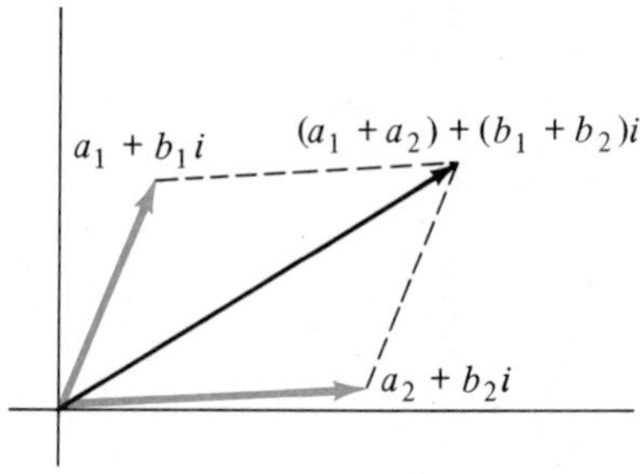

Figure III.2 Addition of complex numbers corresponds to vector addition.

Since points (a, b) in the plane may be regarded as position vectors $\langle a, b\rangle$ originating at the origin (0, 0), we may interpret the complex number $a + bi$ as the position vector $\langle a, b\rangle$. As Figure III.2 illustrates, the definition of addition for complex numbers agrees with the definition of vector addition. You can verify that the corresponding definitions of scalar multiplication agree as well. The vector interpretation leads directly to the concept of the *modulus* (also called the length, or absolute value) of a complex number.

DEFINITION 2

The **modulus of the complex number** $z = a + bi$ is the real number $|z| = \sqrt{a^2 + b^2}$.

Note that the modulus of a complex number is just its absolute value when the number is real. A concept related to the modulus is the *conjugate* of a complex number.

DEFINITION 3

The **conjugate of the complex number** $z = a + bi$ is the complex number $\bar{z} = a - bi$.

The relationship between the modulus and the conjugate is that

$$z\bar{z} = (a + bi)(a - bi) = a^2 + b^2 = |z|^2.$$

Figure III.3 shows that the conjugate $\bar{z}$ is just the reflection in the real axis of the complex number z.

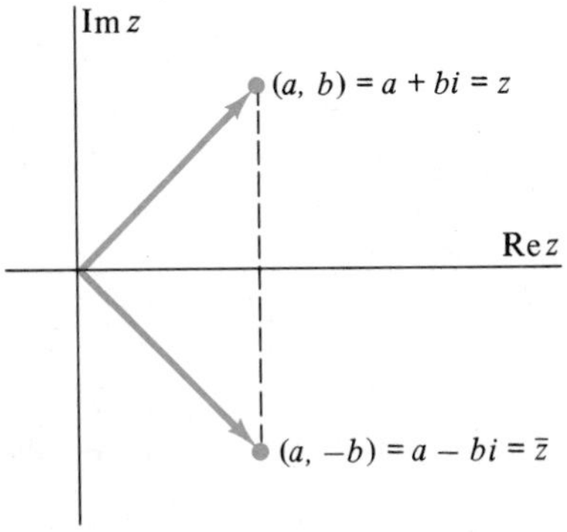

Figure III.3 The conjugate $\bar{z}$ of z is the reflection of z in the real axis.

Division for Complex Numbers

The concepts of modulus and conjugate allow us to define the operation of division. Since

$$z\bar{z} = |z|^2$$

it follows that

$$\left(\frac{1}{|z|^2}\right)(z\bar{z}) = 1,$$

so the reciprocal $\dfrac{1}{z}$, $z \neq 0$, is defined as

$$\frac{1}{z} = \frac{\bar{z}}{|z|^2}, \qquad z \neq 0.$$

Division is then defined as multiplication by a reciprocal.

DEFINITION 4

Let z_1 and z_2 be complex numbers with $z_2 \neq 0$. The quotient $\dfrac{z_1}{z_2}$ is defined to be

$$\frac{z_1}{z_2} = z_1\left(\frac{1}{z_2}\right) = \frac{z_1\bar{z}_2}{|z_2|^2}.$$

Again, this definition agrees with the definition of division in $\mathbb{R}$ when z_1 and z_2 are real.

Example 2 Let $z_1 = 1 + 3i$ and $z_2 = 2 - 4i$. Then

(i) $\bar{z}_1 = 1 - 3i$.
(ii) $z_1\bar{z}_1 = (1 + 3i)(1 - 3i) = (1 + 9) + (3 - 3)i = 10 = |z_1|^2$.
(iii) $\dfrac{1}{z_1} = \dfrac{\bar{z}_1}{|z_1|^2} = \dfrac{1}{10}(1 - 3i) = \dfrac{1}{10} - \dfrac{3}{10}i.$
(iv)
$$\begin{aligned}\frac{z_1}{z_2} = z_1\left(\frac{1}{z_2}\right) = \frac{z_1\bar{z}_2}{|z_2|^2} &= \frac{1}{20}(1 + 3i)(2 + 4i) \\ &= \frac{1}{20}[(2 - 12) + (6 + 4)i] \\ &= -\frac{1}{2} + \frac{1}{2}i.\end{aligned}$$ ■

Solutions of Quadratic Equations

We conclude by recalling that the solutions of the quadratic equation

$$ax^2 + bx + c = 0$$

are given by the quadratic formula as

$$x = -\frac{b}{2a} \pm \frac{\sqrt{b^2 - 4ac}}{2a}$$

when $b^2 - 4ac \geq 0$. Using complex numbers we may now write the roots as

$$x = -\frac{b}{2a} \pm \left(\frac{\sqrt{4ac - b^2}}{2a}\right)i$$

when $b^2 - 4ac < 0$.

Example 3 For the quadratic equation

$$x^2 - 2x + 3 = 0$$

the solutions are

$$\begin{aligned}x &= -\frac{(-2)}{2} \pm \frac{\sqrt{(-2)^2 - 4(1)(3)}}{2} \\ &= 1 \pm \frac{\sqrt{-8}}{2} \\ &= 1 \pm \sqrt{2}\, i.\end{aligned}$$ ■

APPENDIX IV

TABLES OF TRANSCENDENTAL FUNCTIONS

Trigonometric functions (x in radians)

x	$\sin x$	$\cos x$	$\tan x$
0.0	.00000	1.00000	0.00000
0.1	.09983	.99500	.10033
0.2	.19867	.98007	.20271
0.3	.29552	.95534	.30934
0.4	.38942	.92106	.42279
0.5	.47943	.87758	.54630
0.6	.56464	.82534	.68414
0.7	.64422	.76484	.84229
0.8	.71736	.69671	1.02964
0.9	.78333	.62161	1.26016
1.0	.84147	.54030	1.55741
1.1	.89121	.45360	1.96476
1.2	.93204	.36236	2.57215
1.3	.96356	.26750	3.60210
1.4	.98545	.16997	5.79788
1.5	.99749	.07074	14.10142
$\pi/2$	1.00000	.00000	∞
1.6	.99957	−.02920	−34.23254
1.7	.99166	−.12884	−7.69660
1.8	.97385	−.22720	−4.28626
1.9	.94630	−.32329	−2.92710
2.0	.90930	−.41615	−2.18504
2.1	.86321	−.50485	−1.70985
2.2	.80850	−.58850	−1.37382
2.3	.74571	−.66628	−1.11921
2.4	.67546	−.73739	−.91601
2.5	.59847	−.80114	−.74702
2.6	.51550	−.85689	−.60160
2.7	.42738	−.90407	−.47273
2.8	.33499	−.94222	−.35553
2.9	.23925	−.97096	−.24641
3.0	.14112	−.98999	−.14255
3.1	.04158	−.99914	−.04162
π	.00000	−1.00000	.00000
3.2	−.05837	−.99829	.05847
3.3	−.15775	−.98748	.15975

x	$\sin x$	$\cos x$	$\tan x$
3.4	−.25554	−.96680	.26432
3.5	−.35078	−.93646	.37459
3.6	−.44252	−.89676	.49347
3.7	−.52984	−.84810	.62473
3.8	−.61186	−.79097	.77356
3.9	−.68777	−.72593	.94742
4.0	−.75680	−.65364	1.15782
4.1	−.81828	−.57482	1.42353
4.2	−.87158	−.49026	1.77778
4.3	−.91617	−.40080	2.28585
4.4	−.95160	−.30733	3.09632
4.5	−.97753	−.21080	4.63733
4.6	−.99369	−.11215	8.86017
4.7	−.99992	−.01239	80.71271
$3\pi/2$	−1.00000	.00000	$-\infty$
4.8	−.99616	.08750	−11.38487
4.9	−.98245	.18651	−5.26749
5.0	−.95892	.28366	−3.38051
5.1	−.92581	.37798	−2.44939
5.2	−.88345	.46852	−1.88564
5.3	−.83227	.55437	−1.50127
5.4	−.77276	.63469	−1.21754
5.5	−.70554	.70867	−.99558
5.6	−.63127	.77557	−.81394
5.7	−.55069	.83471	−.65973
5.8	−.46460	.88552	−.52467
5.9	−.37388	.92748	−.40311
6.0	−.27942	.96017	−.29101
6.1	−.18216	.98327	−.18526
6.2	−.08309	.99654	−.08338
2π	.00000	1.00000	.00000
6.3	.01681	.99986	.01682
6.4	.11655	.99318	.11735
6.5	.21512	.97659	.22028

Exponential functions

x	e^x	e^{-x}
0.00	1.00000	1.00000
.05	1.05127	.95123
.10	1.10517	.90484
.15	1.16183	.86071
.20	1.22140	.81873
.25	1.28403	.77880
.30	1.34986	.74082
.35	1.41907	.70469
.40	1.49182	.67032
.45	1.56831	.63763
.50	1.64872	.60653
.55	1.73325	.57695
.60	1.82212	.54881
.65	1.91554	.52205
.70	2.01375	.49659

x	e^x	e^{-x}
.75	2.11700	.47237
.80	2.22554	.44933
.85	2.33965	.42741
.90	2.45960	.40657
.95	2.58571	.38674
1.00	2.71828	.36788
2.00	7.38906	.13534
3.00	20.08554	.04979
4.00	54.59815	.01832
5.00	148.41316	.00674
6.00	403.42879	.00248
7.00	1096.63316	.00091
8.00	2980.95799	.00034
9.00	8103.08393	.00012
10.00	22026.46579	.00005

Note: $e^{a+x} = e^a e^x$

Natural logarithms

x	$\ln x$	x	$\ln x$
.1	−2.30258	2.6	.95551
.2	−1.60943	2.7	.99325
.3	−1.20396	2.8	1.02962
.4	−.91628	2.9	1.06471
.5	−.69314	3.0	1.09861
.6	−.51082	3.1	1.13140
.7	−.35666	3.2	1.16315
.8	−.22313	3.3	1.19392
.9	−.10535	3.4	1.22378
1.0	0.00000	3.5	1.25276
1.1	.09531	3.6	1.28093
1.2	.18232	3.7	1.30833
1.3	.26236	3.8	1.33500
1.4	.33647	3.9	1.36098
1.5	.40547	4.0	1.38629
1.6	.47000	4.1	1.41099
1.7	.53063	4.2	1.43508
1.8	.58779	4.3	1.45862
1.9	.64185	4.4	1.48160
2.0	.69315	4.5	1.50408
2.1	.74194	4.6	1.52606
2.2	.78846	4.7	1.54756
2.3	.83291	4.8	1.56862
2.4	.87547	4.9	1.58924
2.5	.91629	5.0	1.60944

x	$\ln x$	x	$\ln x$
5.1	1.62924	7.6	2.02815
5.2	1.64866	7.7	2.04122
5.3	1.66771	7.8	2.05412
5.4	1.68640	7.9	2.06686
5.5	1.70475	8.0	2.07944
5.6	1.72277	8.1	2.09186
5.7	1.74047	8.2	2.10413
5.8	1.75786	8.3	2.11626
5.9	1.77495	8.4	2.12823
6.0	1.79176	8.5	2.14007
6.1	1.80829	8.6	2.15176
6.2	1.82455	8.7	2.16332
6.3	1.84055	8.8	2.17475
6.4	1.85630	8.9	2.18605
6.5	1.87180	9.0	2.19722
6.6	1.88707	9.1	2.20827
6.7	1.90211	9.2	2.21920
6.8	1.91692	9.3	2.23001
6.9	1.93152	9.4	2.24071
7.0	1.94591	9.5	2.25129
7.1	1.96009	9.6	2.26176
7.2	1.97408	9.7	2.27213
7.3	1.98787	9.8	2.28238
7.4	2.00148	9.9	2.29253
7.5	2.01490	10.0	2.30259

Note: $\ln 10x = \ln x + \ln 10$

APPENDIX V

GEOMETRY FORMULAS

(A = area, C = circumference, V = volume, S = surface area, r = radius, b = base, h = height)

Triangle

$$A = \frac{1}{2}bh$$

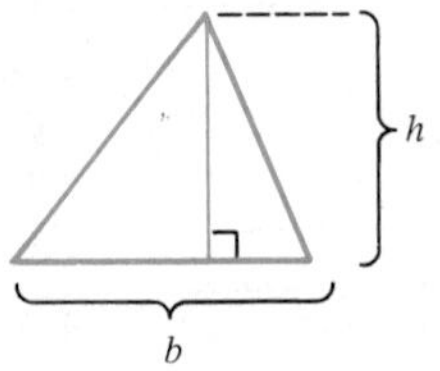

Rectangle

$$A = bh$$

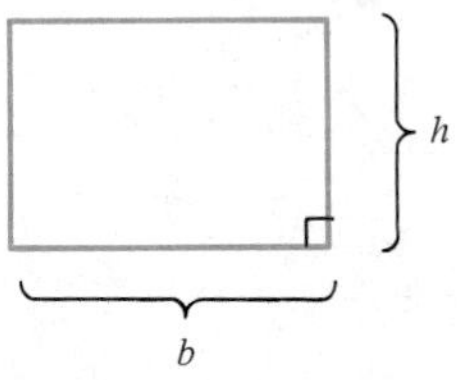

Circle

$$A = \pi r^2$$

$$C = \pi d = 2\pi r$$

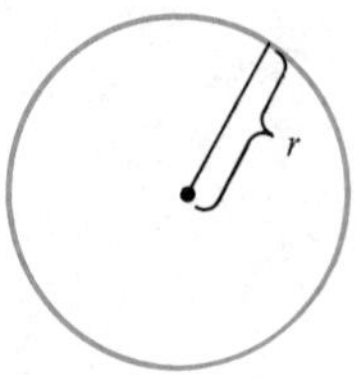

Parallelogram

$$A = bh$$

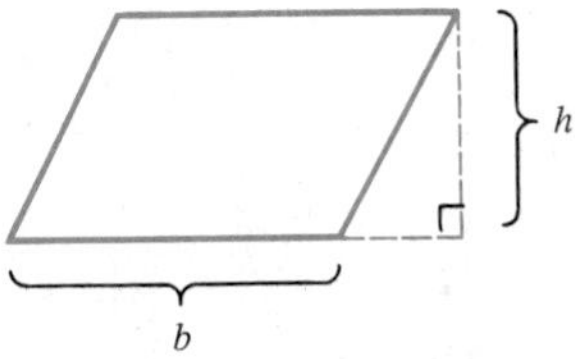

Trapezoid

$$A = \frac{1}{2}(b_1 + b_2)h$$

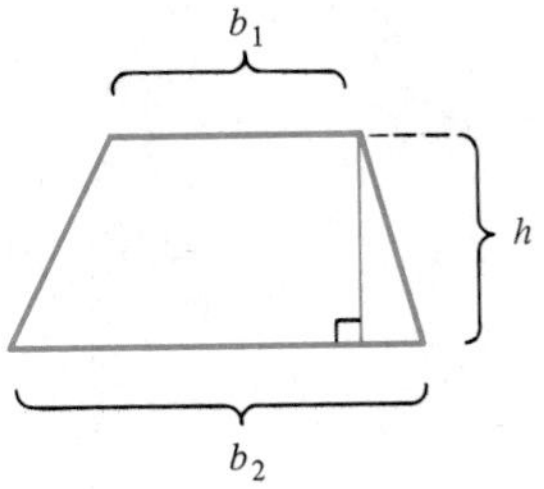

Cylinder with Parallel Bases

$$V = Bh$$

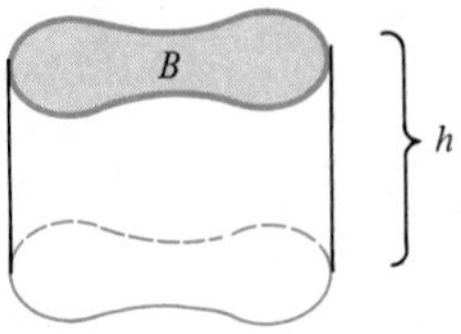

Right Circular Cylinder

$V = \pi r^2 h$

$S = 2\pi rh + 2\pi r^2$

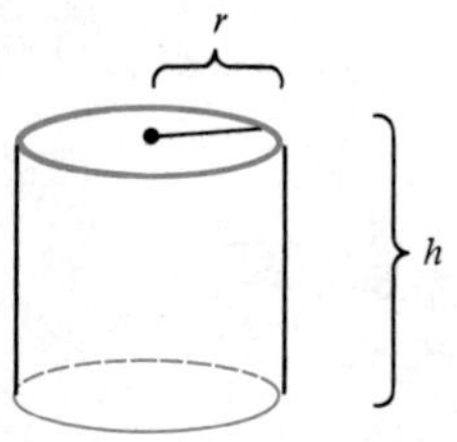

Right Circular Cone

$$V = \frac{1}{3}\pi r^2 h$$

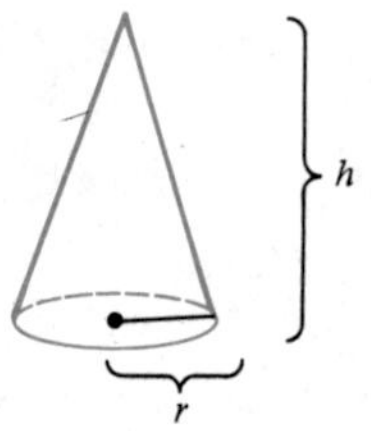

Sphere

$$V = \frac{4}{3}\pi r^3$$

$S = 4\pi r^2$

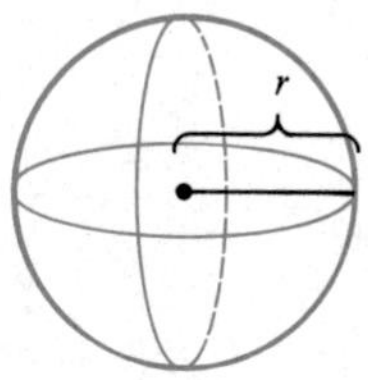

ANSWERS TO ODD NUMBERED EXERCISES

CHAPTER 1

Exercise Set 1.2

1. rational

3. rational

5. integer

7. integer

9. irrational

11. $(-1, 5]$

13. $[-2, 0)$

15. $(-1, 0)$

17. True, true, true

19. $x \geq 2 - \pi/2$

21. $x \geq -7$

23. $-1 < x < 3$

25. $x < -4$ or $x > -2$

27. $x < 6$

29. $-2 < x < 0$ or $x > 1$

31. $-3 < x < 3$ and $x \neq 0$
Equivalently $-3 < x < 0$ or $0 < x < 3$

33. 27

35. 8

37. 9/2

39. $\sqrt{x}\, y^{-1}$

41. 4

43. $x^{-7/2}y^{-1/3}z^{5/2}$

45. $x = 2/5$

47. $x = 4$ or $x = -10/3$

49. $x = 1$ or $x = -1$

51. $x = 2$ or $x = -4$

53. $x = 11/2$

55. $1 \leq x \leq 5$

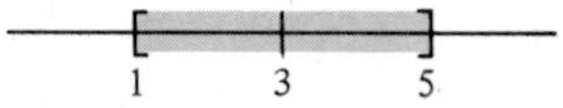

57. $x > -1$ or $x < -3$

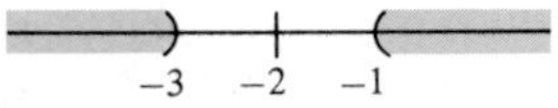

59. $2 \leq x \leq 5$

61. $x \geq 1$

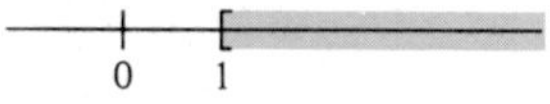

63. x is more than 2 units from 5.

65. The distance from x to 1 is twice the distance from x to 3.

67. $|x + 5| > 4$

69.

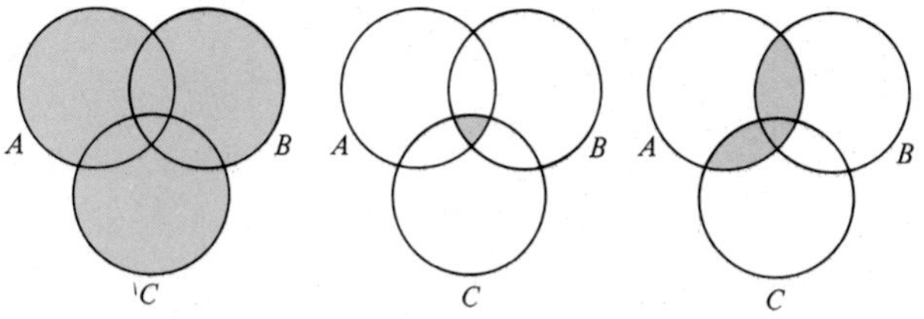

75. $-13 < x < 1$

79. False

83. **a.** $x = 1340/999$

b. $x = \dfrac{a_1a_2 \cdots a_n}{10^n - 1}$

85. $x < \dfrac{-1 - \sqrt{13}}{2}$ or $x > \dfrac{-1 + \sqrt{13}}{2}$

87. $\dfrac{-1 - \sqrt{17}}{2} < x < \dfrac{-1 + \sqrt{17}}{2}$

Exercise Set 1.3

1.

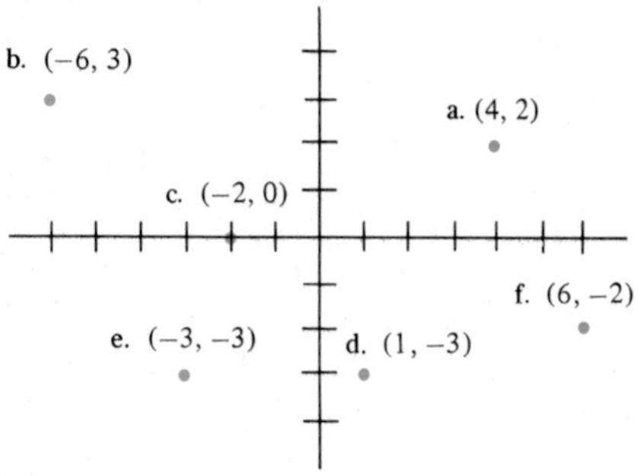

3. (5, 6) and (5, 0)

5. True

7. $b = \sqrt{3}$ or $b = -\sqrt{3}$

9. $8x - 6y - 21 = 0$

11. **a.** (0, 3) **b.** (3, 4)
c. (5/2, 3) **d.** (−3, −1)

13.

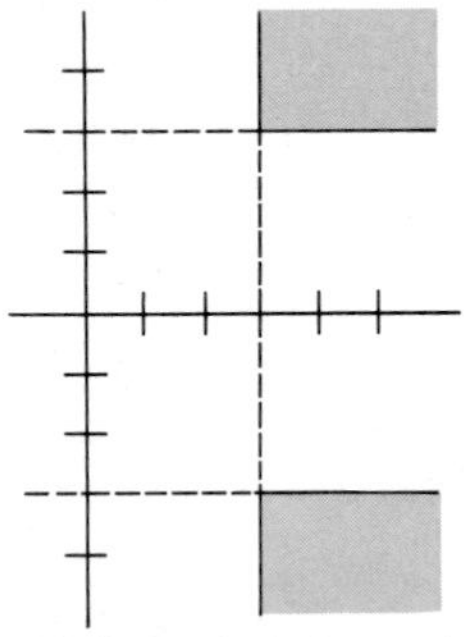

$\{(x, y) \mid x \geqslant 3 \text{ and } y \geqslant 3 \text{ or } y \leqslant -3\}$

15.

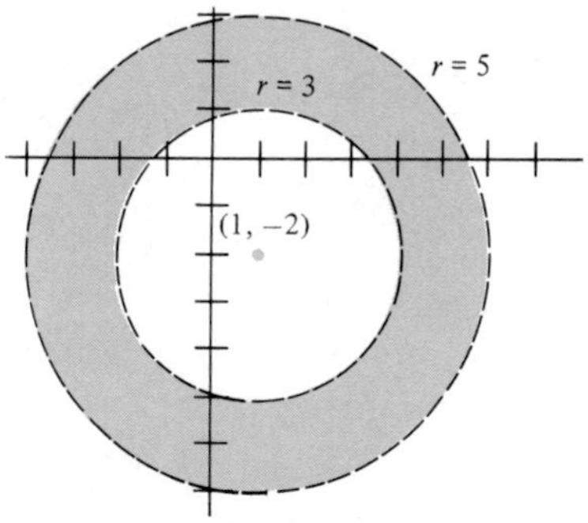

$\{(x, y) \mid 3 < \sqrt{(x-1)^2 + (y+2)^2} < 5\}$

17. $x^2 + y^2 = 25$

19. $(x + 2)^2 + (y - 4)^2 = 9$

21. $(x - 2)^2 + (y - 3)^2 = 16$

23. $\left(x - \frac{3}{2}\right)^2 - \frac{9}{4}$

25. $3\left(x - \frac{3}{2}\right)^2 - \frac{27}{4}$

27. $a\left(x + \frac{b}{2a}\right)^2 - \frac{b^2}{4a}$

29. center (1, 0), radius 3

31. center (−2, −1), radius 4

33. center (3, 2), radius $\sqrt{5}$

35. center $(a, -2a)$, radius 1

37. $(x - 3)^2 + (y - 2)^2 = 9$

39. False. If the three points are collinear they do not determine a circle.

41. $(x - 1)^2 + (y - 1)^2 = 9$

Exercise Set 1.4

1. **a.** 0 **b.** −1/4 **c.** −1/7
d. 3 **e.** −1 **f.** 1

3. False

5. True

7. $b = 4$

9. $y = -2x + 5$

11. $6x - 5y = -36$

13. $y = -3x + 6$

15. $x = -3$

17. $y = -x + 2$

19. $x - 3y + 11 = 0$

21. $x - 3y = -8$

23. False

25. $m = -1$, x-intercept $= 7$, y-intercept $= 7$

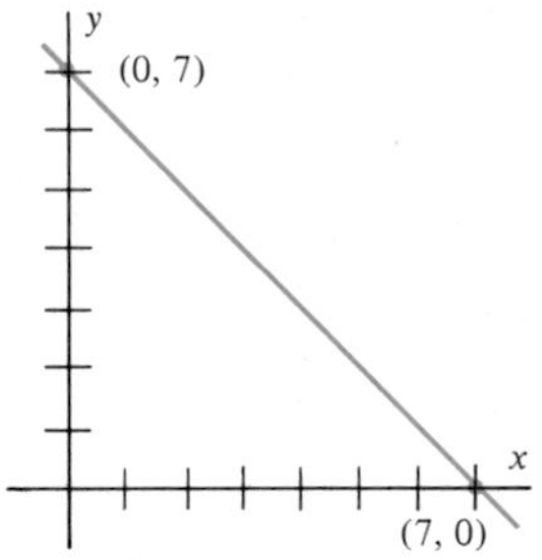

27. $m = -1$, x-intercept $= -3$, y-intercept $= -3$

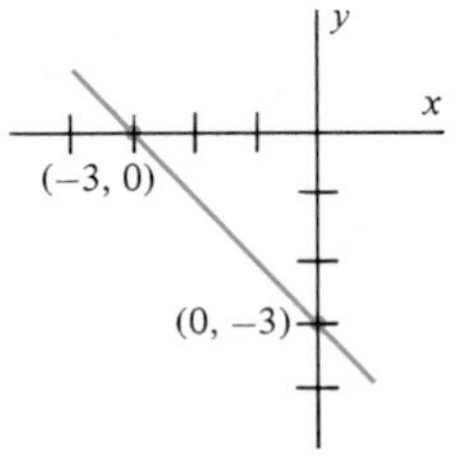

29. $m = 0$, no x-intercept, y-intercept $= 5$

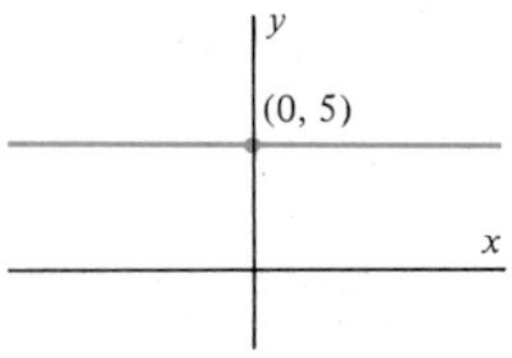

31. no slope, x-intercept = 4, no y-intercept

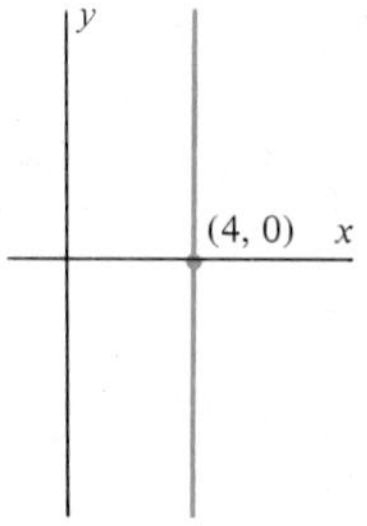

33. $m = \pi/3$, x-intercept $= -4/\pi$, y-intercept $= 4/3$

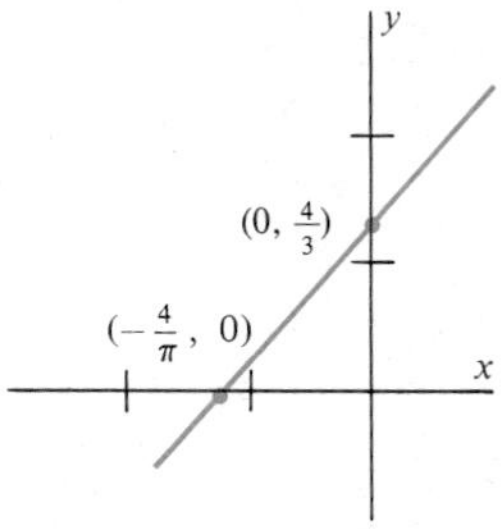

35. (1, 2) **37.** (−3, 0) **39.** no intersection

41. $y = -\dfrac{a}{b}x - \dfrac{c}{b}$, if $b = 0$, $x = -\dfrac{c}{a}$

43. a. yes **b.** yes **c.** no

45. a. 4 **b.** −4, 6, 16, 0

47. $d = -.15P + 15$
$P = \$100$

49. $F = \dfrac{9}{5}C + 32$

51. a. $t = \dfrac{1}{2}T - 11$ **b.** 72, 92

53. a. 1/25,000

55. $y = 3x - 1$

Exercise Set 1.5

1.

$f(x)$	$f(-2)$	$f(0)$	$f(4)$	$f(5)$	$\lvert f(3)\rvert$
$1 - 3x^2$	−11	1	−47	−74	26
$\dfrac{1}{x+2}$	undefined	$\dfrac{1}{2}$	$\dfrac{1}{6}$	$\dfrac{1}{7}$	$\dfrac{1}{5}$
$\dfrac{(x-3)^2}{x^2+1}$	5	9	$\dfrac{1}{17}$	$\dfrac{2}{13}$	0
$\sqrt{x+4}$	$\sqrt{2}$	2	$2\sqrt{2}$	3	$\sqrt{7}$
$\dfrac{1}{\sqrt{16-x^2}}$	$\dfrac{1}{2\sqrt{3}}$	$\dfrac{1}{4}$	undefined	undefined	$\dfrac{1}{\sqrt{7}}$
$\begin{cases} 1-x, & x < -3 \\ x-1, & x > 1 \end{cases}$	undefined	undefined	3	4	2

3. Yes

5. No

7. Yes

9. $x \in (-\infty, \infty)$

11. $x \geq -4$

13. $t \in [0, \infty)$

15. $-4 \leq s \leq 4$

17. $x \neq 0$

19. $x \leq -2$ or $0 \leq x < 2$ or $2 < x$

21. False

23. $y = \dfrac{3}{2}x^2 - 4$

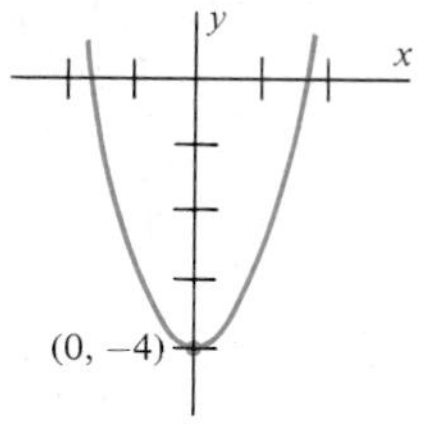

25. $y = 3\left(x - \dfrac{1}{6}\right)^2 + \dfrac{5}{12}$

27. $y = -4(x - 3)^2 + 2$

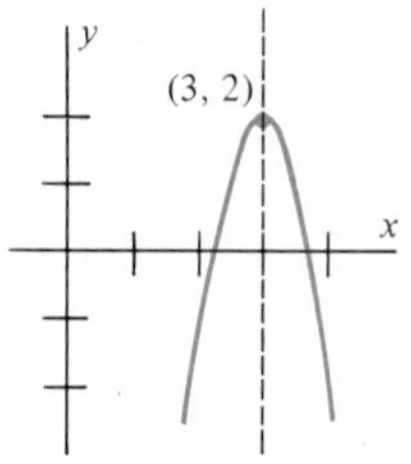

29. $x = -y^2 + 2$

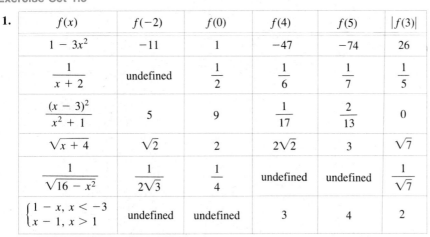

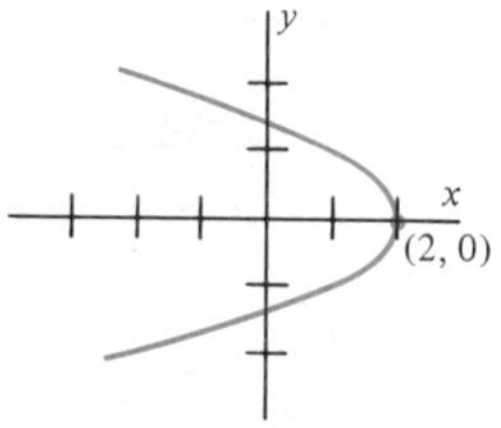

31. $x = -\frac{2}{3}y^2 + 1$

33. $x = 2(y - 1)^2$

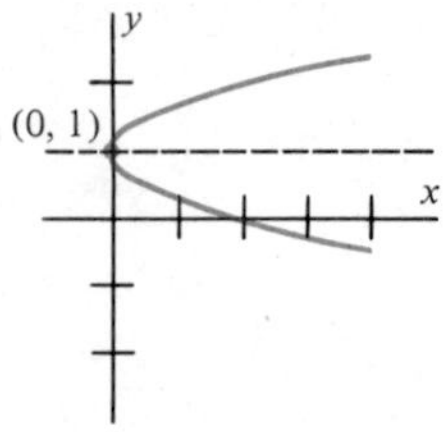

35. $(-3, -5)$, $(2, 0)$

37.

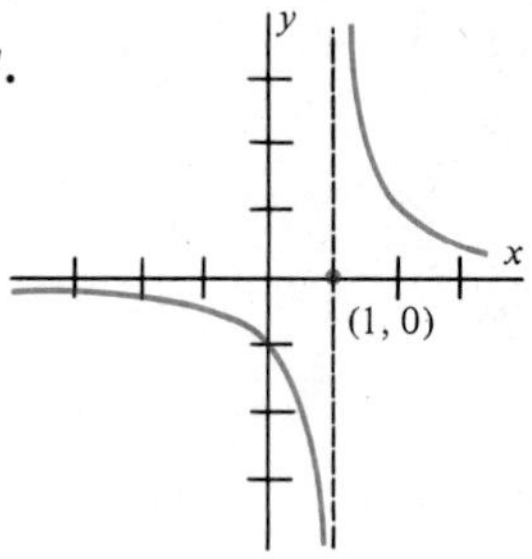

41.

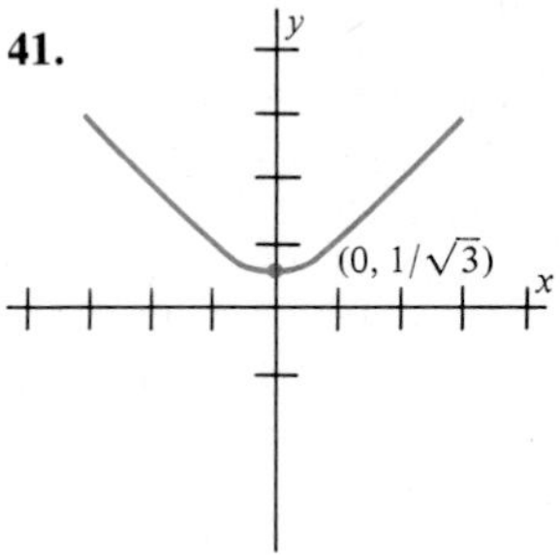

45.

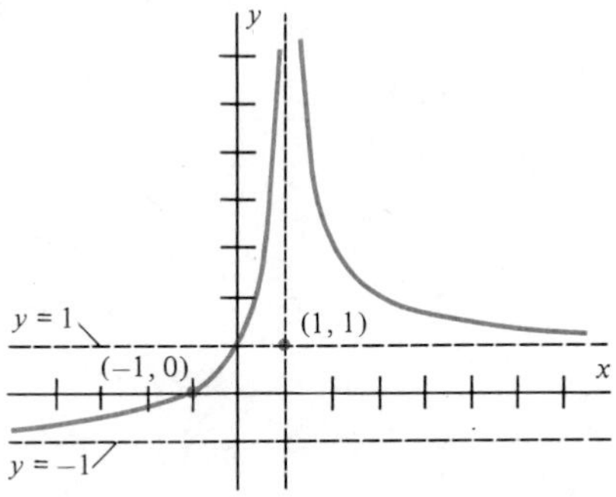

49. **a.** even **b.** even
c. odd **d.** neither
e. odd **f.** even
g. even **h.** neither

51. $3\sqrt{x} + 1$

53. $(3x + 1)^3$

55. $\frac{1}{1 + \sqrt{x}}$

57. $3x^{3/2} + 1$

59. True

61. $f(h) = V = \pi r^2 h$

63. $f(e) = v = e^3$

65. $a(z - b)^2$ is a parabola with vertex $(b, 0)$.

69. **a.** $R = [0, \infty)$ **b.** $R = [1, \infty)$
c. $R = (0, \infty)$ **d.** $R = [0, \sqrt{6}]$
e. $R = [0, \infty)$

Exercise Set 1.6

1. **a.** $\pi/6$ **b.** $5\pi/12$ **c.** $-\pi/12$
d. $7\pi/4$ **e.** $\frac{2\pi}{360} \cdot (x + 30)$
f. $97\pi/6$

3. $\theta_r = \frac{\pi}{180}\theta_d$
$m = \pi/180$

5. **a.** $x = 0, \pi$ **b.** $x = \pi/4, 5\pi/4$
c. $x = \pi/4, 7\pi/4$ **d.** $x = 3\pi/4, 7\pi/4$
e. $x = \pi/4, 3\pi/4$
f. $x = 0, \pi/3, 2\pi/3, \pi, 4\pi/3, 5\pi/3$
g. $x = \frac{3\pi}{4}, \frac{7\pi}{4}$
h. $x = \frac{\pi}{2}, \frac{\pi}{6}, \frac{5\pi}{6}$

7. $t = 2\pi/3$

9. $t = 0$ or $\pi/2 < t \leq \pi$

11. $\pi/2 < t < \pi$

13.

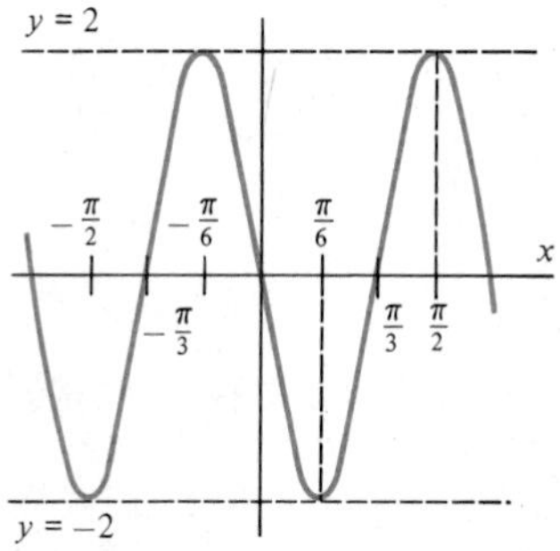

17.

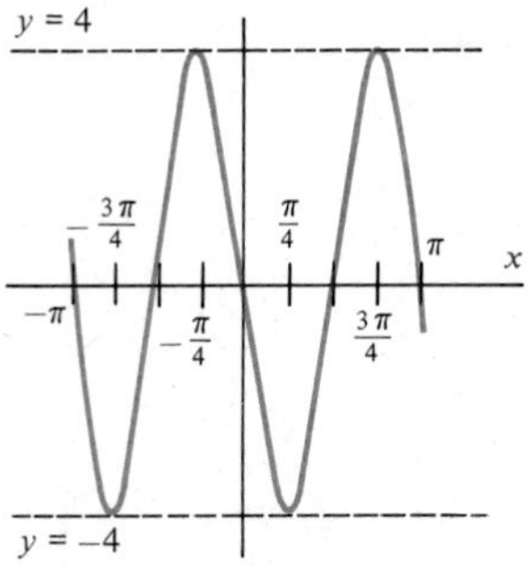

21.

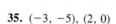

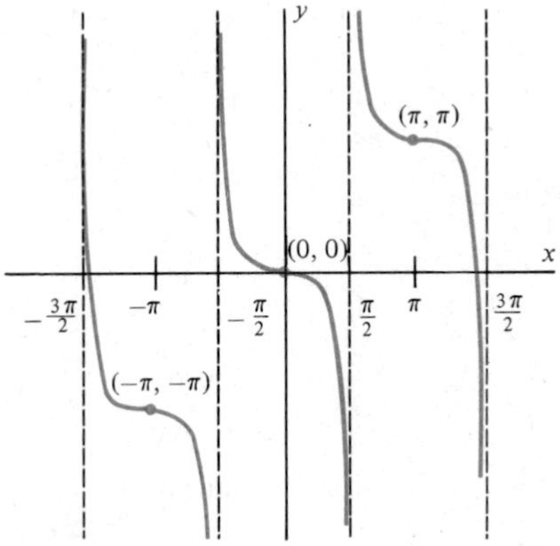

25. 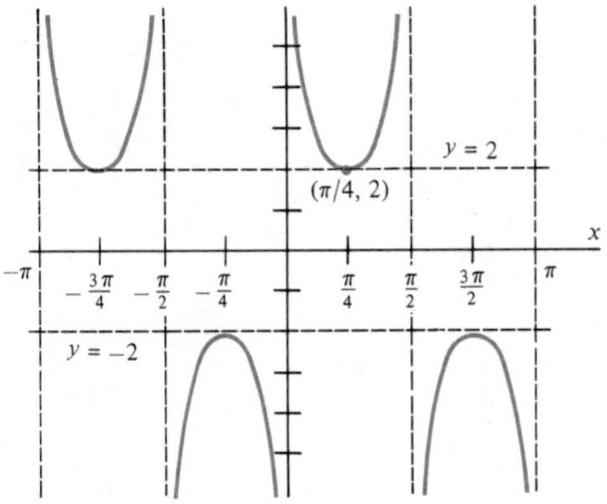

27. odd

29. odd

31. even

33. True

35. $A = \frac{1}{2} b \cdot \text{height}$, $b = x$, height $= h \sin\theta$

37. yes; π; yes; 2π

39. $4\pi/3$ meters

41. $50\sqrt{3}$ meters

43. 30 meters

45. $a = \frac{1}{2}$

Review Exercises–Chapter 1

1. the interval $(0, 1)$

3. a. $[2, 9)$
b. $(3, 7)$

5. $\varnothing$, $\{1\}$, $\{7\}$, $\{19\}$, $\{\pi\}$, $\{1, 7\}$, $\{1, 19\}$, $\{1, \pi\}$, $\{7, 19\}$, $\{7, \pi\}$, $\{19, \pi\}$, $\{1, 7, 19\}$, $\{1, 7, \pi\}$, $\{1, 19, \pi\}$, $\{7, 19, \pi\}$, $\{1, 7, 19, \pi\}$

7. $x \geq 8$

9. a. $[-3, -2] \cup [2, 3]$
b. $(-\infty, \infty)$
c. $\{-4, -3, -2, 2, 3, 4\}$
d. $\{-3, -2, -1, 0, 1, 2, 3\}$
e. $(-\infty, -2] \cup \{-1, 0, 1\} \cup [2, \infty)$
f. $[-3, -2] \cup [2, 3] \cup \{-4, 4\}$

11. $-5 \leq x \leq 19$

13. $A \cup B = A$
$A \cap B = B$

15. $-6 \leq x \leq 5$

17. $-2 \leq x \leq 2$

19. $x \geq 1$ or $x \leq -1$

21. $-2 \leq x \leq -1$ or $x \geq 0$

23. 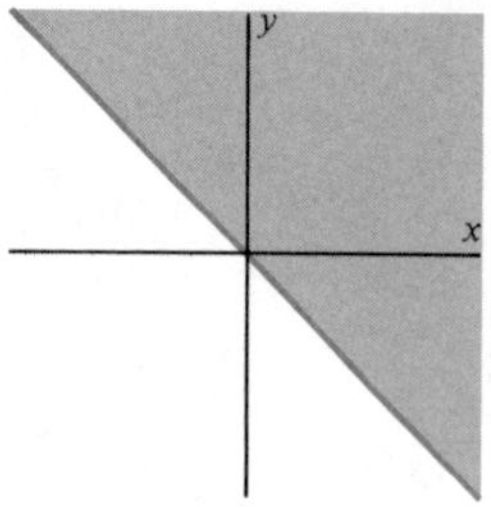

25. d

27. a. $a = 2$ **b.** $a = 10$

29. $x^2 + y^2 = 9$

31. $(0, 0)$; 7

33. $(1, -3)$; 1

35. 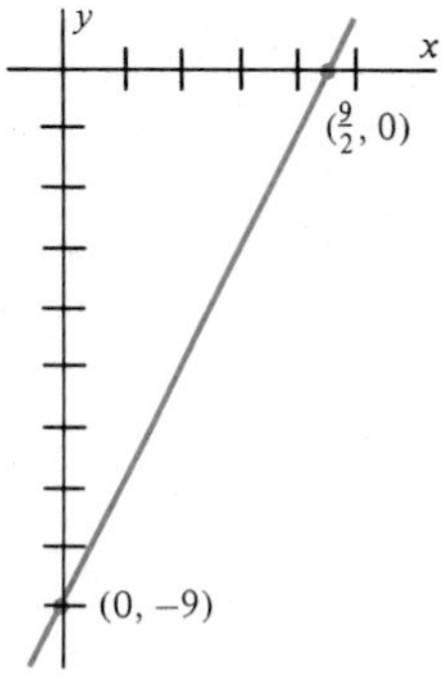

	Slope	y-intercept	x-intercept
37.	1/3	4/3	−4
39.	1	3	−3
41.	none	none	2

43. $y = 6x + 7$

45. $a = 2$

47. $7x + 5y - 8 = 0$

49. $a = \frac{-7}{2}$, $b = 9$

51. 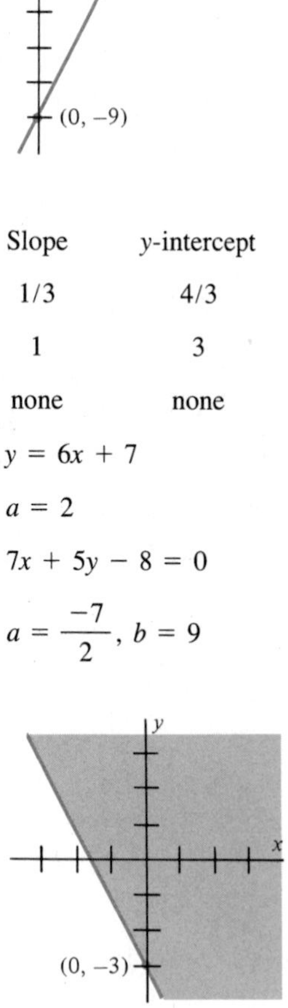

53. 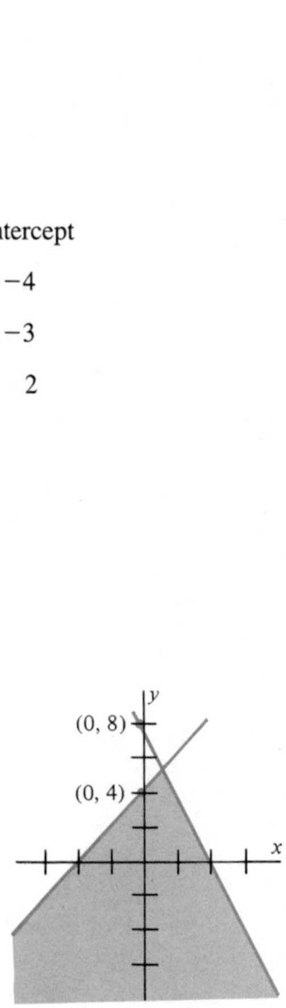

55.

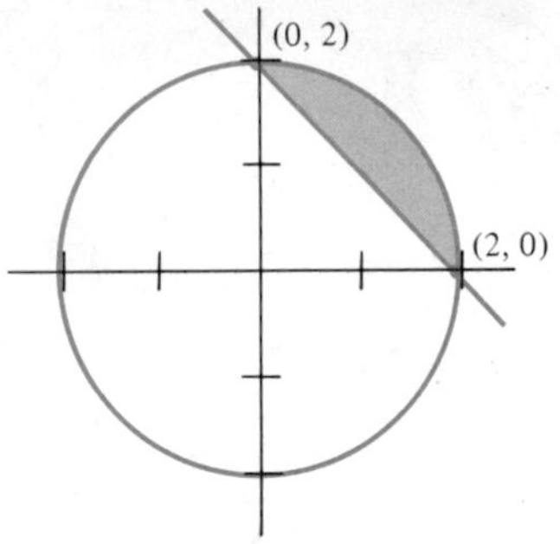

57. $-6208/999$

59. $(3, 1)$

61. $c = \mathrm{K} - 273$

63. $\left(\frac{5}{2}, \frac{-1}{2}\right)$

65. $x \le -1$ or $x \ge 1$

67. all x

69. $x \ne \frac{\pi}{2} + n\pi$

71. even

73. even

75. $y = 2\left(x - \frac{1}{2}\right)^2 + 3$

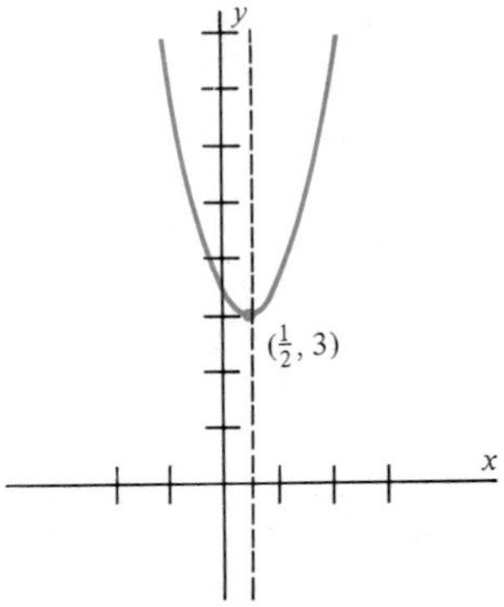

77. $x = -\frac{1}{2}\left(y - \frac{1}{2}\right)^2 - \frac{1}{4}$

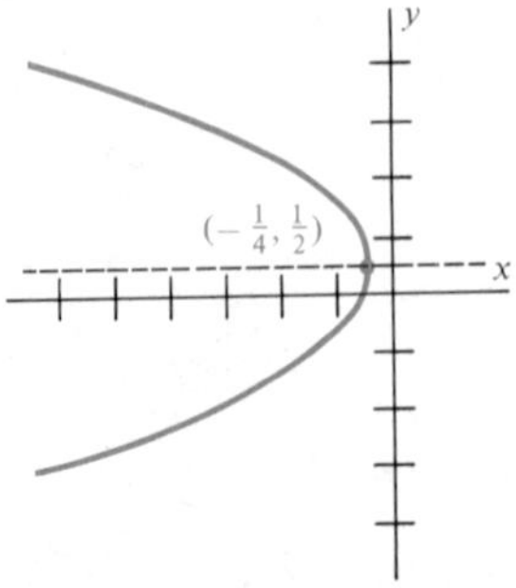

	Sin	Cos	Tan	Sec	Csc	Cot
79.	$\sqrt{2}/2$	$-\sqrt{2}/2$	-1	$-\sqrt{2}$	$\sqrt{2}$	-1
81.	$-\sqrt{3}/2$	$1/2$	$-\sqrt{3}$	2	$-2\sqrt{3}/3$	$-3\sqrt{3}/3$

83. **a.** $2\pi/3$ **b.** $-\pi/4$ **c.** $\pi/3$ or $-\pi/3$

85.

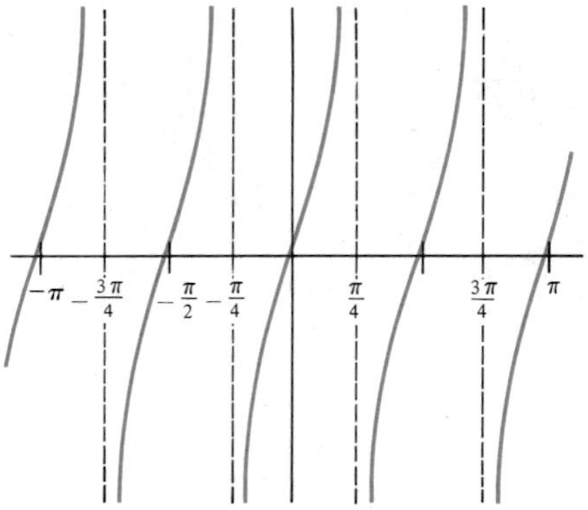

87.

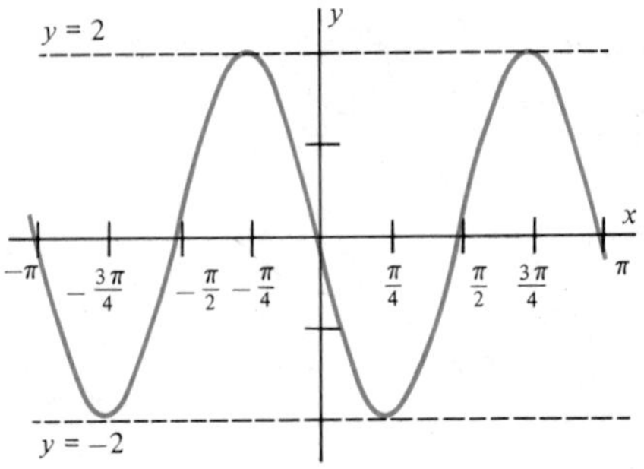

89.

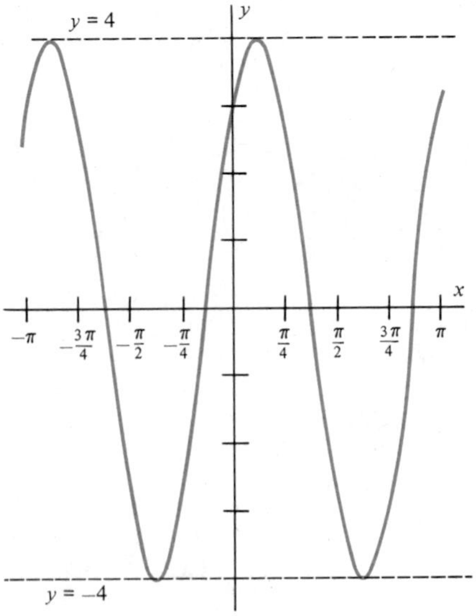

91. $\frac{1}{1 - \sin x}$ **93.** $\sin\left(\frac{1}{1 - x}\right)$ **95.** $\frac{1}{1 - \sin^3 x + \sin x}$

CHAPTER 2

Exercise Set 2.2

1. a. iv; **b.** v; **c.** iii; **d.** i; **e.** ii

3. 1

5. -3

7. $18x$

9. $2x$

11. $6x$

13. $2x + 6$

15. $6x^2 + 1$

17. $4x^3$

19. $\dfrac{1}{2\sqrt{x+4}}$

21. $-\dfrac{1}{x^2}$

23. $\dfrac{-1}{2\sqrt{x^3}}$

25. 8

27. 4

29. 1/4

31. $y = 8x - 8$

33. $y = 4x + 9$

35. $4y - x - 8 = 0$

37. 9; yes

39. 0.707

41. $\left(\dfrac{-\sqrt{3}}{2}, \dfrac{1}{4}\right)$; $\left(\dfrac{\sqrt{3}}{2}, \dfrac{1}{4}\right)$

43. False

45.

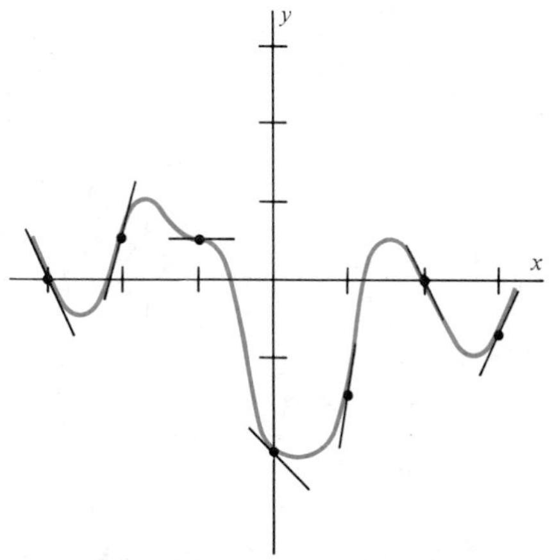

Exercise Set 2.3

1. i

3. i

5. iii

7. 5

9. $\sqrt{2}/2$

11. 6

13. -5

15. 14

17. 32

19. 4

21. 6

23. 1

25. 0

27. -2

29. x^2; 1

31. x^3; 3

33. $x^3 + x$; 2

35. False

37. a. -1 **b.** -5

39. $y = 6x + 1$

41. $y = \dfrac{1}{2}x - \dfrac{1}{2}$

43. $y = 32x - 16$

47. a. 1/3 **b.** 0

Exercise Set 2.4

1. 2

3. -10

5. $-8\sqrt{3}$

7. 0

9. 2

11. 1/10

13. 6/17

15. 9/2

17. $-3/8$

19. 4

21. -4

23. 9

25. 2

27. -18

29. 1

31. $-2/7$

33. 3

35. 3/5

37. 3

39. 0

41. 0

43. 1

Exercise Set 2.5

1. a. -1 **b.** 2
c. 2 **d.** 0

3. a. 2 **b.** does not exist **c.** 0

5. 0

7. 0

9. does not exist

11. 0

13. 4

15. 0

17. 1

19. 0

21. 0

23. 0

25. 3/2

27. a. 1 **b.** 1 **c.** yes

29. $\lim_{x\to 1^+} f(x) = 1 = \lim_{x\to 1^-} f(x)$

31. $c = 1$

Exercise Set 2.6

1. $x = 0$

3. $x = n\pi$; n an integer

5. $x = 1$

7. $x = -1$, or 2

9. $x = 2$

11. $x = -1$

13. $(-\infty, 6)$, $(6, \infty)$

15. all $(n\pi, n\pi + \pi)$; n an integer

17. $(-2, 0)$, $(2, \infty)$

19. $(-\infty, \infty)$

21. all $\left[n\pi, n\pi + \frac{\pi}{2}\right)$; n an integer

23. $\left(n\pi + \frac{\pi}{2}, n\pi + \frac{3}{2}\pi\right)$; $n = 1, \pm 2, \pm 3, \ldots$ also $\left(-\frac{\pi}{2}, -1\right), \left(-1, \frac{\pi}{2}\right), \left(\frac{\pi}{2}, 2\right), \left(2, \frac{3\pi}{2}\right)$

25. $(-\infty, \infty)$

27. $f(a) = 2$

29. $f(a) = 0$

31. $k = 3$

33. $k = -1$

35. **a.** yes **b.** yes **c.** no

37. 1

39. -2

41. $-\sqrt{3}$

43. False

Exercise Set 2.7

1. **a.** 1 **b.** 0.2 **c.** 0.025

3. **a.** 2/3 **b.** 1/3 **c.** 0.1

5. **a.** 0.3 **b.** 0.2 **c.** 0.14

7. **a.** 1 **b.** 0.5 **c.** 0.13

Review Exercises–Chapter 2

1. 5

3. 77

5. 6

7. -3

9. 4/3

11. $\sqrt{2}$

13. 1

15. 4

17. 1/3

19. $-27/7$

21. 3

23. does not exist

25. does not exist

27. 1

29. 1

31. 0

33. $(-\infty, 2), (2, \infty)$

35. $(-\infty, 0), (0, \infty)$

37. $(-\infty, 0), (0, \infty)$

39. $(-\infty, \infty)$

41. $(-\infty, -2), (-2, \infty)$

45. 5

47. 2

49. 0

51. $\pi/2$

53. 1

CHAPTER 3

Exercise Set 3.1

1. 3

3. $6x^2$

5. $3ax^2 + 2bx + c$

7. $-2/(2x + 3)^2$

9. $1/2\sqrt{x + 1}$

11. $-\frac{1}{2}(x + 1)^{-3/2}$

13. $-2x^{-3}$

15. $3(x + 3)^2$

17. $-\frac{1}{2}(x + 5)^{-3/2}$

19. $-6(x - 1)^{-3}$

21. $2x + 9y - 3 = 0$

23. $6y + x - 15 = 0$

25. $(0, 4)$

27. $a = -1, b = 3$

Exercise Set 3.2

1. $24x^2 - 2x$

3. $3ax^2 + b$

5. x

7. $2x + 1$

9. $12x^3 - 24x^2 + 12x - 16$

11. $6x^5 - 8x^3 + 2x$

13. $-5x^4 + 12x^2 - 3$

15. $(3 - x)^{-2}(-x^2 + 6x - 6)$

17. $(1 - x^3)^{-2}(-x^6 + 12x^3 + 12x^2 + 4)$

19. $(x^2 - 3x)^{-2}(x^4 - 6x^3 - 13x^2 - 28x + 42)$

21. $2x - 2 + 2x^{-2}$

23. $-6x^{-2} - 18x^{-3}$

25. $(t^2 + t + 1)^{-2}(1 - t^2)$

27. $(x - 1)^{-3}(-4x - 4)$

29. $(\theta + 3)^{-2}(2\theta^3 + 12\theta^2 + 18\theta + 6)$

31. $(cx^2 + d)^{-2}(-acx^2 - 2bcx + ad)$

33. $(x^2 - 10x + 21)^{-2}$ $(6x^5 - 90x^4 + 252x^3 + 140x^2 - 540x - 240)$

35. $2x + 1$

37. $4x^3$

39. $21x^6 + 88x^3 + 27x^2 + 7$

41. $-2(x^2 + 3x + 2)^{-2}(2x + 3)$

43. $4u^3 - 2u - 2$

47. $3x^2 + 6x + 2$

49. $30t^9 - 168t^7 - 189t^6 + 6t^5 + 945t^4 - 28t^3 - 27t^2 + 63$

51. $6u(u^2 - 4)^2$

55. 1

57. -2

59. -2

61. 16/189

63. 444

65. $9x - y - 13 = 0$

67. $107x + y - 172 = 0$

69. $13x + 100y - 36 = 0$

71. $a = 3$

73. $\frac{1}{3}x^{-2/3}$

Exercise Set 3.3

1. a. 3 **b.** never **c.** $[0, \infty)$

3. a. $-1(1 + t)^{-2}$ **b.** never **c.** never

5. a. $3(t - 2)(t - 4)$ **b.** 2, 4 **c.** $[0, 2) \cup (4, \infty)$

7. a. $-2(1 + t)^{-2}$ **b.** never **c.** $[0, \infty)$

9. a. $3(t - 3)(t - 1)$ **b.** 1, 3 **c.** $[0, 1) \cup (3, \infty)$

11. 13/16

13. 49/4 *m*

Exercise Set 3.4

1. $3(x + 4)^2$

3. $6(x^2 - 7x)^5(2x - 7)$

5. $(x^4 - 5)^2(13x^4 - 5)$

7. $-6x(x^2 - 9)^{-4}$

9. $6x^{-1/2}(3\sqrt{x} - 2)^3$

11. $24(x - 3)^3(x + 3)^{-5}$

13. $40x(4x^2 + 7)^4$

15. $\{(1 - x)\cdot(x^3 + 1)^2\cdot 9x^2 + [(x^3 + 1)^3 + 1]\}/(x - 1)^2$

17. $-6(x^2 + x + 1)^{-7}(2x + 1)$

19. $(2x^3 + x)^{-4}(x^4 - 2x^2)^4(22x^6 + 13x^4 - 14x^2)$

21. $3(ad - bc)(ax + b)^2(cx + d)^{-4}$

23. $3[(x^4 - 4x^2)(2x + 5)]^2\{2(x^4 - 4x^2) + (4x^3 - 8x)(2x + 5)\}$

25. $\frac{3}{2}x^{-5/2}(x^2 - 1)(x - 1)$

27. $(-2x^2 + 5x - 3)/(\sqrt{x^2 + 1}(x + 5)^4)$

29. $(\sqrt{x} + 7)^2(9 - x^2)^{-3/2}\left[\frac{3}{2}\frac{1}{\sqrt{x}}(9 - x^2) + x\sqrt{x} + 7x\right]$

31. $-6x(1 - x^2)^2$

33. $3x^{-4} - x^{-2}$

35. $2(5x^4 + 5)(x^5 + 5x + 5)$

37. $\dfrac{-1}{4\sqrt{x}\sqrt{10 - \sqrt{x}}}$

39. $-3(6 - x^3)^2(x^{-4})(6 + 2x^3)$

41. $(x + 2)/(x + 3)$

43. $1/(2 + 2\sqrt{x})$

45. $60x^3(x^4 + 1)^2[(x^4 + 1)^3 + 1]^4$

47. $2b(bx + c)^3/\sqrt{a + (bx + c)^4}$

49. $8x(x^2 + 1)[(x^2 + 1)^2 + 1]$

51. $1/(4\sqrt{x + x\sqrt{x}})$

53. $-\frac{1}{4}(x + 3)^{-5/4}$

55. $y = x$

57. $y = 0$

59. $x + y - 1 = 0$

61. 300

63. $a = 3$
$b = 2$

Exercise Set 3.5

1. $3 \cos 3x$

3. $\sin\left(\frac{\pi}{2} - x\right)$

5. $\dfrac{x \cos\sqrt{1 + x^2}}{\sqrt{1 + x^2}}$

7. $-2 \sin 2x - 3 \cos 3x$

9. $\cos^2 x - 2x \cos x \sin x$

11. $\sec^2 x \tan x + \sec^3 x$

13. 0

15. $-\sin x$

17. $2 \sec 2x \tan 2x$

19. $-4x \sin x^2 \cos x^2$

21. $x \sin(\sin\sqrt{1 - x^2}) \cos(\sqrt{1 - x^2})(1 - x^2)^{-1/2}$

23. $-\frac{3}{2}\csc^2\left(\frac{x}{2}\right)\cot^2\left(\frac{x}{2}\right)$

25. $\sin\left(\frac{1}{x}\right) - \frac{1}{x}\cos\left(\frac{1}{x}\right)$

27. $3x^2 \sec x^3 \tan x^3$

29. $\dfrac{3 \sec^3\sqrt{x + 1}\tan\sqrt{x + 1}}{2\sqrt{x + 1}}$

31. $(4 \cos 2x + 8 \sin 2x + 2)/2\sqrt{4 + \sin 2x}\,(2 + \cos 2x)^{3/2}$

33. $\left\{\sec x \tan^2 x\left[\frac{1}{2\sqrt{x}}\cot\left(3x - \frac{\pi}{4}\right) - 3\sqrt{x}\csc^2\left(3x - \frac{\pi}{4}\right)\right] - \sqrt{x}\cot\left(3x - \frac{\pi}{4}\right)[\sec x \tan^3 x + 2 \tan x \sec^3 x]\right\}$
$\cdot(\sec^{-2} x \tan^{-4} x)$

35. $-4 \sec^4(x^2 - \pi)\tan(x^2 - \pi)(2x)[1 + \sec^4(x^2 - \pi)]^{-2}$

37. $\frac{1}{2}\left(3x \cos x \sin^2 x - \sin x \cos^2 x - x \cos x\right)$
$(x \sin x \cos^2 x)^{-3/2}$

39. $y - \frac{\sqrt{3}}{2} = x - \frac{\pi}{6}$

41. 4/49

43. a. $\frac{\pi}{6} + \frac{n\pi}{3}$ **b.** 18 **c.** 6

45. $y - 1 = \frac{-4}{\pi}\left(x - \frac{\pi^2}{16}\right)$

47. They differ by a constant.

49. $m = -\dfrac{1}{2}$

51. $\cot x$: $x \neq n\pi$

$\sec x$: $x \neq \dfrac{\pi}{2} + n\pi$

$\csc x$: $x \neq n\pi$

53.

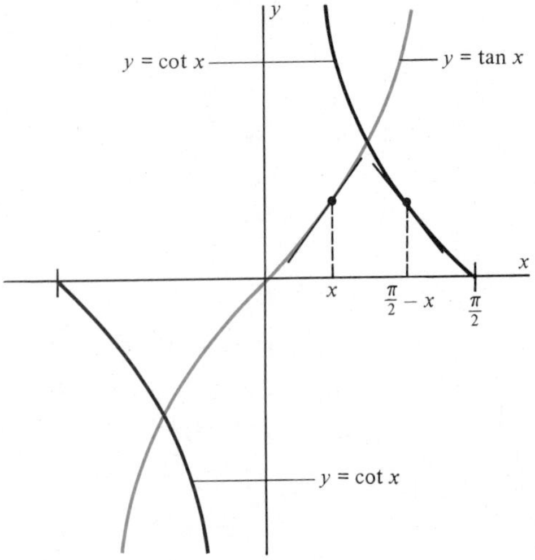

Exercise Set 3.6

1. $12x$

3. $2(1 + x)^{-3}$

5. $(x^2 + 7)^{-1/2} - x^2(x^2 + 7)^{-3/2}$

7. $2\cos(2x)$

9. $2(1 + \cos x)^{-3}\sin^2 x + (1 + \cos x)^{-2}\cos x$

11. $-2\sec(1 - x^2)\tan(1 - x^2)$
$+ 4x^2\sec(1 - x^2)\tan^2(1 - x^2) + 4x^2\sec^3(1 - x^2)$

13. $\dfrac{3x}{\sqrt{x^3 + 1}} - \dfrac{9x^4}{4(x^3 + 1)^{3/2}}$

15. $kx(1 - x^2)^{-1/2} + kx^3(1 - x^2)^{-3/2}$

17. $\sec x\tan^2 x + \sec^3 x$

19. $\dfrac{-\sec\sqrt{1 - x^2}\tan\sqrt{1 - x^2}}{1 - x^2} - \dfrac{x^2\sec\sqrt{1 - x^2}\tan\sqrt{1 - x^2}}{(1 - x^2)^{3/2}}$
$+ \dfrac{x^2\sec^2\sqrt{1 - x^2}(\tan^2\sqrt{1 - x^2} - \sec\sqrt{1 - x^2})}{(1 - x^2)^{3/2}}$

21. $-\dfrac{2}{9}x^{-4/3} + \dfrac{40}{9}x^{2/3}$

23. $2(x + 2)^{-3}$

25. $11880x^8 - 12096x^5 + 2160x^2$

27. $A\sin 2x + B\cos 2x$

29. $y = A\sin x + B\cos x$

33. a.

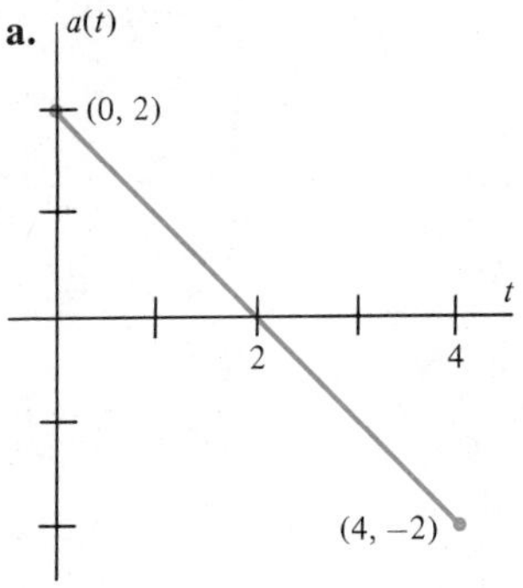

b. $v(1) > \dfrac{3}{2}$ **c.** $v(4) < -\dfrac{3}{2}$ **d.** No

e. Let $s(t) = t^2 - \frac{1}{6}t^3 + 100t$

35. $(-1)^n 2^{-(n+1)}[2(n - 1) - 1][2(n - 2) - 1] \cdots [1]x^{-[n-1/2]}$

37. $a = 3$

Exercise Set 3.7

1. $-x/y$

3. $-\sqrt{y}/\sqrt{x}$

5. $\cos^2 y$

7. $\dfrac{\sin y + y\sin x}{\cos x - x\cos y}$

9. $\dfrac{\sqrt{y}(y - 1)}{2\sqrt{x}(1 - 2\sqrt{xy})}$

11. $\dfrac{-3x^2 - 2xy - y^2}{x^2 + 2xy + 3y^2}$

13. $\dfrac{2(y - 1)\sqrt{x + y} - 1}{1 - 2x\sqrt{x + y}}$

15. $\dfrac{6x - \csc^2(x + y)}{\csc^2(x + y) - \csc^2 y}$

17. $-y/x$

19. $5x^4/4y^3$

21. $-4y^{-3}$

23. $(2y^2 - 2xy)(2y - x)^{-3}$

25. $x + y - 6 = 0$

27. $x + y - 4 = 0$

29. $3x - 5y = 0$

33. $x = \sqrt{2}$
$x = -\sqrt{2}$

35. $x = 1$
$x = -1$

37. $y = -1$
$y = -3$
$x = 0$
$x = -2$

41. $y - \dfrac{3\sqrt{3}}{2} = \dfrac{4\sqrt{3}}{3}(x - 2)$

Exercise 3.8

1. $\frac{4}{3}(x+2)^{1/3}$

3. $\frac{1}{2}x^{-1/2} - \frac{1}{2}x^{-3/2}$

5. $\frac{2}{3}x^{-1/3} - \frac{2}{3}x^{-5/3}$

7. $-4x(x^2+1)^2(x^2-1)^{-3} + 4x(x^2+1)(x^2-1)^{-2}$

9. $1/(6x^{2/3}\sqrt{1+x^{1/3}})$

11. $\dfrac{-2x^{1/3}}{3(x^{1/3}+x)^2}$

13. $\frac{1}{2}x^{-1/2} + \frac{1}{4}x^{-3/4} + \frac{1}{8}x^{-7/8}$

15. $\dfrac{4x(1-x^2)(x^2+1)^{-1/3} + 8x(x^2+1)^{2/3}}{3(1-x^2)^{7/3}}$

17. $\frac{1}{2}\left(\frac{ax^2+b}{cx+d}\right)^{-1/2}\left\{\frac{acx^2+2adx-bx}{(cx+d)^2}\right\}$

19. $(\sin^3 x - \sqrt{x})^{-2/3}\left\{(\sin^3 x - \sqrt{x})^{1/3}\left(\frac{19}{12}x^{7/12} - \frac{5}{4}x^{1/4}\right)\right.$

$\left. -(x^{19/12} - x^{5/4})\,\frac{1}{3}(\sin^3 x - \sqrt{x})^{-2/3}\left[3\sin^2 x\cos x - \frac{1}{2\sqrt{x}}\right]\right\}$

21. $5\sin(x^{2/3}+x^{-4/3})^{5/2}\cos(x^{2/3}+x^{-4/3})^{5/2}(x^{2/3}+x^{-4/3})^{3/2}$
$\cdot\left(\frac{2}{3}x^{-1/3} - \frac{4}{3}x^{-7/3}\right)$

23. $1/(3a^2+1)$

25. $3(1+a)^{-2/3}$

27. $x-1$

29. $3x/(1-x)$

31. x^3-2

33. $2x+y-8=0$

35. $x=0$

41. a, c, d

Exercise Set 3.9

1. $73/12 = 6.0833$

3. $31/128 = 0.2143$

5. $\pi/90 = .0349$

7. $3004/300 = 10.0133$

9. $49/250 = 0.1960$

11.

Exercise	Value	Relative Error	% Error
1	6.0827	4.93×10^{-5}	4.93×10^{-3}
2	6.9282	5.71×10^{-5}	5.71×10^{-3}
3	0.2425	.1163	11.63
4	4.9866	2×10^{-5}	2×10^{-3}
5	0.0349	0	0
6	0.0349	.4986	49.86
7	10.0133	0	0
8	7.8490	1.27×10^{-4}	1.27×10^{-2}
9	0.1961	5.1×10^{-4}	5.1×10^{-2}
10	0.5174	1.93×10^{-4}	1.93×10^{-2}

13. $2x\sin(1-x^2)\,dx$

15. $[\sec(\pi-x) - x\sec(\pi-x)\tan(\pi-x)]\,dx$

17. $(2+x)^{-3}(-x^3-6x^2-2)\,dx$

19. 8

21. **a.** 75π
b. 6π

23. $.0225\pi = 9\%$

25. .0025

27. ±1 meter

29. $.24 = 24\%$

Review Exercises–Chapter 3

1. $-7(3x-7)^{-2}$

3. $(3x^4+x^2)^{-2}(-36x^5+9x^4-12x^3+x^2-2x)$

5. $\frac{1}{2\sqrt{t}}\sin t + \sqrt{t}\cos t$

7. $(x^3+x^2+x)^{-2}(-3x^2-2x-1)$

9. $-(t^{-2}-t^{-3})^{-2}(-2t^{-3}+3t^{-4})$

11. $2x\sin x^2 + 2x^3\cos x^2$

13. $-3(x^4+4x^2)^{-4}(4x^3+8x)$

15. $(2-7x)(x^2+1)^{-3/2}$

17. $30x(x^2+1)^2[(x^2+1)^3-7]^4$

19. $(-3x^2+18x+9)/(2\sqrt{3x-9}(x^2+3)^{3/2})$

21. $-3\cot^2 s\csc^2 s$

23. $(\cos^2 x\sqrt{1+\sin x})^{-1}\left(\frac{1}{2}\cos^2 x + \sin x + \sin^2 x\right)$

25. $\frac{9}{2}t^{7/2} - 15t^{3/2}$

27. $\frac{1}{3}x^{-2/3}(-17x^{-6}+44x^{-4}-32x^{-3}-20x^{-2}+16x^{-1}+4)$

29. $f''(x) = 2\cos x - x\sin x$
$f'''(x) = -3\sin x - x\cos x$

31. $f''(x) = \sec x\tan^3 x + 5\sec^3 x\tan x$
$f'''(x) = \sec x\tan^4 x + 18\sec^3 x\tan^2 x + 5\sec^5 x$

33. $v = \left(1 - \frac{2}{\sqrt{t}}\right)/2(1+\sqrt{t})^{5/2}$

$a = \frac{1}{2}t^{-3/2}(1+\sqrt{t})^{-5/2} - \frac{5}{4}(1-2t^{-1/2})(1+\sqrt{t})^{-7/2}\left(\frac{1}{2\sqrt{t}}\right)$

35. $v = \dfrac{(t^2+1)^{3/2}}{2\sqrt{t}} + 3t^{3/2}\sqrt{t^2+1}$

$a = 6\sqrt{t}\sqrt{t^2+1} - 4\left(t + \dfrac{1}{t}\right)^{3/2} + \dfrac{3t^{5/2}}{\sqrt{t^2+1}}$

37. 6

39. $1/(2\sqrt{x+1})$

41. $(12x - y)/(x + 8y)$

43. $1/(1 + 2y + 3y^2)$

45. $-(y \csc^2(xy) + y)/(x \csc^2(xy) + x)$

47. $\sin y/(1 - x \cos y)$

49. $-(y^2 + x^2)/y^3$

51. $x + y - 1 = 0$

53. $2x + y + 2 - \dfrac{\pi}{4} = 0$

55. $(2\sqrt{2} + 2)x - y + \sqrt{2} - \dfrac{\pi}{4}(2 + 2\sqrt{2}) = 0$

57. $(\sin x + x \cos x)\, dx$

59. $x(1 + x^2)^{-1/2}\, dx$

61. $\dfrac{-\sin\sqrt{x}}{2\sqrt{x}}\, dx$

63. 4.0667

65. .88

67. $2/(x - 3)$

69. $(x/a)^{1/3}$

71. differentiable

73. not differentiable

75. .075 = 7.5%

77. 2.5

79. −1.4%

81. $-\sqrt{3}/3$

85. $t = 1$

CHAPTER 4

Exercise Set 4.2

1. 7; $2 + 2t$

3. 0; $\dfrac{1 - t^3}{2\sqrt{t}} - 3t^2\sqrt{t}$; $f'(0)$ not defined

5. $\dfrac{17}{3}$; $\dfrac{5}{2}t^{3/2} - 3t^{1/2}$

7. $4\sqrt{3}$; $\pi \sec^2 \pi t$; $k'(n\pi)$ not defined

9. $-\dfrac{4}{3\pi}$; $-4 \sec^2 t \tan t\,(1 + \sec^2 t)^{-4}$

11. 1200π cm^3/sec

13. 40π m^2/sec

15. 40 units/sec

17. 64 km/hr

19. −0.75 m/sec

21. 0.2 units/min

23. 22.5 cm^2/sec

25. $\dfrac{20}{3\pi} \approx 2.12$ m/min

27. 1.5 cm/sec

29. 0.0288 rad/sec

31. $480kS^2$

33. $k/500$

35. $-\alpha r^3/2Gm_1m_2$

37. a. $3/(1 - E)$ deg/min
b. $-\dfrac{0.02ET_c}{(1 - E)^2}$ deg/hr

39. 67 km/hr

Exercise Set 4.3

Critical numbers	Minimum	Maximum
1. $0, \frac{2}{3}$	$f\left(\frac{2}{3}\right) = -\frac{4}{27}$	$f(3) = 18$
3. 2	$f(1) = f(3) = -\frac{1}{3}$	$f(2) = -\frac{1}{4}$
5. 0, 2	$f(-1) = -4$	$f(2) = 4$
7. 0	$f\left(-\frac{\pi}{4}\right) = f\left(\frac{\pi}{4}\right) = 0$	$f(0) = 1$
9. 1	$f\left(\frac{1}{2}\right) = -3$	$f(2) = \frac{3}{4}$
11. 0	$f(0) = 2$	$f(-2) = 2 + \sqrt[3]{4} \approx 3.587$
13. −1	$f(-1) = 0$	$f(0) = 1$

Critical numbers	Minimum	Maximum
15. 0	$f(0) = 0$	$f(2) = \frac{4}{3}$
17. 0	$f(0) = 0$	$f\left(\frac{\pi}{2}\right) = f\left(-\frac{\pi}{2}\right) = \frac{\pi}{2}$
19. $-1, 0, 1$	$f(-2) = -56$	$f(2) = 56$
21. 0	$f(0) = 0$	$f(-1) = f(8) = 1$
23. 0, 1	$f(0) = 0$	$f(1) = \frac{1}{2}$
25. $\frac{6 - \sqrt{116}}{10} \approx -0.477$	$f(0) = \frac{1}{4}$	$f(-0.477) \approx 0.269$
27. 0	$f(0) = 0$	$f(8) = \frac{\sqrt{8}}{3}$

29. No. True if both maxima occur at the same value of x. Maximum of $h(x) \le L + M$.

31. $a = -\frac{1}{4}$, $b = -3$

33. **a.** same **b.** translated by c
c. multiplied by c (In **b** and **c,** $x + c$ and cx must be in $[a, b]$).

35. **d.** The critical numbers for $|h(x)|$ are the critical numbers for $h(x)$ and the zeros of $h(x)$.

37. $\min f\left(\frac{5}{3}\right) = 0$, $\max f(4) = 7$

39. $\min f\left(\frac{\pi}{2}\right) = 0$, $\max f(\pi) = 1$

Exercise Set 4.4

1. **a.** 0, 10 **b.** 5, 5 **c.** 0, 10

3. 5 m × 10 m

5. $\ell = 10$ m, $w = 20$ m

7. $\frac{40}{3}$ cm × $\frac{40}{3}$ cm × $\frac{10}{3}$ cm

9. 6 cm × 6 cm × 2 cm

11. 3 cm wide × 2 cm high

13. $\frac{4}{\sqrt{3}}$ units wide × $\frac{8}{3}$ units high

15. $r = \frac{P}{4}$, $\theta = 2$ rad

17. 4 trees

19. $4\sqrt{2} \times 4\sqrt{2} = 32$

21. 24 cm × 48 cm

23. $r = 2$, $h = \frac{5}{3}$

25. min at $t = 3, 9$; max at $t = 0, 6$

29. $\left(\frac{4}{3}, \frac{\sqrt{5}}{3}\right)$

31. $w = \frac{50}{3}$, $h = \frac{50}{3}$, $\ell = \frac{100}{3}$

33. $20/(1 + \sqrt[3]{2}) \approx 8.85$ m from the weaker source

35. $x = 75$ m

37. $x = 5$ m

39. $r = \frac{\sqrt[3]{192}}{\sqrt[6]{2}}$, $h = \sqrt[3]{384}$

41. $8\sqrt{2}$ ft

43. $3\sqrt{1 + 2^{2/3}} + 6\sqrt{1 + 2^{-2/3}} \approx 12.49$ ft

Exercise Set 4.5

1. $x = 192$

3. $x = 5$

5. Companies 1, 2, and 4

7. 6 times/yr

9. 8 times/yr

11. $x = 20$

13. **a.** $R(x) = x(600 - 3x)$
b. $P(x) = 450x - 3.5x^2 - 4000$
c. $MR(x) = 600 - 6x$, $MC(x) = 150 + x$
d. $x = 64.3$; $x = 64$ or 65, max profit for $x = 64$
e. $MR = MC$

15. $MC(x) = A(x)$

19. 69 dishwashers

21. **a.** $Q(101) = 99$ **b.** $E(101) = \frac{101}{99}$ (elastic)

23. Elastic for all $p > 0$

25. Inelastic for all $p > 1$

Exercise Set 4.6

1. 0.2293

3. 0.618

5. 2.303

7. −1.495

9. 2.303

Review Exercises–Chapter 4

	Minimum	Maximum
1.	$f(-1) = f(1) = -\frac{1}{3}$	$f(0) = -\frac{1}{4}$
3.	$f(-1) = -3$	$f(0) = 1$
5.	$f(-1) = 0$	$f(1) = 2$
7.	$f\left(-\frac{1}{\sqrt{2}}\right) = -\sqrt{2}$	$f(1) = 1$
9.	$f(-1) = -5$	$f(-2) = -3$
11.	$f(-1) = f(1) = -1$	$f(-2) = f(2) = 8$
13.	$f(-\pi) = f(\pi) = \frac{1}{3}$	$f(0) = 3$
15.	$f(2) = \frac{4}{\sqrt{5}}$	$f\left(\frac{1}{2}\right) = \frac{5}{\sqrt{5}}$

17. $a = \frac{1}{9}$

19. $60\sqrt{3}$ cm²/sec

23. 1.6180

25. 5.593

27. $\frac{1}{5\pi}$ cm/sec

29. $-\frac{V}{324\pi a^2}$ m/sec

31. 40π m²/sec

33. 5 wide × $5\sqrt{3}/2$ high

35. $-\frac{1}{8}$ rad/sec

37. $\left(\frac{1}{4}, \frac{\sqrt{17}}{4}\right), \left(\frac{1}{4}, \frac{-\sqrt{17}}{4}\right)$

39. 60 km/hr

41. depth/width = $\sqrt{2}$

43. 0.3 m/sec

45. 7 times/year

49. 1.3242

51. 127 items/week

53. $4\sqrt[3]{4} \times 8\sqrt[3]{4} \times 8\sqrt[3]{4}$

CHAPTER 5

Exercise Set 5.2

1. No

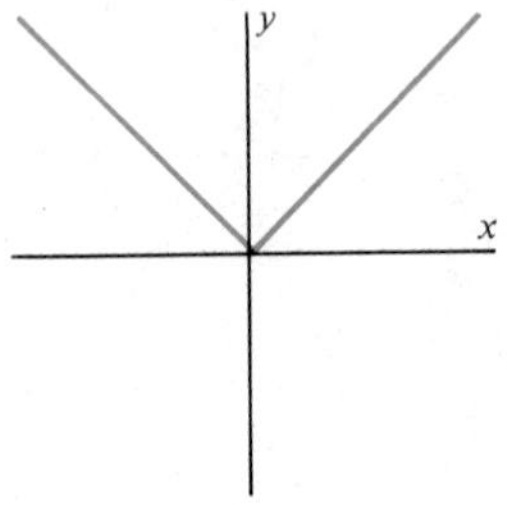

9. MVT does not apply; $f(x)$ not differentiable at $x = 0$.

11. $c = 0$

13. $c = \sqrt{3}$

19. No

23. $c \approx 2.789$

25. $c \approx 1.406$

Exercise Set 5.3

	Increasing	Decreasing
1.	$[1, \infty)$	$(-\infty, 1]$
3.	$(-\infty, -1), (-1, \infty)$	
5.	$\left[\frac{4n-1}{2}\pi, \frac{4n+1}{2}\pi\right]$	$\left[\frac{4n+1}{2}\pi, \frac{4n+3}{2}\pi\right]$
7.	$(-\infty, 0]$	$[0, \infty)$
9.	$(-\infty, \infty)$	
11.	$(-\infty, -1], [7, \infty)$	$[-1, 7]$
13.	$(-\infty, 1), (1, \infty)$	
15.	$(-\infty, -8), (-8, 8), (8, \infty)$	
17.		$(-\infty, \infty)$

19. $\left[\frac{7}{3}, \infty\right)$ $\left(-\infty, \frac{7}{3}\right]$

21. $a = 3$

23. $(-\infty, 25)$

25. Any constant function, $f(x) = c$

Exercise Set 5.4

1. $x = \frac{3}{2}$, rel. min.

3. $x = 7$, rel. min.; no rel. max.

5. $x = 0$, rel. max.; $x = 2$, rel. min.

7. $x = -2$, rel. max.; $x = \frac{4}{3}$, rel. min.

9. $x = (2n + 1)\pi$, stationary points

11. No critical numbers; $f(-1)$ undefined

13. $x = 0$, rel. min.

15. $x = -3, +3$, rel. minima; $x = 0$, rel. max.

17. $x = -2, +2$, critical numbers; no extrema

19. $x = 0$, rel. max.; $x = 2$, rel. min.

21. $x = 1 - \sqrt{2}$, rel. max.; $x = 1 + \sqrt{2}$, rel. min.

23. No relative extrema

25. At most $n - 1$ relative extrema; yes.

27. **a.** $a = 3$, $b = -9$
b. $x = -3$, rel. max.; $x = 1$, rel. min.

29. $a = 0$

31. $r = 5\sqrt[3]{\frac{2}{\pi}}$, $h = 10\sqrt[3]{\frac{2}{\pi}}$

35. $(-1, 1)$

Exercise Set 5.5

1. Concave down on $(-\infty, 0)$, concave up on $(0, +\infty)$

3. Concave down on $(-\infty, 4)$, concave up on $(4, +\infty)$

5. Concave up on $(-\infty, -2)$, concave down on $(-2, +\infty)$; infl. pt. $(-2, 0)$

7. Concave down on $\left(-\infty, -\frac{1}{2}\right)$, concave up on $\left(-\frac{1}{2}, +\infty\right)$; infl. pt. $\left(-\frac{1}{2}, 0\right)$

9. Concave down on $\left(-\infty, \frac{1}{2}\right)$, concave up on $\left(\frac{1}{2}, \infty\right)$; infl. pt. $\left(\frac{1}{2}, -\frac{7}{2}\right)$

11. Concave up on $\left(n\pi, \left(n + \frac{1}{2}\right)\pi\right)$, concave down on $\left(\left(n - \frac{1}{2}\right)\pi, n\pi\right)$; infl. pts. $(n\pi, 0)$

13. Concave down on $\left(\frac{4n + 1}{4}\pi, \frac{4n + 3}{4}\pi\right)$, concave up on $\left(\frac{4n + 3}{4}\pi, \frac{4n + 5}{4}\pi\right)$; infl. pts. $\left(\frac{2n + 1}{4}\pi, \frac{1}{2}\right)$

15. Concave up on $(-\infty, -3)$ and $(2, \infty)$, concave down on $(-3, 2)$; infl. pts. $(-3, -375)$ and $(2, -70)$

17. Concave up on $(-\infty, -1)$ and $(3, \infty)$, concave down on $(-1, 3)$; infl. pts. $(-1, -89)$ and $(3, -73)$

19. Concave up on $\left(\frac{-5 - 3\sqrt{5}}{10}, 0\right)$ and $\left(\frac{-5 + 3\sqrt{5}}{10}, 1\right)$. Concave down on $\left(-\infty, \frac{-5 - 3\sqrt{5}}{10}\right)$, $\left(0, \frac{-5 + 3\sqrt{5}}{10}\right)$, and $(1, \infty)$. Infl. pts. $\left(\frac{-5 - 3\sqrt{5}}{10}, -\frac{5 - \sqrt{5}}{60} \cdot \sqrt[3]{100(5 + 3\sqrt{5})}\right)$, $\left(\frac{-5 + 3\sqrt{5}}{10}, \frac{5 + \sqrt{5}}{60} \cdot \sqrt[3]{100(-5 + 3\sqrt{5})}\right)$, and $(0, 0)$.

21. rel. max.

23. rel. max.

25. rel. min.

27. rel. max.

29. $a > 0$

31. Any constant function, $f(x) = c$, for any b.

33. None; one; $k - 2$

35. $f(x) = \frac{x + 2}{x}$ is an example.

Exercise Set 5.6

1. 3/10

3. −1/3

5. 1/4

7. 0

9. 0

11. 0

13. 0

15. $+\infty$

17. $+\infty$

19. $+\infty$

21. −1

23. $+\infty$

25. $-\infty$

27. $y = 0$

29. $y = 2$

31. $y = 0$

33. $y = 3$

35. $a = -3$

37. $x = -2$

39. $x = 2$ and $x = -2$

41. $x = 0$

43. $r = 2/3$, $a = 6$

45. 5

47. 0

Exercise Set 5.7

1.

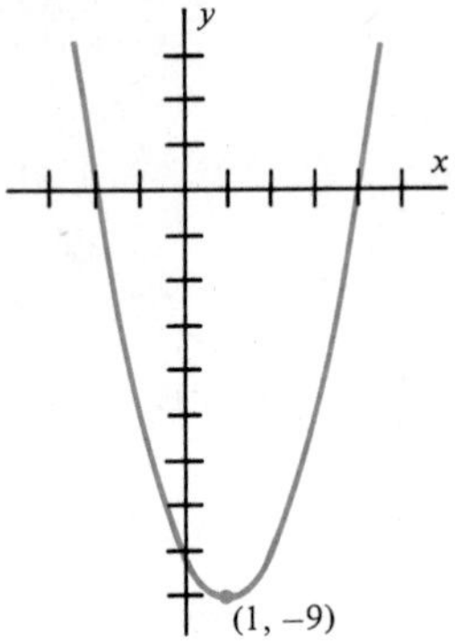

5.

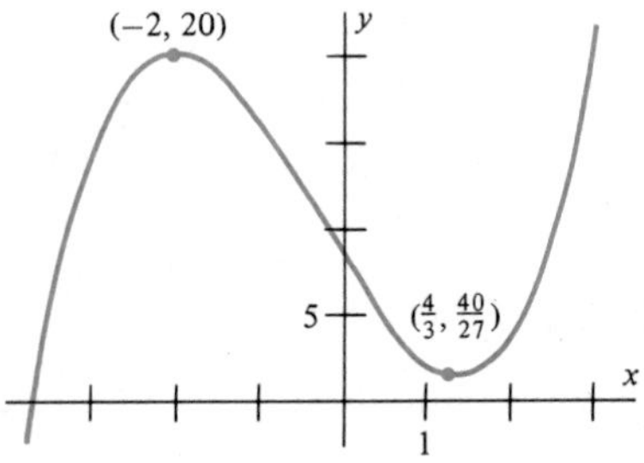

9.

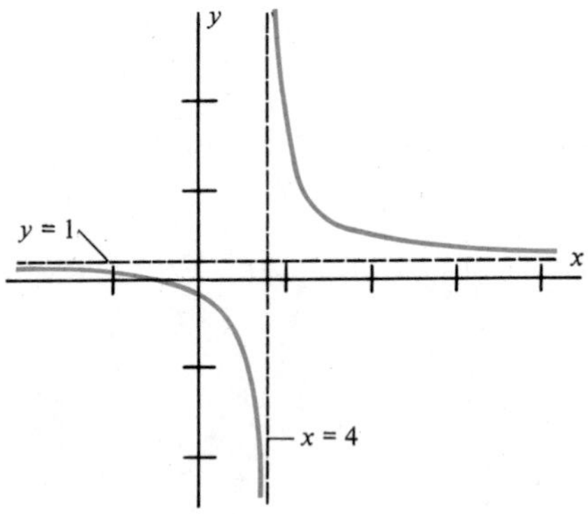

13.

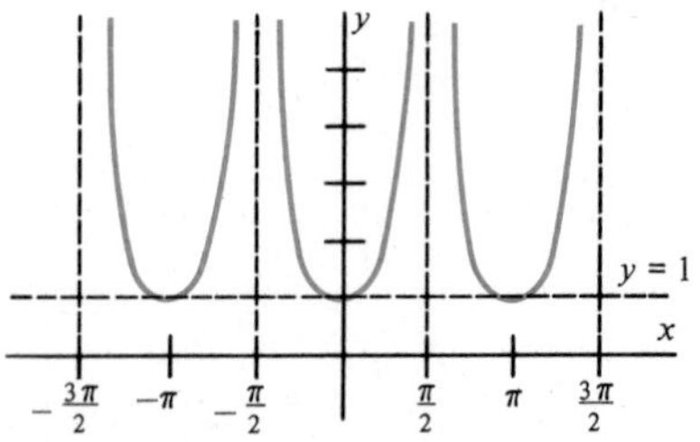

17.

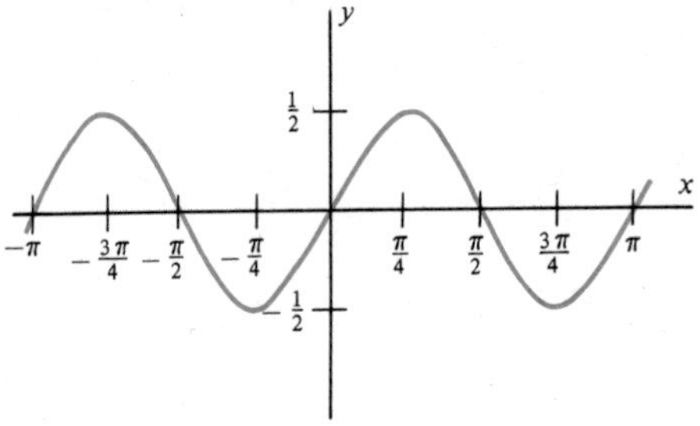

21.

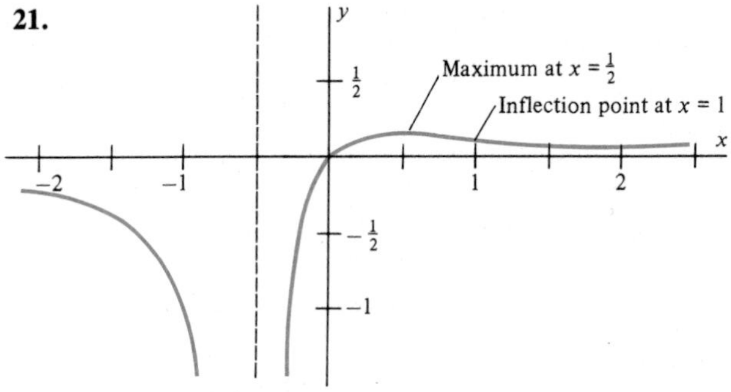

25.

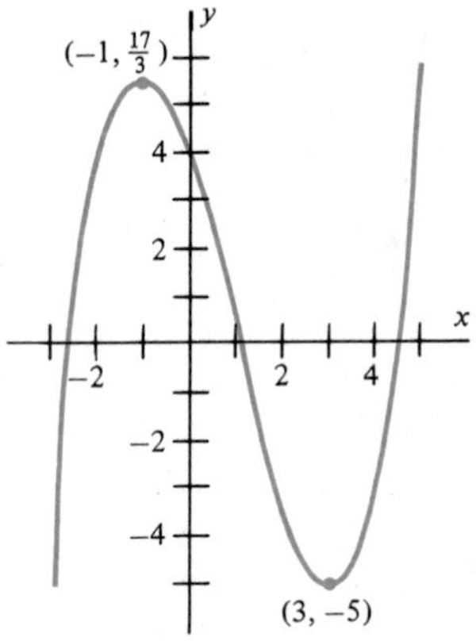

29.

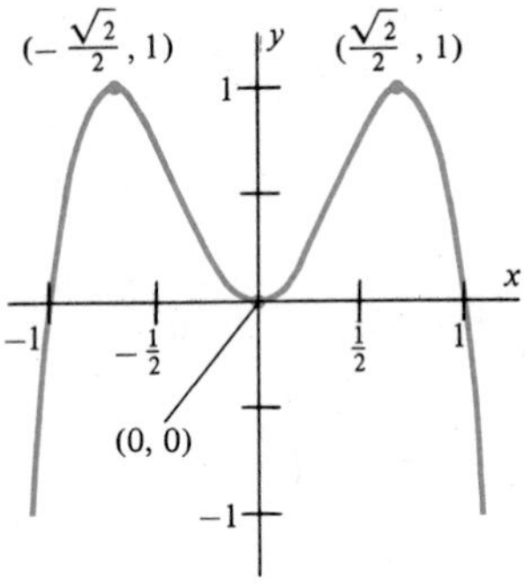

33.

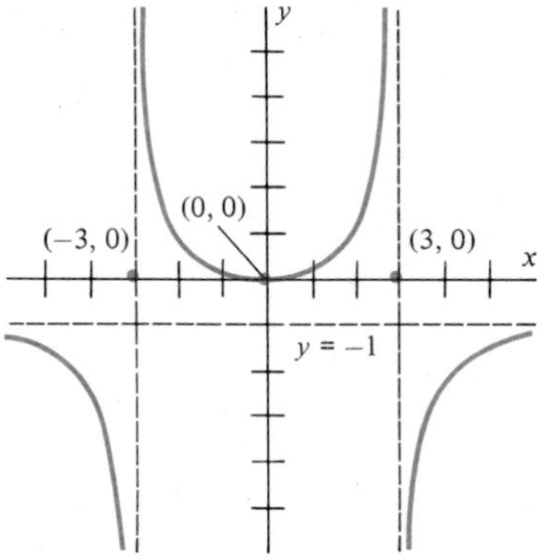

37.

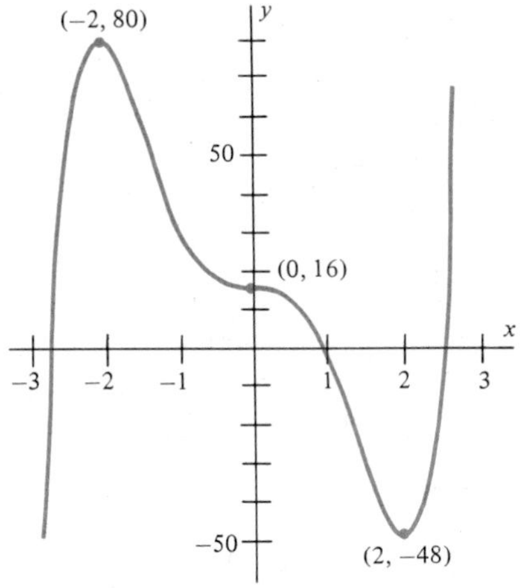

Exercise Set 5.8

1. 6

3. 0

5. 1/2

7. 1

9. 2

11. 1

13. $+\infty$

15. $+\infty$

17. 1/6

19. 0

21. $+\infty$

23. 1/2

25. 0

27. $\dfrac{3\sqrt{2}}{10} \approx 0.424$

29. 1

31. 3

33. 1

35. 3/4

37. 0

39. 1

41. 1

43. 1

45. 0

49. 0

51. $+\infty$

Exercise Set 5.9

1. a. ii **b.** i **c.** iv **d.** iii

3.

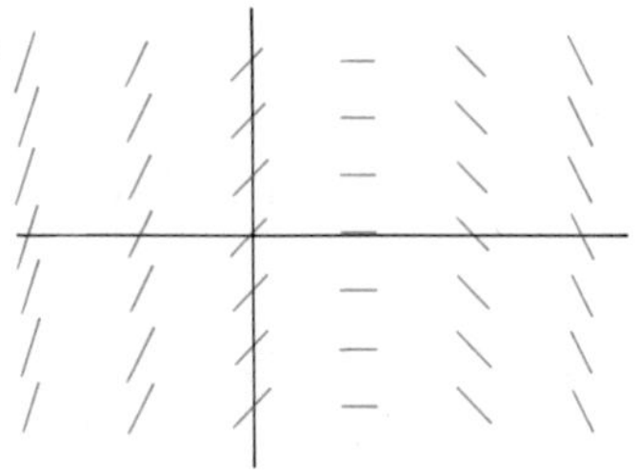

5.

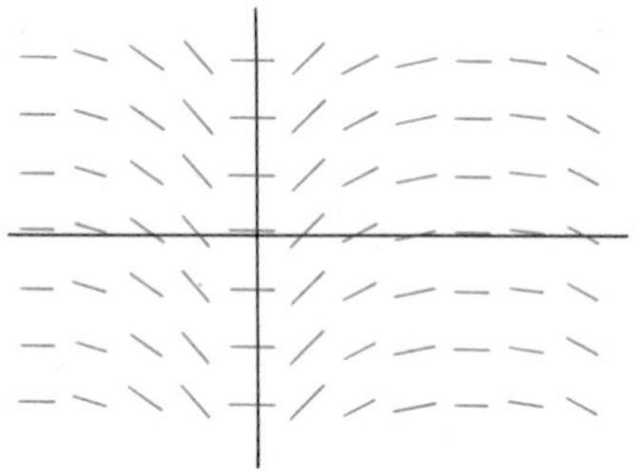

7.

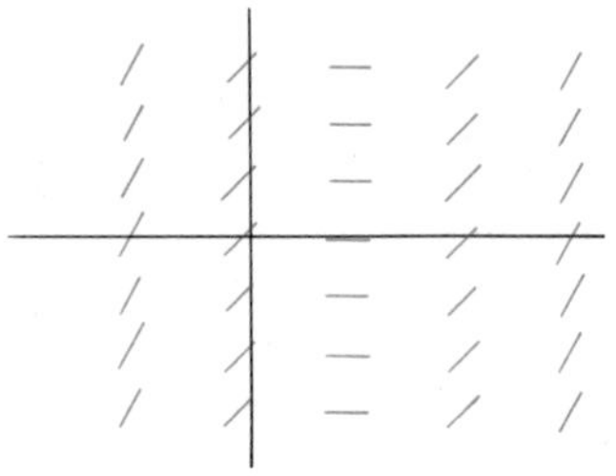

9.

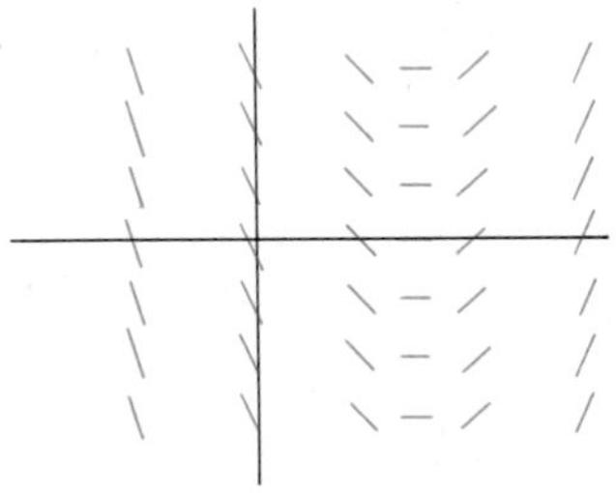

11. $x - \dfrac{1}{4}x^4 + C$

13. $\dfrac{2}{3}(x + 2)^{3/2} + C$

15. $2 \sin t + \cos t + C$

17. $\dfrac{2}{7}t^{7/2} + \dfrac{2}{3}t^{3/2} + C$

19. $\dfrac{t^4}{4} + \dfrac{2t^3}{3} + \dfrac{t^2}{2} + 2t + C$

21. $\dfrac{1}{3}(x^2 + 1)^{3/2} + C$

23. $\dfrac{1}{2}\sec(2\theta) + C$

25. $-\dfrac{1}{2}\cot(x^2) + C$

27. $-\dfrac{2}{27}(3t^3 - 9t + 9)^{3/2} + C$

29. $\tan(x - \pi) + C$

31. $\frac{1}{3}(2\sqrt{x}+3)^3 + C$

33. $\frac{1}{3}(x^2-1)^{3/2} + C$

35. $-\frac{1}{2}\sec(\pi - x^2) + C$

37. $\frac{1}{21}(x^6 - 3x^4)^{7/2} + C$

39. $s(t) = t + t^2$

41. $s(t) = \frac{1}{3}t^3 + t^2 + 2$

43. $s(t) = \frac{1}{2}(\sqrt{t}+3)^4 - \frac{81}{2}$

45. $v(t) = 2t + 3,\ s(t) = t^2 + 3t$

47. $v(t) = \frac{3}{2}t^2,\ s(t) = \frac{1}{2}t^3 + 20$

49. $v(t) = 2t^2 + 4t + 8,\ s(t) = \frac{2}{3}t^3 + 2t^2 + 8t + 12$

51. $v(t) = -\cos t + 1,\ s(t) = -\sin t + t + 2$

53. $45/9.8 \approx 4.6$ sec

55. **a.** $\sqrt{150/4.9} \approx 5.5$ sec
b. $9.8\sqrt{150/4.9} \approx 54$ m/sec (downward)

57. $a = -2$ m/sec^2

Exercise Set 5.10

1. $y = \frac{1}{2}x^2 - x + C$

3. $y = \frac{1}{3}x^3 + x^{-1} + C$

5. $y = \frac{1}{2}x^2 - \frac{2}{3}x^{3/2} + C$

7. $y^2 = x^2 + C$

9. $y = \dfrac{1}{2x^2 + C}$

11. $y^2 = x^2 + 15$

13. $y = \dfrac{-1}{3(1+x^3)} + \dfrac{5}{6}$

15. $y = \dfrac{-1}{x + \dfrac{x^3}{3} - 1}$

17. $y = \frac{1}{12}x^4 + 4x + 2$

19. $f(x) = \frac{1}{2}x^2 - \frac{2}{3}x^{3/2} + \frac{13}{6}$

Review Exercises–Chapter 5

1.

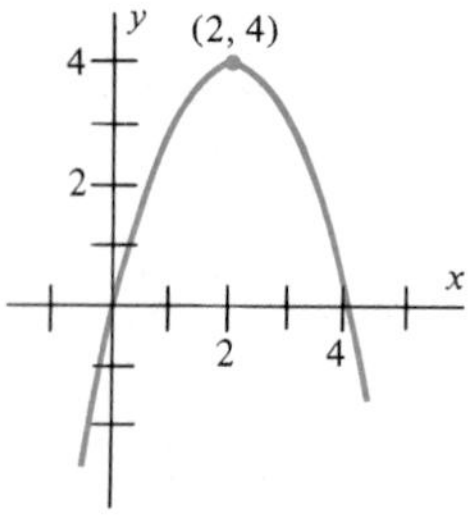

5.

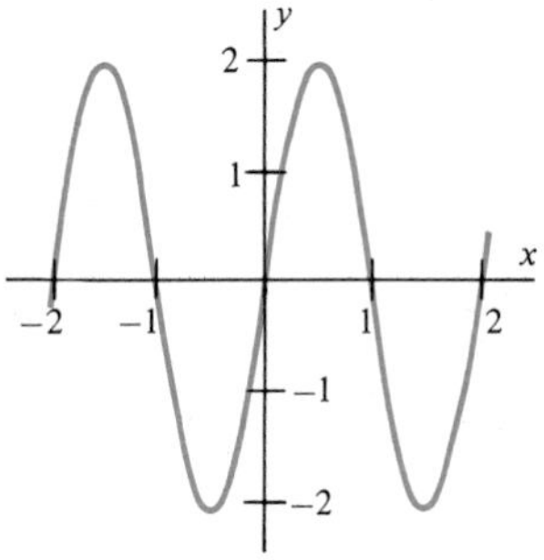

9.

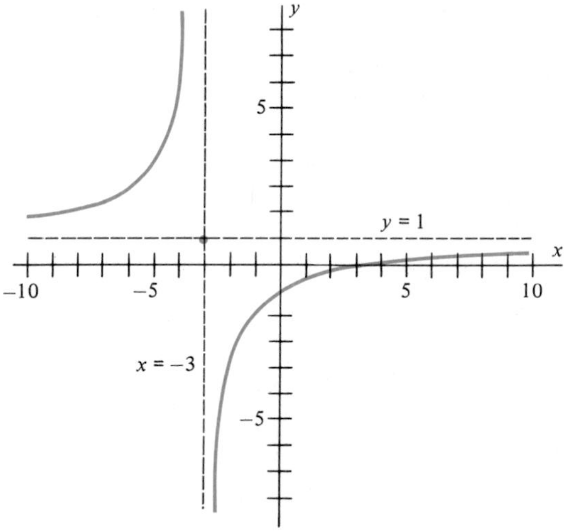

13.

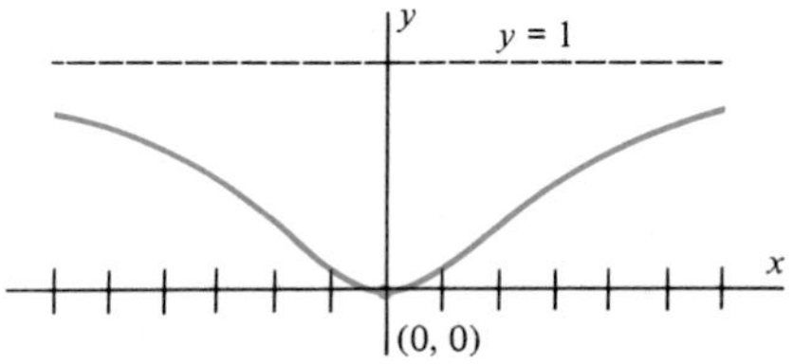

19.

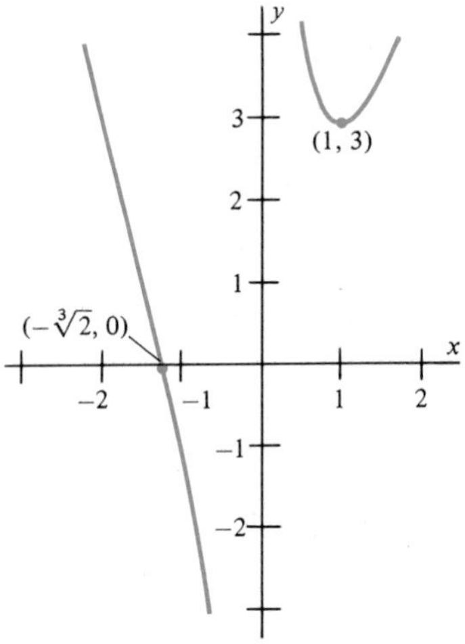

23.

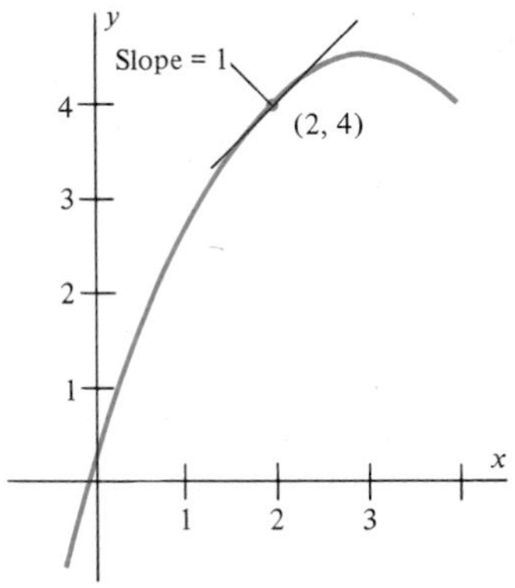

25. **a.** $s(t) = t^2 + \dfrac{1}{t+1} - 1$ **b.** $a(t) = 2 + \dfrac{2}{(t+1)^3}$

27. $(1 + \sqrt[3]{9})^{3/2} \approx 5.4$ meters **33.** $c > b^2/4$

31. True **35.** -1

37. 2

39. $\dfrac{9 - \sqrt{3}}{3 - \sqrt{3}}$

41. 0

43. 2

45. $\dfrac{1}{5}x^5 - 3x^4 + 12x^3 + C$

47. $\dfrac{3}{4}x^4 + C$

49. $\dfrac{1}{4}t^4 + \dfrac{6}{7}t^{7/2} + t^3 + \dfrac{2}{5}t^{5/2} + C$

51. $-\dfrac{1}{3}(9 - s^2)^{3/2} + C$

53. $\dfrac{1}{3}x^3 - \dfrac{1}{2}x^2 + x + C$

55. $\sin x + C$

57. $\dfrac{-1}{8x^2 + 4} + C$

59. $\sqrt{1 + x^2} + 3$

61. $-\sin x + \cos x + 4x - 1$

63. $y^2 = x^2 + 1$

65. $y = \tan x + 1$

67. $y = \dfrac{-1}{2 + x^2}$

CHAPTER 6

Exercise Set 6.1

7. Larger. The rectangles extend outside the triangle.

Exercise Set 6.2

1. 6

3. 40

5. $\dfrac{\sqrt{2}}{4}\pi$

7. 12

9. 85

11. $(2 + \sqrt{2})\dfrac{\pi}{4}$

13. **a.** $\dfrac{17}{2}$

b. $10 - \dfrac{6}{n}$

c. 10

d. $10 + \dfrac{6}{n}$

e. 10, 0
f. 10

15. 26

17. **a.** $\dfrac{65}{8}$

b. $\dfrac{17}{2} - \dfrac{3}{2n}$

c. $\dfrac{17}{2}$

d. $\dfrac{17}{2} + \dfrac{3}{2n}$

e. $\dfrac{17}{2}$, 0

f. $\dfrac{17}{2}$

19. 7

21. 6

23. **a.** $y = \sqrt{1 - x^2}$
b. 0.56
c. 0.68
d. $\pi/4 \approx 0.785$

25. $\overline{S}_3 \approx 0.896$, $A \approx 3.58$; $\overline{S}_6 \approx 0.85$, $A \approx 3.40$

Exercise Set 6.3

1. **a.** -3 **b.** 1 **c.** 3
d. 6 **e.** -1 **f.** -3

3. 30

5. 6

7. -20

9. $\dfrac{-10}{3}$

11. $\dfrac{11}{6}$

13. $\dfrac{112}{3}$

15. -99

17. $A = 8\dfrac{1}{2}$

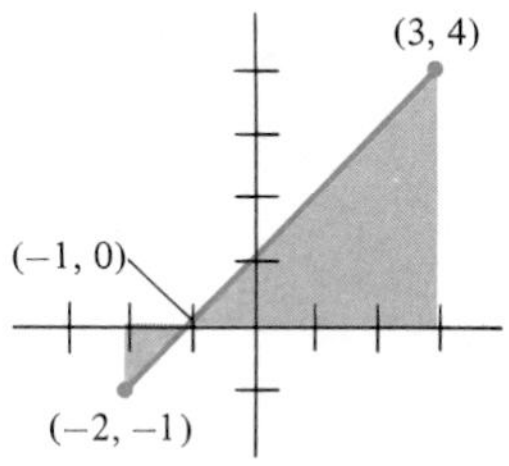

19. $A = 5$

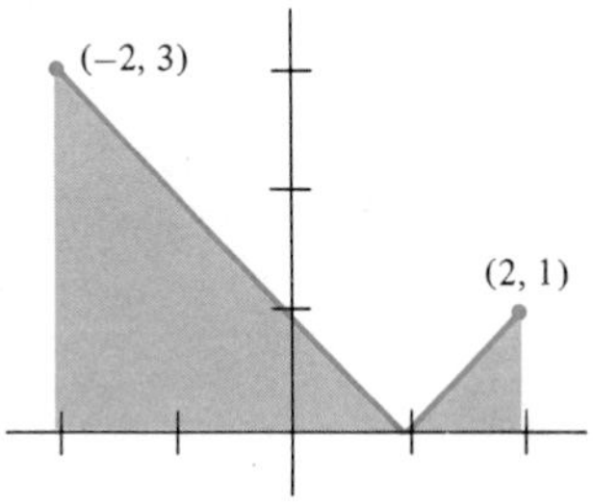

21. $2\pi + 4$

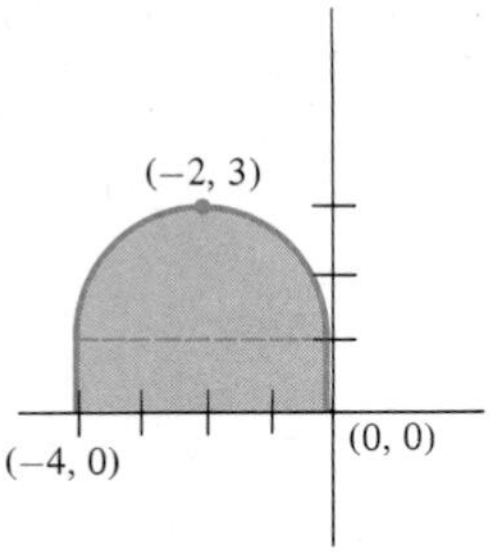

23. $18\pi + 18$

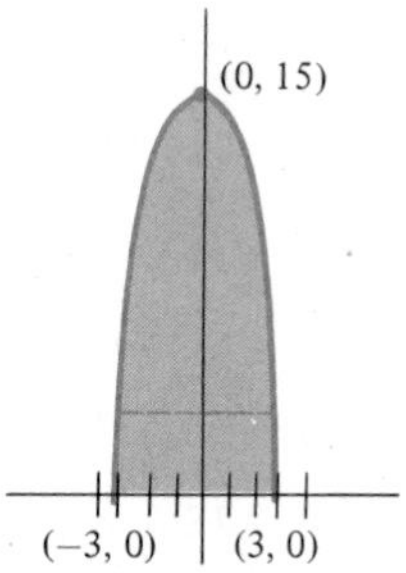

25. 22.97

27. 0.345

29. 265.64

31. 24.529
23.517
23.384

Exercise Set 6.4

1. -4

3. $\dfrac{10 - 8\sqrt{2}}{5}$

5. $\dfrac{3}{4}$

7. $\dfrac{\pi}{4} - \dfrac{1}{2}$

9. $\dfrac{7}{3}$

11. 5

13. $\sqrt{3}$

15. $\dfrac{1366}{15}$

17. $\dfrac{-1}{8}$

19. $\dfrac{1}{2} - \dfrac{\sqrt{2}}{4}$

21. 0

23. $\dfrac{1}{3a}((a + b)\sqrt{a + b} - b\sqrt{b})$

25. 0

27. $2 + 2 \cos \dfrac{\pi}{\sqrt{2}}$

29. $\dfrac{6}{49}$

31. $\dfrac{64}{3}$

33. $\dfrac{1}{12}$

35. $\dfrac{\pi}{8} - \dfrac{1}{4}$

37. 2

39. **a.** 2
b. 4
c. 0
d. $\dfrac{3}{2}$
e. -4
f. $\dfrac{3}{2}$

41. 4

43. $\dfrac{117}{3}$

45. $8\sqrt{6} - 9$

47. $\dfrac{56}{3}$

49. $\dfrac{29}{6}$

51. 26

53. 3

55. $\dfrac{53}{3}$

57. $\dfrac{5}{16}$

Exercise Set 6.5

1. 18

3. 2

5. $\dfrac{155}{3}$

7. $\dfrac{2}{\pi}$

9. $\dfrac{1}{3}$

11. 10

13. 1

15. $\dfrac{1}{3}$

17. 130

19. **a.** 2.2
b. 2.225, 2.257, 2.262

Exercise Set 6.6

1. $\dfrac{x^2}{2} - \dfrac{1}{2}$

3. $\sin x$

5. $\dfrac{a}{3}x^3 + \dfrac{b}{2}x^2 + cx$

7. $x^2 \sin x^2$

9. $-x^3 \cos^2 x$

11. $-2x \sin(x^8 + 1)$

13. $xe^{-x^2} + \displaystyle\int_3^x e^{-t^2}\,dt$

15. 1

17. $\dfrac{27}{2}$

19. $n > 160$

21. 1.4227

Exercise Set 6.7

1. 28

3. $\dfrac{\pi + \sqrt{2}\pi}{4}$

5. 1.3253

7. 1.1051

9. $\dfrac{\pi}{9}(4 + \sqrt{3})$

11. 0.6649

13. 6.5998

15. 0.6658

17. 6.5846

19.

n	5	20	100	250
s	0.6645375	0.6656965	0.6657707	0.6657731

21. **a.** 2.4565
b. 1.1713

Review Exercises–Chapter 6

1. **a.** $\frac{33}{4}$
b. $\frac{51}{4}$

3. **a.** 1.3949
b. 1.7282

5. **a.** 0.6046
b. 2.4184

7. 8

9. 18

11. $\frac{-1}{15}$

13. $\frac{9}{20}$

15. $\frac{7}{2}$

17. $\frac{200}{3}$

19. $\frac{14}{3}$

21. $\frac{8}{3}$

23. $\frac{196}{3}$

25. $1 - \frac{\sqrt{2}}{2}$

27. $\frac{1}{2}$

29. 28,542,630

31. 2

33. $\frac{3}{4}$

35. 1

37. $\frac{1}{30}$

39. -286.2381

41. $\frac{-2}{9}$

43. $\frac{10}{3}$

45. $\frac{16}{3}$

47. $\frac{83}{15}$

49. $\frac{7}{3}$

51. $\frac{23}{3}$

53. $\frac{26}{3}$

55. $\frac{4}{3}\sqrt{2}$

57. $\frac{243}{5}a^2 + 54a + 27$

59. $\frac{32}{3}$

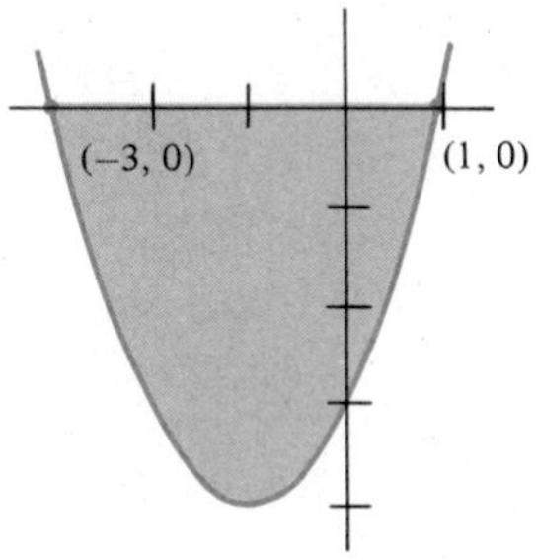

61. 8

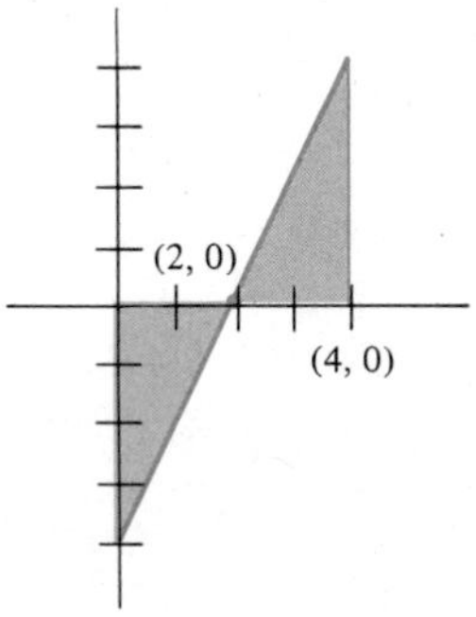

63. **a.** Object is located 11 units to the left of the origin.
b. 16
c. to the left

65. $\frac{64}{27}$

67. **a.** 0
b. $x^2\sqrt{1 + x}$
c. 18
d. $8x^2\sqrt{1 + 2x}$

69. False

71. True

73. $-2\sec(2x)$

75. 18

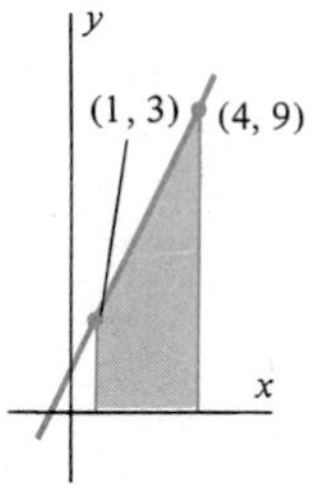

77. $\frac{9\pi}{4} + 3$

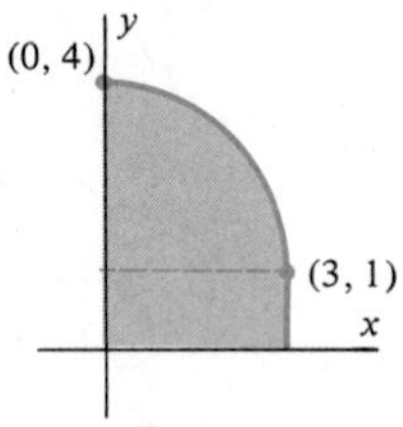

79. 4π

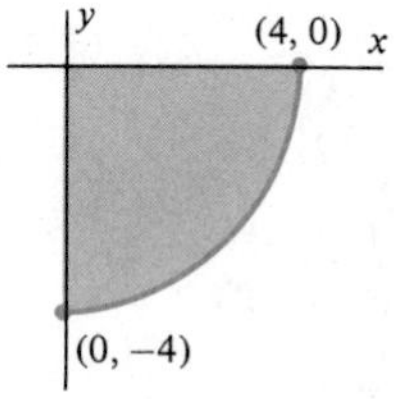

81. 2.2

83. 1.4054

85. 0.8660

CHAPTER 7

Exercise Set 7.2

1. 6

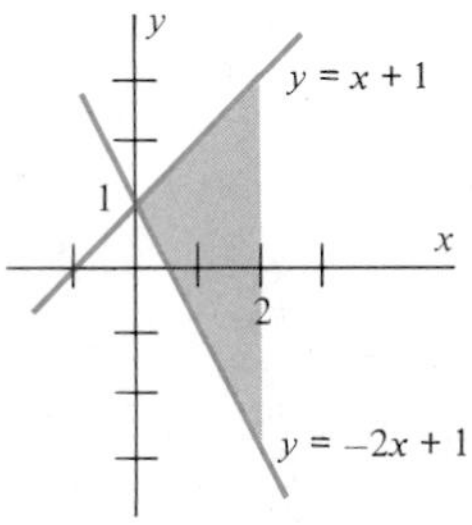

3. $\frac{80}{3}$

5. $4\sqrt{2}$

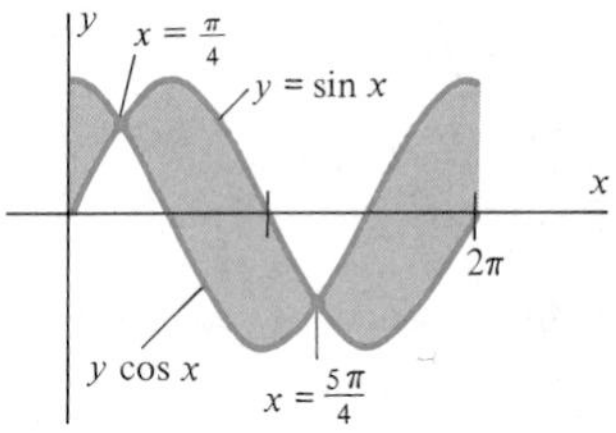

7. $\frac{52}{3}$

9. $\frac{44}{3}$

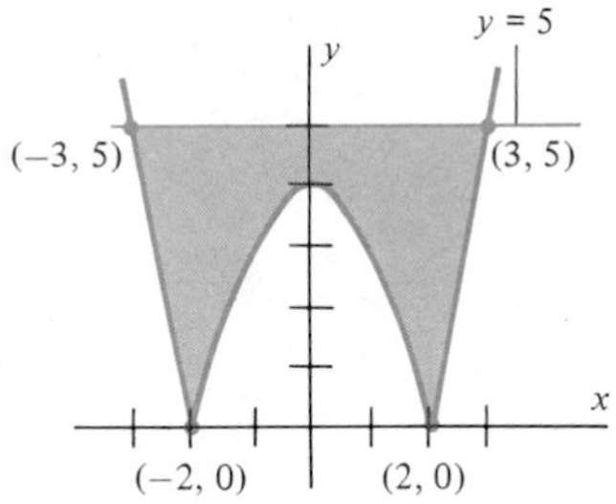

11. 1

13. $\frac{125}{6}$

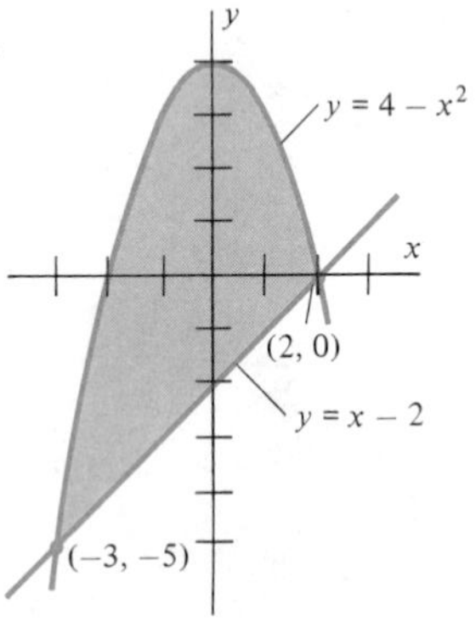

15. $\frac{1}{12}$

17. $\frac{125}{6}$

19. 27

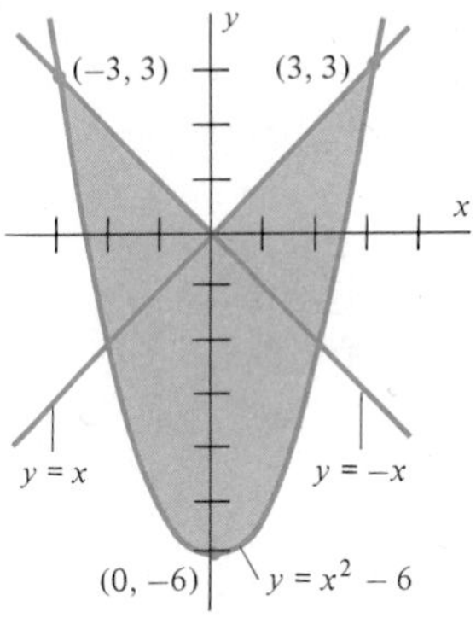

21. $\frac{1}{12}$

23. $\frac{32}{15}$

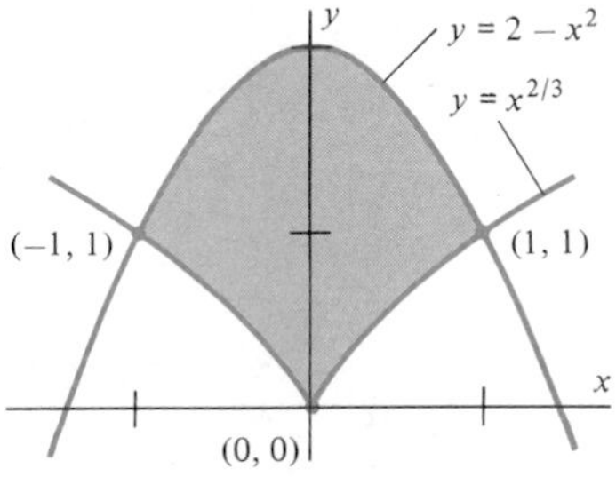

25. $\frac{4}{5}$

27. 3

29. 57.292

31. $a = 4 - 4^{2/3}$

33. 2.095 ($n = 20$)

35. 37.67 ($n = 30$)

Exercise Set 7.3

1. 117π

3. $\frac{\pi^2}{2}$

5. $\pi\left(1 - \frac{\pi}{4}\right)$

7. $\frac{7\pi}{8}$

9. $\frac{128\pi}{15}$

11. $\pi(\pi + 6)/4$

13. 8π

15. $\frac{64\pi}{5}$

17. $\frac{3\pi}{2}$

19. $\frac{4\pi r^3}{3}$

21. $\frac{512}{3}$

23. $\frac{2048}{15}$

25. $\frac{1}{3}\pi h(R^2 + Rr + r^2)$

27. $\frac{5}{42\pi}$

29. $\frac{136\pi}{15}$

31. $\frac{225}{2}\pi$

33. $V = \frac{4}{3}\pi a^2 b$

35. $\frac{2}{3} r^3 \tan\theta$

37. ≈ 4.16

39. ≈ 118.24

Exercise Set 7.4

1. $\frac{\pi}{3}$

3. $\frac{484}{5}\pi$

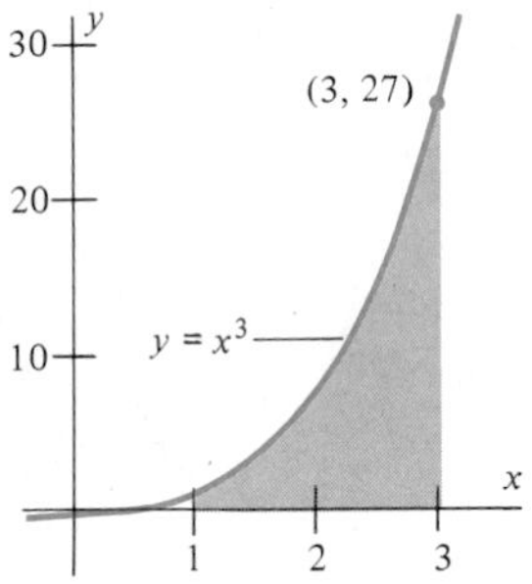

5. $\frac{1024}{15}\pi$

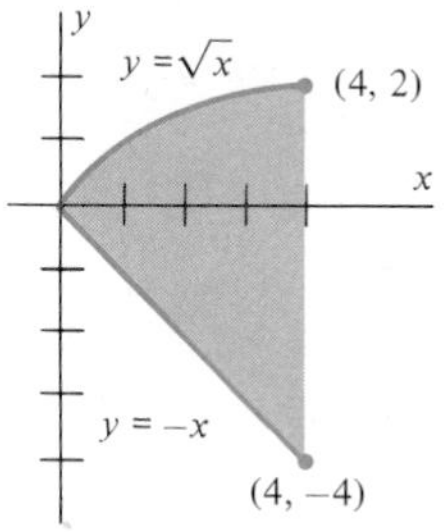

7. 9π

9. $\frac{\pi^2}{2} - \pi$

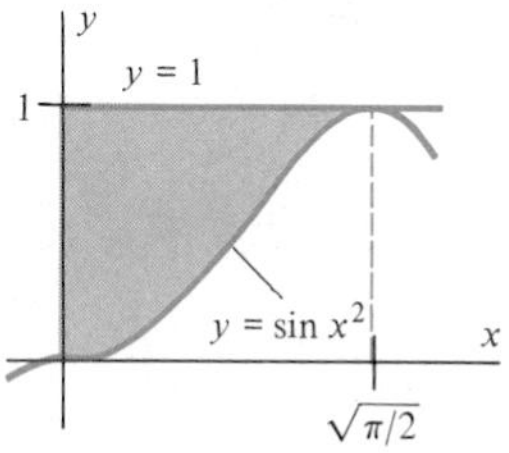

11. 18π

13. $24\pi - \frac{16\sqrt{2}}{9}\pi$

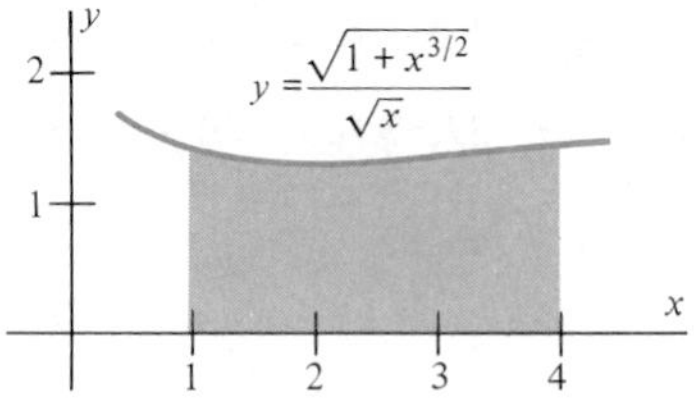

15. $\frac{1088}{15}\pi$

17. $\frac{49}{30}\pi$

19. $\frac{8}{15}\pi$

21. 56π

23. $4\pi^2$

25. 19.74

Exercise Set 7.5

1. $5\sqrt{5}$

3. $\frac{8}{27}(10\sqrt{10} - 1)$

5. $\frac{2}{3}(5\sqrt{5} - 2\sqrt{2})$

7. 12

9. $\frac{87}{8}$

11. $\frac{14}{3}$

13. $\frac{\pi}{27}(730\sqrt{730} - 1)$

15. $\frac{8\pi}{3}(5\sqrt{5} - 2\sqrt{2})$

17. $\frac{112}{3}\pi$

19. 151.63

21. $n = 4$; $S \approx 40.35$

23. 2.9579

Review Exercises–Chapter 7

1. 8

3. $\frac{32}{3}$

5. $\frac{125}{6}$

7. $2\sqrt{2} - 2$

9. $\frac{9}{2}$

11. $\frac{8}{15}$

13. $\frac{4}{3}$

15. 8π

17. 4

19. $\frac{\pi\sqrt{3}}{2}$

21. $\pi\left(1 - \frac{\sqrt{2}}{2}\right)$

23. $\frac{213}{64}\pi$

25. $\left(\frac{9}{5}\sqrt[3]{9} + 9\sqrt[3]{3}\right)\pi$

27. $\frac{13\sqrt{13}}{27} - \frac{8}{27}$

29. $\frac{33}{16}$

31. **a.** $\frac{64}{3}$ **b.** $\frac{16\sqrt{3}}{3}$

33. $\frac{4}{3}$

35. 5

37. 3.82

CHAPTER 8

Exercise Set 8.2

1. 421.2 lb

3. 166.4 lb

5. 341,000 lb

7. 332.8 lb

9. 571.9 lb

11. 3 lb

13. Equal force acting on opposite side of rectangular plate

15. 157.083 lb

17. 2995.2 lb

Exercise Set 8.3

1. 5 ft-lb

3. 15 ft-lb

5. 25 ft-lb

7. True

9. **a.** 5 lb/ft
b. 5 lb
c. 22.5 ft-lb

11. $74{,}880\pi$ ft-lb

13. $1{,}248{,}000\pi$ ft-lb

15. $74{,}880\pi$ ft-lb

17. 3833.9π ft-lb

19. 400 ft-lb

21. **a.** $\frac{160}{q^2}$ N-m
b. 40 N-m

23. 0.758 K

Exercise Set 8.4

1. $\frac{100}{39}$

3. 4.8 ft

9. $x_3 = 2$, $y_3 = \frac{-5}{3}$

11. $\left(\frac{3}{5}, \frac{12}{35}\right)$

13. $\left(\frac{4}{3}, \frac{4}{3}\right)$

15. $\left(0, \frac{4}{3\pi}\right)$

19. $\left(\frac{4}{5}, \frac{2}{7}\right)$

21. $\left(\frac{8}{3}, 0\right)$

23. $\left(0, \frac{4}{\pi}\right)$

25. $\left(\frac{49}{22}, \frac{933}{220}\right)$

27. $\frac{115}{18}$

29. $\left(0, 0, \frac{5}{2}\right)$

33. $\left(0, \frac{54}{55}\right)$

Exercise Set 8.5

1. $90\pi^2$
3. $\dfrac{3\pi}{10}$
5. $8\pi A$
7. 19,968 lb
9. 1291.65 lb
11. $4\pi^2$
23. $(b - a)\overline{f(x)}$ = area bounded by $f(x)$
25. False
27. False

Exercise Set 8.6

1. 9.125°
3. a. −3.5°
   b. −3.5°
   c. −3.58°
5. 5
7. −1
9. $\dfrac{22}{3}$
11. $\dfrac{2 - \sqrt{2}}{2\sqrt{\pi}}$
13. 0
15. $\dfrac{15°}{2}$ C
17. 10
21. ±3

Review Exercises–Chapter 8

1. 18.75 ft-lb
3. 652,400 ft-lb
5. 4000 ft-lb
7. 36.4 lb
11. 12 cm
13. 1.64
15. 58.70 cm
17. $\left(\dfrac{4}{3}, \dfrac{2}{3}\right)$
19. $\left(\dfrac{1}{2}, \dfrac{6}{15}\right)$
21. $\bar{x} = \bar{y} = 0$
23. $\dfrac{128\pi}{3}$
25. 0
27. 0
29. $\dfrac{17}{12}$

CHAPTER 9

Exercise Set 9.1

1. 3
3. 1/3
5. −3
7. 3/2
9. 0

11.

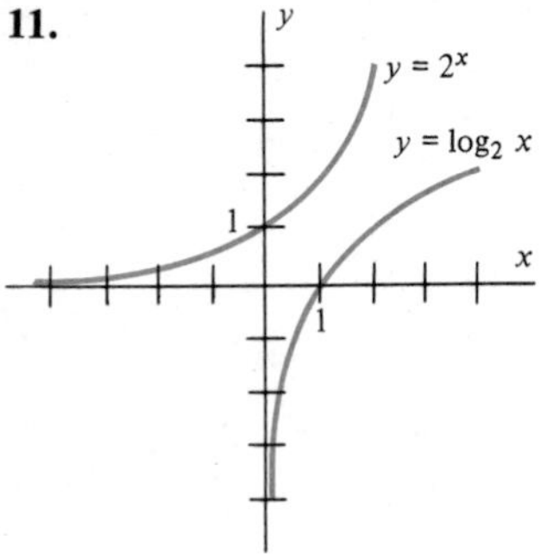

13.

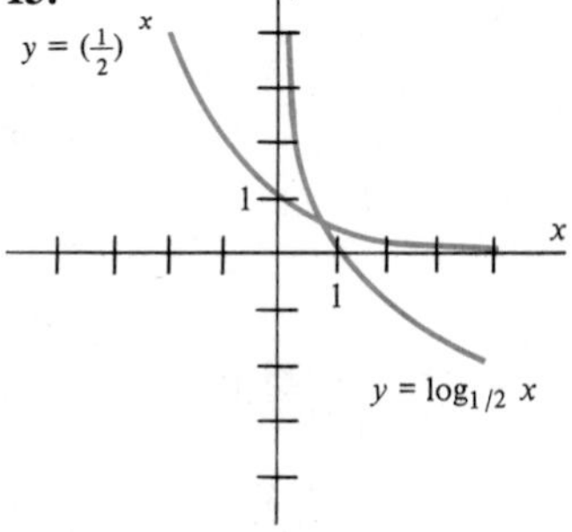

15. 2

23. True

Exercise Set 9.2

1. a. 1/2 b. 2 c. $\sqrt{2}$ d. 0
   e. e^2 f. 2 g. e^2 h. 2
3. True
5. $\ln x + 1$
7. $\dfrac{1}{2} \cdot \dfrac{(3x^2 - 1)}{(x^3 - x)}$
9. $\dfrac{\cos(\ln x)}{x}$
11. $\dfrac{\ln x}{(1 + \ln x)^2}$
13. $\dfrac{162(\ln\sqrt{x})^3}{x}$
15. $\dfrac{\dfrac{b}{a + bt}\ln(c + dt) - \dfrac{d}{c + dt}\ln(a + bt)}{\ln^2(c + dt)}$
17. $-\tan x$
19. $\dfrac{(\ln t - 1)^2}{(1 + \ln^2 t)^2}$
21. $\dfrac{x}{y}$
23. $\dfrac{y(x - y - x\ln y)}{x(x + y\ln x)}$
25. $2x\ln(2x^2)$
27.

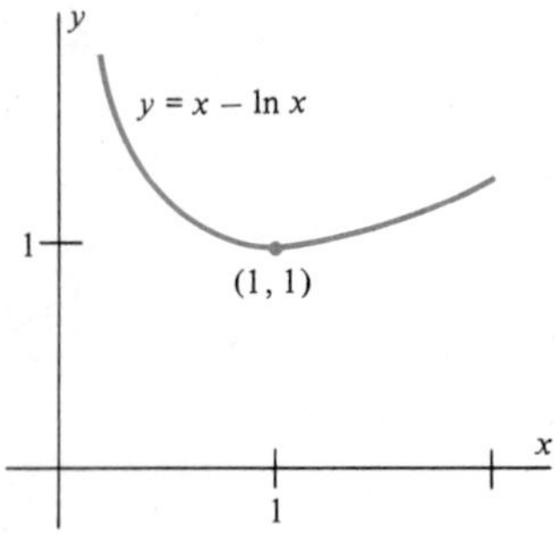

29.

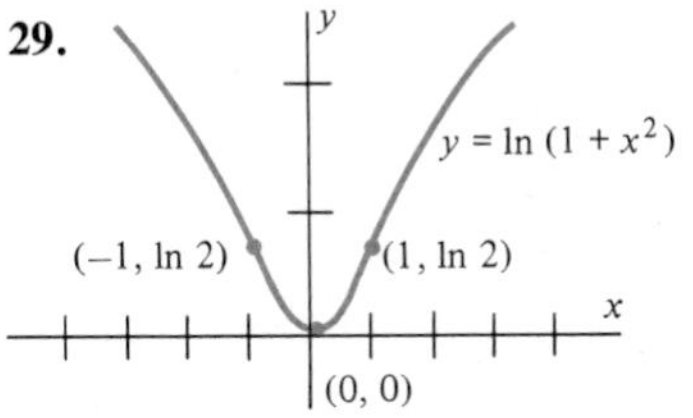

31.

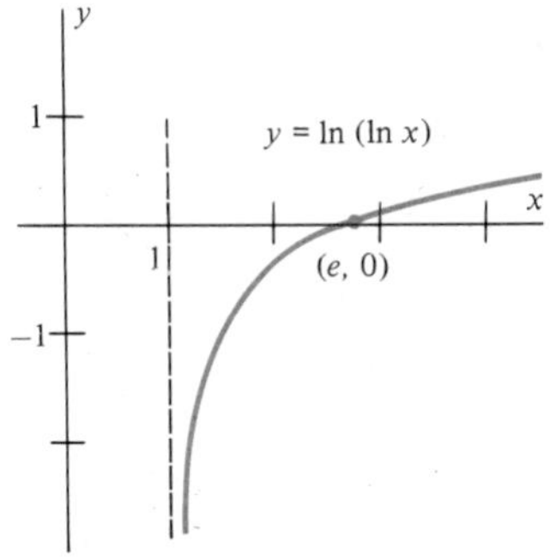

33. $-\ln|1 - x| + c$

35. $= \frac{1}{2} \ln|x^2 - 2x| + c$

37. $\frac{1}{3} \ln|x^3 + 9x| + c$

39. $\ln|\ln x| + c$

41. $\frac{1}{3}(\ln|x|)^3 + c$

43. $-2 \ln|1 - \sqrt{x}| + c$

45. $\frac{x^4}{4} - \frac{x^3}{3} + 2x^2 - 3x + 4 \ln|x + 1| + c$

47. $\ln 2$

49. $\left[\frac{1}{2} \ln|\ln x^2|\right]_e^{e^2} = \frac{1}{2} \ln 2$

51. $2 \ln 3 - 3 \ln 2$

53. $\int_3^4 \frac{dx}{x - 2} = \ln 2$

55. a. $\ln(1 + \epsilon) \approx \ln 1 + \frac{\epsilon}{1} = \epsilon$

b. $\ln(0.8) = \ln(1 + (-0.2)) \approx -0.2$
$\ln(0.95) = \ln(1 - 0.05) \approx -0.05$
$\ln(1.05) = \ln(1 + 0.05) \approx 0.05$
$\ln(1.2) = \ln(1 + 0.2) \approx 0.2$

57. $\pi \int_3^5 \frac{2}{x - 1} dx = 2\pi \ln 2$

59. $\frac{1}{e - 1}$

61. c. 5%

65. $x \approx 0.01866$

67.

n	$\sim\ln 2$	$\sim\ln 4$	$\sim\ln 10$
10	0.6931503	1.386520	2.311978
20	0.6931473	1.386310	2.303565
50	0.6931471	1.386295	2.302618
200	0.6931468	1.386294	2.302585

Exercise Set 9.3

1. a. 2 **b.** $\frac{1}{4}$ **c.** $\frac{x}{y}$ **d.** $-x^2$

e. $\sqrt{x}$ **f.** 2^x **g.** $\frac{1}{x}$ **h.** x^4

i. $x^2 + \ln x$ **j.** $\frac{1}{x} \cdot e^x$

3. $y = x - \ln|x + 3|$

5. $3e^{3x}$

7. $\frac{e^{\sqrt{t}}}{2\sqrt{t}}$

9. $(2x - 1)e^{x^2 - x}$

11. $e^x(\sin x + \cos x)$

13. $\frac{e^x}{e^x + 1} - \frac{1}{x + 1}$

15. $-6x \cdot e^{x^2} \cdot (2 - e^{x^2})^2$

17. $\frac{1}{2} \cdot (e^x - e^{-x})$

19. $\frac{1}{2} \frac{e^{\sqrt{x}}}{\sqrt{x}} \cdot \left(\ln\sqrt{x} + \frac{1}{\sqrt{x}}\right)$

21. $\frac{1 - xy}{x^2}$

23. $\frac{1}{e^y(x + 2y) - 2}$

25. True

27. True

29. $-e^{-x} + c$

31. $\frac{1}{2} e^{x^2 + 3} + c$

33. $\frac{1}{8}(1 + e^{2x})^4 + c$

35. $-e^{1/x} + c$

37. $\ln(1 + e^x) + c$

39. $\frac{-1}{(3 + e^x)} + c$

41. $2e^{\sqrt{x}} + e^{2\sqrt{x}} + c$

43. $\frac{1}{e} - \frac{1}{4}$

45. $e - 1$

47. 2

49. $e^2 - 1$

51. ∞

53. 0

55. e

57. 1

59. 1

61. $(-1, 1)$; $(0, 0)$; $(1, 1)$

63. $(-1, -1/e)$

65. $\int_0^2 \pi \cdot (e^{2x})^2 \, dx = \frac{\pi}{4}(e^8 - 1)$

67. 0

Exercise Set 9.4

1. a. $e^{\ln 2 + \ln x}$ **b.** $e^{3 \ln \pi}$ **c.** $e^{\ln 7 \cdot \ln x}$
d. $e^{\sqrt{2} \ln 4}$ **e.** $e^{\ln 3 \cdot \sin x}$ **f.** $e^{x \ln 2 + (1 - x)\ln 4}$

3. $10^a = e^{a \ln(10)}$, $e^b = 10^{b \log_{10}(e)}$

5. $y' = (\ln 3) \cdot 3^x$

7. $\dfrac{2}{(2x-1)\ln 10}$

9. $\dfrac{1}{x\cdot\ln 10\cdot\ln x}$

11. $2t\log_2 t + \dfrac{t}{\ln 2}$

13. $\ln\pi\cdot\pi^x + \pi x^{\pi-1}$

15. $x^x\cdot(\ln x + 1)$

17. $x^{\sqrt{x}-1/2}\left(1+\dfrac{\ln x}{2}\right)$

19. $\cos x^{\sin x}\left[\cos x\cdot\ln|\cos x| + \dfrac{\sin^2 x}{\cos x}\right]$

21. $\dfrac{5^x}{\ln 5}+c$

23. $\dfrac{2}{\ln\pi}\pi^{\sqrt{x}}+c$

25. $\dfrac{1}{1+2\ln a}\cdot x^{(2\cdot\ln a+1)}+c$

27. $\dfrac{40}{\ln 3}$

29. True

33. e^2

35. $\dfrac{3}{4}$

37. $r=\ln(1.10)$

39. $\left[\dfrac{1}{x}+\dfrac{1}{x+1}+\dfrac{1}{x+2}-\dfrac{1}{x+3}-\dfrac{1}{x+4}-\dfrac{1}{x+5}\right]\cdot\dfrac{x(x+1)(x+2)}{(x+3)(x+4)(x+5)}$

41. $\left[\dfrac{1}{3}\cdot\dfrac{2x}{x^2+1}-\dfrac{1}{x+1}-\dfrac{2}{x+2}\right]\cdot\dfrac{3\sqrt{x^2+1}}{(x+1)(x+2)^2}$

43. $\dfrac{7}{8}x^{-1/8}$

Exercise Set 9.5

1. e^{2t}

3. $e^{5(t-1)}$

5. e^{-2t}

7. False; $P'(t)=\ln(1.05)P(t)$

9. $e^{1/2}$

11. 1

13. e^{-1}

15. $\dfrac{2500}{e^{0.8}}\approx 1123.32$

17. $\ln(1.05)\approx 0.04879$

19. False

21. a. $v(t)=v_0e^{(-c/m)t}$
b. $v(t)=10e^{-4t}$
c. 0

23. \$606.53; \$620.92

25. 3.4×10^{-5} atm

27. Sell after 0.83 years. The original price does not effect this answer.

29. 23.3 hr

31. 8.1 μg

33. 1.69 years

Exercise Set 9.6

1. $y=2e^t-1$

3. $y=\dfrac{-\pi}{2}e^{-t}+\pi$

5. $r^2-1=0;\ C_1e^x,\ C_2e^{-x}$

7. $r^2+2r-3=0;\ C_1e^x,\ C_2e^{-3x}$

9. $r^2+4r+1=0;\ Ce^{(-2\pm\sqrt{3})x}$

13. $y=\dfrac{1}{a}(Ce^{-at}+b)$

15. $y=Ce^{-1/t}$

17. $y=Ce^{t^2}-\dfrac{3}{2}$

19. $y=2t+Ct^{-2}$

21. $y=e^{-t}+Ce^t$

23. $y=e^{x^2/2}$

25. 13.6°C

27. $I=\dfrac{-E}{R}e^{-(R/L)t}+\dfrac{E}{R}$ and $\lim\limits_{t\to\infty}I=\dfrac{E}{R}$

Review Exercises Chapter 9

1. a. 4 b. 1/7776 c. 32 d. $\dfrac{1}{(\sqrt{2})^3}$

3. a. $y=x-2$ b. $y=\sqrt{x}$
c. $y=\dfrac{1}{2}(x-1)$ d. $y=x^2$

5. $2x\ln(x-a)+\dfrac{x^2}{x-a}$

7. $\dfrac{2}{x\ln x}$

9. $2e^{x^2}(x\tan 2x+\sec^2 2x)$

11. $\dfrac{(1+t+e^t)e^t}{(1+e^t)^2}$

13. $\dfrac{(t^2+b)\cdot\ln(t^2+b)+2t(t-a)\ln(t-a)}{(t-a)(t^2+b)[\ln(t^2+b)]^2}$

15. $\left(\dfrac{\sqrt{t}-2}{2t}\right)e^{\sqrt{t}-\ln t}$

17. $\dfrac{-1}{3x^2}-\dfrac{1}{2(1-2x)}-\dfrac{1}{8x}$

19. $\dfrac{-(2xy\ln y+y^2)}{(2x^2+xy\ln x)}$

21. $\dfrac{2x+ye^{xy}}{2y-xe^{xy}}$

23. $\dfrac{-y}{x}$

25. $\dfrac{4\sqrt{x}e^{2x}+\cos\sqrt{x}}{4\sqrt{x}(e^{2x}+\sin\sqrt{x})}$

27. $\dfrac{-1}{2}\ln|1-x^2|+c$

29. $\dfrac{1}{4}\ln|4x+2x^2|+c$

31. $2e^{x/2}+c$

33. $2(\sqrt{2}-1)$

35. $\dfrac{5}{6}-2\ln 2$

37. $\dfrac{e^4+3}{2}$

39. $e^x-e^{3x}+\dfrac{3}{5}e^{5x}-\dfrac{1}{7}e^{7x}+c$

41. $\ln|x^3 + x^2 - 7| + c$

43. $\frac{-1}{2} \ln 2$

45. $\frac{1}{2}(\ln x + 2)x^{\sqrt{x}-1/2}$

47. $\left(\frac{1}{\sqrt{e}}, -\frac{1}{2e}\right)$

49. $(-1, 1)$

51. $\frac{15}{4} - \ln 16$

53. $2\pi \ln 2$

55.

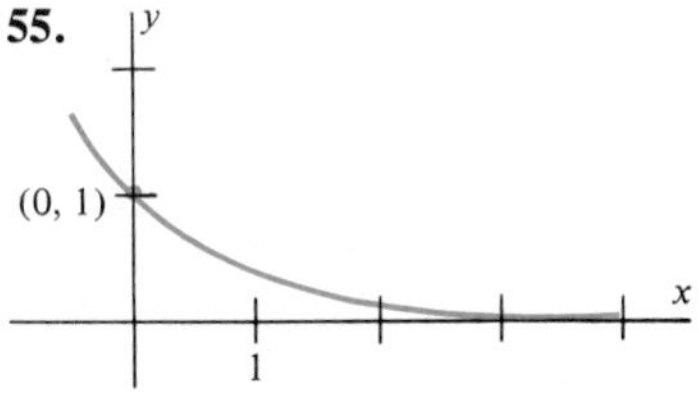

57.

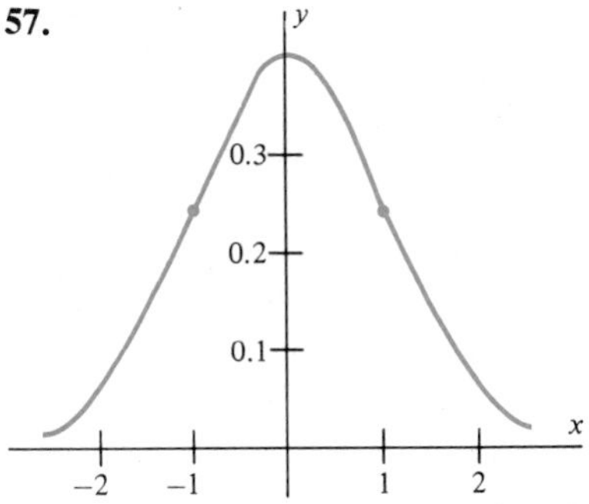

59. $e^{\sqrt{4.1}} \approx \left(\frac{41}{40}\right)e^2$

61. **a.** $\frac{1}{2} - \frac{1}{t+2}$
b. $5 - \ln 6$

63. $r = \ln 2$

65. $\frac{\ln^2 t}{36\pi}$

67. $x \approx 0.158594$, $x \approx 3.146193$

69. $y = 4e^{-x}$

71. $y = 2 - e^{-x}$

73. $y = e^{-x}$

75. $x \approx -0.351734$

77. **a.** 14.46 hr
b. 48 hr

79. e^3

81. $\frac{1}{e^3}$

83. **a.** $\frac{dr}{dt} = k$
b. $r(t) = k_1 t + k_2$
c. $r(t) = 10 - \frac{t}{10}$
d. 4 cm

85. $y = 30 + Ce^{kt}$
$C = 70$
$K = \frac{\ln 5 - \ln 7}{3}$
so $y(10) = 52.8°\text{C}$

CHAPTER 10

Exercise Set 10.2

1. $\frac{1}{3} \sin 3x + C$

3. $\frac{1}{2} \sec 2x + C$

5. $\tan x + C$

7. $\frac{1}{2} \ln|\sec x^2 + \tan x^2| + C$

9. $\frac{1}{6} \ln|\sec(3x^2 - 1)| + C$

11. $\frac{1}{4} \tan^4 x + C$

13. $e^{\tan x} + C$

15. $\ln|x + \tan x| + C$

17. $\frac{1}{2} \cot(1 - x^2) + \frac{1}{2}(1 - x^2) + C$

19. $\ln|\sec x + \tan x| + \ln|\sec x| + C$

21. $\frac{1}{-2} \ln|\sec(x^2 - 2x) + \tan(x^2 - 2x)| + C$

23. $e^{\sin x} + C$

25. $-2 \cot\sqrt{x} + 2 \csc\sqrt{x} + C$

27. $\frac{1}{2}\sqrt{1 + \sin 4x} + C$

29. $2\sqrt{\tan \theta + 1} + C$

31. $\frac{1}{2} \ln|2 + \sqrt{3}|$

33. $\frac{6}{\pi} \ln\left|\frac{1}{2\sqrt{3} - 3}\right|$

35. $\frac{1}{2} \tan^2 x + C$

37. $\ln\left|\frac{1}{2\sqrt{3} - 3}\right|$

39. $\frac{\pi^2}{2} - \frac{4}{3}\pi$

41. $\frac{2\pi}{3}(1 + \sqrt{7}) \approx 7.6356$

Exercise Set 10.3

1. $\frac{1}{2}x - \frac{1}{8}\sin 4x + C$

3. $-\frac{1}{3}\cos^3 x + C$

5. $-\cos x + \frac{2}{3}\cos^3 x - \frac{1}{5}\cos^5 x + C$

7. $\frac{2}{15}$

9. $\frac{1}{5}\sec^5 x - \frac{1}{3}\sec^3 x + C$

11. $\frac{5\pi}{16}$

13. $\frac{-1}{2}\cot^2 x - \ln|\sin x| + C$

15. $\frac{1}{4}$

17. $x - \frac{1}{2}\cos 2x + C$

19. $\frac{1}{5}\tan^5 x + \frac{1}{7}\tan^7 x + C$

21. $\frac{1}{7}\sec^7 x - \frac{2}{5}\sec^5 x + \frac{1}{3}\sec^3 x + C$

23. $\frac{2}{3}\sin^3 x + C$

25. $2(\sin x)^{1/2} - \frac{2}{5}(\sin x)^{5/2} + C$

27. $\frac{-1}{3(1 + \tan x)^3} + C$

29. $\frac{1}{16}\sin 8x + \frac{1}{4}\sin 2x + C$

31. $\frac{1}{2}\cos^2\theta - \ln|\cos\theta| + C$

33. $\frac{2}{3}(\tan x)^{3/2} + C$

35. $\frac{-1}{4}\frac{\cos(a+b+c)x}{a+b+c} - \frac{1}{4}\frac{\cos(a+b-c)x}{(a+b-c)} - \frac{1}{4}\frac{\cos(a-b+c)x}{(a-b+c)} - \frac{1}{4}\frac{\cos(a-b-c)x}{(a-b-c)} + C$

37. 0 **39.** L

41. 0 **49.** $\frac{\pi}{3}$

51. $\ln|\sqrt{2} + 1|$

53. **a.** $\frac{1}{2}t + \frac{1}{4}\sin 2t$ **b.** $\frac{t^2}{4} - \frac{1}{8}\cos 2t + \frac{33}{8}$

55. $\ln|2 + \sqrt{3}|$

Exercise Set 10.4

1. $\frac{\pi}{6}$ **3.** 0 **5.** π

7. $\frac{\pi}{2}$ **17.** 0 **27.** $\frac{1}{x}$

9. $\frac{1}{\sqrt{3}}$ **19.** $\sqrt{3}$ **29.** $\frac{\sqrt{x^2-1}}{x}$

11. 2 **21.** $\frac{x}{\sqrt{1-x^2}}$ **37.** False

13. $\frac{-1}{2}$ **23.** $2x\sqrt{1-x^2}$ **39.** $\theta = \tan^{-1}\frac{x}{100}$

15. 1 **25.** $\frac{\sqrt{x^2-1}}{x}$ **41.** $\theta = \tan^{-1}\frac{x}{20}$

Exercise Set 10.5

1. $\frac{3}{\sqrt{1-9x^2}}$ **3.** $\frac{1}{2\sqrt{t-t^2}}$ **5.** $\frac{-e^{-x}}{\sqrt{1-e^{-2x}}}$

7. $\frac{-1}{2\sqrt{(\cos^{-1}x)(1-x^2)}}$

9. $\frac{1}{(1+x^2)\tan^{-1}x}$

11. $\frac{-1}{\sqrt{1-x^2}}$

13. $\frac{1-2x\tan^{-1}x}{(1+x^2)^2}$

15. $\frac{1}{2}\tan^{-1}\left(\frac{x}{2}\right) + C$

17. $\frac{1}{2}\sec^{-1}\left(\frac{x}{4}\right) + C$

19. $-\sqrt{1-x^2} + C$

21. $\sec^{-1} e^x + C$

23. $2\tan^{-1} e^{\sqrt{x}} + C$

25. $\frac{1}{3}\tan^{-1} x^3 + C$

27. $\frac{\pi^2}{32}$

29. $\frac{1}{\sqrt{2}}\left(\tan^{-1}(\sqrt{2}) - \tan^{-1}\left(\frac{1}{\sqrt{2}}\right)\right) + C$

31. $\frac{\pi}{8}$ **37.** $\frac{1}{4}$ rad/sec

33. $2\tan\left(2x + \frac{\pi}{4}\right)$ **39.** $\frac{2\sqrt{6}}{3}x + \left(\frac{\pi}{6} - \frac{2}{3}\sqrt{3}\right)$

35. 1200 rad/hr **41.** $\frac{\pi^2}{6}$

Exercise Set 10.6

1. 0

3. 5/4

5. 3/5

7. 17/15

9. $\ln(1 + \sqrt{2})$

11. $2 \cosh 2x$

13. $\cos x \tanh x + \sin x \operatorname{sech}^2 x$

15. $2(\cosh(4x))^{-1/2} \sinh(4x)$

17. $\dfrac{-\sinh x}{\cosh^2 x}$

19. $-2xe^x \operatorname{csch} x^2 \coth x^2 + e^x \operatorname{csch} x^2$

21. $\dfrac{2}{\sqrt{4x^2 + 1}}$

23. $\dfrac{2s}{1 - s^4}$

25. $\dfrac{\pi}{\cosh^{-1} \pi x \cdot \sqrt{\pi^2 x^2 - 1}}$

27. $\sinh x \cdot \cosh x^2 + 2x \cosh x \cdot \sinh x^2$

29. $\dfrac{2 \cosh(\ln x^2)}{x}$

31. $\dfrac{x \operatorname{sech}^2(\sqrt{1 + x^2})}{\sqrt{1 + x^2}}$

33. $\ln\left(\dfrac{2 + \sqrt{5}}{1 + \sqrt{2}}\right)$

35. $x - \dfrac{1}{2} \tanh(2x) + C$

37. $\dfrac{(e^2 - e^{-2})^2}{8}$

39. $\ln|3 + \sqrt{8}|$

41. $\dfrac{1}{3} \ln\left(\dfrac{3}{5} + \dfrac{1}{5} \cdot \sqrt{34}\right)$

43. $\ln|\cosh x| + C$

45. $2(\sinh x)^{1/2} + C$

47. $\dfrac{1}{3} \tanh^3 x + C$

49. False

51. $-\infty$

59. $z_1 = e^{kz}$, $z_2 = e^{-kz}$

$y_1 = \dfrac{z_1 - z_2}{2}$, $y_2 = \dfrac{z_1 + z_2}{2}$.

61. $y_1 = C_1 \sinh k\pi x + C_2 \cosh k\pi x$

$y_2 = C_1 e^{k\pi x} + C_2 e^{-k\pi x}$

63. $\ln \dfrac{25}{17}$

65. No relative extrema.

67. $f\left(\dfrac{1}{2} \ln \dfrac{7}{3}\right) = \dfrac{19\sqrt{21}}{84}$

71. Even: $\cosh x$, $\operatorname{sech} x$ Odd: $\sinh x$, $\tanh x$, $\coth x$, $\operatorname{csch} x$

75. $[0, +\infty)$

77. $\tanh x$: $-\infty < x < \infty$
$\coth x$: $x \in (-\infty, 0) \cup (0, \infty)$
$\operatorname{sech} x$: $x \geq 0$
$\operatorname{csch} x$: $x \in (-\infty, 0) \cup (0, \infty)$

Exercise Set 10.7

1. $A \cos 4t + B \sin 4t$

3. $A \cos 3t + B \sin 3t$

5. $\cos 4t$

7. 0

9. **a.** $R = \sqrt{A^2 + B^2}$ **b.** $\cos(c) = \dfrac{B}{\sqrt{A^2 + B^2}}$

c. $\sin c = \dfrac{A}{\sqrt{A^2 + B^2}}$ **d.** $B \cos \omega t + A \sin \omega t$

13. $2 \cos\left(2t + \dfrac{\pi}{6}\right)$

15. $6\sqrt{2} \cos\left(t + \dfrac{\pi}{4}\right)$

17. $\dfrac{1}{10} \cos 2t$

19. $\dfrac{3}{2\pi}, \dfrac{1}{\pi}, \dfrac{1}{2}, \dfrac{1}{2\pi}$

23. $x(t) = 25$
$y(t) = 16$

25. Eventually both predator and prey will become extinct.

Review Exercises–Chapter 10

1. $\dfrac{\pi}{3}$

3. $\dfrac{\pi}{3}$

5. $\dfrac{\pi}{4}$

7. 1

9. $\frac{3}{5}$

11. $\frac{1}{2\sqrt{x(1-x)}}$

13. $\frac{2t}{1+(1-t^2)^2}$

15. $2x\sinh(1-x) - x^2\cosh(1-x)$

17. $6\csc(\cot 6x)\cot(\cot 6x)\cdot(\csc^2 6x)$

19. $\frac{2}{(2x+1)\sqrt{1-(\ln(2x+1))^2}}$

21. $\frac{1-2x\tan^{-1}x}{(1+x^2)^2}$

23. $\frac{x}{(x^2+4)(\sqrt{x^2+3})}$

25. $-\frac{\operatorname{sech}^2 x}{(\pi+\tanh x)^2}$

27. $\frac{1}{5}(\sinh x + \cos^{-1}2x)^{-4/5}\cdot\left(\cosh x - \frac{2}{\sqrt{1-4x^2}}\right)$

29. $\frac{1}{\sin^{-1}x\sqrt{1-x^2}}$

31. $\tanh^{-1}(\ln x) + \frac{1}{1-(\ln x)^2}$

33. $\frac{x}{\tanh^{-1}(x^2)\cdot(1-x^4)}$

35. $\frac{1}{2}\cosh^{-1}(2x) + C$

37. $\frac{1}{10}\sin^5 2x + C$

39. $\frac{-1}{2} - \frac{\pi}{4}$

41. $\frac{\pi}{2}$

43. $\frac{2}{3}$

45. $\tan^2 2x - x + C$

47. $\frac{1}{6}$

49. $\sin^{-1}\left(\frac{x}{2}\right) + C$

51. $\frac{-1}{2}(\cos^{-1}x)^2 + C$

53. $2\tan^{-1}(e^{\sqrt{x}}) + C$

55. $\frac{2}{3}(\sin^{-1}x)^{3/2} + C$

57. $\frac{1}{6}\left(\sec^3\frac{2\pi}{9} - 1\right)$

59. $2\sqrt{3}$

61. $\frac{-\sqrt{3}}{16}$

63. $\frac{2}{7}\cos^{7/2}x - \frac{2}{3}\cos^{3/2}x + C$

65. $\cos x + \frac{1}{\cos x} + C$

67. $\ln|1+\sinh x| + C$

69. $\frac{1}{3}\cosh^3 x - \cosh x + C$

71. $\frac{\pi}{2}$

73. $\pi\left(1-\frac{\pi}{4}\right)$

75. $\pi\tan^{-1}4$

77. $\frac{2xy}{\sqrt{x^4y^2-1} - x^2}$

79. no relative extrema
Point of inflection at $x = 0$
$y'' < 0$ on $(-\infty, 0)$, concave down
$y'' > 0$ on $(0, \infty)$, concave up

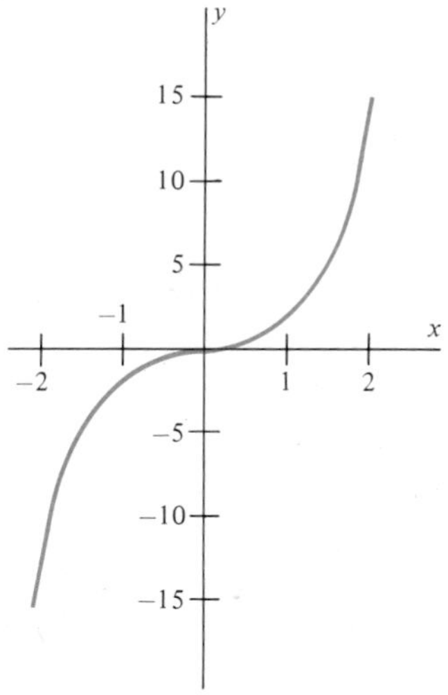

81. 28.845°

83. $\frac{2\sqrt{6}}{3}x + \left(\frac{\pi}{6} - \frac{2\sqrt{3}}{3}\right)$

85. $x = 2,\quad y = \cot^{-1}2$
$x = -2,\quad y = \cot^{-1}(-2)$

87. $\frac{\sqrt{3}}{60}$ radians/sec

91. $x(t) = R\cos(5t - c)$
$x(t) = A\sin 5t + B\cos 5t$

93. $\cos\sqrt{2}\,t$

95. $A\cos 4t + B\sin 4t$

CHAPTER 11

Exercise Set 11.1

1. $\frac{1}{3}(x^2+1)^{3/2}+C$

3. $x - \ln|1+x| + C$

5. $\frac{2}{\pi}$

7. $\frac{-1}{2}\sqrt{1-x^4}+C$

9. $\ln\left|\frac{1+e}{2}\right|$

11. $\frac{1}{2}\ln|x^2+4x+7|+C$

13. $\frac{1}{\pi}(e-1)$

15. $2e-2$

17. $\frac{9}{64}$

19. $\tan^{-1}e^x + C$

21. $\frac{3\pi-8}{32\pi}$

23. 1

25. $\frac{1}{6}$

27. $\frac{(x^2+\pi)}{2}\ln(x^2+\pi) - \frac{1}{2}(x^2+\pi)+C$

29. $\frac{-1}{3}\csc^3 x + C$

31. $\frac{1}{2}\ln|\sec(\pi+x^2)+\tan(\pi+x^2)|+C$

33. $\frac{3^{2x}}{2\ln 3}+\frac{10\cdot 3^x}{\ln 3}+25x+C$

35. $2\sqrt{x}+9x+18x\sqrt{x}+\frac{27}{2}x^2+C$

37. $2(\sqrt{x}+1)-2\ln|\sqrt{x}+1|+\frac{1}{\sqrt{x}+1}+C$

39. $\frac{x}{\tan(x^2)}+C$

41. $\frac{1}{4}(\sin x-\cos x)^4+C$

Exercise Set 11.2

1. $xe^x - e^x + C$

3. $x\sin x + \cos x + C$

5. $x\tan^{-1}x - \frac{1}{2}\ln|1+x^2|+C$

7. $\frac{x}{\pi}\tan\pi x - \frac{1}{\pi^2}\ln|\sec\pi x|+C$

9. $\frac{x}{2}\sin(\ln x) - \frac{x}{2}\cos(\ln x)+C$

11. 1542.28

13. $x(\ln x)^2 - 2x\ln x + 2x + C$

15. $\frac{x}{2}e^{2x} - \frac{1}{4}e^{2x}+C$

17. $\frac{9-e^2}{24}$

19. $x\cosh x - \sinh x + C$

21. $\frac{2}{5}e^{2x}\cos x + \frac{1}{5}e^{2x}\sin x + C$

23. $\left(\frac{x^2}{2}-\frac{1}{4}\right)\sin^{-1}x - \frac{x}{4}\sqrt{1-x^2}+C$

25. $\frac{2}{3}x^{3/2}\ln x - \frac{4}{9}x^{3/2}+C$

27. $x^2\sqrt{1+x^2} - \frac{2}{3}(1+x^2)^{3/2}+C$

29. $x\tanh^{-1}x + \frac{1}{2}\ln|1-x^2|+C$

31. $x\sinh^{-1}x - \sqrt{x^2+1}+C$

33. $\frac{x^2}{2}\sinh^{-1}x - \frac{x}{4}\sqrt{1+x^2} - \frac{1}{4}\ln|x+\sqrt{x^2+1}|+C$

35. $-4x^{3/2}e^{-\sqrt{x}} - 12xe^{-\sqrt{x}} - 24\sqrt{x}\,e^{-\sqrt{x}} - 24e^{-\sqrt{x}}+C$

37. $\frac{1}{\pi}$

39. $\pi(e-2)$

41. $\frac{1}{1+\pi^2}\left(\frac{\pi}{e}+\pi\right)$

43. $\frac{2}{\pi}$ meters

Exercise Set 11.3

1. $-\sqrt{1-x^2}+C$

3. $\sin^{-1}\left(\frac{x}{3}\right)+C$

5. $\frac{x\sqrt{x^2-4}}{2}+2\ln\left|\frac{x+\sqrt{x^2-4}}{2}\right|+C$

7. $\ln\left|\frac{1-\sqrt{1-x^2}}{x}\right|+\sqrt{1-x^2}+C$

9. $\frac{1}{16}\theta + \frac{1}{32}\sin 2\theta + C = \frac{1}{16}\sec^{-1}\left(\frac{x}{2}\right) + \frac{1}{8}\frac{\sqrt{x^2-4}}{x^2} + C$

11. $\frac{1}{128}\tan^{-1}\left(\frac{x}{4}\right) + \frac{x}{32(16+x^2)} + C$

13. $\frac{1}{\sqrt{1-x^2}} + C$

15. $\frac{1}{3}(x^2-1)^{3/2} + C$

17. $\ln\left|\frac{\sqrt{x^2+3}+x}{\sqrt{3}}\right| - \frac{x}{\sqrt{x^2+3}} + C$

19. $\frac{x\sqrt{x^2+a^2}}{2} - \frac{a^2}{2}\ln(x+\sqrt{x^2+a^2}) + C$

21. $\ln\left|\frac{x+\sqrt{x^2-a^2}}{a}\right| - \frac{\sqrt{x^2-a^2}}{x} + C$

23. $\frac{-1}{\sqrt{x^2+a^2}} + C$

25. $\frac{1}{a}\sec^{-1}\left(\frac{x}{a}\right) + C$

27. $\frac{-\sqrt{x^2+a^2}}{x} + |\sqrt{x^2+a^2}+x| + C$

29. $\frac{x\sqrt{x^2+9}}{2} - \frac{3}{2}\ln\left|\frac{\sqrt{x^2+9}+x}{3}\right| + C$

31. $\sqrt{x^2+4} - \frac{3}{2}\ln\left|\frac{\sqrt{x^2+4}-2}{x}\right| + C$

33. $-\sqrt{9-x^2} - 4\sin^{-1}\left(\frac{x}{3}\right) + \frac{5}{3}\ln\left|\frac{3-\sqrt{9-x^2}}{x}\right| + C$

35. $20 - 9\ln 3$

37. 68.86%

39. $\frac{\sqrt{2}}{4} + \frac{1}{4}\ln\left|\frac{2+\sqrt{2}}{\sqrt{2}}\right|$

41. $\sqrt{10} - \sqrt{2} + \ln\left|\frac{(\sqrt{10}-1)}{3(\sqrt{2}-1)}\right|$

Exercise Set 11.4

1. $\frac{-1}{x-2} + C$

3. $\ln\left|\frac{\sqrt{x^2+6x+13}+x+3}{2}\right| + C$

5. $\ln\left|\frac{(x-3)+\sqrt{x^2-6x}}{3}\right| + C$

7. $-\sqrt{6x-x^2} + 3\sin^{-1}\left(\frac{x-3}{3}\right) + C$

9. $2\sqrt{x^2+6x+18} - 5\ln\left|\frac{\sqrt{x^2+6x+18}-(x+3)}{3}\right| + C$

11. $\ln\left|\sqrt{\frac{x^2-8x-7}{23}}\right| + \frac{6\sqrt{23}}{23}\ln\left|\frac{(x-4)-\sqrt{23}}{\sqrt{x^2-8x-7}}\right| + C$

13. $\frac{x^2+2x}{4(x^2+2x+2)} - \frac{1}{2}\tan^{-1}(x+1) - \frac{(x+1)}{2(x^2+2x+2)} + C$

15. $\frac{-\sqrt{7}}{49}\ln\left|\frac{\sqrt{88-18x+x^2}-\sqrt{7}}{x-9}\right| - \frac{1}{7\sqrt{88-18x+x^2}} + C$

17. $\frac{1}{2}\left(x+\frac{5}{2}\right)\sqrt{4x^2+20x+29} - \ln\left|\frac{\sqrt{4x^2+20x+29}}{2} + \left(x+\frac{5}{2}\right)\right| + C$

Exercise Set 11.5

1. $\ln\left|\frac{x}{x+1}\right| + C$

3. $-3\ln|1-x| + \ln|1+x| + C$

5. $\frac{-1}{2}\ln|1-x| + \frac{1}{2}\ln|1+x| + C$

7. $\ln|x| - \frac{1}{2}\ln|x^2+1| + C$

9. $\frac{1}{2}\ln|x+1| - \frac{1}{4}\ln(x^2+1) + \frac{1}{2}\tan^{-1}x + C$

11. $3\ln|x+2| + \frac{1}{2}\ln|x^2+1| - \tan^{-1}x + C$

13. $\ln|x| - \frac{3}{x+1} + C$

15. $2\ln|x| + \frac{1}{x+1} + C$

17. $\frac{1}{2}\ln|x| - \ln|x+1| + \frac{1}{2}\ln|x+2| + C$

19. $\ln|x| + 2\ln|x-3| + \ln|x^2+1| + C$

21. $4 \ln|x| - 2 \ln|x^2 + 1| + C$

23. $\frac{-1}{4a^3} \ln|a - x| + \frac{1}{4a^2(a - x)} + \frac{1}{4a^3} \ln|a + x| - \frac{1}{4a^2(a + x)} + C$

25. $9 \ln|x + 3| + \frac{1}{2} \ln|x^2 + 2x + 6| + C$

27. $3 \ln|x - 1| + \ln|x^2 + x + 1| + C$

29. $2 \ln|x| + 6 \ln|x - 2| + \ln|x + 4| + C$

31. $4 \ln 3 - \ln 2$

33. $3\pi \ln 5 - 4\pi \ln 3$

37. $\frac{100}{1 + \exp(-2t + 100C)}$; C a constant. 100

Exercise Set 11.6

1. $x - 2\sqrt{x} - 2 \ln|1 + \sqrt{x}| + C$

3. $2\sqrt{x} - 6 \sqrt[3]{x} + 24 \sqrt[6]{x} - 48 \ln|\sqrt[6]{x} + 2| + C$

5. $\frac{3}{5}(x + 1)^{5/3} - \frac{3}{2}(x + 1)^{2/3} + C$

7. $\frac{2\sqrt{3}}{3} \tan^{-1}\left(\frac{2 \tan\left(\frac{x}{2}\right) + 1}{\sqrt{3}}\right) + C$

9. $\frac{-1}{\tan\left(\frac{x}{2}\right)} + C$

11. $\frac{3}{8}(1 + x)^{8/3} - \frac{6}{5}(1 + x)^{5/3} + \frac{3}{2}(1 + x)^{2/3} + C$

13. $\frac{\sqrt{2}}{2} \ln\left|\frac{\tan\left(\frac{x}{2}\right) - 1 + \sqrt{2}}{\tan\left(\frac{x}{2}\right) - 1 - \sqrt{2}}\right| + C$

15. $\frac{1}{56}(1 - 2x)^{7/2} - \frac{3}{40}(1 - 2x)^{5/2} + \frac{1}{8}(1 - 2x)^{3/2} - \frac{1}{8}(1 - 2x)^{1/2} + C$

17. $2 \sin^{-1}\left(\frac{\sqrt{1 + x}}{\sqrt{2}}\right) - \sqrt{1 - x^2} + C$

19. $\frac{\sqrt{a}}{a} \ln\left|\frac{\sqrt{a + bx} - \sqrt{a}}{\sqrt{a + bx} + \sqrt{a}}\right| + C$

21. $\frac{2}{3}(2 + x)^{3/2} - 4(2 + x)^{1/2} + C$

Exercise Set 11.7

1. $\frac{1}{5} \ln\left(\frac{e^{5x}}{1 + e^{5x}}\right) + C$

3. $\frac{4}{729} \ln\left(\frac{9 + 2x}{x}\right) - \frac{(9 + 2x) + 2x}{81x(9 + 2x)} + C$

5. $\frac{1}{10} \ln(2 + 5x^2) + \frac{3}{\sqrt{10}} \tan^{-1}\left(\frac{\sqrt{5}x}{\sqrt{2}}\right) + C$

7. $\frac{1}{\pi}\left\{\frac{1}{3} \tan^3 \pi x - \tan \pi x + \pi x\right\} + C$

9. $\frac{\sin 3x}{6} - \frac{\sin 9x}{18} + C$

11. $\frac{-1}{2x} - \frac{3}{4} \ln\left|\frac{2 - 3x}{x}\right| + C$

13. $\frac{9}{20} \ln\left|\frac{2x + 5}{2x - 5}\right| + C$

15. $\frac{1}{10} \tan^2 5x - \frac{1}{5} \ln|\sec 5x| + C$

17. $\frac{4}{5}(3x^2 - 24x + 288)\sqrt{6 + x} + C$

19. $\frac{6x}{14\sqrt{14 - x^2}} + C$

21. $\frac{\pi}{8} \tan 2x - \frac{\pi}{8} \sec 2x + C$

23. $\frac{-3}{4} \ln\left|\frac{4 + \sqrt{x^2 + 16}}{x}\right| + C$

25. $\frac{1}{3} \tan^{-1}(3x + 2) + C$

27. $\frac{-7}{2} \ln\left|\frac{2 + \sqrt{x^2 + 8x + 20}}{x + 4}\right| + C$

29. $\sin^{-1}\left(\frac{x}{4}\right) - x\sqrt{16 - x^2} + C$

Exercise Set 11.8

1. diverges

3. diverges

5. 1

7. $\frac{1}{2}$

9. $\frac{e^4}{2}$

11. 1

13. diverges
17. $\frac{-1}{4}$
21. divergent
15. $\frac{9\pi}{4}$
19. $\frac{2e^3}{9}$
23. $\frac{1}{4}$
25. $\sqrt{\frac{\pi}{2}}$
27. $\frac{1}{4}$
31. diverges, $p \geq -1$
converges, $p < -1$
35. diverges
37. converges
39. diverges
41. converges
45. $\frac{\pi}{2e^2}$, diverges
47. a. 4,121,000 dollars
b. 12.5 million dollars

Review Exercises–Chapter 11

1. $\frac{1}{3}(x^2 + 9)^{3/2} + C$
3. $\frac{x}{2}\sqrt{x^2 + a^2} + \frac{a^2}{2} \ln\left|\frac{x + \sqrt{x^2 + a^2}}{a}\right| + C$
5. $\ln|x(x + 3)| + C$
7. diverges
9. $\frac{x^2}{2} \ln^2 x - \frac{x^2}{4}[2 \ln x - 1] + C$
11. $\frac{\sqrt{5}}{5} \tan^{-1}\left(\frac{x - 2}{\sqrt{5}}\right) + C$
13. $2\sqrt{x + 4} + 2 \ln\left|\frac{\sqrt{x + 4} - 2}{\sqrt{x + 4} + 2}\right| + C$
15. $\frac{32 - 16\sqrt{2}}{15}$
21. $1 - \frac{\pi}{4}$
17. $\sin^{-1}\left(\frac{x}{a}\right) + C$
23. $\frac{2e^3}{9} + \frac{1}{9}$
19. $\frac{-1}{x} + 3 \ln|x + 1| + C$
25. $\sin^{-1}\left(\frac{x - 2}{\sqrt{5}}\right) + C$
27. $-x - 8\sqrt{x} - 16 \ln|2 - \sqrt{x}| + C$
29. diverges
31. $\frac{x}{2}\sqrt{x^2 - a^2} - \frac{a^2}{2} \ln\left|\frac{x + \sqrt{x^2 - a^2}}{a}\right| + C$
33. $\ln|x + 2| + \frac{2\sqrt{3}}{3} \tan^{-1}\left(\frac{2x + 1}{\sqrt{3}}\right) + C$
35. -1
37. $\frac{1}{4}x^3\sqrt{x^2 - 4} - \frac{5}{2}x\sqrt{x^2 - 4} + 6 \ln\left|\frac{x + \sqrt{x^2 - 4}}{2}\right| + C$
39. $2 \ln|x| + \frac{2}{3} \ln|x + 1| + \frac{2\sqrt{3}}{9} \tan^{-1}\left(\frac{2x - 1}{\sqrt{3}}\right) + C$
41. diverges
43. $\frac{2}{3}(1 - e^x)^{3/2} - 2(1 - e^x)^{1/2} + C$
45. $\sqrt{x^2 - 6x + 13} + 5 \ln|(x - 3) + \sqrt{x^2 - 6x + 13}| + C$
47. $\frac{\sqrt{3}}{3} \ln\left|\frac{\tan\left(\frac{x}{2}\right) - 2 - \sqrt{3}}{\tan\left(\frac{x}{2}\right) - 2 + \sqrt{3}}\right| + C$
49. $\frac{x}{2}\sqrt{1 + x^2} - \frac{1}{2} \ln|\sqrt{1 + x^2} + x| + C$
51. 1
53. $\ln\left|\frac{\sqrt{a^2 + x^2}}{a}\right| + C$
55. $3 \ln|x - 1| + \ln|x^2 + 4| + C$
57. $\frac{\pi}{3}$
59. $\sqrt{36 + x^2} + C$
61. $\ln|x| + 2 \ln|x + 2| + 3 \ln|x - 1| + C$
63. 2
65. $\frac{x^2}{2} - 2 \ln|x^2 + 2| - \frac{2}{x^2 + 2} + C$
67. 2
69. $-2 \cos\left(\frac{x}{2}\right) + 2 \sin\left(\frac{x}{2}\right) + C$
71. diverges
73. $\frac{7\sqrt{13}}{13} \tan^{-1}\left(\frac{x}{\sqrt{13}}\right) + C$
75. $\frac{\pi}{6} \ln\left|\frac{6 - \sqrt{36 - x^2}}{x}\right| + C$
77. $\frac{4}{3(7 - 3x)} + C$
79. $\frac{1}{\pi^2} \ln\left|\frac{x}{\pi + 4x}\right| + \frac{1}{\pi(\pi + 4x)} + C$
81. $98 \sin^{-1}\left(\frac{x}{7}\right) - \frac{x}{2}\sqrt{49 - x^2} + C$
83. $\sin^{-1}\left(\frac{2x - 5}{5}\right) + C$
85. $\frac{x^2}{2} \sin^{-1}(2x) + \frac{x}{8}\sqrt{1 - 4x^2} + \frac{1}{16} \sin^{-1}(2x) + C$

CHAPTER 12

Exercise Set 12.2

1. $1 - x + \frac{1}{2}x^2 - \frac{1}{6}x^3 + \frac{1}{24}x^4$

3. $\frac{\sqrt{2}}{2} - \frac{\sqrt{2}}{2}\left(x - \frac{\pi}{4}\right) - \frac{\sqrt{2}}{4}\left(x - \frac{\pi}{4}\right)^2 + \frac{\sqrt{2}}{12}\left(x - \frac{\pi}{4}\right)^3 + \frac{\sqrt{2}}{48}\left(x - \frac{\pi}{4}\right)^4 - \frac{\sqrt{2}}{240}\left(x - \frac{\pi}{4}\right)^5 - \frac{\sqrt{2}}{1440}\left(x - \frac{\pi}{4}\right)^6$

5. $x - \frac{1}{2}x^2 + \frac{1}{3}x^3 - \frac{1}{4}x^4$

7. $x + \frac{1}{3}x^3$

9. $1 + x^2$

11. $1 - \frac{1}{2}x - \frac{1}{8}x^2 - \frac{1}{16}x^3$

13. $\sqrt{2} + \sqrt{2}\left(x - \frac{\pi}{4}\right) + \frac{3\sqrt{2}}{2}\left(x - \frac{\pi}{4}\right)^2 + \frac{11\sqrt{2}}{6}\left(x - \frac{\pi}{4}\right)^3$

15. $\frac{1}{2} - \frac{1}{4}x + \frac{1}{48}x^3$

19. True

Exercise Set 12.3

1. $e^{-x} = 1 - x + \frac{1}{2}x^2 - \frac{1}{6}x^3 + \frac{e^{-c}}{24}x^4$; c is between 0 and x

3. $\cos x = 1 - \frac{1}{2}x^2 + \frac{1}{24}x^4 - \frac{\sin c}{120}x^5$; c is between 0 and x

5. $\tan^{-1} x = x - \frac{1}{3}x^3 + \frac{c - c^3}{(1 + c^2)^4}x^4$; c is between 0 and x

7. $\frac{1}{1 + x^2} = \frac{1}{2} - \frac{1}{2}(x - 1) + \frac{1}{4}(x - 1)^2 + 4\frac{c - c^3}{(1 + c^2)^4}(x - 1)^3$

9. $\sec x = \sqrt{2} + \sqrt{2}\left(x - \frac{\pi}{4}\right) + \frac{3\sqrt{2}}{2}\left(x - \frac{\pi}{4}\right)^2 + \frac{(\sin c)(5 + \sin^2 c)}{6 \cos^4 c}\left(x - \frac{\pi}{4}\right)^3$

11. $\sinh x = x + \frac{1}{6}x^3 + \frac{\cosh c}{5!}x^5$

13. $\cosh x = \frac{5}{4} + \frac{3}{4}(x - \ln 2) + \frac{5}{8}(x - \ln 2)^2 + \frac{1}{8}(x - \ln 2)^3 + \frac{1}{24}(\cosh c)(x - \ln 2)^4$

21. $P_n(x) = 1 - x^2 + x^4 - x^6 + \cdots$, using all terms of degree $\leq n$.

Exercise Set 12.4

Note: in Problems 1–20, there are many ways to handle the R_n and make accuracy estimates.

1. $\ln(1.5) \approx 0.416\overline{6}$

3. $\sqrt{3.91} \approx 1.99737$

5. $\cos 1 \approx 0.540317$

7. $\sqrt[3]{10} \approx 2.152\overline{7}$

9. $|R_2(x)| \leq 0.0000208$

11. $|R_1(x)| \leq 0.00125$

13. $|R_1(x)| < 0.000118$

15. $|R_1(x)| < 0.00005$

17. $n = 6$

19. n is 2.

23. $|R_1(x)| < 0.005483$

27. 0.31

29. 0.9

Review Exercises—Chapter 12

1. $2x - \frac{4}{3}x^3 + \frac{4}{15}x^5$

3. $\frac{\pi\sqrt{2}}{8} + \frac{\sqrt{2}}{2}\left(1 - \frac{\pi}{4}\right)\left(x - \frac{\pi}{4}\right) - \frac{\sqrt{2}}{4}\left(2 + \frac{\pi}{4}\right)\left(x - \frac{\pi}{4}\right)^2 + \frac{\sqrt{2}}{12}\left(\frac{\pi}{4} - 3\right)\left(x - \frac{\pi}{4}\right)^3$

5. $1 + x^2$

7. $\frac{\pi}{4} + \frac{1}{2}(x - 1) - \frac{1}{4}(x - 1)^2 + \frac{1}{12}(x - 1)^3$

9. $1 + (\log 2)x + \frac{(\log 2)^2}{2}x^2 + \frac{(\log 2)^3}{6}x^3$

11. $\left|R_3\left(\frac{17\pi}{90}\right)\right| \leq 9.89775 \cdot 10^{-7}$

13. $|R_2(35)| \le 0.000079$

15. $\left|R_3\left(\frac{1}{4}\right)\right| < 0.000488$

CHAPTER 13

Exercise Set 13.2

1. $\left\{\frac{1}{3}, \frac{2}{5}, \frac{3}{7}, \frac{4}{9}, \ldots\right\}$. Converges.

$\lim_{n\to\infty} \frac{n}{2n+1} = \lim \frac{1}{2+\frac{1}{n}} = \frac{1}{2}$.

3. $\left\{-1, -\frac{1}{3}, -\frac{1}{11}, 0, \ldots\right\}$. Converges to 0.

5. $\left\{\frac{1}{2}, \frac{1}{5}, \frac{1}{10}, \frac{1}{17}, \ldots\right\}$. Converges to 0.

7. $\{\sqrt{5}, \sqrt{5}, \sqrt{5}, \sqrt{5}, \ldots\}$. Converges to $\sqrt{5}$.

9. $\left\{10, \frac{40}{1+\sqrt{2}}, \frac{60}{1+\sqrt{3}}, \frac{80}{3}, \ldots\right\}$. Diverges.

11. $\left\{\frac{2}{3}, \frac{3+\sqrt{2}}{4}, \frac{3-\sqrt{3}}{5}, \frac{5}{6}, \ldots\right\}$. Converges to 0.

13. $\left\{\sqrt{2}, \frac{1}{2}\sqrt{6}, \frac{2}{3}\sqrt{3}, \frac{1}{2}\sqrt{5}, \ldots\right\}$. Converges to 1.

15. $\left\{1, \cos\frac{1}{4}, \cos\frac{2}{9}, \cos\frac{3}{16}, \ldots\right\}$. Converges to $\cos 0 = 1$.

17. $\left\{\frac{3}{2}, \frac{1+\sqrt{2}}{2\sqrt{2}}, \frac{2+3\sqrt{3}}{6\sqrt{3}}, \frac{5}{8}, \ldots\right\}$. Converges to $\frac{1}{2}$.

19. $\left\{\frac{1}{2}, \frac{1}{6}, \frac{1}{12}, \frac{1}{20}, \ldots\right\}$. Converges.

$\lim\left(\frac{1}{n} - \frac{1}{n+1}\right) = \lim\left(\frac{1}{n(n+1)}\right) = 0$.

21. $\{\sqrt{2}-1, \sqrt{3}-2, 2-\sqrt{3}, \sqrt{5}-2, \ldots\}$. Converges to 0.

23. $\left\{\sqrt{3}, \frac{3}{2}, \frac{\sqrt{19}}{3}, \frac{\sqrt{33}}{4}, \ldots\right\}$. Converges to $\sqrt{2}$.

25. $\left\{1, \sqrt{2}, \frac{3}{2}, 4\sin\frac{\pi}{8}, \ldots\right\}$. Converges to $\frac{\pi}{2}$.

27. $\left\{2, \frac{9}{4}, \frac{64}{27}, \frac{625}{256}, \ldots\right\}$. Converges to e.

Exercise Set 13.3

1. 2 **3.** 1 **5.** 0 **7.** Diverges **9.** 1 **11.** 0 **13.** 0 **15.** e^3 **17.** 1

25. $a_n = 2n - 5$.

27. $\{1, -1, 1, -1, 1, -1, \ldots\}$

29. No. An example is

$\left\{1, -\frac{1}{2}, \frac{1}{3}, -\frac{1}{4}, \frac{1}{5}, -\frac{1}{6}, \ldots\right\}$.

Exercise Set 13.4

1. $-\frac{1}{2} + \frac{1}{4} - \frac{1}{8} + \frac{1}{16} - \cdots$

3. $\frac{3}{5} + \frac{5}{11} + \frac{9}{29} + \frac{17}{83} + \cdots$

5. $\ln\frac{1}{2} + \ln\frac{2}{3} + \ln\frac{3}{4} + \ln\frac{4}{5} + \cdots$

7. Converges to $\frac{7}{6}$.

9. Diverges

11. Converges; $\frac{27}{23}$.

13. Converges to $\frac{1}{1-\frac{1}{2+x}} = \frac{x+2}{x+1}$.

15. Converges to $-\frac{1}{2}$.

17. Diverges

19. Converges to $\frac{1}{3}$

21. Diverges

23. Converges to $\frac{3}{3-\sqrt{2}}$.

25. **a.** $\sum_{k=1}^{\infty} \frac{3}{10^k}$ **b.** $\frac{1}{3}$

27. **a.** $\sum_{k=1}^{\infty} \frac{92}{100^k}$
b. $\frac{92}{99}$

29. **a.** $\sum_{k=1}^{\infty} \frac{412}{1000^k}$
b. $\frac{412}{999}$

35. $5h$

37. No, if $c = 0$.

41. $S_n = S_{n-1} + a_n$

43.

S_5	S_{10}	S_{20}	S_{50}	S_{100}
2.736625514	2.96531694	2.999398543	2.999999997	3.

The actual sum is $\frac{1}{1 - \frac{2}{3}} = 3$.

45.

S_5	S_{10}	S_{20}	S_{50}	S_{100}
−.3718688642	−1.128682959	−1.295565042	−1.301540778	−1.301541025

The actual sum ≈ -1.301541025.

Exercise Set 13.5

1. Diverges
3. Diverges
5. Converges
7. Converges
9. Converges
11. Diverges
13. Diverges
15. Converges

NOTE: For exercises 17–46 there are many possible solutions:

17. Diverges
19. Diverges
23. Converges
25. Converges
27. Diverges
29. Diverges
31. Diverges
33. Diverges
35. Diverges
37. Diverges
39. Diverges
41. Converges
43. Diverges
45. Converges
51. $\frac{1}{11} \le (\text{error}) \le \frac{1}{10}$
53. $n > \sqrt{\ln 25} \approx 1.794$

Exercise Set 13.6

1. Diverges; ratio test.
3. Converges; ratio test.
5. Converges; ratio test.
7. Converges; limit comparison test using $\sum \frac{1}{k^2}$.
9. Converges; root test
11. Diverges, by Theorem 7.
13. Diverges; ratio test.
15. Converges; ratio test.
17. Converges; ratio test.
19. Diverges; ratio test.
21. All $x > 0$.
23. Converges for $x < 1$ and diverges for $x \geq 1$.

Exercise Set 13.7

1. **a.** Yes **b.** Yes
3. **a.** Yes **b.** Yes
5. **a.** Yes **b.** Yes
7. **a.** No **b.** Yes
9. **a.** No **b.** Yes
11. **a.** Yes **b.** No
13. **a.** No **b.** No
15. **a.** No **b.** Yes
17. **a.** Yes **b.** Yes
19. **a.** Yes **b.** Yes
21. **a.** No **b.** No
23. **a.** Yes **b.** Yes

25. a. No
b. Yes

27. a. Yes
b. Yes

29. a. Yes
b. Yes

31. all x

33. all x

35. $n \geq 98$

Review Exercises—Chapter 13

1. Diverges
3. Diverges
5. Converges to 0
7. Diverges
9. Diverges
11. $\frac{6}{5}$
13. 2
15. $37 \sum_{k=1}^{\infty} \frac{1}{100^k} = \frac{37}{99}$
17. $\frac{1}{48}$
19. $\frac{10}{9}$
21. Converges
23. Diverges
25. Converges
27. Converges
29. Converges
31. Converges
33. Diverges
35. Converges
37. Diverges
39. Converges
41. Converges
43. Converges
45. Converges
47. Diverges
49. Converges
51. Converges
53. Converges
55. Converges
57. Converges
59. Converges conditionally
61. Diverges
63. Converges absolutely
65. Diverges
67. Converges absolutely
69. $\frac{1}{13}$
71. $\frac{1}{11}$
73. $-1 < c \leq 1$
75. Σb_k must diverge
77. a. 90.6192 meters
b. 110 meters
79. a. Converges
b. Converges
c. Converges

CHAPTER 14

Exercise Set 14.2

1. $[-1, 1)$
3. $(-\infty, \infty)$
5. $(-\infty, \infty)$
7. $[-1, 1)$
9. $(-1, 1]$
11. $[-1, 1]$
13. $(-\infty, \infty)$
15. Converges only for $x = 1$.
17. $\left(\frac{1}{3}, 1\right]$
19. $(-1, 1)$
21. $[2, 4]$
23. $(2 - e, 2 + e)$
25. $\left(-\frac{3}{7}, \frac{1}{7}\right)$
27. $(-1, 1)$
29. $[-2, 4)$
31. $[-e, e)$
33. $(-\infty, \infty)$
35. $[-1, 1]$

Exercise Set 14.3

1. $\Sigma_{k=0}^{\infty} 2^k x^k; \quad r = \frac{1}{2}$
3. $\Sigma_{k=0}^{\infty}(-1)^k 4^k x^{2k}; \quad r = \frac{1}{2}$
5. $\Sigma_{k=0}^{\infty}(-1)^k x^{2k+1}; \quad r = 1$
7. $-1 + \Sigma_{k=1}^{\infty}(-1)^{k+1} 2x^k; \; r = 1$
9. $\Sigma_{k=0}^{\infty} x^{4k}; \quad r = 1$
11. $-2\,\Sigma_{k=1}^{\infty}(-1)^k k x^{k-1}; \quad r = 1$
13. $\Sigma_{k=1}^{\infty} k(-1)^{k+1} x^{2k-1}; \quad r = 1$
15. $\Sigma_{k=0}^{\infty}(2k + 1)x^{2k}; \quad r = 1$
17. $\Sigma_{k=1}^{\infty} 2k(-1)^{k+1} 4^k x^{2k-1}; \quad r = \frac{1}{2}$
19. $\Sigma_{k=1}^{\infty} 2k(-1)^{k-1} x^{k-1}; \quad r = 1$
21. $-\sum_{k=0}^{\infty} \frac{x^{k+1}}{k+1}; \quad r = 1$
23. $2 \sum_{k=0}^{\infty} \frac{(-1)^k 4^k x^{2k+1}}{2k+1}; \quad r = \frac{1}{2}$
25. $\sum_{k=0}^{\infty} \frac{(-1)^k x^{k+1}}{4^{k+1}(k+1)} + \ln(4); \; r = 4$
27. $\sum_{k=0}^{\infty} (-1)^k \frac{x^{2k+2}}{k+1}; \quad r = 1$

Exercise Set 14.4

1. $\sum_{k=0}^{\infty} \frac{2^k x^k}{k!}$, all x

3. $\sum_{k=0}^{\infty} \frac{\sqrt{2}(-1)^{k(k+1)/2}\left(x - \frac{\pi}{4}\right)^k}{2(k!)}$; all x

5. $5 + 4(x - 2) + (x - 2)^2$; all x

7. $\sum_{k=1}^{\infty} (-1)^{k-1} \frac{x^k}{k3^k} + \ln(3)$; $-3 < x \le 3$

9. $\sum_{k=0}^{\infty} \frac{x^k(\ln 2)^k}{k!}$; all x

11. $1 + \sum_{k=1}^{\infty} \frac{\frac{3}{2}\left(\frac{3}{2} - 1\right) \cdots \left(\frac{3}{2} - k + 1\right)x^k}{k!}$; $|x| < 1$

13. See Example 8, Section 14.4.

15. $\sum_{k=0}^{\infty} \frac{(-1)^k x^{2k}}{(2k + 1)!}$; all x

17. $\sum_{k=1}^{\infty} \frac{\sqrt{2}}{2}\left(x - \frac{\pi}{4}\right)^k \left[\frac{(-1)^{(k-1)(k-2)/2}}{(k - 1)!} + \frac{\pi}{4}\frac{(-1)^{(k-1)k/2}}{k!}\right] + \frac{\pi\sqrt{2}}{8}$

19. $1 + \sum_{k=1}^{\infty} \frac{2^{2k-1}x^{2k}(-1)^k}{(2k)!}$; all x

21. $1 + x^2$. Use $\frac{d}{dx}(\tan x) = \sec^2 x$

23. $1 + \sum_{k=0}^{\infty} \frac{x^{4k+4}}{(2k + 2)!}(-1)^{k+1}$

25. $\sum_{k=0}^{\infty} \frac{x^{2k+1}}{(2k)!}$

27. 0.035 **29.** 0.095 **31.** 0.485

33. $1 + \sum_{k=1}^{\infty} \frac{2^k x^k\left(\frac{1}{2}\right)\left(\frac{1}{2} - 1\right) \cdots \left(\frac{1}{2} - k + 1\right)}{k!}$; $|x| < \frac{1}{2}$

35. $27 + \sum_{k=1}^{\infty} \frac{\frac{3}{2}\left(\frac{3}{2} - 1\right) \cdots \left(\frac{3}{2} - k + 1\right)x^k}{3^{k-3}}$; $|x| < 3$

Exercise Set 14.5

1. $y = a_0 \sum_{k=0}^{\infty} \frac{(-1)^k x^k}{k!} = a_0 e^{-x}$

3. $y = a_0 \sum_{k=0}^{\infty} \frac{6^k x^k}{k!} = a_0 e^{6x}$

5. $y = a_0 \sum_{k=0}^{\infty} \frac{x^{2k}}{k!} = a_0 e^{x^2}$

7. $y = \sum_{k=0}^{\infty} \frac{(-1)^k 4^k x^k}{k!} = e^{-4x}$

9. $y = a_1 \sum_{k=1}^{\infty} \frac{(-1)^k x^k}{k[(k - 1)!]^2}$

11. $y = \sum_{k=0}^{\infty} \frac{x^{2k}(-1)^k}{(2k)!} = \cos x$

Exercise Set 14.6

1. $-2 \pm \sqrt{3}i$

3. $3 \pm 2i$

5. $0, -1, 5$

7. $Ae^x + Be^{-3x}$

9. $e^{-2x}(A \cos\sqrt{3}x + B \sin \sqrt{3}x)$

11. $Ae^{-5x} + Be^{2x}$

13. $A \sin 3x + B \cos 3x$

15. $e^{3x}(A + xB)$

17. $e^{2x}(A + xB)$

19. $e^x(A \cos 2x + B \sin 2x)$

21. $a > 0$

Review Exercises—Chapter 14

1. $(2, 4)$ **3.** $[1, 3)$ **5.** $(-1, 1]$

7. Replace $k - 1$ by k on the right

9. $a_n = \frac{c^n a_0}{n!}(-1)^n$

11. $\Sigma_{k=0}^{\infty}(-1)^k x^{4k}$; $|x| < 1$

13. $1 + \sum_{k=1}^{\infty} \frac{\frac{1}{3}\left(\frac{1}{3} - 1\right) \cdots \left(\frac{1}{3} - k + 1\right)x^{2k}}{k!}$

15. $\sin\sqrt{x} = -\frac{1}{2\pi^2}\left(x - \frac{\pi^2}{4}\right) + \cdots$

17. $\sum_{k=0}^{\infty} \frac{x^{2k+2}}{k!}$

19. $e^{1/2} \sim 1.649$

21. $y = a_1 \sum_{k=1}^{\infty} \frac{x^k}{k[(k - 1)!]^2}$

23. $x - x^3$

25. $e^{2x} \sin 2x$

CHAPTER 15

Exercise Set 15.2

	Vertex	Focal Length	Axis	Focus
1.	(0, 0)	$\frac{1}{8}$	$x = 0$	$\left(0, \frac{1}{8}\right)$
3.	(0, 2)	$-\frac{1}{8}$	$x = 0$	$\left(0, \frac{15}{8}\right)$
5.	(1, 0)	$\frac{1}{12}$	$y = 0$	$\left(\frac{13}{12}, 0\right)$
7.	$\left(0, \frac{1}{4}\right)$	-2	$x = 0$	$\left(0, -\frac{7}{4}\right)$
9.	(2, 1)	2	$x = 2$	(2, 3)
11.	(−3, 2)	$\frac{3}{2}$	$y = 2$	$\left(-\frac{3}{2}, 2\right)$
13.	(−2, 1)	1	$y = 1$	(−1, 1)
15.	(−1, −2)	$\frac{3}{2}$	$x = -1$	$\left(-1, -\frac{1}{2}\right)$
17.	(−3, 1)	$\frac{5}{2}$	$x = -3$	$\left(-3, \frac{7}{2}\right)$
19.	(1, −2)	$\frac{1}{2}$	$y = -2$	$\left(\frac{3}{2}, -2\right)$

1.

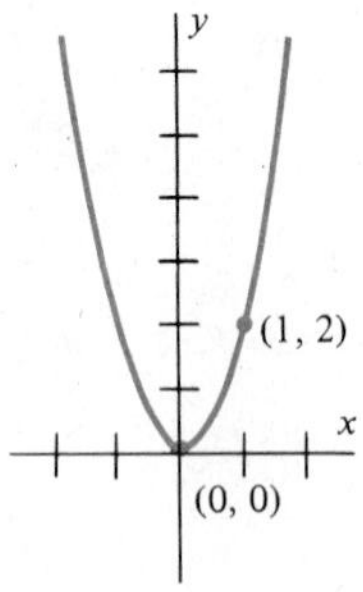

5.

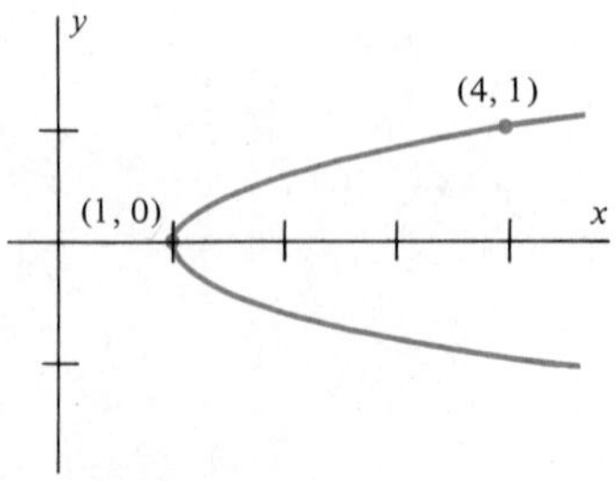

9.

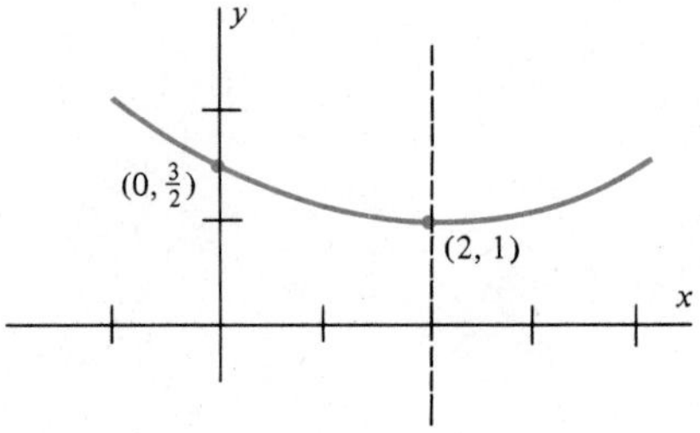

13.

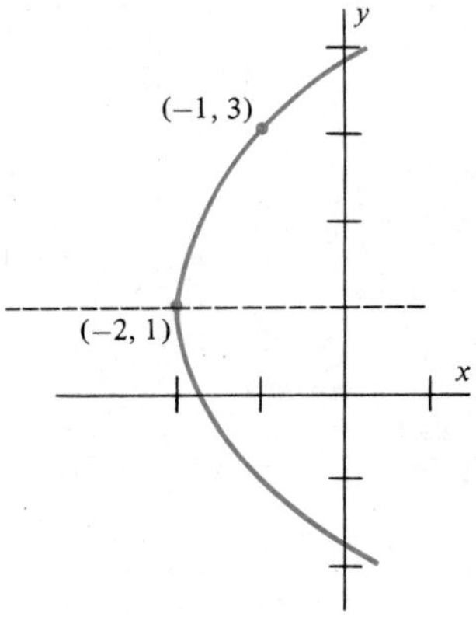

17.

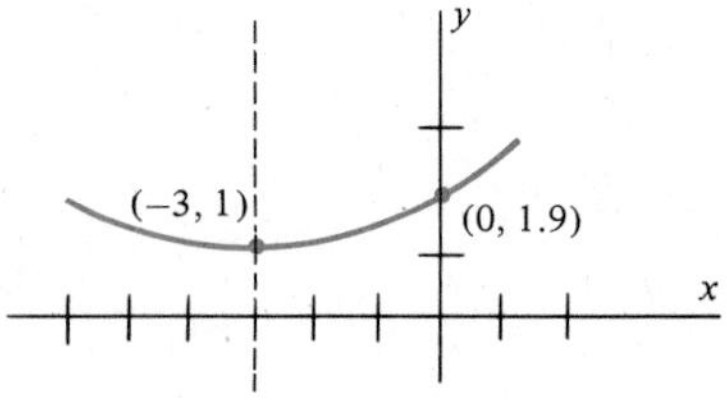

21. $x^2 = 12y$

23. $(x - 1)^2 = 32(y - 3)$

25. $(y + 6)^2 = 16(x + 1)$

27. $x^2 = 16(y - 1)$

33. $\frac{8}{3}$

35. $\frac{32}{3}$

37. -0.85

Exercise Set 15.3

	Center	Major Axis	Minor Axis	Foci
1.	(0, 0)	3	1	$(\pm 2\sqrt{2}, 0)$
3.	(0, 0)	4	2	$(\pm 2\sqrt{3}, 0)$
5.	(0, 0)	5	3	$(0, \pm 4)$
7.	(2, 1)	2	$\sqrt{3}$	(2, 0), (2, 2)
9.	(−3, 1)	4	2	$(-3 \pm 2\sqrt{3}, 1)$
11.	(1, −1)	3	2	$(1, -1 \pm \sqrt{5})$
13.	(−3, 1)	5	3	(1, 1), (−7, 1)
15.	(1, −1)	$\sqrt{3}$	$\sqrt{2}$	(0, −1), (2, −1)

1.

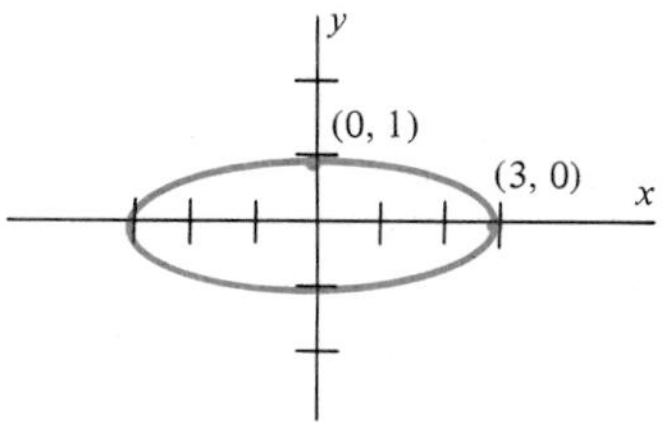

5.

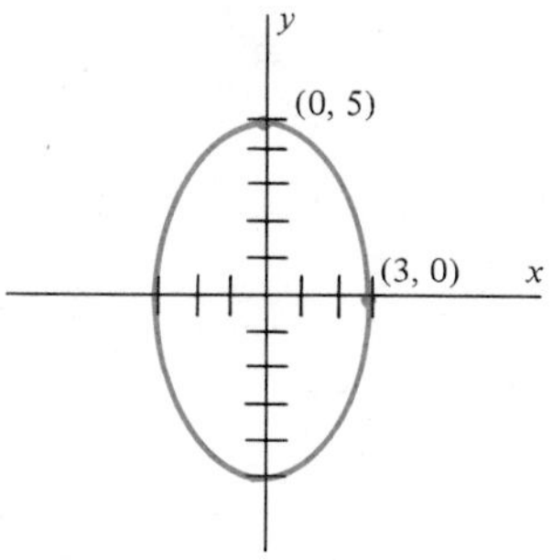

9.

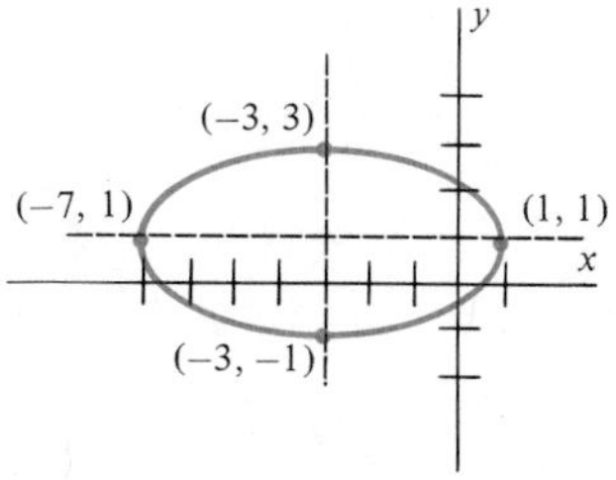

13.

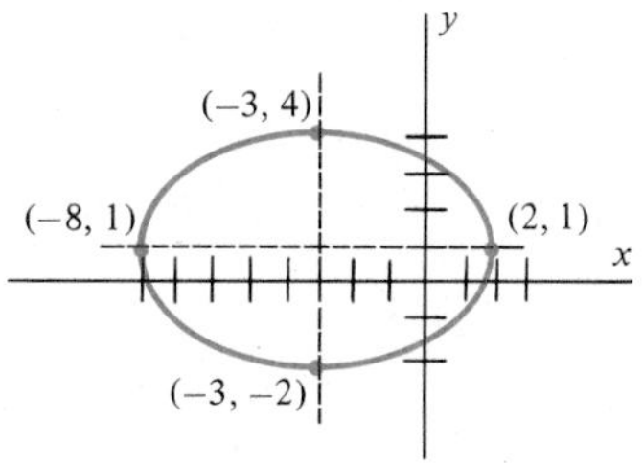

17. $\dfrac{x^2}{16} + \dfrac{y^2}{9} = 1$ **19.** $\dfrac{x^2}{25} + \dfrac{y^2}{9} = 1$

21. $\dfrac{(x-4)^2}{16} + \dfrac{(y-2)^2}{25} = 1$

23. $\dfrac{(x+2)^2}{16} + \dfrac{(y-6)^2}{25} = 1$

25. 50π **27.** $\dfrac{2b^2}{a}$ **29.** True

33. $\dfrac{(x+1)^2}{36} + \dfrac{(y-3)^2}{27} = 1$

35. 48π **37.** ≈ 181,000 ft/lb

Exercise Set 15.4

	Center	Asymptotes
1.	(0, 0)	$y = \pm\dfrac{3}{2}x$
3.	(0, 0)	$y = \pm\dfrac{3}{\sqrt{5}}x$
5.	(0, 0)	$y = \pm 2x$
7.	(0, 0)	$y = \pm\dfrac{\sqrt{7}}{2}x$
9.	(−3, 4)	$y - 4 = \pm\dfrac{3}{2}(x + 3)$
11.	(2, −3)	$y + 3 = \pm\dfrac{3}{2}(x - 2)$
13.	(0, 0)	$y = \pm\dfrac{1}{2}x$

1.

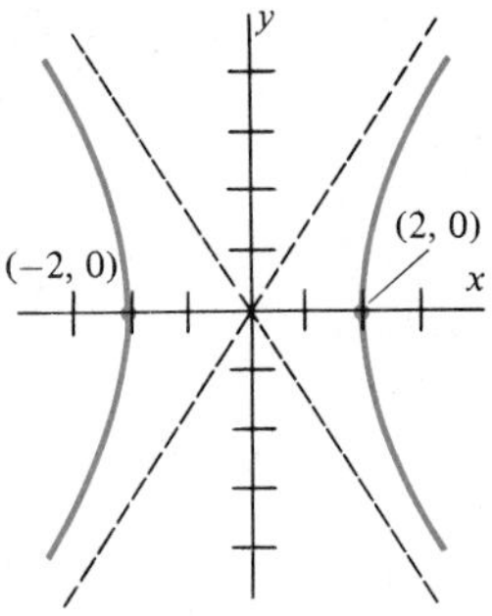

7.

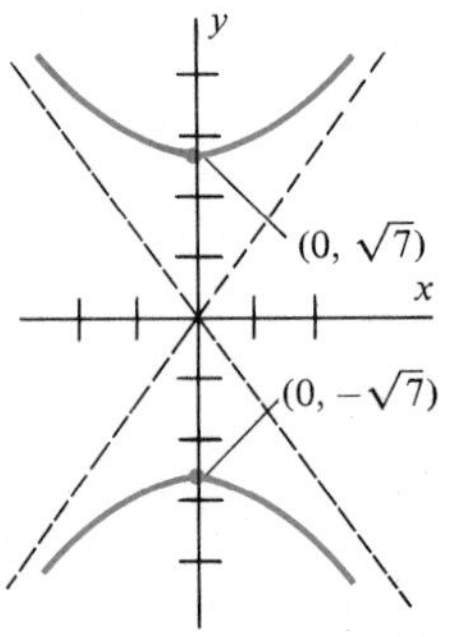

9.

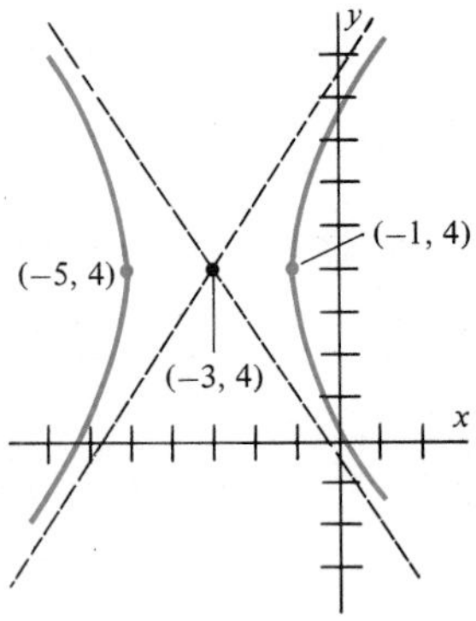

15.

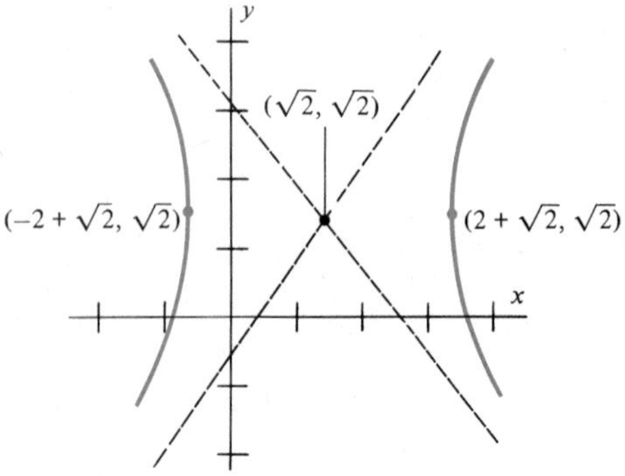

	Center	Asymptotes
15.	$(\sqrt{2}, \sqrt{2})$	$y - \sqrt{2} = \pm\sqrt{2}(x - \sqrt{2})$
17.	$(-2, 3)$	$y - 3 = \pm\frac{4}{5}(x + 2)$
19.	$(-5, 7)$	$y - 7 = \pm\frac{4}{3}(x + 5)$

17.

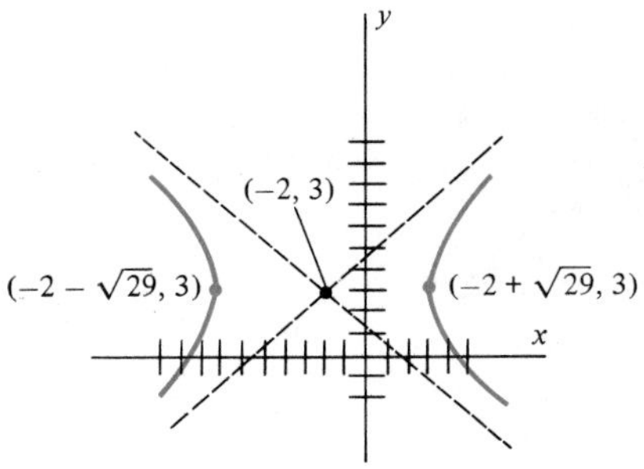

21. $x^2 - \dfrac{y^2}{3} = 1$

23. $\dfrac{(y-2)^2}{4} - x^2 = 1$

25. $\dfrac{(x+1)^2}{16} - \dfrac{(y-2)^2}{64} = \dfrac{1}{5}$

27. $\dfrac{y^2}{16} - \dfrac{x^2}{9} = 1$

29. They have the same asymptotes: $y = \pm\dfrac{b}{a}x$.

31. $y - y_0 = \dfrac{b^2 x_0}{a^2 y_0}(x - x_0)$

Exercise Set 15.5

1. $\theta = \dfrac{\pi}{8}$ **3.** $\theta = \dfrac{\pi}{4}$

5. $\theta = \dfrac{1}{2}\tan^{-1}\dfrac{4}{3}$

7. $\theta = \dfrac{\pi}{4}$, $\dfrac{x'^2}{4} + \dfrac{y'^2}{16} = 1$

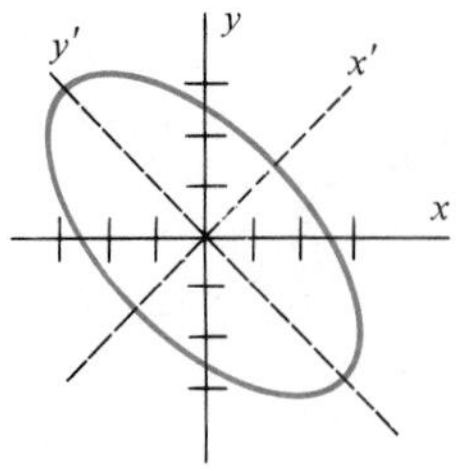

9. $\theta = \dfrac{\pi}{6}$, $x'^2 = -(y' + 4)$

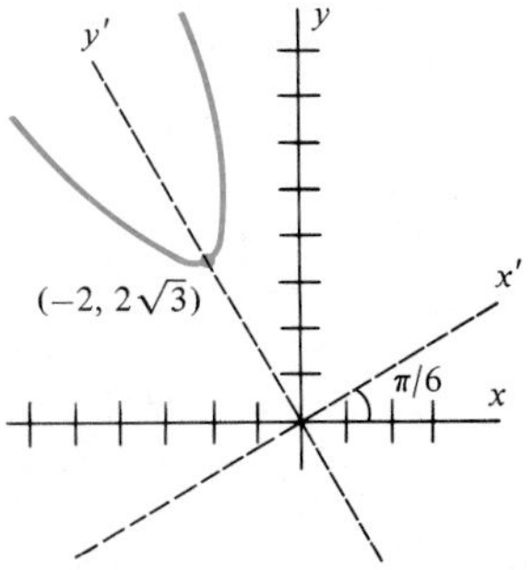

11. $\theta = \dfrac{1}{2} \arcsin \dfrac{24}{25}$, $y'^2 = 2(x' - 2)$

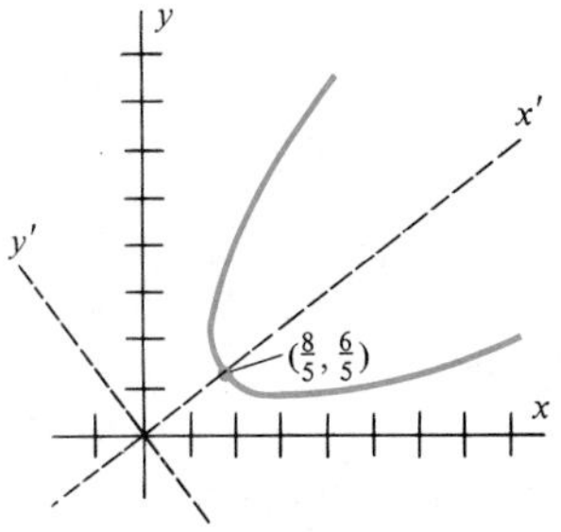

13. $\theta = -\dfrac{\pi}{6}$, $\dfrac{x'^2}{2} + \dfrac{y'^2}{4} = 1$

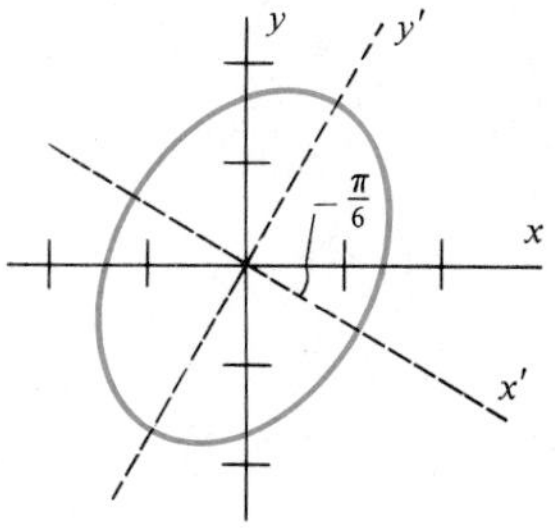

15. $\theta = \dfrac{\pi}{4}$, $x'^2 = y'$

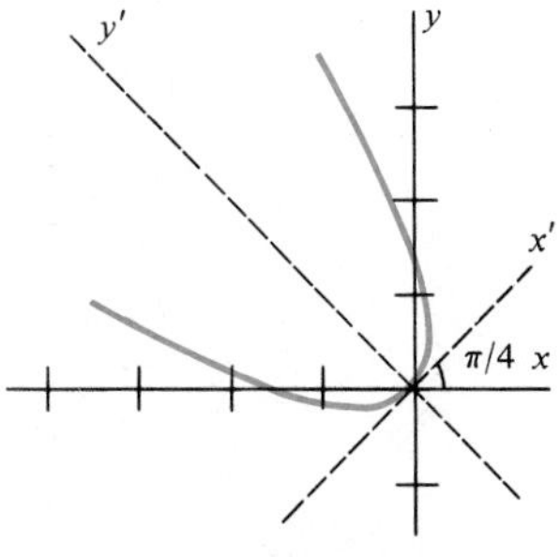

17. $\left(\mp\dfrac{3}{2}, \pm\dfrac{3\sqrt{3}}{2}\right)$

19. $\left(\pm\dfrac{3}{2}, \pm\dfrac{3\sqrt{3}}{2}\right)$

Review Exercises–Chapter 15

1. $(x - 3)^2 + (y + 4)^2 = 36$

3. $(x - 2)^2 + (y - 3)^2 = 16$

5. $(y - 4)^2 = 8x$

7. $y = 3x^2 - 6x + 5$

9. $\dfrac{(x + 1)^2}{9} + \dfrac{(y - 1)^2}{5} = 1$

11. Outside

13. $(-3, -5)$ and $(2, 0)$

15. $(2, 2)$ and $(-2, 2)$

17. $(x - 1)^2 + (y + 3)^2 = 8$, circle

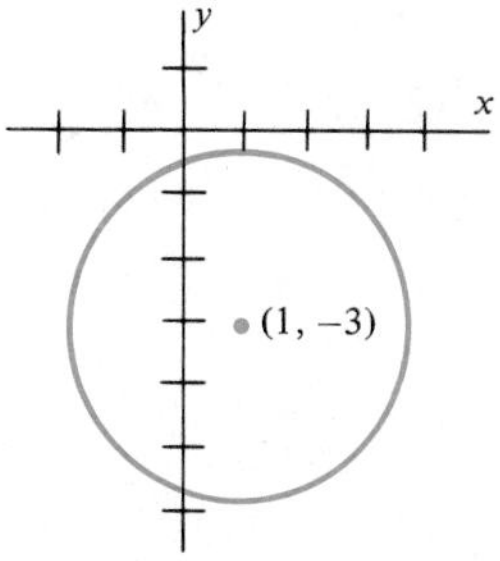

19. $\dfrac{y^2}{4} - \dfrac{x^2}{2} = 1$, hyperbola

21. $\dfrac{x^2}{12} + \dfrac{y^2}{16} = 1$, ellipse

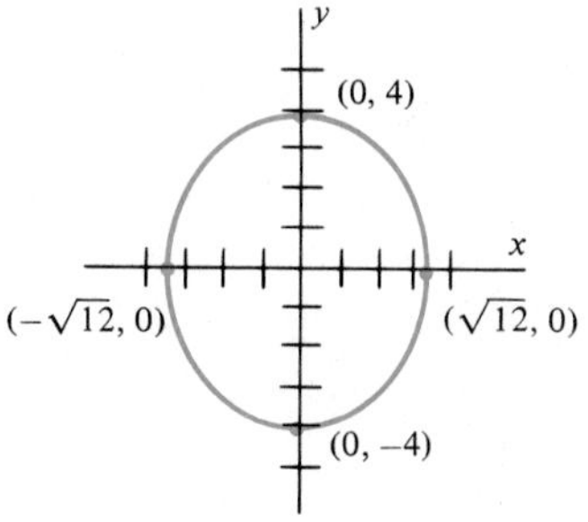

23. $\dfrac{x^2}{4} + \dfrac{y^2}{9} = 1$, ellipse

25. $\dfrac{x^2}{9} - \dfrac{y^2}{4} = 1$, hyperbola

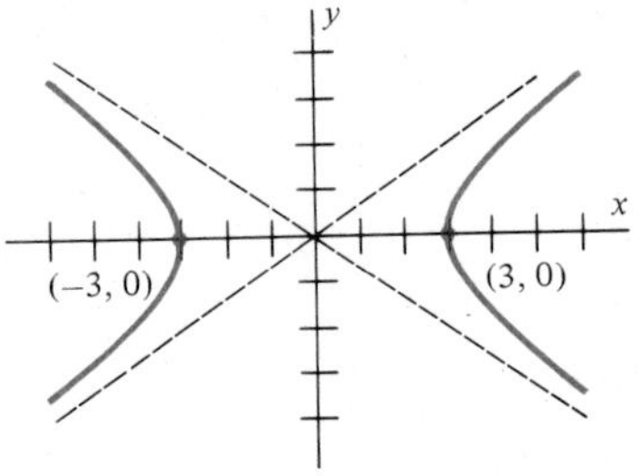

27. $y^2 = -(x - 2)$, parabola

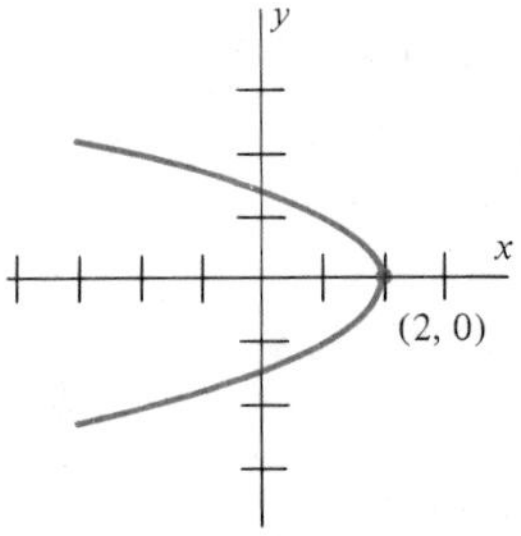

29. $(x + 7)^2 + (y - 5)^2 = 4$, circle

31. $\dfrac{x^2}{8} - \dfrac{y^2}{9} = 1$, hyperbola

33. $\dfrac{x^2}{25} + \dfrac{y^2}{4} = 1$, ellipse

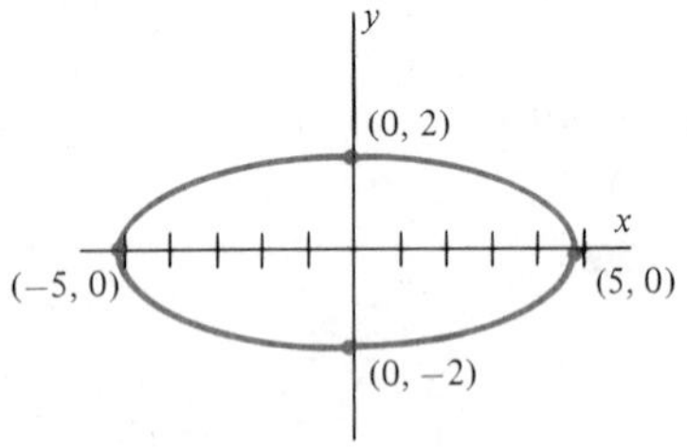

35. $\dfrac{y^2}{4} - \dfrac{x^2}{4} = 1$, hyperbola

37. $(x - a)^2 + (y + 2a)^2 = 1$, circle (Assume $a > 0$)

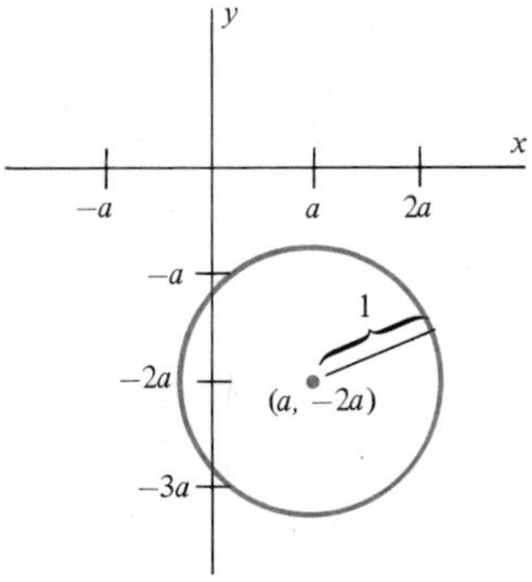

39. $y^2 - x^2 = 2$, hyperbola

41. $(x - 1)^2 + (y - 3)^2 = 7$, circle

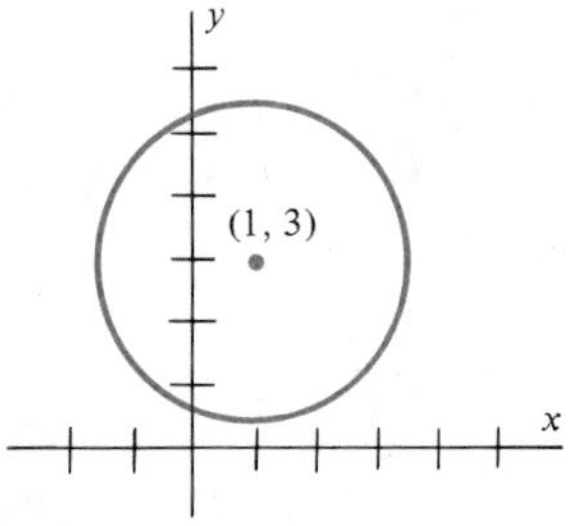

43. $\dfrac{y^2}{25} - \dfrac{x^2}{9} = 1$, hyperbola

45. $y^2 = \dfrac{1}{8}(x - 6)$, parabola

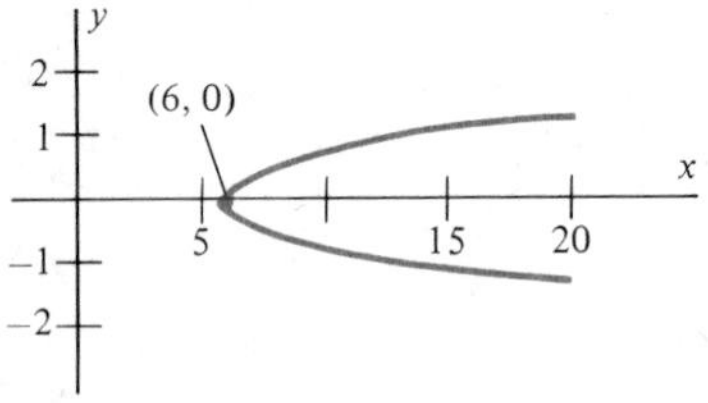

CHAPTER 16

Exercise Set 16.2

1. $P = (0, 1)$

3. $P = (0, 0)$

5. $P = (1, -1)$

7. $P = (3, 0)$

9. $P = \left(\sqrt{2}, \dfrac{\pi}{4}\right)$

11. $P = (-3, 0)$

13. $P = \left(-2, \dfrac{2\pi}{3}\right)$

15. $P = \left(-2\sqrt{2}, \dfrac{15\pi}{4}\right)$

17. symmetric about the x-axis

19. symmetric about the y-axis

21. x-axis, origin, y-axis

23. x-axis, y-axis, origin

25. $r = 2$

27. $r^2 = \dfrac{4}{1 + 3\cos^2\theta}$

29. $r = \dfrac{6}{\cos\theta} = 6\sec\theta$

31. $r = -2\sin\theta$

33. $x^2 + y^2 - 4y = 0$

35. $x^4 + x^2y^2 - y^2 = 0$

37. $x = 4$

39. $y^2 = 1 + 2x$

41.

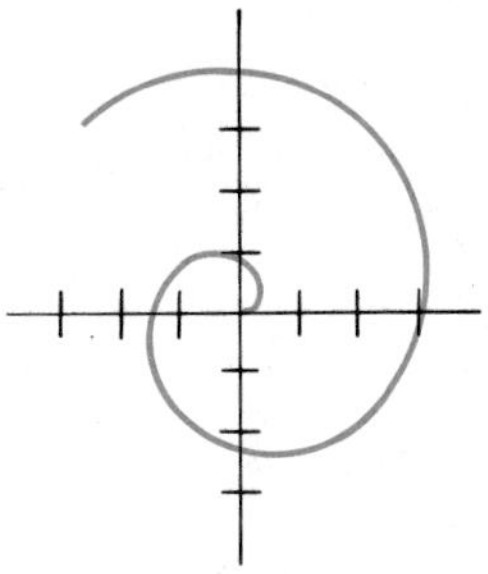

43.

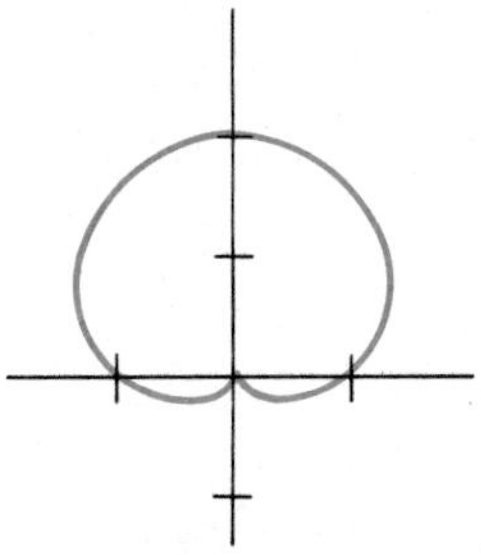

45.

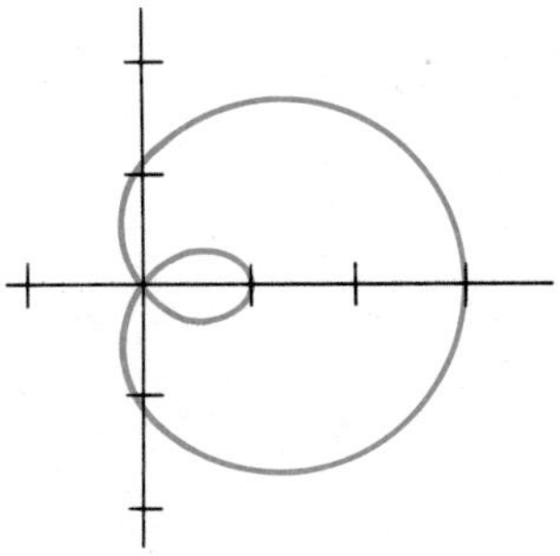

47.

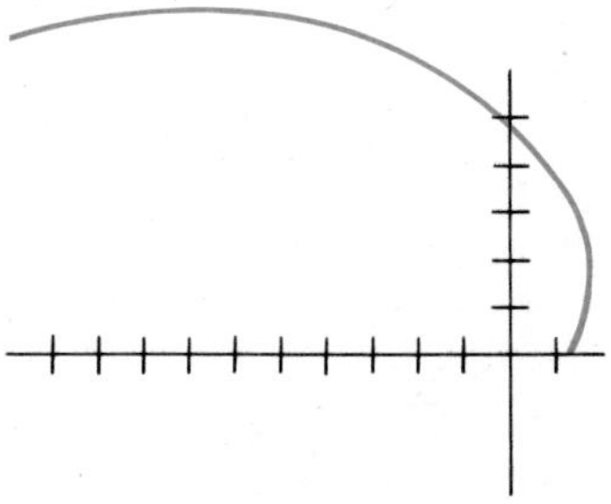

49.

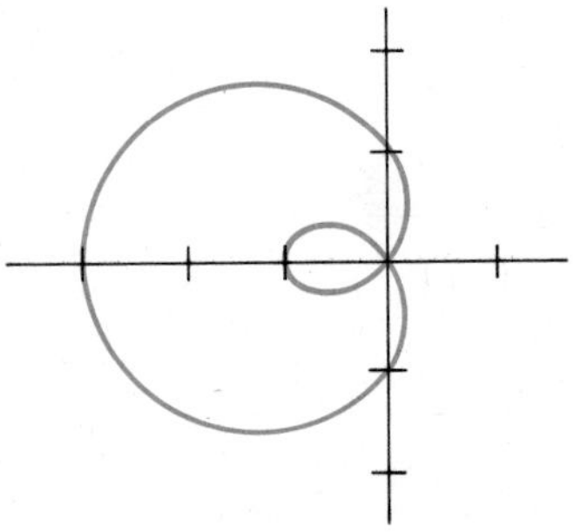

Exercise Set 16.3

1.

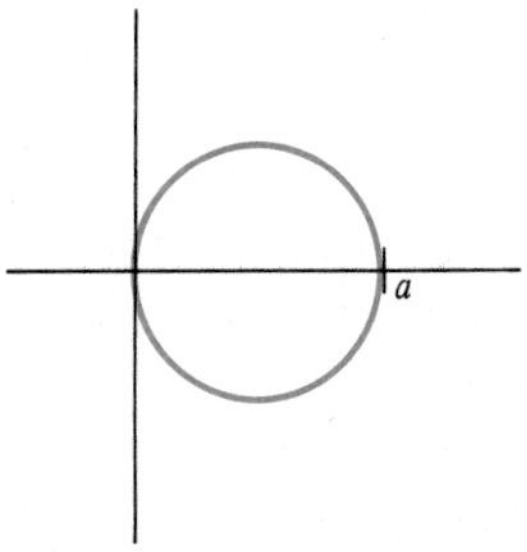

3.

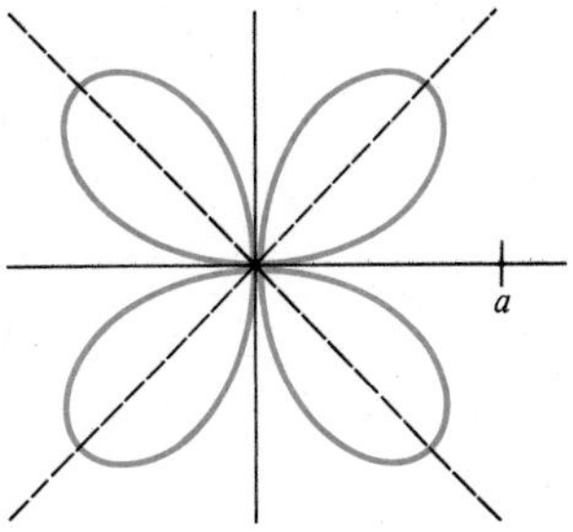

5.

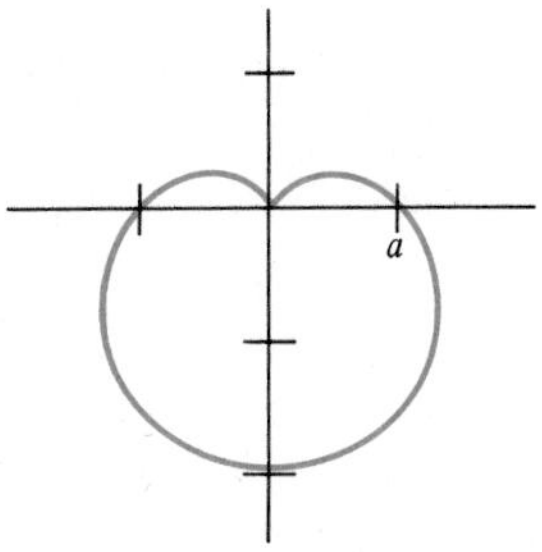

7.

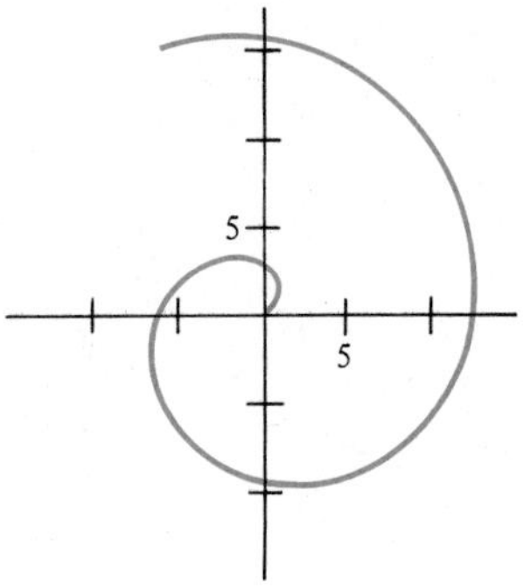

9.

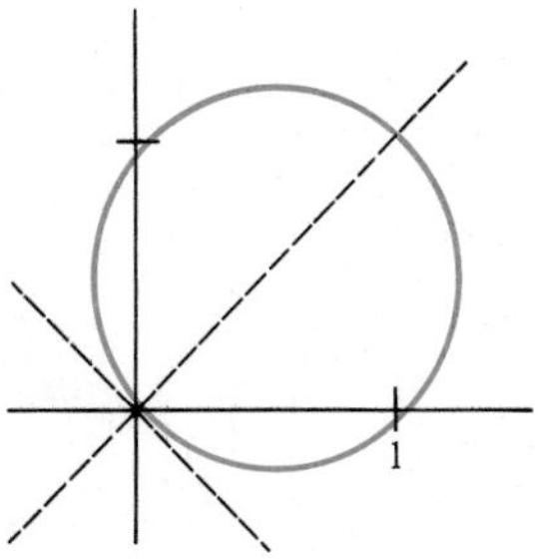

11. $\left(\sqrt{2}, \dfrac{\pi}{4}\right), \left(-\sqrt{2}, \dfrac{5\pi}{4}\right), (0, \theta)$

13. $\left(\dfrac{3}{2}, \dfrac{\pi}{6}\right), \left(\dfrac{3}{2}, \dfrac{5\pi}{6}\right), (0, \theta)$

15. $\left(\dfrac{3a}{2}, \dfrac{\pi}{3}\right), \left(\dfrac{3a}{2}, \dfrac{5\pi}{3}\right), (0, \theta)$

17. $\left(a, \dfrac{\pi}{2}\right), \left(a, \dfrac{3\pi}{2}\right), (0, \theta)$

19. $\left(\dfrac{4}{\pi}, \dfrac{\pi}{4}\right)$

21. $y^2 = a^2 \pm 2ax$ and $x^2 = a^2 \mp 2ay$

23. $y^2 = 1 - 2x$

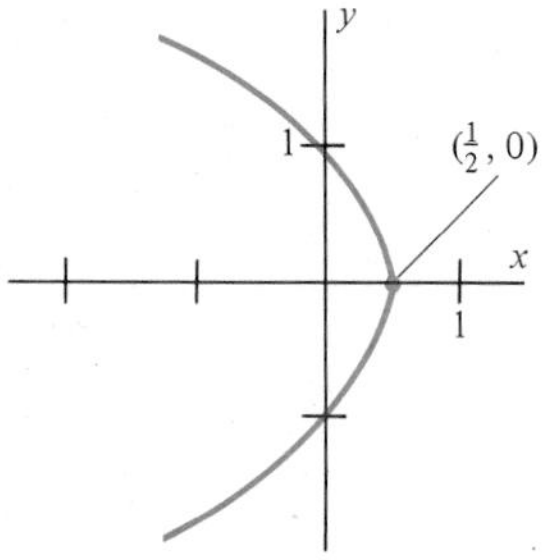

25. $\dfrac{\left(x + \dfrac{4}{3}\right)^2}{\left(\dfrac{8}{3}\right)^2} + \dfrac{y^2}{\left(\dfrac{4}{\sqrt{3}}\right)^2} = 1$ (ellipse)

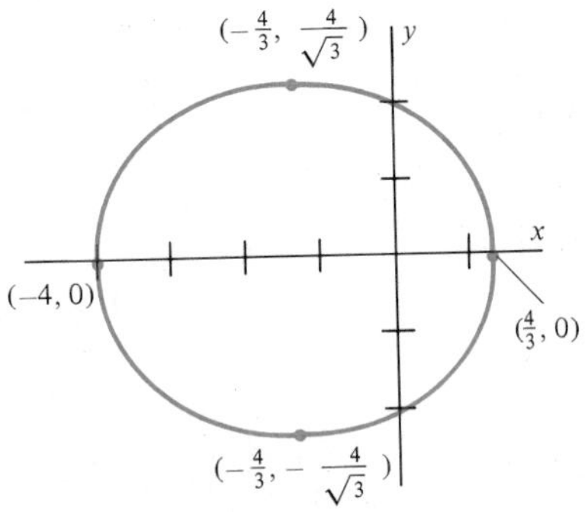

27. $y^2 = 4x + 4$ (parabola)

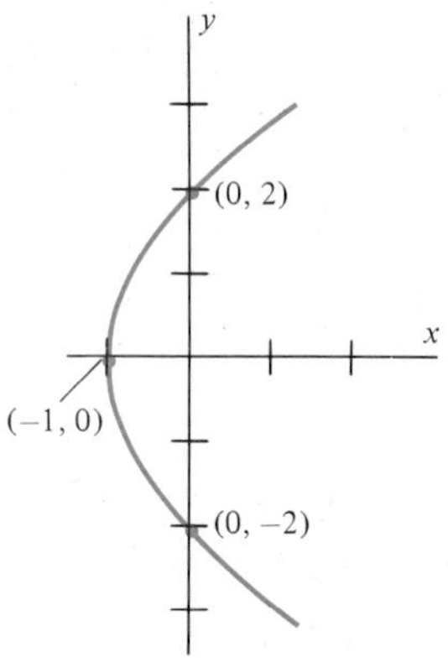

Exercise Set 16.4

1. $\dfrac{3\pi}{4}$

3. π

5. $\dfrac{9}{4}\pi + 4$

7. $\dfrac{33\pi}{8} + 4\sqrt{2} + \dfrac{1}{4}$

9. π

11. $\dfrac{\pi a^2}{2}$

13. $\dfrac{a^2\pi}{4}$

15. $\dfrac{\pi}{2}$

17. $\dfrac{3a^2\pi}{2} - 4a^2$

19. $\dfrac{\pi - 2}{2}$

21. $\dfrac{1}{4}e^{2\pi} - \dfrac{\pi^3}{6} - \dfrac{1}{4}$

23. $2\pi + 3\sqrt{3}$

25. π

27.

	$n = 10$	$n = 100$	$n = 200$
a.	2.587376	2.494899	2.489281
b.	4.885851	4.964704	4.966249
c.	0.138355	0.091575	0.089359

Exercise Set 16.5

1. $y = x + 6$

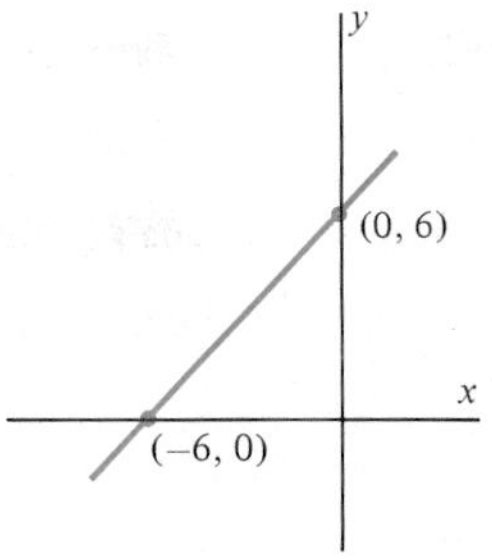

3. $y = \dfrac{2}{3}x + \dfrac{4}{3}$

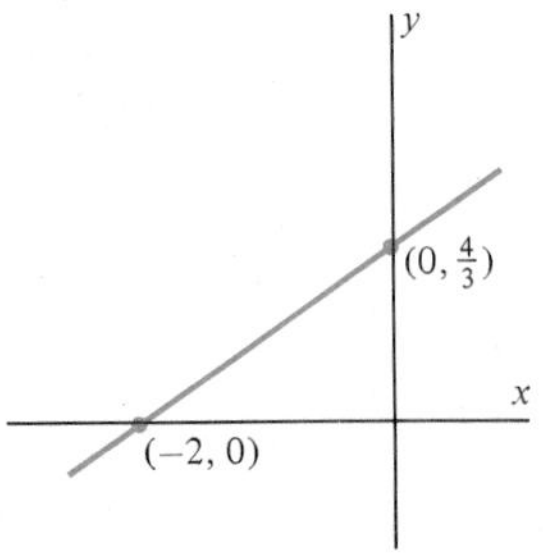

5. $y = x - 2;\quad x \geq 1$

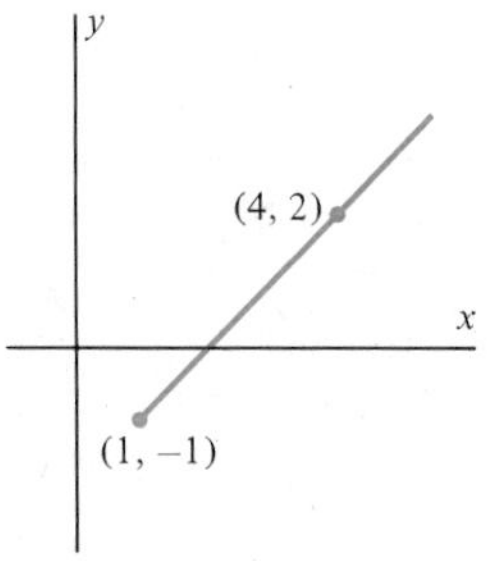

7. $y = x^3$

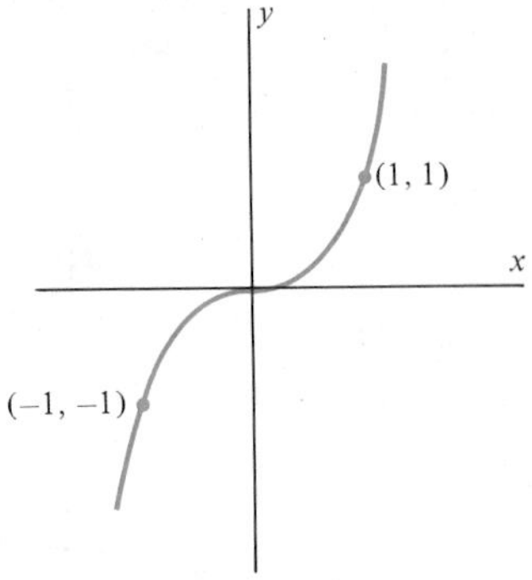

9. $x^2 - y^2 = 1$

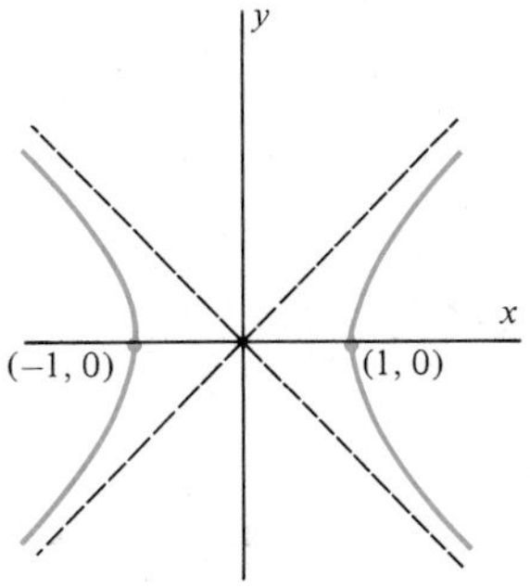

11. $y = 2x\sqrt{1 - x^2}; \qquad |x| \leq 1$

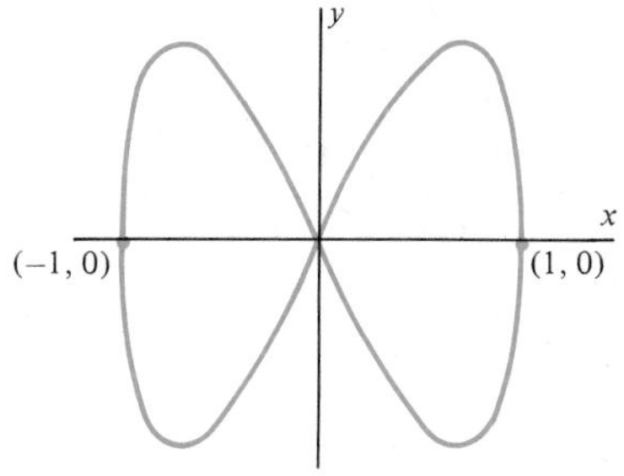

13. slope $= 4, \quad y = 4x - 3$

15. slope $= -\sqrt{3}, \quad y = -\sqrt{3}x + 2$

17. slope $= -48, \quad y = -48x + 29$

19. slope $= -1, \quad y = -x + 2$

21. 0 **23.** $\dfrac{5}{\sqrt{3}}$ **25.** $\dfrac{-2}{\pi}$

27. vertical tangent: $(1, 0)$, $(-1, 0)$
horizontal tangent: $(0, 1)$, $(0, -1)$

29. vertical tangent at $(4, 2)$
horizontal tangent at $(5, -1)$

31. vertical tangent at $(6, 0)$
horizontal tangent at $\left(\dfrac{27}{4}, \dfrac{1}{4}\right)$

33. $(1, 5)$

35. $x(\theta) = f(\theta)\cos\theta = a\sin 3\theta \cdot \cos\theta$
$y(\theta) = f(\theta)\sin\theta = a\sin 3\theta \cdot \sin\theta$

37. $(0, 0), \left(\dfrac{3}{2}, \dfrac{\pi}{6}\right), \left(\dfrac{3}{2}, \dfrac{5\pi}{6}\right)$

39. vertical tangent at $(2, 0)$, $\left(\dfrac{1}{2}, \dfrac{2\pi}{3}\right)$, $\left(\dfrac{1}{2}, \dfrac{4\pi}{3}\right)$

41. **a.** $y = -(x^2 - 8x + 8)$
$y = x + 2$
b. Yes. At $(2, 4)$ and $(5, 7)$
c. The particles collide at $t = -2$ and at $t = 1$ at the points $(2, 4)$ and $(5, 7)$

Exercise Set 16.6

1. $\dfrac{8}{27}(2^{3/2} - 1)$ **3.** $2(3^{3/2} - 1)$ **5.** 36

7. $(5^{3/2} - 8^{3/2}) + (12\sqrt{2} - 3\sqrt{5} + 12\ln(\sqrt{5} - 1) - 12\ln(\sqrt{8} - 2))$

9. 3 **13.** $\dfrac{16\pi}{3}(8^{3/2} - 8)$

11. $\sqrt{2}(e^{\pi} - 1)$ **15.** $\dfrac{\pi}{6}[27 - 5^{3/2}]$

17. $$24\pi\left[4\sqrt{3} - \frac{1}{4}\cdot 5^{3/2} + \frac{1}{8}\sqrt{5} + \frac{1}{16}\ln(2 + \sqrt{5}) - \frac{2^{3/2}}{6} + \frac{1}{8}\cdot 2^{3/2} - \frac{1}{16}\sqrt{2} - \frac{1}{16}\ln(1 + \sqrt{2})\right]$$

19. 4 **23.** 4π **27.** $\sqrt{2}(e^{2\pi} - 1)$

21. $\dfrac{\sqrt{5}}{2}[e^{2\pi} - 1]$ **25.** 3π **29.** $\sqrt{2}\pi(e^{2\pi} - 1)$

31. $S = \displaystyle\int_a^b 2\pi x(t)\sqrt{(x'(t))^2 + (y'(t))^2}\,dt, \qquad a \leq t \leq b$

33. $\dfrac{13\pi}{4}$

Review Exercises–Chapter 16

1. $x^2 + y^2 = 25$

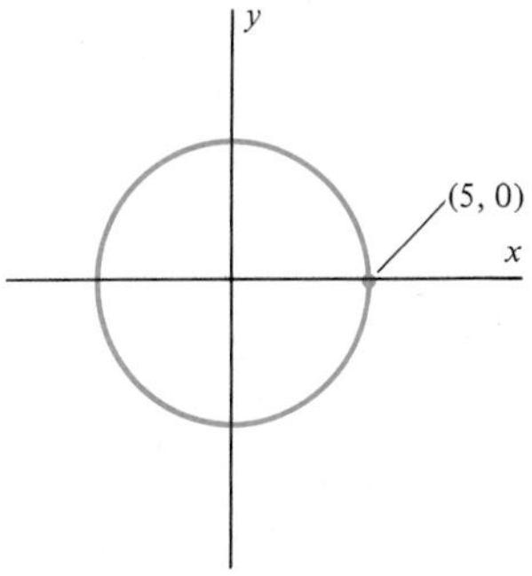

3. $\sqrt{x^2 + y^2} = t$

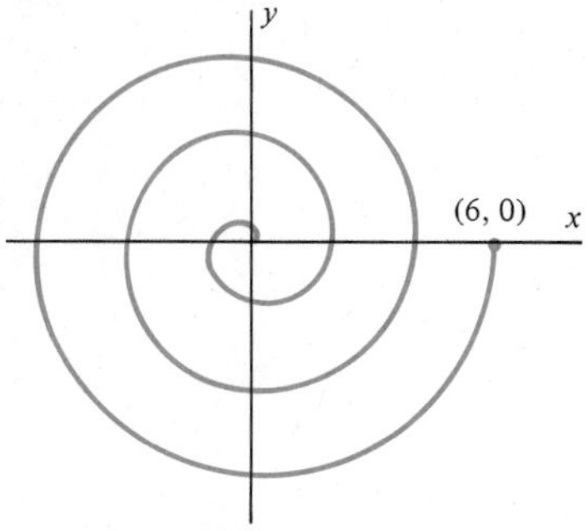

5. $y = 2x\sqrt{1 - x^2}$

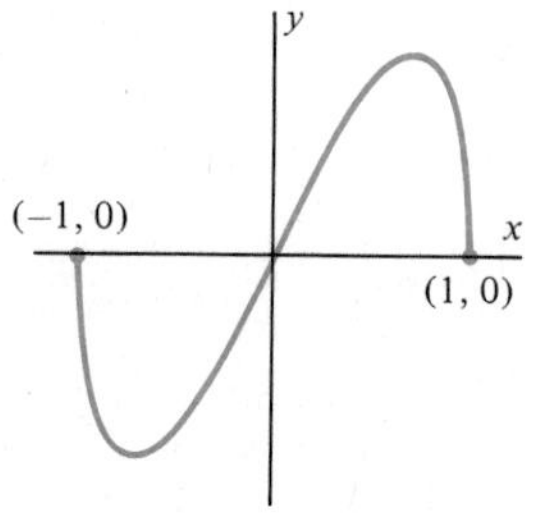

7. $y = x \tan \pi\sqrt{x^2 + y^2}$

9. $y = 2x\sqrt{1 - x^2}$

11. $x(t) = 3 \cos t$
$y(t) = 2 \sin t$

13. $x(t) = 2t^2 + 4t + 5$
$y(t) = t$

15.

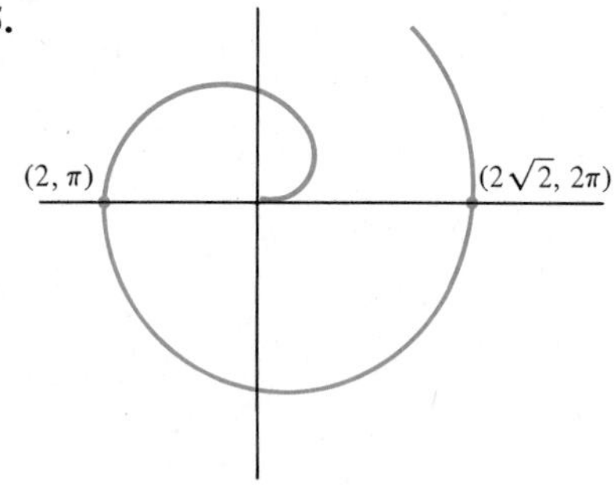

17. $x = \dfrac{y^2}{9} - 1$ (parabola)

21.

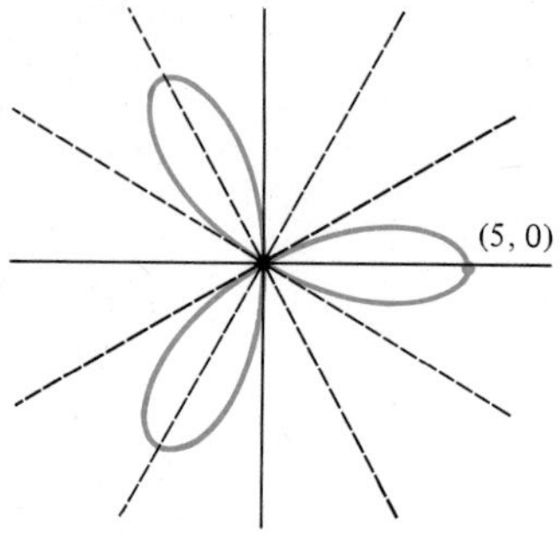

23. $r = 6 \sin \theta$

25.

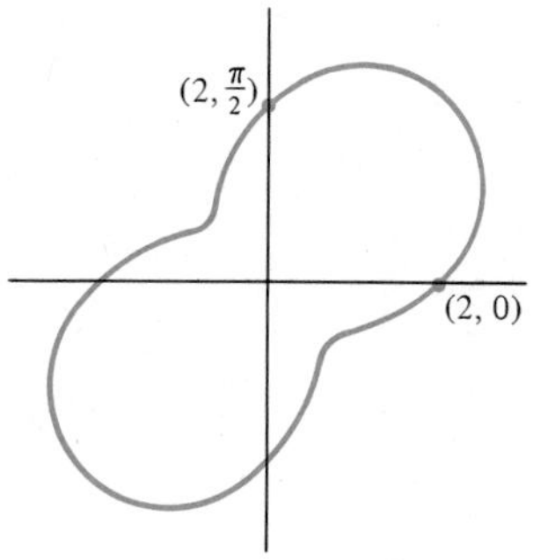

29. 6π

31. $\dfrac{3a^2\pi}{2}$

33. 8π

37. 3.27

39. π

41. $\dfrac{\pi}{2}$

43. 2.83

45. $\dfrac{\pi}{8} - \dfrac{1}{4}$

47. 4π

49. $\dfrac{8\pi}{3} - 2\sqrt{3}$

CHAPTER 17

Exercise Set 17.2

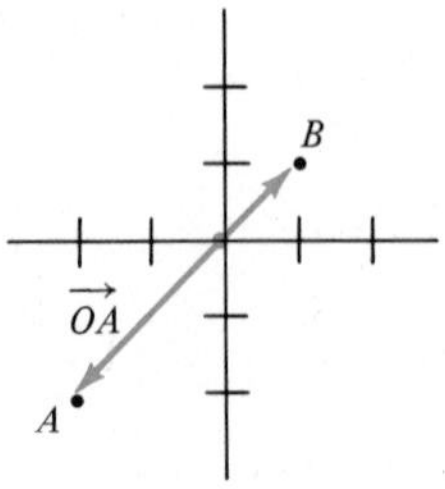

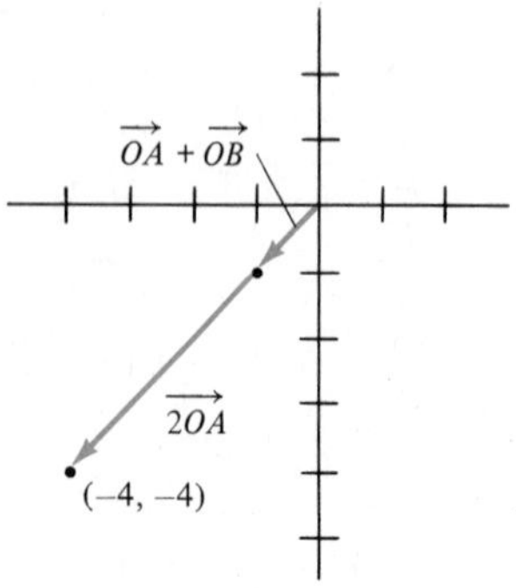

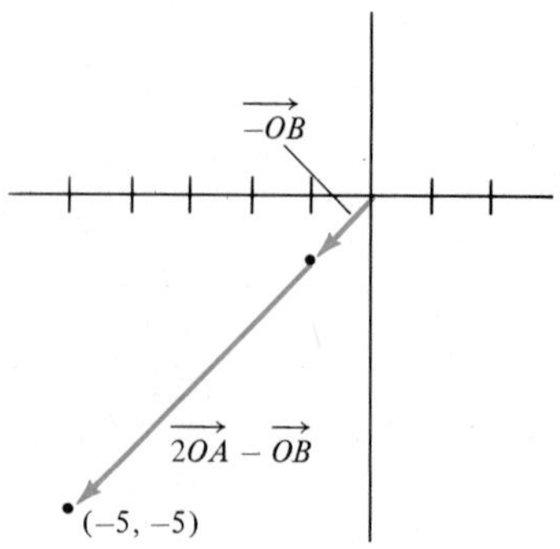

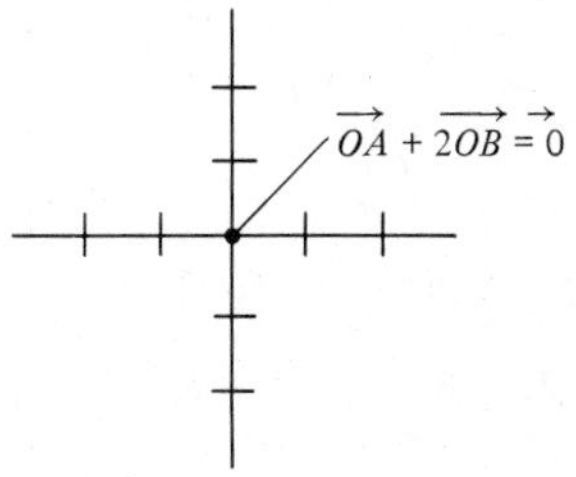

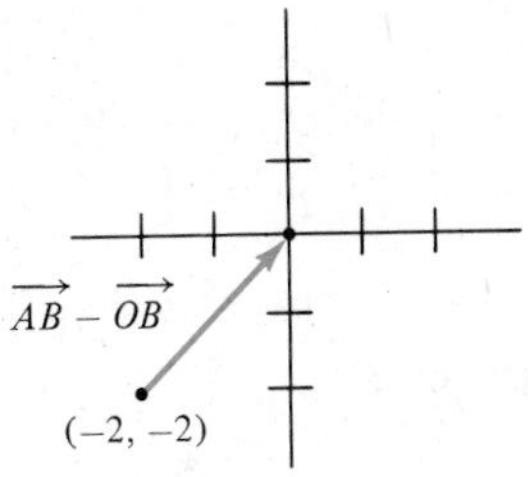

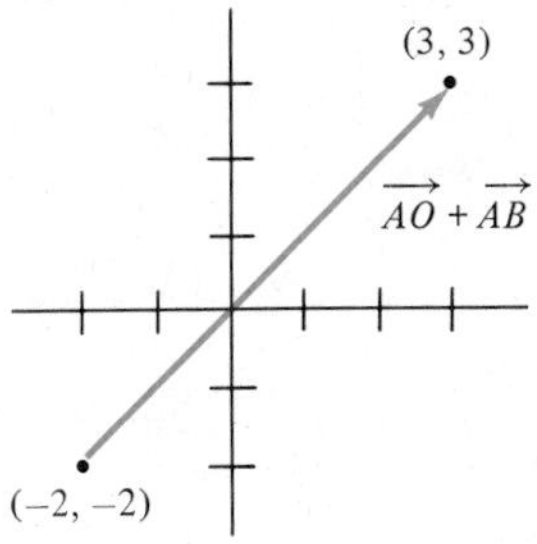

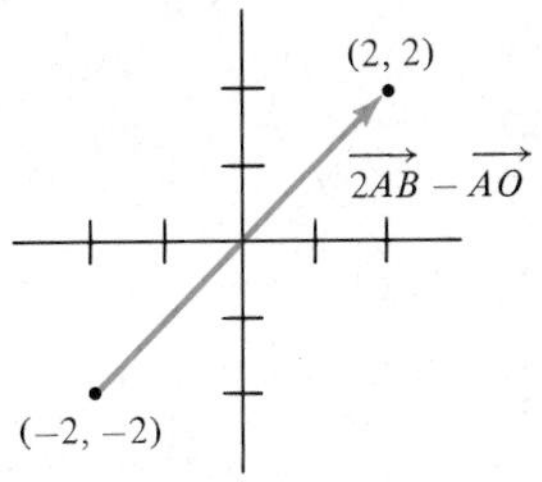

3. **a.** $D = (2, 3)$
b. $D = (10, 3)$ or $(0, -1)$

5. $2\sqrt{2}$ **7.** 90 m **9.** $\theta = \sin^{-1}\dfrac{3}{5}$

11. **a.** 60 km east of P and 400 km to the north
b. 404.5 km
c. $\theta = \sin^{-1} 0.15$

13. $50\sqrt{2}$ lb

Exercise Set 17.3

1.

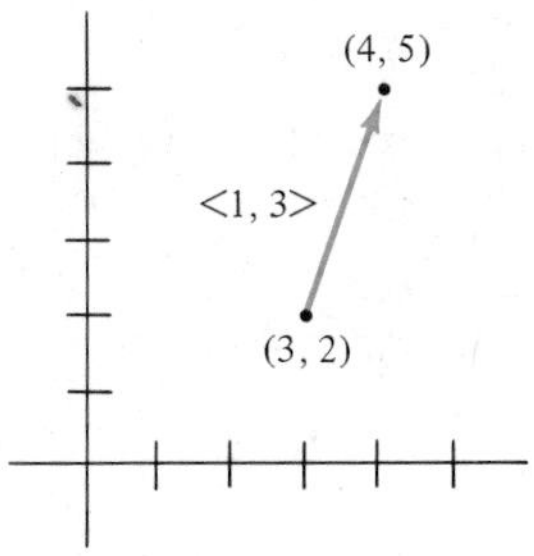

3.

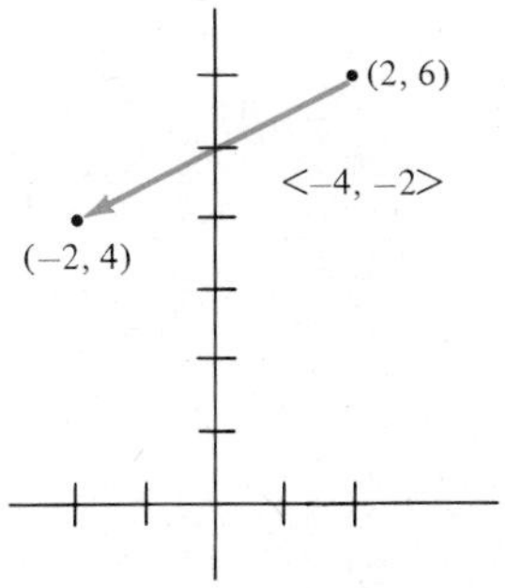

5.

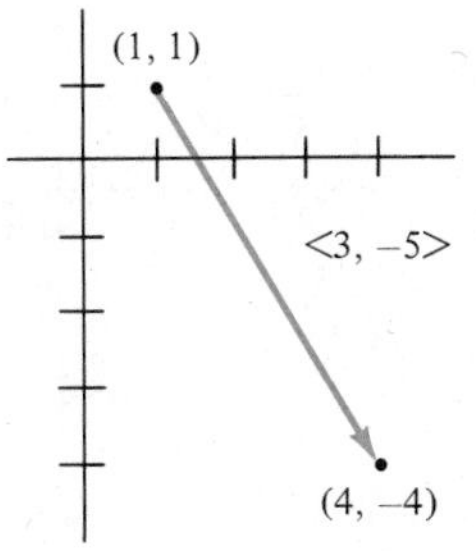

7. $\langle -1, 2\rangle$

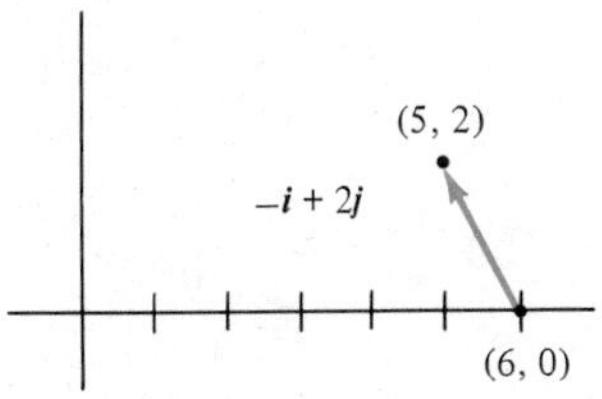

9. $\langle -3, 2\rangle$

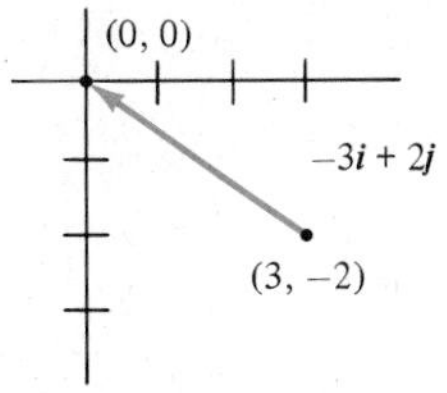

11. $\langle -3, -3\rangle$
13. $\langle -3, -14\rangle$
15. $\langle -6, -6\rangle$
17. $\langle -1, 9\rangle$
19. $\langle 10, 14\rangle$
21. $\langle -1, 5\rangle$
23. $\langle 7, 11\rangle$
25. $\langle 4, 6\rangle$
27. $\langle 19, 19\rangle$
29. $\langle 12, 18\rangle$
31. $\dfrac{\pi}{2}$
33. $\dfrac{5\pi}{6}$
35. $\dfrac{5\pi}{3}$
37. $\sqrt{37}$
39. $|\alpha\sqrt{1 + \alpha^2}|$
41. $\sqrt{2}$
43. $3\sqrt{5}$
45. $-i$

47. $\left\langle -\dfrac{1}{2}, \dfrac{\sqrt{3}}{2} \right\rangle$

49. v originating at $(1, 1)$ and terminating at $\left(1 + \dfrac{1}{\sqrt{10}}, 1 + \dfrac{3}{\sqrt{10}}\right)$ and $-v$ originating at $(1, 1)$ and terminating at $\left(1 - \dfrac{1}{\sqrt{10}}, 1 - \dfrac{3}{\sqrt{10}}\right)$.

Exercise Set 17.4

1. 1
3. −28
5. −15
7. $-\dfrac{7}{\sqrt{170}}$

9. **a.** 14
b. −28
c. −14
d. 42
e. 78
f. −56

g. $\frac{14}{\sqrt{34}}$

h. $-\frac{1}{\sqrt{13}}$

i. $\left\langle -\frac{7}{5}, \frac{14}{5} \right\rangle$

j. $\left\langle \frac{16}{13}, \frac{24}{13} \right\rangle$

11. a. $\frac{7}{\sqrt{26}}$

b. 2

c. $\frac{7}{\sqrt{13}}$

d. $\left\langle \frac{35}{26}, \frac{7}{26} \right\rangle$

e. $\left\langle \frac{17}{26}, -\frac{85}{26} \right\rangle$

f. $\left\langle \frac{51}{13}, \frac{34}{13} \right\rangle$

21. $4\sqrt{2}$

23. $1000\sqrt{2}$ ft-lb

25. $\boldsymbol{v}$ and $\boldsymbol{w}$ being parallel and of same direction.

Exercise Set 17.5

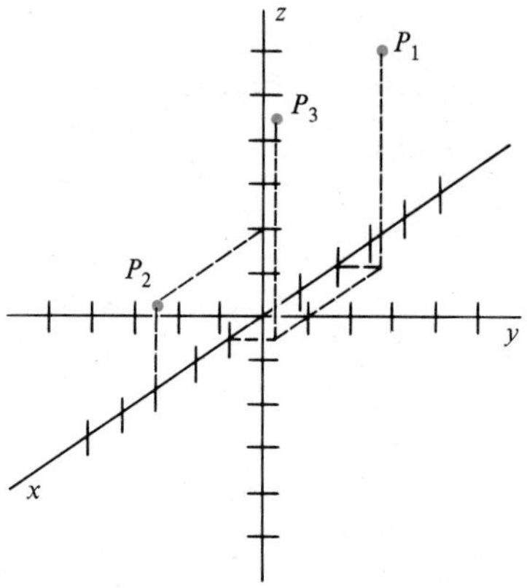

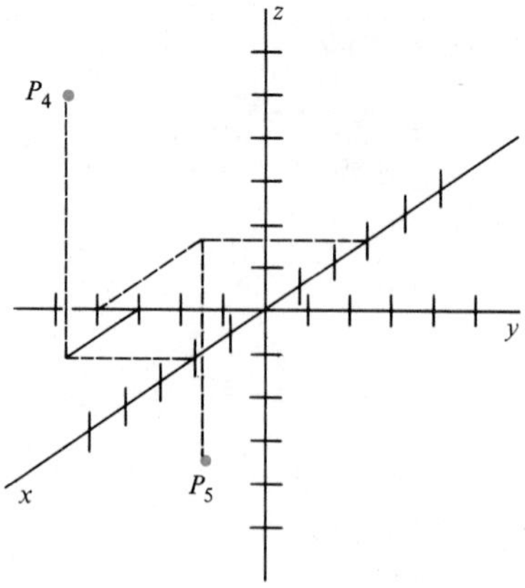

3. a. $2\sqrt{2}$ **b.** $6\sqrt{2}$ **c.** $\sqrt{22}$ **d.** $\sqrt{a^2 + b^2 + c^2}$

5. $(-3, 3, -3)$

7. center $(0, 0, 2)$ and radius 3

9. center $(2, -1, 3)$ and radius 4

11. center $(3, -1, -2)$ and radius 5

13. $\langle -16, 29, 5 \rangle$

15. $\langle 12, -13, 15 \rangle$

17. $9\sqrt{6}$

19. 1

21. $\langle -12, 1, -7 \rangle$

23. -6

25. $\left\langle \frac{3}{\sqrt{17}} + \frac{5}{\sqrt{35}}, \frac{-2}{\sqrt{17}} - \frac{1}{\sqrt{35}}, \frac{2}{\sqrt{17}} + \frac{3}{\sqrt{35}} \right\rangle$

27. $\left\langle \frac{23}{7}, \frac{-23}{35}, \frac{69}{35} \right\rangle$

29. $\frac{-3}{\sqrt{6}}$

31. a. $\langle 2, 1, 2 \rangle$ **b.** $\langle 4, -1, -5 \rangle$ **c.** $\langle 6, 3, -1 \rangle$

33. a. $\cos\alpha = \frac{3}{\sqrt{14}}$, $\cos\beta = \frac{-1}{\sqrt{14}}$, $\cos\gamma = \frac{2}{\sqrt{14}}$

b. $\cos\alpha = \frac{6}{\sqrt{41}}$, $\cos\beta = -\frac{2}{\sqrt{41}}$, $\cos\gamma = \frac{1}{\sqrt{41}}$

c. $\cos\alpha = -\frac{1}{\sqrt{21}}$, $\cos\beta = \frac{2}{\sqrt{21}}$, $\cos\gamma = -\frac{4}{\sqrt{21}}$

35. a. $\left\langle \frac{2}{3}, -\frac{4}{3}, \frac{4}{3} \right\rangle$ and $\left\langle -\frac{2}{3}, \frac{4}{3}, -\frac{4}{3} \right\rangle$

b. $a = -3$

c. $-\frac{5}{\sqrt{14}}\boldsymbol{i} + \frac{10}{\sqrt{14}}\boldsymbol{j} - \frac{15}{\sqrt{14}}\boldsymbol{k}$

d. $a = -\frac{3}{2}$, $b = \frac{1}{2}$

37. $t = 2$

Exercise Set 17.6

1. $-7\boldsymbol{i} + 8\boldsymbol{j} + 11\boldsymbol{k}$

3. $5\boldsymbol{i} - 4\boldsymbol{j} - \boldsymbol{k}$

5. 12

7. $3\boldsymbol{i} + 10\boldsymbol{j} + \boldsymbol{k}$

9. $-16\boldsymbol{i} + 8\boldsymbol{k}$

11. $37\boldsymbol{i} + 30\boldsymbol{j} - 11\boldsymbol{k}$

13. 40

15. $5\sqrt{3}$

17. 8

19. $\frac{5}{2}\sqrt{2}$

21. $\frac{5\sqrt{10}}{2}$

23. $\frac{23}{2}$

25. $\boldsymbol{k}$ or $-\boldsymbol{k}$

27. $\frac{\boldsymbol{i}}{\sqrt{3}} - \frac{\boldsymbol{j}}{\sqrt{3}} - \frac{\boldsymbol{k}}{\sqrt{3}}$ and $-\frac{\boldsymbol{i}}{\sqrt{3}} + \frac{\boldsymbol{j}}{\sqrt{3}} + \frac{\boldsymbol{k}}{\sqrt{3}}$

31. 53

Exercise Set 17.7

1. $x(t) = 1 + t$, $y(t) = 2 + t$, $z(t) = 3 - t$

3. $x(t) = 3 + 3t$, $y(t) = -1 - t$, $z(t) = 5 + 5t$

5. $x(t) = -4 + t,\ y(t) = 2 + 3t,\ z(t) = 1 + 2t$

7. $x(t) = 1 - 10t,\ y(t) = 2 + 3t,\ z(t) = t$

9. $\dfrac{x-7}{1} = \dfrac{y+6}{4} = \dfrac{z-3}{-2}$

11. $\dfrac{x-0}{1} = \dfrac{y+6}{8} = \dfrac{z-4}{4}$

13. $\dfrac{x-3}{3} = \dfrac{y+1}{-1} = \dfrac{z-5}{5}$

15. $x(t) = 3t,\ y(t) = 2t,\ z(t) = 5t$

17. $x(t) = 4t - 4,\ y(t) = -2t + 2,\ z(t) = 3t - 3$

19. $\boldsymbol{i} + 4\boldsymbol{j} - 2\boldsymbol{k}$

21. $4\boldsymbol{i} - 2\boldsymbol{j} + 3\boldsymbol{k}$

23. $-\dfrac{\boldsymbol{a} \cdot \boldsymbol{b}}{|\boldsymbol{b}|^2}$

25. $\sqrt{3}$

27. $\dfrac{\sqrt{210}}{3}$

29. $\dfrac{1}{3}\sqrt{\dfrac{254}{5}}$

31. **d.** $\dfrac{16}{\sqrt{53}}$

33. $(2, 1, 2)$

35. $x + 2y - z = 8$

37. $2x - 2y + 3z = 19$

39. $x + 2y + z = 1$

41. $29x - 6y - 15z = -16$

43. $x - y = -2$

45. $2\boldsymbol{i} - 3\boldsymbol{j} + \boldsymbol{k}$

47. $3x - 6y + 2z = 31$

49. $x(t) = \dfrac{27}{6} + \dfrac{7}{6}t,\ y(t) = \dfrac{9}{6} + \dfrac{5}{6}t,\ z(t) = t$

51. $\theta = \cos^{-1}\dfrac{1}{\sqrt{57}}$

53. $\dfrac{27}{\sqrt{38}}$

55. $x(t) = 2 + 2t,\ y(t) = 4 + 3t,\ z(t) = -3 - 7t$

Exercise Set 17.8

1.

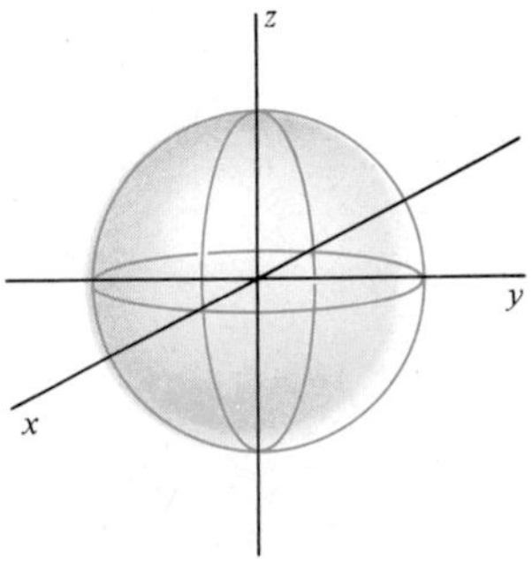

5.

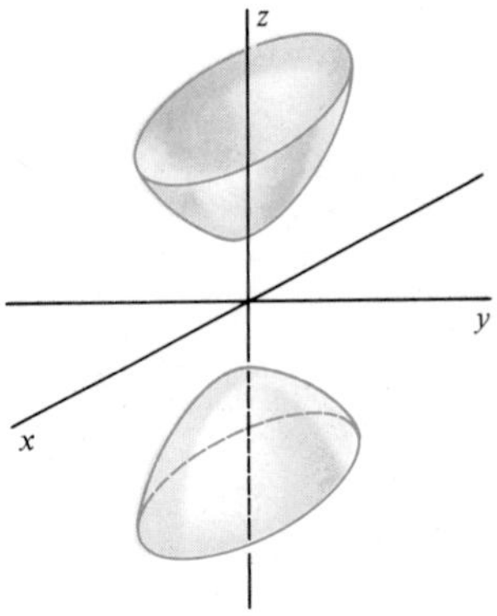

7.

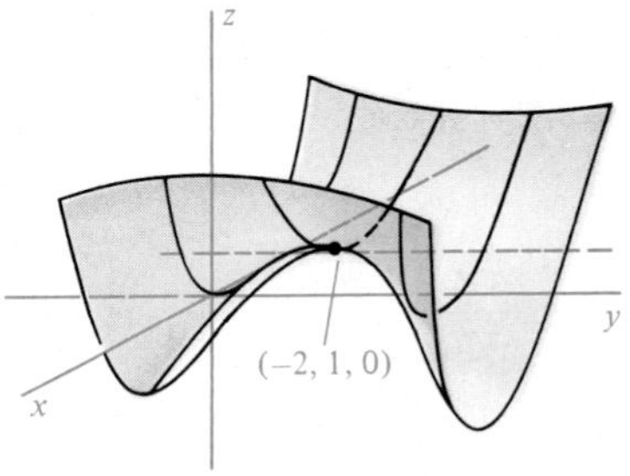

11.

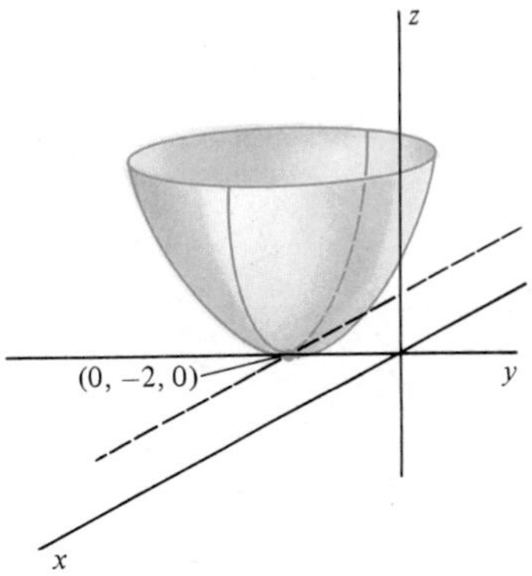

17.

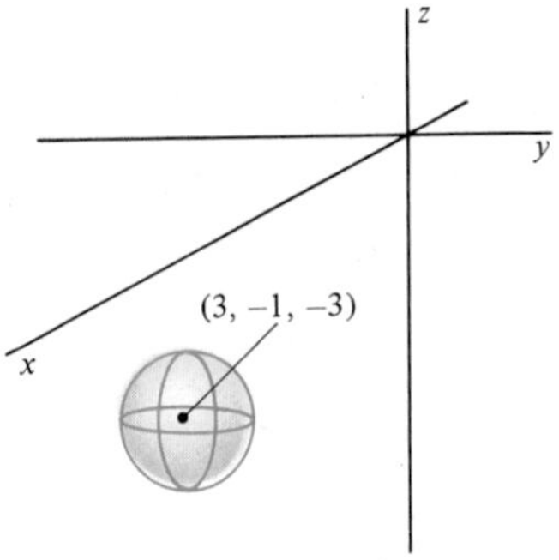

21.

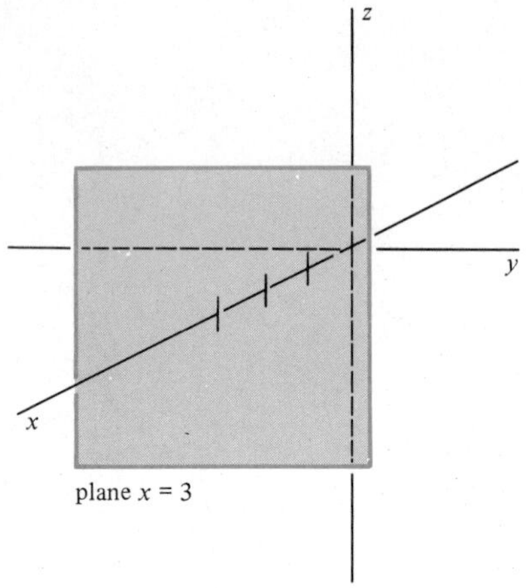

25.

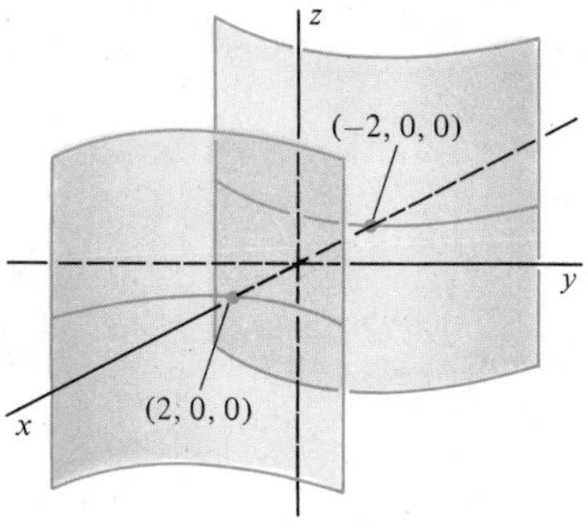

29.

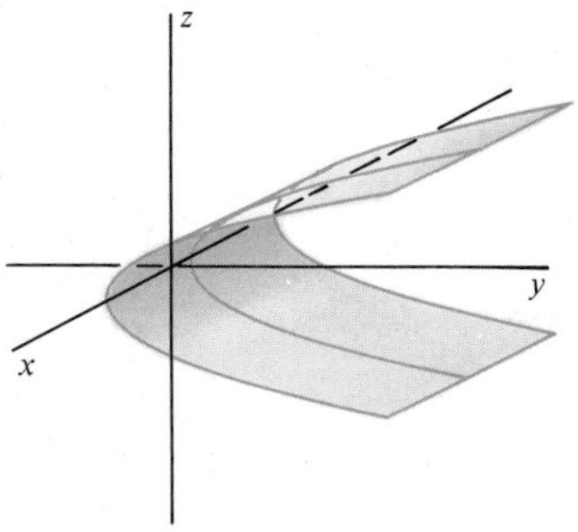

33.

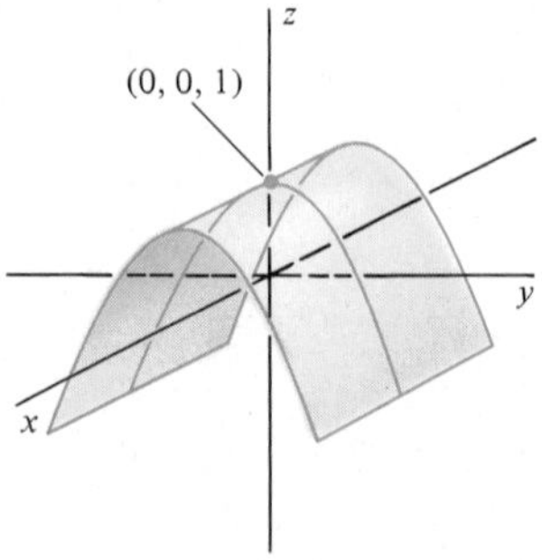

37. $y^2 = 4x^2 + 4z^2$

39. $y = x^2 + z^2$

Exercise Set 17.9

1. a. $\left(\sqrt{2}, \frac{\pi}{4}, 0\right)$ **b.** $\left(2, \frac{\pi}{6}, 3\right)$

c. $\left(\sqrt{2}, \frac{3\pi}{4}, -2\right)$ **d.** $\left(2, \frac{2\pi}{3}, 4\right)$

e. $\left(3, \frac{\pi}{2}, -5\right)$ **f.** $\left(2, \frac{3\pi}{4}, \sqrt{2}\right)$

3. a. $\left(1, 0, \frac{\pi}{2}\right)$ **b.** $\left(2, \frac{\pi}{4}, \frac{\pi}{4}\right)$

c. $\left(2, \frac{7\pi}{4}, \frac{\pi}{4}\right)$ **d.** $\left(2\sqrt{2}, \frac{\pi}{3}, \frac{\pi}{4}\right)$

e. $\left(2\sqrt{2}, \frac{5\pi}{6}, \frac{3\pi}{4}\right)$ **f.** $\left(4, \frac{3\pi}{4}, \frac{\pi}{4}\right)$

5. a. $\left(1, 0, \frac{\pi}{2}\right)$ **b.** $\left(2, -\frac{\pi}{4}, \frac{\pi}{4}\right)$

c. $\left(2\sqrt{2}, \frac{\pi}{3}, \frac{\pi}{4}\right)$ **d.** $\left(2\sqrt{2}, \frac{\pi}{4}, \frac{\pi}{4}\right)$

e. $\left(2, \frac{5\pi}{3}, \frac{\pi}{2}\right)$ **f.** $\left(2\sqrt{2}, \frac{\pi}{6}, \frac{\pi}{4}\right)$

7. $z = 3$

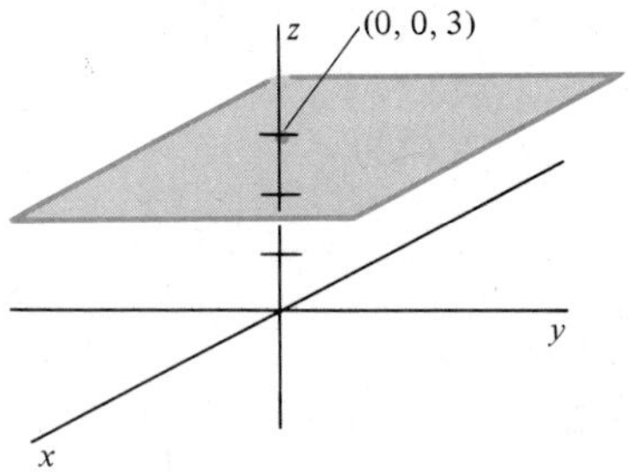

9. $z^2 = 4(x^2 + y^2)$

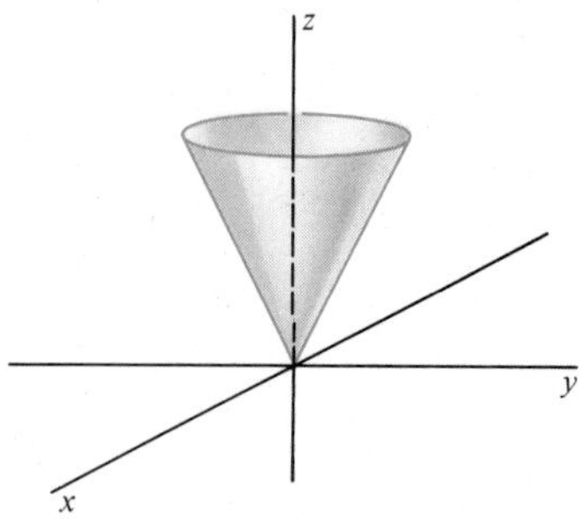

11. $x^2 + y^2 + z^2 = 4$

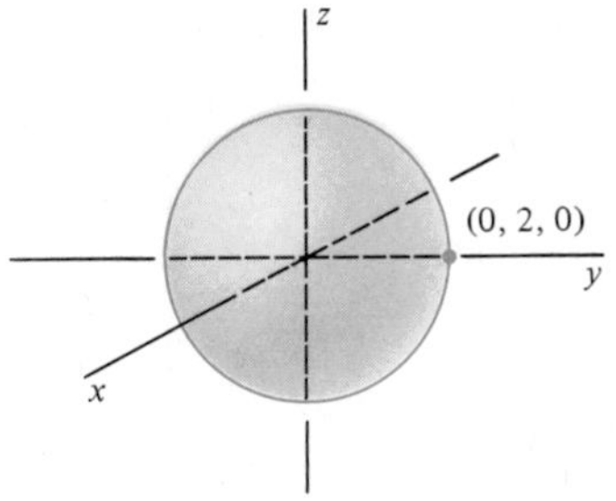

13. $x^2 - y^2 = \alpha^2$

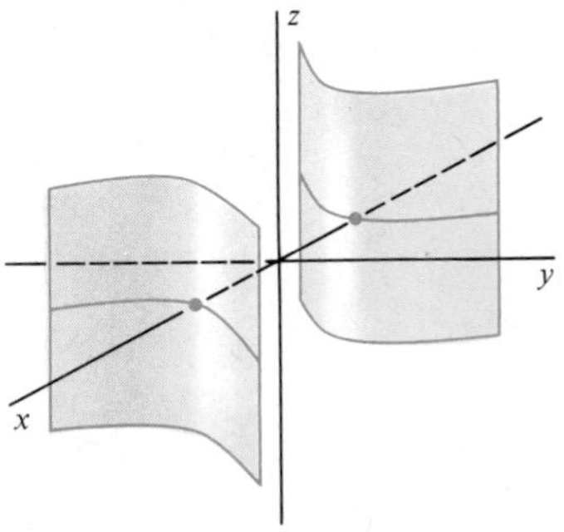

15. $r^2 = 9$

17. $r^2 = 9z$

19. $r^2 \cos^2 \theta + z^2 = 4$

21. $r^2 \cos \theta \sin \theta = 4 + ar \cos \theta$

23. $\rho^2 \sin^2 \phi = 9$

25. $\rho \sin^2 \phi = 9 \cos \phi$

27. $1 - \sin^2 \phi \sin^2 \theta = \dfrac{4}{\rho^2}$

29. $\rho \sin^2 \phi \sin 2\theta - 2a \sin \phi \cos \theta = \dfrac{8}{\rho}$

Review Exercises–Chapter 17

1. $\sqrt{86}$

3. center $(3, -2, 1)$, radius 2

5. $\dfrac{-1}{\sqrt{77}}$

7. $x(t) = 23t + 2,\ y(t) = -t + 1,\ z(t) = -9t - 3$

9. $7x - 9y + 3z = -22$

11. $8\boldsymbol{i} + \boldsymbol{j} - 2\boldsymbol{k}$

13. $\boldsymbol{r}(t) = (2 - 12t)\boldsymbol{i} + 11t\boldsymbol{j} + 9t\boldsymbol{k}$

17.

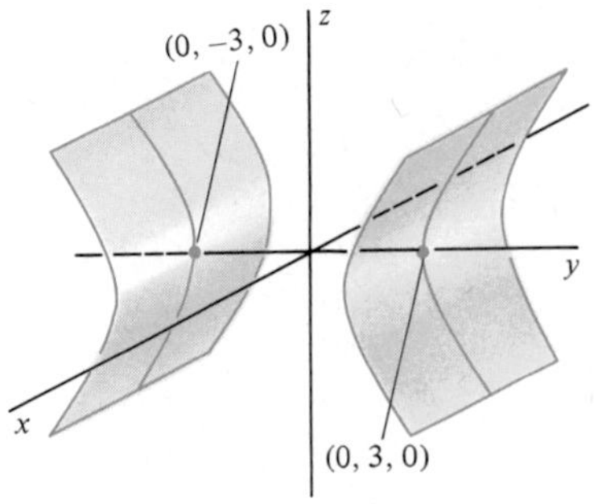

19. $\left(\dfrac{\sqrt{6}}{2}, \dfrac{\sqrt{6}}{2}, 1\right)$

21. $2x - y + 6z = 27$

23. **a.** $\dfrac{22}{5}$ **b.** $\dfrac{66}{5} \cdot \boldsymbol{j} + \dfrac{88}{5}\boldsymbol{k}$

25. $\dfrac{1}{2}\sqrt{118}$

27. $4x + z = 18$

29. $\dfrac{x-1}{3} = \dfrac{y-2}{7} = \dfrac{z-3}{-1}$

31. $z^2 = x^2 + y^2,\ z \geq 0$

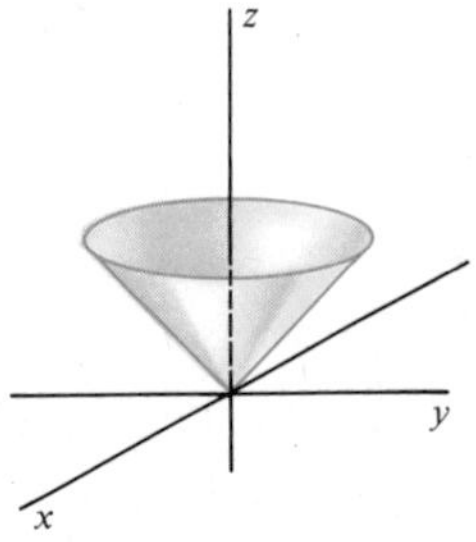

33. $(x - 1)^2 + y^2 = 1$

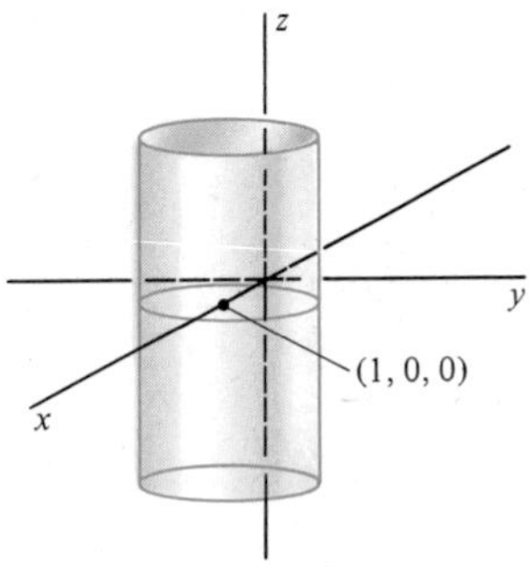

35. **a.** $\dfrac{10}{\sqrt{11}}$ **b.** $\dfrac{10}{7}\boldsymbol{i} + \dfrac{15}{7}\boldsymbol{j} - \dfrac{5}{7}\boldsymbol{k}$

37. $20x + 15y + 12z = 60$

39. $x(t) = 5,\ y(t) = -t + 2,\ z(t) = \dfrac{1}{2}t - 3$

41.

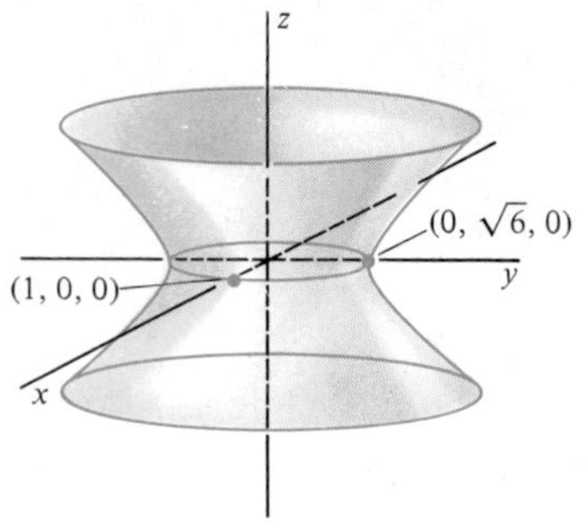

43.

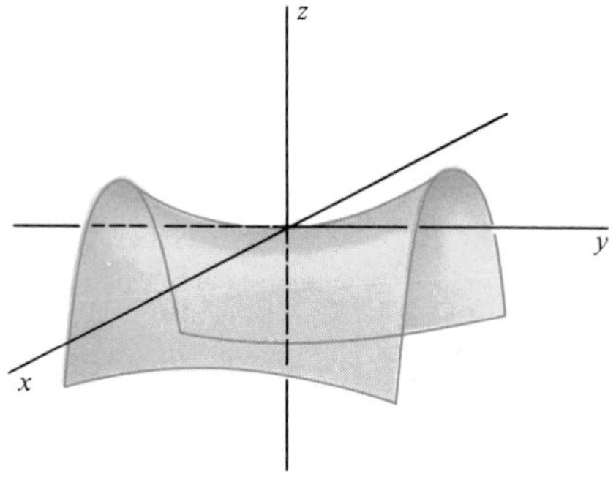

45.

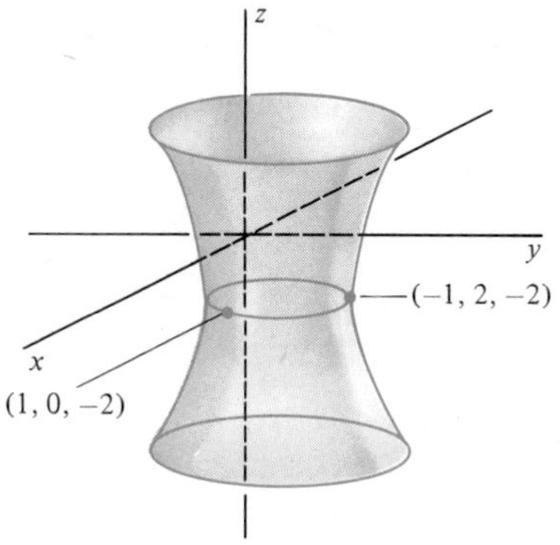

47. $-10\boldsymbol{i} + 14\boldsymbol{j} + 2\boldsymbol{k}$ and $10\boldsymbol{i} - 14\boldsymbol{j} - 2\boldsymbol{k}$

CHAPTER 18

Exercise Set 18.1

1.

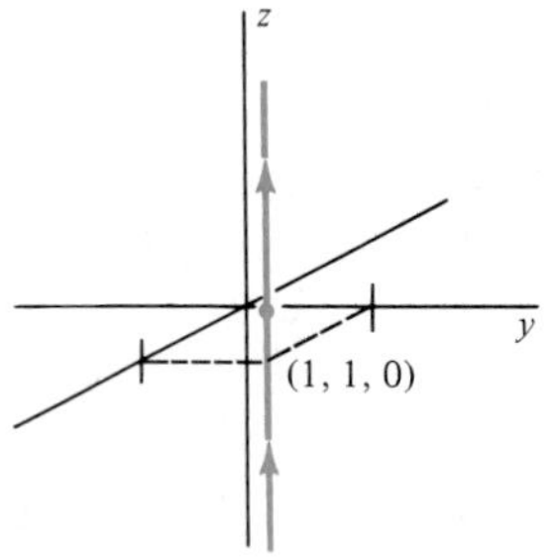

3.

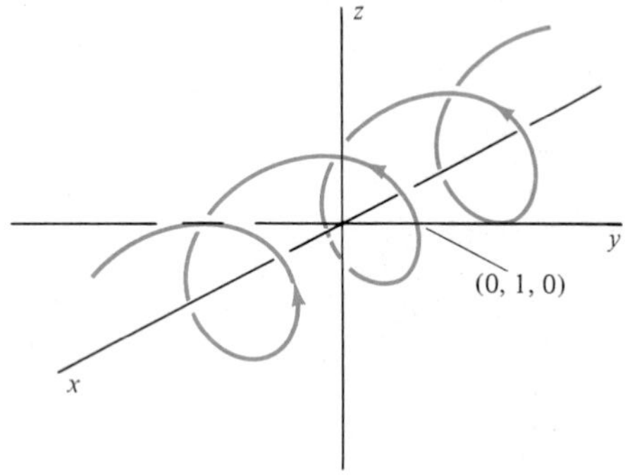

5.

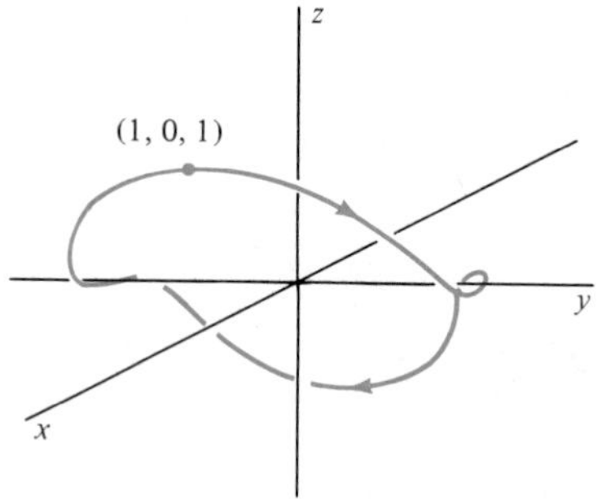

7.

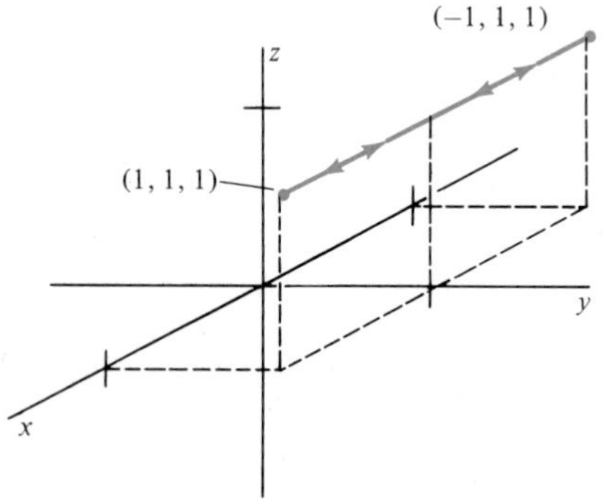

9. $t \geq 0$

11. $t \neq \pm 1$

13. $t \geq 0$ and $t \neq 3$

15. $t \geq 0$

17. a. $f(t) = \cos 10\pi t\boldsymbol{i} + \sin 10\pi t\boldsymbol{j} + 2.5t\boldsymbol{k}$
b. 2.5 mm/sec

Exercise Set 18.2

1. $(t + \sin t)\boldsymbol{i} + (3t + \cos t)\boldsymbol{j} + (t^2 + 1)\boldsymbol{k}$

3. $(3t + 2 \sin t)\boldsymbol{i} + (9t + 2 \cos t)\boldsymbol{j} + (3t^2 + 2)\boldsymbol{k}$

5. $t \sin t + 3t \cos t + t^2$

7. $(1 + t + 3\sqrt{t})\boldsymbol{i} + (4 - 3t - t^2)\boldsymbol{j} + (t + 3e^t)\boldsymbol{k}$

9. $(1 + t - \sqrt{t}) \sin t\boldsymbol{i} + (t - t^2) \sin t\boldsymbol{j} + (t - e^t) \sin t\boldsymbol{k}$

11. $\sqrt{t}(1 + t) + (1 - t)(1 - t^2) + te^t$

13. $\frac{\sqrt{2}}{2}\boldsymbol{i} + \boldsymbol{j} + \frac{\sqrt{2}}{2}\boldsymbol{k}$

15. $\boldsymbol{i} + \boldsymbol{k}$

17. $6\boldsymbol{i} + 7\boldsymbol{j} + 14\boldsymbol{k}$

19. $\left(-\frac{\pi}{2} + n\pi, \frac{\pi}{2} + n\pi\right), n = 0, \pm 1, \ldots$

21. $[0, 1)$ and $(1, \infty)$

23. $[0, 1)$

25. $(-\infty, 1)$ and $(1, \infty)$

29. a. $\boldsymbol{f}(t)$:$(0, \infty)$ $\boldsymbol{g}(t)$:$(0, \infty)$ $\boldsymbol{h}(t)$:all t
b. $\boldsymbol{f}(t)$:$(0, \infty)$ $\boldsymbol{g}(t)$:$(0, \infty)$ $\boldsymbol{d}(t)$:all t

Exercise Set 18.3

1. $(0, \infty)$

3. $(-3, -2)$ and $(-2, \infty)$

5. $\boldsymbol{i} + \frac{1}{2\sqrt{t}}\boldsymbol{j}$

7. $\frac{1}{2\sqrt{t}}\boldsymbol{i} - \frac{3}{2}t^{-5/2}\boldsymbol{j} + \frac{2}{2t - 1}\boldsymbol{k}$

9. $\frac{1}{\sqrt{1 - t^2}}\boldsymbol{i} + \frac{t}{\sqrt{1 + t^2}}\boldsymbol{j} - 3t^2e^{-t^3}\boldsymbol{k}$

11. $-\frac{1}{4}t^{-3/2}\boldsymbol{i} + \frac{15}{4}t^{-7/2}\boldsymbol{j} - \frac{4}{(2t - 1)^2}\boldsymbol{k}$

15. $(3 \cos t - 2)\boldsymbol{i} + (6t + 2 \sin t)\boldsymbol{k}$

17. $-\sin t\boldsymbol{i} + (3t^2 + \sin^2 t - \cos^2 t)\boldsymbol{j} - \boldsymbol{k}$

19. $3e^{3t} \cos e^{3t}\boldsymbol{i} + 6e^{6t}\boldsymbol{k}$

23. $\frac{2}{3}t^{3/2}\boldsymbol{i} + \frac{1}{2}e^{2t}\boldsymbol{j} + \ln|t|\boldsymbol{k} + \boldsymbol{C}$

25. $(t \ln t - t)\boldsymbol{i} + \ln|\ln|t||\boldsymbol{j} + \boldsymbol{C}$

27. $(3 + t)\boldsymbol{i} + \left(5 + \frac{t^3}{3}\right)\boldsymbol{j}$

29. $\left(3 + \frac{t}{2} + \frac{1}{4} \sin 2t\right)\boldsymbol{i} + \left(\frac{t}{2} - \frac{1}{4} \sin 2t - 6\right)\boldsymbol{j} + (3 - e^{-t})\boldsymbol{k}$

31. $e^{-4t}\boldsymbol{i} + 4e^{-4t}\boldsymbol{k}$

33. $\left(\frac{b^2 - a^2}{2}\right)\boldsymbol{i} + \left(\frac{b^3 - a^3}{3}\right)\boldsymbol{j} - \left(\frac{b^4 - a^4}{4}\right)\boldsymbol{k}$

35. $\frac{\sqrt{2}}{2}\boldsymbol{i} + \frac{1}{2}\boldsymbol{j} + \left(\frac{\pi}{8} + \frac{1}{4}\right)\boldsymbol{k}$

Exercise Set 18.4

1. $48\boldsymbol{i} + \frac{1}{4}\boldsymbol{j}$

3. $\sqrt{2}\boldsymbol{i} + 2\boldsymbol{j}$

5. $-2\boldsymbol{i} + \boldsymbol{j} + 6\boldsymbol{k}$

7. $\frac{1}{\sqrt{a^2 + b^2}}(-a\boldsymbol{i} + b\boldsymbol{j})$

9. $t = \sqrt{2}$

11. $(64\boldsymbol{i} + 2\boldsymbol{j} + 2\boldsymbol{k}) + t\left(48\boldsymbol{i} + \frac{1}{4}\boldsymbol{j}\right)$

13. $\frac{2 - x}{2} = y + 5 = \frac{z + 6}{6}$

15. a. $\frac{\pi}{4}$
b. $\frac{\pi}{4}$

17. $\frac{\pi}{2}$

19. $\frac{1}{2}\sqrt{6} + \frac{1}{2}\ln(2 + \sqrt{6}) - \frac{1}{4}\ln 2$

21. $\frac{1}{\sqrt{17}}(\boldsymbol{i} + 4\boldsymbol{j})$

23. $\alpha = \frac{\sqrt{2}}{2}$

27. $\frac{\sqrt{2}}{2}\cos t\boldsymbol{i} - \frac{\sqrt{2}}{2}\sin t\boldsymbol{j}$

29. $(4\boldsymbol{i} + \boldsymbol{j} + 8\boldsymbol{k}) + t(2\boldsymbol{i} + 12\boldsymbol{k})$

Exercise Set 18.5

1. $\boldsymbol{v} = 2\boldsymbol{j}, \quad \boldsymbol{a} = 0$

3. $\boldsymbol{v} = -3 \sin 3t\boldsymbol{i} + 3 \cos 3t\boldsymbol{j}, \quad \boldsymbol{a} = -9 \cos 3t\boldsymbol{i} - 9 \sin 3t\boldsymbol{j}$

5. $\boldsymbol{v} = \frac{2}{t}\boldsymbol{i} - 2 \sin 2t\boldsymbol{j} - \frac{1}{t^2}\boldsymbol{k}, \quad \boldsymbol{a} = \frac{-2}{t^2}\boldsymbol{i} - 4 \cos 2t\boldsymbol{j} + \frac{2}{t^3}\boldsymbol{k}$

7. $\boldsymbol{v} = e^t(\cos t - \sin t)\boldsymbol{i} + e^t(\sin t + \cos t)\boldsymbol{j} - \sin t\boldsymbol{k}$,
$\boldsymbol{a} = -2e^t \sin t\boldsymbol{i} + 2e^t \cos t\boldsymbol{j} - \cos t\boldsymbol{k}$

9. $|v| = 2\sqrt{41}$

11. $(3 + \sin t)\boldsymbol{i} + (3 - \cos t)\boldsymbol{j}$

13. $(\tan^{-1} t - 2)\boldsymbol{i} + \left(\frac{1}{2}e^{t^2} + \frac{1}{2}\right)\boldsymbol{j} + 4\boldsymbol{k}$

15. $v = 2t\boldsymbol{i} + t\boldsymbol{k}, \quad r = (t^2 + 3)\boldsymbol{i} - \boldsymbol{j} + \left(\frac{t^2}{2} + 4\right)\boldsymbol{k}$

17. $v = \sin t\boldsymbol{i} + (1 - \cos t)\boldsymbol{j} + \boldsymbol{k},$
$r = (4 - \cos t)\boldsymbol{i} + (-1 + t - \sin t)\boldsymbol{j} + (t + 2)\boldsymbol{k}$

21. Speed $= \alpha \cdot$ radius

23. a. $50\sqrt{3}\, t\boldsymbol{i} + \left(50t - \frac{1}{2}gt^2\right)\boldsymbol{j}$

b. $\frac{1250}{9.8}$ m

c. $\frac{100}{9.8}$ sec

d. $\frac{5000\sqrt{3}}{9.8}$ m

e. 100 m/sec

Exercise Set 18.6

1. $T(s) = \frac{1}{2}\boldsymbol{i} + \frac{\sqrt{3}}{2}\boldsymbol{j}, \quad r''(s) = 0, \quad K(s) = 0$

3. $T(s) = \frac{-\sqrt{2}}{2}\sin(s)\boldsymbol{i} + \frac{\sqrt{2}}{2}\cos(s)\boldsymbol{j} + \frac{\sqrt{2}}{2}\boldsymbol{k},$
$r''(s) = \frac{-\sqrt{2}}{2}\cos(s)\boldsymbol{i} - \frac{\sqrt{2}}{2}\sin(s)\boldsymbol{j}, \quad K(s) = \frac{\sqrt{2}}{2}$

5. $K = 0$

7. $\frac{e^t}{(1 + e^{2t})^{3/2}}$

9. $K = \frac{1}{2}$

11. $K = \frac{2}{(1 + 4x^2)^{3/2}}$

13. $\frac{|6x|}{(1 + 9x^4)^{3/2}}$

15. $\frac{|\cos x|}{(1 + \sin^2 x)^{3/2}}$

17. $\frac{2}{(1 + 4t^2)^{3/2}}$

19. $\frac{2e^t(1 + e^{2t})^{1/2}}{(2e^{2t} + 1)^{3/2}}$

21. $\left(\frac{\pi}{2}, 0\right)$

23. a. $\frac{2}{5^{3/2}}$ **b.** $\left(\frac{7}{2}, -4\right)$ **c.** 2

27. $\frac{3}{2^{3/2}} \frac{1}{(1 + \sin\theta)^{1/2}}$

29. $\frac{3}{2^{3/2}} \frac{1}{(1 - \cos\theta)^{1/2}}$

Review Exercises–Chapter 18

1. $\frac{x^2}{a^2} + \frac{x^2}{b^2} = 1$

3. $[-1, 0)$ and $(0, 1]$

5. $t = 2n\pi, \quad n = 0, \pm 1, \ldots$

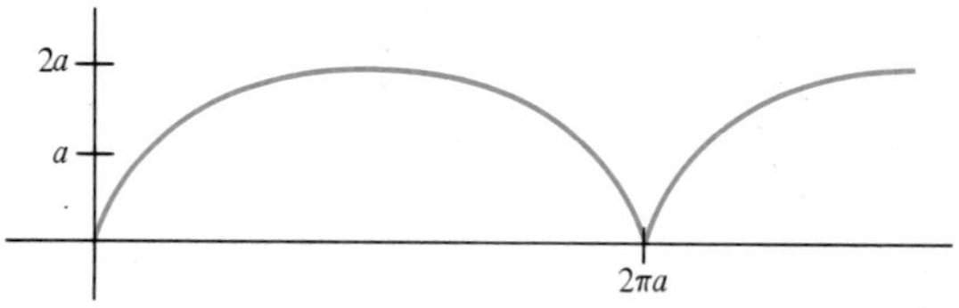

7. $3\boldsymbol{i} + \sqrt{2}\boldsymbol{j} + 2\boldsymbol{k}$

9. $t = \frac{\pi}{2}, \ \frac{3\pi}{2}$

11. $t = \pm 3, \ 2$

13. $r'(t) = e^t(t + 1)\boldsymbol{i} + e^{-t}(1 - t)\boldsymbol{j}$
$r''(t) = e^t(t + 2)\boldsymbol{i} + e^{-t}(t - 2)\boldsymbol{j}$

15. $r'(t) = \frac{1}{t}\boldsymbol{i} + \frac{1}{2\sqrt{t}}e^{\sqrt{t}}\boldsymbol{j}$
$r''(t) = \frac{-1}{t^2}\boldsymbol{i} + \frac{1}{4t^{3/2}}e^{\sqrt{t}}(\sqrt{t} - 1)\boldsymbol{j}$

17. $\boldsymbol{i} + \left(1 - \frac{2}{e}\right)\boldsymbol{j}$

19. $4e^{3t}\boldsymbol{i} + 3e^{3t}\boldsymbol{j}$

21. $r(t) = 5\cos\left(\frac{1}{5}t\right)\boldsymbol{i} + 5\sin\left(\frac{1}{5}t\right)\boldsymbol{j}$

23. a. $K(t) = \frac{1}{2}$

b. $T(t) = \frac{\sqrt{2}}{2}\boldsymbol{i} - \frac{\sqrt{2}}{2}\sin t\boldsymbol{j} + \frac{\sqrt{2}}{2}\cos t\boldsymbol{k}$

25. $3\sqrt{13}$

27. $\frac{12}{(145)^{3/2}}$

33. $x = \frac{-\ln 2}{2}$

35. approximately 4.14 m

CHAPTER 19

Exercise Set 19.1

1. Yes **3.** No **5.** Yes **7.** No

9. $\{(x, y) | (x, y) \neq (0, 0)\}$

11. $\{(x, y) | y \geq x\}$

13. all $(x, y) \in \mathbb{R}^2$

15. $\{(x, y, z) | x^2 + y^2 + z^2 \leq 4\}$

17. $\{(x, y) | x \neq 3, y \neq x\}$

19. $\{(x, y, z) | xyz > 0\}$

21. **a.** $h(x, y) = \sqrt{x + y^2}$
b. $\{(x, y) | x + y^2 \geq 0\}$

23. $h(x, y) = (x - \cos y)^2 + (y + \sin x)^2$

25. $V = \dfrac{1}{3}\pi r^2 h$

27. $T(g, h) = \sqrt{\dfrac{2h}{g}}$

29. **a.** 14
b. $2 \times (1 - 2y)$
c. $x^2(1 + 2y^3)$

31. $(0, \infty)$

Exercise Set 19.2

1. -5

3. -3

5. -7

7. $\dfrac{1}{e}$

9. $\dfrac{\pi}{20}$

11. 0

13.

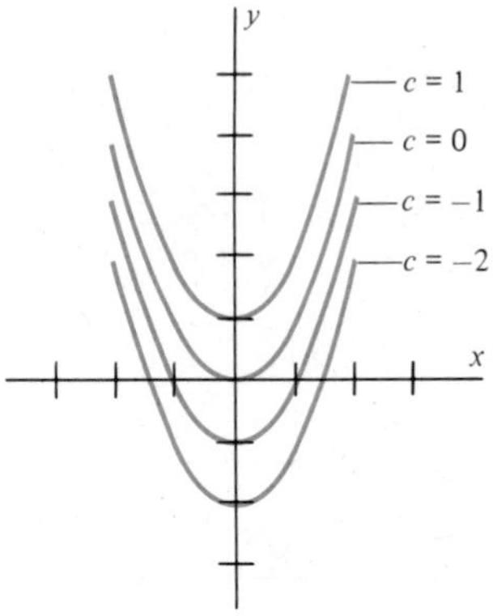

17.

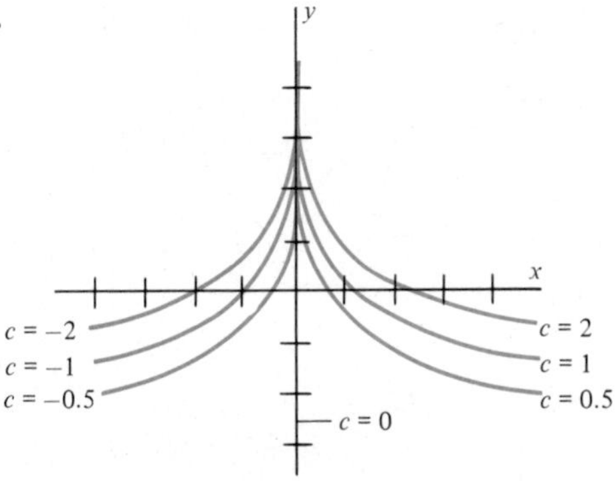

21.

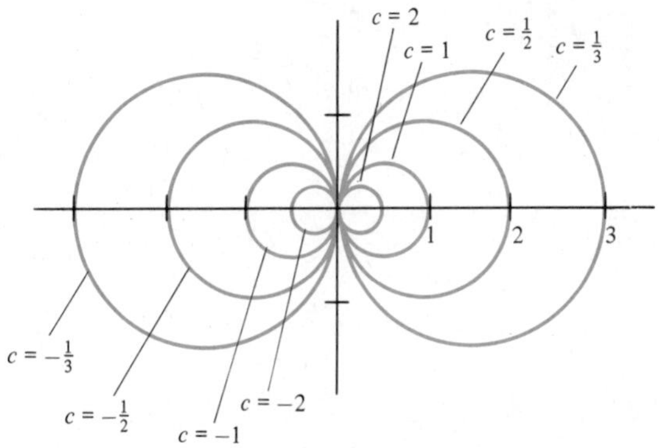

25.

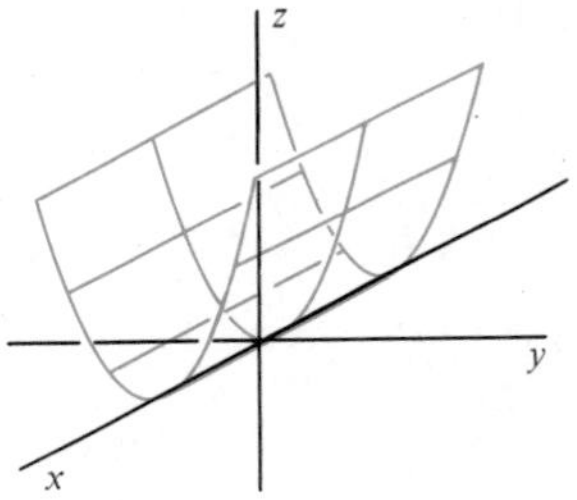

27.

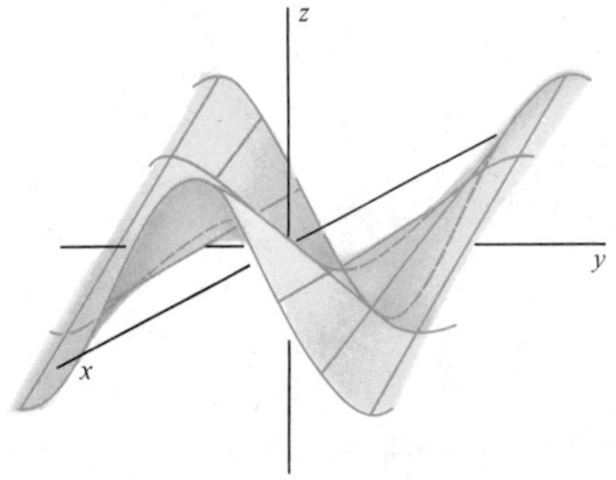

33. open

35. open

37. not open

39. does not exist

41. No

43. Yes

45. 0

47. False

Exercise Set 19.3

1. $\dfrac{\partial f}{\partial x} = y,\ \dfrac{\partial f}{\partial y} = x$

3. $\dfrac{\partial f}{\partial x} = \tan y^2,\ \dfrac{\partial f}{\partial y} = 2xy \sec^2 y^2$

5. $\dfrac{\partial f}{\partial x} = 2xe^{x^2+y^2},\ \dfrac{\partial f}{\partial y} = 2ye^{x^2+y^2}$

7. $\dfrac{\partial f}{\partial r} = \sin\left(\dfrac{\pi}{2} - \theta\right),\ \dfrac{\partial f}{\partial \theta} = -r\cos\left(\dfrac{\pi}{2} - \theta\right)$

9. $\dfrac{\partial f}{\partial x} = \dfrac{y^2+1}{xy^2+x-y}, \dfrac{\partial f}{\partial y} = \dfrac{2xy-1}{xy^2+x-y}$

11. $\dfrac{\partial f}{\partial x} = yx^{y-1}, \dfrac{\partial f}{\partial y} = x^y \ln x$

13. $\dfrac{\partial f}{\partial r} = 2r\cos\theta, \dfrac{\partial f}{\partial \theta} = -r^2\sin\theta$

15. $\dfrac{\partial f}{\partial x} = y^3, \dfrac{\partial f}{\partial y} = 3xy^2 - z^2, \dfrac{\partial f}{\partial z} = -2yz$

17. $\dfrac{\partial f}{\partial x} = \dfrac{2yz}{(x+y)^2}\left(\dfrac{x-y}{x+y}\right)^{z-1},$

$\dfrac{\partial f}{\partial y} = \dfrac{-2xz}{(x+y)^2}\left(\dfrac{x-y}{x+y}\right)^{z-1}$

$\dfrac{\partial f}{\partial z} = \left(\dfrac{x-y}{x+y}\right)^{z}\ln\left(\dfrac{x-y}{x+y}\right)$

19. $\dfrac{\partial f}{\partial r} = \dfrac{\sqrt{s^2-2r+t}\cdot\dfrac{s\ln t}{2\sqrt{r}} + \dfrac{2\sqrt{rs}\ln t}{\sqrt{s^2-2r+t}}}{s^2-2r+t},$

$\dfrac{\partial f}{\partial s} = \dfrac{\sqrt{r}\ln t\sqrt{s^2-2r+t} - \dfrac{\sqrt{r}s^2\ln t}{\sqrt{s^2-2r+t}}}{s^2-2r+t},$

$\dfrac{\partial f}{\partial t} = \dfrac{\left(\dfrac{\sqrt{rs}}{t}\right)\sqrt{s^2-2r+t} - \dfrac{\sqrt{rs}\ln t}{2\sqrt{s^2-2r+t}}}{s^2-2r+t}$

21. 125 **23.** $\dfrac{1}{2\sqrt{3}}$

25. $\dfrac{\partial^2 f}{\partial r^2} = 2\cos\theta, \dfrac{\partial^2 f}{\partial r\,\partial\theta} = -2r\sin\theta,$

$\dfrac{\partial^2 f}{\partial \theta^2} = -r^2\cos\theta$

27. $w_{xx} = \dfrac{-2x^2+2y^2+2z^2}{(x^2+y^2+z^2)^2},$

$w_{yy} = \dfrac{2x^2-2y^2+2z^2}{(x^2+y^2+z^2)^2},$

$w_{zz} = \dfrac{2x^2+2y^2-2z^2}{(x^2+y^2+z^2)^2}$

29. a. $v_2 = \dfrac{A_1v_1}{A_2}$

b. $\dfrac{\partial v_2}{\partial A_2} = \dfrac{-A_1v_1}{(A_2)^2}$

c. $\dfrac{-100}{9}$

d. $\dfrac{3}{20}$

31. $\rho\sin\phi\cos\theta$ **33.** $\sin\phi\cos\theta$ **35.** $-\rho\sin\phi$

37. a. $\cos(x+y)^2 - \cos x^2$
b. $\cos(x+y)^2$

41. a. $\dfrac{\partial v}{\partial a} = \dfrac{a\sigma}{2\epsilon_0\sqrt{a^2+r^2}}$

b. $\dfrac{\partial v}{\partial r} = \dfrac{\sigma}{2\epsilon_0}\left(\dfrac{r}{\sqrt{a^2+r^2}} - 1\right)$

Exercise Set 19.4

1. $2x + 6y - z = 10$

3. $x + 5y + z = -3$

5. $x - 2y - z = 4$

7. $x + y - z = 2$

9. $y + 2z = 2$

11. $y - z = 1$

13. $2x + ez = 0$

15. $\boldsymbol{r}(t) = 3\boldsymbol{i} - \boldsymbol{j} + 4\boldsymbol{k} + t(4\boldsymbol{i} + 4\boldsymbol{j} - \boldsymbol{k})$

17. $\boldsymbol{r}(t) = \boldsymbol{i} - 2\boldsymbol{j} + 2\boldsymbol{k} + t(-\boldsymbol{i} + 2\boldsymbol{j} - 2\boldsymbol{k})$

19. $\boldsymbol{r}(t) = \boldsymbol{i} + \boldsymbol{j} + t(\boldsymbol{j} - \boldsymbol{k})$

21. (2, 3, 20)

25. $18x + 16y - z = 25$

27. $x_0x + y_0y + z_0z = r^2$

29. $-x - 2y + z + \dfrac{\pi}{2} = 0$

Exercise Set 19.5

1. relative minimum: (0, −2, 0)

3. saddle point (−3, 2, 0)

5. saddle point (0, 0, 9)

7. relative minimum: (1, 3, 1)

9. saddle point (0, 0, 0)

11. saddle point (0, 0, 0)

13. saddle point (0, 0, 0)
relative maximum: $\left(-\dfrac{4}{3}, -\dfrac{4}{3}, \dfrac{64}{27}\right)$

15. saddle points $\left(0, \dfrac{\pi}{2} + n\pi, 0\right)$

19. The graph is unbounded.

23. 2 m × 2 m × 4 m

25. $\left(\dfrac{ad}{a^2+b^2+c^2}, \dfrac{bd}{a^2+b^2+c^2}, \dfrac{cd}{a^2+b^2+c^2}\right)$

27. $14'' \times 14'' \times 28''$

29. Absolute minimum: $-2\sqrt{2} + 4$
Absolute maximum: $2\sqrt{2} + 4$

31. Absolute minimum: 3
Absolute maximum: 9

35. 86

37. a. $y = \dfrac{23}{20}x - \dfrac{19}{4}$

b.

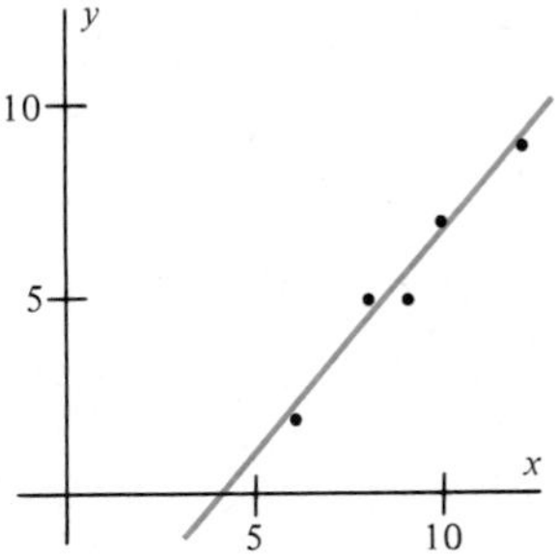

c. $y(4) = \dfrac{-3}{20}$

Exercise Set 19.6

1. $2xy^4\,dx + 4x^2y^3\,dy$

3. $\dfrac{x}{\sqrt{x^2 = y^4}}\,dx + \dfrac{2y^3}{\sqrt{x^2 + y^4}}\,dy$

5. $\dfrac{e^{\sqrt{x}}}{2\sqrt{x}}\cos y\,dx - e^{\sqrt{x}}\sin y\,dy$

7. $2xyz^3\,dx + x^2z^3\,dy + 3x^2yz^2\,dz$

9. $\dfrac{1}{y + 3^z}\,dx - \dfrac{x}{(y + 3^z)^2}\,dy - \dfrac{x3^z \ln 3}{(y + 3^z)^2}\,dz$

11. $\dfrac{y^2 + z^2 - x^2 + 2xy}{(x^2 + y^2 + z^2)^2}\,dx + \dfrac{y^2 - x^2 - z^2 - 2xy}{(x^2 + y^2 + z^2)^2}\,dy + \dfrac{2zy - 2zx}{(x^2 + y^2 + z^2)^2}\,dz$

13. a. $f(2, 3) = 48$
$f(2.1, 3.2) = 57.12$
b. $\Delta f = 9.12$
c. $df = 8.6$
d. $\Delta f - df = 0.52$

15. 5.076

17. 0.7441

19. 26.8

21. 55.44

23. Relative error $= 1.97 \times 10^{-5} = 1.97 \times 10^{-3}$ percent

25. Relative error $= 1.9 \times 10^{-3} = 0.19$ percent

27. Relative error $= 0.0767 = 7.67$ percent

29. Percent change $= \left(\dfrac{0.1\,C + 0.05\,D}{M}\right) \times 100$

31. $\alpha = \beta$

33. 0.1296

Exercise Set 19.7

1. $-16t + 4t^3$

3. $-\sin^3 t + 2 \sin t \cos^2 t$

5. $\dfrac{1}{4\sqrt{t}\sqrt{\sqrt{t} + e^{t^2}}} + \dfrac{te^{t^2}}{\sqrt{\sqrt{t} + e^{t^2}}}$

7. $-6t^2 - 2t + 2 \sin t \cos t$

9. $2e^{-2t}(b^2 - a^2) \sinh t \cosh t + 2e^{-2t}$

11. $-2t^3 \sin^4 t \cos t + 3t^2 \sin^3 t \cos^2 t + 3t^3 \sin^2 t \cos^3 t$

13. a. $2te^{2st} + 2st^2 - 2s + 2t$
b. $2se^{2st} + 2s^2t + 2s - 2t$

15. a. $[y^2 \cos xy^2 - 2xy](2s - t) + [2xy \cos xy^2 - x^2](2t^2s)$
b. $[y^2 \cos xy^2 - 2xy](-s) + [2xy \cos xy^2 - x^2](2s^2t)$

17. a. $y^3z^2 \sin t - 3xy^2z^2t \sin s - 4sxy^3z$
b. $sy^3z^2 \cos t + 3xy^2z^2 \cos s + 4txy^3z$

19. a. $2ue^{u^2}y^2 - 4xye^{2v}$
b. $4xyue^{u^2} - 2x^2e^{2v}$

21. $6rt^2(1 - \cos\theta) + \dfrac{r^2t \sin\theta}{\sqrt{1 + t^2}}$

23. 116π cm^3/sec

25. a. $D(t) = \sqrt{1 + t^2}$
b. $D'(t) = \dfrac{t}{\sqrt{1 + t^2}}$

29. a. $12s^2t^2 - 8y^3 + 48s^2y^2$
b. $2s^4 + 8y^3 + 48t^2y^2$

31. $2s^2t^2e^{rst} - 4s^2t^2e^{2rst} + rs^3t^3e^{rst}$

33. $-4yre^{y^2} \sin\theta \cos\theta + (2x + 4xy^2)re^{y^2} \sin\theta \cos\theta - e^{y^2} \sin\theta + 2xye^{y^2} \cos\theta$

Exercise Set 19.8

1. $6\boldsymbol{i} + 9\boldsymbol{j}$

3. $\dfrac{\sqrt{2}}{2}\left(1 - \dfrac{\pi}{2}\right)\boldsymbol{i} + \dfrac{\sqrt{2}}{4}\pi\boldsymbol{j}$

5. $6\boldsymbol{i} + \boldsymbol{j} + 4\boldsymbol{k}$

7. $5\boldsymbol{i} - 3\boldsymbol{j} - \boldsymbol{k}$

9. $\left(e + \dfrac{1}{e}\right)\boldsymbol{i} + \dfrac{1}{4}\left(e + \dfrac{1}{e}\right)\boldsymbol{j} + 2\left(e - \dfrac{1}{e}\right)\boldsymbol{k}$

11. $-30\sqrt{2}$

13. $\dfrac{2\sqrt{3}-1}{18}$

15. $\dfrac{2}{\sqrt{3}}$

17. $\dfrac{3\sqrt{3}}{\sqrt{11}}$

19. $\dfrac{13\sqrt{2}}{6} + \dfrac{1}{3}\left(\dfrac{1}{e} - e\right) + \dfrac{\pi}{12}\left(\dfrac{1}{e} + e\right)$

21. N: $2\boldsymbol{i} + 3\boldsymbol{j} + t(16\boldsymbol{i} - 6\boldsymbol{j})$
T: $2\boldsymbol{i} + 3\boldsymbol{j} + t(6\boldsymbol{i} + 16\boldsymbol{j})$

23. N: $4\boldsymbol{i} + 4\boldsymbol{j} + t(\boldsymbol{i} + \boldsymbol{j})$
T: $4\boldsymbol{i} + 4\boldsymbol{j} + t(\boldsymbol{i} - \boldsymbol{j})$

25. $\sqrt{5}$ **27.** -1 **29.** $\dfrac{1}{2\sqrt{2}}$

31. **a.** $(3, 1 + \sqrt{2})$, $(3, 1 - \sqrt{2})$
b. $(1, 1)$, $(5, 1)$

37. $2\boldsymbol{i} + \boldsymbol{j}$ **39.** $-\boldsymbol{i} + \boldsymbol{j}$

Exercise Set 19.9

1. N: $-2\boldsymbol{i} + 8\boldsymbol{k}$
$2x - 8z = -18$

3. N: $2\boldsymbol{i} + 8\boldsymbol{j} - 3\boldsymbol{k}$
$2x + 8y - 3z = 16$

5. N: $5\boldsymbol{i} - \boldsymbol{j} - \boldsymbol{k}$
$5x - y - z = 4$

7. N: $-\boldsymbol{j}$
$y = 1$

9. $\dfrac{5}{\sqrt{51}}\boldsymbol{i} + \dfrac{5}{\sqrt{51}}\boldsymbol{j} - \dfrac{1}{\sqrt{51}}\boldsymbol{k}$

13. $x^2 + y^2 = \dfrac{1}{2}$ and $z = \dfrac{\sqrt{2}}{2}$

Exercise Set 19.10

	Maximum	Minimum
1.	$f(0, \pm 1) = 4$	$f(\pm 1, 0) = 2$
3.	none	$f(1, -1) = 2$
5.	$f\left(\dfrac{1}{\sqrt{2}}, \dfrac{1}{\sqrt{2}}\right) = f\left(\dfrac{-1}{\sqrt{2}}, \dfrac{-1}{\sqrt{2}}\right) = \dfrac{1}{2}$	$f\left(\dfrac{1}{\sqrt{2}}, \dfrac{-1}{\sqrt{2}}\right) = f\left(\dfrac{-1}{\sqrt{2}}, \dfrac{1}{\sqrt{2}}\right) = -\dfrac{1}{2}$
7.	$f(2, 0) = 12$	$f(-2, 0) = -4$
9.	$f\left(-\dfrac{2}{3}, \pm\dfrac{\sqrt{5}}{3}\right) = \dfrac{48}{9}$	$f(1, 0) = -3$
11.	$f\left(\dfrac{1}{\sqrt{6}}, \dfrac{2}{\sqrt{6}}, -\dfrac{1}{\sqrt{6}}\right) = \sqrt{6}$	$f\left(-\dfrac{1}{\sqrt{6}}, -\dfrac{2}{\sqrt{6}}, \dfrac{1}{\sqrt{6}}\right) = -\sqrt{6}$
13.	$f(2, 2, 2) = 6$	$f(-2, -2, -2) = -6$

15. $f\left(2, -1 + \dfrac{\sqrt{2}}{2}, \dfrac{3}{2}\right) = f\left(2, -1 - \dfrac{\sqrt{2}}{2}, \dfrac{3}{2}\right) = \dfrac{3}{4}$

17. $\left(3 - 9\dfrac{\sqrt{13}}{13}, -2 + 6\dfrac{\sqrt{13}}{13}\right)$

19. $(1, 0, 1)$

23. $\dfrac{2r}{\sqrt{3}} \times \dfrac{2r}{\sqrt{3}} \times \dfrac{2r}{\sqrt{3}}$

25. $r = \sqrt[3]{\dfrac{1}{\pi}}$, $h = 2\sqrt[3]{\dfrac{1}{\pi}}$

Exercise Set 19.11

1. $x - y + C$

3. None

5. None

7. $x^3 \cos y - x + y^2 + C$

9. $x^2 e^{xy} - \sqrt{y} + C$

11. $x^2 y^2 + y \cos x + C$

13. $x^2 y = C$

15. $\sqrt{x^2 + y^2} = C$

17. Not exact

19. Not exact

21. $x^2 + y^2 = C$

23. $x^2 y = 3$

25. $xy^2 - x^2 y = 2$

Review Exercises–Chapter 19

1. no

3.

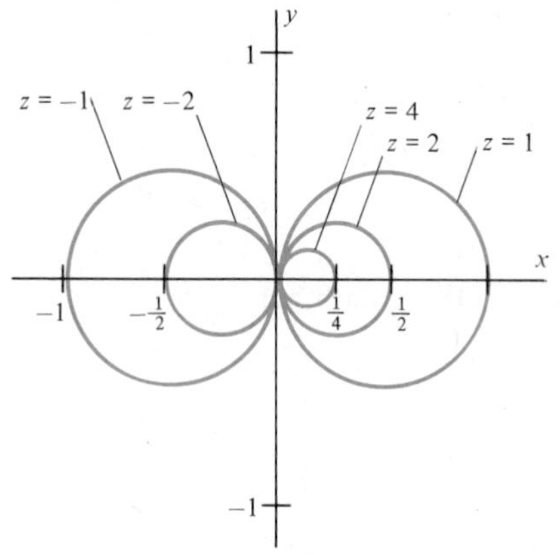

5. $-5\boldsymbol{i} - 8\boldsymbol{j}$

7. $-\boldsymbol{i} - \boldsymbol{k}$

9. $\dfrac{\partial f}{\partial x}(x, y) = \sin\sqrt{x^2 + y^2} + \dfrac{x^2 \cos\sqrt{x^2 + y^2}}{\sqrt{x^2 + y^2}}$,
$\dfrac{\partial f}{\partial y}(x, y) = \dfrac{xy \cos\sqrt{x^2 + y^2}}{\sqrt{x^2 + y^2}}$

11. $\dfrac{784}{15}$

13. $\dfrac{\partial f}{\partial r} = 18x^2 + 4y^2 + 4xy - 2y,$

$\dfrac{\partial f}{\partial s} = 45x^2 + 10y^2 - 16sxy + 8sy$

15. $3x + 4y = 6$

19. $-\dfrac{36}{5}\boldsymbol{i} - \dfrac{12}{5}\boldsymbol{j} - (2 - \tan^{-1} 2)\boldsymbol{k}$

23. $\dfrac{2\sqrt{3}}{3} \times 2\sqrt{3} \times \dfrac{4\sqrt{3}}{3}$

25. (0, 0, 0): saddle point

27. (0, 0, 1): saddle point

29. (0, 0, e): saddle point

31. **a.** 1
b. 3
c. $\boldsymbol{i} + 3\boldsymbol{j}$

33. $f(x, y) = x^2e^y + \cos y + C$

35. $dR = \left(\dfrac{1}{\dfrac{1}{10} + \dfrac{1}{20} + \dfrac{1}{25}}\right)^2\left(\dfrac{1}{(100)(1000)} + \dfrac{1}{400(4000)} + \dfrac{1}{625(6250)}\right)$

39. maximum $5 + 6\sqrt{2}$
minimum $5 - 6\sqrt{2}$

41. no

43. $2x + y + 2z = 6$

51. $(e \tan^{-1} 3)\boldsymbol{i} + \dfrac{e}{10}\boldsymbol{j} + \left(\dfrac{e}{10} - e \tan^{-1} 3\right)\boldsymbol{k}$

53. $4t^2(4x^2 + 4xy - 2)e^{y^2-x^2} + 4txe^{y^2-x^2} + 4t^2(4y^2 + 4xy - 2)e^{y^2-x^2} + 4tye^{y^2-x^2}$

55. $\dfrac{32\sqrt{3}}{3}$

CHAPTER 20

Exercise Set 20.2

1. 1

3. 7

5. $\dfrac{1}{2}$

7. $\dfrac{8}{3}$

9. $\dfrac{1}{4}\left(e - \dfrac{1}{e}\right)$

11. $\dfrac{1}{2}$

13. $\dfrac{4\pi}{3}$

15. $\dfrac{5}{6}$

17. 8

19. $\dfrac{1}{2} \ln \sqrt{2}$

21. $\dfrac{\pi\sqrt{2}}{8} + 1 - \sqrt{2}$

23. 6

25. $\dfrac{1}{2}$

27. $8e(e - 1)$

Exercise Set 20.3

1. $-\dfrac{1}{6}$

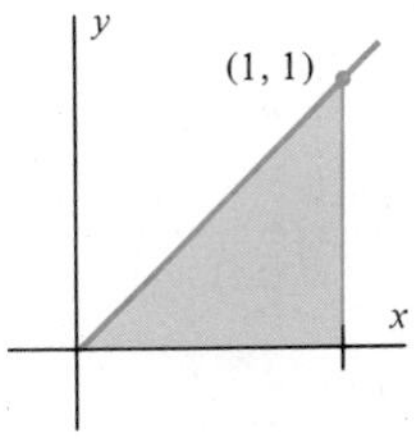

3. $\dfrac{13}{24}$

5. 0

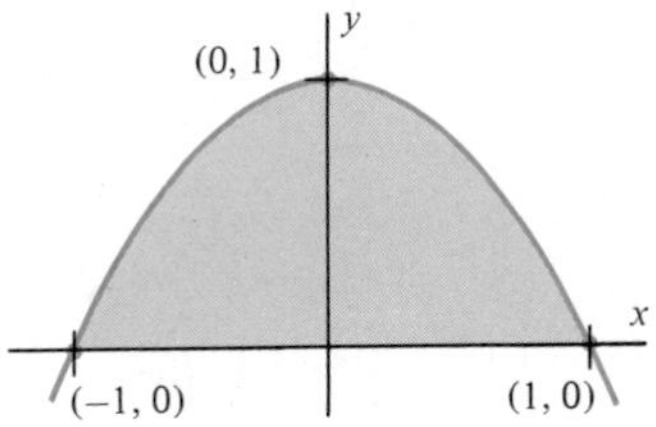

7. $\dfrac{1}{20}$

9. $\dfrac{1}{3}$

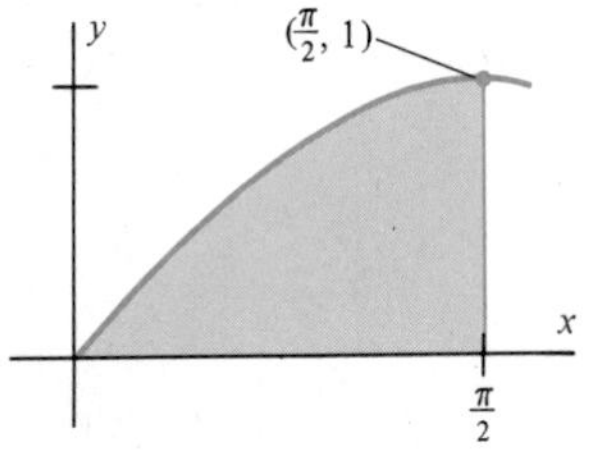

11. $\frac{1}{2}$

13. $\frac{2}{15}$

15. $\frac{1}{3}(1 + e)^{3/2} - \frac{4}{3}2^{3/2} + 2\sqrt{e}\sqrt{1 + e} + 2\ln\left|\frac{\sqrt{e} + \sqrt{1 + e}}{1 + \sqrt{2}}\right|$

17. $\frac{1}{24}$

19. $\frac{3e^2 - e^{-2}}{4}$

21. $\frac{1}{3}$

23. 2

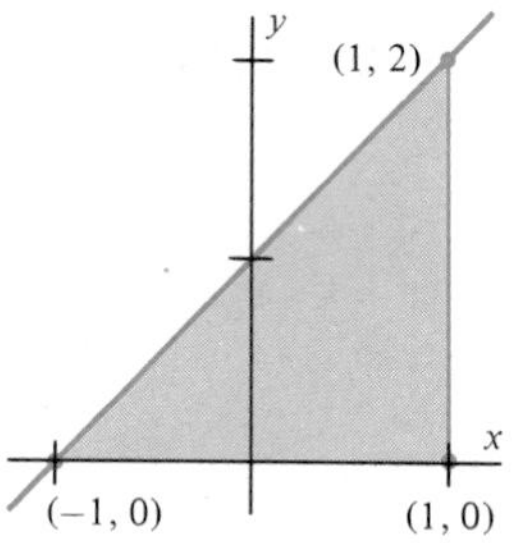

25. $\frac{1}{10}$

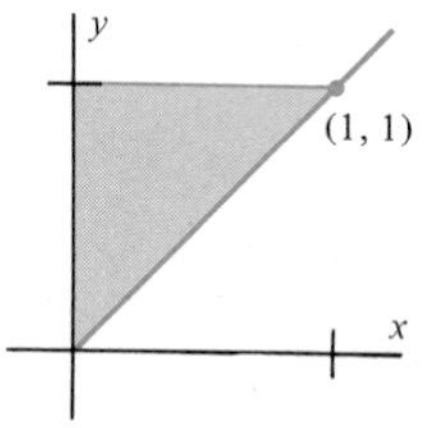

27. $\int_0 \int_{e^y}^{e} f(x, y)\,dx\,dy$

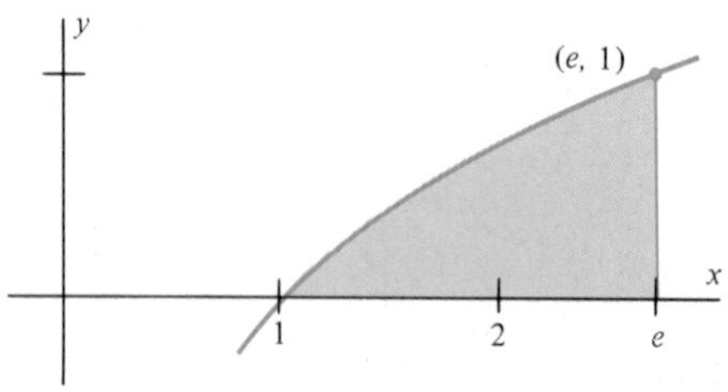

29. $\frac{9}{2}$

31. $\sqrt{2} - 1$

33. $\frac{1}{3}$

35. 6π

37. 10π

39. $\frac{4\pi}{3}(15)^{3/2}$

41. 1

Exercise Set 20.4

1. $\frac{2\pi}{3}$

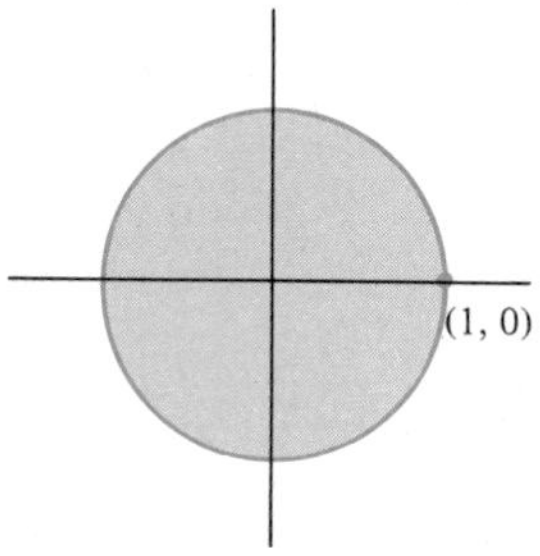

3. $\pi(e^{a^2} - 1)$

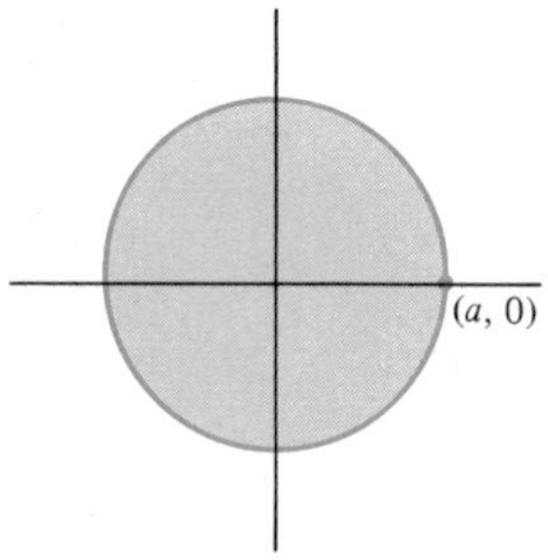

5. $\frac{45\pi}{16}$

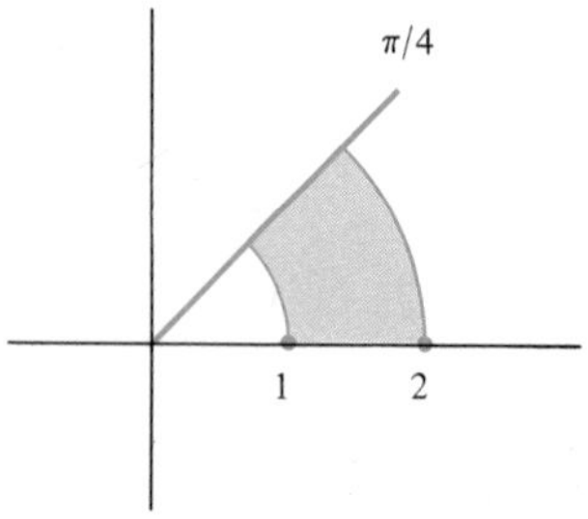

7. $\frac{4}{27}$

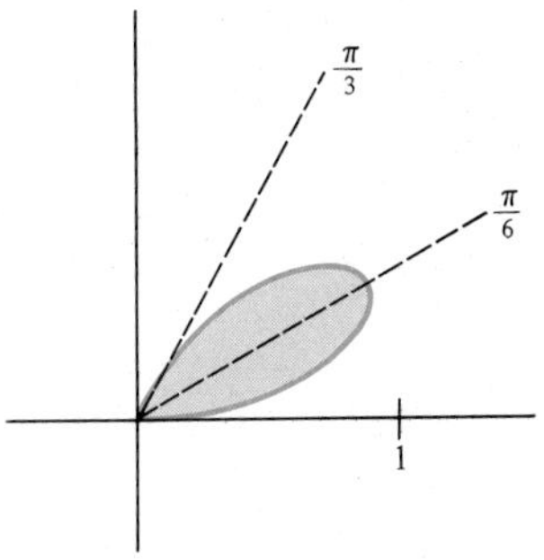

9. $\frac{\pi}{4}$

11. $2 + \frac{\pi}{4}$

13. $\frac{9\pi}{2}$

15. $\frac{\pi}{2}$

17. $\frac{\pi}{4}(e - 1)$

19. 2

21. 0

23. 16π

25. $2\sqrt{3}\,\pi$

27. 3π

29. 16π

31. $\frac{\pi}{4}$

33. $\frac{16\pi}{3}$

Exercise Set 20.5

1. 3

3. 8

5. $\sqrt{2} - 1$

7. $\frac{\pi}{4}$

9. 2π

11. $(\bar{x}, \bar{y}) = \left(\frac{11}{9}, \frac{5}{9}\right)$

13. $(\bar{x}, \bar{y}) = \left(\frac{3}{\pi}, \frac{3}{\pi}\right)$

15. $(\bar{x}, \bar{y}) = \left(\frac{17}{54}, \frac{13}{9}\right)$

17. $(\bar{x}, \bar{y}) = \left(0, \frac{8}{5}\right)$

19. $(\bar{x}, \bar{y}) = (2.6, 1.7)$

21. $(\bar{x}, \bar{y}) = \left(\frac{4a}{3\pi}, \frac{4a}{3\pi}\right)$

23. $(\bar{x}, \bar{y}) = \left(\frac{8}{5}, 2\right)$

25. 4λ

27. $M = \int_c^d \int_{g_\ell(y)}^{f_r(y)} \rho \, dx \, dy$

$= \rho \int_c^d [f_r(y) - g_\ell(y)] \, dy$ as in 4.

29. $(\bar{x}, \bar{y}) = \left(0, \frac{224}{81\pi}\right)$

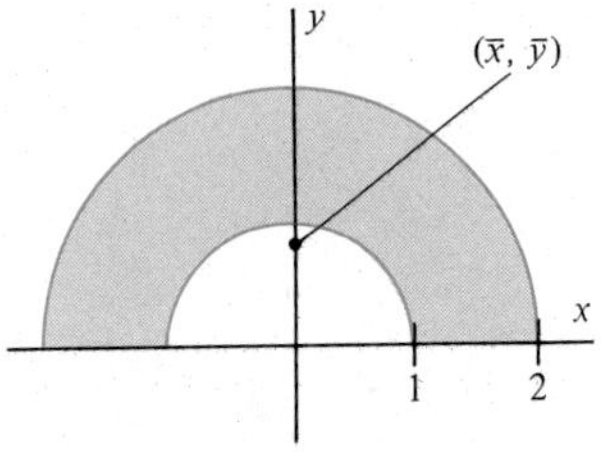

Exercise Set 20.6

1. $\sqrt{3}$

3. $\frac{\pi}{6}(13^{3/2} - 1)$

5. $2\sqrt{3}$

7. $3\sqrt{2}\pi$

9. $\sqrt{2} + \ln(1 + \sqrt{2})$

11. $\frac{5\sqrt{2}}{12} + \frac{1}{4}\ln(3 + 2\sqrt{2})$

13. $\frac{\pi}{6}(7^3 - 29^{3/2})$

15. $4\pi(2 - \sqrt{3})$

17. $A \approx 9.717$

Exercise Set 20.7

1. $\frac{1}{8}$

3. $\frac{1}{2}$

5. $\frac{14}{3}$

7. 0

9. $\int_0^2 \int_{y/2}^1 \int_0^{x+y} f(x, y, z) \, dz \, dx \, dy$

11.

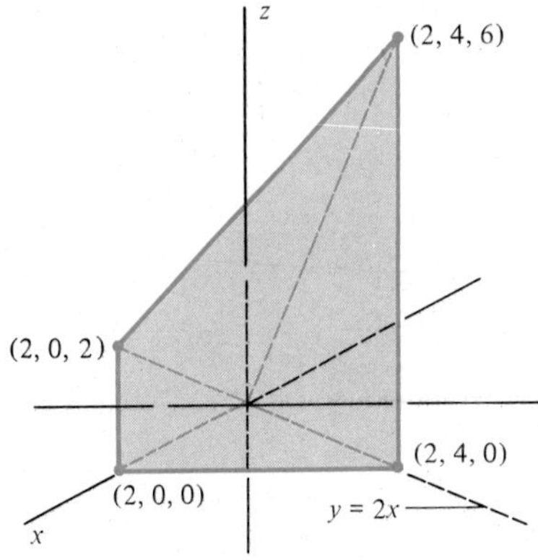

13. $\frac{2\pi}{3}$

15. 27π

17. 0

19. π

21. $\frac{405\pi}{8}$

23. 12

25. $(\bar{x}, \bar{y}, \bar{z}) = \left(0, \frac{3}{4}, \frac{15}{8}\right)$

27. $(\bar{x}, \bar{y}, \bar{z}) = \left(\frac{3}{4}, \frac{3}{4}, \frac{3}{4}\right)$

29. $\int_0^2 \int_0^{1-y/2} \int_0^{4-4x-2y} (x - y + z)\,dz\,dx\,dy$,
$\int_0^1 \int_0^{4-4x} \int_0^{2-2x-z/2} (x - y + z)\,dy\,dz\,dx$,
$\int_0^4 \int_0^{2-z/2} \int_0^{1-y/2-z/4} (x - y + z)\,dx\,dy\,dz$,
$\int_0^2 \int_0^{4-2y} \int_0^{1-y/2-z/4} (x - y + z)\,dx\,dz\,dy$,
$\int_0^4 \int_0^{1-z/4} \int_0^{2-2x-z/2} (x - y + z)\,dy\,dx\,dz$

Exercise Set 20.8

1. $\frac{81\pi}{2}$

3. $(\bar{x}, \bar{y}, \bar{z}) = (0, 0, 6)$

5. 18π

7. $\frac{2\sqrt{2}\,4^4}{3^2} \approx 80$

9. $\frac{32}{9}$

11. 0

13. π

15. $(\bar{x}, \bar{y}, \bar{z}) = \left(0, \frac{3}{8}, \frac{3}{8}\right)$

17. $\frac{16\sqrt{2}\,\pi}{3}$

19. 2π

21. $V = \frac{16}{3}R^3$

23. $9\pi\left(1 - \frac{\sqrt{3}}{3}\right)$

25. $\frac{32\sqrt{2}\,\pi}{5}$

27. $\frac{40\pi}{3}$

29. $\frac{104\pi}{3}\left(1 - \frac{\sqrt{2}}{2}\right)$

Review Exercises–Chapter 20

1. $\frac{1}{15}$

3. $2\left(1 - \frac{1}{\sqrt{e}}\right)$

5. 64

7. $16\left(\frac{\pi}{2} - 1\right)$

9. 3π

11. $\frac{abc}{6}$

13. 36π

15. $(\bar{x}, \bar{y}, \bar{z}) = \left(0, 0, \frac{3}{4}\right)$

17. $\frac{2a^3}{9}(3\pi - 4)$

19. $4\sqrt{17}\,\pi$

21. $\frac{32\pi}{3}(8 - 3\sqrt{3})$

23. $\frac{977}{20}$

25. $(\bar{x}, \bar{y}) = \left(0, \frac{16}{3\pi}\right)$

27. $\frac{\pi^2}{3}(e - 1)$

29. $(\bar{x}, \bar{y}) = \left(\frac{5}{6}, 0\right)$

31. $\frac{81\pi}{4}$

33. $32\pi\rho$

35. $\frac{\pi}{2}(\sqrt{2} - 1)$

37. $\frac{256}{9}$

41. 36π

43. $8\sqrt{3}\,\pi$

CHAPTER 21

Exercise Set 21.1

1.

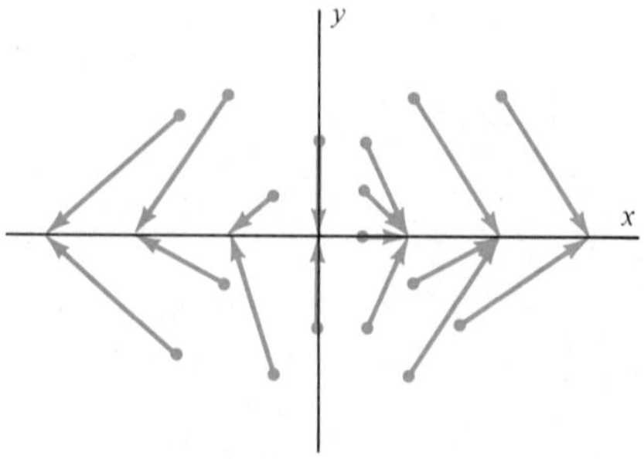

3.

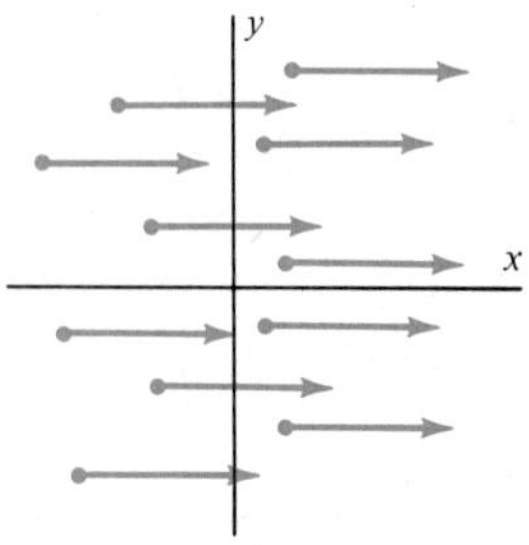

5.

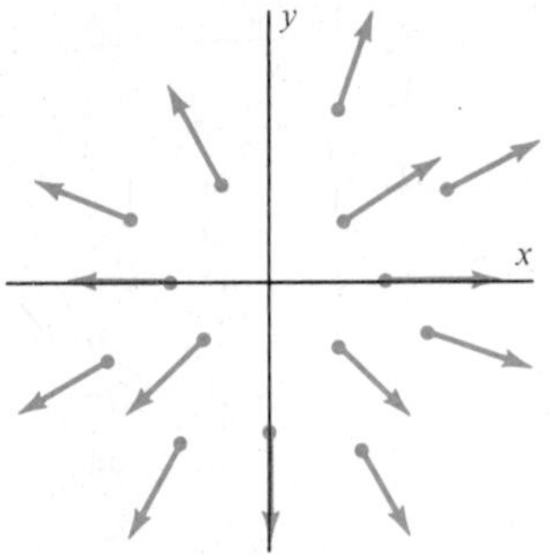

7.

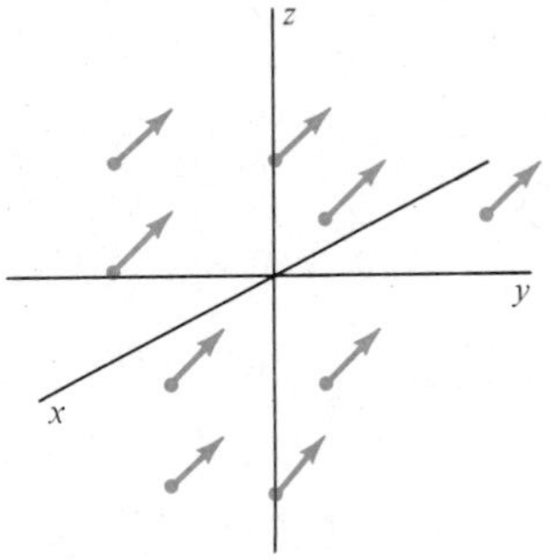

9.

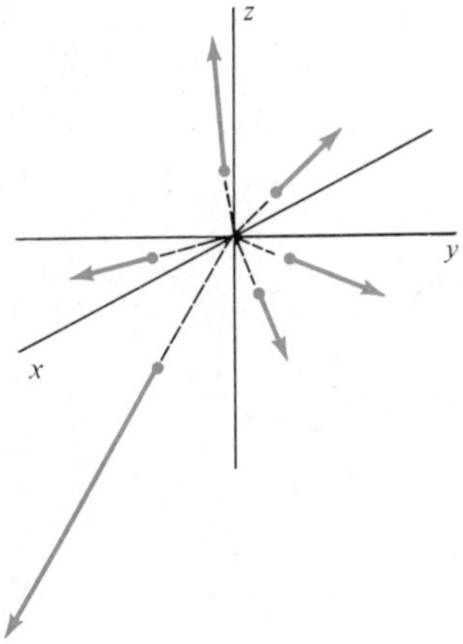

11. 4, 5, 9

13. $F = 2xi - 2yj$

15. $\dfrac{1}{\sqrt{x^2 + y^2 + z^2}}(xi + yj + zk)$

17. $e^{x-y}(zi - zj + k)$

19. not a gradient

21. not a gradient

23. $F = \nabla(e^{xyz})$

27. a. $F(0, 0, 0) = 4i$

b. $F(-2, 0, 0) = -\dfrac{26}{9}i$

c., d.

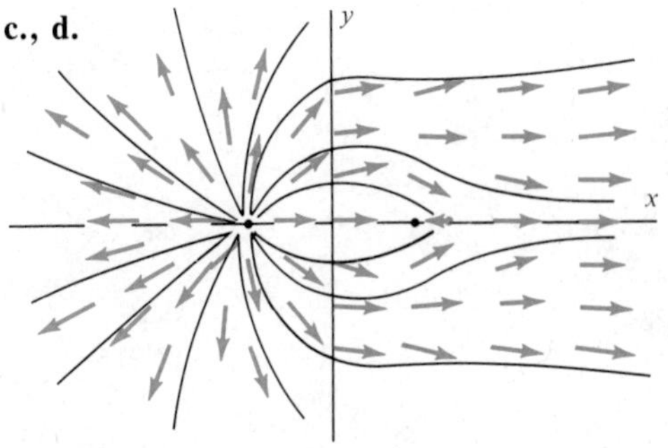

Exercise Set 21.2

1. 16

3. $\dfrac{1708}{3}$

5. $\dfrac{328}{15}$

7. $\dfrac{\pi^2}{4} + \pi$

9. 12

11. $-\pi$

13. -4

15. $\dfrac{\pi}{2}\left(\dfrac{\pi}{2} - 1\right)$

17. 1

19. 3

21. $\dfrac{1}{6} + \dfrac{\pi}{16}$

23. $\dfrac{-442}{3}$

25. $\dfrac{4}{3}$

29. $\dfrac{70\sqrt{5}}{3}$

31. $\dfrac{2}{3}(5^{3/2} - 1)$

33. 5π

Exercise Set 21.3

For problems 1 through 10 $F = \nabla f$ with

	f	$\int_c F \cdot dr$
1.	xy	3
3.	e^{xy}	$e^2 - 1$
5.	$x^2 + y^2$,	0
7.	xe^{y^2},	$2e$
9.	$e^x \sin y$,	$e\dfrac{\sqrt{2}}{2}$

11. $-a$

19. No, the field is not conservative.

Exercise Set 21.4

1. $\dfrac{1}{2}$

3. $\dfrac{2}{3}$

5. $\dfrac{\pi}{8}$

7. $-\dfrac{11}{60}$

9. 0

11. -5π

13. 0

15. $-\frac{9\pi}{2}$

17. $\frac{1}{6}$

19. $\frac{3a^2\pi}{8}$

Exercise Set 21.5

1. $\frac{23\sqrt{6}}{4}$

3. $\frac{-153\sqrt{14}}{5}$

5. 0

7. $4\delta\pi(9 - 3\sqrt{5})$

9. $\frac{32\pi}{3}$

11. 2π

13. 16π

15. $\frac{4\pi a^3}{3}$

17. 3

19. No, not the same: $\frac{4\pi}{r}$

21. $(\bar{x}, \bar{y}, \bar{z}) = \left(0, 0, \frac{a}{2}\right)$

23. $(\bar{x}, \bar{y}, \bar{z}) = \left(\frac{1}{2}, \frac{1}{2}, \frac{1}{2}\right)$

Exercise Set 21.6

1. 0

3. $z\cos(yz)\boldsymbol{i} + xy\boldsymbol{j} + (y\sin xy - xz)\boldsymbol{k}$

5. 0

7. not conservative

9. π

11. $-\frac{2}{3}$

13. 4

15. -2π

17. 0

19. 0

Exercise Set 21.7

1. $2x + 2y + 2z$

3. $1 + x + yx$

5. $-y\sin(xy) + xze^{xyz} + xy\cos(xz)$

7. 0

9. -150

11. $\frac{11}{6}$

13. 0

15. $\frac{512\pi}{15}$

17. $\frac{80\pi}{3}$

Review Exercises–Chapter 21

1. a. $-y\sin x\ln z\boldsymbol{i} + \cos x\ln z\boldsymbol{j} + \frac{y\cos x}{z}\boldsymbol{k}$

b. $[3x^2yz^2 + y\cos(xy)]\boldsymbol{i} + [x^3z^2 + x\cos(xy)]\boldsymbol{j} + 2x^3yz\boldsymbol{k}$

3. $\boldsymbol{F} = \nabla\left(-\tan^{-1}\left(\frac{y}{x}\right)\right)$

5. -2π

7. $\boldsymbol{F} = \nabla(xy^2z)$

9. $\frac{1}{15}$

15. $(a + b + c)V$

17. $-\frac{4}{3}$

19. π

21. $-\frac{3}{2}$

23. $\frac{2}{5}$

25. $\sqrt{6}$

27. $\frac{31\sqrt{2}\,\pi}{5}$

29. $-\frac{15\pi}{2}$

31. 0

33. $8\pi(a + b + c)$

35. $-\frac{100}{3}$

37. 0

39. $W(t) = \frac{\cos^4 t}{4} + t\sin t - \frac{1}{4}$

INDEX

D

T

U

V

W

X

Y

Z